ENCYCLOPEDIA OF PLANT PHYSIOLOGY

EDITED BY

W. RUHLAND

COEDITORS

E. ASHBY · J. BONNER · M. GEIGER-HUBER · W. O. JAMES
A. LANG · D. MÜLLER · M. G. STÅLFELT

VOLUME XVIII

SEXUALITY · REPRODUCTION ALTERNATION OF GENERATIONS

CONTRIBUTORS

L. BAUER · W. DÖPP · K. ESSER · H. ETTL · H. FÖRSTER · C. HARTE · E. HAUSTEIN
M. KAPOOR · K. KÖHLER · M. KROH · H. F. LINSKENS · D. G. MÜLLER
K. NAPP-ZINN · K. NEUMANN · U. NÜRNBERG-KRÜGER · A. NYGREN · J. R. RAPER
F. RESENDE · G. SÖRGEL · H. A. v. STOSCH · J. H. TAYLOR · H. WEBER · W. WEBER

WITH ASSISTANCE OF J. STRAUB SUBEDITED BY

H. F. LINSKENS

WITH 366 FIGURES

SPRINGER-VERLAG
BERLIN · HEIDELBERG · NEW YORK
1967

HANDBUCH DER PFLANZENPHYSIOLOGIE

HERAUSGEGEBEN VON

W. RUHLAND

IN GEMEINSCHAFT MIT

E. ASHBY · J. BONNER · M. GEIGER-HUBER · W. O. JAMES
A. LANG · D. MÜLLER · M. G. STÅLFELT

BAND XVIII

SEXUALITÄT · FORTPFLANZUNG GENERATIONSWECHSEL

BEARBEITET VON

L. BAUER · W. DÖPP · K. ESSER · H. ETTL · H. FÖRSTER · C. HARTE · E. HAUSTEIN
M. KAPOOR · K. KÖHLER · M. KROH · H. F. LINSKENS · D. G. MÜLLER
K. NAPP-ZINN · K. NEUMANN · U. NÜRNBERG-KRÜGER · A. NYGREN · J. R. RAPER
F. RESENDE · G. SÖRGEL · H. A. v. STOSCH · J. H. TAYLOR · H. WEBER · W. WEBER

UNTER MITWIRKUNG VON J. STRAUB REDIGIERT VON

H. F. LINSKENS

MIT 366 ABBILDUNGEN

SPRINGER-VERLAG
BERLIN · HEIDELBERG · NEW YORK
1967

ISBN-13: 978-3-642-95001-8 e-ISBN-13: 978-3-642-95000-1
DOI: 10.1007/978-3-642-95000-1

Softcover reprint of the hardcover 1st edition 1967
Library of Congress Catalog Card Number 56–3135

Titel-Nr. 5684

Inhaltsverzeichnis. — Contents.

I. Einleitung. — Introduction.

Reproduction, sexuality and fertilization. A general concept. By Professor Dr. H. F. Linskens . . . 1
Literature . . . 4

II. Bacterial conjugation.

By Professor Dr. M. Kapoor. With 6 figures . . . 5
I. Introduction . . . 5
II. Sexual differentiation in *E. coli* . . . 6
1. The mating types . . . 6
2. The bacterial surface . . . 9
a) Type I pili . . . 9
b) F-pili . . . 10
c) Sex specific bacteriophages . . . 11
d) The interaction of bacteriophage and F-pili . . . 11
e) The physical and chemical properties of F-pili . . . 12
III. The factors effecting fertility in bacteria . . . 12
1. F and its variants . . . 12
2. Colicinogenic factors . . . 14
3. Resistance factors or multiple drug resistance factor . . . 15
IV. Physiology of mating process in bacteria . . . 16
1. Pairing of compatible cells . . . 16
2. Establishment of cellular contact . . . 17
3. Transfer of genetic material during conjugation . . . 17
a) Kinetics of transfer . . . 17
b) Transfer of sex factors . . . 18
V. Chromosome replication and the mechanism of transfer . . . 19
1. Replication of the bacterial chromosome . . . 19
2. Sex factor and its integration with the chromosome . . . 21
3. The mechanism of chromosome transfer during conjugation . . . 22
VI. Conjugation in bacteria other than *Escherichia coli* . . . 24
1. *Enterobacteriaceae* . . . 24
2. Conjugation in *Pseudomonas aeruginosa* . . . 25
3. Conjugation in *Vibrio cholerae* . . . 25
VII. Concluding remarks . . . 26
Literature . . . 26

III. Physiologie der Geschlechtsbestimmung und Geschlechtsdifferenzierung. — Physiology of sex determination and sex differentiation.

A. Genetik der Geschlechtsbestimmung. — Genetical background of sex determination.

a) Die Geschlechtsfaktoren bei haplogenotypischer Bestimmung. Von Dr. Herbert Förster. Mit 6 Abbildungen . . . 31
Prinzipielles der haplogenotypischen Geschlechtsbestimmung . . . 31
Varianten der Geschlechtsbestimmung . . . 35
Die Funktion der Geschlechtsrealisatoren . . . 38
Geschlechtsbestimmung und Kernphasenwechsel . . . 39

Der Nachweis für haplogenotypische Geschlechtsbestimmung 42
Die Sonderstellung der Pilze. (Das Verhältnis von Sexualität und Inkompatibilität) 43
Literatur 49

b) Die Wirkung (der Geschlechtsfaktoren) bei diplogenotypischer Bestimmung. Von Professor Dr. Ursula Nürnberg-Krüger. Mit 13 Abbildungen 51
Einführung 51
Nachweise für den Homozygotie-Heterozygotie-Mechanismus bei diplogenotypischer Geschlechtsbestimmung 52
I. Kreuzungen zwischen autöcischen und diöcischen Arten 52
a) Correns' *Bryonia*-Versuche 52
b) Weitere Kreuzungen autöcisch und diöcisch 53
II. Certationsversuche 56
III. Geschlechtschromosomen — Heterochromosomen 57
a) XO-Typus 60
b) XY-Typus 61
c) XY_1Y_2-Typus 64
d) Weitere multiple Geschlechtschromosomen-Typen 65
e) Unsichere Heterochromosomen-Nachweise 66
f) Diöcisten mit homomorphen Geschlechtschromosomen 66
IV. Geschlechtsgebundene Vererbung 66
a) *Melandrium* 66
b) *Carica papaya* 68
c) *Cannabis sativa* 69
d) *Spinacia oleracea* 69
Die genetische Zusammensetzung des geschlechtsbestimmenden Mechanismus . . 69
I. Bei Diöcisten 69
a) Der *Drosophila*-Typus 69
b) Der *Melandrium*-Typus 72
II. Bei Subdiöcisten 81
a) Der *Drosophila*-Typus 81
b) Der *Melandrium*-Typus 87
c) Abweichende Formen 90
III. Bei Gynodiöcisten 92
a) Gen-Balance-System 93
b) Kern-Plasma-System 93
IV. Bei autöcischen Pflanzen 94
a) Primärzwitter 94
b) Sekundärzwitter 100
Schluß 101
Literatur 101

B. Phänotypische Geschlechtsbestimmung. — Phenotypic sex determination.

a) Phänotypische Geschlechtsbestimmung bei Algen. Von Professor Dr. Kurt Köhler. Mit 4 Abbildungen 110
A. Chlorophyceen 111
B. Phäophyceen 113
C. Phänotypische „Überraschungen" bei *Callithamnion* und *Chaetomorpha* . . . 114
D. Die Geschlechtsbestimmung bei den Diatomeen 115
1. *Centrales* 115
2. Die pennaten Diatomeen 116
Schlußbetrachtung 118
Literatur 119

b) Phänotypische Geschlechtsbestimmung bei Pilzen. Von Professor Dr. Kurt Köhler. Mit 9 Abbildungen 121
A. Phycomyceten 123
B. Hefen 126

C. *Euascomycetes* 127
D. Hinweis auf sexuelle Differenzierung von Kernen bei Autogamie 130
Schlußbetrachtung 131
Literatur 131

c) **Phänotypische Geschlechtsbestimmung bei Archegoniaten.** Von Professor Dr. LEOPOLD BAUER. Mit 11 Abbildungen 134
I. Einleitung 134
II. Geschlechterverteilung 135
1. Gametophytenstufe 135
2. Sporophytenstufe 138
III. Auslösende Faktoren der Geschlechtsbestimmung 140
1. Genetische Faktoren 140
2. Nichtgenetische Faktoren 142
a) Ernährung 142
b) Wirkstoffe 143
c) Korrelationserscheinungen zwischen den sexuellen Tendenzen 144
IV. Die Stabilität der Geschlechtsdetermination 146
V. Sekundäre Geschlechtsmerkmale 147
a) Physiologische Merkmale 147
b) Morphologische Unterschiede 148
VI. Schlußbemerkung 149
Literatur 150

d) **Modifikative Geschlechtsbestimmung bei Spermatophyten.** Von Professor Dr. KLAUS NAPP-ZINN. Mit 3 Abbildungen 153
I. Modifikative Geschlechtsbestimmung bei Gametophyten 153
1. Verweiblichung männlicher Gametophyten 153
2. Vermännlichung weiblicher Gametophyten 154
II. Modifikative Geschlechtsbestimmung bei Sporophyten 155
1. Pflanzenarten, bei deren Sporophyten modifikative Geschlechtsbestimmung bekannt ist 156
a) Gymnospermae 158
Coniferae S. 158. — Cycadinae S. 159.
b) Monocotyledoneae 160
Scheuchzeriaceae (Juncaginaceae) S. 160. — Alismataceae S. 160. — Hydrocharitaceae S. 160. — Liliaceae S. 160. — Musaceae S. 161. — Gramineae S. 161. — Cyperaceae S. 163. — Araceae S. 164.
c) Dicotyledoneae 164
Ranunculaceae S. 164. — Capparidaceae S. 165. — Cruciferae S. 165. — Crassulaceae S. 166. — Eucommiaceae S. 166. — Rosaceae S. 166. — Oxalidaceae S. 166. — Geraniaceae S. 167. — Rutaceae S. 167. — Anacardiaceae S. 167. — Begoniaceae S. 167. — Caricaceae S. 168. — Aceraceae S. 168. — Hippocastanaceae S. 169. — Vitaceae S. 169. — Halorrhagaceae S. 169. — Onagraceae S. 169. — Caryophyllaceae S. 169. — Nyctaginaceae S. 172. — Chenopodiaceae S. 172. — Polygonaceae S. 173. — Moraceae S. 173. — Cannabinaceae S. 173. — Urticaceae S. 179. — Euphorbiaceae S. 179. — Myricaceae S. 180. — Salicaceae S. 180. — Betulaceae S. 181. — Juglandaceae und Cupuliferae S. 182. — Primulaceae S. 182. — Ebenaceae S. 182. — Plantaginaceae und Borraginaceae S. 182. — Labiatae S. 182. — Solanaceae S. 183. — Scrophulariaceae S. 183. — Gesneriaceae S. 183. — Rubiaceae S. 184. — Dipsacaceae S. 184. — Cucurbitaceae S. 184. — Compositae S. 190.
2. Zusammenfassende Übersicht über die an der modifikativen Geschlechtsbestimmung beteiligten Faktoren 192
a) Edaphische Faktoren 192
b) Luftfeuchtigkeit 193
c) Licht 193
d) Temperatur 194

e) Transplantation, Resektion und andere Verletzungen 194
f) Parasitenbefall 196
g) Stellung der Blüten, Alter und Größe der Pflanzen 196
h) Behandlung mit tierischen Geschlechtshormonen 198
i) Beeinflussung des Wuchsstoffhaushaltes 198
k) Behandlung mit Gibberellinen 199
3. Interpretation der diplomodifikativen Geschlechtsbestimmung 199
Anhang: Physiologische Kastration 202
Literatur 202

C. The role of specific secretions in the induction and development of sexual organs and in the determination of sexual affinity. By Professor Dr. JOHN R. RAPER. With 1 figure 214
The induction and coordinated development of sexual organs 216
The determination of sexual affinity 228
Literature 233

D. Generationswechsel. — Life cycles.
Determination von Gametophyt und Sporophyt. Von Professor Dr. LEOPOLD BAUER. Mit 10 Abbildungen 235
I. Einleitung 235
II. Faktoren bei der Entstehung des Generationswechsels 235
1. Die Qualität der Keimzellen 235
a) Allgemeine Gesichtspunkte 235
b) Vegetative Entwicklung der Gameten 237
c) Sexuelles Verhalten der Sporen 238
d) Unterschiedliche vegetative Determination der Sporen 238
2. Verschiebung der Meiosis 239
a) Bildung einer diploiden Phase 239
b) Bildung einer dikaryotischen Phase 240
3. Morphogenetische Prinzipien 240
a) Plasmodiale Organisation 241
b) Verflechtung von Zellfäden (unechte Gewebebildung) 242
c) Echte Gewebe- und Organbildung 242
d) Gestaltunterschiede zwischen den Generationen 243
III. Determination in der Ontogenese 245
1. Natürlicher Entwicklungsablauf 245
2. Der Stabilitätsgrad der Sporophytendetermination bei Restitutionen 248
3. Experimentell induzierte Apogamie 250
IV. Mischbildungen zwischen Gametophyten und Sporophyten 253
Literatur 254

IV. Physiologie der Fortpflanzung. — Physiology of reproduction.

A. General principles of sexual and asexual reproduction and life cycles. By Professor Dr. FLÁVIO RESENDE. With 13 figures 257
1. General concept of sex 257
A. Historical survey 257
B. Gamones 260
C. About somatic sexuality 260
D. Relative sexuality 261
2. Sex in higher plants 261
3. Life cycles 262
4. Flowering and sex expression 268
A. The inflorescence 268
B. The flower 271
5. Origin of dioecia from monoecia. Possible mechanism of the origin of sex chromosome heteromorphy 275
6. Asexual reproduction 276
A. Vegetative propagation 277

B. Parthenogenesis and apomictosis 277
C. Apogamy and apospory 278
D. Apogonosis and "zygogamosis" 278

Literature 278

B. Geschlechtliche Fortpflanzung. — Sexual reproduction.

a) Thallophyten. — Thallophyta.

1. Die chemischen Grundlagen der Befruchtung (Gamone). Von Professor Dr. KURT KÖHLER. Mit 21 Abbildungen 282

I. Einleitung 282
II. Allgemeine Übersicht über die morphologischen und physiologischen Vorgänge bei der Befruchtung (oder Paarung) 283
III. Die Fernwirkungen 285
1. Chemotaxis (Sirenine) 285
a) *Oedogonium* 286
b) *Cutleria* 287
c) *Fucus* 287
d) *Chlamydomonas* 289
e) *Allomyces* 290
2. Chemotropismus (Sirenine) 291
a) *Achlya* 292
b) Hefen 294
c) Mucorineen 294
d) *Ascobolus* 297
e) *Bombardia* 297
IV. Agglutinations- und Zellkontaktwirkungen 297
1. Isogame Gruppen 298
a) Die Vorgänge bei *Chaetomorpha* und *Chlamydomonas* 298
b) Die Bedeutung der Geißel für die Gruppenbildung 303
c) Weitere Agglutinationsstoffe 306
2. Oogame Formen 307
3. Hefen 310
4. Bakterien 313
V. Schlußbetrachtungen 316

Literatur 317

2. Die Verbreitung der Incompatibilität* bei Thallophyten. Von Professor Dr. KARL ESSER 321

I. Begriffsbestimmungen 321
II. Vorkommen von Incompatibilitätserscheinungen in den verschiedenen Klassen der Thallophyten 322
1. *Cyanophyceae* 322
2. *Flagellatae* 322
3. *Algae* 322
4. *Lichenes* 323
5. *Fungi* 323
III. Verbreitung der homogenischen Incompatibilität bei den Eumycetes . . 324
A. Euascales 325
B. Holobasidiomycetes 327
C. Phragmobasidiomycetes (Protobasidiomycetes) 335
D. Phragmobasidiomycetes (Hemibasidiomycetes) 336

Literatur 338

b) Archegoniaten und Spermatophyten. — Embryophyta.

1. Meiosis. By Professor Dr. J. HERBERT TAYLOR. With 13 figures 344

I. Distinguishing features of meiotic divisions 344
A. Introductory remarks 344

* In Abweichung von der Rechtschreibung wird die Schreibweise mit „c" auf ausdrücklichen Wunsch des Autors in seinem Beitrag beibehalten. Die Redaktion.

B. Typical features of meiotic divisions 346
1. The long prophase. 347
2. The pairing of homologous chromosomes 347
3. The formation of chiasmata and the exchange of genetic material . 350
4. Variations in the coiling cycle 352
5. The unitary behavior of the centromeres of homologous chromosomes 353
C. Synthesis of nucleic acids and proteins 354
1. DNA (deoxyribonucleic acid) synthesis 354
2. RNA (ribonucleic acid) synthesis 356
3. Protein synthesis 357
II. Aberrations of meiosis 359
A. Genetic aberrations 359
B. Culture of excised anthers 359
C. Aberrations induced by ionizing radiations 361
III. Summarizing statement 363
Literature 365

2. Pollen. Von Professor Dr. H. F. LINSKENS. Mit 20 Abbildungen 368
I. Entwicklung der Pollenkörner 368
1. Wandbildung 369
2. Tapetum 369
3. Beeinflussung der Pollenbildung 370
4. Pollenausschüttung 372
5. Pollenkitt 372
II. Morphologie 373
III. Biochemie des Pollens 373
1. Pollenwand (Sporoderm) 373
2. Pollenfarbstoffe 377
a) Lokalisation 377
b) Chemie der Pollenfarbstoffe 378
α) Anthocyane 378
β) Carotinoide 378
γ) Anthoxanthine 379
c) Funktion 379
3. Polleninhaltsstoffe 380
a) Proteine 380
b) Pollenfermente 380
c) Freie Aminosäuren 381
d) Nucleinsäuren 382
e) Kohlenhydrate 382
f) Lipide 383
g) Wirkstoffe 383
h) Anorganische Substanzen 383
IV. Pollenkeimung 384
A. Keimung *in vitro* 384
1. Keimungsbedingungen 384
2. Die Bedeutung des Zuckers und der Borsäure für die Keimung . . 385
3. Gegenseitige Beeinflussung des Pollens bei der Keimung 387
4. Der Pollenschlauch 388
B. Keimung *in vivo* 393
1. Ernährung des Pollenschlauches 393
2. Tropismus der Pollenschläuche 394
V. Lebensfähigkeit des Pollens 395
1. Historisches 395
2. Aufbewahrungsbedingungen 395
a) Luftfeuchtigkeit 395
b) Temperatur 396
c) Der Sauerstoffpartialdruck 396
3. Ursachen der Abnahme der Keimfähigkeit 396
Literatur 399

3. Befruchtung der Archegoniaten und Blütenpflanzen. Von Professor Dr. Erik Haustein. Mit 30 Abbildungen 407

A. Archegoniatae 407
Öffnung der Antheridien 407
Öffnung der Archegonien 409
Verschleimen der Kanalzellen 409
Bau der Spermatozoiden 410
Bau der Eizellen 411
Weg der Spermatozoiden zu den Archegonien 412
Weg der Spermatozoiden im Archegoniumhals 412
Befruchtung 413
a) Moose 413
b) Farne 413

B. Cycadales und Ginkgoales 414
Pollenkammer und Bestäubungstropfen 414
Bestäubung 415
Keimung der Mikrosporen 415
Entstehung und Bau der Spermatozoiden 415
Entwicklung der Eizellen 416
Bau der Archegonien 417
Weg der Spermatozoiden zu den Archegonien 418
Öffnung der Pollenschläuche 418
Öffnung des Archegoniumhalses 419
Befruchtung 419

C. Coniferae 421
Bestäubung 421
Pollenkeimung und Pollenschlauchwachstum 423
Entstehung der Spermazellen 424
Entwicklung der Eizellen 426
Bau der Archegonien 426
Zeitliche Koordinierung 427
Eindringen der Pollenschläuche in die Eizelle 428
Befruchtung 428
Doppelte Befruchtung 430

D. Gnetales 430

E. Angiospermae 430
Auffinden der Samenanlagen durch die Pollenschläuche 431
Der männliche und weibliche Gametophyt 434
Der männliche Gametophyt 434
Der weibliche Gametophyt 435
Eintritt des Pollenschlauches in den Embryosack 436
Chemotropismus der Pollenschläuche 437
Verhalten der Spermazellen im Embryosack 438
Plasmaverschmelzung (Plasmogamie) 438
Kernverschmelzung (Karyogamie) 439
Korrelative Beziehungen 441
Polyspermie 442

Literatur 442

4. Gonenkonkurrenz. Von Professor Dr. Cornelia Harte 447

A. Problemstellung 447

B. Grundlagen der Untersuchungen 448
I. Genetische Beobachtungen 448
II. Ursachen für abnorme Spaltungszahlen 449
a) Allgemeines 449
b) Selektion in der Diplophase 449
c) Selektion bei der Befruchtung 449
d) Selektion der Gameten 450
III. Entwicklungsgeschichtliche Grundlagen der Gonenkonkurrenz 450
a) Entwicklung von den Gonen zu den Gameten im ♂ Geschlecht . . 450
b) Entwicklung von den Gonen zu den Gameten im ♀ Geschlecht . . 451

C. Experimentelle Untersuchungen der Certation 451
I. Genetische Teste 451
a) Überlegungen zur Versuchsplanung 451
b) Veränderung der Pollenmenge 452
c) Abschneiden des Griffels 452
d) Aufteilung des Fruchtknotens 453
e) Erfassung der langsamen Pollenschläuche 453
f) Versuche mit nur einer Samenanlage je Fruchtknoten 454
g) Einfluß von Umweltbedingungen auf die Certation 454
h) Untersuchung von Pollengemischen 456
II. Untersuchung des Pollenschlauchwachstums 457
a) Allgemeines zur Versuchsplanung 457
b) Wachstum im Griffel 457
c) Kulturversuche 458
d) Zur Deutung der Wachstumsversuche 458
III. Identifizierung der gametophytisch wirkenden Gene 459
IV. Genphysiologie der Certation 463
a) Veränderung der Certation unter dem Einfluß des Griffels 463
b) Wechselwirkungen zwischen verschiedenen Genen 466
c) Plasmonwirkung 467
d) Zusammenhang der Certation mit der Selbststerilität 468
e) Das Reaktionsschema der Certation 469
V. Vermutete Zusammenhänge zwischen Pollenschlauchwachstum und Vitalität der nächsten Generation 470
D. Gonenkonkurrenz in der Samenanlage 471
I. Grundlagen 471
II. Einfluß der Genkomplexe bei *Oenothera* 472
a) Variabilität der Gonenkonkurrenz 472
b) Einfluß von Umweltfaktoren 473
c) Einfluß genetischer Faktoren 473
III. Einfluß einzelner Gene auf die Gonenkonkurrenz und genetische Wechselwirkungen 474
IV. Vergleich der Konkurrenz in Pollenschlauch und Samenanlage . . . 474
E. Zusammenfassung 475
Literatur 476

5. **Die selektive Befruchtung.** Von Professor Dr. Erik Haustein. Mit 7 Abbildungen 479
Definition 479
Nachweis der selektiven Befruchtung 479
Reaktionsgeschwindigkeit 483
Affinität zwischen Samenanlagen und Pollenschläuchen 485
Die genetischen Grundlagen der Affinität 489
Der Einfluß des Diplonten 492
Weitere Analyse der Affinität 493
Der unterschiedliche Zustand der Samenanlagen in einem Fruchtknoten . . . 498
Die materiellen Grundlagen der selektiven Befruchtung 500
Chemotropie-Versuche 502
Selektive Befruchtung bei anderen Objekten 503
Literatur 503

6. **Inkompatibilität der Phanerogamen.** Von Professor Dr. H. F. Linskens und Dr. Marianne Kroh. Mit 19 Abbildungen 506
I. Einleitung 506
II. Historisches 507
III. Verbreitung der Inkompatibilität bei den Phanerogamen 507
IV. Selbst-Inkompatibilität 508
1. Hemmung der Pollenkeimung 509
2. Hemmung des Pollenschlauchwachstums 511
a) Pollenschlauchhemmung bei Homostylen 512
b) Hemmung des Pollenschlauchwachstums bei Heterostylen 514

3. Inkompatibilitätsreaktion in der Samenanlage 516
4. Theorien 517
5. Überwindung der Selbst-Inkompatibilität 520
a) Veränderung der genetischen Bedingungen 521
b) Veränderung der Milieubedingungen 521
V. Kreuzungs-Inkompatibilität 523
VI. Interspezifische Inkompatibilität 524
Literatur 525

C. Ungeschlechtliche Fortpflanzung. — Asexual reproduction.

a) Apomixis bei Archegoniaten. Von Professor Dr. W. Döpp. Mit 15 Abbildungen 531
1. Einleitung 531
2. Obligate (habituelle) Apogamie 532
a) In Verbindung mit Restitutionskernbildung. Phasenwechsel (in den Sporenmutterzellen) und Reduktionsteilung vorhanden 532
α) Cytologie, Genetik, Keimungsverhältnisse 532
β) Ursachen und Entstehung der obligaten Apogamie 539
γ) Bedeutung der obligaten Apogamie 542
b) Apogamie verbunden mit Aposporie. Phasenwechsel und Reduktionsteilung nicht vorhanden 544
3. Fakultative (induzierte) Apogamie 545
4. Parthenogenesis 547
5. Apomixis bei *Isoetes* 549
Literatur 549

b) Apomixis in the angiosperms. By Professor Dr. Axel Nygren. With 43 figures 551
Terminology 551
I. Agamospermy 552
A. Diplospory 552
a) Embryo sac formation 552
b) Pollen formation 555
B. Apospory 556
a) Embryo sac formation 556
b) Pollen formation 558
C. Adventitious embryony 558
D. The development of the endosperm in agamospermous species 560
II. Vegetative propagation 561
Biotype, form, and species formation in apomictic species 563
Appendix 569
Literature 583

c) Vegetative Fortpflanzung, Parthenogenese und Apogamie bei Algen. Von Dr. H. Ettl, Dr. D. G. Müller, Dr. K. Neumann, Professor Dr. H. A. v. Stosch und Dr. W. Weber. Mit 110 Abbildungen 597
A. Einleitung 597
a) Vegetative Fortpflanzung 597
b) Apogamie und Parthenogenese 600
Literatur 601
B. Cyanophyta 602
a) Hormogonien 603
1. Entstehung 603
2. Entwicklungsphysiologie 605
b) Nannocyten und Endosporen 606
c) Fortpflanzungskörper mit Dauerorgancharakter 608
1. Hormocysten 608
2. Dauerzellen und Heterocysten, Morphologie und Bildung 608
3. Entwicklungsphysiologie der Entstehung von Dauerzellen und Heterocysten 612

Allgemeine Bedingungen . . . 612
Die Bildung der Dauerzellen S. 612. — Die Heterocysten S. 613.
Zum Mechanismus der Spontandifferenzierung von Heterocysten und Dauerzellen . . . 615
4. Die Keimung von Heterocysten und Dauersporen . . . 617
Literatur . . . 618

C. Euglenophyceae, Chloromonadophyceae und Cryptomonadophyceae . . . 620
a) Varianten der Zellteilung . . . 620
b) Cysten . . . 623
Literatur . . . 624

D. Dinophyta . . . 626
a) Parasitische Dinophyten . . . 626
b) Die monadoide Stufe . . . 627
1. Zellteilung, Gonidienbildung . . . 627
2. Cysten . . . 630
Entwicklungsgeschichtliche Dignität der Cysten . . . 631
Bildungsbedingungen . . . 631
Reifung, Lebensdauer und Keimung der Cysten . . . 632
c) Capsale, coccale und trichale Formen . . . 634
Literatur . . . 635

E. Chrysophyta . . . 637
I. Chrysophyceae . . . 637
a) Euchrysophyceae . . . 637
1. Vegetative Vermehrung . . . 637
Rhizopodiale Formen . . . 637
Flagellaten . . . 638
Kolonievermehrung bei Flagellaten . . . 639
Capsale, coccale und trichale Formen . . . 640
Bedingungen für die vegetative Fortpflanzung . . . 641
2. Dauerstadien, Cysten und Sexualität . . . 642
Pseudocysten, Resistenz . . . 642
Cysten . . . 643
Sexualität S. 644. — Zweikernige Cysten S. 644. — Parthenogenetische Cysten S. 645. — Vegetative Cystenbildung S. 645. — Bedeutung und Bildungsbedingungen der Cysten S. 645. — Cystenkeimung S. 646.
b) Haptophyceae . . . 646
1. Bewegliche und unbewegliche Phase; Generationswechsel . . . 646
Die unbewegliche Phase in Vermehrung und Fortpflanzung . . . 647
Vegetative Fortpflanzung innerhalb der unbeweglichen Phase S. 648.
Die bewegliche Phase, ihre Vermehrung und Fortpflanzung . . . 649
Die Vermehrung in der monadoiden Phase S. 649.
Die Übergänge zwischen den Phasen . . . 650
2. Cysten . . . 650
3. Pleomorphismus . . . 651
4. Der „theoretische" Cyclus von *Coccolithus fragilis* . . . 651
5. Bedingungen für die Phasenübergänge . . . 652
c) Parthenogenese und Apogamie bei Chrysophyceen . . . 653
Literatur . . . 653

II. Diatomeen . . . 657
a) Sprunghafte vegetative Zellgrößenänderungen . . . 657
b) Kolonieteilung . . . 660
c) Dauerzustände und Dauerzellen . . . 661
1. Dauerzustände . . . 661
2. Dauerzellen . . . 661
Bildung und System . . . 662
Modi der Membranbildung bei Dauerzellen S. 662. — Klassifizierung der Dauerzellen S. 664. — Doppelschalen S. 666.

Entwicklungsphysiologie der Dauerzellen 666
Bildung S. 666. — Experimentelles zur Dauersporenentstehung S. 666. — Die Keimung der Dauersporen S. 667.
d) Pleomorphismus 668
1. Abgrenzung 668
2. Vorkommen 669
3. Phasenübergänge 670
4. Entwicklungsphysiologische Dignität 671
5. Ökologisches 671
e) Parthenogenese und Apogamie bei Diatomeen 672
1. Centrales 672
2. Pennatae 673
Typ III Automixis und das Problem der Geschlechtsbestimmung . 674
Apomixis 677
3. Die Bedingungen 678
Literatur 679
III. Heterokontae (Xanthophyceae) 682
a) Die Fortpflanzungsarten 682
1. Die einzelligen und die trichalen Formen 682
2. Heterosiphonales 684
b) Experimentelle Untersuchungen zur Fortpflanzung 686
1. Die Kultur von Heterokonten 686
2. Die Alternative sexuelle-vegetative Fortpflanzung 687
3. Auslösung von Zoosporen und Aplanosporen 688
Zoosporen 688
Aplanosporen 689
4. Periodizität der Sporenentlassung 690
5. Die Keimung der vegetativen Fortpflanzungskörper 690
c) Parthenogenese bei Heterokonten 690
Literatur 691
F. Phaeophyta 692
a) Vegetative Fortpflanzung 692
1. Einleitung 692
2. Die verschiedenen vegetativen Fortpflanzungsweisen 693
3. Plurilokuläre Sporangien 693
4. Unilokuläre Sporangien 694
5. Tetrasporangien 695
6. Monosporangien 695
7. Aplanosporangien 696
8. Hypnocysten, Brutkörper, Adventivthalli 696
9. Haftscheiben, Stolonen 697
10. Vegetative Vermehrung durch Fragmentation 697
11. Einige Bemerkungen zur Physiologie der Fortpflanzung 697
Ausbildung der Sporangien 697
Entleerung der Sporangien 697
Die Schwärmer der Braunalgen 698
Weiterentwicklung der festgesetzten Fortpflanzungszellen 699
Heteroblastie 700
12. Die Bedeutung der vegetativen Fortpflanzung im Generationswechsel der Braunalgen 700
b) Apogamie und Parthenogenese bei Braunalgen 703
1. Parthenogenese isogamer Braunalgen 703
2. Parthenogenese oogamer Braunalgen 703
3. Cutleria 703
4. Apogamie 704
Literatur 704
G. Chlorophyta 705
I. Chlorophyceae 705
a) Bewegliche Fortpflanzungszellen (Zoosporen-Typus) 707
1. Nackte Volvocales (Tochterzellenbildung) 707

2. Behäutete Volvocales 708
3. Celluläre Chlorophyceen 712
Tetrasporales S. 715. — Chlorococcales S. 716. — Ulothrichales S. 717. — Oedogoniales S. 720. — Coenoblastische und siphonoblastische Formen S. 720.
4. Modifizierte Zoosporen 722
Hemizoosporen S. 722. — Amöboide Stadien der Zoosporen S. 722. — Vielfachschwärmer und Synzoosporen S. 723.
b) Unbewegliche Fortpflanzungszellen (Autosporen-Typus) 724
1. Hemiautosporen 724
2. Autosporen 727
2. Autosporine Fadenbildung 727
c) Andere Formen vegetativer Fortpflanzung 728
1. Dauerzustände (Palmellen und Gloeocysten) 729
2. Dauerzellen (Aplanosporen und Akineten) 729
3. Fragmentation 731
d) Auslösung von Fortpflanzungszellen 733
1. Licht-Dunkel-Wechsel 733
2. Temperaturänderung 734
3. Konzentration und Zusammensetzung der Nährlösung 735
4. Acidität 736
5. Wasserströmung und Benetzung 736
e) Anhang: Pedinomonadales 737
f) Parthenogenese und Apogamie bei Chlorophyceen 741
1. Volvocales 741
2. Tetrasporales und Chlorococcales 742
3. Ulothrichales 742
4. Oedogoniales 743
Literatur 743

II. Conjugatophyceae 746
a) Teilung 746
b) Zerfall von Fäden 747
c) Aplanosporen 748
d) Akineten (Chlamydosporen) 749
e) Parthenogenese und Apogamie bei Conjugaten 750
Literatur 752

III. Charophyceae 752
a) Vegetative Vermehrung durch Sproßknoten 753
b) Bulbillen 755
1. Sproßknöllchen 755
2. Wurzelknöllchen 756
c) Parthenogenese 758
Literatur 759

H. Rhodophyta 760
a) Fortpflanzung durch Sporen 760
1. Autosporen 760
2. Monosporen 760
3. Sporen aus mehrsporigen Sporangien 765
Tetrasporen S. 765. — Bisporen S. 770. — Polysporen S. 770.
4. Parasporen 771
b) Sonstige Fortpflanzungskörper 771
c) Vegetative Vermehrung (sensu stricto) 772
d) Parthenogenese und Apogamie 774
Literatur 774

d) Vegetative Fortpflanzung bei Pilzen. Von Professor Dr. Georg Sörgel 777
1. Die Bildung von Sproßmycel 777
2. Die Bildung von Chlamydosporen 780

3. Die Bildung von Sklerotien 781
Temperatur S. 781. — Licht S. 782. — Feuchtigkeit S. 782. — p_H S. 782. — Gase S. 782. — Ernährungsfaktoren S. 783.
4. Die Bildung von Bulbillen 784
Literatur 784

e) **Vegetative Fortpflanzung bei Spermatophyten.** Von Professor Dr. Hans Weber. Mit 12 Abbildungen 787
I. Allgemeines 787
II. Morphologisch-anatomische Grundlagen 788
1. Vegetative Fortpflanzung durch Sproßverzweigung 788
2. Knospenbildung an Blättern 794
3. Sproßbildung an Wurzeln 797
III. Zur Physiologie der vegetativen Fortpflanzung 800
1. Einfluß der Temperatur 800
2. Einfluß des Photoperiodismus 801
3. Wuchsstoffwirkung 802
4. Vegetative Fortpflanzung und Karyologie 804
5. Korrelationen zwischen generativer und vegetativer Fortpflanzung . . . 804
Literatur 805

Namenverzeichnis. — Author Index 809
Sachverzeichnis (Deutsch-Englisch) 837
Subject Index (English-German) 856

Mitarbeiter von Band XVIII. — Contributors to volume XVIII.

LEOPOLD BAUER, Professor Dr., Institut für Allgemeine Botanik der Universität Mainz, 6500 Mainz, Saarstraße 21.

W. DÖPP †.

KARL ESSER, Professor Dr., Institut für Allgemeine Botanik der Ruhr-Universität Bochum, 4630 Bochum-Querenburg, Im Lottental.

HANS ETTL, Dr. rer. nat., Březová, n. Svit. (Tschechoslowakei).

HERBERT FÖRSTER †.

CORNELIA HARTE, Professor Dr., Universität zu Köln, Institut für Entwicklungsphysiologie, 5000 Köln-Lindenthal, Gyrhofstraße 17.

ERIK HAUSTEIN, Professor Dr., Botanisches Institut der Universität Erlangen-Nürnberg, 8520 Erlangen, Schloßgarten 4.

MANJU KAPOOR, Dr. Assistant Professor, Department of Biology, University of Calgary, Alberta (Canada).

KURT KÖHLER, Dr. Professeur associé, Centre National de la Recherche Scientifique, Département de Biochimie Macromoléculaire, 34 Montpellier (Frankreich).

MARIANNE KROH, Dr. phil., Botanisch Laboratorium, Universiteit, Driehuizerweg 200, Nijmegen (Niederlande).

H. F. LINSKENS, Professor Dr. phil., Botanisch Laboratorium, Universiteit, Driehuizerweg 200, Nijmegen (Niederlande).

DIETER G. MÜLLER, Dr., Wissenschaftlicher Assistent, Max Planck-Institut für Züchtungsforschung, Abt. Genetik, 5000 Köln-Vogelsang.

KLAUS NAPP-ZINN, Dr., Professeur à la Faculté des Sciences de Grenoble, Laboratoire de Physiologie Végétale, Domaine Universitaire, 38 St. Martin d'Hères (Frankreich).

KLAUS NEUMANN, Dr. rer. nat., Wissenschaftlicher Assistent, Botanisches Institut der Universität, 3550 Marburg/Lahn, Pilgrimstein 4.

URSULA NÜRNBERG-KRÜGER, Professor Dr., Humboldt-Universität, Institut für Allgemeine Botanik, Abt. Genetik, 1040 Berlin 4, Luisenstraße 53.

AXEL NYGREN, Professor, Department of Genetics and Plant Breeding, Uppsala 7 (Schweden).

JOHN R. RAPER, Professor of Botany, Department of Biology, Harvard University, 16 Divinity Avenue, Cambridge, Massachusetts 02138 (USA).

FLÁVIO RESENDE †.

GEORG SÖRGEL †.

HANS-ADOLF VON STOSCH, o. Professor Dr., Botanisches Institut, 3550 Marburg/Lahn, Pilgrimstein 4.

J. HERBERT TAYLOR, Professor, Institute of Molecular Biophysics, Florida State University, Tallahassee, Florida 32306 (USA).

HANS WEBER, Professor Dr., Institut für Spezielle Botanik der Universität Mainz. 6500 Mainz, Saarstraße 21.

WILHELM WEBER, Dr., Dozent, 7410 Reutlingen, Ringelbachstraße 47.

Reproduction, sexuality and fertilization.

A general concept.

By

H. F. Linskens.

Multiplication seems to be an essential criterion of life. Living systems do not have an unlimited lifespan because of an internal complexity which is not well understood yet. It seems also that living machinery undergoes something like wastage, which destroys balanced cell functions. Furthermore, only multiplication of well-functioning entities guarantees survival of living units. Since they are easily destroyed by external injury, also, only an increase in total population size permits the colonizing of new areas by dispersal.

Multiplication of living systems presents as its first property that the products of multiplication are more or less identical to their predecessors. The *replication unit* we call a *cell*. In unicellular organisms especially, binary or multiple fissions arise from a succession of mitotic cell divisions. Most protistic cells can maintain continuity and increase in population for long periods by many sequences of mitotic division. This form of identical replication which is a form of *reproduction* seems to be very efficient in maintaining cells in a constant, suitable environment.

However the situation changes as soon as environmental conditions are altered and the population of an organism of absolute uniformity must adjust with them. Survival of the fittest requires capacity for adaptation. Adaptation is only possible within a population when there is a system which enables variation to be maintained and stored so that it gives an advantage as soon as changing environment demands it. In this case reproduction therefore implies that changes in the information bearing material are not lost but are involved in a new form of redistribution of nuclear material, called *meiosis*. Or in other words, meiotic division is a manner of adjusting the information bearing genic materials to *adaptation requirements*. Meiosis in reproduction represents a new step forward to a higher degree of fitness in evolution.

Meiotic division can occur at various stages of life in different organisms but generally follows the same pattern in most plants and animals of the eukaryotic type. Important features of meiosis are: (a) the number of chromosomes in the cell and the DNA content of the cell nucleus are halved at the same time; (b) while mitotic division delivers daughter cells with identical genomes, meiotic division results in cells with different genomes; (c) daughter cells differ qualitatively from parent cells in their chromosomal make-up. The latter results from the newly involved mechanism of synapsis which includes the alignment of homologous chromosomes, and the possibility of random exchange of sister chromatic pieces (formation of chiasmata). The result of meiosis is cells — usually followed by a second, mitotic division which gives rise to a tetrad — which are reduced in number of chromosomes and differ by possessing a new combination of chromosomes and/or bearing physically exchanged chromatical material.

The evolution of meiosis has two important consequences: the origin of sexuality and the necessity of fertilization.

Sexuality means capability of recombining genetic material. There are many theories on origin of sex and hence this term is rich in implication. The existence of a general bisexual potency is still under discussion because in microorganisms tetrapolar sexes have also been demonstrated. Sexuality is only a special aspect of higher evolution of reproduction. Terms as male and female are anthropomorphic reprojections and are full of behavioural and psychological aspects. But biophysically seen, there are only two (or more) partly different forms of the same species which allow new combinations of characteristics to arise in organisms, and therefore make possible biophysical evolution to new level.

One should clearly distinguish between *fertility* and sexuality. Sexuality expresses itself by the production of sexual organs or individuals, such as the number of the flowers on a spermatophyte plant; on the other hand, fertility means the production of fully capable gametes, which are able to form fertile zygotes. Male sterility, abortion of zygotes, and reduced seed or embryo production are expressions of reduced fertility in higher plants. Both characters are genetically and physiologically independent.

The sex stage is not a fixed one. It can change during development and may be induced by environmental conditions or be triggered by special genes. In fact sex lability and induction during development is one of the main problems of morphogenesis.

The transfer of genetic information from one individual to another is not a monopoly of sexual reproduction. If one defines sexual reproduction as regular alteration of karyogamy and meiotic division in a life cycle and the linked possibility of recombination, then also other recently detected "genic systems" in a great number of sexual microorganisms give the same possibilities. The "dance of genes", the pool formation and the accumulation of recessive mutants in heterozygotic organisms can also be guaranteed by other mechanisms viz. by *transformation* (e.g. in *Pneumococcus*), *transduction* by bacteriophages (e.g. in *Escherichia coli*), *lysogenesis* (e.g. in *E. coli*), or by *parasexual cycles* (e.g. in *Aspergillus, Penicillium, Fusarium*). They all resemble normal sexual cycles in uniting genic determinants of different cellular origin into one cell. In this way genetic recombination is secured.

The other inherent consequence of meiosis is the necessity of restoration of the chromosome number. This takes place in various ways, especially in animals. In plants the most common means is *fertilization* including *syngamy*. The manner of fertilization shows a great variety of features in detail and has to be discussed for every type of organism. Every fertilization process takes place in three phases namely, plasmogamy, karyogamy and chromosomogamy. These steps of conjugation can be separated in time and place in various groups of organisms and can be distributed throughout the life cycle. Autogamy and endogamy can replace sexual syngamy or the fusion of free gametes. Autogamy in higher plants is analogous to self-fertilization. To prevent inbreeding there are many ecological and physiological mechanisms to form barriers to self-fertilization.

Sexual reproduction in angiosperms and fungi is extensively regulated by *incompatibility*, which determines by genetically regulated biochemical processes the pattern of mating among individual strains. This system may play the same role as dioecy in other organisms by restricting or preventing inbreeding among the members of the same race and increase outbreeding by like alleles in the two mates (homogenic incompatibility). The opposite effect can be caused by restricting or preventing outbreeding between members of the same race in order to promote inbreeding by sexual interaction of unlike alleles (heterogenic incompatibility).

Fertilization as the transition from haploid to diploid phase, and meiosis as the reduction of chromosome number, are complementary events. Sexuality has to be seen in the aspect of rearrangement of genetical material between individuals which assures variability. This guarantees genetic *homeostasis*, that is, optimal adaptation of a species or a race to changing environmental conditions. Later in higher organisms these "regenerating" effects of the sex processes become combined with *multiplication* and *dispersal* aspects.

Thus fertilization and meiosis are closely linked together. Fertilization rebuilds the diploid state, which was disturbed by meiosis. And furthermore, existence of fertilization makes possible *dissemination of genes* within a whole population. It introduces an integrating influence for many individuals, as a generalizing system of gene dispersal. As a consequence of the many problems which are raised in differentiation, approximation and fusion of gametes, new structures and mechanisms are discovered. These are for example synchrony in production, sexual affinity, tropic orientation, contact mechanism, concurrent phenomena in the fusion process, metabolizing of supernumerary sperms and so on. All of them are the fascinating object of sex physiology at all levels of organization.

Alternation of meiotic division with reduction of chromosome numbers and fertilization involving the union of two cells with the immediate effect of reduction in number of individuals involve processes of development which secure the regular alternation of both: thus *the life cycle.*

Change from vegetative to reproductive phase in all diploid tissues, with the exception of the gametes which are haploid, is called *diplontic* life cycle and is observed in some algae. When a separate haploid phase exists between meiotic reduction and diploidisation by fertilization, the life cycle is called *haplodiplontic.* This is very common in higher plants, ferns, many algae and fungi. The products of meiosis do not fuse again immediately, but grow out to haploid *gametophytes,* which produce the gametes. The diploid individual which rises from the zygote and is the place of meiosis (meiospores) is called a *sporophyte.* The kind and size of the meiotic products can be different; if they are little and form sperms (antherozoids, spermatozoids, spermatozoa, microgametes etc.), they are called *microgametophytes*; while if these are larger and form eggs (macrogamete, ovules, oosphere), they are called *macrogametophytes* (megagametophytes).

Within the angiosperm there is a strong *tendency to reduce the gametophyte* and to give up its autonomy. Special physiological problems arise in connection with metabolic exchange between the surrounding sporophytic tissue and the ripening microgametophytes (pollen) and macrogametophytes (egg). This nourishing function is continued to the developing embryo.

As demonstrated to this point, all physiological aspects of reproduction, sexuality and fertilization are strongly linked with genetical aspects. At the highest level of development, a sexual organism has two possibilities to include variations in evolution: rearrangement of genetical information by meiotic crossing-over and accumulation of recessive mutations.

And finally we find on the higher level of evolution a tendency to reduce sexuality. This means reproduction without genetic recombination, called *asexual* or *apomictic reproduction.* Apomixis is found most frequently in unicellular organisms, but also in mosses, ferns and phanerogams. Because in apomictic organisms all descendents are genetically identical to the mother, every mutation remains manifest and a new constant microspecies is born. On the other hand, once a fit organism is evolved it has less chance to lose its adaptiveness by micro-mutation. So asexual or vegetative reproduction permits a rapid increase of an adapted population under favourable environment conditions. Under changing life conditions, the apomictic population can only survive by mutations.

This volume was outlined in 1957 by Prof. STRAUB (Max-Planck-Institut für Züchtungsforschung, Köln-Vogelsang), who also took into contract most of the authors. The articles written in 1958 were set in 1959. Thanks to the cooperative and willing attitude of all the contributors, the articles were checked again and brought up to date in 1966.

Contributors and redaction are very thankful to the publishers who did not spare time and costs to bring out this delayed volume in the best possible way.

Literature.

(For further information on this topic.)

AUSTIN, C. R.: Fertilization. Englewood Cliff, N.Y.: Prentice-Hall Inc. 1965.

COOK, ST. A.: Reproduction, heredity and sexuality. London: McMillan & Co. Ltd 1964.

HARTMANN, M.: Die Sexualität, 2. Aufl., S. 464—480. Stuttgart: Fischer 1956.

JACOB, F., and E. L. WOLLMAN: Sexuality and the genetics of Bacteriae. New York: Acad. Press 1961.

MONROY, A.: Chemistry and physiology of fertilization. New York: Holt, Reinhart and Winston 1965.

WENRICK, D. H. (Ed.): Sex in microorganisms. Washington, D.C.: Am. Ass. Adv. Sci. 1954.

Bacterial conjugation.

By

M. Kapoor.

With 6 figures.

I. Introduction.

The transfer of genetic material from one cell to another is accomplished in a variety of ways. The mode of transfer and the subsequent incorporation of the transferred material is a consequence of the basic structural organization of the genetic material in a given organism. In plants and animals, nucleus, the repository of genetic information is generally delimited by a membrane from the surrounding cytoplasm and the biologically important macromolecules are organized into complex structures, the chromosomes. Such genetic systems comprise the *Eukaryotes*, a group distinct from the simpler organisms referred to as the *Prokaryotes*, characterized by the absence of a conventional nucleus and the nuclear apparatus. The mechanics of both genetic systems, nevertheless, are based on the inherent properties and biochemical activity of a common macromolecule, DNA, which is the predominant carrier of genetic information. The major difference between the two types of systems lies in the presence of additional non-genic material, in the form of basic proteins and histones in the nuclei of *Eukaryotes* and it is this which is largely responsible for the organization of DNA into chromosome threads. The "nuclei" of the *prokaryotes*, including bacteria, possess the functional equivalent of a chromosome in the form of a single double-helical DNA molecule, free of histones and other proteinaceous material. In both systems, genetic exchange mechanisms can be interpreted in terms of the behaviour of a "linkage group". In the following pages the bacterial linkage group will be referred to as the "bacterial chromosome".

Of the three mechanisms of genetic exchange recognized in the bacteria, i.e. transduction, transformation and conjugation, the last named only will be discussed. Transformation and transduction involve the transfer of relatively small segments of DNA from the donor to the recipient cell either through an external agent, which is a virus in the process of transduction, or by means of direct uptake by physiologically mature or competent cells. In conjugation, on the other hand, fairly large segments of DNA are transferred by direct cell-to-cell contact. With the application of the information derived from the study of conjugation it has been possible to elucidate the nature of the bacterial chromosome and the processes of recombination and segregation, as occurring in bacteria. It has been established that in *Escherichia coli*, and most likely many other bacterial genera, the linkage group is present in the form of a circular double helix of DNA. In addition to the chromosome the cell may contain one or several nonchromosomal or extrachromosomal inclusions, probably also present as circular DNA molecules, known as "episomes" or "episomic elements". Episomes can exist as autonomous entities or alternately in a state of integration with the bacterial chromosome. A class of episomes referred to as "sex factors" (including the "fertility" F factor and its variants) confers the property of fertility on a normally infertile or F^- cell, thus converting it into a cell which is able to donate its genetic material to a recipient cell. The sex factor can, thus, be transferred to bacterial cells deficient in it and once it has secured entrance in a cell, it is capable of multiplying at a rate much faster than the bacterial chromosomes and can, therefore, spread rapidly in a population of F^- cells. If, on the other hand, the sex factor were to get attached to the chromosome and exist in an integrated state, then its replication is so regulated that it replicates only in phase with the chromosome. At the same time, integration of the sex factor with the chromosome alters the F^+ cell converting it into a cell which is capable of donating its chromosome or segments thereof to a donor cell. A summary of the information available at present regarding the physiological and biochemical implications of transfer of the sex factor, F, through conjugation, its interaction with the bacterial genome, and the regulation of the activity of the chromosomal and F replicons, is presented in the following pages. Emphasis will be placed on studies conducted, mainly, with *Escherichia coli* but reference will be made to other genera wherever possible.

The subject of bacterial conjugation has been reviewed exhaustively in a number of books and recent publications from which a detailed description of many of the aspects of this subject can be obtained (JACOB and WOLLMAN, 1961, 1962; GROSS, 1964; ADELBERG and PITTARD, 1965).

II. Sexual differentiation in *E. coli*.

1. The mating types.

Genetic recombination in bacteria was first demonstrated by LEDERBERG and TATUM in 1946 by isolation of prototrophs from a mixture of two different auxotrophs of *Escherichia coli* K12, and performing crosses between appropriate genotypes. Most of the crosses were made between strains of *E. coli* K12, originally isolated by TATUM in 1945. All the earlier work was conducted with two strains *A*, which is auxotrophic for methionine and biotin synthesis (Met^- $Biot^-$); and *B*, auxotrophic for threonine, leucine and vitamin B_1 (Thr^- Leu^- B_1^-). On crossing *A* and *B*, recombinant progeny were obtained which showed different permutations and combinations of the characters of the parents. It was observed that the cross

$$A\ (met^-\ Biot^-\ Thr^+\ Leu^+\ B_1^+) \times B\ (Met^+\ Biot^+\ Thr^-\ Leu^-\ B_1^-)$$

gave rise to cell types with genotypes having all conceivable combinations of the characters of the parents, which were detected on plating the mixture on minimal medium supplemented with one nutrient at a time. For example, if the mixture of parental strain was plated on a minimal medium enriched with leucine, Met^+ $Biot^+$ Thr^+ Leu^- B_1^+ and Met^+ $Biot^+$ Thr^+ Leu^+ B_1^+ cells were found to grow and form colonies on the surface of the medium. The proportion of the leu^- recombinants isolated from a given number of parent cells was always in excess of the proportion of Met^+ $Biot^+$ Thr^+ Leu^+ B_1^+ prototrophs. These results were suggestive of a sexual fusion of some of the parental cells, followed by a recombination and segregation of genes.

The first evidence of sexual differentiation in bacteria was presented by LEDERBERG (1951) and HAYES (1952) demonstrating two physiologically distinct mating types. In crosses between strains *A* and *B* carrying the differential marker for streptomycin sensitivity and streptomycin resistance, pretreatment with the antibiotic led to a loss of viability if the recipient strain was streptomycin sensitive, without any impairment of viability on using a sensitive donor strain.

Cross 1)* A, $S^s \times$ B, $S^r \rightarrow$ Recombinants on streptomycin medium
Cross 2) A, $S^r \times$ B, $S^s \rightarrow$ No recombinants

From this experiment HAYES concluded that the two strains involved in this cross played different roles in recombination. One of the strains presumably acted as a donor and was not effected by streptomycin treatment and the other strain was postulated to be the recipient, which, when treated with streptomycin, led to a loss of recombinants.

Further evidence of sexual differentiation was revealed by the differential effect of ultraviolet light on strains *A* and *B*. The frequency of recombination was seen to increase as a result of irradiation of *A*, whereas exposure of *B* to UV led to a decrease in the frequency of recombination. The existence of two physiologically distinct sexual strains was confirmed by the discovery of some mutants, isolated from strain *A* ($Biot^-$ Met^-), which when crossed with standard strain *B* (Thr^- Leu^- vit B_1^-) formed no recombinants. These were classed as F^- mutants and it was noticed that a cross between two F^- mutants never produced any recombinants. The strain which was interfertile with F^- was hence designated F^+ and was regarded as a donor strain, a cross between two donor strains being sterile likewise (LEDERBERG, CAVALLI-SFORZA and LEDERBERG, 1952; CAVALLI-SFORZA, LEDERBERG and LEDERBERG, 1953; HAYES, 1953). HAYES postulated that this distinction between the recipient and donor strains could be attributed to the presence of a factor, symbolised "F", for fertility; so that a cell containing F behaved as a "male" and a cell devoid of F acted like a "female" in a cross. As a consequence of cell contact between F^+ und F^- cells the genes controlling the synthesis of methionine, biotin, threonine, leucine and thiamine (selected markers) underwent recombination. Further, the property of fertility was demonstrated to be transmissible in that it could be readily transferred from F^+ donors to bacteria of the F^- type by incubating a mixture of the two.

In early crosses between F^+ and F^- differing in a number of unselected markers it became evident that the recombinant genotype always resembled the F^- parent in the distribution of unselected characters. In other words, the recombinant progeny always inherited the alleles of unselected markers from the female parent. The donor strain usually appeared to transfer only a part of its genome to the recipient, the contribution of the two parents to the genotype of the progeny being by no means equal (HAYES, 1952, 1953; CLOWES and ROWLEY, 1954). Thus the initial work of HAYES and others established that in *E. coli*, conjugation occurred between two physiologically distinct mating types equivalent to the mating types recognized in *Fungi*, and that *E. coli* could be considered to be a heterothallic species, the contribution of the two parents to the progeny being unequal. When F^+ and F^- strains are mixed, F^- cells become infected with F and it has been demonstrated (JACOB and WOLLMAN, 1955) that the transmission of F is accomplished in 4—5 minutes of contact at 37° C.

A third sexual phenotype was soon described by CAVALLI-SFORZA (1950) and by HAYES (1953b) as Hfr (for high frequency recombination), derived from *E. coli* K12 (met^- $biot^-$ F^+) resistant to nitrogen mustard, on mating with the female (thr^- leu^- vit B_1^-) strain gave rise to approximately a 1000-fold more

* Symbols represent: S^s, streptomycin sensitive; S^r, streptomycin resistant.

recombinants compared to the original standard F^+. The Hfr strain isolated by CAVALLI was designated the symbol Hfr C and that isolated by HAYES as Hfr H. Hfr strains were special male strains which yielded recombinants with a very high degree of efficiency and exhibited several of the properties attributed to the donor strains. In addition, some of the properties exhibited by these strains were unique:

a) treatment with antibiotics like streptomycin did not effect the frequency of recombination (as contrasted with the F^+ strains);

b) the recombinants obtained from crosses with Hfr males always yielded F^- progeny unlike the crosses with F^+ males whose progeny always received the F factor;

c) some characters were transferred with a high frequency and some of the others with a low frequency.

Hfr donor never contributed any transmissible F, unlike the F^+ donors, and it was also impossible for Hfr to acquire F from F^+. Hfr mating types, it was concluded, did not contain any free F. However, Hfr H, HAYES' culture sometimes reverted back to F^+ and simutaneously regained the capacity to transmit F. This led to the suggestion that Hfr types arose as a result of a genetic alteration in F^+, accompanied by the attachment of F factor to the bacterial chromosome thereby converting the F^+ male into a high frequency recombination type. At the same time the replication of free F in the bacterial cell was repressed (Table 1). F^+ and Hfr types represent alternate states of the F factor which functions as an autonomous entity in the F^+ but is completely integrated into the bacterial chromosome in the Hfr type. In the integrated state the F factor behaves as a segment of the bacterial chromosome and its replication is coordinated with it (Table 2). RICHTER (1957) reported a third Hfr type designated Hfr 3 which reverts back to the F^+ state with a high frequency. In the years following the work of HAYES, CAVALLI and others several Hfr types were reported, in each case the attachment of F was presumed to be at a different locus.

Table 1. *Mating types of E. coli K 12.*

Symbol	Type	Sex factor (F)
F^-	recipient	absent
F^+	donor	present in an autonomous state
Hfr	donor	present in an integrated state

Table 2.

Cross	Progeny	Frequency of recombination
$F^+ \times F^-$	F^+	very low
$F^+ \times F^+$	F^+	very low
Hfr $\times F^-$	F^-	very high

CAVALLI-SFORZA and JINKS (1956) and JACOB and WOLLMAN (1957, 1961) determined the linkage relationships and the sequence of linked genes in a variety of Hfr types and characterized each Hfr type in terms of the nature and sequence of the genetic determinants it transferred during conjugation. JACOB and WOLLMAN (1957, 1961) prepared linkage maps for several of these Hfr types (for details *see* JACOB and WOLLMAN, 1961) (Table 3). By standard mapping techniques involving interrupted matings it was seen that the genetic determinants were always transferred in a fixed order for a given Hfr strain. The sequence of characters transferred differed from one Hfr to another. All Hfr types, considered collectively, showed that although the order of gene transfer was unique to each Hfr type, the relative position of the genetic determinants on the linkage maps remained constant. JACOB and WOLLMAN interpreted these data as being demonstrative of the linear and oriented nature of each Hfr chromosome with a fixed leading end (origin) and a tail end. The chromosome or the linkage map of *E. coli* was pictured as a circular structure, each Hfr type representing the attachment of the F factor to the chromosome at a different site (Fig. 3). The circular chromosome was assumed to open up at the point of the attachment of the F, the end farthest away from the F becoming the origin and that with F becoming the distal end. Since F happens to be the last segment of the linear Hfr chromosome, it is seldom transmitted to the recipient cell due to interruption of mating. This also explains the observation that recombinant progeny remain F^- with regard to the sex factor (Cavalli and Jinks, 1953; Jacob and Wollman, 1957, 1958) (Table 2).

Jacob and Wollman (1957, 1958, 1961) have formulated a hypothesis to explain the origin of the different sexual types in *E. coli* according to which the chromosomes of *E. coli* F^- type is circular and the cell is devoid of extra-chromosomal sex factors. The sex factors can be easily acquired by the F^- cell by direct contact with F^+ cells which also have a circular chromosome plus a few to several sex factors, presumably also circular (Fig. 4). As a result of the interaction between the circular chromosome of the F^+ cell and one F episome the F^+ cell is converted to the Hfr type (Fig. 5). The free and the integrated states of F are in a dynamic equilibrium and the reverse reaction, namely, the detachment of the episome, restores the circular chromosome and the free episome. This hypothesis has not been clearly proved, however.

Although several Hfr types have been isolated (Hayes, 1953, 1954; Taylor and Adelberg, 1960; Richter, 1957, 1961; Sneath, 1962) the number of possible F-attachment sites seems to be limited and it has been pointed out that in all probability several of the Hfr mutants reported by different investigators belong to the same attachment site (Matney et al., 1963). Matney et al. (1964) have reported the mapping of seventeen F-attachment sites on the *E. coli* chromosome showing the location of 44 different Hfr mutants.

2. The bacterial surface.

a) Type I pili.

From the surface of many bacterial species including *E. coli*, *Salmonella*, *Shigella flexneri*, *Klebsiella*, *Proteus* and *Chromobacterium prodigiosum*, are seen numerous outgrowths in the

Table 3. *The linkage groups of different Hfr types and the order of transfer of genes.* (From Jacob and Wollman, 1961.)

Types of Hfr	Order of transfer of genetic characters																		
Hfr H	0	T	L	Az	T_1	Pro	Lac	Ad	Gal	Try	H	S-G	Sm	Mal	Xyl	Mtl	Isol	M	B_1
1	0	L	T	B_1	M	Isol	Mtl	Xyl	Mal	Sm	S-G	H	Try	Gal	Ad	Lac	Pro	T_1	Az
2	0	Pro	T_1	Az	L	T	B_1	M	Isol	Mtl	Xyl	Mal	Sm	S-G	H	Try	Gal	Ad	Lac
3	0	Ad	Lac	Pro	T_1	Az	L	T	B_1	M	Isol	Mtl	Xyl	Mal	Sm	S-G	H	Try	Gal
4	0	B_1	M	Isol	Mtl	Xyl	Mal	Sm	S-G	H	Try	Gal	Ad	Lac	Pro	T_1	Az	L	T
5	0	M	B_1	T	L	Az	T_1	Pro	Lac	Ad	Gal	Try	H	S-G	Sm	Mal	Xyl	Mtl	Isol
6	0	Isol	M	B_1	T	L	Az	T_1	Pro	Lac	Ad	Gal	Try	H	S-G	Sm	Mal	Xyl	Mtl
7	0	T_1	Az	L	T	B_1	M	Isol	Mtl	Xyl	Mal	Sm	S-G	H	Try	Gal	Ad	Lac	Pro
AB 311	0	H	Try	Gal	Ad	Lac	Pro	T_1	Az	L	T	B_1	M	Isol	Mtl	Xyl	Mal	Sm	S-G
AB 312	0	Sm	Mal	Xyl	Mtl	Isol	M	B_1	T	L	Az	T_1	Pro	Lac	Ad	Gal	Try	H	S-G
AB 313	0	Mtl	Xyl	Mal	Sm	S-G	H	Try	Gal	Ad	Lac	Pro	T_1	Az	L	T	B_1	M	Isol

form of long, thin, straight, thread-like appendages with a diameter of 250 Å or more and a length varying from 10 to 12 microns (BRINTON et al., 1954). The number of appendages varies from 10 to several thousand per cell. These thread-like appendages are distinct from the flagella and are known to occur both in the flagellated and nonflagellated forms (DUGUID et al., 1955). The surface outgrowths referred to variously as "pili" (BRINTON et al., 1954; BRINTON 1959), "fine threads" (ANDERSON, 1949), "bristles" (WOLLMAN et al., 1956), "fimbriae" (DUGUID et al., 1955, 1957, 1958) and "filaments" (HOUWINK and VAN ITERSON, 1950) are visible only under the electron microscope. The pili were described in detail first by BRINTON et al. (1954), who distinguished between two phases, the piliated and non-piliated phases within a given bacterial strain, differing in electrophoretic mobility of the cells. The electrophoretic mobility of piliated cells was approximately half that of the non-piliated ones. Piliation also gave rise to smaller and smoother colonies in *E. coli*. The piliated strains mutated readily to a non-piliated form (BRINTON, 1957).

The piliated strains (P^+) transmit the property of piliation to non-piliated (P^-) forms through genetic recombination, both in intraspecific and interspecific matings. On mixing piliated (P^+, Hfr) donor strains of *E. coli* and non-piliated stable recipient (F^- P^-) strains of *Salmonella typhosa*, recombinant progeny showed piliation on the surface (BRINTON and BARON, 1960). On rapid mixing in a waring blendor the type I pili were easily removed from the bacterial cell and could be further separated from the remainder of the cell contents by differential centrifugation (BRINTON, 1959) and by free boundary electrophoresis. Pili are proteinaceous in nature and contain several polymerized protein subunits of molecular weight 17,000, forming right handed helices 70 Å in diameter with an axial hole 20—25 Å in diameter.

b) F-pili.

The pili or appendages described above, now known as type I pili, occur independently of the sexual nature of bacterial strain — piliated forms may occur in F^-, F^+ or Hfr strains. A second variety of pili discovered by CRAWFORD and GESTELAND (1964) were found to be sex specific. This type of pili have been shown to be growing exclusively on the surface of male or donor cells, both F^+ and Hfr. Electron micrographs published by CRAWFORD and GESTELAND illustrated the attachment of male specific phage R17 to some of the pili on the surface of F^+ and Hfr but not to those of F^- strain. BRINTON et al. (1964) observed the adsorption of another RNA phage, M12, to special type of pili on Hfr and F^+ males of *E. coli* K12. Due to a close association between these pili and the F factor, BRINTON et al. (1964) proposed the term F-pili to describe all male specific pili. The average number of F-pili is one to two per cell for both F^+ and Hfr. They vary in length from a few tenths of a micron to 10—12 microns.

The relationship between the presence of F-pili and the fertility factor F in the cell was confirmed by growth of *E. coli* K12 F^+ in the presence of acridine orange. Treatment with the latter rendered the cell free of its F factor and the clones derived from the resulting F^- cells invariably showed no development of F-pili and no capacity to adsorb M12 (BRINTON et al., 1964). Type I pili were completely unaffected by the removal of F. The reverse experiment, where the introduction of F into an F^- strain by conjugation with F^+ strain also induced the formation of F-pili and the property of adsorbing M12 in the recombinant progeny, established the relationship between the F-pili and the F factor unequivocally.

Introduction of *E. coli* K 12 F factor into F^- strains of *E. coli* B/r, *Salmonella typhimurium*, *Salmonella typhosa*, *Shigella flexneri*, and *Proteus mirabilis*, normally without F-pili, was followed by an appearance of F-pili. These remarkable observations have firmly established the morphological bases of sexual differentiation in bacteria.

c) Sex specific bacteriophages.

Phenotypic expression of the physiological characteristics of the sexually distinct types in bacteria is afforded by the isolation of bacteriophages, exhibiting the property of differential adsorption on donor or recipient strains. Examples of male specific phages growing on F^+ and Hfr strains of *E. coli* (Table 4) include both RNA containing as well as DNA containing phages. To name a few of them f 2 (LOEB and ZINDER, 1961), M 12 (HOFSCHNEIDER, 1963) and R 17 (CRAWFORD and GESTELAND, 1964) are all RNA phages and f 1 is a DNA phage (ZINDER et al., 1963) belonging to the group of virulent phages isolated by LOEB (1960). Both the male strains, F^+ and Hfr, adsorb these phages whereas the female strains do not.

Table 4. *Some examples of male specific bacteriophages.*

Phage	Response on strains of *E. coli* K 12			Reference
	F^-	F^+	Hfr	
f 1 to f 7	—	+	+	LOEB (1960)
f 2 (RNA)	—	+	+	LOEB and ZINDER (1961)
f 1 (DNA)	—	+	+	ZINDER et al. (1963)
R 17 (RNA)	—	+	+	
M 12 (RNA)	—	+	+	HOFSCHNEIDER (1963)
(phi)	+	—	—	DETTORI et al. (1961)
W-31	+	+ no progeny	+	WATANABE and OKADA (1964)

Female specific phages (HAKURA et al., 1964) of the temperate variety have been reported which form plaques on F^- of *E. coli*. Virulent phages forming clear plaques on F^- and small, turbid ones on F^+ with a low eop (efficiency of plating) (DETTORI et al., 1961; WATANABE and OKADA, 1964) are also known. One of these phages W-31 (WATANABE and OKADA, 1964) adsorbs on to the two male, F^+ and Hfr, and the female strains equally effectively and furthermore it is even able to inject its nucleic acid into the F^+, Hfr and F^- bacterial cells. However, further development of the phage is inhibited in the F^+ cell where the autonomous F interferes with the replication of the phage genome. Removal of F from the host cell by acridine treatment relieves the inhibition and the phage genome is allowed to replicate normally. The phage replication, on the other hand, proceeds normally in *E. coli* Hfr strains, indicating a modification in the properties of the sex factor upon integration with the bacterial chromosome. Only in the free state is the sex factor able to inhibit the multiplication of the phage nucleic acid inside the bacterial cell; in the integrated state the sex factor is inactivated or is unable to function and form a gene product. The product of the sex factor may be a repressor specific for the phage nucleic acid, possibly preventing its replication by interferring with the synthesis of a messenger-like molecule.

d) The interaction of bacteriophage and F-pili.

The kinetics of phage adsorption have been studied by the use of P^{32} labelled coliphage f 2 and cell free suspensions of F-pili, prepared by treatment of F^+ cells in a waring blendor and incubating the two under appropriate conditions

(VALENTINE and STRAND, 1965). Viable phages form complexes with F-pili; the F-pili-phage complex can be separated from the free phage by passing the mixture through membrane filters which retain the complex but not the free phage. The residual radioactivity is used as an index of unreacted phage. This led IPPEN and VALENTINE (1965) to devise an "assay" system for F-pili based on the correlation between phage adsorption and concentration of F-pili. The rate of adsorption appears to be independent of temperature and adsorption seems to occur readily even at 0° C.

F-pili synthesis occurs in coordination with the growth of F^+ bacteria. It is initiated during the lag phase, the rate of synthesis is most rapid during the exponential phase, following which there is a steady decline and a plateau is reached during the stationary phase. During all stages of bacterial growth F-pili can be readily detached from the male cells. The detached F-pili are also able to react with bacteriophage.

e) The physical and chemical properties of F-pili.

1. F-pili are elongated, hollow tubes (VALENTINE and STRAND, 1965), comprised of a number of repeating structural units which bind phage particles.

2. Bouyant density measurements of F-pili by ultracentrifugation on gradients of CsCl indicate that the density of F-pili corresponds to 1.97 (WENDT et al., 1966).

3. Cell free perparations of F-pili are inactivated by treatment with chloroform. F-pili phage complexes are also dissociated by chloroform.

4. Heating to a temperature of 80° C inactivates the F-pili thus impairing their capacity to combine with phage particles.

5. F-pili-phage complexes are resistant to RNAase, whereas free phages are susceptible.

At present more details about the chemical nature of F-pili are not available. They seem to differ from type I pili in containing lipid material. In the absence of any information concerning the fine structure of F-pili it is difficult to speculate upon the nature of bonding with the phage.

III. The factors effecting fertility in bacteria.

Fertility, defined as the ability of bacterial cells to act as genetic donors in a cross, is controlled by the factor F in *E. coli*. Reference has been made to F and its relationship to the mating type in *E. coli*. The wild type F of *E. coli* K12 is generally referred to as F1 in order to distinguish it from the variant forms of F isolated subsequently. All episomes (JACOB and WOLLMAN, 1958) which influence the fertility of the bacterial cell in any way will be collectively termed sex factors.

1. F and its variants.

The main features of F1 are:

a) it can exist in alternative autonomous or fully integrated states within a bacterial cell;

b) in an autonomous state it multiplies at a rate more rapid than the bacterial genome and consequently spreads through the population very rapidly (LEDERBERG, 1958);

c) in F^+ cells the factor F can be eliminated by treatment with acriflavine and acridine orange (HIROTA and IIJIMA, 1957) and in this process the cell becomes F^-, a similar treatment has no effect on the Hfr cell.

In order to understand the interaction between the bacterial chromosome and F let us consider the bacterial chromosome to be one unit of replication or one replicon (Fig. 1) (for a discussion of replicon *see* section V) and F another, the F^- cells contain only the major replicon (bacterial chromosome) or Replicon I (Rep I), F^+ cells, therefore, contain rep I and several F-replicons (Rep II). The Hfr cells can be assumed to have a complex replicon (Rep III) formed as a result of integration between Replicons I and II, k1 and k2 being the rate constants for association and the dissociation of the sex factors from the chromosome.

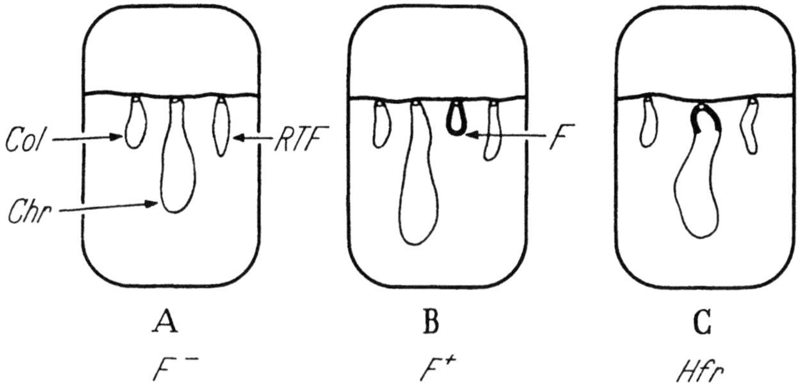

Fig. 1 A—C. The replicons of the bacterial cell. *Chr*, chromosome; *RTF*, resistance transfer factor; *F*, fertility factor F.

The interrelationships of the three replicons are diagrammatically represented below:

Rep I + n Rep II or ⇅ or F^- factor F	$\underset{k2}{\overset{k1}{\rightleftharpoons}}$	Rep III or Hfr
F^+ (Rep I + n Rep II)		

Evidence has been presented to suggest that the F factor is composed largely of DNA. This has been demonstrated by the use of P^{32}-labelled sex factor which is transferred to a non-radioactive female cell of the F^- type (Lavallé and Jacob, 1961). The newly converted F^+ cell loses its sex factor as a function of P^{32} decay. F factor contains nucleic acids and appears to be composed of 5×10^5 nucleotides. The loss of sex factor occurs only if P^{32} has been incorporated into the DNA of the original donor bacterium. Incorporation under conditions inhibitory to DNA synthesis in the donor cell did not lead to a subsequent loss of the factor in the recipient, upon P^{32} disintegration (Driskell and Adelberg, 1961). Size estimations based on P^{32} decay inactivation of the wild type F indicate that the F genome is 1.3—3.7% of the bacterial nucleus (Driskell-Zamenhof and Adelberg, 1963).

Several variants of the original F have been reported by different investigators. These modified factors are termed F′ (F-prime) (Adelberg and Burns, 1960; Hirota, 1959; Jacob et al., 1960; Hirota and Sneath, 1961). Adelberg and Burns reported the selection of a variant of F from cultures of Hfr differing from the wild type *E. coli* K12 F (F1) factor by its capacity to mediate the transfer of large segments of chromosomes. This property was attributed to the occurrence of a chromosomal locus with a high affinity for the sex factor (Table 5). This chromosomal locus which is closely linked to the *lac* locus is designated *sfa* (sex factor affinity) by Adelberg and Burns. The interaction of F′ and the bacterial chromosome is postulated to proceed by an unstable attachment to one preferred site on the chromosome, giving rise to a population with majority of its members behaving as donors. The recombinant progeny of the majority of donors also act as high frequency donors. These modified donors can be visualized as donors intermediate in properties between F+ and Hfr type donors in that they mediate transfer with a high frequency and at the same time give rise to recombinants that act as high frequency donors themselves (Table 6). Adelberg and Burns (1960) suggested that one of the F′ factors (F2) differed from the wild type F1 by incorporation of a larger segment of the bacterial chromosome. According to calculations made from P^{32} decay inactivation data, the F2 genome has been estimated to comprise 2.8—8.1% of the bacterial *E. coli* K12 nucleus (Driskell-Zamenhof and Adelberg, 1963).

Table 5. *Sexuality in E. coli.*

Sexual type	Sex factor	State of sex factor	Affinity for chromosome	Frequency of recombination
F−	absent	—	—	—
F+	present	autonomous	low	very low
Hfr	present	chromosomal	—	very high
F+ (F′)	present	autonomous	high (one site of preference)	high

Table 6. *Properties of donor strains.*

Cell type	Frequency of chromosome transfer	Phenotype of recombinant progeny
F+	very low	donor
Hfr	very high	recipient (very rarely donor)
F′ P4X-1	relatively high	majority donors
F′ AB-1	relatively high	majority donors
F′ F2	relatively high	majority donors

2. Colicinogenic factors.

Colicins comprise a subgroup of a family of antibiotics, known as the bacteriocins, which are produced by bacteria and inhibit the growth of the same or closely related species (for a review of colicins *see*: Fredericq, 1957, 1958). Colicinogenic strains, the strains that produce colicins, are common among *E. coli*, *E. freundii*, *Shigella*, *Paracolobactrum*, *Aerobacter* and *Serratia* (Reeves, 1965). Colicins are macromolecules composed of lipocarbohydrate protein complexes, colicin activity being associated with the protein moiety (Reeves, 1965). Colicinogenic strains synthesize many types of colicins which have been studied extensively and classified by Fredericq (1957, 1963). Colicinogeny and bacteriocinogeny, in general, are genetically determined. The genetic determinants (Col or C-factors), responsible for the synthesis of colicins, are stable, extrachromosomal, genetic elements which behave as independent units of replication (replicons).

Colicinogenic strains (Col+) can transfer the ability to produce specific colicins to noncolicinogenic (Col−) strains by direct cell contact. Following cell contact, the recipient cell also acquires an immunity to the action of the colicin produced by the specific *Col* factor. The recombinant progeny resemble the recipient (Col−) parent in all other characters. Col factors are always transferred independently of the bacterial chromosome and are not known to exist in an integrated state with it. In addition to colicinogeny, col factor also confers fertility on *Salmonella typhimurium* cells (OZEKI and HOWARTH, 1961). Colicinogenic donor strains of *E. coli* are also able to confer fertility on *E. coli* K12 F^- recipients (FREDERICQ, 1963). The physiological factors controlling the synthesis of colicins are not very well understood. In a given population of cells only some cells produce colicins at any one time but the non-producing cells can be induced to produce the antibiotics by treatment with UV (JACOB et al., 1952; AMATI, 1964), Mitomycin C (IIJIMA, 1962) and thymine deprivation (LUZZATI and CHEVALLIER, 1964).

Transmission of the col agent in *E. coli* is a rapid process and may occur within a few minutes of contact between the recipient and the donor strain. The transfer of col factors occurs not only from F^+ cell but also from F^- (CLOWES, 1961; SMITH and STOCKER, 1962). In crosses involving F^+ col$^+$ and F^- col$^-$ strains of *E. coli* K12 the col agent is transferred at a very high frequency. Nearly all the recombinants recovered from such a cross are F^+ col$^+$. An analysis of crosses involving Hfr col$^+$ and F^- col$^-$ was interpreted by AFOLDI et al. (1958) to be indicative of an attachment between the col factor and the bacterial chromosome. In light of later work this interpretation was proven to be wrong and at present it is believed that col agents do not integrate with the bacterial chromosome. Col$^-$ strains of *Salmonella* can also acquire the property of colicinogeny on growth in mixed cultures with col$^+$ strains of *E. coli*. *S. typhimurium* Lt 2 is normally Col$^-$ but can be converted to the col$^+$ state by mixed cultures with *E. coli* col$^+$ strains. Studies of kinetics of transfer of the various col factors to the col$^-$ cell have shown that the col I or col B agent confers the property of fertility upon the cells of *S. typhimurium* LT 2 immediately following transfer, thus making the young col$^+$ cells act as male cells in conjugation with other col$^-$ cells. The donor cells, however, retained the property of colicinogeny (STOCKER et al., 1963) even after transfer. Col I factor not only is able to initiate conjugation but is also able to provide a stimulus to chromosomal transfer (see HAYES, 1965 for a detailed discussion of this point).

The chemical nature of colicinogenic factor E 1 has been determined by using a col$^+$ *E. coli* strain as a donor and transferring the factor to a noncolicinogenic strain of *Proteus mirabilis* and extracting DNA from col$^-$ *Proteus* cells and from col$^+$ cells after the col factor had been received from the donor bacteria. A comparison of bouyant densities of the isolated DNA revealed an additional component in the col$^+$ progeny corresponding to the DNA contributed by the col factor (DEWITT and HELINSKI, 1965). The origin of this new DNA component was verified by using a tritium-labelled colicinogenic donor and a C^{14}-labelled noncolicinogenic recipient. The DNA of the new col$^+$ derivative of *Proteus mirabilis* contained C^{14} whereas the DNA band representing the col factor DNA contained tritium. It was thus demonstrated unequivocally that col factors, like the F factors, were composed largely of DNA.

3. Resistance factors or multiple drug resistance factor.

A third variety of sex factors found in bacteria includes the multiple drug resistance factors (R-factors). R-factor was described in 1959 when it was discovered that multiple drug resistance for streptomycin, chloramphenicol, tetracycline and in some cases sulfonamide could be transferred from resistant *Shigella* strains to sensitive strains of *E. coli* by physical contact between the cells (see WATANABE, 1963 for a complete review). The property of multiple drug resistance was shown to be controlled by an episome like factor known variously as resistance factor R-factor (WATANABE and FUSAKAWA, 1960; NAKAYA et al., 1960).

The R-factors contain another region which promotes conjugation between the *Enterobacteriaceae.*

Multiple drug resistance factors can be transferred to many members of the Enterobacteriaceae including *Proteus* and *Serratia marcescens* (Harada et al., 1960; Lebek, 1963) through conjugation and are known to replicate independently of the bacterial chromosome. The introduction of R-factor alters certain physiological properties of the recipient bacteria. If, for example, *E. coli* K12 F^+ or F'^+ strains are employed as recipients of R-factors, the presence of F inhibits the acceptance of R (Watanabe and Fusakawa, 1960, 1961), and likewise the entry of F and its variants is suppressed if the cell already contains R (Jacob and Adelberg, 1959). This competitive inhibition between the two episomes, R-factors and the F family, is suggestive of a common mechanism of transfer and perhaps the involvement of the same reaction sites. Male strains of *E. coli* K12 which are normally sensitive to bacteriophages f1 and f2 become resistant if they also contain the R-factors (Watanabe et al., 1962).

R-factors like the F factors are also believed to be composed of DNA. This is evident from the fact that the transfer of R-factor to *Proteus mirabilis* is accompanied by the appearance of additional DNA components, apparent by density gradient centrifugation on CsCl. The episomal DNA is distinguished from the normal chromosomal DNA of *Proteus* by differences in buoyant density and base ratios. The removal of the episome by treatment with acridine orange results in the disappearance of the DNA band associated with R-factor (Falkow et al., 1966). The number of R-factors harboured by the bacterial cell differs from species to species. In *E. coli* and *Serratia marcescens* there seems to be only one copy of R per bacterial chromosome but *Proteus mirabilis* contans as many as 10—12 R-factors for each bacterial chromosome (Rownd et al., 1966).

IV. Physiology of mating process in bacteria.

The mating process in bacteria consists of three successive stages:

1. pairing of compatible cells;
2. establishment of cellular contact; and
3. the transfer of genetic material from the male to the female cell.

1. Pairing of compatible cells.

The pairing of cells is brought about by contact between compatible cells. Contact between cells can be effected by mixing cell populations in appropriate proportions and permiting random collisions between cells. Collisions between compatible cells in some instances are expected to lead to effective pairing. The rate at which collisions between cells occur is proportional, up to a limited extent, to the concentration of the two mating types of bacteria (Hayes, 1957). The formation of effective pairs is necessarily a prerequisite to the onset of conjugation. As the methods used in calculating the percentage of conjugation are based on an analysis of the products of recombination, there is no measure available of assessing the efficiency of pairing. A distinction must, however, be made between the number of actual contacts and the proportion of effective pairings for a standardized sample size. It must be realized that the number of effective contacts is far exceeded by the proportion of actual contacts. As to what proportion of actual contacts yield effective pairing would depend on a number of physiological and chemical factors.

2. Establishment of cellular contact.

The establishment of cellular contact follows the formation of effective pairing. Cell contact is an expression of the surface properties of the bacterial strains participating in a cross. The surface of the female bacterium usually has one or more receptive sites corresponding to specific sites on the surface of the male cell. Differences in the surface structure of different mating types had been anticipated for a long time but only recently convincing proof of the presence of specific appendages on the male cells has been obtained (BRINTON et al., 1964). These male specific appendages, F-pili, determine the sites for the point of contact between the sexually compatible cells. The male cells, or more accurately donor cells of all types are susceptible to bacteriophages of a kind which require attachment to the F-pili, the female cells being resistant to attack by the same phages (LOEB, 1960).

It is very likely that F-pili represent the points of contact between the male and the female cell and act as channels for the transport of male chromosome to the female cell. Pairing of cells can hence be pictured as a sequence of events starting with the random contact between compatible cells, orientation of the cells to bring the free end of F-pili in contact with a specific site on the female cell surface and an enzymatic reaction to catalyze the dissolution of the female cell wall at the point of contact with the pilus.

3. Transfer of genetic material during conjugation.

a) Kinetics of transfer.

The kinetics of the transfer of genetic material were first studied by HAYES (1955, 1957). He crossed Hfr donor bacteria of the genetic constitution thr^+ leu^+ met^- $T6^s$ with recipient bacteria of the genotype thr^- leu^- met^+ F^- $T6^r$. Conjugation was initiated by mixing the two types of bacteria under optimal conditions of temperature and pH. Small aliquots were withdrawn from the mixture at regular intervals during conjugation, diluted and plated to select Thr^- leu^+ recombinant progeny. In a parallel set of experiments he treated the mixture with bacteriophage T6 before dilution and plating, thus destroying the donor cells and interrupting the process of mating at regular intervals. These experiments of HAYES and those of JACOB and WOLLMAN established that the transfer of genetic material from the donor to the recipient is a partial oriented, unidirectional process (Fig. 2). Zygotes showing a recombinant genotype were recovered not earlier than 10 minutes had elapsed. This lag period is required for the cell to establish conditions imperative for transfer to occur.

The entire process of the transfer of chromosomal markers of an Hfr strain takes approximately 107 minutes at 37^0 C. The speed of chromosome transfer is believed to remain constant so that the probability of chromosome breakage per unit chromosome length is also constant. A gradient of recombination frequencies is therefore observed, the markers located closest to the origin appearing in recombinants with very high frequencies and distal markers with an extremely low frequency.

The presence of F-pili has been claimed to be necessary for chromosome transfer (BRINTON et al., 1964). Removal of F-pili from the male cell is accompanied by a drastic reduction in the number of recombinant zygotes recovered. The regeneration of F-pili and the appearance of the ability to transfer the chromosome are correlated. Assuming that F-pili have the same or closely similar dimensions to type I pili, it is conceivable that a chromosome could migrate through the axial hole which has a diameter just large enough to accommodate a molecule of double-stranded DNA. BRINTON et al. (1964) have cast serious doubts on the existence of the so called "conjugation bridge" as seen in the earlier electron micrographs of bacterial pairs during mating. According to their results it is likely that the "conjugation bridges" are mere artifact of the techniques of electron microscopy.

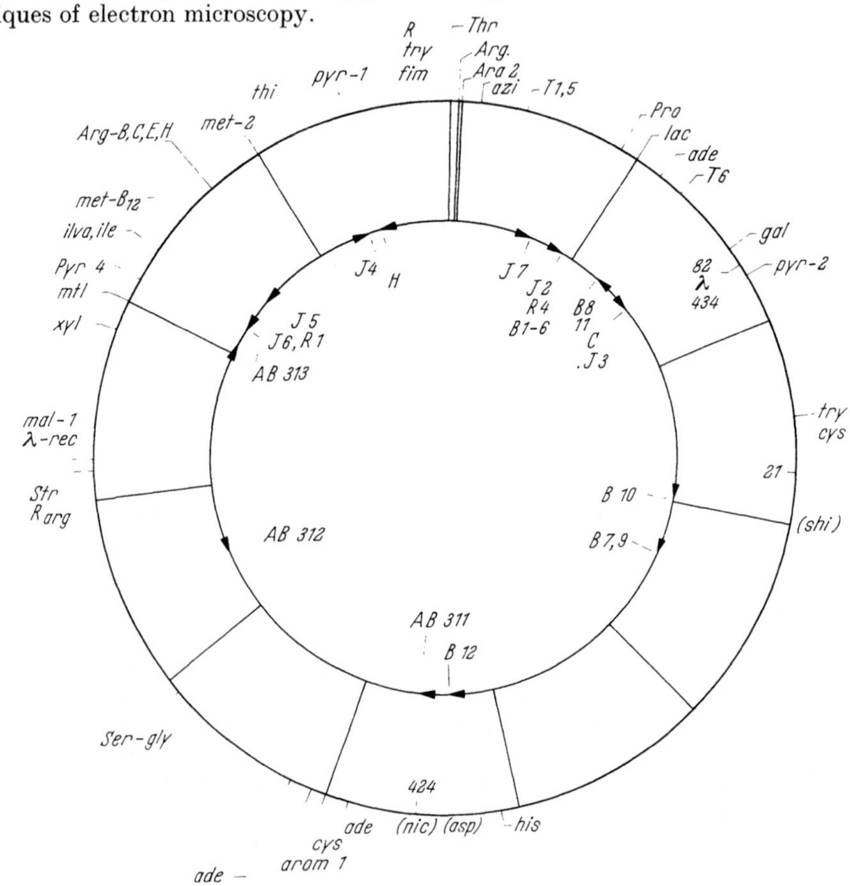

Fig. 2. The circular linkage group of *E. coli* K-12. (After HAYES, 1965.) Symbols represent: *ade*, adenine; *ara*, arabinose; *arg*, arginine; *arom*, aromatic amino acids; *asp*, aspartate; *azi*, sodium azide resistance/sensitivity; *cys*, cysteine; *fim*, fimbriation; *gal*, galactose; *his*, histidine; *ile*, isoleucine; *ilva*, isoleucine + valine requirement; *lac*, lactose; *leu*, leucine; *lys*, lysine; *λ-rec*, receptors for phage λ; *mal*, maltose; *met*, methionine; *met-B_{12}*, methionine + vitamin B_{12} requirement; *mtl*, mannitol; *nic*, nicotinamide; *pyr*, pyrimidine; R_{arg}, regulator locus for arginine pathway; R_{try}, regulator locus for tryptophan pathway; *ser-gly*, serine + glycine requirement; *shi*, shikimic acid; *str*, streptomycin resistance/sensitivity; *T 1, 5*, resistance sensitivity to phages T 1 – T 5; *T 4, T 6*, phage T 4, phage T 6, resistance/sensitivity; *thi*, thiamine (vitamin B_1); *try*, tryptophan; *xyl*, xylose. 82, λ, 434, 21, 424, 18 are the sites of attachment of the respective prophages. Hfr strains: AB 311, AB 312, AB 313, B 1—12, H, C, J 2—6, R, and R 4.

b) Transfer of sex factors.

Data obtained from interrupted mating experiments give an approximation of the time necessary to transfer some of the sex factors. SNEATH and LEDERBERG (1961) confirmed the findings of JACOB and WOLLMAN (1955, 1961) that a minimum time of 4 minutes, at 37° C was required for the transmission of F factor from F^+ to F^- cells. Col I factors appear to require as little as 2.5 minutes (STOEKER et al., 1963) and R factors are transmitted within 5 minutes.

The rate of transfer of F is considerably reduced in the presence of periodate (SNEATH and LEDERBERG, 1961).

V. Chromosome replication and the mechanism of transfer.

1. Replication of the bacterial chromosome.

The bacterial chromosome or the chromosome of F^- and F^+ bacteria consists of a single double-stranded DNA molecule, 700 to 900 μ long without any free ends (CAIRN, 1963). A circular linkage map of F^+ (Fig. 3) had been established as early as 1956 by JACOB and WOLLMAN. Hfr chromosome, on the other hand, is a linear structure expressing itself as an open genetic map. The chromosomes of all the three strains are stable genetical units. The F^+ cell, in addition, possesses many copies of F, which is also a DNA molecule and probably also circular. The bacterial chromosome and F replicate at least once per generation, one copy going to each of the daughter cells. By autoradiography of tritium-labelled *E. coli* chromosomes, CAIRNS (1963) succeeded in elucidating the mechanism of replication of the bacterial chromosome. The main features of the replication process are, that semiconservative replication is initiated at one point only without opening of the circular structure; the two daughter rings are held together at one point of attachment. The occurrence of a single growing point which moves along the chromosome (CAIRNS, 1963; YOSHIKAWA and SUEOKA, 1963) has been verified by other investigators too (NAGATA, 1963; BONHOEFFER and GIERER, 1963). As a result of initiation of replication at any given point both the DNA strands are copied simultaneously, although they are of opposite polarity. The direction of replication is the same in both the strands. This means that *in vivo* two types of DNA polymerases must function simultaneously, one of them catalyzing replication, say, in the $3' \rightarrow 5'$ direction and the other in the $5' \rightarrow 3'$ direction. DNA synthesis is known to occur throughout the entire division cycle (MAALOE, 1961).

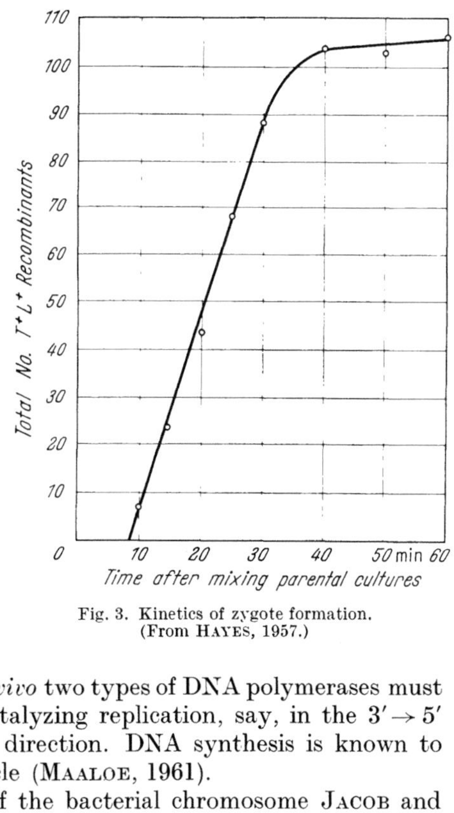

Fig. 3. Kinetics of zygote formation. (From HAYES, 1957.)

To explain the mode of replication of the bacterial chromosome JACOB and BRENNER (1963) invoked the "replicon model". According to the replicon hypothesis, a single genetic element, a chromosome or an episome, which can replicate independently constitutes a *replicon*. A replicon is defined as a unit of replication. Implicit in the hypothesis is the assumption that a replicon contains the necessary control mechanisms for the regulation of its own replicative activity. These controls include an *initiator*, a product of a structural gene, which is responsible for the activation of a specific site, the *replicator*, on the chromosome. Replicator is not only the starting point of DNA synthesis but it also provides a mechanism for opening the chain, as well as the rotation and stepwise unwinding of the unduplicated region of the molecule. The initiator is specific in its reaction with the replicator and it seems to operate by virtue of a built in "site", analogous to the "active site" of an enzyme. The active "site" of an initiator must correspond to a restricted number of replicators on the chromosome. Further, the sex factor constitutes a separate replicon and replicates independently of the bacterial chromosome and presumably has its own initiator and replicators. In an integrated state, in the Hfr chromosome the two replicons are integrated and the sites of attachment of the F factor correspond to homologous regions on the chromosome. Each sex factor carries several different nucleotide sequences homologous to each of the several different sites on the bacterial chromosome. According to the hypothesis proposed by JACOB and BRENNER once the integration of F has occurred, the F replicon loses the capacity for independent replication and

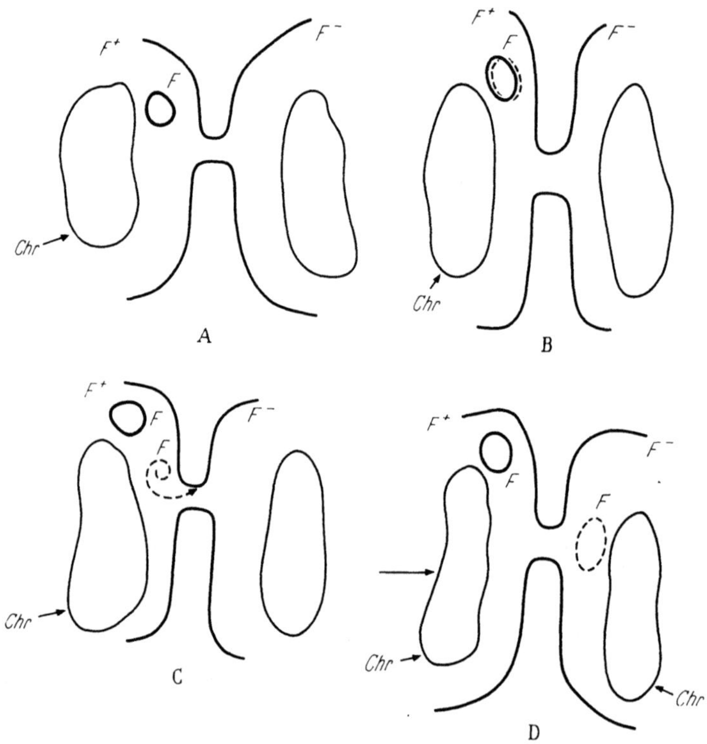

Fig. 4A—D. Transfer of F factor from F+ to F− cell. *Chr*, chromosome. A, F+ and F− cells in contact. B, Replication of F. C, Migration of F to F− recipient. D, Conversion of F− cell to a F+ donor.

becomes incorporated into the bacterial chromosomal replicon. The initiation of the replication cycles is regulated by the chromosomal system. It is further suggested that the individual replicons are attached to a site on the bacterial membrane (Fig. 1). The site of attachment might provide the site for the location of the replicator also, so that the entire chromosome moves through this point during replication. At the end of a replication cycle and before cell division, the attachment site at the membrane also divides and the eventual segregation between the two daughter replicons is completed by synthesis of membrane material in the area between the two new sites. The presence of such an attachment site can also be inferred from the findings of PRITCHARD and LARK (1964) who obtained data from studies on premature initiation of chromosome replication in *E. coli*. They derived evidence for the possible involvement of a regulatory protein, connecting the origin and the end of a chromosome. The association of the newly replicated DNA in *Bacillus subtilis* with a cell membrane has also been confirmed by GANESAN and LEDERBERG (1965).

2. Sex factor and its integration with the chromosome.

The F genome is believed to be a circular replicon like the bacterial chromosome. It is also composed of DNA and probably replicates in a manner similar to the bacterial and other replicons. F can replicate independently or as a part of the bacterial chromosome. It can form stable attachment to homologous regions on the chromosome. Ocassionally F incorporates segments of chromosomal material, adjacent to its site of attachment, resulting in a partial complex of F and bacterial genome. If now the sex factor was to dissociate from the chromosome and attain autonomy, the chromosomal determinant it carries is transferred with it at high frequency, whereas normally it would be transferred at a low frequency. This phenomenon is referred to as F-duction or sex duction (JACOB and WOLLMAN, 1961) and the F-linked chromosomal determinant is

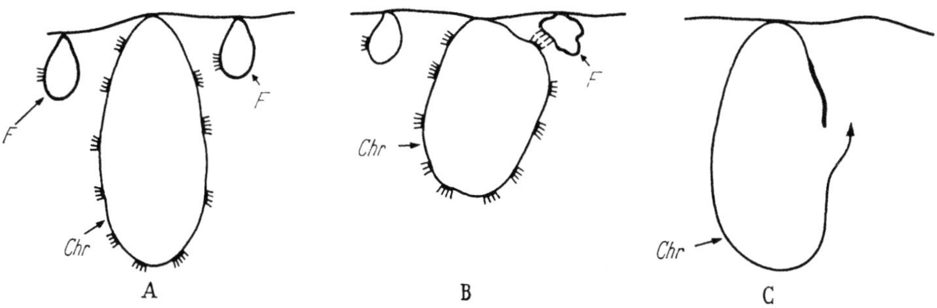

Fig. 5A—C. Integration of bacterial and F replicons. A, Chromosomal and other replicons shown attached to hypothetical attachment sites on the bacterial cell membrane. The chromosomal replicon is seen to contain several attachment sites which are homologous to a site on the F factor. B, Synapsis between the chromosomal and F replicons. C, Interaction between the two replicons.

known as the F-merozygote. Different types of F-merozygotes can arise by interaction of F with different segments of the chromosome. The size and nature of the incorporated segment varies with the site of integration.

FALKOW and CITARELLA (1965) determined the size and composition of a F-merozygote formed of F and the *lac* operon by transferring it from *E. coli* to *Proteus mirabilis* and determining the degree of molecular homology with the host genome and the *E. coli* genome, by the use of molecular hybridization techniques. DNA isolated from *Proteus* with and without the F-merozygote was "hybridized" with *E. coli* DNA. It was inferred from the results of this experiment that the DNA of *Proteus* containing the F-merozygote contained almost 4% more of DNA homologous to that of *E. coli* K12 than did the strains without the merozygote. The F genome was estimated to be equivalent to 1.9% of *E. coli* Hfr genome. The F-merozygote was 2.5% of the *E. coli* genome, indicating a higher degree of homology between the two. The existence of homology between the F factor and the bacterial chromosome provides a means of attachment to the chromosome and the formation of the Hfr chromosome. Assuming that a particular F factor can become linked to any one of a number of loci on the bacterial chromosome, all characterized by regions of homology with the F genome, it is reasonable to postulate that the F and chromosomal genomes react at random and that all the sites are potential reaction sites (Fig. 5). An hypothesis to account for the integration of F and the chromosome based on the incorporation of the F genome by reciprocal crossing over or breakage and reunion between the two genomes has been formulated (see CAMPBELL, 1962).

3. The mechanism of chromosome transfer during conjugation.

The mobilization of the bacterial chromosome preceding the transfer process is a direct consequence of the incorporation of the sex factor into the bacterial chromosome. Chromosomal transfer is a polarized process, beginning at one end and progressing sequentially in the opposite direction (Fig. 6). This is strongly reminiscent of the manner of chromosome replication, which is also polarized moving from one point on the DNA molecule toward the others in a fixed direction. Due to a number of similarities between chromosomal transfer and replication it has generally been recognized that the two processes are closely connected. The process of transfer of the Hfr chromosome has baffled investigators for quite a while.

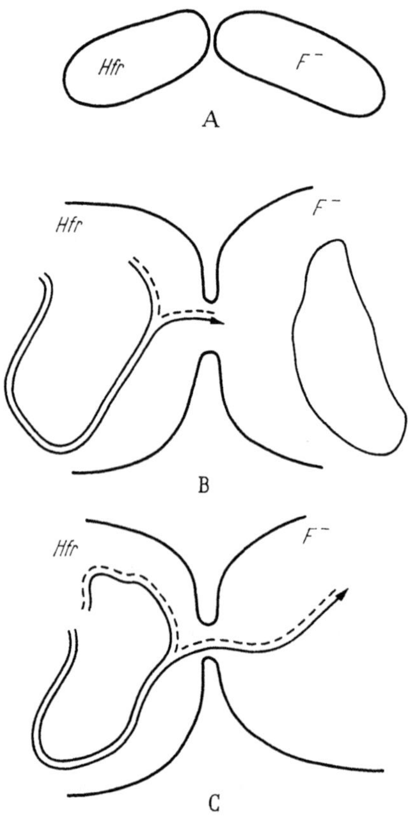

Fig. 6A—C. Transfer of Hfr chromosome during conjugation. A, Pair formation between Hfr and F^- cells. B, Establishment of cell contact and initiation of chromosomal replication. The newly synthesized DNA is represented by a broken line. C, A late stage in chromosomal transfer.

In general, two different kinds of models have been proposed to account for the sequence of events that occur during the transfer of chromosome of the Hfr type of bacteria. The first model proposed by JACOB and BRENNER (1963) is based on the premise that replication is essential to the process of transfer and that replication is initiated in the male cell in response to the stimulus of conjugation, one of the replicas is then transferred to the female cell. This hypothesis presupposes that the force or the energy necessary for the movement of the Hfr chromosome is supplied by the replication process itself. The requirement of some form of energy, by the male cell, to mediate chromosome transfer was recognized by FISHER (1957).

According to the second model proposed by BOUCK and ADELBERG (1963) the transfer process can occur independently of replication and the energy required for chromosomal movement is believed to be derived from a source other than replication. The basic difference between the two models lies in the time relationships between replication and transfer. If the JACOB-BRENNER hypothesis was correct it should be impossible for transfer to proceed under conditions inhibitory to DNA synthesis. If such conditions prevailed immediately preceding and during conjugation, no transfer should occur. However, if normal conditions were restored during conjugation, normal chromosomal transfer should commence. On the other hand, if the BOUCK-ADELBERG model was correct, then inhibition of DNA synthesis immediately preceding conjugation should eliminate subsequent transfer. A reversion to experimental conditions, conducive to replication during conjugation, should not result in a resumption of transfer. The questions which must be settled are, whether DNA synthesis plays any role in the transfer process and if so whether it occurs prior to or concurrently with the transfer process.

The main features of the JACOB-BRENNER model are, a) replication of the Hfr chromosome in a population of nonconjugating cells is initiated at the chromosomal replicator, b) the event of conjugation stimulates replication at a second locus, the F-replicator, c) the conjugation induced replication is such that the origin is duplicated first and d) one of the newly formed daughter replicas moves to the female cell, the energy for this movement being provided by the synthesis of new DNA.

According to the BOUCK-ADELBERG model (1963), on the other hand, replication is postulated to begin always at the F factor and proceed in a direction away from the F and towards the origin. With the commencement of mating in a population, the cycle of replication which had been initiated before the onset of mating, is completed. Following completion of the replication cycle, ring closure does not occur in the new replica and the free end, which is incidently also the origin, begins to migrate into the recipient cell. This situation demands that the youngest part of the DNA molecule be the first to enter the female cell and that the oldest portion of the Hfr daughter replica comprising the F end almost never do so.

JACOB and BRENNER's hypothesis derives strong support from data obtained from an experiment in which Hfr bacteria, labelled with tritiated thymidine and C^{13} were mated with nonlabelled F^{-} bacteria in nonradiactive, C^{12} medium. It was observed that if transfer was permitted to occur, and the transferred DNA analysed, a very large fraction of the DNA appeared to be intermediate in density between the Hfr DNA and that of the entire population. If, however, the cells were permitted to mate but not to transfer their chromosomes, a major proportion of the labelled DNA had density equivalent to that of the Hfr bacteria. This shows that the transferred DNA consisted of DNA molecules which had been formed by replication, after mixing of the male and female populations. Any DNA synthesized after initiation of mating would have to be "light", hence the "hybrid" density of the transferred molecule. Furthermore, "hybrid" DNA could only be formed if replication and transfer were inseparable and proceeded simultaneously.

The most vital test of the accuracy of either of the two models rests on the question as to whether *any* transfer could occur in the absence of DNA replication. The JACOB-BRENNER hypothesis explicitly states that *no* transfer can occur under these circumstances. The BOUCK-ADELBERG hypothesis assumes that transfer could occur independently of DNA synthesis. Phenethyl (PEA) alcohol is an inhibitor of DNA synthesis without any appreciable effect on RNA and protein syntheses (BERRAH and KONETZKA, 1962). In experiments designed to elucidate the relationship between replication and transfer, BOUCK and ADELBERG (1963) interrupted synthesis after initiation of transfer in order to inhibit the transfer of late markers. They did record a small amount of transfer in the absence of DNA synthesis. JACOB, BRENNER and CUZIN (1963), too, observed transfer of late markers after addition of PEA but they attributed this transfer to the occurrence of DNA turnover and residual DNA synthesis.

To further understand the mechanism of PEA inhibition of DNA synthesis ROESER and KONETZKA (1964) treated *E. coli* Hfr cells with PEA to the point when all DNA synthesis was prevented. Now if PEA was removed cells resumed DNA synthesis, but once resumption of synthetic activity had occurred, further addition of PEA failed to prevent DNA synthesis until the previously initiated replication cycle had been completed. Extending these findings to the role of DNA synthesis in transfer of the chromosome they demonstrated that for chromosomal transfer to occur it was essential that a new replication cycle be initiated *after* the establishment of cellular contact. This means that DNA synthesis *per se* is entirely irrelevant and only that replication cycle, which begins in response to the mating stimulus, gives rise to the appropriate replica capable of migrating into the recipient cell. This explains the continuation of transfer even after PEA treatment in the experiments of JACOB et al. (1963), BOUCK and ADELBERG (1963).

Another crucial test of the predictions made by the two models would be to devise experiments to distinguish between the DNA of the male cells, synthesised prior to and during mating. The ratio of the amount of "*old*" and "*new*" DNA donated to the female cell by the Hfr would be the most important deciding factor between the two models. An analysis of chromosome transfer was performed by labelling the Hfr DNA with tritiated thymine prior to mating and/or by labelling DNA made during mating (GROSS and CARO, 1966). To eliminate the possibility of any DNA synthesis by the female cell, an adenine requiring F^- was employed and it was ascertained that the *ade* gene did not pass on to the female cell during the course of the experiment. The Hfr strain used was thymine requiring. A special advantage of using these particular strains was derived from the morphological difference between the Hfr and F^-, under the electron microscope F^- cells appeared ovoid whereas the Hfr strain was rod shaped. This morphological difference was exploited in the subsequent quantitative autoradiography. In one set of experiments DNA was labelled prior to mating and the labelling terminated at the commencement of mating. In the second experiment, labelling was permitted to continue even during mating. The zygotes from the second experiment showed twice as much of radioactivity as did the zygotes from the first experiment, where radioactivity was withheld from the medium during mating. In the latter instance only one strand of the incoming DNA was radioactive and the "new" strand synthesised during conjugation was nonradioactive. However, if mating was also conducted in radioactive medium both "old" and "new" strands were radioactive. This conclusion was verified by a further extension of the same experiment. On performing a cross between "cold" Hfr cells and adding tritiated thymine 20 minutes after mating had ensued, the F^- recipients showed an equal amount of label in the "new" and "old" DNA received following the 20 minute period. All this strengthens the hypothesis that the DNA which is transferred to the female cells consists of one strand synthesised in the Hfr donor prior to the commencement of mating and a second complementary strand synthesised during mating.

At present it is difficult to choose between the JACOB-BRENNER and the BOUCK-ADELBERG models. Overwhelming evidence, however, suggests that in the integrated state (Hfr chromosome) the bacterial or chromosomal replicon is primed by the F replicator and that the chromosomal replicators are nonfunctional. When mating begins a new replication cycle is initiated in the Hfr and the chromosome is replicated semiconservatively, one daughter replica going to the female cell and the other one remaining in the donor. DNA synthesis *per se* is entirely immaterial to the process of transfer but DNA synthesis induced by cell contact is essential for chromosomal transfer.

VI. Conjugation in bacteria other than *Escherichia coli*.

1. *Enterobacteriaceae.*

Conjugation has been reported in bacterial genera other than *Escherichia coli*. Some of the other Enterobacteriaceae, which have been extensively investigated, are *Salmonella* and *Shigella*. The sexual systems encountered in these genera are based on the introduction of F and its variants from *E. coli* K12. The F factors confer the ability to conjugate on these recipient strains, as well as the ability to transfer the F back to the original *E. coli* donor. The F has been transferred from *E. coli* to *Salmonella* species (Makela et al., 1962). Both the wild type F as well as F′ with an attached segment of chromosome can be transferred to *Salmonella*. The F^+ *Salmonella* derivative can now transfer its acquired F both to other *Salmonellas* and to *E. coli*. In the wild state all *Salmonellas* appear to be F^- and are, therefore, unable to initiate conjugation. On acquiring F from *E. coli* the converted *Salmonella* strains behave in a manner very similar to the donor strain.

On transfer of the recombination system of *E. coli* to *Salmonella abony*, Makela (1963) observed that the *E. coli* F established a stable relationship with *Salmonella* chromosome forming several Hfr types, including some with special sex factor affinity loci (sfa), which could be distinguished on basis of the sequence of the transferred markers and the characteristic origins. They yielded recombinants with a frequency of up to 2 percent.

Several of the F variants can also be established in strains of *Salmonella*, *Shigella*, *Serratia* and *Klebsiella* by transfer from *E. coli* and also by transfer between these genera (Falkow et al., 1961). $F_{13\ \mathrm{stable}}$ is able to infect 23 different strains of *Salmonella*, *Shigella*, *Serratia* and *Klebsiella*.

2. Conjugation in *Pseudomonas aeruginosa*.

Conjugation in *Pseudomonas aeruginosa* is controlled by a transmissible factor termed FP (Holloway, 1955; Holloway and Jennings, 1958). The strains possessing FP (FP^+) are very similar in behaviour to the F^+ strains of *E. coli* and are able to donate FP to recipient strains devoid of FP. The FP^+ strains are resistant to the action of acriflavine and acridine orange and there is no conversion of FP^+ to FP^- in the presence of these two dyes. Crosses between two FP^- strains are sterile, those between FP^- and FP^+ and two FP^+ yielded recombinants. Evidence indicates that there is a unidirectional transfer of chromosomal material from FP^+ donor to FP^- recipients and the transfer is only partial as in the case of *E. coli*, for it is observed that the recombinant progeny inherit the unselected markers of the recipient parent. By means of genetic crosses and analysis of recombinant progeny it has been concluded that the linkage group of *Pseudomonas* is organized in the form of a single chromosome, as in *E. coli*.

3. Conjugation in *Vibrio cholerae*.

Strains of *Vibrio cholerae* can be divided into two classes, equivalent to donor and recipient strains of the other bacterial genera, mentioned in the foregoing pages. The sexual determinant, P, plays the role of the conjugation factor and crosses of P^+ and P^- strains yield recombinants while those between two P^- strains are, of course, sterile (Bhaskaran, 1958, 1960; Bhaskaran and Iyer, 1961). The fertility factor also carries a locus responsible for the formation of lytic substances and contact between two compatible cells can lead to the transfer of fertility and the determinant for synthesis of lytic substances. The lytic substances are very similar to the bacteriocins (Bhaskaran, 1960). Transfer of genetic material is unidirectional from P^+ to P^- and incomplete. Details of the linkage group of *Vibrio* are not available currently.

VII. Concluding remarks.

It is apparent from a consideration of the process of conjugation in bacteria that most of the experimental work and investigations into the mechanism of the action of the sex factor, integration with the chromosome and transfer mechanism have been confined to *Escherichia coli* K12. Recombination systems operating in other bacterial species will doubtlessly reveal additional genetic mechanisms. It is more than likely, however, that the *E. coli* system will provide a fundamental pattern liable to be used as a basis for interpretation of other schemes, which could be regarded as variants of the basic *E. coli* scheme. Certain generalizations seem warranted in this connection, namely, that all transferrable sex factors behave in a similar manner; some of them incorporate segments of chromosomal material only if they contain regions in their genome homologous to a few or several points on the bacterial genome; the bacterial chromosome, on the other hand, may contain several potential sites where attachment with the sex factor could occur; all these sites are equivalent and that the attachment and/or genetic exchange with a given sex factor is conditioned by the nucleotide sequence of the sex factor genome. Synapsis occurs between regions of homology on the two genomes, and as a consequence of synapsis genetic exchange may occur by breakage and reunion of DNA strands. Alternatively, the sex factor genome may be incorporated into the chromosome or more accurately the chromosome may be incorporated into the sex factor, depending upon the preferred interpretation of the integration process.

The question of chromosomal transfer between the Hfr and F^- bacteria, though debated at length, still remains unsettled. In view of the data currently available the following picture of conjugation and chromosome transfer emerges:

1. In an Hfr cell chromosomal duplication is controlled by the F replicator so that replication starts at the position of attachment of the F and moves towards the origin.
2. The origin is the last point to be duplicated.
3. Interrupted mating experiments show that origin is the first point to enter the female cell.
4. During conjugation, DNA replication must occur if chromosomal transfer is to proceed. However, DNA synthesis *per se* is irrelevant and only that DNA, whose synthesis is initiated after contact between Hfr and F^-, is crucial to the transfer process.
5. If DNA synthesis is inhibited during conjugation chromosomal transfer is inhibited concurrently.
6. It follows as a corollary of 3 that since the replication of DNA is semiconservative the DNA transferred to the female cell contains one strand of "old" Hfr DNA (preconjugation) and a second complementary strand derived from synthesis during conjugation.

Literature.

Adelberg, E. A., and S. N. Burns: Genetic variation in the sex factor of *Escherichia coli*. J. Bact. **79**, 321—330 (1960). — Adelberg, E. A., and J. Pittard: Chromosome transfer in bacterial conjugation. Bact. Rev. **29**, 161—172 (1965). — Afoldi, L. F., F. Jacob, E. L. Wollman et R. Mazé: Sur le déterminisme génétique de colicinogenie. C. R. Acad. Sci. (Paris) **246**, 3531—3533 (1958). — Amati, P.: Vegetative multiplication of colicinogenic factors after induction by *E. coli*. J. molec. Biol. **8**, 239—246 (1964). — Anderson, T. F.: In: The nature of bacterial surface (eds. A. A. Miles and N. W. Pirie), p. 76. Oxford: Blackwell Sci. Publ. 1949.

Baron, L. S., W. F. Carey, and W. M. Spilman: Characteristics of a high frequency of recombination (Hfr) strain of *Salmonella typhosa* compatible with *Salmonella*, *Shigella* and *Escherichia* species. Proc. nat. Acad. Sci. (Wash.) **45**, 976—983 (1959). — BARON, L. S., and S. FALKOW: Genetic transfer of episomic elements from *Salmonella typhosa* to *Vibrio cholerae*. Rec. Genet. Soc. Amer. **30**, 59 (1961). — BARON, L. S., W. M. SPILMAN, and W. F. CAREY: Diploid heterozygous hybrids from matings between *Escherichia coli* and *Salmonella typhosa*. J. exp. Med. **112**, 361—372 (1960). — BERRAH, G., and W. A. KONETZKA: Selective and reversible inhibition of the synthesis of deoxyribonucleic acid by phenethylalcohol. J. Bact. **83**, 738—744 (1962). — BHASKARAN, K.: Genetic recombination in *Vibrio cholerae*. J. gen. Microbiol. **19**, 71—75 (1958). — Recombination of characters between mutant stocks of *Vibrio cholerae*, strain 162. J. gen. Microbiol. **23**, 47—54 (1960). — BHASKARAN, K., and S. S. IYER: Genetic recombination in *Vibrio cholerae*. Nature (Lond.) **189**, 1030 (1961). — BONHOEFFER, F., and A. GIERER: On the growth mechanism of bacterial chromosome. J. molec. Biol. **7**, 534—540 (1963). — BOUCK, N., and E. A. ADELBERG: The relationship between DNA synthesis and conjugation in *Escherichia coli*. Biochem. biophys. Res. Commun. **11**, 24—27 (1963). — BRINTON, C. C.: Les filaments capsulaires de *Escherichia coli* et leurs manifestations biologiques. Experientia (Basel) **13**, 55 (1957). — BRINTON, C. C., and L. S. BARON: Transfer of piliation from *Escherichia coli* to *Salmonella typhosa* by genetic recombination. Biochim. biophys. Acta (Amst.) **42**, 298—311 (1960). — BRINTON, C. C., A. BUZZELL, and M. A. LAUFFER: Electrophoresis and phage susceptibility studies on a filament-producing variant of the *E. coli* B bacterium. Biochim. biophys. Acta (Amst.) **15**, 533—542 (1954). — BRINTON, C. C., and M. J. STONE: Chemical composition of bacterial pili — a fibrous protein. Bact. Proc. 96 (1961). — BRINTON jr., C. C.: Non-flagellar appendages of bacteria. Nature (Lond.) **183**, 782—786 (1959). — BRINTON jr., C. C., P. GEMSKI jr., and J. CARNAHAN: A new type of bacterial pilus genetically controlled by the fertility factor of *Escherichia coli* K 12 and its role in chromosome transfer. Proc. nat. Acad. Sci. (Wash.) **52**, 776—783 (1964). — BRINTON jr., C. C., P. GEMSKI jr., S. FALKOW, and C. S. BARON: Location of the piliation factor on the chromosome of *Escherichia coli*. Biochem. biophys. Res. Commun. **5**, 293—298 (1961).

CAIRN, J.: The bacterial chromosome and its manner of replication as seen by autoradiography. J. molec. Biol. **6**, 208—213 (1963). — CAMPBELL, A.: Episomes. Advanc. Genet. **11**, 101—146 (1962). — CAVALLI-SFORZA, L. L.: La sessulita nei batteri. Boll. Ist. sieroter. milan. **29**, 281—289 (1950). — CAVALLI-SFORZA, L. L., and J. L. JINKS: Observations on the genetic and mating system of *E. coli* K 12. Proc. 9th Intern. Congr. Genetics, Bellagio 1953, p. 967—969. — CAVALLI-SFORZA, L. L., J. LEDERBERG, and E. M. LEDERBERG: An infective factor controlling sex compatibility in *Bacterium coli*. J. gen. Microbiol. **8**, 89—103 (1953). — CLOWES, R. C.: Colicine factors as fertility factors in bacteria: *Escherichia coli* K-12. Nature (Lond.) **190**, 988—989 (1961). — CLOWES, R. C., and D. ROWLEY: Some observations on linkage effects in genetic recombination in *E. coli* K 12. J. gen. Microbiol. **11**, 250—260 (1954). — CRAWFORD, E. M., and R. F. GESTELAND: The adsorption of bacteriophage R-17. Virology **22**, 165—167 (1964).

DETTORI, R., G. A. MACCACRO, and G. L. PICCININ: Sex-specific bacteriophages of *Escherichia coli* K 12. G. Microbiol. **9**, 141—150 (1961). — DEWITT, W., and D. R. HELINSKI: Characterization of colicinogenic factor E 1 from a non-induced and mitomycine C-induced *Proteus* strain. J. molec. Biol. **13**, 692—703 (1965). — DOVE, W. F., and J. J. WEIGLE: Intracellular state of the chromosome of Bacteriophage Lambda. I. The eclipse of infectivity of the bacteriophage DNA. J. molec. Biol. **12**, 620—629 (1965). — DRISKELL-ZAMENHOF, P. J., and E. A. ADELBERG: Studies on the chemical nature and size of sex factors of *Escherichia coli* K 12. J. molec. Biol. **6**, 483—497 (1963). — DUGUID, J. P., and R. R. GILLIES: Fimbriae and adhesive properties in dysentry bacilli. J. Path. Bact. **74**, 397—411 (1957). — Fimbriae and haemagglutinating activity in *Salmonella, Klebsiella, Proteus* and *Chromobacterium*. J. Path. Bact. **75**, 519—529 (1958). — DUGUID, J. P., I. W. SMITH, G. DEMPSTER, and P. N. EDMONDS: Non-flagellar filamentous appendages ("Fimbriae") and haemagglutinating activity in *Bacterium coli*. J. Path. Bact. **70**, 335—348 (1955).

FALKOW, S., and L. S. BARON: Episomic elements in a strain of *Salmonella typhosa*. J. Bact. **84**, 581—589 (1962). — FALKOW, S., R. V. CITARELLA, J. A. WOHLHIETER, and T. WATANABE: The molecular nature of R-factors. J. molec. Biol. **17**, 102—116 (1966). — FALKOW, S., J. MARMUR, W. F. CAREY, W. M. SPILMAN, and L. S. BARON: Episomic transfer between *Salmonella typhosa* and *Serratia marcescens*. Genetics **46**, 703—706 (1961). — FALKOW, S., J. A. WOHLHIETER, R. V. CITARELLA, and L. S. BARON: Transfer of episomic elements to *Proteus*. I. Transfer of F-linked chromosomal determinants. J. Bact. **87**, 209—219 (1964). — FISHER, K. W.: The role of krebs cycle in conjugation in *Escherichia coli* K-12. J. gen. Microbiol. **16**, 120—135 (1957). — FREDERICQ, P.: The colicins. Ann. Rev. Microbiol. **11**, 7—22 (1957). — The biological replication of macromolecules, p. 104. Cambridge: Cambridge University Press 1958. — On the nature of colicinogenic factors: a review. J. theor. Biol. **4**, 159—165 (1963).

GANESAN, A. T., and G. LEDERBERG: A cell-membrane bound fraction of bacterial DNA. Biochem. biophys. Res. Commun. **18**, 824—834 (1965). — GROSS, J.: Conjugation in bacteria.

In: I. C. Gunsalus and R. Y. Stanier (ed.), The bacteria, vol. 5. New York: Academic Press 1964. — Gross, J. D., and L. Caro: Genetic transfer in the bacterial mating. What mechanism insured the orderly transfer of DNA. Science **150**, 1679—1684 (1965). — DNA transfer in bacterial conjugation. J. molec. Biol. **16**, 269—284 (1966).

Hakura, A., N. Otsuji, and V. Hirota: A temperate phage specific for female strains of *Escherichia coli* K-12. J. gen. Microbiol. **35**, 69—73 (1964). — Harada, K., M. Suzuki, H. Kameda, and S. Mitsuhashi: On the drug resistance of enteric bacteria. 2. Transmission of the drug resistance among Enterobacteriaceae. Jap. J. exp. Med. **30**, 289—299 (1960). — Hayes, W.: Genetic recombination in *E. coli* K 12: Analysis of the stimulating effect of ultraviolet light. Nature (Lond.) **169**, 1017—1018 (1952a). — Genetic recombination in *Bact. coli* K 12: unidirectional transfer of genetic material. Nature (Lond.) **169**, 118—119 (1952b). — Observations on a transmissible agent determining sexual differentiation in *Bact. coli*. J. gen. Microbiol. **8**, 72—88 (1953a). — Mechanism of genetic recombination in *E. coli*. Cold Spr. Harb. Symp. quant. Biol. **18**, 75—93 (1953b). — The kinetics of the mating process in *Escherichia coli*. J. gen. Microbiol. **16**, 97—119 (1957). — Conjugation in *Escherichia coli*. Brit. med. Bull. **18**, 36—40 (1962). — The genetics of bacteria and their viruses. Oxford: Blackwell Sci. Publ. 1965. — Hewitt, R., and D. Billen: Reorientation of chromosome replication after exposure to ultraviolet light of *Escherichia coli*. J. molec. Biol. **13**, 40—53 (1965). — Hirota, Y.: Mutants of the F factor in *Escherichia coli* K-12. Rec. Genet. Soc. Amer. **28**, 75 (1959). — The effect of acridine dyes on mating type factors in *Escherichia coli*. Proc. nat. Acad. Sci. (Wash.) **46**, 57—64 (1960). — Hirota, Y., and T. Iijima: Acriflavine as an effective agent for eliminating F factors in *Escherichia coli* K 12. Nature (Lond.) **180**, 655—656 (1957). — Hirota, Y., K. Nishimura, F. Orskov, and I. Orskov: Effect of drug-resistance factor R on the F properties of *Escherichia coli*. J. Bact. **87**, 341—351 (1964). — Hirota, Y., and P. H. A. Sneath: F′ and F mediated transduction in *E. coli* K-12. Jap. J. Genet. **36**, 307—318 (1961). — Holland, I. B.: A bacteriocin specifically affecting DNA synthesis in *Bacillus megaterium*. J. molec. Biol. **12**, 429—438 (1965). — Holloway, B. W.: Genetic recombination in *Pseudomonas aeruginosa*. J. gen. Microbiol. **13**, 572—581 (1955). — Holloway, B. W., and B. Fargie: Fertility factors and genetic linkage in *Pseudomonas aeruginosa*. J. Bact. **80**, 362—368 (1960). — Holloway, B. W., and P. A. Jennings: An infectious fertility factor for *Pseudomonas aeruginosa*. Nature (Lond.) **181**, 855—856 (1958). — Horiuchi, K., and E. A. Adelberg: Growth of male specific bacteriophage in *Proteus mirabilis* harboring F-genotes derived from *Escherichia coli*. J. Bact. **89**, 1231—1236 (1965). — Houwink, A. L., and W. van Iterson: Macromolecular mono-layer in the cell wall of *Spirillum* spec. Biochim. biophys. Acta (Amst.) **10**, 360—366 (1953).

Iijima, T.: Transfer of a colicinogenic factor with multiple resistance factor in *Escherichia coli* K 12. Jap. J. Genet. **37**, 183—193 (1962). — Ippen, K. A., and R. C. Valentine: An assay for the male substance (F-pili) of *Escherichia coli* K-12. Biochem. biophys. Res. Commun. **21**, 21—27 (1965).

Jacob, F., et E. A. Adelberg: Transfert de charactères génétiques par incorporation au facteur sexuel d'*Escherichia coli*. C. R. Acad. Sci. (Paris) **249**, 189—191 (1959). — Jacob, F., et S. Brenner: Sur la régulation de la synthèse du DNA chez les bactéries l'hypothèse du réplicon. C. R. Acad. Sci. (Paris) **256**, 289—300 (1963). — Jacob, F., S. Brenner, and F. Cuzin: On the regulation of replication in bacteria. Cold Spr. Harb. Symp. quant. Biol. **28**, 329—348 (1963). — Jacob, F., A. Lwoff, A. Siminovitch et E. L. Wollman: Définition de quelques termes relatifs à la lysogénie. Ann. Inst. Pasteur **84**, 222—224 (1953). — Jacob, F., P. Schaeffer, and E. L. Wollman: Episomic elements in bacteria. Symp. Soc. gen. Microbiol. **10**, 67—91 (1960). — In: W. Hayes and R. C. Clowes (ed.), Microbial genetics. Tenth Symp. of the Society for General Microbiology held at the Royal Institution London, April 1960, p. 67. Cambridge: Cambridge University Press 1960. — Jacob, F., L. Siminovitch et E. L. Wollman: Sur la biosynthèse d'une colicine et sur son mode d'action. Ann. Inst. Pasteur **83**, 295—315 (1952). — Jacob, L., et E. L. Wollman: Etapes de la recombinaison génétique chez *E. coli* K12. C. R. Acad. Sci. (Paris) **240**, 2566—2568 (1955). — Recombinaison génétique et mutants de fertilité chez *E. coli* K12. C. R. Acad. Sci. (Paris) **242**, 303—306 (1956a). — Sur les processus de conjugaison et de recombinaison chez *Escherichia coli*. I. L'induction par conjugaison au induction zygotique. Ann. Inst. Pasteur **91**, 486—510 (1956). — Analyse des groupes de liaison génétique de différentes souches donatrices. C. R. Acad. Sci. (Paris) **245**, 1840—1843 (1957). — Les épisomes, éléments génétique ajoutés. C. R. Acad. Sci. (Paris) **247**, 154—156 (1957). — Sexuality and the genetics of bacteria. New York: Academic Press 1961.

Lark, C., and K. G. Lark: Evidence for two distinct aspects of the mechanisms regulating chromosome replication in *Escherichia coli*. J. molec. Biol. **10**, 120—136 (1964). — Lark, K. G., and R. Bird: Premature chromosome replication induced by thymine starvation: Restriction of replication to one of the two partially completed replicas. J. molec. Biol.

13, 607—610 (1965). — LARK, K. G., and C. LARK: Regulation of chromosome replication in *Escherichia coli:* Alternate replication of two chromosomes at slow growth rates. J. molec. Biol. **13**, 105—126 (1965). — LAVALLÉ, R., et F. JACOB: Sur la sensibilité des épisomes sexuel et colicinogène d'*E. coli* K 12 à la désintégration du radiophosphore. C. R. Acad. Sci. (Paris) **252**, 1678—1680 (1961). — LEBEK, G.: Übertragung der Mehrfachresistenz gegen Antibiotika und Chemotherapeutika von *E. coli* auf andere Spezies gramnegativer Bakterien. Zbl. Bakt., I. Abt. Orig. **189**, 213—223 (1963). — LEDERBERG, J.: Prevalence of *E. coli* strains exhibiting genetic recombination. Science **114**, 68—69 (1951). — Extranuclear transmission of the F compatibility factor in *E. coli.* Abstr. 7th Int. Congr. Microbiol., Stockholm 1958, p. 59—60. — LEDERBERG, J., L. L. CAVALLI, and E. M. LEDERBERG: Sex compatibility in *E. coli.* Genetics **37**, 720—730 (1952). — LOEB, T.: Isolation of a bacteriophage specific for the F, F^+ and Hfr mating types of *Escherichia coli* K 12. Science **131**, 932—933 (1960). — LOEB, T., and N. D. ZINDER: A bacteriophage containing RNA. Proc. nat. Acad. Sci. (Wash.) **47**, 282—289 (1961). — LURIA, S. E.: Host induced modifications of bacterial viruses. Cold Spr. Harb. Symp. quant. Biol. **18**, 237—244 (1953). — LURIA, S. E., and J. W. BURROUS: Hybridization between *Escherichia coli* and *Shigella.* J. Bact. **74**, 461—467 (1957). — LUZZATI, D., et M. R. CHEVALLIER: Induction par carence en thymine, de la production de colicine par des bacteries colicogenes thymine-exigeantes. Ann. Inst. Pasteur **107**, 152—162 (1964).

MAALØE, O.: The control of normal DNA replication in bacteria. Cold Spr. Harb. Symp. quant. Biol. **26**, 45—52 (1961). — MACCACARO, G. A.: Cell surface and fertility in *Escherichia coli.* Nature (Lond.) **176**, 125—126 (1955). — MACCACARO, G. A., e R. DETTORI: Studi sulle fimbrie batteriche. IV. Metabolismo ossidativo e fermentativo in cellule fimbriate e efimbriate. G. Microbiol. **7**, 52—68 (1959). — MAKELA, P. H.: Hfr males in *Salmonella abony.* Genetics **48**, 423—429 (1963). — MAKELA, P. H., J. LEDERBERG, and E. M. LEDERBERG: Patterns of sexual recombination in enteric bacteria. Genetics **47**, 1427—1439 (1962). — MATNEY, T. S., E. P. GOLDSMITH, N. S. ERWIN, and R. N. SCROGG: A preliminary map of genomic sites for F-attachment in *Escherichia coli* K 12. Biochem. biophys. Res. Commun. **17**, 278—281 (1964). — MORSE, M. L., E. M. LEDERBERG, and J. LEDERBERG: Transductional heterogenotes in *Escherichia coli.* Genetics **41**, 758—779 (1956).

NAKAYA, R., A. NAKAMURA, and Y. MURATA: Resistance transfer agents in *Shigella.* Biochem. biophys. Res. Commun. **3**, 654—659 (1960). — NISHIMURA, Y., and Y. HIROTA: Genetic analyses of conjugal fertility (F-mating) and mutants of sex factor in *Escherichia coli* K 12. Jap. J. Genet. **37**, 403 (1962).

ØRSKOV, I., and F. ØRSKOV: An antigen termed F^+ occurring in F^+ *E. coli* strains. Acta path. microbiol. scand. **48**, 37—46 (1960). — OZEKI, H.: Abortive transduction in purine requiring mutants of *Salmonella typhimurium.* Genetic studies with bacteria. Carnegie Inst. Wash. Publ. **612**, 97—106 (1956). — OZEKI, H., and S. HOWARTH: Colicine factors as fertility factors in bacteria. Nature (Lond.) **190**, 986—988 (1961).

PITTARD, J., and E. A. ADELBERG: Gene transfer by F′ strains of *Escherichia coli* K 12. III. An analysis of the recombination events occurring in the F′ male and in the zygote. Genetics **49**, 995 (1964). — PITTARD, J., J. S. LOUTIT, and E. A. ADELBERG: Gene transfer by F′ strains of *Escherichia coli* K 12. I. Delay in initiation of chromosome transfer. J. Bact. **85**, 1394—1401 (1963). — PRELL, H. H.: DNA transfer from prophage to phage progeny after zygotic induction. J. molec. Biol. **13**, 329—339 (1965). — PRITCHARD, R. H., and K. G. LARK: Induction of replication by thymine starvation at the chromosome origin in *Escherichia coli.* J. molec. Biol. **9**, 288—307 (1964).

RICHTER, A.: Complementary determinants of an Hfr phenotype in *E. coli* K 12. Genetics **42**, 391 (1957). — Attachment of wild types F to factor to a pecific chromosomal region in a variant strain of *Escherichia coli* K 12: the phenomenon of episome alteration. Genet. Res. **2**, 333—345 (1961). — ROWND, R., R. NAKAYA, and A. NAKAMURA: Molecular nature of the drug resistance factors of the Enterobacteriaceae. J. molec. Biol. **17**, 376—393 (1966).

SCAIFE, J., and D. GROSS: The mechanism of chromosome mobilization by an F-prime factor in *Escherichia coli* K12. Genet. Res. **4**, 328 (1962). — SILVERS, S. D.: The transfer of material during mating in *Escherichia coli.* J. molec. Biol. **6**, 349—360 (1963). — SMITH, S. M., and B. A. D. STOCKER: Colicinogeny and recombination. Brit. med. Bull. **18**, 46—51 (1962). — SNEATH, P. H. A.: Sex factors as episomes. Brit. med. Bull. **18**, 41—45 (1962). — SNEATH, P. H. A., and H. LEDERBERG: Inhibition by periodate of mating in *Escherichia coli.* Proc. nat. Acad. Sci. (Wash.) **47**, 86—90 (1961).

TATUM, E. L.: X-ray induced mutant strain of *E. coli.* Proc. nat. Acad. Sci. (Wash.) **31**, 215—219 (1945). — TAYLOR, A. L., and E. A. ADELBERG: Evidence for a closed linkage group in Hfr males of *Escherichia coli* K12. Biochem. biophys. Res. Commun. **5**, 400—404 (1961).

VALENTINE, R. C., and M. STRAND: Complexes of F-pili and RNA bacteriophage. Science **148**, 511—513 (1965).

WATANABE, T.: Infective heredity of multiple drug resistance in bacteria. Bact. Rev. **27**, 87—115 (1963). — WATANABE, T., and T. FUSAKAWA: "Resistance transfer factor" an episome in Enterobacteriaceae. Biochem. biophys. Res. Commun. **3**, 660—665 (1960). — Episome-mediated transfer of drug resistance in Enterobacteriaceae. IV. Interaction between resistance transfer factor and F-factor in *Escherichia coli* K12. J. Bact. **83**, 727—735 (1962). — WATANABE, T., T. FUSAKAWA, and T. TAKANO: Conversion of male bacteria of *Escherichia coli* K12 to resistance of f phages by infection with the episome "resistance transfer factor". Virology **17**, 218—219 (1962). — WATANABE, T., and M. OKADA: New type of sex factor-specific bacteriophage of *Escherichia coli*. J. Bact. **87**, 727—736 (1964). — WENDT, L. W., K. A. IPPEN, and R. C. VALENTINE: General properties of F-pili. Biochem. biophys. Res. Commun. **23**, 375—380 (1966). — WOHLHIETER, J. A., C. C. BRINTON jr., and L. S. BARON: Utilization of carbohydrates and metabolic intermediates by piliated and nonpiliated bacteria. J. Bact. **84**, 416—421 (1962). — WOHLHIETER, J. A., S. FALKOW, R. V. CITARELLA, and L. S. BARON: Characterization of DNA from a *Proteus* strain harboring an episome. J. molec. Biol. **9**, 576—588 (1964). — WOLLMAN, E. L., et F. JACOB: Sur les processus de conjugaison et de recombinaison chez *E. coli*. V. Le mécanisme du transfert de matérial génétique. Ann. Inst. Pasteur **95**, 641—666 (1958).

YOSHIKAWA, H., and N. SUEOKA: Sequential replication of *Bacillus subtilis* chromosome. I. Comparison of marker frequencies in exponential and stationary growth phases. Proc. nat. Acad. Sci. (Wash.) **49**, 559 (1963).

ZINDER, N. D., R. C. VALENTINE, N. ROGER, and W. STOECKENIUS: f 1, a rod-shaped male-specific bacteriophage that contains DNA. Virology **20**, 638—640 (1963).

Die Geschlechtsfaktoren bei haplogenotypischer Bestimmung.

Von

H. Förster.

Mit 6 Abbildungen.

Prinzipielles der haplogenotypischen Geschlechtsbestimmung.

Die haploide, speziell die haplogenotypische Geschlechtsbestimmung (GB) ist der einfachste bekannte GB-Modus. Eine Geschlechtsdifferenzierung gibt es dabei nur in der haploiden Kernphase, die immer die Kernphase der kopulierenden Partner, der Gameten, Zellen oder Kerne ist. Die diploide Kernphase ist im Gegensatz zu den Verhältnissen bei diplogenotypischer GB nicht sexuell differenziert. — Da die Genetik der Haplonten einfacher ist als die Genetik der Diplonten, so ist auch die haplogenotypische GB einfacher als die diplogenotypische GB. An die Stelle der komplizierteren Homozygotie-Heterozygotie-Mechanismen bei der diplogenotypischen GB tritt die einfache Aufspaltung der geschlechtsbestimmenden Erbfaktoren durch die Reduktionsteilungen. Bei der diplogenotypischen GB wird das Geschlecht dagegen immer bei der Befruchtung bestimmt, dem der Reduktionsteilung entgegengesetzten Vorgang im Kernphasenwechsel.

Die rein haploide sexuelle Differenzierung und GB ist auf niedere Organismen beschränkt. Da die Begriffe Geschlecht und Sexualität nach dem ursprünglichen und gewöhnlichen Sprachgebrauch auf die ausgeprägten morphologischen Differenzierungen von Individuen, Geschlechtsorganen und Geschlechtszellen — dem Geschlechtsdimorphismus von Männchen und Weibchen bei höheren Organismen — abzielen, sind sie für die Anwendung auf niedere Organismen, bei denen mitunter jede morphologische Geschlechtsdifferenzierung fehlt (Isogamie), nicht unwesentlich verallgemeinert worden. Die Erfahrung hat gezeigt, daß auch bei isogamen Formen eine bipolare physiologische Differenzierung, die in Analogie zum Geschlechtsdimorphismus steht, eine Voraussetzung der Kopulation ist. Geschlecht und Sexualität bedeuten also heute die bipolaren, morphologischen oder physiologischen Differenzierungen der Kopulationspartner. Scheinbare Abweichungen von den Regeln der Bipolarität („multipolare Sexualität") sind anders zu erklären. Historisch war diese Verallgemeinerung die Grundlage der heutigen Sexualitätstheorie der Befruchtung von Max Hartmann (1956). Sie wurde durch die erste Entdeckung einer haplogenotypischen GB bei Isogamie an den Zygomyceten *Phycomyces blakesleanus* durch Blakeslee (1904) experimentell begründet. Bei isogamen Formen wird der Geschlechtsunterschied oft durch die Bezeichnungen „*plus*" und „*minus*" statt ♂♂ und ♀♀ gekennzeichnet. Solche Bezeichnungen sind immer rein konventionell und willkürlich. Sie sollen primär nicht einer Homologisierung des Geschlechts von verschiedenen Formen dienen, sondern nur die innerhalb einer Form bestehende sexuelle Bipolarität kennzeichnen.

Abgesehen von einer Reihe genetisch komplizierter Sonderfälle — es handelt sich meist um experimentelle Bastardkreuzungen verschiedener Rassen — erfolgt die genetische GB in der Natur meist nach einem einfachen, monofaktoriellen Mendelschema. Dem steht die Tatsache gegenüber, daß die Geschlechtsunterschiede immer in einer größeren Zahl morphologischer und physiologischer Merkmale bestehen. Es ist unwahrscheinlich, daß sie genphysiologisch alle nur von dem einen Paar genetisch erfaßbarer, geschlechtsbestimmender Erbfaktoren abhängen. Diese Diskrepanz läßt sich mit CORRENS (1920, 1928) und HARTMANN (1923) folgendermaßen erklären: Es gibt 2 Anlagensysteme, A für Männchen und G für Weibchen, die die Grundlage für die entwicklungsphysiologische Ausprägung der beiden Geschlechter sind. Sie bilden einen zusammengehörenden Anlagenkomplex, der als A—G-Komplex bezeichnet wird. Außerdem gibt es einfach mendelnde Faktoren M (für ♂♂) und F (für ♀♀), nach v. WETTSTEIN (1924) werden sie Realisatoren genannt, deren Zusammenspiel und Verteilung darüber entscheidet, welche der beiden Geschlechtsanlagen entfaltet wird.

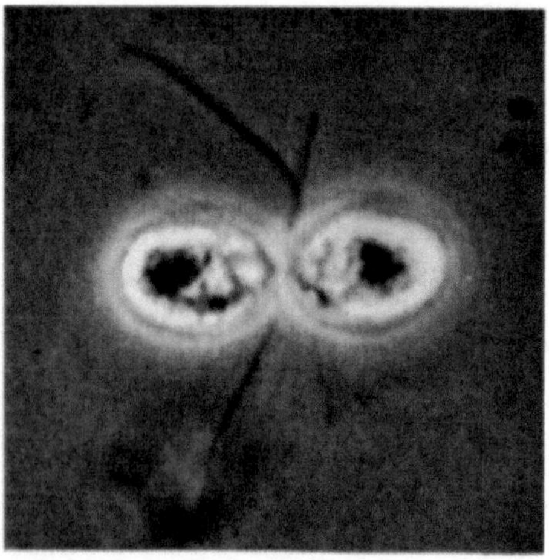

Abb. 1. Ein Kopulationspärchen von *Chlamydomonas moewusi*. Die Gameten sind mit den Vorderenden, an denen die Geißeln ansetzen, verbunden. Lebendaufnahme Phasenkontrast, Elektronenblitz.

Daß es auch bei isogamen Arten eine ganze Reihe sexueller Differenzen geben kann, lehren zwei einfache Beispiele von isogamen Chlamydomonaden. Die Geißeln der Gameten von *Chlamydomonas eugametos* sind mit geschlechtsspezifischen Stoffen besetzt, die den ersten Schritt im Kopulationsvorgang, eine Agglutination oder Gruppenbildung von geschlechtsverschiedenen Gameten bewirken. Diese agglutinierenden Gamone (Gamone = Befruchtungsstoffe) lassen sich von jedem Geschlecht isoliert gewinnen. Sie verursachen eine Gruppenbildung zwischen Gameten des Gegengeschlechtes (Isoagglutination) (FÖRSTER und WIESE 1954). So wird durch Gameten eines Geschlechts der erste Kopulationsschritt imitiert. Alle folgenden Kopulationsschritte unterbleiben aber. Der Grund dafür ist, daß die notwendigen komplementären sexuellen Differenzierungen für die weiteren Kopulationsvorgänge fehlen. Unter anderem kann die Plasmaverbindung zwischen den Papillen nicht hergestellt werden. — Eine weitere bipolare Differenzierung hat LEWIN (1950) an *Chlamydomonas moewusi* genetisch nachgewiesen. Die Kopulationspaare dieser Art schwimmen als Planozygoten noch einige Zeit in der Suspension, bevor sich die Zygoten zur Ruhe setzen. Da die Paare Kopf an Kopf verbunden sind (Abb. 1), muß der eine Partner seine ursprüngliche Bewegungsrichtung verändern. Eine unbewegliche, aber begeißelte und kopulationsfähige Mutante, die geschlechtsunabhängig mendelt, gibt mit der Ausgangsform bewegliche oder unbewegliche Kopulationspaare, je nachdem, mit welchem Geschlecht das mutierte Gen eingekreuzt wird. Daraus ist zu schließen, daß auch die Bewegung normaler Planozygoten geschlechtsspezifisch durch einen Partner bestimmt wird, während der andere Partner nach der Paarbildung seine Geißelbewegung einstellt und zur Bewegung der Zygote nichts beiträgt.

Die beiden Beispiele demonstrieren, daß die sexualspezifischen Differenzierungen der beiden Kopulationspartner meist in zahlreichen physiologischen Einzelheiten genau aufeinander abgestimmt sein müssen, damit eine Kopulation möglich ist. Geht diese Abstimmung oder Komplementarität zwischen den geschlechtsverschiedenen Partnern verloren, so wird die Kopulation unmöglich, bzw. bei kleineren Abweichungen in entsprechendem Maß erschwert. Sind Mutationen, die geschlechtsunabhängig vererbt werden, die Ursache solcher Kopulationshindernisse, so können daraus leicht inkompatible genetische Systeme entstehen, die im Test ein multipolares Kreuzungsschema ergeben. Das rein genetische Experiment läßt es allerdings bei isogamen Arten nicht immer zu, zwischen geschlechtsbestimmenden Faktoren und Inkompatibilitätsfaktoren zu unterscheiden. Genetische Inkompatibilität ist besonders bei Pilzen weit verbreitet (RAPER 1954). Ein multipolares Kreuzungsschema kann außerdem durch geschlechtsunabhängig mendelnde Letalfaktoren verursacht werden.

Auch die eingeschlechtlich bestimmten Individuen besitzen die Anlagensysteme oder den A—G-Komplex für beide Geschlechter. Diese allgemeine Eigenschaft wird als „bisexuelle Potenz" bezeichnet (HARTMANN 1956). Die weite Verbreitung strenger Diöcie bei phänotypischer GB (= modifikatorische GB) ist ein Beispiel dafür. Genetisch sind beide Geschlechter gleich. Durch zufällige Variationen wird jeweils eines der beiden potentiellen Geschlechter realisiert. Sehr oft gibt es inerhalb einer Art Rassen mit phänotypischer und solche mit genotypischer GB. — Experimentelle Geschlechtsumwandlungen durch Hormone sind ebenfalls Nachweise der bisexuellen Potenz. — Bei diplogenotypischer GB nach dem X—O-Typus ist die bisexuelle Potenz unmittelbar zu sehen. Das heterozygote Geschlecht mit dem X-Chromosom hat den gesamten Genbestand des homozygoten Geschlechts, dem nur das zusätzliche X-Chromosom fehlt. Damit hat das heterozygote Geschlecht auch die genetische Potenz des homozygoten Geschlechts. — Bei dem Lebermoos *Sphaerocarpus donnellii* mit haplogenotypischer GB wurden strahleninduzierte Mutanten erhalten, die eine Geschlechtsumkehr von Weibchen in Männchen bewirkten. Cytologisch sind das nach KNAPP und HOFFMANN (1939) zum Teil Chromosomenmutanten, die ein Stück des weibchenbestimmenden X-Chromosoms, wahrscheinlich das Stück mit dem Locus des F-Realisators, verloren haben. Die ♂♂-Potenz muß also in dem restlichen Chromosomenmaterial des genetischen Weibchens vorhanden gewesen sein, sonst könnte der Chromosomenstückverlust nicht die Geschlechtsumwandlung bewirken. — Die bisexuelle Potenz ist die allgemeine Grundlage aller geschlechtlichen Differenzierungen. Sie erklärt zugleich die weite Verbreitung der „bipolaren Zweigeschlechtlichkeit" (HARTMANN 1956).

Fast alle genetischen Experimente durch Bastardierungen, Polyploidisierungen usw. bewirken meist nur Veränderungen der GB-Mechanismen. Die eigentliche Grundlage der sexuellen Differenzierung, die bisexuelle Potenz, hat sich bisher mit wenigen Ausnahmen experimentellen Zugriffen entzogen. Nur wenige Genloci für einzelne morphologische oder physiologische Geschlechtsmerkmale (Gene des A—G-Komplexes) — nicht zu verwechseln mit geschlechtsgebundenen Merkmalen — sind bisher exakt erfaßt. Die umfangreichen Veröffentlichungen von MOEWUS zu diesem Thema über die Genphysiologie der Geschlechtsdifferenzierung, speziell über die Genketten für Termone und Gamone, sind heute aus der Literatur zu streichen. Der größte Teil seiner Angaben ist widerlegt (FÖRSTER und WIESE 1954, 1956; RYAN 1955, LEWIN 1956, FÖRSTER 1957), der Rest bleibt zu bezweifeln bis eventuell Einzelheiten reproduziert werden sollten.

Bei Arten mit diplogenotypischer GB gibt es oft morphologisch differenzierte Chromosomen, sog. Heterochromosomen, durch deren Verteilung und Kombination das Geschlecht bestimmt wird. Heterochromosomen für haplogenotypische GB sind nur bei einigen Moosen bekanntgeworden. Arten mit Heterochromosomen bieten den experimentellen Vorteil, daß genetische Versuche leicht durch cytologische Beobachtungen ergänzt werden können. Für die Natur liegt ihr Wert darin, daß Heterochromosomen nicht konjugieren. Sind innerhalb einer Art oder Rasse Heterochromosomen für die GB ausgebildet, so kann der GB-Mechanismus durch Faktorenaustausch nicht umgeworfen werden. Die Diöcie und das jeweils erwünschte und eingestellte Zahlenverhältnis der Geschlechter wird dadurch gut garantiert. Meist ist ein 1:1-Verhältnis optimal für die Kopulationshäufigkeit. Ähnlich konservativ wie bei Heterochromosomen ist der GB-Mechanismus, wenn die Realisatoren allele Gene in homologen Chromosomen sind. Das scheint die Regel zu sein. Andernfalls müßte der Wechsel Diöcie → Monöcie bzw. genotypische GB → phänotypische GB genau so häufig vorkommen wie die Rekombination anderer Merkmale, weil die F- und M-Faktoren zusammen auf ein Chromosom rekombinieren könnten. Als Ausnahmen von der Regel haben Hüttig (1935) und Greis (1941) bei den Pilzen *Glomerella lycopersici* und *Sordaria fimicola* Realisatorenaustausch beschrieben. Die genetischen Verhältnisse sind in beiden Fällen kompliziert und die Versuchsergebnisse sind eventuell auch anders zu deuten (z. B. Heitz 1945). Die Pilze spielen in der Genetik der GB durch zahlreiche Komplikationen eine Sonderrolle. Moewus hat über eine ganze Reihe von Realisatoraustauschvorgängen bei Algen berichtet. An seinen Stämmen von *Chlamydomonas eugametos* (Förster und Wiese 1954), *Chlamydomonas paradoxa* und *pseudoparadoxa* (Stifter unveröffentlicht) waren seine Angaben nicht zu bestätigen. Der Rest kann daher auch bezweifelt werden. Ein Wechsel von haplogenotypischer Diöcie → Monöcie bei dem Ascomyceten *Hypomyces solani* wurde von Hirsch (1949) cytologisch aufgeklärt. Die Ursache ist Nondisjunktion der homologen Chromosomen mit den geschlechtsbestimmenden Faktoren bei der Meiose. Die beiden homologen Chromosomen gehen in denselben Tochterkern, statt alternativ verteilt zu werden.

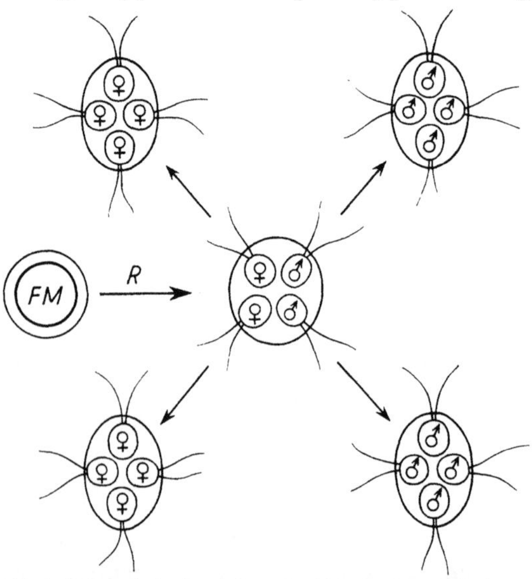

Abb. 2. Schematische Darstellung des Miktohaplonten *Gonium pectorale*. *FM* diploide Zygote; *R* Reduktion zur miktohaplontischen Primärkolonie. Die äußeren 4 Tochterkolonien sind eingeschlechtlich.

Übergänge von Monöcie ⇌ Diöcie kommen oft durch den Übergang Miktohaplont ⇌ reiner Haplont zustande. Miktohaplonten sind nach Kniep (1928) Formen, die 2 Arten haploider Kerne enthalten. Die Volvocinee *Gonium pectorale* veranschaulicht den Sachverhalt (Abb. 2). Wenn die Gonium-Zygote keimt, entstehen durch die Reduktionsteilungen vier haploide Zellen, von denen je zwei den F- und je zwei den M-Realisator besitzen. Die 4 Zellen bilden eine Kolonie. Diese Primärkolonie ist ein Miktohaplont. Die Abkömmlinge dieser Kolonie können untereinander kopulieren. Werden die Nachkommen der 4 Gonen getrennt weitergezüchtet, so gibt es keine intraklonalen Kopulationen mehr. Während bei *Gonium pectorale* der Übergang von der miktohaplontischen Primärkolonie zu eingeschlechtlichen Kolonien regulär im Entwicklungscyclus auftritt, kommen solche Übergänge von Miktohaplonten zu reinen Haplonten bei Pilzen oft accessorisch vor (s. HARTMANN 1956), z. B. bei *Neurospora tetrasperma*. Umgekehrt ist es bei Pilzen mehrfach gelungen, aus zwei reinen Haplonten Miktohaplonten experimentell herzustellen (BURGEFF, s. HARTMANN 1956). Diese Ergebnisse bei Pilzen sind verständlich, wenn man bedenkt, daß der dikaryotische Zustand, eine Paarkernphase, bei ihnen meist eine reguläre Phase im Entwicklungscyclus ist.

Varianten der Geschlechtsbestimmung.

Variationen der genetischen Geschlechtsbestimmung durch Polyploidisierung, durch Bastardierung und Mutation äußern sich hauptsächlich auf 2 Arten und Weisen. Die Realisationsstärken können sich verändern und die Realisatorloci können sich verändern bzw. was für den letzteren Fall wahrscheinlich richtiger ist, neue Genloci können geschlechtsentscheidend werden. Klassisch für den ersten Fall, der im Rahmen der haplogenotypischen GB allein interessiert, sind die Kreuzungsversuche von GOLDSCHMIDT zwischen verschiedenen *Lymantria*-Rassen. Je nach der Stärke von den Realisatoren der Eltern ergab sich eine Nachkommenschaft von verschiedenem Intersexualitätsgrad. Der Grad der Intersexualität ist direkt ein Maß für die Stärke der eingekreuzten F- und M-Faktoren (s. Diplogenotypische Geschlechtsbestimmung). Die meisten Polyploidieversuche, die eine summative oder subtraktive Realisatorstärke zeigen, je nach dem Zahlenverhältnis der Chromosomen, die F- oder M-Faktoren enthalten, sind an Formen mit diplogenotypischer GB gemacht worden. Ähnliche Ergebnisse bei haplogenotypischer GB gibt es in kleinerem Rahmen bei Moosen. Die normalen Gametophyten der getrenntgeschlechtlichen Moose sind haploid. Bei dem Lebermoos *Sphaerocarpus donnellii* sind [7 Autosomen + X] = ♀♀ und [7 Autosomen + Y] = ♂♂. LORBEER erhielt diploide Gametophyten [14 Autosomen + X + Y]. Die Keimlinge waren alle weiblich (LORBEER 1927) bzw. fast alle weiblich (ALLEN 1935). Bei gleicher Valenz der F- und M-Realisatoren im X- und Y-Chromosom hätte man nur Zwitter, bzw. ♂♂ und ♀♀ in annähernd gleicher Zahl erwartet. Der F-Faktor hatte dagegen eine größere Realisationsstärke als der M-Faktor. WETTSTEIN (1924) fand umgekehrt bei diploiden und tetraploiden Gametophyten von dem Laubmoos *Bryum caespiticum* M stärker als F, aber bei triploiden Gametophyten 2 F stärker als 1 M, also das Stärkeverhältnis 2 F > 1 M > 1 F. Es gibt eine Reihe vergleichbarer Ergebnisse bei Moosen (AHRENS 1937, HEITZ 1942, MOUTSCHEN 1954, HOFFMANN 1956, 1957).

Eine unterschiedliche Kopulationshäufigkeit zwischen verschiedenen Stämmen derselben Art oder Rasse, die genetisch bedingt ist, wurde bei einigen Pilzen gefunden. Tabelle 1 gibt ein Beispiel dafür von einer diöcischen Mucorinee aus HARTMANN (1956). Wegen der weiten Verbreitung von Inkompatibilitätsfaktoren bei Pilzen kann die unterschiedliche Kopulationsfähigkeit zwischen einzelnen Stämmen eventuell durch verschiedene Grade der Kompatibilität respektiv Inkompatibilität physiologisch erklärt werden.

Zahlreiche ähnliche Befunde über die verschiedene Valenz von Realisatoren, insbesondere die Tatsache, daß bei gleicher Stärke beider Geschlechtsbestimmer oft Intersexe oder Zwitter auftreten, passen gut in die Theorie von der relativen sexuellen Differenzierung, die HARTMANN aus theoretischen Überlegungen 1909 aufgestellt hatte. Nach dieser Theorie soll die „bisexuelle Potenz" immer die Grundlage aller Geschlechtsdifferenzierungen sein. Außerdem soll die Geschlechtsverschiedenheit nicht absolut qualitativ, mit der einfachen Alternative Männchen oder Weibchen sein, sondern sie soll im Prinzip quantitativ sein und alle Abstufungen von stark männlicher Determination bis zu stark weiblicher Determination, je nach der Realisatorstärke, erlauben. Danach können Grenzfälle vor-

Tabelle 1. *Absidia spec. (diöcische Mucorinee). Kombinationen einer größeren Zahl von Plus- und Minus-Stämmen.* Die Zahlen 1—4 bezeichnen den Stärkegrad der sexuellen Reaktion, gemessen an der Zahl der Zygoten. 0 = keine Reaktion, 4 = stärkste Reaktion. (Nach Blakeslee, Cartledge, Welch und Bergner 1927.)

		♂(−)																	
		400	687	635	807	678	681	634	680	682	402	405	251	403	394	404	398	399	809
(+) ♀	395	3	3	4	4	4	4	2	3	3	3	3	3	2	3	2	3	2	1
	397	3	4	3	4	4	4	3	3	3	2	2	3	3	2	2	2	1	1
	392	4	3	3	3	2	4	2	4	4	3	2	2	3	3	2	2	1	0
	393	3	4	3	4	4	3	3	2	3	2	2	3	2	2	2	3	2	0
	684	4	2	4	2	2	2	3	2	2	3	3	2	3	2	2	2	2	1
	808	4	3	2	2	2	2	3	3	2	3	3	2	3	2	2	1	1	1
	636	4	3	2	2	2	2	2	3	3	2	2	3	2	2	2	2	2	0
	396	2	3	3	3	3	2	2	2	2	2	2	2	2	2	2	2	2	1
	688	4	2	3	3	2	2	2	2	2	3	2	2	2	3	2	1	1	1
	676	3	2	2	2	2	2	3	2	2	2	2	2	2	2	2	2	2	1
	679	3	2	3	2	2	2	2	2	2	2	2	2	2	2	2	1	0	1
	401	2	3	2	3	3	3	2	2	2	1	2	2	1	1	2	0	0	2
	683	3	2	2	2	2	2	2	2	2	2	2	1	2	2	2	2	0	1
	677	2	1	0	0	1	1	2	1	0	1	2	0	0	0	1	1	1	0

kommen, die annähernd in der Mitte des bipolaren Schemas liegen. Bei rein haploider Geschlechtsdifferenzierung können solche intermediären Formen dann unter Umständen sowohl mit stark männlichen Formen als ♀♀, wie mit stark weiblichen Formen als ♂♂ reagieren. Solche Fälle werden als "relative Sexualität" bezeichnet. Es sind im Laufe der Zeit eine ganze Reihe solcher Ausnahmefälle von relativer Sexualität beschrieben worden (Hartmann 1956). Die vielen Fälle, die Moewus berichtet hat, muß man allerdings heute außer acht lassen. Soweit es versucht wurde, sie an seinen eigenen Stämmen von *Chlamydomonas eugametos, paradoxa* und *pseudoparadoxa* zu reproduzieren, ist es auch nicht andeutungsweise gelungen (unveröffentlichte Ergebnisse der Hartmann-Schüler Wiese, Stifter und Förster).

Ausbalancierte Realisatorsysteme bei diploiden Genomen, bzw. geringe Wirkungsstärke des einen Realisators bei haploiden Genomen können unterschiedliche Auswirkungen haben: Es können Zwitter entstehen, bei denen die beiden komplementären Geschlechtscharaktere nebeneinander ausgebildet sind, oder das Geschlecht kann modifikatorisch bestimmt werden, wenn die Organisation den zwittrigen Zustand nicht erlaubt. In diesem Fall ist die Diöcie, im Unterschied zum Geschlecht, letztlich genetisch bedingt. Zwischen diesen beiden Alternativen — zwittrig oder getrenntgeschlechtlich mit modifikatorischer GB — gibt es verschiedene Übergänge und Zwischenformen. So sind z. B. die Intersexe des Schmetterlings *Solenobia triquetrella* nach Seiler (1949, 1951) Mosaikgebilde aus männlichen und weiblichen Zellen und Zellverbänden. Die ganzen intersexuellen Individuen sind „zwittrig", jede einzelne Zelle des sexuellen Mosaiks ist aber entweder rein männlich oder rein weiblich determiniert. Ähnliche, aber weniger eindeutige Mosaikgebilde von sekundären Geschlechtsmerkmalen gibt es bei polyploiden, gemischtgeschlechtlichen Moosgametophyten (Lorbeer 1938, Hoffmann 1957). In anderen Fällen werden die Geschlechtscharaktere nacheinander ausgebildet (Proterandrie oder Protogynie). Diese

Unterschiede sind hauptsächlich für die phänotypische GB von Interesse (s. dort). Von den verschiedenen Übergangsformen ist vor allem die ,,Subdiöcie" für die genotypische GB wichtig. Als subdiöcisch werden in erster Linie Formen bezeichnet, die ein Realisatorsystem für genotypische GB haben, dessen Durchschlagskraft aber gering ist und das daher modifikatorisch überspielt werden kann. Dadurch wird neben dem genetisch bestimmten Geschlecht auch die Ausbildung von Zwittern oder des Gegengeschlechts möglich. Besonders bei den Blütenpflanzen und bei den Anuren ist Subdiöcie weit verbreitet. Bei Rassen von

Abb. 3a—c. *Trichonympha* (polymastiginer Flagellat, phänotypische GB, diöcisch). a Oben normaler ♀-Gamet mit charakteristischen Geschlechtsmerkmalen. Ein dichter Ring von Granula ist um den ,,Empfängnishügel" gesammelt, in den der untere ♂-Gamet eindringt. b Relative Sexualität. Der eindringende Gamet ist deutlich, wenn auch wesentlich schwächer ,,weiblich" differenziert. c Dreierkopulation, der zuletzt eindringende Gamet zeigt fast keine ♀-Differenzierung mehr. Vergr. 500fach. (Nach CLEVELAND 1949.)

Callithamnium (HASSINGER-HUIZINGA 1952) und *Chaetomorpha* (KÖHLER 1956), die normal haplogenotypische GB haben, wurden Unregelmäßigkeiten gefunden, die auf Subdiöcie deuten.

Das Phänomen der relativen Sexualität ist auf Formen beschränkt, bei denen nur die kopulierenden Gameten sexuell differenziert sind. Auch hierbei taucht die Frage auf, ob die einzelnen Gameten selbst beide komplementären, physiologischen Geschlechtsdifferenzierungen aufweisen (Analogie zum Zwitter) oder ob in einer Suspension regulär eingeschlechtlich determinierte Gameten, wenn sie z. B. alle von einer normalerweise streng diöcischen Pflanze stammen, einzelne vom Gegengeschlecht sind (Analogie zur Subdiöcie).

Das eindeutigste Beispiel relativer Sexualität hat CLEVELAND (1949) bei dem hypermastiginen Flagellaten *Trichonympha* gefunden. Die Gameten dieses Flagellaten sind cytologisch deutlich anisogam (Abb. 3) und zeigen oft eine verschieden starke Ausprägung der cytologisch erkennbaren sekundären Geschlechtsmerkmale. Die komplexe Organisation dieser *Trichonympha*-Gameten und die abgebildete Dreierkopulation macht für diesen Fall die erste Deutung wahrscheinlicher (Analogie zum Zwitter). Bei den einfacheren Gameten der Braunalge,

Ectocarpus siliculosus, dem Paradebeispiel für relative Sexualität, ist dagegen die zweite Deutung wahrscheinlicher (Analogie zur Subdiöcie). Sowohl *Trichonympha* wie *Ectocarpus* haben allerdings eine modifikatorische GB. Relative Sexualität bei Arten mit genotypischer Geschlechtsbestimmung ist nur bei einigen Pilzen gefunden worden (HARTMANN 1956). Bei Pilzen gibt es so ziemlich alle erdenkbaren Variationen und Komplikationen der Geschlechtsdifferenzierung und GB (GÄUMANN 1949, RAPER 1954, 1955; BURNETT 1956, HAWKER 1957).

Mit der Theorie von der Relativität der GB und dem umfangreichen Erfahrungsmaterial über verschiedene Realisatorvalenzen sind intermediäre Realisatoren für Zwittrigkeit in guter Übereinstimmung. GALAN (1951) fand bei *Ecballium elaterium* drei allele Realisatoren für männlich, weiblich und zwittrig. Da *Ecballium* diplogenotypische GB hat, konnte durch Kreuzungsversuche das Stärkeverhältnis der Realisatoren festgestellt werden. Der ♂-Realisator ist stärker und der ♀-Realisator ist umgekehrt schwächer als der Faktor für Zwittrigkeit: ♂ > ⚥ > ♀ („>" = dominant). LEUPOLD (1950) fand bei Schyzosaccharomyces Pombe ebenfalls drei allele Realisatoren für die Geschlechter respektiv Paarungstypen *plus*, *zwittrig* und *minus*. LORBEER (1938) konnte den ♀-Realisator im X-Chromosom des Lebermooses *Sphaerocarpus donnellii* durch Röntgenstrahlen in einen Realisator für Zwittrigkeit umwandeln und weiter aus ♀- oder ⚥ -Realisatoren im X-Chromosom, das normalerweise weiblichbestimmend ist, ♂-Realisatoren erhalten. Im Vergleich mit den Ergebnissen von KNAPP am gleichen Objekt (S. 33) liegt es nahe, den Zwitter-Realisator als Schwächung des ♀-Realisators zu einer intermediären Zwischenform aufzufassen und die ♂♂ mit X-Chromosom eventuell als völligen Verlust der Funktion des ♀-Realisators zu deuten. Unabhängig von jeder Deutung bleibt die gefundene Mutationsreihenfolge für den Geschlechtsrealisator im X-Chromosom ♀ → ⚥ → ♂ experimentell gesichert.

HARTMANN (1956) hat bisher das Vorkommen einzelner Gene für Zwittrigkeit stark bezweifelt und immer versucht, das Auftreten von genetischen Zwittern unter getrenntgeschlechtlichen Formen auf Realisatoren-Austausch zurückzuführen. Das umfangreiche Material über Austauschvorgänge zwischen Realisatoren von MOEWUS schien seine Einstellung zu bestätigen. Heute ist dieses Argument entfallen und dem Verfasser scheint das Auftreten von Zwitter-Realisatoren besser in die Sexualitätstheorie von HARTMANN, speziell in die These von der Relativität der sexuellen Differenzierung zu passen, als die Hypothese vom Realisator-Austausch.

Die Funktion der Geschlechtsrealisatoren.

Über die physiologische Funktion der Geschlechtsrealisatoren ist nur wenig bekannt. Am eindeutigsten ist unsere Kenntnis über die Realisatorfunktion bei *Melandrium* (WESTERGAARD 1953, WIESE 1956). Das diöcische *Melandrium* hat eine diplogenotypische GB mit Heterogametie im männlichen Geschlecht: XX = ♀♀, XY = ♂♂. Eine cytologisch definierte Mutation in dem ♂♂-bestimmenden Y-Chromosom führt zu echten Zwittern, bei denen die männlichen und die weiblichen Organe in den Blüten vollständig ausgebildet sind. Bei dieser Mutation ist offenbar das Segment mit dem M-Realisator ausgefallen. Die männliche Potenz in den mutierten Pflanzen blieb trotzdem vollständig erhalten. Das M-Gen gehört augenscheinlich nicht zur genetischen Grundlage für die Entwicklung der männlichen Geschlechtscharaktere, dem A-Komplex. Es wirkt vielmehr nur als ♂-Realisator, indem es die Entfaltung der weiblichen Potenzen unterdrückt. Durch seinen Ausfall entfalten sich daher beide Potenzen und es entstehen die echten Zwitter.

Neben diesem Realisatorsegment wurden cytogenetisch an anderen Mutanten im Y-Chromosom zwei andere Chromosomenabschnitte gefunden, die für die entwicklungsphysiologische Ausprägung der männlichen Organe verantwortlich sind. Der Ausfall des einen Segments hemmt die Pollenbildung und bewirkt dadurch Sterilität. Fällt das zweite Segment aus, so werden die Staubgefäße nicht gebildet. Die genetischen ♂-A-Faktoren in diesen Segmenten wirken aber in keiner Weise als Realisatoren; sie beeinflussen die Ausbildung der weiblichen Potenzen nicht. Das ist einer der wenigen Fälle, in denen Gene des A—G-Komplexes lokalisiert sind. Bemerkenswert ist, daß die bisexuelle Potenz im homozygoten Geschlecht hierbei unvollständig ist, da die Weibchen das Y-Chromosom mit den beiden nachgewiesenen A-Faktoren nicht besitzen. Das ist eine Ausnahme, die nur bei Heterochromosomen vorkommen kann, weil die in ihnen lokalisierten A—G-Faktoren nicht rekombinieren können. Wäre es die Regel, so könnten bei Arten ohne Heterochromosomen die männlichen und weiblichen Geschlechtsanlagen wirr durcheinander rekombinieren.

Ob die Geschlechtsrealisatoren immer vom A—G-Komplex ganz unabhängig sind und nur als Unterdrücker des Gegengeschlechts oder in ähnlicher Weise wirken, läßt sich aus den vereinzelten Kenntnissen allerdings nicht generell schließen. Es wären Fälle denkbar, bei denen die Realisatoren zum A—G-Komplex gehören und selbst genphysiologisch an der Entwicklung des ihnen entsprechenden Geschlechts mitwirken. In solchen Fällen wäre zumindest bei haplogenotypischer GB die bisexuelle Potenz immer unvollständig, da die Realisatoren bei der Reduktionsteilung alternativ verteilt werden. In der Regel kann die alternative Verteilung einzelner A- oder G-Faktoren aber nicht zu klaren Realisatorsystemen und strenger Getrenntgeschlechtlichkeit führen. Es ist vielmehr zu erwarten, daß der Ausfall eines beliebigen ♂ A- oder ♀ G-Faktors meist nur die unvollständige Ausbildung und die Sterilität des zugehörigen Geschlechts bewirkt, wie bei den beiden Melandrium-Mutanten. Es scheint daher wahrscheinlicher zu sein, daß die Funktion der Realisatoren meist ähnlich ist wie bei Melandrium.

Geschlechtsbestimmung und Kernphasenwechsel.

Die Kopulation mit der Karyogamie und die Meiose sind die antagonistischen Vorgänge im Wechsel von haploider und diploider Kernphase. Je nachdem, welche der beiden Kernphasen die eigentliche vegetative Lebensphase der Art ist, bezeichnet man die Individuen als Haplonten oder als Diplonten. Sind beide Kernphasen eigenständige, vegetative Lebensphasen, so reden wir von Diplohaplonten. Diplohaplonten haben danach einen antithetischen oder heterophasischen Generationswechsel, da eine haploide mit einer diploiden Generation, meist obligatorisch, wechselt. Ob die heterophasischen Generationen morphologisch gleich oder verschieden sind, ist ohne Belang. Desgleichen sind Unterschiede in der Lebensdauer der beiden Phasen unwichtig.

Der Sexualvorgang, die Befruchtung oder Kopulation, bewirkt den Übergang von der Haplophase zur Diplophase. Für die Kopulation sind regulär im Entwicklungscyclus spezifische morphologische oder nur physiologische Differenzierungen ausgebildet. Die kopulierenden Partner selbst, im typischen Fall sind es besondere Geschlechtszellen oder Gameten, müssen ebenfalls zueinander komplementäre Oberflächendifferenzierungen haben, damit sie miteinander reagieren können. Die Gesamtheit der im einzelnen sehr variablen komplementären Differenzierungen, die die physiologische Grundlage der Kopulation sind, werden als sexuelle Differenzierungen bezeichnet. Werden die zueinander komplementären Sexualdifferenzierungen nur alternativ ausgebildet, so resultiert daraus Diöcie, im anderen Fall entstehen Zwitter. Die Diöcie als „alternative Reaktionsnorm" kann als solche also genetisch bedingt sein, ohne daß das Geschlecht genetisch bestimmt wird. Genetische Diöcie und genetische GB sind zu unterscheiden.

Die Verteilung der Geschlechtsrealisatoren bei haplogenotypischer GB erfolgt durch die Reduktionsteilungen, die den Übergang von der Diplophase zur Haplophase bewirken. Die Kenntnis, an welcher Stelle im Entwicklungscyclus einer Art die Reduktion erfolgt, ist daher eine Voraussetzung für die Erkenntnis der Art und Weise ihrer GB. Im wesentlichen sind drei verschiedene Fälle ja nach dem Ort der Reduktionsteilungen im Entwicklungscyclus zu unterscheiden.

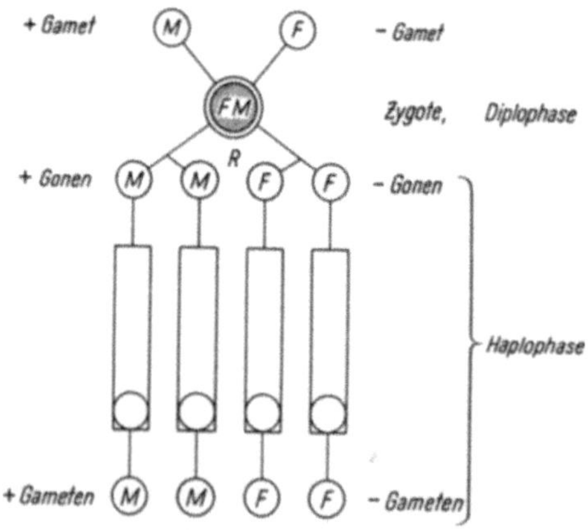

Abb. 4. Schematische Darstellung der genotypischen Geschlechtsbestimmung im Kernphasenwechsel eines Haplonten. *F* und *M* Geschlechtsrealisatoren; *R* Reduktionsteilung, diploide Zygote schraffiert.

Bei Haplonten erfolgt die Reduktion bei der Keimung der diploiden Zygote, die durch die Kopulation von 2 Geschlechtszellen entstanden ist. Durch die zwei meiotischen Teilungen entstehen im einfachsten Fall 4 Keimzellen oder Gonen, die der Ursprung der vegetativen Haplophase sind. Sie können sich im Gegensatz zur diploiden Zygote mitotisch vermehren. Je 2 Gonen erhalten den gleichen Geschlechtsrealisator. Das Geschlecht ihrer Nachkommen bzw. des aus einer Gone entstehenden Individuums ist dadurch festgelegt. Je nach der Art wird dann durch Variationen der Umwelt oder durch endogene Reifungsvorgänge die Sexualität ausgelöst und das erblich festgelegte Geschlecht realisiert. Es entstehen wiederum Gameten, die kopulieren und Zygoten bilden. Damit beginnt der Sexualcyclus von neuem. Abb. 4 ist ein Schema für den Sexualcyclus eines Haplonten mit genetischer GB. R bezeichnet die Reduktion, die bei Haplonten immer „zygotisch" ist, d. h. bei der Zygotenkeimung stattfindet.

Die eindeutigsten Beispiele für reine Haplonten mit genotypischer GB gibt es unter den Volvocineen, so bei *Chlamydomonas*, *Gonium*, *Chlorogonium*, *Chlamydobotrys*, *Eudorina*, *Dunaliella* und außerdem bei *Oedogonium*. Zu bemerken ist, daß nicht bei allen kolonialen Volvocineen miktohaplontische Kolonien wie bei Gonium vorkommen. Bei *Eudorina* wird z. B. nur aus einer der 4 Gonen, die durch die Reduktionsteilung aus der Zygote entstehen, eine neue Kolonie. Die restlichen 3 Gonen gehen zugrunde. Daher ist bereits die erste Kolonie, die aus der Zygote hervorgeht, entweder *plus*- oder *minus*-geschlechtlich.

Bei Diplohaplonten hat auch die Diplophase ein eigenständiges vegetatives Dasein, sie wird nicht nur durch die Zygote repräsentiert. Die Zygote keimt vielmehr zu einem diploiden Sporophyten aus, der keine Sexualorgane oder Geschlechtszellen bildet. Der Sporophyt erzeugt nur ungeschlechtliche Zoosporen oder Agameten, die den Gameten morphologisch oft sehr ähnlich sind, die aber nicht kopulieren können. Aus ihnen entwickelt sich der Gametophyt. Die Meiose erfolgt bei der Zoosporenbildung. Dabei werden die Geschlechtsrealisatoren alternativ auf die Zoosporen verteilt. Die Gametophyten sind daher haploid und ihr Geschlecht ist determiniert. Durch endogene oder exogene Faktoren wird die Sexualität ausgelöst und jeder Gametophyt erzeugt, entsprechend seinem festgelegten Geschlecht, Gameten, die miteinander kopulieren können und wieder Zygoten bilden. Damit beginnt der Cyclus von vorn. Abb. 5 gibt den Lebenscyclus eines Diplohaplonten mit genotypischer GB schematisch wieder.

Diplohaplonten mit haplogenotypischer GB sind besonders bei vielen marinen Algen weit verbreitet. Bei verschiedenen Arten der Grünalgen *Cladophora, Ulva, Enteromorpha, Chaetomorpha,* der Braunalgen *Cutleria, Laminaria, Dictiota* und der Rotalgen *Ceramia, Polysiphonia, Callithamnium* ist genotypische GB gefunden worden. Alle Laubmoose und alle Lebermoose sind Diplohaplonten, viele haben genotypische GB.

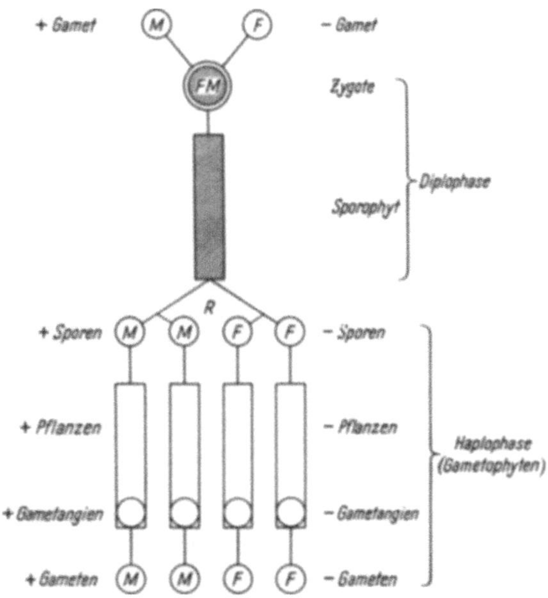

Abb. 5. Schematische Darstellung der genetischen Geschlechtsbestimmung im Kernphasenwechsel eines Diplohaplonten. *F* und *M* Geschlechtsrealisatoren, Diplophase gestrichelt; *R* Reduktionsteilung.

Bei reinen Diplonten wäre eine haplogenotypische GB prinzipiell ebenfalls möglich, wenn die kopulationsfähigen Gameten zugleich mit den Reduktionsteilungen gebildet werden. Das ist z. B. nach GEITLER (1928) bei der Diatomee *Nitzschia subtilis* der Fall. Den sicheren Nachweis haplogenotypischer GB eines Diplonten gibt es allerdings bisher nicht. Es bleibt immer die Möglichkeit offen, daß zugleich mit der Meiose die Sexualität ausgelöst wird, sonst aber Monöcie respektiv phänotypische GB vorliegt.

FRIEDMANN (1959) fand, daß bei der Grünalge *Prasiola stipitata* die Reduktion erst kurz vor der Gametenbildung stattfindet. Der diploide Sporophyt bildet diploide Zoosporen, aus denen sich diploide Gametophyten entwickeln. Nach einer gewissen Wachstumsperiode erfolgt in einzelnen Zellagen des Gametophyten die Meiose. Bei der ersten meiotischen Teilung werden die Geschlechtsrealisatoren getrennt. Durch folgende Mitosen entstehen aus den reduzierten Zellen haploide Zellareale auf dem sonst diploiden Gametophyten, deren Geschlecht determiniert ist. Die geschlechtsverschiedenen Areale sind an ihrer verschiedenen Farbe morphologisch erkennbar. Aus jedem Areal entstehen Gameten von einem Geschlecht. Der ganze Gametophyt ist zwittrig, jedes haploide Zellareal ist aber durch die Reduktion eingeschlechtlich determiniert. Die intermediäre Reduktion ist bei dieser Art fast bis zu einer gametischen Reduktion verschoben.

Eine Komplikation in das allgemeine einfache Schema des Kernphasenwechsels bringen die Pilze. Bei vielen Pilzen folgt auf die Plasmogamie nicht sofort die Kernverschmelzung zur eigentlichen Diplophase, sondern es schiebt sich eine „dikaryotische“ Phase dazwischen. Die Elternkerne behalten dabei ihre Individualität. Die Paarkerne teilen sich in der Regel synchron weiter. Die Karyogamie wird meist bis kurz vor die Reduktionsteilung hinausgeschoben. Für den genetischen Mechanismus der GB ist dieses absonderliche Geschehen ohne Einfluß. Im Prinzip ist nichts wesentlich anders als bei normalen Diplohaplonten. In anderer Hinsicht resultieren daraus sonst nicht bekannte und nicht mögliche

Variationen im Sexualverhalten. Ein bereits befruchtetes dikaryotisches Mycel kann z. B. ein noch haploides Mycel befruchten, indem es einen seiner Paarkerne abgibt. Das dikaryotische Mycel wird dadurch wieder haploid. Die Abb. 6 veranschaulicht den Vorgang. Ein solcher Vorgang, nach seinem Entdecker das „Bullersche Phänomen" genannt, ist ein Beispiel für relative Sexualität. Es ist immer nur eine genetische Sorte von Kernen, die aus dem dikaryotischen Mycel in das haploide Mycel hinüberwandert. Offenbar haben die Kerne, die ihren Partner wechseln, auf Grund der verschiedenen genetischen Konstitution eine größere physiologische „Affinität" zu den Kernen des neuen Haplomycels.

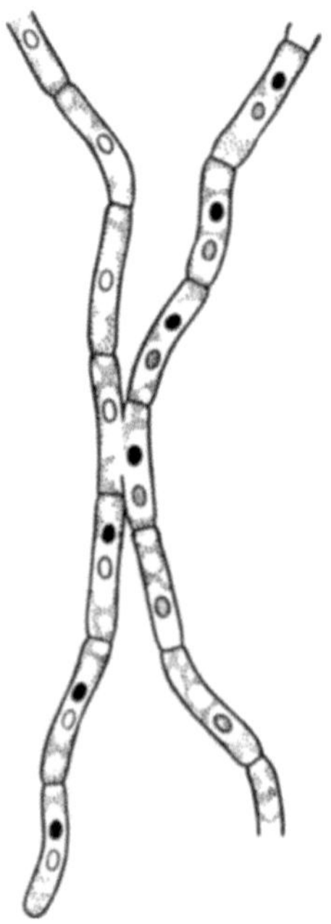

Abb. 6. Schema zur Erläuterung des Bullerschen Phänomens. ● ⊙ ○ drei genetisch verschiedene Kernsorten. Der linke Faden mit dem Dikaryon ● ⊙ (unterer Teil) gibt den ●-Kern an den rechten Faden, der im unteren Teil noch haploid ist. Im oberen Teil ist der rechte Faden bereits dikaryotisch ● ○ und der linke wieder haploid ⊙. (Nach HARTMANN 1956).

Der Nachweis für haplogenotypische Geschlechtsbestimmung.

Bei solchen Haplonten, die sich einerseits beliebig lange rein vegetativ züchten lassen und bei denen andererseits willkürlich, durch kontrollierte Bedingungen, die Sexualität ausgelöst werden kann, ist die Art ihrer GB am einfachsten festzustellen. Wenn genetisch reine Kulturen (Klonkulturen) über längere Zeit immer nur dasselbe Geschlecht, männlich oder weiblich, respektiv *plus* oder *minus*, realisieren, so darf man sicher sein, daß das Geschlecht genetisch determiniert ist. Es dürfen keine intraklonalen Kopulationen auftreten, und die Kombination verschiedener Klone muß immer das gleiche Kreuzungsschema ergeben, also entweder immer positive Sexualreaktion oder keine Reaktion zwischen den gleichen Testklonen.

Schwieriger wird die Entscheidung, wenn die sexuelle Periode obligatorisch im Entwicklungscyclus eingebaut ist, also die haploide Phase sich nicht beliebig lange rein vegetativ vermehren kann und die Sexualität nicht willkürlich auszulösen ist. Mitunter ist es möglich, das Geschlecht der Gameten zu identifizieren, ohne sie kopulieren zu lassen. Diese Möglichkeit bietet sich vor allem bei Gameten, die anisogam sind und deren Geschlecht an morphologischen Merkmalen zu erkennen ist. Manchmal gelingt es, die Gameten „parthenogenetisch" wieder zur Entwicklung zu bringen. Die Nachkommen bzw. die Pflanzen, die aus geschlechtlich bestimmten Gameten parthenogenetisch entstehen, müssen in der nächsten und allen folgenden Sexualperioden wieder das gleiche Geschlecht zeigen. — Aus der Tatsache, daß ein Gametophyt eingeschlechtlich ist und nur ♂♂- oder ♀♀-Gameten (*plus*- oder *minus*-Gameten) erzeugt, kann allein noch nicht auf genetische GB geschlossen werden. Der Gametophyt ist dann zwar nicht zwittrig, weil er nicht beide Gametensorten produziert, sondern diöcisch, aber sein Geschlecht kann trotzdem nur modifikatorisch bedingt sein, wie das z. B. bei Rassen von *Ectocarpus* der Fall ist. Erst die parthenogenetische Aufzucht solcher Gameten von eingeschlechtlichen Gametophyten kann Aufschluß über die Art der GB geben.

Am klarsten und einwandfrei läßt sich die genetische GB durch Gonen- oder Tetradenanalyse nachweisen. Die Keimlinge einer Zygote oder eines Sporangiums, die durch Reduktion entstehen, werden einzeln isoliert und getrennt aufgezogen. Bei genotypischer GB muß das Zahlenverhältnis der Geschlechter 1:1 sein. Am deutlichsten wird die Analyse, wenn nur 4 Gonen durch die zwei meiotischen Teilungen gebildet werden und keine weiteren mitotischen Teilungen folgen, die das klare Spaltungsergebnis verwischen. Unregelmäßigkeiten, die durch zusätzliche Mitosen oder das Absterben einzelner Gonen auftreten, lassen sich aber oft cytologisch nachprüfen.

Die Sonderstellung der Pilze. (Das Verhältnis von Sexualität und Inkompatibilität.)

Die Sexualitätstheorie der Befruchtung nach Correns und Hartmann ist aus der vergleichenden Betrachtung der vielgestaltigen Kopulationsvorgänge im gesamten Organismenreich erwachsen. Im wesentlichen ist die Kopulation überall durch bipolare morphologische und physiologische Mechanismen, die letztlich genetisch begründet sind, verursacht und bedingt. Immer sind es spezifisch komplementär aufeinander abgestimmte Differenzierungen, die obligatorisch oder accessorisch im Entwicklungsgang einer Art auftreten und die Kopulation und damit den Kernphasenwechsel ermöglichen. Je nach den Verhältnissen gibt es bisexuelle oder zwittrige Formen und alternativ differenzierte oder eingeschlechtliche Formen. In dieses klare Bild sind an einigen Stellen im Reich des Lebendigen Variationen getreten, die scheinbar von der Regel abweichen. Die Ausnahmen haben manchmal dazu verleitet, die Regel zu bestreiten, während es immer der bessere Weg ist, erst zu versuchen, die Ausnahmen auf die Regel zurückzuführen.

Bei zwittrigen Blütenpflanzen kennt man genetische Sterilitätsfaktoren. Im einfachsten Fall bewirken sie, daß Pflanzen vom Genotyp Ss nur mit solchen vom Genotyp ss kopulieren können oder fertil kreuzbar sind. Das führt zur Selbststerilität der Zwitterpflanzen. Wegen der klaren morphologischen Homologie der Geschlechtsorgane, die sich eindeutig als männlich oder weiblich charakterisieren lassen, haben die Sterilitätsfaktoren der Blütenpflanzen nie das Grundkonzept der Sexualitätstheorie gestört. Die Inkompatibilitätsfaktoren geben im Gegenteil wichtige Ansatzpunkte für eine genauere physiologische Analyse der Sexualvorgänge, wenn sie den normalen physiologischen Verlauf der Befruchtung an definierten Stellen blockieren oder morphologische Differenzierungen der Geschlechtsorgane bewirken, die die Befruchtung verhindern. Sie müssen dann Variationen der genetischen Grundlage für die Geschlechtsdifferenzierung, d. h. Variationen im A—G-System sein. Damit gewinnen sie für das Sexualitätsproblem, speziell für die Physiologie der Kopulation und die Genphysiologie der Geschlechtsausbildung größeres Interesse, als die formal-genetische Analyse der Realisatorsysteme.

Bei den *Ciliaten* sind Verhältnisse bekanntgeworden, die sich schwerer in das Schema der Sexualitätstheorie einordnen lassen. Bei *Paramaecium*, dem typischen Vertreter unter isogamen Ciliaten, paaren sich 2 Individuen, die sich aneinanderlegen und 2 Kopulationskerne austauschen. Die beiden (genetisch gleichen) Kerne verhalten sich verschieden, der eine ist ein „Wanderkern" und wandert zum „stationären" Kern des Konjugationspartners und umgekehrt. Die konjugierenden Gamonten wurden früher immer als Zwitter angesehen. Als Sexualunterschied galt nur das verschiedene Verhalten der beiden Kerne. Seit einiger Zeit hat man zunehmend viele Arten gefunden, die nicht intraklonal konjugieren können. Zur Konjugation sind vielmehr zwei verschiedene „Paarungstypen" notwendig. Die Paarungstypen können phänotypisch oder genotypisch bestimmt sein. Bei *Paramaecium bursaria* fanden Jennings und Chen ganze Serien von

Tabelle 2. *Paarungstypsystem bei Paramaecium bursaria.* + Konjugation; − keine Konjugation; (+) sterile Konjugation. (Nach JENNINGS und CHEN, aus SONNEBORN 1947.)

Varietät		I	II	III	IV	V	VI
	Paarungstyp	A B C D	E F G H I K L M	N O P Q	R S	T	U V W X
I	A	−+++	−−−−−−−−	−−−−	−−	−	−−−−
	B	+−++	−−−−−−−−	−−−−	−−	−	−−−−
	C	++−+	−−−−−−−−	−−−−	−−	−	−−−−
	D	+++−	−−−−−−−−	−−−−	−−	−	−−−−
II	E		−+++++++	−−−−	(+)−	−	−−−−
	F		+−++++++	−−−−	−−	−	−−−−
	G		++−+++++	−−−−	−−	−	−−−−
	H		+++−++++	−−−−	−−	−	−−−−
	I		++++−+++	−−−−	−−	−	−−−−
	K		+++++−++	−−−−	(+)−	−	−−−−
	L		++++++−+	−−−−	(+)−	−	−−−−
	M		+++++++−	−−−−	(+)−	−	−−−−
III	N			−+++	−−	−	−−−−
	O			+−++	−−	−	−−−−
	P			++−+	−−	−	−−−−
	Q			+++−	−−	−	−−−−
IV	R				−+	−	−−−−
	S				+−	−	−−−−
V	T					−	−−−−
VI	U						−+++
	V						+−++
	W						++−+
	X						+++−

Paarungstypen, die sich auf einzelne kreuzbare Gruppen oder Varietäten verteilen. Ihre Kombination ergibt allerdings kein Zweierschema, wie es bei einer klaren Getrenntgeschlechtlichkeit auftritt, sondern jeder Gamont von einem Paarungstyp kann mit jedem anderen Paarungstyp der gleichen Varietät konjugieren. Nur die „homozygote" Konjugation ist ausgeschlossen. Die Tabelle 2 veranschaulicht den Sachverhalt (SONNEBORN 1947). Zum Kontrast ist ein bipolares Kreuzungsschema von 8 Klonen von *Chlamydomonas eugametos*, die aus acht isolierten Gonen gezüchtet wurden, zur Seite gestellt (Tabelle 3). Die verschiedenen Paarungstypen beweisen, daß die Konjugationspartner keine kompletten Zwitter sind, wie man früher annahm, sonst müßten sie intraklonal konjugieren können. Es können auch nicht streng eingeschlechtliche Zellen sein, denn dann müßte ein bipolares Zweierschema auftreten. Offenbar fehlt jedem Paarungstyp ein spezifischer physiologischer Charakter, der für die Interaktion der Zellen unter sich notwendig ist. Das kann eventuell ein Befruchtungsstoff ähnlicher Art sein, wie sie KIMBALL (1941) bei *Euplotes patella* gefunden hat. Jeder Paarungstyp hat die Lücke an einer anderen

Tabelle 3. *Zweierschema von 8 Klonen von Chlamydomonas eugametos.*

		♂				♀			
		1	2	3	4	5	6	7	8
♂	1	−	−	−	−	+	+	+	+
	2	−	−	−	−	+	+	+	+
	3	−	−	−	−	+	+	+	+
	4	−	−	−	−	+	+	+	+
♀	5	+	+	+	+	−	−	−	−
	6	+	+	+	+	−	−	−	−
	7	+	+	+	+	−	−	−	−
	8	+	+	+	+	−	−	−	−

Stelle im Differenzierungsmuster für die Sexualreaktion. Die verschiedenen Paarungstypen können sich daher komplementär ergänzen. Zwischen den Varietäten muß dagegen der Unterschied im Differenzierungsmuster so groß sein, daß die Sexualreaktion physiologisch unmöglich wird. Man kann solche unvollständigen Zwitter als eine Übergangsform von Zwittrigkeit zur Eingeschlechtlichkeit ansehen, wie sie die anisogamen Peritrichen zeigen.

Zu dieser Deutung ist generell folgendes zu sagen. Die Klassifikation der GB ist in der Hauptsache auf drei verschiedene Alternativen gegründet, eingeschlechtlich-zwittrig, genotypische-phänotypische GB, haploide-diploide GB und Geschlechtsdifferenzierung. Es ist erstaunlich, daß sich bei den meisten Arten ihr GB-Modus zu den vier resultierenden Haupttypen haplophänotypisch, diplophänotypisch, haplogenotypisch und diplogenotypisch eindeutig zuordnen läßt. Dessen ungeachtet muß man aber erwarten, daß die Typen nicht ohne Übergänge und Zwischenformen nebeneinanderstehen. Die morphologischen und physiologischen Geschlechtsdifferenzierungen sind außerordentlich vielgestaltig und variieren von Art zu Art. Es ist unwahrscheinlich, daß sich mit einem einfachen, auf wenige Alternativen gegründeten Begriffssystem die Mannigfaltigkeit der Natur vollkommen adäquat beschreiben läßt. In einzelnen Fällen sind Zwischenformen lange bekannt. So hat die Subdiöcie eine Mittelstellung zwischen genotypischer und phänotypischer GB. Die oft diskutierte polyfaktiorelle GB ist ebenfalls ein ähnlicher Grenzfall. Wenn die Nachkommen einer Kreuzung ein unrationales und schwankendes Zahlenverhältnis der Geschlechter zeigen, so ist nicht zu unterscheiden, ob das Geschlecht jedes einzelnen Individuums genetisch polyfaktoriell oder phänotypisch bestimmt ist. Es können unter solchen Bastarden beide Fälle nebeneinander vorkommen, indem in einzelnen Rekombinationen irgendein Gen geschlechtsentscheidend wirkt und andere phänotypisch bestimmt werden, weil eine vollständige Balance der geschlechtsbestimmenden Gene herrscht. Besonders bei Fischen wurden solche Komplikationen gefunden (Winge 1934).

Die häufigen Intersexe lassen sich ebenfalls als Grenzfälle zwischen Zwittrigkeit und Eingeschlechtlichkeit deuten. Übergänge zwischen haploider und diploider Geschlechtsdifferenzierung sind dagegen bisher kaum gefunden worden. Gerade hier liegt eine besondere Komplikation vor, weil bei haploider GB nur die Haplophase sexuell determiniert ist, bei diploider GB aber beide Kernphasen sexuell determiniert und differenziert sein müssen. Die besonderen Verhältnisse der Ciliaten — eine Übersicht gibt Grell (1956) — sind eventuell solche Übergänge. Unter den Holotrichen gibt es rein zwittrige Formen ohne Komplikationen durch Paarungstypen, wie *Paramaecium caudatum*, das intraklonal konjugieren kann. Sie bilden je zwei sexuell verschiedene, haploide Kopulationskerne. Dann gibt es ganze Reihen von Formen mit verschiedenen Paarungstypsystemen, bei denen die Konjuganten nicht mehr vollständige Zwitter sind, aber noch beide sexuellen Kernsorten bilden. Die verwandten Peritrichen sind dagegen ganz eingeschlechtlich. Sie kopulieren anisogam und haben phänotypisch GB. Jeder Konjugant bildet auch nur noch Wanderkerne oder stationäre Kerne.

Unabhängig von jeder speziellen Deutung können die 4 Haupttypen der GB nicht unvermittelt nebeneinanderstehen. Es ist vielmehr eine Variabilität der GB-Modi vorauszusetzen. Einzelne Varianten, aber keineswegs alle, könnten Übergangsformen zwischen den Haupttypen sein.

Die größte Variabilität der GB ist unter Pilzen zu finden. Parallel damit ist bei ihnen auch die Geschlechtsdifferenzierung außerordentlich plastisch. Wahrscheinlich hängt die große Variabilität in beiderlei Hinsicht ursächlich zusammen.

Es gibt unter den Pilzen Formen mit ausgeprägter Anisogamie bis zu Formen mit reiner „Somatogamie", bei denen morphologisch keine spezifischen Kopulationsorgane vorhanden sind, sondern morphologisch somatische Zellen kopulieren. Reine Gametenkopulation ist selten. Oft kopulieren die Gametangien. Sie sind mitunter morphologisch als „Oogonien" und „Antheridien" zu unterscheiden. Bei vielen Ascomyceten wird nur noch eine Sorte von Gametangien gebildet, die durch somatische Zellen befruchtet werden. Einen ziemlich vollständigen Überblick über die progressiven, sexualmorphologischen Rückbildungen der verschiedenen Arten gibt GÄUMANN (1949). Die meisten Pilze sind fraglos monöcisch oder zwittrig.

Bei einer Reihe von *Ascomyceten* ergibt die Kombination von Einspormycelien ein klares bipolares Zweierschema. Auf Grund dieser Kreuzungsergebnisse, bei denen zunächst nur auf die Perithezienbildung geachtet wurde, schien es sich um Getrenntgeschlechtlichkeit mit genotypischer GB zu handeln. Genauere Untersuchungen von AMES, DODGE, DRAYTON, ZICKLER u. a. (HARTMANN 1956) zeigten, daß diese scheinbaren Diöcisten in morphologischer Hinsicht Zwitter sind, die Oogonien und Antheridien bilden. Es kann aber immer nur ein Mycel vom genetischen Reaktionstyp A mit dem Reaktionstyp a kopulieren. Homozygote Paarung ist nicht möglich. Dies ist die Erklärung für die scheinbare Diöcie. Morphologisch sind es einwandfrei Zwitter mit Inkompatibilitätsfaktoren A, a, die wie Geschlechtsrealisatoren alternativ spalten.

Diese Entdeckung führte zu beachtlichen Komplikationen bei isogamen Arten, die keine morphologisch verschiedenen Kopulationsorgane bilden, aber bei Kreuzungen ein Zweierschema ergeben. Hier taucht die Frage auf, ob es sich um „echte" Diöcie durch alternativ verteilte Realisatoren oder ebenfalls um „scheinbare" Diöcie von Zwittern mit Inkompatibilitätsfaktoren handelt. *Phycomyces blakesleanus*, bei dem zuerst eine genetische Getrenntgeschlechtlichkeit bei Isogamie gefunden wurde, gehört unter vielen anderen zu diesen strittigen Grenzfällen. Voraussetzung für die Entscheidung wäre die klare Kenntnis über die physiologische Funktion von Realisatorgenen und Inkompatibilitätsgenen. Es ist fraglich, ob sich zwischen ihnen überhaupt immer klare Grenzen ziehen lassen. Die generelle Problematik der physiologischen Realisatorwirkung wurde an dem Melandriumbeispiel bereits diskutiert.

Bei vielen *Basidiomyceten* und *Hymenomyceten* kommt als weitere Komplikation hinzu, daß das Kopulationsverhalten nicht nur durch zwei allele Faktoren A, a bestimmt wird, sondern daß es oft ganze Serien multipler Allele gibt A^1, A^2, A^3 ...... A^n. Jede heterozygote Kombination führt zur Befruchtung, die homozygoten nicht. Das ist im Prinzip dasselbe komplementäre Schema wie bei den erwähnten Paarungstypen der Ciliaten.

Noch eine weitere Komplikation bringt der sog. tetrapolare Kreuzungstyp. Hier entscheidet die Kombination von Allelen zweier verschiedener Loci darüber, ob eine Kreuzung möglich ist oder nicht. Im einfachsten Fall von 2 Allelen für jeden Locus Aa, Bb werden 4 Sorten von Sporen gebildet: AB, Ab, aB, ab. Vollkompatible Reaktionen gibt es nur bei Kombinationen, in denen beide Loci heterozygot sind. Von den 16 möglichen Kombinationen sind also vier kompatibel, die anderen nicht. Tabelle 4 gibt ein Beispiel von einem Hymenomyceten nach KNIEP. In den meisten Fällen von Tetrapolarität gibt es für beide Loci mehr als 2 Allele. Seltener sind auch mehr als 2 Loci gefunden worden, die das Kreuzungsverhalten kontrollieren.

Es ist sehr viel Mühe darauf verwandt worden, in diese verwirrenden genetischen Verhältnisse Ordnung zu bringen. HARTMANN hat auf Grund der Versuche von GREIS an dem Basidiomyceten *Solenia anomala* und von QUINTANILHA an dem

Tabelle 4. *Aleurodiscus polygonius (Hymenomycet)*. Reaktion von 15 Haplonten eines Hymeniums, tetrapolarer Typus. (Nach KNIEP 1928.)

		I AB				II ab			III Ab					IV aB		
		1	2	4	11	3	10	13	5	6	8	12	15	7	9	14
I AB	1	o	o	o	o	Z	Z	Z	o	o	o	o	o	o	o	o
	2	o	o	o	o	Z	Z	Z	o	o	o	o	o	o	o	o
	4	o	o	o	o	Z	Z	Z	o	o	o	o	o	o	o	o
	11	o	o	o	o	Z	Z	Z	o	o	o	o	o	o	o	o
II ab	3	Z	Z	Z	Z	o	o	o	o	o	o	o	o	o	o	o
	10	Z	Z	Z	Z	o	o	o	o	o	o	o	o	o	o	o
	13	Z	Z	Z	Z	o	o	o	o	o	o	o	o	o	o	o
III Ab	5	o	o	o	o	o	o	o	o	o	o	o	o	Z	Z	Z
	6	o	o	o	o	o	o	o	o	o	o	o	o	Z	Z	Z
	8	o	o	o	o	o	o	o	o	o	o	o	o	Z	Z	Z
	12	o	o	o	o	o	o	o	o	o	o	o	o	Z	Z	Z
	15	o	o	o	o	o	o	o	o	o	o	o	o	Z	Z	Z
IV aB	7	o	o	o	o	o	o	o	Z	Z	Z	Z	Z	o	o	o
	9	o	o	o	o	o	o	o	Z	Z	Z	Z	Z	o	o	o
	14	o	o	o	o	o	o	o	Z	Z	Z	Z	Z	o	o	o

Hymenomyceten *Coprinus fimetarius*, die beide vom tetrapolaren Typ sind, den einen Locus als echten Realisator und den anderen als Sterilitätsfaktor gedeutet. GREIS fand Sterilitätsfaktoren von sehr verschiedener Wirkung, die homozygot entweder die Plasmogamie, die Karyogamie, die Fruchtkörperbildung oder die Basidienbildung verhinderten. QUINTANILHA fand, daß die verschiedenen A-Allele bei seinem Objekt eine verschiedene Affinität zueinander haben. Das zeigte der Austausch von Kernen zwischen haploiden und dikaryotischen Mycelien. Bei den A-Faktoren könnte es sich also um echte Realisatoren von verschiedener Valenz handeln. Generalisieren läßt sich diese Deutung nicht auf alle Fälle.

Ähnliche unterschiedliche Auswirkungen wie bei Solenia wurden bei *Schizophyllum commune* abhängig von der heterokaryotischen Kombination der A- und B-Faktoren gefunden (RAPER 1955). Bei *Glomerella* wurden zahlreiche Mutanten verschiedener Loci gefunden, die homozygot bestimmte Schritte im Sexualcyclus verhindern (WHEELER 1954). Heterozygot können sich die mutierten und außer Funktion gesetzten Gene komplementär ergänzen. Dadurch gibt es alle Übergänge von selbstfertilen, monöcischen Mycelien bis zum multipolaren Kreuzungsverhalten, je nach der Rekombination verschiedener Verlustmutanten oder monöcischer Wildtypallele.

Genetische Faktoren, die homozygot Sterilität bedingen, weil sie primär vegetative Funktionen, wie das Wachstum stören, sind selbstverständlich keine Sexualfaktoren. Wenn solche Faktoren ein bipolares Kreuzungsschema bewirken, so hat das nichts mit Getrenntgeschlechtlichkeit zu tun.

Genetische Faktoren, die aber spezifisch bestimmte Sexualvorgänge, wie die Plasmogamie oder die Karyogamie blockieren, sind dagegen Sexualfaktoren. Die Frage bleibt nur, ob sie als „Realisatoren“ angesehen werden können, wenn sie ein bipolares Kreuzungsschema bewirken. Es sind nicht eigentlich geschlechtsbestimmende Gene, sondern Loci, die einzelne sexualphysiologische Vorgänge kontrollieren. In das allgemeine Schema der GB, das zwischen A—G-Komplex einerseits und F—M-Realisatoren andererseits unterscheidet, lassen sich diese Verhältnisse nicht ohne weiteres einordnen. Trotzdem lassen sie sich aber vom Standpunkt der Sexualitätstheorie sehr einfach erklären.

Arten, die differenzierte Kopulationsorgane, Antheridien und Oogonien bilden, sind in den meisten Fällen in morphologischer Hinsicht Zwitter, da beide Sorten von Geschlechtsorganen von einem Haplomycel gebildet werden. Sie besitzen das volle sexuelle Anlagensystem, den kompletten A—G-Komplex. Der progressive Abbau morphologischer Sexualcharaktere ist durch den progressiven Verlust der zugrunde liegenden genetischen Potenzen verursacht. Das A—G-System ist auf rein physiologische Potenzen für die Kopulation zusammengeschrumpft, wenn die Befruchtung nur noch rein somatogam erfolgt. Es ist zu erwarten, daß der progressive Abbau des A—G-Systems nicht einseitig auf morphologische Merkmale beschränkt ist, sondern auch A—G-Gene für physiologische Funktionen Hand in Hand damit sukzessive ausfallen. Solche Formen sind dann imkomplette Zwitter. Sie können nur noch mit Typen kopulieren, die den Ausfall in ihrem eigenen Differenzierungsmuster kompensieren. Dadurch können komplexe genetische Paarungstypsysteme entstehen. Ein formales Beispiel soll das veranschaulichen.

Es seien A^+, B^+, C^+, D^+, E^+, F^+ 6 Loci eines selbstfertilen (monöcischen) Zwitters, die die Sexualvorgänge kontrollieren. Sie symbolisieren die volle bisexuelle Potenz oder das komplette A—G-System. Eine Einteilung der 6 Loci in A-Faktoren und G-Faktoren ist willkürlich.

Eine Mutation im B-Locus zu B_1^- bewirkt Inkompatibilität der Mutanten, da die zugehörige Sexualfunktion ausfällt, wenn B_1^- homozygot kombiniert wird. Die Mutante B_1^- kann nur noch mit der zwittrigen Ausgangsform oder mit Mutanten A^-, C^-, D^-, E^-, F^-, mit denen sie sich diploid oder dikaryotisch funktionell ergänzt, kopulieren. So können durch Mutation verschiedener Loci Zwei-Locus-Systeme entstehen, die im Kreuzungsverhalten Diöcie oder „Heterothallie" ergeben.

Besteht jeder physiologische Locus A, B, C, D, E, F aus einer Reihe genetischer Untereinheiten, z. B. B = r, s t, u ..., die getrennt mutieren können, so resultieren daraus physiologisch verschiedenen Mutanten B_1^-, B_2^-, B_3^- ..., die homozygot ebenfalls den Ausfall der sexualphysiologischen B-Funktion bewirken und daher inkompatibel sind. Sie sind fertil nur mit Genotypen, die das Gen B^+ haben wie der Ausgangszwitter oder in heterozygoter Kombination mit anderen B^--Allelen, die sich komplementär ergänzen zum funktionellen B-Locus. Daraus resultiert ein Paarungsverhalten, das durch multiple Allelserien kontrolliert wird. Die Untereinheiten vom B-Locus r, s, t, u ... können nicht nur isoliert mutieren, sondern sie können auch untereinander rekombinieren, so daß aus 2 Inkompatibilitätsallelen B_1^-, B_2^- neue Allele B_3^-, B_4^- entstehen können. Daß neue allele Inkompatibilitätsfaktoren nicht nur durch Mutationen, sondern auch durch Faktorenaustausch von Untereinheiten entstehen können, hat Papazian an *Schizophyllum commune* gezeigt (Papazian 1950, 1951; Raper 1955).

Sind solche funktionellen Minus-Mutanten im A—G-System die Ursache für ein diöcisches Paarungsverhalten, so sind natürlich auch Rekombinationen von heterothallischen plus- und minus-Paarungstypen zu Zwittern zu erwarten. In diesen Fällen wird die Diöcie nicht durch Verteilung von F- und M-Faktoren bei komplettem A—G-System verursacht, wie es im Idealfall echter genotypischer GB anzunehmen ist, sondern durch die alternative Verteilung von A- und G-Minus-Mutanten. Die A- und G-Faktoren können natürlich nicht zueinander allel sein, da sie im haploiden Chromosomensatz gemäß dem Grundprinzip der „bisexuellen Potenz" vollständig vorhanden sein müssen. Die gerade bei Pilzen gefundenen Übergänge von Diöcie zu Monöcie durch Austauschvorgänge sind auf dieser Basis gut zu verstehen.

Die komplexe „physiologische Heterothallie" (WHITEHOUSE 1949) der Pilze und ihr kompliziertes genetisches Bestimmungsmuster von Paarungstypen erklärt die dargelegte Hypothese durch Variationen im A—G-System. Sie führt die sexualmorphologischen Rückbildungen und die damit parallel laufende, zunehmend komplizierter werdende Heterothallie der Pilze auf die gleiche Ursache zurück, den progressiven Verlust von A—G-Potenzen. Dieser Weg der genetischen „GB" unterscheidet sich also durch das Auftreten zahlreicher Komplikationen von dem allgemeinen Modus der genotypischen GB durch F- und M-Realisatoren, die zusätzlich zum A—G-Komplex wirken. Gemeinsam ist beiden GB-Modi die Basis der bisexuellen Potenz oder der zwittrige Ausgangszustand.

Im allgemeinen wird bei der begrifflichen Einteilung der Heterothallie der Pilze nur das genetische Bestimmungsmuster der Paarungstypen berücksichtigt. Ein Verständnis wird aber erst durch die vergleichende Betrachtung der morphologischen, physiologischen und genetischen Kopulationsfaktoren gewonnen.

Literatur.

ALLEN, C. E.: The occurrence of polyploidy in *Sphaerocarpus*. Amer. J. Bot. **22**, 635—644 (1935).

BLAKESLEE, A. F.: Sexual reproduction in the *Mucorineae*. Proc. Amer. Acad. Arts Sci. **40**, 205—319 (1904). — BLAKESLEE, A. F., J. L. CARTLEDGE, D. S. WELCH and A. D. BERGNER: Sexual dimorphism in Mucorales. I. Intraspecific reactions. Bot. Gaz. **84**, 27—50 (1927). — BURNETT, J. H.: The mating system of fungi. New Phytologist **55**, 50—90 (1956).

CLEVELAND, L. R.: Hormone-induced sexual cycles of flagellates. I. Gametogenesis, fertilization, and meiosis in *Trichonympha*. J. Morph. **85**, 197—296 (1949). — CORRENS, C.: Die geschlechtliche Tendenz der Keimzellen gemischtgeschlechtlicher Pflanzen. Z. Bot. **12**, 49, Ges.Abh. 1051 (1920). — Bestimmung, Vererbung und Verteilung des Geschlechts bei höheren Pflanzen, Bd. 2, S. 1—138. Berlin: Gebr. Bornträger 1928.

DREW, K. M., and I. FRIEDMANN: Occurrence of motile gametes in *Prasiola stipitata* Suhr. Nature (Lond.) **180**, 557—558 (1957).

FÖRSTER, H.: Das Wirkungsspektrum der Kopulation von *Chlamydomonas eugametos*. Z. Naturforschung. **12**b, 765—770 (1957). — FÖRSTER, H., u. L. WIESE: Gamonwirkungen bei *Chlamydomonas eugametos*. Z. Naturforsch. **9**b, 548 (1954). — FÖRSTER, H., L. WIESE u. G. BRAUNITZER: Über das agglutinierend wirkende Gynogamon von *Chlamydomonas eugametos*. Z. Naturforsch. **11**b, 316—317 (1956). — FRIEDMANN, I.: Structure, life history and sex determination of *Prasiola stipitata* Suhr. Ann. Bot., N.S. **23**, 571—594 (1959).

GÄUMANN, E.: Die Pilze. Grundzüge ihrer Entwicklungsgeschichte und Morphologie. Basel: Birkhäuser 1949. — GALAN, F.: Analyse génétique de la monoecie et de la dioecie zygotiques et de leur différence dans *Ecballium elaterium*. Act. Salmanticensia Sec. Biol. 7—15 (1951). — GEITLER, L.: Kopulation und Geschlechtsverteilung bei einer *Nitzschia*-Art. Arch. Protistenk. **59**, 506—549 (1928). — GREIS, H.: Mutations- und Isolationsversuche zur Beeinflussung des Geschlechts von *Sordaria fimicola* (Rob.). (Ein Beitrag zur Frage nach der Stabilität und Labilität bei Monözisten und Diözisten.) Z. Bot. **37**, 1—116 (1941).— GRELL, K. G.: Protozoologie. Berlin-Göttingen-Heidelberg: Springer 1956.

HARTMANN, M.: Über sexuelle Differenzierung und relative Sexualität. Studia Mendeliana, Brünn 203—228 (1923). — Die Befruchtungsvorgänge der hypermastiginen Flagellaten und ihre theoretische Bedeutung. Arch. Protistenk. **99**, 328—358 (1954). — Die Sexualität, 2. Aufl. Stuttgart: Gustav Fischer 1956. — HASSINGER-HUIZINGA, H.: Generationswechsel und Geschlechtsbestimmung bei *Callithamnion corymbosum* (SM) Lyngb. Arch. Protistenk. **98**, 91—125 (1952). — HEITZ, E.: Über Beziehungen zwischen Polyploidie und Gemischtgeschlechtlichkeit bei Moosen. 2. Jber. schweiz. Ges. Vererb. **17** (1942). — Kritische Bemerkungen betreffend den Nachweis von Realisatorgenen bei *Sordaria fimicola*. Vjschr. naturforsch. Ges. Zürich **40** (1945). — HIRSCH, H. E.: The cytogenetics of sex in *Hypomyces solani* f. *cucurbitetae*. Amer. J. Bot. **36**, 113—121 (1949). — HÜTTIG, W.: Die Sexualität bei *Glomerella lycopersici* Krüger und ihre Vererbung. Biol. Zbl. **55**, 74—83 (1935).

KIMBALL, R. F.: The nature and inheritance of mating types in *Euplotes patella*. Genetics **27** (1941). — KNAPP, E., u. I. HOFFMANN: Geschlechtsumwandlung bei *Sphaerocarpus* durch Verlust eines Stückes des X-Chromosoms. Chromosoma (Berl.) **1**, 130—146 (1939). — KNIEP, H.: Die Sexualität der niederen Pflanzen. Jena: Gustav Fischer 1928. — KÖHLER, K.: Entwicklungsgeschichte, Geschlechtsbestimmung und Befruchtung bei *Chaetomorpha*. Arch. Protistenk. **101**, 223—269 (1956).

LEUPOLD, U.: Die Vererbung von Homothallie und Heterothallie bei *Schizosaccharomyces Pombe*. C. R. Lab. Carlsberg, Sér. physiol. **24**, 27 (1950). — LEWIN, R. A.: Control of sexual activity in *Chlamydomonas* by light. J. gen. Microbiol. **15**, 170 (1956). — Gamete behavior in *Chlamydomonas*. Nature (Lond.) **166**, 76 (1950). — LORBEER, G.: Untersuchungen über Reduktionsteilung und Geschlechtsbestimmung bei Lebermoosen. Z. indukt. Abstamm.- u. Vererb.-Lehre **44** (1927). — Über das Vorkommen von drei verschiedenen Geschlechtsrealisatoren bei den Lebermoosen. Planta (Berl.) **27**, 708—717 (1938).

PAPAZIAN, H. P.: The genetics and physiology of the incompatibility alleles and some related genes in *Schizophyllum commune*. Doctor's thesis Univ. Chicago 1950. — Physiology of the incompatibility factors in *Schizophyllum commune*. Bot. Gaz. **112**, 143—163 (1950). — The incompatibility factors and a related gene in *Schizophyllum commune*. Genetics **36**, 441—459 (1951).

QUINTANILHA, A.: Étude génétique du phénomène de Buller. Bol. Soc. broteriana, Sér. II **14**, 17—46 (1939).

RAPER, J. R.: Life cycles, sexuality, and sexual mechanisms in the fungi. Sex in microorganisms. Amer. Assoc. for the Advancement of Sci. 1954, pp. 42—81. — Some problems of specificity in the sexuality of plants. Biological specificity and growth. 12. Symposium, Princeton: Princeton University Press 1955.

SEILER, J.: Das Intersexualitätsphänomen. Zusammenfassende Darstellung des Beobachtungsmaterials an *Solenobia triquetrella* und Deutungsversuch. Experientia (Basel) **5**, 425—428 (1949). — SEILER, J., et al.: Analyse des intersexuellen Fühlers von *Solenobia triquetrella* (*Psychidae, Lepid.*). Rev. suisse Zool. **58**, 489—495 (1951). — SONNEBORN, T. M.: Recent advances in the Genetics of *Paramecium* and *Euplotes*. Advanc. Genet. **1** (1947). — STARR, R. C.: Heterothallism in *Cosmarium Botrytis Var. Subtumidum*. Amer. J. Bot. **41**, 601—607 (1954).

WETTSTEIN, F. v.: Morphologie und Physiologie des Formwechsels der Moose auf genetischer Grundlage. I. Z. indukt. Abstamm.- u. Vererb.-Lehre **33**, 1—236 (1924). — Über Fragen der Geschlechtsbestimmung bei Pflanzen. Naturwiss. **12**, 761—768 (1924). — WESTERGAARD, M.: Über den Mechanismus der Geschlechtsbestimmung bei *Melandrium album*. Naturwiss. **40**, 253—260 (1953). — WHEELER, H. E.: Genetics and evolution of heterothallism in *Glomerella*. Phytopathology **44**, 342—345 (1954). — WHITEHOUSE, A. L. K.: Multiple-allelomorph heterothallism in the fungi. New Phytologist **48**, 212—244 (1949). — WIESE, L.: Die diplogenotypische Geschlechtsbestimmung. Fortschr. Zool. **10**, 391—439 (1956). — WINGE, Ö.: The experimental alternation of sex-chromosomes into autosomes and vice versa, as illustrated by *Lebistes*. C. R. Lab. Carlsberg, Sér. physiol. **21** (1934).

Die Wirkung (der Geschlechtsfaktoren) bei diplogenotypischer Bestimmung.

Von

U. Nürnberg-Krüger.

Mit 13 Abbildungen.

Einführung.

Die Erscheinung der diplogenotypischen Geschlechtsbestimmung (G.B.) ist bisher nur bei höher entwickelten Organismen gefunden worden, bei Blütenpflanzen und Metazoen. Die Diplophase ist vom Augenblick der Zygotenbildung an sexuell determiniert. Die genetische sexuelle Tendenz der Haplophase wird durch die Reduktionsteilung (R.T.) bestimmt, ihre physiologische Sexualfunktion dagegen ist durch den Geschlechtscharakter der die Gameten bildenden Organe der Diplophase festgelegt. Bei den Blütenpflanzen kommen mehrere Typen der geschlechtlichen Ausbildung der Diplophase vor: Bisher am besten untersucht sind die Verhältnisse bei einigen streng unisexuellen diöcischen Arten; sie haben ♂ und ♀ Individuen. Bei der Mehrzahl der Arten liegt aber monöcische Verteilung der Sexualorgane (in getrennten Blüten auf demselben Individuum) oder zwittrige Geschlechtsausbildung vor (mit ♂ *und* ♀ Organen in sämtlichen Blüten). Zwischen diesen beiden Typen gibt es viele Zwischenformen, wie Subdiöcisten, bei denen ein Geschlecht, meist das weibliche, rein ausgebildet wird, das andere jedoch ± fließende Übergänge vom rein ♂ zum stark monöcischen oder zwittrigen Zustand zeigt. Bei den Polyöcisten treten zwei oder mehr verschieden-geschlechtliche Formen in einer Fortpflanzungsgemeinschaft auf, von denen in der Regel eine autöcisch (Correns) ist, d. h. monöcisch oder zwittrig. So kommen Arten mit Gynodiöcie vor (♀♀ + ⚥⚥), mit Androdiöcie (♂♂ + ⚥⚥) oder Triöcie (♀♀ + ⚥⚥ + ♂♂).

Die phylogenetische Entwicklung der verschiedenen sexuellen Fortpflanzungsmöglichkeiten ist allem Anschein nach ausgegangen von haploiden Organismen mit der genetischen Grundlage für eine allgemeine bisexuelle Potenz, die entweder die Ausbildung haplomonöcischer Arten oder haplodiöcischer Arten mit phänotypischer G.B. ermöglicht. Der nächste Schritt der Entwicklung führte zu Formen, bei denen nicht Außenfaktoren die Realisierung eines der beiden Geschlechtsfaktoren-Komplexe bestimmten, sondern ein Genpaar bzw. ein Einzelgen, der Realisator (nach v. Wettstein). Nach Correns hat sich aus dem haplomonöcischen Zustand ohne Realisatoren der diploautöcische Typus entwickelt. Er besitzt die gleiche allgemeine bisexuelle Potenz, die Fähigkeit zur Ausbildung ♂ und ♀ Geschlechtsorgane und Gameten. Die beiden genetischen Geschlechtsfaktoren-Komplexe (für ♂ und ♀ Ausbildung) faßt Correns als AG-Komplex zusammen.

Von diesem Grundtypus der Blütenpflanzen ausgehend, ist es einmal durch phänotypische G.B. der Diplophase zur Diöcie gekommen. Bei ihr wird durch Wirkung von Außenfaktoren metagam die Realisierung einer der beiden sexuellen Potenzen entschieden. Zum anderen wurde Diöcie durch *diplogenotypische G.B.*

erreicht. Die G.B. erfolgt bei dieser phylogenetisch hochentwickelten Form nicht wie bei der haplogenotypischen G.B. bei der R.T. endgültig, sondern bei der Rekombination der aus der R.T. hervorgegangenen Gameten in der Befruchtung. Ein Geschlechtstyp, meist der weibliche, bildet in der R.T. bei homozygot vorhandenem genetischem Realisator Gameten gleicher Geschlechtstendenz. Der zweite Geschlechtstyp, meist der männliche, erzeugt bei Heterozygotie für den Realisator Gameten, von denen die eine Hälfte die gleiche, die andere die entgegengesetzte Geschlechtstendenz besitzt wie die des ersten Geschlechts. Die Realisatoren können auf bestimmten, in einigen Fällen morphologisch besonders gestalteten sog. Geschlechtschromosomen lokalisiert sein. Man bezeichnet dann das Geschlechtschromosom des homogameten Geschlechts als X-Chromosom, das nur im heterogameten Geschlecht vorkommende als Y-Chromosom. Die physiologische Funktion der Gameten wird bestimmt durch das genetische Geschlecht der Diplophase. Aus der Tatsache der an vielen Stellen des Systems erfolgten Entwicklung zur genetischen G.B. bei Diöcie folgt, daß die G.B.-Mechanismen nicht identisch sein müssen, und also verschiedene Typen beobachtet werden können. In etwa 75% der *Angiospermen*-Familien ist Diöcie beobachtet worden, die vorwiegend genetisch gesteuert ist (Yampolsky und Yampolsky 1922).

Die Feststellung, daß diplogenotypische G.B. bei einem Diöcisten vorliegt, kann nach Correns erfolgen:

1. Durch den Nachweis der Heterogametie eines Geschlechts,
2. durch Verschiebung des Geschlechtsverhältnisses bei Beeinflussung der Heterogameten (Konkurrenzversuche),
3. durch heteromorphe Geschlechtschromosomen,
4. durch das Vorkommen geschlechtsgekoppelter Merkmale und schließlich auch
5. durch Konstanz der Geschlechtsausbildung bei vegetativer Vermehrung unter verschiedenen Umweltbedingungen.

Die noch sehr geringen Kenntnisse über die genotypische G.B. bei zwittrigen Pflanzen gründen sich hauptsächlich auf Untersuchungen an Diöcisten und Subdiöcisten.

Nachweise für den Homozygotie-Heterozygotie-Mechanismus bei diplogenotypischer Geschlechtsbestimmung.

I. Kreuzungen zwischen autöcischen und diöcischen Arten.

Das zahlenmäßige Verhältnis der Geschlechter bei Diöcisten ist in den meisten Fällen annähernd ein 1:1-Verhältnis. Ein derart regelmäßiges Auftreten zweier gleich großer Gruppen ist für einfach mendelnde genetische Faktoren bei Rückkreuzungen mit dem recessiven Elter immer wieder beobachtet worden. Correns hat nachweisen können, daß ein entsprechender Mechanismus auch das konstante Auftreten zweier sexuell verschieden determinierter Individuenklassen bei Diöcisten bewirkt.

a) Correns' *Bryonia*-Versuche.

Bei der Kreuzung der getrennt-geschlechtlichen *Bryonia dioica* mit der gemischt-geschlechtlichen *Bryonia alba* ging Correns (1907) von der Annahme aus, daß gemischt-geschlechtliche Individuen weibliche und männliche Keimzellen mit gleichartiger gemischt-geschlechtlicher Tendenz bilden. Es mußte sich dann bei Kombination der ♀ Keimzelle einer ♀ Pflanze mit Keimzellen von gemischtgeschlechtlicher Tendenz ein anderes Geschlecht der Zygoten ergeben als bei Kombination derselben mit ♂ Keimzellen einer ♂ Pflanze. Aus der Geschlechtsausbildung der Nachkommen sollte sich die geschlechtliche Tendenz der Keimzellen der diöcischen Art ablesen lassen. Voraussetzung war, daß die Zygoten nicht wieder autöcische Pflanzen ergaben.

CORRENS erhielt aus der Kreuzung *Bryonia dioica* ♀ × *Br. alba* ⚥ 589 rein weibliche Nachkommen. Das ♀ der *Bryonia dioica* bildet also nur Keimzellen mit einer weiblichen Tendenz. Sie ist stärker als die zwittrige Tendenz der Pollen von *Bryonia alba*. Die reziproke Kreuzung *Bryonia alba* ⚥ × *Br. dioica* ♂ ergab unter 358 Bastarden 171 rein männliche und 187 rein weibliche. Demnach können die männlichen Keimzellen verschiedene Geschlechtstendenz haben, sowohl weibliche wie auch männliche. Von beiden Typen wurden etwa gleich viele Keimzellen gebildet. Es prävaliert in der Zygote sowohl die weibliche wie auch die männliche Tendenz über die zwittrige.

Symbolisiert man das homogamete ♀ mit zwei gleichen Buchstaben XX und das heterogamete ♂ mit zwei verschiedenen, XY, so ergibt sich bei der Fortpflanzung der diöcischen Art folgender Verlauf:

	♀♀	♂♂
P	xx	xy
Verteilung auf die Keimzellen in der R.T.	x x	x y
F_1-Kombinationen, die durch die Befruchtung entstehen	xx xx	xy xy

= 50% xx (♀♀) + 50% xy (♂♂)

Es werden also ständig in etwa gleicher Häufigkeit Weibchen und Männchen gebildet. Daraus folgt die Homogametie der Weibchen der *Bryonia dioica* und die Heterogametie der Männchen. Außerdem zeigte sich, daß die Geschlechtstendenz eine charakteristische Stärke oder Valenz besitzt. Die durch y symbolisierte ♂ Tendenz ist dominant über die ♀ Tendenz x. xx ist also der homozygot-recessive Typus. CORRENS gibt den Geschlechtsgenotyp der ♀♀ mit AG AG + $\gamma\gamma$ wieder, wobei AG den AG-Komplex eines haploiden Gameten bedeutet und γ den Realisator für Weiblichkeit; der Genotyp der ♂♂ ist dann AG AG $\gamma\,\alpha$, wobei α den Realisator für Männlichkeit darstellt.

b) Weitere Kreuzungen autöcisch und diöcisch.

1. *Bryonia*. CORRENS' Bastarde zwischen *Br. dioica* und *Br. alba* waren vollständig steril. HEILBRONN und seine Mitarbeiter erhielten einen *dioica* × *alba*-Bastard, der im Stambuler Klima vom 2. Jahr ab fertil war (HEILBRONN und BASARMAN 1942, HEILBRONN 1948). Der reziproke Bastard war auch dort steril. Sie bezogen die monöcische *Bryonia macrostylis* und die diöcische *Br. multiflora* in die Kreuzungen ein. Dabei verhält sich *macrostylis* wie *alba* in der Kombination mit *dioica*. In den wenig umfangreichen Rückkreuzungen des *dioica* × *alba*-Bastards treten neben nichtblühenden neutralen Pflanzen ♀♀ und ♂♂ auf. Eine Ausnahme bilden in der 4. Rückkreuzung mit *alba* zwei monöcische Pflanzen. Die beiden diöcischen Arten unterscheiden sich jedoch in ihrem G.B.-Mechanismus, denn es ergab zwar *multiflora* ♀ × *dioica* ♂: 50% ♀♀ und 50% ♂♂, die reziproke Kreuzung aber 17 Neutra und 17 monöcische (♀♂) Pflanzen (HEILBRONN 1953). Ein ähnliches Ergebnis brachte die Kreuzung der beiden monöcischen Arten *macrostylis* × *alba* = 12 N : 12 (♀♂), rez. 2 N : 1 ♀ : 1 ♂ (BILGE 1955). Es wird ein Valenzunterschied der Realisatoren angenommen und ein Valenzunterschied der hier im Plasma verankerten bisexuellen Potenz, die beide bei *dioica* stärker sein müssen.

Neuere Versuche, die Sterilität der *dioica* × *alba*-Bastarde durch Polyploidisierung zu überwinden (BRABEC 1962), führten an den mit Acenaphthen behandelten weiblichen Bastarden *Br. dioica* × *alba* und *Br. alba* × *dioica* zur Bildung fertiler weiblicher Blüten. Auch der Pollen männlicher Blüten, die an behandelten Pflanzen entstanden, sah zu ca. 75% funktionsfähig aus. Bestäubungen der Bastarde mit *dioica*-Pollen ergaben an beiden reziproken Bastarden triploide Samen. Die Geschlechterverteilung der triploiden partiell fertilen Nachkommenschaft mit 14♀♀:20♂♂ ist bei einem P von 47,6% mit dem 1:1-Verhältnis vereinbar. Monöcische Pflanzen traten nicht auf. Entweder wird die geschlechtsbestimmende Plasmawirkung von *Br. alba* durch zwei *dioica*-Genome überwunden, oder es verhalten sich in den Valenzen die *alba*-Sippen der beiden Autoren unterschiedlich.

2. *Ecballium.* Bei Bastardierungen einer nordspanischen *var. monoicum* mit der südspanischen *var. dioicum* der Art *Ecballium elaterium* hat GALÁN (1951, 1953/54) die in der Tabelle dargestellten Ergebnisse erhalten, die er durch eine Allelenserie der geschlechtsbestimmenden Faktoren erklärt (Tabelle 1).

Tabelle 1. *Ergebnisse der Kreuzungen zwischen Ecballium elaterium monoicum (m) und Ecballium elaterium dioicum (d).* (Aus GALÁN 1951.)

Kombination	♀	⚥	♂
m⚥ × m⚥		22	
d♀ × d♂	61		78
d♀ × m⚥		881	
m⚥ × d♂	4	378	423
(d × m) × (d × m)⚥	10	18	
(m × d) × (m × d)⚥	6	11	1
(m × d)⚥ × (d × m)⚥		3	1
	16	32	2
(m⚥ × d♂)⚥ × (m⚥ × d♂)♂	2	44	35
(d♀ × m⚥)⚥ × (m⚥ × d♂)♂	5	58	70
	7	102	105
(m⚥ × d♂)⚥ × d♂	35	30	48
m⚥ × (m⚥ × d♂)♂	3	547	483
m⚥ × (m⚥ × d♂)⚥	1	58	

E. el. monoicum (♀♂) = a a
E. el. dioicum ♂ = a^d a^D
E. el. dioicum ♀ = a^d a^d

Er nimmt also an, daß ein Genlocus der primär monöcischen Form durch zwei Mutationen zum Realisator geworden ist.

In diese Vorstellung fügt sich die Beobachtung, daß sich nach Neutronenbestrahlung an einem ♂ der *var. dioicum* ein Sektor mit völlig normalen weiblichen Blüten entwickelte (GOMEZ-CAMPO und CASAS-BUILLA 1965). Der Zweig war steril und bildete parthenokarpe Früchte. Es wird auf eine Chromosomenmutation geschlossen, etwa eine Deletion, die das Gen a^D einschließt.

3. *Acnida und Amaranthus.* In der Familie der *Amaranthaceae* untersuchte MURRAY (1940) drei rein diöcische Arten der Gattung *Acnida: A. tamariscina, A. cuspidata, A. tuberculata,* die jede normales 1:1-Geschlechtsverhältnis haben. Eine Reihe von Hybriden verhielten sich ebenso. Einige Kombinationen brachten ein 1♀:2♂♂-Verhältnis, das durch ein geschlechtsgebundenes Letalgen im X-Chromosom erklärt wird. Alle Kombinationen der diöcischen *Acnida cuspidata* und *A. tuberculata* mit bestimmten monöcischen Arten erzeugten bei diöcischem ♀ nur ♀♀, bei diöcischem ♂ eine 1:1-Spaltung in ♀♀ und ♂♂. In den entsprechenden Kombinationen mit *Acnida tamariscina* als ♂ traten stärkere Abweichungen vom 1:1-Geschlechtsverhältnis auf und außerdem ein größerer Anteil an "neuter plants", Pflanzen, die zwar Inflorescenzanlagen, aber keine Blüten zeigen. In den Kreuzungen mit einer anderen monöcischen Art, *Amaranthus spinosus,* treten praktisch nur ♂♂ auf, einige Monöcisten und ein bestimmter Anteil an Neutra.

Danach ist nicht wie bei *Bryonia* auf jeden Fall die Geschlechtstendenz der diöcischen Pflanzen stärker als die der monöcischen, sondern es liegt hier eine verschiedene Valenz der Geschlechtstendenzen vor. Es wirkt zwar der männliche Realisator der diöcischen Art geschlechtsbestimmend auf das Genom der monöcischen Art, aber die weibliche Tendenz der diöcischen Form wird von den männlichen Tendenzen des monöcischen Genoms überdeckt. Wie bei *Bryonia* ist bei *Acnida* das ♂ heterogametisch.

Schema der Geschlechtsbildung:

d = ♀ Genom d. diöc. Art	m_1 Genom d. monöc. Art 1. Typs
D = ♂ Genom d. diöc. Art	m_2 Genom d. monöc. Art 2. Typs
diöcische reine Arten:	reine monöcische Arten:
♀ = dd	1. Typs = m_1m_1
♂ = Dd	2. Typs = m_2m_2

Bastarde: d m_1 = ♀
D m_1 = ♂
d m_2 = ♂
D m_2 = ♂
m_1m_2 = Monöcist 1. Typ (manchmal intermediär).

Für die Herausbildung verschiedener Genomtypen im Verlauf der Evolution spricht auch die Tatsache, daß sowohl bei den monöcischen als auch bei den diöcischen Arten solche mit n = 16 und n = 17 als Grundzahl vorkommen (GRANT 1958).

4. *Fragaria.* Kreuzungen zwischen diöcischen und monöcischen bzw. zwittrigen Arten wurden auch in der Gattung *Fragaria* durchgeführt (CORRENS 1928; LILIENFELD 1933, 1936; SCHIEMANN 1931, 1937, 1943; STAUDT 1952, 1954). Bei dieser Gattung als bisher einziger der höheren Pflanzen ist Heterogametie für das weibliche Geschlecht nachgewiesen worden (RICHARDSON 1914, VALLEAU 1923). Nach Kreuzungsergebnissen mit den oktoploiden *Fr. virginiana* und *Fr. chiloënsis* haben MORGAN, STURTEVANT und BRIDGES (1915) bereits auf diese Möglichkeit hingewiesen.

Ergebnisse:

RICHARDSON	*virg.* ♀ × *vesca* od. *grandifl.* ⚥	: ♀♀ + ⚥⚥ (zum Teil zwittrige ♂♂)
	chil. ⚥ × *grandifl.* ⚥	: ⚥⚥ + ♂♂
	chil. ♀ × ⚥	: 12 ♀♀ + 12 ♂♂
CORRENS	*grandifl.* ⚥ × *elatior* ♂	: nur ♂♂
LILIENFELD (1933)	2n *nipponica* ⚥ × 6n *elatior* ♂	: nur ♂♂
	6n *elatior* ♀ × 2n *nipponica* ⚥	: 15 ♀♀ + 20 ♂♂
SCHIEMANN (1937)	6n *elatior* ♀ × 2n *collina* ⚥	: 38 ♀♀ + 29 ♂♂

Entgegen den Angaben von PETROV und TUKAN (1937) wies STAUDT (1952, 1954) auch für die tetraploide diöcische *Fr. orientalis* in Kreuzungen mit der zwittrigen *Fr. vesca* die weibliche Heterogametie nach. Bei *Fragaria* zeigte sich häufig im männlichen Geschlecht eine geringe Tendenz zur Zwittrigkeit, sowohl in den reinen Arten wie bei den Bastarden (STAUDT 1961); in der Regel sind die an ♂♂ gebildeten Samen jedoch nicht keimfähig.

5. *Thalictrum.* KUHN (1930b) berichtet über Kreuzungen zwischen der streng getrennt-geschlechtlichen Art *Thalictrum Fendleri*, die ein konstantes 1:1-Verhältnis der Geschlechter aufweist, mit drei zwittrigen Arten. Die reziproken Kombinationen gelangen nicht.

Thalictrum Fendleri ♀ × *Th. foetidum* ⚥ = sterile ⚥⚥
Thalictrum Fendleri ♀ × *Th. Delavayi* ⚥ = 26 sterile ⚥⚥ + 2 ± ♀♀
Thalictrum Fendleri ♀ × *Th. aquilegifolium* ⚥ = 1 ♂ + 2 ⚥⚥ (± ♂)

Diese Ergebnisse lassen sich wieder mit der Annahme weiblicher Homogametie vereinbaren.

In anderen Kreuzungen, z. B. *Lychnis floscuculi* ⚥ × *Melandrium rubrum* ♂ (NYGREN 1957c) oder *Melandrium furcatum subsp. angustiflorum* ⚥ × *M. rubrum* ♂ (NYGREN 1957b) erwies sich der zwittrige Zustand dominant über den getrenntgeschlechtlichen.

Durch die im Vorhergehenden geschilderten Versuche konnte für eine Reihe diöcischer Arten nachgewiesen werden, daß 1. die G.B. genetisch erfolgt, 2. bei der R.T. 2 Sorten von sexuell determinierenden Gameten vom einen Geschlecht gebildet werden, und zwar meist vom männlichen Geschlecht, 3. bei der R.T. im anderen Geschlecht nur eine Gametensorte entsteht, 4., daß die Geschlechtstendenz des heterogameten Geschlechts dominiert.

II. Certationsversuche.

Ein weiterer Beweis für das Vorhandensein eines homogameten und eines heterogameten Geschlechts in bezug auf die Geschlechtsbestimmungs-Faktoren bei Diöcisten sind die Certationsversuche, die von Correns (1917, 1918, 1921a,b) durchgeführt wurden. Es mußten eventuell vorhandene Differenzen der beiden Gametentypen des heterogamen Geschlechts, z. B. unterschiedliche Resistenz gegen bestimmte Außenbedingungen, zu einer Veränderung des mechanischen (Correns) 1:1-Geschlechtsverhältnisses führen. Da Kombinationen verschiedener Einzelpflanzen z. B. bei *Melandrium dioicum* unterschiedliche Geschlechtsverhältnisse ergaben (Shull schon 1910), konnten richtige Zahlen nur bei Behandlung gleicher Einzelpflanzen-Kombinationen erhalten werden. Veränderung der Konkurrenz-Verhältnisse der befruchtenden Pollen bei *Melandrium* und bei

Tabelle 2a. *Einfluß der Pollenmenge auf das Zahlenverhältnis der Geschlechter.* (Nach Correns 1929.)

Versuchspflanze	I. Sehr viel Pollen				II. Wenig Pollen				Differenz II—I in %
	n	♀	♂	% ♂	n	♀	♂	% ♂	
					A. Melandrium				
15d	1144	754	390	34,09	1169	660	509	43,54	9,45
21a III	525	373	152	28,95	708	378	330	46,61	17,66
22b III	167	111	56	33,53	210	131	79	37,62	4,09
25b I	420	304	116	27,62	290	165	125	43,10	15,48
zusammen	2256	1542	714	31,65	2377	1334	1043	43,78	12,13
					B. *Rumex acetosa*				
A♀+D♂	1175	1027	148	12,60 ± 0,97	1152	818	334	28,99 ± 1,34	+ 16,39 ± 1,65
B♀+E♂	1145	1057	88	7,68 ± 0,79	1118	753	365	32,65 ± 1,42	+ 24,97 ± 1,61
C♀+E♂	1109	1039	70	6,31 ± 0,65	1140	786	354	31,05 ± 1,39	+ 24,74 ± 1,52

Tabelle 2b. *Melandrium, Wirkung einer Unterbrechung der Leitungsbahnen für die Pollenschläuche.* (Nach Correns 1921.)

♀ Versuchspflanze	I. Kontrolle			II. Unterbrechungsversuch			Differenz II—I
	n	♂	% ♂	n	♂	% ♂	
37 F	866	344	39,72 ± 1,66	853	268	31,42 ± 1,59	8,30 ± 2,30
40 B	1746	664	38,03 ± 1,16	1065	347	32,58 ± 1,44	5,45 ± 1,85
66 A	2445	1066	43,60 ± 1,00	714	179	25,85 ± 1,64	17,75 ± 1,71
72	772	286	37,05 ± 1,74	449	120	26,73 ± 2,09	10,32 ± 2,72

Rumex acetosa ergaben ein Sinken der Männchenzahlen bei erhöhter Konkurrenz durch Vermehrung der Pollenmenge (Tabelle 2). Eine Unterbrechung der Leitungsbahnen für die ♂ Gametenkerne und Kontrolle der Lage der befruchteten Samenanlagen im Fruchtknoten von *Melandrium* zeigten, daß männlich bestimmender Pollen langsamer wächst (Abb. 1). Behandlung des Pollens mit Alkoholdämpfen (Correns 1921, 1922, 1924) und Alternlassen des Pollens führte bei

Melandrium zu einer Steigerung der Männchenzahlen um 10—21% bzw. bis fast auf 100%. Demgegenüber blieb das Alter der Eizellen ohne Einfluß auf das Geschlechtsverhältnis. Diese Tatsachen sprechen für die aus den *Bryonia*-Versuchen gefolgerte Heterogametie der Männchen[1]. Das Fehlen einer Verschiebung durch Konkurrenz bei *Fragaria* (Correns 1928) läßt die Möglichkeit noch offen, daß hier nicht das männliche Geschlecht heterogamet ist. *Humulus japonicus* ergibt nach Versuchen von Kihara und Hirayoshi (1932) bei reichlicher Bestäubung eine Vermehrung des Weibchenanteils in der Nachkommenschaft, weist also auch männliche Heterogametie auf.

Certationsversuche an *Cannabis sativa* führte Laskowska (1961) durch. Während das Alter

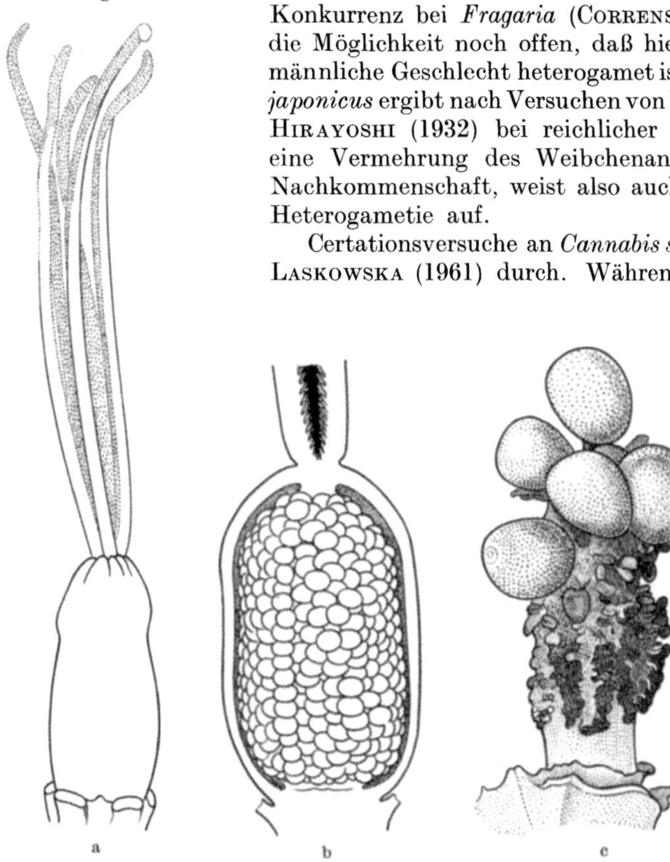

Abb. 1a—c. Aus einer weiblichen Blüte von *Melandrium*. a Fruchtknoten mit Stempel, b Lage der vielen Samenanlagen an der langgestreckten mittelständigen Placenta, c Entwicklung weniger Samenanlagen am oberen Ende der Placenta nach Bestäubung mit sehr wenig Pollen. (Nach Correns 1918 und 1928.)

der Narben das Geschlechtsverhältnis nicht beeinflußt, ergibt sich eine Erhöhung des ♀♀-Anteils bei Verwendung ganz frischen oder 14—17 Tage alten Pollens. Es sind also männliche Gametenklassen mit unterschiedlichem Entwicklungsrhythmus vorhanden.

III. Geschlechtschromosomen — Heterochromosomen.

Durch Untersuchungen von Wilson (1910) wurde der Zusammenhang des mechanischen Geschlechtsverhältnisses 1:1 mit dem schon längere Zeit bei Insekten and anderen Arthropoden beobachteten Vorkommen sog. akzessorischer

[1] Hartshorne (1963) fand keine Verschiebung des Geschlechtsverhältnisses bei *M. dioicum*, wogegen Lawrence (1963, 1964) einen Effekt sowohl bei *M. album* wie auch bei *M. dioicum* zeigen konnte. Bei *M. dioicum* differieren verschiedene X-Chromosomen in der Wirkung (van Nigtevecht 1966): Ein bestimmtes X-Chromosom verhinderte durch seinen hohen positiven Selektionswert die Entstehung ♂ Nachkommen. Alterungsversuche am Pollen führten zunächst zu keiner Veränderung des Geschlechtsverhältnisses, später eher zu einer Steigerung des ♀ Anteils.

Chromosomen und Heterochromosomen aufgeklärt. Nachdem bei Pflanzen zuerst bei *Sphaerocarpus donnelli* von ALLEN (1917) in weiblichen Exemplaren ein großes X-Chromosom, bei männlichen ein diesem entsprechendes kleines Y-Chromosom beobachtet war, wurden von BLACKBURN (1923) bei *Populus tremula* und *Salix*, von WINGE (1923) bei *Melandrium* und *Humulus*, von BLACKBURN und HARRISON (1924) bei *Salix viminalis* und *Populus tremula*, von KIHARA und ONO (1923a, b) bei *Rumex acetosa* und von SANTOS (1923) bei *Elodea* Heterochromosomen im männlichen Geschlecht entdeckt. Nicht alle Heterochromosomen haben Bedeutung für die Geschlechtsbestimmung. Sehr häufig aber haben sie sich durch ihre Verteilung in der Meiose auf verschieden-geschlechtlich bestimmende Gametenzellen als ,,Geschlechtschromosomen" identifizieren lassen. Sie zeichnen sich oft durch starken Heterochromatin-Gehalt aus, der eventuell mit besonderer genetischer Stabilität in Zusammenhang steht (KNAPP 1935). Zahlreiche Untersuchungen haben für eine Reihe diöcischer Blütenpflanzen heteromorphe Geschlechtschromosomen ergeben. Eine ganze Anzahl von diöcischen

Tabelle 3a. *Gut gesicherte Fälle heteromorpher Geschlechtschromosomen bei höheren Pflanzen.* (Nach WESTERGAARD 1958, ergänzt.)

Art	2n ♀	2n ♂	Autoren
Moraceae			
Cannabis sativa	18+XX	18+XY	HIRATA (1929); MACKAY (1939); YAMADA (1943)
Humulus lupulus	18+XX	18+XY	JACOBSEN (1957)
Humulus lupulus	$X_1Y_1X_2Y_2$	nicht vorhanden	SINOTO (1929); NEVE (1958)
Humulus lupulus var. *cordifolius*	$16+X_1X_1X_2X_2$	$16+X_1Y_1X_2Y_2$	ONO (1937)
Humulus japonicus	14+XX	$14+XY_1Y_2$	KIHARA (1929a); KIHARA und HIRAYOSHI (1932); JACOBSEN (1957)
Polygonaceae			
Rumex subgenus *acetosella*			
Rumex angiocarpus	12+XX	12+XY	Á. LÖVE (1943)
Rumex tenuifolius	24+(XX)XX	24+(XX)XY	Á. LÖVE (1943)
Rumex acetosella s. str.	36+(XXXX)XX	36+(XXXX)XY	Á. LÖVE (1943)
Rumex graminifolius	48+(XXXXXX)XX	48+(XXXXXX)XY	Á. LÖVE (1943)
Rumex subgenus *acetosa*			
Rumex hastatulus	6+XX	$6+XY_1Y_2$	SMITH (1955)
Rumex hastatulus (Herkunft Nordkarolina)	6+XX	$6+XY_1Y_2$	BEN W. SMITH (1963)
(Herkunft Süd- und Mitteltexas)	8+XX	8+XY	BEN W. SMITH (1963)
Rumex acetosa und verwandte Arten	12+XX	$12+XY_1Y_2$	ONO (1930, 1935); YAMAMOTO (1938)
Rumex paucifolius	24+(XX)XX	24+(XX)XY	Á. LÖVE und SARKAR (1956)
Rumex paucifolius (diploide Rasse)	12+XX nicht heteromorph	12+XY nicht heteromorph	BEN W. SMITH (1958)
Caryophyllaceae			
Melandrium album	22+XX	22+XY	ONO (1939b); WESTERGAARD (1940); WARMKE (1946a—c)
Melandrium rubrum	22+XX	22+XY	D. LÖVE (1940, 1944)

Tabelle 3b. *Fragliche oder ungenügend gesicherte Fälle heteromorpher Geschlechtschromosomen bei höheren Pflanzen.* (Nach WESTERGAARD 1958, ergänzt.)

Art	Autoren
Cycadaceae	
Cycas pectinata	Vorhanden: ABRAHAM und MATHEW (1962)
Hydrocharitaceae	
Elodea canadensis	SANTOS (1924)
Elodea gigantea	SANTOS (1923)
Hydrilla verticillata	SINOTO und KIYOHARA (1928); SINOTO (1929b)
Palmae	
Trachycarpus excelsus	SINOTO (1929b)
Trachycarpus fortunei	SINOTO (1929b)
Liliaceae	
Asparagus officinalis	Nicht vorhanden: KAMO (1929); FLORY (1932) Vorhanden: REIMANN-PHILIPP (1959); REIMANN-PHILIPP, ZILM und ERESEN (1960)
Smilax-Arten	Vorhanden: NAKAJIMA (1937, 1942) Nicht vorhanden: LINDSAY (1930)
Dioscoreaceae	
Dioscorea-Arten	NAKAJIMA (1937, 1942); SMITH (1937) (♀:XX, ♂:XY oder XO)
Salicaceae	
Salix *Populus* *Toisusu*	Vorhanden: BLACKBURN und HARRISON (1924); MEURMAN (1925); HARRISON (1926); BLACKBURN (1929); ERLANSON und HERMANN (1927); SINOTO (1929b); NAKAJIMA (1937, 1942); JOHNSON (1940); SMITH (1943); SUTO (1944) Manchmal gefunden: VAN BUIJTENEN und EINSPAHR (1959) Nicht vorhanden: PETO (1938); HÅKANSSON (1938); MÜNTZING (1936); JOHNSSON (1940); WILKINSSON (1944)
Moraceae	
Cudrania triloba	SINOTO (1929b)
Morus bombycis	SINOTO (1929b)
Urticaceae	
Urtica dioica	MEURMAN (1925) (zweifelhaft, cf. Å. und D. LÖVE, 1942)
Santalaceae	
Buckleya joan	NAKAJIMA (1937)
Loranthaceae	
Phoradendron flavescens	BILLINGS (1932, 1933) (♀:2A, ♂:2A + X)
Phoradendron villosum	BILLINGS (1932, 1933)
Chenopodiaceae	
Atriplex hymenelytra	BILLINGS (1934) (♀:$2A + X_1X_1X_2X_2$, ♂:$2A + X_1X_2Y$) Nicht vorhanden: LORZ (1937)
Spinacia oleracea	Nicht vorhanden: STAMPS (1910); WINGE (1923); SINOTO (1929); HAGA (1935); LORZ (1937); BEMIS und WILSON (1953); JANICK und STEVENSON (1955); ELLIS und JANICK (1960); SUTO und SUGIYAMA (1961a, b); IIZUKA und JANICK (1962) Vorhanden: DRESSLER (1958); BOSE und JANICK (1961)
Spinacia tetrandra	ARARATJAN (1939)
Spinacia oleracea	Vorhanden: ZOSCHKE (1956) Nicht vorhanden: HAGA (1935); ARARATJAN (1939); BEMIS und WILSON (1953)
Caryophyllaceae	
Silene otites	FAVARGER (1946)
Silene densiflora	FAVARGER (1946)
Cercidiphyllaceae	
Cercidiphyllum japonicum	NAKAJIMA (1942)
Menispermaceae	
Coccolus trilobus	NAKAJIMA (1937)

Tabelle 3b. (Fortsetzung.)

Art	Autoren
Rosaceae	
Fragaria elatior	Kihara (1930) (♀:XY, ♂:XX)
Rutaceae	
Zanthoxylum piperitum	Sinoto (1929b); Nakajima (1937)
Daphniphyllaceae	
Daphniphyllum macropodum	Sinoto (1929b)
Aquifoliaceae	
Ilex serrata var.	Nakajima (1942)
Aceraceae	
Acer negundo	Vorhanden: Sinoto (1929b) Nicht vorhanden: Foster (1933)
Simarubeceae	
Picrassima quassioides	Nakajima (1942)
Dilleniaceae	
Actinidia kolomicta	Nakajima (1942)
Actinidia polygama	Nakajima (1942)
Theaceae	
Eurreya japonica	Nakajima (1942)
Datiscaceae	
Datisca cannabina	Sinoto (1929b)
Valerianaceae	
Valeriana dioica	Meurman (1925)
Cucurbitaceae	
Trichosanthes cucumeroides	Nakajima (1937)
Trichosanthes japonica	Nakajima (1937)
Trichosanthes dioica	Patel (1952)
Trichosanthes multiloba	Kurita (1939)
Coccinea indica	Kumar *et al.* (1952)

Arten besitzt aber auch Geschlechtschromosomen, die nicht heteromorph sind und zum Teil auch nicht von den übrigen als „Autosomen" bezeichneten Chromosomen zu unterscheiden sind (Lindsay 1930).

Die Ausbildung von heteromorphen Geschlechtschromosomen ist offenbar eine Sekundärerscheinung. Da zumindest das Y-Chromosom im einen Geschlecht fehlt, kann es keine allgemein lebenswichtigen Gene mehr enthalten und konnte infolgedessen morphologische Veränderungen erfahren, die keiner Selektion unterworfen wurden. Bei den meisten Arten liegen diese Änderungen in Form von Substanzverlust vor, so daß in der Regel das Y-Chromosom das kleinere der Geschlechtschromosomen ist. Westergaard (1958) hat die gut fundierten Fälle des Vorkommens von Heterochromosomen und die weniger sicheren zusammengestellt (Tabelle 3a, b).

Der Versuch, homomorphe Geschlechtschromosomen nur nach ihrem cytologischen, von dem der Autosomen abweichenden Verhalten zu erkennen, ist verschiedentlich gemacht worden (Kumar u. a. 1945). Ohne gleichzeitigen genetischen Beweis ist die Identifizierung jedoch nicht einwandfrei (Storey 1953).

a) XO-Typus.

Der XO- oder Protenor-Typus (Wilson) zeigt im heterogameten Geschlecht ein Chromosom weniger; d. h. die morphologische Veränderung ging im Laufe der phylogenetischen Entwicklung so weit, daß Y schließlich ausfiel, also auch keinerlei für das betreffende Geschlecht wichtigen Gene, z. B. Fertilitätsgene, mehr enthielt. Dieser Typus der Geschlechtschromosomen ist bei Tieren verschiedentlich bekannt, bei Pflanzen wurde er zunächst von WINGE (1923) für *Vallisneria gigantea* beschrieben, 1927 aber von ihm selber und von JØRGENSEN widerlegt. SMITH (1937) hat ihn für *Dioscorea*-Arten beschrieben. Seine Befunde sind aber nicht ganz eindeutig, so daß dieser Typus vermutlich nur bei Tieren besteht. Das spricht auch für die jeweils phylogenetisch junge Entstehung der Diöcie bei Pflanzen aus dem ursprünglich zwittrigen Zustand (CORRENS 1928).

b) XY-Typus.

Der XY- oder Lygaeus-Typus (WILSON) ist der am häufigsten gefundene Typ der Geschlechtschromosomen. Bei ihm liegt im homogameten Geschlecht ein homologes Geschlechts-Chromosomenpaar vor (XX), im heterogameten Geschlecht ist ein Geschlechtschromosom, das Y-Chromosom, nur zum Teil homolog (XY). SANTOS (1924) hat die Verteilung der unterschiedlichen Chromosomen in der Meiose

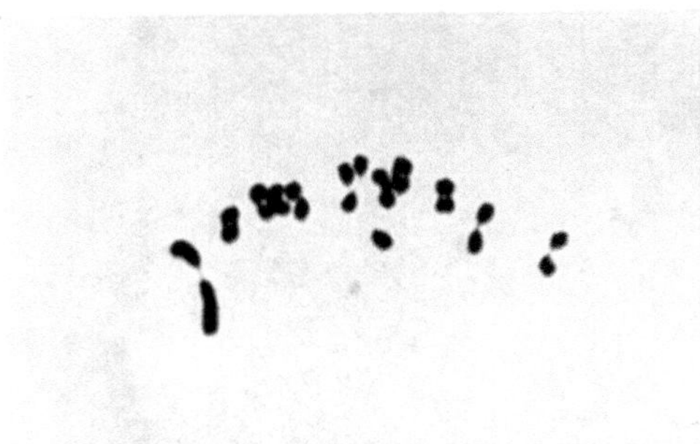

Abb. 2.

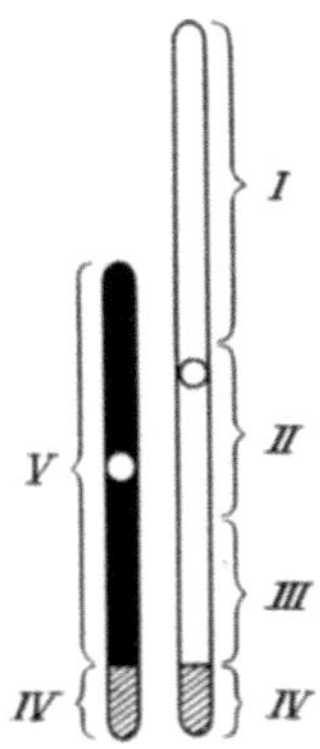

Abb. 3.

Abb. 2. Meiosis in einem diploiden Männchen, 11 Paare Autosomen und das XY-Geschlechtschromosomenpaar links. Mikrophoto aus einem Orcein-Streichpräparat. Vergrößerung 2400 mal. (Aus WESTERGAARD 1953.)

Abb. 3. Der Geschlechtsbestimmungsmechanismus bei *Melandrium*. Schema *I—II—III*: Das differentielle Segment von Y; *IV* die homologen Segmente; *V* differentielles Segment von X. Segment *I* enthält die Gene (oder das Gen), die die Entwicklung der weiblichen Geschlechtsorgane unterdrücken. Die Gene in *II* kontrollieren die ersten Stadien der Staubfädenentwicklung. Segment *III* kontrolliert die letzten Stadien der Staubfädenentwicklung. Vgl. den Text. (Aus WESTERGAARD 1948.)

der ♂♂ von *Elodea canadensis* und *E. gigantea* beobachtet, die in jeder Tetrade zur Bildung von zwei kleineren und zwei etwas größeren Pollen führte. Allerdings ging der Unterschied später wieder verloren. Bei *Cycas pectinata* ist das X-Chromosom mit einem Satelliten gekennzeichnet, der dem Y fehlt. ABRAHAM und MATHEW (1962) konnten die Verteilung in der Meiose verfolgen: nur die Hälfte der Pollen erhielt ein SAT-Chromosom.

1. *Melandrium.* Eingehend untersucht ist das Hetero-Chromosomenpaar bei *Melandrium* (Abb. 2). Nachdem BLACKBURN (1923) ganz richtig das Y als das größere des Paares bezeichnet hatte, wurde von anderen Autoren (WINGE 1923, MEURMAN 1925, BRESLAWETZ 1929) dasselbe für das X-Chromosom gehalten. Dieser Ansicht schloß sich BLACKBURN 1929 an, und erst 1939 konnte durch WARMKE und BLAKESLEE direkt an Polyploiden gezeigt werden, daß Y hier größer ist. Es waren autotetraploide ♀ × ♂ gekreuzt worden, ♀ XXXX × ♂ XXYY und ergaben viele XXXY-Männchen. Da aus dieser Kreuzung keine Pflanzen mit 3 Y entstehen konnten, mußte sich in einer Zelle direkt die Größe von X und Y messen lassen: X ist etwa $^2/_3$ so lang wie Y mit submedianer SPA. Beide zeigen in der Meiose ein terminales Chiasma, während im Weibchen beide X mit zwei terminalen Chiasmata paaren. Das deutet auf mangelnde Homologie des einen

Armes von Y mit dem kürzeren Arm von X, während die beiden anderen Arme homolog zu sein scheinen. Später hat Westergaard (1948) zeigen können, daß nur die terminalen Enden dieser Arme homolog sind (Abb. 3). Es wird also durch paarungsfähige Teile die geregelte Bildung von X- und Y-Gameten gewährleistet, gleichzeitig aber durch die Struktur-Heterozygotie (Darlington 1932) der Austausch von auf X und Y lokalisierten Geschlechtsrealisatoren verhindert. In polyploiden Formen mit mehreren Y ist die Paarungstendenz zwischen Y und Y erheblich größer als zwischen X und Y, so daß XXYY ♂ vorwiegend XY-Keimzellen bilden. Warmke und Blakeslee (1939b) haben außer Bivalenten auch häufig Quadrivalente beobachtet mit folgenden Konfigurationen:

Paarungsfigur:	X Y / | | / X Y	Y X / ⟍ ⟍ / X Y	X Y / <>< / X Y	Y Y / | | / X—X
Gameten:	37,5% XY	40% XY	8,75% XY	8,75% XX, YY

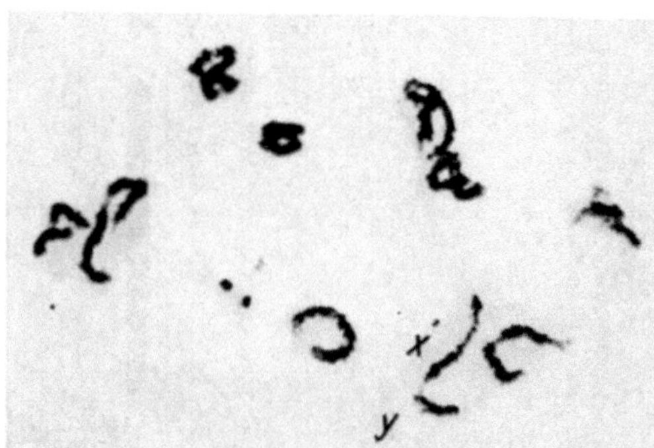

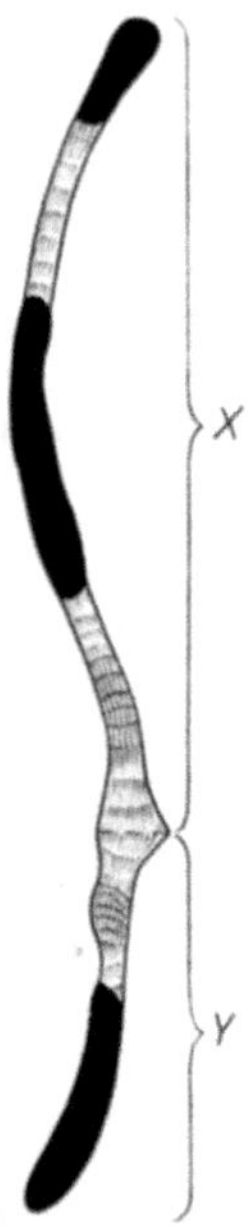

a b

Abb. 4a u. b. *Humulus lupulus.* Meiosis beim Männchen, end-to-end-Paarung von X und Y im Diplotän. (Aus Jacobsen 1957.)

2. *Humulus lupulus.* Das schon 1923 von Winge entdeckte XY-Chromosomenpaar der Männchen von *Humulus lupulus* wurde eingehend von Jacobsen (1957) untersucht. Das Hetero-Chromosomenpaar konnte im Pachytän nicht identifiziert werden, ist aber im Diplotän erkennbar in end-to-end-Paarung mit deutlich strukturellen Unterschieden, die in der Diakinese verschwinden. Der Größenunterschied bleibt jedoch, ebenso in der Metaphase I und Anaphase I, entgegen den Befunden von Dark (1952) und Derenne (1954). Das X-Paar war in der Meiose der ♀♀ nicht deutlich diagnostizierbar, so daß nicht geklärt werden konnte, welches Chromosom X und welches Y ist. In der Mitose der ♂♂ ließen sich die Geschlechtschromosomen auf Grund der morphologischen Differenzen in der Prophase identifizieren. Es wurde ein mittelgroßes Chromosom gefunden, das in der Mitte des kurzen Armes eine weniger färbbare Sektion besitzt. Im ♀ finden sich zwei solche Chromosomen, sind also X; das Y hat einen kürzeren, diffus endenden Arm. Die homologen Segmente umfassen den kurzen Arm des Y (etwa $^1/_4$ des Chromosoms), im X 2 Segmente nahe der SPA (etwa $^1/_6$ des ganzen Chromosoms). Y ist stark heterochromatisch (Abb. 4). Die Deutung von Winge (1932), daß Y zweimal so groß wie X ist und mit starker medianer Einschnürung durch Zusammenlagerung von 2 X entstanden zu denken sei, dürfte wohl zu sehr von den Befunden von *Humulus japonicus* beeinflußt sein (s. dort).

3. *Rumex.* Innerhalb des Subgenus *Acetosella* liegt eine Polyploidenreihe vor aus vier diöcischen perennierenden Arten mit *R. angiocarpus* (2n = 14), *R. tenui-*

folius (4n = 28), *R. acetosella s. str.* (6n = 42) und *R. graminifolius* (8n = 56). *R. acetosella* wurde von MEURMAN (1925) im ♂ mit 2n = 41 festgestellt. Er bezeichnete eine in der Meiose auftretende Kette von 3 Chromosomen in Analogie zu *R. acetosa* als Geschlechtschromosomen (s. S. 65). KIHARA (1925, 1929) stellte für ♀ 2n = 42 fest, für ♂ 2n = 41—43. Er beobachtete in fast allen Pflanzen Quadrivalente und hält die Kettenbildung nicht für ausreichend zur Identifizierung der Geschlechtschromosomen. Erst die Untersuchung der diploiden Art *R. angiocarpus* brachte die Aufklärung des Geschlechtschromosomen-Mechanismus. A. LÖVE (1944) fand in der Metaphase I der Meiose im ♂ regelmäßig ein Hetero-Chromosomenpaar, meist am Rande der Platte, das manchmal auch in der Prophase identifiziert werden konnte und früher kontrahierte als die Autosomen. Im ♀ wurden in der Mitose 2 Chromosomen gefunden, die etwas größer waren als alle anderen mit submedianer SPA 1,4:1,6 mit 3,0 μ gegen die längsten Autosomen mit 2,7 μ. In ♂ Mitosen war das Hetero-Chromosomenpaar mit X = 3,0 μ und dem Partner = 4,4 μ = Y leicht zu finden. Also sind wie bei *Melandrium* bei *Rumex angiocarpus* im ♂ XY vorhanden mit größerem Y, 2,0 μ:2,4 μ mit submedianer SPA.

In dem tetraploiden *R. tenuifolius* mußte nach dem normalen Geschlechtsverhältnis von ♀ und ♂ ebenfalls eine einfache Struktur-Heterozygotie (DARLINGTON) in dem einen Geschlecht vorliegen, die aus dem Status der diploiden Art hervorgegangen sein konnte: ♀ = XXXX und ♂ XXXY. Im ♂ wurde in der Meiose in 298 Zellen an 16 Individuen 245mal die Paarung XX und XY gefunden, 27mal eine Kette von 4 Chromosomen mit Y an einem Ende, 26mal eine Kette von 3 Chromosomen XXY und ein X-Univalent. Die große Häufigkeit der bivalenten Paarung (82%) wird von LÖVE (1944) auf verringerte Homologie eines X mit den übrigen X zurückgeführt. Es ist also eines der Geschlechts-Chromosomenpaare im Begriff, seine Bedeutung für den Geschlechtsbestimmungs-Mechanismus zu verlieren, d. h. autosom zu werden (WESTERGAARD 1940).

Wird *Rumex acetosella s. str.* als autohexaploid angesehen (LÖVE 1941b), so muß die Geschlechtschromosomen-Konstitution ♀ = XXXXXX und ♂ = XXXXXY erwartet werden. Es wurden beobachtet in der Meiose des ♂ Bivalente, Ketten aus 4 und Ketten aus 6 Chromosomen, von denen stets eines in der Größe von den anderen abwich, dieses jeweils nur in Endbindung mit einem oder zwei anderen. Konfigurationen mit univalenten Y waren nicht zu sehen. In den meisten Zellen mit XY-Bivalent traten auch die übrigen 4 X als Bivalente auf. Aus der geringen Frequenz der Tetravalentbildung schließt LÖVE (1944) darauf, daß in einem Paar der X-Chromosomen die Differenzierung zu Autosomen bereits weiter fortgeschritten ist als beim zweiten Paar.

4. *Rumex paucifolius.* Ähnlich wie bei *R. tenuifolius* sind bei *R. paucifolius* aus dem Subgenus *Acetosa* die heteromorphen Geschlechtschromosomen im ♂ XXXY und im ♀ XXXX (LÖVE und SARKAR 1956). Allerdings zeigen sich hier die X im ♀ und ♂ deutlich länger als die Autosomen, während im ♂ Y als einzelnes sehr kleines Chromosom vorhanden ist.

In einer diploiden Rasse konnten allerdings von SMITH (1958) keine Heterochromosomen gefunden werden.

5. *Cannabis sativa.* Beim Hanf ist nach TAKENAKA (1953) Y wie bei *Melandrium* das größte Chromosom überhaupt, während HOFFMANN (1941) Y als kleineres zweigliedriges Chromosom neben dem größeren dreigliedrigen X beschreibt. TAKENAKA konnte an triploiden XXY ♂♂ zeigen, daß X so groß ist wie das größte Autosomenpaar, Y aber größer und im Trivalent in der Meiose in Endstellung auftritt: X—X—Y. Die Paarung und Verteilung von X und Y erfolgt in der Meiose hauptsächlich so, daß XY-Gameten entstehen. Es wurde 77mal

Y—Y, 29mal X—Y, 26mal Y—X—X—Y und 8mal Y—X—Y als Paarungsbild beobachtet. Demnach ist auch bei *Cannabis* die Affinität zwischen YY größer als zwischen XY.

6. *Asparagus officinalis.* Die Heteromorphie von X und Y ist beim Spargel nicht so groß. Nach Reimann-Philipp (1959, 1960) fällt in der Diakinese der Pollenmutterzellen bei ♂ und Andromonöcisten eines der 10 Bivalente auf durch ungenügende Paarung und geringe Längendifferenz. Die Aufspaltung in der Selbstungsnachkommenschaft von Andromonöcisten brachte die erwarteten Zahlenverhältnisse und parallel mit dem Geschlechtstyp die Verteilung der Chromosomen dieses Bivalents.

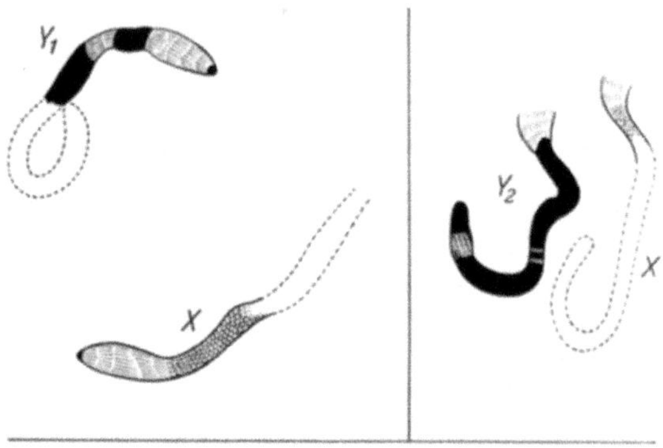

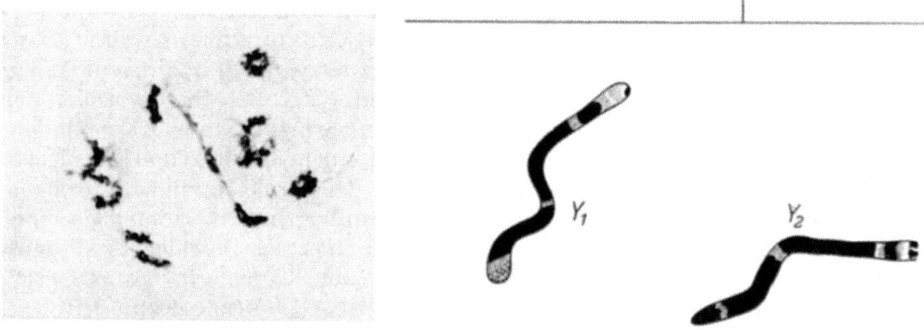

a b

Abb. 5 a u. b. *Humulus japonicus.* Meiosis im Männchen, Diakinese mit den heterochromatischen Y in end-to-end-Paarung an beiden Enden von X. (Aus Jacobsen 1957.)

c) XY_1Y_2-Typus.

1. *Humulus japonicus.* Ein neuer Geschlechtschromosomen-Mechanismus, der im ♂ XYY aufweist, wurde bei *Humulus japonicus* von Kihara (1929b) aufgedeckt, der den dreiteiligen Geschlechtschromosomenkomplex der ♂ als XY_1Y_2 deutete, entgegen der älteren Auffassung von Winge (1923, 1929), daß 3 X vorhanden seien. Beide, X und Y, sind größer als die Autosomen. Die Y-Chromosomen haben eine unterschiedliche Struktur und sind ausgesprochen heterochromatisch, noch im Ruhekern sichtbar. Y_1 besitzt submedian die SPA, Y_2 median. Alle 4 Enden sind morphologisch charakterisiert, Y_2 hat ein nicht heterochromatisches Ende am 1. Arm, der strukturelle Ähnlichkeit zu dem einen Arm von X zeigt und mit ihm end-to-end paart. In Y_1 sind beide Enden nicht heterochromatisch; das Ende des langen Arms ist nach der Ähnlichkeit homolog dem Ende des kurzen Arms von X. Die homologen Stücke betragen weniger als $^1/_6$ je Chromosom (Jacobsen 1957, Abb. 5). Kihara beobachtete als häufigste Konfiguration Y_1—X—Y_2, die in der Meiose X- und Y_1Y_2-Gameten ergab (1932). Aus Winges Auffassung des dreifachen Chromosomenkomplexes als 3 X resultierte seine Deutung bei *Humulus lupulus*, das größere Chromosom des Geschlechtschromosomenpaares im ♂ sei Y, entstanden durch Fusion zweier end-to-end gepaarter X. Diese Annahme ist aber von Darlington und Westergaard widerlegt worden.

SKOVSTEDTs Versuche (1953), die Geschlechtschromosomen auch in somatischem Gewebe zu identifizieren, sind *nicht* gelungen. Im Gegensatz zu WINGE, den Japanern und JACOBSEN findet er im ♂ ein dreifach vorhandenes Chromosom als kleinstes des ganzen Satzes. Die Angaben JACOBSENs differieren mit denen von KIHARA und HIRAYOSHI nur in bezug auf die mediane Einschnürung von X und Y_2. Die Anwesenheit der stark heterochromatischen, wahrscheinlich teilweise inerten Y-Chromosomen wird als Anzeichen einer langen unabhängigen Entwicklung dieses Typs gedacht (JACOBSEN 1957).

2. *Rumex acetosa.* Ein ganz ähnlicher Typus wurde (1923, 1925) von KIHARA und ONO für *Rumex acetosa L.* nachgewiesen. Im ♂ konnten 6 Autosomenpaare und Y_1XY_2 als dreiteiliger Komplex beobachtet werden, beim ♀ 6 Autosomenpaare, die alle kleiner sind, und außerdem ein großes XX-Paar. Nach ONO (1926, 1930) ist Y_1 größer als Y_2, X gleichschenklig, Y_1 ähnlich, aber etwas kleiner, Y_2 dagegen deutlich ungleichschenklig, noch kleiner. Die Autosomen sind alle ähnlich stäbchenförmig, zum Teil mit Köpfchen. Eingehende Untersuchungen liegen vor von KIHARA und ONO (1923, 1925), von SINOTO (1924, 1929), KIHARA und YAMAMOTO (1931). Die gleichen Verhältnisse für die Geschlechtschromosomen wurden von ONO (1930) und TAKENAKA (1930) für *Rumex montanus* gefunden. Bei 2 X + Y_1 + Y_2-Formen zeigte sich XX + Y_1 + Y_2 als der häufigste Paarungstyp, aber auch Y_1XY_2 + X ist beobachtet worden. Bei einer anderen Pflanze mit 2 Y_1 + Y_2 + X lag 87mal Y_1XY_2 + Y_1- und 113mal XY_2 + Y_1Y_1-Konstellation vor, also etwas stärkere Bindungsaffinität der beiden Y_1 untereinander als zum homologen Segment von X. Auch in den gewöhnlichen diploiden Männchen wird die Verbindung zwischen X und Y_1 als locker angegeben (KIHARA und ONO 1925, Abb. 6). Den gleichen Geschlechtschromosomen-Typus fanden KIHARA und ONO auch für *Rumex thyrsiflorus*, *R. arifolius* und *R. nivalis*. Nach LÖVE (1944) gehören alle diöcischen Arten der Sektion *Euacetosa* des Subgenus *Acetosa* zum gleichen Geschlechtschromosomen-Typus mit 2n = 14 bzw. 15.

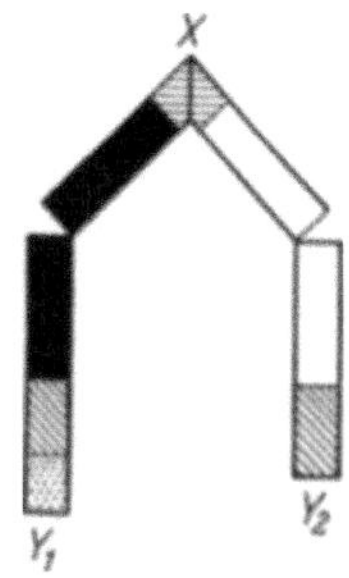

Abb. 6. Schematische Darstellung der Homologieverhältnisse der Geschlechtschromosomen von *Rumex acetosa*. (Aus ONO 1935.)

Die Arten der Sektion *Hastati* sind alle stark gynodiöcisch mit 2n = 8. SMITH und SMITH (1947) haben für *Rumex hastatulus* mit der halben Chromosomenzahl wie *R. acetosa* im ♂ ebenfalls $Y_1X\,Y_2$-Paarung angegeben, in gleicher Weise wie bei *R. acetosa.* X ist aber in somatischen Zellen trotz der geringen Chromosomenzahl schwer von den größten Autosomen zu unterscheiden. Y_1 und Y_2 sind ungleich groß. Das würde bedeuten, daß die *R. acetosa*-Gruppe als tetraploid angesehen werden muß. Ihr Geschlechtsbestimmungs-Mechanismus ist beim Übergang von der diploiden Stufe zur tetraploiden durch Verlust eines Geschlechts-chromosomen-Satzes erhalten geblieben (SMITH 1955).

Neuerliche Befunde von SMITH (1963, 1964) über das Vorliegen eines zweiten Karyotyps innerhalb der Art *R. hastatulus* bei einer Herkunft aus Mitteltexas mit 2n = 10 bei 8 + XX im ♀ und 8 + XY im ♂ deuten auf die abgeleitete Natur des XY_1Y_2-Typus. Seine Entstehung wäre denkbar auf dem Wege über zwei reziproke Translokationen zwischen den Heterosomen und einem Autosomenpaar.

d) Weitere multiple Geschlechtschromosomen-Typen.

1. Eine abweichende Konstellation wird von NAKAJIMA (1937) für *Humulus lupulus L. var. cordifolius* (2n = 20) angegeben, die in der Meiose im ♂ 8 Bivalente und eine Kette aus 2 Bivalenten in end-to-end-Bindung zeigt, wie sie auch

Sinoto (1929) bereits gesehen hat. Die mittleren beiden Chromosomen der Kette sind größer, die Anaphasen-Verteilung erfolgt in der Regel abwechselnd, gelegentlich gehen alle 4 Chromosomen an einen Pol. Nakajima faßt die Kette als $X_1Y_1X_2Y_2$ auf, so daß X_1X_2- und Y_1Y_2-Gameten gebildet werden. Dieser vierteilige Geschlechtschromosomen-Komplex wird gedeutet als entstanden durch reziproke Translokation zwischen XY und einem Paar Autosomen (Westergaard 1958). Beim ♀ hat Ono (1937) nur Bivalentbildung beobachtet.

2. Für *Atriplex hymenelytra* hat Billings (1934) im ♂ einen dreiteiligen Geschlechtschromosomen-Komplex beschrieben, der aus 2 X und 1 Y besteht, X_1X_2Y. Alle 3 Chromosomen sind gleich groß. Im ♀ ist 1 Chromosom mehr vorhanden, $X_1X_2X_1X_2$. Es handelt sich aber eventuell auch bei den X_1X_2-Paaren um jeweils ein großes eingeschnürtes X, das wie zwei in end-to-end-Bindung erscheint.

e) Unsichere Heterochromosomen-Nachweise.

Bei einer Reihe anderer diöcischer Pflanzen sind von einigen Autoren Heterochromosomen angegeben, von anderen wurden sie nicht beobachtet. Als nicht sicher gelten auch Befunde, die Heterochromosomen nur in einem Geschlecht gezeigt haben, ohne Nachweis eines homologen Chromosomenpaares im anderen Geschlecht. Dabei bleibt ungewiß, ob es sich tatsächlich um Geschlechtschromosomen handelt (s. Tabelle 3b).

f) Diöcisten mit homomorphen Geschlechtschromosomen.

Bei den diöcischen Blütenpflanzen, bei denen keine heteromorphen Geschlechtschromosomen gefunden worden sind, hat man bisher Geschlechtschromosomen außer bei *Spinacia oleracea* noch nicht identifizieren können. Die Fälle geschlechtsgebundener Vererbung sind so selten bei Pflanzen beobachtet worden, daß mit ihrer Hilfe der Nachweis auch nur bei dieser Art gelang.

1960 konnten Ellis und Janick ein Karyogramm dieser Art aufstellen, nach dem sich alle 6 Chromosomen eindeutig identifizieren lassen. Aus Kreuzungen zwischen triploiden und diploiden Pflanzen ließen sich für alle 6 Chromosomen Trisome herstellen. Der trisome Typ „reflex“ konnte als derjenige identifiziert werden, der Chromosom 1 dreimal enthielt. Aus den Nachkommen ließ sich ableiten, daß dieses Chromosom Träger der Geschlechtsrealisatoren sein mußte. 1963 wurden mit Hilfe spontan aufgetretener und induzierter Translokationstypen diese Befunde bestätigt (Iizuka und Janick 1963).

IV. Geschlechtsgebundene Vererbung.

Einen weiteren Beweis für den Homo-Heterozygotie-Mechanismus der G.B. ergab die geschlechtsgebundene Vererbung von Merkmalen, die an sich keinen Zusammenhang mit der Ausbildung des Geschlechts haben. Die einzigen Objekte, an denen sie bei Pflanzen beobachtet wurde, sind *Melandrium*, *Carica papaya*, *Cannabis sativa* und bedingt *Spinacia olereacea*.

a) *Melandrium*.

1. Der „*angustifolia*“-Faktor. Baur beschrieb 1912 bei *Melandrium album* eine grasartig schmalblättrige Mutante unter 100 normalblättrigen Pflanzen. Außer den Laubblättern waren auch Kelch- und Blütenblätter schmal. Er erhielt aus der Kreuzung normales ♀ × schmalbl. ♂ in F_1 normale ♀♀ und normale ♂♂, in F_2 waren alle schmalblättrigen Individuen ♂♂ (39). Von den 112 normalblättrigen Pflanzen wurden 20 untersucht auf ihr Geschlecht: 12 waren ♀, 8 ♂. Shull (1914)

führte die Untersuchung weiter. Die normale Breitblättrigkeit wird hervorgerufen durch das Zusammenwirken eines B-Faktors mit dem übrigen Genotyp von XX-Pflanzen. SHULL erhielt in der gleichen Kreuzung in F_2 44 normal: 6 *angustifolia*. Daraus folgt, daß die ♀ homozygot, die ♂ heterozygot für den Faktor B sind.

44 normal : 6 *angustifolia*
32 ♀ + 11 ♂ + 6 ♂
2 : 1 : 1

Ist bezüglich der Geschlechtsrealisatoren das ♀ FF, wobei F den Weiblichkeits-Realisator bedeutet, das ♂ Ff, so ist der Faktor B mit dem Realisator gekoppelt: ♀ = $\widehat{BF}\,\widehat{BF}$, ♂ = $\widehat{BF}\,\widehat{Bf}$ oder aber $\widehat{BF}\,\widehat{bf}$. Das mutierte ♂ hatte die Konstitution $\widehat{bF}\,\widehat{bf}$. Die F_1 mußte dann $\widehat{bF}\,\widehat{BF}$ ♀ und $\widehat{bf}\,\widehat{BF}$ ♂ enthalten, also alle normalblättrig.

F_2 = 25% ♀ $\widehat{bF}\,\widehat{BF}$ + 25% ♀ $\widehat{BF}\,\widehat{BF}$ + 25% ♂ $\widehat{BF}\,\widehat{bf}$ + 25% ♂ *angustifolia* $\widehat{bF}\,\widehat{bf}$.
50% norm. ♀ : 25% norm. ♂ : 25% *angustif.* ♂
2 : 1 : 1

Die Ergebnisse waren völlig denen parallel, die man bei *Drosophila* für die Vererbung der roten und weißen Augenfarbe gefunden hatte. Demnach waren die ♀ der Normalsippe homozygot, die ♂ heterozygot für den *angustifolia*-Faktor.

SHULL hatte damals zwar ein besonders großes Chromosomenpaar bei *Melandrium* beobachtet, ohne aber die Heteromorphie zu konstatieren, so daß er nicht die Lokalisation des Faktors in den Geschlechtschromosomen cytologisch verfolgen konnte. Die Prüfkreuzungen brachten jedoch stets zu wenig ♀. Die Tatsache, daß schmalblättrige ♀ gar nicht in den entsprechenden Kreuzungen gefunden wurden, bedurfte einer besonderen Klärung, die zuerst v. UBISCH (1933) versuchte mit der Annahme einer unterschiedlichen Keimungs- und Pollenschlauch-Wachstumsgeschwindigkeit, wobei die $\widehat{BF}$-Pollen am schnellsten zur Samenanlage gelangen sollten, $\widehat{bf}$ etwas langsamer und $\widehat{bF}$ bedeutend langsamer, etwa mit $^1/_{10}$ Geschwindigkeit von $\widehat{BF}$.

Eine etwas andere Erklärung hat WINGE (1927) gegeben, der annimmt, daß der Faktor b im X-Chromosom gewöhnlich im Pollen letal wirkt, nicht aber in der Eizelle.

1966 berichtet VAN NIGTEVECHT erneut über ein Gen f für Schmalblättrigkeit in seinem Material von *M. album*, das in gleicher Weise die Laubblätter, besonders die Rosettenblätter, beeinflußt und die Form der Petalen verändert. Die Kreuzungsergebnisse zeigen eindeutig den X-gebunden recessiven Charakter des Gens. Der Autor hält für möglich, daß es sich nicht um das *angustifolia*-Gen BAURs handelt, weil in den Kreuzungen auch ♀ Pflanzen mit schmalen Blättern auftraten. Auf der anderen Seite zeigten schmalblättrige Pflanzen herabgesetzte Fertilität sowohl der X_f- als auch der Y-Pollen.

2. Der "*aurea*"-Faktor. Einen zweiten geschlechtsgebundenen Faktor hat WINGE (1927) entdeckt, ebenfalls bei *Melandrium* in der F_2 einer Kreuzung *Melandrium dioica* × *M. album*, in der eine Gruppe gelblicher ♂ herausspaltete, die später als "*aurea*" bezeichnet wurden. In Nachkommenschaften von "*aurea*"-♂ traten grüne ♀, grüne ♂, einige grüne ⚥ und "*aurea*"-♂ auf. Das Fehlen homozygoter "*aurea*"-♀ in entsprechenden Kreuzungen erklärte WINGE mit der Lage des recessiven "*aurea*"-Gens im Y-Chromosom. Später (1931) nach Kontrolle weiterer Generationen kam er zu dem Schluß, es sei X-gekoppelt und homozygot letal, so daß auf diese Weise heterozygote ♀♀ × "*aurea*"-♂ keine "*aurea*"-♀♀ ergeben konnten. In einigen Kreuzungen wurde eine Spaltung von 15 grün: 1 "*aurea*" gefunden, wobei alle "*aurea*" ♂♂ waren, woraus sich ergab, daß noch ein zweiter autosomaler Faktor an der Merkmalsbildung beteiligt sein mußte. Den "*aurea*"-♂♂ gibt WINGE die Formel aa $X_{au}Y$. Sie spalten aus ♀♀ mit AaX_{au}X, AaX_{au} X_{au} und aaX_{au}X heraus. Das Manko an "*aurea*" ♂♂, das meist gefunden wurde, erklärt sich durch geringere Vitalität der "*aurea*"-Pflanzen, außerdem durch den natürlich vorhandenen ♀-Überschuß bei *Melandrium*.

3. Der „*abnormal*“-Typus. Später fand WINGE in der Kreuzung *Melandrium album* ♀ × *rubrum* ♂ noch einen Typus „abnormal“, blasse Pflanzen, die den Kelch nicht öffnen, so daß die Petalen nicht freiwerden. Homozygote n n-Pflanzen waren weniger vital. X_nX_n und X_nY_n-Pflanzen zeigten das Merkmal. Es wurden 3 autosomale Gene gefunden: G, H und I,

die die Wirkung von n verhindern. Es gelang, sie durch Selektion zu entfernen. Die Pflanzen waren in einer F_2 aufgetreten: 116 normale ♀♀ und 60 normale ♂♂ und 7 nn ♀♀, 0 nn ♂♂. Rückkreuzung eines F_1 ♂ mit der Mutter *Melandrium album* ergab 736 normale ♀♀, 353 normale ♂♂. 39 nn ♀♀, 0 nn ♂♂. In anderen Rückkreuzungen der reziproken Kombination waren alle nn-Pflanzen ♂. Die Ausgangspflanzen hatten nach Winge folgende Konstitution: *album* ♀ Gg hh ii X_nX_n, *album* ♂ Gg HH ii X_nY_n, *rubrum* ♀ gg Hh JJ $X_{au}X_N$, *rubrum* ♂ gg Hh Ji X_NY_N. *rubrum* ♀ gg hh JJ X_NX_N.

Für das gleiche Gen fand van Nigtevecht (1966) sowohl bei *M. album* als auch bei *M. dioicum* (syn. *M. rubrum*) eine sehr unterschiedliche Expressivität.

4. Der "*variegata*"-Typus. Als weiteren Fall beschrieb Winge (1931) das "*variegata*"-Merkmal bei *Melandrium*, das nur an ♀ auftrat. Offenbar liegt hier ein System von autosomalen und X-gebundenen Genen vor, wobei nicht zu unterscheiden ist, ob ein Gen für "*variegata*" in den Autosomen und ein Hemmungsgen im X liegt oder das "*variegata*"-Gen selber im X-Chromosom.

5. "*Hermaphroditic males.*" Einen sehr komplizierten Erbgang, der jedoch ebenfalls eine Koppelung mit einem Geschlechtschromosom zeigt, hat das Merkmal "*hermaphroditic males*". Die Aufhebung des X-gebundenen Charakters bei Kreuzungen mit nicht verwandten Typen steht in engem Zusammenhang mit der später zu besprechenden Verteilung der einzelnen Geschlechtsgene bei *Melandrium*.

6. Blütenfärbung. Kunze berichtet 1955 über ein weiteres Merkmal bei *Melandrium*, das geschlechtsgebunden vererbt wird und einen erneuten Beweis für die Heterogametie-Hypothese des männlichen Geschlechts bildet: einen Anthocyan-Faktor, der hauptsächlich in den Blüten wirksam wird. Eine weiße ♂ Form von *Melandrium rubrum* ergab mit normal roten ♀♀ gekreuzt: rote ♀♀, weiße ♀♀ und rote ♂♂; rotes F_1 ♀ × weißes ♂ gab weiße ♀♀, sowie rote und weiße ♂♂. Diese Ergebnisse konnten durch autosomale Gene erklärt werden in der Annahme, daß die Weibchenbestimmer der weißen Form auf der Narbe der roten Form beschleunigtes Wachstum zeigen. Aber die Kreuzung eines weißen Weibchens aus der Rückkreuzung mit dem roten Original ♂ brachte nur 36 rote ♀♀ und 30 weiße ♂♂; die F_2 aus einem dieser roten ♀ und einem roten ♂ der ersten Kreuzung ergab 102 rote ♀♀, 18 rote ♂♂ und 15 weiße ♂♂. Es zeigte sich wieder die *criss-cross*-Vererbung, wie sie für die rote und weiße Augenfarbe bei *Drosophila* bekannt ist. Nach der Analyse von Kunze sind die Genotypen für

rubrum rot = ♀ $\widehat{XR}$ $\widehat{XR}$ AA; ♂ $\widehat{XR}$ Y AA
rubrum weiß = ♀ $\widehat{Xr}$ $\widehat{Xr}$ AA; ♂ $\widehat{Xr}$ Y AA
album = ♀ $\widehat{XR}$ $\widehat{XR}$ aa; ♂ $\widehat{XR}$ Y aa

und nur Vorhandensein des autosomalen A mit dem X-gekoppelten R ergibt Rotfärbung.

Bei *Melandrium* ist eine Reihe von Vererbungserscheinungen beschrieben, bei denen nicht ganz klar ist, ob es sich um geschlechtsgebundenen oder geschlechtsbeeinflußten Erbgang handelt. Im letzteren Fall handelt es sich bei der Merkmalsbildung um physiologische Koppelung mit dem Mechanismus der Geschlechtsausbildung im Sinne der Entwicklung eines Geschlechtsdimorphismus.

Schulz (1890) konstatierte bei *M. album* ♀ kleinere Petalen als bei den ♂. Löve (1944) stellte bei dieser Art größere Blüten an den ♂ Pflanzen fest. Das gleiche beobachtete Baker (1951), während bei *M. dioicum* umgekehrt die ♀ Pflanzen größere Blüten besitzen. Löve deutete diese Tatsache als geschlechtsgebundenen Erbgang. Die genannten Autoren ebenso wie Winge (1931) stellten unterschiedliche Intensität in der Blütenfärbung bei den beiden Geschlechtern fest. Winge vermutet als Ursache geschlechtsgebundene Gene; Löve nimmt an, daß eben diese Gene die Acidität in den Petalen betreffen. Unterschiedliche Ausbildung des Besatzes mit Drüsenhaaren bei ♂ und ♀ Pflanzen wird von van Nigtevecht (1966) ebenfalls so gedeutet, daß die grundlegenden Gene in ihrer Wirksamkeit durch die Geschlechtsausbildung beeinflußt werden.

b) *Carica papaya.*

Geschlechtsgebundene Vererbung bei *Carica papaya* beschreibt Hofmeyr (1939a). Er fand einen Blüten-Farbfaktor (Y), gelb-weiß und einen Faktor für Stammfärbung (M), die beide X—Y gekoppelt vorkommen und sicher die Heterogametie der ♂♂ zeigen (Westergaard 1958).

c) *Cannabis sativa.*

Schließlich wird von IMAI 1938 (zit. nach ALLEN 1940) für Hanf das Auftreten zweier Mutanten „gelb-grün" und „gelb-gefleckt" nach Röntgenbestrahlung des Pollens beschrieben. Beide Charaktere traten nur bei ♂♂ auf und wurden vom Verf. als im X-Chromosom lokalisiert aufgefaßt.

d) *Spinacia oleracea.*

In einer japanischen Sorte wurden von SUTO und SUGIYAMA (1960) drei Wuchstypen unterschieden, zwei männliche und ein weiblicher. „B ♂" besitzt brakteenähnliche Blätter am Blütenschaft, in deren Achseln Knäuel von sitzenden Blüten gebildet werden; „L ♂" hat stärker gestreckten Blütenschaft mit voll entwickelten Laubblättern (leafy). Die Blüten sind kurz gestielt. Die „♀" zeigen persistierende Brakteolen, die die Samen dann verhärtet umgeben, „Prickly Seed". Der L- und B-Faktor zeigen Y-gebundenen Erbgang. (Hierzu auch ROSA 1925, SUGIMOTO 1947, KATAYAMA u. SHIDA 1956, sowie EGUCHI u. ICHIKAWA 1940.)

Aus den verschiedenen Möglichkeiten der geschlechtsgebundenen Vererbung und den cytologischen Beobachtungen an Heterochromosomen ergeben sich 3 Möglichkeiten der Lokalisation der gekoppelten Faktoren in den Geschlechtschromosomen: 1. kann ein solcher Faktor im differentiellen Segment von X lokalisiert sein und ist dann mit dem im X liegenden Realisator absolut gekoppelt; 2. kann der Faktor im differentiellen Segment von Y liegen, ist dann mit dem Y-Realisator absolut gekoppelt; 3. kann er im homologen Segment der Geschlechtschromosomen liegen und dann dementsprechend Austausch zeigen, also sowohl mit X als auch mit Y gekoppelt auftreten.

Die genetische Zusammensetzung des geschlechtsbestimmenden Mechanismus.

Zur näheren Analyse des geschlechtsbestimmenden Mechanismus ergaben sich drei verschiedene Untersuchungsmethoden: das Studium α) von Autopolyploiden, β) von Aneuploiden und γ) von Formen mit X- oder Y-Fragmenten. Dabei stellte sich heraus, daß mindestens zwei sich grundlegend unterscheidende Typen von Mechanismen vorliegen, von denen der zweite erst 1958 auch bei Tieren analysiert ist.

I. Bei Diöcisten.

a) Der *Drosophila*-Typus.

BRIDGES (1921, 1926) hatte festgestellt, daß bei *Drosophila* das Geschlecht eines triploiden Tieres (oder auch polyploider oder heteroploider Tiere) nicht durch Zusammenwirken von X und Y bestimmt wird, sondern gegeben ist durch das Verhältnis von X (mit ♀-Bestimmer) zu den Autosomen (mit ♂-Bestimmer), so daß bei X/A[1] = 1 ♀♀, bei X/A = 0,5 ♂♂ entstehen. Aus allen dazwischenliegenden Relationen entstehen Intersexe verschiedener Grade.

1. *Rumex acetosa.* Bei den Blütenpflanzen ist *Rumex* das bestuntersuchte Objekt dieses Typus.

CORRENS hatte mit seinen Certationsversuchen 1922 das ♂ bei *Rumex acetosa* als heterogamet nachgewiesen. Auffällig ist die starke Verschiebung des Geschlechtsverhältnisses zugunsten der ♀♀, die offenbar auf gametisch und zygotisch geringerer Konkurrenzfähigkeit der ♂♂ beruht. Bei einigen Sippen wurde jedoch weitgehende Annäherung an 1:1 beobachtet (CORRENS 1922). Durch KIHARA und ONO (1923, 1925) waren zytologisch die ♀♀ als 2n = 14 (12a[1] + 2 X) nachgewiesen, die ♂♂ als 2n = 15 (12a + Y_1XY_2).

[1] A = haploider Autosomensatz; *a* = einzelnes Autosom.

α) Ono untersuchte seit 1926 eine große Zahl von natürlich-vorkommenden Intersexen, die sich vorwiegend als Triploide erwiesen. Diese Intersexe sind gewöhnlich trimonöcisch, haben also ♂, ♀ und ⚥ Blüten. In den ⚥ Blüten sind die Geschlechtsorgane häufig schlecht entwickelt. Der Anteil dieser ⚥ Blüten wechselt von Pflanze zu Pflanze. Die Variabilität des Intersexualitätsgrades entspricht der von Goldschmidt (1931) für triploide Intersexe bei Insekten angegebenen. Er kann leicht durch Umwelteinflüsse verschoben werden. Aus der Zusammenstellung ergibt sich, daß der ♀ Realisator F im X und der ♂ Realisator in den Autosomen liegt, also Y_1 und Y_2 ohne Bedeutung sind (Yamamoto 1934b, Ono 1935), und das Verhältnis von X:A über das Geschlecht entscheidet:

	F:M[1]	Geschl.-Index	Geschl.
3n = 20 = 2 X + 18a	200:180	1,11	⚥
21 = 3 X + 18a	300:180	1,67	♀
21 = X + 2 Y + 18a	100:180	0,56	♂
22 = 2 X + 2 Y + 18a	200:180	1,11	⚥

Tetraploide ♂♂ fand Ono (1932, 1935) mit den Formeln: 2 X—2 Y—24a; 2 X—3 Y—24a; 2 X—4 Y—24a. Er erhielt außerdem tetraploide Intersexe, wie sie auch schon von Ono und Shimotomai (1928) beschrieben wurden, die bessere Fertilität als die 3n-Intersexe aufweisen:

Chromosomenkonstitution	F:M	Geschl.-Index	Geschl.
4 X + 24a	400:240	1,67	♀
2 X + 2 Y + 24a	200:240	0,83	♂
2 X + 3 Y + 24a	200:240	0,83	♂
2 X + 4 Y + 24a	200:240	0,83	♂
3 X + 2 Y + 24a	300:240	1,25	⚥

Für diploide ♀♀ ergibt sich dann

14 = 2 X + 12a   200:120   1,67

und für diploide ♂♂

15 = X + 2 Y + 12a   100:120   0,83.

Ein pentaploides Weibchen aus einer F_2 aus Bastarden, zwar sehr klein und mit wenig Blüten, aber deutlich ♀, hatte 35 Chromosomen

35 = 5 X + 30a   500:300   1,67.

Eine 44 chromosomige Pflanze

44 = 4 X + 4 Y + 36a   400:360   1,11   war ⚥.

β) Die Frage nach dem Männlichkeitsbestimmer in den Autosomen konnte durch Untersuchung *aneuploider* Formen bearbeitet werden, die in den Nachkommenschaften Triploider häufiger auftreten (fast alle Typen mit 2n = 15—21) und im Gegensatz zu *Drosophila* lebensfähig sind. Ono berichtet 1935 darüber, daß die Pflanzen dann intersex werden, wenn ein oder mehr Autosomen in X + 2 Y + 12a Typen zusätzlich vorhanden sind oder zwei oder mehr Autosomen bei 2 X + 12a. Das bedeutet, daß in den Autosomensätzen nicht nur männlich wirkende Faktoren, sondern auch weiblich wirkende vorhanden sind. Nach Chromosomenstudien von Ono (1930b, 1935), Kihara und Yamamoto (1931) und Yamamoto (1933b) sind die Autosomen unterscheidbar. Es wurden 8 Karyotypen beschrieben mit unterschiedlicher geographischer Verbreitung. Zwei Autosomen stehen offenbar in enger Beziehung zur Männlichkeitsbestimmung. (2 X + 2 A) + 2a können je nach den a ♀♀ oder auch ⚥⚥ sein. Pflanzen

[1] Man kann nach Goldschmidt die Stärke der Realisatoren durch Zahlenwerte ausdrücken; z. B. F = 100, 6a enthalten 1 M = 60, also M < F < MM.

mit X + 2Y + 2A + a_2 + a_5 werden nicht rein ♂, sondern ⚥. Andere zusätzliche Autosomen, auch nicht a_2 oder a_5 allein, rufen keine Geschlechtsveränderung hervor. Die polymorphen Chromosomen a_2 und a_5 werden daher als „*Gyno-Autosomen*" (YAMAMOTO) bezeichnet und sollen eine Netto-Weiblichkeitstendenz besitzen. Die i-förmigen a_1, a_4 und a_6 sind dagegen mit Netto-Männlichkeitstendenz (YAMAMOTO) ausgestattet und wirken als „*Andro-Autosomen*". a_3 scheint keine bemerkenswerte Geschlechtstendenz zu besitzen. Sind 3 a einer Gruppe überzählig, so wird das Geschlecht verändert, ist die Verteilung 2:1, so bleibt das Geschlecht gleich. Bei 2A + 2X + 4a *(Androautos.)* werden stets ⚥⚥ erhalten. Bei 1 X-Pflanzen wurde keine Änderung beobachtet. 2A + 5a + 2X ergibt ⚥, bei 1 X entstehen Übermännchen.

Manche der Autosomen scheinen wirkungslos zu sein. 70 von YAMAMOTO (1938) beobachtete Pflanzen mit X + 2Y + 2A + a zeigten keine Geschlechtsänderung. Er hält den von ONO beobachteten ⚥ für gleichartig mit dem Intersextyp, der bei normalen 2A ♂♂ auftreten kann. Es scheint ein Chromosom entweder in ♂ oder in ♀ Richtung in seiner Valenz unterschwellig zu sein, so daß keine Geschlechtsverschiebung auftreten kann.

Es wurden auch Intersexe gefunden, die einen normal erscheinenden diploiden ♂ oder ♀ Chromosomensatz besaßen. Die ♂ Intersexe hatten guten Pollen, die ♀ waren im Gynäceum gut fertil. YAMAMOTO (1938) und TAKENAKA (1935) vermuten, daß es sich um Sippenbastarde handelt analog zu den *Lymantria*-Bastarden von GOLDSCHMIDT. YAMAMOTO erhielt nach R-Bestrahlung diploide Intersexe ähnlich den diploiden Intersexen, die STURTEVANT (1920) und LEBEDEFF (1934) bei *Drosophila*-Arten fanden, und die beide auf ein recessives autosomales Gen zurückführen. Jedenfalls wird aus den Ergebnissen bei *Rumex acetosa* geschlossen, daß die Geschlechtsausbildung durch die Mitwirkung verschiedener geschlechtsbestimmender Faktoren zustande kommt, daß aber durch die Summierung weiblicher Faktoren im X die alternative Vererbung bewirkt wird. Den gleichen G.B.-Typus haben offenbar alle Arten der Sektion *Euacetosa* des Subgenus *Acetosa* bei *Rumex*.

2. *Rumex hastatulus*. Von *R. hastatulus* ist bisher der in Nordkarolina verbreitete Karyotyp mit XY_1Y_2 im ♂ im polyploiden Zustand untersucht worden.

α) Der Genotypus $3A+2X+Y_1Y_2$ erwies sich als rein ♂; die Kombination $4A+3X+Y_1Y_2$ ergab ♂♂ und ♂ Intersexe. 3A+3X-Individuen waren ♀. Daraus ist auf das X/A-Verhältnis als geschlechtsentscheidend zu schließen mit dem kritischen Wert von 0,75. Die Ergebnisse schließen aber eine Beeinflussung der Geschlechtsbestimmung durch die Y-Chromosomen noch nicht aus (SMITH 1958).

β) Untersuchungen an Aneuploiden sind noch ohne Aussagekraft.

γ) Eine Pflanze wurde gefunden, die zu den normalen 10 Chromosomen ein telozentrisches Fragment enthielt. Es bestand aus einem einzelnen Arm, der halb so lang war wie das Chromosom 3 der Nomenklatur von SMITH (1963). Das Centromer funktionierte normal. Alle Nachkommen dieser Pflanze, die 9 Chromosomen und das Fragment enthielten, waren ♂ und morphologisch von normalen ♂♂ nicht zu unterscheiden. Alle Pflanzen der Konstitution 2A+XX+f bildeten ⚥-Blüten. Dieses Fragment muß also einen Faktor für männliche Ausbildung enthalten. Der Effekt auf triploider Ebene war ähnlich: 3A+3X+f waren Intersexe. In einigen Zellen in der Diakinese von 9+f Pflanzen wurde es in Assoziation mit dem distalen Ende eines der Y-Chromosomen gefunden. Es ist zu schließen, daß dieses Fragment Homologie zu einem Y-Arm zeigt. Damit wäre zwar X/A entscheidend für die G.B., aber Y ist nicht inert, sondern enthält Gene, die die männliche Ausbildung fördern. Es enthält aber keine Faktoren, welche die ♀ Ausbildung verhindern.

♀	2A+2X*	3A+3X
⚥	2A+2X (1 Pflanze)	
	2A+2X+f	3A+3X+f (2 Pflanzen)
	2A+XY_1Y_2 (7,7%)	3A+XXY_1Y_2 (3 Pflanzen)
		3A+XXY_1Y_2+f (1 Pflanze)
♂	2A+XY_1Y_2 (92,3%)	3A+XXY_1Y_2 (5 Pflanzen)
	2A+XY_1Y_2+f	

* Ist keine Zahl angegeben, immer große Pflanzenanzahl.

3. *Cannabis sativa.* Die Wildformen des Hanfes sind reine Diöcisten, während die hauptsächlich genetisch untersuchten Kultursorten sich zu Subdiözisten entwickelt haben. Aus den an ihnen gewonnenen Ergebnissen folgt, daß der ursprüngliche diöcische Hanf den gleichen G.B.-Modus besitzt wie *Rumex acetosa.* Näheres über die Analyse siehe bei den Subdiöcisten unter IIa.

b) Der *Melandrium*-Typus.

Nach Ableitungen von Muller (1925) u. a. wirkt der G.B.-Modus bei *Drosophila* gegen eine Entstehung natürlicher polyploider Rassen oder Arten mit diöcischer Geschlechtsverteilung. Das Auffinden normal eingeschlechtiger Pflanzen bei polyploidem *Melandrium* führte zur Aufklärung eines zweiten Typs der G.B., der von Warmke und Blakeslee in Amerika und von Westergaard in Dänemark weitgehend analysiert wurde.

1. *Melandrium.* Der XX—XY-Mechanismus bei *Melandrium album (Lychnis alba)* und *Melandrium dioicum (L. dioica)* war bereits von Correns (1907) festgestellt worden durch Kreuzungen von *(L. alba × dioica)* ♀♀ mit *Silene viscosa* (Hermaphrodit mit Neigung zur Gynomonöcie) als ♂♂. Sie ergab nur ♀♀, allerdings mit etwas weiter entwickelten Stamina-Rudimenten als die Mutter sie hatte. Die reziproken Bastardsamen waren nicht keimfähig. Die gleichartige Kreuzung *Melandrium album* ♀ × *Silene viscosa* ⚥ = ∞ ♀♀ war schon von Kölreuter durchgeführt worden. Der cytologische Nachweis der männlichen Heterogametie gelang Blackburn (1923, s. S. 61). Die Konkurrenzversuche hatten die männliche Heterogametie ebenfalls klar erwiesen, ebenso die Kreuzungen von Shull (1910b, 1911, 1914a), Hertwig (1922) und Winge (1931).

α) Durch Colchicin-Behandlung hergestellte tetraploide *Melandrium* erwiesen sich als normal diöcisch. Aus der überwiegend gefundenen Autosyndese der Geschlechtschromosomen im ♂ XXYY schloß Westergaard bereits 1938 auf vorwiegende Bildung von XY-Gameten (vgl. S. 62). Sie wurden von Warmke und Blakeslee (1939) mit etwa 90% festgestellt, so daß man statt der zufallsgemäßen Verteilung in F_1 ♀ XXXX × ♂ XXYY = 1 XXXX + 4 XXXY + 1 XXYY einen erheblich stärkeren Anteil an XXXY-Typen erwartete. Gefunden wurden: 3 ♀♀ XXXX + 78 ♂♂ + mask. ⚥⚥ XXXY + 2 ♂♂ XXYY. Die Kreuzungen zwischen tetraploiden und diploiden Pflanzen ergaben über 100 3n und 4n-Pflanzen:

2n	2A + XX	17 ♀
	2A + XY	20 ♂
3n	3A + XXX	1 ♀
	3A + XXY	4 ♂; 1 ± ⚥
4n—1	4A + XXX	1 ♀
	4A + XXY	1 ♂
4n	4A + XXXX	3 ♀
	4A + XXXY	65 ♂; 3 ± ⚥
	4A + XXYY	2 ♂
4n+1	4A + XXXXY	1 ⚥
	4A + XXXYY	3 ♂

Aus diesen Zahlen ging eindeutig hervor, daß Bildung männlicher Blüten nur bei Vorhandensein des Y-Chromosoms auftritt (Abb. 7). Da Pflanzen mit 4A + XXXY = ♂ und 4A + XXXX = ♀ sich nur durch das Y unterschieden, muß das Y-Chromosom im Gegensatz zu *Drosophila* und *Rumex acetosa* männlich bestimmende Elemente tragen, die über 2X oder 3X prävalieren können. Y kann

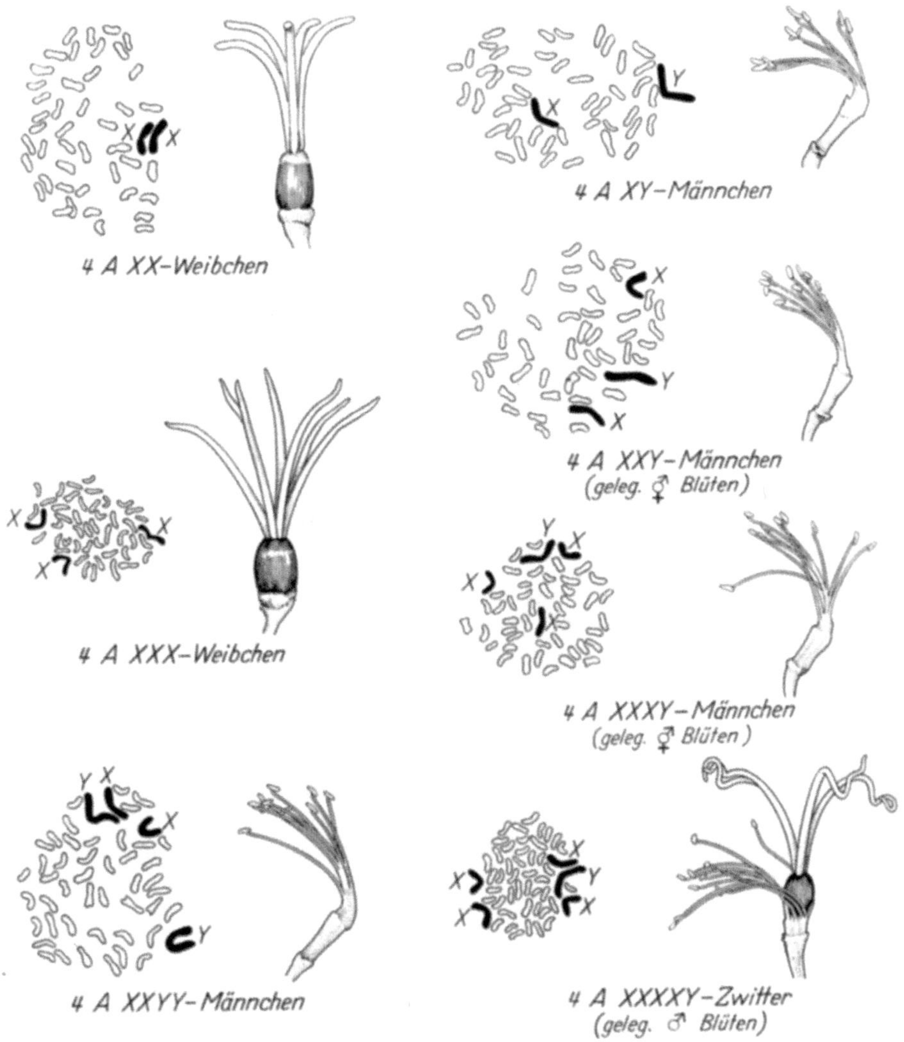

Abb. 7. Geschlechtsausbildung verschiedener experimentell erhaltener Typen von *Melandrium*. (Nach WARMKE 1946.)

aber nicht völlig 4X ausbalancieren, da 4A + XXXXY = ⚥. Diese tetraploiden Pflanzen des 2. G.B.-Typus zeigten die Möglichkeit der Bildung konstanter polyploider diöcischer Sippen: ♀ XXXX × ♂ XXXY = 1 ♀ XXXX : 1 ♂ XXXY. Die XXXY ♂ unterschieden sich meist nicht von den XXYY ♂, nur an 10% der Pflanzen kommen einige ⚥ Blüten vor.

β) Auch 4A XXY-Pflanzen waren ♂ ausgebildet, gelegentlich mit ⚥-Blüten. Stark zwittrig zeigte sich dagegen die Pflanze 4A + XXXXY, während eine Pflanze der gleichen Konstitution bei WESTERGAARD aus europäischen Herkünften rein ♂ war. Selbstung des 4A XXXXY ⚥, der rein normales, nicht fragmentarisches Y besaß, ergab: 7 ♀♀ (6 XXXX + 1 XXXXX), 5 ♂♂ (2 XXXY, 2 XXXYY, 1 XXXXYY) und 8 ⚥⚥ (XXXXY).

Aus der Kreuzung XXXXY ⚥ × XXXY ♂ gingen 9 ♂♂ (1 XXYY, 5 XXXY, 3 XXXYY) und 4 ⚥⚥ (XXXXY) hervor. Ohne Ausnahme ergab also 1 Y gegen 4 X stark zwittrige Pflanzen, aber gegen 3 X ♂ (keine schwach zwittrigen Pflanzen darunter). Daraus folgt auch, daß das Y normal gebaut war. Pflanzen 2A XXY erwiesen sich als ♂, selten mit 1 ⚥ Blüte.

Die weiblich bestimmende Wirkung der X-Chromosomen konnte aus folgender Reihe abgelesen werden, bei der sich nur die Anzahl der in der Pflanze vorhandenen X-Chromosomen änderte:

4A XY = ♂
4A XXY = ♂ (gelegentlich ⚥ Blüten)
4A XXXY = ♂ (gelegentlich ⚥ Blüten)
4A XXXXY = ⚥ (gelegentlich ♂ Blüten)
2A XY = ♂
2A XXY = ♂ (gelegentlich ⚥ Blüten)

In den 2A XXY-Pflanzen waren zum Teil X-Fragmente vorhanden, die dann mit dem X in der Meiose paarten, also aus dem differentiellen Arm des X bestanden, der offenbar die weiblich bestimmenden Elemente enthält. Die Ergebnisse von ONO (1939, 1940) stimmen damit überein. Eine Veränderung der Zahl der Autosomensätze brachte gegenüber gleichbleibendem XY-Verhältnis keine Geschlechtsveränderung.

2A XXY = ♂ → 3A XXY = ♂ → 4A XXY = ♂
(m. geleg. ⚥-Bl.) (m. geleg. ⚥-Bl.) (m. geleg. ⚥-Bl.)

Sie enthalten nach WARMKE und BLAKESLEE keine wichtigen weiblichen Faktoren. *Aneuploide* Formen aus Kreuzung triploider Pflanzen waren bei WARMKE und BLAKESLEE (1940a, b, c) stets rein ♀ oder rein ♂, also fehlende oder zusätzliche Autosomen für die G.B. ohne Einfluß. Das Verhältnis von X/A ist wirkungslos, denn 4A XX hat die gleiche Relation wie 4A XXYY, ist aber rein ♀. Die weiblich bestimmende Wirkung des X muß im amerikanischen Material stärker sein. Für die rassenverschiedene Valenz der Realisatoren spricht auch das Fehlen von Intersexen in der Nachkommenschaft triploider Pflanzen des amerikanischen Materials, während sie im dänischen häufig waren. Die meisten dieser Pflanzen hatten mehr als 33 Chromosomen: eine der Zwitterpflanzen, ein Euhermaphrodit im Sinne CORRENS', zeigte cytologisch 4X-Chromosomen, deren Paarungsbilder aber die gleichen Konfigurationen wie bei XXXY-Typen aufwiesen, so daß WESTERGAARD 3X + 1Y mit terminaler Defizienz annimmt. Eine zweite androhermaphrodite Pflanze war 4n—1 = 43 + XXXY. WARMKE (1946c) fand unter 36 einfach trisomischen Pflanzen (2A + a + XX) keine Geschlechtsveränderung. Trisome ♂♂ wurden nicht beobachtet. In ähnlicher Richtung liegen die Ergebnisse von ONO (1940), der bei triploiden (XXX + 3A + 1 bis 4a) keine Geschlechtsänderung beobachtet hat. Im Gegensatz dazu erhielt WESTERGAARD unter 46 trisomen Pflanzen 10 ± hermaphrodite neben 21 ♀♀ und 15 ♂♂ (Tabelle 4).

Die Geschlechtsausbildung der Selbstungs-Nachkommenschaften dieser ⚥⚥ läßt sich nur verstehen, wenn X-Chromosomen zusammen mit bestimmten Autosomen-Kombinationen der stark männlichen Wirkung des Y die Waage halten können. Unter 205 aneuploiden XXXY-Pflanzen waren 72 ♂♂ und 133 ♀♀. Diese Unterschiede führt WESTERGAARD auf autosomale Wirkung zurück. Die Zwitter waren vorwiegend „*Androhermaphroditen*" mit nur einigen primären zwittrigen Blüten.

Die *aneuploiden* Zwitter unterschieden sich phänotypisch von den *Fragmentations-Zwittern*. Bei ihnen sind nur die ersten Blüten ganz zwittrig, die letzten jedoch rein männlich. Durch Auslese kann der Zwittertyp in ⚥ Richtung verschoben werden, sie wirkt aber nur auf bessere Ausbildung der ♀ Organe. ♂ Organe werden nicht rückgebildet. Aus Selbstung *aneuploider* Zwitter können ♂ und ⚥ oder ♀ hervorgehen, während das Auftreten von ♂ bei *Fragmentations-Zwittern* nicht möglich ist. Da Pflanzen mit XXY—XXYY—XXXY—XXXYY entweder ♂ oder ⚥ werden können und die einzigen variablen Faktoren die Autosomen sind, muß der Ausschlag von diesen erfolgen. Offenbar wirken einige Autosomen ♀-bestimmend, andere ♂-bestimmend. Dafür sprechen auch die (3A + XXY + 2a) und (4A + XXXY + a)-Pflanzen, die einige ♂ Blüten zeigten (WESTERGAARD 1940).

γ) Sowohl WARMKE (1946a, b) als auch WESTERGAARD (1946) erhielten unabhängig voneinander Pflanzen mit fragmentierten Y-Chromosomen. WARMKE beobachtete in einer 2A + XXY-Pflanze 5% spontane Brüche bei der Meiose in Selbstung einer hoch ingezüchteten Linie, die praktisch homozygot war. Und zwar zeigte auch der differentielle Arm des Y Brüche. Es wurde mit Hilfe dieser Defizienzen ein Gen in der Nähe des Zentromers festgestellt, das Anlage der Antheren und Entwicklung bis kurz nach der Meiose ermöglicht. Fehlt es, werden gar keine männlichen Strukturen angelegt. Nahe dem Ende des differentiellen Armes liegt ein Gen (oder Genkomplex), das die Vollendung der ♂ Entwicklung ermöglicht. Eine dritte Genregion im terminalen Viertel des mit X homologen Armes von Y unterdrückt die gesamte ♀ Entwicklung. Duplikationen einzelner Regionen können nicht andere fehlende Stücke ersetzen.

Tabelle 4. *Chromosomen-Konstitution und Geschlecht in triploiden Nachkommenschaften von Melandrium.* (Nach WESTERGAARD 1958.)

Geschlechts-chromosomen-Konstitution	Zahl der X-Chromosomen	♂♂	⚥⚥	Summe	% ♂♂
XY	1	15	0	15	100
XXY	2	25	3	28	
XXYY		37	5	42	89
XXYYY		1	0	1	
XXXY	3	72	133	205	
XXXYY		7	5	12	36
XXXXY	4	0	8	8	
XXXXYY		0	1	1	0
Summe		157	155	312	

Ebenfalls durch Defizienzen am Y hat WESTERGAARD die Lage der geschlechtsentscheidenden Gene im Y feststellen können, fand jedoch für sein Material eine etwas andere Lagerung. Der Weiblichkeits-Unterdrücker liegt hier im distalen Ende des differentiellen Armes, während der Hauptgenkomplex für die Auslösung der Antherenbildung sich im differentiellen Stück zwischen SPA und homologem Ende befindet. Der dritte Faktor, der Fertilität der Antheren hervorruft, ist in der Nähe des homologen Segmentes, eventuell im homologen Segment selber lokalisiert. WESTERGAARD empfiehlt, nur Gene im differentiellen Segment der Geschlechtschromosomen als „Geschlechtsgene" zu bezeichnen. Der Unterschied der Faktorenlagerung ließe sich durch eine zentrale Inversion bei der Entwicklung der amerikanischen Rassen erklären (Abb. 8).

Der Y^1-Typ zeigt eine Defizienz des differentiellen Armes von Y. Alle Pflanzen, die Y^1 neben X und den Autosomen enthalten, sind Euhermaphroditen mit ausschließlich zwittrigen Blüten mit vollentwickelten weiblichen und männlichen Geschlechtsorganen. Solche Pflanzen ergeben, mit normalen ♀♀ gekreuzt oder geselbstet, nur ♀♀ oder ⚥⚥, niemals ♂♂. Da das Fehlen dieses Stückes keinen Einfluß auf die normale Entwicklung der männlichen Organe oder vegetativen

Charaktere hat, ist unwahrscheinlich, daß noch weitere Gene in dieser Region liegen (WESTERGAARD 1958). Das Auftreten von Intersexen in triploiden Nachkommenschaften zeigt, daß die Männlichkeits-Gene nicht absolut epistatisch über die Weiblichkeitsbestimmer sind, wohl aber wirken sie ungehindert neben den Weiblichkeitsgenen. Das zeigen auch Pflanzen, die in Nachkommenschaften N_2O-behandelter polyploider Pflanzen auftraten (4A + 3X + Y = ♂, 3A + 3X und 3A + 2X + Y = ♀; 4A + 3X + Y = ⚥) (NYGREN 1957).

Beim Typus Y^3, dem das homologe Segment und ein Teil des differentiellen Segmentes fehlt, sind die Pflanzen männlich steril. Pflanzen mit einem normalen Y + Y^3 werden normale fertile ♂♂. WESTERGAARD gibt für den Mechanismus bei *Melandrium* 1953 den Vergleich mit einer Maschine mit zwei potentiellen Funktionen: männliche Entwicklung und weibliche Entwicklung. Es gibt 2 Möglichkeiten einer Blockierung der Maschine: Einmal durch Ansetzen einer Bremse, in diesem Fall Funktionieren des Weiblichkeits-Unterdrückers im Y, der nur ein Arbeiten in männlicher Entwicklungsrichtung zuläßt, zum anderen durch Entfernen wichtiger Maschinenteile, der beiden Genkomplexe im Y, die für die Antherenbildung unerläßlich sind, so daß nur die weibliche Entwicklung ablaufen kann. Durch absolute Koppelung beider Systeme im differentiellen Teil des Y ist die einwandfreie Weichenstellung des G.B.-Mechanismus gewährleistet. So erklärt sich, daß alle bei polyploiden Pflanzen beobachteten Intersexe ein Y im Chromosomenbestand enthielten. Einen Hinweis auf die Funktionsweise der beiden männlich wirkenden Genkomplexe im Y kann vielleicht einmal die Analyse der Antherenbildung an XX-Pflanzen durch *Ustilago violacea* geben. VAN NIGTEVECHT (1966) vermutet, daß die Wirkung von Y über Veränderung des Auxinspiegels der Pflanzen verläuft. Aus dem Verhalten der geschlechtsgebunden vererbten Faktoren bei *Melandrium* ergibt sich auch ihre ungefähre Lage zu den Geschlechtsgenen (WESTERGAARD 1953): die Gene *angustifolia* und *chlorina* müssen, da sie X-gebunden auftreten, im differentiellen Segment von X liegen, das von WINGE beschriebene Hemmungsgen für Buntblättrigkeit im differentiellen Teil von Y und das X—Y-gebunden vererbte Gen „*abnormal*" im homologen Segment (Abb. 8).

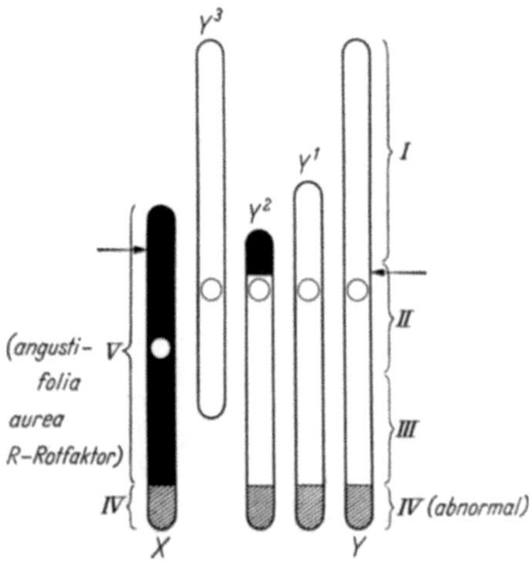

Abb. 8. Schema des Geschlechtsbestimmungsmechanismus bei *Melandrium*. Segment *I* enthält das Gen (oder die Gene), das die Entwicklung der weiblichen Geschlechtsorgane unterdrückt. Die Gene in *II* kontrollieren die ersten Stadien der Staubfädenentwicklung. Segment *III* kontrolliert die letzten Stadien der Staubfädenentwicklung. *IV* die homologen Segmente. *V* differentielles Segment von *X*. Y^1 und Y^3 durch Defizienz entstandene *Y*-Mutanten, Y^2 durch Translokation entstanden (die vermutlichen Bruchstellen bei *X* und *Y* sind durch Pfeile gekennzeichnet). In Klammern die in den entsprechenden Segmenten lokalisierten geschlechtsgebundenen Gene. (Nach WESTERGAARD 1953, verändert.)

An Pflanzen mit Defizienz am differentiellen Arm von X konnte WARMKE (1946c) zeigen, daß Weiblichkeitsfaktoren in diesem Teil vorhanden sind.

Aus diesen an polyploiden Zwittern gewonnenen Einsichten lassen sich auch die länger bekannten diploiden Zwitter besser verstehen, die von G. SHULL, P. HERTWIG u. a. beschrieben wurden. SHULL (1910, 1911) fand in normal diöcischen Kulturen von *Lychnis dioica* einige Hermaphroditen (etwa 1:1000), die bei Selbstung ♀♀ und ⚥⚥ (in der Minderzahl) ergaben und sehr wenige ♂♂. Aus Kreuzungen mit normalen ♀♀ gingen ebenfalls sowohl ♀♀ und ⚥⚥ als auch wenige ♂♂ hervor. Das Ergebnis einer Kreuzung ⚥ × ♂ waren 29 ♀♀, 12 ♂♂ und 2 ⚥⚥. Weibliche Nachkommen dieser Zwitter verhielten sich mit normalen Männchen wie normale Weibchen. Shull schloß daraus auf Mutation entweder eines ♂ oder eines ♀ Geschlechtsgenes, die zur Zwitterbildung beim ♂ führt und nicht durch Eizellen übertragbar ist. Er fand in der Nachkommenschaft genetischer Zwitter einzelne normale ♂♂, 11 unter 5467, die er als durch Rückmutation entstanden ansieht (1911, 1914). Es könnte sich hier um Veränderung des Weiblichkeits-Unterdrückers im Y gehandelt haben.

G. und P. Hertwig (1922) berichten über die Nachkommenschaften zweier Zwitter der *Lychnis dioica*. Sie erhielten:

♀ × ⚥ = ♀ + ⚥
♀ × ♂ = 263 ♀♀ + 246 ♂♂ + 3 ⚥⚥
⚥ × ⚥ = 187 ♀♀ + 158 ⚥⚥
⚥ × ♂ = 14 ♀♀ + 14 ♂♂ + 14 ⚥⚥ (weniger stark ♀ als die Ausgangszwitter)
♀ (aus ⚥) × ♂ = 450 ♀♀ + 330 ♂♂ + 93 ⚥⚥.

Die Zwittrigkeit wurde also hier abweichend von Shulls Zwittern auch durch Eizellen übertragen. Hier muß außer einer schwachen Änderung des Weiblichkeits-Unterdrückers eine Verstärkung der Weiblichkeitsfaktoren im X vorliegen.

Eine Kreuzung von ♀ (aus ⚥) × ⚥ brachte 103 ♀♀, 5 ♂♂ (wahrscheinlich ⚥⚥) und 73 ⚥⚥. Diese Ergebnisse wurden schon von den Autoren durch mutative Änderung der Geschlechtspotenzen im Männchen in weiblicher Richtung erklärt, so daß die Zwitter hier auch aus Männchen entstanden. Die bei Selbstung der XY-Zwitter erwarteten homozygoten YY—♂♂ traten nicht auf, sie sind eventuell nicht vital, wofür auch die große Zahl verkümmerter Samen spricht. Alle von Belar (1925) untersuchten ⚥⚥ hatten den Chromosomenbestand der Männchen 22a + XY, so daß Heitz' Erwägungen über nondisjunction gegenstandslos werden.

Correns (1928) fand Zwitter aus Bestäubungen mit überaltertem Pollen, die er auch durch Valenzänderungen der männchenbestimmenden Pollen erklärt. Eventuell sind sie auf erhöhte Fragmentierung von Y in altem Pollen zurückzuführen.

Im Prinzip ähnlich sind die Ergebnisse Winges (1931), die er mit XY-Hermaphroditen aus der Kreuzung *Lychnis dioica* × *alba* erhielt. Er schließt daraus auf zahlreiche entweder in männlicher oder weiblicher Richtung wirkende Gene sowohl in X und Y als auch in den Autosomen. Bei gestörter Balance der Netto-Tendenzen resultieren Zwitter. Er identifizierte ein stark männlich wirkendes autosomales Gen, das den Hermaphroditismus aufhebt und fand außerdem einen recessiven Letalfaktor im X, der homozygot die Entstehung von ♀♀ verhindert. Selbstung von Zwittern, die ♂ und ⚥-Blüten auf der gleichen Pflanze zeigen und ± fertil sind, ergab ♀♀ und ⚥⚥ und 1 ♂ (D. Löve 1942). Die ♂♂ zeigen Heterochromosomen, die auf eine Translokation zwischen X und Y hinweisen. Es wurde die Spaltung erwartet: 1 normales ♀ (XX):2 ⚥⚥ (XX_Y):1 X_YX_Y-Individuum. Ob die letzten überleben würden, war fraglich. Die tatsächlichen Geschlechtsverhältnisse sind dem Verfasser nicht bekanntgeworden. X_YX_Y-Individuen zeigten größte Sensibilität gegen Behandlung mit Geschlechtshormonen. 1940 berichtet D. Löve von 8 Intersexen, die zum Teil XY, zum Teil XX sein sollten, und schließt daraus, daß auch andere genetische und Umweltbedingungen einen Einfluß auf die Geschlechtsausbildung haben können. Besonders viele Intersexe wurden nach heißen Sommerperioden beobachtet. Allerdings konnte Löve die X in den somatischen Platten nicht identifizieren.

Ein einzelner Fall eines Zwitters, der für XX-Typ gehalten und von Åkerlund (1927) beobachtet wurde, war nicht fertil. Nach Westergaard ist aus dem Paarungsverhalten der Geschlechtschromosomen auf ein fragmentiertes Y + X zu schließen, so daß auch hier ursprünglich XY-Kombination angenommen werden kann. Die XX-⚥⚥ D. Löves könnten ähnlich zustande gekommen sein.

Leman (1960) untersuchte die Meiose einer Anzahl von XY-Intersexen und ihrer Nachkommen und fand häufig ein sich aberrant verhaltendes Autosomenpaar, während X und Y normal paarten. Da am Y keine Veränderung festgestellt werden konnte, schließt er daraus auf eine Beteiligung der Autosomen an der Geschlechtsausbildung.

1966 berichtet van Nigtevecht über eine hermaphrodite diploide Pflanze, die nach zwei Generationen Inzucht in einer *M. album*-Familie auftrat. In allen Blüten waren Stamina und Pistill ausgebildet, jedoch nur 3—4 Griffel anstelle der 5 beim ♀. Die Pflanze war ♀ schwach fertil, der Pollen gut fertil, obwohl die Antheren etwas schwächer entwickelt waren als beim ♂. In Kreuzungen mit 2n- und 4n-Pflanzen von *M. album* wurden nur ♀ und ⚥ beobachtet, ebenso in den Folgegenerationen. Die Ausbildung der Pistille variierte von kleinen mit 2 Griffeln bis zu größeren mit 4 Griffeln; selten wurden 1- oder 5-grifflige Blüten gefunden. In den 3n-Nachkommen lag die gleiche Variation vor, nur mit etwas größeren Organen. Van Nigtevecht vermutet eine Mutation des Weiblichkeitssuppressors im Y zu völligem Ausfall oder in einen weniger effektiven Status zu Ya. Cytologisch zeigt Ya keine Differenz zu Y.

Einen zweiten Hermaphroditentyp erhielt er nach Alternlassen des Pollens im Exsiccator mit gut entwickelten ♂ und ♀ Organen. Kreuzungen dieser Pflanze mit 2n und 4n ♀♀ ergaben nur ♀ Nachkommen, wie sie Correns (1924) mit seinen Hermaphroditen erhalten hatte. Im Gegensatz zu dessen Ergebnissen brachten jedoch Kreuzungen zwischen ♀ und ♂ Nachkommen auch nur ♀ Pflanzen. Kreuzungen zwischen ♀ und ♂ Pflanzen aus Bestäubung mit altem Pollen spalteten in ♂♂ und ♀♀ und einige wenige ⚥⚥ (Androhermaphroditen). Weitere Alterungsversuche mit Pollen sowohl von *M. album* wie *M. dioicum* führten zur Entstehung von 11 Hermaphroditen. Selbstung oder Kreuzung mit ♀ Pflanzen ergab rein ♀ Nachkommenschaften. Ein Certationseffekt lag nicht vor. Die Analyse des Y-Chromosoms zeigte hier eine Defizienz, die nicht mit Y_1, Y_2 oder Y_3 von Westergaard identisch ist. Ein dritter ⚥-Typ als Androhermaphrodit trat auf in einer Kreuzung zwischen verschiedenen *dioicum*-Populationen neben den normalen ♂♂ und ♀♀. Viele Blüten dieser Pflanze waren rein ♂, andere hatten ein rudimentäres ♀ Organ; in einigen Blüten waren kleine Pistille mit einem oder zwei Griffeln ausgebildet. Nach Selbstung spalteten ♀♀ und ⚥⚥ heraus, nach Kreuzung mit einer ♀ Pflanze 35♀♀, 14 ♂♂ und 23 ⚥⚥. Die Expressivität des Zwittercharakters unterscheidet diesen Typ deutlich von den beiden erstgenannten. In dieser Familie war Selektion auf die zwittrige Ausbildung erfolgreich bis zur ausschließlichen Bildung ⚥ Blüten an den Pflanzen mit gleichzeitiger Vergrößerung der Pistille und Steigerung der Griffelzahl auf 5. Zeigten die ♀ Organe die Größe wie an XX-Pflanzen, wies das Androeceum eine Reduktion auf. Van Nigtevecht konnte zeigen, daß weder X noch Y für diese ⚥ Ausbildung verantwortlich sind. Dieser Typus des autosomal bedingten Hermaphroditen wurde nur in *M. dioicum* gefunden, nicht bei *M. album*, obwohl 60000 XY-Pflanzen untersucht wurden.

2. *Silene otites.* Den *Melandrium*-Typus der G.B. zeigt auch *Silene otites* nach Untersuchungen polyploider Pflanzen durch Warmke (1942). Die untersuchte Sippe war diöcisch, jedoch ohne Heterochromosomen. Aus der Kreuzung XXXX ♀ × XXYY ♂ wurden ungefähr erhalten 5♂♂:1♀, d. h. es müssen 2 Klassen von Männchen gebildet werden, von den zu erwartenden Typen 1 XXXX:4 XXXY:1 XXYY entweder XXYY + XXXY oder XXXX + XXXY, je nachdem, ob ♂ oder ♀ Heterogametie vorliegt. Kreuzung der ♀♀ mit diploiden ♂♂ muß bei XXXX = ♀ eine triploide Population mit 1:1-Geschlechtsverhältnis ergeben, bei XXYY = ♀ wieder 5 ♂♂:1 ♀. Das Ergebnis 1:1 beweist die Homogametie der ♀♀ und den gleichen Modus wie bei *Melandrium*, d. h. stark ♂ Wirkung von Y.

Ältere Untersuchungen von Correns lassen eventuelle männliche Homogametie anderer Sippen vermuten. Correns (1928) erhielt aus Selbstungen von ♂♂ mit einigen Zwitterblüten nur ♂♂ oder intersexe ♂♂. Sansome (1938) hält ebenfalls die ♀♀ für heterogamet, während bei *Silene pseudotites* das ♂ heterogamet ist.

3. *Populus tremula.* 1936 berichtet Nilsson-Ehle über einen triploiden männlichen Klon von *Populus tremula* (3n = 57) mit 44% normalem Pollen (Müntzing 1936). Später sind 9 spontane 3n-Formen gefunden worden, von denen 3 reine ♀♀ und 4 reine ♂♂ waren; von zweien ist das Geschlecht unbekannt. Nach Müntzing können die Pflanzen XXX oder XXY sein. Da die beiden reinen Geschlechter gefunden sind, liegen die Verhältnisse wie bei *Melandrium*, wenn XXY = ♂ ist. Blackburn und Harrison (1924) nehmen 1 Paar Heterochromosomen im ♂ an, deren Unterschied aber nicht sehr groß sein kann; sie wurden von anderen Autoren nicht gesehen (van Buijtenen u. Einspahr 1959). Über die Geschlechtsverhältnisse künstlicher tetraploider und aneuploider Formen aus 4n × 4n liegen keine Angaben vor (Johnsson 1942, 1945). Zwitter wurden ebenfalls gefunden (Bail 1869, 1877), in neuerer Zeit einer von Runquist (1951), der vorwiegend ♂, aber auch ⚥ und vereinzelt ♀ Blüten trug. Die Selbstungsnachkommen sind noch nicht analysiert. Zwei neue Funde zeigen gyno-intersexuellen Charakter, innerhalb jedes Kätzchens variiert der Blütenbau in der Zahl der Antheren; rudimentäre Antheren wurden nicht beobachtet. Es handelt sich eventuell um Polyploide, denn die Pollengröße variiert sehr stark. Lester (1963) untersuchte in Connecticut 138 Bäume und fand bei 38% unterschiedliche Bi-

sexualitätsstufen. Stets ist aber ein Geschlecht vorherrschend. Ein besonderes Verteilungsmuster der abweichenden Blüten wurde nicht festgestellt. Ein genetischer Einfluß auf die Variabilität der Geschlechtsausbildung wird vermutet. Charakteristisch ist die Umwandlungsrichtung von ♂ zu ♀. Nach PETO (1938) kommen bei Artbastarden von *Populus tremuloides* mit *alba* und *P. alba* × *grandidentata* Zwitter vor. Ein diploider Zwitterklon zeigte normale Meiose in den Antheren, bildete aber zum Teil unreduzierte Pollen, so daß in der Nachkommenschaft Triploide gefunden wurden (SEITZ 1954). Diese sehr fragmentarischen Befunde sprechen für Vorliegen des *Melandrium*-Typus.

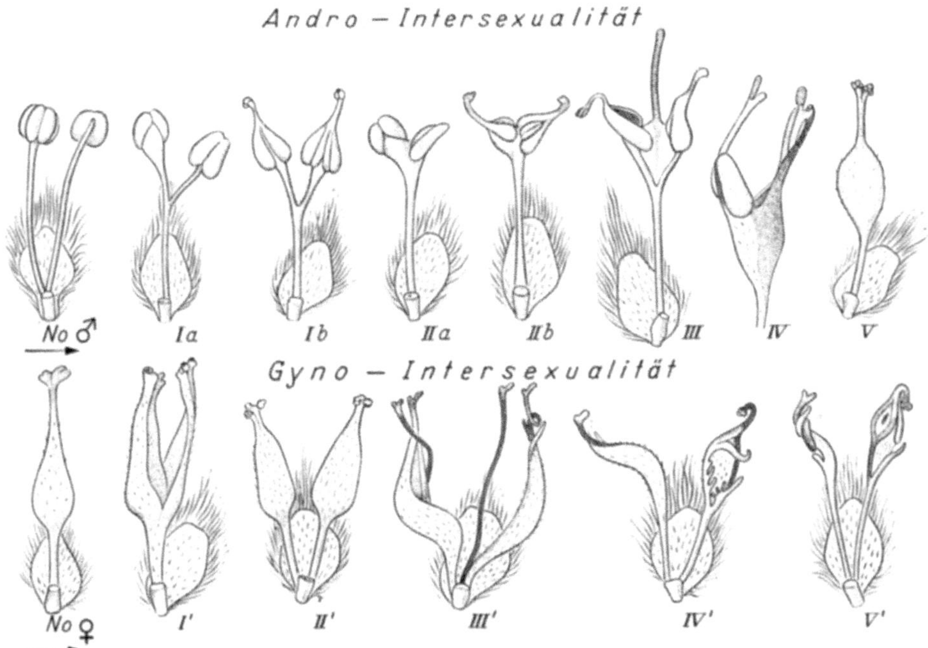

Abb. 9. Übergangsformen zwischen normalen Blüten des einen Geschlechts zu normalen Blüten des anderen Geschlechts bei *Salix*. (Nach RAINIO 1927.)

4. *Salix*. Bei den diöcischen *Salix*-Arten hat HERIBERT-NILSSON 1918 über umfangreiche Beobachtungen der Geschlechtsausbildung berichtet. Bei einigen Arten liegt ein 1:1-Verhältnis vor, bei anderen stark abweichend ein Überwiegen der ♀♀. Bei allen Geschlechts-Aufspaltungen, die einen höheren Weibchenanteil als 10:1 aufweisen, ist *Salix cinerea* an der Kreuzung beteiligt. *Cinerea* ♀ × ♂ ergab nur 24 ♀♀. Lediglich in 2 Kreuzungen: *Salix viminalis* × *daphnoides* und *(repens* × *viminalis)* × *aurita* zeigten sich Intersexe, und zwar mit Mosaikcharakter. Daneben traten normale ♀♀ und ♂♂ auf. RAINIO (1927) trennt die intersexen Typen in *Iso-Intersexe* (alle Blüten gleiche Intersexualitätsklasse) und *Hetero-Intersexe* (mit verschiedenen Blütentypen). Er unterscheidet 5 Klassen Andro-Intersexe und 5 Klassen Gyno-Intersexe (Abb. 9). HERIBERT-NILSSON nimmt einen Geschlechtsfaktor und einen Faktor für Diöcie an, so daß in einer Kreuzung von 2 Pflanzen, die heterozygot für D waren, 3 ♀♀:3 ♂♂:2 Monöcisten erwartet werden müssen. Das Untersuchungsergebnis von 12 ♀♀:11 ♂♂:9 ⚥⚥ zeigte damit weitgehende Übereinstimmung.

Das Vorliegen einer Polyploidenserie bei den *Salix*-Arten (*S. lucida*, *S. aurita*, *S. cinerea* = 4n, *S. Andersoniana* = 6n) spricht für einen *Melandrium*-ähnlichen

Mechanismus der G.B. Blackburn und Harrison (1924) und Harrison (1926) haben in den ♂♂ der diploiden *Salix caprea* und *S. viminalis* und den genannten Polyploiden stets nur 1 Hetero-Chromosomenpaar gefunden. Bei *Salix caprea* aufgetretene ⚥⚥ konnten sie als genetische ♂♂ identifizieren. Für das Vorliegen des *Melandrium*-Mechanismus sprechen die Befunde Håkanssons (1938), der aus der Kreuzung *Salix viminalis* × *caprea* in F_2 3 triploide reine ♂♂ erhielt und 2 reine ♀♀, keine Intersexe. *Salix viminalis* × *(viminalis* × *caprea tripl.)* ergab 2 triploide ♂♂ und 1 aneuploides ♂♂. Aus der Kreuzung *Salix viminalis* × *daphnoides* gingen häufig Intersexe hervor, die oft fehlende Bindung in der Meiose zeigten. Die Intersexualität wurde verursacht durch zwei fehlende oder durch zusätzliche Chromosomen. Ausschlaggebend für die G.B. ist folglich das Verhältnis von Y zu X + Autosomen wie bei *Melandrium*.

5. *Coccinea indica*. Kumar und Vishveshwaralah berichten 1952, daß eine gynodiöcische Form sich als genetisch weiblich mit 22A + XX erwiesen hat, allerdings hochgradig steril ist; es tritt keine normale Pollenbildung ein. Danach können zwar männliche Organe bei Abwesenheit von Y gebildet werden, seine Anwesenheit ist jedoch für die Entstehung männlicher Gameten unerläßlich. Eine triploide Pflanze mit einem Trivalent aus 2 X und dem deutlich größeren Y war ein reines ♂, also auch hier ein Typus wie bei *Melandrium*.

6. *Acnida*. Murray erhielt 1940 durch Colchicinierung 2 tetraploide reine ♀ und 1 ♂. Aus der Paarung der beiden Typen bei *Acnida tamariscina* gingen 131 ♀♀ und 1633 ♂♂ hervor (1:12). Die diploiden Pflanzen ergaben normales 1:1-Verhältnis. Von den ♂♂ waren analog den Ergebnissen bei *Melandrium* 1502 vom XXXY-Typ und 131 vom XXYY-Typ wie der Vater. Die Konstitution der ♂♂ konnte genetisch in Kreuzungen mit dem gleichen ♀ geprüft werden. Die beiden ♂♂ sind phänotypisch nicht zu unterscheiden (♀ XXXX × ♂ XXYY = 1 ♀:12 ♂♂; ♀ XXXX × ♂ XXXY = 1 ♀:1 ♂). XXYY ♂♂ bilden etwa zu 85% XY-Gameten. Sexuelle Abweichungen sind von Murray nicht beobachtet worden. Also ist auch hier wie bei *Melandrium* eine tetraploide konstante diöcische Rasse möglich. Triploide aus den Kreuzungen 4n ♀ × 2n ♂ und 2n ♀ × 4n ♂ ergaben 32 ♀♀ + 27 ♂♂ + 1 ⚥. Dieser einzige Zwitter ist phänotypisch ein umgewandeltes ♂ mit ♂-, ♀- und ⚥-Blüten. Selbstung ergab keine Samen. An 2n ♂ wurden unter abnormen Außenbedingungen Zwitterblüten beobachtet. Es liegen im Y die männlich-bestimmenden Gene wie bei *Melandrium*, und zwar auch so starke, daß 3X + 4A durch 1 Y zum ♂ bestimmt werden. Y scheint außer den Geschlechtsgenen ein volles Komplement von Allelomorphen zu enthalten, da bei 10000 Nachkommen von bestrahlten Pollen keine geschlechtsgebundene Mutation beobachtet werden konnte (Murray 1940b). Autosomale Geschlechtsgene haben sich nicht feststellen lassen. Wenn sie vorhanden sind, können es nur wenige im homozygoten Zustand sein. Auch bei Artkreuzungen wurden keine Intersexe erhalten, auch nicht in F_2-Spaltungsgenerationen. Spontan aufgetretene 3n und 4n Pflanzen waren ebenfalls entweder ♂ oder ♀ (Grant 1958).

7. *Rumex acetosella*. Als ebenfalls zum *Melandrium*-Typus gehörend hat sich in der Gattung *Rumex* der ganze Subgenus *Acetosella* erwiesen (Löve und Löve 1940; Löve und Sarkar 1956; Smith 1955; Löve 1957). Die diploiden Arten haben XX:XY-Mechanismus, *Rumex atlanticus* Coss. und *R. angiocarpus* Merb. mit identischen Geschlechtschromosomen in beiden Arten, aber sehr differenten Autosomen. Künstliche Tetraploide von *Rumex angiocarpus* sind dem tetraploiden *R. tenuifolius* sehr ähnlich, bei dem ein stabiler 4X:3X + Y-Mechanismus der G.B. vorliegt, der auch wieder stark männliche Wirkungen des Y demonstriert. Obwohl in der Meiose der Männchen auch Quadrivalente gebildet werden,

sind XY-Bivalente viel häufiger. Der hexaploide *Rumex acetosella* s. str., der offenbar erst verhältnismäßig spät durch Allopolyploidie aus dem diploiden *R. angiocarpus* und dem tetraploiden *R. tenuifolius* entstanden ist und im ♀ 6X, im ♂ 5X + Y besitzt, zeigt ebenfalls häufiger Bivalent-Bildung von XY als Multivalent-Bildung (Löve 1943). Der oktoploide *Rumex graminifolius* Lam. ist als Autotetraploid von *R. tenuifolius* oder eben als autooktoploid anzusehen, und die G.B. erfolgt durch 8X:7X + Y. Erst hier werden hin und wieder leicht intersexe Individuen gebildet, die sich als Formen mit 2n = 57 oder 58 erwiesen. Mindestens eines der zusätzlichen Chromosomen war ein X, so daß erst bei 1Y gegen 8X + 8A die Stärke der Männchenbestimmer ihre Grenze erreicht. Den gleichen Mechanismus zeigt auch *Rumex paucifolius* im Subgenus *Acetosa*, jedoch ist der männliche Determinant schwächer. Das Y-Chromosom, das bei *R. acetosella* das größte ist, ist hier wahrscheinlich das kleinste. Die Art ist tetraploid mit 2n = 28. Das ♀ enthält vier besonders große Chromosomen, das ♂ nur drei dieser großen Chromosomen und ein besonders kleines, wahrscheinlich das Y (Löve und Sarkar 1956). Die Tatsache, daß in einigen Populationen rein ♂ und hermaphrodite Pflanzen vorkommen, zeigt an, daß die männlichen Gene in den Autosomen so stark sein können wie die weiblichen Gene in den Autosomen und X-Chromosomen zusammen. 1958 berichtet Smith über eine diploide diöcische Rasse von *R. paucifolius* mit 2n = 14, bei der bisher keine heteromorphen Geschlechtschromosomen gefunden wurden. Kreuzungen zwischen *R. hastatulus* und *R. paucifolius* (2n = 28) ergaben reziprok eine diöcische F_1. Es traten keine Intersexe auf, jedoch 6 von 51 Pflanzen waren „Neuter" ohne Blüten.

Bei den Diöcisten haben sich also 2 Typen der Wirkung der Geschlechtsfaktoren feststellen lassen, der *Drosophila*-Typus und der *Melandrium*-Typus. Im ersten Falle liegen im X-Chromosom starke Weiblichkeitsfaktoren, in den Autosomen Männlichkeits- und schwache Weiblichkeitsfaktoren. Es entscheidet das quantitative Verhältnis von X/A alternativ über das Geschlecht. Bei unterschwelligen Valenzunterschieden, z. B. bei Bastardierungen oder Polyploidie, resultiert Intersexualität. Im zweiten Fall liegen im Y starke Männlichkeitsfaktoren, entweder ein Weiblichkeits-Unterdrücker allein oder gekoppelt mit Männlichkeitsgenen. Im X befinden sich Weiblichkeitsgene, in den Autosomen können Weiblichkeits- und Männlichkeitsgene vorhanden sein. Die alternative Geschlechtsentscheidung erfolgt dann durch das quantitative Verhältnis von Y/X + A. Je nach der Stärke der männlichen Valenz von Y tritt Intersexualität erst bei höheren Polyploidiestufen auf, wenn X + A einen Grenzwert gegenüber der Valenz von Y erreichen.

II. Bei Subdiöcisten.

Außer diesen mehr oder weniger eingehend analysierten rein diöcischen Arten sind auch eine größere Zahl sog. Subdiöcisten auf ihren G.B.-Modus hin untersucht worden. Nach Correns handelt es sich bei diesen Pflanzen um genetisch getrenntgeschlechtliche Formen, bei denen die Trennung nicht scharf ausgeprägt ist, so daß neben den Blüten des eigentlichen Geschlechts auch zwittrige Blüten und Blüten des anderen Geschlechts auftreten. Meist ist diese Erscheinung auf eines der Geschlechter beschränkt.

a) Der *Drosophila*-Typus.

1. *Cannabis sativa.* Der Hanf, dessen wilde Formen zwar weitgehend rein diöcisch sind, dessen Kultursorten jedoch züchterisch verstärkte Neigung zur Monöcie aufweisen, zeigt den gleichen G.B.-Typus wie *Rumex acetosa*, obwohl

er sehr viel komplizierter erscheint. Die Analyse des Geschlechtsvererbungsmodus war erforderlich für die züchterische Bearbeitung des Hanfes, dessen ♂ Pflanzen mit ihrem Habitus und Wachstumsrhythmus für Kultursorten unerwünscht sind. Es liegen zahlreiche ältere Beobachtungen über das Geschlechtsverhältnis und die Beeinflußbarkeit der Geschlechtsausbildung durch Umweltfaktoren vor. McPhee (1924) fand 115 ♂♂ auf 100 ♀♀. In der Regel überwiegen jedoch die ♀♀. In einigen Fällen war das Verhältnis 100 ♂♂ auf 200 ♀♀, bei Belowitzkaja (1939) sogar nur 100 ♂♂ auf 300 ♀♀. Nach Hoffmann (1946/47) kommen auf 100 ♂♂ etwa 110—120 ♀♀. Durch mehrjährige Selektion ließen sich aus der Sorte *Carmaguola* mit dem Geschlechtsverhältnis 100 ♂♂ auf 113,7 ♀♀ vier Rassen züchten mit konstanten unterschiedlichen Relationen (Crescini 1955). Es liegt sexueller Dimorphismus des Habitus und der Blüten vor, der unter ungünstigen Bedingungen verstärkt wird. Im Gewächshaus unter schlechten Lichtverhältnissen tritt die Bildung von ⚥-Blüten und Blüten des entgegengesetzten Geschlechts auf, am häufigsten Bildung von ⚥-Blüten an ♂♂ (Petit 1952). Viele dieser abnormen Blüten bleiben allerdings funktionsunfähig. Ähnliche Geschlechtsumwandlung kommt nach Entfernung der ursprünglich gebildeten Blüten vor (McPhee 1924 und v. Sengbusch 1952), hier jedoch vorwiegend an ♀ Pflanzen. Die Geschlechtsverschiebung ist leichter erzielbar bei spontan aufgetretenen Monöcisten, die schon lange bekannt sind (Holuby 1878). Diese Monöcisten hatten in der Regel den ♀-Wuchstypus und die ♂-Blüten traten gleichmäßig zwischen den ♀ auf.

Aus Kreuzungen mit Intersexen erhielt McPhee (1925):

	♀	♂
Intersex geselbstet	71	—
♀ × Intersex	56	—
♀ × Intersex	31	3
♀ × ♂	46	42

Die 3 ♂ der 3. Kreuzung sind wahrscheinlich aus Fremdbefruchtung entstanden. Frei abgeblühte ♂ Intersexe ergaben nur 4 Pflanzen: 3 ♂ und 1 ♀.

Diese Ergebnisse sprechen für ♀ Homozygotie der Geschlechtsfaktoren: Geschlechtschromosomen sind von McPhee wie von Strasburger (1910) nicht gefunden worden. Sie wurden jedoch von Hirata (1929) und Driga (zit. nach Hoffmann), Mackay (1939), Yamada (1943) und Hoffmann (1952) beobachtet.

Hirata (1928) hat an Intersexen der Varietät Karafuto folgende Ergebnisse:

♀ Intersex geselbstet	47 ♀:24 ♀ Intersex:	3 ♂	1924
	22 ♀:13 ♀ Intersex:	—	1925
	9 ♀: 8 ♀ Intersex:	—	1926

Die gleichen Verhältnisse erhielt er bei Kreuzung von ♀ mit ♀ Intersexen, dagegen in Kreuzungen mit ♂ Intersexen folgende Zahlen:

reine ♀ × ♂ Intersex = 28 ♀♀ + 6 ♂♂ + 22 ♂ Intersexe
♀ Intersex × ♂ Intersex = 13 ♀♀ + 9 ♂♂
♂ Intersex geselbstet = 6 ♀♀ + 2 ♂♂

Aus der Tatsache, daß in Kreuzungen mit ♂ oder ♂ Intersexen eine Geschlechtsaufspaltung eintritt, geht hervor, daß die ♂ heterogamet, also XY sind. Die ♀ Intersexe und die Intersexe McPhees müssen XX sein. Hirata schließt in seiner Erklärung einen Kompromiß zwischen der Deutung der Verhältnisse bei *Lymantria* nach der Analyse durch Goldschmidt und bei *Drosophila*: die geschlechtsbestimmenden Faktoren liegen in X und Y, im X ist jedoch die Valenz der *weiblichen* Faktoren höher, im Y die der *männlichen*. Y ist stärker als X, so daß XY = ♂ ist. Er nimmt zusätzlich für praktisch alle Individuen Differenzen in den Valenzen der Geschlechtsfaktoren in den Geschlechtschromosomen an, so daß in der relativen Balance Unterschiede auftreten, die zur Intersexbildung führen. Dafür spricht, daß bei den Individuen hoher Intersexualitätsgrade alle Nachkommen unter normalen Bedingungen Intersexualität zeigen. Die verschiedenen Valenzstufen sind mutativ entstanden zu denken. Eine sehr ähnliche Auffassung vertritt v. SENGBUSCH. Er nimmt an, daß die ganze Reihe verschiedener Geschlechtsstufen (Abb. 10) mit weiblichem Wuchstypus eigentlich genetisch ♀, d. h. XX ist, und die Reihe mit männlichem Wuchstypus XY. Die Monöcisten sind je nach ihrer Geschlechsausbildung verschieden bezeichnet:

1 + Monöcisten überwiegend männliche Blüten,
2 + Monöcisten ideale Ausprägung (halb männlich, halb weiblich),
3 + Monöcisten ideale Ausprägung (annähernd halb männlich, halb weiblich),
4 + Monöcisten wenige männliche Blüten,
5 + Monöcisten nur vereinzelte männliche Blüten.

Nach seiner Arbeitshypothese (v. SENGBUSCH 1942) ist das Realisatorallel mit dem höheren Valenzwert weitgehend dominant:

XY $F_{50}f$ = 50 = reine ♂,
XX $F_{50}F_{50}$ = 100 = reine ♀,
XX $F_{40}F_{40}$ = 80 = monöc. ⚥, stark weiblich blühend,
XX $F_{30}F_{30}$ = 60 = monöc. ⚥, mehr männlich blühend,
XX $F_{25}F_{25}$ = 50 = maskulinisiert. ♀ (Geschlechtsumkehr),
XX $F_{50}F_{50}$ oder $F_{50}F_{40}$ oder $F_{50}F_{30}$ oder $F_{50}F_{25}$ = reines ♀.

Die Selektion auf gleichzeitig reifende Typen, die von v. SENGBUSCH, NEUER, JORDAN, HUHNKE (1950); HOFFMANN (1941); GRISCHKO (1937); BELOWITZKAJA und GRESTSCHUCHIN (1939) durchgeführt wurde, brachte ähnliche Ergebnisse. HOFFMANNs gleichzeitig reifender Hanf enthält etwa 70—76% reine ♀♀, 6 bis 8% Monöcisten ♀ Typs, 7—12% ♂ blühende Pflanzen mit weiblichem Habitus, 4—6% normale ♂♂ und sehr selten weiblich blühende Pflanzen mit männlichem Habitus. Dieselben Autoren versuchten einen monöcischen Hanf durch Selektion zu erzielen, bekamen aber trotz strenger Auslese in hohem Prozentsatz sog. feminisierte ♂♂ (d. h. männlich blühende Pflanzen mit weiblichem Habitus) und reine ♂♂. Jedenfalls ist aber eine Auslese nach verschiedenen Typen der Monöcie möglich (GRISCHKO 1935; KÖHLER 1958; BÓCSA 1958; BREDEMANN et al. 1961), und zwar ergeben Kombinationen von ⚥ (1 Strich, nach v. SENGBUSCH) Monöcisten untereinander eine Verschiebung in der Nachkommenschaft in männlicher Richtung, Kombinationen von ⚥ (2 Strich) Monöcisten untereinander eine Verschiebung in weiblicher Richtung. Aus den Ergebnissen der verschiedenen Kombinationen:

1. ♀ × Monöc. versch. Auspräg. = prakt. nur ♀♀
2. ♀ (aus monöc. Nachkommensch.) × ⚥ und ⚥ = etwa 50% reine ♀♀ und 50% Monöcisten
3. ⚥ × ⚥ (× ⚥) = 17% ♀♀ und rund 80% Monöcisten.
4. ⚥ × ⚥ (× ⚥) (× ⚥ = wenig ♀♀ und viel Monöcisten und sehr wenig ♂ blühende Pflanzen
5. ⚥ × ⚥ = prakt. keine ♀♀, meist fem. ♂♂

zieht v. SENGBUSCH folgende Schlüsse auf den G.B.-Mechanismus: Seine Monöcisten sind genetisch ♀♀, XX, die Diöcie ist fast vollständig dominant über die Monöcie. ♀♀, die aus der Nachkommenschaft von Monöcisten stammen, müssen sich genetisch von reinen ♀♀ der diöcischen Sorten unterscheiden. Auch die fem. ♂ müssen XX sein, also eigentlich maskulinisierte ♀♀. Aus Nachkommenschafts-Prüfungen zeigten sich genetische Differenzen zwischen verschiedenen Monöcisten. Daher nimmt v. SENGBUSCH eine Serie von allelen Realisatoren an.

Abb. 10a—e. Verschiedene Kombinationen von Habitus und Geschlechtsausbildung: a weibliche Pflanze, b monöcische Pflanze mit weiblichem Habitus, c feminisierte männliche Pflanze, d männliche Pflanze, e maskulinisierte weibliche Pflanze.

Das reine ♂ XY mit F_{50}f hat die Valenzsumme 50, ebenso das maskul. ♀ mit XX $F_{25}F_{25}$, so daß XY-Pflanzen männlich blühen können mit männlichem Habitus und XX-Pflanzen mit weiblichem Habitus. Aus dem Auftreten von Pflanzen mit nur ⚥ Blüten oder zum Teil ⚥ Blüten schließt v. SENGBUSCH auf Realisatoren für Zwittrigkeit ebenfalls in Allelenserien außer den Realisatoren für Monöcie. Dadurch erklärt er das gemischte oder getrennte Vorkommen beider Geschlechtstypen. Modifikatorische Geschlechtsverschiebung findet sich leichter bei den Monöcisten als bei den rein diöcischen ♂♂. Der Realisator für Monöcie hat auf die Geschlechtsausbildung selber keinen Einfluß, aus den Kreuzungszahlen kann nicht entschieden werden, ob er im X oder in den Autosomen liegt. v. SENGBUSCH hat also unter den Monöcisten nur Pflanzen mit XX-Konstitution angenommen. Typen mit männlichem Wuchstypus und ♀ und ♂ oder nur ♀ Blüten erklärt er durch Mutation eines Genes "*locker*", das bei normalen ♀♀ oder Monöcisten den Blütenstandstyp dem männlichen angleicht.

Abb. 10c.

BREDEMANN et al. berichten 1961 über Monöcisten-Nachkommenschaften, die in 50% reine ♀♀ und 50% ♀♀ mit ♂ Habitus spalten. Sie vermuten bei den Monöcisten mit ♂ Habitus XY-Konstitution mit Ausfall der Y-Eizellen, so daß die 1:1-Spaltung zustande kommt.

HOFFMANN dagegen ist der Ansicht, daß Monöcisten verschiedener Grade sowohl XX als auch XY oder sogar YY sein können, hervorgerufen durch das Eingreifen autosomaler Gene in die G.B. Er hat bei einigen Monöcisten den XY-Charakter cytologisch kontrolliert. Infolgedessen gibt er auch eine abweichende Interpretation für die im Prinzip mit gleichem Ergebnis verlaufenden Kreuzungen. Die Pflanzen mit männlichem Habitus in Kombinationen von ♀♀ und Monöcisten, die v. SENGBUSCH durch Fremdeinkreuzungen erklärt, bezieht HOFFMANN (1952) in seine Deutung ein. WESTERGAARD (1958) beurteilt die in fast allen Versuchsserien angewandte räumliche Isolierung beim Hanf als Windbestäuber mit Skepsis und hält HOFFMANNs Argumente für nicht ganz stichhaltig.

KÖHLER (1958) stellte fest, daß subgynöcische Pflanzen mit ♀ Habitus geselbstet nur wieder Pflanzen ergeben mit ♀ oder ♀+♂ Blüten und mit ♀ Habitus. Daraus wird auf Homozygotie der Faktoren für den ♀ Wuchstyp geschlossen, und die Monöcie wird als genetisch bedingte Variation der Gynöcie aufgefaßt. Echte ♂♂ wären dann für Geschlecht wie Wuchstyp heterozygot. Er nimmt zwei nicht allele und nicht gekoppelte Gene für Monöcie an. Diese Monöcieallele sollen ihrer Wirkung nach schwache Männlichkeitsrealisatoren sein, die sich nur auf das Geschlecht der Pflanzen mit ♀ Habitus auswirken. Sie werden nach seinen Ergebnissen unabhängig von dem XY-Realisatorsystem des diöcischen Hanfes verteilt:

1. XY aabb — XY AABB Männchen — männlicher Habitus
2. XX aabb
3. XX aabB
4. XX aAbb
5. XX aaBB
6. XX AAbb
7. XX aAbB
8. XX aABB
9. XX AAbB
10. XX AABB Weibchen

(3.–9., ↑: Zunehmende Tendenz, anstelle von weiblichen männliche Blüten auszubilden; 2.–9.: weiblicher Habitus)

HOFFMANN nimmt genetische Verschiedenheit der typisch männlichen Pflanzen aus den monöcischen Kreuzungen von den ♂♂ aus diöcischen Sippen an auf Grund des seiner Ansicht nach vorhandenen Unterschiedes in der Zahl männlicher Pflanzen in ihren Nachkommenschaften. WESTERGAARD findet diesen Unterschied nicht in den Zahlen. Nach HOFFMANN haben die Geschlechtschromosomen an Bedeutung für die G.B. eingebüßt in den monöcischen und gleichzeitig reifenden Sorten des Hanfes. Ihre Funktion ist von Autosomen übernommen worden, wie WINGE es für *Lebistes* beschrieben hat (WINGE 1934). HOFFMANN geht davon aus, daß männliche Pflanzen im Geschlecht leichter beeinflußbar sind als weibliche. Monöcisten sind ebenfalls leicht verschiebbar, deshalb nimmt er an, daß männlich fertile Formen mit weiblichem Wuchstypus auch XY sind (1946/47). Sinngemäß bezeichnet er solche Formen als feminisiert, wogegen maskulinisierte Weibchen XX sind. Die allgemein schlechtere Fertilität bei maskulinisierten Monöcisten auch in den männlichen Blüten stützt diese Auffassung.

Die sekundären Geschlechtsmerkmale wie der Wuchstypus werden sicher polyfaktoriell vererbt, werden aber offenbar durch den X—Y-Mechanismus so gesteuert, daß immer mit den ♂ Blüten der männliche Wuchstyp und vice versa eintritt. Da ein fließender Übergang zwischen den verschiedenen Geschlechtstypen vorhanden ist, kann angenommen werden, daß die Faktoren für den Wuchstypus die Blütenbildung beeinflussen können bzw. wesentliche Faktoren für die Blütenausbildung mit solchen für den Wuchstypus identisch sind oder eng gekoppelt, also pleiotrope Wirkung vorliegt. Die polymeren Faktoren kann man sich über alle Autosomen verteilt denken. GRISCHKO (1937) glaubt die Geschlechtsfaktoren von den Wuchsfaktoren trennen zu können; seine Vermutungen über die Entstehung der feminisierten und maskulinisierten Formen entsprechen aber nach HOFFMANN (1946/47) nicht den Ergebnissen.

Nach GRISCHKOs Auffassung erfolgt die G.B. beim Hanf nicht durch Umkombination einzelner Chromosomen, sondern die sich stets wiederholende Kombination ganzer Genome bei rassisch unterschiedlicher Dominanz der Staubblatt-Bildung oder Fruchtblatt-Bildung.

Für gleichzeitig reifenden Hanf gilt das Schema:

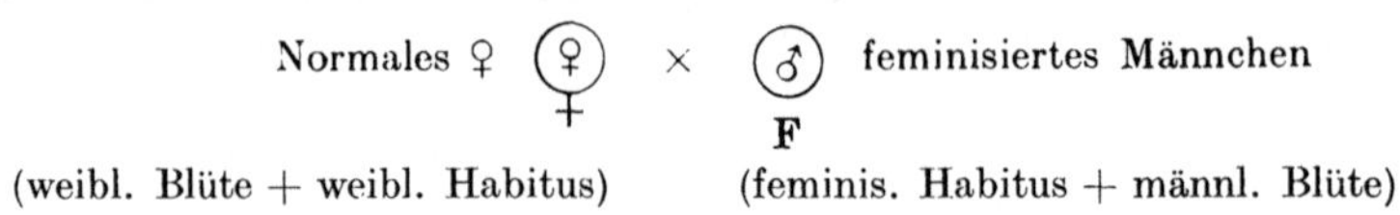

Das Männchen bildet die Geschlechtszellen F → Ga und F → Ag (F = Genkomplex der sekundären weiblichen Merkmale, Ga = Dominanz der weiblichen Tendenz, Ag = Dominanz der männlichen Tendenz der Geschlechtsausbildung.) In F_1 entstehen dann verschiedene Typen:

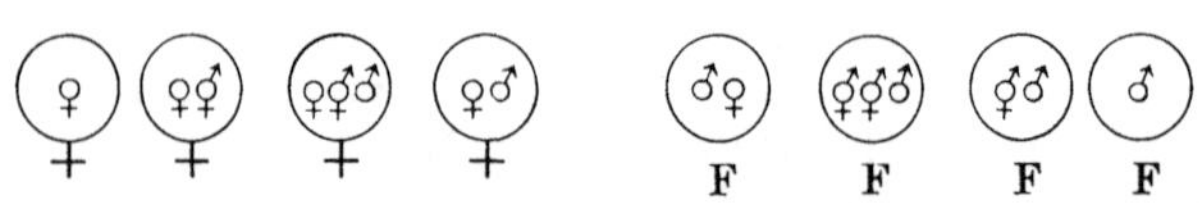

Außerdem vermutet GRISCHKO Plasmonwirkung, die HOFFMANN als unwahrscheinlich, wenigstens in größerem Ausmaß, wegen fehlender Unterschiede zwischen reziproken Bastarden ablehnt.

HOFFMANN hält für möglich, daß 2 X mit den Autosomen ein Übergewicht an weiblichen Faktoren haben, 1 X mit den Autosomen ein Übergewicht an männlichen Faktoren. Dann müßte Y leer sein und die Autosomen würden überwiegend männlich wirken. Das Realisator-Gen im X könnte eine ähnliche Wirkungsweise wie die autosomalen Gene haben. Seine Sonderstellung wäre nur durch die Lage im X gegeben.

Andere Vorstellungen wie Lage des weiblichen Realisators im X, des männlichen im Y oder des weiblichen Realisators in den Autosomen, des männlichen im Y stoßen auf Schwierigkeiten bei der Erklärung, warum bei der Kreuzung von ♀♀ aus diöcischen Rassen mit feminisierten Typen fast nur ♀♀ entstehen. Die Zahl der an der G.B. beteiligten Faktoren wird auf sicher mehr als zwei oder drei geschätzt. Durch Kreuzung und Selektion zeigten sich aus der diöcischen XX—XY-Ausgangsform erreichbar:

Rassen mit einheitlichem Habitus:

1. Mit monofaktorieller G.B.
2. Mit XX ♀ und XY ⚥ und feminisierten XY♂.
3. Mit XX ♀ und XY ⚥.
4. Mit YY ⚥ mit polyfaktoriell vererbter Ausprägung der Monöcie.

Es sollten sich auch Rassen ohne sekundäre Geschlechtsmerkmale mit XY ♀ und YY feminisierten ♂ erreichen lassen, die dann das Vorkommen ♀ und ♂ Heterogametie bei nahe verwandten Arten verstehen ließen, wie es bei Mais in den verschiedenen diöcischen Rassen von Emerson und Jones gezeigt wurde.

Die G.B. bei *Cannabis sativa* ist noch nicht als geklärt anzusehen. Moutschen und Dahmen haben aus bestrahltem Samen eine erhöhte Anzahl von intersexen Pflanzen erhalten (37,5%), und zwar etwa gleich viel ♂ und ♀ Intersexe, bei nur 12,5% ♂♂ und 50% ♀♀ entsprechend den Kontrollversuchen. Die Intersexualität führen die Autoren auf erhöhte Strahlenempfindlichkeit des heterochromatischen X zurück. Ein aus Selbstung dieser Zwitter erhaltener Haplont mit 1 X + 1 A blühte zuerst rein männlich, hatte später auch einige zwittrige Blüten, so daß auch hier eine genetische Veränderung des einen X angenommen werden muß. Künstlich tetraploide Pflanzen wurden von Blakeslee und Warmke (1938, 1939a) untersucht. Sie fanden in der 2. Generation 4 reine ♂♂, 28 reine ♀♀. Die Abweichung vom 1:1-Geschlechtsverhältnis und das Fehlen der Intersexe läßt vermuten, daß XXXX- und XXXY-Pflanzen ♀♀ sind, XXYY-Pflanzen ♂♂, bei denen in der Meiose X + X und Y + Y paaren. Außerdem wurde ein triploides Intersex gefunden, das wahrscheinlich XXY war, woraus auch nur eine im Höchstfall schwache Beteiligung des Y an der G.B. folgt. Takenaka (1953) erhielt aus Kreuzung des ♀ 36a + 4 X × ♂ 18a + X + Y ein triploides ♂ mit dem Chromosomenbestand 27a + X + X + Y. Er fand aber auch mehrere Intersexe der gleichen Konstitution. Durch Colchicin polyploid gemachte monöcische Pflanzen waren bei Hoffmann (1961) zum Teil wieder monöcisch, zeigten in ihren Nachkommenschaften aber sehr variable Geschlechtsausbildung.

b) Der *Melandrium*-Typus.

1. *Mercurialis.* Den *Melandrium*-Typus vertreten auch unter den Subdiöcisten eine ganze Reihe von Arten. Schon früh Anlaß zu Untersuchungen gaben die Verhältnisse bei *Mercurialis annua.* Es sind verwiegend die ♀♀ gemischt-geschlechtlich. Bei ihnen liegen Selbstungsversuche vor (Bitter 1909, Strasburger 1910, Yampolsky 1910). Bitter erhielt in 8 Versuchen 732 ♀♀:21 ♂♂, gibt jedoch keine Erklärung. Strasburger und Yampolsky sowie Kuhn (1936, 1939) erhielten bei Selbstung von ♀♀ stets nur ♀♀, Kuhn bekam aus 43 ♀♀ → 6369 ♀ Nachkommen. Das bedeutet, daß der Pollen von homogameten XX ♀ nur X sein konnte, also nur XX-Individuen entstehen konnten. Die wenigen Männchen Bitters müssen als Fremdbefruchtungen gedeutet werden. Selbstungsversuche an männlichen Pflanzen mit Zwitterblüten liegen von Strasburger und Yampolsky nur in sehr kleinen Zahlen vor und hatten nur ♂♂ ergeben. Kuhn erhielt, wie theoretisch bei männlicher Heterogametie erwartet, eine Spaltung von 3 ♂♂:1 ♀. Unter den ♂♂ traten 2 Klassen auf: 16 pollenreiche, die mit normalen ♀ gekreuzt ein 1:1-Geschlechtsverhältnis aufwiesen, und 8 pollenarme, von denen 2 mit ♀♀ gekreuzt über 80% ♂♂ ergaben, also als YY-Typen aufzufassen sind.

X hat offenbar keine Bedeutung für die Ausbildung der Männchen, ebenso nicht das Verhältnis von X:A oder Y:A; sowohl Y:A = 1:2 und 1:1 sind voll ausgebildete Männchen. Daraus geht hervor, daß *Mercurialis* im Prinzip dem *Melandrium* folgt. Das veränderte Fertilitätsmaß des Pollens der YY-♂♂ beruht eventuell auf dem Fehlen bestimmter im X liegender Gene (WESTERGAARD 1958).

THOMAS (1958) stellte fest, daß *M. annua s. str.* diploid ist, die Varietät *ambigua* hexaploid. Er nimmt für die 6n ♂♂ die Konstitution XXXXXY an und für die monöcischen Pflanzen XXXXXX. DURAND (1957) gibt *M. ambigua* mit 2n = 48 als monöcisch an, wie auch eine tetraploide Form (2n = 32) mit Artrang. Geschlechts-Chromosomen konnten nicht identifiziert werden. Tetraploide F_1-Bastarde *M. annua* × *ambigua* sind fast sterile ♂♂ und sterile ♀♀. 27 monöcische Pflanzen der Nachkommenschaften erwiesen sich als hexaploid. Die meisten ♂♂, die gefunden wurden, waren XXXXXY, einige auch XXXXYY, wie in Kreuzungen festgestellt wurde. Die Zahl der Y hatte keinen Einfluß auf die männliche Sterilität. Es wird angenommen, daß in den monöcischen Sippen die Valenz des F-Faktors so gering ist, daß der hypothetische Weiblichkeitsstoff nur zur Bildung je einer ♀ Blüte in jeder Blattachsel ausreicht und die übrigen Blüten ♂ werden. Ferner soll Y einen Hemmungsfaktor für die Bildung dieses Stoffes tragen, so daß bei seiner Gegenwart nur noch ♂ Blüten gebildet werden können. Dieser Realisator ist wie bei *Melandrium* so stark, daß 4—5 X + 6 A zu ♂♂ bestimmt werden. Die Valenz des F-Realisators im X oder A scheint aber wohl so gering zu sein, daß die ♀♀ nicht immer reingeschlechtlich sind. Als Hemm- bzw. Wirkstoffe kommen vermutlich Wuchsstoffe in Frage (s. auch VAN NIGTEVECHT (1966). Untersuchungen von HESLOP-HARRISON (1957) zeigten eine starke Reduktion der ♂ und eine leichte Zunahme der ♀ Blüten unter CO-Atmosphäre, wobei die Wirkung eventuell über die IES-Oxydase erfolgt.

2. *Spinacia oleracea.* Alle diöcischen Handelssorten des Spinat enthalten einen geringen Prozentsatz verschiedengradiger Monöcisten, für deren Bildung nicht Umwelteinflüsse ausschlaggebend sind (BEMIS und WILSON 1953). Aus stark monöcischen Kultursorten lassen sich diöcische Linien, reinerbige intersexe Linien und spaltende intersexe Linien erhalten. Das Auftreten der Intersexe wird nach SUGIMOTO (1948) durch einen zum XY-Mechanismus zusätzlichen autosomalen Faktor, nach BEMIS und WILSON durch Austausch zwischen zwei sonst enggekoppelten autosomalen Genen hervorgerufen, während JANICK und STEVENSON (1955a) mutative Veränderung eines X annehmen. Umfangreiche Kreuzungen von ♂ und ♀ diöcischen Typs und Monöcisten verschiedener Grade zeigten, daß Monöcisten XX-, XY- und YY-Typus haben können (DRESSLER 1958). Durch das Auftreten homozygoter XX- oder YY-Monöcisten wird polyfaktorielle Geschlechtsvererbung als bewiesen angesehen. Jedoch entstehen bei Fehlen von X nur schwach weibliche Monöcisten, keine reinen ♀♀, bei Fehlen von Y niemals reine ♂♂. In diöcischen Sippen wirkt Y männlich-bestimmend, ohne jedoch Grundgene der männlichen Organbildung zu enthalten, wie die XX-Monöcisten beweisen. Künstlich hergestellte Tetraploide (NOGUCHI 1943, JANICK und STEVENSON 1955, DRESSLER 1958, MAHONEY, JANICK und STEVENSON 1959) zeigten, daß einfache Chromosomen-Verdoppelung keine Geschlechtsveränderung hervorruft. 4n ♀ + 4n ♂ ergab 63 ♂♂:10 ♀♀ (5:1). In der Meiose der ♂♂ wurde überwiegend Bivalent-Paarung beobachtet, bei Triploiden Trivalent-Bildung. Aus Kreuzung 4n ♀ × 2n ♂ entstanden 1 ♀ oder ⚥:1 ♂, aus 2n ♀ × 4n ♂ = 5 ♂:1 ♀ bzw. 2♂:1 ♀. Danach sind XXY-Pflanzen ♂♂, während XXX-Pflanzen ♀ oder ⚥ sein können auf Grund modifikatorischer Faktoren in den Autosomen. Da 1 Y gegen 3 X epistatisch wirkt, sind die männlich-bestimmenden Potenzen sehr stark. Im X scheinen grundlegende weibliche Gene zu liegen, denn XY-Pflanzen

brachten bei JANICK und STEVENSON niemals Samen hervor. DRESSLER fand folgende Beziehung zwischen Geschlechts-Chromosomen und Geschlecht an euploiden Formen:

	2n	3n	4n
♀	= XX, XY	XXX, XXY, XYY	XXXX, XXXY, XXYY, XYYY,
Monöcisten	= XX, XY, YY	XXX, XXY, XYY, YYY	XXXX, XXXY, XXYY, XYYY, YYYY
♂	= XY, YY	XXY, XYY, YYY	XXXY, XXYY, XYYY, YYYY

Daraus folgt, daß der vollständige AG-Komplex in den Autosomen und den homologen Segmenten der Geschlechts-Chromosomen lokalisiert ist, denn XXXX- und YYYY-Pflanzen können monöcisch sein. Y muß einen Unterdrücker für die weibliche Ausbildung enthalten, denn nur Typen mit mindestens einem Y können reine Männchen werden. X muß einen ähnlichen Faktor für die männliche Ausbildung tragen. Das genetische Gesamtmilieu beeinflußt jedoch die Durchschlagskraft der Realisatoren. DRESSLERs Deutungen beruhen zum großen Teil auf der cytologischen Diagnose der Genotypen in bezug auf die Geschlechtschromosomenkonstitution. Die von ihm als Geschlechtschromosomen gedeuteten heteromorphen Chromosomen seines Materials sind nicht identisch mit den von ELLIS und JANICK (1960) und BOSE und JANICK (1961) sowie IIZUKA und JANICK (1962, 1963) als Träger der Realisatoren ermittelten Chromosomen. AKKOÇ (1965) hat das Material von DRESSLER erneut bearbeitet und wie die vorher genannten Autoren Chromosom 1 als Geschlechtschromosom bestätigt. Während nach DRESSLER X und Y keine geschlechtsproduzierenden Gene enthalten sollen, ist AKKOÇ der Meinung, daß sie doch vorhanden sein müssen, weil ohne X kein ♀ und ohne Y kein ♂ gefunden wurde (s. auch JANICK 1956, SUTÔ u. SUGIYAMA 1957, KUCKUCK 1960). IIZUKA u. JANICK (1962, 1963) gelang es, anhand eines Translokationstypus von Chromosom 1 mit zusätzlichem Segment von Chromosom 2 die Lage der Allele für die Geschlechtsbestimmung annähernd durch die

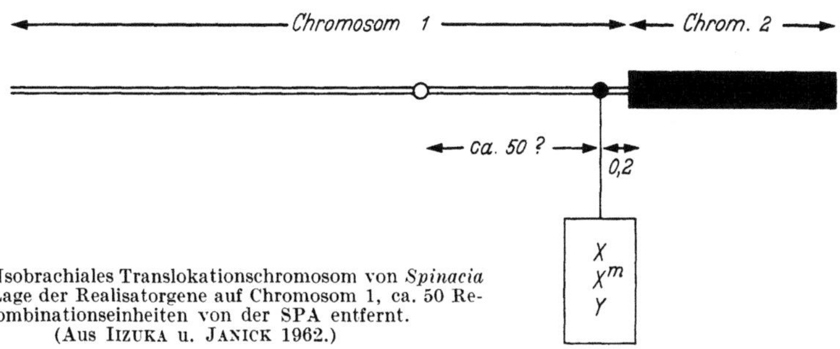

Abb. 11. Isobrachiales Translokationschromosom von *Spinacia oleracea*, Lage der Realisatorgene auf Chromosom 1, ca. 50 Rekombinationseinheiten von der SPA entfernt. (Aus IIZUKA u. JANICK 1962.)

Rekombinationshäufigkeiten festzulegen (Abb. 11). An trisomen Pflanzen konnte die Wirkung der einzelnen Chromosomen verfolgt werden (TABUSKI 1958, IIZUKA u. JANICK 1962).

Bei Spinat zeigte sich, daß verschiedene Rassen in bezug auf die Chromosomenstruktur auftreten. Es liegt hier wahrscheinlich ein Fall vor, in dem die Entwicklung zur Heteromorphie der Geschlechtschromosomen noch auf einer frühen Stufe steht. Derartige Translokationstypen führen zu einem Polymorphismus, wie er bei Insekten mehrfach beobachtet wurde (MAINX 1962).

3. *Vitis.* Die Wildreben der Gattung *Vitis* sind offenbar rein diöcisch. REINER berichtet 1910 von einem spontan aufgetretenen Zwitter. In den Kultursorten ist

züchterisch der Anteil an Zwittern erheblich gesteigert (Breider und Scheu 1938). Die meisten Kultursorten sind heute gemischt-geschlechtlich. Bei *Vitis vinifera* und *V. riparia* scheinen die männlichen Blüten leichter zur Zwittrigkeit durch Außeneinflüsse verschoben zu werden als bei *Vitis rupestris*. Valleau und Negrul (1936) halten die ♂♂ für heterogamet und fanden 1:1-Verhältnis. Zschokke beobachtete bei *Vitis vinifera ssp. Silvaner* 37 ♂♂:9 ♀♀. Selbstung zwittriger Sorten brachte 75% ⚥:25% ♀. Das bedeutet also, daß die Zwitter heterozygot sind. Dieses 3:1-Verhältnis wäre nur zu verstehen, wenn YY = ⚥. Sie dürften nach Selbstung und Kreuzung mit XY ⚥ nur ⚥ ergeben, wie es Ziegler (1923) mitteilt. Breider und Scheu fanden bei *Riesling*-Selbstung und bei F_1 (*Silvaner* × *Riesling*)-Selbstung nur ⚥⚥. Die meisten *Riesling*stämme erwiesen sich als YY-⚥, es gibt aber auch XY-⚥-Stämme, die in Kreuzung mit XY-*Silvaner* 3 ⚥:1 ♀ aufwiesen. Aus Kreuzung von *vinifera* YY ⚥ × *riparia* ♂ entstanden 1 ⚥:1 ♂. Die ⚥-Bastarde sind heterozygot, in F_2 spalten sie in 935 ♀♀ XriXri und 1724 ⚥⚥ XviYvi. Da innerhalb der *vinifera* die YY ⚥⚥ sind, können sie auf Grund der veränderten autosomalen Reaktionsbasis, der sie gegenüberstehen, hier nur ♂♂ sein. In der Rückkreuzungsnachkommenschaft eines Bastards einer *Euvitis*-Form (2n = 38) mit *V. rotundifolia* (2n = 40) — Untergattung *Muscadinia* — mit *rotundifolia* fand Dunstan (1964) bei 35 Pflanzen 1 ♀ : 7 ⚥. Nach Wagner (1960) bilden die ♀♀ zwar auch Pollen, die Pollenkörner unterscheiden sich aber von denen der ⚥⚥ durch das Fehlen der Keimsporen und Keimfurchen. Alle Ergebnisse sind nur verständlich, wenn sowohl die männliche als auch die weibliche Entwicklung fördernde Faktoren in den Autosomen angenommen werden und im Y ein männlich wirkender Realisator. Somit kann auch bei *Vitis* der *Melandrium*-Typ vorausgesetzt werden und nur bei Fehlen von Y können ♀♀ entstehen (Husfeld 1961).

4. *Asparagus officinalis* zeigt ab und zu an männlichen Pflanzen Zwitterblüten. Aus Selbstung dieser Blüten an ♂♂ erhielten Rick und Hanna (1943) 155 ♂♂:43 ♀♀. 8 getestete ♂♂ ergaben mit normalen ♀♀ rein männliche Nachkommenschaft, 17 eine 1:1-Spaltung. Ähnlich wie bei *Mercurialis* wiesen YY-♂ schlechtere Pollenfertilität auf. Das gleiche Ergebnis hatten Robbins und Jones (1928). Y muß hier Träger wichtiger Sexualgene sein (Kappert 1941, Sneep 1953a, Reimann-Philipp et al. 1960, Huyskes u. Sneep 1962). Peirce u. Currence (1962) kommen aufgrund ihrer Beobachtungen an Inzuchtlinien, F_1, F_2 und Rückkreuzungspopulationen zu dem Schluß, daß drei dominante Komplementärgene nötig seien für die Ausbildung von Hermaphroditen. Wenigstens zwei dieser Gene könnten auf dem X liegen, getrennt durch einen Austauschabstand von 30—40 Einheiten. Aus bestimmten Kreuzungen vermuten sie auch eine Interaktion zwischen ♀ Cytoplasma und dem Y-Chromosom.

c) Abweichende Formen.

Nicht alle untersuchten Arten der Subdiöcisten lassen sich einem der beiden bei Diöcisten bekannten Typen zuordnen. Es sind z. B. auch multiple Allelie der Geschlechtsfaktoren (vgl. auch S. 83/84 v. Sengbusch) und das Auftreten von Letalgenen beobachtet worden.

1. *Carica papaya* hat ♂, ♀ und ⚥-Pflanzen. Aus verschiedenen Kombinationen entstehen folgende Geschlechtstypen (Hofmeyr 1938, 1953 und Storey 1938, 1953):

1. ♀ × ♂	1 ♀: – ⚥: 1 ♂
2. ♀ × ⚥	1 ♀: 1 ⚥: – ♂
3. ♂ geselbstet	1 ♀: – ⚥: 2 ♂
4. ⚥ geselbstet	1 ♀: 2 ⚥: – ♂
5. ⚥ × ♂ (♂ × ⚥)	1 ♀: 1 ⚥: 1 ♂

Beide Autoren nehmen ein Gen an mit 3 Allelen, von denen M_1 ♂, M_2 ⚥ und m ♀ wirken. STOREY hält alle doppelt dominanten Formen für zygotisch letal: M_1M_1, M_1M_2 und M_2M_2. ♂ und ⚥ sind also Geschlechts-Heterozygoten, die 25% nicht lebensfähige Samen bringen.

M_1m = ♂, M_2m = ⚥, mm = ♀, so daß die genannten Kreuzungen folgende Konstitution hätten:

1. mm × M_1m → 1 mm : 1 M_1m
2. mm × M_2m → 1 mm : 1 M_2m
3. M_1m geselbstet → 1 mm : 2 M_1m : 1 (M_1M_1)
4. M_2m geselbstet → 1 mm : 2 M_2m : 1 (M_2M_2)
5. M_2m × M_1m } M_1m × M_2m } → 1 mm : 1 M_2m : 1 M_1m : 1 (M_1M_2)

HOFMEYR neigt für die Wirkungsweise der Gene zur Balance-Theorie von BRIDGES und nimmt für die Gesamtheit der Autosomen einen Wert für Männlichkeit und für die 3 Allele bestimmte Werte für Weiblichkeit an, so daß z. B. M_2m einen Wert von 0 bekommt und ⚥ ist. STOREY fordert dagegen einen in den Geschlechts-Chromosomen lokalisierten Komplex von Genen. Und zwar soll m einen recessiven Unterdrücker für das Andröceum, M_1 einen Faktor für die Ausbildung der ♂ Blütenstandsform, einen recessiven zygotischen Letalfaktor, einen cross-over-suppressor enthalten, M_2 den Letalfaktor, den cross-over-suppressor und einen recessiven Weiblichkeits-Unterdrücker. Die Zahlenverhältnisse stimmen bei genetischer Männlichkeit der Zwitter XY mit dem bei *Melandrium* überein, bei dem YY letal ist und ♂♂ oder ⚥⚥ geselbstet ein 1 ♀ : 2 ♂♂-Verhältnis ergeben. Die Deutung durch 3 Allele erscheint STOREY (1953) und WESTERGAARD (1958) zu einfach, um eine befriedigende Erklärung zu geben. HOFMEYR (1945) erhielt aus der Kreuzung künstlich tetraploider ♀ × ♂ = 51 ♂♂ : 21 ♀♀ (3 : 1). Die XXYY- und XXXY-Pflanzen sind danach subandröcische ♂♂, wobei die XXXY-Typen morphologisch stärker weibchenähnlich sind. Aus Selbstung einer XXYY-Pflanze wurden 12 ♂♂ und 2 ♀♀ erhalten. Y hat also eine männlich-bestimmende Bedeutung. Der Faktor M_1 ist demnach so stark männlich wirkend, daß er dominant über m ist, M_2 dagegen hat eine erheblich schwächere Valenz mit intermediärer Wirkung gegen m. Geschlechtsverschiebung an ⚥ Pflanzen ist auch hier leicht zu erreichen in ♀ Richtung durch Polyäthylen und in ⚥ Richtung durch Reduktion der Blattfläche (LANGE 1961).

2. Bei *Rubus idaeus* haben CRANE und LAWRENCE (1931) einen auf 2 Genpaaren basierenden Mechanismus festgestellt. MF = ⚥, Mf = ♂, mF = ♀ und mf = Neuter. Es konnten verschiedene Sorten gefunden werden. Die Sorte Lloyd George ist reinerbig zwittrig, also MM FF. Eine andere zwittrige Sorte ergab gekreuzt mit ♂♂ eine Aufspaltung in ⚥⚥ und ♂♂: Mm Ff ⚥ × MM ff ♂ = 71 ⚥⚥ + 69 ♂♂. Eine dritte Sorte brachte bei Selbstung 128 ⚥⚥ + 30 ♂♂, war also MM Ff ⚥.

3. *Antennaria dioica* ist im ♂ XY (v. UBISCH 1932). Die auftretenden Zwitter sind genetische ♀♀ und genetische ♂♂. v. UBISCH nimmt auch Valenzänderung der Realisatoren auf mutativem Wege an und erhielt nach Selbstung subandröcischer ♂♂ 71 ♀♀ + 51 ♂♂ + 99 ⚥⚥, was einem 1♀ : 2♂ (⚥)-Verhältnis entspricht (v. UBISCH 1936, WESTERGAARD 1958), also Letalität der YY vermuten läßt wie bei *Carica papaya*.

4. *Urtica*. In der Gattung *Urtica* sind neben der ziemlich rein diöcischen *U. dioica* auch mehr oder weniger subdiöcische Arten bekannt. Bei *U. dioica* hatte STRASBURGER (1910) bei Selbstung von ♀♀ mit gelegentlich männlichen Blüten rein weibliche Nachkommenschaften erhalten. NEGODI 1929 (zit. nach ALLEN 1940) untersuchte *U. caudata*, bei der andröcische und gynodiöcische Varietäten (vgl. S. 51) vorkommen und nimmt veränderte Potenzen des einen oder

anderen Geschlechtsfaktors an. Bei *U. cannabina*, die weibliche und monöcische Pflanzen bildet, erhielt er aus Kreuzung von ♀ × ⚥ = 19 ♀♀ + 17 ⚥⚥. In einem anderen Stamm fand er bei Kreuzungen verschieden stark weiblicher ⚥⚥ miteinander unterschiedliche Ergebnisse. Die stärker weiblichen ⚥⚥ gaben bei Selbstung nur 3% monöcische Pflanzen, in der Kreuzung 2—7%. Er nimmt beim ⚥ eine Änderung von F in F_1 an, so daß MM FF = ♀ und MM FF_1 = ⚥ ist. Die verschiedene Zusammensetzung der Nachkommenschaften der Monöcisten wird vom Autor durch unterschiedliche Produktion funktionsfähiger MF_1- (Monöcie bedingender) Pollen erklärt.

5. Bei *Thalictrum* fand KUHN aus Kreuzungen der ziemlich streng diöcischen Art *Thalictrum Fendleri* mit gemischt-geschlechtlichen Arten, daß polymere Faktoren in den Autosomen an der G.B. mitwirken können (Tabelle 5). Es wurden subdiöcische Individuen gefunden mit fließenden Übergängen bis zum ± ♂ (mit voller Karpellzahl). Aus Kreuzungen dieser verschiedenen Stufen von Subdiöcisten mit ♀♀ folgte, daß die Stufen durch besondere polymere Faktoren bedingt sind und nicht durch multiple Allele der Realisatoren. Aus Selbstung der Subandröcisten gingen 481 ♂♂ + 197 ♀♀ hervor, die ♂♂ ergaben wie bei *Mercurialis* zu $^2/_3$ mit normalen ♀♀ eine 1:1-Spaltung, zu $^1/_3$ nur ♂♂. Diese waren also YY und zeigten vorwiegend rein männliche, die heterozygoten ♂♂ häufig subandröcische oder ± ♂ Geschlechtsausbildung. Bei *Thalictrum dayscarpum* und *Th. polyganum* fand KUHN 1933 ♂ und subgynöcische Individuen im Verhältnis 1:1. Die ♀♀ und subgynöcischen Pflanzen erwiesen sich als XX, die Subandröcisten beider Arten spalteten in Selbstungen 3 ♂:1 ♀. KUHN stellt die Arbeitshypothese auf, daß ein plasmatischer Männlichkeitsfaktor gegen $\gamma\gamma$ (homozygote Weiblichkeitsfaktoren im X-Chromosom) beim ♀ hypostatisch ist, dagegen beim ♂ gegen $\alpha\gamma$ (YX) mit den Autosomen zusammen über γ dominiert.

Tabelle 5. *Kreuzungen zwischen dem diöcischen Thalictrum fendleri und zwittrigen Arten.* (Nach WESTERGAARD 1958.)

	N	♀♀	♂♂	⚥♂	Zahl der Karpelle je Blüte
Thalictrum fendleri ♀ × *T. foetidum*	10	0	0	10	3,7 ± 0,11
Thalictrum fendleri ♀ × *T. delawayi*	14	0	0	14	3,7 ± 0,05
Thalictrum fendleri ♀ × *T. aquilegifolium*	3	0	0	3	0,5 ± 0,07

III. Bei Gynodiöcisten.

Ein Typus der Geschlechtsverteilung, der zunächst gewisse Ähnlichkeit mit dem der Subdiöcisten zu haben scheint, jedoch wohl etwas anderes darstellt, ist der der Polyöcisten, von denen bisher nur einige Gynodiöcisten näher auf ihre G.B. untersucht wurden. Während bei einem Subdiöcisten z. B. ♀♀ und genetisch männliche Pflanzen mit ± starker Verschiebung zur Zwittrigkeit vorhanden sein können (also mit ♂ und ♀ oder ⚥ Blüten), sind bei einem Gynodiöcisten ♀♀ und ⚥⚥ ausgebildet. Und zwar stellen diese ⚥⚥ stets Primärzwitter dar, also nicht Reduktionsformen eines diöcischen Status. Während bei dem ± Monöcisten der subdiöcischen Arten Geschlechts-Realisatoren mit verhältnismäßig schwacher Valenz vorhanden sind, besitzen die ⚥⚥ oder Monöcisten der gynodiöcischen Arten keine Realisatoren für das männliche Geschlecht. Es lassen sich bisher zwei grundsätzlich verschiedene Mechanismen unterscheiden. Die gleichmäßige Entstehung weiblicher und autöcischer Individuen in einer Art kann entweder durch ein Gen-System erreicht werden oder aber durch eine Balance zwischen Kerngenen und Plasmon.

a) Gen-Balance-System.

1. Für *Origanum vulgare* fanden LEWIS und CROWE (1956) einen Weibchenanteil von 30—50% neben den zwittrigen Pflanzen. Alle ⚥⚥ ergaben nach Selbstung ⚥⚥. ♀♀ — bis auf eine kleine Nachkommenschaft — spalteten nach Bestäubung durch ⚥⚥ in 227 ♀♀ + 339 ⚥⚥. Die ⚥⚥ der F_1 erwiesen sich zum Teil als reinerbig, zum Teil spalteten sie. Es wird Kontrolle der Geschlechtsausbildung durch 2 Gene angenommen: F veranlaßt dominant den Ausfall der Antheren, H unterdrückt dominant die Wirkung von F. So kann ein ♀ FFhh oder Ffhh sein; beide spalten bei Kreuzung mit ⚥⚥, die sowohl ffHh sein können und dann bei Selbstung reinerbig sind, als auch FfHh und dann bei Selbstung ♀ und ⚥ ergeben. Da bei einer solchen Population mit ♀♀ und ⚥⚥ die ⚥⚥ dreimal so oft ihr Genom in die Nachkommenschaft liefern wie die ♀♀, so müssen die ♀♀, um erhalten zu bleiben, entweder einen doppelt so großen Selektionswert haben wie die ⚥⚥, oder es müssen durch selektive Eliminierung in ähnlicher Weise wie durch Selbst-Sterilitätsfaktoren ⚥-Gameten ausgeschlossen werden. Theoretisch mögliche Homozygote ffhh wurden nicht gefunden, sind eventuell letal. Der HH-Genotyp trat mit 25% der erwarteten Häufigkeit auf, er ist selbststeril. Die Ergebnisse lassen vermuten, daß das gynodiöcische System bei *Origanum* sich aus einem Selbst-Sterilitäts-System entwickelt hat mit sporophytischer Pollenkontrolle und Befruchtungsverhinderung im Embryosack.

2. Bei *Salvia nemorosa* hat LINNERT (1958) eine ebenfalls faktoriell begründete Gynodiöcie festgestellt. Die ♀♀ ergaben ♀ und ⚥ Nachkommen, die ⚥⚥ dagegen nur ⚥. Die Bestäubung kann natürlich immer nur durch ⚥⚥ erfolgen. Auch diejenigen ⚥⚥, die von ♀-Pflanzen abstammen, bringen nur ⚥⚥ hervor. Einzelne ♀♀ brachten rein weibliche Nachkommenschaften, die meisten spalteten jedoch in ♀♀ und ⚥⚥, manchmal 1:1, manchmal 3:1 oder 5:1. Am häufigsten wurde ein 1:1-Verhältnis beobachtet. LINNERT nimmt ein dominantes Gen F an für Weiblichkeit, das homozygot recessiv ff Zwittrigkeit hervorruft. Zusätzlich zu dem Faktorenpaar Ff werden zygotische Letal- und Subletalfaktoren vermutet, um das Geschlechtsverhältnis zu erklären. Die Nachkommenschaften der ⚥⚥ waren meist klein, auch die ♀♀ ergaben oft zu wenig ⚥⚥; auf der anderen Seite wurden viel taube Samen gefunden. Es wird mit mehreren Letalfaktoren gerechnet, von denen einer gekoppelt ist mit f, so daß f let/f let Individuen ausfallen. Die Letalwirkung dieses Faktors soll abhängig vom genetischen Gesamtmilieu sein, so daß in bestimmten Fällen diese Kombination doch überlebt. Weitere Subletalfaktoren setzen die Fertilität der ⚥⚥ herab. Versuche, andere Gene aufzufinden, die mit F und dem Hauptletalfaktor gekoppelt sind, waren bisher erfolglos.

b) Kern-Plasma-System.

Bei Gynodiöcisten hat sich noch ein anderer Typus der Lokalisation und Wirkung von Geschlechtsfaktoren feststellen lassen. Aus Untersuchungen von CORRENS (1904, 1908, 1926) ergab sich 1. für *Cirsium oleraceum* und 2. für *Satureia hortensis* stets eine muttergleiche Ausbildung des Geschlechts. Beide Arten enthalten rein weibliche Pflanzen neben normal zwittrig ausgebildeten. Bei *Satureia* zeigte sich, daß Selbstungen von ⚥⚥ stets wieder ⚥⚥ hervorbrachten, Bestäubungen von ♀♀ mit ⚥⚥ stets nur ♀♀. In gleicher Weise verhielten sich die ♀♀ bei *Cirsium oleraceum*. Bei *C. palustre*, das ebenfalls aus ♀♀ und ⚥⚥ besteht, erwiesen sich die ⚥⚥ als heterozygot; sie spalten bei Selbstung ♀♀ ab. Kreuzungen ♀ × ⚥ ergaben 50% ♀♀ und 50% ⚥⚥. *Cirsium arvense* ist subdiöcisch, hat ♀♀ und ♂♂, die gelegentlich zwittrigen Charakter tragen. Kreuzungen ♀ × ♂ ergaben normal ♀ + ♂; ♂ × ♂ = ♂ + ♀. *C. canum* ist dagegen normal zwittrig. In Kreuzungen *C.*

oleraceum ♀ × *C. canum* ⚥ wurden ebenfalls nur ♀♀ erhalten. Aber auch weitere Rückkreuzung mit *canum* ergab keine Änderung der Geschlechtsausbildung. Nach achtmaliger Kreuzung mit *canum* und morphologisch schon völliger Gleichheit mit *C. canum* wurden nur geschlechtlich muttergleiche, also weibliche Pflanzen erhalten. Das zeigt, daß in diesen beiden Fällen die G.B. durch Zusammenwirken des Plasmons mit genetischen Faktoren erfolgt (v. Wettstein 1924).

3. Ähnliche Geschlechtsverhältnisse liegen bei der von Kappert (1943) beobachteten gynodiöcischen *Aquilegia vulgaris* vor, bei der spontan plasmatisch bedingte Pollensterilität auftrat, so daß die physiologischen ♀♀ nur ♀♀, die ⚥⚥ nur ⚥⚥ ergaben. Eine solche Lokalisation von Geschlechtsfaktoren im Plasma hatte Goldschmidt (1938) für die G.B. bei *Lymantria* als einem diöcischen Organismus abgeleitet.

Plasmatisch bedingte Pollensterilität wurde inzwischen bei einer Reihe von Arten bekannt und bei Kulturpflanzen züchterisch ausgenutzt, z.B. bei *Zea mays* (Duvick 1958, Sprague u. Tavčar 1959), *Solanum lycopersicum* (Currence 1962) und *Beta vulgaris* (Savitsky 1958).

Diese plasmatisch gesteuerte Form der Gynodiöcie kann entweder durch Plasmon-Mutation oder durch Bastardierung zustande kommen. Sie gestattet jedoch anscheinend nicht eine Weiterdifferenzierung zur vollständigen Diöcie.

IV. Bei autöcischen Pflanzen.

Am schwierigsten anzugreifen und daher am wenigsten bearbeitet sind die Verhältnisse bei normal zwittrigen Pflanzen. Es ist gezeigt worden, daß aus Kreuzungen zwischen zwittrigen oder monöcischen Arten und diöcischen Arten sowohl Dominanz der Diöcie über die Monöcie *(Bryonia, Lychnis, Acnida* usw.) als auch Dominanz der Monöcie über die Diöcie beobachtet wurde *(Ecballium, Bryonia)*. Offenbar ist der phylogenetisch ältere Zustand der zwittrige oder monöcische, aus dem sich über subdiöcische Zwischenstufen der diöcische Zustand als der phylogenetisch jüngere entwickelt hat (Correns 1912, Hartmann 1956, Lewis 1941). Es gibt aber auch die umgekehrte Entwicklung zu sog. Sekundärzwittern (s. die Kulturformen bei *Fragaria, Vitis, Cannabis, Spinacia*), die auch wieder über subdiöcische Zwischenstufen sich entwickeln konnten oder können. Während bei den Subdiöcisten die Entscheidung darüber, ob sie eine Stufe auf dem Wege zur Diöcie oder von der Diöcie zur Autöcie darstellen, allenfalls aus den Geschlechtsverhältnissen systematisch verwandter Gruppen zu erschließen ist, müssen sich nach der Auffassung von Correns primäre und sekundäre ⚥⚥ dadurch unterscheiden, daß die ersten niemals Geschlechts-Realisatoren besitzen, sondern nur die Geschlechts-Grundgene (Westergaard 1958). Es sind Autöcie und Diöcie ausbalancierte Zustände, bei denen jede mutative Veränderung zwangsläufig zur Abänderung in Richtung auf den anderen Zustand hin führen muß (Allen 1940).

a) Primärzwitter.

1. *Mit Gen-Balance.* Murray hat in den Kreuzungen innerhalb der *Amaranthaceae* 2 Typen von zwittrigen Arten festgestellt. *Amaranthus retroflexus, A. hybridus, A. caudatus* und *A. powelii* sind Monöcisten des 1. Typs; bei ihnen ist die erste Blüte jeder Traube ♂, alle anderen ♀ Dieser Typus ergibt, mit diöcischen *Acnida*-Arten gekreuzt, mit diöcischem ♀ als Mutter lauter ♀♀, mit diöcischem ♂ als Vater 50% ♀♀ und 50% ♂♂, zeigt also das gleiche Verhalten wie *Bryonia alba* zu *Br. dioica*. *Amaranthus spinosus* gehört zum 2. Typ, bei dem alle Blüten einer Traube vom gleichen Geschlecht sind, die männlichen Trauben jedoch ter-

minal an den Haupt- und Nebenästen sitzen. Dieser Typus ergab mit diöcischen *Acnida* in beiden Kreuzungsrichtungen nur ♂♂ und einzelne monöcische Pflanzen neben etwa ebenso vielen Neuter-Pflanzen, die gar keine Blütenbildung zeigten. Demnach muß zwischen den G.B.-Genen der beiden *Amaranthus*-Typen ein Unterschied bestehen, zumindest in den Epistasie-Verhältnissen gegen die Gene der diöcischen Arten: Beim 2. Typ sind die männlichen Geschlechtsgene der Monöcisten epistatisch über die weiblichen Gene der diöcischen Art, während die männlichen Gene der diöcischen Art sich epistatisch zeigen über die weiblichen der monöcischen Art. Für den geringen Anteil monöcischer Pflanzen, der sowohl in Rückkreuzung mit dem monöcischen Elter als auch mit dem diöcischen Elter auftrat, ist noch keine Erklärung gegeben.

Das Auftreten der „Neutra" konnte auf ein recessives Gen zurückgeführt werden, das in bestimmten Stämmen von *Acnida tamariscina* homozygot oder in anderen heterozygot gefunden wurde und sich nur in bestimmten Kreuzungen des obigen Typs manifestiert.

Interessant ist ferner das Ergebnis der Kreuzung zwischen den beiden monöcischen *Bryonia*-Arten *alba* ♀ × *macrostylis* ♂, die 1 ♀ und 1 ♂ ergab (BILGE 1955). Umgekehrt gingen aus der Kreuzung der beiden diöcischen Arten *dioica* ♀ × *multiflora* ♂ nur monöcische Pflanzen hervor und Neuter-Pflanzen, aus der reziproken 50% ♀♀ und 50% ♂♂ (ARSAN, zit. nach HEILBRONN 1953, und HEILBRONN 1953). Nicht geklärt ist das Fehlen von ⚥⚥ in den Rückkreuzungen von *(dioica* × *alba)*-Bastarden mit *alba*, in denen Pflanzen ohne die Geschlechts-Chromosomen von *dioica* herausspalten müssen. HEILBRONN hält hier eine Wirkung des *dioica*-Plasmons für möglich. Bei *Bryonia* hat jedenfalls der G.B.-Apparat sowohl bei den diöcischen als auch bei den monöcischen Arten eine unterschiedliche Struktur. Über die Art der Unterschiede läßt sich allerdings noch nichts sagen.

Aus seinen Kreuzungen mit einer monöcischen und einer diöcischen Varietät von *Ecballium elaterium* hat GALAN (1951) auf das Vorhandensein eines homozygoten Allels zum Realisator in der monöcischen Sippe geschlossen. Auch hier treten wieder in den Kreuzungen Neuter-Pflanzen zum Teil in großer Zahl auf. Über den Charakter des Realisators der diöcischen Form ist bei *Ecballium* noch nichts Näheres bekannt.

KUHN (1930) erhielt bei Kreuzung von *Thalictrum Fendleri* mit den diöcischen Arten *foetidum*, *Delavayi* und *aquilegifolium* stets ⚥⚥, die je nach der Art einen anderen Grad der Verschiebung der Geschlechtsausbildung in ♂ Richtung zeigten. Diese Ergebnisse zeigen ebenfalls Unterschiede im genetischen System der Geschlechtsbildung an.

Es liegt eine ganze Reihe von Beobachtungen über *Sterilitätsfaktoren* vor, die in den meisten Fällen männliche Sterilität auf Grund einfacher Mutationen bedingen. Diese Sterilität kann an den verschiedensten Entwicklungsstufen von der ersten Anlage eines männlichen Organs bis zur Bildung funktionsunfähigen Pollens angreifen. (Eine ältere Zusammenfassung s. ALLEN 1940, ferner RAEBER u. BOLTON 1955, AR-RUSHDI 1956, FEJER 1958, JOHNSON 1958, LESLEY u. LELSEY 1958, PALIWAL u. HYDE 1958.) Daß durch solche Sterilitätsfaktoren ein funktionierender XX:XY-Geschlechtsmechanismus aufgebaut werden kann, zeigten die Befunde von JONES (1932, 1934) an *Zea Mays*. ALLEN (1940) gibt eine Übersicht der bis dahin identifizierten Sterilitätsfaktoren beim Mais. Einige Gene haben die bestimmte Tendenz, die Bildung ♀ Blüten zu fördern und die ♂ Entwicklung zu unterdrücken. Andere haben die umgekehrte Wirkung. Diese Gene müssen wohl als spezifische Geschlechtsgene bezeichnet werden. "Silkless" sk ist ein ♀ Sterilitätsfaktor. "Tassel-seed" ts_2 kann die Wirkung von sk völlig aufheben. sk sk ts_2 ts_2-Pflanzen erzeugen in allen Blütenständen Samen und sind funktionelle

♀♀. Wenn sie mit sk sk Ts_2 ts_2, heterozygoten "tassel-seed"-Pflanzen, die funktionelle ♂♂ sind, gekreuzt werden, resultiert in der Nachkommenschaft ein 1:1-Verhältnis von ♀ zu ♂, das konstant erhalten bleibt. Pflanzen, die für beide Gene heterozygot sind, spalten in 9 normale Monöcisten:3 silkless:4 tassel-seed (36:17:30). Einige tassel-seed-Pflanzen haben etwas Pollen. Ihre Nachkommenschaft mit einer ♀ Pflanze besteht nur aus tassel-seed-Pflanzen. Dadurch wurde erwiesen, daß die ♀♀ sk sk ts_2 ts_2 sind und die ♂♂ sk sk Ts_2 ts_2. Nach 3 Generationen war das Geschlechtsverhältnis 1,24:1 (also 44,7% ♂♂) geworden. ts ist also ein ausgesprochen weiblichkeitfördernder Faktor, silkless spezifisch die Männlichkeit fördernd und gleichzeitig die weibliche Tendenz unterdrückend. Die bei Diöcisten häufiger beobachteten sekundären Geschlechtsmerkmale und das Balance-Verhältnis zwischen Geschlechts-Chromosomen und Autosomen sind nach JONES in der phylogenetischen Entwicklung auftretende spätere Verfeinerungen eines schon arbeitenden Apparates.

Mit Hilfe solcher Sterilitätsfaktoren hat sich auch ein XY:XX-Apparat mit weiblicher Heterogametie aufbauen lassen (EMERSON 1932).

Bei der bisher nur als monöcisch beschriebenen Art *Bryophyllum daigremontianum* wurden von RESENDE und LINSKENS (1961) Geschlechtstyp-Mutanten gefunden. Sie zeigten gleichzeitig mit der Atrophie der Stamina abnehmende Größe und schwächere Ausfärbung der Korolle. Kreuzungen der ♀ Klone mit normalen ⚥⚥ brachten eine Spaltung in 18 ♀ : 8 (♀ → ⚥) : 32 ⚥ Pflanzen. Phänokopien der Mutanten konnten erzielt werden durch chemische Agentien wie auch geringe Lichtintensität: das Androeceum wurde reduziert, Korollengröße und -ausfärbung ebenfalls. Zwei Pigmente (Delphinidinglykoside) werden in ⚥⚥ im Dunkeln nicht gebildet. Bei direkter Belichtung läuft ihre Synthese normal und erfolgt auch bei den ♀ Blüten. Die Ausbildung des Androeceums bei den ♀ Mutanten kann jedoch durch Licht nicht beeinflußt werden. Demnach liegt bei *Bryophyllum* eine enge genetische Korrelation vor zwischen Androeceum-Korolle einerseits und Gynaeceum-Kalyx andererseits. Von der Art des Mutationsschrittes muß es abhängen, ob die Entwicklung in gynödiöcischer Richtung oder in diöcischer führt. Die Autoren entwickeln eine Möglichkeit, wie mit *einem* Mutationsschritt aus einer monöcischen eine diöcische Rasse mit XY-Mechanismus entstehen könnte. Ein solcher Vorgang hätte in der Evolution eine wesentlich größere Chance, zur Diöcie zu führen als der zwei Mutationsschritte erfordernde Weg der Bildung einer diöcischen Rasse, wie er beim Mais dargestellt wurde.

2. *Kern-Plasma-Balance.* Daß Sterilitätsfaktoren nicht nur Kerngene sind, sondern auch im Plasmon liegen und dann zur Subdiöcie führen können, zeigte das Auftreten plasmatisch bedingter Pollensterilität bei Mais, über die RHOADES (1950, 1954) berichtete. Solche vom Plasmon zusammen mit bestimmten genetischen Faktoren hervorgerufene Pollensterilität fanden schon BATESON und GAIRDNER (1921) bei *Linum*. v. WETTSTEIN (1946) konnte modifizierende Gene im Kern nachweisen, die diese plasmatische Sterilität graduell mildern können. SCHWARTZ (zit. nach RHOADES 1954) fand bei Mais einen Fall, bei dem die Pollensterilität nur auftrat, wenn ein bestimmtes Plasma mit einem dominanten Gen in Kombination mit dem recessiven Allel eines Unterdrücker-Locus vorlag. RHOADES hat mehrere verschiedene plasmasterile Typen erhalten, die nicht identisch sind. Er konnte auch Entmischungsvorgänge beobachten, die den Entmischungsvorgängen bei gestörten *Epilobium*-Bastarden (MICHAELIS 1951) entsprachen.

Die Beteiligung des Plasmons an der G.B. zeigten sehr eindrucksvoll die Artkreuzungen innerhalb der Gattung *Streptocarpus*, die OEHLKERS durchführte.

Diese Untersuchungen dürften wohl neben denen bei Mais die gründlichsten bezüglich der Geschlechtsausbildung bei autöcischen Pflanzen sein.

OEHLKERS kreuzte zunächst die zwittrigen Arten *Streptocarpus Wendlandii* × *St. Rexii* (OEHLKERS 1938, 1940, 1956). R ♀ × W ♂ ergab Pflanzen mit normal ge-

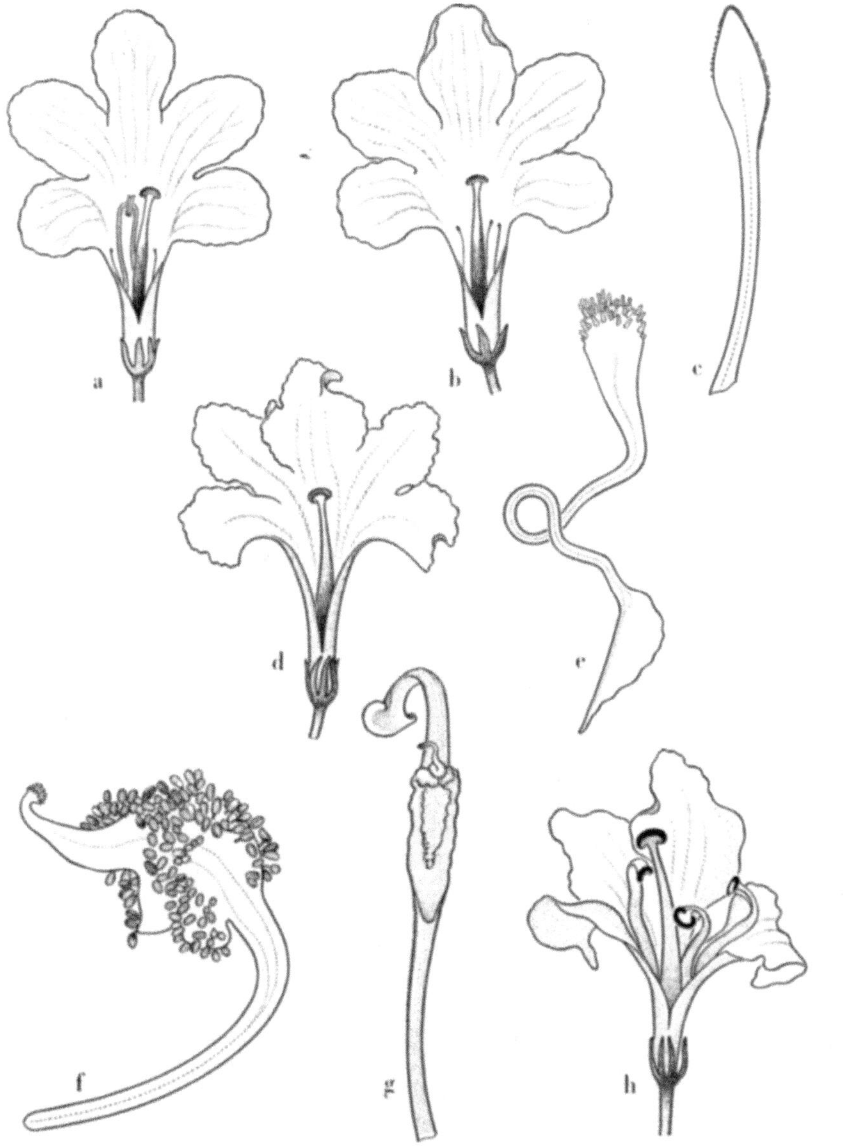

Abb. 12a—h. Fortschreitende Verweiblichung der Blüten von *Streptocarpus*-Bastarden. a—d Reduktion der Antheren, e—h Umbildung der Antherenanlagen zu Karpellen. (Aus OEHLKERS 1935.)

bildeten ⚥-Blüten, aber herabgesetzter ♀ Fertilität, also einer schwachen Verschiebung zur Männlichkeit. Eine Rückkreuzung des Bastards mit dem Vater ergab eine 1:1-Spaltung von normal fertilen Pflanzen zu solchen mit dem Fertilitätsgrad der F_1. Der reziproke Bastard W ♀ × R ♂ zeigt völlige ♂ Sterilität, die zwei sonst fertilen Staubblätter sind wie die übrigen nur staminodial entwickelt.

Es liegen also funktionelle ♀♀ vor. Bei der Rückkreuzung dieses Bastards (WR) ♀ × R ♂ tritt bei der Hälfte der Pflanzen eine Verstärkung der Verweiblichung ein, d. h. die Staminodien fehlen entweder völlig oder sie werden zu Nebenfruchtknoten umgebildet; die Pflanzen sind dann Überweibchen. Nach weiteren Rückkreuzungen und Selektion auf *Rexii*-Charaktere entstehen fast völlig *Rexii*-ähnliche Pflanzen, die reine ♀♀ sind und statt der Antherenanlagen Fruchtblätter enthalten. Die F_2 (W × R) × (R × W) zeigt eine Aufspaltung mit allen Übergängen von normal entwickeltem Andröceum über Staminodialwerden der An-

Tabelle 6.

Normales Androeceum	Antheren fehlen, meist Staminodien vorhanden	Verweiblichtes Androeceum. Staminodien mit Griffelpapillen und Samenanlagen oder Nebenfruchtknoten
Alle bisher untersuchten spontanen Arten und Gartenformen		
Rexii × *R.* Gartenform und reziprok		
Rexii × *grandis* und *grandis* weiß, und reziprok		
Rexii × *solenanthus* und reziprok		
Wendlandii × *Comptonii*		
Wendlandii × *polyanthus* und reziprok sowie 75% der F_2-Pflanzen und 50% der Rückkreuzung *(Wendl.* × *polyanthus)* × *polyanthus*	25% der F_2 *Wendl.* × *polyanthus* und 50% der Rückkreuzung *(Wendl.* × *polyanthus)* × *polyanthus*	
polyanthus × *Rexii* und *Rexii* Gartenform		
Rexii × *Wendlandii* und *Rexii* Gartenform	*Wendlandii* × *Rexii* und *Rexii* Gartenform	*(Wendl.* × *Rexii)* × *Rexii*, ebenso *Rexii* Gartenform
grandis × *Wendlandii*	*Wendlandii* × *grandis* und *grandis* weiß	1% der Blüten *Wendl.* × *grandis*, ferner *(Wendl.* × *grandis)* × *grandis*
solenanthus × *Wendlandii*	*Wendlandii* × *solenanthus*	10% der Blüten *Wendl.* × *solenanthus*, ferner *(Wendl.* × *solenanthus)* × *solenanthus*
	Comptonii × *Rexii* und *Rexii* Gartenform	*(Comptonii* × *Rexii)* × *Rexii*
solenanthus × *Comptonii*	*Comptonii* × *solenanthus*	

theren bis zu ihrem Fehlen und schließlich an ihrer Stelle Ausbildung offener Karpelle, Nebenfruchtknoten (Abb. 12). Die Rückkreuzung (W × R) ♀ × W ♂ ergibt eine 1:1-Spaltung von normal fertil zum Zustand der F_1.

Oehlkers schloß hieraus auf das Vorhandensein eines AG-Komplexes im Genom, der bei *Streptocarpus Rexii* eine Netto-Weiblichkeits-Tendenz besitzt, welcher im Plasmon eine ebenso starke Netto-Männlichkeits-Tendenz entgegenwirkt, so daß in der reinen Art *St. Rexii* beide Tendenzen ausbalanciert sind und also beide Geschlechter in den Blüten normal entwickelt werden. Bei *St. Wendlandii* muß danach umgekehrt im Genom eine Netto-Männlichkeits-Tendenz und im Plasmon eine Netto-Weiblichkeits-Tendenz vorliegen. Durch Einbeziehung anderer Arten der Gattung in seine Versuche konnte Oehlkers (1952, 1958, 1961, 1964) zeigen, daß stufenweise Differenzierungen in den geschlechtsbestimmenden Elementen innerhalb der Gattung vorliegen müssen. *St. Comptonii* verhält sich wie *Wendlandii*; *St. Rexii*, *St. Grandis* und *St. solenanthus* bilden eine Reihe mit einer potentiellen Steigerung der Netto-Geschlechtstendenzen des

Genoms in der genannten Reihenfolge, wobei jedoch zwischen *Rexii* und *Grandis* eine ganze Stufe, zwischen *Grandis* und *solenanthus* nur eine halbe Stufe liegt (Tabelle 6). GOLDSCHMIDT (1938) gibt den Ergebnissen in Analogie zu seinen *Lymantria*-Befunden eine andere Deutung, indem er jeweils für Weiblichkeit und Männlichkeit nur einen Faktor annimmt. Sie sollen hier bei den einzelnen Arten ver-

Abb. 13a—d. Verweiblichung der Blüten von *Solanum*-Bastarden. a—c Reduktion des Androeceums, d Bildung zusätzlicher Karpelle. (Aus KOOPMANS 1951.)

schiedene Valenzen besitzen und einer im Genom, einer im Plasmon verankert sein. HARTMANN (1956) akzeptiert wohl den AG-Komplex von CORRENS, schließt sich jedoch nicht seiner Auflösung in Einzelgene an, wie sie WESTERGAARD für notwendig hält. Außerdem vertritt HARTMANN den Standpunkt, daß außer dem stets potenzlosen AG-Komplex bei Primärzwittern ausbalancierte Realisatoren vorhanden sind, die in den verschiedenen Arten verschiedene Potenzen haben können.

Marquardt (1963, 1964) versucht eine Deutung aufgrund des Jacob-Monod-Modells von Operon- und Regulatorsystem.

Reziprok unterschiedliche Beeinflussung der Geschlechtsausbildung, die nicht nur in Sterilität des einen Geschlechts, sondern in aktiver Verschiebung resultiert, zeigten auch die *Solanum*-Artkreuzungen von Koopmans (1951, 1955, 1959). Sie erhielt in F_2 von *Sol. rybinii* × *chacoënse* eine Variationsreihe, beginnend mit normalem Androeceum über Reduktion der Antheren, Zahlenreduktion der Antheren-Rudimente bis zu Umbildung der Antherenanlagen zu Nebenfruchtknoten (Abb. 13). In der Kreuzung *Sol. rybinii* × *commersonii* trat eine Verschiebung des Geschlechts zur Männlichkeit auf. Rückkreuzungsfamilien zeigten einen bestimmten Grad der Geschlechtsausbildung in Abhängigkeit vom Umwandlungsgrad der Mutter. Alle Rückkreuzungen von *rybinii* × *chacoënse* mit *rybinii* als Vater ergaben normale Blütenbildung. Die reziproke Kreuzung *chacoënse* × *rybinii* brachte eine ähnliche F_1. Ein Stamm von *chacoënse* zeigte jedoch abweichende Ergebnisse. Eine größere Zahl von F_1-Pflanzen besaß abnorme Antheren, anders gestaltet als bei der F_2 der 1. Kreuzung. Hauptsächlich betroffen war die Länge der Antheren; die Ausbildung variierte von etwas reduziert bis rudimentär oder völlig fehlend. Diese Typen kehren in F_2 wieder. Rückkreuzungen mit *rybinii* brachten viele Pflanzen mit abnormen Antheren, solche von abnormen F_1-Pflanzen in höherem Anteil als solche von normalen. Weitere Rückkreuzung mit *rybinii* ergab eine weitere Steigerung des Prozentsatzes der Pflanzen mit abnormem Androeceum. Demnach wirkt auch in dieser zwittrigen Gattung an der Ausbildung des Geschlechts ein balanciertes System, das sich aus mehreren Kerngenen und Faktoren des Plasmons zusammensetzt. In gleiche Richtung weisen die Befunde an Bastarden zwischen zwittrigen *Fragaria*-Arten (Nürnberg-Krüger 1958) und Kreuzungen von *Digitalis purpurea* × *D. lutea* (Heilbronn und Yakar 1942). Auch die Analyse der Hemmungssysteme bei *Godetia* (Hiorth 1948a, b) scheint Aufklärung über ähnliche Kern-Plasma-Balance zu geben.

b) Sekundärzwitter.

Die Untersuchungen an Subdiöcisten haben gezeigt, daß eine Reihe von Arten, die in der Natur normale Diöcie aufweisen, sekundär mehr oder weniger zur Monöcie oder Zwittrigkeit übergehen können (*Fragaria*, *Cannabis* usw.). In diesen Fällen müssen die bei der diöcischen Form geschlechtsentscheidend wirkenden Gene ihre Bedeutung verloren haben. Entweder ist ihre Valenz unter ein Wirkungsmaximum gesunken oder es hat eine Umlagerung der beide Geschlechter bestimmenden Gene stattgefunden, so daß wieder ein ausbalancierter Zustand erreicht wird. Die Tatsache, daß alle bisher bekannten ⚥⚥ dieser Art Kulturpflanzen sind, spricht dafür, daß dieser Vorgang unter dem Einfluß einer Selektion erfolgt ist. Die Verhältnisse bei Hanf und Spinat sind bei den Subdiöcisten dargestellt worden. Für *Fragaria* hatten Richardson (1914) und Mangelsdorf und East (1927) festgestellt, daß bei Kreuzung der oktoploiden *Fr. virginiana* ♀ mit der oktoploiden *Fr. chiloensis* ♂ 1:1 ♀♀ und (♂♂ und zwittrige ♂♂) entstehen. Diese zwittrigen ♂♂ ergaben im Durchschnitt etwa 17% Samenansatz. Vermutlich sind durch hier einsetzende Selektion die heutigen zwittrigen Gartensorten entstanden, also alle eigentlich genetische ♂♂ mit 66% Samenansatz. *Fr. grandiflora* erscheint nach Kuhn (1930a) heute gynodiöcisch mit heterogameten ♀♀ neben sekundären ⚥⚥. Die Kombination ♀ × ⚥ ergibt ♀ + ⚥, d. h. die ♀♀ unterscheiden sich von den ursprünglichen ♀♀ der Art durch ihre Nachkommenschaft. Die einfachste Erklärung ergibt die Annahme einer Valenzschwächung des männlichen Realisators, die schließlich in vollfertilen ⚥⚥ zum Wert 0 führt. Der ♀ Realisator hat dagegen seine Valenz behalten und ist immer dominant.

Schluß.

Es stellt sich heute die diplogenotypische G. B. als ein sehr vielgestaltiger Vorgang dar. Eine Reihe älterer Auffassungen haben Korrekturen erfahren müssen; es kann weder die Theorie GOLDSCHMIDTs noch das Schema von CORRENS alle Fälle, die heute bekannt sind, zureichend erklären. Ausgehend vom phylogenetisch-primären Zustand bei den Blütenpflanzen besitzt jedes Individuum eine allgemeine bisexuelle Potenz (HARTMANN 1909, 1956) die die Fähigkeit bedingt, beiderlei Sexualorgane und Sexualzellen zu bilden. Von einigen Autoren als zusätzlich vorhanden, von anderen als gleichbedeutend damit angesehen, liegt in jedem Individuum eine nach CORRENS als AG-Komplex bezeichnete Summe von Genen, die die Ausbildung der beiderlei Organe und Keimzellen ermöglicht, also eigentlich aus 2 Komplexen besteht. Die Analyse dieser beiden Gengruppen, die sich aus karyotischen und plasmonischen Elementen zusammensetzen, wurde erst in sehr wenigen Fällen in einzelnen Punkten erreicht. Durch das Auftreten von Hemmungsfaktoren für einen der beiden Komplexe wie auch durch Sortierung weiblich und männlich wirkender Gene und ihre Lokalisation in bestimmten, keinen Gen-Austausch zeigenden Abschnitten der Geschlechts-Chromosomen ist vielfach im Laufe der Evolution ein übergeordneter Mechanismus entwickelt worden. Er führte im Extrem durch bestimmte quantitative Relationen von AG und Realisator-Genen zur alternativen Unterdrückung eines der beiden Gen-Komplexe und so zur Bildung getrennt-geschlechtlicher Organismen. Es gibt aus Versuchen zur Geschlechtsumwandlung bei Monöcisten eine Reihe von Hinweisen, daß Hormone aus der Gruppe der Auxine und Gibberelline eine Rolle spielen als Substanzen, über die die Gene wirksam werden (SHIFRISS 1956, 1961, YUICHI 1957, HESLOP-HARRISON u. HESLOP-HARRISON 1958a, b, RESENDE u. VIANA 1959, PETERSON u. LA MOINE D. ANHDER 1960 GALUN 1961, MITCHELL u. WITTWER 1962, KÖHLER 1964, SHIFRISS, GEORGE jr. u. QUIÑONES 1964).

Dieser Realisatoren-Mechanismus kann sehr verschiedenartig in seiner Struktur und seiner Wirkung sein, jedoch sorgt stets ein einfacher Rückkreuzungsvorgang für die Erhaltung beider Geschlechter in etwa gleichen Zahlen. Jede weitere Veränderung dieses ausbalancierten G.B.-Vorganges muß dann wieder über ähnliche Zwischenstufen wie bei seiner Entstehung zu einer Abweichung in Richtung auf den Ausgangszustand der Zwittrigkeit oder Monöcie führen.

Literatur.

ABRAHAM, A., and P. M. MATHEW: Cytological studies in the cycads: Sex chromosomes in *Cycas*. Ann. Bot. **26**, 261—266 (1962). — ÅKERLUND, E.: Ein *Melandrium*-Hermaphrodit mit weiblichem Chromosomenbestand. Hereditas (Lund) **10**, 153—159 (1927). — AKKOÇ, E.: Zytogenetische Beiträge zur Geschlechtsvererbung beim Spinat *(Spinacia oleracea)* im Hinblick auf die Züchtung triploider Hybridsorten. Z. Pflanzenzüchtg **53**, 226—246 (1965). — ALLEN, C. E.: A chromosome difference correlated with sex differences in *Sphaerocarpus*. Science, N. s. **46**, 466—467 (1917). — The genotypic basis of sex expression in angiosperms. Bot. Rev. **6**, 229—300 (1940). — ARARATJAN, A. G.: Heterochromosomes in the wild spinach. Dokl. Akad. Nauk SSSR **24**, 56—57 (1939). — AR-RUSHDI, ABBAS H.: Inheritance in *Nicotiana tabacum*. XXVI. Sterility genes from *Tomentosae* species. J. Genet. **54**, 9—22 (1956).

BAIL, TH.: Über androgyne Blüthenstände bei solchen Monöcisten und Diöcisten, bei denen Trennung der Blüthenstände Regel ist. Schr. naturforsch. Ges. Danzig **2**, 3—5 (1869). — Neuere Beobachtungen der Androgynie bei Salicinen. Schr. kgl. phys. ökonom. Ges. Königsberg, I. Abh. **18**, 94—95 (1877). — BAKER, H. G.: The inheritance of certain characters in crosses between *Melandrium dioicum* and *M. album*. Genetica **25**, 126—156 (1951). — BATESON, W., and A. F. GAIRDNER: Male sterility in flax, subject to two types of segragation. J. Genet. **11**, 269 (1921). — BAUR, E.: Ein Fall von geschlechtsbegrenzter Vererbung bei *Melandrium album*. Z. indukt. Abstamm.- u. Vererb.-Lehre **8**, 335—336 (1912). — BEERMANN, W.: Geschlechtsbestimmung und Evolution der genetischen Y-Chromosomen bei *Chironomus*. Biol. Zbl. **74**, 525—544 (1955). — BELAR, C.:

Der Chromosomenbestand der *Melandrium*-Zwitter. Z. indukt. Abstamm.- u. Vererb.-Lehre **39**, 184—190 (1925). — Belowitzkaja, N. A., i E. J. Grestschuchin: Ob odnowremennoi konople. Izv. Akad. Nauk SSR., Ser. Biol. **1939**, Nr 3 [Russisch]. — Bemis, W. P., and G. B. Wilson: A new hypothesis explaining the genetics of sex determination in *Spinacia oleracea* L. J. Hered. **44**, 91—95 (1953). — Bilge, E.: Recherches morphologiques, anatomiques et génétiques sur *Bryonia macrostylis*. Rev. Fac. Sci. Istanbul, Sér. B **20**, 121—146 (1955). — Billings, F. H.: Male gametophyte of *Atriplex hymenelytra*. Bot. Gaz. **95**, 477—484 (1934). — Bitter, G.: Zur Frage der Geschlechtsbestimmung von *Mercurialis annua* durch Isolation weiblicher Pflanzen. Ber. dtsch. bot. Ges. **27**, 120 (1909). — Blackburn, K. B.: Sex chromosomes in plants. Nature (Lond.) **112**, 687—688 (1923). — On the occurrence of sex chromosomes in flowering plants with some suggestions as to their origin. Proc. intern. Congr. Plant. Sci. **1**, 299—306 (1929). — Blackburn, K. B., and J. W. Heslop-Harrison: A preliminary account of the chromosomes and chromosome behaviour in *Salicaceae*. Ann. Bot. **38**, 361—378 (1924). — Blakeslee, A. F., and H. E. Warmke: Size of seed and other criteria of polyploids. Science **88**, 440 (1938). — Effect of polyploidy upon the sex mechanism in dioecious plants. Genetics **24**, 88—89 (1939). — Bocsa, I.: Beiträge zur Züchtung eines ungarischen monözischen Hanfes und zur Kenntnis der Inzuchterscheinungen beim Hanf (*Cannabis sativa* L.). Z. Pflanzenzüchtg **39**, 11—34 (1958). — Bose, S., and J. Janick: Karyo-races in *Spinacia oleracea*. Amer. J. Bot. **48**, 238—241 (1961). — Brabec, Fr.: Über triploide fertile *Bryonia*-Bastarde. Biol. Zbl. **81**, 291—302 (1962). — Bredemann, G., K. Garber, W. Huhnke u. R. v. Sengbusch: Die Züchtung von monözischen und diözischen fasererträgreichen Hanfsorten (Fibrimon und Fibrida). Z. Pflanzenzüchtg **46**, 235 (1961). — Breider, H., u. H. Scheu: Die Bestimmung und Vererbung des Geschlechts innerhalb der Gattung *Vitis*. Gartenbauwiss. **11**, 627—674 (1938). — Breslavetz, L.: Zytologische Studien über *Melandrium album* L. Planta (Berl.) **7**, 444—460 (1929). — Bridges, C. B.: Triploid intersexes in *Drosophila melanogaster*. Science **54**, 252—254 (1921). — Buijtenen, J. P. van, and D. W. Einspahr: Note on the presence of sex chromosomes in *Populus tremuloides*. Bot. Gaz. **121**, 60—61 (1959).

Caspari, E.: Cytoplasmic inheritance. Advanc. Genet. **2**, 1—66 (1948). — Correns, C.: Experimentelle Untersuchungen übre die Gynodioecie. Ein typisch spaltender Bastard zwischen einer einjährigen und einer zweijährigen Sippe des *Hyoscyamus niger*. Ber. dtsch. bot. Ges. **22**, 506—524 (1904). — Die Bestimmung und Vererbung des Geschlechts, nach neuen Versuchen mit höheren Pflanzen. Berlin 1907. — Weitere Untersuchungen über die Geschlechtsformen polygamer Blütenpflanzen und ihre Beeinflußbarkeit. Jb. wiss. Bot. **45**, 661—700 (1908a). — Die Rolle der männlichen Keimzellen bei der Geschlechtsbestimmung der gynodiöcischen Pflanzen. Ber. dtsch. bot. Ges. **26**, 686—701 (1908b). — Vererbung und Bestimmung des Geschlechts. Naturwiss. Rdsch. **27**, 557—560, 572—576 (1912). — Untersuchungen über Geschlechtsbestimmung bei Distelarten. S.-B. Akad. Wiss. Berlin **20**, 448 (1916). — Ein Fall experimenteller Verschiebung des Geschlechtsverhältnisses. S.-B. Akad. Wiss. Berlin **51**, 685 (1917). — Fortsetzung der Versuche zur experimentellen Verschiebung der Geschlechtsverhältnisse. S.-B. Akad. **50**, 1175 (1918). — Versuche, bei Pflanzen das Geschlechtsverhältnis zu verschieben. Hereditas **2**, 1 (1921a). — Zweite Fortsetzung der Versuche zur experimentellen Verschiebung des Geschlechtsverhältnisses. S.-B. Akad. **18**, 330 (1921b). — Alkohol und Zahlenverhältnis der Geschlechter bei einer getrenntgeschlechtigen Pflanze *(Melandrium)*. Naturwissenschaften **10**, H. 49, 1049 (1922a). — Geschlechtsbestimmung und Zahlenverhältnisse der Geschlechter bei Sauerampfer *(Rumex acetosa)*. Biol. Zbl. **42**, 465 (1922b). — Über den Einfluß des Alters der Keimzellen. I. Dritte Fortsetzung der Versuche zur experimentellen Verschiebung des Geschlechtsverhältnisses. S.-B. preuß. Akad. Wiss. IX, 70—104 (1924). — Über Fragen der Geschlechtsbestimmung bei höheren Pflanzen. Z. indukt. Abstamm.- u. Vererb.-Lehre **41**, 5—40 (1926). — Handbuch der Vererbungswissenschaften. II. Bestimmung, Verteilung und Vererbung des Geschlechts. 1928. — Crane, M. B., and W. J. C. Lawrence: Inheritance of sex, colour and hairiness in the raspberry, *Rubus idaeus*. J. Genet. **24**, 243—255 (1931). — Crescini, Fr.: Selezione artificiale usw. (Künstliche Selektion und Geschlechtsverhältnis bei Hanf.) Caryologia **7**, 415—419 (1955). — Crew, F. A. E.: Sex-determination. 3. rev. edit. London: Methuen & Co.; New York: John Wiley & Sons 1954. — Currence, T. M.: Tomato breeding. In: Handbuch der Pflanzenzüchtung, 2. Aufl., Bd. VI, S. 351ff. (Hrsg. Roemer-Rudorf). Berlin u. Hamburg: Paul Parey 1962.

Dark, S. O. S.: The use of polyploidy in hop breeding. Ann. Rept. Wye Coll, Dept. Hop. Res. 1952, 34—42 (1952). — Darlington, C. D.: Recent Advances in Cytology. London: Churchill 1932. — Dean, H. L.: Variations in style number and other gynoecial structures of *Lycheris alba*. Science **130**, 42—43 (1959). — Derenne, P.: Au sujet des chromosomes sexuels du Houblon. Bull. Inst. Agro. Gemblaux **22**, No 1—2 (1954). — Dressler, O.: Zytogenetische Untersuchungen an diploidem und polyploidem Spinat (*Spinacia oleracea* L.) unter besonderer Berücksichtigung der Geschlechtsvererbung als Grundlage einer Inzucht-Heterosis-Züchtung. Z. Pflanzenzüchtg **40**, 385—424 (1958). — Driga, J. E.: Cytological investigation of hemp. (Preliminary communication.) In: GRISKO et al., Genetics and breeding of hemp, p. 65—72. Voesojnanyi naucno-issled. Inst. Konopli, Trudy, Moskau u. Leningrad 1937. — DUNSTAN, R. T.: Hybridization of *Euvitis* and *Vitis rotundifolia* backcrosses to *Muscadine*. Proc. Amer. Soc. Hort. Sci. **84**, 238 (1964). — DURAND, B.: Polymorphisme, polyploidie et répartition des sexes chez les *Mercuriales annuelles*. C. R. Acad. Sci. (Paris) **244**, 1249—1251 (1957). — DUVICK, D. N.: Allelism and comparative genetics of fertility restoration of cytoplasmatically pollen sterile maize. Genetics **41**, 544 (1956). — The regional localization of pollen sterility in partially sterile cytoplasmatically pollen sterile maize. Proc. internat. Congr. of Genet. Montreal **II**, 72 (1958).

EGUCHI, T., and H. ICHIKAWA: Studies on the flower bud differentiation and bolting in spinach [Jap.]. J. Hort. Ass. Jap. **11**, 13—56 (1940). — ELLIS, J. R., and J. JANICK: The chromosomes of *Spinacia oleracea*. Amer. J. Bot. **47**, 210—214 (1960). — EMERSON, R. A.: The present status of maize genetics. Proc. 6. Internat. Congr. Genetics **1**, 141—167 (1932). — ERNST, A.: Austausch und Mutation im Komplex-Gen für Blütenplastik und Inkompatibilität bei *Primula*. Z. indukt. Abstamm.- u. Vererb.-Lehre **88**, 517—599 (1957).

FARMER, R. E.: Sex-ratio and sex-related characteristics in eastern cottonwood. Silvae Genetica **13**, 116 (1964). — FEJÈR, S. O.: A gene for femaleness in hermaphrodite *Lolium perenne* L. Proc. internat. Congr. of Genet. Montreal **2**, 81—82 (1958).

GALÁN, F.: Analyse génétique de la monoecie et de la dioecie zygotiques et de leur différence dans *Ecballium elaterium*. Acta Salmanticenia, Ciencias, Sec. Biol. I, 7—15 (1951). — Genética y feno genética del sexo en *Ecballium elaterium*. Atti 9. Congr. internaz. genet. Bellagio (Como), Parte 2, 1954, 942—944 (1953/54). — GALUN, E.: Study of the inheritance of sex expression in the cucumber. The interaction of major genes with modifying genetic and non-genetic factors. Genetica **32**, 134—163 (1961). — GATES, R. R.: Polyploidy and the sex chromosomes. Acta biotheor. (Leiden) **11**, 27—44 (1953). — GOLDSCHMIDT, R.: Eine merkwürdige Beziehung zwischen Geschlechtsgenen, Entwicklungsgeschwindigkeit und Zahlenverhältnis der Geschlechter. Naturwissenschaften **19**, 735—736 (1931a). — Die sexuellen Zwischenstufen. Monogr. Gesamtgeb. Pflanzen, Tiere, Bd. 23, 528 S. Berlin: Springer 1931b. — A *Lymantria*-like case of intersexuality in plants (OEHLKERS' Werk) usw. J. Genet. **36**, 531—535 (1938). — GOMEZ-CAMPO, C., and U. CASAS-BUILLA: Radiation-induced sex reversal in *Ecballium elaterium*. Nature (Lond.) **208**, 805 (1965). — GOTTSCHALK, W.: Über strahlenbedingte erbliche Sterilitätserscheinungen bei pflanzlichen Versuchsobjekten. Atompraxis **6**, 2, 341—347 (1960). — Über die zeitliche Aufeinanderfolge bestimmter Gestaltungsprozesse bei der Ausdifferenzierung von Vegetationskegeln zu Blüten. Naturwissenschaften **48**, 59 (1961). — Der Funktionswechsel von Blütenorganen als Folge der Wirkung mutierter Gene. Ber. dtsch. bot. Ges. **75**, 77 (1962). — GRANT, W. F.: Cytological aspects of sex determination and phylogenetic trends in dioecious species of *Amaranthus*. Proc. internat. Congr. of Genet. Montreal **2**, 103 (1958). — GRISCHKO, N. N.: Die Züchtung des einhäusigen Hanfes und von Hanf mit gleichzeitig ausreifenden männlichen und weiblichen Pflanzen. Allruss. Akad. Wiss. Landw., Ser. 3,I (1935) (zit. bei BOCSA 1958). — Gleichzeitig reifender Hanf. [Russisch]. Novoje selskom chosaistewe Wipusk **5** (1937).

HAGA, T.: Sex and chromosomes in *Spinacia oleracea* L. Jap. J. Genet. **10**, 218—222 (1935). — HAGERUP, O.: *Empetrum hermaphroditum* (Lge.) Hagerup. A new tetraploid dioecious species. Dansk. bot. Ark. **5**, 1—17 (1927). — HÅKANSSON, A.: Zytologische Studien an *Salix*-Bastarden. Hereditas (Lund) **24**, 1—32 (1938). — HARTMANN, M.: Die Sexualität, 2. Aufl. Stuttgart: Gustav Fischer 1956. — HARTSHORNE, J. W.: The heterogametic sex in dioecious flowering plants. Genetics today. Proc. XI. Intern. Congr. Genetics I, 232. Oxford: Pergamon Press 1963. — HAUSTEIN, E.: Eine androgyne haploide *Oenothera scabra*. Planta (Berl.) **56**, 475—478 (1961). — HEILBRONN, A.: Über die Genetik von Monözie und Getrenntgeschlechtigkeit bei *Bryonia*-Bastarden. Proc. 8. Internat. Congr. Genet. Hereditas (Lund), Suppl.-Vol. 34, 590—591 (1948). — Über die Rolle des Plasmas bei der Geschlechtsbestimmung der Bryonien. Rev. Fac. Sci. Istanbul, Sér. B **18**, 205—206 (1953). — HEILBRONN, A., u. M. BASARMAN: Über die F_2 der *Bryonia*-Bastarde und ihre Bedeutung für das Problem der Geschlechtsrealisation. Rev. Fac. Sci. Istanbul, Sér. B **7**, 138—144 (1942). — HEILBRONN, A., u. N. YAKAR: Gynodiöcie und Intersexualität als Folge der Bastardierung zwittriger Species. Rev. Fac. Sci. Istanbul, Sér. B **7**, 317—322 (1942). — HERIBERT-NILSSON, M.: Experimentelle Studien über Variabilität, Spaltung, Artbildung und Evolution in der Gattung *Salix*. Lunds Univ. Årsskr., N.F. Avd. **2**, 14, Nr 28 (1918). — HERICH, R.: Nucleole and sex differentiation. Caryologia **14**, 375—381 (1961). — Geschlechtsausprägung und Größe der Zellkerne sowie der Pollenkörner bei monözischem Hanf verschiedener Formen. Züchter **31**, 48—51 (1961). — HERTWIG, G., u. P. HERTWIG: Die Vererbung des Hermaphroditismus bei *Melandrium*. Z. indukt. Abstamm.- u. Vererb.-Lehre **28**, 259—294 (1922). — HESLOP-HARRISON, J. W.: Heterochromosomes and polyploidy. Nature (Lond.) **117**, 50 (1926). — HESLOP-HARRISON, J.: Auxin and sexuality in *Cannabis sativa*. Physiol. Plantarum (Kbh.) **9**, 588—597 (1956). — HESLOP-HARRISON, J. and Y.:

The effect of carbon monoxide on sexuality in *Mercurialis ambigua* L. fils. New Phytologist **56**, 352—355 (1957). — Long-day and auxin induced male sterility in *Silene pendula* L. Port. Acta biol. A **5**, 79—94 (1958a). — Photoperiod, auxin and sex balance in a long-day plant. Nature (Lond.) **181**, 100—102 (1958b). — HIORTH, G.: Über Hemmungssysteme bei *Godetia Whitneyi*. Z. indukt. Abstamm.- u. Vererb.-Lehre **82**, 12—63 (1948a). — Über Hemmungssysteme bei *Godetia Whitneyi*. II. Z. indukt. Abstamm.- u. Vererb.-Lehre **82**, 277—330 (1948b). — HIRATA, K.: Sex determination in hemp, *Cannabis sativa* L. J. Genet. **19**, 65 (1928). — Cytological basis of the sex determination in *Cannabis sativa*. Jap. J. Genet. **4**, 198 (1929). — HOFFMANN, W.: Gleichzeitig reifender Hanf. Züchter **13**, 277—283 (1941). — Die Vererbung der Geschlechtsformen des Hanfes (*Cannabis sativa* L.). I. Züchter **17/18**, 257—277 (1946/47). — Die Vererbung der Geschlechtsformen des Hanfes (*Cannabis sativa* L.). II. Züchter **22**, 147—158 (1952). — Hanf, *Cannabis sativa* L. In: Handbuch der Pflanzenzüchtung, 2. Aufl. (Hrsg. ROEMER-RUDORF), Bd. V, S. 204ff. Berlin u. Hamburg: Paul Parey 1961. — HOFFMANN, W., u. E. KNAPP: Röntgenbestrahlungen beim Hanf. Züchter **12**, 1—9 (1940). — HOFMEYR, J. D. J.: Genetical studies of *Carica papaya* L. I.—II. Un. S. Afr. Dept. Agr. and Forestry Sci. Bull. **187**, 5—64 (1938). — Sex-linked inheritance in *Carica papaya* L. S. Afr. J. Sci. **36**, 283—285 (1939a). — Some suggestions on the mechanism of sex determination in *Carica papaya*. S. Afr. J. Sci. **36**, 288—290 (1939b). — Sex reversal as a means of solving breeding problems of *Carica papaya*. S. Afr. J. Sci. **49**, 228—232 (1953). — HOFMEYR, J. D. J., and H. VAN ELDEN: Tetraploidy in *Carica papaya* L. induced by colchicine. S. Afr. J. Sci. **38**, 181—185 (1942). — HOLUBY, J. I.: *Cannabis sativa monoica*. Öst. bot. Z. **28**, 367—369 (1878). — HUHNKE, W., CHR. JORDAN, H. NEUER u. R. v. SENGBUSCH: Grundlagen für die Züchtung eines monöcischen Hanfes. Z. Pflanzenzüchtg **29**, 55—75 (1950). — HUSFELD, B.: Reben. In: Handbuch der Pflanzenzüchtung (Hrsg. ROEMER-RUDORF), 2. Aufl., Bd. VI, S. 723ff. Berlin u. Hamburg: Paul Parey 1962. — HUYSKES, J. A., u. J. SNEEP: Sproß- und Blattgemüse: Spargel. In: Handbuch der Pflanzenzüchtung (Hrsg. ROEMER-RUDORF), 2. Aufl., Bd. VI, S. 131ff. Berlin u. Hamburg: Paul Parey 1962.

IIZUKA, M., and J. JANICK: Cytogenetic analysis of sex determination in *Spinacia oleracea*. Genetics **47**, 1225—1241 (1962). — Sex chromosomes translocations in *Spinacia oleracea*. Genetics **48**, 273—282 (1963). — IMAI, Y.: Sex-linked mutant characters in the hemp, *Cannabis sativa*. J. Genet. **35**, 431—432 (1938).

JACOBSEN, P.: The sex chromosomes in *Humulus* L. Hereditas (Lund) **43**, 357—370 (1957). — JANICK, J.: Inheritance of sex in tetraploid spinach. Proc. Amer. Soc. Hort. Sci. **66**, 361—363 (1956). — JANICK, J., and M. IIZUKA: Sex determination in a spinach line containing a heteromorphic sex chromosome. Genetics **46**, 873—874 (1961) (Abstr.). — JANICK, J., and E. C. STEVENSON: A genetic study of the heterogametic nature of the staminate plant in spinach (*Spinacia oleracea* L.). Proc. Amer. Soc. Hort. Sci. **63**, 444—446 (1954). — Genetics of the monoecious character in spinach. Genetics **40**, 429—437 (1955a). — Environmental influences on sex expression in monoecious lines of spinach. Proc. Amer. Soc. Hort. Sci. **65**, 416—422 (1955b). — The effects of polyploidy on sex expression in spinach. J. Hered. **46**, 151—156 (1955c). — JOHNSON, A. G.: Male sterility in *Brassica*. Nature (Lond.) **182**, 1523ff. (1958). — JOHNSSON, H.: Cytological studies of diploid and triploid *Populus tremula* and of crosses between them. Hereditas (Lund) **26**, 321—352 (1940). — JONES, D. F.: The interaction of specific genes determining sex in dioecious maize. Proc. 6. Internat. Congr. Genet. **2**, 104—107 (1932). — Unisexual maize plants and their bearing on sex differentiation in other plants and animals. Genetics **19**, 552—567 (1934). — JØRGENSEN, C. A.: Chromosomes and sex in *Vallisneria*. J. Genet. **18**, 63—75 (1927).

KAPPERT, H.: Willkürliche Änderung des Geschlechtsverhältnisses und ihre Ausnützung in der Pflanzenzüchtung. Umschau **49**, 772 (1941). — Untersuchungen über die Plasmonwirkungen bei *Aquilegia* (Gynodiözie-Heterosis-Gestalt des Sporus). Flora (Jena) **37**, 95—105 (1943). — KATAYAMA, Y., and S. SHIDA: Development of intersexual flowers and their location on the stalk in spinach [Jap.]. Jap. J. Breed. **6**, 19—22 (1956). — KENG, H.: Androdioecism in the flowers of *Trochodendron Aralioides* (Sieb. et Zucc.). J. Arnold Arboret. (Harvard Univ.) **40**, 158 (1959). — KIHARA, H.: Chromosomes of *Rumex acetosella*. Bot. Mag. (Tokyo) **39**, 353—360 (1925). — The sex chromosomes of *Humulus japonicus*. Jap. J. Genet. **4**, 55—63 (1929a). — Sex-chromosomes in plants. Jap. J. Genet. **4**, 90—101 (1929b). — A case of linkage of sex-chromosomes with autosomes in the pollen mother cell of *Humulus japonicus*. Jap. J. Genet. **5**, 73—80 (1929c). — Karyologische Studien an *Fragaria* mit besonderer Berücksichtigung der Geschlechtschromosomen. Cytologia (Tokyo) **1**, 345—357 (1930). — KIHARA, H., u. I. HIRAYOSHI: Die Geschlechtschromosomen von *Humulus japonicus* Sieb. et Zucc. 8. Congr. Jap. Ars. Sci. S. 363—368, 1932. — KIHARA, H., and T. ONO: Cytological studies on *Rumex* L. I. Chromosomes of *Rumex acetosa* L. Bot. Mag. (Tokyo) **37**, 36 (84, 90) (1923a). — Cytological studies on *Rumex* L. II. On the relation of chromosome number and sexes in *Rumex acetosa* L. Bot. Mag. (Tokyo) **37**,

147—149 (1923b). — The sex-chromosomes of *Rumex acetosa*. Z. indukt. Abstamm.- u. Vererb.-Lehre **39**, 1—7 (1925). — KIHARA, H., u. Y. YAMAMOTO: Karyomorphologische Untersuchungen an *Rumex acetosa* L. und *R. montanus* Desf. Cytologia (Tokyo) **3**, 84—118 (1931). — KNAPP, E.: Zur Frage der genetischen Aktivität des Heterochromatins, nach Untersuchungen am X-Chromosom von *Sphaerocarpus Donnellii*. Ber. dtsch. bot. Ges. **53**, 751—760 (1935). — KÖHLER, D.: Zur Vererbung der Monözie beim Hanf. Z. Vererbungsl. **89**, 437—447 (1958). — Die Vererbung des Blühtermins bei *Cannabis sativa*. Z. Pflanzenzüchtg **42**, 339 (1960). — Veränderung des Geschlechts von *Cannabis sativa* durch Gibberellinsäure. Ber. dtsch. bot. Ges. **74**, 275—278 (1964a). — Geschlechtsbestimmung bei Blütenpflanzen. Ergebn. Biol. **27**, 98—115 (1964b). — KOOPMANS, A.: Cytogenetic studies on *Solanum tuberosum* L. and some of its relatives. Genetica **25**, 193—337 (1951). — Changes in sex in the flower of the hybrid *Solanum rybinii* × *S. chacoense*. Genetica **26**, 359—380 (1952). — Changes in sex in the flower of the hybrid *Solanum rybinii* × *S. chacoense*. II. Plasmatic influence upon gene action. Genetica **27**, 273—285 (1954). — Changes in sex in the flower of the hybrid *Solanum rybinii* × *S. chacoense*. III. Data about the reciprocal cross *Solanum chacoense* × *S. rybinii*. Genetica **27**, 465—471 (1955). — Changes in sex in the flower of the hybrid *Solanum rybinii* × *S. chacoense*. IV. Further data from the reciprocal cross *S. chacoense* × *S. rybinii*. Genetica **30**, 384—390 (1959). — KUCKUCK, H.: Problems and results in the breeding of agricultural crops in Germany (Abstr.). Jap. J. Breed. **10**, 114—115 (1960). — KUHN, E.: Die Geschlechtsformen bei *Fragaria* und ihre Vererbung. Züchter **2**, 1—11 (1930a). — Über Kreuzungen des getrenntgeschlechtigen *Thalictrum Fendleri* mit gemischtgeschlechtigen Arten. Biol. Zbl. **50**, 79—102 (1930b). — Untersuchungen über die Vererbung des Geschlechts bei subdiözischen Blütenpflanzen. Ber. dtsch. bot. Ges. **54**, 9—10 (1936). — Selbstbestäubungen subdiözischer Blütenpflanzen, ein neuer Beweis für die genetische Theorie der Geschlechtsbestimmung. Planta (Berl.) **30**, 457—470 (1939). — KUMAR, L. S. S., A. ABRAHAM and V. K. SRINIVASAN: The cytology of *Carica papaya* Linn. Indian J. Agric. Sci. **15**, 242—253 (1945). — KUMAR, L. S. S., and S. VISHVESHWARALAH: Sex mechanism in *Coccinea indica* Wight and Arw. Nature (Lond.) **170**, 330—331 (1952). — KUNZE, G.: Ein Fall geschlechtsgebundener Vererbung bei *Melandrium* (Garcke). Ber. dtsch. bot. Ges. **68**, 249—256 (1955).

LACZYNSKA-HULEWICZOWA, T.: Ein Beitrag zum Einfluß der Außenfaktoren auf das Geschlecht des di- und tetraploiden Hanfes. Acta Soc. bot. pol. **26**, 515—549 (1957) [Polnisch]. — LANGE, A. H.: Factors affecting sex changes in the flowers of *Carica papaya* L. Proc. Amer. Soc. Hort. Sci. **77**, 252—264 (1961). — LASKOWSKA, R.: Influence of the age of pollen and stigmas on sex determination in hemp. Nature (Lond.) **192**, 147—148 (1961). — LAWRENCE, W. J. C.: Studies on *Streptocarpus* Lindl. V. Speciation and gene systems. Heredity **12**, 333—356 (1958). — LEBEDEFF, G. A.: Genetics of hermaphroditism in *Drosophila viridis*. Proc. nat. Acad. Sci. (Wash.) **20**, 613—616 (1934). — LEMAN, A.: Über Unregelmäßigkeiten der Meiose bei Zwittern von *Melandrium rubrum*. Ber. dtsch. bot. Ges. **83**, 167 (1960). — LESLEY, M. M., and J. W.: Male-sterile mutants in the tomato from seed treated with the radioactive isotope P^{32}. Proc. Amer. Soc. Hort. Sci. **71**, 339—343 (1958). — LESTER, D. T.: Variation in sex expression in *Populus tremuloides* Michx. Silvae Genetica **12**, 141—151 (1963). — LEWIS, D.: Male sterility in natural populations of hermaphrodite plants. New Phytologist **40**, 50—63 (1941). — The evolution of sex in flowering plants. Biol. Rev. Cambridge philos. Soc. **17**, 46—47 (1942). — LEWIS, D., and L. K. CROWE: The genetics and evolution of gynodioecy. Evolution **10**, 115—125 (1956). — LILIENFELD, F.: Karyologische und genetische Studien an *Fragaria*. I. Ein tetraploider fertiler Bastard zwischen *F. nipponica* (n = 7) und *F. elatior* (n = 21). Jap. J. Bot. **6**, 425—458 (1933). — III. Geschlechtsverhältnisse in den F_2- und Folgegenerationen des Bastards zwischen der getrenntgeschlechtigen *F. elatior* und der zwittrigen *F. nipponica*. Mem. Coll. Agric. Kyoto Univ. **38**, 1—58 (1936). — LIMBERK, J.: Versuche zur Änderung des Geschlechts beim Hopfen *(Humulus lupulus)*. Čs. Biol. **2**, 306—313 (1953) [Russisch]. — Änderung des Geschlechts beim Hopfen *(Humulus lupulus)*. Čs. Biol. **3**, 252—255 (1954) [Russisch]. — LINDSAY, R. H.: The chromosomes of some dioecious angiosperms. Proc. nat. Acad. Sci. (Wash.) **15**, 611—613 (1929). — The chromosomes of some dioecious angiosperms. Amer. J. Bot. **17**, 152—174 (1930). — LINNERT, G.: Geschlechtsvererbung. Eine Übersicht über die wesentlichen genetischen Grundlagen. Z. Bot. **45**, 77—96 (1957). — Kerngesteuerte Gynodiözie bei *Salvia nemorosa*. Z. Vererb.-Lehre **89**, 36—51 (1958). — LÖVE, Á.: Études cytogénétiques des *Rumex*. II. Bot. Not. (Lund), 1941, 155—172 (1941b). — Cytogenetic studies on *Rumex subgenus Acetosella*. Hereditas (Lund) **30**, 1—136 (1944). — Sex determination in *Rumex*. Proc. Genet. Soc. Canada **2**, 31—36 (1957). — LÖVE, D.: A tetraploid, dioecious *Melandrium* found in nature. Hereditas (Lund) **28**, 241—242 (1942a). — Intersexuality in *Melandrium rubrum* probably caused by a translocation between the sex chromosomes. Hereditas (Lund) **28**, 497—498 (1942b). — Cytogenetic studies on dioecious *Melandrium*. Bot. Not. (Lund) 1944, 125—213 (1944). — LÖVE, Á., and D. LÖVE: Experimental sex reversal in plants. Svensk bot. T. **34**, 248—252 (1940). — LÖVE, Á., and N. SARKAR: Cytotaxonomy and sex determination in *Rumex paucifolius*. Canad. J. Bot. **34**, 261—268 (1956). — LOOMIS, N. H., C. F. WILLIAMS and M. M. MURPHY: Inheritance of flower types in muscadine grapes. Proc. Amer. Soc. horticult. Sci. **64**, 279—283 (1954). — LORBEER, G.: Die Distanzkonjugation der total heterochromatischen Geschlechtschromosomen im triploiden Sporogon von *Sphaerocarpus Donnellii*. Ber. dtsch. bot. Ges. **54**, 98—123 (1936). — LORZ, A. P.: Cytological investigations on five chenopodiacious genera with special emphasis on chromosome morphology and somatic doubling in *Spinacia*. Cytologia (Tokyo) 8, 241—276 (1937). — LÜNING, K. G.: Studies on induced mutations in male and female germ line of *Drosophila melanogaster*. Hereditas (Lund) **42**, 483—486 (1956).

MACKAY, E.: Sex chromosomes of *Cannabis sativa*. Amer. J. Bot. **26**, 707—708 (1939) (zit. bei KÖHLER 1958). — MAEKAWA, T.: Widerstands- und Selbstregulierungsvermögen gegen Geschlechtsänderung bei Hanfpflanzen und seine Beziehung zur Theorie der Geschlechtsbestimmung. Jb. wiss. Bot. **70**, 512—564 (1929) (zit. bei KÖHLER 1958). — MAHONEY, D. L., J. JANICK, and E. C. STEVENSON: Sex determination in diploid-triploid crosses of *Spinacia oleracea*. Amer. J. Bot. **46** (5), 372—375 (1959). — MAINX, F.: Ein neuer Modus der genotypischen Geschlechtsbestimmung. Biol. Zbl. **81**, 335—340 (1962). — MAILING, H.: The chromosome number of *Honckenya peploides* (L.) Ehrh. with a note on its mode of sex determination. Hereditas (Lund) **43**, 517—524 (1957). — MANGELSDORF, A. J., and E. M. EAST: Studies on the genetics of *Fragaria*. Genetics **12**, 307—339 (1927). — MARQUARDT, H.: Die Deutung eines Falles plasmatischer Vererbung bei *Streptocarpus* als Kern-Plasma-Operonsystem. 1. Vorl.-Notiz. 2. Hauptarbeit. Biol. Zbl. **82**, 489—490 (1963); **83**, 1—18 (1964). — MCCLUNG, C. E.: The accessory chromosome — sex-determinant? Biol. Bull. **3**, 43—84 (1902). — MCPHEE, H. C.: The influence of environmental on sex in hemp, *Cannabis sativa* L. J. agric. Res. **28**, 1067—1080 (1924). — The genetics of sex in hemp. J. agric. Res. **31**, 935—943 (1925). — MEURMAN, O.: The chromosome behaviour of some dioecious plants and their relatives with special reference to the sex chromosomes. Soc. Sci. fenn., Comm. Biol. **2**, 1—104 (1925). — MICHAELIS, P.: Interactions between genes and cytoplasm in *Epilobium*. Cold Spr. Harb. Symp. quant. Biol. **16**, 121—129 (1951). — MITCHELL, W. D., and S. H. WITTWER: Chemical regulation of flower sex expression and vegetative growth in *Cucumis sativus* L. Science **136**, 2, 880 (1962). — MOSTOVOJ, K.: Konstitutionssterilität bei der Paprika (*Capsicum annuum* L.). Rozn.**16** (3), 308—316, Sbornik (1941) [Tschechisch]. — MOUTSCHEN, J., et M. DAHMEN: Intersexualité induite par les rayons gamma chez *Cannabis sativa* L. Bull. Soc. roy. bot. Belg. **87**, 21—28 (1955). — MÜNTZING, A.: The chromosome of a giant *Populus tremula*. Hereditas (Lund) **21**, 383—393 (1936). — MULLER, H. J.: Why polyploidy is rarer in animals than in plants. Amer. Naturalist **59**, 346—353 (1925). — MURRAY, M. J.: An F_2 study of interspecific characters. Genetics **25**, 128 (1940a). — Colchicine-induced tetraploids in dioecious and monoecious species of *Amaranthaceae*. J. Hered. **31**, 477—485 (1940b). — The genetics of sex determination in the family *Amaranthaceae*. Genetics **25**, 409—431 (1940c).

NAKAJIMA, G.: Cytological studies in some dioecious plants. Cytologia (Tokyo) (Fujii Jub.-Vol.) Part I, 282—292 (1937). — NEGODI, G.: Ricerche sulla distribuzione e transmissione dei sessi in *Urtica caudata* Vahl. Nuovo Giorn. bot. ital. **36**, 60—126 (1929). — NEGRUL, A. M.: The genetic basis of grape breeding. Bull. Appl. Bot., Genet. Plant Breed. (URSS.) 8, 1—149 (1936). — NEUER, H., u. R. v. SENGBUSCH: Die Geschlechtsvererbung bei Hanf und die Züchtung eines monözischen Hanfes. Züchter **15**, 49—62 (1943). — NEVE, R. A.: Sex chromosomes in the hop *Humulus lupulus*. Nature (Lond.) **181**, 1084—1085 (1958). — NICOLAISEN, N., u. R. HANOW: Bestimmung der Geschlechtsverhältnisse bei Spinat. Z. Pflanzenzüchtg **23**, 476—484 (1941). — NIGTEVECHT, G. VAN: Genetic studies in dioecious *Melandrium*. I. Sex-linked and sex-influenced inheritance in *Melandrium album* and *Melandrium dioicum*. Genetica (s'Grav.) **37**, 281—306 (1966a). — Genetic studies in dioecious *Melandrium*. II. Sex determination in *Melandrium album* and *Melandrium dioicum*. Genetica (s'Grav.) **37**, 307—344 (1966b). — NILSSON-EHLE, H.: Über eine in der Natur gefundene Gigasform von *Populus tremula*. Hereditas (Lund) **21**, 379—382 (1936). — NOGUCHI, Y.: On polyploidy in spinach. Jap. J. Genet. **19**, 106—108 (Pl.Br.Abs. 21, 3076) (1943). — NÜRNBERG-KRÜGER, U.: Genetische Untersuchungen an diploiden *Fragaria*-Arten. I. Die Kreuzung zwischen *Fr. vesca* und *Fr. nilgerrensis*, ihre Auswirkung auf die Morphologie der Pflanzen und die Ausbildung des Geschlechts. Z. Vererb.-Lehre **89**, 747—773 (1958). — NYGREN, A.: Polyploids in *Melandrium* induced by nitrous oxide. Hereditas (Lund) **41**, 287—290 (1955). — Studies in polyploid *Melandrium album* produced by nitrous oxide, N_2O. Kgl. Lantbrukshögsk. Ann. **23**, 393—404 (1957a). — Experimental studies in Scandinavian alpine plants. III. Intra- and interspecific crosses in *Melandrium furcatum* (Ref.) Hultén subsp. *angustifolium* (Ruprecht) Walpers. Kgl. Lantbrukshögsk. Ann. **23**, 405—411 (1957b). — A fertile hybrid *Lychnis floscuculi* and *Melandrium rubrum* and its sex segregating offspring. Kgl. Lantbrukshögsk. Ann. **23**, 413—421 (1957c).

OEHLKERS, F.: Bastardierungsversuche in der Gattung *Streptocarpus* Lindley. I. Plasmatische Vererbung und die Geschlechtsbestimmung von Zwitterpflanzen. Z. Bot. **32**, 305—393 (1938). — III. Neue Ergebnisse über die Genetik von Wuchsgestalt und Geschlechtsbestimmung. Ber. dtsch. bot. Ges. **58**, 76—91 (1940). — IV. Weitere Untersuchungen über Plasmavererbung und Geschlechtsbestimmung. Z. Bot. **37**, 158—182 (1941). — Neue Überlegungen zum Problem der außerkaryotischen Vererbung. Z. indukt. Abstamm.- u. Vererb.-Lehre **84**, 213—250 (1952). — Bastardierungsversuche in der Gattung *Streptocarpus* Lindl. V. Phänotypische und genotypische Geschlechtsbestimmung bei dem Bastard *Streptocarpus* (Wendlandii × Rexii). Z. indukt. Abstamm.- u. Vererb.-Lehre **87**, 722—734 (1956). — IX. Neue Untersuchungen über die Geschlechtsbestimmung bei *Streptocarpus*. Z. Vererbungsl. **92**, 445—456 (1961). — Cytoplasmic inheritance in the genus *Streptocarpus* Lindl. Advanc. Genet. **12**, 329—370 (1964). — OEHLKERS, F., u. U. EBELL: VII. Die genetischen und plasmonischen Grundlagen für die Pollensterilität verschiedener *Streptocarpus*-Bastarde. Z. Vererbungsl. **89**, 559—586 (1958). — ONO, T.: Größenverhältnisse der Geschlechtschromosomen von *Rumex acetosa* L. Sci. Rep. Tôhoku Imp. Univ. **4**, 159—160 (1926). — Further investigations on the cytology of *Rumex*. I.—V. Bot. Mag. (Tokyo) **44**, 168—176 (1930a). — Chromosomenmorphologie von *Rumex acetosa*. Sci. Rep. Tôhoku Imp. Univ., 4. Ser. **5**, 415—422 (1930b). — Polyploidy in *Rumex acetosa*. Bot. Mag. (Tokyo) **46**, 259—260, 321—326 (1932). — Chromosomen und Sexualität von *Rumex acetosa*. Sci. Rep. Tôhoku Imp. Univ. **4**, 10, 41—210 (1935). — On sex-chromosomes in wild hops. Bot. Mag. (Tokyo) **51**, 110—115 (1937). — Polyploidy and sex determination in *Melandrium*. Bot. Mag. (Tokyo) **53**, 549—556 (1939). — Polyploidy and sex determination in *Melandrium*. II. Bot. Mag. (Tokyo) **54**, 225—230 (1940). — ONO, T., and N. SHIMOTOMAI: Triploid and tetraploid intersexes of *Rumex acetosa* L. Bot. Mag. (Tokyo) **42**, 266—270 (1928).

PALIWAL, R. L., and B. B. HYDE: The association of a single B-chromosome with male sterility in *Plantago*. Proc. internat. Congr. Genet. Montreal **2**, 212 (1958). — PARSONS, P. A.: Evolution of sex in the flowering plants of South Australia. Nature (Lond.) **181**, 1673—1674 (1958). — PEIRCE, L. C., and T. M. CURRENCE: The inheritance of hermaphroditism in *Asparagus officinalis* L. Proc. Amer. Soc. Hort. Sci. **80**, 368—376 (1962). — PETERSON, C. E., and LA MOINE D. ANHDER: Induction of staminate flowers on gynoecious cucumbers with gibberellin A_3. Science **131**, 1673—1674 (1960). — PETIT, J.: Sur le détermination du sexe chez *Cannabis sativa* dans les conditions experimentales du phytotron de Liège. Bull. Soc. roy. Sci. Liège **21**, 464—476 (1952). — PETO, F. H.: Cytology of poplar species and natural hybrids. Canad. J. Res., Sect. C **16**, 445—455 (1938). — PETROV, D. F., and D. A. TUKAN: On the heterozygosity of the male sex in *Fragaria orientalis* Los. C. R. Acad. Sci. URSS, N. s. **16**, 459—460 (1937).

RAEBER, J. G., and A. BOLTON: A new form of male sterility in *Nicotiana tabacum* L. Nature (Lond.) **176**, 314—315 (1955). — RAINIO, A. J.: Über die Intersexualität bei der Gattung *Salix*. Ann. Soc. zool. bot. fenn. Vanamo **5**, 165—275 (1927). — REIMANN- PHILIPP, R.: Cytologische Untersuchungen zur Identifizierung der Geschlechtschromosomen bei Spargel (*Asparagus off.*). Bericht 2. Kongr. „Eucarpia" 1959. — REIMANN-PHILIPP, R., B. ZILM u. H. ERESEN: Cytologische Untersuchungen zur Identifizierung der Geschlechtschromosomen bei *Asparagus officinalis* L. in Verbindung mit der Züchtung rein männlicher Sorten. Z. Pflanzenzüchtg **42**, 295—304 (1960). — REINER: Zit. nach H. BREIDER u. H. SCHEU, Gartenbauwiss. **11** (1938) (1910). — RESENDE, F., and H. F. LINSKENS: On hermaphrodite and female plants in *Bryophyllum daigremontianum* and its segregation. I. Port. Acta biol. A **6**, 155—178 (1961). — RESENDE, F., and M. J. VIANA: Gibberellin and sex expression. Port. Acta biol. A **6**, 77—98 (1959). — RHOADES, M. M.: Meiosis in maize. Heredity **41**, 59—67 (1950). — Chromosomes, mutations and cytoplasm in maize. Science **120**, 115—120 (1954). — RICHARDSON, C. W.: A preliminary note on the genetics of *Fragaria*. J. Genet. **3**, 171—177 (1914). — RICK, C. M., and G. C. HANNA: Determination of sex in *Asparagus officinale* L. Amer. J. Bot. **30**, 711—714 (1943). — ROBBINS, W. W., and H. A. JONES: Secondary sex characters in *Asparagus officinalis* L. Hilgardia **1**, 183—202 (1925). — ROSA, J. T.: Sex expression in spinach. Hilgardia **1**, 258—274 (1925). — RUNQUIST, E. W.: Ett fall av androgyna hängen hos *Populus tremula* L. Bot. Not. (Lund) 1951, 188—191 (1951).

SAEZ, F.: Gradient of the heterochromatization in the evolution of the sexual system. "Neo-X—Neo-Y". Port. Acta biol. A **7**, 111 (1963). — SANSOME, F. W.: Sex determination in *Silene otites* and related species. J. Genet. **35**, 387—396 (1938). — SANTOS, J. K.: Differentiation among chromosomes in *Elodea*. Bot. Gaz. **75**, 42—59 (1923). — SAVITSKY, V. F.: Differences in interaction of cytoplasm-genes in different male-sterile races in *Beta vulgaris* L. USA, Utah, Salt Lake City, Beet Sugar Development Foundations and US Department of Agriculture. Proc. internat. Congr. Genet. Montreal **2**, 251 (1958). — SCHAFFNER, J. H.: Observations on the sexual state of various plants. Ohio J. Sci. **23**, 149—159 (1923). — SCHIEMANN, E.: Geschlechts- und Artkreuzungsfragen bei *Fragaria*. Jena:

Fischer 1931. — Artkreuzungen bei *Fragaria*. II. Z. indukt. Abstamm.- u. Vererb.-Lehre **73**, 375—390 (1937). — Artkreuzungen bei *Fragaria*. III. Die *vesca*-Bastarde. Flora (Jena), N. F. **37**, 166—192 (1943). — Schulz, A.: Beiträge zur Kenntnis der Bestäubungseinrichtungen und Geschlechtsverteilung bei den Pflanzen. II. Bibl. Bot. **17**, 182—196 (1890). — Schwartz, D. (zit. bei Rhoades): The interaction of nuclear and cytoplasmic factors in the inheritance of male sterility in maize. Genetics **36**, 676—696 (1951). — Seitz, W.. Über das Auftreten von Triploiden nach der Selbstung anomaler Zwitterblüten einer Graupappelform. Z. Forstgenetik u. Forstpflanzenzüchtg **3**, 1—6 (1954). — Sengbusch, R. v.: Beitrag zum Geschlechtsproblem bei *Cannabis sativa*. Vorl. Mitt. Z. Vererb.-Lehre **80**, 616—618 (1942). — Ein weiterer Beitrag zur Vererbung des Geschlechts bei Hanf als Grundlage für die Züchtung eines monözischen Hanfes. Z. Pflanzenzüchtg **31**, 319—338 (1952). — Shifriss, O.: Sex instability in *Ricinus*. Genetics **41**, 265—280 (1956). — Sex expression in *Ricinus*. Genetics **42**, 394 (1957). — Conventional and unconventional systems controlling sex variations in *Ricinus*. J. Genet. **57**, 361—388 (1961). — Shifriss, O., W. L. George jr., and J. A. Quinones: Gynodioecism in cucumber. Genetics **49**, 285—291 (1964). — Shull, G. H.: Inheritance of sex in *Lychnis*. Bot. Gaz. **49**, 110—125 (1910b). — Reversible sex-mutants in *Lychnis dioica*. Bot. Gaz. **52**, 329—368 (1911). — Sexlimited inheritance in *Lychnis dioica* L. Z. indukt. Abstamm.- u. Vererb.-Lehre **12**, 265—302 (1914a). — Sinoto, Y: On chromosome behaviour and sex determination in *Rumex acetosa* L. Bot. Mag. (Tokyo) **38**, 153—162 (1924). — On the tetrapartite chromosome in *Humulus lupulus*. Proc. Imp. Acad. (Tokyo) **5**, 46—47 (1929a). — Chromosome studies in some dioecious plants, with special reference to the allosomes. Cytologia (Tokyo) **1**, 109—191 (1929b). — Skovstedt, A.: Some colchicine experiments with hops. Hereditas (Lund) **39**, 156—160 (1953). — Smith, B. W.: Notes on the cytology and distribution of *Dioscoreaceae*. Bull. Torrey bot. Club. **64**, 189—197 (1937). — Sex chromosomes and natural polyploidy in dioecious *Rumex*. J. Hered. **46**, 226—232 (1955). — Chromosome behaviour and sex expression in induced polyploids of dioecious *Rumex hastatulus*. Genetics **42**, 345—396 (1957). — Polyploidy hybridization and sexuality in dioecious *Rumex hastatulus* and *R. paucifolius*. Proc. internat. Congr. Genet. Montreal **2**, 264 (1958). — The mechanism of sex determination in *Rumex hastatulus*. Genetics **48**, 1265 (1963). — The evolving karyotype of *Rumex hastatulus*. Evolution (Lawrence, Kansas) **18**, 93—104 (1964). — Smith, B. W., and M. Th. Smith: Sex chromosomes in *Rumex hastatulus* Baldw. with XY_1Y_2 pairing (Abstr.). Genetics **32**, 104—105 (1947). — Sneep, J.: The significance of andromonoecy for the breeding of *Asparagus officinalis* L. Euphytia **2**, 89—95 (1953). Zit. bei Reimann-Philipp 1962. — Spinat. In: Handbuch der Pflanzenzüchtung (Hrsg. Roemer-Rudorf), 2. Aufl., Bd. VI, S. 227ff. Berlin u. Hamburg: Paul Parey 1962. — Sprague, G. F., u. A. Tavcar: Mais *(Zea Mays)*. In: Handbuch der Pflanzenzüchtung (Hrsg. Roemer-Rudorf), 2. Aufl., Bd. II, S. 103ff. Berlin u. Hamburg: Paul Parey 1959. — Staudt, G.: Cytogenetische Untersuchungen an *Fragaria orientalis* Los. und ihre Bedeutung für Artbildung und Geschlechtsdifferenzierung in der Gattung *Fragaria* L. Z. indukt. Abstamm.- u. Vererb.-Lehre **84**, 361—416 (1952). — Die Vererbung des Geschlechts bei der tetraploiden diözischen *Fragaria orientalis* Los. Vorl. Mitt. Ber. dtsch. bot. Ges. **67**, 385—387 (1954). — Die Entstehung und Geschichte der großfrüchtigen Gartenerdbeeren *Fragaria* × *ananassa* Duch. Züchter **31**, 212—218 (1961). — Taxonomic studies in the genus *Fragaria*. Typification of *Fragaria* species known at the time of Linnaeus. Canad. J. Bot. **40**, 869—886 (1962). — Storey, W. B.: Segregation of sex types in *Solo papaya* and their application to the selection of seed. Proc. Amer. Soc. horticult. Sci. **35**, 83—85 (1938). — Genetics of the *papaya*. J. Hered. **44**, 70—78 (1953). — Strasburger, E.: Über geschlechtsbestimmende Ursachen. Jb. wiss. Bot. **48**, 427—520 (1910). — Sturtevant, A. H.: A gene in *Drosophila melanogaster* that transforms females into males. Genetics **30**, 297—299 (1945).— Sugimoto, Y.: Studies on the breeding of spinach. 1. On the sex expression [Japanese]. J. Hort. Ass. Jap. **16**, 203—210 (1947). — Studies on the breeding of spinach. 2. Sex expression and genetical explanation. Hort. Ass. Jap. J. **17**, 77—83 (1948). — Suto, T., and S. Sugiyama: On the genetic modification on the sex ratio in spinach [Japanese Abstr.]. Jap. J. Breed **7**, (Suppl.), 11 (1957a). — Sex inheritance in spinach, with special reference to the existence of YY males [Japanese Abstr.]. Jap. J. Genet. **33**, 325 (1958). — Sex expression and determination in spinach. 1. Growth habit and its sex-limited inheritance. Jap. J. Bot. **17**, 163—176 (1960). — Sex chromosomes of Japanese spinach. Agr. and Hort. (Tokyo) **36**, 407—408 (1961a) [Japanisch]. — 2. Inheritance and breeding of intersexuality. Jap. J. Breed. **10**, 215—222 (1961b). — 3. Inheritance and breeding of intersexuality (continued). Jap. J. Breed. **11**, 29—36 (1961c). — Sex determination and expression in spinach. Chromosome morphology and behaviour in diploid plants. Chromosoma (Berl.) **51/52**, 1718—1733 (1962).

Tabushi, J.: Trisomics of spinach. Biol. Abstr. **34**, 6161 (1958). — Takenaka, Y.: On the sex chromosomes of *Rumex montanus* Desf. Bot. Mag. (Tokyo) **44**, 176—184 (1930). — Cytogenetic studies on sex in *Cannabis sativa* L. II. Meiosis in diploid, triploid and tetraploid males. Ann. Rep. Nat. Inst. Genet. Jap. **3**, 48—49 (1953). — THOMAS, R. G.: Sexuality in diploid and hexaploid races of *Mercurialis annua* L. Ann. Bot. **22**, 55 (1958).

UBISCH, G. v.: Abweichungen vom mechanischen Geschlechtsverhältnis bei *Melandrium dioicum*. Biol. Zbl. **42**, 112—118 (1922). — Selbstfertilität und Geschlechtsverhältnis bei *Antennaria dioica* (Gaertn.) Biol. Zbl. **52**, 307—312 (1932). — Genetic studies on the nature of hermaphrodite plants in *Antennaria dioica* (L.). Gaertn. Genetics **21**, 282—294 (1936).

VALLEAU, W. D.: Inheritance of sex in the grape. Amer. Naturalist **50**, 554—564 (1916). — The inheritance of flower types and fertility in the strawberry. Amer. J. Bot. **10**, 259—274 (1923).

WAGNER, E.: Die Feststellung des Geschlechts bei Rebensämlingen. Vitis (Landau) **2**, 190—197 (1960). — WARMKE, H. E.: A new method for determining the sex heterozygote with morphologically undifferentiated sex chromosomes and its application to *Silene otites*. Genetics **27**, 174 (1942). — Sex determination and sex balance in *Melandrium*. Amer. J. Bot. **33**, 648—660 (1946a). — An analysis of male development in *Melandrium* by means of Y-chromosome deficiencies. Genetics **31**, 234—235 (1946b). — A study of spontaneous breakage of the Y-chromosome in *Melandrium*. Amer. J. Bot. **33**, 224 (1946c). — WARMKE, H. E., and A. F. BLAKESLEE: Effect of polyploidy upon the sex mechanism in dioecious plants. Genetics **24**. 88 (1939a). — Sex mechanism in polyploids of *Melandrium*. Science **89**, 391—392 (1939b). — The establishment of an 4n dioecious race in *Melandrium*. Amer. J. Bot. **27**, 751—762 (1940a). — Polyploidy and the sex mechanism in *Melandrium*. Genetics **25**, 139 (1940b). — Further differences in the determination of sex in *Melandrium* and *Drosophila*. Collecting Net **15**, 174 (1940c). — WESTERGAARD, M.: Induced tetraploidy in *Melandrium album*. Nature (Lond.) **142**, 917 (1938). — Studies on cytology and sex determination in polyploid form of *Melandrium album*. Dansk. bot. Ark. **10**, 1—131 (1940). — Structural changes of the Y chromosome in the offspring of polyploid *Melandrium*. Hereditas (Lund) **32**, 60—64 (1946a). — Aberrant Y chromosomes and sex expression in *Melandrium album*. Hereditas (Lund) **32**, 419—443 (1946b). — The relation between chromosome constitution and sex in the offspring of triploid *Melandrium*. Hereditas (Lund) **34**, 257—279 (1948). — Über den Mechanismus der Geschlechtsbestimmung bei *Melandrium album*. Naturwiss. **40**, 253—260 (1953). — The mechanism of sex determination in dioecious flowering plants. Advanc. Genetics **9**, 217—281 (1958). — WESTON, E. W.: Changes in sex in the hop caused by plant growth substances. Nature (Lond.) **188**, 81—82 (1960). — WETTSTEIN, F. v.: Über Fragen der Geschlechtsbestimmung bei Pflanzen. Naturwiss. **12**, 761—768 (1924). — Untersuchungen zur plasmatischen Vererbung. I. *Linum*. Biol. Zbl. **65**, 149 (1946). — WILSON, E. B.: Recent researches on the determination and heredity of sex. Science, N. s. **29**, 53—70 (1909). — The chromosomes in relation to the determination of sex. Sci. Progr. **16**, 570 (1910). — WINGE, Ø.: On sex chromosomes, sex determination and preponderance of females in some dioecious plants. C. R. Lab. Carlsberg **15**, 1—26 (1923). — On a Y-linked gene in *Melandrium*. Hereditas (Lund) **9**, 274—284 (1927). — On the nature of the sex chromosomes in *Humulus*. Hereditas (Lund) **12**, 53—63 (1929a). — Critical remarks to Y. SINOTO's paper on a tetrapartite sex chromosome complex in *Humulus*. Hereditas (Lund) **12**, 269—270 (1929b). — X- and Y-linked inheritance in *Melandrium*. Hereditas (Lund) **15**, 127—165 (1931). — The nature of sex chromosome. Proc. 6. Internat. Congr. Gen. **1**, 343—355 (1932). — The experimental alteration of sex chromosomes into autosomes and vize versa, as illustrated by *Lebistes*. C. R. Lab. Carlsberg **21**, 1—49 (1934). — WINGE, Ø., and O. LAUSTSEN: On a cytoplasmatic effect of inbreeding in homozygous yeast. C. R. Lab. Carlsberg (Ser. physiol.) **23**, 17—40 (1940). — WITSCHI, E.: Genic and morphologic structure of sex chromosomes. Proc. internat. Genet. Symp. 1956 (Suppl.-Vol. zu Cytologia) 133—138 (1957). — Genetic and postgenetic sex determination. Experientia (Basel) **16**, 274—278 (1960).

YAMADA, I.: The sex-chromosomes of *Cannabis sativa* L. Rep. Kihara Inst. Biol. Res. **2**, 64—68 (1943). — YAMAMOTO, Y.: Karyotypes in *Rumex acetosa* L. and their geographical distribution. Jap. J. Gen. 8, 264—274 (1933b). — Karyogenetische Untersuchungen bei der Gattung *Rumex*. I. Hetero- und Euploidie bei *Rumex acetosa* L. Cytologia (Tokyo) **5**, 317—336 (1934b). — Geschlechtsbestimmung bei eu- und aneuploiden Pflanzen. Mem. Coll. Agric. Kyoto Imp. Univ. **43**, 1—59 (1938). — YAMPOLSKY, C.: Inheritance of sex in *Mercurialis annua*. Amer. J. Bot. **4**, 410 (1919). — YAMPOLSKY, C., and H. YAMPOLSKY: Distribution of sex forms in the phanerogamic flora. Bibl. genet. **3**, 1—63 (1922). — YUICHI, S.: Artificial control of sex differentiation in Japanese red pine and black pine strobiles. J. Fac. Agric., Tottori Univ. **3**, 1—29 (1957).

ZIEGLER, A.: Die Rebenzüchtung in Bayern. Bretten: Fr. Esser 1923. — ZOSCHKE, U.: Untersuchungen über die Bestimmung des Geschlechts beim Spinat (*Spinacea oleracea* L.) unter besonderer Berücksichtigung der Züchtung eines monözischen oder gleichzeitig schossenden Spinats. Z. Pflanzenzüchtg **35**, 257—296 (1956).

Phänotypische Geschlechtsbestimmung bei Algen.

Von

Kurt Köhler.

Mit 4 Abbildungen.

Bei einer großen Anzahl von Organismen im Tier und Pflanzenreich ist die Ausprägung des Geschlechtes offenbar nicht durch genetische Faktoren, sondern durch andere physiologische Verhältnisse oder (ökologische) Außenbedingungen verursacht. Bestimmter ausgedrückt würde das bedeuten, daß hier nicht die im genetischen Experiment erfaßbaren Geschlechtsrealisatoren, wie sie der haplogenotypischen Geschlechtsbestimmung zugrunde gelegt wurden, in der Meiose auf die Gonen verteilt werden, sondern in allen Stufen des Entwicklungscyclus gegenwärtig sind, die Organismen daher als genetische Zwitter angesehen werden müssen (s. FÖRSTER, S. 31 ff., in diesem Band). Es würde demnach hier gefordert werden, daß bei den hier zu betrachtenden Algen, die in der Haplo- oder Diplophase die Potenz besitzen, beide Sorten von Geschlechtsorganen oder Geschlechtszellen zu bilden, jeweils beide Geschlechtsfaktoren vorhanden seien. Diese Art der Geschlechtsbestimmung wird als phänotypische oder modifikatorische Geschlechtsbestimmung bezeichnet. Die entscheidenden Faktoren, die einen ganzen Organismus, einen Teil desselben oder auch nur eine einzelne Zelle in weiblicher oder männlicher (bzw. Plus- oder Minus-) Richtung differenzieren, können Umwelts- und Kulturbedingungen oder innere physiologische Zustände wie Alter oder Lage eines Teiles im Gesamtorganismus usw. sein. Während bei den Metazoen und den höheren Pflanzen genauere genetische und physiologische Analysen vorliegen, sind unsere Kenntnisse bei den Algen und Pilzen bescheiden.

Am besten wird die phänotypische Geschlechtsbestimmung zwar durch die gemischtgeschlechtlichen oder monöcischen Algen repräsentiert. Darüber hinaus wurde aber auch bei getrenntgeschlechtlichen Algen, bei denen man zunächst eine genotypische Determination vermutet hatte, phänotypische Geschlechtsbestimmung experimentell nachgewiesen. Entscheidend für die sekundäre Ausprägung nur eines Geschlechtes ist vor allem der Zeitpunkt in der Entwicklung, an dem durch entsprechende Außenbedingungen die Weiche gestellt wird. Dies kann sehr spät geschehen, so daß z. B. erst die Gametangien bzw. die Geschlechtszellen in einer oder der anderen Richtung differenziert sind. Eine weitere Möglichkeit ist aber, daß bereits die keimende Spore oder die junge Pflanze beeinflußt würde. Schließlich gäbe es noch den Weg der progamen Geschlechtsbestimmung, bei dem z. B. schon die unbefruchteten Eier entsprechend ihrer Größe zu Weibchen- bzw. Männcheneiern determiniert sind (solche Fälle sind bisher nur bei Metazoen bekannt). HARTMANN hat in seinem Buch „Die Sexualität" die Grundlagen der phänotypischen Geschlechtsbestimmung erläutert und durch zahlreiche Beispiele ausführlich dargestellt. Eine Übersicht über die Sexualitätsphänomene bei den Algen hatte schon Ende der 20er Jahre KNIEP (1928) geschaffen. Es wird hier der Versuch gemacht werden, diese Frage aus dem Blickwinkel der Pflanzenphysiologie erneut zu betrachten.

A. Chlorophyceen.

Besonders übersichtlich und leicht zu entscheiden sind die Verhältnisse bei den reinen Haplonten, vor allem bei den einzelligen *Volvocales* und den Conjugaten. Im allgemeinen hat dort jede Zelle die Fähigkeit, zum Gameten zu werden. Die Weiche wird also für die einzelne Geschlechtszelle alternativ gestellt. Die Reduktionsteilung findet in der Zygote statt. Für den Modus der Geschlechtsbestimmung ist daher der entscheidende Nachweis aus einer Kultur der Gonen bzw. der „keimenden Zygoten" zu erbringen. Experimentell ergeben sich daher folgende Möglichkeiten: 1. Aus dem einen (nach der Reduktionsteilung überlebenden Gonen-)Keimling entwicklen sich gemischtgeschlechtliche Pflanzen oder Klone. 2. Werden 4 oder mehr Gonen aus der Zygote gebildet, so sollte ihr Geschlecht nicht nach der Regel festgelegt sein und möglichst vom Verhältnis 2:2 abweichen. 3. Eine Aufzucht von Gameten auf parthenogenetischem Wege zu Pflanzen oder Klonen sollte keine Geschlechtskonstanz aufweisen. Die Frage wird hier also durch die Nachkommen beantwortet. Es sollen im folgenden von den Algen, bei denen offenbar haplophänotypische Geschlechtsbestimmung vorzuliegen scheint, einige Beispiele kurz genannt werden; nur der Gattung *Vaucheria* soll anschließend ein etwas größerer Raum gewidmet werden. Unter den Grünalgen beobachtete SCHULZE (1927) bei der Volvocinee *Stephanosphaera pluvialis* Kopulationen in seinen Klon-Kulturen. Jede von seinen 14 geprüften Kulturen erbrachte gleiche Ergebnisse. Obwohl aus der Gattung *Chlamydomonas* überwiegend diöcische Arten bekannt sind, wurde vereinzelt auf Ausnahmen aufmerksam gemacht. Eine von STREHLOW (1929) aufgefundene und kultivierte monöcische *Chlamydomonas (= monoica)* zeigte in 12 Klonen Kopulationen bzw. Zygoten. Von den anderen von STREHLOW untersuchten Arten sind *Chlamydobotrys gracilis* und *stellata* sowie *Polytoma ulvella* monöcisch. Diese Reihe ließe sich für die einzelligen *Volvocales* beliebig verlängern; genetische oder physiologische Versuche liegen jedoch nicht vor, die nähere Beweise für nicht genotypische Geschlechtsbestimmung bringen würden.

In der Gattung *Volvox*, deren Arten allgemein diöcisch sind, galt *V. globator* als der offenbar einzige monöcische Vertreter. Jedoch sind nach MAINX (1929) auch bei *V. aureus* Kolonien mit Oogonien *und* Antheridien nicht selten, womit er zufällige Beobachtungen einiger älterer Autoren bestätigt. Auch bei *Eudorina* kommen beide Geschlechtsorgane gelegentlich im gleichen Coenobium vor. TIFFANI (1935) fand *Pleodorina californica* während des frühen Sommers getrenntgeschlechtlich vor; später nach Ansteigen der Temperatur und erhöhtem p_H traten Zwitter auf. *Hydrodictyon*, das Wassernetz ist ebenfalls, wie schon KLEBS (1896) gefunden hatte, gemischtgeschlechtlich (MAINX 1931 a). Monöcisten finden sich weiter bei der Gattung *Oedogonium* (MAINX 1931 b), bei den Characeen und den Konjugaten. HARTMANN (1913, unveröffentlicht) konnte in einem Druck- und Temperaturexperiment bei einem zwittrigen *Oedogonium* die Ausbildung von Oogonien unterdrücken. Bei *Oe. kurzii* erscheinen die Antheridien gewöhnlich einen Tag später als die Oogonien, so daß man hier bei flüchtiger Beobachtung neben zwittrigen rein weibliche Pflanzen vorfinden würde (SPESSARD 1930). Man hat nun vielfach für solche Spirogyren, bei denen leiterförmige Kopulation erfolgt, genotypische Getrenntgeschlechtlichkeit vermutet. Durch Kulturversuche hat CZURDA (1925) jedoch gezeigt, daß „Klon"-Kulturen wieder zu einem gemischtgeschlechtlichen Faden heranwachsen. Er hatte Einzelzellen, die bei der Kopulation zweier Fäden keinen Partner gefunden hatten, getrennt weiterkultiviert. In 2 Kulturen von je einer männlichen Zelle entstanden Fäden, die nach 10 bis 20 Tagen Leiter-Kopulationen ergaben. Das gleiche wurde in einem Falle für eine von einer weiblichen Zelle ausgehenden Kultur gefunden.

Alle diese Beobachtungen, zu denen noch viele weitere zugefügt werden könnten, sind als Hinweise auf die Möglichkeit phänotypischer Geschlechtsbestimmung zu werten.

Ein offenbar günstiges Objekt für physiologische Untersuchungen ist *Vaucheria*. Schon KLEBS (1896) hatte gefunden, daß bei *V. repens*, wo in der Regel ein Oogon und ein Antheridium paarweise zusammenstehen, in Temperatur- und Druckexperimenten nur Antheridien angelegt werden (Abb. 1). Die bisexuelle Potenz von schon abgetrennten Oogonien und Antheridien konnte v. WETTSTEIN bei *V. hamata* nachweisen. Durch Anstechen brachte er die isolierten Geschlechtsorgane zum Auswachsen und erhielt Regenerate mit männlichen und weiblichen Geschlechtsorganen. Diese gemischtgeschlechtliche Tendenz, die hier selbst bei differenzierten Geschlechtsorganen in Erscheinung tritt, ist bei allen

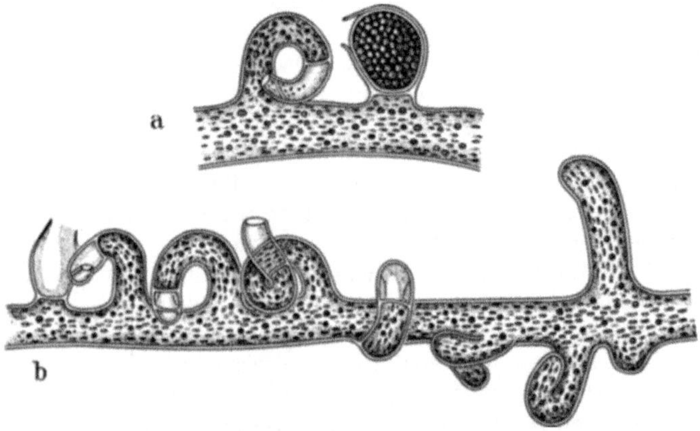

Abb. 1a u. b. *Vaucheria repens*. a Normale Verteilung der Oogone und Antheridien; b Faden nur mit Antheridien. (Nach KLEBS 1896, aus HARTMANN 1956.)

Arten der Gattung *Vaucheria* vorhanden. Neuerdings berichtet RIETH, daß auch die bisher als diöcisch angesehene Art *V. dichotoma* in Wirklichkeit monöcisch ist, wie er durch Einsporkulturen finden konnte. Auch einige andere Autoren waren zu der Ansicht gelangt, obwohl von ihnen keine Kulturen vorliegen. Bei *V. dichotoma* sind häufig ganze Zweige rein getrenntgeschlechtlich, so wie daß auch bei den morphologisch so analog gebauten Saprolegniaceen unter den Pilzen vorkommt. Es würde also das Geschlecht hier, wie bei manchen Achlya-Arten, in einem frühen Entwicklungsstadium wohl durch innere Faktoren bestimmt. Auch wenn es sich hier nur, wie sich RIETH vorsichtig ausdrückt, um eine monöcische Rasse von *V. dichotoma* eines einzelnen Standortes handelt, wird man nicht fehlgehen, bei *Vaucheria* genotypische Geschlechtsbestimmung auszuschließen. Auf die Frage, von welcher Art solche inneren Faktoren sein können, die bei der Geschlechtsbestimmung möglicherweise eine Rolle spielen, haben HUSTEDE (1957) (s. auch v. DENFFER und HUSTEDE 1955) in gewisser Weise eine Antwort gegeben. Bei *V. sessilis* fördert die Zugabe von Tryptophan und β-Indolylessigsäure (IES) die Synzoosporenbildung, also die ungeschlechtliche Fortpflanzung, der Zusatz von β-Indolcarbonsäure die Entwicklung der Sexualorgane. β-Indolcarbonsäure entsteht auch durch Photolyse und fermentativen Abbau aus Tryptophan oder IES. Besonders auffallend ist dabei, daß nicht nur die sexuelle Tendenz, sondern besonders die Entwicklung weiblicher Organe gefördert wird. Die Tabelle 1 zeigt, in welcher Weise diese Faktoren aus der Tryptophanreihe geschlechtsmodifizierend wirken.

Tabelle 1. *Vaucheria sessilis. Zusammenstellung der beobachteten Gametangiengruppierungen aus den Versuchen mit Tryptophan-Abkömmlingen.* (Nach HUSTEDE 1957.)

Anordnung	% ♀	% ♂	Vorkommen
♂	0	100	Kontrollkulturen
♀ ♂ ♂ ♂	25	75	
♀ ♂ ♂	33	66	
♀ ♂ ♂ ♂ ♂ ♀	33	66	
♀ ♂ ♂ ♂ ♀	40	60	
♀ ♂ ♂ ♀	50	50	Kontrollkulturen, β-Indol-Aldehyd, β-Indol-Carbons., Indol (10^{-6}—10^{-8} g/cm³)
♀ ♂	50	50	
♀ ♂ ♀	66	33	
♀ ♂ ♂ ♀ ♀	60	40	β-Indol-Carbons. 10^{-6} g/cm³
♀ ♂ ♀ ♀	75	25	
♀	100	0	

Unter den Grünalgen sind vermutlich nur die Codiaceen und die Dasycladaceen Diplonten. Bei ihnen findet die Reduktionsteilung in den Gametangien statt (WILLIAMS 1925, SCHULZE 1938). Einige Arten (oder Rassen) der Gattung *Codium* tragen männliche *und* weibliche Gametangien auf der gleichen Pflanze und häufig auf der gleichen Blase. Neben älteren Forschern haben HARTMANN und HÄMMERLING im Laufe von 20 Jahren 2 *Codium*arten in Neapel untersucht und dabei sehr wechselnde Verhältnisse vorgefunden. Dies wurde durch längere Kultur im Herbst 1948 durch HARTMANN erneut bestätigt. Die Pflanzen von *Codium elongatum* waren anfangs rein weiblich und wandelten sich darauf über ein zwittriges Stadium in Männchen um. Bei *Dasycladus (Dasycladacae)* fand JOLLOS einmal im Herbst 1926 zwittrige Pflanzen, während ihm solche während des ganzen Sommers nicht zu Gesicht gekommen waren. Da BERTHOLD (1880), HARTMANN (1925, in litteris), HÄMMERLING (1940) und KÖHLER (1957) in all diesen Jahren nur getrenntgeschlechtliche Pflanzen von *Dasycladus clavaeformis* gefunden haben, kann die Frage nach der Geschlechtsbestimmung hier nicht sicher beantwortet werden. *Acetabularia mediterranea* und *A. wettsteinii* scheinen dagegen sehr häufig monöcisch zu sein. Da die Kopulation nur zwischen Gameten erfolgt, die aus verschiedenen Cysten stammen, müßte die Geschlechtsbestimmung in der Entwicklung der Cysten erfolgen (HÄMMERLING 1934). Dies hat im Prinzip schon STRASBURGER (1882) beobachtet. Es liegt die Vermutung nahe, daß die Geschlechtsbestimmung phänotypisch erfolgt.

B. Phäophyceen.

Unter den Braunalgen sind eine Reihe gemischtgeschlechtlicher Arten bekannt geworden, wie die Untersuchungen von Exemplaren natürlicher Standorte gezeigt haben. Das fällt besonders dann ins Auge, wenn die Gameten von verschiedener Größe sind, also Anisogamie bzw. Oogamie vorliegt. Von einigem Interesse ist hier *Ectocarpus siliculosus*. Diese isogame Alge ist nach KUCKUCK (1912) in Helgoland zwittrig. Dagegen haben eine Reihe von Beobachtern in Neapel immer wieder Getrenntgeschlechtlichkeit nachgewiesen (BERTHOLD 1881, OLTMANNS 1897, HARTMANN 1925, 1934, 1937). HARTMANN gelang es, männliche sowie weibliche Gameten parthenogenetisch heranzuziehen. Eine Kultur aus männlichen Gameten ergab getrenntgeschlechtliche Gametophyten beider Geschlechter, die untereinander kopulierten. (In der Regel sind jedoch Parthenogametenzuchten steril.) Offenbar erfolgt also die Geschlechtsdifferenzierung bei *Ectocarpus* in einem sehr frühen Entwicklungsstadium. Ein weiterer Hinweis auf phänotypische Geschlechtsbestimmung ergibt sich daraus, daß Pflanzen, die auf der gleichen Unterlage (der Braunalge *Scytosiphon*) gewachsen waren, gleiches Geschlecht hatten. Diese Regel wurde in 10 von 14 geprüften Fällen eingehalten. In 2 weiteren Fällen zeigten die Ausnahmepflanzen sog. relative Sexualität. Diese Ausnahmekopulationen wurden insgesamt 37mal von HARTMANN beobachtet.

Als Diplonten sind unter den Braunalgen nur die *Fucales* bekannt: auch bei ihnen sind neben rein getrenntgeschlechtlichen Arten (z. B. *Fucus vesiculosus*, *F. serratus*) eine Reihe von Zwittern gefunden worden (*Fucus spiralis*, *Cystosira barbata*), bei denen Antheridienstände und Oogonien im gleichen Konzeptakel stehen. Bei anderen Arten finden sich auf dem gleichen Thallus rein weibliche und rein männliche Konzeptakeln. Auch gynomonöcische, andromonöcische und trimonöcische Organe kommen vor. Die Verhältnisse wurden bisher noch nicht experimentell verfolgt.

C. Phänotypische „Überraschungen“ bei *Callithamnion* und *Chaetomorpha*.

Statt einer Liste von Naturbeobachtungen an Rotalgenspecies soll hier ein Fall, der in jahrelanger Kultur näher untersucht wurde, aufgeführt werden: im übrigen sind unter den Rotalgen Monöcisten wie Diöcisten bekannt. Von *Callithamnion corymbosum* war nun durch Aufzucht von Tetrasporenpflanzen ein guter Hinweis auf genotypische Geschlechtsbestimmung gefunden worden. Aus den Gonen eines Tetrasporangiums entwickeln sich in der Regel getrenntgeschlechtliche Gametophyten mit dem Geschlechtsverhältnis wie 2:2. Vereinzelt wuchsen aber aus Tetrasporen erneut diploide Tetrasporenpflanzen heran; offenbar war hier die Reduktionsteilung ausgefallen. Einige solcher Tetrasporenpflanzen trugen überdies Äste mit Antheridien, andere Äste mit Karpogonen. Als Ursache dieser Störungen wies HASSINGER-HUIZINGA (1952) Genommutationen nach, bei denen überzählige Chromosomen geschlechtsdeterminierende oder -modifizierende Gene enthalten dürften. Von diesen Tetrasporen der zweiten Generation ergab sich in der Regel ein Geschlechtsverhältnis der Gonen von 2:2; daneben traten einige zwittrige haploide Geschlechtspflanzen auf (Tabelle 2).

Tabelle 2. *Callithamnion corymbosum. Gonenanalyse von 4 Tetraden zweier Tetrasporenpflanzen mit Antheridienzweigen.* (Nach HASSINGER-HUIZINGA aus HARTMANN 1956.)

Elternpflanze	Tetrade	Geschlecht der Gametophyten aus den 4 Sporen			
7a Tetr. + ♂	618	♂	steril	♀	⚥
730 Tetr. + ♂	688	♂	♂	♀	♀ mit ♂ Ästen
730 Tetr. + ♀	1420	♂	♂	♀ mit ♂ Ästen	♀ mit ♂ Ästen
730 Tetr. + ♀	1429	♂	—	♀ mit ♂ Ästen	♀ mit ♂ Ästen

Solche zwittrige Ausnahmefälle, die bei den Blütenpflanzen als Subdiöcisten bezeichnet werden, scheinen mir aber bei *Callithamnion* nicht auf einheitlicher Grundlage zu entstehen; so ist sicher ein Teil durch Genommutationen, ein anderer Teil durch Außenbedingungen und wieder andere möglicherweise durch Genmutationen entstanden. Dieser Fall zeigt besonders deutlich, wie letzten Endes ein fließender Übergang zwischen genotypischer und phänotypischer Geschlechtsbestimmung besteht, abhängig davon, wie stark determinierend die postulierten Realisatorgene das Genom selber oder die Außenbedingungen beherrschen. Durch leichte Genommutationen oder durch Aktivierung von Geschlechtsmodifikatoren sind solche Abweichungen formal zu erklären. Freilich ist noch kein Fall ganz gesichert.

Ähnlich verhält sich — und sei darum an dieser Stelle nachgetragen — *Chaetomorpha aerea* (Chlorophycee). Hier hatte KÖHLER (1956) bei Untersuchungen über die Geschlechtsbestimmung Gameten parthenogenetisch aufgezogen und zunächst Sicherheit für eine genotypische Determinierung erhalten. Vereinzelt entwickelten sich jedoch aus den unbefruchteten Gameten über eine Zoosporenpflanzengeneration diöcische Geschlechtspflanzen beider Geschlechter. Da zur näheren Klärung dieses Phänomens keine genetischen und physiologischen Versuche angestellt wurden, kann man das Ergebnis nur hinnehmen. Es mag darauf hinweisen, wie die Verhältnisse komplizierter werden, sobald man eine große Anzahl von Pflanzen über einen längeren Zeitraum kultiviert. Jedenfalls ist auch in vielen offenbar sicheren Fällen außer einem streng genetischen Mechanismus noch genügend Raum für andere Faktoren gelassen.

D. Die Geschlechtsbestimmung bei den Diatomeen.

1. *Centrales.*

Die centralen Diatomeen wurden in neuerer Zeit durch v. STOSCH eingehender untersucht. Bisher haben sich alle Arten als monöcisch erwiesen. Das ergab sich aus Klonkulturen von *Melosira moniliformis*, *Stephanopyxis*, *Biddulphia mobiliensis*, *Lithodesmium* und *Guinardia*, ebenso wie bei der Durchmusterung von Zellkolonien, die als natürliche Klone anzusehen sind: *Melosira varians*, *Streptotheca*, *Bellarochea* und *Biddulphia rhombus*. Die aus einer Auxospore sich zunächst vegetativ entwickelnde Population zeigt später nach Beginn der sexuellen Phase nur Weibchen, darauf eine zwittrige und schließlich eine männliche Phase.

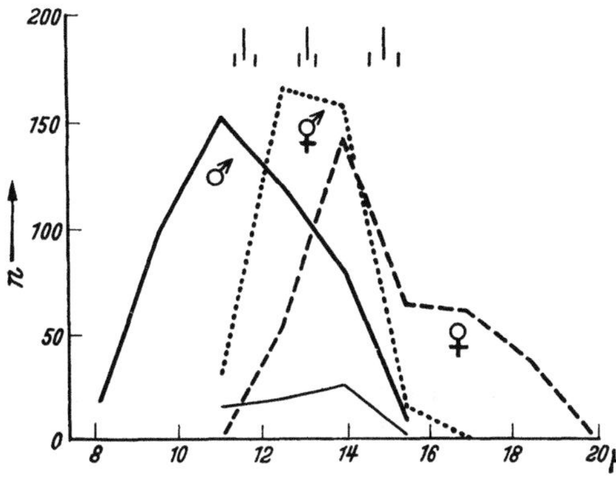

Abb. 2. *Melosira varians*. Breitenvariationskurven der männlichen, weiblichen und zwittrigen Fäden. (Nach v. STOSCH 1951.)

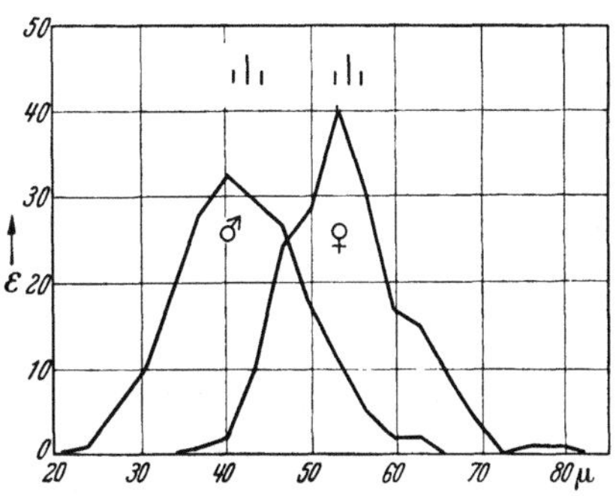

Abb. 3. *Biddulphia rhombus*. Breitenvariationskurven der männlichen und weiblichen Zellen. (Nach v. STOSCH 1956.)

Dies ist besonders bei fädigen Formen unmittelbar sichtbar. Bei *Melosira varians* (Abb. 2) und noch besser bei *Biddulphia rhombus* (Abb. 3) besteht ein Zusammenhang zwischen Zellgröße und Geschlecht für alle Zellen, die in die Meiose eintreten. Bei *B. rhombus* setzt die sexuelle Phase dann ein, wenn eine bestimmte Zellgröße (78 μ) erreicht ist; zunächst werden nur Oogonien gebildet. Von einer Zellbreite von 63 μ ab treten vereinzelt Spermatogonien auf, ab 37 μ können die Zellen nur noch männliche Schwärmer bilden. Es liegt hier somit diplophänotypische Geschlechtsbestimmung vor. Zusätzlich seien in Abb. 4 zwei Schwesterzellen wiedergegeben, die noch mit ihren Hörnern beieinandergehalten sind; die untere Zelle befindet sich im Pachytän der Eientwicklung, die obere (halbe Zelle) mit 2 (in der ganzen Zelle 4!) Spermatogonien ist ebenfalls in der Meiose.

Die Schalengröße scheint auch bei anderen Formen der wesentliche geschlechtsdeterminierende Faktor zu sein; unterhalb einer gewissen Zellgröße werden offenbar nur Spermatogonen gebildet. In der zwittrigen Phase dürften zusätzlich weitere Faktoren wirksam sein. So ergab Dauerbelichtung von *Lithodesmium* fast ausschließlich Oogonien, schwaches Licht in Hell-Dunkelperioden nur männliche Zellen. Dagegen bildeten sich bei starker Belichtung im Tag-Nachtrhythmus etwa gleiche Anzahlen von weiblichen und männlichen Gametangien (v. STOSCH 1954).

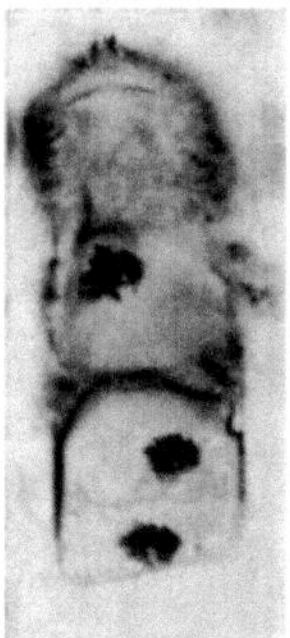

Abb. 4. *Biddulphia rhombus*. Zwei Schwesterzellen in der Meiose; die untere Zelle wird zum Oogon, die obere (nur die eine Hälfte sichtbar) mit Spermatogonen. (Nach v. STOSCH 1956.)

2. Die pennaten Diatomeen.

Weitaus kompliziertere Verhältnisse liegen bei den pennaten Diatomeen vor, bei denen *vor* der Vereinigung der Gameten in der Regel eine Paarung der Mutterzellen erfolgt. Die autogamen Vorgänge sollen hier nicht näher betrachtet werden. Die sexuelle Fortpflanzung ist während der letzten 30 Jahre durch GEITLER eingehend untersucht und zusammenfassend beschrieben worden (1932, 1957). Die neuere Literatur wurde überdies durch v. STOSCH (1950) referiert und im Hinblick auf die Frage nach der Geschlechtsbestimmung diskutiert. „Zusammenfassend muß festgestellt werden, daß ein Fortschritt in der Klärung der grundsätzlichen Fragen gegenüber dem Stand von 1932 nicht erreicht wurde. Das beträchtlich vergrößerte Tatsachenmaterial und seine Analyse haben vielmehr ergeben, daß eine befriedigende Lösung noch schwieriger geworden ist." (GEITLER 1957, S. 283.) Ein großzügiger Vergleich mit den Ciliaten und den „suprasexuellen" Erscheinungen wie der Inkompatibilität bei Pilzen und Blütenpflanzen könnte eines Tages doch zu einem befriedigenden, wenigstens formalen Verständnis der Paarungs- und Sexualitätsverhältnisse bei den Diatomeen führen. Man hat grundsätzlich 2 Möglichkeiten in Erwägung zu ziehen: Nimmt man die sich paarenden Mutterzellen als sexuell (bipolar) differenziert an, so hätte man eine Geschlechtsbestimmung in der Diplophase. Nimmt man dagegen die im Normalfall gebildeten 2 Gameten jeder Mutterzelle als verschiedengeschlechtlich an, so muß man der Paarung der Mutterzellen, die man dann als zwittrig anzusehen hätte, eine suprasexuelle Ursache zubilligen. Zusätzlich wird man vor zwei weitere Entscheidungen gestellt: Man kann jeden einzelnen und speziellen Fall bei jeder Art und Varietät gesondert deuten, was formal einfacher ist. Andererseits wird man den Wunsch haben, ein allgemein gültiges vergleichendes Bild der Paarungs- und Sexualitätsverhältnisse bei den Diatomeen entwerfen zu wollen.

GEITLER unterscheidet im Anschluß an ältere Autoren „auf Grund der Zahl der Gameten einerseits und der Ausprägung der sexuellen Fortpflanzung als Allogamie, Automixis oder Apomixis andererseits" folgende 4 Typen von Vorgängen, die zu Auxosporenbildung führen:

„I. Der ‚Normaltypus': Zwei gepaarte Mutterzellen bilden je 2 Gameten und nach allogamer Kopulation 2 Zygoten bzw. Auxosporen; dabei können die Gameten in verschiedener Weise sich iso- oder anisogam verhalten.

II. Abgeleiteter Typus: Zwei gepaarte Mutterzellen bilden je einen Gameten und allogam eine Zygote bzw. Auxospore; die Gameten verhalten sich iso- oder anisogam.

III. Abgeleiteter Typus von I, entstanden durch Automixis: Eine ungepaarte Mutterzelle bildet 2 Gameten, die pädogam miteinander kopulieren; oder die Gameten als solche sind rückgebildet und die Mutterzelle bildet nur 2 Geschlechtskerne, die autogam fusionieren. In beiden Fällen entsteht aus einer Mutterzelle eine Zygote bzw. Auxospore.

IV. Abgeleiteter Typus von II (vielleicht auch von I und III ?): Unter Apomixis, aber unter Pseudomeiose entsteht aus einer Mutterzelle eine Azygote bzw. eine Auxospore; es handelt sich um diploide Parthenogenese; als Rudiment der Allogamie kann dabei auch Paarung zweier Mutterzellen erfolgen (Beispiel *Cocconeis*)."

Zunächst seien die einfacheren Verhältnisse bei Typ II näher betrachtet. Diatomeen, die je Mutterzelle nur einen Gameten bilden, verhalten sich folgendermaßen: Zur Paarung legen sich zwei diploide Zellen aneinander und ihre Kerne treten gleichzeitig in die Meiose ein, auf die eine mehr oder weniger deutliche inäquale Zellteilung folgt; die kleinere Zelle abortiert, die größere wird zum Gameten; hierauf erfolgt die Kopulation. Es wäre leicht aus solchem Verhalten zu schließen, daß bereits die Mutterzellen sexuell determiniert sind, wenn sichere Kriterien für *Geschlechtsunterschiede* vorhanden wären. In 2 Fällen hat nun GEITLER offenbar derartige Geschlechtsunterschiede gefunden. Bei *Cocconeis placentula var. pseudolineata* findet man eine physiologische „Anisogamie" in der Weise, daß ein aktiver, als männlich angesehener Gamet aus der Mutterschale entlassen wird und in die Schale des passiven weiblichen Gameten hinüberwandert. Es ließ sich meistens finden, daß der Wandergamet überdies aus einer kleineren Mutterzelle entstanden war. Wenn auch der morphologische Unterschied nicht ausschließlich mit dem physiologischen gekoppelt ist, so stammen doch 81% der Wandergameten (aus 207 gezählten Paaren) aus kleineren Mutterzellen. Bei *Navicula seminulum* ist das entsprechende Verhältnis 76% (GEITLER 1932, 1957). Es ist also möglich, in solchen Fällen diplophänotypische Geschlechtsbestimmung anzunehmen, wenngleich bei etwa 20% der Partner die Zellgröße für die Ausprägung des Geschlechtes keine Rolle spielt. Fast man die Mutterzellen dagegen trotzdem als diploide Zwitter auf, so wäre natürlich auch haplophänotypische oder sogar eine Geschlechtsbestimmung über einen Chromosomenmechanismus möglich (Beispiel bei den Metazoen u. a., SEILER 1920). Gar keine Schlüsse lassen sich aus dem Verhalten von *Cocconeis placentula var. klinoraphis* ziehen, wo die Kopulation „isogam" vor sich geht (GEITLER 1932, S. 115).

Werden 2 Gameten je Zelle gebildet („Normalfall", Typ I), so liegt es nahe, besonders bei Ausprägung eines Ruhe- und eines Wandergameten, den Gameten verschiedenes Geschlecht zuzuweisen und Selbststerilität vorauszusetzen. Die Mutterzellen sind daher zwangsläufig als Zwitter anzusehen. Das trifft für die meisten Arten und Rassen des Normaltyps zu. Nun werden aber bei *Navicula halophila* (SUBRAHMANYAN 1947) und *Synedra ulna* (GEITLER 1939a, b) beide Wandergameten bzw. beide Ruhegameten jeweils in einer Mutterzelle gebildet, es fällt also schwer, hier die Mutterzellen als zwittrig anzusehen und das Wandern bzw. das Nichtwandern der Gameten gleichzeitig als Geschlechtsmerkmal zu werten. Noch schwieriger werden die Verhältnisse dadurch, daß bei *S. ulna* gelegentlich die beiden Gameten einer Mutterzelle miteinander kopulieren können (Typ III); hier kann nur die Hypothese der relativen Sexualität, die ja von HARTMANN zur Erklärung autogamer Vorgänge aufgestellt wurde (1909), die gedankliche Ordnung wiederherstellen, oder man gibt das Vorurteil, der Wandergamet sei männlich, der Ruhegamet sei weiblich determiniert, auf. Bei sessilen Arten wie *Gomphonema*, *Rhoicosphenia* und *Anomoeoneis* sucht häufig ein kleinerer beweglicher Partner die sessile Mutterzelle auf, hier könnten nun nochmal dieselben Betrachtungen angefügt werden wie oben bei *Navicula seminulum*.

Man würde hier ein ähnliche — wenn auch nicht befriedigend aufgeklärte — Reihe aufstellen können, wie sie sich bei den Ciliaten von *Paramaecium* über

Vorticella zu den Suctorien ergibt (MÜGGE 1957). Es sei daher zusammenfassend die Ansicht GEITLERs, der beste Kenner dieser Vorgänge, wiedergegeben: „Sicher ist für die bisher in Klonkulturen oder an natürlichen Klonen untersuchten Arten allein, daß die Geschlechtsbestimmung nicht diplogenotypisch vor sich gehen kann: denn es erfolgt Paarung und Kopulation innerhalb der Klone. Selbst wenn die Mutterzellen der „zweigametigen" Arten sexuell determiniert wären, könnte es sich also nur um diplophänotypische Bestimmung handeln. Sind die Mutterzellen aber Zwitter, so erfolgt die Geschlechtsbestimmung bei der Gametogenese, also während oder nach der Meiose, und somit haplophasisch (wobei aber, schon rein gedanklich, eine Unterscheidung zwischen haplogeno- und phänotypisch schwer fällt). Völlig klar ist dies aber der Fall bei den obligat automiktischen Typen: hier sind sicher die Mutterzellen sexuell neutral und erst die Gameten sexuell determiniert.

Die autogamen Vorgänge (Typ III) tragen nichts zum Verständnis der Sexualverhältnisse bei. Sind die Mutterzellen zwittrig, so können die beiden Gameten z. B. bei *Gomphonema constrictum var.* (1952a), *Cymbella aspera* (1956) oder *Synedra ulna* oder die zwei überlebenden Gonenkerne z. B. bei *Denticula tenuis* (1953) genotypisch oder phänotypisch, eventuell durch differentielle Zellteilung bipolar differenziert sein. Es sei erwähnt, daß die Deutung ähnlicher automiktischer Befruchtungserscheinungen durch die Sexualitätstheorie HARTMANNs zur Hypothese der relativen Sexualität führte. „War einmal gefordert, daß die Ursache der Befruchtung eine sexuelle Differenzierung in zwei polare, miteinander reaktionsfähige Gametensorten oder Kerne wäre, so ergab sich eben die logische Notwendigkeit, bei der Paarung von Schwesternkernen, im extremsten Falle bei der Parthenogamie, eine verschiedene sexuelle Stärke der Kerne, eine Polarität vorauszusetzen" (KÖHLER 1957).

Schlußbetrachtung.

Die hier versuchte, auf typische Fälle beschränkte Übersicht über die Erscheinungen bei phänotypischer und modifikatorischer Geschlechtsbestimmung fußt hauptsächlich auf morphologischen Untersuchungen und Beobachtungen von Entwicklungscyclen in Algenkulturen. Man möchte sagen: leider! Ihrer Formulierung nach ist die modifikatorische Art der Geschlechtsbestimmung genetisch begründet, indem die Anwesenheit je eines Allels für männliche und weibliche Differenzierung, Realisatorgene genannt (v. WETTSTEIN 1924), im haploiden Zustand vorausgesetzt werden; der doppelte Satz entsprechend bei den Diplonten. Diese Formulierung sollte in erster Linie eine genetische Untersuchung bedingen mit dem Hauptgegenstand, die Natur solcher Gene bzw. mit ihnen gekoppelter Merkmalsfaktoren aufzuklären. Das ist bei Algen bisher nur durch MOEWUS geschehen (1935a, b, 1936, 1937, s. dazu: RENNER 1958). Die zweite nicht minder interessante Seite dieses Problems ist seine Bedeutung für die Entwicklungsphysiologie; „die moderne Forschung fragt natürlich auch nach den Eigenschaften der Nachkommen" (OLTMANNS S. 162). Darüber hinaus gelang es, bei anderen Organismen Näheres über die Bedingungen für die Entwicklung der Geschlechtsorgane zu erfahren (s. RAPER in diesem Band). Zweifellos hat die morphologische und entwicklungsgeschichtliche Forschung eine Menge interessanten Materials angesammelt und auf vergleichender Grundlage zu Deutungsversuchen geführt, die experimentelle Bearbeitung der Grundlagen der phänotypischen Geschlechtsbestimmung steckt noch in den Anfängen.

Literatur.

Berthold, G.: Die geschlechtliche Fortpflanzung von *Dasycladus clavaeformis AG.* Bot. Ztg **38**, 648—651 (1880). — Bemerkungen zu der Abhandlung von Fr. Oltmanns „Über Scheinkopulationen bei Ectocarpeen und anderen Algen". Flora (Jena) **83**, 415—425 (1897).

Czurda, V.: II. Zur Kenntnis der Geschlechtsverhältnisse bei *Spirogyra*. Ber. dtsch. bot. Ges. **42**, 441—444 (1925).

Denffer, D. v., u. H. Hustede: Wuchsstoffbedingte Umstimmung von der sexuellen zur vegetativen Entwicklung bei *Vaucheria sessilis*. Flora (Jena) **142**, 489 (1955).

Geitler, L.: Der Formwechsel der pennaten Diatomeen. Arch. Protistenk. **78**, 1—226 (1932). — Die Auxosporenbildung von *Synedra ulna*. Ber. dtsch. bot. Ges. **57**, 432—600 (1939a). — Gameten- und Auxosporenbildung von *Synedra ulna* im Vergleich mit anderen Diatomeen. Planta (Berl.) **30**, 551—566 (1939b). — Untersuchungen über Kopulation und Auxosporenbildung pennater Diatomeen. I. Automixis bei *Gomphonema constrictum var. capitata*. Öst. bot. Z. **99**, 376—384 (1952a). — Die Auxosporenbildung von *Rhoicosphenia curvata*. Öst. bot. Z. **99**, 78—88 (1952e). — Über differentielle Teilung und einen im Zellbau begründeten, kopulationsbegrenzenden Faktor bei der Diatomee *Cocconeis*. Z. Naturforsch. **7b**, 411—414 (1952f). — Allogamie und Autogamie bei der Diatomee *Denticula tenuis* und die Geschlechtsbestimmung der Diatomeen. Öst. bot. Z. **100**, 331—352 (1953). — Lebendbeobachtung der Gametenfusion bei *Cymbella*. Öst. bot. Z. **101**, 74—80 (1954a). — Paarbildung und Gametenfusion bei *Anomoeoneis exilis* und einigen anderen pennaten Diatomeen. Öst. bot. Z. **101**, 441—452 (1954b). — Automixis, Geschlechtsbestimmung und Pyknose von Gonenkernen bei *Cymbella aspera*. Planta (Berl.) **47**, 359—373 (1956). — Über die Paarung bei *Navicula seminulum* und die Ausprägung der Geschlechtsmerkmale bei pennaten Diatomeen. Ber. dtsch. bot. Ges. **70**, 45—48 (1957a). — Die sexuelle Fortpflanzung der pennaten Diatomeen. Biol. Rev. **32**, 261—295 (1957b).

Hämmerling, J.: Über die Geschlechtsverhältnisse von *Acetabularia mediterranea* und *Acetabularia Wettsteinii*. Arch. Protistenk. **83**, 57—97 (1934). — Zur Lebensweise, Fortpflanzung und Entwicklung verschiedener Dasycladaceen. Arch. Protistenk. **97**, 7—56 (1944). — Hartmann, M.: Autogamie bei Protisten und ihre Bedeutung für das Befruchtungsproblem. Arch. Protistenk. **14**, 264 (1909). — Über die relative Sexualität bei *Ectocarpus siliculosus*. Ein experimenteller Beitrag zur Sexualitätshypothese der Befruchtung. Naturwiss. **13**, 975—980 (1925). — Untersuchungen über die Sexualität von *Ectocarpus siliculosus*. Arch. Protistenk. **83**, 110—153 (1934). — Ergänzende Untersuchungen über die Sexualität von *Ectocarpus siliculosus*. Arch. Protistenk. **89**, 382—392 (1937). — Die Sexualität, 2. Aufl. Stuttgart: Gustav Fischer 1956. — Hartmann, M., u. J. Hämmerling: Beiträge zur Kenntnis der Befruchtung und Sexualität mariner Algen. II. Über phänotypische Geschlechtsbestimmung von *Codium elongatum* und *tomentosum*. Pubbl. Staz. zool. Napoli **22**, 1—9 (1950). — Hassinger-Huizinga, H.: Generationswechsel und Geschlechtsbestimmung bei *Callithamnion corymbosum*. Arch. Protistenk. **98**, 91—124 (1952). — Hustede, H.: Untersuchungen über die stoffliche Beeinflussung der Entwicklung von *Stigeoclonium falklandicum* und *Vaucheria sessilis* durch Tryptophanabkömmlinge. Biol. Zbl. **76**, 555—595 (1957).

Jollos, V.: Untersuchungen über die Sexualitätsverhältnisse von *Dasycladus clavaeformis*. Biol. Zbl. **46**, 279—295 (1926).

Klebs, G.: Die Bedingungen der Fortpflanzung bei einigen Algen und Pilzen. Jena 1896. 543 S. — Kniep, H.: Die Sexualität der niederen Pflanzen. Jena: Gustav Fischer 1928. — Köhler, K.: Entwicklungsgeschichte, Geschlechtsbestimmung und Befruchtung bei *Chaetomorpha*. Arch. Protistenk. **101**, 224—268 (1956). — Neue Untersuchungen zur Sexualität von *Dasycladus*. Arch. Protistenk. **102**, 209—218 (1957). — Kuckuck, P.: Beiträge zur Kenntnis der Meeresalgen. II. Die Fortpflanzung der Phaerosporeen. Wiss. Meeresuntersuch., N. F. **5**, Abt. Helgoland 155—188 (1912).

Mainx, F.: Über die Geschlechterverteilung bei *Volvox aureus*. Arch. Protistenk. **67**, 205—214 (1929). — Physiologische und genetische Untersuchungen an Oedogonien I. Mitt. Z. Bot. **24**, 481—527 (1931). — Gametenkopulation und Zygotenkeimung bei *Hydrodictyon reticulatum*. Arch. Protistenk. **75**, 502—516 (1931). — Moewus, F.: Über den Einfluß äußerer Faktoren auf die Geschlechtsbestimmung bei *Protosiphon*. Biol. Zbl. **55**, 293—309 (1935a). — Über die Vererbung des Geschlechts bei *Polytoma Pascheri* und bei *Polytoma uvella*. Z. indukt. Abstamm.- u. Vererb.-Lehre **69**, 374—417 (1935a). — Faktorenaustausch, insbesondere der Realisatoren bei *Chlamydomonas*-Kreuzungen. Ber. dtsch. bot. Ges. **54**, 45—57 (1936). — Methodik und Nachtrag zu den Kreuzungen zwischen *Polytoma*-Arten und zwischen *Protosiphon*-Rassen. Z. indukt. Abstamm.- u. Vererb.-Lehre **73**, 63—107 (1937). — Mügge, E.: Die Konjugation von *Vorticella campanula* (Ehrbg.). Arch. Protistenk. **102**, 165—208 (1957).

Oltmanns, Fr.: Über die Sexualität der Ectocarpeen. Flora (Jena) **86**, 86—99 (1899). — Morphologie und Biologie der Algen, 2. Aufl., Bd. 3. Jena 1923.

RENNER, O.: Auch etwas über FRANZ MOEWUS, *Forsythia* und *Chlamydomonas*. Z. Naturforsch. **13**b, 399—403 (1958). — RIETH, A.: Beobachtungen zur Entwicklungsgeschichte einer *Vaucheria* der Sektion *Woroninia*. Flora (Jena) **142**, 156 (1955).

SCHULZE, B.: Zur Kenntnis einiger Volvocales *(Chlorogonium, Haematococcus, Stephanosphaera, Spondylomoraceae* und *Chlorobrachis)*. Arch. Protistenk. **58**, 508—576 (1927). — SCHULZE, K. L.: Cytologische Untersuchungen an *Acetabularia mediterranea* und *Acetabularia Wettsteinii*. Arch. Protistenk. **92**, 179—225 (1939). — SEILER, J.: Geschlechtschromosomen-Untersuchungen an Psychiden. I—III. Arch. Zellforsch. **15** (1920). — SPESSARD, E. A.: Fertilization in a living Oedogonium. Bot. Gaz. **89**, 385—394 (1930). — STOSCH, H. A. v.: Entwicklungsgeschichtliche Untersuchungen an zentrischen Diatomeen. I. Die Auxosporenbildung von *Melosira varians*. Arch. Mikrobiol. **16**, 101—135 (1951). — Entwicklungsgeschichtliche Untersuchungen an zentrischen Diatomeen. II. Geschlechtszellenreifung, Befruchtung und Auxosporenbildung einiger grundbewohnender Biddulphiaceen der Nordsee. Arch. Mikrobiol. **23**, 327—365 (1956). — Die Oogamie von *Biddulphia mobiliensis* und die bisher bekannten Auxosporenbildungen bei den *Centrales*. VIII. Congr. Intern. Bot. Sect. **17**, 58—68 (1954). — Pennate Diatomeen. (Referat.) Z. Bot. **43**, 89—99 (1955). — STRASBURGER, E.: Histologische Beiträge. IV. Über das Verhalten des Pollens und der Befruchtungsvorgänge bei den Gymnospermen. — Schwärmsporen, Gameten, pflanzliche Spermatozoiden und das Wesen der Befruchtung. Jena: Gustav Fischer 1892. — STREHLOW, K.: Über die Sexualität einiger *Volvocales*. Bot. Z. **21**, 625—692 (1929). — SUBRAHMANYAN, R.: On somatic division, reduction division, auxospore formation and sex differentiation in *Navicula halophila*. J. Ind. Bot. Soc. Iyengar Commem. Vol., pp. 239—266. 1948.

TIFFANI, L. H.: Homothallism and other variations in *Pleodorina californica* Shaw. Arch. Protistenk. **85**, 140—144 (1935).

WETTSTEIN, F. VON: Künstliche haploide Parthenogenese bei *Vaucheria* und die geschlechtliche Tendenz ihrer Keimzellen. Ber. dtsch. bot. Ges. **38**, 260—266 (1920). — Über Fragen der Geschlechtsbestimmung bei Pflanzen. Naturwiss. **12**, 761 (1924). — WILLIAMS, M. M.: Cytology and phylogeny of siphonaceous algae. I. The cytology of the gametangia of *Codium tomentosum* (Stackh.). Proc. Linn. Soc. N.S. Wales **50**, 98—111 (1925).

Phänotypische Geschlechtsbestimmung bei Pilzen.

Von

Kurt Köhler.

Mit 9 Abbildungen.

Während im vorausgegangenen Kapitel über die Geschlechtsbestimmung bei den Algen im allgemeinen kein Zweifel darüber bestand, was man als Sexualvorgang anzusehen hätte, bereiten besonders die höheren Pilze, die keine Sexualorgane aufweisen, Schwierigkeiten in bezug auf die Frage, wie und wo sich eine *sexuelle* Bipolarität manifestiert. Jedoch schon die Betrachtung der Diatomeen unter den Algen hatte gezeigt, daß eine befriedigende Erklärung der Gamonten- und Gametenpaarung noch nicht gefunden ist. Um hier bei den Pilzen der Mannigfaltigkeit der Paarungsvorgänge besser gerecht zu werden, sei von vornherein zwischen 3 Gruppen von Phänomenen unterschieden:

A. Supra- oder extrasexuelle Vorgänge: die Inkompatibilität. Vergleichbare Erscheinungen bei anderen Organismen dürften die Paarungstypen der Ciliaten und die Gamontenpaarung der Diatomeen sein (Esser u. Kuenen 1965).

B. Sexuelle Vorgänge: Phänotypische Geschlechtsbestimmung würde hier beispielsweise durch Monöcisten repräsentiert werden, gleichgültig, ob diese selbstfertil oder selbststeril sind. Als sexuelle Differenzierungen sind auch die Paarungstypen der Hefen anzusehen. Eine dritte Gruppe bilden die

C. *Parasexuellen Erscheinungen.* Mit dem Abnehmen sexueller Differenzierung tritt die Bedeutung der Kerne stärker in den Vordergrund. Daher sollten Vorgänge wie Autogamie, Parthenogamie und die Folgen von Hyphenfusionen im Bullerschen und im Dualphänomen gesondert betrachtet werden, ebenso wie die „Aufregulierung" parthenogenetisch entwickelter Algengameten oder die endomitotische Diploidisierung bei dem hypermastiginen Flagellaten *Barbulanympha* (Cleveland 1954) etwas Besonderes darstellen. Eine Liste von Beispielen solcher Alternativen zur sexuellen Fortpflanzung findet man bei Esser und Kuenen (1965, S. 98).

Solche Vorgänge wurden teilweise von Köhler (1956) „Attrappen" der Befruchtung genannt und ihnen „Attrappen" für Meiosis gegenübergestellt. Auch wenn man diese Einteilung nicht allgemein übernehmen will, erleichtert sie die Darstellung des Folgenden. Nur bei wirklich sexuellen Vorgängen (B) wird man daher nach phänotypischer Geschlechtsbestimmung sensu stricto suchen können. Inwieweit man infrasexuelle Erscheinungen dabei berücksichtigen will, ist weitgehend vom Einzelfall und von der subjektiven Ansicht des Betrachters abhängig. Phylogenetisch gesehen, sind wenigstens ein Teil der parasexuellen Erscheinungen von Sexualvorgängen abzuleiten.

Es wird also zunächst die Frage zu stellen sein, wo das, was der Sexualität bei anderen Organismen vergleichbar ist, bei den Pilzen sichtbar wird. Zwar sind unter den meisten Phycomyceten die Dinge völlig eindeutig; jedoch schon bei den Mucorineen meint Whitehouse verschiedene Stämme nicht als sexuell, sondern durch Inkompatibilitätsfaktoren differenziert ansehen zu müssen. Ebenso geben einige Autoren bei Hefen dem allgemeinen Ausdruck „Paarungstyp" den Vorzug. Esser und Kuenen (1965) haben die Fortpflanzungsbegriffe neuerdings geordnet; sie weichen dabei stark von Whitehouse und Burnett ab (Abb. 1). Wie z. B. verschiedene Vorschläge zustande kommen, wird durch einen Blick auf Abb. 2 (Raper 1954) klar. Der Begriff homo- oder heterothallisch ist verwendbar,

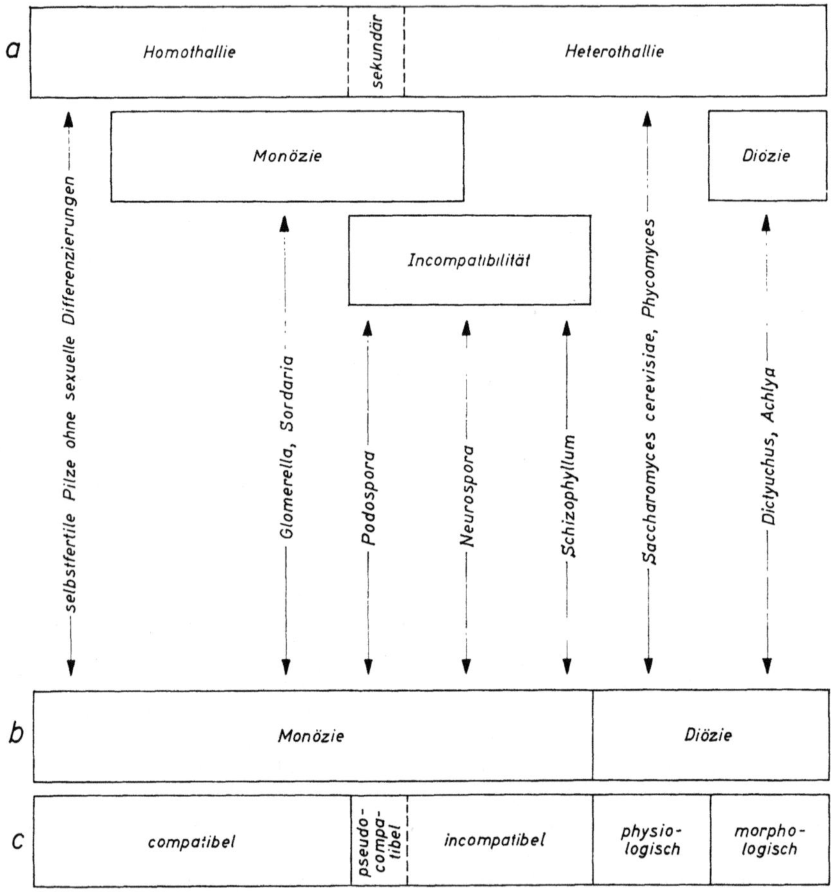

Abb. 1a—c. Fortpflanzungs-Systeme bei den Pilzen. a Zusammenstellung der bisher verwendeten Einteilungsprinzipien unter Berücksichtigung der generellen Unterteilung in homo- und heterothallische Arten. b Vorgeschlagene neue Einteilung nach einer erweiterten Fassung der Begriffe Monöcie und Diöcie. c Unterteilung der Monöcisten und Diöcisten in verschiedene Fortpflanzungs-Systeme. Die Länge der einzelnen Blocks steht in keinem Zusammenhang mit der zahlenmäßigen Verteilung der einzelnen Systeme in der Natur. Zur besseren Veranschaulichung der von uns vorgeschlagenen Einteilung haben wir einige der bekannteren Pilze in das Schema eingezeichnet. (Esser und Kuenen 1965.)

wenn man die nach der Meiose gebildeten Sporen aufzieht und die sich ergebenden *Stämme* testet. Diese Stämme können untereinander gleich sein (homothallisch) oder bipolar und tetrapolar differenziert. Das hat nur teilweise etwas mit Sexualität zu tun, denn die Stämme selbst können getrenntgeschlechtlich oder zwittrig sein, je nach den Sexualorganen, die sie tragen. Bei einer Diskussion der phänotypischen Geschlechtsbestimmung richten wir den Blick dorthin, wo sich *Geschlecht* manifestiert. In der unteren Hälfte der Spalte 0, II und III in Abb. 2 werden zwittrige Stämme als repräsentative Vertreter der phänotypischen Geschlechtsbestimmung sichtbar. In diesen 3 Spalten wird man vorzugsweise die

Organismen zu suchen haben, mit denen man genetische und physiologische Untersuchungen zur Geschlechtsbestimmung ausführen kann.

Die morphologischen und entwicklungsgeschichtlichen Grundlagen der Pilze, besonders auch ihrer sexuellen Vorgänge wurden durch GÄUMANN eingehend dargestellt. Eine Reihe von Aufsätzen RAPERs zeigt sein erfolgreiches Bemühen, gedankliche Ordnung in die Mannigfaltigkeit der bekannten Tatsachen zu bringen. Den gleichen Versuch haben WHITEHOUSE und BURNETT in jüngster Zeit gemacht. Eine das gesamte Tier- und Pflanzenreich umfassende Übersicht über die Probleme der Geschlechtsbestimmung hat HARTMANN in seinem Werk

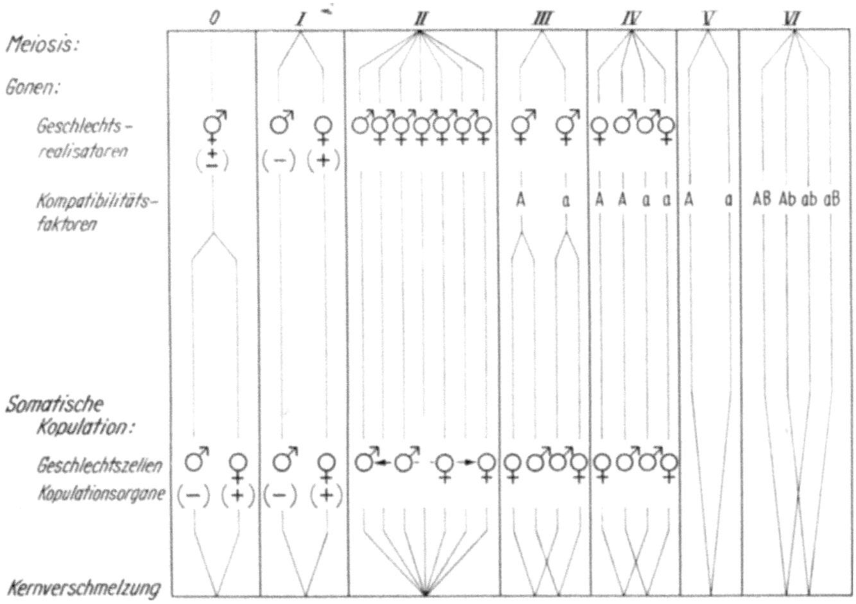

Abb. 2. Schemata der verschiedenen Sexualvorgänge bei Pilzen. *0* homothallisch; *I—VI* heterothallisch; *0, II, III* monöcisch; *I, IV* diöcisch; *V* diöcische Hefen und Mucorineen; *III, IV*, (*V*), *VI* mit Imkompatibilitätsfaktoren. (Nach RAPER 1954.)

„Die Sexualität" gegeben. Eine gründliche Behandlung aller dieser Themen findet man neuerdings in dem Buch von ESSER und KUENEN (1965) „Genetik der Pilze". Dort ist vor allem die Literatur der letzten Jahre vollständig berücksichtigt. Im folgenden werden hier nur typische Beispiele ausgewählt.

A. Phycomyceten.

Innerhalb der Ordnungen der *Chytridiales*, *Monoblepharidales* und der Oomyceten finden sich reine Haplonten (oder Diplohaplonten, wie einige Arten der *Blastocladiales*). Die weitaus meisten sind zwittrig. So gibt KNIEP (1928) für die Saprolegniaceen unter 94 untersuchten Arten 93 monöcisch, nur eine als diöcisch an. PETHYBRIDGE und MURPHY (1913) haben Conidien von eingeschlechtlichen Zweigen von *Phytophthora infestans* anfgezogen und Mycelien mit beiderlei Geschlechtsorganen erhalten. Das gleiche hat MURPHY (1918) aus isolierten Hyphenabschnitten für die dikline *Ph. erythroseptica* nachgewiesen. Unter den *Blastocladiales* hat KNIEP (1928, 1930) zuerst weibliche Gameten von *Allomyces* gelegentlich zu parthenogenetischer Entwicklung bringen können; desselbe gelang in neuerer Zeit EMERSON (1938, 1941) bei *A. crovanicus* und *A. macrogynus*. Aus solchen Versuchen ergab sich die bisexuelle Potenz der *Allomyces*-Gameten und es konnte nicht gezweifelt werden, daß bei den *Blastocladiales* die Geschlechtsbestimmung phänotypisch erfolgt, bis neuerdings au⸱ interspezifischen Kreuzungen von *A. macrogynus* und *A. arbuscula* diöcische Nachkommen erhalten wurden (EMERSON (1954).

Wie innerhalb einer Verwandtschaftsgruppe Fälle von offenbarer Monöcie bis zu scheinbarer Diöcie ausgebildet sein können, illustrieren die Verhältnisse bei den *Saprolegniales*. So können z. B. bei *Achlya colorata* die Antheridien direkt unter dem Oogon aus dem Gametangienstiel oder aus dem Oogon selber herauswachsen (Abb. 3a). Bei anderen *Achlya*-Arten befinden sich die weiblichen und männlichen Organe oft getrennt an verschiedenen Seitenzweigen; auch die Haupthyphen können bereits sexuell differenziert sein. Da außerdem noch alle möglichen

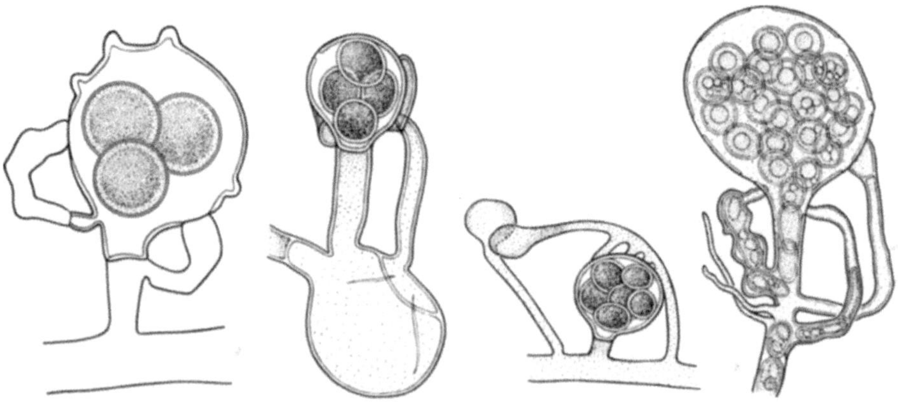

Abb. 3a—d. Besonderheiten bei der Differenzierung von Geschlechtsorganen bei Saprolegniaceen. a *Achlya colorata:* Antheridien aus dem Oogon sprossend. b *A. americana:* Aus einem Oogon sprossen ein weiteres Oogon und ein Antheridium. c *A. americana:* Antheridienschläuche, die sich am Ende zu Oogonien umdifferenzieren. d *Saprolegnia paradoxa:* Antheridien mit Eiern. a Nach COKER (1923); b und c nach HUMPHREY (1892), d nach MAURIZIO (1899). (b—d aus HARTMANN 1956.)

Zwischenstufen vorkommen, wie z. B., daß Antheridien in einer Region des Mycels an rein männlichen Zweigen entstehen, in anderen Regionen ebenso in der Nachbarschaft von Oogonien, so wird man nicht fehlgehen, bei dieser Gruppe phänotypische Geschlechtsbestimmung anzunehmen. Ältere Beobachter wie HUMPHREY (1892), MAURIZIO (1899) und COKER (1923) bilden Fälle ab, bei denen sich Oogonien oder deren Anlagen in Antheridien umwandeln oder umgekehrt (Abb. 3b—d). Weiterhin konnten aus unbefruchteten Eiern von *Achlya* gemischtgeschlechtliche Pflanzen herangezogen werden. In mehreren zwittrigen Arten, z. B. bei *Thraustotheca clavata* und *A. americana*, konnte durch Sexualhormone von stark eingeschlechtlichen Pflanzen verwandter Arten Oogonienbildung bzw. Antheridienentwicklung hervorgerufen werden. Nach COKER (1923, zit. HAWKER 1957) soll die Antheridienbildung stärker vom Phosphatgehalt abhängen als die Bildung von Oogonien. Die bisexuelle Potenz und gemischtgeschlechtliche Tendenz dieser Pilze begünstigt physiologische Untersuchungen. So hat RAPER bei *Achlya* ein ganzes Hormonsystem aufzeigen können; in ähnlicher Weise konnte BISHOP (1940) bei *Sapromyces reinschii* Antheridienentwicklung durch Filtrate weiblicher Hyphen induzieren.

S. reinschii wurde zunächst genau wie *Dictyuchus monosporus* als rein diöcisch angesehen. Von dem letzten berichten jedoch COUCH (1926) und NAGAI (1931) von einem weiblichen Stamm, der eine latente männliche Tendenz zeigte. Das ist ein interessantes Phänomen, denn auch bei *Achlya bisexualis* findet man neben rein männlichen und rein weiblichen auch gynomonöcische Stämme (RAPER 1936). In gleicher Weise zeigt *Sapromyces reinschii* die Tendenz, in weiblichen Stämmen gelegentlich Antheridien zu bilden. Der männliche Stamm 5 entwickelt auf normalem Nährsubstrat reichlich (ungeschlechtliche) Zoosporangien und kurze Antheridien, bei Anwesenheit von weiblichem Mycel ein langes, dichtes Netz männlicher Hyphen. Weibliche Stämme bilden auf Nähragar nur vereinzelt

♂ - - - - ♂ und/oder ♀ - - - ♀
E 87 → 11,159 → 89 → 107,190 → 88 → 155 → 78,80,184
England Illinois

Abb. 4. *Achlya ambisexualis* (s. Abb. 1, Spalte II). Sexuelles Verhalten von 11 Stämmen untereinander. (Näheres s. Text; aus RAPER 1954.)

Zoosporangien und gewöhnlich abortive Oogenien, gelegentlich legen sie zusätzlich Antheridien an und führen Selbstbefruchtung durch. BISHOP (1940) vermerkt ausdrücklich, daß zwar die Weibchen die Tendenz zur Anlage von männlichen Organen besitzen, nie dagegen die Männchen Oogonien anlegen. Damit liegen offenbar die Weibchen am weitesten auf der sexuellen Seite der Entwicklung, die Männchen mehr auf der vegetativen. Eine solche lineare Tendenz hat HUSTEDE (1957) schon an der so analog gebauten Grünalge *Vaucheria* gezeigt (s. S. 112 in diesem Band). Diese Ansicht, die sich hier durch entwicklungsgeschichtliche Untersuchungen ergab, könnte neue Anreize zu physiologischen Experimenten bei den Oomyceten geben. Vielleicht würde damit auch eine neue Grundlage für eine Diskussion über relative Sexualität oder Subdiöcie, einen von den Blütenpflanzen entlehnten Begriff, gefunden werden. Ein weiterer Beitrag hierfür ist ein von RAPER (1954) unter der Überschrift „Multiple alternate sexual factors" erwähnter Fall. Aus mehr als 100 isolierten Stämmen von *Achlya ambisexualis* waren 11 von besonderer Bedeutung. Diese Stämme, die alle rein diöcisch auftreten und untereinander fertil sind, lassen sich in 7 Gruppen linear anordnen. Je nach der Stellung des beigegebenen Partners im

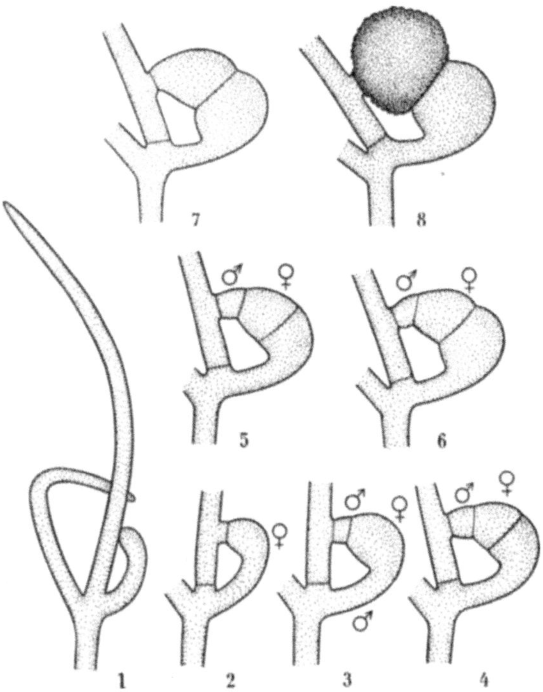

Abb. 5. *Zygorrhynchus heterogamus*. Gametangienentwicklung bei einem Monöcisten (*1—6*), Kopulation (*7*) und Zygotenbildung (*8*). (Nach BLAKESLEE 1913, aus HARTMANN 1956.)

Schema kann jeder Stamm entweder männlich oder weiblich reagieren, abgesehen von den beiden eingeschlechtlichen Endgliedern: der Stamm E 87 ist rein männlich, die Stämme 78, 80 und 184 reagieren ausschließlich weiblich, wenn immer sie mit den anderen Stämmen zusammengebracht werden. Dagegen treten die Stämme 11 und 159 gegenüber E 87 als Weibchen auf, gegenüber 89, 107, 190 usw. benehmen sie sich männlich (Abb. 4, RAPER 1947, 1954).

Solcher Erscheinungen wegen sind die niederen Pilze besonders geeignete Objekte für physiologische Untersuchungen nach dem Wesen der phänotypischen Geschlechtsbestimmung.

Unter den Mucorineen, für deren meiste Vertreter haplogenotypische Geschlechtsbestimmung angenommen wird (BLAKESLEE, seit 1904), sind einige monöcische Arten, z. B. *Zygorrhynchus heterogamus* und *Absidia spinosa* bekanntgeworden (Abb. 5). Gelegentlich treten auch bei sonst getrenntgeschlechtlichen Formen sog. neutrale Mycelien auf, die beide Sorten von Gametangien tragen, untereinander kopulieren und Zygoten bilden können. BURGEFF (1915) hat nun bei *Phycomyces* nachgewiesen, daß derartige Pflanzen Miktohaplonten sind, also Kerne beiden Geschlechtes enthalten. Er konnte dies experimentell nachmachen, indem er zwei getrenntgeschlechtliche Hyphenabschnitte aufeinanderpfropfte und so deren Kerne mischte. Die Gemischtgeschlechtlichkeit wird hier nur durch eine Population getrenntgeschlechtlicher Kerne vorgetäuscht.

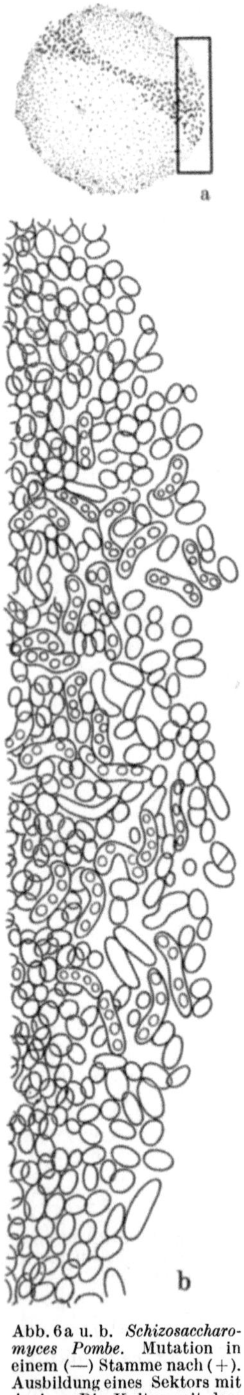

Abb. 6a u. b. *Schizosaccharomyces Pombe*. Mutation in einem (—) Stamme nach (+). Ausbildung eines Sektors mit Asci. a Die Kultur mit dem mutierten Sektor, eine Reaktionslinie zeigend. 60×. b Der Kulturrand an der Sektorengrenze mit Ascusbildung. 700×. (Nach LEUPOLD 1950, aus HARTMANN 1956.)

B. Hefen.

Die Genetik der Hefen wurde besonders in 2 Arbeitskreisen, bei WINGE und bei LINDEGREN, eingehend studiert. Sie fanden, daß ein großer Teil der Hefen getrenntgeschlechtlich ist (LINDEGRENs Paarungstypen a und α). Trotzdem kommen in vielen solchen Kulturen gelegentlich sog. illegitime Kopulationen vor. Die sich in der Folge entwickelnden Asci enthalten Sporen, welche nicht oder nur bedingt lebensfähig sind. Die Gattung *Zygosaccharomyces* stellt offenbar ein Sammelbecken für solche Saccharomyces-Stämme dar, die im Verlauf der Entwicklung „soweit gelangt sind", daß aus illegitimen Kopulationen regelmäßig lebensfähige Sporen erzeugt werden (LINDEGREN und LINDEGREN 1943). Man kann aber auch mit WINGE und ROBERTS einfach annehmen, daß es sich um monöcische Stämme handelt. Es gibt demnach neben rein getrenntgeschlechtlichen Arten andere, die eine mehr oder minder starke Tendenz zur Monöcie haben. Die Frage nach der Geschlechtsbestimmung bei Hefen hat LEUPOLD (1950) näher untersucht. Er fand zunächst, daß seine Kultur von *Schizosaccharomyces pombe* aus einer Mischung von zwei monöcischen, zwei getrenntgeschlechtlichen und einem sterilen Stamm besteht. Genetische Experimente führten diese Erscheinungen auf 2 Gene für Monöcie und zwei Geschlechtsfaktoren zurück. Ein weiteres Beispiel für eine gewisse „sexuelle Instabilität" ergibt sich aus einem mehrmonatigem Kulturversuch eines zwittrigen Stammes. Der Prozentsatz der Sporenbildung ging im Laufe der Zeit zurück. Durch Isolierung von Einzelzell- und Einzelsporkulturen ließ sich nachweisen, daß ein Teil der Zellen

Plus- oder in anderen Fällen zum Minus-Paarungstyp mutiert, also diöcisch geworden war. Umgekehrt traten gelegentlich in Kulturen von reinen Plus- bzw. Minuszellen Gruppen oder Sektoren mit Asci auf (Abb. 6). Falls man sich nicht unbedingt der Deutung LEUPOLDs anschließt und Mutationen als Ursache für den Wechsel von Monöcie und Diöcie ansieht, bleiben weitere Möglichkeiten offen. So hält z. B. HARTMANN (1956) ein bestimmtes Allel für Zwittrigkeit für unwahrscheinlich. Man könnte die Erscheinungen ebenso wie die illegitimen Kopulationen auch als Fälle von relativer Sexualität ansehen.

Eine besondere Art der Kopulation (= Pädogamie) ist bei der Gattung *Nadsonia* bekannt. Eine größere Mutterzelle (Makrogamet) verschmilzt mit ihrer Sproßzelle (Mikrogamet). Die Kerne paaren sich und wandern darauf in

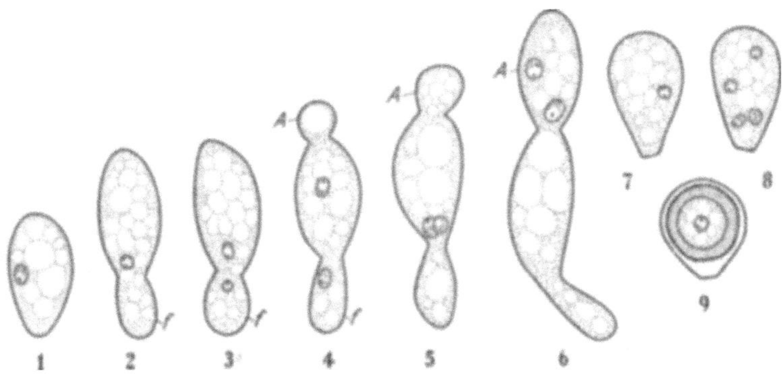

Abb. 7. *Nadsonia*. Pädogame Kopulation. *1—3* Mutterzelle (Makrogamet mit Sproßzelle (Mikrogamet); *5—7* Kopulation der beiden Kerne und ihre Wanderung in den Ascus (*A*); *8—9* Sporenbildung. 1350×. (Nach NADSON und KONOKOTIN 1926, aus GÄUMANN 1952.)

eine zweite Sproßzelle an der gegenüberliegenden Seite, wo die Befruchtung erfolgt und die einzige Ascospore gebildet wird (Abb. 7). Durch Bestrahlung konnten folgende Abweichungen vom normalen Sexualvorgang erhalten werden: 1. Kopulationen eines Makrogameten mit einer fremden Sproßzelle, 2. Kopulationen von zwei Makrogameten, 3. Kopulationen zwischen einem Makrogameten und zwei Mikrogematen. Offenbar fördert Bestrahlung eine Tendenz zur Allogamie (NADSON und PHILIPPOV 1926).

C. *Euascomycetes.*

Der größte Teil der Euascomyceten dürfte als zwittrig anzusehen sein, mit Ausnahme einiger *Laboulbemales* (s. KNIEP 1928, TAXTER 1896, 1908; BENJAMIN und SHANOR 1950), *Taphrinales* und dem Discomyceten *Stromatinia narcissi* (DRAYTON et al. 1952). Im Normalfall werden Ascogone und Antheridien auf dem gleichen Mycel gebildet. Daneben zeigen die Ascomyceten in den verschiedenen Ordnungen eine ausgeprägte Tendenz zur Rückbildung oder zum völligen Verlust der Sexualorgane; so übernehmen verschiedentlich Conidien (Spermatien) oder schließlich vegetative Zellen die Rolle der männlichen Geschlechtszellen. Bei den Arten, wo beide Sexualorgane funktionslos sind oder überhaupt nicht mehr ausgebildet werden, wird das Dikaryon durch autogame Vorgänge hergestellt oder Mycelien mit verschiedenen Inkompatibilitätsfaktoren kopulieren somatogam. Da die durch Inkompatibilitätsfaktoren bedingte Bipolarität oder Tetrapolarität bei den somatogamen Ascomyceten und den Basidiomyceten keine unmittelbare Aussagen über die Sexualität zuläßt, wurden sie hier nicht berücksichtigt.

Ein besonderer Typ phänotypischer Geschlechtsbestimmung bei den Ascomyceten und den Rostpilzen ist dadurch gegeben, daß die monöcischen Mycelien selbststeril sind; diese Selbststerilität wird durch einen genetischen Inkompatibilitätsfaktor an einem Locus gesteuert. Beispiele hierfür wären *Ascobolus magnificus*, *Sclerotina gladioli*, *Neurospora sitophila*. Diese komplizierten Verhältnisse seien hier an der von AMES (1934) untersuchten *Podospora anserina* näher erläutert. Zunächst hatte sich ergeben, daß aus allen der vier gebildeten Ascosporen dieses Pilzes monöcische selbst*fertile* Mycelien entstehen. Die cytologische Untersuchung zeigte, daß solche Sporen normalerweise zwei Kerne besitzen. AMES konnte nun nachweisen, daß verschiedene monöcische Äste in solchen Kulturen selbststeril sind und sich untereinander in zwei Kompatibilitätsgruppen einordnen lassen. Nun kommen neben den normalen allerlei große (mit mehr als zwei Kernen) und kleine einkernige Sporen vor (Abb. 8). Kulturen dieser einkernigen Sporen wachsen zu Mycelien heran, die Ascogone und Spermatien ausbilden, aber selbst *steril* sind. *P. anserina* ist also monöcisch, die Geschlechtsbestimmung erfolgt daher phänotypisch. Das Paarungssystem wird jedoch genetisch durch suprasexuelle Faktoren gesteuert. Weitere neuere Arbeiten über *Podospora* sind bei ESSER und KUENEN (1965) besprochen.

Wie eingangs, erwähnt wird eine Übersicht über die *Sexualitäts*verhältnisse erschwert durch die Verwendung verschiedener Begriffssysteme. WHITEHOUSE (1949), RAPER (1954), BURNETT (1956) u. a. haben versucht, eine gewisse Ordnung zu schaffen (Abb. 1 und 2). Es ist jedoch, wie oben ersichtlich, nicht möglich, den Begriff Homothallie mit Monöcie und Heterothallie mit Diöcie gleichzusetzen. Der Fall *Podospora* zeigt, daß ein heterothallischer Pilz sehr wohl monöcisch sein kann. Der Begriff physiologische Heterothallie — im Gegensatz zu morphologischer Heterothallie, welche wohl mit Diöcie gleichzusetzen wäre — wird einer Unterscheidung zwischen sexuellen und suprasexuellen Faktoren nicht gerecht. In einem Aufsatz über phänotypische Geschlechtsbestimmung muß notwendigerweise auf solche Faktoren Bezug genommen werden, die sich phänotypisch als Geschlechtsmerkmale manifestieren. Die Schwierigkeit tritt erst dort auf, wo solche Merkmale nicht vorhanden sind. In Abb. 9 wird daher allen Varianten des Schemas eine bisexuelle Differenzierung der „Gameten"kerne zugrunde gelegt, Kernen, die a) in Sexualorganen getrennt heranreifen können und deren plasmatische Umgebung (die Thalli) b) durch suprasexuelle Faktoren differenziert sein können, es nicht unbedingt auch sein müssen. Eine Einteilung der Pilze nach ihren Paarungs- und Sexualreaktionen wird eine vergleichend morphologische Betrachtung gebührend berücksichtigen müssen. Das gilt besonders für die Ascomyceten, bei denen innerhalb enger Verwandtschaftskreise Gametangienbefruchtung, Selbststerilität, Somatogamie und Autogamie in mannigfaltigen Variationen vorkommen. Das dürfte sehr einleuchten bei einer Durchsicht der Familien der *Aspergillaceae*, *Sordariaceae* oder *Pyronemaceae*, etwa an Hand von GÄUMANNs Werk „Die Pilze".

Der Begriff phänotypische Geschlechtsbestimmung ist seiner Auffassung nach der genotypischen Geschlechtsbestimmung gegenübergestellt und infolgedessen genetisch begründet in der Weise, daß beide genetische Geschlechtsfaktoren (M = männlich; F = weiblich) in jedem Abschnitt der Entwicklung, ob Haplo- oder Diplophase, vorhanden sein sollen. Damit erhebt sich die Forderung, diese beiden Faktoren in genetischen Zwittern tatsächlich nachzuweisen. Solche Möglichkeiten ergeben sich dort, wo man eingeschlechtliche Mutanten erhalten konnte. So wurden durch HÜTTIG (1935) an *Glomerella lycopersici* und GREIS (1941) an *Sordaria fimicola* genetische Experimente über Faktorenaustausch und Vererbung von Monöcie und Diöcie durchgeführt. Ein neuerer Fall soll noch angefügt werden: Aus der hermaphroditen, selbststerilen Wildform des Ascomyceten *Hypomyces solani f. cucurbitae* haben HANSEN und SNYDER (1943,

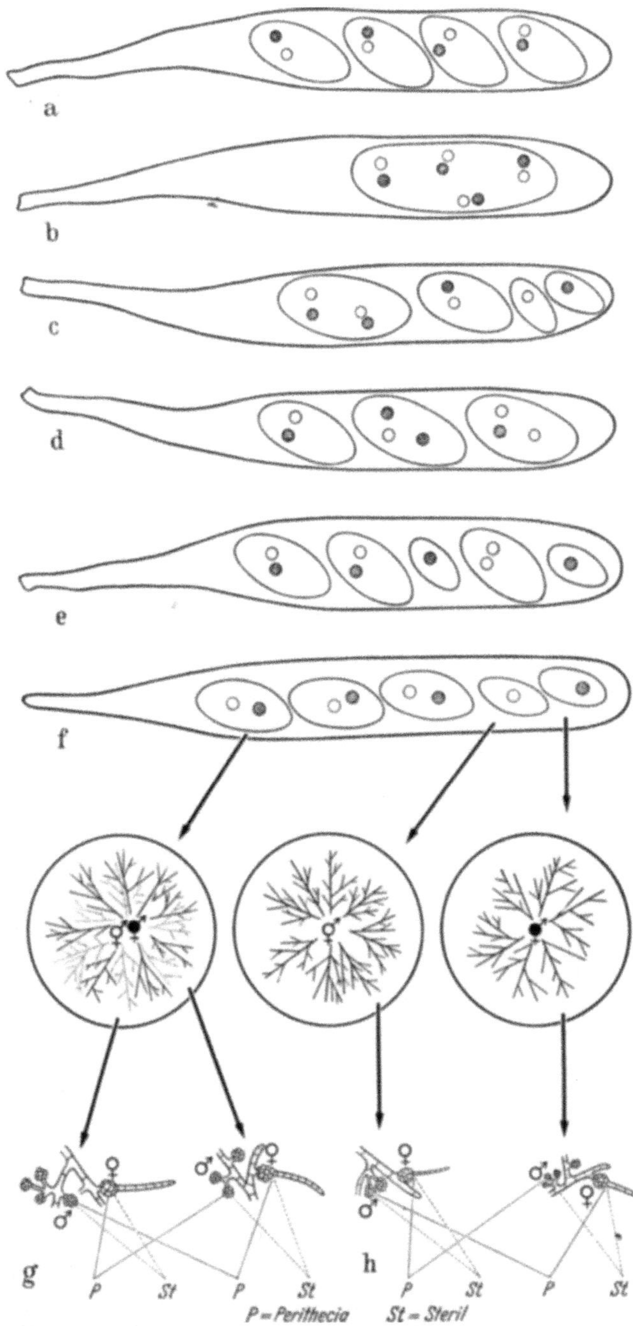

Abb. 8a—h. *Sordaria anserina*. Sexualität- und Selbststerilität. a Normaler Ascus mit 4 zweikernigen Sporen, die alle zu Miktohaplonten auskeimen (g); b—f unregelmäßige Sporenbildung mit einkernigen Sporen in c, e und f. Diese gehören verschiedenen Paarungstypen an, sind selbststeril, obwohl sie fertile Geschlechtszellen besitzen (h). (Schematisch, nach AMES 1934.)

1946) Mutanten erhalten, die keine Protoperithezien anlegen, dagegen befruchtungsfähige Spermatien lieferten, die also als rein männliche Pflanzen anzusehen sind. Weiter wurden Mutanten gewonnen, die zwar Protoperithezien und Conidien liefern, deren Conidien aber nicht als Spermatien brauchbar sind, was also zu physiologisch eingeschlechtlichen weiblichen Stämmen führte. Da diese Mutationen in beiden Kompatibilitätsgruppen auftreten, stehen vier eingeschlechtliche Stämme zur Verfügung (Abb. 2, Spalte IV). Kreuzungen zwischen rein weiblichen und rein männlichen Stämmen ergaben für 200 isolierte Ascosporen: 64 ♀, 84 ♂, 24 ⚥, 28 sterile Mycelien. Die cytologischen Untersuchungen erbrachten, daß die Mutanten durch den Verlust des zweiten bzw. dritten Chromosoms entstanden wären (HIRSCH 1949). Dieser Befund ist aber neuerdings durch EL-ANI (1954) widerlegt worden.

D. Hinweis auf sexuelle Differenzierung von Kernen bei Autogamie.

Die Basidiomyceten und hier besonders die Hymenomyceten sind in der Regel heterothallisch, das Dikaryon wird über somatogame Fusion von Hyphen verschiedener Mycelien hergestellt. Diese Art der Kopulation ist auch vereinzelt unter den höheren Ascomyceten bekannt (z. B. *Penicillium Brefeldianum*, *Sordaria Brefeldii*, *Humaria rutilans*); ebenso könnte man die Paarung der niederen, den Hefen ähnlichen Ascomyceten (*Endomycetales*, *Taphrinales*) unter dem Begriff Somatogamie zusammenfassen. Nimmt man nun an, daß bei solchen Organismen auf Grund unisexueller Differenzierung (und) oder durch Inkompatibilität der Kerne ein Dikaryon innerhalb eines Mycels nicht zustande kommen kann, so mag es erlaubt sein, die homothallischen Arten, bei denen ein Dikaryon „autogam“ aus irgendwelchen Tochterkernen der Gonen gebildet wird, als phänotypisch differenzierte Vertreter hier anzuführen. So keimt z. B. die Basidiospore von *Coprinus sterquilinus* zunächst zu einem siphonalen (vielkernigen, querwandlosen) Mycel aus. Nach gewisser Zeit erfolgt eine differenzierende Kernteilung, deren Tochterkerne von nun an offenbar als Dikaryon zusammenbleiben, also als bipolar differenziert anzusehen sind. Damit ist die Basis für das typische dikaryotische Mycel mit Schnallen- und Querwandbildung gelegt (BULLER 1931, 1958). Wird durch Mikromanipulation einer der Kerne des Dikaryons im Verlauf der Schnallenbildung abgetötet, so entsteht zwar zunächst vorübergehend ein haploides Mycel, das aber später spontan wieder ein Dikaryon herstellt (HARDER 1926). Ähnlich wie bei *C. sterquilinus* liegen die Verhältnisse bei *C. radians* (VANDENDRIES 1925), *C. rostrupianus* (NEWTON 1936) und *C. micaceus* (VANDENDRIES 1927), ebenso bei dem Gastromyceten *Sphaerobolus stellatus* (PILLAY 1923). Gelegentlich ist auch bei anderen Arten in Einsporkulturen Schnallenbildung beobachtet worden (BRUNSWICK 1924 a u. b, VANDENDRIES u. a.), ohne daß über die karyologischen Verhältnisse Näheres bekannt wäre.

Bei *Corticium terrestre* wird nach KNIEP (1913) der dikaryotische Zustand bereits in der Basidiospore hergestellt, in dem die Tochterkerne des Gonenkernes bereits zusammenbleiben; das gleiche wurde bei *Coniophora cerebella* beobachtet (KEMPER 1937). Diesen Beispielen lassen sich noch weitere aus der Klasse der Ascomyceten anfügen.

Bei *Taphrina potentillae*, einem Parasiten auf den Blättern von *Potentilla canadensis*, werden in ähnlicher Weise die Ascosporen zweikernig, so daß das ganze Sproßmycel dikaryotisch ist (ebenso: *T. Johansonii* und *T. coryli*, MARTIN 1924, 1927, 1936). Bei *Morchella elata* (GREIS 1940) und *Sclerotina trifoliorum* (BJÖRLING 1942) wird das Dikaryon in der Perithezium-Mutterzelle durch autogame

Paarung von Kernen gebildet. Damit soll die Aufzählung weiterer Beispiele abgebrochen werden; eine Besprechung aller Variationen von Autogamie würde den Rahmen des Themas sprengen.

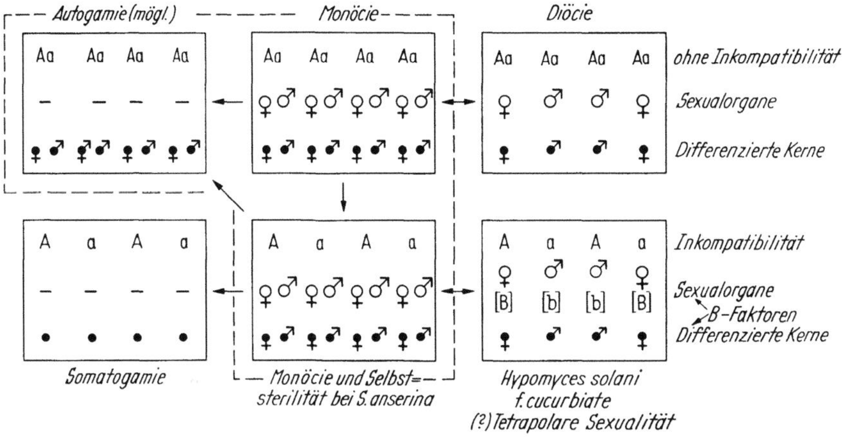

Abb. 9. Versuch einer Gliederung der Sexualitätserscheinungen bei Pilzen unter Berücksichtigung von Inkompatibilitätsfaktoren, Ausbildung von Sexual- bzw. Kopulationsorganen oder Gameten und den mußmaßlichen Kerndifferenzierungen. — — — Bereich der in diesem Artikel berücksichtigten gemischtgeschlechtlichen Fälle mit offenbarer phänotypischer Geschlechtsbestimmung.

Schlußbetrachtung.

Die Fragestellung nach der Ausprägung der Sexualität und damit auch nach dem Modus der Geschlechtsbestimmung führte bei den niederen Pilzen, den Phycomyceten, zu eindeutigen und entwicklungsphysiologisch interessanten Antworten. So ist z. B. bei den Saprolegniaceen und vermutlich auch bei anderen Pilzen (s. Raper in diesem Band) ein Hormonsystem nachgewiesen worden, das bei Monöcisten für eine geregelte Entwicklung der Sexualorgane und den Ablauf der Befruchtung verantwortlich ist. In vielen Ordnungen der Ascomyceten und bei den Basidiomyceten macht sich die Tendenz bemerkbar, die Sexualorgane zurückzubilden und sie durch andere suprasexuelle Faktoren zu überlagern. Um hier jedoch zum Kern der Probleme vorstoßen zu können, müßten besonders zwei Fragen weiter verfolgt werden: 1. Welches sind die physiologischen und stofflichen Grundlagen der Inkompatibilität einerseits und der Sexualität andererseits? 2. Welche Bedingungen bewirken im dikaryotischen System das Wandern, die Anziehung und das Nichtverschmelzen der Kerne in den entsprechenden Zeitabschnitten? Eine Antwort auf die zweite Frage würde darüber hinaus zu einem besseren Verständnis der mannigfaltigen autogamen Befruchtungsweisen und der Apomixis führen.

Literatur[1].

Ames, L. M.: Hermaphroditism involving self-sterility and crossfertility in the Ascomycete *Pleurage anserina*. Mycologia **26**, 392—414 (1934).

Benjamin, R. K., and L. Shanor: The development of male and female individuals in the dioecious species *Laboulbenia formicarum*. Amer. J. Bot. **37**, 471—476 (1950). — Bishop, H.: A study of sexuality in *Sapromyces Reinschii*. Mycologia **32**, 505—529 (1940). — Björling, K.: Undersökningar Rörande Klöverrötan. II. Studier av Utvecklingshistoria och variation hos *Sclerotina trifoliorum*. Statens Vöxlskyddsanst., Stockholm Medd. **37**, 154 S. (1942). — Blakeslee, A. F.: Sexual reproduction in the Mucorineae. Proc. Amer. Acad. Arts Sci. **40**, 205—319 (1904). — Brunswik, H.: Untersuchungen über die Geschlechts- und Kern-

[1] Die neuere Literatur wurde ausführlich bei Esser und Kuenen (1965) berücksichtigt.

verhältnisse bei der Hymenomycetengattung *Coprinus*. Bot. Abh. **5**, 1—152 (1924a). — Neuere Untersuchungen über die Sexualitätsverhältnisse bei den Pilzen (Sammelreferat). Z. indukt. Abstamm.- u. Vererb.-Lehre **24**, 214—228 (1924b). — Buller, A. H. R.: Researches on fungi IV. Further observations on the *Coprini* together with some invetigations on social organisation and sex in the *Hymenomycetes*. New York: Hafner Publ. Co. 1958 (frühere Ausgabe: London 1931). — Burgeff, H.: Untersuchungen über Variabilität, Sexualität und Erblichkeit bei *Phycomyces nitens* Kunze. Flora (Jena), N. F. **108**, 353—448 (1915). — Burnett, J. H.: The mating systems of Fungi. I. New Phytologist **55**, 50—90 (1956).

Cleveland, L. R.: Hormone-induced sexual cycles of flagellates. X. Autogamy and endomitosis in *Barbulanympha* resulting from interruption of haploid gametogenesis. J. Morph. **95**, 189—212 (1954). — Coker, W. C.: The *Saprolegniaceae*, with notes on other water molds Chapel Hill: Univ. North Carolina Press 1923. 201 S. — Couch, J. N.: Heterothallism in *Dictyuchus*, a genus of the water molds. Ann. Bot. **40**, 849—881 (1926).

Drayton, F. D., and J. W. Groves: *Stromatinia narcissi*, a new sexually dimorphic Discomycete. Mycologia **44**, 119—140 (1952).

El-Ani, A. S.: The genetics of sex in *Hypomyces solani f. cucurbitae*. Amer. J. Bot. **41**, 110—113 (1954). — Emerson, R.: A new life cycle involving cyst formation in *Allomyces*. Mycologia **30**, 120—132 (1938). — An experimental study of the life cycles and taxonomy of *Allomyces*. Lloydia **4**, 77—144 (1941). — Emerson, R., and C. M. Wilson: Interspecific hybrids and the cytogenetics and cytotaxonomy of *Euallomyces*. Mycologia **46**, 396—434 (1954). — Esser, K., u. R. Kuenen: Genetik der Pilze. Berlin-Heidelberg-New York: Springer 1965.

Gäumann, E.: The Fungi a discription of their morphological features and evolutionary development. New York and London: Hafner Pubbl. Co. 1952. (Deutsche Ausgabe: Basel, Birkhäuser.). — Greis, H.: Befruchtungsarten bei *Morchella*. Jb. wiss. bot. **87**, 244—266 (1940). — Mutations- und Isolationsversuche zur Beeinflussung des Geschlechtes von *Sordaria fimicola* (Rob.). (Ein Beitrag zur Frage nach der Stabilität und Labilität der Sexualreaktionen bei Monözisten und Diözisten.) Z. Bot. **37**, 1—116 (1941).

Hansen, H. N., and W. C. Snyder: The dualphenomenon and sex in *Hypomyces solani f. cucurbitae*. Amer. J. Bot. **30**, 419—422 (1943). — Inheritans of sex in fungi. Proc. nat. Acad. Sci. (Wash.) **32**, 272—273 (1946). — Harder, R.: Mikrochirurgische Untersuchungen über die geschlechtliche Tendenz der Paarkerne des homothallischen *Coprinus sterquilinus* Fries. Planta (Berl.) **2**, 446—453 (1926). — Hartmann, M.: Die Sexualität. Das Wesen und die Grundgesetzlichkeiten des Geschlechts und der Geschlechtsbestimmung im Tier- und Pflanzenreich, 2. Aufl. Stuttgart: Gustav Fischer 1956. — Hawker, E. L.: The physiology of reproduction in Fungi. Cambridge: Cambrigde Univ. Press 1957. — Hirsch, H. E.: The cytogenetics of sex in *Hypomyces solani f. curcurbitae*. Amer. J. Bot. **36**, 113—121 (1949). — Hüttig, W.: Die Sexualität bei *Glomerella lycopersici* Krüger und ihre Vererbung. Biol. Zbl. **55**, 74—83 (1931). — Humphrey, J. E.: The *Saprolegniaceae* of the United States with notes on other species. Trans. Amer. Phil. Soc. **17**, 63—148 (1892). — Hustede, H.: Untersuchungen über die stoffliche Beeinflussung der Entwicklung von *Stigeoclonium falklandicum* und *Vaucheria sessilis* durch Tryptophanabkömmlinge. Biol. Zbl. **76**, 555—595 (1957).

Kemper, W.: Zur Morphologie und Zytologie der Gattung *Coniophora*, insbesondere des sogenannten Kellerschwamms. Zbl. Bakt., II. Abt. **97**, 100—124 (1937). — Kniep, H.: Beiträge zur Kenntnis der Hymenomyceten. I, II. Z. Bot. **5**, 593—637 (1913). — Die Sexualität der niederen Pflanzen. Jena: Gustav Fischer 1928. 544 S. — Über den Generationswechse- von *Allomyces*. Z. Bot. **22**, 433—441 (1930). — Köhler, K.: Entwicklungsgeschichte, Geschlechtsbestimmung und Befruchtung bei *Chaetomorpha*. Arch. Protistenk. **101**, 223—268 (1956).

Leupold, U.: Die Vererbung von Homothallie und Heterothallie bei *Schizosaccharomyces Pombe*. C. R. Lab. Carlsberg, Sér. physiol. **24**, Nr 27 (1950). — Lindegren, C. C.: Yeast genetics: life cycles, cytology hybrydization, vitamin synthesis, and adaptive enzymes. Bact. Rev. **9**, 111—170 (1945). — Lindegren, C. C., and G. Lindegren: Segregation, mutation, and copulation in *Saccharomyces cerevisiae*. Ann. Mo. bot. Gard. **30**, 453—464 (1943).

Martin, E. M.: Cytological studies of *Taphrina coryli*. Trans. Wiss. Acad. Sci. **21**, 345—356 (1924). — Cytological studies of the *Exoascaceae*. *Taphrina Johansonii* and *Taphrina deformans*. J. Elisha Mitchell Sci. Soc. **43**, 15—16 (1927). — Morphological and cultural studies of *Taphrina potentillae*. Bot. Gaz. **98**, 339—347 (1936). — Maurizio, A.: Beiträge zur Biologie der Saprolegnieen. Z. Fischerei- u. Hilfswiss. **7**, (2) 1—66 (1899). — Murphi, P. A.: The Morphology and Cytology of the sexual organs of *Phytophthora erythroseptica* Pethyb. Ann. Bot. **32**, 115—153 (1918).

Nagai, N.: Studies on the Japanese Saprolegniaceae. J. Fac. Agr. Hokkaido Imp. Univ. **32**, 1—43 (1931). — Nadson, G., and A. Konokotine: Étude cytologique sur les levures a copulation heterogamique du genre *Nadsonia* Syd. Ann. Sci. nat. Bot., Sér. X **8**, 165—183 (1926). — Nadson, G. A., et G. S. Philippov: Sur des anomalis de sexualite chez la levure

Nadsonia fulvescens provoquee par les rayons x. C. R. Soc. Biol. (Paris) **95**, 433—434 (1926). — NEWTON, D. E.: The bisexuality of individual strains of *Coprinus rostrupianus*. Ann. Bot. **40**, 105—128 (1926).

PETHYBRIDGE, G. H., and P. A. MURPHY: On pure cultures of *Phytophthora infestans* De Bary and the development of oospores. Sci. Proc. roy. Dublin Soc. **13**, 566—588 (1913). — PILLAY, T. P.: Zur Entwicklungsgeschichte von *Sphaerobolus stellatus*. Jb. Phil. Fak. Univ. Bern **3**, 197—219 (1923).

RAPER, J. R.: Heterothallism and sterility in *Achlya* and observations on the cytology of *Achlya bisexualis*. J. Elisha Mitchell Soc. **52**, 274—289 (1936). — On the distribution and sexuality of *Achlya ambisexualis*. Amer. J. Bot. **34** (Abstr.) 609 (1947). — Sexual hormons in *Achlya*. VII. The hormonal mechanism in homothallic species. Bot. Gaz. **112**, 1—24 (1950). — Life cycles, sexuality, and sexual mechanisms in the Fungi. Aus: Sex in Microorganism. Washington Amer. Ass. Adv. Sci. **1954**, 42—81. — Some problems of the specificity in the sexuality of plants. In: Biological specificity and growth, 12. Symp. Soc. study of developm. and growth. Princeton: Univ. Press 1955.

THAXTER, R.: Contribution towards a monograph of the *Laboulbeniaceae*. I. Mem. Amer. Acad. Arts Sci. **12**, 195—419 (1896).

VANDENDRIES, R.: I. L'hetero-homothallisme dans le genre *Coprinus*. Bull. Soc. roy. bot. Belg. **57**, 8 (1925). — II. Les mutations sexuelles des *Basidiomycetes*. Bull. Soc. roy. bot. Belg. **58**, 10 (1925). — Nouvelles recherches experimentales sur le comportement sexuel de *Coprinus micaceus*. Mém. Acad. roy. Belg., Cl. Sci., Ser. II **9**, 1—128 (1927).

WHITEHOUSE, H. L. K.: Heterothallism and sex in the fungi. Biol. Rev. **24**, 411—447 (1949). — WINGE, Ö., and C. ROBERTS: Causes of deviations from 2:2 segregations in the tetrads of monohybrid yeast. C. R. Lab. Carlsberg, Sèr. physiol. **25**, 285 (1954b).

Phänotypische Geschlechtsbestimmung bei Archegoniaten.

Von

Leopold Bauer.

Mit 11 Abbildungen.

I. Einleitung.

Für die Mehrzahl der daraufhin untersuchten Archegoniaten ist phänotypische Geschlechtsbestimmung nachgewiesen; abgesehen von einigen noch nicht restlos geklärten Fällen (s. unten) besteht lediglich in der primitivsten Archegoniatengruppe (Bryophyten) neben phänotypischer auch eindeutig genotypische Geschlechtsbestimmung.

Die phänotypische Geschlechtsbestimmung tritt uns in zwei Typen entgegen:

1. Die sexuelle Determination erfolgt während der Entwicklung des gemischtgeschlechtigen *Gametophyten* (normalerweise in der Haplophase), also frühestens *nach vollzogener Reduktionsteilung*. Alle Sporen haben daher (mindestens anfangs) gemischtgeschlechtige Tendenz. Wir sprechen von haplophänotypischer (CORRENS 1934) oder von haplomodifikatorischer (HARTMANN 1956) Geschlechtsbestimmung. Sie stellt den ursprünglichsten Modus unter den höheren Pflanzen dar (CORRENS 1928; Stufe I), von dem aus die phylogenetische Weiterentwicklung bis zu den Blütenpflanzen (über Stufe III zu Stufe IV nach CORRENS) führt, von dem sich aber auch (als Seitenentwicklung) die haplogenotypische Geschlechtsbestimmung eines Teiles der Bryophyten durch Neuerwerbung von Realisatorgenen ableitet (Stufe II nach CORRENS).

Neben *primärer* Gemischtgeschlechtigkeit unterscheidet man noch eine *sekundäre* (CORRENS 1928), welche sich von der haplogenotypischen *Getrennt*geschlechtigkeit ableitet. Sie entsteht durch Regeneration aus dem gemischtgeschlechtigen Sporophyten oder durch Nichtreduktion bei der Sporenbildung; die Gametophyten sind dann diploid. In selteneren Fällen führt auch ein Realisatoraustausch zur Gemischgeschlechtigkeit; die Gametophyten bleiben dann haploid. Die sekundäre Gemischtgeschlechtigkeit hat vor allem für die Bryophyten Bedeutung (s. unten).

Durch Auszählung der Moose Deutschlands (vgl. LIMPRICHT 1895, 1904) hat man versucht, einen Überblick über das Zahlenverhältnis von phänotypischer zu genotypischer Geschlechtsbestimmung bei den Bryophyten zu erhalten (z. B. CORRENS 1899; GREGUSS 1925; HARTMANN 1956). So sollen etwas mehr als die Hälfte getrenntgeschlechtig sein; doch sind diese Zahlen mit der größten Unsicherheit behaftet (vgl. auch KNIEP 1929), denn für die Systematiker sind echte (genotypisch) diöcische Formen von rhizautöcischen bzw. „phänotypisch diöcischen" (s. unten) meist nicht zu trennen; sie erscheinen alle als „diöcisch". Da erst in relativ wenigen Fällen genotypische Getrenntgeschlechtigkeit exakt bewiesen worden ist, bleibt das wirkliche Zahlenverhältnis noch völlig offen.

2. Die Entscheidung über das Geschlecht fällt zu einem bestimmten Zeitpunkt im gemischtgeschlechtigen *Sporophyten* (diplophänotypisch, CORRENS bzw. diplomodifikatorisch, HARTMANN), also *vor der Reduktionsteilung*. Sporen und Gameto-

phyt sind daher (phänotypisch) diöcisch. Dieser Modus (Stufe III nach CORRENS) hat sich zweifellos aus der haplophänotypischen Geschlechtsbestimmung entwickelt.

Tabelle. *Verteilung der einzelnen Archegoniatengruppen auf die verschiedenen Typen der Geschlechtsbestimmung.* (Stufen nach CORRENS.) (Gesperrt = ausschließlich fossile Formen.)[1]

Stufe II *Sphaerocarpus*-Typ isospor und heterospor	← Stufe I *Funaria*- oder *Pteris*-Typ isospor →	Stufe III *Selaginella*-Typ heterospor
Bryophyten (Teil)	**Bryophyten** (Teil)	
	Pteridophyten:	**Pteridophyten:**
	Psilophytinae	—
	Psilotinae	—
	Lycopodiinae:	*Lycopodiinae:*
	Lycopodiales	—
	—	*Selaginellales*
	—	Lepidophytales
	—	Isoetales
	Articulatae:	*Articulatae:*
	Protoarticulatales	—
	Sphenophyllales (Teil)	Sphenophyllales (Teil)
	Equisetales (rezent alle, fossil Teil)	Equisetales (Teil)
	Filicinae:	*Filicinae:*
	Primofilices	—
	Eusporangiatae	—
	Leptosporangiatae: Filicales	Leptosporangiatae: Hydropteridales

[1] Bei den fossilen Formen konnte naturgemäß nur die Unterscheidung zwischen Isosporie und Heterosporie getroffen werden. Die Zuordnung der isosporen Formen zum *Pteris*-Typ hat nur Wahrscheinlichkeitswert; das Vorkommen von genotypischer Haplodiöcie wäre an sich möglich.

II. Geschlechterverteilung.

1. Gametophytenstufe.

Bei den *Bryophyten* besteht eine große Mannigfaltigkeit hinsichtlich der räumlichen Verteilung der Geschlechtsorgane und des Zeitpunktes der Geschlechtertrennung während der Individualentwicklung des Gametophyten.

Wenn die Trennung der sexuellen Tendenzen gegen Ende der vegetativen Phase direkt in der Scheitelregion des Moospflänzchens erfolgt, werden Archegonien und Antheridien in einem Gametangienstand vereinigt. Die Systematiker sprechen (nach der Nomenklatur von LINDBERG) von „*synöcischen*" (zwittrigen) Moosen, wenn diese Art der Verteilung erblich festgelegt ist *(Pellia calycina, Bryum bimum)*. GOEBEL (1928, S. 158) hält dieses Verhalten (und zwar Archegonien nach bzw. über den Antheridien) für das ursprünglichste (vgl. dagegen SCHELLENBERG 1920). Oft wird noch eine „paröcische" Anordnung unterschieden, wenn am gleichen Stämmchen die Gametangienstände des einen Geschlechtes etwas vom anderen, terminalen abrücken *(Webera nutans)*.

Eine noch weitere räumliche Trennung erfolgt, wenn die verschiedengeschlechtigen Gametangienstände außer am Sproßgipfel an verschiedenen Seitenästen ausgebildet werden. In diesem Fall wird von „*monöcischer*" oder auch „*autöcischer*" Verteilung gesprochen (*Funaria hygrometrica*; vgl. auch *Splachnum ampullaceum*, Abb. 10). Wenn wir jedoch mit CORRENS (1934) den Begriff „autöcisch" ganz allgemein für den „gemischtgeschlechtigen Zustand" verwenden (haploautöcisch = gemischtgeschlechtig in der Haplophase; diploautöcisch = gemischtgeschlechtig in der Diplo-(Sporophyten-)phase; vgl. auch HARTMANN 1956, S. 19), wäre es wohl besser, für diesen Verteilungstyp den Ausdruck „monöcisch" zu reservieren (obwohl er ursprünglich für Blütenpflanzen [Sporophytenstufe!] definiert wurde).

„*Rhizautöcisch*" schließlich sind solche Moose, die rein ♀ und ♂ Pflanzen zwar voneinander getrennt, aber am gleichen Protonema bilden (Abb. 1).

Theoretisch wäre außerdem „phänotypische Diöcie" möglich, wie sie auch bei isosporen Farnen und *Equisetum* vorkommt (s. unten), wenn die Determination auf sehr frühen Keimstadien erfolgen würde. Doch ist bei Bryophyten eine solche Geschlechterverteilung (als Regel) noch nicht beschrieben worden; besonders bei Lebermoosen, welche aus einer Spore fast stets nur einen einzigen geschlechtsreifen Gametophyten bilden, wäre danach zu suchen.

Abb. 1. *Ephemerum serratum*. Rhizautöcische Verteilung der ♂ und ♀ Pflanzen; größere ♀ Pflanze bereits mit Sporogon. (Nach GOEBEL 1910.)

Sowohl *Protandrie (Funaria hygrometrica)* als auch *Protogynie (Tayloria lingulata)* sind beobachtet worden. Die Reihenfolge der Geschlechter ist meist erblich festgelegt, doch ist ausnahmsweise eine zeitweilige Umstimmung möglich (s. unten).

Nicht immer sind diese Termini von systematischem Wert. Bei einigen Objekten ist nachgewiesen worden, daß alle gemischtgeschlechtigen Verteilungstypen auch an ein und derselben Sippe verwirklicht sein können [dipl. Sporogonregeneratsippen von *Bryum capillare* (HOFFMANN 1956) und *Splachnum luteum* (BAUER 1959); das haploautöcische *Splachnum ampullaceum* (BAUER, noch unveröffentlicht)]. Ein solches Verhalten bezeichnen die Systematiker als „*heteröcisch*"[1]. Auch für andere Arten wird (ohne daß sich die Angaben auf Beobachtungen an ein und derselben Sippe beziehen) „Heteröcie" angegeben (z. B. *Bryum pendulum*, zit. nach CORRENS 1934). Es werden sogar Arten aufgeführt, welche zwittrige bzw. monöcische und diöcische Verteilung aufweisen (z. B. *Bryum pallescens* und *Webera cruda* bzw. *Leptobryum piriforme*; zit. nach CORRENS 1928 und 1934). Experimentelle und cytologische Daten über solche Arten fehlen jedoch noch. Zur endgültigen Beurteilung jedes Einzelfalles muß streng zwischen *primärer* und *sekundärer* Gemischtgeschlechtigkeit unterschieden werden (CORRENS, S. 22f.; vgl. auch III, 1). Ferner könnte manches dieser Moose auch heteröcisch sein, wobei „Diöcie" versteckte Rhizautöcie (oder „phänotypische Diöcie"?) wäre.

Die räumliche Verteilung der Geschlechtsorgane sagt selbstverständlich noch nichts Verbindliches über den Zeitpunkt der phänotypischen Geschlechtertrennung aus. So treten in zwittrigen Gametangienständen beide Geschlechter fast nie völlig gleichzeitig auf, sie können sogar (beobachtet bei gemischtgeschlechtigen Splachnaceen; BAUER 1959) monatelang rein eingeschlechtig sein. Ebenso besagt die monöcische Verteilung nicht, daß die Entscheidung über das Geschlecht bei der Anlegung der Seitenäste erfolgt; sie fällt in korrelativer Abhängigkeit vom Geschlecht der Achse höherer Ordnung (vgl. unten) erst zu

---

[1] CORRENS (1938, S. 2, Fußnote) hatte diesen Terminus in Abänderung seines bisherigen Sinnes für „getrenntgeschlechtig" vorgeschlagen, später allerdings (1934) zugunsten von „diöcisch" (das ursprünglich eine engere Bedeutung hatte) wieder aufgegeben.

einem viel späteren Zeitpunkt. Lediglich bei Rhizautöcie wird man mit Recht sagen können, daß die Determination auf einem frühen Stadium der Individualentwicklung (spätestens bei der Anlegung der Stämmchenknospen) erfolgt sein muß.

Lycopodium. Experimentaluntersuchungen liegen wegen der schwierigen Kulturbedingungen nicht vor. BRUCHMANN (1898) fand an gesammeltem Material die Prothallien von *L. annotinum, clavatum* und *selago* alle monöcisch, wobei in der Regel auf dem zentralen Teil der generativen Oberfläche zuerst Antheridien auftreten. Bei *L. complanatum* finden sich an vielen Prothallien, besonders wenn

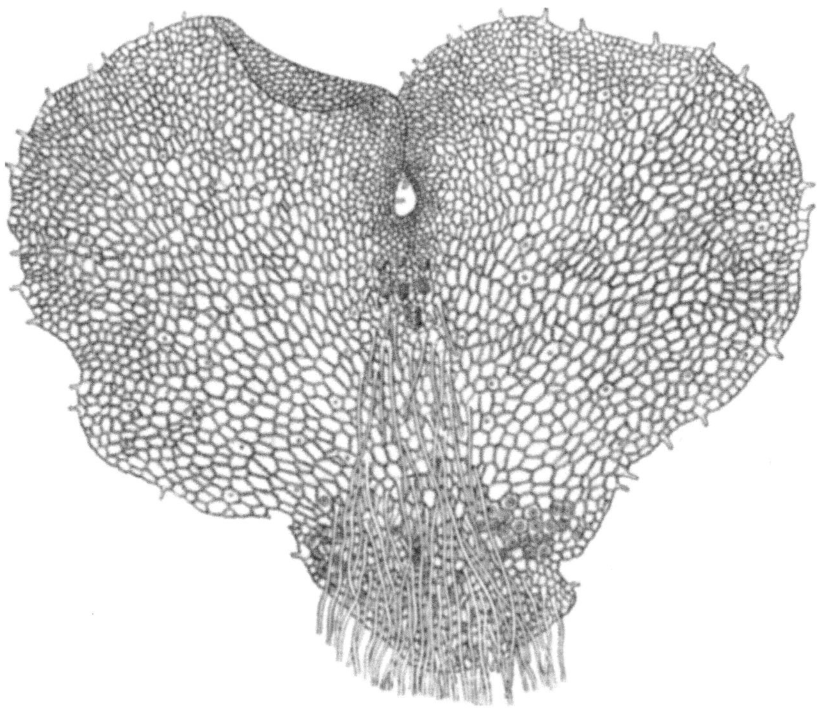

Abb. 2. Prothallium eines leptosporangiaten Farnes *(Dryopteris)* von der Unterseite gesehen; Archegonien im oberen, Antheridien im unteren Teil. (Nach KNY aus WETTSTEIN 1935.)

sie in großer Tiefe wachsen, nur Antheridien. Doch ist hier nur durch ungünstige Außenbedingungen die Manifestation des ♀ Geschlechtes unterdrückt. Rein ♀ Prothallien wurden dagegen nie beobachtet.

Auch bei den tropischen Formen trägt der generative Teil beiderlei Geschlechtsorgane (TREUB 1884, 1890).

Isospore Farne. Bei „typischer" Ausbildung, wie sie in den Lehrbüchern abgebildet ist, trägt das Prothallium an der Unterseite Gruppen von Archegonien (in der Nähe des Meristems) und — teils räumlich von ihnen getrennt (Abb. 2), teils mit ihnen vermischt (DÖPP 1927) — Antheridien. Doch sind Abweichungen von diesem Modus in Abhängigkeit von den Ernährungsbedingungen (s. unten) sehr häufig und für zahlreiche Arten der Normalfall. Die Determination erfolgt dann auf sehr frühen Keimstadien und für das gesamte heranwachsende Prothallium einheitlich, so daß rein ♀ *(Blechnum brasiliense, Gynogramme chrysophylla)* bzw. ☿ *(Ceratopteris thalictroides, Gynogramme sulfurea)* neben rein ♂ Individuen entstehen. CZAJA (1921), der diese Verhältnisse bei einer größeren Zahl von Farnen geklärt hat, spricht von „ökologischer" (fakultativer) Diöcie im Gegensatz zur „habituellen" (obligaten); Correns (1928) gebraucht auch den Ausdruck „phänotypische" Diöcie, den ich für diesen Bericht bevorzuge.

Equisetum. Wir haben hier offenbar eine besonders starke Neigung zu „phänotypischer" Diöcie, d. h. es besteht bereits auf sehr frühen Stadien eine starke Bereitschaft, die Entscheidung über die spätere Geschlechtsausprägung (vor allem des ♂ Geschlechtes!) festzulegen. Wollersheim (1957) diskutiert als frühestmöglichen Determinationszeitpunkt die Reifephase der Sporen. Obwohl auf diese Weise fast immer eingeschlechtige ♂ und ♀ (neben ☿) Individuen auftreten (Abb. 3), neigen alle neueren Untersucher (z. B. Ludwigs 1911; Kashyap 1915, 1917; Schratz 1928; Walker 1931; Wollersheim 1957) dazu, phänotypische Geschlechtsbestimmung anzunehmen. An den ♀ Prothallien läßt sich immer die ♂ Tendenz zum Vorschein bringen, an den ♂ Prothallien jedoch nicht immer die ♀. Obwohl die Zahlenverhältnisse der Geschlechter sehr stark schwanken, beansprucht das sich gelegentlich (Schratz, Wollersheim, Orth) einstellende Verhältnis ♂ : ♀ (bzw. ☿) von etwa 1:1 Aufmerksamkeit, da sich dahinter eine gewisse, jedoch schwach entwickelte genotypische Diöcie verstecken könnte (vgl. auch Correns 1928 und Mäckel 1924).

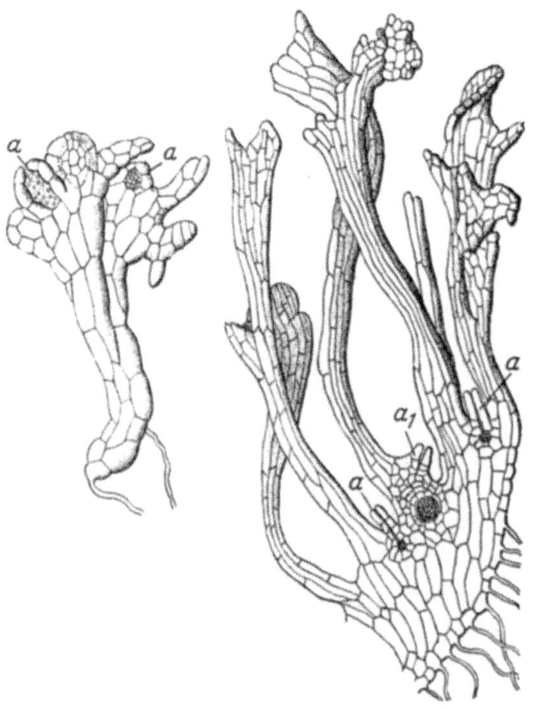

Abb. 3. Phänotypische Diöcie und Dimorphismus bei *Equisetum arvense*. (Nach Hofmeister aus Wettstein 1935).

2. Sporophytenstufe.

Die diplophänotypische Geschlechtsbestimmung findet sich bei den Bryophyten noch nicht, sie hat sich aber in den verschiedenen Verwandtschaftskreisen der höheren Archegoniaten mehrmals unabhängig aus der haplophänotypischen entwickelt; man könnte sich vorstellen, daß die Determinierungsbereitschaft stufenweise in immer frühere Entwicklungsphasen verlegt wurde (s. *Equisetum*) und sogar die Reduktionsteilung übersprang.

Bei *Selaginella* tritt die Geschlechtsdeterminierung spätestens bei der Anlage der Sporangien oder im Laufe ihrer frühen Entwicklungsstadien ein (Goebel 1901): Sie sind entweder Mikro- oder Makrosporangien. Dabei wird deutlich, daß die Mikrosporangien und -sporen sich gegenüber der isosporen Ausgangsstufe am wenigsten verändert haben (Goebel 1901, 1930). Bei den „oligomakrosporangiaten" Selaginellablüten (*S. elegans hort.*) finden wir aber auch — deutlicher als bei den anderen Typen — auffallende Größen- und Gestaltsunterschiede zwischen Mikro- und Makrosporophyllen (Abb. 4).

Die Feststellung von CORRENS (1928, S. 23), daß bei den Selaginellen die Entscheidung, ob ♂ oder ♀ Sporen gebildet werden, bereits bei der Anlage der Sporophylle fällt, bedarf einer kurzen Erläuterung. Bei den Selaginellen (wie übrigens auch bei den Psilotaceen) sind die Sporangien gar keine Blattprodukte (wie üblicherweise), sondern sie entspringen *oberhalb* der Sporophyllanlage (GOEBEL 1930, S. 1232); ursprünglich sind aber wahrscheinlich auch die *Selaginella*-Sporangien einmal blattbürtig gewesen, wie der Vergleich mit *Lycopodium* nahelegt.

Die Verteilung der Makro- und Mikrosporangien ist nicht einheitlich. Bei Formen mit aufrechten, radiären Blüten findet man in der Regel im unteren Abschnitt Makro- und im oberen Mikrosporangien (bei den „Oligomakrosporangiaten" an der Basis der Blüte nur ein einziges sehr großes Makrosporangium; GOEBEL 1928; vgl. Abb. 4). — Als Seltenheit ist gynomonöcische bzw. gynodiöcische Verteilung (*S. pectinata* bzw. *S. Martensii*) oder sogar Triöcie (*S. substipata*) beschrieben worden (nach CORRENS 1934).

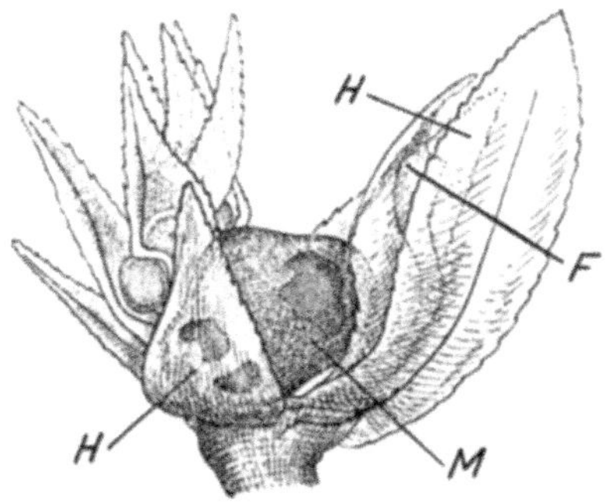

Abb. 4. Blüte von *Selaginella elgans hort.* mit einem Makrosporangium, dessen Sporophyll einen deutlichen Flügel (*F*) hat, und mehrere Mikrosporangien. *H* sterile, das Makrosporangium teilweise umhüllende Blätter. (Nach GOEBEL 1928.)

Bei *Isoetes* gehen in den radiären Sprossen normalerweise ebenfalls die Makroden Mikrosporophyllen voraus (BRAUN 1863, GLÜCK 1924, GOEBEL 1928). Bei schwächlichen Exemplaren von *I. tegulensis* kann die Zahl der Makrosporangien

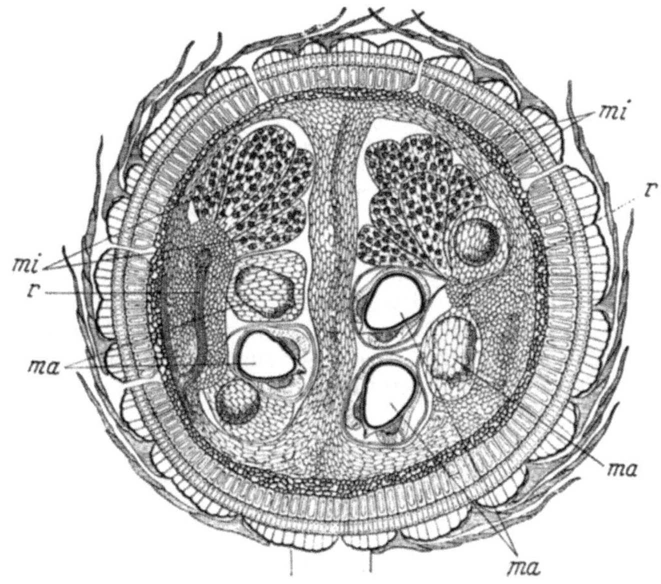

Abb. 5. *Pilularia globulifera.* Längsschnitt durch ein Sporokarp. Zwei Sori mit Makrosporangien (*ma*) und Mikrosporangien (*mi*); Receptaculum (*r*). (Nach LUERSSEN aus R. WETTSTEIN 1935.)

auf 1 oder sogar 0 reduziert sein. Auch *I. melanopodium* bildet teils eingeschlechtige, teils gemischtgeschlechtige Blüten aus; für *I. butleri* wird sogar (wahrscheinlich phänotypische) Diöcie angegeben (nach CORRENS 1934).

Bei den *heterosporen Farnen (Hydropteridales)* sind mehrere Sporangien in Sori vereinigt. Ursprünglich kommen Makro- und Mikrosporangien in einem Sorus gemeinsam vor (GOEBEL 1930). Die sexuelle Determination erfolgt also bei der Anlage der Sporangien. So ist es bei den *Marsiliaceen* (Marschall 1925); vgl. Abb. 5. — Bei *Azolla* finden wir bereits einen Übergang zu eingeschlechtigen Sori. Zwar enthält ein Sorus entweder nur ausgebildete Mikrosporangien oder ein Makrosporangium (Abb. 6), aber die Anlagen der andersgeschlechtigen Sporangien sind in beiden Fällen stets deutlich nachweisbar (Strasburger 1899, Goebel 1930). Auch hier liegt der Determinationszeitpunkt etwa bei der Anlage der Sporangien, doch bekommt jeweils das eine Geschlecht so stark das Übergewicht, daß es die entgegengesetzten Anlagen im Wachstum unterdrückt. —

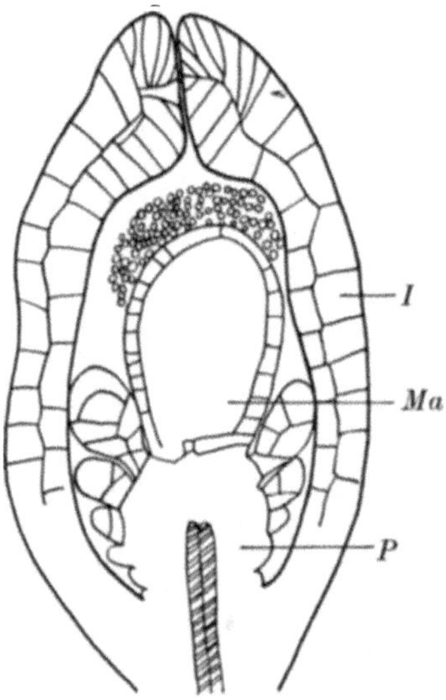

Abb. 6.

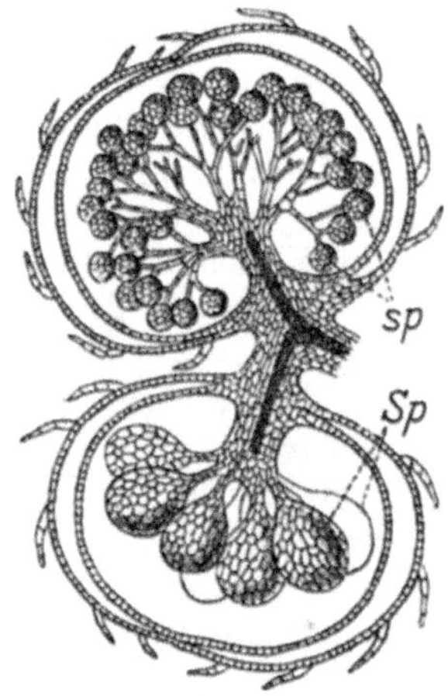

Abb. 7.

Abb. 6. *Azolla filiculoides*. Längsschnitt durch einen Makrosorus. *I* Indusium; *Ma* Makrosporangium; *P* Placenta. Oberhalb des Makrosporangiums sind *Anabaena*-Zellfäden sichtbar, unterhalb des Makrosporangiums Anlagen von Mikrosporangien, welche verkümmern. (Nach Goebel 1930.)

Abb. 7. *Salvinia natans*. Längsschnitt durch einen Sorus mit Mikrosporangien (*sp*) und einen solchen mit Makrosporangien (*Sp*). (Nach Luerssen aus Wettstein 1935.)

Am stärksten abgeleitet ist *Salvinia*, bei der bereits mit der Bildung des Sorus die Entscheidung über das Geschlecht fällt: Es werden normalerweise nur Makrosporangien (Makrosori) oder Mikrosporangien (Mikrosori) gebildet (Abb. 7); gelegentlich findet man allerdings auch Sori mit Makro- und Mikrosporangien zusammen (Mettenius 1846; Heinricher 1882).

III. Auslösende Faktoren der Geschlechtsbestimmung.

1. Genetische Faktoren.

Der AG-Komplex (Correns 1928), dem Hartmann neuerdings (1956) noch eine ganz allgemeine und unbestimmte bisexuelle Potenz BP voranstellt, braucht hier nicht weiter erörtert zu werden (vgl. Beitrag Resende). Darüber hinaus hat Correns noch „Z-Gene" unterschieden, welche die Reihenfolge bestimmen, in der die eigentlichen Geschlechtsgene (A- oder G-Gene seines AG-Komplexes) in Tätigkeit treten und unter anderem auch den Determinationszeitpunkt kontrollieren. Sie würden also dafür verantwortlich sein, daß die Geschlechtsverteilungstypen der Synöcie, Monöcie, Rhizautöcie und phänotypischen Diöcie vielfach wenig variabel sind und in der Systematik als Artkennzeichnung verwendet werden können. — Bei den heterosporen Pteridophyten ist, je nach

Verwandtschaftskreis, der Zeitpunkt der Sporophytenentwicklung, zu dem die Entscheidung über das Geschlecht fällt, erblich festgelegt.

Es soll ferner diskutiert werden, wieweit mit der Möglichkeit zu rechnen ist, daß auch bei phänotypischer Geschlechtsbestimmung (ausbalanzierte) F- und M-Realisatorgene eine Rolle spielen. CORRENS hat die Frage klar verneint — sowohl für haplo- als auch für diplophänotypische Geschlechtsbestimmung (Ausnahmen sekundäre Zwitter). Für ihn treten die Realisatorgene erst auf einer phylogenetisch höheren Stufe als Neuerwerbung auf. HARTMANN nimmt neuerdings (1956) den entgegengesetzten Standpunkt ein (vgl. auch v. WETTSTEIN 1936). Unter dem Eindruck gewisser Experimentalbefunde bei Algen (MOEWUS 1935) und Pilzen (HÜTTIG 1935 und GREIS 1941) über Austausch der Realisatoren zweifelt er, ob es haploide Primärzwitter ohne F und M *überhaupt* gibt. Die Frage kann zur Zeit nicht eindeutig beantwortet werden.

Doch bezieht sich das Problem auch auf die *Sekundärzwitter*, die ja definitionsgemäß mit Realisatorgenen ausgerüstet sind, und so verstanden lautet die Frage. welchen mengenmäßigen Anteil sie an den natürlich vorkommenden gemischtgeschlechtigen Arten haben. Auch darüber wissen wir nichts Sicheres. HEITZ (1942) wies nach, daß gemischtgeschlechtige Laub- und Lebermoose gegenüber ihren getrenntgeschlechtigen näheren Verwandten häufig verdoppelte Chromosomenzahlen besitzen („Polyploidie-Monöcie-Relation"), so in den Gattungen *Metzgeria, Alicularia, Calypogeia, Radula, Mnium.* Auch bei *Bryum caespititium* (im typischen Fall haploautöcisch) sind natürlich vorkommende gemischtgeschlechtige Sippen mit der doppelten Chromosomenzahl der haploautöcischen bekannt geworden (v. WETTSTEIN 1942). Obwohl solche Moose meist keine Gigasmerkmale aufweisen, wird ihre Entstehung aus getrenntgeschlechtigen Arten mit Sicherheit angenommen werden dürfen.

Bei den Laubmoosen sind Sekundärzwitter durch Sporogonregeneration experimentell leicht herstellbar (vgl. vor allem das aus *Bryum caespititium* hervorgegangene *Bryum corrensii*, dessen Entstehung — allmählicher Verlust der Gigasmerkmale! — einen Einblick in die Entstehung der natürlich vorkommenden Sekundärzwitter gestattet). Bei den Lebermoosen *Pellia Neesiana* (SHOWALTER 1927) und *Sphaerocarpus donellii* (LORBEER 1927) wurden diploide Sekundärzwitter gefunden, die mindestens im zweiten Beispiel durch Nichtreduktion bei der Sporenbildung entstanden waren.

Die Wirkung der Realisatoren, bzw. mit ihnen gekoppelter Gene, auf die Geschlechtsbestimmung experimentell hergestellter Sekundärzwitter beleuchtet der Vergleich diploider (♀♂ = protandrisch), triploider (♀♀♂ = protogyn) und tetraploider (♀♀♂♂ = ausgeprägter protandrisch als die diploiden) Pflanzen von *Bryum caespititium* (v. WETTSTEIN 1924). Triploide (♂♂♀) Pflanzen von *Splachnum luteum* sind demgegenüber trotz quantitativen Überwiegens ♂ bestimmender Gene noch immer protogyn, gehen allerdings früher in den zwittrigen Zustand über und bilden leichter „konstante" Männchen als die diploiden (BAUER, noch unveröffentlicht).

Von Bedeutung für den angeschnittenen Fragenkomplex sind auch Schlußfolgerungen, die HEITZ (1927) aus cytologischen Beobachtungen an zwei nahe verwandten Lebermoosarten (*Pellia fabroniana*, haplodiöcisch und *Pellia epiphylla*, haploautöcisch) zieht, von denen die erstere die Chromosomenzahl $n = 8 + 1$ Geschlechtschromosom, die zweite $n = 9$, also *nicht* die doppelte Chromosomenzahl der getrenntgeschlechtigen Verwandten hat! Die Autosomen der getrenntgeschlechtigen Art lassen sich nach ihrer Gestalt — die genau definiert werden kann — mit den entsprechenden Chromosomen der gemischtgeschlechtigen homologisieren; das übrige *Epiphylla*-Chromosom müßte demnach einem

Geschlechtschromosom homolog sein (es ist dem X-Chromosom ähnlicher als dem Y-Chromosom). Wenn die Schlußfolgerung richtig ist, könnte man mindestens bei einem Teil der haploautöcischen Moose mit dem Vorhandensein von realisatorwirksamen Genen rechnen, wobei natürlich größte Vorsicht bei einer Verallgemeinerung geboten ist. Ein Befund v. WETTSTEINs (1942) gibt die Möglichkeit, auch haploautöcische Moose, die offensichtlich „Geschlechtschromosomen" haben, als „Sekundärzwitter" zu deuten: Bei der Aufzucht von Sporen eines Sporogons von *Bryum caespititium* (haplodiöcisch) fand er einzelne Sippen, welche ihrem ganzen Verhalten nach als *haplo*autöcisch aufzufassen waren (der cytologische Beweis konnte nicht geführt werden). Hier wird der seltene Fall eines Realisatoraustausches vermutet, entsprechend den Befunden an Algen (MOEWUS) und Pilzen (GREIS und HÜTTIG).

2. Nichtgenetische Faktoren.

a) Ernährung.

Die Untersuchungen gerade dieses für die Physiologie der phänotypischen Geschlechtsbestimmung wichtigen Gebietes sind noch wenig über erste Beobachtungen hinausgekommen. Man kann die Ergebnisse aller einschlägigen Untersuchungen zusammenfassen und sagen, daß *bei allen Archegoniaten die Pflanzen zur Manifestation der ♀ Tendenz größere Ansprüche an die vegetative Entfaltungsmöglichkeit stellen als zur Manifestation der ♂ Tendenz.*

Bei *Funaria hygrometrica* (z. B. JANSEN 1909, SCHELLENBERG 1921, v. WETTSTEIN 1924) werden unterhalb einer bestimmten Lichtintensität, wenn überhaupt Geschlechtsreife eintritt, an Haupt- und Seitenästen nur Antheridien gebildet, während unter optimalen Bedingungen auf einen Antheridienstand ein Archegonienstand zu kommen pflegt. Man kann durch Steuerung der Beleuchtungsstärke (die hier als Ernährungsfaktor zu verstehen ist) die Ausbildung der Archegonien sehr weit hinausschieben oder aber in anderen Fällen (Rassenverschiedenheiten?) eine mehr gleitende Umstimmung erzielen, so daß an den Seitenästen 1. und höherer Ordnung in allmählich steigender Zahl Archegonien (also ☿ Stände, BROWN 1919) gebildet werden, ehe reine Archegonienstände auftreten (vgl. auch BARTHELMESS an einer *Physcomitrium piriforme*-Mutante 1938). Für *Bryum capillare* (dipl. Regeneratsippe) fand HOFFMANN, daß es nur zur Bildung von ♀ Ästen kommt, wenn die Pflänzchen sich längere Zeit kräftig vegetativ entwickeln konnten. Bei Mutanten, wie der von ERNST (1958) untersuchten *Funaria*-Mutante, kann dieses Bild der unterschiedlichen Ernährungsansprüche noch verstärkt sein: Während „normalerweise" nur Antheridien produziert werden, kann die Archegonienbildung durch Zuckerfütterung ausgelöst werden. — Die allmähliche Vermännlichung der protogynen diploiden Sporogonregeneratsippe von *Splachnum luteum* ist ein weiteres Beispiel (BAUER 1959): Unter den sehr schlechten Bedingungen alter Reagensglaskulturen kann sich die ♀ Tendenz nicht mehr genügend durchsetzen und ermöglicht so das allmähliche Aufkommen der ♂ Tendenz. Hier spielen aber offenbar auch Faktoren eine Rolle, die mit dem Alterungsprozeß zu tun haben (jedoch von der Ernährung nicht exakt zu trennen sind) und die eine „Erschöpfung" der ♀ Tendenz (bzw. deren Manifestationskraft) zur Folge haben.

Zu den Ernährungsfaktoren gehört auch der p_H-Wert des Kulturmediums. BORNHAGEN (1930) hat bei *Splachnum ampullaceum* das Geschlechtsverhältnis im sauren und alkalischen Milieu verglichen; es ergaben sich aber keine reproduzierbaren Werte.

Bei den Prothallien isosporer Farne (CZAJA) und von *Equisetum* (PRANTL 1881; BUCHTIEN 1887; LUDWIGS 1911; KLEBS 1915, 1917; NAGAI 1914; KASHYAP

1914, 1917; SCHRATZ 1928; WALKER 1931; WOLLERSHEIM 1957) kommt der Einfluß des Ernährungsfaktors noch extremer zum Ausdruck: Hier wirkt sich Qualität und Quantität (Mangel) der Nährstoffe bzw. somatische Schädigung (Röntgenbestrahlung, WOLLERSHEIM) stets bereits auf Entwicklungsstadien aus, die weit vor der sexuellen Phase liegen. Völliger Nährstoffmangel im Medium ermöglicht z. B. bei *Ceratopteris thalictroides* nur Antheridienbildung; dazu reichen die Reservestoffe der Spore gerade noch aus (PRANTL). Verhinderung der Archegonienbildung wurde sowohl bei mangelnder C- als auch N- (PRANTL, NAGAI), P- (NOLL nach SCHULTZE 1903), Ca- und S-Versorgung (LUDWIGS 1911) festgestellt. Andererseits kann Glucosefütterung bei *Equisetum limosum* bis zu 90% ♀ Prothallien hervorrufen (WOLLERSHEIM). Oft wird der Ernährungsfaktor nur durch Dicht- und Weitsaat definiert.

Die wenigen diesbezüglichen Beobachtungen bei den heterosporen Pteridophyten gehen in die gleiche Richtung (vgl. GOEBEL 1928). Die Makrosporangien von *Selaginella* stellen größere Ernährungsansprüche und können unter ungünstigen Bedingungen, z. B. während eines trüben, lichtarmen Winters, gelegentlich ganz unterdrückt werden (*S. stenophylla* u. a.), so daß man nur rein ♀ Blüten findet (vgl. auch oben *Isoetes*). — Bemerkenswert ist, daß bei plagiotropen Blüten (auch bei invers dorsiventralen) die Makrosporangien sich an der *Unterseite* (am vegetativen Sproß die geförderte Seite = größere Blätter) befinden. Bei kümmerlich wachsenden Pflanzen (*S. pouzoliana*) tragen Ober- und Unterseite nur Mikrosporangien. Ob die Schwerkraft in irgendeiner Weise an dieser Sporangienverteilung beteiligt ist, wie es HIERONYMUS (zit. nach GOEBEL 1928) für möglich hält, ist experimentell nicht untersucht. Immerhin sind auch bei Selaginellen mit hängenden Blüten oder nach unten gekehrter Spitze nicht selten an der Spitze Makrosporangien. — An schwächlichen Seitenzweigen von *Salvinia* (z. B. am Ende der Vegetationsperiode) treten als Anomalie nur Mikrosori auf (GOEBEL 1928).

b) Wirkstoffe.

Bei den Moosen sind nur wenige diesbezügliche Beobachtungen veröffentlicht. IES und Kinetin haben bei *Splachnum luteum* weder die Ausbildung der Geschlechtsorgane aufgehalten noch eine Geschlechtsverschiebung bewirkt (BAUER 1959, 1963). Bei *Equisetum limosum* fand WOLLERSHEIM (1956) keinen eindeutigen Einfluß von IES auf die Verteilung der Geschlechtsorgane.

DÖPP (1950, 1959) und NÄF (1958) berichten übereinstimmend, daß Extrakte von ♂, mehr sogar noch von ♀ Prothallien von *Pteridium aquilinum* die Bildung von Antheridien fördern. Nur junge Stadien (bis etwa zur Bildung der Herzform) sind für die antheridienfördernde Substanz (A-Substanz, DÖPP) empfindlich. Außer einer früheren Geschlechtsreife kommt es unter ihrem Einfluß auch zu einer Verschiebung des Geschlechtsverhältnisses; sie spielt also eine Rolle bei der sexuellen Determination. — DÖPP errechnet etwa 0,003% als Mindestmenge der A-Substanz in der Versuchslösung, während NÄF aus seinen Experimenten eine Wirkungsgrenze von $1{,}6 \times 10^{-9}$ ableitet. Daraus ergibt sich, daß es sich nicht um eine Antheridienförderung auf dem Umweg einer unspezifischen Hemmung handeln kann. Die A-Substanz wird auch von unverletzten Pflanzen in das Medium abgegeben und es ist nicht ausgeschlossen, daß der „Dichtsaateffekt" nicht lediglich als Nahrungsmangel sondern auch — durch Anreicherung im Nährboden oder durch Direktübertragung auf das Nachbarprothallium — teilweise als A-Substanz-Wirkung aufzufassen ist.

Die Wirkung ist nicht auf *Pteridium aquilinum* beschränkt, sondern ist (obwohl sie auch bei einer Reihe von Objekten bisher negativ war) sogar besonders deutlich z. B. bei *Dryopteris filix mas*, der normalerweise eine lange Reifezeit hat, und bei *Onoclea sensibilis*, der in Kultur keine Antheridien bildet. Überzeugend sind vor allem die Wirkungen auf *Nothlaena sinuata*, einen apogamen Farn, dem normalerweise beiderlei Geschlechtsorgane überhaupt fehlen, der aber unter dem Einfluß der A-Substanz Antheridien bildet (Abb. 8).

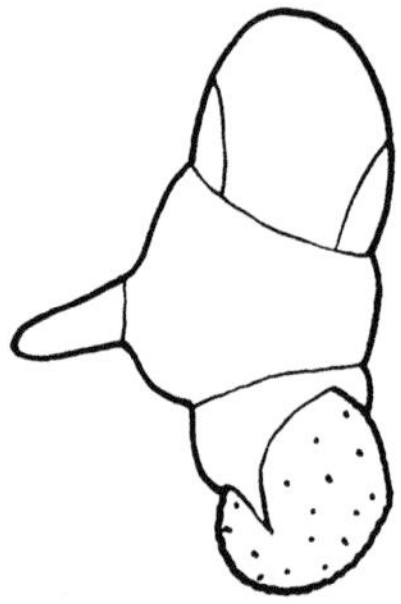

Abb. 8. *Notholaena sinuata.* Wenigzelliger Vorkeim mit jungem Antheridium (A-Substanz-Wirkung). (Nach DÖPP 1959.)

Die A-Substanz ist gut in Wasser, weniger gut in Äther und Aceton löslich, wärmestabil, an Kohle adsorbierbar; sie wird zerstört durch heiße HCl, heißes H_2O_2, nicht aber heißes NH_4OH (DÖPP; NÄF).

Das Unempfindlichwerden älterer Stadien gegen die A-Substanz (keine Antheridienbildung mehr auch bei 15000fach überhöhter Konzentration; NÄF 1958) erklärt DÖPP durch eine antheridienhemmende „(H-)Substanz", deren Bildung er in lebhaft sich teilenden Meristemzellen vermutet (vgl. dazu unten die mit der Entfaltung der ♀ Tendenz gekoppelte „vegetative" Tendenz). SCHRAUDOLF (1962, 1966) konnte bei Schizaeaceen mit Gibberellinen ebenfalls Antheridienbildung auslösen; eine Identität der A-Substanz mit einem Gibberellin scheint aber nicht vorzuliegen (VOELLER 1964).

Prothallien von *Equisetum arvense* behandelte ORTH (1934) mit Follikelhormon. Er fand keine direkte Wirkung auf die Geschlechtsverteilung, wohl aber eine Förderung der vegetativen Entwicklung, die eine Vermehrung der ♀ ausgeprägten Prothallien (bis auf 50%) zur Folge hatte.

c) Korrelationserscheinungen zwischen den sexuellen Tendenzen.

Wenn sich die sexuellen Determinierungsvorgänge zur Zeit der Geschlechtsreife abspielen (Befunde vor allem an Laubmoosen), zeigt sich eine deutliche Abhängigkeit der Manifestation des einen Geschlechts von der vorangegangenen Geschlechtsreife des anderen. Zwei Möglichkeiten der Korrelation sind besonders auffallend:

Der eine Modus kann beispielsweise an den lange bekannten Verhältnissen bei den Funariaceen erläutert werden (z. B. SCHELLENBERG, v. WETTSTEIN 1924). Die Bildung der Archegonien ist hier an das *vorherige* Auftreten von Antheridien gebunden (Protandrie)[1]. Entfernung des Hauptgipfels führt im Seitenast anstelle des eigentlich fälligen ♀ Geschlechtes wieder zu Antheridien. Das läßt sich bis zu Seitenachsen höherer Ordnung fortsetzen. Natürlich wird durch die Entfernung der Gipfelblättchen die Assimilationsleistung des Moosstämmchens herabgesetzt, und man könnte auch einen relativen Nährstoffmangel als Erklärung für den Ersatz des ♀ durch das „anspruchslosere" (♂) Geschlecht mit heranziehen. Doch ist bei mehrfacher Verzweigung trotz Verlust der jeweiligen Astspitze die Assimilationsleistung des Gesamtstämmchens sicher größer als im ungestörten „Normalstämmchen", das nur aus ♂ Haupt- und ♀ Seitensproß besteht. Auch bei Lebermoosen kann man ähnliche Zusammenhänge finden (KACHROO 1956 bei *Mannia indica* und *Asterella pathankotensis*).

Etwas komplizierter sind die Beziehungen beim zweiten Modus, der bei einigen Splachnaceen anzutreffen ist (BAUER 1959, 1961a, 1961b, 1963). In scheinbarer Umkehr der Verhältnisse bei der Protandrie ist hier das Sichtbarwerden der ♂ Tendenz an die vorherige Produktion von Archegonien gebunden (Protogynie). Es

[1] GOEBEL (1930, S. 991) teilt den seltenen Fall mit, daß bei *Funaria* der *erste* Sproß mit einem Archegonium abschloß.

ist aber keine einfache Umkehr (hervorgerufen etwa durch entgegengesetzte Dominanzverhältnisse der verantwortlichen geschlechtsbestimmenden Gene), sondern hier spielt ein neuer, das Grundverhalten überlagernder Faktor eine wesentliche Rolle: er ist an die ♀ Tendenz gekoppelt und soll seinem Verhalten nach vorläufig mit „vegetativer Tendenz des ♀ Geschlechtes" (im folgenden kurz „vegetative Tendenz") umschrieben werden (vgl. BAUER 1959). Er verzögert bzw. hemmt die Reifeprozesse beider Geschlechter (♂ > ♀); unter bestimmten Kulturbedingungen (Dauerlicht von Tageslichtleuchtstofflampen, 1000—2000 Lux, T = 20—23° C) wird bei *Splachnum luteum* (dipl.) nur die Manifestation des ♂ Geschlechtes, bei *Spl. ampullaceum* jedoch die von beiden Geschlechtern (noch unveröffentlicht) unterdrückt. Die Zurückdrängung dieser „vegetativen Tendenz" ist Voraussetzung für die Ausbildung der Geschlechtsorgane, insbesondere der männlichen. Sie ist bei guten Wachstumsbedingungen sehr stabil, aber offenbar empfindlich gegen ungünstige Vegetationsverhältnisse. Ihre Regeneration kann nach einmal eingetretener Reife im ♂ Geschlecht (vielleicht unter deren Einfluß) sehr verzögert und schließlich auch völlig blockiert werden. Da sie ein wesentlicher Teil (bzw. ein Merkmal) der ♀ Tendenz ist, kann sich das ♀ Geschlecht in diesem Extremfall überhaupt nicht mehr manifestieren; die ♂ Determination ist stabilisiert (*Spl. luteum*, dipl.; *Spl. rubrum*, dipl.; *Spl. ampullaceum*, hapl. und dipl.; s. auch unten).

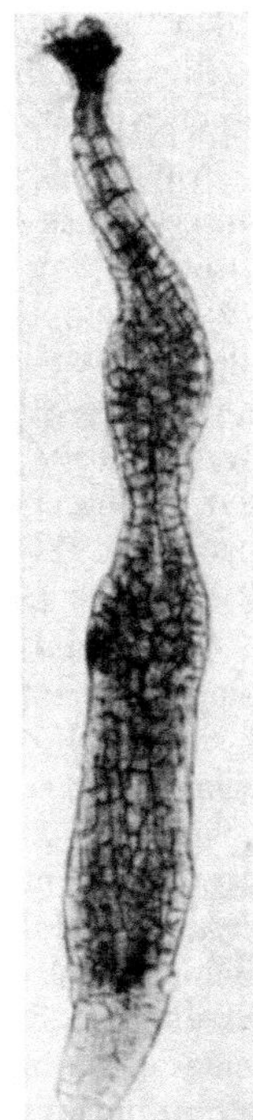

Abb. 9. Doppelbildung von Archegonium und Antheridium bei *Bryum capillare* L. (Nach HOFFMANN 1956.)

Wenn sich die „vegetative Tendenz" nur sehr langsam regeneriert, kann der Fall eintreten, daß eine sonst streng protogyne Sippe (*Spl. ampullaceum*, dipl.; noch unveröffentlicht) über einige vegetative Passagen zu einem protandrischen Verhalten umgestimmt wird. Dabei zeigt sich interessanterweise, daß die bestehende Reife im ♂ Geschlecht die Zeit bis zur Produktion von Archegonien auf wenige Tage verkürzt (normale Dauer unter für die Bildung günstigen Bedingungen mehrere Wochen) — sogar unter Verhältnissen (s. oben), unter denen im „protogynen Zustand" keine Geschlechtsorgane angelegt werden. Man wird annehmen dürfen, daß die Reifeprozesse in beiden Geschlechtern nicht sehr spezifisch, sondern nahe verwandt sind, und daß bei einer Umdeterminierung der Reifezustand vom anderen Geschlecht „übernommen" werden kann. — Wahrscheinlich sind auch die gar nicht so selten zu beobachtenden Mittelbildungen zwischen Archegonien und Antheridien (HOLFERTY 1904; WILSON 1915; HAUPT 1926; v. WETTSTEIN 1942; HAMANT 1943; MOUTSCHEN 1954; HOFFMANN 1956, 1957; BAUER 1959; vgl. Abb. 9) — abgesehen von ihren genetischen Aspekten — so zu interpretieren, daß bei einer sexuellen Umdifferenzierung im Stadium der (relativ unspezifischen) Geschlechtsreife sofort der andersgeschlechtige Organcharakter zum Vorschein kommt. Mittelformen zwischen Archegonien und Antheridien sind übrigens auch bei Pteridophyten anzutreffen (z. B. *Lycopodium complanatum*, BRUCHMANN 1898).

Bei den heterosporen Pteridophyten ist protogynes Verhalten die Regel (vgl. GOEBEL 1928, S. 165ff.). SHATTUCK (1910) konnte bei *Marsilea quadrifolia* nach Zerstörung der zuerst auftretenden Makrosporangien in den Mikrosporangien den Entwicklungsgang in Richtung auf die Bildung von Makrosporen abändern.

IV. Die Stabilität der Geschlechtsdetermination.

Man hat mehrfach durch Regenerationsexperimente an sexuell determinierten Zellen (Antheridien- und Archegonienwand, CORRENS 1920; Bauchkanalzelle, v. WETTSTEIN 1924; beide an *Funaria hygrometrica*) oder an Perichätialblättern eingeschlechtiger Gametangienstände (*Amblystegium*, MARCHAL 1909; *Physcomitrium*, CORRENS 1913) den Beweis dafür angetreten, daß die einseitige sexuelle Determination leicht wieder rückgängig gemacht werden kann, daß also der Determination keine Veränderung im Erbgut zugrunde liegt. Bei Rückführung in den vegetativen Zustand stellt sich das Verhältnis und die Reihenfolge der Geschlechter im Wechselspiel von genetischen (Z-Gene) und Außenfaktoren immer wieder in etwa gleicher Weise ein.

Prinzipiell ähnlich verhalten sich die isosporen Farne — auch jene mit ,,phänotypisch diöcischen" Prothallien und *Equisetum*. GOEBEL (1928, S. 1060) beobachtete an den Halszellen von Archegonien (an alternden Prothallien von *Hemionitis palmata* und *Lygodium japonicum*) kleine Regenerationsprodukte mit Antheridien. Doch ist bei den ♂ determinierten Prothallien das Hervorbringen des anderen Geschlechtes schwieriger als umgekehrt und gelingt nicht einmal immer (CZAJA, SCHRATZ, WOLLERSHEIM); aber auch im ♀ Geschlecht findet sich — wie oben schon erwähnt — eine deutliche Neigung, den Geschlechtscharakter zu bewahren. Wenn keine Befruchtung erfolgt, kann z. B. bei *Blechnum brasiliense* (CZAJA) der ♀ Zustand jahrelang erhalten bleiben.

Auch bei Laubmoosen (Splachnaceen) wurde in letzter Zeit eine bei diesen bisher noch nicht beobachtete Stabilität der sexuellen Determination gefunden (BAUER; vgl. oben, Abschnitt III). Die ♀ Ausprägung läßt sich bei vegetativer Vermehrung unbegrenzt lange aufrechterhalten, andererseits besitzt auch die ♂ Tendenz ein starkes Beharrungsvermögen, wenn sie einmal in Erscheinung getreten ist; im Extremfall wird die ♂ Determination völlig konstant (*Spl. luteum, rubrum, ampullaceum*). Zur Zeit der Geschlechtsreife kann sich in verschiedenen Teilen des Moosstämmchens (nicht nur in der sexuellen Region!) ein unterschiedliches ,,Stärkeverhältnis" der beiden sexuellen Tendenzen herausbilden und wegen der (im Gegensatz zu *Funaria*) relativen Beständigkeit des sexuellen Determinationszustandes sexuell verschiedene Regenerate bedingen.

Während man bei den isosporen Farnen und *Equisetum* noch einwenden könnte, daß die an sich kümmerlichen Regenerate der ♂ Pflanzen eben das anspruchslosere (♂) Geschlecht produzieren, lassen die Ergebnisse an den Splachnaceen diese Deutung nicht zu. Ist die ♂ Ausprägung einmal stabil geworden, läßt sie sich auch durch optimale Kulturbedingungen nicht mehr umstimmen. Da aber der Stabilisierungsvorgang gleitend, über instabile Zwischenstufen, verläuft, dürfte wohl auch hier eine Änderung im Erbgut auszuschließen sein; er trägt eher den Charakter einer Dauermodifikation (BAUER 1959, 1963)). Das Wesen dieser Dauermodifikation dürfte wahrscheinlich mit dem Verhalten jenes Faktors aus dem ♀ bestimmenden Komplex zusammenhängen, der mit ,,vegetativer Tendenz" charakterisiert wurde. Sein Überwiegen ermöglicht die ♀ Ausprägung, seine Zurückdrängung (durch noch unbekannte Faktoren) läßt die ♂ Ausprägung zu[1]. Die relative Beständigkeit *beider* Geschlechter läßt sich also durch den *gleichen* Mechanismus erklären und das oben erwähnte verschiedene ,,Stärkeverhältnis" der Geschlechter als ein verschieden starker Hemmungsgrad der ,,vegetativen Tendenz", und damit der ♀ Tendenz insgesamt, auffassen.

[1] Nach prinzipiell dem gleichen Schema erklärt BAUER (1963) das Auftreten von Zwergmännchen am haploiden Weibchen von *Splachnum rubrum* (genotypische Geschlechtsbestimmung!).

Dieses Geschlechtsverhalten ist bei den Moosen offenbar nicht auf die Splachnaceen (vgl. auch SCHWEIZER 1924; BORNHAGEN 1930; ARENS 1939) beschränkt. Es scheint, daß die bivalens-Sippe von *Barbula unguiculata* (HOFFMANN 1957) auch zu diesem Typ gehört. Ferner hat die einseitig ♂ Ausprägung der Sporogonregeneratsippe von *Eurhynchium stokesii* (MOUTSCHEN 1954) ihre Parallele in den „konstanten" Männchen der Splachnaceen, insbesondere einiger Sporogonregeneratsippen von *Spl. ampullaceum*, welche (obwohl die durch Colchicinbehandlung gewonnene diploide Sippe protogynes Verhalten zeigt!) auch in zahlreichen Subkulturen rein ♂ ausgeprägt sind (BAUER 1961a, 1961b, 1963, vgl. dazu BORNHAGEN 1930). Einige nicht weiter analysierte eigene Beobachtungen seien in diesem Zusammenhang noch angefügt: Von *Bryum caespititium* wurde eine Sporogonregeneratsippe gewonnen, welche nur Antheridien bildete (vgl. dazu v. WETTSTEIN 1924); *Anthoceros crispulus*, das längere Zeit als synöcischer Klon gezogen wurde, verlor in der Kultur irreversibel die Fähigkeit, das ♀ Geschlecht zu manifestieren.

Obwohl die Untersuchungen — vor allem bei den Pteridophyten — noch sehr lückenhaft sind, sieht es doch so aus, als ob sich innerhalb der verschiedenen Verwandtschaftskreise der Archegoniaten dieser besondere Modus des Geschlechtsverhaltens in gleicher Weise herausgebildet hat. Die Gemeinsamkeit besteht darin, daß die geschlechtliche Determination dauermodifikatorische Züge aufweist (vgl. CORRENS 1934), und es wird vermutet, daß auch der Mechanismus der gleiche ist. Als Arbeitshypothese soll von einem *Splachnum-Blechnum-Equisetum*-Typ gesprochen werden. Es ist nicht ausgeschlossen, daß die „H-Substanz" in *Pteridium aquilinum* (DÖPP 1959) etwas mit der „vegetativen Tendenz" zu tun hat.

Bei den heterosporen Pteridophyten konnte die Stabilität der sexuellen Determination noch nicht durch Regenerationsexperimente an den Gametophyten geprüft werden.

V. Sekundäre Geschlechtsmerkmale.

a) Physiologische Merkmale.

Bei den *Bryophyten* sind die meisten bekannten Geschlechtsdifferenzen genotypisch bedingt (vgl. dazu Beitrag BOPP, dieses Handbuch). Aber auch bei phänotypischer Geschlechtsbestimmung (z. B. *Splachnum ampullaceum*, haploautöcisch) treten — schon in der vegetativen Phase — Unterschiede auf: Die Bereitschaft, Stämmchenknospen zu bilden, ist im ♀ Geschlecht größer als im ♂, wenn man die Weibchen mit den stabilisierten Männchen vergleicht (BAUER, noch unveröffentlicht).

Allen Archegoniaten ist gemeinsam, daß das ♀ Geschlecht *größere Ansprüche an die vegetative Entfaltungsmöglichkeit* stellt als das ♂ (längere Reifezeiten), andererseits muß auch auf die *stärkere Entfaltungsbereitschaft* („vegetative Tendenz"), die in gewissen Archegoniatengruppen besonders ausgeprägt ist, hingewiesen werden. Vermutlich beziehen sich beide Gesichtspunkte auf den gleichen physiologischen Faktor, der erste auf seine Bildung (Zusammenhang mit dem Determinationsvorgang), der zweite auf seine Auswirkungen nach erfolgter Determination. Möglicherweise spielt die „vegetative Tendenz" auch bei protandrischen Gametophyten eine gewisse Rolle bei den längeren Reifezeiten im ♀ Geschlecht, aber mit schwächerer Ausprägung. — Das anspruchslosere Geschlecht ist das ♂; es hat daher eine kürzere Reifezeit. Dieser Satz gilt mit der Einschränkung, daß bei protogynen Gametophyten die „vegetative Tendenz" die ♂ Geschlechtsreife hemmt. Fällt dieser überlagernde Faktor weg (stabilisierte Männchen bei Splachnaceen, Männchen phänotypisch diöcischer Farne und von *Equisetum*), dann gilt auch hier, daß die Männchen früher geschlechtsreif werden.

Ob die gelblichere Farbe der ♂ *Equisetum*-Prothallien (SCHRATZ 1928; WOLLERSHEIM 1957) nur eine Folge des schwächlicheren Wuchses der Männchen oder ein echter Ausdruck des geschlechtsspezifischen Stoffwechsels ist, muß noch offenbleiben.

An dieser Stelle seien auch die Befunde von JOYET-LAVERGNE (1926a, b; 1927) erwähnt, nach denen die Sporen von *Equisetum arvense, limosum* und *maximum* sich in solche mit stärkerem und solche mit schwächerem Redoxpotential teilen lassen. Da bei zwittrigen Blütenpflanzen die Samenanlagen stärker reduzieren als die Pollenkörner, glaubt er, auch bei *Equisetum* sexuell verschieden determinierte Sporen vor sich zu haben, wobei in Analogie zu den Blütenpflanzen das ♀ Geschlecht die stärkere Reduktionskraft hätte. SCHRATZ (1928) konnte jedoch die Befunde an seinem *Equisetum*-Material nicht reproduzieren. Versuche dieser Art (vgl. auch MANOILOFF 1922/23, 1924; SATINA und BLAKESLEE 1926; SCHRATZ 1926) entspringen der Suche nach einer für alle geschlechtlich sich fortpflanzenden Organismen identischen biochemischen Reaktion, welche das Wesen des Gegensatzes der Geschlechter charakterisiert.

b) Morphologische Unterschiede.

Der Dimorphismus ist nicht auf genotypische Geschlechtsbestimmung beschränkt. Auch Autöcie läßt nicht selten sehr ausgeprägte Unterschiede zu. — Geschlechtsspezifische Faktoren modifizieren häufig (Laubmoose) sehr deutlich die vegetative Region in der näheren Umgebung der Geschlechtsorgane bzw. der Gametangienstände. In den Antheridienständen der Laubmoose finden sich beispielsweise häufig typisch gestaltete Paraphysen, die in reinen Archegonienständen fehlen (z. B. Funariaceen; vgl. v. WETTSTEIN 1942). Die Gestalt der ♂ Paraphysen ist bei *Splachnum* so charakteristisch (SCHWEIZER, ARENS, BORNHAGEN, BAUER), daß man in den ersten, zwischen den Archegonien erscheinenden Paraphysen (welche vor den Antheridien erscheinen) das allmähliche Hervortreten der ♂ Tendenz erkennen kann.

Abb. 10. *Splachnum ampullaceum*; sexueller Dimorphismus bei monöcischer Verteilung der Geschlechtsorgane. (Nach BORNHAGEN 1930.)

Auch die Perichätialblätter sind häufig mehr oder weniger abgeändert. Bei *Funaria* sind sie der Größe nach etwas verschieden (♀>♂), an den ♂ Ästen scheibenförmig ausgebreitet, im ♀ Geschlecht knospenförmig geschlossen. Extrem verschieden sind sie aber beispielsweise bei der Gattung *Splachnum*, bei der die gemischtgeschlechtigen (auch haploautöcischen) Sippen die gleichen charakteristischen Merkmale aufweisen wie die haplodiöcischen (Abb. 10). Da die Perichätialblätter vor den Gametangien angelegt werden, dürfte die spezifische Gestaltung weniger auf einer „Ausstrahlung" geschlechtsspezifischer Stoffe aus dem generativen Scheitel beruhen als vielmehr darauf, daß bereits in den letzten vegetativen Abschnitten der Stämmchenentwicklung in zunehmendem Maße sexuelle Reifungsprozesse anlaufen. Man könnte sie in Parallele zur „Blühreife" der höheren Pflanzen setzen (HOFFMANN 1956); wenn sie während der Ausbildung der Perichätialblätter stark genug ist, findet man nicht selten an deren Spitzen mehr oder weniger gut ausgebildete Geschlechtsorgane, je nach der vorherrschenden sexuellen Tendenz im Gametangienstand (ELSMANN 1922; SPRINGER 1935; HOFFMANN 1956, 1957; BAUER noch unveröffentlicht). Bei *Splachnum luteum* ist die Neigung, Geschlechtsorgane an den Perichätialblättern zu bilden, im ♂ Geschlecht (dipl.>hapl.) stärker ausgeprägt als im ♀ (BAUER).

Die kürzer dauernde vegetative Phase im ♂ Geschlecht bedingt oft eine geringere Größe der ♂ Pflanzen, und man findet nicht selten bei den isosporen Farnen mit phänotypischer Diöcie und bei *Equisetum* ausgesprochene Zwergmännchen (vgl. Abb. 3). Die kleineren ♂ Prothallien wachsen ohne Meristem, die größeren ♀ Prothallien haben eine deutliche meristematische Region (z. B. PRANTL 1881). Auch bei den Moosen sind auffällige Größenunterschiede geschlechtsverschiedener (rhizautöcischer) Pflanzen anzutreffen, z.B. bei *Ephemerum serratum* (Abb. 1).

Die *heterosporen* Pteridophyten weisen bereits am Sporophyten mehr oder weniger starke Geschlechtsunterschiede auf (Sporophylle: *Selaginella elgans* hort.; Sporangien: *Selaginella erythropus*, GOEBEL 1934).

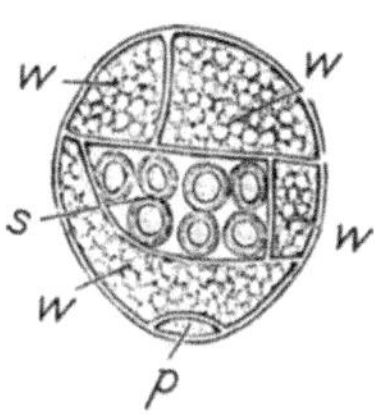

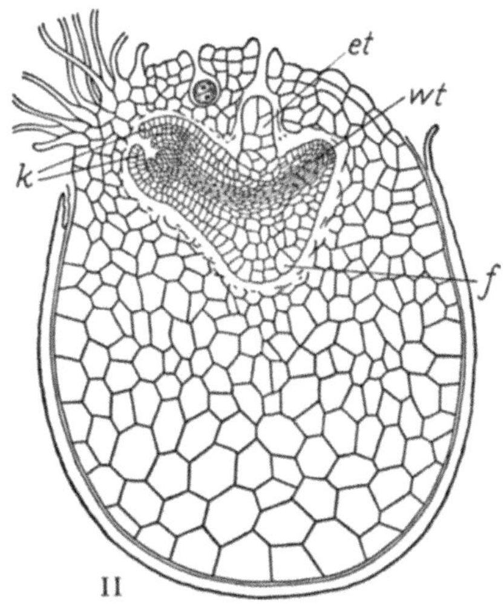

Abb. 11. Gekeimte Mikrospore (*Selaginella stolonifera*) und gekeimte Makrospore (*Selaginella Martensii*); Längsschnitt. *p* Prothallium-(Rhizoid-)Zelle; *w* Antheridiumwandzellen; *s* spermatogene Zellen. (Kombiniert nach BELAJEFF und BRUCHMANN, aus BOWER 1934.)

Besonders ausgeprägt ist jedoch der Dimorphismus in der Größe der Sporen (Mikrosporen ♂; Makrosporen ♀) und der Gametophyten. Diese haben in beiden Geschlechtern eine Verminderung in der Größe und der Ausdifferenzierung gegenüber der isosporen Stufe erfahren, das ♂ Geschlecht stärker als das ♀ (Abb. 11). Damit ist eine Entwicklung eingeleitet, welche durch stufenweise Verstärkung der Gametophytenreduktion zu den Blütenpflanzen überleitet.

VI. Schlußbemerkung.

Angesichts der Tatsache, daß bereits bei den Bryophyten der Weg zur genotypischen Geschlechtsbestimmung eingeschlagen worden ist, darf man fragen, ob nicht auch bei den höheren Archegoniaten wenigstens Ansätze in dieser Richtung vorhanden sind. In der Tat lassen einige noch nicht restlos geklärte Fälle diese Möglichkeit offen, und es soll auf die Problematik ausdrücklich noch einmal hingewiesen werden. Es handelt sich hierbei (wie oben schon angedeutet) um diöcische Formen, für die zum Teil mit Sicherheit „phänotypische" Diöcie nachgewiesen worden ist (CZAJA, SCHRATZ, WOLLERSHEIM). So weist DÖPP (1937, 1950) darauf hin, daß *Pteridium aquilinum* auch unter günstigsten Umständen und bei Dünnsaat kleine ♂ Prothallien bildet und nimmt daher eine „gewisse, schwach entwickelte Diöcie" (d. h. genotypisch) an. Auch der Befund von WOLLERSHEIM, daß Glucosefütterung bis zu 90% ♀ *Equisetum*-Prothallien erzeugt, schließt eine solche Möglichkeit nicht aus, insbesondere, da sich der Prozentsatz nicht ausdrücklich auf die Zahl der ausgesäten Sporen, sondern offenbar nur auf die geschlechtsreif gewordenen Pflanzen bezieht. Noch immer gilt also für diese Formen der Vorbehalt, den CORRENS (1928, S. 22; vgl. dazu

MÄCKEL 1924) geltend macht. Er diskutiert die Möglichkeit, „daß es doch zweierlei Sporen gäbe, solche mit einer *schwachen* männlichen Tendenz und solche mit einer *schwachen* weiblichen, indem zwar zweierlei Realisatoren vorhanden wären, wie bei einem diöcischen Moose, ihre Wirkung aber gegenüber der phylogenetisch zugrunde liegenden, gemischtgeschlechtigen Tendenz und den äußeren Einflüssen nur schwach wäre".

Literatur.

ARENS, P.: Die Geschlechterverteilung bei einigen Splachnaceen. Ber. dtsch. bot. Ges. **57**, 486—494 (1939).

BARTHELMESS, A.: Mutationsversuche mit einem Laubmoos *Physcomitrium piriforme*. I. Phänoanalyse der Mutanten. Z. indukt. Abstamm.- u. Vererb.-Lehre **74**, 479—518 (1938). — BAUER, L.: Zur Physiologie der Geschlechtsdifferenzierung bei *Splachnum luteum*. Planta (Berl.) **53**, 628—646 (1959). — Über die Stabilisierung der ♂ Geschlechtstendenz bei Splachnaceen. Naturwissenschaften **48**, 58 (1961a). — Über die Geschlechtsausprägung der Sporogonregeneratsippen von *Splachnum ampullaceum*. Naturwissenschaften **48**, 58—59 (1961b). — On the stabilization of the male sexual tendency in Musci. J. Linn. Soc. (Bot.) **58**, 337—342 (1963). — BOODLE, C.: The monoecism of *Funaria hygrometrica* Sibth. Ann. Bot. **20**, 293—299 (1906). — BORNHAGEN, H.: Geschlechtsverteilung und Geschlechtsdimorphismus bei *Splachnum ampullaceum* L. und *Splachnum sphaericum* (L. f.) Swartz. Beih. bot. Zbl., I. Abt. **46**, 407—434 (1930). — BOWER, F. O.: Fortpflanzung der Gewächse (Archegoniaten: Farne). In Handwörterbuch der Naturwissenschaften, 2. Aufl., Bd. 4. Jena 1934. — BRAUN, A.: Über die *Isoetes*arten der Insel Sardinien. S.-B. phys.-math. Kl. Berl. Akad., 7. Dez., S. 563 (1863). — BUCHTIEN, O.: Entwicklungsgeschichte des Prothalliums von *Equisetum*. Bibliotheca Bot. H. 8 (1887). — BRUCHMANN, H.: Über die Prothallien und die Keimpflanzen mehrerer europäischer Lycopodien. Gotha 1898.

CAMPBELL, D. H.: The development of the ostrich fern, *Onoclea Struthiopteris*. Mem. Boston Soc. Nat. Hist. **4**, No 2 (1887). — The structure and development of mosses and ferns, 3. Aufl. New York 1918. — CORRENS, C.: Untersuchungen über die Vermehrung der Laubmoose durch Brutorgane und Stecklinge. Jena 1899. — Geschlechterverteilung und Geschlechtsbestimmung (bei Pflanzen). In Handwörterbuch der Naturwissenschaften, Bd. 4. Jena 1913. — Die geschlechtliche Tendenz der Keimzellen gemischtgeschlechtiger Pflanzen. Z. Bot. **12**, 49—60 (1920). — Bestimmung, Vererbung und Verteilung des Geschlechtes bei den höheren Pflanzen. In Handbuch der Vererbungswissenschaft, herausgeg von E. BAUR u. M. HARTMANN, Bd. II. Berlin 1928. — Geschlechtsbestimmung und -verteilung bei Pflanzen. In Handwörterbuch der Naturwissenschaften, 2. Aufl., Bd. 4. Jena 1934. — CZAJA, A. TH.: Über Befruchtung, Bastardierung und Geschlechtertrennung bei Prothallien homosporer Farne. Z. Bot. **13**, 545—587 (1921). — Zur Frage der habituellen Diözie bei *Onoclea Struthiopteris*. Ber. dtsch. bot. Ges. **42**, 300—304 (1924).

DÖPP, W.: Untersuchungen über die Entwicklung von Prothallien einheimischer Polypodiaceen. Pflanzenforsch., H. 8, 1—58 (1927). — Gestaltung und Organbildung innerhalb der Gametophytengeneration der Polypodiaceen unter besonderer Berücksichtigung genetischer Gesichtspunkte. Beitr. Biol. Pflanz. **24**, 201—238 (1936). — Eine die Antheridienbildung bei Farnen fördernde Substanz in den Prothallien von *Pteridium aquilinum* (L.) Kuhn. Ber. dtsch. bot. Ges. **63**, 139—147 (1950). — Über eine hemmende und eine fördernde Substanz bei der Antheridienbildung in den Prothallien von *Pteridium aquilinum*. Ber. dtsch. bot. Ges. **72**, 11—24 (1959).

GERHARDT, E.: Untersuchungen über die Vorkeimentwicklung einiger einheimischer Farne. Diss. Marburg 1927. — GLÜCK, H.: Biologische und morphologische Untersuchungen über Wasser- und Sumpfgewächse. IV. Jena 1934. — GOEBEL, K.: Archegoniatenstudien. IX. Flora (Jena) 88, 207—228 (1901). — X. Flora (Jena)**96**, 1—202 (1906). — Über sexuellen Dimorphismus bei Pflanzen. Biol. Zbl. **30**, 657—679 (1910). — Organographie der Pflanzen, 3. Aufl., Teil 1: Allgemeine Organographie. Jena 1928; Teil 2: Bryophyten-Pteridophyten. Jena 1930. — GREGUSS, P.: Die Sporenverschiedenheit der Musci. Bot. Archiv **12**, 473—480 (1925). — GREIS, H.: Mutations- und Isolationsversuche zur Beeinflussung des Geschlechts von *Sordaria fimicola* (Rob.). Ein Beitrag zur Frage nach der Stabilität und Labilität der Sexualreaktionen bei Monözisten. Z. Bot. **37**, 1—116 (1941).

HARTMANN, M.: Die Sexualität, 2. Aufl. Stuttgart 1956. — HEINRICHER, E.: Die näheren Vorgänge bei der Sporenbildung der *Salvinia natans*. S.-B. ksl. Akad. Wiss. Wien **85**, 492—522 (1882). — HEITZ, E.: Geschlechtschromosomen bei *Pellia Fabbroniana* (diözisch) und *P. epiphylla* (monözisch). Ber. dtsch. bot. Ges. **45**, 607—610 (1927). — Über die Beziehung zwischen Polyploidie und Gemischtgeschlechtigkeit bei Moosen. 2. Jber. schweiz. Ges. Vererb. **17**,

444—448 (1942). — HOLFERTY, G. M.: The archegonium of *Mnium cuspidatum*. Bot. Gaz. **27**, 106—127 (1904). — HÜTTIG, W.: Die Sexualität bei *Glomerella lycopersici* Krüger und ihre Vererbung. Biol. Zbl. **55**, 74—83 (1935).

JACHIMSKY, H.: Beitrag zur Kenntnis von Geschlechtschromosomen und Heterochromatin bei Moosen. Jb. wiss. Bot. **81**, 203—238 (1935). — JOYET-LAVERGNE, PH.: Sur les différences des potentiels d'oxydation réduction dans les spores d'une Prêle: *Equisetum arvense*. C. R. Acad. Sci. (Paris) **182**, 980—982 (1926a). — L'heterogamie des spores des Prêles et les caractères de sexualisation du cytoplasma. C. R. Acad. Sci. (Paris) **182**, 1555—1556 (1926b). — Sur les caractères physico-chimiques de la sexualité dans les spores de *Equisetum maximum*. C. R. Soc. Biol. (Paris) **116**, 1217—1218 (1927).

KACHROO, P.: Probable factors responsible for formation of sporogonia in *Mannia indica* St. and *Asterella pathankotensis* Kash. J. Indian bot. Soc. **35**, 120—122 (1956). — KASHYAP, S. R.: The structure and development of the prothallus of *Equisetum debile* Roxb. Ann. Botany **28**, 163—181 (1914). — Notes on *Equisetum debile* Roxb. Ann. Botany **31**, 439—445 (1917). — KLEBS, G.: Zur Entwicklungsphysiologie der Farnprothallien. I.—III. S.-B. heidelberg. Akad. Wiss., math.-nat. Kl., Abt. B **3**, 3—138; **4**, 3—82 (1916); **7**, 3—104 (1917). — KNIEP, H., u. R. BAUCH: Geschlechtsverteilung bei den Pflanzen. Tabul. biol. ('s-Grav.) **5** (1929).

LAGERBERG, T.: Morphologisch biologische Bemerkungen über die Gametophyten einiger schwedischer Farne. Svensk. bot. T. **2**, 229—276 (1908). — LIMPRICHT, K. G.: Die Laubmoose Deutschlands, Österreichs und der Schweiz. In RABENHORST, Kryptogamenflora, Bd. IV, S. 1—3. 1895—1904. — LORBEER, G.: Untersuchungen über Reduktionsteilung und Geschlechtsbestimmung bei Lebermoosen. Z. indukt. Abstamm.- u. Vererb.-Lehre **44**, 1—109 (1927). — LUDWIGS, K.: Untersuchungen zur Biologie der Equiseten. Flora (Jena) **103**, 385—439 (1911).

MÄCKEL, H. G.: Zur Kenntnis der späteren Entwicklungsstadien der Prothallien von *Equisetum arvense*. Beih. Rep. spec. nov. reg. veget. **28**, 1—36 (1924). — MANOILOFF, E. O.: Über eine chemische Reaktion bei der Geschlechtsbestimmung der zweihäusigen Pflanzen. Bull. Appl. Bot. and Plantbr. **13**, 505 (1922/23). — Weitere Erfahrungen über meine Blutreaktion zur Geschlechtsbestimmung bei Menschen und Tieren und durch Chlorophyll bei Pflanzen. Münch. med. Wschr. **71**, 1784 (1924). — MARCHAL, EL., et EM. MARCHAL: Aposporie et sexualité chez les mousses. II. Bull. Acad. roy. Belg., Cl. Sci. **12**, 1247—1288 (1909). — MARSCHALL, C. C.: Differentiation of sporangia in *Marsilea quadrifolia*. Bot. Gaz. **79**, 85—94 (1925). — METTENIUS, G.: Beiträge zur Kenntnis der Rhizocarpeen. Frankfurt 1846. — MOTTIER, D. M.: Notes on the sex of the gametophyte of *Onoclea struthiopteris*. Bot. Gaz. **50**, 209—213 (1910). — Behaviour of certain fernprothallia under prolonged cultivation. Bot. Gaz. **83**, 244—266 (1927). — MOEWUS, F.: Über die Vererbung des Geschlechtes bei *Polytoma Pascheri* und bei *Polytoma uvella*. Z. indukt. Abstamm.- u. Vererb.-Lehre **69**, 374—417 (1935). — Faktorenaustausch, insbesondere der Realisatoren bei *Chlamydomonas*-Kreuzungen. Ber. dtsch. bot. Ges. **54**, 45—57 (1936).

NÄF, U.: The demonstration of a factor concerned with the initiation of antheridia in polypodiaceous ferns. Growth **20**, 91—105 (1956). — On the physiology of antheridium formation in the bracken fern [*Pteridium aquilinum* (L.) Kuhn]. Physiol. Plantarum (Cph.) **11**, 728—746 (1958). — NAGAI, I.: Physiologische Untersuchungen über Farnprothallien. Flora (Jena) **106**, 281—330 (1914). — NEGER, F. W.: Die Sporenausstreuung bei *Selaginella helvetica* und *S. spinulosa*. Flora (Jena) **101**, 74—77 (1911).

ORTH, H.: Die Wirkung von Follikelhormon auf die Entwicklung der Pflanze. Z. Bot. **27**, 565—607 (1934). — Morphologische und physiologische Untersuchungen an Farnprothallien. Planta (Berl.) **25**, 104—150 (1936).

PRANTL, K.: Beobachtungen über die Ernährung der Farnprothallien und die Verteilung der Sexualorgane. Bot. Ztg **39**, 753—758, 770—775 (1881).

RUMBERG, I.: Entwicklungsgeschichte der Prothallien von *Equisetum silvaticum* und *E. palustre*. Planta (Berl.) **15**, 1—42 (1932).

SATINA, S., and A. F. BLAKESLEE: Biochemical differences between sexes in green plants. Proc. nat. Acad. Sci. (N.Y.) **12**, 191, 197 (1926). — SCHELLENBERG, G.: Über die Verteilung der Geschlechtsorgane bei den Bryophyten. Beih. bot. Zbl., I. Abt. **37**, 115—153 (1920). — SCHMELZEISEN, W.: Beiträge zur Entwicklungsgeschichte der Prothallien einiger Marattiaceen, Cyatheaceen und Polypodiaceen. Flora (Jena) **127**, 46—80 (1933). — SCHRAUDOLF, H.: Die Wirkung von Phytohormonen auf Keimung und Entwicklung von Farnprothallien. I. Auslösung der Antheridienbildung und Dunkelkeimung bei Schizaeaceen durch Gibberellinsäure. Biol. Zbl. **81**, 731—740 (1962). — II. Analyse der Wechselbeziehung zwischen Gibberellinsäurekonzentration, Antheridienbildung und physiologischem Alter der Prothalliumzellen in *Anemia Phyllitidis*. Planta (Berl.) **68**, 335—352 (1966).— SCHRATZ, E.: Zur Frage der Geschlechtsdiagnose auf Grund chemischer Reaktionen. Biol. Zbl. **46**, 727—741 (1926). — Über die Geschlechterverteilung bei *Equisetum*. Biol. Zbl. **48**, 617—639 (1928). —

SCHULTZE, O.: Zur Frage der geschlechtsbildenden Ursachen. Arch. mikr. Anat. **63**, 197 (1903). — SHATTUCK, C. H.: The origin of heterospory in *Marsilia*. Bot. Gaz. **49**, 19—40 (1910). — SHOWALTER, A. M.: Hermaphroditism in dioicious hepatic. Proc. nat. Acad. Sci. (N.Y.) **13**, 369 (1927). — STRASBURGER, E.: Über geschlechtsbestimmende Ursachen. Jb. wiss. Bot. **48**, 427—520 (1910).

TREUB, M.: Études sur les Lycopodiacées. Ann. Jard. bot. Buitenzorg **4**, **5**, **7** (1884 bis 1890).

VOELLER, B.: Antheridogens in ferns. In: Régulateurs naturels de la croissance végétale. Gif s/Yvette, Paris 1964.

WALKER, E.: The gametophyte of *Equisetum laevigatum*. Bot. Gaz. **71**, 370—391 (1921). — The gametophytes of 3 species of *Equisetum*. Bot. Gaz. **92**, 1—22 (1931). — The gametophyte of *Equisetum scirpoides*. Amer. J. Bot. **24**, 40—43 (1937). — WETTSTEIN, F. v.: Über Fragen der Geschlechtsbestimmung bei Pflanzen. Naturwissenschaften **12**, 761 (1924). — Morphologie und Physiologie des Formwechsels der Moose auf genetischer Grundlage. I. Z. indukt. Abstamm.- u. Vererb.-Lehre **33**, 1—236 (1924). — Gesichertes und Problematisches zur Geschlechtsbestimmung. Ber. dtsch. bot. Ges. **54**, 23—38 (1936). — Über einige Beobachtungen und experimentelle Befunde bei Laubmoosen. Ber. dtsch. bot. Ges. **60**, 394—414 (1942). — WETTSTEIN, R.: Handbuch der systematischen Botanik, 4. Aufl. Leipzig u. Wien 1935. — WILSON, M.: Sex determination in *Mnium hornum*. Ann. Botany **29**, 433—440 (1915). — WUIST, E. D.: The physiological conditions for the development of monoecious prothallia in *Onoclea Struthiopteris*. Bot. Gaz. **49**, 216—219 (1910).

Modifikative Geschlechtsbestimmung bei Spermatophyten.

Von

Klaus Napp-Zinn.

Mit 3 Abbildungen.

Bei Spermatophyten kann man ebenso wie bei niederen Pflanzen in Anlehnung an M. Hartmann (1956 und früher) zwischen haplomodifikativer und diplomodifikativer Geschlechtsbestimmung[1] unterscheiden, je nachdem, ob die Modifikation den Gameto- oder den Sporophyten betrifft. Für beides, besonders aber für den zweiten Fall, findet man zahlreiche Beispiele in der teratologischen Literatur, die sich begreiflicherweise meist auf die bloße Deskription beschränkt und deren erschöpfende Wiedergabe in einem Handbuch der Pflanzen*physiologie* sich von selbst verbietet. Viel spärlicher fließen die Quellen der Information über die physiologischen Grundlagen der Geschlechtsausprägung. Dies gilt insbesondere für die

I. Modifikative Geschlechtsbestimmung bei Gametophyten.

Hier sind die Verweiblichung männlicher und die Vermännlichung weiblicher Gametophyten in Betracht zu ziehen. Beide Male handelt es sich darum, daß aus einem Staubblatt oder einem Fruchtblatt Gametophyten desjenigen Geschlechtes hervorgehen, welches man auf Grund der morphologischen Natur des betreffenden Blütenorgans gerade *nicht* erwarten sollte. Solche Phänomene erscheinen aber verständlich, wenn man sich vergegenwärtigt, daß Pollenkorn und Eizelle über „genau die gleichen Anlagen (Potenzen) im gleichen (genotypischen!) Zustand" verfügen (Correns 1928, S. 24).

1. Verweiblichung männlicher Gametophyten.

Die Entwicklung von Pollenkörnern zu embryosackartigen Gebilden wurde namentlich bei *Hyacinthus orientalis* wiederholt festgestellt, so bereits von Němec (1898) bei petaloiden Antheren, später auch von Stow (1930a, b, 1933), Naithani (1937) und de Mol (1921, 1923, 1933, 1934, 1938). Nach Stow (1930a, b) veranlaßte erhöhte Gewächshaustemperatur einen Teil der Pollenkörner, sich zu Embryosäcken auszugestalten (wogegen der größte Teil des Pollens zugrunde ging). Dabei sprang die Exine, die Intine wuchs aus, und durch eine Reihe von Mitosen entstand jeweils ein achtkerniger Embryosack, dessen Eizelle vielleicht sogar befruchtet werden konnte (Stow 1933). — Auch in Versuchen von Naithani wurden durch Wärmebehandlung „unreifer" Zwiebeln (5 Wochen bei

[1] Die Ausdrücke „haplo-" und „diplomodifikativ" können zwar als Hybridwörter unter sprachlichen Gesichtspunkten nicht befriedigen; die Kombination der Zahlwörter mit dem Begriff „phänotypisch" (mit dem man die modifikative Geschlechtsbestimmung vielfach bezeichnet) scheint mir aber aus sachlichen Gründen ungeeignet, und auf die Bildung neuer Termini möchte ich verzichten, um die Übersicht nicht unnötig zu erschweren.

80—82° F = etwa 27° C) in vielen Pollenkörnern weitere Kernteilungen ausgelöst, die zu typischen, achtkernigen Embryosäcken führten. Wenn auch NAITHANI daneben 60% tote Pollenkörner fand, so lehnte er doch STOWs Auffassung ab, derzufolge die sterbenden Pollenkörner durch Abscheidung von Nekrohormonen die Geschlechtsumstimmung der anderen Pollenkörner herbeigeführt haben sollten[1]. — Schließlich konnte DE MOL (1938) zeigen, daß nicht nur Wärme, sondern auch Kälte die Geschlechtsänderung herbeizuführen vermag: „Reif" ausgegrabene Hyazinthenzwiebeln wurden 6—42 Wochen lang verschiedenen, zum Teil auch wechselnden Wärmegraden ausgesetzt. Herrschte zur Zeit der Blütenanlegung eine Temperatur von < 18,3° C oder > 26,7° C, so kam es zu Blütenmißbildungen; < 4,4° C und > 26,7° C zur Zeit der Meiose bewirkten die Ausbildung embryosackähnlicher Pollenkörner.

GEITLER (1941) fand Embryosäcke in Pollenkörnern, die aus kontabeszenten Antheren einiger Exemplare von *Ornithogalum nutans* hervorgegangen waren, aber wieder nur aus solchen Antheren, in denen ein Teil der Pollenkörner (meist < 50% gegenüber 60—100% bei *Hyacinthus*) zugrunde gegangen war. In einigen embryosackartigen Pollenkörnern zählte GEITLER bis zu 16 Kerne. Vergleichbare Erscheinungen fand MOFFETT (1932) bei *Kniphofia*.

Ob der von SUTÔ (1955, S. 144, Abb. 5—7) dargestellten embryosackähnlichen Entwicklung von Pollenmutterzellen von *Zea Mays* mit 3, 5 bzw. 6 teils di-, teils tetra-, teils octoploiden Kernen eine Verweiblichung zugrunde liegt, sei dahingestellt. Dies gilt auch für die von JOHRI und TIAGI (1952) beschriebenen mehrkernigen Pollenkörner von *Cuscuta reflexa*.

Relativ selten scheint sexuelle Umstimmung von Gametophyten bei Dikotylen entdeckt worden zu sein. YAMPOLSKY (1934) fand bei *Mercurialis annua* verschiedene Übergangsbildungen zwischen ♂ und ♀ Sporophyllen (vgl. S. 179), darunter auch Antheren, die neben Pollenkörnern Embryosäcke enthielten. — BRESLAVEC (1935) berichtete über Umwandlung von Pollenmutterzellen in embryosackähnliche Gebilde bei *Cannabis sativa*; auch beschrieb er (1937) hier freie Samenanlagen, die aus dem sporogenen Gewebe von Antheren bzw. aus dem Tapetum entstanden waren. — BERGER (1953) untersuchte 10000 Antheren von zwei *Rubus*-Arten und vier *Rubus*-Bastarden; in 48 Fällen beobachtete er die Entstehung von Embryosäcken (deren acht Kerne aber im Zentrum blieben) aus dem Tapetum bzw. aus der Antherenepidermis. — Endlich entwickeln sich Mikrosporenmutterzellen der Santalacee *Leptomeria Billardierii* zu embryosackähnlichen Gebilden mit zunächst vier Kernen an beiden Polen. Es kommt dann zur Bildung von Eizellapparat und Antipoden und zur Wanderung zweier Kerne in die Mitte (RAM 1959). Niemals war normale Pollenbildung (oder auch nur Meiosis) bei Staubblättern dieser Species zu beobachten, weshalb RAM Apogamie vermutet. Unklar bleibt freilich, inwieweit diese Verweiblichung männlicher Gametophyten gen- oder umweltbedingt ist.

Es muß offenbleiben, ob auch die Entwicklung von Samenanlagen an der Außenseite von Staubblättern bei dem kultivierten *Sempervivum tectorum* (laut CORRENS 1928 schon ALBRECHT V. HALLER bekannt) sowie bei *Papaver orientale* hierher gehört. Dies gilt auch für die Samenanlagen in und an verweiblichten Antheren von *Cannabis sativa* (BRESLAVEC und ZAUROV 1937 u. a.), *Argemone mexicana* (SACHAR 1955), *Digitalis lutea* × *D. purpurea* (HEILBRONN und YAKAR 1942) und *Digitalis lanata* (SCHWERDTFEGER 1954).

2. Vermännlichung weiblicher Gametophyten.

Der umgekehrte Fall — Entwicklung von Zellen des Embryosackapparates zu Pollenkörnern — scheint bislang noch nicht mit Sicherheit beobachtet worden zu sein. Immerhin kann auf Pollenbildung in Samenanlagen von *Begonia Veit-*

[1] NAITHANIS Versuche waren übrigens nur bei einer Sorte (Yellow Hammer) erfolgreich.

chiana hybrida hort. (im Nucellus von Samenanlagen an überzähligen Petalen gefüllter Blüten, GOEBEL 1886; ähnlich SANDT 1921 bei *B. tuberhybrida*) und von *Petunia hybrida* (MOLLIARD 1893, 1896) hingewiesen werden. Auch R. P. SINGHs Abb. 5 (1956), die sich auf *Trewia nudiflora* bezieht, erweckt den Anschein, als ob hier eine — freilich unvollkommene — Samenanlage einen Pollensack umschließe. Weiter fand SINGH (1956) an der Basis des Fruchtknotens mancher ♀ Blüten dieser Euphorbiacee akzessorische Fruchtknoten, die an der Oberfläche Pollensäcke ausdifferenzierten, welche meist einkernige, dreiporige, wenigstens zum Teil keimfähige Pollenkörner entließen. Andere vergleichbare Fälle wurden bei *Salix* (CHAMBERLAIN 1897, HAGERUP 1938), *Mercurialis annua* (STRASBURGER 1910, C. YAMPOLSKY 1934), *Dianthus* (RAO 1940), *Rosa arvensis* (MASTERS 1867), *Passiflora palmata* (SALTER 1863), *Citrus medica* var. *Limonia* (FAROOQ 1952), *Cuscuta reflexa* (JOHRI und TIAGI 1952), *Morus alba* (S. P. SINGH 1954), *Argemone mexicana* (SACHAR 1955) und *Crotalaria spectabilis* (BISWAS 1956) bekannt. Bei *Zelkowa crenata* beobachtete FINN (1935) Mikrosporenbildung in der Wand des Fruchtknotens; möglicherweise war diese Anomalie durch ein Insekt hervorgerufen worden, dessen Larven wiederholt in mißgebildeten Fruchtknoten dieser Art gefunden wurden.

II. Modifikative Geschlechtsbestimmung bei Sporophyten.

Einiges mehr weiß man über die Beeinflußbarkeit des Geschlechts der Sporophyten, genauer: ihrer Blüten. Die weitaus meisten Spermatophyten sind nämlich Zwitter, und nur bei relativ wenigen gibt es ♂ und ♀ Individuen, deren Blüten aber gleichwohl im einen oder anderen Sinne umgestimmt werden können. Überhaupt ist bei allen Spermatophyten eine modifikative Änderung des Geschlechts der Blüten denkbar, die bei gemischtgeschlechtigen Blüten darin bestehen kann, daß ein Geschlecht mehr oder weniger unterdrückt wird.

Dies ist bei vielen eingeschlechtig erscheinenden Blüten schon normalerweise der Fall. Hier können unter Umständen die gewöhnlich verkümmerten Organe des „anderen" Geschlechts durch bestimmte experimentelle Maßnahmen zu voller Entwicklung gebracht werden. Die Zahl der Sporophylle erscheint dann erhöht. Beispiele hierfür trifft man vornehmlich in solchen Verwandtschaftskreisen an, in denen ohnehin gemischtgeschlechtige Blüten die Regel sind. Umgekehrt kann hier aber auch dadurch eine Geschlechtsänderung zustande kommen, daß in normalerweise gemischtgeschlechtigen Blüten die Organe des einen Geschlechtes aus was für Gründen auch immer an ihrer Ausbildung und Entfaltung gehindert werden (vgl. Abb. 1). Endlich sind hier solche Taxa wie gewisse Ranunculaceen-Gattungen in Betracht zu ziehen, bei denen die Zahl der Sporophylle sowieso nicht streng festgelegt ist, bei denen sich deshalb eine Änderung der Geschlechtsverhältnisse auch schon in einer mehr oder minder deutlichen Verschiebung des prozentualen Anteils der Staub- und der Fruchtblätter an der Gesamtzahl der Sporophylle manifestieren kann (s. S. 164). „Verweiblichung" kann deshalb ebensogut Vermehrung oder bessere Entwicklung der Fruchtblätter wie Reduktion der Staubblätter bedeuten.

In allen soeben besprochenen Fällen kann die Hemmung oder Ausbildung der Organe eines Geschlechtes ganz verschiedene Grade erreichen. Fassen wir z. B. die Staubblätter ins Auge: Das eine Extrem besteht darin, daß sie gar nicht in Erscheinung treten, d. h. sich nicht über das Primordialstadium hinaus entwickeln. Das gegenteilige Extrem (d. h. der geringste Grad der Hemmung) ist darin zu sehen, daß die Staubblätter zwar äußerlich normal erscheinen, aber keinen keimfähigen Pollen liefern. In unserer Darstellung werden wir uns im wesentlichen auf solche Fälle beschränken, in denen die Hemmung oder Förderung des einen oder anderen Geschlechtes bereits makroskopisch festzustellen ist, und die zuletzt erwähnten Phänomene nur anhangsweise kurz streifen (vgl. S. 202).

Daß solchen graduellen Unterschieden nicht immer genug Rechnung getragen wurde, mag die Ursache für ein gut Teil der Widersprüche in der einschlägigen Literatur gewesen sein. Als Quelle weiterer Widersprüche (und dies gilt auch für die im nachfolgenden Absatz erörterten Fälle) kommt der Umstand in Betracht, daß — abgesehen von dem als maßgeblich betrachteten Faktor — die Außenbedingungen, unter denen die Versuchspflanzen aufwuchsen, differierten oder sich änderten, sofern sich diese Pflanzen nicht schon selbst hinsichtlich ihres Alters, ihres Habitus oder ihrer genetischen Konstitution unterschieden.

Eine andere Möglichkeit findet man besonders bei Haplantheen verwirklicht, bei denen eingeschlechtige Blüten vorherrschen: Im Falle modifikativer Geschlechtsumstimmung treten hier intersexuelle Sporophylle auf (z. B. Staubblätter, die an ihrer Spitze einen Narbenlappen und am Konnektiv Samenanlagen entwickelt haben); die Zahl der Sporophylle kann hierbei unverändert bleiben (vgl. Abb. 2).

Unter den bisherigen Veröffentlichungen, die einen Überblick über Fälle von modifikativer Geschlechtsbestimmung bei Spermatophyten bieten, sind in erster Linie jene von Correns (1928) und J. Heslop-Harrison (1957) zu nennen. Gerade in den letzten Jahren hat diese Forschungsrichtung aber lebhaften Auftrieb erhalten — einerseits dadurch, daß sich auch die Gibberelline im Hinblick auf die Geschlechtsbestimmung als wirksam erwiesen haben (vgl. S. 199), andererseits durch das Aufkommen praktischer, in erster Linie züchterischer und pflanzenbaulicher Gesichtspunkte (vgl. z. B. die Abschnitte über Gymnospermen, S. 158/59, und über „physiologische Kastration", S. 202). Eine Neudarstellung erscheint daher unumgänglich notwendig. — Weitere Übersichtsreferate, die sich großenteils auf einen oder wenige Aspekte oder auf bestimmte Spermatophyten-Taxa beschränken, verdanken wir H. v. Mohl (1837, rein morphologisch; diskutiert Fälle von Geschlechtsumkehr einzelner Sporophylle im Hinblick auf den morphologischen Wert von Blütenorganen), Düsing (1884, sehr spekulativ, aber mit zahlreichen Hinweisen auf ältere Veröffentlichungen), C. und H. Yampolsky (1922, morphologisch-systematisch), Loehwing (1938), Minina (1952), Resende (1957b), Lang (1961, S. 937), J. Heslop-Harrison (1963) und Köhler (1964a).

Im folgenden wird zunächst eine Übersicht über diejenigen Pflanzenarten[1] gegeben, bei denen Möglichkeiten der modifikativen Geschlechtsbestimmung bekannt geworden sind. Ein Überblick über die modifizierenden Faktoren schließt sich an.

1. Pflanzenarten, bei deren Sporophyten modifikative Geschlechtsbestimmung bekannt ist.

Einleitend mögen einige Begriffe in Erinnerung gebracht werden, die zur Kennzeichnung der Verteilung der Geschlechter verwendet werden. Dabei wird weitgehend die von Correns (1928, S. 29) gegebene Übersicht zugrunde gelegt. Zwar sind noch mehr Begriffe in Gebrauch, doch mag für unsere Zwecke die folgende Liste genügen.

Monoklin (gemischtgeschlechtig) sind Blüten, die Staub- *und* Fruchtblätter enthalten.

Diklin (eingeschlechtig) sind Blüten, die nur Staub- oder nur Fruchtblätter enthalten.

Hermaphroditisch (zwittrig) sind Pflanzen, bei denen alle Blüten gemischtgeschlechtig sind.

Gynomonöcisch sind Pflanzen, deren Blüten teils weiblich, teils gemischtgeschlechtig sind.

Andromonöcisch sind Pflanzen, deren Blüten teils männlich, teils gemischtgeschlechtig sind.

Trimonöcisch sind Pflanzen, deren Blüten teils weiblich, teils männlich, teils gemischtgeschlechtig sind.

Monöcisch sind Pflanzen, deren Blüten teils weiblich, teils männlich sind.

Gynöcisch sind Pflanzen, die nur weibliche Blüten tragen.

Andröcisch sind Pflanzen, die nur männliche Blüten tragen.

Monomorph sind Arten, deren Individuen bezüglich der Geschlechtsausprägung übereinstimmen.

Polymorph sind Arten, deren Individuen bezüglich der Geschlechtsausprägung differieren.

Diöcisch sind Arten, deren Individuen teils gynöcisch, teils andröcisch sind.

Polyöcisch sind Arten, deren Individuen teils nur monokline, teils nur dikline Blüten tragen.

[1] Die Gültigkeit der verwendeten Artnamen wurde nicht überprüft.

Abb. 1. *Silene pendula*. Sporophylle von Pflanzen, die unter dem Einfluß einer Langtagbehandlung gynomonöcisch geworden waren. Die Antheren der beiden linken Blüten sind gänzlich steril. Die mittlere Blüte entwickelte fertilen Pollen in den Antheren der drei längsten der insgesamt zehn Staubblätter. Die beiden rechten Blüten sind funktionell gemischtgeschlechtig, zeigen aber verschiedene Grade der Entwicklung des Gynoeceum zur Zeit der Anthese. (Aufnahme Prof. Dr. J. Heslop-Harrison.)

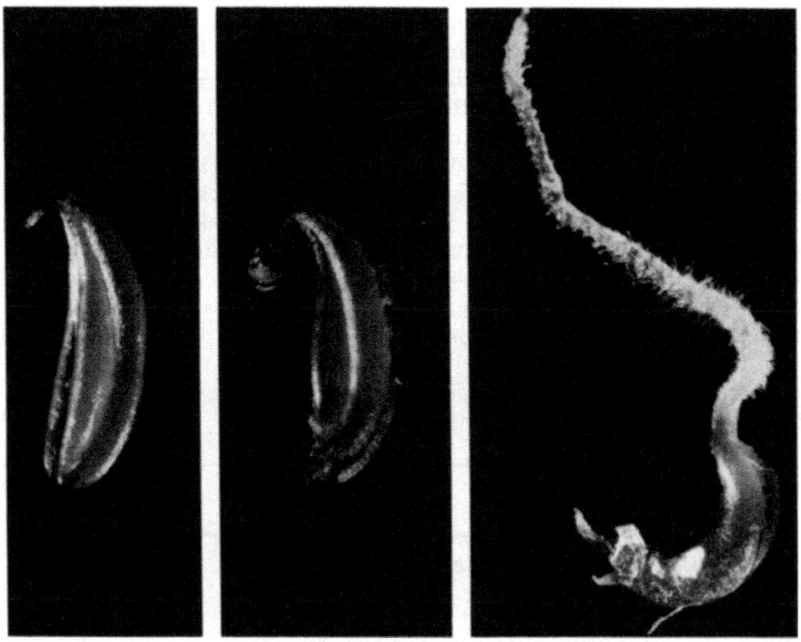

Abb. 2. *Cannabis sativa*. Intersexuelle Sporophylle einer genetisch ♂ Pflanze, von links nach rechts zunehmende Grade der Verweiblichung zeigend. (Aufnahme Prof. Dr. J. Heslop-Harrison.)

Gynodiöcisch sind Arten, deren Individuen teils gynöcisch, teils zwittrig sind.

Androdiöcisch sind Arten, deren Individuen teils andröcisch, teils zwittrig sind.

Beispiele für die selteneren Typen werden bei Correns (1928) genannt. — Bei den Versuchen zur Beeinflussung des Geschlechts stehen Experimente mit monöcischen und diöcischen Pflanzen bzw. Arten im Vordergrund. Das Schwergewicht liegt dabei begreiflicherweise auf Kulturpflanzen (Nutz- und Zierpflanzen), wie sich sogleich zeigen wird.

a) Gymnospermae.

Coniferae.

Die meisten auf Gymnospermen bezüglichen Mitteilungen gehen kaum über die Beschreibung der normalen Anordnung von ♂ und ♀ Blüten innerhalb des Sproßsystems hinaus. Individuen verschiedener (monöcischer) *Pinus*-Arten sind in der Jugend zunächst steril, später ♀ und schließlich — in zur Spitze hin zunehmendem Maße — ♂ (Prat 1930, 1936)[1]. Vorzeitig (d.h. im Alter von weniger als etwa 8 Jahren) blühende Kiefern bilden gewöhnlich zuerst ♀ Zapfen. In einem von Herrmann (1951) beschriebenen Fall erschienen jedoch zunächst ♂ Blüten. Es handelte sich dabei um einen erst 16 Monate alten, noch im kurztrieblosen Primärstadium befindlichen Sämling von *Pinus montana*, der auf einen gleichaltrigen Keimling von *Picea excelsa* gepfropft worden war. — Wareing (1953) konnte junge, etwa 12jährige Exemplare von *Pinus silvestris* durch Entfernung der Endknospe zur vorzeitigen Entwicklung männlicher Blüten veranlassen. Auch wenn die Endknospe durch Insekten beschädigt wird, kommt es zu verfrühter Bildung von ♂ Blüten. — Weitere Beobachtungen über den Einfluß von Resektion auf das Geschlecht von *Pinus*-Blüten verdanken wir Y. Saito (1957), Melchior und Heitmüller (1961) sowie Melchior (1962); bei deren Versuchen standen forstgenetische (bzw. züchterische) Gesichtspunkte im Vordergrund, die eine Abkürzung der Generationsdauer als wünschenswert erscheinen ließen. Was zunächst noch *P. silvestris* betrifft, so reagierten 3jährige Pfropflinge unterschiedlich: Bei manchen Klonen trat eine Vermehrung der ♀ Zapfen ein, bei anderen eine solche der ♂ Blüten. Nach Rückschnitt der Seiten- und Gipfeltriebe 5jähriger Pfropflinge Ende Februar (= am wirksamsten) bis Ende April war die Zahl der ♂ Blüten (pro Pfropfling) erhöht, nicht dagegen nach Resektion im Mai oder Juni. Die Zahl der ♀ Zapfen erfuhr dabei keine wesentliche Änderung (Melchior und Heitmüller 1961). Bei 6jährigen Pfropflingen entsprach der Vermehrung der ♂ Blüten (nach Rückschnitt) eine Abnahme der ♀ (vgl. Tabelle 1). Natürlich ist es nicht immer leicht, mit Sicherheit zu sagen, ob tatsächlich eine Umstimmung von Knospen stattgefunden hat — oder ob bei der Resektion ♀ Zapfenanlagen nur einfach abgeschnitten und dadurch ♂ Blüten zu vorzeitiger Entwicklung veranlaßt worden sind. Um eine Umstimmung scheint es sich jedenfalls in der Regel bei der (wohl oft durch Para-

Tabelle 1. *Mittlere Zahl der ♂ Blüten und der ♀ Zapfen an 6jährigen Pfropflingen von Pinus silvestris nach Rückschnitt.* (Nach Melchior 1962; die von diesem Autor als signifikant betrachteten Unterschiede sind durch Klammern gekennzeichnet.)

Behandlung	♂ Blüten (pro Pfropfling)	♀ Zapfen (pro Pfropfling)
Rückschnitt aller Seitentriebe und des Gipfeltriebes um 1/3 am 22. 2. 60	46,2	0,02
Rückschnitt sämtlicher Seitentriebe um 1/3 am 25. 3. 60	12,2	0,6
Rückschnitt sämtlicher Seitentriebe mit Ausnahme des jüngsten Quirles um 1/3 am 25. 3. 60	2,1	0,8
Nicht zurückgeschnitten (Kontrolle)	0,3	1,5

[1] Schon Meehan (1878b) hatte übrigens bei monöcischen Coniferen (*Pinus*-, *Larix*- und *Abies*-Arten) festgestellt, daß Zweige, die einmal fruktifiziert hatten, später nur noch ♂ Blüten (natürlich an Seitentrieben nächsthöherer Ordnung) produzierten; er machte dafür verschlechterte Ernährung (infolge zunehmender Beschattung oder anderer Faktoren) verantwortlich, merkte aber selbst, daß diese Erklärung z. B. für die diöcische *Juniperus virginiana* nicht ausreichen könne.

siten hervorgerufenen) sog. Zapfensucht von Kiefern und anderen Nadelhölzern sowie bei manchen Befunden von Y. SAITO zu handeln. An Stellen, die normalerweise von ♂ Blüten eingenommen werden, konnte SAITO durch Resektion bestimmter Triebe — bei *Pinus densiflora* im Februar, März und vorzugsweise im April, bei *P. Thunbergii* in der Zeit von Ende Dezember bis Anfang März — die Bildung von ♀ Zapfen hervorrufen. Auch durch rechtzeitiges Besprühen mit α-Naphthylessigsäure konnte Y. SAITO noch nicht ausdifferenzierte Knospen dieser beiden *Pinus*-Arten, die sich ihrer Stellung nach zu ♂ Blüten hätten entwickeln müssen, zu ♀ Zapfen umstimmen. — Laut KARSCHON (1951) wird bei *P. silvestris* das Geschlecht der ersten Blüte durch photo- und thermoperiodische Einflüsse mitbestimmt, wobei geographische Breite und Meereshöhe des Standorts eine Rolle spielen.

Andere Versuche von HEITMÜLLER und MELCHIOR (1960) sowie von MELCHIOR (1960, 1961) zielten darauf ab, bei Lärchen-Arten *(Larix decidua* und *L. leptolepis)* durch „mechanische" Eingriffe vorzeitige Blütenbildung zu erzielen und das Geschlecht der Blüten im einen oder anderen Sinne zu beeinflussen. Die Ergebnisse eines solchen Versuches veranschaulicht Tabelle 2. Die verschiedenen

Tabelle 2. *Einfluß verschiedener mechanischer Eingriffe auf Zahl und Geschlecht der Blüten 4jähriger Pfropflinge von Larix leptolepis.* (Nach MELCHIOR 1961.)

Behandlung	Zahl der untersuchten Pfropflinge	♂ Blüten	♀ Zapfen	Verhältnis ♂ Blüten : ♀ Zapfen
		pro Pfropfling		
Gipfelkrümmen	60	20,9	0,11	180:1
Strangulation	56	5,9	0,16	37:1
Ringelung	37	88,5	12,6	7:1
Unbehandelt (Kontrolle) . . .	77	4,7	0,14	33:1

Behandlungen waren im Mai vorgenommen worden, die Blüten wurden im folgenden Frühjahr ausgezählt. Wie man sieht, hat Gipfelkrümmen nur eine Vermehrung der ♂ Blüten, Ringelung aber daneben eine relativ noch stärkere Vermehrung der ♀ Zapfen und damit, aufs ganze gesehen, eine Verweiblichung bewirkt. Übrigens scheint das Zahlenverhältnis der ♂ Blüten und der ♀ Zapfen bei *Larix* auch mit der von Jahr zu Jahr schwankenden „Blühwilligkeit" zusammenhängen.

Eine ähnliche Beziehung zwischen steriler, ♂ und ♀ Phase besteht nach COURTOT und BAILLAUD (1955) auch bei *Chamaecyparis nootkatensis.* Die Geschlechterverteilung ist hier bei jedem einzelnen Zweig streng fixiert: Die Spitze ist steril, proximal folgt eine ♀ Zone und noch weiter basalwärts findet man schließlich ♂ Blüten tragende Seitenzweige nächsthöherer Ordnung. Abb. 2 in der Notiz von COURTOT und BAILLAUD zeigt sehr schön, wie sich sterile, ♀ und ♂ Zone in einem Zweigsystem von Jahr zu Jahr spitzenwärts weiterschieben. — Bei *Thuja* gelang es MCVEY (1961), durch Behandlung mit Gibberellinsäure die Bildung von ♀ Zapfen zu fördern.

Cycadinae.

Ein Exemplar von *Cycas revoluta* wurde mit der Säge längs halbiert; die eine Hälfte blieb in Mentone und blühte ♀, die andere kam nach Cap Martin und blühte aus unbekannten Gründen ♂; ihr Pollen wurde zur Bestäubung der erstgenannten benutzt (SCHUSTER 1932, S. 43/44).

b) Monocotyledoneae.

Scheuchzeriaceae (Juncaginaceae).

Bei *Lilaea* finden sich an der Basis des ährenförmigen Blütenstandes einige ♀ Blüten mit einem langgriffeligen Fruchtblatt. Weiter oben folgen zahlreiche ⚥ Blüten mit je einem Staub- und Fruchtblatt. Den Abschluß der Inflorescenz bilden einige ♂ Blüten, die nur je ein Staubblatt enthalten (ENGLER und GILG 1919).

Alismataceae.

Mitteilungen über die Geschlechtsverhältnisse von *Sagittaria*-Arten haben HILDEBRAND (1893) und SCHAFFNER (1929) veröffentlicht. Bei *S. sagittifolia* umfaßt die Inflorescenz normalerweise zwei untere Wirtel von je drei ♀ Blüten und darüber fünf bis sechs Wirtel von je drei ♂ Blüten, wobei erstere rudimentäre Stamina und letztere einen Schopf unvollkommener Fruchtblätter enthalten. In einem von HILDEBRAND beschriebenen abnormen Fall war eine ♀ Blüte des untersten Wirtels durch einen Trieb mit vier Blütenwirteln ersetzt, von denen der unterste zwei annähernd monokline, die übrigen aber nur ♂ Blüten besaßen. — SCHAFFNER glaubt, daß die Umschlagstelle von ♀ zu ♂ Blüten in dem grundsätzlich ebenso aufgebauten Blütenstand von *S. latifolia* verschiebbar sei und ihre Lage von einem physiologischen Gleichgewicht abhänge. Er stützt sich dabei allerdings auf keinerlei Experimente, sondern allein auf die Untersuchung der Blütenverhältnisse von je 100 Individuen dreier Lokalpopulationen, bei denen 43—57% der Blüten ♀ waren, ohne daß sich hierbei eine Beziehung zur Gesamtzahl der Blüten gezeigt hätte.

Hydrocharitaceae.

Bei *Stratiotes aloides* dehnt sich das Areal der ♀ Exemplare in Europa um 13° weiter nach Norden aus als das der ♂ (68° gegenüber 55° n. Br.), was von GEDDES und THOMPSON (1889) nicht auf modifikative Geschlechtsbestimmung, sondern auf unterschiedliche Elimination beider Geschlechter, also auf ein sekundäres Geschlechtsmerkmal, und von GRAEBNER (1908) auf ausschließlich vegetative Fortpflanzung in weiten Gebieten zurückgeführt wird. (In anderen Fällen können bei diöcischen Pflanzen Abweichungen vom normalen Geschlechterverhältnis — 1:1 — auf verschiedener Häufigkeit ♂- und ♀-bestimmender Befruchtungen beruhen, wodurch ebenso wie durch die unterschiedliche Überlebensrate, Augenfälligkeit und Blütezeit der ♂ und ♀ Individuen modifikative Geschlechtsbestimmung vorgetäuscht werden kann, vgl. CORRENS 1928 und HESLOP-HARRISON 1957.)

Liliaceae.

Nach Versuchen von LEWITZKY (1925) begünstigt schlechte Ernährung bei dem andromonöcischen *Veratrum nigrum* die Entwicklung von rein ♂ Blüten. Ähnliches vermutete schon H. MÜLLER (1881) in bezug auf *V. album*. Durch Einschnitte in den Stengel konnte LEWITZKY die Entwicklung monokliner statt männlicher Blüten — und umgekehrt — hervorrufen. — *Asparagus officinalis* gilt normalerweise als diöcisch. Die Blüten werden jedoch monoklin angelegt, und ein Geschlecht wird später unterdrückt. Nur selten wird in Blüten von ♂ Pflanzen das Gynoeceum voll entwickelt (RICK und HANNA 1943). Auch VAZART (1959) betrachtet anscheinend Individuen mit vollentwickelten monoklinen Blüten im allgemeinen als genetisch ♂. Auf einem Spargelfeld verhält sich nämlich die Zahl der ♀ Pflanzen zu der der ♂ und zwittrigen Stöcke gewöhnlich etwa wie 1:1. Allerdings fand er auch Stöcke, welche Triebe mit gemischtgeschlechtigen und andere mit rein ♀ Blüten hervorbrachten. MALHOTRA (1930) berichtete, daß durch Behandlung des zur Bestäubung verwendeten Pollens mit ätherischem Öl der normale ♀♀-Überschuß bei *Asparagus officinalis* in einen ♂♂-Überschuß umgekehrt worden sei, wobei es sich jedoch um die Steigerung der Befruchtungsaussichten ♂-bestimmender Pollenkörner, nicht also um modifikative Geschlechtsbestimmung im eigentlichen Sinne handelte. — Bei *Muscari comosum* werden die untersten, gewöhnlich sterilen Blüten durch eine Infektion durch *Ustilago Vaillantii* zur vollen Entwicklung der Antheren veranlaßt (MAGNIN 1892). — Bei den von ŠAVROV (1961) erwähnten Übergangsgebilden zwischen Staub- und Fruchtblättern in Blüten von *Lilium Martagon* und *L. dahuricum* bleibt offen, ob extreme Standortsbedingungen eine Rolle gespielt haben könnten; die betreffenden Pflanzen waren in dem polaren Alpengarten von Kirovsk in den Chibiny-Bergen auf der Halbinsel Kola aufgewachsen.

Musaceae.

Musa paradisiaca hat zwar morphologisch monokline, funktionell aber — durch Abort der Samenanlagen oder Verkümmerung der Staubblätter — eingeschlechtige Blüten (JOSHI 1939). In jeder Inflorescenz entstehen zuerst ♀, dann ♂ Blüten. Durch Amputation basaler Früchte oder ♀ Blüten konnten Primordien, aus denen sonst ♂ Blüten hervorgegangen wären, dazu veranlaßt werden, sich zu fertilen ♀ Blüten weiterzuentwickeln, was auf reichlichere Nährstoffversorgung zurückgeführt wird. So konnte man auch Blütenstände erzielen, in denen ♂ und ♀ Blüten mehrmals etagenweise miteinander abwechselten.

Gramineae.

Eine rein deskriptive Übersicht über die Verteilung von ♂ und ⚥ Blüten in den Ährchen verschiedener Gräser gibt CORRENS (1928, S. 31). Weitere Einzelheiten sind einschlägigen systematischen Werken zu entnehmen.

Bei der sardinischen Weizensorte „Denti di cani" wurden bei Dauerlicht und hoher Temperatur (20—25° C) die Antheren zu karpellähnlichen Gebilden umgestaltet; bei Dauerlicht und niedrigerer Temperatur (<20° C) waren die Stamina nur rückgebildet, und die Pollenbildung unterblieb. Zeitstufenversuche lehrten, daß das in dieser Hinsicht empfindlichste Stadium gewöhnlich 20—30 Tage nach der Keimung durchlaufen wird (MELETTI 1961). — Über vergleichbare Erscheinungen bei Blüten von *Hordeum sativum*, die in dem trocken-heißen Sommer 1947 an verschiedenen Stellen in Großbritannien beobachtet wurden, berichten GREGORY und PURVIS (1947), freilich ohne auf mögliche Ursachen einzugehen. Es handelte sich dabei um Blüten mit 1—3 überzähligen Karpellen, die offensichtlich anstelle von Staubblättern entstanden waren und auch Frucht ansetzten; daneben wurden gelegentlich auch Staubblätter gefunden, die an der Spitze Narbenäste trugen. — Relativ späte Infektion der Inflorescenz mit Sporen von *Ustilago longissima* bewirkt bei *Glyceria maxima* vegetative Durchwachsung des Fruchtknotens, wogegen die Stamina kaum beeinflußt werden (LAMBERT 1947). — Weiter sei auf *Coix Lacryma-Jobi* hingewiesen. bei der in dem trockenen Sommer 1959 in der Pariser Gegend spontane Entwicklung monokliner anstelle von ♂ Blüten an der Triebspitze zu beobachten war. Dasselbe Phänomen trat auch bei experimenteller Trockenkultur (verknüpft mit etwas höherer Luftfeuchtigkeit und geringerer Lichtintensität) auf (JACQUES-FÉLIX 1959) und war auch schon von ARBER (1934) erwähnt worden.

Zahlreiche Arbeiten beschäftigen sich mit den Geschlechtsverhältnissen von *Zea Mays*. In den Blattachseln des mittleren Sproßbereichs finden sich bekanntlich in Kolben vereinigte ♀ Blüten, während ♂ Blütenstände die Sproßspitze einnehmen. Abweichungen von dieser Norm können einerseits rein genetische Ursachen haben, was hier außer Betracht bleibt. Andererseits sind aber auch modifikative Abweichungen von der normalen Anordnung der ♂ und ♀ Blütenstände möglich. Sie bestehen am häufigsten darin, daß in der terminalen, gewöhnlich ♂ Inflorescenz — ausschließlich oder neben ♂ — ♀ Blüten auftreten (vgl. Abb. 3). Dergleichen beobachteten z. B. K. MÜLLER (1858) und BLARINGHEM (1906a) nach Resektion des Haupttriebes: Eine in Höhe des Bodens abgeschnittene Pflanze regenerierte eine Anzahl basaler Triebe, von denen einer mit einer verbänderten Rispe abschloß, deren ♂ Blüten in ♀ „metamorphosiert" waren. HESLOP-HARRISON (1957) weist darauf hin, daß ein photoperiodischer Effekt hierbei nicht ausgeschlossen ist, da die Regeneration im Spätsommer bei abnehmender Tageslänge erfolgte[1]. — Auf die Bedeutung der Tageslänge hat in diesem Zusammenhang vor allem SCHAFFNER immer wieder (1925a, b, 1927, 1930, 1933, 1935) aufmerksam gemacht: Bei zwei Mais-Varietäten zeigten 21 von 28 bzw. 44 von 55 Pflanzen insofern im 9—10 Std-Tag eine Geschlechtsumstimmung, als in der terminalen Rispe (meist an ihrer Basis) fertile ♀ Blüten entstanden; bei „mittleren" Tageslängen fehlten dort, besonders an der Spitze, Blüten jeglichen Geschlechts. RICHEY und SPRAGUE (1932) zufolge wird die kurztagbedingte Geschlechtsumkehr (♂ → ♀) in der endständigen Inflorescenz durch Kälte begünstigt; unter bestimmten Bedingungen wirkt dabei eine Temperatursenkung um 5,5° C (von 80 auf 70° F) ähnlich wie eine Verkürzung der Photoperiode um $4^1/_2$ Std, wogegen die Lichtintensität keine nennenswerte Rolle spielt. Im einzelnen herrschen beträchtliche Sortenunterschiede; eine größere Tendenz zur Geschlechtsänderung wird durch ein recessives Hauptgen und eine Reihe weniger wirksamer Zusatzgene bedingt. Nach CHOUDHRI und KRISHAN (1946) wird durch 18—24 Std-Langtag die Ausprägung des ♀ Geschlechts an der ganzen Pflanze unterdrückt. Was noch die Temperaturwirkung anlangt, so scheint es im Hinblick auf eine Verweiblichung der ♂ Rispen in erster Linie auf die *Nacht*temperatur anzukommen; liegt sie unterhalb von 13° C, so entstehen ♀ Blüten (WENT 1957). J. HESLOP-HARRISON (1961) stellt fest, daß sich die Wirkungen des Kurztages und der niederen Nachttemperatur addieren können und daß eine Verminderung oder gar eine vollständige Beseitigung der ♂ Fertilität durch diese beiden Faktoren selbst dann noch möglich ist, wenn diese schon nicht mehr die vollständige Ausbildung der Fruchtblätter zu bewirken vermögen; die ♂ Blüten an der Hauptachse der Rispe sowie jene an der Basis ihrer Seitenachsen sind übrigens die ersten, die gegen den verweiblichenden Einfluß von Kurztag und Kälte immun werden.

Abb. 3. *Zea Mays*. ♀ Blüten in der gewöhnlich rein ♂ endständigen Inflorescenz. (Aufnahme R. SCHRAMM.)

[1] Da dieselbe Erscheinung auch in der Nachkommenschaft dieses Triebes auftrat, erscheint mir eine Knospenmutation nicht ausgeschlossen (genotypisch bedingte Androgynie wurde beim Mais schon oft beschrieben); dies dürfte auch für die Fälle von erblicher Vermännlichung der ♀ Blüten in seitlichen Kolben bei anderen Pflanzen derselben Nachkommenschaft (BLARINGHEM 1906b) gelten.

Weiterhin kann auch eine Infektion durch *Ustilago maydis* die Verweiblichung von Blüten der terminalen Rispe (CHIFFLOT 1909, ILTIS 1911) sowie die Verlaubung der Fruchtblätter und die Proliferation von Blüten auslösen (L. R. und CH. TULASNE 1847, ILTIS 1910).

In den letzten Jahren wurde wiederholt Modifikation der Geschlechtsverhältnisse von *Zea Mays* nach verschiedenartigen Wuchsstoffbehandlungen beobachtet. BUIS (1956) tauchte den Stumpf zurückgeschnittener Blätter in Cumarin-Lösungen (10—100 p.p.m.) und fand später an Seitensprossen der betreffenden Pflanzen solche Kolben, die an der Basis ♀, an der Spitze ♂ Blüten trugen, wogegen die Hauptachse zuweilen an der Spitze eine kleine ♀ Ähre aufwies. Durch Maleinsäurehydrazid (R. H. MOORE 1950, A. W. NAYLOR 1950) sowie durch Gibberellinsäure (P. M. NELSON und ROSSMAN 1958) konnte man partielle oder vollständige Sterilität der ♂ Rispen hervorrufen, im Falle der Gibberellinsäure vielfach verbunden mit dem Auftreten von ♀ Blüten in diesen Rispen (NICKERSON 1959). NICKERSON und HEWITSON (1963) befassen sich mit dem Einfluß von α-Naphthylessigsäure und β-Indolylbuttersäure auf die Geschlechtsausprägung der Mutante *Vestigial (Vg)*, PETRŮ und ŘETOVSKÝ (1957) sowie PETRŮ und ŠESTÁK (1961) mit den Geschlechtsverhältnissen von Maispflanzen, die aus isolierten Sproßspitzen hervorgingen, welche man zunächst auf synthetischen Medien kultiviert hatte[1].

HEYER (1884, 1886), RIEDE (1922), SABININ (1937), MININA (1938) sowie CHOUDHRI und KRISHAN (1946) beschäftigten sich mit dem Einfluß mineralischer Düngung auf die Geschlechtsausbildung beim Mais. Als Regel mag gelten, daß zusätzliche N-Düngung ebenso wie K- oder P-Mangel und Stickoxyd-Behandlung Verweiblichung, dagegen K-Düngung Vermännlichung begünstigt. Dichtsaat verzögert die Entwicklung ganz allgemein und damit insbesondere auch die der ♀ Kolben (EDEL'ŠTEJN und PAPONOV 1964).

Abschließend sei auf Angaben von GOEBEL (1910), MILLER (1920), RIEDE (1922), BONNETT (1940, 1948) und USTINOVA (1956) hingewiesen, die auf die morphologischen Grundlagen der Geschlechtsausprägung beim Mais Licht werfen. Wie bei *Asparagus* und *Musa* (s. oben) werden hier die Blüten monoklin angelegt; aber ein Geschlecht wird später gewöhnlich unterdrückt. Im Herbst kann in ♂ Rispen der Ablast der ♀ Organe unterbleiben, womit oft die kolbenförmige Ausbildung eines oder mehrerer Rispenäste verbunden ist. Volle Entwicklung der Antheren in seitlichen, sonst rein ♀ Blütenständen und deren Umwandlung zu Rispen beobachtete USTINOVA seltener.

Cyperaceae.

Wie bei den Gramineen können auch bei den Cyperaceen, zumal auch bei der Gattung *Carex*, Blüten verschiedenen Geschlechtes in unterschiedlicher Weise auf ein und demselben oder auf verschiedenen Individuen angeordnet sein (vgl. z. B. ARNAL 1952b, 1954), ohne daß man viel über die modifikative Beeinflussung dieser Verhältnisse wüßte. STRASBURGER (1900) fand an sonst ♂ *Carex*-Ähren nicht selten ♀ Blüten. MARTENS (1939) beobachtete bei der in der Regel diöcischen *Carex picta*, daß ♂ Exemplare in späteren Jahren vereinzelt auch ♀ Ähren trugen. Bei *Scirpus caespitosus* pflegen von den vier bis acht Blüten einer Ähre die ein bis vier obersten ♀ zu sein, und auch bei verschiedenen Arten der Gattung *Eriophorum*, die sonst durch gemischtgeschlechtige Blüten charakterisiert ist, sind statt dessen rein ♀ Blüten anzutreffen (SCHULZ 1892).

[1] Eine einschlägige Arbeit von PREDA und CRĂCIUN (1960) hat mir nicht vorgelegen.

Araceae.

Auch bei den zahlreichen Araceen findet man ♀ und ♂ Blüten in wechselnder Anordnung, die oft von systematischem Wert und für bestimmte Taxa charakteristisch sein kann. Nennenswertes Interesse im Hinblick auf die Frage nach der experimentellen Beeinflußbarkeit scheinen jedoch allein die Geschlechtsverhältnisse der Gattung *Arisaema* gefunden zu haben. Bei *A. japonica* beispielsweise, die man früher fälschlich für diöcisch gehalten hatte, folgt einem etwa zweijährigen blütenlosen Jugendstadium eine ein- bis zweijährige ♂ Phase; den Rest des Lebens füllt die ♀ Phase aus. Kultur auf trockenem Sand verzögert den Übergang zum ♀ Zustand und kann ♀ Pflanzen zur Rückkehr zum ♂ Stadium veranlassen (Maekawa 1924). — Umgekehrt hatte bei ♂ Exemplaren von *Arisaema triphyllum* Verpflanzung in nährstoffreichen Boden die Bildung von ♀ Infloreseenzen zur Folge (v. Treskow 1895, Atkinson 1898, Schaffner 1922); derselbe Effekt konnte auch durch Entblätterung und Kultur in trockenem Boden erzielt werden (Schaffner 1925a, b, 1926). — Bei *A. dracontium* gibt es ♂ und monöcische Individuen. Erstere werden in nährstoffreichem Boden monöcisch, letztere in armem Boden oder nach Entfernung der Blätter ♂ oder steril (Schaffner). — Eine wichtige Rolle spielt schließlich auch die Größe des Cormus, die mit den oben erwähnten Geschlechtsphasen deutlich korreliert ist (Maekawa 1927 bei *A. japonica*, Camp 1932 bei *A. triphyllum*): Radikale Resektion von ♀ Individuen führt zurück zur ♂ Phase (Atkinson 1898 bei *A. triphyllum*, Maekawa 1927 bei *A. japonica*). Maekawa macht hierfür die verfügbare Assimilatmenge, Heslop-Harrison (1957) dagegen ein hormonales Agens verantwortlich. Übrigens hat Maekawa (1927) bei *A. japonica* ganz vereinzelt auch monokline Blüten gefunden.

c) Dicotyledoneae.

Ranunculaceae.

In gemischtgeschlechtigen Blüten kann sich eine Tendenz zur Vermännlichung oder zur Verweiblichung auf zweierlei Weise manifestieren: einmal dadurch, daß die Organe des einen Geschlechts wohl ausgebildet werden, die des anderen aber mehr oder weniger verkümmern, was bei *Musa* und *Zea*, wie wir sahen, zur funktionellen Monöcie führt, — ein andermal dadurch, daß das Zahlenverhältnis der ♂ und ♀ Sporophylle in der einen oder anderen Richtung verschoben wird. Diese zweite Möglichkeit kommt für Arten mit variablen Sporophyllzahlen in Betracht, wie sie unter den Ranunculaceen reichlich zu finden sind. Ein hübsches Beispiel hierfür schildert Laibach (1952): Bei *Helleborus foetidus* nimmt der Prozentsatz der Karpelle unter den Sporophyllen einer Blüte im basalen und im apikalen Sproßabschnitt mit der Ordnungszahl des Seitenzweiges zu, den die betreffende Blüte abschließt (diese Zunahme entspricht der Aufblühfolge); nur bei Seitentrieben im mittleren Sproßbereich sinkt der Prozentsatz der Karpelle anfangs (von der I.—III. Ordnung), um dann bis zur VI. Ordnung wieder anzusteigen (Tabellen l. c., S. 133). Aus Zählungen zu verschiedenen Zeitpunkten ergab sich für diesen Prozentsatz eine ähnliche Kurve wie jene, die Correns (1907) für den Prozentsatz von ♀ Blüten bei gynomonöcischen Exemplaren von *Satureia hortensis* (s. unten) ermittelt hatte (Laibach 1952, Abb. 4 und 5).

Mit modifikativer Geschlechtsbestimmung bei *Thalictrum dioicum* befaßte sich Schaffner (1923c, 1925a); J. Heslop-Harrison (1957) bezeichnet seine Darstellung als "somewhat too confused". ♂ und ♀ Exemplare von *Th. dasycarpum* sollen sich laut Schaffner (1925b) nach Verpflanzen von nassen Standorten in trockenen Lehmboden zu zwittrigen entwickelt haben.

Capparidaceae.

Bei der monöcischen *Cleome spinosa* ist wieder die erste der beiden in der Einleitung zu den Ranunculaceen aufgezeigten Möglichkeiten verwirklicht: Die Blüten werden gemischtgeschlechtig angelegt, ein Geschlecht pflegt aber unterdrückt zu werden; in jedem Blütenstand erfolgt innerhalb weniger Wochen ein mehrmaliger Übergang von einem Geschlecht zum anderen (STOUT 1923). Die Tendenz zur Geschlechtsumkehr innerhalb der Inflorescenz konnte durch Variation der N-Ernährung nicht geändert werden, doch steigerte N-Düngung das Gesamtwachstum und damit zugleich den Anteil der als ♀ fungierenden Blüten. Die Amputation von ♀ Blüten oder Früchten in unteren Sproßabschnitten bewirkte, daß im distalen Bereich an Stelle der Griffel die Staubblätter abortierten, so daß funktionell ♀ Blüten an Stelle von ♂ entstanden (MURNEEK 1927).

Cruciferae.

Bei *Capsella Bursa-pastoris* stellt man häufig fest, daß die untersten Schötchen einer Fruktificenz verhältnismäßig klein bleiben und keinen Samen ansetzen. Die Ursache hierfür ist darin zu suchen, daß die betreffenden Blüten keine fertilen Staubblätter besessen hatten, so daß zur Zeit der Empfängnisbereitschaft keine Selbstbestäubung erfolgen konnte. Da für alle Pflanzen einer Population weithin dasselbe gilt, kam auch Fremdbestäubung nicht in Betracht. Schon frühzeitig machte man die jahreszeitlichen klimatischen Bedingungen für dieses Verhalten verantwortlich. So fand WILLIS (1893) — in England — am 12. März nur Individuen mit ♀ Blüten, am 16. April solche, bei denen die zuerst entwickelten Blüten ganz abortierte, die später entfalteten fertile, aber zu kurze Staubblätter aufwiesen; im weiteren Jahreslauf wurden normale ⚥ Blüten und im November wieder rein ♀ Blüten gebildet. BURKILL (1897) beobachtete, daß 1895 mehrere 100 nach strengen Januar- und Februarfrösten erblühte Pflanzen nur rudimentäre Stamina trugen und ⚥ Blüten erst ab 9. April auftraten. Nach dem milden Winter 1895/96 waren die ersten Blüten gleichfalls ♀; ⚥ Blüten wurden jetzt aber schon ab Ende März gefunden. LAIBACH (1951) bestätigte diese Beobachtungen und stellte fest, daß die verkümmerten Staubblätter von ♀ Erstlingsblüten nur 0,6—1,0 mm lang waren, während die fertilen Stamina gemischtgeschlechtiger Blüten 3 mm maßen. Magerer oder guter Boden, schattiger oder sonniger Standort spielten für die Geschlechtsverhältnisse keine Rolle. Im Frühjahr 1918 blühten Exemplare, die frostfrei im Kalthaus überwintert hatten, rein ⚥; Individuen gleicher Herkunft bildeten dagegen nach Frostexposition zunächst ♀ Blüten. Auf welches Entwicklungsstadium der Frost eingewirkt hatte, ist nicht ganz klar; jedenfalls werden durch den Frost hervorgerufene Welkungserscheinungen für die Verkümmerung der Staubblätter verantwortlich gemacht. Auch nach Hitzebehandlungen (einige Stunden 40^{0} C) produzierten vorher ⚥ blühende Endflorescenzen (nach einer Reihe von verkümmerten Blüten) zunächst bis zu 20 rein ♀ Blüten und erst dann wieder monokline. Starb die Endflorescenz infolge dieser Behandlung ab, so trieben Conflorescenzen aus, die aber nur ausnahmsweise ♀ Blüten entwickelten. Auch wenn man Pflanzen ausriß (wobei das Wurzelsystem beschädigt wurde), welken ließ und wieder einpflanzte, traten ähnliche Effekte auf (bis zu 6 ♀ Blüten).

Annähernd gleiches Verhalten beobachtete LAIBACH (1951) bei *Arabis albida*, *Arabidopsis Thaliana* sowie (nur einmal) bei *Thlaspi arvense*, niemals aber bei alpinen Cruciferen oder bei anderen Frühjahrsblühern aus dieser Familie.

Crassulaceae.

Entsprechend der für diese Familie gültigen Regel sind auch die *Bryophyllum*-Blüten gewöhnlich gemischtgeschlechtig. Durch stufenweise Herabsetzung der Lichtintensität erzielte RESENDE (1946, 1949, 1950) jedoch bei *B. Daigremontianum* an Stelle von ⚥ Blüten zunächst ♀ Blüten mit Krone, dann ♀ Blüten ohne Krone und schließlich ganz vergrünte Blüten. Er folgerte daraus eine Reihe abnehmender Vegetativität: Kelch — Fruchtknoten — Krone — Staubblätter. Auch einer Behandlung mit β-Indolylessigsäure oder mit 2,3,5-Trijodbenzoesäure fielen bei *B. Daigremontianum* als erstes die Staubblätter zum Opfer (RESENDE 1951, RESENDE und VIANA 1952). Eine bescheidenere Reduktion des Androeceum (Verkürzung der Filamente) wurde — wohl bei *Bryophyllum crenatum* × *B. Daigremontianum* „var. mut. B" — durch Temperaturen unterhalb von 13° C erreicht (RESENDE 1957a). — Bei einem subgynöcischen Klon der Langkurztagpflanze *B. proliferum*, der gewöhnlich fast ausschließlich ♀ Blüten (mit abortierter Krone) und nur vereinzelt ⚥ Blüten (mit voll entwickelter Krone) trug, untersuchten RESENDE und VIANA (1959) den Einfluß von Gibberellinsäure (GA_3) auf die Geschlechtsverhältnisse. Eine Behandlung mit 0,005%iger GA_3 im Kurztag begünstigte die Bildung von ⚥ Blüten, bewirkte also Vermännlichung. Im späteren Verlauf des Versuches wurde die GA_3-Konzentration verdoppelt; es entstanden dann aber wieder „weiblichere" Blüten (mit vier oder mehr Griffeln und reduziertem Androeceum oder ohne Stamina, z. T. mit mißgebildeten Sporophyllen). — Endlich besprühte CATARINO (1964) Inflorescenzen von *B. crenatum* mit einer 0,005%igen Kinetinlösung und erzielte dadurch Verweiblichung (üppigere Entwicklung von Kelch und Gynoeceum, Reduktion von Krone und Androeceum); im einzelnen hing das Ausmaß der Kinetinwirkung davon ab, in welcher Phase ihrer Entwicklung die einzelne Blüte dem Einfluß von Kinetin ausgesetzt worden war.

KLEBS (1905; hier auch Hinweise auf ältere Literatur, besonders bezüglich *Sempervivum tectorum*) fand bei 10 von 287 daraufhin untersuchten Blüten von *Sempervivum Funkii* je ein Staubblatt, das teilweise in ein Fruchtblatt umgewandelt war und neben Pollensäcken auch Samenanlagen trug. In zwei Fällen stand umgekehrt im Kreis der Fruchtblätter je ein Staubblatt, das als metamorphosiertes Karpell interpretiert wurde.

Eucommiaceae.

Durch reichliche Stecklingsentnahme 2—3mal im Jahr sowie durch Verwundungen an Zweigen und Rinde sollen ältere und jüngere ♂ Exemplare von *Eucommia ulmoides* zur Bildung von ♀ Blüten und Früchten veranlaßt worden sein (BOSSÉ 1937).

Rosaceae.

Bei *Geum intermedium* sind die ersten Blüten laut CORRENS (1907, 1908, 1928) ♂, die späteren ⚥. Im ersten Vegetationsjahr werden mehr eingeschlechtige Blüten gebildet als — nach Erstarkung — im zweiten Jahr. Über Möglichkeiten, diesen Übergang durch Außenfaktoren zu beeinflussen, scheint nichts bekannt zu sein.

Oxalidaceae.

v. UBISCH (1935) arbeitete mit *Oxalis stricta*, Herkunft Zürich. Griffel und Antheren erreichten hier gewöhnlich annähernd gleiche Länge. An manchen Individuen wurden aber besonders langgriffelige Blüten gefunden, deren abnormes Griffelwachstum auf eine Infektion mit *Ustilago oxalidis* zurückgeführt werden konnte. *Ustilago* enthielt Auxin und Biosstoffe, die das Längenwachstum der Griffel ausgelöst haben könnten. Diese Beobachtung verdient auch im Hinblick auf Fragen der Heterostylie Beachtung.

Geraniaceae.

Nach Correns (1907, 1908) sind bei den gynomonöcischen *Geranium sylvaticum* und *G. pratense* die sich später öffnenden Blüten zu einem steigenden Prozentsatz gemischtgeschlechtig. Im Laufe der Blühperiode eines Exemplars von *G. pratense* hatte die Kurve des Prozentsatzes tauglicher Antheren unter den theoretisch überhaupt möglichen Antheren folgende Extremwerte: 22. Mai 0% — 30. Mai 95% — 1. Juni 83% — 3./4. Juni 100% — 7. Juni 50%.

Rutaceae.

An einem Exemplar von *Citrus trifoliata* fand Baccarani (1938) 47% vollentwickelte gemischtgeschlechtige Blüten, 5,2% mit leicht reduziertem Gynoeceum und 47,8% funktionell ♂ Blüten mit atrophiertem Gynoeceum, das gleichwohl noch schwach entwickelte Samenanlagen enthielt. Wie in so vielen anderen Fällen bleiben auch hier die Ursachen der Geschlechtsverhältnisse ungewiß. Dies gilt auch für die Beobachtungen von Uphof (1932) über das Auftreten von ♂ Blüten bei *Citrus Limonia*, *C. aurantifolia*, *C. medica*, *C. sinensis*, *C. paradisi*, *C. nobilis* var. *deliciosa* sowie *Microcitrus australasica*.

Anacardiaceae.

Die Geschlechtsverhältnisse der Blüten eines *Rhus Cotinus*-Strauches können sich im Lauf der Jahre ändern (Schulz 1892). — Bei der Mangopflaume (*Mangifera indica*), Sorte Kalapadi, kann der Anteil der ♀ Blüten an der Gesamtzahl der Blüten durch mehrmaliges rechtzeitiges Besprühen der ganzen Pflanze mit α-Naphthylessigsäure (50 oder 100 p.p.m.) besonders bei der Februar-März-Blühperiode deutlich gesteigert werden; normalerweise herrschen die ♂ Blüten vor (Mallik, Sahay und D. L. Singh 1959).

Begoniaceae.

In einer viel zitierten Arbeit befaßt sich Irmscher (1924) mit der Anordnung von Blüten beiderlei Geschlechts in den vierblütigen Wickeln von *Begonia Wallichiana*. Fast immer folgt auf 3 ♂ eine ♀ Blüte, und nur ausnahmsweise werden zunächst 2 ♂, dann 2 ♀ Blüten gebildet. Die dann anstelle einer ♂ entstandene ♀ Blüte unterscheidet sich jedoch dadurch von gewöhnlichen ♀ Blüten, daß das Perianth (wie sonst bei ♂ Blüten) vier- und nicht fünfzählig, das Gynoeceum aber zwei- anstatt dreizählig ist. Irmschers Versuche, diese Erscheinung auch künstlich hervorzurufen, schlugen fehl.

Im Gegensatz hierzu kann bei *Begonia semperflorens* das Zahlenverhältnis der ♂ zu den ♀ Blüten in verschiedenen Blütenständen eines oder verschiedener Individuen stark variieren. An Exemplaren, die auf N-gedüngtem, tonigem Lehm gezogen worden waren, ermittelte Matzke dieses Zahlenverhältnis zu 2,83:1, an solchen in Sandkultur dagegen zu 4,48:1. Reichliche Ernährung hatte nicht nur die absolute Blütenzahl versiebenfacht, sondern die Entwicklung von ♀ Blüten stärker begünstigt als die Bildung von ♂ Blüten.

Neuerdings ist R. Noack (1962) den Ursachen für die Entstehung gemischtgeschlechtiger Blüten in den Infloreszenzen von *Begonia Cathayana* nachgegangen. Diese monoklinen Blüten nehmen den Platz von ♂ Blüten ein, und ihr Fruchtknoten ist *ober*ständig; die „echten" ♀ Blüten finden sich — wie bei *B. Wallichiana* — immer an den Wickelästen letzter Ordnung und weisen einen *unter*ständigen Fruchtknoten auf. Zwar bleibt der Fruchtknoten der gemischtgeschlechtigen Blüten immer steril; mit zunehmender Verweiblichung dieser Blüten geht aber auch ihre Pollenfertilität zurück. Was nun die für diese Art Verweiblichung maßgeblichen Außenfaktoren betrifft, so hat Noack in zwei aufeinanderfolgenden Jahren eine positive Korrelation zwischen der Zahl monokliner Blüten und der täglichen Sonnenscheindauer feststellen können.

Caricaceae.

In den letzten Jahren hat auch wieder die Frage einiges Interesse geweckt, ob man das Geschlecht der Blüten von *Carica Papaya* modifikativ beeinflussen könne, die als normalerweise diöcisch gilt. Das diesbezügliche Schrifttum findet sich großenteils bei Lange (1961a) referiert. Schon Bordage (1898), Higgins und Holt (1914) sowie Higgins (1916) hatten berichtet, daß zurückgeschnittene ♂ Exemplare in der Folge ♀ Blüten bilden, wogegen bei ♀ Bäumen Resektion in dieser Hinsicht wirkungslos sei. Dikshit (1956) beobachtete nach künstlicher Reduktion der Blattfläche gesteigerte Parthenokarpie und deutete diese Erscheinung als Verstärkung der ♀ Tendenz.

Lange (1961a) selbst konnte jedoch bei zwittrigen Exemplaren durch Rückschnitt keine Verweiblichung erzielen und vermutet, daß wenigstens bei manchen der soeben erwähnten Befunde früherer Autoren eher jahreszeitliche Veränderungen bestimmter Faktoren eine Rolle gespielt haben könnten. Deren Bedeutung kommt unter anderem auch bei Storey, Jones, Daris und Holdaway (1941), Awada (1958) sowie R. N. Singh, Majumder und Sharma (1963) zur Sprache. Nach Storey u. Mitarb. soll Winterwetter die Bildung von Früchten, Spätsommer- und Herbstwetter die Bildung von Blüten ohne funktionsfähigen Fruchtknoten begünstigen. Dementsprechend fand Awada die größte Zahl ♀-steriler (d. h. funktionell ♂) Blüten in den Monaten Oktober, November und Dezember, insbesondere bei denjenigen Pflanzen, die an dem wärmsten von drei berücksichtigten Standorten wuchsen.

Aber auch die jahreszeitlichen Wetteränderungen sind nicht alleine für die Geschlechtsausprägung von *Carica Papaya* verantwortlich; vielmehr zeigt sich, daß die Bildung von ♀ Blüten ganz allgemein mit der Wüchsigkeit („vigor") korreliert ist, und alles, was die Wüchsigkeit steigert, erhöht zugleich auch die Zahl der ♀ Blüten, so z. B. stärkere Bewässerung und ausreichende N-Düngung (Awada und Ikeda 1953, Awada 1957). So stellt auch Lange (1961a) fest, daß kräftige junge Bäume (gemessen an Wuchshöhe und Umfang) früher blühen und mehr zum ♀ Geschlecht tendierende Blüten entwickeln als schwächere Exemplare gleichen Alters; jede Verminderung der Wüchsigkeit, z. B. infolge Wassermangels oder Alterung, infolge Erkrankung oder Resektion der Wurzeln oder (in Widerspruch zu Dikshit, s. oben) infolge Reduktion der Blattfläche um 50 oder 90%, führt zur Vermännlichung.

Andererseits fand Lange, daß die Entstehung von ♀ Blüten durch niedrige Wärmegrade sowie durch große Tag-Nacht-Differenzen der Temperatur gefördert wird. Derselbe Effekt wurde übrigens erzielt, indem man ein Polyäthylen-Zelt über sonst zwittrigen Pflanzen errichtete (Lange vermutet, daß es hierbei in erster Linie auf die Ansammlung gasförmiger Stoffwechselprodukte und nicht auf die Verminderung der Lichtintensität ankam) — oder indem man zwittrige Zwerg-Papayas oder rein ♂ blühende Exemplare mit 2,3-Dichlorisobutyrat und 2,2-Dichlorpropionat besprühte (Lange 1961b).

Aceraceae.

Bei *Acer Pseudoplatanus*, *A. campestre* und *A. platanoides* herrscht keine feste Regel bezüglich der Anordnung männlicher und weiblicher Blüten innerhalb einer Inflorescenz (Wittrock 1886, Schulz 1888/90; vgl. auch Correns 1928, Abb. 35 [nach Wittrock]). Vereinzelt kommen bei *A. platanoides* auch eingeschlechtige Exemplare vor. — In letzter

Zeit hat sich SCHOLZ (1960) mit der Geschlechtsausprägung bei *A. Pseudoplatanus* und *A. platanoides* beschäftigt; bei manchen Individuen blieben die Geschlechtsverhältnisse während des ganzen Beobachtungszeitraumes konstant, während sie bei anderen von Jahr zu Jahr wechselten. Es handelte sich dabei stets um gemischtgeschlechtige und ± ♂ Blüten (bei denen die ♀ Organe ± reduziert waren). — Bei sonst ♀ Exemplaren des (ebenso wie *A. Negundo*) diöcischen *A. dasycarpum* fand übrigens schon MEEHAN (1878a) öfters Zweige mit ♂ Blüten (aber nie das Umgekehrte!); irgendeine Beziehung dieser Erscheinung zur Größe des Zweiges, zur Beblätterung oder zu anderen Faktoren konnte MEEHAN nicht entdecken.

Hippocastanaceae.

Wie jedermann leicht feststellen kann, sind die Blütenstände von *Aesculus Hippocastanum* andromonöcisch; Früchte werden gewöhnlich nur im basalen Bereich der Inflorescenz angesetzt. Sämlinge blühen allerdings zuerst rein männlich (HOFFMANN 1871). — PORSCH (1937) fand an zwei Blütenständen von *Ae. macrostachya* unmittelbar unter einer Verletzung der Inflorescenzachse monokline anstelle von ♂ Blüten, was er als Hinweis auf den Einfluß der Ernährungsverhältnisse auf die Fruchtknotenentwicklung verstanden wissen möchte. (*Der Anlage nach* sind die *Aesculus*-Blüten immer gemischtgeschlechtig.)

Vitaceae.

Bei *Vitis vinifera* unterscheidet BREIDER (1962) drei Typen der Blütenverlaubung. Einer davon — der durch Vergrünung der Staubblätter bei unverminderter Fertilität des Fruchtknotens gekennzeichnete „Sylvaner-Typ" — kann auch modifikativ auftreten, nämlich „unter dem Einfluß von abnormen Umweltbedingungen wie Witterung, Düngung ...".

Halorrhagaceae.

Bei trimonöcischen *Myriophyllum*-Arten werden — ähnlich den Verhältnissen bei *Lilaea* (S. 160) — zuerst ♀, später ⚥ und schließlich ♂ Blüten entwickelt (PETERSEN 1898, S. 234).

Onagraceae.

Fuchsia procumbens gilt als triöcisch; man unterscheidet 1. langgriffelig-♂ sterile, 2. kurzgriffelig-♀ sterile und 3. mittelgriffelig-zwittrige Formen. Bei den zuletzt genannten handelt es sich um die normalen, gewöhnlich selbstfertilen Gartenformen; diese können durch Kälte in Verbindung mit Kurztag zur Reduktion des Griffels veranlaßt werden, in deren Gefolge Sterilität der ♀ Organe eintritt (HOLDSWORTH 1959).

Caryophyllaceae.

Die zweite der bei den Ranunculaceen skizzierten Möglichkeiten der Ausprägung von „Geschlechtstendenzen" ist gelegentlich auch bei Caryophyllaceen realisiert. Auch hier kann die Zahl der Staubblätter beträchtlich schwanken, wie CORRENS (1928, S. 33, Tabelle 4) an Hand von *Dianthus sylvestris* zeigte. Selbst bei ein und demselben Exemplar dieser Art unterlag die Zahl der Stamina pro Blüte von Jahr zu Jahr auffälligen Variationen, deren Ursachen wiederum im Dunkeln blieben.

Versuche über die modifikative Geschlechtsbestimmung bei *Silene noctiflora* stellte schon GAERTNER (1844) an: Bei drei fruchtenden Individuen dieser (gynodiöcischen) Art, die bis dahin nur ⚥ Blüten entwickelt hatten, wurden alle Kapseln weggeschnitten; die in den folgenden 2—3 Wochen neu gebildeten Blüten hatten größtenteils kontabeszierte Antheren, die späteren wieder normale. CORRENS (1907) machte für diese Erscheinung Ernährungsfaktoren verantwortlich.

Gynomonöcische Stöcke von *Silene inflata* verhalten sich nach CORRENS (1905) wie *Geranium pratense* (s. dort): Der Anteil der ♀ Blüten ist unter den zuerst entfalteten am größten.

Die Wirkung verschiedener Tageslängen auf die Geschlechtsverhältnisse zweier Sorten der Langtagpflanze *Silene pendula* (die normalerweise ⚥ Blüten entwickelt) war Gegenstand einiger Experimente von J. und Y. HESLOP-HARRISON (1958a). Die Versuchspflanzen wurden in zwei Gruppen eingeteilt. Gruppe I erhielt von der Keimung an 40 21—22 Std-Langtage und anschließend 50 8 Std-Kurztage. Am Ende dieser Periode zeigten die Pflanzen der Gruppe I noch rosettigen Wuchs; nach Rückübertragung in Langtagbedingungen gingen sie sofort zur Blütenbildung über. Gruppe II erhielt ständig Langtag; nach 90 Tagen waren Blütenanlagen sichtbar. Eine Untersuchung der Antheren 130 Tage nach der Keimung ergab, daß bei den Pflanzen der Gruppe I etwa 1%, bei denen der Gruppe II etwa 50% der Antheren steril waren (vgl. Abb. 1, S. 157). Mit der Antherensterilität waren meist besonders starkes Wachstum des Fruchtknotens und Verkürzung der Kronblätter verbunden. Dieselben Erscheinungen, jedoch bis zu 100% sterile Antheren, konnten bei Pflanzen der Gruppe I durch Wuchsstoffbehandlung (0,5% α-Naphthylessigsäure [NES] in Lanolinpaste auf die Blätter aufgetragen) erzielt werden; nicht behandelte Seitenzweige derselben Individuen zeigten normale Antherenfertilität. Die überhaupt fertile Antheren enthaltenden Blüten an NES-behandelten Trieben waren proterogyn, während im Normalfall Proterandrie herrscht. — J. und Y. HESLOP-HARRISON führen auf Grund dieser Ergebnisse die photoperiodische Beeinflussung der Geschlechtsausprägung auf eine Änderung des Wuchsstoffhaushaltes zurück. Da aber bei *S. pendula* unter dem Einfluß von NES nur die Weiterentwicklung bereits angelegter Staubblätter gehemmt, aber offenbar niemals ein Stamen in ein Karpell umgewandelt wird, glaubt J. HESLOP-HARRISON (1959) mit gutem Grund, zumindest in diesem Falle Wuchsstoffe eher als geschlechtsregulierende denn als geschlechtsbestimmende Agentien betrachten zu müssen.

Der durch Langtag oder NES induzierten Antherensterilität scheint meist eine Verkümmerung des Tapetums zugrunde zu liegen, die eine mangelhafte Ausbildung des Pollens zur Folge hat (Pollenkörner vacuolisiert, Kern an der Wand liegend). In manchen Fällen (bei ständiger Langtageinwirkung) gelangte die Antherenentwicklung allerdings noch vor der Ausdifferenzierung des Archespors zum Stillstand (J. und Y. HESLOP-HARRISON 1958b).

Einige der Hauptobjekte bei der Erforschung der Grundlagen der Geschlechtsbestimmung gehören der Gattung *Melandrium* an. Hier gibt es wie bei der klassischen *Bryonia* Arten mit unterschiedlicher Geschlechterverteilung, die nicht nur miteinander bastardiert werden können, sondern auch Gegenstand zahlreicher Untersuchungen über modifikative Geschlechtsbestimmung waren. Diese liefern ein Beispiel dafür, daß auch bei Diöcisten nicht allein die genotypische Geschlechtsbestimmung eine Rolle spielt (vgl. v. WETTSTEIN 1936). Hierbei stehen jene Untersuchungen im Vordergrund, die die Vermännlichung von ♀ Exemplaren von *M. album* und *M. rubrum (M. dioecum, Lychnis dioica, L. vespertina)* infolge von Infektionen durch *Ustilago violacea (U. antherarum)* betreffen (auch bei ♀ Individuen der gynodiöcischen *Silene dichotoma* und der diöcischen *S. Otites* beobachtet; CORRENS 1928). Diese durch *Ustilago* induzierte Vermännlichung besteht darin, daß die normalerweise rudimentären Staubblätter — bei gleichzeitiger Reduktion des Gynoeceum — zu voller Entwicklung gelangen (*Deuterohermaphrodismus*, CORRENS) und zur Reifezeit *Ustilago*-Sporen enthalten. Diese Umstimmung erfolgt nur an den von *Ustilago* befallenen Teilen einer Pflanze; die übrigen bleiben ♀ (L. R. und CH. TULASNE 1847, MAGNIN 1889 [hier auch eine Übersicht über die bis dahin erschienenen einschlägigen Veröffentlichungen], VUILLEMIN 1891, STRASBURGER 1900). In nur schwach infizierten Inflorescenzen entstehen gelegentlich Zwitterblüten, in denen auch die Samenanlagen wohl entwickelt sind (BAKER 1947b). Daß die Geschlechtsumstimmung von ♀ *Melandrium* schon allein auf Grund der *Ustilago*-Infektion erfolgen kann, hält BAKER (1947a) für ein Argument gegen einen qualitativen Unterschied zwischen „♀-Suppressoren“ und „♂-Promotoren“, wie ihn WESTERGAARD (1946a, b, 1948, 1953) fordert, der hierzu selbst schreibt: “It is in fact surprising that external agents like the smut fungus *Ustilago* are able to neutralize this (postulated) female suppressor which is absolutely epistatic in the unaffected females" (1946b, S. 438). Näher liegt für BAKER (1957) die Annahme, daß die Infektion auf dem Wege über eine Veränderung des Wuchsstoffspiegels die Geschlechtsausprägung beeinflußt. Neue Befunde von GARAY und SÁGI (1960) sprechen allerdings nicht unbedingt zugunsten dieser Auffassung. In infizierten und nicht infizierten Kelchblättern männlicher Blüten wurde eine größere IES-Oxydase-Aktivität als in Sepalen weiblicher Blüten ermittelt; eine *Ustilago*-Infektion glich die Kelchblätter weiblicher Blüten zwar bezüglich des Anthocyan-Gehaltes, nicht aber hinsichtlich der IES-Oxydase-Aktivität den Verhältnissen bei ♂ Blüten an. — Die für die Vermännlichung maßgebende Infektion erfolgt übrigens gewöhnlich durch Narben offener Blüten und läßt sich — entgegen älterer Ansicht — nicht durch Samen auf die nächste Generation übertragen (BAKER 1947b).

Die von SCHOPFER (1940a, 1941) durchgeführte Kreuzung eines gesunden ♀ Exemplars von *M. album* mit einem *Ustilago*-infizierten ⚥ Exemplar von *M. dioecum* (= *M. rubrum*) hatte mehrdeutige Ergebnisse, so daß sie bezüglich der modifikativen Geschlechtsbestimmung keine sicheren Aufschlüsse gab (vgl. BAKER 1950). Auch bei der von WERTH (1912) geschilderten Verweiblichung männlicher Individuen von *M. album* bleibt unklar, inwieweit genetische oder Außenfaktoren beteiligt sind.

ERLENMEYER und GEIGER-HUBER (1935) konnten übrigens im Kulturmedium von *Ustilago violacea* sowie in (mit HCl behandelten) Ätherextrakten *kein* im Hahnenkammtest wirksames ♂ Sexualhormon feststellen.

Von größerem Interesse sind auch die Versuche von A. und D. LÖVE (1940, 1945), denen bei *Melandrium rubrum* (= *M. dioecum*) die Geschlechtsumstimmung mit Hilfe tierischer Sexualhormone gelang, nachdem eine größere Zahl von gleichartigen Versuchen anderer Autoren mit anderen Objekten in den 1930er Jahren fehlgeschlagen war (weshalb sie hier auch übergangen werden). Bei ♀, ♂ und Übergangsformen wurde die Sproßspitze samt Blütenknospen amputiert. Nach Dekapitation der Hauptachse wurde in die Achseln des obersten Laubblattpaares 1 g Lanolinpaste mit 0,0005—0,1% Oestron, Oestradiol, Oestradiolbenzoat, Testosteron oder Testosteronpropionat aufgetragen. Bei ♀ Pflanzen kam es nach Behandlung mit Testosteron zur Bildung reduzierter Antheren, bei ♂ Pflanzen nach Oestron-Behandlung zur Bildung kleiner Fruchtknoten an den Blüten der durch die Lanolinpaste hindurchwachsenden Regenerate. Die größten Effekte in beiden Richtungen wurden bei Übergangsformen erzielt (vgl. die bei J. HESLOP-HARRISON 1957, S. 67, abgedruckte Tabelle): Unter Oestroneinfluß wurden Gynohermaphroditen (deren Blüten praktisch ♀ sind, aber funktionslose Staubblätter enthalten) rein ♀, während Androhermaphroditen Blüten mit größerem Fruchtknoten und ohne Staubblätter entwickelten. Erwähnung verdient noch, daß in jedem Falle etwa 50% der Versuchspflanzen vorzeitig zugrunde gingen, daß bei den Überlebenden das Wachstum zunächst 14 Tage lang gehemmt war und daß auch von A. und D. LÖVE unter Benutzung anderer Methoden durchgeführte Versuche mit *M. rubrum* mißlangen.

Durch Pfropfungen konnten bislang bei *Melandrium* keine Änderungen der Geschlechtsverhältnisse herbeigeführt werden. Dies gilt sowohl für die Vereinigung von verschiedengeschlechtigen Sprossen derselben Art als auch für die Transplantation von zwittrigen auf eingeschlechtige Sprosse (und umgekehrt) verschiedener Arten (SCHOPFER und BLUMER 1939, SCHOPFER 1943, Á. und D. LÖVE 1945, WESTERGAARD 1953). Ebensowenig besteht bei *M. album* und *M. rubrum* ein Zusammenhang zwischen der Insertionshöhe der Früchte und dem Geschlecht der aus ihren Samen hervorgehenden Pflanzen (NOLL 1907).

Eine Reihe von Untersuchungen (über die MATZKE 1930 referiert) betrifft die Zahl der Staubblätter in den Blüten von *Stellaria media* und damit natürlich auch den Grad ihrer Monoklinie bzw. Männlichkeit: Bei 5684 von BURKILL (1895) analysierten Blüten variierte die Anzahl der Stamina zwischen 0 und 10, jüngere Pflanzen hatten mehr Staubblätter pro Blüte als ältere, die Bodenqualität schien k ine Rolle zu spielen. REINÖHL (1903) kam an Hand von insgesamt 74491 Blüten zu dem Resultat, daß die Zahl der Stamina pro Blüte anfangs steige, dann wieder sinke; die meisten Blüten hatten drei, die zweitmeisten fünf Staubblätter; wachstumsfördernde Düngungs- und Beleuchtungsverhältnisse sollen den Mittelwert steigern, schlechte ihn senken. BÉGUINOT (1910—1921) erwähnt die Abhängigkeit der Staubblattzahl von Sippe und Alter. MATZKE (1930) selbst endlich bestätigt die anfängliche Zunahme der Staubblattzahl (auf etwa 4, etwa Ende Januar) und ihre spätere Abnahme (auf etwa 3, Ende Mai). Die Amputation von Blüten nach der Anthese, von Inflorescenzen oder von distalen Teilen älterer Zweige steigert den Mittelwert; bei panaschierten Pflanzen liegt dieser anfangs niedriger als bei rein grünen, worin MATZKE einen Ernährungseffekt sieht. Im übrigen wiesen 0,12% der von MATZKE untersuchten Blüten abnorme Fruchtblattzahlen (zwischen 0 und 5) auf.

Bei der Langtagpflanze *Cerastium tomentosum* setzten BOCQUET und STROUN (1960) durch Kurztag die Länge der Stamina herab und störten die Pollenbildung; nach Übertragung in Langtag trat Normalisierung ein.

Nyctaginaceae.

LILLO (1924) schilderte ein Exemplar von *Pisonia ambigua*, das 35 Jahre lang nur ♂ Blüten, fortan aber aus unbekannten Gründen nur ♀ Blüten bildete.

Chenopodiaceae.

Bei *Spinacia oleracea* gibt es ♂, ♀ und monöcische Individuen. Untersuchungen von THOMPSON (1955) hatten ganz allgemein das Ergebnis, daß sich die Anteile der ♀♀ und der Monöcisten an den Versuchspflanzen-Populationen zu annähernd 50% addierten; dies sprach dafür, daß die monöcischen Pflanzen genetisch ♀ sind. Im einzelnen betrafen THOMPSONs Versuche den Einfluß der N-Ernährung, der Tageslänge und der Temperatur auf die Geschlechtsausprägung. Der Einfluß der N-Düngung war in Gewächshaus- und Feldversuchen gering oder gleich Null: Von 130 N-arm gezogenen Pflanzen waren unter gewissen Bedingungen 45,4% ♂, von 132 N-reich gezogenen 58,3%; monöcisch waren in beiden Fällen 2,3%. Bei anderen Versuchen erfolgte die Anzucht 5 Wochen lang bei 10—16° C im 15—16 Std-Tag. Dann wurden die Pflanzen auf drei Temperaturbereiche (10—16, 16—21 und 21—27° C) aufgeteilt und innerhalb jedes Temperaturbereichs nochmals auf drei Tageslängen (9—10 Std-Kurztag, 15—16 Std-Langtag und Dauerlicht). Der Anteil der Monöcisten, d. h. der vermännlichten ♀♀, war bei allen Temperaturen am höchsten im Kurztag[1] und bei gleicher Tageslänge am größten unter der Einwirkung der höchsten Temperatur. Einige beispielhafte Ergebnisse lauteten

bei 21—27° im 15—16 Std-Tag 49,4% ♂, 2,2% monöcisch, 48,4% ♀;
bei 21—27° im 9—10 Std-Tag 44,6% ♂, 27,1% monöcisch, 28,3% ♀;
bei 10—16° im 9—10 Std-Tag 50,6% ♂, 7,7% monöcisch, 41,7% ♀.

Ähnliche Ergebnisse hatten fast gleichzeitig durchgeführte Versuche von JANICK und STEVENSON (1955), die mit je einer di- und monöcischen Linie experimentierten. Rein ♂ Pflanzen blieben (wie oben) durch verschiedene Temperaturen und Tageslängen unbeeinflußt. Bei monöcischen und ♀ Exemplaren bewirkte hohe Temperatur (27° C) Vermännlichung; bei monöcischen Individuen

[1] In bezug auf die Blütenbildung ist Spinat nach ROSA (1925) und KNOTT (1932) eine Langtagpflanze.

hatte Steigerung der Tageslänge von 12 auf 18 Std (bei 27° C) Verweiblichung zur Folge. — Nach diesen Befunden überrascht es nicht, daß NICOLAISEN und HANOW (1941) sowie ZOSCHKE (1956) eine Abhängigkeit der Geschlechterverteilung vom Aussaattermin feststellten.

Reichliche N-Düngung soll nach LAURENT (1903) bei Spinat Vermännlichung, starke P-Düngung das Gegenteil bewirken. LAURENTS Material war allerdings nicht sehr umfangreich; monöcische Spinatpflanzen hielt er für transformierte ♂♂.

Versuche von HYLMÖ (1940), die Geschlechtsverhältnisse bei zurückgeschnittenen Spinatpflanzen, Sorte „Nobel" („Matador"), durch tierische Sexualhormone zu beeinflussen, hatten kein eindeutiges Ergebnis, da auch bei Kontrollpflanzen Geschlechtsumkehr beobachtet wurde, wie sie häufig auftritt, wenn Pflanzen zurückgeschnitten wurden[1].

Polygonaceae.

Auch Experimente von A. und D. LÖVE (1945), die auf Geschlechtsumstimmung bei *Rumex Acetosa*, *R. Acetosella* und *R. tenuifolius* durch tierische Sexualhormone abzielten, schlugen fehl. Ebensowenig konnte SPRECHER (1913) bei *R. Acetosa* eine Beeinflussung der Geschlechtsverhältnisse durch Düngung feststellen. Abweichungen vom Verhältnis 1:1 führte er auf verschiedene Überlebensraten zurück. Bei über 6000 kultivierten Exemplaren hatte er ebenso wie bei wild wachsenden Individuen nur etwa 30% ♀ und 70% ♂ Pflanzen gefunden.

Moraceae.

Zehn 4jährige Maulbeerbäume *(Morus alba)* blühten erstmals 1936, davon 5 rein ♀, die anderen rein ♂. Im nächsten Jahr trugen von den erstgenannten 3 nur ♀, 2 neben ♀ auch einige ♂ Blüten. Von den 5 anfangs ♂ Bäumen entwickelten im Folgejahr 3 nur ♂, einer neben ♂ auch einige ♀ Blüten und einer schließlich fast nur ♀ Blüten (SCHAFFNER 1937).

Bei *Artocarpus integrifolia*, dem Brotfruchtbaum, der gewöhnlich ♂ und ♀ Blütenstände trägt, fand VENKATARAYAN (1949) abnorme ♂ Infloreszenzen mit einseitigen basalen ♀ Blüten.

Cannabinaceae.

Diese Familie umfaßt drei diöcische Arten, die bei der Erforschung der modifikativen Geschlechtsbestimmung sämtlich eine mehr oder weniger bedeutende Rolle gespielt haben. Einen gewissen Überblick über den jeweiligen Stand der Untersuchungen an Hanf vermitteln SPRECHER (1913), BESSEY (1933, sehr dürftig!) und KÖHLER (1964a).

Versuche, das Geschlecht von Blüten des Hanfes *(Cannabis sativa)* zu verändern, sollen — laut BRESLAVEC (1936, 1937) — schon SPALLANZANI 1787 und GIROU DE BUZAREINGUES 1827 unternommen haben, aber erst seit MOLLIARD (1897) waren Dutzende von Experimentatoren in dieser Hinsicht erfolgreich[2]. Anders als in weiter oben besprochenen Fällen geht hier die Geschlechtsumstimmung in der Weise vor sich, daß Sporophyllanlagen, die sich ihrer genetischen Bestimmung nach z. B. zu Staubblättern entwickeln müßten, den Charakter von Fruchtblättern annehmen. Diese Umstimmung kann begreiflicherweise nur auf einem gewissen Stadium erfolgen (bei älteren Primordien ist das Geschlecht schon endgültig festgelegt, und bei jüngeren setzt sich das genetisch bestimmte Geschlecht durch). Gelegentlich — vermutlich wenn der Stimulus in ungenügender Stärke oder am „Rande" des entscheidenden Stadiums einwirkt — kommt es zur Entstehung von Übergangsbildungen, die Staub- und Fruchtblattmerkmale in sich vereinigen können (*Intersexe*, CORRENS; vgl. Abb. 2, S. 157). Antheren weisen dann narbenförmige Spitzen auf, Samenanlagen, zuweilen mit befruchtungs- und entwicklungsfähiger Eizelle, findet man in Lokulamenten oder auf dem Konnektiv, und benachbarte Staubblätter können miteinander verwachsen (SCHAFFNER 1921, BRESLAVEC 1936, 1937, BRESLAVEC und

---

[1] Eine ausführlichere Studie von SUGIYAMA und SUTÔ (1964), die sich auch mit modifikativer Geschlechtsbestimmung bei Spinat befaßt, hat mir noch nicht vorgelegen.

[2] STRASBURGER (1900) zufolge hat allerdings schon MAUZ im Jahre 1822 Verweiblichung bei ♂ Hanfpflanzen beobachtet, die zurückgeschnitten oder im Winter aufgezogen worden waren.

Zaurov 1937, J. und Y. Heslop-Harrison 1957c; anatomische Details bei F. E. Rudenko 1956).

Zunächst sei der Einfluß der *Düngung* auf die Geschlechtsverhältnisse des Hanfes besprochen. *Keinen* solchen Einfluß — wenigstens keinen eindeutigen — konnten Girou de Buzareingues (1828), F. Haberlandt (1877)[1], Fisch (1887), Laurent (1903), Sprecher (1913) und Black (1945) feststellen. Molliard (1897) fand zwar bei schlechter Ernährung ein Verhältnis von 119 ♀ : 28 ♂, was aber auch andere Ursachen haben kann, da die Versuche im Winter bei Kurztag und geringer Lichtintensität durchgeführt worden waren (vgl. unten). Halsted (1901) stellte auf ungedüngten Böden ein Geschlechterverhältnis von 1:1, auf gut gedüngten dagegen ein Überwiegen der ♀♀ fest[2]. Schaffner (1919, 1925c, 1927) glaubte, daß nährstoffreicher Boden die durch Kurztag induzierte Umstimmung (♂ → ♀, vgl. unten) begünstige. Durch Kultur in Knopscher Nährlösung mit achtfacher N-Menge wird nach Tibeau (1936) die Verweiblichung gefördert, wogegen N-Mangel-Kultur (Nitrate durch Chloride ersetzt) Vermännlichung hervorrufen soll. Die Versuche wurden allerdings nur mit je sechs Pflanzen pro Gruppe durchgeführt; die Ergebnisse sind daher wohl kaum signifikant[3]. Inzwischen war auch Maekawa (1929) zu der Ansicht gelangt, daß nährstoffarmer Boden die Vermännlichung begünstige, die laut Schaffner (1928) sogar durch Tabakrauch hervorgerufen werden kann!

Aus Unterschieden im N-, K- und P-Gehalt bei verschiedengeschlechtigen Individuen schlossen Erdelsky und Herich (1956), daß N und P ♂-, K ♀-bestimmend wirken und eine entsprechende Düngung die Geschlechtsbestimmung beeinflussen könnte. Merkwürdig ist auch die Mitteilung von Herich (1956), daß bei polyploiden Hanfsorten (bei denen intersexuelle Formen auftreten) durch Quellung von Samen in $MnSO_4$-, KBr-, KJ- und H_2O_2-Lösungen der Prozentsatz der ♂ und der intersexuellen Individuen auf Kosten der ♀ erhöht wurde, während bei gewöhnlichen Hanfsorten umgekehrt der Anteil der ♂ auf Kosten der ♀ Exemplare gesteigert worden sein soll. Neuerdings befaßt sich Hulewicz (1964) wieder mit dem Einfluß der N-Versorgung (sowie der Temperatur und der Photoperiode) auf das Geschlecht von Hanfpflanzen.

In älterer Zeit suchte man auch öfters nach Beziehungen zwischen der Insertionshöhe der Früchte, der Größe und dem Alter der Samen einerseits und dem Geschlecht der aus ihnen hervorgehenden Pflanzen andererseits (Girou de Buzareingues 1828, 1831, F. Haberlandt 1869, 1874, Saccardo 1879, Fisch 1887, Noll 1907, Sprecher 1913). In der Mehrzahl der Fälle wurden keine solchen Beziehungen gefunden; einzelne gegenteilige Angaben scheinen der Kritik nicht standzuhalten. Hingegen stimmen die meisten eben genannten Autoren in der Ansicht überein, daß das Geschlecht der Hanfpflanzen bereits im Samen festgelegt sei und selbst durch extreme Außenbedingungen nicht verändert werden könne (vgl. vor allem Noll). Etwa seit der letzten Jahrhundertwende trat jedoch hinsichtlich dieser Auffassung ein radikaler Wandel ein:

Besonders zahlreich sind die Beobachtungen über die Bedeutung der *Tageslänge* für die Geschlechtsbestimmung beim Hanf. Molliard (1898a) fand Inter-

[1] F. Haberlandt hatte zunächst (1869) überhaupt keinen Einfluß von Düngung und Aussaatzeit auf das Geschlechterverhältnis von Hanfpflanzen feststellen können; später (1874) schienen ihm Lichtmangel und reichlichere Düngung die Entwicklung des ♀ Geschlechtes zu begünstigen; schließlich (1877) zeigte sich aber, daß eine Variation des Geschlechterverhältnisses durch die mehr oder weniger hohe Sterblichkeitsrate der ♂ Individuen hervorgerufen werden kann.

[2] Heslop-Harrison (1957) findet in der Arbeit von Halsted keinen konkreten Hinweis darauf, daß tatsächlich eine Umstimmung von ♂ Pflanzen erfolgt war. Übrigens glaubte auch schon Leydecker (1870), daß man auf reichem Boden mehr ♀, auf armem mehr ♂ Hanfpflanzen antreffe, wofür sich natürlich F. Haberlandts Erklärung (s. vorstehende Fußnote) anbietet.

[3] In Versuchen von Borthwick und Scully (1954) beeinflußte der N-Gehalt der Nährlösung nicht die Zahl der ♂ Blüten auf ♀ Pflanzen, sondern nur ihre „Entwicklungsrate".

sexualität besonders bei zuerst ♂ blühenden Individuen, die im Winter gezogen worden waren. Bei einer Aussaat waren von 147 Pflanzen 119 ♀ und 28 ♂, bei einer anderen von 172 Pflanzen 148 ♀ und 24 ♂; von den ♂ Individuen zeigten im ersten Fall 21, im zweiten alle Exemplare Übergänge zum ♀ Geschlecht. MOLLIARD führte dies wohl fälschlich auf die geringe Lichtintensität statt auf die Kurztagverhältnisse zurück. STRASBURGER (1900) konnte die Befunde von MOLLIARD nicht bestätigen, da er bei ähnlichen Versuchen durch Gewächshausschattierung den sommerlichen Langtag nicht ausgeschaltet hatte. Wohl aber wurden MOLLIARDs Ergebnisse von TOURNOIS (1911, 1912) reproduziert, der im Winter durch natürlichen Kurztag, nicht aber im Sommer durch künstlichen 6 Std-Kurztag bei ♂ Hanfpflanzen ausgiebige Verweiblichung erzielte. Die Versuche von TOURNOIS wurden von SCHAFFNER (1921a, 1923a, 1925a, b, 1931) erfolgreich wiederholt. Bei Sippen, die im Sommer im Freiland nur rein ♀ und rein ♂ Exemplare hervorbrachten, wurden im Winter im Gewächshaus bis zu 80% verweiblichte ♂♂ und bis zu 88% vermännlichte ♀♀ erzielt, deren eigentliches Geschlecht aber noch an der Blattform erkannt werden konnte. Die höchsten Prozentsätze verweiblichter ♂♂ ermittelte SCHAFFNER (1931) nach Novemberaussaat. In Ergänzung zu STRASBURGERs Mißerfolg stellte SCHAFFNER (1923a) fest, daß — im Gegensatz zu Kurztag — Schattierung unter Langtagbedingungen selbst dann keine Verweiblichung von ♂ Individuen zur Folge hat, wenn das Wachstum infolge der geringen Lichtintensität (im Vergleich zu voll beleuchteten Kontrollpflanzen) um mehr als die Hälfte zurückbleibt. — MCPHEE (1924) konnte durch Kurztag (im Sommer durch Rolläden arrangiert) allein keine Intersexualität erreichen, wohl aber bei 14 von 15 ♂ Pflanzen durch 7 Std-Kurztag und ständige Entfernung von Seitenknospen (die unter Langtagverhältnissen wirkungslos war) sowie bei 53% ♂ Pflanzen durch Winteraussaat ohne Entknospung. Auch bei ♀ Pflanzen steigerte Kurztag den Prozentsatz intersexueller Blüten (MCPHEE 1925). — CORRENS (1928) kultivierte im Winter im Gewächshaus eine Hanfsippe, deren ♀♀ unverändert blieben, während die ♂♂ verweiblicht waren (Blüten monströs, kein Fruchtansatz). — Auch BRESLAVEC (1936, 1937) sowie BRESLAVEC und ZAUROV (1937) erzielten Intersexe durch Kurztag, desgleichen durch Röntgenstrahlen. — Von Natur aus monöcische Hanfrassen fanden anscheinend erstmals in einschlägigen Versuchen von HUHNKE, JORDAN, NEUER und V. SENGBUSCH (1951) Verwendung. Je später im Jahr (von Ende Juni ab!) die Aussaat erfolgte, desto weiblicher wurden die Monöcisten. Bei extremem Kurztag und geringer Lichtintensität (z. B. nach Herbst- oder Winteraussaat) bildeten bis zu 80% der Pflanzen solcher monöcischer Rassen ausschließlich ♀ Blüten. — PETIT (1952) scheint als erster Versuche über die photoperiodische Geschlechtsbestimmung beim Hanf in Klimakammern unter konstanten Bedingungen (20° C, 80% relative Luftfeuchtigkeit, 8 bzw. 16 Std-Tag, 5000 bis 6000 Lux) durchgeführt zu haben. Im Langtag betrugen das Geschlechterverhältnis 1:1, die Wuchshöhe 150—200 cm (wie normalerweise im Freiland) und das geringste Blühalter 1—$1^1/_2$ Monate. Im Kurztag wurden ein Mindestblühalter von 3—4 Wochen und eine Wuchshöhe von nur etwa 40 cm erzielt; die ♂ Blüten waren ± verweiblicht, es wurden daher ♀ und ⚥ Pflanzen im Verhältnis 1:1 gefunden. Gingen der Langtagbehandlung 10 oder 20 Kurztage voraus, so war das Resultat das gleiche wie nach ständiger Langtageinwirkung. 41 Kurztage und anschließender Langtagaufenthalt hatten zur Folge, daß die schon blühenden Pflanzen zu lebhaftem vegetativem Wachstum übergingen; bildeten sie schließlich erneut Blüten, so entstanden an den bis dahin ⚥ Pflanzen nur ♂ Blüten. 30 Langtage und weiterhin Kurztagbehandlung bewirkten Sistierung des Wachstums bei etwa 80 cm Wuchshöhe; ein Teil der Pflanzen bildete zuerst ♂, dann ⚥ Blüten. Kurztag führte also stets zu Wachstumshemmung, Blühbeschleunigung und Verweiblichung. — BORTHWICK und SCULLY (1954) berichteten — wie schon SCHAFFNER (1921) — über die photoperiodisch induzierte Vermännlichung weiblicher Hanfpflanzen. Im winterlichen Kurztag produzierten ♀ Exemplare bei voller Lichtintensität mehr ♂ Blüten als bei einer auf 63% reduzierten Lichtintensität. Zu weiteren Versuchen wurden ♀ Pflanzen benutzt, die aus Selbstungen der soeben erwähnten hervorgegangen waren. Photoperioden, erheblich kürzer als die kritische Tageslänge der Blütenbildung, induzierten ♂ Blüten bei 25—45% der (♀) Versuchspflanzen, wogegen Photoperioden knapp unterhalb der kritischen Tageslänge *keine* Vermännlichung bewirkten. Auch niedere Nachttemperatur vor oder während der photoperiodischen Induktion löste die Bildung ♂ Blüten aus. — In Versuchen von THOMAS (1956) bildeten ♀ Hanfpflanzen nach 45tägiger Anzucht in Dauerlicht bei tags 21—26, nachts 15,5—21° C und anschließender 8 Std-Kurztag-Behandlung ♂ Blütenstände am 7. oder 8. (dem ersten inflorescenztragenden) Knoten. THOMAS vermutet daher, daß monöcische und ♀ Pflanzen eine ♂ Phase durchlaufen, die sich bei ♀ Exemplaren normalerweise infolge Blütenlosigkeit nicht zu manifestieren pflege, die aber im vorliegenden Versuch über den Beginn der Blütenbildung hinaus verlängert worden sei. — ŁĄCZYŃSKA-HULEWICZOWA (1957) konnte im Gegensatz zu den vorstehend referierten Befunden keinen Einfluß der Tageslänge auf die Geschlechtsverteilung bei der diploiden diöcischen Hanfsorte LKCSD feststellen, wohl aber bei diploiden monöcischen (Sorte Bredemanns) und tetraploiden diöcischen Hanfsorten. Diese zeigten im 10 Std-Kurztag starke Verweiblichung, verbunden mit Entwicklungshemmung der ♂ Blüten (verminderte Zahl und Fertilität der Pollenkörner). Die Tageslängenabhängigkeit manifestierte sich auch im Verlauf der normalen Ontogenese: Eine in der Jugend unter Kurztagbedingungen ♀ erscheinende Pflanze nimmt im Alter unter Langtagverhältnissen männlichen Charakter an. — Endlich hat sich auch KÖHLER (1958) mit monöcischen Hanfpflanzen befaßt. ♂ Blüten fanden sich oberhalb großer Blätter, ♀ Blüten oberhalb kleiner Blätter; Blattgröße und -form waren in komplizierter Weise von der Tageslänge und anderen Faktoren abhängig. Eine der „blühinduzierenden" Kurztagexposition vorausgehende Langtagperiode ruft bei monöcischen Hanfpflanzen die Bildung einer um so größeren Zahl von ♂ Blüten hervor, je länger diese Langtagperiode andauert und je stärkeres Licht dabei angewandt wird (KÖHLER 1961, 1964a).

Die Rolle der *Temperatur* wurde merkwürdigerweise viel seltener untersucht. Die diesbezüglichen Beobachtungen von BORTHWICK und SCULLY wurden bereits im vorigen Absatz erwähnt. In Versuchen von MAEKAWA (1929) trugen bei Frühjahrsaufzucht im Gewächshaus 57—100% der ♂ Individuen gemischtgeschlechtige Blüten; durch Kalthauskultur wurde jedoch — verglichen mit Warmhauskultur — die Tendenz zur Verweiblichung von ♂ Exemplaren vermindert, die Wuchshöhe gesteigert. Nach NELSON (1944) herrschten bei 15°-Kultur die ♀, bei 30°-Kultur die ♂ Blüten vor.

Durch starkes *Zurückschneiden* brachte PRITCHARD (1916) im Jahre 1914 alle so behandelten ♀♀ und 14—20% der ♂♂ dazu, Blüten des anderen Geschlechtes (neben solchen des bisherigen) zu bilden; bei einem 1909 durchgeführten gleichartigen Versuch war der Erfolg jedoch geringer. MAEKAWA (1929) beobachtete weiterhin nach Dekapitation von ♂ Hanfpflanzen Bildung von intersexuellen Blüten an Seitenzweigen; später erfolgte die Rückkehr zum ursprünglichen Geschlecht. ♀ Exemplare blieben ihrem Geschlecht jedoch (im Gegensatz zu den Beobachtungen von PRITCHARD) weitgehend treu. SAVELLI und SOSTER (1929) beschnitten 1812 Hanfpflanzen und konnten auch bei den männlichen keine Geschlechtsänderung feststellen. — Monöcische Exemplare, die sonst an jedem

Seitenzweig ♂ und ♀ Blüten trugen, blühten nach Dekapitation und Ausgeizen der ♀ Infloreszenzen im oberen Teil rein ♀, im unteren rein ♂ (NEUER und v. SENGBUSCH 1943), wie es auch schon BRAUN (1873) gelegentlich (neben anderen Abweichungen) bei ♀ Individuen beobachtet hatte. Bei F_1-Pflanzen, die aus der Kreuzung von ♀ Pflanzen mit Monöcisten hervorgegangen waren, wurde durch Ausgeizen weiblicher die Bildung männlicher Blüten begünstigt (HUHNKE, JORDAN, NEUER und v. SENGBUSCH 1951, vgl. Tabelle 3). — Erst in jüngster Zeit rief CH. JU. RUDENKO (1963) wieder Übergänge zu ♀ Blüten, die Bildung von Samenanlagen an oder in Antheren und die Entwicklung von Makro- und Mikrosporen in ein und demselben Sporophyll hervor, indem er Blütenstände männlicher Pflanzen beschädigte.

Tabelle 3. *Vermännlichung von Hanfpflanzen (vgl. Text) durch Ausgeizen.* (Nach HUHNKE, JORDAN, NEUER und v. SENGBUSCH 1951.) *Bei den Zahlen handelt es sich um die prozentuale Häufigkeit der drei Phänotypen.*

Zeitpunkt der Bonitierung	Andröcisten	Monöcisten	Gynöcisten
Zur Zeit der Blüte	1,1	8,8	90,1
Nach dem Ausgeizen und Wiederaustreiben . .	1,1	28,6	70,3

Mittels *Pfropfung* verschiedengeschlechtiger Partner konnte das Geschlecht des Reises oder der Unterlage in keinem Fall[1] beeinflußt werden (STRASBURGER 1900, MCPHEE 1924, BRESLAVETZ 1935, KUHN 1941, LIMBERK 1959; vgl. aber J. und Y. HESLOP-HARRISON 1957c, s. unten). KUHN vermutet daher, daß die geschlechtsbestimmenden Stoffe auf die Einzelzelle beschränkt seien.

Von größerem Interesse sind die Untersuchungen von J. bzw. J. und Y. HESLOP-HARRISON über den Einfluß von α-Naphthylessigsäure (NES), 2,3,5-Trijodbenzoesäure (TJBS) und Kohlenmonoxyd auf das Geschlecht des Hanfes, die anschließend besprochen werden mögen. In einer ersten Versuchsreihe (J. HESLOP-HARRISON 1956) wurden Pflanzen (zunächst unbekannten Geschlechtes) anfangs 20 Tage unter Langtagbedingungen bei $> 20^0$ C aufgezogen und dann — zwecks Induktion der Blütenanlegung — 10 Kurztagen ausgesetzt. Nun wurde bei einem Teil der Pflanzen auf die Unterseite je eines Blattes 0,25 g einer Lanolinpaste aufgetragen, die 0,5%ige NES enthielt. Bei der jetzt folgenden Langtagkultur manifestierte sich die NES-Wirkung, von der sogleich die Rede sein wird. Endlich erhielten die Pflanzen vom 70. Tag ab wieder Kurztag; durch diese „zweite Induktion" wurden sie veranlaßt, ihr genetisch festgelegtes Geschlecht zu offenbaren. Dabei zeigte sich, daß NES die ♀ Individuen bezüglich des Geschlechtes unbeeinflußt gelassen hatte, während die ♂ Exemplare durchweg zur zeitweise ausschließlichen Bildung von ♀ Blüten veranlaßt worden waren. Monokline Blüten wurden in diesem Versuch nicht gefunden. Daß in ständigem Langtag keine Blütenentfaltung und nur bei ♀ Exemplaren vereinzelt Blütenanlegung erfolgte, spricht dafür, daß sich das hier verwendete Pflanzenmaterial von dem von PETIT (1952, s. oben) benutzten genetisch unterschied. — Kohlenmonoxydbehandlung, die auch schon in anderen Fällen wuchsstoffähnliche Wirkungen zeitigte, führte bei *Cannabis* zu ähnlichen Resultaten wie die Applikation von NES. ♀ Pflanzen wurden durch CO ebensowenig wie durch NES beeinflußt. Wurden ♂ Individuen einmal 24 oder 48 Std einer 1%igen CO-Atmosphäre ausgesetzt, so traten intersexuelle Blüten auf, ♀ Blüten freilich erst nach mehrmaliger CO-Exposition. Im Gegensatz zu den oben erwähnten erfolglosen Transplantationen ließ sich der CO-Effekt in zwei erfolgreichen Pfropfungen übertragen:

[1] Nur SAVINE soll in einem — anscheinend unveröffentlichten — Vortrag über positive Ergebnisse berichtet haben (BRESLAVETZ 1935).

Einmal 48 Std CO-exponierte, 1,3 cm lange, entblätterte Pfropfreiser wurden einer unbehandelten Unterlage aufgesetzt, die hierauf mit der Bildung von intersexuellen Blüten reagierte (J. und Y. Heslop-Harrison 1957c). Da jedoch keine Kontrollpfropfungen erwähnt werden, erscheint ein auf Resektion der Unterlage beruhender Effekt nicht ausgeschlossen.

Führt also, wie wir gesehen haben, die Applikation eines Wuchsstoffes wie NES zur Verweiblichung von ♂ Individuen, so erscheint es denkbar, daß Behandlung von ♀ Exemplaren mit einem Wuchsstoffantagonisten wie TJBS umgekehrt Vermännlichung bewirkt. Nach J. und Y. Heslop-Harrison (1957b) ist dies jedoch nicht der Fall. Ob TJBS die durch NES oder CO induzierten Umstimmungsvorgänge blockieren oder aufheben kann, wurde allerdings noch nicht erforscht.

Schließlich untersuchte man in letzter Zeit verschiedentlich den Einfluß von *Gibberellinsäure* (GA_3) auf das Geschlecht von Hanfblüten. In einem Versuch von Atal (1959) wurden fünf noch vegetative ♀ Exemplare in vier aufeinanderfolgenden Wochen je einmal mit GA_3 (100 p.p.m.) besprüht. Bei vier dieser Pflanzen wurden daraufhin zunächst ♂, dann intersexuelle und zuletzt weibliche Blüten entwickelt. Bei einer schon blühenden ♀ Pflanze führte GA_3-Behandlung gleichfalls zu vorübergehender Bildung von ♂ Blüten. — Auch Köhler (1964b) berichtet über Vermännlichung von *Cannabis* unter dem Einfluß von Gibberellinsäure, wenigstens bei heterozygoten ♀♀ (*XXm*, Subgynöcisten)[1]. Dagegen schildert Herich (1960) einen Fall von mutmaßlicher Verweiblichung durch GA_3: Samen der Sorte Rastislavická wurden 24 Std lang in GA_3-Lösung (5, 10, 25 oder 100 p.p.m.) eingequollen. In zwei Wiederholungen wurden je 400—500 Pflanzen pro Variante aufgezogen. Allein nach Vorbehandlung mit 10 p.p.m. GA_3 wich der Quotient ♂:♀ Pflanzen (0,82:1) merklich, aber nicht überzeugend, von dem Quotienten der Kontrollpflanzen (1,07:1) ab.

Ähnliche Befunde wie bei *Cannabis* wurden von Tournois (1909, 1911, 1912, 1914) bei *Humulus japonicus* erhoben. Pflanzen dieser (ebenfalls diöcischen) Art wurden durch winterliche Gewächshauskultur unter natürlichen Kurztagbedingungen oder durch 6 Std-Kurztage im Sommer (in einer Dunkelkammer) zu vorzeitigem Blühen veranlaßt[2]. ♀ Individuen zeigten keine Besonderheiten, ♂ waren dagegen zum Teil verweiblicht: Im übrigen normale Antheren hatten eine narbenartige Spitze, im Extremfall waren Staubblätter zu Karpellen umgewandelt und bildeten Fruchtknoten. In anderen Fällen wurde in einzelnen Blüten inmitten von Staubblättern ein zentraler Fruchtknoten mit zwei oder mehr Narben entwickelt. Die kurztaginduzierte Verweiblichung konnte durch Erhöhung der Luftfeuchtigkeit noch gesteigert werden. Die meisten intersexuellen Exemplare kehrten mit zunehmender Tageslänge zur rein ♂ Form zurück. Schaffner (1923b, 1925a, b) konnte Tournois' Ergebnisse bestätigen[3]. Ähnliche Erscheinungen hatte auch schon Figdor (1911) beobachtet; die Ursache sah er in Nahrungsmangel, Lufttrockenheit, Kälte und einer „bestimmten chemischen Lichtintensität".

Bei *Humulus Lupulus* wurde die Verweiblichung ♂ Exemplare durch Dekapitation erzielt (Limberk 1953). Weiterhin berichtete Limberk (1954) über zwei Fälle von Geschlechtsbeeinflussung durch Pfropfung: ♀ Triebe des Saazer Kulturhopfens, seit 800 Jahren vegetativ vermehrt, wurden als Pfropfreis, ♂ Wildhopfen-Sämlinge als Unterlage verwendet. In beiden Fällen ging aus der Unterlage

---

[1] Bei normalen Hanf-♀♀ war GA_3 in Köhlers Versuchen freilich wirkungslos.

[2] Es handelt sich hier um die Darstellung eines Falles von Photoperiodismus — ein Jahrzehnt vor der als „klassisch" bekannten Veröffentlichung von Garner und Allard (1920).

[3] Schaffner (1927b) erwähnt übrigens Geschlechtsumkehr in *beiden* Richtungen unter dem Einfluß von Kurztag und guter Stickstoffversorgung.

ein Trieb hervor, der ♀ „Zapfen" trug, die nach Form und Größe von denen des Saazer Hopfens abwichen. Auch hier ist natürlich an eine verweiblichende Wirkung der vorangegangenen Resektion der Unterlage zu denken. — Neuerdings gelang WESTON (1960) die Vermännlichung von genetisch ♀ Exemplaren von *H. Lupulus* durch Behandlung mit α-(2-Chlorphenylthio)propionsäure in 0,05%-iger Lösung. Erwähnt sei schließlich noch, daß gelegentlich spontan (LEPLAE) oder bei Stecklingsvermehrung (NYPELS 1897) aufgetretene monöcische Individuen bekannt geworden sind.

Urticaceae.

Bei *Urtica dioica* findet man zuweilen monöcische Exemplare, bei denen verschiedengeschlechtige Sproßabschnitte in verschiedener Weise miteinander abwechseln (♀ → ♂, ♂ → ♀, ♀ → ♂ → ♀), ohne daß über die Ursachen etwas bekannt zu sein scheint (HILDEBRAND 1893, LAIBACH 1952). PREDA und GHISA (1957) behandelten *Urtica dioica*-Pflanzen vom Keimlingsstadium bis zum Sichtbarwerden von Blütenanlagen mit gonadotropem Hormon (Prolan) und glauben, außer einer Beschleunigung des Wachstums auch Verweiblichung erzielt zu haben: In einem ersten Versuch blühten von 81 Pflanzen 74 ♀ und 7 ♂, in einem zweiten Versuch von 44 Pflanzen 40 ♀ und 4 ♂.

Euphorbiaceae.

Obwohl diese Familie die umfangreiche monöcische Gattung *Euphorbia*[1] einschließt, sind Versuche unter unserem Gesichtspunkt lange Zeit offenbar nur mit der diöcischen *Mercurialis annua* sowie mit *M. ambigua* durchgeführt worden, bei der es andröcische und monöcische Individuen gibt. Erst in den letzten Jahren sind einige Untersuchungen an den monöcischen Arten *Ricinus communis* und *Manihot utilissima* hinzugekommen. Die morphologischen Grundlagen der Geschlechtsumstimmung sind hier (zumindest bei ♂ Individuen von *Mercurialis*) dieselben, wie sie oben für *Cannabis* geschildert wurden (YAMPOLSKY 1920b, 1934).

Wasser- und N-reiche Böden sollen nach GIARD (1898) bei ♂ Exemplaren von *M. annua* den Übergang zur Monöcie begünstigen. LAURENT (1903) konnte demgegenüber keinen Einfluß der mineralischen Ernährung auf die Geschlechtsbestimmung feststellen. MOLLIARD (1898b) fand bei Frühjahrskultur (ab 18. April) ein Verhältnis von 1894 ♂:1637 ♀, bei Sommerkultur (ab 25. Juni) ein solches von 2019 ♂:1999 ♀, wofür er die im Durchschnitt um 6,5° C höhere Temperatur verantwortlich machte, was aber kaum zwingend erscheint.

Im Gegensatz zu STRASBURGER (1910), der bei 200 nach starker Beschneidung regenerierenden ♂ Individuen nur an einem einzigen Exemplar ein paar ♀ Blüten gefunden hatte (die vielleicht auch ohne Beschneidung aufgetreten wären), erzielte YAMPOLSKY (1930a) an 11 von 18 kräftigen ♂ Exemplaren nach derselben Behandlung ♀ und intersexuelle Blüten; später kehrten diese Pflanzen zur Bildung von ♂ Blüten zurück (vgl. auch YAMPOLSKY 1920c). 18 nicht beschnittene Kontrollpflanzen blieben ständig ♂. — Pfropfungen, auch Dreifachpfropfungen, waren bei *M. annua* wirkungslos (VÖCHTING 1892, STRASBURGER 1900, YAMPOLSKY 1930b, 1957, KUHN 1941).

Bei *M. ambigua* stellte THOMAS (1956) fest, daß Dauerlicht im Vergleich mit 8 Std-Kurztag (bei tags 26—32° C, nachts 17—21° C) die Ausbildung von ♀ Blüten verzögert; das gleiche gilt (bei Dauerlicht) für eine Durchschnittstemperatur von 24° C, verglichen mit einer solchen von 15,5° C.

J. und Y. HESLOP-HARRISON (1957a) bedienten sich zur Anzucht des Pflanzenmaterials für ihre Versuche über den Einfluß von CO auf die Geschlechtsausprä-

[1] Soweit ich sehe, bezieht sich auf modifikative Geschlechtsbestimmung bei *Euphorbia* nur eine Beobachtung von MOLLIARD (1895): Bei von *Uromyces scutellatus* befallenen Exemplaren von *Euphorbia Cyparissias* wird die ♀ Blüte im ersten Cyathium jedes Strahles der Dolde entwickelt, während sie bei gesunden Individuen unterdrückt ist.

gung von *M. ambigua* derselben den Eintritt der ♀ Phase verzögernden Bedingungen (> 20 Std-Langtag und Temperatur > 22° C). 7—10 Tage alte Pflanzen wurden zweimal 48 Std lang einer 1%igen CO-Atmosphäre ausgesetzt. Im Vergleich mit sonst ebenso behandelten „Luft-Kontrollen" war bei monöcischen Exemplaren nach dieser Behandlung (an deren Ende noch keine Blütenanlagen zu sehen waren) an jedem einzelnen Knoten die Zahl der ♂ Blüten gesenkt, die (viel geringere) Zahl der ♀ Blüten erhöht. An den fünf ersten Knoten oberhalb der Keimblätter wurde daher bei CO-behandelten Individuen ein Verhältnis von 5,7 ♂:1 ♀ Blüte gegenüber einem solchen von 26,6 ♂:1 ♀ Blüte bei „Luft-Kontrollen" ermittelt.

MUKERJI (1936) berichtete über ein Vorherrschen von ♀ Exemplaren der *M. perennis* an schattigen Standorten auf würzigen (?; engl. "mull") Böden. Ob dabei modifikative Geschlechtsbestimmung zugrunde lag, erscheint zweifelhaft.

Bei *Ricinus communis* steigert Kurztag die ♂ Tendenz, während Langtag (bzw. zunehmende Tageslänge) sie vermindert; dementsprechend erschienen die von SHIFRISS (1956) untersuchten Individuen im Frühjahr am „weiblichsten", im Winter am „männlichsten". — Später erforschte SHIFRISS (1961) den Einfluß von Gibberellinsäure (GA_3) auf die Geschlechtsverhältnisse von *Ricinus*-Blüten. Eine *vorzeitige* GA_3-Behandlung (vor Beginn der Differenzierung von Blütenknospen) verzögert die Blütenbildung erheblich; ein spezifischer Gibberellin-Effekt auf die Blütenbildung bleibt dann aus. Erfolgt dagegen die Behandlung mit GA_3 (als K-Gibberellat, 250—1500 mg/Liter) erst auf dem 2-, 3- und 4-Blatt-Stadium, so ist die Blütenbildung nur schwach verzögert, zugleich aber die Entwicklung von ♀ Blüten stark gefördert[1]. α-Naphthylessigsäure, die in anderen Fällen verweiblichend wirkt, hatte bei *Ricinus* keinerlei Effekt. Übrigens sind die Geschlechtsverhältnisse der *Ricinus*-Blütenstände auch mit ihrer Insertionshöhe korreliert: Bei unbehandelten Pflanzen, die unter konstanten Bedingungen gehalten wurden, enthielten die zweite und die dritte Infloreszenz eine geringere Zahl von ♀ Blüten als die erste und die vierte.

Bei *Manihot utilissima* gibt es Rassen, die zur Bildung von Staminodien in sonst ♀ Blüten neigen. Solche Staminodien können zum Teil auch Pollen produzieren. Ihre Anzahl pro Blüte (maximal 10) hängt unter den Freilandbedingungen der Elfenbeinküste von den jahreszeitlichen Verhältnissen ab: In einem von MIÈGE (1959) untersuchten Falle wiesen im Mai alle ♀ Blüten Staminodien auf (durchschnittlich 3,9 pro Blüte), im September aber nur 6,8% der ♀ Blüten (durchschnittlich nur 0,082 Staminodien pro Blüte).

Myricaceae.

Die Ursachen der von DAVEY und GIBSON (1917) erwähnten unterschiedlichen Geschlechtsverhältnisse von *Myrica Gale* an verschiedenen Standorten sind nicht ohne weiteres klar. An trockeneren Stellen fanden diese Autoren etwa 75% ♂, 10% ♀ und 25% monöcische Individuen, an feuchteren etwa 56% ♂, 38% ♀ und 6% monöcische und zwittrige Exemplare. Im einzelnen beschrieben DAVEY und GIBSON monöcische Sträucher mit ♀ und ♂ Kätzchen, Individuen mit monöcischen Kätzchen (unten meist ♂, oben ♀, Mittelzone ⚥) und Sträucher mit ⚥ Kätzchen. Ein und derselbe Strauch schwankte hinsichtlich seiner Geschlechtsausprägung von Jahr zu Jahr: 1913 erschien er ♀, 1914 monöcisch, 1915 fast rein ♂ und 1916 wieder mehr ♀.

Salicaceae.

SCHLENKER (1953) fand in einem Bestand 5- und 6jähriger F_1-Bastarde von *Populus tremula* und *P. alba* zahlreiche vorzeitig blühende Sämlinge, darunter viele mit androgynen Kätzchen und monoklinen Blüten. Reziproke Bastarde differierten deutlich.

[1] Schon hier sei darauf hingewiesen, daß GA_3 andererseits bei *Cucumis sativus* Vermännichung bewirkt.

Bei *Salix*-Arten sind monöcische Exemplare, androgyne Kätzchen und gemischtgeschlechtige Blüten gang und gäbe. So schilderte bereits BRAUN (1853, S. 28) ein in Schwetzingen befindliches, ♂ und ♀ Blüten tragendes Individuum der sonst fast nur in ♀ Exemplaren kultivierten *Salix babylonica*; vgl. auch ARNAL (1952a) bezüglich *Salix sepulcralis*. Über ähnliche Erscheinungen bei 9 amerikanischen Weidenarten berichtete SMITH (1940); ein Strauch von *Salix brachycarpa* hatte nur abnorme Blüten. Gelegentlich wurden Staubblätter gefunden, die distal eine Narbe trugen. Über die physiologischen Ursachen solcher Bildungsabweichungen scheint nichts bekannt zu sein. Als Stecklinge behandelte Zweige ♀ Exemplare von *Salix fragilis* blühten zuerst meist ♂ (RIEDE 1925). In der Regel scheinen aber Weiden bei Stecklingskultur monöcisch zu werden und relativ selten allein Blüten des dem ursprünglichen entgegengesetzten Geschlechtes zu tragen (SCHULZ 1892). Als Ursache für die Bildung intersexueller Einzelblüten kommen übrigens bei *Salix*-Arten der *Caprea*-Gruppe parasitäre Milben, insbesondere *Epitrimerus salicobius*, in Betracht (J. W. HESLOP-HARRISON 1924).

Betulaceae.

Bei Birken (*Betula* spec.) zählte HEYER (1884, 1886) um so mehr ♀ Blüten, je besser die Kulturbedingungen waren. Nach Untersuchungen von RIEDE (1925) an *B. verrucosa*, deren Sämlinge anfangs rein ♂ blühen, spielen auch Lichtintensität und Bewurzelungsverhältnisse eine Rolle. — Auf die Bedeutung der Lichtverhältnisse für die Geschlechtsausprägung von Birkenblüten kommen auch in neuerer Zeit wieder einige Autoren zu sprechen. So säten LONGMAN und WAREING (1959) im Februar 1957 Samen von *B. verrucosa* aus und kultivierten die Keimlinge im Gewächshaus bei 15—25° C und Langtag oder Dauerlicht; während die blütenlose Jugendphase gewöhnlich mindestens 5 Jahre andauert, zeigten diese Sämlinge bereits im November des Aussaatjahres die ersten Inflorescenzanlagen; trotz ihrer Stellung an Seitenknospen entwickelten sie sich zu ♂ Kätzchen; später folgten auch ♀ und monöcische (androgyne) Kätzchen. Solche androgynen Kätzchen traten gleichfalls in Versuchen von v. DELLINGSHAUSEN und STERN (1958) mit *Betula*-Klonpflanzen auf, und zwar vorzugsweise dann, wenn die vorausgegangene Vegetationsperiode durch Gewächshausaufenthalt künstlich verlängert worden war (die Blütenknospen für das nächste Jahr werden dann z. B. schon im April statt im Juni angelegt). NEKRASOV, VARTAZAROVA und BORODINA (1963) glauben, die Androgynie von Kätzchen eines Exemplares von *B. japonica* (das in Moskau aus einem aus Leningrad erhaltenen Samen hervorgegangen war) darauf zurückführen zu können, daß proximale und distale Hälfte dieser Kätzchen während des Herbstes und Winters unterschiedlichen Licht- und Feuchtigkeitsverhältnissen ausgesetzt waren; die Knospenschuppen hatten nämlich nur den basalen Abschnitt umhüllt. Dieselbe Anomalie hatte auch schon FROLOVA (1958) an *B. Rezniczenkoana* entdeckt; die androgynen Kätzchen wiesen an der Spitze mehr ♂, an der Basis mehr ♀ Blüten auf. Endlich sei noch erwähnt, daß SCHULZ (1892) an der Basis von ♂ wie von ♀ *Betula*-Kätzchen auch monokline Blüten gefunden hat.

Sämlinge von *Corylus Avellana* bilden im Gegensatz zu solchen von *Betula* zunächst nur ♀ Blüten (HILDEBRAND 1881). Sexuell voll entwickelte Zweige älterer Sträucher tragen an Basis und Spitze ♀, in der Mitte ♂ Inflorescenzen und in den Übergangszonen androgyne Zweige (ARNAL 1954). Gemischtgeschlechtige Blüten trifft man nach SCHULZ (1892) bei *Corylus* nur selten an.

Demselben Autor zufolge trägt wohl jeder ♂ *Alnus glutinosa*-Baum mehr oder minder zahlreiche ♀ Kätzchen, die an der Basis monokline Blüten aufweisen.

Juglandaceae und Cupuliferae.

Wie schon HILDEBRAND (1881) angibt, entwickeln Keimlinge von *Juglans regia* und *Quercus rubra* in den ersten Jahren der Blütenbildung nur ♀ Blüten. In dem feuchtheißen August 1890 stellte HILDEBRAND (1893) fest, daß Kätzchen von *Qu. ilicifolia* rein ♀ blühten, obwohl sie im Hinblick auf ihre Stellung in den Blattachseln vorjähriger Triebe hätten ♂ blühen müssen. HILDEBRAND blieb keineswegs im Zweifel über die morphologische Grundlage dieser mutmaßlichen Umstimmung, hatte doch SCHULZ (1892) gerade erst entdeckt, daß ♂ *Quercus*-Blüten in der Regel sichtbare Anlagen der ♀ Geschlechtsorgane enthalten. — Eine Notiz von PAIXÃO-CORREIA (1955), welcher durch 0,25% NES in Lanolinpaste die vorzeitige Bildung von ♂ Kätzchen an 2—3jährigen Zweigen 10—15jähriger Exemplare von *Qu. Suber* ausgelöst zu haben glaubt, hält im Hinblick auf unzureichende Kontrollen der Kritik nicht stand. Über das Auftreten von ♂ und gemischtgeschlechtigen Kätzchen an sonst ♀ Exemplaren von *Castanea sativa* und „physiologische Zweihäusigkeit" der Eßkastanie referiert BENČAT' (1964 [hier auch Angaben über ältere Beobachtungen an *Castanea*], 1965).

Primulaceae.

AYMARD (1960) berichtet über die Entstehung von Pollensäcken an Fruchtknoten von *Anagallis arvensis* nach Behandlung mit 2,4-D, doch ist dabei nicht ausgeschlossen, daß es sich auch nur um die Verwachsung von Staub- und Fruchtblättern handeln könnte.

Ebenaceae.

Diospyros Kaki, die gewöhnlich monöcische, zuweilen auch gynomonöcische Kakipflaume, bildet in ihren oberen Teilen vorwiegend ♀ Blüten (MININA 1949). Wie in vielen anderen Fällen werden auch hier die Blüten monoklin angelegt, doch ist auch noch in adulten, funktionell eingeschlechtigen Blüten das unterdrückte Geschlecht durch Staminodien bzw. Pistillodien vertreten (vgl. die praktisch rein deskriptive Arbeit von NAMIKAWA, SISA und ASAI 1932).

Plantaginaceae und Borraginaceae.

In gynomonöcischen Ähren von *Plantago lanceolata* sind stets die letzten, oft auch die allerersten Blüten ♀. Ähnlich verhält sich auch *Echium vulgare* (CORRENS 1907). STOUT (1919) gab eine langatmige, rein deskriptive Darstellung von intersexuellen Bildungen bei der erstgenannten Art.

Labiatae.

CORRENS (1907, 1908) stellte umfangreiche Untersuchungen über die Geschlechtsverhältnisse bei gynomonöcischen Exemplaren von *Satureia hortensis* an (bei der es auch ♀ Individuen gibt, die in CORRENS' Versuchen einer modifikativen Geschlechtsänderung unzugänglich waren). Sowohl in den einzelnen Partialinflorescenzen als auch an der ganzen Pflanze wurden zunächst monokline Blüten, dann solche mit (einigen oder nur) kontabeszenten Antheren und zuletzt rein ♀ Blüten gebildet. Der Prozentsatz vollkommen zwittriger Blüten unter den neu aufblühenden stieg im Verlauf einer Beobachtungsreihe vom 29. Juni bis 31. Juli von 78 auf 95% und sank dann bis zum 4. September auf 1%; der Anteil der rein ♀ Blüten stieg vom 29. Juni—4. September von 1 auf 86%. Die restlichen Blüten hatten ± kontabeszente Antheren; ihr Anteil erreichte Maxima in der Zeit vom 29. Juni—9. Juli (21%) und am 28. August (40%). Die genannten Prozentsätze geben freilich nur ungefähre Hinweise, da etliche Blüten, deren Krone nach einem Tag noch nicht abgefallen war, doppelt oder öfter gezählt worden waren (CORRENS 1908). — Neben dem jahreszeitlichen Ablauf der Blühphase hatten auch die Kulturbedingungen maßgeblichen Einfluß auf die Geschlechtsverhältnisse. Je vier Exemplare bildeten: bei Kultur unter vollem Licht in großen Töpfen mit nährstoffreicher Erde 1187 Blüten, darunter 14,2% ♀; bei 3 Wochen lang auf $^1/_{60}$ verminderter Lichtintensität 56,8% ♀ Blüten unter 2185; in kleinen Töpfen mit stark sandiger Erde bei voller Beleuchtung 378 Blüten, davon 27,2% ♀; im letzteren Fall entwickelten vier Exemplare bei 3 Wochen lang auf $^1/_{60}$ verringerter Lichtintensität 465 Blüten, darunter

78,7% ♀. Auch die Wuchsdichte scheint von Bedeutung zu sein: In einer Versuchsreihe von CORRENS (1906) traten unter 315 pikierten Pflanzen keine ♀, dagegen unter 965, die zu bis zu 260 in Aussaatschalen von 17 cm Durchmesser verblieben, 920 monöcische und 45 rein ♀ Exemplare auf. — Wenn alle Blüten bald nach ihrer Entfaltung abgeschnitten wurden, war anfangs der Prozentsatz der ♀ Blüten, später aber der Anteil der ⚥ Blüten (im Vergleich zu heilen Pflanzen) erhöht; der Eintritt der rein ♀ Phase war verzögert oder ganz unterbunden (CORRENS 1907).

Solanaceae.

Versuche von BURK (1930) und HOWLETT (1936, 1939) mit *Lycopersicum esculentum* lassen sich dahingehend zusammenfassen, daß die absolute Länge der Fruchtblätter und das Verhältnis Fruchtblattlänge : Staubblattlänge im Kurztag, bei geringer Lichtintensität und bei reichlicher N-Versorgung am größten waren, während unter entgegengesetzten Bedingungen der gegenteilige Effekt eintrat. Mit der Reduktion der Staubblattlänge, die im Extremfall zur partiellen Unterdrückung der Staubblattbildung führte, gingen eine Hemmung der Meiose und eine Verminderung der Pollenfertilität einher. — Durch Besprühen von Tomatenpflanzen mit TJBS, NES oder Maleinhydrazid erzielte REHM (1952) Sterilität im ♂ Geschlecht.

Bei *Hyoscyamus niger* hatte Heteroauxin-Behandlung nach photoperiodischer Induktion teilweisen oder vollständigen Ablast der Krone und des Androeceum mancher Blüten zur Folge (RESENDE 1954). — Denselben Effekt — manchmal verbunden mit gleichzeitiger Reduktion des Gynoeceum, nicht aber mit Verkümmerung der Krone — konnte man bei *H. niger* durch monatelange Injektion von K-Gibberellat (in 0,01%iger Lösung) in die Sproßachse erzielen (RESENDE und VIANA 1959).

Bei *Solanum nigrum* gelang es, die Antherenentwicklung durch Applikation von TJBS (KIERMAYER 1958, 1961) oder von Phthalsäuremono-α-naphthylamid (KIERMAYER 1960) zu unterdrücken.

Scrophulariaceae.

VÖCHTING (1893) bemerkte, daß bei *Mimulus Tilingi* in Zimmerkultur im Sommer bei zunehmendem Abstand vom Fenster die Antherenentwicklung gehemmt wurde, wobei aber noch normaler Pollen entstand. Die Griffel erreichten zwar die übliche Größe, doch trat — selbst nach Bestäubung mit Pollen von normalen Blüten — kein Samenansatz ein. Die durch die verringerte Lichtintensität ausgelösten Hemmungserscheinungen betrafen also in morphologischer Hinsicht mehr die Stamina, in funktioneller Hinsicht mehr den Fruchtknoten.

Gesneriaceae.

In Versuchen von OEHLKERS (1956) verschob sich die Geschlechtsausprägung bei den Bastarden *Streptocarpus Wendlandii* × *S. Rexii* und *S. Wendlandii* × *S. cyaneus* mit zunehmendem Alter unter konstanten Temperaturbedingungen nach der ♂ Seite. Hohe Temperatur (etwa 20° C) begünstigte die Ausbildung des ♀, niedere (10—15° C) die des ♂ Geschlechts. Die Rückkreuzung *S.* ([*W.* × *R.*] × *R.*) ergab eine 1:1-Spaltung zwischen F_1-Typen und stark verweiblichten Formen (mit Nebenfruchtknoten bei den Nachkommen von vermännlichten Kältepflanzen); letzteres wird auf die Veränderung des *Wendlandii*-Plasmons unter dem Einfluß der Kälte zurückgeführt. Weitere, mehr beiläufige Studien über modifikative Geschlechtsbestimmung bei *S. Wendlandii* verdanken wir RESENDE und VIANA (1952) sowie HESS (1959): Hatten erstere auch bei dieser Gesneriacee durch Behandlung mit IES oder mit TJBS die übliche, mit dem Androeceum beginnende Reduktionsreihe hervorrufen können, so gelang es umgekehrt HESS, durch Anwendung von 2-Thiouracil (200 γ pro Pflanze in der Woche nach 8wöchiger Blühinduktion) die Entwicklung von einem oder zwei überzähligen Staubblättern pro Blüte und somit ein bescheidenes Maß von Vermännlichung zu erzielen.

Rubiaceae.

Bei *Galium cruciatum* sind die ersten Blüten der axillären Blütenstände im mittleren Sproßabschnitt gemischtgeschlechtig, die folgenden ♂. Auch bei *Asperula taurina* sind in den Dichasien die Endblüten erster Ordnung monoklin, die späteren ♂ (SCHULZ 1888, CORRENS 1928).

Dipsacaceae.

Eine Infektion durch *Ustilago succisae* führt bei *Succisa pratensis* zum Abort des Gynoeceum (BAKER 1957). Nach MOLLIARD (1895), der allerdings von „*Peronospora violacea*" und von „*Scabiosa Succisa*" spricht, gilt dies nur bei *stärkerem* Befall; eine schwächere Infektion hat dagegen bei *S. Succisa, S. Columbaria* und *Knautia arvensis* allein Reduktion und Atrophie der Staubblätter zur Folge (wie sie bei stärkerer Infektion *außer* der Verkümmerung des Gynoeceum eintritt).

Cucurbitaceae.

Zu dieser Familie gehört eine beträchtliche Anzahl monöcischer Arten, unter denen die wirtschaftlich wichtigeren schon frühzeitig Anlaß zu Versuchen über die modifikative Geschlechtsbestimmung gegeben haben. Die ersten, freilich noch unvollkommenen diesbezüglichen Angaben hat man anscheinend KNIGHT (1819) zu verdanken, dessen Wassermelonen *(Citrullus vulgaris)* bei besonders hoher Gewächshaustemperatur nur ♂ Blüten bildeten. HEYER (1884, 1886) stellte ausgedehnte Beobachtungen über den Einfluß der Kulturbedingungen auf das Geschlecht der Blüten von *Cucurbita Pepo* und *Cucumis sativa* an; einige Ergebnisse sind in Tabelle 4 zusammengestellt (ausführlichere Wiedergabe bei CORRENS 1928, Tabelle 2 und 3).

Tabelle 4. *Gesamtzahl der Blüten und Anteil der männlichen Blüten bei Cucumis sativus und Cucurbita Pepo unter verschiedenen Kulturbedingungen.* (Nach HEYER 1884, S. 66, aus CORRENS 1928, S. 33, vereinfacht.)

Standort	*Cucumis sativus*			*Cucurbita Pepo*		
	Zahl der Versuchspflanzen	Mittlere Zahl der Blüten pro Pflanze	Prozentsatz der ♂ Blüten	Zahl der Versuchspflanzen	Mittlere Zahl der Blüten pro Pflanze	Prozentsatz der ♂ Blüten
Warmhaus . .	10	55,4	66,8	6	18,5	71,2
Gartenboden .	17	53,7	81,0	8	133,0	76,4
Sandboden . .	22	29,2	86,8	13	141,7	80,6

Bevor wir die Ergebnisse neuerer Versuche für die einzelnen Arten getrennt wiedergeben, müssen wir noch kurz auf die morphologischen Grundlagen der Geschlechtsausprägung bei Cucurbitaceen eingehen. Zwar glaubte noch J. HESLOP-HARRISON (1959), die Blüten der monöcischen Cucurbitaceen zu den "basically unisexual flowers" rechnen zu müssen. Hiergegen sprachen aber schon Befunde von WHITAKER (1931), der bei *Cucumis sativus, Citrullus vulgaris, Cucurbita moschata* und *C. Pepo* vereinzelt gemischtgeschlechtige Blüten entdeckt hatte, sowie der Umstand, daß *Cucumis Melo* neben ♂ auch monokline Blüten trägt (vgl. BRANTLEY und WARREN 1960). Untersuchungen von GALUN u. Mitarb. schafften nun endgültig Klarheit darüber, daß bei *Cucumis sativus* — ähnlich wie beim Mais (s. oben, S. 163) — *alle* Blüten in der Jugend ein bisexuelles Stadium durchlaufen, das in bestimmten Fällen bis zur Floration anhält, wogegen gewöhnlich die Organe des einen oder des anderen Geschlechtes eine Entwicklungshemmung erfahren, so daß diese Blüten schließlich eingeschlechtig erscheinen (ATSMON und GALUN 1960, GALUN und ATSMON 1960). Dabei besteht nun eine bemerkenswerte Beziehung zwischen der Größe einer Blütenknospe und der Größe bzw. dem Alter des Tragblattes: In der Achsel von Blättern gynöcischer und zwittriger Exemplare finden sich größere, weiterentwickelte Blütenknospen als in der Achsel gleichgroßer Blätter monöcischer und andromonöcischer Exemplare; mit anderen Worten: Die sich entwickelnde Blütenknospe erreicht den Wendepunkt bezüglich der Geschlechtsbestimmung auf der ♀ Pflanze nahe einem jüngeren (wuchsstoffreicheren), auf der ♂ Pflanze nahe einem älteren (wuchsstoffärmeren) Blatt (GALUN und ATSMON 1960). Später zeigte sich ganz allgemein, daß ♀ Blüten schneller, d. h. in einem geringeren Abstand vom Vegetationskegel, entfaltet werden als männliche; diese Entwicklungsbeschleunigung war am größten unter „verweiblichenden" Bedingungen (Kälte, Kurztag; vgl. unten): Alles was das Geschlecht der Blüten beeinflußt, beeinflußt offenbar auch die Beziehung zwischen der Größe der Blütenknospe und dem Alter des zugehörigen Deckblattes (ATSMON und GALUN 1962, ATSMON, GALUN und JAKOB 1965).

Hinsichtlich der Rolle *edaphischer Faktoren* bei der Bestimmung des Geschlechtes von Gurkenblüten *(Cucumis sativus)* kam TIEDJENS (1928) zu ähnlichen Resultaten wie HEYER (s. oben). $NaNO_3$-Düngung steigerte (im Vergleich mit N-armem, ungedüngtem Boden) bei *Cucumis sativus* im Langtag die Zahl der ♂ Blüten um 45,5%, die der ♀ Blüten um 55,1%, im Kurztag die der ♂ Blüten um 3,1%, die der ♀ Blüten um 20,7% (diesen Angaben liegen Untersuchungen an je 20 Pflanzen zugrunde). Nach N-Mangelkultur und anschließender N-Düngung wurden im Kurztag reichlich ♀ Blüten gebildet, selbst an den unteren Knoten, an denen die Blätter schon abgefallen waren. — Auch in Versuchen von MININA (1938) wurde bei *Cucumis sativus* und *C. Melo* durch periodische N-Gaben zur Zeit der Keimung, des Blühbeginns und der Vollblüte (verglichen mit solchen Pflanzen, die nur bei Versuchsbeginn die volle mineralische Nährstoffgarnitur erhalten hatten) bei gleichzeitigem P-Mangel die Zahl der ♀ Blüten — parallel mit dem Zuckergehalt — gesteigert, die der ♂ aber gesenkt. So wurden in einem Versuch bei N-gedüngten Exemplaren von *C. sativus* 39,3 ♂ : 5,6 ♀ Blüten I. Ordnung und 7,6 ♂ : 7,3 ♀ Blüten II. Ordnung gefunden, bei den Kontrollpflanzen dagegen 58,5 ♂ : 4,1 ♀ Blüten I. Ordnung und 15,8 ♂:4,0 ♀ Blüten II. Ordnung. ATSMON und GALUN (1962) konnten allerdings *keinen* Einfluß der N-Ernährung auf das Geschlecht von Gurkenblüten feststellen. Vielleicht reagieren verschiedene Sorten in dieser Hinsicht unterschiedlich.

TIEDJENS (1928) befaßte sich auch schon mit der Rolle der *Tageslänge* und der *Lichtintensität*. Bei Kultur auf N-reichem Boden vom 8. März—21. Juni war das Geschlechterverhältnis der *Cucumis*-Blüten

bei natürlicher Tageslänge und voller Lichtintensität . . .	14,2 ♂ : 1 ♀,
bei natürlicher Tageslänge und verminderter Lichtintensität	26,5 ♂ : 1 ♀,
bei nächtlichem Zusatzlicht (ab 1 Uhr früh)	17 ♂ : 1 ♀.

In einem anderen Versuch wurden Individuen einer „♀-betonten" Sorte vom 1. Dezember—1. März kultiviert. Das Geschlechterverhältnis der Blüten betrug in diesem Fall

bei natürlicher Tageslänge (etwa 9—11 Std Licht) 1,3 ♂ : 1 ♀ bei 12—13 ♀ Blüten pro Pflanze,
bei Zusatzlicht (ab 2 Uhr früh) 2,9 ♂ : 1 ♀ bei 12—13 ♀ Blüten pro Pflanze,
bei 6 Std-Kurztag (Kümmerwuchs) 1,1 ♂ : 1 ♀ bei 3,7 ♀ Blüten pro Pflanze.

Unter reduziertem Licht bildeten einige Sorten nur ♀ Blüten. Eine „♂-betonte" Linie entwickelte im Langtag 64 ♂ : 1 ♀ Blüte, im Kurztag 25,5 ♂ : 1 ♀ Blüte. Die Amputation von ♀ Blüten steigerte die Tendenz zur Anlegung von neuen ♀ Blüten, so daß im Kurztag fast nur noch ♀ Blüten entstanden, wogegen reifende Früchte bewirkten, daß bereits angelegte ♀ Blüten abfielen, ohne sich geöffnet zu haben. — Nach TIEDJENS' Befunden verwundert es nicht, daß auch EDMOND (1930) bei verschiedenen Gurkensorten eine Abhängigkeit des Geschlechterverhältnisses der Blüten von der Jahreszeit feststellte (vgl. Tabelle 5). Daß die Pflanzen hier im

Tabelle 5. *Abhängigkeit des Geschlechterverhältnisses bei Blüten verschiedener Gurkensorten von der Jahreszeit.* (Nach EDMOND 1930 aus HESLOP-HARRISON 1957, S. 55/56, verändert.)

	Kulturperiode		
	27. 6.—6. 9.	15. 12.—15. 4.	17. 2.—25. 5.
Sorten insgesamt	8,6—39,3 ♂ : 1 ♀	0,0005—0,06 ♂ : 1 ♀	0,012—0,345 ♂ : 1 ♀
Sorte "Extra Long White Spiny"	21,2 ♂ : 1 ♀	0,00715 ♂ : 1 ♀	0,0127 ♂ : 1 ♀

ganzen „weiblicher“ erschienen als in TIEDJENS' Versuchen, mag teils auf genetischen Ursachen, teil auf edaphischen Faktoren und geringerer Nachttemperatur beruhen. — Weiterhin konnten NITSCH, KURTZ, LIVERMAN und WENT (1952) bei *Cucumis sativus* (Sorte Boston Pickling) feststellen, daß bei allen angewandten Temperaturen außer dauernd 30° C an den ersten 25 Knoten im 8 Std-Kurztag mehr ♀ Blüten gebildet werden als im 16 Std-Langtag. Ebenso bewirkte in Versuchen von GALUN (1959b) eine Steigerung der Tageslänge von 13 auf 18 Std Vermännlichung, und zwar sowohl bei einer monöcischen Linie als auch bei einer gynöcischen, deren Individuen nur an der Basis der Hauptachse ein paar ♂ Blüten zu tragen pflegten (vgl. Tabelle 6). Nur DANIELSON (1944) scheint dagegen den größten Anteil der ♂ Blüten an der Gesamtzahl der Blüten gerade bei Kurztagkultur (8 Std- gegenüber 12 oder gar 16 Std-Tag) erzielt zu haben.

Tabelle 6. *Einfluß der Tageslänge auf die Zahl der ♂ und der ♀ Blüten bei je einer monöcischen und gynöcischen Linie von Cucumis sativus.* (Nach GALUN 1959b.)

	Monöcische Linie		Gynöcische Linie	
	13 Std-Tag	18 Std-Tag	13 Std-Tag	18 Std-Tag
Zahl der ♂ Blüten pro Pflanze . .	115,7	108,2	2,33	13,5
Zahl der ♀ Blüten pro Pflanze . .	2,0	1,0	13,33	9,08
Quotient ♀:♂ Blüten	0,0171	0,0093	5,71	0,67
Zahl der Knoten unter der ersten ♀ Blüte	13,54	25,55	1,77	4,16

Weiter gelang es GALUN (1956) bei einer von zwei untersuchten Sorten, durch *Kältebehandlung* (3 Tage bei +3° C) von 18 Std vorgequollenen Samen die Ordnungszahl des ersten Knotens mit ♀ Blüte herabzusetzen. (Nach SHIFRISS und GALUN [1956] ist diese Ordnungszahl mit dem Geschlechterverhältnis der Blüten korreliert).

In anderen Versuchen ging GALUN (1959b) den Wirkungen der *Amputation* bestimmter Blätter nach. Handelte es sich dabei um junge Blätter, so wurde bei gynöcischen Exemplaren die „♂ Tendenz“ verstärkt; wurden ältere Blätter entfernt, so trat in manchen Fällen Verweiblichung ein. Über Ergebnisse weiterer Resektionsversuche berichtet KRIBBEN (1963).

Bemerkenswert sind auch verschiedene Untersuchungen über den Einfluß von *Wuchsstoffen* und Wuchsstoffantagonisten auf das Geschlecht von Gurkenblüten. In Versuchen von MININA (1938; vgl. auch MININA und TYLKINA 1947, MININA und KUŠNERENKO 1949, MININA 1949, 1952) wurden Gurkenkeimlinge dadurch verweiblicht, daß sie auf dem 3—4 Blatt-Stadium für 11—12 Std in ein abgeschlossenes Gewächshaus kamen, in das Holzofenabgase[1] geleitet wurden, welche nennenswerte Mengen von CO enthielten, dessen wuchsstoffartige Effekte seit langem bekannt sind. Bei solcherweise behandelten Pflanzen entfielen 16 Tage später 2,17 ♂ auf 1 ♀ Blüte, nach 30 Tagen 4,34 ♂ auf 1 ♀ Blüte, bei den nicht mit CO behandelten Kontrollen demgegenüber 16 Tage später 25 ♂ auf 1 ♀ Blüte und 30 Tage später 12,5 ♂ auf 1 ♀ Blüte. Überstieg (in anderen Versuchen) der CO-Gehalt der Luft 1%, so wurden nur noch ♀ Blüten entwickelt[2]. In wieder anderen Fällen wurde die verweiblichende Wirkung der CO-Begasung durch Dekapitation gesteigert[3]. — LAIBACH und KRIBBEN (1949/50, 1950a, b) sowie LAIBACH (1950/51) bearbeiteten 16—18 Tage alte Gurkenkeimlinge der Sorte Delikateß mit verschiedenen Wuchsstoffen (α-Naphthylessigsäure [NES], β-Indolylessigsäure, 2,4-Dichlorphenoxyessigsäure), die in 0,01—0,5%iger Konzentration in Lanolinpaste auf den Stielstumpf des *zweiten* Blattes oder auf die Unterseite des Primärblattes aufgetragen wurden. Gemessen an der Zahl der ♀ Blüten in den ersten 7 oder 10 Blattachseln, war — besonders nach NES-Applikation — eine ausgesprochene Verweiblichung festzustellen; gleichzeitig waren die Zahl der ♂ Blüten und die Gesamtzahl der Blüten vermindert, die Seitentriebentwicklung gefördert. Nach NES-Behandlung des Stielstumpfes des *ersten* Blattes trat eigenartigerweise keine geschlechtsmodifizierende Wirkung ein. — Demgegenüber beruhte die von WITTWER und HILLYER (1954) bei feldmäßig angebauten Gurken- und Kürbisrassen durch Besprühen im Primärblattstadium mit Maleinsäurehydrazid (MH; 250—350 p.p.m.), in geringerem Maße auch durch NES und durch 2,3,5-Trijodbenzoesäure (TJBS) erzielte Verweiblichung anscheinend nicht auf einer Vermehrung der ♀ Blüten, sondern auf einer Hemmung der Staubblattentwicklung in ♂ Blüten; die Fig. 1D in der genannten Notiz von WITTWER und HILLYER läßt allerdings eher an Verlaubung von ♂ Blüten denken. — Die von GALUN (1956, s. oben) durch Saatgutvernalisation erzielte Verweiblichung wurde durch Quellung der Samen in Wuchsstofflösungen noch geringfügig gesteigert, und die sonst durch Entfernung junger Blätter bewirkte Vermännlichung (s. oben) konnte GALUN (1959b) durch Behandlung der Blattstümpfe mit NES verhindern. Auch Behandlung von Samen mit 0,03%iger Methylenblaulösung förderte später die Entwicklung von ♀ Blüten (NAUGOL'NYCH 1955). Neuerdings rief auch Allyltrimethylammoniumbromid (als Zusatz zur Nährlösung von Gurkenpflanzen) Verweiblichung hervor (MITCHELL und WITTWER 1962).

Zahlreiche Untersuchungen haben in den letzten Jahren dem Einfluß von *Gibberellinen*, insbesondere von Gibberellinsäure (GA_3), auf das Geschlecht von Gurkenblüten gegolten. Ganz allgemein, auch bei sonst rein ♀ erscheinenden Individuen, wurde durch GA_3 unter den verschiedensten Kulturbedingungen und unabhängig von der Anwendungsform Vermännlichung in dem Sinne erzielt, daß der Prozentsatz der ♂ Blüten und die Ordnungszahl des ersten Knotens mit ♀ Blüte gesteigert wurden. Die praktische Bedeutung dieses Befundes ist wohl offenkundig, wird doch auf diese Weise die Selbstung in der gärtnerischen Praxis erwünschter gynöcischer Linien ungemein erleichtert. Die ersten diesbezüglichen Beobachtungen rühren anscheinend von WITTWER und BUKOVAC (1957, 1958) her; gleichsinnige Mitteilungen verdanken wir unter anderem GALUN

[1] Es handelt sich hierbei um die sog. Klin-Methode, die in Rußland stellenweise (besonders bei Klin, im Moskauer Gebiet) in der gärtnerischen Praxis angewandt wird (DUBROVINA 1958).

[2] Als am wirksamsten erwies sich in dieser Beziehung die Begasung (50—200 Std unter einer Glasglocke) von solchen Pflanzen, die zwei oder drei Blätter entfaltet hatten und etwa 20—22 Tage alt waren (MININA und TYLKINA 1947).

[3] Ähnliche Effekte wie die eben genannten Autoren durch Kohlenmonoxyd konnte MECHANIK (1958) durch Acetylen erzielen.

(1959a, 1961a, b), PETERSON und ANHDER (1960) sowie MITCHELL und WITTWER (1960, 1962). GALUN (1959a) konnte den GA_3-Effekt durch NES (die ja auch schon allein Verweiblichung bewirkt) weitgehend annullieren; in späteren Versuchen dieses Autors (1961a) war die Gibberellin-Wirkung (gemessen an der Ordnungszahl des ersten Knotens mit ♀ Blüte) dann am größten, wenn sich die Blüten am 9.—12. Knoten zur Zeit der GA_3-Behandlung im morphologisch bisexuellen Stadium befanden. Auf diese Weise konnte auch die Wirkung ♂-bestimmender modifizierender Gene nachgeahmt werden (GALUN 1961b). — BUKOVAC und WITTWER (1960a, b) sowie WITTWER und BUKOVAC (1962a) verglichen den vermännlichenden Effekt von GA_3 mit demjenigen anderer Gibberelline. In der zuletzt erwähnten Veröffentlichung wird ein Versuch geschildert, bei dem die Gibberelline A_8, A_5, A_6, A_1, A_3, A_9, A_4, A_3 (muß wohl heißen: A_2) und A_7 in der Reihenfolge ihrer Nennung bei einer gynöcischen Gurkenlinie zunehmend stärkere Vermännlichung hervorriefen, gemessen an der Zahl der vermännlichten unter 10 Pflanzen, an der Zahl der ♂ Blüten pro Pflanze sowie an der Zahl der ♂ Blüten tragenden Knoten pro Pflanze. Die starke Wirksamkeit von A_9, A_2, A_4 und A_7 wird mit dem Fehlen einer C/D-Hydroxylgruppe in Zusammenhang gebracht. — Bei in vitro-Kultur von Gurken-Blütenknospen (auf Whyte's Medium, unter Zusatz von Tryptophan, Kinetin, Caseinhydrolysat und *Cocos*-Milch) zeigten sich übrigens dieselben (gegensätzlichen) Wirkungen von GA_3 und IES, wie wir sie bereits bei ganzen Pflanzen kennengelernt hatten (GALUN, JUNG und LANG 1962, 1963). Eigentümlicherweise bleiben ⚥ Blütenknospen (von hermaphroditen Pflanzen) praktisch unbeeinflußt.

VAN KOOT und VAN DORST (1958) berichten über verschiedene *Cucumis*-Virusstämme, von denen einer Vermännlichung, die beiden anderen (weniger ausgeprägte) Verweiblichung auslösen; sie denken an einen primären Einfluß der Virusinfektion auf den Auxinspiegel der Pflanze[1].

Bei *Cucumis Melo* untersuchten BRANTLEY und WARREN (1960) den Einfluß von Tageslänge, NES und N-Ernährung auf den Anteil monokliner Blüten an der Gesamtzahl der Blüten. Bei der Tageslänge handelte es sich um 8 Std-Kurztag bzw. 16 Std-Langtag (Tageslicht + 200 f.c. Zusatzlicht, Temperatur tagsüber 80—85, nachts 65—70° F). NES wurde in der Weise angewandt, daß ein Viertel des zu $^3/_4$ entfalteten ersten Blattes abgeschnitten und die restliche Spreite 1 Woche lang in NES-Lösung (0, 0,01, 0,1, 1,0 p.p.m.) getaucht wurde. Die Pflanze wurde in Withrow-Nährlösung kultiviert, die 20, 100 oder 300 p.p.m. Stickstoff enthielt. Die wichtigsten Ergebnisse sind in Tabelle 7 zusammengefaßt.

Tabelle 7. *Einfluß von Tageslänge, Stickstoff-Versorgung und NES-Behandlung auf Anzahl und Zahlenverhältnis der ♂ und der ⚥ Blüten von Cucumis Melo.* (Nach BRANTLEY und WARREN 1960.)

	Langtag			Kurztag		
	♂ Blüten	⚥ Blüten	♂:⚥	♂ Blüten	⚥ Blüten	♂:⚥
N 20 p.p.m.	49	5	9,8	29	2	14,5
100 p.p.m.	69	11	6,3	42	11	3,8
300 p.p.m.	138	37	3,7	33	5	6,6
NES 0 p.p.m.	129	38	3,4	44	5	8,8
0,01 p.p.m.	79	28	2,8	62	11	5,6
0,1 p.p.m.	67	27	2,5	53	6	8,8
1,0 p.p.m.	61	28	2,2	37	4	9,3

[1] Eine Artikelserie von ITO u. Mitarb. (ITO und KATO 1953, ITO, KATO, HASHIMOTO und SAITO 1954, ITO und SAITO 1956a, b, c, 1957a, b, 1958, 1960) über Geschlechtsbestimmung bei japanischen Gurken, unter anderem durch Temperatur-, Lichtdauer- und Wuchsstoffeinflüsse, war mir leider unzugänglich.

Mit *Cucumis Anguria* experimentierten HALL (1949) sowie NITSCH, KURTZ, LIVERMAN und WENT (1952). Ersterer untersuchte die Bedeutung der N-Ernährung (Düngung mit NH_4-, Ca- und K-Nitrat) und der Tageslänge (8 Std-Kurztag bei natürlichem Tageslicht bzw. 16 Std-Langtag, zusammengesetzt aus natürlichem Tageslicht und 200 f.c.-Zusatzlicht). Bei N-reicher Aufzucht kamen im Kurztag 4,58 ♂, im Langtag 4,87 ♂ Blüten auf 1 ♀, bei N-Mangel in beiden Tageslängen 6,0 ♂ auf 1 ♀ Blüte. Ein signifikanter Tageslängeneinfluß auf das Geschlecht konnte also nicht festgestellt werden, wohl aber wurden im Kurztag etwa 50% mehr Blüten produziert als im Langtag. Nach NITSCH und Mitarbeitern wird hingegen die Bildung von ♂ Blüten durch Langtag und hohe Temperaturen, die Entwicklung von ♀ Blüten durch niedere Temperaturen und Kurztag begünstigt.

Für *Cucurbita maxima* var. *turbaniformis*, Sorte Acorn (squash, Turbankürbis), gilt nach NITSCH und Mitarbeitern dasselbe: Individuen, die dauernd bei 30° C im 16 Std-Langtag gehalten wurden, bildeten binnen dreier Monate keine ♀ Blüte; im 8 Std-Kurztag bei ständig 30° C entwickelte sich die erste ♀ Blüte am 58. Knoten, bei tags 20 und nachts 10° C stand die erste ♀ Blüte bei Langtagexemplaren am 22., bei Kurztagexemplaren am 16. Knoten. Besprühung von zweiblättrigen Individuen, im 14 Std-Langtag aufgezogen, mit 0,01%iger NES senkte die Ordnungszahl des ersten Knotens mit ♀ Blüte von $>$ 20 auf 9[1], während 24stündige Einwirkung einer 0,1% Äthylen enthaltenden Atmosphäre bezüglich der Geschlechtsausprägung wirkungslos war. Unabhängig von ihrer Verteilung längs der Sproßachse sind ♂ und ♀ Blüten in Abhängigkeit von der Insertionshöhe verschieden entwickelt: Die ersten ♂ Blüten sind unterentwickelt, die späteren normal, die letzten gehemmt; die ersten ♀ Blüten sind normal, die späteren haben einen besonders großen Fruchtknoten, und die letzten neigen — zumal unter Kurztag- und Kälteeinwirkung — zu Parthenokarpie, was als extreme Weiblichkeit gedeutet wird. Nicht der Ablauf, nur die Dauer dieser Phasen konnte experimentell modifiziert werden. — Das Geschlecht jeder einzelnen Blüte soll nach LAIBACH und KRIBBEN (1951), die sich hierbei auf die Ergebnisse ihrer mit Gurken durchgeführten Wuchsstoffversuche stützen, vom Auxinspiegel in der Tragblattachsel zur Zeit der Blütenanlegung abhängen. Den bei Kürbispflanzen feststellbaren Rhythmus, demzufolge zwischen 2 ♀ 3—5 ♂ Blüten eingeschaltet sind (tabellarisch für 10 Exemplare von *C. Pepo* dargestellt), interpretieren LAIBACH und KRIBBEN durch die Annahme, daß bei der Bildung einer ♀ Blüte so viel Wuchsstoff verbraucht werde, daß der Rest eine Zeitlang nur noch für ♂ Blüten reiche. — Weitere Mitteilungen über modifizierende Einflüsse auf das Geschlecht von Kürbisblüten stammen von SCOTT (1933) und von SAVELLI (1960). SCOTT befaßt sich mit der Rolle des Schattens, SAVELLI mit jener des Samengewichts. Der letztgenannte Autor hatte von einer unterernährten Pflanze 36 Samen geerntet, deren Gewicht nur etwa $^1/_3$ des normalen betrug. Von diesen Samen keimten 31 und lieferten Pflanzen, bei denen das ♀ Geschlecht unterdrückt war, während die ♂ Blüten hypertrophierten und mancherlei Mißbildungen zeigten.

Wurden junge Exemplare von *Citrullus vulgaris* mit TJBS-Lösungen besprüht, so war — wie bei der Gurke (s. oben) — das Geschlechterverhältnis zugunsten der ♀ Blüten modifiziert; ältere Individuen, die schon Blüten angelegt hatten, reagierten auf TJBS-Behandlung mit Pollensterilität (REHM 1952). In Versuchen von HENSZ und MOHR (1959) erwies sich TJBS (ebenso wie MH und NES) jedoch

[1] Ähnlich verhielt sich der Turbankürbis in Wuchsstoffversuchen von WITTWER und HILLYER (1954); die Ergebnisse glichen den von diesen Autoren bei *Cucumis Anguria* erzielten (s. oben).

als wirkungslos; Behandlung mit b-dichlorisobuttersaurem Natrium (0,5 oder 0,25%, mit Tween 20 als Benetzungsmittel) führte aber zu Verweiblichung — reife ♂ Blütenknospen öffneten sich nicht trotz normaler Pollenbildung —. während GA_3 das Geschlechterverhältnis der Blüten (wie bei Gurken) zugunsten der männlichen verschob.

Bei *Luffa acutangula* führte wiederholtes Besprühen mit NES, 2,4-D oder p-Chlorphenoxyessigsäure (CPES) zu einer Verzögerung der Blütenbildung und zu einer Verminderung der Gesamtzahl der Blüten, zugleich aber zu einer Steigerung des Anteils der ♀ an der Gesamtzahl der Blüten von 2,9% bei den Kontrollen auf maximal 9,1%; die *absolute* Zahl der ♀ Blüten (und damit auch, wenigstens in bestimmten Fällen, der Früchte) wurde aber nur durch NES und 2,4-D erhöht, durch CPES dagegen gesenkt (SATYANARAYANA und RANGASWAMI 1959).

Echinocystis trägt, wie viele andere Cucurbitaceen, in den Blattachseln ♂ Blüten in Trauben und einzeln stehende ♀ Blüten; eine NES-Behandlung von Keimlingen führt dazu, daß später die ♀ Blüten gefördert, die ♂ aber in ihrer Entwicklung gehemmt werden (J. HESLOP-HARRISON 1959).

Bei *Bryonia alba* folgen auf die ersten, sterilen Achseln solche mit ♂ und später solche mit ♀ Inflorescenzen (CORRENS 1928). Stecklinge, die der ♂ blühenden Region entnommen wurden, und solche aus der ♀ blühenden Region verhalten sich im wesentlichen genau so wie Sämlinge: Immer bilden sie zunächst ♂ und dann erst ♀ Inflorescenzen. Aus diesen Beobachtungen (und solchen an Laubmoosen) folgerte CORRENS (1920), daß bei Zwittern und Monöcisten *jede* Zelle *beiderlei* Geschlechtspotenz besitze, daß also eine derartige Pflanze im Hinblick auf das Geschlecht ihrer Blüten kein genetisches Mosaik darstelle[1].

Bei *Thladiantha dubia* soll BLAVET (1896) Geschlechtsumkehr durch Pfropfung erzielt haben.

Abschließend mögen Befunde von HILDEBRAND (1893) an *Ecballium elaterium* Erwähnung finden: In der Hauptblütezeit (d. h. im Frühsommer) sind hier ♂ und ♀ Blüten ebenso angeordnet, wie wir es weiter oben für *Echinocystis* geschildert haben. Auf den Herbst zu abortieren jedoch die ♂ Blütenstände zunehmend, und nur noch die einzelnstehenden ♀ Blüten werden voll entwickelt. HILDEBRAND möchte hierfür jedoch nicht (wie es für uns heute naheliegen würde) das Herbstwetter verantwortlich machen, sondern gibt eine teleologische, blütenbiologische Erklärung: *E. elaterium* sei protogyn, spätere ♂ Blüten wären daher funktionslos; Umbelliferen seien umgekehrt protandrisch, die letzten Blüten der Inflorescenz darum rein ♂, denn sonst stünde den letzten ♀ Organen kein Pollen mehr zur Verfügung.

Compositae.

Gemischtgeschlechtige, ♂, ♀ und geschlechtslose Blüten sind bei vielen Compositen in charakteristischer Weise in einem Köpfchen vereinigt, so daß ihrer Anordnung systematische Bedeutung beigemessen wird. Die verschiedenartigen Blüten sind in konzentrischen Zonen gruppiert (Strahlen- und Scheibenblüten unterscheiden sich meist auch in geschlechtlicher Hinsicht); ihr Anteil an der Gesamtzahl der Blüten eines Köpfchens wird in erster Linie durch Faktoren modifiziert, die die Größe des Köpfchens verändern, wächst doch der Umfang (die Peripherie: Strahlenblüten) proportional dem Radius, der Flächeninhalt (Scheibenblüten) proportional dem Quadrat des Radius[2]. In solchen Fällen kann

[1] Im Gegensatz zu diesen Befunden von CORRENS an *Bryonia* hängt bei der Euphorbiacee *Manihot utilissima* die Zahl der (zum Teil fertilen) Staminodien pro ♀ Blüte unter anderem auch davon ab, welcher Region der Mutterpflanze das betreffende Exemplar als Steckling entnommen worden war (MIÈGE 1959).

[2] SZYMKIEWICZ (1934) fand allerdings bei *Solidago Virga-aurea* und *Arnica montana* eine *lineare* (und keine exponentielle) Beziehung zwischen der Zahl der Scheiben- und jener der Strahlenblüten. Die Gerade, welche diese Relation im Diagramm darstellt, hatte bei zwei Rassen von *S. Virga-aurea* und bei terminalen und lateralen *Arnica*-Köpfchen einen unterschiedlichen Steigungswinkel; bei *Arnica* würde demnach das Zahlenverhältnis der verschiedenartigen Blüten in einem Köpfchen durch dessen Position im Sproßsystem bestimmt.

wohl kaum von modifikativer Geschlechtsänderung gesprochen werden. Eher wäre dies bei einer Änderung der Breitenverhältnisse der einzelnen Zonen möglich, die beispielsweise zu gefüllten Köpfchen führen kann. Die „Füllung“ von Compositen-Köpfchen scheint aber vor allem genotypische Ursachen zu haben.

Gründlicher untersucht wurden die Möglichkeiten modifikativer Geschlechtsbeeinflussung bisher anscheinend nur bei einigen Arten, die zu den Ambrosiinen gehören, einer gelegentlich zur selbständigen Familie (Ambrosiaceae) erhobenen Subtribus der Heliantheae. Jedes Köpfchen enthält hier entweder nur ♀ oder nur ♂ Blüten.

Nach Beobachtungen von MEEHAN (1880) entwickelt *Ambrosia artemisifolia* L. bei dichtem Stand fast ausschließlich ♂, bei vereinzelter Stellung auf gutem Boden vorwiegend ♀ Köpfchen. *A. artemisifolia* L. bildet während eines einmonatigen Aufenthaltes bei 7 Std-Kurztag (zwecks Auslösung der Blütenbildung) ♂ und ♀ Köpfchen, nach Rückübertragung in sommerlichen Langtag aber fast nur noch ♀ Köpfchen (GARNER und ALLARD 1920). Auch in Versuchen von JONES (1936) mit diesem Objekt (var. *elatior*) wurden durch Verlängerung der Dauer der täglichen Belichtung die ♀ Köpfchen auf Kosten der ♂ vermehrt. Schließlich konnte JONES (1936) durch Dekapitation bei Sämlingen aus der Nachkommenschaft (freilich nicht reinerbiger) monöcischer Individuen den Prozentsatz der rein ♀ Exemplare steigern.

Ähnlich liegen die Verhältnisse bei *Ambrosia trifida*. Auch hier entstanden bei sommerlicher Langtag-Kultur, unterbrochen durch einige „blüteninduzierende“ Kurztage, ♀ Köpfchen an solchen Stellen, an denen sonst ♂ zu erwarten waren, und zwar in um so größerer Zahl, je mehr Kurztage geboten worden waren (verglichen wurden die Wirkungen von 2, 10 und 20 Kurztagen); 6 Std-Tag erwies sich in dieser Hinsicht als wirksamer denn Tageslängen von 12 und 15 Std (MANN 1942). Nach JONES (1947) wird die Ausbildung von ♂ Köpfchen bei *A. trifida* und *A. elatior* durch hohe Temperaturen (35° C während der Dunkelperiode) und Langtag gefördert. Der Grad der genetischen Determination scheint bei *A. trifida* geringer als bei *A. elatior* zu sein (MANN 1942).

Einen Monat alte Exemplare von *Xanthium pennsylvanicum* reagierten auf sieben 8 Std-Kurztage und anschließenden ständigen 17—18 Std-Langtag-Aufenthalt mit langsamer Blütenbildung, in deren Verlauf viele ♂ Köpfchen entstanden; dauernder 8 Std-Kurztag führte dagegen zu rascher Blütenbildung, wobei weniger oder keine ♂, dafür aber mehr (allerdings großenteils taube) ♀ Köpfchen auftraten. Die Pollenfertilität war etwa der Zahl der ♂ Köpfchen proportional: Unter bestimmten Ernährungsverhältnissen betrug sie im ersten Fall 95,4%, im zweiten nur 28,6% (NEIDLE 1938). — Versuche von F. L. NAYLOR (1941) am selben Objekt hatten im wesentlichen ähnliche Ergebnisse. — Reichlichere N-Versorgung erhöhte die Zahl der ♂ Köpfchen (NEIDLE 1938).

v. WITSCH (1961) untersuchte bei *Xanthium Strumarium* „X 7“, einer fast tagneutralen Rasse (bei der die Blütenbildung geringfügig durch Kurztag gefördert wird), den Einfluß von Tageslänge und K-Gibberellat auf die Bildung von ♂ und ♀ Köpfchen. Langtag erhöhte relativ und absolut die Zahl der ♂ Inflorescenzen pro Pflanze und verzögerte das erste Auftreten und die weitere Entwicklung von ♀ Köpfchen. K-Gibberellat schob in beiderlei Tageslängen den Termin der Blütenbildung hinaus und verminderte im Kurztag die Zahl der ♂ und ♀ Köpfchen pro Pflanze; im Kurz- wie im Langtag veränderte es das Geschlechterverhältnis zugunsten der ♀ Köpfchen.

KIERMAYER (1959) gab jungen Exemplaren von *Helianthus annuus* eine Zeitlang alle 4 Tage 1 cm^3 einer wäßrigen alkalischen TJBS-Lösung ($5 \cdot 10^{-4}$) auf die Blätter. Später zeigte sich bei diesen Pflanzen eine deutliche Verweiblichung: Die Staubblätter waren im allgemeinen verkümmert, die Narbenlappen oft vermehrt; in den gewöhnlich sterilen Randblüten wurden häufig normale Griffel sowie petaloide Stamina entwickelt.

Eine Untersuchung von CORRENS (1916) galt subandröcischen Stöcken des subdiöcischen *Cirsium arvense*: Die *durchschnittliche* Zahl von Achänen in Köpfchen „männlicher" Exemplare schwankte von Standort zu Standort zwischen 0 und 1, wobei aber genetische Unterschiede nicht ausgeschlossen scheinen. Im Einzelfall wurden bis zu 115 Früchte in „männlichen" Köpfchen gezählt.

2. Zusammenfassende Übersicht über die an der modifikativen Geschlechtsbestimmung beteiligten Faktoren.

Bevor im folgenden die Wirkung einzelner modifizierender Faktoren auf die Geschlechtsausprägung besprochen wird, muß daran erinnert werden, daß bei Pflanzen, die einer Geschlechtsbeeinflussung durch Außenfaktoren zugänglich sind, auch *genetische* Faktoren in diesem Zusammenhang eine Rolle spielen. Das manifestiert sich z. B. darin, daß Individuen verschiedener Sorten oder Linien ein und derselben monöcischen Pflanzenart unter gleichen Umweltbedingungen einen verschieden starken Zug zur Ausprägung des einen oder des anderen Geschlechtes zeigen (vgl. EDMOND 1930 bei *Cucumis sativus*, RICHEY und SPRAGUE 1932 bei *Zea Mays*). Die strenge Determination der Stellung männlicher und weiblicher Blüten innerhalb der Inflorescenz bei vielen monöcischen Pflanzen weist in dieselbe Richtung (viele Compositen und Araceen; *Begonia Wallichiana* nach IRMSCHER 1924). Keinesfalls trifft (wie wir schon gesehen haben) die Annahme zu, daß die Geschlechtsbestimmung bei Diöcisten i. w. S. immer nur genotypisch, bei Monöcisten i. w. S. ausschließlich modifikativ erfolge. Vielmehr spielen in beiden Fällen Erb- und Umweltfaktoren eine bald größere, bald geringere Rolle (so auch schon HILDEBRAND 1893)[1].

Unter den wirksamen Außenfaktoren seien als erste

a) Edaphische Faktoren

in Betracht gezogen. *Nährstoffreiche Böden* (im Vergleich zu sandigen), lockerer Stand der Pflanzen und *reichliche Wasserversorgung* pflegen die Verweiblichung von einhäusigen und ♂ Pflanzen zu begünstigen, während für nährstoffarme Böden, Dichtsaat und Trockenheit in der Regel das Umgekehrte gilt (HEYER 1884, 1886 bei *Zea, Betula, Cucumis sativus* und *Cucurbita Pepo*; EDEL'ŠTEJN und PAPONOV 1964 bei *Zea Mays* und *Cucumis sativus*; v. TRESKOW 1895, ATKINSON 1898, SCHAFFNER 1922, 1925b und MAEKAWA 1924 bei *Arisaema*; LEYDECKER 1870, F. HABERLANDT 1874, HALSTED 1901, SCHAFFNER 1925c und MAEKAWA 1929 bei *Cannabis sativa*; H. MÜLLER 1881 und LEWITZKY 1925 bei *Veratrum*; ferner MEEHAN 1880, DÜSING 1884, S. 722, LUDWIG 1885, S. 234, HOFFMANN 1885, KLEBS 1894, BLARINGHEM 1923 und SCHAFFNER 1923c). Insbesondere fördern reiche N-Gaben die Ausprägung des ♀ Geschlechts (GIARD 1898 und TIBEAU 1936 bei *Cannabis*; TIEDJENS 1928 bei *Cucumis sativus*; SABININ 1937 sowie CHOUDHRI und KRISHAN 1946 bei *Zea Mays*; MININA 1938 bei *Zea Mays, Cucumis sativus* und *C. Melo* [N-Düngung verbunden mit P-Mangel][2]; BRANTLEY und WARREN 1960 bei *Cucumis Melo*; HALL 1949 bei *Cucumis Anguria*; MURNEEK 1927 bei *Cleome spinosa*; HOWLETT 1936, 1939 bei *Lycopersicum esculentum*; MATZKE 1938 bei *Begonia semperflorens*). Speziell über Verweiblichung durch bessere Wasser-

[1] Bei der Aufstellung der nachfolgenden Übersicht war es unerläßlich, die berücksichtigten Befunde bestimmten Typen zuzuordnen. Im Hinblick auf Einzelheiten, die dabei vielfach außer acht bleiben mußten, sei deshalb auf den systematischen Teil (S. 158ff.) zurückverwiesen.

[2] Es handelt sich hierbei um Bedingungen, die der Blütenbildung im allgemeinen abträglich sind.

versorgung bzw. Vermännlichung bei Wassermangel berichten DAVEY und GIBSON (1917: *Myrica Gale*), AWADA und IKEDA (1953), AWADA (1957) sowie LANGE (1961a: sämtlich *Carica Papaya:* mittelbare Wirkung durch Beeinflussung der Wüchsigkeit).

Keinen signifikanten Einfluß der Bodenqualität bzw. der Düngung auf die Geschlechtsverhältnisse fanden F. HABERLANDT (1869, 1877), FISCH (1887), LAURENT (1903), SPRECHER (1913) und BLACK (1945) bei *Cannabis*; LAURENT (1903) bei *Mercurialis annua*; BURKILL (1895) bei *Stellaria media*; SPRECHER (1913) bei *Rumex Acetosa* und THOMPSON (1955) bei *Spinacia oleracea*. Nach SCHAFFNER (1925b bzw. 1927) führte bei *Thalictrum dasycarpum* Verpflanzen von einem nassen Standort auf trockenen Lehmboden, bei *Cannabis sativa* reichlichere N-Ernährung zu Änderungen in beiden Richtungen. — Bei in vitro-Kultur von Blütenknospen von *Cucumis sativus* hatte die N-Ernährung keinen Einfluß auf die Geschlechtsausprägung (ATSMON und GALUN 1962).

Entgegen der am Beginn dieses Abschnittes wiedergegebenen Regel berichteten LAURENT (1903) über Vermännlichung von *Spinacia* und NEIDLE (1938) über Vermännlichung von *Xanthium* nach reichlicher N-Düngung, CORRENS (1908) über Unterdrückung der Entwicklung zwittriger Blüten bei auf sandigem Substrat gezogenen gynomonöcischen Exemplaren von *Satureia hortensis* und SCHAFFNER (1926) über Verweiblichung von *Arisaema triphyllum* bei Kultur auf trockenem Boden; bei *Cannabis sativa* führten größerer Standraum und höhere Bodenqualität zu üppigerem Wuchs und damit zu „mehr ♂ Habitus" (HUHNKE, JORDAN, NEUER und v. SENGBUSCH 1951), und bei *Stellaria media* soll wuchsfördernde Düngung die mittlere Zahl der Staubblätter pro Blüte erhöhen (REINÖHL 1903). Bodentrockenheit scheint bei *Coix Lacryma-Jobi* Verweiblichung zu bewirken (JACQUES-FÉLIX 1959).

b) Luftfeuchtigkeit.

Über die Rolle der Luftfeuchtigkeit bei der modifikativen Geschlechtsbestimmung liegen nur wenige und zudem großenteils recht ungenaue Mitteilungen vor. HILDEBRAND (1893) machte für bestimmte Verweiblichungsphänomene bei *Quercus ilicifolia* feuchtheißes Wetter verantwortlich, und TOURNOIS (1914) erzielte bei *Humulus japonicus* Verweiblichung durch erhöhte Luftfeuchtigkeit (in Verbindung mit reduzierter Beleuchtung). Weiter entstanden in Versuchen von JACQUES-FÉLIX (1959) mit *Coix Lacryma-Jobi* bei höherer Luftfeuchtigkeit (verknüpft mit verminderter Lichtintensität) monokline Blüten anstelle von männlichen.

c) Licht.

Beim Licht spielen zeitliche Dosierung und Intensität eine Rolle hinsichtlich der Geschlechtsbestimmung. Der Frage nach der Wirksamkeit bestimmter Spektralbereiche ist man in diesem Zusammenhang offenbar noch nicht nachgegangen.

Geringe *Lichtintensität* hemmt oder unterdrückt die Antherenentwicklung in gemischtgeschlechtigen Blüten (CORRENS 1907, 1908 bei *Satureia hortensis*, HOWLETT bei *Lycopersicum esculentum*, RESENDE 1946, 1949, 1950 bei *Bryophyllum Daigremontianum*), begünstigt aber in manchen Fällen die Vermännlichung monöcischer Pflanzen (TIEDJENS bei *Cucumis sativus*, RIEDE bei *Betula verrucosa*); bei gynomonöcischen Exemplaren von *Satureia hortensis* (CORRENS 1908) sowie bei ♀ bzw. monöcischen Individuen von *Cannabis sativa* hatte höhere Lichtintensität einen vermännlichenden (BORTHWICK und SCULLY 1954, KÖHLER 1961, 1964a), Lichtmangel dagegen einen verweiblichenden Effekt (F. HABERLANDT 1874, HUHNKE, JORDAN, NEUER und v. SENGBUSCH 1951, LIMBERK 1959); HEYER (1884) scheint es dagegen bei *Cannabis* ebensowenig wie bei *Mercurialis annua*, *Cucumis sativus* und *Cucurbita Pepo* gelungen zu sein, das Geschlechterverhältnis durch Schattierung zu beeinflussen. Bei *Mimulus Tilingi* wurde durch verminderte Lichtintensität im Androeceum wohl das Wachstum der Stamina, nicht aber die Pollenentwicklung gehemmt, wogegen im Gynoeceum nicht das Griffelwachstum, wohl aber die Fertilität herabgesetzt war (VÖCHTING 1893).

Wie schon J. HESLOP-HARRISON (1957) hervorhebt, scheint bei photoperiodisch empfindlichen Pflanzen jeweils die für die Blütenbildung optimale *Tageslänge* in der Regel die Verweiblichung zu fördern. Im einzelnen bewirkte Kurztag

± ausgeprägte (zumindest relative) Verweiblichung bei *Zea Mays* (Schaffner 1925a, 1930, Richey und Sprague 1932, Choudhri und Krishan 1946, J. Heslop-Harrison 1961), *Humulus japonicus* (Tournois 1909, 1911, 1912, 1914, Schaffner 1923b, 1925a), *Cannabis sativa* (Molliard 1898, Tournois 1911, 1912, Schaffner 1925a, Correns 1928, Breslavec 1936, 1937, Breslavec und Zaurov 1937, Huhnke, Jordan, Neuer und v. Sengbusch 1951, Petit 1952, Łączyńska-Hulewiczowa 1957 [nur monöcischer Hanf]), *Mercurialis ambigua* (Thomas 1956), *Manihot utilissima* (Miège 1959), *Lycopersicum esculentum* (Burk 1930, Howlett 1939), *Cucumis sativus* (Tiedjens 1928, Edmond 1930, Nitsch, Kurtz, Liverman und Went 1952, Galun 1959b, 1961b), *Cucumis Anguria* und *Cucurbita Pepo* (Nitsch et al. 1952), *Ambrosia trifida* (Mann 1942, Jones 1947), *Xanthium Strumarium* (v. Witsch 1961) und *Xanthium pennsylvanicum* (Neidle 1938, Naylor 1941). Ob sich die verweiblichende Wirkung einer Kurztagbehandlung durch nächtliches Störlicht aufheben läßt, scheint noch nicht untersucht zu sein. Jedenfalls führte auch bei der Langtagpflanze *Cerastium tomentosum* Kurztag zu einem gewissen Maß von Verweiblichung (Bocquet und Stroun 1960).

Keinen Einfluß der Tageslänge hatten Hall (1949) bei *Cucumis Anguria* und Łączyńska-Hulewiczowa (1957) bei diöcischer *Cannabis sativa* feststellen können, wogegen Schaffner (1921a) sowie Borthwick und Scully (1954) bei der letztgenannten Art gerade durch Kurztag auch Vermännlichung erzielten und McPhee (1925) bei ♂ wie bei ♀ Hanfpflanzen nach Kurztagexposition eine Steigerung des Prozentsatzes intersexueller Blüten bemerkte; Schaffner (1927) berichtet in bezug auf *Cannabis sativa* und *Humulus japonicus* über Geschlechtsumkehr in *beiden* Richtungen nach Kurztageinwirkung. — Über *Langtag*-induzierte Verweiblichung (bzw. Unterdrückung des ♂ Geschlechts) berichteten Meletti (1961) bei *Triticum vulgare* (Sorte Denti di cani), J. und Y. Heslop-Harrison (1958a) bei *Silene pendula*, Janick und Stevenson (1955) und Thompson (1955) bei *Spinacia oleracea*, Shifriss (1956) bei *Ricinus communis*, Noack (1962) bei *Begonia Cathayana*, Brantley und Warren (1960) bei *Cucumis Melo* sowie Garner und Allard (1920) und Jones (1936) bei *Ambrosia artemisifolia* L.

d) Temperatur.

Verweiblichung durch Kälte oder Vermännlichung durch hohe Temperaturen wurden beobachtet bei *Zea Mays* (Richey und Sprague 1932, Went 1957, J. Heslop-Harrison 1961), *Bryophyllum* (Resende 1957a), *Spinacia oleracea* (Janick und Stevenson 1955, Thompson 1955), *Cannabis sativa* (Nelson 1944), *Mercurialis ambigua* (Thomas 1956), *Carica Papaya* (Awada 1958, Lange 1961a), *Citrullus vulgaris* (Knight 1819), *Cucurbita Pepo* und *Cucumis Anguria* (Nitsch et al. 1952), *Cucumis sativus* (Galun 1956: Samen) sowie bei *Ambrosia trifida* (Jones 1947). Zu gegenteiligen Ergebnissen gelangten Holdsworth (1959) bei *Fuchsia procumbens*, Borthwick und Scully (1954) bei *Cannabis sativa* und Oehlkers (1956) bei *Streptocarpus*-Bastarden. Bei *Capsella Bursa-pastoris* erschienen sowohl nach Einwirkung von Frösten als auch nach Hitzeexposition ♀ Blüten an sonst ⚥ Exemplaren, wofür aber nicht die betreffenden Temperaturen als solche, sondern die in ihrem Gefolge aufgetretenen Welkungserscheinungen verantwortlich gemacht wurden (Laibach 1951).

e) Transplantation, Resektion und andere Verletzungen.

Transplantation hatte anscheinend bislang nur in zwei oder drei Fällen möglicherweise Einfluß auf das Geschlecht eines Pfropfpartners, und zwar zunächst bei der Cucurbitacee *Thladiantha dubia* (Blavet 1896), sodann in einem Versuch von

LIMBERK (1954) mit *Humulus Lupulus*: 2 ♂ Unterlagen bildeten nach Aufpfropfung eines ♀ Reises je einen ♀ Trieb. Es ist aber offenbar nicht ausgeschlossen, daß diese Erscheinung auf die der Pfropfung vorausgehende Resektion des ♂ Partners zurückzuführen ist (vgl. LIMBERK 1953). — J. und Y. HESLOP-HARRISON (1957c) erwähnen zwei erfolgreiche Transplantationen bei *Cannabis sativa*; hier wurde die durch CO-Behandlung induzierte Verweiblichung von je einem (genetisch ♂) Pfropfreis der (genetisch gleichfalls ♂) Unterlage mitgeteilt, die daraufhin ebenfalls zur Bildung intersexueller Blüten überging. Auch hier ist nicht ausgeschlossen, daß die Verweiblichung der ♂ Unterlage auf Resektion beruhte.

In allen anderen Pfropfungsversuchen erfolgte keine Geschlechtsänderung (SANSOME 1930, SCHOPFER und BLUMER 1939a, SCHOPFER 1943, A. und D. LÖVE 1940 und WESTERGAARD 1953 bei *Melandrium*; STRASBURGER 1900, MCPHEE 1924, BRESLAVEC 1935 und KUHN 1941 bei *Cannabis sativa*; VÖCHTING 1892, STRASBURGER 1900, YAMPOLSKY 1930b, 1957 und KUHN 1941 bei *Mercurialis annua*; STRASBURGER 1900 bei *Bryonia dioica* und *Ginkgo biloba*; CORRENS 1928 bezüglich *Aucuba japonica*). AUTENRIETH zufolge berichtete schon BORY DE ST. VINCENT von fehlgeschlagenen Versuchen HUBERTs auf den Mascarenen, bei *Myristica fragrans* das Geschlecht durch Pfropfung (♂ auf ♀) zu verändern.

Dagegen konnte vielfach bei andröcischen und monöcischen Pflanzen durch *Dekapitation*, weitgehendes Zurückschneiden, Stecklingsentnahme (bei Bäumen), Amputation weiblicher oder monokliner Blüten und ähnliche Maßnahmen mehr oder minder ausgeprägte *Verweiblichung* (bzw. Unterdrückung des ♂ Geschlechts) hervorgerufen werden (HEITMÜLLER und MELCHIOR 1960 durch Strangulation bei *Larix leptolepis*; MELCHIOR 1960, 1961 durch Ringelung bei *L. leptolepis* und *L. decidua*; Y. SAITO 1957 durch Resektion bei *Pinus densiflora*; MELCHIOR und HEITMÜLLER 1961 vereinzelt bei *P. silvestris*; K. MÜLLER 1858, BLARINGHEM 1906a, b bei *Zea Mays* [im Spätsommer, Tageslängeneffekt und Knospenmutation nicht ausgeschlossen]; SCHAFFNER 1922 bei *Arisaema* [durch Entblätterung]; JOSHI 1939 bei *Musa paradisiaca*; MURNEEK 1927 bei *Cleome spinosa*; GAERTNER 1844 bei *Silene noctiflora*; PORSCH 1937 bei *Aesculus macrostachya*; CH. JU. RUDENKO 1963 bei *Cannabis sativa*; LIMBERK 1953 bei *Humulus Lupulus*; YAMPOLSKY 1920c, 1930a bei *Mercurialis annua*; BORDAGE 1898, HIGGINS und HOLT 1914, HIGGINS 1916 und DIKSHIT 1956 bei *Carica Papaya* [vgl. aber CORRENS 1928]; BOSSÉ 1927 bei *Eucommia ulmoides*; CORRENS 1907, 1908 bei *Satureia hortensis*; TIEDJENS 1928 sowie MININA und KUŠNERENKO 1949 bei *Cucumis sativus* und JONES 1936 bei *Ambrosia artemisifolia* var. *elatior*).

Das Umgekehrte — *Vermännlichung* nach Dekapitation, Resektion usw. — wurde seltener beobachtet, so von MELCHIOR (1961) bei *Larix leptolepis* (nach Gipfelkrümmen), von WAREING (1953), MELCHIOR und HEITMÜLLER (1961) sowie MELCHIOR (1962) bei sonst zuerst ♀ blühenden Sämlingen von *Pinus silvestris*[1], von ATKINSON (1898) bei *Arisaema triphyllum*, von MAEKAWA (1927) bei *Arisaema japonica*, von MATZKE (1930) bei *Stellaria media*, von HUHNKE, JORDAN, NEUER und v. SENGBUSCH (1951) bei *Cannabis sativa*, von LANGE (1961a) bei *Carica Papaya*, von RIEDE (1925) bei *Salix fragilis* und von GALUN (1959b) bei *Cucumis sativus*.

Veränderungen in *beiden* Richtungen wurden nach derartigen Eingriffen bei *Spinacia oleracea* und bei *Cannabis sativa* gefunden (PRITCHARD 1916, MAEKAWA 1929, NEUER und v. SENGBUSCH 1943). — Keine geschlechtsbeeinflussende Wirkung einer Beschneidung konnten HEYER (1884) und STRASBURGER (1910) bei *Mercurialis annua* sowie SAVELLI und SOSTER (1929) bei *Cannabis sativa* feststellen.

[1] Hierher dürfte auch der von HERRMANN (1951) geschilderte, im Alter von 16 Monaten ♂ blühende Sämling von *Pinus montana* zu rechnen sein.

f) Parasitenbefall.

Auf die Rolle von Pilzinfektionen bei der modifikativen Geschlechtsbestimmung hat zuletzt BAKER (1957) nachdrücklich hingewiesen. Vorher befaßten sich mit der durch *Ustilago violacea* (Pers.) Fuckel ausgelösten Vermännlichung von *Melandrium*-Arten L. R. und CH. TULASNE (1847), BECKER (1869), CORNU (1869), HEYER (1884), GIARD (1888), MAGNIN (1888, 1889 [hier auch Übersicht über die ältere einschlägige Literatur], 1892), VUILLEMIN (1891), STRASBURGER (1900), ERLENMEYER und GEIGER-HUBER (1935), SCHOPFER und BLUMER (1939b), SCHOPFER (1940b, 1941), BAKER (1947a, b) und WESTERGAARD (1953). Vermännlichung wird weiterhin hervorgerufen durch andere Sippen von *U. violacea* bei *Silene inflata* (GIARD 1888), *S. dichotoma* und *S. Otites* (CORRENS 1928), durch *U. succisae* Magnus bei *Succisa pratensis* (BAKER 1957), durch *U. Vaillantii* bei *Muscari comosum* (MAGNIN 1892), durch *U. longissima* (Sow.) Mayen bei *Glyceria maxima* (LAMBERT 1947) und Verweiblichung durch *U. maydis* bei *Zea Mays* (ILTIS 1910, 1911; bei den zwei letztgenannten Phanerogamen auch Verlaubung des Fruchtknotens).

Bei *Oxalis stricta* bewirkte eine Infektion durch *U. oxalidis* auffallende Langgriffeligkeit (v. UBISCH 1935). MOLLIARD (1895) berichtet über Reduktion und Atrophie der Staubblätter in den (⚥) Blüten von *Knautia arvensis* Coult., *Scabiosa Columbaria* L. und *S. Succisa* L. nach Befall durch „*Peronospora violacea* Berk.“; gelegentlich nehmen die Staubblätter dann auch petaloide Gestalt an; bei stärkerer Infektion verkümmert auch das Gynoeceum. Nach *Uromyces scutellatus*-Befall wurden bei *Euphorbia Cyparissias* sonst abortierte ♀ Blüten voll ausgebildet (MOLLIARD 1895). Ferner bewirken Verweiblichung männlicher Blüten: *Tilletia Buchloëana* bei *Buchloë americana*, *Ustilago andropoginis* bei *Andropogon provincialis* und *A. Halleri* (MAGNIN 1891) sowie *U. caricis* bei *Carex praecox* (ROZE 1888).

In einzelnen Fällen können auch tierische Parasiten sowie Virusbefall eine Geschlechtsumstimmung auslösen. So ruft die Milbe *Epitrimerus salicobius* bei *Salix*-Arten der *Caprea*-Gruppe die Bildung intersexueller Einzelblüten hervor (J. W. HESLOP-HARRISON 1924). *Cucumis sativus* kann unter Viruseinfluß mehr ♂ oder mehr ♀ Blüten hervorbringen (VAN KOOT und VAN DORST 1958). EDWARDSON und CORBETT (1961) ziehen auch bei *Petunia* die Möglichkeit virusbedingter ♂-Sterilität in Betracht.

g) Stellung der Blüten, Alter und Größe der Pflanzen.

Ob eine Pflanze in einem bestimmten Alter an einer bestimmten Stelle eine ♂, ♀ oder ⚥ Blüte entwickelt, hängt — wie man seit KLEBS zu sagen pflegt — von den „inneren Bedingungen“ ab. Deren Änderung im Lauf des Pflanzenlebens wird einerseits in mehr oder weniger ausgeprägter Weise durch die „spezifische Struktur“, die Gesamtheit der Erbanlagen, kontrolliert (vgl. Beitrag NÜRNBERG-KRÜGER) und ist andererseits der Einwirkung von Außenfaktoren zugänglich, wie wir sie in den vorangegangenen sechs Abschnitten besprochen haben. In welcher Weise diese Außenfaktoren (einschließlich der Eingriffe nach Art der Resektion) auf die gerade für die Geschlechtsausprägung verantwortlichen inneren Bedingungen Einfluß nehmen, ist noch weitgehend unbekannt, und erst recht entzieht sich die Beschaffenheit der inneren Bedingungen selbst unserer Kenntnis. Bis zu einem gewissen Grade können in dieser Hinsicht die Reaktionen auf „gezielte“ Eingriffe Aufschluß geben, von denen in den drei nächsten Abschnitten die Rede sein wird. Hier jedoch haben wir uns mit den Geschlechtsverhältnissen zu befassen, wie sie sich bei quasi sich selbst überlassenen (monöcischen i. w. S.)

Pflanzen aus dem im Laufe ihres Lebens oder während einer Vegetationsperiode erfolgenden Wechsel der inneren Bedingungen ergeben.

In sehr vielen Fällen hält die genetische Struktur diesen Wechsel der inneren Bedingungen offensichtlich in exakt vorgezeichneten Bahnen. So bilden die Sämlinge vieler monöcischer Bäume anfangs ♀ und erst in späteren Jahren ♂ Blüten. Als Beispiele hierfür haben wir *Corylus Avellana, Juglans regia, Quercus rubra* (HILDEBRAND 1881) sowie *Pinus montana* und andere Coniferen (PRAT 1936, HERRMANN 1951, WAREING 1953) zu nennen. Auch bei *Galium cruciatum* (SCHULZ 1888), *Asperula taurina* und *Bryonia alba* (CORRENS 1928), *Cucumis sativus* und *Diospyros Kaki* (MININA 1949), *Geranium silvaticum* und *G. pratense* (CORRENS 1907, 1908), *Silene inflata* (CORRENS 1905), *Streptocarpus*-Bastarden (OEHLKERS 1956), *Musa paradisiaca* (JOSHI 1939), *Lilaea* (ENGLER und GILG 1919) und *Myriophyllum* (PETERSEN 1898) werden an der Pflanze im ganzen oder in einzelnen Blütenständen zuerst ♀, später ⚥ oder ♂ Blüten entwickelt.

Für viele andere Gewächse gilt das Gegenteil: Die Keimlinge von *Aesculus Hippocastanum* entwickeln zunächst nur ♂ Blüten (HOFFMANN 1871) und erst an älteren Exemplaren treten im basalen Bereich der Inflorescenzen auch gemischtgeschlechtige Blüten auf. Vom Übergang aus der ♂ zur ♀ Phase bei einem 35jährigen *Pisonia*-Baum (LILLO 1924) wurde oben auch schon berichtet, desgleichen von der Verweiblichung von *Arisaema triphyllum* in Abhängigkeit von der mit dem Alter zunehmenden Cormus-Größe (CAMP 1932). Auch in den Blütenständen anderer krautiger Pflanzen werden zuerst ♂ bzw. ⚥ und dann ⚥ bzw. ♀ Blüten gebildet, so bei *Begonia Wallichiana* (IRMSCHER 1924), *Satureia hortensis* (CORRENS 1907) und *Geum intermedium* (CORRENS 1928). *Helleborus foetidus* (mit dem Alter wachsender Prozentsatz der Karpelle unter den Sporophyllen; LAIBACH 1952) mag hier gleichfalls genannt werden. Schließlich sei auch an die Fixierung der Geschlechterverteilung im Sproßsystem von *Chamaecyparis nootkatensis* (COURTOT und BAILLAUD 1955) und von *Corylus Avellana* (ARNAL 1954) erinnert. Von vielen Compositen, Dipsacaceen, Araceen, Gramineen, Cyperaceen usw., die an dieser Stelle noch zu nennen wären, sei hier — wie schon in der nach Familien geordneten Übersicht — abgesehen.

Bei wieder anderen Pflanzen kann von einer exakt festgelegten Reihenfolge der verschiedengeschlechtigen Blüten nicht die Rede sein, und zwar weder im Hinblick auf eine einzelne Vegetationsperiode noch hinsichtlich aufeinanderfolgender Jahre. Dies gilt etwa für monöcische Individuen von *Urtica dioica* (LAIBACH 1952), für *Dianthus sylvestris* (CORRENS 1928), *Myrica Gale* (DAVEY und GIBSON 1916/17), *Salix sepulcralis* (ARNAL 1954), *Morus alba* (SCHAFFNER 1937), *Acer Pseudoplatanus* (WITTROCK 1886) und *A. platanoides* (SCHULZ 1888/90).

Abschließend sei noch auf eine Anzahl Veröffentlichungen hingewiesen, die großenteils schon in unserer systematischen Übersicht erwähnt worden waren und die sich alle in irgendeiner Form mit den Beziehungen zwischen dem Geschlecht von Blüten und ihrer Stellung an der Pflanze, deren Alter oder Größe beschäftigen: J. HESLOP-HARRISON (1961: *Zea Mays*), MAEKAWA (1927: *Arisaema japonica*), VAZART (1959: *Asparagus officinalis*), NOACK (1962: *Begonia Cathayana*), BURKILL (1895: *Stellaria media*), BÉGUINOT (1910—1921: *Stellaria media*), BRUNOTTE (1905: *Humulus Lupulus*), TOURNOIS (1914: *H. Lupulus* und *H. japonicus*), HUHNKE, JORDAN, NEUER und v. SENGBUSCH (1951: *Cannabis sativa*), SHIFRISS (1961: *Ricinus communis*), LONGMAN und WAREING (1959: *Betula verrucosa*) und CORRENS (1920: *Bryonia alba*).

h) Behandlung mit tierischen Geschlechtshormonen.

Im folgenden sind drei Gruppen von Versuchen zu besprechen, die eine „gezielte“ Veränderung der „inneren Bedingungen“ bezwecken. Die Versuche der ersten Gruppe (Abschnitt h) bestehen in der Applikation tierischer Geschlechtshormone, die der zweiten (Abschnitt i) in einer mehr oder minder direkten Beeinflussung des Wuchsstoffhaushaltes und die der dritten (Abschnitt k) in der Anwendung von Gibberellinen.

Loewe und Spohr (1926) sowie Dohrn, Faure, Poll und Blotevogel (1926) und Skarżyński (1933) hatten aus ♀ Pflanzenorganen (z. B. aus Fruchtknoten von *Nuphar luteum* und *Salix Caprea*), ferner aus Kartoffelknollen, Petersilienwurzeln, Hefezellen und anderem Pflanzenmaterial Substanzen mit oestrogener Wirkung isoliert. Seitdem Butenandt und Jacobi (1933) eine solche aus Palmkernextrakten gewonnene Substanz mit einem tierischen Geschlechtshormon (α-Follikelhormon) identifiziert hatten (vgl. auch Butenandt 1933), wurde eine große Zahl von Versuchen unternommen, durch derartige Hormone die sexuelle Entwicklung von Pflanzen zu beeinflussen. Diese Versuche schlugen aber sämtlich fehl, oder ihre Ergebnisse konnten doch wenigstens von anderen Experimentatoren nicht reproduziert werden. Eine Ausnahme machen die sehr sorgfältig durchgeführten Untersuchungen von A. und D. Löve (1940, 1945; vgl. S. 171) an *Melandrium*. Nur hier scheint eine einwandfreie Geschlechtsverschiebung durch Sexualhormone in beiden Richtungen gelungen zu sein. Es darf daher bezweifelt werden, daß diese Stoffe bei Blütenpflanzen im Normalfall an der Geschlechtsausprägung wesentlich beteiligt sind[1]. Als viel erfolgreicher erwiesen sich Versuche zur Geschlechtsbestimmung auf dem Wege über eine

i) Beeinflussung des Wuchsstoffhaushaltes.

Die weitaus meisten einschlägigen Versuche betrafen monöcische Cucurbitaceen, zumal *Cucumis sativus*. Bei Pflanzen dieser Art induzierten *Kohlenmonoxyd* und Acetylen (deren auxinartige Wirkungen auch schon in anderem Zusammenhang bekannt wurden) weitgehende Verweiblichung (Minina 1938, Minina und Tylkina 1947, Minina und Kušnerenko 1949 bzw. Mechanik 1958). Bei demselben Objekt gelang es wiederholt, durch *Wuchsstoffe* (β-Indolylessigsäure [IES], 2,4-Dichlorphenoxyessigsäure [2,4-D] und vor allem α-Naphthylessigsäure [NES]) den Anteil der ♀ an der Gesamtzahl der Blüten wesentlich zu steigern (Laibach und Kribben 1949/50, 1950a, b, 1951, Laibach 1950/51, 1952, Wittwer und Hillyer 1954, Galun 1956, 1959b, Galun, Jung und Lang 1962, 1963 [in vitro-Kultur isolierter Blütenknospen], Kribben 1963 [Regenerate an zurückgeschnittenen Keimlingen]). Das gleiche gilt für *Cucurbita maxima* var. *turbaniformis* (Nitsch, Kurtz, Liverman und Went 1952, Wittwer und Hillyer 1954) und einige weitere Cucurbitaceen (J. Heslop-Harrison 1959: *Echinocystis*, Satyanarayana und Rangaswami 1959: *Luffa acutangula*), ferner für *Pinus densiflora* und *P. Thunbergii* (Y. Saito 1957), *Zea Mays* (J. Heslop-Harrison 1961, Nickerson und Hewitson 1963) und *Mangifera indica* (Mallik, Sahay und D. L. Singh 1959). Ähnlich verhalten sich monöcische Exemplare von *Mercurialis ambigua* (J. und Y. Heslop-Harrison 1957a; CO-Exposition). — Auch bei Pflanzen wie *Bryophyllum Daigremontianum*, *Streptocarpus Wendlandii*, *Hyoscyamus niger* und *Silene pendula*, die im Normalfall gemischtgeschlechtige Blüten tragen, konnte durch IES bzw. NES Verweiblichung hervorgerufen werden; die Entwicklung der Stamina war dann mehr oder weniger stark gehemmt

[1] In den letzten Jahren berichteten anscheinend nur noch Preda und Ghisa (1957) über Versuche, bei denen Placentarhormon (bei *Urtica dioica*) zu Verweiblichung geführt haben soll.

(RESENDE 1951, RESENDE und VIANA 1952, RESENDE 1954 bzw. J. und Y. HESLOP-HARRISON 1958a). — Endlich können auch ♂ *Cannabis*-Individuen durch CO- oder NES-Einwirkung zur Entwicklung von intersexuellen oder gar ♀ Blüten veranlaßt werden (J. HESLOP-HARRISON 1956, J. und Y. HESLOP-HARRISON 1957c). Auch durch einige weniger gebräuchliche wuchsstoffartige Substanzen konnte man in den letzten Jahren Verweiblichung erzielen, so durch Allyltrimethylammoniumbromid bei *Cucumis sativus*, Sorte Marketer (MITCHELL und WITTWER 1962), durch 2,3-Dichlorisobuttersäure und 2,2-Dichlorpropionsäure bei *Carica Papaya* (LANGE 1961b) und *Citrullus vulgaris* (HENSZ und MOHR 1959) sowie durch Phthalsäuremono-α-naphthylamid bei *Solanum nigrum* (KIERMAYER 1960). Allein durch α-(2-Chlorphenylthio)propionsäure scheint (bei *Humulus Lupulus*) Vermännlichung von genetisch ♀ Individuen induziert worden zu sein (WESTON 1960).

Führt also Wuchsstoffbehandlung oder CO-Exposition in fast allen daraufhin untersuchten Fällen[1] — bei Hermaphroditen ebenso wie bei Monöcisten und Diöcisten — zur Verweiblichung, so sollte man erwarten, daß durch Einwirkung von *Wuchsstoffantagonisten* umgekehrt Vermännlichung ausgelöst werden könnte. Dies ist aber bemerkenswerterweise *nicht* der Fall. Bei *Cannabis sativa* hat Trijodbenzoesäure (TJBS) überhaupt keinen Einfluß auf die Geschlechtsausprägung, sondern induziert nur im vegetativen Bereich einige Mißbildungen (J. und Y. HESLOP-HARRISON 1957b); bei *Cucumis sativus* und *Cucurbita Pepo* hatte 2,3,5-TJBS sogar dieselbe verweiblichende Wirkung wie NES (WITTWER und HILLYER 1954; vgl. auch KIERMAYER 1958, 1961 bei *Solanum nigrum*, KIERMAYER 1959 bei *Helianthus annuus* und REHM 1952 bei *Citrullus vulgaris*). Ob sich die Effekte von NES und TJBS addieren oder einander aufheben können, scheint noch nicht untersucht zu sein.

k) Behandlung mit Gibberellinen.

In den 10 Jahren, die seit der Einführung von Gibberellinen in die Versuche zur Physiologie der Blütenbildung vergangen sind, hat sich auch schon eine ganze Reihe von Forschern mit der Wirkung dieser Stoffe auf die Geschlechtsausprägung beschäftigt. Die Ergebnisse dieser Experimente lassen sich dahingehend zusammenfassen, daß Gibberelline bei Cucurbitaceen durchweg Vermännlichung, sonst aber fast immer Verweiblichung bewirken. Die Cucurbitaceen-Studien beziehen sich größtenteils auf *Cucumis sativus* (WITTWER und BUKOVAC 1957, 1958, 1962a, b, GALUN 1959a, 1961a, b, BUKOVAC und WITTWER 1960a, b, MITCHELL und WITTWER 1960, 1962, PETERSON und ANHDER 1960, GALUN, JUNG und LANG 1962, 1963), daneben auch auf *Citrullus vulgaris* (HENSZ und MOHR 1959). Verweiblichung unter dem Einfluß von Gibberellinen wurde im einzelnen beobachtet bei *Thuja* spec. (MCVEIGH 1961), *Zea Mays* (NELSON und ROSSMAN 1958, NICKERSON 1959), *Hyoscyamus niger* (RESENDE und VIANA 1959), *Ricinus communis* (SHIFRISS 1961) und *Xanthium Strumarium* (v. WITSCH 1961). Nur in bezug auf *Bryophyllum proliferum* (RESENDE und VIANA 1959) und *Cannabis sativa* liegen widersprüchliche Befunde vor: ATAL (1959) und KÖHLER (1964b) berichten über Vermännlichung, HERICH (1960) über Verweiblichung von Hanfpflanzen.

3. Interpretation der diplomodifikativen Geschlechtsbestimmung.

Will man den Phänomenen der diplomodifikativen Geschlechtsbestimmung eine Deutung angedeihen lassen, so bieten sich in erster Linie zwei Möglichkeiten dar:

[1] Nur in Versuchen von HENSZ und MOHR (1959) erwies sich NES bei *Citrullus vulgaris* als wirkungslos.

einerseits die der teleologischen Interpretation (vgl. Hildebrand 1893), die wir hier außer acht lassen möchten, — andererseits die der kausalanalytischen. Was diese zweite Möglichkeit betrifft, so dachte man im 19. Jahrhundert in der Regel wohl an einen ± direkten Zusammenhang zwischen Ernährungs- und Geschlechtsverhältnissen; vgl. z. B. Düsing (1884), dem sich Klebs (1894) anschließt.

Einen Schritt weiter geht Tournois (1914), der bei Cannabinaceen eine Beziehung zwischen Geschlecht und osmotischem Wert zu erkennen glaubt: Wird durch Verminderung der Lichtintensität und Steigerung der Luftfeuchtigkeit der osmotische Wert gesenkt, so tritt Verweiblichung ein, während einer Erhöhung des osmotischen Wertes der gegenteilige Effekt entspricht. — In den ersten auf Tournois folgenden Jahrzehnten bemühte man sich eher um den Nachweis, *daß* im einen Falle die Wirkung bestimmter Gene, in einem anderen die bestimmter Außenfaktoren bei der Geschlechtsbestimmung im Vordergrund stehe, als um die Aufklärung der einzelnen Zwischenglieder zwischen Ursache und Endeffekt.

Arnal (1952a, b, 1954) machte für die ein-, zwei- oder mehrmalige Geschlechtsumkehr bei Monöcisten innerhalb einer Vegetationsperiode einfache, doppelte bzw. „multinodale“ Gradienten verantwortlich, über deren physiologisch-chemisches Substrat gar nichts Konkretes gesagt wird. So führen diese Überlegungen nicht weiter, sondern stellen nur sozusagen eine gehobenere Art der Beschreibung dar.

Ernster ist die in den 1950er Jahren vorherrschende Deutung zu nehmen, derzufolge in jedem Fall der Wuchsstoffhaushalt für die modifikative Geschlechtsbestimmung verantwortlich sein soll, eine Auffassung, die sich auf viele erfolgreiche Versuche stützt, über die im Abschnitt II 2 i berichtet worden ist (Laibach und Kribben, J. Heslop-Harrison u. a.). So sollen — nach Laibach — auch in den von Heyer (1884) und Correns (1908) geschilderten Versuchen die Ernährungsverhältnisse auf dem Wege über eine Änderung des Auxingehaltes die Geschlechtsausprägung beeinflußt haben. Aus seinen mit Kribben durchgeführten Versuchen und denen von Resende (1950) zieht Laibach den Schluß, daß für die Ausbildung von Staubblättern oder ♂ Blüten ein relativ niederer Auxinspiegel genüge, für die Entwicklung von Fruchtblättern oder ♀ Blüten ein etwas höherer und für die Bildung von vegetativen Organen ein noch höherer Wuchsstoffgehalt erforderlich sei. — Auch J. Heslop-Harrison (1957) vertritt im wesentlichen diese Ansicht, deren weitere Begründung wir hier nicht im einzelnen wiedergeben möchten. Wuchsstoffkonzentrationen sind bekanntlich für viele Entwicklungsprozesse von großer Bedeutung, — warum also nicht auch bei der Geschlechtsbestimmung?

An die Vorstellungen von J. Heslop-Harrison (1957) knüpft Köhler (1964a, S. 109) mit seinem Versuch einer neuen, umfassenden Hypothese an, die sowohl die genotypische wie die modifikative Geschlechtsbestimmung zu deuten beabsichtigt[1]. Hiernach soll für die Geschlechtsverhältnisse der Blüten je ein ♂- und ♀-bestimmendes Enzym verantwortlich sein, deren Aktivität von der Konzentration einer (vermutlich mit Auxin identischen) Substanz A abhängt; überschneiden sich die Aktivitätsbereiche dieser beiden Enzyme, so kommt es (immer nach Köhler) zur Bildung von monoklinen Blüten, andernfalls nur zur Entstehung von ♂ und (bzw. oder) ♀ Blüten. Eine weitere Modifizierung kann durch andere Substanzen erfolgen, die z. B. die Konzentration der Substrate der beiden erwähnten hypothetischen Enzyme beeinflussen mögen[2].

[1] Dies letztere hervorzuheben ist insofern wichtig, als z. B. Noack (1962) im Hinblick auf erhebliche morphologische Unterschiede zwischen (überwiegend genotypisch bestimmten) ♀ und (modifikativ) verweiblichten ♂ Blüten von *Begonia Cathayana* die Möglichkeit in Betracht zieht, daß hier zwei *verschiedene* Mechanismen im Spiele sind.

[2] So deutete Köhler (schon 1961) die Maskulinisierung monöcischer Hanfpflanzen unter dem Einfluß von Langtag und hoher Lichtintensität (die in der präfloralen Phase geboten wurden) als Folge der Zufuhr unspezifischer Baustoffe zu den blütenbildenden Meristemen.

KÖHLERs Hypothese gibt m. E. keine ausreichende Antwort auf die Frage, was geschieht, wenn sich die Aktivitätsbereiche der beiden Enzyme nicht nur nicht überschneiden, sondern nicht einmal berühren. Darüber hinaus ist aber in den letzten Jahren eine Reihe von Befunden bekannt geworden, welche die (auch von KÖHLERs Hypothese vorausgesetzte) entscheidende Rolle von Wuchsstoffen zumindest zweifelhaft erscheinen lassen. So stellt schon J. HESLOP-HARRISON (1959, S. 273) im Hinblick auf (wenigstens primär) monokline Blüten wie jene von *Silene pendula* fest, daß unter dem Einfluß von Wuchsstoffen nur das eine oder das andere Geschlecht unterdrückt, aber niemals ein Staubblatt in ein Karpell umgewandelt werde. Er folgert daraus, daß Auxine hier eher als Regulatoren denn als Determinatoren fungieren; die eigentlichen Determinatoren seien noch unbekannt.

Auf eine mehr indirekte Weise gehen GARAY und SÁGI (1960) die Beziehungen zwischen Wuchsstoffhaushalt und (*Ustilago*-induzierter) Vermännlichung weiblicher *Melandrium*-Blüten an: Die IES-Oxydase-Aktivität ist in den Sepalen maskulinisierter ♀ Blüten nicht auf das Maß erhöht, das für gesunde und kranke Kelchblätter von genetisch ♂ Blüten charakteristisch ist.

Wenig später berichten CONRAD und MOTHES (1961) über Unterschiede im Auxingehalt von ♂ und ♀ Exemplaren von *Cannabis*[1]; bald darauf zeigt sich aber, daß das „Wuchsstoffübergewicht" der ♀♀ in den ersten Versuchen durch den Wuchsstoffreichtum der Früchte bedingt war; der Auxingehalt der vegetativen Teile hingegen gab keinen Anlaß zu der Vermutung, daß er etwas mit der Geschlechtsausprägung zu tun haben könne (CONRAD 1962).

In ähnlichem Sinne äußert sich auch J. HESLOP-HARRISON (1964) in seinem jüngsten Übersichtsbericht: Zwar könne man das Geschlecht von Blüten in vielen Fällen durch Auxine und Gibberelline beeinflussen, doch lassen andröcische und gynöcische *Cucumis*-Linien sowie unter ♂- oder ♀-begünstigenden Bedingungen gezogene Vegetationskegel von *Zea Mays* keine nennenswerten Unterschiede im Gehalt an solchen Substanzen erkennen. HESLOP-HARRISON denkt deshalb — auch im Hinblick auf die unabänderliche Reihenfolge Kelch—Krone—Androeceum—Gynoeceum in den Blüten — nun eher an eine Serie physiologischer Relais: «Chaque étape commanderait la suivante par le déclanchement de certaines potentialités génétiques sous l'effet de régulateurs à très court rayon d'action, différents des substances de croissance connues actuellement.» HESLOP-HARRISON zieht denn auch in seiner Argumentation viele Beispiele heran, welche die entscheidende Rolle von Wuchsstoffen und Gibberellinen bei der Geschlechtsbestimmung in Frage stellen.

So finden wir uns heute, nachdem man sich in der „Wuchsstoff-Periode" der Entwicklungsphysiologie der Lösung des Problems der Geschlechtsbestimmung schon nahe geglaubt hatte, von einer befriedigenden Antwort auf die Frage nach dem biochemischen Mechanismus der Geschlechtsbestimmung so weit entfernt wie eh und je. Daß diese im allgemeinen etwas mit der Synthese einer Ribonucleinsäure zu tun habe, erscheint zwar nach den Thiouracil-Versuchen von HESS (1959) an *Streptocarpus* als denkbar, nach jenen von J. HESLOP-HARRISON (1960) an *Cannabis* aber als auch nicht sehr wahrscheinlich.

---

[1] KÖHLER (1964a) glaubt, diese Unterschiede auf den allgemeinen Geschlechtsdimorphismus zurückführen zu können; denselben Gesichtspunkt kann man aber wohl auch gegenüber dem Befund von KÖHLER und LANG (1963) geltend machen, denen zufolge Extrakte aus Inflorescenzen von ♀♀ und feminisierten ♂♂ größere Mengen eines Gibberellinantagonisten enthalten als solche aus Blütenständen normaler ♂♂.

Anhang: Physiologische Kastration.

Zum Schluß sei noch kurz von der sog. physiologischen Kastration die Rede. Unter diesem Sammelbegriff versteht man alle jene Methoden, durch die man in größerem Umfang die Bildung fertilen Pollens (und damit Selbstbefruchtung) verhindern kann, ohne die Antheren operativ aus den Blüten(-Knospen) zu entfernen. Hierbei kann es sich (1.) darum handeln, daß infolge bestimmter Maßnahmen überhaupt keine Staubblätter gebildet werden (und dies ist der Grund dafür, daß die physiologische Kastration in diesem Handbuchbeitrag erwähnt wird); in anderen Fällen (2.) werden zwar Staubblätter zunächst entwickelt, dann aber — vor der Pollenbildung — selektiv gehemmt oder abgetötet; in wieder anderen Fällen (3.) wird verhindert, daß ♂ Blütenknospen sich öffnen; in abermals anderen (4.) entsteht befruchtungsunfähiger Pollen. Es bedarf wohl keines besonderen Hinweises darauf, daß bestimmte Formen der physiologischen Kastration eine gewisse Bedeutung für Pflanzenbau und -züchtung erlangt haben.

Folgende Beispiele für die erstgenannte Art von physiologischer Kastration seien genannt: Bei *Zea Mays* verhindert Besprühen mit einer 0,025%igen Lösung von Maleinsäurehydrazid die Bildung von Antheren in der terminalen Rispe (R. H. Moore 1950, A. W. Naylor 1950); bei *Triticum vulgare* können sich die Antheren unter dem Einfluß von Langtag und hoher Temperatur (20—25° C) zu karpellartigen Gebilden entwickeln (Meletti 1961).

Im Sinne der zweiten Art physiologischer Kastration wirken in bestimmten Fällen Wärme (Hovin 1957, 1958: *Poa annua*, 20—29° C zur Zeit der Rispenentfaltung), Kälte (Suneson 1937: *Triticum vulgare*, 27—36° F; Youngner 1961: *Pennisetum clandestinum*, 10° C), Dauerlicht (Meletti 1961: *Triticum vulgare*; Youngner 1961: *Pennisetum clandestinum*) und Dunkelheit (Grinvald, Mavromati und Gutenmacher 1962/63: *Hordeum sativum*, 48—72stündige Verdunkelung).

Ein Beispiel der dritten Art stellt die Verhinderung der Entfaltung von ♂ Blüten bei *Citrullus vulgaris* unter dem Einfluß von 2,3-dichlorisobuttersaurem Natrium dar (Hensz und Mohr 1959).

Pollensterilität wird durch die letztgenannte Substanz bei *Lycopersicon esculentum* (J. F. Moore 1959) und bei *Gossypium* spec. (Eaton 1957 sowie Pate und Duncan 1960) hervorgerufen. — Weitere Beispiele für verschiedene Arten von physiologischer Kastration finden sich im systematischen Teil der vorliegenden Übersicht.

Literatur.

+ = zitiert nach Correns 1928; * = zitiert nach J. Heslop-Harrison 1957. — Quellenhinweise beziehen sich in der Regel auf den *Inhalt* der betreffenden Arbeiten; die bibliographischen Angaben sind zuweilen anderen Quellen entnommen.

Arber, A.: The Gramineae. Cambridge: University Press 1934. — Arnal, C.: Remarques sur le déterminisme du sexe à propos d'un Saule androgyne. Ann. Univ. Saraviensis **1**, 80—87 (1952a). — Essai sur la répartition des sexes chez les *Carex*. Ann. Univ. Saraviensis **1**, 102—114 (1952b). — Gradients sexuels chez les Phanérogames. VIIIe Congr. Internat. de Botanique, Paris, Rapp. et Comm., Sect. 7 et 8, p. 291—292, 1954. — Atal, C. K.: Sex reversal in hemp by application of gibberellin. Curr. Sci. **28**, 408—409 (1959). Zit. nach Ber. wiss. Biol. **145**, 288. — Atkinson, G. F.: Experiments on the morphology of *Arisaema*. Bot. Gaz. **25**, 114 (1898)* — Atsmon, D., and E. Galun: A morphogenetic study of staminate, pistillate and hermaphrodite flowers in *Cucumis sativus* (L.). Phytomorphology **10**, 110—115 (1960). — Physiology of sex in *Cucumis sativus* (L.). Leaf age patterns and sexual differentiation of floral buds. Ann. of Bot., N. S. **26**, 137—146 (1962). — Atsmon, D., E. Galun, and K. M. Jakob: Relative time of anthesis in pistillate and staminate cucumber flowers. Ann. of Bot., N. S. **29**, 277—282 (1965). — Autenrieth, H. F.: Disquisitio quaestionis academicae de discrimini sexuali iam in seminibus plantarum dioïcarum apparente. Tubingae 1821+. — Awada, M.: Effects of water and nitrogen application on composition, growth, sugars in fruits, yield, and sex expression of the papaya plants (*Carica papaya* L.). Techn. Bull. Hawaii agric. Exp. Station **33** (1957). Zit. nach LANGE 1961a. — Relationships of minimum temperature and growth rate with sex expression of papaya plants (*Carica papaya* L.). Techn. Bull. Hawaii agric. Exp. Station **38** (1958). Zit. nach LANGE 1961a. — AWADA, M., and W. IKEDA: Effects of moisture on yield and sex expression of papaya plants (*Carica papaya* L.). Progress Notes Hawaii agric. Exp. Station **97** (1953). Zit. nach LANGE 1961a. — AYMARD, M.: Coalescences florales chez *Anagallis arvensis* L. traité à l'acide 2,4-dichlorophénoxyacétique. Bull. Soc. bot. France **107**, 1—3 (1960).

BACCARANI, M.: Osservazioni sulla distribuzione dei sessi nel *Citrus trifoliata* L. Atti Soc. Nat. e Mat., Modena **69**, 97—102 (1938). Zit. nach Ber. wiss. Biol. **54**, 531. — BAKER, H. G.: Sex in *Melandrium*. Nature (Lond.) **159**, 34 (1947a). — Infection of species of *Melandrium* by *Ustilago violacea* (Pers.) Fuckel and the transmission of the resultant disease. Ann. of Bot., N. S. **11**, 333—348 (1947b). — The inheritance of certain characters in crosses between *Melandrium dioicum* and *M. album*. Genetica **25**, 126—156 (1950). Zit. nach BAKER 1957. — Expression of sex in flowering plants. Nature (Lond.) **180**, 614—615 (1957). — BECKER: Atheneum **1869**, 342+. — BÉGUINOT, A.: Ricerche intorno al polimorfismo della *Stellaria media* (L.) Cyr. N. G. bot. ital., N. S. **17**, 299—326, 348—390 (1910); Teil 3/1 Padova 1920; Teil 3/2 Ferrara 1921. Zit. nach MATZKE 1930. — BENČAT', F.: Morphologische und physiologische Zweihäusigkeit der Edelkastanie *Castanea sativa* Mill. unter den Bedingungen der Slowakei. Biológia (Bratislava) **19**, 912—919 (1964) [Slowakisch mit russ. u. dtsch. Zus.-fass.]. — Einige seltene biologisch-morphologische Erscheinungen der Edelkastanie *Castanea sativa* Mill. unter den Bedingungen der Slowakei. Biológia (Bratislava) **20**, 25—30 (1965) [Slowakisch mit russ. u. dtsch. Zus.fass.]. — BERGER, X.: Intersexe Bildungen in den Antheren partiell apomiktischer *Rubus*arten und *Rubus*bastarde. Z. Zellforsch. **38**, 472—474 (1953). — BESSEY, E. A.: Sex problems in hemp. Quart. Rev. Biol. 8, 194—200 (1933). — BISWAS, K. C.: Staminate pistil in *Crotalaria spectabilis* Roth. Sci. and Culture **21**, 372—373 (1956). Zit. nach R. P. SINGH 1956. — BLACK, C. A.: Effect of commercial fertilizers on the sex expression of hemp. Bot. Gaz. **107**, 114—120 (1945). — BLARINGHEM, L.: Production d'une espèce élémentaire nouvelle de Maïs par traumatisme. C. R. Acad. Sci. (Paris) **143**, 245—247 (1906a). — Production par traumatisme et fixation d'une variété nouvelle de Maïs, le *Zea Mays* var. *pseudo-androgyna*. C. R. Acad. Sci. (Paris) **143**, 1252—1254 (1906b). — Études sur le polymorphisme floral. IV. Sexualité et métamorphose des épis de *Plantago lanceolata* L. Bull. Soc. bot. France **70**, 717—725 (1923). — BLAVET, M. A.: Intermédiaire de l'A.F.A.S. **1** (1896). Zit. nach GIARD 1898. — BOCQUET, G., et M. STROUN: Photopériodisme et formation de l'androcée chez le *Cerastium tomentosum* L. et le *Melandrium diurnum* (Miller) Garcke (Caryophyllaceae). Ber. schweiz. bot. Ges. **70**, 163—170 (1960). — BONNETT, O. T.: Development of the staminate and pistillate inflorescences of sweet corn. J. agric. Res. **60**, 25—37 (1940). — Ear and tassel development in maize. Ann. Missouri bot. Gard. **35**, 269—280 (1948). Zit. nach J. HESLOP-HARRISON 1961. — BORDAGE, E.: Variation sexuelle consécutive à une mutilation chez le Papaya commun. C. R. Soc. Biol. (Paris) **5**, 708 (1898)*. — BORTHWICK, H. A., and N. J. SCULLY: Photoperiodic responses of hemp. Bot. Gaz. **116**, 14—29 (1954). — BOSSÉ, G. G.: Willkürliche Geschlechtsumänderung bei *Eucommia*. Izv. Akad. Nauk SSSR., Otd. mat. i estestv. Nauk, Ser. biol. **1937**, 607—618 [Russisch mit engl. Zus.fass.]. Zit. nach Ber. wiss. Biol. **46**, 485. — BRANTLEY, B. B., and G. F. WARREN: Sex expression and growth in muskmelon. Plant Physiol. **35**, 741—745 (1960). — BRAUN, A.: Das Individuum der Pflanze in seinem Verhältniß zur Species. Generationsfolge, Generationswechsel und Generationstheilung der Pflanze. Abh. kgl. Akad. Wiss., Berlin **1853**, Physik. Abh., 19—122 (1854). — Monöcische Formen des Hanfes (*Cannabis sativa*). S.-B. Ges. naturforsch. Freunde Berlin, 19. Nov. 1872, und Bot. Ztg **31**, 168 (1873). Zit. nach BRESLAVEC und ZAUROV 1937. — BREIDER, H.: Spontane Blütenmutationen bei der Rebe. Züchter **32**, 100—102 (1962). — BRESLAVEC (BRESLAVETZ), L. P.: Abnormal development of pollen in different races and grafts of hemp. Genetica **17**, 154—169 (1935). — Untersuchung der Entwicklung von Hanfblüten, die ihr Geschlecht durch photoperiodische Wirkung verändert haben. Bjull. moskov. obšč. Ispyt. Prir., Otd. biol. **45**, 182—189 (1936) [Russisch mit franz. Zus.fass.]. Zit. nach Ber. wiss. Biol. **34**, 279. — Researches on development of the flower in hemp whose sex has been changed under the influence of photoperiodism. Genetica **19**, 393—412 (1937). — BRESLAVEC, L. P., and E. ZAUROV: A study of hermaphrodite hemp found in field crops. Dokl. Akad. Nauk SSSR., N. S. **16**, 285—288 (1937). — BRUNOTTE, C.: Sur une liane de Houblon (*Humulus Lupulus* L.) hermaphrodite. Rev. gén. Bot. **17**, 109—115 (1905). — BUIS, R.: Action de faibles concentrations de coumarine et d'hydrazide maléique sur le développement du Maïs. Rev. gén. Bot. **63**, 29—37 (1956). — BUKOVAC, M. J., and S. H. WITTWER: Comparative biological activities of gibberellins A_1, A_2, A_3, A_4 and some of their derivatives. In: Plant Growth Regulation. Ames, Iowa: Iowa State College Press 1960a. Zit. nach WITTWER und BUKOVAC 1962b. — Gibberellin modification of flower sex expression in *Cucumis sativus* L. Symp. on Gibberellins, Amer. chem.

Soc., 138th ann. Meeting, New York City, Sept. 14, 15, 1960b. Zit. nach Wittwer und Bukovac 1962b. — Burk, E. F.: The rôle of pistil length in the development of forcing tomatoes. Proc. Amer. Soc. hortic. Sci. **26**, 239—240 (1930)*. — Burkill, J. H.: On some variations in the number of stamens and carpels. J. Linn. Soc. London, Bot. **31**, 216—245 (1895). Zit. nach Matzke 1930. — Fertilisation of spring flowers on the Yorkshire coast. J.of Bot. **35**, 184—189 (1897). Zit. nach Laibach 1951. — Butenandt, A.: Zur Biologie und Chemie der Sexualhormone. Naturwiss. **21**, 49—54 (1933). — Butenandt, A., u. H. Jacobi: Über die Darstellung eines krystallisierten pflanzlichen Tokokinins (Thelykinins) und seine Identifizierung mit dem α-Follikelhormon. Untersuchungen über das weibliche Sexualhormon. X. Hoppe-Seylers Z. physiol. Chem. **218**, 104—112 (1933).

Camp, W. H.: Sex in *Arisaema triphyllum*. Ohio J. Sci. **32**, 147—151 (1932)*. — Catarino, F. M.: Some effects of kinetin on sex expression in *Bryophyllum crenatum* Bak. (LSDP). Port. Acta biol. A **8**, 267—284 (1964). — Chamberlain, C. J.: Contribution to the life history of *Salix*. Bot. Gaz. **23**, 147—179 (1897). Zit. nach R. P. Singh 1956. — Chifflot, M.: Sur la castration thelygène chez *Zea Mays* L. var. *tunicata*, produite par l'*Ustilago Maydis* D. C. (Corda). C. R. Acad. Sci. (Paris) **148**, 426—429 (1909). — Choudhri, R. S., and R. Krishan: Sex differentiation in *Zea mays*. Sci. and Culture **11**, 472—476 (1946)*. — Conrad, K.: Über geschlechtsgebundene Unterschiede im Wuchsstoffgehalt männlicher und weiblicher Hanfpflanzen. Flora (Jena) **152**, 68—73 (1962). — Conrad, K., u. K. Mothes: Über geschlechtsgebundene Unterschiede im Auxin-Gehalt diözischer Hanfpflanzen. Naturwiss. **48**, 26—27 (1961). — Cornu, M.: Note sur le *Melandrium dioicum* attaqué par un *Ustilago*. Bull. Soc. bot. France **16**, 213—214 (1869). — Correns, C.: Weitere Untersuchungen über die Gynodioecie. Ber. dtsch. bot. Ges. **23**, 452—463 (1905). — Die Vererbung der Geschlechtsformen bei den gynodiöcischen Pflanzen. Ber. dtsch. bot. Ges. **24**, 459—474 (1906). — Zur Kenntnis der Geschlechtsformen polygamer Blütenpflanzen und ihrer Beeinflußbarkeit. Jb. wiss. Bot. **44**, 124—173 (1907). — Weitere Untersuchungen über die Geschlechtsformen polygamer Blütenpflanzen und ihre Beeinflußbarkeit. Jb. wiss. Bot. **45**, 661—700 (1908). — Untersuchungen über Geschlechtsbestimmung bei Distelarten. S.-B. kgl. preuß. Akad. Wiss. **1916**, 448—477. — Fortsetzung der Versuche zur experimentellen Verschiebung des Geschlechtsverhältnisses. S.-B. preuß. Akad. Wiss. **1918**, 1175—1200. — Die geschlechtliche Tendenz der Keimzellen gemischtgeschlechtiger Pflanzen. Z. Bot. **12**, 49—60 (1920). — Bestimmung, Vererbung und Verteilung des Geschlechtes bei den höheren Pflanzen. In: Handbuch der Vererbungswissenschaft, herausgeg. von E. Baur u. M. Hartmann, Bd. II, Teil C, 138 S. Berlin: Gebrüder Borntraeger 1928. — Courtot, V., et L. Baillaud: Sur la répartition des sexes chez un *Chamaecyparis*. Ann. Sci. Univ. Besançon, Sér. II, Bot. **6**, 75—81 (1955). — Currence, T. M.: Nodal sequence of flower type in cucumber. Proc. Amer. Soc. hortic. Sci. **29**, 447 (1932)*.

Danielson, L. L.: Effect of daylength on growth and reproduction of the cucumber. Plant Physiol. **19**, 638—648 (1944). — Davey, A. J., and C. M. Gibson: On the distribution of monoecious plants and the occurrence of hermaphrodite flowers in *Myrica Gale*, with observations of variations of sex. Proc. Linn. Soc. London **1916/1917**, 6—7+. — Note on the distribution of the sexes in *Myrica Gale*. New Phytologist **16**, 147—151 (1917)*. — Dellingshausen, M. v., u. K. Stern: Über einige besondere Blütenformen der Birke. Silvae Genetica **7**, 181—188 (1958). — Dikshit, N. N.: Some aspects of parthenocarpy in fruit plants. South Indian Hortic. **4**, 64—69 (1956). Zit. nach Lange 1961a. — Dohrn, M.: W. Faure, H. Poll u. W. Blotevogel: Tokokinine, Stoffe mit sexualhormonartiger Wirkung aus Pflanzenzellen. Med. Klin. **22**, 1417—1419 (1926). — Dubrovina, A. V.: Presowing carbon monoxide fumigation treatment of cucumber seeds. Fiziol. Rastenij **5**, 16—23 (1958) [Russisch mit engl. Zus.fass.]. Zit. nach Ber. wiss. Biol. **128**, 173. — Düsing, C.: Die Regulierung des Geschlechtsverhältnisses bei der Vermehrung der Menschen, Tiere und Pflanzen. Jenai. Z. Naturwiss. **17** (N. F. **10**), 593—940 (1884).

Eaton, F. M.: Selective gametocide opens way to hybrid cotton. Science **126**, 1174—1175 (1957). — Edel'štejn, V. I., i A. N. Paponov: Einfluß der Nährfläche und der Bodenfruchtbarkeit auf Entwicklung und Geschlechtsbildung einiger monöcischer Pflanzen. Izv. Timirjazev Sel'skochozjajstv. Akad. **1964**, No. 2, 138—143 [Russisch]. Zit. nach Bull. signal. **26**—17—907. — Edmond, J. B.: Seasonal variation in sex expression of certain cucumber varieties. Proc. Amer. Soc. hortic. Sci. **27**, 329—332 (1930)*. — Edwardson, J. R., and M. K. Corbett: Asexual transmission of cytoplasmic male sterility. Proc. nat. Acad. Sci. (Wash.) **47**, 390—396 (1961). — Engler, A., u. E. Gilg: Syllabus der Pflanzenfamilien, 8. Aufl. Berlin: Gebrüder Borntraeger 1919. — Erdelský, K., u. R. Herich: Beitrag zur chemischen Differenzierung des Geschlechts von Hanf (*Cannabis sativa* L.). II. Biológia (Bratislava) **11**, 111—115 (1956) [Tschechisch mit dtsch. Zus.fass.]. Zit. nach Ber. wiss. Biol. **113**, 43. — Erlenmeyer, H., u. M. Geiger-Huber: Notiz über die durch einen Brandpilz verursachte Geschlechtsumstimmung bei *Melandrium album*. Helv. chim. Acta **18**, 921—923 (1935).

FAROOQ, M.: An exceptional gynoeceum of *Citrus medica* var. *Limon* L. showing adherent pollen chambers and extraovarian ovules. Curr. Sci. **21**, 72—73 (1952). — FIGDOR, W.: Übergangsbildungen von Pollen- zu Fruchtblättern bei *Humulus japonicus* Sieb. et Zucc. und deren Ursachen. S.-B. Akad. Wiss. Wien, math.-nat. Kl. **120**, Abt. I, 689—707 (1911). — FINN, W.: Ein Fall von eigenartiger „Nacktsamigkeit" bei Angiospermen. Ber. dtsch. bot. Ges. **53**, 178—185 (1935). — FISCH, C.: Über die Zahlenverhältnisse der Geschlechter beim Hanf. Ber. dtsch. bot. Ges. **5**, 136—146 (1887). — FROLOVA, G. D.: Anomalien im Aufbau der Infloreszenz von *Betula rezniczenkoana* (Litw.) B. Schischk. Bot. Ž. **43**, 1311—1314 (1958) [Russisch].

GAERTNER, C. F.: Versuche und Beobachtungen über die Befruchtungsorgane der vollkommeneren Gewächse. Stuttgart 1844. Zit. nach CORRENS 1907. — GALUN, E.: Effect of seed treatment on sex expression in the cucumber. Experientia (Basel) **12**, 218—219 (1956). — Effects of gibberellic acid and naphthaleneacetic acid on sex expression and some morphological characters in the cucumber plant. Phyton (Vicente López) **13**, 1—8 (1959a). — The role of auxins in the sex expression of the cucumber. Physiol. Plantarum (Cph.) **12**, 48—61 (1959b). — Gibberellic acid as a tool for the estimation of the time interval between physiological and morphological bisexuality of cucumber floral buds. Phyton (Vicente López) **16**, 57—62 (1961a). — Study of the inheritance of sex expression in the cucumber. The interaction of major genes with modifying genetic and non-genetic factors. Genetica **32**, 134—163 (1961b). — GALUN, E., and D. ATSMON: The leaf-floral bud relationship of genetic sexual types in the cucumber plant. Bull. Res. Counc. Israel, Ser. D **9**, 43—50 (1960). — GALUN, E., Y. JUNG and A. LANG: Culture and sex modification of male cucumber buds in vitro. Nature (Lond.) **194**, 596—598 (1962). — Morphogenesis of floral buds of cucumber cultured in vitro. Developmental Biol. **6**, 370—387 (1963). — GARAY, A., u. F. SÁGI: Untersuchungen über die Geschlechtsumwandlung bei *Melandrium album* Mill. nach Infektion mit *Ustilago violacea* (Pers.) Fckl. unter besonderer Berücksichtigung der Auxinoxydase und Flavonoide. Phytopath. Z. **38**, 201—208 (1960). — GARNER, W. W., and H. A. ALLARD: Effect of the relative length of day and night and other factors of the environment on growth and reproduction in plants. J. agric. Res. **18**, 553—606 (1920). — GEDDES, P., and J. A. THOMPSON: The evolution of sex. London 1889*. — GEITLER, L.: Embryosäcke aus Pollenkörnern bei *Ornithogalum*. Ber. dtsch. bot. Ges. **59**, 419—423 (1941). — GIARD, A.: Sur la castration parasitaire du *Lichnis dioica* L. par l'*Ustilago antherarum* Fr. C. R. Acad. Sci. (Paris) **107**, 757—759 (1888). — Les variations de la sexualité chez les végétaux. C. R. Soc. Biol. (Paris) **10**, 730—731 (1898). — GIROU DE BUZAREINGUES, C.: Expériences sur la génération des plantes. Ann. des Sci. natur., 1ère Sér. **16**, 140—143 (1828)+. — Suite des expériences sur la génération des plantes. Ann. des Sci. natur., 1ère Sér. **24**, 138—148 (1831). Zit. nach SPRECHER 1913. — GOEBEL, K.: Beiträge zur Kenntniss gefüllter Blüthen. Jb. wiss. Bot. **17**, 207—296 (1886). — Über sexuellen Dimorphismus bei Pflanzen. Biol. Zbl. **30**, 657—679, 692—718, 721—737 (1910). — GRÄBNER, P.: Hydrocharitaceae. In: O. v. KIRCHNER, E. LOEW u. C. SCHRÖTER, Lebensgeschichte der Blütenpflanzen Mitteleuropas, Bd. I, Abt. 1, S. 665 bis 714. Stuttgart: E. Ulmer 1908. — GREGORY, F. G., and O. N. PURVIS: Abnormal flower development in barley involving sex reversal. Nature (Lond.) **160**, 221—222 (1947). — GRINVALD, C., E. MAVROMATI et I. GUTENMACHER: Influenţa lipsei de lumina in formarea organelor florale la orz. An. Inst. centr. Cerc. agric., Ser. C **30**, 157—167 (1962/63) [Rumänisch mit russ. u. franz. Zus.fass.]. Zit. nach Bull. signal. **25**—17—6605.

HABERLANDT, F.: Wiener landw. Ztg **1869**, No. 3. Zit. nach SPRECHER 1913. — Welche Einflüsse bedingen das Geschlecht der Hanfpflanzen? Fühlings landw. Ztg **1874**, 920. Zit. nach SPRECHER 1913. — Welche Einflüsse bedingen das Geschlecht der Hanfpflanzen? Fühlings landw. Ztg **1877**, 881+. — HAGERUP, O. K.: On the origin of some Angiosperms through the Gnetales and the Coniferae. III. The gynoeceum of *Salix cinerea*. Danske Vidensk. Selsk. Biol. Medd. **14**, 4—34 (1938). Zit. nach R. P. SINGH 1956. — HALL, W. C.: Effects of photoperiod and nitrogen supply on growth and reproduction in the gherkin. Plant Physiol. **24**, 753—769 (1949). — HALSTED, B. D.: Experiments with hemp. Rep. N. J. agric. Res. Stat. **21**, 455 (1901)*. — HARTMANN, M.: Die Sexualität. Das Wesen und die Grundgesetzlichkeiten des Geschlechts und der Geschlechtsbestimmung im Tier- und Pflanzenreich, 2. Aufl. Stuttgart: Gustav Fischer 1956. — HEILBRONN, A., u. N. YAKAR: Gynodiöcie und Intersexualität als Folge der Bastardierung zwittriger Species. Rev. Fac. Sci. Univ. Istanbul **7**, 317—322 (1942). Zit. nach SCHWERDTFEGER 1954. — HEITMÜLLER, H.-H., u. G. H. MELCHIOR: Über die blühfördernde Wirkung des Wurzelschnitts, des Zweigkrümmens und der Strangulation an japanischer Lärche *(Larix leptolepis)*. Silvae Genetica **9**, 65—72 (1960). Zit. nach MELCHIOR 1962. — HENSZ, R. A., and H. C. MOHR: Functional male sterility in watermelon induced by chemical treatment. Proc. Amer. Soc. hortic. Sci. **74**, 552—554 (1959). — HERICH, R.: Zum Problem der Differenzierung des Geschlechts von Pflanzen. Biológia (Bratislava) **11**, 220—224 (1956) [Tschechisch mit dtsch. Zus.fass.]. Zit. nach Ber. wiss. Biol. **113**, 43. — Gibberellin and sex differentiation of flowering plants. Nature (Lond.)

188, 599—600 (1960). — Herrmann, S.: Im Primärstadium blühende Kiefer. Naturwiss. 38, 381—382 (1951). — Heslop-Harrison, J.: Auxin and sexuality in *Cannabis sativa*. Physiol. Plantarum (Cph.) 9, 588—597 (1956). — The experimental modification of sex expression in flowering plants. Biol. Rev. 32, 38—90 (1957). — Growth substances and flower morphogenesis. J. Linn. Soc. London, Bot. 56, 269—281 (1959). — Suppressive effects of 2-thiouracil on differentiation and flowering in *Cannabis sativa*. Science 132, 1943—1944 (1960). — The experimental control of sexuality and inflorescence structure in *Zea Mays* L. Proc. Linn. Soc. London 172, 108—123 (1961). — Sex expression in flowering plants. In: Meristems and Differentiation. Brookhaven Symp. in Biol. 16, 109—125 (1963). — The control of flower differentiation and sex expression. Coll. intern. C.N.R.S. 123 (Régulateurs naturels de la croissance végétale), 649—664 (1964). — Heslop-Harrison, J., and Y. Heslop-Harrison: The effect of carbon monoxide on sexuality in *Mercurialis ambigua* L. fils. New Phytologist 56, 352—355 (1957a). — Studies on flowering-plant growth and organogenesis. I. Morphogenetic effects of 2,3,5-triiodobenzoic acid on *Cannabis sativa*. Proc. roy. Soc. Edinb. B 66, 409—423 (1957b). — Studies on flowering-plant growth and organogenesis. II. The modification of sex expression in *Cannabis sativa* by carbon monoxide. Proc. roy. Soc. Edinb. B 66, 424—434 (1957c). — Photoperiod, auxin and sex balance in a long-day plant. Nature (Lond.) 181, 100—102 (1958a). — Long-day and auxin induced male sterility in *Silene pendula* L. Port. Acta biol. A 5, 79—94 (1958b). — Heslop-Harrison, J. W.: Sex in the Salicaceae and its modification by eriophyid mites and other influences. Brit. J. exp. Biol. 1, 445—472 (1924). Zit. nach Bot. Zbl. 146, 412. — Hess, D.: Die selektive Blockierung eines an der Blühinduktion beteiligten Ribosenucleinsäure-Eiweißsystems durch 2-Thiouracil. (Untersuchungen an *Streptocarpus Wendlandii*.) Planta (Berl.) 54, 74—94 (1959). — Heyer, F.: Untersuchungen über das Verhältniss des Geschlechtes bei einhäusigen und zweihäusigen Pflanzen unter Berücksichtigung des Geschlechtsverhältnisses bei Thieren und Menschen. Ber. physiol. Labor. landw. Inst. Univ. Halle 5, 1—165 (1884)*. — Das Zahlenverhältniss der Geschlechter. Dtsch. landw. Presse 13, 163 (1886)+. — Higgins, J. E.: Growing melons on trees. J. Heredity 7, 208—220 (1916). Zit. nach Bessey 1933. — Higgins, J. E., and V. Holt: The papaya in Hawaii. Bull. Hawaii agric. Exp. Station 32, 25—28 (1914). Zit. nach Lange 1961a. — Hildebrand, F.: Die Lebensdauer und Vegetationsweise der Pflanzen, ihre Ursachen und ihre Entwickelung. Bot. Jb. 2, 51—135 (1881). — Über einige Fälle von Abweichungen in der Ausbildung der Geschlechter bei Pflanzen. Bot. Ztg 51, 27—35 (1893). — Hoffmann, H.: Zur Geschlechtsbestimmung. Bot. Ztg 29. 81, 97—109 (1871)+. — Über Sexualität. Bot. Ztg 43, 145—154, 161—169 (1885). — Holdsworth, M.: The production of female-sterile flowers by hermaphrodite plants of *Fuchsia procumbens*. Trans. roy. Soc. New Zealand 86, 105—111 (1959). — Hovin, A. W.: Bulk emasculation by high temperatures in annual blue-grass, *Poa annua* L. Abstr. Golden Anniv. Meeting Amer. Soc. Agron., Atlanta, Georgia, 1957. Zit. nach Meletti 1961. — Reduction of self-pollination by high night temperature in naturally self-fertilized *Poa annua* L. Agronomy J. 50, 369—371 (1958). — Howlett, F. S.: The effect of carbohydrate and of nitrogen deficiency upon microsporogenesis and the development of the male gametophyte in the tomato, *Lycopersicum esculentum* Mill. Ann. of Bot. 50, 767—803 (1936). — The modification of flower structure by environment in varieties of *Lycopersicum esculentum*. J. agric. Res. 58, 79—117 (1939). — Huhnke, W., Ch. Jordan, H. Neuer u. R. v. Sengbusch: Grundlagen für die Züchtung eines monözischen Hanfes. Z. Pflanzenzüchtg 29, 55—75 (1951). — Hulewicz, D.: Badania nad wpływem ezynników zewnęt. nych na modyfikacje płci konopi jednopiennych. Roczn. Nauk Roln., A 88, 193—210 (1964). Zit. nach Bull. signal. 25—17—6609. — Hylmö, B.: Einwirkung durch Hormonbehandlung auf das Geschlecht der *Spinacia*. Bot. Not. (Lund) 1940, 389—394*.

Iltis, H.: Über eine durch Maisbrand verursachte intracarpellare Prolifikation bei *Zea Mays* L. S.-B. Akad. Wiss. Wien, math.-nat. Kl., 119, Abt. I, 331—345 (1910). — Über einige bei *Zea Mays* L. beobachtete Atavismen, ihre Verursachung durch den Maisbrand, *Ustilago Maydis* D. C. (Corda), und über die Stellung der Gattung *Zea* im System. Z. indukt. Abstamm.- u. Vererb.-Lehre 5, 38—57 (1911). — Irmscher, E.: Über eine Abänderung des Zahlenverhältnisses zwischen männlichen und weiblichen Blüten bei der monoecischen *Begonia Wallichiana*. Mitt. Inst. allg. Bot. Hamburg 6, 149—158 (1924). — Ito, H., and T. Kato: Factors responsible for the sex expression of the Japanese cucumber. J. hortic. Ass. Japan 22, 138—144 (1953). Zit. nach Galun 1961b. — Ito, H., T. Kato. K. Hashimoto and T. Saito: Factors responsible for the sex expression of the Japanese cucumber. II. J. hortic. Ass. Japan 23, 65—70 (1954). Zit. nach Galun 1961b. — Ito, H., and T. Saito: Factors responsible for the sex expression of the Japanese cucumber. III. J. hortic. Ass. Japan 25, 101—110 (1956a). Zit. nach Galun 1961b. — Factors responsible for the sex expression of the Japanese cucumber. IV. The role of auxin on the plant growth and sex expression. J. hortic. Ass. Japan 25, 141—151 (1956b). Zit. nach Satyanarayana and Rangaswami 1959. — Factors responsible for the sex expression of the Japanese cucumber.

V. J. hortic. Ass. Japan **25**, 213—220 (1956c). Zit. nach GALUN 1961b. — Factors responsible for the sex expression of the Japanese cucumber. VI. J. hortic. Ass. Japan **26**, 1—8 (1957a). Zit. nach GALUN 1961b. — Factors responsible for the sex expression of the Japanese cucumber. VIII. J. hortic. Ass. Japan **26**, 209—214 (1957b). Zit. nach GALUN 1961b. — Factors responsible for the sex expression of the Japanese cucumber. IX. J. hortic. Ass. Japan **27**, 11—20 (1958). Zit. nach GALUN 1961b. — Factors responsible for the sex expression of the Japanese cucumber. XII. Tohoku J. agric. Res. **11**, 287—308 (1960). Zit. nach ATSMON, GALUN und JAKOB 1965.

JACQUES-FÉLIX, H.: Conditions de milieu et modifications morphologiques chez le *Coix lachryma-Jobii* L. C. R. Acad. Sci. (Paris) **249**, 1929—1931 (1959). — JANICK, J., and E. C. STEVENSON: Environmental influences on sex expression in monoecious lines of spinach. Proc. Amer. Soc. hortic. Sci. **65**, 416—422 (1955). — JOHRI, B. M., and B. TIAGI: Floral morphology and seed formation in *Cuscuta reflexa* Roxb. Phytomorphology **2**, 162—180 (1952). — JONES, K. L.: Studies on *Ambrosia*. II. Effect of certain environmental factors on floral development of *Ambrosia elatior*. Bot. Gaz. **98**, 296—306 (1936). — Studies on *Ambrosia*. IV. Effects of short photoperiod and temperature on sex expression. Amer. J. Bot. **34**, 371—377 (1947). — JOSHI, A. C.: A note on the determination of sex of plantain flowers. J. Genet. **38**, 353—356 (1939).

KARSCHON, R.: Photopériodicité et thermopériodicité chez le Pin sylvestre. J. forest. Suisse **102**, 268—275 (1951). Zit. nach Biol. Abstr. **25**, 821. — KIERMAYER, O.: Induktion von Eingeschlechtigkeit, sowie Gamophyllie an Blüten durch 2,3,5-Trijodbenzoesäure. Planta (Berl.) **52**, 393—404 (1958). — Induktion männlich-steriler Blüten bei *Helianthus annuus* durch 2,3,5-Trijodbenzoesäure (TIBA). Naturwiss. **46**, 457 (1959). — Induktion von Ringfasciation bei *Lycopersicon esculentum* durch Phthalsäuremono-α-naphthylamid (PNA). Naturwiss. **47**, 190—191 (1960). — Über die formative Wirksamkeit der 2,3,5-Trijodbenzoesäure (TIBA) im vegetativen und generativen Bereich von *Solanum nigrum*. Österr. bot. Z. **108**, 101—156 (1961). — KLEBS, G.: Über das Verhältnis des männlichen und weiblichen Geschlechts in der Natur. Jena: G. Fischer 1894. Zit. nach SPRECHER 1913. — Über Variationen der Blüten. Jb. wiss. Bot. **42**, 155—320 (1905). — KNIGHT, TH. A.: Upon the effects of very high temperature on some species of plants. Trans. hortic. Soc. London **3**, 459 (1819)+. — KNOTT, J. E.: Rapidity of response to change in photoperiod. Plant Physiol. **7**, 125—130 (1932). — KÖHLER, D.: Die Entwicklung von *Cannabis sativa* unter dem Einfluß verschiedener Tageslängen. Physiol. Plantarum (Cph.) **11**, 249—259 (1958). — Ein Beitrag zur Physiologie und Genetik der Geschlechtsausprägung von *Cannabis sativa*. Planta (Berl.) **56**, 150—173 (1961). — Geschlechtsbestimmung bei Blütenpflanzen. Erg. Biol. **27**, 98—115 (1964a). — Veränderung des Geschlechts von *Cannabis sativa* durch Gibberellinsäure. Ber. dtsch. bot. Ges. **77**, 275—278 (1964b). — KÖHLER, D., and A. LANG: Evidence for substances in higher plants interfering with response of dwarf peas to gibberellin. Plant Physiol. **38**, 555—560 (1963). — KOOT, Y. VAN, and H. J. M. VAN DORST: The influence of some virus diseases on flower formation in cucumber. T. Planteziekt. **64**, 432—439 (1958) [Holländisch mit engl. Zus.fass.]. Zit. nach Ber. wiss. Biol. **138**, 83. — KRIBBEN, F. J.: Die Entstehung von Blüten aus Kallusgewebe bei der Gurke (*Cucumis sativus* L.). Naturwiss. **50**, 101—102 (1963). — KUHN, E.: Untersuchungen zur Frage einer hormonalen oder zellulären Geschlechtsdifferenzierung bei Blütenpflanzen. (Pfropfungen bei zweihäusigen Arten.) Planta (Berl.) **32**, 286—342 (1941).

ŁĄCZYŃSKA-HULEWICZOWA, T.: Ein Beitrag zum Einfluß der Außenfaktoren auf das Geschlecht des di- und tetraploiden Hanfes. Acta Soc. Bot. Poloniae **26**, 515—549 (1957) [Polnisch mit dtsch. Zus.fass.] — LAIBACH, F., berichtet über die neuesten Ergebnisse der mit seinem Mitarbeiter F. J. KRIBBEN durchgeführten Versuche über die Geschlechtsbeeinflussung von Gurkenblüten durch Wuchsstoffe. Ber. dtsch. bot. Ges. **63** (32) (1950/51). — Die Gynomonözie von *Capsella bursa-pastoris* (L.) Medik. und ihre Ursache. Ber. dtsch. bot. Ges. **64**, 135—141 (1951). — Wuchsstoff und Blütenbildung. Beitr. Biol. Pflanzen **29**, 129—141 (1952). — LAIBACH, F., u. F. J. KRIBBEN: Der Einfluß von Wuchsstoff auf die Bildung männlicher und weiblicher Blüten bei einer monözischen Pflanze (*Cucumis sativus* L.). Ber. dtsch. bot. Ges. **62**, 53—55 (1949/50). — Über die Bedeutung der β-Indolylessigsäure für die Blütenbildung. Ber. dtsch. bot. Ges. **63**, 119—120 (1950a). — Der Einfluß von Wuchsstoff auf die Blütenbildung der Gurke. Naturwiss. **37**, 114—115 (1950b). — Der Einfluß von Wuchsstoff auf das Geschlecht der Blüten bei einer monözischen Pflanze. Beitr. Biol. Pflanzen **28**, 64—67 (1950c). — Die Bedeutung des Wuchsstoffs für die Bildung und Geschlechtsbestimmung der Blüten. Beitr. Biol. Pflanzen **28**, 131—144 (1951). — LAMBERT, J.: Biological flora of the British isles. *Glyceria maxima* (Hartm.) Holmb. J. Ecology **34**, 310—344 (1947). — LANG, A.: Auxins in flowering. Handbuch der Pflanzenphysiologie, Bd. 14, S. 909 bis 950. Berlin-Göttingen-Heidelberg: Springer 1961. — LANGE, A. H.: Factors affecting sex changes in the flowers of *Carica Papaya* L. Proc. Amer. Soc. hortic. Sci. **77**, 252—264 (1961a). — The effect of 2,3-dichloroisobutyrate and 2,2-dichloropropionate on the sex expression of *Carica Papaya* L. Proc. Amer. Soc. hortic. Sci. **78**, 218—224 (1961b). — LAURENT, É.: De l'influence de l'alimentation minérale sur la production des sexes chez les plantes dioïques. C. R. Acad. Sci. (Paris) **137**, 689—692 (1903). — LEPLAE, E.: Bull. Agric. Belg. **10** (partie non officielle), 236. Zit. nach GIARD 1898. — LEWITZKY, G. A.: On natural and voluntary changes in the flowers of *Veratrum nigrum* L. Bull. appl. Bot. Plant Breeding **14**, No. 2, 97—112 (1925)+. — LEYDECKER: Landw. Wochenbl. (Wien) **1870**, 209. Zit. nach SPRECHER 1913. — LILLO, M.: Un cambio curioso de sexualidad. Darwinia **1**, 163—164 (1924)+. — LIMBERK, J.: Versuche zur Änderung des Geschlechtes beim Hopfen *(Humulus lupulus)*. Československ. Biol. **2**, 306—314 (1953) [Russisch mit dtsch. Zus.fass.]. Zit. nach Ber. wiss. Biol. **90**, 334. — Änderung des Geschlechtes beim Hopfen *(Humulus lupulus)*. Českoslov. Biol. **3**, 252—255 (1954) [Russisch mit dtsch. Zus.fass.]. Zit. nach Ber. wiss. Biol. **97**, 76. — The influence of photoperiodicity on the sexual index in hemp (*Cannabis sativa* L.). Biol. Plantarum (Praha) **1**, 176—186 (1959). Zit. nach Ber. wiss. Biol. **144**, 182. — LOEHWING, W. F.: Physiological aspects of sex in angiosperms. Bot. Rev. **4**, 581—625 (1938). — LÖVE, Á., and D. LÖVE: Experimental sex reversal in plants. Svensk bot. Tidskr. **34**, 248—252 (1940). — Experiments on the effects of animal sex hormones on dioecious plants. Ark. Bot. **32**a, 1—60 (1945)*. — LOEWE, S., u. E. SPOHR: Nachweis und Gehaltsbestimmung des weiblichen Brunsthormons in weiblichen Organen des Pflanzenreiches. Anz. Akad. Wiss. Wien **63**, math.-naturw. Kl., 167—169 (1926). — LONGMAN, K. A., and P. F. WAREING: Early induction of flowering in birch seedlings. Nature (Lond.) **184**, 2037—2038 (1959). — LUDWIG, (F.): Die verschiedenen Blütenformen an Pflanzen der nämlichen Art. Biol. Zbl. **4**, 225—234 (1885).

MAEKAWA, T.: On the phenomena of sex transition in *Arisaema japonica* Bl. J. Coll. Agric. Hokkaido Imp. Univ. Sapporo **13**, 217—305 (1924)*. — On intersexualism in *Arisaema japonica* Bl. Jap. J. Bot. **3**, 205—216 (1927)*. — Widerstands- und Selbstregulierungsvermögen gegen Geschlechtsänderung bei Hanfpflanzen und seine Beziehung zur Theorie der Geschlechtsbestimmung. Jb. wiss. Bot. **70**, 512—564 (1929). — MAGNIN, A.: Sur l'hermaphrodisme du *Lichnis dioica* atteint d'*Ustilago*. C. R. Acad. Sci. (Paris) **107**, 663—665 (1888). — Recherches sur le polymorphisme floral, la sexualité et l'hermaphrodisme parasitaire du *Lichnis vespertina* Sbtp. Ann. Soc. bot. Lyon **16**, 203—231 (1889). — Observations sur le parasitisme et la castration chez les Anémones et les Euphorbes. Bull. sci. France et Belg. **23**, 412—436 (1891). Zit. nach FISCHER u. GÄUMANN **1929**. — Nouvelles observations sur la sexualité et la castration parasitaire. C. R. Acad. Sci. (Paris) **115**, 675—678 (1892). — MALHOTRA, R. C.: The sex ratio in *Asparagus officinalis* L. and its artificial modification. J. Genet. **23**, 157—172 (1930). — MALLIK, P. C., R. K. SAHAY and D. L. SINGH: Effect of hormones on the sex-ratio in mango. Curr. Sci. **28**, 410 (1959). Zit. nach Ber. wiss. Biol. **146**, 63. — MANN, L. K.: Effects of photoperiod on sex expression in *Ambrosia trifida*. Bot. Gaz. **103**, 780—787 (1942). — MARTENS, J. L.: Some observations on sexual dimorphism in *Carex picta*. Amer. J. Bot. **26**, 78—88 (1939). — MASTERS, M. T.: On polliniferous ovules in a rose *(Rosa arvensis)*. J. of Bot. **5**, 318 (1867). Zit. nach STRASBURGER 1910. — MATZKE, E. B.: Der Einfluß einiger Bedingungen, besonders der Buntblättrigkeit, auf die Zahl der Staubblätter bei *Stellaria media* (L.) Cyr. Planta (Berl.) **9**, 776—791 (1930). — Inflorescence patterns and sexual expression in *Begonia semperflorens*. Amer. J. Bot. **25**, 465—478 (1938). — McPHEE, H. C.: Influence of environment on sex in hemp. J. agric. Res. **28**, 1067—1080 (1924)+, *. — The genetics of sex in hemp. J. agric. Res. **31**, 935—943 (1925). — McVEY, G. R.: The modifying influences of gibberellin and temperature on the photoperiodic responses of certain woody plants with special reference to *Catalpa speciosa*. Ph. D. Thesis, Michigan State Univ., East Lansing, Michigan, 1961. Zit. nach WITTWER und BUKOVAC 1962b. — MECHANIK, F. JA.: Die Acetylenbehandlung als eine Methode zur Steigerung der Bildung fruchtansetzender weiblicher Blüten bei Gurken. Dokl. Vsesoju. Akad. Sel'sk. Nauk **21**, No. 11, 20—33 (1958) [Russisch]. Zit. nach LANGE 1961a. — MEEHAN, TH.: Notes on *Acer rubrum*. Proc. Acad. nat. Sci. Philad. **30**, 122—123 (1878a). — The law governing sex. Proc. Acad. nat. Sci. Philad. **30**, 267—268 (1878b). — Changes of flowers normally of one sex to the other. Proc. Acad. nat. Sci. Philad. **32**, 353—354 (1880). — MELCHIOR, G. H.: Ringelungsversuche zur Steigerung der Blühwilligkeit an japanischer Lärche *(Larix leptolepis)* und an europäischer Lärche *(Larix decidua)*. Silvae Genetica **9**, 105—111 (1960). Zit. nach MELCHIOR 1962. — Versuche zur Förderung der Blühwilligkeit an japanischen Lärchen-Pfropflingen *(Larix leptolepis)*. Silvae Genetica **10**, 20—27 (1961). — Weitere Untersuchungen zur Förderung der Blütenbildung an Kiefern durch Rückschnitt. Silvae Genetica **11**, 11—15 (1962). — MELCHIOR, G. H., u. H.-H. HEITMÜLLER: Erhöhung der Zahl der männlichen Blüten an *Pinus silvestris*-Pfropflingen durch Rückschnitt. Silvae Genetica **10**, 180—186 (1961). — MELETTI, P.: Induzione sperimentale della maschio-sterilità in *Triticum*. N. G. bot. ital., N. S. **68**, 299—307 (1961). — MIÈGE, J.: La staminodie chez le Manioc en Côte-d'Ivoire. Rev. Cytol. Biol. vég. **20**, 161—185 (1959). — MILLER, E. C.: Development of the pistillate spikelet and fertilization in *Zea Mays* L. J. agric. Res. **18**, 255—266 (1920). —

MININA, E. G.: On the phenotypical modification of sexual characters in higher plants under the influence of the conditions of nutrition and other external factors. Dokl. Akad. Nauk SSSR., N. S. **21**, 298—301 (1938). Zit. nach Ber. wiss. Biol. **50**, 645. — Bedeutung des Alters für die Geschlechtsbestimmung bei Pflanzen. Dokl. Akad. Nauk SSSR., N. S. **69**, 93—96 (1949) [Russisch]. — Modifikation des Geschlechtes bei Pflanzen unter dem Einfluß von Außenfaktoren. [Russisch] Moskau 1952. Zit. nach J. HESLOP-HARRISON 1961. — MININA, E. G., i S. V. KUŠNERENKO: Die Rolle der Blätter bei der Geschlechtsbestimmung der Pflanzen. Dokl. Akad. Nauk SSSR., N. S. **64**, 261—264 (1949) [Russisch]. — MININA, E. G., and L. G. TYLKINA: Physiological study of the effect of gases upon sex differentiation in plants. Dokl. Akad. Nauk SSSR., N. S. **55**, 165—168 (1947). — MITCHELL, W. D., and S. H. WITTWER: Regulation of flower sex expression in gynoecious and monoecious cucumbers with gibberellin and allyltrimethyl-ammonium-chloride. Abstr. No. 35, 55th ann. Meeting Amer. Soc. hortic. Sci., Oklahoma State Univ., Aug. 28—31, 1960. Zit. nach WITTWER and BUKOVAC 1962b. — Chemical regulation of flower sex expression and vegetative growth in *Cucumis sativus* L. Science **136**, 880—881 (1962). — MOFFETT, A. A.: Studies on the formation of multinuclear giant pollen grains in *Kniphofia*. J. Genet. **25**, 315—337 (1932). Zit. nach STOW 1933. — MOHL, H. v.: Beobachtungen über die Umwandlung von Antheren in Carpelle. Diss. 1836. In: Vermischte Schriften botanischen Inhalts, S. 28—44. Tübingen: Fues 1845. Auch französisch: Sur la métamorphose des anthères en carpelles. Ann. des Sci. natur., 2e Sér., Bot. **8**, 50—75 (1837). — MOL, W. E. DE: On the influence of circumstances of culture on the habitus and partial sterility of the pollen-grains of *Hyacinthus orientalis*. Proc. kon. ned. Akad. Wet. **23**, 1289—1302 (1921). Zit. nach NAITHANI 1937. — Duplication of generative nuclei by means of physiological stimuli and its significance. Genetica **5**, 225—272 (1923). — Die Entstehungsweise anormaler Pollenkörner bei Hyazinthen, Tulpen und Narzissen. Cytologia **5**, 31—65 (1933). — Näheres über das Vorfinden nebst dem experimentellen Hervorrufen mehrchromosomiger und embryosackartiger Pollenkörner bei diploiden und heteroploiden holländischen Hyazinthenvarietäten. Cytologia **5**, 204—229 (1934). — Teilungsdisharmonien durch niedrige Temperaturen, die Ursache von Modifikationen, Mutationen und anormalen Pollenkörnern. Gartenbauwiss. **11**, 431—456 (1938). — MOLLIARD, M.: Note sur les particularités que présentent les fleurs doubles du *Petunia hybrida*. Bull. Soc. bot. France **40**, 330—333 (1893). — Recherches sur les cécidies florales. Ann. des Sci. natur., 8e Sér., Bot. **1**, 67—245 (1895). — Sur la formation du pollen dans les ovules du *Petunia hybrida*. Rev. gén. Bot. **8**, 49—58 (1896). — Sur la détermination du sexe chez le Chanvre. C. R. Acad. Sci. (Paris) **125**, 792—794 (1897). — De l'hermaphrodisme chez la Mercuriale et le Chanvre. Rev. gén. Bot. **10**, 321—334 (1898a). — De l'influence de la température sur la détermination du sexe. C. R. Acad. Sci. (Paris) **127**, 669—671 (1898b). — MOORE, J. F.: Male sterility induced in tomato by sodium 2,3-dichloroisobutyrate. Science **129**, 1738—1740 (1959). — MOORE, R. H.: Several effects of maleic hydrazide on plants. Science **112**, 52—53 (1950). — MÜLLER, H.: Alpenblumen, ihre Befruchtung durch Insekten und ihre Anpassungen an dieselben. Leipzig 1881. Zit. nach CORRENS 1907. — MÜLLER, K.: Zur Kenntniss der Umwandlung männlicher Blüthen in weibliche beim Mais. Bot. Ztg **16**, 297—299 (1858). Zit. nach ILTIS 1911. — MUKERJI, S. K.: Contributions to the autecology of *Mercurialis perennis* L. J. Ecology **24**, 38—81 (1936). — MURNEEK, A. E.: Physiology of reproduction in horticultural plants. II. The physiological basis of intermittent sterility, with special reference to the spider flower. Res. Bull. Missouri agric. Exp. Stat. **106** (1927)*.

NAITHANI, S. P.: Chromosome studies in *Hyacinthus orientalis* L. III. Reversal of sexual state in the anthers of *Hyacinthus orientalis* L., var. yellow Hammer. Ann. of Bot., N. S. **1**, 369—377 (1937). — NAMIKAWA, I., M. SISA and K. ASAI: On the flower types of *Diospyros Kaki* L. f. Jap. J. Bot. **6**, 139—172 (1932). — NAUGOL'NYCH, V. N.: Veränderung der Geschlechtsmerkmale bei Gurken zwecks Ertragssteigerung. Bot. Ž. **40**, 715—719 (1955) [Russisch]. — NAYLOR, A. W.: Observations on the effects of maleic hydrazide on flowering of tobacco, maize and cocklebur. Proc. nat. Acad. Sci. (Wash.) **36**, 230—232 (1950). — NAYLOR, F. L.: Effect of length of induction period on floral development of *Xanthium pennsylvanicum*. Bot. Gaz. **103**, 146—154 (1941). Zit. nach Ber. wiss. Biol. **62**, 580. — NEIDLE, E. K.: Nitrogen nutrition in relation to photoperiodism in *Xanthium pennsylvanicum*. Bot. Gaz. **100**, 607—618 (1938). — NEKRASOV, V. I., L. S. VARTAZAROVA i N. A. BORODINA: The occurrence of androgynous inflorescences in *Betula japonica* Sieb. under the conditions of introduction. Bot. Ž. **48**, 436—440 (1963) [Russisch mit engl. Zus.fass.]. — NELSON, C. H.: Growth responses of hemp to differential soil and air temperatures. Plant Physiol. **19**, 294—309 (1944). — NELSON, P. M., and E. C. ROSSMAN: Chemical induction of male sterility in inbred maize by use of gibberellins. Science **127**, 1500—1501 (1958). — NĚMEC, B.: Über den Bau der petaloiden Antheren von *Hyacinthus orientalis*. Bull. intern. Acad. Sci. Bohème **5**, 17—23 (1898). Zit. nach NAITHANI (1937). — NEUER, H., u. R. v. SENGBUSCH: Die Geschlechtsvererbung bei Hanf und die Züchtung eines monöcischen Hanfes. Züchter **15**, 49—62 (1943). —

Nickerson, N. H.: Sustained treatment with gibberellic acid of five different kinds of maize. Ann. Missouri bot. Gard. **46**, 19—37 (1959). — Nickerson, N. H., and W. M. Hewitson: Morphological and anatomical differences in maize induced by treatment with growth-regulating substances. Amer. J. Bot. **50**, 622 (1963). — Nicolaisen, N., u. R. Hanow: Bestimmung der Geschlechtsverhältnisse bei Spinat. Z. Pflanzenzüchtg **23**, 476—484 (1941). — Nitsch, J. P., E. B. Kurtz, J. L. Liverman and F. W. Went: The development of sex expression in cucurbit flowers. Amer. J. Bot. **39**, 32—43 (1952). — Noack, R.: Die zwittrigen und eingeschlechtlichen Blüten von *Begonia cathayana*. II. Der Einfluß äußerer Faktoren auf die Zwitterblütenbildung und deren Fruchtbarkeit. Z. Bot. **50**, 22—33 (1962). — Noll, F.: Vorläufiger Abschluß der Versuche über die Bestimmung des Geschlechts bei dioecischen Pflanzen. S.-B. niederrh. Ges. Natur- u. Heilkde Bonn, naturwiss. Abt. **1907**, 68—91. — Nypels: Notes pathologiques. Bull. Soc. roy. bot. Belg. **36**, 274 (1897). Zit. nach Giard 1898.

Oehlkers, F.: Bastardierungsversuche in der Gattung *Streptocarpus* Lindl. V. Phänotypische und genotypische Geschlechtsbestimmung bei dem Bastard *Str. (Wendlandii* × *Rexii)*. Z. indukt. Abstamm.- u. Vererb.-Lehre **87**, 722—734 (1956).

Paixão-Correia, C. A. da: Sobre a formação de gomos florais adventícios em estacas de *Quercus Suber* L. tratadas com os ácidos I-naftaleno-acético e indol-3-acético. Port. Acta biol. A **4**, 249—263 (1955). — Pate, J. B., and E. N. Duncan: Evaluation of sodium 2.3-dichloroisobutyrate as a selective male gametocide in cotton. Agron. J. **52**, 506—508 (1960). — Petersen, O. G.: Halorrhagidaceae. In: A. Engler u. K. Prantl, Die natürlichen Pflanzenfamilien, Teil III, Abt. 7, S. 226—237. Leipzig: W. Engelmann 1898. — Peterson, C. E., and L. D. Anhder: Induction of staminate flowers on gynoecious cucumbers with gibberellin A_3. Science **131**, 1673—1674 (1960). — Petit, J.: Sur la détermination du sexe chez *Cannabis sativa* dans les conditions expérimentales du phytotron de Liège. Bull. Soc. roy. Sci. Liège **21**, 464—476 (1952). — Petrů, E., and R. Řetovský: The metamorphosis of a monoecious corn plant into a female plant during ontogenesis. Folia biol. **3**, 319—320 (1957). — Petrů, E., and Z. Šesták: Desinfection of kernels and its influence on the growth and photosynthesis of maize seedlings. Biol. Plantarum (Praha) **3**, 148—155 (1961). — Porsch, O.: Zur Lebensgeschichte der *Aesculus*-Blüte. Biologia generalis **12**, 591—618 (1937). — Prat, H.: Note sur les pousses courtes du Pin maritime. Bull. Stat. Biol. Arcachon **27**, 49—67 (1930). Zit. nach Courtot u. Baillaud 1955. — Sur la correspondance entre la structure des pousses de Pins et les cycles saisonniers. Vol. Jubil. Pr. Daniel, pp. 439—457. Rennes 1936. Zit. nach Courtot u. Baillaud 1955. — Preda, V., et O. Crăciun: Contributions à la différenciation sexuelle dirigée chez les plantes monoïques *(Zea mays)*. Stud. Cercet. Biol. (Cluj) **11**, 369—374 (1960) [Rumänisch mit franz. Zus.fass.]. Zit. nach Ber. wiss. Biol. **169**, 277. — Preda, V., et A. Ghisa: Détermination dirigée des sexes chez les plantes dioïques *(Urtica dioica)*. Comun. Acad. Republ. Popul. Romîne **7**, 663—668 (1957) [Rumänisch mit franz. Zus.fass.]. Zit. nach Ber. wiss. Biol. **127**, 168. — Pritchard, F. J.: Change of sex in hemp. J. Heredity **7**, 325—329 (1916). Zit. nach Bessey 1933.

Ram, M.: Occurrence of embryo-saclike structures in the microsporangia of *Leptomeria billardierii* R. Br. Nature (Lond.) **184**, 914—915 (1959). — Rao, V. S.: The occurrence of pollen grains in the ovary wall of *Dianthus*. Curr. Sci. **9**, 77—78 (1940). Zit. nach R. P. Singh 1956. — Rehm, S.: Male sterile plants by chemical treatment. Nature (Lond.) **170**, 38—39 (1952). — Reinöhl, F.: Die Variation im Androeceum der *Stellaria media* Cyr. Bot. Ztg **61**, 159—200 (1903). Zit. nach Matzke 1930. — Resende, F.: Suculentas Africanas. V. The influence of the intensity of illumination on the formation of adventitious roots in *Bryophyllum* and *Kalanchoë*. Port. Acta biol. **1**, 391 (1946)*. — Changing of the ⚥ flowers into ♀ flowers by the action of weak light *(Bryophyllum)*. Port. Acta biol. A **2**, 365—366 (1949). — Contribution to the physiology of development of the inflorescence and of the single flower *(Bryophyllum* and *Kalanchoë)*. Port. Acta biol. A, R. B. Goldschmidt Vol., 729—784 (1950). — As auxinas e o sexo. Bol. Soc. port. Ci. nat., Ser. II **3**, 267 (1951). Zit. nach Resende 1957b. — Auxina e sexo em *Hyoscyamus* L. Bol. Soc. port. Ci. nat., Ser. II **4**, 248—251 (1953). — Híbridos intergenéricos e interspecíficos em Kalanchoideae. II. Rev. Biol. (Lisboa) **1**, 157—168 (1957a). — Phenotypically and genotypically determined variation between monoecious and dioecious strains (flowering plants). Publ. U.I.S.B., Sér. B **34**, 25—45 (1957b). — Resende, F., and M. J. Viana: Some effects of synthetic auxins and antiauxins on the floral state. Bol. Soc. port. Ci. nat., Ser. II **4**, 74—88 (1952). — Gibberellin and sex expression. Port. Acta biol. A **6**, 77—98 (1959). — Richey, F. D., and G. F. Sprague: Some factors affecting the reversal of sex expression in the tassels of maize. Amer. Naturalist **66**, 433—443 (1932). — Rick, Ch. M., and G. C. Hanna: Determination of sex in *Asparagus officinalis*. Amer. J. Bot. **30**, 711—714 (1943). — Riede, W.: Die Abhängigkeit des Geschlechtes von den Außenbedingungen. Flora (Jena) **115**, 259—272 (1922). — Beiträge zum Geschlechts- und Anpassungsproblem. Flora (Jena) **118/119**, 421—452 (1925). — Roberts, R. H., and B. E. Struckmeyer: Further studies of the effects of temperature and other environmental factors upon the photoperiodic responses of plants. J. agric. Res. **59**, 699—709 (1939). —

Rosa, J. T.: Sex expression in spinach. Hilgardia **1**, 259—274 (1925)*. — Roze, E.: L'*Ustilago Caricis* Fuckel (*U. urceolorum* Tul.) aux environs de Paris. Bull. Soc. bot. France **35**, 277—278 (1888). — Rudenko, Ch. Ju.: Embryological investigation of change in sex in hemp. Ukrain. bot. Ž. **20**, 14—20 (1963) [Russisch mit engl. Zus.fass.]. Zit. nach Biol. Abstr. **44**, 7874. — Rudenko, F. E.: Änderung des Geschlechtes beim Hanf. Bot. Ž. **41**, 868—876 (1956) [Russisch].

Sabinin, D. A.: La nutrition minérale comme facteur de morphogenèse. Bull. Soc. Nat. Moscou **46**, 251—253 (1937)*. — Saccardo, P. A.: Sulle cause determinanti la sessualità nella Canape. Bull. Soc. Veneto-Trentina Sci. nat. Padova **1**, 16 (1879). Zit. nach Sprecher 1913. — Sachar, R. C.: The embryology of *Argemone mexicana* L. — A reinvestigation. Phytomorphology **5**, 200—218 (1955). — Saito, Y.: Artificial control of sex differentiation in Japanese red pine and black pine strobiles. J. Fak. Agric. Tottori Univ. **3**, 1—29 (1957). Zit. nach Melchior und Heitmüller 1961. — Salter, S. J.: On a sexual monstrosity, consisting in the development of polliniferous ovules. Trans. Linn. Soc. (Lond.) **24**, 143—150 (1863). Zit. nach Strasburger 1910. — Sandt, W.: Beiträge zur Kenntnis der Begoniaceen. Flora (Jena) **114**, 329—384 (1921). — Sansome, F. W.: Some remarks on the physiology of sexuality in plants. 2nd Intern. Congr. Sex Res. London 1930, p. 248—257. Zit. nach Kuhn 1941. — Satyanarayana, G., and G. Rangaswami: The effect of plant regulators on sex expression in ribbed gourd (*Luffa acutangula* Roxbg.). Curr. Sci. **28**, 376—377 (1959). — Savelli, R.: Manifestazioni di pedogenesi ed effetti di denutrizione in *Cucurbita Pepo*. Atti Accad. naz. Lincei, Ser. VIII **28**, 27—31 (1960). — Savelli, R., e N. Soster: Provocazione della monofillia in *Cannabis sativa* mediante traumi. Atti Accad. naz. Lincei, Ser. VI, Rc. Cl. Sci. fis. mat. nat. **10**, 604—609 (1929). — Šavrov, L. A.: Some peculiar features of intraspecies variability of plants induced to a polar-alpine botanical garden. Izv. Akad. Nauk SSSR., Ser. Biol. **26**, 582—590 (1961) [Russisch mit engl. Zus.fass.]. Zit. nach Ber. wiss. Biol. **170**, 285. — Schaffner, J. H.: Complete reversal of sex in hemp. Science **50**, 311—312 (1919). Zit. nach Schaffner 1927b. — Influence of environment on sexual expression in hemp. Bot. Gaz. **71**, 197—219 (1921). — Control of sexual state in *Arisaema triphyllum* and *Arisaema dracontium*. Amer. J. Bot. **9**, 72—78 (1922). — The influence of relative length of daylight on the reversal of sex in hemp. Ecology **4**, 323—334 (1923a). Zit. nach Jones 1947. — Sex reversal in the Japanese hop. Bull. Torrey bot. Club **50**, 73—79 (1923b)*. — Observations on the sexual state of various plants. Ohio J. Sci. **23**, 149—159 (1923c)*. — Experiments with various plants to produce change of sex in the individual. Bull. Torrey bot. Club **52**, 35—47 (1925a)*. — Sex determination and sex differentiation in the higher plants. Amer. Naturalist **59**, 115—127 (1925b)*. — The influence of the substratum on the percentage of sex reversal in winter-grown hemp. Ohio J. Sci. **25**, 172—176 (1925c)*. — Siamese twins of *Arisaema triphyllum* of opposite sex experimentally induced. Ohio J. Sci. **26**, 276—280 (1926)*. — Control of sex reversal in the tassel of Indian corn. Bot. Gaz. **84**, 440—449 (1927a). — Sex and sex-determination in the light of observations and experiments on dioecious plants. Amer. Naturalist **61**, 319—332 (1927b). — Further experiments in repeated rejuvenations in hemp and their bearing on the general problem of sex. Amer. J. Bot. **15**, 77—85 (1928). — Fluctuation of the point of sex reversal in *Sagittaria latifolia*. Amer. J. Bot. **16**, 191—195 (1929). — Sex reversal and the experimental production of neutral tassels in *Zea Mays*. Bot. Gaz. **90**, 279—298 (1930). — The fluctuation curve of sex reversal in staminate hemp plants induced by photoperiodicity. Amer. J. Bot. **18**, 424—430 (1931). — The production of vestigial and sterile sex organs through sex reversal and neutral sex states. Bull. Torrey bot. Club **60**, 89—97 (1933)*. — Observations and experiments on sex in plants. Bull. Torrey bot. Club **62**, 387—401 (1935)*. — Stability and instability of sexual conditions in *Morus alba*. J. Hered. **28**, 426—427 (1937). — Schlenker, G.: Beobachtungen über die Geschlechtsverhältnisse bei jungen Graupappeln und Aspen. Z. Forstgenetik **2**, 102—104 (1953). Zit. nach Ber. wiss. Biol. **87**, 83. — Scholz, E.: Blütenmorphologische und -biologische Untersuchungen bei *Acer pseudoplatanus* L. und *Acer platanoides* L. Züchter **30**, 11—16 (1960). — Schopfer, W. H.: Recherches sur l'hérédité de l'hermaphrodisme mâle chez *Melandrium*. Verh. schweiz. naturf. Ges. **121**, 159—160 (1940a). Zit. nach Ber. wiss. Biol. **58**, 37. — Recherches sur la phénologie de *Melandrium album* (Miller) Gke parasité par *Ustilago violacea* (Pers.) Fuck. C. R. Acad. Sci. (Paris) **210**, 703—705 (1940b). — Recherches sur les plantes mâles de *Melandrium album* (Miller) Garcke et *dioecum* (L.) Schinz et Thellung. L'action du parasite *Ustilago violacea*. C. R. Soc. Phys. Hist. nat., Genève (Suppl. aux Arch. Sci. physiques etc. 23) **58**, 112—116 (1941). Zit. nach Ber. wiss. Biol. **59**, 128. — Recherches sur les greffes chez *Melandrium* entre espèces hermaphrodite et dioïque. Boissiera **7**, 219 (1943)*. — Schopfer, W. H., et S. Blumer: Recherches sur les greffes sexuelles homo- et hétéroplastiques chez *Melandrium album* (Miller) Garcke et *M. dioecum* (L.) Schinz et Thellung. Verh. schweiz. naturforsch. Ges. **120**, 61—62 (1939a). Zit. nach Kuhn 1941. — Les hermaphrodites de *Melandrium album* (Miller) Garcke et *dioecum* (L.) Simonkai (*M. rubrum* Garcke). Ber. schweiz. bot. Ges. **49**, 414 (1939b). — Schulz, A.:

Beiträge zur Kenntnis der Bestäubungseinrichtungen und Geschlechtsverteilung bei den Pflanzen. I. Bibl. bot. **10** (1888). II. Bibl. bot. **17** (1890)+. — Beiträge zur Morphologie und Biologie der Blüthen. I. II. Ber. dtsch. bot. Ges. **10**, 303—313. 395—409 (1892). — Schuster, J.: Cycadaceae. In: Das Pflanzenreich. Regni vegetabilis conspectus. hersg. von A. Engler, fortgesetzt von L. Diels. Leipzig: W. Engelmann 1932. — Schwerdtfeger, G.: Abweichungen im Blütenbau bei *Digitalis lanata*. Ber. dtsch. bot. Ges. **67**, 248—252 (1954). — Scott, G. W.: Sex ratios and fruit production studies in bush pumpkin. Proc. Amer. Soc. hortic. Sci. **30**, 520—525 (1933). Zit. nach Lange 1961a. — Shifriss, O.: Sex instability in *Ricinus*. Genetics **41**, 265—280 (1956). — Gibberellin as sex regulator in *Ricinus communis*. Science **133**, 2061—2062 (1961). — Shifriss, O., and E. Galun: Sex expression in the cucumber. Proc. Amer. Soc. hortic. Sci. **67**, 479—486 (1956). — Singh, R. N., P. K. Majumder and D. K. Sharma: Seasonal variation in the sex expression of papaya (*Carica papaya* L.). Indian J. agric. Sci. **33**, 261—267 (1963). Zit. nach Bull. signal. **25**—17—6745. — Singh, R. P.: Occurrence of pollen sacs in the gynoeceum of *Trewia nudiflora* Linn. Sci. and Culture **22**, 165—166 (1956). — Singh, S. P.: Morphological studies in *Morus* spp. Agra Univ. J. Res. Sci. **3**, 233—341 (1954). Zit. nach R. P. Singh 1956. — Skarżyński, B.: An oestrogenic substance from plant material. Nature (Lond.) **131**, 766 (1933). — Smith, E. C.: Sex expression in willows. Bot. Gaz. **101**, 851—861 (1940). — Sprecher, A.: Recherches sur la variabilité des sexes chez *Cannabis sativa* L. et *Rumex Acetosa* L. Ann. des Sci. natur.. 9e Sér.. Bot. **17**. 255—352 (1913). — Storey, W. B.. W. W. Jones. G. K. Daris and F. G. Holdaway: Papaya production in the Hawaiian Islands. Bull. Hawaii agric. Exp. Station **87**, 5—22 (1941). Zit. nach Lange 1961a. — Stout, A. B.: Intersexes in *Plantago lanceolata*. Bot. Gaz. **68**. 109—133 (1919). — Alternation of the sexes and intermittent production of fruit in the spider flower *(Cleome spinosa)*. Amer. J. Bot. **10**, 57—66 (1923). — Stow, I.: On the embryosac-like pollen grain produced by change of the external conditions. Jap. J. Gen. **5**, 145—146 (1930a) [Japanisch]. Zit. nach Sutô 1955. — Experimental studies on the formation of the embryosac-like giant pollen grain in the anther of *Hyacinthus orientalis*. Cytologia **1**, 417—439 (1930b). — On the female tendencies of the embryosac-like giant pollen grain of *Hyacinthus orientalis*. Cytologia **5**, 88—108 (1933). — Strasburger, E.: Versuche mit diöcischen Pflanzen in Rücksicht auf Geschlechtsverteilung. Biol. Zbl. **20**, 657—665, 689—698, 721—731, 753—785 (1900). — Über geschlechtsbestimmende Ursachen. Jb. wiss. Bot. **48**, 427—520 (1910). — Sugiyama, S., and T. Sutô: Investigations on sex determination in *Spinacia oleracea* L. Bull. nat. Inst. Agric. Sci., Ser. D **11**, 211—329 (1964) [Japanisch mit engl. Zus.fass.]. Zit. nach Bull. signal. **26**—17—2009. — Suneson, C. A.: Emasculation of wheat by chilling. J. Amer. Soc. Agron. **29**, 247—249 (1937). — Sutô, T.: Cytogenetical studies of male sterility in maize. I. On a case of genic semi-sterility. Bull. nat. Inst. Agric. Sci. Tokyo, Ser. D **5**, 137—178 (1955). — Szymkiewicz, D.: Observations biométriques. IV. Corrélation entre les nombres de fleurs tubuleuses et ligulées dans les capitules des Composées à ligules unisériées. Acta Soc. Bot. Poloniae **11**, Suppl., 153—167 (1934).

Thomas, R. G.: Effects of temperature and length of day on the sex expression of monoecious and dioecious angiosperms. Nature (Lond.) **178**, 552—553 (1956). — Thompson, A. E.: Methods of producing first-generation hybrid seed in spinach. Mem. Cornell agric. Exp. Stat. **336** (1955)*. — Tibeau, M. E.: Time factor in utilization of mineral nutrients by hemp. Plant Physiol. **11**, 731—747 (1936). — Tiedjens, V. A.: Sex ratios in cucumber flowers as affected by different conditions of soil and light. J. agric. Res. **36**, 721—746 (1928). — Tournois, J.: Sur quelques anomalies florales de *Humulus japonicus*. Bull. Mus. Hist. natur. **16**, 331—332 (1909). Zit. nach Tournois 1911. — Anomalies florales du Houblon japonais et du Chanvre déterminées par des semis hâtifs. C. R. Acad. Sci. (Paris) **153**, 1017—1020 (1911). — Influence de la lumière sur la floraison du Houblon japonais et du Chanvre. C. R. Acad. Sci. (Paris) **155**, 297—300 (1912). — Études sur la sexualité du Houblon. Ann. des Sci. natur., 9e Sér., Bot. **19**, 49—191 (1914). — Treskow, M. v.: Über das Vorkommen mehrerer Sporne an den Blüthen von *Viola uliginosa*. Verh. bot. Ver. Prov. Brandenburg **37**, VI—X (1895)+. — Tulasne, L. R., et Ch. Tulasne: Mémoire sur les Ustilaginées comparées aux Urédinées. Ann. des Sci. natur., 3e Sér., Bot. **7**. 12—127 (1847).

Ubisch, G. v.: Beeinflussung der Heterostyliemerkmale von *Oxalis stricta* durch *Ustilago Oxalidis*. Proc. kon. ned. Akad. Wet., Sect. Sci., **38**, 93—100 (1935). — Uphof, J. C. Th.: Wissenschaftliche Beobachtungen und Versuche an Agrumen. IV. Der polygamische Zustand einiger *Citrus*arten. Gartenbauwiss. **7**, 121—141 (1932). — Ustinova, E. I.: Einige Eigentümlichkeiten der Formierung der Fortpflanzungsorgane in den Blütenständen des Maises. Bot. Ž. **41**, 864—867 (1956) [Russisch].

Vazart, B.: Biologie florale de l'Asperge, *Asparagus officinalis* L. Rev. gén. Bot. **66**, 405—418 (1959). — Venkatarayan, S. V.: Sex reversal (heterogamy) in the jack-fruit tree (*Artocarpus integrifolia* Linn.). Curr. Sci. **18**, 14—15 (1949). Zit. nach Ber. wiss. Biol. **69**, 362. — Vöchting, H.: Über Transplantation am Pflanzenkörper. Untersuchungen zur Physiologie und

Pathologie. Tübingen: H. Laupp 1892. Zit. nach KUHN 1941. — Über den Einfluß des Lichtes auf die Gestaltung und Anlage der Blüthen. Jb. wiss. Bot. **25**, 149—208 (1893). — VUILLEMIN, P.: Sur les effets du parasitisme de l'*Ustilago antherarum*. C. R. Acad. Sci. (Paris) **113**, 662—665 (1891).

WAREING, P. F.: Experimental induction of male cones in *Pinus sylvestris*. Nature (Lond.) **171**, 47 (1953). — WENT, F. W.: The experimental control of plant growth. New York: Ronald Press Co. 1957. — WERTH, E.: Weitere Infektionsversuche mit *Ustilago antherarum*. Mitt. kais. biol. Anst. Land- u. Forstw. **12**, 18 (1912). Zit. nach BAKER 1957. — WESTERGAARD, M.: Structural changes of the Y chromosome in the offspring of polyploid *Melandrium*. Hereditas (Lund) **32**, 60—64 (1946a). — Aberrant Y chromosomes and sex expression in *Melandrium album*. Hereditas (Lund) **32**, 419—443 (1946b). — The relation between chromosome constitution and sex in the offspring of triploid *Melandrium*. Hereditas (Lund) **34**, 257—279 (1948). — Über den Mechanismus der Geschlechtsbestimmung bei *Melandrium album*. Naturwiss. **40**, 253—260 (1953). — WESTON, E. W.: Changes in sex in the hop caused by plant growth substances. Nature (Lond.) **188**, 81—82 (1960). — WETTSTEIN, F. v.: Gesichertes und Problematisches zur Geschlechtsbestimmung. Ber. dtsch. bot. Ges. **54**, (23)—(38) (1936). — WHITAKER, T. W.: Sex ratio and sex expression in the cultivated cucurbits. Amer. J. Bot. **18**, 359—366 (1931). — WILLIS, I. C.: On gynodioecism (Third paper). With a preliminary note upon the origin of this and similar phenomena. Proc. Cambridge Philos. Soc. **8**, 129—133 (1893). Zit. nach CORRENS 1904. — WITSCH, H. v.: Die Entwicklung männlicher und weiblicher Infloreszenzen bei *Xanthium* unter dem Einfluß von Tageslänge und Gibberellin. Planta (Berl.) **57**, 357—369 (1961). — WITTROCK, V. B.: Über die Geschlechtsverteilung bei *Acer platanoides* L. und einigen anderen *Acer*-Arten. Bot. Cbl. **25**, 55—68 (1886). — WITTWER, S. H., and M. J. BUKOVAC: Gibberellin and higher plants. X. Field observations with certain vegetable crops. Quart. Bull. Michigan agric. Exp. Station **40**, 352—364 (1957). Zit. nach WITTWER und BUKOVAC 1962b. — The effect of gibberellin on economic crops. Econ. Bot. **12**, 213—255 (1958). Zit. nach ATSMON und GALUN 1960. — Staminate flower formation on gynoecious cucumbers as influenced by various gibberellins. Naturwiss. **49**, 305—306 (1962a). — Some gibberellin effects on flowering of plants. In: Eigenschaften und Wirkungen der Gibberelline, hersg. von R. KNAPP, S. 68—73. Berlin-Göttingen-Heidelberg: Springer 1962b. — WITTWER, S. H., and I. G. HILLYER: Chemical induction of male sterility in cucurbits. Science **120**, 893—894 (1954).

YAMPOLSKY, C.: The occurrence and inheritance of sex intergradation in plants. Amer. J. Bot. **7**, 21—38 (1920a). — Sex intergradation in the flowers of *Mercurialis annua*. Amer. J. Bot. **7**, 95—100 (1920b). — Further observations on sex in *Mercurialis annua*. Amer. Naturalist **54**, 280 (1920c)+. — Induced alteration of sex in the male plant of *Mercurialis annua*. Bull. Torrey bot. Club **57**, 51—58 (1930)*. — The cytology of the intersexual flowers of *Mercurialis annua*. A morphogenetic study. Amer. J. Bot. **21**, 651—672 (1934). — Further experiments with male and female grafts of *Mercurialis annua*. Bull. Torrey bot. Club **84**, 1—8 (1957). Zit. nach Ber. wiss. Biol. **115**, 281. — YAMPOLSKY, C., and H. YAMPOLSKY: Distribution of sex forms in the phanerogamic flora. Biblioth. genet. **3**, 1 (1922)*. — YOUNGNER, V. B.: Low temperature induced male sterility in male-fertile *Pennisetum clandestinum*. Science **133**, 577—578 (1961).

ZOSCHKE, U.: Untersuchungen über die Bestimmung des Geschlechts beim Spinat (*Spinacia oleracea* L.) unter besonderer Berücksichtigung der Züchtung eines monözischen oder gleichzeitig schossenden Spinates. Z. Pflanzenzüchtg **35**, 257—296 (1956).

Nachtrag.

Während der Drucklegung wurden dem Verfasser noch folgende Veröffentlichungen bekannt, die im Text nicht mehr berücksichtigt werden konnten:

GALUN, E., SH. IZHAR and D. ATSMON: Determination of relative auxin content in hermaphrodite and andromonoecious *Cucumis sativus* L. Plant Physiol. **40**, 321—326 (1965).

IYER, C. P. A., and G. S. RANDHAWA: Chemical induction of pollen sterility in grapes. Curr. Sci. **34**, 411—412 (1965). Zit. nach Bull. sign. **27**—17—1614.

JANA, M. K.: X-ray induced female sterility in *Phaseolus mungo*. Naturwissenschaften **52**, 646 (1965).

LECOCQ, M.: Variations morphologiques des fleurs doubles chez le *Begonia tuberhybrida* Voss. Bull. Soc. bot. Fr. **113**, 161—173 (1966).

NUNES, M. A.: Sex expression abnormalities in rice plants suffering from "straighthead". An attempt to hormonal interpretation. Port. Acta biol. A **8**, 285—300 (1964).

PHATAK, S. C., S. H. WITTWER, S. HONMA and M. J. BUKOVAC: Gibberellin-induced anther and pollen development in a stamenless tomato mutant. Nature (Lond.) **209**, 635—636 (1966).

ROBERTS, R. H.: Sex hormones in plants. Plant Physiol. **40**, XLV—XLVI (1965).

SCARAMELLA PETRI, P., e P. STRIGOLI: I fattori ecologici e la sessualità dei fiori del *Citrus trifoliata*. Frutticolt. **20**, 53—58 (1958).

The role of specific secretions in the induction and development of sexual organs and in the determination of sexual affinity.

By

John R. Raper.

With 1 figure.

The probable role of specific chemical agents in the initiation and regulation of sexual processes was recognized by SACHS as early as 1880. The first demonstration of the actual occurrence of such specific agents in plants, however, was given by BURGEFF, in 1924, for *Mucor mucedo*, a member of the common group of phycomycetous fungi, the "black bread molds". Since that time, comparable evidence for specific sexual secretions has appeared for various fungi and algae. The precise manifestations for chemical control, however, vary widely in different groups, thereby reflecting directly or indirectly the diversity of basic sexual processes in the many phylogenetic groupings. Furthermore, in every case for which adequate information is available, the specific activities demonstrably attributable to sexual secretions contribute to one or more of several distinct functions requisite to the achievement of sexual union: the determination of sexual sign, the induction of sexual organs, the spatial orientation of sexual organs, the differentiation of gametangia and/or gametes, and finally, the fusion of differentiated sexual elements. The present chapter deals with those determinative, inductive, and coordinative activities of specific sexual secretions that have been described as operating prior to the copulatory union of gametangia or gametes. The regulatory role of specific secretions in the copulatory process *per se* are treated in Chapter III D. Sexual systems dependent upon external regulation by means of diffusible substances are known only among the lower plants, the fungi and algae; the present account is accordingly restricted to a consideration of these groups of primitive plants.

Although it is very probable that the initiation and regulation of sexual reproduction in all plants are mediated by specific chemical agents, the forms in which such regulatory mechanisms have been demonstrated to date reflect the coincidence of three controlling features: (a) a genetic device for sexual or incompatibility differentiation that imposes a pattern of cross-mating between individuals, (b) an essentially aquatic environment that provides a continuous medium for the diffusion of substances of relatively small molecular size, and (c) a simple form lacking structural elaborations that tend to insulate the organism from the environment. Under these conditions, the interplay of metabolic products between individuals would have high selective advantage, whereas in more highly elaborated and in all truly terrestrial forms the requisite control must have developed along other lines. The evolution and survival of sexual coordinative mechanisms dependent upon external diffusible agents is thus restricted to those primitive forms which live in an actual or an essentially aquatic habitat. In higher plants, the initation and regulation of the early stages of the sexual process are strictly internal functions without direct coordination between the individuals concerned, although through responses to common external stimuli, *e.g.*, *via* photoperiod, their development may be beautifully synchronized.

Externally operating regulatory mechanisms, however, are not strictly limited to morphologically simple, aquatic, self-sterile forms. Similar — perhaps even identical — correlative mechanisms sometimes appear to operate throughout a closely related group in spite of the homothallic nature of a majority of the members, in which internally active agents could unquestionably achieve the necessary regulation more economically. Thus to those cases in which external regulation is indispensable may be added a number of cases in which common regulatory devices reflect close phylogenetic relationships.

The genetic devices that impose self-sterility and obligatory cross-mating permit in many cases of greater flexibility in sexual reactions than would be provided by a strict opposition of ♂ and ♀ potentialities. For example, in the biflagellated Phycomycetes, a series of sexual intergrades and an apparent lack of rigid species-specificity allow fertile matings to occur in a wide variety of combinations, and in many of the higher Ascomycetes, strict incompatibility differentiation is mitigated by reciprocally functional hermaphroditism. A high degree of plasticity in sexual reproductive processes has accordingly evolved among the members of many groups of fungi and algae, and secreted sexual substances are often precisely determining in the kind of sexual activity elicited and in the success of sexual exchange.

The functions of sexual determination, organ induction, and developmental coordination are not as distinct in most cases as the rather wide and inexact usage of the terms *termones* and *gamones* imply. These terms were introduced by KUHN, MOEWUS and JERCHEL (1938) to denote sex-determining and copulatory substances, respectively, secreted by differentiated gametes, and as applied to the cases in the algae where these activities have been demonstrated to reside in specific secretions of differentiated gametes, the use of the terms appear entirely appropriate. Among the fungi, however, demonstrated sexual secretions with only a single exception to date occur and elicit their responses during earlier stages of the sexual progression, and their rare activity as determinants of sexual affinities is shared with and incidental to their activities in organ-induction and developmental regulation. The exceptional case in fungi is *Allomyces*, in which *sirenin*, a substance secreted by the mature ♀ gametangia and gametes, chemotactically attracts ♂ gametes (MACHLIS 1958a, b); whether sirenin is implicated in the actual copulatory process, however, is uncertain. To designate the gametangial primordia-inducing secretions as termones (or even gamones: PLEMPEL 1957) and the gametangial orienting and differentiating secretions as gamones (HARTMANN 1943, 1955) tends both to discredit otherwise useful terms by confusing their meaning and, more importantly, to suggest a common means of regulation as applicable to all sexual processes, however diverse their mechanistic manifestations and their developmental histories. The present author accordingly urges the future restrictions of the terms gamones and termones to secretions of differentiated gametes as consistent with the original definitions, and to apply the more general term *sexual hormones*, and when necessary, newly defined and appropriate terms, to specific sexual secretions in other situations. The usage of the term sexual hormones has both historical and biological justification (RAPER 1952), and the introduction of new specific terms for gametangial-inducing, gametangial-orienting, and gametogenesis-inducing substances would be far preferable to the continued use of previous terms in inappropriate context.

The induction and coordinated development of sexual organs.

Of the several activities that have been attributed to sexual hormones in the lower plants, the *de novo* induction of sexual organs is certainly the most spectacular and the best documented. Furthermore, in each case in which such activity has been found, the secreted agents constitute the sole means of sexual activation; their role is accordingly one of critical significance in the forms in which they occur. Yet within the wide assortment of lower plants in which external initiating and regulating mechanisms might be expected to have evolved and survived, they have been found to date only in three phylogenetically diverse groups, all belonging to the fungi: the Saprolegniales and Mucorales of the Phycomycetes, and the Pezizales of the Ascomycetes. There are many forms among the algae, especially among fresh-water, filamentous species, in which diffusible substances regulating gametangial differentiation and gametogenesis would be expected on theoretical grounds, but to date only a single case has been reported in any detail (RAWITSCHER-KUNKEL and MACHLIS 1962). This system, however, lacks the features of initiation of the sexual process and the induction of sexual organs (see below). Sexually active secretions are better known among the algae as products of differentiated gametes and regulators of mating behavior between gametes.

The specific, organ inducing hormones and the correlative mechanisms of which they are essential parts have somewhat greater biological significance than the detailed elucidation of particular developmental histories: they provide the most clear-cut demonstration, in plants or animals, of the induction of normally occurring organs through the agency of self-secreted specific chemical entities, albeit of unknown identities as yet.

From an historical point of view, the Mucorales have been of particular importance in the development of present concepts of sexual reproduction in lower plants. Both obligatory cross-mating (BLAKESLEE 1904) and diffusible sexual secretions (BURGEFF 1924) were first demonstrated with members of the order, and various of its species remain as often favored material for basic research.

BURGEFF's classical experiment (1924) demonstrating the activity of diffusible secretions in the initiation of the sexual reaction, the induction of sexual organs, and their coordinated development was a logical outgrowth of acute observations of mating behavior between (+) and (—) strains of *Mucor mucedo*. In matings of two compatible plants, marked regions of inhibited growth, "restraint areas", occurred at the line of contact between the two plants, and into this region only a few hyphae penetrated. These occasional intruding hyphae became somewhat enlarged at their tips to form a pregametangial stage, the "corolloid swelling". Since this differentiation from the vegetative state preceded actual contact between elements of the two plants, the activity of stimulatory diffusible secretions was strongly suspected. A simple experimental device provided the first proof of sexual hormones in plants: a permeable colloidin membrane was laid upon the surface of a (—) mycelium growing upon agar, and upon the membrane was placed a block of agar containing (+) mycelium. This procedure established an impenetrable physical barrier between the two mates yet maintained liquid continuity over the short distance between them. As the (+) mycelium became established upon the surface of the membrane, the typical initial stage of sexual activity was elicited upon the adjacent portions of the (—) mycelium across the membrane barrier. This activation at a distance due to diffusible substances was termed "telomorphosis". BURGEFF was also able to demonstrate mutually attractive growth responses in sexually differentiated hyphae, "zygophores" of the two sexual strains on opposite sides of the membrane, a response which he termed "zygotropism".

BURGEFF spoke of probable telomorphic induction of sexual organ primordia in other heterothallic species of the Mucoraceae, including *Absidia glauca*, *Rhizopus nigricans*, and *Mucor hiemalis*, although he cites no specific experimentation to confirm these views except for the case of *Phycomyces nitens*, in which species telomorphic effects were not found. He also stated that zygophores of *Mucor hiemalis* were telomorphically induced through the air, but again he gave no experimental confirmation. This latter effect had earlier been suspected in *Rhizopus nigricans* by BLAKESLEE (1904), who, unable to confirm experimentally the suspected phenomenon, concluded that activation occurred only upon direct physical contact between the two compatible plants.

During the decade following the publication of BURGEFF'S work, a number of other workers, VERKAIK (1930), KÖHLER (1935), RONSDORF (1931), KEHL (1937), and KRAFCZYK (1931, 1935), successfully repeated his major findings and extended the demonstration of the occurrence and activity of sexual substances to a number of other species of the Mucoraceae. The work of KRAFCZYK (1931, 1935) on *Pilobolus crystallinus* is of particular interest, for in this species he was able to provide unequivocal evidence for the response of both (+) and (—) strains to secretions of their opposite numbers, and he gave convincing evidence, though not rigorous proof, that the sexual reaction was under hormonal control throughout early stages until the differentiation of the gametangia.

More recently, BANBURY (1954, 1955) and PLEMPEL (1957) have considerably enlarged the understanding of the role of diffusible hormones in the sexual reaction of *Mucor mucedo*. BANBURY (1954) was unable to demonstrate initial sexual activation, the induction of zygophores, on the mycelium of either strain with a filtrate of the other, but he was able to elicit a definite response in the (+) when treated with a) a filtrate of a mixed culture of (+) and (—) mycelia or b) a filtrate successively passed through (+) and (—) mycelia. BANBURY'S work strongly indicates the necessity of two sequential steps in the interaction prior to any morphological expression: a) the secretion by the (+) mycelium of a substance required for activation of the (—); and b) the secretion by activated (—) mycelium of a substance required to induce zygophores on the (+). This pattern of initiating interaction has been fully confirmed and more elaborately worked out by PLEMPEL (1957), who was able to demonstrate an identical but reciprocal pattern of zygophore induction in the (—) strain, *i.e.*, (—) → (+) → (—) zygophores. PLEMPEL concludes that a distinct substance ("progamone") and an organ-inducing substance ("gamone") must be sequentially secreted by each strain, (+) and (—).

Both BANBURY and PLEMPEL have also attempted to analyse the mutual attraction between (+) and (—) zygophores, the "zygotropism" of BURGEFF, but with rather divergent results. BANBURY (1955) described a mutual attraction *in air* of (+) and (—) zygophores through a distance of 2 mm. and a mutual repulsion *in air* of zygophores of the same sign, either (+) or (—), through a distance of 1 mm. Opaque and transparent physical barriers interposed between zygophores prevented both attraction and repulsion, but when a barrier having a single tiny pore was interposed, zygophores of opposite sign on either side of the barrier grew toward the hole. A gentle current of warm, moist air blown over the zygophores also prevented interaction. BANBURY concluded a) that volatile substances, at least one from the zygophores of each strain, were necessary to account for the observed responses and b) that if only a single volatile substance is emitted by each, stimulation of growth of zygophores of the same sign and inhibition of growth of zygophores of opposite sign would be consistent with cross-attraction and autorepulsion.

In his first paper on the subject, PLEMPEL (1957) was unable to find any evidence of air-borne zygotropic agents and attributed the mutual attraction of zygophores of opposite signs to the gradients of the nonvolatile zygophore-inducing agents present in the substrate. Later studies by PLEMPEL (1960a) and PLEMPEL and DAWID (1961), however, showed the functions of zygophore-induction and zygophore-attraction to be due to distinct agents and clearly demonstrated the presence and activity of gaseous zygotropic substances. The results of this more recent work essentially confirmed and extended the earlier observations of BANBURY; the interpretation of the activity of the zygotropic agents, however, was somewhat different. PLEMPEL and DAWID concluded that the zygotropic agent of either strain stimulated the growth of zygophores of the opposite strain and inhibited the growth of zygophores of the same strain. The mutual tropic growth of compatible zygophores was considered to be regulated by the concentration gradients of the two agents, which were postulated to be highly unstable in air in order to maintain effective gradients.

A minimum of six specific secretions thus account for initiation and regulation of the sexual process in *Mucor mucedo* prior to the time of gametangial differentiation. Activating agent ("progamone"), zygophore-inducing agent ("gamone"), and zygotropic agent are secreted in turn by either strain, each agent eliciting a response only in the mate of the opposite strain. In either strain, each stage in the process is directly dependent upon the last preceding stage in the opposite strain; the hormonal system thus imposes simultaneous and reciprocal development in the two mates.

Concurrently with the work on the biological aspects of the hormonal system in *M. mucedo*, PLEMPEL (1960b, 1963a, 1963b) and PLEMPEL and BRAUNITZER (1958) carried out extensive studies in the chemistry of the zygophore-inducing agents. The zygophore-inducing substances of (+) and (—) strains are apparently very similar structurally, as no chromatographic separation has so far been possible. Filtrates of mixed cultures of (+) and (—) strains, which served as the source material for their work, however, contain far more of the (+) substance, and the identification of this agent has been their primary concern. A detailed account of isolation procedures, etc., are given by PLEMPEL (1963b). Extraction of 250 liters of culture fluid by organic solvents and enrichment by column- chromatography provided an active fraction, which, when treated with benzyol chloride, yielded 67 mg of benzoic acid ester. Hydrolysis of the ester with 0.5 N KOH yielded 29.9 mg of a clear, golden oil having high zygophore-inducing activity. The molecular weight of the isolated "gamone" was determined by diffusion rate as 314 in acetone and 374 in chloroform, and analysis showed the compound to contain carbon (67.72%), hydrogen (7.41%), and oxygen (24.57%). The molecular weight and elemental proportions are consistent with an empirical formula of $C_{20}H_{25}O_5$.

Although the structural formula of the active agent has yet to be determined, its chemical and physical properties are well known, and progress toward its specific identification is comparable to that for *sirenen* of *Allomyces* (MACHLIS *et al.* 1966) and hormone A of *Achlya* (BARKSDALE, pers. communication).

The Saprolegniales, of the aquatic, biflagellate series of the Phycomycetes, have also for a long time intrigued investigators of sexual processes in the lower plants. DE BARY, in 1881, recognized the sequential and spatial relationships of the several stages in the development and differentiation of the ♂ and ♀ sexual organs in a homothallic species of *Saprolegnia*. He further postulated specific secretions of the developing oogonium as the agents responsible for the induction of antheridia and for their directed growth to the oogonium. Later students of the group largely accepted the rationality of this hypothesis, and KAUFFMAN (1908) interpreted the role of phosphates and nitrates in promoting the production of antheridia in *Saprolegnia hypogena*, commonly a parthenogenetic species, as affecting the elaboration of antheridial inducing secretions of the oogonium. Shortly after the publication of BURGEFF's work on *Mucor*, COUCH (1926) presented a very concise account of the sexual interaction between ♂ and ♀ plants of *Dictyuchus monosporus*, the first described heterothallic species of the Saprolegniales. As a result of his analysis of the sexual interaction, COUCH strongly suspected the initiative and regulatory activities of specific secretions, and he performed a number of carefully designed experiments in an attempt to demonstrate their presence and specific activities. That he was unable to do so can only be partially explained in terms of more recent findings in related forms.

The first unequivocal evidence of sexual regulations by diffusible agents in this group of plants was given by BISHOP (1937, 1940) for *Sapromyces reinschii*, a heterothallic species of the Leptomitaceae. In agar matings of ♂ and ♀ plants, antheridial hyphae were regularly produced prior to the actual contact of the mated mycelia; in rare cases, a few oogonial initials were also produced before any hyphal contact, whereupon the growth of antheridial hyphae became oriented over considerable distances toward the newly formed ♀ organs. Proof of diffusible agents was provided by further experimentation: when treated with cell-free filtrates of ♀ plants, grown sterilely in water, numerous antheridial hyphae were regularly induced on ♂ plants, grown upon agar. Filtrates from the ♂ had no comparable effect upon the ♀. Production of antheridial hyphae was thus established as the initial stage in the sexual reaction in the Saprolegniales, a role attributed to oogonial initials since the time of DE BARY. BISHOP's observations and experimental findings have been repeatedly confirmed in subsequent work on related forms.

A hormonal coordinating mechanism has been worked out in considerably greater detail in two heterothallic species of *Achlya*, a genus of the Saprolegniaceae by RAPER (1952 et prior.). In *A. bisexualis* and *A. ambisexualis*, several lines of evidence strongly indicated a highly effective mechanism for the coordination of the sexual reaction, and diffusible chemical substances were early recognized as the most probable agents of the coordination: a) The stages of the sexual reaction invariably follow a rigid sequence as follows:

1. Production of antheridial hyphae.
2. Production of oogonial initials.
3. Attraction of antheridial hyphae to oogonial initials.
4. Delimitation of antheridium at end of antheridial hypha and in contact with oogonial wall.
5. Delimitation of oogonium by septation at base of oogonial initial.
6. Reorganization of oogonial contents to form ♀ gametes (oospheres, or "eggs").
7. Growth of fertilization tubes from antheridium to each gamete.
8. Fertilization.
9. Maturation of zygotes (oospores).

The time course of this progression can be considerably altered by experimental means; the order, never. b) Each stage in the reaction subsequent to the formation of antheridial hyphae is closely correlated, both in the time of appearance and in the intensity of expression, with the last preceding stage. c) By altering the composition of the medium, the sexual reaction can be disrupted at any stage, and all subsequent stages fail to materialize. d) The pattern of sexual interactions in interspecific and intergeneric matings (see below) range from no response through all degrees of partial compatibility to complete compatibility, thus indicating in homothallic as well as heterothallic species a common basic coordinating mechanism in which occur minor specific differences, probably chemical in nature (RAPER 1939, 1950b, BARKSDALE 1960, 1963, 1966b).

Both conclusive evidence for the indicated hormonal coordinating mechanism and considerable information concerning the details of its operation have been obtained by experimental procedures that impose spatial separation between parts of the reacting system. Telomorphic reactions can be induced at several stages of the reaction in a variety of ways. a) In matings of strong ♂ and ♀ plants on an agar medium, induction of antheridial hyphae, induction of oogonial initials, and attraction of antheridial hyphae can commonly be observed prior to any contact between the two mates. b) By interposing a permeable Cellophane membrane between the reacting plants, all of the stages listed in a) plus an additional stage, the delimitation of antheridia appressed to the membrane, have routinely been seen, and in *A. bisexualis*, the delimitation of small oogonia and their reorganization to form eggs have been observed on rare occasions. It is thus clear that the entire sexual reaction up to the ingrowth of fertilization tubes into the oogonium is under the control of diffusible substances. c) The use of filtrates and filtrate fractions, however, has permitted a partial dissection of the reaction, has provided the greater part of the information relating to the number of hormones involved, the sites and/or circumstances of their secretion, and has made it possible to quantify to a considerable degree the initial stage of the reaction, the production of antheridial hyphae (RAPER 1940a, 1942a, b, 1950a).

Early in the work of *Achlya*, an adaptation of the use of filtrates was employed to determine the order and origin of the major types of hormones. A slowly flowing stream of water was successively passed via capillaries through four microaquaria, in which ♂ and ♀ plants were places in two different sequences. The reactions observed in the two cases were as follows:

Direction of flow 〉	------------	------------	------------	------------→
Microaquaria	1	2	3	4
Series 1	♂	♀	♂	♀
Reactions	None	None	Antheridial initials	Oogonial initials
Series 2	♀	♂	♀	♂
Reactions	None	Antheridial initials	Oogonial initials	Antheridial initials and their attraction to capillary

These results clearly showed that a) the initiation of the sexual reaction depends upon secretion(s) of the ♀ plant which induce antheridial hyphae on the ♂, b) oogonial initial production depends upon secretion(s) of the ♂ plant, but only after the induction of antheridial hyphae in response to secretion(s) of the vegetative ♀ plant. A third conclusion drawn from these results now appears to be in doubt (see below): antheridial hyphal attraction depends upon secretion(s) of the ♀, but only after the induction of oogonial initials in response to secretion(s) of the sexually activated ♂, *etc.* (RAPER 1940a).

From these and other lines of evidence, it has been possible to formulate the minimal hormonal mechanism that coordinates the sexual reaction in *Achlya*. The following summary account[1] and the accompanying diagram[2] (Fig. 1) are taken from previous accounts published by the author (RAPER 1951, 1955).

[1] Quoted from American Scientist. By permission.

[2] Reprinted from Biological Specificity and Growth, Princeton University Press. By permission.

"The initial stage of the sexual reaction, the production of antheridial hyphae on the ♂ plant, is induced and quantitatively regulated by four distinct hormones, collectively designated the A-complex. The A-complex comprises hormones A and A^2, secreted by the vegetative ♀ plant, and A^1 and A^3 secreted by the vegetative ♂.

The production of antheridial hyphae on the ♂ constitutes a sensitive and reliable biological assay for the A hormones. When four-day-old ♂ plants, growing upon halved hemp seed, are placed in water containing the appropriate hormones in the proper concentrations,

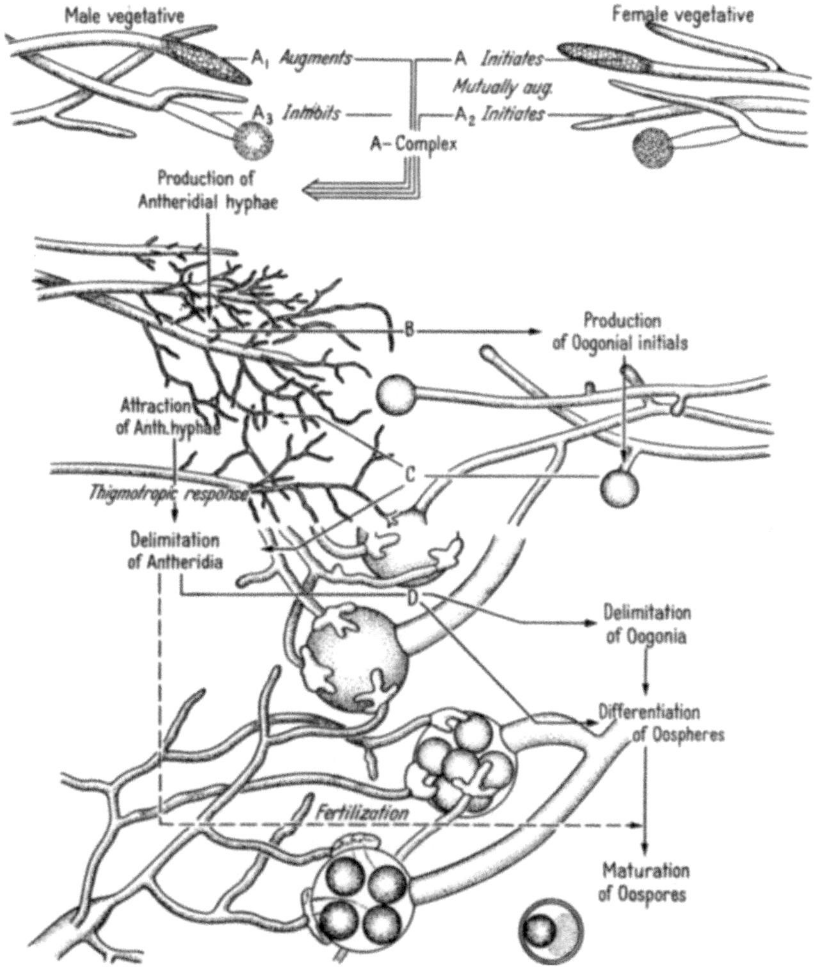

Fig. 1. A semidiagrammatic representation of the sexual progression in heterothallic species of *Achlya* relating the sequence of morphological developments to the origins and specific activities of the several sexual hormones. (From RAPER 1955.)

a discernible response occurs within an hour. On the vegetative hyphae there appear, *de nova*, numerous small lateral bumps, each of which, by elongating rapidly, develops into a characteristic antheridial hypha. The antheridial hyphae, three to four hours after the addition of the hormone(s) commonly attain a length of 0.2—0.5 millimeter. The average number of antheridial hyphae per vegetative hypha (3-millimeter treminal segment) has been adopted as the index of reaction intensity. The average number of antheridial hyphae formed on each vegetative hypha can be varied at will from 0 to about 75 by alteration of hormonal content.

Hormones A and A^3 are acetone-soluble and water-insoluble, whereas hormones A^1 and A^2 are acetone-insoluble and water-soluble. The two hormones secreted by plants of each sex are thereby readily separable by chemical means and may be tested separately or in any desired combination.

The qualitative activities of the hormones of the A-complex are as follows: hormones A and A^2, secreted by the ♀, singly or together, induce the formation of antheridial hyphae; hormone A^1, secreted by the ♂, cannot induce the reaction but significantly augments the activities of hormones A and A^2; hormone A^3, also secreted by the ♂, acts to depress the response induced by the other hormones.

The pattern of quantitative control of antheridial hyphal production by the four hormones of the A-complex is very complicated. Each of the four hormones affects the response in a mathematically characteristic manner and their effects in combination are not always predictable from their separate activities. These relationships have been determined in considerable detail.

The antheridial hyphae secrete, during the period of their formation and continuing growth, a substance, hormone B, which specifically induces the formation of oogonial initials on the ♀. Although the quantitative control of oogonial initial production has not yet been worked out, it is quite apparent that such control is operative: the relative abundance of oogonial initials on the ♀ varies directly with the intensity of the preceding reaction in the ♂.

The oogonial initials secrete hormone C during the period of their rapid growth and for some hours after maximal size is attained. Hormone C induces two specific activities in the antheridial hyphae of the sexually activated ♂: it causes the positive tropic growth of antheridial hyphae along its concentration gradient to the source of secretion, the oogonial initial, and it evokes the formation and delimitation of the swollen, branched antheridium when an antheridial hypha comes into contact with the oogonial initial.

The antheridia, following their delimitation, secrete yet another sexual substance, hormone D, which is required for the final development of the oogonium. Hormone D induces two specific reactions: the delimitation of the oogonium and the reorganization of the protoplasmic contents of the oogonium to form a number of uninucleate oospheres, or eggs.

The formation and directional growth of fertilization tubes from the antheridia to the ripened eggs may also be hormonally initiated and controlled, but as yet no means of experimentation has been devised to test this possibility.

The dual roles of hormones C and D suggest the likelihood that each of these secretions comprises a hormonal complex consisting of two or more specific substances.

The initiation and correlation of the sexual processes in heterothallic species of *Achlya* are thus seen to depend upon a minimum of seven distinct and specific hormones, four secreted by the ♂ and three secreted by the ♀. By means of these chemical 'messengers' each stage in the sexual progression is directly dependent upon the immediately preceding stage. Thus the consummation of the entire sexual interaction, the formation of (fertilized) zygotes, is prevented by a failure of any one of the several component stages. By a proper choice of experimental procedure, the sexual reaction of these plants can be disrupted at any desired stage.

The chemical identities of the various hormones are unknown."

The results of more recent work by BARKSDALE (1960, 1963, 1966b) make necessary some revision of the description of the hormonal system in *Achlya* as delineated in 1951 and considerably extends available information about the nature and activities of the initiating hormone of the sexual progression, hormone A. BARKSDALE has shown that hormone A is removed from the culture fluid by mycelia of various types (♂, intergrade, and ♀ strains of heterothallic species and mycelia of homothallic species) in direct correlation to their ability to produce antheridial hyphae in response to the hormone. Highly reactive ♂'s or homothallic strains thus rapidly take up the hormone, whereas ♀'s and non-reactive homothallic strains remove it very slowly if at all. This study was prompted by the observation that mixed cultures of ♂ and ♀ strains accumulated no hormone A. Withdrawal of the hormone from the medium appears to be no simple absorptive phenomenon, as the hormonal activity cannot be recovered by extraction of the ♂ mycelium with acetone, in which the hormone is readily soluble; inactivation or incorporation into a nonsoluble complex is indicated.

The functions earlier attributed to hormone C, chemotropic attraction of antheridial hyphae and the delimitation and maturation of their tips as antheridia, have been shown to be elicited by hormone A alone. Hormone A absorbed on polystyrene or polyvinyl plastic particles served both to induce antheridial hyphae and to direct their growth to the "mock oogonia", and delimitation of antheridia is regularly induced in the absence of oogonial initials by highly enriched hormone A alone. Comparable attraction of antheridial hyphae to vegetative ♀ hyphae was earlier described by RAPER (1939) for a cross between what was considered to be strains of different species, although in this case no antheridial delimitation was observed. Concentration effects of hormone A could have been decisive here, as BARKSDALE reports the concentration of hormone A required for antheridial delimitation in the ♂ of *A. ambisexualis* to be about a hundred times that required for the induction of antheridial hyphae. In view of the likelihood that these organisms are diploid (MULLINS and RAPER 1965, BARKSDALE 1966a) and that meiosis occurs in the antheridia and oogonia (SANSOME 1961, 1963, 1965, BARKSDALE, unpublished), BARKSDALE suggests that hormone A, in its role in gametangial differentiation and maturation, may serve as an inducer of meiosis.

Although BARKSDALE's demonstrations of these activities of hormone A previously attributed to hormone C are quite definitive, and hormone C has been declared to be superfluous (MACHLIS 1966), this summary dismissal may be premature. More detailed information is needed about (a) the kinetics of the production of hormone A by activated vs vegetative ♀ plants and (b) the characteristics of the hormonal gradient required to elicit the chemotropic response. If, after their induction by hormone B, the oogonial initials serve as loci for the intensive secretion of hormone A, its contunial removal by the ♂ plant would establish the required gradients for antheridial attraction, and the oogonial initials, as the primary (or even sole) source of the hormone and the foci of its gradients, would be the natural locations for antheridial delimitation and maturation. Since these possibilities have not been explored, however, the alternative possibility of the secretion by the oogonial initials of a second hormone, possibly closely related chemically to hormone A and active in attraction and delimitation at lower concentrations, cannot now be rigorously discounted.

Isolation and preliminary characterization of hormone A of *A. bisexualis* have been achieved in recent work by BARKSDALE and McMORRIS. The details of this *tour de force* are as yet unpublished, but, in response to a request for information about the current status of the work, BARKSDALE (personal communication) has written, Aug. 9, 1966: "I can now provide you with a statement by my collaborator, Dr. TREVOR McMORRIS, summarizing what is known of the chemistry of hormone A. Crystalline hormone A (m.p. 250—255^0) isolated from the culture filtrate of *Achlya bisexualis* T5 (female) has a molecular formula of $C_{29}H_{42}O_5$ (M.W. = 470), as determined by high resolution mass spectroscopy, and evidence to date suggests that the hormone is a steroid possessing a stigmasterol carbon skeleton. Hormone A is inactivated by 0.01 M NaOH and by steam distillation. We have found only a single compound with hormone A activity."

The chemotropic activity of a mixture of a minimum of 5—7 L-amino acids, each inactive alone, has been shown to obtain for vegetative hyphae, germ tubes, and antheridial hyphae in many species of the Saprolegniaceae by FISCHER and WERNER (1955). Both *A. bisexualis* (♂ and ♀) and *A. ambisexualis* (♂ and ♀) are listed in a category of somewhat indifferently reactive species and as more responsive to casein hydrolysate than to the amino acid mixture. No specific mention is made of any tests with sexually activated ♂ plants of either species; the relevance of this chemotropic effect to the hormonal mechanism in *Achlya* is accordingly uncertain.

Although an early, unsuccessful attempt was made to demonstrate sexual hormones in *Dictyuchus* (COUCH 1926), it is only in the last year that definitive evidence has been found for hormonal activity in the sexual interaction in this genus, a close relative of *Achlya*. SHERWOOD (1966) reported a number of preliminary experiments in which telomorphic responses could rationally be attributed to specific sexual secretions, but, in these, the responses were quite delayed, and the possibility of contamination by zoospores could not be excluded. One experiment, however, gave definitive proof of hormonal activity: when a slow, continuous stream of water was passed over a mixed culture of ♂ and ♀ plants, through a millipore filter, and over a ♀ test plant, oogonial initials were induced on the ♂ test plant in 5 of 8 tests. No discernible response was seen when other combinations of source and test plants were used, *i.e.*, ♀→♂, ♂→♀, and ♂+♀→♂. If the reasonable assumption is made that the hormonal system in *Dictyuchus* is similar to that of *Achlya*, the hormonal activity demonstrated by SHERWOOD would correspond to hormone B of *Achlya*, and the failure of antheridial induction on the ♂ in filtrate from ♂+♀ might be accounted for by the uptake of ♀-secreted hormones by the ♂ of the mixed culture. It is difficult to rationalize the lack of antheridial hyphae on the ♂ in ♀ filtrate except in terms of a lower concentration of hormone-A like hormone than that required to elicit the response — unless, of course, the hormonal mechanism of *Dictyuchus* differs in a fundamental way from that of *Achlya*.

A common pattern of sexuality in the higher Ascomycetes superposes 1-locus, 2-allele incompatibility upon hermaphroditism, and in most of the heterothallic species, such as *Neurospora crassa*, *N. sitophila*, *Bombardia lunata*, *Venturia inequalis*, *etc.*, ascogonia, the ♀ gametangia, are formed on the mycelia of both incompatibility classes when grown in isolation. Two heterothallic species of *Ascobolus*, *A. magnificus* and *A. stercorarius*, are exceptional in that gametangia — ♂ and ♀ in the former, only ♀ in the latter — are usually formed only upon physical contact of plants of the two opposite incompatibility classes (DODGE 1920, BISTIS 1956).

In a preliminary study of the sexual reaction of *A. stercorarius*, BISTIS (1956, 1957) provided a clear demonstration of a multistage hormonal coordinating mechanism (RAPER 1957). Subsequent work (BISTIS and RAPER 1963, BISTIS 1965) has amply confirmed the earlier findings and added much significant detail. A simple, straightforward account of the sexual reaction in *A. stercorarius* is hampered by the ability of any vegetative structure, oidium, hypha, or germinating ascospore, to become differentiated as a ♂ gametangium, and although the general course of differentiation is similar in all cases, significant differences in detail are evident. The following brief account of the sexual reaction and the evidence for hormonal coordination traces the sequence of interactions that occur between oidia and vegetative mycelia.

Oidia, multinucleate, sausage-shaped cells, are produced in profusion by mycelia of both incompatibility classes *A* and *a*, and serve the dual roles of asexual spores and detached, ♂ gametangia. Transferred to fresh medium, they are capable of germination within 3—5 hours; in the presence of any mycelium, however, their germination is indefinitely suppressed. Removal of the oidia from the presence of the mycelia to fresh medium allows their germination, but the time of germination differs markedly when the oidia have been exposed for about 4 hours to mycelia of the same incompatibility class and to mycelia of the opposite, compatible class: in the former case, germination occurs within 3—5 hours, whereas in the latter, the process of germination is delayed by 8—10 hours. This disparate behavior reflects "sexual activation", the process of gametangial differentiation, that the oidia undergo only when exposed to compatible mycelia.

Interposition of a permeable Cellophane membrane between oidia and mycelia does not alter the course of events either in compatible or incompatible contrasts: both inhibition of germination and sexual activation are determined by diffusible substance(s). That the mechanism of regulation may be rather complex, however, is indicated by differences in the details of interaction in the two reciprocal, compatible contrasts. Oidia of class *A* exposed to mycelia of class *a* are more quickly activated and, when activated, germinate more tardily than do oidia in the reciprocal *a*♂/*A*♀ contrast. Furthermore, *A* oidia can be activated by filtrates of *a* mycelia, but no activation of *a* oidia has been achieved by filtrate of *A* mycelia. There are even more striking differences in the induction and differentiation of attached ♂ gametangia in the two crosses, *A*♂/*a*♀ and *a*♂/*A*♀, and these differences probably play a decisive role in the biology of the organism in nature (Bistis 1965). The characteristic differences between the two reciprocal contrasts seem to be directly attributable to the *A* and *a* incompatibility alleles; crosses (by oidiation or hyphal transfer) among numerous strains were consistent in displaying the two patterns of development dependent upon the incompatibility class of the mycelia from which the ♂ fertilizing elements originated.

Oidia placed upon young, growing mycelia of the opposite incompatibility class induce within a very few hours a marked local proliferation of hyphae, the ascogonial hyphae, one (or a very few) of which proximal to each oidium elongates rapidly into a tighly coiled, cylindrical organ, the ♀ gametangium or ascogonium; septation results in a number of multinucleate cells, of which the terminal member elongates and becomes differentiated as the trichogyne. Again, a permeable membrane interposed between oidia and mycelium does not alter the reaction; induction of ascogonial hyphae and differentiation of the ascogonium are determined by diffusible hormones.

The trichogyne grows directly to the inducing oidium and fuses with it. Prior to actual fusion of the trichogyne with the oidium, however, any relocation of the oidium is quickly reflected (within 10 minutes) in reoriented growth in the new direction; chemotropic control of the growth of the trichogyne is clearly indicated. Furthermore, the trichogyne unerringly discriminates between activated and nonactivated oidia, no response being elicited by the latter. The trichogyne, however, makes no discrimination whatever between activated oidia of the two incompatibility classes: it responds chemotropically to and will readily fuse with an activated oidium of the same or of the opposite incompatibility class. Following fusion of trichogyne and oidium of the same incompatibility class, there is an initial development of the ascogonial apparatus to the stage of the initiation of ascogenous hyphae, but at this point, all development ceases, and no spore-bearing apothecium is formed.

During the latter stages of ascogonial development and following fertilization, lateral hyphal branches arise from the basal cells of the ascogonium and from vegetative hyphae in the immediate vicinity. These hyphae, through rapid, interlacing growth, form the enveloping sheath of the spore-producing fruiting body, the apothecium. An incompatibility-class specificity, probably mediated by diffusible substances, obtains in the process of apothecial development and maturation (Bistis 1965, Bistis and L. S. Olive, unpublished). Apothecial initials, initiated and fertilized by single oidia, removed from the mycelium of their origin and transferred to a mycelium of the same incompatibility class will continue development and mature normally. They will not develop further if transferred to a mycelium of the opposite class. This distinction was used to show that practically all of the apothecia developing in a mating between *A* and *a* vegetative mycelia originated from the *a* parent, i.e., only one, *A*♂/*a*♀, of the two reciprocal matings, *A*♂/*a*♀ and *a*♂/*A*♀, effectively occurs, and the mating essentially displays a sexual dimorphism, with the *A* mycelium acting as ♂ and *a* as female.

To account for the developmental sequence and telomorphic reactions described by BISTIS (1956, 1957) a minimal hormonal coordinating mechanism was postulated by RAPER (1957). The postulated mechanism, revised to account for subsequent findings, is as follows:

1. A class-specific or non-specific inhibitor of vegetative function of oidia.
2. A class-specific activator of oidia. (If *1* is class-specific, it may also function as *2*).
3. A class-specific substance responsible for ascogonial initiation and differentiation.
4. A non-class-specific substance that chemotropically attracts the trichogyne.
5. A class-specific inductor for sheath hyphae.

Substances *2*, *3* and *5* are rigidly class-specific, and to account for the complete reciprocity between the two incompatibility classes, *A* and *a*, these two substances must accordingly exist in complementary duplication in the two classes. Seven or eight classes of specific agents are therefore required to coordinate the demonstrated events of the sexual interaction in *Ascobolus*. Even this postulated array, however, does not adequately explain a number of perplexing features, such as the lack of exact reciprocity in interclass reactions and the minute details of antheridial and ascogonial development. Also, it should be emphasized that the various hormones, with a single exception, are known only by inference from interactions between physically separated elements. The one exception is the agent present in filtrate of the *a* mycelium responsible for activation of *A* oidia; the rapid loss of activity of this agent suggests that it is either volatile or highly unstable.

Chemotropic control of trichogynal growth, similar to that in *Ascobolus*, has been reported in two other species of Ascomycetes. BACKUS (1939) described for *Neurospora sitophila* the branching of trichogynes in close proximity to micro- or macroconidia of the opposite incompatibility class and the subsequent growth of the branches to the inducing conidia. ZICKLER (1953) observed in *Bombardia lunata* two different habits of growth of trichogynes in the presence of extruded spermatia of compatible and of noncompatible origins; in the former, directed growth of the trichogyne to masses of spermatia was evident, and the same response was achieved by the introduction, *via* capillary or agar block, of hot-water extracts of spermatia. No diffusible sexual hormones appear to be involved at earlier stages in the sexual histories of these species, since both ♂ and ♀ elements are fully differentiated in each species without any cross-stimulation between compatible plants.

Among the algae, there is no report of a system that is strictly comparable to those described above in that both initiation of the entire sexual process and its subsequent coordination result from hormonal activity. In *Oedogonium* sp. (RAWITSCHER-KUNKEL and MACHLIS 1962) the former feature is lacking, but the coordinative roles of sexual hormones parallel those of the fungal systems. *Oedogonium* sp. is a heterothallic, nannandrous form, and androsporangia and oogonial mother cells are formed on ♂ and ♀ filaments in isolated culture. If a ♂ filament bearing androsporangia is placed in a culture of ♀ filaments bearing oogonial mother cells, however, a hormonal system obviously becomes operative to determine the coordinated completion of the sexual process.

Androspores, liberated from the sporangia, are strongly attracted to the oogonial mother cells, upon which they become attached in large numbers.

a slight attachment occurring on vegetative cells of the ♀ filaments when a dense population of androspores is present. Encasing the ♀ filament in a thin sheath of agar prevents neither attraction of androspores nor their attachment to the oogonial mother cells; diffusible secretions are clearly responsible for these responses. The androspores then begins to germinate to form short filaments, which are oriented in a very precise manner in relation to the oogonial mother cell, the growth of those attached upon the anterior two-thirds of the oogonial mother cell being directed toward the apex of the cell and rather closely appressed to it, those attached in a posterior position, growing in the opposite direction toward the base of the ♀ cell. Differentiation of the dwarf male or nannandrium, which consists of an elongate cell with a basal holdfast and a terminal, 2-sperm gametangium, proceeds to completion within 10—15 hours after attachment. In the sequence of germination and differentiation, however, only the orientation of the developing dwarf males is critically influenced by the ♀ cell: (a) androspores attached to an agar sheath about the oogonial mother cell are not oriented as upon the normal ♀ cell but grow out normal to the surface to which they are attached; (b) germinating androspores can develop into dwarf male plants in cultures of ♂ filaments in the absence of oogonial mother cells or other ♀ elements.

Following attachment of androspores and during the late stages of the differentiation of dwarf males, the oogonial mother cell initiates a process of differentiation by the formation of a cell-wall ring near its apex. A few hours later, the cell elongates apically and divides into an apical ♀ gametangium, the oogonium, within which a single, large egg is formed, and a basal suffultory cell. This process is dependent upon the presence of *attached* dwarf male plants, as no oogonial differentiation occurred if an agar sheath was present, although dwarf males were present in abundance. It was also shown by comparison with sheathed cells with attached dwarf males, that the presence of the agar sheath did not interfere with the differentiative process.

At maturity, the oogonium and the tips of the dwarf males are surrounded by a heavy layer of mucilagenous material, and into this the ♂ gametes are discharged. A single gamete enters the ripe oogonium through a tiny pore in the oogonial wall.

The pattern of sexual development in *Oedogonium* provides considerable indicative evidence of a multi-stage hormonal coordinating mechanism, although only the agent for the chemotropic attraction of androspores has been shown to be diffusible; all other stages are dependent upon contact of interacting ♂ and ♀ elements and possibly under control by other means — suspected protoplasmic connections between the holdfasts of dwarf males and oogonial mother cells could not be observed. Also, the competence of the androspore to germinate and to differentiate as a dwarf male in complete isolation does not negate its integration in the interaction, as one critical feature of its effective development, its orientation, is under rigid control of the ♀.

Although there are marked differences between the coordinative mechanisms and in the specific roles performed by individual sexual hormones in *Mucor*, *Achlya*, *Ascobolus* and *Oedogonium*, two important points of basic similarity deserve brief mention. a) It is now abundantly clear that the sexual reaction of each of these plants is a true interaction between individuals. In three of the four cases, even the initiation of the sexual reaction depends upon metabolic products of both partners. b) The initiating and regulating mechanism in each is a multistage progression, in which each successive stage is dependent upon the achievement (organ formation, differentiation, *etc.*) of the last preceding stage. In most of the individual stage-to-stage relationships, this sequential dependence has been demonstrated to depend upon diffusible agents. These basic features provide a pattern of control that limits the expenditure of metabolic potential on sexual specialization only to circumstances that afford the maximal opportunity for the culmination of the sexual process.

Beyond the cases described above in which elaborate hormonal coordinating systems have been delineated, initiation of the sexual process by diffusible substances has been reported for *Saccharomyces* of the Hemiascomycetes and *Tremella* of the Heterobasidiomycetes. In a preliminary study of the process of conjugation in *S. cerevisiae*, LEVI (1956) found, when microcolonies of (+) and (—) strains were established in close proximity on agar-films, that conjugation tubes were formed prior to actual contact between the compatible colonies. The telomorphic induction, however, obtained only in the (—) strain, although cells of both incompatibility classes formed conjugation tubes in mixed culture. Minus cells inoculated onto agar-films from which (+) cells had been removed also responded by the formation of conjugation tubes, but no response occurred in (+) cells when the sequence was reversed. Only one of several attempts to demonstrate the induction of conjugation tubes across a membrane barrier was successful, and, in this, only the (—) cells responded.

A more detailed account of a comparable induction of conjugation tubes has been given by BANDONI (1963, 1965) for *Tremella mesenterica*. In this and related forms and in the smuts, basidiospores germinate to produce colonies of yeast-like unicells, which conjugate in compatible pairs to establish filamentous, dikaryotic mycelia. Mating competence in *Tremella* is primarily determined by two alleles, *A* and *a*, at a single incompatibility locus, and conjugation occurs whenever *A* and *a* cells are present together. Because this interaction is frequently seen between cells initially at some distance from each other, diffusible inducing agents were naturally suspect, and BANDONI demonstrated that the induction of conjugation tubes in cells of either incompatibility class is due to the specific secretion of cells of the opposite class. Cultures of cells of either class were autoclaved, and the agar-medium allowed to resolidify; cells of both classes were then streaked onto these media and their subsequent development observed. In every case, cells of the opposite incompatibility class to that of the original culture responded by the production of conjugation tubes, whereas cells of the same class only grew vegetatively. Two class-specific inducing substances are clearly involved. Delicate, short-range adjustments in growth of one or of both of the conjugating tubes to result always in their end-to-end meeting was interpreted as indicative of an additional diffusible agent(s), although this function could also be elicited by the inducing agents at higher concentration near the sources of their secretion. Conjugation tube formation and fusion are dependent only upon heteroallelism for *A* and *a*, but the pattern of subsequent development suggests that the establishment of the dikaryon is further dependent upon heteroallelism for an additional, multiple allelic *B* factor.

The same pattern of conjugation was earlier described in *Ustilago bromivora* by BAUCH (1925) to occur between basidiospores or between unicells derived from basidiospores, and he postulated the involvement of diffusible inducing agents but gave no experimental evidence of their presence. Indicative evidence for such agents in *Tilletia* was also reported more recently by HOLTON (1953)

The determination of sexual affinity.

The activity of specific substances in the determination of a given individual as ♂ or as ♀ can occur only within certain limitations as discussed above. Either an underlying genetic differentiation must allow a directed choice between two different possible sexual expressions or, less likely, the absence of any differentiation must permit the secretive activities of certain individuals to dictate the sexual expression of others. In either case, the resulting exploitation of opportunity afforded by chance proximity of compatible individuals greatly enhances the probability of sexual interaction and the attainment of its consequent biological benefits.

Among the algae there have been described a number of instances in which the pattern of sexual differentiation is such that there may be regular or sporadic fusion between gametes of the same sexual sign as well as between gametes of opposite sexual signs, *e.g.*, *Ectocarpus siliculosus* (HARTMANN 1925), *Dasycladus clavaeformis* (JOLLOS 1926), and *Enteromorpha ramulosa* (HARTMANN 1929). In each of these cases, some correlative mechanism must be operative to force upon differentiated gametes a pattern of sexual behavior diametrically opposite that of their primary physiological conditioning. Specific sexual secretions are in all probability responsible for such cases of sexual reversion (= relative sexuality, HARTMANN 1943 *etc.*). Such determinative activity, however, has been rarely demonstrated.

JOLLOS, in 1926, reported for *Dasycladus clavaeformis* that the ultimate sexual affinity of the gametes of certain weakly reacting strains could be altered by exposure to cell-free filtrates of gametes of strongly reacting strains. Individuals of this species could be divided into two classes, (+) and (—), within each of which there was marked differences of sexual potency. When gametes of a weak (—) strain were suspended for two hours or longer in the filtrate of gametes of a strong (—) strain, the mating response of the weak (—) gametes with (+) gametes was markedly intensified, *i.e.*, they now behaved as strong (—) gametes. Furthermore, the exposure to filtrates of strong (+) gametes rendered weak (—) gametes incapable of mating with (+) gametes and fully fertile with strong (—) gametes. In each case, exposure of differentiated gametes to filtrates of other strains caused an alteration of the original sexual affinity to correspond with that of the gametes from which the filtrate was taken. JOLLOS made no claim that these effects were dependent upon specific secretions of the gametes rather than upon nonspecific changes in the medium. It is only recently that any further report dealing with this problem has appeared. A recent critical attempt by KÖHLER (1957) to confirm the results reported by JOLLOS for *D. clavaeformis* has been unsuccessful.

Comparable but considerably more elaborate sexual determinative effects were later described by MOEWUS and associates (*cf.* MOEWUS 1951 for review and list of pertinent papers) in the unicellular green alga, *Chlamydomonas eugametos*. To specifically identified substances, termones, secreted by ♂ gametes (androtermones: *peonin* and *4-oxy-β-cyclocitral*) and by ♀ gametes (gynotermones: *isorhamnetin* and an *unidentified derivative of picrocrocin*) was ascribed responsibility for predetermining the sexual affinity of all gametes of undifferentiated (monoecious) clones and of a portion of the gametes of partially differentiated (subdioecious) clones to correspond to that of the secreting strain. Recent attempts to confirm various aspects of the work on *Chlamydomonas* have failed to substantiate these effects: RYAN (1955), in colloboration with MOEWUS, was unable to find any effect of peonin and isorhamnetin or of filtrates from ♂ and ♀ gametes on gametes of a monoecious clone, and FÖRSTER and WIESE (1954a, b, and in HARTMANN 1955) were unable to find in filtrates of fully motile and sexually active gametes any trace of the flagellar hormone, *Crocin*, a hydrolytic product originating from the same precursor as picrocrocin, which in turn hydrolysed to form 4-oxy-β-cyclocitral.

The effects described for *Dasycladus* and *Chlamydomonas* occur in species comprising differentiated sexual strains and at a stage in the sexual process following differentiation of the gametes. In a number of algae and fungi characterized by gametic copulation and in which no sexual differentiation is known, a type of pseudo-determinative gametic behavior has been repeatedly observed. In the unicellular alga, *Chlamydomonas paupera* (PASCHER 1931), in the filamentous alga, *Hydrodictyon reticulatum* (MAINX 1931), and in the chytridaceous fungus, *Synchitrium fulgens* (KUSANO 1930), the gametes upon emergence are highly active and are capable of gametic fusion with sedentary mates: as the gametes age, however, they in turn become sedentary and attract to themselves more active (*i.e.*, younger) mates with which they fuse. This pattern of behavior has commonly been interpreted as an age-dependent progression from maleness to femaleness. Whether any actual alteration or determination of sexual affinity is involved here is unknown, but whatever the basic correlative mechanism may be, it effectively discourages intraclonal inbreeding in favour of interclonal outbreeding. Specific sexual secretions have been suspected at one or at both stages in the progression; none has been experimentally demonstrated.

There have been no reports of sexual determination by hormones among the primitive phycomycetous fungi that reproduce by planogametes, and there are no known instances among the algae of the determination of sexual affinity by specific substances at earlier stages of development. The remaining reports deal with certain members of two of the three phylogenetically diverse groups of fungi in which sexual hormonal mechanisms have been described, the Saprolegniaceae and Ascobolaceae. In the heterothallic members of these families, sexual organs are formed only following contact or close proximity of compatible plants. and the basic patterns of sexual determination allow for a limited range of sexual expression.

In two heterothallic species of *Achlya*, *A. bisexualis* and *A. ambisexualis*, there are numerous self-sterile, cross-fertile sexual intergrades that range in predominant sexual affinity from strong ♂ to strong ♀ (RAPER 1940b, 1954), and the sexual reaction elicited in any given intergrade strain depends upon its position in the series relative to that of its mate. A predominant-♂ thus reacts as a ♂ with plants of all strains more ♀ than itself but reacts as ♀ to all stronger ♂'s. The ambivalence of sexual expression in these intergrade strains is entirely explicable in terms of the sexual hormonal coordinating mechanism. The intergrade strains apparently have the capacity to secrete as well as to react to the full complement of sexual substances, ♂ and ♀, though probably in different quantities and with differing sensitivities depending upon the balance of ♂ and ♀ potentialities. Since the sexual reaction is initiated by hormones secreted by the vegetative ♀ plant, in the chance juxtoposition of two potentially bisexual plants, the secretions of the more strongly ♀ partner determines the subsequent role of its mate by inducing antheridial hyphae upon the adjacent portions of its mycelium. Once this initial stage is achieved, the sexual roles of both partners are fully determined, because each of the later stages in the sexual progression is initiated and controlled by secretions of the organs formed or differentiated in the last preceding stage. In *A. ambisexualis*, there are five or more of these intergrade strains, linearly ordered in respect to maleness-femaleness, bridging the interval between pure-♂ and pure-♀. The effective balance between the secretive and sensitivity capacities must accordingly be highly critical in view of the constancy of interstrain reactions. The determination of sexual affinity established by the specific hormones, however, affects only those portions of the mycelium in close proximity to the mate, since plants of all intergrade strains will react simultaneously and vigorously as ♂ and ♀ in different portions of their thalli when mated with strong ♀ *and* strong ♂ partners. As a device to assure cross-breeding in nature, this pattern of ambivalent sexual interaction achieves high effectiveness and accomplishes through multistrain sexual differentiation a degree of outbreeding approaching that provided by multiple alternate incompatibility factors in the higher fungi.

Another type of determination of sexual affinity or competence occurs between ♂ or ♀ strains of the heterothallic species of *Achlya* and numerous homothallic species of *Achlya* and the closely related genus, *Thraustotheca*. The results of interspecific matings, both between homothallic plants and between heterothallic and homothallic plants, range from no interaction to apparent complete compatibility with the production of normal, matured zygotes (RAPER 1950b). Reactions in matings between heterothallic and homothallic plants were shown to depend upon sexual hormones, and the wide range of responses has been interpreted as most probably the result of hormonal specificities (RAPER 1955). In addition to numerous cross- reactive steps that occur in the interaction between heterothallic and homothallic plants to integrate their sexual activities, two specific effects were demonstrated to determine the sexual competence of the homothallic mate — each in a manner that is decicely detrimental to the homothallic partner. a) Hormone A, secreted by ♀ plants of the heterothallic species, causes in certain homothallic species an intense production of antheridial hyphae and the proliferation of immature oogonia to form antheridial hyphae, *i.e.*, redifferentiation. b) Hormone A^1, secreted by heterothallic ♂'s, completely inhibits the production of oogonial initials in a number of homothallic species. In both cases, there is a decided interference with the sexual potential of the homothallic partner: reduction to unisexual ♂ activity in the first instance, and to sexual impotence in the second.

More recently, BARKSDALE (1960, 1966b) has made a more extensive study of crosses between heterothallic and homothallic mycelia. She confirmed and extended the earlier observations on the induction by hormone A of antheridial hyphae on many homothallic species. She also reported a marked, though temporary, inhibition of the formation of oogonial initials in homothallic plants competent to produce antheridial hyphae in response to the hormone and showed that the plant's sensitivity to oogonial inhibition paralleled that to produce antheridial hyphae. Release of the inhibition occurred within a day or so, by which time the removal of hormone by the reactive plant presumably lowered the concentration below inhibitory levels. With a continuing source of hormone A, as in proximity to a ♀ plant, the inhibition of oogonial initials would obtain throughout the period of sexual competence of the homothallic partner.

A type of sexual determination comparable to that between integrade strains of *Achlya* appears to be a normal feature of the heterothallic Ascomycete, *Ascobolus stercorarius* (BISTIS 1965). Plants of both incompatibility classes are functionally hermaphroditic, i.e., plants of opposite incompatibility classes are reciprocally cross-fertile, and the reciprocal crosses proceed in close but not exact parallel whenever a small element of one class — be it a single oidium or a detached hyphal fragment — is transferred to a preformed mycelium of the other. When two relatively massive vegetative mycelia of opposite sign grow together, however, the resulting interaction is almost exclusively of one kind, the *A* mycelium providing the fertilizing elements for ♀ elements on the *a* mycelium. BISTIS (1965) interprets this nonreciprocal determination of sexual expression to be due to the initially more rapid and vigorous development of the *A*/*a* progression than of its reciprocal, *a*/*A*.

The few cases among the algae and fungi in which the determination of sexual affinity is known to result from the activity of diffusible secretions further emphasize the extreme flexibility of sexual systems in these primitive groups of plants. Such determinative activity appreciably extends the range of mating possibilities within a species, and since in each case the extension favors interclonal outbreeding, a significant selective advantage must accrue through the evolution of sexual coordinative mechanisms that permit a directed choice of sexual expressions in different associations. Sexual determinative substances would therefore be expected to be very common throughout those groups of plants in which external regulatory mechanisms could have evolved. That only a few instances have been demonstrated is perhaps surprising; there are quite a number of cases known, however, in which sexual reversals commonly or even routinely occur, and in these, diffusible secretions are properly suspect as the agents of specific determination. Future work will undoubtedly add many new cases of determinative activity of specific sexual secretions to the list now known.

The relationships of the various functions performed by diffusible sexual substances in the lower plants deserve brief examination. It was pointed out above that the demonstrated functions, determination of sexual sign, organ induction, organ orientation, organ differentiation, *etc.*, could not be individually ascribed to separate and distinct substances. The examination of the activity of a single excretion and the consequences of this activity under different circumstances illustrates this plurality of function. In *Achlya*, hormone A, secreted by the ♀ plants of heterothallic species, induces antheridial hyphae on the ♂ plant; hormone A is therefore an organ-inducing agent. In the mating of two sexual intergrade strains, however, the accumulating ♀ secretions of both plants sets off antheridial hyphal production on the mate having the stronger ♂ potency and thereby determines the subsequent sexual activity of both partners; hormone A is accordingly a sex-determining agent. When applied to plants of certain homothallic species, the consequence of the single specific activity of hormone A goes one step further: in causing the already differentiated oogonial initial to proliferate as antheridial hyphae, a previous sexual differentiation is negated and a redetermination of sexual affinity is manifest. The secretion(s) of the sexually activated oidium of *Ascobolus* likewise plays, through the single activity of ascogonial induction, the dual role of local determination of the mycelium as ♀ and the production of sexual organs. Among the algae, comparable overlapping effects are known only in the single case of *Oedogonium* (RAWITSCHER-KUNKEL and MACHLIS 1962), in which the secretion(s) of the oogonial mother cell chemotropically attract the androspores and determine their orientation prior to their differentiation as dwarf male plants. The author can no longer consider the reports of MOEWUS on *Chlamydomonas* as relevant or significant in this connection.

The chemical identity has been determined for none of the active substances considered here, but three, the zygophore-inducing agent of *Mucor*, hormone A of *Achlya*, and sirenin of *Allomyces*, have been isolated and characterized, and continuing work is directed toward their identification. Chemical studies with these substances have in the past presented very real difficulties — the active agents occur and elicit their specific responses in infinitesimal quantities, and the conditions requisite to strong sexual activity rarely coincide with mass culture methods. The possibilities of the successful chemical delineation of these coordinating systems, however, are continually enhanced as additional methodologies of high resolving power become available. Certainly, each of these so imperfectly known chemically regulating mechanisms stands as a challenge to the biologist or the biochemist who would consider the chemical aspects of cellular differentiation and interaction as basic and fundamental biological problems.

Literature.

BACKUS, M. P.: The mechanics of conidial fertilization in *Neurospora sitophila*. Bull. Torrey Bot. Club **66**, 63—76 (1939). — BANBURY, G. H.: Processes controlling zygophore formation and zygotropism in *Mucor mucedo*. Nature (Lond.) **173**, 499 (1954). — Physiological studies in the Mucorales. III. The zygotropism of zygophores of *Mucor mucedo*. J. Exp. Bot. **6**, 235—244 (1955). — BANDONI, R. J.: Conjugation in *Tremella mesenterica*. Canad. J. Bot. **41**, 467—474 (1963). — Secondary control of conjugation in *Tremella mesenterica*. Canad. J. Bot. **43**, 627—630 (1965). — BARKSDALE, A. W.: Inter-thallic sexual reactions in *Achlya*, a genus of the aquatic fungi. Amer. J. Bot. **47**, 14—23 (1960). — The role of hormone A during sexual conjugation in *Achlya ambisexualis*. Mycologia **55**, 627—632 (1963). — Segregation of sex in the progeny of a selfed heterozygote of *Achlya bisexualis*. Mycologia **58**, 802—804 (1966). — The sexual hormones of the fungus *Achlya*. Ann. N.Y. Acad. Sci. (in Press). — BAUCH, R.: Untersuchungen über die Entwicklungsgeschichte und Sexualphysiologie der *Ustilago bromivora* und *Ustilago grandis*. Z. Bot. **17**, 129—177 (1925). — BISHOP, H.: Sexuality in *Sapromyces reinschii*. Diss. Harvard Univ. 1937. — A study of sexuality in *Sapromyces reinschii*. Mycologia **32**, 505—529 (1940). — BISTIS, G.: Sexuality in *Ascobolus stercorarius*. I. Morphology of the ascogonium; plasmogamy; evidence for a sexual hormonal mechanism. Amer. J. Bot. **43**, 389—394 (1956). — Sexuality in *Ascobolus stercorarius*. II. Preliminary experiments on various aspects of the sexual process. Amer. J. Bot. **44**, 436—443 (1957). — The function of the mating type locus in filamentous Ascomycetes. In: Incompatibility in Fungi (K. ESSER and J. R. RAPER, eds.), p. 23—31. Berlin-Heidelberg-New York: Springer 1965. — BISTIS, G., and J. R. RAPER: Heterothallism and sexuality in *Ascobolus stercorarius*. Amer. J. Bot. **50**, 880—891 (1963). — BLAKESLEE, A. F.: Sexual reproduction in the Mucorineae. Proc. Amer. Acad. Sci. **40**, 206—319 (1904). — BURGEFF, H.: Untersuchungen über Sexualität und Parasitismus bei Mucorineen. I. Bot. Abh. (K. GOEBEL), H. 4, 5—135 (1924).

COUCH, J. N.: Heterothallism in *Dictyuchus*, a genus of the water molds. Ann. Bot. **40**, 848—881 (1926).

DE BARY, A.: Untersuchungen über Peronosporeen und Saprolegnieen. Beitr. Morph. u. Physiol. Pilze, Reihe IV (1881). — DODGE, B. O.: Life history of *Ascobolus magnificus*. Origin of the archicarp from two strains. Mycologia **12**, 115—134 (1920).

FISCHER, F. G., u. G. WERNER: Eine Analyse des Chemotropismus einiger Pilze, insbesondere der Saprolegniaceen. Hoppe-Seylers Z. physiol. Chem. **300**, 211—236 (1955). — FÖRSTER, H., u. L. WIESE: Untersuchungen zur Kopulationsfähigkeit von *Chlamydomonas eugametos*. Z. Naturforsch. **9**b, 470—471 (1954a). — Gamonwirkungen bei *Chlamydomonas eugametos*. Z. Naturforsch. **9**b, 548—550 (1954b).

HARTMANN, M.: Über relative Sexualität bei *Ectocarpus siliculosus*. Ein experimenteller Beitrag zur Sexualitätshypothese der Befruchtung. Naturwissenschaften **13**, 975—980 (1925). — Untersuchungen über die Sexualität und Entwicklung von Algen. III. Über die Sexualität und den Generationswechsel von *Chaetomorpha* und *Enteromorpha*. Ber. dtsch. bot. Ges. **47**, 485—494 (1929). — Die Sexualität. Jena 1943. — Sex problems in algae, fungi, and protozoa. Amer. Naturalist **89**, 321—346 (1955). — HOLTON, C. S.: Fusion between secondary sporidia in culture as an index of sex compatibility in *Tilletia* species. Phytopathology **43**, 322—323 (1953).

JOLLOS, V.: Untersuchungen über die Sexualitätsverhältnisse von *Dasycladus clavaeformis*. Biol. Zbl. **46**, 279—295 (1926).

KAUFFMAN, C. H.: A contribution to the physiology of the Saprolegniaceae with special reference to the variations of the sexual organs. Ann. Bot. **22**, 361—388 (1908). — KEHL, H.: Ein Beitrag zur Morphologie und Physiologie der Zygophoren von *Mucor mucedo*. Arch. Mikrobiol. **8**, 379—406 (1937). — KÖHLER, F.: Beitrag zur Kenntnis der Sexualreaktionen von *Mucor mucedo*. Planta (Berl.) **23**, 358—378 (1935). — KÖHLER, K.: Neue Untersuchungen über die Sexualität bei *Dasycladus* und *Chaetomorpha*. Arch. Protistenk. **102**, 209—217 (1957). — KRAFCZYK, H.: Die Zygosporenbildung bei *Pilobolus crystallinus*. Ber. dtsch. bot. Ges. **49**, 141—146 (1931). — Die Bildung und Keimung der Zygosporen von *Pilobolus crystallinus* und sein heterokaryotisches Myzel. Beitr. Biol. Pflanzen **23**, 349—396 (1935). — KUHN, R., F. MOEWUS u. D. JERCHEL: Über die chemische Natur der Stoffe, welche die Kopulation der männlichen und weiblichen Gameten von *Chlamydomonas eugametos* im Lichte bewirken. Ber. dtsch. chem. Ges. **71**, 1541—1547 (1938). — KUSANO, S.: Life history and physiology of *Synchytrium fulgens*, with special reference to its sexuality. Jap. J. Bot. **5**, 35—112 (1930).

LEVI, J. D.: Mating reaction in yeast. Nature (Lond.) **177**, 753—754 (1956).

MACHLIS, L.: Evidence for a sexual hormone in *Allomyces*. Physiol. Plantarum (Cph.) **11**, 181—192 (1958a). — A procedure for the purification of sirenin. Nature (Lond.) **181**, 1790—1791 (1958b). — Sex hormones in fungi. In: The fungi. vol. II (G. C. AINSWORTH and A. S. SUSSMAN, eds.), p. 415—433. New York: Academic Press 1966. — MACHLIS, L., W. H. NUTTING, M. W. WILLIAMS, and H. RAPOPORT: Production, isolation, and characterization of sirenin. Biochemistry **5**, 2147—2153 (1966). — MAINX, F.: Gametenkopulation und Zygotenkeimung bei *Hydrodictyon reticulatum*. Arch. Protistenk. **75**, 502—516 (1931). — MOEWUS, F.: Die Sexualstoffe von *Chlamydomonas eugametos*. Erg. Enzymforsch. **12**, 1—32 (1951). — MULLINS, J. T., and J. R. RAPER: Heterothallism in biflagellate aquatic fungi: preliminary genetic analysis. Science **26**, 1174—1175 (1965).

PASCHER, A.: Über Gruppenbildung und Geschlechtswechsel bei den Gameten einer Chlamydomonadine ... Jb. wiss. Bot. **75**, 551—580 (1931). — PLEMPEL, M.: Die Sexualstoffe der Mucoraceae, ihre Abtrennung und die Erklärung ihrer Funktion. Arch. Mikrobiol. **28**, 151—174 (1957). — Die zygotropische Reaktion bei Mucorineen. I. Mitt. Planta (Berl.) **55**, 254—258 (1960a). — Die Darstellung eines kristallinen Benzoesäure-Esters der Sexualstoffe von *Mucor mucedo*. Naturwissenschaften **47**, 472—473 (1960b). — Die Mucorineen-Gamone. Naturwissenschaften **50**, 226 (1963a). — Die chemischen Grundlagen der Sexualreaktion bei Zygomyceten. Planta (Berl.) **59**, 492—508 (1963b). — PLEMPEL, M., u. G. BRAUNITZER: Die Isolierung der Mucorineen-Sexualstoffe. I. Z. Naturforsch. **13**b, 302—305 (1958). — PLEMPEL, M., u. W. DAWID: Die zygotropische Reaktion bei Mucorineen. II. Mitt. Planta (Berl.) **56**, 438—446 (1961).

RAPER, J. R.: Sexual hormones in *Achlya*. I. Indicative evidence for a hormonal coordinating mechanism. Amer. J. Bot. **26**, 639—650 (1939). — Sexual hormones in *Achlya*. II. Distance reactions, conclusive evidence for a hormonal coordinating mechanism. Amer. J. Bot. **27**, 162—173 (1940a). — Sexuality in *Achlya ambisexualis*. Mycologia **32**, 710—727 (1940b). — Sexual hormones in *Achlya*. III. Hormone A and the initial male reaction. Amer. J. Bot. **29**, 159—166 (1942a). — Sexual hormones in *Achlya*. V. Hormone A[1], a male secreted augmenter or activator of hormone A. Proc. Nat. Acad. Sci. (Wash.) **28**, 509—515 (1942b). — Sexual hormones in *Achlya*. VI. The hormones of the A-complex. Proc. Nat. Acad. Sci. (Wash.) **36**, 524—533 (1950a). — Sexual hormones in *Achlya*. VII. The hormonal mechanism in homothallic species. Bot. Gaz. **112**, 1—24 (1950b). — Sexual hormones in *Achlya*. Amer. Scientist **39**, 110—120 (1951). — Chemical regulation of sexual processes in the Thallophytes. Bot. Rev. **18**, 447—545 (1952). — Life cycles, sexuality, and sexual mechanisms in fungi. AAAS Symposium "Sex in Microorganisms" 42—81 (1954). — Some problems of specificity in the sexuality of plants, pp. 119—140. Biological Specificity and Growth, Ed. E. G. BUTLER. Princeton: University Press 1955. — Hormones and sexuality in lower plants. Biological Action of Growth Substances. Symp. Soc. Exp. Biol. No XI. 143—165 (1957). — RAWITSCHER-KUNKEL, E., and L. MACHLIS: Hormonal integration of sexual reproduction in *Oedogonium*. Amer. J. Bot. **49**. 177—183 (1962). — RONSDORF, L.: Über die chemischen Bedingungen von Wachstum und Zygotenbildung bei *Phycomyces blakesleeanus*. Planta (Berl.) **14**, 482—514 (1931). — RYAN, F. J.: An attempt to reproduce with MOEWUS' collaboration some of his experiments on *Chlamydomonas* and *Polytoma*. Science **122**, 470 (1955).

SACHS, J.: Stoff und Form der Pflanzenorgane. Arb. bot. Inst. Würzburg **2**, 452—488, 689—718 (1880). — SANSOME, E.: Meiosis in the oogonium and antheridium of *Pythium debaryanum* Hesse. Nature (Lond.) **191**, 827—828 (1961). — Meiosis in *Pythium debaryanum* Hesse and its significance in the life history of the Biflagellate. Trans. Brit. Myc. Soc. **46**. 63—72 (1963). — Meiosis in diploid and polyploid sex organs of *Phytophthora* and *Achlya*. Cytologia (Tokyo) **30**, 103—117 (1965). — SHERWOOD, W. A.: Evidence for a sexual hormone in the water mold *Dictyuchus*. Mycologia **58**, 215—221 (1966).

VERKAIK, C.: Über das Entstehen von Zygophoren von *Mucor mucedo* (+) unter Beeinflussung eines von *Mucor mucedo* (—) abgeschiedenen Stoffes. Proc. kon. ned. Akad. Wet. (Amst.) **33**, 656—658 (1930).

ZICKLER, H.: Zur Entwicklungsgeschichte des Ascomyceten *Bombardia lunata*. Arch. Protistenk. **98**, 1—70 (1953).

Determination von Gametophyt und Sporophyt.

Von

Leopold Bauer.

Mit 10 Abbildungen.

I. Einleitung.

Der Begriff des Generationswechsels — ursprünglich für die Archegoniaten aufgestellt (HOFMEISTER 1851) — hat durch die Vermehrung unserer Kenntnisse über die Entwicklungsgeschichte niederer Pflanzen allmählich eine Auflockerung und Erweiterung erfahren. Die Gemeinsamkeit aller Generationswechseltypen besteht darin, daß sich ein ontogenetischer Kreislauf aus einer Folge von Einzelindividuen zusammensetzt, die über die Bildung von verschiedenen Keimzellen miteinander im Zusammenhang stehen.

Wir unterscheiden hinsichtlich der Kernphase zwischen *homophasischem* und *hetero-* bzw. *diphasischem* und hinsichtlich der Gestaltverhältnisse zwischen *isomorphem* und *heteromorphem* Generationswechsel (vgl. FRITSCH 1945a; HARTMANN 1956). Auch *fakultative* Generationswechseltypen gehören in unseren Problemkreis. Die Erweiterung des Generationswechselbegriffes bedingt, daß wir in die Betrachtung Phänomene einbeziehen müssen, die man noch nicht als „typischen" Generationswechsel, wohl aber als „Vorläufer" oder „Anlauf" bezeichnen kann. Ferner müssen wir auf Faktoren hinweisen, die zwar als „Konstruktionselemente" im Generationswechsel wichtig sind, die aber auch ohne Bindung an einen Generationswechsel eine entwicklungsphysiologische Rolle spielen. Die Erörterung der Determination des Gametophyten und Sporophyten muß also den Rahmen bewußt weiter abstecken, als er durch die Definition des Generationswechsels gegeben ist.

Wir werden im folgenden auch manche hypothetischen Gesichtspunkte zu erörtern haben und damit dartun, daß die Hauptaufgabe der experimentellen Bearbeitung der Generationswechselphänomene noch vor uns liegt.

Dabei ergibt sich zunächst ein phylogenetischer Aspekt; d.h. wir wollen einige der Faktoren herausstellen, die für Zustandekommen und Ausbau eines Generationswechsels bedeutsam sind. Weiterhin sind die in der Ontogenese wirkenden Determinationsfaktoren zu betrachten, soweit wir darüber schon Aussagen machen können.

II. Faktoren bei der Entstehung des Generationswechsels.

1. Die Qualität der Keimzellen.

a) Allgemeine Gesichtspunkte.

Die als Generationen bezeichneten Ontogenieabschnitte werden durch die Qualität der Keimzellen definiert, die sie zu bilden vermögen und mit denen sie zugleich abschließen (vgl. WIDDER 1951). Bei einer Alternanz von sexuell und asexuell ausgeprägten Keimzellen spricht man von einer „geschlechtlichen" und einer „ungeschlechtlichen" Generation. Doch können sich in gewissen Fällen in einem Kreislauf auch Sporen verschiedener Gestalt bzw. Determination ablösen.

Der Wechsel in der Bildung von Gameten und Sporen an einem Haplonten ist noch kein Generationswechsel. Jedoch wird beispielsweise bei *Ulothrix zonata* (Schussnig 1960) je nach Jahreszeit Zoosporen- oder Gametenbildung begünstigt, so daß zunächst — oft in längerer Folge — Pflanzen gebildet werden, die nur Zoosporen produzieren; diese werden dann von Individuen mit Gameten

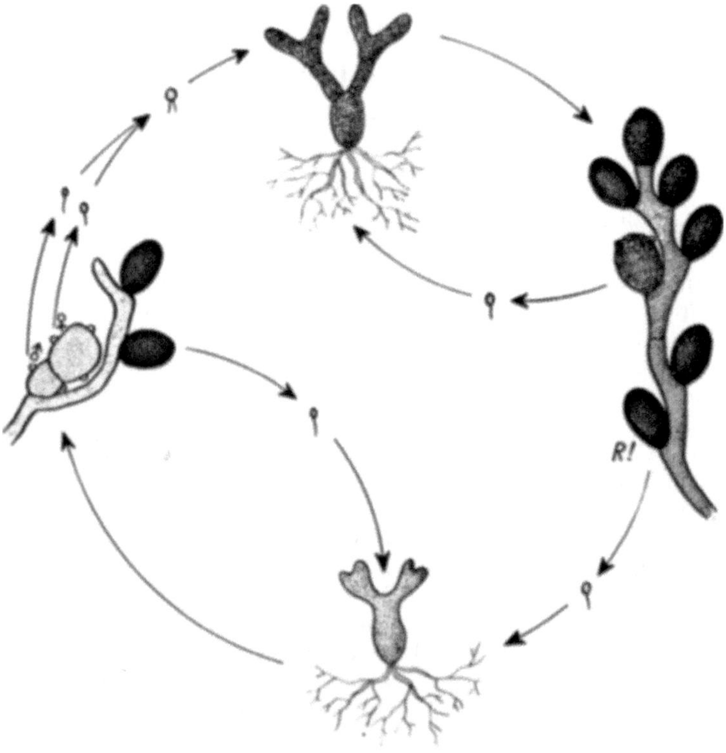

Abb. 1. Entwicklungscyclus einer Mutante von *Allomyces arbuscula* mit Dauersporangien am Gametophyten, halbschematisch. Großer Cyclus: Zygote-Sporophyt (Reduktion)-Gametophyt-Zygote. Kleine Cyclen: 1. Sporophyt (Mitose)-Diplospore-Sporophyt; 2. Dauersporangium am Gametophyten (*keine* Reduktion)-Haplospore-Gametophyt. (Nach Stumm 1958.)

(eventuell zusätzlich mit Sporen) abgelöst. In *Cladophora glomerata* liegt ein entsprechendes Beispiel für einen Diplonten vor (List 1930). Die auseinandergezogenen vegetativen und geschlechtlichen Zustände können in gewissem Sinne als fakultative Generationen angesprochen werden.

Dieser „Anlauf" ist aber wohl nicht ohne weiteres als Vorstufe für einen heterophasischen Generationswechsel anzusehen. Dieser hat vielmehr seine Wurzel in einem veränderten Verhalten der Zygote (vgl. unten S. 239). Wenn jedoch die Entwicklung in der Diplophase nach dem gleichen Determinationsmuster wie in der Haplophase abläuft, könnten die Faktoren der Keimzellbildung für die Diplophase insofern von Bedeutung werden, als der Zeitpunkt, zu dem die Meiosis nachgeholt wird, nicht willkürlich, sondern determiniert ist. Für die Dictyotaceen und Rhodophyceen möchte Schussnig (1960) sogar annehmen, daß die Vierzahl der Tetrasporen nicht erst durch den Teilungsmodus der Meiosis geprägt worden ist; vielmehr soll eine allmähliche Reduktion der Zahl der ungeschlechtlichen Keimzellen (Gonidien) bis herab zur Vierzahl bereits in der Haplophase erfolgt sein. Jedenfalls kommen bei den primitiven Rhodophyceen (Chantrasioideen) in der Haplophase neben merogonen auch tetragone Sporangien (Gonidangien) vor. Bei einer Verschiebung der Reduktionsteilung von der Zygote weg könnte daher der für die Bildung der Tetrasporen bereits in der Haplophase entwickelte Zustand ausgenutzt worden sein (SCHUSSNIG 1960).

Einen interessanten Aspekt in dieser Hinsicht liefern genetische Untersuchungen über das Fortpflanzungsverhalten von *Allomyces arbuscula*. Bei diesem Phycomyceten kommen in der Haplophase gelegentlich neben den Gametangien auch Dauersporangien vor, welche eigentlich die für den Sporophyten typischen Keimzellbehälter darstellen (SOERGEL 1937). Durch eine spontane Ein-Gen-Mutation wurde diese „Mischmycel"-Eigenschaft erbfest, so daß nunmehr (Sippe MX) nicht nur der Sporophyt, sondern auch der Gametophyt über einen Nebenkreislauf verfügt (Abb. 1: STUMM 1958). Durch die Mutation hat also der Gametophyt eine Eigenschaft erhalten (zurückerhalten ?), wie sie ein Haplont vom Typ der *Ulothrix* besitzt: die Fähigkeit, Gameten *und* Sporen zu bilden. Wenn *Allomyces* von einem in dieser Weise ausgestatteten Vorfahren abstammen würde, dann hätte sich im Zuge der heterophasischen Generationswechselentstehung die hinausgeschobene Meiosis in der Tat in den schon vorhandenen Sporangien festgesetzt.

Häufig kommt auch eine *Umdetermination* des Keimzellcharakters — sowohl bei Haplonten als auch bei Haplodiplonten in beiden Kernphasen — vor. Dadurch kann sich ein zusätzlicher Vermehrungsmodus (Nebenkreislauf) einstellen; es können aber auch Bedingungen für den Ausfall einer Generation gegeben sein.

b) Vegetative Entwicklung der Gameten.

In den unteren Stufen der verschiedenen Verwandtschaftskreise ist die sexuelle und asexuelle Determination noch wenig scharf fixiert. Eine vegetative Entwicklung von Gameten ist daher weit verbreitet. Bei *Chlamydomonas eugametos* kann sich z.B. jede Zelle entweder als Gamet oder Agamet verhalten (HARTMANN 1956). Eine Entwicklung von Parthenogameten zu Gametophyten ist unter anderem bei Ulotrichales (SCHILLER 1907; CARTER 1926; BLIDING 1933; KORNMANN 1962), Ectocarpales (BERTHOLD 1881) und *Cutleria* (THURET 1850; HARTMANN 1950) beobachtet worden. Die Fortpflanzung durch Parthenogameten ist zuweilen (*Enteromorpha compressa*, BLIDING 1933), *Ectocarpus virescens*, KNIEP 1928) die vorherrschende Vermehrungsart, so daß im Extremfall durch sekundäre Betonung der Haplophase der Generationswechsel wieder aufgehoben wird.

Es kommt auch eine parthenogenetische Entwicklung zu Pflanzen mit Sporophytenwuchs vor, z.B. bei *Cladophora suhriana* (FÖYN 1934), *Monostroma grevellii* (KORNMANN 1962), Ectocarpales, Sporochnales (SAUVAGEAU 1926; FRITSCH 1945b; SMITH 1951), *Cutleria* (CHURCH 1898; SAUVAGEAU 1908; SCHLÖSSER 1935; HARTMANN 1950). Wieweit damit eine Diploidisierung verbunden war, ist nicht in allen Fällen nachzuprüfen gewesen. Die Bildung haploider *Cutleria*-Pflanzen vom *Aglaozonia*-Typus beschreibt HARTMANN (1950). Andererseits wies FÖYN (1934) Diploidie der parthenogenetisch entstandenen *Cladophora*-Sporophyten nach. Bei *Cutleria* scheinen außer einer individuellen Variabilität auch genetische Unterschiede im Verhalten zu bestehen. Die Gameten von THURETs Pflanzen entwickelten sich zu Gametophyten, während diejenigen von CHURCH *Aglaozonia*-Wuchs annahmen. Andererseits schließt SAUVAGEAU (1908) aus seinen Beobachtungen am natürlichen Standort, sowie HARTMANN (1950) aus Kulturversuchen, daß sich die Eier von Pflanzen der gleichen Population vegetativ in verschiedener

Richtung entwickeln können. Diese Schlußfolgerung wird durch Experimente von SCHLÖSSER (1935) gestützt, in denen er durch Zentrifugieren und chemische Beeinflussung unbefruchtete Eier zur Entwicklung anregen konnte: es ergaben sich Geschlechtspflanzen. Wenn jedoch während der ersten Zellteilung solcher Eier starke Zentrifugalkräfte einwirkten, erzielte er *Aglaozonia*-Stadien. Ob durch das Zentrifugieren die Chromosomenzahl beeinflußt wurde, konnte nicht geklärt werden.

Als Besonderheit zahlreicher diöcischer Oedogonien werden nur die ♂ Gameten vegetativ (Androsporen) und entwickeln sich als „Zwergmännchen" (PRINGSHEIM 1858; PASCHER 1906) zu einer eigenen Generation.

Bei den Ceramiales (vgl. SMITH 1951) sind Sexualorgane an Tetrasporenpflanzen beobachtet worden. Für viele dieser Pflanzen muß offenbleiben, ob es sich um haploide Gametophyten oder diploide Tetrasporophyten gehandelt hat (vgl. auch SCHUSSNIG 1960). Wenn aber, wie für *Nitophyllum punctatum* (SVEDELIUS 1914) oder *Spermothamnion turneri* (DREW 1934, 1943) berichtet wird, die diploiden Sexualorgane funktionsfähig sind, kann der Entwicklungsgang durch das Auftreten triploider Pflanzen, wie sie bei *Plumaria elegans* (DREW 1939) tatsächlich am natürlichen Standort gefunden wurden, kompliziert werden.

In der Diplophase von Ectocarpales und Sphacelariales treten pluriloculäre Keimbehälter auf: sie fungieren jedoch nicht als Gametangien, sondern enthalten Sporen. Sie dienen der vegetativen Vermehrung des Sporophyten und können Nebenkreisläufe bedingen. Bei den Ectocarpales, bei denen die Fortpflanzungsverhältnisse sehr variabel (z.B. MÜLLER 1964) und offenbar immer noch im Fluß sind, können sich die Schwärmer aus den pluriloculären Sporangien zu verschieden gestalteten Pflanzen entwickeln (SAUVAGEAU 1928, 1932; s. unten S. 238).

c) Sexuelles Verhalten der Sporen.

Auf der anderen Seite können auch Sporen sexuell determiniert werden, wie z.B. die Schwärmer aus uniloculären Sporangien von verschiedenen Arten von Ectocarpaceen (KNIGHT 1929; SCHUSSNIG und KOTHBAUER 1934; MATHIAS 1935) oder von Sphaceariaceen (CLINT 1927). Dadurch wird fakultativ der Gametophyt ausgeschaltet. FRITSCH (1945b) weist auf eine gleiche Tendenz, den Gametophyten sekundär zu eliminieren, bei den höheren Ectocarpales hin. Ein entwicklungsphysiologisches Zwischenglied zwischen den Cladophoraceen mit isomorphem diphasischen Generationswechsel und der rein diploiden *Cladophora glomerata* könnte z.B. *Cladophora flavescens* sein, bei der gelegentlich die Schwärmer aus den Sporangien nach der Reduktionsteilung fusionieren sollen (HIGGINS 1931).

d) Unterschiedliche vegetative Determination der Sporen.

In gewissen Fällen liegt einer Generationenfolge auch der Wechsel von Sporen verschiedener Qualität zugrunde, d.h. es kann eine ungeschlechtliche Generation durch eine weitere ungeschlechtliche — aber anders gestaltete — Generation abgelöst werden. Zunächst sei als Beispiel der bekannte dreigliedrige Generationswechsel vieler Rotalgen erwähnt. Ferner gehören hierher die Fortpflanzungsverhältnisse von Ectocarpales (SAUVAGEAU 1928, 1932). Aus den Schwärmern der diploiden pluriloculären Sporangien entwickeln sich entweder wieder normale Sporophyten oder aber kleine Zwergpflanzen, welche die Überwinterungsform darstellen. Diese tragen ihrerseits wieder pluriloculäre Sporangien, deren Schwärmer die Zwergform vermehren können. Es entsteht somit ein heteromorpher Nebengenerationswechsel in der Diplophase, der als fakultativ zu gelten hat. Als Determinationsfaktoren sind die jahreszeitlichen Schwankungen, vor allem wohl der Temperatur, anzusehen. Ob die Determination bereits im Sporangium oder erst bei der Keimung erfolgt, ist unbekannt.

Ein weiteres Beispiel finden wir in der Ulotrichale *Codiolum (Urospora) wormskioldii* var. *biflagellatum* (KORNMANN 1961 a), bei der sich die viergeißeligen Schwärmer in scharfer Abhängigkeit von der Temperatur entweder zu Urosporafäden (unter 10° C) oder zu Zwergpflanzen (über 13° C) entwickeln. Für diese wiederum sind zweigeißelige Zoosporen charakteristisch, aus denen *Codiolum*-Zellen und Zwergpflanzen hervorgehen. Es werden also drei heteromorphe, ungeschlechtlich sich vermehrende Stadien zu Gliedern eines gemeinsamen Cyclus verbunden (Abb. 2). Geschlechtliche Fortpflanzung wurde nicht beobachtet. Am natürlichen Standort reguliert diesen Generationswechsel der jahreszeitliche Temperaturwechsel. Bei der Bangiacee *Porphyra leucosticta* (Abb. 3) geht von verschieden determinierten Karposporen, sowie von den Conchosporen der *Conchocelis*-Stadien und von den Monosporen von Zwergpflanzen eine vielgestaltige Entwicklung aus (KORNMANN 1961 b). Die Determinationsfaktoren sind unbekannt.

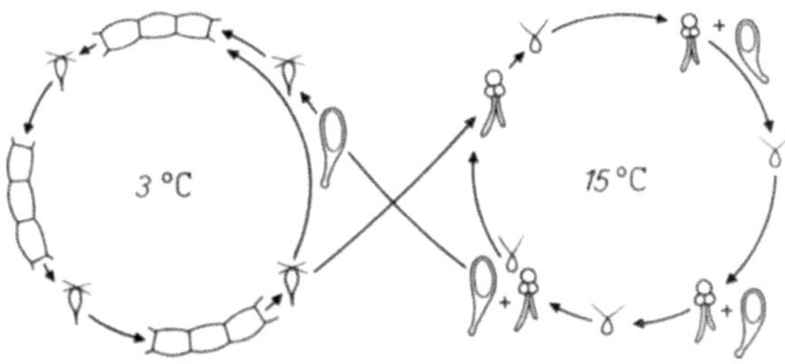

Abb. 2. Entwicklungscyclen von *Codiolum wormskioldii* var. *biflagellatum* in Abhängigkeit von der Temperatur, schematisch. (Nach KORNMANN 1961 a.)

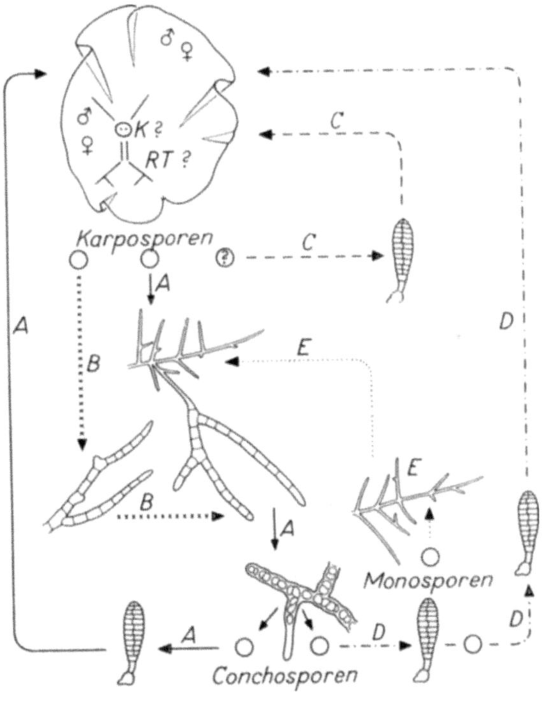

Abb. 3. Gestaltvariationen im Entwicklungscyclus von *Porphyra leucosticta*, schematisch. (Nach KORNMANN 1961 b.)

2. Verschiebung der Meiosis.

a) Bildung einer diploiden Phase.

Für die Entstehung eines isomorphen, diphasischen Generationswechsels genügt formal ein einziger Mutationsschritt, der die Zygote an der Durchführung der Meiosis hindert, indem er sie zu vegetativer Entfaltung nach dem Informationsmuster des Gametophyten zwingt. Die Determination zum Vollzug der Meiosis muß aber in dem vegetativen Zygotenprodukt bestehen bleiben (SMITH 1955; FRITSCH 1945 a).

Es ist auch ein zweiter Weg zur Bildung einer diploiden Generation möglich; wir wissen nicht, inwieweit er bei der Entstehung von Generationswechsel-Kreisläufen eine Rolle gespielt hat. Es könnte nämlich die vegetative Entfaltung der Zygote auch als mehrstufiger Prozeß ablaufen. Anläufe dazu finden wir schon bei den Protisten, z.B. kann bei *Chlamydomonas chlamydogama* und *Chlamydomonas moewusii* (Smith 1951; Lewin 1952, 1953; Schussnig 1960) die Zygote, die sonst häufig ein mehr oder weniger ausgeprägtes Ruhestadium darstellt, eine neue Qualität bekommen. Sie wächst durch photosynthetische Aktivität beträchtlich heran, bevor sie zur Reduktionsteilung kommt, und kann eine Zeitlang ein eigenständiges Dasein führen. Hier wird also die Meiosis nicht durch Zwischenschalten mitotischer Teilungen verzögert, sondern einfach durch Übernahme von Funktionen der vegetativen Phase durch die Zygote selbst. Ein solches Verhalten könnte als physiologische Vorbereitung für die Bildung einer ausgedehnteren diploiden Phase dienen. Bei *Raciborskiella* gehen z.B. nach Wislouch (1925) in der Planozygote mitotische Teilungen vor sich. Für *Ulothrix rorida* wird berichtet (Lind 1932), daß die Zygote direkt zu vegetativen Fäden auswächst, ob fakultativ oder obligatorisch, ist noch nicht sichergestellt. Die Zygote von *Stigeoclonium subspinosum* wächst zu einem wenigzelligen, kleinen Pflänzchen aus, wodurch ein heteromorpher, diphasischer Generationswechsel entsteht (Juller 1937). Man könnte sich vorstellen, daß ein solches Verhalten einen Anfangsschritt für eine stärkere Entfaltung des Sporophyten in isomorpher und heteromorpher Richtung darstellt (vgl. Schussnig 1960).

Bei den Florideen kann die Meiosis bei der Keimung der *Zygote* (*Batrachospermum* u.a.), bei der Bildung von Tetrasporen (cytologische Untersuchungen stehen noch aus) auf dem *Gonimoblasten* (*Liagora tetrasporifera*, *Helminthocladia hudsoni*, Boergensen 1927; Feldmann 1939) und bei der Bildung der Tetrasporen am *Tetrasporophyten* (*Ceramium* u.a.) erfolgen. Man kann in dieser Aufstellung eine progressive Entwicklungsreihe sehen, in der die Meiosis immer weiter hinausgeschoben wird (z.B. Svedelius 1942; Fritsch 1945b; Schussnig 1960) oder aber eine Reduktionsreihe (Goebel 1928), in der die Meiosis sekundär wieder in ein früheres Stadium, im Extremfall wieder zurück zur Zygote, verlegt worden wäre.

b) Bildung einer dikaryotischen Phase.

Beim Vorgang der Befruchtung sind zwei Phasen erkennbar: die *Plasmogamie*, der meist unmittelbar die *Karyogamie* folgt. Beide Teilvorgänge brauchen aber nicht direkt miteinander gekoppelt zu sein. Vielmehr kann sich der dikaryotische Zustand nach erfolgter Plasmogamie mehr oder weniger lange stabil erhalten. Die Ursachen der verzögerten Karyogamie sind noch unbekannt.

Bei der Myxomonadine *Pseudospora parasitica*, die in *Spirogyra* parasitiert, bleiben die Gametenkerne während der ganzen Umwandlungsvorgänge bis zur Dauercyste getrennt und nehmen die Karyogamie wahrscheinlich erst in der reifen Dauerspore vor (Schussnig 1960). Wir sehen bereits hier ein entwicklungsphysiologisches Prinzip, das bei den Eumyceten, insbesondere bei den Basidiomyceten, zu einer ausgedehnten dikaryotischen Phase geführt hat. Wir nennen sie ebenfalls Sporophyt; doch streng genommen hat sie einen anderen entwicklungsphysiologischen Status als die Diplophase der übrigen Generationswechseltypen; die Sporen werden erst nach Erreichen des (nur einzelligen) diploiden Zustandes gebildet.

3. Morphogenetische Prinzipien.

Der Generationswechsel ist häufig mit einem Gestaltwechsel verbunden. Es würde jedoch zu weit führen, die verschiedenen gestaltlichen Ausdrucksmöglichkeiten der Heteromorphie aufzuzählen. Es kann nur auf einige allgemeine morphogenetische „Konstruktionsprinzipien“, deren sich die Gestaltbildung im Generationswechsel bedient, sowie auf einige Faktoren hingewiesen werden, die einen Anstoß zur Gestaltänderung geben könnten.

Die gestaltbildenden Prozesse haben für das Generationswechselgeschehen unter anderem auch die Bedeutung, daß sie meist so eng mit der einen oder anderen Keimzellart gekoppelt sind, daß die Induktion der spezifischen Gestalt zugleich auch das Keimzellverhalten determiniert, also den Status als Gameto- oder Sporophyt einschließt (vgl. unten S. 247).

Ferner ist auch derjenige Gestaltwechsel nicht ohne Bedeutung, der die aufeinanderfolgenden Entwicklungsabschnitte *einer* Generation charakterisiert; dies betrifft vor allem den Wechsel in der Höhe der Organisationsstufe. Die Bedeutung liegt darin, daß sich im Laufe der Evolution die Bereitschaft zur Bildung der jeweiligen Fortpflanzungsorgane einer bestimmten morphologischen Stufe zugeordnet hat. Fast immer ist es die Folgeform, welche in der Regel zugleich die höhere Organisationsstufe repräsentiert. Die Bedingungen ihrer Ausbildung entscheiden somit über die Vollständigkeit des Entwicklungscyclus.

Die Entwicklung höher organisierter Formen aus primitiver gebauten ist ein Grundprinzip der Biologie: es beherrscht die Phylogenese, spielt aber auch eine wichtige Rolle in der Ontogenese („Jugend"- und „Folgeform"; GOEBEL 1928). Der Generationswechsel ist eine höher entwickelte Stufe des Fortpflanzungsverhaltens, und es ist daher natürlich, daß beide Entwicklungstendenzen sich gegenseitig bedingen. Folgende Möglichkeiten, zu höheren morphologischen Einheiten zu gelangen, sind vor allem verwirklicht.

a) Plasmodiale Organisation.

Bereits auf der Flagellatenstufe können — als Konvergenzerscheinungen — in verschiedenen Verwandtschaftskreisen (rhizopodiale Protisten u.a.) durch mitotische Vermehrung der Zellkerne coenokaryotische Zustände entstehen, die als Anlauf einer höheren Organisationsstufe gelten können (SCHUSSNIG 1960). Auf diese Entwicklungsstufe gehört z.B. auch noch *Plasmodiophora brassicae*, der Erreger der Kohlhernie, der aber wegen seines Chitingehaltes in den Membranen bereits zu den primitiven Pilzen gerechnet wird (WETTSTEIN 1935). Seine Schwärmer wandeln sich in der Wirtszelle in eine Amöbe um, deren haploider Kern sich mehrfach mitotisch teilt und so ein „Paraplasmodium" bildet. Von diesem lösen sich einkernige Fragmente ab, die zu Mikrocysten werden und als Gameten anzusprechen sind. Die Zygote wächst unter mehrfachen mitotischen Teilungen wiederum zu einem „Paraplasmodium" heran, dessen Kerne schließlich unter Reduktionsteilung zahlreiche haploide Sporen liefern. Durch Verschiebung der Meiosis hat sich also ein Generationswechsel eingestellt, dessen Gestaltprinzip die Ausbildung mehrkerniger plasmodialer Gebilde in *beiden* Generationen ist.

Das Gestaltprinzip des Plasmodiums finden wir bekanntlich auch bei den Myxomyceten im weiteren Sinne. Hier wird es aber auf andere Weise verwirklicht, und zwar durch mehr oder weniger innige Verschmelzung einzelner Myxamöben. Ein ursprünglicher Typ ist die Bildung von Aggregationsplasmodien bei den Acrasieen (deren verwandtschaftliche Beziehungen zu den übrigen Myxomyceten noch ungeklärt sind). Der erste Schritt der Vereinigung ist eine chemotaktische Anlockung („Acrasin": BONNER 1947; GERISCH 1964). Die höhere Organisationsstufe schließt auch das gemeinsame und einheitliche Reagieren der Myxamöben ein, obwohl sie ihre Individualität nicht aufgeben. Bei den echten Myxomyceten aber fusionieren die Myxamöben so, daß sie ihre Individualität völlig verlieren. Da der Plasmodienbildung eine sexuelle Vereinigung haploider Myxoflagellaten bzw. Myxamöben vorausgeht, ist bei den Myxomyceten der plasmodiale Organisationszustand für die Diplophase allein charakteristisch, die den

Charakter einer Generation besitzt; die gametophytische, oder hier besser sexuelle Generation, wird durch die sich mitotisch teilenden haploiden Einzelflagellaten bzw. -amöben repräsentiert.

b) Verflechtung von Zellfäden (unechte Gewebebildung).

Ein weiteres Prinzip, das zu höher organisierte Gestalteinheiten führt, treffen wir bei den Eumyceten und Florideen an; es lag auch dem Aufbau der fossilen Taxites zugrunde. Der Vegetationskörper besteht aus Zellfäden, die sich nach einem ordnenden Prinzip zusammenlagern und eine physiologische Einheit darstellen. Diesem Konstruktionsprinzip liegen einige Generationswechseltypen zugrunde. Bei den höheren Ascomyceten stellen derartige „Hyphenorganbildungen" das Endstadium der *Haplophase* dar (Apo-, Perithecien), bei den Basidiomyceten sind sie als „Fruchtkörper" der *dikaryotischen* Phase bekannt.

Die gleichen „aggregativen Entwicklungstendenzen" (Schussnig 1960) finden wir aber unter den Ascomyceten auch im Bereich der Conidienvermehrung (Koremien bei *Penicillium*; Conidienstromata in der Gattung *Xylaria*), die in ihrer Organisationshöhe durchaus „Fruchtkörpern" vergleichbar sind.

Die Florideen machen von diesem Gestaltprinzip im ausgebildeten Gametophyten und im Tetrasporophyten Gebrauch, während im Keimling noch häufig gut erkennbar die einfache Fadenform als Jugendform auftritt.

c) Echte Gewebe- und Organbildung.

Das „erfolgreichste" Prinzip für den Aufbau höher organisierter Vegetationskörper ist schließlich die Bildung echter zwei- und dreidimensionaler Gewebe, ausgehend von apikalen oder interkalaren Vegetationspunkten bzw. Initialzonen.

Für den Generationswechsel ist dabei eine Entwicklungstendenz bedeutsam, die von der Stufe der fadenförmigen Algen ausgeht. Der Vegetationskörper differenziert sich in ein plagiotropes (prostrates) und ein orthotropes (erektes) System (Seitenfäden zweiter Ordnung); die prostrate Gestaltungsphase kann auch scheiben- und krustenförmig modifiziert sein. Dieser Typ ist sowohl bei den Chlorophyceen als auch bei den einfachen Braunalgen zu finden und ist übrigens auch der Ausgangspunkt für die Höherentwicklung der Rotalgen. Fritsch (1945a) nennt ihn den „heterotrichen Habitus". In fast allen diesen heterotrichen Gruppen ist nun die Tendenz zum Ausbau der erekten und Reduktion der prostraten Phase festzustellen. Die erekte Phase entwickelt sich zu zum Teil hoch organisierten Zellkörpern. Der heterotriche Habitus ist in diesem erweiterten Sinn also auch das Konstruktionsprinzip der höheren Braun- und Rotalgen. Darüber hinaus dürften dieEntwicklungspotenzen, die in der Grundkonstruktion des heterotrichen Habitus liegen, auch ihre Bedeutung für die Ausgestaltung der beiden alternierenden Generationen höherer Pflanzen haben (Fritsch 1945a). Auf den heterotrichen Habitus der Laubmoosprotonemen haben Allsopp und Mitra (1956) hingewiesen.

Bei heterotricher Ausbildung werden die Fortpflanzungszellen an den Ästen höherer Ordnung angelegt, und ebenso werden die erekten Zellkörper mehr und mehr Träger der Fortpflanzungsorgane. Dies gilt sowohl für den Gameto- als auch für den Sporophyten. Heteromorphie im Generationswechsel basiert daher im allgemeinen nicht auf einem Wechsel von prostrater und erekter Phase, sondern meist auf Gestaltabweichungen innerhalb der erekten Folgeform, welche die größeren Entwicklungspotenzen hat. Bei den höheren Algen tritt das fädige prostrate Stadium immer mehr zurück, und die erekten Äste werden zu hochdifferenzierten, oft mächtigen Gebilden. Das fädige Jugendstadium ist aber meist auch dann noch (auch im Sporophyten) zu erkennen. Doch sei *Laminaria* als ein Typ genannt, bei dem tatsächlich der Gametophyt auf die fädige Form beschränkt ist, während die Folgeform die Gestalt des Sporophyten bestimmt (Abb. 4).

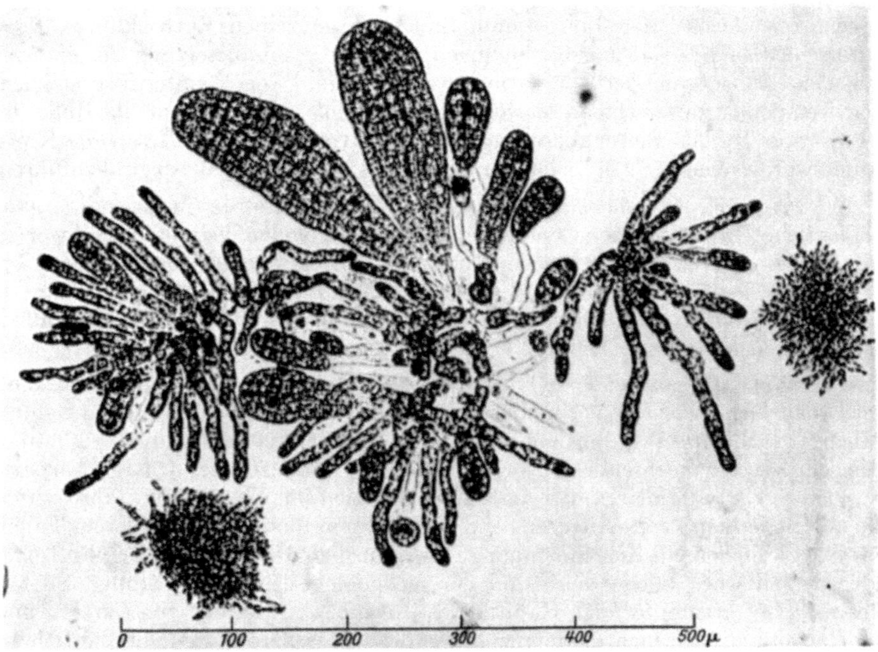

Abb. 4. *Laminaria hyperborea.* ♂ und ♀ Gametophyten mit jungen Sporophyten. (Original KORNMANN.)

d) Gestaltunterschiede zwischen den Generationen.

Ein heteromorpher Generationswechsel kann einmal aus einem ursprünglich isomorphen Generationswechsel entstanden sein (wahrscheinlich unter anderem bei den Phaeophyceen). Er könnte aber auch direkt von einem Haplonten ausgehen, indem der Entwicklungsgang der vegetativ gewordenen Zygote (vgl. oben) von vornherein eigene Wege geht. Weiterhin könnte, wie FRITSCH (1945a) zur Diskussion stellt, in den Fortpflanzungsverhältnissen, wie sie bei *Coleochaete* und *Batrachospermum* vorliegen, ein Ausgangspunkt zur Entwicklung eines heterophasischen Generationswechsels bestehen. Beide Pflanzen ähneln sich darin, daß *nach* der Reduktionsteilung eine kleine, morphologisch abweichende, sporenbildende Zwischengeneration auftritt. Wenn die Meiosis vom Beginn an das Ende dieser Zwerggeneration verlagert würde, wäre ein heteromorpher diphasischer Generationswechsel entstanden, von dem aus die Diplophase weiter ausgebaut werden könnte. Für die Rotalgen könnte man diesen ersten Schritt z.B. in *Liagora tetrasporifera* (s. oben S. 240) als vollzogen ansehen. Eine derartige Entwicklung wird z.B. für die Entstehung des Archegoniaten-Generationswechsels diskutiert (FRITSCH 1945a; SMITH 1955), für den allerdings auch der Weg über einen ursprünglich isomorphen Generationswechsel möglich erscheint (ZIMMERMANN 1958).

Die Heteromorphie kann einmal mit einer Betonung der Haplophase bzw. Reduzierung der Diplophase einhergehen (z.B. *Cutleria* unter den Braunalgen; *Bonnemaissonia* unter den Rotalgen; Bryophyten unter den Archegoniaten), andererseits mit einer Betonung der Diplophase und Reduktion (z.T. bis zur

Elimination) der Haplophase (z. B. *Laminaria* und *Fucus* unter den Braunlagen; höhere Archegoniaten, Samenpflanzen).

Über die *Ursachen* dieser Entwicklungen können wir nur Vermutungen anstellen. Sie dürften sich wohl auch kaum auf einen einheitlichen Nenner bringen lassen. Im Wechsel der Jahreszeiten finden die einzelnen Entwicklungsphasen unterschiedliche Entfaltungsbedingungen, welche die Voraussetzung für eine verschiedene Ausprägung der Generationen sein könnte. Dabei könnten Tageslängen- und Temperaturunterschiede als Selektionsprinzipien wirken. Auf die Rolle der *Temperatur* bei der Einregulierung des fakultativen *Codiolum (Urospora)*-Kreislaufes wurde oben (S. 239) schon hingewiesen, um nur ein Beispiel anzuführen.

Wieweit auch eine *endogene rhythmische* Komponente die (insbesondere morphologische) Alternanz der Generationen begünstigt oder bedingt, ist schwer zu entscheiden. Der Gestaltwechsel im ontogenetischen Cyclus von gewissen Vertretern des Meeresplanktons läßt sich auf dieser Basis interpretieren; denn sie sind als Bewohner tropischer Meere praktisch keinem Rhythmus der Außenbedingungen ausgesetzt. Hierauf hat Oehlkers (1956) hingewiesen.

Gestaltänderungen sind auch oft mit einer Änderung des *Stoffwechsels* und der Ernährungsansprüche verbunden, die ihrerseits einen Milieuwechsel im Gefolge haben. Ursache und Wirkung sind hierbei nicht immer klar auseinander zu halten. Hinsichtlich des Wechsels von Normal- und Zwergpflanzen bei den Ectocarpales (s. oben S. 238) nimmt Knight (1931) an, daß sich im Wechsel der Jahreszeiten die Verfügbarkeit von Nitraten und Phosphaten ändert. In anderen Fällen besteht zweifellos ein Zusammenhang zwischen dem Übergang zu epiphytischer oder parasitischer Lebensweise und der morphologischen Entwicklung. So hat einerseits der krustenförmige Habitus des *Aglaozonia*-Stadiums von *Cutleria* und andererseits die geringe Differenzierung des Moossporophyten eine deutliche Beziehung zur veränderten Lebensweise. Grau (1964) konnte heterotrophe Züge im Stoffwechsel des Laubmoossporophyten feststellen. Veränderte Stoffwechselansprüche der verschiedenen Generationen der Rostpilze haben wohl nicht nur den Wirtswechsel bedingt, sondern auch irgendwie die Morphogenese beeinflußt, die sich allerdings praktisch nur auf die Sporengestalt bezieht.

Der Übergang zum *Landleben* hat sicherlich neue Voraussetzungen für die Morphogenese des Sporophyten der höheren Archegoniaten geschaffen und wohl auch die Tendenz zur Größenverschiebung der Generationen beeinflußt (Svedelius 1921).

Schließlich sei noch anhand der eigenartigen Generationswechselverhältnisse bei den Florideen die Frage aufgeworfen, ob nicht durch *unvollständige Entdetermination* des „Zygotenzustandes" die Morphogenese der Zygotenprodukte beeinflußt werden könnte. Weder die reduzierte (z. B. *Batrachospermum*) noch die unreduzierte Zygote (z. B. *Ceramium*) entwickelt sich zunächst nach dem Informationsmuster des Gameto- bzw. Tetrasporophyten. Goebel spricht hinsichtlich der Karposporenbildung von „fraktionierter Embryobildung". Man hat in der Tat den Eindruck, daß mit der Entwicklung des Karposporophyten als fädiges Gebilde zwar prinzipiell die Jugendform wieder erreicht ist, daß aber offenbar vom „Zygotenzustand" irgend ein Faktor nachwirkt (vgl. unten S. 253), der den normalen Ablauf der Entwicklung blockiert und auch dem Stoffwechsel heterotrophe Züge aufprägt; erst die Karposporenbildung bringt die endgültige Entdeterminierung und ermöglicht den normalen Ablauf der Entwicklung. Auf die mögliche Bedeutung eines derartigen entwicklungsphysiologischen Ausgangspunktes auch für die Entwicklung eines heteromorphen, diphasischen Generationswechsels wurde oben hingewiesen.

III. Determination in der Ontogenese.

1. Natürlicher Entwicklungsablauf.

Für das Zustandekommen des gesamten Generationswechselcyclus sind nicht nur die Determinationsfaktoren entscheidend, die den Übergang von der einen zur anderen Generation kontrollieren, sondern ebenso wichtig ist auch der normale Ablauf der einzelnen Entwicklungsschritte jeder einzelnen Generation selbst, also der Übergang von der sterilen zur fertilen Phase im Gametophyten und die Entwicklung des Sporophyten bis zur Produktion der Sporen. Die einzelnen Entwicklungsschritte vollziehen sich im Zusammenwirken von genetischer Information und äußeren Faktoren, worüber in anderen Bänden dieses Handbuches berichtet wird. Offenbar geht auch von gewissen Entwicklungsschritten eines Abschnittes eine determinierende Wirkung auf den nachfolgenden aus.

Unsere Kenntnisse des Differenzierungsstoffwechsels sind noch sehr lückenhaft. Die aufeinanderfolgenden Entwicklungsstadien unterscheiden sich auffällig in ihren Ansprüchen: z.B. ändert sich ihr Lichtbedürfnis sowohl in quantitativer (GOEBEL 1908, 1932) als auch in qualitativer Hinsicht. *Blaulicht* ist erforderlich zur Bildung der Perithecien von *Pyronema confluens* (CLAUSSEN 1912; HAWKER 1957) und zur Erreichung der flächigen Wuchsform der Farnprothallien (MOHR 1956); *Rotlicht* wird zur Anlegung der Moosknospen benötigt. Auch ihr Wirkstoffbedürfnis ist qualitativen und quantitativen Schwankungen unterworfen (für Kinine als Beispiel s. BAUER 1966a). Unsere mangelhafte Kenntnis dieser Verhältnisse kommt darin zum Ausdruck, daß wir in vielen Fällen nicht in der Lage sind, den gesamten Entwicklungscyclus einer Pflanze in vitro zu reproduzieren bzw. in einem Callus die potentiellen Entwicklungstendenzen zu induzieren.

Von besonderem Interesse ist die Frage, an welcher Stelle des Gesamtkreislaufes und in welcher Weise die Determination des Sporophytencharakters erfolgt bzw. beginnt. Die Diskussion über die Frage, ob die Determination der veränderten Gestalt und des veränderten Verhaltens des Sporophyten direkt etwas mit der Verdoppelung des Chromosomensatzes zu tun habe, hat nur noch historisches Interesse. Bereits GOEBEL (1908) hat sich mit der Frage auseinandergesetzt und sie aufgrund seiner Apogamie- und Aposporieversuche an leptosporangiaten Farnen verneint. Zur gleichen Ansicht sind auch F. v. WETTSTEIN (1942) angesichts der erstmalig bei einem Laubmoos (*Phascum cuspidatum*; SPRINGER 1935) beobachteten Apogamie und SOST (1955) durch seine Colchicinierungsversuche mit *Allomyces* gekommen. Inzwischen sind zahlreiche haploide Sporophyten bei Farnen (vgl. MANTON 1950) und Samenpflanzen (BLAKESLEE u. Mitarb. 1922; IVANOV 1938; KNAPP 1939) bekannt geworden.

Andererseits scheint aber der diploide Zustand doch für die dem Sporophyten zukommende entwicklungsphysiologische Konstellation der „gemäßere" zu sein. Bei Laubmoosen, die zur Apogamie neigen, gehen in der Regel nur diploide Gametophyten dazu über (*Phascum* SPRINGER 1935; *Georgia* BAUER 1956, Abb. 5; *Splachnum*, *Pottia* LAZARENKO 1961, 1963). Allerdings ist Sporogonbildung auch in der Haplophase zu finden: *Desmatodon* (LAZARENKO 1960, 1961), *Physcomitrium* (LAL 1963), *Splachnum*, *Bryum* (BAUER, unpubl.). Bei *Desmatodon randii* handelt es sich aber offenbar um eine allopolyploide Art, und bei *Physcomitrium coorgense*, bei der über 50% der Sporen lebensfähig sind, ist Aufregulierung der Chromosomenzahl nicht ausgeschlossen; in den vom Referenten beobachteten Fällen bilden die haploiden Gametophyten nur gelegentlich, die diploiden dagegen reichlich und oft spontan Sporophytengewebe. — Ferner wäre anzuführen, daß der in der Nachkommenschaft der Röntgenmutante „M70" von *Allomyces arbuscula* (STUMM 1961) aufgetretene haploide Sporophyt nach der fünften Passage seine Ploidiestufe geändert hat; weiterhin wurden parthenogenetisch entstandene Keimlinge von *Cladophora suhriana* (FÖYN 1934) und *Monostroma grevillei* (KORNMANN 1962) offenbar in Verbindung mit spontaner Diploidisierung zu Sporophyten.

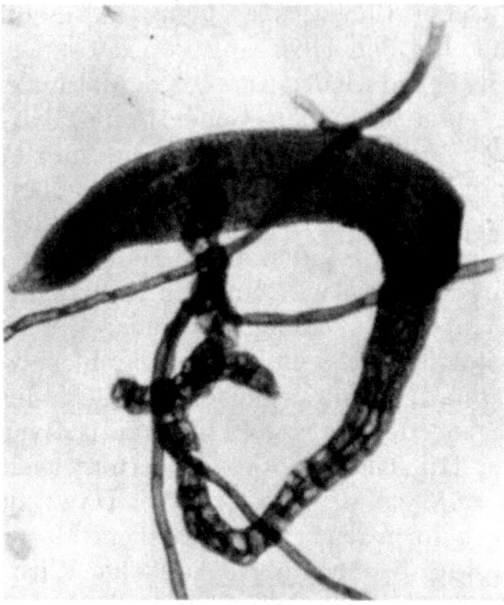

Abb. 5. *Georgia pellucida*. Apogame Sporogonbildung am diploiden Protonema. (Nach BAUER 1956.)

Die Determinierung des Sporophytencharakters ist wahrscheinlich kein einmaliger Prozeß, sondern erfolgt in *Schritten*, zumindest im heteromorphen Generationswechsel. Die *Meiosis* wird bei den Haplonten, als Abschluß des Sexualaktes, im Zusammenhang mit der Befruchtung determiniert. Da wir den heterophasischen Generationswechsel von der Haplontenstufe ableiten, müssen wir annehmen, daß auch die zeitlich verschobene Meiosis ihre Determination zu diesem frühen Zeitpunkt erhalten hat (OEHLKERS 1956). Die zahlreichen Fälle parthenogenetischer Entwicklung der Gameten zu Sporophyten, die bis hinauf zu den Spermatophyten zu beobachten sind, werfen jedoch die Frage auf, ob nicht ein wesentlicher Determinationsschritt zum Ablauf der Meiosis bereits im reifenden ♀ Gametophyten erfolgt. Er könnte in diesen Fällen jedoch an die Potenz zur parthenogenetischen Entwicklung gebunden sein.

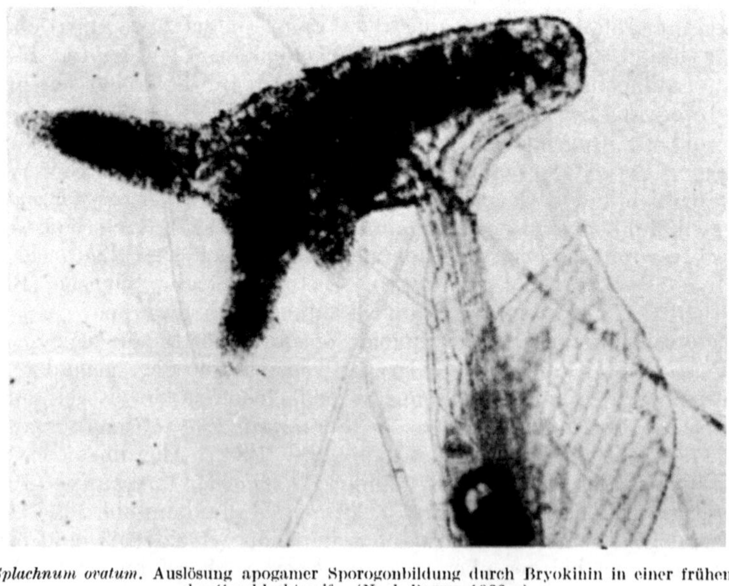

Abb. 6. *Splachnum ovatum*. Auslösung apogamer Sporogonbildung durch Bryokinin in einer frühen Phase der Geschlechtsreife. (Nach BAUER 1966a.)

Im übrigen ist die Meiosisdetermination kein isolierter Schritt. Die Erfahrungen mit der apogamen Sporophytenbildung beweisen, daß eine enge Koppelung und Koordination mit zahlreichen anderen Determinationsfaktoren, insbesondere der Gestaltbildung, besteht. Die apogame Induktion des Sporophytenhabitus schließt also die Voraussetzungen zum Ablauf der Meiosis ein.

Die Vorstellung der schrittweisen Determination der Sporophyten-Charakteristika (insbesondere der morphologischen) ist aus Beobachtungen an Archegoniaten gewonnen worden; eine Verallgemeinerung ist erst nach Erweiterung unserer experimentellen Erfahrungen möglich. — Bei einigen Sippen von *Splachnum pedunculatum* können an den Blattspitzen *alternativ* Sporogone oder Gametangien angelegt werden (BAUER, unpubl.). In anderen Fällen (*Splachnum ovatum*, BAUER 1966a) ist mit Bryokinin nur dann apogame Sporogonbildung auslösbar, wenn wenigstens die ersten Phasen der sexuellen Reifung durchlaufen worden sind (Abb. 6). Gewisse Bedingungen der sexuellen Reifung im Moosgametophyten können also von der apogamen Sporogonbildung ausgenutzt werden, und wenn man von diesen Anomalien auf die Normalentwicklung schließt, ergibt sich als Arbeitshypothese, daß bei den Laubmoosen bereits in der Phase der sexuellen Reifung eine Determination von „Konstruktionselementen" des Sporophyten erfolgt (vgl. BAUER 1963). Eine gemeinsame Voraussetzung für die Bildung der Gametangien und Sporogone der Laubmoose wäre z. B. die Determinierung der keilförmig-zweischneidigen Scheitel-

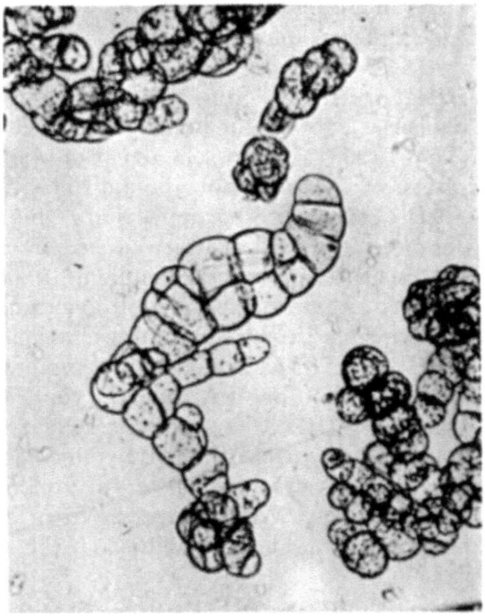

Abb. 7. *Physcomitrium piriforme*. Frühstadium eines Laubmoossporogons, welches sich in einer Calluskultur bildet. (Original BAUER.)

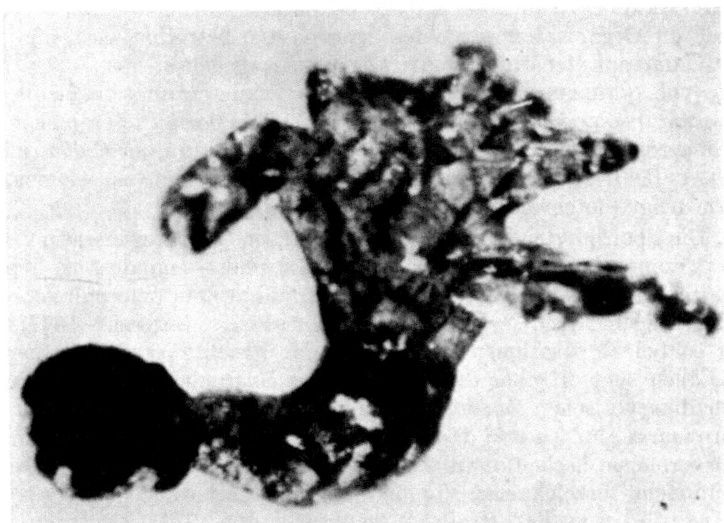

Abb. 8. *Todea barbara*. Prothalloide Bildungen am herauspräparierten und kultivierten Embryo. (Nach DE MAGGIO 1963.)

zelle. Wir wissen nicht, zu welchem Zeitpunkt der Entwicklung die Weichen für diesen Entwicklungsschritt gestellt werden. Bei obiger Annahme müßte aber die Determination über die Bildung der Archegonien hinaus auch in die dort gebildete Eizelle übernommen werden.

Die Vorstellung, daß die Determination des Sporophyten in Teilschritten erfolgt, wird auch durch Beobachtungen am Jugendstadium des Sporophyten nahegelegt. Sporogonanlagen, die sich aus Calluszellen differenzieren (Bauer 1961b, 1963), ähneln sehr den Embryonalstadien des natürlichen Entwicklungsganges. Sie besitzen auf diesem Entwicklungsstadium noch keine zweischneidige Scheitelzelle (Abb. 7) und sind sehr labil. Sie können unter noch näher zu definierenden Bedingungen über ein kurzzelliges, kettenförmiges Stadium in Protonema (vor allem im Dunkeln) übergehen oder zur Calluswuchsform zurückschlagen. Ältere Sporogone mit einer ausgebildeten meristematischen Zone hinter der Scheitelzelle behalten unter den gleichen Bedingungen ihren Differenzierungscharakter; die Determination ist nunmehr stabil.

Auch die Beobachtungen von De Maggio und Wetmore (1961) und De Maggio (1963) an herauspräparierten Embryostadien junger Sporophyten von *Todea barbara* offenbaren eine labile Determination der Frühstadien. Je jünger die Stadien waren, desto größer war die Abweichung von der Normalentwicklung bei Weiterkultur in vitro. Das noch einzellige befruchtete Ei wurde zu einem prothalloiden Embryo (Abb. 8).

2. Der Stabilitätsgrad der Sporophytendetermination bei Restitutionen.

Wenn differenziertes Gewebe zur Regeneration induziert wird, gehen die Regenerationsprodukte meist unter völliger Entdifferenzierung wieder in den Embryonalzustand zurück; dabei wird auch zumeist die Determination zu einem bestimmten Gewebetyp aufgehoben (vgl. Bünning 1955). Es ist die Frage, wieweit dabei auch die Organisationsstufe des Sporophyten betroffen wird, die ja gegenüber der Gametophytenstufe als die abgeleitete zu gelten hat.

Algen (vgl. Schussnig 1960). Die Sporophytendetermination bleibt erhalten bei *Cladophora* (isomorph), *Dictyota* (isomorph), *Laminaria* (heteromorph; Sporophyt > Gametophyt). Doch wird vom *Aglaozonia*-Stadium der *Cutleria* berichtet (Sauvageau 1899; Kuckuck 1899), daß es zum Umschlag zur Gametophytenwuchsform neigt (heteromorph; Sporophyt < Gametophyt).

Pilze. Die Sporophytendetermination ist während der Regeneration stabil bei *Allomyces* (isomorph; Sörgel 1937; Sost 1955) und — mindestens hinsichtlich des dikaryotischen Zustandes — bei den Basidiomyceten (Sporophyt > Gametophyt). Bei einigen Arten (z.B. *Coprinus stercorarius*; Brefeld 1877) kann es sogar zu echten Reparationen im Bereich des Fruchtkörpers kommen. — In anderen Fällen liegt offenbar die Stabilität des Sporophytencharakters lediglich in seinem dikaryotischen Zustand. Man kann experimentell den einen Partner eines Kernpaares abtöten und dadurch wieder den haploiden Zustand und dessen Reaktionsvermögen herbeiführen (Harder 1926, 1927). Auch in dem als „Bullersches Phänomen" bezeichneten Vorgang, bei dem ein haploides und ein dikaryotisches Mycel in „sexuelle" Reaktion treten (Buller 1931), leitet sich von einer ursprünglich dikaryotischen Zelle ein sekundär haploides Mycel ab, während das ursprünglich haploide Mycel dikaryotisiert wird.

Moose. Die Sporogone der Laubmoose sind beim Regenerationsvorgang besonders labil (Sporophyt < Gametophyt). Seit Pringsheim (1876) und Stahl (1876) wissen wir, daß dabei Protonema entsteht; Lebermoos-Sporogone bilden einen thallösen Gametophyten (Lang 1901; Bornhagen 1926; Schwarzenbach 1926; Lorbeer 1934; Burgeff 1937; Bauer 1955). Genauere Untersuchungen (Bauer 1957, 1959a, 1961a, 1963) haben jedoch ergeben, daß in jungen Laub-

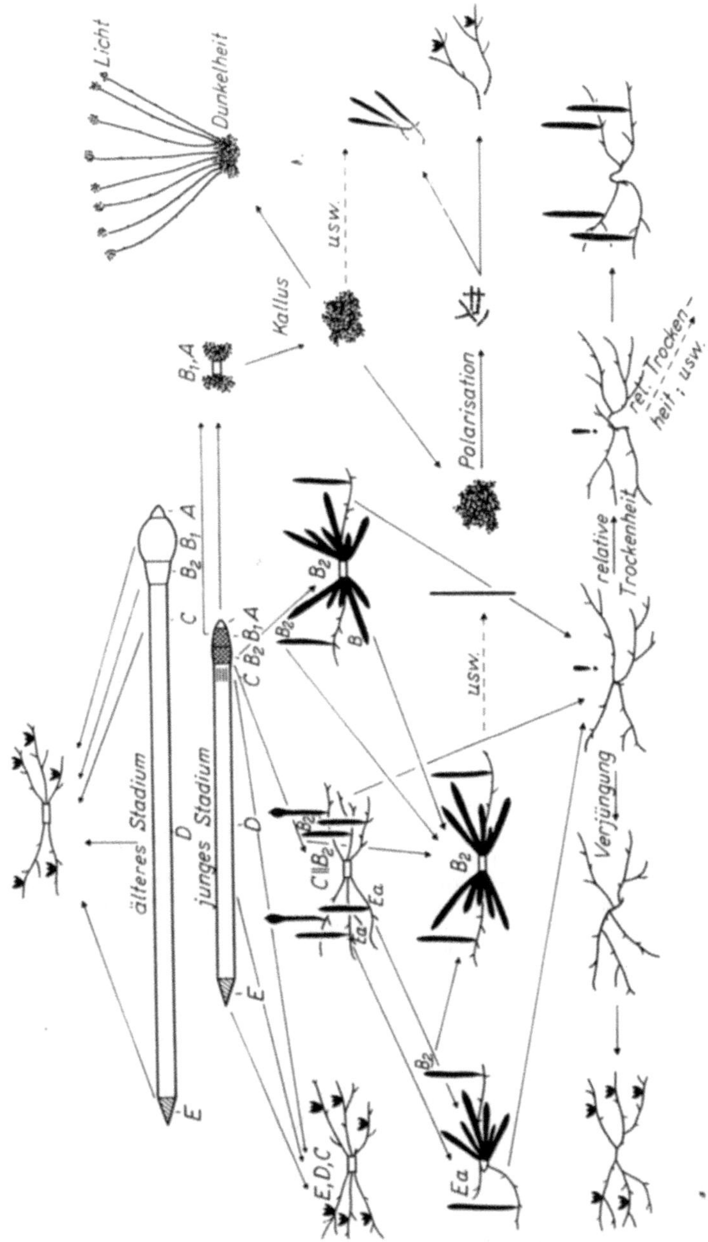

Abb. 9. Regeneration eines Laubmoos-Sporogons (Bastard *Funaria hygrometrica* × *Physcomitrium piriforme*) in Abhängigkeit vom physiologischen Alter des Sporogongewebes: verschiedene Determination der Regenerate, schematisch. *A* Scheitelzellregion; B_1 meristematische Region (im älteren Stadium Kapsel); B_2 meristematische Region (im älteren Stadium Apophyse); *C* Streckungszone; *D* ausdifferenzierte Seta; *E* Setafuß. (Nach Bauer 1963.)

moossporogonen ein entwicklungsphysiologischer Gradient hinsichtlich der Qualität der Regenerate besteht. Im Extremfall (embryonale Spitzenregion) besteht das Regenerationsprodukt wiederum aus Setagewebe bzw. Calluszellen, die jedoch in entwicklungsphysiologischer Hinsicht ebenfalls der Sporophytenstufe zuzurechnen sind; an der Grenze der Spitzenzone zur Streckungszone bleibt im Regenerat (Protonema) eine Restdetermination des Sporophytenstatus erhalten (Abb. 9; vgl. unten S. 253). In gewissen Bastardkombinationen von Funariaceenkreuzungen (BAUER 1961a, 1966b) ist die Sporophytendetermination bei Regenerationen so stabil, daß hier erstmals eine unbegrenzte Vermehrung der Sporogone durch Stecklinge und Setabruchstücke möglich ist. Diese bei Laubmoosen ungewöhnliche Stabilität der Sporophytendetermination dürfte wohl mit der Unausgeglichenheit des genetischen Systems der Bastardkombination zusammenhängen.

Pteridophyten und Spermatophyten. Im allgemeinen wird bei der Regeneration die Sporophytenstufe nicht verlassen, wenn man nicht in die zuweilen fädigen Bildungen im Callusgewebe Gametophytencharaktere hineininterpretieren will (Sporophyt > Gametophyt). Jedoch finden wir bei einer Reihe leptosporangiater Farne eine gewisse Labilität der Sporophytenstufe bei Regenerationen (GOEBEL 1905, 1907, 1908, 1930; BELL und RICHARDS 1958; vgl. auch WETMORE, DE MAGGIO und MOREL 1963). Es gelingt hier, an den Primärblättern junger Keimlinge — und nur an diesen! — außer neuen Sporophytenanlagen auch prothalloide Bildungen zu erzielen (vgl. unten S. 253).

Aus diesem verschiedenen Verhalten kann man ableiten, daß die Festigkeit der Sporophytendetermination vom *relativen Stärkeverhältnis* der Generationen mitbestimmt wird. Im isomorphen Generationswechsel scheint der Sporophyt immer „stark" genug zu sein, um den Sporophytencharakter zu bewahren. Im heteromorphen Generationswechsel spielt es eine Rolle, ob der Gametophyt oder der Sporophyt dominierend ist. Die Modifikation dieser Regel durch das Regenerationsverhalten mehrerer leptosporangiater Farne ist insofern verständlich, als ja ihr Gametophyt, gemessen an vielen Moosgametophyten, durchaus von ansehnlichen Dimensionen und unbegrenzt kultivierbar ist. Außerdem ist hier der besondere entwicklungsphysiologische Status der Primärblätter zu berücksichtigen (vgl. unten).

3. Experimentell induzierte Apogamie.

Die zahlreichen entwicklungsgeschichtlichen Arbeiten über den Generationswechsel der Algen enthalten vielfach Hinweise auf parthenogenetische Entwicklungen. Es existieren auch einige Ansätze zu deren Induzierung (SCHLÖSSER 1935; HARTMANN 1950; vgl. auch oben S. 237). Wichtig ist die Erfahrung, daß sich entwickelnde Eizellen sowohl den gametophytischen als auch den sporophytischen Entwicklungsweg einschlagen können (SAUVAGEAU 1908; SCHLÖSSER 1935; HARTMANN 1950). Die Faktoren sind noch undurchsichtig; da die cytologischen Verhältnisse nicht geklärt wurden, muß mit der Möglichkeit gerechnet werden, daß z.T. Diploidisierungen mit im Spiele sind. Ferner werden gewisse Entwicklungen bei *Ectocarpus* (SVEDELIUS 1921; SCHUSSNIG 1960) ebenfalls als Apogamie gedeutet. Bei den Ascomyceten kommt eine apogame Weiterentwicklung der Ascogone vor (KNIEP 1928). Doch sind alle diese Fälle zu wenig experimentell untersucht.

Bei diploidisierten Sippen von *Laubmoosen* ist Apogamie erst relativ spät beobachtet bzw. induziert worden (SPRINGER 1935; v. WETTSTEIN 1942). Später mehrten sich die Befunde (BAUER 1956, 1959a, 1961a, 1963, 1966; LAZARENKO 1960, 1961, 1963; LAZARENKO, PASHUK und LESNYAK 1961; LAL 1963). Man darf daher wohl annehmen, daß die Voraussetzungen zur Apogamie noch in weiteren Verwandtschaftskreisen gegeben sind; sie können auch bei Laubmoosen in der Haplophase vorliegen (vgl. oben S. 245).

Eine dieser Voraussetzungen liegt in der *genetischen Konstellation*. Sippen von *Phascum cuspidatum* verschiedener Standorte zeigen eine sehr unterschiedliche Neigung zur Apogamie (SPRINGER 1935); das gleiche gilt auch für *Splachnum pedunculatum* (BAUER, unpubl.), bei dem vier skandinavische Sippen bisher nicht zu beeinflussen waren, während drei alpine Sippen eine z.T. sehr ausgeprägte Neigung zur apogamen Sporogonbildung bei in vitro-Kultur zeigten. Bei vorhandener Bereitschaft begünstigt herabgesetzte Hydratur die apogamen Bildungen (SPRINGER 1935; BAUER 1963); dagegen wurde bei *Physcomitrium coorgense* (LAL 1963), das offenbar sehr stark zu spontaner Apogamie neigt, kein Hydratureffekt registriert. Bei gewissen Bastardnachkommen der Kreuzung *Funaria hygrometrica* × *Physcomitrium piriforme* vermag die Zugabe von Chloralhydrat (0,001 M) zum Nährboden apogame Sporogonbildung auszulösen (BAUER 1966b). Da der Chloralhydrateffekt sippenspezifisch ist, müssen die Voraussetzungen hierfür irgendwie in der Unausgeglichenheit des genetischen Systems gesucht werden.

Als weitere determinierende Faktoren für apogame Sporogonbildung sind gewisse *entwicklungsphysiologische Konstellationen* anzusehen, die im normalen Entwicklungsablauf des Gametophyten auftreten, insbesondere die Differenzierungsvoraussetzungen für die keilförmig-zweischneidige Scheitelzelle (vgl. oben S. 247). Es ist auffällig, daß apogam gebildete Sporophyten nur an den Stellen zu finden sind, an denen natürlicherweise sowieso eine zweischneidige Scheitelzelle die Gestaltbildung bestimmt: an den Spitzen junger Moosblättchen (*Phascum*, SPRINGER 1935; *Splachnum*, BAUER unpubl.; *Desmatodon*, LAZARENKO 1960) und anstelle von Gametangienanlagen (*Splachnum*, BAUER 1963, 1966, teilweise unpubl.); auch der Ersatz von Moosstämmchenknospen durch Sporogonknospen gehört hierher, wenn man davon ausgeht, daß der dreiseitig-pyramidalen Scheitelzelle ein mehr oder weniger ausgeprägtes labiles zweischneidiges Stadium vorausgeht.

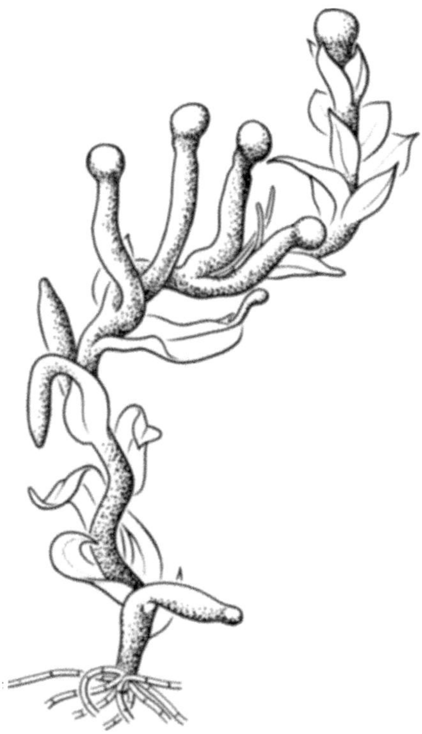

Abb. 10. *Desmatodon ucrainicus* Lazar. Mehrfacher Wechsel zwischen gametophytischer und sporophytischer Wuchsform. (Nach LAZARENKO 1960.)

In der entwicklungsphysiologischen Konstellation kann eine *labile Komponente* enthalten sein, die in Abhängigkeit von Außenfaktoren (Hydratur) ein mehrmaliges ,,Umkippen" von der gametophytischen in die sporophytische Wuchsform, und umgekehrt, erlaubt (F. v. WETTSTEIN 1942; LAZARENKO 1960; Abb. 10).

Die Fähigkeit zur apogamen Sporogonbildung kann auch durch einen *stofflichen Faktor* determiniert sein, der bei der vorausgegangenen aposporen Bildung von Protonema vom Sporogon her übernommen worden ist (BAUER 1959, 1961a, 1963). Dieser Faktor ist labil; er kann — offenbar durch Selbstreproduktion — im sich vermehrenden Protonema ziemlich lange erhalten bleiben. Hierzu sind relativ trockene Kulturbedingungen und Zuckerzusatz zum Nährmedium förderlich (vgl. Abb. 10).

Der von LAL (1963) an *Physcomitrium coorgense*-Gametophyten erzeugte Callus hat eine labilere Determination, als die von BAUER (1961, 1963) an Sporophyten von *Physcomitrium piriforme* und von Funariaceen-Bastarden gewonnenen Calli. Trotzdem darf man ihn wohl nicht ohne weiteres als Gametophytencallus bezeichnen, weil diese Kennzeichnung nicht nur die Herkunft, sondern vielmehr den Determinationscharakter beinhalten sollte. An Gametophyten von gewissen Funariaceen-Bastarden (BAUER, unpubl.) sind ebenfalls Calli induzierbar, welche aber *vorwiegend* zur Sporogondifferenzierung determiniert sind. In entwicklungsphysiologischer Hinsicht ist hier wohl eher die Callusbildung selbst und nicht die Setadifferenzierung aus dem Callus der wesentliche (wenn auch modifizierte) Apogamieschritt. Echte Gametophytencalli haben WARD (1960), D. v. WETTSTEIN (1953) und BAUER (1962) bei Laubmoosen induziert. Die Calli der beiden letzten Autoren sind jedoch nur so lange stabil, wie das induzierende Agens (Chloralhydrat, Acenaphthen) einwirkt, und die aller drei zeigen keinerlei Tendenzen zu sporophytischer Differenzierung.

Wir kennen zahlreiche Beobachtungen über apogame (bzw. parthenogenetische) Entwicklung bei *höheren Archegoniaten* und *Spermatophyten* (vgl. Beitrag DÖPP in diesem Band; ebenso GOEBEL 1930, 1932; STEIL 1939, 1951; MANTON 1951). Als Hauptursache wird Hybridisation angesehen. Doch sind die experimentellen Ansätze zu ihrem kausalen Verständnis ebenfalls noch bescheiden (vgl. auch WETMORE, DE MAGGIO und MOREL 1963).

LANG (1898) induzierte unter anderem bei *Pteridium aquilinum* auf einfache Weise dadurch apogame Sporogonbildung, daß er die Prothallien von der Oberfläche her *trocken* hielt und so eine Befruchtung verhinderte. Außerdem erkannte er eine kräftige *Belichtung* als notwendige Bedingung; dies wurde von DUNCAN (1941, 1943) bestätigt. WHITTIER und STEEVES (1960, 1962) und WHITTIER (1962) nahmen die Experimente LANGs wieder auf und konnten den Belichtungseffekt durch *Zucker*fütterung ersetzen. Die Zahl der gebildeten Sporogone war direkt proportional der Glucosekonzentration (0,8—2,5%). BELL (*Pteridium aquilinum*, 1959) registrierte positive Effekte an *alternden* Kulturen und bei *höheren Konzentrationen* des Mediums, jedoch waren bei ihm Zucker, Auxin und Nucleinsäurebausteine ohne fördernden Einfluß. BROWN (1923) beobachtete bei *Phegopteris polypodioides* apogam gebildete Keimlinge in *erschöpften Nährlösungen*.

Auch bei Farnen ist *Calluswuchs* beobachtet bzw. induziert worden (GOEBEL 1907; BEYERLE 1932; MOREL 1956b; BRISTOW 1962 = Bildung am *Sporophyten*; MOREL und WETMORE 1951; MOREL 1956a; STEEVES, SUSSEX und PARTANEN 1955 = Bildung am *Gametophyten*). Ähnlich wie bei den Laubmooscalli (s. oben) gibt es auch hier Unterschiede in der Differenzierungstendenz (GOEBEL 1907; BEYERLE 1932; STEEVES, SUSSEX und PARTANEN 1955; BRISTOW 1962). BRISTOW konnte zeigen *(Pteris cretica)*, daß in Abwesenheit von Zucker Gametophytendifferenzierung und bei Zuckerfütterung (0,01% und höher) Sporophytendifferenzierung eintritt. — Hinsichtlich der Bezeichnung ,,Gametophytencallus" muß man den gleichen Vorbehalt machen, wie dies oben bei den Mooscalli getan wurde. Entscheidend sind die Differenzierungspotenzen. Wenn ein vom Gametophyten stammender Callus (STEEVES, SUSSEX und PARTANEN 1955) sich bei Abwesenheit von Zucker im Medium zum Gametophyten, bei Zuckergabe aber zum Sporophyten entwickelt, dann ist der zweite Effekt sicher keine reine Apogamie, sondern der Callus ist offenbar — ähnlich wie der ,,Gametophytencallus" von *Physcomitrium coorgense* (LAL 1963) — hinsichtlich seiner Sporophytentendenz vordeterminiert.

Rückblickend kann man feststellen, daß bezüglich der Bedingungen für apogame Bildungen einige wichtige Übereinstimmungen zwischen Moosen und Farnen bestehen. In beiden Pflanzengruppen spielt die *genetische Konstitution* eine wichtige Rolle; ferner wird der Zusammenhang mit den Prozessen der *sexuellen Reifung* deutlich. Die Bedeutung der Hydratur und der Zuckerfütterung für die Induktion wird nicht von allen Autoren einheitlich beurteilt. Dies mag mit der Verschiedenheit der Objekte zu erklären sein.

IV. Mischbildungen zwischen Gametophyten und Sporophyten.

Mischbildungen zwischen den beiden Generationen sind unter anderem insofern von Interesse, als sie zeigen, daß Charaktere einer Phase nicht einen geschlossenen Determinationskomplex darstellen. Die in den Mischbildungen zum Ausdruck kommenden Teildeterminationen könnten ein Modell, vielleicht aber auch die Grundlage für die oben diskutierte schrittweise Determination des spezifischen Charakters der jeweiligen Phase darstellen.

Ob das „Mischmycel" von *Allomyces* (Sörgel 1937), an welchem Gametangien und Dauersporangien (Sporophytenmerkmal) vorkommen, wirklich eine Zwischenform darstellt oder eher als eine Reminiszenz an ein ursprüngliches Verhalten (im phylogenetischen Sinne) aufzufassen ist, sollte man noch offen lassen.

Für unser Problem *nicht* von Bedeutung sind alle jene „Mischbildungen", bei denen durch Umschlagen der Differenzierung Gametophyten- und Sporophytengewebe miteinander verwachsen sind: „Forma Church" der *Aglaozonia* (Sauvageau 1899); „Mischbildungen" nach Art der Abb. 10; apogam gebildete Sporogone an Blattspitzen, in Gametangienständen, am Protonema; das mit Protonema regenerierende Sporogon; das mit Prothallien regenerierende Farnprimärblatt. Der Mischcharakter muß vielmehr im Regenerat selbst liegen.

Bei den Laubmoosen ist in diesem engeren Sinne besonders auf den Fall zu verweisen, in dem während des Regenerationsvorganges ein Determinationsfaktor vom Sporogon in das Regeneratprotonema übergeht (Bauer 1959a, 1963) und dort bis zum Zeitpunkt der Knospenbildung latent bleibt; unter seiner Wirkung entwickeln sich die Moosknospen zu jungen Sporogonen. Seine unverminderte Aktivität über zahlreiche Passagen der Protonemavermehrung schließt eine hormonartige Wirkung eines vom Sporogon abgegebenen Stoffes aus; eher ist an eine Nucleinsäure zu denken. Bemerkenswert ist, daß sich dieser Teilfaktor der Sporogondifferenzierung offenbar auch in der entwicklungsphysiologisch „falschen" Umgebung harmonisch einfügt (vgl. Abb. 9).

Die Calluswuchsform, die sich sowohl vom Gametophyten als auch vom Sporophyten ableiten kann (Moose und Farne), sei hier nur mit Vorbehalt genannt. Sie könnte — soweit sie die Potenz zur Gametophyten- *und* Sporophytendifferenzierung besitzt — nach beiden Richtungen teildeterminiert sein; dann wäre sie eine Mischbildung. Sie könnte sich aber auch — etwa als potentielles Sporophytengewebe — in einer Phase labiler Determination befinden, die durch experimentelle Eingriffe umgestoßen werden kann. In gleicher Weise wären auch die Embryonalstadien junger Sporophyten von Laubmoosen und Farnen (eventuell sogar die Primärblätter junger Farnpflanzen) zu diskutieren; doch hat wohl in diesen Fällen die Annahme einer labilen Determination der reinen Sporophytenstufe die größere Wahrscheinlichkeit für sich. Über den entwicklungsphysiologischen Status der „labilen" und „stabilen" Determination wissen wir nichts.

Eindeutige Mischbildungen sind jedoch zahlreiche durch Regeneration aus Farnprimärblättern gebildete Prothallien, auf die Goebel (1908) aufmerksam gemacht hat. Wir finden an ihnen Sporophytenmerkmale wie Spaltöffnungen und Leitbündel. Da die Regenerate von einer Zelle ausgehen, kann es sich nicht um primäre Chimären handeln.

In diesem Zusammenhang sei auch noch einmal auf die oben dargelegte Auffassung hingewiesen, daß gewisse Informationskomplexe („Scheitelzellcharakter", „Stämmchenform" u.a.) in *beiden* Gestaltphasen zum *normalen* Bestandteil der „Informationsgarnitur" gehören, aber jeweils in verschiedener Kombination verschiedene morphologische Einheiten bedingen (Bauer 1963). Hier handelt es sich zwar nicht um Mischbildungen im obigen Sinne, aber es besteht zweifellos zwischen den beiden Gesichtspunkten eine innere Verwandtschaft. Wir ziehen aus allen diesen Mischphänomenen den gleichen Schluß, den GOEBEL (1908) bereits gezogen hat, „daß die beiden ‚Generationen', in welche sich der Entwicklungsgang ... gliedert, viel weniger voneinander verschieden sind, als man meist angenommen hat".

Literatur.

ALLSOPP, A., and G. C. MITRA: The heterotrichous habit in the protonema of the Bryales. Nature (Lond.) **178**, 1063—1064 (1956).

BAUER, L.: Über die Bildung von Tochterthalli aus den Rhizoiden einer diploiden Sippe von *Anthoceros laevis*. Ber. dtsch. bot. Ges. **68**, 381—384 (1955). — Über vegetative Sporogonbildung bei einer diploiden Sippe von *Georgia pellucida*. Planta (Berl.) **46**, 604—618 (1956). — Regenerationsversuche am Sporogon von *Physcomitrium piriforme* (Brid.). Ber. dtsch. bot. Ges. **70**, 424—432 (1957). — Auslösung apogamer Sporogonbildung am Regenerationsprotonema von Laubmoosen durch einen vom Muttersporogon abgegebenen Faktor. Naturwissenschaften **46**, 154—155 (1959). — Über die Stabilität von Teilfaktoren aus dem Faktorenkomplex der Sporogonbildung von Laubmoosen. Biol. Zbl. **80**, 353—362 (1961a). — Über die Kalluswuchsform des Laubmoossporophyten. Naturwissenschaften **48**, 507—508 (1961b). — Über Wachstum und Differenzierung der Kalluswuchsform der Laubmoose. Ber. dtsch. bot. Ges. **75**, (78) (1962). — On the physiology of sporogonium differentiation in mosses. J. Linn. Soc. (Bot.) **58**, 343—351 (1963). — Isolierung und Testung einer kinetinartigen Substanz aus Kalluszellen von Laubmoossporophyten. J. Pflanzenphysiol. **54**, 241—253 (1966a). — Über die Induktion apogamer Sporogonbildung bei Laubmoosen durch Chloralhydrat. Beitr. Biol. Pflanz. **42**, 113—125 (1966b). — BELL, P. R.: The experimental investigation of the pteridophytic life cycle. J. Linn. Soc. (Bot.) **56**, 188—202 (1959). — BELL, P. R., and B. M. RICHARDS: Induced apospory in polypodiaceous ferns. Nature (Lond.) **182**, 1748—1749 (1958). — BERTHOLD, G.: Die geschlechtliche Fortpflanzung der eigentlichen Phaeosporeen. Mitt. Zool. Stat. Neapel **2**, 401—413 (1881). — BEYERLE, R.: Untersuchungen über die Regeneration von Farnprimärblättern. Planta (Berl.) **16**, 622—665 (1932). — BLAKESLEE, A. F., J. BELLING, M. E. FARNHAM, and A. D. BERGENER: A haploid mutant in the Jimson Weed, *Datura stramonium*. Science **55**, 646—647 (1922). — BLIDING, C.: Über Sexualität und Entwicklung bei der Gattung *Enteromorpha*. Svensk. Bot. Tidskr. **27**, 233—256 (1933). — BONNER, J. T.: Evidence for the formation of cell aggregates by chemotaxis in the development of the slime mold, *Dictyostelium discoideum*. J. exp. Zool. **106**, 1—37 (1947). — BORNHAGEN, H.: Die Regeneration (Aposporie) des Sporophyten von *Anthoceros laevis*. Biol. Zbl. **46**, 578—586 (1926). — BREFELD, O.: Untersuchung über Schimmelpilze, Nr. 3: Basidiomyceten I. Leipzig: O. Felix 1877. — BRISTOW, J. M.: The controlled in vitro differentiation of callus derived from a fern, *Pteris cretica* L., into gametophytic or sporophytic tissues. Develop. Biol. **4**, 361—375 (1962). — BROWN, E. D. W.: Regeneration in *Phegopteris polypodioides*. Bull. Torrey bot. Club **45**, 391—397 (1918). — BÜNNING, E.: Regenerationen bei Pflanzen. In: Handbuch der Allgemeinen Pathologie. Herausgeg. von F. BÜCHNER, E. LETTERER, F. ROULET, Bd. 6/1. Berlin-Göttingen-Heidelberg: Springer 1955. — BULLER, A. H. R.: Researches on Fungi, Vol. IV. London-New York-Toronto: Hafner Publ. Co. 1931. — BURGEFF, H.: Über Polyploidie bei *Marchantia*. Z. indukt. Abstamm.- u. Vererb.-Lehre **73**, 394—403 (1937).

CARTER, N.: An investigation into the cytology and biology of the Ulvaceae. Ann. Bot. **40**, 665—689 (1926). — CHURCH, A. H.: The polymorphy of *Cutleria multifida* (Grev.). Ann. Bot. **12**, 75—109 (1898). — CLAUSSEN, P.: Zur Entwicklungsgeschichte der Ascomyceten. *Pyronema confluens*. Z. Bot. **4**, 1—64 (1912). — CLINT, H. B.: The life history and cytology of *Sphacelaria bipinnata*. Sauv. Publ. Hartley Bot. Lab. Univ. Liverpool **3**, 5—23 (1927).

DE MAGGIO, A. E.: Morphogenetic factors influencing the development of fern embryos. J. Linn. Soc. (Bot.) **58**, 361—376 (1963). — DE MAGGIO, A. E., and R. H. WETMORE: Morphogenetic studies on the fern *Todea barbara*. III. Experimental embryology. Amer. J. Bot. **48**, 551—565 (1961). — DREW, K. M.: Contributions to the cytology of *Spermothamnion turneri* (Mert.) Aresch. 1. The diploid generation. Ann. Bot. **48**, 549—573 (1934). — An investigation of *Plumaria elegans* (Bonnem.) Schmitz with special reference to triploid plants bearing parasporangia. Ann. Bot., N.S. **3**, 347—368 (1939). — Contributions to the cytology of *Spermothamnion turneri* (Mert.) Aresch. 2. The haploid and triploid generations. Ann. Bot., N.S. **7**, 23—30 (1943). — DUNCAN, R. E.: Apogamy in *Doodia caudata*. Amer. J. Bot. **28**, 921—931 (1941). — Origin and development of embryos in certain apogamous ferns of *Dryopteris*. Bot. Gaz. **105**, 202—211 (1943).

FELDMANN, J.: Une Némalionale à carpotétraspores: *Helminthocladia hudsoni* (C. Ag.) J. Ag. Bull. Soc. Hist. Nat. Afrique du Nord **30** (1939). — FÖYN, B. R.: Über den Lebenscyclus einiger Braunalgen. Bergens Mus. Årbok Naturvid. Rekke 1934a. — Lebenszyklus, Cytologie und Sexualität der Chlorophycee *Cladophora suhriana* Kütz. Arch. Protistenk. **83**,

1—56 (1934b). — FRITSCH, F. E.: The structure and reproduction of the algae. Vol. I. Cambridge 1945a. — The structure and reproduction of the algae. Vol. II. Cambridge: Cambridge University Press 1945b.

GERISCH, G.: Die Bildung des Zellverbandes bei *Dictyostelium minutum*. I. Übersicht über die Aggregation und den Funktionswechsel der Zellen. Wilhelm Roux' Arch. Entwickl.-Mech. Org. **155**, 342—357 (1964). — GOEBEL, K.: Kleine Mitteilungen. 3. Aposporie bei *Asplenium dimorphum*. Flora (Jena) **95**, 239—244 (1905). — Experimentell-morphologische Mitteilungen. 1. Künstlich hervorgerufene Aposporie bei Farnen. Sitz.-Ber. bayer. Akad, Wiss. **37**, 119—138 (1907). — Einleitung in die experimentelle Morphologie der Pflanzen. Leipzig u. Berlin 1908. — Organographie der Pflanzen. Teil 1. Allgemeine Organographie. 3. Aufl. Jena: Gustav Fischer 1928. — Organographie der Pflanzen, Teil 2. Bryophyten-Pteridophyten, 3. Aufl. Jena: Gustav Fischer 1930. — GRAU, W.: Stoffwechselvergleich zwischen Gametophyt und Sporophyt eines Laubmooses. Diss. Tübingen 1963.

HARDER, R.: Mikrurgische Untersuchungen über die geschlechtliche Tendenz der Paarkerne des homothallischen *Coprinus sterquilinus* Fr. Planta (Berl.) **2**, 446—453 (1926). — Zur Frage nach der Rolle von Kern und Protoplasma im Zellgeschehen und bei der Übertragung von Eigenschaften. Z. Bot. **19**, 337—407 (1927). — HARTMANN, M.: Beiträge zur Kenntnis der Befruchtung und Sexualität mariner Algen. I. Über die Befruchtung von *Cutleria multifida*. Publ. Staz. Zool. Napoli **22** (1950). — Die Sexualität, 2. Aufl. Stuttgart: Gustav Fischer 1956. — HAWKER, L. E.: The physiology of reproduction in fungi. Cambridge: Cambridge University Press 1957. — HOFFMANN, A.: Untersuchungen über die Geschlechtsdifferenzierung bei haploiden und diploiden Gametophyten von *Bryum capillare* L. Z. indukt. Abstamm.- u. Vererb.-Lehre **87**, 753—768 (1956). — HOFMEISTER, W.: Vergleichende Untersuchungen über die Keimung, Entwicklung und Befruchtung der höheren Kryptogamen und über die Befruchtung der Koniferen. Leipzig: Wilhelm Engelmann 1851.

IVANOV, M. A.: Experimental production of haploids in *Nicotiana rustica* (and a discussion of haploidy in flowering plants). Genetica **20**, 295—397 (1938).

JULLER, E.: Der Generations- und Phasenwechsel bei *Stigeoclonium subspinosum*. Arch. Protistenk. **89**, 55—93 (1937).

KNAPP, E.: Haploide Pflanzen von *Antirrhinum majus*. Ber. dtsch. bot. Ges. **57**, 371—379 (1939). — KNIEP, H.: Die Sexualität der niederen Pflanzen. Jena: Gustav Fischer 1928. — KNIGHT, M.: Studies in the Ectocarpaceae. II. The life-history and cytology of *Ectocarpus siliculosus* Dillw. Transact. Roy. Soc. Edinburgh **56**, 307—332 (1929). — Nuclear phases and alternation in algae. Phaeophyceae. Beih. bot. Zbl. I, **48**, 15—37 (1931). — KORNMANN, P.: Über *Codiolum* und *Urospora*. Helgol. Wiss. Meeresunters. 8, 42—57 (1961a). — Die Entwicklung von *Porphyra leucosticta* im Kulturversuch. Helgol. Wiss. Meeresunters. 8, 167—175 (1961b). — Die Entwicklung von *Monostroma grevillei*. Helgol. Wiss. Meeresunters. 8, 195—202 (1962). — KUCKUCK, P.: Beiträge zur Kenntnis der Meeresalgen. IX. Über den Generationswechsel von *Cutleria multifida* (Engl. Bot.). Grev. Wiss. Meeresunters. Abt. Helgoland, N.F. **3**, 95—116 (1899).

LAL, M.: Experimental induction of apogamy and the control of differentiation in gametophytic callus of the moss *Physcomitrium coorgense* Broth. Plant Tissue and Organ Culture — A Symposium. Herausg. P. MAHESHWARI and N. S. RANGA SWAMY, 363—381 (1963). — LANG, W. H.: On apogamy and the development of sporogonia upon fern prothallia. Philos. Trans. Roy. Soc. B (Lond.) **190**, 187—238 (1889). — On apospory in *Anthoceros laevis*. Ann. Bot. **20**, 503—510 (1901). — LAZARENKO, A. S.: Apogamous formation of sporogonia on the diploid protonema of *Desmatodon*. Dokl. Akad. Nauk USSR. **134**, 1240—1243 (1960). In Russisch mit engl. Zusammenf. — Apogamous sporogonium on the diploid protonema of *Splachnum ovatum* Hedw. Dop. Akad. Nauk Ukrain. SRS. **1**, 99—103 (1961). In Russisch mit engl. Zusammenf. — Apogamous sporogonia in the diplophase of *Pottia intermedia* (Turn.) Fürnr. Dop. Akad. Nauk Ukrain. SRS. **11**, 1524—1526 (1963). — LAZARENKO, A. S., C. T. PASHUK, and E. N. LESNYAK: Apogamy in the haplophase of *Desmatodon randii* (Kenn.) Lazar. Dop. Akad. Nauk Ukrain. SRS. **10**, 1381—1384 (1961). — LEWIN, R. A.: Studies on the flagella of algae. I. General observations on *Chlamydomonas moewusii* Gerloff. Biol. Bull. **102** (1952). — Genetics of *Chlamydomonae moewusii* Gerloff. J. Genet. **51**, 543—560 (1953). — LIND, E. M.: A contribution of the life-history and cytology of two species of *Uhlotrix*. Ann. Bot. **46**, 711—725 (1932). — LIST, H.: Die Entwicklungsgeschichte von *Cladophora glomerata* Kütz. Arch. Protistenk. **72**, 453—481 (1930). — LORBEER, G.: Die Zytologie der Lebermoose mit Berücksichtigung allgemeiner Chromosomenfragen. Jb. wiss. Bot. **80**, 567—818 (1934).

MANTON, I.: Problems of cytology and evolution in the Pteridophyta. Cambridge: Cambridge University Press 1950. — MATHIAS, W. T.: The life-history and cytology of *Phloeospora brachiata* Born. Publ. Hartley Bot. Lab. Univ. Liverpool **13**, 1—23 (1935). — MOHR, H.: Die Abhängigkeit des Protonemawachstums und der Protonemapolarität bei Farnen vom Licht. Planta (Berl.) **47**, 127—158 (1956). — MOREL, M.: Nouvelles méthodes de culture de tissus. Rev. gen. Bot. **63**, 314—324 (1956a). — Prolifération des feuilles d'*Adiantum pedatum* cultivées in vitro. Rev. gen. Bot. **63**, 325—330 (1956b). — MOREL, M. G., and R. H. WETMORE: Fern callus tissue culture. Amer. J. Bot. **38**, 140—143 (1951). — MÜLLER, D. G.: Life cycle of *Ectocarpus siliculosus* from Naples, Italy. Nature (Lond.) **203**, 1402 (1964).

OEHLKERS, F.: Das Leben der Gewächse, Bd. 1. Berlin-Göttingen-Heidelberg: Springer 1956.

PASCHER, A.: II. Über die Zwergmännchen der Oedogoniaceen. Hedwigia **46**, 265—278 (1907). — PRINGSHEIM, N.: Beiträge zur Morphologie und Systematik der Algen. I. Morphologie der Oedogonien. Jb. wiss. Bot. **1**, 1—81 (1858). — Über vegetative Sprossung der Moosfrüchte. Mber. kgl. Akad. Wiss. Berlin **1876**, 425—429.

SAUVAGEAU, C.: Les Cutlériacées et leur alternance de générations. Ann. Sci. Nat. Bot. Sér. VIII, **10**, 265—362 (1899). — Sur la germination parthénogénétique du *Cutleria adspersa*. C. R. Soc. Biol. (Paris) **64**, 698—700 (1908a). — III. Nouvelles observations sur la germination parthénogénétique du *Cutleria adspersa*. C. R. Soc. Biol. (Paris) **65**, 165—167 (1908b). — Sur un nouveau type d'alternance de générations chez les algues brunes; les Sporochnales. C. R. Acad. Sci. (Paris) **182**, 361—364 (1926). — Sur les algues Phéosporées à éclips ou Éclipsiophycées. Rev. Trav. Bot. Néerl. **25**, 260—270 (1928). — Le pléthysmothalle. Bull. Stat. Biol. d'Arcachon **29**, 1—16 (1932). — SCHILLER, J.: Beiträge zur Kenntnis der Entwicklung der Gattung *Ulva*. Sitzber. Akad. Wiss., Math.-nat. Kl. 1 Wien **116**, 1691—1716 (1907). — SCHLÖSSER, L. A.: Zur Entwicklungsphysiologie des Generationswechsels von *Cutleria*. Biol. Zbl. **55**, 198—208 (1935). — SCHUSSNIG, B.: Der Generationswechsel bei den Chlorophyceen im Lichte neuerer Erkenntnisse. Ber. dtsch. bot. Ges. **57**, (148)—(154) (1939). — Handbuch der Protophytenkunde, Bd. 2. Jena 1960. — SCHUSSNIG, B., u. E. KOTHBAUER: Der Phasenwechsel von *Ectocarpus siliculosus*. Österr. bot. Z. **83**, 81—97 (1934). — SCHWARZENBACH, M.: Regeneration und Aposporie bei *Anthoceros*. Arch. Jul.-Klaus-Stift. Vererb.-Forsch. **2**, 91—141 (1926). — SMITH, G. M.: Herausgeber von: Manual of Phycology. Waltham, Mass. USA: Chronica botania 1951. — Cryptogamic botany, Vol. II. New-York-Toronto-London: McGraw-Hill 1955. — SÖRGEL, G.: Untersuchungen über den Generationswechsel von *Allomyces*. Z. Bot. **31**, 401—446 (1937). — SOST, H.: Über die Determination des Generationswechsels von *Allomyces arbuscula* Butl. Arch. Protistenk. **100**, 541—564 (1955). — SPRINGER, E.: Über apogame (vegetativ entstandene) Sporogone an der bivalenten Rasse des Laubmooses *Phascum cuspidatum*. Z. indukt. Abstamm.- u. Vererb.-Lehre **69**, 249—262 (1935). — STAHL, E.: Über künstlich hervorgerufene Protonemabildung an dem Sporogonium der Laubmoose. Bot. Ztg **34**, 689—695 (1876). — STEEVES, T. A., I. M. SUSSEX, and C.R. PARTANEN: In vitro studies on abnormal growth of prothalli of the bracken fern. Amer. J. Bot. **42**, 232—245 (1955). — STEIL, W. N.: Apogamy, apospory and parthenogenesis in the Pteridophytes. Bot. Rev. **5**, 433—453 (1939). — Apogamy, apospory and parthenogenesis in the Pteridophytes. II. Bot. Rev. **17**, 90—104 (1951). — STOSCH, H. A. v.: Ein morphologischer Phasenwechsel bei einer Coccolithophoride. Naturwissenschaften **42**, 423 (1955). — STUMM, C.: Die Analyse von Genmutanten mit geänderten Fortpflanzungseigenschaften bei *Allomyces arbuscula* Butl. Z. Vererb.-Lehre **89**, 521—539 (1958). — SVEDELIUS, N.: Über Sporen an Geschlechtspflanzen von *Nitophyllum punctatum*; ein Beitrag zur Frage des Generationswechsels der Florideen. Ber. dtsch. bot. Ges. **32**, 106—116 (1914). — Einige Bemerkungen über Generationswechsel und Reduktionsteilung. Ber. dtsch. bot. Ges. **39**, 178—187 (1921). — Nuclear phases and alternation in the Rhodophyceae. Beih. bot. Zbl. I, **48**, 38—59 (1931).

THURET, G.: Recherches sur les zoospores des algues et le anthéridies des cryptogames. I. Ann. Sci. Nat. Bot. Sér. III, **14**, 214—260 (1850).

WARD, M.: Callus tissues from the mosses *Polytrichum* and *Atrichum*. Science **132**, 1401—1402 (1960). — Developmental pattern of adventitious sporophytes in *Phlebodium aureum* J. Sm. J. Linn. Soc. (Bot.) **58**, 377—380 (1963). — WETMORE, R. H., A. E. DE MAGGIO, and G. MOREL: A morphogenetic look at the alternation of generations. Maheshwari Comm. Vol., J. Indian bot. Soc. **42** A, 306—320 (1963). — WETTSTEIN, D. v.: Beeinflussung der Polarität und undifferenzierte Gewebebildung aus Moossporen. Z. Bot. **41**, 199—226 (1953). — WETTSTEIN, F. v.: Über einige Beobachtungen und experimentelle Befunde bei Laubmoosen. Ber. dtsch. bot. Ges. **60**, 394—414 (1942). — WETTSTEIN, R.: Handbuch der systematischen Botanik. Leipzig u. Wien: F. Deuticke 1935. — WHITTIER, D. P.: The origin and development of apogamous structures in the gametophyte of *Pteridium* in sterile culture. Phytomorphology **12**, 10—20 (1962). — WHITTIER, D. P., and T. A. STEEVES: The induction of apogamy in the bracken fern. Canad. J. Bot. **38**. 925—930 (1960). — Further studies on induced apogamy in ferns. Canad. J. Bot. **40**, 1525—1531 (1962). — WIDDER, F.: Grundformen des pflanzlichen Generationswechsels. Phyton **3**, 252—272 (1951). — WINKLER, H.: Über den Biontenwechsel und die Abweichungen von seinem normalen Verlauf. Planta (Berl.) **33**, 1—90 (1942/43). — Über Parthenogenesis und Apogamie im Pflanzenreich. Progressus Rei Bot. **2**, 293—444 (1908). — WISLOUCH, S.: Beiträge zur Biologie und Entstehung von Heilschlamm der Salinen der Krim. Acta Soc. Bot. Polon. **2**. 99—129 (1924).

ZIMMERMANN, W.: Wie der Generationswechsel der Archegoniaten und der Phanerogamen entstanden ist. Planta (Berl.) **51**, 511—517 (1958).

General principles of sexual and asexual reproduction and life cycles.

By

Flávio Resende.

With 13 figures.

1. General concept of sex.

A. Historical survey.

It was certainly by witnessing copulation in higher animals, supplemented by observation of primary and secondary sexual characters, that primitive people arrived at the notion of sex: of the male and female which copulate.

Scientific curiosity led to a more detailed study of the distinguishing characteristics of the two sexes and from the very beginning endeavoured to find out the roles of the two progenitors in originating the offspring. This curiosity was concentrated mainly on the human species both from the popular and scientific point of view.

From the scientific view point, the ancient cultures (mainly Greece from the 6th to the 4th centuries B.C.) were fundamentally interested in the different roles of male and female in procreation, especially in the human species. ARISTOTLE for example (460—377 B.C.) dissected many animals for the purpose of gaining knowledge applicable to man.

This same philosopher, more than a century after the elaboration of the relatively well thought out *theory of panspermy* (conceived by the atomists), evolved the mistaken concept of the existence of "seed" only in man *(causa formalis)*. The woman would only provide the material *(causa materialis)* from which the male "seed" would form the new being.

Early interpretations of sex segregation, both popular and scientific, e.g. the differences of heat and strength between the father and mother now seem so very naive. The scientific(!) "theory of the right-left" was dominant for centuries (STUBBE, 1964, p. 20—22).

In fact, up to the second half of the 17th century and the discovery of LEEUWENHOEK'S microscope (1677), the scientific notions of sex and of sexual reproduction and sexual segregation ran closely parallel to popular notions to the point of being coincident with them.

Thus took root the natural and intuitive general notion (which science adopted) that the physiology of sex was expressed by *copulation* of two different beings, male and female. The relation of this act with the origin of the offspring was much discussed by the Greeks, but it was only from the 17th century onwards that it began to be understood that the sexual organs were more than mere waiting chambers for the exit of "seed" arising from the brain, medulla or blood (STUBBE, 1964).

Microscopic observations of sexual secretions made since 1875 (Hertwig, Strasburger, Auerbach, Bütschli) showed that cellular phenomena imitated the macroscopic phenomenon of copulation of multicellular beings. Thus the concept of copulation of higher animals was extended to the cellular world and the copulation[1] of cells began to be called cellular conjugation or fecundation. In unicellular beings copulation is obviously identical to cellular conjugation, which in the majority of cases shows the same physiological differences taking place in male and female copulation: immobility of one cell (female), mobility of the other (male).

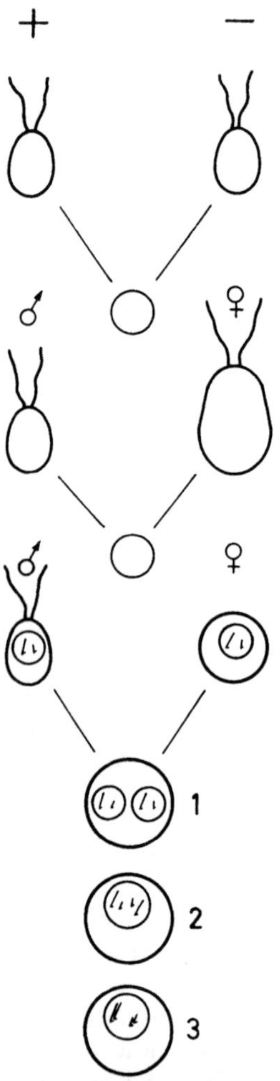

Fig. 1. Schematic representation (from top to bottom) of isogamy, anisogamy and oogamy. By oogamy the 3 steps of cell conjugation are represented: plasmogamy (*1*), karyogamy (*2*) and chromosomogamy (*3*). Two pairs of chromosome are shown. (Original.)

These cells were thus considered as sexual cells and given the Greek name of *gametes*. The mobile male gamete was called the spermatozoid and the female, immobile cell, the ovule. The following symbols were then adopted ♂, □ (masculine) and ♀, ○ (feminine).

The organs of higher animals where the male gametes are produced, the testes, have long been known as the male sexual organs. These were the first sexual organs recognized by man as they were easily observed in the majority of mammals. The female sex organs were seen much later. Herophilos (3th century B.C.) was the first to observe the "female testes", in mammals, which Galen confirmed and corrected regarding the function of the seminal ducts proposed by Herophilos. Only much later did the female organs become known as ovaries and this was certainly due to the original observation by v. Baer (1817) of female sexual cells, the ovules, in the bitch. This was a century and a half after the first observation of spermatozoids in man by Leeuwenhoek in 1677.

After the observation of fertilization of ovules by spermatozoids [cells which are morphophysiologically quite different (oogamy)], the discovery of conjugation between physiologically identical cells which differed morphologically (anisogamy) was made in the case of many lower organisms and also conjugation between "completely" identical cells (isogamy). In the genus *Chlamydomonas (Flagellatae)* for example, we find species showing all three types (Fig. 1). The concept of cellular fecundation was thus extended to include these cases and so a distinction was made between isogamic, anisogamic and oogamic conjugation (Fig. 1).

The existence of different sexes for conjugation, at least in the cases of isogamy, seemed to have become unnecessary! The experimental possibility of crossing

[1] At first sight there seems to be a fundamental difference between *copulation* of higher animals and cellular *conjugation*. In the former there is no fusion whereas in the latter there is the disappearance of the two cellular entities which become one. The cells seem to fuse rather than to copulate. If, however, we look at the essential part of the cell, the chromosomes, we observe that they do not fuse, but come together for a short time and then again separate, i.e. they only copulate (pachytene).

isogamic and anisogamic strains with oogamic strains (BLAKESLEE, MOEWUS) proved, however, that morphological isogamy does not imply physiological isogamy: the two conjugation gametes of an isogamic strains, on being crossed with an oogamic strain, do not conjugate indiscriminately with the spermatozoid or the ovule, but one of them will only conjugate with the ovule, whereas the other will only conjugate with the spermatozoid. It was thus logically deduced that an isogamic strain possesses ♂ and ♀ gametes morphologically indistinguishable at least at the cellular level; the former conjugating only with the ovules and the latter only with the spermatozoa of the oogamic race. And so at the cellular level the same popular criterion of masculinity and femininity was carried on; there is no other test for demonstrating the ♀ or ♂ character of those cells other than the test of the immobile (♀) or mobile (♂) conjugating cells of oogamic strains.

Before carrying out this test the two types of conjugating gametes are simply designated using + and — signs. It is possible to prove experimentally that the + and — gametes do not conjugate with their own kind, but we cannot assign the terms masculine and feminine to + and — (cf. HARTMANN, 1956). The difference between + and — may be genotypically or phenotypically determined.

In the last few decades a still less obvious type of sexuality has been discovered in the *Bacteria* and *Bacteriophages* (cf. WEIDEL, 1964, p. 103). In these cases there is no conjugation at the cellular level but only at the molecular level.

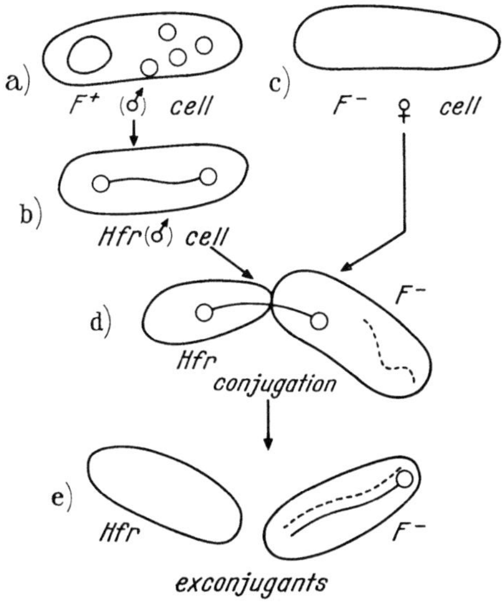

Fig. 2a—e. Conjugation in bacteria. Here there is chromosomogamy only and no cell-fusion. In the resting stage nucleus the chromosome is ring-shaped and by copulation it is opened at one point and goes through the bridge between the two cells (see text). (From WAGNER and MITCHELL, 1964, somewhat modified.)

In electron micrographs of *Escherichia coli* cytoplasmic bridges may be seen between conjugating cells, and there is no cellular fusion but only transfer of the single chromosome of one of the "nuclei" from one cell to another (LEDERBERG, cf. WAGNER and MITCHELL, 1964) (Fig. 2).

As this chromosome consists of only one molecule of DNA (CAIRNS, 1966), we have in these organisms only a *molecular conjugation*. Here again at the molecular level the notion of "maleness" and "femaleness" is adopted bringing into this field the intuitive popular concept of movement and passiveness: the *donor* cell of the chromosome, which moves through the bridge into the other cell is considered as the ♂ cell and the cell which receives the chromosome passively (*recipient* cell) is considered as ♀ (cf. WAGNER and MITCHELL, 1964, p. 122).

In the Bacteriophages the sex phenomena, their "Liebesleben" (WEIDEL, 1964) are deduced indirectly from genetical recombination (cf. Vol. I of this Encyclopaedia), the existence of conjugation and crossing-over phenomena at a molecular level now being accepted.

SERRA (1963) uses the terms *trans-sexual* to describe the sexual behaviour of these submicroorganisms, in contrast to the general case (e.g. mammals) which he calls *eusexual* and the absence of differentiated sexual organs in some fungi and algae, which he describes as *subsexual*.

In this connection we should mention the phenomenon of *recombinative transduction* which consists of pairing between a virus (temperate virus) and a bacterial chromosome giving riso to genetical recombinations (cf. Kühn, 1965a, Fig. 161).

Should these phenomena be considered sexugamy or somatogamy? *Somatogamy* is quite an old expression (cf. Winkler, 1934, p. 453; Weiling, 1965, p. 203) used for designating conjugation of somatic cells (Fig. 3). This possibility of somatic conjugation has been recently used in tissue culture for obtaining intra- and interspecific hybrid cells (Barski, Sorieul and Cornefert, 1960; Ephrussi and Weiss et al., 1965). It seems that the possibility admitted by Noll (1905) and Winkler (cf. Brabec, 1965) of obtaining somatic hybrid-zygotes in grafts is now being carried out in tissue cultures.

Fig. 3. Somatogamy between haploid nuclei of the prothallus of *Lastrea*. In *1* we see one nucleus passing through the cell membrane and one cell with two nuclei and the neighbouring cell without a nucleus. In *2* also there is one cell with no nucleus and the other with two nuclei perhaps in karyogamy. (From Winkler, 1934.)

B. Gamones.

A general characteristic of cell conjugation seems to be the secretion of gamones: male gametes secreting a male gamone, which attracts female gametes, and these in turn secrete female gamones which attract the male gametes. Extracts of female gametes, for example, show an agglutinating action on male gametes and vice-versa (cf. Linskens, 1956, 1957, 1958, 1959, 1960).

This general gamete characteristic is considered to be the best proof of sexual bipolarity. It represents the physiological difference of the two types of gametes, even when there are no morphological differences, i.e. the case of isogamy (Hartmann, 1956).

Are there also gamones responsible for molecular pairing?

If they could be demonstrated at the molecular level we could arrive at a chemical interpretation of the pairing of chromosomes (cf. p. 259) for which, up to present time, only a physico-chemical interpretation has been attempted (Freksa, 1940; Serra, 1963, Kühn, 1965b).

C. About somatic sexuality.

It would be perhaps interesting to ask why is it that in diplonts and haplo-diplonts caryogamy is not followed soon after by chromosomogamy like in the haplonts, the most primitive organisms (Figs. 1, 4).

Since we know that there are genes responsible for pairing of chromosomes (=chromosomogamy), it would not be unreasonable to admit that the delay between caryogamy and chromosomogamy may be due to the activity of repressor genes whose functioning is generally inhibited in the sporangia only.

In exceptional cases (giant chromosomes of diptera, the so-called parasexual cases of Pontecorvo, and other somatogamies) this inhibition takes place in some somatic cells outside the sporangia, and therefore somatogamy occurs. If this inhibition of repressors is due to substances like gamones, we will be dealing with a sexual regulation along the development of a living being: the degree of sexuality is stronger in the sex organs and weaker in the somatic tissues (similar with relative sexuality); accidentally or specifically for some tissues the weakness may shift towards a stronger degree.

Thus the difference between somatic and sexual tissues would be a matter of sexual gradient, and this would be dependent on the interplay between genes for pairing and genes repressing pairing. Thus, these latter genes may be called genes responsible for the somatic stage[1].

There is still, however, no satisfactory explanation for the somatic conjugations between nuclei not possessing homologous chromosomes (between diploid nuclei of different species). These are the only cases in which we should use the expression asexual somatogamy or, better still, asexual fusion of "nuclei and cells". It is not to be expected that we may obtain chromosomogamies.

D. Relative sexuality.

HARTMANN, who formulated the principle of sexual bipolarity, gave us also the principle of relative sexuality: variability in degree of maleness or femaleness to the extent of possibility of conjugation between gametes of the same sex, e.g. female gametes showing weak femininity conjugate with those possessing an enhanced degree of femaleness, similar phenomena taking place within male gametes. HARTMANN and his students claimed to have found such cases in algae and in fungi.

Later on a different interpretation was found for the cases reported in the fungi, and at present relative sexuality in these plants is no longer held (ESSER and KUENEN, 1965).

This type of sexuality, however, may be perhaps invoked to interpret the cases of somatogamy (see above 1, C). This somatic phenomenon imitates most closely sexual conjugation and, like it, originates genetical recombination (STERN, 1931; PONTECORVO, 1954; PRUD'HOMME, 1965). If such cases are considered completely independent of sexuality, then the sex of virions (only inferred from genetical recombination) may also be considered the same nature as an asexual phenomenon of somatogamy.

The pairing-structures (chromosomes), the "cores" (cf. HESS, 1966), could be considered the structures for differentiating somatic pairing from sexual pairing. They have not been found in somatogamies neither in the pairing of the Diptera giant chromosomes. But they have also not been found in the chromosomogamy of *Drosophila* males (MEYER, 1965).

We should rather consider the "core" as "related to" the implementation of crossing-over and not a characteristic of sexual pairing.

2. Sex in higher plants.

The concept of sex developed intuitively from the observation of sexual acts in animals, but primitive tribes even to-day have no notion of sex and sexual reproduction in higher plants. That is, of course, because there is no copulation between male and female in plants. This knowledge is also absent even in regions abundant in unisexual plants but where the country people have received little formal education.

Both in Africa with *Carica papaya* and in S.W. Europe with *Ceratonia siliqua* we have purposely observed that country people do not have the notion of sexuality. It is well known that there are "good trees" which bear fruit (females)

[1] In this relation the recent observation of FELDMAN et al. (1966) of somatic pairing in *Triticum aestivum* is very important.

and "bad trees" which do not bear fruit (males), but the fact that the seed and the fruit are the consequence of fertilization has not occurred to them.

The discovery of sexual organs and sexual reproduction in plants is thus an exclusively scientific acquisition, which owes nothing to popular tradition and it was certainly for this reason that sex was recognized and studied in plants and hybridization carried out much later than in animals (STUBBE, 1964).

Although in Babylon and Syria in 5.000 B.C. the unisexual nature of the date palm was known and THEOPHRASTES (3th—4th century B.C.) had recognised that fertilization in plants was identical to that of fish (external) it is clear that these ideas did not spread. It was only after CAMERARIUS (second half of the 17th century), that scientific circles became interested in demonstrating the existence of sex in higher plants. In fact from this time onwards the subject becomes a hotly debated one in Europe and even in the 18th and 19th centuries the nature of sexual reproduction in plants was still an important problem: this is quite evident from the interest shown by various scientific institutions. The Academies of Sciences of St. Petersburg in 1759, of Prussia in 1819, Haarlem (Holland) in 1830 and of Paris in 1861 offered prizes for work giving the proof of sexuality and of the possibility of obtaining plant hybrids and giving data on their characteristics. The first of these prizes was won by LINNAEUS and the last by NAUDIN.

A similarity between animal and plant sexual cells and organs is thus established. Male organs in plants are however called stamens, whereas the female organs retain the same name, ovaries. In addition, owing to an apparent parallelism, the word ovule is used in plants for what was thought to be the homologue of the female sex cell of animals.

3. Life cycles.

The famous comparative studies by HOFMEISTER (1851—1868) on mosses, ferns and flowering plants, showed that in the majority of cases, plants had a "sui-generis" peculiarity: *alternating generations in their life cycle.* His recognition of this fact in the *Spermatophyta* and the confirmation of this same, easily observed, occurrence in the *Bryophyta* and *Pteridophyta* proved that the so-called ovule of the *Spermatophyta* was in fact a macrosporangium harbouring the whole female prothallus with the macrospore germinating in the macrosporangium (Fig. 4) and hence the previous mistaken ideas in this respect (see above).

Later on it was seen that in any life cycle of an organism reproducing sexually there is a distinction between *alternation of generations*, when present, and *alternation of phases.* These are two different concepts.

We shall define generation in a life cycle as a part of this cycle characterized by an individual, independent or not, which begins with germination of a cell, spore or zygote, and ends with the production of another cell type, either zygote or spore, different from the initial one.

The alternation of generations almost always takes place between a gametophyte generation and one or more sporophytes, these two or three generations completing the life cycle.

In rusts the gametophyte is masked, because of non-differentiation of sex organs. It seems that the life cycle of these plants consists only in alternation of 3 sporophytes: aecidiosporophyte, teleutosporophyte and basidiosporophyte. The aecidiospores are however the result of plasmogamy and the teleutospores the result of caryogamy.

There are, however, many cases of sexual reproduction where it is not possible to distinguish alternation of generations. In such cases (e.g. *Zygnemales, Fucales, Charophytae*, various *Chlorophyta* and almost all the animals) the zygote begins the generation which ends with the production of gametes (gametophyte generation) which on conjugation give rise to a new zygote, and this starts in the following cycle another gametophyte generation. There are *no spores* in

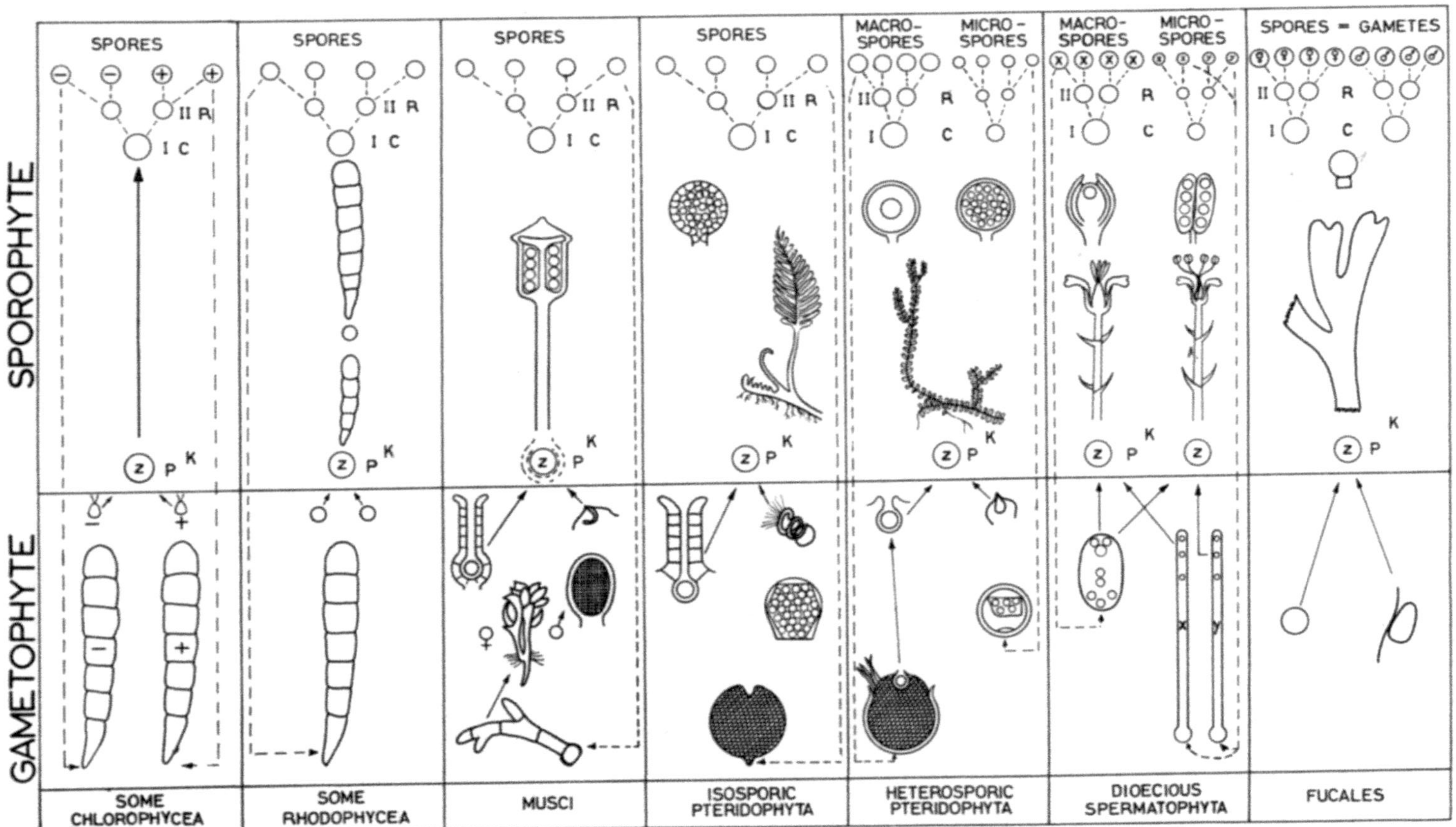

Fig. 4. An illustration of a haplont (far left), a diplont (far-right) and 5 haplo-diplont life-cycles (see text). The second figure from left shows a cycle with 3 generations. The 3 steps of cell-conjugation are indicated by *P* (plasmogamy), *K* (karyogamy) and *C* (chromosomogamy). *I* and *II* show the first and the second mitosis of meiosis. *Z* zygote; *R* chromosomic reduction. (From QUINTANILHA, 1945, modified.)

this cycle). In these cases we have no alternation of generations within a cycle, but successions of cycles which coincide with successions of generations, since each cycle is filled by only one gametophyte generation. The generation is started by a zygote (= fusion of gametes) and will again produce gametes. In all these cases we have, however, alternation of phases *otherwise reproduction would not be sexual* (QUINTANILHA, 1945). In the formation of the zygote there is a change from the gametic phase (1) (n chromosomes) to the zygotic phase (1) (2 n chromosomes) and at this point, which varies with each case, either chromatic reduction takes place immediately after the fusion of gametes *(Zygnemales, Charophytae,* some *Chlorophyta)* or it is pregametic (*Fucales*, animals). In the first instance we have haplonts, in the second diplonts. In nearly every other case where there is alternation between 2 or more generations we have haplo-diplonts. We can, however, have cases of alternation of 2 or 3 generations in haplonts *(Nemaliales, Bonnemaisoniaceae)* and in diplonts (some of the *Chlorophyta*).

The existence of haplonts, diplonts, diplo-haplonts and even "diplont" cells with haploid nuclei *(Ascomycetes* and *Basidiomycetes)* is the result of cellular conjugation not taking place immediately, but showing a variable delay depending on the organism, and being accomplished in steps which are spread out throughout the life cycle in space and time [1].

During conjugation of the two gametes, in the Eucaryontes three stages have been recognised (Fig. 1): (1) plasmogamy, (2) caryogamy and (3) chromosomogamy (pachytene) (RESENDE, 1956a; LINSKENS, 1958, p. 266).

In haplonts these three steps are consecutive. In the other cycles, plasmogamy (1) may be distant from caryogamy (2) and chromosomogamy (3) which are consecutive *(Asco-* and *Basidiomycetes)* or the first two (1, 2) may be consecutive and the third (3) distant from them (all diplonts and haplo-diplonts).

There are also cases of alternation of phases in which these steps occur irregularly during the period of somatic growth (PONTECORVO, 1953, 1954, 1956; PONTECORVO et al., 1958). These cases are different from the normal ones only because the conjugation stages do not take place in the sexual organs (gametangia and sporangia). PONTECORVO called these parasexual cases.

In the first case we have only one type of sex organs (gametangia, similar or dissimilar, ♂♂ and ♀♀). In the second and third there are two types of sex organs: (i), gametangia similar or dissimilar, in relation to plasmogamy *(Eumycetes)* or to plasmo- and caryogamy (e.g. *Musci*, Pteridophytes, Gymnosperms), and (ii), similar sporangia (e.g. *Musci*, isosporic Pteridophytes) or dissimilar ones (heterosporic *Pteridophyta* and *Spermatophyta*) in relation to chromosomogamy (RESENDE, 1956).

Related to this more intimate contact of male and female material in cellular conjugation (chromosomogamy), where even gene exchange may occur (crossing-over), we have the reduction of the chromosome number from 2n to n (R) for which only the second division of meiosis is responsible (RESENDE, 1946, 1947, 1953, 1956c; RESENDE and MANARTE, 1951; LEVAN and MÜNTZING, 1963; STRANDHEDE, 1965).

Usually only the reductional function of meiosis is considered when life cycles are analysed the conjugational function being largely ignored. HARTMANN in his textbook (1956) makes a vague reference to pachytene chromosome conjugation without, however, enlarging on its fundamental role in cellular conjugation.

Consequently, the widespread gross inaccuracy (cf. e.g. the textbook of "STRASBURGER", 1962, p. 650) of considering the life cycles, for example of a thallophyta, a fern and an

[1] These are also known as haploid and diploid phases from which are derived the terms haplont and diplont. This nomenclature is, however, not exact in the polyploid series (LEVAN and MÜNTZING, 1963).

angiosperm as having an alternation of a "sexual" generation (gametophyte) with an "asexual" one (sporophyte) may still be found at the present time. This leads to the absurdity of considering the angiosperm flower, sex organs "par excellence", as belonging to an "asexual" generation while the "sexual" generation is devoid of sex organs! On the other hand sexuality of bacteria and viruses, one of the great advances of molecular biology in the last 15 years, would be considered an illogical designation because here the manifestation of sex — their "Liebes leben" using WEIDEL's expression — is the outcome of chromosomogamic phenomena. There is only *mating* of chromosomes, of DNA molecules, the existence of which in bacteriophage, is inferred from studies of genetical recombination. *There is no other sign of sexuality.* None of these phenomena are homologous with those of the gametophyte ("sexual" generation) but only with the characteristic phenomena of the sporophyte ("asexual" generation)!

A general survey of the plant kingdom (Fig. 4), in the present geological age, shows that only the gymnosperms present clearly differentiated 3 of the 4 possible sex organs: ♀-sporangia (macrosporangia), ♂-sporangia (microsporangia) and ♀-gametangia (archegonia). Of the ♂ gametangia (antheridia) we find *only cells representing the conceptacle* which no longer shows differentiation. In the heterosporic Pteridophyta we unmistakably have the 4 types of sex organs: ♂- and ♀-sporangia and ♂- and ♀-gametangia although the former are largely undifferentiated. In plants lower than the heterosporic Pteridophyta we do not find differentiation of ♂- and ♀-sporangia but only ♂- and ♀-gametangia. In contrast, higher plants such as the Angiosperms, show very markedly differentiated ♂- and ♀-sporangia but no differentiated gametangia.

Thus in the plant world we may discern a progressive differentiation in sex expression of sex organs from the gametophyte to the sporophyte: in *Bryophyta* and isosporic *Pteridophyta* differentiation of sex expression in sex organs is more enhanced in the gametophyte, whereas in the heterosporic *Pteridophyta* and gymnosperms this sex expression shows a gradual shift towards the sporophyte. The gamophyte is so greatly reduced in the Angiosperms that we only see sex organs in the sporophyte (Fig. 4).

The *Thallophyta* offer a broad spectrum of development between gametophyte and sporophyte including the subdivision of the sporophyte into two generations. Restricting our attention to the *Pheophyta*, we may draw a plan of comparative physiology of development (Fig. 5) showing parallel phenomena to those observed on comparing the cycles in *Bryophyta*, *Pteridophyta*, and *Spermatophyta* (Fig. 4). There are, however, within the algae of thr order *Fucales*, a bigger reduction of the gametophyte generation than in the Angiosperms. These algae also show an identical cycle to that found in most animals: the gametophyte "generation" is so greatly reduced that the germinal cell, spore, i.e. the first cell of the gametophyte "generation", is also the last cell of this "generation", the gamete. Thus this cell may be considered to be either a gamete or a spore, and the respective conceptacles may therefore be regarded as gametangia or sporangia. Since the most salient characteristics are those of the gamete these cycles are considered a succession of generations of gametophytes rather than sporophytes. However, from the point of view of homology of generations its sporophyte character is more evident than its gametophyte one (Figs. 4, 5).

In life cycles where only 2 generations alternate (see Figs. 4, 5), the cycle may be carried out by 2, 3, or 4 different individuals. This variation or difference may be pheno- or genotypically determined:

When the cycle is carried out by only 2 individuals, one of the sporophyte, the other of the gametophyte (Fig. 4), the differences between the two are phenotypically determined because the genetical difference is only quantitative. One individual has n and the other $2n$ chromosomes.

In the polyploids the chromosome number in the gametophyte may rise to n = 4x and n = 8x or more [F. v. WETTSTEIN, 1924 (Mosses); KIRCHEISEN, 1960 (Liverworts); PARTANEN, 1965 (Ferns)].

If we now consider the cases of asexuality which show alternation of generations, we shall find that between the 2 individuals of the 2 generations, there is no genetic difference at all, not even a quantitative one. Nevertheless the great qualitative differences hold, for example, between the prothallus and the sporophyte of an isosporic fern. What then, is the cause of this variation between the 2 individuals, one a sporophyte and the other a gametophyte,

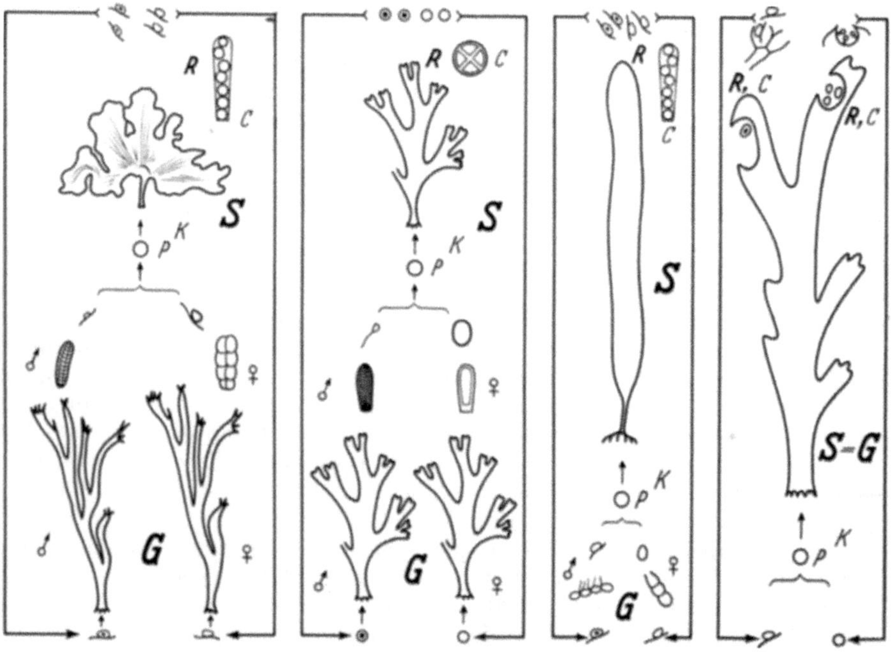

Fig. 5. Schematic representation of the four life-cycles of *Phaeophyta*. Three haplo-diplonts with genetically determined sex and one diplont with phenotypically determined sex. *G* gametophyte; *S* sporophyte (cf. with Fig. 4). (From "STRASBURGER", modified.)

both living in the same environment? Here we are certainly faced with a problem of physiology of development in which the same genetical inheritance, developing in the same external environment, gives rise to different individuals (LINSKENS, 1956, p. 335). On originating the germinal cells (spore and zygote), or the vegetative cells substituting for them (asexuality), must already be the result of a differentiation, i.e. of an impregnation which permanently represses some genetic powers and leaves others (SIGNORET, 1964) with the capacity of becoming active consecutively during growth and development of the respective generation.

On the other hand as there is a great uniformity of appearance in many algae between sporophytes and gamophytes differing only in the type of sex cells they produce, we are led to believe that the previous problem of explaining the great differences in appearance between generations of the same cycle (*Algae*, *Bryophyta* and *Pteridophyta*) may be identical to the well known cases of "juvenile forms". In the same sporophytic generations of many higher plants (e.g. *Eucalyptus*, *Acacia* sp., *Hedra*, certain *Gasteria*, some Kalanchoideae —see PÖLLNITZ, 1938; RESENDE, 1958; RESENDE and WARDEN, 1954; BRINK, 1962) some juvenile aspects differ so much from those of the adult that taxonomists, who do not follow the development of the plants assign them to different species or genera. This is exactly what happened with the sporophyte and gamophyte generations of the same form or species which were not then known to belong to the same life cycle and were classified as if each generation belonged to different families and genera (cf. *Cutleria*, *Bonnemaisoneaceae*, etc.).

Certainly this is a particular instance of the present problems of genetical activity and repression which condition phenotypic variation (differentiation) during development of the same genetical individual. This is intercepted, however, by germinal cells in 2 or 3 physio- and morphologically different individuals, each forming one alternating generation of the same life cycle. Also another salient point to note is the fact that genetically identical individuals (generations)

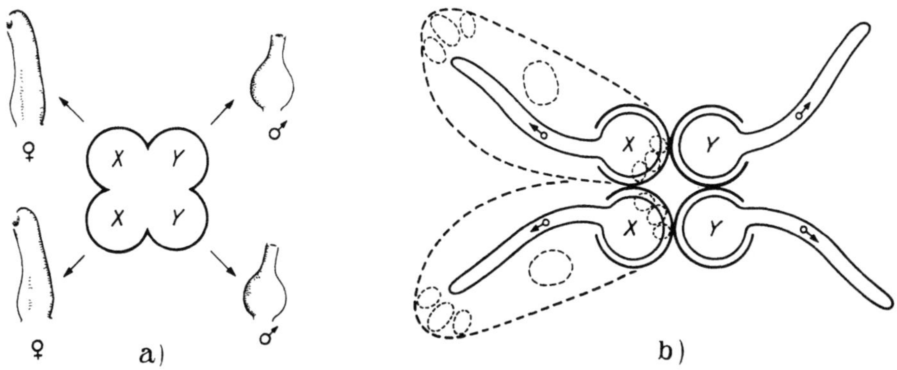

Fig. 6a and b. Representation of the segregation of the spore-tetrad in one dioecious *Thallophyta* or *Bryophyta* (a) and in one dioecious *Spermatophyta* (b); see text. In b is represented with broken lines the ♀-prothallus, the development of which must be expected if the sex-genes had penetrance like in a. *Sphaerocarpus Donnellii* is taken as representative example for *Thallophyta* and *Bryophyta*. (Original.)

may show the same degrees of morphological differences as those between families or even higher taxonomic groups (cf. for experimental differences between gametophytes and sporophytes in Bryophyta, Pteridophyta and Spermatophyta).

When the cycle is formed by three individuals, two in the gametophytic and one in the sporophytic stage (Figs. 4, 5), the interpretation of the morphological variation between individuals is a problem of the same nature as on p. 265. There is, however, a variation from ♂ to ♀ in the gametophyte, which *in all Thallophyta and Bryophyta studied up to the present time*, is always genetically determined with in many cases segregation of spores with heteromorphic sex chromosomes such as in the well known case of *Sphaerocarpus* (liverwort).

Here we have genetical heterosporia, although the spores are morphologically identical. The spore tetrad segregates 2 ♂ and 2 ♀, the former receiving the Y chromosome and the latter the X chromosome (Fig. 6).

In the *heterosporic Pteridophyta and Spermatophyta with sex determined by phenotypical modifications*, this same gametophyte dioecism is phenotypically determined, i.e. its ♂ to ♀ mechanism of differentiation (= variation) is of the same class as, for example, that which distinguishes leaf from root.

Again we have, also for sex, the situation referred to p. 265. In this case the same genetical constitution is shared by 3 different individuals. Besides their enormous morphological differences they are all sexually different: the sporophyte is monoecious (♂ and ♀ sporangia in the same individual) and the two individuals of the gametophyte generation are dioecious, one ♂ and the other ♀, i.e. each producing gametes of one sex. Experiments, carried out mainly with *Bryophyta*, show that any of these cells is able to regenerate a complete plant with all its differentiations.

When the cycle is completed by 4 individuals — *Spermatophyta* with genetically determined sex — there are 2 individuals in the sporophyte, one ♂ and the other ♀, and two others, also one ♂ and another ♀, in the gametophyte. How are we to understand this sexual variation?

Let us consider a case of heteromorphy in sex chromosomes, e.g. the well known and still better studied dioecious species *Melandrium album*. The chromosome set of the zygotic phase (diploid here) differs in the male and in the female. The former show 22 autosomes and one X and one Y chromosome; the females have 22 autosomes and two X chromosomes. The segregation of the tetrad of male spore mother-cells is the same as in *Sphaerocarpus:* two spores receive the X chromosome and other two the Y chromosome. We should therefore expect, as in *Sphaerocarpus*, the production of two ♀ and two ♂ prothalli, i.e. two embryo sacs (X) and two pollen tubes (Y) (Fig. 6). We know, however, that this does not take place but that 4 pollen tubes are formed, i.e. 4 ♂ individuals of the gametophyte generation instead of the expected two ♀ and two ♂ individuals. Again as in *Sphaerocarpus*, there is segregation of the genes responsible for sex variation but in two of the spores *the genotype responsible for the female sex does not possess penetrance.* We have, consequently, two genetic types of spores and therefore of ♂ prothalli and ♂ gametes, i.e. of the 4 male gametes corresponding to the 4 tetrad spores, two are genetically feminine. This "genetical femaleness" will only show its penetrance in the sporophyte generation. This same lack of penetrance is found in cases of female heteromorphy (*Fragaria*, birds, butterflies — HARTMANN, 1956; WESTERGAARD, 1958).

Sporophyte female plants only produce spores and therefore gametes of a single genetic type, the A + X type. Thus we get a "backcross" type of segregation, i.e. 50% of the zygotes show a 2AXX and the other 50% a 2AXY constitution. The zygotes of the first type produce females and those of the second produce males.

Whether there is or is not a pair of heteromorphic chromosomes, this is the general type of sex segregation found in the *Spermatophyta* and the *Fucales*, — if in these Algae dioecism is unequivocally proved to be genetically determined —, and in all higher animals showing genetically determined sex. The difference displayed by the *Fucales* and the animals in relation to the *Spermatophyta* consists only in a greater reduction of the gametophyte here consisting of a single cell, which may be considered to be either a spore or a gamete.

4. Flowering and sex expression.

A. The inflorescence.

The differentiation of sex organs of the sporophyte (sporangia) from the Pteridophyta upwards *(Lycopodium, Selaginella, Equisetum* and *Spermatophyta)* is preceded by a metamorphosis of the vegetative organs, producing the inflorescence. The transition from vegetative growth to sexual reproduction is brought about by what, in the physiology of flowering, is called the "flowering impulse" (cf. Vol. XV, XVI of this Encyclopedia) which is followed by "flower-realisation", i.e. differentiation of the inflorescence. The metamorphosis of vegetative shoots into floral ones generally consists of a decrease in the size of the leaves (bracts) and frequently in complete disappearance of the chlorophyll which may be substituted by petal type coloured pigments. There is also formation of ramifications differing from those of the vegetative state and the formation of flowers.

Between the fully floral state (= 100% floral) and the fully vegetative state, there is a continuous variation (easy to induce in many plants) which may be

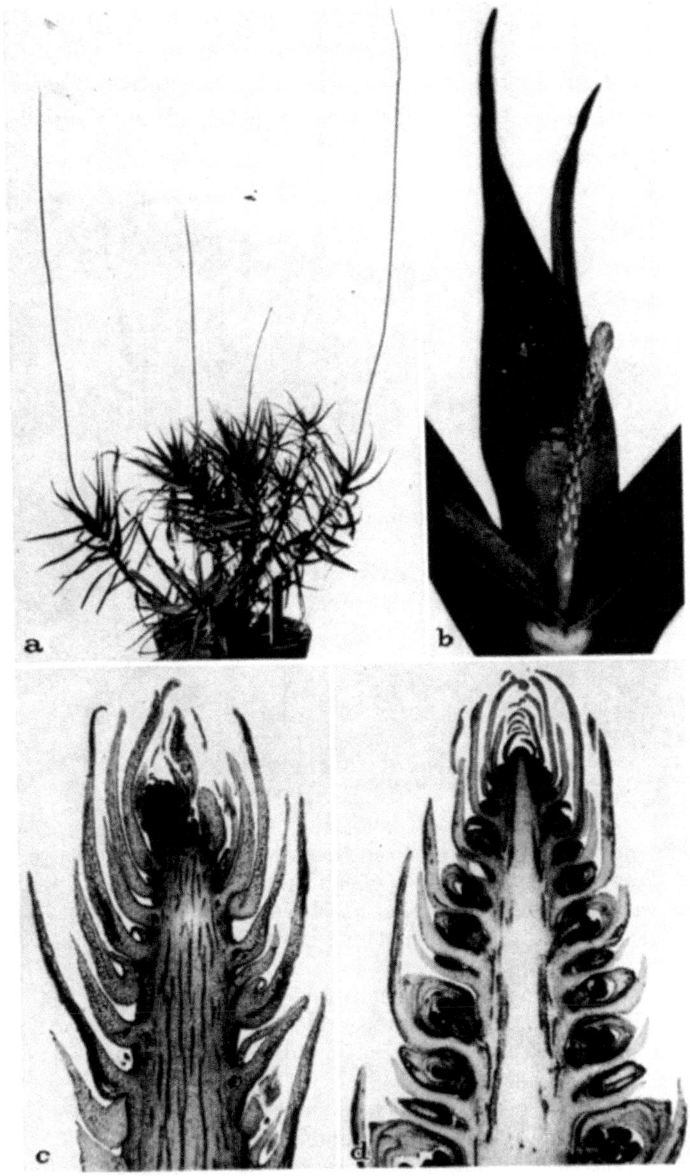

Fig. 7a—d. A spontaneous variant of *Aloë tenuior* × *Aloë striatula* showing inflorescences without flowers. The development of the inflorescence is in all ways like the normal inflorescence but the genes for flower realization are blocked. a, A plant showing 4 inflorescences; b, a magnified inflorescence at the beginning of development in order to show the bracts without flower in the axils; c, d, microscopic picture of the mutated variant (c) and of the normal inflorescence (d). (a, b, RESENDE, unpubl.; c, d, taken from LINSKENS, 1960.)

expressed in degrees of flowering intensity ranging from a floral zero to a floral 100%. The variation can be obtained, for example by direct or indirect modification of *growth regulators* in species of *Kalanchoideae* (cf. Vol. XV/2 of this Encyclopedia, p. 360—362; and RESENDE et al., 1947—1959), in *Coleus* (LAIBACH and Kribben, 1950—1953) and in other plants. The higher the floral degree the smaller the bracts and the greater the number of flowers. Low floral degrees may no longer show differentiated flowers but only large bracts. In the lowest degree no flowers or bracts are seen, but only the ramifications characteristic of the inflorescence.

Harder and his school gave the name of "vegetative inflorescences" to these flowerless inflorescences which may or may not show differentiation of large bracts (cf. Vol. XV/2 of this Encyclopedia, p. 352—360).

Fig. 8. A variant of *Bryophyllum daigremontianum*, with floral branches without flowers or bracts (arrows). (Resende, unpubl.)

All this continuous variation, from "*vegetative*" inflorescences to completely developed "*floral*" inflorescences, is phenotypically determined.

In some cases this same variation has been found to be genotypically determined, i.e. genocopies of the phenotypic variation (cf. for example Resende, 1948).

However, the most interesting genetically determined variant is the one found in the genus *Aloë* (Fig. 7). Here there is a blocking of the genes for flower formation without any correlative modification in the other characters of the inflorescence.

The phenocopy of this process is still unknown and is not to be found in the range of variants of "vegetative" inflorescences.

We have a *flowerless "floral" inflorescence* (Resende, 1956b and unpublished; Linskens, 1960b). This genetic possibility of realizing, in the complete absence of flowers, an inflorescence consisting of only peduncle, raceme and floral bracts allows one to predict that in the gene constellation responsible for complete flowering there may be other genes capable of being blocked.

In 1943 and 1956b we outlined the following steps, for flower development, directed by independent genes but with successive action: a) peduncle, b) raceme, c) bracts and d) flowers (Resende, 1956b, p. 234). The genetical possibility of blocking step d) is shown in Fig. 7.

Later, we observed in *Bryophyllum daigremontianum* a genetically determined variant which showed the process blocked from onwards. As illustrated in Fig. 8 (arrows) there is a peduncle and its ramifications, which may be considered a *raceme*, but no bracts nor flowers.

B. The flower.

Considering now the flower itself, we see that this may also be formed in stages: In a strain of *Br. daigremontianum*, possessing normal (= 100% floral) hermaphrodite flowers with a complete perianth, the author induced, phenotypically, flowers with the following lowering of floral degree: (i), exhibiting only a vegetative calyx (large and succulent), (ii), flowers with only a floral calyx (the size of a normal flower), (iii), flowers with only a floral calyx and gynoecium and (iv), flowers with only calyx, gynoecium and corolla (Fig. 9).

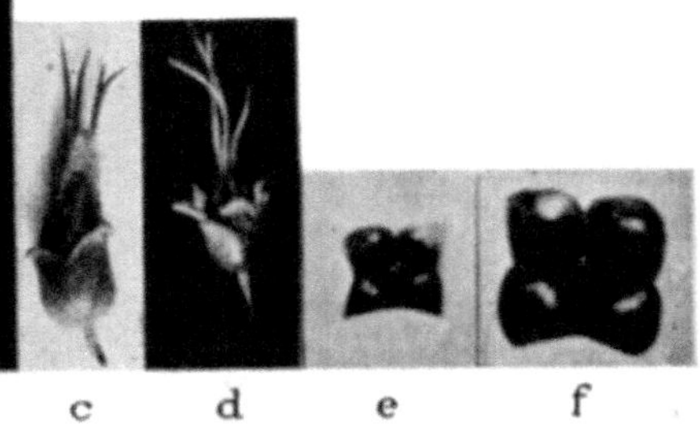

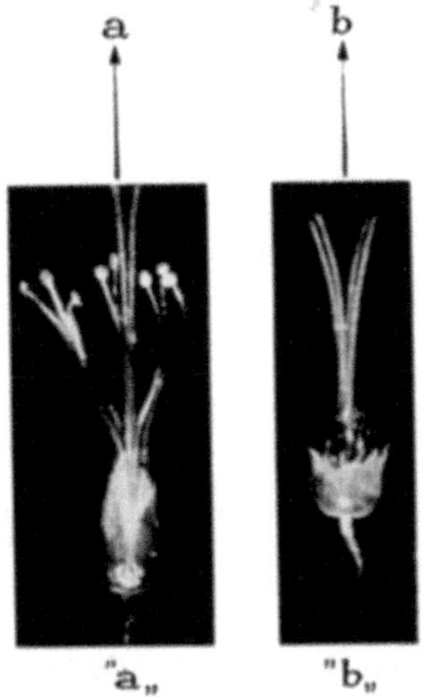

Fig. 9a—f. *Br. daigremontianum.* An illustration of the action of the native auxin level in the leaves on the floral degree of the flower. From a to f we can see that the greater the amount of auxin, in the leaves after flower impulse the more vegetative is the flower. The most vegetative flower is shown in f; here androecium, corolla and gynaeceum do not develop and the calyx is increased and becomes succulent like the vegetative leaves. The 100% floral flower can be seen in a. "a" and "b" show a and b flowers without corolla to see the aspect of androeceum and gynaeceum. (From RESENDE, 1950, Figs. 13 and 19, modified.)

This floral gradient may be produced both by "controlling" floral impulse or by inducing the floral state to regress either completely or partially to the vegetative one with the formation of "inflorescences" and "vegetative flowers" (RESENDE and VIANA, 1959, Fig. 1; LINSKENS and RESENDE, 1966). So as in the case of the inflorescence we may also induce independently in the flower the formation of certain parts or organs and we may obtain in the flower the same flowering gradient as in the inflorescence.

The totally "vegetative flower" in *Bryophyllum* is the one showing only a succulent calyx (Fig. 9g). Between this type of flower and a 100% "floral flower" (Fig. 9a) there is found a continuous variation, phenotypically determined. The internal milieu responsible for this variation — just as in the inflorescence — is directed by growth regulators and this may be expressed by the following ratio $\frac{\text{growth promotors}}{\text{growth inhibitors}} = A$, the higher the values for A the more vegetative the flower will be, i.e. raising the value of A will result in lowering the floral degree of the flower. Very high values for A will lead to total regression to the vegetative state.

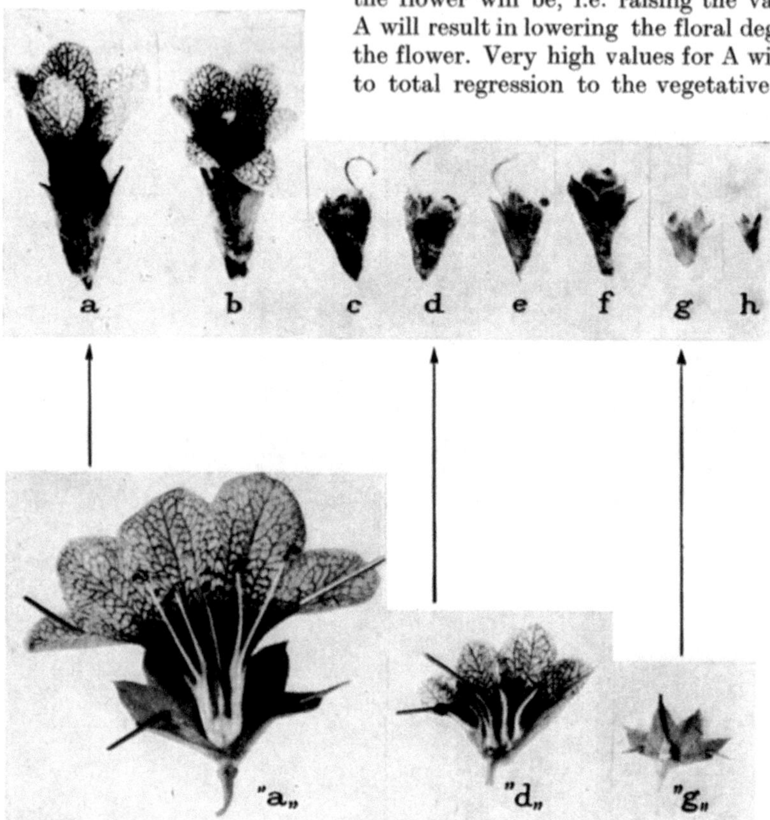

Fig. 10a—h. *Hyoscyamus niger* ⊙. Representation of the same flower variation as in Fig. 9 produced by the action of IAA injected into the plants: in this species, however, neither the calyx nor the gynaeceum increases with the level of auxin, but the flower disappears completely with the increase of growth-promotors in the leaves. In "a" and "d" the corollas have been removed from a and d to show the gynaeceum and androeceum. (From Resende, 1953 and 1957.)

Some of these phenotypical variants have been fixed genetically (genocopies) in *Br. daigremontianum* as may be seen in the continuous variation displayed in Fig. 10, i.e. variants have been obtained which are genetically unlike the wild form in that is a blocking of the genes responsible for the most characteristic floral organs of the flower, the stamens and the corolla (Resende, 1957; Resende and Linskens, 1961).

These experimental data have, since 1950 (Catarino, 1964), led the writer to believe that flower development is carried out polygenically by additive genes controlling native growth regulation $\frac{\Sigma\,\text{FFF}}{\Sigma\,\text{MMM}} = \frac{\text{growth promotors}}{\text{growth inhibitors}}$, Resende, 1950, 1957).

If we now consider the flower sex organs (gynoecium and androecium) and their relative variation (Figs. 9, 11, 12) we may see that, while in *Bryophyllum* and *Hyoscyamus* no clear-cut negative correlation seems to exist (Resende,

1950, 1953) (see ,however, further below) between gynoecium and androecium, in *Silene* this negative correlation does hold. In the case of *Silene* we may admit the existence of a possible epistasy-hypostasy mechanism between genes responsible for the gynoecium and those responsible for the androecium, if a genetically determined mutant (genocopy) of the above type of variation is found to occur.

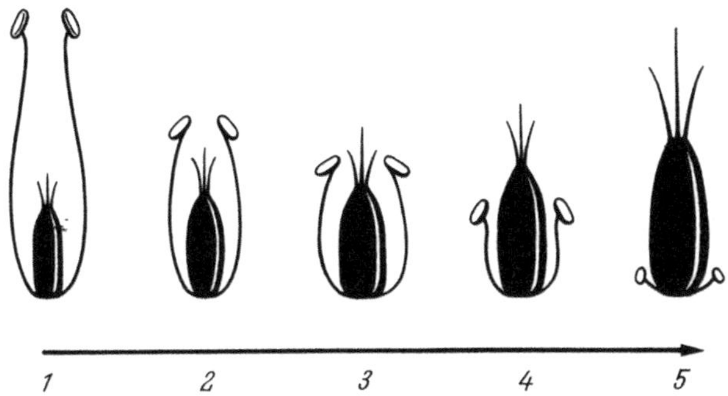

Fig. 11. *Silene pendula.* In this species the action of the growth promotor (NAA) on the leaves is similar to that in *Bryophyllum* and *Hyoscyamus* with the difference, that here the vegetation of the flower occurs by a diminishing of the size of androeceum and corolla and an increase in the size of the calyx and gynaeceum. In this figure only the androeceum and gynaeceum are shown. The vegetation increases from 1→5 (see text). (From J. and Y. HESLOP-HARRISON, 1958b.)

In *Bryophyllum* and *Hyoscyamus*, both in phenotypically (Figs. 9, 11) and genotypically (Fig. 10) determined variation, it is observed that androecium atrophy may even proceed to complete elimination without the gynoecium increasing, at least in relation to ovary development. Ovule number may however increase (CATARINO, 1964) but further on the gynoecium follows the androecium towards complete atrophy as the concentration of growth promotors rises (Fig. 9f, g).

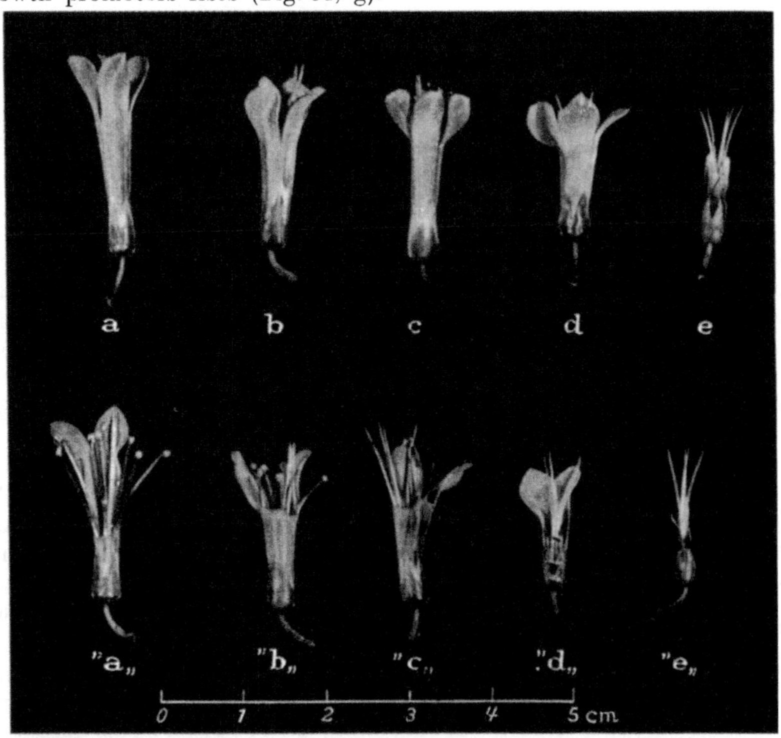

Fig. 12a—e. Flowers of five different strains of *Br. daigremontianum*, differing in the floral degree of the flower. Here we have the same sort of variation as that illustrated in Fig. 9, but genetically determined. d and e are two perfectly genocopies of b and c in Fig. 9. (RESENDE, unpubl.)

All these observations and other more recent ones (cf. NUNES, 1964, and recent references here) seem to point logically to the following hypothesis: sex expression in the Angiosperms is controlled by a group of structural genes, subgroups of which are responsible[1] for the calyx, for the gynoecium, for the corolla and for the androecium and the activation of these genes is governed by the establishment of a specific ratio of growth promotors to growth inhibitors (Fig. 13).

The balance of this ratio of growth *regulators* (= morphoregulators) must be controlled by a system of additive genes, which regulate activity and repression of the structural genes (Fig. 13). Genetic interaction of the epi-hypostasy type may or may not occur depending on the type of material.

This interpretation may be considered, to some extent, as an up date version of that held by CORRENS (HARTMANN, 1956), in which the AG complexes cor-

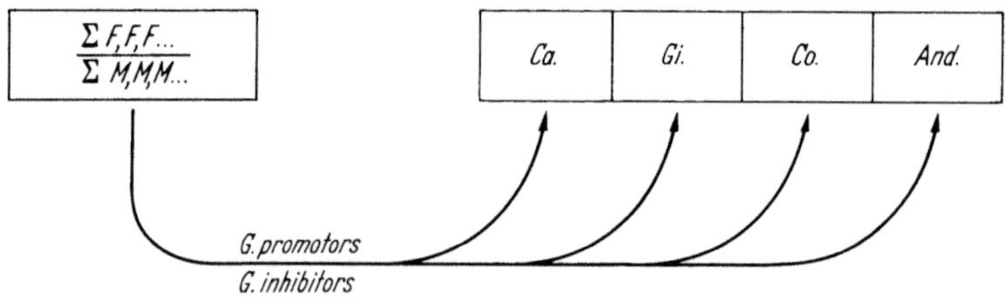

Fig. 13. Schematic representation of two systems of genes which are assumed by the writer to be involved in the formation of the flower (see text). (Original.)

respond to the structural genes responsible for the calyx, gynoecium, corolla and androecium, and the genes governing the ratio of growth regulators correspond to the genes classically called *realizers* (realisators).

What we are advancing differs somewhat from the present day interpretations, which have also sprung from that classical interpretation (GALÁN, 1953—1964; RESENDE, 1957; WESTERGAARD, 1946—1958).

A survey of the observations since CORRENS, of pheno- and genotypically determined variation (cf. above) and of sexual bipolarity of both male and female of dioecious strains, leads us to admit the existence of the two types of genes: the structural genes, with infrequent genetical variation, and those regulator genes (less stable than the former) controlling the balance of the morphogenetically active substances. Sex expression may also be modified by any substances accidentally influencing the action of these genes, or directly affecting the growth regulators they control. This may be the role of certain steroids and of traumatin and olefins, etc. (cf. RESENDE and VIANA, 1959, p. 78—79; HESLOP-HARRISON, 1957; BOPP, 1965).

The inducement of stamen formation observed by CORRENS in female plants of *Melandrium album* by the fungus *Ustilago violacea* is a classical example of the system of regulator genes compensated or substituted by accidentally administered substances affecting the structural genes directly. Here only the blocking of the "genetical regulator" system is substituted or compensated; the "structural genetical" system is kept intact.

If therefore, the former system (structural genes) is a fixed one, all the variation, covering genotypically determined monoecia, dioecia, gynodioecia, gynomonoecia and andromonoecia (WESTERGAARD, 1957) may be explained taking

[1] Here we are not taking into account the genes responsible for flowering impulse and floral realization, a necessary but not sufficient condition for fulfilment of flowers.

into account simply the variation within the additive genes regulating the balance $\frac{\text{growth promotors}}{\text{growth inhibitors}}$ (RESENDE, 1950; 1957; CATARINO, 1964).

This ratio of growth regulators is responsible for the sex gradient not only in the hermaphrodite flower, but also in monoecious plants with unisexual flowers, and the femaleness or maleness of the latter may be augmented by increasing respectively the relative number of female and male flowers (LAIBACH and KRIBBEN, 1950—1953; HESLOP-HARRISON, 1956—1963; GALÁN, 1951—1964; GALUN et al., 1956—1965; CATARINO, 1964; NUNES, 1964; and this Encyclopaedia, 1965, Vol. XV/1, contribution P. NITSCH).

5. Origin of dioecia from monoecia. Possible mechanism of the origin of sex chromosome heteromorphy.

If we are able to fix genetically two independent variants, one with the genetical balance of formula cited above adjusting the ratio $\frac{\text{promotion}}{\text{inhibition}}$ of growth in order that only genes Ca, Gi, Co (genes responsible for calyx, gynoecium and corolla) may be activated, and another with a genetical balance corresponding to a different ratio and resulting in the activation of genes Ca, Co, And. (calyx, corolla, androceum) it should then be possible to change a monoecious strain into a dioecious one. The probability of this change taking place in two genetical steps (one to form the male, another to form the female) is very low; if it is, for example, 10^{-5} for a one step it must be 10^{-10} for a two step change (RESENDE and LINSKENS, 1961, p. 167—171).

We may imagine theoretically a possible mutation from monoccia to dioecia in a single step (RESENDE, 1951, 1957). In monoecia we recognise that the genes represented in formula $\frac{\Sigma F, F \ldots}{\Sigma M, M \ldots}$ are balanced in such a way as to control the hormone ratio responsible for normal development of the hermaphrodite flower or ♂ and ♀ unisexual flowers distributed specifically along a given monoecious plant. If we are able to unbalance the genetical power of the genes ΣF,F... or ΣM,M... and weaken one of these summations, we may arrive at a balance which produces either female or male organs only. If this new genetical balance is originated by the formation of a heterozygoty responsible for one sex the segregation may simply play the role of forming a viable homozygoty responsible for a further new balance which triggers the realization of the other sex.

As an example take the *Drosophila-Rumex* type of sex and, for a primitive monoecious form, imagine a distribution of a certain number of genes of ΣF,F,F..., e.g. six in the same pair of homologues not containing any genes of ΣM,M,M..., and which by an unequal crossing-over results in all these genes appearing in the same chromosome, and no F genes remaining in its homologue. If this positional change of 3 F-genes weakens its genetic power (cf. the gene *bar* in *Drosophila*), transforming *F* into *Fa* (weaker), and producing a viable new combination, we might bring about incapability for ♀ sex organ differentiation and thus originati a male strain. If this male can be crossed with a primitive monoecious form, the possibility will now exist of obtaining in F_3 a combination giving rise to a hormone balance only compatible with female sex organs (nullifying achievement of maleness) and where there may be epistasy of F over M. This epi-hypostatic interplay has been observed in *Melandrium* and *Drosophila* (RESENDE, 1957; RESENDE and LINSKENS, 1961; WESTERGAARD, 1958).

Thus, in a single step of genetic alteration (unequal cross plus position-effect), from two homologous chromosomes, we would have produced two new chromosomes containing different genes which, if the heteromorphy is microscopically visible, would be the X and Y chromosomes. The XX homomorphy would produce the female and the XY heteromorphy the male. The viability of the YY combination is extremely low due to the double deficiency (see Fig. 8 of Resende, 1957).

This same hypothesis may explain the occurrence of the female before the male if the initial inbalance due to crossing-over with genes FF takes place with genes MM. In a similar fashion we may explain female chromosomic heteromorphy.

The only difficulty in this interpretation is that of the cases where there is no epi-hypostatic inter-play between genes FF and MM, as it seems to be in the behaviour of dioecious forms derived from *Bryophyllum* and *Hyoscyamus* type.

The spontaneous occurrence of females genetically different from the monoecious plants can be seen quite frequently in both *Br. daigremontianum* and *Br. crenatum* (Resende and Linskens, 1961; Resende unpubl.) and these could have arisen from a type of mutation producing two new chromosomes (see above) or only one new sex chromosome (Resende and Linskens, 1961, Figs. 2, 3). Up to the present the search for males in a large number of segregating gene rations has been unsuccessful and we are thus inclined towards the second possibility, i.e. the formation of a single new chromosome with deficiencies of genes MM. However, in the absence of epi-hypostatic interplay between genes FF and MM, the first possibility may also occur.

6. Asexual reproduction.

Hans Winkler was probably the botanist who, in the first half of this century, contributed most to problems of asexual reproduction (Winkler, 1908, 1916, 1932, 1934, 1942); his contribution to the problem of the chimaera is also extraordinary (cf. in this Encyclopaedia, Vol. XV/1, contribution Brabec). In the writer's opinion, a very special place should also be given to his logical systematization of asexual processes published in *Planta* in 1942.

We shall summarize here very briefly this systematization, which covers in general all the known cases of asexual reproduction in plants.

Since a definition of sexual reproduction must take into account the life cycle the same applies to asexual reproduction. If sexual reproduction implies a life cycle with alternation of phases, asexual reproduction will be that in which life cycles do not show alternation of phases.

Since the events responsible for alternation of phases are conjugation (plasmo- and karyogamy) and chromatic reduction (chromosomogamy and reductional mitosis[1]), asexual reproduction will appear whenever Nature shows life cycles where such events, or even one of them, are absent.

A zygote, derived from plasmo- or karyogamy, germinates and produces a sporophyte generation. Chromatic reduction gives rise to a spore or an equivalent cell which on germination produces a gametophyte generation. This spore is called gonospore by Winkler. All the processes occurring in nature for substituting these first cells of the two generations, the zygote and the gonospore, are processes of spontaneous asexual reproduction. Man, imitating Nature, frequently carries out provoked asexual reproduction.

[1] The second mitosis of meiosis is the reductional division and hence named "ektomitosis" (Resende, 1956c; Levan and Müntzing, 1963).

The substitution of the zygote is called *Apomixis* and the substitution of the gonospore is called *Aposporosis* (Winkler, 1908, 1942).

Asexual reproduction may completely replace sexual reproduction, or both processes may exist in the same species or strain, or even in the same individual. In the first case we have all the *Spermatophytae* which have lost their genetic capacity for flowering. This is the case with some hybrids (Renner, 1929; Resende, 1943). In the second case which constitutes the majority of plants, the same strain or even the same individual (e.g. *Hieracium*) may show the two types of reproduction, sexual and asexual. Turesson (1926) classifies such organisms as *amphi-apomictic*. All the known cases of asexual reproduction may be divided into 4 different types (see Table 1) taking into account the origin of the sporophyte (substitution of the zygote) and the origin of the gametophyte (substitution of the gonospore).

A. Vegetative propagation.

The fragmentation of a genetical individual into a number of physiologically independent individuals constitutes a normal process in the physiology of development of the mosses where a protonema produces many buds, which separate and give rise to independent plants. The formation of embryos on the leaves of *Bryophyllum* is also a typical normal process of asexual reproduction. It is probably the best studied case of vegetative reproduction (cf. this Encyclopaedia, vol. XV/2, p. 6).

In the case of the Musci we have a vegetative propagation by *Aposporosis* since in the multiplication of the individual the role of the gonospore is eliminated (see Table 1). In *Bryophyllum* vegetative propagation is by *Apomixis* because the "cutting" substitute for the zygote in initiating a new sporophytic individual. Likewise the *direct spore*[1] may be considered either an Aposporosis or an Apomixis according to the generation which is derived from it (see Table 1). In all effects the *direct spore* is a cellular cutting, i.e. a cutting reduced to a cell.

Table 1.

Apomixis	1. Vegetative propagation	1. Vegetative propagation	Aposporosis
	2. Parthenogenesis	2. Apomictosis	
	3. Apogamy	3. Apospory	
	4. Apogonosis	4. Zygogamosis	

The carpellar leaves of many Angiosperms produce adventitious embryos like the vegetative leaves of *Bryophyllum*. Since these embryos result from the vegetative propagation of cells of the ovule nucellus or integument they appear near the place of the zygote-originated embryos, these cases of asexuality also giving rise to seed (nucellar embryony).

Other cases of natural vegetative propagation (Aposporosis) are the *gemmae* (certain *Marchantiales*) and the *soredia* (Lichens), and in higher plants we have multiplication by grafting and by rooting of stems (Apomixis).

B. Parthenogenesis and apomictosis.

If a non-fertilized gamete germinates and so substitutes the zygote producing a sporophyte generation, we have then what is called *parthenogenesis* (Apomixis). If, however, the same non-fertilized gamete germinates and produces a gameto-

[1] This expression means a spore which produces the same generation as that which gave rise to it.

phyte generation the process is designated *Apomictosis*. In either case we may have the ♂ or the ♀ gamete substituting the zygote or the gonospore. When it is the ♂ gamete replacing the zygote or gonospore the process is called *androgenesis* or *andromictosis*, when it is the ♀ gamete it is called *gynogenesis* or *gynomictosis*.

Androgenesis, in the case of oogamy, always implies pseudozygote formation: the gametes conjugate (plasmogamy) and karyogamy does not take place, because there is degeneration of the nucleus of the ♀ gamete.

Only when it is possible to mark genetically the ♂ and the ♀ nuclei can we know indirectly which of them degenerated after plasmogamy.

C. Apogamy and apospory.

Apogamy refers to the process of the substitution of the zygote for a vegetative cell of the gametophyte generation (Apomixis) and apospory the substitution of the gonospore for a vegetative cell of the sporophyte generation (Aposporosis).

In a *Spermatophyta* apogamy will then be the substitution of the zygote by a synergid, mesocist or antipodal cell, and aposporia the substitution of the macrospore by a vegetative cell of the nucellus or integument of the ovule which will produce the embryonic sac.

When these two processes appear in the same life cycle they eliminate alternation of phases but not alternation of generations. They may also, in the same life cycle, alternate parthenogenesis with apospory or with apomeiosis. In this latter case there is a gonospore produced but without a chromosomic reduction, instead of meiosis, the mother cells produce spores by mitosis. In both Pteridophytae and Angiosperms there are various cases of such asexual cycles with alternating generations (on their significance in physiology of development).

D. Apogonosis and "zygogamosis".

When the gonospore loses its function as a "transition spore" (the cell initiating the gametophyte generation, and produces again the sporophyte from which it was derived, the alternation of generations is eliminated from the life cycle. Instead of a sporophyte-gametophyte alternation, a succession of sporophytes takes place. This loss of sexuality by gametophyte elimination is equivalent to a change to reproduction by "direct spores", which we have already classified as a process of vegetative propagation.

Parallel to this type of loss of sexual reproduction in the sporophyte Winkler conceived the process of "zygogamosis" in which the zygote, because it is incapable of originating the sporophyte, gives rise to the gametophyte.

This process is a purely theoretical concept and probably non-existent. Even if it did exist, it would not be a case of loss of sexual reproduction but rather of change of life cycle to the diplont type because, in order for the zygote to exist as such, there should be gametes which must result from chromatic reduction if the cycle is to be a viable one.

Literature.

Aristotle: Opera omnia. Ed. Academia regia Boruss. (ed. I. Becker). — Auerbach, L.: Organologische Studien. Breslau 1874.

Baer, K. E. v.: De ovi mammalium et hominis genesi. Epistola ad Acad. Imp. scientiarium Petropolitanum. Lipsiae, summpt; Leopoldi Vossii 4, 40 (1827). — Barski, G., S. Sorieul et F. Cornefert: Production dans des cultures *in vitro* de deux souches cellulaires en association de cellules de caractère «hybride». C. R. Acad. Sci. (Paris) **251**, 1825 (1960). —

BLAKESLEE, A. F.: Sexual reactions between hermaphroditic and dioecious Mucors. Biol. Bull. **29**, 87—102 (1915). — BOPP, M.: Entwicklungsphysiologie. Fortschr. Bot. **27**, 163—205 (1965). — BRABEC, F.: Pfropfung und Chimären unter besonderer Berücksichtigung der entwicklungsphysiologischen Problematik. In: Handbuch der Pflanzenphysiologie, Bd. XV/2, S. 388—498. Berlin-Heidelberg-New York: Springer 1965. — BRINK, R. A.: Phase-change in higher plants and somatic cell heredity. Quart. Rev. Biol. **37**, 1—22 (1962). — BÜTSCHLI, O.: Vorläufige Mittheilung über Untersuchungen betreffend die ersten Entwicklungsvorgänge im befruchteten Ei von Nematoden und Schnecken. Z. wiss. Zool. **25** (1875a). — Vorläufige Mittheilung einiger Resultate von Studien über die Conjugation der Infusorien und die Zelltheilung. Z. wiss. Zool. **25**, 426—441 (1875b). — Studien über die ersten Entwicklungsvorgänge der Eizelle, die Zelltheilung und die Conjugation der Infusorien. Abh. Senckenberg. naturforsch. Ges. **10**, H. 3, 213—462 (1876).

CAIRNS, J.: The bacterial chromosome. Sci. Amer. **214**, 37—44 (1966). — CAMERARIUS, R. J.: De sexu plantarum epistola. (Translation of MÖBIUS.) Leipzig: Wilhelm Engelmann 1899. — CATARINO, F. M.: Some effects of kinetin on sex expression in *Bryophyllum crenatum* Bak. (LSDP). Port. Acta biol. A **8**, 267—284 (1964).

EPHRUSSI, B., and M. C. WEISS: Interspecific hybridization of somatic cells. Proc. nat. Acad. Sci. (Wash.) **53**, 1040—1042 (1965). — ESSER, K., u. R. KUENEN: Genetik der Pilze. Berlin-Heidelberg-New York: Springer 1965.

FELDMAN, M., T. MELLO-SAMPAYO, and E. R. SEARS: Somatic association in *Triticum aestivum*. Proc. nat. Acad. Sci. (Wash.) **56**, 1192—1199 (1966). — FREKSA, F. H. F.: Bei der Chromosomenkonjugation wirksame Kräfte und ihre Bedeutung für die identische Verdoppelung der Nukleoproteine. Naturwissenschaften **28**, 376—379 (1940).

GALÁN, F.: Analyse génétique de la monoecie et de la dioecie zygotiques et de leur différence dans *Ecballium Elaterium*. Acta Salmanticensia, Ciencias: Sec. Biol. 15 pags. Universidad de Salamanca 1951. — Analyse génétique de la monoecie et de la dioecie zygotiques et de leur différence dans *Ecballium Elaterium*. Proc. 7th Internat. Bot. Congr. Stockholm 1950, 340 (1953). — Genética e fenogenética del sexo en *Ecballium Elaterium*. Atti IX. Congr. Internaz. Genetica, Bellagio (Como) 1953. Caryologia, Suppl., 942—944 (1954). — Teoria genética del sexo zigótico en el caso de *Ecballium Elaterium*. Rev. Biol. (Lisbon) **4**, 187—220 (1964). — GALEN: Galeni opere. Leipzig: Ed. Kühn 1825. — GALUN, E.: Effects of gibberellic acid and naphthalene acetic acid on sex expression and some morphological characters in the Cucumber plant. Phyton (Buenos Aires) **13**, 1—8 (1959a). — The role of auxins in the sex expression in the Cucumber. Physiol. Plantarum (Copenh.) **12**, 48—61 (1959b). — GALUN, E., Y. JUN, and A. LANG: Culture and sex modification of male Cucumber buds *in vitro*. Nature (Lond.) **194**, 596—598 (1962). — Morphogenesis of floral buds of Cucumber cultured in vitro. Develop. Biol. **6**, 370—387 (1963).

HARTMANN, M.: Die Sexualität, 2. Aufl. Stuttgart: Gustav Fischer 1956. — HEROPHILOS: Cit. by STUBBE 1965, p. 34. — HERTWIG, D.: Beiträge zur Kenntnis der Bildung, Befruchtung und Teilung des tierischen Eies. 1. Abh. Morph. Jb. **1**, 347—434 (1875). — HESLOP-HARRISON, J.: Auxin and sexuality in *Cannabis sativa*. Physiol. Plantarum (Copenh.) **9**, 588—597 (1956). — The experimental modification of sex expression in flowering plants. Biol. Rev. **32**, 38—90 (1957a). — The sexuality of flowers. New Biol. **23**, 8—28 (1957b). — HESLOP-HARRISON, J., and Y. HESLOP-HARRISON: Studies on flowering plant growth and organogenesis. I and II. Proc. roy. Soc. Edinb. B **66**, part IV, 21—28 (1957a). — The effect of carbon monoxide on sexuality in *Mercurialis ambigua* L. fils. New Phytologist **56**, 352—355 (1957b). — Photoperiod, auxin and sex-balance in a long-day plant. Nature (Lond.) **181**, 100—102 (1958a). — Long day and auxin induced male sterility in *Silene pendula* L. Port. Acta biol. A **5**, 79—94 (1958b). — Studies on flowering-plant growth and organogenesis. III. Proc. roy. Irish Acad. B **59**, 257—283 (1958c). — Studies on flowering-plant growth and organogenesis. IV. Proc. roy. Irish. Acad. B **61**, 219—231 (1961). — HESS, O.: Strukturelle und funktionelle Organisation der Chromosomen vielzelliger Organismen. Naturw. Rdsch. **19**, 126—184 (1966). — HOFMEISTER, W.: Vergleichende Untersuchung über die Keimung und Fruchtbildung höherer Kryptogamen und die Samenbildung der Koniferen. Leipzig 1851a. — Die Entstehung des Embryos der Phanerogamen. Abh. kgl. sächs. Ges. Wiss. **6** u. **7** (1851b). — Die Lehre von der Pflanzenzelle. In: Handbuch der physiologischen Botanik, Bd. 1. Leipzig: Wilhelm Engelmann 1867. — Allgemeine Morphologie der Gewächse. In: Handbuch der physiologischen Botanik, Bd. 1, 1. Abt. Leipzig: Wilhelm Engelmann 1868.

KIRCHEISEN, G.: Die Bildung von Dyaden und Monaden sowie das Geschlecht diploider Gametophyten bei *Sphaerocarpus donnellii*. Z. Vererbungsl. **91**, 106—113 (1960). — KÜHN, A.: Grundriß der Vererbungslehre, 4. Aufl. Heidelberg: Quelle & Meyer 1965a. — Vorlesungen über Entwicklungsphysiologie, 2. Aufl. Berlin-Heidelberg-New York: Springer 1965b.

LAIBACH, F.: Wuchsstoff und Blütenbildung. Beitr. Biol. Pflanz. **29**, 129—141 (1952). — LAIBACH, F., u. F. J. KRIBBEN: Der Einfluß von Wuchsstoff auf die Bildung männlicher und weiblicher Blüten bei einer monözischen Pflanze *(Cucumis sativus)*. Ber. dtsch. bot.

Ges. **62**, 53—55 (1950a). — Über die Bedeutung der B-Indolylessigsäure für die Blütenbildung. Ber. dtsch. bot. Ges. **63**, 119—120 (1950b). — Der Einfluß von Wuchsstoff auf die Blütenbildung der Gurke. Naturwissenschaften **37**, 114—115 (1950c). — Ein weiterer Beitrag zur Frage nach der Bedeutung des Wuchsstoffs für die Blütenbildung. Beitr. Biol. Pflanz. **29**, 339—352 (1953). — Leeuwenhoek, A. van: Letter to Viscount Brouncker, November 1677. Phil. Trans. **12**, 1040—1043 (1677) — Lederberg, J : Gene recombination and linked segregations in *Escherichia Coli*. Genetics **32**, 505—525 (1947). — Sibling recombinants in zygote pedigrees of *Escherichia Coli*. Proc. nat. Acad. Sci. (Wash.) **43**, 1060—1065 (1957). — Levan, A., and A. Müntzing: Terminology of chromosome numbers. Port. Acta biol. A **7**, 1—16 (1963). — Linskens, H. F.: Physiologie der Fortpflanzung und Sexualität. Fortschr. Bot. **18—22** (1956—1960). — Beitrag zur biochemischen Charakterisierung einer *Aloë*-Mutante mit blütenloser Infloreszenz. Port. Acta biol. A **6**, 125—130 (1959—1962). — Linskens, H. F., and F. Resende: Beitrag zur Characterisierung des Langtag- und Kurztag-Habitus von *Bryophyllum crenatum* Bak. Port. Acta biol. A **6**, 141—154 (1961). — About the action of temperature on the floral impulse in *Bryophyllum* (LSDP). Port. Acta biol. A **9**. 305—307 (1966).

Meyer, G. F.: A possible correlation between the submicroscopic structure of meiotic chromosomes and crossing-over. Proc. Eur. Reg. Conf. on Electron Microscopy **3**, 461—462 (1965). — Moewus, F.: Methodik und Nachträge zu den Kreuzungen zwischen *Polytoma*-Arten und zwischen *Pratosiphon*-Rassen. Z. indukt. Abstamm.- u. Vererb.-Lehre **73**, 63—107 (1937). — Die Sexualität und der Generationswechsel der Ulvaceen und Untersuchungen über die Parthenogenese der Gameten. Arch. Protistenk. **91**, 357—441 (1938).

Naudin, Ch.: Réflexion sur l'hybridation dans les végétaux. Rev. Horticole, Paris Sér. **4**, 351—354 (1855). — Nitsch, J. P.: Physiology of flower and fruit development. In: Handbuch der Pflanzenphysiologie, Bd. XV/1, S. 1537—1547. Berlin-Göttingen-Heidelberg: Springer 1965. — Noll, F.: Die Pfropf-Bastarde von Bronvaux. S.-B. Niederrhein. Ges. Natur- u. Heilkunde Bonn 1905, S. 1—34. — Nunes, M. A.: Sex expression abnormalities in rice plants suffering from "Straighthead". An attempt to hormonal interpretation. Port. Acta Biol. A 8, 284—300 (1964).

Partanen, C. R.: On the chromosomal basis for cellular differentiation. Amer. J. Bot. **52**, 204—209 (1965). — Pöllnitz, K. v.: *Gasteria staynerii* sp. nov. Cactus J. **7**, 1—2 (1938). — Pontecorvo, G.: The genetics of *Aspergillus nidulans*. Advanc. Genet. **5**, 141—238 (1953). — Pontecorvo, K.: Mitotic recombination in the genetic systems of filamentous fungi. Caryologia, Suppl. **6**, 192—200 (1954). — The parasexual cycle in fungi. Ann. Rev. Microbiol. **10**, 393—400 (1956). — Pontecorvo, K., and E. Käfer: Genetic analysis based on mitotic recombination. Advanc. Genet. **9**, 71—104 (1958). — Prud'homme, N.: Somatic recombination in the basidiomycete *Coprinus radiatus*. In: Incompatibility in fungi (ed. K. Esser and J. R. Raper), p. 48—52. Berlin-Heidelberg-New York: Springer 1965.

Quintanilha, A.: Os fundamentos científicos da sexualidade, 2. ed. Lisboa: Cosmos 1945.

Renner, O.: Zur Terminologie des pflanzlichen Generationswechsels. Biol. Zbl. **36**, 337—374 (1916). — Artbastarde bei Pflanzen. In: Handbuch der Vererbungswissenschaften, Bd. 2. (Lief. 7). Berlin: Borntraeger 1929. — Resende, F.: Suculentas Africanas. III. Contribuição para o estudo da morfologia, da fisiologia da floração e da geno-sistemática das Aloinae. Mem. Soc. Brot. A **2**, 1—108 (1943). — Cariocinese Lisboa: Sociedade Portuguesa de Biologia (1946). — Kariokinesis. Port. Acta biol. A **2**, 1—24 (1947a). — Suculentas Africanas. VIII, A inflorescência de *Bryophyllum Daigremontianum* (R. Hamet et Perrier) Berg. (Planta de Dia Curto) em condições extremas de Dia Curto e em condições de Dia Curto ou Longo com luz de fraca intensidade luminosa. Bol. Soc. port. Cie. nat. **15**, 117—122 (1947b). — Luz e floração, inibidores rizogénicos. Bol. Soc. port. Cie. nat. **15**, 188—191 (1948). — Auxin and antiauxin, the hormones responsible for the change of the vegetative into floral phenotypes. Bol. Soc. port. Cie. nat. (2. Ser., **17**, 174—188 (1949). — Contribution to the physiologie of development of the inflorescence and of the single flower *(Bryophyllum* and *Kalanchoë)*. Port. Acta biol. A (R. B. Goldschmidt Vol.), 729—784 (1950). — A determinação de sexo: monoicia, dioicia e subdioicia nos animais e plantas. Bol. Soc. port. Cie. nat. 2. Ser., **3**, 270—272 (1951a). — As auxins e o sexo. Bol. Soc. port. Cie. nat., 2. Ser., **3**, 267 (1951b). — Consequências cito-genéticas do "kinetochoro" difuso. Bol. Soc. port. Cie. nat., 2. Ser., **4**, 252—260 (1953a). — Auxina e sexo em "Hyoscyamus" L. Bol. Soc. port. Cie. nat., 2. Ser,. **4**, 248—251 (1953b). — Conceito de geração "assexuada" num ciclo de vida sexuado. Bol. Soc. port. Cie. nat., 2. Ser., **6**, 257—261 (1956a). — "Inflorescências florais" sem flores (Aloë). Bol. Soc. port. Cie. nat., 2. Ser., **6**, 227—236 (1956b). — Endomitose e exomitose. Rev. Biol. (Lisboa) **1**, 93—95 (1956c). — Phenotypically and genotypically determined variation between monoecious and dioecious strains (flowering plants). Colloque Int. Photo-Thermopériodisme Parma, Publ. No 34, Ser. B de l'U.L.S.B. 25 (1957). — Dominanzwechsel und Epistasiewechsel bei der Entwicklung eines Individuums bzw. eines Organs. Ber. dtsch. bot. Ges. **70** (32) (1958). — On the transmission of the "floral state" through grafting from

LSDP or SDP-donors to LSDP-acceptors in LD and SD. Evidence for the "separation" of growth-hormones from anti-growth-hormones in this transmission. Port. Acta biol. A **6**, 1—17 (1959). — RESENDE, F., and H. F. LINSKENS: On hermaphrodite and female plants in *Bryophyllum daigremontianum* and its segregation. I. Port. Acta biol. A **6**, 155—178 (1961). — RESENDE, F., and M. MANARTE: Physiological changes as the cause of mutations. Port. Acta biol. A **3**, 219—333 (1951). — RESENDE, F., and M. J. VIANA: Suculentas Africanas. IX. Port. Acta biol. A **2**, 211—226 (1948). — Some effects of synthetic auxins and antiauxins on the floral state. Bol. Soc. port. Cie. nat., 2. Ser., **4**, 74—88 (1952). — Gibberellin and sex expression. Port. Acta biol. A **6**, 77—98 (1959). — RESENDE, F., and J. W. WARDEN: As formas juvenis na sistemática e na genética. Bol. Soc. port. Cie. nat., 2. Ser., **5**, 78—80 (1954).

SERRA, J. A.: Evidence on the mechanism of chromosome pairing. Rev. port. zool. Biol. Geral **3**, 79—94 (1961). — Novas interpretações na determinação e hereditariedade do sexo. Rev. port. zool. Biol. Geral **4**, 37—73 (1963). — SIGNORET, J.: Le probléme de la differentiation cellulaire et son abord expérimentel. Ann. Biol. **3**, 163—182 (1964). — STERN, C.: Somatic crossing-over and segregation in *Drosophila melanogaster*. Genetics **21**, 625—730 (1936). — STRANDHEDE, SVEN-OLOV: Chromosome studies in *Eleocharis*, subser. *Palustris*. VI. A possible case of an extra, reductional division giving rise to hemi-haploid pollen nuclei. Bot. Not. (Lund) **118**, 243—253 (1965). — STRASBURGER, ED.: Über Zellbildung und Zelltheilung. Jena: H. Dabis 1875. — Lehrbuch der Botanik für Hochschulen, 28. Aufl. Jena: VEB Gustav Fischer 1962. — STUBBE, H.: Kurze Geschichte der Genetik bis zur Wiederentdeckung der Vererbungsregeln Gregor Mendels, 2. Aufl. Jena: VEB Gustav Fischer 1965.

THEOPHRASTUS: Tom I, Historia plantarum. Tom II, De causis plantarum, hrsg. von WIMMER. Leipzig: Teubner 1854. — TURESSON, K.: Studien über *Festuca* L. I. Hereditas (Lund) 8, 161—206 (1926).

WAGNER, R. P., and H. K. MITCHELL: Genetics and metabolism, 2. ed. New York-London-Sydney: John Wiley & Sons, Inc. 1964. — WEIDEL, W.: Virus und Molekularbiologie (Heidelberger Taschenbücher). Berlin-Göttingen-Heidelberg: Springer 1964. — WEILING, F.: Licht- und elektronenmikroskopische Beobachtung zum Problem der Cytomixis sowie ihrer möglichen Beziehung zur Photocytose. Untersuchungen bei *Cucurbita*-Arten und *Lycopersicum esculentum*. Planta (Berl.) **67**, 182—212 (1965). — WESTERGAARD, M.: Aberrant Y chromosomes and sex expression in *Melandrium album*. Hereditas (Lund) **32**, 419—443 (1946). — The relation between chromosome constitution and sex in the offspring of triploid *Melandrium*. Hereditas (Lund) **34**, 257—279 (1948). — Über den Mechanismus der Geschlechtsbestimmung bei *Melandrium album*. Naturwissenschaften **40**, 253—260 (1953). — The mechanism of sex determination in dioecious flowering plants. Adv. Genet. **9**, 217—281 (1958). — WETTSTEIN, F. v.: Morphologie und Physiologie des Formwechsels der Moose auf genetischer Grundlage. I. Z. indukt. Abstamm.- u. Vererb.-Lehre **33**, 1—236 (1924). — Über Fragen der Geschlechtsbestimmung bei Pflanzen. Naturwissenschaften **12**, 761 (1924). — Gesichertes und Problematisches zur Geschlechtsbestimmung. Ber. dtsch. bot. Ges. **54**, 23—38 (1936). — WINKLER, H.: Parthenogenesis und Apogamie im Pflanzenreich. Jena: Gustav Fischer 1908. — Über die experimentelle Erzeugung von Pflanzen mit abweichenden Chromosomenzahlen. Z. Bot. 8, 417—531 (1916). — Entwicklungsmechanik oder Entwicklungsphysiologie der Pflanzen. In: Handwörterbuch der Naturwissenschaften, Bd. 3, 620—629 2. Jena: Gustav Fischer 1932. — Fortpflanzung der Gewächse. In: Handwörterbuch der Naturwissenschaften, 2. Aufl., Bd. 4. Jena: Gustav Fischer 1934. — WITSCH, H. v.: Vergrünung und Verlaubung. In: Handbuch der Pflanzenphysiologie, Bd. XV/2, S. 352—364. Berlin-Heidelberg-New York: Springer 1965.

Die chemischen Grundlagen der Befruchtung (Gamone).

Von

Kurt Köhler.

Mit 21 Abbildungen.

I. Einleitung.

Im Laufe der letzten Jahrzehnte ist es gelungen, viele biologische Vorgänge auf ihre biochemischen und physikalischen Grundreaktionen zurückzuführen. Wenig weiß man jedoch noch immer von den Vorgängen, die einen Kontakt zwischen Zellen und die Fusion von Zellen bewirken (WEISS 1958), also jenen Tatsachen, die im speziellen Falle, nämlich als Vereinigung einer männlichen mit einer weiblichen Geschlechtszelle, auch typische Ereignisse bei der Befruchtung sind. Nach HARTMANN (seit 1909) ist die *Ursache* der Befruchtung die Sexualität, eine biologische bipolare Zweigeschlechtlichkeit, die, wie man vermuten und wünschen möchte, auf einer chemischen und (oder) physikalischen Polarität beruhen müßte. Ergebnis und *Ziel* der Befruchtung ist die Vereinigung zweier haploider Kerne zu einem diploiden, in dem später die meiotischen Prozesse ablaufen. Obgleich es vorübergehend so aussah, als sei wenigstens für einen Organismus, die einzellige Grünalge *Chlamydomonas*, die chemische Ursache der sexuellen Polarität und der Befruchtung gefunden, wissen wir nach Überprüfung der Ergebnisse und der Publikationen von MOEWUS, daß der von ihm beschrittene Pfad ein Irrweg war (Überblick: MOEWUS 1951; Kritik: unter anderen FÖRSTER und WIESE 1954, RENNER 1957, 1958). Während mehrere Arbeitskreise durch Untersuchungen an tierischen Eiern, vor allem am Seeigelei, besonders in den letzten 20 Jahren reichlich Tatsachenmaterial über die stofflichen Vorgänge der Befruchtung gesammelt haben (Überblick: METZ 1957), sind im Bereich der pflanzlichen Befruchtungserscheinungen nur wenige eingehende physiologische und biochemische Arbeiten bekanntgeworden. Ein Überblick über die verstreuten Veröffentlichungen wird somit manche einmalige und zufällige Beobachtung miteinschließen müssen, um damit zu fruchtbarer Weiterarbeit anzuregen. Einige umfassende Überblicke über das hier zu betrachtende Gebiet, die viele Einzelergebnisse und Gedanken über die chemische Regulierung der Sexualvorgänge zusammentragen, stammen von RAPER (1952, 1957 und in diesem Band) und LEWIN (1954).

Dieser Artikel, der mit Absicht als Überschrift das Wort „Gamone“ trägt, soll enger begrenzt, nur die chemischen Prozesse im Auge haben, die der Vereinigung von Gameten oder deren Äquivalenten zugrunde liegen; also solcher Zellen, Organe oder Organismen, die sich vereinigen müssen, um zwei haploide Kerne von verschiedenem Geschlecht in ein gemeinsames Plasma zu bringen. Von den folgenden Prozessen, dem Heterokaryon, der Kernfusion und der Chromosomenpaarung sind bisher gegenseitige stoffliche Beeinflussungen noch nicht bekanntgeworden.

Der Begriff „Gamon" wurde von HARTMANN, KUHN und SCHARTAU auf Grund ihrer Untersuchungen über die Befruchtungsstoffe der Seeigel geprägt. Er wurde nicht allgemein übernommen und dient vorwiegend im deutschen Sprachgebrauch zur Bezeichnung von Stoffen, die für den Mechanismus der Befruchtung von Bedeutung sind. Der Begriff *Gamon* soll hier in dem Sinne gebraucht werden, wie er ursprünglich vorgeschlagen worden war (HARTMANN und SCHARTAU 1939, S. 587): „Die Wirkstoffe, um die es sich hier handelt, sind spezifische Stoffe, welche die chemische Wechselwirkung zwischen weiblichen und männlichen Geschlechtszellen bedingen und zu deren Vereinigung führen." „Wir schlagen den Namen Gamone (Gametenhormone) vor. Dieser Name ist so kurz, daß er leicht Zusammensetzungen erlaubt (z. B. Phyto-Gamone, Andro-Gamone, Gyno-Gamone usw.). Man muß ihn auch nicht direkt unbedingt auf Gamet zurückführen, sondern auf γάμος (Hochzeit), so daß die wörtliche Übersetzung Hochzeitsstoffe = Befruchtungsstoffe wäre." „Eine solche Namengebung kann nicht auf chemischer Grundlage vorgenommen werden, da nur biologische Gesichtspunkte entscheidend sind." Es kann mit Recht betont werden, daß eine allgemein gültige chemische Grundlage für die Befruchtung oder die Sexualität aller Lebewesen wohl nicht gefunden werden kann (etwa im Sinne MANOILOFFs), sondern daß jedem Einzelschritt der mannigfaltigen Möglichkeiten, die zur Vereinigung der Kerne oder deren Äquivalente, sei es von räumlich getrennten, verschieden geschlechtlichen Pflanzen, Tieren oder Bakterien, oder sei es bei autogamen Vorgängen, führen, eine chemische Reaktions*kette* zugrunde liegt. Es bleibt daher niemandem erspart, sich zuerst gründlich über die morphologischen und cytologischen Abläufe klar zu sein, ehe molekulare Reaktionen in Betracht gezogen werden.

II. Allgemeine Übersicht über die morphologischen und physiologischen Vorgänge bei der Befruchtung (oder Paarung).

Da nun die Algen und die Pilze große Mannigfaltigkeiten bei der sexuellen Vereinigung aufweisen, soll hier notwendigerweise eine Gliederung auf grob morphologischer Basis vorangestellt werden.

1. Bei der Gametenkopulation wird die Befruchtung durch Vereinigung einzelliger haploider Geschlechtszellen vollzogen. Für experimentelles Arbeiten mit Gameten ist Vorausbedingung, daß man sie zu gewisser Zeit und in größeren Mengen erhalten kann; das bedeutet, daß *getrenntgeschlechtliche* Pflanzen bevorzugt für die Untersuchungen von Gamonen in Frage kommen. Die Gameten vieler Algen und niederer Pilze sind begeißelt. Die Tendenz zur Ausbildung größerer, unbeweglicher und sich spät festsetzender Eier läßt sich besonders bei den Braunalgen gut verfolgen (s. unten). Die männlichen Geschlechtszellen der Rotalgen sind unbegeißelt und vor dem Festsetzen ohne aktive Beweglichkeit.

Bei Algen mit begeißelten Gameten sind zwei Arten von Gruppenbildung näher bekanntgeworden:

a) Die Gameten können in beiden Geschlechtern gleich groß sein. Während der Paarung können sich beide Gameten gleich verhalten. Mischt man z. B. von *Chlamydomonas eugametos* Gameten beider Geschlechter, so tritt Gruppenbildung oder „clumping" auf, indem viele Gameten beider Geschlechter sich gegenseitig mit den Geißeln einzufangen versuchen (Typ: Isogamie).

b) Mischt man dagegen von *Ectocarpus siliculosus* Gameten beider Geschlechter, die ebenfalls von gleicher Größe sind, so erfolgt zunächst keine Gruppenbildung. Erst nachdem sich die Gameten des einen Geschlechtes festgesetzt haben, werden diese wie Eier von Gameten des anderen Geschlechtes umschwärmt, auf ganz ähnliche Weise wie die großen unbegeißelten *Fucus*-Eier. Es ist demnach durchaus vertretbar, auch die großen beweglichen Gynogameten von *Cutleria* als „begeißelte Eier" anzusehen, da auch sie erst später, nachdem sie die Geißeln abgeworfen haben, oogam befruchtet werden.

Auf dieser Grundlage sollen daher im folgenden nur mehr die physiologischen Begriffe Isogamie und Oogamie angewandt werden. Diese Unterscheidung basiert dann nicht mehr auf Größe und Form der Gameten, sondern vielmehr darauf, an welcher Stelle bei diesen Gameten das Gamon lokalisiert ist oder sezerniert wird.

2. Bei der Gamontenkopulation paaren sich Mutterzellen, die als Haplonten durch mitotische Zellteilung oder als Diplonten nach der Meiose Gameten oder Kerne bilden, die die Befruchtung durchführen. Für die Diatomeen nimmt Geitler (1958) auf Grund sorgfältiger Beobachtungen an, daß die einzelnen Schritte bei der Paarung ein gegenseitiges Einverständnis voraussetzen. Unter den „tierischen" Protozoen sind bei *Paramaecium* Paarungstypsubstanzen von Proteincharakter bekanntgeworden (Metz 1954, Miyake 1964, Cohen u. Siegel 1963).

3. Gametangien-Befruchtungen kommen bei den Phycomyceten und den Ascomyceten in wenigstens drei Variationen vor:

a) als Eibefruchtung durch Antheridienhyphen,

b) als Kopulation zweier Gametangien,

c) als Befruchtung des Ascogons durch „vegetative" Zellen.

Für jede Variation ist wenigstens ein Beispiel von Gamonwirkungen bekanntgeworden.

Bei den *Oomycetales (Saprolegnia, Achlya)* werden in den Oogonien einkernige Eier gebildet. Die Antheridien wachsen als Schläuche auf die Eier zu, so daß das männliche Gametangium wohl nur den Kern zur Befruchtung beisteuert. Bei den Mucorales und den typischen Ascomyceten müssen beide Gametangien zueinander finden. Bei anderen Ascomyceten schließlich kann das männliche Gametangium zurückgebildet sein und Sporen und Coniden übernehmen als Spermatien die Rolle der Geschlechtszellen. Diese werden häufig durch die Trichogyne, der typischen Kopulationshyphe der Ascomyceten aufgesucht. Die männlichen Kerne wandern durch die Trichogyne in das Ascogon.

4. Hyphenfusionen als sexuelle Vorgänge finden sich bei den Ascomyceten *(Tuberales)* und den Hymenomyceten. Von einer chemischen Regelung dieser Vorgänge sind bisher keine Untersuchungen bekannt geworden.

Obwohl nun die Befruchtungen in den genannten Gruppen ganz verschieden ablaufen, lassen sich die Vorgänge bei den Algen wie bei den Pilzen in einzelne biologische Grundreaktionen zerlegen, die ihrerseits nicht einmal spezifisch für Befruchtungsreaktionen sein müssen.

Im Ausgangsstadium liegen die Kerne verschiedenen Geschlechtes, die sich vereinigen sollen — wenn man von autogamen Vorgängen absieht — in räumlich voneinander getrennten Organismen oder Organen. Als erste Reaktion ist daher eine Entfernungsüberbrückung passiver oder aktiver Art nötig, passiv durch Luft oder Wasserströmung, passiv auch als ungerichtete, dem Organismus eigene natürliche Bewegung (= zufällige Berührung), aktiv im Sinne einer gerichteten Suchbewegung, als Chemotaxis oder Chemotropismus, oft verbunden mit aktivierter Beweglichkeit.

Der zweite Akt wäre ein spezifischer Kontakt oder ein Verkleben der Zellen miteinander, der die Zellen (oder Organe) solange beieinander hält, bis es gelingt, aus der Gruppe der Bewerber die Partner herauszulösen, oder sie in die richtige Lage zueinander zu bringen und schließlich eine Zellverschmelzung zu ermöglichen.

Als weitere Schritte wären das Di- oder Heterokaryon, die Kernfusion, und als letzter Schritt die Paarung von Chromosomen in der Meiose zu nennen. Tabelle 1 und 2 geben einen Überblick über die einzelnen Schritte.

Tabelle 1. *Vergleichende Übersicht über die Befruchtungsvorgänge bei Algen, bei Paramaecium und bei Metazoen.* (Nach KÖHLER 1956.)

Zahl der Partner	Reaktionsschritt	*Chaetomorpha Chlamydomonas*	Braunalgen	Ciliaten	Metazoen
∞:∞ ∞:1	1. Fernwirkung	Gamon? zufällige Berührung?	Chemotaxis bei *Cutleria*	zufällige Berührung	Gynogamon I (Chemotaxis)
	2. Agglutination	an den Geißelspitzen (Gamon)	an der Geißelspitze; Sonderbildung an der Geißel und Proboscis	an den Cilienspitzen (P. T.-Subst.)	zwischen Acrosomfaden und Eihülle (Gynogamon II, Androgamon II)
1:1	3. Isolierung der Paare	Geißelspiralisierung		Holdfast-Region	Befruchtungswelle Befruchtungsmembran
	4. Zellverschmelzung	Plasmastrang bei *Chlamydomonas*		paroral cone (Plasmaaustausch möglich)	
	5. Wanderung der Kerne				
	6. Karyogamie				

III. Die Fernwirkungen.

1. Chemotaxis (Sirenine).

Ältere Forscher gingen von der Theorie aus, daß die Befruchtungsvorgänge und die Gruppenbildung maßgeblich durch Chemotaxis verursacht seien. Manche dieser Beobachtungen und Experimente zielen daher auch allein darauf hinaus, Chemotaxis zu entdecken. Vielleicht könnte einiges davon nach heutigen Erfahrungen eher als Hinweis auf Agglutinations- oder Fallenreaktionen gewertet werden (s. unten KOTTE 1923, PASCHER 1931). Meist wurde Chemotaxis in der Weise nachgewiesen, daß man die beweglichen Gameten — oder vegetativen Organismen — in eine Capillare einwandern ließ. Spermatozoen, die durch Substanzen bei gewisser niedriger Konzentration aktiviert, bei höheren Konzentrationen aber narkotisiert oder getötet werden, werden zwar verstärkt in die Capillare einwandern, nicht aber wieder heraus, wie es normale Spermatozoen vielleicht täten. Sie werden in der Capillare wie in einer *Falle gefangen*. Leider läßt sich nicht immer nachträglich eindeutig feststellen, ob es sich, wo immer die Methode angewandt wurde, um eine wirkliche Anlockung oder eine solche *Fallenreaktion* handelt.

Das Problem der Chemotaxis ist nicht nur für die Befruchtung von Bedeutung. Für die Fruchtkörperbildung bei den Acrasieen und für das Auffinden von Wirts- und Commensalenorganismus (z. B. für *Saprolegnia*-Zoosporen, FISCHER und WERNER 1956) wurde Chemotaxis näher untersucht. Die Bedeutung der Chemotaxis bei Archegoniatenspermatozoiden für die Befruchtung wird an anderer Stelle dieses Handbuches (s. Band XVII/2) besprochen.

Tabelle 2. *Vergleichende Übersicht über die Befruchtungsvorgänge bei Pilzen (Dimictic mating systems).* (Nach BURNETT 1956.)[1]

Process	Mode of operation	*Mucor*	*Achlya*	*Ascobolus magnificus*	*Humaria granulata*	*Neurospora*	Rusts Smuts
Tropistic effects due to extra-mycelial agents, probably diffusable	Zygotropism	+	+	+	+	+	+
	Telemorphosis	+	+	—	—	—	—
	Thigmotropism	+	+	+	±	—	—
Morphological differentiation of special regions in connection with reproduction	None, physiological differentiation only	+	—	—	—	—	—
	As the result of tropistic stimulation	—	+	—	—	—	—
	Localized before commencement of the mating reaction	—	—	+	+	+	±
Plasmogamy	Restricted to differentiated organs or regions of fusion	+	+	+	—	—	—
	Preferential restriction to specialized conjugant regions but fusion can occur elsewhere	—	—	—	+	+	±
	Unrestricted	—	—	—	—	—	+
Nuclear migration	Restricted to region of fusion or potential zygote	+	+	+	—	—	—
	Limited migration through hyphae	—	—	—	+	+	—
	Unlimited migration through hyphae	—	—	—	—	—	+
Nuclear association	Only immediately prior to fusion	+	+	+	+	+	—
	Dicaryon of limited extent or duration	—	—	(+)	(+)	(+)	±

[1] Um Diskussionen über Definitionen zu vermeiden, wird Tabelle 2 in der Originalsprache wiedergegeben.

a) *Oedogonium.*

MAINX (1931) beobachtete bei seinen Untersuchungen an der Grünalge *Oedogonium* in einem mikroskopischen Präparat, daß Spermatozoide, denen durch ein Haar der Zugang zu einem Algenfaden mit Oogonien verwehrt war, sich an diesem Haar in nächster Nähe des Oogons ansammelten, was er auf chemotaktische Anlockung zurückführte. Sobald es einem der Schwärmer gelungen war, das Ei zu befruchten, hörte diese offenbar auf. Das Problem wurde nicht weiter verfolgt.

Ebenfalls aus Beobachtungen wurde ein System verschiedener gegenseitiger Beeinflussung bei einer nannandrischen *Oedogonium*-Art (der männliche Gametophyt ist ein sog. Zwergmännchen) erschlossen (Abb. 1). Die von den männlichen Fäden entlassenen Androsporen werden von einer Substanz, die von der Oogonium-Mutterzelle sezerniert wird, chemotaktisch angezogen. Dieses Agens wird solange abgegeben bis sich eine Androspore festgesetzt hat. Die oberhalb oder unterhalb des Oogons zur Ruhe gekommene Androspore entwickelt sich zum Zwergmännchen, dem männlichen Gametophyten mit dem Antheridium. Die Zwergmännchen wachsen gerichtet (also offenbar chemotropisch) auf die Oogonium-Mutterzelle zu. Das sich entwickelnde Zwergmännchen induziert die Teilung der Oogonium-Mutterzelle in Oogonium und Stützzelle. Das Oogon umgibt darauf sich und das Zwergmännchen mit einer Gallerte, die als Spermiumfalle funktioniert. Die Spermien werden vom Porus des Oogons chemotaktisch angezogen. Während freischwimmende Spermien makandrischer, diöcischer Formen

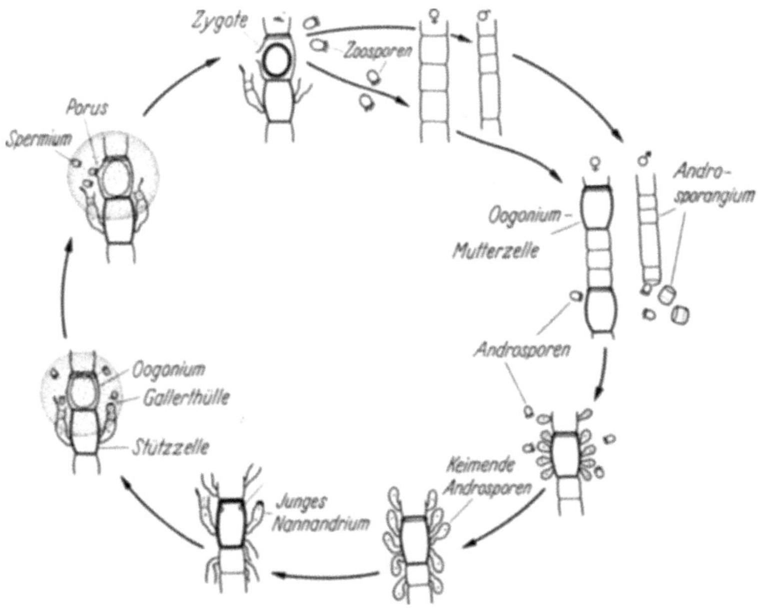

Abb. 1. Befruchtungsvorgang bei einer *Oedogonium*-Art mit Zwergmännchen. (Schematisch nach RAWITSCHER-KUNKEL u. MACHLIS, verändert.)

Chemotaxis vergleichbar der der Androsporen zeigen, kriechen die Spermien der nannandrischen in der Gallerte des Oogons. Auch ein in Agar eingebettetes isoliertes Oogon wirkt chemotaktisch anziehend auf Spermien (RAWITSCHER-KUNKEL u. MACHLIS 1962, HOFFMAN 1960, 1961).

b) *Cutleria.*

Bei der Braunalge *Cutleria* gelang es HARTMANN (1947), Chemotaxis nachzuweisen. Gibt man zu einer Suspension der großen weiblichen, zweigeißligen Gameten einige Tropfen der kleinen Spermatozoide, so erfolgt zunächst, solange die „Eier" beweglich sind, weder Gruppenbildung noch Befruchtung. Erst wenn diese sich festsetzen, sammeln sich reichlich Spermatozoide um sie herum an. Da die Spermatozoide auch in Tropfen von „Eiwasser" einwandern, dürfte Chemotaxis die Ursache hierfür sein.

c) *Fucus.*

Für *Fucus* liegen sorgfältige Untersuchungen über die chemotaktische Anlockung der Spermatozoide durch das Ei vor. Die ersten frühen Experimente von KOTTE ergaben, daß die Reaktion phobische Chemotaxis sei und ein Konzentrationsgefälle voraussetzt. Daß der Schleim, der sich häufig mit den Eiern

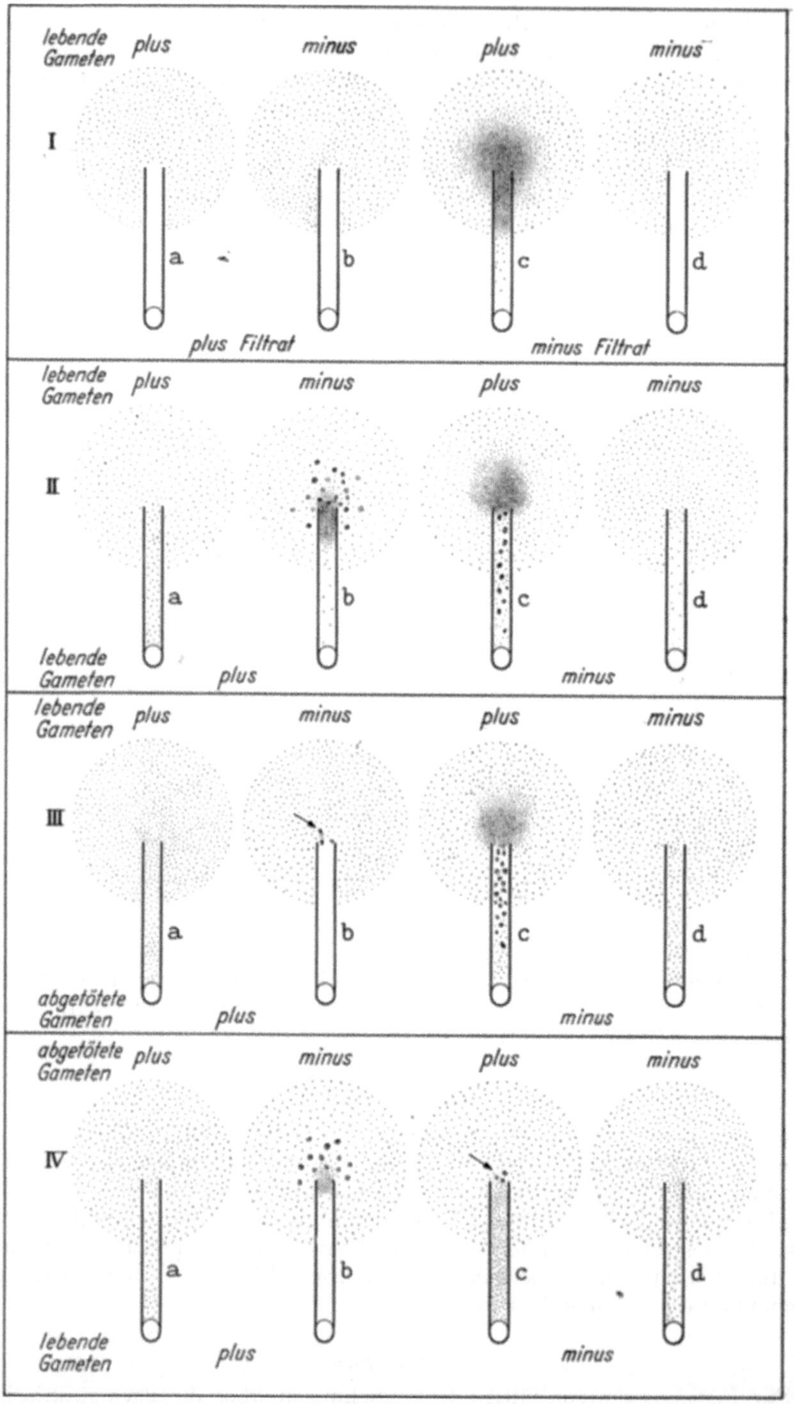

Abb. 2. *Chlamydomonas sp. 24 Tsubo.* Chemotaktische Anlockung von Gameten an der Öffnung einer Capillare. I Anlockung von Plus- und Minus-Gameten durch Filtrate. Nur die Plus-Gameten sammeln sich vor der Capillare mit Minus-„Sirenin" (s. MACHLIS) und wandern in diese ein. II Anlockung von Gameten durch lebende Gameten. Die Plus-Gameten in Capillare 6 wandern aus. Gruppenbildung nur außerhalb. Die Plus-Gameten in c außerhalb

findet, für diese Wirkung verantwortlich sei, dürfte ein Fehlschluß gewesen sein. COOK und ELVIDGE prüften mit Hilfe der Capillarmethode Filtrate von Eiern der Arten *Fucus serratus* Linn. (dioec.), *Fucus vesiculosus* Linn. (dioec.) und *Fucus spiralis* Linn. (monoec.), die 24 Std bei 0° C gestanden hatten und erhielten gute chemotaktische Anlockung. Versuche, die Substanzen zu isolieren, ergaben, daß es sich um flüchtige Stoffe handeln müsse. So können die aktiven Substanzen z. B. mit inerten Gasen aus den Filtraten entfernt und in einer Kühlfalle quantitativ wieder abgefangen werden. Die Autoren kamen dann beim Versuch, die Substanzen mit Petroläther zu extrahieren, scheinbar der Frage näher. Sie fanden, daß Spuren von Petroläther in Seewasser bereits chemotaktische Wirkung auf *Fucus*-Spermatozoide haben. Darauf wurden eine Reihe verschiedener Petroläther und hochgereinigter Kohlenwasserstoffe geprüft. Von ihnen war n-Hexan bei größten Verdünnungen wirksam. Massenspektrographische Untersuchungen ergaben jedoch, daß das Spektrum des Hexans von dem des chemotaktischen Stoffes abweicht. Dessen Molekulargewicht ließ sich mit etwa 74 berechnen. Höhere Titer zeigen auch eine Reihe von anderen flüchtigen Kohlenwasserstoffverbindungen mit dem Molekulargewicht von etwa 74, wie sec. Butylalkohol, Methylbutenyläther, Diäthyläther, Methylamylketon, Essigsäurepropylester, n-Buttersäuremethylester und die Propionsäuremethyl- und -äthylester. Obwohl das Spektrum des Äthyläthers dem des chemotaktischen Stoffes am nächsten kommt, scheint Äther, da er erst in höherer Konzentration wirksam ist, nicht der fragliche Stoff zu sein. Das Problem ist bis jetzt ungelöst. Zusammenfassend läßt sich soviel sagen, daß es sich um eine chemisch inaktive Substanz mit einem Molekulargewicht zwischen 70 und 120 handeln muß, die aus einer geraden Kette aus 5 oder 6 C-Atomen besteht.

d) *Chlamydomonas.*

Auch für *Chlamydomonas (sp. 24) Tsubo* wurde mit Hilfe der Capillarmethode Chemotaxis nachgewiesen (TSUBO 1957). Zellfreie Filtrate von Gameten eines (—)Klons locken (+)Gameten an. Filtrate von (+)Zellen sind dagegen wirkungslos. Es wurde weiterhin die Wirkung toter und lebender Zellen aufeinander geprüft; die Ergebnisse sind in Abb. 2 zusammengefaßt. Werden lebende (+) Zellen in einer Capillare auf (—)Gameten geprüft, so wandern ++ aus der Capillare aus. Gruppenbildung erfolgt nur *außerhalb* der Capillare. Umgekehrt wird eine solche mit (—)Gameten gegen einen Tropfen (+) Zellen geprüft, so wandern diese ein, Gruppenbildung ist nur im *Inneren* der Capillare sichtbar. Das gleiche läßt sich mit toten (+) und lebenden (—) (und umgekehrt) zeigen. Da tote (+) nicht wandern, erfolgt zwischen ihnen und lebenden (—) kaum eine Reaktion, während umgekehrt lebende (+) in die Capillare mit toten (—) einwandern oder aus einer Capillare zu den außerhalb liegenden toten (—) gehen und Gruppen bilden. Nur lebende, kopulationsfähige (—)Gameten geben das chemotaktische Prinzip ab, abgetötete und gut gewaschene dagegen nicht mehr, obwohl tote (—) mit lebenden (+)Gameten Gruppen bilden. Das zeigt deutlich, daß hier die chemotaktische Anlockung und die Agglutination zwei verschiedene Stufen sind. Die chemische Natur des chemotaktischen Prinzips ist noch unbekannt; Untersuchungen mit Conway-Gefäßen ergaben, daß es sich um eine flüchtige Substanz handelt. Andere flüchtige bzw. gasförmige Stoffe wie Äthylen,

der Capillare wandern in diese ein. Gruppenbildung nur innerhalb. III und IV Reaktionen zwischen lebenden und toten Gameten. In IVc werden Plus-Gameten außerhalb der Capillare von den toten im Inneren angelockt. Gruppenbildung innerhalb. Dagegen ist nur schwache Gruppenbildung im umgekehrten Falle an der Öffnung der Capillare sichtbar. In IVb wandern die lebenden Plus-Gameten aus der Capillare aus. Gruppenbildung außerhalb mit toten Minus-Gameten.

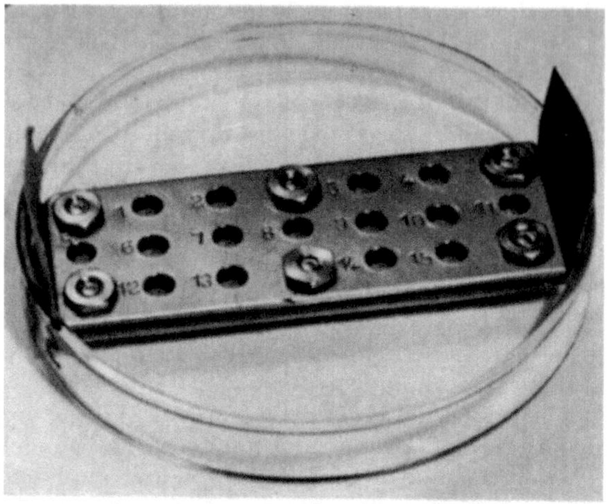

a

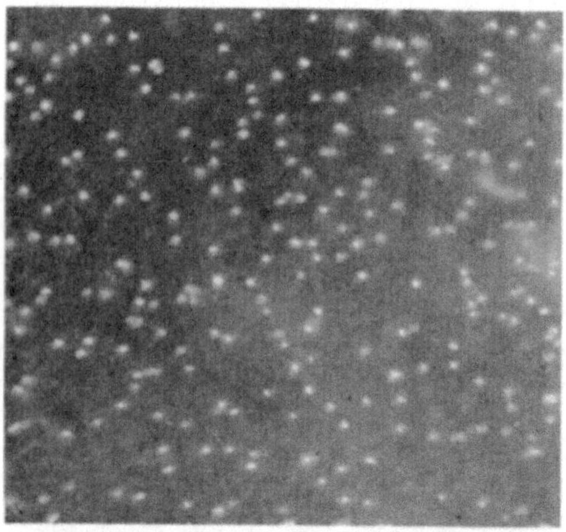

b

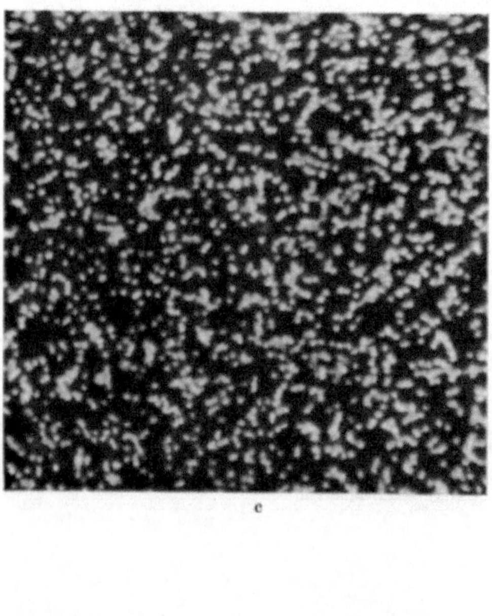

c

Abb. 3a—c. *Allomyces*. Aufnahme der durch das Hormon an die Membran gelockten Gameten. a Der von MACHLIS konstruierte Apparat für die Chemotaxisversuche mit *Allomyces*-Gameten. Gesamtansicht (in normaler Petrischale). b Kontrolle ohne Hormon. c In Hormonlösung.

Acetylen, Kohlenmonoxyd und Leuchtgas reagieren in gleicher Weise. Außer den (+)Gameten von *Chlamydomonas sp. 24* werden durch das (—)Filtrat wie durch die genannten Gase noch beide Geschlechter von *Chl. eugametos*, *Chl. moewusii* und *Chl. sp. 50 Tsubo* chemotaktisch angelockt, von *Chl. reinhardi* dagegen nicht. In keinem Filtrat von Gameten dieser vier Arten konnte irgendein weiteres chemotaktisches Prinzip nachgewiesen werden (TSUBO 1958, 1961).

e) *Allomyces.*

Nachdem es gelungen war, bei *Allomyces* aus interspezifischen Kreuzungen der monöcischen Eltern getrenntgeschlechtliche Nachkommen zu isolieren, war es möglich, auch bei einem Pilz mit reiner Gametenkopulation Sexualstoffe nachzuweisen *(Allomyces macrogynus Burma 1 Da* ♂ und ♀ *A. arbuscula Ceylon 1)*. Bei monöcischen Ausgangspflanzen versammeln sich gewöhnlich die männlichen Gameten um das weibliche Gametangium, ehe die weiblichen Gameten entlassen werden. Werden weibliche Gametangien isoliert und in warmen Agar eingebettet, so sammeln sich auch um die Agarklümpchen männliche Gameten an. Diese Beobachtungen ließen vermuten, daß von den Oogonien diffusible Stoffe in Lösung gehen, die die Gameten anziehen. Die verwendete Methode soll hier näher beschrieben und für weitere Untersuchungen empfohlen werden, da sie möglicherweise Besseres leistet als die vorwiegend angewandte Capillarmethode.

In zwei nichtrostende Stahlplatten und eine Gummischeibe werden 5 mm-Löcher gebohrt, die genau übereinander passen. Auf die untere Stahlplatte wird eine Cellophanmembran gelegt, darüber kommen die Gummischeibe und die zweite Stahlplatte (Abb. 3a). Die ganze Einrichtung wird in eine Petri-Schale mit einer Suspension männlicher Gameten gebracht; von oben können dann in die Reihe von Bohrungen verschiedene Testsubstanzen gegeben werden. Nach 5 min beginnen sich die Gameten am Fenster zu sammeln (Abb. 3b und c). Die Methode bietet den Vorteil, daß gelähmte oder getötete Gameten nicht an der Membran gefangen werden, sondern absinken, während bei der Capillarmethode eine positive Chemotaxis und eine Fallenreaktion weniger gut unterschieden werden können.

Der fragliche Stoff, „Sirenin" genannt, ist hitzebeständig und offenbar nicht flüchtig. Im stark sauren und basischen Bereich geht seine Wirksamkeit verloren. Es ließ sich daher vermuten, daß es sich um einen Ester handeln könnte. Die wäßrigen Extrakte konnten gereinigt werden. Nach Absorption an Holzkohle wird mit Chloroform eluiert und mit einer Kieselsäuresäule fraktioniert. 10 γ/l sind im biologischen Test noch wirksam.

Das „Sirenin" wird als neutrales Molekül nicht an Dowex 2 und 50 Ionenaustauscher absorbiert. Das Infrarotspektrum enthält bei 5,63 μ und 6,85 μ zwei sich überlappende Banden, die auf einen Laktonring und Aldehyd- und (oder) Ketongruppen hinweisen würden. Das Massenspektrum zeigt Banden bei m/e 31, 44 und 58. Das höchste Molekulargewicht nach diesen Daten liegt bei 386. Das Molekül enthält keinen aromatischen Ring und zeigt keine ausgesprochenen Absorptionen im UV-Bereich. Die Elementaranalyse ergab 61,6% ($\pm$1,5) C, 8,8% ($\pm$0,3) H, 3,4% ($\pm$0,3) N und 26,2% O. Das kleinste Molekulargewicht der sich daraus ergebenden Formel $C_{21}H_{36}O_7N$ liegt bei 414. Das gereinigte Sirenin (MACHLIS, WILLIAMS u. RAPPOPORT unveröff.) ist noch bei einer 10^{-6} molaren Konzentration auf männliche Gameten wirksam (CARLILE u. MACHLIS 1965a und b). Weibliche Gametangien sind die einzige nachgewiesene Quelle für das Sirenin. Weibliche Gameten zeigen keine Chemotaxis. Männliche Gameten reagieren nur auf Sirenin, nicht dagegen, wie andere Schwärmer z.B. auf Aminosäuren: Zygoten, Mito- und Meiosporen von *Allomyces* reagieren chemotaktisch ähnlich wie die Zoosporen der Saprolegniaceen (FISCHER u. WERNER 1957; s. unten), *Pythium aphanidermatum* (ROYLE u. HICKMAN 1964) und *Phytophtora* (DUKES u. APPLE 1961) auf Caseinhydrolysat.

Das „Sirenin" (der Name wurde von MACHLIS und VEGIS vorgeschlagen) gehört offenbar einer Gruppe sog. sekundärer Pflanzenstoffe an, über deren Rolle im Stoffwechsel und deren Biosynthese noch wenig bekannt ist. Es zeigt sich aber, daß *Fucus*-Stoffe, *Chlamydomonas*-Stoffe, *Allomyces*-Sirenin und *Mucor*-Gamone eine von den Chemotaktika, die für Archegoniatenspermatozoide und *Saprolegnia*-Sporen aufgefunden wurden, abweichende Rolle und Funktion haben müssen.

Der Name „Sirenin", an jene (♀) Fabelwesen der Antike anknüpfend, die die (♂) Seefahrer auf jene vor der Amalfiküste liegenden Klippen lockten, erscheint mir treffend. Ich schlage vor, ihn auf die ganze Gruppe hier behandelter Stoffe, die vom Ei abgegeben werden und Spermatozoide „sexuell" anlocken, auszudehnen.

2. Chemotropismus (Sirenine).

Bei Pilzen, deren Befruchtung als Gametangienkopulationen, Hyphenfusionen usw. ablaufen, wo also nicht freibewegliche Gameten, sondern die

Pflanze selbst oder besondere Organe Suchbewegungen ausführen, werden diese als Chemotropismus beschrieben. Dieser Begriff besagt, daß es sich im Falle einer sexuellen Reaktion um Stoffe handeln muß, die von dem einen Geschlecht in die Kulturflüssigkeit, in das Gel oder als flüchtige Substanzen abgegeben werden und eine positive Richtungsänderung beim anderen Geschlecht induzieren. Von den Suchbewegungen sind zwei ausgeprägte Möglichkeiten näher analysiert worden: In dem einen Falle wachsen die männlichen Antheridien aus und führen die Such- und Wachstumsbewegungen aus: im zweiten Falle bilden die weiblichen Organe eine lange Trichogyne aus, die, oft mehrzellig werdend, die männlichen Geschlechtszellen aufsucht. Es sind folgende Fälle näher bekanntgeworden:

1. *Achlya* }
2. Hefe } mit Antheridienhyphen,
3. *Mucor* }
4. *Bombardia* } mit Trichogynen.
5. *Ascobolus* }

a) *Achlya*.

Bei einem der diöcischen, im Wasser lebenden Phycomyceten, *Achlya ambisexualis* hat Raper nachgewiesen, daß die Induktion der Geschlechtsorgane und die Gametangienkopulation durch eine Reihe von Gamonen gesteuert wird (s. Beitrag Raper in diesem Band) (Abb. 4). Das Gamon C und nach Barksdale auch das Gamon A ist für das Wachstum der antheridialen Hyphen in Richtung auf die Oogonanlagen verantwortlich. Das Gamon A wird von den weiblichen Hyphen abgegeben und findet sich in meßbarer Menge in der Kulturflüssigkeit, während Kulturen mit nur männlichen Hyphen wenig oder kein Gamon A enthalten (Raper 1940, 1942, Raper u. Haagen-Smit 1942). In Kulturen mit beiden Geschlechtern findet man nur geringe Mengen Gamon A (Abb. 5). Barksdale (1963a und b) zeigte, daß das Gamon A von den männlichen Hyphen aufgenommen und somit der Kultur entzogen wird. Die männlichen Stämme von *A. ambisexualis* E 87, 9 und 10, nehmen das Gamon A in großen Mengen auf und produzieren Antheridien, während z.B. der weibliche Stamm 734 kein Gamon aufnimmt. Das Gamon A kann mit Chloroform aus dem Medium extrahiert und durch Gegenstromverteilung angereichert werden. Es läßt sich an Polystyrol oder Polyvinylpartikel adsorbieren. Die antheridialen Hyphen wachsen auf solche Gamonpartikel zu, umfassen sie wie Oogonien und schnüren Antheridien ab (Abb. 6). Raper hatte die beiden Vorgänge, die chemotropische Anziehung der antheridialen Hyphen und die Abschnürung des Antheridiums einem Gamon C zugeschrieben (Abb. 4). Es ist nicht klar, inwieweit A und C übereinstimmen. Eine vergleichende Untersuchung von Gamon A und dem Sirenin von *Allomyces* ergab, daß beide Stoffe nur artspezifisch wirksam sind (Barksdale, Carlyle u. Machlis 1965).

Im Rahmen einer Analyse des Chemotropismus vegetativer Hyphen bei den Saprolegniaceen fanden Fischer und Werner (1955) Aminosäuren und Hydrolysate von Eiweißen für wirksam. „Mindestens 5—7 Aminosäuren sind erforderlich, um alle Erscheinungen des Chemotropismus (Umkehrkrümmungen der Hyphen, Wuchs bis zur Diffusionsquelle, zusätzliche Verzweigung, gerichtete Sprossung) deutlich hervorzurufen: keine Aminosäure ist für die Auslösung der tropischen Effekte unentbehrlich." Keiner bestimmten Aminosäure konnte jedoch eine spezifische Wirksamkeit zugeschrieben werden. Leucin, Glutaminsäure, Methionin und Cystein zeigen beste Resultate. Die unnatürlichen D-Aminosäuren sind wirkungslos. Diese Ergebnisse sind insofern interessant, als sich auch Antheridienhyphen entsprechend zu verhalten scheinen.

Agarklümpchen mit Caseinhydrolysat wurden in die Nähe von Antheridien von *Achlya polyandra* de Bary gebracht. Eine Anzahl positiver Ergebnisse

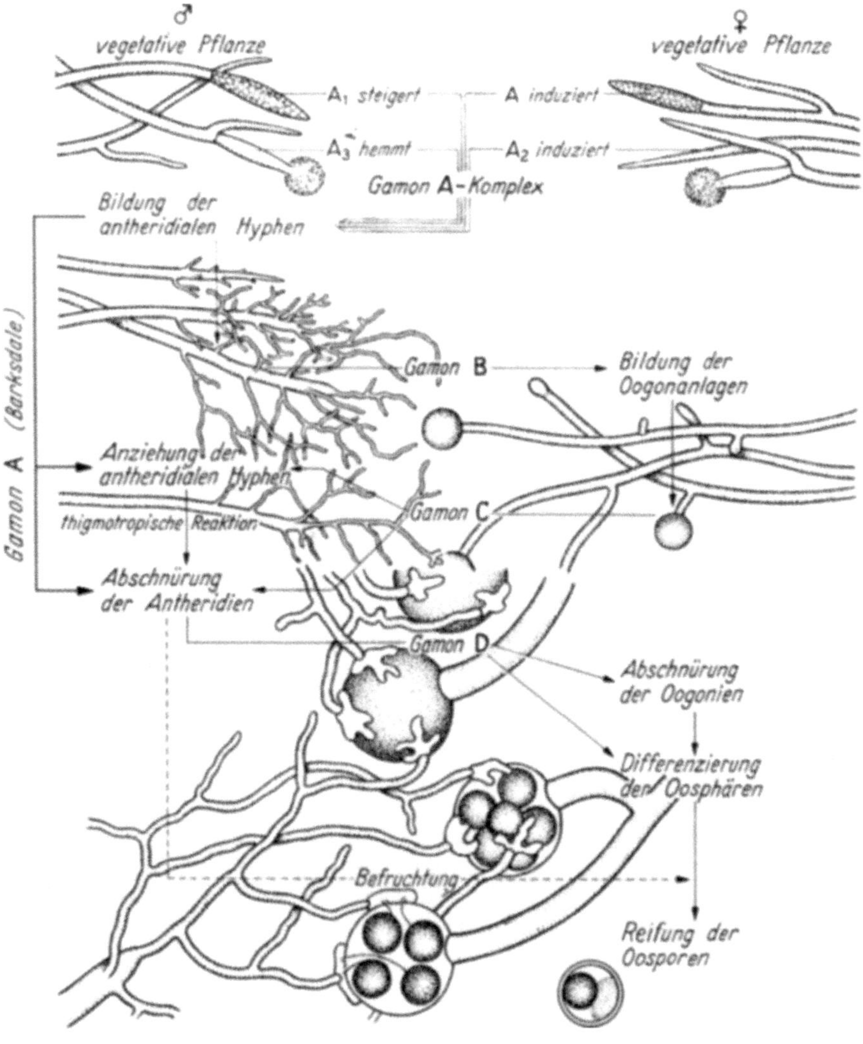

Abb. 4. Halbschematische Darstellung des durch Gamone kontrollierten Sexualvorganges zwischen männlichen und weiblichen Stämmen von *Achlya ambisexualis*. Jede mit einem Buchstaben gekennzeichnete Linie zeigt den Ursprung und die Wirkung eines bestimmten Gamonkomplexes an. Die gestrichelt gezeichnete Linie (Befruchtung) beruht auf noch nicht mit Sicherheit erwiesenen Versuchsdaten. (Nach RAPER, aus ESSER u. KUENEN 1965.)

scheint darauf hinzuweisen, daß das gerichtete Wachstum der Antheridienhyphen offenbar durch Aminosäuren bedingt wird. Leider wurden keine Oogonienextrakte geprüft. Ungeschlechtliche Schwärmer aber, die auf Aminosäuren (und Alkalichloride) reagieren (FISCHER und WERNER 1957), werden ebenso von reifen Oogonien angelockt. Daher könnte man annehmen, daß bei der Oosporenbildung, mit der zugleich überzählige Kerne und Plasmateile degenerieren, aus dem Oogon

Proteinhydrolysate heraus diffundieren, die den Chemotropismus der Antheridienhyphen bedingen.

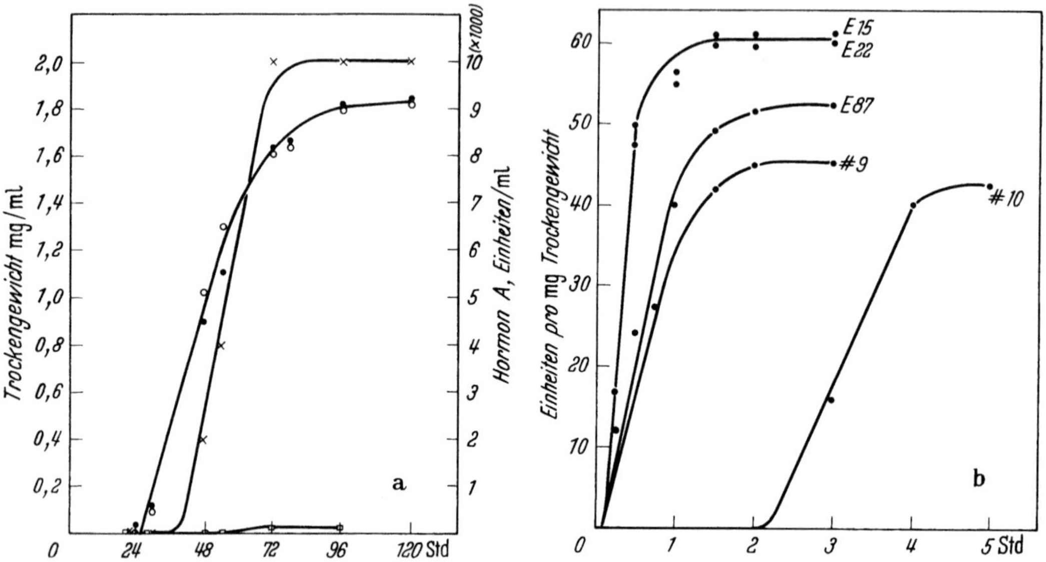

Abb. 5a u. b. a Wachstum und Hormonproduktion einer weiblichen Kultur in Abwesenheit und in Gegenwart männlicher Mycelien. Hormone A im Medium: × = 734 ♀, ○ = 734 ♀ × E 87 ♂. Wachstumskurven: ● = 734 ♀, ○ = 734 ♀ × 87 ♂. b Aufnahme des Hormons A durch männliche (E 15, E 22 und E 87) und durch monöcische Stämme (# 9 und #10). Hormonkonzentration 10^5 Einheiten/ml. (Nach BARKSDALE 1963.)

b) Hefen.

Eine chemotropische Wirkung wurde bei zwei haploiden verschieden geschlechtlichen Hefestämmen beobachtet (LEVI 1956). Unter geeigneten Bedingungen kopulieren (+) und (—)Zellen auf Agar, wobei nur das (—)Geschlecht eine gerichtete Kopulationshyphe ausbildet, ehe die Zellen miteinander Kontakt haben. Agarplatten, von denen kopulierende Zellen entfernt worden waren, induzieren ebenfalls bei (—)Zellen Kopulationshyphen, nicht dagegen bei (+) Zellen. In nur einem einzigen Falle wurde über eine (+)Kultur eine Kollodiumfolie und darüber eine (—)Kultur gelegt. Da es gelang, bei (—)Zellen Kopulationshyphen zu induzieren, dürfte die chemotropische Wirkung und die Induktion der Kopulationshyphen durch ein niedermolekulares Gamon oder Hormon verursacht sein.

c) Mucorineen.

Von den Organismen mit Gametangienkopulationen wurden die Mucorineen näher untersucht. Werden zwei verschiedengeschlechtliche Mycelien von *Mucor mucedo* auf der gleichen Agarplatte gezogen, so bilden sich in der „Kopulationszone", wo sich die beiden Hyphenfilze berühren, die Kopulationsorgane, die (+) und (—)Zygophoren aus. Sobald sich die Mycelien auf 1—2 mm genähert haben, färben sich die Spitzen der Hyphen der Kopulationszone gelb. Danach entstehen in beiden Geschlechtern im Abstand von 2—3 mm von der Hyphenspitze die Geschlechtsorgane, die Zygophoren wachsen aus dem Agar heraus und krümmen sich aufeinander zu, so daß sich im Idealfalle die Spitzen berühren und dort die Zygote gebildet wird. Allerdings verhalten sich die (+)Zygophoren weitgehend passiv, so daß es häufig zu Spitzen-Flanken-Kopulationen kommt, die nicht zur Zygotenbildung führen. Die (+)Zygophoren sind

2—3mal so dick wie die (—)Zygophoren, wesentlich kürzer und geringer an Zahl. Es kommt daher gelegentlich zu sternartigen Figuren, indem sich mehrere (—)Zygophoren an eine (+)Zygophore anlegen. Jedoch wird nur eine

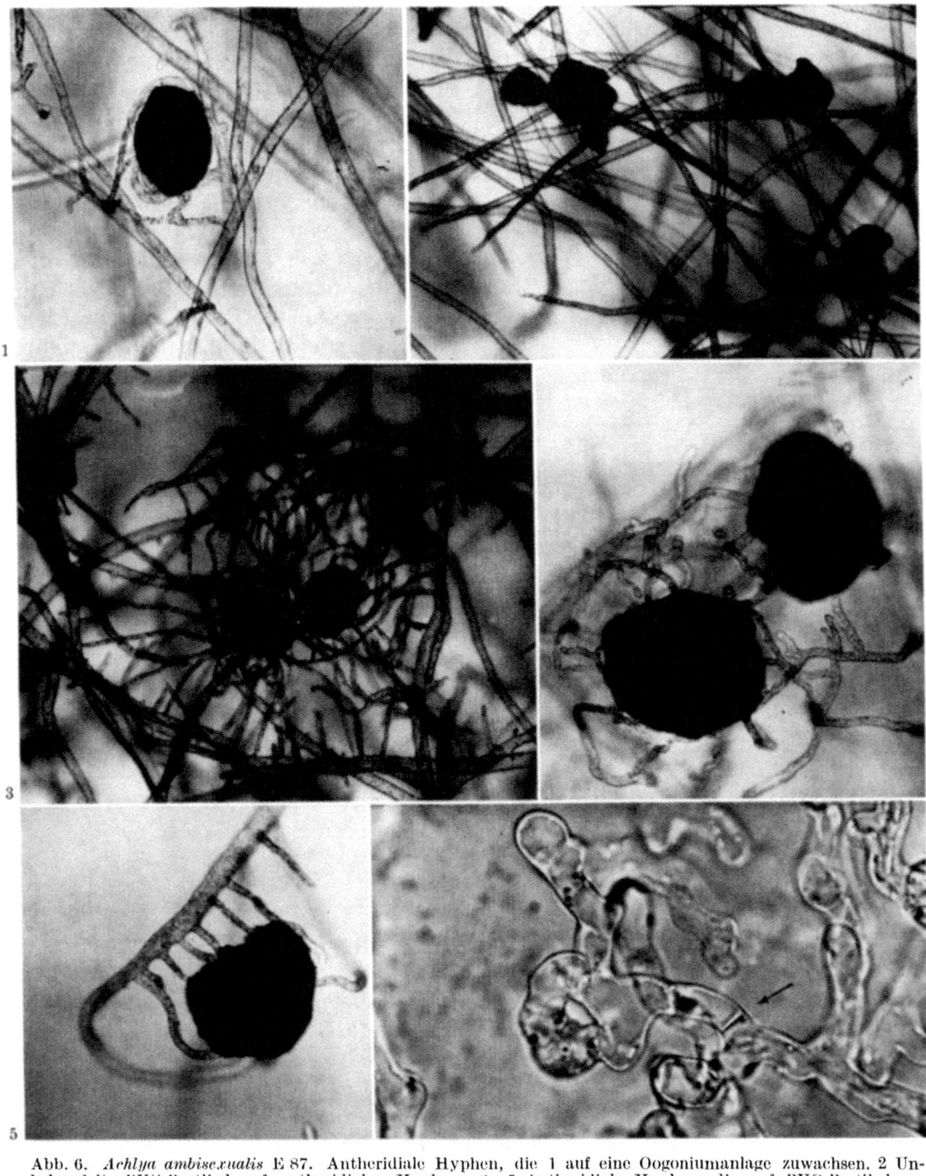

1 2 3 4 5 6

Abb. 6. *Achlya ambisexualis* E 87. Antheridiale Hyphen, die 1 auf eine Oogoniumanlage zuwachsen. 2 Unbehandelte PVC-Partikel auf antheridialen Hyphen. 3—5 Antheridiale Hyphen, die auf PVC-Partikel zuwachsen, die mit Gamon A vorbehandelt wurden. 6 Abschnürung eines Antheridiums unter dem Einfluß von Gamon A. (Aus BARKSDALE 1963.)

Zygote gebildet (BURGEFF 1924, PLEMPEL 1957). Die Sexualreaktionen werden durch spezifische Substanzen ausgelöst. BURGEFF (1924) führte folgenden Versuch hierfür durch: Auf eine Agarplatte wurde eine Celloidinfolie so gelegt, daß sie zu einem Drittel ein (—)Mycel bedeckt und zu zwei Dritteln freie Agarfläche. Darauf wurde ein Agarblock mit (+)Mycel gelegt. Die Hyphenspitzen beider Mycelien wachsen in gleicher Richtung oberhalb und unterhalb der Membran weiter. Die (+)Hyphen bilden am Rande des Agarblocks Zygophoren aus, die sich mit ihren Spitzen nach unten auf die Membran neigen. Das Untermycel gibt also offenbar durch die Folie einen Stoff ab, der beim obengelegenen (+)Mycel die Bildung von Zygophoren induziert und diese *chemotropisch* reizt.

PLEMPEL hat in den letzten Jahren zwei Stoffe, die die Zygophoren in beiden Geschlechtern induzieren, isoliert und sie chemisch charakterisiert (PLEMPEL 1957, 1959, PLEMPEL und BRAUNITZER 1958). Sie wurden als (+) und (—)Gamon bezeichnet. Da dieser Aufsatz nur Reaktionen beschreiben soll, die zur Vereinigung der Geschlechtszellen führen, sei auf eine Zusammenfassung der Arbeiten über Gamone, die die Zygophoren induzieren, von PLEMPEL (1963) hingewiesen. (Abb. 7). Die Induktionsstoffe und eine weitere Stoffgruppe, vielleicht eine Vorstufe, die Progamone, werden in flüssigem Medium gebildet, wenn immer genügend Sauerstoff vorhanden ist (PLEMPEL 1960). Aus dem Rohgamon gewinnt man ein gelbliches, charakteristisch riechendes Öl, das sich in organischen Lösungsmitteln und auch in Wasser löst. Das Massenspektrum ergibt ein Molekulargewicht von 306. Die Summenformel ist $C_{18}H_{26}O_4$. Diese Fraktion hat Gamonaktivität gegenüber (+) und (—)Mycel. Neben Hydrierungs- und oxydativen Spaltungsversuchen weist schon die vermutliche Strukturformel (nach PLEMPEL) auf den Polyencharakter dieser Stoffe hin:

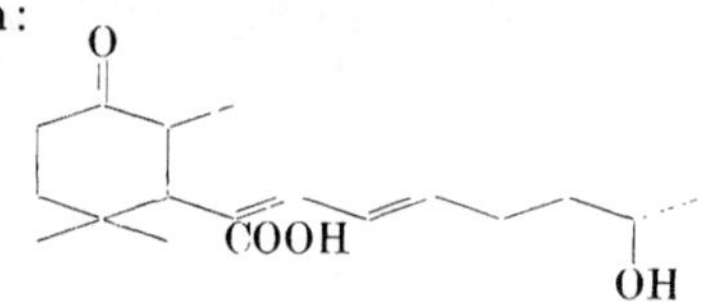

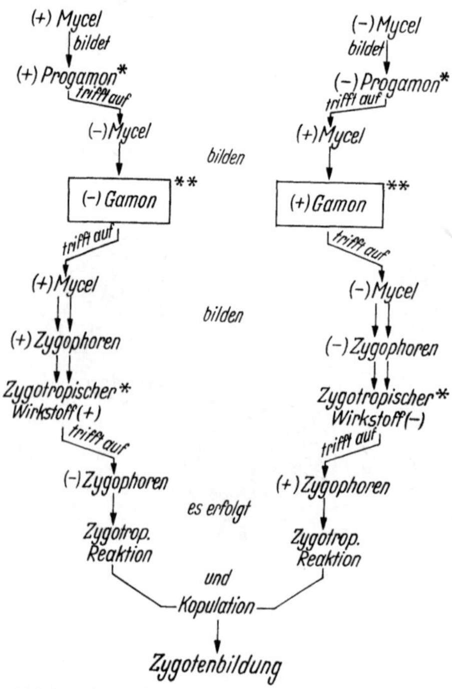

Abb. 7. Schema der Sexualreaktion von *Mucor mucedo*, * im biologischen Test nachgewiesen, ** in kristalliner Form isoliert. (Nach PLEMPEL 1960, verändert aus ESSER u. KUENEN 1965.)

Es konnte in weiteren Versuchen bestätigt werden, daß zwischen dem Auftreten von Carotinoiden und Gamonen Zusammenhänge bzw. gemeinsame Synthesewege bestehen (PLEMPEL 1965). Das bestätigt trotz verschiedentlicher Einwände (REICHEL u. WALLIS 1956, 1957, 1958 u. BURNETT 1956) die Vermutungen von BURGEFF u.a. (BLAKESLEE 1913, 1920, BARNETT u. Mitarb. 1956, SATINA u. BLAKESLEE 1929, HESSELTINE u. ANDERSON 1957).

BANBURY (1954, 1955) ist es gelungen die Induktion der Zygophoren von einer zweiten Reaktion, dem Zygotropismus, abzutrennen. Er zeigte, daß sich die Zygophoren verschiedenen Geschlechtes im Luftraum über der Agarplatte über eine Entfernung von 2 mm anziehen. Die Zygophoren gleichen Geschlechtes halten dagegen in der Regel einen Abstand von 1 mm voneinander. Zygotropische Stoffe hatte schon BURGEFF (1924) gefordert. Durch PLEMPEL (1960) wurden diese Wirkstoffe näher untersucht. Sie sind gasförmig und darüber hinaus organ- und geschlechtsspezifisch. Sie werden erst sezerniert, nachdem die Zygophoren durch die Gamone induziert worden sind (Abb. 7) und sich über das Substrat erheben. Beide Stoffe werden an die Luft abgegeben, dort jedoch sehr schnell oxydiert und unwirksam. Der für das gerichtete Wachstum notwendige Wirkstoffgradient ergibt sich durch die oxydative Veränderung der zygotropischen Substanzen. Durch die Wirkstoffwolke des anderen Geschlechts wird die Zygophore zu verstärktem Wachstum angeregt, das zu Kontakt und Kopulation von (+) und (—)Zygophoren führt (PLEMPEL 1960, 1962, PLEMPEL und DAWID 1961).

d) *Ascobolus.*

Eine weitere Variation der (Gameten-)Gametangienkopulation bei den Ascomyceten ist der *Ascobolus carbonarius*-Typ. Normal entwickelte Ascogone suchen mit Hilfe einer Trichogyne vegetative Propagationszellen, Conidien oder Oidien auf. Dabei zeigt die Trichogyne als Teil des weiblichen Gametangiums chemotropisches Wachstum in Richtung auf die als männliche Spermatien fungierenden Zellen zu. Für *Ascobolus stercorarius* (Bull.) Schröt. ließ sich durch fortgesetztes Verlagern eines Spermatiums ein gerichtetes Wachstum der Trichogyne nachweisen (BISTIS 1956). Beim Wechsel der Richtung krümmt sich die Trichogyne nicht, es wird vielmehr ein neuer „Vegetationspunkt“ an der dem Spermatium zugewandten Seite induziert. In gleicher Weise wie das Spermatium können beliebige Protoplasmaeinheiten wie Keimschläuche, Ascosporen oder vegetative Hyphenspitzen als männliche Organe dienen. Über die Natur dieser diffundierenden Substanzen ist nichts bekannt.

e) *Bombardia.*

Bei einem anderen Ascomyceten *Bombardia lunata* Zckl. reagieren die Trichogynen nur mit Spermatien oder Spermogonien des anderen Paarungstyps. Sie werden chemotaktisch angelockt, wobei sie oft „scharfe Wendungen“ machen, sobald sie in ein wirksames Konzentrationsgefälle kommen. Filtrate von Spermatiensuspensionen wurden in eine Capillare gefüllt und in eine Agarplatte mit weiblichem Mycel gesteckt. Die Trichogynen änderten ihre Wachstumsrichtung auf die Capillare hin und wuchsen, sich typisch schlängelnd, schließlich in diese hinein. Dieselbe Wirkung hatten mit Spermatienlösung getränkte Agarwürfel, auch dann, wenn die Lösung erhitzt worden war. Es ist anzunehmen, daß es sich bei den chemotropisch wirkenden Substanzen ebenfalls um niedermolekulare Stoffe handelt (ZICKLER 1953).

Aus den obengenannten Ergebnissen über die stoffliche Wirkung, welche zur Vereinigung von Gameten über Entfernungen hinweg wirken, können wir kein einheitliches Bild ablesen. Selbst wenn die Struktur jener diffundierenden Stoffe bekannt wäre, wüßten wir nichts über die Art, wie sie von dem anderen Geschlecht aufgenommen und dort wirken würden. Eine allgemeine kausale Erklärung auf chemischer Basis ist weder für den Chemotropismus noch für die sicher kompliziertere Chemotaxis bisher bekannt geworden. Nicht bei allen Organismen haben sich stoffliche Wirkungen über Entfernungen nachweisen lassen.

IV. Agglutinations- und Zellkontaktwirkungen.

Das eigentliche Befruchtungsproblem begänne damit, daß die kerntragenden Zellen oder Organe in Kontakt miteinander kommen und zwar mit den richtigen Wirkgruppen, in der richtigen molekularen Lage. Die Kontakt- oder Agglutinationsreaktion wird daher als erster Schritt in einer Kette von Reaktionen angesehen werden müssen. Durch Aneinanderhaften bestimmter Orte der Zelloberfläche werden erstens die Zellen oder Organe zunächst über eine gewisse Zeit beieinandergehalten, und zweitens werden sie in die richtige Lage gedreht, die für die Verschmelzung und den Kernaustausch günstig ist. Durch diese Bewegung wird aus der „Gruppe“ vieler Bewerber das Paar heraus isoliert. Es wird drittens ein enzymatischer oder lytischer Mechanismus in der Zelle ausgelöst werden müssen, der die verschiedenen Hüllen durchbricht. So mannigfaltig die morphologischen Befruchtungsvorgänge sein mögen, so verschieden werden der zweite und der dritte Schritt aussehen. Vorausgehend wird ihnen eine Adsorptions- oder Agglutinationsreaktion, die die Partner in Kontakt bringt und für gewisse Zeit beieinander hält.

1. Isogame Gruppen.

a) Die Vorgänge bei *Chaetomorpha* und *Chlamydomonas*.

Es sei zuerst der bei isogamer Befruchtung bei Algen und niederen Pilzen ablaufende Vorgang am Beispiel von einigen Grünalgen wie *Chaetomorpha*, *Dasycladus* und *Chlamydomonas* näher beschrieben. Vereinzelt wurde in älterer wie in neuerer Literatur die Meinung vertreten, daß Gameten fast regellos miteinander verschmelzen. Auch bei manchen mitgelieferten Zeichnungen ist man nicht sicher, ob es sich tatsächlich um Zygoten handelt und nicht um mißgebildete Gameten, bei denen die letzte Zellteilung ganz oder teilweise verhindert wurde, oder um „zufällige“ zusammenklebende Zellen. Untersuchungen an *Chaetomorpha aerea* (*Chlorophyceae*, *Chladophorales*) führten dagegen zu dem Schluß, daß die Kopulation nach einem Schema verläuft, dessen Einzelreaktionen von allgemein-

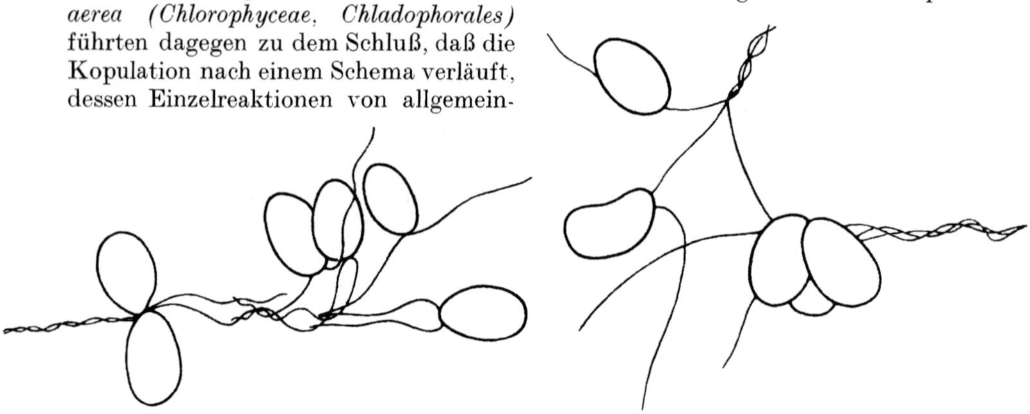

Abb. 8. *Chaetomorpha aerea*. Gruppenbildung 10 min nach Zusammengeben beider Geschlechter mit Lugolscher Lösung. Fixiert. (Nach KÖHLER 1956.)

biologischem Interesse sind (KÖHLER 1956). Gibt man je einen Tropfen Gameten verschiedenen Geschlechtes zusammen, so tritt augenblicklich Gruppenbildung (clumping) ein. Wird eine solche Gruppe sofort (nach 10 sec) fixiert, so erhält man bereits alle wichtigen Informationen (Abb. 8). Offenbar sind es nicht die Körper der Gameten, die den ersten Kontakt untereinander haben, sondern die Geißeln, die auch die zitternden Bewegungen der Gruppen verursachen. Es wird weiter klar, daß vorwiegend die Spitzen der Geißeln miteinander verkleben. Später, etwa nach 30 sec, sieht man viele Paare, deren Geißeln paarweise eng spiralisiert sind (Abb. 9). Dadurch werden die Paare einander soweit genähert, daß an ihren Flanken die Zellverschmelzung beginnt. Anschließend entspiralisieren sich die Geißeln und dienen als lokomotorische Organe für die Zygote. Diese drei Reaktionsschritte konnten näher analysiert werden. Es gelang mit 0,5 % Na- oder K-Citrat in der Kulturlösung die Beweglichkeit der Gameten

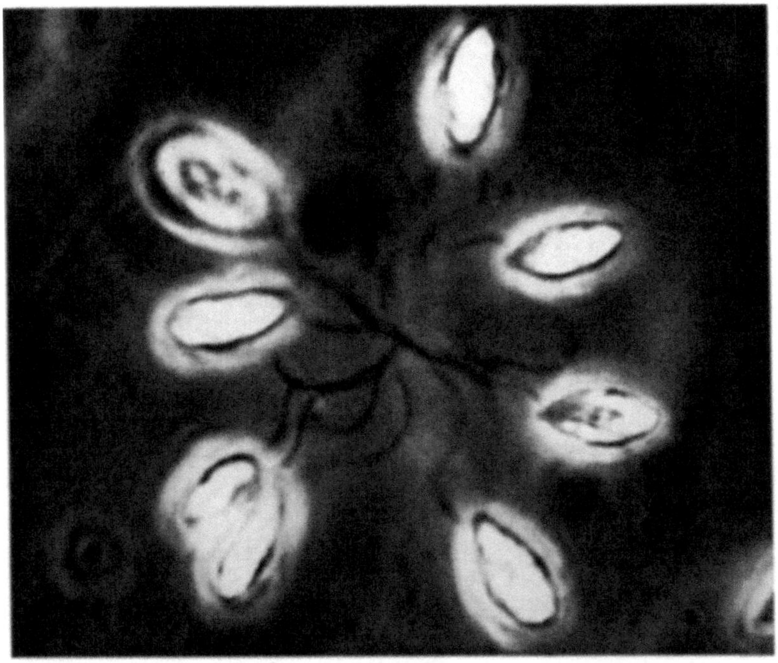

Abb. 10. *Chaetomorpha aerea*. Gruppenbildung von Gameten, deren Geißelbewegung durch Citrat beeinträchtigt ist. Lebend, Phasenkontrastaufnahme mit Elektronenblitz. (Nach KÖHLER 1956.)

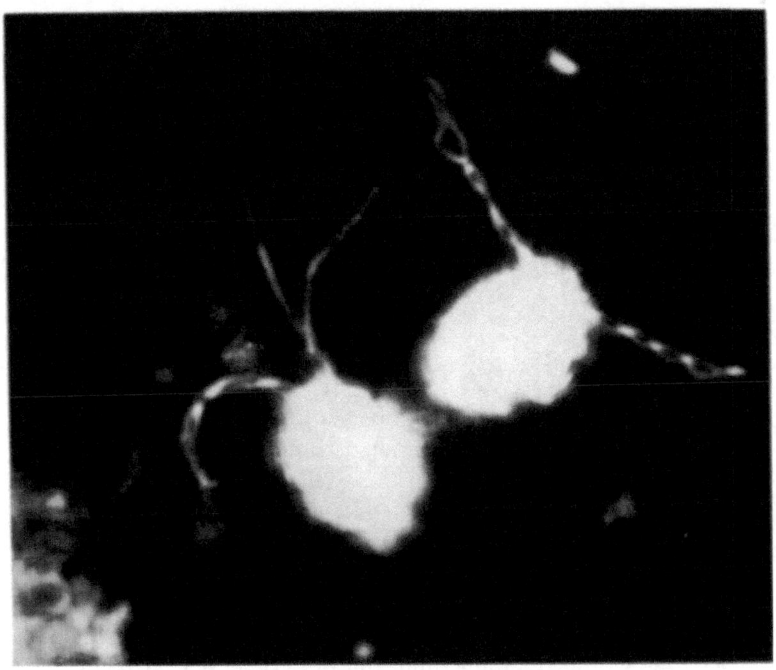

Abb. 9. *Chaetomorpha aerea*. Verschmolzene Gameten. Die Geißeln sind kurz nach der Paarung eng spiralisiert. Fixiert, Dunkelfeldaufnahme. (Nach KÖHLER 1956.)

so zu lähmen, daß die Geißeln steif werden und sich nur langsam bewegen. Mischt man so behandelte Gameten beider Geschlechter, so treten nur während der ersten Minute „stern"- und „döldchen"-artige Gruppen auf (Abb. 10). Die Geißeln

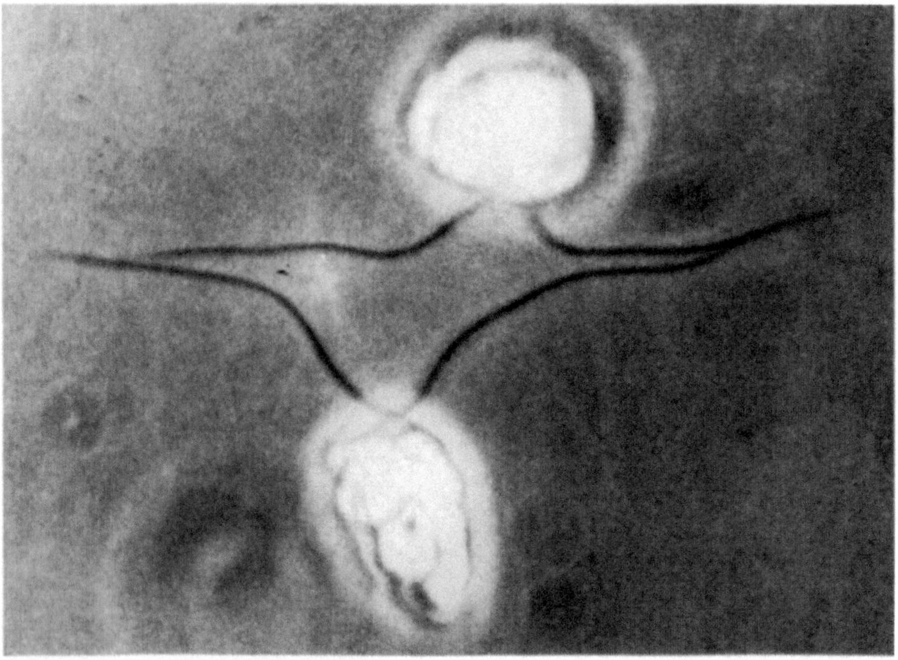

a

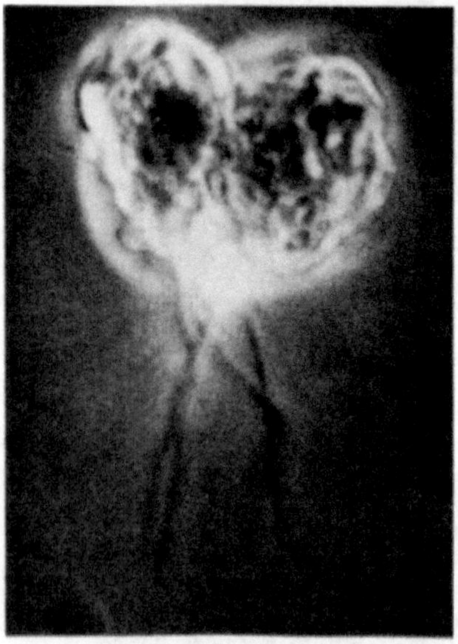

b

werden an ihrer Spitze an einem Agglutinationspunkt festgehalten, wobei der *Kontakt* vorübergehend sehr *fest* sein kann (s. S. **306**, *Pandorina*). Dabei gelingt es den Geißeln nicht sich aufzuspiralisieren, und es treten folglich keine Zygoten auf. Es läßt sich hieraus der Schluß ziehen, daß an der Spitze der Geißel bei beiden Geschlechtern ein geschlechtsspezifischer Stoff oder eine entsprechende Struktur liegt, die die eigentliche chemische Differenzierung der Sexualität darstellt. Ihrem Wesen nach muß sie als Agglutination bezeichnet werden. Der dritte Schritt, die Verschmelzung der Zellkörper, konnte durch Proteinaseeinwirkung (Subtilisin) auf die Zelloberfläche verhindert werden. Obwohl weder die Beweglichkeit der Gameten noch die Gruppenbildung und Spiralisierung der Geißeln gehemmt ist, werden dennoch keine Zygoten gebildet.

Abb. 11a u. b. *Dasycladus clavaeformis*. a Gameten bei der Paarung. Beachte die Geißelspiralisierung. Fixiert, Phasenkontrastaufnahme. b Kopulierende Gameten in Ammonium-sulfatlösung. Lebend, Phasenkontrastaufnahme aus einer Serie.

Daß in der Tat die Geißelspitze für die Gamonwirkung, die Agglutination, verantwortlich ist, könnten auch FÖRSTER und WIESE bei *Chlamydomonas* zeigen, wo durch konzentrierte Filtrate von Gameten eines Geschlechtes beim anderen

Isoagglutination ausgelöst wurde (s. unten). Zum gleichen Schluß führten Untersuchungen an *Dasycladus (Chlorophyceae, Dasyclaceae)*. Ausgehend von den Untersuchungen von JOLLOS (1926) (s. auch Beitrag RAPER, in diesem Band), ergab sich schließlich, daß die Kopulation nach dem gleichen Schema verläuft, wie bei *Chaetomorpha*, so daß dies möglicherweise als allgemeingültig für isogame Grünalgen angesehen werden kann (Abb. 11). Bei *Dasycladus* werden die Geißeln in 1,25% Ammonsulfatlösung gelähmt, was im Kopulationsversuch zu sternoder reigenartigen Gruppen führt. In 1,25% Ammonacetatlösung bildet sich an der Spitze der Geißel eine Blase, die möglicherweise aus gefälltem Gamon besteht. Schließlich ergaben elektronenmikroskopische Untersuchugen in der Tat einen Tropfen an der Spitze der Geißeln (Abb. 12). Wenn man auch aus diesen morphologischen Beobachtungen nur beschränkte Schlüsse ziehen kann, so ergibt sich aus der Menge der Hinweise doch eine Vorstellung über den Befruchtungsvorgang bei diesen Algen.

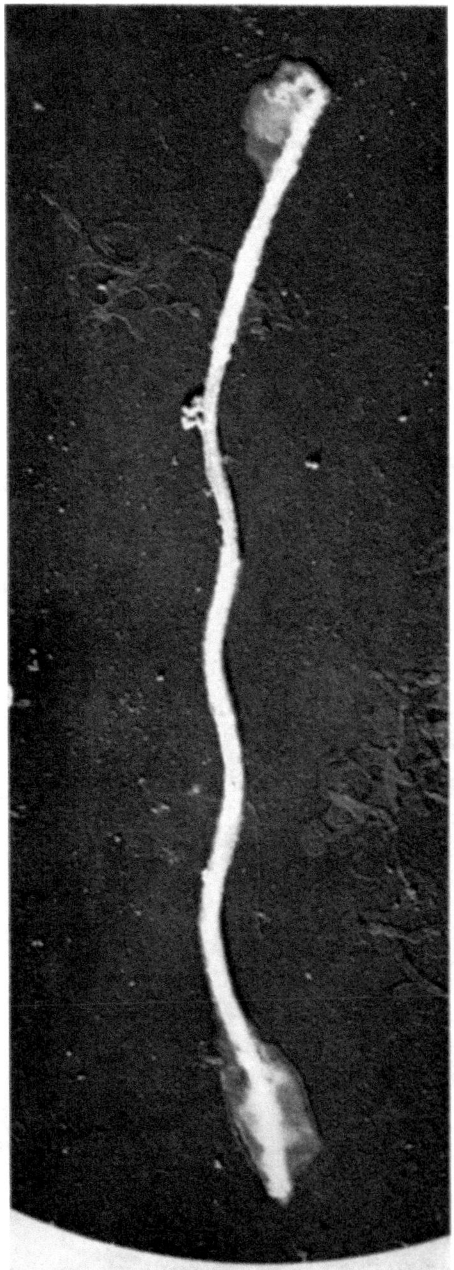

Abb. 12 *Chaetomorpha aerea*. Geißel eines Gameten mit Tropfenbildung an der Spitze. Elektronenmikroskopische Aufnahme Pt Rh bedampft.

Eine genauere Information über die Chemie dieser Substanzen am distalen Ende der Geißel erbrachten die Untersuchungen von FÖRSTER, WIESE und BRAUNITZER (1956). *Chl. eugametos* kann als Palmellastadium auf Agar gezogen werden und vermehrt sich vegetativ reichlich. Überträgt man sie in flüssiges Medium, so bilden sie Geißeln aus und werden beweglich und kopulationsfähig. Diese Gameten kopulieren unter typischer Gruppenbildung. Dabei zeigte sich eigenartigerweise, daß der Agglutinationsstoff an der Geißel in das Medium abgegeben werden kann. Zentrifugate oder Filtrate von kopulationsfähigen Gameten rufen Gruppenbildung beim anderen Geschlecht, sog. Isoagglutination hervor (Abbildung 13; MOEWUS 1933, FÖRSTER und WIESE 1954a, b). Diese agglutinierende Wirkung wird durch zwei geschlechts- und artspezifische hochmolekulare Substanzen verursacht. Sie sind nicht dialysierbar und werden bei höherer

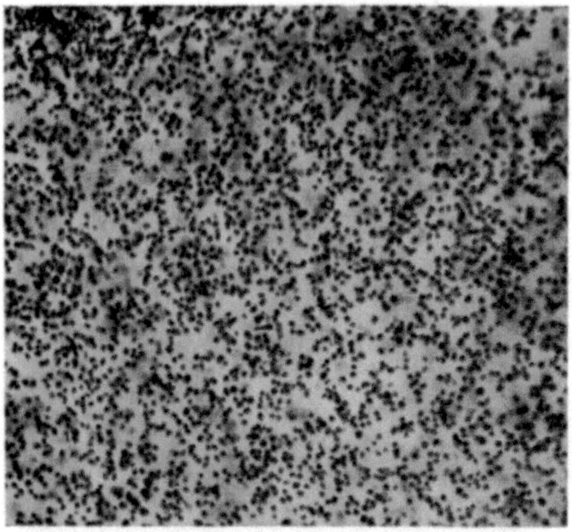

a

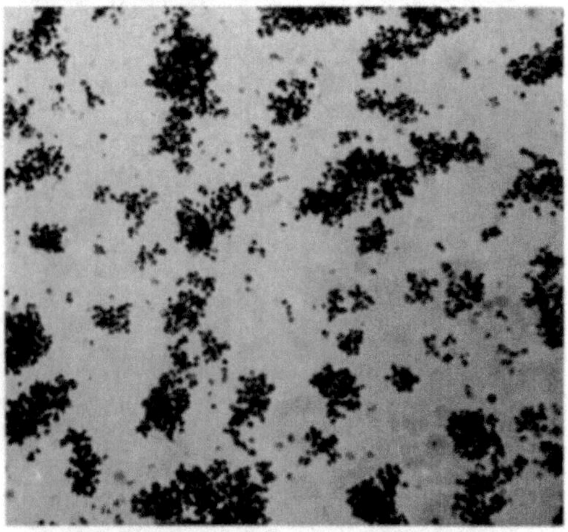

b

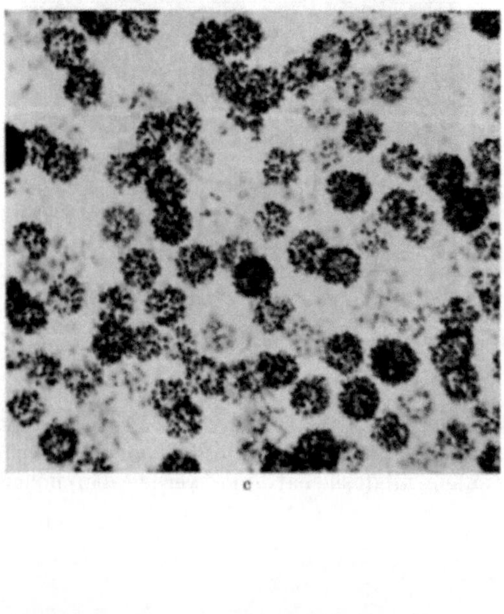

c

Temperatur zerstört, das männliche Gamon nach 10 min bei 50°, das weibliche Gamon bei etwa 90°. Die chemischen und physikalischen Eigenschaften der beiden Gamone wurden näher untersucht (FÖRSTER, WIESE und BRAUNITZER 1956, HARTMANN 1956b, FÖRSTER und WIESE 1958), nachdem sie durch Ammoniumsulfat u. a. ausgefällt und angereichert worden waren. Das Gynogamon ist ein Glykoproteid mit 5,86% N und 3,1% S. Es enthält mindestens 12 Aminosäuren, 2 Hexosen und 2 Pentosen, von denen folgende papierchromatographisch identifiziert wurden:

Glycin, Alanin, Serin, Leucin/Isoleucin, Threonin, Asparaginsäure, Glutaminsäure, Lysin, Arginin, Phenylalanin, Prolin, Oxyprolin; von den *Pentosen:* Xylose und Arabinose; von den *Hexosen:* Rhamnose und eine nicht näher bestimmte Aldohexose. *Mannose* wurde nicht nachgewiesen.

Es hat eine Wirksamkeit bis zu einer Verdünnung von 1 γ/ml. Auf der Ultrazentrifuge sedimentiert es mit einheitlicher Bande. Diese Bande ist mit dem Gamon identisch. Als Beweis für die Einheitlichkeit der Bande wurden in einem Schichtenversuch 40% der Bande sedimentiert. Der Überstand und das Sediment enthielten quantitativ entsprechende Teile der Aktivität. Bei Annahme einer Kugelform ergab sich ein Molekulargewicht 10^8. Partikel solcher Größe ließen sich daher, wie vermutet, auch im Elektronenmikroskop nachweisen. Das Gamon hat in Lösnng als rundes Partikel einen Durchmesser von etwa 0,1 μ. Es wird von der

Abb. 13 a—d. *Chlamydomonas eugametos.* a Suspension kopulationsbereiter männlicher Zellen. b Normale Gruppenbildung beim Zusammengeben beider Geschlechter. c Die männlichen Zellen nach Zugabe konzentrierten Gynogamons. d Isoagglutination von weiblichen Zellen durch männliches Gamon. Lebend, Phasen kontrastaufnahme. Lebendaufnahme mit Elektronenblitz. Vergr. 90 ×. (Nach FÖRSTER, WIESE und BRAUNITZER 1956.)

Teil leicht in destilliertes Wasser Geißel besonders am vordersten oder in sehr verdünnte Salzlösungen abgegeben.

Die Untersuchungen über das Androgamon sind noch nicht abgeschlossen. Es ist ein Glykoproteid von ähnlichen Eigenschaften wie das Gynogamon und besitzt etwa gleichen N-Gehalt.

Durch solche konzentrierte Gamonlösungen konnten bei *Chlamydomonas* die charakteristischen Gruppen erhalten werden, bei denen die Gameten dicht aneinanderliegend eine Kugelschale bilden und mit ihren Geißelspitzen im Kugelmittelpunkt verklebt sind (Abb. 13d).

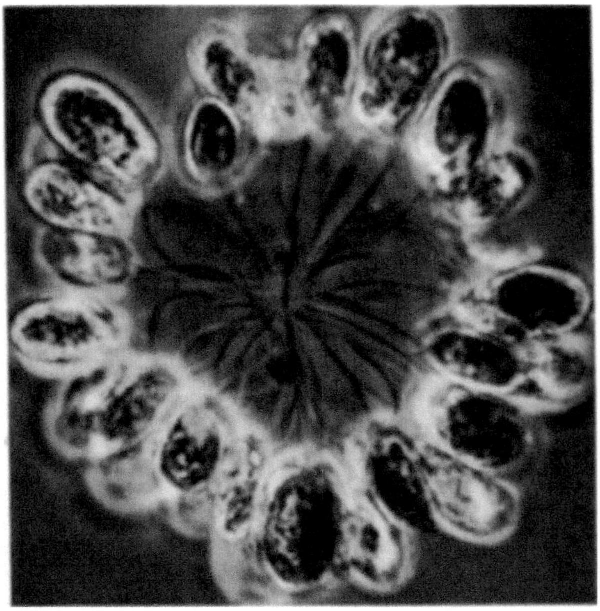

Abb. 13 d.

Die Gruppenbildung findet nur in Gegenwart von Ca^{++}- und Mg^{++}-Ionen statt und wird weder durch Laurylsulfat noch durch Mercaptoäthanol oder andere SH-Gruppen-Reagentien gehemmt. Dagegen wird die Verschmelzung der Zellen bei höheren Konzentrationen dieser Stoffe gehemmt. Eine Reihe anderer Chemikalien bewirken das sofortige Abwerfen der Geißeln. Die abgeworfenen Geißeln des weiblichen Geschlechts von *Chl. eugametos*, des (—) Stammes von *Chl. moewusii* und des Stammes (b) von *Chl. sp. Tsubo* agglutinieren mit Gameten des anderen Geschlechts. Dagegen reagieren mit Osmiumsäure fixierte Gameten beiden Geschlechtes mit lebenden Partnern (Wiese und Jones 1963). Hieraus wird ersichtlich, daß auch bei *Chlamydomonas* die Gruppenbildung und die Zellvereinigung durch verschiedene Stoffe bewirkt wird. Schein-Kopulationen zwischen lebenden und abgetöteten Partnern wurden auch von Hutner u. Provasoli (1951) und von Tsubo (1961) beschrieben.

Die Regulation der Gametogenese wurde von K. Hartmann eingehend bearbeitet. Während weibliche Gameten auch im Dunkeln quantitativ beweglich und kopulationsfähig werden, also Gamon bilden können, benötigen die männlichen Schwärmer Licht, um zu Gameten zu werden. Dagegen brauchen bei der mit *eugametos* sehr nahe verwandten und kreuzbaren *Chl. moewusii* beide Geschlechter Licht, um kopulationsfähig zu werden (Lewin 1956). Temperatur- und Lichtversuche von Förster (1957, 1959), die zur Bestimmung des Wirkungsspektrums führten, lassen offenbar darüber hinaus Schlüsse auf Vorstufen der Gamone zu. Damit würde sich eine Möglichkeit eröffnen, Näheres über deren Biosyntheseweg zu erfahren.

b) Die Bedeutung der Geißel für die Gruppenbildung.

Nachdem an den obengenannten Beispielen die einzelnen Reaktionsschritte und die daran beteiligten Stoffsysteme im Zusammenhang beschrieben worden sind, sollen diese noch durch weitere Einzelbefunde ergänzt werden.

Die Bedeutung der Geißel für die Gruppenbildung und die Befruchtung wurde schon früher von GERLOFF (1940) und LEWIN (1952) erkannt. LEWIN hält eine spezifische „stickiness" der Geißel für die Voraussetzung für die Paarung. Die ersten Versuche, durch Antiserum Gruppenbildung zu induzieren, wurden im Labor von LEVINE und durch WIESE (beide unveröffentlicht) ausgeführt. Allerdings sind die erhaltenen Antikörper nicht geschlechtsspezifisch.

Ebenfalls von LEWIN (und MEINHART 1955) stammen die ersten zufälligen Aufnahmen von Geißeln mit anhängenden Partikeln (*Chlamydomonas moewusii*).

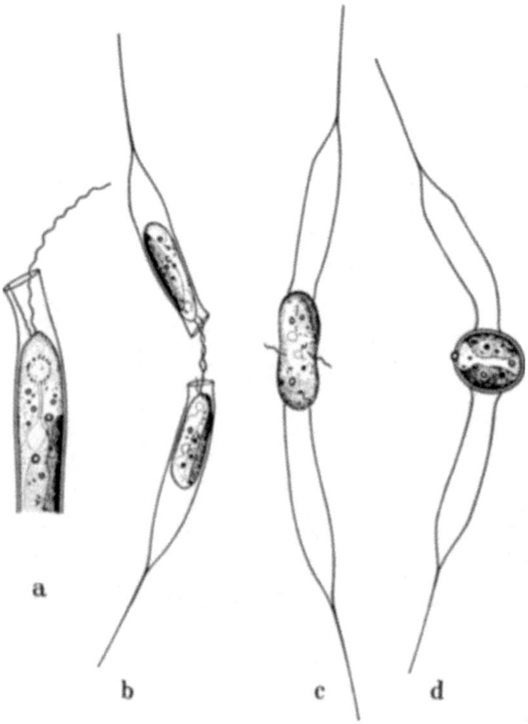

Abb. 14a—d. *Dinobryon borgei*. Kopulation zweier Gameten. a Teil eines Flagellaten. b Erstes Kopulationsstadium. c Zellverschmelzung. d Dauerzygote.

Alle bisher bekannten elektronenoptischen Aufnahmen sind zwar keine entscheidenden Beweise dafür, daß das Gamon von der Geißel abgegeben wird und in Lösung geht, oder daß es nicht abgegeben wird und als Tropfen sichtbar ist. Sie bilden jedoch eine gute Ergänzung zu unseren Vorstellungen (Abb. 12; KÖHLER 1956, HARTMANN 1956).

Ein treffendes Beispiel dafür, daß die Geißeln bei der Kopulation für den Kontakt und Agglutination verantwortlich sind, können wir unter anderem einer Abbildung von SKUJA entnehmen (Abb. 14). Sie zeigt, daß bei der Kopulation der in den Bechern sitzenden *Dinobryon*-Gameten notwendigerweise die Geißeln den ersten Kontakt bekommen. Wie später bei den Braunalgen noch gezeigt werden wird, stellt auch bei ihnen die Spitze einer Geißel den ersten Kontakt mit der Eigallerte her. Ein weiteres Beispiel für die Beteiligung der Fortbewegungsorganelle für die Paarung ist bei den Ciliaten bekannt geworden. Nach METZ (1954) sind die Cilien die Träger der Paarungstypsubstanzen bei *Paramaecium*.

Die entscheidende Bedeutung der Geißel für die Agglutination wird weiterhin durch UV-Mutanten von *Chlamydomonas moewusii* aufgezeigt (LEWIN 1952, 1954). Sie lassen sich in drei Gruppen einordnen:

a) „Lazy", deren Geißel unregelmäßig und langsam schlägt, ohne daß sich die Zellen im Medium fortbewegen.

b) „Paralyzed", bei denen das distale Drittel der Geißel nur zu gelegentlicher Bewegung fähig ist. Solche Zellen können — wie Wildtypzellen — „kriechen" (LEWIN 1952).

c) Mutanten mit verkürzten Geißeln von 2—6 μ Länge.

d) Geißellose Mutanten ohne Bewegung.

Eine Übersicht über eine Reihe von Mutanten vermittelt Tabelle 3. Gelähmte Mutanten (lazy, paralyzed) kopulieren nur spärlich, durch Schütteln und Rühren jedoch kann man passiv die Geißeln der beiden Geschlechter häufiger in Kontakt

Tabelle 3. *Übersicht über die UV-Mutanten von Chlamydomonas moewusii.* (Nach LEWIN.)

Mutanten	Anzahl	Länge der Geißeln mm	Beweglichkeit	Gruppenbildung	Paarung
Geißellos	5	—	—	—	—
Stummelgeißlig	2	2—3	—	+	—
Stummelgeißlig	1	2—3	—	+	+
Paralyzed	2	12—15	—	—	—
Paralyzed	2	12—15	—	+	—
Paralyzed	15	12—15	—	+	+
Beweglich, asexuell	3	12—15	+	—	—
Beweglich, keine Paarung	noch nicht aufgefunden	12—15	+	+	—
Wildtyp		12—15	+	+	+

miteinander bringen, wodurch die Paarungsreaktion ausgelöst wird und somit mehr Zygoten entstehen. Geißellose Mutanten kopulieren unter keinen Bedingungen.

Eine weitere Gruppe von Stoffen mit gerichteter Wirkung sind die übertragbaren Beweglichkeits- und Hemmstoffe bei dem frühen Zygotenpärchen. Zygoten des Wildtyps *(Chlamydomonas eugametos, moewusii* und *sp. 24 Tsubo)* schwimmen als Paare, die durch eine Plasmabrücke miteinander verbunden sind, noch Stunden umher, ehe die Zellen ganz verschmelzen und (im Licht) die Dauerzygote bilden. In solchen Paaren sind nur die (+)Partner für die Vorwärtsbewegung verantwortlich (LEWIN 1950, TSUBO 1956). Die Geißeln

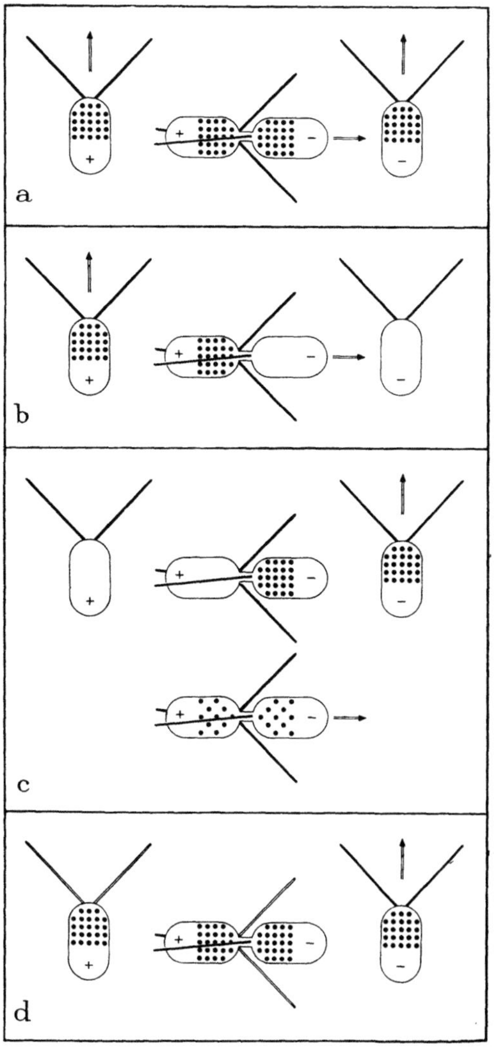

Abb. 15a—d Kreuzung gelähmter Mutanten von *Chlamydomonas moewusii.* Deutung der Erholungsperiode (recovery) bei gepaarten Gameten durch Übertragung eines Faktors durch den Plasmastrang. a Kopulation zwischen Plus und Minus-Wildtypen. Das Paar ist sofort beweglich. Der Plus-Partner bestimmt die Bewegungsrichtung. b Wildtyp Plus × „paralyzed" Minus. Das Paar ist sofort beweglich. Der Plus-Partner bestimmt die Bewegungsrichtung. c „paralyzed" Plus × Wildtyp Minus. Das Paar wird nach 6—8 Std beweglich. Nach der Erholung (recovery) übernimmt der vorher gelähmte Partner die Bewegung. d „paralyzed" Plus × Wildtyp Minus. Da die Mutante irreversibel geschädigt ist („not recoversible") werden die Paare nicht beweglich. Der hypothetische Beweglichkeitsfaktor ist durch schwarze Punkte dargestellt.

der (—)Zellen hören auf zu schlagen, sobald die Plasmabrücke geformt ist. Werden daher gelähmte (—)Zellen mit (+)Wildtyp gekreuzt, so sind bei manchen Mutanten die Paare bald beweglich (Abb. 15). Umgekehrt ergibt die Kreuzung Wildtyp(—) × paralyzed(+) unbewegliche Paare. Bei einigen Kreuzungen von wild(—) × paralyzed(+) beginnen die Paare nach einer gewissen Dunkelperiode zu schwimmen. Diese Erholung (recovery) läßt sich so deuten, daß — sofern man das Paar als Heterokaryon in einem gemeinsamen Plasma betrachtet — ein Beweglichkeitsfaktor durch die Plasmabrücke gelangt und in dem gelähmten (+) die Geißelbewegung induziert. Durch markierte Zellen konnte in der Tat nachgewiesen werden, daß die Geißeln des mutierten (+)Partners nach der Erholungsperiode beweglich werden, nicht die des (—)Wildtyps. Paare aus Kreuzungen gelähmter Mutanten untereinander sind meist unbeweglich. Jedoch bleibt zu fragen, inwieweit solche Faktoren geschlechtsspezifisch sind und ob sie für das Befruchtungsproblem im engeren Sinne Bedeutung haben.

c) Weitere Agglutinationsstoffe.

Historisch betrachtet erbrachte Geitler (1931) die ersten Nachweise für Agglutinationsstoffe, die an das Medium abgegeben werden, bei *Tetraspora*. Gameten eines Geschlechtes, die durch Hitze getötet worden waren, verursachten noch Gruppenbildung, wenn lebende Gameten des anderen Geschlechtes zugegeben wurden. Waren die Gameten jedoch durch die Hitzebehandlung zu sehr „desorganisiert", so umschwärmten die lebenden des anderen Geschlechtes jene toten nur vereinzelt. Dagegen bildeten sie *untereinander Gruppen*. Die ersten Isoagglutinationen wurden auch mit Überstand von Gametenzentrifugaten erhalten. Es ist gleich, ob der Überstand von toten oder lebenden Gameten verwandt wird.

Während der 30er Jahre wurden noch bei folgenden Arten Filtrate von Algengameten geprüft, die Isoagglutinationen bedingen: *Protosiphon botryoides* (Moewus 1923), *Stephanosphaera pluvialis* (Moewus 1933), *Dunaliella salina* (Lerche 1937). Bei allen von Moewus beschriebenen Arten sollte die Zugabe von Bakterien erst die Isoagglutination hervorrufen. Nachdem das für *Chlamydomonas* nach Förster und Wiese nicht zutrifft, mögen die Befunde durch Moewus falsch gedeutet sein. Aus den spärlichen Angaben läßt sich nachträglich keine befriedigende Deutung ableiten. Die gleichen Gamonwirkungen wie für *Chlamydomonas eugametos* fanden Förster und Wiese (1955) bei *Chl. reinhardi* und *Chl. moewusii* (1958) sowie für einen Stamm von der Insel Yap von *moewusii*. An *Pandorina* wurden in jüngster Zeit ebenfalls Gamonwirkungen nachgewiesen (Wilbois 1958). Hier — wie bei *Astrephomene gubernaculifera* und *Volvulina steinii* (Stein 1958) — wurde zunächst gefunden, daß bei diesen kolonialen Formen die Agglutination nicht auf die Gameten beschränkt ist, sondern daß sich die Kolonien bereits paaren, wenn man „reife" Kolonien beider Geschlechter zusammengibt. Erst nach und nach werden die Gameten entlassen, ja zum Teil werden solche, die noch in der gallertigen Kolonie festsitzen, von schon freigesetzten des anderen Geschlechtes herausgerissen. Dies mag darauf hinweisen, wie fest in der Tat die Agglutination ist. Die freien Gameten kopulieren unter Gruppenbildung wie *Chlamydomonas* u. a. Die Filtrate von *Pandorina*-Kolonien beider Geschlechter zeigen je zwei verschiedene Wirkungen: Sie bewirken einmal das Entlassen von Gameten aus der Kolonie und zweitens deren Isoagglutination. Die hierfür verantwortlichen Stoffe sind thermostabil und nicht dialisierbar. Unerklärlich ist, daß sie nicht streng geschlechtsspezifisch sein sollen.

Alle anderen aufgefundenen agglutinierenden Gamone bei Grünalgen sind geschlechts- und artspezifisch. Ihre chemische Natur ist bisher allein bei *Chlamy-*

domonas eugametos aufgeklärt worden. Daß die Gruppenbildung bei den begeißelten, nicht oogamen Grünalgen der Paarung und der Befruchtung vorausgeht, ist allgemein anerkannt.

In 3 Fällen sind Reaktionen zwischen abgetöteten Gameten des einen Geschlechtes und lebenden des anderen beobachtet worden, so bei *Tetraspora* (GEITLER 1931), bei *Chlamydomonas moewusii* (HUTNER und PROVASOLI 1951) und bei *Chlamydomonas* sp. 24 (TSUBO 1956). Von Interesse dürften einige nicht näher analysierte Versuche sein, in denen Gruppenbildung verhindert oder extrem gefördert werden kann. Werden Gameten von *Ulva lactuca* mit Detergent vorbehandelt, so verliert nur ein Geschlecht seine Kopulationsfähigkeit. (Die Gameten werden dabei — wie normalerweise die Zygoten — negativ phototaktisch). Das andere Geschlecht wird durch das Detergent nicht beeinflußt (LEVRING 1955). Schon KLEBS berichtet, daß die Kopulation von *Protosiphon*-Gameten durch 0,5—1,0% Knop-Lösung verhindert wird; das gleiche fand PASCHER für Gameten von Chaetophoracaeen in Morphiumlösung. Umgekehrt konnte die Kopulation durch Mg^{++} Ionen ausgelöst und stark gefördert werden (s. OLTMANNS 1923, S. 159—160). Welche chemischen Veränderungen das Gamon (oder die Zelloberfläche) in diesen Fällen erleidet, ist unbekannt.

2. Oogame Formen.

Obwohl bei vielen Braunalgen, besonders den Ectocarpales, die Gameten beider Geschlechter gleiche Größe haben, erfolgt die Befruchtung in der Regel anisogam bzw. oogam. Die erste Beobachtung über Gruppenbildung bei *Ectocarpus siliculosus* stammt von BERTHOLD. Der weibliche Gamet wird, solange er freibeweglich ist, nicht befruchtet. Erst nachdem er sich nach einiger Zeit festgesetzt und seine Geißeln verloren hat, tritt eine für die Braunalgen und die Eibefruchtung allgemein typische Gruppenbildung auf. Ein in der Mitte der Gruppe befindliches Ei ist von vielen männlichen Gameten oder Spermatozoiden umgeben, die mit der Spitze der längeren Geißel an der „Eigallerte" festgeheftet sind. Die Spitze der männlichen Geißel und die Eimembran oder Eigallerte sind die morphologischen Orte, wo Agglutination erfolgt. Alle Agglutinationsstoffe sind also dort zu suchen. Wenn bisher keine wirksamen Filtrate oder Zentrifugate erhalten worden sind, so möglicherweise deshalb, weil diese normalerweise in Seewasser nicht in Lösung gehen. (Abb. 16a und b).

Genauere Untersuchungen der Gruppenbildung bei *Cutleria* zeigte, daß die Eier mit einer Art Gallerte oder hyalinen Hülle umgeben sind und daß während der Gruppenbildung auch hier die Spermatozoide sich mit der Spitze der *längeren* Geißel am Ei festsetzen, sie daher im Abstand einer Geißellänge um das Ei herum stehen. Die Gruppenbildung dauert etwa 3 min. Dann vereinigt sich plötzlich eines der Spermatozoide mit dem Ei. Damit hört jegliche Anlockung (HARTMANN 1950) und Gruppenbildung auf. Isolierte Eigallerte agglutiniert ebenfalls Spermien (KÖHLER unveröff.). Auch für *Dictyota* berichtet WILLIAMS, daß die Spermatozoide mit der einzigen Geißel sich zunächst an der Eioberfläche festsetzen, ehe die Befruchtung erfolgt. (Abb. 16c, d und e).

Bei *Fucus* werden große unbegeißelte Eier durch kleine Spermatozoide befruchtet. Hier heftet sich offenbar die kurze („vordere") Geißel an der Eioberfläche an (s. FRITSCH II, 1945, S. 377, dort Angaben über Originalliteratur).

Morphologisch ist das *Fucus*ei dem Seeigelei sehr ähnlich gebaut. Da auch ähnliche Veränderungen nach der Befruchtung eintreten, ist man geneigt, auch die stofflichen Vorgänge zu vergleichen. Die Oberfläche des *Fucus*-Eis (ebenso bei *Ascophyllum, Hormosyra*) ist von einer Eigallerte überzogen. Unter ihr befindet sich eine Eimembran, eine Lipoproteidschicht, darunter die Rindengranula, die nach Befruchtung wie beim Seeigel zum Aufbau einer Befruchtungsmembran aufgebraucht werden (LEVRING 1947, 1949, 1952). Die Eimembran kann durch Trypsin abgebaut werden. Das Ei bleibt dadurch zwar befruchtungsfähig, eine Befruchtungsmembran kann jedoch nicht mehr gebildet werden.

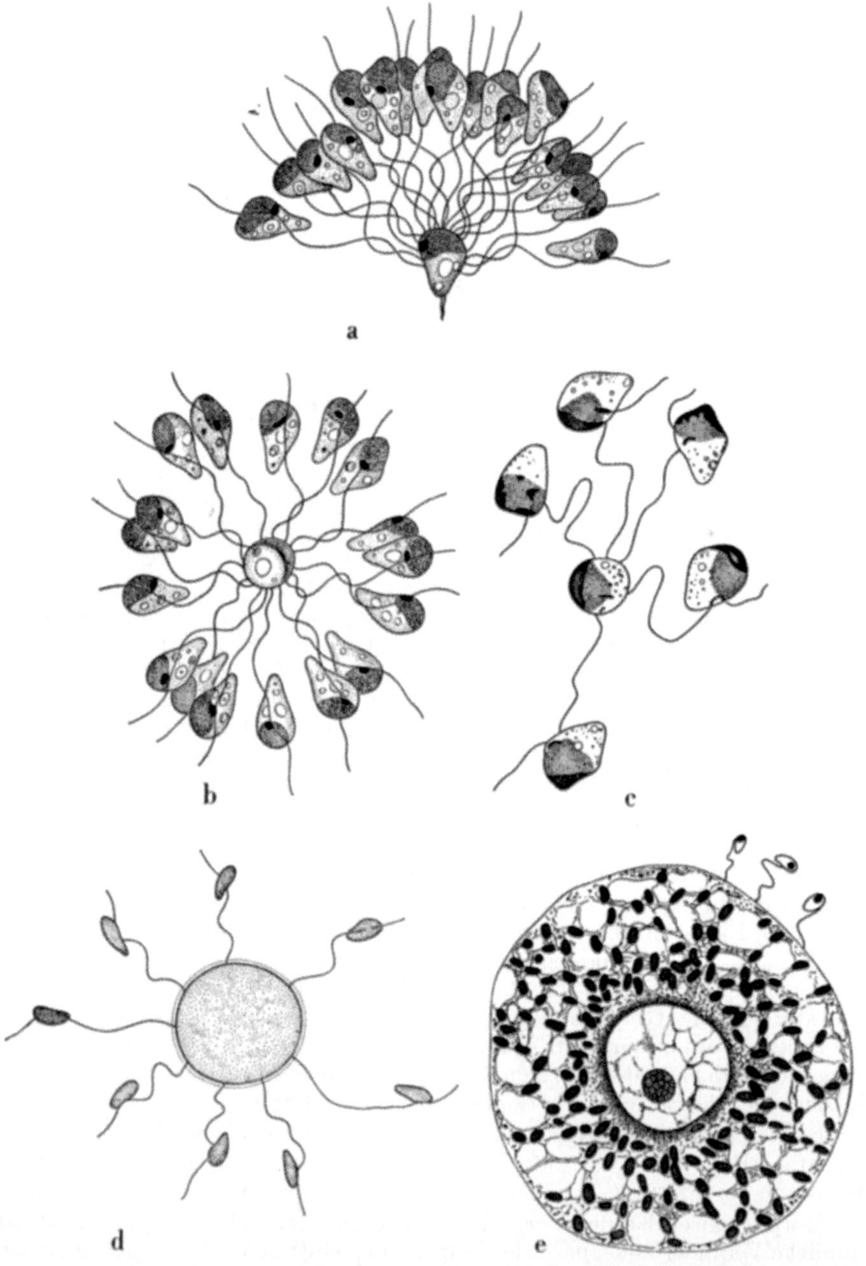

Abb. 16a—e. Gruppenbildung bei den Phaeophyceen. a u. b *Ectocarpus siliculosus.* c *Cladostephus spongiosus,* d *Cutleria multifida.* e *Dictyota dichotoma.* (a u. b nach OLTMANNS, c nach SCHREIBER 1931, d KÖHLER, Original e nach WILLIAMS.)

Das Festheften des männlichen Gameten mit der *einen* Geißel an der Oberfläche des Eies vor der Befruchtung wurde bisher bei folgenden Braunalgen beobachtet (zusammengestellt bei FRIEDMANN 1961).

	Ectocarpales.
Ectocarpus siliculosus (Dillw.) Lingb.	BERTHOLD (1881) SAUVAGEAU (1896b) SAUVAGEAU (1896c) HARTMANN (1937)
Ectocarpus secundus Kütz.	SAUVAGEAU (1896a) SAUVAGEAU (1896c)
Scytosiphon lomentaria (Lyngb.) Ag.	BERTHOLD (1881)
Colpomenia sinuosa (Roth) Derb. et Sol.	KUNIEDA und SUTO (1938)
	Sphacelariales.
Cladostephus spongiosus C. Ag.	SCHREIBER (1931)
	Dictyotales.
Dictyota dichotoma (Huds.) Lamour	WILLIAMS (1904)
	Fucales.
Fucus serratus L.	THURET (1857) THURET und BORNET (1878)
Fucus spiralis L. var. *platycarpus* (Thur.) Batt.	STRASBURGER (1884)
Fucus vesiculosus L.	STRASBURGER (1884) LEVRING (1947)
Halidrys siliquosa (L.) Lyngb.	FARMER und WILLIAMS (1898)
Cystoseira barbata J. Ag.	KNAPP (1931)
Pelvetia canaliculata (L.) Decsne. et Thur.	SUBRAHMANYAN (1958)

Diese Art der Befruchtung beobachtete FRIEDMANN (1961, 1962) auch bei *Prasiola*, einer Grünalge (Abb. 17). Obwohl hier die Geißeln gleichlang sind, berührt nur eine von ihnen die Eizelle und wird im Verlauf der Kopulation zuerst in das Ei aufgenommen. Dieser Vorgang hat große Ähnlichkeit mit der Befruchtung tierischer Eier, wo das Acrosom den ersten Kontakt mit der Eioberfläche herstellt; während die „zweite" Geißel allein als Fortbewegungsorgan dient.

Den Überstand einer Suspension von *Fucus*-Eiern hat KOTTE zuerst untersucht. Er hatte, wie viele ältere Forscher, eigentlich im Sinne, Chemotaxis nachzuweisen und glaubte daher, daß die viscose Substanz seines Eiwassers Spermatozoide chemotaktisch anlocke. Ein Tropfen Eiwasser wurde mit einer Capillarpipette in eine Spermatozoiden-Suspension gebracht. Im Bereich des Tropfens sammelten sich schon nach wenigen Minuten Spermatozoide an, nach $^1/_2$ Std hatte sich (auf dem dunkelgestellten Objektträger) ein gelber Fleck aus Spermatozoiden gebildet. Das Eiwasser enthält eine schleimige Substanz, die sich in Fäden ausziehen läßt und an der sich Spermatozoiden sammeln, die schließlich einen gelben Strich bilden. Die Aktivität des Überstandes geht bei mäßigem Erhitzen nicht verloren (1 min bei 70°). Da wir aber nach den Ergebnissen von COOK und Mitarbeitern wissen, daß die chemotaktischen Stoffe bei *Fucus* flüchtig sind, ist man versucht, KOTTEs Beschreibungen anders zu deuten. Vergleicht man seine Reaktionen mit der Agglutination von Seeigelspermien durch Eiwasser, so ergibt sich eine verblüffende Ähnlichkeit. Es würde sich hier also um eine Agglutination in einem hochmolekularen Stoff handeln. Diese Befunde wären dann in der Tat der früheste Nachweis eines wirklichen Gamons. LEVRING stellte weiter fest, daß Spermien, die mit Dupunol vorbehandelt wurden, Eier nicht mehr befruchten können. Es ist anzunehmen, daß mit Hilfe dieses Detergents Receptorgruppen an der Spermienoberfläche verändert oder in Lösung gebracht wurden. Da bei normaler Befruchtung die sonst noch um das Ei herum sich befindenden Spermien bald unbeweglich werden, schließt LEVRING, daß auch die Eier nach der Befruchtung einen Stoff abgeben, der Polyspermie verhindert und vielleicht Detergentcharakter hat.

In die gleiche Richtung deutet eine Beobachtung von FARMER und WILLIAMS (1898). Spermatozoide von *Halidrys*, die ein Ei umschwärmen, sterben in der Regel ab, sobald es einem von ihnen gelungen ist, das Ei zu befruchten.

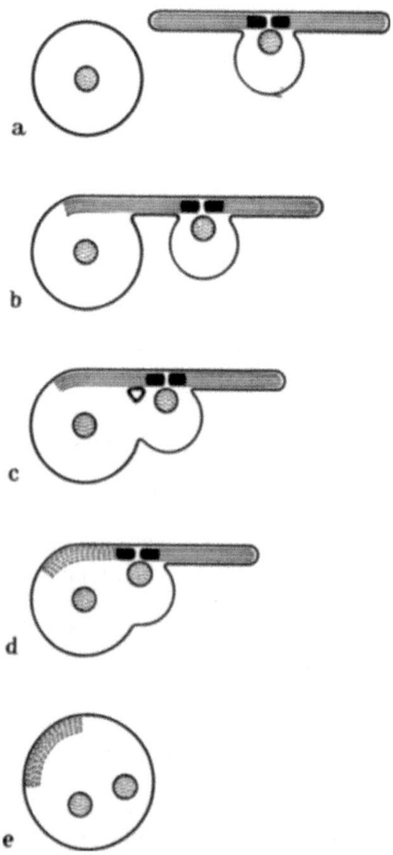

Abb. 17a—e. *Prasiola stipitata*. Oogame Befruchtung. a—c Verschmelzen des unbegeißelten Eies mit dem Spermium beginnend von der Spitze der einen Geißel. d Bewegliche, e festsitzende Zygote. (FRIEDMANN 1962.)

Ebenso wie KOTTE glaubte auch PASCHER (1930), daß die Chemotaxis die bedeutende Rolle für das Zustandekommen der Befruchtung spiele. So „stürzen sich" bei *Sphaeroplea* die männlichen Gameten dieser oogamen Grünalge, sobald sie in die Nähe geöffneter Zellen mit Eiern kommen, „durch die kleinen Öffnungen ins Innere der Oogonien und umschwärmen die Eier .... Wurde ein kleiner Tropfen des Wassers, in dem sich einige weibliche Fadenabschnitte mit geöffneten Oogonien gefunden hatten, in einen spermatozoidhaltigen Tropfen gegeben, so sammelten sich die Spermatozoiden sofort in schnürenartigen Banden oder unregelmäßigen Haufen an, ohne daß es aber zu ausgesprochenen Gruppenbildungen kam. Nach einiger Zeit verteilten sich die Spermatozoiden wieder gleichmäßig". Baumwollfäden, die in Eiwasser gelegen und dieses absorbiert hatten, wurden ebenfalls, wenn auch schwächer „angeschwärmt", was bis zu 1 oder 2 Std andauern konnte. Diese Versuche scheinen mehr auf eine Agglutinationsreaktion als auf Chemotaxis hinzuweisen, wie PASCHER glaubte. Da *Sphaeroplea* ein geeignetes Objekt für weitere Untersuchungen sein könnte, sollte die kleine Beobachtung PASCHERs an dieser Stelle nicht übergangen werden.

3. Hefen.

Bei Hefen sind Agglutinationen bekannt, die in der Praxis große Bedeutung haben. Diese sind zum Teil erwünschte, zum Teil unerwünschte Erscheinungen. In der Brauerei erleichtert die Agglutination die Entfernung der Zellen vom Bier; Weinkellereien bevorzugen ebenfalls agglutinierende Stämme von *Saccharomyces cerevisiae* u. a., Brennereien und Bäckereien dagegen nicht. Trotzdem reichlich Interesse von seiten der Praxis und der Genetik für diese Erscheinung vorliegt, wurde der Zusammenhang von Agglutination und Zygotenbildung offenbar erst durch WICKERHAM (1956) erkannt.

Bei *Hansenula wingei*, einer von WICKERHAM neu beschriebenen und isolierten Hefeart, kommen in der Natur haploide und diploide Zellen vegetativ vor. Einige haploide, getrenntgeschlechtliche Stämme (5, 20, 21) agglutinieren mit Zellen des anderen Geschlechtes unmittelbar innerhalb weniger Sekunden. Die Agglutination ist von solcher Stärke, daß die Zellen sich zu einem festen Ball zusammenrollen und sich in Wasser nicht mehr suspendieren lassen, selbst durch kräftiges Rühren nicht, sie setzen sich im Gegenteil fest am Rührer an (Abb. 18). Agglutinierte Massen zeigen (bei 28° C) nach 55—60 min gepaarte Zellen. Die fertigen Zygoten vermehren sich diploid durch Knospung vegetativ ohne besondere Neigung zu Ascusbildung (nur etwa 6%). Mit vollzogener Paarung wird die Agglutination schwächer und die Zygoten lassen sich schließlich wieder suspendieren.

Zwei andere haploide Stämme (73 und 72) agglutinieren nicht untereinander, konnten jedoch mit den agglutinierenden Stämmen gekreuzt werden. Die

a b c d

Abb. 18a—d. *Hansenula wingei.* a Zellen verschiedenen Geschlechtes wurden auf Agarplatten gemischt und zu einem Ball zusammengerollt. d Werden diese in Wasser gebracht, so bleiben die Zellen agglutiniert. b u. c zeigten als Kontrolle eine Suspension, wenn die Zellen beider Geschlechter getrennt in Wasser gebracht werden. (Nach WICKERHAM 1956.)

Haplonten der F_1 zeigen gegenüber den Teststämmen eine abgestufte Reaktionsfähigkeit, ein Teil agglutiniert sofort, ein anderer Teil „latent", d. h. nach 3 Std Inkubationszeit bei 28° C, ein Rest schließlich agglutiniert gar nicht. Für die „latent" agglutinierenden wird angenommen, daß die Agglutination erst durch gegenseitige Stimulierung, vergleichbar mit adaptiver Enzymsynthese, erfolgt, wohingegen eine sofortige Reaktion mit einem konstitutiven Enzymsystem vergleichbar wäre (WICKERHAM). Die Prüfung vieler Haplonten der F_1 untereinander und mit den Teststämmen ergaben übereinstimmend, daß unmittelbar und „latend" agglutinierende Stämme mehr Zygoten ergeben als nicht agglutinierende. Wenn WICKERHAM auf Grund seiner Ergebnisse glaubt, „daß Agglutination bei Hefen, bei *Saccharomyces* wie bei der auf Agar gezogenen *Hansenula wingei,* die Aufgabe habe, 1. die vegetativen Zellen so dicht aneinander zu bringen, daß die gegenseitige sexuelle Auslösung verstärkt wird und 2. die Zellen fest zusammenzuhalten bis die Paarung erfolgen kann", so darf man die dafür verantwortlichen Stoffe oder Strukturen als Befruchtungsstoffe oder Gamone bezeichnen. Außer *Hansenula* zeigen noch *Citeromyces matritensis (= Torulopsis globosa), Saccharomyces kluyveri* und eine noch nicht benannte Art „*NRRL Y-2408*" „sexuelle" Agglutinationen (WICKERHAM 1958).

Nach quantitativer Untersuchung von Brock (1958) ist die Agglutination der Zellen geschlechts- und artspezifisch. Sie ist am stärksten, wenn man die Zellen zweier verschiedengeschlechtlicher Stämme etwa im Verhältnis 1:1 mischt. Markiert man ein Geschlecht mit Kongorot, so enthält man, wie man auch immer das Verhältnis der beiden Geschlechter in der Suspension ansetzt, ein Sediment agglutinierter Zellen, mit dem Geschlechtsverhältnis 1:1. Die Agglutinationskräfte sind außerordentlich stark. Sie deformieren die Zellwände so, daß kein intercellulärer Raum sichtbar bleibt (Abb. 19).

Durch zweimaliges Waschen der Zellen geht die agglutinierende Wirkung verloren, wird aber durch Cofaktoren, wie Salze, Protaminsulfat oder Gelatine

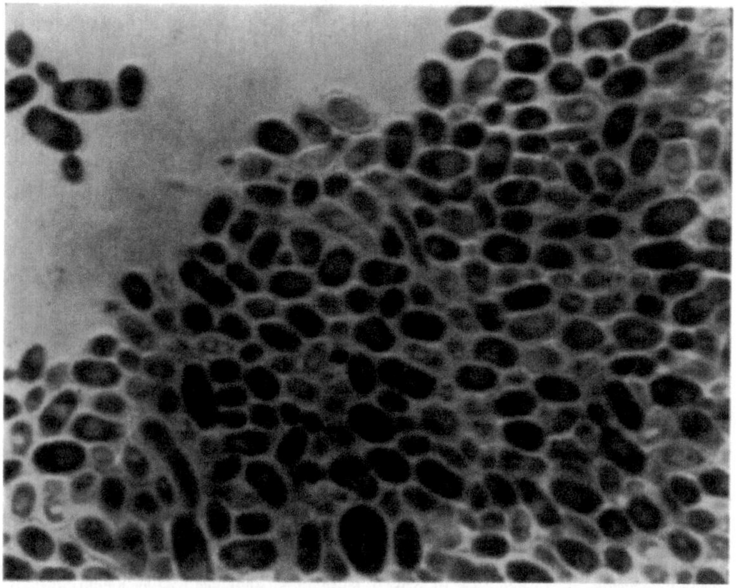

Abb. 19. *Hansenula wingei.* Agglutinierte Zellen. Die Zellwände sind deformiert, kein intercellulärer Raum ist sichtbar. Mit Methylenblau vital gefärbtes Ausstrichpräparat. 1500 ×. (Nach Brock 1958a.)

wieder hergestellt. Beim Erhitzen auf 100° dagegen bleibt die Agglutinationsfähigkeit erhalten. Längeres oder wiederholtes Erhitzen verstärkt sie soweit, daß selbst bei nachfolgendem Waschen die Zellen noch stark agglutinieren. Offenbar werden durch das Erhitzen inaktive Oberflächenkomponenten, die die aktiven Orte überdecken, abgespalten.

Das Gamon ist an die Zellmembran gebunden und kann mit Hilfe von Weinbergschneckenenzym in aktiver Form abgespalten werden (Taylor 1964a, 1965). Die Gamone beider Geschlechter enthalten Proteinkomponenten, die sich spezifisch von Stamm zu Stamm unterscheiden. Während das Gamon des Stammes 21 durch Proteinasen inaktiviert wird, scheint es keine aktiven S—S-Gruppen zu haben. Der Stamm 26 des gleichen Geschlechtes und die Stämm 3 und 5 des anderen Geschlechtes enthalten Gamon, das durch Mercaptoäthanol und andere S—S-Bindungen-spaltende Stoffe inaktiviert wird (Taylor 1964b).

Das geschlechtsspezifische Material des Stammes 5 (= Gamon 5) ist in der Zelle im Cytoplasma in einer aktiven Vorstufe vorhanden, dessen kleinste Einheit eine Sedimentationskonstante von 3,5 S hat (Brock 1965). Es besitzt die gleiche biologische Aktivität und die gleichen chemischen Eigenschaften wie das Gamon an der Zelloberfläche. Es läßt sich an Zellen des Stammes 21 adsorbieren und von ihnen wieder eluieren, und somit reinigen und anreichern. Das gereinigte 3,5 S-Material besteht zu gleichen Teilen aus Protein und Mannan. Die biologische Aktivität wird durch Proteasen und Mercaptoäthanol zerstört. Aus dem Stamm 21 und aus diploiden Zellen kann kein Gamon 5 gewonnen werden.

Trotz der Geschlechtsspezifität der Gamone konnten BROOKBANK u. HEISLER (1963) keine serologischen Unterschiede nachweisen. Daß auch Steroide für die Kopulation eine Rolle spielen, sei kurz erwähnt (HUNT u. CARPENTER 1963).

Mannane spielen auch bei der Bruchbildung der Brauerei-Hefen eine große Rolle. Eine Reihe interessanter Arbeiten von EDDY (1955, 1958) über Agglutinationen bei Saccharomyces-Arten würde die Frage nach ihrer chemischen Natur noch weiter gefördert haben, wenn bekannt wäre, ob diese Agglutinationen, die von technischer Bedeutung in der Brauerei sind, wenigstens teilweise (EDDY 1958) mit Sexualität bzw. Paarung und Ascusbildung in Beziehung stehen. Die Ergebnisse erscheinen mir so wichtig, daß ich sie, auch auf die Gefahr hin, einen nicht sexuellen Vorgang zu mißdeuten, eingehend beschreibe. Schlimmstenfalls würden diese Erscheinungen, bei den unten genannten Modellfällen einzuordnen sein. Unter den verschiedensten Agglutinationen und Aggregationen von Hefezellen sind besonders die sog. Selbstagglutination oder die gegenseitige Agglutination beim Mischen zweier (verschiedengeschlechtlicher?) Stämme wichtig. Zellwände einer Reihe solcher Stämme von *Saccharomyces cerevisiae* und *Saccharomyces carlsbergensis* wurden dargestellt und auf ihre chemische Zusammensetzung untersucht (EDDY und RUDIN 1958). Verschiedene Aufarbeitungen führten zu Protein-, Glukan- und Mannankomponenten. Mit Papain verdaute Membranen ergaben eine Kohlenhydratverbindung (15% der gesamten Zellwand), die gemäß weiterer Auftrennung durch Elektrophorese und Papierchromatographie aus einem Proteinanteil und einem Kohlenhydrat mit nur einem Zucker, Mannose, besteht. Dieser Komplex (von Stamm 3) enthält 85% Kohlenhydrat, 1,2% N und 0,37% P.

Mit Papain inkubierte lebende Zellen verlieren die Fähigkeit zu agglutinieren und geben ein Mannan in den Überstand ab. Daß nun ein Mannose-Polysaccharid tatsächlich die eine agglutinierende Substanz ist, wird weiterhin dadurch klar, daß der monomere Zucker bei gewissen Stämmen die Agglutination hindert, also haptenähnliche Eigenschaften hat (EDDY 1955).

Nur kurz hingewiesen sei auf die Rolle von Ca^{++}-Ionen und Äthanol für die Agglutination bei einigen Hefestämmen, wobei die Bedeutung solch niedermolekularer Stoffe wie Äthanol und der oben beschriebenen chemotaktischen und aktivierenden Substanzen im ganz neuen Lichte erscheinen würden. Der interessante Befund der „latenten" Agglutination (WICKERHAM 1956) unterstreicht a) die Möglichkeit einer sexuellen Beeinflussung *vor* der Agglutination und eröffnet b) Möglichkeiten Vorstufen, der Befruchtungsstoffe zu untersuchen. Die Untersuchung von Hefen in dieser Beziehung erscheint deshalb besonders hoffnungsvoll, weil sie bisher für genetische Experimente leichter zugänglich waren als Algen.

4. Bakterien.

Die Ergebnisse der Bakteriengenetik und der Virusforschung verleiten uns, den Begriff *Befruchtung* noch viel weiter zu fassen, nämlich darin einen Akt zu sehen, der einer Übertragung genetischer Substanz vorausgeht. So wird bei der Bakterienkonjugation zwischen Donor und Empfänger (also ♂ und ♀) unterschieden, je nachdem wer genetisches Material gibt oder empfängt. Zwar werden nicht immer ganze „Kerne" oder besser, ein ganzes „Chromosom" übertragen, sondern nur bestimmte Stücke (Meromixis). Auch bei Virusinfektionen, bei Transduktionen und bei Transformationen wird genetisches Material von außen in das Genom der Bakterienzelle aufgenommen. Befruchtung und Infektion tragen somit in der Tat sehr weitgehende, gemeinsame, *bipolare* Züge (Weidel 1958). „Jedoch sind Adsorption und Eindringen eines Phagenpartikels nur ein Beispiel dafür, wie die Natur ein überall auftauchendes Problem löst: Für bio-

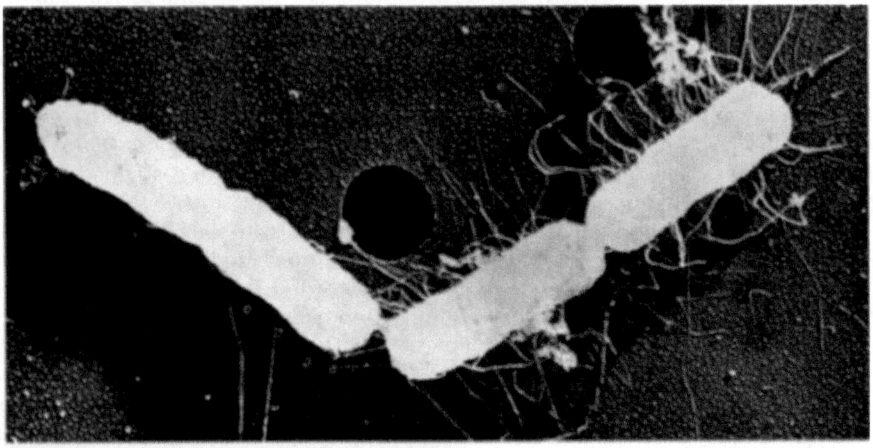

a

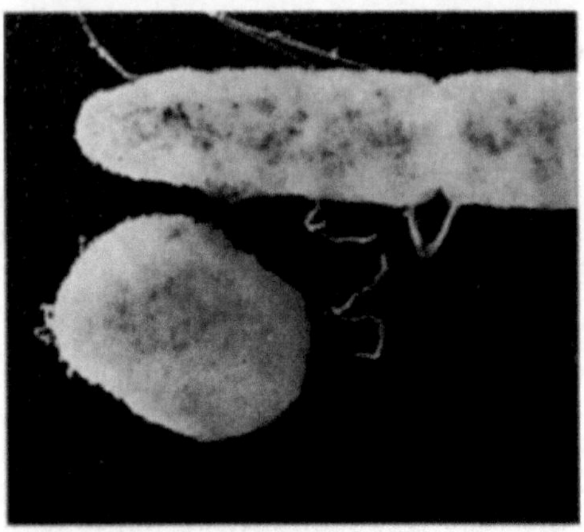

b

Abb. 20a u. b. Elektronenoptische Aufnahme von kopulierenden Bakterien. a Der Partner *E. coli* K-12 Hfr, ist markiert durch abgetötete Bakteriophagen, der andere Partner K-12 $F^-/2$ ist durch die „Geißeln" gekennzeichnet. b K-12 Hfr, eine lange Stäbchenform, kopuliert mit einer kurzen dicken CF^--Zelle. Vergrößerung a 40000×, b 60000×. [Aus Anderson, Wollman und Jacob (1957); und Wollman, Jacob und Hayes (1956) mit Erlaubnis des Biological Laboratory, Cold Spring Harbor, N.Y.]

logische Einheiten aller Art ist es wesentlich mit anderen in spezifischen und dauerhaften Oberflächenkontakt zu kommen, sei es zum einseitigen oder zum gegenseitigen Austausch von Material oder sei es, um vollständig zu verschmelzen. Der Mechanismus der Phageninfektion trägt somit die Züge solcher Vorgänge, die in deutlicheren Fällen als typische Sexualvorgänge bezeichnet wurden."

Tatum und Lederberg (1947), Lederberg (1947) entdeckten in gemischten Kulturen gewisser Mutanten von *Escherichia coli* K 12 Rekombinationen, die sinnvollerweise — verglichen mit sexuellen Vorgängen bei höheren Organismen — eine Paarung der Elternzelle voraussetzten. Kinetische Untersuchungen ergaben, daß dieser Rekombinationsvorgang die Eigenschaften einer Reaktion zweiter Ordnung aufweist, offenbar also einen Kontakt zwischen den teilnehmenden Zellen benötigt (Nelson 1951). Untersucht man die Kinetik des effektiven

Kontaktes, so zeigt sich, daß die Paarung sofort nach dem Zusammengeben der Geschlechter eintritt. Es ergibt sich eine Kurve vom Nullpunkt stetig zunehmend, die eine Sättigung bei etwa 60 min erreicht, zu der Zeit, wo alle kopulationsfähigen Zellen den Partner gefunden haben.

Wir wissen heute, daß — um nur das günstigste Beispiel zu wählen — dabei von einem Donor-Stamm Hfr genetisches Material auf einen Empfängerstamm F⁻ übertragen wird. Dabei gelangt ein Teil des Bakterienchromosoms mit einer gewissen Anzahl von Genen in die Empfängerzelle. Die genetischen Grundlagen hierfür sind durch Wollman, Jacob und Hayes (1956) zusammenhängend berichtet worden. Diese Vorgänge zeigen eine deutliche Parallele zur Befruchtung bei Algen und Pilzen (s. Tabelle 1 und 2), so daß eine Besprechung der Konjugation bei *E. coli* K 12 hier gerechtfertigt ist. Mit Hilfe eines Mikromanipulators wurden die Partner einzelner Paare getrennt und deren Nachkommen getestet (Lederberg 1956, Anderson 1958, Anderson und Mazé 1957, Anderson et al. 1957), so daß kein Zweifel besteht, daß es sich um ein Ergebnis zwischen zwei Zellen handelt. Elektronenmikroskopische Stereoaufnahmen zeigen deutliche Brücken zwischen kopulierenden Bakterien, deren Partner morphologisch unterscheidbar oder durch Phagen markiert waren (Abb. 15). Anderson gibt für den Kopulationsvorgang folgenden Augenzeugenbericht: „Kurz nach dem Mischen beider Kulturen treten viele Pärchen auf, die offenbar durch nicht sichtbare Stränge miteinander verbunden sind. Die Verbindung, mit deren Hilfe „Hfr“-Bakterien die F⁻ durch das Medium schleppen, ist außerordentlich fest; die Bakterien lassen sich z.B. nicht trennen, indem man sie mit einer Mikropipette aufsaugt und ausspritzt. Die Paare nähern sich schließlich einander und verschmelzen durch eine schmale Brücke zu einer Einheit. Der Übergang des genetischen Materials dauert etwa 50 min. In diesem Zustand lassen sich die Bakterien nur mit einem Blendor voneinander trennen“, wovon Wollman u. a. Gebrauch machten, um „partielle Befruchtung“, Meromixis, zu demonstrieren (Wollman, Jacob u. Hayes 1956). Normalerweise trennen sich nach etwa 1 Std die kopulierenden Zellen wiederum.

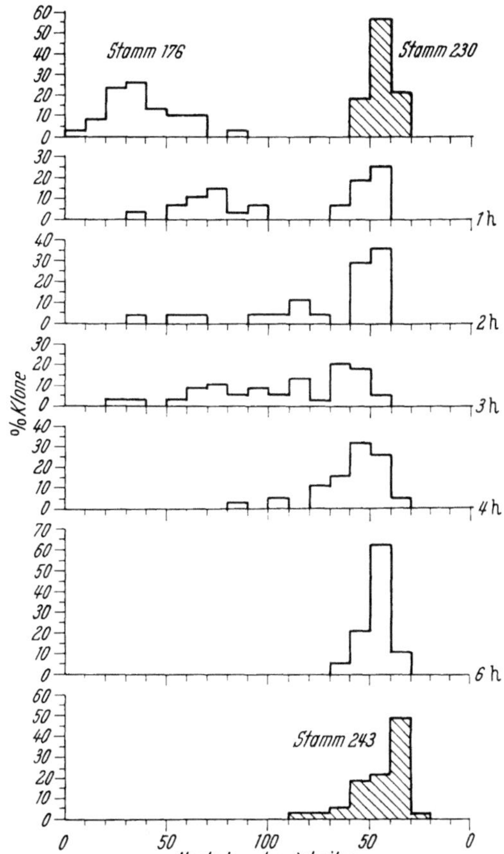

Abb. 21. Veränderung der Zelloberfläche nach Transfer des F⁺-Genes von Stamm 230 auf Stamm 176. (Der Rekombinant wird als Stamm 243 bezeichnet.) Ordinate: % der Klone. Abszisse: Populationsdichte als Nephelometereinheiten (Zeiss-Stufenphotometer). (Weitere Einzelheiten s. bei Maccacaro und Comolli 1956.)

Es ist offenbar, daß die Zelloberfläche eine gewisse physiologische Bereitschaft (Kompetenz) aufweisen muß, sagen wir, daß Gamone in aktiver Form da sein muß. Es wurden daher Versuche angesetzt mit der Absicht, die Zelloberflächen beider Geschlechter näher zu charakterisieren (Maccacaro 1955, Maccacaro u. Comolli 1956). Sie benutzten *E. coli* K 12 F^+ und F^-. Das F^+-Material ist auf F^- übertragbar, so daß z. B. aus der Mischung von Stamm 230 (F^+) mit Stamm 176 (F^-) ein Stamm 243 mit F^+-Merkmalen hervorgeht. Es war zu erwarten, daß Stamm 176 als Empfänger mit dem Geschlecht auch seine Oberflächeneigenschaften ändert. Von diesen sind bisher nur unterschiedliche elektrostatische Ladungen charakterisiert worden, und zwar mit Hilfe von drei Methoden: a) Bestimmung des pH-Wertes, bei dem Selbstagglutination der Zellen eintritt (F^+: 3,331 $\pm$ 0,059; F^-: 2,913 $\pm$ 0,035). b) Farbstoffbindungsvermögen und c) die Stabilität in Nährmedien. Alle drei Methoden zeigen übereinstimmend, daß F^- stärker negativ geladen sind als F^+. Wird daher die oben genannte Kreuzung durchgeführt, so bildet sich im Verlaufe von 6 Std eine *neue* Zelloberfläche aus (Abb. 21). Ob nun diese Merkmale zugleich Eigenschaften des Gamons (oder Receptorsubstanz) sind oder nur „sekundäre" Geschlechtsmerkmale ist bis jetzt nicht klar.

Nach Le Minor (in Wollman, Jacob u. Hayes 1956) und Ørskov u. Ørskov (1960) bestehen auch serologische Unterschiede zwischen Donor und Empfänger; der F^+-Stamm trägt ein neues spezifisches Antigen an seiner Oberfläche. Sneath u. Lederberg (1961) fanden, daß der F^+-Stamm ein mit Perjodat (bzw. Persulfat oder Perborat) reagierendes Kohlenhydrat besitzt. Die mit Perjodat inaktivierten „devirilisierten" Stämme können sich wie weibliche Stämme verhalten. F^--Stämme sind nicht empfindlich gegen Perjodat.

Die männliche Bakterienzelle mit einem F^+-Faktor trägt an ihrer Oberfläche charakteristische Fortsätze, sog. pili (s. Beitrag Kapoor in diesem Band, S. 9ff.). Auf ihnen befinden sich Regionen, an die spezifisch Phagen adsorbiert werden. Ob diese pili auch für die Rekombination von Bedeutung sind, ist noch nicht klar (Caro u. Schnös 1966).

V. Schlußbetrachtungen.

Als Gamone dürften möglicherweise bei den beiden hier behandelten Phyla, den Algen und den Pilzen, einmal die *Anlockungsstoffe*, Sirenine, und zweitens die hochspezifischen *Agglutinations-* und *Receptorstoffe* bezeichnet werden. Der Begriff „Gamon" ist dann streng genommen nicht mit „Hormon" gleichzusetzen. Er wird im Sinne von „Hochzeits-" oder „Paarungsstoffen" allen Wirkungen bis zur Chromosomenpaarung gerecht, wenn man auch nicht in die Kulturflüssigkeit abgegebene Stoffe darin einschließen will: der Ablauf der Befruchtung wird sowohl durch diffusible wie durch unlösliche Stoffe gesteuert.

Der ersten Gruppe, der vorwiegend niedermolekularen Stoffe würde eine vorbereitende Rolle zukommen, Vorbereitung in dem Sinne, daß entweder die Agglutination der Zellen oder allgemeiner die Syngamie ermöglicht oder beschleunigt wird. Notwendigerweise muß die Kinetik einer Paarungsreaktion bei Zugabe oder Entzug von Gamonen untersucht werden. Darauf hat in anderem Zusammenhang Nelson (1951) hingewiesen. Bis jetzt ist keiner der hier erwähnten Stoffe bei Pflanzen auf seine stimulierende Wirkung bei der Paarung hin quantitativ untersucht worden. Damit ist auch die Frage nach der Rolle der Chemotaxis als Mittel zur Beschleunigung des Kontaktes der Geschlechtspartner erkannt. Theoretisch gesehen, müßte es daher auch Anti-Gamone geben.

Die chemische Polarität der Sexualpartner wird vor allem durch die zweite Gruppe der Agglutinations- und Zellwandstoffe manifestiert. Ob dabei Glykoproteide allgemein die entscheidenden Stoffe sind, läßt sich jetzt noch nicht entscheiden.

Literatur.

ANDERSON, TH. F.: Recombination and segregation in *Escherichia coli*. Cold Spring Harb. Symp. quant. Biol. **23**, 47—58 (1958). — ANDERSON, TH. F., et R. MAZÉ: Analyse de la descendance de zygotes formes par conjugaison chez *Escherichia coli* K-12. Ann. Inst. Pasteur **93**, 194—198 (1957). — ANDERSON, TH. F., E. L. WOLLMAN et F. JACOB: Sur les processus de conjugaison et de recombinaison génétique chez *E. coli*. III. Aspects morphologiques en microscopie électronique. Ann. Inst. Pasteur **93**, 450—455 (1957).

BANBURY, G. H.: Processes controlling zygophore formation and zygotropism in *Mucor mucedo* Brefeld. Nature (Lond.) **173**, 499—500 (1954). — Physiological studies in the mucorales. III. The zygotropism of zygophores of *Mucor mucedo* Brefeld. J. exp. Bot. **6**, 235—244 (1955). — BARKSDALE, A. W.: The uptake of exogenous hormone A by certain strains of *Achlya*. Mycologia (N.Y.) **55**, 164—171 (1963). — The role of hormone A during sexual conjugation in *Achlya ambisexualis*. Mycologia (N.Y.) **55**, 627—632 (1963). — BARKSDALE, A. W., M. J. CARLILE, and L. MACHLIS: A comparative study of hormone A and sirenin. Mycologia (N.Y.) **57**, 138—140 (1965). — BARNETT, H. G., V. G. LILLY, and R. F. KRAUSE: Increased production of carotene by mixed (+) and (—) cultures of *Choanephora cucurbitarum*. Science **123**, 141 (1956). — BERTHOLD, G.: Die geschlechtliche Fortpflanzung der eigentlichen Phaeosporeen. Mitth. zool. Stat. Neapel **2**, 401—413 (1881). — BISTIS, G.: Sexuality in *Ascobolus stercorarius*. I. Morphology of the ascogonium; plasmogamy; evidence for a sexual hormonal mechanism. Amer. J. Bot. **43**, 389—394 (1956). — Sexuality in *Ascobolus stercorarius*. II. Preliminary experiments on various aspects of the sexual process. Amer. J. Bot. **44**, 436—443 ((1957). — BLAKESLEE, A. F.: A possible means of identification the sex of (+) and (—) races in the mucors. Science **37**, 880—881 (1913). — Sexuality in mucors. Science **51**, 375—382 (1920). — BROCK, D. T.: Mating reaction in the yeast *Hansenula wingei*. J. Bact. **75**, 697—701 (1958a). — Protein as a specific cell surface component in the mating reaction of *Hansenula wingei*. J. Bact. **76**, 334—335 (1958b). — Purification and characterization of intracellular sex specific mannan protein from yeast. Proc. nat. Acad. Sci. **54**, 1104—1112 (1965). — BROOKBANK, J. W., and M. R. HEISLER: Immunology of the yeast *Hansenula wingei*. J. Bact. **85**, 509—515 (1963). — BURGEFF, H.: Untersuchungen über Sexualität und Parasitismus bei Mucorineen. I. Bot. Abh. **4**, 5—135 (1924). — BURNETT, H. J.: Carotene and sexuality in Mucoraceae, especially in *Phycomyces Blakesleeanus*. New Phytologist **55**, 45—49 (1956). — The mating systems of fungi. I. New Phytologist **55**, 50—90 (1956).

CARLILE, M. J., and L. MACHLIS: The response of male gametes of *Allomyces* to the sexual hormone sirenin. Amer. J. Bot. **52**, 478—483 (1965a). — A comparative study of the chemotaxis of the motile phases of *Allomyces*. Amer. J. Bot. **52**, 484—486 (1965b). — CARO, L. G., and M. SCHNÖS: The attachment of the male-specific bacteriophage F 1 to sensitive strains of *Escherichia coli*. Proc. Nat. Acad. Sci. (Wash.) **56**, 126—132 (1966). — COHEN, L. W., and R. W. SIEGEL: The mating-type substances of *Paramecium bursaria*. Genet. Res. **4**, 143—150 (1963). — COOK, A. H., and I. A. ELVIDGE: Fertilization in the *Fucaceae:* Investigations on the nature of the chemotactic substance produced by eggs of *Fucus serratus* and *F. vesiculosus*. Proc. roy. Soc. (Lond.) B **138**, 97—114 (1951). — COOK, A. H., I. A. ELVIDGE, and Sir I. HEILBRONN: Fertilization, including chemotactic phenomena in the *Fucaceae*. Proc. roy. Soc. (Lond.) B **135**, 293—301 (1948).

DUKES, P. D., and J. L. APPLE: Chemotaxis of zoospores of *Phytophthora parasitica* var. *nicotianae* by plant roots and certain chemical solutions. Phytopathology **51**, 195—197 (1961).

EDDY, A. A.: Flocculation characteristics of yeasts. I. Comparative survey of various strains of *Saccharomyces cerevisiae*. J. Inst. Brewing **61**, 307—312 (1955). — Flocculation characteristics of yeasts. II. Sugars as dispersing agents. J. Inst. Brewing **61**, 313—317 (1955). — Flocculation characteristics of yeasts. III. General role of flocculating agents and special characteristics of a yeast flocculated by alcohol. J. Inst. Brewing **61**, 318—320 (1955). — Composite nature of the flocculation process of top and bottom strains of *Saccharomyces*. J. Inst. Brewing **64**, 143—151 (1958). — EDDY, A. A., and A. D. RUDIN: Part of the yeast surface apparently involved in flocculation. J. Inst. Brewing **64**, 19—21 (1957). — ESSER, K., u. R. KUENEN: Genetik der Pilze. Berlin-Heidelberg-New York: Springer 1965.

FARMER, J. B., and J. L. WILLIAMS: Contributions to our knowledge of the *Fucaceae*. Their life-history and cytology. Phil. Trans. Bot. Soc. Lond., Ser. B **190**, 623—645 (1898). — FISCHER, F. G., u. G. WERNER: Eine Analyse des Chemotropismus einiger Pilze, insbesondere der Saprolegniaceen. Hoppe-Seylers Z. physiol. Chem. **300**, 211—236 (1955). — Die Chemotaxis der Schwärmsporen von Wasserpilzen (Saprolegniaceen). Hoppe-Seylers Z. physiol. Chem. **310**, 65—91 (1958). — FÖRSTER, H.: Das Wirkungsspektrum der Kopulation von *Chlamydomonas eugametos*. Z. Naturforsch. **12b**, 765—770 (1957). — Die Wirkungsstärken einiger Wellenlängen zum Auslösen der Kopulation von *Chlamydomonas moewusii*. Z. Naturforsch. **14**, 479—480 (1959). — FÖRSTER, H., u. L. WIESE: Untersuchungen zur Kopulationsfähigkeit von *Chlamydomonas eugametos*. Z. Naturforsch. **9**b, 470—471 (1954). — Gamonwirkungen bei *Chlamydomonas eugametos*. Z. Naturforsch. **9**b, 548—550 (1954). — Gamonwirkung bei *Chlamydomonas reinhardi*. Z. Naturforsch. **10**b, 551 (1955). — FÖRSTER, H., L. WIESE u. G. BRAUNITZER: Über das agglutinierend wirkende Gynogamon von *Chlamydomonas eugametos*. Z. Naturforsch. **11**b, 315—317 (1956). — FRIEDMANN, I.: Cinemicrography of spermatozoids and fertilization in *Fucales*. Bull. Res. Coun. Israel D **10**, 73—83 (1961). — Cell membrane fusion and the fertilization mechanism in plants and animals. Science **136**, 711—712 (1962). — FRITSCH, F. E.: The structure and reproduction of algae. II. Cambridge: Univ. Press 1945; s. S. 377.

GEITLER, L.: Untersuchungen über das sexuelle Verhalten von *Tetraspora tubrica*. Biol. Zbl. **51**, 173—187 (1931). — Selektive Paarung und gegenseitige Beeinflussung der Kopulationspartner bei der Diatomee *Cocconeis*. Planta (Berl.) **51**, 584—599 (1958). — GERLOFF, J.: Beiträge zur Kenntnis der Variabilität und Systematik der Gattung *Chlamydomonas*. Arch. Protistenk. **94**, 311—502 (1940).

HARTMANN, K. M.: Die Regulation der Gametogenese von *Chlamydomonas eugametos* und *Chlamydomonas moewusii* durch exogene und endogene Faktoren. Vergleichend morphologische, physiologische und biophysikalische Untersuchungen. Diss. Tübingen 1962. — HARTMANN, M.: Autogamie bei Protisten und ihre Bedeutung für das Befruchtungsproblem. Arch. Protistenk. **14**, 264 (1909). — Ergänzende Untersuchungen über die Sexualität von *Ectocarpus siliculosus*. Arch. Protistenk. **89**, 382—392 (1937). — Beiträge zur Kenntnis der Befruchtung und Sexualität mariner Algen. I. Über die Befruchtung von *Cutleria multifida*. Pubbl. Staz. zool. Napoli **22**, 120—128 (1950). — Die Sexualität, 2. Aufl. Stuttgart: Gustav Fischer 1956. — Neue Befunde über Befruchtungsstoffe (Gamone) und ihre theoretische Bedeutung. Nach Untersuchungen von H. FÖRSTER und L. WIESE an *Chlamydomonas eugametos* und K. KÖHLER an *Chaetomorpha aerea*. Naturwiss. **43**, 313—317 (1956). — HARTMANN, M., u. O. SCHARTAU: Untersuchungen über die Befruchtungsstoffe der Seeigel. I. Biol. Zbl. **59**, 571—587 (1939). — HESSELTINE, C. W., and R. F. ANDERSON: Microbial production of carotenoids. I. Mycologia (N.Y.) **49**, 449—452 (1957). — HIRST, G. K.: Adsorption of influenza hemagglutinins and virus by red blood cells. J. exp. Med. **76**, 185—209 (1942). — HOFFMAN, L. R.: Chemotaxis of *Oedogonium* sperms. Southwestern Nat. **5**, 111—116 (1960). — Studies of the morphology, cytology and reproduction of *Oedogonium* and *Oedocladium*. Ph. D. Diss. University of Texas, Austin 1961, 197 p. — HUNT, D. E., and P. L. CARPENTER: Sterols as components in the mating reaction of *Hansenula wingei*. J. Bact. **86**, 845, 847 (1963). — HUTNER, S. H., and L. PROVASOLI: The phytoflagellates. In A. LWOFF, Biochemistry and physiology of Protozoa, p. 27—128. New York: Academic Press 1951.

JOLLOS, V.: Untersuchungen über die Sexualitätsverhältnisse von *Dasycladus clavaeformis*. Biol. Zbl. **46**, 279—295 (1926).

KNAPP, E.: Zur Kenntnis der Polarität der Eier von *Cytoseira barbata*. Planta (Berl.) **14**, 731—751 (1931). — KÖHLER, K.: Entwicklungsgeschichte, Geschlechtsbestimmung und Befruchtung bei *Chaetamorpha*. Arch. Protistenk. **101**, 224—268 (1956). — Neue Untersuchungen zur Sexualität von *Dasycladus*. Arch. Protistenk. **102**, 209—218 (1957). — KUNIEDA, H., and S. SUTO: The life-history of *Colpomenia sinuosa* (Scytosiphonaceae) with special reference to the conjugation of anisogametes. [In Japanese, with English summary.] Bot. Mag. (Tokyo) **52** (622), 539—546 (1938).

LEDERBERG, J.: Gene recombination and linked segregations in *E. coli*. Genetics **32**, 505—525 (1947). — Conjugal pairing in *Escherichia coli*. J. Bact. **71**, 497—499 (1956). — LERCHE, W.: Untersuchungen über Entwicklung und Fortpflanzung der Gattung *Dunaliella*. Arch. Protistenk. **88**, 236—268 (1937). — LEVI, J. D.: Mating reaction in yeast. Nature (Lond.) **177**, 753—754 (1956). — LEVRING, T. R.: Remarks on the surface layers and the formation of the fertilization membrane in *Fucus* eggs. Meddel. Göteborgs Bot. trädg. **17**, 97—105 (1947). — Fertilization experiments with *Hormosira banksii* (Turn.) Dcne. Physiol. Plantarum (Cph.) **2**, 45—55 (1949). — Remarks on the submicroscopical structure of eggs and spermatozoids of *Fucus* and related genera. Physiol. Plantarum (Cph.) **5**, 528—533 (1952). — Some remarks on the structure of gametes of *Ulva lactuca*. Bot. Not. **108**, 40—45 (1955). — LEWIN, R. A.: Gamete behaviour in *Chlamydomonas*. Nature (Lond.) **166**, 76 (1950). — Studies on the flagella of algae. I. General observations on *Chlamydomonas moewusii* Gerloff. Biol. Bull. **103**, 74—79 (1952). — Mutants of *Chlamydomonas moewusii* with impaired motility. J. gen. Microbiol. **11**, 358—363 (1954). — Sex in unicellular algae. In: Sex in microorganisms, p. 100. Washington, D. C.: Amer. Ass. Adv. Sci. 1954. — Control of sexual activity in *Chlamydomonas* by light. J. gen. Microbiol. **15**, 170—185 (1956). — LEWIN, R. A., and J. O. MEINHART: Studies on the flagella of algae. III. Electron micrographs of *Chlamydomonas moewusii*. Canad. J. Bot. **31**, 711—717 (1955).

MACCACARO, G. A.: Cell surface and fertility in *Escherichia coli*. Nature (Lond.) **176**, 125 (1955). — MACCACARO, G. A., and R. COMOLLI: Surface properties correlated with sex compatibility in *Escherichia coli*. J. gen. Microbiol. **15**, 121—132 (1956). — MACHLIS, L.: Evidence for a sexual hormone in *Allomyces*. Physiol. Plantarum (Cph.) **11**, 181—192 (1958). — A procedure for the purification of sirenin. Nature (Lond.) **181**, 1790—1791 (1958). — MAINX, F.: Physiologische und genetische Untersuchungen an Oedogonien. I. Mitt. Z. Bot. **24**, 481—527 (1931). — MANOILOFF, E. O.: Chemical reaction of blood for definition of sex in man, animals, and dioecious plants. Third modification. Amer. J. phys. Anthrop. **13**, No 1 (1929). — METZ, CH. B.: Mating substances and the physiology of fertilization in ciliates. In: Sex in microorganism, p. 284—334. Washington, D.C.: Amer. Ass. Adv. Sci. 1954. — Specific egg and sperm substances and activation of the egg. In: The beginning of embryonic development. Washington, D.C.: Amer. Ass. Adv. Sci. 1957. — MIYAKE, A.: Induction of conjugation by cell-free preparations in *Paramecium multimicronucleatum*. Science **146**, 1583—1585 (1964). — MOEWUS, F.: Untersuchungen über die Sexualität und Entwicklung von Chlorophyceen. Arch. Protistenk. **80**, 469—520 (1933). — Die Sexualstoffe von *Chlamydomonas eugametos*. Ergebn. Enzymforsch. **12**, 173—206 (1951).

NELSON, T. C.: Kinetics of genetic recombination in *Escherichia coli*. Genetics **36**, 162 (1951).

OLTMANNS, F.: Morphologie und Biologie der Algen, 2. Aufl., Bd. 3. Jena: Gustav Fischer 1923. — ØRSKOV, I., and F. ØRSKOV: An antigen termed f^+ occurring in F^+ *E. coli* strains. Acta path. microbiol. scand. **48**, 37—39 (1960).

PASCHER, A.: Über Gruppenbildung und ,,Geschlechtswechsel" bei den Gameten einer Chlamydomonadine (*Chlamydomonas paupera*). Studien und Beobachtungen über die geschlechtliche Fortpflanzung und den Generationswechsel der Grünalgen. I. Jb. wiss. Bot. **75**, 551—580, s. S. 575 (1931). — PLEMPEL, M.: Die Sexualstoffe der *Mucoraceae*. Ihre Abtrennung und die Erklärung ihrer Funktion. Arch. Mikrobiol. **26**, 151—174 (1957). — Die Darstellung eines kristallinen Benzoesäureesters der Gamone von *Mucor mucedo*. Naturwiss. **47**, 227 (1959). — Die zygotropische Reaktion bei Mucorineen. I. Planta (Berl.) **55**, 254—258 (1960). — Die zygotropische Reaktion bei Mucorineen. III. Planta (Berl.) **58**, 509—520 (1962). — Die chemischen Grundlagen der Sexualreaktion bei Zygomyceten. Planta (Berl.) **59**, 492—508 (1963). — Sexualreaktion und Carotin-Synthese bei Zygomyceten. Planta (Berl.) **65**, 225—231 (1965). — PLEMPEL, M., u. G. BRAUNITZER: Die Isolierung der Mucorineen-Sexualstoffe. I. Z. Naturforsch. **13**b, 302—304 (1958). — PLEMPEL, M., u. W. DAWID: Die zygotropische Reaktion bei Mucorineen. II. Planta (Berl.) **56**, 438—446 (1961).

RAPER, J.: Chemical regulation of sexual processes. Bot. Rev. **18**, 447—545 (1952). — The biological action of growth substances. Symp. Soc. exp. Biol. **11**, 143—165 (1957). — RAPER, J. R.: Sexual hormones in *Achlya*. II. Distance reactions, conclusive evidence for a hormonal coordinating mechanism. Amer. J. Bot. **27**, 162—173 (1940). — Sexual hormones in *Achlya*. III. Hormone A and the initial male reaction. Amer. J. Bot. **29**, 159—166 (1942). — RAPER, J. R., and A. J. HAAGEN-SMIT: Sexual hormones in *Achlya*. IV. Properties of hormone A of *A. bisexualis*. J. biol. Chem. **143**, 311—320 (1942). — RAWITSCHER-KUNKEL, E., and L. MACHLIS: The hormonal integration of sexual reproduction in *Oedogonium*. Amer. J. Bot. **49**, 177—183 (1962). — REICHEL, L., u. M. WALLIS: Über die Biosynthese des β-Carotins. Angew. Chem. **68**, 181 (1956). — Über die Biosynthese des β-Carotins. Naturwiss. **44**, 234 (1957). — Zur Biosynthese des β-Carotins. Naturwiss. **45**, 130 (1958). — RENNER, O.: Besprechung: HARTMANN, M.: Die Sexualität. Z. Bot. **45**, 317—352 (1957). — Auch etwas über FRANZ MOEWUS, *Forsythia* und *Chlamydomonas*. Z. Naturforsch. **13**b, 399—403 (1958). — ROYLE, D. J., and C. J. HICKMAN: Analysis of factors governing in vitro accumulation of zoospores of *Pythium aphanidermatum* on roots. 1. Behavior of zoospores. Canad. J. Bot. **10**, 151—162 (1964a). — Analysis of factors governing in vitro accumulation of zoospores of *Pythium aphanidermatum* on roots. 2. Substances causing response. Canad. J. Bot. **10**, 201—219 (1964b).

SATINA, S., and A. F. BLAKESLEE: Criteria of male and female in bread moulds. Proc. nat. Acad. Sci. (Wash.) **15**, 735—740 (1929). — SAUVAGEAU, C.: Sur la fécondation hétérogamique d'une algue phéosporée. C. R. Acad. Sci. (Paris) **123**, 360—362 (1896a). — Sur la conjugaison des zoospores de l'*Ectocarpus siliculosus*. C. R. Acad. Sci. (Paris) **123**, 431—433 (1896b). — Observations relatives à la sexualité des Phéosporées. J. Bot. **10**, 357—367 (1896c). — SCHAUDINN, F.: Neuere Forschungen über die Befruchtung bei Protozoen. Verh. dtsch. zool. Ges. 1905. — SCHREIBER, E.: Über die geschlechtliche Fortpflanzung der Sphacelariales. Ber. dtsch. bot. Ges. **49**, 235—240 (1931). — SKUJA, H.: Körperbau und Reproduktion bei *Dinobryon Borgei*. Sv. bot. Tidskr. **44**, 1 (1950). — SNEATH, P. H., and J. LEDERBERG: Inhibition by periodate of mating in *Escherichia coli* K-12. Proc. nat. Acad. Sci. (Wash.) **47**, 86—91 (1961). — STEIN, R. J.: A morphological study of *Astrephomene gubernaculifera* and *Volvulina steinii*. Amer. J. Bot. **45**, 388—396 (1958). — STRASBURGER, E.: Das botanische Practicum. Jena 1884. — SUBRAHMANYAN, R.: Observations on the anatomy, cytology, development of the reproductive structures, fertilization and embryology of *Pelvetia canaliculata* Dcne. et Thur. III. The liberation of reproductive bodies, fertilization and embryology. J. Indian bot. Soc. **36**, 373—395 (1958).

TATUM, E. L., and J. LEDERBERG: Gene recombination in the bacterium *Escherichia coli*. J. Bact. **53**, 673—684 (1947). — TAYLOR, N. W.: Specific soluble factor involved in sexual agglutination of the yeast *Hansenula wingei*. J. Bact. **87**, 863—866 (1964). — Inactivation of sexual agglutination in *Hansenula wingei* and *Saccharomyces kluyveri* by disulfide-cleaving agents. J. Bact. **88**, 929—936 (1964). — Purification of sexual-agglutination factor from the yeast *Hansenula wingei* by chromatography and gradient sedimentation. Arch. Biochem. **111**, 181—186 (1965). — THURET, G.: Deuxième note sur la fécondation des Fucacées. Mem. Soc. Sci. nat. Cherbourg **5** (1857). — THURET, G., et E. BORNET: Études phycologiques. Paris 1878. — TSUBO, Y.: On the mating of a *Chlamydomonas*, with special references to clumping and chemotaxis. Bot. Mag. (Tokyo) **70**, 327—334 (1958). — Chemotaxis and sexual behavior in *Chlamydomonas*. J. Protozool. **8**, 114—121 (1961).

WEIDEL, W.: Bacterial viruses (with particular reference to adsorption/penetration). Ann. Rev. Microbiol. **12**, 27—48 (1958). — WEISS, P.: Cell contact. Int. Rev. Cytol. **7**, 391—424 (1958). — WICKERHAM, L. J.: Influence of agglutination on zygote formation in *Hansenula wingei*, a new species of yeast. C. R. Lab. Carlsberg, Sér. physiol. **26**, 423—443 (1956). — Sexual agglutination of heterothallic yeasts in divers taxonomic areas. Science **128**, 1504—1505 (1958). — WIESE, L., and R. F. JONES: Studies on gamete copulation in heterothallic *Chlamydomonads*. J. cell. comp. Physiol. **61**, 265—274 (1963). — WILBOIS, A.: Diss. Bloomington, Ind. (USA); s. auch A. COLEMAN. — WILLIAMS, L. J.: Studies in the *Dictyotaceae*. II. The cytology of the gametophyte generation. Amer. J. Bot. **18**, 183—204 (1904); s. S. 193. — WOLLMAN, E. L., F. JACOB, and W. HAYES: Conjugation and genetic recombination in *E. coli* K-12. Cold Spring Harb. Symp. quant. Biol. **21**, 141—162 (1956).

ZICKLER, H.: Zur Entwicklungsgeschichte des Ascomyzeten *Bombardia lunata* Zckl. Arch. Protistenk. **98**, 1—70 (1953).

Die Verbreitung der Incompatibilität bei Thallophyten*.

Von

Karl Esser.

I. Begriffsbestimmungen.

Als *Incompatibilität* oder sexuelle Unverträglichkeit bezeichnet man *jede Hemmung der Befruchtung innerhalb eines sexuellen Fortpflanzungs-Systems hermaphroditer Lebewesen, die nicht auf Defekten der Gameten oder Gametenkerne beruht.* Unter Befruchtung verstehen wir (nach Hartmann 1956) die Verschmelzung zweier geschlechtlich verschiedener Zellen oder Kerne (Gameten oder Gametenkerne). Diese Fusion trägt aber nur den Charakter einer Befruchtung, wenn die damit verbundene Karyogamie entweder unmittelbar oder nach einer Reihe von vegetativen Kernteilungen zu einer Reifungsteilung (Meiosis) führt. Im Gegensatz zur sexuellen Differenzierung, die nicht bei allen Organismen genotypisch ausgelöst wird, werden alle bisher analysierten Incompatibilitäts-Phänomene bei Pflanzen durch Gene gesteuert. Dies schließt jedoch nicht aus, daß die Reaktionsnorm der Incompatibilitäts-Gene durch Umweltfaktoren beeinflußbar ist.

Wenn die Kreuzungsbeziehungen innerhalb einer hermaphroditen Art oder Rasse durch ein Incompatibilitäts-System bestimmt werden, kann dadurch sowohl bei jedem Individuum die Selbstbefruchtung als auch die Kreuzungsbefruchtung zwischen bestimmten Individuen verhindert werden. Im ersten Falle spricht man von Selbst-Incompatibilität und im zweiten von Kreuzungs-Incompatibilität. Da beide Erscheinungsformen der Incompatibilität vielfach miteinander verknüpft sind, genügt es meist, das Vorliegen von sexueller Unverträglichkeit nur mit der Bezeichnung Incompatibilität zu charakterisieren.

Die Grenzen zwischen Incompatibilität und Sexualität, d. h. die Abgrenzung der Incompatibilität monöcischer Pflanzen gegenüber der Getrenntgeschlechtigkeit diöcischer ist bei Blütenpflanzen, die stets klar erkennbare Sexualorgane ausgebildet haben, sehr einfach. Für Thallophyten dagegen ist eine klare Unterscheidung beider Phänomene oft sehr schwierig. Dies trifft für die Arten mit verschiedenen Kreuzungstypen zu, deren hermaphroditer Charakter infolge von Isogamie *(Algae)* oder Fehlen von Geschlechtsorganen *(Holobasidiomycetes)* nicht morphologisch erkennbar ist.

Da die Zwittrigkeit eines Organismus jedoch eine unbedingte Voraussetzung für das Vorkommen von Incompatibilitäts-Erscheinungen ist, bedarf es zur Entscheidung, ob das Kreuzungsverhalten solcher Arten durch Incompatibilität oder Diöcie bestimmt wird, einer Präzisierung des Begriffs „hermaphrodit“: *Ein isogamer Thallophyt oder ein solcher, der keine Geschlechtsorgane ausbildet, ist hermaphrodit, wenn jede Pflanze in der Lage ist, bei der Zygotenbildung als Kern-Donator (männliche Potenz) oder als Kern-Receptor (weibliche Potenz) zu dienen. Er ist selbst-incompatibel, wenn er in mindestens 2 Kreuzungstypen auftritt, von denen jeder*

* Mit Unterstützung der Deutschen Forschungsgemeinschaft.

einzelne sowohl eine männliche als auch eine weibliche Potenz besitzt. Wenn ein Kreuzungstyp nur die Funktion des Donators oder des Receptors aufweist, liegt Diöcie vor.

Auf der Grundlage dieser von Lewis (1954) vorgeschlagenen Definition soll im folgenden in einem summarischen Überblick gezeigt werden, in welchen Klassen der Thallophyten Incompatibilitäts-Erscheinungen vorkommen.

II. Vorkommen von Incompatibilitätserscheinungen in den verschiedenen Klassen der Thallophyten.

1. *Cyanophyceae.*

Da bei diesen Organismen noch nicht mit Sicherheit bekannt ist, ob eine sexuelle Vermehrung vorliegt, kann über Incompatibilität nichts ausgesagt werden.

2. *Flagellatae.*

Für anisogame und oogame Arten ist über das Vorkommen von Incompatibilität nichts bekannt. Die isogamen Arten besitzen meist 2 Kreuzungstypen. Diese scheinen jedoch bei *Chlamydomonas*, der am meisten untersuchten Gattung, die Manifestierung einer morphologisch nicht erkennbaren, jedoch physiologisch vorhandenen Diöcie darzustellen, denn bei Kreuzungen zwischen der isogamen *C. eugametos* und *C. oogama*, einer anisogamen Art, erfolgt die Zygotenbildung nur zwischen +-Individuen und den Makrogameten bzw. zwischen —-Individuen und Mikrogameten (zit. nach Kühn 1950). Die Zellen des +-Kreuzungstyps zeigen also männliches und die des —-Typs weibliches Verhalten.

In die gleiche Richtung weisen auch die Befunde, die Sager (1954, 1955) über die extrachromosomal vererbte Resistenz von *C. reinhardi* gegen Streptomycin erhielt. Sie konnte nämlich zeigen, daß dieses Merkmal nur durch den +-Kreuzungstyp übertragen wird. Leider ist der Mechanismus dieser an einen bestimmten Kreuzungstyp gebundenen Merkmalsübertragung nicht bekannt. Da jedoch die toxische Wirkung des Streptomycins in erster Linie die Plastiden betrifft (Lwoff und Schaeffer 1949), lassen sich die Sagerschen Befunde erklären, wenn man annimmt, daß zur Zygotenbildung nur einer der beiden Kreuzungspartner die Plastide liefert, und zwar im vorliegenden Falle der +-Stamm. Ein Zutreffen dieser Annahme würde gleichfalls ein Beweis für die sexuelle Basis der Existenz von 2 Kreuzungstypen bei der isogamen *Chlamydomonas* sein, denn der Plastidenlieferant (+) würde als der weibliche Partner und der —-Kreuzungstyp als der männliche anzusehen sein.

3. *Algae.*

Auch bei anisogamen und oogamen Algen liegen keinerlei Angaben über Incompatibilitäts-Erscheinungen vor. Die isogamen Arten weisen meist 2 Kreuzungstypen auf (s. die zusammenfassende Darstellung bei Hartmann 1956). Da die Zygotenbildung dieser Algen durch Gametenfusion eingeleitet wird, muß es gemäß unserer Definition offenbleiben, ob die Ausbildung verschiedener Kreuzungstypen eine Erscheinungsform der Diöcie oder der Incompatibilität ist. Auch die bei zahlreichen Formen gefundenen physiologischen und stofflichen Unterschiede (Gamone) (s. bei Hartmann 1956) zwischen den beiden Kreuzungstypen können sowohl eine Charakterisierung der Getrenntgeschlechtigkeit als auch der Incompatibilität sein.

Nach älteren entwicklungsgeschichtlichen Untersuchungen an Desmidiaceen und Diatomeen (s. bei Brieger 1930) können Konjuganten je 2 Isogameten enthalten, die nicht miteinander fusionieren. Nach Übertritt je eines Gameten in den Konjugationspartner entsteht aus jeder Zelle eine Zygote. Diese Befunde scheinen für einen monöcischen Charakter der betreffenden Algen zu sprechen, die als selbst-incompatibel angesehen werden könnten. Ob auch innerhalb dieser Arten eine Kreuzungs-Incompatibilität vorkommt, d. h. ob mehrere Kreuzungstypen existieren, ist nicht bekannt. Infolge des Fehlens genetischer Grundlagen liefern diese Beobachtungen uns leider keine Entscheidung der Fragestellung: Incompatibilität oder Diöcie.

4. Lichenes.

Da bei den Flechten sich nur die Pilzkomponente sexuell vermehrt und die meisten Flechten Euascomyceten enthalten, dürften hier dieselben Incompatibilitäts-Verhältnisse vorliegen wie in dieser Gruppe. Entsprechende Untersuchungen sind allerdings nicht bekannt.

5. Fungi.

Bei den *Myxomycetes* sind isogame Arten mit 2 Kreuzungstypen bekannt. Ob es sich dabei um Diöcie oder Incompatibilität handelt, ist unklar.

Innerhalb der *Phycomycetes* existieren Arten mit verschiedenen Kreuzungstypen in der Hauptsache in der Familie der *Mucoraceae.* Die Zygotenbildung erfolgt bei diesen durch Gametangiogamie. Es muß daher offenbleiben, ob diese Reaktion zwischen morphologisch nicht als weiblich oder männlich differenzierten Sexualorganen[1] durch ein Incompatibilitäts-System gelenkt wird oder der Ausdruck einer Diöcie ist. Diese Unklarheit kann auch nicht durch den Nachweis von Gamonen (BURGEFF 1924, PLEMPEL 1956, 1963, PLEMPEL und BRAUNITZER 1958) behoben werden, denn diese können sowohl „Sexualstoffe" als auch „Incompatibilitätsstoffe" sein.

Wenn man die Gamone der Mucoraceae als Incompatibilitäts-Stoffe auffaßt, läßt sich das Zustandekommen der Sexualreaktion zwischen den beiden Kreuzungstypen als Folgeerscheinung einer sich ergänzenden Wirkung der +- und —-Stoffe entsprechend dem Schlüssel-Schloß-System erklären.

Die gleiche Unklarheit zwischen Sexualität und Incompatibilität besteht auch bei den *Protoascomycetes,* die zahlreiche Arten mit 2 Kreuzungstypen aufweisen. Die Zygotenbildung erfolgt durch Fusion von morphologisch nicht unterscheidbaren Zellen.

Zwar fand LEUPOLD (1950), daß die +-Zellen von *Schizosaccharomyces pombe* stets mit einem Kopulationsfortsatz auf die —-Zellen zuwachsen. Da aber die +- und —-Klone der cytologischen Unterscheidung wegen unter verschiedenen Kulturbedingungen angezogen wurden, kann man in diesem Falle [wie auch schon von HARTMANN (1956) zum Ausdruck gebracht wurde] nicht mit Sicherheit sagen, daß die Bildung der Keimschläuche der Ausdruck einer sexuellen Differenzierung der +- und —-Zellen ist. Ähnliche Beobachtungen von LEVI (1956) an *Saccharomyces cerevisiae* sind ebenfalls nicht eindeutig zu interpretieren. LEVI fand, daß auf Agarmedien nur die —-Zellen Kopulationsfortsätze bilden. In flüssigen Nährmedien bilden jedoch beide Kreuzungstypen diese Organellen.

Dagegen findet man bei den *Euascomycetes* in den meisten Fällen eine deutlich ausgeprägte Monöcie[2]. An Einspormycelien entstehen weibliche Geschlechtsorgane (Ascogone usw.), die durch männliche Gametangien, Gameten oder Gametenkerne befruchtet werden; Einzelheiten über die unterschiedliche morphologische Ausprägung der Geschlechter bei MARTENS 1946 und GÄUMANN 1949. Für das Vorkommen von mehreren Kreuzungstypen unter diesen Zwittern sind Incompatibilitäts-Systeme verantwortlich.

Die *Holobasidiomycetes* tragen an ihren Mycelien, die aus haploiden einkernigen Basidiosporen hervorgehen, keinerlei Differenzierungsgebilde, die als Geschlechtsorgane anzusprechen sind. Bei Arten mit mehreren Kreuzungstypen wird die sexuelle Vermehrung durch Hyphenfusionen zwischen 2 Kreuzungstypen eingeleitet, die einen beidseitigen Kernaustausch ermöglichen (BULLER 1931,

[1] Auf Grund morphologischer Unterschiede der Sexualorgane bezeichnen PLEMPEL und BRAUNITZER (1958) bei *Mucor mucedo* den —-Kreuzungstyp als das weibliche und den +-Typ als das männliche Geschlecht. Diese Annahme bedarf jedoch einer genetischen Überprüfung durch Kreuzungsanalysen, deren Resultate beweisen müssen, daß die morphologischen Differenzen wirklich durch den +/—-Locus und nicht etwa durch andere Genorte bestimmt werden und damit vom Kreuzungstyp nicht abhängig wären.

[2] Ein eindeutiger Nachweis für das Vorliegen von Diöcie konnte in dieser Gruppe bisher nur für drei Arten erbracht werden (Literatur bei ESSER u. KUENEN 1965, S. 83ff.).

Quintanilha 1939). Beide Kreuzungspartner bilden heterokaryotische Hyphen, deren Zellen je einen der beiden Kerne enthalten (Schnallenbildung). An beiden Dikaryen entstehen Fruchtkörper. Jeder Kreuzungstyp ist daher als ein Zwitter anzusehen, der als Kern-Donator und als Kern-Receptor fungieren kann. Die Kreuzungsbeziehungen werden durch Incompatibilitäts-Systeme gesteuert.

Das gleiche gilt auch für die *Ustilaginales*, denen ebenfalls die Sexualorgane fehlen und deren Gonen, die Sporidien, einen verschiedenen Kreuzungstyp besitzen. Jedes Sporidium kann bei der paarweisen Fusion sowohl seinen Kern in den Partner abgeben als auch dessen Kern aufnehmen (Holton 1942).

Die Sexualitätsverhältnisse der *Uredinales* sind mit denen der Euascomycetes vergleichbar. Die Pyknosporen (= Spermatien) sind als männliche Gameten und die Aecidienanlagen als weibliche Geschlechtsorgane aufzufassen (Lamb 1935, Mitter 1936). Da Einspormycelien stets zwittrig sind (Allen 1932), liegt bei Vorhandensein von mehreren Kreuzungstypen Incompatibilität vor.

Entsprechend den genetischen Voraussetzungen, die zur sexuellen Unverträglichkeit führen, kennt man bei Pilzen zwei verschiedene Systeme, für die wir die Bezeichnung „homogenische Incompatibilität“ und „heterogenische Incompatibilität“ eingeführt haben (Esser 1959, 1962, ausführliche Besprechung bei Esser und Kuenen 1965, S. 53ff.). Bei homogenischer Incompatibilität tritt die Zygotenbildung nicht ein, wenn die Kreuzungspartner die gleichen Incompatibilitäts-Gene besitzen. Tragen jedoch zwei sexuell unverträgliche Individuen an allen Incompatibilitäts-Loci verschiedene Allele, so liegt eine heterogenische Incompatibilität vor. Heterogenische Incompatibilität ist bisher nur zwischen den Rassen von 7 Arten (3 Ascomyceten und 4 Basidiomyceten) beobachtet worden. Alle übrigen bisher als Incompatibilität bezeichneten Erscheinungen der sexuellen Unverträglichkeit gehören zum homogenischen System (s. die folgenden Tabellen). Bei diesem System gibt es zwei verschiedene Mechanismen, je nachdem, ob die Incompatibilität durch einen oder zwei Faktoren gesteuert wird. Da im ersten Falle mindestens zwei und im zweiten Falle mindestens vier verschiedene Kreuzungstypen vorhanden sind, hat man diese Mechanismen als bipolar bzw. tetrapolar (Raper 1953) bezeichnet.

Mit eindeutiger Sicherheit sind Incompatibilitätserscheinungen nur bei den *Eumycetes* nachgewiesen, und zwar gehören diese fast ausschließlich zum homogenischen System.

Da die Genetik und Physiologie der Incompatibilität bei Pilzen an anderer Stelle in ausführlicher Form behandelt wurde (s. K. Esser und R. Kuenen 1965, Kapitel „Ontogenese“ und „Fortpflanzung“, sowie Esser und Raper 1965), beschränken wir uns hier auf eine tabellarische Übersicht zur Verbreitung der homogenischen Incompatibilität.

III. Verbreitung der homogenischen Incompatibilität bei den Eumycetes.

Vorbemerkung zu den Tabellen:

Die folgenden Tabellen sind aus den Veröffentlichungen von Whitehouse (1949a, *Euascales*; 1949b, *Holobasisdiomycetes*; 1951, *Ustilaginales*), Quintanilha und Pinto-Lopes (1950, *Holobasidiomycetes*), Craigie (1942, *Uredinales*) und Nobles *et al.* (1957, *Holobasidiomycetes*) zusammengestellt und durch Angaben aus neueren Veröffentlichungen vielfach ergänzt worden.

Der systematischen Gliederung der Tabellen liegt für die *Euascales* die von Gäumann (1949) und für die *Basidiomycetes* die von Wettstein (1935) benutzte Klassifizierung zugrunde.

Da innerhalb der *Euascales* Incompatibilitäts-Erscheinungen nur relativ wenig verbreitet sind, werden in den Tabellen zu jeder Ordnung nur die Familien angegeben, in denen incompatible Arten bekannt sind. Bei den *Basidiomycetes*, deren Arten in der Mehrzahl incompatibel sind, werden dagegen alle Familien angeführt, und zwar in ihrer systematischen Ordnung.

Alle pseudocompatiblen Arten sind mit einer hochgestellten „1" bezeichnet. Unter Pseudocompatibilität verstehen wir die Erscheinung, daß die Fruchtkörperbildung an Einspormycelien bei manchen Arten nicht auf Selbstcompatibilität beruht, sondern durch das Vorhandensein von mehreren Kernen mit verschiedenen Kreuzungstypen in derselben Asco- oder Basidiospore bedingt ist. Ein Fragezeichen in der Spalte „Incompatibilitäts-Mechanismus" und „multiple Allele" bedeutet, daß das Incompatibilitäts-System bzw. die Existenz von multiplen Allelen der betreffenden Art nicht mit Sicherheit bekannt ist. Die Literaturangaben beziehen sich auf den Autor, der zuerst die Incompatibilität der Art beschrieben hat. Werden mehrere Autoren angegeben, so haben diese die Angaben des ersten Autors entweder ergänzt oder berichtigt. Bei den *Ustilaginales*, deren Arten vielfach durch Synonyma bezeichnet werden (FISCHER 1943) und deren Incompatibilitäts-Mechanismen teilweise noch nicht klar sind, werden alle Autoren angegeben, welche die Incompatibilitäts-Erscheinungen der einzelnen Arten untersucht haben.

A. Euascales.

Die Kreuzungsbeziehungen aller incompatiblen Arten werden durch den bipolaren Mechanismus bestimmt. Multiple Allele der Incompatibilitäts-Gene sind nicht bekannt.

Gattung	Art	Autoren
	1. Ordnung. Plectascales.	
	Familie: Aspergillaceae.	
Penicillium	luteum	DERX 1925
(eine Nachuntersuchung zeigte jedoch, daß P. luteum selbstcompatibel ist		RAPER und FENELL 1952
	Familie: Ophiostomaceae.	
Ceratostomella	coerulea Münch.	MITTMANN 1932
(syn. Ophiostoma)	fimbriata (Ell. & Hals.) Ell.	OLSON 1949
	multiannulata Hedgcock & Davids.	ANDRUS 1936
	paradoxa Dade	DADE 1928
	pluriannulata Hedgcock	GREGOR 1932
	quercus Georgewitsch	MITTMANN 1932
	radicicola Bliss	BLISS 1941
	ulmi Buism.	BUISMAN 1932
Chaetomium	elatium Kunze & Schmidt	FOX 1953
	oliviaccum Cook & Ellis	FOX 1953
	2. Ordnung. Perisporales.	
	Familie: Erysyphaceae.	
Erisyphe	cichoracearum DC.	YARWOOD 1935, ALLEN 1936
	3. Ordnung. Myrangiales, incompatible Arten sind nicht bekannt.	
	4. Ordnung. Pseudosphaeriales.	
	Familie: Pseudosphaeriaceae.	
Ophiobolus	graminis Sacc. (syn. O. cariceti B. & Br.)	KIRBY 1923
	sativus (P., K. et B.) Ito & Kurib. (perfekte Form von Helminthosporium sativum P., K. et B.)	TINLINE 1951
Venturia	inaequalis (Cooke) Wint.	KEITT und PALMITER 1938
	pirinia Aderh.	LANGFORD und KEITT 1940
	Familie: Mycosphaerellaceae.	
Mycosphaerella	musicola	STOVER 1963

Gattung	Art	Autoren
5. Ordnung. Hemisphaeriales, incompatible Arten sind nicht bekannt.		
	6. Ordnung. Sphaeriales.	
	Familie: Sordariaceae.	
Bombardia	lunata Zckl.	ZICKLER 1934
Gelasinospora	autosteira nov. spec.	ALEXOPOULOS und SUN 1950
	tetrasperma Dowding[1]	DOWDING 1933
Neurospora	crassa Sh. & D.	SHEAR und DODGE 1927
	sitophila Sh. & D.	SHEAR und DODGE 1927
	tetrasperma Sh. & D.[1]	SHEAR und DODGE 1927
Podospora	anserina (Ces.) Rehm.[1] (syn. Sordaria anserina Rabenh.; syn. Pleurage anserina (Ces.) Kuntze)	DOWDING 1931 a
	curvula	RIZET 1941 a
	minuta Wint.[1] (syn. Sordaria minuta Fuckel)	PAGE 1936
	minuta (achtsporige Form)	RIZET 1940
	neglecta Niessl	RIZET 1941 b
Sordaria	brevicollis spec. nov.	OLIVE und FANTANI 1961
	heterothallis spec. nov.	FIELDS und MANIOTIS 1963
	Familie: Hypocreaceae *.	
Gibberella	cyanogena (Desm.) Sacc.	GORDON 1954
	moniliformis (Sheldon) Wineland	WINELAND 1923
Hypomyces	rosellus Alb. & Schw.	ZYCHA 1935
	solani Rke. & Berth. f. cucurbitae (syn. H. ipomeae (Hals.) Wr.)	DIMOCK 1937
Sphaeronemella	fimicola (Marchal) Pease	PEASE 1948
	7. Ordnung. Diaporthales.	
	Familie: Diaporthaceae.	
Diaporthe	perniciosa March.	CAYLEY 1931
	phaseolorum (C. & E.) Sacc. var. sojae (Lehmann) Wehmeyer	WELCH und GILMAN 1948
8. Ordnung. Clavicipitales, incompatible Arten sind nicht bekannt.		
	9. Ordnung. Pezizales.	
	Familie: Pyronemaceae.	
Ascobolus	carbonarius Karst.	BETTS 1926
	geophilus Saver	BETTS und MEYER 1938
	immersus Pers.	RIZET 1939
	magnificus Dodge	DODGE 1920
	stercorarius (Bull.) Schroet.	DOWDING 1931 b
Ascophanus	granulatus (Bull.) Speg. (syn. Humaria granulata Quél.)	GWYNNE-VAUGHAN und WILLIAMSON 1930
Patella	melaloma (Alb. & Schwein.) Seaver (syn. Lachnea melaloma [Alb. & Schwein] Sacc.)	GWYNNE-VAUGHAN 1937
	10. Ordnung. Helotiales.	
	Familie: Helotiaceae.	
Gloeotinia	temulenta (Prill. & Delacr.) Wilson & Gray.	GRIFFITHS 1958
Martinia	panamensis Whetzel	WHETZEL 1942
Sclerotinia	gladioli Massey	DRAYTON 1932
(syn. Stromatinia)	narcissi	DRAYTON und GROVES 1948
13. und 14. Ordnung. Tuberales, Laboulbeniales, incompatible Arten sind nicht bekannt.		

* *Anmerkung bei der Korrektur:* Von REICHLE und SNIDER [Phytopathology **54**, 1297 (1964)] werden von der in diese Familie gehörenden *Calonectria (Fusarium) rigidiuscula* (Berk. and Br.) 5 incompatible und 2 selbstcompatible Rassen beschrieben.

B. Holobasidiomycetes.

Gattung	Art	Incompatibilitäts-Mechanismus: unbekannt	Incompatibilitäts-Mechanismus: bipolar	Incompatibilitäts-Mechanismus: tetrapolar	Multiple Allelie	Autoren
	1. Ordnung. Hymenomycetes.					
	1. Familie: Thelephoraceae.					
Aleurodiscus	canadensis Skolko		+[1]			Skolko 1944
	polygonius v. Höhn & Litsch.			+		Kniep 1922
Corticum	bombycinum (Sommerf.) Bres.			+		Kniep 1920
	Brinkmanni Bres.		+			Terra 1953
	calceum Burt.	+				Biggs 1938
	centrifugum (Lev.) Bres.			+	+	Butler 1930
	conigenum		+			Shear und Davidson 1944
	coronilla v. Höhn & Litsch Gruppe IV		+			Biggs 1937
	coronilla v. Höhn & Litsch Gruppe I		+		+	
	coronilla v. Höhn & Litsch Gruppe III			+	+	
Corticum	effuscatum Cooke & Ellis			+	+	Nobles 1942
	evolens Schnizl.			+		Robak 1942
	hydnans (Schw.) Burt.	+				Biggs 1938
	incrustans v. Höhn & Litsch		+			Nobles 1937
	lavae Pers.	+				Biggs 1938
	polygonium Pers.	+				Biggs 1938
	porosum (B. & G.) Wakef.	+				Biggs 1938
	radiosum Fr.	+				Biggs 1938
	serum Fr.			+		Kniep 1920
	stramineum Bres.	+				Biggs 1938
	varians Kniep	+				Kniep 1922
Cytidia	floculenta Fr.			+		Terra 1953
	salicina (Fr.) Burt.			+		Biggs 1938
Gloeocystidium	roseo-cremeum Kuhn (s. Peniphora roseo-cremeum)					
Peniophora	affinis Burt.		+			Biggs 1938
	allescheri Bres.		+			Nobles 1935
	candida (Pers.) Lyman			+		Biggs 1938
	ciliata (Fr.) Bres.	+				Kniep 1922
	cinera (Pers.) Cooke			+		Vandendries 1937a
	corticalis (Bull.) Bres.	+				Kniep 1920
	farinosa Bres.			+		Biggs 1938
	incarnata (Pers.) Karst			+		Biggs 1938
	laevis (Fr.) Burt.	+				Biggs 1938
	ludoviciana Burt.		+			Biggs 1938
	merismoides Fr.		+			Nobles et al. 1957
	pubera (Fr.) Sacc.		+			Biggs 1938
	roseo-cremeum Bres.		+			Vandendries 1937a
	setigera D. C.		+			Terra 1953
	tenuis Pat.		+			Terra 1953
	versata Burt.	+				Biggs 1938
	violacea-livida (Sommf.) Bres.	+				Biggs 1938
Solenia	anomala Pat.		+			Greis 1938
	poriaeformis DC (Fr.)	+				Kniep 1922
Stereum	chailletii Pers.			+		Terra 1953
	murrayi (Berk. & Curt) Burt.	+				Biggs 1938
	purpureum (Pers.) Fr.	+				Kniep 1920
	rufum Fr.			+		Biggs 1938
	spadiceum		+			Yen 1948

Gattung	Art	Incompatibilitäts-Mechanismus			Multiple Allelie	Autoren
		unbekannt	bipolar	tetrapolar		
	2. Familie: Clavariaceae.					
Sparassis	crispa (Fr.) Wulf.		+			FRIES 1948
Typhula	erythropus Fr.			+		KNIEP 1920
	gyrans (Batsch) Fr.	+				MACDONALD 1934
	trifolii Rostr.	+				NOBLE 1937
	3. Familie: Hydnaceae.					
Hydnum	auriscalpium L. ex. Fr.			+	+	FRIES 1941
	coralloides Scop. ex. Fr.			+	+	FRIES 1941
	corrugatum Fr.			+	+	FRIES 1941
	strigosum Sw. ex Fr.			+		FRIES 1941
Irpex	pendulus (A. & S.) Fr.			+		FRIES 1941
	fuscoviolaceus Fr.			+	+	FRIES 1941
Odontia	bicolor (A. & S.) Bres.	+				BROWN 1935
	fusco-atra (Fr.) Bres.	+				BIGGS 1938
	separans (Peck.) Brown	+				BROWN 1935
	setigera (Fr.) Mill.		+			BIGGS 1938
	sudans Burt.			+		BIGGS 1938
	uda (Fr.) Bres.		+			NOBLES et al. 1957
Phlebia	merismoides Fr.		+			NOBLES et al. 1957
	strigozonata (Schw.) Burt.		+			BIGGS 1938
Radulum	aurantica Sow.		+			TERRA 1953
	casearium (Morgan) Lloyd		+			NOBLES et al. 1957
	orbiculare Fr.		+			TERRA 1953
	radula (Fr.) Nannf. (syn. R. orbiculare Grev. ex Fr.)		+		+	FRIES 1941
	4. Familie: Polyporaceae.					
Boletus	variegatus (Fr.) Sw.		?			FRIES 1948
Coriolus	(s. Polystictus)					
Favolus	alveolaris (DC. ex Fr.) Quél.			+		NOBLES et al. 1957
Fomes	applanata			+		AOSHIMA 1953
	ellisanus And.			+		NOBLES et al. 1957
	fraxineus			+		MONTGOMERY 1936
	fraxinophilus (Peck.) Sacc.			+		NOBLES et al. 1957
	ignarius (L.) Gill.			+	+	VERALL 1937
	pinicola (Sw.) Cooke		+		+	MOUNCE und MACRAE 1938
	roseus (A. & S.) Cooke		+		+	MOUNCE und MACRAE 1937
	subroseus (Weir.) Overh.		+		+	MOUNCE und MACRAE 1937
Ganoderma	applanatum			+		AOSHIMA 1953
Lenzites	abietina (Bull. ex Fr.) Fr.		+			NOBLES et al. 1957
	betulina (L.) Fr.			+	+	VANDENDRIES und BRODIE 1933a
	malaecensis Sacc. & Cub.	+				BOSE 1930
	saepiaria (Wulf.) Fr.		+		+	FRIES 1936a
	trabea (Pers.) Fr. (syn. L. termophila Falck.)		+		+	MOUNCE und MACRAE 1936
Leptoporus	(s. Polyporus)					
Leucoporus	(s. Polyporus)					
Melanopus	(s. Polyporus)					

Gattung	Art	Incompatibilitäts-Mechanismus unbekannt	bipolar	tetrapolar	Multiple Allelie	Autoren
Merulius	americanus			+		HWANG 1955
	rufus	+				YEN 1947
Polyporus	abietinus (s. Polystictus abietinus)					
	albellus Peck.			+		NOBLES et al. 1957
	adustus (Willd.) Fr.		+		+	VANDENDRIES 1934d, 1936b
	anceps Peck.		+			MOUNCE 1930
	arcularius (Batsch.) Fr.			+		VANDENDRIES 1934d, 1936c
	betulinus (Bull.) Fr.		+			MACDONALD 1937
	borealis (Wahlenb.) Fr.			+		ROBAK 1932
	brumalis (Pers.) Fr.			+		VANDENDRIES 1934d, 1936d
	compactus Overh.			+		NOBLES et al. 1957
	fumosus Pers. ex Fr.			+		NOBLES et al. 1957
	galactinus Berk.		+			NOBLES et al. 1957
	imberbis (Bull.) Quél.		+			VANDENDRIES 1933a
	obtusus Berk.		+		+	EGGERTSON 1953
	ostreiformis		+			BOSE 1934
	palustris Berk.		+		+	NOBLES 1943
	paragmenus Fr.			+		NOBLES et al. 1957
	pubescens Schum. ex Fr.			+		NOBLES et al. 1957
	resinosus Schrad. ex Fr.		+			NOBLES et al. 1957
	sanguineus			+		KAWAMURA 1942
	spumeus			+		YEN 1947
	squamosus (Huds.) Fr.	+			+	VANDENDRIES 1934d, 1936e
	supinus Sw. ex Fr.			+		NOBLES et al. 1957
	tephroleucus		+			NOBLES 1958
	tuberaster Jacq. ex Fries			+ +		VANTERPOOL und MACRAE 1951
	tukahoe	+				MOUNCE 1930
	volvatus Peck.			+		NOBLES et al. 1957
Polystictus	abietinus (Dicks.) Fr. (syn. Polyporus abietinus)			+	+	FRIES und JONASSON 1941; ROBAK 1936, 1942
	hirsutus (Wulf.) Fr.		+			BOSE 1934
	versicolor (L.) Fr.			+		KNIEP 1920
	xanthopus Fr.			+		BANERJEE und SAMADDER 1957
	zonatus Fr.			+		VANDENDRIES und BRODIE 1933b
Poria	albipellucida Baxter		+			NOBLES et al. 1957
	cinerascens (Bres.) Sacc.		+			NOBLES et al. 1957
	eupora (Karst.) Cooke			+		NOBLES et al. 1957
	microspora Overh.		+		+	NOBLES 1943
	mucida (Pers.) sensu Bourdot			+		TERRA 1953
	notata Overh.		+			NOBLES et al. 1957
	tenuis v. pulchella (Schw.) Lowe			+		NOBLES et al. 1957
Trametes	americana Overh.		+		+	MOUNCE und MACRAE 1936
	cinnabarina (Jacq.) Fr.			+	+	VANDENDRIES 1934a, FRIES 1936a
	heteromorpha (Fr.) Bres.		+			NOBLES et al. 1957
	hispida Bagl.			+		VANDENDRIES 1934a

Gattung	Art	Incompatibilitäts-Mechanismus unbekannt	bipolar	tetrapolar	Multiple Allelie	Autoren
Trametes	odorata		+			Mounce 1930
	serialis Fr.		+		+	Robak 1936, 1942; Fries 1936a, Nobles 1943
	suaveolens (L.) Fr.			+	+	Vandendries 1933a
	variiformis Peck		+			Nobles et al. 1957
	5. Familie: Agaricaceae.					
Agrocybe	(s. Naucoria, Pholiata)					
Anellaria	separata (L.) Karst.		+			Vandendries 1923a, b
	separata			+		Routien 1940
Armillaria	mucida (Schrad.) Fr.			+		Kniep 1920
Clitocybe	aurantiaca (Fr.) Wulf.			+	+	Fries 1948
	cyathiformis (Bull.) Fr.		+			Vandendries 1936e
	dicolor Pers.			+		Terra 1953
	gallinacea (Scop.) Fr.			+		Vandendries 1936e
	gigantea Sow.			+		Terra 1953
	infundibuliformis (Schaeff.) Fr.	+				Kniep 1920
	lituus (Fr.)			+		Kühner 1954a
	radicellata (syn. Collybia rhizophora)	+				Yen 1948
Collybia	asema Fr.	+				Kniep 1920
	butyracea (Bull.) Fr.	+				Kniep 1920
	cirrhata (Schum.) Fr.			+	+	Arnold 1935
	cookei (Bres.) Arn.			+		Arnold 1935
	dryophila			+		Yen 1948
	myosurus	+				Yen 1948
	tenacella Pers.			+		Yen 1948
	tuberosa (Bull.) Fr.			+	+	Kniep 1920, Arnold 1935
	velutipes (Curt.) Fr. (syn. Flammulina velutypes)			+	+	Kniep 1920, Vandendries 1923a, Zattler 1924, Funke 1924, Newton 1926a, Heldmaier 1929, Yen 1947
Conocybe	coprophila	?				Yen 1947
	pubescens Gill. (s. Galera pubescens)					
	spicula			+		Yen 1947
Coprinus	atramentarius Fr. ex Bull.	+				Chow 1934
	auricomus Pat.			+		Quintanilha et al. 1941
	boudieri Quél.		+			Quintanilha et al. 1941
	clavatus (Batt.) Quél.	+				Lendner 1920
	comatus Fr.		+		+	Mounce 1921, Brunswik 1924
	cothurnatus God.		+			Lange 1948
	curtus Kalchbr.		+			Brunswik 1924
	curtus Kalchbr.			+		Newton 1926a
	deliquescens Fr.		+			Brunswik 1924
	dissemninatus (Pers.) Quél. (s. Psathyrella disseminata)					
	domesticus (Pers.) Fr.		+			Chow 1934

Gattung	Art	Incompatibilitäts-Mechanismus unbekannt	bipolar	tetrapolar	Multiple Allelie	Autoren
Coprinus	ephemerus (Bull.) Fr.		+			Brunswik 1924
	ephemerus f. bisporus Sass		+[1]			Sass 1929
	erethistes Heim			+		Quintanilha et al. 1941
	extinctorius (Bull.) Fr.			+		Quintanilha 1941, 1944
	fimetarius (L.) Fr. (s. C. lagopus Fr.)			+	+	Bensaude 1918, Kniep 1922, Quintanilha 1933, 1935, 1941
	fimetarius Fr. var. macrorhizus Pers. (s. C. macrorhizus)					
	friesii Quél.		+		+	Brunswik 1924
Coprinus	funarium Métrod.			+		Quintanilha et al. 1941
	gonophyllus		+			Yen 1947
	hiascens Fr.		+			Quintanilha 1941, 1944
	impatiens		+			Yen 1947
	lagopus Fr. sensu Buller (syn. C. fimetarius (L.) Fr.)			+	+	Mounce 1922, Hanna 1925, 1928, Newton 1926b, Buller 1931
	macrorhizus (Pers.) Rea.			+	+	Dickson 1936
	macrorhizus (Pers.) Rea.		+			Routien 1940
	micaceus (Bull.) Fr.			+	+	Brunswik 1924, Vandendries 1926 1927a—c, 1929a, b, Vandendries und Robyn 1929, Routien 1940
	niveus (Pers.) Fr.		+	+	+	Mounce 1922, Brunswik 1924
	papillatus Fr. ex Batsch.					Brunswik 1923
	picaceus (Bull.) Fr.			+	+	Brunswik 1924, Chow 1934
	plagioporus	+[1]				Lamoure 1957
	plicatilis Fr.			+		Routien 1940
	radians (Desm.) Fr.		+			Brunswik 1924, Vandendries 1924, 1925a—d
	radians (Desm.) Fr.			+		Quintanilha et al. 1941
	rostrupianus Hans.		+		+	Newton 1926a
	semilantanus Peck.			+		Routien 1940
	sphaerosporus Dicks.		+			Dickson 1934
	stercorarius (Bull.) Fr.	+				Kniep 1920
	tergiversans Fr.		+			Vandendries 1930
	urticaecola (Berk. & Br.) Bull.		+			Hanna 1939
	velaris Fr.		+			Brunswik 1924
Cortinellus	berkeleyana Ito & Imai			+		Nisikado und Yamauti 1935
	shiitake P. Henn.		+			Oikawa 1935
Crinipellis	stipitarius Fr.			+		Terra 1953
Deconia	atrorufa			+		Pinto-Lopes 1948
	coprophila (s. Psilocybe coprophila Fr.)					

Gattung	Art	Incompatibilitäts-Mechanismus: unbekannt	Incompatibilitäts-Mechanismus: bipolar	Incompatibilitäts-Mechanismus: tetrapolar	Multiple Allelie	Autoren
Deconia	crobula (Fr.) Romagn. (s. Naucoria crobulus (Fr.) Quél.)					
	inquilina (Fr.) Pat. (s. Naucoria inquilina (Fr.) Quél.)					
Drosophila	(s. Hypholoma)					
Flammula	carbonaria Fr.			+		QUINTANILHA et al. 1941
	conissans sensu Lange			+		YEN 1947
	gummosa			+		YEN 1947
	lenta Pers.			+		TERRA 1953
Galera	mycenopsis			+		KÜHNER 1950
	pubescens Gill.			+		VANDENDRIES 1936e
	pubescens f. bispora	+[1]				VANDENDRIES 1936f, 1937b
	sphagnorum Pers.			+		TERRA 1953
	tenera (Schaeff.) Fr.			+	+	OIKAWA 1939
	tenera f. semiglobata Kühner	+				VANDENDRIES 1936f
	tenera f. bispora Sass.	+[1]				SASS 1929
Geopetalum	carbonarium (A. et S.) Pat. (syn. Cantharellus carbonarium)			+		KÜHNER 1954b
Hebeloma	circinans			+		YEN 1947
	truncata sensu Lange			+		TERRA 1953
Hypholoma	appendiculata Fr.			+		QUINTANILHA et al. 1941
	atomata Fr.	+			+	QUINTANILHA 1941, 1944
	badia Romagn.			+	+	QUINTANILHA 1941, 1944
	candolleanum Fr.			+		KNIEP 1920, VANDENDRIES und BRODIE 1933b
	capnoides Fr.			+		FUNKE 1924
	caudata Fr.			+		QUINTANILHA 1941, 1944
	dispersum			+		YEN 1947
	exalbicans Romagn.			+		QUINTANILHA et al. 1941
	fasciculare (Huds,) Fr.			+		KNIEP 1920
	fibrillosa (Pers.) Fr.			+		QUINTANILHA et al. 1941
	gracilis (Pers.) Fr.			+		QUINTANILHA et al. 1941
	hydrophilum (Bull.) Fr.			+		KNIEP 1920
	macrescibilis Britz.			+		QUINTANILHA et al. 1941
	melantinum Fr.			+		VANDENDRIES und BRODIE 1933b
	prona Fr.			+	+	QUINTANILHA 1941, 1944
	semivestita Bk.		+			QUINTANILHA 1941, 1944
	subatrata (Bats.) Fr.		+			QUINTANILHA et al. 1941
	sublateritium (Schaeff.) Fr.			+	+	VANDENDRIES 1934c
	subtilis Fr.			+	+	QUINTANILHA 1941, 1944

Gattung	Art	Incompatibilitäts-Mechanismus			Multiple Allelie	Autoren
		unbekannt	bipolar	tetrapolar		
Hypholoma	trepida Fr.	+			+	QUINTANILHA 1941, 1944
	virginea Lange		+			QUINTANILHA 1941, 1944
Lentinellus	ursinus Fr.			+		TERRA 1953
Lentinus	degener Kalchbr.		+			TERRA 1953
	triginus (Bull.) Fr.			+		QUINTANILHA et al. 1941
Marasmius	alliaceus Jacq.	+				KNIEP 1920
	androsaceus L.			+		TERRA 1953
Marasmius	cohaerens Pers.		+			VANDENDRIES 1936f
	elongatipes Peck.			+	+	ARNOLD 1935
	epiphyllus Pers.			+		TERRA 1953
	oreades (Bolt.) Fr.	+				BULLER 1931
	peronatus (Bolt.) Fr.			+		KNIEP 1928
	rotula Scop.			+		TERRA 1953
	scorodonius Fr.			+		TERRA 1953
Mycena	aetites			+		YEN 1947
	amicita Fr.			+		TERRA 1953
	avanacea			?		YEN 1947
	calopus Fr.	+				KNIEP 1920
	citrinomarginata Gill.			+		TERRA 1953
	crocata Schrad.			+		TERRA 1953
	elegans Pers.			+		TERRA 1953
	epipterygia			?		YEN 1947
	flavoalba Fr.			+		TERRA 1953
	galericulata Fr. ex Scop.	+				VANDENDRIES 1937a, b
	galopus Pers.			+		BOTHE 1935
	haematopus Pers.			+		TERRA 1953
	maculata			+		YEN 1947
	parabolica Fr.	+				KNIEP 1920
	polygramma Bull.			+		BOTHE 1935
	rubromarginata Fr.			+		VANDENDRIES 1936f
	tenella Fries			+[1]		KÜHNER und TERRA 1955
	vitilis Fr. f. tetraspora	+				VANDENDRIES 1937a, b
	vulgaris			+		YEN 1947
	zephyrus			+		YEN 1948
Naucoria	crobolus (Fr.) Quél. (syn. Deconia crobola)		+			QUINTANILHA et al. 1941
	inquilina (Fr.) Quél. (syn. Deconia inquilina)			+		QUINTANILHA et al. 1941
	pediades Fr.		+			VANDENDRIES 1936f
	semiorbicularis (Bull.) Fr. (syn. N. pediades Fr. sec. Quél.)		+		+	ROUTIEN 1940, QUINTANILHA et al. 1941
	semiorbiculares (Bull.) Fr. f. bispora Sass	+[1]				SASS 1929
Omphalia	mairei			?		YEN 1947
	maura Fr.		+		+	VANDENDRIES 1937a, b
Panaeolina	foenisecii (Fr. ex. Pers.) Maire (s. Psilicybe foenisecii)					
Panaeolus	accuminatus ssRick.		+			YEN 1947
	campanulatus (Fr.) Quél.			+	+	VANDENDRIES 1933b
	fimicola Fr.	+				VANDENDRIES 1923a
	papilionaceus (Fr.) Quél.			+		VANDENDRIES 1931
	retirugis Fr.			+		QUINTANILHA 1944

Gattung	Art	Incompatibilitäts-Mechanismus unbekannt	bipolar	tetrapolar	Multiple Allelie	Autoren
Panaeolus	separatus (L.) Fr. (s. Anellaria separata)					
	subalteatus (Bull.) Fr.			+		Brodie 1935a
Panellus	mitis (syn. Pleurotus mitis)			+		Yen 1948
Panus	stipticus (Bull.) Fr.			+	+	Vandendries 1936a, Macrae 1937, 1942
Pholiota	adiposa Fr.		+		+	Mounce 1926
	aurivella (Batsch.) Fr.		+			Martens und Vandendries 1933
Pholiota	cylindracea			+		Vandendries 1934d
	dura (Bolt.) Fr.			+		Quintanilha 1941, 1944
	mutabilis (Schaeff.) Fr. (syn. Kueneromyces mutabilis)			+	+	Vandendries und Brodie 1933b
	polychroa (Berk.) Smith & Brodie			+		Smith und Brodie 1935
	praecox	+				Kniep 1920
	sphaleromorpha (Bull.) Fr.		+			Quintanilha et al. 1941
Pleurotus	columbinus Quél.			+		Vandendries 1932, 1933a
	cornucopioides P.			+		Vandendries 1937a, b
	corticatus Fr.			+		Kaufert 1935, 1936
	nidulans Pers.			+		Vandendries 1933a
	ostreatus (Jacq.) Fr.			+		Vandendries 1933b
	pinsitus Fr.			+		Vandendries 1934b
Psathyra	pennata Fr.			+		Vandendries 1936f
Psathyrella	bifrons			+		Yen 1947
	disseminata (Pers.) Fr. (syn. Pseudocoprinus disseminatus (Fr. ex. Pers.) Kühner		+		+	Quintanilha 1941, 1944
	fragilissima			+		Yen 1947
	microrhiza			+		Yen 1947
	pseudobifrons Romagn.			+		Yen 1947
Psilocybe	coprophila Fr.		+			Gilmore 1926, Brodie 1935b
	coprophila Fr.			+		Vandendries 1937a, b
	crobula	+				Yen 1947
	foenisecii (Pers.) Fr.			+		Quintanilha et al. 1941
	semilanceata Fr.	+				Terra 1953
	subviscida	+				Routien 1940
Ripartites	tricholoma A. et S.			+		Terra 1953
Rhodopaxillus	nudus					Yen 1947
Schizophyllum	commune Fr.			+	+	Kniep 1920, 1922 1923, 1928, 1930; Zattler 1924, Heldmaier 1929, Harnack 1931
	fasciatum Pat.			+	+	Raper 1960
Stropharia	merdaria Fr.		+			Quintanilha et al. 1941
	semiglobata Batsch.		+		+	Oikawa 1939
	umbonatescens Peck.		+[1]			Terra 1955

Gattung	Art	Incompatibilitäts-Mechanismus: unbekannt	bipolar	tetrapolar	Multiple Allelie	Autoren
Trogia	crispa Pers.	+				KNIEP 1922
Xeromorphalina	campanella Batsch.			+		TERRA 1953

2. Ordnung: Gasteromycetes.

1. Unterordnung: Plectobasidiales.

Gattung	Art	unbekannt	bipolar	tetrapolar	Multiple Allelie	Autoren
Scleroderma	aurantium Pers.	+				FRIES 1948
Sphaerobolus	cyclophorus	?				LORENZ 1933
	grandis	?				LORENZ 1933
Sphaerobolus	javanensis	?				LORENZ 1933
	stellatus (Pers.) Tode.			+		FRIES 1948
	tubulosus	?				LORENZ 1933

2. Unterordnung: Eugasteromycetes.

1.—5. Familie. Hymenogastraceae, Hysterangiaceae, Secotiaceae, Podaxaceae, Lycoperdaceae; incompatible Arten sind nicht bekannt.

6. Familie: Nidulariaceae.

Gattung	Art	unbekannt	bipolar	tetrapolar	Multiple Allelie	Autoren
Crucibulum	vulgare Tul.			+	+	FRIES 1936b, 1943
Cyathus	stercorarius			+		BRODIE 1948
	striatus Pers.			+	+	FRIES 1936b, 1940
	vernicosus			+		BRODIE 1949
Nidularia	candida			+		BRODIE 1951
	farcta (Pers.) Fr.			+		FRIES 1948
	microcarpa			+		BRODIE 1951
Mycocalia	denudata (Fr.) J. T. Palmer *		+		+	BURNETT und BOULTER 1963

7. und 8. Familie: Clathraceae, Phallaceae; incompatible Arten sind nicht bekannt.

3. Ordnung: Exobasidiales; incompatible Arten sind nicht bekannt.

4. Ordnung: Darcryomycetales.

1. Familie: Caloceraceae.

Gattung	Art	unbekannt	bipolar	tetrapolar	Multiple Allelie	Autoren
Calocera	cornea Batsch.	+[1]				VANDENDRIES 1937a, b
Femsjonia	luteo-alba	+				YEN 1947

2. Familie: Dacryomycetaceae.

Gattung	Art	unbekannt	bipolar	tetrapolar	Multiple Allelie	Autoren
Dacryomyces	deliquescens	+				YEN 1947

5. Ordnung: Tulasnellales; incompatible Arten sind nicht bekannt.

C. Phragmobasidiomycetes (Protobasidiomycetes).

1. Ordnung: Tremellales.

Familie: Tremellaceae.

Gattung	Art	unbekannt	bipolar	tetrapolar	Multiple Allelie	Autoren
Exidia	glandulosa (Bull.) Fr.		+		+	BARNETT 1937
	nucleata (Schw.) Burt.		+			BARNETT 1937
	recisa (Ditm.) Fr.		+		+	BARNETT 1937
	saccharina Fr.		+		+	BARNETT 1937
Sebacina	calcea (Pers.) Bres.	+				KNIEP 1922
Tremella	mesenterica Fr.			+	?	BANDONI 1963

* Fakultativ pseudocompatibel.

Gattung	Art	Incompatibilitäts-Mechanismus unbekannt	bipolar	tetrapolar	Multiple Allelie	Autoren
	2. Ordnung: Auriculariales.					
	Familie: Auriculariales.					
Auricularia	auriculae-judae (L. ex Fr.) Schroet.		+		+	BARNETT 1937
	auriculae-judae (L. ex Fr.) Schroet.			+		BANERJEE 1956
Pilacre	faginea (Fr.) B. & Br. [syn. Phleogena faginea (Fr.) Link.]	+	?			SHEAR und DODGE 1925, BECKWIRTH 1929

C. Phragmobasidiomycetes (Hemibasidiomycetes).

Gattung	Art	Autoren
	3. Ordnung: Uredinales.	
	Die Kreuzungsbeziehungen aller incompatiblen Arten werden durch den bipolaren Mechanismus bestimmt. Multiple Allele der Incompatibilitäts-Gene sind nicht bekannt.	
	1. Familie: Pucciniaceae.	
Endophyllum	sempervivi (A. & S.) de Bary	ASHWORTH 1931
Gymnosporangium	globosum Farl.	MILLAR 1932
	haraeanum Syd.	KAWAMURA 1934
	juniperi-viginianae Schw.	MILLAR 1932
Phragmidium	speciosum (Fr.) Cooke	BROWN 1940
Puccinia	anomala Rostr.	D'OLIVEIRA 1939
	coronata Corda	ALLEN 1930
	graminis Pers.	CRAIGIE 1927 b
	helianthi Schw.	CRAIGIE 1927 a
	phragmitis (Schum.) Korn.	LAMB 1935
	sorghi Schw.	CUMMINS 1932
	triticina Erikss.	Allen 1931
Uromyces	appendiculatus (Pers.) Fries	ANDRUS 1931
	fabae (Pers.) de Bary	BROWN 1940
	graminis (Niessl.) Diet.	D'OLIVEIRA 1938
	trifolii hybridi (W. H. Davis) Arth.	BROWN 1940
	vignae Barclay	ANDRUS 1931
	2. Familie: Endophyllaceae: incompatible Arten sind nicht bekannt.	
	3. Familie: Melampsoraceae.	
Cronartium	ribicola J. C. Fischer	PIERSON 1933
Melampsora	lini (Pers.) Lev.	ALLEN 1934

Gattung	Art	Incompatibilitäts-Mechanismus unbekannt	bipolar	tetrapolar	Multiple Allelie am Locus A	B	Autoren
	4. Ordnung. Ustilaginales.						
	1. Familie. Ustilaginaceae.						
Sphacelotheca	cruenta (Kühn) Potter		+				RODENHISER 1932, 1934
	destruens (Schlecht.) Stev. & A. G. Johns. (syn. S. panimiliacei (Pers.) Bubák		+				Bauch 1934

Gattung	Art	Incompatibilitäts-Mechanismus: unbekannt	Incompatibilitäts-Mechanismus: bipolar	Incompatibilitäts-Mechanismus: tetrapolar	Multiple Allelie am Locus: A	Multiple Allelie am Locus: B	Autoren
	reiliana (Kühn) Clint. (syn. Sorosporium reilianum (Kühn) McAlpine			+			HANNA 1929, TYLER und SHUMWAY 1935, VAHEEDUDDIN 1936
	schweinfurthiana (Thüm.) Sacc.			+			BAUCH 1932a
	sorghi (Link.) Clint.		?	?			RODENHISER 1932, 1934, ISENBECK 1935, TYLER 1938
Ustilago	anomala J. Kunze		+				KNIEP 1926
	avenae (Pers.) Rostr. (syn. U. perennans Rostr., syn. U. nigra Tapke)		+				KNIEP 1926, KNIEP 1928, HANNA und POPP 1930, HOLTON 1931, HOLTON und FISCHER 1941
	bromovira (Tul.) Fisch. v. Waldh. (syn. U. bullata Berk.)		+				BAUCH 1925, FISCHER 1940b
	cardui Fisch. v. Waldb.		+				KNIEP 1926
	decipiens		+				HÜTTIG 1933
	grandis Fr.		+				KNIEP 1926
	hordei (Pers.) Lagerh. (syn. U. kolleri Wille., syn. U. levis (Kellerm. & Swingle) Magn.)		+				KNIEP 1926, DICKINSON 1927, 1931; HANNA und POPP 1930, HOLTON 1931, POPP u, HANNA 1935
	hypodytes (Schlecht.) Fr.		?	?	?		BAUCH 1934
	longissima (Schlecht.) Tul. longissima var. macrospora Davies			+	?	+	BAUCH 1923, 1930, 1931, 1934; KNIEP 1926, KÄMMERLING 1929
Ustilago	maydis (DC.) Corda			+		+	STAKMAN und CHRISTENSEN 1927, HANNA 1929, EDDINS 1929, CHRISTENSEN 1929, 1931; SLEUMER 1932, BAUCH 1932b, 1934, ROWELL, 1955
	scabiosae (Sow.) Wint.		+				KNIEP 1921
	scorzonerae Schroet.		+				BAUCH 1932b
	striaeformis (Westd.) Niessl.		+				FISCHER 1940a
	succisae Magn.		+				KNIEP 1928
	tragoponis-pratensis (Pers.) Roussel.		+				KNIEP 1926
	tritici (Pers.) Rostr. (syn. U. nuda (Jens.) Rostr.)		+				KNIEP 1926, THREN 1937, TAPKE 1943
	utriculosa (Nees.) Tul.		+				KNIEP 1926
	vinosa Tul.		+				KNIEP 1926
	violacea (Pers.) Fuckel		+				KNIEP 1919, ZILLIG 1921, BAUCH 1922, GOLDSCHMIDT 1928, RODENHISER 1934
	2. Familie: Tilletiaceae.						
Tilletia	caries (DC.) Tul. (syn. T. tritici (Bjerk,) Wint.)		?	?			FLOR 1932, HANNA 1934, BECKER 1936
	foetida (Wallr.) Liro. (syn. T. levis Kühn)		?	?			FLOR 1932

Literatur.

Alexopoulos, C. J., and S. H. Sun: A new species of *Gelasinospora.* Mycologia (N. Y.) **42**, 723–734 (1950). — Allen, C. E.: Sex-inheritence and sex-determination. Amer. Naturalist **66**, 97—107 (1932). — Allen, R. F.: Heterothallism in *Puccinia coronata,* Science **72**, 536 (1930). — Heterothallism in *Puccinia triticina.* Science **74**, 462—463 (1931). — A cytological study of heterothallism in flax rust. J. agricult. Res. **49**, 765—791 (1934). — A cytological study of *Erisyphe polygoni* on *Delphinium.* J. agricult. Res. **53**, 801—818 (1936). — Andrus, C. F.: The mechanisms of sex in *Uromyces appendiculatus* and *U. vignae.* J. agricult. Res. **42**, 559—587 (1931). — Cell relation in the perithecium of *Ceratostomella multiannulata.* Mycologia (N. Y.) **28**, 133—153 (1936). — Aoshima, K.: Sexuality of *Elfvingia applanata (Fomes applanata).* Ngaoa **3**, 4—11 (1953). — Arnold, J. D.: A comperative study of certain species of *Marasmius* and *Collybia* in culture. Mycologia (N. Y.) **27**, 388—417 (1935). — Ashworth, D.: *Puccinia malvacearum* in monospoidial culture. Trans. Brit. mycol. Soc. **16**, 177—202 (1931).

Bandoni, R. J.: Conjugation in *Tremella mesenterica.* Canad. J. Bot. **41**, 467—474 (1963). — Banerjee, S.: Heterothallism in *Auricularia auriculajudae* (Linn.) Schroet. Sci. and Culture **21**, 549—550 (1956). — Banerjee, S., and K. R. Samadder: Studies on heterothallism. II. *Polystictus xanthopus* Fr. Sci. and Culture **22**, 578—579 (1957). — Barnett, H. L.: Studies in the sexuality of the Heterobasidiae. Mycologia (N. Y.) **29**, 626—649 (1937). — Bauch, R.: Kopulationsbedingungen und sekundäre Geschlechtsmerkmale bei *Ustilago violacea.* Biol. Zbl. **42**, 9—38 (1922). — Über *Ustilago longissima* und ihre Varietät *macrospora.* Z. Bot. **15**, 241—279 (1923). — Untersuchungen über die Entwicklungsgeschichte und die Sexualphysiologie der *Ustilago bromivora* und *Ustilago grandis.* Z. Bot. **17**, 129—177 (1925). — Über multipolare Sexualität bei *Ustilago longissima.* Arch. Protistenk. **70**, 417—466 (1930). — Geographische Verteilung und funktionelle Differenzierung der Faktoren bei der multipolaren Sexualität von *Ustilago longissima.* Arch. Protistenk. **75**, 101—132 (1931). — *Sphacelotheca Schweinfurthiana,* ein neuer multipolar sexueller Brandpilz. Ber. dtsch. bot. Ges. **50**, 17—24 (1932a). — Die Sexualität von *Ustilago Scorzonerae* und *Ustilago Zeae.* Phytopath. Z. **5**, 315—321 (1932b). — Über Kreuzungen zwischen bipolar und multipolar sexuellen Brandpilzarten. Z. indukt. Abstamm.- u. Vererb.-Lehre **67**, 242—245 (1934). — Becker, T.: Untersuchungen über Sexualität bei *Tilletia tritici* (Bjerk.) Wint. im Rahmen der Immunitätszüchtung. Phytopath. Z. **9**, 187—228 (1936). — Beckwirth, A. M.: *Pilacre faginea* proves to be a heterothallic fungus. Bull. Torrey bot. Club **56**, 359—360 (1929). — Bensaude, M.: Recherches sur le cyle évolutif et la sexualité chez les Basidiomycetes. Thesis Nemours (Paris) 156 S. 1918. — Betts, E. M.: Heterothallism in *Ascobolus carbonarius.* Amer. J. Bot. **13**, 427—432 (1926). — Betts, E. M., and S. L. Meyer: Heterothallism in *Ascobolus geophilus* Seaver. Amer. J. Bot. **25** Suppl. to No. 10 (1938). — Biggs, R.: The species concept in *Corticum coronilla.* Mycologia (N. Y.) **29**, 686—706 (1937). — Cultural studies in the Theleophoraceae and related fungi. Mycologia (N. Y.) **30**, 64—78 (1938). — Bliss, D. E.: A new species of *Ceratostomella* on the date palm. Mycologia (N. Y.) **33**, 468—482 (1941). — Bose, S. R.: Biology of wood-rotting fungi common in forest areas. Linnean Soc. J. Bot. **98** (323), 417—438 (1930). — Sexuality of *Polyporus ostreiformis* Berk. and *Polystictus hirsutus* Fr. Cellule **42**, 249—266 (1934). — Bothe, F.: Genetische Untersuchungen über die Lichtentwicklung der Hutpilze. I. Arch. Protistenk. **85**, 369—383 (1935). — Brieger, F.: Selbststerilität und Kreuzungssterilität im Pflanzenreich und Tierreich. Berlin: Springer 1930. — Brodie, H. J.: The heterothallism of *Panaeolus subbalteatus* Berk., a sclerotium-producing agaric. Canad. J. Res. **12**, 657—660 (1935a). — The oidia of *Psilocybe coprophila* and the pairing reactions of monosporous mycelia. Canad. J. Res. **12**, 661—667 (1935b). — Tetrapolarity and unilateral diploidisation in the bird's nest fungus *Cyathus stercoreus.* Amer. J. Bot. **35**, 312—320 (1948). — *Cyathus vernicosus,* another tetrapolar bird's nest fungus. Mycologia (N. Y.) **41**, 652—659 (1949). — Two heterothallic species of the genus Nidula. Mycologia (N. Y.) **43**, 329—337 (1951). — Brown, A. M.: The sexual behavior of several plant rusts. Canad. J. Res., C **18**, 18—25 (1940). — Brown, C. A.: Morphology and biology of some species of *Odontia.* Bot. Gaz. **96**, 640—675 (1935). — Brunswik, H.: Über die Sexualitätsverhältnisse bei den Basidiomyceten. Verh. zool.-bot. Ges. Wien **73**, 153—154 (1923). — Untersuchungen über die Geschlechts- und Kernverhältnisse bei der Hymenomycetengattung *Coprinus.* Bot. Abh. **5**, 152 S. (1924). — Buisman, C.: *Ceratostomella ulmi,* de geslachtelijke vorm van *Graphium ulmi* Schwarz. Tijdschr. Pl.-Ziekt. **38**, 1—8 (1932). — Buller, A. H. R.: Researches on Fungi. Vol. IV. London 1931. — Burgeff, H.: Untersuchungen über Sexualität und Parasitismus bei Mucorineen. I. Bot. Abh. **4**, 135 S. (1924). — Burnett, J. H., and M. E. Boulter: The mating systems of fungi. II. Mating systems of the Gasteromycetes *Mycocalia denudata* and *M. duriaena.* New Phytol. **62**, 217—236 (1963). — Butler, L. F.: *Corticium centrifugum,* a heterothallic pathogene of apples. J. agricult. Res. **41**, 269—294 (1930).

Cayley, D. M.: The inheritance of the capacity for showing mutual aversion between monospore-mycelia of *Diaporthe perniciosa* (Marchal.). J. Genet. **24**, 1—63 (1931). — Chow, C. H.: Contribution á l'étude du développpement des Cor pins. Botaniste **26**, 89—234 (1934).— Christensen, J. J.: Mutation and hybridisation in *Ustilago zeae*. Part II. Hybridisation. Techn. Bull. Minn. Agric. Exp. Sta. No **65**, 89—115 (1929). — Studies in the genetics of *Ustilago zeae*. Phytopath. Z. **4**, 129—188 (1931). — Craigie, J. N.: Experiments on sex in rust fungi. Nature (Lond.) **120**, 116—117 (1927a). — Discovery of the function of pycnia in the rust fungi. Nature (Lond.) **120**, 765—767 (1927b). — Heterothallism in the rust fungi and its significance. Trans. roy. Soc. Can., Sect. V, **36**, 19—40 (1942). — Cummins, G. B.: Heterothallism in corn rust and effect of filtering the pycnial exsudate. Phytopathology **22**, 751—753 (1932).

Dade, H. A.: *Ceratostomella paradoxa*, the perfect stage of *Thielaviopsis paradoxa* (de Seynes) von Höhnel. Trans. brit. mycol. Soc. **13**, 184—194 (1928). — Derx, H. G.: L'héterothallie dans le genre *Penicillium* (note préliminaire). Bull. Soc. mycol. France **41**, 375—381 (1925). — Dickinson, S.: Experiments on the physiology and genetics of smut fungi. Hyphal-fusion. Proc. roy. Soc. B **101**, 126—136 (1927). — Experiments on the physiology and genetics of smut fungi. Cultural characters. II. The effect of certain external conditions on their segregation. Proc. roy. Soc. B **108**, 395—423 (1931). — Dickson, H.: Studies in *Coprinus sphaerosporus*. I. The pairing behavior and the characteristics of various haploid and diploid strains. Ann. of Bot. **48**, 527—547 (1934). — Observations on inheritance in *Coprinus macrorhizus* (Pers.) Rea. Ann. of Bot. **50**, 719—734 (1936). — Dimock, A. W.: Observations on sexual relations in *Hypomyces Ipomeae*. Mycologia (N. Y.) **29**, 116—127 (1937). — Dodge, B. O.: The life history of *Ascobolus magnificus*. Mycologia (N. Y.) **12**, 115—134 (1920). — D'Oliveira, B.: New hosts for the aecidial stage of *Uromyces graminis* (Niessl.) Diet. Bol. Soc. Brot. Sér. **2**, 13, 81—89 (1938). — Studies on *Puccinia anomala* Rost. I. Physiologic races on cultivated barleys. Ann. appl. Biol. **26**, 56—82 (1939). — Dowding, E. S.: The sexuality of the normal, giant and dwarf spores of *Pleurage anserina* (Ces.) Kuntze. Ann. of Bot. **45**, 1—14 (1931a). — The sexuality of *Ascobolus stercorarius* and the transportation of the oidia by mites and flies. Ann. of Bot. **45**, 621—637 (1931b). — *Gelasinospora*, a new genus of Pyrenomycetes with pitted spores. Canad. J. Res. **9**, 294—304 (1933). — Drayton, F. L.: The sexual function of the microconidia in certain Discomycetes, Mycologia (N. Y.) **24**, 345—348 (1932). — Drayton, F. L., and J. W. Groves: A new heterothallic *Stromatinia*. Trans. Brit. mycol. Soc. **31**, 356 (1948).

Eddins, A. H.: Pathogenicity and cultural behavior of *Ustilago zeae* (Beckm.). Ung. from different localities. Phytopathology **19**, 885—916 (1929). — Eggertson, E.: An estimate of the number of alleles at the loci for heterothallism in a local concentration of *Polyporus obtusus* Berk. Canad. J. Bot. **31**, 750—759 (1953). — Esser, K.: Die Incompatibilitätsbeziehungen zwischen geographischen Rassen von *Podospora anserina* (Ces.) Rehm. III. Untersuchungen zur Genphysiologie der Barragebildung und Semi-Incompatibilität. Z. Vererb.-Lehre **90**, 445—456 (1959). — Die Genetik der sexuellen Fortpflanzung bei den Pilzen. Biol. Zbl. **81**, 161—172 (1962). — Esser, K., u. R. Kuenen: Genetik der Pilze. Berlin-Heidelberg-New York: Springer 1965. — Esser, K., and J. R. Raper: Incompatibility in fungi. Berlin-Heidelberg-New York: Springer 1965.

Fields, W. G., and J. Maniotis: Some cultural and genetic aspects of a new heterothallic *Sordaria*. Amer. J. Bot. **50**, 80—85 (1963). — Fischer, G. W.: Fundamental studies of the stripe smut of grasses (*Ustilago striaeformis*) in the Pacific North-west. Phytopathology **30**, 93—118 (1940a). — Two cases of haplo-lethal deficiency in *Ustilago bullata* operative against saprophytism. Mycologia (N. Y.) **32**, 275—289 (1940b). — Some evident synonymous relationships in certain graminicolous smut fungi. Mycologia (N. Y.) **35**, 610—619 (1943).— Flor, H. H.: Heterothallism and hybridisation in *Tilletia tritici* and *T. levis*. J. agricult. Res. **44**, 49—58 (1932). — Fox, R. A.: Heterothallism in *Chaetomium*. Nature (Lond.) **172**, 165—166 (1953). — Fries, N.: Über die Sexualität einiger Polyporaceen. Svensk. bot. Tidskr. **30**, 355—361 (1936a). — *Crucibulum vulgare* Tul. und *Cyathus striatus* Pers., zwei Gasteromyceten mit tetrapolarer Geschlechtsverteilung. Bot. Notiser **1936** (1936b). — Researches into the multipolar sexuality of *Cyathus striatus* Pers. Symb. bot. Upsal. **4**, 1—39 (1940). — Über die Sexualität einiger Hydnaceen. Bot. Notiser **1941**, 285—300 (1941). — Über das Vorkommen von geographischen Rassen bei *Crucibulum vulgare* Tul. Arch. Mikrobiol. **13**, 182—190 (1943). — Heterothallism in some Gasteromycetes and Hymenomycetes. Svensk. bot. Tidskr. **42**, 158—168 (1948). — Fries, N., u. L. Jonasson: Über die Interfertilität verschiedener Stämme von *Polyporus abietinus* (Dicks.)| Fr. Svensk. bot. Tidskr. **35**, 177—193 (1941). — Funke, G. L.: Über die Isolierung von Basidiosporen mit dem Mikromanipulator nach Janse und Péterfi. Z. Bot. **16**, 619—623 (1924).

Gäumann, E.: Die Pilze, Grundzüge ihrer Entwicklungsgeschichte und Morphologie. Basel 1949. — Gilmore, K. A.: Culture studies of *Psilocybe coprophila*. Bot. Gaz. **81**, 419—433 (1926). — Goldschmidt, V.: Vererbungsversuche mit den biologischen Arten des Antherenbrandes (*Ustilago violacea* Pers.). Z. Bot. **21**, 1—90 (1928). — Gordon, W. L.: Geographical distribution of mating types in *Giberella cyanogena* (Desm.) Sacc. Nature (Lond.) **173**, 505—506 (1954). — Gregor, M. J. F.: A study of heterothallism in *Ceratostomella pluriannulata* Hedgcock. Ann. mycol. (Berl.) **30**, 1—9 (1932). — Greis, H.: Entwicklungsgeschichtliche Untersuchungen an Basidiomyzeten. IV. Entwicklungsgeschichte von *Solenia anomala* Pat. Jb. wiss. Bot. **87**, 244—266 (1938). — Griffiths, E.: Sexual reproduction and variation in *Gloeotinia temulenta* (Prill & Delacr.) Wilson & Gray. Trans. Brit. mycol. Soc. **41**, 461—482 (1958). — Gwynne-Vaughan, H. C. I.: Contributions to the study of *Lachnea melaloma*. Ann. of Bot., N.S. **1**, 99—106 (1937). — Gwynne-Vaughan, H. C. I., and H. S. Williamson: Contributions to the study of *Humaria granulata* Quel. Ann. of Bot. **44**, 127—145 (1930).

Hanna, W. F.: The problem of sex in *Coprinus lagopus*. Ann. of Bot. **39**, 431—457 (1925). — Sexual stability in monosporous mycelia of *Coprinus lagopus*. Ann. of Bot. **42**, 379—389 (1928). — Studies in the physiology and cytology of *Ustilago zeae* and *Sorosporium reilianum*. Phytopathology **19**, 415—442 (1929). — The physiology of the fungi causing bunt of wheat. Proc. V. Pacif. Sci. Congr. 1934, S. 3195—3204. — *Coprinus urticaecola* on stems of marquis wheat. Mycologia (N. Y.) **31**, 250—257 (1939). — Hanna, W. F., and W. Popp: Relationship of the oat smuts. Nature (Lond.) **126**, 843—844 (1930). — Harnack, W.: Die Entstehung des Paarkernmycels bei *Collybia tuberosa* Bull. und *Schizophyllum commune* Fr. Z. Bot. **24**, 353—380 (1931). — Hartmann, M.: Die Sexualität, 1. Aufl., Jena 1943; 2. Aufl., Stuttgart 1956. — Heldmaier, C.: Über die Beeinflußbarkeit der Sexualität von *Schizophyllum commune* Fr. und *Collybia velutipes* Curt. Z. Bot. **22**, 161—220 (1929). — Holton, C. S.: Hybridisation and segregation in the oat smuts. Phytopathology **21**, 835—842 (1931).— Extent of pathogenicity of hybrids of *Tilletia tritici* and *T. levis*. J. agricult. Res. **65**, 555—563 (1942). — Holton, C. S., and G. W. Fischer: Hybridisation between *Ustilago Avenae* and *U. perennans*. J. agricult. Res. **62**, 121—128 (1941). — Hüttig, W.: Über physikalische und chemische Beeinflussungen des Zeitpunktes der Chromosomenreduktion bei Brandpilzen. Z. Bot. **26**, 1—26 (1933). — Hwang, S. W.: Compatibility and variability in *Merulius americanus*. Mycologia (N. Y.) **47**, 317—328 (1955).

Isenbeck, K.: Untersuchungen über die Physiologie von *Sphacelotheca sorghi*, den gedeckten Körnerbrand von *Sorghum*. Phytopath. Z. **8**, 165—182 (1935).

Kämmerling, H.: Über Geschlechtsverteilung und Bastardierung von *Ustilago longissima* und ihrer Varietät macrospora. Z. Bot. **22**, 113—142 (1929). — Kaufert, F.: The production of asexual spores by *Pleurotus corticatus*. Mycologia (N. Y.) **27**, 333—346 (1935). — The biology of *Pleurotus corticatus* Fries. Tech. Bull. Minn. agric. Exp. Stat. No **114**, 35 S. (1936).— Kawamura, E.: On the function of the spermatia of *Gymnosporangium Haraeanum* Syd. Agr. and Hort. **9**, 1299—1300 (1934). — Studies on the sex of Hymenomycetous fungi. Rev. Appl. Mycol. **21**, 273 (1942). — Keitt, G. W., and D. H. Palmiter: Heterothallism and variability in *Venturia inaequalis*. Amer. J. Bot. **25**, 338—345 (1938). — Kirby, R. S.: Heterothallism in *Ophiobolus cariceti*. Phytopathology **13**, 35 (1923). — Kniep, H.: Untersuchungen über den Antherenbrand (*Ustilago violacea* Pers.) Z. Bot. **11**, 257—284 (1919).— Über morphologische und physiologische Geschlechtsdifferenzierung. Verh. phys.-med. Ges., N.F. **46**, 1—18 (1920). — Über *Urocystis Anemones* (Pers.) Winter. Z. Bot. **13**, 289—309 (1921). — Über Geschlechtsbestimmung und Reduktionsteilung. Verh. phys.-med. Ges., N. F. **47**, 1—28 (1922). — Über erbliche Änderungen von Geschlechtsfaktoren bei Pilzen. Z. indukt. Abstamm.- u. Vererb.-Lehre **31**, 170—183 (1923). — Über Artkreuzungen bei Brandpilzen. Z. Pilzk. **5**, 217—247 (1926). — Die Sexualität der niederen Pflanzen. Jena: Gustav Fischer 1928. — Über Selektionswirkungen in fortlaufenden Massenaussaaten von *Schizophyllum*. Z. Bot. **23**, 510—536 (1930). — Kühn, A.: Grundriß der Vererbungslehre. Heidelberg: Quelle & Meyer 1950. — Kühner, R.: Caractères myceliens et polarité sexuelle de *Galera mycenopsis*. Bull. Soc. mycol. France **65**, 155—167 (1950). — L'amphithallie et ses causes dans la forme bisporique tétrapolaire de *Clitocybe lituus* (Fr.). Bull. Soc. mycol. France **69**, 307—325 (1954a). — Germination, caractères mycéliens et polarité de *Geopetalum (Cantharellus) carbonarium* (A. et S.) Pat. Étude cytologique des copulations illégitimes. Ann. Univ. Lyon., Sér. III, Sect. C **8**, 5—20 (1954b). — Kühner, R., et P. Terra: La race bisporique de *Mycena tenella* Fries est amphithalle tétrapolaire. Rev. Mycol. **20**, 161—177 (1955).

Lamb, I. M.: The initiation of the dicaryophase in *Puccinia phragmitis* (Schum.) Korn. Ann. Bot. **49**, 403—438 (1935). — Lamoure, D.: Hétérocaryose chez les Basidiomycetes amphitalles. C. R. Acad. Sci. (Paris) **244**, 2841—2843 (1957). — Lange, M.: Two species of *coprinus* with notes on their cultural characters. Mycologia (N. Y.) **40**, 739—747 (1948). — Langford, M. H., and G. W. Keitt: Heterothallism in *Venturia pirina*. Phytopathology **30**, 452 (1940). — Lendner, A.: A propos de l'hétérothallisme des Coprins. Bull. Soc. bot. Genève **12**, 337—352 (1920). — Leupold, U.: Die Vererbung von Homothallie und Heterothallie bei *Schizosaccharomyces pombe*. C. R. Lab. Carlsberg, Sér. physiol. **24**, 381—480

(1950). — LEVI, J. D.: Mating reaction in yeast. Nature (Lond.) **177**, 753—754 (1956). — LEWIS, D.: Comparative incompatibility in angiosperms and fungi. Adv. Genet. **6**, 235—285 (1954). — LORENZ, F.: Beiträge zur Entwicklungsgeschichte von *Sphaerobolus*. Arch. Protistenk. **81**, 361—398 (1933). — LWOFF, A., et P. SCHAEFFER: Le décoloration *d'Euglena gracilis* par la streptomycine. C. R. Acad. Sci. (Paris) **228**, 779—781 (1949).

MACDONALD, J. A.: The life history and cultural characteristics of *Typhula gyrans* (Batsch) Fries. Ann. appl. Biol. **21**, 590—613 (1934). — A study of *Polyporus betulinus* (Bull.) Fries. Ann. appl. Biol. **24**, 289—310 (1937). — MACRAE, R.: Interfertility phenomena of the American and European forms of *Panus stypticus* (Bull) Fries. Nature (Lond.) **139**, 674 (1937). — Interfertility studies and inheritance of luminosity in *Panus stypticus*. Canad. J. Res., C **20**, 411—434 (1942). — MARTENS, P.: Cyle de dévelopement et sexualité des ascomycètes. Cellule **50**, 125—310 (1946). — MARTENS, P., et R. VANDENDRIES: Le cycle conidien haploide et diploide chez *Pholiota aurivella*. Cellule **41**, 337—388 (1933). — MILLAR, P. R.: Pathogenicity of three red cedar rusts that occur on apple. Phytopathology **22**, 723—740 (1932). — MITTER, J. H.: Some contributions to our knowledge of heterothallism in fungi. J. Indian bot. Soc. **15**, 318—192 (1936). — MITTMANN, G.: Kulturversuche mit Einsporstämmen und zytologische Untersuchungen in der Gattung *Ceratostomella*. Jb. wiss. Bot. **77**, 185—219 (1932). — MONTGOMERY, H. B. S.: A study of *Fomes fraxineus* and its effect on ashwood. Ann. appl. Biol. **23**, 465—486 (1936). — MOUNCE, I.: Homothallism and the production of fruit-bodies by monosporus mycelia in the genus *Coprinus*. Trans. Brit. mycol. Soc. **7**, 198—217 (1921). — Homothallism and heterothallism in the genus *Coprinus*. Trans. Brit. mycol. Soc. **7**, 256—269 (1922). — A preliminary note on *Fomes pinicola* Fr. and *Pholiota adiposa* Fr. — two heterothallic species of wood-destroying fungi. Phytopathology **16**, 757—758 (1926). — Notes on sexuality in *Fomes pinicola* (Sw.) Cooks, *Fomes roseus* (Fr.) Cke., *Polyporus Tukahoe* (Güssow) Sacc. et Trott., *P. resinosus* (Schrad.) Fr., *P. anceps* Peck., *Lenzites saepiaria* Fr., *Trametes protracta* Fr. and *T. suaveolens* (L.) Fr. Phytopathology Soc. Proc. 1929 (1930). — MOUNCE, I., and R. MACRAE: The behavior of paired monosporous mycelia of *Lenzites saepiaria* (Wulf.) Fr., *L. trabea* (Pers.) Fr., *L. thermophila* Falck. and *Trametes americana* Overh. Canad. J. Res., C **14**, 215—221 (1936). — The behavior of paired monosporous mycelia of *Fomes roseus* (Alb. & Schw.) Cooke and *Fomes subroseus* (Weir.) Overh. Canad. J. Res., C **15**, 154—161 (1937). — Interfertility phenomena in *Fomes penicola*. Canad. J. Res.. C **16**, 354—376 (1938).

NEWTON, D. E.: The bisexuality of individual strains of *Coprinus Rostrupianus*. Ann. Bot. **40**, 105—128 (1926a). — The distribution of spores of diverse sex on the hymenium of *Coprinus lagopus*. Ann. Bot. **40**, 891—917 (1926b). — NISIKADO, Y., and K. YAMAUTI: Studies on the heterothallism of *Cortinellus Berkeleyana* Ito and Imai, an economically important edible mushroom in Japan. Ber. Ohara Inst. **7**, 115—128 (1935). — NOBLE, M.: The morphology and cytology of *Typhula trifolii* Rostr. Ann. Bot., N. S. **1**, 67—98 (1937). — NOBLES, M. K.: Conidial formation, mutation and hybridisation in *Peniophora Allescheri*. Mycologia (N. Y.) **27**, 286—301 (1935). — Production of conidia by *Corticum incrustans*. Mycologia (N. Y.) **29**, 557—566 (1937). — Secondary spores in *Corticum effuscatum*. Canad. J. Res., C **20**, 347—357 (1942). — A contribution toward a clarification of the *Trametes serialis* complex. Canad. J. Res., C **21**, 211—234 (1943). — Cultural characters as a guide to the taxonomy and phylogeny of the Polyporaceae. Canad. J. Bot. **36**, 883—926 (1958). — NOBLES, M. K., R. MACRAE and B. P. TOMLIN: Results of interfertility tests on some species of Hymenomycetes. Canad. J. Bot. **35**, 377—387 (1957).

OIKAWA, K.: Sex in *Cortinellus Shiitake*. Bot. Mag. (Tokyo) **49**, 453—455 (1935). — Diploidisation and fruit-body formation in the Hymenomycetes. Sci. Rep. Tohoku Univ. **14**, 245—260 (1939). — OLIVE, L. S., and A. A. FANTANI: A new heterothallic species of *Sordaria*. Amer. J. Bot. **48**, 124—128 (1961). — OLSON, O. E.: Genetics of *Ceratostomella*. I. Strains in *Ceratostomella fimbriata* (Ell. & Hals.) Ell. from sweet potatoes. Phytopathology **39**, 548—561 (1949).

PAGE, W. M.: Note on abnormal spores in *Podospora minuta*. Trans. Brit. mycol. Soc. **20**, 186—189 (1936). — PEASE, D.: *Sphaeronaemella fimicola* (Marchal) emend.: some characteristics in culture. Mycologia (N. Y.) **40**, 114—124 (1948). — PIERSON, R. H.: Fusion of pycniospores with filamentous hyphae in the pycnium of the white pine blister rust. Nature (Lond.) **131**, 728 (1933). — PINTO-LOPES, J.: The tetrapolarity in *Deconia atrorufa*. Port. Acta biol. A **2**, 249—250 (1948). — PLEMPEL, M.: Die Sexualstoffe der Mucoraceae, ihre Abtrennung und die Erklärung ihrer Funktion. Arch. Mikrobiol. **26**, 151—174 (1956). — Die chemischen Grundlagen der Sexualreaktion bei Zygomyceten. Planta (Berl.) **39**, 492—508 (1963). — PLEMPEL, M., u. G. BRAUNITZER: Die Isolierung der Mucorineen Sexualstoffe I. Z. Naturforsch. **13**b, 302—305 (1958). — POPP, A., and W. F. HANNA: Studies on the physiology of the oat smuts. Sci. Agricult. **15**, 424—433 (1935).

QUINTANILHA, A.: Le problème de la sexualité chez les champignons. Bol. Soc. Brot. **8**, 1—99 (1933). — Cytologie et génétique de la sexualité chez les Hymnomycètes. Bol. Soc.

Brot. **10**, 289—332 (1935). — Étude génétique du phénomène de Buller. Bol. Soc. Brot. **13**, 425—486 (1939). — Doze anos de citologia e génetica dos fungos. Agronomia Lusitana **3**, 241—306 (1941). — La conduite sexuelle de quelques espèces d'Agaricacées. Bol. Soc. Brot. **19**, 27—65 (1944). — Quintanilha, A., et J. Pinto-Lopes: Apercu sur l'état actuel de nos connaissances concernent la "conduite sexuelle" des espèces d'Hyménomycetes. I. Bol. Soc. Brot. **24**, 115—290 (1950). — Quintanilha, A., L. Quintanilha et A. Vasermanis: La conduite sexuelle et la systématique des Hyménomycetes. Rev. Mycol. **6**, 1—48 (1941).

Raper, J. R.: Tetrapolar sexuality. Quart. Rev. Biol. **28**, 233—259 (1953). — Tetrapolarity in *Schizophyllum fasciatum*. Mycologia (N.Y.) **52**, 334—336 (1960). — Raper, K. B., and D. J. Fenell: Homothallism. vs. heterothallism in the *Penicillium luteum* series. Mycologia (N. Y.) **44**, 101—111 (1952). — Rizet, G.: Sur les spores dimorphes et l'héridité de leurs caractères chez un nouvel *Ascobolus hétérothallique*. C. R. Acad. Sci. (Paris) **208**, 1669—1671 (1939). — Sur les divers formes de la distribution des "sexes" chez le *Podospora minuta*. C. R. Soc. Biol. (Paris) **123**, 31 (1940). — Sur la biologie et sexualité du *Podospora curvula*. Bull. Soc. Lin. Norm. **2**, 23—26 (1941a). — A propos de répartition des perithèces dans les cultures mixtes d'ascomycètes héterothalliques. Bull. Soc. Lin. Norm. **2**, 27—30 (1941b). — Robak, H.: Sexuality of *Polyporus borealis* (Wahlenb.) Fr. Svensk. bot. Tidskr. **26**, 266—270 (1932). — Studies in the biology of wood-destroying Hymenomycetes. I. Contribution to the knowledge of homothally and heterothally in some species of Telephoraceae and Polyporaceae. Nyt. Mag. Naturv. **76**, 5—20 (1936). — Cultural studies in some Norwegian wood-destroying fungi. Medd. Vestl. Forsöksst. **25**, 248 S. (1942). — Rodenhiser, H. A.: Heterothallism and hybridisation in *Sphacelotheca sorghi* and *S. cruenta*. J. agricult. Res. **45**, 287—296 (1932). — Studies on the possible origin of physiologic forms of *Sphacelotheca sorghi* and *S. cruenta*. J. agricult. Res. **49**, 1069—1086 (1934). — Routien, J. B.: Cultural and genetical studies of certain Agarics. Mycologia (N. Y.) **32**, 97—103 (1940). — Rowell, J. B.: Functional role of compatibility factors and an in vitro test for sexual compatibility with haploid lines of *Ustilago zeae*. Phytopathology **45**, 370—374 (1955).

Sager, R.: Mendelian and non-mendelian inheritance of streptomycin resistence in *Chlamydomonas Reinhardi*. Proc. nat. Acad. Sci. (Wash.) **40**, 356—363 (1954). — Non-mendelian inheritance in *Chlamydomonas*. Genetics **40**, 594 (1955). — Sass, J. E.: The cytological basis for homothallism and heterothallism in the Agaricaceae. Amer. J. Bot. **16**, 663—701 (1929). — Shear, C. L., and R. W. Davidson: An undiscribed *Corticum* with conidia. Mycologia (N. Y.) **36**, 294—299 (1944). — Shear, C. L., and B. O. Dodge: The life history of *Pilacre faginea* (Fr.) B. & Br. J. agricult. Res. **30**, 407—417 (1925). — Life histories and heterothallism of the red bred mould fungi of the *Monilia sitiphila* group. J. agricult. Res. **34**, 1019—1042 (1927). — Skolko, A. J.: A cultural and cytological investigation of a two-spored Basidiomycete, *Aleurodiscus canadensis* n. sp. Canad. J. Res., C **22**, 251—271 (1944).— Sleumer, H. O.: Über Sexualität und Zytologie von *Ustilago zeae* (Beckm.) Unger. Z. Bot. **25**, 209—263 (1932). — Smith, A. H., and H. J. Brodie: Cultural characters and pairing reactions of monosporous mycelia and development of the fruit body of *Pholiota (Flammula) polychroa*. Bot. Gaz. **96**, 533—546 (1935). — Stakman, E. C., and J. J. Christensen: Heterothallism in *Ustilago zeae*. Phytopathology **17**, 827—834 (1927). — Stover, R. H.: Sexuality and heterothallism in *Mycospaerella musicola*. Canad, J. Bot. **41**,153 1—1532 (1963).

Tapke, V. F.: Occurence, identification, and species validity of the barley loose smuts, *Ustilago nuda*, *U. nigra*, and *U. medians*. Phytopathology **33**, 194—209 (1943). — Terra, P.: Détermination de la polarité sexuelle de trente espèces de Basidiomycètes saprophytes. C. R. Acad. Sci. (Paris) **236**, 115—117 (1953). — Amphithallie et hétérocaryose chez un Basidiomycète bipolaire, *Stropharia umbonatescens* Peck. C. R. Acad. Sci. (Paris) **241**, 1334—1336 (1955). — Thren, R.: Gewinnung und Kultur von monokaryotischem und dikaryotischem Myzel. Z. Bot. **31**, 337—391 (1937). — Tinline, R. D.: Studies on the perfect stage of *Helminthosporium sativum*. Canad. J. Bot. **29**, 467—478 (1951). — Tyler, L. J.: Variation in *Sphacelotheca sorghi* (Link.) Clinton. Tech. Bull. Minn. agric. Exp. Stat. **133**, 1—48 (1938). — Tyler, L. J., and C. P. Shumway: Hybridisation between *Sphacelotheca sorghi* and *Sorosporium reilianum*. Phytopathology **25**, 375—376 (1935).

Vaheeduddin, S.: Hybridisation between *Sphacelotheca cruenta* and *Sorosporium reilianum*. Phytopathology **26**, 111 (1936). — Vandendries, R.: Recherches sur le déterminisme sexúel des Basidiomycètes. Mém. Acad. roy. Belg., Cl. Sci. **5**, 1—98 (1923a). — Nouvelles recherches sur la sexualité Basidiomycètes. Bull. Soc. bot. Belg. **56**, 73—97 (1923b). — Recherches expérimentales sur la bipolarité des sexuelle des Basidiomycètes. Bull. Soc. bot. Belg. **57**, 75—78 (1924). — Contribution nouvelles à l'étude de la sexualité des Basidiomycètes. Cellule **35**, 129—157 (1925a). — L'hétéro-homothallisme dans le genre *Coprinus*. Boll. Soc. bot. Belg. **57**, 139—146 (1925b). — Les mutations sexuelles des Basidiomycètes. Bull. Soc. bot. Belg. **58**, 28—37 (1925c). — Recherches experimentales prouvant la fixité du sexe dans *Coprinus radians* Desm. Bull. Soc. mycol. France **41**, 358—374 (1925d). — La tétrapolarité sexuelle des Coprins. Bull. Soc. bot. Belg. **58**, 180—186 (1926). — Les mutations sexuelles, l'hétérohomothalisme et la stérilité entre races géographiques de *Coprinus micaceaus*. Mém. Acad. roy. Belg., Cl. Sci. **9**, 50 S. (1927a). — Le comportement sexuel du *Coprin micacé*, dans ces rapports avec la dispersion de l'espèce. Bull. Soc. bot. Belg. **60**, 62—65 (1927b). — Nouvelies recherches expérimentales sur le comportement sexuel de *Coprinus micaceus*. Mém. Acad. roy. Belg., Cl. Sci. **9**, 128 S. (1927c). — Comment résoudre le problème sexuel du *Coprin micacé*? Bull. Soc. bot. Belg. **61**, 123—135 (1929a). — Les relations entre souches étrangères expliquées par les aptitudes sexuelles des individus pathénogéniques chez *Coprinus micaceus*. Bull. Soc. mycol. France **45**, 216—248 (1929b). — Les polarités sexuelles de *Coprinus tergiversans* Fr. Bull. Soc. mycol. France **47**, 36—43 (1930). — Les aptitudes et les mutations sexuelles chez *Panaeolus papilionaceus*. Rec. trav. crypt. déd. à L. Mangin. Paris 1931. — La tétrapolarité sexuelle de *Pleurotus columbinus*. Cellule **41**, 267—278 (1932). — Nouvelles investigations dans la domaine sexuel des Hymenomycetès. Bull. Soc. mycol. France **49**, 130—165 (1933a). — De la valeur du barrage sexuel comme critérium dans l'analyse d'une sporée tétrapolaire de Basidiomycete *Pleuritus ostreatus*. Genetica **15**, 202—212 (1933b). — Contribution a l'étude de la sexualité dans la genre *Trametes*. Bull. Soc. mycol. France **50**, 98—110 (1934a). — La polarité sexuelle et la régime conidien chez *Pleurotus pinsitus*. Bull. Soc. mycol. France **50**, 203—212 (1934b). — Les affinités sexuelles de *Hypholoma sublateritium* Fr. Bull. Soc. mycol. France **50**, 213—270 (1934c). — Les polarités sexuelles dans le genre *Pholiota*. Bull. Soc. mycol. France **50**, 270—277 (1934d). — La polarité sexuelle de *Panus stipticus* (Bull.) Fr. Bull. Soc. mycol. France **52**, 54—56 (1936a). — Les tendences sexuelles chez les polyporés. I. *Leptoporus adustus* (Fr. ex Willd.) Quél. Rev. Mycol. **1**, 85—92 (1936b). — Les tendences sexuelles chez les Polyporés. II. *Leucoporus arcularius* (Batsch.) Quél. Rev. Mycol. **1**, 181—190 (1936c). — Les tendences sexuelles chez les Polyporés. III. *Leucoporus brumalis* (Fr. ex Pers.) Quél. Rev. Mycol. **1**, 294—302 (1936d). — Les tendences sexuelles des Polyporés. IV. *Melanopus squamosus* (Huds.) Pat. Bull. Soc. mycol. France **52**, 351—362 (1936e). — Sur la sexualité des Basidiomycètes. C. R. Acad. Sci. (Paris) **203**, 1284—1286 (1936f). — Nouveaux aperçus sur la sexualité des Basidiomycetes. C. R. Acad. Sci. (Paris) **204**, 1084—1086 (1937a). — Les modalités sexuelles des Basidiomycètes. Bull. Soc. bot. Belg. **70**, 66—85 (1937b). — Vandendries, R., et H. J. Brodie: Manifestation de barrages sexuels dans le champignon tétrapolaire *Lenzites betulina* (L.) Fr. Bull. Soc. bot. Belg. **65**, 109—111 (1933a). — Nouvelles investigations dans le domaine de la sexualité des Basidiomycètes et étude expérimentale des barrages sexuels. Cellule **42**, 165—210 (1933b). — Vandendries, R., et G. Robyn: Nouvelles recherches éxpérimentales sur le comportement sexuel de *Coprinus micaceus*. Deuxième Partie. Mém. Acad. roy. Belg., Cl. Sci. **9**, 117 S. (1929). — Vanterpool, T. C., and R. Macrae: Notes on the *Canadian tukahoe*, its occurrence in Canada and the interfertility of its perfect stage, *Polyporus tuberaster* Jacq. ex Fries. Canad J. Bot. **29**, 147—157 (1951). — Verall, A. F.: Variation in *Fomes igniarius* (L.) Gill. Tech. Bull. Minn. agric. Exp. Stat. **117**, 41 S. (1937).

Welch, A. W., and J. C. Gilman: Hetero- and homothallic types of *Diaporthe* on soybeans. Phytopathology **38**, 628—637 (1948). — Wettstein, R.: Handbuch der systematischen Botanik, 4. Aufl. Leipzig u. Wien 1935. — Whetzel, H. H.: A new genus and new species of brown-spored inoperculate Discomycetes from Panama. Mycologia (N. Y.) **34**, 584—591 (1942). — Whitehouse, H. L. K.: Heterothallism and sex in the fungi. Biol. Rev. **24**, 411—447 (1949a). — Multiple-allelomorph heterothallism in the fungi. New Phytol. **48**, 212—244 (1949b). — A survey of heterothallism in the Ustilaginales. Trans. Brit. mycol. Soc. **34**, 340—355 (1951). — Wineland, G. O.: The production in culture of ascigerous stage of *Fusarium moniliforme*. Phytopathology **13**, 51 (1923).

Yarwood, C. E.: Heterothallism of sunflower powdery mildew. Science **82**, 417—418 (1935). — Yen, H. C.: Note préliminaire sur la polarité sexuelle et sur les caractères du mycelium haploide de plusieurs Homobasidiomycètes. C. R. Acad. Sci. (Paris) **224**, 1239—1240 (1947). — Deuxième note préliminaire sur la sexualité et sur les caractères du mycelium de quelques Holobasidiomycètes. C. R. Acad. Sci. (Paris) **226**, 1214—1215 (1948).

Zattler, F.: Vererbungserscheinungen an Hutpilzen (Basidiomyceten). Z. Bot. **16**, 433—499 (1924). — Zickler, H.: Genetische Untersuchungen an einem heterothallischen Askomyceten (*Bombardia lunata* nov. spev.). Planta (Berl.) **22**, 573—613 (1934). — Zillig, H.: Über spezialisierte Formen beim Antherenbrand *Ustilago violacea* (Pers.) Fuck. Zbl. Bakt., II. Abt. **53**, 33—74 (1921). — Zycha, H.: Über *Hypomyces rosellus* (Alb. & Schw.) Tul. Zbl. Bakt., II. Abt. **92**, 449—459 (1935).

Meiosis.

By

J. Herbert Taylor*.

With 13 figures.

I. Distinguishing features of meiotic divisions.

A. Introductory remarks.

In these groups of plants meiosis occurs in the formation of the spores, microspores and megaspores, if both types exist. However, much more detailed information is available concerning the divisions of microsporocytes because of the abundance of cells available for study and the facility with which preparations of microsporocytes are made. The essential features of meiosis are quite uniform throughout the group when the process goes to completion. The aberrations that have evolved usually either alter the timing and sequence of events or allow the process to proceed without one or more of its characteristic features. Very few anomalous new features have been introduced as has occurred in the insects and related forms, for example.

Meiosis typically consists of two nuclear divisions, and therefore, gives rise to four nuclei. From the standpoint of evolution we must think of the process evolving by changes in the regular somatic divisions. Although little can be said about its evolutionary development, the description and analysis of the process will continually be compared to regular mitosis in this treatment.

The various stages of meiosis will be classified as follows:

Division I,

- prophase I (subdivided into five stages)
  - leptotene, elongated unpaired chromosomes.
  - zygotene, active stage of pairing.
  - pachytene, the stable stage of pairing (Fig. 1).
  - diplotene, the stage of separation of homologous chromosomes except at chiasmata.
  - diakinesis, a stage of extreme contraction of chromosomes prior to metaphase (Fig. 2).
- metaphase I (Fig. 3)
- anaphase I (Fig. 4) and
- telophase I.

Interphase (or interkinesis).

---

* Recently in order to take into account some of the work published since the preparation of the original manuscript, several new paragraphs have been added in collaboration with Dr. MARGARET Y. MENZEL, Department of Biological Science, Florida State University.

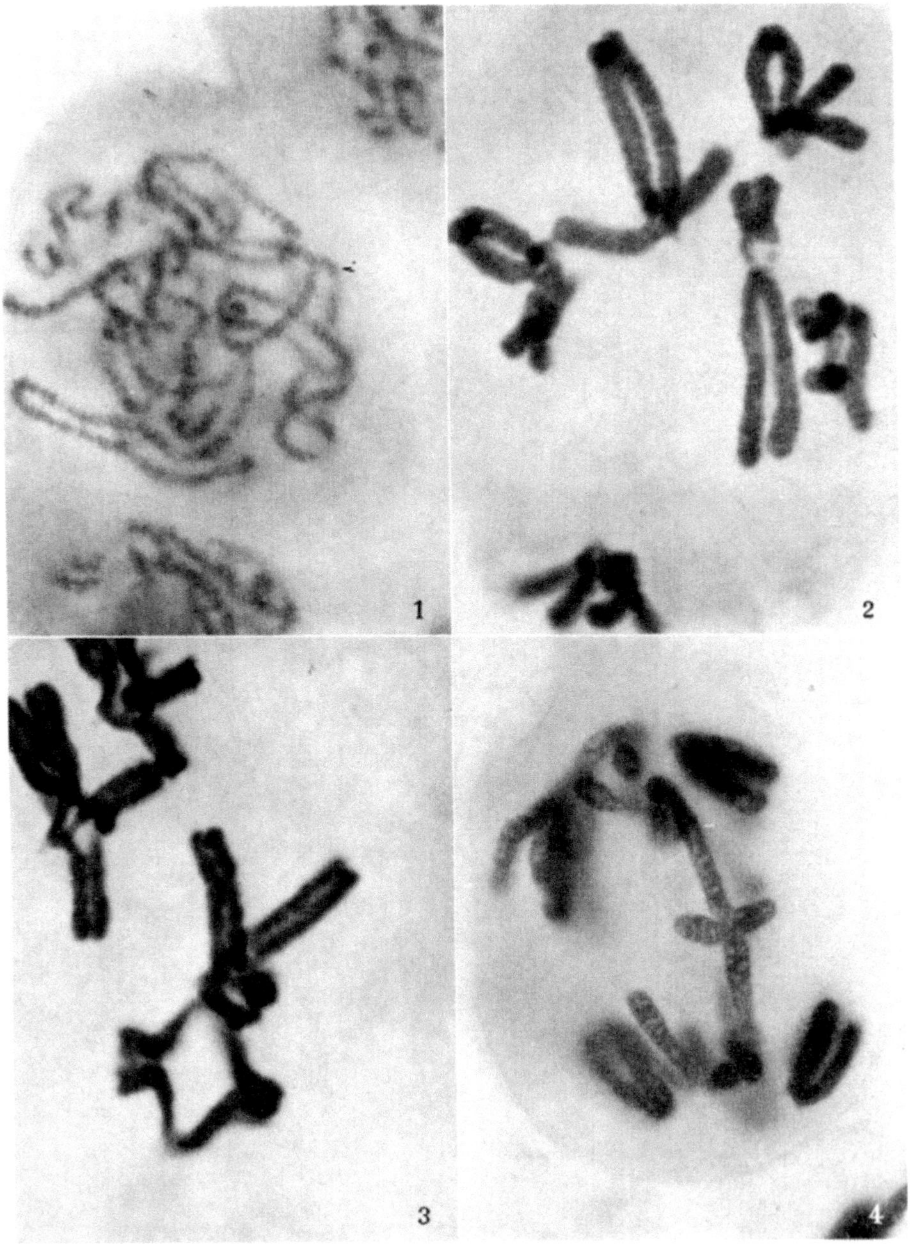

Figs. 1—4. Stages of meiosis in *Tulbaghia violacea*. Fig. 1. Late pachytene. Fig. 2. Diakinesis. Fig. 3. Metaphase I. Fig. 4. Anaphase 3 hours after 15 r of x-radiation showing a typical subchromatid bridge. One arm of each diad is attached at one point with no free fragment.

Division II,
- prophase II.
- metaphase II (Fig. 6).
- anaphase II (Fig. 7).
- telophase II.

Quartet (tetrad) of spores.

The anaphase complement of chromosomes in a microspore division (Fig. 8) is shown for comparison.

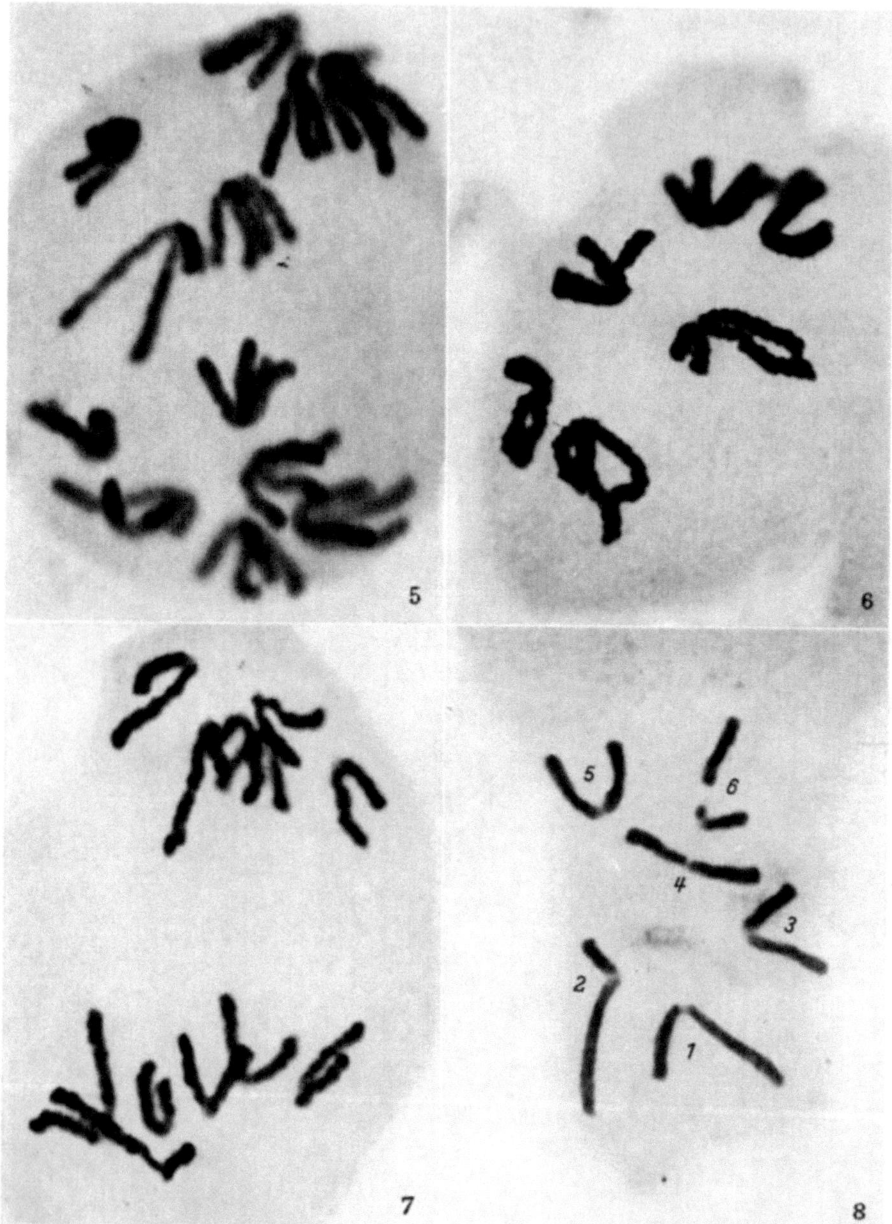

Fig. 5—8. Stages of meiosis and of first mitosis after meiosis in *Tulbaghia violacea.*

Fig. 5. Normal anaphase — notice that each diad has four arms that are well separated. Fig. 6. Metaphase II. Fig. 7. Anaphase II. Fig. 8. Anaphase group of chromosome at the microspore division (first mitosis after meiosis). Chromosomes are numbered in order of size. Chromosome 1 and 2 are nearly the same size but the position of the centromere distinguishes them. Chromosome 3 is only slightly longer than 4 and 5, but the position of the centromere distinguishes it. Chromosome 4 and 5 are morphologically indistinguishable. Chromosome 6 is the shortest one in meiosis but the long secondary constriction which does not show in meiosis makes it easy to recognize in mitosis.

B. Typical features of meiotic divisions.

Among the several characteristic features that distinguish the first division of meiosis from regular somatic divisions are (1) the very long prophase stage,

(2) the pairing of homologous chromosomes, (3) the formation of chiasmata and exchange of genetic material, (4) the variations in the coiling cycle, and (5) the unitary behavior of the centromeres of each homolog.

1. The long prophase.

The length of prophase I varies from several weeks to a few days in microsporocytes. The length of the stages is less well known in megasporocytes, but is probably as long or longer. This period contrasts with a few hours or less for prophase in most somatic divisions. If the few examples which have been investigated are typical, the long prophase is a period of relatively slow metabolic activity (see section C below).

CHAMBERLAIN (1935) reported that microsporocyte stages of *Pinus laricio* last from early October to late April. The exact stages involved are not clear but the pre-prophase or early prophase stages last until January. His drawings clearly show pachytene stages on April 4 and division II on May 3.

SPARROW and SPARROW (1949) report the duration of stages in microsporogenesis of *Trillium erectum* at 4—6° C. Leptotene-zygotene stages require about 30 days, pachytene 18—24 days, diplotene-diakinesis 12—16 days and first and second division are completed in 10—18 days.

In *Lilium longiflorum* prophase I lasts about 7 days at 23° C (TAYLOR and McMASTER 1954 and TAYLOR, unpublished). Leptotene lasts about 2 days, zygotene about one day, pachytene 3 days and diplotene-diakinesis about 1 day. At 20° C. the prophase lasts about 10—11 days.

Prophase I requires about 6 days in *Tulbaghia violacea* (TAYLOR 1958b) and about 5—6 days in *Tradescantia paludosa* (SAX and EDMONDS 1933 and MOSES and TAYLOR 1955).

2. The pairing of homologous chromosomes.

The suggestion has been made that the long prophase together with the elongation of the chromosomes allows pairing of homologs (SAX and SAX 1935). Another factor may be the lower viscosity of the nuclear fluids which becomes apparent toward the end of leptotene and reaches a minimum during zygotene. The nuclei become susceptible to rupture or the extrusion of globules of nuclear material into the cytoplasm or even into adjacent cells (COOPER 1952, TAKATS 1958). The extrusions may occur during fixation and probably also occur as a result of osmotic shock under growing conditions. The results are micronuclei which persist for several days in cells of anthers placed in culture during the sensitive period. The chromosomes or parts of chromosomes in these micronuclei appear very much like the typical pachytene chromosomes after a few days in culture (TAYLOR, unpublished).

One of the most characteristic features of the pairing stage is the collapse of the chromosomes into a rather compact mass when most fixatives are used. This "synizetic knot" of chromatin appears to be largely an artifact of fixation, but the asymmetrical distribution of chromatin is frequently evident when fresh microsporocytes are squeezed from anthers and observed immediately in the surrounding fluids. The condition is less extreme in cells of anthers quick frozen in isopentane and dehydrated in absolute ethanol by the technique described by WOODS and POLLISTER (1955). Another indication of the high content of unbound water is the greater tendency to form ice crystals when anthers are quick frozen at this stage rather than at premeiotic interphase or leptotene.

Toward the end of leptotene in *Lilium longiflorum* the 3—6 nucleoli of each nucleus fuse into one spherical body. Shortly after this occurs the ends of homologous chromosomes frequently are seen lying parallel to each other at the periphery of the nucleus. In some other species, *Tradescantia* for example, a rather typical bouquet is formed with all of the chromosome ends aggregated near one point (KAUFMANN 1925, TAYLOR 1949a). In many other plants including *Lilium* and *Zea* there is no apparent orientation of the chromosomes associated with zygotene pairing. Although pairing frequently may be seen beginning at the ends of chromosomes, there is a limited amount of evidence on the point along the chromosome at which pairing begins and the time required for any two homologous chromosomes to complete the process. Once a contact is made between homologs the pairing (synapsis) appears to proceed in a "zipper" fashion. However, the complete process has never been observed in living material. Conclusions based on cells fixed at various stages of pairing indicate that pairing proceeds relatively slowly and in *Lilium* each homologous pair may require from 12 to 24 hours to reach the stable stage of pairing typical of pachytene.

Hope that the high resolving power of the electron microscope might elucidate questions of chromosome organization has usually been frustrated by the membraneless, microfibrillar nature of the chromatin. However, MOSES (1956) and FAWCETT (1956) discovered peculiar, highly characteristic structures in meiotic prophase nuclei of primary spermatocytes in crayfish and human respectively. MOSES termed these structures "synaptinemal complexes" and showed (MOSES 1958) that they were located in nuclear regions which contained paired prophase chromosomes. In sagittal section, a complex is composed of a fine central, electron-dense axis, the "*central element*", flanked at some distance by dense "*lateral elements*" and connected to them by very fine transverse fibrils. These components constitute the "*cores*" or "*axial elements*". Outside the cores are wide bands of less dense fibrillar material, not sharply demarcated from the granular-fibrillar background. Cytochemical studies of the complexes (NEBEL and COULON 1962a, b, COLEMAN and MOSES 1964) suggest that most of the deoxyribonucleic acid is located in the outer fibrils and that the electron-dense core elements are not disrupted by nucleases.

Synaptinemal complexes have subsequently been demonstrated in a variety of animals and plants (MOSES 1964). Their role is still conjectural. MEYER (1961) found them in oocytes of *Drosophila melanogaster* but not in spermatocytes. Since crossing over occurs in oogenesis, but not in spermatogenesis in this fly, MEYER speculated that the complexes had some connection with crossing over. Triploid oocytes had some chromosomes involved in complexes and some excluded (MEYER 1964). However, MENZEL and PRICE (1966) found evidence that the complexes are associated with synapsis regardless of whether crossing over is to ensue. They examined fine structure of pachytene nuclei in pollen mother cells of *Lycopersicon esculentum* (tomato), *Solanum lycopersicoides*, their diploid F_1 hybrid, and a tomato haploid. In the two species, pairing was homologous and followed by chiasma formation in all pairs. In their diploid hybrid, the pairing though complete, was homeologous and many pairs failed to form chiasmata. In the haploid, an average of about one region per nucleus was paired non-homologously and no chiasmata were formed. Nevertheless typical synaptinemal complexes were found in all four materials, regardless of whether synapsis was homologous, homeologous, or non-homologous.

That pairing is a consequence of incomplete reproduction of chromosomes in meiotic prophase is an old idea proposed by DARLINGTON (1932). He suggested later (1955) that pairing is a substitute for reproduction, but did not consider that pairing involves the DNA specifically. The first suggestion that the specificity of chromosome pairing might be related to base sequences in single stranded

DNA was probably expressed by GOLDSCHMIDT (1955) on the basis of the WATSON-CRICK (1953) model for DNA structure. He expressed the "wild idea" that somehow the DNA chains get separated in preleptotene or postleptotene chromosomes, which then attract the homologs because of the tendancy of DNA to form two stranded structures when complementary chains are available. He did not speculate on what might keep complementary chains of the homologs from reassociating since these should be close together and have a much better chance of finding partners than homologous regions.

In what sense the chromosome may be incomplete is not clear for the DNA appears to be replicated before meiotic prophase (see section I, C.) and histones are also synthesized at the same time (DE 1961). However, some recent reports indicate that a small but detectable synthesis of DNA may occur during meiotic prophase (WIMBER and PRENSKY 1963 and HOTTA et al. 1966). Although it is not clear that this DNA is a part of the chronosomes, there remai ns the possibility that a delayed synthesis of a very small part of the total DNA remained undetected in the original studies.

A hypothesis to explain the specificity of pairing based on the occurrence of single stranded DNA has been proposed by TAYLOR (1967). The idea is based on the concept that a chromosome is composed of relatively short units of replication *(replicons)* in which the two chains replicate out of sequence. After the replication of one chain some block was assumed to occur at random in the replicons so that a random sample was left single stranded in each nucleus. At zygotene when the nuclear viscosity drops, these regions would find complementary regions if they exist in the same nucleus and the specificity of pairing could be explained without the assumption of any special forces except those known to operate in renaturation of DNA. The replicated half of each replicon involved would be two stranded already and would prevent sister chromatids from pairing. Pairing would cease when the block in DNA synthesis was released later in prophase (presumably late pachytene). The testing of the hypothesis requires more elaborate studies of DNA replication than have yet been devised.

To be applicable to angiosperms, any general theory of chromosome pairing at meiosis must take into account the behavior of chromosomes in interspecific hybrids and in allopolyploids, since the latter constitute perhaps 50 per cent of angiosperm species (DAWSON 1962). Chromosome pairing in interspecific hybrids may range from virtually none to virtually complete; the degree of pairing is taken as an index of the amount of homology between the genomes of the parental species. Allopolyploids resulting from doubling the chromosome number of the primary hybrids may exhibit only bivalents, only multivalents, or various mixtures; often univalents are also found even though each chromosome now has an exact homologue present. Theoretically, each pair of chromosomes in the primary hybrid could be represented by an association of four chromosomes (quadrivalent) in the allopolyploid, each trivalent by a hexavalent, etc. The observed multivalent frequency often falls far short of the predicted frequency even though all the chromosomes are paired. This is usually interpreted and can sometimes be shown by genetic segregation ratio (e.g. GERSTEL 1956, GERSTEL and PHILLIPS 1957, 1958, PHILLIPS and GERSTEL 1959, SARVELLA 1958) to mean that homologues from the same ancestral genome tend to pair with each other to the partial or complete exclusion of their counterparts (homeologues) from other genomes. This phenomenon was termed "*differential affinity*" by DARLINGTON (1937); the term "*preferential pairing*" is also in common use today. Its existence implies that, given a choice, chromosomes will tend to pair with the most nearly identical partner available. If no partner with any degree of homology is present, pairing of manifestly non-homologous chromosome regions (different chromosomes or regions of the same chromosome) may occur (McCLINTOCK 1933, RHOADES and McCLINTOCK 1935).

Interpretation of preferential pairing is complicated by the fact that as a matter of expediency it is read at metaphase I, whereas the critical events occurred earlier, either at the time of synapsis or of chiasma formation or both. MENZEL (1964) has shown that in a allotetraploid of *Lycopersicon esculentum* × *Solanum lycopersicoides*, which exhibits very strong preferential pairing, it is synapsis which is preferential. Probably examples of other types of preferential pairing exist, e.g. non-preferential synapsis followed by preferential chiasma formation; this would probably result in both multivalents and univalents, a common finding in newly arisen allopolyploids.

Preferential pairing seems to be under genetic control. The most striking example is in tetraploid and hexaploid wheats. A region (single locus ?) on the long arm of chromosome 5B has been shown to completely suppress homeologous pairing, so that only homologous bivalents are formed (SEARS and OKAMOTO 1958, RILEY and CHAPMAN 1958, RILEY 1960, RILEY, CHAPMAN and KIMBER 1960). Plants nullisomic for this arm have multivalent frequencies approaching those in experimental hybrids involving putative wild ancestors of cultivated polyploid wheats. FELDMAN (1966) has shown that doses of 5 BL in excess of four depress chiasma formation in homologous pairs; six doses lead to asynapsis. FELDMAN (1966) and FELDMAN, MELLO-SAMPAYO and SEARS (1966) believe that the mode of action is primarily upon pre-meiotic association of homologues during mitotic cycles preceding meiotic prophase.

3. The formation of chiasmata and the exchange of genetic material.

When the homologous chromosomes begin to separate at diplotene there are normally one or several places along each pair at which separation fails to occur. In favorable preparations of sporocytes of some species, this failure of separation can be seen to be due to the exchange of chromatid partners at these points, chiasmata. The unresolved question is whether during diplotene separation, the sister chromatids of each homolog always act as a unit and separate from the other two homologous chromatids — the one-plane hypothesis (JANSSENS 1909, 1924, DARLINGTON 1937). If this is the mode of separation, chiasmata would be the result of exchanges of chromatid segments between homologs (crossovers) which occurred prior to diplotene. On the other hand if the four chromatids begin separation at different places, sisters together at some points and non-sisters at other places — two-plane separation (SAX 1932) — the chiasmata would be formed whether crossing over had occurred or not.

Genetic evidence from tetrad analysis indicates that separation at the centromere is regularly reductional, *i.e.*, sister chromatids remain together at the centromere during diplotene separation. Topological evidence from bivalent interlocking (MATHER 1938) and from chiasma formation in trivalents (DARLINGTON 1930) strongly support the one-plane hypothesis. However, the evidence is not unequivocal and a final decision should not be made without more evidence. In species with terminalized chiasmata the non-sister and sister chromatids lie together at different points along the chromosome at late prophase I and metaphase I even though the original separation is one plane. Since they appear to have the same proximity as chromatids in bivalents without terminalization non-sister chromatids must have the same affinity for each other as sisters in these cases.

Recent autoradiographic studies (TAYLOR 1965) demonstrated that segregation of chromatids is regularly reductional at the centromere regions in the spermatocytes of the grasshopper, *Romalea*. However, a critical study for the whole length of the chromosomes at diplotene remains to be done.

BROWN and ZOHARY (1955) studied the frequency of chiasmata and the pattern of segregation of a pair of heteromorphic chromosomes in *Lilium formosanum*. The frequency of first and second division segregation, i.e., the occurrence of two equal chromatids or a long and short one at anaphase I, was in accord with the hypothesis that each chiasmata seen at diplotene or diakinesis represents a crossover, if segregation of centromeres is regularly reductional. The diplotene configurations produced by a paracentric inversion in *Lilium* and a pericentric inversion in maize (ZOHARY 1955) were also in accord with the same idea. More recent papers by NODA (1960), KAYANO (1960), JAIN and BASAK (1963) and ZEN (1964) supplement the evidence for the chiasmatype hypothesis presented earlier by DARLINGTON (1935, 1937) and MATHER (1938).

However, all of these studies including the classic demonstration of new chromosome arrangements associated with crossing-over (STERN 1931 and CREIGHTON and McCLINTOCK 1931) could be explained by what is now called copy choice in the duplication of chromosomes (BELLING 1928). Finally, however a study of exchanges among tritium-labeled chromatids in spermatocytes of the grasshopper shows that breakage and exchange of homologous chromatids actually occurs in meiosis (TAYLOR 1965). Moreover the frequency of the exchanges between labeled and unlabeled chromatids is exactly that which would be predicted by the chiasmata frequency.

The classic studies of CREIGHTON and McCLINTOCK on *Zea Mays* (1931) and STERN (1931) on *Drosophila* finally established the correlation between genetic crossing over and the exchange of chromosomal segments. While there may be exceptional types of non-reciprocal recombinations that do not involve exchange of large segments (LINDEGREN 1956, MITCHELL 1957), the validity of the classical hypothesis of reciprocal exchange between homologous chromosomes has not been seriously challenged. There is, however, a paucity of information concerning the time at which these exchanges occur and the mechanism involved.

Since the chiasmata appear at the transition between pachytene and diplotene, this has often been assumed to be the period at which crossing over occurs. However, it is quite possible that crossing over is not limited to any one period of meiosis, but occurs at any time at which homologous chromosomes are paired. This would appear to restrict crossing over to the period between zygotene and anaphase I, but there is the possibility that regions of homologous chromosomes are paired temporarily before zygotene. SMITH (1942a, b) has presented evidence that pairing is initiated as early as late telophase of the mitosis preceding meiosis in certain insects. However, there is no evidence that this behavior is typical of any of the higher plants.

The fact that two parental and two crossover types are recovered from a single tetrad indicates that crossing over occurs at a four strand stage. Additional evidence is the observation that the percentage of recombinations between any two loci rarely exceeds 50 percent regardless of their genetic map distance. Since exceptions occur, crossing over at the two strand stage remains a possibility. However, its regular occurrence, as suggested by MOEWUS (1940) on the basis of the recovery of only ditype segregation in *Chlamydomonas* at temperatures below 5° C, was not supported by similar experiments on another species of this alga (EBERSOLD 1956). PERKINS (1955) has reviewed the data available from tetrad analysis and pointed out its usefulness in solving some of the problems of meiosis.

Since chromosomes become effectively double before meiotic prophase (TAYLOR 1957, 1958b MITRA 1958 and SAUERLAND 1956) the knowledge that crossing-over usually occurs at a four strand stage does not help in locating the period of crossing over. Experiments on the effect of various environmental factors on chiasmata frequency suggest that crossing over may be affected by treatments as late as early pachytene, but not in late pachytene and diplotene (STRAUB 1937, 1938). Heat shocks reduced chiasma frequency when applied at early pachytene; milder shocks were effective if applied during leptotene. TAYLOR (1949b) reported effects on the frequency of interlocked bivalents from osmotic shock produced by placing anthers in culture. Since the greatest effect was obtained by making the change at late pachytene, a clue to the time of crossing over may be at hand (see section on "Aberrations produced by culture of excised anthers" below).

Recent reports tend to confirm the hypothesis that crossing over occurs in prophase after the major synthesis of DNA is completed. ABEL (1965) showed that intergenic recombination, which could be influenced by temperature, occurred after the sporocytes of *Sphaerocarpus* were in leptotene and therefore following the period of chromosome duplication. HENDERSON (1966) showed that heat treatments could reduce chiasmata frequency in grasshopper spermatocytes when applied 2—3 days after premeiotic DNA synthesis. DNA synthesis was timed in relation to the heat treatment in the same individuals by labeling with H^3-thymidine just before treatment and noting the sequence in arrival of labeled cells and cells with reduced chiasmata at metaphase. However, a similar study with *Drosophila* females (GRELL and CHANDLEY 1965) indicated that the time of recombination coincided with DNA replication.

4. Variations in the coiling cycle.

The coiling characteristic of meiotic chromosomes has been described many times in various species with large chromosomes (see reviews by KAUFMANN 1949, and MANTON 1950). It is now recognized that, at least in forms with large chromosomes, coils exist in the leptotene chromosomes (RIS 1945 and OEHLKERS 1957). This means that coiled chromosomes pair during zygotene. Behavior of this type does not appear to present any greater difficulties than the events during later meiotic stages that lead to the separation of the chromatids which are coiled together. The gyres of the coils evidently mesh with each other and the really significant pairing probably occurs at the molecular level anyway.

The gyres increase in diameter during pachytene and the chromosomes shorten accordingly. During diplotene another set of coils develop. The chromosomes at diakinesis and metaphase I have two sets of coils, major and minor. The major coils are fewer in number and larger in gyre diameter than those in metaphase chromosomes of somatic divisions. The difference may be, in part, related to the fact that each two chromatids are coiled together as one unit rather than separately as in regular divisions. The major and minor coils persist through anaphase and telophase. The behavior of coils during the interkinesis is of three types according to MANTON (1950). In one type there is no essential change and no reorganization of the chromosomes occurs before the next division. *Trillium* and *Vicia* are examples. A second and third type have the chromosomes relatively uncoiled during the interphase between the first and second divisions of meiosis. At the second division the coils are more like those at a somatic mitosis. MANTON (1950) distinguished between the behavior of *Osmunda* and *Tradescantia*. *Osmunda* represents a type in which the coiling at the second division is intermediate between that typical of the first division and a somatic mitosis with gyre numbers in the ratio of 4:8:16 for the three types of divisions. Although the second divisions chromosome of *Tradescantia*, *Rhoeo* and *Lilium* are more like those of a somatic mitosis than those of *Osmunda*, they are not indistinguishable. Considerable variation occurs depending on the environmental conditions, but the chromosomes of the second division of meiosis can regularly be distinguished from those of somatic tissues on the basis of their coiling pattern.

There is no general agreement on a universal pattern of coiling. Actual differences between various species may exist. A number of authors have described a coiling pattern in which many coils of small gyre diameter exist in prophase of both meiotic and mitosis. As prophase progresses the number of gyres is reduced by some process of gyre elimination so that a much smaller number of large gyres remain at metaphase (Sparrow 1942, Manton 1950 and Swanson 1957). On the other side Cleveland (1949) maintains that the change involves no reduction in gyre number, but that the major coil arises as kinks in the once-coiled early prophase chromosomes. These irregular kinks gradually transform into the helix within helix seen at metaphase.

Observations and speculations on the mechanism of coiling have provoked a number of hypotheses. A torsional mechanism is almost certainly involved although several authors who have studied coiling in the past have attributed it to contraction of a matrix (Sax and Humphrey 1934) or to the elongation of the chromonemata within a matrix (Huskins 1941). A number of features of coiling make it difficult to explain in terms of an internally directed torsion. The coiling is random in direction for any given segment or for the whole chromosome; the direction may reverse at any point and there is no necessary quantitative relation between the number of right and left turns in any one chromosome. As pointed out by Huskins (1941) the mechanism is probably only indirectly if at all related to internal molecular coiling as postulated by Darlington (1935). A final solution of the problem requires a better knowledge of the fine structure of chromosomes. However, a model that would explain the mechanism of coiling and can account for all of the known features of the cycle has been suggested by Taylor *et al.* (1957) and Taylor (1958a).

5. The unitary behavior of the centromeres of homologous chromosomes.

The unitary behavior of the centromeres of dyads accounts for the segregation which occurs at the first anaphase. The centromere of each dyad, whether divided or single, acts as a unit in orientation on the spindle and movement to the poles. In this way two chromatids of each bivalent (tetrad of some authors) are moved to opposite poles. During the late metaphase or the early anaphase movement (Fig. 3) the chromatid arms of each dyad, which have been coiled together during all of the previous stages, separate freely indicating that the coils were essentially paranemic. The two chromatids remain closely associated only at the centromere region. During the prophase and metaphase of the second division the chromatids are often united only at the centromere (Fig. 5 and 6) unlike somatic divisions in which chromatids lie side by side until anaphase. Lima-de Faria (1953) believes that the chromatids are attached not at the centromere, but at loci on each side of it. As a result of crossing-over parts of the chromatids of each dyad are homologous rather than identical as in regular divisions. Therefore the second division is essential to complete the segregation of homologous chromosomes. It has long been recognized that the second division is in many respects more like a regular somatic mitosis than the first division. It is distinguished, in addition to the above mentioned differences, by the absence of chromosome duplication in the preceding interphase or interkinesis between the two divisions. There are some who still entertain the idea that the segregation of the centromeres at the first division may not be regularly reductional (however, see section I, B, 4, above). The best direct evidence on this point is based on tetrad analysis which is rarely possible in the higher plants. The evidence from bivalent interlocking also strongly favors the idea that separation is regularly reductional. The non-homologous chromosomes pass through the loop containing the centromere in a high percentage of the cases. A non-homologous chromosome has never been observed to be trapped between two chromatids of the same dyad.

C. Synthesis of nucleic acids and proteins.

1. DNA (deoxyribonucleic acid) synthesis.

Synthesis of DNA in higher plants is usually completed before zygotene pairing begins. The evidence comes from two sources. Microphotometric measurements of the Feulgen-dye content of individual nuclei (SWIFT 1950, 1953) in maize, *Lilium* and *Tradescantia* show that the amount of DNA doubles before leptotene except in *Tradescantia* where the synthesis extends into leptotene. Autoradiographic studies of incorporation of ^{32}P into individual nuclei showed that DNA-phosphorus was incorporated only during this same period in *Lilium longiflorum* and *Tradescantia* (TAYLOR 1953, TAYLOR and MCMASTER 1954, MOSES and TAYLOR 1955) and in *Lilium Henryi* (PLAUT 1953). As shown in Fig. 9 the incorporation of ^{32}P into DNA of microsporocytes of *Tulbaghia violacea* also occurs prior to leptotene (TAYLOR 1958b). The incorporation of C^{14}-orotic acid into DNA of *Lilium longiflorum* (Fig. 10) confirms the data from studies with P^{32} (TAYLOR 1959). In none of these studies was there any indication of synthesis of DNA during any stage of meiotic prophase, with the exception noted for *Tradescantia*. An exception has been reported for *Trillium* as a result of the microchemical analysis of large groups of isolated pollen mother cells. DNA appeared to increase during most of prophase and reached a maximum during diplotene (SPARROW, MOSES and STEELE 1952). However, these results have not been confirmed by studies of synthesis in individual cells. The first division of meiosis, interkinesis, and the second division are all completed without additional synthesis of DNA.

The two exceptions mentioned in section I, B, 2 should be kept in mind. WIMBER and PRENSKY (1963) reported an incorporation of H^3-thymidine into spermatocytes of the newt, *Trituris*, at late pachytene. The amount was very small compared to that which occurs during regular DNA replication preceding meiotic prophase. However, a recent attempt to verify this report (TAYLOR and CALLAN, unpublished) indicates that most if not all of the incorporation is into a cytoplasmic fraction. If any H^3-thymidine is incorporated into the chromosomes at late pachytene and early diplotene, it probably can never be demonstrated autoradiographically because of the cytoplasmic incorporation. The report of an incorporation of P^{32} and H^3-thymidine into DNA of microsporocytes of *Lilium* and *Trillium* (HOTTA et al. 1966) is much more convincing. However, the amount is less than 0.3 of one percent of that which occurs in regular premeiotic replication. They took several precautions to avoid contamination from the tapetal nuclei which rapidly synthesize DNA during most of meiotic prophase. They also isolated nuclei of spermatocytes to show that the incorporation was not cytoplasmic. However, the DNA had peculiar density characteristics in a cesium chloride gradient and a base ratio unlike a random sample of the

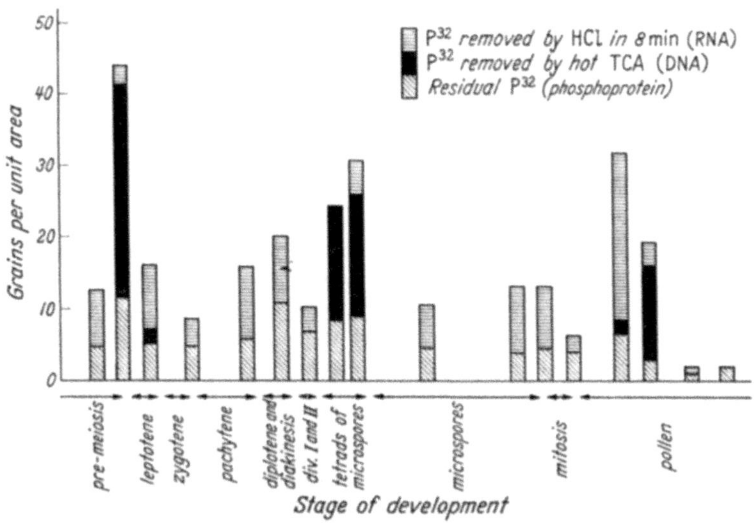

Fig. 9. Graph showing the incorporation of P^{32} into nucleic acids and proteins determined by autoradiography: Incorporation for 24 hour-periods during the 16—20 days required for microgametogenesis in *Tulbaghia violacea* are shown. (After TAYLOR 1958b.)

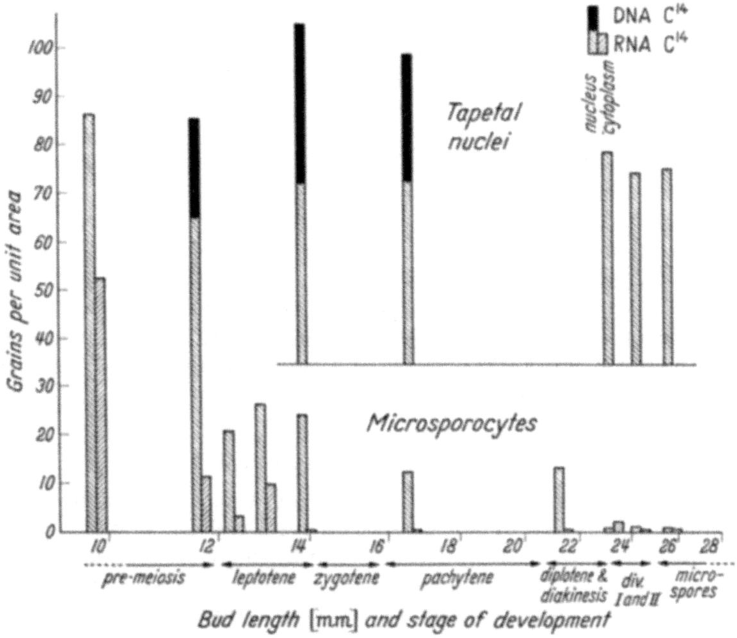

Fig. 10. Graph showing the incorporation of C^{14} from orotic acid-2-C^{14} into nuclei of *Lilium longiflorum*. Incorporation was allowed during periods of about 24 hours at 29° before anthers were fixed and prepared for autoradiography. At this temperature meiotic prophase lasts about 5—6 days. (After TAYLOR 1959.)

DNA. Whether its production is related to genetic recombination must remain an open question although this idea is an interesting possibility.

The behavior of tapetal cells during meiosis has been given considerable attention. In *Lilium*, *Tradescantia* and *Tulbaghia*, these nuclei undergo two divisions during the period when the microsporocytes are in meiotic prophase. The first mitosis occurs when the microsporocytes are in zygotene or early pachytene and the second one when they are in late pachytene or diplotene. The mitoses are usually not accompanied by cytokinesis and the spindles frequently fuse during the second mitosis with the result that tetraploid nuclei occur in some of the multinucleate cells. DNA content of each nucleus is doubled prior to each division (Fig. 10). No further synthesis of DNA occurs after the second division (TAYLOR and McMASTER 1954, TAYLOR 1958b).

The nuclei of microsporocytes and tapetal cells become susceptible to extrusion of globules of DNA from the cell at the beginning of zygotene. The extrusion frequently occurs during fixation, but may be induced by various osmotic or mechanical disturbances of anthers during these stages. Transfer of DNA containing globules into the cytoplasm and between cells has been inferred from observing these extrusions (SPARROW and HAMMOND 1947 ,COOPER 1952). Recently TAKATS (1959) has made an extended study of the phenomena and his work supports the idea that the globules of chromatin are extrusions which do not represent a normal metabolic process in anthers. The studies with isotopic tracers show that there is normally no transfer of DNA between cells during meiosis by extrusion or any other method. The function of the tapetum appears to begin late in meiosis or perhaps not until the spore walls begin to form after meiosis.

With respect to DNA metabolism, the cell differentiating into a microsporocyte is not appreciably different from any other cell. In *Lilium* and *Tulbaghia* the asynchronized synthesis of DNA, which normally precedes each mitosis in the sporogenous tissue, stops a few days before leptotene. Cells with the doubled amount of DNA continue to divide asynchronously until all cells have the amount of DNA typical of early interphase nuclei. After a day or two in this state synthesis of DNA begins in the microsporocytes at the top of each anther at about the same time. The wave of synthesis appears to proceed downward in a very regular manner so that the cells at the base of the anther may be just entering the period of synthesis when those at the top have completed the process. Whether the synchrony is imposed at the period just preceding DNA synthesis or whether the gradient in development occurs earlier in the differentiation of the anther is not clear. The gradient, which is evident when stages of division are examined along the length of the anther, is maintained during the whole of meiosis.

2. RNA (ribonucleic acid) synthesis.

Both preceding and following DNA synthesis in sporogenous cells (Fig. 9 and 10), there is a period of cell growth during which RNA synthesis reaches a peak (MOSES and TAYLOR 1955, TAYLOR 1958b, 1959). This is shown by the incorporation of P^{32} and C^{14}-orotic acid into a fraction removed by ribonuclease. The incorporation occurs in both nucleus and cytoplasm. By the time DNA synthesis begins the rate of this incorporation has dropped. During DNA synthesis the rate remains relatively low, but rises again in early leptotene. By zygotene incorporation reaches a very low level in both cytoplasm and nucleus. There is additional synthesis during early pachytene, but this is confined largely to the nucleolus in lilies. In *Tradescantia* the nucleolus disappears during this period and incorporation of label into RNA is at a low level until diplotene-diakinesis when the chromosomes revert to a "diffuse", semi-interphase condition. The synthesis of RNA as measured by incorporation of P^{32} reaches a maximum at this period, but ceases before the first metaphase. Synthesis is low or absent during the remainder of meiosis. The behavior in *Lilium* differs in that the incorporation into the cytoplasm is almost absent from leptotene until divisions are completed. However, incorporation into the nucleus continues at a relatively low rate up until diakinesis. *Tulbaghia* shows more incorporation than *Lilium* but relatively less than *Tradescantia*. There appears to be a positive correlation between the diffuseness of the diplotene chromosomes and the rate of RNA synthesis.

Synthesis of RNA in the tapetal cells occurs during most of meiosis. It continues at a high rate during the period after the last mitosis. This apparently is a growth period during which the tapetum is differentiated for its functions during the early microspore stages and perhaps beginning in the late stages of meiosis.

3. Protein synthesis.

Not much information is available concerning protein synthesis during meiosis in higher plants. A study of incorporation of S^{35} when inflorescences were grown in labeled sulfate showed that the largest amounts of sulfur were bound by the nucleus during the period of DNA synthesis in *Tradescantia* (TAYLOR and TAYLOR 1953). A more extensive study with glycine-2-C^{14} in *Lilium* revealed a more

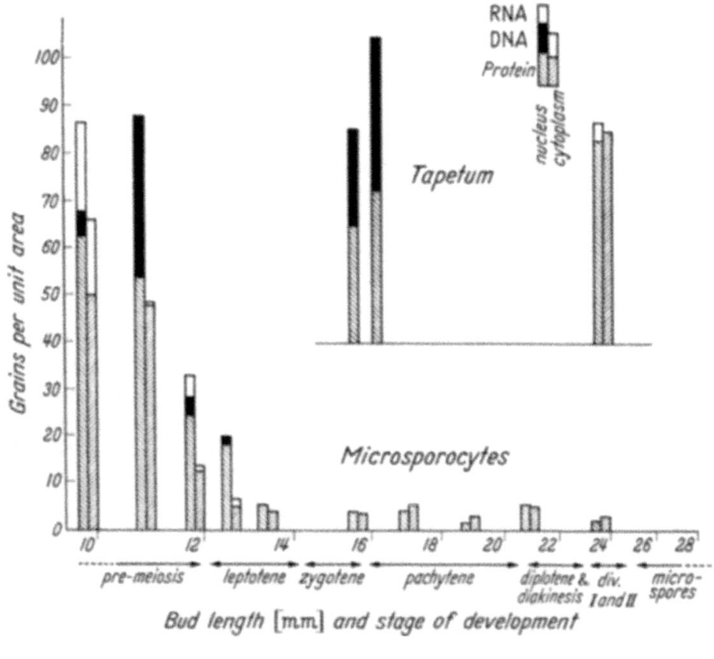

Fig. 11. Graph showing the incorporation of C^{14} into microsporocytes and tapetal cells of *Lilium longiflorum* from glycine-2-C^{14} during periods of about 28 hours at 20°. Autoradiographs were made of replicate slides after fractional removal of the isotope by the use of ribonuclease and deoxyribonuclease. (From TAYLOR 1959.)

complete picture of protein and nucleic acid synthesis as measured by the incorporation of C^{14} from glycine (TAYLOR 1959). Incorporation is rapid in the microsporocytes at the growth period preceding DNA synthesis (Fig. 11). Labeled proteins accumulate rapidly in both nucleus and cytoplasm, with the concentration about 20—25 percent higher in the nucleus. During this period RNA is also rapidly labeled by the glycine-C^{14} in both nucleus and cytoplasm. During

DNA synthesis the incorporation into nuclear and cytoplasmic proteins continues at about equal rates and only slightly reduced as compared to the period preceding DNA synthesis. However, the rate of synthesis of proteins diminishes rapidly during leptotene and by zygotene is quite low in the nucleus and cytoplasm. Incorporation continues at about the same rate throughout the remainder of the meiotic prophase. No additional incorporation occurs during spindle formation. Therefore, the spindle proteins must be synthesized before the spindle makes its appearance. In the early stages while the sporogenous cells are most active in incorporation of glycine-C^{14}, the tapetal cells are relatively inactive. They begin their rapid uptake a day or more preceding the mitotic division which occurs when the microsporocytes are in zygotene. Protein and DNA synthesis either occurs simultaneously or the exposure to the isotope was long enough so that they appear to coincide (Fig. 11). In the two groups of nuclei selected for counting when the incorporation was at a peak, there was no detectable RNA synthesis (Fig. 11). Both these examples are at the stage preceding the first mitosis. A similar incorporation precedes the second mitosis and the tapetal cells are heavily labeled while the adjacent microsporocytes in pachytene have only a small amount of label.

At the late stages of meiosis the tapetum also incorporates glycine-C^{14} rapidly into proteins but DNA is not labeled at all. No labeled RNA was detected in the cytoplasm when glycine-C^{14} was the source of label, but a small amount accumulated in the nuclei. The incorporation of orotic acid-C^{14} into the RNA demonstrates that synthesis continues in the tapetum until meiosis is complete. However, the rate of incorporation slows down in the late stage of meiosis and reaches a low level by the time the microspores are formed. It may be noted that there is no evidence for transfer of labeled proteins or nucleic acids from tapetum to the adjacent microsporocytes which are labeled very lightly at this stage.

However, in studies with sulfate-S^{35} the tapetum appears to transfer a labeled component which is incorporated into the wall of the microsporocyte at late stages of meiosis and later more extensively into the wall of the developing microspore. The concentration of the substance is high in the tapetal cytoplasm and also accumulates in droplets along the inner walls of the tapetal cells adjacent to the microspores. The microspore walls accumulate by far the highest concentration of this material. Sulfate is bound during the formation of cell walls at other times but not in the concentration observed here when the thick walls are laid down. Part of the sulfate can be released by digestion with pectinase (TAYLOR and TAYLOR 1953). Perhaps the sulfate is bound to mucopolysaccharide, but no further attempts have been made to identify the material. Since glycine is not incorporated extensively at this period, the substance must not be proteinaceous.

Since this article was originally written, DE (1961) studied the synthesis of histones in microsporocytes of *Tradescantia* by means of autoradiography with H^3-arginine. Most of the histone-like proteins were synthesized during DNA replication at premeiotic interphase and early leptotene, but some synthesis continued during the post-DNA synthetic period and some histone-like proteins were labeled in the nucleolus. HOTTA and STERN (1963) studied the effects of various inhibitors of protein synthesis on the course of meiosis in cultured anthers of *Trillium*. They found a low, but significant synthesis of proteins through most of meiotic prophase. However, the rate of incorporation of C^{14}-leucine was greatest during late pachytene and diplotene. Inhibitors which stopped protein synthesis also blocked development of microsporocytes.

II. Aberrations of meiosis.

A. Genetic aberrations.

Perhaps the most frequent changes in meiosis are those that cause it to revert toward mitosis in one or several features. In various species that have apomictic forms, meiosis may fail completely, and be replaced by mitotic divisions which result in unreduced gametes. In embryo sacs of some forms these cells may function to produce an embryo without fertilization. Various intermediate types have been found in which meiosis proceeds in various ways. DARLINGTON (1937) has reviewed some of the instances. One of the best examples is *Artemisia nitida* which has various kinds of apomixis. In some forms pairing occurs, in others no stage equivalent to pachytene is seen. In one with a pachytene stage meiotic-like contraction of chromosomes occurs but the chromosomes at diakinesis are unpaired. Two diploid megaspores are formed.

In various species normal sporogenesis may be interrupted at almost any point by changes that have occurred during evolution. Some of these changes lead to inviable cells, but frequently they result in various types of apomixis. These are all genetic aberrations, although only a few have been demonstrated to be due to specific Mendelian factors. Asynapsis in *Zea Mays*, *Crepis*, *Datura*, and tomato are examples known to be due to single factors. The mutant producing asynapsis in *Zea* was described by BEADLE (1933). Although pachytene pairing occurs the chromosomes separate during diplotene and diakinesis and by metaphase I a variable but generally low number of chromosomes remain paired. Enough viable male gametes are formed to permit a study of crossing over in homozygous asynaptic plants. Rather unexpectedly RHOADES (1947) in a study of the frequency of crossing over in the chromosomes from both reduced and unreduced gametes found a higher than normal frequency. An asynaptic triploid race of *Allium* has been described in which zygotene and pachytene pairing appear quite normal, but the number of chiasmata observed at metaphase was extremely low compared to that which occurs in any of the above mentioned cases. The frequency was about 100th percent of the expected, but two trivalents were observed among many cells examined. The random distribution of chiasmata could hardly have produced these trivalents. Similar indications of non-random distribution of crossing over and chiasmata were observed by RHOADES and BEADLE in the asynaptic *Zea*.

A gene "el" described by RHOADES (unpublished) affects the pattern of coiling and produces very elongated chromosomes, particularly noticeable in anaphase II. Another locus or group of loci in chromosome 10 induce knobs to act as precocious centric centers that move to the poles ahead of the normal centromeres (RHOADES 1955). These result in preferential segregation of certain linked loci in megasporogenesis.

The list of genetic changes affecting meiosis could be made very long, but only a few of them have been studied in enough detail to yield information on the physiology of the process. Many of the changes can also be induced by varying the environmental conditions. Perhaps the best way to accomplish these changes is by culture of anthers *in vitro* which is considered in the next section.

B. Culture of excised anthers.

The ideal method for the study of the physiology of meiosis in higher plants would be the culture of sporocytes isolated from the plant. Although this has often been attempted, success has been limited. The cells survive only a few hours. SHIMAKURA (1934) reported the survival of isolated *Tradescantia* microsporocytes from metaphase I through both divisions in sucrose solution. Survival for longer periods has not been reported. Unfortunately most of the interesting and significant events of meiosis that distinguish it from other nuclear divisions occur in the prophase of division I. Survival at this stage is of short duration when microspores are isolated. Recently Hotta et al. (1966) report more success in culturing isolated microsporocytes of *Lilium* and *Trillium*, but details were not given.

The next most promising technique for experiments on meiosis in higher plants is the culture of excised whole anthers. Here again success has been limited, but there may be hope for improvement. Gregory (1940) reported experiments on the culture of anthers removed from flowers at various stages of development. With *Lilium longiflorum* he was able to get anthers with microsporocytes at diplotene to survive until meiosis was complete. Microsporocytes in anthers removed at earlier stages did not reach division, but after a number of days the sporogenous tissue changed to a parenchyma-like undifferentiated tissue. On the basis of these experiments, and grafting experiments with tomato buds in which he was able to get meiosis to continue, he advanced the hypothesis that for continued survial anthers required some essential substances other than sugar and minerals supplied from the remainder of the plant.

In an attempt to supply these hypothetical growth factors Taylor (1950b) added several substances to a basic medium used by Bonner (1943) for growing excised roots. Improvement in survival was indicated for complex mixtures such as casein hydrolysate and coconut milk, but the percentage of anthers which reached division stages was too low to make experimental studies practical without further improvement in the technique of culture.

Excised anthers of different species appear to have rather different potentials for surviving. Lima-de Faria (1950) reported the survival of anthers of barley from meiotic prophase through the pollen mitosis without any additions to White's (1943) basic medium. St. Gimesi *et al.* (1949) reported the production of pollen by anthers of *Lilium regale* excised at meiosis and cultivated in a mineral solution to which was added only glutamic acid or valine. The stage of meiosis at excision and the viability of the pollen were not mentioned. Attempts to repeat the experiments with *Lilium longiflorum* (Taylor, unpublished) were not very successful. A little improvement in the survival at meiotic prophase stages was indicated with glutamic acid. Microsporocytes sometimes survived from late leptotene through both divisions, but the divisions were frequently rather abnormal. No microspores ever reached the division stages in these cultures. The thick wall of the microspores did not develop in anthers which were excised with microsporocytes at early meiotic prophase. Excision at late stages of meiosis resulted in thick-walled microspores without any ability to divide. Sparrow *et al.* (1955) reported survival of *Trillium* anthers in various modifications of the medium used by Taylor (1950b). Survival from late stages of meiosis through microspore division was common. Recently, Vasil (1957) reported considerable improvements in the survival of *Allium* anthers cultivated in a simple medium to which kinetin and gibberellic acid had been added. A high percentage of cells were reported to survive from early meiotic prophase through both divisions. However, in our laboratory addition of these substances to the medium used for culturing anthers of *Lilium longiflorum* failed to make any noticeable improvement (Kosan, unpublished).

Microsporocytes in cultured anthers show many of the aberrations which are known to be induced by genetic changes in other instances. Various degrees of asynapsis occur (Taylor 1950b) depending on how early the anthers are placed in culture. Cells that have begun to differentiate into microsporocytes may change and undergo rather typical mitotic divisions. This is very striking in *Podophyllum* (TAYLOR 1950a). Anthers were placed in culture about the time meiotic prophase began. Five or six days later many of the anthers had hundreds of cells in synchronized mitotic-like divisions. Some of the cells had typical mitotic chromosomes with two separate chromatids. In others the two chromatids remained coiled together as they typically are in meiotic prophase. Intermediate conditions existed. After division the cells formed a parenchymatous tissue which failed to differentiate further. Samples excised at earlier stages showed similar cells in prophase stages. Pairing failed to occur, but many of the cells had the two chromatids of each chromosome closely associated as in typical meiotic prophase. Other cells exhibited typical mitotic prophases. Tapetal cells and microsporocytes were in some instances not distinguishable in smear preparations.

In cultures of *Tradescantia* anthers tapetal cells were found with chromosomes having an organization similar to meiotic chromosomes. At metaphases the two chromatids of each chromosome were closely associated and coiled into large coils nearly like the typical major coils of the meiotic chromosomes. Intermediate conditions and typical mitotic chromosomes were also observed.

These observations suggest that there is no trigger mechanism that sets the meiotic and tapetal cells, for example, on different paths from which they may not turn back. Differentiation appears to be a continuing process and cells may achieve various degrees of meiotic behavior. Likewise they may retain most of the characteristics and lose or gain only a few traits of the adjacent tissue. The mechanism of differentiation can probably be studied with profit in meiotic cells for these are one of the few types of cells with a potential genetic future which undergo considerable differentiation in higher plants. So far the results have been rather unpromising but with the application of proper techniques much may be learned.

Another result was obtained from cultures of excised anthers of *Tradescantia* which may throw light on the time at which crossing over occurs. A large increase in the percentage of interlocked bivalents occurred when the anthers were placed in culture at very late pachytene, about 18—20 hours before metaphase I (TAYLOR 1949b). Although the suggestion was made that the interlocks might be associated with desynapsis and secondary pairing of the ends of homologous chromosome, later studies (unpublished) gave no support for this hypothesis. The studies showed that the effect of culture was probably in the nature of osmotic shock. The same results could be produced with far greater reproducibility by immersing the cut stems of inflorescences in strong sucrose solutions to produce severe wilting. When the stems were put back into mineral solution without sucrose the plants recovered and meiotic prophase was not interrupted. Chiasma frequency at metaphase was not reduced, but the number of interlocked bivalents increased as much as in the cultures.

The most likely explanations for the increae in interlocking is that the treatments cause chiasma to form closer to the ends of the chromosomes and thereby retain a higher percentage of the chromosomes trapped between paired homologs. These remain as interlocked ring bivalents until anaphase I. If this interpretation is correct the time of crossing over which has often been equated with chiasma formation may occur at the transition between pachytene and diplotene as several hypotheses predict. (see section I, B, 3).

C. Aberrations induced by ionizing radiations.

Meiosis appears to be more sensitive to ionizing radiation than most other stages in the life cycle. Chromosome breaks are induced at a high frequency

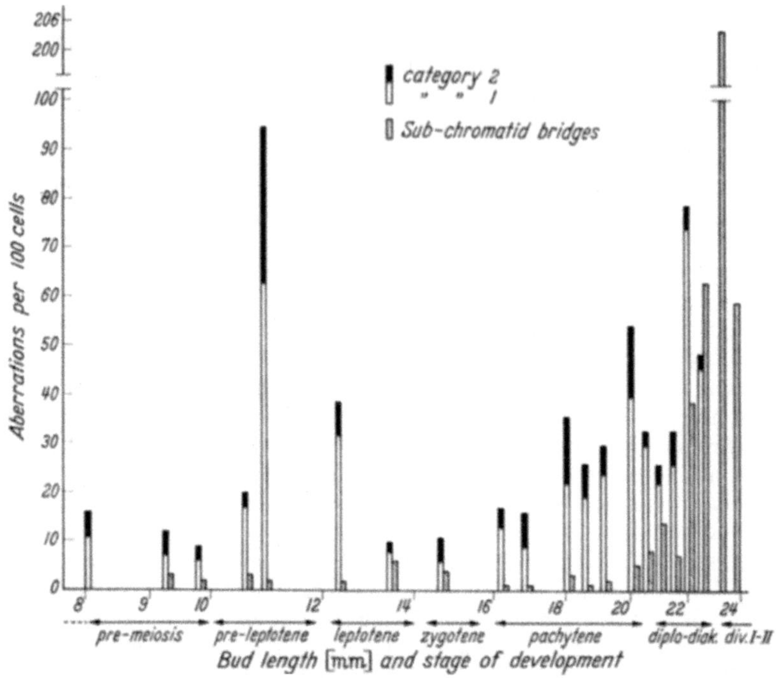

Fig. 12. The frequency and type of chromosomal aberrations induced by 30 r of x-radiation applied at various stages of meiosis and pre-meiosis. The category 1 aberrations were those produced by two breaks while those in category 2 could not be classified in this respect. (After MITRA 1958.)

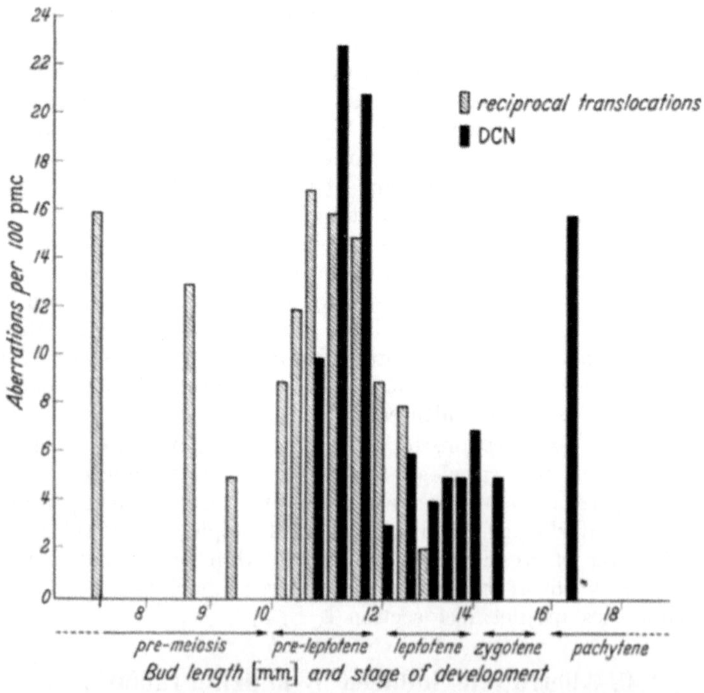

Fig. 13. Graph showing the period of transition from chromosome type aberrations to chromatid type in premeiotic interphase and early meiotic prophase of *Lilium longiflorum*. The chromosomes first act as if composed of two chromatids at late premeiotic interphase which appears to be the period of chromosome duplication. (After MITRA 1958.)

by low doses of x-rays. In *Trillium* the most sensitive period is diplotene and diakinesis (SPARROW *et al.* 1952). In *Lilium* (Fig. 12) the same period of sensitivity was found and in addition another sensitive period occurs at late premeiotic interphase (pre-leptotene) when DNA synthesis is occurring (MITRA 1958).

MITRA (1956, 1958) has made a study of the time at which the chromosomes of *Lilium* become double to x-ray breakage and rejoin in a manner which shows independent behavior of the two chromatid (Fig. 13). Buds were irradiated at various stages which could be estimated from the length of the buds (TAYLOR and McMASTER 1954). Only chromosome type aberrations were produced by irradiation at stages up to late premeiotic interphase. X-rays applied at late premeiotic interphase and later produce a type of aberration which can only be explained on the assumption that two chromatids were present in the chromosomes at the time of reunion. The aberration is a chromatid bridge between non-homologous centromeres (DCN) which either go to the same or opposite poles at anaphase I.

A new type of aberration begins to appear when meiotic chromosomes are irradiated at late pachytene or diplotene. This new type of aberration (Fig. 4) appears to be a reunion at a sub-chromatid level. Bridges are formed at anaphase without free fragments. The chromatids (homologous or non-homologous) adhere at a single point. The aberration has been interpreted as a break and reunion between half chromatids (CROUSE 1954, SAX and KING 1955). Since these aberrations are not produced until the diplotene separation begins at late pachytene, they may coincide with some little understood reorganization of the chromosome. The sensitivity to this type of aberration increases after pachytene and reaches a peak in late diplotene. A similar result was obtained by SPARROW *et al.* (1957) from studies with *Trillium.*

These data confirm the older observations which occasionally indicated that the leptotene chromosomes are already double. Synthesis of DNA and the major part of the protein of the nucleus occurs simultaneously in premeiotic interphase. This period must be the time of chromosome duplication.

III. Summarizing statement.

The hypothesis of meiosis which attracted the most attention in the 1930's was DARLINGTON's precocity theory (DARLINGTON 1932, 1937). The basic assumption of this hypothesis was that the nuclei entered the meiotic prophase before the chromosomes were duplicated. As a result the chromosomes pair because of the natural tendency of homologous or identical chromatids to lie together in pairs. In regular divisions the duplicated chromosomes with two chromatids each satisfy this tendency and no pairing occurs. During pachytene the chromosomes were supposed to duplicate. Simultaneously, with the pairing tendency satisfied by the two sister chromatids, the homologs began to separate. Crossing-over according to DARLINGTON's hypothesis resulted from torsions that developed when relational coiling of chromosomes occurred. When the strain broke a chromatid at any one point the additional stress placed on the other three frequently resulted in the breakage of another at the same locus. Rotation and reunion frequently produced exchanges between homologous chromatids.

BELLING (1928) also proposed a hypothesis of crossing-over based on the assumption that chromosomes duplicate after pairing. According to his hypothesis no breakage was required. Chromomeres were assumed to form identical chromomeres along side of themselves during duplication. These were then connected by interchromomeric fibers. Since connections grew to the nearest loose chromomere, connections were likely to link chromomeres produced by homologous chromosomes where a sharp twist occurred along the paired chromosomes.

Unfortunately we must now reject both of these hypotheses in their original form. Many morphological studies have indicated that the chromosomes are doubled at earliest meiotic prophase. Moreover the studies of time of synthesis of DNA and protein show that in most species the chromosomal DNA and most histones are produced before zygotene pairing begins. In addition the chromosomes are functionally composed of two chromatids before leptotene in *Lilium* at least (MITRA 1958, SAUERLAND 1956). However, the recent reports which indicate that a small amount of DNA synthesis occurs during meiotic prophase may provide a way out of the apparent anomalous situation. The hypothesis of a delayed synthesis of one strand of the DNA in a few replicons (TAYLOR 1967) might be regarded as a modification of DARLINGTON'S precocity theory. But even this remains a theory with very little evidence to support it. The most widely discussed hypothesis of recombination is based on the demonstration of repair synthesis of DNA in microorganism. The hypotheses of WHITEHOUSE (1963) and HOLLIDAY (1964) depend heavily on these hypothetical mechanisms of repair, which remain to be demonstrated as possible in higher organisms and then to be associated with recombination. The only clearly demonstrated events associated with crossing over is the breakage and exchange of segments of DNA in lambda phages (MESELSON and WEIGLE 1961, KELLENBERGER et al. 1961) and the breakage and exchange of segments between homologous chromatids which occurs in meiosis (TAYLOR 1965). The events which many be associated with these exchanges, particularly the aberrant segregation of markers (OLIVE 1959, LISSOUBA et al. 1962 and others), which has sometimes been called gene conversion can be accounted for, in theory at least, by the filling of gaps left in one or both strands of the DNA near the point of reunion (ANRAKU and TOMIZAWA 1965). Although these subtle changes may be demonstrated in phage DNA where the systems lend themselves to very precise analysis, it seems doubtful that presently available techniques will ever allow the analysis to be extended to the complex systems of higher cells.

The suggestions of PONTECORVO (1958) and others that a significant part of crossing over occurs before zygotene pairing has lost some of its former appeal with the demonstration that events as late as pachytene can influence genetic recombination and chiasmata formation. In addition, the study of ROSSEN and WESTERGAARD, which shows that DNA replication in the ascomycete *Neotiella* occurs before fertilization when the chromosomes are in different nuclei, is difficult to reconcile with this idea. However, as long as the possibility remains that part of the DNA is single stranded because of a block in replication or for any other reason, the idea of prezygotene pairing of at least part of the DNA is possible. For example, if a few long loops, similar to lampbrush loops, extended out from a chromosome and paired with homologous regions, presumably single stranded, the contraction of these loops would produce zygotene pairing by pulling together homologs. The phenomenon, if could be watched under a microscope, would look like a long range force, because the connections would be invisible.

In the original manuscript written in 1958 the statement was made that many of the basic problems in meiosis must for the present be considered unanswered. In the seven years since that statement was made some significant progress in experimental techniques has been made. In addition, much more speculation, and hopefully, some constructive thinking has been given the problem by more people than seemed interested at that time. The optimism expressed seven years ago that tools and techniques were available still holds and progress in understanding allows one to predict that progress in experimental work will follow from the changing concepts.

Literature.

ABEL, W. O.: Über den Zeitpunkt des crossing-over und der Chromosomenverdopplung bei *Sphaerocarpus*. Z. Vererbungsl. **96**, 228—233 (1965). — ANRAKU, N., and J. TOMIZAWA: Molecular mechanisms of genetic recombination in bacteriophage. III. Joining of parental polynucleotides of phage T 4 in the presence of 5-fluorodeoxyuridine. J. molec. Biol. **11**, 501—508 (1965).

BEADLE, G. W.: Further studies in asynaptic maize. Cytologia **4**, 269—287 (1933). — BELLING, J.: A working hypothesis for segmental interchange between homologous chromosomes in flowering plants. Univ. Calif. Publ. Bot. **14**, 283—291 (1928). — BONNER, J.: Further experiments on the nutrition of isolated tomato roots. Bull. Torrey bot. Club **70**, 184—189 (1943). — BROWN, S. W., and D. ZOHARY: The relationship of chiasmata and crossing-over in *Lilium formosanum*. Genetics **40**, 850—873 (1955).

CHAMBERLAIN, C. J.: Gymnosperms, structure and evolution, pp. 286—287. Chicago: Chicago Univ. Press 1935. — CLEVELAND, L. R.: The whole life cycle of chromosomes and their coiling systems. Trans. Amer. phil. Soc., N. S. **39**, 1—100 (1949). — COLEMAN, J. R., and M. J. MOSES: DNA and the fine structure of synaptic chromosomes in the domestic rooster *(Gallus domesticus)*. J. Cell Biol. **23**, 63—78 (1964). — COOPER, D. C.: The transfer of desoxyribose nucleic acid from the tapetum to the microsporocytes at the onset of meiosis. Amer. Naturalist **86**, 219—229 (1952). — CREIGHTON, H. B., and B. MCCLINTOCK: A correlation of cytological and genetical crossing-over in *Zea Mays*. Proc. nat. Acad. Sci. (Wash.) **17**, 492—497 (1931). — CROUSE, H. V.: X-ray breakage of lily chromosomes at first meiotic metaphase. Science **119**, 485—487 (1954).

DARLINGTON, C. D.: A cytological demonstration of genetic crossing over. Proc. roy. Soc. B **107**, 50—59 (1930). — Recent advances in cytology, 1st edit. London 1932. — The internal mechanics of the chromosomes. I., II. and III. Proc. roy. Soc. B **118**, 33—96 (1935).— Recent advances in cytology. London: Churchill Ltd. 1937. — The chromosome as a physico-chemical entity. Nature (Lond.) **176**, 1139—1144 (1955). — DAWSON, G. W. P.: An introduction to the cytogenetics of polyploids. Philadelphia: F. A. Davis Co. 1962. — DE, D. N.: Autoradiographic studies of nucleoprotein metabolism during the division cycle. Nucleus **4**, 1—24 (1961).

EBERSOLD, W. T.: Crossing over in *Chlamydomonas reinhardi*. Amer. J. Bot. **43**, 408—410 (1956).

FAWCETT, D. W.: The fine structure of chromosomes in the meiotic prophase of vertebrate spermatocytes. J. biophys. biochem. Cytol. **2**, 204—406 (1956). — FELDMAN, M.: The effect of chromosomes 5B, 5D and 5A on chromosomal pairing in *Triticum aestivum*. Proc. nat. Acad. Sci. (Wash.) **55**, 1447—1453 (1966a). — FELDMAN, M., T. MELLO-SAMPAYO, and E. R. SEARS: Somatic association in *Triticum aestivum*. Proc. nat. Acad. Sci. (Wash.) **56**, 1192—1199 (1966b).

GERSTEL, D. U.: Segregation in new allopolyploids of *Gossypium*. I. The R_1 locus in certain new world-wide American hexaploids. Genetics **41**, 31—44 (1956). — GERSTEL, D. U., and L. L. PHILLIPS: Segregation in new allopolyploids of *Gossypium*. II. Tetraploid combinations. Genetics **42**, 783—797 (1957). — Segregation of synthetic amphiploids in *Gossypium* and *Nicotiana*. Cold Spring Harb. Symp. quant. Biol. **23**, 225—237 (1958). — GOLDSCHMIDT, R. B.: Theoretical genetics. Berkeley: University of California Press 1955. GREGORY, W. C.: Experimental studies on the culture of excised anthers in nutrient solution. Amer. J. Bot. **27**, 687—692 (1940). — GRELL, R. F., and A. C. CHANDLEY: Evidence bearing on the coincidence of exchange and DNA replication in the oocyte of *Drosophila melanogaster*. Proc. nat. Acad. Sci. (Wash.) **53**, 1340—1346 (1965).

HENDERSON, S. A.: Time of chiasma formation in relation to the time of deoxyribonucleic acid synthesis. Nature (Lond.) **211**, 1043—1047 (1966). — HOLLIDAY, R.: A mechanism for gene conversion in fungi. Genet. Res. **5**, 282—304 (1964). — HOTTA, Y., M. ITO, and H. STERN: Synthesis of DNA during meiosis. Proc. nat. Acad. Sci. (Wash.) **56**, 1184—1191 (1966). — HOTTA, Y., and H. STERN: Inhibition of protein synthesis during meiosis and its bearing on intracellular regulation. J. Cell Biol. **16**, 259—279 (1963). — HUSKINS, C. L.: The coiling of chromonemata. Cold Spr. Harb. Symp. quant. Biol. **9**, 13—17 (1941).

JAIN, H. K., and S. L. BASAK: Genetic interpretation of chiasmata in *Delphinium*. Genetics **48**, 329—339 (1963). — JANSSENS, F. A.: Spermatogénèse dans les Batraciens. V. La theorie de la chiasmatypie, nouvelle interpretation des cinèses de maturation. Cellule **25**, 387—411 (1909). — La chiasmatypie dans les Insectes. Cellule **34**, 135—359 (1924).

KAUFMANN, B. P.: The existence of double spiral chromatin bands and a "bouquet" stage in *Tradescantia pilosa* Lehm. Amer. Naturalist **59**, 190 (1925). — Chromosome structure in relation to the chromosome cycle. II. Bot. Rev. **14**, 57—109 (1948). — KAYANO, H.: Chiasma studies in structural hybrids. III. Reductional and equational separation in *Disporum sessile*. Cytologia (Tokyo) **25**, 461—467 (1960). — KELLENBERGER, G., M. L. ZICHICHI, and J. J. WEIGLE: Exchange of DNA in the recombination of bacteriophage lambda. Proc. nat. Acad. Sci. (Wash.) **47**, 869—878 (1961).

LEVAN, A.: The cytology of *Allium amplectens* and the occurrence in nature of its asynapsis. Hereditas (Lund) **26**, 353—394 (1940). — LIMA-DE FARIA, A.: Meiosis and pollen mitosis in rye under controlled conditions. Hereditas (Lund) **36**, 106—109 (1950). — The regions of special cycle of division of *Agapanthus* chromosomes. Chromosoma (Berl.) **6**, 33—44 (1953). — LINDEGREN, C. C.: Stability of the gene. Science **121**, 26—27 (1956). — LISSOUBA, P., J. MOUSSEAU, G. RIZET, and J. L. ROSSIGNOL: Fine structure of the genes in the ascomycete *Ascobolus immersus*. Advanc. Genet. **11**, 343—380 (1962).

MANTON, I.: The spiral structure of chromosomes. Biol. Rev. **25**, 486—508 (1950). — MATHER, K.: Crossing over. Biol. Rev. **13**, 252—292 (1938). — MCCLINTOCK, B.: The association of non-homologous parts in chromosomes in the mid-prophase of *Zea mays*. Z. Zellforsch. **19**, 191—237 (1933). — MENZEL, M. Y.: Preferential chromosome pairing in allotetraploid *Lycopersicon esculentum-Solanum lycopersicoides*. Genetics **50**, 855—862 (1964). — MENZEL, M. Y., and J. M. PRICE: Fine structure of synapsed chromosomes in F_1 *Lycopersicon esculentum-Solanum lycopersicoides* and its parents. Amer. J. Bot. **53** (in press) (1966). — MESELSON, M., and J. J. WEIGLE: Chromosome breakage accompanying genetic recombination in bacteriophage. Proc. nat. Acad. Sci. (Wash.) **47**, 857—868 (1961). — MEYER, G.: Fine structure of spermatocyte nuclei of *Drosophila melanogaster*. Proc. European Regional Conf. Electron Microscopy, Delft, 1960, 951—954 (1961). — A possible correlation between the submicroscopic structure of meiotic chromosomes and crossing-over. Proc. Third European Regional Conf. Electron Microscopy, Prague, 1964, 461—462 (1964). — MITCHELL, H. K.: Crossing over and gene conversion in *Neurospora*. In: Symp. on the chemical basis of heredity, edit. by WM. D. MCELROY and B. GLASS. Baltimore: Johns Hopkins Press 1957. — MITRA, S.: Types of aberrations produced by irradiation of premeiotic and meiotic chromosomes of *Lilium*. Genetics **41**, 654 (1956). — Effects of x-rays on chromosomes of *Lilium longiflorum* during meiosis. Genetics **43**, 771—789 (1958). — MOEWUS, F.: Die Analyse von 42 erblichen Eigenschaften der *Chlamydomonas eugametos*-Gruppe. III. Teil: Die 10 Koppelungsgruppen. Z. indukt. Abstamm.- u. Vererb.-Lehre **78**, 501—522 (1940). — MOSES, M. J.: Chromosomal structures in crayfish spermatocytes. J. biophys. biochem. Cytol. **2**, 215—218 (1956). — The relation between the axial complex of meiotic prophase chromosomes and chromosome pairing in a salamander *(Plethodon cinereus)*. J. biophys. biochem. Cytol. **4**, 633—638 (1958). — The nucleus and chromosomes: a cytological perspective. In: Cytology and cell physiology, 3rd edit. (G. H. BOURNE, ed.). New York: Academic Press 1964. — MOSES, M. J., and J. H. TAYLOR: Desoxypentose nucleic acid synthesis during microsporogenesis in *Tradescantia*. Exp. Cell Res. **9**, 474—488 (1955).

NEBEL, B. R., and E. M. COULON: The fine structure of chromosomes in pigeon spermatocytes. Chromosoma (Berl.) **13**, 272—291 (1962a). — Enzyme effects on pachytene chromosomes of the male pigeon evaluated with the electron microscope. Chromosoma (Berl.) **13**, 292—299 (1962b). — NODA, S.: Chiasma studies in structural hybrids. II. Reciprocal translocation in *Lilium Maximowiczii*. Cytologia (Tokyo) **25**, 456—460 (1960).

OEHLKERS, F., u. PAUL EBERLE: Spiralen und Chromomeren in der Meiosis von *Bellevalia romana*. Chromosoma (Berl.) 8, 351—363 (1957). — OLIVE, L. S.: Aberrant tetrads in *Sordaria fimicola*. Proc. nat. Acad. Sci. (Wash.) **45**, 727—732 (1959).

PERKINS, D. D.: Tetrads and crossing over. J. cell. comp. Physiol. **45** (Suppl. 2), 119—149 (1955). — PHILLIPS, L. L., and D. U. GERSTEL: Segregation of new allopolyploids of *Gossypium*. III. Leaf shape segregation in hexaploid hybrids of New World cottons. J. Hered. **50**, 103—108 (1959). — PLAUT, W. S.: DNA synthesis in the microsporocytes of *Lilium henryi*. Hereditas (Lund) **39**, 439—444 (1953). — PONTECORVO, G.: Trends in genetic analysis. New York: Columbia Univ. Press 1958.

RHOADES, M. M.: Crossover chromosomes in unreduced gametes of asynaptic maize. Genetics **32**, 101 (1947). — The cytogenetics of maize. In: Corn and corn improvement, edit. G. F. SPRAGUE. New York: Academic Press 1955. — RHOADES, M. M., and B. MCCLINTOCK: The cytogenetics of maize. Bot. Rev. **1**, 292—325 (1935). — RILEY, R.: The diploidisation of polyploid wheat. Heredity **15**, 407—429 (1960). — RILEY, R., and V. CHAPMAN: Genetic control of the cytologically diploid behavior of hexaploid wheat. Nature (Lond.) **182**, 713—715 (1958). — RILEY, R., V. CHAPMAN, and G. KIMBER: Position of the gene determining the diploid-like meiotic behavior of wheat. Nature (Lond.) **186**, 259—260 (1960). RIS, H.: The structure of meiotic chromosomes in the grasshopper and its bearing on the nature of chromomeres and lampbrush chromosomes. Biol. Bull. **89**, 242—257 (1945).

SARVELLA, P.: Multivalent formation and genetic segregation in some allopolyploid *Gossypium* hybrids. Genetics **43**, 601—619 (1958). — SAUERLAND, HEINZ: Quantitative Untersuchungen von Röntgeneffekten nach Bestrahlung verschiedener Meiosisstadien bei *Lilium candidum* L. Chromosoma (Berl.) **7**, 627—654 (1956). — SAX, H. J., and K. SAX: Chromosome structure and behavior in mitosis and meiosis. J. Arn. Arboretum **16**, 423—439 (1935). — SAX, K.: The cytological mechanism of crossing over. J. Arn. Arboretum **13**, 180—212 (1932). — SAX, K., and H. W. EDMONDS: Development of the male gametophyte in *Tradescantia*. Bot. Gaz. **95**, 156—163 (1933). — SAX, K., and L. M. HUMPHREY: Structure of meiotic chromosomes in microsporogenesis of *Tradescantia*. Bot. Gaz. **96**, 353—362 (1934). — SAX, K., and E. D. KING: An x-ray analysis of chromosome duplication. Proc. nat. Acad. Sci. (Wash.) **41**, 150—155 (1955). — SEARS, E. R., and M. OKAMOTO: Intergenomic chromosome relationships in hexaploid wheat. Proc. X. Internat. Congr. Genetics (Montreal) **2**, 258—259 (1958). — SHIMAKURA, K.: The capability of continuing divisions of *Tradescantia* pollen mother-cells in saccharose solution. Cytologia **5**, 363—372 (1934). — SMITH, S. G.: Polarization and progression in pairing-pachytene observations in *Neodiprion* (Hymenoptera). Canad. J. Res. **20**, 368—379 (1942a). — Polarization and progression in pairing. II. Premeiotic orientation and the initiation of pairing. Canad. J. Res., D **20**, 221—229 (1942b). — SPARROW, A. H.: The structure and development of the chromosome spirals in microspores of *Trillium*. Canad. J. Res., D **20**, 257—266 (1942). — SPARROW, A. H., and M. R. HAMMOND: Desoxyribonucleic-acid-containing bodies in the cytoplasm of pollen mother cells at early meiotic prophase. Genetics **32**, 107 (1947). — SPARROW, A. H., M. J. MOSES and R. STEELE: A cytological and cytochemical approach to an understanding of radiation damage in dividing cells. Brit. J. Radiol. **25**, 182—188 (1952). — SPARROW, A. H., V. POND and S. KOJAN: Microsporogenesis in excised anthers of *Trillium erectum* grown on sterile media. Amer. J. Bot. **42**, 384—394 (1955). —SPARROW, A. H., and RHODA SPARROW: Treatment of *Trillium erectum* prior to and during mass production of permanent smear preparations. Stain Technol. **24**, 47—55 (1949). — ST. GIMESI, W. FRENYO and G. L. FARKAS: Experiments in the cultivation of stamens in vitro. Hung. Acta biol. **1**, 37—39 (1949). — STERN, C.: Zytologisch-genetische Untersuchungen als Beweise für die Morgansche Theorie des Faktorenaustauschs. Biol. Zbl. **51**, 547—587 (1931). — STRAUB, J.: Die Wirkung von Temperaturstößen auf die Reduktionsteilung. Entwicklungsphysiologische Untersuchungen über die Meiosis von *Gasteria trigona*. Ber. dtsch. bot. Ges. **55**, 160—166 (1937). — Untersuchungen zur Physiologie der Meiosis. VII. Die Abhängigkeit der Chiasmabildung bei *Vicia Faba* und *Campanula persicifolia* von äußeren Bedingungen. Z. Bot. **32**, 225—266 (1938). — SWANSON, C. P.: Cytology and cytogenetics. Englewood Cliffs, N. J.: Prentice Hall 1957. — SWIFT, H.: The constancy of DNA in plant nuclei. Proc. nat. Acad. Sci. (Wash.) **36**, 643—654 (1950). — Quantitative aspects of nuclear nucleoproteins. Int. Rev. Cytol. **2**, 1—76 (1943).

TAKATS, STEPHEN T.: Chromatin extrusion and DNA transfer during microsporogenesis. Chromosoma (Berl.) **10**, 430—453 (1959). — TAYLOR, J. H.: Chromosomes from cultures of excised anthers. J. Hered. **40**, 86—88 (1949a). — Increase in bivalent interlocking and its bearing on the chiasma hypothesis of metaphase pairing. J. Hered. **40**, 65—69 (1949b).— Transition of meiosis to mitosis and visa versa in cultures of excised anthers. Genetics **35**, 136—137 (Abstr.) (1950a). — The duration of differentiation in excised anthers. Amer. J. Bot. **37**, 137—143 (1950b). — Autoradiographic determinations of incorporation of P^{32} into chromosomes during mitosis and meiosis. Exp. Cell. Res. **4**, 164—173 (1953). — The time and mode of duplication of chromosomes. Amer. Naturalist **91**, 290—221 (1957). — Autoradiographic studies of the organization and mode of duplication of chromosomes. In: Current Activities in Molecular Biology, edit. R. E. ZIRKLE. Chicago: Chicago Univ. Press 1958a. — Incorporation of phosphorus-32 into nucleic acids and proteins during microgametogenesis of *Tulbaghia*. Amer. J. Bot. **45**, 123—130 (1958b). — Distribution of tritium-labeled DNA among chromosomes during meiosis. I. Spermatogenesis in the grasshopper. J. Cell Biol. **25**, 57—67 (1965). — Patterns and mechanisms of genetic recombination. In: Molecular genetics, part II (edit. J. H. TAYLOR). New York: Academic Press 1967. — TAYLOR, J. H., and R. D. MCMASTER: Autoradiographic and microphotometric studies of desoxyribose nucleic acid during microgametogenesis in *Lilium longiflorum*. Chromosoma (Berl.) **6**, 489—521 (1954). — TAYLOR, J. H., and S. H. TAYLOR: The autoradiograph — a tool for cytogenetics. J. Hered. **44**, 128—132 (1953). — TAYLOR, J. H., P. S. WOODS, and W. S. HUGHES: The organization and duplication of chromosomes as revealed by autoradiographic studies using tritium-labeled thymidine. Proc. nat. Acad. Sci. (Wash.) **43**, 122—128 (1957).

VASIL, I. K.: Effect of kinetin and gibberellic acid on excised anthers of *Allium cepa*. Science **126**, 1294—1295 (1957).

WATSON, J. D., and F. H. C. CRICK: Molecular structure of nucleic acids. Nature (Lond.) **171**, 737—740 (1953). — WHITEHOUSE, H. L. K.: A theory of crossing-over by means of hybrid deoxyribonucleic acid. Nature (Lond.) **199**, 1034—1040 (1963). — WIMBER, D. E., and W. PRENSKY: Autoradiography with meiotic chromosomes of the male newt *(Triturus viridescens)* using H^3-thymidine. Genetics **48**, 1731—1738 (1963).

ZEN, S.: Chiasma studies in structural hybrids. VI. Heteromorphic bivalent and reciprocal translocation in *Allium fistulosum*. Cytologia (Tokyo) **26**, 67—73 (1961). — ZOHARY, D.: Chiasmata in a pericentric inversion in *Zea mays*. Genetics **40**, 874—877 (1955).

Pollen.

Von

H. F. Linskens.

Mit 20 Abbildungen.

Der Pollen entsteht in den Antheren (Mikrosporophyllen) aus den *Gonen*. Die Gonen ihrerseits sind hervorgegangen aus den *Gonotokonten* (Pollenmutterzellen, Mikrosporocyten) und haben die Meiose durchlaufen. Die Theken der Antheren können als *Mikrosporangien* angesehen werden, wenn die Pollenkörner auch keine vollkommene Analogie zu den Mikrosporen aufweisen.

Die Monographie von Schnarf (1929) faßt die zur damaligen Zeit bekannten physiologischen Gesichtspunkte ebenfalls zusammen. Neuere Übersichten über die Kenntnisse vom männlichen Gametophyten haben Maheshwari (1949), Johri und Vasil (1960, 1961), Linskens (1964a, b) und Stanley (1964) gegeben.

I. Entwicklung der Pollenkörner.

Die Entwicklung des Pollens erfolgt bei den Angiospermen nach verschiedenen Typen (Wulff 1939). Die meisten Arten der Angiospermen bilden ihre Pollenkörner nach dem Normaltypus (Tabelle 1); dieser zeichnet sich dadurch aus, daß der Teilung des primären Pollenkerns ein starkes Wachstum der Mikrosporen, Vacuolisierung und Bildung der Exine vorausgehen.

Tabelle 1. *Die Typen der Pollenkorn-Entwicklung.* (Nach Wulff 1939.)

Normal-Typ	*Juncus*-Typ	*Triglochin*-Typ
Meiosis	Meiosis	Meiosis
Starkes Wachstum der Mikrosporen Auftreten von Vacuolen Bildung der Exine		Geringes Wachstum der Mikrosporen Anfänge der Exinebildung
Teilung des primären Pollenkornkerns	Teilung des primären Pollenkornkerns	Teilung des primären Pollenkornkerns
	Wachstum der Pollenzelle(n) Auftreten von Vacuolen Bildung der Exine	Hauptwachstum des Pollenkorns Auftreten von Vacuolen Weitere Ausbildung der Exine
[Teilung des generativen Kerns Weiteres Wachstum des Pollenkorns]	Teilung des generativen Kerns Weiteres Wachstum des Pollenkorns	[Teilung des generativen Kerns Weiteres Wachstum des Pollenkorns]
Vorkommen: Bei der überwiegenden Mehrzahl der Monokotylen und Dikotylen, sowohl bei Arten mit zweikernigem als auch bei solchen mit dreikernigem Pollen	Vorkommen: Juncaceen und Cyperaceen, in Verbindung mit der Bildung von Pollentetraden, dreikernige Pollenzellen	Vorkommen: *Najas*, *Lilaea*, *Ceratophyllum*, *Ruppia*, *Aponogeton*, *Triglochin* (*Zostera*?), zwei- oder dreikernige Pollenkörner

1. Wandbildung.

Die Aufteilung der Gonen (Tetradenbildung) erfolgt bei den *Dikotylen* in den meisten Fällen durch *simultane* Teilung, d.h. es gehen zunächst aus der Meiose 4 Gonenkerne hervor, die sodann durch vier gleichzeitig entstehende Wände ausgekammert werden (Oehlkers 1956). Die Wandbildung erfolgt hingegen bei den *Monokotylen* meist *succedan:* es werden zunächst 2 Zellen und durch einen weiteren Teilungsschritt daraus insgesamt 4 Zellen gebildet. Die Wand des Gonotokonten löst sich auf und die 4 Gonen werden frei. Die Bildung des Pollenkornes aus den Gonen in der Anthere beruht auf der frühzeitigen Ausbildung des männlichen Gametophyten und der besonderen Ausgestaltung der Pollenwand.

Die bei der Bildung der Pollen-Tetraden entstehenden Zellwände werden durch Entmischung der Plasmakolloide gebildet, die zur Entstehung von *Cytoplasmaplatten* führt; sie sind von kurzer Lebensdauer. Unmittelbar nach der Anlage der Cytoplasmaplatten setzt die Abscheidung von Callose ein, die zuerst von Mangin (1889) entdeckt wurde. Diese wird jedoch wieder aufgelöst und für die Ausgestaltung der Oberfläche der Exine und zum Wachstum der Pollenkörner verwendet (Eschrich 1964). Die Calloseschichten erweisen sich hier also als Reservestoff, dessen Verflüssigung zunächst die PMZ-Wand, sodann die Mikrosporenkammerwand und zuletzt die Mittellamellen erfaßt. Trotz ihres kurzen Daseins sind die Tetradenwände als strukturell normale Zellwände anzusehen (Mühldorf 1939, 1941). Die von den Pollenmutterzellen zu Beginn der meiotischen Prophase synthetisierte Callosewand kann als Molekularsieb interpretiert werden, das den Eintritt größerer Moleküle aus dem umgebenden Plasma ausschließt. Dadurch wird verhindert, daß die Fähigkeit des haploiden Pollenkerns, sich in seinem eigenen Cytoplasma autonom zu manifestieren, beeinträchtigt wird (Heslop-Harrison 1963, 1964, 1966).

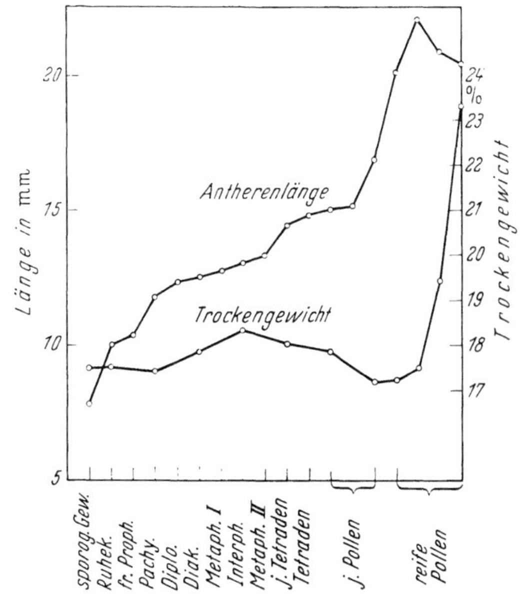

Abb. 1. Antheren-Länge und Trockengewicht während der Pollenentwicklung von *Petunia hybrida*. (Nach Linskens 1956a, verändert.)

Der Entwicklungsprozeß der Pollenkörner ist äußerlich kenntlich am *Wachstum der Gesamtanthere* (Abb. 1). Dieses vollzieht sich offensichtlich in Schüben (Erickson 1948, Linskens 1956a) und geht mit physiologischen Veränderungen des Tapetumbreies einher. Der Zeitpunkt der Meiose ist durch einen merklichen Rückgang der Atmungsintensität gekennzeichnet. Das Antherentrockengewicht erreicht zu diesem Zeitpunkt ein erstes Maximum. Ein zweiter Atmungsrückgang tritt zur Zeit der Pollenmitose ein (Erickson 1948). Im Verlauf der Karyokinese kommt es zu periodischen Schwankungen der freien elektrischen Ladungen an der Oberfläche und in dem Inneren der Zelle (Yamaha 1938b).

2. Tapetum.

Im Verlaufe der Entwicklung des Pollens spielt das Tapetum eine wichtige ernährungsphysiologische Rolle. Es geht im allgemeinen aus der parietalen

Schicht des Archespors hervor; in einigen Fällen kann es auch aus den sporogenen Zellen entstehen (SCHNARF 1929). Nach dem physiologischen Verhalten der Tapetumzellen bei der Ernährung des Pollens unterscheidet man das *Sekretionstapetum* und das *amoeboide Tapetum* (SCHNARF 1923). Beim amoeboiden Tapetum-Typus (Periplasmodium) lösen sich die Zellwände vorzeitig auf, so daß sich der Inhalt der Tapetumzellen zwischen die in Ausbildung befindlichen Pollenkörner ergießt. Beim Sekretionstapetum werden die Zellwände nicht aufgelöst; vielmehr besitzen die Tapetum-Zellen den Charakter eines Drüsengewebes: Mehrkernigkeit, Endopolyploidie (BROWN 1949, MECHELKE 1952, WUNDERLICH 1954, CARNIEL 1963), Plasmareichtum, Fehlen von Vacuolen.

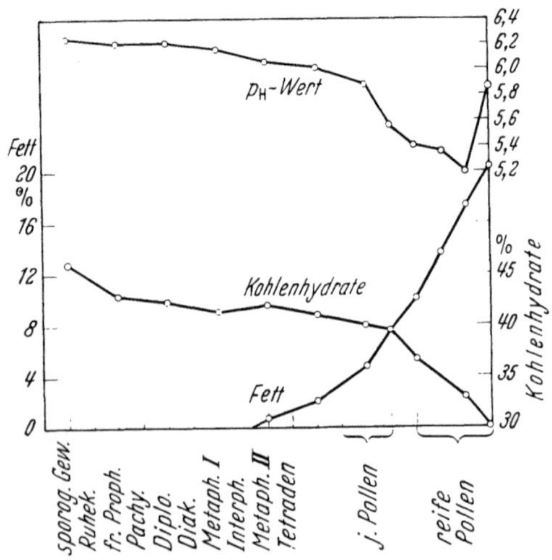

Abb. 2. Änderung des pH-Wertes des Tapetumbreies sowie Umbau der Kohlenhydrate und Fette während der Pollenentwicklung von *Petunia hybrida*. (Nach LINSKENS 1956a, verändert.)

Die Zellen des Archespors haben sich schon vor dem Eintritt der Meiose aus dem Gewebeverband gelöst und sind gegeneinander abgerundet. Sie befinden sich also in einer Art „*Nährlösung*", die zunächst von den Tapetumzellen sezerniert wird bzw. in die hinein sie sich schließlich auflösen. Die gesamten Stoffbewegungen, die für die in den Gonotokonten ablaufenden Wachstums- und Entwicklungsvorgänge notwendig sind, können also nur noch auf dem Wege der *Diffusion* erfolgen. Besonders erstaunlich ist die in vielen Fällen vorhandene exakte *Synchronisierung der Meiose*, die in einem Pollenfach gleichzeitig abläuft (NEUMANN 1963, ERICKSON 1964). Auch dieses sicherlich komplizierte Steuerungssystem beruht also auf einem biochemischen Mechanismus des Tapetumbreies. Um so erstaunlicher ist die noch geringe Kenntnis von den chemischen Abläufen während der Pollen-Entwicklung. Mikrochemisch kann der Übergang von Fetttröpfchen aus den Tapetumzellen in das Archespor und deren Aufnahme durch den jungen Pollen nachgewiesen werden (*Petunia*, WELZEL 1953). Bei vielen Arten verändert sich die Färbung des Antherensaftes mit fortschreitender Auflösung der Pektinlamellen der Pollenmutterzellen (DORSEY 1914). Aus den Veränderungen des p_H-Wertes im Tapetum-Brei (Abb. 2) lassen sich Rückschlüsse auf *intensiven Stoffaustausch und -umbau* ziehen (YAMAHA 1938a, LINSKENS 1956). Während der Pollenreifung läßt sich ein starker Anstieg der Auxin-Produktion feststellen (WEINLAND 1941). COOPER (1952) konnte zeigen, daß feulgenpositive Körperchen durch die Wandporen des jungen Pollens ins Innere wandern. Es scheint die Annahme gerechtfertigt, daß auf diese Weise ein Transfer von Nucleinsäurebausteinen der zerfallenden Tapetumzellen und -kerne in die Gonotokonten stattfindet (LINSKENS 1958). Eine weitere entscheidende Funktion des Tapetums ist die Beteiligung an der Ausbildung der Pollenmembran (Sporoderm) (ROWLEY 1962), insbesondere an der Modellierung des Reliefs der Exine (BIOURGE 1892).

3. Beeinflussung der Pollenbildung.

Die Ausbildung des Pollens ist durch innere und äußere Faktoren beeinflußbar und gesteuert. Es sind zahlreiche *Gene* bekannt, deren Veränderungen (Mutationen) sich durch Störungen im Ablauf der meiotischen Teilungen und der

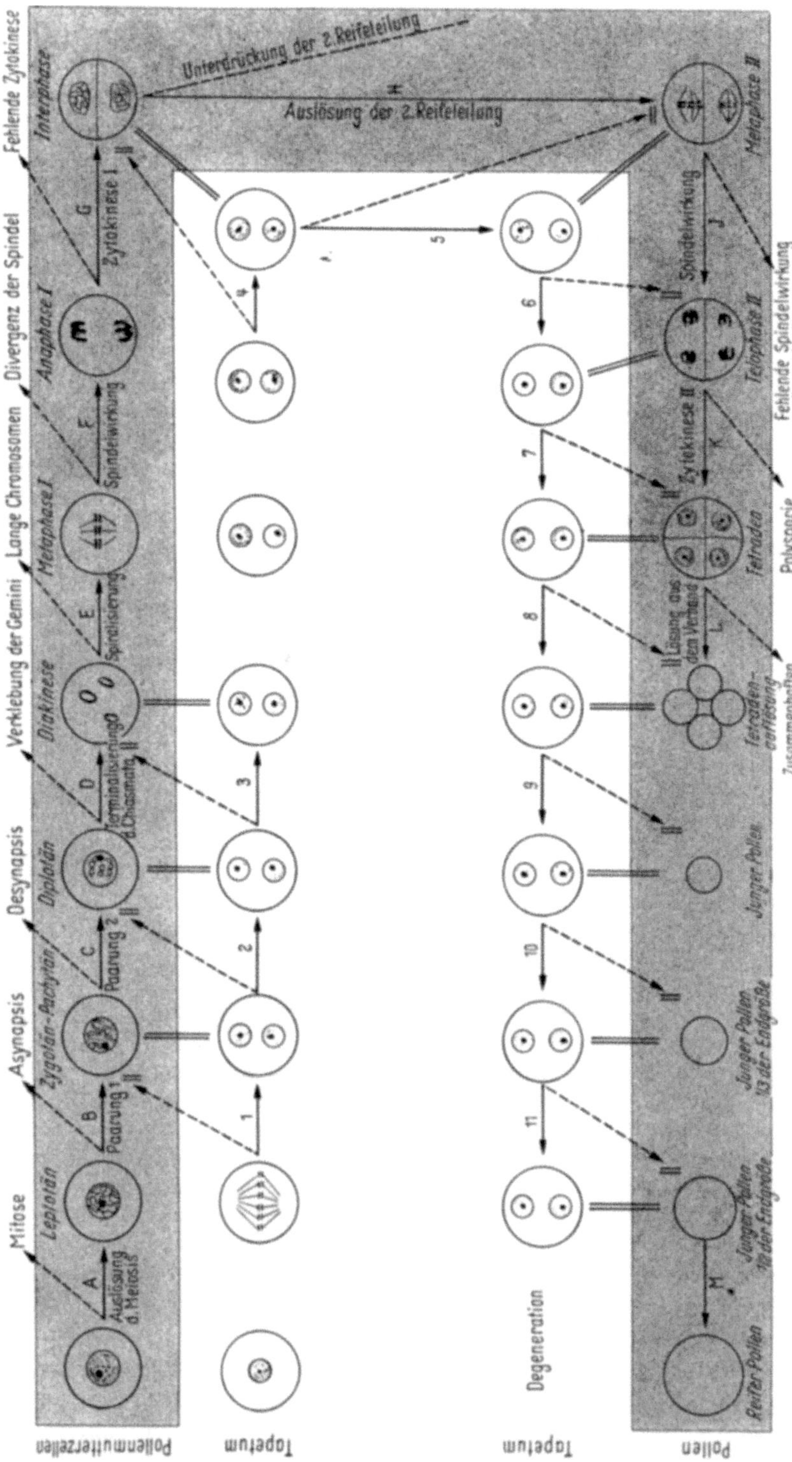

Abb. 3. Schematische Darstellung der Pollenentwicklung und der Tapetumzellen unter Berücksichtigung der steuernden Gene. Buchstaben: Pollengene, Zahlen: Tapetumgene. ⟶ = normale Genwirkung; ---→ = mutierte Genwirkung; ═ = Blockierung einer Tapetumsekretion. (Nach WELZEL 1953, verändert.)

postmeiotischen Entwicklungen bemerkbar machen (vgl. Beadle und McClintock 1928, Blakeslee 1928, Avery 1929, Kostoff 1930, Oehlkers 1937, Clark 1939, Welzel 1953). Welzel hat versucht, die bekannten *Tapetumgene* in ihrer Wirkung auf den Entwicklungsprozeß in einem Schema zusammenzustellen (Abb. 3). Diese Gene wirken sowohl auf den cytologischen Prozeß, als auch auf die ernährungsphysiologischen Vorgänge im Tapetum während der Gonenbildung.

Durch die *Milieufaktoren* wird die Entwicklung der Pollenkörner ebenfalls beeinflußt. Es ist dabei erstaunlich, daß die Modifizierbarkeit der Pollengröße durch veränderte Wasserhaushalts- und Ernährungsbedingungen der Pflanze verhältnismäßig gering ist (Wagenitz 1955). Wassermangel im Boden macht sich zwar durch Verringerung der Pollengröße und Verzögerung der Reifeteilung bemerkbar (Schoch-Bodmer 1940, Anikiev und Goroščenko 1950). Hohe Lufttemperaturen wirken ebenfalls verkleinernd (Mikkelsen 1949). Dabei spielen offensichtlich die Nachttemperaturen während der Pollenentwicklung eine besondere Rolle: wiederholte Abkühlung der Blütenknospen führt zur Bildung von abnormen Pollenkörnern mit einer variablen Anzahl von Keimporen. Der größte Pollendurchmesser wird bei Tomaten und *Xanthium* bei einer Nachttemperatur um 17° C erreicht. Der Exinecharakter wird kaum beeinflußt (Kurtz-Liverman 1958). Verdunkelung während der Ausbildung der Tetraden führt zur Sterilität infolge Degeneration des jungen Pollens (Kudrjavcev 1954). Außer durch Polyploidisierung ist die Pollengröße aber eine nur schwer veränderliche Maßgröße (Schwanitz 1950). Die Pollengröße hat jedenfalls keinen Einfluß auf die Befruchtungsfähigkeit unter normalen Kompatibilitätsverhältnissen (Wettstein und Niklas 1955).

4. Pollenausschüttung.

Das Aufspringen der Anthere kann bei allen Antheren einer Blüte gleichzeitig erfolgen *(Alisma, Aquilegia)* oder nacheinander innerhalb eines längeren Zeitraumes (*Helleborus* bis zu 26 Tage). Die Öffnung des Staubbeutels und die Entlassung des Pollens weist bei den einzelnen Species sowohl in der Tageszeit des Auswurfs, als auch in der Menge bedeutende Unterschiede auf: Die meisten Arten produzieren Pollen tagsüber von 7—17 Uhr. Die längste Darbietung pro Tag wurde bei *Helleborus abchasicus* während Tag *und* Nacht gefunden, die kürzeste (maximal 5 Std) bei *Alisma plantago*. Die *Rosaceen* präsentieren Pollen hauptsächlich am Nachmittag. Bei den *Monokotylen* hingegen ist die Pollen-Produktion gleichmäßig zwischen Vor- und Nachmittag verteilt (Alltagsgruppe). Auch die *Gesamtpollenmenge* einer Blüte schwankt in weiten Grenzen von 54,7 mg *(Cucurbita pepo)* bis 0,005 mg *(Sonchus oleraceus)* pro Blüte und Tag.

Der *Mechanismus* kann durch tiefe Temperatur und durch Wasser an der Anthere verzögert werden. Die untere Temperaturgrenze für das Aufplatzen der Antheren liegt zwischen 4,5 und 5° C bei starker Beleuchtung; bei bedecktem Himmel erhöht sich die Grenztemperatur um 4° C. Hohe Außentemperaturen und hohe Lichtintensität beschleunigen nicht nur das Öffnen der Blüten, sondern auch das Aufplatzen der Antheren (Percival 1955). Die meteorologischen Bedingungen haben einen entscheidenden Einfluß auf Anthese und Pollenausschüttung (Emecz 1962).

5. Pollenkitt.

Die eingetrockneten Reste des Tapetumsaftes kleben manchmal an der Oberfläche der reifen, entlassenen Pollenkörner als *Pollenkitt* (Fette, Viscin), der ein Verklumpen der Pollenmassen bedingen kann (Knoll 1930). Später bei der

Quellung löst sich der Pollenkitt sofort nach der Benetzung auf (TROLL 1928, STEFFEN 1953). *Funktionell* soll es sich beim Pollenkitt also um einen Überzug der Pollenkörner handeln, der das Anheften an der Oberfläche der Narbe bewirkt (MEYEN 1893, TROLL 1928, POHL 1930a und b).

Über die *chemische Beschaffenheit* des Pollenkittes bestehen immer noch keine Klarheiten. Ältere Angaben stammen von LOPRIORE 1905, KÜSTENMACHER 1911, RICHTER 1925, POHL 1933. Offensichtlich ist die chemische Grundlage des Stoffgemisches von öliger, harzartiger Beschaffenheit von Pflanzenart zu Pflanzenart verschieden. Die Tatsache, daß in manchen Fällen starke Flüchtigkeit nachzuweisen war, deutet auf ätherische Öle hin (HILDEBRAND 1897, ZETSCHKE 1930, KNOLL 1930). Typisch ist die Anfärbbarkeit mit Fettfarbstoffen wie Sudan III, Rosanilin und Nilblausulfat, sowie die Schwärzbarkeit durch Osmiumsäure (PANKOW 1958). Offensichtlich handelt es sich in den meisten Fällen um fettige Öle, in denen Farbstoffe wie Carotinoide und Antoxanthine gelöst sind. Die Öltröpfchen der Antherenflüssigkeit sind mit dem Pollenkitt identisch; sie werden während der Reifung des Pollens in das Hohlraumsystem der Exine eingelagert (PANKOW 1958).

II. Morphologie.

Die Morphologie der reifen Pollenkörner erfreut sich einer besonders guten und systematischen Bearbeitung. Dafür ist vor allem das Interesse der Pollenstratigraphie (Pollenanalyse) und der medizinischen Allergieforschung entscheidend gewesen. Für die Morphologie ist neben *Größe*, *Polarität* und *Symmetrie* vor allem Lage und Form der *Keimporen* sowie die Ausgestaltung der *Pollenwand* (Sporoderm) von Interesse. Zusammenfassende Darstellungen darüber liegen vor von FISCHER (1890), WODEHOUSE (1953), ERDTMAN (1952, 1954).

Für die Untersuchung spielt heute neben der Licht- und Elektronenmikroskopie (MÜHLETHALER 1953, SITTE 1953) auch die LO-Analyse nach ERDTMAN (1956) eine bedeutende Rolle. Über Nomenklatur-Probleme besteht keine Einigkeit.

Die *Größe der Pollenkörner* variiert von 0,1765 mm ⌀ bei *Mirabilis Wrightii*, 0,168 mm ⌀ bei *Cucurbita ficifoli* bis zu 0,015 mm ⌀ bei *Philadelphus acuminatus* (AMELUNG 1893). Bei der Größenbestimmung spielt das Einbettungsmedium eine wesentliche Rolle (SCHOCH-BODMER 1936).

III. Biochemie des Pollens.

Zusammenfassungen über die Ergebnisse der modernen Pollenchemie hat LUNDÉN (1954, 1956) gegeben. Vergleiche auch die jährlichen Berichte über neue Inhaltsstoffe des Pollens in den „Fortschritten der Botanik" (LINSKENS 1955b, 1956b, 1957, 1958b). Unter den organischen Naturprodukten erfreut sich der Blütenstaub seit alters her besonderer Aufmerksamkeit. Einmal ist er ein wichtiges Bienen-Nahrungsmittel, zum andern ist er für das Zustandekommen des Heufiebers bei allergischen Personen verantwortlich. Sicher spielt eine wichtige Rolle, daß sich große Mengen einheitlichen Materials verhältnismäßig leicht gewinnen lassen (HELLSTRÖM 1956).

1. Pollenwand (Sporoderm).

E. STRASBURGER (1886) hat als erster den Bau und das Wachstum der Pollenwand eingehend beschrieben. Man hat seitdem auf Grund lichtmikroskopischer Beobachtungen eine innere Schicht *(Intine)* und eine äußere Wand *(Exine)* unterschieden, wobei die letztere mit einem Netzwerk von Stäbchen versehen ist (Abb. 4). Die älteren Untersuchungen über die Chemie des Sporoderms sind bei CZAPEK (1913) zu finden. Die wichtigsten Formelemente der Pollenwand und ihre Bezeichnungen nach heutiger Auffassung sind in Abb. 5 (MÜHLETHALER 1955) zusammengestellt. Detaillierte Beschreibungen finden sich bei ERDTMAN (1952, 1954). Infolge der Bedeutung der Membranstruktur für die Pollenanalyse werden eine Vielfalt teilweise synonymer Nomenklatur-Systeme benutzt, die durch elektronenmikroskopische Untersuchungen dauernd mehr detailliert werden können (WITTMANN u. WALKER 1965).

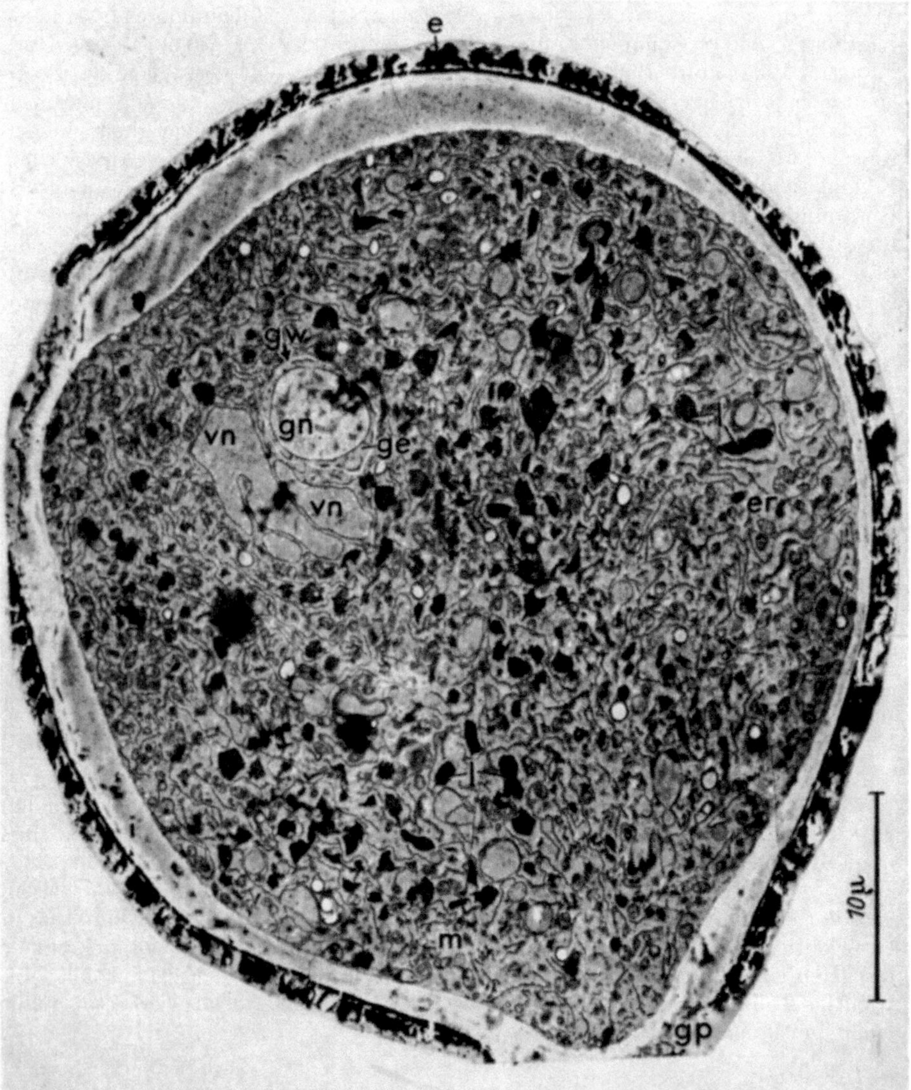

Abb. 4. Querschnitt durch ein Pollenkorn von *Petunia*. *e* Exine; *er* endoplasmatisches Reticulum; *ge* generative Zelle; *gn* generativer Kern; *gp* Keim-Pore; *l* Lipoid-Vacuolen; *vn* vegetativer Kern, stark gelappt und mehrfach angeschnitten; *i* Intine. (Elektronenmikroskopische Aufnahme von M. M. A. SASSEN.)

Zur Frage der Entwicklung des Sporoderms hat die Elektronenmikroskopie wichtige Beiträge in neuer Zeit geliefert (FERNANDEZ-MORÁN und DAHL 1952, AFZELIUS 1954, 1955, 1956, AFZELIUS, ERDTMAN und SJÖSTRAND 1954). Die Entwicklung der Exine beginnt bereits während der meiotischen Teilung. Die endgültige Form erhält das Pollenkorn jedoch erst dann, wenn die Tapetumzellen sich zum Tapetumbrei aufgelöst haben.

Die große Mannigfaltigkeit in der Formgestaltung der Pollenwand, die für die einzelne Pflanzenart soweit spezifisch ist, daß eine Determinierung rezenter und fossiler Pollen danach möglich ist, läßt sich auf eine — allerdings hypothetische — Grundform zurückführen, die bereits alle Elemente des Sporoderms enthält.

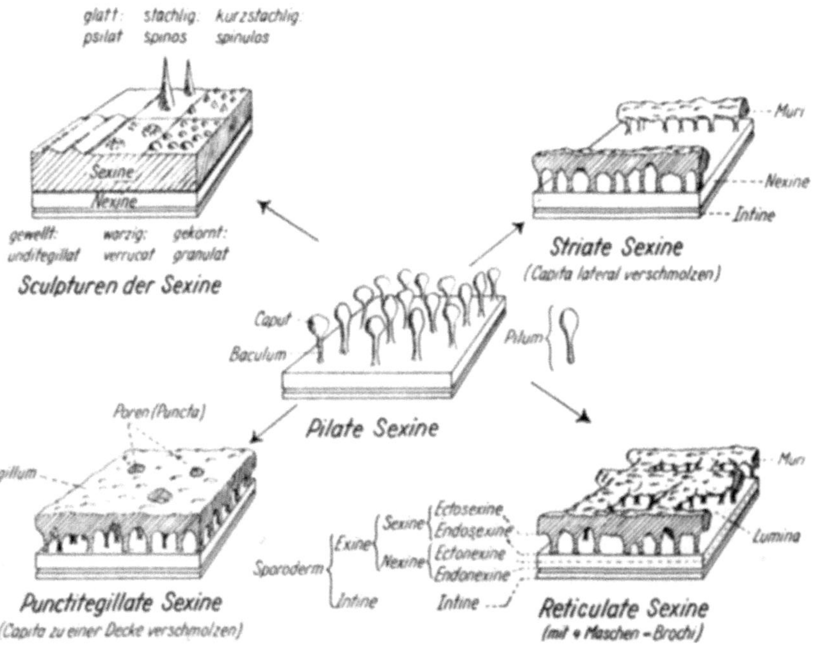

Abb. 5. Die Formelemente des Sporoderms und ihre Nomenklatur. In der Mitte: hypothetische Ausgangsform. Links oben: Skulpturen der Sexine. (Nach MÜHLETHALER 1955.)

Es ist dies die pilate Sexine, von der sich striate, punctitegillate und reticulate Sexine ableiten lassen (Abb. 5). Alle Pollenmembranen lassen das gleiche *Bauprinzip* erkennen (MÜHLETHALER 1953): Auf der im Schnitt strukturlosen, dünnen pektinhaltigen Intine liegt die kontrastreiche, dicke Exine, die aus *Sporopollenin* (hochmolekulare Terpene) besteht (ZETSCHKE und HUGGLER 1928; ZETSCHKE und VICARI 1931). Die Exine besteht aus einer zusammenhängenden Basalschicht, aus der zahlreiche radial nach außen gerichtete Stäbchen auswachsen, die am Ende keulenförmig verdickt sind und verwachsen können (WAGENITZ 1955). So entsteht in der Aufsicht ein Netzwerk mit zusammenhängendem Hohlraumsystem. Die unterschiedliche stoffliche Zusammensetzung der beiden Schichten äußert sich auch in dem verschiedenen Streuvermögen im EM-Bild (Abb. 6). Genauere chemische Untersuchungen der Pollenmembran fehlen noch immer; die Natur der Sporopollenine (ZETSCHKE 1932), die einen hohen Anteil an der Trockensubstanz der Pollen darstellen (Tabelle 2), ist unbekannt. Im EM erscheinen diese Stoffe amorph. Nach den Untersuchungen von SITTE (1953) ist

Tabelle 2. *Gehalte einiger Pollen an Sporopollenin und Cellulose in Prozent des Trockengewichtes.* (Nach ZETSCHKE 1932.)

	Cellulose in % Trocken-substanz	Sporo-pollenin in % Trocken-substanz
Pinus silvestris . . .	2,0	21,9
Picea orientalis . . .	2,2	20,0
Corylus avellana . .	1,1	7,3
Papaver spec. . . .	0,6	5,0
Phönix dactylifera .	0,6	4,3

in ihnen ein Fibrillengeflecht aus Cellulose vorhanden, wie es für Primärwände typisch ist. Dieses ist jedoch sehr stark inkrustiert bzw. adkrustiert. Auf der Außenfläche besitzt die Exine häufig leisten- oder stäbchenförmige Auswüchse (Abb. 7). Nach STRASBURGER (1886) erfolgt deren Bildung durch Plasma, das aus dem Zellinnern in die Membran einwandert. Die Oberflächenbeschaffenheit

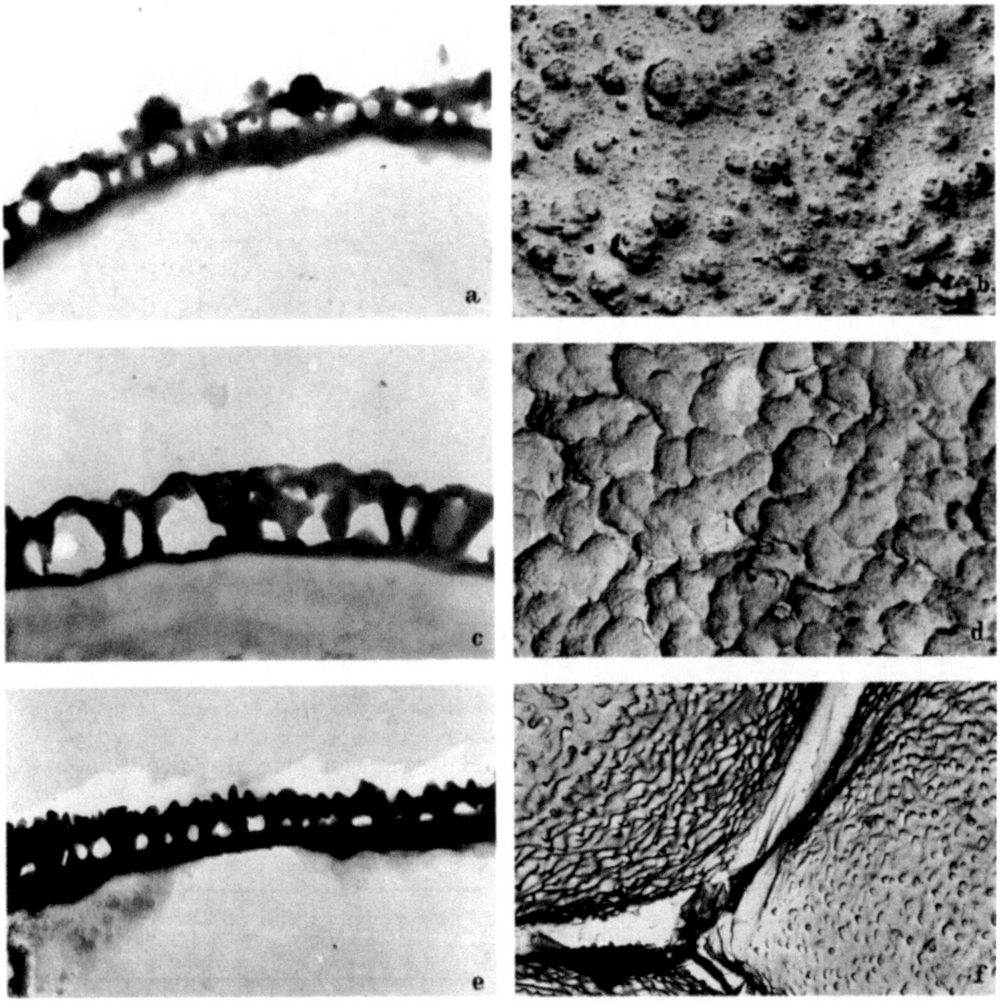

Abb. 6a—f. Elektronenmikroskopische Aufnahmen von Pollen. Linke Reihe (a, c, e) Querschnitte durch Pollenwand, rechte Reihe (b, d, f) Oberflächenabdrücke des Sporoderms. a *Quercus robur* 10000×; b *Quercus robur* 10000×; c *Pinus nigra* 9000×; d *Pinus nigra* 8000×; e *Fagus silvatica* 8000×; f *Fagus silvatica* 5000×. (Nach MÜHLETHALER 1955.)

des Pollens ist im allgemeinen granulär. Diese Granulierung kommt wahrscheinlich dadurch zustande, daß die Pollenkörner auch nach der Fertigstellung der Exine noch wachsen (Abb. 8). Dabei wird das Sporoderm offenbar plastisch gedehnt und an der Außenseite granulär. Daß dabei häufig isodiametrische Felder entstehen, erklärt sich aus dem amorphen Zustand der Sporopollenine (SITTE 1953).

Die morphologische Ausgestaltung des Sporoderms, einschließlich der strukturellen Elemente, welche eine große taxonomische Bedeutung haben, muß als Aus-

Abb. 7. Schnitt durch Pollen-Membran von *Cucurbita pepo* 7000×. (Nach MÜHLETHALER 1955.)

druck des Genotypus der haploiden Spore angesehen werden — und *nicht* als etwas, das aus dem umgebenden Milieu des Tapetum aufgeprägt wird. Die hauptsächlichen Kennzeichen der Pollenwand werden zu einem Zeitpunkt determiniert, wenn die Sporen noch eingeschlossen sind von der Pollenmutterzell-Wand und so wirksam vom Protoplasma des Tapetums isoliert sind (HESLOP-HARRION 1963a, b, 1964).

Die Tapetumzellen sind also am Aufbau der Wand nicht *unmittelbar* beteiligt. Sie bedingen jedoch die häufig vorhandene Einlagerung von fettig-öligen Substanzen in die Hohlräume der Exine. Der Verschluß der Keimpore erfolgt an der Austrittsöffnung durch einen „Zwischenkörper", der eine stark quellbare, linsenförmige Verdickung der Intine darstellt. Pflanzen mit Windbestäubung haben eine starke Intine, die bei *Carex gracilis* fast das ganze Lumen des Pollenkornes erfüllt (MÜHLETHALER 1953).

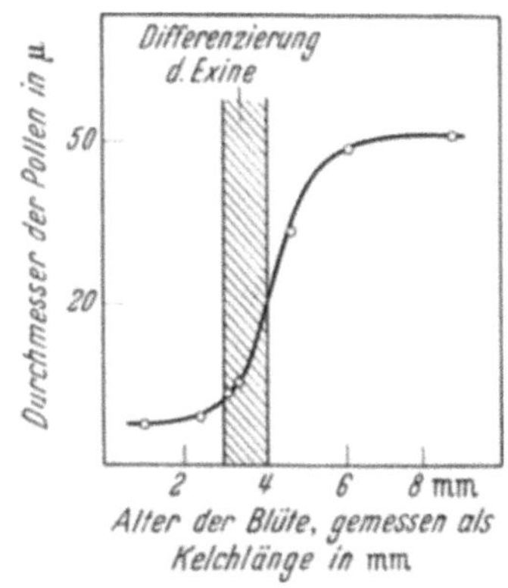

Abb. 8. Abhängigkeit der Pollenkorngröße vom Alter der Blüten bei *Epilobium angustifolium*. (Nach SITTE 1953.)

2. Pollenfarbstoffe.

a) Lokalisation.

Bei der Betrachtung geöffneter Antheren fällt die häufig intensive Färbung der Pollenkörner auf, die je nach Art der Pflanze verschieden ist. Auch beim Auspressen des noch unreifen Anthereninhaltes weist dieser reichen Farbstoffgehalt auf. FERGUSON-HUNT (1934) fand allein bei *Petunia* Färbungen in 38 Abstufungen, die sich in die Gruppen blau, gelb, grün und grau einteilen ließen. Die Frage entsteht: *Wo* ist der Pollen gefärbt? MÖBIUS (1923) gibt an, daß 80% der von ihm untersuchten Pollenarten ihre Färbung durch die Exine erhalten, in etwa 20% der Fälle durch ein gelbes Öl, das an deren Oberfläche hafte und durch die Tapetumzellen produziert werde. Danach ist also die Färbung des Sporoderm der ursprüngliche Zustand. Nur in wenigen Fällen nimmt er Färbung durch den Polleninhalt an. ZETSCHKE und VICARI (1931) bezeichnen die Farbstoffe des Pollens als teilweise unabtrennbare, teils extrahierbare Begleiter der Membran. Es wird auch häufig beobachtet, daß Pollenarten bei längerem Lagern, besonders am Sonnenlicht, ausbleichen, während andere ihre intensive Färbung behalten. Dies führt zu der Annahme, daß die Membransubstanz an sich farblos ist, aber mehr oder minder dauerhaft angefärbt ist, wobei der Farbstoff teilweise in den polymeren Verband der Membransubstanz übergegangen ist.

Der wesentliche Anteil der Pollenfärbung kommt wohl durch *Pigmentanlagerung* zustande. Auch weist steriler und normaler Pollen derselben Anthere die gleiche Färbung auf; es ist daher anzunehmen, daß die Farbstoffe nicht im Polleninnern, sondern in der Wand lokalisiert sind (SCHOCH-BODMER 1939). Dies scheint sich nach neueren Untersuchungen (MÜHLETHALER 1953) zu bestätigen: bei *Narcissus* sind die Hohlräume der Exine mit einer gelben Flüssigkeit gefüllt, die von den Tapetumzellen, die in das Antherenfach eingewandert sind, gebildet worden ist.

b) Chemie der Pollenfarbstoffe.

Die bisher in Pollen untersuchten Farbstoffe gehören zu den Gruppen der Anthocyane, Antoxanthine und Carotinoide.

α) Anthocyane. Anthocyane als Pollenfarbstoffe werden von MÖBIUS (1923) erwähnt und zwar für das Innere des Pollens von *Papaver bracteatum* und der roten *Tulpe*. Im Pollen von *Lythrum salicaria* sind neben Flavonen auch Anthocyane vorhanden (SCHOCH-BODMER 1938). Bei Untersuchungen an *Petunia*-Pollen fand FERGUSON-HUNT (1934), daß die Pollenkörner diploider und polyploider Rassen tiefgefärbte Strukturen aufweisen, die „Farbkörperchen" genannt werden. Diese wurden bei solchem Pollen gefunden, der bereits während der Entwicklung sein Wachstum einstellte. Er weist stark verdickte Wände auf und im Innern eine konzentrierte Anthocyan-Lösung.

β) Carotinoide. Die Carotinoide sind als stark ungesättigte Kohlenwasserstoffe vom Isopren abzuleiten und werden wegen ihrer guten Löslichkeit in Fetten und Fettlösungsmitteln auch als Lipochrome bezeichnet. v. EULER und Mitarbeiter (1945) untersuchten zahlreiche Pollenarten auf ihren Carotingehalt. In Blütenstaub von *Populus tremula*, *P. nigra*, *Betula pubescens*, *Pinus montana*, *Typha latifolia* und *Campanula persicifolia* konnten keine Carotinoide nachgewiesen werden, während der Pollen von *Salix alba*, *S. nigra*, *Lilium candidum* und *Taraxacum officinale* Carotin enthielt. Bei *Lilium candidum* und *Taraxacum* wurden außer β-Carotin auch noch Lycopin, Xanthophyll und kleinere Mengen Zeaxanthin und Crocetin nachgewiesen. Da diese Carotine auch im Bienenpollen gefunden wurden, schloß man, daß ein grundlegender Unterschied zwischen anemophilen und entomophilen Blütenpflanzen bestehen müsse. Dies wurde durch die Resultate von VIVINO und PALMER (1944) unterstützt, die in bienengesammeltem Pollen stets sowohl Carotin als auch Xanthophyll finden konnten. β-Carotin und Lycopin wurden von KARRER und LEUMANN (1951) im Pollen von *Cyclamen persicum* gefunden; zwei weitere Carotinoide konnten nicht mit Sicherheit identifiziert werden, doch handelt es sich wahrscheinlich um β-dihydro-Carotin und um Neurosporin. In dem Pollen von *Asphodelus albus* fand TAPPI (1952) β-Carotin, Xanthophyll, Xanthophyll-Epoxyd und Flavoxanthin, bei *Acacia dealbata* ließen sich Epoxyd-α-Carotin, Epoxyd-Xanthophyll, α- und β-Carotin, sowie Flavoxanthin und Xanthophyll nachweisen. Pigmente, deren Absorptionsmaxima zum Rot hin verschoben sind, wie die des epiphasischen Carotinoids und des Hydroxycarotinoids, wurden im Pollen bisher nicht gefunden; sie sind überhaupt in höheren Pflanzen selten. Zusammenstellung der in neuerer Zeit gefundenen Pollenfarbstoffe s. Tabelle 3.

γ) Anthoxanthine. Unter den Pollenfarbstoffen haben die Anthoxanthine die weiteste Verbreitung. Sie lassen sich in mehrere Gruppen aufteilen, von denen die Flavone und Flavonole am häufigsten im Pflanzenreich vertreten sind. Sie kommen meist als farblose oder gelbliche Kristalle und in glykosidischer Bindung vor. Als Zucker treten sowohl Monosen als auch Saccharide auf. Im Rahmen der Anamneseuntersuchungen des Heuschnupfens hat zuerst HEYL (1919) im Blütenstaub von *Ambrosia* und *Artemisia* Flavonolglykoside nachgewiesen. Weiter haben v. EULER und Mitarbeiter (1945), v. EULER (1948), MÖBIUS (1923)

Tabelle 3. *Pollenfarbstoffe in verschiedenen Pflanzenarten.* (Nach Untersuchungen von TAPPI und Mitarbeitern 1948—1955.)

	Leucojum vernum	*Acacia dealbata*	*Lilium mantchiuricum*	*Lilium candidum*	*Anemone coronaria*
α-Carotin	×	×			
β-Carotin	×	×	×		
Xanthophyll	×	×			
Xanthophyll-Epoxyd	×	×			
Flavon-Glucoside	×	×			
α-Carotin-Epoxyd		×		×	
Flavoxanthin		×			
Violaxanthin			×		
Capsanthin			×		
Capsorubin			×		
Antheraxanthin			×	×	
Delphinidin					×
Malvidin					×
Cyanidin					×
Pelargonidin					×
Quercetin		×			
Acetylnaringenin		×			
Isorhamnetin				×	
Phytocosan				×	
Phytofluen				×	
γ-Sitosterol				×	

sowie FERGUSON-HUNT (1934) in anderen Pollenarten nicht näher identifizierte Flavone aufgefunden. KUHN und LÖW (1944) fanden im *Crocus*-Pollen Isorhamnetin-diglycosid. In neuerer Zeit haben TAPPI und Mitarbeiter zahlreiche weitere Verbindungen isolieren und identifizieren können (Tabelle 3).

c) Funktion.

Die Vermutung von KUHN (1943) und MOEWUS (1950), daß die Flavonole und deren Glykoside teilhaben an einer Form der Inkompatibilitätsreaktion, hat sich nicht bestätigen lassen (ESSER und STRAUB 1954, LEWIS 1954, VISSER 1956, REZNIK 1957). Allerdings berichtet SCHWARZENBACH (1951) von *Cyclamen*-Pollen, daß er daraus 2 Carotinoide gewinnen konnte, die hemmend auf die Pollenkeimung wirken. Die *Hemmwirkung* soll sich durch unbefruchtete Samenanlagen aufheben lassen. Hier wird also ein physiologischer Mechanismus angedeutet, der im Zusammenhang steht mit der Befruchtungsreaktion. Von PFEIFFER (1955) wurde die Vermutung ausgesprochen, daß die Veränderungen der Pigmentausstattung der Pollenmembran bei der *Alterung* mit dem Verlust der Keimfähigkeit in Zusammenhang stehen.

Eine wesentliche Funktion der Pollenmembran und der in ihr lokalisierten Farbstoffe ist im *Strahlenschutz*, also der Abschirmung des gegenüber Ultraviolett hochempfindlichen genetischen Materials der männlichen Geschlechtszelle, zu suchen. Nachdem bekannt ist, daß alle untersuchten Pollensorten schon nach

8stündiger Sonnenbestrahlung (Werfft 1951) einer deutlichen Reduktion ihrer Keimkraft unterworfen sind, erscheinen die Pigmente als *Sonnen-UV-Filter* von besonderer Bedeutung (Asbeck 1954a, b, 1955). Dabei hat der Pollen 2 Funktionen in sich vereint: den Leichtbau und den Schutz gegen UV-B (280 bis 320 mμ), indem er das 2 μ dicke Filter auf dem Prinzip verstärkter chromatischer Absorption aufgebaut hat: Die Umwandlung schädlicher Wellenlängen in langwelliges Licht mittels Fluorescenz. Auch findet eine *Abwandlung der Exine-Färbung* entsprechend der UV-Belastung statt. In stark exponiertem Pollen variiert die Farbe bis orange-rotorange-rot. Außerdem hat offensichtlich die Temperatur und die Einstrahlung während der Pollenentwicklung einen Einfluß auf die Färbung des Pollens (Schoch-Bodmer 1939).

3. Polleninhaltsstoffe.

a) Proteine.

Der Eiweißgehalt verschiedener Pollenarten schwankt zwischen 12 und 30% der Trockensubstanz (Czapek 1920), (vgl. auch Tabelle 4). Etwa 70—80% des Gesamtstickstoffes besteht aus Protein-N (Kiesel und Rubin 1929, Freiberg und Clark 1955). Heyl (1917, 1919, 1920) isolierte aus Pollen Proteosen, Gluteline und Albumine. Augustin (1953) fand Globuline und Albumine als Fraktionen der Pollenproteine. Mittels Elektrophorese ließen sich Pollenproteine fraktionieren (Linskens 1955a, Bingham *et al.* 1964). Bei *Nicotiana, Petunia* und *Salpiglossis* wiesen die Proteine des Pollen gleiche *Aminosäurenzusammensetzung* im Totalhydrolysat auf (Bellartz 1956). Auch die allergieaktiven Komponenten werden noch immer unter den Pollen-Proteinen gesucht (Loveless und Timasheff 1955).

Tabelle 4. *Gesamtanalysen einiger Pollenarten.* (Nach Elser und Ganzmüller 1930.)

	Alnus spec.	*Pinus spec.*	*Corylus avellana*
H_2O %	7,11	7,45	2,33
Trockensubstanz	92,89	92,55	97,77
Fett	23,55	14,59	8,81
Eiweiß	23,44	12,55	16,19
Invertzucker	4,60	1,61	6,43
Rohrzucker	0	5,82	3,95
Asche gr.	3,78	2,88	4,53
(SiO_2) unlöslich	0,90	0,20	1,01
K_2O	1,91	0,94	2,46
CaO	0,343	0,259	0,360
Fe_2O_3	0,315	0,040	0,114
P_2O_5	1,13	1,39	1,37
Mg	0	0	0
Cl	0	0	0

Während der Pollenentwicklung zeigt das Proteinmuster eine phasenspezifische Veränderung, so daß jedem Stadium ein spezifisches Proteinmuster zugeordnet werden kann, das einem phasenspezifischen Enzym- und Isoenzym-Muster entspricht (Nasatir *et al.* 1960; Linskens u. Schrauwen 1966). Die Proteinzusammensetzung ändert sich jedoch schon mit dem Altern des Pollens (Bingham *et al.* 1964).

Das vegetative Plasma mancher Pollenarten ist mit Wasser mischbar und nicht plasmolysierbar, da es keine Grenzschicht besitzt. Aber auch plasmolysierbare Pollensorten weisen keine dem Plasma-Lemma gleichwertige Membran auf. Dem entspricht die extrem hohe Permeabilität der Pollen für Anelektrolyte, insbesondere Traubenzucker. Nur die generative Zelle des Pollens besitzt ein echtes Plasma-Lemma (Hofmeister 1956). Möglicherweise ist sie sogar von einer echten Zellwand umgeben (Sassen 1964).

b) Pollenfermente.

Die mannigfache und komplexe physiologische Aktivität, welche die Pollenkörner als Geschlechtszellen beim Befruchtungsvorgang ausüben müssen, kommt in ihrer reichen Fermentgarnitur zum Ausdruck. Es wurden in neuerer Zeit im Pollen folgende Fermente mit Sicherheit nachgewiesen: Katalase, Amylase, Saccharase (VINSON 1927, ELSER und GANZMÜLLER 1930, SOSA-BOURDOUIL 1939, HAECKEL 1951), Cozymase, Dehydrogenase (v. EULER 1948, v. EULER und v. EULER 1948, PALUMBO 1953), Phosphatase (NAKAMURA und BECKER 1951, HAECKEL 1951, PALUMBO 1953), Proteinase, Phosphorylase (BELLARTZ 1956), Diaphorase, Cytochromoxidase (OKONUKI 1942, 1943, v. EULER, HELLER und HÖGBERG 1948), Pektinase (PATON 1921, VINSON 1927), Adenosintriphosphatase (PALUMBO 1953), Folsäure-Konjugase (NIELSEN-HOLMSTRÖM 1957), Cutinase (LINSKENS und HEINEN 1962) Hydrolasen (GÓRSKA-BRYLAS 1965).

Eine ältere Zusammenstellung aus verschiedenen Ordnungen des Systems (Tabelle 5) zeigt, daß die *Fermentausrüstung* bei den einzelnen Arten *unterschiedlich* ist. Sie ist mit den speziellen Gegebenheiten des Schlauchwachstums

Tabelle 5. *Fermentausrüstung des reifen Pollens einiger Pflanzenarten.* (Nach PATON 1921.)

	Amylase	Invertase	Lipase	Proteinase	Katalase	Nuclease	Tyrosinase	Cellulase	Pektinase
Lilium rubrum	+	+	·	·	+	+	—	·	·
Acer rubrum	+	+	·	·	+	+	—	·	+
Pyrus malus silvestris	+	+	·	+	+	+	—	+	+
Pinus austriaca	+	+	+	+	+	+	—	+	+
Magnolia grandiflora	+	+	·	+	+	+	—	—	+
Taraxacum officinale	+	+	·	+	+	+	—	—	·
Bellis perennis	+	·	—	+	+	+	—	—	+
Rumex spec.	+	·	+	+	+	+	·	|	|
Ulmus spec.	+	·	·	+	+	+	—	—	·
Solidago virgaurea	+	·	—	+	+	+	—	—	+
Secale cereale	+	·	+	+	+	+	—	+	+
Phleum pratense	+	·	—	+	+	+	—	+	+
Zea mays	+	·	·	+	+	+	—	+	·

+ positiv; — negativ; · nicht getestet.

in Verbindung zu bringen: mittels der polleneigenen enzymatischen Werkzeuge muß der Schlauch in der Lage sein, die mechanischen Widerstände beim Durchwachsen des Leitgewebes zu beseitigen und gleichzeitig die Ernährung und den Aufbau der Schlauchwand sicherzustellen. Man beobachtet daher während der Keimung im allgemeinen ein starkes *Anwachsen der fermentativen Aktivität* (STANLEY u. LINSKENS 1964, 1965; LINSKENS u. HEINEN 1962). Bei *Dicotylen*, die hauptsächlich ein kompaktes Griffel-Leitgewebe besitzen, herrscht Pollen mit hoher Phosphatase und geringer Amylase- und Invertase-Aktivität vor; hingegen findet sich bei *Monokotylen*, die meist mit einem Griffelkanal versehen sind, stets niedrige Phosphatase-Aktivität bei hoher Amylase- und Invertase-Aktivität. Beziehungen zwischen der Höhe der Phosphatase-Aktivität und der Griffellänge bzw. der Wachstumsgeschwindigkeit ließen sich nicht nachweisen. Die Fermentaktivität sinkt mit dem Altern des Pollens sowie mit der hohen Stellung der Blüten im Blütenstand ab (HAECKEL 1951).

c) Freie Aminosäuren.

Es liegen zahlreiche Untersuchungen über den Gehalt an freien Aminosäuren im Pollen vor, die meist mit Hilfe papierchromatographischer Methoden durchgeführt wurden (AUCLAIR und JAMIESON 1948, SARKAR, WITTWER, LUECKE und SELL 1949, MARQUARDT und VOGG 1952, BATHURST 1954, HATANO 1955, NIELSEN,

GRÖMMER und LUNDEN 1955, BELLARTZ 1956, LINSKENS und TUPÝ 1966). Die meisten untersuchten Aminosäuren treten sowohl in freier als auch in gebundener Form auf. Von allen Autoren werden Arginin, Histidin, Leucin, Methionin, Threonin, Lysin und Valin angegeben. Bei allen untersuchten Pollenarten fällt der *hohe Gehalt an freiem Prolin* auf (Tabelle 6), der bis zu 1,65% des Trockengewichtes erreicht (BATHURST 1954, LINSKENS u. TUPÝ 1966). Mengenmäßig an zweiter Stelle treten Asparaginsäure und Glutaminsäure mit den zugehörigen Amiden häufig auf. Nur als freie Aminosäuren werden angegeben: Serin, Methionin und Threonin.

Tabelle 6. *Freie Aminosäuren im Pollen.* Menge in γ/1 mg lufttrockenen Pollens; + = sicher nachgewiesen, aber nur in Spuren, (+) = nicht regelmäßig gefunden. (Nach BELLARTZ 1956.)

Aminosäure	*Nicotiana*	*Petunia*	*Salpiglossis*
α-Alanin	0,2	1,1	2,3
β-Alanin	+	(+)	+
γ-Aminobuttersäure	0,6	1,3	1,1
Arginin	—	2,6	2,4
Asparagin	6,2	10,5	8,8
Asparaginsäure	1,3	0,7	1,0
Cystein/Cystin	+	+	+
Glutamin	1,4	3,4	3,4
Glutaminsäure	3,1	3,2	2,6
Glycin	+	+	0,2
Leucin	(+)	(+)	(+)
Lysin	(+)	(+)	3,0
Ornithin	+	0,7	0,7
Pipecolinsäure (?)	+	+	+
Prolin	12,2	14,2	14,5
Serin	0,4	0,8	0,9
Threonin	+	+	+
Tyrosin	(+)	(+)	+
Valin	(+)	(+)	+

d) Nucleinsäuren.

Bei der Bestimmung der DNS im reifen Pollen wurden gefunden: in *Betula pubescens* 0,453 %, *Pinus montana* 0,186 %, *Populus tremula* 0,31% (v. EULER, AHLSTRÖM, HÖGBERG und PETTERSON 1945), *Lilium henryi* 0,8% (LINSKENS 1958b). RNS in freier Form wurde ebenfalls nachgewiesen (v. EULER, HELLER und HÖGBERG 1948).

In den keimenden Pollenschläuchen von *Nicotiana* in vitro wird eine neue Form von RNS synthetisiert, deren Basenzusammensetzung keine Ähnlichkeit mit der im Pollenkorn vorhandenen RNS hat (TANO u. TAKAHASHI 1964). Die Frage, ob diese Pollenschlauch-RNS Information trägt, ist noch nicht geklärt.

Die Basen-Zusammenstellung der RNS in den subcellulären Zellfraktionen haben STANLEY und YEE (1966) untersucht.

e) Kohlenhydrate.

Die älteste bekannte Zuckeranalyse in Pollen gibt HEYL (1917) von *Artemisia* mit 7,3% Pentosanen, 2,1% Dextrinen, 0,4% Saccharose und 1,6% Glucose. Es werden von anderen Untersuchern (KIESEL und RUBIN 1929, ELSER und GANZMÜLLER 1930, TODD und BETHERIK 1942, ROBBINS, SAMUELS und MOSKO 1948, WEYGANG und HOFMANN 1950, LINSKENS 1955a, BELLARTZ 1956) gefunden: Glucose, Fructose, Ribose, Saccharose. Maltose und Ribose treten stets nur in Spuren auf. Interessant ist der Nachweis von Lactose (KUHN und LÖW 1949) und das Vorhandensein von Uronsäure (ROBBINS, SAMUELS und MOSKO 1948). Der Gehalt an Kohlenhydraten

Tabelle 7. *Kohlenhydratverteilung von drei verschiedenen Rassen von Zea mays.* Menge in Prozent des Trockengewichtes. (Nach ANDERSON und KULP 1922.)

Kohlenhydrate	Yellow dent corn %	White flint corn %	Popcorn %
Stärke	11,07	19,04	18,03
Rohfaser	5,45	—	—
Reduzierende Zucker	3,5	5,38	4,95
Saccharose	9,09	2,97	14,18
Pentosane	10,60	—	—

ist starken Schwankungen ausgesetzt, sowohl innerhalb einer Art in verschiedenen Jahren (KIESEL und RUBIN 1929), als auch zwischen den Rassen (Tabelle 7) und Klonen einer Art (ANDERSON und KULP 1922, LINSKENS 1955a).

f) Lipide.

Die Analysen der Pollen-Lipoide und -Fette sind bisher nur recht summarisch, ohne daß auf deren Zusammensetzung näher eingegangen wird (RIDI und WAFA 1950). Im Alkoholextrakt von *Artemisia* wurden 10,8% Fett und 0,75% Lecithin gefunden (HEYL 1922), bei *Senf*-Arten schwankt der Fettgehalt zwischen 8,6 und 13,1% (SOSA und SOSA-BOURDOUIL 1952). Weiterhin wurden nachgewiesen: Phosphatide (ANDERSON und KULP 1922), Cetylalkohol (HEYL 1922), Phytosterole (Anderson 1923, KOESSLER 1918, HEYL 1922), Sterole, Palmitinsäure, Laurinsäure, Trikosan (SOSA und SOSA-BOURDOUIL 1952).

g) Wirkstoffe.

Der *Vitamingehalt* des Pollens ist unter dem Gesichtspunkt des Nährwertes untersucht. Es wurden gefunden: Thiamin, Riboflavin, Nicotinsäure, Ascorbinsäure (HEYDACK und PALMER 1940, KITZES, SCHUETTE und ELVEHJEM 1943, NIELSEN, GRÖMMER und LUNDEN 1955, VIVINO und PALMER 1944, RIDI und WAFA 1947, SAGROMSKY 1947, WEYGAND und HOFMANN 1950, NIELSEN 1956), Nicotinsäure, Panthotensäure (VIVINO und PALMER 1944, RIDI und WAFA 1947), die Vitamine A, D, E und K (VIVINO und PALMER 1944). Einige neuere Analysenergebnisse für verschiedene Pollenarten sind in Tabelle 8 wiedergegeben.

Tabelle 8. *Vitamine in verschiedenem Pollen.* Menge in µg/g Pollen Trockengewicht. (Nach NIELSEN, GRÖMMER und LUNDÉN 1955.)

	Zea Mays	*Alnus glutinosa*	*Alnus incana*	*Pinus montana*
Riboflavin . . .	5,7	11,2	12,1	5,6
Nicotinsäure . .	40,7	82,7	82,3	79,8
Pantothensäure .	14,2	4,2	5,0	7,8
Pyridoxin . . .	5,9	5,7	6,8	3,1
Biotin	0,52	0,65	0,69	0,62
Inositol (mg/g) .	30,0	3,0	3,5	9,0

Nach NIELSEN (1956) ändert sich der Vitamingehalt während des Alterungsprozesses des Pollens.

Die älteren Mitteilungen über *Wuchsstoffgehalt* im Pollen sind von EULER, AHLSTRÖM, HÖGBERG und PETTERSON (1945) zusammengefaßt worden. Wuchsstoffe im Pollen haben weiterhin nachgewiesen: MUIR (1947), YAKUSHKINA (1947), LARSEN und SIN MIN TUNG (1950), REDEMANN (1950), KÜHLWEIN und ANHÄUSSER (1951), LAIBACH (1932). Der ruhende Pollen enthält sowohl Indolylessigsäure als auch Indolacetonitril (LUND 1956, MÜLLER 1953). Besonders reich sollen die Pollinien der *Orchideen* an Wuchsstoff sein. Doch handelt es sich dabei wohl um die an der Außenseite der Pollenmasse anhaftende Substanz. Auch sollen im Blütenstaub der *Dattelpalme* oestrogene Substanzen nachweisbar sein (RIDI und WAFA 1947, HASSAN und WAFA 1947).

h) Anorganische Substanzen.

Der Aschegehalt des Pollens beträgt rund 4%, er schwankt zwischen 2,07% bei bienengesammeltem Mischpollen (TODD und BETHERIK 1942) und 10,6% bei *Ambrosia artemisifolia* und *A. trifida* (KOESSLER 1918). Einige Einzelanalysen sind in Tabelle 4 wiedergegeben (ELSER und GANZMÜLLER 1930). Auffallend ist, daß kein Magnesium gefunden wurde, ein Befund, der auch für andere Pollenarten bestätigt wird (BERTRAND 1940). Die Ergebnisse einer aschenanalytischen Untersuchung sind in Tabelle 9 zusammengestellt (nach ANDERSON und KULP 1922, v. EULER, AHLSTRÖM, HÖGBERG und PETTERSON 1945, FAMINTZIN und PRZYBYTEK 1889, RIDI und WAFA 1947, TODD und BETHERIK 1942, VIVINO und PALMER 1944).

Tabelle 9. *Aschenanalyse verschiedener Pollen-Arten.* Mengenangaben in Prozent der Gesamtasche.

Asche	*Zea mays* %	*Typha latifolia* %	*Dactylifera palmae* %	*Typha latifolia* %	Pollen-gemisch %
K	35,58	1,244	—	0,97	—
Ca	1,02	0,097	0,55	0,30	0,29
Mg	4,60	0,283	—	0,24	0,24
Fe	0,25	0,025	0,025	0,0038	0,55
Cl	0,80	0,510	—	—	—
SiO_2	3,76	—	—	—	—
Al	0,22	—	—	—	—
Na	0,69	0,127	—	—	—
Cu	—	—	—	—	1,57
P	18,92	0,406	0,65	0,49	0,43
S	0,69	0,240	—	—	—
Mn	—	0,0069	—	—	—

IV. Pollenkeimung.

Pollenkörner sind in reifem Zustand ruhende Pflanzenorgane. Um ihre Funktionen, die Ausbildung eines Pollenschlauches und Durchführung des Befruchtungsaktes, ausführen zu können, müssen sie aus dem Zustand latenten Lebens zurückkehren zur vollen Aktionsbereitschaft: quellen, keimen und wachsen.

Die Rolle des Pollens beim Befruchtungsprozeß hat zuerst R. J. CAMERARIUS (1694) erkannt. Trotz der wichtigen Untersuchungen von KOELREUTER (1761/66) und SPRENGEL (1793) ist dieses Wissen erst Anfang des 19. Jahrhunderts Allgemeingut der Botanik geworden. AMICI (1824, 1830) hat als erster das Auskeimen der Pollenkörner beobachtet und das gerichtete Wachstum der Pollenschläuche zu den Samenanlagen beschrieben. HUGO VON MOHL (1834) sah, daß die Pollen auch in Wasser unabhängig von der Narbe auskeimen können. Seitdem haben sich zahlreiche Forscher mit der künstlichen Kultur des männlichen Gametophyten von Samenpflanzen beschäftigt. Ziel solcher *in vitro*-Untersuchungen ist es, die Kenntnis vom Verhalten der Pollenschläuche im natürlichen Milieu der weiblichen Blütenorgane zu fördern.

A. Keimung *in vitro.*

1. Keimungsbedingungen.

Untersuchungen über das Pollenschlauchwachstum *in vitro* sind in großer Zahl durchgeführt worden, Literaturzusammenstellung bei BRINK 1924 und BRANSCHEIDT 1930. Hinsichtlich der *Keimungsbedingungen* werden nach JOST (1905) drei verschiedene Typen unterschieden:

I. Der Pollen benötigt zur Keimung nur *flüssiges Wasser* (LIDFORSS 1896, 1899a, b; ADAMS 1916, KNIGHT 1917, PATON 1921, SCHOCH-BODMER 1936, RICHTER 1939, COOPER 1939, EHLERS 1951, SCHWARZENBACH 1953). Zwar sind die erzielten Keimprozentsätze häufig ziemlich niedrig; aber prinzipiell war der Pollen in der Lage, einen Schlauch zu bilden, ohne Energiezufuhr von außen.

II. Die Pollenkörner bedürfen zur Keimung außer Wasser einer ganz *bestimmten chemischen Substanz,* die *in vivo* in der Narbe enthalten ist. Als solche Reizstoffe werden in einigen Fällen Zucker (BURCK 1900), in anderen organische Säuren (MOLISCH 1893) angesehen.

III. Der Pollen keimt *nur in Zuckerlösungen* ganz bestimmter Konzentration. Dabei zeigt sich, daß im allgemeinen die Konzentrationen für bestimmte Species spezifisch sind (Zusammenstellung bei VISSER 1955). Andere Arten hingegen keimen unabhängig von der Höhe der Konzentration in 1—40%igen Saccharose-Lösungen (MOLISCH 1893, PFUNDT 1910). Hier erhebt sich natürlich sogleich die Frage nach der Bedeutung der löslichen Kohlenhydrate im Keimmedium: Wirkt der Zucker hier *ernährend* oder durch seine *osmotischen* Eigenschaften?

Zu den von JOST beschriebenen Keimungsbedingungen kann noch eine vierte hinzugefügt werden:

IV. Der Pollen keimt nur auf einem *festen Substrat* aus. Als solches kann feuchtes Fließpapier (JOST 1907), Deckglaskultur bei begrenzter Luftfeuchtigkeit (FIRBAS 1922), eingetrockneter Narbenschleim (RENNER 1919), Gelatine- oder Agar-Substrat (WALDERDORF 1924, KUHN 1937) dienen. Das wesentliche Prinzip dabei scheint zu sein, daß es sich um Bedingungen mit begrenzter Wasserzufuhr handeln muß.

Die Keimkraft des Pollens wird weiterhin beeinflußt durch das *Alter und die Reife des Pollens* (HAYASE und HIRAIZUMI 1955), sowie die Temperatur vor und nach der Anthese (HAYASE 1955).

2. Die Bedeutung des Zuckers und der Borsäure für die Keimung.

Seit BRINK (1924) die Bedeutung des Keim-Mediums sowohl hinsichtlich der osmotischen Funktion als auch in bezug auf die Nährstoffversorgung untersucht hat, ist die Frage der Bedeutung des Zuckers für den Keimvorgang noch immer offen. VISSER (1955) stellt die Literatur über die beiden Möglichkeiten vergleichend gegenüber. Danach muß man annehmen, daß fehlende Zuckerversorgung nicht für das Platzen der Pollenschläuche bei *in vitro*-Kultur verantwortlich ist. Vielmehr kann die osmotische Funktion bzw. die begrenzte Nachlieferung von Wasser auch von anderen Stoffen kolloidaler Natur durch Diffusionshemmung übernommen werden (KUHN 1937, TISCHLER 1917, ESSER 1953, MOLISCH 1893) (Abb. 9). Aus den Untersuchungen von EHLERS (1951) ist ersichtlich, daß zahlreiche Pollensorten in einem Medium ohne Zucker, jedoch unter Zusatz von Borsäure Schlauchlängen produzieren können, die zum Erreichen der Samenanlage *in vivo* genügen.

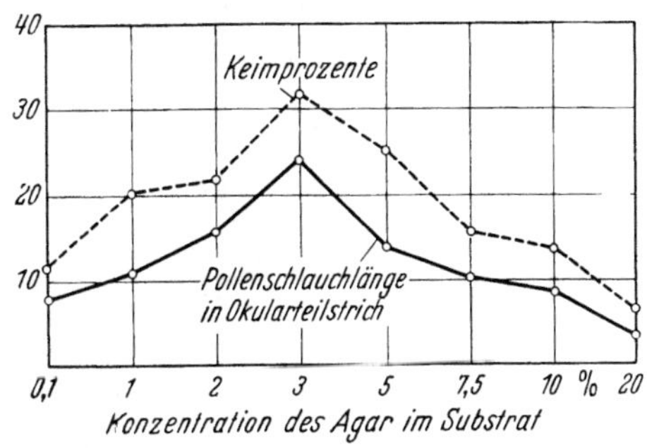

Abb. 9. Abhängigkeit der Pollenkeimung und des Pollenschlauch-Wachstums bei *Petunia hybrida* von der Agar-Konzentration, *in vitro* ohne Zusatz von Zucker und Borsäure. (Unveröffentlicht.)

Eine zentrale Funktion für das erfolgreiche Keimen und Wachsen der Pollenschläuche hat die Borsäure, deren Bedeutung SCHMUCKER (1933, 1935) entdeckt hat. Es ist noch immer unklar, in welcher Form diese im Narbenschleim vorliegt und in welche Prozesse sie beim Keimungsvorgang eingreift (GÄRTEL 1956). Nachdem ECHOLS und MERGEN (1956) bei *Pinus* zeigen konnten, daß die höchsten Keimprozente im zucker- und borfreien Medium sich bei Zusatz eines chelatbildenden Eisenkomplexes erzielen lassen, liegt die Vermutung nahe, daß Zucker und Borsäure nicht unmittelbar als für die Ernährung und Entwicklung notwendige Stoffe anzusehen sind. Jedenfalls steht die spezifische Bedeutung des Bors in keiner Beziehung zur Sauerstoff- und Zuckeraufnahme (O'KELLEY 1957). Wohl aber kann das Bor für den Transport der Zucker als Boratkomplex (VASIL 1964) und die Mobilisierung der Reservestoffe, sowie die Aktivitätserhöhung der ATP eine Rolle spielen (DANIEL-VÁRÓCZY 1957, VASIL 1958).

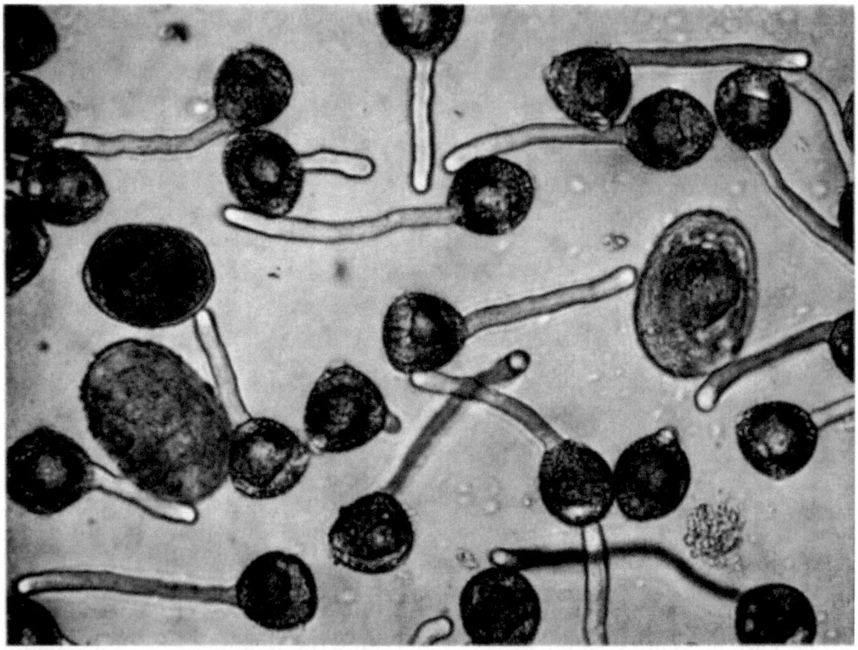

Abb. 10. *Forsythia*-Pollen in Kombination mit Pollen von *Gagea lutea*: positive Beeinflussung zur Keimung.

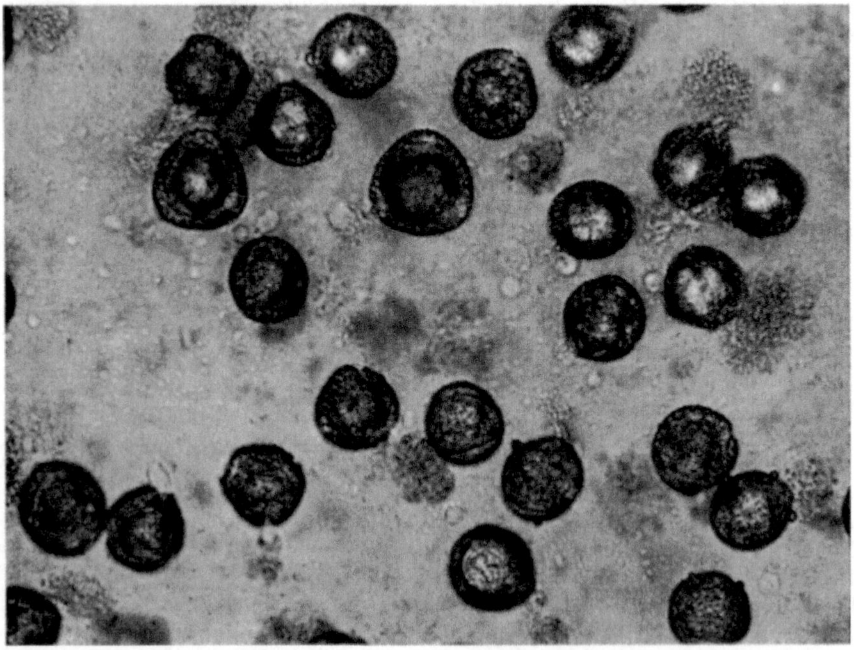

Abb. 11. Pollen von *Forsythia suspensa* in 15% Saccharose-Lösung. Keine Keimung.

Neben der Rolle, die Bor als Enzym-Effektor bei der Saccharosesynthese (FÄHNRICH 1964) hat, dürfte es eine entscheidende Rolle in der Pektinsynthese des wachsenden Pollenschlauches spielen (STANLEY u. LOEWUS 1964).

Außer Zucker und Borsäure sind natürlich zahlreiche andere physiologische Substanzen im Pollenkeimungstest untersucht worden, insbesondere *Wuchsstoffe* (ADDICOTT 1943, SMITH 1939, ANHÄUSSER 1952, POHL 1951, SCHWARZENBACH 1953), *Vitamine* (RAGHAVAN und BARUAH 1956, COOPER 1939, ECHOLS und MERGEN 1956, BECK und JOLY 1941, DANDLIKAR, COOPER und TRAUB 1938, SMITH 1942) und *Spurenelemente* (RAGHAVAN und BARUAH 1956). Die Resultate lassen nur bedingt Rückschlüsse auf die Verhältnisse *in vivo* zu.

3. Gegenseitige Beeinflussung des Pollens bei der Keimung.

Bei der Keimung spielt die wechselseitige Beeinflussung der Pollenkörner im Keimsubstrat eine wichtige Rolle (Zusammenstellung bei GRÜMMER 1955). Umfangreiche Beobachtungen über das Verhalten des Pollens in Mischkulturen stammen von BRANSCHEID (1930). Zunächst fand er, daß Pollen von *Hamamelis virginica* in Zuckerlösungen, in denen bereits 1× oder 2× Pollen von *Corylus avellana* gekeimt war, gegenüber frischen Lösungen geförderte Keimprozente bei gehemmtem Wachstum ergab. In Mischkulturen waren je nach Kombination der Arten fördernde oder hemmende Einflüsse zu beobachten. Bei Mischungen aus Pollen von kurz- und langgriffligen Formen von *Primula vulgaris* war die Keimfähigkeit stets um 50% reduziert (ZANONI 1930). Andere Untersuchungen stammten von GOLUBINSKY (1946, 1951), der Steigerung der Keimung und Förderungen des Schlauchwachstums bei Pollengemischen aus Pflanzengesellschaften mit gleicher Blütezeit zu finden glaubt. In gleicher Richtung liegen auch die Resultate von BORRISS und KROLOP (1955), die Beziehungen zur systematischen Stellung der einzelnen Partner bei der Hemmung zu erkennen glauben (Abb. 10 u. 11). Die *Mutualreaktion* zwischen keimenden Pollenkörnern ist jedoch nicht nur bei Art-Kombinationen zu finden, sondern auch innerhalb einer Species. Darüber liegen zahlreiche Beobachtungen vor (unter anderen BRINK 1924, KUHN 1937, SCHMUCKER 1938, SAVELLI 1940, ADDICOTT 1943, VISSER 1951). Da diese wechselseitigen Stimulationen bei allen untersuchten Arten gefunden wurden, dürften sie allgemein verbreitet sein, wenngleich die *in vitro*-Beobachtungen nicht kritiklos auf die Verhältnisse *in vivo* zu übertragen sind. Immerhin hat LOBANOV (1951) nachgewiesen, daß bei reichlicher Bestäubung intervarietale Kreuzungen gefördert werden können. Die Stärke der mutualen Förderung ist von der Zahl der Pollenkörner im Volumen des Keimmediums abhängig (SAVELLI und CARUSO 1940, VISSER 1955) (Abb. 12).

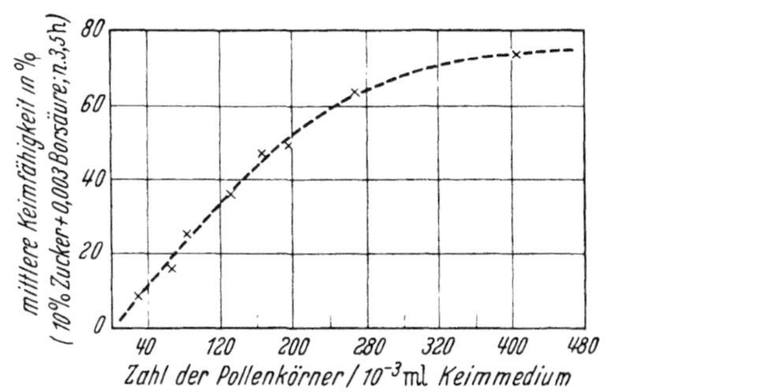

Abb. 12. Abhängigkeit der Keimfähigkeit von Pollen von der Zahl der Pollenkörner im gleichen Volumen des Keimmediums. *Pyrus malus* var. *Précoce de Trévoux.*

Über die *chemische Natur* der Stoffe, die für die gegenseitige Beeinflussung der Pollenkörner verantwortlich sind, lassen sich noch keine Angaben machen. Das hemmende Prinzip im *Compositen*pollen ließ sich durch Äther nicht extrahieren, während der Pollenkitt sich im Äther leicht löste. Außen an der Pollenwand anhaftende Reste des Pollenkittes dürften also nicht mit dem Hemmstoff identisch sein (BORRISS und KROLOP 1955). Lange Zeit neigte man der Auffassung zu, daß das „Pollen-Hormon" mit Heteroauxin identisch sei (LAIBACH 1932,

JAKUSCHKINA 1947, KÜHLWEIN 1948, LARSEN und TUNG 1950, ANHÄUSSER 1952), zumal POHL (1951) bei *Petunia* zeigen konnte, daß die Wachstumsgeschwindig-

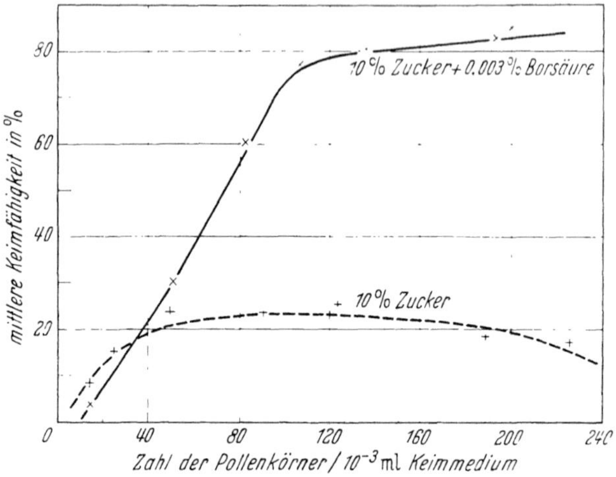

Abb. 13. Die Abhängigkeit der Pollenkeimung von der Zahl der Pollenkörner im gleichen Volumen des Keimmediums. *Pyrus communis* var. *Conférence,* Test mit Pollen von Zweigen in Wasser.

keit der Pollenschläuche durch β-Indolylessigsäure gesteigert werden kann. Die optimale Konzentration war davon abhängig, welche Strecke die Schläuche bereits im Griffelgewebe zurückgelegt hatten. Es muß also eine weitere physiologische Komponente im Spiel sein. Aus Keimversuchen unter Berücksichtigung der Dichte der Pollenzahl (Abb. 13, 14) ergibt sich, daß Borsäuremangel durch steigende Anzahl der Pollenkörner im Medium kompensiert werden kann. Mithin ist sowohl für den Boreffekt als auch die Mutual-Stimulation das gleiche stoffliche Prinzip verantwortlich. Die Hypothese, daß es sich um einen Borsäure-Komplex handelt, kann durch verschiedene Gesichtspunkte gestützt werden (VISSER 1955), ist jedoch noch nicht bewiesen.

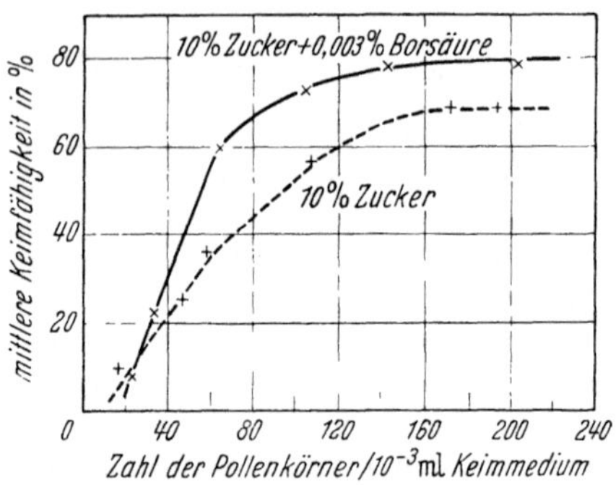

Abb. 14. Wie Abb. 13, jedoch Test mit Pollen von Zweigen, die 0,01 % Borsäurenlösung während der Pollenentwicklung zugeführt erhielten, durchgeführt. (Nach VISSER 1955.)

4. Der Pollenschlauch.

Für den normalen Keimvorgang der Phanerogamen-Pollen haben Keimfalten und -poren in mannigfaltiger Ausbildung die Funktion eines Keimapparates, durch dessen Betätigung erst das Pollenschlauchwachstum möglich wird. Daneben gibt es zahlreiche Pollen ohne speziellen Keimapparat, nonaperturate Pollen, bei denen sich der Pollenschlauch durch lochförmige, schlitzförmige Öffnungen oder durch Fragmentierung der Exine den Weg bahnt (MÜLLER-STOLL 1956). Der Widerstand der Membran wird durch *Quellung der Pollenkolloide* (LICHTE 1957) überwunden. Dabei kommt es unter Dehnung der Intine zur Ausstülpung an der Stelle des geringsten Widerstandes. Der Schlauch ist also ein Abkömmling der Intine (Abb. 4). Die Saugkraft des Pollens wurde im Augenblick des Keimungsbeginns mit etwa 45 Atm. berechnet. Im

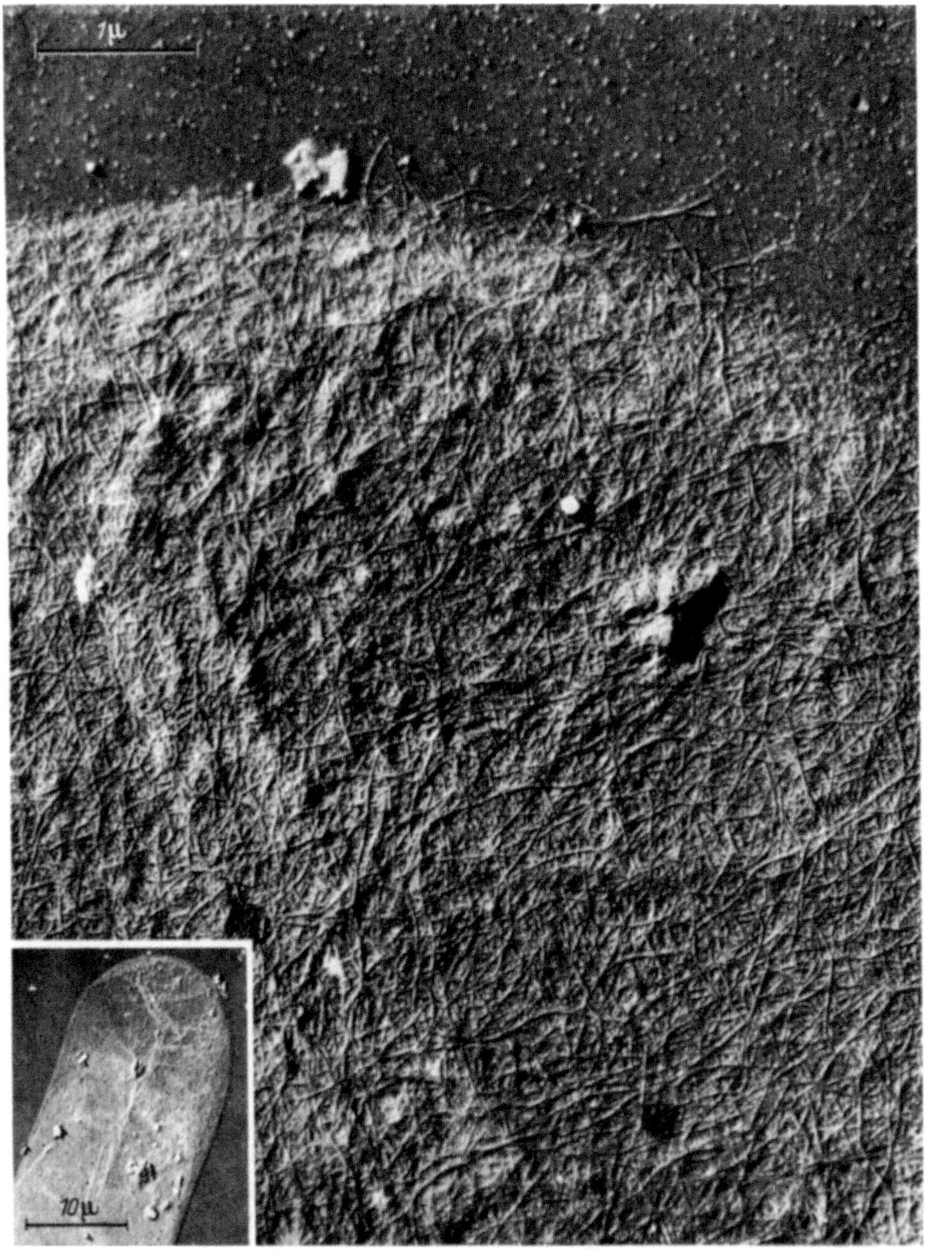

Abb. 15 A. Spitze eines Pollenschlauches von *Petunia*. Einsatz: Übersichtsbild. Der Pollenschlauch wurde mit Äthanolamin behandelt, um das Cellulose-Skelet sichtbar zu machen. Die Mikrofibrillen an der Schlauchspitze bilden ein Netzwerk, das keine Vorzugsorientierung zeigt. (Elektronenmikroskopische Aufnahme von M. M. A. SASSEN.)

wachsenden Pollenschlauch dürfte sie etwas geringer sein, da durch die Membranbildung lösliche Substanzen laufend verbraucht werden. Andererseits entstehen durch *Transformierung der Reservestoffe* wieder neue, osmotisch wirksame Substanzen (SCHOCH-BODMER 1936). Der Pollenschlauch besteht hauptsächlich aus Cellulose (BIOURGE 1892). Diese erscheint jedoch stark durch nichtcellulose Stoffe maskiert zu sein (Abb. 15). Als primäre Konstituente ist im Pollenschlauch, vor allem in der Spitzenregion, Pektin vorhanden. m-Inosit wird als Pektinvorstufe in die Schlauchwand eingebaut (STANLEY u. LOEWUS 1964). Solange sich der Schlauch in lebhaftem Wachstum befindet, ist die Spitze weniger stark inkrustiert (MÜHLETHALER und LINSKENS 1956). Der Pollenschlauch hat *Primärwandstruktur* (O'KELLEY und CARR 1954, MÜHLETHALER und LINSKENS

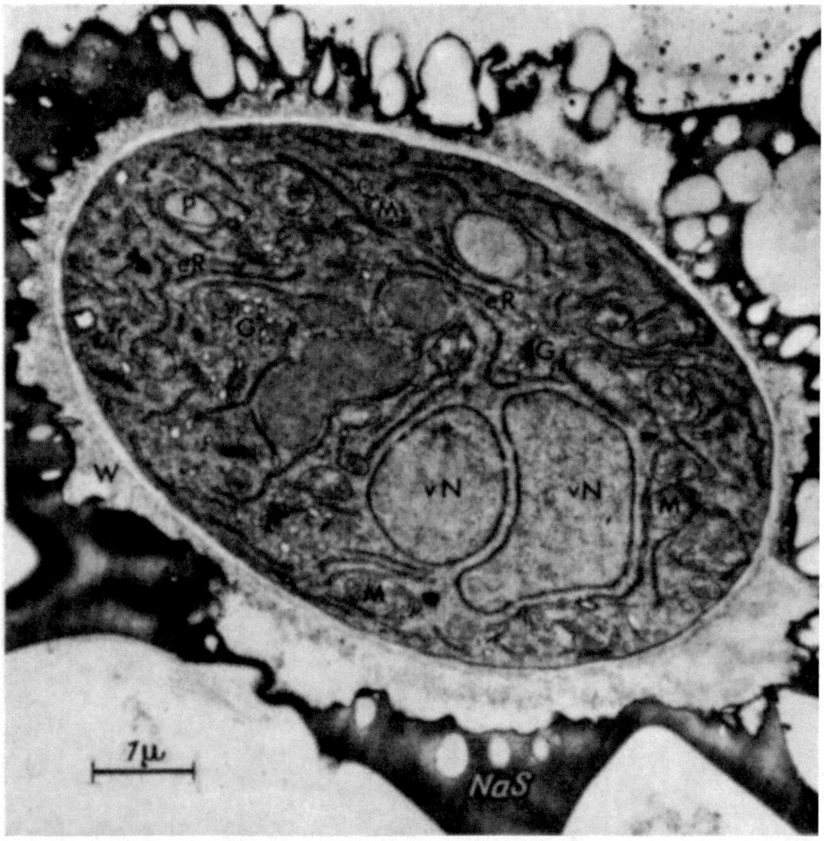

Abb. 15 B. Querschnitt durch einen *Petunia*-Pollenschlauch, ca. 2½ Std nach Bestäubung fixiert. Aufnahme: M. KROH. *G* Golgi-Material; *M* Mitochondrien; *vN* gelappter vegetativer Kern; *p* Plastid; *eR* endoplasmatisches Reticulum; *W* Pollenschlauchwand; *NaS* Narbensekret, in dem der Pollen gekeimt ist.

1956, FREY-WYSSLING 1957, SASSEN 1964b) und ist wahrscheinlich an der Spitze entweder offen oder jedenfalls nur von sehr lockerer Struktur (Abb. 16c, d); damit hat er Ähnlichkeit mit wachsenden Haaren (HOUWINK und ROELOFSEN 1954). Nach hinten wird er durch Anlagerung neuer Stränge verstärkt. Da das Flächenwachstum auf die ersten 4—7 μ hinter der Spitze beschränkt ist, erfolgt eine nachträgliche Dehnung der Membran und Umorientierung der Fibrillen nicht mehr (SCHOCH-BODMER 1945, 1947).

In den Pollenschläuchen entstehen auch bei künstlicher Kultur im Grenzgebiet zwischen dem vorderen, noch dicht mit fein granuliertem Plasma angefüllten Teil der Spitze und dem basalen Teil *Kallose-Pfropfen* (ESCHRICH 1956, MÜLLER-STOLL und LERCH 1957a, LINSKENS und ESSER 1957). Nach neueren Untersuchungen (MÜLLER-STOLL und LERCH 1957b) entsteht die Callose nach besonderen Reizvorgängen im Cytoplasma im Kontakt mit den submikroskopischen Wandunebenheiten und wird unter degenerativen Bedingungen lokal angehäuft. Das elektronenmikroskopische Bild zeigt den Kallosepfropfen im

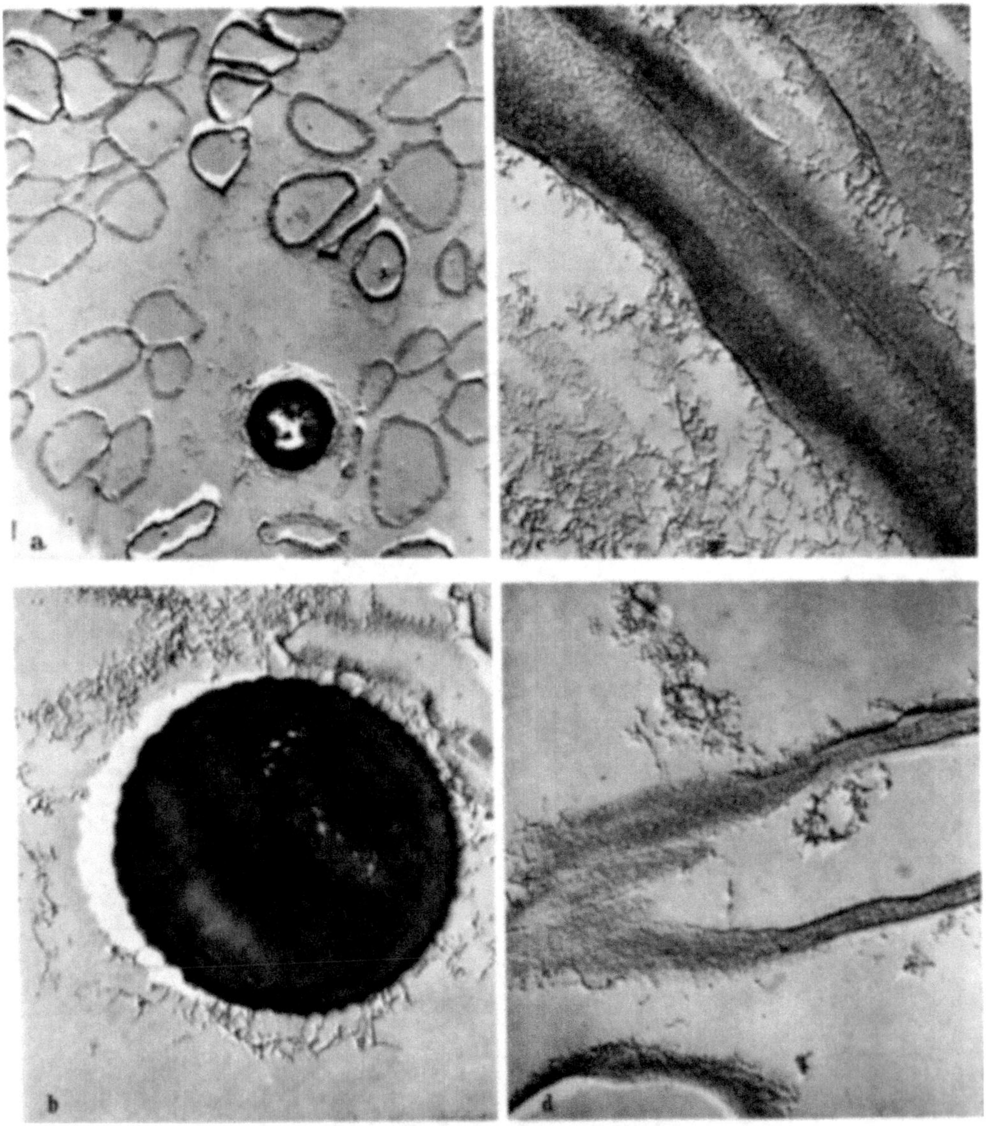

Abb. 16a—d. Elektronenmikroskopische Aufnahmen des Pollenschlauches von *Petunia hybrida*. a Querschnitt durch einen Griffel mit Pollenschläuchen und Leitgewebszellen (Pektin weggelöst). In der Mitte ein angeschnittener Kallosepfropfen. 2000×. b Querschnitt durch Pollenschlauch im Bereich des Kallosepfropfens, 7000×. c Mittelstück eines Pollenschlauches, 7000×. d Längsschnitt durch einen Pollenschlauch in der Spitzenregion, 7000×. (Aufnahmen: MÜHLETHALER und LINSKENS, unveröffentlicht.)

Pollenschlauch als ein sehr stark absorbierendes, amorphes Gebilde, das mit Poren durchsetzt ist (Abb. 16b). BOBILIOFF-PREISSER (1917) beobachtete, daß die lebhafte Plasmaströmung durch diese submikroskopischen Öffnungen hindurch verlief.

Abb. 17—19. (Legenden s. S. 73).

Die plasmatische Feinstruktur des Pollenschlauches zeigt alle Kennzeichen einer wachsenden Zelle (Abb. 15 B). Der Protoplast ist von einer Membran umgeben, das mit dem endoplasmatischen Reticulum verbunden ist, Mitochondrien, Plastiden, Fettvacuolen und Dictyosomen sind zahlreich anwesend. Zahlreiche Golgi-Vesikel sind besonders in der Spitzen-Region vorhanden (Abb. 15 B, 17, 19). Der generative Kern ist von einem eigenständigen Plasma, sowie einer zellwandähnlichen Struktur (Abb. 4, *gw*; SASSEN 1964a, b) umgeben, so daß man sicherlich von einer generativen Zelle bzw. nach der Mitose von Sperma-*Zellen* zu sprechen hat.

Die Formenmannigfaltigkeit der Pollenschläuche bei künstlicher Kultur im hängenden Tropfen ist sehr groß. Dabei treten zahlreiche *Formabweichungen* auf, die unter natürlichen Wachstumsbedingungen im weiblichen Griffel in dieser Weise nicht oder selten beobachtet wurden. Neben blasiger Auftreibung der Schlauchspitze, Gabelwuchs, Krümmungen und Spiralisierung treten durch überoptimale Substratkonzentration keulige und kuglige Schläuche auf (HARTMANN-DICK und MÜLLER-STOLL 1955).

Die *Wachstumsleistung* verschiedener Pollenarten hat KÜSTER (1951) zusammengestellt. Nach RENNER (1919) ist eine hohe Wachstumsgeschwindigkeit stets mit kurzer Wachstumsdauer verbunden. Bei *Impatiens* wurde die Wachstumsleistung in bezug auf die Ausdehnung der dabei tätigen Zone mit einer Streckung derselben um 220% in der Minute errechnet (WENT 1933). Der Pollenschlauch vermag auch bei Energiezufuhr von außen (O'KELLEY 1955) nicht unbegrenzt zu wachsen. Nicht entschieden ist die Frage, ob die Limitierung auf inneren Bedingungen beruht oder durch die Wirkung blokkierender Stoffwechselausscheidungen des wachsenden Schlauches in das Substrat bedingt ist.

B. Keimung *in vivo*.

1. Ernährung des Pollenschlauches.

Auf das Problem der Ernährung des Pollenschlauches während seines Wachstums durch das Leitgewebe des Griffels hat zuerst AMICI (1830) hingewiesen. Die meist reichliche Reservestoffausrüstung des reifen Pollens legt die Vermutung nahe, daß die Pollenschläuche aus diesen aufgebaut werden können. Den stärksten Hinweis für eine *Resorption von Griffelmaterial* durch den wachsenden Pollenschlauch haben jedoch SCHOCH-BODMER und HUBER (1947) mitgeteilt: Sie finden, daß nach dem Durchwachsen die Schläuche mehr als 40% des kollenchymatischen Leitgewebsvolumen einnehmen; die ernährungsphysiologische Funktion der Leitzellen kann also histophysiologisch als bewiesen gelten. Auch aus *in vitro*-Kulturen ergeben sich Hinweise in diesem Sinne: Läßt man Pollenschläuche in konzentrierten Zuckerlösungen wachsen, so kann sekundäre Stärkebildung im Innern derselben nachgewiesen werden (IWANAMI 1956, TANAKA 1955). Und schließlich haben HELLMERS und MACHLIS

Abb. 17. Querschnitt durch einen Pollenschlauch von *Petunia* in der Spitzen-Region, hergestellt nach der Gefrier-Ätz-Technik von H. MOOR; Aufnahme: M.M.A. SASSEN. W_1 Äußerer Teil der Zellwand mit Cellulose Mikrofibrillen; W_2 innerer Teil der Zellwand von granulärer Struktur, vorwiegend aus Pektin bestehend; *PM* Plasmamembran; *GV* Golgi-Vesikel; *G* Golgi-Material.

Abb. 18. Detail des Plasmas in einem Pollen-Korn von *Petunia*; hergestellt mittels Gefrier-Ätz-Technik von H. MOOR; Aufnahme: M.M.A. SASSEN. *NMI* Innere Kernmembran, von innen her gesehen; *NMA* äußere Kernmembran, von innen her gesehen; *L* Lipid-Granula; *V* Vacuolen; *ER* endoplasmatisches Reticulum; *G* Golgi-Material.

Abb. 19. Golgi-Apparat aus dem Cytoplasma eines Pollenschlauches (Gefrier-Ätz-Technik von M. MOOR; Aufnahme: M.M.A. SASSEN). G_1 Golgi-Material im Querschnitt; G_2 Golgi-Material in Aufsicht.

(1956) aus Atmungsmessungen Indizienbeweise für eine Ernährung der Pollenschläuche aus dem Substrat des Leitgewebes gegeben. Der gegenwärtige Stand der Kenntnisse muß wohl so charakterisiert werden, daß eine *Wasseraufnahme* und *Substraterschließung* durch den Schlauch bei zahlreichen Arten angenommen werden kann. Dies gilt insbesondere für Species mit langem Griffel und jene Fälle, in denen eine lange Ruheperiode zwischen Pollenkeimung und Befruchtung liegt (STANLEY 1957, 1958). Der direkte Beweis für eine Substanzaufnahme der Pollenschläuche aus dem Leitgewebe des Griffels (Zucker, Aminosäuren) ließ sich durch radioaktive Markierung des weiblichen Partners zeigen (LINSKENS und ESSER 1959). Besonders wichtig ist die Tatsache, daß die Pollenschläuche in der Lage sind, Kohlendioxyd in organische Säuren, Aminosäuren und Proteinfraktionen einzubauen. Die Energie für Reduktion und Einbau wird wahrscheinlich durch Oxydation von Kohlenhydraten geliefert, die aus dem weiblichen Gametophyten bezogen werden muß (STANLEY, YOUNG und GRAHAM 1958). Es findet also schon vor der eigentlichen Befruchtung ein intensiver Stoffaustausch zwischen den Geschlechtspartnern statt.

2. Tropismus der Pollenschläuche.

Die Richtungsweisung der eben gekeimten Pollenschläuche in das Innere des Narbengewebes erfolgt sicherlich durch *Hygrotropismus*; darauf weist auch die ungeordnete Wachstumsrichtung von Pollen auf der Narbe hin, wenn man diese in eine feuchte Kammer einschließt. Die Gravitation ist wirkungslos. Sodann spielt die *mechanische Leitung* auf den kontinuierlichen Schleimbahnen im Griffel bei zahlreichen Arten eine wichtige Rolle (KAIENBURG 1950, STEFFEN 1951, IWANAMI 1953). Zwischen der Narbe und dem Ovar besteht eine *elektrische Potentialdifferenz*, die ebenfalls für das gerichtete Wachstum der Schläuche verantwortlich gemacht werden kann (BÜNNING 1953, ZEIJLEMAKER 1956). *In vitro* läßt sich zeigen, daß Pollenschläuche *(Narcissus)* stets in Richtung auf die Kathode zu wachsen und gegenüber der Anode bereits bei einer Potentialdifferenz von 0,5 V/cm gehemmt sind (ZEIJLEMAKER 1956). Für den letzten Abschnitt des Weges muß man eine *chemotropische Anlockung* durch die Samenanlagen annehmen. Diese ist jedoch nicht bei allen Arten vorhanden (KAIENBURG 1950, TSAO 1949). Die positiven Befunde von BRINK (1924) konnten von SCHNEIDER (1956) *in vitro* reproduziert und statistisch gesichert werden. Die chemische Natur der chemotropisch wirkenden Stoffe ist unbekannt. In einigen Fällen konnte gefunden werden, daß sie hitzelabil und von relativ geringer Molekülgröße sind (MIKI 1954, 1955, 1964). Wesentlich ist jedoch, daß sie von hoher Spezifität sein müssen, da sie für den Mechanismus der selektiven Befruchtung verantwortlich sind.

Wie SCHILDKNECHT und BENONI (1963a, b) gezeigt haben, können auch Gemische von Salzen, Aminosäuren und Zuckern in bestimmter Zusammenstellung die chemotropische Anziehung simulieren. Die sexuelle Affinität wird jedenfalls durch die Samenanlagen gesteuert (GLENK 1964).

Als chemotropisch wirksamer Faktor ist das Calcium-Ion wieder aufgefunden worden (MASCARENHAS u. MACHLIS 1962, MACHLIS u. RAWITSCHER-KUNKEL 1963, BREWBAKER u. KWAK 1964). Es kann jedoch kaum als universales chemotropisches Agens angesehen werden (ROSEN 1964). Vielmehr scheint das Calcium mit der Überwindung des Populationseffektes zusammenzuhängen. Es fördert auch in Anwesenheit von toxischen Kationen, metabolischen und osmotischen Inhibitoren noch das Schlauchwachstum.

V. Lebensfähigkeit des Pollens.

1. Historisches.

Die ältesten Angaben über den Umgang mit Pollen werden vom Blütenstaub der Dattelpalmen gemacht. In den Handelsverträgen der Hammurabi-Periode um 2000 v. Chr. (ZIRKLE 1935) werden die männlichen Inflorescenzen der *Dattelpalmen* schon als wichtiges Importprodukt behandelt. Die Erfahrung, daß bei Aufbewahrung an einem dunklen und trockenen Platz die Fähigkeit zur Befruchtung bis zur neuen Vegetationsperiode erhalten bleibt, ist uralt (KÄMPFER 1719). Eine mehr oder weniger systematische Erforschung der Lebensdauer des Pollens setzte am Ende des 19. Jahrhunderts ein. Angaben über die Lebensfähigkeit des Blütenstaubes von über 80 Pflanzenarten, die bei niedriger Luftfeuchtigkeit gelagert wurden, werden von MANGIN (1886), RITTINGHAUS (1886) und MOLISCH (1893) gemacht. GOFF (1901), SANDSTEN (1909) und ROEMER (1915) fanden, daß Pollen verschiedener Species bei niedriger Aufbewahrungstemperatur länger seine Lebensfähigkeit behält als bei hoher Lagerungstemperatur.

Eine Zusammenstellung der Literatur über Pollenlagerung in alter Zeit gibt VISSER (1955), in neuerer Zeit JOHRI u. VASIL (1960), LINSKENS (1964a).

Die Lebensfähigkeit der Pollenkörner wird entweder an deren Keimfähigkeit oder ihrer Fähigkeit, zum Fruchtansatz zu führen, getestet. Beide Eigenschaften brauchen sich nicht zu entsprechen (PASSECKER 1930). Für eine Reihe von Arten ist in Tabelle 10 die Lebensfähigkeit unter verschiedenartigen Lagerungsbedingungen zusammengestellt. Entscheidend ist die Frage: Wieweit kann man die physiologische Aktivität während der Lagerung reduzieren, ohne die Lebensfähigkeit der Pollenkörner zu schädigen?

2. Aufbewahrungsbedingungen.

Daraus ergibt sich, daß für die Erhaltung der Keimkraft die Aufbewahrungsbedingungen von entscheidender Bedeutung sind. Die *Milieufaktoren* lassen sich in 3 Gruppen aufgliedern:

a) Luftfeuchtigkeit.

Es ist nicht überraschend, daß die Lebensfähigkeit des Pollens mit abnehmender relativer Luftfeuchtigkeit während der Lagerung zunimmt. Dennoch kann bei vielen Pollen der Wassergehalt nicht ohne Schädigung der Vitalität unter ein bestimmtes Niveau reduziert werden. Es besteht also in diesen Fällen eine bestimmte optimale Feuchtigkeit für maximale Lebensdauer. Bei Erhöhung der Feuchtigkeit wird infolge der Möglichkeit größerer physiologischer Aktivität die Lebensdauer verkürzt (Abb. 20). Andere Pollenarten erhalten ihre Lebensfähigkeit bei einer relativen Feuchtigkeit nahe um 0%. Es muß angenommen werden, daß die wasserhaltende Kapazität ihrer Kolloide so groß ist, daß sie selbst bei extremer Trockenheit dieses nicht abgeben. PFUNDT (1910) sowie HOLMANN und BRUBAKER (1926) fanden bei 200 verschiedenen Pollenarten, daß ungefähr bei der Hälfte derselben bei Aufbewahrung über konzentrierter Schwefelsäure (± 0,005% relative Feuchtigkeit) die Lebensfähigkeit besser erhalten bleibt als bei 30% Luftfeuchtigkeit.

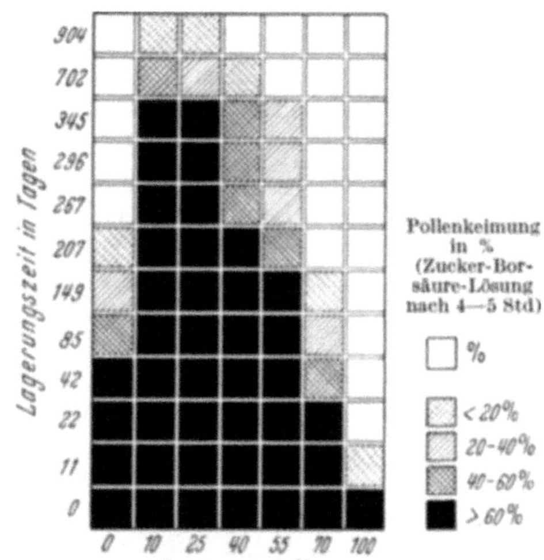

Abb. 20. Lebensdauer des Pollens von *Pyrus communis* var. Clapp's *Liebling* bei Aufbewahrung unter verschiedenen Feuchtigkeitsbedingungen bei 2—4° C. (Nach VISSER 1955.)

Eine kleine Gruppe von Pollen, insbesondere von *Gramineen*, verliert unter trockenen Bedingungen ihre Lebensfähigkeit sehr viel schneller als bei 80 bis 100% Feuchtigkeit. Dieser Pollen verliert auch bei niedrigen Lagerungstemperaturen seine Lebensfähigkeit. Die kurze Lebensdauer ist zweifellos durch seine schwache und wenig widerstandsfähige Struktur bedingt.

b) Temperatur.

Ein weiterer bedeutender Faktor für die Lebensfähigkeit des Pollens ist die Temperatur. Aus der Fülle des Beobachtungsmaterials ergibt sich, daß durch Erniedrigung der Temperatur in der Nähe des Null-Punktes die Lebensfähigkeit verlängert werden kann (McGUIRE 1952, MORGANDO 1949). Jedoch scheint der Faktor Wassergehalt ein integrierender zu sein. OLMO (1942) fand nämlich, daß ein Pollen bei Aufbewahrung von —12° C seine Vitalität bei 28% relativer Feuchtigkeit besser behält als bei 56% relativer Feuchtigkeit.

Die Trocknung ist also für die Aufbewahrung bei Temperaturen unter 0° C unerläßlich. In dieser Beziehung ist es interessant, darauf hinzuweisen, daß *Gramineen*-Pollen nicht unter dem Null-Punkt gelagert werden können; offensichtlich führt die Höhe des Wassergehaltes der Pollen zu Frostschäden. Von besonderem Interesse sind natürlich Lagerungsversuche bei extrem tiefen Temperaturen. Es scheint so, daß die Lebensfähigkeit der Pollen mit abnehmender Temperatur zunimmt. Es erhebt sich jedoch die Frage, was die Temperaturerniedrigung z. B. auf —200° C für die Verlängerung des Lebens der Pollen bedeutet. Theoretisch kann angenommen werden, daß bei solchen niedrigen Temperaturen alle physiologischen Aktivitäten auf 0 reduziert sind. Das würde bedeuten, daß bei solchen Extremtemperaturen dem gelagerten Pollen die Keimfähigkeit für eine unbestimmte Zeit unverändert erhalten bleibt. In der Tat scheinen die Ergebnisse von BREDEMANN und Mitarbeitern (1947) bei *Lupinus* diese theoretische Erwartung zu bestätigen. Die Autoren schließen aus ihren Resultaten, daß Lupinen-Pollen, der bei —180° C aufbewahrt würde, für 1000000 Jahre unverändert lebensfähig bliebe. Es scheint jedoch, daß sich solcher Pollen auch durch eine Trockenresistenz, d.h. maximale Lebensfähigkeit bei niedriger Feuchtigkeit, auszeichnen müsse.

c) Der Sauerstoffpartialdruck.

Neben der Anwendung niedriger Luftfeuchtigkeit hat man durch Erniedrigung des O_2-Spiegels und Erhöhung des CO_2-Gehaltes versucht, die Lebensfähigkeit des Pollens zu verlängern. Wie aus Tabelle 10 hervorgeht, vermag in manchen Fällen eine Verringerung des Sauerstoffgehaltes der Luft die Lebensfähigkeit zu verlängern. Dennoch scheint dieser Faktor nicht für alle Pollenarten von großer Bedeutung zu sein. Durch Vakuum-Lagerungsbedingungen werden andere Pollen in ihrer Keimfähigkeit reduziert.

3. Ursachen der Abnahme der Keimfähigkeit.

Die primären Ursachen für die Abnahme der Keim- und Lebensfähigkeit des Pollens unter normalen physiologischen Bedingungen muß in einem Verbrauch von Atmungsmaterial gesucht werden. Die Ergebnisse von NIELSEN (1956) deuten daraufhin, daß auch ein Verbrauch an Fermenten und Wuchsstoffen stattfindet. Außerdem ist an die Anhäufung von sekundären Atmungsprodukten und deren Blockierungseinfluß zu denken. PFEIFFER (1955) bringt den Verlust der Keimfähigkeit mit der Veränderung der Ölanlagerungen in der Pollenmembran in Zusammenhang. Es liegen auch Andeutungen vor, die auf die

Tabelle 10. *Lebensfähigkeit verschiedener Pollenarten unter unterschiedlichen Lagerungsbedingungen.*

Art	Aufbewahrungsbedingungen			Lagerungszeit in Tagen	Keimung in %		Autor
	Temperatur in °C	Relative Luftfeuchtigkeit in %	Luftdruck		vor Aufbewahrung	nach Lagerung	
Azalea mollis	+2	10	—	168	80	2	VISSER (1955)
Betula lutea	~+18	25	—	30	40	3	JOHNSON (1943)
Betula verrucosa	+5	0	Vakuum	920	60	20	JENSEN (1964)
Cinchona ledgeriana	+10	30—50	—	365	45—65	5—10	PFEIFFER (1944)
Cucurbita moschata	—17	0	Vakuum	30	98	98	GRIGGS *et al.* (1950)
Cydonia spec.	+2 bis +7	25	—	550	50—60	46	KING *et al.* (1938)
Ginkgo biloba	+5	0	—	700	90	35—45	TULECKE (1954)
Haemanthus katherinae	+20	100	—	1	80	60	PRUZSINSKI (1960)
Haemanthus katherinae	+20	15	—	1	80	15	PRUZSINSKI (1960)
Haemanthus katherinae	+20	100	—	3	80	8	PRUZSINSKI (1960)
Haemanthus katherinae	+20	48	—	3	80	60	PRUZSINSKI (1960)
Hevea spec.	+6	70—80	—	19	90	37	DIJKMAN (1938)
Hordeum vulgare	+10	30, 60—90	—	1	40	0	ANTHONY *et al.* (1920)
Lilium henryi	+0,5	35	—	194	69	53	PFEIFFER (1955)
Lupinus polyphyllus	—190	30—70	—	93	78	78	BREDEMANN (1947)
Lycopersicum esculentum	+2 bis +4	10	—	252	47	10	VISSER (1955)
Medicago sativa	—17	0	Vakuum	34	88	73	GRIGGS *et al.* (1953)
Olea spec.	—17	0	Vakuum	374	33	32	GRIGGS *et al.* 1953)
Pinus nigra	+5	0	Vakuum	920	80	20	JENSEN (1964)
Pinus nigra	+8	0	Vakuum	920	80	15	JENSEN (1964)
Pinus nigra	+5	60—80	—	378	80	10	JENSEN (1964)
Pinus strobus, *Pinus retinosa*	~+18	25	—	413	93	20	DUFFIELD *et al.* (1941)
	+4	25	—	413	93	70	DUFFIELD *et al.* (1941)
	0	50	—	413	93	91	DUFFIELD *et al.* (1941)
Pinus strobus, *Pinus retinosa*, *P. sylvestris*, *Pinus banksiana*	~+18	15—35	—	365	90	70	JOHNSON (1943)
	+2	25—75	—	365	90	90	JOHNSON (1943)
Prunus armeniaca	+2 bis +8	50	—	912	60—80	20—30	NEBEL *et al.* (1936)
Prunus avium (cerasus)	—17	0	Vakuum	410	53	37	GRIGGS *et al.* (1953)
Prunus avium (cerasus)	0 bis +2	25	—	550	57	55	KING *et al.* (1938)
Prunus avium (cerasus)	+2 bis +8	50	—	1460	60	20	NEBEL *et al.* (1936)
Prunus communis	—17	0	Vakuum	346	91	82	GRIGGS *et al.* (1953)
Prunus communis	—17	0	Vakuum	1130	91	24	GRIGGS *et al.* (1953)
Prunus communis	+2	0	—	550	70	53	KING *et al.* (1938)

Tabelle 10. (Fortsetzung.)

Art	Aufbewahrungsbedingungen: Temperatur in °C	Aufbewahrungsbedingungen: Relative Luftfeuchtigkeit in %	Aufbewahrungsbedingungen: Luftdruck	Lagerungszeit in Tagen	Keimung in %: vor Aufbewahrung	Keimung in %: nach Lagerung	Autor
Prunus domestica (insititia)	—17	0	Vakuum	435	69	60	GRIGGS *et al.* (1953)
Prunus domestica (insititia)	+10 bis +30	0	—	400	76	5	MANARESI (1924)
Prunus persica	0	50	—	550	85	42	KING *et al.* (1938)
Pyrus salicina	—17	0	Vakuum	439	40	35	GRIGGS *et al.* (1953)
Pyrus communis	—17	0	Vakuum	419	77	65	GRIGGS *et al.* (1953)
Pyrus communis	+10 bis —30	0	—	400	55	3	MANARESI (1924)
Pyrus communis	+2 bis —4	10	—	662	66	42	VISSER (1955)
Pyrus communis	+2 bis —4	0	Vakuum	1032	64	15	VISSER (1955)
Pyrus communis	—20	0	Vakuum	1032	64	50	VISSER (1955)
Pyrus communis	—190	0	Vakuum	662	64	50	VISSER (1955)
Pyrus communis	—17 —37	5	—	3287	—	1	USHIROZAWA (1951)
Pyrus malus	—17	0	Vakuum	385	92	64	GRIGGS et al. (1953)
Pyrus malus	+10 bis +30	0	—	400	93	7	MANARESI (1924)
Pyrus malus	+2 bis +8	50	—	1461	70—80	20	NEBEL (1939)
Pyrus malus	+2 bis +8	10	Vakuum	673	76	70	VISSER (1955)
Pyrus malus	—20	30—70	Vakuum	673	76	63	VISSER (1955)
Rhododendron catawbiense	+2 bis +4	30—70	—	252	30	2	VISSER (1955)
Rhododendron catawbiense	—20	30—70	—	662	30	25	VISSER (1955)
Rhododendron catawbiense	—190	30—70	—	662	30	25	VISSER (1955)
Rosa damascena	+18	0	—	50	85	5	ZOLOVIVH *et al.* (1963)
Saccharum spontaneum	+4	90—100	—	8	90	70—90	SARTORIS (1942)
Tulipa hybrida	+20	99	—	4	60	10	PRUZSINSKI (1960)
Tulipa hybrida	+20	95	—	4	60	35	PRUZSINSKI (1960)
Tulipa hybrida	+20	82	—	4	60	15	PRUZSINSKI (1960)
Tulipa hybrida	+20	60	—	4	60	5	PRUZSINSKI (1960)
Tulipa hybrida	+20	90	—	10	40	15	PRUZSINSKI (1960)
Tulipa hybrida	+20	48	—	10	40	2	PRUZSINSKI (1960)
Trollius europaeus	+20	71	—	1	80	75	PRUZSINSKI (1960)
Trollius europaeus	+20	95	—	3	90	2	PRUZSINSKI (1960)
Trollius europaeus	+20	6	—	3	6	70	PRUZSINSKI (1960)
Vitis vinifera	+10	25	—	365	43	10	OLMO (1942)
Vitis vinifera	+2	25	—	730	43	7	OLMO (1942)
Vitis vinifera	—12	28	—	1461	43	12	OLMO (1942)
Zea mays	+4	90—100	—	8	90	60—70	SARTORIS (1942)

Wichtigkeit der mineralischen Ernährung während des Pollenbildungsprozesses für die Lebensfähigkeit hinweisen (GREBINSKIJ und ROLIK 1949). Außerdem spielen sicherlich die bakterielle Kontamination während der Lagerung (TULECKE 1954) und die destruktive Wirkung des UV-Anteils bei Sonnenbestrahlung eine wesentliche Rolle bei der Verminderung der Vitalität (WERFFT 1951). Die Alterung des Pollens von virusinfizierten Blüten erfolgt bedeutend schneller als normalerweise, obgleich ein Nachweis des Virus im reifen Pollenkorn bisher noch nicht gelang. Möglicherweise wirkt das Virus auf den Alterungsprozeß durch kolloidchemische Veränderungen und Blockierung der Amylase- und Invertase-Aktivität (SCHADE 1957).

Literatur.

ADAMS, J.: On the germination of the pollen grains of apple and other fruit trees. Bot. Gaz. **61**, 131–147 (1916). — ADDICOTT, F. T.: Pollen germination and pollen tube growth as influenced by pure growth substances. Plant Physiol. **18**, 270—279 (1943). — AFZELIUS, B. M.: La structure du sporoderme vue au microscope électronique. Rapp. VIIe Congr. Intern. Bot., S. 241, 1954. — On the fine structure of the pollen wall in Clivia miniata. Bot. Not. (Lund) **108**, 141–143 (1955). — Electron-microscope investigation into exine-stratification. Grana palynolog. **1**, 22—37 (1956). — AFZELIUS, B. M., G. ERDTMAN and F. S. SJÖSTRAND: On the fine structure of the outer part of spore wall. Svensk bot. Tidskr. **48**, 155—161 (1954). — AMELUNG, B.: Flora (Jena) **77**, 176 (1893). Zit. in Tabul. biol. ('s-Grav.) **5**, 230—231 (1929). — AMICI, J. B.: Observations microscopiques sur diverses espèces de plantes. III. Du pollen. Ann. Sci. natur. **3**, 65—70 (1824). — Note sur la mode d'action du pollen sur le stigmate (Extrait d'une lettre de M. Amici à M. Mirbel). Ann. Sci. natur. **21**, 329—332 (1830). — ANDERSON, R. J.: Composition of corn pollen. J. biol. Chem. **55**, 611 (1923). — ANDERSON, R. J., and W. L. KULP: Analysis and composition of corn pollen. J. biol. Chem. **50**, 433 (1922). — ANHÄUSSER, H.: Keimung und Schlauchwachstum des Gymnospermenpollens unter besonderer Berücksichtigung des Wuchsstoffproblems. Beitr. Biol. Pflanz. **29**, 297—338 (1952). — ANTHONY, S., and H. V. HARLAN: Germination of barley pollen. J. agric. Res. **18**, 525–536 (1920). — ANIKIEV, V. V., u. E. N. GOROŠČENKO: Die Wirkung von Wassermangel im Boden auf Gerstenpollen im Lichtstadium der Entwicklung. Dokl. Akad. Nauk SSSR. **74**, 373—376 (1950). — ASBECK, F.: Sonnenlicht und Biogenese. Strahlentherapie **93**, 602—609 (1954a). — Strahlenschutz des Pollens. Proc. intern. Photobiol. Congr., biol. sect., S. 328—331. 1954b. — Fluorescierender Blütenstaub. Naturwiss. **42**, 632 (1955). — AUCLAIR, J. L., and C. A. JAMIESON: A quantitative analysis of acids in pollen collected by bees. Science **108**, 357 (1948). — AUGUSTIN, R.: Chemical, immunological and electrophoretic data of biologically active protein compounds of grasspollen extracts. Biochem. J. **54**, x (1953). — AVERY, P.: Chromosome number and morphology in Nicotiana. IV. The nature and effects of chromosomal irregularities in N. alata var. grandiflora. Univ. Calif. Publ. Bot. **11**, 265 (1929).

BATHURST, N. O.: The amino-acids of grass pollen. J. exp. Bot. **5**, 252—256 (1954). — BEADLE, G. W., and B. MCCLINTOCK: A genic disturbance of meiosis in Zea mays. Science **68**, 433 (1928). — BECK, W. A., and R. A. JOLY: Some growth phenomena in cultured pollen tubes. Trans. Amer. micr. Soc. **60**, 149—162 (1941). — BELLARTZ, S.: Das Pollenschlauchwachstum nach arteigner und artfremder Bestäubung einiger Solanaceen und die Inhaltsstoffe ihres Pollens und ihrer Griffel. Planta (Berl.) **47**, 588—612 (1956). — BERTRAND, G.: Sur la presence contestée du Mg dans le pollen. C. R. Acad. Sci. (Paris) **210**, 683 (1940). — BINGHAM, W. E., S. L. KRUGMAN, and E. F. ESTERMANN: Acrylamide electrophoresis of pine pollen protein. Nature (Lond.) **202**, 923—924 (1964). — BIOURGE, P.: Recherches morphologiques et chimiques sur les grains de pollen. Cellule **8**, 44—80 (1892). — BLAKESLEE, A. F.: Genetics of Datura. Z. Vererb.-Lehre Suppl. **1**, 117 (1928). — BOBILOFF-PREISSER, W.: Zur Physiologie des Pollens. Beih. bot. Zbl., Abt. I **34**, 459—492 (1917). — BORRISS, H., u. H. KROLOP: Über physiologische Wechselwirkungen zwischen Pollen verschiedener Pflanzenarten. Naturwiss. **42**, 301—302 (1955).— BRANSCHEIDT, P.: Zur Physiologie der Pollenkeimung und ihrer experimentellen Beeinflussung. Planta (Berl.) **11**, 368—456 (1930). — BREDEMANN, G., K. GARBER, P. HARTECK u. K. A. SUHR: Die Temperaturabhängigkeit der Lebensdauer von Blütenpollen. Naturwiss. **34**, 279—280 (1947). — BREWBAKER, J. L., and B. H. KWAK: The calcium ion and substances influencing pollen growth. In: Pollen Physiology and Fertilization (ed. H. F. LINSKENS), p. 145—151. Amsterdam: North Holland Publ. Co. 1964. — BRINK, R. A.: The physiology of pollen. I. The requirements for growth. Amer. J. Bot. **11**, 218—228 (1924a). — The physiology of pollen. IV. Chemotropism; effects on growth of grouping grains; formation and function of callose plugs. Amer. J. Bot. **11**, 417—436 (1924b). — BROWN, S. W.: Endomitosis in the tapetum of tomato. Amer. J. Bot. **36**, 703

(1949). — BÜNNING, E.: Entwicklungs- und Bewegungsphysiologie der Pflanze, 3. Aufl., S. 487. Berlin-Göttingen-Heidelberg 1953. — BURCK, W.: Preservatives on the stigma against the germination of foreign pollen. Proc. kon. ned. Akad. Wet. **3**, 264—274 (1900).

CAMERARIUS, R. J.: Epistola ad D. Mich. Bern. Valentini de sexu plantanum. Tuebingae 1694. (Abgedruckt in: Opuscula botanici argumenti collegit et edidit Johann Christian Mikan. Praguae 1797.) — CARNIEL, C.: Das Antherentapetum. Öst. bot. Z. **110**, 145—176 (1963). — CLARK, F. J.: Cytogenetic studies of divergent meiotic spindle formation in Zea mays. Amer. J. Bot. **27**, 547 (1939). — COOPER, D. C.: The transfer of DNA from the tapetum to the microsporocyten at the onset of meiosis. Amer. Naturalist **86**, 219—229 (1952). — COOPER, W. C.: Vitamins and the germination of pollen grains and fungous spores. Bot. Gaz. **100**, 844—852 (1939). — CZAPEK, F.: Biochemie der Pflanzen, Bd. 1, S. 702. 1913; Bd. 2, S. 289. 1920.

DANDLIKAR, W., W. C. COOPER and H. P. TRAUB: Vitamin B_1 and germination of pollen. Science **88**, 622 (1938). — DANIEL, L., and E. VÁRÓCZY: Pollenphysiological studies: II. The effect of boron on the germination of pollen grains and on the growth of pollen tubes. Növénytermelés **6**, 309—330 (1957). — DORSEY, M. J.: Pollen development in the grape, with special reference to sterility. Univ. Minnesota agric. Exp. Stat. Bull. **144**, 1—60 (1914). — DUFFIELD, J. W., and A. G. SNOW: Pollen longevity of Pinus strobus and P. retinosa as controlled by humidity and temperature. Amer. J. Bot. **28**, 175—177 (1941). — DIJKMAN, M. J.: Voorlopige gegevens over het bewaren van Hevea stuifmeel. Arch. Rubbercult. **22**, 239—255 (1938).

ECHOLS, R. M., and F. MERGEN: Germination of slash pine pollen in vitro. Forest Sci. **2**, 321—327 (1956). — EHLERS, H.: Untersuchungen zur Ernährungsphysiologie der Pollenschläuche. Biol. Zbl. **70**, 432—451 (1951). — ELSER, E., u. J. GANZMÜLLER: Die chemische Zusammensetzung einiger Blütenstaubarten. Hoppe-Seylers Z. physiol. Chem. **194**, 21—32 (1930). — EMECZ, T. I.: The effect of meteorological conditions on anthesis in agricultural grasses. Ann. Bot., N.S. **26**, 159—172 (1962). — ERDTMAN, G.: Pollen Morphology and Plant Taxonomy. Stockholm 1952. — An Introduction to Pollen Analysis. Chron. bot., Waltham **12** (1954). — "LO-Analysis" and "Welcker's Rule", a centenary. Svensk bot. Tidskr. **50**, 135—141 (1956). — ERICKSON, E. O.: Synchronous cell and nuclear division in tissue of higher plants. In: Synchrony in cell division and growth (ed. E. ZEUTHEN), p. 11—38. New York: Interscience 1964. — ERICKSON, R. O.: Cytological and growth correlations in the flower bud and anther of Lilium longiflorum. Amer. J. Bot. **35**, 729—739 (1948). — ESCHRICH, W.: Kallose. Protoplasma **47**, 487—530 (1956). — Die Callosesynthese bei Pollenmutterzellen von Cucurbita ficifolia. In: Pollen Physiology and Fertilization (ed. H. F. LINSKENS), p. 48—51. Amsterdam 1964. — ESSER, K.: Genomverdoppelung und Pollenschlauchwachstum. Z. indukt. Abstamm.- u. Vererb.-Lehre **85**, 28—51 (1953). — ESSER, K., u. J. STRAUB: Das Pollenschlauchwachstum bei Forsythia, eine Stellungnahme zu der Moewus-schen Hemmstoff-Ferment-Hypothese. Biol. Zbl. **73**, 449—455 (1954). — EULER, H. v.: Biochemische Untersuchungen von diploiden und triploiden Espen aus normalen und röntgenbestrahlten Pollen. Ark. Kemi, Mineral. Geol. A **26**, No 30 (1948). — EULER, H. v., L. AHLSTRÖM, B. HÖGBERG u. J. PETTERSON: Coenzyme, Enzyme, Wuchsstoffe und Reizstoffe in Pflanzenorganen. Ark. Kemi, Mineral. Geol. A **19**, No 4 (1945). — EULER, H. v., u. J. VON EULER: Über die katalatischen Wirkungen pflanzlicher Zellen. Ark. Kemi. Mineral. Geol. A **26**, No 22 (1948). — EULER, H. v., L. HELLER u. K. G. HÖGBERG: Nukleoproteide aus Pollen von Birke. Ark. Kemi. Mineral. Geol. A **26**, No 21 (1948).

FÄHNRICH, P.: Untersuchungen über den Einfluß des Bors bei der Pollenkeimung und beim Pollenschlauchwachstum. In: Pollen Physiology and Fertilization (ed. H. F. LINSKENS), p. 120—127. Amsterdam: North Holland Publ. Co. 1964. — FAMINTZIN, A., u. D. S. PRZYBYTEK: Aschenanalysen des Pollens von Pinus sylvestris. Bull. Acad. Imper. Sci. St. Peterbourg **30**, 358 (1889). — FERGUSON-HUNT, F.: Studies on Petunia. Bot. Gaz. **96**, 342 (1934). — FERNANDES-MORÁN, H., and A. O. DAHL: Electron microscopy of ultra thin frozen sections of pollen grains. Science **116**, 465—467 (1952). — FIRBAS, H.: Über die künstliche Keimung des Roggen- und Weizenpollen und seine Haltbarkeit. Z. Pflanzenzüchtg **8**, 70—73 (1922). — FISCHER, H.: Beiträge zur vergleichenden Morphologie der Pollenkörner. Breslau 1890. — FREIBERG, S. R., and H. E. CLARK: Changes in nitrogen fractions and proteolytic enzymes of soybean plants treated with 2,4-D. Plant Physiol. **30**, 39 (1953). — FREY-WYSSLING, A.: Formation and patterns of cellulose fibrils. J. cell comp. Physiol. **49**, Suppl. **1**, 63—70 (1957).

GÄRTEL, W.: Untersuchungen über die Bedeutung des Bors für die Rebe unter besonderer Berücksichtigung der Befruchtung. Weinberg u. Keller **3**, 132—139, 185—192, 233—241 (1956). — GOFF, E. S.: A study of certain conditions affecting the setting of fruits. Wiscon. agr. Exp. Stat. Rep. **18**, 289—303 (1901). — GOLUBINSKIJ, I. N.: Über den wechselseitigen Einfluß der Pollenkörner verschiedener Species beim gemeinsamen Auskeimen in künstlichen Medien. [Russisch.] Dokl. Akad. Nauk SSSR. **53**, 71—74 (1946). — Der Einfluß von Beimengungen des Pollens benachbarter Pflanzen auf die Pollenkeimung. [Russisch.] Dokl.

Akad. Nauk SSSR. **76**, 591—592 (1951). — Górska-Brylas, A.: Hydrolases in pollen grains and pollen tubes. Acta Soc. bot. Polon. **34**, 589—604 (1965). — Grebinskij, S. O., u. R. P. Rolik: Der Einfluß mineralischer Düngemittel auf die Lebensfähigkeit und die Dimensionen des Pollens. Dokl. Akad. Nauk SSSR. **68**, 1109—1112 (1949). — Griggs, W. H. G. H, Vansell and B. T. Iwakiri: Pollen storage. Cal. agr. Exp. Stat. **7**, 12 (1953). — Griggs, W. H., G. H. Vansell and J. F. Reinhardt: The germinating ability of quick-frozen bee-collected apple pollen stored in a dry ice container. J. econ. Entomol. **43**, 549 (1950). — Grümmer, G.: Die gegenseitige Beeinflussung höherer Pflanzen — Allelopathie. Jena 1955.

Haeckel, A.: Beitrag zur Kenntnis der Pollenfermente. Planta (Berl.) **39**, 431 (1951). — Hartmann-Dick, U., u. W. Müller-Stoll: Zytomorphologische Studien über das normale und pathologische Verhalten von Pollenschläuchen in künstlicher Kultur. Öst. bot. Z. **102**, 273—300 (1955). — Hassan, A., and M. H. Aboul Wafa: An oestrogenic substance in palm-pollen grains of data tree. Nature (Lond.) **159**, 140 (1947). — Hatano, K.: The amino acids in the pollen and female flower of Pinus Thunbergii. Bull. Tokyo Univ. Forests **48**, 149—151 (1955). — Hayase, H.: Effect of temperature before and after anthesis on pollen germinating power in cucumber and egg plant. Jap. J. Breeding **5**, 143—148 (1955). — Hayase, H., and Y. Hiraizumi: Relationship between pollen age and the optimum conditions (saccharose concentration and p_H value) of the artificial media. Jap. J. Breeding **5**, 51—60 (1955). — Hellström, N.: Determination of the purity of pollen grains collected in large scale quantities. Grana palynolog. **1**, 3—21 (1956). — Heslop-Harrison, J.: An ultrastructural study of pollen wall ontogeny in Silena pendula. Grana palynolog. **4**, 7—24 (1963a). — Ultrastructural aspects in differentiation in sporogenous tissue. Sympos. Soc. exp. Biol. **17**, 315—430 (1963b). — Cells walls, cell membranes and protoplasmic connections during meiosis and pollen development. In: Pollen Physiology and Fertilization (ed. H. F. Linskens), p. 39—47. Amsterdam: North Holland Publ. Co. 1964. — Cytoplasmic continuities during spore formation in flowering plants. Endeavour **25**, 65—72 (1966). — Heydak, M. H., and L. S. Palmer: Vitamin content of bee food. J. econ. Entomol. **33**, 396 (1940). — Heyl, F. W.: Analysis of ragweed pollen. J. Amer. chem. Soc. **39** (1917). — The yellow coloring substances of ragweed pollen. J. Amer. chem. Soc. **41**, 1285 (1919). — The ragweed pollen proteins. J. Amer. chem. Soc. **42** (1920). — The phytosterols of ragweed pollen. J. Amer. chem. Soc. **44** (1922). — Hofmeister, L.: Über die Plasmagrenzschichten im Pollenkorn. Protoplasma **46**, 367 379 (1956). — Holman, R. M., and F. Brubaker: On the longevity of pollen. Univ. Calif. Publ. Not. **13**, 179—204 (1926). — Houwink, A. L., and P. A. Roelofsen: Fibrillar architecture of growing plant cell walls. Acta bot. neerl. **3**, 385—395 (1954).

Iwanami, Y.: On the inductive tissue and the growth of the pollentube in the style. Bot. Mag. (Tokyo) **66**, 189—196 (1953). — The starch and the sugar in the pollen grain. Bot. Mag. (Tokyo) **69**, 91—95 (1956).

Jakuschkina, N. L.: Über Wuchsstoffe im Blütenstaub. [Russisch.] Dokl. Akad. Nauk SSSR. **56**, 549—552 (1947). — Johnson, L. P. V.: The storage and artificial germination of forest tree pollens. Canad. J. Res., C **21**, 332—342 (1943). — Johri, B. M., and I. K. Vasil: The pollen and pollen tube. Ergebn. Biol. **23**, 1—13 (1960). — Physiology of pollen. Bot. Rev. **27**, 325—383 (1961). — Jost, L.: Zur Physiologie des Pollens. Ber. dtsch. bot. Ges. **23**, 504—515 (1905). — Über die Selbststerilität einiger Blüten. Bot. Ztg **65**, 77—117 (1907).

Kämpfer, E.: Amoenitas, exotica, Lemgovia 1712, S. 1—108. — Kaienburg, A. L.: Zur Kenntnis der Pollenplastiden und der Pollenschlauchleitung einiger Oenotheraceen. Planta (Berl.) **38**, 377—430 (1950). — Karrer, P., and E. Leumann: Carotinoids from the pollen of Cyclamen persicum. Helv. chim. Acta **34**, 1412 (1951). — Kiesel, A., u. B. Rubin: Untersuchungen über pflanzliche Fortpflanzungszellen. II. Beitrag zur Kenntnis der Bestandteile der Pollenkörner der Zuckerrübe. Hoppe-Seylers Z. physiol. Chem. **182**, 241 (1929). — King, J. R., and C. O. Hesse: Pollen longevity studies with deciduous fruits. Proc. Amer. Soc. hort. Sci. **36**, 310—313 (1938). — Kitzes, G., H. A. Schuette and C. A. Elvehjem: The B vitamins in honey. J. Nutr. **26**, 241 (1943). — Knight, L. I.: Physiological aspects of the self-sterility of the apple. Proc. Amer. Soc. hort. Sci. **14**, 101—105 (1917). — Knoll, F.: Über Pollenkitt und Bestäubungsart. Z. Bot. **23**, 609—675 (1930). — Koelreuter, J. G.: Vorläufige Nachricht von einigen das Geschlecht der Pflanzen betreffenden Versuchen und Beobachtungen, nebst Fortsetzungen 1, 2 und 3, 1761/66. (Ostwalds Klassiker, Nr 46, Leipzig, Engelmann.) — Koessler, J. H.: Studies on pollen and pollen diseases. The chemical composition of rangweed pollen. J. biol. Chem. **35**, 415 (1918). — Kostoff, D.: A chromosomal chimera in Tobacco. Somatic non-disjunction and doubling of chromosomes in a Nicotiana hybrid. J. Hered. **21**, 445 (1930). — Kudrjavcev, V. A.: Zur Bedeutung der Lichtintensität bei der Entwicklung der Fortpflanzungsorgane der Gerste. Dokl. Akad. Nauk SSSR., N.S. **97**, 349—353 (1954). — Kühlwein, H.: Über keimungsfördernde Substanzen in Pollen und Narben. Planta (Berl.) **35**, 528—535 (1948). — Kühlwein, H., u. H. Anhäusser: Veränderungen des Gymnospermenpollens durch Lagerung. Planta (Berl.) **39**, 476 (1951). — KÜSTENMACHER, M.: Propolis. Ber. dtsch. pharm. Ges. **31**, 65—92 (1911). — KÜSTER, E.: Die Pflanzenzelle, S. 722. Jena 1951. — KUHN, E.: Physiologie der Pollenkeimung bei Matthiola. Planta (Berl.) **27**, 304—333 (1937). — KUHN, R.: Über die biologische Bedeutung der Borsäure. Wien. Chem. Ztg **46**, 1—9 (1943). — KUHN, R., u. I. LÖW: Über das Flavonolglycosid aus Crocus-Pollen. Ber. dtsch. chem. Ges. **77**, 196 (1944). — Über das Vorkommen von Milchzucker im Pflanzenreich. Chem. Ber. **82**, 474 (1949). — KURTZ jr., E. B., and J. L. LIVERMAN: Some effects of temperature on pollen characters. Bull. Torrey bot. Club **85**, 136—138 (1958).

LAIBACH, F.: Pollenhormon und Wuchsstoff. Ber. dtsch. bot. Ges. **50**, 383—390 (1932); **77**, 196 (1944). — LARSEN, P., and S. M. TUNG: Growth-promoting and growth-retarding substances in pollen from diploid and triploid apple varieties. Bot. Gaz. **111**, 436—447 (1950). — LEWIS, D.: Comparative incompatibility in Angiosperms and Fungi. Advanc. Genet. **6**, 235—285 (1954). — LICHTE, H. F.: Über die Physiologie von Angiospermenpollen und ihre Bedeutung für die Pflanzenzüchtung. Angew. Bot. **31**, 1—28 (1957). — LIDFORSS, B.: Zur Biologie des Pollens. Jb. Bot. **29**, 1—38 (1896). — Über den Chemotropismus der Pollenschläuche. Ber. dtsch. bot. Ges. **17**, 236—242 (1899a). — Weitere Beiträge zur Biologie des Pollens. Jb. Bot. **33**, 232—312 (1899b). — LINSKENS, H. F.: Physiologische Untersuchungen der Pollenschlauch-Hemmung selbststeriler Petunien. Z. Bot. **43**, 1—44 (1955a). —Physiologie der Fortpflanzung und Sexualität. Fortschr. Bot. **17**, 791—818 (1955b). — Physiologische Untersuchungen zur Reifeteilung. I. Über die Änderung einiger physiologischer Zustandsgrößen während der Pollenmeiose und Pollenentwicklung von Lilium henryi. Ber. dtsch. bot. Ges. **69**, 353—360 (1956a). — Physiologie der Fortpflanzung und Sexualität. Fortschr. Bot. **18**, 329—346 (1956b); **19**, 385—398 (1957). — Physiologische Untersuchungen zur Reifeteilung. II. Mitt. Über die Änderung der Nukleinsäuıengehaltes während der Pollenmeiose und Pollenentwicklung von Lilium henryi. Acta bot. neerl. **7**, 61—68 (1958a). — Physiologie der Fortpflanzung und Sexualität. Fortschr. Bot. **20**, 266—280 (1958b). — Pollen physiology. Ann. Rev. Plant Physiol. **15**, 255—270 (1964a). — Die Änderung des Protein- und Enzym-Musters während der Pollenmeiose und Pollenentwicklung. Planta (Berl.) **69**, 79—91 (1966). — LINSKENS, H. F. (ed.): Pollen Physiology and Fertilization. Amsterdam: North Holland Publ. Co. 1964b. — LINSKENS, H. F., u. Kl. ESSER: Über eine spezifische Anfärbung der Pollenschläuche im Griffel und die Zahl der Kallosepfropfen nach Selbstung und Fremdung. Naturwiss. **44**, 10 (1957). — Stoffaufnahme der Pollenschläuche aus dem Leitgewebe des Griffels. Proc. kon. ned. Acad. Wet., Ser. C **62**, 150—154 (1959). — LINSKENS, H. F., u. W. HEINEN: Cutinase in Pollen. Z. Bot. **50**, 338—347 (1962). — LINSKENS, H. F., and J. TUPÝ: The amino-acids pool in the style of self-incompatible strains of Petunia after self- and cross-pollination. Genetics and Breeding Res. **36**, 151—158 (1966). — LOBANOV, G. A.: The effect of different quantities of pollen upon fertilization results. Hort. Abstr. No 1374, **21** (1951). — LOPRIORE, G.: Über die Vielkernigkeit der Pollenkörner und Pollenschläuche von *Auraucaria Biwillii* Hock. Ber. dtsch. bot. Ges. **23**, 335—346 (1905). — LOVELESS, M. H., and S. N. TIMASHEFF: Fraction of ragweed extract by semicontinuous electrophoretisis-convection. Arch. Biochem. **58**, 298—317 (1955). — LUND, H. A.: The biosynthesis of IEA in the styles and ovaries of Tabacco preliminary to the setting of fruit. Plant Physiol. **31**, 334—339 (1956). — LUNDÉN, R.: A short introduction to the literature on pollen chemistry. Sv. kem. Tidskr. **66**, 201—213 (1954). — Literature on pollen chemistry. Grana palynolog. **1**, 3—21 (1956).

MACHLIS, L., and E. RAWITSCHER-KUNKEL: Mechanisms of gametic approach in plants. Int. Rev. Cytol. **15**, 97—138 (1963). — MAHESHWARI, P.: The male gametophyt of angiosperms. Bot. Rev. **15**, 1—75 (1949). — MANARESI, A.: Ricerce sulla longevita del polline di alcune piante da frutto. Stat. Sper. Agr. Ital. **48**, 33—55 (1924). — MANGIN, L.: Recherches sur le pollen. Bull. Soc. bot. France **33**, 512—517 (1886). — MARQUARDT, P., u. G. VOGG: Pharmakologische und chemische Untersuchungen über Wirkstoffe im Bienenpollen. Arzneimittel.-Forsch. **2**, 267—271 (1952). — MASCARENHAS, J. P., and L. MACHLIS: The hormonal control of the directed growth of pollen tubes. Vitam. and Horm. **20**, 347—372 (1962). — MCGUIRE, D. C.: Storage of tomato pollen. Proc. Amer. Soc. hort. Sci. **60**, 419—425 (1952). — MECHELKE, F.: Die Entstehung der polyploiden Zellkerne des Antherentapetums bei Antirrhinum majus L. Chromosoma (Wien) **5**, 199 (1952). — MEYEN, F. J. F.: Neues System der Pflanzenphysiologie, Bd. 3. Berlin 1893. — MIKI, H.: A study of tropism of pollen tubes to the pistil. I. Tropism in Lilium. Bot. Mag. (Tokyo) **67**, 143—147 (1954), — A study of tropism of pollen tubes to the pistil. II. Tropism in Camellia sinensis. Bot. Mag. (Tokyo) **68**, 293—298 (1955). — MIKI-HIROSIGE, H.: Tropism of pollen tubes to the pistils. In: Pollen Physiology and Fertilization (ed. H. F. LINSKENS), p. 152—158. Amsterdam: North Holland Publ. Co. 1964. — MIKKELSEN, V. M.: Has temperature any influence on pollengrainsize? Physiol. Plantarum (Cph.) **2**, 323—324 (1949). — MOEBIUS, M.: Über die Färbung der Antheren und des Pollens. Ber. dtsch. bot. Ges. **41**, 12 (1923). — MOEWUS, F.: Zur Physiologie und

Biochemie der Selbststerilität bei Forsythien. Biol. Zbl. **69**, 181—197 (1950). — MOHL, H. v.: Über den Bau und die Formen der Pollenkörner. Bern 1834. — MOLISCH, H.: Zur Physiologie des Pollens, mit besonderer Rücksicht auf die chemotropischen Bewegungen der Pollenschläuche. S.-B. Akad. Wiss. Wien, math.-nat. Kl. **102**, 423—449 (1893). — MORGANDO, A.: Indagini sulla biologia fiorale des pomodore (Solanum Lycopersicum). Sci. genet. (Torino) **3**, 183—193 (1949). — MÜHLDORF, A.: Über die Bildung und Auflösung der Wände bei den Pollentetraden angiospermer Pflanzen. Beih. bot. Zbl. I **60**, 557—640 (1941). — MÜHLETHALER, K.: Untersuchungen über die Struktur der Pollenmembran. Mikroskopie **8**, 103 bis 110 (1953). — Die Struktur einiger Pollenmembranen. Planta (Berl.) **46**, 1—13 (1955). — MÜHLETHALER, K., u. H. F. LINSKENS: Elektronen-mikroskopische Aufnahmen von Pollenschläuchen. Experientia (Basel) **12**, 253—254 (1956). — MÜLLER, R.: Zur quantitativen Bestimmung von Indolylessigsäure mittels Papierchromatographie und Papierelektrophorese. Beitr. Biol. Pflanz. **30**, 1—32 (1953). — MÜLLER-STOLL, W.: Über das Verhalten der Exine bei nonaperturaten Angiospermen-Pollen bei Quellung und Keimung. Grana palynolog. (Uppsala) **1**, 38—58 (1956). — MÜLLER-STOLL, W. R., u. G. LERCH: Über Nachweis, Entstehung und Eigenschaften der Kallosebildungen in Pollenschläuchen. Flora (Jena) **144**, 297—334 (1957a). — Über den physiologischen Charakter der Kallosebildung in Siebröhren und Pollenschläuchen und die Beziehung zu den Zellvorgängen. Biol. Zbl. **76**, 595—612 (1957b). — MUIR, R. M.: The relations of growth hormones and fruit development. Proc. nat. Acad. Sci. (Wash.) **33**, 303 (1947).

NAKAMURA, G. R., and E. L. BECKER: Phosphate of ragweed pollen. Arch. Biochem. **33**, 78 (1951). — NASATIR, M., A. M. BRYAN, and S. D. RODENBERG: Changes in soluble proteins of developing lily anthers. Science **132**, 897—898 (1960). — NEBEL, B. R.: Longevity of pollen in apple, pear, plum, peach, apricot and sour cherry. Proc. Amer. Soc. hort. Sci. **37**, 130—132 (1939). — NEBEL, B. R., and M. L. RUTTLE: Storage experiments with pollen of cultivated fruit trees. J. Pomol. **14**, 347—359 (1936). — NEUMANN, K.: Meiotische Teilungswellen in Antheren und ihre morphologische Bedingtheit. Biol. Zbl. **82**, 665—719 (1963). — NIELSEN, N.: The vitamin content of pollen after storing. Acta chem. scand. **10**, 332—333 (1956). — NIELSEN, N., J. GRÖMMER and R. LUNDÉN: Investigations on the chemical composition of pollen from some plants. Acta chem. scand. **9**, 1100—1106 (1955). — NIELSEN, N., and B. HOLMSTRÖM: On the occurence of folic acid, folic acid conjugates and folic acid conjugases in pollen. Acta chem. scand. **11**, 101—104 (1957).

OEHLKERS, F.: Neue Versuche über cytologisch-genetische Probleme (Physiologie der Meiosis). Biol. Zbl. **57**, 126 (1937). — Das Leben der Gewächse, Bd. I. Berlin-Göttingen-Heidelberg 1956. — O'KELLEY, J.: External carbohydrates in growth and respiration of pollen tubes in vitro. Amer. J. Bot. **42**, 322—327 (1955). — O'KELLEY, J., and P. H. CARR: An electron micrographic study of the cell walls of elongating cotton fibers, root hairs and pollen tubes. Amer. J. Bot. **41**, 261—264 (1954). — O'KELLEY, J. C.: Boron effects on growth, oxygen uptake and sugar absorption by germinating pollen. Amer. J. Bot. **44**, 239—244 (1957). — OKONUKI, K.: Coenzymes in pollen. Science **12**, 221 (1942). — The respiration of pollen. Acta phytochim. (Japan) **13**, 93 (1943). — OLMO, H. P.: Storage of grape pollen. Proc. Amer. Soc. hort. Sci. **41**, 219—224 (1942).

PALUMBO, R. F.: A cytochemical investigation of enzymes activities in the developing pollen of Tradescantia paludosa and Lilium longiflorum car. Croft. Diss. Abstr. **13**, 966 (1953).— PANKOW, H.: Über den Pollenkitt bei *Galanthus nivalis*. Flora **146**, 240—253 (1958). — PASSECKER, F.: Kann man aus der Keimfähigkeit des Pollens in Zuckerlösung auf dessen Tauglichkeit zur Befruchtung schließen? Gartenbauwiss. **3**, 201—237 (1930). — PATON, J. B.: Pollen and pollen enzymes. Amer. J. Bot. 8, 471—501 (1921). — PERCIVAL, M. S.: The presentation of pollen in certain angiosperms and its collection by Apis mellifica. New Phytologist **54**, 353—368 (1955). — PFEIFFER, N. E.: Prolonging the life of Cinchona pollen by storage under controlled conditions of temperature and humidity. Contr. Boyce Thompson Inst. **13**, 281—295 (1944). — Effect of lyophilization on the viability of Lilium pollen. Contr. Boyce Thompson Inst. **18**, 153—158 (1955). — PFUNDT, M.: Der Einfluß der Luftfeuchtigkeit auf die Lebensdauer des Blütenstaubes. Jb. wiss. Bot. **47**, 1—40 (1910). — POHL, F.: Beziehungen zwischen Pollenbeschaffenheit, Bestäubungsart und Fruchtknotenbau. Beih. Bot. Zbl. A **46**, 247—285 (1930a). — Kittstoffreste auf der Pollenoberfläche windblütiger Pflanzen. Beih. Bot. Zbl. A **46**, 286—305 (1930b). — Herkunft der Pollenklebstoffe und Funktion der Antherenhaare bei *Cucurbita pepo* L. Beih. Bot. Zbl. **50**, I, 138—149 (1933). — POHL, R.: Die Wirkung von Wuchsstoff und Hemmstoff auf das Wachstum der Pollenschläuche von Petunia. Biol. Zbl. **70**, 119—128 (1951). — PRUZSINSKY, S.: Über Trocken- und Feuchtluftresistenz des Pollens. S.-B. Öst. Akad. Wiss., math.-nat. Kl., Abt. I, **169**, 43—100 (1960).

RAGHAVAN, V., and H. K. BARUAH: Germination of pollen grains and growth of pollen tubes under the influence of certain auxins, vitamins and trace elements. Phyton (Argent.) **7**, 77—88 (1956). — REDEMANN, C. T.: Biochemical studies of pollen from Zea Mays. Microfilm Abstr. **10**, No 1, 3 (1950). — RENNER, O.: Zur Biologie und Morphologie des männlichen

Haplonten einiger Oenotheren. Z. Bot. **11**, 305—380 (1919). — REZNIK, H.: Über die Pigmentausstattung der Pollen Heterostyler. Biol. Zbl. **76**, 351—359 (1957). — RICHTER, F. I.: A simple method of making germinating tests of pine-pollen. J. Forestry **37**, 574—576 (1939). — RICHTER, S.: Über den Öffnungsmechanismus der Antheren bei einigen Vertretern der Angiospermen. Planta (Berl.) **8**, 154—184 (1929). — RIDI, M. S., and A. WAFA: An oestrogenic substance in palm pollen grains of the date palm. J. roy. egypt. med. Ass. **30**, 124 (1947). — General composition and vitamin content of the pollen grain of the date palm. J. roy. egypt. med. Ass. **33**, 168 (1950). — RITTINGHAUS, P.: Der Einfluß der Luftfeuchtigkeit auf die Lebensdauer des Blütenstaubes. Verh. naturwiss. Ver. Rheinland **43**. 123—166 (1886). — ROBBINS, K. C., A. SAMUEL and M. MOSKO: Chemical studies on a skinrective fraction from short ragweed pollen. J. Allergy **19**, 35 (1948). — ROEMER, TH.: Zur Pollenaufbewahrung. Z. Pflanzenzücht. **2**, 83—86 (1915). — ROSEN, W. G.: Chemotropism and fine structure of pollen tubes. In: Pollen Physiology and Fertilization (ed. H. F. LINSKENS), p. 159—166. Amsterdam: North Holland Publ. Co. 1964. — ROWLEY, J. R.: Stranded arrangement of sporopollenin in the exine of microspores of *Poa annua*. Science **137**, 526—528 (1962).

SAGROMSKY, H.: Über einige Bestimmungen des Vitamin B_1-Gehaltes in Pollen mit Hilfe des Phycomycestestes. Biol. Zbl. **66**, 140 (1947). — SANDSTEN, E. P.: Some conditions which influence the germination and fertility of pollen. Univ. Wisconsin agr. Exp. Stat. Res. Bull. **4**, 149—172 (1909). — SARKAR, B. C. R., S. H. WITTWER, R. W. LUECKE and H. M. SELL: Quantitative estimation of some amino acids in sweet corn pollen. Arch. Biochem. **22**, 353 (1949). — SARTORIS, G. B.: Longevity of sugar cane and corn pollen. A method for long distance shipment of sugar cane pollen by airplane. Amer. J. Bot. **29**, 395—400 (1942). — SASSEN, M. M. A.: Fine structure of germinated petunia pollen. In: Pollen Physiology and Fertilization (ed. H. F. LINSKENS), p. 167—169. Amsterdam: North Holland Publ. Co. 1964a. — Fine structure of pollen tubes of Petunia. Acta Bot. Neerl. **13**. 175—181 (1964b). — SAVELLI, R.: Sur le méchanism de la stimulation mutuelle des grains de pollen germants en collectivité. C. R. Acad. Sci. (Paris) **210**, 546—548 (1940). — SAVELLI, R., et C. CARUSO: Stimulation mutuelle dans la germination des grains de pollen de Nicotiana. C. R. Acad. Sci. (Paris) **210**, 184—186 (1940). — SCHADE, C.: Keimfähigkeit und Lebensdauer des Pollens von Kartoffel-X-Virus-infizierter *Nicotiana acuminata*. Phytopath. Z. **30**, 225—236 (1957). — SCHILDKNECHT, H., u. H. BENONI: Über die Chemie der Anziehung von Pollenschläuchen durch die Samenanlagen von Oenotheren. Z. Naturforsch. **18**b, 45—54 (1963a). — Versuche zur Aufklärung des Pollenschlauch-Chemotropismus von Narcissen. Z. Naturforsch. **18**b, 656—661 (1963b). — SCHMUCKER, TH.: Zur Blütenbiologie tropischer Nymphaea-Arten. Planta (Berl.) **18**, 641—650 (1933). — Über den Einfluß von Borsäure auf Pflanzen, insbesondere keimende Pollenkörner. Planta (Berl.) **23**, 264—289 (1935). — SCHNARF, K.: Über das Verhalten des Antherentapetums einiger Pflanzen. Österr. bot. Z. **73**, 242—245 (1923). — Embryologie der Angiospermen. In Handbuch der Pflanzenanatomie, Abt. II, Teil 2, Bd. X/2. Berlin 1929. — SCHNEIDER, G.: Wachstum und Chemotropismus von Pollenschläuchen. Z. Bot. **44**, 175—205 (1956). — SCHOCH-BODMER, H.: Zur Methodik der Größenbestimmung von Pollen, mit besonderer Berücksichtigung von Corylus avellana. Ber. schweiz. bot. Ges. **45**, 62—70 (1936). — Zur Physiologie der Pollenkeimung bei Corylus avellana: Pollen- und Narbensaugkräfte, Quellungserscheinungen der Kolloide des Pollens. Protoplasma **25**, 337—371 (1936). — Farbumschlag des Pollenanthocyans durch saure Narbensekrete bei Lythrum salicaria. Ver. schweiz. naturforsch. Ges. Chur **1938**, 179. — Zur Physiologie der Pollenfärbung von Lythrum salicaria. Vjschr. naturforsch. Ges. Zürich **84**, 225—235 (1939). — The influence of nutrition upon pollen grain size in *Lythrum salicaria*. J. Genet. **40**, 393—402 (1940). — Über das Spitzenwachstum der Pollenschläuche. Ber. schweiz. bot. Ges. **55**, 154—168 (1945). — SCHOCH-BODMER, H., u. P. HUBER: Die Ernährung der Pollenschläuche durch das Leitgewebe. Vjschr. naturforsch. Ges. Zürich **92**, 43—48 (1947). — SCHWANITZ, F.: Untersuchungen an polyploiden Pflanzen. VI. Pollengröße und Zellkerngröße bei diploiden und autotetraploiden Pflanzen. Züchter **20**, 53—57 (1950). — SCHWARZENBACH, F. H.: Die Wirkung von Carotinoiden auf die Keimung von Cyclamenpollen. Helv. chem. Acta **34**, 1064—1069 (1951). — Carotinoide als Wirkstoffe der Fortpflanzungsphysiologie von Cyclamen persicum. Vjschr. naturforsch. Ges. Zürich **98**, Beih. 1 (1953). — SITTE, P.: Untersuchungen zur submikroskopischen Morphologie der Pollen- und Sporenmembranen. Mikroskopie (Wien) **8**, 290—299 (1953). — SMITH, P. F.: The influence of 3-indoleacetic acid on pollen germination. Science **90**, 163—164 (1939). — Studies an the growth of pollen with respect to temperature, auxins, colchicine and vitamin B_1. Amer. J. Bot. **29**, 56—66 (1942). — SOSA, A., et C. SOSA-BOURDOUIL: Études de quelques constituents chimique du pollen de Corylus avellana L. C. R. Acad. Sci. (Paris) **235**, 971—973 (1952). — SOSA-BOURDOUIL, C.: Notes on pollen. Rev. bot. appl. et agr. trop. **19**, 495 (1939). — SPRENGEL, C. K.: Das entdeckte Geheimnis der Natur im Bau und in der Befruchtung der Blumen. Berlin: Fr. Vieweg d. Ältere 1793. (Oswald's Klassiker Nr 48, 49, 50, 51.) — STANLEY, R. G.: Certain aspects of a study of the physiology of flowering in Pine. Symp. Tree Physiol. Harvard Univers., pp. 12—13, 1957. — Methods and concepts applied to a study of flowering in pine. In: The physiology of forest trees (ed. K. V. THIMANN), p. 583—599. 1958. — Physiology of pollen and pistil. Sci. Progr. **52**, 122—132 (1964). — STANLEY, R. G., and H. F. LINSKENS: Enzyme activation in germinating Petunia pollen. Nature (Lond.) **203**, 542—544 (1964). — Protein diffusion from germinating pollen. Physiol. Plantarum (Copenh.) **18**, 47—53 (1965). — STANLEY, R. G., and F. LOEWUS: Boron and myo-inositol in pollen pectin biosynthesis. In: Pollen Physiology and Fertilization (ed. H. F. LINSKENS), p. 128—136. Amsterdam: North Holland Publ. Co. 1964. — STANLEY, R. G., and A. W. G. YEE: Ribonucleic acid base composition in subcellular fractions of pine pollen. Nature (Lond.) **210**, 181—183 (1966). — STANLEY, R. G., L. C. T. YOUNG, and J. S. D. GRAHAM: Carbon dioxide fixation in germinating pine pollen. Nature (Lond.) **182**, 1462—1463 (1958). — STEFFEN, K.: Zur Kenntnis des Befruchtungsvorganges bei Impatiens glanduligera. Planta (Berl.) **39**, 175—244 (1951). — Zytologische Untersuchungen an Pollenkorn und -schlauch. I. Phasenkontrastoptische Lebensuntersuchungen an Pollenschläuchen von Galanthus nivalis. Flora (Jena) **140**, 140—174 (1953). — STRASBURGER, E.: Über das Wachstum vegetabilischer Zellhäute. Histolog. Beitr. **2**, 36 (1889).

TANAKA, K.: The pollen germination and pollen tube development in Pinus densiflora. I. The effect of storage, temperature and sugars. Sci. Rep. Tokohu Univ. **4**, Ser. Biol. **21**, 185—198 (1955) — TANO, S., and H. TAKAHASHI: Nucleic acid synthesis in growing pollen tubes. J. Biochem. (Tokyo) **56**, 578—580 (1964). — TAPPI, G.: Sui pigmenti del Leucojum vernum. Atti Accad. Sci. Torino **83**, 1—8 (1948/49). — Sui pigmenti del polline di Asphodelus albus. Gazz. chim. ital. **81**, 621—624 (1951). — TAPPI, G., e E. MENZIANI: Sui pigmenti del polline di Lilium mantchiuricum. Atti Soc. Nat. Math. Modena **85**, 28—31 (1954). — Sul pigmento flavonolico del polline di Lilium candidum L. Gazz. chim. ital. **85**, 694—702 (1955). — TAPPI, G., e A. MONZANI: Sui pigmenti del polline di Anemone coronaria. Gazz. chim. ital. **85**, 732—736 (1955). — Sui constituenti del cemento pollinico di Lilium candidum L. Gazz. chim. ital. **85**, 725—731 (1955). — TAPPI, G., A. SPADA e R. CAMERONI: Sui pigmenti flavonici del polline di Acacia dealbata (var. Le Gaulois). Gazz. chim. ital. **85**, 703—713 (1955). — TISCHLER, G.: Pollenbiologische Studien. Z. Bot. **9**, 417—489 (1917). — TODD, F. E., and O. BETHERIK: The composition of pollens. J. econ. Entomol. **35**, 312 (1942). — TROLL, W.: Über Antherenbau, Pollen und Pollination bei Galanthus L. Flora (Jena) **123**, 321—343 (1928). — TSAO, T. H.: A study of chemotropism of pollen tubes in vitro. Plant Physiol. **24**, 494—504 (1949). — TULECKE, W. R.: Preservation and germination of the pollen of Gingko under sterile conditions. Bull. Torrey bot. Club. **81**, 509—512 (1954).

USHIROZAWA, K., and J. SHIBUKAWA: Studies on the germination and fertilization of long preserved apple pollen. Aomori Apple Exp. Stat. Bull. **1951**.

VASIL, I. K.: Studies on pollen germination. Proc. Delhi Univers. Seminar on Mod. Development in Plant Physiol. 1958, p. 123—126. — Effect of boron on pollen germination and pollen tube growth. In: Pollen Physiology and Fertilization (ed. H. F. LINSKENS), p. 107—119. Amsterdam: North Holland Publ. Co. 1964. — VINSON, C. G.: Some nitrogenous constituents of corn pollen. J. agric. Res. **35**, 261 (1927). — VISSER, T.: Bloembiologie en kruisingstechniek bij appel en peer. Med. Dir. Tuinb. **14**, 707—726 (1951). — Germination and storage of pollen. Meded. Landbouwhogesch. Wageningen **55**, 1—68 (1955). — Germination and pollination experiments with Forsythia. Proc. kon. ned. Acad. Wet., Ser. C **59**, 685—693 (1956). — VIVINO, A. E., and L. S. PALMER: The chemical composition and nutritional value of pollens collected by bees. Arch. Biochem. **4**, 129 (1944).

WAGENITZ, G.: Über die Änderung der Pollengröße von Getreiden durch verschiedene Ernährungsbedingungen. Ber. dtsch. bot. Ges. **68**, 297—302 (1955). — Pollenmorphologie und Systematik in der Gattung Centaurea. Flora (Jena) **142**, 213—279 (1955). — WALDERDORF, M.: Über Kultur von Pollenschläuchen und Pilzmycelien auf festem Substrat bei verschiedener Luftfeuchtigkeit. Bot. Arch. **6**, 84—110 (1924). — WATANABE, K.: Studies on the germination of grass pollen. I. Liquid exudation of the pollen on the stigma before germination. Bot. Mag. (Tokyo) **68**, 40—44 (1955). — WEINLAND, H.: Das Wachstum der Hypanthien bei den Oenotheren. Z. Bot. **36**, 401—430 (1941). — WELZEL, G.: Entwicklungsgeschichtlich-genetische Untersuchungen an pollensterilen Mutanten von Petunia. Z. indukt. Abstamm.- u. Vererb.-Lehre **86**, 35—53 (1954). — WETTSTEIN, W., u. L. NIKLAS: Vergleichende Pollenuntersuchungen an verschiedenen Lärchenrassen. Österr. bot. Z. **102**, 520—523 (1955). — WENT, F. A. F. C.: Lehrbuch der allgemeinen Botanik. Jena 1933. — WERFFT, R.: Über die Lebensdauer der Pollenkörner in der freien Atmosphäre (Sonnen- und Ultraviolettbestrahlung). Biol. Zbl. **70**, 354—367 (1951). — WEYGAND, F., u. H. HOFMANN: Polleninhaltsstoffe. I. Mitt.: Zucker, Folinsäure und Ascorbinsäure. Chem. Ber. **83**, 405—415 (1950). — WITTMANN, G., and D. WALKER: Towards simplification in sporoderm description. Pollen et Spores **7**, 443—456 (1965). — WODEHOUSE, R. P.: Pollen grains, their structure, identification and significance in science and medicine. New York 1935. — WULFF, H. D.: Die

Entwicklung der Pollenkörner von Triglochin palustris L. Jb. wiss. Bot. 88, 161 (1939). — Wunderlich, R.: Über das Antherentapetum mit besonderer Berücksichtigung seiner Kernzahl. Eine kritische Zusammenfassung. Österr. bot. Z. **101**, 1 (1954).

Yakushkina, N. I.: The growth substance in pollen. Dokl. Akad. Nauk SSSR. **56**, 549 (1947). — Yamaha, G.: Über die Ionenkonzentration des Antherenschleimes bei Lilium-Arten. Proc. Imp. Acad. Sci. (Tokyo) **14**, 81—82 (1938a). — Potentialmessungen an den sich teilenden Pollenmutterzellen von Lilium speciosum. Proc. Imp. Acad. Sci. (Tokyo) **14**, 83—85 (1938b).

Zanoni, G.: Antagonismo pollonico. Riv. Biol. **12**, 126—135 (1930). — Zeijlemaker, F. C. J.: Growth of pollen tubes in vitro and their reaction on potential differences. Acta bot. neerl. **5**, 179—186 (1956). — Zetschke, F.: Sporopollenine. In Klein, Handbuch der Pflanzenanalyse, Bd. 3, S. 205. Wien 1932. — Zetschke, F., u. K. Huggler: Untersuchungen über die Membran der Sporen und Pollen. Justus Liebigs Ann. Chem. **461**, 89—108 (1928). — Zetschke, F., u. H. Vicari: Untersuchungen über die Membran der Sporen und Pollen. Helv. chem. Acta **14**, 62—67 (1931). — Zirkle, C.: The beginning of plant hybridization, p. 231. Philadelphia: Univers. Pensilv. Press 1935.

Befruchtung der Archegoniaten und Blütenpflanzen.

Von

Erik Haustein.

Mit 30 Abbildungen.

A. Archegoniatae.

Die Befruchtung der Archegoniaten (Bryophyten und Pteridophyten) ist dadurch charakterisiert, daß sich die befruchtungsfähigen, frei beweglichen männlichen Gameten (Spermatozoiden) in einem die Antheridien und Archegonien bedeckenden wäßrigen Medium, in das sie von den Antheridien entleert wurden, aktiv zu den Archegonien — chemotaktisch von diesen angelockt — bewegen. Durch den von Schleim erfüllten Halsteil des Archegoniums gelangen sie in unmittelbare Nähe der Eizelle, worauf die Vereinigung der männlichen und weiblichen Fortpflanzungszellen und die Verschmelzung der männlichen und weiblichen Kerne die Befruchtung abschließt.

Bevor die eigentliche Befruchtung erfolgen kann, müssen einige die Befruchtung vorbereitende Vorgänge ablaufen. Diese sind die Öffnung der Antheridien und Archegonien sowie die Verschleimung der Kanalzellen. Erst der Abschluß dieser vorbereitenden Stadien schafft die Voraussetzungen für den Ablauf der Befruchtung.

Öffnung der Antheridien.

Die Öffnung der Antheridien ist ein aktiver Vorgang; sie erfolgt bei Moosen und Farnen grundsätzlich in gleicher Weise mit Hilfe schleimhaltiger, quellungsfähiger Wandzellen. Die mit der Quellung des Schleimes verbundene starke Volumenzunahme führt an einer bestimmten Stelle zum Aufreißen und Entleeren des Antheridiums. Im einzelnen bestehen jedoch eine Reihe von Unterschieden in der Ausbildung und Anordnung der an dem Öffnungsmechanismus beteiligten Wandzellen.

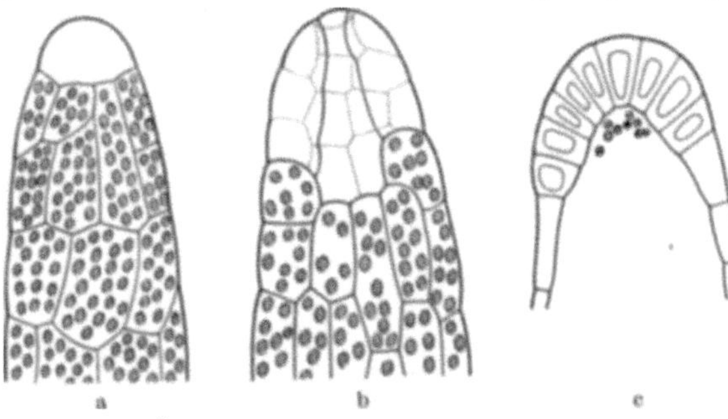

Abb. 1a—c. Öffnungskappe von Laubmoos-Antheridien. a *Funaria hygrometrica* in Außenansicht, die Öffnungskappe besteht aus einer Zelle; b entleertes Antheridium von *Polytrichum*, Öffnung halbiert gedacht; c Spitze eines Antheridiums von *Atrichum undulatum* im Längsschnitt, die Zellen der Öffnungskappe treten durch die Schleimablagerung auf den Zellwänden hervor. (Nach GOEBEL.)

Bei den Laubmoosen wird eine von den übrigen Wandzellen sich scharf abhebende Scheitelkappe aus einer größeren *(Polytrichum, Atrichum)* oder geringeren Zahl (1—2 bei *Funaria hygrometrica*) fast isodiametrischer Zellen mit stark verdickten Wandungen gebildet (Abb. 1). Diese

Wandverdickungen werden erst im Laufe der Entwicklung auf Kosten der in diesen Zellen reichlich vorhandenen Stärke abgelagert; sie sind außerordentlich quellungsfähig, so daß sie als Schleim angesprochen werden können. Der Druck bei der Verquellung dieser Zellen führt zu einer starken Spannung und schließlich zum Platzen der das Antheridium überziehenden Cuticula. Die entstehende Öffnung wird von den Zellresten der Öffnungskappe begrenzt. Eine Öffnung des Antheridiums erfolgt nur bei Zufuhr von Wasser, das auch nach dem Austreten der Spermatozoiden als Medium für deren Bewegung lebensnotwendig ist. Abweichend von der Mehrzahl der Laubmoose verhält sich *Sphagnum*, bei dem eine scharf abgegrenzte Öffnungskappe fehlt; an deren Stelle ist eine größere, unbestimmte Zahl von Wandzellen an der Öffnung beteiligt.

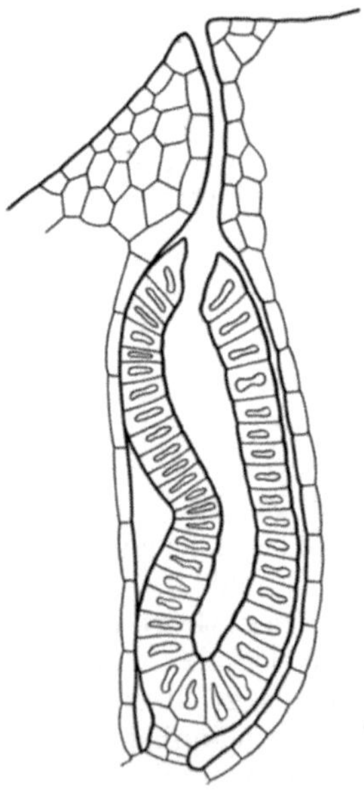

Abb. 2. *Marchantia polymorpha*, entleertes Antheridium, die Wandzellen sind schleimreich. Die Gruben, in denen die Antheridien stehen, haben enge Mündungen. (Nach GOEBEL.)

Diese Ausbildung des Öffnungsapparates gilt allgemein für die Lebermoose. Die durch die Quellung ausgelöste Volumenzunahme ist dabei oft so groß, daß sich die die Öffnung bildenden Wandzellen wurstartig nach außen krümmen *(Jungermaniaceae)*. Bei den tief in den Thallus eingesenkten Antheridien der Marchantiaceen kommt für die Öffnung weniger die Spannung gegen die Cuticula des Antheridiums (Abb. 2), als vielmehr die Spannung gegen die Wand der Grube in Betracht, in der das Antheridium sitzt. Bei manchen Formen, deren Grubenmündung in eine über die Oberfläche hervorragende Spitze ausgezogen ist, findet ein regelrechtes Aussprühen der Spermatozoiden statt, das sich bis auf eine Entfernung von einem halben Meter erstrecken kann. Der Druck auf die Antheridien kann noch verstärkt werden, indem bei

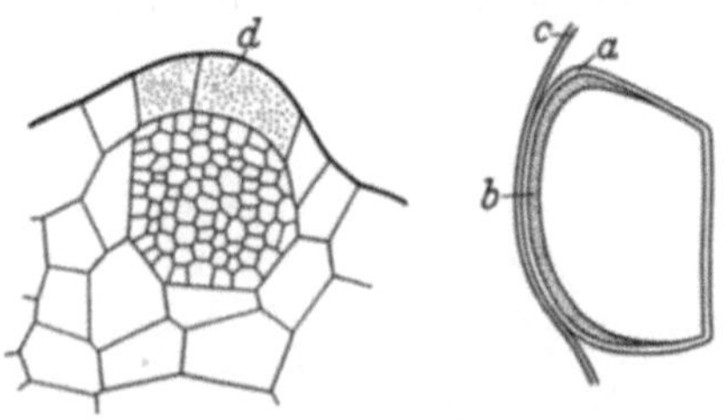

Abb. 3.

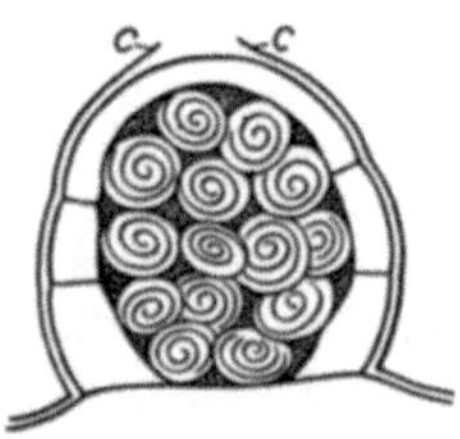

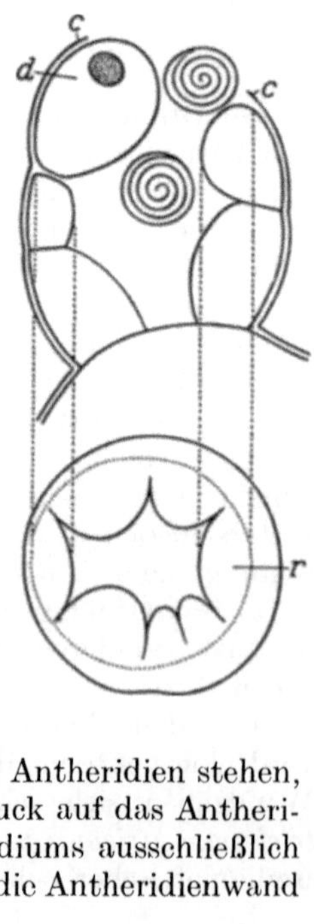

Abb. 4.

Abb. 3. *Equisetum palustre.* Links Längsschnitt durch ein ganz eingesenktes Antheridium, *d* Deckschicht. (Nach GOEBEL.) Rechts Schnitt durch eine Deckelzelle eines ungeöffneten Antheridiums. *c* Cuticula; *a* Zellwand; *b* Schleimschicht. (Nach LUDWIGS.)

Abb. 4. *Woodsia ilvensis*, Antheridien im Längsschnitt. Links reifes Antheridium, die Cuticula ist geplatzt; rechts entleertes Antheridium, unten in Aufsicht. *c* Cuticula; *d* Deckelzelle; *r* Ringzelle. (Nach SCHLUMBERGER.)

manchen Arten auch die Wandzellen der Grube, in der die Antheridien stehen, schleimhaltig sind und nach Quellung einen zusätzlichen Druck auf das Antheridium ausüben. Schließlich kann die Entleerung des Antheridiums ausschließlich durch den Druck der Grubenwandung bedingt sein, während die Antheridienwand selbst passiv bleibt. Eine Sonderstellung nehmen die Antheridien von *Anthoceros* insofern ein, als sie ohne Verbindung nach außen im Innern des Thallus eingeschlossen sind. In diesen Fällen sind die das Dach der Grube bildenden Thalluszellen mit Schleim erfüllt und öffnen sich durch Zurückschlagen; jetzt erst öffnen sich auch die Antheridien, aber im Gegensatz zu den übrigen Lebermoosen mit Hilfe einer an die Laubmoose erinnernden Öffnungskappe (GOEBEL 1930).

Bei den Pteridophyten mit eingesenkten Antheridien (*Equisetum*, *Lycopodium*, *Ophioglossaceae*, *Marattiaceae*) erfolgt die Öffnung dadurch, daß eine oder mehrere, durch ihre helle Färbung ausgezeichnete Deckschichtzellen mit stark quellbarer Schleimschicht bei Wasseraufnahme sich nach außen krümmen und eine Öffnung entstehen lassen (Abb. 3). Die Quellung des Antheridieninhalts wirkt mit, einen Druck auf die Wandschicht auszuüben, so daß nach dem Öffnen durch Zusammenziehen der gespannten Wände der schleimige Inhalt ausgepreßt wird. Bei den Arten mit nicht eingesenkten Antheridien (leptosporangiate Farne) enthalten alle Wandzellen, vor allem die Deckelzelle einen wasserhellen Schleim und sind an der Öffnung beteiligt. Die Deckelzelle, deren Verbindung mit den übrigen Zellen schon vorher gelockert war, löst sich los, die stehenbleibenden Wandzellen dehnen sich stark aus und pressen den Antheridieninhalt heraus (Abb. 4).

Öffnung der Archegonien.

Die Öffnung der Archegonien erfolgt bei den Moosen ähnlich wie bei den Antheridien mit Hilfe einer aus den obersten Halswandzellen gebildeten Öffnungskappe, deren Zellen auf Kosten von Stärke Schleim entwickeln. Auf den durch die Quellung erzeugten Druck hin trennen sich die Zellen der leicht keulig aufgetriebenen Spitze voneinander — gelegentlich können sie sich auch ganz loslösen — und krümmen sich mehr oder weniger trichterförmig nach außen (Abb. 5).

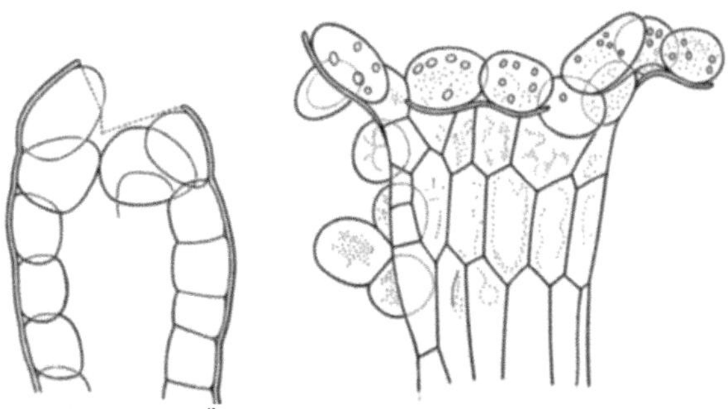

Abb. 5. *Mnium undulatum.* Links halb, rechts ganz geöffnetes Archegonium. (Nach ZIELINSKI.)

Bei den Pteridophyten ist über die Öffnung der Archegonien weniger bekannt. Bei *Equisetum* und *Selaginella* liegen an der Spitze des Archegoniums vier lange, große, durch ihren schleimhaltigen Inhalt auffallende Zellen, die sich beim Öffnen nach außen krümmen. Bei den leptosporangiaten Farnen sind an der Öffnung auch tiefer gelegene Zellen der Wand des Archegoniums beteiligt.

Verschleimen der Kanalzellen.

Noch bevor die Öffnung der Archegonien vollendet ist, sind — bei Moosen und Farnen in gleicher Weise — die Halskanalzellen schleimig verquollen. In seltenen Fällen können sich auch 2 oder 3 der obersten Wandungszellen daran beteiligen. Häufig kann man in diesem Schleim noch zwei Bestandteile unterscheiden; der eine, graue körnige, entspricht dem ehemaligen Plasma nebst den Kernen, der andere, homogene, stark lichtbrechende, den verquollenen Zellwänden. Der Verschleimung der Halskanalzellen voraus geht die Verschleimung der Bauchkanalzelle (*Lygodium palmatum:* ROGERS 1927).

Bau der Spermatozoiden.

Allgemein dürfte gelten, daß es sich auch bei den männlichen Fortpflanzungszellen der Moose und Farne um wirkliche Zellen handelt. Wie die Beobachtungen an günstigen Objekten gezeigt haben, ist sowohl bei den Spermatozoiden der Moose als auch bei denen der Farne der Kern von einer dünnen Plasmaschicht umgeben.

Trotz der wenigen direkten Beobachtungen an Moosen scheint wohl allgemein der Kern des funktionsfähigen Moos-Spermatozoids von einer dünnen, meist allerdings nicht feststellbaren Plasmahülle umgeben zu sein. Deutlich ist dagegen das Plasmastück am hinteren Ende des Spermatozoids, das vielfach dem Spermatozoid-Ende etwas seitlich angeheftet ist. Die Verbindung zwischen diesem Plasmastück und dem übrigen Körper des Spermatozoids ist jedoch ziemlich lose und das Plasmastück scheint in der Mehrzahl der Fälle während des Vordringens bis zur Eizelle vom übrigen Körper abgelöst zu werden (Abb. 6).

Abb. 6.

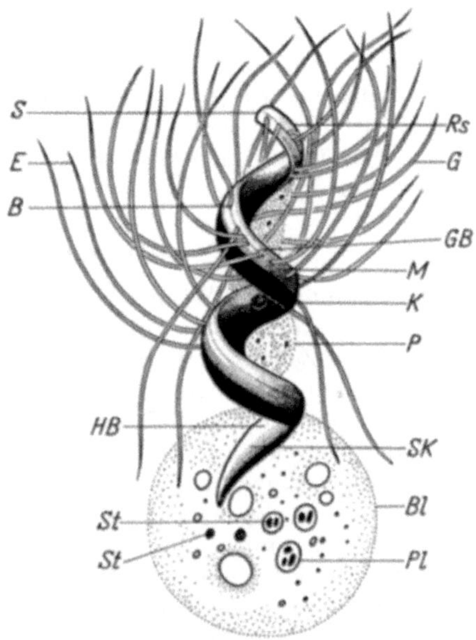

Abb. 7.

Abb. 6. Spermatozoiden von *Riccardia pinguis* (links) und *Marchantia polymorpha* (rechts) bei derselben Vergrößerung. (Nach STEIL.)

Abb. 7. Morphologie eines Polypodiaceen-Spermatozoids, schematisch. *B* Basalkörper der Geißeln; *Rs* Randsaum; *G* Geißel; *GB* Geißelband; *E* Endstück der Geißeln; *Bl* Endbläschen; *S* Endsaum; *SK* heller gefärbte Fortsetzung des Geißelbandes; *M* Mittellinie im Geißelband; *K* Zellkern; *HB* hinteres Ende des Geißelbandes; *P* Plasmateil; *Pl* Plastide; *St* Stärkekörner. (Nach YUASA.)

Hinsichtlich der Plasmaeinschlüsse liegen verschiedene Beobachtungen vor. Bei *Mnium* befinden sich an dem schon im Antheridium abgestreiften Plasmastück Plastid und Chondriosom (HAMANT 1954), bei *Atrichum undulatum* wird die Zahl der Chondriosomen während der letzten Teilungen der Spermatogenese ständig vermindert und bereits in den Spermatidzellen lassen sich keine Chondriosomen mehr feststellen (EYMÉ 1954). Dagegen besitzen die Spermatozoiden von *Atrichum undulatum*, ebenso diejenigen von *Sphagnum cymbifolium* und *Funaria hygrometrica* zwei Chloroplasten (EYMÉ und CAPOT 1954). Der eine bildet zusammen mit dem Kern den Apikalkörper des Spermatozoids und kann also mit diesem in die Eizelle übertragen werden; der zweite befindet sich in dem apikalen Plasmastück und wird vor dem Eindringen des Spermatozoids in das Archegonium abgestoßen. Bei den Lebermoosen *Marchantia* und *Dumortiera* befindet sich ein Chloroplast am hinteren Ende des Spermatozoids, das bei der Befruchtung übertragen werden soll, eine Angabe, die wohl der Nachprüfung bedarf (YUASA 1954).

Nach mehrfachen, gesicherten Beobachtungen ist auch das schraubig gedrehte Kernstück der Farn-Spermatozoiden von einer dünnen Plasmahülle umgeben. Die Fortsetzung dieser Plasmahülle bildet am Hinterende des Spermatozoids das spitz auslaufende Plasmastück, dessen allgemeine Verbreitung jedoch nicht bekannt ist. Außer dieser dem Kern eng anliegenden plasmatischen Hülle hängt dem Schraubenband aber noch weiter ein ziemlich großer, sog. Plasmarest an, der in vielen Fällen von der letzten Schraubenwindung festgehalten erscheint (Abb. 7). In diesem Plasmarest sind wiederum Plastiden, Stärkekörner, Mitochondrien und Fetttröpfchen enthalten. Die Zahl der Plastiden beträgt dabei für *Selaginella* drei, bei den *Eufilicinae* sind es stets mehrere Plastiden. Wie bei den Moosen wird auch bei den Farnen der Plasmarest vor der Befruchtung abgestoßen, so daß — jedenfalls für *Selaginella* und die *Filicinae* — eine rein mütterliche Plastidenvererbung vorliegt. Bemerkenswert im Hinblick auf die Geißelaktivität ist die Ansammlung von Mitochondrien in der Nähe der Geißelinsertionen bei *Pteridium aquilinum* (MANTON 1959).

Bau der Eizellen.

Die kugeligen, häufig auch eiförmigen und dann in Richtung des Archegoniums gestreckten Eizellen der Moose lassen im allgemeinen keinen bipolaren Bau erkennen. Bei vielen Arten löst sich das Cytoplasma der reifen Eizelle von seiner Wand ab, so daß anzunehmen ist, daß es sich bei der Zusammenballung des Eiprotoplasten um eine allgemeine Erscheinung bei den Moosen handelt. Dazu kann noch eine Vergrößerung des Hohlraums durch Wachstum des Archegoniumbauches kommen. Der Raum zwischen der reifen Eizelle und der Archegoniumwand ist von Schleim erfüllt. Die Begrenzung der reifen Eizelle ist dann wohl nur eine Plasmahaut. Die Eizellen sind verhältnismäßig arm an Inhaltsstoffen; regelmäßig finden sich Plastiden, Stärke und Mitochondrien, gelegentlich auch Öltröpfchen.

Während bei den meisten Arten zur Zeit der Befruchtung Hals- und Bauchkanalzellen verschleimen, findet sich mit großer Häufigkeit bei *Anthoceros* und *Sphagnum* eine persistierende Bauchkanalzelle, die sich in vieler Hinsicht ähnlich wie die Eizelle verhält. So ballt sich die Bauchkanalzelle vor der Befruchtung in gleicher Weise zusammen und löst sich unter Bildung einer schleimigen Umhüllung von ihrer Wandung ab wie die Eizelle; die Bauchkanalzelle kann auch als Eizelle fungieren, d. h. also befruchtet werden, ein Verhalten, das insofern nicht überrascht, als beide Tochterzellen der Zentralzelle sind.

Auch die Eizellen der Farne besitzen zur Zeit der Befruchtung in vielen Fällen keinen cytologisch erkennbaren bipolaren Bau. Ein ausgeprägt bipolarer Bau findet sich dagegen bei den großen Eizellen der *Marsileaceae*, indem deutliche Unterschiede in der Beschaffenheit des Cytoplasmas zwischen der oberen und unteren Hälfte der Eizelle bestehen: während die untere Hälfte zahlreiche Stärkekörner und größere und kleinere Vacuolen enthält, besteht die obere, dem Halsteil zugekehrte Hälfte aus feinkörnigem Protoplasma mit einer feinfaserigen Struktur, wobei der Kern an der Grenze der beiden Hälften liegt. Im Gegensatz zu den Moosen löst sich die Eizelle der Farne jedoch nicht von der Archegoniumwand ab; lediglich die feste Membran zwischen Bauchkanal- und Eizelle verschwindet bei der Reife der Eizelle, so daß diese gegen den Halsteil zu nur von einer Plasmahaut begrenzt ist. Abweichend davon scheint sich *Equisetum* zu verhalten, wo sich die Eizelle wie bei den Moosen vor der Befruchtung von der Wand ablöst und in einer schleimigen Substanz liegt. Eine Nachuntersuchung dieser Verhältnisse dürfte vielleicht wünschenswert sein. Das Vorhandensein eines sog. „Empfängnisflecks" scheint jedoch mindestens fraglich. Häufig erscheint ein solcher dadurch vorgetäuscht, daß die Spitze der Eizelle nach innen zu gewölbt ist. Etwas abweichend sind die Verhältnisse bei *Marsilea*; hier besitzt die vom Rande gegen die Mitte zu fortschreitend verdickte Zellwand zwischen Bauchkanal- und Eizelle ein großes Loch bzw. einen Tüpfel, durch den die Spermatozoiden zur Eizelle hindurchtreten können. An Inhaltskörpern finden sich in der Eizelle wiederum Plastiden, Stärkekörner und Mitochondrien. Nach MÜHLETHALER und BELL sollen die Plastiden und Mitochondrien in der Eizelle von *Pteridium aquilinum* jedoch regelmäßig degenerieren und aus der Kernmembran wieder neu entstehen (MÜHLETHALER u. BELL 1962; BELL u. MÜHLETHALER 1964; BELL 1966). Demgegenüber konnte für die Eizelle von *Dryopteris filix mas* (MENKE u. FRICKE 1964) sowie für die Eizelle des Lebermooses *Sphaerocarpus donnellii* (DIERS 1964, 1965a, b, 1966), wo die Verhältnisse wohl nicht allzu unähnlich sein dürften, weder eine Degeneration von Plastiden und Mitochondrien, noch deren Neuentstehung aus der Kernmembran beobachtet werden. Wenn diese letztere Möglichkeit zwar auch nicht mit völliger Sicherheit ausgeschlossen werden konnte, so darf sie jedoch zumindest als sehr unwahrscheinlich betrachtet werden.

Weg der Spermatozoiden zu den Archegonien.

Das zur Öffnung der Antheridien und Archegonien notwendige Wasser dient nach dem Öffnen der Antheridien und Entleeren der Spermatozoiden als Medium auf ihrem Weg zu den Archegonien, von denen sie chemotaktisch angelockt werden. Daß von den geöffneten Archegonien tatsächlich eine chemotaktische Anlockung ausgeht, konnte im Wahlversuch zwischen reifen und unreifen Archegonien von *Pteridium aquilinum* (WILKIE 1954) gezeigt werden: die Spermatozoiden sammeln sich am reifen Archegonium in größerer Zahl an, als nach der Wahrscheinlichkeit zu erwarten wäre. Da auch verletzte Prothalliumzellen und Sporophytenzellen, ebenso auch Blattzellen eine gerichtete Anlockung hervorrufen, dürfte die chemotaktisch wirksame Substanz eine einfache organische Säure sein. Diese dürfte bei den Moos- und Farn-Archegonien wohl von den durch die Schleimbildung zerstörten Hals- und Bauchkanalzellen abgegeben werden. Die zahlreichen und eingehenden Untersuchungen zur Chemotaxis der Moos- und Farnspermatozoiden, vor allem von PFEFFER und seinen Schülern, werden an anderer Stelle dieses Handbuches besprochen (s. Band XVII/2).

Weg der Spermatozoiden im Archegoniumhals.

Als Folge der chemotaktischen Anlockung dringen im allgemeinen die Spermatozoiden stets in größerer Zahl in den Hals des Archegoniums ein. Wie sich jedoch die Spermatozoiden im Hals gegen die Eizelle bewegen, ist nicht geklärt. Daß eine Bewegung durch Geißeln ausscheidet, wird von allen Untersuchern angegeben. Sobald die Spermatozoiden in den Schleim des Halsteils gelangt sind, verlieren sie irreversibel ihre Beweglichkeit (*Pteridium aquilinum*, WILKIE 1954). Ob diese Unbeweglichkeit gleichzeitig an einen Verlust der Geißeln gebunden ist, erscheint jedoch unwahrscheinlich, weil in manchen Fällen auch im Bauchteil der Archegonien begeißelte Spermatozoiden beobachtet wurden (*Lygodium palmatum*, ROGERS 1927).

Befruchtung.

a) Moose.

Von den zahlreichen in den Halsteil eingedrungenen und die Eizelle auf allen Seiten umgebenden Spermatozoiden, die — wie schon erwähnt — ihr Plasmastück verloren haben, tritt ein, gelegentlich auch mehrere (Polyspermie) Spermatozoiden in die Eizelle ein. Dabei scheint keine Stelle der zur Zeit der Befruchtung nur aus einer dünnen Plasmahaut bestehenden Oberfläche der Eizelle bevorzugt zu sein. Ob beim Eindringen der Spermatozoiden auch immer die Geißeln mit eindringen oder ob in manchen Fällen das Bewegungsstück vor dem Eindringen abgeworfen wird, ist nicht klar. Sicher wird aber die dünne, den Kern umgebende Plasmahaut des Spermatozoids mit in die Eizelle aufgenommen. Die Aufnahme der Spermatozoiden scheint in verschiedener Weise vor sich zu gehen: entweder legt sich das Spermatozoid in seiner ganzen Länge an die Eizelle an und wird von dieser gleichzeitig aufgenommen, oder das Spermatozoid bohrt sich mit seiner Spitze in die Eizelle ein. Vielfach schwillt das aufgenommene Spermatozoid im Protoplasma der Eizelle deutlich an oder nimmt eine eiförmige Gestalt an (*Sphaerocarpus donnellii*). Bei *Funaria flavicans* und *Riccardia pinguis* scheint das Eindringen eines Spermatozoids auch ein Wachstum der Eizelle auszulösen.

Nach einer für die verschiedenen Arten verschieden langen Zeit nach dem Eindringen (10—36 Std) erfolgt die Kernverschmelzung. Auch bei der Kernverschmelzung können sich die beiden Kerne entweder aneinanderlegen und nach Verschwinden der Trennwand miteinander verschmelzen oder der Spermakern kann sich (bei *Riccardia pinguis* unter starkem Anschwellen der eingedrungenen Spitze) in den Eikern einbohren. Bei Polyspermie scheinen die überzähligen Spermatozoiden im Eiplasma zugrunde zu gehen. Daß bei *Sphagnum* gelegentlich auch beide Eizellen befruchtet werden können, berichtet LYON (1905).

b) Farne.

Dem Eindringen der Spermatozoiden steht bei den Farnen nur das an den Halsteil angrenzende Stück der Eizelloberfläche („Empfängnisfleck“ ?) zur Verfügung. Im allgemeinen dringt nur ein Spermatozoid (das zuerst angekommene ?, STRASBURGER) in die Eizelle ein. Die bei Moosen verhältnismäßig häufig beobachtete Polyspermie kommt bei Farnen nur als seltener Ausnahmefall vor, was vielleicht mit den verschieden großen „freien Oberflächen“ der befruchtungsfähigen Eizellen zusammenhängen mag.

Alle Beobachter stimmen darin überein, daß die Spermatozoiden mit ihrem Vorderende in die Eizelle eindringen, und zwar nach den Beobachtungen STRASBURGERs in verhältnismäßig kurzer Zeit (3—4 min). Nach der Aufnahme bleibt das Spermatozoid einige Zeit anscheinend unverändert im Eiplasma liegen. Inwieweit das ganze Spermatozoid oder nur Teile desselben in die Eizelle aufgenommen werden, ist nicht klar; vielleicht besteht auch keine Einheitlichkeit des Verhaltens. Bei *Lygodium* wird das ganze Spermatozoid mit seiner Plasmahaut (aber ohne den Plasmarest) und den Geißeln aufgenommen, wobei die Plasmahaut als helle Zone um den Kern herum sichtbar bleibt. Nach Beobachtungen YUASAs an Polypodiaceen sollen die Geißeln vor dem Eintritt in die Eizelle zerstört werden. Eine Übertragung männlicher Plastiden und Mitochondrien ist äußerst unwahrscheinlich, da der diese Organellen enthaltende Plasmarest in allen Fällen vor der Befruchtung abgestoßen wird. Ähnlich wie beim Eindringen des ganzen Spermatozoids in die Eizelle erfolgt auch die Kernverschmelzung in der Weise, daß das Vorderende des Spermatozoidkerns in den

Eikern eindringt. Der anfänglich homogen erscheinende Spermatozoidkern läßt nach der Kernverschmelzung ein Chromatinnetzwerk erkennen, das sich mit dem des Eikerns vereinigt und in ihm aufgeht.

B. Cycadales und Ginkgoales.

Gegenüber den Archegoniaten sind bei den *Cycadales* und *Ginkgoales* zwei Änderungen eingetreten, die auf den Gesamtablauf der Befruchtung von Bedeutung sind: einmal liegt der die Archegonien bildende Gametophyt im Innern eines geschlossenen Gewebes, des Nucellus; zweitens wird die Befruchtung eingeleitet durch die Übertragung von ganzen Mikrosporen. Die Entwicklung der Spermatozoiden in der zu einem schlauchförmigen Haustorium auswachsenden Mikrospore erfolgt erst im Innern einer im Nucellus entstehenden Höhlung, der Pollenkammer, und der Weg der Spermatozoiden zu den Archegonien wird erst durch Auflösung der dazwischenliegenden Nucelluszellen frei. Die Befruchtung ist vom Wasser unabhängig geworden.

Die die Befruchtung vorbereitenden Stadien umfassen daher folgende Teilprozesse: Entstehung der Pollenkammer im Nucellus und Ausscheidung eines Bestäubungstropfens zum Auffangen der durch den Wind übertragenen Mikrosporen; Befestigung der keimenden Mikrosporen in der Wand der Pollenkammer mit Hilfe eines Haustoriums; Ausbildung der Spermatozoiden und Entlassen der befruchtungsfähigen Spermatozoiden nach Auflösen der zwischen Pollenkammer und Archegonien liegenden Nucelluszellen. Auf der weiblichen Seite: Entwicklung der Makroprothallien auf den von den keimenden Mikrosporen ausgehenden Reiz hin und Öffnung der Archegonien.

Pollenkammer und Bestäubungstropfen.

Die Entstehung der Pollenkammer erfolgt bei den Cycadeen und *Ginkgo* in weitgehend übereinstimmender Weise. Gleichzeitig mit der beginnenden Meiose der Makrosporenmutterzelle im Nucellus beginnen die unterhalb der Nucellusspitze gelegenen Zellen zu degenerieren und mit der Fertigstellung der Makrospore ist durch völlige Auflösung der degenerierten Zellen ein allseitig geschlossener Hohlraum entstanden, der an der Spitze von einer einschichtigen Zellage bedeckt ist (Abb. 8). Bei *Ginkgo* ist dabei der Ablauf der beiden Vorgänge so streng koordiniert, daß beide von einer gemeinsamen Ursache abzuhängen scheinen (FAVRE-DUCHARTRE 1956, SLOOVER-COLINET 1963). Mit der beginnenden Prothalliumbildung schließlich sterben dann auch die Zellen der einschichtigen Deckschicht ab und die Pollenkammer öffnet sich direkt in die Mikropyle.

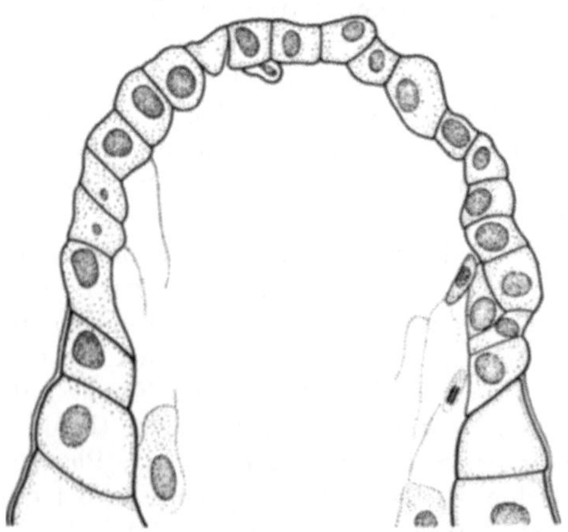

Abb. 8. *Ginkgo biloba*, obere Begrenzung der Pollenkammer vor der Öffnung. (Nach FAVRE-DUCHARTRE.)

Der zur Zeit der Bestäubung aus der Mikropyle austretende, leicht klebrige Bestäubungstropfen, der sich nach Entfernen — bei *Ginkgo* bis zu dreimal innerhalb von 6 Tagen — wieder ergänzen kann und dessen Volumen das Volumen der Pollenkammer um ein Mehrfaches übertrifft, entsteht wohl durch Sekretion der Nucelluszellen. Die Annahme, daß der Bestäubungstropfen lediglich durch die bei der Bildung der Pollenkammer aufgelösten Zellen entsteht, dürfte unzutreffend sein.

Bestäubung.

Die aus den Mikrosporangien entlassenen Mikrosporen, die bei den Cycadeen dreizellig (Prothalliumzelle, generative und vegetative Zelle), bei *Ginkgo* vierzellig sind (zwei Prothalliumzellen, von denen die erste aber bald degeneriert, generative und vegetative Zelle), werden durch den Wind übertragen. Eine vielfach behauptete Übertragung durch die Curculioniden *Phloeophagus* und *Derelomus* bei einigen südafrikanischen *Encephalartus*-Arten wird von CHAMBERLAIN äußerst skeptisch beurteilt. Die auf den Bestäubungstropfen aufgetroffenen Mikrosporen sinken in diesem der Schwerkraft folgend bald auf den Grund der Pollenkammer oder sie gelangen mit dem Eintrocknen des Bestäubungstropfens in dieselbe (Abb. 9). Bei *Ginkgo* nehmen sie dabei reichlich Wasser auf und die zunächst spindelförmigen Mikrosporen nehmen rasch eine kugelige Form an. Mit dem Eintrocknen des Bestäubungstropfens wird gleichzeitig auch die Pollenkammer nach außen wieder fest verschlossen. Von der bei *Bowenia* in eine engere obere und eine weitere untere Höhlung unterteilten Pollenkammer schrumpft die obere ein und nur die untere bleibt erhalten.

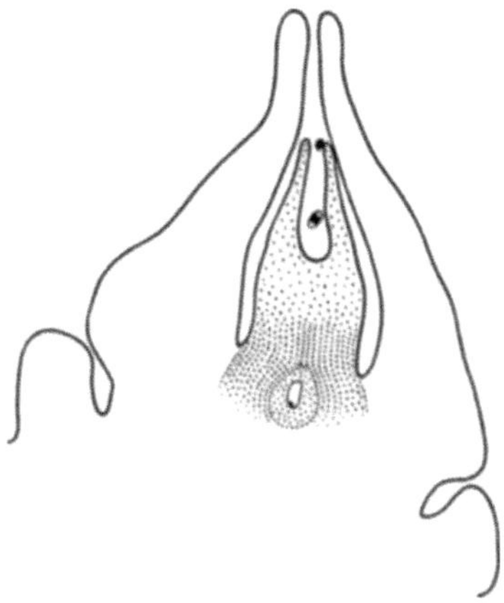

Abb. 9. *Ginkgo biloba*, geöffnete Pollenkammer mit zwei Pollenkörnern (das obere von *Ginkgo*, das untere von einer Conifere.) (Nach FAVRE-DUCHARTRE.)

Keimung der Mikrosporen.

Unmittelbar nach dem Eindringen in die Pollenkammer oder erst nach einer kürzeren Ruhezeit beginnen die Mikrosporen zu keimen, indem sich die Intine an der von der Exine nicht bedeckten Stelle der Mikrospore auszudehnen beginnt und die vegetative Zelle schlauchartig auswächst und zwischen die Nucelluszellen der Pollenkammerwand eindringt; zwischen den Nucelluszellen bilden sich dann reichlich rhizoidenähnliche Verzweigungen. Es entsteht ein typisches Haustorium, das die Mikrospore mechanisch befestigt und seine Ernährung ermöglicht. Die von der Exine umschlossene Spitze der Mikrospore mit der Prothalliumzelle und der generativen Zelle ragt frei in das Innere der Pollenkammer hinein (Abb. 10).

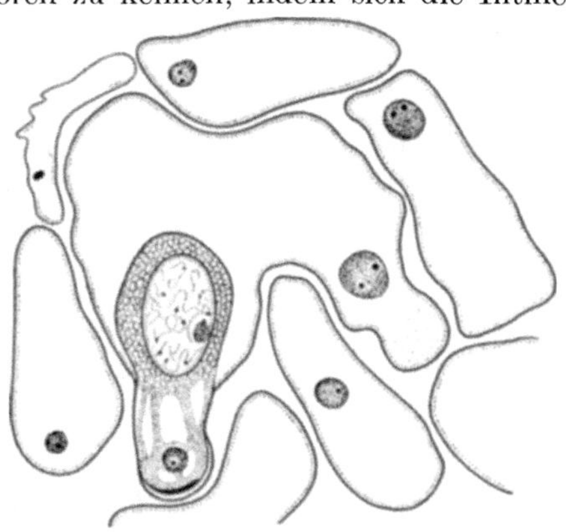

Abb. 10. *Ginkgo biloba*, beginnendes Einwachsen der haustorialen Zelle zwischen die Zellen der Pollenkammerwand. (Nach FAVRE-DUCHARTRE.)

Entstehung und Bau der Spermatozoiden.

An der Spitze des schlauchförmig (Cycadeen) bzw. blasig *(Ginkgo)* angeschwollenen freien Endes der Mikrosporen wird die generative Zelle nach Abgabe einer sterilen Stielzelle zur spermatogenen Zelle; der Kern der spermatogenen Zelle teilt sich und die beiden nackten Tochterzellen werden — unter Mitwirkung der Blepharoplasten — zu den Spermatozoiden, wobei der ganze Inhalt der spermatogenen Zelle zur Bildung der Spermatozoiden aufgebraucht wird. Abweichend verhält sich *Microcycas calocoma*, indem die von der generativen Zelle abgegebene Stielzelle sich wiederholt teilt und bei jedem Teilungsschritt eine weitere spermatogene Zelle abgibt, so daß schließlich 8, gelegentlich sogar 11 spermatogene Zellen entstehen, deren jede in normaler Weise zwei Spermatozoiden ausbildet (Abb. 11).

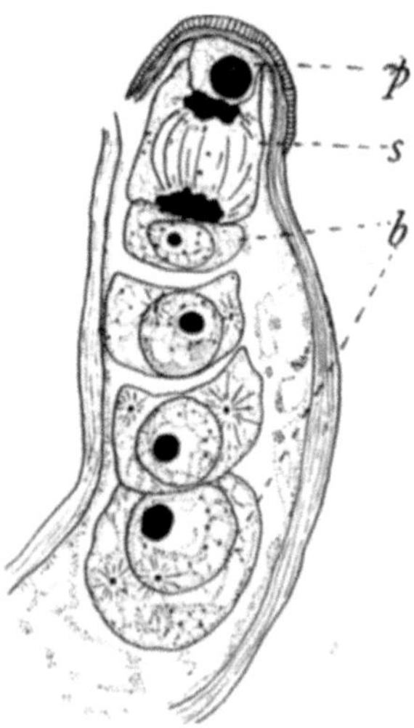

Abb. 11.

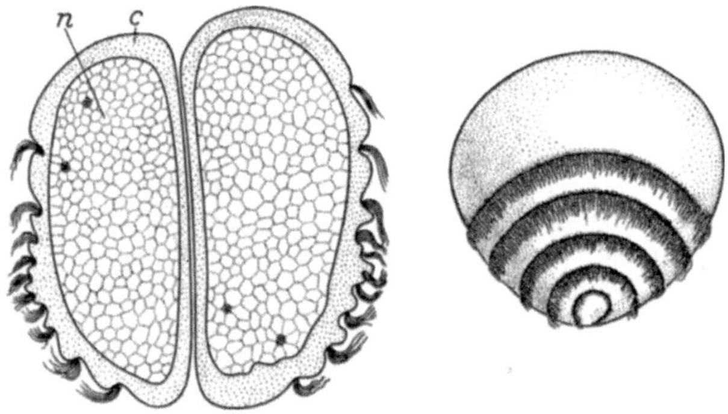

Abb. 12.

Abb. 11. *Microcycas calocoma*, Pollenschlauch mit Prothalliumzelle (*p*), spermatogener Zelle (*s*) in Prophase und Körperzellen (*b*). (Nach DOWNIE.)

Abb. 12. *Zamia muricata*, Spermatozoiden. Links die beiden Spermatozoiden nahezu reif im optischen Längsschnitt. *n* Kern; *c* Cytoplasma. Rechts reifes Spermatozoid. (Nach WEBBER.)

Die bei den Cycadeen mit freiem Auge sichtbaren Spermatozoiden und die kleineren von *Ginkgo* zeichnen sich durch ihre außerordentliche Größe (*Dioon edule* 230 μ Durchmesser, 300 μ Länge; *Ginkgo* 50—85 μ:80—110 μ) wie durch die Größe des Zellkerns und den Bau ihres Bewegungsapparates aus. Die frei beweglichen Spermatozoiden haben eine annähernd kugelige Gestalt. Den größten Teil nimmt dabei aber der Kern ein, während die ihn umgebende Plasmaschicht im Verhältnis dazu dünn ist (Abb. 12). Ein bei den Bryophyten und Pteridophyten vor der Befruchtung abgestoßener Plasmarest ist nicht vorhanden. Etwas unter der Hautschicht des Protoplasten liegt das Wimperband, das bei den Cycadeen 5—6 Umläufe bildet (bei *Cycas* nach rechts, bei *Zamia* und *Dioon* nach links gedreht), bei *Ginkgo* $2^1/_2$ Umläufe nach rechts gedreht ist. Besonders auffallend ist die Tatsache, daß der Kern der Spermatozoiden total Feulgen-negativ ist. Das Protoplasma der Spermatozoiden besitzt ein granuläres Chondriom (*Ginkgo*; FAVRE-DUCHARTRE 1956), ist aber frei von Reservestoffen.

Entwicklung der Eizellen.

Die Entwicklung des weiblichen Prothalliums beginnt im Anschluß an die Makrosporenbildung; zur Zeit der Bestäubung besteht es erst aus wenigen freien Kernen. Eine normale Weiterentwicklung des Makroprothalliums ist jedoch sowohl bei den Cycadeen als auch bei *Ginkgo* gebunden an die Keimung der Mikrosporen in der Pollenkammer. Unterbleibt die Mikrosporenkeimung, dann fallen die Samenanlagen ohne weitere Entwicklung ab. Daß es sich dabei um einen von den keimenden Mikrosporen ausgehenden stofflichen Reiz handelt, der für die Weiterentwicklung notwendig ist, scheint aus der Tatsache hervorzugehen, daß durch Einbringen einiger Kriställchen von IES oder NES in den Bestäubungstropfen wenigstens das Abfallen der Samenanlagen verhindert werden kann; zu einer Prothalliumentwicklung kommt es jedoch auch in diesem Falle nicht (Favre-Duchartre 1956). Beide Vorgänge, Mikrosporenkeimung und Weiterentwicklung des Makroprothalliums sind dabei zeitlich eng aufeinander abgestimmt; eine künstlich hinausgezögerte Bestäubung ist nur eine begrenzte Zeit lang wirksam (Favre-Duchartre 1956).

Von den zahlreichen Archegonienanlagen entwickelt sich nur eine verhältnismäßig geringe Anzahl zu befruchtungsfähigen Archegonien weiter. Durchschnittlich beträgt ihre Anzahl bei den Cycadeen 3; bis zu 10 werden für *Dioon* angegeben; außergewöhnlich groß ist die Zahl bei *Microcycas*, wo einige Hundert auf der ganzen Oberfläche des Makroprothalliums, auch auf der basalen Seite, verteilt sind. Aber auch in diesem Fall wird der letzte Entwicklungsschritt, nämlich die Teilung des Kerns der Zentralzelle in den Eikern und den Bauchkanalkern, nur von der Archegoniengruppe am mikropylaren Ende durchgeführt; ebenso entsteht eine Pollenkammer nur in Verbindung mit dieser Archegoniengruppe, Die Normalzahl der Archegonien bei *Ginkgo* beträgt 2; sie kann seltener 3—5, ausnahmsweise auch nur 1 betragen.

Bau der Archegonien.

Die befruchtungsfähigen Archegonien bestehen aus der großen Eizelle, dem kurz vor der Befruchtung von der Zentralzelle abgegebenen Bauchkanalkern (Cycadeen) bzw. der Bauchkanalzelle *(Ginkgo)* und dem aus 4 Zellen bestehenden Halsteil, dessen Zellen bei *Ginkgo* vor der Befruchtung paarweise übereinandergeschoben sind (der zweite Teilungsschritt bei der Ausbildung der Halszellen erfolgt in der Regel so spät, daß vielfach nur zwei Halszellen beobachtet wurden). Der bei den Cycadeen im allgemeinen bald degenerierende Bauchkanalkern bleibt jedoch z.B. bei *Zamia umbrosa* erhalten und kann sogar — wenn auch nicht allzu häufig — Aussehen (und Funktion ?) eines zweiten Eizellkerns annehmen (Bryan und Evans 1957). Die an die Eizelle angrenzenden Zellen des Makroprothalliums dienen der Nährstoffzufuhr in die Eizelle und sind in besonderer Weise ausgebildet (Deckschicht, jacket layer). Sie sind reich an Plasma und Inhaltsstoffen und enthalten bei den Cycadeen vielfach 2—4 große Zellkerne. Bei *Cycas* ist weiterhin die Wand zwischen der Eizelle und den Zellen der Deckschicht bis zu 150 μ dick und mit zusammengesetzten Tüpfeln versehen. Der Kern der Eizelle wandert gegen die Zellmitte, sein Volumen nimmt vor der Befruchtung ganz auffallend und rasch zu (*Macrozamia reidlei* $300 \times 500\,\mu$); bei *Ginkgo* sind in ihm die völlig entspiralisierten Chromonemata zu erkennen (Favre-Duchartre). Gegenüber der Zentralzelle zeigt das Plasma der Eizelle nur wenig Unterschiede. Die Vacuolen bilden sich weitgehend zurück. An Reservestoffen enthält es Eiweißvacuolen, ferner (bei *Ginkgo*) Eiweißgranula von einer Größe von 2—5 μ, die von den ersteren nach Entstehung und Entwicklung deutlich zu unterscheiden sind. Sie erscheinen erst spät im Plasma der Zentralzelle, sind sicherlich nicht chondriosomaler Herkunft und scheinen auch nicht in Vacuolen zu entstehen. Nach Favre-Duchartre sollen sie nucleärer Herkunft sein. Ein charakteristisches Kennzeichen der Eizell-Reservestoffe ist ihre relativ späte Bildung erst kurz vor der Befruchtung.

Zwischen den Eiweißvacuolen und vor allem in dem dichteren Plasma um den Zellkern finden sich die Chondriosomen. Ein vielfach beobachteter Hohlraum zwischen dem eigentlichen Eiplasma und den Halszellen („chambre receptive“ nach HIRASÉ) wird von FAVRE-DUCHARTRE (1956) als Zeichen einer beginnenden Degeneration betrachtet.

Weg der Spermatozoiden zu den Archegonien.

Vor der Teilung des Kerns der Zentralzelle erfährt das Prothalliumgewebe um die Archegonien bei den Cycadeen ein gesteigertes Wachstum, wodurch die Archegonien in eine Vertiefung, die Archegonienkammer, versenkt werden, während bei *Ginkgo* die Archegonienkammer ringförmig ausgebildet ist, da der zwischen den beiden Archegonien gelegene Teil des Nucellus kurz säulenförmig erhöht ist. Gleichzeitig mit der Entwicklung der Archegonienkammer reißt auch die das ganze Prothallium umhüllende Makrosporenmembran, so daß die Archegonien nicht mehr von ihr bedeckt sind. Ebenso haben sich auch die die untere Begrenzung der Pollenkammer bildenden Nucelluszellen fortschreitend aufgelöst, so daß schließlich Pollenkammer und Archegonienkammer einen gemeinsamen, aber nicht von Flüssigkeit erfüllten Hohlraum bilden, in den die freien schlauch- bzw. blasenförmigen Enden der Mikrosporen hineinhängen bzw. sich infolge eines chemotropischen Reizes in die Richtung auf die Archegonien zu krümmen (GOEBEL 1932). Ältere Angaben, wonach bei *Ginkgo* die Pollenkammer schon vor dem Öffnen der Pollenschläuche zum Teil von Flüssigkeit erfüllt sein soll (Sekretion von der zentralen Prothalliumspitze bei *Ginkgo*?), bedürfen nach CHAMBERLAIN (1935) einer Überprüfung.

Öffnung der Pollenschläuche.

Die Öffnung der Pollenschläuche erfolgt bei den Cycadeen und bei *Ginkgo* an der Spitze; Cycadeen und *Ginkgo* sind also in der Bezeichnung GOEBELs acrogam (Abb. 13). Hinsichtlich des Öffnungsmechanismus vermutet GOEBEL, daß die an der Spitze liegende Prothallium- und Stielzelle als „Öffnungsapparat“ des Pollenschlauches wirksam ist. Durch Zunahme des Turgors in diesen beiden Zellen (vorübergehende Stärkespeicherung), die an der Spitze des Pollenschlauches fest angeheftet sind, wo noch die cuticularisierte Exine vorhanden ist, wird diese am wenigsten dehnungsfähige Stelle zerrissen und der Inhalt des Pollenschlauches wird ausgepreßt. Zusammen mit dem flüssigen Inhalt des Pollenschlauches ergießen sich die beiden Spermatozoiden, gelegentlich auch die Reste der Stiel- und Prothalliumzelle, in die Archegoniumkammer.

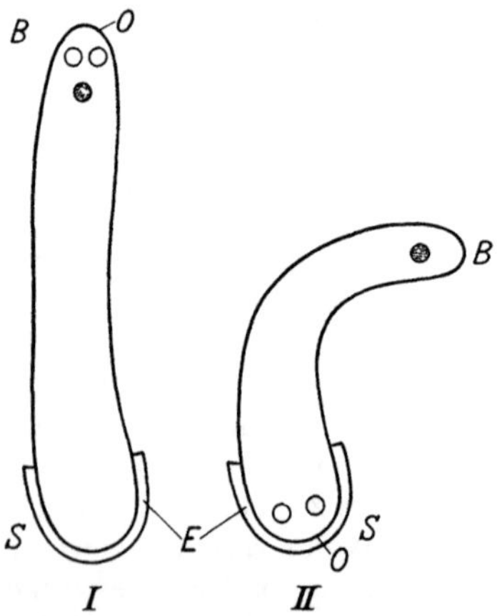

Abb. 13. Schema für akrogame (*II*) und basigame (*I*) Pollenschläuche. *S* Spitze; *B* Basis; *E* äußere Hülle der Mikrospore; *O* Öffnungsstelle des Pollenschlauchs. Die generativen Kerne sind hell, die Schlauchkerne dunkel gehalten. (Nach GOEBEL.)

Schon bei der Keimung der Mikrosporen wurde auf die Abhängigkeit der Weiterentwicklung der ganzen Samenanlage von der erfolgten Keimung der Mikrospore hingewiesen. Ähnliche Wechselbeziehungen liegen auch bei der Entwicklung der ♂ und ♀ Gametophyten vor, die streng synchron verläuft und vor allem kurz vor der Befruchtung sehr rasch abläuft. Da aber nach FAVRE-DUCHARTRE die Kernteilungen in den ♂ und ♀ Gametophyten stets gleichzeitig erfolgen, erscheint es möglich, daß beide durch einen gemeinsamen (hormonalen)

Faktor ausgelöst werden. Das gleiche gilt wohl auch für die Reifung der verschiedenen Archegonien eines Prothalliums, die ebenfalls streng gleichzeitig erfolgt.

Öffnung des Archegoniumhalses.

Über die Öffnung des vor der Befruchtung wohl allgemein vierzelligen Archegonienhalses besteht bei den verschiedenen Autoren keine Übereinstimmung. Zeitlich erfolgt sie erst, wenn der Grund der Archegonienkammer und die nur wenig darüber hinausragenden Halszellen von der Pollenschlauchflüssigkeit bedeckt sind. CHAMBERLAIN nimmt für die Cycadeen osmotische Kräfte in Anspruch: durch den konzentrierten Pollenschlauchinhalt sollen die zunächst voll turgescenten Halszellen ihren Turgor verlieren und den Durchtritt der Spermatozoiden ermöglichen, die von der Eizelle geradezu hineingezogen werden sollen. Nach den Untersuchungen von BROUGH-TAYLOR (1940) an *Macrozamia* sollen ebenfalls kleinste Schwankungen in der Turgescenz der Halszellen im Verhältnis zu dem Druck, der durch eine erhöhte Turgescenz der Eizelle erzeugt wird, genügen, um die Halszellen voneinander zu trennen. Demgegenüber nimmt LEE bei *Ginkgo* eine mechanische Öffnung des Archegonienhalses von der Eizelle aus an: Wenn die Spermatozoiden in die Pollenkammer entlassen sind, stülpt sich das Eiplasma durch die — degenerierte — Bauchkanalzelle hindurch gegen die Halszellen vor und preßt diese dadurch auseinander, wobei gleichzeitig durch die so gebildete Öffnung eine chemotaktisch wirksame Flüssigkeit ausgeschieden werden soll. Andere Beobachter sprechen von aufgelösten Halszellen (nach SCHUSTER 1932), WEBBER (1897) nimmt ein aktives Eindringen der Spermatozoiden durch die zerrissenen Halszellen an.

Auffallenderweise wird die Möglichkeit eines aktiven Öffnens des Archegonienhalses — ähnlich wie bei den Pteridophyten und Bryophyten — nie in Erwägung gezogen, obwohl für die Halskanalzellen wiederholt eine besondere (drüsige) Ausbildung angegeben wird. Lediglich FAVRE-DUCHARTRE beschreibt für *Ginkgo* die Öffnung so, daß sich die vier Halskanalzellen verlängern, paarweise übereinanderschieben, sich aufblähen und so den Zusammenhang an der Überschiebungsstelle lösen.

Für das Auffinden der Archegonien durch die Spermatozoiden wird eine chemotaktische Anlockung von CHAMBERLAIN (1935) gänzlich in Abrede gestellt. Wenn sie angenommen wird, besteht keine Klarheit darüber, wovon diese Anlockung ausgehen soll (Halskanalzellen ?). Möglicherweise wird auch bei den Cycadeen und *Ginkgo* die chemotaktische Anlockung erst nach dem Öffnen der Archegonien wirksam, wobei wie bei den Bryophyten und Pteridophyten die Anlockung von der unmittelbar vor der Befruchtung degenerierten Bauchkanalzelle bzw. dem Bauchkanalkern ausgehen könnte. Damit sind die vorbereitenden Stadien der Befruchtung abgeschlossen und die eigentliche Befruchtung kann erfolgen.

Befruchtung.

Das Eindringen der Spermatozoiden in die Eizelle ist zweifellos ein aktiver Vorgang, wobei den Spermatozoiden, deren Größe die Öffnungsweite des Halses um ein Vielfaches übertrifft, ihre Elastizität und Weichheit zustatten kommt, mit deren Hilfe sie ihre Gestalt auf Druck hin weitgehend verändern können. Da in der Regel eine ganze Anzahl von Pollenschläuchen gleichzeitig ihre Spermatozoiden in die Archegonienkammer entleeren, ist es nicht verwunderlich, wenn ähnlich wie bei den Bryophyten und Pteridophyten mehrere Spermatozoiden durch den Archegonienhals in die Eizelle eindringen können, was wieder für eine chemotaktische Anlockung spricht. Bei *Stangeria paradoxa* beobachtete CHAMBERLAIN in einem Falle sogar 7 überzählige Spermatozoiden in einer Eizelle. Von allen eingedrungenen Spermatozoiden gelangt aber nur eines, wohl das zuerst eingedrungene, zur Befruchtung. Unter Zurücklassung der plasmatischen Hülle und des Bewegungsapparates, die in der Nähe des Archegonienhalses liegen bleiben, bewegt sich der Spermakern auf den zur Zeit der Befruchtung in der Zellmitte gelegenen Eikern zu und verschmilzt mit diesem (Abb. 14). Für die Bewegung des Spermakerns zum Eikern scheidet eine chemotaktische Anlockung als Ursache aus, da sie durch die vollzogene Verschmelzung für die übrigen Spermatozoiden unwirksam wird. Die überzähligen Spermatozoiden bleiben an der Peripherie der Eizelle liegen; ihre Kerne verlassen die Plasmahülle nicht und sie degenerieren bald. Ausnahmsweise wurde in einem Falle bei *Ginkgo* Dispermie beobachtet (FAVRE-DUCHARTRE 1956). Das Plasma, das von dem die Befruchtung ausführenden Spermakern zurückgelassen wird — kenntlich an seinem granulären Chondriom und dem Fehlen von Reserve-

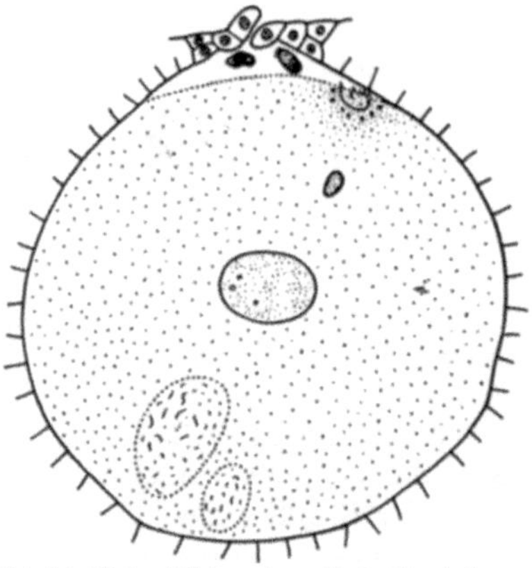

Abb. 14. *Ginkgo biloba*, schematische Darstellung der ersten Phase der Befruchtung. Der ♂ Kern auf dem Wege zum Eikern, Geißelband und ♂ Cytoplasma am Rande der Eizelle. Ein überzähliges Spermatozoid und die degenerierte Bauchkanalzelle unter dem Archegoniumhals. (Nach FAVRE-DUCHARTRE.)

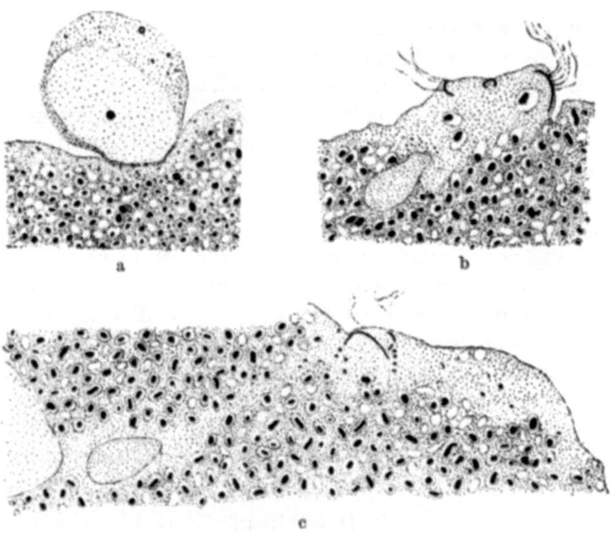

Abb. 15a—c. *Ginkgo biloba*, Kernverschmelzung. a Das Spermatozoid in Berührung mit der Eizelle; b beginnende Wanderung des ♂ Kerns unter Zurücklassung des ♂ Cytoplasmas; c ♂ Kern unmittelbar vor Erreichen des Eikerns (am linken Bildrand); am rechten Rand die Reste des ♂ Cytoplasmas und des Geißelbandes. (Nach FAVRE-DUCHARTRE.)

stoffen — ist noch kurze Zeit vom Eiplasma zu unterscheiden, während das Wimperband noch längere Zeit erkennbar bleibt, später, nach vollzogener Kernverschmelzung aber ebenfalls nicht mehr nachweisbar ist.

Auffallend sind die vielfach zu beobachtenden Größenunterschiede zwischen den beiden verschmelzenden Kernen. So ist das Verhältnis von ♀ zum ♂ Kern bei *Bowenia* 3:1, bei *Ginkgo* sogar 10:1. Während die älteren Autoren bei der Kernverschmelzung dem ♀ Kern eine ausgesprochen aktive Rolle zuteilten, in der Weise, daß dieser eine Mulde bildet und den ♂ Kern fortschreitend umhüllt, scheint es sich nach den neueren Untersuchungen vielmehr um eine Einstülpung des ♀ Kerns zu handeln, die durch den aktiv eindringenden ♂ Kern hervorgerufen wird. Der vor der Befruchtung Feulgen-negative ♂ Kern wird während der Verschmelzung bis zur Auflösung der trennenden Kernmembranen in zunehmendem Maße wieder Feulgen-positiv. Nach allen Beobachtungen sind die verschmelzenden Kerne im Ruhezustand. Mit der ersten folgenden Mitose dürfte — entgegen früheren Beobachtungen — die volle Kernverschmelzung vollzogen sein (Abb. 15). Eine gelegentliche Embryoentwicklung ohne Bestäubung und Spermatozoidenbefruchtung wird auf Grund von cytologischen und Freilandbeobachtungen an zwei Arten von *Encephalartos* von CHAMBERLAIN (1912) und SEDGWICK (1924) für möglich und wahrscheinlich gehalten; in diesen Fällen soll an die Stelle einer Spermatozoidenbefruchtung die Verschmelzung des Eizellkerns mit dem persistierenden Bauchkanalkern treten. Neben Spaltungs-Polyembryonie ist für *Cycas circinalis* und andere *Cycas*-Arten auch einfache Polyembryonie durch Befruchtung zweier Archegonien beobachtet worden (RAO 1964).

C. Coniferae.

Bei weitgehend übereinstimmendem Bau der Mikrosporen (Pollenkörner) und Samenanlagen unterscheiden sich die Coniferen wesentlich dadurch von den Cycadeen und *Ginkgo*, daß an die Stelle beweglicher Spermatozoiden unbewegliche Spermazellen getreten sind. Damit ist ein grundsätzlich anderer Übertragungsmechanismus verbunden, indem diese unbeweglichen Spermazellen an der Spitze der schlauchartig auswachsenden vegetativen Zelle des Pollenkorns, des Pollenschlauchs, durch das nicht aufgelöste Nucellusgewebe hindurch zu den Archegonien gebracht und durch die sich nicht mehr öffnenden Zellen des Archegoniumhalses in die Eizelle entlassen werden.

Folgende Teilprozesse bereiten die Befruchtung vor: Ausscheidung des Befruchtungstropfens zum Auffangen der durch den Wind übertragenen Pollenkörner und aktiver Verschluß der Mikropyle nach erfolgter Bestäubung; Pollenkeimung und Pollenschlauchwachstum durch das Gewebe des Nucellus und Ausbildung der Spermazellen; Entwicklung des Makroprothalliums mit den Archegonien auf den durch die Bestäubung ausgelösten Reiz und Eindringen des Pollenschlauchs in die Eizelle.

Bestäubung.

Die durch den Wind verbreiteten Pollenkörner der Coniferen gelangen in der Mehrzahl der Fälle zunächst auf die Mikropyle, die einen Bestäubungstropfen ausscheidet, durch den die Pollenkörner festgehalten werden und auf den Nucellus gelangen. Eine Pollenkammer wie bei den Cycadeen und *Ginkgo* ist nur noch andeutungsweise bei einigen Coniferen zu erkennen. Besonders ausgeprägt ist der Bestäubungstropfen bei *Taxus* und *Cephalotaxus*, wo er auch nach Abschütteln mehrmals aufs neue innerhalb von 10 Std ergänzt wird. Die Flüssigkeit wird dabei aus dem Nucellusgipfel ausgeschieden. Gelangt ein Pollenkorn in den Bestäubungstropfen, so wird die Exine durch Quellen der Intine gesprengt und schließlich von dem auf das Doppelte seiner ursprünglichen Größe gequollenen Pollenkorn ganz abgeworfen. Durch seine Schwere sinkt es in dem

Tropfen unter bzw. gelangt bei hängender Samenanlage durch Eintrocknen des Tropfens auf den Nucellus. Bei den Arten, deren Samenanlagen nicht frei über die Schuppenblätter des Zapfens herausragen, wie bei *Taxus*, *Cephalotaxus* und den *Cupressaceae*, rücken zur Zeit der Bestäubung die Zapfenschuppen durch Streckung der Internodien der Zapfenachse auseinander *(Pinaceae)* und die Pollenkörner werden vielfach durch Kiele oder Rinnen an den Deckschuppen zu den Mikropylen geleitet. Nach der Bestäubung wird der Zapfen durch starkes Heranwachsen der Zapfenschuppen wieder geschlossen.

Mit Abschluß der Bestäubung wird die Mikropyle jeder Samenanlage verschlossen. Dieser Mikropylenverschluß kommt zustande durch das Heranwachsen der Epidermiszellen im Innern der Mikropylarröhre *(Phyllocladus*, *Podocarpus*, *Cupressaceae)* oder er geht in der Weise vor sich, daß sich die zwischen der äußeren und inneren Epidermis gelegenen Integumentzellen stark vergrößern und teilen *(Pinaceae*, *Torreya)*. Interessant ist dabei die Beobachtung NORÉNs (1907) an *Juniperus*, daß der Verschluß der Mikropyle durch einen von den Pollen ausgehenden Reiz ausgelöst wird; ein Verschluß tritt nämlich immer nur dann ein, wenn arteigener Pollen in die Mikropyle gelangt ist; fremder Pollen löst keinen Verschluß aus.

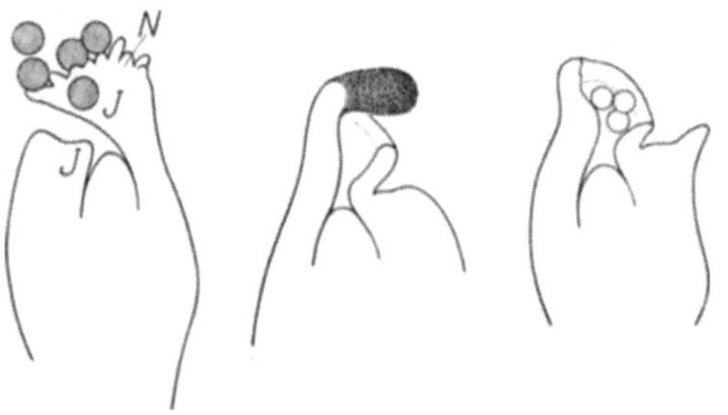

Abb. 16. *Larix decidua*. Links Längsschnitt durch ein Makrosporangium, dessen Mikropylarnarbe (*N*) mit Sporen (schraffiert) bedeckt ist; Mitte Schema für die Veränderung der Mikropylarnarbe; der schrumpfende Teil schraffiert; rechts Längsschnitt durch ein Makrosporangium, bei dem die Mikrosporen durch Schrumpfen der Mikropylarnarbe in die Mikropyle gelangt sind. (Nach GOEBEL.)

Von besonderem Interesse ist die verschiedene Ausbildung der Mikropyle, wo wir drei Typen unterscheiden können: die Röhren-, die Zangen- und die Narben-Mikropyle. Die schon von den Cycadeen und *Ginkgo* bekannte röhrenförmig ausgebildete Mikropyle findet sich unter den Coniferen bei *Taxus* und den *Cupressaceae*. Eine Zangenmikropyle liegt bei den meisten *Pinaceae* vor. Bei dieser ist die Mikropyle ungleichseitig ausgebildet, indem die der Zapfenachse zugekehrte Seite höher und rechts und links in zwei lange, dünne Fortsätze ausgezogen ist, die wohl am Festhalten des Bestäubungstropfens beteiligt sind. Bei der Narbenmikropyle *(Larix*, *Pseudotsuga)* schließlich ist der der Achse zugekehrte Teil noch stärker ausgebildet, aber nicht in zwei seitliche Auswüchse verlängert, sondern wie die Narbe vieler Angiospermen mit „Papillen" versehen, welche die Pollenkörner festhalten. Die Mikropyle selbst ist zur Zeit der Bestäubung dadurch vollständig zugedeckt. Durch Einschrumpfen der Innenseite dieses narbenähnlichen Mikropylarlappens krümmt sich dieser nach innen und jetzt erst gelangen die Pollenkörner in den Mikropylarkanal (Abb. 16).

Ganz abweichend von diesem verbreitetsten Bestäubungsmodus verhalten sich die *Araucariaceae* und *Saxegothea*, bei denen der Nucellus so weit aus dem Integument herausragt, daß eine Bestäubung der Mikropyle ausgeschlossen ist. Die herausragende Nucellusspitze bildet weiter wie die Narbe vieler Angiospermen Papillen. Während bei *Saxegothea* die Pollenkörner wohl direkt auf die fleischig angeschwollene Nucellarspitze gelangen, die zur Zeit der Bestäubung von einer schleimigen Flüssigkeitsausscheidung bedeckt ist, die die Pollenkörner festzuhalten vermag, fallen sie bei *Araucaria* und *Agathis* von der Nucellusspitze entfernt auf die Karpelle oder auf die Ligularschuppe. Hier bereits wachsen sie schlauchartig aus.

Pollenkeimung und Pollenschlauchwachstum.

Wenn wir zunächst absehen von den eben erwähnten Sonderfällen von *Araucaria* und *Agathis*, wo der Pollen weit entfernt von den Samenanlagen auf der Fruchtschuppe seine Entwicklung beginnt, und *Saxegothea*, wo die Pollenkörner direkt auf die papillöse Nucellusspitze zu liegen kommen, gelangt der Pollen erst von der Mikropyle aus auf den Nucellusscheitel. Eine Pollenkammer, wie bei den Cycadeen und *Ginkgo*, ist nur noch andeutungsweise bei einigen Coniferen zu erkennen, indem es an der Spitze des Nucellus zur Auflösung einer Zellpartie, allenfalls zur Bildung einer seichten muldenförmigen Vertiefung kommt. Dagegen fällt in manchen Fällen eine besondere Ausbildung eines Zellkomplexes an der Nucellusspitze auf; dieser als Pollenkissen bezeichnete Zellkomplex hebt sich durch sein Aussehen sehr auffallend von dem übrigen Nucellusgewebe ab und dürfte mit der Keimung der Pollenkörner in Zusammenhang stehen. Bei *Sciadopitys verticillata*, wo es besonders deutlich ausgebildet ist, besteht es gegenüber dem kleinzelligen übrigen Gewebe aus einer Gruppe von großen Zellen mit Intercellularräumen und dünnen Wänden, die an der Oberfläche eine durchsichtige Flüssigkeit ausscheiden.

Die Zahl der Pollenkörner, die auf den Nucellusscheitel kommen, ist sehr verschieden, zum Teil ist ihre Anzahl recht beträchtlich; so wurden bei *Phyllocladus alpinus* z.B. gegen 15 Pollenschläuche gezählt, bei *Taxus* sogar 22 in einer einzigen Samenanlage.

Die Pollenkörner beginnen gleich oder kurze Zeit nach der Bestäubung die Pollenschläuche zu bilden *(Pinus silvestris)* oder es vergeht einige Zeit bis zur Keimung (*Pseudotsuga douglasii* bis 10 Tage, *Larix decidua* einige Wochen). Ebenso verschieden ist auch die weitere Entwicklung der Pollenschläuche im Nucellus, indem sie gleichmäßig von der Keimung des Pollenkorns bis zur Befruchtung abläuft oder durch eine längere Ruheperiode (Winterruhe, z.B. bei der Mehrzahl der untersuchten *Pinus*-Arten) in zwei Wachstumsperioden unterteilt ist. Sind mehrere Pollenschläuche auf einem Nucellusscheitel, so kann das Wachstum der einzelnen Pollenschläuche doch sehr verschieden verlaufen; bei *Torreya taxifolia* z.B. wachsen manche Pollenschläuche rasch zum Prothallium und erreichen dieses noch in einem Zustand mit wenigen (16—64) freien Kernen, andere wachsen langsamer und erreichen das Prothallium erst, wenn schon Archegonien angelegt sind.

Während des Wachstums nimmt der Durchmesser der Pollenschläuche vielfach beträchtlich zu (z.B. *Taxus*, *Podocarpus*); insbesondere erweitert sich die Pollenschlauchspitze beim Auftreffen auf das Prothallium. Die Wand der Pollenschläuche, die aus der Intine hervorgeht, ist im allgemeinen dünn; auffällig dickwandige Pollenschläuche wurden bei *Podocarpus macrophyllus* und *Acmopyle pancheri* beobachtet, wo sie doppelt so dick wie die Makrosporenmembran waren.

Sehr variabel ist die Wachstumsrichtung der Pollenschläuche in bezug auf das ♀ Prothallium. Manche Pollenschläuche wachsen geradewegs von der Nucellusspitze in der Achse des Nucellus auf den Scheitel des Prothalliums zu, andere (bei derselben Art) schlagen recht abweichende Bahnen ein. Neben den Pollenschläuchen, die mehr oder weniger direkt von der Nucellusspitze auf das ♀ Prothallium zuwachsen, wachsen die Pollenschläuche anderer Arten an der Seite des Nucellus in die Tiefe hinab. Bei *Athrotaxis selaginoides* etwa wächst der Pollenschlauch vom Scheitel des Nucellus aus zwischen Nucellus und Integument hinab, dringt auf halber Höhe des Nucellus in diesen ein und wächst dann im Kontakt mit der Makrosporenmembran weiter. Auch bei *Pherosphaera* wächst der Pollenschlauch seitlich an der Prothalliumspitze weiter, bis er die seitlich gelegenen Halszellen eines Archegoniums erreicht hat.

Während ihres Wachstums im Nucellus können sich die Pollenschläuche gelegentlich auch verzweigen, so bei *Pinus*; der Inhalt des Pollenschlauchs fließt jedoch nur in einen dieser Zweige hinein, der andere entwickelt sich nicht weiter. Ganz abnorm in dieser Hinsicht ist das Wachstum der Pollenschläuche von *Araucaria* und *Agathis*, die weit entfernt von der Nucellusspitze bereits auf der Fruchtschuppe zu keimen beginnen. Bei *Araucaria* gehen dabei von dem mehr oder weniger geradlinig auf den Nucellus gerichteten Hauptast Seitenäste aus, die in allen Richtungen über die Oberfläche der benachbarten Karpelle hinkriechen und diese wie ein Spinngewebe überziehen. Viele dieser Verzweigungen enthalten überhaupt keine Zellkerne. Bei *Agathis* können Verästelungen des Pollenschlauchs sogar in die Rinde der Zapfenachse eindringen und in unregelmäßigem Verlauf bis in das Xylem vordringen. In die Samenanlage treten die Pollenschläuche im allgemeinen durch den aus der Mikropyle herausragenden Nucellusscheitel, der narbenartig ausgebreitet ist und eine Flüssigkeit sezerniert. Bei *Agathis* verzweigen sich die Pollenschläuche auch nach ihrem Eintritt in den Nucellus, sie können aber auch direkt auf das Prothallium zuwachsen. Die Pollenschläuche treten dabei so zahlreich in den Nucellus ein, daß das Gewebe dadurch fast ganz zerstört wird.

Entstehung der Spermazellen.

Die Pollenkörner der Gymnospermen zeigen zur Zeit der Bestäubung eine große Mannigfaltigkeit ihres inneren Baues. Allgemein kommt ihnen eine ausgesprochene Polarität zu, wobei die Lage der Hauptachse durch die Orientierung der Pollenkörner in der Tetrade bestimmt ist. An den noch zu Tetraden zusammengefaßten Pollenkörnern können wir eine nach außen gerichtete Basis und eine nach innen gerichtete Spitze unterscheiden. Bei *Pinus* konnte Goebel feststellen, daß sich die Prothalliumzellen an der Spitze bilden, während das basale Ende zum Pollenschlauch auswächst (Abb. 17). Eine ähnliche Beobachtung hatte Warming (1879) auch für Cycadeen gemacht.

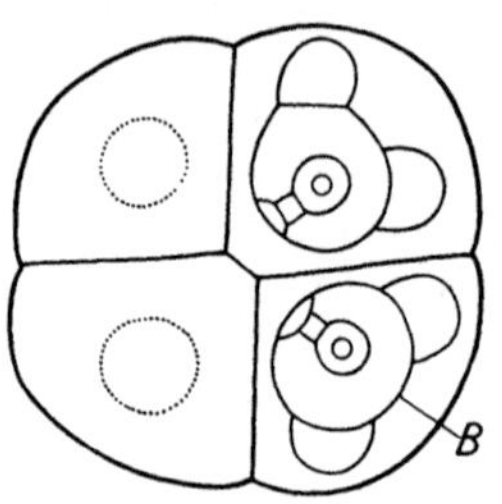

Abb. 17. *Pinus montana*, Pollentetrade. Schema für die Lage der Mikrosporen innerhalb einer Sporenmutterzelle (alle vier Tochterzellen sichtbar). *B* Basis. (Nach Goebel.)

Zur Zeit der Bestäubung sind bei den *Pinaceae* neben dem vegetativen (Pollenschlauch-) Kern zwei Prothalliumzellen und die antheridiale Zelle enthalten. Davon weichen einmal die *Podocarpaceae* (ohne *Pherosphaera*) und die *Araucariaceae* ab, bei denen die beiden primären Prothalliumzellen weitere sekundäre Teilungen erfahren, während andererseits die *Taxaceae*, *Cephalotaxaceae*, *Taxodiaceae*, *Cupressaceae* und *Pherosphaera* überhaupt keine Prothalliumzellen bilden und zum Teil überhaupt einkernig (*Taxus*, *Cupressaceae* zum Teil) oder zweikernig (antheridiale und Pollenschlauchzelle) sind.

Soweit Prothalliumzellen vorhanden sind, sind sie nirgends bleibend; bei den *Pinaceae* werden sie frühzeitig zusammengedrückt und selbst bei *Araucaria* und *Podocarpus*, wo sie die größte Entwicklung erreichen, werden ihre Zellwände aufgelöst und die Kerne wandern in den Pollenschlauch ein.

In den Fällen, wo das reife Pollenkorn einkernig ist, erfolgt die Teilung in die antheridiale und die Pollenschlauchzelle noch vor der Bildung des Pollenschlauchs *(Juniperus)* oder (10—12 Tage nach der Bestäubung) in dem sich bereits zur Bildung des Pollenschlauchs streckenden Pollenkorn *(Taxus)*.

Die folgende Entwicklung vollzieht sich nun einheitlich bei allen Arten ohne grundsätzliche Abweichung im Pollenschlauch. Zunächst teilt sich die antheridiale

Zelle in die Stielzelle und die spermatogene Zelle, die anfänglich beide noch in der Spitze des Pollenkorns liegen. Die spermatogene Zelle wandert nun in den Pollenschlauch ein, wobei nach GOEBEL die Loslösung der spermatogenen Zelle durch Platzen der Stielzelle erfolgt bzw. bei *Araucaria* und *Podocarpus* mit der Auflösung der Zellwände der Prothalliumzellen auch die spermatogene Zelle frei wird. Sie wird dabei wesentlich größer und auch der Kern vergrößert sich.

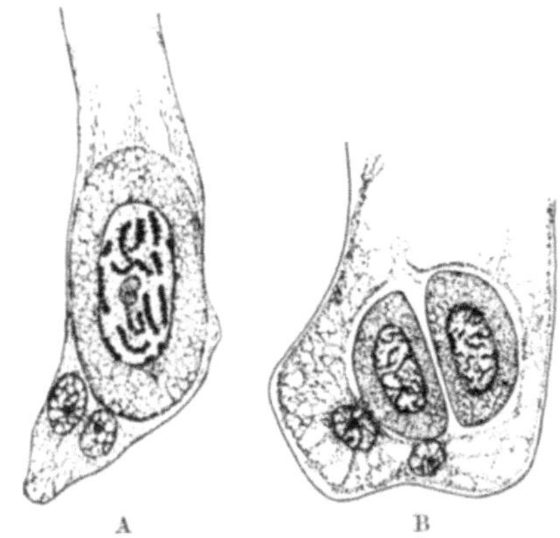

Abb. 18 A u. B. *Thuja occidentalis*, Pollenschlauchspitze. A Körperzelle mit Stiel- und Schlauchkern; B die zwei Spermazellen. (Nach LAND.)

Unabhängig von der im einzelnen vorliegenden Mannigfaltigkeit findet sich schließlich an der Spitze des wachsenden Pollenschlauchs immer wieder die für die Gymnospermen charakteristische Dreier-Gruppe, bestehend aus der spermatogenen Zelle und zwei freien Kernen (Pollenschlauch- und Stielkern) (Abb. 18).

Der letzte Entwicklungsschritt, die Teilung der spermatogenen Zelle in die zwei Spermazellen erfolgt — oft nach einem vorausgegangenen Ruhestadium — fast ausnahmslos unmittelbar vor der Befruchtung. Auffallend weichen lediglich *Abies balsamea* und *A. veitchii* ab, bei denen die Spermazellen bereits im Pollenkorn vor dem Stäuben gebildet werden (SMITH 1923). Auch bei *Araucaria brasiliana* erfolgt die Teilung einen Monat oder mehr vor der Befruchtung außerhalb der Samenanlage (BURLINGAME 1915). Bemerkenswert ist ferner das Verhalten von *Taxodium* und *Juniperus*, wo die Teilung der spermatogenen Zelle stets gleichzeitig mit der Teilung des Zentralkerns im Archegonium erfolgt.

Hinsichtlich des Baues der Spermazellen können wir zwei Gruppen unterscheiden, je nachdem, ob durch die Teilung der spermatogenen Zelle zwei distinkte Spermazellen entstehen *(Taxaceae, Taxodiaceae, Cephalotaxaceae, Cupressaceae)* oder ob bei der letzten Teilung die Cytokinese mehr oder weniger unvollständig ist und die beiden Spermakerne in einem gemeinsamen Cytoplasma liegen *(Pinaceae)* oder durch eine ganz zarte Plasmamembran voneinander getrennt sind *(Araucariaceae* und *Podocarpaceae)*.

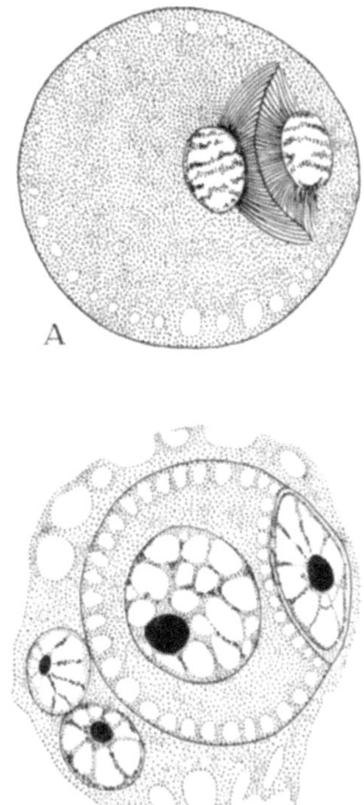

Abb. 19 A u. B. *Taxus canadensis*, A sich teilende Körperzelle mit der Bildung zweier ungleicher Spermazellen; B die beiden Spermazellen. (Nach DUPLER.)

In den Fällen, wo zwei ungleich große Spermazellen bzw. Kerne vorliegen, können die beiden Tochterzellen bzw. Kerne ursprünglich gleich sein und es kann erst im weiteren Verlauf der Entwicklung der eine Tochterkern degenerieren (z. B. *Araucaria*), wobei die Größenunterschiede zwischen den beiden Kernen sehr variabel sein können. Auf der anderen Seite kann die Ungleichheit der beiden Spermazellen von Anfang an die Folge einer inäqualen Teilung sein; so verlagert sich bei *Taxus canadensis* (DUPLER 1917, FAVRE-DUCHARTRE 1960) der spermatogene Kern vor der Teilung an die Peripherie der Zelle und durch eine linsenförmige Wand wird eine kleinere von einer größeren Zelle getrennt (Abb. 19). Auffallenderweise hat BRANDSCHEIDT bei künstlicher Pollenschlauchkultur von *Taxus baccata* beobachtet, daß die beiden Spermazellen ursprünglich gleich groß sind und die Größenunterschiede erst später sichtbar werden. Die kleinere Spermazelle kann im Verlauf der weiteren Entwicklung völlig degenerieren (*Araucaria brasiliana:* BURLINGAME 1915, oder *Podocarpus coriacea:* COKER 1902), so daß nur eine einzige funktionsfähige Spermazelle vorhanden ist.

An Inhaltsstoffen der Spermazellen ist vielfach eine dichte Stärkehülle um den Spermakern vorhanden (z.B. bei *Cunninghamia sinensis, Taxodium distichum*), in anderen Fällen finden sich reichlich Mitochondrien (*Cephalotaxus fortunei:* FAVRE-DUCHARTRE 1958, *Pinus maritima:* MANGENOT 1938).

Entwicklung der Eizellen.

Wie bei den Cycadeen und *Ginkgo* wird auch bei den Coniferen die Weiterentwicklung des Makroprothalliums durch den von den keimenden Pollenkörnern ausgehenden Reiz angeregt; unbestäubte Samenanlagen entwickeln sich nicht weiter und fallen ab. Versuche, durch künstliche Zufuhr von Wuchsstoffen — ähnlich wie bei *Ginkgo* — hatten jedoch bei *Cephalotaxus fortunei* keinen Erfolg; es genügte jedoch die Bestäubung nur einer einzigen Samenanlage, um das Abfallen des ganzen ♀ Zapfens zu verhindern (FAVRE-DUCHARTRE 1956).

In dem zum Teil erst lange Zeit nach der Bestäubung ausgebildeten Prothallium — bei *Cephalotaxus fortunei* beginnt die Prothalliumbildung erst 11 Monate nach der Bestäubung — entstehen die Archegonien in wechselnder Zahl und Anordnung. In der Mehrzahl der Fälle ist die Anzahl der Archegonien verhältnismäßig gering; sie beträgt bei *Taxus* 5—8, *Abies alba* 2—3, *Picea excelsa* 2—7. Sie kann aber in einigen Fällen auch sehr beträchtlich sein, so bei *Actinostrobus pyramidalis* 15—30 und bei *Widdringtonia cupressoides* sogar 30—100, was mit der großen Oberflächenentwicklung der langgestreckten Prothallien parallel geht.

Hinsichtlich der Verteilung der Archegonien ist die häufigste Anordnung eine polare an dem der Mikropyle zugekehrten Ende, wobei im wesentlichen zwei verschiedene Typen zu unterscheiden sind. Beim Pinaceen-Typ stehen die Archegonien einzeln, durch Prothalliumgewebe voneinander getrennt; jedes Archegonium besitzt seine eigene Deckschicht. Beim Cupressaceen-Typ sind die Archegonien zu Gruppen vereinigt, die Eizellen der Archegonien grenzen unmittelbar aneinander und die ganze Gruppe ist von einer gemeinsamen Deckschicht umgeben. In den wenigen Fällen einer apolaren oder zerstreuten Anordnung *(Sequoia* und *Widdringtonia)* liegen die Archegonien im oberen Teil des Prothalliums zerstreut wie beim Pinaceen-Typ, gegen die Mitte zu sind sie wie bei den *Cupressaceae* zu Gruppen angeordnet.

Sowohl die einzelnen Archegonien wie auch die Archegoniengruppen erscheinen bei voller Ausbildung gewöhnlich in das Prothallium eingesenkt, indem das die Archegonien umgebende Gewebe stärker heranwächst. Es kommt dadurch zur Entstehung von trichter- oder muldenförmigen Vertiefungen, den Archegonienkammern.

Bau der Archegonien.

Die befruchtungsfähigen Archegonien bestehen aus der großen Eizelle, der kurz vor der Befruchtung abgegebenen, meist kurzlebigen Bauchkanalzelle *(Pinaceae)* bzw. in allen anderen Fällen dem bald degenerierenden Bauchkanalkern — bei *Sciadopitys verticillata* beginnt die Degeneration, bevor noch ein abgegrenzter Kern entstanden ist —, sowie aus dem Halsteil. Der Halsteil ist im Gegensatz zu den Cycadeen und *Ginkgo* aus einer größeren Zahl von Zellen aufgebaut; ihre Zahl ist auch innerhalb einer Art sehr variabel; bei *Cephalotaxus drupacea* z.B. besteht er aus 2—5 Zellen in sehr wechselnder Anordnung. Bisweilen liegen die Halszellen sogar in zwei oder mehr Schichten übereinander.

Mit der Reife des Archegoniums hat auch die das einzelne Archegonium bzw. die Archegoniengruppe umgebende Schicht des Prothalliumgewebes, die Deckschicht, ihre endgültige Ausgestaltung erfahren. Ihre Innenwände sind beträchtlich verdickt und mit Tüpfeln versehen, die zu kleineren Gruppen vereinigt sind, die wieder zusammen ein größeres Tüpfelfeld bilden. Ihre Kerne sind groß, die Zellen sind stets mehrkernig und zeigen einen ausgesprochen drüsigen Bau, der ihrer Funktion, der Ernährung der Eizelle, entspricht.

Gleichzeitig mit dieser Entwicklung der Deckschicht ist auch — zeitlich beginnend mit der Teilung des Zentralkerns in Ei- und Bauchkanalkern — eine zunehmende Einlagerung von Inhaltsstoffen zu beobachten, während die große zentrale Vacuole, mit Ausnahme der *Cupressaceae*, fast völlig verschwunden ist. Zunächst treten in erheblicher Menge Stärkekörner auf, später bilden sich Eiweißvacuolen, deren Auftreten kurz vor der Befruchtung seinen Höhepunkt erreicht. Vor allem bei den *Pinaceae* sind sie besonders stark und auffallend ausgebildet. Es scheint sich dabei um dichtere Cytoplasmaportionen mit Mitochondrien zu handeln, die durch eine Membran gegen das übrige Cytoplasma abgegrenzt sind (Takao 1959, 1960; Camefort 1959, 1960). An weiteren wichtigen Inhaltskörpern der reifen Eizelle sind dann noch die Mitochondrien zu nennen.

Zeitliche Koordinierung.

Wie schon erwähnt, löst erst die Keimung der Pollenkörner allgemein für alle Coniferen die Weiterentwicklung des Prothalliums aus. Darüber hinaus bestehen in einigen Fällen auch bei der weiteren Differenzierung enge Wechselbeziehungen zwischen den Pollenschläuchen und der Ausbildung der Archegonien. So liegen bei *Sequoia*, *Widdringtonia*, *Actinostrobus* u.a. die befruchtungsfähigen Archegonienkomplexe stets in unmittelbarer Nähe der Pollenschläuche. Die Archegonienbildung wird dabei aber nicht durch die Pollenschläuche erst ausgelöst; die Archegonium-Initialen entstehen vielmehr unabhängig von den Pollenschläuchen auf dem Prothallium. Der zufällige Verlauf der Pollenschläuche bestimmt lediglich, welche dieser Archegoniumanlagen sich zu befruchtungsfähigen Archegonien weiterentwickeln.

Eine weitere Pollenschlauch-Archegonium-Beziehung scheint dann vorzuliegen in dem vielfach weitgehenden Zusammentreffen der Teilungen der antheridialen Zelle und der Zentralzelle, die beide mit wenigen Ausnahmen „kurz vor der Befruchtung“ erfolgen.

Was schließlich die Beziehungen zwischen den einzelnen Archegonien eines Prothalliums untereinander betrifft, so war bei *Cycas* und *Ginkgo* die Entwicklung der Archegonien eines Prothalliums stets streng gleichzeitig. Fast gleichzeitig dürfte auch bei den Coniferen, deren Archegonien in Gruppen innerhalb einer gemeinsamen Deckschicht liegen (*Juniperus communis*, *J. virginianus*, *Cunninghamia sinensis*) die Entwicklung, gemessen an der Teilung des Zentralkerns, ablaufen. Für *Cephalotaxus drupacea*, wo die Archegonien voneinander getrennt und je von einer eigenen Deckschicht umgeben sind, gibt Favre-Duchartre an, daß die Teilungen der Zentralzellen bei den Archegonien eines

Prothalliums nie streng gleichzeitig ablaufen, sondern daß die Teilungsstadien von einem Archegonium ausgehend auf die räumlich benachbarten fortzuschreiten scheinen, so als ob von einem Archegonium aus ein teilungsauslösender Reiz ausginge.

Eindringen der Pollenschläuche in die Eizelle.

Über das Eindringen der Pollenschläuche in die Eizelle liegen nur wenige genauere, beschreibende Angaben vor, über die daran beteiligten Faktoren herrscht eine weitgehende Unklarheit. Im einzelnen scheint der in Frage stehende Vorgang ziemlich variabel zu sein. Was zunächst den häufigsten Fall isolierter Archegonien betrifft, so dringt hier der Pollenschlauch stets aktiv durch den Hals, gelegentlich, wenn noch eine Bauchkanalzelle vorhanden ist, auch durch diese bis zum Plasma der Eizelle vor. Und zwar erfolgt das Eindringen durch den Hals rein mechanisch unter Zerstörung der Halszellen oder, indem der Pollenschlauch seitlich zwischen Hals- und Prothalliumzellen hindurch wächst, ohne die Halszellen zu zerstören. Ein Eindringen des Pollenschlauchs in das Cytoplasma der Eizelle wurde nie beobachtet; bei Berührung der Eizelle öffnet sich der Pollenschlauch und sein Inhalt ergießt sich in die Eizelle. Ein Öffnungsapparat — wie etwa bei den akrogamen Cycadeen — fehlt bei den basigamen Coniferen. Die Öffnung erfolgt durch „Auflösen" der Schlauchspitze. Daß die Eizelle im Augenblick des Eindringens des Pollenschlauches ihr Volumen nicht vergrößert, sie also nicht unter einer elastischen Spannung steht, könnte vielleicht mit dem weitgehenden Abbau der Vacuolen unmittelbar vor der Befruchtung in Zusammenhang stehen. Wenn auch gelegentlich von einem „heftigen" Eindringen des Pollenschlauchinhalts in die Eizelle gesprochen wird (*Araucaria brasiliana*, BURLINGAME 1915), so dürfte im allgemeinen, wohl schon aus räumlichen Gründen, nur eine begrenzte Menge des Pollenschlauchinhalts ohne irgendwelchen größeren mechanischen Druck in die Eizelle übertreten.

Wesentlich unklarer scheinen die Verhältnisse bei den *Taxodiaceae* und *Cupressaceae* zu liegen, wo die Archegonien in Gruppen zusammenliegen und die breiten Pollenschlauchenden im allgemeinen mehrere Archegonien überdecken. In diesen Fällen soll der Inhalt des Pollenschlauchs in die unmittelbar über den Archegonien gelegene gemeinsame Archegonialkammer entleert werden, von wo dann — nach vorheriger Zerstörung der Halszellen ? — die Spermazellen zu den Eizellen gelangen könnten. Die damit gegebene Möglichkeit der Befruchtung zweier benachbarter Archegonien durch einen einzigen Pollenschlauch wurde mehrfach beschrieben, wozu wegen der Ausbildung von zwei gleich großen distinkten Spermazellen in diesen Fällen die dafür notwendigen Voraussetzungen gegeben sind. Ein bei *Podocarpus* und *Cephalotaxus* beobachtetes Austreten von Eiplasma aus dem Halsteil wird von FAVRE-DUCHARTRE als Kunstprodukt betrachtet.

Befruchtung.

Hinsichtlich der Elemente, die aus dem Pollenschlauch in die Eizelle übertreten und an der Befruchtung teilnehmen, besteht eine ziemliche Variabilität. In jedem Fall wird außer dem befruchtenden Spermakern auch noch das Plasma der Spermazelle übertragen. Vielfach wird auch der zweite Spermakern, außerdem der Pollenschlauch- und Stielkern und auch das Plasma des Pollenschlauchs in die Eizelle aufgenommen. Allerdings liegen gerade bei den „überzähligen" Elementen des Pollenschlauchinhalts auch widersprechende Angaben der Autoren vor. Wie weit in diesen Fällen auch unrichtige Beobachtungen oder eine natürliche Variabilität vorliegen, ist schwer zu entscheiden. Daß sich das Plasma der Spermazelle (bzw. Zellen) und des Pollenschlauchs mit dem Eiplasma vereinigen, ist in allen Fällen sicher. Das weitere Schicksal, insbesondere eine Beteiligung des ♂ Plasmas am Aufbau des Proembryos, ist jedoch unterschiedlich. In manchen Fällen schlüpft nämlich der Spermakern aus seiner plasmatischen Hülle aus, die in der Nähe des Halses liegenbleibt und mit ihren ganzen Inhaltskörpern, auch dem ♂ Chondriom, allmählich resorbiert wird und wandert allein bzw. höchstens von einer dünnen Plasmahaut umgeben zum Eikern. In anderen Fällen wieder wandert mit dem Spermakern auch dessen Plasma zum Eikern und umgibt dort — nachweisbar an den mitgeführten Inhaltskörpern (Stärke, Chondriosomen) — den Verschmelzungskern. Bei *Cephalotaxus drupacea* soll sogar das Chondriom beider Spermazellen (und damit deren Plasma) den Fusionskern umgeben (FAVRE-DUCHARTRE).

Auf seinem Wege zur Eizelle (aktive Bewegung?) erfährt der Spermakern, von wenigen Ausnahmen abgesehen, eine beträchtliche und überdies sehr rasche Größenzunahme. Er erreicht allerdings nur in wenigen Fällen (*Juniperus*, *Sequoia*) die Größe des Eikerns; im allgemeinen ist der Spermakern im Augenblick der Kernverschmelzung nur $^1/_3$ so groß wie der Eikern. Die auf die Annäherung der beiden Kerne folgende Phase der Kernverschmelzung verläuft nun im wesentlichen auf zwei verschiedene Arten, die allerdings nicht scharf voneinander geschieden sind, vor allem auch innerhalb eines Objekts selbst variieren können. Entweder dringt nämlich der ♂ Kern ohne wesentliche Formänderung gleichmäßig in den ♀ Kern ein, bis er schließlich fast völlig von diesem umgeben ist und die Membran des Fusionskerns überwiegend von der Membran des Eikerns gebildet wird (vgl. dazu die Aufnahmen von MCWILLIAM und MERGEN 1958) (Abb. 20). Oder die beiden Kerne lagern sich unter gegenseitiger Abplattung aneinander, so daß die Membran des Fusionskerns aus zwei ursprünglich verschiedenen Teilen zusammengesetzt ist. Bei dem Fehlen eines hinreichend großen Beobachtungsmaterials kann über die Ursachen dieses verschiedenartigen Verhaltens nichts ausgesagt werden. Erwähnung verdient noch die Beobachtung, daß eine zwischen den beiden verschmelzenden Kernen gelegene dünne Plasmaschicht vielfach in den Fusionskern mit einbezogen werden kann (FERGUSON 1904, LAWSON 1909, HAUPT 1941).

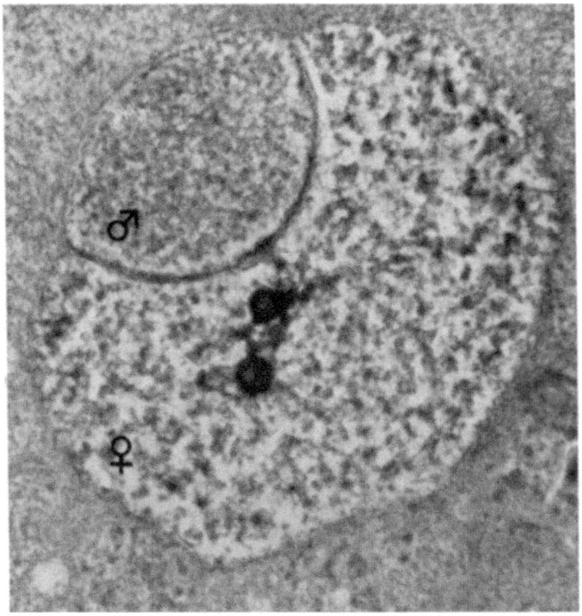

Abb. 20. *Pinus nigra*, Kernverschmelzung. Kerne unmittelbar nach der Verschmelzung, getrennt durch die doppelte Membran; zwei Nucleoli im Eikern. (Nach MCWILLIAMS und MERGEN.)

Aus der Tatsache, daß bei der ersten Mitose des Zygotenkerns, die unmittelbar nach der Auflösung der trennenden Kernmembran einsetzt, abgesehen von *Sequoia* und *Cryptomeria*, noch getrennte väterliche und mütterliche Spindeln auftreten, die sich allerdings bis zur Metaphase vereinigen, ist zu schließen, daß die beiden verschmolzenen Kerne auch in dem äußerlich einheitlich erscheinenden Fusionskern wenigstens zunächst noch ihre Individualität beibehalten. Erst mit Abschluß der ersten Mitose findet diese ihr Ende und streng genommen können wir erst dann von einem Verschmelzungskern sprechen.

Polyembryonie durch Spaltung sowie einfache Polyembryonie durch Befruchtung mehrerer Archegonien (bei *Pinus* durchschnittlich 2—3) ist bei den Coniferen häufig zu beobachten (Zusammenstellung bei CHAMBERLAIN 1935, S. 348—350). Doch gelangt nur jeweils ein Embryo zur vollen Reife.

Doppelte Befruchtung.

Die außer dem befruchtenden Spermakern in die Eizelle eingedrungenen Kerne bleiben an der Spitze der Eizelle liegen, ohne ihre Lage zu verändern und werden mehr oder weniger rasch desorganisiert. Wenn gelegentlich der zweite Spermakern länger erhalten bleibt und sich sogar gleichzeitig mit dem Zygotenkern noch teilen kann (*Taxus*, *Cephalotaxus*, FAVRE-DUCHARETR 1957, 1961), so zeigt das, daß der zweite Spermakern doch auch eine gewisse Aktivität besitzen kann. Wenn dann gleichzeitig auch der Schwesterkern des Eikerns, der Bauchkanalkern einen gametischen Charakter annimmt, kann es — als Ausnahmeerscheinung — sogar zu einer Verschmelzung des zweiten Spermakerns mit dem Bauchkanalkern, also zu einer „doppelten Befruchtung" kommen. Diese Erscheinung wurde schon mehrfach beobachtet und an ihrer Realität kann nicht gezweifelt werden; unter dem Gesichtspunkt einer Differenzierungslabilität ist sie durchaus verständlich. Eine doppelte Befruchtung wurde bisher beschrieben von *Pinus laricio* (CHAMBERLAIN 1899), *Tsuga* (MURRILL 1900), *Thuja* (LAND 1902) und *Abies balsamea* (HUTCHINSON 1915) (vgl. dazu auch HERIBERT-NIELSSON 1941 und SRIVASTAVA 1953).

D. Gnetales.

Das zunehmende Interesse, das insbesondere die Gattung *Gnetum* in den letzten Jahren gefunden hat, erstreckt sich naturgemäß in erster Linie auf entwicklungsgeschichtliche Fragen (KHAN 1941, 1943; FAGERLIND 1941; MEHRA 1950a, b; BATTAGLIA 1951, 1958; WATERKEYN 1954, 1959, 1960; NARANG 1955a, b; NEGI und LATA 1957; VASIL 1959; MAHESHWARI und VASIL 1961; SANWAL 1962). Unsere Kenntnisse über die Befruchtungsverhältnisse dieser heterogenen Gruppe sind daher noch immer ziemlich begrenzt. Da die *Gnetales* überdies bei ihrer isolierten Stellung überwiegend von phylogenetischem Interesse sind, darf auf die bisherigen zusammenfassenden Darstellungen verwiesen werden.

E. Angiospermae.

Gegenüber den Gymnospermen, wo die Pollenkörner — abgesehen von *Araucaria* und *Agathis* — direkt oder durch Vermittlung eines Bestäubungstropfens auf den Nucellus gelangen und dort zum Pollenschlauch auswachsen, sind bei den Angiospermen die Samenanlagen nicht mehr frei zugänglich. Die Pollenkörner gelangen vielmehr zunächst auf die papillöse Narbe des mit dem Fruchtknoten in direkter Verbindung stehenden Griffels und keimen dort zum Pollenschlauch aus. Erst die Spitze des Pollenschlauches erreicht die im Innern des Fruchtknotens eingeschlossenen Samenanlagen. Eine wesentliche Veränderung hat auch der ♀ Gametophyt erfahren, indem an die Stelle eines Makroprothalliums mit Archegonien der Embryosack getreten ist. Die Befruchtung selbst erfolgt jedoch auch bei den Angiospermen wieder durch Vereinigung einer Eizelle mit einer Spermazelle. Neu hinzu tritt jedoch die doppelte Befruchtung, die zur Entstehung eines Nährgewebes führt.

Auffinden der Samenanlagen durch die Pollenschläuche.

Von den die Befruchtung einleitenden Prozessen sind die Bestäubung, die Pollenkeimung und das Pollenschlauchwachstum im Griffel bereits besprochen; wir können uns also gleich dem nächsten Teilprozeß zuwenden, nämlich dem Auffinden der Samenanlagen durch die Pollenschläuche, die in der Höhlung des Fruchtknotens angelangt sind. Hier scheint innerhalb der Angiospermen eine ziemliche Mannigfaltigkeit zu herrschen.

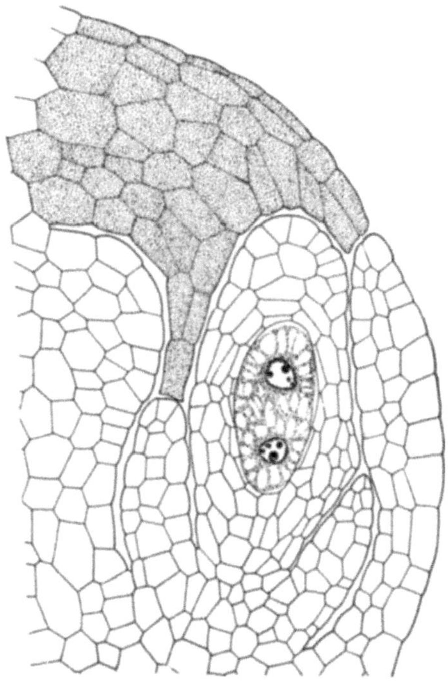

Abb. 21. *Acalypha indica*, Obturator (schraffiert) aus Placentagewebe über dem Nucellus liegend. (Nach MAHESHWARI und JOHRI.)

In allen genauer untersuchten Fällen besteht eine kontinuierliche Bahn für den Pollenschlauch von der Narbe durch den Griffel und schließlich auch von der Placenta bis zur Mikropyle. Der letzte Schritt kann dabei in der Mehrzahl der Fälle auf einer oberflächlichen Schleimbahn verlaufen oder aber auch weiterhin im Innern von Geweben erfolgen. Eine Überwindung größerer Lufträume scheint ausgeschlossen. Wo solche überhaupt vorliegen, ist der zu überwindende Abstand immer klein und wohl stets von Schleim erfüllt, oder sie sind durch besondere Gewebebildungen überbrückt.

Wenn wir zunächst den ersten Fall einer oberflächlichen Leitung auf einer Schleimbahn betrachten, so haben die Untersuchungen von STEFFEN an *Impatiens glandulifera* gezeigt, wie genau dieser Weg vorgeschrieben sein kann, so daß die Leitung der Pollenschläuche ganz zwangsläufig erfolgt. Ein Pollenschlauch, der auf seinem Weg nach abwärts auf einen Funiculus trifft, folgt dessen Lauf; der genaue Weg ist dabei dadurch gegeben, daß der Funiculus auf der nach außen gerichteten Wand eine schleimführende Epidermis besitzt, während die Epidermiszellen der abgekehrten Seite gerbstoffhaltig sind. Weiter ist an der Übergangsstelle vom Funiculus zur Samenanlage eine von besonders viel Schleim erfüllte Ringfurche ausgebildet, die den Pollenschlauch direkt zur Mikropyle hinleitet. Wasserspalten auf der inneren Fruchtblattepidermis sorgen dabei für eine gleichmäßig feuchte Atmosphäre. Die Leitung der Pollenschläuche erfolgt wahrscheinlich rein mechanisch; für eine chemotropische Anziehung liegen bei diesem Objekt keine Anzeichen vor. Dafür spricht auch das weitere Verhalten der Pollenschläuche von der Mikropyle bis zur Spitze des Embryosacks, die zwar in der Regel geradlinig in die Mikropyle eindringen, gelegentlich aber auch zwischen die beiden Integumente hineinwachsen können. Da ein Wachstum auch in abortierte Samenanlagen erfolgt, sowie in Samenanlagen, deren Synergiden durch vorausgegangene Pollenschläuche bereits zerstört sind, ist für dieses

Objekt ein (von den Synergiden ausgehender?) Chemotropismus unwahrscheinlich. Ganz anders verläuft der Weg des Pollenschlauchs bei dem völlig endotropen Wachstum, wie es erstmals für *Casuarina* nachgewiesen wurde. Bei *Casuarina* dringt der Pollenschlauch durch die unmittelbare Fortsetzung des Griffels bis zum Gefäßbündel des Fußes hinunter, wo er sich, abgelenkt durch das feste Gewebe, dem von den schlauchartigen Makrosporen aufgelockerten Gewebe zuwendet und zwischen diesen bzw. in diesen bis zum Eiapparat emporsteigt. Ähnlich liegen die Verhältnisse bei einer Reihe von Amentaceen. Im Gegensatz zu den mehr oder weniger klar begrenzten Bahnen beim ektotropen

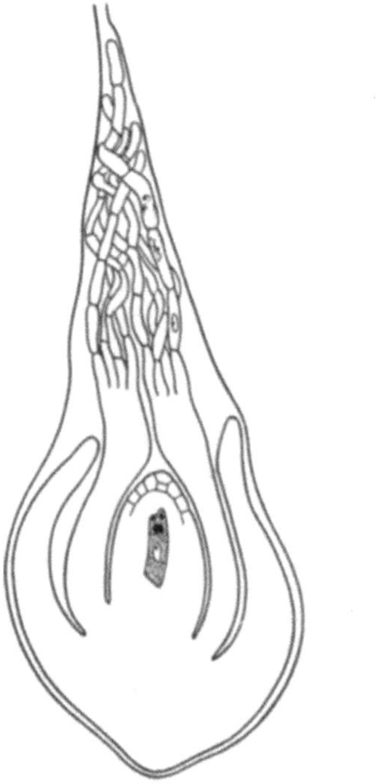

Abb. 22. *Myriocarpa longipes*, Obturator aus Zellen des inneren Integuments. (Nach FAGERLIND.)

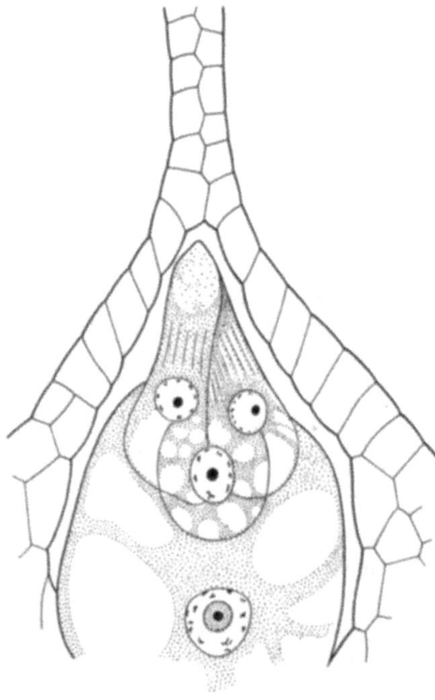

Abb. 23. *Polygonum persicaria*, reifer Embryosack mit einem Teil der verlängerten Nucellusspitze. (Nach SOUÈGES.)

Wachstum der Pollenschläuche auf einer Schleimbahn ist für die Chalazogamie bei dem Fehlen eines ausgesprochenen Leitgewebes eine starke Variabilität im Lauf der Pollenschläuche charakteristisch.

Wegen der weiteren zahlreichen Einzelheiten muß auf die Handbücher verwiesen werden (neuere Zusammenfassung bei STEFFEN in MAHESHWARI 1963, S. 105—133). Hier sei nur auf einen Punkt noch hingewiesen. So tiefgreifend die Unterschiede zwischen diesen beiden Wachstumsweisen der Pollenschläuche erscheinen, so darf aber nicht übersehen werden, daß bei manchen Objekten ektotropes und endotropes Pollenschlauchwachstum nebeneinander vorkommen können, etwa bei *Brassica oleracea* (THOMPSON 1933). Gewöhnlich dringt der Pollenschlauch bei *Brassica* durch die Mikropyle ein; gelegentlich kann er aber auch von der Chalaza aus den Embryosack erreichen (weitere Beispiele siehe MAHESHWARI 1950, S. 186).

Eine weitere Möglichkeit, den Übergang von dem Leitgewebe zur Mikropyle herzustellen, besteht in der Ausbildung von besonderen Geweben. Als erstes ist hier zu nennen der Obturator, der eine Wucherung der Placenta darstellt (Abb. 21).

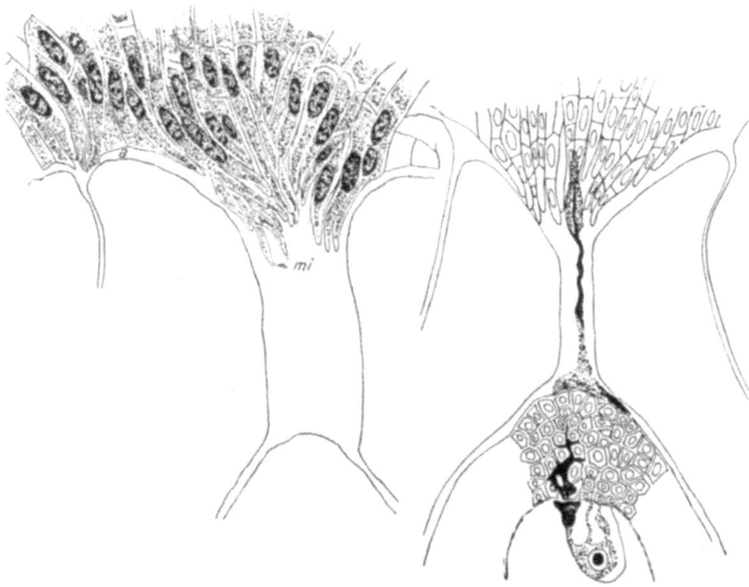

Abb. 24. *Daphne laureola.* Links Teil des Fruchtknotens mit Zellen des Obturators, die sich nach abwärts in die Mikropyle einschieben; rechts weiter fortgeschrittenes Stadium mit Pollenschlauchverlauf. *mi* Mikropyle. (Nach FUCHS.)

Das Gewebe des Obturators schiebt sich dabei als endotropes Leitgewebe zwischen Placenta und Nucellus ein, so daß beide in direkte Berührung treten, wobei entweder der Obturator in die Mikropyle einwächst oder der Nucellus dem Obturator entgegenwächst *(Euphorbiaceae).* Ein derart zwischengeschaltetes Leitgewebe kann aber auch ausgehen von der Samenanlage und zwar, indem sich die Zellen des inneren Integuments nach oben verlängern und als lockeres Leitgewebe in den Griffelkanal eindringen *(Myriocarpa)* (Abb. 22). Noch eigentümlicher ist die Ausbildung eines Leitgewebes bei *Polygonum*-Arten, wo die Spitze des Nucellus in einen langen, aus zwei Zellreihen gebildeten Fortsatz auswächst, der aus der Mikropyle heraus- und ziemlich weit in den Griffelkanal hineinragt Abb. 23).

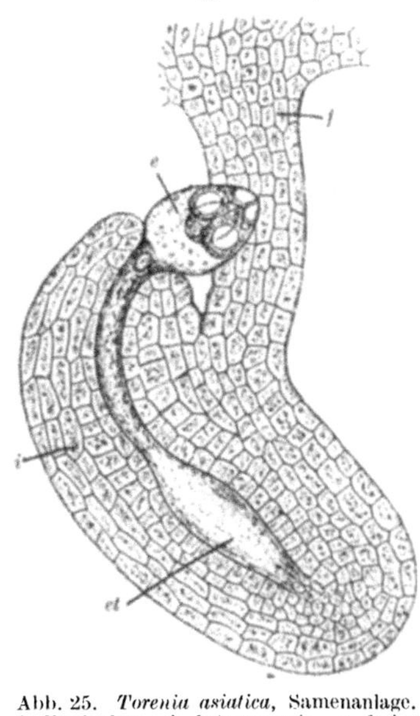

Abb. 25. *Torenia asiatica,* Samenanlage. *f* Funiculus; *i* Integument; *e* freier Embryosackscheitel; *et* dessen im Innern der Samenanlage erweiterter Teil. (Nach ENGLER.)

Ein Leitgewebe kann aber auch in der entgegengesetzten Richtung ausgehen von der Basis des Griffelkanals, der ein dichtes Büschel papilllöser Zellen durch die Mikropyle dem Nucellus entgegenwachsen läßt *(Thymelaeaceae)* (Abb. 24). Hierher gehören auch die Haarbildungen am Funiculus, die häufig auf der der Mikropyle zugekehrten Seite des Funiculus auftreten und auf die Mikropyle zugerichtet sind (*Didierea, Ruta graveolens* usw.).

Alle bisher erwähnten Einrichtungen erleichtern den Pollenschläuchen den Weg zu den Samenanlagen bzw. Embryosäcken. Eine gänzlich andere Möglichkeit, das Auffinden von Pollenschläuchen und Embryosäcken zu ermöglichen, besteht schließlich darin, daß die Embryosäcke den Pollenschläuchen entgegenwachsen. Ein schrittweises Auswachsen der Embryosäcke in Richtung auf die Mikropyle zu ist bei einer ganzen Reihe von Objekten bekannt (MAHESHWARI, S. 142). Bei *Torenia asiatica* z. B. wächst der Embryosack aus der Mikropyle heraus, so daß das bauchig angeschwollene Ende des Embryosacks mit den Synergiden und der Eizelle offen dem Pollenschlauch zugänglich ist (Abb. 25). Bei *Galium lucidum* liegt der Embryosack vollständig außerhalb der Mikropyle, bei *Vitis pallida* tritt der Embryosack aus der Mikropyle heraus, preßt sich der Fruchtknotenwand an und dringt sogar in deren Epidermis ein. Noch weiter gesteigert ist dieses Verhalten bei den *Santalaceae* und *Loranthaceae*, etwa bei *Scurrula atropurpurea* und *Dendrophthoe pentandra*. Bei diesen verlängern sich die Embryosäcke während ihrer Entwicklung stark nach unten und oben; während sie am unteren Ende aber durch eine Kollenchymschicht aufgehalten werden, dringen sie nach oben mit ihrer Spitze aus dem Fruchtknoten bis in das Griffelgewebe hinein. Bei *Helixanthera ligustrina* schließlich gelangen die Embryosäcke zum Teil sogar bis unter die Narbenepidermis. Der normal gebaute Eiapparat kommt dabei erst im Griffelkanal zur Ausbildung.

Der männliche und weibliche Gametophyt.

Die Entwicklung und der Bau des ♂ und ♀ Gametophyten sind wiederholt zusammenfassend dargestellt worden (zuletzt bei VAZART 1958; STEFFEN in MAHESHWARI 1963, S. 15—40; JOHRI ebenda, S. 69—103), so daß darauf verwiesen werden kann. Nur die für den Zusammenhang wichtigsten Punkte seien in knappster Form zusammengestellt.

Der männliche Gametophyt.

Die Entwicklung des ♂ Gametophyten verläuft innerhalb der Angiospermen außerordentlich einförmig. Die aus der Meiose hervorgehende Gone erfährt stets zwei Teilungen: Durch die erste Teilung entsteht die vegetative und die generative Zelle, im zweiten Teilungsschritt, der entweder noch im Pollenkorn oder erst im Pollenschlauch stattfindet, teilt sich die generative Zelle in die zwei Spermazellen. Zu Beginn der ersten Teilung wird der anfänglich zentral gelegene Kern einseitig verlagert, so daß in einer inäqualen Teilung zwei ungleich große Zellen entstehen, die kleinere, linsenförmige, generative Zelle und die größere vegetative Zelle, deren Kerne sich ebenfalls deutlich in ihrer Größe und Struktur voneinander unterscheiden. Die inäquale erste Pollenkornmitose führt zwar auch zu einer zahlenmäßig inäqualen Verteilung der

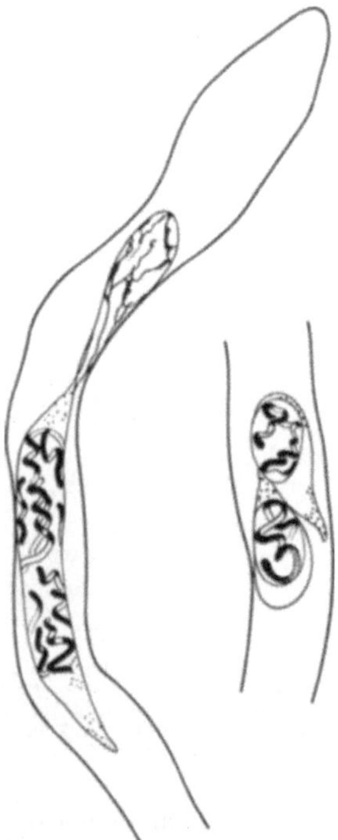

Abb. 26. *Galanthus nivalis*, lebend beobachtete Pollenschläuche mit Verteilung der Tröpfchenscheide bei der Kernteilung. Links generativer Kern in Prophase, Tröpfchenscheide in Form einer Doppelkappe; die im Bild untere Kappe läßt eine Verlagerung der Tröpfchen erkennen. Vor (im Bild oberhalb) der generativen Zelle der vegetative Kern. Rechts Spermazellen kurz nach der Durchschnürung des vegetativen Plasmas. Die im Bild obenliegende Spermazelle hat die untere passiert. Ihre Tröpfchenscheide befindet sich gerade in Verteilung, die größere Tröpfchenzahl liegt noch im schwanzartig ausgezogenen Teil, dem früheren Ende der generativen Zelle. (Nach STEFFEN.)

Plasmapartikel; die Verteilung der Plasmapartikel pro Plasmaeinheit wird jedoch nicht geändert und auch das Verhältnis Proplastiden: Chondriosomen + Sphärosomen innerhalb der generativen Zelle entspricht dem in der vegetativen Zelle (STEFFEN und LANDMANN 1958a, b).

Die zunächst noch an der Peripherie des Pollenkorns gelegene generative Zelle wird nun in das Innere des Pollenkorns, d.h. in das Innere der vegetativen Zelle verlagert; sie nimmt dabei spindelförmige Gestalt an und ist durch eine deutliche Plasmahaut scharf begrenzt (BOPP-HASSENKAMP 1960). Noch innerhalb des Pollenkorns oder erst im Pollenschlauch teilt sich die generative Zelle in zwei gleich große Spermazellen. An wesentlichen Einschlüssen sind in den Spermazellen regelmäßig Plastiden und Chondriosomen vorhanden (STEFFEN und LANDMANN 1958a, b; RICHTER-LANDMANN 1959; BOPP-HASSENKAMP 1960; DIERS 1963); bei manchen Arten findet sich sowohl bei der generativen Zelle als auch bei den Spermazellen eine sog. Tröpfchenscheide (Abb. 26). Sie stellt ein Vacuom dar und kann osmotisch Wasser aufnehmen. Ihre Bedeutung wird von STEFFEN (1953) in einer osmo-regulatorischen Funktion zur Erhaltung des kolloidalen Zustandes des generativen Plasmas und in einer Stabilisierung beim Eintritt der Spermazellen in den Embryosack gesehen.

Der weibliche Gametophyt.

Trotz der verschiedenartigen Entstehung des Embryosacks (mono-, di- oder tetraspor) ist der Bau des befruchtungsfähigen Embryosacks bei der Mehrzahl der Angiospermen auffallend einheitlich. Wenn wir von den wenigen Fällen von

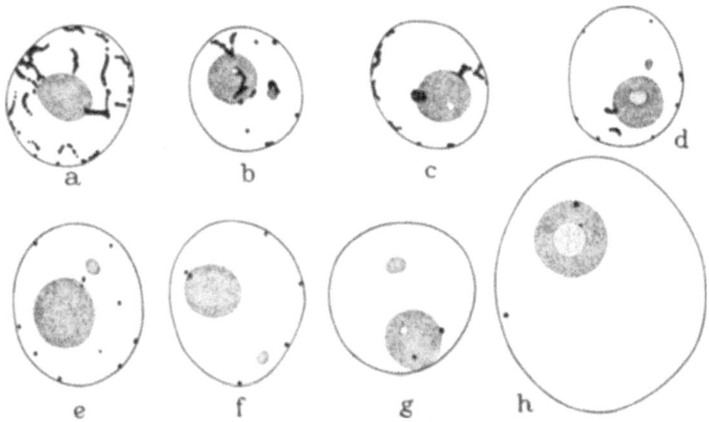

Abb. 27a—h. *Impatiens glandulifera*, Reifung des Eikerns. Die Eikerne sind gleichsinnig orientiert (obere Seite = mikropylare Seite). Feulgensche Nuclealreaktion. a Letztes Stadium der Kernwachstumsphase; Streckung der Chromatinfäden und Auftreten von Chromomeren; b—h Nuclealitätswechsel; c Stadium der 9 Restchromomeren; g typischer befruchtungsreifer Kern mit Restnuclealität; h hypertrophierter Kern. (Nach STEFFEN.)

Peperomia, *Plumbago*, *Plumbagella* und *Acalypha indica* absehen, wo der Eiapparat nur aus der Eizelle bzw. aus der Eizelle und einer Synergide besteht, haben wir in allen anderen Fällen wenigstens einen normalen dreizelligen Eiapparat, meist ist der Embryosack einheitlich aus dem dreizelligen Eiapparat, den drei Antipoden und den zwei Polkernen aufgebaut.

Der Eiapparat besteht aus der Eizelle und den beiden Synergiden; alle drei Zellen sind von einer festen Membran begrenzt. In der gewöhnlich weiter in den Embryosack hineinragenden Eizelle liegt der Kern mit seinem Plasma am unteren Ende, an der Spitze liegt die große Vacuole. Bei den Synergiden umgekehrt befindet sich eine große Vacuole am unteren Ende, das mikropylare Ende wird vom sog. Fadenapparat (Synergidenkappe) eingenommen. An einer Reihe von Objekten wurde beobachtet, daß Eikern und Polkerne vor der Befruchtung auffallende Veränderungen erfahren („Reifungsprozeß"), die zu einer fortschreitenden Nuclealitätsverminderung bzw. sogar zu völliger Anuclealität führen können; doch scheint diese nur dadurch bedingt zu sein, daß die Konzentration der DNS unterhalb der Nachweisgrenze der normalen Feulgen-Reaktion liegt (Abb. 27).

An Inhaltskörpern finden wir in der Eizelle (wie auch im Embryosack) Plastiden, die vielfach auch Stärke bilden können sowie Mitochondrien, so daß also diese beiden Elemente von Anfang an in die Zygote gelangen.

Eintritt des Pollenschlauchs in den Embryosack.

Die genaue Art des Eintritts des Pollenschlauchs in den Embryosack bereitet der Beobachtung gewisse Schwierigkeiten, so daß vielfach nur allgemeine Beobachtungen über den Ort des Eindringens, weniger aber über den wirklichen Vorgang selbst vorliegen. Allgemein gilt zunächst, daß der Pollenschlauch niemals direkt in die Eizelle eindringt, sondern sich stets in den Embryosack ergießt. Der Weg des Pollenschlauchs in den Embryosack kann nun sehr verschieden sein, wobei der wesentliche Punkt die Beziehung zwischen Pollenschlauch und Synergiden ist.

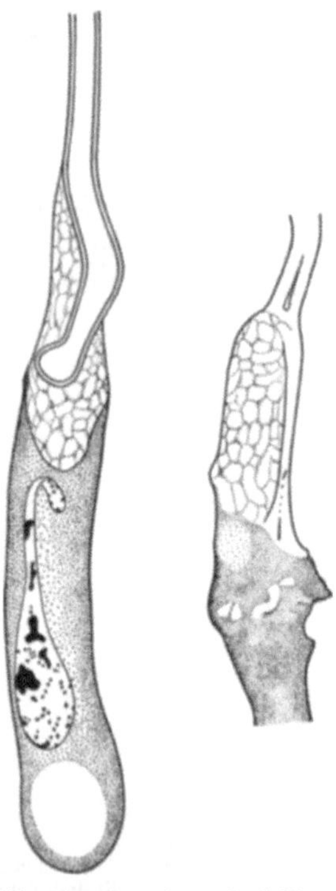

Abb. 28. *Impatiens glandulifera*, Eindringen des Pollenschlauchs in eine Synergide. Links Pollenschlauch dringt in den Fadenapparat ein; rechts mikropylarer Teil der Synergide nach Ergießen des Pollenschlauchplasmas, Pollenschlauch durch Callose abgeriegelt. (Nach STEFFEN.)

In vielen Fällen nimmt der Pollenschlauch seinen Weg zwangsläufig auf die Synergiden zu, wenn nämlich die Mikropyle die Form eines engen Kanals hat, dessen unteres Ende von den gegen die Mikropyle zu vorgewölbten Synergidenkappen verschlossen ist, wie etwa bei *Impatiens glandulifera* (STEFFEN 1951). Der auf den Fadenapparat einer Synergide auftreffende Pollenschlauch durchdringt diesen, bis er mit dem eigentlichen Synergidenplasma unterhalb des Fadenapparates in direkte Berührung kommt. Jetzt erst öffnet sich die Pollenschlauchspitze und ergießt ihren Inhalt in die Synergide, was stets ein Kollabieren des Synergidenkerns zur Folge hat (Abb. 28). Das Pollenschlauchwachstum ist damit beendet und nun übernimmt die Synergide die Weiterleitung der Gameten, indem nun ihrerseits die Synergide an ihrem vacuoligen, gegen den Embryosack gerichteten Ende platzt und mit mehr oder weniger großer Vehemenz die ♂ Gameten in den Raum zwischen Eizelle und Polkern schleudert. Unter der Annahme, daß die osmotischen Verhältnisse im Pollenschlauch und in der Synergide voneinander verschieden sind, kann das Öffnen des Pollenschlauchs in der Synergide als Plasmoptyse durch zu rasche Wasseraufnahme aufgefaßt werden; durch die Übernahme des Pollenschlauchinhalts können nun aber auch die ursprünglichen osmotischen Verhältnisse der Synergide gegenüber denen des Embryosacks im gleichen Sinne verändert werden, so daß jetzt auch die Synergide durch rasche Wasseraufnahme platzt und ihren Inhalt in den Embryosack ergießt.

Daß die Synergiden wesentlich für das Öffnen des Pollenschlauchs sind, ergibt sich daraus, daß nachfolgende Pollenschläuche durch bereits zerstörte Synergiden hindurch in den Embryosack einwachsen können, ohne sich zu öffnen, was allerdings die weitere Annahme notwendig macht, daß nunmehr im Embryosack andere osmotische Verhältnisse herrschen als vor der ersten Befruchtung. Erscheint in diesen Fällen die Funktion der Synergiden, das Öffnen und Entleeren der Pollenschläuche in den Embryosack zu bewirken, plausibel, so läßt sich dieser Befund jedoch nicht verallgemeinern. Denn wir kennen auch zahlreiche Fälle, wo die Synergiden an der Öffnung des Pollenschlauchs nicht beteiligt sind, wenn wir ganz absehen von den Fällen, wo der Embryosack überhaupt keine Synergiden besitzt, wie beim *Plumbagella*- und *Plumbago*-Typ. So tritt der Pollenschlauch bei *Fagopyrum* zwischen Eizelle und Synergiden in den Embryosack, ohne daß die Synergiden verletzt werden, oder zwischen Embryosackwand und den Synergiden bei *Cardiospermum*. Aber auch wenn die Synergiden direkt gar nicht beteiligt erscheinen, können sie trotzdem in manchen Fällen durch den Eintritt des Pollenschlauchs beeinflußt werden; unter entsprechenden räumlichen Verhältnissen könnte man allerdings auch nur an einen rein mechanischen Druck des Pollenschlauchs denken, der die Eizelle jedoch nicht erfaßt, weil diese tiefer liegt. Häufig zeigen nämlich die Synergiden nach dem Pollenschlauchdurchtritt eine geänderte Färbbarkeit und beschleunigte Degeneration des Kerns.

Zu der Variabilität innerhalb der verschiedenen Arten kommt dann aber auch noch eine gewisse Variabilität innerhalb ein und derselben Art, wie sie für *Coffea* beschrieben wurde und vielleicht bei hinreichendem Material auch bei anderen Objekten zu finden sein wird. Wenn wir schließlich dazu noch die Fälle nehmen, wo zur Zeit der Befruchtung die Synergiden bereits degeneriert sind, dann macht es diese Variabilität in den Beziehungen zwischen Pollenschlauch und Synergiden schwer, die Rolle der Synergiden bei der Befruchtung eindeutig zu klären. Vielleicht kann aber diese Frage nach der Bedeutung der Synergiden ganz allgemein gar nicht gestellt werden, indem ihnen eine solche Bedeutung gar nicht (mehr ?) in allen Fällen zukommt.

Zur Öffnung der Pollenschläuche ist noch zu erwähnen, daß sich in einigen Fällen der Pollenschlauch in zwei Äste gabelt, die gegen die Eizelle bzw. gegen die Polkerne zu gerichtet sind und je eine Spermazelle entlassen (*Peperomia:* FAGERLIND 1939; *Portulaca, Phryma:* COOPER 1940, 1941), während bei *Coffea* (MENDES 1941) zwei getrennte subterminale Öffnungen des Pollenschlauchs vorliegen.

Chemotropismus der Pollenschläuche.

Die Frage nach der Rolle der Synergiden bei der Befruchtung hängt eng zusammen mit der Frage nach einer chemotropischen Anlockung der Pollenschläuche durch die Samenanlagen, womit zunächst aber nur ganz allgemein die Tatsache ausgedrückt werden soll, daß die Pollenschläuche die Samenanlagen bzw. die Eizellen auffinden. Obgleich nämlich in vielen Fällen eine Reihe von Vorrichtungen eine rein mechanische Leitung der Pollenschläuche zu den Samenanlagen bzw. durch die Mikropyle zum Embryosack gewährleisten, so gibt es andererseits eine Reihe von Fällen — es mag genügen, auf die endotrop wachsenden Pollenschläuche hinzuweisen —, wo eine rein mechanische Leitung allein unzureichend erscheint und deshalb eine chemotropische Anlockung nicht nur vom Embryosack als ganzem, sondern von der Spitze des Embryosacks — und damit vielleicht von den Synergiden — gefordert werden muß.

Irgendwelche gesicherten experimentellen Untersuchungen, die als einwandfreier Beweis für diese Annahme gelten könnten, liegen jedoch bis heute noch nicht vor. Auch die in vitro-Versuche von Tsao (1949) lassen wohl nicht mehr als die Möglichkeit eines Chemotropismus zu. Wesentliche Hinweise zu dieser Frage aus neuerer Zeit stammen dagegen von genetischer Seite. Der Ausfall bestimmter Kreuzungen zwischen Oenotheren der Sektion *Raimannia* lassen sich zwanglos erklären unter der Annahme, daß Samenanlagen mit genetisch verschiedenen Embryosäcken Pollenschläuche bestimmter genetischer Konstitution in stark wechselndem Maße anziehen (vgl. dazu den Beitrag „Selektive Befruchtung"). Aussagen über die Art der wirksamen Stoffe und den Ort ihrer Bildung lassen sich aus diesen Untersuchungen naturgemäß nicht gewinnen. Die bisherigen Versuche zur Identifizierung der für eine chemotropische Anziehung möglicherweise verantwortlichen Stoffe (Schildknecht und Benoni 1963) können wegen der Unzulänglichkeit des biologischen Tests nicht als beweisend angesehen werden.

Verhalten der Spermazellen im Embryosack.

Die mit dem Übergang in ein anderes Medium verbundene Schockwirkung hat vielfach eine Störung in der Struktur der Spermakerne beim Eintritt in den Embryosack zur Folge, die sich in mehr oder weniger starken Krümmungen des Kerns ausdrückt, sich aber allmählich wieder normalisiert. Daß diese Strukturänderungen und die nachfolgende Normalisierung rein physiologisch bedingt sind, ergibt sich auch daraus, daß nicht nur die zur Befruchtung gelangenden Spermazellen diese Erscheinung zeigen, sondern auch die in manchen Fällen übertragenen überzähligen Spermazellen eines nachfolgenden Pollenschlauchs (Steffen 1951).

Eine auffallende und anscheinend allgemein verbreitete Erscheinung ist dann die Größenzunahme der Spermakerne nach ihrem Eintritt in den Embryosack. Irgendwelche Unterschiede zwischen den beiden Spermazellen bzw. deren Kernen sind jedoch nicht festzustellen. Da beide zunächst stets gepaarten Spermazellen sich nach ihrem Eintritt in den Embryosack in verschiedenen Richtungen bewegen, der gelegentlich mit entleerte vegetative Kern keine Lageänderung zeigt, ist der Schluß auf eine Eigenbeweglichkeit der Spermazellen berechtigt. Wieweit diese durch eine Plasmaströmung unterstützt oder sogar ersetzt werden kann, bleibt aber offen (Steffen 1951, 1953). Auf alle Fälle erfolgt die Bewegung der Spermazellen sehr rasch, da nur selten Spermazellen auf dem Wege zu den ♀ Partnern beobachtet werden. Daß die ♂ Gameten im Embryosack tatsächlich stets ganze Zellen sind, dürfte außer Zweifel sein.

Plasmaverschmelzung (Plasmogamie).

Ebenso wie die Bewegung der Spermazellen zur Eizelle bzw. zu den Polkernen, hat sich auch die Vereinigung der Spermazelle mit der Eizelle (und bis zu einem gewissen Grad die Plasmogamie zwischen der zweiten Spermazelle und dem Embryosackplasma) bisher der Beobachtung entzogen, wiederum wohl wegen der Geschwindigkeit, mit der sich dieser Vorgang abspielt. Die beiden einzigen Fälle, bei denen der Durchtritt beobachtet wurde (*Lilium martagon:* Heimans 1928 und *Antirrhinum majus:* Anderson 1936), scheinen zweifelhaft (Steffen 1951). Über den genauen Ablauf und den Mechanismus der Syngamie bestehen daher nur Vermutungen. Einleuchtend erscheint die Vorstellung von Wylie (1941), wonach die Berührungsmembran der beiden nackten Protoplasten plötzlich zusammenbricht und damit die Vereinigung in einem einzigen Akt hergestellt ist. Der Nachweis des ♂ Plasmas in der Eizelle wird dabei davon abhängen, wie rasch die anschließende Angleichung der beiden Cytoplasmen vor sich geht; sie wird vor allem dann möglich sein, wenn aus irgendwelchen Gründen der Viscositätsausgleich verlangsamt ist. Tatsächlich gelang es STEFFEN bei *Impatiens glandulifera* erstmals, die Übertragung des ♂ Cytoplasmas auch cytologisch nachzuweisen, nachdem aus genetischen Gründen diese Tatsache schon längst selbstverständlich war. Im Gegensatz dazu beobachtet RICHTER-LANDMANN (1958, 1959), ebenfalls an *Impatiens glandulifera* das „Durchtreten" intakter Spermazellen in die Eizelle, die auch im Eizellplasma intakt erhalten bleiben. Erst nachdem sich die Spermazelle eng an den Eikern angelegt hat und die Kernverschmelzung beendet ist, erfolgt die Plasmaangleichung.

Grundsätzlich die gleichen Faktoren werden auch für die Plasmogamie der zweiten Spermazelle mit dem Embryosackplasma maßgebend sein, lediglich mit dem Unterschied, daß die Angleichung der beiden Cytoplasmen wesentlich langsamer erfolgt; denn wir beobachten ja beide Spermazellen mit ihrer Membran im Embryosack, bis sich die erste Spermazelle mit der Eizelle vereinigt hat. Erst mit der vollzogenen Angleichung der beiden Cytoplasmen verschwindet die Begrenzung der zweiten Spermazelle und damit hat auch das Plasma der zweiten Spermazelle seine Individualität verloren. Sowohl in der Eizelle als auch im Embryosack liegen mit Abschluß der Plasmogamie die Sexualkerne in einem physiologisch einheitlichen Cytoplasma, in dem nun die Kernverschmelzung stattfindet bzw. bereits stattgefunden hat (RICHTER-LANDMANN 1959).

Kernverschmelzung (Karyogamie).

Die Spermakerne sind im allgemeinen viel kleiner als die Eikerne; sie vergrößern sich jedoch oft ganz beträchtlich auf ihrem Weg zum Eikern. Sie sind in diesem Stadium von sehr verschiedener Form; häufig eiförmig bis nahezu kugelig, in anderen Fällen wieder langgestreckt und dann oft wurmförmig gekrümmt. Doch wird man auf Unterschiede in der Form der Spermakerne kein allzu großes Gewicht legen dürfen. Besonders auffallend und wichtig scheinen aber gewisse Strukturänderungen der Spermakerne zu sein, die völlig den Vorgängen entsprechen, die sich bei der Bildung des befruchtungsfähigen Eikerns abgespielt haben. Mit der Anlagerung des Spermakerns an den Eikern beobachten wir nämlich auch im Spermakern einen Verlust der Nuclealität, der jedoch bei *Impatiens glandulfiera*, wo diese Verhältnisse besonders eingehend untersucht wurden (STEFFEN 1951), nicht ganz soweit geht, wie im Eikern, indem nämlich alle 9 Rest-Chromozentren erhalten bleiben. Daß dieser „Reifungsprozeß" aufs engste mit der Befruchtung zusammenhängt, ergibt sich daraus, daß überzählige Spermakerne, die nicht mit einem ♀ Kern in Berührung kommen, diesen Prozeß nicht durchlaufen, sondern ihre Nuclealität behalten.

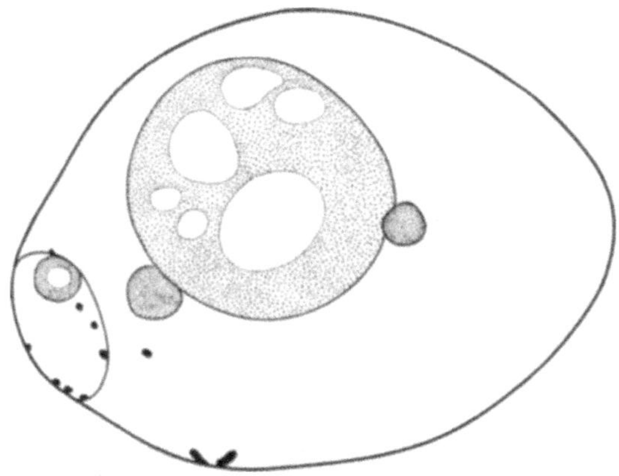

Abb. 29. *Impatiens glandulifera*, Befruchtung des sekundären Embryosackkerns. Der Spermakern ist eingesunken; Kernmembran noch sichtbar, Chromomerenstruktur deutlich. Nucleolus in Bildung begriffen. Im sekundären Embryosackkern beginnt (unten links) die Regeneration der Thymonucleinsäure. (Nach STEFFEN.)

Erst jetzt erfolgt die Karyogamie. Die Verschmelzung beginnt damit, daß der bei der Anlagerung abgeplattete Spermakern unter Erhaltenbleiben der beiden einander berührenden Kernmembranen in den Eikern einsinkt. Nach RICHTER-LANDMANN (1958) beginnt die Karyogamie bei *Impatiens glandulifera* mit der Anlagerung der intakten Spermazelle an den zu befruchtenden Kern. In noch ungeklärter Weise tritt der in Richtung auf den ♀ Kern verlagerte Spermakern durch die ♂ Zellwand hindurch und sinkt allmählich — ohne ♀ Plasma berührt zu haben — in den ♀ Kern ein.

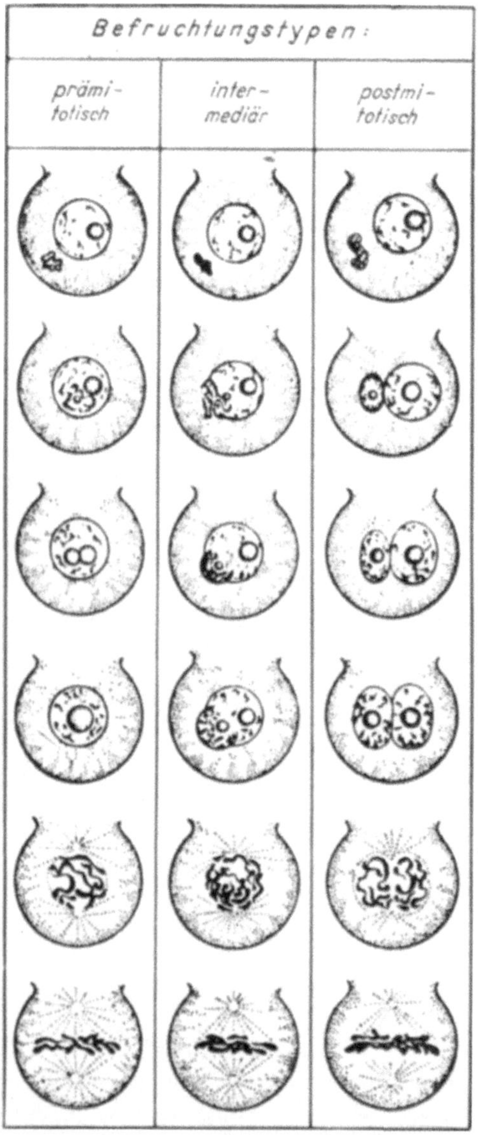

Abb. 30. Befruchtungstypen. (Nach GERASSIMOVA-NAVASHINA 1961.)

Die nun folgenden Vorgänge spielen sich so rasch ab, daß sie im einzelnen schwer zu verfolgen sind. Die Teilbeobachtungen fügen sich jedoch gut in die Vorstellung ein, daß die einander berührenden Membranen von einer Stelle ausgehend sich auflösen und daß diese so entstandene Öffnung sich rasch erweitert und dadurch der ♂ und ♀ Kern in direkte Berührung miteinander treten. Das aufgelöste Teilstück der Eikernmembran wird also gewissermaßen durch ein entsprechendes Stück der Spermakernmembran ersetzt. Mit dem vollzogenen Abbau der trennenden Membranen wird dann auch die anfängliche Vorwölbung des Spermakerns eingeebnet. Keinesfalls erfolgt jedoch die Karyogamie in der Weise, daß der ♂ Kern die Membran des Eikerns durchbricht und direkt in den Eikern eindringt.

Nach elektronenmikroskopischen Beobachtungen dagegen (JENSEN 1964) verläuft die Karyogamie in der Weise, daß zunächst Teile des endoplasmatischen Reticulums, die mit den Membranen der beiden einander genäherten Kerne in Verbindung stehen, sich vereinigen, die so entstandenen Membranbrücken sich verkürzen, bis die inneren Kernmembranen miteinander in Berührung kommen und verschmelzen und eine direkte Verbindung zwischen beiden Kernen hergestellt ist. Schließlich vergrößern sich diese Kernbrücken und verdrängen dabei das noch zwischen den beiden Kernen liegende Cytoplasma mit seinen Organellen.

Eine gewisse Lagebeziehung zwischen dem eindringenden Spermakern und dem Nucleolus des Eikerns, die in manchen Fällen zu beobachten ist, deutet wohl nicht das Vorhandensein einer bestimmten Eindringstelle an, sondern ist umgekehrt als Reaktion des Nucleolus auf das Eindringen des Spermakerns zu deuten, der sich vor diesem in die entgegengesetzte Richtung zurückzieht.

Mit dem Abschluß der Karyogamie durchlaufen nun sowohl der Eikern als auch der Spermakern, jetzt aber in umgekehrter Richtung einen Nuclealitätswechsel (Abb. 29), indem jetzt die DNS wieder mit der Feulgen-Reaktion nachweisbar wird (nähere Einzelheiten s. VAZART 1958). Dabei ist auch zu beobachten, daß keine völlige Durchmischung der beiden Kerne stattfindet. In grundsätzlich der gleichen Weise erfolgt auch die Karyogamie zwischen den Polkernen und dem zweiten Spermakern; jedoch verläuft hier der ganze Prozeß mit größerer Geschwindigkeit als in der Eizelle und ist eher abgeschlossen, obwohl der Eikern als erster vom Spermakern erreicht wird.

In den letzten Jahren hat sich jedoch, vor allem durch die Arbeiten von GERASSIMOVA-NAVASHINA gezeigt, daß die Verhältnisse hinsichtlich der Karyogamie insofern nicht einheitlich liegen, als wir zwischen einem — weit verbreiteten — „prämitotischen" Typ *(Compositae, Gramineae)*, wo die Kernverschmelzung unmittelbar nach der Berührung der beiden Sexualkerne erfolgt, und einem „postmitotischen". Typ *(Liliaceae)* zu unterscheiden haben, wo die Kernfusion erst dann stattfindet, wenn die beiden Sexualkerne die Mitose begonnen haben, die dann nach der Vereinigung als erste Zygotenmitose weitergeführt wird. Zwischen beiden vermittelt ein „intermediärer" Typ, wo der Spermakern zwar nach der Berührung mit dem Eikern in diesen aufgenommen wird, die endgültige Vereinigung der beiden Geschlechtskerne aber erst mit der ersten Zygotenmitose abgeschlossen ist (GERASSIMOVA-NAVASHINA 1960, 1961) (Abb. 30).

Korrelative Beziehungen.

Für abnorme Entwicklungsgänge, wie Pseudogamie, Adventivembryonie usw. ist eine Entwicklungsanregung durch die Bestäubung in vielen Fällen bekannt. So wird z.B. durch das Pollenschlauchwachstum bei *Alchemilla argyrophylla* (GENTCHEFF und GUSTAFSSON 1940) die Bildung aposporer Embryosäcke stimuliert, womit deren Aufgabe abgeschlossen ist; die weitere Entwicklung erfolgt autonom.

Weniger bekannt bzw. untersucht sind solche korrelativen Beziehungen zwischen Bestäubung bzw. Pollenschlauchwachstum und der „normalen" Entwicklung der ♀ Fortpflanzungsorgane. Wie bei den Gymnospermen scheint auch bei manchen Angiospermen die Anlegung bzw. die Weiterentwicklung der Samenanlagen von dem durch die Bestäubung ausgeübten Reiz abzuhängen. So sind bei *Quercus*, *Fagus* und *Corylus* zur Zeit der Bestäubung überhaupt noch keine Samenanlagen vorhanden. Für *Corylus* ist bekannt, daß die Weiterentwicklung des Fruchtknotens unterbleibt, wenn bei verfrühtem Stäuben der ♂ Kätzchen die Narben der ♀ Blüten noch nicht entwickelt sind (GOEBEL 1901). Experimentelle Untersuchungen darüber liegen jedoch nicht vor. Auch bei Orchideen sind zur Zeit der Bestäubung die Samenanlagen noch ganz rudimentär und ihre Weiterentwicklung hängt von einer erfolgreichen Bestäubung ab. Ähnlich verhalten sich auch einige Dikotylen, z.B. *Fraxinus*, *Forsythia* und *Syringa dubia*.

Daß die Möglichkeiten solcher korrelativen Beziehungen noch viel weiter gehen können, zeigt die Beobachtung an *Coffea arabica* (MENDES 1941). Der Befruchtung geht bei *Coffea* immer die Verschmelzung der Polkerne voraus.

Diese Verschmelzung scheint nun durch die Bestäubung beschleunigt zu werden: denn in unbestäubten Blüten wurden sogar 15 Tage nach der Blütenöffnung die Kerne noch getrennt gesehen.

Polyspermie.

Im Gegensatz zu allen bisherigen cytologischen und genetischen Befunden wurde in den letzten Jahren (im wesentlichen seit 1948) vor allem von russischen Forschern auf das Vorkommen von Polyspermie, und zwar als Norm jeder Befruchtung hingewiesen und der Nachweis für die Richtigkeit dieser Vorstellungen durch eine Reihe von Untersuchungen über erfolgreiche Bestäubungen mit Pollengemischen genetisch verschiedener Pollen zu erbringen versucht. Folgende Vorstellungen wurden dabei entwickelt: 1. In eine Eizelle können mehrere Spermien eindringen; ein Spermakern verschmilzt mit dem Eikern, die übrigen Spermien werden von der Eizelle assimiliert. Die hierdurch hervorgerufenen Veränderungen ihrer physikalischen und chemischen Eigenschaften können sich auch auf die Erbanlagen des heranwachsenden Embryos auswirken („Polyspermie"). 2. Spermien können auch in das somatische Fruchtblattgewebe eindringen; dadurch ergeben sich Veränderungen des für die Ernährung des Embryos maßgebenden Gewebes, was dann ebenfalls die Merkmalsbildung beeinflussen kann („somatische Befruchtung").

Sorgfältige genetische Untersuchungen über das Vorliegen von Polyspermie bei *Salpiglossis variabilis* (HIEMENZ 1955) und *Lupinus angustifolius* und *L. luteus* (ZACHOW 1958) haben jedoch einwandfrei ergeben, daß bei diesen drei Objekten immer nur die Merkmale *einer* der im Pollengemisch vorhandenen Vaterformen in der Nachkommenschaft eines Kreuzungskorns beobachtet werden konnten.

Auf mögliche Ursachen für die Befunde der russischen Autoren weist KAPPERT (1948) hin: Versuchsfehler (Möglichkeit der Fremdbestäubung bei Mais) und mangelnde genetische Reinheit des Versuchsmaterials. HIEMENZ schließt aus den mitgeteilten Kreuzungsergebnissen ebenfalls auf Heterozygotie des Ausgangsmaterials.

Durch Verwendung markierter (^{32}P und ^{35}S) fremder Pollenschläuche, die nicht zur Befruchtung gelangten, wurde gezeigt, daß der ^{32}P bzw. ^{35}S trotzdem in die Embryonen eingegangen waren; und zwar in einer Menge, die aus etwa 7 nicht an der Befruchtung beteiligten Pollenschläuchen stammt (EKLUNDH-EHRENBERG, v. EULER und HEVESY 1947; POLJAKOW und DMITRIEVA 1955; POLJAKOW 1955), woraus ebenfalls auf die Teilnahme mehrerer Pollenschläuche an der Befruchtung geschlossen wird. Solange jedoch der Beweis nicht erbracht ist, daß Stoffe aus dem Pollenschlauch auf keinem anderen Weg als durch direkten Transport in andere Zellen bzw. Gewebe übergehen können, kommt diesen Versuchen keine Beweiskraft zu.

Nach wie vor besteht kein Anlaß, die bisherigen Vorstellungen über die Befruchtungsvorgänge bei den Angiospermen einer grundsätzlichen Revision unterziehen zu müssen.

Literatur.

Zusammenfassende Darstellungen mit Literaturnachweisen.

MAHESHWARI, P.: An introduction to the embryology of angiosperms. New York-Toronto-London: McGraw-Hill 1950. — MAHESHWARI, P. (ed.): Recent advances in the embryology of angiosperms. Intern. Soc. of Plant Morphologists. University of Delhi 1963. — MODILEVSKY, J. S.: Cytoembryologie der Höheren Pflanzen. Kiew: Verl. d. Akad. d. Wiss. d. Ukrain. SSR. 1963 [Russisch].

PODDUBNAJA-ARNOLDI, V. A.: Allgemeine Embryologie der bedecktsamigen Pflanzen. Moskau: Verl. d. Akad. d. Wiss. d. UdSSR 1964 [Russisch].

SCHNARF, KARL: Embryologie der Angiospermen. In K. LINSBAUERS Handbuch der Pflanzenanatomie, Bd. X/2. Berlin: Gebrüder Bornträger 1929. — Embryologie der Gymnospermen. In K. LINSBAUERS Handbuch der Pflanzenanatomie, Bd. X/2. Berlin: Gebrüder Bornträger 1933. — Vergleichende Cytologie des Geschlechtsapparates der Kormophyten. Monographien zur vergleichenden Cytologie, Bd. 1. Berlin: Gebrüder Bornträger 1941.

VAZART, BERNARD: Differenciation des cellules sexuelles et fecondation chez les phanerogames. In: Protoplasmatologia, Handbuch der Protoplasmaforschung, herausgeg. von HEILBRUNN u. WEBER, Bd. VII/3a. Wien: Springer 1958.

Im Text erwähnte bzw. in den obigen Handbüchern noch nicht enthaltene Literatur.

ANDERSON, L. E.: Mitochondria in the life cycles of certain higher plants. Amer. J. Bot. **23**, 490—500 (1936).

BATTAGLIA, E.: The male and female gametophytes of angiosperms. An interpretation. Phytomorphology **1**, 87—116 (1951). — L'evoluzione del gametofito maschile dalle Gimnosperme alle Angiosperme sulla base di recenti osservazioni in *Gnetum*. Caryologia (Firenze) **11**, 217—228 (1958). — BELL, P. R.: Organelle in der Eizelle von *Dryopteris filix-mas* (L.) Schott. Naturwissenschaften **53**, 232 (1966). — BELL, P. R., and K. MÜHLETHALER: The degeneration and reappearence of mitochondria in the egg cells of a plant. J. Cell Biol. **20**, 235—248 (1964). — BOPP-HASSENKAMP, G.: Elektronenmikroskopische Untersuchungen an Pollenschläuchen zweier Liliaceen. Z. Naturforsch. **15**b, 91—94 (1960). — BRANSCHEIDT, P.: Befruchtungsphysiologische Untersuchungen an *Taxus baccata* L. Ber. dtsch. bot. Ges. **57**, 495—505 (1939). — BRENMAN, M. S. J., and J. DOYLE: The gametophytes and embryogeny of *Athrotaxis*. Sci. Proc. roy. Dublin Soc. **27**, 193—252 (1956). — BROUGH, P., and M. H. TAYLOR: An investigation of the life cycle of *Macrozamia spiralis* Miq. Proc. linn. Soc. N.S.Wales **65**, 494—524 (1940). — BRYAN, G. S., and R. J. EVANS: Types of development from the central nucleus of *Zamia umbrosa*. Amer. J. Bot. **44**, 404—415 (1957). — BURLINGAME, L. L.: The morphology of *Araucaria brasiliensis*. III. Fertilization, the embryo and the seed. Bot. Gaz. **59**, 1—39 (1915).

CAMEFORT, H.: L'acide ribonucléique dans l'oosphère et le proembryo du *Pinus Laricio* var. *austriaca*. C. R. Acad. Sci. (Paris) **243**, 2134—2137 (1956). — Sur la nature cytoplasmique des inclusions dites «vitellines» de l'oosphère du *Pinus laricio* (var. *austriaca*): étude en microscopie électronique. C. R. Acad. Sci. (Paris) **248**, 1568—1570 (1959). — Évolution de l'organisation du cytoplasme dans la cellule centrale et l'oosphère du *Pinus Laricio* Poir. (var. *austriaca*). C. R. Acad. Sci. (Paris) **250**, 3707—3709 (1960). — CHAMBERLAIN, C. J.: Oogenesis in *Pinus Laricio*. Bot. Gaz. **27**, 268—280 (1899). — Morphology of *Ceratozamia*. Bot. Gaz. **53**, 1—19 (1912). — Gymnosperms, structure and evolution. Chicago, Ill.: Chicago: University Press 1935. — COKER, W. C.: Notes on the gametophytes and embryo of *Podocarpus*. Bot. Gaz. **33**, 89—107 (1902). — COOPER, D. C.: Macrosporogenesis and embryology of *Portulaca oleracea*. Amer. J. Bot. **27**, 326—330 (1940). — Macrosporogenesis and the development of the seed of *Phryma leptostachya*. Amer. J. Bot. **28**, 755—761 (1941).

DANGEARD, P., et J. EYMÉ: Les plastes et les mitochondries dans la cellule apicale de quelques Muscinées. C. R. Acad. Sci. (Paris) **222**, 335—336 (1946). — DIERS, L.: Elektronenmikroskopische Beobachtungen an der generativen Zelle von *Oenothera hookeri* Torr. et Gray. Z. Naturforsch. **18**b, 562—566 (1963). — Bilden sich während der Oogenese bei Moosen und Farnen die Mitochondrien und Plastiden aus dem Kern? Ber. dtsch. bot. Ges. **77**, 369—371 (1964). — Elektronenmikroskopische Beobachtungen zur Archegoniumentwicklung des Lebermooses *Sphaerocarpus donnellii* Aust. Die Entwicklung des jungen Archegons bis zum Stadium der fertig ausgebildeten sekundären Zentralzelle. Planta (Berl.) **66**, 165—190 (1965a). — Elektronenmikroskopische Untersuchungen über die Eizellbildung und Eizellreifung des Lebermooses *Sphaerocarpus donnellii* Aust. Z. Naturforsch. **20**b, 795—801 (1965b). — On the plastids, mitochondria, and other cell constituents during oögenesis of a plant. J. Cell Biol. **28**, 527—543 (1966). — DOWNIE, D. G.: Male gametophyte of *Microcycas calocoma*. Bot. Gaz. **85**, 437—450 (1928). — DUPLER, A. W.: The gametophytes of *Taxus canadensis* Marsh. Bot. Gaz. **64**, 115—136 (1917).

ENGLER, A.: In A. ENGLER u. K. PRANTL, Die natürlichen Pflanzenfamilien. Nachträge zum II.—IV. Teil. Leipzig: Wilhelm Engelmann 1897. — EYMÉ, J.: Observations sur la spermatogenèse des Mousses. C. R. Acad. Sci. (Paris) **234**, 657—659 (1952). — Recherches cytologiques sur les mousses. Botaniste (Paris), Sér. **38**, 1—166 (1954). — EYMÉ, J., et L. CAPOT: Sur la structure de l'anthérozoide de *Sphagnum cymbifolium* Ehrh. examiné au microscope électronique. C. R. Acad. Sci. (Paris) **238**, 1336—1338 (1954).

FAGERLIND, F.: Die Entwicklung des Embryosackes bei *Peperomia pellucida*. Ark. Bot. A **29**, 1—15 (1939). — Bau und Entwicklung der *Gnetum*-Gametophyten. Kgl. svenska

Vet.-Akad. Handl., III. ser. **19**, 55 S. (1941). — Die Samenbildung und die Zytologie bei agamospermischen und sexuellen Arten von *Elatostemma* und einigen nahestehenden Gattungen nebst Beleuchtung einiger damit zusammenhängender Probleme. Kgl. svenska Vet.-Akad. Handl., III. ser. **21**, 1—130 (1944). — FAVRE-DUCHARTRE, M.: Contribution à l'étude de la reproduction chez le *Ginkgo biloba*. Rev. Cytol. et Biol. végét. **17**, 1—216 (1956). — Contribution à l'étude de la fécondation chez *Cephalotaxus*. C. R. Acad. Sci. (Paris) **243**, 1349—1352 (1956). — Contribution à l'étude de la reproduction chez *Cephalotaxus drupacea*. Rev. Cytol. et Biol. végét. **18**, 305—343 (1957). — Contribution à l'étude des spermatozoides de *Taxus baccata*. Rev. Cytol. et Biol. végét. **21**, 329—337 (1960). — Comportement des noyaux spermiques surnuméraires chez quelques Gymnospermes sensu lato. Bull. Soc. bot. France **108**, 196—199 (1961). — FERGUSON, M. C.: Contributions to the knowledge of the life history of *Pinus* with special reference to sporogenesis of the gametophytes and fertilization. Proc. Wash. Acad. Sci. **6**, 1—153 (1904). — FUCHS, A.: Beiträge zur Embryologie der *Thymelaeaceae*. Öst. bot. Z. **87**, 1—41 (1938).

GENTCHEFF, G., and A. GUSTAFSSON: Parthenogenesis and pseudogamy in *Potentilla*. Bot. Not. (Lund) 109—132 (1940). — GERASSIMOVA-NAVASHINA, H.: A contribution to the cytology of fertilization in flowering plants. The Nucleus **3**, 111—120 (1960). — Cytologische Fragen über die doppelte Befruchtung. In: Morphogenese der Pflanzen. Bd. 2. Moskau: Staatl. Univ. 1961 [Russisch]. — GOEBEL, K.: Organographie der Pflanzen. Jena: Gustav Fischer 1898—1901; 3. Aufl. 1928—1933.

HAMANT, CL.: Remarques sur la spermatogenèse des Mousses. Congr. Intern. Bot. Paris. Rapp. et Commun., Sect. **9**, 51—52 (1954). — HAUPT, A. W.: Oogenesis and fertilization in *Pinus Lambertiana* and *P. monophylla*. Bot. Gaz. **102**, 482—498 (1941). — HEIMANS, J.: Chromosomen und Befruchtung bei *Lilium martagon*. Rec. Trav. bot. néerl. A **25**, 138—167 (1928). — HUTCHINSON, A. H.: Fertilization in *Abies balsamea*. Bot. Gaz. **60**, 457—472 (1915).

JENSEN, W. A.: Observations on the fusion of nuclei in plants. J. Cell Biol. **23**, 669—672 (1964).

KHAN, R.: A note on "double fertilization" in *Ephedra foliata*. Curr. Sci. **9**, 323—324 (1940). — Contributions to the morphology of *Ephedra foliata* Boiss. II. Fertilization and embryogeny. Proc. nat. Acad. Sci. (Wash.) **13**, 357—375 (1943).

LAND, W. J. G.: A morphological study of *Thuja*. Bot. Gaz. **34**, 249—259 (1902). — LAWSON, A. A.: The gametophytes and embryo of *Pseudotsuga Douglasii*. Ann. Bot. **23**, 163—180 (1909). — LEE, C. L.: Fertilization in *Ginkgo biloba*. Bot. Gaz. **117**, 79—100 (1955).— LUDWIGS, K.: Untersuchungen zur Biologie der Equiseten. Flora (Jena) **103**, 385—440 (1911). — LYON, H. L.: Polyembryony in *Sphagnum*. Bot. Gaz. **39**, 365—366 (1905).

MAHESHWARI, P., and B. M. JOHRI: The embryo sac of *Acalypha indica* L. Beih. bot. Zbl. A **61**, 125—136 (1941). — MAHESHWARI, P., and V. VASIL: *Gnetum*. Botanical monograph No 1. Council of Scientific and Industrial Research, New Delhi 1961. — MANGENOT, G.: Sur les oosphères, les tubes polliniques et la fécondation chez le Pin maritime. C. R. Acad. Sci. (Paris) **206**, 364—366 (1938). — MANTON, I.: Observations on the microanatomy of the spermatozoid of the bracken fern (*Pteridium aquilinum*). J. biophys. biochem. Cytol. **6**, 413—418 (1959). — MCWILLIAMS, J. R., and F. MERGEN: Cytology of fertilization in *Pinus*. Bot. Gaz. **119**, 246—249 (1958). — MEHRA, P. N.: Occurrence of hermaphrodite flowers and the development of female gametophyte in *Ephedra intermedia* Shrenk et Mey. Ann. Bot., N. s. **14**, 165—180 (1950a). — Inequality in size of the male nuclei in the genus *Ephedra*. Ann. Bot., N. s. **14**, 331—339 (1950b). — MENDES, A. J. T.: Cytological observations in *Coffea*. VI. Embryo and endosperm development in *Coffea arabica* L. Amer. J. Bot. **28**. 784—789 (1941). — MENKE, W., u. B. FRICKE: Beobachtungen über die Entwicklung der Archegonien von *Dryopteris filix mas*. Z. Naturforsch. **19**b, 520—524 (1964). — MÜHLETHALER, K., u. P. R. BELL: Untersuchungen über die Kontinuität von Plastiden und Mitochondrien in der Eizelle von *Pteridium aquilinum* (L.) Kuhn. Naturwissenschaften **49**, 63—64 (1962). — MURRILL, W. A.: The development of the archegonium and fertilization in the Hemlock Spruce (*Tsuga canadensis* Carr.). Ann. Bot. **14**, 583—607 (1900).

NARANG, N.: Contributions to the life history of *Ephedra campylopoda*. I. Male and female gametophytes. Proc. Indian Sci. Congr. No 3, p. 223—224, 1955a. — Contributions to the life history of *Ephedra campylopoda*. II. Fertilization and embryogeny. Proc. Indian Sci. Congr. No 3, p. 224, 1955b. — NEGI, V., and M. LATA: Male gametophyte and megasporogenesis in *Gnetum*. Phytomorphology **7**, 230—236 (1957). — NILSSON, H.: Die Homologie des Angiospermen-Embryosackes. Bot. Not. (Lund) 50—58 (1941). — NORÉN, C. O.: Zur Entwicklungsgeschichte des *Juniperus communis*. Uppsala Univ. Arsskr. Mat.- Nat. **1**, 64 S. (1907).

RICHTER-LANDMANN, W.: Quantitative Untersuchungen über die Zahl der plasmatischen Organelle von Sperma- und Eizelle und der Befruchtungsvorgang von *Impatiens glandulifera* Royle. Naturwiss. **45**, 554—555 (1958). — Der Befruchtungsvorgang bei *Impatiens* *glandulifera* Royle unter Berücksichtigung der plasmatischen Organelle von Spermazelle, Eizelle und Zygote. Planta (Berl.) **53**, 162—177 (1959). — ROGERS, L. M.: Development of the archegone and studies in fertilization. Cellule **37**, 325—352 (1927).

SANWAL, M.: Morphology and embryology of *Gnetum gnemon* L. Phytomorphology **12**, 243—264 (1962). — SCHILDKNECHT, H., u. H. BENONI: Über die Chemie der Anziehung von Pollenschläuchen durch die Samenanlagen von Oenotheren. Z. Naturforsch. **18b**, 45—54 (1963). — SCHLUMBERGER, O.: Familienmerkmale der Cyatheaceen und Polypodiaceen und die Beziehungen der Gattung *Woodsia* und verwandter Arten zu beiden Familien. Flora (Jena) **102**, 383—414 (1911). — SCHUSTER, J.: Cycadaceae. In A. ENGLER, Das Pflanzenreich, Heft 99. Leipzig: Wilhelm Engelmann 1932. — SEDGWICK, P. G.: Life history of *Encephalartos*. Bot. Gaz. **77**, 300—310 (1924). — SLOOVER-COLINET, A. DE: Chambre pollinique et gamétophyte mâle chez *Ginkgo biloba*. Cellule **64**, 127—145 (1963). — SMITH, R. W.: Life history of *Cedrus atlantica*. Bot. Gaz. **75**, 203—208 (1923). — SOUÈGES, E. C. R.: Recherches sur l'embryogénie des Polygonacées. Bull. Soc. bot. France **66**, 168—199; **67**, 1—11, 75—85 (1919). — SRIVASTAVA, R. K.: The evolutionary concept of the morphology of endosperm. Bot. Mag. (Tokyo) **66**, 88—97 (1953). — STEFFEN, K.: Zur Kenntnis des Befruchtungsvorganges bei *Impatiens glanduligera* Lindl. Cytologische Studien am Embryosack der Balsamineen. Planta (Berl.) **39**, 175—244 (1951). — Zytologische Untersuchungen an Pollenkorn und -schlauch. 1. Phasenkontrast-optische Lebenduntersuchungen an Pollenschläuchen von *Galanthus nivalis*. Flora (Jena) **140**, 140—174 (1953). — STEFFEN, K., u. W. LANDMANN: Chondriosomen-, Sphärosomen- und Proplastidenzahlen in Beziehung zu den Differenzierungsvorgängen bei der Antherenentwicklung. Planta (Berl.) **51**, 30—48 (1958b). — STEFFEN, K., u. W. RICHTER-LANDMANN: Entwicklungsgeschichtliche und cytologische Untersuchungen am Balkentapetum von *Gentiana cruciata* L. und *Impatiens glandulifera* Royle. Planta (Berl.) **50**, 423—460 (1958a). — STEIL, W. N.: The antherozoid of the Genus *Riccardia*. Bull. Torrey bot. Club **50**, 197—201 (1923). — STRASBURGER, E.: Die Befruchtung bei den Farnkräutern. Jb. wiss. Bot. **7**, 390—408 (1870).

TAKAO, A.: A cytochemical study on the proteid vacuoles in the egg of *Pinus thunbergii* Parl. Bot. Mag. Tokyo **72**, 289—297 (1959). — Histochemical studies on embryogenesis of *Pinus thunbergii* Parl. Bot. Mag. Tokyo **73**, 379—388 (1960). — THOMPSON, R. C.: A morphological study of flower and seed development in cabbage. J. agricult. Res. **47**, 215—232 (1933). — TSAO, TSUNG-HSUN: A study of chemotropism of pollen tubes in vitro. Plant Physiol. **24**, 494—504 (1949).

VASIL, V.: Morphology and embryology of *Gnetum ula* Brongn. Phytomorphology **9**, 167—215 (1959). — VAZART, J.: Étude comparée de la spermatogenèse chez les fougeres et les *Equisetum*. Congr. Intern. Bot. Paris, Rapp. et Commun., Sect. 9—10, 8, 55—56 (1954).

WATERKEYN, L.: Études sur les *Gnétales*. I. Le strobile ♀, l'ovule et la graine. Cellule **56**, 105—147 (1954). — Études sur les Gnétales. II. Le strobile mâle, la microsporogenèse et le gamétophyte mâle de *Gnetum africanum* Welw. Cellule **60**, 1—78 (1959). — Études sur les Gnétales. IV. Le tube micropylaire et la chambre pollinique de *Gnetum africanum* Welw. Cellule **61**, 81—95 (1960). — WEBBER, H. J.: The development of the antherozoids of *Zamia*. Bot. Gaz. **24**, 16—22 (1897). — Spermatogenesis and fecundation of *Zamia*. U. S. Dept. Agricult., Bur. Pl. Ind. Bull. **2**, 110 S. (1901). — WILKIE, D.: The movements of spermatozoa of bracken (*Pteridium aquilinum*). Exp. Cell Res. **6**, 384—391 (1954). — WYLIE, R. B.: Some aspects of fertilization in *Vallisneria*. Amer. J. Bot. **28**, 169—174 (1941).

YUASA, A.: Studies in the cytology of *Pteridophyta*. XXXIII. Cytomorphological study on the plant spermatozoid, with special reference to its phylogenetic meaning. Bot. Mag. (Tokyo) **67**, 6—14 (1954). — Does the spermatozoid accompany the plastid in fertilization? Congr. Intern. Bot. Paris, Rapp. et Commun., Sect. 9—10, 8, 57—58 (1954).

ZIELINSKI, F.: Beiträge zur Biologie des Archegoniums und der Haube der Laubmoose. Flora (Jena) **100**, 1—36 (1910).

Literatur zu Polyspermie.

Über den Befruchtungsprozeß bei Pflanzen und Tieren. Neue sowjetische Arbeiten. Folge 1. 31. Beiheft zur „Sowjetwissenschaft". Berlin, 1952. Mit Arbeiten von: AWAKIAN u. JASTREB; ELLENHORN u. SWETOSAROWA; FEIGINSON; KOLSOW; PODDUBNAJA-ARNOLDI; PRESENT; SAIKOWSKAJA; TURBIN; WASSILZOWA.

BOGDANOWA, E. N.: Erzeugung hybrider Pflanzen des Winterweizens mit den Merkmalen zweier Vaterformen. Nachr. der Leningrader Universität 1952, Nr 1.

EKLUNDH-EHRENBERG, C., H. v. EULER and G. HEVESY: Note on the number of pollen grains identified in the fruit of the aspen. Ark. Kemi B **23**, H. 6, Nr 5, 1—5 (1947).

GERASSIMOVA-NAVASHINA, H., and T. B. BATYGINA: Fertilization in *Scilla sibirica* Andr. Bot. Ž. **43**, 959—988 mit engl. Zus.fass. (1958) [Russisch]. — GLUSCHTSCHENKO, I. E.: Die Erscheinung der Polybefruchtung bei Pflanzen. Agrobiologie 3—18 (1957). — The phenomenon of polyfertilization in plants. Proc. Int. Genet. Symp., 1956, Suppl. zu Cytologia, 1957, p. 32—34.

HIEMENZ, G.: Untersuchungen an *Salpiglossis variabilis* über Gonenkonkurrenz und selektive Befruchtung und ihre Auswirkung auf die Nachkommenschaft. Biol. Zbl. **74**, 337—370 (1955).

KAPPERT, H.: Referat über FEIGINSON, 1948. Züchter **19**, 120 (1948).

NAWASCHIN, M. S., J. N. GERASSIMOWA-NAWASCHINA u. M. S. JAKOWLEW: Über die Rolle der nichtzellularen lebenden Materie im Reproduktionsprozeß der Pflanzen. Nachr. Akad. Wiss. UdSSR., Biol. Ser. **5**, 7—32 (1952). Deutsch in: Sowjetwiss. **7**, 725—752 (1954).

PODDUBNAJA-ARNOLDI, V. A.: Investigation of the process of fertilization in angiosperms in vivo. Bot. Ž. **43**, 178—193 mit engl. Zus.fass. (1958) [Russisch]. — Allgemeine Embryologie der bedecktsamigen Pflanzen. S. 183ff. Moskau: Verl. d. Akad. d. Wiss. d. UdSSR 1964 [Russisch]. — POLJAKOW, I. M.: Use of radioactive isotopes in studies of plant fertilization. Conf. Akad. Sci. USSR. on the peaceful uses of atomic energy. Div. Biol. Sci. 139—144 (1955). — Contemporary state of the problem of selective fertilization in plants. Ž. obšč. Biol. **19**, 177—186 mit engl. Zus.fass. (1958) [Russisch]. — POLJAKOW, I. M., u. A. M. DMITRIEVA: Neue Wege zur Untersuchung der Befruchtungsprozesse höherer Pflanzen mittels radioaktiver Isotopen. Ž. obšč. Biol. **16**, 3—22 (1955).

TER-AWANESJAN, D. W.: Über die Vererbung der Merkmale zweier Vaterformen bei der Hybridisation der Baumwolle. Agrobiologie 105—109 (1949).

ZACHOW, F.: Ergebnisse der Bestäubung mit Pollengemischen bei *Lupinus angustifolius* und *Lupinus luteus*. Beitrag zur Frage der Polyspermie. Züchter **28**, 241—252 (1958).

Gonenkonkurrenz.

Von

Cornelia Harte.

A. Problemstellung.

Bei allen Lebewesen läßt sich beobachten, daß mehr Fortpflanzungszellen gebildet werden, als nötig sind. Nur ein Teil der Keimzellen kommt zur Fortentwicklung. Es findet also immer eine Konkurrenz statt zwischen den Fortpflanzungszellen um die Möglichkeit, sich weiter entwickeln zu können. Das Stadium, auf dem dieser Konkurrenzkampf sich abspielt, liegt an verschiedenen Stellen des Entwicklungscyclus, je nachdem, ob es sich um Haplo-, Diplobionten oder Organismen mit einem antithetischen Generationswechsel handelt. Bei den meisten Arten hat es den Anschein, als ob der Ausgang dieser Konkurrenz nur durch Zufallseinflüsse entschieden würde. In wenigen Fällen gibt es aber Anhaltspunkte dafür, daß die Zellen, zwischen denen sich die Konkurrenz abspielt, nicht von gleichen Startbedingungen ausgehen, sondern daß ihre genetische Konstitution dabei eine Rolle spielt. Die einzigen exakten und experimentell unterbauten Beobachtungen zu diesem Problem liegen an höheren Pflanzen vor. Hierbei standen zunächst die genetischen Folgen der Konkurrenz im Vordergrund des Interesses, und erst von dort aus wurde nach den entwicklungsgeschichtlichen Grundlagen gefragt. Eine entwicklungsphysiologische Analyse dieser Erscheinungen wurde erst in wenigen Fällen versucht. Der gebräuchliche Ausdruck „*Gonenkonkurrenz*" bedeutet also im weitesten Sinne nichts anderes als eine Beschreibung der Tatsache, daß bei höheren Pflanzen und Tieren zwischen den Produkten der Meiosis, den Gonen, oder ihren direkten Abkömmlingen eine Konkurrenz stattfindet um die Möglichkeit, die nächste Generation zu bilden. Er wird aber in diesem Sinn nie verwendet, sondern nur dann, wenn die Konkurrenz nicht zufallsgemäß, sondern unter dem Einfluß genetischer Faktoren entschieden wird. Bei höheren Pflanzen werden dabei auch Selektionsvorgänge in der aus den Gonen entstandenen haploiden Generation einbegriffen.

Zur Terminologie ist zu erwähnen, daß in den ersten Untersuchungen von CORRENS u. a. zunächst einfach von Konkurrenz gesprochen wird. In der englischen Literatur wurde der Ausdruck „*selective fertilization*" (BRIEGER 1926) verwendet. Im gleichen Sinn wird auch „differential fertilisation" gebraucht (EMERSON 1934). Für die Konkurrenz zwischen den wachsenden Pollenschläuchen wurde dann von HERIBERT-NILSSON (1920) die Bezeichnung „*Zertation*" eingeführt, abgeleitet von lat. „*certatio*" = Wettlauf, um diese Form der Konkurrenz von anderen Möglichkeiten zu unterscheiden. Als Certationskreuzung wird seitdem die Rückkreuzung von Pollen einer Heterozygoten, der aus konkurrierenden Genotypen besteht, auf die Elternformen bezeichnet. Von SIRKS (1926) wurde die Schreibweise in „*Certation*" abgewandelt. BRINK (1925) gebrauchte die Umschreibung „*differential pollentube growth*" und grenzte diese ab gegen „*selective fertilization*". Kurz danach übernahm HARLAND (1926) beide Bezeichnungen für deutlich unterschiedene Tatbestände, da er für abweichende Spaltungszahlen, ohne auf

Einzelheiten einzugehen, „*differential pollentube growth or selective fertilization*" als mögliche Ursachen diskutiert. Von RHOADES (1942) wurde „preferential segregation" eingeführt als allgemeine Bezeichnung für das Verhalten von Nachkommenschaften, in denen eine Abweichung von den zu erwartenden Mendelschen Spaltungsverhältnissen auftritt.

Bei der Verwendung der älteren Literatur ergab sich zunächst die Schwierigkeit, daß die variationsstatistische Auswertung dem damaligen Stand entsprechend durchgeführt wurde und heutigen Ansprüchen nicht mehr genügt. Hierbei erweist es sich jedoch als ein unschätzbarer Vorteil, daß fast alle Untersucher ihre Befunde ausführlich protokolliert und die Kreuzungsergebnisse in sehr detaillierten Tabellen dargestellt haben, so daß eine erneute statistische Auswertung mit modernen Methoden ohne weiteres möglich war und in vielen Fällen interessante Aufschlüsse geben konnte über Fragen, die damals nicht zur Diskussion standen, heute aber in einem anderen Zusammenhang wichtig sind.

B. Grundlagen der Untersuchungen.

I. Genetische Beobachtungen.

Die Feststellung von Erwartungswerten für das Zahlenverhältnis der verschiedenen Genotypen in der Nachkommenschaft von Bastarden beruht auf der Annahme, daß in den zur Befruchtung gelangenden Gameten alle Genotypen in demselben Verhältnis vorkommen, in dem sie nach der Meiosis in den Gonen vertreten sind, d. h. daß die beiden Allele eines Gens, für das ein diploides Individuum heterozygot ist, in je der Hälfte der von diesem Individuum gebildeten Keimzellen enthalten sind. Es darf also zwischen dem Augenblick, in dem die Gonen aus der Meiosis hervorgehen, und dem Zeitpunkt, in dem die aus den Gonen direkt oder durch Zwischenschaltung einer haploiden Generation gebildeten Gameten zur Befruchtung gelangen, kein Vorgang eingeschaltet sein, der durch eine Ausmerze oder eine Vermehrung einzelner Genotypen eine Verschiebung der Zahlenverhältnisse zur Folge hat. Auf dieser Voraussetzung beruhen alle Methoden, bei denen aus gefundenen Spaltungsverhältnissen in F_2 und Rückkreuzungsgenerationen Rückschlüsse gezogen werden auf die Anzahl der an der Vererbung eines bestimmten Merkmals beteiligten Gene oder auf Koppelungsbeziehungen zwischen verschiedenen Loci.

Meist ist diese Annahme auch zutreffend, aber schon bald nach der Wiederentdeckung der Spaltungsregeln im Jahre 1900 traten die ersten Zweifel an der Allgemeingültigkeit dieser Voraussetzung auf. Wiederholt wurden in Fällen, die an dem Vorliegen eines monohybriden Erbganges keinen Zweifel ließen, doch Spaltungszahlen beobachtet, die von der Erwartung abwichen. Die ersten Befunde hierzu teilte CORRENS bereits 1902 mit, Endospermmerkmale beim Mais betreffend (sugary). Es folgten bald weitere Fälle, so die Abweichung vom mechanischen Geschlechtsverhältnis 1:1 bei Pflanzen (*Melandrium:* CORRENS 1917, KUNZE 1955) und Tieren (Zusammenfassung bei SCHLEIP 1912 und LENZ 1912), und die Rotnervigkeit bei *Oenothera* (HERIBERT-NILSSON 1916).

Schon bei den ersten Beobachtungen wurde gefolgert, daß eine mögliche Ursache für die unerwarteten Spaltungsergebnisse darin zu sehen sei, daß zwischen der Entstehung der Gonen und dem Befruchtungsvorgang der Gameten sich auf irgendeine Weise die Zahlenverhältnisse verschoben hätten. Die Annahme einer Reduplikation einzelner, bevorzugter Gametensorten (unter anderen HERIBERT-NILSSON 1916, 1924) wurde sehr schnell verworfen und die Ursache für die vermutete ungleiche Häufigkeit von Gameten mit verschiedenen Allelen in Selektionsvorgängen gesucht, die in den Entwicklungsgang zwischen Gone und Gamet eingreifen (CORRENS l. c., RENNER 1917, 1919, HERIBERT-NILSSON 1920, 1925).

II. Ursachen für abnorme Spaltungszahlen.

a) Allgemeines.

Für die Interpretation abweichender Spaltungszahlen sind jedoch grundsätzlich eine ganze Reihe verschiedener Möglichkeiten in Betracht zu ziehen. Zwischen der Entstehung der Gonen, der Befruchtung und der Auszählung der Spaltungszahlen an den aus den Zygoten hervorgegangenen Individuen liegen jeweils längere Zeiträume, in denen viele Entwicklungsschritte ablaufen. Wenn diese genetisch gesteuert werden, so können unter dem Einfluß verschiedener Allele hier Unterschiede auftreten, die dazu führen, daß auf einzelnen Entwicklungsstadien die verschiedenen Genotypen eine Differenz der Entwicklungsgeschwindigkeit oder eine verschiedene Anfälligkeit gegen äußere Einflüsse zeigen. Das Ergebnis ist ein verstärkter Ausfall einzelner Genotypen und eine Verschiebung der später feststellbaren Spaltungszahlen, wobei die Größe dieser Verschiebung von der Stärke des Selektionsvorgangs abhängig ist. Der Schluß auf Gonenkonkurrenz als Ursache für abweichende Spaltungszahlen setzt also voraus, daß die anderen Möglichkeiten bekannt sind und durch die Versuchsanstellung als Erklärung für den vorliegenden Einzelfall ausgeschaltet werden.

Die vielfältigen Ansätze für Selektionsvorgänge lassen sich in wenigen Gruppen zusammenfassen, in denen jeweils der Zeitpunkt des Eingriffs gemeinsam ist, unabhängig von der Art des Wirkungsmechanismus. Diese Angriffspunkte sind:

1. Die Gonen.
2. Die aus den Gonen entstandenen Haplonten (Gametophyten).
3. Die Gameten.
4. Der Befruchtungsvorgang.
5. Die Zygoten nach der Befruchtung und die anschließende embryonale und postembryonale Entwicklung.

Abweichende Spaltungszahlen in den Eizellen (Embryosäcken) können auch ohne Selektion durch einen Vorgang, der von Rudloff (1933) als „Polarisation der Meiose“ bezeichnet wurde, zustande kommen. Cytologische Untersuchungen zu diesem Problem liegen am Mais vor (Rhoades 1942, 1952). Bei Pflanzen, die neben einem normalen Chromosom 10 ein abnormes Chromosom 10 enthalten, tritt dieses in etwa 70% der Embryosäcke auf anstelle der erwarteten 50%. Die Untersuchung von 200 MMZ (Makrosporenmutterzellen) ergab, daß das Bivalent 10 sich in der Meiose bevorzugt so orientiert, daß das abnorme Chromosom 10 in die basale Gone gelangt, die immer den Embryosack bildet. Die Folge sind Störungen der erwarteten Spaltungszahlen für alle Loci der 10. Koppelungsgruppe. Weitere genetische und cytologische Daten hierzu wurden von Longley (1945) und Emmerling (1959) gebracht.

b) Selektion in der Diplophase.

Die Feststellung darüber, ob eine Selektion in der Diplophase vor der Auszählung stattgefunden hat, ist am einfachsten, da ausfallende Zygoten durch Untersuchung der Samen und Embryonen, der Keimfähigkeit und der Absterberate nach der Keimung leicht festgestellt werden können. Es wird nicht mehr gefordert als genaue Beobachtung und Protokollführung über die Entwicklung der Kulturen. Bereits von Heribert-Nilsson (1916) wurde betont, daß eine solche, sich auf alle Einzelheiten erstreckende Kontrolle über die Ursachen der Ausfälle eine Voraussetzung für jede Untersuchung zur Gonenkonkurrenz ist, um eine herabgesetzte Vitalität einzelner diploider Genotypen als Quelle möglicher Fehlschlüsse auszuschalten. Nur wenn bei Verminderung dieser Ausfälle durch sorgfältige Kulturbedingungen keine Normalisierung der Spaltungszahlen erfolgt, ist die Ursache in früheren Stadien zu suchen.

c) Selektion bei der Befruchtung.

Eine Störung der zufallsgemäßen Spaltung durch eine Differenz der Befruchtungsfähigkeit der verschiedenen Pollensorten ist einfach zu prüfen, indem der Samenansatz verglichen wird, der nach Selbstung der beiden Elternformen und der Heterozygoten oder nach Bestäubung verschiedener Blüten einer dieser Formen oder einer dritten mit allen 3 Pollensorten festzustellen ist. In allen bisher auf diese Weise geprüften Fällen, insbesondere beim Mais, ergab sich dabei keine Benachteiligung einer der Pollensorten. Dies heißt aber, daß bei Ausschaltung der Konkurrenz (nur x_1- oder nur x_2-Pollen auf der Narbe) keine verminderte Funktionstüchtigkeit einer Sorte festzustellen war.

Bei der Verschmelzung der Gameten kann eine Störung der Spaltungszahlen durch selektive Befruchtung entstehen. Mit diesem Begriff wird der Tatbestand bezeichnet, daß die Befruchtung einer Eizelle durch eine Spermazelle, bzw. bei höheren Pflanzen das der Befruchtung vorangehende Eindringen des Pollenschlauches in die Samenanlage, nicht zufallsgemäß erfolgt, sondern unter dem Einfluß der genetischen Konstitution der beiden Gameten bzw. Haplonten steht. Dabei ist die Verschmelzung bestimmter Gametentypen häufiger als die anderer und einzelne Kombinationen sind sogar völlig unmöglich. Die Ausschließung dieser Möglichkeit erfordert den Nachweis, daß einmal in den befruchtungsfähigen Samenanlagen alle zu erwartenden haploiden Genotypen in einer relativen Häufigkeit vertreten sind, die einer zufallsgemäßen Auswahl der Gonen entspricht, und daß zum anderen keine Unterschiede in der Entwicklung der Pollenschläuche in Abhängigkeit von ihrer genetischen Konstitution bestehen. Die Methoden hierfür wurden von SCHWEMMLE und Mitarbeitern ausgearbeitet (vgl. den Beitrag „Selektive Befruchtung" in diesem Handbuch).

Die selektive Befruchtung kann auch als ein Sonderfall der Gonenkonkurrenz betrachtet werden, wobei die Konkurrenz erst im letzten Augenblick vor der Befruchtung stattfindet und die Überlegenheit einer Gametensorte nicht von vornherein gegeben ist und nur von den genetischen Eigenschaften des betreffenden Haplonten bestimmt wird, sondern erst in jedem Einzelfall entsteht im Zusammenwirken mit und in Abhängigkeit von der genetischen Konstitution des jeweiligen Partners.

d) Selektion der Gameten.

Funktionsausfälle der Gameten machen sich durch histologisch feststellbare, vor der Befruchtung abortierende Eizellen oder durch trotz Eindringen eines Pollenschlauches nicht entwicklungsfähige Samenanlagen bemerkbar. Es liegen aber bisher bei höheren Pflanzen keine Beobachtungen darüber vor, daß genetisch bedingte Funktionsausfälle der Gameten, d. h. der Eizellen in histologisch normalen Embryosäcken oder der Spermazellen in normalen Pollenschläuchen, vorkommen.

III. Entwicklungsgeschichtliche Grundlagen der Gonenkonkurrenz.

a) Entwicklung von den Gonen zu den Gameten im ♂ Geschlecht.

Wenn aus abweichenden Spaltungszahlen durch Ausschluß der anderen Möglichkeiten der Verdacht auf Gonenkonkurrenz als ihre Ursache gegeben ist, so ist eine nähere Untersuchung und möglichst eine experimentelle Prüfung dieser Vermutung nötig. Voraussetzung dafür ist eine genaue Kenntnis der Entwicklungsgeschichte, um die Punkte zu finden, an denen eine Selektion der Genotypen durch Konkurrenz möglich ist.

Bei den Spermien der Tiere werden die Gonen ohne weitere Zellteilungen zu Gameten, so daß hier aus der ungleichen Geschlechtsverteilung auf einen Unterschied in der Befruchtungswahrscheinlichkeit geschlossen wurde. [Für eine Kritik dieser Schlußfolgerung vgl. jedoch LUDWIG u. BOOST (1943, 1947).] Bei den höheren Pflanzen ist dagegen zwischen der Entstehung der Gonen und der Ausbildung der Gameten eine ganze Generation eingeschaltet, der Gametophyt, der zwar stark reduziert ist und nur in Abhängigkeit vom Sporophyten lebens- und funktionsfähig ist, aber doch selbständige Entwicklungsvorgänge durchmacht, die Ansatzpunkte für Selektionsvorgänge bieten.

Im ♂ Geschlecht wandeln sich bei den höheren Pflanzen alle Gonen zu Pollenkörnern um, die bereits das erste Entwicklungsstadium des Gametophyten darstellen. Nach der ersten Kernteilung gehen sie in einen Ruhezustand ein, der bei der Keimung auf der Narbe aufgehoben wird. Der Pollenschlauch wächst dann zwischen den Narbenpapillen in den Griffel ein und dort durch das Leitgewebe zum Fruchtknoten. Während dieser Wachstumsphase läuft die zweite Mitose ab, die zur Bildung der beiden Spermazellen führt. Eine Konkurrenz zwischen verschiedenen Genotypen kann auf allen Stadien dieser Entwicklung stattfinden. Zunächst besteht die Möglichkeit, daß zwischen Gonen bzw. jungen Pollenkörnern in der Anthere eine echte Konkurrenz um die Verteilung der in begrenzter Menge vorhandenen Nährstoffe vorliegt (HARTE 1952), die zu einer Benachteiligung bestimmter Genotypen führt. Der nächste Schritt, bei dem sich genetisch bedingte Unterschiede geltend machen können, ist die Keimung auf der Narbe, wo sowohl Keimfähigkeit wie Keimgeschwindigkeit ausschlaggebend sein können. Völliger Funktionsausfall auf diesem Stadium, der zur Elimination bestimmter Allele im ♂ Geschlecht führt, ist mehrfach nachgewiesen, so von Renner (1917) für die inaktiven Pollenkörner von *Oenothera*, kann aber nicht als Konkurrenz bezeichnet werden. Am häufigsten wurde bisher ein Wettbewerb während der Wachstumsphase der Pollenschläuche postuliert. Eine Benachteiligung beim Eintritt in die Samenanlage wird als selektive Befruchtung gesondert behandelt.

b) Entwicklung von den Gonen zu den Gameten im ♀ Geschlecht.

Bei der Entwicklung des ♀ Gametophyten gehen aus der Makrosporen-Mutterzelle (MMZ) 4 Gonen hervor, von denen sich im Normalfall eine zum Embryosack umbildet. Wenn hierbei Zufallseinflüsse oder Bedingungen, die von der genetischen Konstitution der Gonen unabhängig sind, z. B. ihre Lage innerhalb der linear angeordneten Tetrade, darüber entscheiden, aus welcher Gone der Embryosack hervorgeht, so liegt hier zwar eine Konkurrenz der Gonen vor, aber es kann dadurch keine Verschiebung der genetischen Zahlenverhältnisse in der Nachkommenschaft zustandekommen, sondern nur zufallsgemäße Schwankungen in der Häufigkeit der beiden Allele in den Keimzellen. Eine Verschiebung der Zahlenverhältnisse kann aber auf 2 Wegen zustandekommen. Einmal kann die genetische Konstitution der Gonen ein Faktor sein, der mit darüber entscheidet, welche sich zum Embryosack entwickeln wird, so daß dabei bestimmte haploide Genotypen bevorzugt, andere benachteiligt sind. Zum anderen können während der Entwicklung und nach Fertigstellung des Embryosackes sich Vitalitätsunterschiede der einzelnen Haplonten bemerkbar machen, die eine unterschiedliche Absterberate bedingen, so daß unter den abortierenden Embryosäcken bestimmte Genotypen häufiger vertreten sind als andere. Im ersten Fall kann von einer Konkurrenz zwischen den Gonen gesprochen werden, während im zweiten Fall dieser Ausdruck nur bedingt zu verwenden ist, vielmehr eine Konkurrenz zwischen den Samenanlagen unter dem Einfluß der in ihnen enthaltenen Haplonten vorliegt.

Die Kenntnis des Typs der Embryosackentwicklung ist also erste Voraussetzung, wenn bei einem Objekt eine Aussage über die Möglichkeiten der Gonenkonkurrenz im ♀ Geschlecht gemacht werden soll. Auf denkbare Komplikationen bei abweichenden Entwicklungstypen soll jedoch hier nicht eingegangen werden, weil eine solche Erörterung in Anbetracht der wenigen Arten, bei denen bisher Angaben über Gonenkonkurrenz im Embryosack vorliegen, nur hypothetischen Charakter haben könnte.

Wegen dieser verschiedenartigen Ansatzpunkte für eine Konkurrenz in beiden Geschlechtern müssen die Verhältnisse für Embryosack und Pollenschlauch zunächst getrennt behandelt werden.

C. Experimentelle Untersuchungen der Certation.

I. Genetische Teste.

a) Überlegungen zur Versuchsplanung.

Die ersten Vermutungen über genetisch gesteuerte Unterschiede im Pollenschlauchwachstum wurden auf Grund von Kreuzungsergebnissen geäußert. Dementsprechend basieren auch die meisten Teste, die zur Untersuchung derartiger Hypothesen ausgearbeitet wurden, auf der Prüfung von Nachkommenschaften aus Kreuzungen mit heterogenem Pollen. Das Pollenschlauchwachstum zeigt, wie alle biologischen Vorgänge, auch bei genetisch einheitlichem Material eine Variabilität. Wenn nun durch genetische Einflüsse diese Variabilität vergrößert wird, so wird die Genotypenverteilung in der Gruppe der „schnellen" Pollenschläuche eine andere sein als unter den langsam Wachsenden. Die Versuchsanordnung besteht also darin, daß man versucht, einmal alle Pollenschläuche zur Befruchtung zu bringen und in einem anderen Versuch dafür sorgt, daß die Befruchtung nur durch die schneller wachsende Gruppe ausgeführt werden kann (Correns 1917). Die Pollenkörner müssen dabei durch verschiedene Allele eines Locus oder mehrerer Loci, „Testgene", die einen phänotypischen Effekt in der

Diplophase haben, markiert sein, so daß aus dem Phänotyp der Individuen in der Nachkommenschaft die genetische Konstitution der zur Befruchtung gelangenden Pollenkörner erschlossen werden kann.

In manchen Fällen, für die das Vorliegen von Gonenkonkurrenz behauptet wird, zeigt sich jedoch, daß bei Spaltungsverhältnissen, die von der Erwartung bei monogener Vererbung abweichen, Gonenkonkurrenz als Ursache angenommen wird, ohne daß eine Prüfung dieser Hypothese durch darauf ausgerichtete Versuche erfolgt (z. B. SMITH 1956, LUIG 1960, LUIG, MCWHIRTER u. BAKER 1958).

b) Veränderung der Pollenmenge.

Griffel gleicher genetischer Konstitution, meist die der homozygot-recessiven Form, werden mit Pollen der Heterozygoten bestäubt, dabei aber die Menge des für die Bestäubung verwendeten Pollens variiert. In einem Versuch wird bestäubt mit einer abgezählten Anzahl von Pollenkörnern oder wenigstens mit einer sehr geringen Menge, wobei auf jeden Fall die Anzahl der Pollenkörner geringer sein muß als die der im Fruchtknoten zur Verfügung stehenden Samenanlagen. Dadurch ist gewährleistet, daß alle Pollenschläuche, deren Wachstumsleistung ausreicht, die Fruchtknotenhöhlung zu erreichen, dort noch unbefruchtete Samenanlagen vorfinden und somit alle zur Befruchtung gelangen. Die Nachkommen aus einer solchen Kreuzung mit spärlicher Bestäubung geben also die genaue Verteilung der Genotypen in den befruchtungsfähigen Pollenschläuchen wieder. Der Gegenversuch besteht in der Bestäubung mit reichlich Pollen, so daß viel mehr Pollenschläuche als befruchtungsfähige Samenanlagen vorhanden sind. Dabei werden nur die zuerst ankommenden Pollenschläuche zur Befruchtung gelangen, während die langsam wachsenden keine unbefruchteten Samenanlagen mehr vorfinden und funktionslos bleiben. In dieser Nachkommenschaft werden also die Genotypen, die den schneller wachsenden Pollenschläuchen (PS) entsprechen, häufiger vertreten sein als diejenigen, die aus den langsam wachsenden hervorgehen. Die Verschiebung der Spaltungsverhältnisse gegenüber dem Parallelversuch mit spärlicher Bestäubung zeigt also, ob die PS, die sich in ihrer Wachstumsgeschwindigkeit unterscheiden, auch genetisch verschieden sind.

Diese Methode wurde fast gleichzeitig von CORRENS (1917, 1918, 1919) an *Melandrium* und RENNER (1919) an *Oenothera* entwickelt. Die Versuche von RENNER wurden zwar später publiziert, aber bereits 1917 durchgeführt, noch vor der ersten Veröffentlichung von CORRENS. Später wurde sie von HERIBERT-NILSSON (1925) am R-r-Gen von *Oenothera*, von SIRKS (1926) an *Datura* und von OEHLKERS (1923) und HARTE (1953 und unveröffentlicht) ebenfalls an *Oenothera* angewendet.

c) Abschneiden des Griffels.

Die genaue entwicklungsgeschichtliche Untersuchung des Pollenschlauchwachstums führte noch zu einem weiteren experimentellen Ansatz mit dem Ziel, die langsamen Schläuche überhaupt von der Befruchtung auszuschließen. Zunächst muß festgestellt werden, nach welcher Zeit die ersten PS den Griffel durchwachsen haben, wie lange es dauert, bis die Hauptmasse der schnellen PS den Fruchtknoten erreicht hat und in welchem zeitlichen Abstand die langsamen Schläuche folgen. Wenn reichlich bestäubt und dann zwischen diesen beiden Zeitpunkten der Griffel abgeschnitten wird, so ist damit sichergestellt, daß ausschließlich PS der schnellen Gruppe die Befruchtung vollziehen können. Die Untersuchung der Nachkommenschaft und vor allem der Vergleich zwischen Gruppen, bei denen spärlich bestäubt und dann der Zeitpunkt des Abschneidens variiert wurde, ergibt Aufschluß über die genetische Konstitution der schnell wachsenden PS und damit eine Beantwortung der Frage, ob eine genetisch bedingte Pollenschlauchkonkurrenz vorliegt.

Solche Abschneideversuche wurden von CORRENS (1921) an *Melandrium*, HERIBERT-NILSSON (1920, 1923) und HARTE (unveröffentlicht) an *Oenothera* und SIRKS (1926) an *Datura* durchgeführt. Eine Bestäubung an der Spitze der Griffel, verglichen mit Bestäubung an der Basis und kombiniert mit nachfolgendem Abschneiden der Griffel, ergab bei *Melandrium* eine deutliche Verschiebung im Geschlechtsverhältnis, die mit einer schärferen Konkurrenz und starker Benachteiligung der ♂-bestimmenden Pollenschläuche bei längerem Weg zu erklären ist [Bestäubung an Spitze = 25% ♂, an Basis = 43,6% ♂ (CORRENS 1921)].

d) Aufteilung des Fruchtknotens.

Eine dritte Möglichkeit, die Nachkommen aus den beiden Gruppen der schnell und langsam wachsenden PS zu trennen, ist gegeben durch die entwicklungsgeschichtliche Beobachtung, daß die Befruchtung der Samenanlagen durch die Pollenschläuche nur selten rein zufallsgemäß erfolgt, sondern daß hierbei eine bestimmte Reihenfolge mehr oder weniger streng eingehalten wird.

Bei *Melandrium* befruchten die ersten PS vorzugsweise, aber nicht ausschließlich die oberen Samenanlagen, so daß die später ankommenden PS nur wenige unbefruchtete Samenanlagen im oberen Teil des Fruchtknotens, dagegen viele im unteren Teil vorfinden. Wenn ein Fruchtknoten bei der Ernte durch Querschnitte zerlegt wird und die Samen aus den einzelnen Regionen getrennt ausgesät werden, wird man aus dem oberen Teil fast nur Nachkommen der schnellen, im unteren Viertel vorwiegend Nachkommen der langsamen Schläuche erhalten. Ein Vergleich der Spaltungszahlen im oberen und unteren Teil der Frucht gibt daher ebenfalls Auskunft über genotypische Unterschiede zwischen den beiden Gruppen von PS. Die erste Anwendung dieser Methode wurde von CORRENS (1918, 1921) an *Melandrium* durchgeführt, später von OEHLKERS (1923) an *Oenothera*. Bei diesem Objekt befruchten allerdings die ersten PS die unteren Samenanlagen, während die oberen Samenanlagen erst von den letzten PS befruchtet werden (OEHLKERS 1923). In jedem Einzelfall ist also eine entwicklungsgeschichtliche Klärung des Verhaltens der PS vor der Planung und Deutung derartiger Versuche notwendig.

e) Erfassung der langsamen Pollenschläuche.

Theoretisch müßte es möglich sein, auch die schnell wachsenden PS auszuschalten und die Befruchtung nur durch die langsamen vollziehen zu lassen. Die Möglichkeit hierzu ist gegeben, wenn es gelingt, die Bestäubung so frühzeitig vorzunehmen, daß die zuerst ankommenden PS noch keine fertigen Samenanlagen vorfinden und zugrunde gehen, während die langsamer wachsenden PS dann im Fruchtknoten anlangen, wenn die Samenanlagen befruchtungsfähig geworden sind. Die meisten Nachkommen aus einer derartigen Kreuzung müßten dem Genotyp entsprechen, der bei normaler Bestäubung stark benachteiligt ist.

Bisher wurden keine Versuche eigens mit dieser Anordnung veröffentlicht, jedoch liegen Beobachtungen vor über die Kreuzung *Oenothera Hookeri sulfurea* × *(biennis* × *Hookeri)*, die keine andere Deutung zulassen. Der Pollenelter ist heterozygot für $s/+^{s}$. Bei normaler Bestäubung ist s sehr stark benachteiligt (OEHLKERS 1926, EMERSON 1931), während in der genannten Kreuzung, für die sehr junge Blüten mit noch kaum sezernierenden Narben als ♀ Kreuzungspartner verwendet wurden, die *sulfurea*-Nachkommen sehr stark überwiegen, so daß unerwartet viele Samenanlagen von s-Pollenschläuchen befruchtet sein müssen (HARTE, unveröffentlicht).

Eine andere Möglichkeit ergab sich beim Mais. Das Gen sp_2 (small pollen) zeigt nach Selbstung und Rückkreuzung der Heterozygoten gestörte Spaltungsverhältnisse. Aus den Daten ist zu ersehen, daß nur etwa 3,3% sp_2-Pollenschläuche zur Befruchtung gelangen anstelle von 50%. Durch den Größenunterschied zwischen den kleinen sp_2-Pollenkörnern und den großen $+^{sp_2}$-Körnern ist eine mechanische Trennung der beiden Genotypen möglich. Durch Bestäubung mit einem Pollengemisch, das auf diese Weise mit sp_2-Körnern angereichert war, ließ sich nachweisen, daß die Pollenschläuche dieses Genotyps voll funktionsfähig sind, aber relativ langsamer wachsen als solche mit dem $+^{sp_2}$-Allel (MANGELSDORF 1931, 1932; RHOADES u. RHOADES 1939).

f) Versuche mit nur einer Samenanlage je Fruchtknoten.

Bei Pflanzen, die nur eine Samenanlage je Fruchtknoten enthalten, ist die Durchführung solcher Teste erschwert, weil in jedem Fall nur der zuerst angekommene PS die Befruchtung ausführt. Von JONES (1922) und BRINK (1925) wurde aber für Mais eine Modifikation ausgearbeitet. Die Grundlage bildet die Beobachtung, daß die Variationsbereiche der schnellen und langsamen PS sich überschneiden, die Unterschiede sich aber um so deutlicher manifestieren, je größer die zu durchwachsende Strecke ist.

Bei einer Vielzahl von befruchteten Blüten werden unter den zuerst ankommenden PS um so mehr der bevorzugten Klasse sein, je länger die Strecke zwischen dem Ort der Keimung der Pollenkörner und der Samenanlage ist. Es besteht also die Möglichkeit, einmal die natürlich vorhandenen Unterschiede der Narbenlänge für den Test auszunutzen und zum anderen die Narbenlänge künstlich zu verändern. Beim Mais sind die Narben durchschnittlich um so länger, je weiter unten an der Inflorescenz die betreffende Blüte sitzt. Es ist daher zu erwarten, daß bei den unteren Blüten die Konkurrenz zwischen genetisch verschiedenartigen PS schärfer hervortritt als bei den oberen. Die Folge davon wird sein, daß im unteren Teil der Inflorescenz eine Verschiebung der Spaltungszahlen erfolgt zugunsten derjenigen Genotypen, die aus der Befruchtung durch die schnell wachsenden PS hervorgehen. Für das Gen *waxy* wurde mit dieser Versuchsanordnung festgestellt, daß die Wx-Pollenschläuche schneller wachsen und eine größere durchschnittliche Länge erreichen als die wx.

Etwas abgewandelt wurde dieser Versuch von BRINK (1925). Es wurden bei einer Serie von Mais-Inflorescenzen die Narben gestutzt, bei einer anderen durch Einbeuteln zu verstärktem Wachstum gebracht und dann die verschieden langen Narbenfäden mit Pollen von Heterozygoten *Wx/wx* bestäubt. Es wurde erwartet, daß bei den längeren Narben die Konkurrenz stärker wurde, die Verschiebung der Spaltungszahlen gegenüber der Erwartung also deutlicher war als bei kurzen Narben, was auch tatsächlich gefunden wurde. Ähnliche Versuche führte auch MANGELSDORF (1929) durch.

g) Einfluß von Umweltbedingungen auf die Certation.

Mehrfach wurde der Versuch unternommen, durch Einwirkung von verschiedenartigen Außenbedingungen eine Änderung der Certation zu erzielen.

Kreuzungen, die STRASBURGER (1900) durchführte, ergaben an *Melandrium* nach Bestäubung mit reifem Pollen mehr ♂ als wenn unreifer Pollen verwendet wurde. Die Versuchsplanung war aber wegen der noch unklaren Fragestellung nicht exakt genug, um daraus weitere Schlüsse zu ziehen. Bei Verwendung von altem Pollen, der 10—20 Tage im Exsiccator aufbewahrt war, erhielt CORRENS (1918) am gleichen Objekt mehr ♀ als bei frischem Pollen. Nach extremer Alterung (110 Tage) erfolgte nur noch eine geringe Befruchtung, ergab dann aber 90% ♂. Diese stärkere Resistenz soll vielleicht in Zusammenhang mit der normalerweise langsameren Entwicklung der ♂-bestimmenden Pollenkörner stehen. Der Widerspruch zum ersten Versuch wird durch die verbesserte Versuchstechnik erklärt (CORRENS 1921b). Ein ähnlicher Versuch mit altem Pollen bei *Hyoscyamus niger* × *pallida* wurde von CORRENS (1921a) dahingehend gedeutet, daß der Pollen mit dem Allel *pal* resistenter sein sollte als derjenige mit $+^{pal}$, aber die statistische Nachprüfung (χ^2-Zerlegung) ergab eindeutig, daß in der Rückkreuzung zwar eine gesicherte Abweichung vom erwarteten 1:1-Verhältnis besteht, daß aber zwischen den einzelnen Bestäubungen und den beiden Gruppen völlige Homogenität gegeben ist (l. c. Tabelle 1: für 1:1 ist $P < 0{,}01$; für Homogenität in Gruppen $0{,}1 > P > 0{,}05$; für Homogenität zwischen Gruppen $0{,}2 > P > 0{,}1$). Die Werte könnten also als eine leichte Certation mit Bevorzugung von *pal* gedeutet werden, auf die aber das Altern des Pollens ohne Einfluß ist.

Nachdem bei *Melandrium* die Certation zwischen ♂- und ♀-bestimmenden Pollenschläuchen sichergestellt war, konnte CORRENS (1922) eine Verschiebung des Geschlechtsverhältnisses durch relative Zunahme der ♂ erzielen, wenn er den Pollen mit Alkoholdämpfen behandelte. Dies wurde von ihm gedeutet als stärkere Resistenz der ♂-bestimmenden Pollenkörner gegen äußere Schädigung. Es bleibt aber offen, ob tatsächlich Pollenkörner selektiv zum Absterben gebracht wurden oder ob sich als Folge der Behandlung die Konkurrenzverhältnisse verändert haben.

Ein eindeutiger Versuch, in dem eine Verschiebung der Certation im Zusammenhang mit Umwelteinflüssen anzunehmen ist, wurde von HERIBERT-NILSSON (1920) mitgeteilt. Bei *Oenothera Lamarckiana* besteht zwischen Pollenschläuchen mit den Allelen R und r eine deutliche Konkurrenz, die sich mit der

Jahreszeit verschiebt. Bei Rückkreuzung mit Rr auf rr im Juli ist die Abweichung von der Erwartung 1:1 mit 1,2 R:1 r sehr gering. Sie verschiebt sich dann Ende Juli auf 2 R:1 r und bis Ende August auf 2,7 R:1 r. Die statistische Nachprüfung ergab, daß das Material ausreicht, um diese Veränderung zu sichern. Es wird angenommen, daß bei niedrigen Temperaturen die an sich schon langsameren r-Schläuche stärker gehemmt sind als R, so daß die Differenz der Wachstumsgeschwindigkeit zunimmt. Diese Deutung läßt sich nach HERIBERT-NILSSON auch auf die Versuche von DE VRIES (1915) über die Vererbung des Faktors *nanella* anwenden.

Für diese Beobachtung ist aber noch eine andere Deutung möglich. Bei Antirrhinum wurde für das Gen *caeca* eine Bevorzugung von *cae* gegenüber $+^{cae}$ festgestellt. Die Aufgliederung des Materials in einzelne Nachkommenschaften zeigt gesicherte Differenzen der Certation, wobei regelmäßig zu erkennen war, daß eine Korrelation zwischen der Stellung der Blüte an der Inflorescenz und der Höhe des *cae*-Überschusses gegeben war. Alle aus dem Pollen einer Blüte stammenden Familien ergaben jeweils den gleichen Wert. Andere Variationsursachen, wie etwa Witterungseinflüsse, konnten ausgeschaltet werden, und es blieb nur die Annahme übrig, daß die Ernährung der Blüten während der Ausbildung der Pollenkörner ein entscheidender Faktor sei, und daß bei mangelhafter Zufuhr von Nährstoffen an die oberen Blüten der Inflorescenz die *cae*-Pollen diese bevorzugt aufnehmen und dadurch ein besseres Wachstum der Pollenschläuche ergeben (HARTE 1952). Da in den Versuchen von HERIBERT-NILSSON und DE VRIES ebenfalls Blüten von Pflanzen verschiedenen Alters und daher unterschiedlicher Stellung an der Inflorescenz verwendet wurden, ist die für *Antirrhinum* gegebene Deutung vielleicht auch auf *Oenothera* zu übertragen. Eine Entscheidung ist nicht möglich, weil in diesem Fall beide Variablen, Stellung der Pollenblüte an der Pflanze und jahreszeitliche Witterung, zusammenfallen.

Ähnliche Beobachtungen liegen für *Vigna sinensis* vor. An Pflanzen, die für den Faktor X/x (Anthocyanfleck in den Blattachseln) heterozygot waren, wurden die Blüten nach der Aufblühfolge geselbstet und die Nachkommenschaften danach geordnet. Zu Beginn besteht ein gesicherter Überschuß an Recessiven, der gegen Mitte der Vegetationsperiode verschwindet und dann in einen Überschuß an Dominanten übergeht (HARLAND, HAIGH und LOCHRIE 1926). Auch diese Daten lassen sich am einfachsten erklären durch eine Konkurrenz zwischen den Gonen mit X und x, die sich mit zunehmendem Alter der Pflanze gesetzmäßig ändert.

In Versuchen über die Vererbung des Merkmals „*glutinous*“ beim Reis *(Oryza sativa)*, das dem Gen *waxy* beim Mais entspricht, wurde ein eindeutiges Defizit an Recessiven gefunden, das von CHAO (1927) auf Pollenschlauchkonkurrenz zurückgeführt wurde. Nachkommenschaften einer Pflanze aus verschiedenen Ernten sind inhomogen in bezug auf dieses Defizit. Eine hohe Sterilitätsrate, bedingt durch den Abort von Embryosäcken vor der Befruchtung bei ungünstigen Umweltbedingungen, geht parallel mit einem Überschuß an Recessiven. Diese Erscheinung, ebenso wie die Diskrepanzen zwischen verschiedenen Autoren, das Spaltungsverhältnis für gl betreffend, wurden erklärt durch die Annahme, daß die Wirkung von gl auf das Pollenschlauchwachstum durch Umwelteinflüsse variiert wird, so daß sowohl eine Benachteiligung wie eine Bevorzugung von gl gegenüber +gl entstehen kann.

In fast allen Certationsversuchen, die mit größerem Material durchgeführt wurden und für die die Versuchsdaten ausführlich aufgegliedert in den Tabellen publiziert sind, ist festzustellen, daß die Werte für die einzelnen Versuchsglieder manchmal weit auseinanderfallen und die Spaltungszahlen stärker variieren, als dies zufällig möglich wäre. Hierin ist ein Hinweis darauf zu sehen, daß verschiedenartige Einflüsse die Konkurrenz verändern können, auch wenn für den einzelnen Versuch nicht bekannt ist, welcher Art diese Einwirkung gerade war.

Bei Gerste *(Hordeum vulgare)* bleibt für die abweichenden Spaltungszahlen des Faktors „*Zwerg*“ (HALLQVIST 1923) nur Gonenkonkurrenz als Erklärung übrig, und auch hier bestehen sowohl zwischen Familien wie in Familien gesicherte Unterschiede in der Benachteiligung von „*Zwerg*“ im Pollen (χ^2-Zerlegung: $P_{hom} < 0{,}01$ für beide Vergleiche). Die Berechnung der

Varianzen für die beiden Komponenten zeigt jedoch, daß diese gleich sind ($F = 1,1$; n_1 (in Familien) $= 267$; n_2 (zwischen Familien) $= 24$; $P > 0,05$). Eine genetische Ursache für die Unterschiede zwischen Familien kann daher nicht gegeben sein. Es bleibt als einfachste Erklärung die Annahme von Umweltwirkungen, die die Gonenkonkurrenz verändern und deren Variabilität unabhängig von genetischen Differenzen der einzelnen Familien ist und daher alle gleichmäßig betrifft.

In Versuchen über die Vererbung des *grandiflora*-Merkmals von *Petunia* wurde für den Locus G eine Certation festgestellt, bei der die Blüten einer Pflanze in bezug auf das Ausmaß der Benachteiligung von G-Pollenschläuchen nicht homogen sind (REIMANN-PHILIPP 1961). Auch hier bietet sich als Erklärung nur die Wirkung von Umweltfaktoren an.

h) Untersuchung von Pollengemischen.

Alle diese Versuche geben nur Aufschluß über die relativen Wachstumsunterschiede zwischen den Pollenschläuchen aus den beiden Pollensorten, die von einer heterozygoten Pflanze gebildet werden. Daneben erhebt sich jedoch die Frage: Sind die PS mit x_1 immer gegenüber x_2 benachteiligt, unabhängig von ihrer Herkunft, oder läßt sich dieser Unterschied nur feststellen zwischen Pollenkörnern, die auf einer Heterozygoten, also in Konkurrenz miteinander, gebildet wurden? Es ist nämlich denkbar, daß die geringere Wachstumsgeschwindigkeit eine weitere Ursache in einer Differenz der Nährstoffreserven hat, die zurückgeht auf eine unterschiedliche Aufnahme der Reservestoffe durch die genetisch verschiedenen Gonen in der Anthere. Auch genetische Wechselwirkungen zwischen den Gonen und der Pflanze, auf der sie entstehen, sind möglich. Der nächste Versuch ist also die Bestäubung mit einem Pollengemisch aus möglichst gleichen Anteilen der beiden Elternformen.

Ein Vergleich mit absoluten Erwartungswerten, wie bei Bestäubung mit Pollen einer Heterozygoten, ist hierbei nicht möglich, weil die Mischung nie genau ein vorgegebenes Verhältnis der beiden Sorten haben wird, aber die relative Differenz zwischen den beiden Gruppen müßte deutlich werden. Wenn der Unterschied zwischen x_1 und x_2 im Pollen erst während des Wachstums wirksam wird, so muß er auch an den Pollenmischungen feststellbar sein. Ist er dagegen durch eine Konkurrenz zwischen beiden Sorten von Gonen in der Anthere während der Pollenreifung bestimmt, so dürfen beide Sorten im Gemisch keinen Unterschied erkennen lassen.

JONES (1920, 1922) führte am Mais derartige Untersuchungen durch. Wenn das Gemisch auf die Narben der beiden Elternformen gebracht wird, werden, wie schon beschrieben, die schnelleren PS bevorzugt die Blüten mit den längeren Narben, also an der Basis der Inflorescenz, befruchten, während der Anteil der langsamen PS bei Befruchtung der Spitzenblüten zunehmen sollte. Die beiden Elternformen unterschieden sich durch 2 Gene, wobei jede je ein recessives und ein dominantes Allel enthielt. Durch diese doppelte Markierung mit *su* und *white* war es möglich, die Kreuzungen mit einem Pollengemisch auf beide Elternformen durchzuführen und jeweils am Kolben die Selbstungs- und Kreuzungskörner zu erkennen. Das Ergebnis war überraschend. Es zeigte sich nämlich, daß in jedem Fall an der Basis der Inflorescenz relativ mehr Selbstungen vorgekommen waren als an der Spitze. Auf jeder der beiden Elternformen wies also der eigene Pollen das schnellere Wachstum auf. Auch bei *Gossypium* soll nach Bestäubung mit Pollenmischungen nur sehr selten Bastardsamen entstehen, was ebenfalls auf eine Bevorzugung des eigenen Pollens deutet (BALLS 1919, zit. nach JONES 1922).

Damit ist zugleich gezeigt, daß bei diesen Objekten die genetische Konstitution des Sporophyten keinen Einfluß auf die Konkurrenzfähigkeit der von ihm gebildeten Pollenkörner hat. Weiter ergibt sich, daß die Ursache des unterschiedlichen Wachstums hier nicht in einer Konkurrenz zwischen den Gonen in der Anthere liegen kann, sondern daß die Wachstumsreaktion erst im Zusammenwirken mit dem Griffelgewebe zustande kommt.

In anderen Fällen jedoch zeigt sich bei Bestäubung mit Pollengemischen eine deutliche Überlegenheit des Fremdpollens, so bei *Cheiranthus* (Bateman 1956). In der Gattung *Streptocarpus* kommen bei Artkreuzungen alle Übergänge vor, von einer Überlegenheit des Selbstpollens bis zu einer weitgehenden Verdrängung durch den Fremdpollen, je nach der untersuchten Artkombination (Lawrence 1958). Diese Kreuzungen wurden aber nicht individuen-reziprok ausgeführt, wie die älteren von Jones. Aus den Schwankungen der Konkurrenzfähigkeit können keine allgemeineren Schlüsse gezogen werden, weil keine Möglichkeit besteht, den individuellen Einfluß vom Arteinfluß zu trennen.

Welche Vorsicht bei Interpretation der Bestäubung mit Pollengemischen gewahrt werden muß, zeigen Versuche von Schwemmle (1951, 1952a) an *Oenotheren* der *Raimannia*-Gruppe. Wenn Griffel der *Oe. odorata* (Komplexkonstitution v · I in *odorata*-Plasma) bestäubt werden mit Pollen, der nur den Komplex I enthält und nach einiger Zeit (9—18 Std) mit Pollen von *Oe. scabra (hsc · hsc)* nachbestäubt wird, so überholen die *hsc*-Pollenschläuche zum Teil die I-Schläuche und befruchten einen erheblichen Teil der Samenanlagen. Werden dagegen beide Pollensorten gleichzeitig auf die Narbe gebracht, so tritt eine Hemmung des *hsc*-Pollens ein, der diese Schläuche weitgehend von der Befruchtung ausschließt. Es besteht also nicht nur eine Wechselwirkung zwischen dem Griffelgewebe und den Pollenschläuchen, sondern auch noch zwischen den verschiedenen gleichzeitig wachsenden Pollenschläuchen. Diese muß an den Vorgang des Wachstums gebunden sein, weil die relative Hemmung der *hsc*-Schläuche sich nur bemerkbar macht, wenn gleichzeitig wachsende Schläuche anderer genetischer Konstitution vorhanden sind, während die Anwesenheit älterer Schläuche keine Hemmung bewirkt.

II. Untersuchung des Pollenschlauchwachstums.

a) Allgemeines zur Versuchsplanung.

Neben der genetischen Prüfung kann ein grundsätzlich anderer Weg eingeschlagen werden, nämlich die direkte Beobachtung der wachsenden Pollenschläuche im natürlichen Medium, dem Leitgewebe des Griffels oder als Pollenschlauchkultur auf einem künstlichen Substrat. Die Aussagekraft beider Versuchsanordnungen ist sehr verschieden. Sie ergänzen sich aber in wichtigen Punkten, und die Grundlagen für die Versuchsanstellung und die Deutung sind für beide gleich.

Wenn zwischen 2 Genotypen im Pollen einer heterozygoten Pflanze die Konkurrenz auf einem Unterschied in der Keimung oder in der Wachstumsgeschwindigkeit beruht, so muß sich dies äußern in der Weise, daß sich statt der normalen, eingipfligen Verteilungskurve für Keimung der Körner und Wachstum der Schläuche eine zweigipflige Kurve ergibt, wobei sich die Variabilitätsbereiche mehr oder weniger überschneiden können. Im Extremfall ist die Zusammensetzung aus 2 Kurven nur an einem gegenüber den Homozygoten verbreiterten Maximum und der vergrößerten Streuung feststellen.

b) Wachstum im Griffel.

Wichtige Aufschlüsse können erhalten werden, wenn die Wachstumskurven von genetisch homogenem und heterogenem Pollen im Griffelgewebe miteinander verglichen werden. Hierzu werden Narben mit dem Pollen bestäubt und in verschiedenen Zeitabständen danach die Griffel präpariert und die Länge der Pollenschläuche gemessen. Es ergibt sich bei einheitlichem Pollen eine eingipflige Verteilung, bei heterogenem Wachstum ein Zweigipfligkeit, wobei die beiden Gruppen der schnell und langsam wachsenden Schläuche mehr oder weniger scharf getrennt sind. Meist kommt der Unterschied um so deutlicher zum Ausdruck, je größer der zeitliche Abstand zwischen Bestäubung und Messung ist. Es sind Kontrollen durch gleichartige Untersuchungen nach Bestäubung von Griffeln gleicher genetischer Konstitution mit genetisch einheitlichem Pollen der beiden Homozygoten nötig.

Von BUCHHOLZ und BLAKESLEE (1922, 1927, 1936) wurden solche Untersuchungen an *Datura* durchgeführt. Die trisome Mutante *Globe* zeigt ein deutliches Defizit an Trisomen nach Selbstung und Kreuzung. Die Pollenschlauchlängen ergeben bei histologischer Untersuchung des Griffels eine zweigipflige Kurve, wobei die beiden Gruppen um so weiter auseinander liegen, je länger die Entwicklungszeit war. Es findet hier eine deutliche Konkurrenz zwischen den normalen Pollenschläuchen und denen mit dem überzähligen Chromosom statt. Pollen von normalen Pflanzen liefert eine eingipflige Verteilung der Pollenschlauchlänge, die bei vergleichbaren Entwicklungsstadien mit dem ersten Gipfel des *Globe*-Pollens übereinstimmt (BUCHHOLZ und BLAKESLEE 1922). Bei einer Genmutante *tricarpel* (*tc*), die einen Recessivenausfall zeigt, wurden die 3 Genotypen (*TcTc, Tctc, tctc*) in allen 9 Kombinationen miteinander gekreuzt (BUCHHOLZ und BLAKESLEE 1927). Sobald *tc* im Pollen enthalten ist, finden sich sehr viele abnorme Pollenschläuche, die anschwellen und platzen. In den befruchtenden Pollenschläuchen ist *tc* dadurch seltener vertreten als in den Pollenkörnern, aber eine Benachteiligung durch Konkurrenz ist nicht vorhanden. Wenn *tc*-Pollenschläuche überhaupt den Griffel durchwachsen, so benötigen sie hierzu durchschnittlich dieselbe Zeit wie die normalen. Dieser Fall zeigt deutlich die Bedeutung derartiger entwicklungsgeschichtlicher Untersuchungen zur Abgrenzung der Konkurrenz von anderen Störungsursachen.

Einen besonders günstigen Fall untersuchte RENNER (1919). Es wurde eine Pollenmischung von *Oenothera Lamarckiana* und *Oe. muricata* auf Blüten der ersteren Art gekreuzt und das Wachstum der Pollenschläuche untersucht. In der zweigipfligen Verteilung mit einem deutlichen Abstand zwischen beiden Gruppen war es möglich, jeden Pollenschlauch eindeutig seiner Herkunft nach zu erkennen, da die Pollenkörner der beiden Arten sich durch die Form der Stärkekörner unterscheiden. Die schnellwachsenden Schläuche wurden als von *Oe. Lamarckiana*, die langsamen als von *Oe. muricata* stammend identifiziert.

Bei Heterozygoten von *Oenothera* der Konstitution *velans*·^{h}*franciscana* und *flavens*·^{h}*Hookeri* ließ sich die zweigipflige Verteilung für die Wachstumsgeschwindigkeit der Pollenschläuche in verschiedenen Griffeln nachweisen, während die Schläuche nach Bestäubung mit Pollen von Homozygoten eine eingipflige Verteilung ergaben. Künstliche Pollengemische ergeben jedoch im Gegensatz zum Verhalten von Pollen von Heterozygoten einheitliche Wachstumskurven der Pollenschläuche. Eine unterschiedliche Wachstumsgeschwindigkeit genetisch verschiedener Pollensorten ist nur feststellbar, wenn diese zusammen auf einer heterozygoten Pflanze gebildet wurden (HARTE u. GOTTER, unveröffentlicht).

c) Kulturversuche.

Kulturversuche auf künstlichem Substrat gestatten eine direkte Beobachtung der Keimung und der ersten Wachstumsstadien, lassen Unterschiede aber nur erkennen, wenn diese bereits bei der ersten Entwicklung der Haplonten auftreten. In diesem Fall wäre die Möglichkeit gegeben, die Abhängigkeit der Wachstumsreaktion von der Ernährung der PS zu prüfen. Dieser Versuch wurde bisher noch nicht zur Klärung der Pollenschlauchkonkurrenz herangezogen. Durch Keimungsversuche konnten aber RENNER und CLELAND (1933) zeigen, daß der Komplex *excellens* aus *Oenothera chicaginensis* im Pollen nicht zu inaktiven Körnern führt. Die mangelnde Übertragung durch den Pollen muß hier, im Gegensatz zu den Eizellenkomplexen anderer *Oenothera*-Arten, durch eine extreme Unterlegenheit der Pollenschläuche gegenüber allen übrigen pollenaktiven Komplexen zustandekommen.

d) Zur Deutung der Wachstumsversuche.

Die Wachstumsuntersuchungen gestatten also eine Aussage darüber, ob bei einer beobachteten Abweichung der Spaltungszahlen, für deren Ursache keine andere Deutung als eine Selektion in der ♂ Haplophase übrigbleibt, diese vielleicht durch eine Keimungshemmung oder eine abnorme Entwicklung zu erklären ist. Die Kulturversuche können auch zeigen, ob eine Benachteiligung durch verzögerte Keimung und eine Konkurrenz in der ersten Wachstumsphase vorliegt. Die

Untersuchungen am Griffel bieten darüber hinaus die Möglichkeit, eine unterschiedliche Entwicklung im Laufe des gesamten Pollenschlauchwachstums aufzufinden.

Diese Teste zeigen jedoch, wenn sie positiv ausfallen, nur an, daß aus dem verwendeten Pollen 2 Gruppen von PS entstehen, die sich in ihrer Wachstumsgeschwindigkeit unterscheiden, während sie keine direkten Rückschlüsse zulassen in bezug auf die Frage, worin die Ursache hierfür zu suchen ist, insbesondere darüber, ob diese beiden Gruppen sich in dem genetischen Faktor unterscheiden, für den die veränderten Spaltungszahlen gefunden wurden. Die Kreuzungsuntersuchungen mit experimenteller Trennung der beiden Gruppen müssen dann darüber entscheiden, ob die schnell und die langsam wachsenden PS sich auch genetisch unterscheiden in dem Faktor, für den die Konkurrenz vermutet wird.

Bei negativem Ausfall der Versuche, wenn also keine zweigipflige Verteilungskurve gefunden wurde, können entweder die Variationsbereiche sich soweit überschneiden, daß mit dieser Methode keine Trennung möglich ist, oder die Konkurrenz kann erst in den Spätstadien des Wachstums im Fruchtknoten oder bei der Befruchtung zur Auswirkung gelangen. Wenn auch unter Verwendung von Vergleichskulturen mit homogenem Pollen keine Differenzen festgestellt werden können, so besagt dies nur, daß sich die Keimung und das erste Wachstum der genetisch verschiedenen Pollen nicht unterscheiden, läßt aber keine Rückschlüsse auf spätere Entwicklungsstadien oder auf das Verhalten im Griffel zu. Auch wenn die Unterschiede überhaupt erst durch die Konkurrenz, d. h. das gemeinsame Wachstum im Griffelgewebe zustande kommen, liefern die Kulturversuche keine Entscheidung. Wird also eine eingipflige Kurve der Keimungs- und Wachstumsgeschwindigkeit der Pollenschläuche in der Kultur erhalten, die sich nicht vom Kurvenverlauf bei Verwendung von genetisch einheitlichem Pollen unterscheidet, so ist dies kein Beweis gegen das Vorliegen einer Konkurrenz zwischen den Pollenschläuchen im Griffel. Bei positivem Ausfall dieses Testes ist dagegen durch den Vergleich mit den beiden Homozygoten zu zeigen, daß die gefundene Zweigipfligkeit ihre Ursache tatsächlich in den zur Diskussion stehenden genetischen Unterschieden hat und nicht auf andere Wirkungen phänotypischer oder genotypischer Art zurückzuführen ist. Kulturversuche auf künstlichen Nährböden sind also immer nur als zusätzliche Informationsquelle zu verwerten, können allein aber keine Entscheidung liefern. Falls dieser Test jedoch positiv ausfällt, ist damit zugleich gezeigt, daß eine unterschiedliche Wachstumsgeschwindigkeit genetisch verschiedener Pollenschläuche von diesen selbst bestimmt wird, ohne weitere Einwirkungen von außen her.

III. Identifizierung der gametophytisch wirkenden Gene.

Aus der Feststellung der Tatsache, daß Pollenschläuche verschiedener genetischer Konstitution eine unterschiedliche Wachstumsgeschwindigkeit aufweisen, ergibt sich die Frage nach der Ursache dieser Erscheinung. Der ♂ Gametophyt ist zwar stark reduziert und bietet somit nur wenig Möglichkeiten für eine Genwirkung. Der Pollenschlauch ist aber, wenn auch kurzlebig, doch ein Zentrum starker vegetativer Aktivität, die wahrscheinlich kontrolliert wird durch den vegetativen Kern und damit unter dem Einfluß der in diesem Kern enthaltenen Gene steht. Zwischen Pollenschläuchen, deren Kerne sich durch verschiedene Allele eines Gens unterscheiden, können demnach Unterschiede der vegetativen Entwicklung erwartet werden, die sich, da praktisch keine Möglichkeit einer Beeinflussung der Formbildung besteht, hauptsächlich in Differenzen der Entwicklungsgeschwindigkeit äußern werden.

Eine Gonenkonkurrenz wurde bisher vor allem festgestellt für Gene, die auf Kohlenhydrate und Farbstoffe einwirken, so *Wx/wx*, *Su/su* beim Mais, ähnlich wirkende Gene beim Reis (PARNELL 1921, CHAO 1927), Genkomplexe bei *Oenothera*, die sich durch die Form der Stärke im Pollen unterscheiden, für eine Allelenserie bei *Salpiglossis*, die die Blütenfarben beeinflußt (HIEMENZ 1957), $s/+^{s}$ bei *Oenothera*, den Rotnervfaktor derselben Art und *P/p* bei *Datura* (SIRKS 1926). Andererseits tritt Gonenkonkurrenz auf bei den Genen, die für die Geschlechtsausprägung verantwortlich sind, wobei die ♂- und ♀-bestimmenden Gameten sich durch die Geschlechtschromosomen, also durch den Chromatingehalt, unterscheiden. Dasselbe trifft für die Trisomen zu. Es ist aber gleich zu ergänzen, daß für eine Reihe anderer Gene, deren physiologische Wirkung auf den Diplonten von anderer Art ist, ebenfalls Gonenkonkurrenz gefunden wurde [*Co/co* (LANGENDORFF 1928) und $fr/+^{fr}$ bei *Oenothera* (HARTE 1948), $serp/+^{serp}$ bei *Antirrhinum* (HARTE 1952), *s/S* bei *Datura* (SIRKS 1926) u. a.].

Mehrfach wurde die Annahme geäußert, daß die verschiedene Befruchtungswahrscheinlichkeit, insbesondere der Überlegenheit der ♂-bestimmenden Spermien damit zusammenhängt, daß sie durch das Fehlen des X-Chromosoms leichter seien und daher eine größere Beweglichkeit aufweisen würden (SCHLEIP 1912, LENZ 1914). Bisher wurde jedoch noch kein Beweis dafür geliefert, daß diese Parallelität von Fehlen des X-Chromosoms und größerer Befruchtungswahrscheinlichkeit auf einer größeren Beweglichkeit beruht und zwischen beiden Erscheinungen ein ursächlicher Zusammenhang besteht. Die gleiche Annahme über eine mögliche Ursache wurde auch für das unterschiedliche Wachstum der ♂- und ♀-bestimmenden Pollenschläuche bei *Melandrium* gemacht (CORRENS 1921), weil die mit dem kleineren X-Chromosom versehenen Schläuche schneller wachsen als die mit dem größeren Y-Chromosom. Da hier aber die Kerne nur einen relativ kleinen Teil des Pollenschlauches ausmachen und zudem ihr Wachstum und nicht die Beweglichkeit der Gameten den Ausschlag in der Konkurrenz gibt, ist es unwahrscheinlich, daß hier ein so mechanischer Zusammenhang gegeben ist. Darüber hinaus gilt noch der Einwand, daß aus der Parallelität zweier Beobachtungen nicht auf eine ursächliche Beziehung geschlossen werden darf. Nachdem diese Ansicht aber einmal vorhanden war, wurde sie ohne allzu große Bedenken auch auf andere Fälle übertragen, in denen Abweichungen vom mechanischen Geschlechtsverhältnis gefunden wurden, so für *Rumex* (CORRENS 1924). Die spätere Nachprüfung der genetischen Grundlagen der Geschlechtsbestimmung bei diesen Formen zeigte aber hier ganz andere Ursachen für das Abweichen vom Verhältnis 1:1 für ♂:♀ auf.

Für die Abweichungen vom mechanischen Geschlechtsverhältnis bei Tieren wurde wegen der Unzulänglichkeit der Vorstellung einer rein mechanischen Beziehung zwischen Chromatingehalt und Befruchtungswahrscheinlichkeit schon bald eine physiologische Erklärung gesucht (Zusammenfassung bei LUDWIG und BOOST 1947).

Die Feststellung von Gonenkonkurrenz bei Genen, die auf den Kohlenhydratstoffwechsel einwirken, führte zu der Annahme, daß das ungleiche Wachstum genetisch verschiedener PS dadurch zustande kommt, daß sie über verschiedenartige KH-Reserven verfügen oder die gleiche Reservestoffmenge unterschiedlich schnell mobilisieren können (BRINK und MCGILLIVRAY 1924). So naheliegend diese Annahme zunächst erscheinen mag, so verlor sie doch bald an Gewicht durch die Untersuchung des Stoffwechsels wachsender PS. Diese führte nämlich zu der Feststellung, daß die Pollen auch ohne nennenswerte Reservestoffe noch keimen und wachsen können, andererseits im Normalfall ihre Stärkereserven selten ganz verbrauchen und schon nach kurzer Wachstumszeit KH aus den Griffelgeweben aufnehmen (Zusammenfassung der älteren Literatur bei BRINK 1924). Dies macht es unwahrscheinlich, daß ein direkter Zusammenhang zwischen den Genen, die auf den KH-Stoffwechsel einwirken, und dem Pollenschlauchwachstum besteht. Das gleiche gilt für Gene, die auf die Farbstoffbildung, besonders der Anthocyane und Flavone, einwirken und damit indirekt in den KH-Haushalt eingreifen.

Die Untersuchungen am Mais führen hier weiter. Nachdem zunächst für *sugary* die Certation beobachtet war, traten neue Beobachtungen über *waxy* hinzu (KEMPTON 1919, COLLINS und KEMPTON 1911, BRINK und MCGILLIVRAY 1924). Die Ergebnisse bei Kombination der beiden Loci lassen sich nicht durch eine einfache Addition der Wirkung erklären. Bei Kreuzung von Pollen einer Pflanze, die für *Su/su* heterozygot ist, auf einen Griffel, der *Su* enthält (homo- oder heterozygot), entsteht, wie schon erwähnt, eine Benachteiligung der Pollenschläuche mit *su* (JONES 1920, 1922). Dies gilt aber nur, solange nur die F_1 betrachtet wird. Wenn aus der F_2 einzelne heterozygote Pflanzen geselbstet werden, entstehen 3 Typen: normal spaltende, Pflanzen mit niedriger Recessivenrate in der Nachkommenschaft wie die F_1 und Pflanzen mit extrem hoher Recessivenrate und einem Defizit an Dominanten. Wenn *su/Su* selber die Konkurrenz bestimmt, ist dieser Umschlag nicht zu erklären, und vor allem nicht die Heterogenität der einzelnen F_2-Nachkommenschaften. Es entspricht aber der Erwartung, wenn *Su* und *su* nicht direkt für die Konkurrenz verantwortlich sind, sondern eine Koppelung mit einem nur gametophytisch wirksamen Gen *Ga/ga* vorliegt, so daß durch crossing-over neue Kombinationen entstehen können (EMERSON 1925). Die Pflanzen mit niedriger Recessivenrate müßten in ihrer genetischen Konstitution der F_1 entsprechen *(Ga Su/ga su)*, die Pflanzen mit hoher Recessivenrate müßten die Kombination *Ga su/ga Su* besitzen und diejenigen mit normaler Spaltung für *Ga* oder *ga* homozygot sein. In der gleichen Kreuzung war noch der embryonale Letalfaktor de_1 enthalten, der derselben Koppelungsgruppe angehört und dessen Spaltungszahlen der Erwartung bei Zutreffen der oben gestellten Annahme entsprechen (MANGELSDORF und JONES 1926).

In den Untersuchungen über das Gen *waxy* ist es zunächst durch die verschiedenen Stärkeformen im Pollen möglich, die Spaltung 1:1 dort zu kontrollieren. In den Pollenschläuchen, die zur Befruchtung gelangen, besteht sowohl nach Selbstung wie nach Rückkreuzung auf *wx/wx* im Durchschnitt ein Defizit an *wx*, aber es lassen sich normal spaltende Familien und andere mit einem sehr starken Defizit auffinden (KIESSELBACH und PETERSEN 1926, KEMPTON 1919, KEMPTON und COLLINS 1911).

Die Berechnung der Homogenität zwischen den einzelnen Familien zeigt in allen Tabellen gesicherte Unterschiede (χ^2-Test, $P_{hom} < 0{,}001$) zwischen den einzelnen Nachkommenschaften an. Bei der Prüfung der Pflanzen, für die mehrere Kolben ausgewertet wurden (Pflanzen mit 2 Kolben aus Tabelle 6 von KIESSELBACH und PETERSEN), könnte die Gesamtspaltung *Wx:wx* etwa 1:1 sein, aber die χ^2-Zerlegung zeigt deutlich, daß die Kolben auf jeder Pflanze gut in ihren Spaltungsverhältnissen übereinstimmen, während die Unterschiede zwischen den Pflanzen sehr stark gesichert sind und die scheinbar normale Spaltung im Gesamtmaterial nur durch die Kombination extremer Typen vorgetäuscht wird (Homogenität zwischen Pflanzen: $\chi^2 = 30{,}47$; $n = 8$; $P < 0{,}001$; Homogenität in Pflanzen: $\chi^2 = 2{,}99$; $n = 9$; $P > 0{,}9$).

Die Autoren fanden diese Streuung zunächst unerklärlich, aber der von MANGELSDORF und JONES (1926) vorgenommene Vergleich mit ihren eigenen Daten führte zu der Annahme, daß *wx* ebenfalls, wie *su*, mit einem gametophytisch wirkenden Faktor gekoppelt ist. Der verwendete Stamm *waxy* war für diesen heterozygot, und außerdem war von den ersten Bearbeitern nicht vergleichbares Material aus verschiedenen Stämmen und mehreren Generationen zusammengeworfen worden. Wenn man auf die Originaltabellen von KEMPTON zurückgreift, so ist zu erkennen, daß die F_1-Pflanzen, die nach Selbstung eine normale Spaltung ergeben, auch als Pollenelter ein Verhältnis *Wx:wx* = 1:1 hervorbringen. Wenn dagegen nach Selbstung ein Defizit an *wx* vorliegt, so ist dasselbe auch nach Rückkreuzung dieser Pflanze als Pollenelter zu beobachten. Auf der gleichen Basis sind auch die Resultate von BRINK (1925) zu interpretieren.

Der gametophytische Faktor des *wx*-Stammes zeigt allerdings einige Abweichungen gegenüber *Ga*. Das ursprünglich mit *wx* gekoppelte Allel verlangsamt das Pollenschlauchwachstum und wirkt dabei unabhängig von der genetischen Zusammensetzung des Griffelgewebes. Für die meisten Gene beim Mais, für die Certation festgestellt wurde, wird danach angenommen, daß dieser Einfluß nicht direkt von den betreffenden Loci ausgeht, sondern von mit ihnen gekoppelten, nur gametophytisch wirksamen Genen. Diese sind daran zu erkennen, daß nach Selbstung von F_2-Pflanzen, also bei Einbeziehung einer weiteren Generation, eine Veränderung des Certationsverhältnisses zu beobachten ist. Es treten dann Familien mit hoher, normaler und niedriger Recessivenrente nebeneinander auf. Diese Annahme schließt nicht aus, daß es neben diesen gametophytischen Faktoren noch andere Gene gibt, die sowohl im Sporophyt wie im Gametophyt eine Wirkung entfalten.

Nachdem damit die Methode zur Auffindung gametophytisch wirkender Gene ausgearbeitet war, wurde in vielen anderen Fällen versucht, im Analogieschluß Störungen der Spaltungszahlen gegenüber der Erwartung auf die Wirkung derartiger Gene zurückzuführen, ohne daß in jedem Fall alle Kriterien geprüft wurden [BEMIS (1959) bei Limabohnen; beim Mais: BURNHAM (1936) in der Koppelungsgruppe 5, LONGLEY (1961) ein anderer Locus in derselben Koppelungsgruppe, SCHWARTZ (1960) in der Koppelungsgruppe, die ein Gen für die Bildung von Esterasen enthält, RHOADES (1943, 1948) in der 3. Koppelungsgruppe, sowie BURNHAM (1958), KIKUDOME (1958), RHOADES (1955, 1958) und RHOADES u. DEMPSEY (1957) an verschiedenen Loci].

Der nächste Schritt ist, das crossing-over zwischen diesen Genen und den Testloci zu untersuchen und sie in ihrer Koppelungsgruppe zu lokalisieren. Vor allem beim Mais konnten auf diesem Wege in mehreren Koppelungsgruppen *Ga*-Gene (von EMERSON 1925 eingeführte Abkürzung für gametophytisch wirksame Gene) gefunden werden. Die Schwierigkeit bei der Koppelungsuntersuchung ist, daß in die Berechnung zwei Unbekannte eingehen, die Benachteiligung von *ga* gegenüber *Ga*-Pollenschläuchen und das crossing-over zwischen *Ga* und dem Testlocus. Der Austausch kann nur berechnet werden, wenn ein Schätzwert für die Konkurrenz zugrunde gelegt wird, und umgekehrt. Näherungsmethoden, die auf der gleichzeitigen Verwendung von zwei Testloci derselben Koppelungsgruppe beruhen, wurden zunächst auf graphischem Wege von MANGELSDORF u. JONES (1926), dann mit exakteren Formeln von EMERSON (1934), BRIEGER (1937) und BRIEGER, TIDBURY u. TSENG (1938) ausgearbeitet und für die Lokalisierung der *Ga*-Gene in der 4. und 5. Koppelungsgruppe beim Mais angewendet. EMERSON (1934) untersuchte das zuerst gefundene, mit *su* gekoppelte Gen (Gruppe 4) zum ersten Mal in einem exakten, orthogonalen Versuchsplan. Die vier möglichen Bastarde, die heterozygot für *Su/su* und gleichzeitig homozygot für *Ga/Ga* oder *ga/ga* oder heterozygot *Ga/ga* im Koppelungs- oder Abstoßungsfall mit *Su* sind, wurden sowohl als Pollen- wie als Eizellenelter mit den beiden Homozygoten *Ga su* und *ga su* rückgekreuzt. Durch Kombination der Ergebnisse dieses Versuchs mit weiteren, unter Hinzuziehung anderer Loci (de_1, Ts_5, de_{16}, *Tu*) in gleicher Weise gewonnenen Daten konnte die Lokalisation von *Ga* durchgeführt werden. In ähnlicher Weise wurde das *Ga*-Gen in der 5. Koppelungsgruppe lokalisiert (LONGLEY 1961) und ein *Ga*-Gen bei der Gerste (TABATA 1961).

Neuere Untersuchungen an *Oenothera* ergaben, daß die schon beschriebene Konkurrenz der Pollenschläuche, die am Verhalten der Testloci *s*, *fr* und *de* festgestellt wurde, nicht auf eine direkte Wirkung dieser Gene zurückzuführen ist, sondern auf ein zur gleichen Koppelungsgruppe gehörendes, gametophytisch wirksames Gen, das wahrscheinlich proximal im Chromosomenende 3, nahe dem Translokationspunkt und dem Locus *s* liegt (HARTE 1961). Die Fortsetzung dieser Untersuchungen führte zu einem orthogonalen Versuchsplan, in dem die Komplexe *flavens* und ^{h}*Hookeri* mit den Allelen *s* und $+^{s}$ kombiniert wurden, so daß alle möglichen Kombinationen vorhanden waren, mit $de/+^{de}$ als zusätzlichem Testlocus, und in Rückkrezungen mit den verschiedenen Homo- und Heterozygoten geprüft werden konnten. Auf diesem Wege ließ sich nachweisen, daß die Certation unter dem Einfluß dieses gametophytischen Locus unabhängig vom Plasmon und der Anwesenheit der verschiedenen Allele der beiden Testloci und auch unabhängig von der genetischen Konstitution des Griffels ist, aber durch unbekannte Umwelteinflüsse variiert werden kann. Frühere Befunde, die als Wirkung des Plasmons oder des Griffelgewebes gedeutet wurden, sind durch die Umweltvariabilität zu erklären (HARTE 1965).

CHAO (1927) lehnt beim Reis die Wirkung eines gametophytischen Faktors ab und nimmt die direkte Wirkung des *gl*-Gens auf das Pollenschlauchwachstum an. Die Durchsicht der F_2-Daten zeigt aber, daß ihre Interpretation auch durch die Annahme eines mit *gl* gekoppelten *ga*-Locus mit relativ hohem crossing-over zwischen beiden Loci möglich ist.

IV. Genphysiologie der Certation.

a) Veränderung der Certation unter dem Einfluß des Griffels.

Neben der phänotypischen Variabilität, die sich zeigt in Schwankungen der Benachteiligung einer Pollensorte nach Kreuzung mit dem Pollen aus verschiedenen Blüten einer Pflanze oder bei Wiederholung desselben Versuchs zu verschiedenen Zeiten, sind noch andere Einflüsse zu vermuten. Bei vergleichbaren Versuchen zeigte es sich schon sehr früh, daß es vielfach nicht gleichgültig ist, welche Kreuzungspartner verwendet wurden. CORRENS (1921b) mußte feststellen, daß alle Angaben über Durchschnittswerte größerer Versuchsserien bei *Melandrium* deswegen ohne Informationswert sind.

Die Verschiebung des Geschlechtsverhältnisses beim Vergleich von spärlicher und reichlicher Bestäubung ist z. B. sehr verschieden groß, je nach dem Elternpaar, das verwendet wurde. Bei der Zusammenfassung von Versuchen, in denen der Pollen für beide Serien von verschiedenen Pflanzen entnommen wurde, kann dadurch ein vollkommen falscher Eindruck entstehen. Die Nachrechnung seiner Daten (CORRENS 1921a) zeigt darüber hinaus, daß Pollen von verschiedenen ♂ bei Kreuzung mit dem gleichen ♀ das gleiche Geschlechtsverhältnis ergibt, während bei Kreuzung von Pollen eines ♂ auf mehrere ♀ der Unterschied in der Certation zwischen ♂- und ♀-bestimmendem Pollen gut gesichert ist. Die verschiedenen ♀ Pflanzen üben also einen unterschiedlichen Einfluß auf das Wachstum der beiden Sorten von Pollenschläuchen aus. Bei *Antirrhinum* ließ sich dagegen mit Sicherheit ein Einfluß der ♀ Kreuzungspartner in der Rückkreuzung $cae/cae \times cae/+^{cae}$ ausschalten (HARTE 1952). Allerdings wurde dabei im Gegensatz zu *Melandrium* mit Inzuchtlinien gearbeitet, für die eine weitgehende Übereinstimmung in allen übrigen Genen angenommen werden kann.

Zu besonders interessanten Ergebnissen kam die Analyse bei einigen Genen für Endospermmerkmale beim Mais. Die erste Untersuchung von CORRENS (1902) zeigt für das Gen *sugary* bei Selbstung der Heterozygoten ein deutliches Überwiegen von *Su* über *su* im Pollen.

Die χ^2-Zerlegung läßt erkennen, daß zwischen Pflanzen einer Familie und zwischen Familien (F_1 aus verschiedenen ♀ *sugary*-Pflanzen stammend) keine Unterschiede in der Häufigkeit von *su* in der F_2 bestehen (l. c., Zusammenfassung von Tabelle 1 und 2: χ^2 für Übereinstimmung mit der Erwartung 3:1 gibt $P < 0{,}001$; für Homogenität zwischen Familien $\chi^2 = 4{,}2$; $n = 1$; $0{,}05 > P > 0{,}02$; für Homogenität in Familien $\chi^2 = 21{,}3$; $n = 22$; $0{,}7 > P > 0{,}5$). Ebenso verhalten sich mehrere Kolben an einer Pflanze gleich (l. c., Tabelle 1: 8 Pflanzen mit je 2—3 Kolben: für Homogenität zwischen Pflanzen ist $\chi^2 = 8{,}2$; $n = 7$; $0{,}5 > P > 0{,}3$; für Homogenität in Pflanzen $\chi^2 = 22{,}0$, $n = 9$; $0{,}02 > P > 0{,}01$; $F = s^2$ in Pflanzen/s^2 zwischen Pflanzen $= 1{,}9$; $P > 0{,}05$).

Die Rückkreuzung von *Su/su* ♂ auf *su/su* ♀ zeigt jedoch einen unerwarteten Überschuß von *su*-Körnern, der von CORRENS infolge eines Rechenfehlers in seiner Tabelle 4 nicht bemerkt wurde. Jones (1924) bestätigte bei Fortführung der Versuche mit *su* das Defizit von *sugary*-Körnern nach Selbstung und fand den Überschuß bei Rückkreuzung auf die recessive Elternform.

Mit dieser Beobachtung kommt ein ganz neuer Gesichtspunkt in das Problem hinein. Um die Untersuchungen zu vereinfachen, wurde meist mit Selbstung der Heterozygoten oder mit Rückkreuzung auf den recessiven Elter gearbeitet. Diese Versuche können aber ausgedehnt werden durch Rückkreuzung auf beide Elterntypen, also auch auf die dominante Form. Dabei ist natürlich die Prüfung einer weiteren Generation nötig, um festzustellen, welcher Anteil der Nachkommen aus der Rückkreuzung jeweils durch den Pollen das dominante bzw. das recessive Allel erhalten hat, also homozygot oder heterozygot ist, um dann Rückschlüsse auf die Befruchtungshäufigkeit der beiden Sorten von Pollenschläuchen zu ziehen. Wegen dieser zusätzlichen Arbeit ist die Anzahl der untersuchten Fälle und der Umfang dieser Untersuchungen relativ gering, aber es ist in einigen dieser Fälle mit Sicherheit zu erkennen, daß bei Rückkreuzung auf die beiden Elternformen mit Pollen einer heterozygoten Pflanze jeweils verschiedene Pollenschläuche bevorzugt, sind und zwar meist diejenigen, die das Allel enthalten, das auch im Griffelgewebe vorkommt.

Von Jones (l. c.) wurde beim Mais diese Rückkreuzung auf den dominanten Elter durchgeführt (*Su*/*Su* × *Su*/*su*). Die Prüfung der Nachkommenschaft zeigte einen unerwartet hohen Überschuß an Homozygoten, die also aus dem Pollen das dominante Allel erhalten haben mußten (*Su*:*su* = 0,8:0,2). Aus der F_2 läßt sich auf Grund der nachgewiesenen normalen Spaltung in den Eizellen berechnen, daß unter den zur Befruchtung kommenden Pollenschläuchen das Verhältnis *Su*:*su* ~ 0,67:0,33 ist. Bei Selbstung der Heterozygoten ist also die Benachteiligung von *Su* geringer als bei Rückkreuzung auf die dominante Elternform. Ein versteckter plasmatischer Einfluß ist auszuschalten, da die Zahlenwerte für die Nachkommen der beiden reziproken F_1-Bastarde homogen sind. Die Konkurrenz zwischen den Pollenschläuchen mit *Su* und *su* ist nicht immer gleich groß, sondern hängt ab von der genetischen Konstitution des Griffelgewebes, in dem die Pollenschläuche wachsen. Bei Kreuzung auf homozygote ♀ sind immer die Pollen bevorzugt, die dasselbe Allel enthalten wie das Griffelgewebe, besonders deutlich bei *Su*, während in der Heterozygoten *Su*/*su* in bezug auf diese Wirkung eine unvollständige Dominanz von *Su* festzustellen ist.

Da es sich hier, wie von anderen nachgewiesen wurde, um die Wirkung des gametophytischen Faktors *ga* handelt, ergibt sich ein Widerspruch zu den späteren Untersuchungen. Von Jones wurde die Existenz derartiger Gene noch nicht berücksichtigt, die Planung der im folgenden erwähnten Versuche war aber gerade darauf abgestellt. Durch Auswertung eines orthogonalen Versuchs, in dem eine Anzahl verschiedener Genotypen in verschiedenen Rückkrezungen geprüft wurden, konnte Emerson (1934) nachweisen, daß Pollenschläuche mit dem recessiven Allel *ga* des gametophytisch wirksamen, mit *su* gekoppelten Gens des Chromosoms 4 auf Griffeln, die für *Ga* homo- oder heterozygot sind, relativ langsamer wachsen als solche mit *Ga*, während auf *ga*/*ga*-Griffeln keine Certation nachzuweisen ist.

In diesen Zusammenhang gehören auch die Fälle der nicht-reziproken Kreuzungssterilität, die beim Mais untersucht wurden (Demerec 1929, Nelson 1952, Schwartz 1950). Diese konnte auf die Wirkung einer Serie multipler Allele am erwähnten *Ga*-Locus im Chromosom 4 zurückgeführt werden. Pollenschläuche der Konstitution *ga* sind völlig gehemmt, wenn sie auf einem Griffel wachsen, der Ga^s/Ga^s oder Ga^s/Ga ist. Die Herstellung der F_1 gelingt nur, wenn *ga*/*ga* der ♀ Kreuzungspartner ist. Die Prüfung der Nachkommen aus Pollen der F_1 zeigt normale Spaltung bei Rückkreuzung auf *ga*/*ga* ♀ und Ausfall der *ga*-Pollen bei

Rückkreuzung auf Griffel, die Ga^s enthalten, also auch bei Selbstung. Das Allel *Ga* nimmt, wie oben erwähnt, eine Mittelstellung ein, indem *Ga*-Pollen auf allen Griffeln wachsen können und *ga*-Pollen auf *Ga/Ga*- und *Ga/ga*-Griffeln zwar relativ zu Ga^s- und *Ga*-Pollen gehemmt, aber nicht vollständig an der Befruchtung gehindert werden. Der Vergleich dieser Versuche mit denen von JONES läßt vermuten, daß in den letzteren vielleicht zufällig für die verschiedenen Rückkreuzungen Pflanzen mit unterschiedlicher Kombination der *ga*-Allele mit denen des *su*-Locus verwendet wurden, wodurch die Umkehrung der Certation erklärt wird.

Aus den Daten des umfangreichen Materials, das HERIBERT-NILSSON (1916) für den Rotnervfaktor von *Oenothera Lamarckiana* mitteilt, läßt sich entnehmen, daß dort die gleichen Verhältnisse vorliegen. In den Rückkreuzungen *rr* ♀ × *Rr* ♂ ist unter den zur Befruchtung gelangenden Pollenschläuchen das Verhältnis der Allele $R:r = 0{,}6:0{,}4$. Bei Selbstung, also Kreuzung des Pollengemisches $R + r$ auf Griffel der Konstitution *Rr*, läßt sich dieses Verhältnis aus den Daten desselben Autors berechnen als $R:r = 0{,}9:0{,}1$. Die Rückkreuzung auf die dominante Form *RR* ist wegen der Letalität dieser Homozygoten nicht durchzuführen. Aus den Daten ist zu erkennen, daß die Überlegenheit der Pollenschläuche mit *R* über *r* auf einem Griffel, der *R* enthält, viel größer ist als auf *rr*.

Die Verschiebung der Konkurrenz in Abhängigkeit vom Griffelgewebe kommt noch besonders deutlich zum Ausdruck in Versuchen mit *Oenothera* (RENNER 1919). Pollen des Bastards *Oe. Lamarckiana* × *muricata*, der die Komplexe *velans* und *curvans* enthält, wurde auf verschiedene Arten gekreuzt. Bei *Oe. Lamarckiana* und *suaveolens* wurde in der Konkurrenz mit *velans* praktisch keine Befruchtung durch *curvans* erzielt, während Kreuzungen mit *Oe. muricata* bei diesen Arten durchaus möglich sind. Auf *Oe. biennis* variierte das Verhältnis *velans*: *curvans* zwischen 30:1 und 12:1, auf *Oe. muricata* betrug es 3:1. Bei isolierter Betrachtung dieser letzten Kreuzung könnte der Eindruck entstehen, daß der aus *Oe. muricata* stammende Komplex *curvans* auf den Griffeln dieser Art stark benachteiligt wäre, aber der Vergleich mit den anderen Ergebnissen zeigt doch, daß *curvans*-Pollen auf dem Griffel der eigenen Art relativ am schnellsten wächst.

Daß es auch gametophytisch wirkende Gene gibt, die ihren Einfluß nur im Pollenschlauch ausüben und seine Wachstumsleistung bestimmen, ohne erkennbaren Zusammenhang mit der Konstitution des Griffelgewebes, konnten BUCHHOLZ und BLAKESLEE (1936) an *Datura* zeigen sowie JIMÉNEZ u. NELSON (1965) am Locus Ga^m 9 beim Mais.

Bei *Oenothera* sind in den Chromosomenenden 1 bis 4 mehrere Gene bekannt, die als Testloci für eine Untersuchung der Pollenschlauchkonkurrenz verwendet werden können. Es sind dies *s*, *fr*, *de* und der Translokationspunkt (HARTE 1948, 1953). In umfangreichen Versuchen, in denen die Allele dieser Loci in verschiedener Weise kombiniert waren und Pollen derselben heterozygoten Blüte jeweils auf mehrere Griffel verschiedener genetischer Konstitution verteilt wurde, gelang der Beweis, daß die genetische Konstitution des Griffels ohne Bedeutung ist. Die entscheidende Reaktion muß vielmehr entweder zwischen den Gonen in der Anthere der Heterozygoten oder zwischen den Gonen und der Pflanze, auf der sie entstehen, erfolgen (HARTE, unveröffentlicht).

Bei Selbstung einer Heterozygoten findet jeder Pollenschlauch im Griffel das entsprechende Allel vor und es könnte erwartet werden, daß bei Fehlen von Dominanz beide Sorten von PS in gleicher Weise reagieren, eine in den Rückkreuzungen feststellbare Konkurrenz bei Selbstung also aufgehoben wäre. Dies ist jedoch allem Anschein nach meist nicht der Fall, wenn auch keine allgemeine Aussage darüber möglich ist und es durchaus denkbar erscheint, daß es für einzelne Gene zutrifft.

Zur Klärung dieser Frage wäre es nötig, den Erbgang von vielen Genen, die nach Selbstung in der F_2 eine normale Spaltung mit ungestörter Vitalität aller Phänotypen zeigen, genau zu untersuchen in Rückkreuzungen auf beide Eltern mit Prüfung der Nachkommenschaft dieser F_1R-Pflanzen in ausreichendem Umfang und mit einer Protokollführung, die es gestattet, alle anderen Störungsursachen auszuschalten. Bisher liegt in der Literatur kein Fall vor, der für diese Fragestellung ausgewertet werden könnte.

b) Wechselwirkungen zwischen verschiedenen Genen.

Weitere Aufschlüsse geben hierzu die Untersuchungen über das Zusammenwirken mehrerer Gene bei der Certation. Ausführliche Daten liegen allerdings nur für 2 Objekte vor, Mais und *Oenothera*.

Die Ergebnisse der Kombination von *wx* und *Su* sind durch die Koppelung mit gametophytischen Faktoren noch nicht völlig geklärt. Pflanzen, die für *waxy* heterozygot, für *Su/Su* homozygot sind, weisen nur ein geringes Defizit an *wx* nach Selbstung auf (23% statt 25%). Ist die Pflanze dagegen homozygot für *su/su*, so wird das Defizit größer (10—20%). Die Rückkreuzungen zeigen, daß *Su* oder *su* im Griffelgewebe ohne Einfluß auf die Konkurrenz zwischen *Wx* und *wx* ist. Bei den doppelt-heterozygoten *Su/su Wx/wx* (Loci nicht gekoppelt) werden alle 4 Kombinationen zunächst in gleicher Häufigkeit gebildet. Das Defizit an *wx* ist unter den *Su*- und *su*-Nachkommen gleich groß.

Die Anwesenheit von *Su* oder *su* (oder des damit gekoppelten gametophytischen Gens *Ga/ga*) im gleichen Pollenkorn hat also keinen Einfluß auf die Certation von *wx*, sondern ein Unterschied läßt sich nur feststellen in Abhängigkeit von *Su* oder *su* in der Pflanze, auf der der Pollen gebildet wurde. Die wahrscheinlichste Deutung ist hier diese, daß bei einer Entwicklung der Pollen die Körner, die *wx* (oder den mit *wx* gekoppelten gametophytischen Faktor) enthalten, in Konkurrenz mit *Wx* auf einem *su/su*-Elter bereits benachteiligt sind und die genetische Konstitution des Sporophyten so auf die ersten Entwicklungsstadien der jungen Pollen einwirkt, daß davon die spätere Wachstumsreaktion beeinflußt wird (BRINK und BURNHAM 1927).

Ein Zusammenwirken zweier Gene wurde von SIRKS (1926) bei *Datura* untersucht. Bevorzugt sind die Allele *p* und *S*, gegenüber *P* und *S*. Der Einfluß von *p* ist dabei stärker als von *S*, denn *ps* ist überlegen gegenüber *Ps*. Ein weiteres Gen *B/b* zeigt keine Certation und beeinflußt auch nicht die Konkurrenz der beiden anderen Gene. Bei zwei weiteren Genen von *Oenothera*, *fr* und *s*, ist bei Selbstung und Rückkreuzung auf die recessive Elternform jeweils das recessive Allel unterlegen. In der Kombination addieren sich die Wirkungen, wobei sich zeigt, daß von *fr* der stärkere Einfluß ausgeht, so daß die Reihe entsteht: $+^{fr}+^{s} > +^{fr}s > fr\,+^{s} > fr\,s$ (HARTE 1948, 1953).

Es handelt sich hier um die Wirkung eines zu dieser Koppelungsgruppe gehörenden gametophytischen Gens, das wahrscheinlich mit dem in der Kombination *flavens*·*ʰHookeri* näher untersuchten identisch ist und dessen Wirkung durch den genetischen Hintergrund und Umwelteinflüsse stark variiert werden kann (HARTE 1961). Die Konkurrenzfähigkeit der durch das Allel *fr* markierten *stringens*-Chromosomen hängt im Pollen weitgehend von der Komplexverbindung ab, die jeweils vorliegt, und geht von fast völliger Ausschaltung von *fr* im Pollen von *ʰHookeri*·*stringens* über normale Spaltung von $fr/+^{fr}$ und $s/+^{s}$ in *flavens*·*stringens* bis zu deutlicher Überlegenheit in der Kombination *ᵇalbicans* $+^{fr}$·*stringens fr*. Die verschiedenen Kreuzungen dieses Vergleichs wurden alle auf *fr/fr*-Pflanzen derselben Herkunft ausgeführt. Für diese Unterschiede kann die Ursache also nicht im Griffelgewebe liegen, sondern muß durch das Zusammenwirken des gametophytischen, eng mit *fr* gekoppelten Gens mit anderen Genen bestimmt sein, die im einzelnen nicht bekannt sind (HARTE 1953, 1961). Eine andere Möglichkeit der Deutung dieser Befunde liegt in der Annahme einer Serie multipler Allele am gametophytisch wirksamen Locus. Die bisher vorliegenden experimentellen Daten gestatten keine Unterscheidung zwischen diesen Hypothesen.

Weiter wurden noch einige interessante Umkehrreaktionen beobachtet. Wie schon erwähnt, ist bei *Oenothera Lamarckiana* in der Pollenschlauchkonkurrenz im allgemeinen *R* gegen *r* überlegen, ebenso der Komplex *velans* über *gaudens*. Die Wirkung dieser beiden Komponenten addiert sich aber nicht einfach, sondern es ergibt sich die Reihe *R-velans* > *r-velans* > *r-gaudens* > *R-gaudens*. Die relative Wirkung der Komplexe bleibt sich also gleich (*velans* > *gaudens*), unabhängig von der Anwesenheit von *R* oder *r*, aber die relative Wirkung dieser beiden Allele, ob sie also fördernd oder hemmend wirken, hängt von der Kombination ab, in der sie auftreten (HIORTH 1926). In weiteren Versuchen von DE VRIES (1924) sind die Pollenschläuche, die einen abgeänderten *gaudens*-Komplex ohne den Letalfaktor enthalten, schneller als die mit dem unveränderten Komplex. Es erscheint daher möglich, daß die Unterlegenheit von *gaudens* mit der Anwesenheit des Letalfaktors verknüpft ist. Diese Versuche sagen jedoch nichts darüber aus, ob die Wirkung vom Letalfaktor direkt ausgeht oder von anderen, im gleichen Chromosomenende lokalisierten Genen bestimmt wird. Ein anderer Fall, der am besten durch eine Umkehrung der Konkurrenzfähigkeit bei Kombination von verschiedenen Genen zu erklären ist, wurde bei *Melandrium* gefunden (SHULL 1914, v. UBISCH 1922). Ein schmalblättriges ♂ ergab in den Nachzuchten Spaltungszahlen, aus denen hervorgeht, daß der Schmalblattfaktor in der Certation mit dem ♀—♂-Mechanismus zusammenwirkt, so daß die Reihe ♀B > ♂b > ♀b > (♂B) entsteht. Die letzte Kombination kam nie zur Befruchtung.

LONGLEY (1961) fand einen anderen gametophytischen Faktor beim Mais im Chromosom 5, bei dem die Pollenschläuche mit dem recessiven Allel benachteiligt sind, wenn der Griffel einen anderen, nicht zu dieser Koppelungsgruppe gehörenden Faktor enthält. Die Ausschaltung der Recessiven ist nicht vollständig. Die Stärke der Hemmung wird beeinflußt durch einen oder mehrere modifizierende Faktoren, für die nicht geklärt werden konnte, ob sie im Griffel oder im Pollenschlauch wirken, sowie durch nicht näher zu fassende Umweltbedingungen.

Es ist allgemein also sowohl ein Vorherrschen der Wirkung eines Gens, eine Addition der Wirkung verschiedener Gene als auch eine komplizierte, nicht aus der Reaktion der einzelnen Gene voraussagbare Wechselwirkung möglich.

c) Plasmonwirkung.

Die Versuche mit Oenotheren der Sect. *Raimannia* brachten noch weitere Komplikationen zutage (SCHWEMMLE 1951, 1952a, b, 1953, SCHWEMMLE und KOEPCHEN 1953).

Bei Selbstung von *Oe. Berteriana* (Komplexe B·1) besteht eine Konkurrenz zwischen den beiden Pollensorten mit deutlicher Überlegenheit von B über 1. Bei Selbstung der *Oe. odorata* werden Pollen mit deren Komplexen *v* und I gebildet. Die Pollenschläuche mit v wachsen bei Selbstung viel schneller als diejenigen mit I. Durch mehrfache Kreuzungen können die Komplexe beider Arten ausgetauscht werden, so daß Kerne der Konstitution v·I (= *Oe. odorata*) kombiniert sind mit dem Plasma von *Oe. Berteriana*. Die Pollenkörner enthalten dann v und I, aber im Plasma einer anderen Art. Bei Selbstung dieser Form ist die Konkurrenz zwischen v und I ebenfalls feststellbar, aber die Differenz ist deutlich geringer als in der *Oe. odorata*. Werden Narben dieser *Oe.* (v · I + *Berteriana*-Plasma) mit der ursprünglichen *Oe. odorata* bestäubt, dann zeigt sich eine Umkehrung der Konkurrenz, indem jetzt die I-Schläuche schneller wachsen als die v-Schläuche. Dasselbe tritt ein, wenn als Griffel die ursprüngliche *Oe. Berteriana* verwendet wird, die ganz andere Komplexe im Kern aufweist. Wenn Pollen der Konstitution (I + *odorata*-Plasma) und (I + *Berteriana*-Plasma) zusammen auf die Narben von *Oe. odorata* gebracht wird, zeigt sich der letztere als überlegen. Die Umkehrung kommt also dadurch zustande, daß die Hemmung von I im Griffel der *Oe. odorata* in eine Förderung umschlägt, sobald I mit einem anderen Plasmon kombiniert ist.

Hiermit vergleichbar sind die Ergebnisse über die Gonenkonkurrenz unter dem Einfluß einzelner Gene bei anderen *Oenothera*-Arten. Wenn die Komplexkombination *flavens* · *stringens* im Plasma von *Oe. suaveolens* realisiert ist, zeigt sie normalerweise fast keine Konkurrenz zwischen den Allelen *s* und $+^s$. Wenn jedoch dieselbe Kombination im Plasma der *Oe. strigosa* vorliegt, so ist die Konkurrenz deutlich feststellbar mit Unterlegenheit von *s* (HARTE 1948).

Das Wachstum der Pollenschläuche wird hier zwar von den Genen im Pollenkern gelenkt, aber diese reagieren im Zusammenwirken mit dem Plasmon, mit dem sie kombiniert sind, mit Stoffen, die im Griffelgewebe in Abhängigkeit vom Plasmon dieser Zellen gebildet werden, ohne daß die Beteiligung von Genwirkstoffen des Sporophyten angenommen werden muß.

Bei *Epilobium* wurde für die Spaltung des Faktors *Cr/cr* (cruciate Petalen) in Rassenkreuzungen der Art *E. hirsutum* ein geringer Überschuß an Dominanten gefunden, in der Artkreuzung *E. parviflorum* × *hirsutum cruciat* ein schwaches Defizit, woraus auf eine unterschiedliche Pollenschlauchkonkurrenz mit Überlegenheit von *Cr* in den *hirsutum*-Bastarden, von *cr* in den *parviflorum*-Bastarden geschlossen wurde (ESCHENBERGER 1956). Bei der nachgewiesenen Bedeutung des Plasmons in der Gattung *Epilobium* könnte hieraus eine Plasmonabhängigkeit der Certation abgeleitet werden. Die Nachprüfung der Spaltungszahlen ergibt aber, daß beide Einzelspaltungen noch gut mit der Erwartung 3:1 übereinstimmen, ebenso die Gesamtspaltung, und daß beide Kreuzungsgruppen nicht stärker voneinander abweichen, als bei zufälliger Probenentnahme erwartet werden kann (für 3:1 ist $\chi^2 = 1{,}63$; $n = 1$; $0{,}3 > P > 0{,}2$; für Homogenität: $\chi^2 = 5{,}01$; $n = 1$; $0{,}05 > P > 0{,}02$). Für diesen Fall kann also weder die Certation im allgemeinen noch ihre Umkehrung in verschiedenen Bastarden als bewiesen angesehen werden.

d) Zusammenhang der Certation mit der Selbststerilität.

Die Beobachtungen darüber, daß die Konkurrenz und die Überlegenheit eines bestimmten Allels beeinflußt wird von der genetischen Konstitution des Griffelgewebes, machen es wahrscheinlich, daß der Beeinflussung des Pollenschlauchwachstums, die zur Konkurrenz führt, wenigstens zum Teil eine Reaktion zugrunde liegt, die der Immunitätsreaktion bei selbststerilen Pflanzen ähnlich ist (vgl. Kapitel „Selbststerilität" dieses Handbuches).

Bereits JONES (1920) kam zu der Ansicht, daß eine physiologische Wechselwirkung zwischen Pollenschlauch und Griffelgewebe auf genetischer Grundlage für die Differenzen der Wachstumsgeschwindigkeit verantwortlich sein könnte. Bei Selbstung ist das Pollenschlauchplasma gleich dem Plasma der Griffelzellen, da beide von derselben Pflanze geliefert wurden. Diese Verwandtschaft scheint die Entwicklung zu begünstigen und schon geringe Differenzen wirken sich hier aus. JONES vergleicht darüber hinaus das Verhältnis zwischen Pollenschlauch und Griffel demjenigen bei einer Pfropfung, bei der die Verwachsung um so besser gelingt, je näher verwandt Reis und Unterlage sind, und nimmt an, daß der Hemmung eine Reaktion zugrunde liegt, die dem toxischen Effekt von Fremdproteinen bei Tieren verwandt ist.

Diese Auffassung der Bedeutung der plasmatischen Verwandtschaft stößt aber auf Schwierigkeiten. Einmal sind mehrere Fälle bekannt, in denen der Fremdpollen gefördert, der Selbstpollen gehemmt ist (vgl. Abschn. C IV b). Zum anderen kann diese Deutung nicht zutreffen für den Pollen von Heterozygoten. Hier werden zwei Sorten von Pollenkörnern auf derselben Pflanze gebildet und zeigen trotzdem ein unterschiedliches Wachstum. Dies spricht nicht dagegen, daß bei der Wachstumsreaktion Stoffe beteiligt sind, die im Cytoplasma der Pollenschläuche enthalten sind und im erwähnten Sinne wirken, aber es läßt sich ausschließen, daß diese mit dem genetisch definierten Plasmon zusammenhängen und ebenso, daß es sich um eine Prädetermination der Gametophyten im Sinne einer Nachwirkung der genetischen Konstitution des Sporophyten handelt, die alle auf einer Heterozygoten gebildeten Pollenkörner gleichartig betreffen müßte.

Bei Bestäubungen mit Pollenmischungen ist aber damit zu rechnen, daß beide Pollensorten auf Pflanzen verschiedener genetischer Konstitution ausgebildet wurden, so daß dort im Sinne einer Prädetermination eine Nachwirkung vorliegen könnte und die Reaktion in diesen Versuchen in Wirklichkeit nicht zwischen den Genprodukten des Griffelgewebes und denen des Pollenschlauchkernes, sondern denen der pollenbildenden Pflanze erfolgt, die als Reservestoffe im Pollen enthalten sind. Dieser Einwand kann ausgeschaltet werden, indem das natürliche Pollengemisch, das auf einer Heterozygoten entsteht, zum Vergleich verwendet wird. Hierbei standen beiden Sorten von Pollenkörnern die gleichen Nährstoffe zur Verfügung, auch wenn sie diese vielleicht in unterschiedlichem Maß aufgenommen haben. In einigen der beim Mais untersuchten Fälle verhielt sich das künstliche Gemisch genau wie das natürliche, während bei *Oenothera* ein entsprechender Versuch den Beweis dafür lieferte, daß sich die Konkurrenz zwischen den sich entwickelnden Pollenkörnern in der Anthere abspielt (vgl. C II b).

Diese Befunde lassen die Deutung zu, daß vielfach das Pollenschlauchwachstum nicht ausschließlich von den Genen des vegetativen Kernes bestimmt wird, sondern von diesem in Wechselwirkung mit der genetischen Konstitution des Griffelgewebes, in dem der Pollenschlauch wächst. Damit ist das Problem der Pollenschlauchkonkurrenz mit dem der Selbststerilität verbunden. Während bei Selbststerilität oder Selbstinkompatibilität diese Wechselwirkung zwischen Griffel und Pollenschlauch zu einer Hemmung führt, wenn beide die gleichen Allele enthalten, ist auch die umgekehrte Reaktion möglich, die einer Beschleunigung bei genetischer Gleichheit bzw. Hemmung bei genetischer Verschiedenheit. Die Fälle, in denen bei genetischer Gleichheit nur eine relative Verzögerung des Wachstums, aber keine absolute Hemmung eintritt, wären dann als Zwischenglieder anzusehen. Es wäre besonderer Untersuchung wert, diesen Reaktionen nachzugehen, die eine Begünstigung des Pollenschlauches bei genetischer Übereinstimmung bzw. eine Benachteiligung bei genetischen Unterschieden gegenüber dem Griffelgewebe bewirken, um festzustellen, ob hier grundsätzlich derselbe oder ein ganz anderer Mechanismus als bei Hemmung infolge genetischer Übereinstimmung vorliegt. Bei *Oenothera* findet wahrscheinlich eine entsprechende Inkompatibilitätsreaktion zwischen den genetisch verschiedenen, jungen Pollenkörnern oder zwischen ihnen und dem Antherengewebe statt.

e) Das Reaktionsschema der Certation.

Obgleich die bisherigen Untersuchungen über die Certation nicht darauf abgestellt waren, Reaktionstypen zu suchen, sondern sich vorwiegend damit befaßten, Feststellungen über das Vorkommen einer Konkurrenz zu treffen, so ist doch aus der Zusammenfassung aller Beobachtungen zu entnehmen, daß es sich um ein ausnehmend komplexes Geschehen handelt, das nicht auf einen Nenner zu bringen ist, sondern bei dem sehr verschiedenartige Reaktionen zusammenwirken.

Die Konkurrenz zwischen zwei genetisch verschiedenen Pollensorten kann auf verschiedenen Entwicklungsstadien stattfinden:

a) Während der Pollenentwicklung in der Anthere.
b) Bei der Keimung.
c) Beim Wachstum im Leitgewebe des Griffels.
d) Kurz vor der Befruchtung im Fruchtknoten.

Für diese Konkurrenz lassen sich verschiedenartige Einflüsse aufzeigen. Von Bedeutung sind folgende:

1. Die genetische Konstitution des Sporophyten kann eine Wirkung auf die Konkurrenz zwischen den genetisch verschiedenen Pollenkörnern in den Antheren haben, die beim Wachstum der Pollenschläuche nachwirkt: sporophytisch-genetische Prädetermination (*Oenothera;* vielleicht Mais; Wechselwirkung zwischen den mit *su* und *wx* gekoppelten gametophytischen Genen).

2. Der physiologische Zustand des Sporophyten während der Entwicklung der Pollenkörner in der Anthere hat einen Einfluß auf die Konkurrenz, so daß die genetisch verschiedenen Pollen dadurch unterschiedliche Startbedingungen erhalten, auf dem Weg über Ernährungseinflüsse: sporophytisch-phänotypische Prädetermination (z. B. *Antirrhinum,* Testgen: *caeca*).

3. Die genetische Konstitution des Gametophyten wirkt direkt auf das Wachstum (Mais: mit *wx* gekoppelter gametophytischer Faktor).

4. Die genetische Konstitution des Gametophyten tritt in Wechselwirkung mit der genetischen Konstitution des Griffelgewebes (z. B. Mais: *Ga*; vermutlicher Wirkungstyp der Certation bei verschiedenen Objekten). Im letzteren Fall kann diese Wechselwirkung auf homozygoten Griffeln entweder zu einer Förderung der

genetisch gleichartigen Pollenschläuche, oder zu einer Förderung der Pollenschläuche mit dem abweichenden Allel führen. In heterozygoten Griffeln kann völlige Dominanz oder intermediäre Wirkung gegeben sein. Förderung und Hemmung sind in vielen Fällen nicht absolut, sondern relativ zur Vergleichsgruppe.

5. Das Plasmon, mit dem die Pollenschlauchkerne kombiniert sind, beeinflußt die Reaktion (Komplexe v und I im Plasma von *Oe. odorata* und *Berteriana* auf Griffel der *Oe.* v · I + *Berteriana*-Plasma).

6. Das Plasmon des Griffelgewebes verändert die Reaktion der Pollenschläuche bestimmter genetisch-plasmonischer Konstitution (Komplexe v und I + *odorata*-Plasma auf Griffel mit Plasma von *Oe. odorata* oder *Berteriana*).

Aus der Zusammenfassung dieser Gesichtspunkte ergibt sich folgende Reaktionskette beim Pollenschlauchwachstum. Unter der Wirkung einer Reihe von Genen des vegetativen Kernes ist die Ausgangsgröße der Wachstumsgeschwindigkeit des Pollenschlauches bestimmt. Diese Anfangssituation kann variiert werden durch den Einfluß der genetischen Konstitution des Sporophyten, auf dem die Pollenkörner entstehen oder von den Pollenkörnern aufeinander, und von Umweltfaktoren, die über den physiologischen Zustand des Sporophyten indirekt auf die Pollenkörner während ihrer Entwicklung einwirken. Wenn die Pollenkörner mit dieser verschiedenen Ausgangssituation auf die Narbe gelangen und keimen, so kann zusätzlich im Pollenschlauch eine Wechselwirkung zwischen den Genen, die während des Wachstums wirksam werden, unter dem Einfluß des Plasmons einerseits und der genetisch-plasmonischen Konstitution des Griffelgewebes, in dem der Pollenschlauch wächst, andererseits einsetzen. Wahrscheinlich erfolgt dies nach Art einer Immunitätsreaktion, die eine relative Hemmung bzw. Förderung bestimmter Genotypen bewirkt. Hierdurch wird während der Wachstumsphase selbst die endgültige Geschwindigkeit dieses Vorgangs bestimmt, die demnach für die einzelnen Genotypen verschieden ist. Die Folge ist eine Konkurrenz der Pollenschläuche in Abhängigkeit von ihrem Genotyp, die bei einer begrenzten Anzahl von befruchtungsfähigen Samenanlagen im Fruchtknoten eine unterschiedliche Befruchtungswahrscheinlichkeit bedingt.

Die Beweisführung für das Zutreffen oder den Ausschluß eines bestimmten Einflusses in einem gegebenen Fall ist zwar theoretisch sehr einfach und die dazu notwendigen Versuchspläne sind leicht aufzustellen, aber in der Praxis entstehen durch den erforderlichen Umfang der Kreuzungen, die Aufzucht der Nachkommenschaften und die für die Lösung mancher Fragen notwendige genetische Prüfung jeder einzelnen Pflanze durch Testkreuzungen und Untersuchung ihrer Nachkommenschaft so große Schwierigkeiten, daß die Analyse bisher nur in sehr wenigen Fällen über das erste Stadium hinaus vorangetrieben wurde.

V. Vermutete Zusammenhänge zwischen Pollenschlauchwachstum und Vitalität der nächsten Generation.

Die größere Wachstumsgeschwindigkeit der PS wurde zunächst als größere Vitalität gedeutet und daraus geschlossen, daß die durch solche Pollenschläuche übertragenen Gameten auch diese stärkere Vitalität vererben würden und die aus den ersten Bestäubungen entstehenden Pflanzen dementsprechend kräftiger sein müßten. Diese Vorstellung von einem Zusammenhang der Vitalität der haploiden Generation und der daraus entstehenden Diplonten wurde zunächst von CORRENS an *Mirabilis* (1900) als Hypothese entwickelt, dann von HERIBERT-NILSSON (1920) als sicher hingestellt. Diese Ansicht wurde auch beibehalten, obwohl verschiedenartige Teste, die diese Autoren zur Stützung ihrer eigenen Ansichten durchgeführt hatten, keinen solchen Zusammenhang erweisen konnten. Erst eine exakte Nachprüfung durch HIEMENZ (1955) zeigte die Haltlosigkeit dieser Behauptung.

Im Gegenteil sind Samen von *Salpiglossis*, die aus spärlicher Bestäubung stammen, meist größer und ergeben kräftigere Sämlinge als solche aus dicht gefüllten Früchten nach reichlicher Bestäubung, bei der nur die schnellsten PS zur Befruchtung gelangten. Die Ursache hierfür liegt darin, daß bei nur wenigen Samen die Embryonen besser ernährt werden und die Konkurrenz der Embryonen um die Nährstoffe und den Platz in der jungen Frucht wegfällt. Dieser Vorsprung der Sämlinge auf Grund ihrer besseren Ausstattung mit Reservestoffen wird jedoch im Lauf der weiteren Entwicklung allmählich ausgeglichen, insbesondere bei Pflanzen mit langer Vegetationsdauer, bis schließlich kein Unterschied mehr festzustellen ist. Es ist also mit Sicherheit kein Einfluß vorhanden, der gedeutet werden könnte im Sinne der Entstehung von kräftigen Embryonen und Pflanzen aus kräftigen Pollenschläuchen.

Die Konkurrenz zwischen den Haplonten hat keinen selektiven Einfluß auf die Vitalitätseigenschaften der diploiden Generation, sondern jede ist ihren eigenen Entwicklungsgesetzen unterworfen. Der Einfluß bestimmter Allele auf das Wachstum der Haplonten läßt keinen Rückschluß auf ihre Wirkung im Diplonten zu, noch ist von den genetisch bedingten Eigenschaften der 2n-Generation aus eine Voraussage auf das Verhalten der Haplonten mit den entsprechenden Allelen möglich, wie die einwandfreie Übertragung von letalen und subletalen Allelen beweist.

Die statistische Nachprüfung der Differenzen zwischen Samen aus spärlicher und reichlicher Bestäubung bei *Mirabilis* (CORRENS 1900) zeigte zusätzlich, daß die Bedeutung der geringen Differenzen überschätzt wurde. Unter Berücksichtigung der Unterschiede zwischen gleichartigen Versuchen liegen alle im Bereich der Zufallsstreuung ($P = 0{,}03$ für die Differenz des Pflanzengewichts nach Bestäubung mit ein oder mit vielen Pollenkörnern).

D. Gonenkonkurrenz in der Samenanlage.

I. Grundlagen.

Für die Gonenkonkurrenz in den Samenanlagen sind ganz andere entwicklungsgeschichtliche Grundlagen gegeben als bei den Pollenschläuchen. Wie schon dargelegt wurde, entsteht im Normalfall in jeder Samenanlage aus der Makrosporenmutterzelle durch die Meiosis eine einzige Tetrade, in der die Gonen linear angeordnet sind. Eine Gone entwickelt sich dann zum Embryosack weiter. Von den 4 Gonen werden, in Abhängigkeit von der genetischen Konstitution der Mutterpflanze, je zwei gleichartig oder, bei starker Heterozygotie und hohem crossing-over, alle vier verschieden sein, aber für jedes Allel werden sie paarweise übereinstimmen. Die Konkurrenz findet nicht, wie bei den Pollenschläuchen, zwischen großen Mengen von Gonen bzw. Haplonten statt, sondern zwischen diesen 4 Zellen, von denen nur eine gewinnen kann.

Bei den meisten Objekten ist bekannt, daß die Entscheidung darüber, welche Gone bevorzugt sein wird, durch die Lage in der Tetrade gefällt wird. Nur bei wenigen Arten entwickelt sich vorwiegend diejenige, die an der Mikropyle liegt (z. B. bei *Oenothera*), meist dagegen diejenige am anderen Ende der Tetrade. Es findet also eine Konkurrenz statt, die aber entschieden wird auf Grund des Vorteils, der sich für die betreffende Gone aus ihrer Lage in der Samenanlage ergibt. Die Frage, ob daneben noch andere Einflüsse, so besonders die genetische Struktur der verschiedenen Gonen bei Heterozygotie der Mutterpflanze, mitwirken bei der Entscheidung darüber, ob auch eine der anderen Gonen zur Entwicklung eines Embryosackes gelangen kann, ist bisher nur an *Oenothera* genauer bearbeitet. Diese Frage ist aber interessant in bezug darauf, ob bei der Entwicklung des ♀ Gametophyten Genwirkungen feststellbar sind, oder ob in diesem durch seine enge Verbindung mit dem Sporophyten keine selbständige Steuerung der Entwicklungsvorgänge möglich ist. Der erste Nachweis einer selbständigen Genwirkung im Embryosack wurde von BRINK (1925) geführt durch den Nachweis, daß auf einer heterozygoten Maispflanze *Wx/wx* die beiden Typen von Embryosäcken der Konstitution *Wx* bzw. *wx* an ihren Kohlenhydraten eindeutig zu unterscheiden sind.

Der Nachweis einer Gonenkonkurrenz in der Samenanlage bei der Bildung des Embryosackes stellt an die Versuchsplanung dieselben Ansprüche wie die Untersuchung der Pollenschlauchkonkurrenz, nämlich Ausschluß aller anderen Faktoren, durch die eine Störung der Spaltungszahlen hervorgerufen werden könnte. Eine gesicherte Abweichung von der Erwartung einer Spaltung 1:1 in der Rückkreuzung der Eizellen einer Heterozygoten mit Pollen der homozygot-recessiven Elternform gibt dann Hinweise auf das Vorkommen einer Gonenkonkurrenz. In vielen Versuchen über die Certation wurden gleichzeitig die Spaltungszahlen bei Verwendung der Eizellen geprüft. In den meisten Fällen konnten keine Abweichungen gefunden werden. Obgleich *Oenothera* das einzige Objekt ist, für das Gonenkonkurrenz in der Samenanlage bisher nachgewiesen wurde, besagt dies nicht, daß bei anderen Pflanzen eine solche nicht vorkommt, aber es wurde noch nicht systematisch danach gesucht. Mit Sicherheit kann daher nur gesagt werden, daß einer Konkurrenz der ♂ Haplonten nicht unbedingt eine der ♀ Gonen oder Haplonten zu entsprechen braucht. Es ist zur Zeit keine Aussage darüber möglich, ob es etwa auch Fälle gibt, in denen eine Konkurrenz der ♀ Haplonten vorliegt, aber normale Spaltung bei Rückkreuzung mit dem Pollen der Heterozygoten.

In den schon erwähnten Fällen von „preferential segregation" beim Mais ließ sich Konkurrenz als Ursache durch cytologische und embryologische Untersuchungen ausschalten.

II. Einfluß der Genkomplexe bei *Oenothera*.

a) Variabilität der Gonenkonkurrenz.

Die ersten Beobachtungen über das Vorkommen von abweichenden Spaltungszahlen in genetischen Versuchen bei *Oenothera*, die durch eine Gonenkonkurrenz bei der Bildung des Embryosacks ihre Erklärung finden können, wurden von RENNER (1917, 1918, 1921) gemacht, und zwar zunächst in bezug auf ganze Genkomplexe.

Bei einigen der komplex-heterozygotischen *Oenothera*-Arten sind beide Komplexe gleichmäßig in den Eizellen vertreten (z. B. *velans* und *gaudens* in *Oe. Lamarckiana*), während bei anderen, heterogamen Arten der Komplex, der im Pollen aktiv ist, nicht in der erwarteten Häufigkeit, d. h. 50%, in den Eizellen auftritt (z. B. *curvans* in *Oe. muricata*, *rubens* in *Oe. biennis*). Diese Komplexe kommen äußerst selten in den Eizellen vor, aber sie fehlen nicht vollständig. Andererseits zeigten die histologischen Untersuchungen, daß abortierende Samenanlagen nicht häufiger sind als bei den homozygoten Arten. Es kann daher die Wirkung von haplophasischen Letalfaktoren ausgeschlossen werden. Als Erklärung wurde angenommen, daß hier nicht, wie bei den homozygoten und den isogamen Arten, die Lage der Gonen in der Tetrade darüber entscheidet, welche sich zum Embryosack entwickeln wird, sondern daß daneben die genetische Konstitution der Gonen mitspricht.

Der nächste Schritt war die Untersuchung der sog. diplarrhenen Verbindungen, in denen 2 Pollenkomplexe vereinigt sind. RENNER (1940) stellte fest, daß die Pollenkomplexe allgemein eine „konstitutionelle Schwäche" aufweisen, wodurch die Fähigkeit dieser Gonen, einen Embryosack zu bilden, stark vermindert wird. Die dadurch frei werdenden Plätze würden in den Samenanlagen durch eine andere Gone, die dann den antagonistischen Komplex enthält, eingenommen. Weitere Untersuchungen zeigten aber, daß die Konkurrenzfähigkeit bestimmter Komplexe weitgehend von dem Antagonisten bestimmt wird, mit dem sie jeweils kombiniert sind. So ist *gaudens* in *Oe. Lamarckiana* in etwa 50% der Eizellen vertreten, wird dagegen in der Kombination mit *^{h}Hookeri* fast völlig verdrängt (RENNER 1925). Ebenso kann der Pollenkomplex *punctulans* sich gegenüber *flavens* stärker in den Eizellen durchsetzen als gegenüber *excellens* in seiner Stammart *Oe. chicaginensis* (RENNER und CLELAND 1933).

Die quantitative Untersuchung dieser Frage führte zu weiteren Einsichten in denMechanismus, der die Gonenkonkurrenz bestimmt. Die genetisch-entwicklungsgeschichtlichen Grundlagen der Versuchsplanung für die Bearbeitung dieser Fragestellung wurden bereits ausführlich dargestellt, so daß hier darauf verwiesen werden kann (HARTE 1954, 1955, 1958a). Eine große Anzahl von Eizellen- und Pollenkomplexen wurde in vielen verschiedenen Kombinationen zusammengebracht und mit Hilfe der Markierung der Komplexe mit Letalfaktoren geprüft, in welchem Umfang die beiden miteinander in dem betreffenden Bastard vereinigten Komplexe in der Eizelle auftraten und wieweit dieses Verhältnis variieren kann.

Überraschenderweise stellte sich heraus, daß die Variabilität der Konkurrenz zwischen zwei bestimmten Komplexen sehr groß ist. Sowohl innerhalb der Pflanzen wie zwischen verschiedenen Geschwisterpflanzen der gleichen Aufzucht bestehen erhebliche, statistisch gesicherte Differenzen in der Häufigkeit, mit der der Pollenkomplex in der Eizelle auftritt. Die Konkurrenzfähigkeit eines Pollenkomplexes in der Kombination mit einem bestimmten Eizellenkomplex liegt nicht unveränderlich fest, sondern ist einer starken Umweltvariabilität unterworfen, die aber nicht in erkennbarer Abhängigkeit vom Entwicklungszustand der Pflanze oder von bekannten Umwelteinflüssen steht (HARTE 1958a).

b) Einfluß von Umweltfaktoren.

Extreme Veränderungen der Umwelt, wie Unterschiede zwischen Witterungsbedingungen in verschiedenen Jahren oder Mangelkultur, bedingen in der Hauptsache eine Vergrößerung der Variabilität zwischen den Pflanzen. Es gibt also Umweltfaktoren, die die Konkurrenz zwischen 2 Komplexen in den Samenanlagen im Sinne einer Förderung eines der beiden und einer Hemmung des anderen verändern, aber durch die untersuchten Extrembedingungen wird vor allem eine erhöhte Labilität der Pflanzen erzielt, so daß diese mit verstärkten Ausschlägen auf die Einwirkung der übrigen, im einzelnen noch unbekannten Faktoren reagieren (HARTE 1958b).

c) Einfluß genetischer Faktoren.

Der Vergleich der verschiedenen Bastarde, in denen derselbe Pollenkomplex mit verschiedenen Eizellenkomplexen, bzw. derselbe Eizellenkomplex mit verschiedenen Pollenkomplexen verbunden war, zeigt an, daß vom Pollenkomplex der stärkere Einfluß auf die Konkurrenz zwischen den Gonen um die Bildung des Embryosackes ausgeht. Ein Pollenkomplex reagiert aber sehr verschiedenartig auf die einzelnen Kombinationen. Der Eizellenkomplex folgt nicht nur passiv, sondern wirkt aktiv an der Konkurrenz mit, so daß die Häufigkeit, mit der ein Pollenkomplex in einer gegebenen Kombination in der Eizelle auftritt, wesentlich vom jeweiligen Partnerkomplex mitbestimmt wird. Dabei ist es aus der Kenntnis des Verhaltens der beiden Komplexe mit anderen Kombinationen nicht vorauszusagen, wie in einer bestimmten, neuen Kombination das Ergebnis sein wird, denn es bestehen deutliche Wechselwirkungen zwischen beiden Partnern von der Art, daß mit einem Pollenkomplex der Partner einer Kombination stark, der einer anderen schwach sein kann, während dieselben Eizellenkomplexe mit einem anderen Pollenkomplex das umgekehrte Stärkeverhältnis erkennen lassen (HARTE 1958c).

Wenn 2 Eizellenkomplexe miteinander vereinigt werden, so sollte erwartet werden, daß beide in der Hälfte der Eizellen auftreten, oder daß einer stärker ist und den anderen zurückdrängt. OEHLKERS (1943) gelang eine derartige Verbindung zwischen *albicans* (aus *Oe. biennis*) und *curtans* (aus *Oe. Cockerelli*). Hierbei zeigte sich die gleiche Erscheinung wie bei den anderen Artkreuzungen: Es trat weder die eine noch die andere Erwartung ein, sondern eine extreme, nicht durch zufällige Schwankungen erklärliche Variabilität im Vorkommen der beiden Komplexe in den Eizellen. Die Häufigkeit von *curtans* schwankte zwischen 10,7, 57,6 und 65,5% der Samenanlagen (χ^2-Test für Homogenität: $P < 0{,}001$).

Diese genetischen Faktoren wirken jedoch mit den Umwelteinflüssen zusammen. Sie bestimmen im wesentlichen die Variationsbreite, innerhalb derer in einer gegebenen Kombination die Umwelt variierend wirken kann. Besonders auffallend ist die Feststellung, daß dieser Variationsbereich für die Kombinationen, die in den natürlich vorkommenden Arten gegeben sind, relativ eng ist, während er für Bastarde, in deutlicher Abhängigkeit von der Verwandtschaft der miteinander gekreuzten Arten, stark vergrößert ist. In den Arten müssen die Einflüsse, die von den beiden Komplexen ausgehen, so gegeneinander ausbalanciert sein, daß ein gegen Veränderungen der Umwelt gut gepuffertes System entsteht. In den Bastarden wird dieses Gleichgewicht gestört. Dadurch werden die Pflanzen anfällig gegen variierende Einflüsse, die auf den Vorgang der Gonenkonkurrenz einwirken. Für den Ausgang der Konkurrenz in den Samenanlagen ist somit ein Zusammenspiel von genetischen und physiologischen Faktoren maßgebend (HARTE 1958c).

III. Einfluß einzelner Gene auf die Gonenkonkurrenz und genetische Wechselwirkungen.

Auch über den Einfluß einzelner Gene auf die Gonenkonkurrenz in der Samenanlage liegen einige Beobachtungen an *Oenothera* vor.

Für den Komplex *stringens* von *Oenothera strigosa* konnte festgestellt werden, daß die Unterlegenheit in der Konkurrenz in der Samenanlage nicht auf einem einzelnen Gen beruht, sondern wahrscheinlich durch das Zusammenspiel sehr vieler Faktoren, die in verschiedenen Chromosomen lokalisiert sind, zustande kommt (Harte 1943, 1948). In spaltenden Bastarden, die diesen Komplex enthalten, sind eine ganze Anzahl von Genen gefunden, die in normalen Zahlenverhältnissen in den Eizellen realisiert werden können (Oehlkers 1923, Harte 1942, 1948). Andererseits wurde eine *lutescens*-Form gefunden, die einen abgeänderten *flavens*-Komplex enthält, der nicht in den Eizellen auftreten kann. Da diese Form aus einer Kreuzung zwischen *Oe. suaveolens* und *strigosa* hervorgegangen war, ist zu vermuten, daß hier Faktoren, die die Unterlegenheit von *stringens* wesentlich mitbestimmen, durch crossing-over auf den Komplex *flavens* übergegangen sind (Oehlkers 1923). Der Komplex *elongans* aus *Oe. Cockerelli* kann sich in Kreuzungen so verändern, daß er auch in den Eizellen auftreten kann (Oehlkers 1943). Der Faktor R/r (Rotnervigkeit), der im Pollen eine ausgeprägte Certation erkennen läßt, zeigt in den Eizellen normale Spaltung (Heribert-Nilsson 1916ff.).

Für $s/+^{s}$ als Testgen ist die Gonenkonkurrenz schon länger bekannt (Emerson 1931, Oehlkers 1933), später wurde sie auch für $fr/+^{fr}$ nachgewiesen. Dabei sind in den Eizellen bei der Konkurrenz dieselben Allele benachteiligt, die auch im Pollen des betreffenden Bastards unterlegen sind (Harte 1948, 1953).

Die Konkurrenz um die Bildung des Embryosacks in der Samenanlage konnte, ebenso wie die Certation, auf die Wirkung eines gametophytischen Gens in der Koppelungsgruppe mit den Chromosomenenden 1 bis 4 zurückgeführt werden, das entweder mit dem bei der Pollenentwicklung wirkenden Gen identisch oder sehr eng mit ihm gekoppelt ist. In der Koppelungsgruppe der Chromosomenenden 5—6—13—14 ließ sich ein weiterer, gametophytisch im ♀ Geschlecht wirkender Faktor nachweisen, der einen völligen Ausfall der Makrosporen, die die *stringens*-Chromosomen 5·14, 6·13 enthalten, bewirkt, aber im Pollen unwirksam ist, so daß diese Chromosomen dort ungehindert übertragen werden (Harte 1961).

Im Zusammenwirken dieser beiden Loci zeigt sich, daß die Häufigkeit, mit der sie einzeln, bzw. in der Doppelheterozygoten in den verschiedenen Kombinationen in den Eizellen auftreten, wesentlich von der Komplexkombination abhängt. Die Konkurrenz wird in Wechselwirkung mit dem gesamten genetischen Hintergrund bestimmt. Die Variabilität geht dabei von fast normaler Spaltung bis zu fast völligem Fehlen der Recessiven in den Eizellen.

Im Zusammenwirken mehrerer Gene kann entweder ein Gen bzw. eine Chromosomengruppe entscheidend sein oder es kann eine Kombination der Wirkung verschiedener Gene stattfinden (Beispiele für beide Fälle bei Harte 1953, 1961).

IV. Vergleich der Konkurrenz in Pollenschlauch und Samenanlage.

Die Koppelungsgruppe der Chromosomenenden 1 bis 4 ist bisher die einzige, für die eine Konkurrenz sowohl im Pollenschlauch wie in den Samenanlagen festgestellt wurde. Auffallend ist, daß die Konkurrenzverhältnisse in jedem Fall stark variieren und daß die Variationsreihe, die sich ergibt, wenn die Komplexkombinationen geordnet werden nach der relativen Häufigkeit, mit der die Keimzellen mit den entsprechenden Allelen zur Befruchtung gelangen, für Pollen und Eizellen gleich ist. Diese Gleichheit der Reaktion wurde dadurch erklärt, daß es sich sowohl bei der Bildung der Pollenschläuche wie bei der Entstehung des Embryosackes aus der Makrospore um Wachstumsvorgänge handelt, bei denen eine genetische Steuerung in gleicher Weise denkbar ist (Harte 1953). Inzwischen wurde festgestellt daß bei *Oenothera* der Certation eine Wechselwirkung zwischen der genetischen Struktur des Pollen und derjenigen der Heterozygoten, auf der er gebildet wird, oder zwischen den Gonen zugrunde liegt. Die Gonen befinden sich nun in der Samenanlage in einem Kontakt miteinander und mit dem Gewebe der Mutterpflanze, der noch enger sein dürfte als dies zwischen Pollen und Antherengewebe der Fall ist. Es ist daher denkbar, daß auch hier die Konkurrenz durch eine Wechselwirkung, die entweder direkt zwischen den Gonen oder zwischen ihnen und dem diploiden, heterozygoten Gewebe, in das sie eingebettet sind, besteht, beeinflußt wird.

Im Gegensatz zur Untersuchung der Certation ist hier keine Möglichkeit gegeben, die genetische Umwelt, d. h. den Sporophyten, auf dem die Konkurrenz stattfindet, zu verändern, da dieser immer heterozygot sein muß, wenn die genetische Grundlage der Konkurrenz zwischen den Gonen geprüft werden soll.

Der Ausgang der Konkurrenz zwischen den 4 Gonen ist daher bestimmt als Resultante aus verschiedenartigen Einwirkungen. Diese sind: die Lage der Gone in der Tetrade, wechselnde Außeneinflüsse, die durch die Mutterpflanze vermittelt werden oder direkt auf die Gonen einwirken, die genetische Struktur der Gonen und den genannten Wechselwirkungen.

Beim Vergleich der vorliegenden Daten fällt auf, daß die gametophytisch wirkenden Gene, für die eine Wechselwirkung zwischen den wachsenden Pollenschläuchen und dem Griffelgewebe nachgewiesen wurde, in den Eizellen eine normale Spaltung ergeben. Sie sind also im ♀ Gametophyten unwirksam. In den Fällen, in denen die für die Certation verantwortliche Reaktion in der Anthere bei der Entwicklung der Pollenkörner abläuft, findet sich eine gleichartige Konkurrenz in beiden Geschlechtern. Es ist in diesem Fall nicht geklärt, ob es sich um die Wirkung eines Gens handelt oder um zwei sehr eng gekoppelte Gene, von denen eines in den Pollenkörnern, das andere in den Gonen der Samenanlage wirkt. Der ersten Möglichkeit, *ein* Gen, kommt dabei die größere Wahrscheinlichkeit zu, da sowohl bei einer gegenseitigen Beeinflussung der benachbarten Gonen einer Tetrade oder in einer Anthere als auch bei einer Abhängigkeit der Gonen von ihrer Mutterpflanze in beiden Geschlechtern derselbe Reaktionsmechanismus angenommen werden kann. Es hat demnach den Anschein, daß es zwei verschiedene Mechanismen der Gonenkonkurrenz gibt.

E. Zusammenfassung.

Der Grund dafür, daß bisher die Analyse nur an wenigen Objekten vorgenommen wurde, liegt darin, daß in jedem Fall zunächst die theoretisch bestehenden Möglichkeiten durchdacht werden müssen und dann ein diesen angepaßter Versuchsplan entwickelt werden muß, der eine Entscheidung über die Struktur des vorliegenden Systems gestattet. Bei der Vielzahl der zu beachtenden Komplikationen läßt sich dies nur in einer auf dieses Ziel hin ausgerichteten Untersuchung erreichen. Die vielen noch offenen Fragen rühren daher, daß in manchen Fällen die referierten Befunde Nebenergebnisse der Untersuchung anderer Probleme sind, bei denen der Aufbau des Versuchs nicht für eine tiefgehende Analyse der Gonenkonkurrenz geeignet war.

Die Zusammenfassung der vorliegenden Untersuchungen über die Gonenkonkurrenz bei höheren Pflanzen zwischen den Pollenschläuchen und in der Samenanlage zeigt, daß diese so einfach aussehenden Vorgänge wie das Wachstum einer Zelle, des Pollenschlauchs oder die Vergrößerung einer Gone zum Embryosack vom entwicklungsphysiologischen Standpunkt aus extrem komplizierte Vorgänge sind. Die primäre Ursache für Differenzen der Entwicklungsgeschwindigkeit bzw. der Konkurrenzfähigkeit liegt in der genetischen Konstitution der Haplonten. Auf dieser Grundlage reagiert aber der Haplont auf genetisch gesteuerte und von der Umwelt her bestimmte Einflüsse, die ihn treffen von der

Pflanze her, auf der er entsteht, und von der Pflanze her, auf der er die Wachstumsvorgänge durchmacht. Für die Gonen, die zum Embryosack werden, sind diese beiden identisch, aber für den Pollenschlauch komplizieren sich die Vorgänge noch weiter dadurch, daß der ♂ Haplont während seiner Entwicklung unter dem Einfluß von zwei verschiedenen Sporophyten steht. Der Pollen wird auf einer Pflanze gebildet und der Pollenschlauch wächst auf einem Griffel, der zu einer anderen Pflanze gehören kann. Aus dieser Konstellation sind viele Möglichkeiten einer Beeinflussung der Wachstumsvorgänge auf genetischer und phänotypischer Grundlage abzuleiten, von denen für die meisten Fälle bisher bereits Beispiele gefunden wurden. Solange nur Einzelfälle betrachtet wurden, schienen in bezug auf die Abhängigkeit der Gonenkonkurrenz von verschiedenen Faktoren manche Widersprüche vorhanden zu sein, aber die Zusammenfassung läßt die Verzahnung der einzelnen Vorgänge erkennen und zeigt neben den Komplikationen auch die Möglichkeiten, allgemeine Gesetzmäßigkeiten zu finden. Damit sind jetzt, in Verbindung mit den Befunden über Selbststerilität und selektive Befruchtung, die Grundlagen gegeben für eine genetisch fundierte, entwicklungsphysiologische Analyse der haploiden Generation der höheren Pflanzen.

Literatur.

BALLS, W. L.: The cotton plant in Egypt. London 1919. — BATEMAN, A. J.: Cryptic self-incompatibility in the wallflower: *Cheiranthus cheiri* L. Heredity **10**, 257—261 (1956). — BEMIS, W. P.: Selective fertilization in lima beans. Genetics **44**, 555—562 (1959). — BRIEGER, F. G.: Mendelian factors producing selective fertilization. Amer. Naturalist **60**, 183—191 (1926). — Genetic control of gametophyte development in maize. I. A gametophyte character in chromosome five. J. Genet. **34**, 57—80 (1937). — BRIEGER, F. G., G. E. TIDBURY and H. P. TSENG: Genetic control of gametophyte development in maize. II. The quarter test. J. Genet. **36**, 17—38 (1938). — BRINK, R. A.: The physiology of pollen. I., II., II., IV. Amer. J. Bot. **11**, 218—228, 283—294, 351—364, 417—436 (1924). — Mendelian ratios and the gametophyte generation in angiospermes. Genetics **10**, 359—394 (1925). — BRINK, R. A., and C. R. BURNHAM: Nucleus and cytoplasm in relation to differential pollen-tube growth. Proc. nat. Acad. Sci. (Wash.) **13**, 238—242 (1927). — BRINK, R. A., and J. H. MACGILLIVRAY: Segregation for the waxy character in maize pollen and differential development of the male gametophyte. Amer. J. Bot. **11**, 465—469 (1924). — BUCHHOLZ, J. T., and A. F. Blakeslee: Studies of the pollen tubes and abortive ovules of the globe mutant of *Datura*. Science **55**, 597—599. (1922). — Abnormalities in pollen-tubes growth in *Datura* due to the gene "Tricarpel". Proc. nat. Acad. Sci. (Wash.) **13** 242—249 (1927). — Genes from radium treatment affecting pollen-tube growth in *Datura*. Genetics **21**, 731—751 (1936). — BURNHAM, C. R.: Differential fertilization in the Bt Pr linkage group of maize. J. Amer. Soc. Agron. **28**, 968—975 (1936).

CHAO, L. F.: The disturbing effect of the glutinous gene in rice on a mendelian ratio. Genetics **13**, 191—225 (1928). — COLLINS, G. N., and J. H. KEMPTON: Inheritance of waxy endosperm in hybrids of Chinese maize. 4th Confer. Internat. Gen. (Paris), 1911, S. 347—356. — CORRENS, C.: Über den Einfluß, welchen die Zahl der zur Bestäubung verwendeten Pollenkörner auf die Nachkommenschaft hat. Ber. dtsch. bot. Ges. **18**, 422—435 (1900). — Scheinbare Ausnahmen von der Mendelschen Spaltungsregel für Bastarde. Ber. dtsch. bot. Ges. **20**, 159—172 (1902). — Ein Fall experimenteller Verschiebung des Geschlechtsverhältnisses. S.-B. kgl. preuß. Akad. Wiss. **51**, 685—717 (1917). — Fortsetzung der Versuche zur experimentellen Verschiebung des Geschlechtsverhältnisses. S.-B. kgl. preuß. Akad. Wiss. **50**, 1175—1200 (1918). — Versuche, bei Pflanzen das Geschlechtsverhältnis zu verschieben. Hereditas (Lund) **2**, 1—24 (1921). — Zweite Fortsetzung der Versuche zur experimentellen Verschiebung des Geschlechtsverhältnisses. S.-B. preuß. Akad. Wiss. **18**, 330—354 (1921). — Der Einfluß des Alterns der Keimzellen auf das Zahlenverhältnis spaltender Bastarde. Naturwiss. **9**, 313—315 (1921). — Alkohol und Zahlenverhältnis der Geschlechter bei einer getrenntgeschlechtigen Pflanze (*Melandrium*). Naturwiss. **10**, 1049—1052 (1922).

DEMEREÇ, M.: Cross sterility in maize. Z. Vererb.-Lehre **50**, 281—291 (1929).

EMERSON, R. A.: A possible case of selective fertilization in maize hybrids. Anat. Rec. **29**, 136 (1925). — Relation of the differential fertilization genes, Ga/ga, to certain other genes of the Su-Tu linkage group in maize. Genetics **19**, 137—156 (1934). — EMERSON, S.: Genetic and cytological studies on *Oenothera*. II. Z. Vererb.-Lehre **49**, 381 (1931). — EMMERLING, M. H.: Preferential segregation of structuraly modified chromosomes in maize. Genetics **44**,

625—645 (1959). — ESCHENBERGER, F.: Über den Erbgang der Sepaloidie bei *Epilobium*. Z. Bot. **44**, 89—108 (1956).

HALLQVIST, C.: Gametenelimination bei der Spaltung einer zwerghaften und chlorophylldefekten Gerstensippe. Hereditas (Lund) **4**, 190—205 (1923). — HARLAND, S. C., J. C. HAIGH and J. L. LOCHRIE: A progressive variation with age of a simple Mendelian ratio in the cowpea. Genetica **8**, 507—512 (1926). — HARTE, C.: Meiosis und crossing-over. Weitere Beiträge zur Cytogenetik von *Oenothera*. Z. Bot. **38**, 65—137 (1942). — Über die Aufhebung des Gonen- und Zygotenausfalls bei *Oenothera*. II. Flora N. F. **37**, 120—124 (1943). — Cytologisch-genetische Untersuchungen an spaltenden *Oenothera*-Bastarden. Z. Vererb.-Lehre **82**, 495—640 (1948). — Untersuchungen über die Nachkommenschaft von Heterozygoten der graminifolia-Koppelungsgruppe von *Antirrhinum majus*. Z. Vererb.-Lehre **84**, 480—507 (1952). — Untersuchungen über Gonenkonkurrenz und crossing-over bei spaltenden *Oenothera*-Bastarden. Z. Vererb.-Lehre **85**, 97—117 (1953). — Untersuchungen zur Gonenkonkurrenz bei der Entwicklung des Embryosackes von *Oenothera*. C. R. Acad. Sci. (Paris), Sect. 8, 178—179 (1954). — Gonenkonkurrenz im Embryosack bei *Oenothera*. Ber. dtsch. bot. Ges. **68**, (17)—(18) (1955). — Untersuchungen über die Gonenkonkurrenz in der Samenanlage bei *Oenothera* unter Verwendung der Letalfaktoren als Markierungsgene. I., II., III. Z. Vererb.-Lehre **89**, 473—496. 497—507, 715—728 (1958). — Untersuchungen über die Gonenkonkurrenz bei *Oenothera* unter Verwendung der Testloci fr, s und de. Z. Vererb.-Lehre **92**, 142—164 (1961). — Untersuchungen über die Pollenschlauchkonkurrenz bei *Oenothera* in der Komplexkombination *flavens*·h*Hookeri*. (In Vorbereitung.) — HERIBERT-NILSSON, N.: Die Spaltungserscheinungen der *Oenothera Lamarckiana*. Acta Univ. Lund. **12**, 1—131 (1916). — Zuwachsgeschwindigkeit der Pollenschläuche und gestörte Mendelzahlen bei *Oenothera Lamarckiana*. Hereditas (Lund) **1**, 41—67 (1920). — Zertationsversuche mit Durchtrennung des Griffels bei *Oenothera Lamarckiana*. Hereditas (Lund) **4**, 177—190 (1923). — Multiple monofaktorielle Reduplikation als der Ausdruck partialer Heterogamie bei *Oenothera fallax*. Hereditas (Lund) **5**, 1—13 (1924). — Das Ausbleiben der dominanten Heterozygoten in bezug auf die Nervenfarbe bei *Oenothera Lamarckiana*. Hereditas (Lund) **6**, 387—391 (1925). — HIEMENZ, G.: Untersuchungen an *Salpiglossis variabilis* über Gonenkonkurrenz und selektive Befruchtung und ihre Auswirkung auf die Nachkommenschaft. Biol. Zbl. **74**, 337—370 (1955). — HIORTH, G.: Zur Kenntnis der Homozygoten-Eliminierung und der Pollenschlauchkonkurrenz bei *Oenothera*. Z. Vererb.-Lehre **43**, 171—237 (1926).

JIMÉNEZ, J. R., and O. E. NELSON: A new fourth chromosome gametophyte locus in maize. J. Hered. **56**, 259—263 (1965). — JONES, D. F.: Selective fertilization in pollen mixtures. Biol. Bull. **38**, 251—289 (1920). — Selective fertilization and the rate of pollen-tube growth. Biol. Bull. **43**, 167—174 (1922). — Selective fertilization among the gametes from the same individuals. Proc. nat. Acad. Sci. (Wash.) **10**, 218—221 (1924).

KEMPTON, J. H.: Inheritance of waxy endosperm in maize. U. S. Dept. Agric. Bull. **754**, 1—99 (1919). — KIESSELBACH, T. A., and F. PETERSON: The segregation of carbohydrates in crosses between waxy and starchy types of maize. Genetics **11**, 407—422 (1926). — KUNZE, G.: Ein Fall von geschlechtsgebundener Vererbung bei *Melandrium* (Garcke). Ber. dtsch. Bot. Ges. **68**, 249—256 (1955).

LANGENDORFF, H.: Zur Kenntnis der Genetik und Entwicklungsgeschichte von *Oenothera fallax, rubirigida* und *Hookeri-albata*. Bot. Archiv **29**, 474 (1930). — LAWRENCE, W. J. C.: Studies on *Streptocarpus* LINDL. V. Heredity **12**, 333—356 (1958). — LENZ, F.: Über die krankhaften Erbanlagen des Mannes und die Bestimmung des Geschlechts beim Menschen. Jena 1912. — LONGLEY, A. E.: Abnormal segregation during megasporogenesis in maize. Genetics **30**, 11—113 (1945). — A gametophyte factor on chromosome 5 of corn. Genetics **46**, 641—647 (1961). — LUDWIG, W., u. C. BOOST: Die Abhängigkeit des menschlichen Geschlechtsverhältnisses von Erb- und Umwelteinflüssen, insbesondere vom Kriege. Klin. Wschr. **22**, 189—194 (1943). — Über Certation und Spermiendimorphismus bei Tier und Mensch. Z. Naturforsch. **2**b, 222—226 (1947). — LUIG, N. H.: Differential transmission of gametes in wheat. Nature (Lond.) **185**, 636—637 (1960). — LUIG, N. H., K. S. MCWHIRTER and E. P. BAKER: Proc. Linn. Soc. N.S.W. **83**, 340 (1958).

MANGELSDORF, P. C.: The relation between length of styles and Mendelian segregation in a maize cross. Amer. Naturalist **63**, 139—150 (1929). — Modification of Mendelian ratios in maize by mechanical separation of gametes. Proc. nat. Acad. Sci. (Wash.) **17**, 698—700 (1931). — Mechanical separation of gametes in maize. J. Hered. **23**, 289—295 (1932). — MANGELSDORF, P. C., and D. F. JONES: The expression of Mendelian factors in the gametophyte of maize. Genetics **11**, 423—455 (1926).

NELSON jr., O. E.: Non-reciprocal cross-sterility in maize. Genetics **37**, 101—124 (1952).

OEHLKERS, F.: Vererbungsversuche an *Oenothera*. III. Z. Vererb.-Lehre **31**, 201—260 (1923). — Erblichkeit und Cytologie einiger Kreuzungen mit *Oenothera strigosa*. Jb. wiss. Bot. **65**, 411—446 (1926). — Crossing-over bei *Oenotheren*. Z. Bot. **26**, 385—430 (1933). — Über die Aufhebung des Gonen- und Zygotenausfalls bei *Oenothera*. I. Flora, N. F. **37**, 106—119 (1943).

PARNELL, F. R.: Note on the determination of segregation by examination of the pollen of rice. J. Genet. **11**, 209—212 (1921).

REIMANN-PHILIPP, R.: Untersuchungen über die Vererbung des *grandiflora*-Merkmals bei *Petunia* × *hybrida* Vilm. Z. Pflanzenzücht. **48**, 143—176 (1962). — RENNER, O.: Versuche über die gametische Konstitution der Oenotheren. Z. Vererb.-Lehre **18**, 121—294 (1917). — Weitere Vererbungsstudien an Oenotheren. Flora, N.F. **11**, 641—667 (1918). — Zur Biologie und Morphologie der männlichen Haplonten einiger Oenotheren. Z. Bot. **11**, 305—380 (1919). — Heterogamie im weiblichen Geschlecht und Embryosackentwicklung bei den Oenotheren. Z. Bot. **13**, 609—621 (1921). — Untersuchungen über die faktorielle Konstitution einiger komplexheterozygotischer Oenotheren. Berlin: Gebr. Bornträger 1925. — Kurze Mitteilungen über *Oenothera*. IV. Flora, N. F. **34**, 145—158 (1940). — RENNER, O., u. R. E. CLELAND: Zur Genetik und Cytologie der *Oenothera chicaginensis* und ihrer Abkömmlinge. Z. Vererb.-Lehre **66**, 275—318 (1933). — RHOADES, M. M.: Preferential segregation in maize. Genetics **27**, 395—407 (1942). — Preferential segregation in maize. In: Heterosis, ed. J. W. GOWEN, p. 66—80. Ames/Iowa: Iowa State College Press 1952. — RHOADES, M. M., and V. H. RHOADES: Genetic studies with factors in the tenth chromosome in maize. Genetics **24**, 302—314 (1939). — RUDLOFF, F. C.: Zur Polarisation in der Reduktionsteilung heterogamer Oenotheren. Z. Vererb.-Lehre **65**, 147—179 (1933).

SCHLEIP, W.: Geschlechtsbestimmende Ursachen im Tierreich. Fortschr. Zool. **3**, 165—328 (1912. — SCHWARTZ, D.: The analysis of a case of cross-sterility in maize. Proc. nat. Acad. Sci. (Wash.) **36**, 719—724 (1950). — Genetic studies on mutant enzymes in maize: Synthesis of hybrid enzymes by heterozygotes. Proc. nat. Acad. Sci. (Wash.) **46**, 1210—1215 (1960). — SCHWEMMLE, J.: Gibt es eine selektive Befruchtung? II. Biol. Zbl. **70**, 193—252 (1951). — Gibt es eine selektive Befruchtung? III. Biol. Zbl. **71**, 152—183 (1952a). — Selektive Befruchtung als Erklärung unerwarteter Kreuzungsergebnisse. Biol. Zbl. **71**, 353—384 (1952). — Selektive Befruchtung bei der *Oenothera odorata*. Biol. Zbl. **72**, 405—424 (1953). — SCHWEMMLE, J., u. W. KOEPCHEN: Weitere Untersuchungen zur selektiven Befruchtung. Z. Vererb.-Lehre **85**, 307—346 (1953). — SHULL, G. H.: Sex-limited inheritance in *Lychnis dioica* L. Z. Vererb.-Lehre **12**, 265—302 (1914). — SIRKS, M. J.: Mendelian factors in *Datura*. I., II. Genetica 8, 485—500, 518—524 (1926). — SMITH, F. L.: Inheritance of three seed-coat color genes in *Vigna sinensis* Savi. Hilgardia (Berkeley/Calif.) **24**, 279—296 (1956). — STRASBURGER, E.: Versuche mit diözischen Pflanzen in Rücksicht auf Geschlechtsverteilung. Biol. Zbl. **20**, 657—665, 689—698, 721—731, 753—785 (1900).

TABATA, M.: Studies of a gametophyte factor in barley. Jap. J. Genet. **36**, 157—167 (1961).

UBISCH, G. v.: Abweichung vom mechanischen Geschlechtsverhältnis bei *Melandrium dioicum*. Biol. Zbl. **42**, 112—118 (1922).

VRIES, H. DE: Über amphikline Bastarde. Ber. dtsch. bot. Ges. **33**, 461 (1915) — Preferential fertilization in *Oenothera Lamarckiana*. Bot. Gaz. **77**, 73—79 (1924).

Anhang zum Literaturverzeichnis

Folgende Artikel wurden aus „Maize Genetic Cooperation News Letter“ zitiert:

BURNHAM, C. R.: Test of Ga specificity. **32**, 92—93 (1958). — KIKUDOME, G.: Preferential segregation in chromosome 9. **32**, 81—87 (1958). — RHOADES, M. M.: Gametophyte factor in chromosome 3. **17**, 7 (1943); **22**, 9—10 (1948). — Neocentromere formation as a cause of preferential segregation. **29**, 49 (1955). — Relation of crossing-over to preferential segregation. **32**, 71—72 (1958). — RHOADES, M. M., and E. DEMPSEY: Further studies on preferential segregation. **31**, 77—80 (1957).

Die selektive Befruchtung.

Von

Erik Haustein.

Mit 7 Abbildungen.

Definition.

Von selektiver Befruchtung sprechen wir dann, wenn durch irgendwelche Vorgänge oder Faktoren, die während der letzten prägamen Phase wirksam werden, die freie Kombination von Samenanlagen und Pollenschläuchen gestört wird. Die letzte prägame Phase umfaßt dabei den Abschnitt der Befruchtung, in dem die in der Fruchtknotenhöhle angelangten Pollenschläuche auf die Samenanlagen zuwachsen.

Nachweis der selektiven Befruchtung.

Der erste gesicherte Fall von selektiver Befruchtung wurde von Schwemmle (Schwemmle und Mitarbeiter 1938) bei der Kreuzung zwischen *Oenothera Berteriana* und *Oenothera odorata* beschrieben. Beide Arten sind Komplexheterozygoten; *Oe. Berteriana* mit den Renner-Komplexen B und *l*, *Oe. odorata* mit den Renner-Komplexen v und I. Sie bilden also zur Hälfte Fortpflanzungszellen mit dem Komplex B, zur Hälfte Fortpflanzungszellen mit dem Komplex *l (Oe. Berteriana)*, bzw. zur Hälfte Fortpflanzungszellen mit dem Komplex v, zur Hälfte Fortpflanzungszellen mit dem Komplex I *(Oe. odorata)*.

Aus der Kreuzung dieser beiden Arten können also folgende vier neuen Komplexverbindungen entstehen: B·v, B·I, *l*·v und *l*·I, die nun ihrerseits wieder als Kreuzungspartner dienen können. Bei allen diesen Kreuzungen hat sich nun die überraschende Tatsache ergeben, daß die Komplexverbindung B·v regelmäßig, wenn auch nicht allzu zahlreich auftritt, wenn der B-Komplex durch den Vater übertragen wird, daß sie dagegen praktisch vollkommen fehlt, wenn der B-Komplex von der Mutter stammt: In 128 Kreuzungen mit insgesamt 15224 Pflanzen, in denen der B-Komplex von der Mutter stammte, waren nur 24 = 0,15% B·v aufgetreten; in 118 Kreuzungen, wo der B-Komplex vom Vater übertragen wurde, waren unter insgesamt 10093 Pflanzen 953 = 9,4% B·v, also fast 63mal so viel. Da alle anderen Erklärungsmöglichkeiten ausgeschlossen werden konnten, schloß Schwemmle (1938, S. 374) „mit aller Vorsicht" auf selektive Befruchtung: Die B-Samenanlagen werden von den v-Pollenschläuchen nicht befruchtet.

In drei weiteren Arbeiten (Schwemmle 1949, 1951; Loertzer 1954) konnte dann die Frage nach dem Verbleib der B·v mit aller Sicherheit im Sinne einer selektiven Befruchtung entschieden werden. Wird nämlich die Narbe der B·*l* nur mit v-Pollen belegt (solche Formen, die im Pollen nur den v-Komplex übertragen, gibt es), dann müssen — vorausgesetzt, daß die Annahme einer selektiven Befruchtung richtig ist — die B-Samenanlagen unbefruchtet bleiben und nur die *l*-Samenanlagen befruchtet werden. Der bei diesen Bestäubungen erzielte Samengehalt wird also nur halb so groß sein wie bei optimaler Befruchtung aller Samenanlagen, etwa nach Freibestäubung (unter der weiteren Annahme, daß gleichviel B- und *l*-Samenanlagen befruchtet werden). Wird nun eine mit v-Pollen vorbestäubte Blüte einige Zeit (14—19 Std) später mit dem Pollen einer Homozygoten, etwa der *Oe. argentinea* (ha·ha), die einheitlich nur ha-Pollen überträgt, nachbestäubt, dann werden die ha-Schläuche im Fruchtknoten nur noch B-Samenanlagen vorfinden und befruchten; der Samenansatz muß dann wesentlich höher sein. In den vollständigen Aufzuchten müssen im ersten Falle nur *l*·v, im zweiten Fall neben den *l*·v auch noch B·ha auftreten. Bei den reziproken Kreuzungen, wo also die v·I nur mit B-Pollen bestäubt wird, wird ein derartiger Unterschied nicht vorhanden sein. Entsprechendes gilt dann natürlich auch für Kreuzungen mit B·I, B·v bzw. v·I oder *l*·v als Mutter. Die Versuchsergebnisse entsprachen voll den Erwartungen: So war etwa (Schwemmle 1949, S. 212) in der Kreuzung B·*l* × v·I der Ansatz 36,5% (im Durchschnitt aus 4 Kapseln, bezogen auf den gesamten Kapselinhalt). in der Kreuzung B·*l* × Typ A, der im Pollen allein den v-Komplex überträgt, dagegen nur 16,4%. Wurden aber die Narben der B·*l* zuerst mit dem Pollen von Typ A bestäubt und wurde 14—19 Std später mit Pollen der *Oe. argentinea* nachbestäubt, dann stieg der Samenansatz auf 25,5%. Wie die Aufzuchten ergaben, kam die Zunahme an gesunden Samen allerdings nicht ausschließlich durch die B-Samenanlagen zustande; zu einem bestimmten Prozentsatz waren auch noch *l*-Samenanlagen von den v-Pollenschläuchen der Vorbestäubung ausgelassen worden, die durch die später nachfolgenden ha-Schläuche ebenfalls noch befruchtet werden konnten. Es ergab sich aber eindeutig, ebenso auch aus den zahlreichen anderen Kreuzungen, daß trotz ausreichender v-Pollenschläuche die B-Samenanlagen ausgespart bleiben, die dann von den später nachfolgenden ha-Schläuchen noch befruchtet werden können. Eine nochmalige Wiederholung der Versuche durch Loertzer (1954) mit B·I und *l*·v als Mutter konnte die Ergebnisse von Schwemmle noch einmal bestätigen.

Auch wenn die gefundenen Zahlen nicht an die theoretisch zu erwartenden herankommen und noch manche Teilfragen offenbleiben, so besteht aber kein Zweifel, daß das Vorkommen selektiver Befruchtung für die B·v damit eindeutig bewiesen ist.

Der Gedanke an eine selektive Befruchtung ist beinahe so alt wie die moderne Vererbungsforschung. Bereits 1901 (S. 85 Anm., G.A. S. 278) gibt Correns allgemein als Möglichkeit für das Zustandekommen abweichender Spaltungszahlen an, daß nicht alle drei möglichen Verbindungen (A + A, A + a, a + a) gleich gut gelingen. Ein Jahr später (1902a, S. 166, G.A. S. 294, 1902b, S. 75, G.A. S. 308) erklärt er dann das Defizit an rezessiven Zuckermaispflanzen aus der Kreuzung zwischen Zucker- und Stärkemais damit, daß die eine Kombination schwerer gelingt als die andere. Wie Renner (1917, S. 152) dazu bemerkt, „scheint er dabei an selektive Befruchtung gedacht zu haben". Auch Baur (1911, S. 116) schließt sich dieser Correnssehen Vermutung an, wenn er schreibt: „... bleibt nur noch die Annahme übrig, daß von den theoretisch möglichen 4 Kombinationen der beiderlei Sexualzellen nicht alle sich gleich leicht verwirklichen, daß z. B. zwischen Sexualzellen, die beide den Faktor für Stärkebildung nicht enthalten, überhaupt nicht oder nur selten ein Sexualakt zustande käme..." Johannsen (1913, S. 519) dagegen denkt bei der Besprechung dieses Falles an eine schnellere Keimung oder *sonstige schnellere Wirkung* der Pollenkörner mit dem Stärkebildung bedingenden Gene. Daran hat übrigens auch Correns selbst bereits in seiner ersten Mitteilung des Maisfalles neben der Möglichkeit selektiver Befruchtung gedacht: „...das eine Korn bilde seinen Schlauch rascher als das andere" und er scheint später dieser Möglichkeit zeitweise den Vorzug gegeben zu haben. So 1917 (G.A. 42, S. 852); wo er bei Besprechung der Pollenschlauchkonkurrenz zwischen männchen- und weibchenbestimmenden Pollenschläuchen auf den „ersten derartigen Fall" (nämlich den Maisfall von 1902) verweist und von „funktionstauglicheren" Keimzellen spricht; S. 877 l. c. spricht er direkt von rascher wachsenden Pollenschläuchen (bei *Melandrium*). Im Zusammenhang mit der Besprechung der Ergebnisse von Heribert-Nilsson und von Renner (1917) betont er dann (1921, G.A. 52, S. 1101) nochmals: „Ich selbst hatte schon 1902 eine ungleiche Schnelligkeit der beiderlei

Pollensorten als einen der Wege bezeichnet, auf dem bei einem einfachen mendelnden Bastarde ein Zahlenverhältnis der Nachkommen erklärt werden könne, das stark von der Erwartung abwich." Offenbar war sich CORRENS aber doch nicht völlig sicher, denn er kommt im gleichen Jahr (1921, G.A. 54, S. 1133) nochmals auf den Maisfall zurück und meint, die Entscheidung zwischen den beiden Möglichkeiten, nämlich dem leichteren Gelingen der einen Verbindung bzw. dem schnelleren Pollenschlauchwachstum ,,ist für den Fall auch jetzt noch unsicher". Tatsächlich wurde der Maisfall später, in erster Linie durch MANGELSDORF (1929) auf experimentellem Wege zugunsten der Pollenschlauchkonkurrenz entschieden, indem er die Abhängigkeit des Spaltungsverhältnisses von der Entfernung der Bestäubungsstelle von den Samenanlagen nachwies.

Der zweite Fall einer abweichenden Spaltung, zu dessen Erklärung selektive Befruchtung herangezogen wurde, war dann das Rotnervenmerkmal der Oenotheren. Bei *Oe. Lamarckiana* sind alle untersuchten Rotnerven heterozygot im R-Faktor, sie spalten immer Weißnerven ab. Nachdem HERIBERT-NILSSON zunächst (1915a) Elimination nach der Zygotenbildung annimmt, schließt er dann (1915b) auf selektive Befruchtung: Die R-Samenanlagen sollen nur von r-Pollenschläuchen befruchtet werden, die R-Schläuche sollen die Befruchtung der R-Embryosäcke nicht auszuführen vermögen und in der Mikropyle steckenbleiben, so daß nachfolgende r-Schläuche noch kopulieren können. Die R-Pollenschläuche werden von den R-Samenanlagen abgestoßen, an der Befruchtung gehindert. An Stelle einer Elimination lebensunfähiger RR-Zygoten soll eine ,,Prohibition" treten. Auf Grund seiner Versuchsergebnisse lehnt jedoch RENNER (1917) die Ansicht HERIBERT-NILSSONs scharf ab; trotzdem hält dieser aber noch 1920 seine Meinung aufrecht. Erst als RENNER nochmals (1921) eindeutig nachweisen kann, daß der Rotnervenfaktor bei allen bekannten Oenotheren-Sippen, nicht bloß bei *Oe. Lamarckiana* im homozygoten Zustand letal wirkt, die RR-Homozygoten also in den tauben Samen stecken, schließt sich HERIBERT-NILSSON (1925) ebenfalls dieser Anschauung an. Schließlich wurde das Rotnerven-Problem nochmals zusammenfassend durch einen Schüler RENNERs, HIORTH (1922, S. 206ff.) besprochen und der Verbleib der RR-Homozygoten auch cytologisch untersucht und geklärt.

Trotz dieses negativen Befundes im Rotnervenfall hat RENNER aber trotzdem noch wiederholt die Möglichkeit einer selektiven Befruchtung zugelassen; so spricht er in einem Vortrag vor der Deutschen Gesellschaft für Vererbungswissenschaft (1924, S. 328) im Zusammenhang mit der Prohibitionstheorie von HERIBERT-NILSSON: ,,Die Möglichkeit solcher selektiven Befruchtung ist natürlich nicht zu bestreiten." Und 1925 (Anm. S. 140) bei der Besprechung abweichender Spaltungszahlen: ,,Auch auf selektive Befruchtung wird geprüft werden müssen." Über den inzwischen beschriebenen dritten Fall, wenn auch nicht einer selektiven Befruchtung im bisherigen Sinne einer völligen Unterdrückung der Befruchtung, sondern einer "preferential fertilization" von DE VRIES (1924) wollte er (an der gleichen Stelle) noch kein Urteil wagen. DE VRIES hatte den abweichenden Ausfall bestimmter Selbstungen bzw. Kreuzungen (auf die später noch einzugehen sein wird) damit erklärt, daß die eine der beiden Pollensorten rascher im Fruchtknoten ankommt als die andere und daß diese weiterhin von der einen Eizellsorte ,,kräftiger" angezogen wird, so daß zunächst fast ausschließlich diese bestimmte Kombination zustande kommt und erst später die übrigen noch möglichen Kombinationen gebildet werden. Insofern geht dieser Deutungsversuch über die bisherigen Fälle selektiver Befruchtung hinaus, als eine bestimmte Kombination nicht mehr gänzlich ausgeschlossen ist, sondern nur erschwert bzw. eine andere bevorzugt ist.

Schließlich ist noch der Vollständigkeit halber die Arbeit von COBB-BLANCHARD (1926) zu erwähnen, in der sie ebenfalls selektive Befruchtung als eine Erklärungsmöglichkeit ihrer Kreuzungen mit *Oe. pratincola* heranzieht. Die ohne Kenntnis der Rennerschen Arbeiten geschriebene Arbeit kann natürlich keinerlei positiven Beitrag zur Frage der selektiven Befruchtung bringen.

Es ist daher nicht verwunderlich, wenn RENNER in seinem Handbuchartikel (1929) die Frage einer selektiven Befruchtung endgültig für erledigt erklärt. Zu dem gleichen Schluß kommt auch JONES (1928, S. 3): "If we limit the definition of fertilization to the fusion of two cells, the gametes, to form the zygote, there is no clear case known of a selective action within the species where the individuals are all compatible with each other." Auch BRIEGER (1930, S. 160) schreibt: ,,es muß bis auf weiteres als eine Regel gelten, daß bei Sippenkreuzungen an sich funktionsfähige Pollenschläuche, die bis in die Fruchtknotenhöhle vorgedrungen sind, die noch unbefruchteten, aber befruchtungsfähigen Samenanlagen befruchten können."

Die zahlreichen anderen Arbeiten über ,,selektive Befruchtung" brauchen in diesem Zusammenhang nicht erwähnt zu werden, da bei ihnen der Begriff der selektiven Befruchtung viel weiter gefaßt ist und alle beschriebenen Fälle im wesentlichen auf Pollenschlauchkonkurrenz zurückgeführt werden können. Soweit sich wirkliche Hinweise auf die Möglichkeit einer selektiven Befruchtung in unserem Sinne darunter finden, sollen sie später noch besprochen werden.

War mit der B·v der erste sichere Fall selektiver Befruchtung bekanntgeworden, so führte ein zweites Problem wiederum zu einem neuen Fall von selektiver Befruchtung. Es war die Frage nach den Homozygoten der *Oe. Berteriana.* Bei der Selbstung der *Oe. Berteriana* (B·*l* × B·*l*) sind theoretisch 3 Kombinationen zu erwarten: die Komplexheterozygote B·*l* und die beiden Homozygoten B·B und *l*·*l* und zwar im Verhältnis 1:1. Tatsächlich treten aber bei den Aufzuchten immer nur B·*l* auf. Bei den Oenotheren der *biennis*-Gruppe, wo grundsätzlich dieselben Verhältnisse vorliegen, hatten RENNER (1914, 1916, 1917) sowie HIORTH (1926) zeigen können, daß die Homozygoten als absterbende Embryonen zu den tauben Samen werden. In sorgfältigen embryologischen Untersuchungen versuchte nun SCHÜTZ, ein Schüler von SCHWEMMLE (1938), den Nachweis der B·B- und *l*·*l*-Homozygoten auch für die *Oe. Berteriana* zu führen. Daß diese nicht in den prallen Samen stecken konnten, ergab der hohe Keimprozentsatz von 95%. Weiter ergab die embryologische Untersuchung des gesamten Kapselinhaltes, daß neben den prallen Samen mit vollentwickeltem Embryo noch weiter unterschieden werden konnten: Große Samen mit kopfförmigem Embryo; sog. Pulversamen ohne Embryo, aber mit eingewachsenem Pollenschlauch; Pulversamen aus unbefruchteten Samenanlagen und schließlich noch Pulversamen aus sterilen Samenanlagen ohne Embryosack, mit verfallenen Tetraden.

Da die großen Samen zu 95% keimen und reine B·*l* ergeben und da etwa 95% der großen Samen einen vollentwickelten Embryo besitzen, können in diesen die Homozygoten nicht stecken. Für diese kommen nur die folgenden 2 Klassen in Frage, die überhaupt auf eine Befruchtung zurückgehen: Die großen Samen mit kopfförmigem Embryo und die Pulversamen ohne Embryo, aber mit eingewachsenem Pollenschlauch. Selbst wenn nun diese beiden Klassen die Homozygoten darstellen würden, ergibt sich dann sofort ein neues Problem: Das Verhältnis der Heterozygoten zu den Homozygoten ist etwa 14:1 an Stelle von 1:1. Auf jeden Fall werden also die Homozygoten in viel zu geringer Anzahl gebildet, wenn sie überhaupt in diesen beiden Klassen von Samen stecken.

Eine Pollenschlauchkonkurrenz als Erklärung scheidet aus, weil dadurch ja lediglich das Verhältnis der Homozygoten B·B und *l*·*l* untereinander verschoben würde. Da weiterhin aus den Kreuzungen SCHWEMMLES mit Sicherheit bekannt ist, daß in beiden Geschlechtern die B- und *l*-Fortpflanzungszellen in gleicher Häufigkeit gebildet werden, bleibt für das Fehlen der Homozygoten einzig und allein der Schluß, daß sie, wenn überhaupt, in viel zu geringer Häufigkeit gebildet werden. SCHÜTZ erklärte diese Tatsache wiederum durch selektive Befruchtung, indem er annahm (SCHÜTZ 1938, S. 496): „...daß der B- und *l*-Komplex eine größere Affinität zueinander besitzen als jeder von beiden zu seinesgleichen." Doch konnte auch durch die embryologische Untersuchung des Kapselinhalts aus den Kreuzungen B·*l* × B·(II) (im Pollen nur der B-Komplex übertragen) bzw. B·*l* × *l*·(II) (im Pollen nur der *l*-Komplex übertragen) keine völlige Klarheit erzielt werden.

Diese konnte erst aus vielfach wiederholten und abgewandelten umfangreichen Kreuzungsserien wieder mit Hilfe von Vor- und Nachbestäubungen gewonnen werden, durch die ein wesentlich reicheres Material zur Verfügung stand als bei den mühsamen embryologischen Untersuchungen von SCHÜTZ (SCHWEMMLE 1951, 1952, 1953). Unter Verwendung der beiden Komplexverbindungen B·(II) und *l*·(II) als Vater, die also im Pollen nur den B- bzw. nur den *l*-Komplex übertragen, ergab sich zunächst, daß sich die beiden Homozygoten B·B und *l*·*l* ganz verschieden verhalten. Wird nämlich die B·*l* bestäubt mit Pollen der *l*·(II), dann liegen die Verhältnisse fast genau so wie bei den B·v in der Kreuzung B·*l* × v·I: Die *l*-Pollenschläuche befruchten fast ausschließlich die B-Samenanlagen; daraus entstehen die B·*l*-Heterozygoten; nur zu einem ganz geringen Prozentsatz befruchten sie auch die *l*-Samenanlagen. Trotz ausreichender *l*-Samenanlagen und ausreichender *l*-Pollenschläuche werden nur ganz wenige *l*·*l*-Homozygoten gebildet. Wie für die B·v liegt auch für die *l*·*l* selektive Befruchtung vor.

Ganz anders liegen aber die Verhältnisse in der Kreuzung B·*l* × B·(II). Zwar entstehen wiederum reichlich die *l*·B (die Samenanlage zuerst geschrieben), im Gegensatz zu der vorigen Kreuzung aber ebenso reichlich die Homozygoten B·B, die dann allerdings infolge eines Letalfaktors auf frühen Entwicklungsstadien absterben und zu tauben Samen werden. Für die B-Samenanlagen und die B-Pollenschläuche gibt es also keine selektive Befruchtung, die B·B können völlig normal gebildet werden.

Wenn also bei den Selbstungen der B·*l*, wo also B- und *l*-Pollenschläuche gemeinsam befruchten können, keine *l*·*l*-Homozygoten entstehen, so ist das wegen der selektiven Befruchtung klar. Wenn aber auch keine B·B-Homozygoten auftreten (oder höchstens ganz vereinzelt), dann muß notwendigerweise gefolgert werden, daß die B-Samenanlagen bereits von den *l*-Pollenschläuchen befruchtet sind, ehe die B-Pollenschläuche dazu die Möglichkeit haben bzw. dazu imstande sind. Die B·B-Homozygoten fehlen also deshalb, weil die B-Pollenschläuche den *l*-Pollenschläuchen gegenüber in der Konkurrenz um die Befruchtung der B-Samenanlagen unterliegen.

Selektive Befruchtung auf der einen und Pollenschlauchkonkurrenz auf der anderen Seite sind also die beiden Faktoren, deren geordnetes Zusammenspiel zu dem bemerkenswerten Resultat führt, daß trotz des Vorhandenseins zweier Samenanlagen- und zweier Pollensorten alle befruchteten Samenanlagen nur die Komplexheterozygote B·*l* darstellen.

Reaktionsgeschwindigkeit.

Bei der Aufklärung der Befruchtungsverhältnisse der B·*l* hat sich außer der selektiven Befruchtung bei *l*·*l* noch ein weiteres Problem ergeben, das ebenfalls für die Frage der selektiven Befruchtung von Bedeutung ist. Bereits Schütz hatte bei seinen embryologischen Untersuchungen gefunden, daß in der Kreuzung B·*l* × B·(II) das Verhältnis der gebildeten B·*l*- zu den B·B-Embryonen von der zur Verfügung stehenden Pollenmenge abhängt; während bei reichlicher Bestäubung die B·*l* und B·B etwa gleich häufig sind, werden bei spärlicher Bestäubung die B·B in viel geringerer Häufigkeit gebildet als die B·*l*, ja sie können überhaupt vollkommen fehlen. Das gleiche Ergebnis fand auch Schwemmle (1953a, S. 131): Wurden 2 oder 3 von den 4 Narben der B·*l* mit Pollen der B·(II) bestäubt (und später mit Pollen der lebensfähigen Homozygoten I·I nachbestäubt), dann entstanden aus der Vorbestäubung 14,4% B·B und 13,8% B·*l*. Wurde dagegen nur ein Schenkel spärlich mit Pollen der B·(II) vorbestäubt, dann entstanden daraus nur 1,3% B·B gegenüber 9,1% B·*l*. Während also die Häufigkeit der B·*l* — als Folge der geringeren Pollenmenge — von 13,8% auf 9,1% absinkt, sinken die B·B von 14,4% auf 1,3% ab. Bei spärlicher Bestäubung befruchten also die B-Pollenschläuche *vorzugsweise* die *l*-Samenanlagen.

Das gleiche beobachten wir bei der Kreuzungsserie, wo die B·II als Mutter zunächst mit B·(II) vorbestäubt und dann mit *l*·(II) nachbestäubt wurde, wenn wir nur die aus der Vorbestäubung stammenden B·II und B·B (die in den tauben Samen stecken) betrachten (Schwemmle 1953a, S. 131). Bei gutem Samenansatz von 40,4% ist das Verhältnis B·B:B·II = 1:1,2. Je schlechter der Samenansatz wird, um so mehr verschiebt sich das Verhältnis zugunsten der B·II. Bei einem Ansatz von 20,8% ist das Verhältnis B·B:B·II = 1:7,1. Mit abnehmender Menge an Pollenschläuchen wird also wieder *bevorzugt* die Komplexverbindung B·II gebildet.

Ein dritter hierher gehöriger Fall liegt schließlich bei der *Oe. odorata* vor. Die Verhältnisse scheinen zunächst ähnlich zu liegen wie bei der *Oe. Berteriana*. Die prallen Samen keimen zu durchschnittlich 80%, gelegentlich auch 90% und geben reine v·I. In ganz verschwindender Häufigkeit treten einige (lebensfähige) v·v-Homozygoten auf, ganz selten kommen sogar die offenbar kräftigsten der im odorata-Plasma an sich nicht lebensfähigen I·I-Homozygoten zur Keimung, um dann allerdings sofort abzusterben. Weitere Homozygoten werden zum Teil noch in den nicht keimenden prallen Samen und den geschrumpften Samen vorliegen. Wenn also auch einige Homozygoten gebildet werden, dann wieder wie bei der *Oe. Berteriana* in einer viel zu geringen Häufigkeit.

In 3 Arbeiten (SCHWEMMLE 1949, 1953b; SCHWEMMLE und KOEPCHEN 1953) konnte SCHWEMMLE, wieder mit der Methode der Vor- und Nachbestäubungen, den Verbleib der Homozygoten aufklären. Kreuzt man die v·I mit der aus der *Oe. odorata* abgeleiteten Trisomen Typ A, die im Pollen nur den v-Komplex überträgt, dann entstehen in dieser Kreuzung 15,3% v·v und 26,2% I·v (SCHWEMMLE 1951, S. 234). In der Kreuzung v·I × I·I andererseits entstehen neben 13,7%, v·I noch 19,1% I·I (l.c., S. 241). Das heißt also, daß an sich die beiden Homozygoten v·v und I·I in nicht unbeträchtlicher Menge gebildet werden können, wenn nur mit einerlei Pollen bestäubt wird (auf die Bedeutung der verschiedenen Häufigkeiten, mit der die einzelnen Komplexverbindungen auftreten, soll erst im nächsten Abschnitt eingegangen werden; für unsere augenblicklichen Betrachtungen können sie noch außer acht gelassen werden). Die Verhältnisse liegen also insofern anders als bei der *Oe. Berteriana*, wo ja die eine Homozygote, nämlich *l·l*, infolge selektiver Befruchtung in keinem Fall gebildet werden kann.

Daß nun trotzdem bei den Selbstungen der v·I, wenn also v-Pollen und I-Pollen gleichzeitig in die Fruchtknotenhöhle hinunterwachsen, keine oder nur ganz wenige v·v- und I·I-Homozygoten gebildet werden, erklärt SCHWEMMLE (l. c., S. 245) damit, daß die v-Schläuche schneller hinunterwachsen und „infolge ihrer größeren Affinität zu den I-Samenanlagen" diese *in erster Linie* und *nur zögernd* die v-Samenanlagen befruchten; die wenig hintendreinkommenden I-Schläuche finden dann praktisch nur mehr die nach der Bildung weniger v·v noch reichlich vorhandenen v-Samenanlagen und befruchten diese.

Diese Erklärung ist in dieser Form wohl nicht ganz zutreffend. Was zunächst die Unterschiede in der Wachstumsgeschwindigkeit der v- und I-Pollenschläuche im Griffel der v·I betrifft, so wurde diese von SCHNEIDER (1956, S. 186) bestimmt; zur Durchwachsung eines Griffels von 3 cm Länge benötigen die v-Pollenschläuche 4 Std 40 min, die I-Pollenschläuche 4 Std 50 min (in Wirklichkeit sind das die Zeiten, nach denen bei reichlicher Bestäubung der Narben je die ersten 10 Pollenschläuche an der unteren Schnittfläche angekommen waren). Daß sich bei so geringen Differenzen der schnellsten Pollenschläuche die Gesamtkurven für die beiderlei Pollenschläuche weitgehend überschneiden werden, ist wohl selbstverständlich. Es genügt weiter, auf die Überlegungen hinzuweisen, die CORRENS (1921, G. A. 59, S. 1094ff.) gerade zu diesem Problem anstellt, um einzusehen, daß einfache Pollenschlauchkonkurrenz, zudem mit so geringen Unterschieden, das Fehlen der Homozygoten nicht erklären kann. Entscheidend an der Erklärung SCHWEMMLEs ist also lediglich die „bevorzugte" bzw. „zögernde" Befruchtung der wechselseitigen Fortpflanzungszellen.

Wir haben also in allen drei besprochenen Fällen dieselbe Beobachtung. Bei B·*l* × B·(II) befruchten die B-Schläuche vorzugsweise die *l*-Samenanlagen und nur zögernd die B-Samenanlagen; bei B·II × B·(II) befruchten die B-Schläuche bevorzugt die II-Samenanlagen und bei v·I × v·I schließlich kann die viel zu geringe Häufigkeit der Homozygoten v·v und I·I nur erklärt werden durch die Annahme, daß die v-Schläuche zuerst und in erster Linie die I-Samenanlagen, dann erst die I-Schläuche die v-Samenanlagen befruchten.

Um diese ganz allgemeinen Begriffe einer „bevorzugten" bzw. einer „zögernden" Entstehung bestimmter Komplexverbindungen zu präzisieren, führte HAUSTEIN (1955, S. 259) den Begriff der Reaktionsgeschwindigkeit ein: Wir verstehen darunter die Zeit, die verstreicht, bis es zur Auslösung einer positiven Reaktion zwischen Pollenschlauch und Samenanlage kommt. Bei einer „bevorzugten" Verbindung ist diese Zeit kurz, bei einer „zögernden" mehr oder weniger lang. Zu jeder Befruchtung gehört also auch noch der Begriff der Reaktionsgeschwindigkeit, mit der diese Befruchtung durchgeführt wird. Ob also eine bestimmte Komplexverbindung zustande kommt, hängt nicht nur ab von der genetischen Konstitution der beiden Partner, sondern unter Umständen auch noch von der Reaktionsgeschwindigkeit, wie etwa das Beispiel der Kreuzung B·*l* × B·(II) zeigt: Die Reaktion zwischen den *l*-Samenanlagen und den B-Pollenschläuchen erfolgt schneller als die Reaktion zwischen den B-Samenanlagen und den B-Pollenschläuchen. Bei Mangel an B-Schläuchen werden diese rasch mit den *l*-Samenanlagen reagieren und für die B-Samenanlagen sind dann nur mehr wenige B-Schläuche übrig.

Daß der aus den wenigen, oben angeführten Beispielen erschlossene Begriff der Reaktionsgeschwindigkeit tatsächlich seine Berechtigung hat, haben die speziell darauf gerichteten Untersuchungen von LECHNER (1964) an einem zahlenmäßig umfangreichen Material eindeutig ergeben. Gleichzeitig zeigten sie aber auch, daß eine unterschiedliche Reaktionsgeschwindigkeit durchaus nicht immer vorzuliegen braucht; so war z.B. in den Kreuzungen v·I × I·I, B·*l* × I·I und B·*l* × hsc·hsc kein unterschiedliches Verhalten der I- bzw. hsc-Pollenschläuche gegenüber den genetisch verschiedenen Samenanlagen zu erkennen. Wie dann weiter die Arbeit von SCHWEMMLE und LOHMEYER (1963) gezeigt hat, ist das Vorkommen unterschiedlicher Reaktionsgeschwindigkeiten keineswegs nur auf die Sekt. *Raimannia* beschränkt. Auch bei der zur Sekt. *Renneria* gehörenden *Oe. campylocalyx* befruchten die (allein pollenaktiven) c_k-Pollenschläuche „bevorzugt" die c_l-Samenanlagen und erst dann „zögernd" die c_k-Samenanlagen (SCHWEMMLE und LOHMEYER S. 418ff.).

Unter bestimmten Bedingungen, z. B. wenn die Pollenmenge nicht zur Befruchtung sämtlicher Samenanlagen ausreicht, oder bei Konkurrenz zweier Pollenschläuche kann auch eine unterschiedliche Reaktionsgeschwindigkeit zu einem Abweichen von der reinen Zufallsverteilung, also zu selektiver Befruchtung führen.

Affinität zwischen Samenanlagen und Pollenschläuchen.

Mit der Lösung der beiden Ausgangsprobleme, die zur Entdeckung einer selektiven Befruchtung führten, nämlich 1. der Frage nach dem Verbleib der B·v in der Kreuzung B·*l* × v·I und 2. der Frage nach den Homozygoten B·B und *l*·*l* bei den Selbstungen der *Oe. Berteriana* haben sich, eigentlich nur nebenbei, weitere Tatsachen ergeben, die aber ebenfalls von allgemeiner Bedeutung für die Befruchtungsphysiologie und auch speziell für das Problem der selektiven Befruchtung sind.

Bei den zahllosen Kreuzungen, bei denen nur mit einer einheitlichen Pollensorte bestäubt wurde, hatte sich nämlich die ganz auffallende Tatsache ergeben, daß einmal von der Gesamtheit der fertilen Samenanlagen immer nur ein ganz bestimmter Bruchteil durch die Pollenschläuche befruchtet wird und 2. daß die Häufigkeit, mit der eine bestimmte Komplexverbindung gebildet wird, von der genetischen Konstitution der beiden Partner abhängt und für die einzelnen Komplexverbindungen ganz spezifisch ist. So zeigt etwa Tabelle 1 den Ausfall einiger Kreuzungen der *Oe. Berteriana* bzw. der *Oe. odorata,* nach reichlicher Bestäubung mit nur einer Pollensorte, wenn also jede Pollenschlauchkonkurrenz und jeder Unterschied in der Reaktionsgeschwindigkeit ausgeschlossen ist.

Die Abhängigkeit von der genetischen Konstitution der beiden Partner ist dabei aber nicht so zu verstehen, daß es gleichgültig ist, welcher Komplex durch die Eizelle, welcher durch den Pollen übertragen wird. Das zeigten ja bereits die Verhältnisse bei der Entstehung der B · v; als weiteres Beispiel sei aus der Tabelle 1 noch auf die Entstehung der B · *l* hingewiesen: B ♀ · *l* ♂ = 26,0%, *l* ♀ · B ♂ = 15,2%.

Tabelle 1. Erklärung im Text. (Nach Schwemmle 1951.)

B · *l* × hsc · hsc	25,8% B · hsc	20,9% *l* · hsc
B · *l* × I · I	20,4% B · I	19,0% *l* · *I*
B · *l* × ha · ha	14,0% B · ha	20,4% *l* · ha
B · *l* × B · (II)	15,2% B · B	16,8% *l* · B
B · *l* × *l* · (II)	24,2% B · *l*	1,2% *l* · *l*
v · I × hsc · hsc	8,8% v · hsc	14,4% I · hsc
v · I × ha · ha	14,3% v · ha	13,2% I · ha
v · I × Typ A (v)	15,3% v · v	26,2% I · v

Um die dadurch aufgezeigten Gesetzmäßigkeiten klarer erfassen zu können, hat Schwemmle den Begriff der „Affinität" eingeführt (Schwemmle 1951, S. 207; 1952, S. 379): „Die Häufigkeit, mit der bestimmte Komplexverbindungen entstehen, ist sowohl von der genetischen Konstitution der Eizelle in den fertilen Samenanlagen wie der Pollenschläuche abhängig. Diese bezeichnen wir als Affinität. Ein Maß für diese ist der prozentuale Anteil der verschiedenen möglichen Kombinationen in einer Kreuzung bezogen auf den gesamten Kapselinhalt."

Diese Bedeutung hatte der Affinitätsbegriff nicht von Anfang an. Als erster spricht Schütz wiederholt von „Affinität", aber in einem ganz allgemeinen Sinne, z. B. (Schütz, S. 498): „...dann wird diese vorzugsweise *l*-Samenanlagen befruchten, weil ja der B-Komplex eine größere Affinität zum *l*-Komplex als zu seinesgleichen besitzt". In einem ähnlich allgemeinen Sinne spricht auch Schwemmle (1949, S. 205): „Es wäre aber auch denkbar, daß zwar die B-Schläuche auf die v-Samenanlagen zuwachsen, daß aber die Anziehung gering ist oder die B-Schläuche nicht stark genug reagieren, kurz gesagt, *die Affinität zu gering ist,* so daß die Befruchtung unterbleibt." An der gleichen Stelle (l. c., S. 229) spricht er auch schon von „Reihen abgestufter Affinität" zwischen bestimmten Samenanlagen und Pollensorten. Erst als sich bei den weiteren Untersuchungen herausstellte, daß die verschiedenen Komplexverbindungen (bei fehlender Konkurrenz) in ganz bestimmten, aber untereinander stark verschiedenen Häufigkeiten gebildet werden, erhielt der Affinitätsbegriff einen ganz neuen Inhalt, indem er jetzt direkt mit der Befruchtungshäufigkeit gleichgesetzt wurde und der zahlenmäßige Wert, mit dem eine Komplexverbindung gebildet wird (ausgedrückt in Prozent des gesamten Kapselinhalts) direkt als Maß für die Affinität genommen wurde. Obwohl damit der Affinitätsbegriff einen wesentlichen Bedeutungswandel erfahren hatte und nunmehr als Befruchtungshäufigkeit klar definiert und seine Größe eindeutig zahlenmäßig bestimmbar schien, wurde er trotzdem auch noch weiterhin in seiner anfänglichen allgemeinen qualitativen Bedeutung weiter verwendet, was sehr rasch zu logischen Schwierigkeiten führen mußte, die dann dazu führten, daß durch Haustein der als reine Häufigkeit definierte (neue) Affinitätsbegriff noch ergänzt wurde durch die Einführung des Begriffs der Reaktionsgeschwindigkeit. Mit Hilfe dieser beiden eindeutig definierten Begriffe ist es jetzt möglich, jeden Befruchtungsvorgang, soweit er sich in der letzten prägamen Phase abspielt, eindeutig in Abhängigkeit von den jeweiligen Voraussetzungen zu beschreiben, z. B. reichliche oder spärliche Bestäubung, einheitliche Pollensorte oder Pollenmischung usw.

Es ist vielleicht nicht uninteressant, darauf hinzuweisen, daß bereits vor mehr als 100 Jahren ganz ähnliche Gedanken, und zwar ebenfalls auf Grund experimenteller Unterlagen, von GÄRTNER geäußert wurden (GÄRTNER 1849, S. 188): „Unter *Wahlverwandtschaft* unter den Pflanzen verstehen wir demnach die größere oder geringere Neigung verschiedener reiner Arten, sich durch Bastardbefruchtung zu einem neuen Produkt zu verbinden.“ Auch hat GÄRTNER bereits versucht, diese Beziehungen, also den Grad der Wahlverwandtschaft zahlenmäßig zu erfassen, indem er ihn ausdrückte durch den Gehalt an Samen aus einer Artkreuzung bezogen auf den Samengehalt bei Selbstung gleich 1. Als Beispiel sei auf Tabelle 2 verwiesen (GÄRTNER, S. 216/217).

Tabelle 2. Erklärung im Text. (Nach GÄRTNER.)

♀ Verbascum Lychnitis fl. albo		♀ Dianthus barbatus	
proprio polline	1,0000	♂ proprio polline	1,0000
♂ Var. fl. luteo	0,9081	superbus	0,8111
phoeniceum	0,8061	japonicus	0,6666
nigrum	0,6336	Armeria	0,5333
Blattaria fl. albo	0,6224	barbato-carthusianorum	0,3111
— — luteo	0,4387	chinensis	0,2600
thapsiforme	0,4081	collinus	0,2333
austriacum	0,3877	deltoides	0,2222
macranthum	0,2653	chinensis latifol. (Schraderi)	0,1354
Thapsus	0,2142	carthusianorum	0,1111
pyramidatum	0,0306	prolifer	0,0333
		virgineus	0,0111
		pulchellus	0,0096
		arenarius	0,0084
		diutinus	0,0033

GÄRTNER ging dabei von der Annahme aus (S. 206), „daß aus einer solchen hybriden Verbindung zwar mehr oder weniger gute Samen hervorgehen, je nach der Gunst oder Ungunst der Nebenumstände, daß aber jede solche hybride Verbindung niemals über ein gewisses *Maximum* von keimungsfähigen Samen zu erzeugen vermag; welches Maximum jeder derartigen Verbindung eigen ist.“ So auffällig die Parallelen zu den wechselnden Affinitätswerten bei den oben angeführten Oenotheren-Kreuzungen sind, so können sie doch nicht als Vergleichsmaterial herangezogen werden; es könnte ja sein, daß der Grund für den abnehmenden Samenansatz bei den Gärtnerschen Kreuzungen in Wirklichkeit auf einer entsprechenden Abnahme der Pollenkeimung beruht, um nur eine Möglichkeit zu erwähnen.

Wesentlich größeres Gewicht könnte jedoch den bereits S. 481 kurz erwähnten Versuchen von DE VRIES (1924) zukommen, wenn auch hier nicht wieder einige Einwände gemacht werden müßten, die den Wert der mitgeteilten Zahlen stark beeinträchtigen. Interessant sind hier in erster Linie die Kreuzungen, in denen *Oe. lucida* (=gaudens. subvelans in der Rennerschen Bezeichnung) mit einer Reihe von anderen Formen gekreuzt wurde, die aber alle nur eine einheitliche Pollensorte übertrugen. Aus der Tabelle 3 in der de Vriesschen Bezeichnung der Komplexe (DE VRIES 1924, S. 77) ist die auffallende Übereinstimmung mit den *Oe. Berteriana*- bzw. *Oe. odorata*-Kreuzungen nach Bestäubung mit einheitlichem Pollen zu ersehen. Leider finden sich aber bei DE VRIES keinerlei Angaben über Zahl und Ausbildung des ganzen Kapselinhalts, Keimung der Samen, Zahl der aufgezogenen Keimlinge usw., so daß dadurch kaum ein Urteil möglich ist; es genügt z. B., nur auf die ungleiche Keimgeschwindigkeit der verschiedenen Komplexverbindungen hinzuweisen.

Tabelle 3. Erklärung im Text. (Nach DE VRIES.)

Cross	Percentage	
	laeta' × mut.	deserens' × mut.
Oe. lucida × albida	5 Lamarckiana	95 rubrinervis
Oe. lucida × elongata	14 empty seeds	86 lucida
Oe. lucida × blandina	33 Lamarckiana	67 subrobusta
Oe. lucida × deserens	42 lucida	58 deserens
Average	22	78

Wenn wir uns nochmals der Definition der Affinität als der Häufigkeit einer gebildeten Komplexverbindung, bezogen auf den gesamten Kapselinhalt, zuwenden, so ist dieses Maß ganz speziell auf die vorliegenden Oenotheren zugeschnitten. Als natürliches Maß erschiene selbstverständlich der Anteil der befruchteten Samenanlagen, bezogen auf die Gesamtzahl der befruchtungsfähigen Samenanlagen einheitlicher genetischer Konstitution. Damit sind aber schon die Schwierigkeiten gegeben; denn diese Zahl ist bei den Oenotheren gar nicht ohne weiteres feststellbar. Von den im Rahmen der selektiven Befruchtung untersuchten Oenotheren der Section *Raimannia* ist lediglich eine einzige Art, *Oe. argentinea* dadurch ausgezeichnet, daß ihre sämtlichen Samenanlagen fertil sind und bei Selbstbestäubung auch zu 100% befruchtet werden. Andere Arten, sowohl die komplexheterozygotischen, wie die *Oe. Berteriana, Oe. odorata,* als auch die homozygotischen, z. B. *Oe. longiflora, Oe. scabra* besitzen einen mehr oder weniger großen Prozentsatz steriler Samenanlagen, in denen überhaupt kein befruchtungsfähiger Embryosack vorhanden ist. Die Bestimmung der Zahl der fertilen Samenanlagen durch Bestimmung des Samenansatzes ist, wie wir gesehen haben, wegen der für die einzelnen Samenanlagen und Pollenschläuche stark wechselnden Affinitäten nicht mit Sicherheit möglich. Diese kann höchstens Schätzungswerte ergeben. Eine genaue Feststellung müßte für jede Art durch mühsame cytologische Untersuchung getroffen werden, wobei noch weiter erschwerend hinzukommt, daß unter Umständen dieser Wert in Abhängigkeit von geringfügigen genetischen Änderungen (z. B. in Abhängigkeit vom Tupfungsfaktor u. ä.) sich ebenfalls ändern kann. Die genauesten Bestimmungen liegen bisher vor für *Oe. Berteriana.* Unter Verwendung von 2 Zählungen von Weidner-Rauh (S. 430 und 433) sowie von 14 Zählungen von Schütz (Tabelle 1—3, S. 493, 503, 516) ergeben sich folgende 16 Werte für die Zahl der sterilen Samenanlagen bei der *Oe. Berteriana* (zusammengestellt bei Schwemmle 1951, S. 197): 29,6%, 29,7%, 35,7%, 38,4%, 40,1%, 40,7%, 41,1%, 42,2%, 42,8%, 43,6%, 45,9%, 46,7%, 48,1%, 49,1%, 50,0%, 50,3%.

Die Werte schwanken also sehr stark zwischen 30 und 50%, wobei jedoch nur 4 Werte zwischen 30 und 40%, 12 dagegen zwischen 40 und 50% liegen, so daß man also mit Werten in diesem Bereich am ehesten wird rechnen müssen.

Weitere Bestimmungen liegen noch vor für *Oe. longiflora* (Weidner-Rauh, S. 443/444, 448/449), *Oe. odorata* (v. Zitek, S. 12), B·I (Huber, S. 473), *l*·I (Leuchtmann, S. 487), 4 Komplexverbindungen mit ^{h}Hookeri (Kistner, S. 542). Aus allen diesen Bestimmungen, sowie auch aus den durch den besten Ansatz bei Kreuzungen gewonnenen Schätzungswerten ergibt sich nun die starke Schwankung des Gehalts an sterilen Samenanlagen bei den verschiedenen Oenotheren bzw. neu hergestellten Komplexverbindungen (z. B. B·I und *l*·I). Andererseits ist damit aber auch eine gewisse Grundlage für die Berechnung der tatsächlichen Affinitäten (als Anteil der befruchteten bezogen auf die befruchtungsfähigen Samenanlagen) gegeben, was unerläßlich ist, wenn Affinitäten verglichen werden sollen, die aus Kreuzungen mit verschiedenen Müttern stammen. Solche Vergleiche wurden durchgeführt in kleinerem Maße von Loertzer (1954, S. 44) für einige Homozygoten-Kreuzungen, ausführlicher von Schwemmle und Simon (1956, S. 562) für Kreuzungen von Oenotheren der Section *Raimannia* mit *Oe. Hookeri,* die in der Tabelle 4 wiedergegeben sind.

So notwendig also die Umrechnung der Affinitätswerte auf die tatsächlich vorhandene Anzahl der fertilen Samenanlagen bei Vergleichen wie den eben angeführten ist, so ist andererseits wegen der Ungenauigkeit der Bestimmung der fertilen Samenanlagen wie vor allem wegen der großen Schwankungsbreite des Anteils der fertilen Samenanlagen diese Umrechnung wohl auf die unumgänglich notwendigen Fälle zu beschränken, während es in allen anderen Fällen aus praktischen Gesichtspunkten doch zweckmäßiger ist, den an sich völlig falschen, aber mit weitgehender Sicherheit zu bestimmenden Wert der befruchteten Samenanlagen, bezogen auf den gesamten Kapselinhalt, also fertile + sterile Samenanlagen zusammen, als Maß für die Affinität zu nehmen. Eine Korrektur an diesen „praktischen" Affinitätswerten wäre allerdings insofern angebracht, als sie lediglich nach dem Ausfall der Aufzuchten bestimmt werden, also durch

Tabelle 4. Erklärung im Text. (Nach SCHWEMMLE und SIMON.)

♀	Fertile Sa.-Anl. %	Affinitäten bezogen auf		
			Ges. Ka.-Inh.	Fertile Sa.-Anlagen
B·*l*	50—60	B — hHook	4,3	17,2—14,3
		l — hHook	7,4	29,6—24,7
v·I	50—60	v — hHook	12,4	49,6—41,3
		I — hHook	7,2	28,8—24,0
v·I	50—60	v — hHook	3,5	14,0—11,7
(Bert. Pl.)		I — hHook	3,2	12,8—10,7
l·v	50—60	*l* — hHook	14,9	59,6—49,7
(Bert. Pl.)		v — hHook	18,7	74,8—62,3
l·v	50—60	*l* — hHook	15,5	62,0—51,7
(od. Pl.)		v — hHook	15,2	60,8—50,7
$l_{str} \cdot v_{str}$	40—50	l_{str} — hHook	5,8	29,0—23,2
		v_{str} — hHook	7,7	38,5—30,8
$l_{Se} \cdot v_{Se}$	40—50	l_{Se} — hHook	7,6	38,0—30,4
		v_{Se} — hHook	4,4	22,0—17,6
mk·ml	40—60	ml — hHook	1,8	9,0—6,0
		mk — hHook	0,0	0,0
$k_{oak} \cdot l_{oak}$	40—60	k_{oak} — hHook	10,1	50,5—33,7
		l_{oak} — hHook	4,8	24,0—16,0
ck·cl		ck — hHook	0,0	0,0
		cl — hHook	0,0	0,0
hsc·hsc	70—80	hsc — hHook	66,1	94,4—82,6
I·I	60—80	I — hHook	24,3	40,5—30,4
ha·ha	100	ha — hHook	19,9	27,5—14,1
hl·hl	70—80	hl — hHook	14,8	21,1—18,5

die wechselnden Werte der Keimung entsprechend beeinflußt werden. Diese Schwankungen könnten aber in einfacher Weise dadurch kompensiert werden, daß man die beobachteten Werte auf eine theoretisch 100%ige Keimung umrechnet.

Die genetischen Grundlagen der Affinität.

Mit der Tatsache, daß die Häufigkeit, mit der eine bestimmte Komplexverbindung gebildet wird, von der genetischen Konstitution der Eizelle und des Pollenschlauches abhängt, ist gleichzeitig die genetische Bedingtheit des ganzen hier in Frage kommenden Prozesses ausgedrückt. Diese genetische Grundlage der Affinität weiter aufzulösen, bereitet aber bei dem vorliegenden Untersuchungsobjekt, den Oenotheren, wegen der Komplexheterozygotie mit der starren Bindung des Anlagenbestandes kaum zu überwindende Schwierigkeiten. Die Möglichkeiten beschränken sich bei den vorliegenden Formen im wesentlichen auf die Bildung von Austauschkomplexen bzw. Austausch von sich paarenden Chromosomen bei bestimmten Komplexverbindungen, die in der Diakinese mehrere unabhängige Ringe bilden, auf das Auftreten 8-chromosomiger Komplexe bei Trisomen, und auf die Verbindung bestimmter Komplexe mit fremdem Plasma und fremden Plastiden, was durch entsprechende Kreuzungen und auf dem Umweg über Chimärenbildung möglich ist. Über die dabei gewonnenen Ergebnisse soll an Hand einiger typischer Beispiele in aller Kürze zusammenfassend berichtet werden.

Die Komplexverbindung ($B \cdot l \times v \cdot I$) $l \cdot I$ besitzt die Diakinese-Anordnung $8 + 4 + 2$. Durch Austausch des 4-Ringes entsteht aus dem I-Komplex der mit II bezeichnete Austauschkomplex, der sich also aus 5 I-Chromosomen und 2 l-Chromosomen zusammensetzt. Vergleichen wir nun die Affinitäten der 3 Komplexe: I(7 I-Chromosomen), II (5 I-Chromosomen + 2 l-Chromosomen) und l (7 l-Chromosomen) als ♀ in bezug auf die 3 Komplexe I, l und B als ♂, so ergeben sich die in Tabelle 5 zusammengestellten Werte (SCHWEMMLE 1952b, S. 381). Da außerdem auch noch der Austausch des freien Paares erfaßt werden kann, da auf dem l-Chromosom des Paares der Tupfungsfaktor T lokalisiert ist, entstehen weiterhin alle 3 Komplexe auch getupft und ungetupft. Aus der Tabelle ersehen wir nun einmal, daß durch Übernahme von 2 bzw. 3 l-Chromosomen die Affinität I—l sehr stark absinkt, fast bis auf den Wert der Affinität l—l. Weniger ausgeprägt zeigt sich der Einfluß der übernommenen Chromosomen bei den Affinitäten zum B- bzw. I-Komplex; doch liegt auch hier wieder die Affinität des II-Komplexes zwischen den entsprechenden Affinitäten des I-Komplexes und l-Komplexes zum B- bzw. I-Komplex. Weiter ist daraus aber zu sehen, daß sich der durch den Austausch des Paarchromosoms bedingte Unterschied auf die Affinität nicht auszuwirken scheint.

Tabelle 5. Erklärung im Text. (Nach SCHWEMMLE 1952b.)

	Affinitäten		
	I	l	B ♂
♀ l_T l_t	23,5	2	23
II_T II_t	20	5	21
I_T I_t	17	22,7	20,8

Ein solcher Einfluß des Tupfungsfaktors bzw. des den Tupfungsfaktor tragenden l-Chromosoms ließ sich jedoch nachweisen in Vergleichskreuzungen mit $v \cdot I$ und $v \cdot I_T$ als ♀ (GRESS 1957, S. 79), indem durch die Tupfung die Affinität des v- bzw. I-Komplexes der ♀ zum v- bzw. I-Komplex des ♂ in den meisten Fällen erniedrigt wurde. Durch Verwendung der Mutante Rinnig (Ri liegt auf demselben I-Chromosom wie t) ließ sich, wieder aus Kreuzungen mit $v \cdot I_{Ri}$ zeigen, daß sogar allein der Besitz des Faktors Ri ausreicht, um die Affinität zu beeinflussen, und zwar wiederum im Sinne einer Herabsetzung der Affinitätswerte.

Noch eindeutiger zeigt den möglichen Einfluß eines einzigen Gens auf die Affinität der durch crossing-over austauschbare Hellgrünfaktor h (SCHWEMMLE 1963a, b). So wird z.B. die Affinität der v_h und I_h-Samenanlagen zu den v- und zu den hsc-Pollenschläuchen im Vergleich zu den v- und I-Samenanlagen verringert (SCHWEMMLE 1963a, S. 142). Ähnliches gilt für die Affinität der B_h- und l_h-Samenanlagen zu den verschiedenen Pollenschläuchen im Vergleich zu den B- und l-Samenanlagen. Doch hängt das Ausmaß der Affinitätserniedrigung sehr stark von der genetischen Konstitution der Pollenschläuche ab (SCHWEMMLE 1963b, S. 231). Der Einfluß des Hellgrünfaktors zeigt sich aber auch auf die Reaktionsfähigkeit der Pollenschläuche, indem diese z.B. für die B- und v-Pollenschläuche durch den Faktor h erniedrigt wird. Doch erschweren die großen Schwankungen der gleichen Kreuzungen in verschiedenen Jahren, ja sogar ein direkt entgegengesetztes Verhalten (SCHWEMMLE 1963b, S. 231) eine gesicherte Aussage außerordentlich. Eine weitere, nicht unerhebliche Schwierigkeit gerade bei der Wirkung eines einzigen Gens, ist dann dadurch gegeben, daß die miteinander verglichenen Komplexe, jeweils mit und ohne Hellgrünfaktor, in ihrem übrigen Genbestand mit Sicherheit nicht identisch sind.

Auf die Möglichkeit, daß bei den 8-chromosomigen Eizellen der Trisomen, wo zu einem chromosomalen Austausch zwischen 2 Komplexen auch noch eine zahlenmäßige Vermehrung der Chromosomen hinzukommt, die Affinitäten ebenfalls stark beeinflußt werden können, haben bereits BUCHNER (1956, S. 547) und DITTMANN (1956, S. 581) in ihren Arbeiten über die 3 Trisomen Typ A, Typ B und Typ C (von der Konstitution $v \cdot I_8$) hingewiesen. Genauere Untersuchungen über die Affinitäten 8-chromosomiger Eizellen wurden dann vorgenommen von SCHWEMMLE an der Trisomen *Oe. Berteriana var. angusta* (vom Typ $B \cdot l_8$) (SCHWEMMLE 1958a) und an der Trisomen *Oe. odorata var. tenella* (vom Typ $v_8 \cdot I$) (SCHWEMMLE 1958b). Dabei ergab sich, daß die l_8-Samenanlagen der *Oe. Berteriana var. angusta* unabhängig von den verwendeten Pollenschläuchen, wenn auch in einem für die einzelnen Pollensorten verschiedenen Ausmaß, weniger häufig befruchtet werden als die *l*-Samenanlagen der *Oe. Berteriana* ($B \cdot l$). Das gleiche gilt für die v_8-Samenanlagen der *Oe. odorata var. tenella* in bezug auf die v-Samenanlagen der $v \cdot I$. Es kommt aber wieder auf die genetische Konstitution der befruchtenden Pollenschläuche an, wie groß der Unterschied ist; die ha-Schläuche der *Oe. argentinea* befruchten die v_8-Samenanlagen sogar genau so häufig wie die v-Samenanlagen.

Tabelle 6. Erklärung im Text. (Nach SCHWEMMLE 1957d.)
Die Affinitäten v—v und I—v.

♀		v (od. Pl.)	v (Bert. Pl.) ♂
v	od. Pl. + od. Pld.	17,7	14,7
	od. Pl. + Bert. Pld.	12,9	13,7
	Bert. Pl. + od. Pld.	16,9	18,3
	Bert. Pl. + Bert. Pld.	18,6	18,0
	arg. Pl. + arg. Pld.	16,1	17,5
I	od. Pl. + od. Pld.	29,8	27,2
	od. Pl. + Bert. Pld.	24,7	25,1
	Bert. Pl. + od. Pld.	24,1	24,4
	Bert. Pl. + Bert. Pld.	31,8	33,1
	arg. Pl. + arg. Pld.	28,1	27,3

Tabelle 7. Erklärung im Text. (Nach SCHWEMMLE 1957d.)
Die Affinitäten v—I und I—I.

♀		I (od. Pl.)	I (Bert. Pl.) ♂
v	od. Pl. + od. Pld.	25,0	29,5
	od. Pl. + Bert. Pld.	20,6	21,6
	Bert. Pl. + od. Pld.	21,6	24,1
	Bert. Pl. + Bert. Pld.	25,4	28,0
	arg. Pl. + arg. Pld.	22,8	25,0
I	od. Pl. + od. Pld.	26,8	27,1
	od. Pl. + Bert. Pld.	25,4	27,3
	Bert. Pl. + od. Pld.	26,3	27,0
	Bert. Pl. + Bert. Pld.	35,0	35,2
	arg. Pl. + arg. Pld.	26,8	28,3

In zahlreichen Arbeiten konnte dann weiter der Einfluß des Plasmas und der Plastiden auf die Affinität nachgewiesen werden (SCHWEMMLE 1952, 1957c, d; SCHWEMMLE und KOEPCHEN 1953, GRESS 1957). Als besonders eindrucksvolles Beispiel sollen hier die Untersuchungen von SCHWEMMLE (1957d) angeführt werden, wo als ♀♀ die $v \cdot I$ mit fünf verschiedenen Plasma- und Plastiden-Kombinationen Verwendung fand (od. Pl. + od. Pld; od. Pl. + Bert. Pld; Bert. Pl. + od. Pld.; Bert. Pl. + Bert. Pld.; argentinea Pl. + arg. Pld.) und als ♂♂ Formen, die nur eine Pollensorte übertrugen, nämlich die Komplexe I (od. Pl. + od Pld.), I (Bert. Pl. + Bert. Pld.), v (od. Pl. + od. Pld.), v (Bert. Pl. + Bert. Pld.). In den beiden Tabellen 6 und 7 (SCHWEMMLE 1957d, S. 532, 539) bzw. Abb. 1 und 2 (l. c., S. 547/548) sind die Ergebnisse zusammengefaßt, die deutlich den starken Einfluß auch der außerkaryotischen Komponenten auf die Höhe der Affinitäten erkennen lassen.

Zusammenfassend läßt sich also über die bisherigen Ergebnisse zur Frage der genetischen Bedingtheit der Affinität sagen, daß jede Änderung der genetischen Gesamtkonstitution des ♂ oder ♀ Partners, ob es sich nun um große chromosomale

Unterschiede oder nur um scheinbar geringfügige Plasma-Unterschiede handeln mag, die Höhe der Affinität beeinflussen kann.

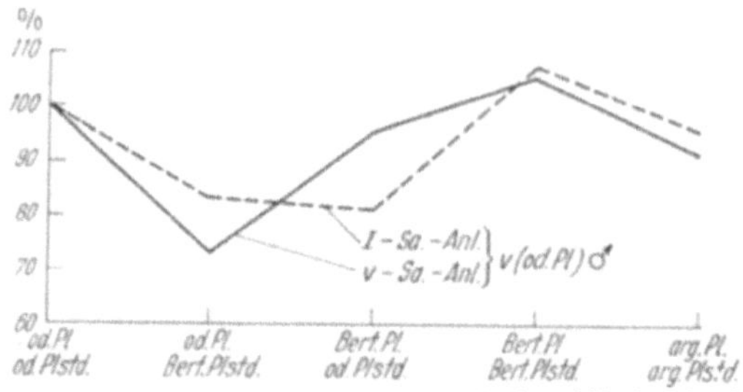

Abb. 1. Relative Affinitäten der v- und I-Samenanlagen zu v-Pollenschläuchen in Abhängigkeit von Plasma und Plastiden. (Nach Schwemmle 1957d.)

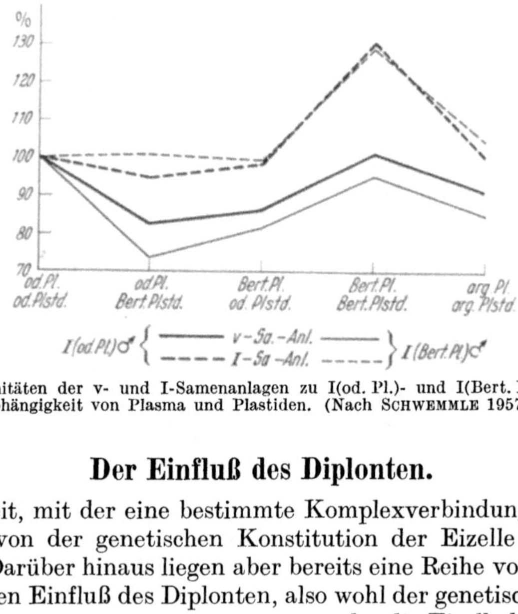

Abb. 2. Relative Affinitäten der v- und I-Samenanlagen zu I(od. Pl.)- und I(Bert. Pl.)-Pollenschläuchen in Abhängigkeit von Plasma und Plastiden. (Nach Schwemmle 1957d.)

Der Einfluß des Diplonten.

Die Häufigkeit, mit der eine bestimmte Komplexverbindung entsteht, hängt in erster Linie von der genetischen Konstitution der Eizelle und des Pollenschlauches ab. Darüber hinaus liegen aber bereits eine Reihe von Beobachtungen vor, die auch einen Einfluß des Diplonten, also wohl der genetischen Konstitution der die Eizelle bzw. den Embryosack beherbergenden Samenanlage erkennen lassen.

Tabelle 8. Erklärung im Text. (Nach Schwemmle und Simon.)

Mutterpflanze	Samenanlage	„echte" Affinität zu ʰHook %
B·*l* (Bert. Pl.)	*l* (Bert. Pl.)	29,6—24,7
l·v (Bert. Pl.)	*l* (Bert. Pl.)	59,6—49,7
v·I (Bert. Pl.)	v (Bert. Pl.)	14,0—11,7
l·v (Bert. Pl.)	v (Bert. Pl.)	74,8—62,3

Als erstes Beispiel wollen wir einige Kreuzungen mit *Oe. Hookeri* ♂ betrachten. Vergleichen wir die Häufigkeiten, mit der *l*- bzw. v-Samenanlagen von den ʰHookeri-Pollenschläuchen befruchtet werden, dann ergeben sich in Abhängigkeit von der Mutterpflanze beträchtliche Unterschiede zwischen an sich identischen Verbindungen. Die Zusammenstellung (Tabelle 8) zeigt in aller Deutlichkeit — ausgedrückt durch die auf die Zahl der fertilen Samenanlagen bezogene „echte" Affinität —, daß die *l*-Samenanlagen der *l*·v viel häufiger befruchtet werden als die Samenanlagen mit dem gleichen *l*-Komplex in der B·*l*. Noch stärker sind die Unterschiede zwischen den v-Samenanlagen der $l \cdot$v und den v-Samenanlagen der v·I. „Offenbar ist es nicht gleichgültig, ob die gleich konstituierten Samenanlagen im Fruchtknoten dieses oder jenes Diplonten enthalten sind" (SCHWEMMLE und SIMON, S. 563).

Ähnliche Beobachtungen, wenn auch mit weniger ausgeprägten Unterschieden, machten auch RICHTER und GRESS: So werden in der Kreuzung B·l_{Ri} × B·(II) die B-Samenanlagen von den B-Pollenschläuchen weniger häufig befruchtet (11,3%) als in den Kreuzungen B·l_{T} × B·(II) (14,6%) und B·l_{t} × B·(II) (15,3%) (RICHTER, S. 386). In der Kreuzung v·I_{Ri} (Bert. Pl.) × Typ A (Bert. Pl.) entstehen die v·v nur zu 10,6% gegenüber 15,3% v·v in der Kreuzung v·I (Bert.Pl.) × Typ A (Bert. Pl.) (GRESS, S. 29/30). Auf einen Einfluß des Diplonten schließt SCHWEMMLE (1958c, d) auch bei den Kreuzungen mit der trisomen *Oe. Berteriana var. angusta* (B·l_8) und der trisomen *Oe. odorata var. tenella* (v_8·I): Die B-Samenanlagen der B·l_8 werden von den hsc-Pollenschläuchen der Homozygoten *Oe. scabra* weniger häufig befruchtet als die B-Samenanlagen der B·l_{t}; umgekehrt werden die I-Samenanlagen der v_8·I durch die B-, hsc- und ha-Pollenschläuche viel häufiger als die I-Samenanlagen der ursprünglichen v·I befruchtet.

Zu den die Affinität bestimmenden Faktoren der genetischen Konstitution von Eizelle und Pollenschlauch im weitesten Sinne kann also noch weiter modifizierend der Einfluß des Diplonten treten, in dem sich die Eizelle befindet.

Weitere Analyse der Affinität.

Die zahllosen Affinitätsbestimmungen, die seit 1943 unter den verschiedensten Gesichtspunkten durchgeführt wurden, haben schon bald eine Tatsache erkennen lassen, die einer Erklärung zunächst große Schwierigkeiten entgegensetzte. Die Affinitätswerte in den aufeinanderfolgenden Jahren, aber auch zu verschiedenen Zeiten eines Jahres, sogar die von verschiedenen Untersuchern gleichzeitig durchgeführten Bestimmungen an völlig identischen Kreuzungen zeigten ganz auffallende Verschiedenheiten. In der Tabelle 9 sind einige Werte für die Kreuzungen B·l × I·I, B·l × ha·ha und B·l × hsc·hsc zusammengestellt (nach SCHWEMMLE 1957a, S. 136).

Daß die „praktischen" Affinitätswerte (also bezogen auf den gesamten Kapselinhalt) starke Schwankungen aufweisen können, ist als selbstverständlich zu erwarten, da ja auch der Gehalt an sterilen Samenanlagen, wie die Untersuchungen von WEIDNER-RAUH und SCHÜTZ gezeigt haben, beträchtlichen Schwankungen unterliegt. Wieweit dazu auch noch eine Variabilität der Affinität selbst kommt, läßt sich nicht sagen. Wir dürfen aber mit Sicherheit annehmen, daß die durch Schwankungen im Gehalt an sterilen Samenanlagen oder durch Schwankungen der Affinität selbst variierenden praktischen Affinitätswerte eine für biologische Vorgänge charakteristische Häufigkeitsverteilung zeigen werden, so daß eine mit zunehmender Zahl der Einzelbestimmungen immer genauere Bestimmung der Affinität möglich ist. Daß aber diese Mittelwerte wiederum so stark schwanken sollen, scheint bei der außerordentlich großen genetischen Einheitlichkeit des vorliegenden Materials schwer verständlich. Jedenfalls wurden aus diesem Grunde bei den weiteren Untersuchungen, soweit irgendmöglich, alle für einen Vergleich benötigten Kreuzungen möglichst gleichzeitig hergestellt, um diese Einflüsse auszuschalten. Trotzdem blieb dieses Problem bestehen.

In erster Linie konnte man natürlich als Ursache an die Außenbedingungen, Temperatur, Feuchtigkeit usw. denken. Das ist zweifellos auch der Fall. So konnte RICHTER nachweisen, daß bei völlig gleichartiger Durchführung der Kreuzungen die Ansätze zu verschiedenen Zeiten während eines Jahres ganz

Tabelle 9. Erklärung im Text. (Nach SCHWEMMLE 1957a.)

	Jahr	Zahl der Kreuzungen	Zahl der Aufzuchten	große Sa. %	B·I %	l·I %	
B·l × I·I	1944	10	2	48,8	21,1	18,9	
	1949	12	3	51,9	20,3	23,4	
	1949	12	3	55,6	22,3	27,3	2 Narbenschenkel bestäubt
	1949	10	3	52,5	20,3	25,2	1 Narbenschenkel bestäubt
	1950	17	7	58,9	24,1	22,7	
	1952	20	5	60,5	28,6	27,4	HUBER
	1953	18	4	39,9	15,2	20,8	2 Narbenschenkel bestäubt
	1953	20	3	56,3	19,7	23,0	RICHTER
					B·ha	l·ha	
B·l × ha·ha	1943		2	34,3	12,6	20,4	
	1944	10	3	34,2	12,4	15,0	
	1953	30	4	41,4	16,1	22,5	RICHTER
					B·hsc	l·hsc	
B·l × hsc·hsc	1944	10	4	52,9	25,8	20,9	
	1949	12	3	43,5	16,3	23,9	
	1949	12	3	56,3	22,9	29,5	2 Narbenschenkel bestäubt
	1949	12	3	51,5	21,0	27,7	1 Narbenschenkel bestäubt
	1953	11	5	38,2	17,1	19,6	2 Narbenschenkel bestäubt 48 Std nach der Kastration
	1953	24	8	56,6	24,5	26,6	RICHTER

verschieden sein können. Als Beispiel seien etwa die Kreuzungen B·l × I·I angeführt, die, wie Abb. 3 zeigt, am 11. 8. 53 wesentlich besser ansetzten als nur 8 Tage später am 19. 8. 53. Außerdem zeigten diese Kreuzungen, daß die Ände-

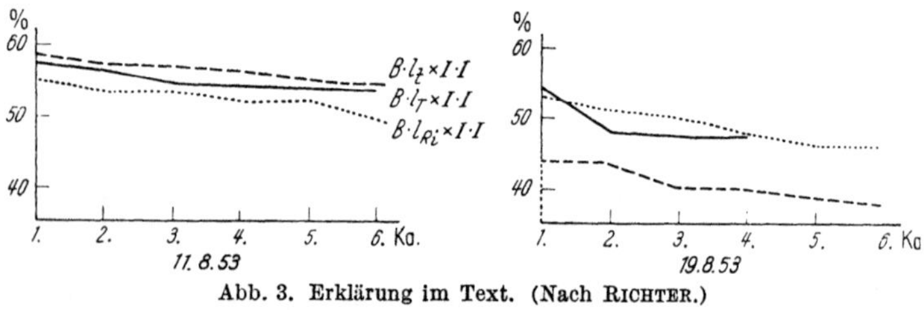

Abb. 3. Erklärung im Text. (Nach RICHTER.)

rungen im Samenansatz in Abhängigkeit von der verschiedenen genetischen Konstitution der Mutterpflanzen durchaus nicht gleichsinnig zu erfolgen brauchen. Sie sind z. B. bei B·l_t ♀ viel größer als bei B·l_T bzw. B·l_{Ri} ♀♀.

Als zweite viel wesentlichere Ursache für die verschiedenen Werte ein und derselben Affinität ergab sich dann aber der Zeitpunkt der Bestäubung (SCHWEMMLE 1957a, b, 1958a, b). Nachdem schon früher Kreuzungen mit Nachbestäubungen durchgeführt wurden und sich dabei ergeben hatte, daß sogar bei Bestäubung bis 50—60 Std nach der Kastration noch ein Ansatz erfolgt, wurden nun systematisch Vergleichskreuzungen mit Frühbestäubungen und Spätbestäubungen hergestellt. Als Beispiel seien zunächst die Kreuzungen B·l × hsc·hsc angeführt (SCHWEMMLE 1957a, S. 138). Bestäubt wurde früh: 12 Std nach der Kastration;

mittel: 48 Std nach der Kastration, und sehr spät: 80 Std nach der Kastration, also zu einem Zeitpunkt, wo die Petalen schon rotbraun verfärbt waren. Den Ausfall der Kreuzungen zeigt Abb. 4. Die Durchschnittswerte von jeweils 10 (bzw. 9) Kreuzungen sind für die frühen Bestäubungen 50,4%, für die mittleren 43,2% und für die sehr späten 20,2%.

Es war nun weiterhin interessant zu wissen, ob die aus dieser Kreuzung entstehenden B·hsc und *l*·hsc in gleicher Weise an der Abnahme des Samenansatzes beteiligt waren oder ob sich hier etwa Unterschiede zeigten. Wie die Aufzuchten ergaben, war tatsächlich das letztere der Fall. In den Frühbestäubungen sind die B·hsc und die *l*·hsc ungefähr gleich häufig, in den Spätbestäubungen dagegen sind die B·hsc deutlich viel weniger häufig als die *l*·hsc.

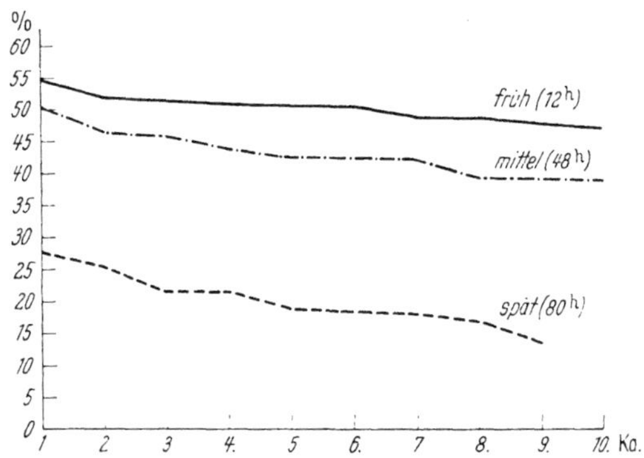

Abb. 4. Samenansatz der Kreuzung B · l_T × hsc · hsc in Abhängigkeit von der Bestäubungszeit. (Nach SCHWEMMLE 1957a.)

Diese — und zahlreiche weitere — Kreuzungen zeigen also einmal, daß die Häufigkeit, mit der die Samenanlagen befruchtet werden, mit zunehmendem Alter der Blüte abnimmt, und zweitens, daß diese Abnahme für die beiderlei Samenanlagen (offenbar durch deren genetische Konstitution bedingt) ungleich erfolgt, indem die B-Samenanlagen rascher „altern" als die *l*-Samenanlagen.

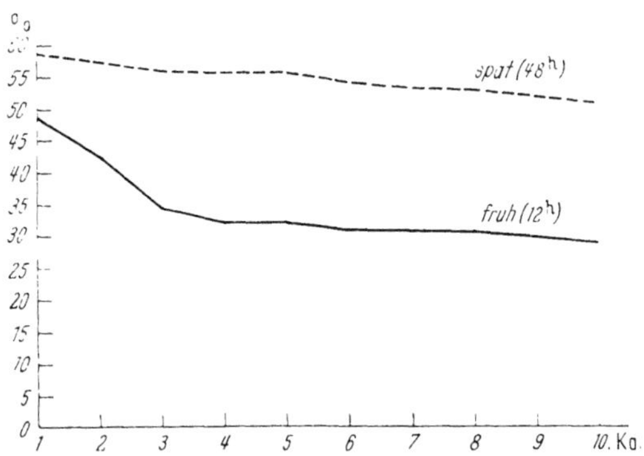

Abb. 5. Samenansatz der Kreuzung B · l_T × v · I *tenella* in Abhängigkeit von der Bestäubungszeit. (Nach SCHWEMMLE 1957a.)

Es scheint sehr naheliegend, daß der Samenansatz mit zunehmendem Alter einer Blüte zum Zeitpunkt der Bestäubung schlechter wird. Daß das aber nicht selbstverständlich ist, zeigen die folgenden Kreuzungen: Stellen wir die Kreuzung B·*l* × v·I *tenella* (die nur den I-Komplex überträgt) früh und spät her, dann haben wir gerade das Umgekehrte (Abb. 5). Die Durchschnittswerte des Samenansatzes für 10 Kapseln bei Frühbestäubung (12 Std) sind 33,9%, die für Spätbestäubung (48 Std) aber 54,4%. Die Spätbestäubungen setzen also jetzt wesentlich besser an als die Frühbestäubungen. Die Aufzuchten zeigen ferner, daß bei den Frühbestäubungen die B·I durchweg weniger häufig sind als die *l*·I, daß jedoch bei den Spätbestäubungen die B·I meistens häufiger sind als die *l*·I.

Als Ergebnis der Kreuzungen mit Früh- und Spätbestäubung mit B·*l* ♀ können wir also feststellen, daß die Befruchtungshäufigkeit sehr stark vom Zeitpunkt der Bestäubung abhängt, also vom Alter der Samenanlagen. Die B·*l* × hsc·hsc-Kreuzungen lassen sich so deuten, daß die Samenanlagen allmählich ihre Fähigkeit, die hsc-Pollenschläuche anzuziehen, verlieren; die Samenanlagen altern. Für die genetisch verschiedenen Samenanlagen (in unserem Falle die B- und die *l*-Samenanlagen) scheint dieses Altern aber durchaus verschieden schnell vor sich zu gehen: Die B-Samenanlagen altern rascher als die *l*-Samenanlagen.

Andererseits scheinen die Kreuzungen der B·*l* mit v·I *tenella* zu zeigen, daß die jungen Samenanlagen noch nicht ihre optimale Anziehungskraft erreicht haben; diese erreichen sie offenbar erst einige Zeit nach der Blütenöffnung. Die Samenanlagen müssen erst „reifen". Zu einem ähnlichen Schluß war auch schon LOERTZER (1954, S. 14) bei ihren Nachbestäubungskreuzungen B·I × Typ A × hsc·hsc (bzw. nachbestäubt × *l*·(II), × ha·ha und × I·I) gekommen: „Vielleicht werden die Samenanlagen der B·I (Bert. Pl.) später befruchtungsreif, als es das volle Erblühen anzeigt, vielleicht ist dieser Zeitpunkt der Nachbestäubung (ca. 15 h nach dem Aufblühen) der Zeitpunkt, zu dem eine Bestäubung erst voll wirksam sein würde."

Die auffallenden Unterschiede zwischen den B·*l* × hsc·hsc- und den B·*l* × v·I *tenella*-Kreuzungen sucht nun SCHWEMMLE zu erklären durch die Annahmen, daß 1. die Konzentration der die Anziehung der Pollenschläuche bewirkenden Samenanlagenstoffe mit dem Altern der Samenanlagen immer mehr zunimmt, und 2. daß die Anziehung der Pollenschläuche mit steigender Konzentration zunächst besser, dann aber nach Überschreitung eines Optimums wieder schlechter wird. Naheliegender und einfacher läßt sich vielleicht folgende Vorstellung bilden: Die, letzten Endes die Affinität bedingende, von den Samenanlagen ausgehende Anziehung der Pollenschläuche nimmt auch noch nach dem Aufblühen der Blüte bis zu einem maximalen Wert zu, um später mit dem fortschreitenden Alter der Blüte wieder abzunehmen. Die I-Pollenschläuche reagieren nun nach ihrem Eintritt in die Fruchtknotenhöhle verhältnismäßig rasch, sie treffen also bei Frühbestäubung die Samenanlagen noch im aufsteigenden Ast, der Samenansatz ist gering; bei Spätbestäubung treffen sie die Samenanlagen noch beim maximalen Wert, der Samenansatz ist besser. Die Reaktion zwischen den hsc-Pollenschläuchen und den B- bzw. *l*-Samenanlagen erfolgt im Vergleich zur Reaktion der I-Pollenschläuche verzögert; bei den Frühbestäubungen erfolgt die Reaktion der Pollenschläuche erst dann, wenn die Samenanlagen bereits ihren maximalen Wert erreicht haben, daher der gute Samenansatz; die Spätbestäubungspollenschläuche treffen dagegen bereits den absteigenden Ast, daher der schlechtere Ansatz.

Für diese Vorstellung spricht auch der Ausfall der nächsten Kreuzung (SCHWEMMLE 1957a, S. 157) B·*l* × Typ A (Abb. 6): Der Ansatz ist bei Früh-

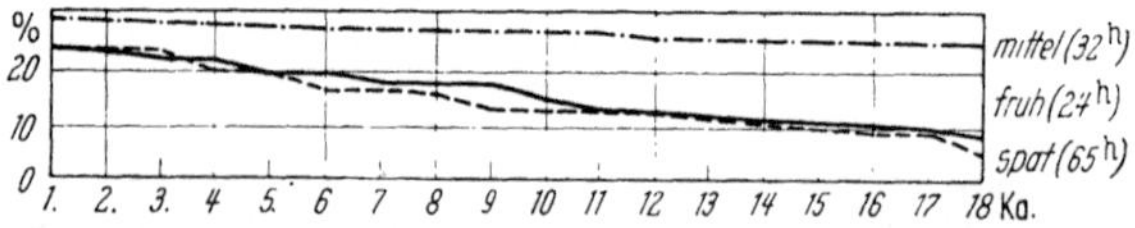

Abb. 6. Samenansatz der Kreuzung B·*l* × Typ A in Abhängigkeit von der Bestäubungszeit. (Nach SCHWEMMLE 1957a.)

bestäubung (24 Std) (im Durchschnitt der 10 besten Kapseln) 19,6%, bei mittlerer Bestäubung (32 Std) 27,9% und bei später Bestäubung (65 Std) 18,5%. Früh- und Spätbestäubungen setzen also schlechter an als eine Bestäubung zu einem mittleren Zeitpunkt. Auch dieses Ergebnis paßt gut zu unseren oben entwickelten Vorstellungen: Die Reaktion der v-Pollenschläuche erfolgt verhältnismäßig schnell; daher treffen die Frühbestäubungen in den aufsteigenden Ast. Die mittleren Bestäubungen treffen in das Maximum und die Spätbestäubungen in den absteigenden Ast.

Der entscheidende Faktor bei diesen Überlegungen ist also die Vermutung, daß die Pollenschläuche nach ihrer Ankunft in der Fruchtknotenhöhle nicht unmittelbar auf die Samenanlagen zuwachsen, sondern daß bis zum Eintritt dieser Reaktion eine mehr oder weniger lange Zeit verstreichen kann, deren Dauer von der genetischen Konstitution der betreffenden Pollenschläuche abhängen kann (hsc—I). Es handelt sich also um die gleiche Erscheinung, die schon früher zur Aufstellung des Begriffs der Reaktionsgeschwindigkeit führte; leider liegen aber gerade über diesen Abschnitt der Befruchtung für die vorliegenden Oenotheren keine Angaben vor. Lediglich WEIDNER-RAUH (1939, S. 429) erwähnt bei einer Kapsel von B·*l*, die 24 Std nach der Bestäubung mit I·I untersucht wurde, daß in einen Teil der Samenanlagen bereits ein Pollenschlauch eingewachsen war. „Die Befruchtung der Eizellen findet in der Regel also wohl 24—48 Std nach der Bestäubung statt.“ Wenn man bedenkt, daß die Pollenschläuche in wenigen Stunden den Griffel durchwachsen können (SCHNEIDER 1956), dann vergeht doch

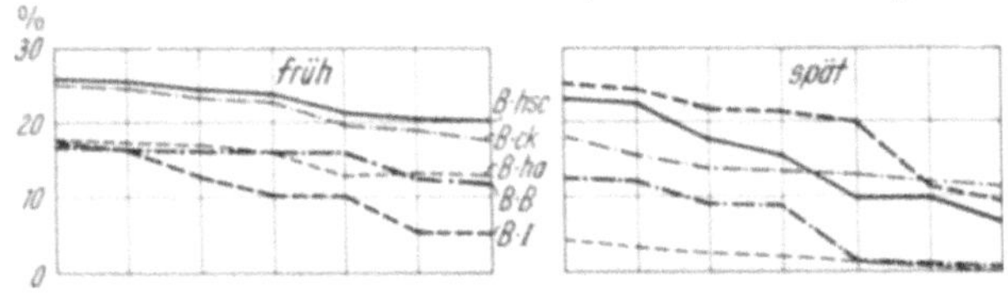

Abb. 7. Vergleich der Häufigkeit der Formen mit dem B-Komplex aus den Kreuzungen mit B·l_T ♀. (Nach SCHWEMMLE 1957a.)

noch eine ganz beträchtliche Zeit, bis die Befruchtung tatsächlich vollzogen ist. Eine Aufklärung über die Zeit, die vom Eintritt der Pollenschläuche in den Fruchtknoten wenigstens bis zum Eindringen in die Mikropyle einer Samenanlage verstreicht, läßt sich allerdings nur durch mühsame Untersuchungen gewinnen. Unter den erwähnten Gesichtspunkten wäre eine eingehende Kenntnis darüber aber äußerst notwendig.

Diese Tatsache, daß zwischen der Ankunft der Pollenschläuche im Fruchtknoten und der eigentlichen Befruchtung unter Umständen eine ziemlich lange Zeitspanne liegen kann, würde dann auch die an den Nachbestäubungskreuzungen immer wieder gemachte Beobachtung verständlich machen, daß nämlich die Pollenschläuche aus der Nachbestäubung die Pollenschläuche aus der etwa 24 Std vorher erfolgten Vorbestäubung noch „überholen“ können (SCHWEMMLE 1957b, S. 176).

Die Abhängigkeit der Affinitätswerte vom Zeitpunkt der Bestäubung ist natürlich entscheidend für die Frage nach den relativen Affinitäten etwa zweier Samenanlagen zu einer Pollenschlauchsorte. Das erhellt am besten aus der Abb. 7, wo für Früh- und Spätbestäubung der B·*l* mit verschiedenen Pollensorten die Häufigkeiten, mit der die B-Samenanlagen von den jeweiligen Pollensorten befruchtet werden, eingetragen sind. Zunächst ergibt sich daraus eine erneute Bestätigung der Tatsache, daß die B-Samenanlagen die verschiedenen Pollenschläuche verschieden gut anziehen; aber es ergibt sich weiter, daß für einen Vergleich von Affinitätswerten der Zeitpunkt der Bestäubung von ganz entscheidender Bedeutung sein kann. Stellt man nämlich die Affinitäten der B-Samenanlagen zu einer Reihe von Pollensorten nach fallenden Werten für Früh- und für Spätbestäubung zusammen, dann haben wir folgende Verhältnisse (SCHWEMMLE 1957a, S. 163):

In den Frühbestäubungen werden die Komplexverbindungen in der Reihenfolge gebildet

$$B\cdot hsc \approx B\cdot ck > B\cdot B \approx B\cdot ha > B\cdot I.$$

In den Spätbestäubungen dagegen in folgender Reihenfolge

$$B\cdot I > B\cdot hsc > B\cdot ck > B\cdot B > B\cdot ha.$$

Gleiches gilt auch für die Affinitäten der *l*-Samenanlagen (Schwemmle 1957a, S. 165), die je nach dem Zeitpunkt der Bestäubung wieder ganz verschieden ausfallen.

Die schon bald nach Erkennen der ersten Unregelmäßigkeiten getroffene Maßnahme, nur solche Kreuzungen miteinander zu vergleichen, die — soweit möglich — genau zur gleichen Zeit hergestellt wurden, hat damit ihre volle Rechtfertigung erfahren; sie ist tatsächlich die einzige Möglichkeit, korrekte Werte zu erhalten.

Was nun den Begriff der Affinität betrifft, so hat dieser natürlich nach wie vor seinen Sinn. Die anfängliche Annahme eines für jede Samenanlagen-Pollenschlauch-Kombination festen Affinitätswertes muß jedoch aufgegeben werden. Ein solcher Wert existiert nicht. Denn auch unter völlig konstanten Außenbedingungen ist die Affinität eine Funktion des Alters der Samenanlagen; sie steigt von einem Anfangswert zu einem Maximum, das aber nicht notwendigerweise mit dem Zeitpunkt des Aufblühens zusammenfallen muß, um dann mit zunehmendem Alter der Blüte wieder abzusinken.

Der unterschiedliche Zustand der Samenanlagen in einem Fruchtknoten.

Weiteren Aufschluß über das Reifen und Altern der Samenanlagen im Fruchtknoten ergaben dann Untersuchungen, die zunächst eigentlich einer ganz anderen Frage galten, nämlich der Frage, wie die Nachbestäubungspollenschläuche die noch nicht befruchteten Samenanlagen befruchten. Dazu wurden die Kapseln nach der Samenreife in drei Teile zerlegt, diese Drittel getrennt ausgezählt und zum Keimen ausgelegt. Dabei ergab sich zunächst, daß die Nachbestäubungspollenschläuche tatsächlich die von der Vorbestäubung ausgesparten bzw. noch nicht befruchteten Samenanlagen durch den ganzen Fruchtknoten von unten bis oben hin befruchten können, daneben aber noch ein weiteres, was für uns im Zusammenhang mit dem Altern der Samenanlagen von Bedeutung ist: Innerhalb eines Fruchtknotens befinden sich die Samenanlagen nicht alle im gleichen „Reifezustand". Die Samenanlagen „reifen" und „altern" von unten fortschreitend nach oben. Je nach dem Zeitpunkt der Bestäubung werden sich die Samenanlagen in den einzelnen Abschnitten eines Fruchtknotens also ganz verschieden verhalten.

Das zeigen deutlich die nächsten Kreuzungen. Wird die B·*l* × Typ A früh bestäubt (22 Std nach der Kastration), dann haben die einzelnen Drittel von oben nach unten folgende Werte des Samenansatzes: 18,9%—22,2%—23,0%; für die Kreuzung B·*l* × *l*·(II) (24 Std) ergaben sich die Werte: 13,6%—12,4%—15,5% (Schwemmle 1957b, S. 183 bzw. 197). Bei Spätbestäubungen jedoch ist der Ansatz völlig ungleich auf die einzelnen Abschnitte verteilt; der Ansatz im oberen Drittel ist um vieles höher als der im mittleren und unteren Drittel: In der Kreuzung B·*l* × I·I (47 Std) haben wir die Werte: 62,1%—38,6%—14,8%; in der Kreuzung B·*l* × hsc·hsc die Werte: 69,9%—31,3%—10,1% (Schwemmle 1957b, S. 183 bzw. 197).

Wenn auch die 4 Kreuzungen streng genommen nicht verglichen werden können, so sind die Unterschiede zwischen den Früh- und den Spätbestäubungen doch völlig gesichert. Da wir aus den früher besprochenen Kreuzungen mit Früh- und Spätbestäubungen wissen, daß die Samenanlagen eines Fruchtknotens je nach dem Zeitpunkt der Bestäubung verschieden gut die Pollenschläuche anziehen — sie reifen, erreichen ein Maximum und altern wieder —, ist die Erklärung ohne weiteres nahegelegt: Die untersten Samenanlagen sind zuerst imstande, die Pollenschläuche anzulocken; je höher im Fruchtknoten die Samenanlagen liegen, um so später erreichen sie den zur Befruchtung notwendigen Reifezustand. Die „Reifung“ der Samenanlagen schreitet von unten nach oben fort. Ebenso werden aber auch die untersten Samenanlagen auch die ersten sein, die ihre Fähigkeit, einen Pollenschlauch anzulocken, verlieren, sie altern auch als erste und dieses Altern der Samenanlagen schreitet wieder von unten nach oben fort. Zu einem Zeitpunkt, wo die untersten Samenanlagen bereits deutlich gealtert sein können, können also die obersten gerade noch den maximalen Wert der Anziehung besitzen. Das dürfte z. B. bei den beiden Spätbestäubungen B·*l* × I·I und B·*l* × hsc·hsc der Fall gewesen sein.

Vergleichen wir dann aber diese beiden Kreuzungen noch etwas näher miteinander, dann sehen wir, daß bei den B·*l* × I·I-Kreuzungen der Ansatz im untersten und auch im mittleren Drittel besser, im obersten dagegen etwas schlechter ist als bei den B·*l* × hsc·hsc-Kreuzungen. Das würde aber wieder ganz gut zu unseren früheren Erklärungen passen, daß nämlich die Reaktion der I-Pollenschläuche rascher erfolgt als die Reaktion der hsc-Pollenschläuche; die I-Pollenschläuche erreichen daher die untersten Samenanlagen noch in einem günstigeren Zustand als die hsc-Schläuche. Umgekehrt erreichen die I-Schläuche die obersten Samenanlagen in dem noch nicht ganz maximalen Zustand wie die später in Reaktion tretenden hsc-Schläuche.

Die Beurteilung der oben gegebenen Zahlenwerte für die einzelnen Drittel einer Kapsel und die daraus gezogenen Schlüsse über das Reifen und Altern der Samenanlagen beruhen natürlich auf der Annahme, daß die fertilen und sterilen Samenanlagen über den ganzen Fruchtknoten hinweg gleichmäßig verteilt sind. Tatsächlich dürfte diese Voraussetzung aber nicht immer zutreffen. Wenn z.B. bei *Oe. Berteriana* in einem Fall im oberen Drittel 83,6% große Samen gefunden wurden, bezogen auf den ganzen Inhalt des Drittels (Schwemmle 1964, S. 419), und wir einen Wert von 60% für die fertilen Samenanlagen, bezogen auf den ganzen Fruchtknoten zugrunde legen, dann zeigt das eindeutig, „daß bei der *Oenothera berteriana* die fertilen Samenanlagen nicht gleichmäßig über den Fruchtknoten verteilt sind“. Darauf weist auch Landspersky (1963) schon hin. Der Vergleich der einzelnen Drittel gibt also in Wirklichkeit ein sehr verzerrtes Bild, das bei der Beurteilung des Reifens und Alterns korrekterweise berücksichtigt werden müßte.

Daß diese ganzen Erscheinungen des Reifens und Alterns der Samenanlagen nicht auf die B·*l* allein beschränkt sind, zeigen die zahlreichen Kreuzungen von v·I mit *Oe. argentinea* (ha·ha), *Oe. scabra* (hsc·hsc), *Oe. campylocalyx* (ck·cl) und B·(II), jeweils mit Früh- und Spätbestäubung (Schwemmle 1958a, b). Ohne auf weitere spezielle Einzelbeispiele einzugehen, ergibt sich aus diesen Kreuzungen wieder, daß der Zeitpunkt der Bestäubung für den Samenansatz von entscheidender Bedeutung ist und daß die Affinitätswerte zwischen einer Samenanlagensorte und einer Pollensorte in Abhängigkeit von dem Reifezustand der Samenanlagen erheblich variieren können. Wenn im übrigen bei einer Wiederholung der Versuche durch Landspersky (1963) die Befunde von Schwemmle nur zum Teil bestätigt werden konnten, zeigt das nur, wie schwer die mit dem Reifen und Altern der Samenanlagen zusammenhängenden Prozesse einer exakten Analyse zugänglich sind.

Durch die Verwendung von v·I mit verschiedenen Plasma- und Plastiden-Kombinationen (od. Pl. + od. Pld., Bert. Pl. + od. Pld., Bert. Pl. + Bert. Pld.) ließ sich dann auch hier wieder ein Einfluß des Plasmas und der Plastiden auf die Affinität feststellen, wobei der Gedanke sehr naheliegt, daß die primäre Ursache für die beobachteten Unterschiede vielleicht eher in einem Einfluß des

Plasmas und der Plastiden auf den Vorgang des Reifens und Alterns der Samenanlagen als in einem Einfluß dieser beiden Faktoren auf die Höhe der Affinität selbst beruht. Gerade aus diesen Ergebnissen wird besonders deutlich, wie sehr das Manifestwerden von Plasma- und Plastiden-Einflüssen auf die Affinität vom Zeitpunkt der Bestäubung abhängen kann, so daß dadurch die früheren Befunde, wonach in manchen Jahren Plasma- und Plastiden-Einflüsse eindeutig nachgewiesen werden konnten, in anderen Jahren aber überhaupt nicht vorhanden waren, ihre natürliche Erklärung finden.

Was schließlich noch die zu Beginn angeschnittene Frage nach dem Weg der Pollenschläuche bei der Befruchtung betrifft, so suchte Schwemmle zunächst (Schwemmle 1957) seine Befunde mit der Annahme in Übereinstimmung zu bringen: „Nun werden bei den Oenotheren die unten im Fruchtknoten befindlichen Samenanlagen zuerst befruchtet", obwohl Kaienburg (1950) für *Raimannia* nachgewiesen hatte, daß die den Fruchtknoten zuerst erreichenden Schläuche sich in den äußersten Teilen des Leitgewebes halten und die obersten Samenanlagen befruchten (S. 409). Erst in einer weiteren Arbeit kommt auch Schwemmle zu dem Schluß: „Die bislang besprochenen Kreuzungen lassen die Annahme berechtigt erscheinen, daß die Pollenschläuche oben im Fruchtknoten zu befruchten beginnen (Schwemmle 1964, S. 422).

Auf Grund seiner anatomischen Untersuchungen über den Weg der Pollenschläuche bei der Befruchtung kommt Glenk (1964) allerdings zu dem Schluß, daß bereits das gerichtete Wachstum der Pollenschläuche vom Leitgewebe auf die Oberfläche der Placenten unter Abbiegen aus der Längsrichtung durch chemotropische Anziehung von den Samenanlagen bedingt ist, und daß die Reihenfolge der Befruchtung daher durch den *optimalen* Reifezustand der Samenanlagen bestimmt ist. Die Überlegungen Schwemmles werden dadurch jedoch wohl kaum hinfällig; denn für das Wachstum der Pollenschläuche auf die Placenta-Oberfläche und zu den Samenanlagen wird der Reifezustand der Samenanlagen wohl nur unter den extremen Bedingungen sehr früher oder sehr später Bestäubungen der bestimmende Faktor sein, während bei einer Bestäubung unter „normalen" Bedingungen wohl alle Samenanlagen im Fruchtknoten hinreichend reif sein dürften, um die Pollenschläuche anziehen zu können.

Die materiellen Grundlagen der selektiven Befruchtung.

An der Tatsache, daß das gerichtete Wachstum der Pollenschläuche auf die Samenanlagen eine chemotropische Reaktion der Pollenschläuche auf spezifische, in den Samenanlagen gebildete Anlockungsstoffe darstellt, wird wohl kaum ein berechtigter Zweifel bestehen. Allerdings ist zu bedenken, daß Renner in seiner Arbeit über den Weg der Pollenschläuche im Fruchtknoten der Oenotheren (Renner und Herzog 1943, S. 220) zunächst nur die Möglichkeit einer solchen Reaktion erwägt: „Wieweit chemotropische Reizbarkeit an der Lenkung der Schläuche zu den Mikropylen beteiligt ist, bleibt zu untersuchen." Nach den vorliegenden Ergebnissen, die im Rahmen der Untersuchungen von Schwemmle und seinen Schülern in einer solchen Fülle gemacht wurden, erscheint es aber doch berechtigt, an Stelle nur einer Möglichkeit doch mit der Tatsache einer chemotropischen Reaktion zu rechnen. Es wäre schwer verständlich, wie sonst die im einzelnen geschilderten quantitativen und qualitativen Beziehungen zustande kommen sollten. Es erübrigt sich, in diesem Zusammenhang auf die früheren Ansichten zu diesem Problem im einzelnen einzugehen: es genügt, auf die Zusammenstellung bei Brieger (1930) zu verweisen.

Wenn wir uns im folgenden auf die Oenotheren der Section *Raimannia* beschränken, dürfen wir wohl annehmen, daß die Samenanlagen in Abhängigkeit von der genetischen Konstitution des Embryosacks bzw. der Eizelle, die sie enthalten, spezifische, chemotropisch wirksame Stoffe besitzen, auf die die Pollenschläuche, wiederum in Abhängigkeit von ihrer genetischen Konstitution, mit einem gerichteten Wachstum reagieren können. Diese Samenanlagenstoffe sind weiter in einer bestimmten Menge vorhanden, die wiederum in Abhängigkeit vom Alter der Samenanlagen variieren kann, indem sie zunächst bis zu einem Maximum zunimmt, um später wieder abzusinken: Die Samenanlagen „reifen" und „altern". Entsprechend der damit zeitlich sich ändernden Zahl von Samenanlagen mit einer für das Eintreten einer erfolgreichen Reaktion notwendigen Mindestmenge dieses Anlockungsstoffes ändert sich auch die Häufigkeit, mit der die Samenanlagen von einer gegebenen Pollenschlauchsorte befruchtet werden, also der Wert der Affinität.

Durch eine Reihe von Tatsachen wird nahegelegt, daß an der Bildung dieser Anlockungsstoffe der Embryosack wesentlich beteiligt ist. Der wichtigste Punkt ist die Abhängigkeit dieser Stoffe von der genetischen Konstitution des Embryosacks, der ja von der genetischen Konstitution des übrigen Gewebes der Samenanlagen durchaus verschieden sein kann. Weiter spricht dafür die Tatsache, daß in die sterilen Samenanlagen der Oenotheren (die also keinen Embryosack enthalten) nach WEIDNER-RAUH (1939, S. 432, 446) nie ein Pollenschlauch einwächst. Das gleiche hatte auch schon GEERTS (1909, S. 177) an *Oe. Lamarckiana* beobachtet. Auch SCHNARF (1929, S. 286) zog diesen Schluß, BRIEGER (1930, S. 50) erscheint er jedoch nicht unbedingt zwingend. Noch weitergehende Schlüsse, etwa auf bestimmte Zellen des Embryosacks als Produzenten des Anlockungsstoffes dürften wohl rein hypothetisch sein und wir müssen uns wohl noch immer der Ansicht von BRIEGER (S. 51) anschließen: „Wir wissen nichts Genaues darüber, wo die chemotropisch wirksamen Stoffe produziert und ausgeschieden werden." Daß aber der Embryosack nicht unbedingt allein für die Anlockung entscheidend zu sein braucht, zeigen andererseits die Beobachtungen über den modifizierenden Einfluß des Diplonten, also wohl des Gewebes der Samenanlagen auf die primär sicher irgendwie vom Embryosack aus gesteuerte Produktion von Anlockungsstoffen.

Die zweite Frage ist dann die nach der Ursache der verschiedenen Reaktionen zwischen Pollenschläuchen und Samenanlagen in Abhängigkeit von der genetischen Konstitution der *beiden* Partner. Zwei Möglichkeiten sind gegeben. 1. Die in den genetisch verschiedenen Samenanlagen vorliegenden Anlockungsstoffe sind miteinander identisch, jedoch in Abhängigkeit von der genetischen Konstitution quantitativ verschieden; die unterschiedlichen Reaktionen beruhen auf einer verschiedenen Reaktionsfähigkeit der Pollenschläuche. 2. Die Anlockungsstoffe sind qualitativ verschieden, also spezifisch für die genetisch verschiedenen Embryosäcke bzw. die entsprechenden Samenanlagen. Nach einer von ARNOLD (1958, S. 88) angestellten Überlegung läßt sich diese Frage wohl eindeutig zugunsten der zweiten Alternative entscheiden: Wir dürfen zunächst wohl voraussetzen, daß für den Eintritt einer Reaktion zwischen Pollenschlauch und Samenanlage eine untere Grenze für die Menge des Anlockungsstoffes existiert, unterhalb der keine Reaktion mehr erfolgt. Weiter dürfen wir annehmen, daß die in den einzelnen Samenanlagen vorhandenen Anlockungsstoffe um einen mittleren Wert variieren. Vergleichen wir nun — unter der Annahme, daß sich die in verschiedenen Samenanlagen gebildeten Anlockungsstoffe nur quantitativ voneinander unterscheiden — etwa die Affinitäten B ♀ — v ♂ = 0 und B ♀ — *l* ♂ = 24,2, dann müssen wir schließen, daß die Konzentration des B-Anlockungsstoffes unter der für die v-Pollenschläuche charakteristischen Reizschwelle liegt, nicht aber für die *l*-Pollenschläuche. Aus den Affinitätswerten *l* ♀—v ♂ = 20,0 und *l* ♀—*l* ♂ = 1,2 folgt aber gerade das Gegenteil. Unsere Annahme von nur quantitativen Unterschieden führt also zu einem offenen Widerspruch.

Auf die Möglichkeit, daß von den Samenanlagen in Abhängigkeit von ihrer genetischen Konstitution ungleiche Mengen von Anlockungsstoffen entstehen, hat bereits DE VRIES in seiner schon früher erwähnten Arbeit über bevorzugte Befruchtung hingewiesen (1924, S. 77): "Perhaps the amount of attracting substances secreted in the micropyle is different in the two cases." (Unter Bezugnahme auf die von der *Oe. lucida* gebildeten beiden Samenanlagen deserens' bzw. laeta' in der de Vriesschen Bezeichnung).

Einen Versuch zur direkten Bestimmung der von den Samenanlagen gebildeten Anlockungsstoffe unternahmen BENONI und SCHILDKNECHT (1963). In Extrakten aus Samenanlagen von fünf verschiedenen Oenotheren (*Oe. argentinea*, *Oe. longiflora*, *Oe. scabra*, *Oe. affinis* und I·I) konnten sie sieben Aminosäuren, fünf Peptide, ein primäres aliphatisches Amin, eine Indolverbindung, zwei nicht mit Ninhydrin reagierende Amine und fünf Zucker nachweisen. Im biologischen Test (s. Chemotropieversuche) mit Pollen der *Oe. longiflora* werden die *longiflora*-Pollenschläuche durch diese einzelnen Verbindungen in einer jeweils charakteristischen Verdünnung mehr oder weniger stark angezogen; durch Zugabe von Kaliumionen oder durch Gemische von Aminosäuren mit Zuckern kann die Wirkung gesteigert werden. Da die genetisch verschiedenen Samenanlagen alle die gleichen Stoffe abgeben, wird die Spezifität auf quantitative Unterschiede in der Zusammensetzung der Gemische zurückgeführt. Wegen der methodischen Unzulänglichkeit des biologischen Tests mit überdies nur angedeuteten Unterschieden gegenüber den Kontrollen kann jedoch dem Versuch zur Aufklärung der chemischen Grundlagen der Affinität kaum eine Beweiskraft zugesprochen werden.

Chemotropie-Versuche.

Es war ein naheliegender Gedanke, die durch die Kreuzungen nachgewiesene stark wechselnde Anlockung von Pollenschläuchen durch die Samenanlagen durch direkte Versuche zu verifizieren. In einer Reihe von Arbeiten wurde dieser Versuch unternommen (SCHWEMMLE und KOEPCHEN 1953, LOERTZER 1954, SCHNEIDER 1956, SCHWEMMLE und SIMON 1956). Die Untersuchungen von KOEPCHEN, LOERTZER und SIMON zu dieser Frage dürften wegen methodischer Mängel für eine Beurteilung ausscheiden. In der von SCHNEIDER angegebenen Versuchsanordnung sind die wesentlichsten Mängel zwar behoben, aber trotz der Parallelen zwischen den Versuchsergebnissen und den aus den Kreuzungen ermittelten Affinitätswerten können die Befunde nicht überzeugen. Solange es nicht gelingt, ein stets reproduzierbares, über einen längeren Zeitraum sich erstreckendes Wachstum der Pollenschläuche zu erzielen (die Versuche wurden alle nur mit frisch keimenden Pollenkörnern durchgeführt) und solange nicht die Möglichkeit gegeben ist, eine Störung der ganzen Reaktion durch allenfalls von der Wundstelle der isolierten Samenanlagen ausgehende Stoffe mit Sicherheit auszuschließen, kann von einem positiven Nachweis wohl kaum gesprochen werden. Hinzu kommen noch Schwierigkeiten, die sich aus dem Objekt selbst ergeben. Außer der *Oe. argentinea* enthalten alle anderen untersuchten Arten zu einem ziemlich hohen Prozentsatz sterile Samenanlagen, die äußerlich nicht kenntlich sind und die fertilen Samenanlagen sind bei den komplexheterozygotischen Arten ebenfalls nicht nach Komplexen zu trennen. Es scheint noch verfrüht, ohne ausreichende methodische Vorarbeiten an diese Frage heranzugehen.

Selektive Befruchtung bei anderen Objekten.

Wir hatten schon bei dem geschichtlichen Rückblick über die selektive Befruchtung darauf hingewiesen, daß die vor den Arbeiten von SCHWEMMLE diskutierten Fälle von selektiver Befruchtung sich schließlich alle auf andere Weise erklären ließen. Wenn wir nun noch die übrigen Fälle betrachten, für die ebenfalls selektive Befruchtung angenommen wurde, dann müssen wir sagen, daß eine selektive Befruchtung im Sinne der eingangs gegebenen Definition bis heute erst in einem einzigen Fall wirklich streng bewiesen wurde, nämlich bei *Salpiglossis* (HIEMENZ 1955, S. 352): Werden heterozygote Mütter mit homozygoten Vätern rückgekreuzt, dann gibt reichliche Bestäubung mit dem Inhalt einer ganzen Anthere ein klares 1:1-Verhältnis, Bestäubung mit nur 25 Pollentetraden dagegen in vielen Fällen eine deutliche Bevorzugung einer bestimmten Eizellsorte. Diese Verhältnisse entsprechen also ganz der früher beschriebenen Kreuzung $B \cdot l \times B \cdot (II)$ bei reichlicher und spärlicher Bestäubung und wie dort dürfte auch bei *Salpiglossis* die Ursache für die beobachtete selektive Befruchtung in einer unterschiedlichen Reaktionsgeschwindigkeit der Pollenschläuche zu den beiderlei genetisch verschiedenen Samenanlagen liegen. Bemerkenswert ist der *Salpiglossis*-Fall deshalb, weil sich hier die beiden Samenanlagensorten lediglich durch ein einziges, bekanntes Genpaar unterschieden.

Von den übrigen noch beschriebenen Fällen einer selektiven Befruchtung ist aber keiner wirklich streng bewiesen; allerdings ist selektive Befruchtung als Möglichkeit durchaus gegeben. Hierher gehören die Tabak-Kreuzungen von POLJAKOW und MICHAILOWA (1950, zitiert nach ARNOLD, S. 91) und die Beobachtungen von LAMPRECHT an *Pisum* (1954). Die Möglichkeit einer selektiven Befruchtung könnte vielleicht auch noch vorliegen bei den *Gossypium*-Kreuzungen von KEARNEY und HARRISON (1924) sowie bei der *aurea*-Mutante von *Pisum* (LAMPRECHT 1952). In beiden Fällen gehen jedoch die Autoren auf diese Möglichkeit zur Erklärung des Ausfalls ihrer Kreuzungen gar nicht ein. Ein endgültiges Urteil über diese letzten Fälle kann aber wohl nur eine erneute, kritische Nachuntersuchung erbringen.

Literatur.

ARNOLD, C.-G.: Selektive Befruchtung. Ergebn. Biol. **20**, 67—96 (1958). — Zum Problem der selektiven Befruchtung. Z. Bot. **50**, 113—126 (1962).

BAUR, E.: Einführung in die experimentelle Vererbungslehre. Berlin: Borntraeger 1911. — BENONI, H.: Über die Chemie der Anziehung von Pollenschläuchen durch Samenanlagen bei Oenotheren und Narzissen. Diss. Erlangen 1962. — BRIEGER, F.: Selbststerilität und Kreuzungssterilität. Berlin: Springer 1930. — BUCHNER, E.: Genetische und zytologische Untersuchungen an trisomen Mutanten der *Oenothera odorata* I. Flora (Jena) **143**, 543—574 (1956).

COBB-BLANCHARD, F.: Heterogametic and homogametic hybrids between two mutations of *Oenothera pratincola*. Pap. Michigan Acad. Sci. **6**, 133—180 (1926). — CORRENS, CARL: Die Ergebnisse der neuesten Bastardforschungen für die Vererbungslehre. Ber. dtsch. bot. Ges. **19**, (71)—(94) (1901). Gesammelte Abhandlungen zur Vererbungswissenschaft aus periodischen Schriften 1899—1924. Berlin: Springer 1924 (G.A.) 8, 264—286. — Scheinbare Ausnahmen von der Mendel'schen Spaltungsregel für Bastarde. Ber. dtsch. bot. Ges. **20**, 159—172 (1902a). G.A. 9, 287—299. — Über den Modus und den Zeitpunkt der Spaltung der Anlagen bei den Bastarden vom Erbsen-Typus. Bot. Ztg **60**, Abt. II, 65—82 (1902b). G.A. 10, 300—314. — Ein Fall experimenteller Verschiebung des Geschlechtsverhältnisses. S.-B. kgl. preuß. Akad. Wiss. **51**, 685—717 (1917). G.A. 42, 849—879. — Versuche bei Pflanzen das Geschlechtsverhältnis zu verschieben. Hereditas (Lund) **2**, 1—24 (1921a). G. A. 52, 1088—1108. — Der Einfluß des Alterns der Keimzellen auf das Zahlenverhältnis spaltender Bastarde. Naturwissenschaften **9**, 313—315 (1921b). G.A. 54, 1133—1136.

DITTMANN, CHR.: Genetische und zytologische Untersuchungen an trisomen Mutanten der *Oenothera odorata* II. Flora (Jena) **143**, 575—602 (1956).

GÄRTNER, C. F. VON: Versuche und Beobachtungen über die Bastarderzeugung im Pflanzenreich. Stuttgart: K. F. Hering & Co. 1849. — GEERTS, J. M.: Beiträge zur Kenntnis der Cytologie und partiellen Sterilität von *Oenothera Lamarckiana*. Rec. Trav. bot. néerl. **5**, 93—206 (1909). — GLENK, H.-O.: Untersuchungen über die sexuelle Affinität bei Oenotheren. In H. F. LINSKENS: Pollen physiology and fertilization. Amsterdam: North-Holland Pub. Co. 1964, 170—181. — GRESS, M.: Der Einfluß des Plasmas und einer geringen Änderung des Genoms auf die Affinität zwischen Samenanlagen und Pollenschläuchen bei *Oenothera odorata*. Diss. Erlangen 1957.

HAUSTEIN, E.: Neuere Arbeiten zur Befruchtungsphysiologie. Z. Bot. **43**, 253—261 (1955). — „Selektive Befruchtung" im Pflanzenreich. Umschau **57**, 432—435 (1957). — HERIBERT-NILSSON, N.: Eliminierung der positiven Homozygoten bezüglich der Rotnervigkeit bei *Oenothera Lamarckiana*. Bot. Not. (Lund) **1915**, 23—25 (1915a). — Die Spaltungserscheinungen der *Oenothera Lamarckiana*. Lunds Univ. Årsskr. **12**, Nr 1 (132 S.) (1915b). — Zuwachsgeschwindigkeit der Pollenschläuche und gestörte Mendelzahlen bei *Oenothera Lamarckiana*. Hereditas (Lund) **1**, 41—67 (1920). — Das Ausbleiben der dominanten Homozygoten in Bezug auf die Nervenfarbe bei *Oenothera Lamarckiana*. Hereditas (Lund) **6**, 387—391 (1925). — HIEMENZ, G.: Untersuchungen an *Salpiglossis variabilis* über Gonenkonkurrenz und selektive Befruchtung und ihre Auswirkung auf die Nachkommenschaft. Biol. Zbl. **74**, 337—370 (1955). — HIORTH, G.: Zur Kenntnis der Homozygoten-Eliminierung und der Pollenschlauchkonkurrenz bei Oenotheren. Z. indukt. Abstamm.- u. Vererb.-Lehre **43**, 171—237 (1927). — HUBER, H.: Die Eliminierung von Austauschgonen bei Oenotheren. I. Z. indukt. Abstamm.- u. Vererb.-Lehre **86**, 459—479 (1955).

JOHANNSEN, W.: Elemente der exakten Erblichkeitslehre, 2. Aufl. Jena: Gustav Fischer 1913. — JONES, D. F.: Selective Fertilization. Univ. of Chicago Sci. Ser. Chicago: The University of Chicago Press 1928.

KAIENBURG, A.-L.: Zur Kenntnis der Pollenplastiden und der Pollenschlauchleitung bei einigen Oenotheraceen. Planta (Berl.) **38**, 377—430 (1950). — KEARNEY, T. H., and G. J. HARRISON: Selective Fertilization in cotton. J. agric. Res. **27**, 329—340 (1924).

LAMPRECHT, H.: Über Chlorophyllmutanten bei *Pisum* und die Vererbung einer neuen, goldgelben Mutante. Agri hort. genet. (Landskrona) **10**, 1—18 (1952). — Selektive Befruchtung im Lichte des Verhaltens interspezifischer Gene in Linien und Kreuzungen. Agri hort. genet. (Landskrona) **12**, 1—37 (1954). — LANDSPERSKY, N.: Das unterschiedliche Reifen und Altern der Samenanlagen in den verschiedenen Abschnitten des Fruchtknotens der *Oenothera Berteriana*. Biol. Zbl. **82**, 9—29 (1963). — LECHNER, K.: Selektive Befruchtung auf Grund unterschiedlicher Reaktionsgeschwindigkeiten zwischen Pollenschlauch- und Eizellsorten. Biol. Zbl. **83**, 541—560 (1964). — LEUCHTMANN, G.: Die Eliminierung von Austauschgonen bei Oenotheren. II. Z. indukt. Abstamm.- u. Vererb.-Lehre **86**, 480—497 (1955). — LOERTZER, B.: Weitere Untersuchungen zur selektiven Befruchtung. II. Diss. Erlangen 1954.

MANGELSDORF, P. C.: The relation between length of style and Mendelian segregation in a maize cross. Amer. Naturalist **63**, 139—150 (1929).

POLJAKOW, I. M., u. P. W. MICHAILOWA: Der Einfluß der quantitativen Wechselbeziehungen von Blütenstaub verschiedener Arten in Blütenstaubmischungen auf die Wahlbefruchtung bei Tabak und Machorka. Ž. obšč. Biol. **11**, 73—90 (1950) [Russisch].

RENNER, O.: Befruchtung und Embryobildung bei *Oenothera Lamarckiana* und einigen verwandten Arten. Flora (Jena) **107**, 115—150 (1914). — Die tauben Samen der Önotheren. Ber. dtsch. bot. Ges. **34**, 858—869 (1916). — Versuche über die gametische Konstitution der Önotheren. Z. indukt. Abstamm.- u. Vererb.-Lehre **18**, 121—294 (1917). — Das Rotnervenmerkmal der Önotheren. Ber. dtsch. bot. Ges. **39**, 264—270 (1921). — Vererbung bei Artbastarden. Z. indukt. Abstamm-. u. Vererb.-Lehre **33**, 317—347 (1924). — Untersuchungen über die faktorielle Konstitution einiger komplexheterozygotischer Önotheren. Bibl. genet. (Lpz.) **9**, 168 S. (1925). — Artbastarde bei Pflanzen. In: Handbuch der Vererbungswissenschaft, Bd. II. Berlin: Borntraeger 1929. — RENNER, O., u. G. PREUSS-HERZOG: Der Weg der Pollenschläuche im Fruchtknoten der Oenotheren. Flora (Jena) **136**, 215—222 (1943). — RICHTER, C.-M.: Untersuchungen zur Physiologie der Befruchtung bei Oenotheren. Z. Bot. **44**, 377—407 (1956).

SCHILDKNECHT, H., u. H. BENONI: Über die Chemie der Anziehung von Pollenschläuchen durch die Samenanlagen von Oenotheren. Z. Naturforsch. **18b**, 45—54 (1963). — SCHNARF, K.: Embryologie der Angiospermen. In: K. LINSBAUER, Handbuch der Pflanzenanatomie, Abt. II, Teil 2, Bd. X/2. Berlin: Borntraeger 1929. — SCHNEIDER, G.: Wachstum und Chemotropismus von Pollenschläuchen. Z. Bot. **44**, 175—205 (1956). — SCHÜTZ, GEORG: Genetische und zytologische Untersuchungen an Eu-Oenotheren. Selektive Befruchtung bei der Komplexheterozygote *Oenothera Berteriana*. Z. Bot. **33**, 481—525 (1938). — SCHWEMMLE, J.: Gibt es eine selektive Befruchtung? Biol. Zbl. **68**, 195—231 (1949). — Gibt es eine selektive Befruchtung? II. Biol. Zbl. **70**, 193—252 (1951). — Gibt es eine selektive Befruchtung? III. Biol. Zbl. **71**, 152—183 (1952a). — Selektive Befruchtung als Erklärung unerwarteter Kreuzungsergebnisse. Biol. Zbl. **71**, 353—384 (1952b). — Der Einfluß des Plasmas auf die Affinität zwischen Samenanlagen und Pollenschläuchen. Biol. Zbl. **71**, 487—499 (1952c). — Selektive Befruchtung bei der *Oenothera Berteriana*. Biol. Zbl. **72**, 129—146 (1953a). — Selektive Befruchtung bei der *Oenothera odorata*. Biol. Zbl. **72**, 405—424 (1953b). — Selektive Befruchtung bei *Oenothera Selowii* und *Oenothera stricta*. Biol. Zbl. **75**, 563—575 (1956a). — Über Vergleichskreuzungen mit ähnlich konstituierten Oenotheren. Biol. Zbl. **75**, 643—664 (1956b). — Der Samenansatz bei Kreuzungen mit der *Oenothera Berteriana* ♀ nach Früh- und Spätbestäubung. Planta (Berl.) **49**, 135—167 (1957a). — Das unterschiedliche Reifen und Altern der Samenanlagen in den verschiedenen Abschnitten des Fruchtknotens der *Oenothera Berteriana*. Planta (Berl.) **49**, 168—204 (1957b). — Der Einfluß des Plasmas auf die Affinität zwischen Samenanlagen und Pollenschläuchen. II. Biol. Zbl. **76**, 443—453 (1957c). — Der Einfluß des Plasmas und der Plastiden auf die Affinität zwischen Samenanlagen und Pollenschläuchen. Biol. Zbl. **76**, 529—549 (1957d). — Über Kreuzungen mit verschiedenen Formen der *Oenothera odorata* bei Früh- und Spätbestäubung. I. Planta (Berl.) **51**, 223—248 (1958a). — Über Kreuzungen mit verschiedenen Formen der *Oenothera odorata* bei Früh- und Spätbestäubung. II. Planta (Berl.) **51**, 249—287 (1958b). — Der Samenansatz bei Kreuzungen mit der trisomen *Oenothera Berteriana var. angusta*. Biol. Zbl. **77**, 153—165 (1958c). — Über Kreuzungen mit der trisomen *Oenothera odorata var. tenella*. Biol. Zbl. **77**, 329—347 (1958d). — Der Einfluß des Faktors h auf die Affinität zwischen Samenanlagen und Pollenschläuchen. I. Z. Vererb.-Lehre **94**, 133—142 (1963a). — Der Einfluß des Faktors h auf die Affinität zwischen Samenanlagen und Pollenschläuchen. II. Z. Bot. **51**, 217—232 (1963b). — Weitere Untersuchungen zur Physiologie der Befruchtung bei *Oenothera berteriana*. Biol. Zbl. **83**, 409—425 (1964). — Schwemmle, J., u. Mitarb.: Genetische und zytologische Untersuchungen an Eu-Oenotheren. Z. indukt. Abstamm.- u. Vererb.-Lehre **75**, 358—800 (1938). — Schwemmle, J., u. W. Koepchen: Weitere Untersuchungen zur selektiven Befruchtung. Z. indukt. Abstamm.- u. Vererb.-Lehre **85**, 307—346 (1953). — Schwemmle, J., u. H. Lohmeyer: Über Versuche mit *Oenothera campylocalyx*. Z. Vererb.-Lehre **94**, 412—426 (1963). — Schwemmle, J., u. R. Simon: Der Samenansatz bei Oenotherenkreuzungen. Planta (Berl.) **46**, 552—568 (1956).

Vries, H. de: Preferential fertilization in *Oenothera Lamarckiana*. Bot. Gaz. **77**, 73—79 (1924).

Weidner-Rauh, E.: Untersuchungen über die partielle Sterilität der Oenotheren: Das Pulver bei Eu-Oenotheren. Z. indukt. Abstamm.- u. Vererb.-Lehre **76**, 422—486 (1939).

Zitek, R. v.: Der Nachweis der v·v- und I·I-Homozygoten der *Oenothera odorata*. Diss. Erlangen 1954.

Inkompatibilität der Phanerogamen.

Von

H. F. Linskens und **Marianne Kroh.**

Mit 19 Abbildungen.

I. Einleitung.

Die Inkompatibilität ist ein gengesteuertes Phänomen. Die Physiologie der Inkompatibilität wird also im strengen Sinne Gen-Physiologie sein. Zur Untersuchung der physiologischen Mechanismen sind daher nur solche Pflanzen geeignet, bei denen die genetische Grundlage für den Inkompatibilitätsprozeß bekannt ist. Weiterhin wird man wünschen, daß sich die Versuchspflanzen zur Gewinnung eines größeren Versuchsmaterials vegetativ vermehren lassen, also die Herstellung von Klonen gestatten.

Entwicklungsphysiologisch gesehen gibt es für die Phanerogamen 2 Formen der „*Unverträglichkeit*" beim Befruchtungsvorgang: entweder bei *inter*spezifischer (zwischen verschiedenen Arten) oder bei *intra*spezifischer (innerhalb einer Art oder Rasse) Kreuzung.

Interspezifische Inkompatibilitätserscheinungen werden als allgemein mit dem Artbegriff verbunden betrachtet. Man muß annehmen, daß sie fundiert sind in den physikalischen und chemischen Organisationsmerkmalen, welche species-spezifisch sind.

Bei der *intraspezifischen* Inkompatibilität unterscheidet man zwischen einer Behinderung des Befruchtungsvorganges zwischen den Keimzellen eines Individuums einerseits und den Keimzellen der Nachkommen eines Elternpaares (Stout 1923) anderseits: Liegt eine Unverträglichkeit bei Selbstbestäubung vor, so spricht man von *Selbst-Inkompatibilität* (self-incompatibility, Selbststerilität); tritt diese bei Fremdbestäubung auf, so spricht man von *Kreuzungs-Inkompatibilität* (cross-incompatibility, Kreuzungssterilität). In beiden Fällen handelt es sich um Entwicklungsstörungen im oder unmittelbar nach dem Kopulationsvorgang. Da wir hier die Störungen, welche sich für die Übertragung des Pollens auf die Narbe aus dem Blütenbau durch herkogame und dichogame Einrichtungen ergeben, außer Betracht lassen, ist die Inkompatibilität der Phanerogamen also stets eine *physiologische Abwehrreaktion* während des Befruchtungsprozesses (Linskens 1957).

Bereits Sirks (1917, 1927) hat darauf hingewiesen, daß Inkompatibilität in strengem Sinne *keine* Sterilität ist. Man kann sie eher nach dem Vorschlag von Brieger (1930) als Parasterilität bezeichnen. Eine saubere begriffliche Unterscheidung zwischen Sterilität und Inkompatibilität erscheint um so wichtiger, als beide Phänomene durch voneinander unabhängige genetische Mechanismen gesteuert werden und gleichzeitig nebeneinander auftreten können (z.B. bei *Lotus corniculatus*) (Bubar 1959).

Als erster hat wohl Stout (1917) für das Phänomen der sexuellen Unverträglichkeit den Ausdruck der "incompatibility" angewandt. Aber auch dieser

Begriff hat sich nicht allgemein durchgesetzt. So werden nach wie vor auch in der modernen Literatur Inkompatibilität und Sterilität als Synonyme gebraucht.

II. Historisches.

Durch das grundlegende Werk CHARLES DARWINS (1. Aufl. 1877) ist das Problem der Inkompatibilität der Blütenpflanzen in seinem ganzen Umfang in das Bewußtsein der Botaniker getreten. Jedoch bereits früher wurden von J. G. KÖLREUTER (1761, 1766, 1809), CHRISTIAN KONRAD SPRENGEL (1793), ANDREW KNIGHT (1799), FRITZ MÜLLER (1866), F. HILDEBRAND (1867) und HERMANN MÜLLER (1873) Beobachtungen gemacht, die darauf hindeuteten, daß die Übertragung von Blütenstaub von fremden Blüten auf die Narben „nicht ohne Nutzen" sein könne. Diese führten sodann zur Formulierung des "*lege di dichogama*" durch DELPINO (1868—1874), wonach „*es die Natur nicht haben wolle, daß irgendeine Blume durch ihren eigenen Staub befruchtet werde*" (SPRENGEL 1793). Der eigentliche Entdecker der Selbststerilität der Pflanzen scheint KÖLREUTER (1764) gewesen zu sein, der beobachtete, daß 3 Pflanzen von *Verbascum phoeniceum* in zwei aufeinanderfolgenden Jahren nach Bestäubung mit eigenem, gutem Pollen keinen Samenansatz brachten. Auf die weite Verbreitung von Selbststerilitätserscheinungen hat bereits HERBERT (1837) hingewiesen.

Aber erst DARWIN wurde durch seine Beobachtungen an *Orchideen*keimlingen aus kreuzbestäubten Blüten, welche stets den selbstbestäubten überlegen waren, zu Versuchen an *Eschholzia californica, Abutilon Darwinii, Senecio cruentus, Reseda odorata* und *R. lutea* angeregt, welche die ersten systematischen Experimente zur Inkompatibilitätsforschung darstellen. Diese führten zu einer allgemeineren Formulierung, welche auch heute noch als Darwin-Knightsches Gesetz bekannt geblieben ist; es lautet: "*Nature tells us in the most emphatic manner that she abhorrors perpetual selffertilization.*" Ausführliche Zusammenstellungen der historisch bedeutsamen Arbeiten haben LEHMANN (1927), EAST (1929) und BRIEGER (1930) gegeben. Rezente Arbeiten sind bei LINSKENS (1964) besprochen.

III. Verbreitung der Inkompatibilität bei den Phanerogamen.

Die Inkompatibilität ist unter den Phanerogamen sehr weit verbreitet. Nach DARWIN (1877) hat zuerst KNUTH (1898) eine Zusammenstellung der Verbreitung der Selbst-Inkompatibilität im System gegeben, die jedoch keinen Anspruch auf Vollständigkeit erheben kann. Die letzte große Zusammenstellung stammt von EAST (1940) und wurde post mortem von SAX herausgegeben. Bei über 3000 Arten aus mindestens 68 Familien wurde Selbst-Inkompatibilität nachgewiesen (vgl. auch BREWBAKER 1959), vorherrschend jedoch bei krautigen Gewächsen der *Cruciferae, Saxifragaceae, Rosaceae, Solanaceae, Scrophulariaceae, Compositae* und *Liliaceae*. In diesen Gattungen treten aber selbstfertile und inkompatible Arten nebeneinander auf. Innerhalb der *Cruciferae* wurden 102 selbst-kompatible und 80 selbst-inkompatible Arten gefunden (BATEMAN 1951, 1955). Polyphyletischer Ursprung des Phänomens ist daher anzunehmen (BATEMAN 1952, HUF-BRINCKMANN und LINSKENS 1961). Selbst-Inkompatibilität scheint jedoch eine relativ alte „Erfindung" der sexuellen Fortpflanzung zu sein. Die meisten Blütenpflanzen haben entweder früher ein Selbst-Inkompatibilitäts-System besessen oder haben es heute noch. Rezente Arten, welche selbst-kompatibel sind, sind wohl von selbst-inkompatiblen Vorfahren abzuleiten (LEWIS 1955, LEWIS und CROWE 1958, CROWE 1964). Für den Evolutionsprozeß der Blütenpflanzen spielt die Selbst-Inkompatibilität durch Erhaltung des Bastardcharakters und Förderung der Variabilität jedoch eine große Rolle.

Auf interessante Korrelationen zwischen Pollencytologie und Inkompatibilitätstypus bei Homostylen hat BREWBAKER (1957) aufmerksam gemacht: Formen mit gametophytischer Determination haben vorwiegend binucleaten Pollen (Ausnahme: *Gramineae*) und weisen eine Hemmung des Pollenschlauchwachstums im Griffel auf. Sporophytisch determinierte Arten besitzen dagegen trinucleaten Pollen und ihre Inkompatibilitätsreaktion ist auf der Narbenoberfläche lokalisiert.

Eine Deutung kann in der Weise erfolgen, daß man annimmt, daß die letzteren relativ ärmer an Zucker und anderen notwendigen Metaboliten sind, da sie diese Reservestoffe bei der zweiten zusätzlichen Mitose verbraucht haben.

IV. Selbst-Inkompatibilität.

Die Selbst-Inkompatibilitäts-Reaktionen bei höheren Pflanzen lassen sich nach verschiedenen Gesichtspunkten klassifizieren (Tabelle 1).

Tabelle 1. *Einteilung der Inkompatibilität bei Blütenpflanzen nach Kriterien aus Morphologie, Genetik und Physiologie.* (Nach Lewis 1954.)

Es bedeuten: sporoph. = sporophytische Kontrolle, gametoph. = gametophytische Kontrolle, Dominanz = die Allele zeigen Dominanz, Indiv. = die Allele wirken einzeln individuell.

Morphologie	Genetik			Physiologie
	Allele	Kontrolle		
		des Pollen	des Griffel	
Heteromorphie	2	sporoph. Dominanz	Dominanz	Komplementäre Förderung oder oppositionelle Hemmung
		sporoph. Dominanz od. Indiv.	Dominanz od. Indiv.	
Homomorphie	zahlreiche	gametoph. Indiv.	Indiv.	Ausschließlich oppositionelle Hemmung

Theoretisch sind für den physiologischen Mechanismus zwei Möglichkeiten gegeben: a) eine Hemmung als Ursache für die Inkompatibilität (Sterilität), b) eine Stimulation als Ursache für die Kompatibilität (Fertilität) (Tabelle 2). Das physiologische Problem ist: Wie wird der Pollenschlauch daran gehindert, die Eizelle zu erreichen und eine erfolgreiche Befruchtung zu vollziehen? Physiologisch betrachtet umfaßt also die Inkompatibilität alle Prozesse, die zwischen Bestäubung und Zygotenentwicklung liegen.

Tabelle 2. *Beziehungen zwischen den genetischen Reaktionen und den verschiedenen Möglichkeiten einer biochemischen Steuerung der Inkompatibilität.* (Nach Lewis 1954, ergänzt.)

I. Oppositionelle Hemmung		⟶		*Inkompatibilität.*
Reaktion:				Mechanismen:
zwischen gleichen Allelen (z. B. $S_1 \times S_1$)	⟶	Hemmung des Pollenschlauches	⟶	1. Hemmung von Fermenten oder Wuchsstoffen für das Schlauchwachstum. 2. Hemmung von Fermenten, die für das Eindringen der Schlauchspitze in das Narben- und Griffelgewebe nötig sind. 3. Blockierung der Pollenschlauchmembran.
II. Komplementäre Förderung		⟶		*Kompatibilität*
Reaktion:				
zwischen ungleichen Genen (z. B. $S \times s$)	⟶	Förderung des Pollenschlauches	⟶	1. Stimulation der Pollenschläuche. 2. Abfangen bzw. Beseitigen eines Hemmstoffes, der normalerweise im Pollen vorhanden ist.

Nach dem Ort resp. Zeitpunkt der Reaktion läßt sich folgende Einteilung vornehmen (Sears 1937):

1. Die Hemmung erfolgt vor oder bei der Pollenkeimung, also auf der Narbenoberfläche.

2. Der Pollenschlauch wird während seines Wachstumsprozesses im Leitgewebe des Griffels verlangsamt bzw. gestoppt.

3. Die Inkompatibilitätsreaktion erfolgt bei Eintreffen der Pollenschläuche in der Samenanlage. In diesem Falle scheinen die Integumente für die Reaktion eine Rolle zu spielen.

4. Als Folge der Inkompatibilitätsreaktion tritt nach der Karyogamie Abortion der befruchteten Eizelle oder des jungen Embryos ein.

1. Hemmung der Pollenkeimung.

Diese Form der Selbststerilitätsreaktion ist bei *Cruciferen* (BEATUS 1929, 1931, 1934; KAKIZAKI 1930, SEARS 1937, TATEBE 1939, BATEMAN 1954, 1955; KROH 1956, OELKE 1957, CHRIST 1958 u.a.), *Compositen* (GERSTEL 1950, HUGHES und BABCOCK 1950, CROWE 1954), sowie bei *Pelargonium*, *Secale* (SEARS 1937) und *Coffea* (MEDINA und MENDES CONAGIN 1959, DEVREUX, VALLAEYS, POCHET und GILLES 1959) verbreitet. Ein großer Teil der Pollenkörner keimt nach Selbstbestäubung nicht aus. Die wenigen Schläuche, welche gebildet werden, bleiben in der oberflächlichen Papillenschicht der Narbe stecken (RILEY 1936). Wesentlich ist, daß die Sterilitätsreaktion auf die oberste Narbenregion beschränkt ist; tieferliegendes Narbengewebe und der Griffel vermögen keine Hemmung auf die Pollenkeimung und das Pollenschlauchwachstum mehr auszuüben. Dies konnte in eindrucksvollen Versuchen an *Raphanus* vor allem KROH (1955) nachweisen (Abb. 1). Daraus ergab sich weiterhin, daß im Griffel kein Hemmungsgefälle besteht, das hemmende Prinzip also auch im Narbengewebe produziert oder aktiviert werden muß. Neben einer Reduktion der Pollenkeimungsrate wird auch eine extreme Verzögerung der Keimungsgeschwindigkeit des Pollen bei inkompatibler Bestäubung beobachtet.

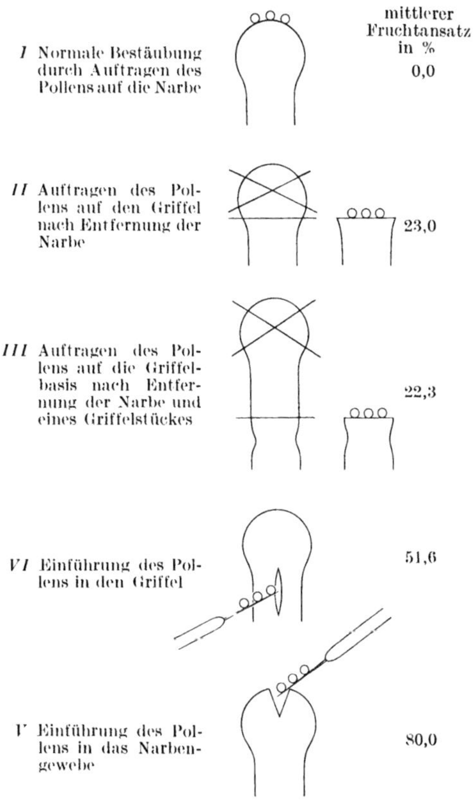

Abb. 1. Fruchtansatz von *Raphanus raphanistrum* bei verschiedenen Bestäubungsmethoden. (Nach KROH 1956.)

Die Erscheinung, daß bei Pflanzen, deren Narben mit einer intakten Cuticula bedeckt sind, die Pollenkeimung unterbleibt, wurde mehrfach beschrieben. Für einige *Papilionaceen*-Arten konnte nachgewiesen werden, daß die Narben erst nach Verletzung empfängnisfähig werden (MÜLLER 1873, JOST 1907, KIRK und STEVENSON 1931, DRAYNER 1959, HOLDEN und BOND 1960).

OELKE (1957) und TATEBE (1964) sehen in der Änderung der Transpirationsgröße, des osmotischen Wertes und der Permeabilität im Laufe der Narbenentwicklung, sowie entwicklungsbedingten pH-Änderungen wesentliche Faktoren für das Zustandekommen der Inkompatibilitätsreaktion bei *Cruciferen*. Durch Erhöhung der Luftfeuchtigkeit läßt sich nämlich die Hemmung der Pollenkeimung und des Pollenschlauchwachstums aufheben. Dennoch kommt es nicht zur Befruchtung, da die Selbstungsschläuche nicht in die Wände der Narbenpapillen, die mit einer Cuticula überzogen sind, einzudringen vermögen (Abb. 2a und b).

CHRIST (1959) entwickelte folgende Vorstellung von dem Inkompatibilitätsmechanismus. Nach einer Bestäubung steht dem Pollen auf den Narbenpapillen nur so viel Wasser zur Verfügung, daß er quellen, jedoch nicht keimen kann.

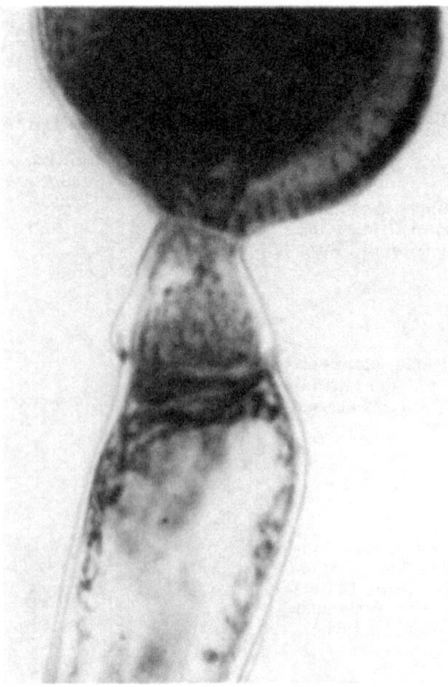

Abb. 2a.

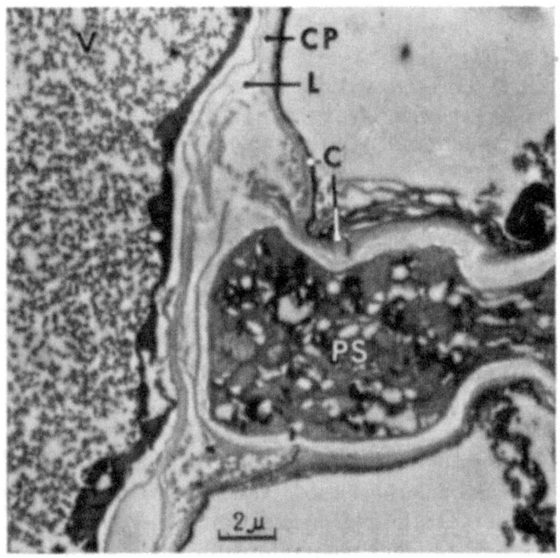

Abb. 2b.

Abb. 2a. Bei Fremdung dringt der Pollenschlauch von *Cardamine pratensis* in die Narbenpapille (unten) ein. Die Schlauchspitze sitzt der Papille mit einem Wulst auf. Zum Durchdringen benötigt der Schlauch 5—10 min. Die Wand der wulstigen Schlauchspitze verschmilzt mit der Papillenwand. 1800×. (Nach CHRIST 1959.)

Abb. 2b. Der Pollenschlauch (*PS*) von *Brassica nigra* hat nach Fremdung an der Berührungsstelle mit der Papillenwand die Cuticula (*C*) aufgelöst und ist in die Cellulose-Pektinschicht (*CP*) eingedrungen. (*L* Celluloselamellen; *V* Vacuole.) (Nach KROH 1964.)

Im Falle einer Fremdbestäubung macht der Pollen durch Abgabe und Wirksamwerden eines cutinspaltenden Fermentes die Narbenpapillen permeabel und erhält dadurch die zum Keimen notwendige Feuchtigkeit. Nach Selbstbestäubung wird das im Pollen vorhandene Fermentsystem infolge der Inkompatibilitätsreaktion inaktiviert. Der Selbstungspollen kann aus Wassermangel und infolge der Unfähigkeit, in die Zellwände des Narbengewebes einzuwachsen, seine Befruchtungsfunktion nicht ausführen.

Die Hypothese von CHRIST wird unterstützt durch die beiden Befunde von LINSKENS und HEINEN (1961, 1962), daß cutinspaltende Fermentsysteme ausschließlich im Pollen von Pflanzen vorkommen, deren Narben mit einer Cuticula bedeckt sind, und daß von *Cruciferen* isolierte Narbencuticula durch eine aus Pilzen gewonnene Cutinase, abgebaut werden kann.

Auf Grund von Experimenten, in denen das Verhalten der Pollenkörner nach einem Narbenwechsel (= Veränderung des S-Gen-Milieus) beobachtet wurde, erfahren die Pollenkörner nach einer Fremdbestäubung eine irreversible Aktivierung ihres cutinspaltenden Fermentsystems, wodurch auch ein Eindringen des Pollens in eine „Selbstungs"-Narbe ermöglicht werden kann (KROH 1966). Die spezifische Inkompatibilitätsreaktion soll nach Selbstbestäubung durch eine Kombination am „S-Allel-Areal" (SAMPSON 1962) eintreten und eine Blockierung der Cutinase-Aktivierung bewirken.

2. Hemmung des Pollenschlauchwachstums.

Am häufigsten äußert sich die Inkompatibilitätsreaktion in Form einer Hemmung des Pollenschlauchwachstums im Leitgewebe des Griffels, während die

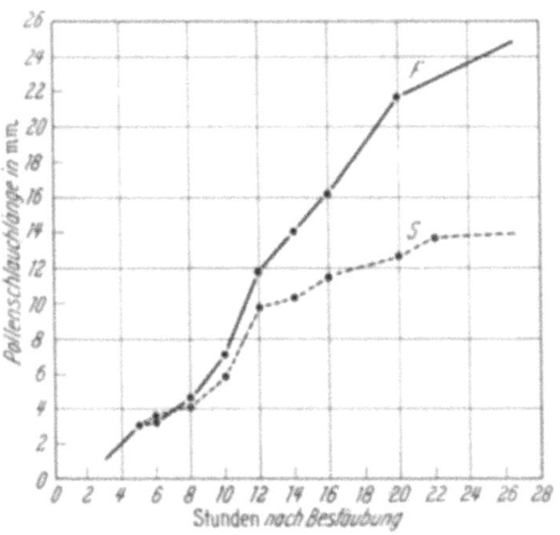

Abb. 3. Wachstumskurven von Pollenschläuchen in vivo bei Selbst- (*S*) und Fremdbestäubung (*F*). (Nach STRAUB 1946, bei *Petunia hybrida*.)

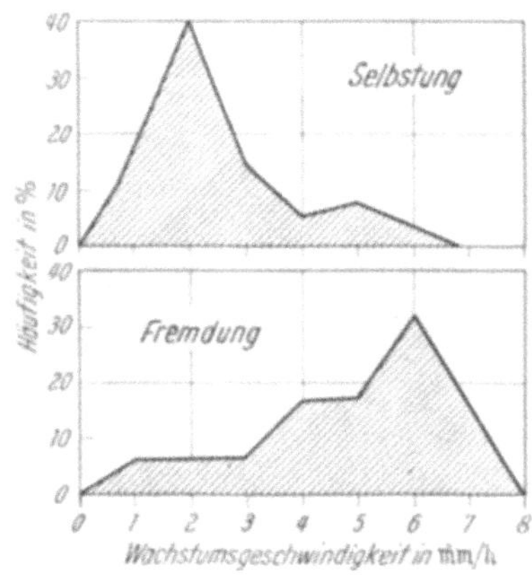

Abb. 4. Wachstumsgeschwindigkeit der Pollenschläuche von *Oenothera* nach Selbst- (oben) und Fremdbestäubung (unten). (Nach EMERSON 1940.)

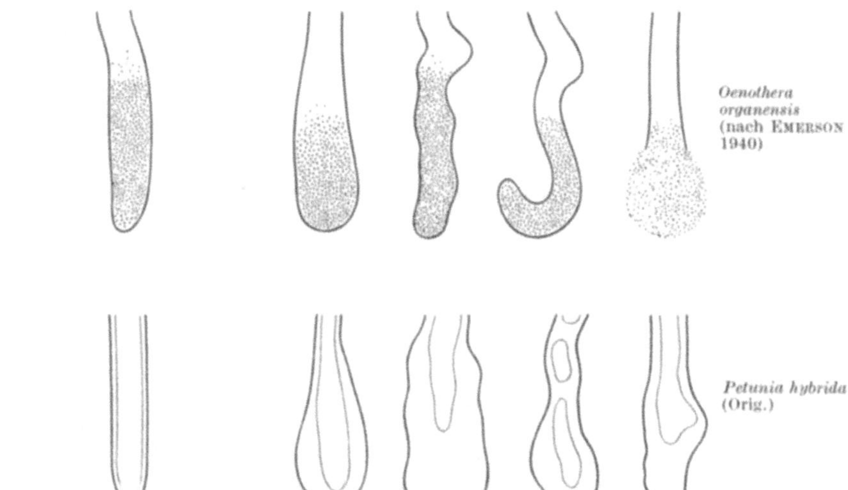

Abb. 5. Morphologie der Pollenschlauchspitzen bei normalem Wachstum nach kompatibler und bei gehemmtem Wachstum nach inkompatibler Bestäubung. (Nach EMERSON 1940 und Orig.)

Schläuche auf ihrem Weg zur Eizelle mit diesem in Kontakt stehen. Sie kann sich unmittelbar unter dem Narbengewebe, in der Mitte des Griffels oder an dessen Basis abspielen. Das Wachstum der Masse der Selbstschläuche wird gewöhnlich verlangsamt und schließlich sistiert, während die Fremdschläuche mit mehr oder weniger gleichbleibender Wachstumsgeschwindigkeit die Samenanlage erreichen (Abb. 3). Man muß sich jedoch darüber klar sein, daß nicht alle Pollenschläuche in gleichem Maße gehemmt werden. Es kommt vielmehr, trotz einheitlicher genetischer Konstitution, zu einer Häufigkeitsverteilung (Abb. 4), in der das schnellere Wachstum der Masse der Fremdschläuche zum Ausdruck kommt (EMERSON 1940).

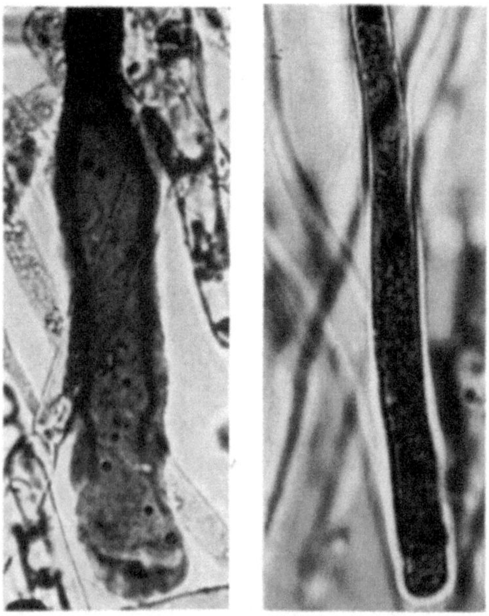

Abb. 6. Pollenschlauchspitzen von *Petunia* von einem Selbstungsschlauch (links) und einem Fremdungsschlauch (rechts).

Die gehemmten Pollenschläuche inkompatibler Bestäubungen verändern während des verlangsamten und schließlich sistierten Wachstums ihre Gestalt: Ihre Spitzen schwellen keulenförmig an, krümmen sich und können ihren Inhalt vorzeitig in das Leitgewebe durch Platzen entleeren (Abb. 5, 6). Dabei kommt es meist auch zu starken Cellulose- und Calloseanlagerungen im Bereich der Pollenschlauchspitze (YASUDA 1934, EMERSON 1940, SCHLÖSSER 1961). Häufig bleibt in inkompatiblen Pollenschläuchen die mitotische Teilung des generativen Kerns aus oder setzt gegenüber dem kompatiblen Schlauch verspätet ein, während der vegetative Kern schon in den ersten Stunden nach der Keimung verschwindet (SCHLÖSSER 1961).

a) Pollenschlauchhemmung bei Homostylen.

Auch bei normalem, ungehemmtem Wachstum der Pollenschläuche besteht eine intensive physiologische Wechselwirkung zwischen Pollenschlauch und Leitgewebe (LINSKENS und ESSER 1959, TUPÝ 1961).

Für den Wachstumsprozeß von Selbst- und Fremdschläuchen lassen sich zahlreiche physiologische Unterschiede aufzeigen, die Hinweise zum Verständnis eines Mechanismus geben. Die Atmung der Selbstschläuche (Abb. 7) ist im ersten Abschnitt des Wachstums trotz reduzierter Geschwindigkeit erhöht. Dies deutet auf einen erhöhten Energieverbrauch während der Hemmreaktion hin (LINSKENS 1953, 1954, 1955). Durch Beeinflussung der Syntheseprozesse mittels Belichtung und Verdunklung ließ sich die Intensität der Hemmreaktion stören (EMERSON 1940, LINSKENS 1955). Es bestehen also *Zusammenhänge zwischen der Selbst-Inkompatibilitätsreaktion und der Ernährung der Pollenschläuche.* Für eine Störung des normalen Kohlenhydratstoffwechsels sprechen auch die Beobachtungen, daß inkompatible Pollenschläuche gegenüber kompatiblen eine größere Anzahl von Callose-Pfropfen per Schlauchlänge (LINSKENS und ESSER 1957, TUPÝ 1959, SCHLÖSSER 1961), sowie ein dichteres Gerüst von Cellulosefibrillen in der Wand der Pollenschlauchspitze aufweisen (MÜHLETHALER und LINSKENS 1957). Demnach scheint eine superoptimale Festlegung von Kohlenhydraten vorzuliegen, welche bei kompatibler Bestäubung für den weiteren Aufbau der Pollenschlauchwand auf dem Wege zur Samenanlage benötigt werden. Doch dürften diese Phänomene sekundärer Natur sein.

Während des Pollenschlauchwachstums tritt bei gleichzeitigem Zustrom von Kohlenhydraten eine Senkung des Zuckerspiegels im Griffel auf, die gepaart ist mit einer Anhäufung von Asparagin. Dieser Prozeß läuft in fremdbestäubten Griffeln schneller ab als in selbstbestäubten und läßt auf einen intensiven Proteinverbrauch durch die Atmungsprozesse schließen. Hierauf weist auch die nach Selbstbestäubung auftretende starke Umsetzung von Prolin, Glutamin und Glutaminsäure, von denen die beiden letztgenannten auf Grund ihrer Ketogruppe direkt mit dem Respirationsprozeß in Verbindung stehen (Tupý 1961, 1964, Linskens und Tupý 1966). Die Protein- und Nucleinsäurefraktionen von Extrakten unbestäubter, selbst- und kreuzbestäubter Griffel zeigen deutliche Unterschiede. Charakteristische Fluorescenzmuster waren nach chromatographischer Trennung von Griffelextrakten zu beobachten. So erscheint in selbstbestäubten Griffeln eine noch nicht identifizierte Verbindung, die einen hemmenden Effekt auf das Pollenschlauchwachstum ausübt (Linskens 1965). Eine zunehmende Aktivität von „saurer Phosphatase" und Cytochromoxidase kann im Verlauf der Inkompatibilitätsreaktion beobachtet werden (Schlösser 1961).

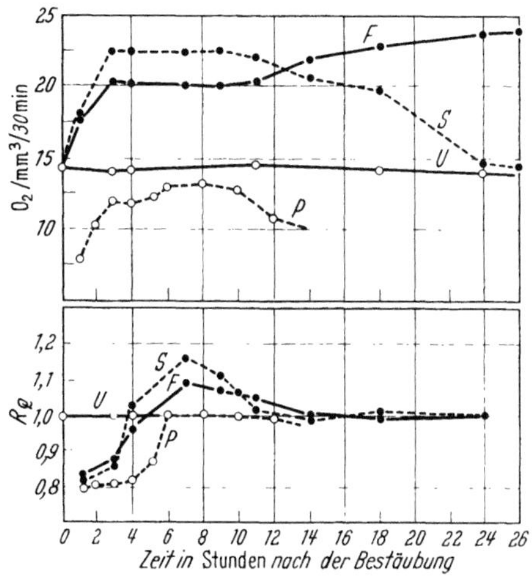

Abb. 7. Atmungsverhalten bei Griffeln von *Petunia*. Oben: Atmung als O_2-Verbrauch, unten: respiratorischer Quotient. *P* keimende Pollen in vitro ohne Griffel; *U* unbestäubte Griffel; *F* fremdbestäubte Griffel; *S* selbstbestäubte Griffel. (Nach Linskens 1955.)

Für den Primäreffekt im Sinne der Eastschen Immunitätstheorie der Selbst-Inkompatibilität geben die serologischen Untersuchungen von Lewis (1953) interessante Aufschlüsse (Abb. 8). Nach intravenöser und intraperitonealer Injektion von Pollenextrakten bestimmter Inkompatibilitätsklone von *Oenothera* in Kaninchen können zwischen den entstehenden Antiseren und den homologen Antigenen Präcipitatreaktionen gefunden werden, die gegenüber den halbhomologen und heterologen wesentlich stärker sind. Diskussion der Methode und der Ergebnisse s. Moritz (1958).

Serologische Teste mit Leitgewebs- und Pollenextrakten von *Petunia* stimmen in ihren Resultaten mit denen von Lewis überein (Linskens 1960). Nach vorheriger Trennung der Extrakte durch Immunelektrophorese und anschließender Reaktion auf das Antiserum sind in Abhängigkeit von der S-Allel-Zusammenstellung der Reaktionspartner charakteristische Niederschlagszonen zu beobachten (Linskens 1965). Untersucht man die Eiweißreaktionen von Leitgeweben von *Petunia* nach Selbst- und Fremdbestäubung, so sind Unterschiede in den Fraktionen zu registrieren (Abb. 9). Man findet nämlich im geselbsteten Griffel zwei neue Komponenten (X, Y), die sich nach Hydrolyse als Glucoproteide erweisen, während nach Fremdung lediglich eine glucoproteidische Fraktion (Z) gefunden wird. Dabei ist dieser Befund unabhängig von den zur Bestäubung verwendeten S-Allelen (Linskens 1954, 1955).

Am Zustandekommen dieser Fraktionen sind sowohl Pollenschlauch wie Leitgewebe beteiligt (Linskens 1958a, b, 1959).

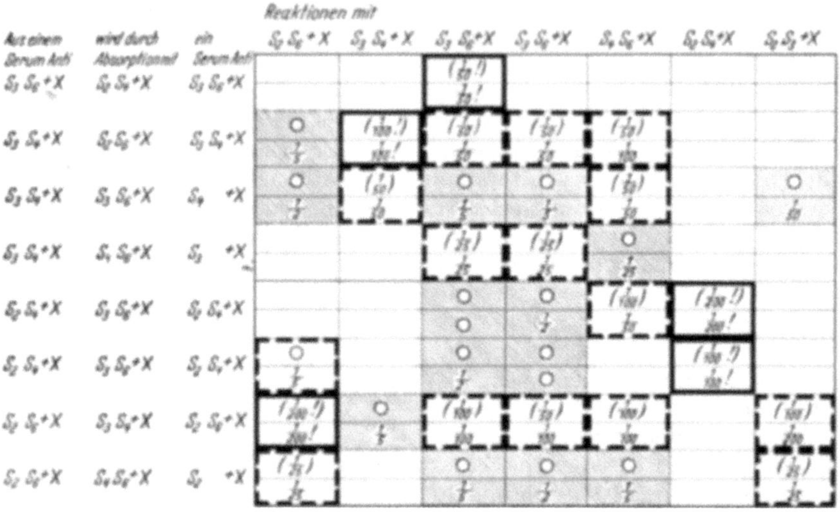

Abb. 8. Serologische Teste von Pollenextrakten nach Kaninchen-Passage. Es bedeuten: Die dick umrandeten Felder: homologe Reaktionen, gebrochen umrandete Felder: halbhomologe Reaktionen, schraffierte Felder: heterologe Reaktionen. Eingeklammerte Zahlen: Sollwerte, übrige Zahlen: Versuchswerte. Leere Felder: nicht ausgeführte Reaktionen. (Nach LEWIS 1952, hier in der veränderten Form nach MORITZ 1958.)

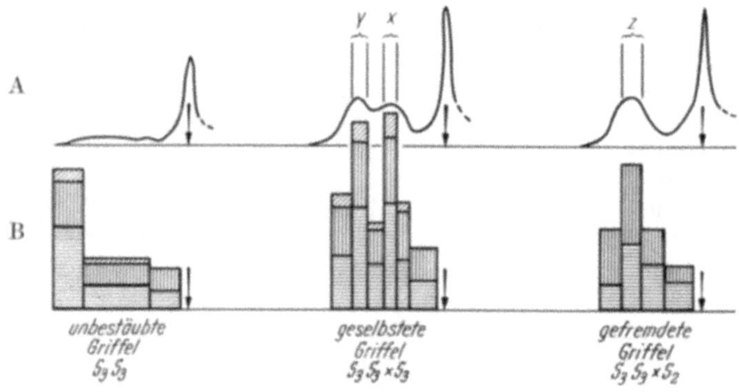

Abb. 9A u. B. Elektrophoretische Auftrennung von Leitgewebsextrakten von unbestäubten (links), selbstbestäubten (Mitte) und fremdbestäubten (rechts) *Petunia*-Griffeln. Obere Reihe (A): Extinktion der Pherogramme nach Proteinfärbung. Untere Reihe (B): Quantitative Bestimmung der Zucker nach 10 min. Hydrolyse der verschiedenen Pherogramm-Abschnitte. Flächen sind mengenproportional. ↓ Startpunkt. (Nach LINSKENS 1955.)

b) Hemmung des Pollenschlauchwachstums bei Heterostylen.

Heterostyle Arten sind dadurch gekennzeichnet, daß man innerhalb der Species Gruppen unterscheiden kann, die sich durch die Länge der Griffel in ihren Blüten unterscheiden. Meist ist Heterostylie verknüpft mit Heterantherie, einer verschiedenen Stellung der Staubbeutel. Man unterscheidet gewöhnlich 2 Typen von Heterostylie: *dimorphe*, wobei zwei verschiedene Griffel- bzw. Antheren-Höhen zu beobachten sind, und *trimorphe* Heterostylie, wobei drei verschiedene Griffel- bzw. Antherenlängen zu unterscheiden sind.

Mit den Heterostylie-Merkmalen sind häufig noch andere primäre und sekundäre Merkmale und Unterschiede gekoppelt (osmotischer Wert der Narbenpapille, Größe der Pollenkörner usw.) (Abb. 10). Die Größe der Pollenkörner und die Antherenstellung sind jedoch durchaus voneinander genetisch unabhängig (ERNST 1953, 1962). Zusammenfassung bei BRIEGER 1930.

Heterostylie ist in der Regel mit Inkompatibilität verknüpft; d. h. die Griffel sind nur jeweils mit Pollen aus gleich langen Staubbeuteln erfolgreich zu bestäuben (*legitime* Bestäubung), während alle anderen Bestäubungsarten keinen Samen-

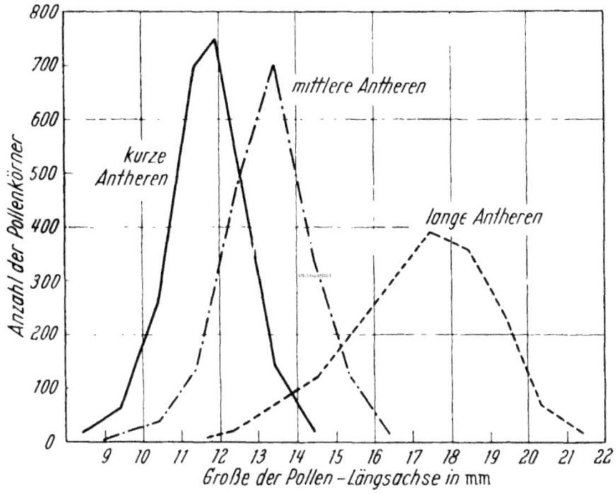

Abb. 10. Unterschiedliche Pollenkorn-Größen von *Lythrum salicaria* je nach der Antherenlänge. Je Klasse wurden 1800 Pollenkörner gemessen. (Nach BODMER 1927.)

ansatz geben (*illegitime* Bestäubungen) (Abb. 11 und 12). Die Inkompatibilität der Heterostylen kommt vorwiegend durch eine Hemmung des Pollenschlauchwachstums im Narbenkopf oder im Griffelleitgewebe zustande (STRASBURGER

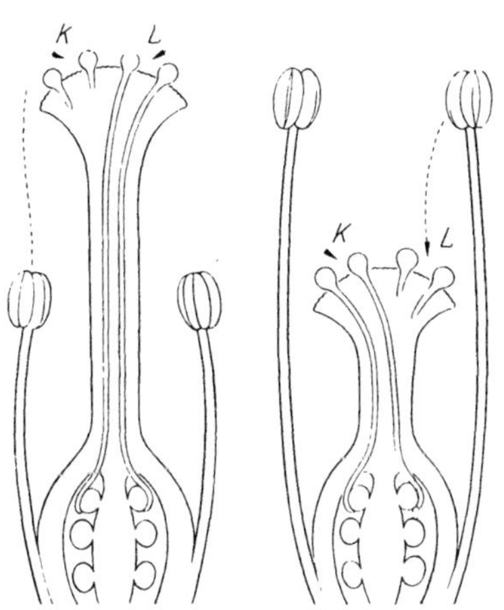

a b

Abb. 11.

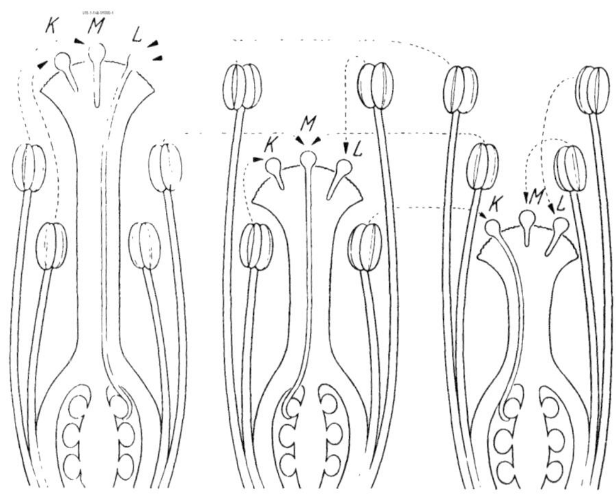

a b c

Abb. 12.

Abb. 11. Schema der Bestäubungsmöglichkeiten und des resultierenden Pollenschlauchverhaltens bei einer dimorph-heterostyl-heteranthеren Pflanze. (Nach BRIEGER 1930.)

Abb. 12. Schema der Bestäubungsmöglichkeiten und des resultierenden Pollenschlauchverhaltens bei einer trimorph-heterostyl-heteranthеren Pflanze. (Nach BRIEGER 1930.)
Es bedeuten: *K* Pollen aus kurzen, *M* Pollen aus Mittel-, *L* Pollen aus langen Antheren.

1886, ESSER 1953) (Abb. 13). Eine Ausnahme bildet die distyle *Narcissus tazetta*, bei der die Inkompatibilität, welche genetisch nicht mit Distylie gekoppelt ist, einen Zusammenbruch der Samenanlagen nach dem Eindringen der Pollenschläuche in den Embryosack bedingt (DULBERGER 1964).

Für die Heterostylen sind durchaus einfache Prinzipien der Hemmung inkompatibler Bestäubungen denkbar, z. B. mechanische Hindernisse im Narben- und Griffelgewebe. So konnte Lewis (1943) für *Linum* zeigen, daß verschiedene osmotische Werte der beiden Pollensorten bzw. der Narbenpapillen über Hemmung oder Nicht-Hemmung entscheiden (Abb. 14). Für andere Arten (*Lythrum*, *Fagopyrum*) wies hingegen Schoch-Bodmer (1945) nach, daß bei den 3 Typen im anatomischen Bau der Griffel keinerlei Unterschiede vorliegen.

Auf Grund von Untersuchungen an *Forsythia* glaubte Moewus (1950a und b), einen komplementären Förderungsmechanismus mit einer „Hemmstoff - Ferment - Hypothese" fordern zu müssen. Nachuntersuchungen (Esser 1953, Lewis 1954, Esser-Straub 1954, Visser 1956, Reznik 1957) konnten jedoch einen Zusammenhang zwischen Pollenflavonfarbstoffen und Inkompatibilitätsverhalten nicht stützen.

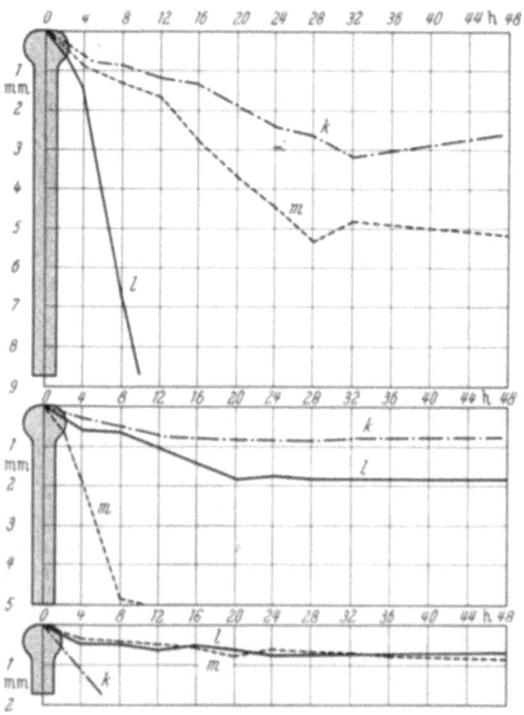

Abb. 13.

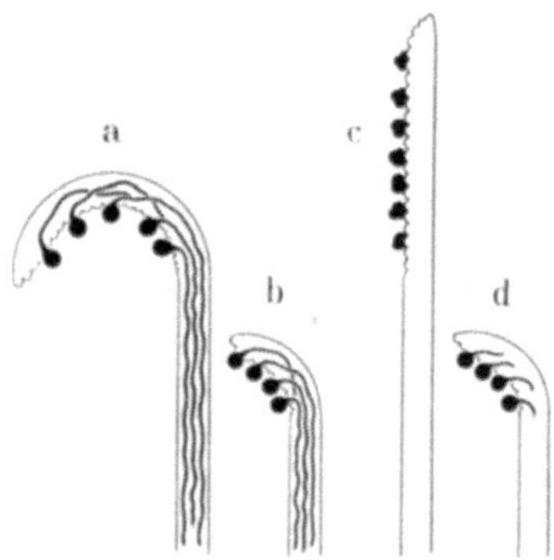

Abb. 14a--d.

Abb. 13. Pollenschlauchwachstum in Lang-, Mittel- und Kurz-Griffeln von *Lythrum salicaria*. Ordinate: Griffellänge bzw. Pollenschlauchlänge in Millimeter, Abscisse: Zeit in Stunden. Es bedeuten: *l* Pollen aus langen Antheren, *m* Pollen aus Mittel-Antheren; *k* Pollen aus kurzen Antheren. (Nach Esser 1953, verändert.)

Abb. 14a--d. Beziehungen zwischen dem osmotischen Druck und dem Verhalten von Pollen und Griffel bei *Linum grandiflorum*. Wenn der Quotient Pollen/Griffel unter 4/1 liegt, so keimt der Pollen nicht (c); liegt er höher, so platzen die Pollenschläuche nach kurzem Wachstum (d). Ist der Quotient des osmotischen Druckes Pollen/Griffel genau 4:1, so ist Befruchtung möglich (a, b). (Nach Lewis 1943.)

Man darf annehmen, daß das Inkompatibilitätsverhalten von Heterostylen in ähnlicher Weise wie bei den Homostylen durch ein Zusammenwirken von Pollenschlauch und Griffel erklärt werden kann.

3. Inkompatibilitätsreaktion in der Samenanlage.

Der Fall, daß sich die Inkompatibilitätsreaktion erst in der Samenanlage, zum Teil nach der Befruchtung, vollzieht, ist der am wenigsten realisierte. Er ist beschrieben bei *Gasteria* (Sears 1937), *Hemerocallis* (Stout und Chandler 1933), *Antirrhinum* (Straub 1941), *Narcissus tazetta* (Dulberger 1964), *Liquidambar styraciflua* (Schmitt und Perry 1964) und *Theobroma cacao* (Knight und Rogers 1955, Cope 1958, 1962; Bennet und Cope 1959, Naundorf 1959, Pandey 1960, Glendinning 1960). Während das Pollenschlauchwachstum im Griffel ungehindert erfolgen kann, degeneriert die befruchtete Eizelle häufig unmittelbar und gleichzeitig mit den nicht befruchteten. Der männliche Partner scheint also nicht in der Lage zu sein, die Entwicklung des Embryos einzuleiten. Nach NAUNDORF (1959) sind für das Auftreten der Inkompatibilität beim Kakao Hemmstoffe verantwortlich, die nach inkompatibler Bestäubung im Fruchtknoten gebildet werden. Extrahiert und in ein Glucose-Agar-Medium gebracht, hemmen sie das Pollenschlauchwachstum selbststeriler, fördern dagegen das Schlauchwachstum selbstfertiler Bäume. Auf Grund cytologischer Beobachtungen kommen hingegen BENNET und COPE (1959) zu folgender Auffassung: In Gegenwart eines dominanten S-Allels wird im Cytoplasma des Spermakerns eine spezifische Substanz adsorbiert. Diese soll beim Zusammentreffen von Spermakern und Polkernen der gleichen dominanten Allelenausrüstung in das die Polkerne umgebende Cytoplasma diffundieren, mit spezifischen „sites" an der Oberfläche der beiden Polkerne reagieren und diese gegen eine Fusion abschirmen. Da beim Kakao das Wachstum der Pollenschläuche im Griffel und Embryosack ungehemmt verläuft, versucht PANDEY (1960) das Inkompatibilitätsystem durch postmeiotisch entstehende Vorstufen spezifischer Inkompatibilitätsstoffe zu erklären, die erst im Pollenschlauch und Embryosack in spezifische, zur Inkompatibilitätsreaktion befähigte Substanzen umgewandelt werden.

Wenn dieser Inkompatibilitätstyp auch noch zahlreiche ungeklärte Probleme aufweist, wird man doch auch für ihn eine wechselseitige Reaktion zwischen dem Cytoplasma von Eizelle resp. Embryosack und des Pollens nach Art einer Immunitätsreaktion annehmen dürfen.

4. Theorien.

DARWIN (1877) gab, dem Zeitgeist entsprechend, der Selbst-Inkompatibilität eine teleologische Deutung. Einmal aufgewiesen, haben immer wieder aufs neue Untersucher nach dem Wesen des Phänomens gefragt. Schon STRASBURGER (1886) sah in der Inkompatibilität nicht nur einen Mangel an „*sexueller Affinität*", sondern glaubte, daß fehlende oder mangelhafte Pollenschlauchbildung im Gefolge einer Selbstbestäubung auf einer „*Summierung bestimmter, in beiden Protoplasmen übereinstimmend vertretener Stoffe*" beruhe. „Durch diese Summierung würde das Maß überschritten, bis zu welchem der betreffende Stoff ohne Nachteil von den Pollenschläuchen, eventuell auch von der Narbe ertragen werden kann." Schon bei dieser älteren Interpretation fällt auf, daß nicht gegenseitige strukturelle Beeinflussung von Pollenschlauch und Griffel für den Hemmeffekt verantwortlich gemacht werden, sondern stoffliche Prinzipien. JOST (1907) hat auf Grund zahlreicher Pollenkeimungsversuche die Vermutung geäußert, „*daß qualitative chemische Differenzen in den Leitgeweben der verschiedenen Individuen für das Zustandekommen der Hemmung des Pollenschlauches im Griffel*" verantwortlich seien. Während STRASBURGER (1886) den individuellen Stoffen die Bedeutung von Giften beilegte, sah JOST in ihnen „gleichgültige Stoffe", die „wachstumsanregende Stimulantia" sind. Diese Jostschen „*Individualstoffe*" sind nicht als plasmatische Unterschiede aufzufassen, sondern als lösliche und diffusible Substanzen, Träger individueller Differenz. CORRENS (1913, 1919) führte diese Vorstellungen weiter. Da die hemmenden Prinzipien vererbt werden und die Anlagen dazu dem Mendelschen Spaltungsgesetz folgen, sind die stofflichen Agentien als „*Linienstoffe*" anzusehen, deren bestimmte Kombination für jedes Individuum charakteristisch ist. EAST und MANGELSDORF (1926) hielten es zunächst für wahrscheinlich, daß eine Verschiedenheit der Linienstoffe ein beschleunigtes Wachstum bedingt. Nach BRIEGER (1927) war es jedoch wahrscheinlicher, daß die Inkompatibilität auf einer Herabsetzung der Wachstumsgeschwindigkeit beruhe.

Die Ausbildung jedes dieser stofflichen Prinzipien beruht nach CORRENS auf einer Anlage, die wie andere Anlagen über die Keimzellen vererbt wird. Die Kombination der Anlagen und damit die der Hemmstoffe ist das Individuelle; die Ausbildung jedes einzelnen Stoffes ist spezifisch. In die Terminologie der modernen Genetik übersetzt bedeutet dies: Die Kombination der gengesteuerten Prozesse führt im Individuum zur Ausprägung eines individuellen Merkmales. Bei CORRENS wurde somit erstmalig eine Verbindung zwischen dem entwicklungsphysiologischen Prozeß und dessen erblicher Bedingtheit hergestellt. In umfangreichen Untersuchungen hat YASUDA in den Jahren 1927—1934 (Zusammenfassung: YASUDA 1934) an *Petunia violacea* die Natur des hemmenden Agens bei der Selbststerilitätsreaktion zwischen Pollenschlauch und Griffelgewebe zu ergründen versucht. Er sieht die Ursache der Selbst-Inkompatibilität in einem „*Ovarialstoff*", der hauptsächlich von der Placenta gebildet wird und sich durch Diffusion akropetal verbreitet. Dieses genabhängige, wasserlösliche, stoffliche Prinzip beschleunigt bei Fremdbestäubung und hemmt bei Selbstbestäubung das Pollenschlauchwachstum. Knospen- und Pseudofertilität werden von YASUDA so erklärt, daß der Ovarialstoff noch nicht bzw. in unterschwelliger Konzentration gebildet wird. Leider ließen sich nicht alle von YASUDA gemachten Beobachtungen reproduzieren. Im Griffel findet sich nämlich kein Gradient der Hemmwirkung; diese ist vielmehr in allen Zonen des Leitgewebes gleich groß (STRAUB 1946). Der Yasudaschen Theorie mangelt es vor allem daran, daß sie die Hemmwirkung nicht als Ausdruck einer Wechselwirkung ansieht. Sie verkennt, daß ein bestimmter Pollenschlauch auf eine bestimmte Griffeleigenschaft hin mit einer spezifischen Reaktion antworten muß. Es war von großer Bedeutung, daß JOST bereits in 1907 mit wenigen Sätzen auf die Ähnlichkeit der für die Inkompatibilität im Pflanzenreich gefundenen Verhältnisse mit Tatsachen aus dem Gebiete der Immunitätslehre hinwies. Der damit in die Diskussion eingeführte Gedankengang wurde von EAST (1929) auf Grund der Befunde eigener Untersuchungen und der seiner Schule zur „Immunitätstheorie" der Selbst-Inkompatibilität ausgebaut. EAST sieht in einer Reaktion zwischen Griffelgewebe und Pollenschlauch eine Wechselbeziehung, die ihr Analogon in Immunitätsreaktionen hat. Er kommt zu der Spekulation, daß sich Griffel und Schlauch wie intercellulärer Parasit und Wirt zueinander verhalten, wobei die in beiden vorhandenen gleichartigen S-Gene über eine Art Antigen-Antikörper-Reaktion das Wachstum hemmend beeinflussen (Abb. 15). Die Eastsche Immunitätstheorie besticht durch ihre Einfachheit und die damit gegebene Möglichkeit, experimentelle Befunde der Inkompatibilitätsforschung zu deuten. Sie bleibt jedoch rein spekulativ, solange noch „kein Licht in die Chemie des Verhaltens selbststeriler Pflanzen geworfen wird" (EAST 1929). Auf Grund neuerlicher Untersuchungen des Pollen-

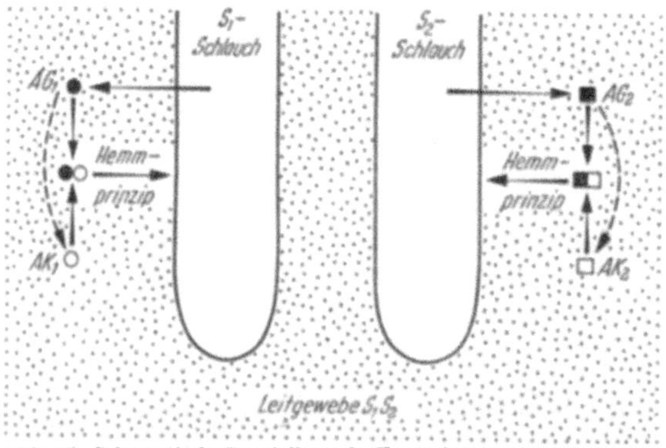

Abb. 15. Schematische Darstellung der Eastschen Immunitätstheorie der Pollenschlauchhemmung bei Vorliegen von Selbst-Inkompatibilität. Die spezifischen Antigene (*AG*) werden vom Pollenschlauch gebildet und abgegeben; sie lösen im Leitgewebe (punktiert) die Bildung spezifischer Antikörper (*AK*) aus (gestrichelter Pfeil). *AG* und *AK* vereinigen sich zu dem hemmenden Prinzip. (Verändert nach STRAUB 1948.)

schlauchwachstums von *Petunia* lehnte STRAUB (1946, 1947, 1948) zunächst eine Hemmstoffbildung als Grundlage für das Sistieren des Pollenschlauchwachstums bei inkompatibler Kombination ab. Vielmehr nahm er an, daß das S-Gen im Pollen die Bildung eines spezifischen Stoffes *(PS-Stoff)* in bestimmter Menge bewirkt, der zum Wachstum des Schlauches benötigt wird. Im Leitgewebe wird bei Anwesenheit des gleichen S-Genes ein spezifisches System gebildet *(LG-Stoff)*, das den PS-Stoff seiner eigentlichen Funktion entzieht, ihn inaktiviert. Bei Vorliegen verschiedener S-Gene in Pollen- und Griffel-Zellen, d. h. also bei Fremdbestäubung, wird der PS-Stoff in seiner Wirkung hingegen nicht beeinträchtigt. Selbstung und Fremdung würden sich also nach STRAUB lediglich durch unterschiedlichen Verbrauch eines gengesteuerten Wachstumsstoffes auf der gleichen vom Pollenschlauch zu durchwachsenden Wegeinheit unterscheiden. Diese „*Verbrauchstheorie*" erweitert demnach die Vorstellung der Immunitätstheorie auf die Ernährung des Pollenschlauches während des Wachstums im Griffel. Bereits LANG (1949) hat darauf hingewiesen, daß ein prinzipieller Unterschied zwischen der Immunitätstheorie und der Verbrauchstheorie nicht besteht. Auf Grund neuer experimenteller Daten hat STRAUB jedoch seine „Verbrauchstheorie" modifiziert (STRAUB 1954) und sich weitgehend den Gedankengängen der nunmehr schon klassischen Immunitätstheorie angeschlossen (STRAUB 1958).

Ausgehend von der Genabhängigkeit der Inkompatibilitätsreaktion nehmen also alle Theorien eine stoffliche Grundlage für die Unverträglichkeitsreaktion an. Dabei müssen sie der Tatsache gerecht werden daß es sich dabei um eine Wechselwirkung zwischen Pollenschlauch und Leitgewebe (Mutualreaktion) handelt, die von großer phylogenetischer Bedeutung ist.

Es ist natürlich weitgehend eine Definitionsfrage ob man die Inkompatibilitätsreaktion als Antigen-Antikörper-Reaktion im Sinne der Immunitätstheorie (EAST 1939) ansehen will. *Für* eine solche Erklärung sprechen:

1. Die sehr hohe genetische Spezifität derselben,
2. die serologische Spezifität der Pollen- und Griffelextrakte nach den Versuchen von LEWIS (1952) (Abb. 7) und LINSKENS (1960, 1965),
3. die wahrscheinliche Beteiligung von Proteinkörpern an ihrem Zustandekommen (LINSKENS 1955) (Abb. 8),
4. die Tatsache daß das Wachstum von Selbstungspollen nach vorhergehender Passage von Pollen gleicher genetischer Konstitution (Doppelbestäubung) gefördert ist (STRAUB 1948). Man kann dabei an die teilweise Absättigung eines vorhandenen Antikörpers-Bestandes denken,
5. die Tatsache daß bei höheren Temperaturen nicht eine Beschleunigung des Wachstums der Schläuche zu beobachten ist, sondern eine verstärkte Hemmung (LEWIS 1942, 1953b, STRAUB 1958, HAGMAN 1962) (Abb. 17 und 18),
6. die Feststellung, daß die Inkompatibilitätsreaktion ein endothermer Prozeß ist, der seine Energie aus erhöhter Atmung bezieht (Abb. 6),
7. die Beteiligung beider Geschlechter am Zustandekommen der beobachteten „Abwehrkörper" von Glucoproteid-Natur (LINSKENS 1958, 1959).

Eine ausführliche Diskussion über die Kriterien einer Antigen-Antikörper-Reaktion im Zusammenhang mit der Selbst-Inkompatibilität hat MORITZ (1958) gegeben. Danach wäre vor allem wegen der relativ kurzen „Inkubationszeit" für das Hemmstoffsystem am ehesten der Begriff der "*Isoantikörper*" zutreffend.

Auf Grund der physiologischen und biochemischen Resultate sowie des genetischen Hintergrunds (Zusammenfassung bei LUNDQUIST 1965) stellt LEWIS (1965) folgende „*Dimer*"-Hypothese für das gametophytische Inkompatibilitätssystem zur Diskussion. Nach ihr liefert der S-Gen-Komplex ein für jedes S-Allel spezifisches Polypeptid, dessen Molekül in Pollen und Griffel identisch gebaut ist.

Die Polypeptide polymerisieren im Pollen und Griffel zu einem „*Dimer*". Die Inkompatibilitätsreaktion wird eingeleitet durch die Kombination identischer „*Dimere*" zu einem „*Tetramer*" mit Hilfe eines allosterischen Moleküls (z. B. Glucose). Erst das „*Tetramer*" wirkt als Regulator, der entweder die Synthese eines unspezifischen Inhibitors für das Schlauchwachstum induziert oder die Synthese eines Auxins für das Schlauchwachstum unterdrückt. Molekulargewichtsbestimmungen der S-Proteine von Pollen, Griffel und Inkompatibilitätskomplex müssen die Richtigkeit dieser Hypothese erweisen.

Die Fülle der experimentellen Befunde kann jedoch auch noch unter weiteren Gesichtspunkten diskutiert werden: So läßt sich die *Klon-Selektions-Theorie* der Immunität auch auf das Inkompatibilitätsphänomen anwenden. Danach müßte die Selektion der spezifischen Antikörper gegen die Pollen-Antigene bereits während der Pollenentwicklung stattfinden, also während der nahen Stoffwechsel-Beziehung in der Anthere. Das würde bedeuten, daß der Pollenschlauch bei seinem Wachstumsprozeß durch den Griffel bereits auf die selektierten, komplementären Antikörper-Klone trifft, wodurch auch die Schnelligkeit der Reaktion erklärt würde. Experimentelle Befunde, wie Kastrationsexperimente in frühen Knospenstadien (LINSKENS 1964) und Zusammenbruch der Hemmreaktion nach Bestrahlung des Griffels vor der Bestäubung (LINSKENS et al. 1960, KUMAR und HECHT 1965), stützen in gewisser Hinsicht diese Deutung (LINSKENS 1962).

Schließlich kann der biochemische Mechanismus der Inkompatibilitätsreaktion auch mit den Modellen der Enzym-Induktion, der Rückkoppelung und End-Produkt-Hemmung der Regulator-Operator-Hypothese erklärt werden. Die *molekularbiologische Deutung* bietet zudem einen neuen Ansatz für experimentelle Arbeiten, die jedoch noch im Gange sind (LINSKENS 1964).

Zum Abschluß der theoretischen Erörterungen sei noch auf den *Zeitpunktt der genetischen Determinierung* und auf die *Lokalisation des S-Proteins* eingegangen. Nach PANDEY (1960) liegt der Zeitpunkt der S-Gen-Wirkung bei den heteromorph-sporophytischen Arten prämeiotisch, bei den homomorph-sporophytischen meiotisch (nach Anaphase II, vor Trennung der 4 Mikrosporen oder nach der Cytokinese, aber vor der Exineausbildung), bei den homomorph-gametophytischen Arten postmeiotisch (nach der Cytokinese oder nach der Exineentwicklung). Beim sporophytischen System sind also die für die Inkompatibilität verantwortlichen spezifischen Substanzen bereits im cytoplasmatischen Material vorhanden, das am Aufbau der Pollenwand beteiligt ist (PANDEY 1958). Damit wäre auch zu erklären, daß die Inkompatibilitätsreaktion bei sporophytisch determinierten Arten direkt nach dem Kontakt der Pollenkörner mit der Narbe zustande kommt. Für gametophytisch determinierte Arten kann man auf Grund des Nachweises einer S-Protein-Diffusion aus intaktem Pollen annehmen, daß der Ort der Wirksamkeit des S-Proteins beim Wachstum der Pollenschläuche durch den Griffel in der Nähe oder an der Schlauchwandoberfläche liegt (MÄKINEN und LEWIS 1962, VAN DER PLUIJM und LINSKENS 1966). Für letztere Annahme spricht der Befund, daß gleichzeitig den Griffel durchwachsende kompatible und inkompatible Pollenschläuche sich gegenseitig nicht beeinflussen. Weitere Experimente sind auch hier notwendig, um dieses Problem endgültig zu klären.

5. Überwindung der Selbst-Inkompatibilität.

Zur entwicklungsphysiologischen Untersuchung der Selbst-Inkompatibilität ist es notwendig, homozygote Formen selbststeriler Pflanzen in die Hand zu bekommen. Die bei der Überwindung des genphysiologischen Mechanismus anwendbaren Methoden ergeben darüber hinaus wichtige Aufschlüsse über dessen biochemische Grundlagen (Zusammenstellung bei STRAUB 1958). Die Ausschaltungsmethoden gliedern sich in:

a) Veränderung der genetischen Bedingungen.

1. *Modifikator-Gene* verändern oder heben die Wirkung der S-Gene auf (BRIEGER 1927, EAST 1932, EUE 1953, BREWBAKER 1960, MOSIG 1960).

2. *S_f-Allele* der gleichen S-Serie unterbinden die Inkompatibilitätsreaktion (EAST und YARNELL 1929, EUE 1953, MOSIG 1960, SURIKOV 1961).

3. Der Grad der Hemmung ist von dem quantitativen Verhältnis der S-Gene abhängig. Daher kann durch *Polyploidisierung* bei sinkendem Anteil eines bestimmten S-Genes Fruchtansatz erreicht werden (STRAUB 1941, STOUT-CHANDLER 1941, 1942; LEWIS 1943, STECHER). Den gleichen Effekt im Griffel erreicht man ebenfalls mit diploiden Pollenkörnern (Abb. 16). Auf diese Weise lassen sich selbst bei Heterostylen im Falle illegitimer Bestäubungen Befruchtungen erzielen (ESSER 1953.)

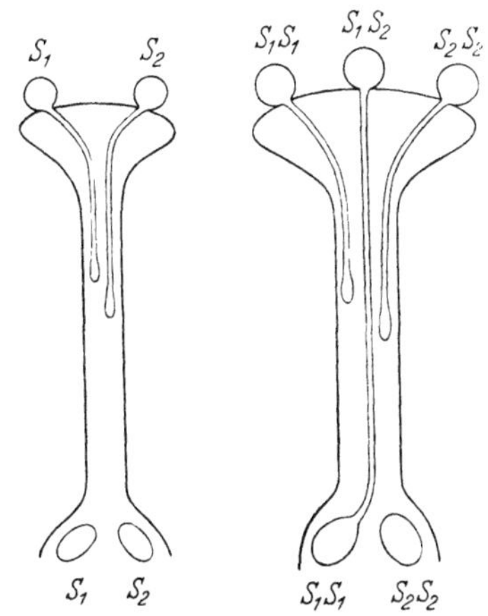

Abb. 16. Schematische Darstellung der Pollenschlauchhemmung in di- (links) und tetraploiden (rechts) *Petunien*. Es liegen in beiden Fällen die gleichen S-Allele vor. Der Pollenschlauch mit den beiden verschiedenen S-Allelen ist ungehemmt und erreicht die Eizelle. (Nach STRAUB 1948.)

4. Bei selbststerilen Pflanzen mit gametophytischer Determination des Pollencharakters kann es darüber hinaus in polyploiden Formen zu einer "*competition*" der S-Gene in heterogenen, diploiden Pollenkörnern kommen (LEWIS 1947, 1954; BREWBAKER 1954, BREWBAKER und SHAPIRO 1959, BREWBAKER und NATARAJAN 1960, PANDEY 1962). S-heterogener Pollen mit "*competition*" kann auch als Folge meiotischer Störungen auftreten, z. B. durch Translokation oder anomales *crossing-over*, wodurch die beiden homologen S-Allele auf einem Chromosom vereinigt werden (BREWBAKER 1960, BREUER 1961).

5. Spontane oder röntgeninduzierte *Mutationen* führen im allgemeinen zum einseitigen Verlust der S-Aktivität des Pollens (LEWIS 1951, 1954; LUNDQUIST 1958, 1962). Eine Ausnahme bildet die Beobachtung einer Inaktivierung des gesamten S-Gens (des Pollen- und Griffel-"*site*") bei der amphidiploiden *Trifolium repens* (PANDEY 1956).

b) Veränderung der Milieubedingungen.

Ohne Änderung der genetischen Grundlagen der Inkompatibilität läßt sich die Reaktion zwischen männlichem und weiblichem Partner so verändern, daß Fertilität die Folge ist:

1. Bei denjenigen Selbststerilen, bei denen die Hemmung auf einen bestimmten Gewebebezirk, z. B. die *Narbenzone lokalisiert* und begrenzt ist, kann durch Entfernen desselben (Abb. 1) Befruchtung erzielt werden (KAKIZAKI 1930, TATEBE 1939, 1955; SWAMINATHAN 1955, KROH 1956, CHRIST 1958). Werden jedoch die Pollenschläuche auch während ihres Wachstums gehemmt, so kann durch Verkürzen der Wegstrecke das Erreichen der Samenanlage ermöglicht werden (STRAUB 1947).

2. Auch lassen sich bei Verwendung von unreifen Griffeln *(Knospenbestäubung)* Befruchtungen erzielen (EAST 1923). Die Ursachen liegen zum Teil in der kürzeren

Weglänge, welche die noch nicht gestreckten Griffel darstellen. Aus Griffelpfropfungen geht jedoch hervor, daß auch der physiologische Hemmungsmechanismus nicht voll funktioniert (Kakizaki-Kasai 1933, Straub 1958). Zwar kann die anfängliche Wachstumsgeschwindigkeit der Schläuche in Knospengriffeln zunächst geringer sein, als in reifen Griffeln (Abb. 17). Doch wird die Gesamtwachstumsstrecke eben durch den Ausfall der Hemmreaktion zum Erreichen der Eizelle ausreichen. Die Ursache des langsameren Wachstums ist in der noch unzureichenden Nährstoffversorgung des unreifen Knospengriffels zu suchen (East 1934).

Bei *Nicotiana* läßt sich die Samenproduktion nach Knospenbestäubung steigern, wenn vor der Bestäubung auf die sekretfreien Knospennarben Sekret von reifen Narben appliziert wird (Pandey 1963).

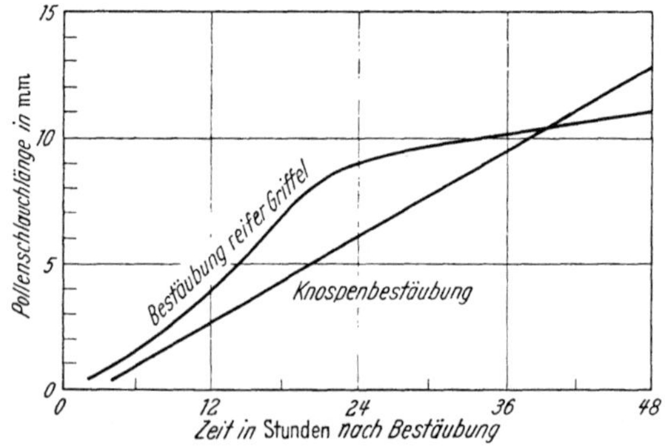

Abb. 17. Verlauf des Pollenschlauchwachstums bei Knospenbestäubung (24 Std vor der Anthese) und in reifen Griffeln von *Oenothera organensis*. (Nach East 1934, verändert.)

3. Schon Darwin (1866) war es bekannt geworden, daß gegen Ende der Vegetationsperiode in genetisch inkompatiblen Rassen schwache Fertilität auftritt *(end-season-fertility, pseudo-fertility)*. Diese Erscheinung ist nicht allgemein verbreitet (East 1923, Yasuda 1930).

4. Durch Infiltration der selbstbestäubten Griffel mit Wasser oder wäßrigen Lösungen kann die Inkompatibilitätsreaktion bei *Petunia* ebenfalls soweit geschwächt werden, daß die Masse der Pollenschläuche die Eizellen erreichen kann (Linskens 1955). Offensichtlich liegt hier ein *Verdünnungseffekt* an wasserlöslichen Hemmprinzipien vor.

5. Das Pollenschlauchwachstum ist *temperaturabhängig*. Lewis (1942) hat gefunden (Abb. 18 und 19), daß höhere Temperaturen weniger die Wachstumsgeschwindigkeit der Schläuche als die Hemmreaktion beeinflussen. Infolgedessen kann es durch tiefe Temperaturen zu einer Schwächung der Hemmreaktion kommen (vgl. Straub 1958, Hagman 1962, Kwack 1965). Werden hingegen Griffel und Narbe *vor* Selbstbestäubung einer kurzfristigen Warmwasserbehandlung (50° C) unterzogen, so kann bei *Oenothera* normales Schlauchwachstum beobachtet werden. Dieser Befund wird auf eine Zerstörung oder Inaktivierung der für die Inkompatibilitätsreaktion verantwortlichen Substanzen zurückgeführt (Bali 1963, 1964; Hecht 1964).

6. Bei *doppelter Bestäubung* mit einem zeitlichen Abstand von 3—4 Std kann erreicht werden, daß der 2. Schub der Pollenschläuche näher an die Samenanlagen heranwächst. Dabei erreichen einige Schläuche die Eizellen. Man muß sich vorstellen, daß die im Griffel vorliegenden Hemmstoffe durch die erste Bestäubung

mengenmäßig so stark reduziert werden, daß die Hemmung der Pollenschläuche aus der 2. Bestäubung abgeschwächt wird (STRAUB 1958).

7. Durch *Applikation von Wuchsstoffen* (EMSWELLER und STUART 1948, EMSWELLER und URING 1965) oder *Calcium-Ionen* (KWACK 1965) kann die Inkompatibilitätsreaktion völlig oder teilweise überwunden werden. Die Wuchsstoffe — möglicherweise auch die Calcium-Ionen — bewirken eine Beschleunigung des Pollenschlauchwachstums, wobei der Effekt von Calcium auf einer Veränderung der Pollenschlauchwand beruhen soll.

8. Auch die Ernährungsbedingungen spielen in manchen Fällen eine Rolle für die Stärke der Ausprägung der Inkompatibilitätsreaktion (TANAKA 1952). *Unterdrückung des vegetativen Wachstums* durch mangelhafte Düngung kann erhöhte Fertilität bei an sich inkompatiblen Pflanzen hervorrufen (STOUT 1923).

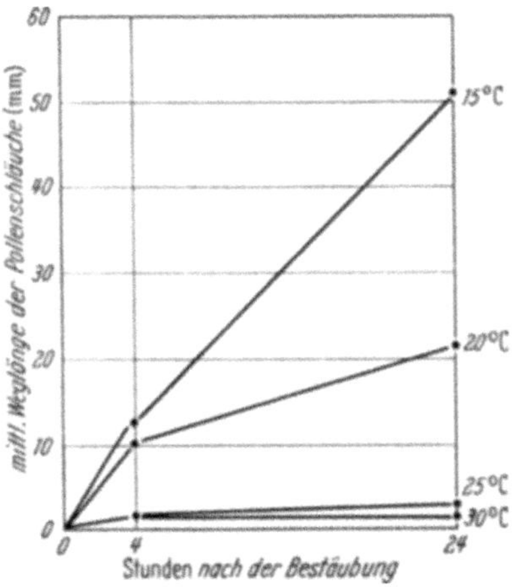

Abb. 18. Pollenschlauchwachstum nach Selbstbestäubung von *Oenothera organensis* unter verschiedenen Temperaturbedingungen. (Nach LEWIS 1942, verändert.)

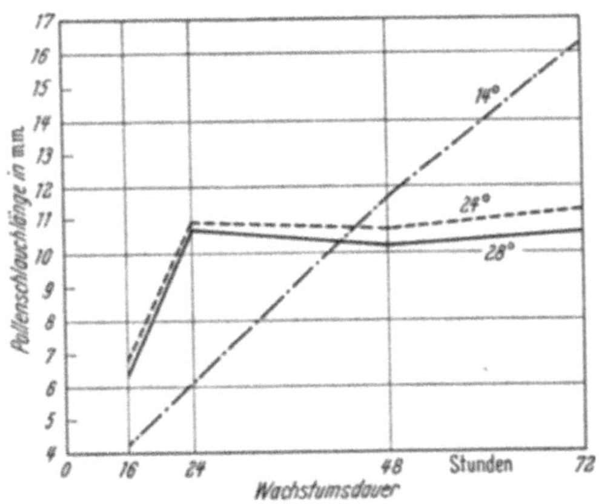

Abb. 19. Pollenschlauchwachstum von *Petunia* nach Selbstbestäubung ($S_4S_4 \times S_4$) unter verschiedenen Temperaturbedingungen. (Nach STRAUB 1958.)

9. *Röntgen-* (LINSKENS, SCHRAUWEN, VAN DER DONK 1960) sowie *UV-Bestrahlung* (KUMAR und HECHT 1965) des Griffelgewebes vor der Passage der Schläuche führt ebenfalls zu einem teilweisen Zusammenbruch der Selbstinkompatibilität. Ihr Einfluß dürfte in erster Linie auf einer Störung des intracellulären Immunitätsmechanismus beruhen.

10. Schließlich kann die Selbstinkompatibilität durch *Chimärenbildung* (selbstinkompatibles-selbstkompatibles Gewebe) aufgehoben werden (GÜNTHER 1961, 1964).

V. Kreuzungs-Inkompatibilität.

Während bei der Selbst-Inkompatibilität ein funktionsfähiger Pollen bestimmter genetischer Konstitution auf dem weiblichen Organ gleicher Konstitution nicht zur erfolgreichen Befruchtung kommt, kommt es bei der Kreuzungs-Inkompatibilität zu Entwicklungs-Hemmungen, wenn Unterschiede in der erblichen Veranlagung bestehen, während gleichzeitig bei gleicher genetischer Determination des Pollens und der als Partner benutzten weiblichen Pflanze volle Fertilität herrscht. Für die heterostylen Arten kennzeichnet Kreuzungssterilität alle Bestäubungskombinationen mit Übertragung kleiner Pollen auf die Narben langgriffliger, sowie die von großen Pollen auf die Narbe kurzgriffliger Blüten (ERNST 1953, 1962).

Der klassische Fall der Kreuz-Inkompatibilität liegt beim *Mais* vor (Demerec 1929, Whiteley-Thomas-Johnson 1957). Weitere Beispiele sind von *Datura* (Rappaport-Satina-Blakeslee 1950, Rietsema-Satina-Blakeslee 1954, Rietsema-Blondel-Satina-Blakeslee 1955) und *Primula*-Bastarden (Ernst 1953, 1955) beschrieben.

Im Gegensatz zu den Verhältnissen bei der Selbst-Inkompatibilität sind jedoch hier sowohl die genetischen als auch die physiologischen Ursachen noch weitgehend unklar. Eine nahe Verwandtschaft zwischen den beiden Inkompatibilitätserscheinungen ist anzunehmen (Lewis 1956). Bei *Solanum* konnte das gehemmte Pollenschlauchwachstum durch Entfernen der Narbe vor der Bestäubung und Aufbringen einer künstlichen Narbe aus Agar-Gelatine mit Zucker überwunden werden, so daß Befruchtung erreicht wurde (Swaminathan 1955). Bei *Datura* kommt es zwar zur Verschmelzung von Plasma und Kernen, die Embryonen gehen jedoch zugrunde und die Samenanlage wird durch eine tumorartige Wucherung ausgefüllt. Aus den Ovarialtumoren ließen sich wirksame Extrakte gewinnen, deren Hemmungsprinzip mit Indolylessigsäure zu identifizieren war (Rietsema-Satina-Blakeslee 1954). Im Falle der Pollenschlauchhemmung während des Wachstums im Griffel ließ sich ein Zusammenhang mit den Inhaltsstoffen andeuten: mangelhafte Zuckerreserven lassen bei Kreuzbestäubung das Pollenschlauchwachstum frühzeitig zum Erlahmen kommen (Bellartz 1956). Dies läßt sich auch durch die Beobachtung stützen, daß Verkürzung der Griffellänge unter Umständen zum Durchwachsen der Schläuche führen kann (Gardella 1950). Daneben dürften jedoch letztlich für die Kreuzungs-Inkompatibilität auch plasmatische Inkompatibilität, möglicherweise nach Art einer Immunitätsreaktion eine Rolle spielen. Hier bietet sich dem Physiologen noch ein reiches Arbeitsfeld.

VI. Interspezifische Inkompatibilität.

Wenig untersucht sind bisher die physiologischen Mechanismen, welche den Inkompatibilitätserscheinungen bei Kreuzungen zwischen weit voneinander entfernten Species, Gattungen und Familien zugrunde liegen. Hier läßt sich die Untersuchung ja nicht mit Hilfe von Hybridengewinnung betreiben, da keine Fertilität gegeben ist. Bei der interspezifischen Inkompatibilität besteht häufig Einseitigkeit, d. h. der Pollen von selbst-kompatiblen Arten ist im Griffel von selbst-inkompatiblen Arten gehemmt, während bei reziproker Kreuzung keine Hemmung vorkommt *(unilaterale interspezifische Inkompatibilität)* (Lewis und Crowe 1958).

Zwischen intra- und interspezifischer Inkompatibilität finden sich zahlreiche Analogien. Als Kreuzungsbarrieren können gehemmte Pollenkeimung (Röbbelen 1960), verzögertes Pollenschlauchwachstum (McWilliam 1959, Röbbelen 1960, Batygina, Dolgova und Korobova 1961, Evans 1962, Hagman 1964) oder gestörte Embryoentwicklung auftreten (Datta und Sen 1959, Davies und Wall 1960, Patel und Datta 1960, Hagman und Mikkola 1963). Die in ihrem Wachstum gestörten Pollenschläuche weisen einen erhöhten Sauerstoffverbrauch auf (Roggen 1963) und zeigen die gleichen morphologischen Veränderungen wie die Selbstschläuche (Roggen 1963, Zych 1965). Die Inkompatibilität kann bei einigen Artkreuzungen durch Veränderung der Milieubedingungen abgeschwächt oder aufgehoben werden (Davies und Wall 1960, Röbbelen 1960). Auf Grund dieser Übereinstimmungen, sowie des Befundes, daß die Hemmreaktion gegenüber Temperaturerhöhungen die gleiche Verstärkung wie innerhalb einer Art zeigt (Lewis und Crowe 1958), darf man annehmen, daß die Natur der Hemmreaktion des Pollenschlauchwachstums bei interspezifischer Inkompatibilität die gleiche ist wie bei intraspezifischer. Bei zahlreichen Art-Kombinationen ist die Hemmung des Schlauchwachstums stärker oder tritt früher in Aktion als bei reiner Selbstung (BATEMAN 1943, MCGUIRE und RICK 1954). Bei einigen *Pinus*-Artkreuzungen dürfte die Inkompatibilität in erster Linie darauf beruhen, daß die Pollenschläuche unfähig sind, das Ovar mit einem Faktor von Wuchsstoffcharakter zu versorgen, der die vollständige Entwicklung des Gametophyten gewährleistet (MCWILLIAM 1959). Die bei Artkreuzungen häufig auftretende gehemmte Embryoentwicklung nach normaler Befruchtung kann in einigen Fällen auf eine Störung der Endospermentwicklung zurückgeführt werden (MANGELSDORF und REEVES 1931, COOPER 1951, DATTA und BANERJI 1960, KROH 1962). Für die Überwindung der interspezifischen Inkompatibilität kommt der Embryokultur eine große Bedeutung zu (vgl. BRINK und COOPER 1947, GRANT, BULLEN und NETTANCOURT 1961, KANTA, RANGA SWAMY und MAHESHWARI 1962).

Auf Grund der Kreuzungsresultate von Cruciferengattungen diskutiert SAMPSON (1962) einen physiologischen Mechanismus, mit dessen Hilfe kompatible Pollenkörner und Narben einander „erkennen" können. Dieser beruht auf der Möglichkeit einer molekularen Kombination an 2 Arealen komplementärer Strukturen, dem „S-*Allel-Areal*" und dem „*Species-Areal*". Intraspezifische Inkompatibilität resultiert aus der Kombination am „S-*Allel-Areal*"; interspezifische Inkompatibilität wird durch den Ausfall der Kombination am „*Species-Areal*" bedingt. Kombination am „Species-Areal" führt zur Kompatibilität.

Da bislang nur Ansätze biochemischer Untersuchungen vorliegen (HAGMAN 1964), lassen sich die Befunde vorläufig weitgehend mit Erweiterungen einer Antigen-Antikörper-Hypothese befriedigend erklären.

Literatur.

BALI, P. N.: Some experimental studies on the self-incompatibility of *Oenothera rhombipetala*. Phyton (Argent.) **20**, 97—103 (1963). — Effect of heat on the self-incompatibility reaction of *Oenothera rhombipetala* Nutt. Sci. and Cult. (Calcutta) **30**, 294—295 (1964). — BATEMAN, A. J.: Specific differences in *Petunia*. II. Pollen growth. J. Genet. **45**, 236—242 (1943). — Ann. Report John Innes Horticult. Inst. 1951, 19—20. — Self-incompatibility systems in Angiosperms. I. Theory. Heredity **6**, 285—310 (1952). — Self-incompatibility systems in Angiosperms. II. *Iberis amara*. Heredity **8**, 305—332 (1954). — Self-incompatibility systems in Angiosperms. III. Cruciferae. Heredity **9**, 53—68 (1955). — BATYGINA, T. P., O. A. DOLGOVA, and S. N. KOROBOVA: Behaviour of pollen tubes in intra- and interspecific hybridization. Dokl. Akad. Nauk SSSR **136**, 1482—1485 (1961). — BEATUS, R.: Zur Kenntnis der Selbststerilität von *Cardamine pratensis*. Z. indukt. Abstamm.- u. Vererb.-Lehre **59**, 285—348 (1931). — Die Selbststerilität von *Cardamine pratensis*. Jb. wiss. Bot. **80**, 457—503 (1934). — Über die Selbststerilität von *Cardamine pratensis*. Ber. dtsch. bot. Ges. **47**, 189—199 (1929). — BELLARTZ, S.: Das Pollenschlauchwachstum nach arteigener und artfremder Bestäubung einiger Solanaceen und die Inhaltsstoffe ihres Pollens und ihrer Griffel. Planta (Berl.) **47**, 588—612 (1956). — BENNETT, M. C., and F. W. COPE: Nuclear fusion and non-fusion in *Theobroma cacao* L. Nature (Lond.) **183**, 1540 (1959). — BHATTACHARJYA, S. S., and H. F. LINSKENS: Recent advances in the physiology of self-sterility in plants. Sci. and Cult. (Calcutta) **20**, 370—373 (1955). — BODMER, H.: Beiträge zum Heterostylie-Problem bei *Lythrum salicaria* L. Flora (Jena) **22**, 306—341 (1927). — BREUER, K.: Selbstfertile Diploide aus polyploiden selbststerilen Petunien; die Duplikation des S-Locus. Z. Vererbungsl. **92**, 253—260 (1961). — BREWBAKER, J. L.: Incompatibility in autotetraploid *Trifolium repens*. I. Competition and self-compatibility. Genetics **39**, 307—316 (1954). — Pollencytology and self-incompatibility systems in plants. J. Hered. **48**, 271—277 (1957). — Incompatibility and the pollen grain. Rec. Adv. in Bot. **2**, 46 (1959). — Induced mutations of incompatibility alleles and the structure of the gene. J. Agric. and Hort. **35**, 109—113 (1960) [Jap.]. — BREWBAKER, J. L., and S. T. NATARAJAN: Centric fragments and pollen part mutation of incompatibility alleles in *Petunia*. Genetics **45**, 699—704 (1960). — BREWBAKER, J. L., and N. SHAPIRO: Homozygosity and S gene mutation. Nature (Lond.) **183**, 1209—1210 (1959). — BRIEGER, F.: Über die Genetik und Physiologie der

Selbststerilität. Naturwiss. **15**, 734—740 (1927). — Über genetische Pseudofertilität bei der selbststerilen *Nicotiana Sanderae hort.* Biol. Zbl. **47**, 122—128 (1927). — Selbststerilität und Kreuzungssterilität im Pflanzenreich und Tierreich. Monographien Physiol. **21** (1930). — BRINK, R. A., and D. C. COOPER: The endosperm in seed development. Bot. Rev. **13**, 423—541 (1947). — BUBAR, J. S.: An association between variability in ovule development within ovaries and self-incompatibility in *Lotus* (Leguminosae). Canad. J. Bot. **36**, 65—72 (1958). — Differences between self-incompatibility and self-sterility. Nature (Lond.) **183**, 411—412 (1959).

CHRIST, B.: Entwicklungsgeschichtliche und physiologische Untersuchungen über die Selbststerilität von *Cardamine pratensis* L. Z. Bot. **47**, 88—112 (1959). — COOPER, D. C.: Caryopsis development following matings between diploid and tetraploid strains in *Zea mays*. Amer. J. Bot. **38**, 702—708 (1951). — COPE, F. W.: Incompatibility in *Theobroma cacao*. Nature (Lond.) **181**, 279 (1958). — The mechanism of pollen incompatibility in *Theobroma cacao* L. Heredity **17**, 157—182 (1962). — CORRENS, C.: Selbststerilität und Individualstoffe. Festschr. 84. Verslg Dtsch. Naturforsch. u. Ärzte. Münster i. Westf., S. 186—217. 1912 [auch abgedruckt in: Biol. Zbl. **33**, 389—423 (1913)]. — Individuen und Individualstoffe. Naturwiss. **4**, 183—187, 193—198, 210—213 (1916). — CRANE, M. B., and W. J. C. LAWRENCE: Genetical and cytological aspects of incompatibility and sterility in cultivated plants. J. Pomology **7**, 276—301 (1929). — CROWE, L. K.: Incompatibility in *Cosmos bipinnatus*. Heredity **8**, 1—11 (1954). — The evolution of incompatibility in species of *Oenothera*. Heredity **9**, 293—322 (1955). — The evolution of outbreeding in plants. I. The angiosperms. Heredity **19**, 435—457 (1964).

DARWIN, C.: The effects of cross and self fertilization in the vegetable kingdom, 1st edit. London 1877, 2nd edit. 1900. — DATTA, R. M., S. K. DANA, and S. N. BANERJI: Investigations on the interspecific hybridization between the autotetraploids of the cultivated Jute species (*Corchorus olitorius* Linn. and *C. capsularis* Linn.) and on the failure of viable seed formation in them. Genet. iberica **12**, 139—165 (1960). — DATTA, R. M., and S. K. SEN: On the case of seed failure or seed setting in 4n wild *Corchorus capsularis* Linn. × 2n wild *C. olitorius* L. and their reciprocals through cytogenetical studies. Delpinoa (Genova) **1**, 129—149 (1959). — DAVIES, D. R., and E. T. WALL: Effect of gammaradiation on interspecific incompatibility within the genus *Brassica*. Z. Vererbungsl. **91**, 45—51 (1960). — DELPINO, F.: Ulteriori osservazione sulla dicogamie nel regno vegetale. Milano 1868—1875 [auch in: Atti Soc. ital. Sci. nat. **11—17** (1869—1874]). — Sugli apparechi della fecondazione della piante autocarpee. Firenze 1867. — DEMEREC, M.: Cross-sterility in maize. Z. indukt. Abstamm.- u. Vererb.-Lehre **50**, 281—291 (1929). — DEVREUX, M., G. VALLAEYS, P. POCHET et A. GILLES: Recherches sur l'autostérilité du caféier robusta *(Coffea canephora)*. Publ. Inst. Nat. Agron. Congo Belge, Ser. Sci. **44**, 87 (1959). — DRAYNER, J. M.: Self- and cross-fertility in field beans (*Vicia faba* Linn.). J. Agric. Sci. **53**, 387—403 (1959). — DULBERGER, R.: Flower dimorphism and self-incompatibility in *Narcissus tazetta*. Evolution (Lawrence, Kansas) **18**, 361—363 (1964).

EAST, E. M.: An interpretation of self-sterility. Proc. nat. Acad. Sci. (Wash.) **1**, 95—100 (1915). — Genetical aspects of self- and cross-sterility. Amer. J. Bot. **10**, 468—473 (1923). — Self-sterility. Bibliogr. genet. **5**, 331—368 (1929). — Studies on self-sterility. IX. The behavior of crosses between self-sterile and self-fertile plants. Genetics **17**, 175 (1932). — The reaction of the stigmatic tissue against pollentube growth in selfed self-sterile plants. Proc. nat. Acad. Sci. (Wash.) **20**, 364—368 (1934). — The distribution of self-sterility in the flowering plants (edit. K. SAX). Proc. Amer. phil. Soc. (Philadelphia) **82**, 449—518 (1940). — EAST, E. M., and A. J. MANGELSDORF: Studies on self-sterility. VII. Heredity and selective pollen-tube growth. Genetics **11**, 466—481 (1928). — EAST, E. M., and J. B. PARK: Studies on selfsterility. I. The behavior of self-sterile plants. Genetics **2**, 505—607 (1917). — Sammelreferat über die älteren Eastschen Arbeiten s. E. LEHMANN, Z. indukt. Abstamm.- u. Vererb.-Lehre **27**, 141—152 (1921). — EAST, E. M., and S. H. YARNELL: Studies on self-sterility. VIII. Self-sterility allolemorphs. Genetics **14**, 455 (1929). — EMERSON, S.: Growth of incompatible pollen tubes in *Oenothera organensis*. Bot. Gaz. **101**, 890—911 (1940). — EMSWELLER, S. L., and N. W. STUART: Use of growth regulating substances to overcome incompatibilities in *Lilium*. Proc. Amer. Soc. horticult. Sci. **51**, 581—589 (1948). — ERNST, A.: Die Relation Antherenstellung/Pollenkorngröße bei Blütendimorphismus und das Kompatibilitätsproblem. Planta (Berl.) **42**, 81—128 (1953). — Self-fertility in monomorphic *Primulas*. Genetica **27**, 391—448 (1955). — Stammesgeschichtliche Untersuchungen zum Heterostylieproblem. VII. Stand des Nachweises monomorpher Arten, homostyler Sippen und anderen genetisch bedingten Abweichungen vom „klassischen" Blütendimorphismus in den Sektionen der Gattung *Primula*. Arch. Klaus-Stift. Vererb.-Forsch. **37**, 1—127 (1962). — ESSER, K.: Genomverdoppelung und Pollenschlauchwachstum bei Heterostylen. Z. indukt. Abstamm.- u. Vererb.-Lehre **85**, 28—44 (1953). — ESSER, K., u. J. STRAUB: Das Pollenschlauchwachstum bei *Forsythia*, eine Stellungnahme zu der Moewusschen Hemmstoff-Ferment-Hypothese. Biol. Zbl. **73**, 449—455 (1954). — Eue, L.: Über die entwicklungsphysiologischen Grundlagen der Selbststerilität bei der Mutante defecta von *Petunia*. Z. indukt. Abstamm.- u. Vererb.-Lehre **85**, 423—428 (1953). — Evans, A. M.: Species hybridization in *Trifolium*. I. Methods of overcoming species incompatibility. Euphytica **11**, 164—176 (1962). — Species hybridization in *Trifolium*. II. Investigating the pre-fertilization barriers to compatibility. Euphytica **11**, 256—262 (1962).

Gardella, C.: Overcoming barriers to crossability due to the style length. Amer. J. Bot. **37**, 219—224 (1950). — Gerstel, D. U.: Self-incompatibility studies in *Guayule*. II. Inheritance. Genetics **35**, 482—506 (1950). — Glendinning, D. R.: Selfing of self-incompatible *Cocoa*. Nature (Lond.) **187**, 170 (1960). — Grant, W. F., M. R. Bullen, and D. de Nettancourt: Embryocultures of interspecific diploid hybrids in *Lotus*. Amer. J. Bot. **48**, 533—537 (1961). — Günther, E.: Durch Chimärenbildung verursachte Aufhebung der Selbstinkompatibilität von *Lycopersicon peruvianum*. Ber. dtsch. bot. Ges. **74**, 333—336 (1961). — Reziproke Bastarde zwischen *Lycopersicon esculentum* Mill. und *Lycopersicon peruvianum* (L.) Mill. Naturwissenschaften **51**, 443—444 (1964).

Hagman, M.: Ann. Rep. Forest Res. Inst. of Finland **2**, 1—42 (1962). — The use of disc electrophoresis and serological reactions in the study of pollen and style relationships. In: H. F. Linskens (Hrsg.), Pollen physiology and fertilization, p. 244—250. Amsterdam: North Holland Publ. Co. 1964. — Incompatibility in *Pinus*. Hereditas (Lund) **52**, 245 (1964). — Hagman, M., and L. Mikkola: Observations on cross-, self-, and interspecific pollinations in *Pinus peuce* Griseb. Silvae Genet. (Frankfurt a. M.) **12**, 73—79 (1963). — Hecht, A.: In: H. F. Linskens (Hrsg.), Pollen physiology and fertilization, p. 237—243. Amsterdam: North Holland Publ. Co. 1964. — Heinen, W., and H. F. Linskens: Enzymic breakdown of stigmatic cuticula of flowers. Nature (Lond.) **191**, 1416 (1961). — Herbert, W.: Amaryllidaceae, with a treatise on cross-bred vegetables. London 1837. — Hildebrand, F.: Die Geschlechtverteilung bei den Pflanzen, und das Gesetz der vermiedenen und unvorteilhaften stetigen Selbstbefruchtung. Leipzig 1867. — Holden, H. H. W., and D. A. Bond: Studies on the breeding system of the field bean, *Vicia faba* (L.). Heredity **15**, 175—192 (1960). — Huf-Brinckmann, M., and H. F. Linskens: Observations on the presence of self-incompatibility in the genus *Aloë*. Rev. Biol. (Lisboa) **2**, 151—153 (1961). — Hughes, M. B., and E. B. Babcock: Self-incompatibility in *Crepis foetida* (L.) subsp. *rhoeadifolia* (Bieb.) Schinz et Keller. Genetics **35**, 570—588 (1950).

Jost, L.: Zur Physiologie des Pollens. Ber. dtsch. bot. Ges. **23**, 504—515 (1905). — Über die Selbststerilität einiger Blüten. Bot. Ztg **65**, 77—117 (1907).

Kakizaki, Y.: Studies in the genetics and physiology of self- and cross-incompatibility in the common cabbage (*Brassica oleracea* L. var. *capitata* L.). Jap. J. Bot. **5**, 133—208 (1930). — Kakizaki, Y., and T. Kasai: Bud pollination in cabbage and radish. Some examples of conspicuous "pseudo-fertility" in normal self-incompatible plants. J. Hered. **24**, 359—360 (1933). — Kana, K., N. S. Ranga Swamy, and P. Maheshwari: Test-tube fertilization in a flowering plant. Nature (Lond.) **194**, 1214—1217 (1962). — Kirk, L. E., and T. M. Stevenson: Factors which influence spontaneous self-fertilization in sweet clover *Melilotus*. Canad. J. Res. **5**, 313—326 (1931). — Knight, A.: Phil. Trans. **1799**, 202. — Knight, R., and H. H. Rogers: Incompatibility in *Theobroma cacao*. Heredity **9**, 69—77 (1955). — Knuth, P.: Handbuch der Blütenbiologie. Bd. 1, Leipzig 1898. — Kolreuter, J. G.: Vorläufige Nachricht von einigen das Geschlecht der Pflanzen betreffenden Versuchen und Beobachtungen, nebst Fortsetzungen 1, 2 und 3. 1761/66 (auch in: Oswaldts Klassiker Bd. 46). — Mem. Acad. St. Petersbourg **3**, 197 (1809). — Kroh, M.: Genetische und entwicklungsphysiologische Untersuchungen über die Selbststerilität von *Raphanus raphanistrum*. Z. indukt. Abstamm.- u. Vererb.-Lehre **87**, 365—384 (1956). — Vergleichende Untersuchungen an *Phaseolus coccineus*-Selbstungen und Kreuzungen zwischen *Ph. vulgaris* und *Ph. coccineus*. Z. Pflanzenzüchtg **47**, 201—216 (1962). — An electron microscopic study of the behavior of Cruciferae pollen after pollination. In: H. F. Linskens (Hrsg.), Pollen physiology and fertilization, p. 221—224. Amsterdam: North Holland Publ. Co. 1964. — Reaction of pollen after transfer from one stigma to another. (Contribution to characterization of the incompatibility mechanism with *Cruciferae*.) Züchter (Genetics and Breeding Res.) **36**, 185—189 (1966). — Kumar, S., and A. Hecht: Inactivation of incompatibility in *Oenothera organensis* following ultraviolet irradiation. Naturwissenschaften **52**, 398—399 (1965). — Kwack, B. H.: Stylar culture of pollen and physiological studies of self-incompatibility in *Oenothera organensis*. Physiol. Plantarum (Kbh.) **18**, 297—305 (1965).

Lang, A.: Entwicklungsphysiologie. Fortschr. Bot. **12**, 432—435 (1949). — Lehmann, E.: Selbststerilität, Heterostylie. In: Handbuch der Vererbungswissenschaften, Bd. 2, Lfg 4, S. 1—43. Berlin 1927. — Lewis, D.: The physiology of incompatibility in plants. I. The effect of temperature. Proc. roy. Soc. B **131**, 13—26 (1942). — The incompatibility sieve for producing polyploids. J. Genet. **45**, 261—264 (1943). — The physiology of incompatibility in plants. II. *Linum grandiflorum*. Ann. Bot., N.S. **7**, 115—122 (1943). — Physiology of incompatibility in plants. III. Autopolyploids. J. Genet. **45**, 171—185 (1943). — Incompatibility in plants, its genetical and physiological synthesis. Nature (Lond.) **153**, 575—578 (1944). — Chemical control of fruit formation. J. Pomology **22**, 175—183 (1946). — Competition and dominance of incompatibility alleles in diploid pollen. Heredity **1**, 85—108 (1947). — Incompatibility in flowering plants. Biol. Rev. **24**, 472—496 (1949). — Structure of the incompatibility gene. III. Types of spontaneous and induced mutation. Heredity **5**, 399—414 (1951). — Serological reactions of incompatibility substances. Proc. roy. Soc. B **140**, 127—135 (1952). — Competition and dominance of incompatibility alleles in diploid pollen. Heredity **1**, 85—108 (1952). — Comparative incompatibility in angiosperms and fungi. Advanc. Genet. **6**, 235—285 (1954). — Incompatibility, a summing up. C. r. des Seances VIIIe Congr. Intern. de Bot. Paris, Sect. **10**, 81—85 (1954). — Sexual incompatibility. Sci. Progr. **172**, 593—603 (1955). — Incompatibility and plant breeding. Brookhaven Symp. exp. Biol. **9**, 89—100 (1956). — A protein dimer hypothesis on incompatibility. Genetics Today **3**, 657—663 (1965), London: Pergamon Press. — LEWIS, D., and L. K. CROWE: Unilateral interspecific incompatibility in flowering plants. Heredity **12**, 233—256 (1958). — LINSKENS, H. F.: Physiologische und chemische Unterschiede zwischen selbst- und fremdbestäubten Petunien-Griffeln. Naturwiss. **40**, 28—29 (1953). — Biochemical studies on the incompatibility-reaction in the style of *Petunia*. Rapp. et Commun. VIIIe Congr. Intern. de Bot. Paris, Sect. **10**, 146—147 (1954). — Physiologische Untersuchungen zur Pollenschlauchhemmung selbststeriler Petunien. Z. Bot. **43**, 1—44 (1955). — Die Abwehrreaktionen der Pflanzen. Utrecht-Nijmegen 1957. — Zur Frage der Entstehung der Abwehr-Körper bei der Inkompatibilitätsreaktion von *Petunia*. I. Mitt.: Versuche zur Markierung der Griffel mit P^{32}- und C^{14}-Verbindungen. Ber. dtsch. bot. Ges. **71**, 3—10 (1958a). — La participation des deux sexes à la formation des «corps de resistance» chez la reaction d'incompatibilité de *Petunia*. Abstr. IV. Intern. Congr. Biochemistry Wien **15**, 189—190 (1958b). — Zur Frage der Entstehung der Abwehrkörper bei der Inkompatibilitätsreaktion von *Petunia*. II. Mitt.: Versuche zur Markierung des Pollens mit P^{32}- und C^{14}-Verbindungen. Ber. dtsch. bot. Ges. **72**, 84—92 (1959). — Zur Frage der Entstehung der Abwehrkörper bei der Inkompatibilitätsreaktion von *Petunia*. III. Mitt.: Serologische Teste mit Leitgewebs- und Pollenextrakten. Z. Bot. **48**, 126—135 (1960). — Die Anwendung der „clonal selection theory" auf die Erscheinung der Selbstinkompatibilität der Blütenpflanzen. Port. Acta biol. A **6**, 251—258 (1962). — Pollen physiology. Ann. Rev. Plant Physiol. **15**, 255—270 (1964). — Biochemische aspecten van incompatibiliteits-verschijnselen bij de befruchting der bloemplanten. Biol. Jaarb. (Dodonaea) **33**, 35—40 (1965). — Biochemistry of incompatibility. Genetics Today **3**, 629—635 (1965), London: Pergamon Press. — LINSKENS, H. F., u. K. ESSER: Über eine spezifische Anfärbung der Pollenschläuche im Griffel und die Zahl der Kallosepfropfen nach Selbstung und Fremdung. Naturwiss. **44**, 16 (1957). — Stoffaufnahme der Pollenschläuche aus dem Leitgewebe des Griffels. Proc. Kon. Ned. Akad. v. Wetensch. **62**, 150—154 (1959). — LINSKENS, H. F., u. W. HEINEN: Cutinase-Nachweis in Pollen. Z. Bot. **50**, 338—347 (1962). — LINSKENS, H. F., J. A. M. SCHRAUWEN u. M. VAN DER DONK: Überwindung der Selbstinkompatibilität durch Röntgenbestrahlung des Griffels. Naturwissenschaften **47**, 547 (1960). — LINSKENS, H. F., and J. TUPÝ: The amino acids pool in the style of self-incompatible strains of *Petunia* after self- and cross-pollination. Züchter (Genetics and Breeding Res.) **36**, 151—158 (1966). — LUNDQUIST, A.: Self-incompatibility in rye. IV. Factors related to self-seeding. Hereditas (Lund) **44**, 193—256 (1958). — The nature of the two-loci incompatibility system in grasses. I. The hypothesis of a duplicative origin. Hereditas (Lund) **48**, 153—168 (1962). — The genetics of incompatibility. Genetics Today **3**, 637—647 (1965), London: Pergamon Press.

MÄKINEN, Y. L. A., and D. LEWIS: Immunological analysis of incompatibility S proteins and of cross-reacting material in a self-compatible mutant of *Oenothera organensis*. Genet. Res. **3**, 352—363 (1962). — MANGELSDORF, P. C., and R. G. REEVES: Hybridization of Maize, *Tripsacum*, and *Euchlaena*. J. Hered. **22**, 329—343 (1931). — MATHER, K.: Specific differences in *Petunia*. I. Incompatibility. J. Genet. **45**, 215—235 (1943). — MCGUIRE, D. C., and C. M. RICK: Self-incompatibility in species of *Lycopersicum* sect. *Eriopersicum* and hybrids with *L. esculentum*. Hilgardia **23**, 101—124 (1954). — MCWILLIAM, J. R.: Interspecific incompatibility in *Pinus*. Amer. J. Bot. **46**, 425—433 (1959). — MEDINA, D. M., e C. H. T. MENDES CONAGIN: Self-incompatibility in *Coffea dewevrei*. Bragantia (São Paulo) **18**, 283—293 (1959). — MOEWUS, F.: Zur Physiologie und Biochemie der Selbststerilität bei *Forsythia*. Biol. Zbl. **69**, 181—197 (1950a). — Die Bedeutung von Farbstoffen bei den Sexualprozessen der Algen und Blütenpflanzen. Z. angew. Chem. **62**, 496—502 (1950b). — MORITZ, O.: Die Serologie der pflanzlichen Eiweißkörper. In: Handbuch der Pflanzenphysiologie. Bd. 8. S. 356—414. 1958. — MOSIG, G.: Zur Genetik von *Petunia hybrida*. I. Die Selbststerilität. Z. Vererbungsl. **91**, 158—163 (1960). — MÜHLETHALER, K., u. H. F. LINSKENS: Elektronenmikroskopische Aufnahmen von Pollenschläuchen. Experientia (Basel) **12**, 253—254 (1956). — MÜLLER, F.: Selbststerilität bei Orchideen 1866/67 [auch in:

Ges. Abh. 2, 99—122 Jena (1915)]. — Bestäubungsversuche an *Abutilon*-Arten. Jena. Z. Med, u. Naturw. Ges. Abh. **4**, 22, 144 (1873). — Müller, H.: Die Befruchtung der Blumen durch Insekten und die gegenseitige Anpassung beider. Leipzig 1873.

Naundorf, G.: Beitrag zur Frage der Selbststerilität bei *Theobroma Cacao* L. Gordian **59**, Nr 1405, 7—10, Nr 1406, 8—11, Nr 1407, 8—11 (1959).

Oelke, J.: Zur Physiologie der Selbst- und Kreuzungssterilität beim Radieschen. Züchter **27**, 358—369 (1957).

Pandey, K. K.: Mutations of self-incompatibility alleles in *Trifolium pratense* and *T. repens*. Genetics **41**, 327—343 (1956). — Time of the S allele action. Nature (Lond.) **181**. 1220—1221 (1958). — Microsporogenesis, time of S allele action, pollen cytology and place of inhibitions and their relationship with the evolution of gametophytic and sporophytic systems of self-incompatibility. Proc. 10th Int. Gen. Congr. (Montreal) **2**, 212—213 (1958). — Evolution of gametophytic and sporophytic systems of self-incompatibility in angiosperms. Evolution (Lancaster, Pa.) **14**, 98—115 (1960). — Interspecific incompatibility in *Solanum* species. Amer. J. Bot. **49**, 874—882 (1962). — Stigmatic secretion and bud pollinations in self- and cross-incompatible plants. Naturwissenschaften **50**, 408—409 (1963). — Patel, G. I., and R. M. Datta: Interspecific hybridisation between *Corchorus olitorius* and *C. capsularis* L. and the cytogenetical basis of incompatibility between them. Euphytica **9**, 89—110 (1960). — van der Pluijm, J. E., u. H. F. Linskens: Feinstruktur der Pollenschläuche im Griffel von *Petunia*. Züchter (Genetics and Breeding Res.) **36**, 220—224 (1966).

Rappaport, J., S. Satina and A. F. Blakeslee: Ovular tumors and inhibition of embryo growth in incompatible crosses of *Datura*. Science **111**, 276—277 (1950). — Reznik, H.: Über die Pigmentausstattung der Pollen Heterostyler. Biol. Zbl. **76**, 351—359 (1957). — Rietsema, J., B. Blondel, S. Satina and A. F. Blakeslee: Studies on ovule and embryo growth in *Datura*. I. A growth analysis. Amer. J. Bot. **42**, 449—455 (1955). — Rietsema, J., S. Satina and A. F. Blakeslee: On the nature of the embryo inhibitor on ovular tumors of *Datura*. Proc. nat. Acad. Sci. (Wash.) **40**, 424—431 (1954). — Riley, H. P.: The genetics and physiology of self-sterility in the genus *Capsella*. Genetics **21**, 24—39 (1936). — Rittinghaus, P.: Einige Beobachtungen über das Eindringen der Pollenschläuche ins Leitgewebe. Verh. naturhist. Ver. preuß. Rheinlande **43**, 105 (1886). — Röbbelen, G.: Über die Kreuzungsunverträglichkeit verschiedener *Brassica*-Arten als Folge eines gehemmten Pollenschlauchwachstums. Züchter **30**, 300—312 (1960). — Roggen, H. P. J. R.: Enkele fysiologische bepalingen bij de incompatibele kruising tussen *Salpiglossis sinuata* en *Petunia hybrida*. Nijmegen (unpubl.) (1963).

Sampson, D. R.: Intergeneric pollen-stigma incompatibility in the *Cruciferae*. Canad. J. Genet. Cytol. **4**, 38—49 (1962). — Schlösser, K.: Cytologische und cytochemische Untersuchungen über das Pollenschlauchwachstum selbststeriler Petunien. Z. Bot. **49**, 266—288 (1961). — Schmitt, D., and T. O. Perry: Self-sterility in sweet-gum. Forest Sci. **10**, 302—305 (1964). — Schoch-Bodmer, H.: Zur Frage der „Hemmungsstoffe" bei Heterostylen. Arch. Klaus-Stift. Vererb.-Forsch., Erg.-Bd. zu **20**, 403—416 (1945). — Sears, E. R.: Cytological phenomena connected with self-sterility in the flowering plants. Genetics **22**, 131—181 (1937)— Sirks, M. J.: Stérilité, auto-inconceptibilité et différenciation sexuelle physiologique. Arch. néerl. Sci. exact. nat., Sér. B **3**, 205—234 (1917). — De verklaring der zelfsteriliteit als erfelijk verschijnsel. Landbouwk. T. **39**, 1—19 (1927). — Sprengel, C. K.: Das entdeckte Geheimnis der Natur im Bau und in der Befruchtung der Blumen. Berlin 1793 (auch in: Oswaldts Klassiker Bd. 48—51). — Stecher, M.: Zit. nach Straub 1958. — Stout, A. B.: Fertility in *Cichorium intybus*. The sporadic occure of self-fertile plants among the progeny of self-sterile plants. Amer. J. Bot. **4**, 375—395 (1917). — The physiology of incompatibilities. Amer. J. Bot. **10**, 459—461 (1923). — Stout, A. B., and C. Chandler: Pollentube behavior in *Hemerocallis* with special reference to incompatibilities. Bull. Torrey bot. Club **60**, 397—414 (1933). — Change from self-incompatibility to self-compatibility accompanying change from diploidy to tetraploidy. Science **94**, 118 (1941). — Hereditary transmission of induced tetraploidy and compatibility in fertilization. Science **96**, 257—258 (1942). — Strasburger, E.: Über fremdartige Bestäubung. Jb. wiss. Bot. **17**, 50 (1886). — Straub, J.: Die Beseitigung der Selbststerilität durch Polyploidisierung. Ber. dtsch. bot. Ges. **59**, 296—304 (1941). — Zur Entwicklungsphysiologie der Selbststerilität von *Petunia*. Z. Naturforsch. **1**, 287—291 (1946). — Zur Entwicklungsphysiologie der Selbststerilität von *Petunia*. II. Das Prinzip des Hemmungsmechanismus. Z. Naturforsch. **2**b, 433—444 (1947). — Neue Ergebnisse der Selbststerilitätsforschung. Naturwiss. **35**, 23—26 (1948). — Die Physiologie der Selbststerilität. Rap. et Comm. VIIIe Congr. Intern. de Bot. Paris, Sect. **10**, 148—151 (1954). — Das Überwinden der Selbststerilität. Z. Bot. **46**, 98—111 (1958). — Surikov, I. M.: Variation of the incompatibility reaction in rye. Ref. Zhur. Biol. No 10G274 (1961). — Swaminathan, M. S.: Overcoming cross-incompatibility among some mexican diploid species of *Solanum*. Nature (Lond.) **176**, 887—888 (1955).

TANAKA, R.: Vergleichende Untersuchungen über die Fertilität bei Selbstbestäubung in der Polyploidie bei *Chrysanthemum*. Jap. J. Genet. **27**, 1—2 (1952). — TATEBE, T.: The effect of the mutilation of stigma on self-fertility in the japanese radish. J. Horticult. All. Japan **10**, 62—65 (1939). — Studies on the self-incompatibility of the common cabbage. II. On inhibiting substance. J. Horticult. Ass. Japan **24**, 167—174 (1955). — Studies on the genetics of self- and cross-incompatibility in the japanese radish. II. J. Horticult. All. Japan **25**, 157—162 (1956a). — Physiological researches on the fertility of the buckwheat. V. Further studies on the behavior of pollen and pollentube. Jap. J. Breeding **6**, 156—162 (1956b). — Studies on the physiological mechanism of self-incompatibility in Japanese radish. I. Influence of air humidity on self-fertility. J. Jap. Soc. Hort. Sci. **33**, 62—66 (1964). — THOMPSON, K. F.: Self-incompatibility in marrow-stem kale, *Brassica oleracea* var. *asphala*. J. Genet. **55**, 45—60 (1957). — TUPÝ, J.: Calloseformation in pollen tubes and incompatibility Biol. Plant. (Praha) **1**, 192—198 (1959). — Changes in glucose and fructose level in *Nicotiana alata* styles and ovaries accompanying compatible and incompatible pollen tube growth. Biol. Plant. (Praha) **3**, 1—14 (1961). — Investigation of free amino-acids in cross-, self-, and non-pollinated pistils of *Nicotiana alata*. Biol. Plant. (Praha) **3**, 47—64 (1961). — Metabolism of proline in styles and pollen tubes of *Nicotiana alata*. In: H. F. LINSKENS (Hrsg.), Pollen physiology and fertilization, p. 86—94. Amsterdam: North Holland Publ. Co. 1964.

VISSER, T.: Germination and pollination experiments with *Forsythia*. Proc. kon. ned. Akad. Wet., Ser. C **59**, 685—693 (1956).

WHITELEY, J. R., W. I. THOMAS, and I. J. JOHNSON: Cross-incompatibility in Maize. Agronomy J. **49**, 513—518 (1957).

YASUDA, S.: Physiological researches of the fertility in *Petunia violacea*. VI. Growth of the pollen tubes in the style. Bot. Mag. (Tokyo) **43**, 156—169 (1929). — On the special substance that inhibits self-fertilization. Physiological consideration on its nature, based on the results of the experiments on the fertility of *Petunia*. Bot. Mag. (Tokyo) **46**, 225 (1932). — Physiological research on self-incompatibility in *Petunia violacea*. Bull. Imp. Coll. Agricult. and Forest. Morioka Nr 20 (1934).

ZYCH, C. C.: Incompatibility in crosses of red, black, and purple raspberries. Proc. Amer. Soc. Hort. Sci. **86**, 307—312 (1965).

Apomixis bei Archegoniaten.

Von

W. Döpp.

Mit 15 Abbildungen.

1. Einleitung.

Apogamie ist nach WINKLER (1908) die Entstehung eines Sporophyten aus einer oder mehreren vegetativen Zellen des Gametophyten ohne einen dabei erfolgenden Befruchtungsvorgang, Parthenogenese die Entwicklung eines Sporophyten aus einer unbefruchteten Eizelle. Beide Vorgänge zusammen mit der vegetativen Propagation bezeichnet WINKLER (1908) als Apomixis. Die Chromosomenzahl ist bei Apogamie und Parthenogenesis im Gametophyten und Sporophyten die gleiche. Diese Fortpflanzungsweise ist bei Pteridophyten häufig, jedoch nicht immer mit einem vollständigen Verlust der Fähigkeit zur geschlechtlichen Fortpflanzung verbunden.

WINKLER (1908) unterscheidet generative und somatische Apogamie und Parthenogenesis, je nachdem, ob die vegetative Zelle oder die Eizelle, die den Sporophyten liefert, die haploide oder die diploide Chromosomenzahl besitzt. Diese Bezeichnung hat HARTMANN (1909) durch die Begriffe „haploid" für generativ und „diploid" für somatisch ersetzt. Gegenüber den Bedenken, die hinsichtlich dieser Definition vorgebracht werden können, schlage ich vor, die Apogamie oder Parthenogenesis entsprechend der Zahl der Chromosomensätze durch die Bezeichnungen „monoploid", „diploid", „triploid", „tetraploid" usw. zu charakterisieren, soweit dies möglich ist. Dann hätte ein haploider Sporophyt mit nur einem Chromosomensatz — ein Fall, dem besondere Bedeutung zukommt — eine eindeutige Bezeichnung, die nur für diesen zutrifft, während bei Anwendung der Begriffe „generative Apogamie" (WINKLER) und „haploide Apogamie" (HARTMANN) der Sporophyt einen oder zwei oder auch mehr Chromosomensätze besitzen kann.

Apogamie und Parthenogenesis kommen bei Moosen, soweit bekannt ist, nicht vor. Bei den Pteridophyten gibt es zahlreiche Fälle von Apogamie und auch einige von Parthenogenesis. Apogamie ist in den Tropen prozentual stärker vertreten als in Europa; sie findet sich bei zahlreichen Gattungen und Arten, jedoch nicht in allen Familien der *Filices* (Tabelle 1), Parthenogenesis bei einigen *Marsilia*-, *Salvinia*- und *Selaginella*-Arten.

Bei den *Psilophytinae* und *Equisetinae* sowie bei *Lycopodium* und *Isoetes* sind beide Fortpflanzungsarten unbekannt. Unter den Eusporangiatae ist *Marattia sambucina* apogam (STOKEY 1942).

Seit der Entdeckung der Apogamie durch DE BARY (1878) und FARLOW (1878) waren Apogamie und Parthenogenesis bei Pteridophyten sehr häufig Gegenstand von Untersuchungen. Zusammenfassungen und ältere Literatur findet man bei SADEBECK (1902), WINKLER (1908), ERNST (1908), F. O. BOWER (1923), GOEBEL (1930), ROSENBERG (1930), DÖPP (1937), IRENE MANTON (1950), STEIL (1939, 1951).

Die Apogamie bei Farnpflanzen kann obligat (habituell) oder fakultativ sein. Im ersteren Fall ist sie erblich, im letzteren spricht man auch von induzierter Apogamie.

Tabelle 1. *Gattungen, in denen (nicht bei allen Arten) Apogamie auftritt.* Familien nach Copeland (1947). Nur einige neuere Arbeiten sind angeführt.

	Hymenophyllaceae.
Trichomanes (Stokey)	
	Pteridaceae.
Pteris (Manton 1950, Walker)	*Actiniopteris* (Stokey)
Pellaea (Steil, Tryon)	*Hemionitis* (Manton u. Sledge)
Notholaena (Cheilanthes)	*Adiantum* (Mehra; Roy u. Sinha)
	Davalliaceae.
Humata (Manton u. Sledge)	
	Aspidiaceae.
Dryopteris (Steil, Döpp, Manton 1950)	*Athyrium* (Mehra u. Verma)
Cyrtomium (Manton 1950)	*Diplazium* (Manton u. Sledge)
Polystichum (Patterson)	
	Aspleniaceae.
Asplenium (Manton u. Sledge)	

2. Obligate (habituelle) Apogamie.

a) In Verbindung mit Restitutionskernbildung. Phasenwechsel (in den Sporenmutterzellen) und Reduktionsteilung vorhanden.

α) Cytologie, Genetik, Keimungsverhältnisse.

Diese Art der Fortpflanzung ist in der Familie der Polypodiaceen (im Sinne von Diels) so häufig, nicht etwa nur als Abnormität bei Gartenvarietäten, sondern auch bei vielen wildwachsenden Arten, daß man sie nicht als abnorm ansehen kann, zumal die apogamen den geschlechtlichen Farnen an Vitalität nicht nachstehen (Döpp, Steil, Manton u. a.). Archegonien treten selten auf, wohl aber fast immer Antheridien mit normal funktionsfähigen Spermatozoiden, so daß in der Natur wie auch im Experiment die Möglichkeit der Kreuzung mit verwandten sexuellen Formen gegeben ist. Als Reduktionserscheinung infolge von Apogamie kann das Fehlen der Antheridien unter gewöhnlichen Kulturbedingungen bei *Notholaena sinuata* angesehen werden (Woronin); nur mit Hilfe eines Extraktes aus Prothallien von *Pteridium aquilinum* war es möglich, Antheridien zu erhalten (Döpp 1958). Der apogame Sporophyt entsteht

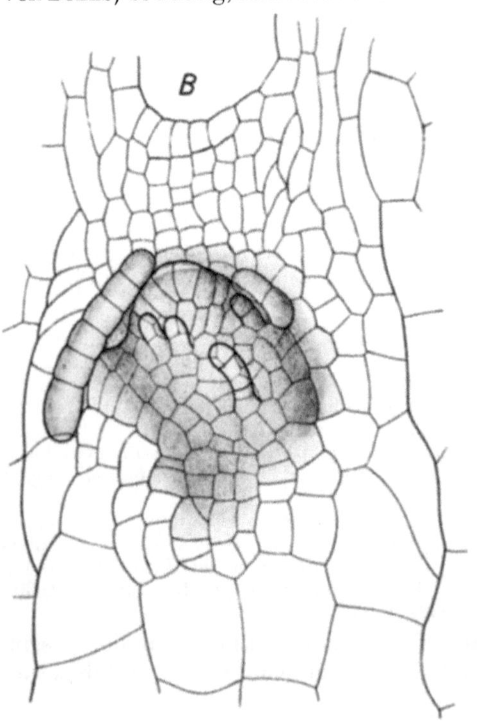

Abb. 1. Apogamie bei *Pteris cretica*. Anlage des 1. Blattes auf der Unterseite des Prothalliums. *B* Scheitelbucht. (Nach de Bary.)

entweder hinter der Scheitelbucht des Prothalliums (Abb. 1, 2) auf der dem Licht abgewandten Seite, da, wo sonst die Archegonien sitzen, oder an einem aus der Scheitelbucht herauswachsenden Fortsatz *(Notholaena)*, den Woronin als erstes verkümmertes Blatt deutet, da er aus einer Zelle des Randmeristems mit zweischneidiger Scheitelzelle herauswächst. Bei *Notholaena* und *Pellaea* können schon sehr kleine Prothallien eine apogame Sprossung bilden. Meist

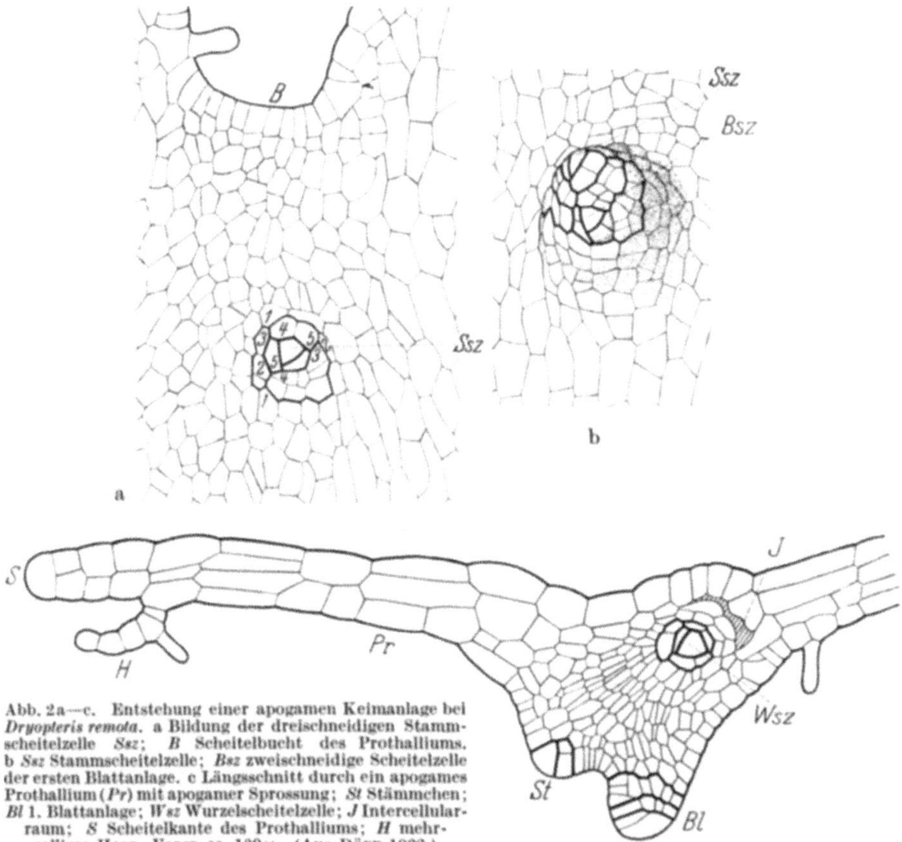

Abb. 2a—c. Entstehung einer apogamen Keimanlage bei *Dryopteris remota*. a Bildung der dreischneidigen Stammscheitelzelle *Ssz*; *B* Scheitelbucht des Prothalliums. b *Ssz* Stammscheitelzelle; *Bsz* zweischneidige Scheitelzelle der ersten Blattanlage. c Längsschnitt durch ein apogames Prothallium (*Pr*) mit apogamer Sprossung; *St* Stämmchen; *Bl* 1. Blattanlage; *Wsz* Wurzelscheitelzelle; *J* Intercellularraum; *S* Scheitelkante des Prothalliums; *H* mehrzelliges Haar. Vergr. ca. 130×. (Aus Döpp 1932.)

entsteht zuerst die Blattanlage, aber auch der Sproßscheitel kann zuerst auftreten. Auch andere Abweichungen werden oft beobachtet, über deren Ursache Leitgeb berichtet.

Obwohl beim Übergang vom Gametophyten zum Sporophyten kein Phasenwechsel erfolgt, liegt eine normale Reduktionsteilung vor. Diese wird dadurch ermöglicht, daß im 8-zelligen sporogenen Gewebe eine vollständige Kernteilung (Restitutionskernbildung[1]) stattfindet (Abb. 3, *R*): Die Chromosomen teilen sich, bleiben aber zu einem Kern vereinigt (Abb. 4c, d). Es bilden sich somit 8 Sporenmutterzellen vom Typus I. Vielfach angelegte unvollständige Wände lassen diese Sporenmutterzellen in jüngeren Stadien wie eingeschnürt erscheinen (Abb. 4a). In der darauffolgenden Diakinese (Abb. 4e) konjugiert jedes Chromosom mit seinem Schwesterchromosom. Die 8 Sporenmutterzellen liefern 32 Sporen,

[1] Dieser von Rosenberg geprägte Begriff ist hier in erweitertem Sinne gebraucht (vgl. Döpp 1955).

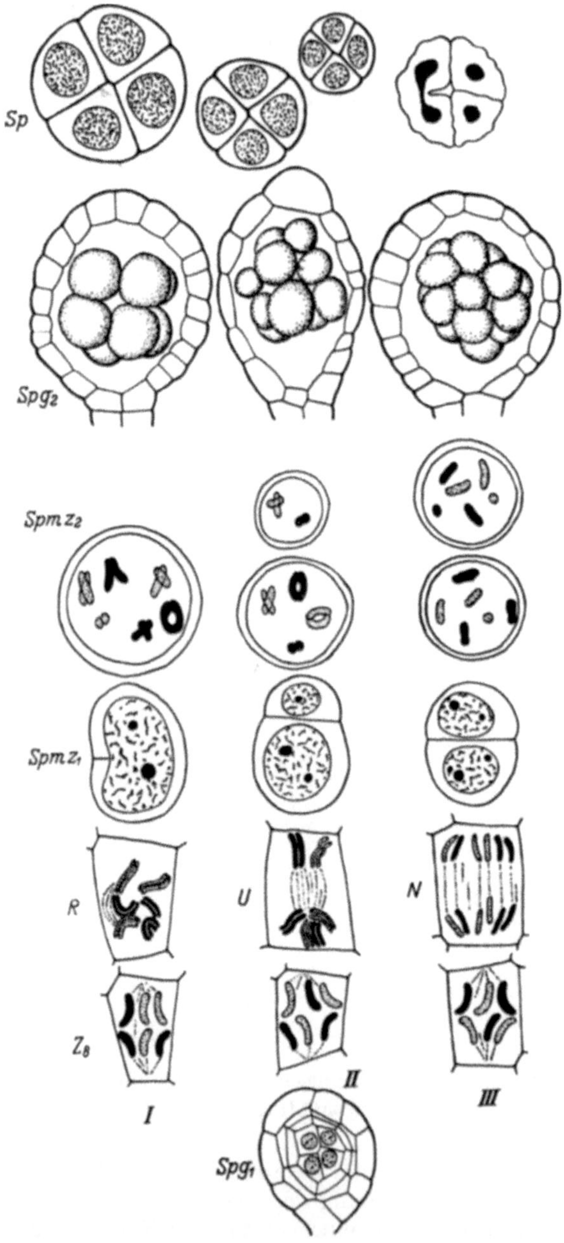

Abb. 3. Schematische Darstellung der Vorgänge in den Sporangien einer apogamen *Dryopteris*. Spg_1 Sporangium mit 8-zelligem sporogenem Gewebe; Z_8 Einzelzellen aus dem 8-zelligen sporogenen Gewebe in Metaphase aus 3 verschiedenen Sporangien; *I* Entwicklung einer Sporenmutterzelle vom Typus I, entsprechend II und III; *R* Restitutionskernteilung; *U* unregelmäßige Kernteilung, durch die in Verbindung mit unregelmäßiger Zellteilung zwei meist ungleich große Sporenmutterzellen vom Typus II entstehen; *N* normale Kernteilung, die mit nachfolgender normaler Zellteilung zwei gleich große Sporenmutterzellen vom Typus III liefert; $Spmz_1$ junge Sporenmutterzellen; $Spmz_2$ ältere Sporenmutterzelle in Diakinese, bei I und II mit Gemini, bei III mit Univalenten; Spg_2 Sporangien mit abgerundeten Sporenmutterzellen; *Sp* Sporentetraden. In den sporogenen Zellen sind 2 Chromosomensätze zu je 3 Chromosomen angenommen, von denen der eine schwarz, der andere punktiert ist. (DÖPP 1955.)

während bei einem sexuellen Farn 16 Sporenmutterzellen und dementsprechend 64 Sporen gebildet werden. Man kann daher die Sporenzahl 32 nahezu als Charakteristikum apogamer Farne ansehen (DÖPP 1932, 1939, 1955, MANTON 1950, TRYON u. BRITTON 1958).

Zum Unterschied zum Typus I erfolgt in den Sporenmutterzellen vom Typus III keine Restitutionskernbildung; die Teilung im 8-zelligen sporogenen Gewebe ist normal und liefert 16 Sporenmutterzellen wie bei einem geschlechtlichen Farn. Die Diakinese (Abb. 4f) läßt jedoch entweder nur ungepaarte Chromosomen oder zum mindesten zahlreiche Univalente erkennen; die Reifeteilungen sind gestört und dementsprechend die Sporen steril, abgesehen von besonderen, weiter unten zu besprechenden Ausnahmefällen.

Für den Entwicklungsgang ist offenbar im allgemeinen auch bedeutungslos ein weiterer Typ von Sporenmutterzellen (Typus II): Auf dem 8-Zellstadium des sporogenen Gewebes kann die Masse der gespaltenen Chromosomen durchgeschnürt werden und somit die Bildung zweier meist ungleich großer Kerne erfolgen. Dementsprechend werden meist zwei verschieden große Sporenmutterzellen gebildet, zuweilen solche mit nur 10—15 Gemini. Sind es noch weniger, so stirbt die Zelle ab. Wahrscheinlich können auch einzelne Tochterchromosomen von einander getrennt und auf die beiden Kerne verteilt werden, so daß in diesen nicht nur Paare, sondern auch einige einzelne Chromosomen vorhanden sind. Eine solche Teilung findet in einer bis mehreren, selten in allen 8 sporogenen Zellen eines Sporangiums anstelle der Bildung von Restitutionskernen statt. Im Anschluß daran entstehen normal aussehende,

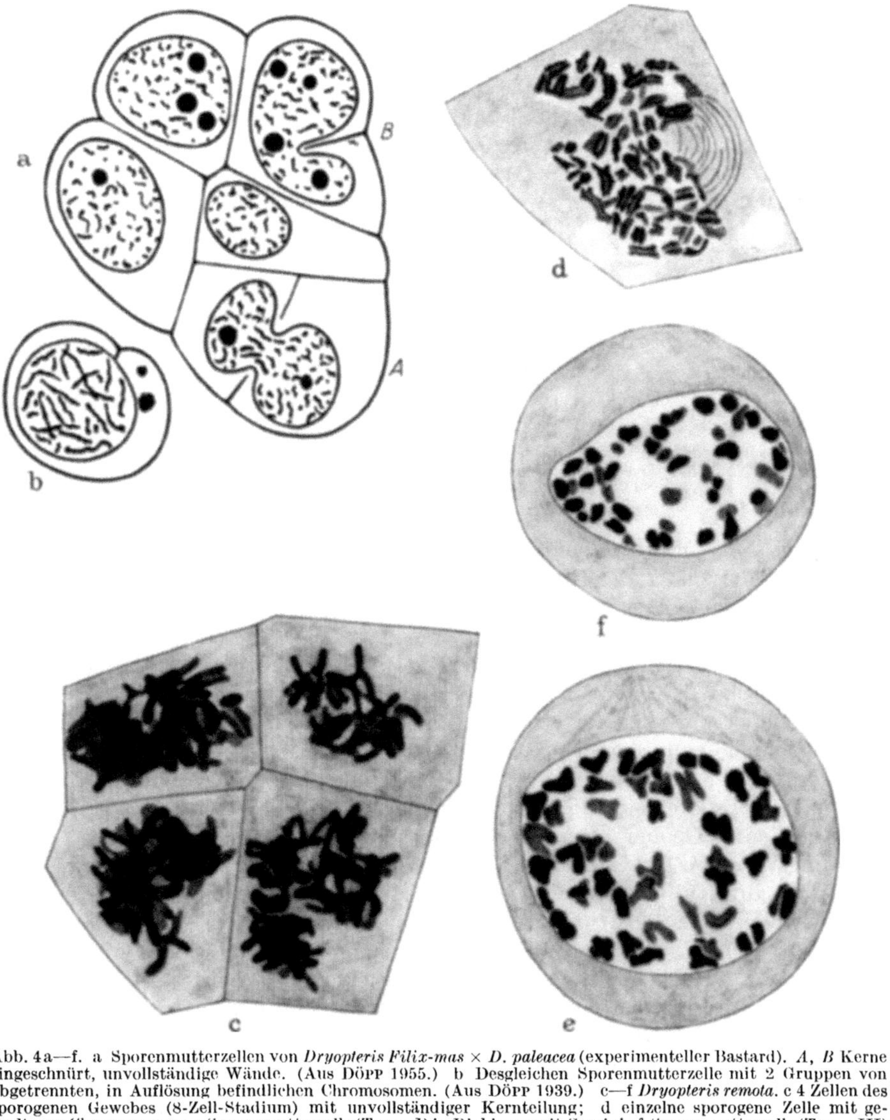

Abb. 4a—f. a Sporenmutterzellen von *Dryopteris Filix-mas* × *D. paleacea* (experimenteller Bastard). *A*, *B* Kerne eingeschnürt, unvollständige Wände. (Aus DÖPP 1955.) b Desgleichen Sporenmutterzelle mit 2 Gruppen von abgetrennten, in Auflösung befindlichen Chromosomen. (Aus DÖPP 1939.) c—f *Dryopteris remota*. c 4 Zellen des sporogenen Gewebes (8-Zell-Stadium) mit unvollständiger Kernteilung; d einzelne sporogene Zelle mit gespaltenen Chromosomen; e Sporenmutterzelle (Typus I) in Diakinese mit Gemini; f Sporenmutterzelle (Typus III) in Diakinese mit Univalenten. Vergr. ca. 1900×. (c—e aus DÖPP 1932.)

kleinere Sporen mit unvollständigen Genomen. Ob diese keimfähig sind, erscheint auf Grund von Beobachtungen möglich, sofern sie nicht zu klein sind. Aber selbst wenn Keimung erfolgt, ist noch nicht sicher, ob sich lebensfähige Pflanzen aus Sporen vom Typus II[1] entwickeln können.

[1] Im folgenden ist der Einfachheit wegen von Sporen vom Typus I usw. die Rede.

Daß die apogame Pflanze fertil ist, wird nach dem Vorhergehenden durch die Restitutionskernbildung bewirkt. Aus den cytologischen Befunden ergibt sich ferner, daß die Nachkommenschaft auch bei apogamen Bastarden konstant sein muß. Die Restitutionskernbildung mitsamt der anschließenden ersten Reifeteilung hat nämlich keinen anderen Endeffekt als den einer gewöhnlichen Mitose; eine Aufspaltung kann somit nicht stattfinden. Mit anderen Worten, die Nachkommenschaft ist konstant, wie auch die Beobachtungen bestätigen.

Einschränkend muß allerdings gesagt werden, daß eine gewisse Variabilität theoretisch möglich erscheint. In den Fällen von Apogamie in der Gattung *Dryopteris* wurde nämlich Chromosomenelimination beobachtet (Döpp 1932, 1939, 1955): Zuweilen sondern sich bei der Restitutionskernteilung einige Chromosomen aus der Hauptmasse ab und werden, ohne durch eine Wand von den übrigen Chromosomen abgetrennt zu werden, resorbiert (Abb. 4b). Ob nun Sporen mit etwas verminderter Chromosomenzahl keimfähig sind und lebensfähige Pflanzen liefern können, kann mit Sicherheit ebensowenig gesagt werden wie bei Sporen vom Typus II. Ebensowenig ist erwiesen, ob die Chromosomenelimination eine Bedeutung im Hinblick auf Variabilität und Evolution hat.

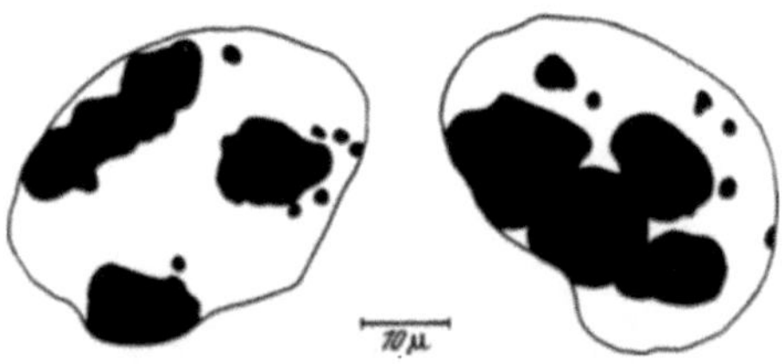

Abb. 5. Vielkernige Sporen aus Sporenmutterzellen von Typus III von *Pteris jumbrosa var. iuvanti*. (Aus Schifferdecker 1957.)

In den Sporenmutterzellen vom Typus III kann anstelle der 1. oder 2. Reifeteilung eine unvollständige Kernteilung erfolgen, so daß „eingeschnürte Sporen" entstehen (Döpp 1932, *Dryopteris remota*). Ausbleiben der Zellteilung ergibt 4- oder mehrkernige Sporen (Abb. 5); sind dabei eine oder mehr Kernteilungen unvollständig, so resultieren 3-, 2- oder 1-kernige Sporen (Schifferdecker 1957, apogame *Pteris*-Arten). Wenigerkernige Sporen können aber auch durch Kernfusion gebildet werden. Schifferdecker diskutiert ausführlich die Möglichkeit einer Keimung von Sporen aus 16-Zellen-Sporangien. Wenn im Anschluß an die Metaphase I eine unvollständige Kernteilung erfolgt, so enthält die entstehende Großspore (Monade oder homöotypisch gebildete Dyade) vermutlich das vollständige Genom der Mutterpflanze, müßte also keimfähig sein, was durch Keimungsversuche mit Monaden von *Pteris Rivertoniana* erwiesen wird. Prämeiotische Restitutionskernbildung und in den Sporenmutterzellen während der Meiosis ablaufende Restitutionskernbildung haben also denselben Effekt. Die Tetradensporen dürften allerdings auch hier kaum keimfähig sein. Aber gerade die Frage der Möglichkeit einer Weiterentwicklung solcher Sporen wäre von besonderem Interesse, da sie ebenso wie Sporen vom Typus II nicht das vollständige Genom besitzen.

Einen Hinweis bezüglich der Keimfähigkeit von Sporen des Typus III versucht Manton (1950, S. 194) zu geben. Nach ihren Beobachtungen ergaben Sporen einer pentaploiden Pflanze aus dem Formenkreis von *Dryopteris paleacea* einige Prothallien mit Archegonien. Die pentaploide Pflanze enthält als Bastard außer Chromosomensätzen von *Dryopteris paleacea* einen oder mehrere der sexuell sich fortpflanzenden *Dryopteris Filix-mas* oder einer verwandten sexuellen Form, und so wäre denkbar, daß die für Apogamie verantwortlichen Gene ganz oder teilweise abgetrennt worden sind, so daß archegontragende Prothallien entstanden. Sporophyten konnten jedoch aus diesen nicht herangezogen werden. Es könnte sich aber meines Erachtens auch um Typus II-Sporen handeln.

Bei *Dryopteris paleacea* erhielt Döpp (1950) nach Chromosomenverdoppelung durch Colchicinbehandlung keimfähige Sporen vom Typus III außer größeren, ebenfalls keimfähigen Typus I-Sporen. Tryon und Britton (1958) weisen auf einen nicht näher beschriebenen Fall hin, wo auf der gleichen apogamen Pflanze von *Pellaea intermedia* Sporangien mit 32 Sporen und Sporangien mit 64 nicht abortierten Sporen vorkommen. Da in den bekannten Fällen von Apogamie die III-Sporen, falls vorhanden, im allgemeinen steril sind, ist anzunehmen, daß Autopolyploidie hier nicht vorliegt.

Die Keimungsverhältnisse der apogamen Arten sind vielfach schlechter als bei den sexuellen. Bei *Dryopteris remota* keimen 50—70% der Sporen, bei *Dryopteris Filix-mas, D. austriaca* etwa 88—96%. SCHIFFERDECKER (1957) fand bei sexuellen *Pteris*-Arten Keimprozente zwischen 90,3 und 56,9, bei apogamen von 66,4 bis herunter zu 2,5. Dieser Unterschied dürfte hauptsächlich in dem Vorhandensein von nichtkeimfähigen Sporen des Typus II und III begründet sein. Auf Grund eigener Beobachtungen sowie nach MANTON (1950) und SCHIFFERDECKER ist die Keimzahl um so höher, je mehr Typus I vertreten ist. So ist in der polyploiden Reihe der natürlichen *Dryopteris paleacea* und *D. Tavelii* (MANTON 1950) und bei den experimentell hergestellten Bastarden zwischen *Dryopteris Filix-mas* und *D. paleacea* (DÖPP 1955) mit steigendem Polyploidiegrad die Zahl der Sporenmutterzellen vom Typus I und die Sporenproduktion geringer, der Grad der Unregelmäßigkeit der Ausbildung größer und die Keimzahl geringer.

In der eben genannten polyploiden Reihe des *Dryopteris paleacea-D. Tavelii*-Formenkreises ist der höhere Polyploidiegrad entstanden durch Bastardierung (Allopolyploidie), so daß der Schluß gezogen werden kann, daß die Kreuzung mit einer sexuellen Form das erhöhte Auftreten vom Typus III bewirkt (DÖPP 1939, 1955; MANTON 1950). Dafür spricht auch, daß der experimentell hergestellte Bastard *Dryopteris Filix-mas* × *D. Tavelii* infolge fast ausschließlichen Vorkommens von Typus III hochgradig steril ist (DÖPP, unveröffentlicht). Ferner möchte ich annehmen, daß Typus III, vielleicht auch Typus II, dann nicht ausgebildet wird, wenn der apogame Farn kein Bastard ist. Dies könnte der Fall sein bei *Pteris cretica var. albo-lineata* mit 32 Chromosomen (HEUN 1929), die nur Sporangien mit 8 Sporenmutterzellen besitzt, ebenso bei *Pteris arguta* (SCHIFFERDECKER 1957). Die apogame *Notholaena distans* hat sehr gleichmäßig ausgebildete Sporen. *Phegopteris polyploidioides* weist fast nur Typus I auf. Demgemäß hat folgende Vorstellung große Wahrscheinlichkeit: Typus III ist ebensowenig ein Charakteristikum eines apogamen Farnes wie ungleichmäßige Ausbildung der Sporen, sondern beides ist bedingt durch Bastardierung mit einem geschlechtlichen, 16 Sporenmutterzellen führenden Farn; wird ein apogamer Farn, der nur Typus I ausbildet, mit einem sexuellen Farn gekreuzt, so bilden die F_1 und die folgenden Generationen nicht einen intermediären Sporangientyp aus, sondern außer Typus I auch Typus III, ferner auch entsprechend einem mehr oder weniger intermediären Charakter des Spindelapparates Sporenmutterzellen vom Typus II in wechselnder Anzahl.

Innerhalb einer Art oder eines Individuums ist das Verhältnis Typus III: Typus I nahezu konstant, jedoch kommen auch hier Unterschiede vor: Es kann z. B. bei jungen Pflanzen das Verhältnis von III:I erhöht sein. Auch ungünstige Ernährungsverhältnisse und Verpflanzungen scheinen eine Verschiebung zugunsten von III zu bewirken.

Zu den 3 Typen von Sporenmutterzellen fügt MANTON noch einen vierten hinzu, den Verf. seitdem gelegentlich auch beobachtet hat: Restitutionskernbildung erfolgt 2mal und zwar auf dem 4- und auf dem 8-Zellstadium des sporogenen Gewebes. Die Sporenmutterzellen haben daher die 4fache Chromosomenzahl und bilden nach normal verlaufender Meiosis 4 Riesensporen, im ganzen 16 in einem Sporangium. Quadrivalente wurden jedoch ebensowenig gebildet wie nach Polyploidisierung durch Colchicin bei *Dryopteris Filix-mas* (ROSENDAHL

1940). Es ist möglich, daß aus solchen Sporen gelegentlich autoploide apogame Pflanzen, z. B. bei *Pteris cretica*, entstehen. Auch bei *Asplenium cheilosorum* kommen vermutlich zwei aufeinanderfolgende Restitutionskernbildungen vor (Manton und Sledge 1954).

Ob Restitutionskernbildung anstelle der 2. Reifeteilung nur dann erfolgt, wenn wie in II, III kein vollständiges Genom vorliegt, oder auch in Fällen mit vollständigem Genom (Typus I), konnte nicht festgestellt werden. Im letzteren Fall wäre eine weitere Möglichkeit zur Chromosomenverdopplung gegeben. Bei *Dryopteris paleacea* und *Dryopteris remota* keimen nach eigenen Beobachtungen „eingeschnürte" Sporen, allerdings nur die größeren.

Ein neuer Typ der Aposporie wurde bei Farnen des *Asplenium aethiopicum*-Komplexes gefunden. Hier zeigen die Sporenmutterzellen bei der Diakinese vollständige Asynapsis. Nach der ersten meiotischen Teilung wird ein Restitutionskern gebildet, der sich normal teilt und dann Dyaden von Diplosporen liefert (Braithwaite 1964).

Die Paarungsverhältnisse der Chromosomen in den 16-er Sporenmutterzellen können wichtige Aufschlüsse über die Herkunft einzelner Formen geben, wie besonders Manton (1950) an verschiedenen Beispielen eingehend gezeigt hat. Es können nur Uni-, aber außerdem auch Bi- und Multivalente auftreten. Die Zahl derselben ist bei den einzelnen Arten nahezu konstant. Bei der diploiden *Dryopteris paleacea* wurde z. B. keine Paarung beobachtet, wohl aber bei der tetra- und pentaploiden, jedoch bei der tetraploiden Form nicht in dem Maße, um auf Autopolyploide schließen zu können, so daß hybrider Ursprung sehr wahrscheinlich erscheint.

Apogame Farne, die durch Kreuzung einer apogamen mit einer sexuellen Form entstanden sind, besitzen zweifellos eine oder mehrere Anlagen für Archegonbildung. Wir wissen aber nicht, ob in allen Fällen von Apogamie Anlagen für Archegonien vorhanden sind oder ob sie auch gänzlich fehlen können. Es gelang mir bisher nicht, durch stoffliche Beeinflussung apogame Prothallien von Bastarden, von denen anzunehmen ist, daß sie die Anlage für Archegonien enthalten, zur Archegonbildung zu veranlassen. Von besonderem Interesse sind in dieser Hinsicht Farne, deren Prothallien apogame Fortpflanzung haben, außerdem aber — allerdings nicht funktionierende — Archegonien aufweisen [*Lastrea pseudo-mas* = *D. paleacea* var. *polydactyla* Dadds (Farmer und Digby 1907); *Pteris cretica var. albo-lineata* (Heun 1925); *Actiniopteris australis* (Stokey 1948)]. Zum Teil könnten solche Fälle durch Kreuzung mit sexuellen Formen, zum Teil durch die Annahme einer stufenweisen Entstehung der Apogamie auf Grund multipler Allelie erklärt werden:

Sexuell ⟶ Apogam mit Archegonien ⟶ Apogam ohne Archegonien

Pseudomixis (Farmer und Digby 1907), d. h. Ersatz der normalen Befruchtung durch Verschmelzung der Kerne benachbarter vegetativer Prothallienzellen und Entstehung eines Sporophyten aus einer auf diese Weise diploid gewordenen Zelle, gibt es bei Farnen nicht (Döpp 1939, Manton 1950, S. 159). Den zweikernigen Zellen, die in den Prothallien von *Pellaea atropurpuera* auftreten, schreibt Hayes (1924) Ernährungsfunktion für den apogamen Embryo zu.

Über die physiologischen Ursachen der Restitutionskernteilung ist nichts bekannt. Da es möglich ist, durch Colchicinbehandlung von Blättern, die junge Sporangien tragen, Sporenmutterzellen und Sporen mit erhöhter Chromosomenzahl zu erhalten (Rosendahl 1940, Döpp unveröffentlicht), liegt es nahe, an stoffliche Wirkungen zu denken. Solche sollen nach Steil (1951) von den Tapetenzellen ausgehen können.

β) Ursachen und Entstehung der obligaten Apogamie.

Über die Ursachen und die Entstehung der obligaten Apogamie herrscht wenig Klarheit. Einige Autoren heben hervor, daß bei apogamen vorwiegend hohe, bei sexuellen niedrigere Chromosomenzahlen vorkommen (MANTON 1950, SCHIFFERDECKER 1957 u. a.). In Ceylon sind die meisten apogamen Arten polyploid (Tabelle 2).

Die von ALICE F. TRYON und D. M. BRITTON (1958) untersuchten tri- und tetraploiden 5 Arten oder Varietäten der Gattung *Pellaea* sind mit einer Ausnahme apogam; von 6 sexuellen Formen waren 5 diploid, nur eine tetraploid (Grundzahl 29). Von 5 untersuchten sexuellen *Pteris*-Arten erwiesen sich 4 als diploid, eine als tetraploid; unter 11 apogamen *Pteris*-Arten waren eine diploid, 4 triploid, 5 tetraploid, 1 pentaploid (SCHIFFERDECKER 1957).

Es wäre aber nicht zwingend, daraus zu schließen, daß Auto- oder Allo-Polyploidie irgendwie die Ursache der Apogamie sein müßte, ebensowenig wie Diploidie einer Eizelle die Fähigkeit zur parthenogenetischen Entwicklung bedingt. Durch Aposporie erzeugte tetraploide Prothallien sind nicht apogam, nur einige apospor entstandene Prothallien von *Aspidium marginale* trugen Sporangien (LAWTON 1932 u. a.). Vielmehr könnte umgekehrt Apogamie die Ursache dafür sein, daß eine neu auftretende polyploide Rasse oder Form sich als solche ohne Schwierigkeit fortpflanzen kann. Eine polyploid gewordene apogame Form dürfte nämlich mit größerer Wahrscheinlichkeit erhalten bleiben als eine solche mit geschlechtlicher Fortpflanzung. Denn da bei sexuellen, nicht aber bei apogamen Prothallien für die Bildung eines Sporophyten das Vorhandensein von Antheridien, unter Umständen auf einem anderen Prothallium, und im richtigen Zeitpunkt erfolgende Benetzung Voraussetzung sind, da ferner selbst bei günstigen Bedingungen nicht immer Befruchtung erfolgt und mit zunehmender Chromosomenzahl offenbar größere Schwierigkeiten für die Befruchtung auftreten, ist es leichter möglich, daß aus einem apogamen Prothallium ein Sporophyt entsteht als aus einem geschlechtlichen. Hinzu kommt, daß die Apogamen nicht aufspalten und infolgedessen als solche in der Nachkommenschaft konstant erhalten bleiben. Triploide sexuelle Formen spalten auf und ihre Nachkommen haben verschiedene Chromosomenzahlen, wie MANTON (1950) für *Osmunda regalis* gezeigt hat. Auch die Nachkommen von autotetraploider *Osmunda regalis* ($4n = 88$) haben verschiedene Chromosomenzahlen zwischen 39 und 48 (MANTON 1950). Ein durch Kreuzung einer diploiden sexuellen und einer diploiden apogamen Form erzeugter Triploider ist apogam und hat daher eine gleichförmige Nachkommenschaft von nur Triploiden (WALKER 1958). Entsprechendes gilt auch für Tetraploide, Pentaploide usw.

Tabelle 2. *Polyploidiegrad der in Ceylon aufgefundenen 14 apogamen Farnarten.* (Nach MANTON 1953 und MANTON und SLEDGE 1954.)

Diploid	*Tetraploid*
Pteris biaurita	*Hemionitis arifolia*
P. linearis	
	Pentaploid
Triploid	*Diplazium silvaticum*
Pellaea boivini	*Athyrium pinnatum*
Adiantum caudatum	
A. philippinenuse	*Hexaploid*
Humata repens	*Adiantum hispidulum*
Dryopteris paleacea	
Diplazium dilatatum	
Asplenium cheilosorum	
Athyrium maximum	

ERNST (1918) und andere Autoren nehmen Bastardierung als Ursache der Apogamie an. Dies würde bedeuten, daß 2 sexuelle Elternteile einen apogamen Bastard liefern könnten. Nun bedeutet aber die Apogamie ja nicht nur einen Verlust, nämlich der geschlechtlichen Fortpflanzung und der Fähigkeit zur Archegonbildung, sondern auch das Auftreten neuer Fähigkeiten, z. B. zur Restitutionskernbildung, was man sich schwer durch Kreuzung erklären kann. Dagegen würden auch die Fälle von Apogamie bei niedrigen, monoploiden Chromosomenzahlen sprechen [*Pteris cretica var. albo-lineata* mit 32 Chromosomen (Heun 1925),

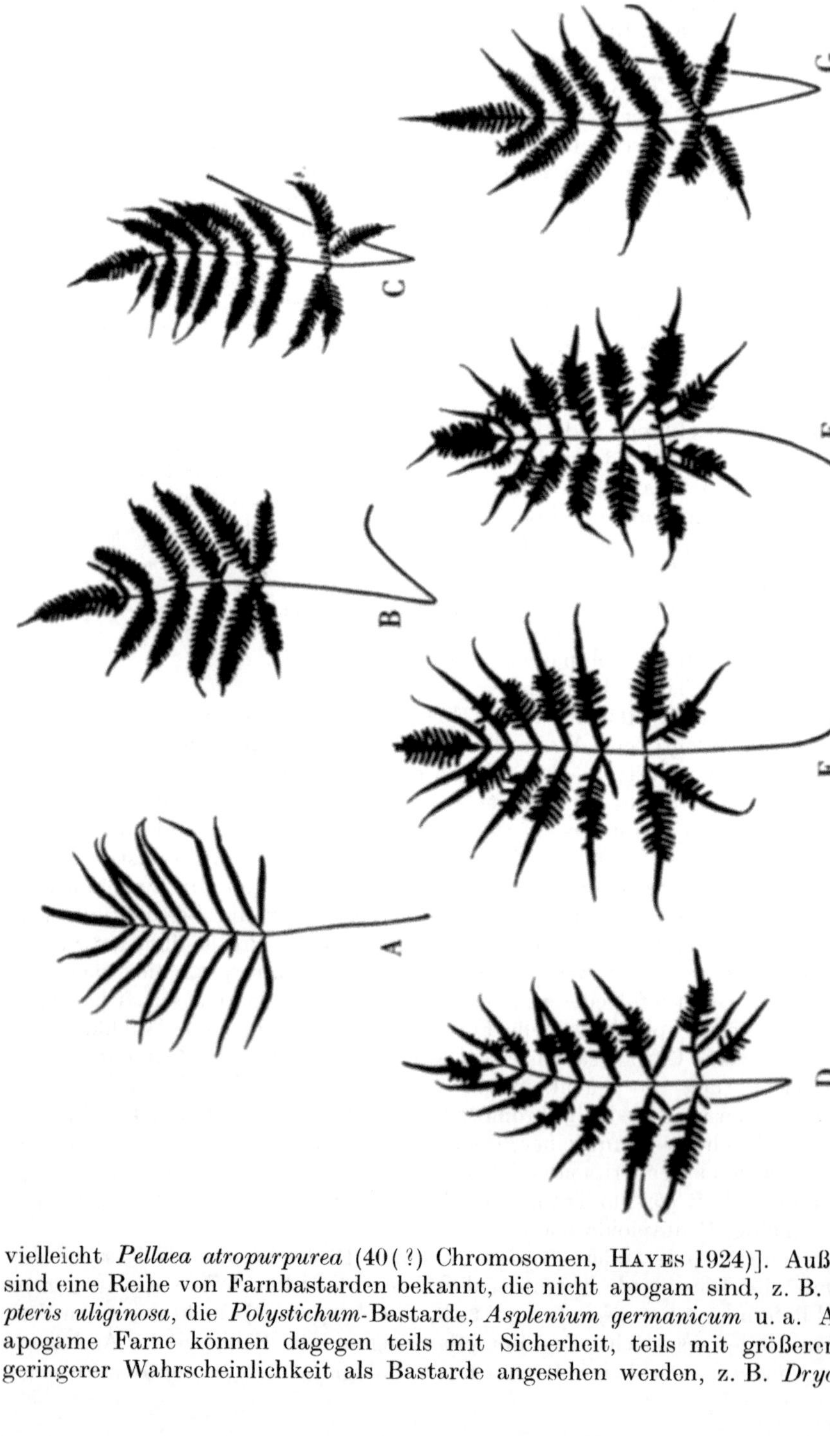

Abb. 6A—G. Obere Reihe: A *Pteris multiaurita* (sexuell), B *P. confusa* (apogam), C *P. quadriaurita* (sexuell). Untere Reihe: Experimentell hergestellte triploide apogame Bastarde: D *P. multiaurita* × *P. confusa*, E (*P. multiaurita* × *P. quadriaurita*) × *confusa*, F (*P. multiaurita* × *P. quadriaurita*) × *P. confusa*, G *P. quadriaurita* × *P. confusa*. Verkleinerung auf $^{1}/_{8}$×. (Nach Walker 1958.)

vielleicht *Pellaea atropurpurea* (40(?) Chromosomen, Hayes 1924)]. Außerdem sind eine Reihe von Farnbastarden bekannt, die nicht apogam sind, z. B. *Dryopteris uliginosa*, die *Polystichum*-Bastarde, *Asplenium germanicum* u. a. Andere apogame Farne können dagegen teils mit Sicherheit, teils mit größerer oder geringerer Wahrscheinlichkeit als Bastarde angesehen werden, z. B. *Dryopteris remota*, *Dryopteris Tavelii*, *Pteris*-Bastarde (Abb. 6). Experimentell hergestellte Bastarde zwischen sexuellen und apogamen Formen sind wiederum apogam (DÖPP 1939, 1955, WALKER 1958). Auch die diploide *Dryopteris paleacea* dürfte zwei verschiedene Genome besitzen und daher ein amphidiploider Bastard sein, da sie in den 16-er Sporenmutterzellen (Typus III) keine Bivalente aufweist. Diese Hybriden sind jedoch nicht infolge der Bastardierung apogam geworden, sondern die Apogamie stammt von einem schon apogamen Elternteil. Ein solcher apogamer Bastard bleibt konstant erhalten, ein Bastard zweier sexueller Formen dagegen spaltet entweder auf oder ist steril, falls nicht Amphidiploidie eintritt.

GOEBEL (s. WORONIN 1908, S. 160) ist der Meinung, daß Apogamie durch Mutation entstanden sei. Diese Theorie hat den Vorzug der Einfachheit und Klarheit. Ein Hinweis dafür wäre dann gegeben, wenn wenigstens in einigen Fällen zu einer apogamen Form innerhalb der gleichen Art eine morphologisch

Tabelle 3.

	Sexuelle Form	*Apogame Form*
Dryopteris hirtipes . . .	diploid 2n = 82 MANTON u. SLEDGE (1954)	triploid 3n = 123 (*D. atrata*) MANTON (1950)
Hemionitis arifolia . . .	? MEHRA u. CHOWDHURY (1954)	tetraploid MANTON u. SLEDGE (1954)
Adiantum caudatum . . .	diploid 2n = 60 (Indien) ROY u. SINHA (1957) tetraploid 4n = 120 (Malaya) MANTON u. SLEDGE (1954)	triploid 3n = 90 (Ceylon) MANTON u. SLEDGE (1954)

sehr nahestehende sexuelle Form mit der gleichen Chromosomenzahl existieren würde. Dies scheint nach unserer jetzigen Kenntnis höchstens bei *Pellaea ovata* der Fall zu sein, bei der eine diploide sexuelle und eine diploide apogame Varietät existieren (TRYON und BRITTON); im übrigen liegen nur einige Beispiele mit verschiedener Chromosomenzahl vor (Tabelle 3).

Über die Genetik der Apogamie und der Restitutionskernbildung ist im übrigen kaum etwas bekannt. Es scheint keine Aussicht zu bestehen, ein geeignetes Objekt für die genetische Untersuchung in Form eines Bastardes zu finden, bei dem etwa eine Mendelspaltung in „apogam“ und „sexuell“ möglich wäre. Durch Mutation konnten die Fälle von Apogamie bei niedrigchromosomigen Formen, die nicht als Bastarde anzusehen sind, erklärt werden. Es fehlt jedoch bisher jegliches Tatsachenmaterial, insbesondere auch experimentelle Belege, aus denen hervorginge, daß die Apogamie etwa auf ein oder einige mutierte Gene zurückzuführen sei. Die Schwierigkeiten für eine genetische Untersuchung liegen in dem Nichtspalten der Bastarde infolge Restitutionskernbildung und der fehlenden Fertilität der Typus III-Sporen.

Bei *Pteris cretica* hält auch I. MANTON (1950) eine Mutation nicht für ausgeschlossen. Die Autorin schließt auch aus dem Verhalten eines pentaploiden *Dryopteris*-Bastards aus dem Formenkreis von *D. Tavelii*, dessen Sporen außer apogamen Prothallien eine sehr geringe Anzahl von Prothallien mit Archegonien liefern, daß eine Spaltung sexuell-apogam sehr wohl möglich, dagegen eine plasmatische Vererbung ausgeschlossen sei. Im übrigen ist die Autorin aber der Auffassung, daß die Apogamie nicht bedingt sei durch einen einzelnen dominanten oder rezessiven Erbfaktor, vielmehr scheine sie das Resultat einer durch Heterozygotie bedingten Unausgeglichenheit vieler genetischer Faktoren zu sein; diese sei vermutlich bei Triploiden besonders groß, und gerade unter den apogamen Farnen komme ja auch besonders häufig Triploidie vor. Sie kommt zu dem Schluß, daß die Bastardierung nicht nur die Gelegenheit, sondern die Ursache zur Entstehung der Apogamie sei.

Von besonderer Bedeutung für die Induktion der Apogamie scheint die Zucker-Konzentration in einem sterilen Medium zu sein (Whittier 1962). So konnten Whittier und Steeves (1960, 1962) bei zahlreichen Farnen durch Zusatz von 2,5% Glucose in normalen Gametophyten Apogamie induzieren. Bei einer höheren Zuckerkonzentration (4%) fördern Naphthylessigsäure und Gibberellinsäure, wenn sie während der Induktionsphase des Gametophyten gegeben werden, die apogame Sporophytenbildung; durch Behandlung mit Kinetin wird sie hingegen reduziert oder gar unterdrückt (Whittier 1964a). Die Induktion der Apogamie erwies sich bei *Pteridium*-Gametophyten als lichtabhängig; die Entwicklung des apogamen Sporophyten ist an die Anwesenheit von Licht oder die Gegenwart von Bernsteinsäure oder Zucker gebunden (Whittier 1964b).

Einige Autoren nehmen als Ursache für die Apogamie äußere Bedingungen an, besonders Trockenheit (Woronin 1908, Stokey 1948); so sei z. B. bei Farnen trockener Felsen (*Notholaena*-Arten) und bei der xerophytischen *Actiniopteris australis* die Apogamie entstanden. Jedoch ist zu sagen, daß solche Bedingungen nur als auslesende Faktoren wirkend gedacht werden können.

γ) Bedeutung der obligaten Apogamie.

Die Bedeutung der obligaten Apogamie (sowie auch Parthenogenesis) bei Farnen ist zunächst darin zu sehen, daß bei ihnen trotz Verlustes der geschlechtlichen Fortpflanzung die Erhaltung der Art, auch nach Polyploidwerden, gesichert

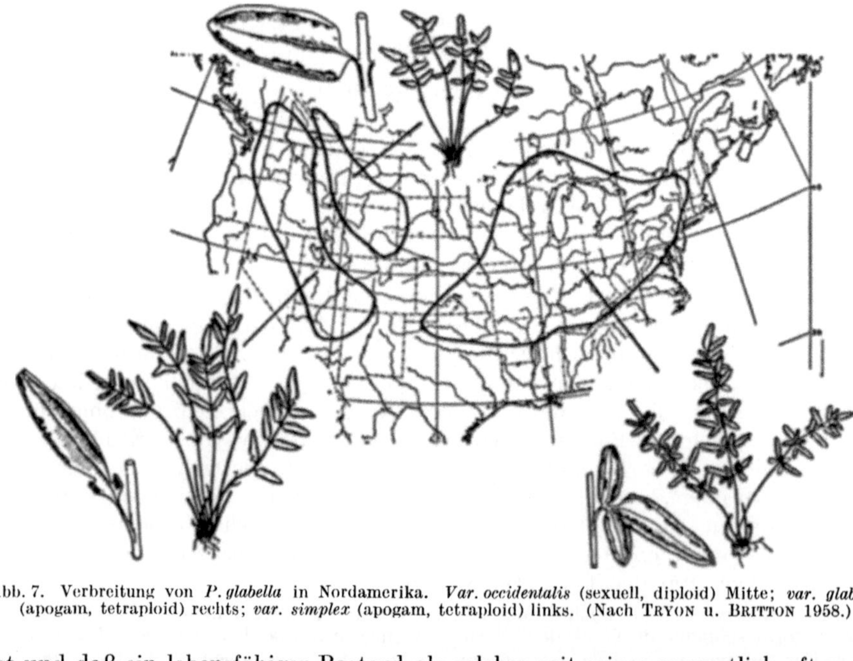

Abb. 7. Verbreitung von *P. glabella* in Nordamerika. *Var. occidentalis* (sexuell, diploid) Mitte; *var. glabella* (apogam, tetraploid) rechts; *var. simplex* (apogam, tetraploid) links. (Nach Tryon u. Britton 1958.)

ist und daß ein lebensfähiger Bastard als solcher mit seiner vermutlich oft recht günstigen Kombination von Eigenschaften erhalten bleiben kann. Ferner führt die Apogamie, zum Teil nach Kreuzung mit sexuellen Formen, zur Entstehung neuer, konstanter Arten, die sich ein besonderes, von dem der verwandten sexuellen Formen verschiedenes Verbreitungsgebiet erobern (Abb. 7, 8). NORDHAGEN (1947) nimmt Verbreitung von *Dryopteris paleacea* über die ganze Erde an; aller-

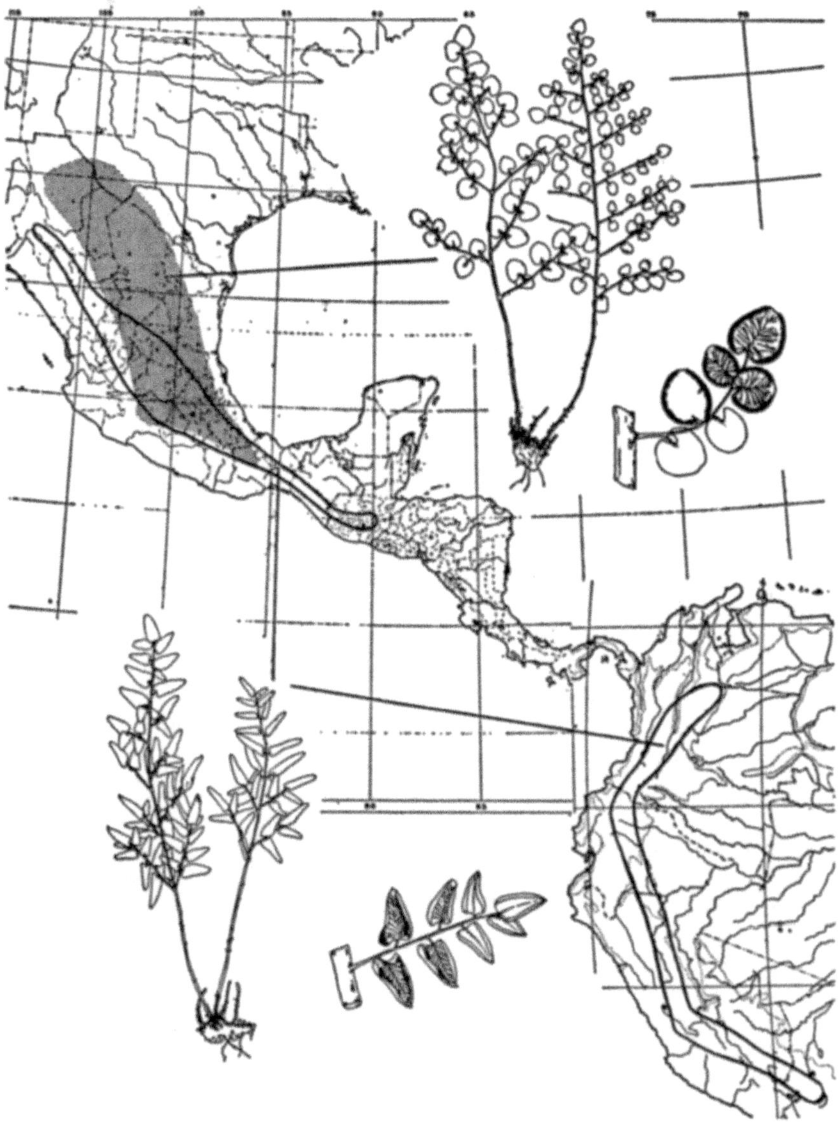

Abb. 8. Verbreitung von *Pellaea sagittata*. *Var. cordata* (sexuell, diploid) oben; *var. sagittata* (apogam, triploid) unten. Stellenweise zusammen mit *var. cordata* kommt ein apogamer Stamm von *P. ovata* vor, aus deren Kreuzung die apogame *var. sagittata* hervorgegangen sein dürfte, die nach Südamerika eingewandert ist. (Nach TRYON u. BRITTON 1958.)

dings spricht sich ALSTON (1957) für zwei verschiedene Formen aus, eine amerikanische [*D. parallelogramma* (Kunze) Alston] und eine asiatische [*D. Wallichiana* (Spreng) Hyl.].

Außerdem sieht Verfasser die Bedeutung der Apogamie in folgendem: In zahlreichen Prothallienkulturversuchen zeigte sich eine Überlegenheit apogamer

Farne hinsichtlich der Zahl der aus den Prothallien gebildeten Sporophyten und ihres früheren Auftretens gegenüber nahe verwandten sexuellen Formen (Döpp 1939, 1955). Trotz günstigster Bedingungen während eines längeren Zeitraumes lieferten aus nicht bekannten Ursachen die Prothallien verschiedener sexueller Arten keine Sporophyten. Bei sexuellen Prothallien ist die Sporophytbildung außer an Archegonien gebunden an das Vorhandensein von Antheridien; letztere werden aber nicht immer auf jedem Prothallium angelegt. Dann ist die Anwesenheit eines 2. Prothalliums wie bei den heterosporen Pteridophyten erforderlich. Die Wahrscheinlichkeit dafür ist um so geringer, je größer die Entfernung von der Mutterpflanze ist. Aus diesen Gründen dürfte ein apogamer oder parthenogenetischer Farn bei Hetero-, aber auch in manchen Fällen von Isosporie leichter die Möglichkeit einer Ausbreitung auf weite Entfernungen hin haben. Die zur Befruchtung notwendige Benetzung dürfte in der Natur nicht immer gegeben sein. Somit könnten unter Umständen apogame Prothallien an trockene Standorte besser angepaßt sein als sexuelle.

Ein weiterer Vorteil bei Apogamie liegt darin, daß bei Bastarden die Heterosis in der Nachkommenschaft erhalten bleibt. Hier sei besonders auf *Dryopteris Tavelii* hingewiesen, die in Süd- und Südwestdeutschland und in der Schweiz oft in riesigen Exemplaren vorkommt.

Apogame Farne erscheinen infolge ihrer charakteristischen Eigentümlichkeiten besonders geeignet zur Varietätenbildung. Ist nämlich einmal durch Mutation eine Varietät entstanden, so dürfte sie auch in der Nachkommenschaft als solche erhalten und konstant bleiben, da 1. eine Aufspaltung nicht erfolgt, 2. eine Kreuzung mit einem ebenfalls apogamen, unverändert gebliebenen Individuum der gleichen Art unmöglich ist. Eine nähere Prüfung dieser Frage erscheint jedoch erforderlich.

b) Apogamie verbunden mit Aposporie. Phasenwechsel und Reduktionsteilung nicht vorhanden.

Hier verläuft die Entwicklung wesentlich einfacher. Gewisse Farnarten und -varietäten haben die Fähigkeit zur Sporenbildung verloren, da sie keine oder nur sterile Sporangien ausbilden. Prothallien können bei ihnen nur apospor entstehen, d. h. vegetative Zellen vom Blattrand (Abb. 9), von der Blattfläche oder eine Zelle der Sporangienwand wachsen ohne Kernphasenwechsel zu Prothallien aus. Beispiele sind in England aufgefundene Varietäten [*Lastrea pseudomas* (= *Dryopteris paleacea*) *var. cristata apospora* (ohne Archegonien), *Athyrium filix-femina var. clarissima* Jones (mit Archegonien) (Farmer und Digby 1907). *Osmunda javanica* (Sarbadhikari 1939), *Tectaria trifoliata* (Steil 1951) und *Trichomanes Kaulfussii var. aposporum* (Georgevitsch 1910)]. Bei *Pellaea* beschreibt Woronin einen Fall, in dem aus einem sehr kleinen Prothallium eine verkümmerte Blattanlage entstand, die ihrerseits wieder in ein Prothallium mit Spaltöffnungen und Haaren überging. Antheridien werden in allen Fällen ausgebildet. Falls Archegonien vorhanden sind, üben sie einen chemotaktischen Reiz auf die Spermatozoide aus.

Während die Aposporie häufig fakultativ ist und nur unter Versuchsbedingungen sowie meist nur an Primärblättern auftritt, falls diese abgetrennt und auf feuchtem Substrat bei hoher Luftfeuchtigkeit und schwacher Beleuchtung kultiviert werden, ist sie in den genannten Fällen mehr oder weniger obligat und tritt auch an großen Wedeln auf, oft selbst dann, wenn diese noch aufgerichtet in Verbindung mit der Mutterpflanze sind. Jedenfalls ist ohne Aposporie ein abgeschlossener Entwicklungsgang nicht möglich (Abb. 10).

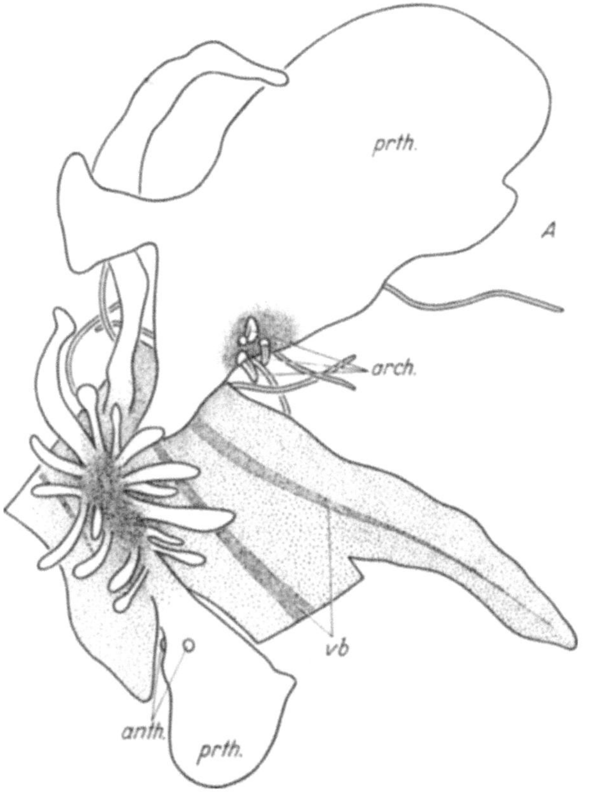

Abb. 9. Aposporie bei *Athyrium Filix-femina var. clarissima* Jones. Teil eines Fiederchens (punktiert) mit Leitbündeln (*vb*) und einem Sorus, aus dem anstelle von Sporangien Prothallien (*prth.*) (nicht punktiert) herauskommen. *anth.* Antheridien; *arch.* Archegonien. (Aus BOWER 1923)

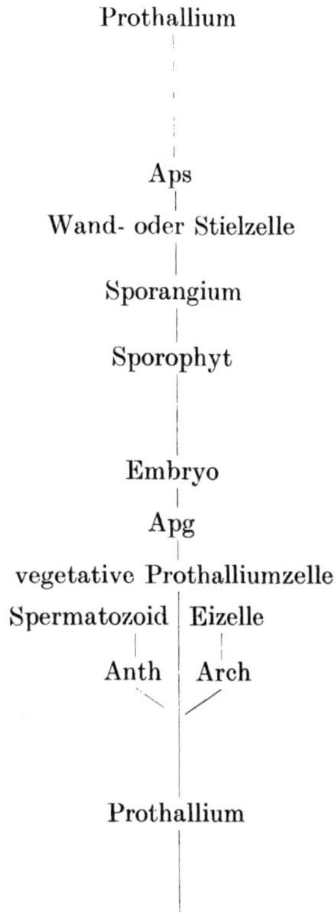

Abb. 10. Apogamie (*Apg*) verbunden mit Aposporie (*Aps*).

3. Fakultative (induzierte) Apogamie.

Werden aus Sporen erhaltene Farnprothallien, insbesondere von Gartenvarietäten, längere Zeit in hellem Licht und bei höherer Temperatur kultiviert und dabei zur Verhinderung der Befruchtung nicht von oben benetzt, sondern nur von unten bewässert, so wachsen sie oft zu beträchtlicher Größe heran und man beobachtet vielfach, jedoch nicht bei allen Arten und nicht an allen Prothallien, Bildungen, die gewöhnlich nur für den Sporophyten charakteristisch sind, z. B. Tracheiden, mehrzellige Haare, Schuppen, Epidermiszellen, Spaltöffnungen oder auch Mittelbildungen zwischen Prothallium und Pflänzchen. Auch Behandlung einzelliger Keimlinge mit Röntgenstrahlen wirkte sich in einem Einzelfall in gleichem Sinne aus (ROTTMANN 1938). Dies braucht noch kein Anzeichen einer Apogamie zu sein. Auch in Kallus-Gewebekulturen von Prothallien von *Osmunda cinnamomea* waren Tracheiden vorhanden; Apogamie wurde jedoch hierbei nicht beobachtet (MOREL und WETMORE 1951). Gelegentlich treten aber auch apogame Sprossungen auf, die sich in manchen Fällen zu einer Pflanze entwickeln (LANG 1898, 1929, MOTTIER 1915, 1927, DUNCAN 1941, MANTON 1950, MANTON und WALKER 1954, MEYER 1952). Die Apogamie ist in diesen Fällen nicht obligat, sie tritt nur unter besonderen, allerdings noch nicht genügend analysierten Kulturbedingungen auf; man sagt, sie ist durch äußere Faktoren und durch längere Kulturdauer induziert. Man spricht auch von fakultativer Apogamie.

Ob nur die äußeren Faktoren oder Dauerkultur, also vielleicht Alterungsvorgänge, möglicherweise auch Einflüsse von seiten der Archegonien mit ihren nicht befruchteten Eizellen die Apogamie hervorrufen können oder ob nicht doch eine gewisse erbliche Veranlagung vorliegt, wobei man an Selektionswirkungen durch die Umweltfaktoren denken könnte, mag dahingestellt bleiben (vgl. auch MEYER 1952, S. 16; MANTON und WALKER 1954, S. 378).

Abb. 11. *Dryopteris Filix-mas*, links Blatt von einer sexuell entstandenen tetraploiden Pflanze, rechts Blatt von einer apogam entstandenen diploiden Pflanze. (Nach MANTON u. WALKER 1954.)

Wie bei obligater Apogamie können auch Wurzeln ohne Blatt- und Sproßanlage entstehen. Ein weiterer extremer Fall von induzierter Apogamie in Verbindung mit Reduktion des Sporophyten und Frühreife liegt vor bei einer Varietät von *Scolopendrium vulgare* und bei *Dryopteris dilatata var. crist. gracilis* (LANG 1898, 1929). Hier entstehen Sporangien auf Prothallien. Die Chromosomenzahl dieser Farne wird nicht angegeben. Unter anderen Bedingungen traten Sporangien auf einigen apospor erzeugten Prothallien von *Aspidium marginale* auf (LAWTON). Einige dieser Prothallien bildeten auch Epidermiszellen, Schließzellen und Tracheiden außer normalen Prothallienzellen. Im Gegensatz zu LANGS Versuchen waren die aposporen Prothallien nicht alt, und die Befruchtung war nicht verhindert worden. Über die Bedingungen der normalen Sporangienbildung wissen wir kaum etwas, erst recht nichts über die Ursachen der Entstehung von Sporangien auf Prothallien.

Natürliche fakultative Apogamie scheint vorzuliegen bei *Cystopteris fragilis f. polyapogama* (HEILBRONN 1910).

Vielfach gehen die apogamen Pflänzchen nicht direkt aus gewöhnlichen Prothallienzellen hervor, sondern die „Umstimmung" erfolgt allmählich: Zunächst bildet sich z. B. ein zylindrischer Fortsatz aus der Scheitelbucht des Prothalliums, oder es entstehen eine oder mehrere Wucherungen oder Höcker aus vegetativen Prothallienzellen oder durch Umbildung einer Archegonanlage.

I. MANTON (1950) untersuchte 2 Fälle von induzierter Apogamie cytogenetisch. Bei *Doodia caudata*, bei der DUNCAN (1941) fakultative Apogamie erhielt, waren die apogam entstandenen jungen Pflanzen zunächst wesentlich kleiner als nach Befruchtung entstandene; jedoch glich sich später der Unterschied weitgehend aus. Die Sporen der apogamen Pflanzen keimten nicht, obwohl sie äußerlich gut ausgebildet waren. In den Sporenmutterzellen fanden sich Uni- und zahlreiche Bivalente. Daraus kann geschlossen werden, daß die untersuchte *Doodia caudata* tetraploid und zwar, da die Paarung nicht vollkommen ist, allotetraploid ist; der apogame Sporophyt dagegen besitzt die diploide Zahl. Demgegenüber erwies sich *Scolopendrium vulgare* (n = 36) als diploide Art ohne Hybridencharakter. Sexuell erzeugte Pflanzen hatten 72, apogam gebildete nur 36 Chromosomen. Die 36 Chromosomen des apogamen Sporophyten blieben in der Diakinese ungepaart, so daß sie mit großer Wahrscheinlichkeit als monoploider Satz und der apogame Sporophyt als haploid aufzufassen sind. Hier handelt es sich also um einen Fall von haploider Apogamie.

Bei *Doodia caudata* konnten am gleichen Prothallium diploide und tetraploide Sporophyten festgestellt werden (MANTON 1950). Fakultative Apogamie erhielt auch MEYER (1952) durch Trockenkultur bei *Asplenium septentrionale*, wo am gleichen Prothallium ebenfalls apogame und sexuelle Sporophyten auftraten, die beide von derselben Spore stammten und wo das gleiche Genom in dem einen einfach, im anderen nach Einzelkultur, d. h. infolge von Selbstbefruchtung doppelt vorhanden war. Gegenüber den sexuell entstandenen Pflänzchen waren die apogamen sehr klein und von gelblichgrüner Farbe.

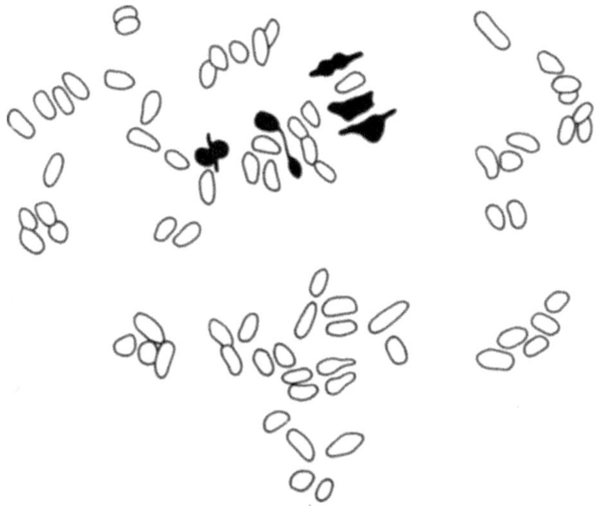

Abb. 12. Meiotische Metaphase von der apogam entstandenen diploiden Pflanze von *Dryopteris Filix-mas* (Abb. 11, rechts). Gepaarte Chromosomen schwarz, Univalente nur im Umriß gezeichnet. (Nach MANTON u. WALKER 1954.)

Besonders schöne Beispiele geben MANTON und WALKER (1954): Apogam entstandene, diploide Pflanzen (2n = 82) von *Dryopteris dilatata* und *D. Filix-mas* (Abb. 11) zeigten in der Meiosis keine bzw. nur geringe Paarung (Abb. 12). Daraus kann mit Sicherheit geschlossen werden, daß die normalen, tetraploiden Pflanzen beider Arten amphidiploid sind, obwohl über die Eltern nichts ausgesagt werden kann.

Die Erscheinungen der Apogamie und Aposporie zeigen, daß die Ausbildung von gametophytischem und sporophytischem Gewebe nicht an die Chromosomenzahl gebunden ist; auch auf dem gleichen Organ können beiderlei Zellen nebeneinander vorkommen. Dies spricht dafür, daß die Ausbildung von Prothallien- oder Sporophytgewebe ein Differenzierungsvorgang ist, entsprechend der Bildung von Parenchym-, Epidermiszellen, Tracheiden, Siebröhren usw. in einer Sproßachse.

4. Parthenogenesis.

Ein klassischer Fall von Parthenogenesis bei Pteridophyten liegt vor bei *Marsilia Drummondi* (STRASBURGER 1907). Das Archegonium bleibt geschlossen (Abb. 13). Allem Anschein nach findet in den Sporangien eine Restitutionskernbildung mit anschließender Reduktionsteilung statt (STRASBURGER, GUSTAFSSON 1935). Mikroprothallien- und Antheridienentwicklung konnte STRASBURGER nicht feststellen. Er glaubt, daß auch Makro- und Mikrosporen mit der reduzierten Chromosomenzahl gebildet werden. Erhöhung der Temperatur des umgebenden Wassers erhöht die Zahl der sich apogam entwickelnden Eizellen. Während STRASSBURGER für die heterotypische Metaphase von *Marsilia Drummondi* und für das Makroprothallium eine Dipoidzahl von 32 Chromosomen angibt, erhielt GUSTAFSSON bei *Marsilia Drummondi* und der Varietät *tomentosa* ungefähr 210 Chromosomen. MAHABALE und D'MELLO (1952) finden in den Mikrosporangien der parthenogenetischen *Salvinia auriculata* auf dem Stadium des 8-zelligen sporogenen Gewebes Restitutionskernbildung und später 32 Mikrosporen. — Wahrscheinlich sind noch weitere *Marsilia*- und *Salvinia*-Art parthenogenenetisch.

FARMER und DIGBY (1907) stellten diploide Parthenogenesis bei *Athyrium Filix-femina var. clarissima* BOLTON (Abb. 14) sowie bei *Scolopendrium vulgare var. crispum Drummondae* fest. Die Sporangien sind bei der *Athyrium*-Varietät steril, bei der Varietät von *Scolopendrium* fehlen sie. Jedoch werden Prothallien apospor gebildet, bei ersterer

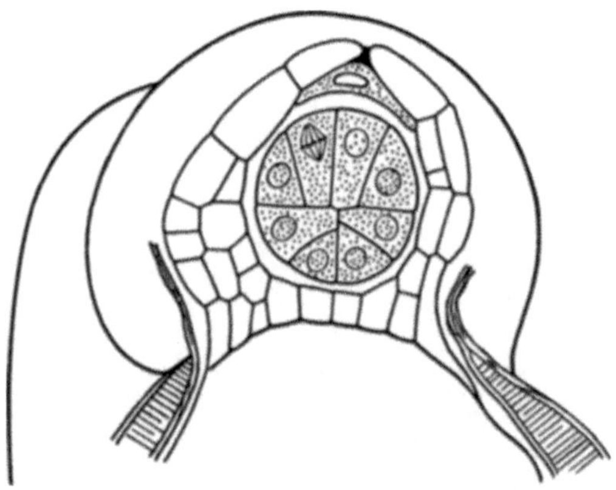

Abb. 13. Apogamie bei *Marsilia Drummondi* (nach STRASBURGER). Vergr. ca. 200×. Archegon geschlossen, Bauchkanalzelle erhalten.

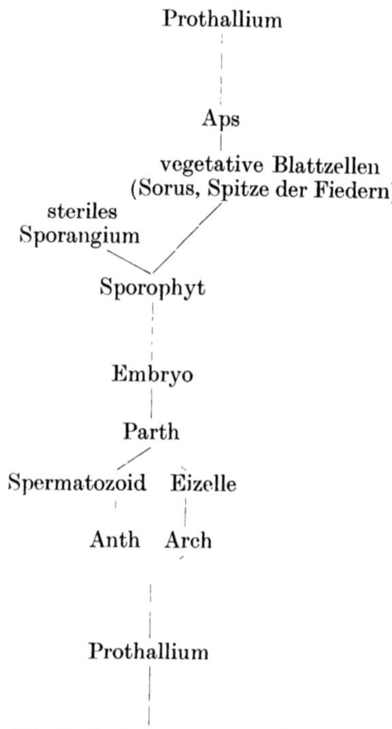

Abb. 14. Parthenogenesis (*Parth*) verbunden mit Aposporie (*Aps*).

Art aus den Spitzen der Fiedern und aus der Sporangienwand, bei letzterer hauptsächlich aus Zellen des Blattrandes. Archegonien und Antheridien sind normal. Bei *Athyrium Filix-femina var. clarissima* BOLTON bewegen sich die Spermatozoiden im Halskanal bis zur Eizelle vor, ohne aber in die Eizelle einzudringen. Es ist unentschieden, ob sie trotzdem auf die Eizelle einen Anreiz zur Weiterentwicklung ausüben. Bevor sich bei *Scolopendrium vulgare var. crisp. Drummondae* der Archegonhals öffnet, umgibt sich die Eizelle mit einer Membran, so daß die Möglichkeit einer Befruchtung ausgeschlossen ist. Die Aposporie parthenogenetischer oder apogamer Formen ist auf den Britischen Inseln offenbar unter dem Einfluß des ozeanischen Klimas entstanden.

Selaginella rubricaulis hat nach BRUCHMANN (1912) auf dem gleichen Prothallium zweierlei Archegonien (Abb. 15). Die einen haben vorgewölbte Halszellen und öffnen sich, Hals- und Bauchkanalzellen verschleimen; die Eizellen gehen jedoch nach dem Öffnen das Archegons zugrunde. Die anderen öffnen sich nicht; Halszellen wölben sich nicht vor, während Hals- und Bauchkanalzellen nicht verschleimen; die Eizelle entwickelt sich ohne Befruchtung.

Mikrosporen und Mikroprothallien werden gebildet; jedoch haben die Spermatozoiden keine Funktion. Auffallend ist die Zahl von Makrosporen. Auch bei

Selaginella selaginoides (= *spinulosa*, BRUCHMANN 1912) und bei *Selaginella anocardia* (GOEBEL 1915) liegt Parthenogenesis vor, ferner nach R. M. TRYON (1954) gebietsweise auch bei *Selaginella rupestris*. Schon HIERONYMUS (1902) hatte bei dieser Art Parthenogenesis vermutet, da ihm bei allen von ihm untersuchten Pflanzen das völlige Fehlen der Mikrosporen aufgefallen war. Mitteilungen über die cytologischen Verhältnisse in den Sporangien liegen nicht vor. Die beschriebenen Fälle von Parthenogenesis bei Pteridophyten sind obligat; experimentell hervorgerufene Parthenogenesis ist nicht bekannt.

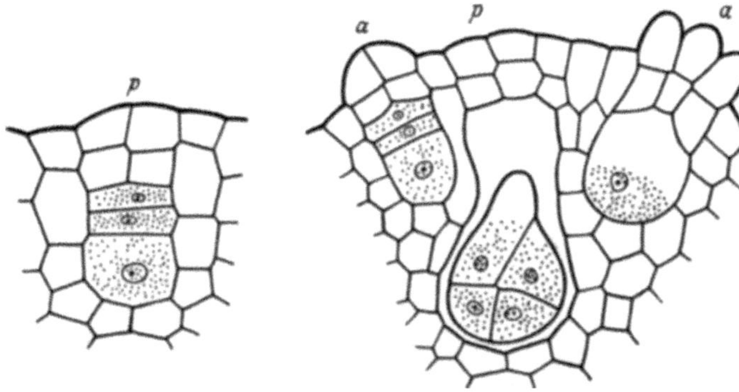

Abb. 15. *Selaginella rubricaulis* (nach BRUCHMANN 1912). *a* Archegonien mit ungehemmter Entwicklung, *p* Archegonien mit geschlossenem, in der Entwicklung gehemmten Hals und parthenogenetischem Embryo. Vergr. ca. 600×.

5. Apomixis bei *Isoetes*.

Bemerkenswert ist, daß ABRAHAM und NINAN (1958) und NINAN (1958) für *Isoetes coromandelina* Apogamie angeben. In den Makrosporenmutterzellen erfolgt keine Reduktionsteilung, sondern Äquationsteilung, da die Paarung der Chromosomen unterbleibt. Es entstehen also Makrosporen mit der unreduzierten Chromosomenzahl. Die Mikrosporen sind nicht lebensfähig infolge von Unregelmäßigkeiten bei der Verteilung der Chromosomen in den Mikrosporenmutterzellen. Welcher Art die Apogamie ist, wird nicht angegeben. Es könnte auch ein anderer apomiktischer Vorgang in Frage kommen.

Literatur.

ABRAHAM, A., and C. A. NINAN: Cytology of *Isoetes*. Curr. Sci. **27**, 60—61 (1958). — ALSTON, A.H.G.: The American fern usually known as *Dryopteris paleacea*. Amer. Fern J. **47**, 91–92 (1957).

BOWER, F. O.: The Ferns. Bd. 1. 1923. — BRAITHWAITE, A. F.: A new type of apogamy in ferns. New Phytologist **63**, 293—305 (1964). — BRUCHMANN, H.: Zur Embryologie der Selaginellaceen. Flora (Jena) **104**, 180—224 (1912).

COPELAND, E. B.: Genera Filicum. Waltham 1947.

DÖPP, W.: Die Apogamie bei *Aspidium remotum* Al. Br. Planta (Berl.) **17**, 86—152 (1932). — Gestaltung und Organbildung innerhalb der Gametophytgeneration der Polydiaceen unter besonderer Berücksichtigung genetischer Gesichtspunkte. Beitr. Biol. Pflanzen **24**, 201—238 (1937). — Cytologische und genetische Untersuchungen innerhalb der Gattung *Dryopteris*. Planta (Berl.) **29**, 481—533 (1939). — Experimentell erzeugte Bastarde zwischen *Dryopteris Filix-mas* (L.) Schott und *D. paleacea* (Sw.) C. Chr. Planta (Berl.) **46**, 70—91 (1955). — Über eine hemmende und eine fördernde Substanz bei der Antheridienbildung in den Prothallien von *Pteridium aquilinum*. Ber. dtsch. bot. Ges. **72**, 11—24 (1959). — DUNCAN, R. E.: Apogamy in *Doodia caudata*. Amer. J. Bot. **28**, 921—931 (1941).

ERNST, A.: Bastardierung als Ursache der Apogamie im Pflanzenreich. Jena 1918.

FARMER, J. B., and L. DIGBY: Studies in apospory and apogamy in Ferns. Ann. of Bot. **21**, 161—199 (1907).

GEORGEVITSCH, P.: Aposporie und Apogamie bei *Trichomanes Kaulfussii* Hk. et Grew. Jb. wiss. Bot. **48**, 155—170 (1910). — GOEBEL, K.: *Selaginella anocardia*, eine neue apogame Art. Flora (Jena) **108**, 324—326 (1915). — Organographie der Pflanzen, 3. Aufl., Teil 2, S. 1133—1138. Jena 1930. — GUSTAFSSON, AKE: Studies on the mechanism of parthenogenesis. Hereditas (Lund) **21**, 1—118 (1935/36).

Hartmann, M.: Autogamie bei Protisten und ihre Bedeutung für das Befruchtungsproblem. Jena 1909. — Hayes, Doris, M.: Some studies of apogamy in *Pellaea atropurpurea*. Trans. Amer. micr. Soc. **43**, 119—135 (1924). — Heilbronn, A.: Apogamie, Bastardierung und Erblichkeitsverhältnisse bei einigen Formen. Flora (Jena) **101**, 1—42 (1910). — Heun, Alphonse, L.: The cytology of apogamy in *Pteris cretica* Linn. var. albo-lineata Hort. Bull. Torrey bot. Club **66**, 549—562 (1939). — Hieronymus, G., u. R. Sadebeck: Selaginellaceen. Natürliche Pflanzenfamilien, Bd. 4, Teil I, S. 621—715. 1900.

Lang, W. H.: Apogamy and the development of sporangia on prothalli. Phil. Trans. B **190**, 187—238 (1898). — On a variety of *Scolopendrium* that bears sporangia on the prothallus. Ann. of Bot. **43**, 355—374 (1929). — Lawton, Elva: Regeneration and induced polyploidy in Ferns. Amer. J. Bot. **19**, 303—333 (1932). — Leitgeb, H.: Die Sproßbildung an apogamen Farnprothallien. Ber. dtsch. bot. Ges. **3**, 169—176 (1885).

Mahabale, T. S., and Joyce D'Mello: Apogamy in *Salvinia*. Curr. Sci. **21**, 227—228 (1952). — Manton, Irene: Problems of cytology and evolution in Pteridophyta. Cambridge 1950. — The cytological evolution of the Fern flora of Ceylon. Symp. Soc. exp. Biol. **7**, 174—185 (1953). — Manton, I., and W. A. Slegde: Observations on the cytology and taxonomy of the Pteridophyte Flora of Ceylon. Phil. Trans. B **238**, 127—185 (1954). — Manton, I., and S. Walker: Induced apogamy in *Dryopteris dilatata* (Hoffm.) A. Gray and *Dryopteris Filix-mas* (L.) Schott *emend.* and its significance for the interpretation of the two species. Ann. of Bot., N. S. **18**, 377—383 (1954). — Mehra, P.: Cytological investigation of apogamy in *Adiantum lunulatum*. Proc. nat. Acad. Sci. India **14**, 189—204 (1949). — Mehra, P. N., and S. C. Verma: The cytology of some *Athyrium* species from Northern India. Ann. of Bot. **21**, 455—459 (1957). — Meyer, D.: Untersuchungen über Bastardierung in der Gattung *Asplenium*. Bibl. bot. **123**, 1—134 (1952). — Morel, G., and R. H. Wetmore: Fern callus tissue culture. Amer. J. Bot. **38**, 141—143 (1951). — Mottier, D. M.: Behaviour of certain Fern prothallia under prolonged cultivation. Bot. Gaz. **83**, 244—266 (1927).

Ninan, C. A.: Studies on the cytology and phylogeny of the Pteridophytes. V. Observations on the Isoetaceae. J. Indian bot. Soc. **37**, 93—103 (1958).

Patterson, Mary Thomasine: Cytology of apogamy in *Polystichum tsussemense*. Bot. Gaz. **104**, 107—114 (1942).

Rosenberg, O.: Apogamie und Parthenogenesis bei Pflanzen. In: Handbuch der Vererbungswissenschaft, Bd. II L. Berlin 1930. — Rosendahl, Gertraud: Versuche zur Erzeugung von Polyploidie bei Farnen durch Colchicinbehandlung sowie Beobachtungen an polyploiden Farnprothallien. Planta (Berl.) **31**, 597—637 (1940/41). — Rottmann, W.: Versuche zur Gewinnung abweichender Formen mit Farnsporen und Gametophyten. Diss. Marburg 1938. — Roy, R. P., and B. M. B. Sinha: Chromosomal types in *Adiantum caudatum* L. Curr. Sci. **26**, 322—323 (1957).

Sadebeck, R.: Pteridophyten. Natürliche Pflanzenfamilien, Bd. I, Teil 4, S. 34—39. Leipzig 1902. — Sarbadhikari, P. C.: Cytology of apogamy and apospory in *Osmunda javanica* Bl. Ann. of Bot. **53**, 137—145 (1939). — Schifferdecker, Ilse: Untersuchungen über die Gattung *Pteris*, im besonderen die Zytologie ihrer apogamen Formen. Z. indukt. Abstamm.- u. Vererb.-Lehre 88, 163—183 (1957). — Steil, W. N.: A study of apogamy in *Nephrodium hirtipes*. Ann. of Bot. **33**, 109—132 (1919). — New cases of apogamy in certain homosporous leptosporangiate Ferns. Bot. Gaz. **95**, 164—167 (1933). — Apogamy, apospory and parthenogenesis in the Pteridophytes. I. Bot. Rev. **5**, 433—453 (1939). II. Bot. Rev. **17**, 90—104 (1951). — Stokey, A.: Gametophytes of *Marattia sambucina* and *Macroglossum Smithii*. Bot. Gaz. **103**, 559—569 (1942). — Reproductive structures of the gametophytes of *Trichomanes* and *Hymenophyllum*. Bot. Gaz. **109**, 363—380 (1948). — The gametophyte of *Actiniopteris australis* (L. fil.) Link. J. Indian bot. Soc. **27**, 40—49 (1948). — Strasburger, E.: Apogamie bei *Marsilia*. Flora (Jena) **97**, 123—191 (1907).

Tryon, Alice F., and D. M. Britton: Cytotaxonomic studies on the Fern genus *Pellaea*. Evolution **12**, 137—145 (1958). — Tryon, Rolla M.: Evolution in the *Selaginella rupestris* group (section *Stachygyanndrum*). 8. Congr. Int. Bot., Rapp. et Comm. av. le Congr., Sect. 4, 20, Paris 1954.

Walker, T. G.: Hybridization in some species of *Pteris* L. Evolution **12**, 82—92 (1958). — Whittier, D. P.: The origin and development of apogamous structures in the gametophyte of *Pteridium* in sterile culture. Phytomorphologie **12**, 10—20 (1962). — The effect of growth factors on apogamy in *Pteridium* gametophytes. Amer. J. Bot. **51**, 663—664 (1964a). — The influence of cultural conditions on the induction of apogamy in *Pteridium* gametophytes. Amer. J. Bot. **51**, 730—736 (1964b). — Whittier, D. P., and T. A. Steeves: The induction of apogamy in the bracken fern. Canad. J. Bot. **38**, 925—930 (1960). — Further studies induced apogamy in ferns. Canad. J. Bot. **40**, 1525—1531 (1962). — Winkler, H.: Parthenogenesis und Apogamie im Pflanzenreich. Jena 1908. — Fortpflanzung der Gewächse. 7. Apomixis. In Handwörterbuch der Naturwiss., 2. Aufl. — Woronin, H.: Apogamie und Aposporie bei einigen Farnen. Flora (Jena) **98**, 101—162 (1908).

Apomixis in the angiosperms.

By

Axel Nygren.

With 43 figures.

Terminology.

In 1908 WINKLER gave the definition of *apomixis* as „den Ersatz der geschlechtlichen Fortpflanzung durch einen anderen, ungeschlechtlichen, nicht mit Kern- und Zellverschmelzung verbundenen Vermehrungsprozeß". Apomixis is a derived stage occuring in species with sexual relatives in the same or related genera. There are two main groups of apomictic propagation in the Angiosperms, *agamospermy* (TÄCKHOLM 1922), *i.e.* the formation of seeds without a sexual process, and vegetative propagation by means of propagules such as bulbils or by runners, stolons and so on.

We distinguish between three different types of agamospermy, 1. *diplospory*, 2. *apospory*, and 3. *adventitious embryony*. We speak of *diplospory* (EDMAN 1931), when the EMC develops into an unreduced ES and the embryo is formed either by the egg cell, = *parthenogenesis* (SIEBOLD 1856), or by any other cell of the gameteophyte, = *apogamety* (RENNER 1916). *Apospory* (DRUERY 1886, and BOWER 1886, 1889) is at hand, when some somatic cell in the nucellus forms the diploid ES, the egg cell of which parthenogenetically develops into an embryo. By chance other cells of the ES in addition to the egg cell may give rise to embryos. In aposporous species the EMC either degenerates at an early stage or forms an ES, which is normally superseded by one or more aposporous ES's. *Adventitious embryony* finally means that the embryos are formed directly from somatic cells in the nucellus or the integuments, while the ES develops in a normal, sexual way. Adventitious embryony normally leads to the development of more than one embryo.

In some rare cases the egg cell of a reduced embryo sac divides directly without fertilization and forms a haploid embryo, this process is called *haploid parthenogenesis*. In a number of diplosporous and aposporous species the central nucleus normally needs fertilization in spite of the fact that the egg cell develops parthenogenetically, = *pseudogamy* (FOCKE 1881).

Vegetative propagation includes a reproduction by bulbils, bulbs, runners, suckers and so on. In the Angiosperms all such organs are produced by the sporophyte only. Bulbils in many plants, particularly belonging to the grass and the lily families, are formed in the floral region, a phenomenon which already by LINNAEUS (1737, 1759) was named *vivipary*. In viviparous plats the bulbils normally germinate before they are spred. The term vivipary earlier has been used even for the situation occurring in some mangrove plants, the seeds of which germinate on the mother plant. For such cases the term *biotechnosis* was coined by MATTFELD in 1920, while SERNANDER (1927) speaks of *fructificative* and *vegetative vivipary*. In this survey vivipary *s.str.*, *i.e.* vegetative vivipary, will be taken up to discussion only.

I. Agamospermy.

A. Diplospory.

a) Embroy sac formation.

There is a rich flora of terms covering what is here called diplospory; the interested reader has to consult the papers by STEBBINS (1941) and BATTAGLIA (1951, 1955, 1963) as well as the work by GUSTAFSSON (1946, 1947). The author follows the terminology given by GUSTAFSSON.

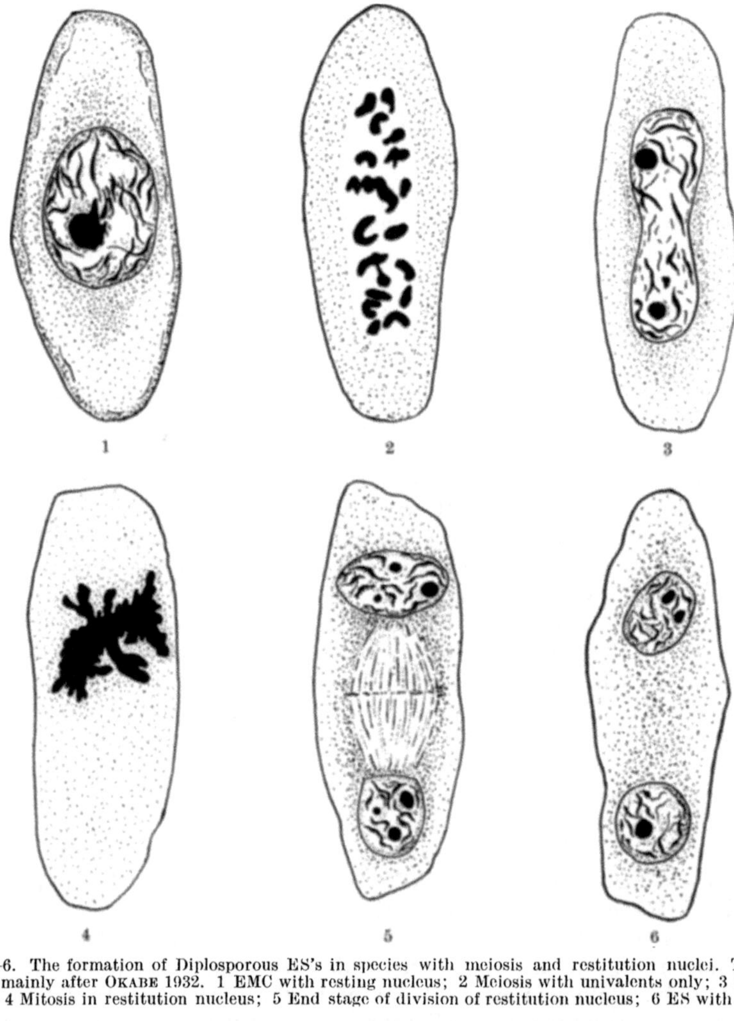

Figs. 1—6. The formation of Diplosporous ES's in species with meiosis and restitution nuclei. Type *Ixeris dentata*, mainly after OKABE 1932. 1 EMC with resting nucleus; 2 Meiosis with univalents only; 3 Restitution nucleus; 4 Mitosis in restitution nucleus; 5 End stage of division of restitution nucleus; 6 ES with two nuclei.

It is typical to diplospory that the EMC gives rise to the ES. This process is more easily studied in species with unicellular archespore than in species with multicellular archespore, and therefore the types given here are taken from species with unicellular archespore. We may divide the diplosporous species into two groups as to the type of division at hand. Either some type of restitution nucleus occurs, the derivatives of which form the ES, *group I*, or the EMC directly gives rise to the ES, *group II*. The first of these groups can be split in two subgroups, the first of which, *I:1*, is characterized by the EMC directly turning into an ES without cell formation after the first division. In the second type, *I:2*, the EMC divides into two cells one of which degenerates, while the other one forms the ES.

Ixeris dentata (Figs. 1—6), studied by OKABE (1932) is a typical representative of the type I:1. The EMC divides by a division only holding univalents, *i.e.* a

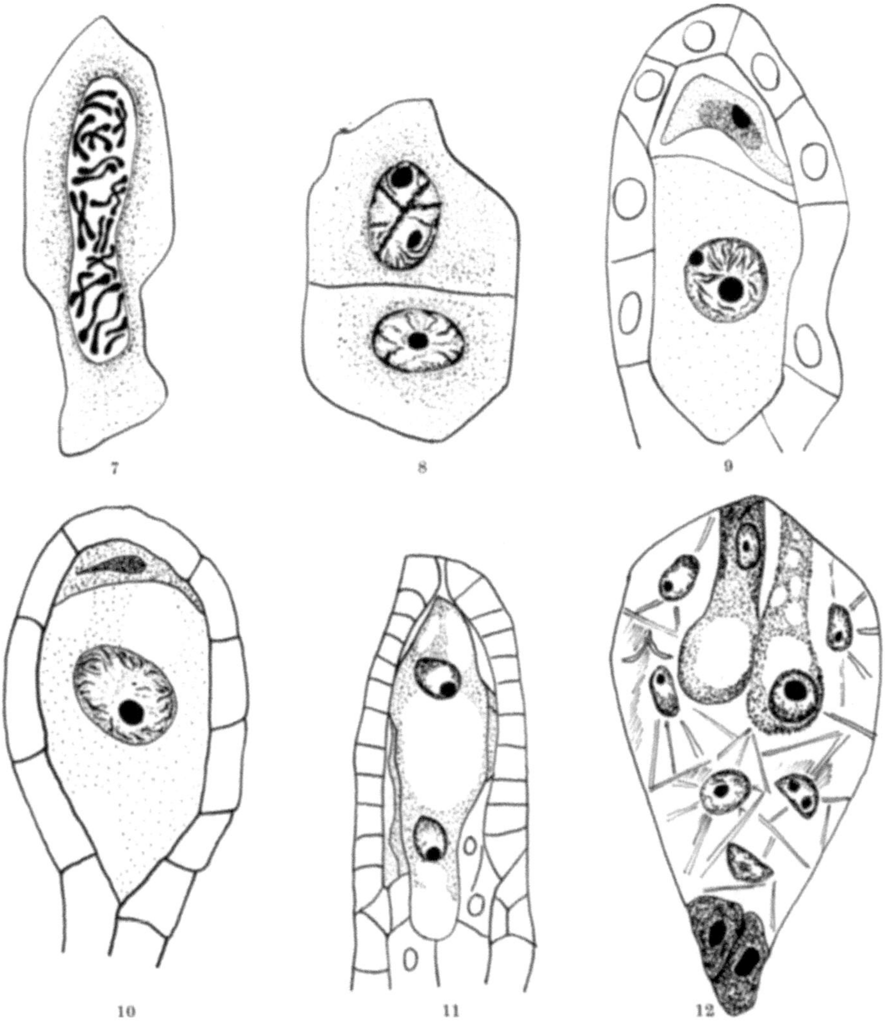

Figs. 7—12. The formation of Diplosporous ES's in species with restitution nucleus and degeneration of one macrospore. Type *Taraxacum*, mainly after OSAWA 1913. 7 Restitution nucleus; 8 Two celled stage as aresult of mitotic division of the restitution nucleus; 9 Degeneration of the upper cell of the dyade; 10 The lower cell develops an ES; 11 ES with two nuclei; 12 Completed ES.

semiheterotypic division according to ROSENBERG (1927), followed by the formation of a restitution nucleus (Fig. 3). The restitution nucleus divides mitotically and forms a dyade which turns into an ES with eight nuclei. The *Ixeris* type also is characteristic to species of the genus *Erigeron* (BERGMAN 1944, FAGERLIND 1947) and *Euhieracium* (BERGMAN 1941).

The apomictic species of the genus *Taraxacum* belong to type I:2 (Figs. 7—12). FAGERLIND (1947) as well as BATTAGLIA (1948) have proved that the EMC in

Taraxacum divides at a very early stage by a division holding unpaired chromosomes only. These chromosomes later on form a restitution nucleus (Fig. 7), which divides by a regular mitotis and form a dyade (Fig. 8). The chalazal part of this dyade gives rise to the ES, while the micropylear part degenerates (Figs. 9—12). The *Taraxacum* type of diplospory is typical to *Balanophora*

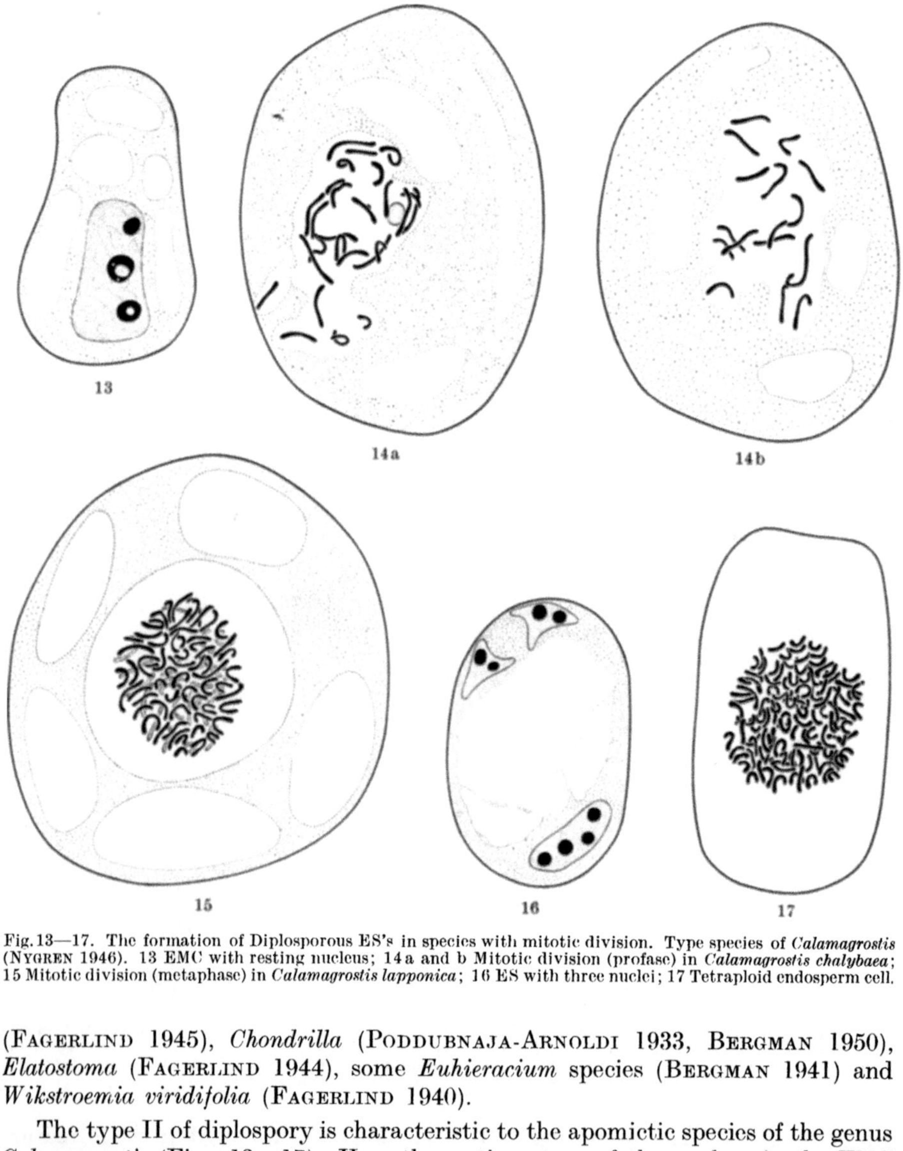

Fig. 13—17. The formation of Diplosporous ES's in species with mitotic division. Type species of *Calamagrostis* (NYGREN 1946). 13 EMC with resting nucleus; 14a and b Mitotic division (profase) in *Calamagrostis chalybaea*; 15 Mitotic division (metaphase) in *Calamagrostis lapponica*; 16 ES with three nuclei; 17 Tetraploid endosperm cell.

(FAGERLIND 1945), *Chondrilla* (PODDUBNAJA-ARNOLDI 1933, BERGMAN 1950), *Elatostoma* (FAGERLIND 1944), some *Euhieracium* species (BERGMAN 1941) and *Wikstroemia viridifolia* (FAGERLIND 1940).

The type II of diplospory is characteristic to the apomictic species of the genus *Calamagrostis* (Figs. 13—17). Here the resting stage of the nucleus in the EMC (Fig. 13) is very long. Normally the meiosis in the PMC's have reached the tetrad stage, when the EMC finally divides. The division is a typic mitosis (Figs. 14, 15), which rests for a very short time, a circumstance which has made it difficult and laborious to find. Therefore it has not been proved to occur in more than a relatively restricted number of species.

Antennaria was the first genus studied in which the *mitotic division* was found (JUEL 1898, 1900, STEBBINS 1932b), and thus type II usually is named the *Antennaria* type. It must be remembered, however, that according to GUSTAFSSON (1946, 1947) the *Antennaria* type also includes the *Ixeris* type. Diplospory with mitotic division is typical to *e.g. Antennaria* sp. (STEBBINS 1932b, BERGMAN 1935c, 1937, 1951; NYGREN 1950b), *Calamagrostis*, 7 sp. (NYGREN 1946, 1949, 1951, 1954, 1958), *Euhieracium* sp. (BERGMAN 1941), *Eupatorium glandulosum* (HOLMGREN 1919), *Poa*, 5 sp. (great number of authors, see Appendix), and *Potentilla* (RUTISHAUSER 1943a, b, c; 1945a, b; 1946, 1947, 1948, 1949; RUTISHAUSER and HUNZIKER 1954).

b) Pollen formation.

It is typical to many diplosporous species to have an irregular meiosis on the male side. In 1917 ROSENBERG distinguished between three types of divisions in *Euhieracium* with regard to the degree of pairing in the meiosis. In the *boreale type* the number of bivalents is relatively high, a considerable number of univalents is allways at hand, however. This type of meiosis shows great similarities to that common to the PMC's of many hybrids. In the *laevigatum type* the chromosomes are meiotic but unpaired, only univalents occur. This type later on was named *semiheterotypic division* by ROSENBERG (1927). The *pseudoillyricum type* finally is characterized by a mitotic division, a circumstance which gives ROSENBERG a reason to make the statement (*l.c.*, p. 193) „. . . stimmt also sehr gut mit dem Teilungsmodus in der Embryosackmutterzelle von z. B. *Antennaria alpina* überein." It might be remembered that the mitotic type of division occurring in the EMC of *Antennaria* was proved first by STEBBINS in 1932b.

The *boreale* and *laevigatum* types of ROSENBERG are the most common ones in diplosporous species, while his *pseudoillyricum* type is very seldom observed. The different types of division sometimes occur together, sometimes alternate, however, which in single cases causes considerable confusement as to the interpretation. In 1946 GUSTAFSSON and NYGREN found that in spite of the fact that *Hieracium robustum* in its natural environment in southern France had quite normal meiosis this division turned over into a mitosis when the species was cultivated in southern Sweden. In this case the change from one type of division to another has been interpreted as a temperature effect. In *Calamagrostis purpurea* on the other hand the type of division depends upon the age of the panicles (Table 1).

Table 1. *Occurrence of meiotic and mitotic division types in one clone of C. purpurea in 1943 and 1944.* (NYGREN 1946, p. 182.)

Fix. date	Fix. nr.	PMC's with		EMC's with		Number of			
		Meiosis %	Mitosis %	Meiosis %	Mitosis %	Florets	Florets with pollen %	EMC's	PMC's
8./9.6.43	135,163	100.0	0	100.0	0	506	100.0	27	6.666
22.6.44	269,270	30.8	69.2	45.5	54.5	344	46.0	66	2.672
22.6.44	244	8.2	91.8	71.4	28.6	134	1,5	7	122
22.6.44	271,272	7.1	92.9	33.3	66.7	67	18.0	12	1.191
22.6.44	273,274	0	100.0	27.5	72.5	402	0.5	80	5.042
						1.562		209	16.052

The fixations made the same day, 22.6.44, in Table 1 are all placed in order according to the age of the panicles. The panicles first produced on the plant in the year give a higher number of meiotic divisions than those panicles which develop later in the season. A similar situation in 1951 was found to occur by Bøcher in *Arabis Hoelboelli*. Sometimes the divisions in the PMC's break down completely or are inhibited, phenomena common to *Calamagrostis purpurea* and *C. chalybaea*, in which species big plasmodia are formed in the loculi, and which only exceptionally produce pollen able to function. Thus there are in diplosporous species a great variation in the way of pollen formation. This may depend upon the fact that only some diplosporous species are pseudogamous, and because of this reason the quality of the pollen is not so important for the production of seeds. On the other hand, the pollen is usually good in pseudogamous species such as *Poa alpina* (Müntzing 1940, 1954).

B. Apospory.

a) Embryo sac formation.

Although apospory was stated in ferns as early as in 1886 by Druery and Bower the discovery in 1907 by Rosenberg that apospory also occurred in the angiosperms is very important. Rosenberg studied species belonging to the subgroup *Pilosella* of the genus *Hieracium* in cooperation with Ostenfeld and his observations have passed half a century without remarks! As type example in this survey we will take the two *Poa* species *arctica* and *pratensis*, however (Figs. 18—26) (Nygren 1950). Just as in diplosporous species there are aposporous ones with a unicellular or a multicellular archespore. In both cases the situation may by very complicated, however. The genus *Poa* has a unicellular archespore. The EMC (Fig. 18, upper cell) normally passes through meiosis (Fig. 19) and may form a triade or a tetrade, the chalazal cell of which generally forms an ES. At a very early stage aposporous initial cells are found in the nucellus, in some cases before the EMC has begun to divide (Fig. 18, lower cell). Sometimes the EMC degenerates at an early stage and therefore the aposporous initial cells only give rise to embryo sacs (Fig. 20). In the two *Poa* species used as type, particularly in *Poa arctica*, the number of aposporous ES's can be high, in a stated case not less than twelve. Irrespectively if the EMC forms a reduced ES or not there will be a competition between ES's in the nucellus. Those cells which are closest situated to the vessels and therefore also to the nutriment will be winners in this competition. The normal ES only wins when it has begun to divide much earlier than the aposporous initials. Anyhow, the situation in the nucellus is often confusing (Figs. 22—24). In *Poa pratensis* it is a common situation that more than one embryo sac is completed, a circumstance which entails the development of more than one seed in one nucellus. This phenomenon not always is kept apart from polyembryony. All studied aposporous *Poa* species are pseudogamous and parthenogenetic. Therefore a situation often observed in the ES is the one given in Fig. 25, where the embryo is well developed at the same time as the two polar nuclei lie close together but do not fuse. Normally the polar nuclei do not form the central nucleus until the sperm nucleus enters the ES, often all three nuclei fuse at the same time. In such cases as no fertilization of the "central nucleus" does occur, the ES degenerates notwithstanding that a multicellular embryo exists. Exceptionally an endosperm may be formed even if no fertilization has taken place (Fig. 26, center), but in these cases we do not know if some stimulation from a sperm nucleus entering the ES has caused the fusing of the polar nuclei and made it possible for them to divide. In aposporous species pseudogamy plays a great role in the species formation, which will be taken up later on (*cf.* p. 565).

Apospory is the most important type of agamospermy known in the Angiosperms, up to now it has *e.g.* been found in such genera as *Alchemilla*, *Crepis*,

Parthenium, *Potentilla*, *Ranunculus* and *Rubus*, all rich in species, in addition to the grass genera already mentioned. A very restricted number of species belonging to these genera have been studied carefully until now, however.

In a number of genera and species belonging to the *Bothriochloidae*-group within the grass family the aposporous ES:s do not hold the same number of

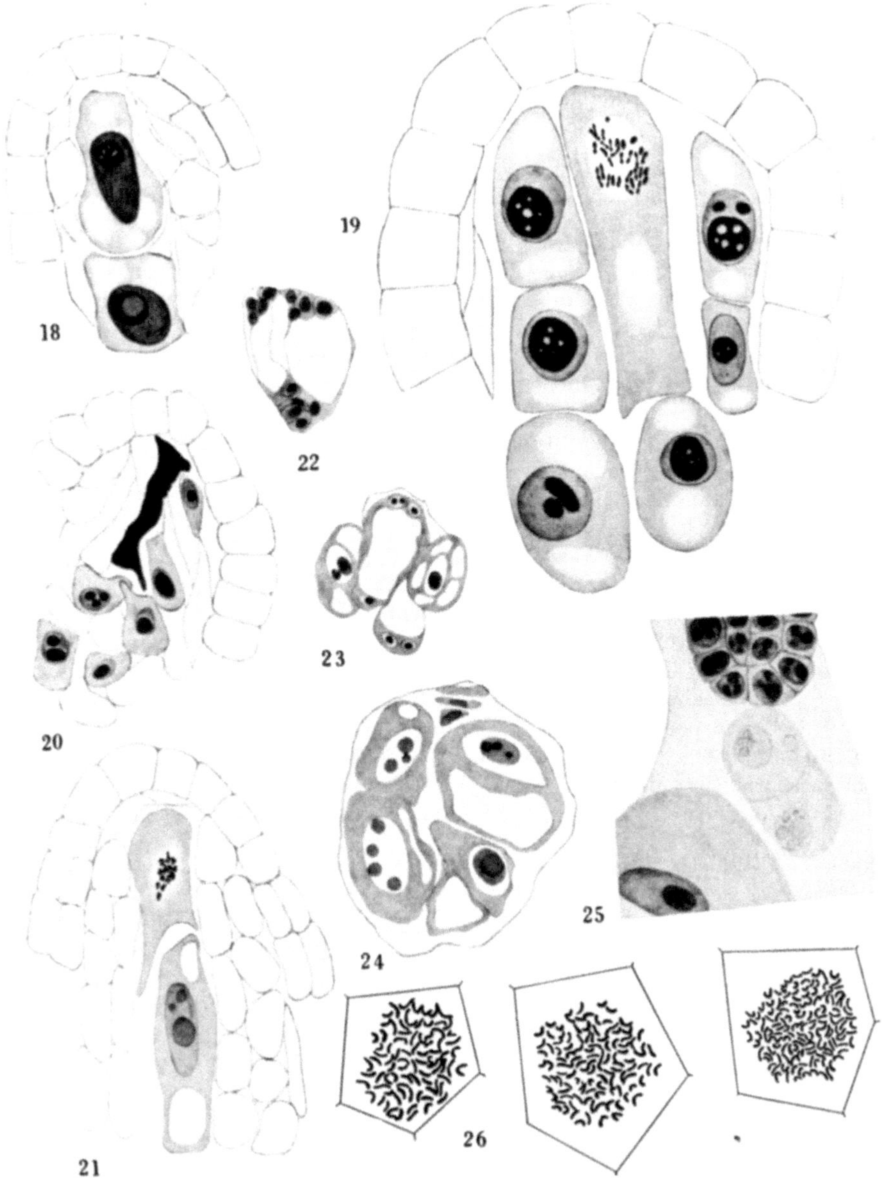

18 19 20 21 22 23 24 25 26

Figs. 18—26. The formation of Aposporous ES's. Type *Poa arctica* and *Poa pratensis* (NYGREN 1950a). 18 EMC and one aposporous initial cell in the nucellus; 19 EMC in meiotic division, six aposporous initial cells; 20 EMC degenerated, six aposporous initial cells; 21 Aposporous cell just under the EMC develops an ES. 22, 23 and 24 Three, four and five aposporous ES's developed in the nucellus; 25 Multicelled embryo and unfused polar nuclei waiting for fertilization (= pseudogamy); 26 Endosperm cells of *Poa arctica caespitans*, from left to right: fertilized reduced ES (triploid endosperm), unfertilized unreduced ES (tetraploid endosperm), fertilized unreduced ES (pentaploid endosperm).

nucleae as the sexual sacs (the antipods are often missing), and therefore it is possible in these species to count the exact number of aposporous and sexual sacs at an early stage of development (*cf.* HARLAN, BROWN, EMERY and coworkers 1957—1966).

b) Pollen formation.

Because of the fact that all species with apospory thoroughly studied are pseudogamous their pollen formation is very regular compared to most diplosporous ones. Undoubtedly this factor plays an important role in evolution. As a contrast may be mentioned that *Poa arctica* subsp. *caespitans* does not produce pollen in spite of the fact that it forms well developed seeds (NANNFELDT 1940). In the area where it grows other species of *Poa* occur, however (NYGREN 1950).

C. Adventitious embryony.

Already in 1841 J. SMITH observed that female plants of the dioecious *Alchornea (Caelebogyne) ilicifolia* belonging to the *Euphorbiaceae* set seed in spite of the fact that no male plants were present. In 1878 STRASBURGER investigated the species and found that the embryos formed originated from adventitious buds in the nucellus. Thus adventitious embryony is the first described type of agamospermy. It is now known in a number of species mainly belonging to the families *Buxaceae*, *Cactaceae*, *Euphorbiaceae*, *Myrtaceae*, *Orchidae* and *Rutaceae*.

The species with adventitious embryony can be divided in two groups, the first of which contain such species as need no fertilization for the formation of an endosperm, while a fertilization is necessary in species of the second group. To the first group belong *e.g. Alchornea ilicifolia (l.c.) Euphorbia dulcis* (CARANO 1925, 1926) and *Opuntia aurantiaca* (ARCHIBALD 1939). In these species the egg cell degenerates at an early stage, while adventitious embryos are formed from nucellus cells close to the ES. The endosperm develops autonomously just as in many diplosporous species. As type example for the second group we will take the genus *Citrus*, in which genus many species form adventitios embryos. The orange, *Citrus aurantium*, was studied by STRASBURGER in 1878, later on by OSAWA in 1912. Other species and varieties among others have been investigated by FROST in 1926 and 1938 and by FURUSATO in 1953, 1954 and 1960. Our figures follow the interpretation given by OSAWA (Figs. 27—30). In *Citrus* the EMC forms an ES in a normal way, the egg cell of which is fertilized (Fig. 27). Close to the egg

Table 2. *The number of embryos per seed in Citrus varieties and the production of hybrid seedlings in crosses.* (FROST 1926, combination of Table 1, p. 370, and Table 2, p. 372.)

Variety	Number of embryos per seed after self-fertilization or open pollination (ten seeds examined of each variety)											Artificial cross-fertilization	
	1	2	3	4	5	6	7	8	9	10	>10	Average	Seedlings per seed	Hybrid seedlings %
King mandarin	10											1.0	1.01	79.8
Sweet lemon	9	1										1.1	1.00	100.0
Lisbon lemon	8	2										1.2	1.06	83.9
Eureka lemon	8	1	1									1.3	1.08	73.9
Ruby orange	3	5	1	1								2.0	1.19	47.6
Mediterranean sweet orange	3	4	2				1					2.4	1.10	61.0
Dancy tangerine	3	1	4		1	1						2.8	1.26	18.5
Imperial grapefruit	1	1	2	3	1		1	1				4.1	1.28	46.7
Owari Satsuma			1		6			2	1			5.8	1.39	21.2
Willow-leaf mandarin			2		2	2	1	1		1	1	6.5	1.27	18.6

apparatus adventitious initial cells arise (Fig. 27, cells with nuclei in the nucellus), which later on grow out to embryos (Figs. 28 and 30). The normal embryo (Fig. 29) may be completed or may degenerate. The number of embryos per seed varies considerably, which is clear from Table. 2. This table also makes clear that the number of hybrids formed after artificial cross fertilization decreases at the same time as the number of formed embryos per seed increases.

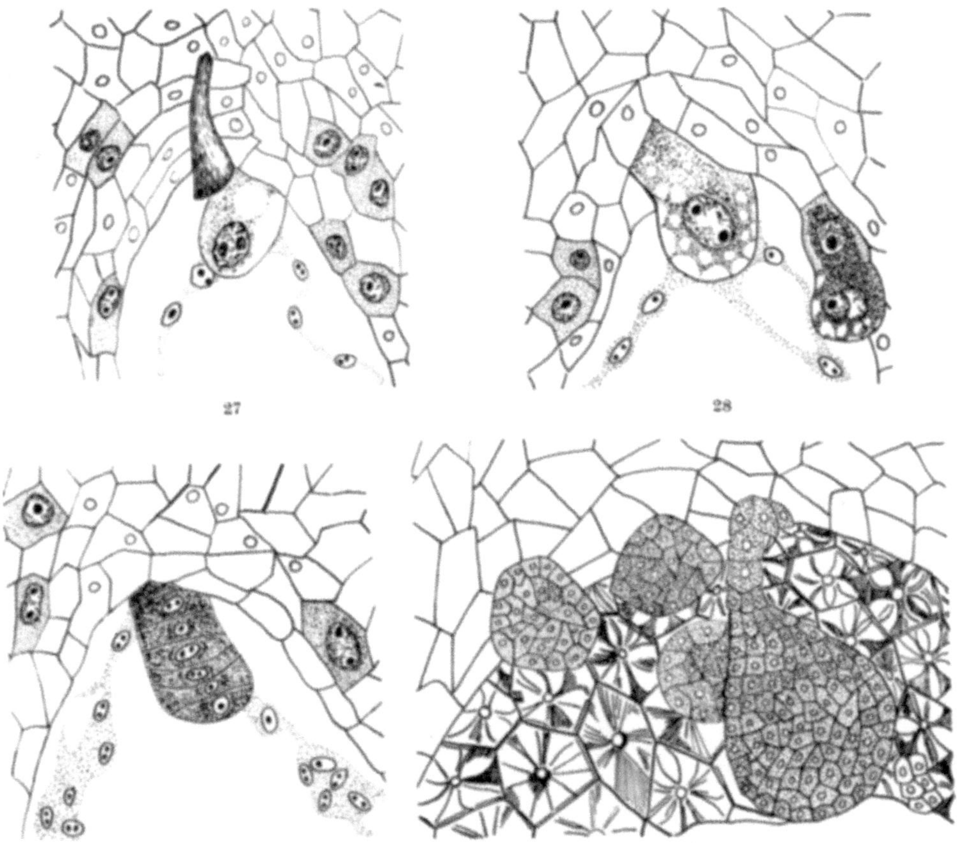

27 28 29 30

Figs. 27—30. Adventitious embryology. Type *Citrus* (OSAWA 1912). 27 Fertilization of egg cell. Adventitious initials in the nucellus; 28 Egg cell and adventitious embryo; 29 The egg cell has developed an embryo, adventitious initial cells; 30 Adventitious embryos surrounded by endosperm cells.

FURUSATO (1954, 1956, 1960) has analyzed the number of embryos per seed in a number of crosses between different species of *Citrus*. He has found that the formation of adventitious embryos begins after fertilization, and is of the opinion that fertilization is neccessary for the development of such embryos, because of the fact that fruits developed after hormone treatment do not form any seeds. The adventitious embryos develop by chance, and various growth stages may be found in the same seed. In the cross *C. Unshu* × *C. Natsudaidai* the average number of embryos is 12.3, with a variation in number between 1 and 29, while in the hybrid *C. Unshu* × *C. leiocarpa* var. *tumida* the mean is 14.3 and finally in *C. Unshu* × *Poncirus trifoliata* 16.8. Sometimes monoembryonic seeds are to be found in species normally producing polyembryonic ones. The trait mono- versus polyembryony is controlled by an allele pair. Polyembryony being the dominant and monoembryony the recessive allele (1960, p. 60). During the last years work on *Citrus* has been taken up with embryo cultures, the first of which have been reported by SWAMY (1958a, b). This work ist still in progress.

D. The development of the endosperm in agamospermous species.

Much confusement is met with in the litterature as to the development of the endosperm in diplosporous and aposporous species. In most diplosporous species the endosperm has the double chromosome number of the mother plant, brought about by an autonomous development of the central nucleus. This is the case *e.g.* in *Antennaria*, *Chondrilla*, *Calamagrostis* and *Taraxacum*. Other diplosporous species such as *Poa alpina* (HÅKANSSON 1943) and *Zephyranthes texana* (PACE 1913) are pseudogamous and need a fertilization of the central

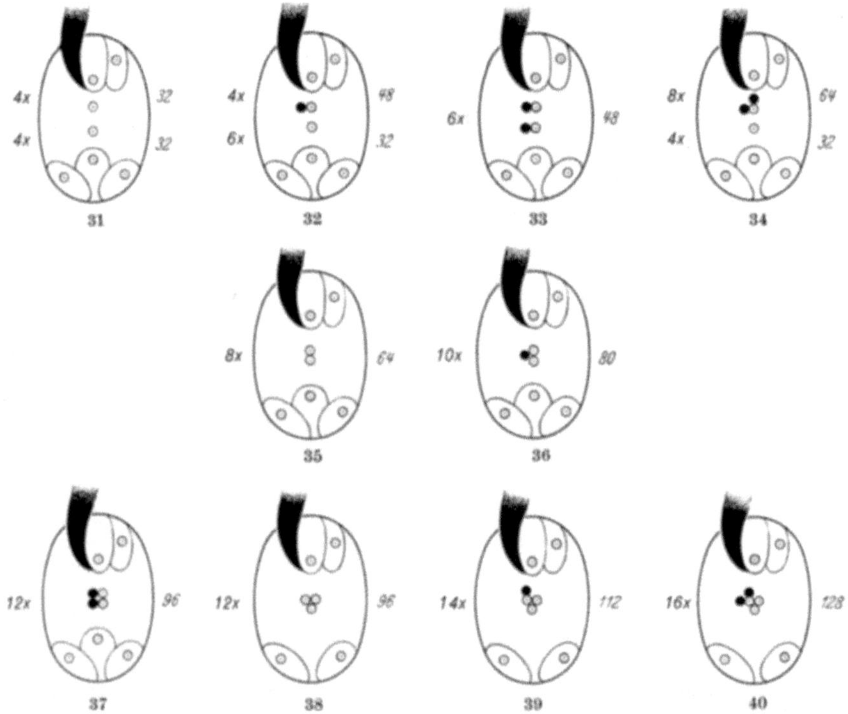

Figs. 31—40. The formation of endosperm in *Ranunculus auricomus* (RUTISHAUSER 1956). The species is aposporous with $2n = 32$ ($= 4n$). 31 The polar nuclei are unfused, endosperm tetraploid; 32 One polar nucleus fertilized, one unfertilized. Endosperm tetraploid and hexaploid; 33 Both polar nuclei fertilized. Endosperm hexaploid; 34 One polar nucleus unfertilized, the other one fertilized by two pollen nuclei. Endosperm tetraploid and octoploid; 35 Both polar nuclei fused. Endosperm octoploid; 36 Both polar nuclei fused and fertilized by one pollen nucleus. Endosperm decaploid; 37 Both polar nuclei fused and fertilized by two pollen nuclei. Endosperm duodecaploid; 38 Three polar nuclei are formed. They fuse and form duodecaploid endosperm; 39 Three polar nuclei, fertilized by one pollen nucleus. Endosperm quattuordecaploid; 40 Three polar nuclei, fertilized by two pollen nuclei. Endosperm sedecimploid.

nucleus to be able to develop an endosperm. Most aposporous species behave in the same way. Our knowledge of the chromosome number in the endosperm of aposporous species has been very incomplete until the 1950th, which has been caused partly by the difficulty to find suitable cells for a chromosome count in microscopic slides, partly by the difficulty to analyze such cells as to their chromosome number. It may be mentioned that until 1950 the chromosome number of three endosperm cells in all had been determined in such a common species as

Poa pratensis! Ten years ago RUTISHAUSER and cooworkers worked out smear methods to study the chromosome numbers of the endosperm in a number of species. His studies in the aposporous *Ranunculus auricomus* (1956) has verified earlier findings by HÅKANSSON in *Poa alpina (l.c.)* and by NOAK in *Hypericum* (1939). In one respect his investigation differs from all earlier studies made, however: it holds a sufficient number of determinations to permit conclusions to be drawn, a circumstance which appears from Table 3 (*l.c.*, p. 327).

Table 3. *Chromosome numbers found in the endosperm of Ranunculus auricomus, n = 8.* (RUTISHAUSER 1956, p. 327.)

Combination	Chromosome number of the endosperm													Sum
	4n	5n	6n	7n	8n	9n	10n	11n	12n	13n	14n	15n	16n	
4n × 4n														
Self-fertilization	2		7		3		38		117		1		2	170
Interspecific crosses . . .							7		67				1	75
4n × 2n (diploid sexual) .		2	1			1	27							31
Number of determinations														276

The results obtained by RUTISHAUSER in tetraploid *Ranunculus auricomus* are brought together in the Figs. 31—40. From this survey it is quite evident that the variation in chromosome number is considerable in a pseudogamous species and that more careful studies than earlier were made are neccessary in most such species until general conclusions can be drawn. In the actual case a tetraploid species with $2n = 32$ has been proved to form endosperm with most chromosome numbers between 32 and 128.

II. Vegetative propagation.

According to GUSTAFSSON (1946, p. 45—66) we may distinguish between three groups of vegetative propagation in higher plants:

A. Propagules formed outside the floral region. No fertilization or seed-dispersion in spite of functioning sexual organs. To this group belong *e.g.* the dioecious *Elodea canadensis* in Europe. The female plant was introduced into Scotland and Ireland in 1840. It spreads with shoots in the autumn. *Agave americana* was introduced to Europe in the 16th century from Mexico. In the Mediterranean region it develops very powerful runners and spreads vegetatively. European species belonging to this group are *e.g. Epipogium aphyllum* and *Malaxis paludosa*.

B. Propagules formed outside the floral region. Sterility phenomena occurring. *Lilium bulbiferum* and *Fritillaria imperialis* are typical representatives of this group. They propagate by bulbils or bulblets. In *Fritillaria imperialis* fertile as well as sterile races are known. The tendency to bulblet formation is much stronger in the sterile than in the fertile races. In *Ranunculus ficaria* bulbils are formed in the axils of the stem leaves. The species is known as diploid, triploid and tetraploid. The tetraploid race forms bulbils, while the diploid one produces seeds (NEVES 1942).

C. Propagules arising instead of or in addition to flowers (vivipary).

Group C is the most important one in apomictic plants. It was divided already by BRAUN (1857) in three subgroups: 1. The bulbils are accessory formations in the floral axils or floral branches. They often prevent fruits and seeds from developing. 2. Bulbils are formed instead of flowers. 3. The inflorescens or parts thereof are transformed into vegetative shoots.

Species belonging to the genera *Agave* and *Fourcroya* are counted to subgroup 1. Various species behave differently according to the ability to produce seeds. Sometimes they produce seeds only, sometimes propagules only are formed and in some cases there are seeds as well as propagules in the same inflorescens.

The second subgroup is exemplified by *Polygonum viviparum*, known already to LINNAEUS. The seedsetting in this species is very poor, it multiplies by bulbils. Another example is *Saxifraga foliolosa* of the mountains in Scandinavia and

41

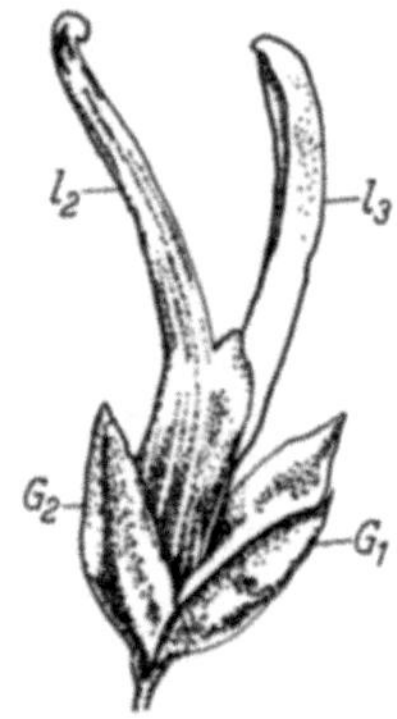

42

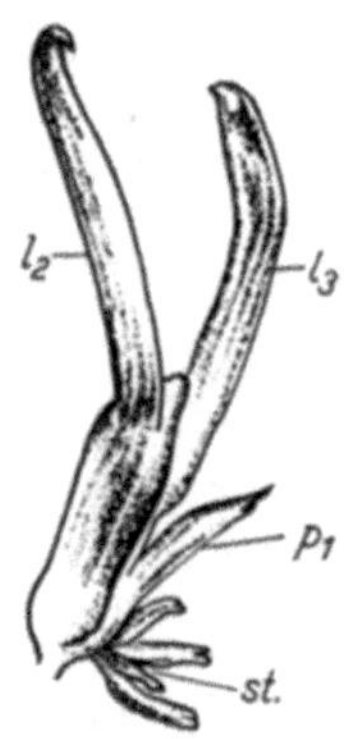

43

Figs. 41—43. Vivipary in *Poa*. 41 Viviparous panicle of *Poa alpina*. 42 and 43 Drawings of single viviparous flowers. *G* glumes; *l* lemma; *p* palea; *st* stamina.

Greenland. Here solitary flowers only occur in the top of the inflorescens, all others are transformed into bulbils. Almost the same situation occurs in *Allium oleraceum* and *carinatum*.

The third subgroup among other species includes the viviparous grasses, which have been studied by a great number of authors. In these species the lemmas have turned into shoots (Figs. 41—43). Vivipary has been particularly well studied in *Deschampsia*, *Festuca* and *Poa*, and in all three genera the tendency of vivipary increases with the chromosome number. Diploid viviparous grasses were unknown until in 1952, when SKALINSKA described a viviparous diploid *Poa alpina* ($2n = 14$) with a very restricted area of distribution in the Tatra district of Poland. FLOVIK has studied triploid and tetraploid *Deschampsia alpina* from Spitzbergen (1938). The triploid had almost no bulbils in the panicles, while the tetraploid was an obligate viviparous type. Diploid *Deschampsia alpina* from Sweden is non viviparous (NYGREN 1949). A very interesting study was made by

LAWRENCE in 1945. He cultivated *Deschampsia caespitosa* from Sweden in California together with North American material of the same species. The Swedish biotypes turned viviparous in spite of the fact that the species in Sweden is non viviparous. The North American material on the other hand did not change. In 1949 the author cultivated *Deschampsia* caespitosa from Abisko in Swedish Lapland (68° N. lat.) at Uppsala (60° N. lat.) under short day treatment. The material turned viviparous just as in the experiment of LAWRENCE. In 1956 SCHWARZENBACH studied Greenland specimens of *Poa alpina* cultivated in Switzerland. In this material there was a transformation of vivipary to non vivipary, however, probably depending upon the difference in the light and temperature conditions between the experimental field and the natural environment. In 1960 YOUNGNER was able to control the initiation of inflorescence, reproductive structures, and proliferations in *Poa bulbosa*. Thus inflorescence initiation was promoted by 16 hr light following vernalization of 1 week or more at 10°C. Normal florets developed under high temperatures (21°—27°C) and long days following initiation. Bulbils developed in place of normal florets at short days (8 hr.) and low temperatures (20°C, or below).

In a recent preliminary study NYGREN and ALMGÅRD have been able to control vivipary in normally viviparous *Poa alpina* and *bulbusa* by varying light and temperature conditions and thus verify the results obtained by YOUNGNER (1960). *Poa Chaixii* which is non viviparous turned viviparous after short day treatment (1962).

Biotype, form, and species formation in apomictic species.

We may distinguish between obligate and facultative apomicts, but experiments have shown that completely obligate apomicts are very rare. It is a fact that the behaviour of an apomict may vary in different climates, and it has even been proved that the sexual tendency in one and the same specimen may change not only between years but also between panicles during one and the same year *(Calamagrostis purpurea)* as well as in different climates *(Deschampsia caespitosa, Poa alpina)*. The first experimental control of apomixis was worked out by KNOX and HESLOP-HARRISON (1963). They made experiments with a tetraploid ($2n = 40$) race of the Australian species *Dichanthium aristatum* (POIR) C. E. HUBBARD, proved to be a facultative, aposporeous apomict. Under continous short days, the plants produced 1.54 ± 0.24 sacs per ovule, the percentage of aposporeous sacs being up to 79%, while with minimum photoperiodic induction (40 short days), 1.17 ± 0.09 sacs were formed per ovule, of which a maximum of c. 47% were apospereous. The differences observed were highly significant, and it may be concluded that the light regime does govern the incidence of apomixis in this tetraploid race, continuous exposure to short days increasing the numbers of aposporous sacs produced. Compare also HJELMQVIST and GRAZI (1964, 1965).

There are three modes of biotype and form formation in apomictic species: 1. By formation of euploid, aneuploid or polyploid aberrants. No pollination occurs. 2. By self-fertilization. 3. By hybridization. Common among pseudogamous species.

1. Formation of euploid, aneuploid or polyploid aberrants without pollination. As type genus we will take *Taraxacum*. The diploid, sexual species of this genus have the chromosome number $2n = 16$, while the apomicts are triploid with $2n = 24$. SØRENSEN and GUDJONSSON in 1946 made a very thorough investigation of the formation of aberrants in a number of Danish *Taraxacum* species. The species studied were all apomicts ($2n = 24$), which frequently formed aberrants

Table 4. *Aberrants formed in apomictic Taraxacum species (2n = 24).* (SØRENSEN and GUDJONSSON 1946, p. 24.)

1. Species	Number of plants	2n = 23 aberrants *elegans* %	*truncata* %	*plumosa* %	*hamosa* %	*pygmaea* %	*tenuis* %	Sum %
T. laciniosifrons	6.000	0.1	0.05	0.3	0.1	0.1	0.05	0.7
T. polyodon	1.650	0.05	0.05	0.1	0.2	0.2	+	0.5

2.	2n = 23 aberrants *olivacea* %	*crassifolia* %	Aberrant plants not belonging to the 2n = 23 series %	Total %	*gigas* %
T. laciniosifrons	—	+	1.4	2.1	0.8
T. polyodon	0.1	+	0.6	1.3	0.8

with 2n = 23, 2n = 26, as well as with some polyploid numbers, all morphologically distinct and distinguishable from the mother types, *cf.* Table 4.

The *gigas* mutants either have 2n = 44 or 2n = 46 and have arisen after chromosome doubling of original plants with 2n = 22 and 2n = 23. In eleven other species studied, SØRENSEN and GUDJONSSON proved that the total number of aberrants varied between 8.0 and 0.1 %. In second and later generations a further differentiation may happen, which is clear from Table 5.

Table 5. *Offspring of aberrants with 2 n = 23 in Taraxacum.* (SØRENSEN and GUDJONSSON *l. c.*, p. 26.)

Primary aberrant, 2n = 23	Number of plants	Secondary aberrant plants (mainly 2n = 22 aberrants)	*gigas* (2n = 46)
T. laciniosifrons aberr. *elegans*	843	c. 11	2.3
truncata . . .	787	c. 2	0.2
plumosa . . .	196	c. 4	0.5
hamosa	126	c. 3	0.8
pygmaea . . .	350	c. 4	3.0
T. polyodon elegans	237	c. 11	0.4
truncata	288	c. 2	2.4
plumosa	154	c. 4	0.7
hamosa	369	c. 3	0.0
pygmaea	410	c. 3	3.4

It has been proved (*l.c.*, p. 38) that there exists eight morphologically different types of chromosome aberrants with 2n = 23, each one of which lacks one of the eight chromosomes in one of the basic chromosome sets of the idiogram of *Taraxacum*. Diploid sexual species have two, apomictic triploids three such basic sets of eight chromosomes.

In addition to the aberrants with aneuploid or polyploid chromosome numbers there also are deviating plants with the normal chromosome number, 2n = 24, in the cultures, however. In one experiment a very big plant was cloned in 300 individuals, three of which turned out to be morphologically deviating aberrants holding twentyfour chromosomes somatically.

Similar results as those found in *Taraxacum* have been obtained in *Poa nervosa* by GRUN (1955). *Poa nervosa* is an obligate apomict of western North America, which is diplosporous with mitotic division in the EMC. No pollen is formed in this species, but in spite of this seed setting is good. The endosperm has been proved to be tetraploid, and thus the species is not pseudogamous. The chromosome numbers vary between 2n = 63 and 2n = 91 in specimens from California and Washington.

2. By self-fertilization. Normally obligate apomictic species may for some time have the ability to sexual and apomictic reproduction, they constitute facultative apomicts. A diplosporous species with such an ability is *Calamagrostis purpurea*. In a study of 673 two years old offspring plants belonging to 40 different clones of *C. purpurea* (2n = 56), which all lacked the capacity of producing pollen in nature, it was revealed (NYGREN 1949) that 31 plants belonging to eight of these clones produced pollen during the first year they flowered. Some of these progeny plants were bagged. The offspring raised from the seed harvested on individual panicles of these bagged plants in some cases had deviating chromosome numbers from different panicles of the same mother plant, which is clear from Table 6.

Table 6. *Offspring of bagged Calamagrostis purpurea with pollen.* (NYGREN 1949, p. 287.)

Mother clone	Chromosome number of mother clone 2n	Number of Ia[1] plant	Chromosome number of (Ia)[1] plant 2n	Chromosome numbers of the Ia[2] plants. Each row represents the offspring of one bagged panicle
J 3—41	56	1—3	56	2n = 70, 80, 84
D 2—41	56	1—3	56	2n = 56
		1—4	56	2n = 56, 56, 56
		2—2	56	2n = 56, 56, 56
		2—3	56	2n = 56, 58, 70
			56	2n = 56, 56
			56	2n = 68, 69, 70, 70
		2—4	56	2n = 56, 56, 56
		3—2	56	2n = 63, 64, 68, 68
S 2—38 A	56	1—1	56	2n = 63, 64, 68, 68
G 1—41 A	56	1—4	56	2n = 65
J 2—41 B	56	1—1	56	2n = 69, 70, 70
		2—4	56	2n = 67, 70
			56	2n = 70, 71, 71
		3—1	56	2n = 56, 56, 56, 56
		3—4	56	2n = 56, 56, 56, 56, 56
R 1—41 B	56	3—1	56	2n = 67
A	56	1—1	56	2n = 66, 66, 67
		4—1	56	2n = 66, 70

The heading "Mother clone" has reference to plants from nature, while Ia[1] denotes descendants of the mother clone raised after bagging. Next generation of these Ia[1] plants, raised after bagging, is designated Ia[2], and so on.

Thus, plants with other chromosome numbers than that of the mother clone have been obtained. Many of these chromosome aberrants were morphologically distinct and deviated from the original clone (NYGREN 1951a).

In a pseudogamous, aposporous spcies, *Poa irrigata*, LÖVE (1952) found a similar variation in chromosome numbers as that of *Calamagrostis purpurea*.

Pseudogamous species need a fertilization of the central nucleus to be able to produce seeds, and therefore it may sometimes happen that even the egg cell will be fertilized. The ability for the egg cell to be fertilized varies in different years and in different climates in *Poa pratensis*, however (NYGREN 1951b).

In 1958 JULÉN has published some interesting results obtained in the second generation from X ray treatment of one single clone of *Poa pratensis*. 286 X_2 families, all containing twelve progeny plants were studied. In 38 cases all, and in 147 cases between one and ten of the progeny plants were aberrants.

In these cases the mother plant obviously has been highly or partially sexual. As a contrast the remaining 101 progenies were all maternal. It is very possible that JULÉN by X ray treatment has been able to increase the number of sexual ES's or to increase the ability for the egg cell in sexual or aposporous ES's to be fertilized. An embryological investigation of JULEN's matherial made by GRAZI, UMAERUS and ÅKERBERG has proved this assumpsion (1961). HANSON and JUSKA who have treated *Poa pratensis* with neutrons (1962) report an increased number

Table 7. *Chromosome numbers of offspring of Poa irrigata from Iceland.* (LÖVE 1952, p. 22—23.)

Chromosome number of mother plants 2n	Chromosome numbers of offspring 80—90	90—100	100—110	110—120	120—130	130—140	140—150	Total
84 (three strains)	125	1		1	14			141
87	42	2	3	1	7	6		61
91		38				1		39
98		42	2				2	46
101		3	39					42
107			34					34
119				35	3			38
126					42			42
129					33	5		38
131					1	36		37
133						38		38
136						34		34
139						30	4	34
140 (two strains)							73	73
143							32	32
147							39	39
Total	167	86	78	37	100	150	150	768

of sexual plants in N_1, while BASHAW and HOFF have found that γ-radiation decreases the fertility of some plants of *Paspalum dilatatum*, while the degree of apospory is not affected by this type of radiation (1962), *cf.* also GUSTAFSON and GADD (1965).

3. By hybridization. Hybridization is a common phenomenon in many apomictic species. As type examples we will take up the two genera *Rubus* and *Poa*.

The apomictic system in European *Rubi* is well known by the classic studies of LIDFORSS (1899, 1914) and GUSTAFSSON (1943, 1946, 1947). In 1951 EINSET also turned some light over the conditions occurring in the blackberries of the new world, a study which was completed by PRATT and EINSET in 1955 and PRATT, EINSET and CLAUSEN (1958) as to the development of the ES in some of these species. The chromosome numbers in North American *Rubi* runs from $2n = 14$ to $2n = 63$, and most of the species can be crossed in any combination with each other. Embryologic studies of *Rubus alleghensis* ($2n = 14$), *R. canadensis* ($2n = 21$), *R. localis* ($2n = 21$), *R. bellobatus* ($2n = 28$), *R. abactus* ($2n = 35$), *R. meracus* ($2n = 49$), and *R. flagellaris* ($2n = 63$) proved that all these species developed aposporous ES's, the diploid *R. alleghensis* very seldom, however. All species are facultative pseudogamous (1955, 1958). Table 8 gives the results of interspecific crosses between maternal species with $2n = 28$ and pollen producing species with various chromosome numbers.

The variation in chromosome number is similar to that earlier discussed in the genera *Calamagrostis* and *Poa*, and just as in these genera morphologically deviating aberrants are common.

Table 8. *Chromosome numbers of seedlings of 28 chromosome plants* (*l. c.*, p. 770).

28-chromosome seed parent	Cross	Pollen parent and 2n numbers	Chromosome numbers of seedlings											
			14	21	28	29	30	32	33	34	35	40	42	49
R. Huttonii	×	Open-pollinated			40								3	
R. allegheniensis (Lowden) . . .	×	Self	2											
	×	*R. flagellaris* (63)											2	
R. Rosa (Eldorado)	×	Self	1		1									
	×	*R. bellobatus* (Erie) (28)	2		70	3								
R. Rosa (Snyder) .	×	Self			2									
R. Hanesii	×	Open-pollinated			37									
R. bellobatus . . .	×	Self	4		61								5	
	×	*R. Wiegandii* (36)	1		5									
	×	*R. meracus* (49)	3		20						1			3
	×	Open-pollinated	4		29									
R. bellobatus (Erie)	×	Self	1		46							1	5	
	×	Kittatinny (21)	2		48						1		5	
	×	*R. Cardianus* (35)	1		31		1	3	1	1	1			1
R. bellobatus (Brewer)	×	Self	5	2	2						1			
	×	Open-pollinated	1		1									
R. bellobatus (Mersereau) . . .	×	Self	2		1									
	×	Open-pollinated			2									
R. bellobatus (Kittatinny) . .	×	Self	1	1	32		1						2	
	×	*R. Cardianus* (35)			7								1	
	×	Open-pollinated	14		157								2	
			43	3	592	3	2	3	1	1	4	1	24	4 = 681

No apomictic genus has been so thoroughly investigated as to the production of interspecific hybrids as the genus *Poa* (Clausen and coworkers, 1940—1962). Particularly *Poa pratensis* and the west North American species *P. ampla* and *P. scabrella* have been combined in a great number of crosses with various European, American, and Australien species (Clausen *et al.* 1962). Just as is known in the genera *Calamagrostis* and *Rubus* the F_1 hybrids very often have deviating chromosome numbers, which is clear from Table 9.

Later reports (Clausen *et al.* 1953, Cox and Clausen 1955) have given some data, which go in the same direction as those of Table 9. The chromosome numbers of some F_2 families do not deviate too much from the figures of Table 9.

The main results obtained by the Clausen group is that it has been proved that almost any combination between these mostly aposporous *Poa* species give rise either to apomictic or sexual F_1 hybrids, which in later generation sooner or later form new apomictic strains again. These highly polyploid *Poa* species have the ability to pick up genes, single chromosomes or whole chromosome sets from related species and in this way constitute new forms, which is of great evolutionary importance. This ability of forming new apomictic forms is well illustrated in Table 10, in which the F_2 and F_3 progenies of some combinations made have been collocated. During the last decade much work has been done in the genera *Bothriochloa*, *Capillipedium* and *Dichantium* by Borgaonkar, Harlan, de Wet, Richardson and cooworkers (1961—1966). Out of *Bothriochloa* and *Dichantium* 300 F_2 families have been analyzed, 25 of which showed sexual tendencies. The main results obtained verify those by the Clausen group.

We have no reason here to discuss the old matter of dispute if apomixis is a dominant or recessive character. Most data point in the direction of recessivity.

Table 9. *Chromosome numbers of Poa hybrids.* (GRUN and TRIPLETT 1953, p. 121.)

Parents			Somatic chromosome numbers		
			Parents		
Female	Male		Female	Male	F_1 hybrids
ampla Albion	× *pratensis*	Mather	64	68	63, 70, 92, 93, 94, 96, 98, 100, 104
Kahlotus	× *pratensis*	Athabasca	64	70	77, 79, *ca.* 83, 84
Heise	× *pratensis*	Newport	70	81	73, 75, *ca.* 79
Condon	× *pratensis*	unknown	63	—	*ca.* 98
Enterprise	× *pratensis*	unknown	63	—	96
Albion	× *pratensis*	Lapland	63	74	63, 68, 70, 72, 73
Albion	× *pratensis*	Athabasca	64	70	75, 103
scabrella Watsonville	× *pratensis*	Athabasca	82	70	71, *ca.* 119
Las Posas	× *pratensis*	Lapland	82—84	74	76, 79, *ca.* 110
Las Posas	× *pratensis*	Mather	82—84	68	64, 67, 68, 70, 71, 74, 75
Las Posas	× *pratensis*	Athabasca	84	70	69, 70, 71, 72, 73, *ca.* 74
Las Posas	× *ampla*	Kahlotus	82—84	63—4	75—76
Las Posas	× *compressa*	Chorsum	84	50	65, 105, *ca.* 115
Las Posas	× *compressa*	Crescent Mills	84	42	63—65, 101
ampla Albion	× *compressa*	Chorsum	64	50	*ca.* 50, *ca.* 62, 87
compressa Chorsum	× *ampla*	Albion	50	64	56, 59
ampla Spokane	× *compressa*	Crescent Mills	63	42	*ca.* 81, 84
Albion	× *arida*	North Platte	—	63	*ca.* 50, 59, 61
arida North Platte	× *ampla*	Albion	63	—	*ca.* 97, 98
scabrella Las Posas	× *arida*	North Platte	84	63	67
arida North Platte	× *scabrella*	Las Posas	63	84	*ca.* 105
Canbyi Blue Mts.	× *pratensis*	Athabasca	84	70	77
gracillima Slate Creek	× *scabrella*	Yager Creek	84	84	86
Falen Leaf L.	× *pratensis*	Athabasca	86	70	*ca.* 72, *ca.* 74, 76
scabrella Lucerne	× *arachnifera*	Stillwater	*ca.* 63	58	57

Table 10. *Number of progenies of interspecific hybrids of Poa.* (CLAUSEN *et al.* 1949, p. 101.)

Hybrid combinations	F_2 progenies			F_2 progenies			Totals
				From aberrants in apomictic F_2s		From sexual F_2's	
	Apomictic	Sexual	Apomictic	Sexual	Apomictic	Sexual	
ampla × *pratensis*	20	15	7	2	6	7	57
× *compressa*	1	5	—	—	4	4	14
× *arida*	1	1	—	—	—	—	2
Canbyi × *pratensis*	—	1	—	—	—	3	4
scabrella × *pratensis*	14	25	—	2	30	23	84
× *compressa*	—	1	—	—	—	—	1
× *arida*	2	1	—	—	—	—	3
× *ampla*	2	2	—	—	—	—	4
× *gracillima*	3	3	—	—	—	—	6
Totals	43	54	7	4	30	37	
	97		11		67		175

Anyhow, experimental studies made in genera rich in apomictic species have proved that even if apomixis may be broken down in a single cross it will sooner or later be restored in later generations of these sexual derivatives (*cf.* however, ALMGÅRD 1966). Facultative apomixis such as it has established itself *e.g.* in *Calamagrostis, Poa and Rubus* is the most successful mechanism for creating of new forms to nature ever existing in higher plants.

Appendix.

Species with known or suspected apomixis in 1966.

(A apospory, AE adventitious embryony, D diplospory, P parthenogenesis, V vivipary.)

Family, genus and species	Type of apomixis	2n number	References
Amarantaceae			
Aerva tomentosa	D+P	—	SACHAR and MURGAI (1958), MURGAI (1959)
Amaryllidaceae			
Agave angustiflora	V	60, 90	BERGER (1915), VIGNOLI (1936, 1937a, b), GRANICK (1944), DARLINGTON and JANAKI-AMMAL (1945)
"candelabrum"	V	90	
cantula	V	60, 90	
fourcroydes	V	ca. 140	
Lespinassei	V	60	
lurida	V	90	
Salmiana var.	V	120	
sisalana	V	(138, 141—)150	
vivipara	V	60	
zapupe	V	(110—)120	
Calostoma Cunninghamii	(AE?)	—	SCHLIMBACH (1924)
purpureum	(AE?)	—	
Cooperia pedunculata	D+P	48	COE (1953)
Fourcroya cubensis	V	—	ERNST (1918), DARLINGTON and JANAKI-AMMAL (1945)
gigantea	V	60	
Zephyranthes texana	D+P	24	PACE (1913), DARLINGTON and JANAKI-AMMAL (1945), BROWN (1951)
Anacardiaceae			
Mangifera indica	AE	40, ca. 52	STRASBURGER (1878), WEBBER (1931), JULIANO and CUEVAS (1932), JULIANO (1934, 1937), DARLINGTON and JANAKI-AMMAL (1945), LEROY (1947), SACHAR and CHOPRA (1957)
Apocynaceae			
Amsonia Tabernaemontani	(AE?)	—	ANDERSSON (1931)
Araceae			
Aglaonema commutatum	A+P	118, 119, 120	PFITZER (1962)
psedobracteatum		60	
roebelinii		60	
Spathiphyllum Patinii	AE	18	SCHÜRHOFF and JÜSSEN (1925), DARLINGTON and JANAKI-AMMAL (1945)

Species with known or suspected apomixis in 1966 (continued).

Family, genus and species	Type of apomixis	2n number	References
Balanophoraceae			
Balanophora globosa	D+P	High	FAGERLIND (1945), DARLINGTON and JANAKI-AMMAL (1945)
japonica	(D+P)	94—112	
Belutaceae			
Alnus rugosa	AE (D+P)	28	WOODWORTH (1930)
Bombacaceae			
Pachira oleagnea	AE	72	BAKER (1960)
Burmanniaceae			
Burmannia coelestis	D+P	64—72	ERNST and BERNARD (1912), SCHOCH (1920)
Buxaceae			
Sarcococca Hookeriana	AE	—	ORR (1923), WIGER (1930, 1935), SIMONET and MIEDZYRZECKI (1932)
pruniformis	AE	28	
ruscifolia	AE	56	
humilis	AE(?)	—	
zeylanica	AE(?)	—	
Cactaceae			
Opuntia aurantiaca	AE	—	BRAUN (1860), GANONG (1898), MONTEMARTINI (1899), HULL (1915), ARCHIBALD (1939)
ficus indica	AE	—	
leucantha	AE	—	
Rafinesquii	AE	—	
glaucophylla	(AE?)	—	
tortispina	(AE?)	—	
Calycanthaceae			
Calycanthus fertilis	(D+P?)	22	PETER (1920), SCHÜRHOFF (1923), DARLINGTON and JANAKI-AMMAL (1945)
floridus	(AE?, D+P?)	22	
occidentalis	(AE?, D+P?)	—	
Campanulaceae			
Adenophora latifolia	—	—	SUKAČEV (1940)
liliifolia	—	—	
Potaninii	—	—	
utriculata	—	—	
Capparidaceae			
Capparis frondosa	AE	—	MAURITZON (1934)
Isomeris arborea	—	17?	BILLINGS (1937)
Casuarinaceae			
Casuarina distyla	D+P	(22), 33, 44	BARLOW (1958, 1959)

Celastraceae			
Celastrus scandens	AE	46	Andersson (1931), Bowden (1940)
Evonymus americanus	(AE?)	64	Braun (1860), Strasburger (1878), Bally (1916), Andersson (1931)
europeus	(AE?)	64	
latifolius	AE	—	
Compositae			
Antennaria alpina	D+P	84	Kerner (1876), Juel (1900), Fernald (1899, 1933, 1936), Porsild (1915), Fries (1919), Ekman (1927), Stebbins (1932, 1935, 1941), Bergman (1935c, 1951), Bøcher (1950), Nygren (1950b), Battaglia (1956, 1963), Urbanska (1959), Urbanska-Worytkiewicz (1961, 1962a, b, 1965)
Brainerdii	D+P	42	
canadensis	D+P	83—86	
carpatica	A, D+P?	40—42?, 56	
fallax	D+P	84	
glabrata	D(?)+P(?)	63	
groenlandica	—	—	
intermedia	D+P	ca. 80	
magellanica	D+P	56	
neodicica	D+P	ca. 52	
occidentalis	D+P	75—85	
Parlinii	D+P	84	
petaloidea	D+P	75—80	
Porsildii	D+P	63	
Arnica alpina	D+P	56, (60)	Afzelius (1936), Flovik (1940), Battaglia (1963)
Chamissonis	D+P	—	
diversifolia	(D+P)	—	
Artemisia nitida	A, D; no P	27	Chiarugi (1926), Battaglia (1963)
Centaurea cyanus	A; no P	24	Bergman (1935a)
Chondrilla acantholepis	D+P	15	Rosenberg (1912), Poddubnaja-Arnoldi (1933), Bergman (1944, 1950, 1952b, c), Battaglio (1949)
brevirostris	D+P	15	
coronifera	(D+P)	15	
graminea	D+P	20	
juncea	D+P	15	
Kouzneezowii	(D+P)	15	
lejosperma	(D+P)	15	
ornata	(D+P)	20	
pauciflora	D+P	15	
Cichorium intybus	A+?	18	Plumier (1960)
Coreopsis bicolor	A; no P	24	Gelin (1934), Battaglia (1963)
Crepis acuminata	A+P	22—88	Babcock and Stebbins (1938), Stebbins and Jenkins (1939), Babcock (1947) I and II, Battaglia (1963)
atribarba	A+P	22—88	
Bakeri	A+P	22—55	
barbigera	A+P	44—88	

Species with known or suspected apomixis in 1966 (continued).

Family, genus and species	Type of apomixis	2n number	References
intermedia	A+P	33—88	
modocensis	A+P	22—88	
monticola	A+P	22—88	
occidentalis	A+P	22—88	
pleurocarpa	A+P	22—88	
Erigeron annuus	D+P	(26), 27	Holmgren (1919), Tahara (1921), Carano (1921), McDonald (1927), Söderberg (1929), Okabe (1932), Gustafsson (1935), Ikeno (1935), Oka (1935), Bergman (1942, 1944), Fagerlind (1947a), Battaglia (1950), Battaglia (1963)
Karwinskianus var. mucronatus	D+P	32—34, 36	
ramosus	D+P	27	
Eupatorium glandulosum	D+P	51	Holmgren (1919), Battaglia (1963)
Hieracium			
Subg. Euhieracium (apomicts)	D+P	18, 27, 36	Ostenfeldt (1912), Rosenberg (1907, 1917, 1927, 1930), Schnarf (1919, 1929), Haberlandt (1921), Alton Gustafsson (1933), Christoff and Popoff (1933), Christoff (1940, 1942), Christoff and Christoff (1948), Bergman (1935b), Gustafsson (1935), Gustafsson and Nygren (1946), Gentcheff and Gustafsson (1940), Gentcheff (1937, 1941), Battaglia (1947c), Heitz (1951), B. and G. Turesson (1960, 1963), Skawinska (1962), Battaglia (1963)
Pilosella (apomicts)	A+P	27, 36, 45, 54, 63	
Ixeris dentata	D+P	21	Okabe (1932), Darlington and Janaki-Ammal (1945), Nisioka (1963)
Leontodon hispidus	A; no P	14	Bergman (1935a)
Leontopodium alpinum	D+P	52	Sokolowska-Kulczycka (1959)
Parthenium argentatum (apomicts)	A, D+P	(36, 37), 54—72—144	Bergner (1944, 1946), Esau (1944, 1946), Gardner (1946a, b), Powers (1945), Powers and Gardner (1945), Powers and Rollins (1945), Rollins (1945a, b, 1949), Catcheside (1947, 1950), Gerstel (1950a, b), Gerstel and Mishance (1950), Gerstel and Riner (1950), Gerstel, Hammond and Kidd (1953)
incanum	A, D+P	54—90	
Picris hieracioides	A; no P	10	Bergman (1935a)
Rudbeckia californica	D+P	76	Battaglia (1945, 1946a, b, 1947a—d, 1952a—c, 1954, 1955), Fagerlind (1946), Battaglia (1963)
Deamii	D+P	76	
laciniata	D+P	76	
speciosa	D+P	76	
Sullivantii	D+P	76	
triloba	D+P	76	

Taraxacum vulgaria apomicts	D + P	24	Gustafsson (1932, 1935), Poddubnaja-Arnoldi and Dianowa (1934), Matsuura and Suto (1935), Poddubnaja-Arnoldi (1939), Turrill (1938), Koroleva (1939), Erlandsson (1939), Sörensen and Gudiónsson (1945), Sörensen (1958), Fagerlind 1947), Heribert-Nilsson (1947), Battaglia (1948), Cooper and Brink (1949) Tschermak-Woess (1949), Haran (1950–1952), Kappert (1954), Fürnkranz (1960, 1961), Malecka (1958, 1962), Hou-Liu (1963), Battaglia (1963), Petrovskaya-Baranova and Zinger (1965), Zinger, Poddubnaja-Arnoldi, Petrovskaya and Polunina (1965)
obligua apomicts	D + P	24, 32	
erythrosperma apomicts	D + P	24, 32	
spectabilis apomicts	D + P	32	
maculigera apomicts	D + P	24, 48	
palustria apomicts	D + P	(24), 32	
arctica apomicts	D + P	40	
ceratophora apomicts	D + P	24, 32	
Other apomicts	D + P	24, 32, 40	
Youngia Sect. Crepidopsis	—	—	Stebbins (1941)
Cruciferae			
Arabis Holboellii	D + P	14, 21	Bøcher (1947, 1951, 1954)
Erysimum inconspicuum	?	54	G. A. Mulligan (1966)
Euphorbiaceae			
Alchornea (Caelobogyne) ilicifolia	AE	—	Strasburger (1878)
Euphorbia dulcis	AE	12, 28	Carano (1925, 1926), Perry (1943)
Gentianaceae			
Cotylanthera tenuis	(D + P ?)	—	Oehler (1927), Schnarf (1929)
Graminae			
Agropyron scaber	A + P	42	Hair (1936)
Agrostis stolonifera	V	—	Åberg (1940)
tenuis	V	—	
Anthephora pubescens	A + P		Brown and Emery (1957a, 1958), Brown (1958)
Bothriochloa decipiens	A + P		Brown and Emery (1957a, b, 1958), Celarier and Harlan (1957), Celarier (1957), Hubbard (1950), Brown (1958), Harlan, Celarier, Richardson, Brooks, Mehra (1958), Harlan and Celarier (1959), Harlan and De Wet (1963), Harlan, Brooks, Borgaonkar and De Wet (1964), Borgaonkar and De Wet (1961, 1963), Borgaonkar (1964), Borgaonkar, Harlan and De Wet (1962), Borgaonkar and De Wet (1964), Dewald and Harlan (1961), Mehra, Subramanyam and Swaminathan (1962), De Wet and Borgaonkar (1963), De Wet, Borgaonkar and Richardson (1963), De Wet and Harlan (1963, 1966), De Wet and Higgins (1963, 1964), De Wet and Richardson (1963), De Wet and Scott (1965), De Wet and Singh (1964), Faruqi (1964a, b)
ewartiana	A + P		
intermedia	A + P	20, 40, 50, 60, 80	
ischaemum	A + P	40, 50, 60	
Brachiaria brizantha	A + P		Brown and Emery (1957a, 1958), Brown (1958), De Wet (1958), Chedda and Harlan (1964)
serrata	A + P	18	

Species with known or suspected apomixis in 1966 (continued).

Family, genus and species	Type of apomixis	2n number	References
Bouteloua curtipendula	D + P	20, 40, 42, 50, 80—102	Fultz (1942), Myers (1947), Harlan (1948, 1949), Harlan and Emery (1958), Bryant (1952), Snyder and Harlan (1953), Brown (1958), Gould (1959), Gould and Kapadia (1964), Kapadia and Gould (1964), Mohamed and Gould (1966)
gracilis	?	20—84	
hirsuta	?	21, 37, 42	
Buchloë dactyloides	A + P) ?		Brown and Emery (1958), Brown (1958)
Calamagrostis canadensis	D + P	42—68	Stenar (1932b), Nygren (1946, 1948a, b, 1949a, 1951b, 1954, 1958, 1962, 1966b)
chalybaea	D + P	42	
crassiglumis	D + P	140	
inexpansa	D + P	84—103	
Langsdorffi	D + P	56	
lapponica	D + P	42—112	
purpurascens	D + P	42—56	
purpurea	D + P	56—91	
Callipedium parviflorum	A + P	20, 40, 50	Brown and Emery (1958), Celarier and Harlan (1957), Brown (1958), Harlan and Celarier (1959), De Wet, Borgaonkar and Richardson (1963)
spicigerum	A + P	40	
Cenchrus setigerus	A + P	34, 36, 44, 54, 56	Fisher, Bashaw and Holt (1954), Darlington and Janaki Ammal (1945), Brown (1958), Bogdan (1961), De Lisle (1964), Jagannath, Ponnaiya, Raman (1964)
ciliaris	A + P	32, 34, 36, 40, 44, 54, 56	
Chloris andropogonoides	A + P	40	Brown and Emery (1957a, 1958), Brown (1958), Anderson (1965)
cucullata	A + P	40	
gayana	A + P	20, 40	
verticillata	A + P	40	
Deschampsia alpina	V	(26)—39—52, 56	Flovik (1938, 1940), Hagerup (1939), Nannfeldt (1940) Lawrence (1945), Nygren (1949b), Wicherley (1953), Hedberg (1958), Kawano (1963)
caespitosa (apomicts)	V	26	
Dichantium annulatum	A + P	(20), 40, 60	Brown and Emery (1957a, 1958), Celarier and Harlan (1957), Brown (1958), Harlan, Celarlier, Richardson, Brooks and Mehra (1958), Harlan and Celarier (1959), Harlan and De Wet (1963), Mehra (1961, 1964a, b), Mehra, Subramanyam and Swaminathan (1962), De Wet, Mehra and Borgaonkar (1961), De Wet, Borgaonkar and Richardson (1963), De Wet and Borgaonkar (1963), Dewald and Harlan (1961), Borgaonkar and Singh (1962), Borgaonkar and De Wet (1960, 1963), Borgaonkar, Harlan and De Wet (1962), Knox and Heslop-Harrison (1963), Singh (1965), Singh and Mehra (1965)
aristatum	A + P	40	
caricosum	A + P	40	
nodosum	A + P	40	

Echinochloa frumentacea occ.	A + P	54	BROWN (1958)
Eleucine coracana (apomicts)	V	36	LI (1950)
Eragrostis chloromelas	(D + P)?	40, 60—63	BROWN and EMERY (1957a, 1958), BROWN (1958), DE WET (1958), STREETMAN (1963)
Heteropogon contortus	A + P	20, 40, 50, 60, 80	BROWN and EMERY (1957a, 1958), BROWN (1958), EMERY and BROWN (1958), ROY, SINGH and GUPTA (1965)
Hierochloe odorata	A + P	56	NORSTOG (1963)
Hyparrhenia hirta	A + P	30, 45	BROWN and EMERY (1957a, 1958), BROWN (1958), DE WET (1954)
Nardus stricta	D + P	26	COULON (1923), RYCHLEWSKI (1961)
Panicum deustum	A + P	36	BROWN and EMERY (1957a, 1958), BROWN (1958), DE WET (1954), SHAMA KUMARI (1960), BASHAW and HUTCHISON (1963), HUTCHISON and BASHAW (1964)
maximum	A + P	18, 32, 36, 48	
obtusum	A + P	20, 40	
virgatum	A + P	18, 21—36, 54, 72, 90, 108	
Paspalum dilatatum	A + P	40	SMITH (1948), BURTON (1948), WYCHERLEY (1953), HAYMAN (1956), SNYDER (1957), BROWN and EMERY (1958), WARMKE (1954), BASHAW and HOLT (1958a, b), BURTON and FORBES (1960, 1961), BURTON (1962), BURTON and HOFF (1962), REUSCH (1961)
hartwegianum	A + P		
malacophyllum	A + P	40	
notatum	A + P	40	
scrobiculatum	A + P	40	
secans	A + P	40	
Pennisetum ciliare	A + P	36	NARAYAN (1951), FISHER, BASHAW and HOLT (1954), SNYDER, HERNANDEZ and WARMKE (1955), CARR and ENG KOK NG (1956), BROWN and EMERY (1957a, 1958), BROWN (1958), GILDENHUYS and BRIX (1959), BASHAW (1961, 1962)
clandestinum	A + P	36	
dubium	A + P	66	
orientale	A + P	36, 45	
purpureum	A + P	28, 56	
ramosum occ.	A + P	10	
setaceum	A + P	27	
villosum	A + P	—	
Poa alpina (apomicts)	V, D + P	14—57	MÜNTZING (1933, 1940, 1954, 1966), MÜNTZING and NYGREN (1955), ÅKERBERG (1941), ÅKERBERG and BINGEFORS (1953), LÖVE and LÖVE (1942), HÅKANSSON (1943), NYGREN (1950a, 1955a, 1957, 1962a), NYGREN and ALMGÅRD (1962), SKALINSKA (1952), WYCHERLEY (1953), SCHWARZENBACH (1956)
ampla	A + P	56, 63	NYGREN (1951a, 1954), COX and CLAUSEN (1955), CLAUSEN, HIESEY and NOBS (1962)
arachnifera	V	54—56, 63	W. L. BROWN (1939), HARTUNG (1945), CLAUSEN et al. (1952), CLAUSEN, HIESEY and NOBS (1962)
arctica			
subsp. caespitans	A + P	56	NANNFELDT (1940), NYGREN (1950a)
subsp. depauperata	A + P	75—79	NANNFELDT (1940), NYGREN (1950a)
subsp. elongata	A + P	68—76	NANNFELDT (1940), NYGREN (1950a)

Species with known or suspected apomixis in 1966 (continued).

Family, genus and species	Type of apomixis	2n number	References
subsp. microglumis	A + P	68—82	NANNFELDT (1940), NYGREN (1950a)
subsp. stricta	V	(38), 39	NANNFELDT (1940), NYGREN (1950a)
var. vivipara	V	56	FLOVIK (1938)
diff. forms	A + P	56—ca. 100	ARMSTRONG (1937), FLOVIK (1938), ENGELBERT (1940, 1941), NYGREN (1950a)
arida	A + P	63	NYGREN (unpublished), TRIPLETT (1953)
bulbosa	V	14, 28, 42, 56	ARMSTRONG (1937), ÅKERBERG (1941), HARTUNG (1946), NYGREN (1957), YOUNGNER (1960), NYGREN and ALMGÅRD (1962)
compressa	A + P	35, 42, 49, 50, 56	ARMSTRONG (1937), TURESSON (1938), ÅKERBERG (1941), BRITTINGHAM (1941, 1943), M. A. CHRISTOFF (1942), HARTUNG (1946), CLAUSEN *et al.* (1952)
glauca	D + P	42, 47, 50, 56, 60 65, 70	KIELLANDER (LÖVE and LÖVE 1942), MÜNTZING (1933), ARMSTRONG (1937), ÅKERBERG (1941)
granitica	A + P	64, 67, 70, 71, 72, 78, 82, 94	SKALINSKA (1959)
herjedalica (= *alpina* × *pratensis*)	V, A + P	47—80	NANNFELDT (1937), NYGREN (1950a)
jemtlandica (= *flexuosa* × *alpina*)	V	37, (39)	NANNFELDT (1937), NYGREN (1950a, 1955b)
nemoralis	(D + P)	28, 29—38, 42 43, ± 56	KIELLANDER (LÖVE and LÖVE, 1942), MÜNTZING (1933), ARMSTRONG (1937), ÅKERBERG (1941)
nervosa	D + P	(28), 63, 76, 80, 81—90	GRUN (1951, 1952, 1955b)
nobilis	V	61—82	SKALINSKA (1955)
palustris	D + P	28, 29, 30, 42	ARMSTRONG (1937), KIELLANDER (LÖVE and LÖVE 1942), HARTUNG (1946), ÅKERBERG and NYGREN (1958)
pratensis subsp. alpigena	A + P	46—92	ÅKERBERG (1941), HARTUNG (1946), CLAUSEN *et al.* (1947, 1957), NYGREN (1950a), NYGREN and ÅKERBERG (1957)
subsp. alpigena var. vivipara	V, A + P	38, 42, 52, 69—81	FLOVIK (1938), WESTERGÅRD and SÖRENSEN (LÖVE and LÖVE 1942), NYGREN (1950a)
subsp. angustifolia	A + P	50, 64, 65	ÅKERBERG (1941), JUHL (1952, 1955)
subsp. eupratensis	A + P	41—124	MÜNTZING (1933, 1940), RANCKEN (1934), ARMSTRONG (1937), BROWN (1939, 1941), TINNEY (1940), ÅKERBERG (1941), M. A. CHRISTOFF (1942, 1944), NIELSEN (1945, 1946, 1957), NIELSEN and SMITH (1951a, b, 1952), HARTUNG (1946), CLAUSEN (1961), CLAUSEN *et al.* (1947, 1957, 1961, 1962), NYGREN (1951a, 1966), GRUN (1951, 1952, 1954, 1955a), GRUN and TRIPLETT (1952), HIESEY (1956), NYGREN and ÅKERBERG (1957), ÅKERBERG and NYGREN (1958), JULÉN (1958, 1960, 1961), JUHL-NOODT

			(1955), Harlan and Celarier (1958), Duich and Musser (1959), Watson and Clausen (1961), Grazi, Umaerus and Åkerberg (1961), Hanson and Juska (1962), Almgård (1960a, b, 1964, 1966),
subsp. irrigata	A + P	84—147	Åkerberg (1941), Löve (1952)
scabrella	A + P	82—84	Nygren (unpublished), Cox and Clausen (1955), Clausen *et al.* (1962)
Schmidtia bulbosa	?	36	Brown and Emery (1957a, 1958), Brown (1958)
kalahariensis	?	—	
Setaria leucopila	A + P	54, 68	Emery (1957)
villosissima	A + P	54	
viridis	V	18	Li (1950)
Themeda quadrivalvis	A + P	40	Brown and Emery (1958)
triandra	A + P	20, 22, 30, 40, 45, 49, 50, 51, 53, 54, 56, 60, 68, 71, 80	Brown and Emery (1957a, b, 1958), Carnahan and Hill (1961)
Tricholaena monachne	A + P	36	Brown and Emery (1957a, 1958)
Tripsacum dactyloides	D + P	36, 45, 54, 72, 90, 108	Farquharson (1955), Brown and Emery (1958), Brown (1958)
Urochloa bulbodes	A + P	36	Brown and Emery (1957a, 1958)
mosambicensis	A + P	28, 30, 32	
pullulans	A + P	28	
trichopus	A + P	30	
Guttiferae			
Garcinia mangostana	AE	—	Sprecher 1919), Horn (1932)
Hypericum perforatum	A + P	32 (60)	Noack (1939), Hoar and Haertl (1932)
Juglandaceae			
Juglans sp.	? ?	—	Zarubin (1949)
Liliaceae			
Allium amplectens	(D + P ?)	24	Levan (1940, 1944)
ascalonicum	V	16	Darlington and Janaki-Ammal (1945)
carinatum	V	16, 24	Levan (1937), Diannelidis (1944), Geitler (1944)
cepa	V	16	Darlington and Janaki-Ammal (1945)
nutans	AE ?	40—42	Håkansson (1951, 1957)
odorum	AE	16, 32	Levan (unpubl.), Darlington and Janaki-Ammal (1945), Håkansson (1951), Håkansson and Levan (1957)
oleraceum	V	32	Levan (1937)
proliferum	V	16	Darlington and Janaki-Ammal (1945)
roseum var. bulbiferum	V	48	Messeri (1930)
sativum	V	16	Darlington and Janaki-Ammal (1945)

Species with known or suspected apomixis in 1966 (continued).

Family, genus and species	Type of apomixis	2n number	References
schoenoprasum	V	16, 24, 27, 29, 32	LEVAN (1935)
scorodoprasum	V	16	DARLINGTON and JANAKI-AMMAL (1945)
vineale	V	32	DARLINGTON and JANAKI-AMMAL (1945)
viviparum	V	16	DARLINGTON and JANAKI-AMMAL (1945)
Dianella tasmanica	?	1[illegible] (76, 80, 84)	CURTIS (1952)
coerulea	?	(16, 32, 48)	
Hosta coerulea	AE	60	STRASBURGER (1878), AKEMINE (1935), SUTO (1936)
Lilium bulbiferum	V	24	TISCHLER (1935—1936)
longiflorum	? + P	48	EMSWELLER and UHRING (1962)
Nothoscordum fragans	AE	16, 18, 19, 24	STRASBURGER (1878), STENAR (1932a), LEVAN and EMSWELLER (1938), DARLINGTON and JANAKI-AMMAL (1945), D'AMATO (1949), HÅKANSSON (1953)
Smilacina racemosa	AE	—	MACALLISTER (1913)
Tulipa Gesneriana	AE	(24, 36)	BAMBACIONI-MEZZETTI (1932), BETH (1938)
Trillium sp.	P ?	—	JEFFREY (1948)
Malphigiaceae			
Hiptage Madablota	A	—	SUBBA RAO (1940)
Musaceae			
Musa Balbisiana × *acuminata*	P ?		DODDS and SIMMONDS (1948)
Myrtaceae			
Eugenia Hockeri	AE	—	V. D. PIJL (1934), JOHNSON (1936), TIWARY (1926, 1929)
Jambos	AE	28, ±42, ±54	
malaccensis	AE	22	
Cumini	(AE ?)	—	
Ochnaceae			
Ochna serrulata	AE	35	CHIARUGI and FRANCINI (1930)
Onagraceae			
Taraxia ovata	(P ?)	—	JOHANSEN (1931)
Orchidaceae			
Bletilla striata var. gebina	Haploid P		MIDUNO (1940)
Cephalanthera damasonium	Haploid P	32	HAGERUP (1947)
longifolia	Haploid P	32	HAGERUP (1947)
Epipactis latifolia	Haploid P	40	HAGERUP (1945)
Habenaria platyphylla	?	—	SWAMY (1946b)
Listera ovata	Haploid P	34, 36	HAGERUP (1947)
Nigritella nigra (apomicts)	AE	ca. 64	AFZELIUS (1928, 1932)
Orchis maculatus	Haploid P	40, 80	HAGERUP (1944)
strictifolius	Haploid P	40	HAGERUP (1947)

Platanthera chlorantha	Haploid P	42	Hagerup (1947)
Zeuxine sulcata	AE	—	Swamy (1946a)
Zygopetalum Mackayi	AE	96	Suessenguth (1923), Darlington and Janaki-Ammal (1945), Afzelius (1959)
Plumbaginaceae			
Statice olaeefolia var. confusa	D+P	27	D'Amato (1940, 1949)
Limonium binervosum	?	32	Baker (1954)
lychnidifolium	?	24	
Polygonaceae			
Acetosa arifolia	?	14	Löve (1942, 1949)
pratensis	?	—	
thyrsiflora	?	—	
Atraphaxis frutescens	A+P	88 (?)	Edman (1931)
Polygonum viviparum	V	83—88, ±100, ±110	Löve and Löve (1942)
Primulaceae			
Primula japonica × chungensis	"automixis"	—	Ernst (1949, 1950, 1951a, b)
× Cockburnia		—	
× pulverulenta	—	—	
Wilsoni × Cockburnia	—	—	
Ranunculaceae			
Ranunculus acris	?	—	Marsden-Jones and Turrill (1935, 1952)
auricomus coll.	A+P	16, 32, 40, 48	Häfliger (1943), Rozanova (1932), Koch (1933, 1939), Sörensen (1938), Marklund (1939), Marklund and Rousi (1961), Darlington and Janaki-Ammal (1945), Rutishauser 1953, 1954, 1954a, b, c, 1956, 1960, 1966), Rousi (1956), Julin (1963, 1964, 1965), Jankun and Izmailow (1965)
bulbosus	?	—	Marsden-Jones and Turill (1945)
Rosaceae			
Alchemilla subg. Aphanes			
arvensis	A+P	48	Böös (1924), Gentcheff and Gustafsson (1940b), Gudjonsson (1941), Hjelmqvist (1959a)
subg. Eualchemilla			
alpina coll.	(A+P)	>100	Murbeck (1901), Gentcheff and Gustafsson (1940b), Turesson (1943, 1956, 1958)
vulgaris coll.	(A+P)		Murbeck (1901), Böcher (1938), Gentcheff and Gustafsson (1940b), Turesson (Löve and Löve 1942), (1957), Ehrenberg (1945)
acutiloba		ca. 100	
glabra		c. 90 ca. 100	
glomerulans		64 (?)	
micans		ca. 93	
pastoralis		101	
subcreanata		ca. 90	

Species with known or suspected apomixis in 1966 (continued).

Family, genus and species	Type of apomixis	2n number	References
Cotoneaster acutifolia var. villosula	A + P	68	SAX (1954), HJELMQVIST (1962), ZEILINGA (1964)
bullata	A + P	68	
obscura	A + P	68	
var. nova	A + P	—	
racemiflora var. soongorica	A + P	—	
rosea	A + P	—	
Crataegus, numerous American apomicts	?	Polyploid	LONGLEY (1924a), PALMER (1932), SAX (1931)
Malus hupehensis	A + P	51	SAX (1932, 1959), DERMEN (1936), OLDÉN (1953), HJELMQVIST (1957, 1959b)
sikkimensis	(A + P)?	51	
sargentii var. rosea	(A + P)?	51	
toringoides	(A + P)?	51	
rockii	A + P	68	
sargentii	A + P	68	
Sieboldii	A + P	68	
toringoides var. macrocarpa	A + P	68	
Potentilla adscharica	P	42	FORENBACHER (1914), MÜNTZING (1928, 1931, 1958a, b, c, d, 1960), A. and G. MÜNTZING (1941, 1945), POPOFF (1935), SHIMOTOMAI (1935), GENTCHEFF and GUSTAFSSON (1940b), CLAUSEN, KECK and HIESEY (1940), CLAUSEN (1951), CHRISTOFF and PAPASOVA (1943), RUTISHAUSER (1943a, b, c, 1945a, b, 1946, 1947, 1948, 1949, 1960), HÅKANSSON (1946), HUNZIKER (1954), SKALINSKA and CZAPIK (1958), CZAPIK (1962a, b), SMITH (1963a, b)
arenaria	D + P, A + P	28, 35, 42, 56	
argentea	A + P	14, 28, 35, 42, 56	
arguta	—	14	
argyrophylla	—	63	
aurea	—	14	
Breweri	—	72—102	
canescens	P	42	
chrysantha	—	42	
collina	P	35, 42	
Crantzii	(D + P)?	28, 42, 42 + f	
dealbata	—	35	
Degeni	—	42	
diversifolia	—	42—101	
dissecta	—	42	
Drummondii	—	±64 — ±108	
flabelliformis	—	56 — ±66	

gelida	—	42	
gracilis	—	56—±109	
hirta	—	28	
intermedia	—	28, 56	
kurdica	—	56	
Levieri	—	42	
multifida	—	28	
nevadensis	—	28	
praecox	A+P	42	
pulcherrima	—	56—±109	
recta	P	28, 42	
verna (including) *P. Tabernaemontani*	A+P, D+P	28, 42, 49, 84	
Rubus subg. Eubatus sect.			
Moriferi			
European apomicts	A+P	21—49	Crane and Thomas (1940), Gustafsson (1943), Christen (1950, 1952), Berger (1953a, b), Haskell (1955, 1957, 1960a, b, c, 1961a, b, 1966), Haskell and Hill (1961), Haskell and Tun (1961), Heslop-Harrison (1953), Watson (1958), Dowrick (1961)
idaeeus	A+P	21	
American apomicts	A+P	(14)—63	Longley (1924b), Crane and Thomas (1940), Einset (1947, 1951), Waldo and Darrow (1948), Pratt and Einset (1955), Pratt, Einset and Clausen (1958)
Sorbus aria coll.	—	68	Moffett (1931); Sax (1931), Liljefors (1934, 1953, 1955a, b)
(obtusifolia)	—	68	
(rupicola)	—	68	
chamaemespilus	A+P	68	
hybrida (=fennica)	A+P	68	
intermedia	—	68	
lancifolia	—	51	
Meinichii	—	68	
minima	—	51	
Mougeoti	—	51	
sudetica	—	51	
Rutaceae			
Citrus			
Numerous species and varieties, as well as related genera	AE	18, 27, 36	Strasburger (1878), Osawa (1912), Frost (1926), Toxopeus (1936), Traub (1938), Weber (1940), Bacchi (1943), Krug (1943), Moreira and Gurgel (1947), Hofmeier and Oberholzer (1948), Furusato (1952, 1953a, b, c, 1954, 1960), Mamporija (1957), Swamy (1958a, b), Majsuradze (1961), Raghuvanshi (1962)

Species with known or suspected apomixis in 1966 (continued).

Family, genus and species	Type of apomixis	2 n number	References
Ptelea trifoliata	(AE ?)	?	MAURITZON (1935)
Skimmia japonica	A + P	—	DESAI (1961)
lavellei	A + P	—	
Triphasia aurantiola	AE	—	MAURITZON (1935)
Xanthoxylum alatum	(AE ?)	—	LONGO (1908), MAURITZON (1935)
Bungei	AE	—	
Saururaceae			
Houttuynia cordata	D + P	(52—56), 94—98, 100—104	SHIBATA and MIYAKE (1908), SÖDERBERG (1927), OKABE (1930)
Saxifragaceae			
Saxifraga cernua	V	50, ca. 66	SKOVSTED (1934), LÖVE and LÖVE (1942), BÖCHER (1938b), HARMSEN (1939), FLOVIK (1940)
foliolosa	V		
Thymeliaceae			
Wikstroemia viridifolia	D + P	27	FAGERLIND (1940)
Urticaceae			
Elatostoma acuminatum	A + P	50	FAGERLIND (1944), KRAUSE (1931)
eurhynchum	D + P	52	
latifolium	A + P	60	
machaerophyllum	D + P	60	
peltifolium	D + P	60	
penniverve	—	—	
repens	—	High	
rostratum	—	—	
sinuatum	D + P	80	
Procris frutescens	—	—	

Literature.

Åberg, E.: Viviparous forms of *Agrostis*. Ann. agric. Coll. Sweden 8, 461—463 (1940). — Afzelius, K.: Embryologische und zytologische Studien in *Senecio* und verwandten Gattungen. Acta Hort. Berg. 8, 123—219 (1924). — Die Embryobildung bei *Nigritella nigra*. Svensk bot. Tidskr. **22**, 82—91 (1928). — Zur Kenntnis der Fortpflanzungsverhältnisse und Chromosomenzahlen bei *Nigritella nigra*. Svensk bot. Tidskr. **26**, 365—369 (1932). — Apomixis in der Gattung *Arnica*. Svensk bot. Tidskr. **30**, 572—579 (1936). — Apomixis and polyembryony in *Zygopetalum Mackayi* Hook. Acta Horti Bergiani **19**, 7—13 (1959). — Akemine, T.: Chromosome studies on *Hosta*. I. The chromosome numbers in various species of *Hosta*. Hokkaido Imp. Univ., J. Fac. Sci. V **5** (1), 25—32 (1935). — Åkerberg, E.: Cytogenetic studies in *Poa pratensis* and its hybrid with *Poa alpina*. Diss. 1941; also in Hereditas (Lund) **28**, 1—126 (1942). — Åkerberg, E., and S. Bingefors: Progeny studies in the hybrid *Poa pratensis* × *Poa alpina*. Hereditas (Lund) **39**, 125—136 (1953). — Åkerberg, E., u. A. Nygren: *Poa pratensis, trivialis, palustris, compressa* und verwandte Arten. In: Handbuch der Pflanzenzüchtung, II. Aufl., S. 392—418. 1959. — Almgård, G.: Experiments with *Poa*. I. Studies of *Poa longifolia* Trin. as a fodder grass and as a component in interspecific hybrids. Ann. roy. Agric. Coll. Sweden **26**, 77—119 (1960a). — Some Swedish results of breeding new *Poa* types. Proc. 8th int. Grassl. Congr. p. 71—73 (1960b). — Experiments with *Poa*. II. Studies of Polyhaploid *Poa*-types. Ann. Agric. Coll. Sweden **30**, 335—373 (1964). — Experiments with *Poa*. III. Further studies of *Poa longifolia* Trin. with special reference to its cross with *Poa pratensis* L. Ann. Agric. Coll. Sweden **32**, 3—65 (1966). — Andersson, A.: Studien über die Embryologie der Familien Celastraceae, Oleaceae und Apocynaceae. Lunds Univ. Årsskr. **27** (1), 1—112 (1931). — Anderson, D.: Introgression in Texas species of Chloris (Graminae). Amer. J. Bot. **52**, 647 (1965). — Archibald, E. E. A.: The development of the ovule and seed of jointed cactus (*Opuntia aurantiaca* Lindley). S. Afr. J. Sci. **36**, 195—211 (1939). — Armstrong, J. M.: A cytological study of the genus *Poa* L. Canad. J. Res. C **15**, 281—297 (1937).

Babcock, E. B.: The genus *Crepis*, Part I, 197 pp. 1947. — Babcock, E. B., and G. L. Stebbins jr.: The American species of *Crepis*. Carnegie Inst. Wash., Publ. **504**, 1—199 (1938). — Bacchi, O.: Cytological observations in citrus. III. Megasporogenesis, fertilization, and polyembryony. Bot. Gaz. **105**, 221—225 (1943). — Baker, H. G.: The agamic complex in *Limonium* (Sections Densiflorae and Dissiflorae Boiss.) Proc. VII. Int. Bot. Congr. 1954, p. 329—330. — Apomixis and polyembryony in *Pachira oleaginea (Bombaceae)*. Amer. J. Bot. **47**, 296—302 (1960). — Bally, W.: Zwei Fälle von Polyembryonie und Parthenokarpie. Verh. schweiz. naturforsch. Ges. **98**, 169—170 (1916). — Bambacioni-Mezzetti, V.: Sullo sviluppo dell'embrione in *Tulipa Gesneriana* L. Ann. Bot. (Roma) **19**, 145—155 (1932). — Barlow, B. A.: A stage in the evolution of apomixis in *Casuarina*. Aust. J. Sci. **21**, 108 (1958). — Polyploidy and apomixis in the *Casuarina distyla* species group. Aust. J. Bot. **7**, 238—251 (1959). — Bashaw, E. C.: Apomixis and sexuality in buffelgrass *(Pennisetum ciliare) (L)* Link. Ann. Meet. Amer. Soc. Agr. Abstr. **1961**, **47** (1961). — Apomixis and sexuality in buffelgrass *(Pinnisetum ciliare)*. Crop Sci. **2**, 412—415 (1962). — Bashaw, E. C., and B. J. Hoff: Effects of irradiation on apomictic common dallisgrass *(Paspalum dilatatum)*. Crop Sci. **2**, 501—504 (1962). — Bashaw, E. C., and E. C. Holt: Megasporogenesis, embryo sac development and embryogenesis in Dallisgrass, *Paspalum dilatatum* Poir. Agron. J. **50**, 753—756 (1958). — Reproduction in dallisgrass, *Paspalum dilatatum* Abstr. Ann. Meet. Amer. Soc. Agron. Lafayette, Ind. **1958**, **43** (1958). — Bashaw, E. C., and D. J. Hutchison: Accessory chromosomes in *Panicum coloratum*. Canad. J. Genet. Cytol. **5**, 281—289 (1963). — Battaglia, E.: Fenomeni cirologici nuovi nella embriogenesi ("semigamia") e nella microsporogenesi ("doppio nucleo di restituzione") di *Rudbeckia laciniata* L. Nuovo Giorn. bot. ital. **52**, 34—38 (1946a). — Ricerche cariologiche ed embryologiche sul genere *Rudbeckia* (Asteracea). I—VIII. Nuovo Giorn. bot. ital. **53**, 1—69, 437—511 (1946b). — La "Semigamia", singolare comportamento del nucleo spermatico nelle uova diploidi delle specie apomittiche del genere *Rudbeckia* (Asteraceae) e consequente embriogenesi di tipi chimerico. R. C. Accad. Naz. Lincei VIII **2** (1), 63—67 (1947a). — Ricerche cariologiche ed embryologiche sul genere *Rudbeckia* (Asteracea). IX—XII. Nuovo Giorn. bot. ital. **54**, 377—431, 531—567 (1947b). — Apomissia in *Hieracium ramosum* Waldst. et Kit. Atti Soc. Toscana Sci. Nat. **54**, 1—11 (1947c). — Sulla terminologia dei processi apomittici. Nuovo Giorn. bot. ital., Nuova Ser. **53**, 23 pp. (1947d). — Ricerche sulla parameiosi restiuzionale nel genere *Taraxacum*. Caryologia **1**, 1—47 (1948). — L'alterazione della meiosi nella reproduzione apomittica di *Chondrilla juncea* L. Caryologia **2**, 23—30 (1949). — L'alterazione della meiosi nella riproduzione apomittica di *Erigeron karwinskianus* DC. var. *mucronatus* DC. (Asteraceae). Caryologia **2**, 165—204 (1950). — Development of Angiosperm embryo sacs with nonhaploid eggs. Amer. J. Bot. **38**, 718—724 (1951). — Nuovi reperti di apomissia e di amfimissia nel genere *Rudbeckia* (Compositae). Atti Soc. Toscana Sci. Nat., Mem., B **59**, 205—209 (1952a). — Ricerche embriologiche su *Rudbeckia columnaris* Sims. Atti Soc. Toscana Sci. Nat., Mem., B **59**, 172—197 (1952b). — Utilizzazione di dati embriologici in funzione di una migliore conoscenza della sistematica del genere *Rudbeckia (Compositae)*. Atti Soc. Toscana Sci. Nat., Mem., B **59**, 217—221 (1952c). — Ricerche embriologiche nel genere *Arnica (Compositae)*. Atti Soc. Toscana Sci. Nat., Mem., B **59**, 210—216 (1952d). — The concepts of spore, sporogenesis and apospory. Phytomorphology **5**, 173—177 (1955). — Unusual cytological features in the apomictic *Rudbeckia sullivantii* BOYNTON et BEADLE. Caryologia **8**, 1—32 (1955). — Do new types of embryo sac development occur in *Antennaria carpatica*? Phytomorphology **6**, 119—123 (1956). — Recent advances in the embryology of Angiosperms. Int. Soc. Plant Morph. Univ. Delhi **1963**, 221—264. — BENNETT, H. W., and E. C. BASHAW: An interspecific hybrid in *Paspalum*. J. Hered. **51**, 81—85 (1960). — BERGER, A.: Die Agaven. 288 pp. 1915. — BERGER, X.: Intersexe Bildungen in den Antheren partiell apomiktischer *Rubusarten* und *Rubus*bastarde. Z. Zellforsch. **38**, 472—474 (1953). — Untersuchungen über die Embryologie partiell apomiktischer *Rubus*bastarde. Ber. schweiz. bot. Ges. **63**, 224—265 (1953). — BERGMAN, B.: Zytologische Studien über die Fortpflanzung bei den Gattungen *Leontodon* und *Picris*. Svensk bot. Tidskr. **29**, 155—301 (1935a). — Zytologische Studien über sexuelles und asexuelles *Hieracium umbellatum*. Hereditas (Lund) **20**, 47—64 (1935b). — Zur Kenntnis der Zytologie der skandinavischen *Antennaria-Arten*. Hereditas (Lund) **20**, 214—226 (1935c). — Eine neue apomiktische *Antennaria*. Svensk bot. Tidskr. **31**, 391—394 (1937). — Zur Embryologie der Gattung *Erigeron*. Svensk bot. Tidskr. **36**, 429—443 (1942). — A contribution to the knowledge of the embryo sac mother cell and its development in two apomicts. Svensk bot. Tidskr. **38**, 249—259 (1944). — Meiosis in two different clones of the apomictic *Chondrilla juncea*. Hereditas (Lund) **36**, 297—320 (1950). — On the formation of reduced and unreduced gametophytes in the females of *Antennaria carpatica*. Hereditas (Lund) **37**, 501—518 (1951). — Asyndesis in macrosporogenesis of diploid, triploid and tetraploid *Chrysanthemum carinatum*. Hereditas (Lund) **38**, 83—90 (1952a). — Chondrilla chondrilloides, a new sexual *Chondrilla* species. Hereditas (Lund) **38**, 367—369 (1952b). — Chromosome morphological studies in *Chondrilla juncea* and some remarks on the microsporogenesis. Hereditas (Lund) **38**, 128—130 (1952c). — BERGNER, A. D.: Guayule with low chromosome numbers. Science **99**, 224—225 (1944). — Polyploidy and eneuploidy in Guayule. U. S. Dept. Agr., Tech. Bull. **918**, 1—36 (1946). — BETH, K.: Untersuchungen über die Auslösung von Adventivembryonie durch Wundreiz. Planta (Berl.) **28**, 296—343 (1938). — BILLINGS, F. H.: Some new features in the reproductive cytology of angiosperms, illustrated by *Isomeris arborea*. New Phytologist **36**, 301—326 (1937). — BØCHER, T. W.: Zur Zytologie einiger arktischen und borealen Blütenpflanzen. Svensk bot. Tidskr. **32**, 346—361 (1938). — Cytological studies of *Arabis Holboellii* Hornem. Hereditas (Lund) **33**, 573 (1947). — Chromosome numbers of some arctic or boreal flowering plants. Medd. om Grønland **147** (6), 32 pp. (1950). — Cytological and embryological studies in the amphiapomictic *Arabis Holboellii* complex. Kon. danske Vid. Selsk. Biol. Skr. **6** (7), 59 pp. (1951). — Experimental taxonomical studies in the *Arabis Holboellii* complex. Svensk. bot. Tidskr. **48**, 31—44 (1954). — BÖÖS, G.: Neue embryologiche Studien über *Alchemilla arvensis* (L.) Scop. Bot. Not. (Lund) **77**, 209—250 (1924). — BOGDAN, A. V.: Breeding behaviour of *Cenchrus ciliaris* in Kenya. E. Afr. Agric. For. J. **26**, 241 (1961). — BORGAONKAR, D. S.: Apomixis and speciation in the Bothriochloninae. Diss. Ohl. St. Coll. Stillwater. U.S.A. 1964, 25, No 64—8898, 762—63 (1964). — BORGAONKAR, D. S., J. R. HARLAN, and J. M. J. DE WET: A study of apomictic mechanisms in the *Bothriochloininae*. Amer. J. Bot. **49**, 674 (1962). — BORGAONKAR, D. S., and A. P. SINGH: Species relationships in *Dichantium*. IV. *D. annulatum* and *D. papillosum* Phyton (Argent.) **19**, 101—107 (1962). — BORGAONKAR, D. S., and J. M. J. DE WET: A cytological of hybrids between *Dichanthium annulatum* and *Dichanthium fecundum*. Phyton (Argent.) **12**, 137—144 (1960). — Segregation among partial apomictic *Bothriochloa* hybrids. Amer. J. Bot. **48**, 545 (1961). — Cytological studies in some segregating population of hybrids within the *Bothriochloininae*. Cytologia (Tokyo) **28**, 54—67 (1963). — Cytogenetical study of hybrids between *Dichanthium annulatum* and *D. fecundum*. III. Phyton (Argent.) **21**, 55—60 (1964). — BOWDEN, W. M.: Diploidy polyploidy, and winter hardiness relationship in the flowering plants. Amer. J. Bot. **27**, 357—371 (1940). — BOWER, F. O.: On apospory in ferns. J. Linn. Soc. Bot. **21**, 360—368 (1886). — Note on attempts to induce aposporous development in ferns. Ann. Bot. **4**, 168—169 (1889). — BRAUN, A.: Über Parthenogenesis bei Pflanzen. Abh. Kgl. Akad. Wiss. Berlin 1856, S. 311 bis 376 (1857). — Über Polyembryonie und Keimung von *Caelebogyne*. Abh. Kgl. Akad. Wiss. Berlin 1859, S. 109—263 (1860). — BRITTINGHAM, W. H.: An artificial hybrid between two species of bluegrass. J. Hered. **32**, 57—63 (1941). — Type of seed formation as indicated by the nature and extent of variation in Kentucky bluegrass, and its practical implications. J. agric. Res. **67**, 225—264 (1943). — BROOKS, M. H.: A comparative study of the patterns of development in certain facultative apomicts. Proc. X. Int. Congr. Genetics, Montreal, 36 (1958). — BROWN, W. L.: Chromosome complements of five species of *Poa* with an analysis of variation in *Poa pratensis*. Amer. J. Bot. **26**, 717—723 (1939). — The cytogenetics of *Poa pratensis*. Ann. Missouri bot. Garden **28**, 493—522 (1941). — BROWN, W. V.: Apomixis in *Zephyranthes texana* Herb. Amer. J. Bot. **38**, 697—701 (1951). — Apomixis as related to geographical distribution in the panicoid grass tribes. J. S. Afr. Bot. **24**, 191–200 (1958). — BROWN, W. V., and W. H. P. EMERY: Some South African apomictic grasses. J. South Afr. Bot. **23**, 123—125 (1957a). — Apomixis in the *Graminae*, tribe *Andropogoneae: Themeda triandra* and *Bothriochloa ischaemum*. Bot. Gaz. **118**, 246—253 (1957b). — Apomixis in the *Gramineae: Panicoideae*. Amer. J. Bot. **45**, 253—263 (1958). — BURTON, G. W.: The method of reproduction in common Bahia grass, *Paspalum notatum*. J. Amer. Soc. Agron. **40**, 443—452 (1948). — Conventional breeding of dallisgrass, *Paspalum dilatatum* POIR. Crop Sci. **2**, 491—494 (1962). — BURTON, G. W., and I. FORBES: The genesis and manipulation of obligate apomixis in common Bahia grass *(Paspalum notatum* FLUGGE*)*. Proc. 8th Int. Grassl. Congr. Reading, 1960, p. 66—71. — BURTON, G. W., and IAN FORBES jr.: Cytology of diploids, natural and induced tetraploids and intraspecies hybrids of Bahiagrass, *Paspalum notatum* FLUGGE. Crop Sci. **1**, 6, 402—406 (1961).

CARANO, E.: Nuovo ricerche sulla embriologia delle Asteraceae. Ann. Bot. (Roma) **15**, 97—196 (1921). — Sul particolare sviluppo del gametofito ♀ di *Euphorbia dulfis* L. R. C. Accad. Lincei, Cl. Sci. fis., mat. e nat., Ser. VIa **1**, 633 (1925). — Ulteriori osservazioni su *Euphorbia dulcis* L., in rapporto col suo comportamento apomittico. Ann. Bot. (Roma) **17**, 50—79 (1926). — CARNAHAN, H. L., and D. HELEN HILL: Cytology and genetics of forage grasses. Bot. Rev. **27**, 1—162 (1961). — CARR, D. J., and ENG KOK NG: Experimental induction of flower formation in Kikuyu grass (*Pennisetum clandestinum* Hochst.). Austr. J. agric. Res. **7**, 1—6 (1956). — CATCHESIDE, D. G.: The B-chromosomes of *Parthenium argentatum*. Heredity **1**, 393—394 (1947). — The B-chromosomes of *Parthenium argentatum*. Genet. Iber **2**, 139—148 + 1 pl. (1950). — CELARIER, R. P.: The cyto-geography of the *Bothriochloa ischaemum* complex. II. Chromosome behaviour. Amer. J. Bot. **44**, 729—738 (1957). — CELARIER, R. P., and J. R. HARLAN: Apomixis in *Bothriochloa, Dichanthium* and *Capillipedium*. Phytomorphology **7**, 93—102 (1957). — CELARIER, R. P., K. L. MEHRA, and M. L. WULF: Cytogeography of the *Dichanthium annulatum* complex. Brittonia **10**, 59—72 (1958). — CESCA, G.: Richerche cariologiche ed embriologiche sulle *Euphorbiaceae*. I — Su alcuni biotipi di *Euphorbia dulcis* L. della Toscana. Caryologia **14**, 79—96 (1961). — CHASE, S. S.: Selection for parthenogenesis and monoploid fertility in maize. Genetics **37**, 573—574 (1952). — CHEDDA, H. R., and J. R. HARLAN: Chromosomal evolution and genome building in *Bothriochloininae*. Niger. agric. J. **1**, 18—21 (1964). — CHIARUGI, A.: Aposporia e apogamia in "*Artemisia nitida*" Bertol. Nuovo Giorn. bot. ital., Nuova Ser. **33**, 501—626 (1926). — CHIARUGI, A., e E. FRANCINI: Apomissia in "Ochna serrulata" Walp. Nuovo Giorn. bot. ital., Nuova Ser. **37**, 1—250 (1930). — CHRISTEN, H. R.: Untersuchungen über die Embryologie pseudogamer und sexueller *Rubus*arten. Ber. schweiz. bot. Ges. **60**, 153—198 (1950). — Die Embryologie von *Rubus idaeus* und von Bastarden zwischen *Rubus caesius* und *Rubus idaeus*. Z. indukt. Abstamm.- u. Vererb.-Lehre **84**, 454—461 (1952). — CHRISTOFF, M.: Über die Fortpflanzungsverhältnisse bei einigen Arten der Gattung *Hieracium* nach einer experimentell induzierten Chromosomenvermehrung. Planta (Berl.) **31**, 73—90 (1940). — Die genetische Grundlage der apomiktischen Fortpflanzung bei *Hieracium aurantiacum*. Z. indukt. Abstamm.- u. Vererb.-Lehre **80**, 103—125 (1942). — CHRISTOFF, M., and M. A. CHRISTOFF: Meiosis in the somatic tissue responsible for the reduction of chromosome number in the progeny of *Hieracium Hoppeanum* Schult. Genetics **33**, 36—42 (1948). — CHRISTOFF, M., u. G. PAPASOVA: Die genetischen Grundlagen der apomiktischen Fortpflanzung in der Gattung *Potentilla*. Z. indukt. Abstamm.- u. Vererb.-Lehre **81**, 1—27 (1943). — CHRISTOFF, M., u. A. POPOFF: Cytologische Studien über die Gattung *Hieracium*. Planta (Berl.) **20**, 440—447 (1933). — CHRISTOFF, M. A.: Embryologische Studien über die Fortpflanzung einiger *Poa*-Arten. Jb. Univ. Heil. Klim. Ochrid, Sofia, Fak. Land.- u. Forstwirtsch. **20**, 169—178 (1942). — Die Polyembryonie als Ursache für die Variabilität in den apomiktischen Arten der Gattung *Poa*. Jb. Univ. Heil. Klim. Ochrid, Sofia, Fak. Land.- u. Forstwirtsch. **22**, 153—179 (1944). — CLAUSEN, J.: Stages in the evolution of plant species. 206 pp. 1951. — Introgression facilitated by apomixis in polyploid *Poas*. Euphytica **10**, 87—94 (1961). — CLAUSEN, J., P. GRUN, W. M. HIESEY and M. NOBS: New *Poa* hybrids. Carnegie Inst. Wash., Year Book **51**, 111—117 (1952). — CLAUSEN, J., P. GRUN, A. NYGREN and M. NOBS: Genetics and evolution of *Poa*. Carnegie Inst. Wash., Year Book **50**, 109—111 (1951). — CLAUSEN, J., W. M. HIESEY, P. GRUN and M. NOBS: Survey of the range-grass program. Carnegie Inst. Wash., Year Book **51**, 107—111 (1952). — CLAUSEN, J., W. M. HIESEY and M. A. NOBS: The *Poa* program. Carnegie Inst. Wash., Year Book **52**, 170—173 (1953). — *Poa* investigations. Carnegie Inst. Wash. Year Book **54**, 172—175 (1955). — Studies in *Poa*. Carnegie Inst. Wash., Year Book **55**, 236—239 (1956). — Contrasting tolerance ranges of apomictic species and hybrids of *Poa*. Carnegie Inst. Wash., Year Book **56**, 293—295 (1957). — Studies in *Poa* hybridization. Carnegie Inst. Wash. Year Book **61**, 325—333 (1962). — CLAUSEN, J., D. KECK and W. M.

HIESEY: Experimental studies on the nature of species. I. Effect of varied environments on western North American plants. Carnegie Inst. Wash. Publ. **520**, 1—452 (1940). — Experimental studies on the nature of species. II. Plant evolution through amphiploidy and autoploidy, with examples from the Madiinae. Carnegie Inst. Wash., Publ. **564**, 1—174 (1945). — Experimental taxonomy. Carnegie Inst. Wash., Year Book **46**, 95—104 (1947). — Experimental taxonomy. Carnegie Inst. Wash., Year Book **47**, 105—110 (1948). — CLAUSEN, J., D. KECK, W. M. HIESEY and P. GRUN: Experimental taxonomy. *Poa* investigations. Carnegie Inst. Wash., Year Book **48**, 97—103 (1949). — Experimental taxonomy. *Poa* investigations. Carnegie Inst. Wash., Year Book **49**, 104—107 (1950). — CLAUSEN, J., D. KECK, W. M. HIESEY, P. GRUN, A. NYGREN and M. NOBS: Climatic tolerances of *Poa* species and hybrids. Carnegie Inst. Wash., Year Book **50**, 105—108 (1951). — CLAUSEN, J. C., M. A. NOBS, and W. M. HIESEY: Studies in *Poa*. Carnegie Inst. Wash. Year Book **60**, July 1, 1960 — June 30, 384—385 (1960/61). — CLIFFORD, H. T.: Vivipary in *Eleusine indica* (L.) GAERTN. Nature (Lond.) **184**, 1888 (1959). — COE, G. E.: Cytology of reproduction in *Cooperia pedunculata*. Amer. J. Bot. **40**, 335—343 (1953). — COOPER, D. C., and R. A. BRINK: The endosperm-embryo relationship in an autonomous apomict, *Taraxacum officinale*. Bot. Gaz. **3**, 139—153 (1949). — COULON, J. DE: *Nardus stricta*. Étude physiologique, anatomique et embryologique. Mém. Soc. Vaud. Sci. Natur. No 6, 247—442 (1923). — COX, L. M., and J. CLAUSEN: Chromosome numbers of hybrid *Poa* lines. Carnegie Inst. Wash., Year Book **54**, 175—177 (1955). — CRANE, M. B., and P. T. THOMAS: Segregation in asexual (apomictic) offspring in *Rubus*. Nature (Lond.) **143**, 684 (1939). — CURTIS, W. M.: Variation in certain Tasmanian plants. New Phytologist **51**, 398—414 (1952). — CZAPIK, R.: Badania embriologiczne nad rodzajem *Potentilla* L. II. *P. arenaria*. Embryological studies in the genus *Potentilla* L. II. *Potentilla arenaria*. Acta biol. cracov. **5**, 29—42 (1962). — Badania embriologiczne nad rodzajem *Potentilla* L. III. Mieszańce miedzy *P. Crantzii* i *P. arenaria*. Embryological studies in *Potentilla* L. III. Hybrids between *P. Crantzii* and *P. arenaria*. Acta biol. cracov. **5**, 43—61 (1961).

D'AMATO, F.: Apomissia in *Statice oleaefolia* Scop. var. *confusa* Goodr. (Plumbaginaceae). Nuovo Giorn. bot. ital. **47**, 504 (1940). — Risultati di una Carioembriologica in una populazione di *Nothoscordum fragrans* Kunth. Caryologie **1**, 194—200 (1949a). — Triploidia e apomissa in *Statice oleaefolia* Scop. var. *confusa* Goodr. Caryologia **2**, 71—84 (1949b). — DARLINGTON, C. D., and E. K. JANAKI-AMMAL: Chromosome atlas of cultivated plants. 397 pp. 1945. — DARLINGTON, C. D., and A. P. WYLIE: Chromosome atlas of flowering plants. 519 pp. London 1955. — DE LISLE, D. G.: Chromosome numbers in Cenchus (Graminae). Amer. J. Bot. **51**, 1133—1134 (1964). — DERMEN, H.: Aposporic parthenogenesis in a triploid apple, *Malus hupehensis*. J. Arn. Arb. **17**, 90—105 (1936). — DESAI, SHARDA: Studies on apospory in *Skimmia*. Phyton (Argent.) **16**, 147—152 (1961). — DEWALD, C. L., and J. R. HARLAN: Stigma removal studies on certain accessions of *Bothriochloa intermedia* and *Dichanthium annulatum*. Crop Sci. **1**, 15—17 (1961). — DE WET, J. M. J.: Chromosome numbers of a few South African grasses. Cytologia (Tokyo) **19**, 97—103 (1954). — Additional chromosome numbers in Transvaal grasses. Cytologia (Tokyo) **23**, 113—118 (1958). — DE WET, J. M. J., and D. S. BORGAONKAR: Aneuploidy and apomixis in *Bothriochloa* and *Dichantium*. Bot. Gaz. **124**, 437—440 (1963). — DE WET, J. M. J., D. S. BORGAONKAR, and W. L. RICHARDSON: Chromosome number and mode of reproduction in the *Bothriochloininae*. Caryologia **16**, 47—55 (1963). — DE WET, J. M. J., and J. R. HARLAN: Role of apomixis in the evolution of the *Bothriochloa-Dichanthium* complex. Crop Sci. **3**, 314—316 (1963). — Morphology of the compilospecies *Bothriochloa intermedia*. Amer. J. Bot. **53**, 94—98 (1966). — DE WET, J. M. J., and M. L. HIGGINS: Species relationships within the *Bothriochloa pertusa* complex. Phyton (Argent.) **20**, 205—211 (1963). — Cytology of the *Bothriochloa pertusa* complex (1964). — DE WET, J. M. J., K. L. MEHRA, and D. S. BORGAONKAR: Chromosome association in *Dichantium* hybrids. Cytologia (Tokyo) **26**, 78—82 (1961). — DE WET, J. M. J., and W. L. RICHARDSON: Morphological variation and species relationships in *Dichantium*. Phyton (Argent.) **20**, 19—28 (1963). — DE WET, J. M. J., and B. D. SCOTT: Essential oils as taxonomic criteria in *Bothriochloa*. Bot. Gaz. **126**, 209—214 (1965). — DE WET, J. M. J., and A. P. SINGH: Species relationships in *Dichantium* V. The diploid species. Caryologia **17**, 153—160 (1964). — DIANNELIDIS, T.: Über das spontane Vorkommen von diploidem *Allium carinatum*. Öst. bot. Z. **93**, 66—72 (1936). — DODDS, K. S.: Polyhaploids in *Solanum demissum*. Nature (Lond.) **166**, 795 (1950). — DODDS, K. S., and N. W. SIMMONDS: Genetical and cytological studies of *Musa*. IX. The origin of an edible diploid and the significance of interspecific hybridization in the banana complex. J. Genet. **48**, 285—292 (1948). — The cytogenetics of bananas. Proc. VIII. Int. Congr. Genet. Hereditas (Lund) Suppl. Vol. 35, 262 (1949). — DOWRICK, G. J.: Biology of reproduction in *Rubus*. Nature (Lond.) **191**, 680—682 (1961). — DRUERY, CH. T.: Observations on a singular mode of development in the Lady-Fern (*Athyrium Filix-foemina*). J. Linn. Soc. Bot. **21**, 354—357 (1886a). — Further notes on a singular mode of reproduction in *Athyrium Filix-foemina*, var. *clarissima*. J. Linn. Soc. Bot. **21**, 358—360

(1886b). — DUARA, B. N., and G. L. STEBBINS, jr.: A polyhaploid obtained from a hybrid derivative of *Sorghum halepense* × *S. vulgare* var. *sudanense*. Genetics **37**, 369—374 (1952). — DUICH, J. M., and H. B. MUSSER: The extent of aberrants produced by Merion Kentucky bluegrass, *Poa pratensis* L., as determined by first and second generation progeny tests. Agronomy J. **51**, 421—424 (1959).

EDMAN, G.: Apomeiosis und Apomixis bei *Atraphaxis frutescens* C. Koch. Acta Horti Berg. **11**, 13—66 (1931). — EHRENBERG, L.: Kromosomtalen hon några kärlväxter. Bot. Not. (Lund) **98**, 430—437 (1945). — EINSET, J.: Chromosome studies in *Rubus*. Gents Herb. **7**, 181—192 (1947). — Apomixis in American polyploid blackberries. Amer. J. Bot. **38**, 768—772 (1951). — EKMAN, E.: Notes on some Greenland Antennariae. Svensk bot. Tidskr. **21**, 49—57 (1927). — EMERY, W. H. P.: A study of reproduction in *Setaria macrostachya* and its relatives in the southwestern United States and northern Mexico. Bull. Torrey Bot. Club **84**, 106—121 (1957). — EMERY, W. H. P., and W. V. BROWN: Apomixis in the *Graminae*. Tribe *Adropogoneae: Heteropogon contortus*. Madrano **14**, 238—246 (1958). — ENGELBERT, V.: Reproduction in some *Poa* species. Canad. J. Res. C **18**, 518—521 (1940). — The development of twin embryo sacs, embryos and endosperm in *Poa arctica* R. Br. Canad. J. Res. C **19**, 135—144 (1941). — ERLANDSSON, S.: The chromosome numbers of some *Taraxacum* species. Bot. Not. (Lund) **92**, 261—264 (1939). — ERNST, A.: Bastardierung als Ursache der Apogamie im Pflanzenreich. 666 pp. 1918. — Addenda, Corrigenda und Desiderata zur Genetik des amphidiploiden Artbastardes *Primula kewensis*. Arch. Klaus-Stift. Vererb.-Forsch. **24**, 1—104 + 2 pl (1949). — Resultate aus Kreuzungen zwischen der tetraploiden. monomorphen *Primula japonica* und diploiden, mono- und dimorphen Arten der Sekton *Candelabra*. Arch. Klaus-Stift. Vererb.-Forsch. **25**, 135—236 + 4 pl (1950). — "Maternal hybrids" nach interspezifischen Bestäubungen in der Gattung *Primula*. 1. Sekton *Candelabra*. Arch. Klaus-Stift. Vererb.-Forsch. **26**, 1—96 + 2 pl. (1951a). — "Maternal hybrids" nach interspezifischen Bestäubungen in der Gattung *Primula*. 2. Sekton *Farinosae*. Arch. Klaus-Stift. Vererb.-Forsch. **26**, 187—322 + 2 pl. (1951b). — ERNST, A., u. CH. BERNHARD: Entwicklungsgeschichte des Embryosackes und des Embryos von *Burmannia candida* Engl. und B. *Championii* Thw. Ann. Jard bot. Buitenzorg II **10**, 161—188 (No 9 in Beiträge zur Kenntnis der Saprophyten Javas) (1912). — ESAU, K.: Apomixis in Guayule. Proc. nat. Acad. Sci. (Wash.) **30**, 352—355 (1944). — Morphology of reproduction in Guayule and certain other species of *Parthenium*. Hilgardia **17**, 61—101 (1946).

FAGERLIND, F.: Embryologische, zytologische und bestäubungsexperimentelle Studien in der Familie Rubiaceae nebst Bemerkungen über einige Polyploiditätsprobleme. Acta Horti Berg. **2**, 195—470 (1937). — Zytologie und Gametophytenbildung in der Gattung *Wikstroemia*. Hereditas (Lund) **26**, 23—50 (1940). — Die Samenbildung und die Zytologie bei agamospermischen und sexuellen Arten von *Elatostoma* und einigen nahestehenden Gattungen nebst Beleuchtung einiger damit zusammenhängender Probleme. Kung. Svensk Vet.-Akad. Handl. **21** (4), 1—130 (1944). — Bildung und Entwicklung des Embryosacks bei sexuellen und agamospermischen *Balanophora*-Arten. Svensk bot. Tidskr. **39**, 65—82 (1945). — Sporogenesis, Embryosackentwicklung und pseudogame Samenbildung bei *Rudbeckia laciniata* L. Acta Horti Berg. **14**, 39—90 (1946). — Macrogametophyte formation in two agamospermous *Erigeron* species. Acta Horti Berg. **14**, 221—247 (1947a). — Makrosporogenese und Embryosackbildung bei agamospermischen *Taraxum*-Biotypen. Svensk bot. Tidskr. **41**, 365—390 (1947b). — FARQUHARSON, L. I.: Apomixis and polyembryony in *Tripsacum dactyloides*. Amer. J. Bot. **42**, 737—743 (1955). — FARUQI, S. A.: The significance of apomixis in evolution and speciation of the *Bothriochloinae*. Biologia (Lahore) **9**, 57—64 (1963). — Cytogenetic studies of the *Bothriochloa intermedia* complex. Cytologia (Tokyo) **29**, 280—297 (1964a). — Cytogeography of the *Bothriochloa intermedia complex*. Diss. Oklah. St. Coll. Stillwater. U.S.A. Abstr. 25, No 64—8915, 774—775 (1964b). — FERNALD, M. L.: Some Antennarias of northern New England. Rhodora **1**, 71—75 (1899). — Recent discoveries in the Newfoundland flora. Rhodora **35** (*Festuca prolifera*, p. 133—136; *Antennaria*, p. 327—346) (1933). — Memoranda on Antennaria. Rhodora **38**, 229—231 (1936). — FISHER, W. D., E. C. BASHAW and D. C. HOLT: Evidence for apomixis in *Pennisetum ciliare* and *Cenchrus setigerus*. Agron. J. **46**, 401—404 (1954). — FLOVIK, K.: Cytological studies of arctic grasses. Hereditas (Lund) **24**, 265—376 (1938). — Chromosome numbers and polyploidy within the flora of Spitzbergen. Hereditas (Lund) **26**, 430—440 (1940). — FOCKE, W. O.: Die Pflanzenmischlinge. 569 pp. Berlin 1881. — FORENBACHER, A.: Die Fortpflanzungsverhältnisse bei der Gattung *Potentilla*. Jugoslav. Akad. Znanosti i unjetnosti, Zagreb Izvjesca o raspravama matem. prirod. Razreda **1**, 86—97 (1914). Cit. after SCHNARF 1929. — FRIES, TH. C. E.: *Antennaria alpina* (L.) Gaertn. och dess skandinaviska elementararter. Svensk bot. Tidskr. **13**, 179—193 (1919). — FROST, H. B.: Polyembryony, heterozygosis and chimeras in citrus. Hilgardia **1**, 365—402 (1926). — FÜRNKRANZ, D.: Cytogenetische Untersuchungen an *Taraxacum* im Raume von Wien. Öst. bot. Z. **107**, 310—350 (1960). — Cytogenetische Untersuchungen an *Taraxacum* im Raume von Wien. 11. Hybriden zwischen *T. officinale* und *T. palustre*.

Öst. bot. Z. **108**, H. 4/5, 408—415 (1961). — FULTZ, J. L.: Somatic chromosome complements in *Bouteloua*. Amer. J. Bot. **29**, 45—55 (1942). — FURUSATO, K.: Genetics and breeding of *Citrus*. Nat. Inst. Genet. (Japan) Ann. Rep. **1951**, 41 (1952). — Tetraploidy in *Citrus*. Nat. Inst. Genet. (Japan) Ann. Rep. No 3, 51—52 (1952). — Abnormal growth of *Citrus* trees caused by deviating chromosome numbers in the grafting stocks. Nat. Inst. Genet. (Japan) Ann. Rep. No 3, 52—53 (1952). — Studies on the polyembryony in *Citrus*. Jap. J. Genet. **28**, 165—166 (1953). — Studies on polyembryony in *Citrus*. Nat. Inst. Genet. (Japan) Ann. Rep. No **4**, 56 (1953). — Inheritance of mono and polyembryony in Citrus. Ann. Rep. Nat. Inst. Genet. **11**, 59—60 (1960).

GANONG, W. F.: Upon polyembryony and its morphology in *Opuntia vulgaris*. Bot. Gaz. **25**, 221—228 (1898). — GARBER, E. D.: Cytotaxonomic studies in the genus Sorghum. Univ. Calif. Publ. Bot. **23** (6), 283—362 + pl. 40—45 (1950). — GARDNER, E. J.: Sexual plants with high chromosome number from an individual plant selection in a natural population of guayule and mariola. Genetics **31**, 117—124 (1946a). — Wind pollination in guayule, *Parthenium argentatum* Gray. Proc. Amer. Soc. Agron. **38**, 264—272 (1946b). — GEITLER, L.: Natürliches diploides *Allium carinatum*. Ber. dtsch. bot. Ges. **61** (1943), 210—211 (1944). — GELIN, O.: Embryologische und cytologische Studien in Heliantheae-Coreopsidinae. Acta Horti Berg. **11**, 99—128 (1934). — GENTCHEFF, G.: Zytologische und embryologische Studien über einige *Hieracium*-Arten. Planta (Berl.) **27**, 165—195 (1937). — Degenerative phenomena in the male gametophyte of *Hieracium* in relation to mitotic behavior of the cell. Univ. St. Kliment, Ochridsky Sofia, Fac. Agr., Year Book **19**, 107—150 (1941). — GENTCHEFF, G., and Å. GUSTAFSSON: The balance system of meiosis in Hieracium. Hereditas (Lund) **26**, 209—249 (1940a). — Parthenogenesis and pseudogamy in Potentilla. Bot. Not. (Lund) **93**, 109—132 (1940b). — GERASSIMOVA-NAVASHINA, E. N.: The effect of temperature conditions on the course of embryological processes in plants. Dokl. Akad. Nauk CSSR **131**, 1—6, 81—84 (1961). — GERASSIMOVA-NAVASHINA, H.: Fertilization and events leading up to fertilization, and their bearing on the origin of angiosperms. Phytomorphology **11**, 1, 2, 139—146 (1961). — GERSTEL, D. U.: Is resistance to Verticillium wilt in guayule related to chromosome number? Agron. J. **42**, 310—311 (1950a). — Self-incompatibility studies in guayule. II. Inheritance. Genetics **35**, 482—506 (1950b). — GERSTEL, D. U., B. L. HAMMOND and C. KIDD: An additional note on the inheritance of apomixis in guayule. Bot. Gaz. **115**, 89—93 (1953). — GERSTEL, D. U., and W. M. MISHANEC: On the inheritance of apomixis in *Parthenium argentatum*. Bot. Gaz. **112**, 96—106 (1950). — GERSTEL, D. U., and MARY E. RINER: Self-incompatibility studies in guayule. I. J. Hered. **41**, 49—55 (1950). — GILDENHUYS, P. J., and K. BRIX: Apomixis in *Pennisetum dubium*. S. Afr. J. Agric. Sci. **2**, 231—245 (1959). — GOULD, F. W.: Notes on apomixis in sidecoats grama. J. Range Mgmt. **12**, 25—28 (1959). — Biosystematic studies in the *Bouteloua curtipendula* complex. III. pollen size as related to chromosome numbers. Amer. J. Bot. **51** (2), 166—172 (1964). — GRANICK, E. B.: A karyosystematic study of the genus *Agave*. Amer. J. Bot. **31**, 283—298 (1944). — GRAZI, F., M. UMAERUS, and E. ÅKERBERG: Observations on the mode of reproduction and the embryology of *Poa pratensis*. Hereditas (Lund) **47**, 489—541 (1961). — GRUN, P.: *Poa nervosa*, an extreme in asexual reproduction. Carnegie Inst. Wash., Year Book **50**, 112—113 (1951). — Apomixis and variation in *Poa nervosa*. Carnegie Inst. Wash., Year Book **51**, 117—119 (1952). — Effect of environment upon chromosomal pairing of some species and hybrids of *Poa*. Amer. J. Bot. **39**, 318—323 (1952). — Cytogenetic studies of *Poa*. I. Chromosome numbers and morphology of interspecific hybrids. Amer. J. Bot. **41**, 671—678 (1954). — Cytogenetic studies of *Poa*. II. The pairing of chromosomes in species and interspecific hybrids. Amer. J. Bot. **42**, 11—18 (1955). — Cytogenetic studies in *Poa*. III. Variation within *Poa nervosa*, an obligate apomict. Amer. J. Bot. **42**, 778—783 (1955). — GRUN, P., and E. L. TRIPLETT: *Poa* cytology. Carnegie Inst. Wash., Year Book **51**, 119—122 (1952). — GUDJÓNSSON, G.: Om *Aphanes arvensis* L. og A. *microcarpa* (Boiss et Reut.) Rothm. og deres Udbredelse i Danmark. Svensk bot. Tidskr. **45**, 352—370 (1941). — GUSTAFSSON, ALTON: Cytological studies in the genus *Hieracium*. Bot. Gaz. **94**, 512—533 (1933). — GUSTAFSSON, Å.: Zytologische und experimentelle Studien in der Gattung Taraxacum. Hereditas (Lund) **16**, 41—62 (1932). — Studies on the mechanism of parthenogenesis. Hereditas (Lund) **21**, 1—112 (1935). — The genesis of the European blackberry flora. Lunds Univ. Årsskr., N. F. **39** (6); 1—200 (1943). — Apomixis in higher plants. Part 1—III. Lunds Univ. Årsskr., N. F. Avd. II **42** (3), **43** (2), **43** (12), 370 pp. (1946/47). — GUSTAFSON, Å., and I. GADD: Mutations and crop improvement. IV. *Poa pratensis* L. *(Gramineae)*. Hereditas (Lund) **53**, 90—102 (1965). — GUSTAFSSON, Å., and A. NYGREN: The temperature effect on pollen formation and meiosis in *Hieracium robustum*. Hereditas (Lund) **32**, 1—14 (1946).

HABERLANDT, G.: Die Entwicklungserregung der Eizellen einiger parthenogenetischer Kompositen. S.-B. preuß. Akad. Wiss. **51**, 861—881 (1921). — HÄFLIGER, E.: Zytologisch-embryologische Untersuchungen pseudogamer Ranunkeln der *Auricomus*-Gruppe. Ber. schweiz. bot. Ges. **53**, 317—382 (1943). — HAGERUP, O.: Studies on the significance of polyploidy III. *Deschampsia* and *Aira*. Hereditas (Lund) **25**, 185—192 (1939). — On fertilization, polyploidy and haploidy in *Orchis maculatus* L. sens. lat. Dansk bot. Ark. **11** (5), 1—25 (1944). — Facultative parthenogenesis and haploidy in *Epipactis latifolia*. Kung. danske Vid. Selsk., Biol. Medd. **19** (11), 1—14 (1945). — The spontaneous formation of haploid, polyploid, and aneuploid embryos in some orchids. Kung. danske Vid, Selsk., Biol. Medd. **20** (9), 22 pp. (1947). — HAIR, J. B.: Subsexual reproduction in Agropyron. Heredity **10**, 129—160 (1956). — HÅKANSSON, A.: Die Entwicklung des Embryosacks und die Befruchtung bei *Poa alpina*. Hereditas (Lund) **29**, 25—61 (1943). — Ergänzende Beiträge zur Embryologie von *Poa alpina*. Bot. Not. (Lund) **97**, 299—311 (1944). — Untersuchungen über die Embryologie einiger *Potentilla*-Formen. Lunds Univ. Årsskr., N. F. Avd. II **42** (5), 1—70 (1946). — Embryology of *Poa alpina* plants with accessory chromosomes. Hereditas (Lund) **34**, 233—247 (1948). — Parthenogenesis in *Allium*. Bot. Not. (Lund) **104**, 143—179 (1951). — Die Samenbildung bei *Nothoscordum fragans*. Bot. Not. (Lund) **110**, 129—139 (1953). — Notes on the giant chromosomes of *Allium nutans*. Bot. Not. (Lund) **106**, 196—200 (1957). — HÅKANSSON, A., and ALBERT LEVAN: Endo-duplicational meiosis in *Allium odorum*. Hereditas (Lund) **43**, 179—200 (1957). — HANSSON, A. A., R. J. CARBER and W. M. MYERS: Yields of individual and combined apomictic strains of Kentucky bluegrass (*Poa pratensis* L.). Agron. J. **44**, 125—128 (1952). — HANSON, A. A., and F. V. JUSKA: Induced mutation in Kentucky bluegrass. Crop Sci. **2**, 369—371 (1962). — HARAN, N.: Apomixis in *Taraxacum cyprium* Lindb. Palestine J. Bot. J Series **5**, 237—247 (1950—1952). — HARLAN, J. R.: Tucson side-oats grama: an improved strain. Oklahoma crops and soils 1947. Research reported at the second annual Oklahoma crops and soils conference. Okl. agric. Exp. Stat., Bull. B **319**, 123—125 (1948). — Apomixis in side-oats grama (*Bouteloua curtipendula* [Michx.] Torr.). Amer. J. Bot. **36**, 495—499 (1949). — HARLAN, J. R., M. H. BROOKS, D. S. BORGAONKAR, and J. M. J. DE WET: Nature and inheritance of apomixis in *Bothriochloa* and *Dichantium*. Bot. Gaz. **125**, 41—46 (1964). — HARLAN, J. R., M. H. BROOKS, and R. P. CELARIER: Can we break the apomictic barrier? Rep. 14th South Past. and Foage Crop Impr. Conf., Kentucky Exp. St., Lexington Kent. sid. 11—12 (1957). — HARLAN, J. R., and R. P. CELARIER: Apomixis and species formation in the *Bothriochloaea Keng*. Rec. Adv. in Bot. (Montreal) **3**, 706—710 (1959). — HARLAN, J. R., R. P. CELARIER, W. L. RICHARDSON, MARGARET HOOVER BROOKS, and K. L. MEHRA: Studies on old world bluestems II. Oklahom. St. Univ. Techn. Bull. T-72, 1—23 (1958). — HARLAN, J. R., and J. M. J. DE WET: Role of apomixis in the evolution of the *Bothriochloa dichanthium* complex. Crop Sci. **3**, 314—316 (1963). — HARMSEN, L.: Studies on the embryology and cytology of *Saxifraga*. Medd. om Grønland **125** (4), 1—15 (1939). — HARTUNG, M. E.: Chromosome numbers in *Poa*, *Agropyron*, and *Elymus*. Amer. J. Bot. **33**, 516—531 (1946). — HASKELL, G.: Analyses of sexual-apomictic blackberry populations, and their ecological consequenses. Cold Spr. Harb. Symp. quant. Biol. **20**, 111—126 (1955). — Phenoploidy and the distribution of European and British blackberries (*Rubus fructicosus*). Heredity **14**, 221 (1960). — Role of the male parent in apomictic *Rubus* species. Heredity **14**, 101—113 (1960a). — Selection, correlated responses and speciation in subsexual *Rubus*. Genetica **30**, 240—260 (1960b). — The genetics of *Rubus* breeding. Phyton (Argent.) **12**, 173—187 (1961). — Genetics and the distribution of British *Rubi*. Genetica **32**, 118—132 (1961). — The history, taxonomy and breeding system of apomictic British *Rubi*. Reprod. Biol. and Taxon. of Vasc. Plants, p. 141—151. B.S.B. I. Conf. Rep. 9, Pergamon Press. (1966). — HASKELL, G., and J. HILL: Segregation in tetraploid Blackberries. Heredity **16**, 3, 354—358 (1961). — HASKELL, G., and N. N. TUN: Developmental sequence of chromosome number in a cytologically unstable *Rubus* hybrid. Genet. Res. **2**, 10—24 (1961). — HASKELL, G. M.: Breeding behaviour of subsexual *Rubus* species. Heredity **11**, 427 (1957). — HEDBERG, O.: Cyto-taxonomic studies in Scottish mountain plants, notably *Deschampsia caespitosa* (L.) PB., *s. lat.* Svensk. bot. Tidskr. **52**, 37—46 (1958). — HEITZ, E.: Embryologischer Nachweis von Agamospermie mittels Simultanmethode. Experientia (Basel) **7**, 456 (1951). — HERIBERT-NILSSON, N.: Totale Inventierung der Mikrotypen eines Minimiareals von *Taraxacum officinale*. Hereditas (Lund) **33**, 119—142 (1947). — HESLOP HARRISON, Y.: Cytological studies in the genus *Rubus* L. 1. Chromosome numbers in the British *Rubus* flora. New Phytologist **52**, 22—39 (1953). — HIESEY, W. M.: Growth studies under controlled temperatures. Carnegie Inst. Wash., Year Book **50**, 99—105 (1951). — Responses of hybrid bluegrass strains to contrasting environments. Proc. 7th Int. Grassland Congr., Massey Agric. Coll., Palmeston North, New Zealand, sid. 1956, p. 212—221. — HJELMQVIST, H.: The apomictic development in *Malus Sieboldii*. Bot. Not. (Lund) **110**, 455—467 (1957). — Studien über Embryologie und Variabilität bei einigen *Aphanes*-Arten. Bot. Not. (Lund) **112**, 17—64 (1959a). — On the embryology of two *Malus* hybrids. Bot. Not. (Lund) **112**, 453—464 (1959b). — The embryo sac development of some Cotoneaster species. Bot. Not. (Lund) **115**, 208—236 (1962). — HJELMQVIST, H., and F. GRAZI: Studies on variation in embryo sac development. Bot. Not. (Lund) **117**, 141—166 (1964). — Studies on variation in embryo sac development. Second part. Bot. Not. (Lund) **118**, 329—360 (1965). —

HOFMEYR, J. D. J., and P. C. J. OBERHOLZER: Genetic aspects associated with the propagation of *Citrus*. Farming S. Afr. **23**, 201—208 (1948). — HOLMGREN, I.: Zytologische Studien über die Fortpflanzung bei den Gattungen *Erigeron* und *Eupatorium*. Kung. Svensk Vet.-Akad. Handl. **59** (7), 1—118 (1919). — HORN, C. L.: Existence of only one variety of cultivated mangosteen explained by asexually formed "seed". Science **92**, 237—238 (1940). — HOU-LIU, S. Y.: The chromosome counts of some taraxacum species. Acta bot. neerl. **12**, 76—83 (1963). — HUBBARD, J.: A study of *Andropogon ischaemum* L. M.S. thesis, A. & M. College of Texas 1950. — HULL, E. D.: Polyembryony in *Opuntia Rafinesquii*. Amer. J. Bot. **21**, 56—57 (1915). — HUNZIKER, H. R.: Beitrag zur Aposporie und ihrer Genetik bei *Potentilla*. Arch. Klaus-Stift. Vererb.-Forsch. **29**, 135—222 (1954). — HUTCHISON, D. J., and E. C. BASHAW: Cytology and reproduction of *Panicum coloratum* and related species. Crop Sci. **4**, 151—153 (1964).

IKENO, S.: Studien über die mutative Entstehung eines hochmutablen Genes bei einer parthenogenetischen Pflanzenart. Z. indukt. Abstamm.- u. Vererb.-Lehre **68**, 517—542 (1935).

JAGANNATH, D. R., B. W. X. PONNAIYA, and W. S. RAMAN: A study of morphological and cytological variation in chromosomal races of two species of *Cenchrus*. Madras Agr. J. **51**, 362 (1964). — JANKUN, A., and R. IZMAILOW: Badania cytotaksonomiczne nad polimorficznym gatunkiem *Ranunculus cassubicus* L. — Cytotaxonomical studies in the polymorphic species *Ranunculus cassubicus* L. Acta biol. cracoviensia, Ser. Bot. **7** (1964). — JEFFREY, E. C.: Hormones in relation to parthenogenesis. Genetics **33**, 615 (1948). — JOHANSEN, D.: Studies on the morphology of the Onagraceae. III. *Taraxia ovata* (Nutt.) Small. Ann. Bot. **45**, 111—124 (1931). — JOHNSON, A.: Polyembryony in *Eugenia Hookeri*. Amer. J. Bot. **23**, 83—85 (1936). — JOHNSON, B. L.: High rubber yielding selections from a natural population of guayule. Agron. J. **42**, 345—350 (1950). — JUEL, O.: Parthenogenesis bei *Antennaria alpina* (L.) Br. Bot. Zbl. **74**, 369—372 (1898). — Vergleichende Untersuchungen über typische und parthenogenetische Fortpflanzung bei der Gattung *Antennaria*. Kung. Svensk Vet. Akad.-Handl. **33** (5), 1—59 (1900). — JUHL, H.: Zytologische Untersuchungen an einigen Formen von *Poa pratensis* L. in Schleswig-Holstein. Flora (Jena) **139**, 462—476 (1952). — JUHL-NOODT, H.: Apomixis und umweltbedingte Variation bei *Poa pratensis* L. Züchter **25**, 3, 80—86 (1955). — JULÉN, GÖSTA: Über die Effekte der Röntgenbestrahlung bei *Poa pratensis*. Züchter **28**, 37—40 (1958). — The effect of X-raying on the apomixis in *Poa pratensis*. Proc. 6th Ann. Meet. Ital. Soc. Agric. Genetics. Genet. agr. **13**, 60—66 (1960). — The effect of X-rays on the apomixis in *Poa pratensis*. Int. Atom Energy Ag., Vienna 1961, p. 527—532. — JULIANO, J. B.: Origin of embryos in the strawberry mango. Philipp. J. Sci. **54**, 553—556 (1934). (Cited after STEBBINS 1941). — Embryos of Carabao Mango, *Mangifera indica* L. Philipp. Agr. **25**, 749—760 (1937). (Cited after STEBBINS 1941.) — JULIANO, I. B., and N. Y. CUEVAS: Floral morphology of the mango (*Mangifera indica* L.) with special reference to the pico variety from the Philippines. Philipp. Agr. **21**, 449—472 (1932). Cited after STEBBINS 1941. — JULIN, E.: Der Formenkreis des *Ranunculus auricomus* L. in Schweden. I. Diagnosen und Fundortsangaben einiger Sippen aus Södermanland. Ark. Bot. (Stockh.) Ser. 2, **6** (1), 1—28 (1963). — Några ord om Sveriges majranunkelflora. Svensk bot. Tidskr. **58**, 27—30 (1964). — Der Formenkreis des *Ranunculus auricomus* L. in Schweden. II. Diagnosen und Fundortsangaben neuer Sippen aus Södermanland. Ark. Bot., Ser. 2, **6** (2), 29—108 (1965).

KAPADIA, Z. L., and F. W. GOULD: Biosystematic studies in the *Bouteloa curtipendula var. caespitosa*. Bull. Torrey Bot. Club **91**, 465—478 (1964). — KAPPERT, H.: Experimentelle Untersuchungen über die Variabilität eines Totalapomikten (*Taraxacum officinale* Web.). Ber. dtsch. bot. Ges. **67**, 325—334 (1954). — KAWANO, S.: Cytogeography and evolution of the *Deschampsia caespitosa* complex. Canad. J. Bot. **41**, 719—742 (1963). — KERNER, A.! Parthenogenesis bei einer angiospermen Pflanze. S.-B. Akad. Wiss. Wien, math.-nat. Kl. I **74**, 469 (1876). — KIELLANDER, C. L.: A subhaploid *Poa pratensis* L. with 18 chromosomes and its progeny. Svensk bot. Tidskr. **36**, 200—220 (1942). — KNOX, R. B., and J. HESLOP-HARRISON: Experimental control of aposporeous apomixis in a grass of the *Andropogoneae*. Bot. Not. (Lund) **116**, 127—141 (1963). — KOCH, W.: Schweizerische Arten aus der Verwandschaft des *Ranunculus auricomus* L. Ber. schweiz. bot. Ges. **42**, 740—753 (1933). — Zweiter Beitrag zur Kenntnis des Formenkreises von *Ranunculus auricomus* L. Ber. schweiz. bot. Ges. **49**, 541—554 (1939). — KOROLEVA, V. A.: Interspecific hybridization in the genus *Taraxacum*. C. R. Acad. Sci. URSS. **24**, 174—176 (1939). — KRAUSE, O.: Zytologische Studien bei den Urticales unter besonderer Berücksichtigung der Gattung *Dorstenia*. Planta (Berl.) **13**, 29—84 (1931). — KRUG, C. A.: Chromosome numbers in the subfamily Aurantioideae, with special reference to the genus *Citrus*. Bot. Gaz. **104**, 602—611 (1943).

LAWRENCE, W. E.: Some ecotypic relations of *Deschampsia caespitosa*. Amer. J. Bot. **32**, 298—314 (1945). — LEROY, J. F.: Sur un "complexe agamique" des manguiers et sur l'origine et la phylogénie des variétés. Rev. Bot. Agr. Trop. **27**, 304—309 (1947). — LEVAN, A.: Zytologische Studien an *Allium schoenoprasum*. Diss. Lund 1935, also in Hereditas (Lund) **22**, 1—128 (1936/37). — Cytological studies in the *Allium paniculatum* group. Hereditas (Lund) **23**,

317—370 (1937). — The cytology of *Allium amplectens* and the occurrence in nature of its asynapsis. Hereditas (Lund) **26**, 353—394 (1940). — LEVAN, A., and S. L. EMSWELLER: Structural hybridity in *Nothoscordum fragrans* and the origin of terminally attached chromosomes. J. Hered. **29**, 291—294 (1938). — LI, LAI-YUNG: Vivipary in some Chinese plants. Bot. Gaz. **112**, 358—359 (1950). — LIDFORSS, B.: Batologiska iakttagelser. I. Öfv. Kung. Svensk Vet.-Akad. Förh. **56** (1), 21—35 (1899). — Batologiska iakttagelser. II. Öfv. Kung. Svensk Vet.-Akad. Förh. **58** (1), 59—90 (1901). — Studier över artbildningen inom släktet *Rubus*. Ark. Bot. **4** (6), 1—41 (1905). — Studier över artbildningen inom släktet *Rubus*. II. Ark. Bot. No. 16, 1—43 (1907a). — Über das Studium polymorpher Gattungen. Bot. Not. (Lund) **60**, 241—261 (1907b). — Resumé seiner Arbeiten über *Rubus*. Z. indukt. Abstamm.- u. Vererb.-Lehre **12**, 1—13 (1914). — LILJEFORS, A.: Über normale und apospore Embryoentwicklung in der Gattung *Sorbus*, nebst einigen Bemerkungen über die Chromosomenzahlen. Svensk bot. Tidskr. **28**, 290—299 (1934). — Studies on propagation, embryology and pollination in *Sorbus*. Acta Horti Berg. **16**, 277—329 (1953). — Cytological studies in *Sorbus*. Acta Horti Berg. **17**, No 4, 47—113 (1955). — Studies on species formation in the genus *Sorbus* in Scandinavia. Stockholm. Diss. 1955. 10 pp. — LINNAEUS, C.: Flora Lapponica. Amstelaedami 1737. — Flora Alpina. Amoenitates Academicae. Vol. **4**. Holmiae 1759. — LÖVE, A.: Cytogenetic studies in *Rumex*. III. Some notes on the Scandinawian species of the genus. Hereditas (Lund) **28**, 289—296 (1942). — Agamospermy in Acetosa. Hereditas (Lund) **35**, 390—393 (1949). — Preparatory studies for breeding Icelandic *Poa irrigata*. Hereditas (Lund) **38**, 11—32 (1952). — LÖVE, A., and D. LÖVE: Chromosome numbers of Scandinavian plant species. Bot. Not. (Lund) **95**, 19—59 (1942). — Chromosome numbers of Northern plant species. Reykjavik 1948, 131 pp. (1948). — LONGLEY, A. E.: Cytological studies in the genus *Crataegus*. Amer. J. Bot. **11**, 295—317 (1924a). — Cytological studies in the genus Rubus. Amer. J. Bot. **11**, 249—282 (1924b). — LONGO, B.: La poliembrionia nello *Xanthozylum Bungei* Planch. senza fecondazione. Bull. Soc. Bot. Ital. **17**, 113—115 (1908).

MACALLISTER, F.: On the cytology and embryology of *Smilacina racemosa*. Trans. Wisc. Acad. Sci., Arts Lett. **17** (1), 599—660 (1913). Cited after SCHNARF 1929. — MAJSURADZE, N. I.: The growth of sexual and nucellar embryos of the orange. Agrobiologija **9**, 309—310 (1961). — MALECKA, J.: Liczby chromosomów niektórych polskich gatunków Taraxacum. Chromosome numbers of some Taraxacum-species in Poland. Acta biol. cracov. Sér. bot. **1**, 55—56 (1958). — Badania cytologiczne nad rodzajem Taraxacum. Cytological studies in the genus Taraxacum. Acta biol. cracov. **5**, 117—136 (1962). — MAMPORIJA, F. D.: The effect of the fertilizing element upon nucellar cells in citrus fruits. Agrobiologija **5**, 119—131 (1957). — MARKLUND, G.: Einige Sippen der *Ranunculus auricomus*-Gruppe. Mem. Soc. Fauna Flora fenn. **16**, 45—53 (1939). — MARKLUND, G., and A. ROUSI: Outlines of evolution in the pseudogamous *Ranunculus auricomus* group in Finland. Evolution (Lawrence, Kansas) **15**, 4, 510—522 (1961). — MARSDEN-JONES, E. M., and W. B. TURRILL: Studies in *Ranunculus*. III. Further experiments concerning sex in *Ranunculus acris*. J. Genet. **31**, 363—378 (1935). — Studies in *Ranunculus*. IV. Additional experiments with *Ranunculus bulbosus* and *R. acris*. J. Genet. **51**, 26—31 (1952). — MATSUURA, H., and T. SUTO: Contributions to the idiogram study in phanerogamous plants. I. J. Fac. Sci. Hokkaido Imp. Univ. V **5** (1), 33—75 (1935). — MATTFELD, J.: Über einen Fall endocarper Keimung bei *Papaver somniferum* L. Verh. bot. Ver. Prov. Brandenburg **62**, 1—8 (1920). — MAURITZON, J.: Die Embryologie einiger Capparidaceen sowie von *Tovaria pendula*. Ark. Bot. A **26** (15), 1—14 (1934). — Über die Embryologie der Familie Retaceae. Svensk bot. Tidskr. **29**, 319—347 (1935). — McDONALD, C. C.: A study of seed development in three species of *Erigeron* with special reference to apogamy. Bull. Torrey bot. Club **54**, 479—497 (1927). — MEHRA, K. L.: Chromosome numbers, geographical distribution and taxonomy of the *Dichanthium annulatum* complex. Phyton (Argent.) 157—166, No 12 (1961). — The *Dichantium annulatum* complex. II. Relationships between the tropical and Mediterranean types. Phyton (Argent.) **21**, 119—126 (1964). — The *Dichantium annulatum* complex. III. Origin and artificial synthesis of *Dichantium papillosum* STAPF. Caryologia **17**, 545—556 (1964). — MEHRA, K. L., and R. P. CELARIER: Cytotaxonomic notes on the *Dichanthium annulatum* complex. Proc. Oklahoma Acad. Sci. **38**, 22—25 (1958). — MEHRA, K. L., K. N. SUBRAMANYAM, and M. S. SWAMINATHAN: Cytogenetical studies in some members of *Andropgoneae*. J. Indian Bot. Soc. **41**, 491—502 (1962). — MENDES, A. J. T.: Parthenogenesis, parthenocarpy and instances of abnormal fertilization in *Coffea*. Bragantia (São Paulo) **6**, 265—273 (1946). — A citologia e o melhoramento do cafeeiro. (Cytology and the improvement of coffee.) Bot. Sup. Serv. Café **22**, 236—240 (1947). — Coffee cytology. Proc. VIII. Int. Congr. Gen. Hereditas (Lund) Suppl. Vol. 628 (1949). — MESSERI, A.: Il numero dei cromosomi dell'*Allium roseum v.* bulbilliferum e dell'*Allium* confr. *odorum* e nuovo esempi di rapporti fra apomissia e polyploidismo. Nuovo Giorn. bot. ital. **37**, 276—277 (1930). — MIDUNO, T.: Chromosomenstudien an Orchidazeen. III. Über das Vorkommen von haploiden Pflanzen bei *Bletilla striata* Reichb. f. var. *gebine* Reichb. f. Cytologia **11** (1940). Cited after Biol. Abstr. **15**, 16193 (1941). — MOFFETT, A. A.: The chromosome constitution of the Pomoideae. Proc. roy. Soc. B **108**, 423—446 (1931). — MOHAMED, A. H., and F. W.

GOULD: Biosystematic studies in the *Bouteloa curtipendula* complex. V. Megasporogenesis and embryo sac development. Amer. J. Bot. **53**, 166—169 (1966). — MONTEMARTINI, L.: Contributo allo studio del'anatomia del frutto e del seme delle Opunzie. Atti Ist. Bot. Univ. Pavia II **5**, 59—68 (1899). Cited after SCHNARF 1929. — MOREIRA, S., J. T. A. GURGEL and L. F. DE ARRUDA: Polyembryony in *Citrus*. Bragantia (São Paulo) **7**, 69—106 (1947). — MÜNTZING, A.: Pseudogamie in der Gattung *Potentilla*. Hereditas (Lund) **11**, 267—283 (1928). — Note on the cytology of some apomictic *Potentilla species*. Hereditas (Lund) **15**, 166—178 (1931). — Apomictic and sexual seed formation in *Poa*. Hereditas (Lund) **17**, 131—154 (1932/33). — Further studies on apomixis and sexuality in *Poa*. Hereditas (Lund) **26**, 115—190 (1940). — Different chromosome numbers in root tips and pollen mother cells in a sexual strain of *Poa alpina*. Hereditas (Lund) **34**, 127—129 (1946). — Accessory chromosomes in *Poa alpina*. Heredity **2**, 49—61 (1948). — Accessory chromosomes in *Secale* and *Poa*. Proc. VIII. Int. Congr. Genet. Hereditas (Lund) Suppl. Vol. **35**, 402—411 (1949). — The cytological basis of polymorphism in *Poa alpina*. Hereditas (Lund) **40**, 459—516 (1954). — Further studies on intraspecific polyploidy in *Potentilla argentea (coll.)*. Bot. Not. (Lund) **111**, 209—227 (1958a). — The balance between sexual and apomictic reproduction in some hybrids of *Potentilla*. Hereditas (Lund) **44**, 145—160 (1958b). — Heteroploidy and polymorphism in some apomictic species of *Potentilla*. Hereditas (Lund) **44**, 280—329 (1958c). — Sublethal species hybrids in *Potentilla*. Hereditas (Lund) **44**, 554—556 (1958d). — Apomixis and sexuality in new material of *Poa alpina* from middle Sweden. Hereditas (Lund)**54**, 314—337 (1966). — MÜNTZING, A., and G. MÜNTZING: Some new results concerning apomixis, sexuality and polymorphism in *Potentilla*. Bot. Not. (Lund) **94**, 237—278 (1941). — Recent results in *Potentilla*. Hereditas (Lund) **28**, 232—235 (1942). — Spontaneous changes in chromosome number in apomictic *Potentilla collina*. Hereditas (Lund) **29**, 451—460 (1943). — A pentaplid F_1 hybrid between two diploid *Potentilla* species. Hereditas (Lund) **30**, 631—638 (1944). — The mode of reproduction of hybrids between sexual and apomictic *Potentilla argentea*. Bot. Not. (Lund) **98**, 49—71 (1945). — MÜNTZING, A., and A. NYGREN: A new diploid variety of *Poa alpina* with two accessory chromosomes at meiosis. Hereditas (Lund) **41**, 405—522 (1955). — MUKHERJEE, S. K.: Origin and distribution of *Saccharum*. Bot. Gaz. **119**, 55—61 (1957). — MULLIGAN, G. A.: Chromosome numbers of the family *Cruciferae*. III. Canad. J. Bot. **44**, 309—319 (1966). — MURBECK, Sv.: Parthenogenetische Embryobildung in der Gattung *Alchemilla*. Lunds Univ. Årsskr. **36**, Afd, 2, No 7, 1—45 (1901). — MURGAI, P.: In vitro experiments on apomictic *Aerva tomentosa* FORSK. Proc. forty-sixth Indian Sci. congr. 1959, part III, p. 326—327. — MYERS, W. M.: Cytology and gentics of forage grasses. Bot. Rev. **13**, 319—421 (1947).

NANNFELDT, J. A.: On *Poa jemtlandica* (Almqu.) Richt., its distribution and possible origin. Bot. Not. (Lund) **90**, 1—27 (1937). — On the polymorphy of *Poa arctica* R. Br. Symb. bot. upsal. **4** (4), 1—85 (1940). — NARAYAN, K. N.: Cytogenetic studies of apomixis in *Pennisetum*. Ph. D. thesis, Univ. Calif 1951. — NIELSEN, E. L.: Cytology and breeding behavior of selected plants of *Poa pratensis*. Bot. Gaz. **106**, 357—382 (1945). — Breeding behavior and chromosome numbers in progenies from twin and triplet plants in *Poa pratensis*. Bot. Gaz. **108**, 26—40 (1946a). — The origin of multiple macrogametophytes in *Poa pratensis*. Bot. Gaz. **108**, 41—50 (1946b). — Developmental sequence of embryo and endosperm in apomictic and sexual forms of *Poa pratensis*. Bot. Gaz. **108**, 531—534 (1947). — NIELSEN, E. L., and D. C. SMITH: Comparisons of clonal isolations of *Poa pratensis* L. from good and poor pastures for vigor, variability and disease reactions. Agronomy J. **43** (4), 214—218 (1951). — Dimorphic panicle formation in *Poa pratensis*. Bot. Gaz. **112**, 4, 534—535 (1951). — Interrelations of selected plant characters in Kentucky bluegrass (*Poa pratensis* L.). Bot. Gaz. **114**, 53—62 (1952). — NISIOKA, T., and N. TAKAHASHI: Studies on the promotion of sprouting and breaking of dormancy in winter-buds of *Cornus controversa*. Phylogenetic study in the *Ixeris dentata* group II. General aspect of the *Ixeris dentata* group. Jap. J. Bot. **18**, 2, 169—225 (1963). — NOACK, K.: Über *Hypericum*-Kreuzungen. VI. Fortpflanzungsverhältnisse und Bastarde von *Hypericum perforatum* L. Z. indukt. Abstamm.- u. Vererb.-Lehre **76**, 569—601 (1939). — NORSTOG, K.: Apomixis and polyembryony in *Hierochloe odorata*. Amer. J. Bot. **50**, 815—821 (1963). — NYGREN, A.: The genesis of some Scandinavian species of *Calamagrostis*. Hereditas (Lund) **32**, 131—262 (1946). — Further studies in spontaneous and synthetic *Calamagrostis purpurea*. Hereditas (Lund) **34**, 113—134 (1948a). — Some interspecific crosses in *Calamagrostis* and their evolutionary consequences. Hereditas (Lund) **34**, 387—413 (1948b). — Apomictic and sexual reproduction in *Calamagrostis purpurea*. Hereditas (Lund) **35**, 285—300 (1949a). — Studies on vivipary in the genus *Deschampsia*. Hereditas (Lund) **35**, 27—32 (1949b). — Cytological and embryological studies in arctic *Poae*. Sym. bot. upsal. **10** (4), 1—64 + 15 pl. (1950a). — A cytological and embryological study of *Antennaria Porsildii*. Hereditas (Lund) **36**, 483—485 (1950b). — Embryology of *Poa*. Carnegie Inst. Wash., Year Book **50**, 113—115 (1951a). — Form and biotype formation in *Calamagrostis purpurea*. Hereditas (Lund) **37**, 519—532 (1951b). — How to breed Kentucky blue grass, *Poa pratensis* L. Hereditas (Lund) **39**, 51—56 (1953). —

Investigations on North American *Calamagrostis.* Hereditas (Lund) **40**, 377—397 (1954a). — Apomixis in the Angiosperms. Bot. Rev. **20**, 577—649 (1954b). — Chromosome studies in the *Poa laxa* group. Ann. roy. agric. Coll. Sweden **22**, 359—368 (1955a). — Further studies in diploid *Poa alpina* without accessory chromosomes at meiosis. Ann. roy. agric. Coll. Sweden **22**, 179—191 (1955b). — *Poa timoleontis* HELDR., a new diploid species of the section *Bolbophorum* A. and GR. with accessory chromosomes only in the meiosis. Ann. roy. agric. Coll. Sweden **23**, 489—495 (1957). — Investigations on North American *Calamagrostis.* II. Ann. roy. agric. Coll. Sweden **24**, 363—368 (1958). — On *Poa badensis, Poa concinna, Poa pumila, Poa xerophila* and the possible origin of *Poa alpina.* Ann. Acad. Reg. Sci. Upps. **6**, 1—29 (1962a). — Artificial and natural hybridization in European *Calamagrostis.* Symb. Bot. Upsal. **17** (3), 1—105 (1962b). — Apomixis (including vegetative reproduction). Apomixis in the Angiosperms, with special reference to *Calamagrostis* and *Poa.* Biol. and taxonom. of vasc. plants. B.S.B.I. Conf. Rep. No **9**, 131—140 (1966). — NYGREN, A., and G. ALMGÅRD: On the experimental control of vivipary in *Poa.* Ann. roy. Agric. Coll. Sweden **28**, 27—36 (1962). — NYGREN, A., and E. ÅKERBERG: Studies in species and hybrid derivatives of *Poa* new to practical work. Acad. Reg. Sci. Upsal. **1**, 53—69 (1957).

OEHLER, E.: Entwicklungsgeschichtlich-zytologische Untersuchungen an einigen saprophytischen Gentianaceen. Planta (Berl.) **3**, 641—733 (1927). — OKA, H. T.: Zytologische Beobachtungen über eine Mutantensippe von *Erigon annuus.* Jap. J. Genet. **10**, 237—241 (1935). — OKABE, S.: Über Parthenogenesis bei *Houttuynia cordata.* Jap. J. Genet. **6**, 14—19 (1930). — Parthenogenesis bei *Ixeris dentata.* Bot. Mag. (Tokyo) **46**, 518—523 (1932). — OLDÉN, E. J.: Sexual and apomictic seed formation in *Malus Sieboldii* Rehd. Bot. Not. (Lund) **106**, 105—128 (1953). — ORR, M. Y.: Polyembryony in *Sarcococca riscifolia* Stapf. Roy. Bot. Gard. Edinburgh, Notes **14**, 21—23 (1923). Cited after SCHNARF 1929. — OSAWA, J.: Cytological and experimental studies in *Citrus.* J. Coll. Agr., Imp. Univ. Tokyo **4**, 83—116 (1912). — Studies on the cytology of some species of *Taraxacum.* Arch. Zellforsch. **10**, 450—469 (1913). — OSTENFELD, C. H.: Experiments on the origin of species in the genus *Hieracium* (apogamy and hybridsism). New Phytologist **11**, 347—354 (1912).

PACE, L.: Apogamy in *Atamosco.* Bot. Gaz. **56**, 376—394 (1913). — PALMER, E. J.: The *Crataegus* problem. J. Arn. Arb. **13**, 342—362 (1932). — PERRY, B. A.: Chromosome numbers and phylogenetic relationships in the Euphorbiaceae. Amer. J. Bot. **30**, 527—543 (1943). — PETER, J.: Zur Entwicklungsgeschichte einiger Calycanthaceen. Beitr. Biol. Pflanzen **14**, 59—84 (1920). — PETROVSKAYA-BARANOVA, T. P., and N. V. ZINGER: Atavistic adaptations to the sexual process in seedbuds of parthenogenetic species of *Taraxacum.* Transact. Moscow Soc. Nat. **13**, 257—266 (1965). — PFITZER, P.: Apomixis und Embryosackentwicklung in der Gattung *Aglaonema* ENGL. Portugal. Acta biol. **6**, 3—4, 279—293 (1962). — PIJL, L. VAN DER: Über die Polyembryonie bei *Eugenia.* Rec. Trav. bot. néerl. **31**, 113—187 (1934). — PLUMIER, W.: Biologie florale et fécondation. (Colloquium of studies on witloof chicory Gembloux.) Ann. Gembloux **66**, 151—156 (1960). — PODDUBNAJA-ARNOLDI, W. A.: Geschlechtliche und ungeschlechtliche Fortpflanzung bei einigen *Chondrilla*-Arten. Planta (Berl.) **19**, 46—86 (1933). — Development of pollen and embryosac in interspecific hybrids of *Taraxacum.* C. R. Acad. Sci. URSS. **24**, 374—377 (1939). — PODDUBNAJA-ARNOLDI, V., u. V. DIANOWA: Eine zytoembryologische Untersuchung einiger Arten der Gattung *Taraxacum.* Planta (Berl.) **23**, 19—46 (1934). — POPOFF, A.: Über die Fortpflanzungsverhältnisse der Gattung *Potentilla.* Planta (Berl.) **24**, 510—522 (1935). — PORSILD, M. P.: On the genus *Antennaria* in Greenland. Medd. om Grønland **51**, 265—281 (1915). — POWERS, L.: Fertilization without reduction in guayule (*Parthenium argentatum* Gray) and a hypothesis as to the evolution of apomixis and polyploidy. Genetics **30**, 323—346 (1945). — POWERS, L., and E. J. GARDNER: Frequency of aborted pollen grains and microcytes in guayule, *Parthenium argentatum* Gray. J. Amer. Soc. Agron. **37**, 184—193 (1945). — POWERS, L., and R. C. ROLLINS: Reproduction and pollination studies on guayule, *Parthenium argentatum* Gray and P. *incanum* H. B. K. J. Amer. Soc. Agron. **37**, 96—112 (1945). — PRATT, C., and J. EINSET: Development of the embryo sac in some American blackberries. Amer. J. Bot. **42**, 637—645 (1955). — PRATT, C., J. EINSET, and T. CLAUSEN: Embryology, breeding behaviour and morphological characteristics of apomictic, triploid *Rubus idaeus* L. Bull. Torrey Bot. Club **85**, 242—254 (1958).

RAGHAVAN, T. S.: The sugarcanes of India. J. Hered. **42**, 199—206 (1951). — RAGHUVANSHI, S. S.: Cytogenetical studies in Genus *Citrus.* IV. Evolution in Genus *Citrus.* Cytologia (Tokyo) **27**, 172—188 (1962). — RANCKEN, G.: Zytologische Untersuchungen an einigen wirtschaftlich wertvollen Wiesengräsern. Acta agral. fenn. **29**, 1—92 (1934). — RENNER, O.: Zur Terminologie des pflanzlichen Generationswechsels. Biol. Zbl. **36**, 337—374 (1916). — REUSCH, J. D. H.: The relationship between reproductive factors and seed set in *Paspalum dilatatum.* S. Afr. J. Agric. Sci. **4**, 513—530 (1961). — ROLLINS, R. C.: Interspecific hybridization in *Parthenium.* I. Crosses between Guayule (P. *argentatum*) and Mariola (P. *incanum*). Amer. J. Bot. **32**, 395—404 (1945a). — Evidence for genetic variation among apomictically produced plants of several F_1 progenies of Guayule (*Parthenium argentatum*) and Mariola (P. *incanum*). Amer. J. Bot. **32**, 554—560 (1945b). — Sources of genetic variation in *Parthenium argentatum*

Gray (Compositae). Evolution **3**, 358—368 (1949). — Rosenberg, O.: Cytological studies on the apogamy in *Hieracium*. Bot. Tidsskr. **28**, 143—170 (1907). — Über die Apogamie bei *Chondrilla juncea*. Svensk bot. Tidsskr. **6**, 915—919 (1912). — Die Reduktionsteilung und ihre Degeneration in *Hieracium*. Svensk bot. Tidskr. **11**, 145—206 (1917). — Die semiheterotypische Teilung und ihre Bedeutung für die Entstehung verdoppelter Chromosomenzahlen. Hereditas (Lund) 8 (1926, 1927), 305—338 (1927). — Apogamie und Parthenogenesis bei Pflanzen. In: Handbuch der Vererbungswissenschaften, Liefg 12, S. 1—66. 1930. — Rousi, A.: Cytotaxonomy and reproduction in the apomictic *Ranunculus auricomus* group. Ann. bot. Soc. zool.-bot. fenn. "Vanamo" **29**, No 2, 1—64 + 1 pl. (1956). — Roy, R. P., A. P. Singh, and P. K. Gupta: A new chromosome race in *Heteropogon contortus*. Current Sci. **34**, 190 (1965). — Rozanova, M. A.: Versuch einer analytischen Monographie der conspecies *Ranunculus auricomus* Korsh. Trav. Inst. Sci. Nat. Peterhof 8,19—148 (1932). — Rutishauser, A.: Konstante Art- und Rassenbastarde in der Gattung *Potentilla*. Mitt. naturforsch. Ges. Schaffhausen **18**, 111—134 (1943a). — Über die Entwicklungsgeschichte pseudogamer Potentillen. Arch. Klaus-Stift. Vererb.-Forsch. **18**, 687—691 (1943b). — Untersuchungen über die Fortpflanzung und Bastardbildung apomiktischer Potentillen. Ber. schweiz. bot. Ges. **53**, 1—83 (1943c). — Über die Fortpflanzung einiger Bastarde von pseudogamen Potentillen. Arch. Klaus-Stift. Vererb.-Forsch., Erg.-Bd. zu **20**, 300—314 (1945a). — Zur Embryologie amphimiktischer Potentillen. Ber. schweiz. bot. Ges. **55**, 19—32 (1945b). — Über Kreuzungsversuche mit pseudogamen Potentillen. 6. Jber. Schweiz. Ges. Vererb. S. S. G. Arch. Klaus-Stift. Vererb.-Forsch. **21**, 469—472 (1946). — Untersuchungen über die Genetik der Aposporie bei pseudogamen Potentillen. Experientia (Basel) **3**, 204 (1947). — Pseudogamie und Polymorphie in der Gattung Potentilla. Arch. Klaus-Stift. Vererb.-Forsch. **23**, 267—424 + 2 pl. (1948). — Untersuchungen über Pseudogamie und Sexualität einiger Potentillen. Ber. schweiz. bot. Ges. **59**, 409—419 (1949). — Die Entwicklungserregung des Endosperms bei pseudogamen *Ranunculus*arten. Mitt. naturforsch. Ges. Schaffhausen **25**, 1—45 + 5 Taf. (1953/54). — Entwicklungserregung der Eizelle bei pseudogamen Arten der Gattung *Ranunculus*. Bull. schweiz. Akad. med. Wiss. **10**, Fasc. 6, 491—512 (1954). — Cytogenetik des Endosperms. Ber. schweiz. bot. Ges. **66**, 318—336 (1956).— Untersuchungen über die Evolution pseudogamer Arten. Ber. Schweiz. Bot. Gesellsch. 70, 113—125 (1960). — Genetik der Pseudogamie bei *Ranunculus auricumus s.* l. W. Koch. Ber. schweiz. bot. Ges. **75**, 157—182 (1966). — Rutishauser, A., u. H. R. Hunziker: Untersuchungen über die Zytologie des Endosperms. Arch. Klaus-Stift. Vererb.-Forsch. **25**, 477—483 (1950). — Weitere Beiträge zur Genetik der Aposporie pseudogamer Potentillen. Arch. Klaus-Stift. Vererb.-Forsch. **29**, 223—233 (1954). — Rychlewski, J.: Studia cytoembriologiczne nad apomiktycznym gatunkiem *Nardus stricta* L. Cytological studies in the apomictic species *Nardus stricta* L. Acta biol. cracov. Ser. Bot. **4**, 1—23 (1961).

Sachar, R. C., and R. N. Chopra: A study of the endosperm and embryo in *Mangifera* L. Indian J. Agric. Sci. **27**, 219—228 (1957). — Sachar, R. C., and P. Murgai: Embryology of *Aerva tomentosa* Forsk. Current Sci. (Delhi) **27**, 105—106 (1958). — Sakai, K.: Studies on the chromosome number in Alpine plants. I. J. Genet. **9**, 226—230 (1934). — Sax, H. J.: Polyploidy and apomixis in *Cotoneaster*. J. Arn. Arb., Bol. **35**, 334—365 (1954). — Sax, K.: The origin and relationships of the Pomoideae. J. Arn. Arb. **12**, 3—22 (1931). — Chromosome relationships in the Pomoideae. J. Arn. Arb. **13**, 363—367 (1932). — The cytogenetics of facultative apomixis in *Malus* species. J. Arnold Arboretum Harvard Univ. **40**, 289—297 (1959). — Schlimbach, H.: Beiträge zur Kenntnis der Samenanlagen und Samen der Amaryllidaceen mit Berücksichtigung des Wassergehaltes der Samen. Flora (Jena) **117**, 41—54 (1924).— Schnarf, K.: Beobachtungen über die Endospermentwicklung von *Hieracium aurantiacum*. S.-B. Akad. Wiss. Wien, math.-nat. Kl. I **128**, 755—771 (1919). — Embryologie der Angiospermen. In: Handbuch der Pflanzenanatomie. 690 pp. 1929. — Schoch, M.: Entwicklungsgeschichtlich-cytologische Untersuchungen über die Pollenbildung und Bestäubung bei einigen *Burmannia*-Arten. Diss. Zürich 1920. 95 pp. — Schürhoff, P. N.: Zur Apogamie von *Calycanthus*. Flora (Jena) **116**, 73—84 (1923). — Schürhoff, P. N., and F. J. Jüssen: Nuzellarembryonie bei *Spathiphyllum Patinii* (Hogg.) N.E. Br. Ber. dtsch. bot. Ges. **43**, 454—456 (1925). — Sernander, R.: Zur Morphologie und Biologie der Diasporen. N. Act. Reg. Soc. Sci. Upsal., Vol. extra ordin. edit. 1927. — Shama Kumari, K.: Cytogenetic investigations in *Paniceae*: Occurrence of apospory in a diploid species of *Panicum-Panicum antidotale* Retz. Current Sci. **29**, 191 (1960). — Shibata, K., u. K. Miyake: Über Parthenogenesis bei *Houttuymia cordata*. Bot. Mag. (Tokyo) **22**, 141—144 (1908). — Shimotomai, N.: Chromosomenzahlen und Phylogenie bei der Gattung *Potentilla*. Hiroshima Univ., J. Sci. B Div. 2, **1**, 1—11 (1930). — Siebold, C. Th. E. v.: Wahre Parthenogenesis bei Schmetterlingen und Bienen. Leipzig 1856. Cited after Braun 1857. — Simonet, M., et Ch. Miedzyrzecki: Étude caryologique de quelques espèces arborescentes ou sarmenteuses d'ornement. C. R. Soc. Biol. (Paris) **111**, 969—973 (1932). — Singh, A. P.: Intergenetic cross of *Dichantium annulatum* with *Bothriochloa ischaemum*. Cytologia (Kyoto) **30**, 54—57 (1965). — Singh, A. P., and K. L. Mehra: Cytological analysis of hybrids within the *Dichantium annu-*

latum complex. Cytologia (Kyoto) **30**, 307—316 (1965). — SKALINSKA, M.: Cyto-ecological studies in *Poa alpina* L. var. *vivipara* L. Bull. Acad. Polon. Sci. et Lett., Sci. Math. et Nat. B Sci. Nat. (1) 253—283 + 1 pl. (1952). — *Poa nobilis* n. sp., a new viviparous species of the High Tatra. Acta Soc. Bot. Pol. **24**, No 4, 749—761 (1955). — Embryological studies in *Poa granitica* BR. BL., an apomictic species of the *Carpathian range*. Acta biol. cracov., Ser. bot. **2**, 91—110 + 7 pl. (1959). — SKALINSKA, M., and R. CZAPIK: Studies in the cytology of the genus *Potentilla*. Acta Biol. cracov., Ser. bot. **1**, 137—149 (1958). — SKAWINSKA, R.: Apomiksja u *Hieracium alpinum* L. Apomixis in *Hieracium alpinum* L. Acta biol. cracov. **5**, 90—96 (1962). — SKOVSTED, A.: Cytological studies in the tribe Saxifrageae. Dansk bot. Ark. **8** (5), 1—52 (1934). — SMITH, B. W.: Hybridity and apomixis in the perennial grass, *Paspalum dilatatum*. Genetics **33**, 628—629 (1948). — SMITH, G. L.: Studies in *Potentilla* L. I. Embryological investigations into the mechanism of agamospermy in British *P. Tabernaemontani Aschers*. New Phytologist **62**, 264—282 (1963). — Studies in *Potentilla* L. II. Cytological aspects of apomixis in *P. Crantzii* (CR.) Beck ex fritisch. New Phytologist **62**, 283—300 (1963). — SMITH, J.: Notice of a plant which produces seeds without any apparent action of pollen. Meeting of June 18th 1839. Trans. Linn. Soc. London **18** (1841). Cited after BRAUN 1857. — SNYDER, L. A.: Apomixis in *Paspalum secans*. Amer. J. Bot. **44**, 318—324 (1957). — SNYDER, L. A., and J. R. HARLAN: A cytological study of *Bouteloua gracilis* from western Texas and eastern New Mexico. Amer. J. Bot. **40**, 702—707 (1953). — SNYDER, L. A., A. R. HERNANDEZ and H. E. WARMKE: The mechanism of apomixis in *Pennisetum ciliare*. Bot. Gaz. **116**, 209—221 (1955). — SÖDERBERG, E.: Über die Chromosomenzahl von *Houttuynia cordata*. Svensk. bot. Tidskr. **21**, 247—250 (1927). — Bemerkungen zur Nomenklatur der parthenogenetischen *Erigeron*-Arten. Svensk bot. Tidsskr. **23**, 261—262 (1929). — SÖRENSEN, TH.: Ranunkelstudier. II. Om variation og nedarvningsforhold hos *Ranunculus auricomus* L. Bot. Tidsskr. **44**, 433—438 (1938). — Sexual chromosome aberrants in triploid apomictic *Taraxaca*. Sv. Bot. Tidskr. **54**, 1—22 (1958). — SÖRENSEN, TH., and G. GUDJÓNSSON: Spontaneous chromosome aberrants in apomictic *Taraxaca*. Kong. danske Vid. Selsk. Biol. **4**(2), 1—48 (1946). — SOKOLOWSKA-KULCZYCKA, A.: Apomiksja u *Leontopodium alpinum* CASS. Apomixis in *Leontopodium alpinum* CASS. Acta biol. cracov., Ser. bot. **2**, 51—63 + 2 pl. (1959). — SPRECHER, M. A.: Étude sur la sémence et la germination du *Garcinia mangostana* L. Rev. gén. Bot. **31**, 513—531, 611—634 (1919). — STEBBINS jr., G. L.: Cytology of *Antennaria*. I. Normal species. Bot. Gaz. **94**, 134—151 (1932a). — Cytology of *Antennaria*. II. Parthenogenetic species. Bot. Gaz. **94**, 322—344 (1932b). — New species of *Antennaria* from the Appalachian region. Rhodora **37**, 229—237 (1935). — Apomixis in the angiosperms. Bot. Rev. **7**, 507—542 (1941). — Variation and evolution in plants. 643 pp. 1950. — STEBBINS, G. L., and J. A. JENKINS: Aposporie development in the North American species of *Crepis*. Genetica **21**, 191—224 (1939). — STENAR, H.: Studien über die Entwicklungsgeschichte von *Nothoscordum fragrans* Kunth und N. *striatum* Kunth. Svensk bot. Tidskr. **26**, 25—44 (1932a). — Parthenogenesis in der Gattung *Calamagrostis*. Ark. Bot. A **25** (6), 1—8 (1932b). — STRASBURGER, ED.: Über Polyembryonie. Jena. Z. Naturw. **12**, 647—670 (1878). — STREETMAN, L. J.: Reproduction of the lovegrasses, the genus Eragrostis-I.E. Chloromelas steud., E. Curvula (Schrad.) Nees, E. Lehmanniana Nees and E. Superba Peyr. Wrightia **3**, 41—51 (1963). — SUBBA RAO, A. M.: Studies in the Malphigiaceae. I. Embryo sac development and embryogeny in the genera *Hiptage*, *Banisteria* and *Stigmatophyllum*. J. Indian bot. Soc. **18**, 145—156 (1940). Cited after Biol. Abstr. **15**, 19328 (1941). — SUESSENGUTH, K.: Über die Pseudogamie bei *Zygopetalum Mackayi* Hook. Ber. dtsch. bot. Ges. **41**, 16—23 (1923). — SUKACEV, V. N.: Polymorphismus und Apomixis bei Arten der Gattung *Adenophora*. J. Bot. URSS. **25**, 297—303 (1940). Cited after Bot. Zbl. **85**, 168 (1941/42). — SUTO, T.: List of chromosome number and idiogram types in Liliaceae and Amaryllidaceae. Jap. J. Genet. **12**, 107—112, 157—162, 221—231 (1936). — SWAMY, G. L.: The embryology of *Zeuxine sulcata* Lindl. New Phytologist **45**, 132—136 (1946a). — Embryology of *Habenaria*. Proc. nat. Inst. Sci. India **12** (8), 413—426 (1946b). — SWAMY, R.: Culture of nucellar tissue of *Citrus in vitro*. Experientia (Basel) **14**, 111—112 (1958). — In vitro culture of nucellus and embryos of *Citrus*. Mod. Develop. Plant Physiol. (P. MAHESHWARI, ed.). University of New Delhi, India, 104—105 (1958).

TÄCKHOLM, G.: Zytologische Studien über die Gattung *Rosa*. Acta Horti Berg. **7**, No 3, 97—381 (1922). — TAHARA, M.: Cytologische Studien an einigen Kompositen. Imp. Univ. Tokyo, J. Coll. Sci. **43** (7), 1—51 (1921). — TINNEY, F. W.: Cytology of parthenogenesis in *Poa pratensis*. J. agric. Res. **60**, 351—360 (1940). — TISCHLER, G.: Pflanzliche Chromosomenzahlen. Nachtrag Nr 2, Teil I and II. Tab. Biol. Period .V/VI, 1935/36. — TIWARY, N. K.: On the occurence of polyembryony in the genus *Eugenie*. J. Indian bot. Soc. **5** (1926). Cited after PIJL 1934. — Further observations on the seeds and seedlings of *Eugenia Jambolana* Lamk. J. Indian bot. Soc. 8 (1929). Cited after PIJL 1934. — TOXOPEUS, H. J.: Die Züchtung von Unterlagen für *Citrus sinensis* Osb. immun gegen *Phytophthora parasitica*, die Ursache der "gumdisease" in Japan. Züchter **8**, 1—10 (1936). — TRAUB, H. P.: Artificial control of nucellar embryony in *Citrus*. Science **83**, 165—166 (1938). — TRIPLETT, E. L.: Cytology of

Poa hybrids. Carnegie Inst. Wash., Year Book **52**, 173—174 (1953). — TSCHERMAK-WOESS, E.: Diploides *Taraxacum vulgare* in Wien und Niederösterreich. Öst. bot. Zeitschr. **96**, 56—63 (1949). — TURESSON, G.: Studien über *Festuca ovina* L. I. Normalgeschlechtliche, halb- und ganzvivipare Typen nordischer Herkunft. Hereditas (Lund) **8**, 161—206 (1926). — Studien über *Festuca ovina* L. II. Chromosomenzahl und Viviparie. Hereditas (Lund) **13**, 177—184 (1930). — Studien über *Festuca ovina* L. III. Weitere Beiträge zur Kenntnis der Chromosomenzahlen viviparer Formen. Hereditas (Lund) **15**, 13—16 (1931). — Chromosome stability in Linnean species. Ann. agric. Coll. Sweden **5**, 405—416 (1938). — Variation in the apomictic microspecies of *Alchemilla vulgaris* L. Bot. Not. (Lund) **96**, 413—427 (1943). — Variation in the apomictic microspecies of *Alchemilla vulgaris* L. II. Progeny tests in agamotypes with regard to morphological characters. Bot. Not. (Lund) **109**, 400—404 (1956). — Variation in the apomictic microspecies of *Alchemilla vulgaris* L. III. Geographical distribution and chromosome number. Bot. Not. (Lund) **110**, 413—422 (1957). — Observations on some clones and clone progenies in *Alchemilla alpina* L. Bot. Not. (Lund) **111**, 159—164 (1958). — TURESSON, G., and B.: Experimental studies in *Hieracium pilosella* L. I. Reproduction, chromosome number and distribution. Hereditas (Lund) **46**, 717—736 (1960). — Observations on chromosome number and reproduction in some piloselloids. Bot. Not. (Lund) **116**, 157—160 (1963). — TURRILL, W. B.: Material for a study of taxonomic problems in *Taraxacum*. Bot. Soc. Exch. Club. Brit. Isl. **11**, 570—589 (1938). — TYSDAL, H. M.: Apomictic interspecific hybrids are promising for rubber production from guayule. Agron. J. **42**, 351—359 (1950).

UHRING, J., and S. L. EMSWELLER: Parthenogenesis in tetraploid *Lilium Longiflorum*. Amer. J. Bot. **49**, 9 (1962). — URBAŃSKA, K.: Cytologia *Antennaria carpatica* i *A. dioica*. Cytology of *Antennaria carpatica* and *A. dioica*. Acta biol. cracov., Ser. bot. **2**, 43—49 (1959). — URBAŃSKA-WORYTKIEWICZ, K.: Badania embriologiczne nad rodzajem *Antennaria* Gaertn. I. Rozwój zalażków *A. carpatica* (Wahlb.) Bl. et Fing. Embryological investigations in *Antennaria* Gaertn. I. Development of the ovules of *A. carpatica* (Wahlb.) Bl. et Fing. Acta biol. cracov., Ser. bot. **4**, 49—64 (1961). — Badania embriologiczne nad rodzajem *Antennaria* Gaertn. II. Rozwój pylku *A. carpatica* Bl. et Fing. Embryological investigations in *Antennaria* Gaertn. Development of the pollen of *A. carpatica* Bl. et Fing. Acta biol. cracov. **5**, 97—102 (1962). — Badania embriologiczne nad rodzajem *Antennaria* Gaertn. III. Mieszańce eksperymentale pomiedzy *A. carpatica* Bl. et Fing. i *A. dioica* (L.) Gaertn. Embryological investigations in *Antennaria* Gaertn. III. Experimental hybrids between *A. carpatica* Bl. et Fing. and *A. dioica* (L.) Gaertn. Acta biol. cracov. **5**, 103—115 (1962). — Badania cytologiczne nad doswiadczalnymi mieszancami pomiedzy *Antennaria carpatica* (Wahlb.) Bl. et Fing. i *A. dioica* (L.) Gaertn. Cytological studies in experimental hybrids between *Antennaria carpatica* (Wahlb.) Bl. et Fing. and *A. dioica* (L.) Garten. Acta biol. cracov. **5**, 51—61 (1965).

VIGNOLI, L.: Cariologia del genere *Agave*. Lav. Ist. Bot. Palermo **7**, 1—44 (1936). — Cariologia del genere *Agave*. Nota II. Lav. Ist. Bot. Palermo **8**, 1—6 (1937a). — Grandezza cellulare e poliploidia in *Agave*. Lav. Ist. Bot. Palermo **8**, 88—106 (1937b).

WALDO, G. F., and G. M. DARROW: Origin of the logan and the mammoth blackberries. J. Hered. **39**, 99—107 (1948). — WARMKE, H. E.: Cytotaxonomic investigations of some varieties of *Panicum maximum* and *P. purpurascens* in Puerto Rico. Agron. J. **43**, 143—491 (1951). — Apomixis in *Panicum maximum*. Amer. J. Bot. **41**, 5—11 (1954). — WATSON, P. J.: The distribution in Britain of diploid and tetraploid races within the *Festuca ovina* group. New Phytologist **57**, 11—18 (1958). — WATSON, P. J., and J. CLAUSEN: Phenotypic responses to contrasting environments in the genus *Poa*. Scott. Plant Breed. St. Pentlandfield, Rep. **1961**, 64—78. — WATSON, W. C. R.: Handbook of the *Rubi* of Great Britain and Ireland. Cambridge 1958. — WEBER, J. M.: Polyembryony. Bot. Rev. **6**, 575—598 (1940). — WERBER, H. J.: The economic importance of apogamy in *Citrus* and *Mangifera*. Proc. Amer. Soc. Hort. Sci. **28**, 57—61 (1931). — WICHERLEY, P. R.: Proliferation of spikelets in British grasses. Watsonia **3**, 41—56 (1953). — WIGER, J.: Ein neuer Fall von autonomer Nuzellarpolyembryonie. Bot. Not. (Lund) **83**, 368—370 (1930). — Embryological studies on the families Buxaceae, Meliaceae, Simarubaceae and Burseraceae. Diss. Lund 1935. 130 pp. — WINKLER, H.: Über Parthenogenesis und Apogamie im Pflanzenreiche. Progr. nei Bot. **2**, 293—454 (1908). — WOODWORTH, R. H.: Cytological studies in the Betulaceae. III. Parthenogenesis and polyembryony in *Alnus rugosa*. Bot. Gaz. **89**, 402—409 (1930).

YONG, D. P.: Autogamous *Epipactis* in Scandinavia. Bot. Not. (Lund) **106**, 233—270 (1953). — YOUNGNER, V. B.: Environmental control of initiation of the inflorescence, reproductive structures, and proliferations in *Poa bulbosa*. Amer. J. Bot. **47**, 753—757 (1960).

ZARUBIN, A. F.: Possibility of parthenogenesis in walnut and black walnut. Priroda No 10, 64—65 (1949). — ZEILINGA, A. E.: Polyploidy in *Cotoneaster*. Bot. Not. (Lund) **117**, 262—278 (1964). — ZINGER, N. V., V. A. PODDUBNAJA-ARNOLDI, T. P. PETROVSKAYA, and N. N. POLUNINA: To the question of the causes of apomixis. Transact. Moscow Soc. Nat. **13**, 201—237 (1965).

Vegetative Fortpflanzung, Parthenogenese und Apogamie bei Algen.

Von

H. Ettl, D. G. Müller, K. Neumann, H. A. v. Stosch und W. Weber*.

Mit 110 Abbildungen.

A. Einleitung.

a) Vegetative Fortpflanzung.

Die beiden Termini der Themenstellung sind sehr heterogenen Inhaltes. Ohne uns mit den möglichen Bedeutungen abzugeben, die der erste von ihnen, die „Vegetative Fortpflanzung“, bei anderen Autoren erhält, beginnen wir damit, ihn für den jetzigen Zweck passend zu umgrenzen. Wir schließen also — wenigstens im Grundsatz — alle jene Vorgänge aus, die mit der Ausbildung von Gameten, ihrer Vereinigung und deren unmittelbaren Folgen, also auch der Zygote und ihrer Keimung, zu tun haben: ferner die Prozesse der Zellvermehrung. So bezeichnen wir diejenigen Vermehrungsweisen der Zelle, welche von der in Frage kommenden Art üblicherweise durchgeführt wird. Tritt also neben vorwiegender Zweiteilung noch Zoo- oder Autosporenbildung auf, so gehört die letztere zur vegetativen Fortpflanzung, fehlt die Teilung, so zur Zellvermehrung. Insgesamt verfahren wir so, wie bei Vielzellern allgemein üblich, wo die Zellteilung als Grundlage des Wachstums von der Fortpflanzung getrennt wird. Bei Einzellern wären Zellteilung oder sie ersetzende Vorgänge diejenigen, welche das „Wachstum“ einer Population ausmachen. Doch wird es bei Einzellern oft nötig sein, auf die Morphologie der vegetativen Vermehrung einzugehen, um die Fortpflanzungsvorgänge abzugrenzen.

Es bleibt dann eine große Mannigfaltigkeit von Vermehrungs- und Fortpflanzungsweisen, von den Gonidien zu den Dauerzellen und Brutkörpern, denen gemeinsam ist, daß sie ungeschlechtlich erfolgen. Diese Mannigfaltigkeit bedarf einer nomenklatorischen Gliederung. Nach dem Erscheinen der Bücher SCHUSSNIGS (1953, 1954, 1960), die eine konsequent durchgeführte einheitliche Terminologie der Fortpflanzungsorgane darbieten, wäre es lockend, diese nunmehr zu verwenden. Doch fürchten wir, dieses in erster Linie physiologischen Fragen gewidmete Handbuch zu belasten, wenn wir eine noch ungebräuchliche Terminologie allein zu Worte kommen lassen, und werden daher von den in der etwas älteren Literatur, etwa bei FRITSCH (1934, 1945) oder in einer repräsentativen Monographie der betreffenden Gruppen üblichen Fachworten Gebrauch machen und bei ihrem erstmaligen Erscheinen in diesem Text den Ausdruck SCHUSSNIGS, soweit abweichend, in spitzen Klammern hinzufügen.

* Bearbeiter: ETTL: G I Chlorophyceae, G II Conjugatophyceae; MÜLLER: F Phaeophyta; NEUMANN: G III Charophyceae; v. STOSCH: A Einleitung, B, C, D, E Niedere Algen; WEBER: H Rhodophyta.

Die zweite Heterogenität bietet der Begriff Algen, der systematisch immerhin die Mehrzahl der existierenden Pflanzenstämme umfaßt und wegen des Inbegriffenseins der akaryonten Cyanophyta mehr grundsätzlich Differierendes enthält als z. B. das gesamte Tierreich, das wie die übrigen Algen und die Cormophyten eukaryont ist. Insgesamt müssen wir mit 7—8 Stämmen rechnen. Die Zahl der Stämme würde sich um einen vermehren, wenn wir die Phäophyten von den Chrysophyten abtrennen, wie es hier aus praktischen Gründen geschah, obwohl dies sachlich nicht mehr gerechtfertigt ist. Da man die Phyla dadurch kennzeichnen möchte, daß sich zwischen ihnen verwandtschaftliche Beziehungen nicht mehr rational begründen lassen, wenn man von den durch die Charakterisierung *eukaryont* vermittelten Gemeinsamkeiten absieht, so folgt, daß auch ähnliche Fortpflanzungsweisen zwischen den Stämmen nicht mehr homolog sind oder zu sein brauchen. Die Autosporen einer Blaualge sind also wahrscheinlich nur konvergent zu denen der Grünalge. A priori dürfen wir also auch nicht postulieren, daß die Bedingungen, welche solche nur analogen Organe auslösen, vergleichbar seien. Das ist die sachliche Begründung dafür, daß wir hier nicht nach Fortpflanzungsformen, sondern nach Pflanzenstämmen gliedern.

Jedoch ist frühzeitig ein bedeutender Versuch unternommen worden, von den experimentellen Befunden her zu allgemeinen, d. h. von der systematischen Stellung der Organismen unabhängigen Aussagen über die Bedingungen zu kommen, welche die Fortpflanzung gegenüber dem Wachstum charakterisieren. Er ist an den Namen Georg Klebs geknüpft, dessen Experimente über die Fortpflanzung von Algen und Pilzen 1896 zusammengestellt und in den nächsten Jahren theoretisch kommentiert das damals revolutionierende Ergebnis brachten, diese Ereignisse seien nicht endogen bedingt, sondern *absolut* durch Außenbedingungen gesteuert. Weniger erfolgreich konnte Klebs der Natur der Dinge nach in der Formulierung allgemeiner Gesetzmäßigkeiten über die Art dieses gegenseitigen Verhältnisses von Wachstum und Fortpflanzung sein. Die vier Sätze darüber in der Arbeit von 1900 über Pilze haben sich, obwohl sehr weit gefaßt, zum Teil als tautologisch, zum anderen als nur bedingt zutreffend erwiesen und sind in dem Artikel im „Handwörterbuch der Naturwissenschaften" von 1913 auf einen einzigen reduziert worden, daß nämlich „der entscheidende Grund für das Auftreten der Fortpflanzungsorgane anstelle des vegetativen Wachstums in quantitativen Veränderungen der für alle Gestaltungsvorgänge wichtigen allgemeinen äußeren Bedingungen liegt". Es kennzeichnet die Situation, daß gleichzeitig Pilze und Algen, die früher von Klebs hinsichtlich der Fortpflanzung gemeinsam behandelt waren, nunmehr getrennt besprochen werden. Der Satz hält gegenüber Einwänden, z. B. von seiten Pfeffers (1907), an der in allen Fällen wesentlichen Rolle der Außenbedingungen bei der Bewirkung der Fortpflanzung fest. Deshalb kann er auch im Hinblick auf den Nachweis endonomer Rhythmiken bei Algen (Bühnemann 1955, Rieth 1959, Bünning und Müller 1961), sowie auf das Verhalten der hapaxanten Algen und der Diatomeen nicht mehr als allgemein zutreffend angesehen werden, sosehr er in vielen Einzelbeispielen gültig ist. Bezüglich der Hapaxanten stellt Klebs (1900) selbst die Frage, was aus einer *Hydrodictyon*-Zelle wohl werde, wenn das Wachstum weitergehe, ohne daß die Außenbedingungen die Fortpflanzung verursacht hätten: entweder Tod oder doch zuletzt Fortpflanzung, was aber erst bewiesen werden müsse. Nun kann man zwar bei der einkernigen hapaxanten *Acetabularia calyculus* unter Verzicht auf eine normale Formbildung das Wachstum so lenken, daß von einem bestimmten Zeitpunkt an die Fortpflanzung unter sonst dazu geeigneten Bedingungen nicht oder nicht mehr normal erfolgt, aber bei *Hydrodictyon* und *Botrydium* scheint es doch eine artspezifische obere Größe der Coeloblasten zu geben, bei deren Annäherung die

Neigung zur Gonidienbildung ansteigt, d.h. diese letztere auf immer geringere Reize hin stattfindet, so daß gleichzeitig das Primat in der Bestimmung, was geschieht, von den Außenbedingungen auf die Innenbedingung Zellgröße überwechselt; und bei gewissen zentrischen Diatomeen (s. S. 667) entscheidet die durch die Breitenabnahme bei der Teilung veränderliche Zellgröße allein darüber, ob bzw. welche Art von Fortpflanzung stattfinden kann. Weiter fragt sich auch, ob in allen Fällen Fortpflanzung und Wachstum notwendig durch verschiedene Bedingungen charakterisiert sind. KLEBS selbst nannte 1913 die Blütenpflanze *Parietaria officinalis* als Beispiel für gleichzeitiges Wachsen und Blühen unter allen Gewächshausbedingungen; ganz ähnlich verhält sich in unseren Kulturen das Männchen von *Antithamnium sarniense* (v. STOSCH), das seit einigen Jahren ununterbrochen wächst und gleichzeitig fruchtet.

Zu einem etwas enger gefaßten Ergebnis kam man in dem Versuch, die sexuelle Fortpflanzung allein gegenüber dem Wachstum oder auch gegenüber der vegetativen Fortpflanzung bedingungsmäßig zu charakterisieren. Gegen das vegetative Wachstum schien sie durch das Zurücktreten der Mineralsalzernährung (nach Versuchen von BENECKE 1908 u.a. handelt es sich dabei insbesondere um den Stickstoff) zugunsten der Kohlenhydratversorgung und alle Umstände, die dieses fördern, gekennzeichnet, während umgekehrt die Zoosporenbildung von einem Zurückbleiben der letzteren begleitet sein kann. Doch zeigte CZURDA (1933) durch kritische Literaturstudien und eigene Versuche an *Spirogyra*-Arten, daß es Fälle von sexueller Fortpflanzung gibt, in denen nicht Stickstoffmangel, sondern die Wasserstoffionenkonzentration die entscheidende Bedingung ist. Ebenso zeigen die Versuche von v. DENFFER und HUSTEDE bei *Vaucheria* (s. S. 687), deren Verhalten sich im großen mit den Klebsschen Formulierungen zu decken schien, daß hier der Übergang in die sexuelle Vermehrung überhaupt nicht trophisch bedingt ist oder bedingt zu sein braucht. Daher ist auch der Zusammenhang zwischen Stickstoff-(bzw. Mineralsalz-)Mangel und sexueller Vermehrung nicht in einfacher Weise von allgemeingültiger Art. Auch die neuerdings in sehr weiter Verbreitung in der Organismenwelt gefundene ,,Innere Uhr" (s. z.B. BÜNNING 1963) ist, soweit überhaupt, keinem von beiden Prozessen eindeutig zugeordnet: bei *Dictyota* (s. S. 695) reguliert sie anscheinend nur die sexuelle, bei *Oedogonium* (s. S. 734) die vegetative und bei *Vaucheria* (s. S. 690) möglicherweise beide Fortpflanzungsarten.

Während die generalisierenden Schlußfolgerungen sowie ein Teil der experimentellen Ergebnisse (CZURDA 1933) von KLEBS der schnellebigen Relativierung physiologischer Arbeit anheimfallen, sind seine Begriffsbestimmungen — in sprachlich modifizierter Form — noch heute in Geltung. Die genotypisch bestimmte *Reaktionsnorm* stellt im Zusammenwirken mit den Außenbedingungen meist nach einigem Wachstum einen Zustand *(innere Bedingungen)* des Organismus her, welcher diesen befähigt, (,,von selbst" oder) auf bestimmte Änderungen der *Außenbedingungen* hin zur Fortpflanzung überzugehen. Die anfängliche Abtrennung besonderer ,,formativer Reize" (HERBST), von KLEBS auch ,,spezifische Bedingungen" genannt, von den allgemeinen Außenbedingungen wurde 1904 aufgegeben.

In dieser Veröffentlichung macht KLEBS auch darauf aufmerksam, daß die eine Fortpflanzung verursachenden Eingriffe von zweierlei Natur sein können[1], und diese Unterscheidung scheint uns wichtig, gerade weil sie in der Folgezeit kaum beachtet wurde. Entweder wirken sie durch die *Änderung* von Außenfaktoren und, wie KLEBS am Beispiel der Auslösung von *Vaucheria repens*-

[1] KLEBS schreibt (S. 454): Der einen Entwicklungsvorgang veranlassende Reiz ist in vielen Fällen nicht bloß für seine Auslösung, sondern auch für seine Ausführung notwendig.

Zoosporen durch Benetzung der vorher trocken kultivierten Pflanze zeigte, vermittels der *Geschwindigkeit* dieser Änderung oder aber unabhängig von einer Änderungsgeschwindigkeit durch den schließlich erreichten *Zustand*. Die erste reizartige Form der Bewirkung, „Auslösung" im eigentlichen Sinne, hat kürzer dauernde Fortpflanzungsstöße zur Folge; sie kann sogar unabhängig von der Richtung der Änderung sein. So löst bei *Vaucheria repens* sowohl Überführung aus verdünnter Nährlösung in Wasser, wie auch u. U. in konzentrierte Nährlösung Zoosporenbildung aus; beides erfolgt im Licht. Einen Schub von Zoosporen erhält man auch bei Minderung der Lichtintensität auf ein geringeres Niveau. Verdunkelt man aber völlig, so wirkt der nunmehrige Zustand als „Dauerreiz", und die Zoosporenbildung hält bis zur Erschöpfung des Thallus an. In die unter den Reizphysiologen schwebende Diskussion darüber, ob man in diesem letzten Falle überhaupt von Reiz sprechen dürfe, soll hier nicht eingetreten werden. Man wird also sagen können, daß im mehrdimensionalen Bedingungsraum der Eintritt in bestimmte Bereiche *unabhängig* vom zeitlichen Verlauf Fortpflanzung nach sich zieht — hierher wären auch Fälle wie die Dauerfertilität von *Antithamnium sarniense* zu rechnen — daß diese aber in anderen Bereichen nur erfolgt auf bestimmte *Bedingungsänderungen* hin, bei denen also der Zeitfaktor wesentlich ist, in noch anderen schließlich *gar nicht*. Und man muß hinzufügen, daß unter die „dimensionalen" Vektoren des Bedingungsraumes auch innere Bedingungen der verschiedensten Art gleichberechtigt eintreten können.

Das oben über die Verschiedenartigkeit der zum gleichen Effekt führenden Einwirkungen Gesagte macht deutlich, wie schwer es gerade bei der vegetativen Fortpflanzung ist, über mehr als eine Art oder Artengruppe hinaus verbindliche Ursachen für das gleiche Ereignis anzugeben, und CZURDA begründet, daß wegen der *ökologischen* Verschiedenheiten der Arten einer Gattung auf der Seite der Außenbedingungen auch nicht viel Gemeinsamkeit erwartet werden könne. Doch muß man mit KLEBS annehmen, daß einem gleichen Fortpflanzungsereignis mindestens ähnliche Innenzustände („spezifische innere Bedingungen") vorausgehen, in welche die Pflanze die verschiedenen Außeneinwirkungen transformierte. KLEBS scheint zuletzt (nach manchen Meinungsänderungen, s. WINKLER 1934) angenommen zu haben, dieser physiologische Zustand stehe mit der Nährsalz-Zucker-Relation in Zusammenhang, die er auch für das Eintreten der sexuellen Fortpflanzung verantwortlich hielt. So skeptisch man der speziellen Hypothese gegenüberstehen wird, so ist es doch eine mögliche und sogar aussichtsreiche Annahme, daß, genetisch vergleichbares Material vorausgesetzt, einem bestimmten Endzustand, also etwa einer Sporulation, auch dann gleichartige Reaktionsketten vorausgehen, wenn die auslösenden Bedingungen verschieden waren (s. auch COLEMAN 1961).

b) Apogamie und Parthenogenese.

Die Termini der Überschrift sind vergleichend entwicklungsgeschichtlich definiert; nur bei einer Art oder in einem engeren Verwandtschaftskreis, in welchem Gamie vorkommt, kann von *Apogamie*, der Entstehung von Zygobionten mit haploidem oder diploidem Chromosomensatz aus vegetativen (also auch aus vegetativen Fortpflanzungs-) Zellen des Gamobionten, und nur, wo es Gameten gibt, von *Parthenogenese*, einer Entwicklung der Geschlechtszelle oder einer dieser homologen Zelle ohne äußere Befruchtung gesprochen werden. Beide Termini sind damit nicht anwendbar in Gruppen, bei denen Sexualität nicht, noch nicht oder nicht hinreichend bekannt ist, wie bei den Blaualgen und den meisten Flagellatenstämmen.

Eine allgemeingültige und logische Durchgliederung der Abweichungen vom normalen Biontenwechsel hat WINKLER (1942) vorgenommen. Sie ist terminologisch gültig, da ein zweiter ähnlich umfassender Versuch nicht besteht, und wird nachstehend wiedergegeben.

A. Apomixis. Abweichende Entstehung des Zygobionten.

1. Vegetative Propagation des Zygobionten.
2. Apogamie: Zygobiont entsteht aus vegetativer Zelle des Gamobionten.
3. Parthenogenesis: Zygobiont entsteht aus unbefruchteten Gameten.
4. Apogonosis: Zygobiont entsteht aus Gonospore.

B. Aposporosis. Abweichende Entstehung des Gamobionten.

1. Vegetative Propagation des Gamobionten.
2. Aposporie: Gamobiont entsteht aus vegetativer Zelle des Zygobionten.
3. Apomiktosis: Gamobiont entsteht aus unbefruchteten Gameten.
4. Zygogamosis: Gamobiont entsteht aus Zygote.

Sowohl Zygo- als auch Gamobiont kann dabei auch einzellig sein.

Von diesen Fortpflanzungen sind A_1 und B_1, die vegetative Propagation von Zygobiont und Gamobiont für die vielzelligen Algen, bereits in der vegetativen Fortpflanzung (s. oben) inbegriffen; ebenso B_2, die Aposporie und A_4, die Apogonosis. B_4, Zygogamosis, kommt mit einer möglichen Ausnahme (*Cutleria adspersa*, SAUVAGEAU 1899) bei Algen nicht vor. Es bleiben die Apogamie, die wieder in der diploiden (Zygo-) oder der haploiden (Gamo-) Phase ablaufen kann und die Gametenentwicklungen ohne Befruchtung. Bei der Parthenogenese liegen die Verhältnisse insofern nicht einfach, als das, was WINKLER als Parthenogenese bezeichnet — aus dem Parthenogameten entsteht ein Zygobiont —, gewöhnlich nicht von der Apomiktosis unterschieden wird, bei der aus dem Parthenogameten die Geschlechtspflanze zurückerhalten wird. Wir werden daher beide Fälle unter Parthenogenesis führen und weiter zwischen Parthenogenese sensu stricto (der Parthenogenese WINKLERs) und Apomiktosis unterscheiden. Letztere findet sich häufig bei Chlorophyten und Phaeophyceen. Bei der Parthenogenese sensu stricto kann 1. Meiosis erfolgen und der Zygobiont haploid bleiben: *generative Parthenogenese* (SUOMALAINEN 1950). Bei einer 2. Form der Parthenogenese wird der Zygobiont diploid (*somatische Parthenogenese*, SUOMALAINEN; die Termini WINKLERs decken sich mit jenen SUOMALAINENs nicht vollständig), indem a) die Meiosis erfolgt und der diploide Zustand durch autogame Kernverschmelzung wieder hergestellt wird (*automiktische Parthenogenese*, SUOMALAINEN) oder b) die Meiosis durch mannigfaltige Abwandlungen gegenüber ihrem normalen Verlauf zur Diploidie führt, oder schließlich durch eine diploide mehr oder weniger typische Mitose ersetzt wird (*apomiktische Parthenogenese*, SUOMALAINEN). Die unter 2. genannten Vorgänge treten im tierischen Ei in sehr zahlreichen Formen auf (SUOMALAINEN; NARBEL-HOFSTETTER 1964, dort Literatur).

Die Vorgänge der Apogamie und Parthenogenese werden im folgenden, soweit überhaupt bekannt, im Anschluß an die vegetative Fortpflanzung der einzelnen Gruppen behandelt; die entsprechende spezielle Literatur ist bei diesen eingereiht (E I Chrysophyceae, E II Diatomeen, E III Heterokontae, F Phaeophyta, G I Chlorophyceae, G II Conjugatophyceae, G III Charophyceae, H Rhodophyta).

Literatur.

In dem Abschnitt „A. Einleitung“ dieser Zusammenstellung sind die wichtigsten Lehr- und Handbücher über Algen sowie die Literatur zur Einleitung aufgeführt. Das Schrifttum der systematischen Gruppen wurde zusammengehalten. In den Einzelbeiträgen wird mit vor die Jahreszahl gesetztem „A“ auf die hier gebrachte Literatur hingewiesen.

Benecke, W.: Über die Ursachen der Periodizität im Auftreten der Algen auf Grund von Versuchen über die Bedingungen der Zygotenbildung bei *Spirogyra communis*. Internat. Rev. ges. Hydrobiol. **1**, 533—552 (1908). — Bühnemann: 1955 s. Lit. Grünalgen. — Bünning, E.: Die physiologische Uhr, 2. Aufl. Berlin-Göttingen-Heidelberg: Springer 1963. — Bünning, E., u. P. Müller: Wie messen Organismen lunare Zyklen? Z. Naturforsch. **16**b, 391—395 (1961).

Chadefaud, M.: Les végétaux non vasculaires. (Vol. 1 par Chadefaud et Emberger. Traité de Botanique systematique.) Paris: Masson & Cie. 1960. — Christensen, T.: Alger. In: T. W. Böcher, M. Lange u. T. Sørensen. Botanik. København: Munksgaard 1962. — Coleman, A.: The role of modern culture methods in the study of algal life cycles. Quart. Rev. Biol. **36**, 247—253 (1961). — Czurda, V.: Experimentelle Analyse der kopulationsauslösenden Bedingungen bei Mikroorganismen. I. Untersuchungen an Algen (*Spirogyra*. *Zygnema* und *Hyalotheca*). Beih. Bot. Zbl. 1. Abt., **51**, 711—757 (1933).

Dangeard, P.: Traité d'algologie. Paris: Paul Lechevalier 1933.

Ehrenberg, S. G.: Die Infusionsthiere als vollkommene Organismen. Berlin u. Leipzig: Voss 1838.

Fott, B.: Algenkunde. Jena: VEB Gustav Fischer 1959. — Fritsch, F. E.: The structure and reproduction of the algae, vol. 1 1934, vol. 2 1945. Cambridge: University Press.

Hartmann, M.: Die Sexualität, 2. Aufl. Stuttgart: Gustav Fischer 1956.

Klebs, G.: Die Bedingungen der Fortpflanzung einiger Algen und Pilze. Jena: Gustav Fischer 1896. — Zur Physiologie der Fortpflanzung einiger Pilze. III. Jb. wiss. Bot. **35**. 1—120 (1900). — Über Probleme der Entwicklung. Biol. Zbl. **24**, 257 (1904). — Fortpflanzung der Gewächse; Physiologie. In: Handwörterbuch der Naturwissenschaften, Bd. 4. S. 276—296. Jena: Gustav Fischer 1913. — Kniep, H.: Die Sexualität der niederen Pflanzen. Jena: Gustav Fischer 1928.

Narbel-Hofstetter, M.: Les altérations de la méiose chez les animaux parthénogénétiques. Protoplasmatologia **6**, 1—163 (1964).

Oltmanns, F.: Morphologie und Biologie der Algen, 2. Aufl. Jena: Gustav Fischer 1922.

Peacock, A. S., and U. Weidmann: Recent work on the cytology of animal parthenogenesis. Przegl. zool. **5**, 5—27, 101—122 (1961). — Pfeffer, W.: Untersuchungen über die Entstehung der Schlafbewegungen. Abh. Kgl. Sächs. Ges. Wiss. **1907**, 257—472 (1907). — Pringsheim, E. G.: Farblose Algen. Stuttgart: Gustav Fischer 1963.

Rieth: 1959 s. Lit. E III Heterokonten.

Sauvageau, C.: Les Cutlériacées et leur alternance de générations. Ann. sci. nat. Bot., Sér. VIII. **10**, 265—362 (1899). — Schussnig, B.: Grundriß der Protophytologie. Jena: VEB Gustav Fischer 1954. — Handbuch der Protophytenkunde. Jena: VEB Gustav Fischer 1953, 1960. — Smith, G. M.: The fresh-water algae of the United States, 2nd ed. New York: McGraw-Hill Book Co. 1950. — Manual of phycology. Waltham, Mass.: Chronica Botanica Co. 1951. — Cryptogamic botany, 2nd ed. New York: McGraw-Hill Book Co. 1955. — Suomalainen, E.: Parthenogenesis in animals. Advanc. Genet. **3**, 193—253 (1950).

White, M. J. D.: Animal cytology and evolution, 2nd ed. Cambridge: University Press 1954. — Winkler, H.: Parthenogenesis und Apogamie im Pflanzenreiche. Jena: Gustav Fischer 1908. — Verbreitung und Ursache der Parthenogenesis im Pflanzen- und Tierreiche. Jena: Gustav Fischer 1920. — Fortpflanzung der Gewächse: Physiologie der Fortpflanzung. In: Handwörterbuch der Naturwissenschaften, 2. Aufl., Bd. 4, S. 435—450. Jena: Gustav Fischer 1934. — Über den Biontenwechsel und die Abweichungen von seinem normalen Verlauf. Planta (Berl.) **33**, 1—90 (1942).

B. Cyanophyta.

Sieht man von einigen wohl nur ungenügend untersuchten Flagellatengruppen ab, so ist dieses der einzige Stamm im Pflanzenreich, in dem neben Zellteilung — wenigstens mit Sicherheit — nur ungeschlechtliche Fortpflanzungsvorgänge nachgewiesen werden konnten (s. aber Kumar 1962). Mit seinen akaryotischen Zellen steht er auch cytologisch von den übrigen Algen ganz isoliert. Für die morphologischen Grundlagen sei auf Geitler (1960) verwiesen.

Die Fortpflanzung der Blaualgen kann geschehen:

a) Durch Freiwerden ganzer Einzelzellen, „*Gonidien*“ (hier wegen Fehlens experimenteller Unterlagen nicht weiter behandelt), oder von fadenförmigen Zellverbänden, *Hormogonien*, aus dem Thallus.

b) Durch *Nannocyten*, welche innerhalb der Membran einer gewöhnlichen vegetativen Zelle als Folge succedaner Vielfachteilungen oder durch *Endosporen* ⟨Endogonidien⟩, die ähnlich, aber in deutlich durch ihre Größe als Sporangien herausgehobenen Zellen entstehen. Wenn zwischen den succedanen Teilungen stärkeres Wachstum der Tochterzellen erfolgt, resultieren Übergänge zur gewöhnlichen Zweiteilung, wenn andererseits die Sporangien nicht mehr deutlich durch Größe abgesetzt sind, solche zwischen Endosporen und Nannocyten, weshalb FRITSCH (1942) auf eine terminologische Trennung dieser beiden Fortpflanzungsarten verzichtet. Von den Endosporen leiten sich vermutlich (GEITLER) die in Ketten gebildeten Exosporen ⟨Exogonidien⟩ von *Chamaesiphon* sowie die Sporenketten von *Stichosiphon* bzw. die Stäbchenreihen von *Cyanophanon* ab. Diese letzten drei Fortpflanzungskörper können hier übergangen werden, da sie die einzige Vermehrungsart ihrer Besitzer darstellen.

c) Durch die Verwandlung von Zellen *(Dauerzellen* und *Heterocysten)* oder Zellreihen *(Hormocysten)* in dickwandige Dauerorgane.

a) Hormogonien.

1. Entstehung.

Die höheren, echte Fadenverbände bildenden Blaualgen sind durch Hormogonien ausgezeichnet, nach welchen die Gruppe benannt ist. Es handelt sich um sehr kurze oder längere, wenig differenzierte, aber in der Form ihrer Zellen bisweilen (z. B. *Nostoc*) von den Thalluszellen abweichende scheidenlose Fadenstücke, denen die Verbreitung der Art zukommt. Häufig sind sie dazu mit Bewegungsvermögen ausgestattet. Ihnen fehlt im typischen Falle eine Zellteilungstätigkeit während der Phase des Herumkriechens (GEITLER), die jedoch nach Einstellen der Bewegung im Vorgang der „Keimung" wiedergewonnen wird. Die Hormogonien von *Nostoc* zeigen jedoch Teilungen in der beweglichen Phase (HARDER 1917a, LAZAROFF und VISHNIAC 1961). Die scheidenlosen Fäden der Gattung *Oscillatoria* können als permanent gewordene Hormogonien angesehen werden, die ebenfalls Bewegung und Teilungstätigkeit vereinigen. Sie können ihrerseits in kürzere Hormogonien zerfallen. Bei *Plectonema* und anderen Hormogonalen wurden auch einzellige Hormogonien beobachtet, die dann als Gonidien zu bezeichnen wären (z. B. SUBA-CLAUS 1965). Die Hormogonien entstehen bei den einfacher gebauten *Hormogonalen* aus beliebigen, bei den höher differenzierten aus den jüngeren Teilen des Thallus (Abb. 1). Sie lösen sich aus dem Fadenverband oder im Falle der Seitenzweighormogonien von *Nostochopsis* und anderen Stigonemataceen vom Hauptast bzw. dessen Lateralen wie bei der eigenartigen *Doliocatenella* (GEITLER 1942). Dieses geschieht 1. direkt, d.h. durch einfache Trennung benachbarter Fadenzellen (z.B. bei *Nostoc:* HARDER 1917a, bei *Gomontiella:* PALIK 1961), oder 2. häufig durch Absterben mehrerer oder einzelner zwischenliegender Zellen — diese als *Nekriden* (Spaltkörper) oft noch lange sichtbar — oder 3. auch durch Abbrechen an interkalaren Heterocyten wie bei *Nostoc* (HARDER 1917a, GLADE 1914, LAZAROFF und VISHNIAC 1961) und anderen Formen (*Anabaenopsis*, VARMA und MITRA 1962). In solcher Art als Trennzellen zu dienen, wurde unter anderem insbesondere von BORZI (1878) als Hauptfunktion der Heterocysten angenommen. Eigentümlich ist das Verhalten von *Gomontiella*, deren Fäden aus aufeinander gestapelten hufeisenförmigen Zellen aufgebaut sind und die Fadentrennungen sowohl durch Quer- als auch durch Längsteilungen erfahren können (CLAUS 1960, PALIK 1961). Schließlich nehmen auch die Keimlinge von Dauerzellen die Eigenschaften von Hormogonien an, wofür HARDER (1917a) eindrucksvolle Schilderungen (Abb. 2) liefert. Die Hormogonien können wie vegetative Fäden Gasvacuolen bilden. Bei *Oscillatoria rubescens* entstehen sie unter Nekridenbildung in sehr alten Kulturen und können dann praktisch farblos sein; die Thylakoide sind mehr oder weniger vollständig verschwunden, werden aber auf frischen Medien aus dem Plasmalemma neugebildet (Jost 1965); dies ist aber wohl ein Ausnahmefall. Nach Harder (1917a) (*N. punctiforme*) sowie nach Lazaroff und Vishniac (1961, 1962, 1964) (*N. muscorum*) spielt sich der

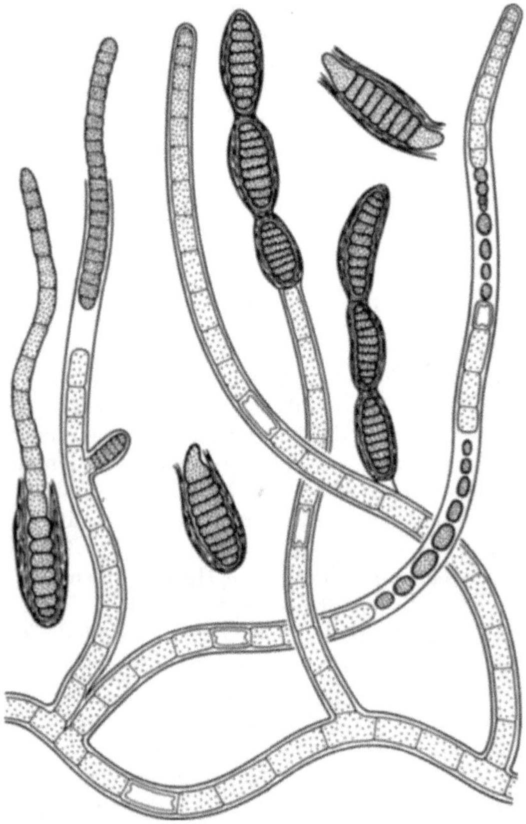

Abb. 1.

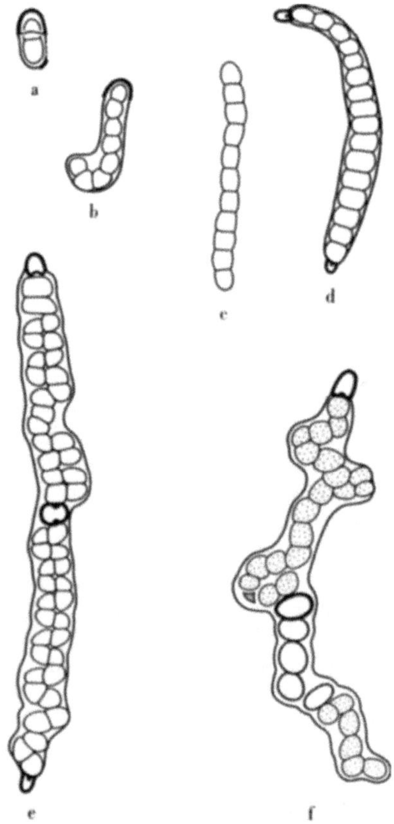

Abb. 2a—f.

Abb. 1. *Westiella lanosa*. Der Thallus gibt links ein Hormogonium ab und entwickelt darunter eine Seitenast-Hormocyste, in der Mitte eine Kette von endständigen und eine weitere von seitenständigen Hormocysten; rechts interkalare Ketten von Dauersporen. Die Heterocysten haben zylindrische Form. Am Rande Keimungsstadien von Hormocysten. (Nach Frémy 1930.)

Abb. 2a—f. Keimung der Dauerspore und Hormogonentwicklung bei *Nostoc punctiforme*. a, b Keimungsstadien der Spore, die ausschlüpfenden Hormogonien (c) kriechen herum, kommen zur Ruhe, entwickeln Terminalheterocysten (d) und treten dann in Teilung ein, wobei sich der Faden durch die Scheide beengt auffaltet (e); in seiner Mitte hat er eine Interkalarheterocyste entwickelt. f Junger Thallus mit Kette von Dauersporen. (Nach Harder 1917a.)

Wachstumscyclus von *Nostoc* in Kultur folgendermaßen ab (Abb. 2): Das Hormogonium kriecht nach seiner Entstehung herum, kommt dann bei gleichzeitiger Ausbildung von Terminalheterocysten (s. S. 608) zur Ruhe (de Puymaly 1957 sieht daher die Heterocysten als den Rhizoiden anderer Algen entsprechende Bildungen an), sezerniert eine Scheide und tritt in ausgiebige Teilungen ein. Dabei liegen die Teilungswände vielfach parallel zur Längsachse des Fadens. Geitler (1921) wies jedoch nach (aus den Figuren Harders geht dieses auch ohne weiteres hervor), daß nicht ein Polaritätswechsel, sondern durch extreme Stauchung des Fadens hervorgerufene Umlagerung der Zellen hierfür die Ursache ist; die Stauchung wiederum ist bedingt durch geringe Dehnbarkeit der Scheide. Diese Stadien sind also sehr wahrscheinlich nicht in Wirklichkeit „aseriat", wie LAZAROFF u. VISHNIAC sie benennen. Später lockert sich die Scheide (Abb. 2f), ein gewellter Fadenverlauf wird deutlich, und es treten interkalare Heterocysten auf; an diesen kann ein Brechen des Fadenzusammenhanges unter Freilegung von sekundären Hormogonien sowie der Heterocysten eintreten. Dieser Cyclus wiederholt sich in der gleichen Kultur u.U. mehrmals (HARDER); das bedeutet jedoch noch nicht, daß es sich bei der geschilderten Abfolge, selbst bei *Nostoc*, um einen Entwicklungscyclus handele, wovon LAZAROFF und VISHNIAC sprechen. Daß eine in einem solchen Sinne zwangsläufige Folge von Ereignissen nicht besteht, geht aus dem Verhalten der Art auf Glucosenährböden im Licht hervor, wo im Gegensatz zum Saccharosemedium der periodische Wechsel von Wachstum und Hormogonbildung ausfällt (LAZAROFF und VISHNIAC 1962, 1964).

2. Entwicklungsphysiologie.

Die Bildungsbedingungen für die Hormogonien sind experimentell wenig untersucht. Jene aus Sporen entstehen bei deren Keimung (s. S. 617). *Nostoc muscorum* liefert Hormogonien anscheinend nach Übertragung von Zellmaterial alter Kulturen auf frisches Medium. Die Hormogonien einer unbestimmten *Nostoc*-Art (BUNT 1961) jedoch sistieren ihre Bewegung nach Übertragung auf neues Medium und wachsen zu Thalli aus.

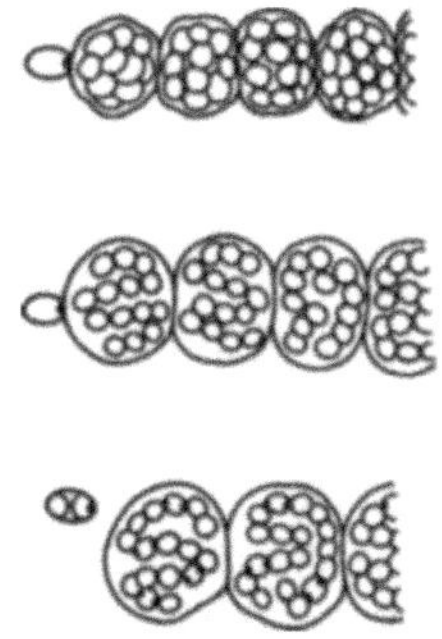

Abb. 3a—c. *Nostoc muscorum*. a Weiterentwicklung von Hormogonien, die Abb. 2d entsprechen würden, im Dunkeln auf Glucose: „Aseriates Stadium", b u. c das gleiche: Weiterentwicklung nach Beleuchtung. (Schema, nach LAZAROFF und VISHNIAC 1961.)

LAZAROFF und VISHNIAC finden bei Kultur von Hormogonien des *Nostoc muscorum* auf Glucose oder Saccharose im Dunkeln, daß jede der zwischen den Endheterocysten liegenden Zellen sich innerhalb ihrer Membran oder Scheide weiterteilt und dabei mächtig und kugelförmig anschwillt. Über diesen „aseriaten" Zustand (Abb. 3a) geht die Entwicklung bei Heterotrophie im Dunkeln nicht hinaus. Bereits geringe Belichtung (ca. 10 Lux) genügt aber, um Verquellung der Scheiden, Faden- und Heterocysten- sowie Hormogonienbildung hervorzurufen. Da Licht dieser Intensität nicht ausreicht, autotrophes Wachstum herbeizuführen, denken die Autoren an eine nicht über die Photosynthese laufende morphogenetische Wirkung. Dem entspricht das Aktionsspektrum (LAZAROFF und SCHIFF 1962). Obwohl Blaualgen in einem weiten Spektralbereich assimilieren, tritt die Wirkung nur durch einen schmalen Rotbereich mit Schwerpunkt bei 650 nm auf; möglicherweise ist das Allophycocyanin der Receptor. Die Rotinduktion kann durch eine ihr folgende Einstrahlung von Licht eines breiten Bereichs um 520 nm gelöscht werden, wobei das Löschlicht im Optimum dem Induktionslicht nur quantengleich zu sein braucht, was auf eine photochemisch einfache Reaktion hinweist. Das Empfängersystem ist aber nicht bekannt. Dunkelrot bleibt ohne Wirkung. Die Belichtung (LAZAROFF und VISHNIAC 1961) kann im Dunkeln durch einen Extrakt von Lichtmaterial der Alge ersetzt werden. Allerdings ließ sich der Kontrollversuch mit Extrakten aus heterotrophen Dunkelzellen aus technischen Gründen noch nicht durchführen.

Schwierig ist es, mit einigen die morphologischen Einzelheiten betreffenden Annahmen dieser Autoren (LAZAROFF und VISHNIAC 1962) einverstanden zu sein. Wurden „aseriate" Dunkelzellen leicht gequetscht, so fand man innerhalb der

Scheide kurze ein- bis vierzellige Fadenstücke; wiederholte man dies nach einem und zwei Tagen Belichtung, so waren die Fadenstücke länger und ihre Zahl geringer (s. auch Abb. 3b u. c). Zuletzt war nur noch ein einziger Faden mit inzwischen entwickelten Heterocysten sichtbar. Die Verfasser schließen, es habe reelle Fragmentation vorgelegen und sie werde während der Entwicklung im Licht durch end to end-Verschmelzung der kurzen Fäden rückgängig gemacht. Das wäre ein sehr ungewöhnlicher Vorgang. Genauso wahrscheinlich oder wahrscheinlicher ist die Fragmentation aber durch das Quetschen herbeigeführt; dieses wirkte dann nach beginnender Quellung der Scheiden im Licht weniger destruktiv.

Weiter ist zu erwähnen: Im Starklicht vorbehandeltes Material kommt nach Dunkelkultur auf Saccharose (meist jedoch nicht auf Glucose) über das „aseriate" Stadium hinaus. Die Scheiden quellen etwas, Heterocysten werden aber nicht angelegt. Dennoch sollen aus der einzelnen Scheide mehrere Hormogonien auskriechen. Nicht genügend belegt erscheint jedoch die Ansicht, daß derartige Hormogonien auf dem Substrat verschmelzen und danach sporenartige Bildungen produzieren. Die dies demonstrierenden Bilder stammen nämlich aus Impfungen, für welche, wie gerade angegeben gewonnenes, aseriates Material homogenisiert wurde, und können leicht dadurch zustande gekommen sein, daß einzelne Zellen durch mechanischen Schock die Fähigkeiten, sich zu teilen, nicht aber diejenige zur Plasmavermehrung einbüßten oder vorübergehend einbüßten („Sporen") oder daß aus ihnen anomal Verzweigungen auswuchsen („Verschmelzungen"). So erübrigt sich auch, auf weitere Vorstellungen der Autoren einzugehen, welche einen regelrechten Sexualcyclus mit Meiosis und Gamie für *Nostoc* konstruieren.

Die Bewegung der Hormogonien von *N. punctiforme* (HARDER 1927a) wird durch Lichtentzug in ihrer Dauer erhöht und entsprechend durch Starklicht wie diejenige von *N. muscorum* (LAZAROFF und VISHNIAC 1961, auf anorganischen Medien, hier bei ca. 100 Lux) unterdrückt. Auf Glucosemedien tritt diese Hemmung erst bei etwa 1000 Lux ein, sie fehlt aber auf Saccharose. Der Effekt der unterschiedlichen Wirkung der beiden Zucker (s. auch weiter oben) für die integrale Thallusmorphologie wird von LAZAROFF und VISHNIAC (1964) dargestellt.

b) Nannocyten und Endosporen.

Einige Blaualgen führen unter entsprechenden Bedingungen die Zweiteilung so durch, daß nicht wie gewöhnlich jedem, sondern nur jeweils dem zweiten, dritten oder höheren Teilungscyclus das zur Ausgangsgröße zurückführende Wachstum zugeordnet ist. GEITLER (1960) nennt als Beispiel für die Zusammenfassung von 2 Teilungen *Merismopedia* und *Gomphosphaeria*, für diejenige von 3 Teilungen *Chroococcus turgidus*. Hier findet auch während der Teilungen der Gruppen begrenztes Wachstum statt; fehlt es ganz, so resultiert die reine Vielfachteilung, welche bei Blaualgen als *Nannocyten*bildung bekannt und bei Chroococcalen verbreitet ist. GEITLER erhielt bei *Aphanothece caldariorum*, einer Art mit zylindrischen Zellen und ausschließlicher Querteilung, beim Übertragen von Standorten im Gewächshaus auf Benecke-Lösung den Zerfall der Zellen in Nannocyten zuletzt kugeliger Form und vom halben Durchmesser des Fadens, so daß hier neben Längs- auch Querteilungen stattgefunden haben müssen. Nannocytenartiger Zerfall kommt auch bei gewissen Wuchsformen der thermobiontischen Hormogonale *Mastigocladus lamellosus* vor (SCHWABE 1962).

Die *Endosporen* der *Chamaesiphonales* sind dadurch definiert, daß sie nach succedanen Vielfachteilungen innerhalb eines besonderen Sporangiums ⟨Aplanogonidangium⟩ und gewöhnlich in hoher Zahl entstehen (Abb. 4). Elektronenmikroskopische Untersuchungen von BECK (1963) an *Pleurocapsa fuliginosa* zeigen, daß bei der Bildung der ersten Furchen kräftige, an die dickwandige Sporangienwand anschließende partielle Septen entstehen, die Sporen bei den weiteren Teilungen eine Zeitlang nackt sind, sich zuletzt aber noch im Sporangium behäuten. Die Muttermembran öffnet sich bei der Reife mit einem Riß. Auch die Endosporen von *Chroococcidiopsis kashaii* sind erst zuletzt behäutet; sie entweichen durch einen Porus in der Sporangienmembran (FRIEDMANN 1961). Auch für die übrigen Endosporen neigt GEITLER (1960) heute dazu, das Vorhandensein von Eigenmembranen bei der Reife anzunehmen.

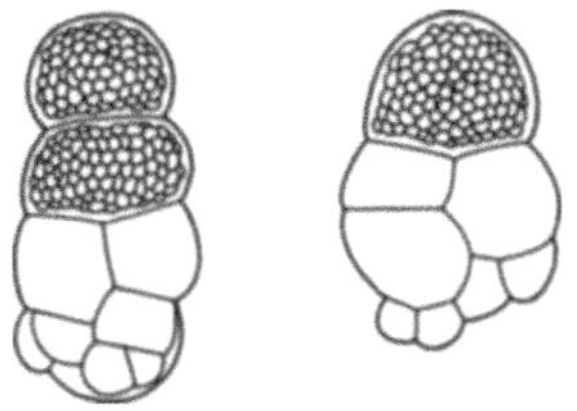

Abb. 4. *Pleurocapsa fuliginosa*. Thalli mit terminalen Endosporangien. (Nach BECK 1963.)

BECK studierte auch die Cytologie der Sporenbildung. Während die Spore nur einen Chromatinapparat enthält, wird ihr Wachstum im Sinne einer „Polyenergidie" von einer Vermehrung dieser Elemente begleitet. Der succedanen Teilung geht dann eine Umordnung der bis dahin im Plasma verteilten Einzelthylakoide und Thylakoidstapel voraus, bis diese die einzelnen Chromatinkörper becherartig umhüllen. Nur im Sporenzustand existiert die klassische Trennung in Centro- und Chromatoplasma. Im Gegensatz dazu scheint bei *Dermocarpa fucicola* bis zur Fortpflanzung ein einziger Chromatinapparat vorhanden zu sein (GARDNER 1906).

Die Endosporen entstehen, um nur Beispiele zu geben, bei *Myxosarcina* (VARMA und MITRA 1962), wo eine Altersordnung der durch Zweiteilung entstehenden Zellen in der räumlichen Kolonie fehlt, zu 1—8 aus allen Zellen etwa gleichzeitig. Bei *Chroococcidiopsis kashaii* teilt sich die Mutterzelle in 2, 4 oder 8 sekundäre Sporangien, die sich behäuten und heranwachsen, worauf ihr Inhalt in Sporen zerfällt. Die Wände der Sekundärsporangien lösen sich nunmehr auf und die Sporen treten aus der Mutterzellmembran durch einen Porus aus. Bei der hochdifferenzierten, mit einer Kriechsohle und aus ihr sich erhebenden erekten Fäden ausgestatteten *Nematoradaisia* (GARDNER 1917) wird immer die Endzelle der aufrechten Fäden in einem von enormer Schwellung begleiteten Reifungsvorgang zum Sporangium bestimmt. Ähnlich verhält sich in der Natur die etwas einfacher gebaute *Pleurocapsa minor*, die aber in Kultur genommen (GEITLER 1925) aus allen Thalluszellen Sporangien liefert. Die Determination zum Sporangium durch die Altersordnung erweist sich dadurch als relativ. Bei *Pleurocapsa fuliginosa* (BECK 1963) entstehen die Sporangien aus den Endzellen mehrreihiger (Abb. 4), bei der eigenartigen Gattung *Endonema* (PASCHER 1929) aus jenen der einreihigen polaren Fäden (Abb. 5).

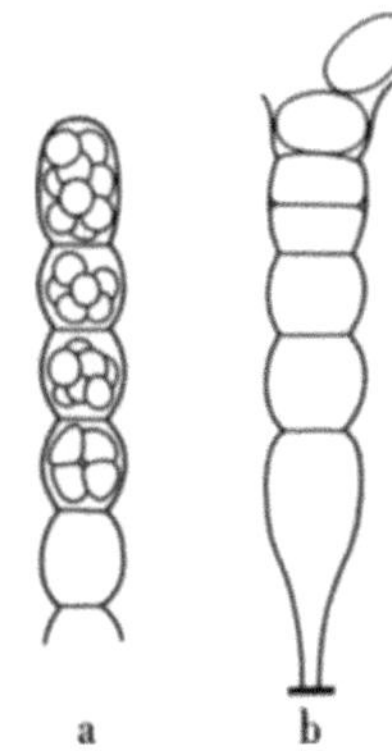

Abb. 5a u. b. *Endonema moniliforme*. a Bildung zahlreicher Endosporen in den Endzellen. b Ganzer Faden, Endzelle hat nur 4 Endosporen gebildet, von denen zwei austreten; die beiden anderen bleiben im Fadenverband. (Nach PASCHER, aus GEITLER 1932.)

Die Gattung *Dermocarpa* besitzt definitionsgemäß nur Endosporangien-Fortpflanzung, keine Zweiteilung. Bei *Pleurocapsa fuliginosa* (BECK) jedoch verwischt sich der Unterschied zu jener Gattung, indem in Seewassernährlösung der *Dermocarpa*-Typus der Entwicklung vorherrscht, in zur Hälfte mit Chu-Lösung, einem Medium für Süßwasseralgen, verdünnter Seewasserlösung aber erst durch Zweiteilungen einer initialenähnlichen Zelle der pleurocapsale Zell„faden" und darauf von dessen Ende proximalwärts fortschreitend Sporangium nach Sporangium entsteht.

c) Fortpflanzungskörper mit Dauerorgancharakter.

Dauerorgane im eigentlichen Sinn existieren nur bei einem Teil der Hormogonalen und bei wenigen Chroococcalen, ganz selten bei Chamaesiphonalen. Doch besitzen auch die vegetativen Zellen zahlreicher Blaualgen die Fähigkeit, ohne Schaden auszutrocknen, Grunderfordernis einer aerophytischen Lebensweise. Literatur darüber sowie über Hitze- und Kälteresistenz bei Fogg (1956) und Hess (1962); Resistenz gegen Gefrieren Whitton (1962), gegen Lyophilisieren Watanabe (1959) und Holm-Hansen (1964).

1. Hormocysten.

Die Hormocysten sind von ganz wenigen Stigonemataceen, Scytonemataceen und Rivulariaceen bekannt und stellen Hormogonien mit Funktion und Bau eines Dauerorgans dar, der letztere in Reservestoffreichtum und dem Besitz einer dicken verfestigten Scheide sich ausdrückend (Abb. 1). Bei der Keimung streckt sich der Faden und beginnt sich zu teilen (Beschreibungen bei Frémy 1930, Cholnoky 1935, Böcher 1946).

Die Paketkolonien der zu den *Pleurocapsaceen* gerechneten *Myxosarcina* können, wie anhangsweise bemerkt sei, als Ganze eine derbe, gefältete Membran ausscheiden und als Dauerorgane fungieren, die nachher mit Endosporen (bzw. Nannocyten) keimen (Varma und Mitra 1962).

2. Dauerzellen und Heterocysten, Morphologie und Bildung.

Sehr viel mehr ist über die einzelligen Dauerzustände bekannt, die beide durch Umwandlung vegetativer Zellen entstehen und von denen die ersteren wesentlich auf Nostocaceen und Rivulariaceen beschränkt sind, die zweiten den meisten Gattungen der fadenbildenden *Hormogonales* zukommen. Während die Dauerzellen klar die Funktion besitzen, die ihr Name angibt, ist die biologische Bedeutung der Heterocysten, welche häufig zusammen mit Dauersporen und gerne in deren Nähe vorkommen, zwar viel diskutiert, aber nur unbefriedigend geklärt worden, wenn eine einheitliche Klärung überhaupt möglich ist; darüber verweisen wir auf Geitler (1960), Tischer (1957) sowie auf die ausführlichen Diskussionen bei Fritsch (1951) und de Puymaly (1957). Da Heterocysten aber wiederholt keimten (Brand 1901, Spratt 1911, Geitler 1921, Hollerbach 1928, Steinecke 1934, Iyengar u. Desikachary 1944, Fogg 1951, Desikachary 1956, Wolk 1965a), seien sie hier mitbehandelt. Bei *Nostoc* und *Tolypothrix* scheinen sie auch wirklich Dauerzellen ersetzen zu können (Geitler, Hollerbach).

Dauerzellen. Die Dauerzellen (akinetes) sind nicht völlig auf die *Hormogonales* beschränkt. So bilden Chroococcalen wie *Gloeocapsa* dickwandige, bei manchen Arten bestachelte, an Dauerzellen erinnernde und von vegetativen Zellen durch ihre angehobene Größe, nicht aber in der Form, verschiedene Zustände aus (Brand 1900, Geitler 1932, Jaag 1945), deren Keimung unter schließlicher Sprengung der Membran von Brand und insbesondere von Jaag geschildert wurde. Die Bildungsbedingungen werden verschieden angegeben. Brand kennt *Gl. alpina* Näg. mit Dauersporen nur von feuchten Standorten, während Jaag bei *Gl. sanguinea* (Ag.) Kütz sens. Jaag, welche *Gl. alpina* Naeg. einschließt sowie bei *Gl. kützingiana* Näg. sens. Jaag gerade relativ trockene Standorte (Tintenstriche) als charakteristisch für dieses Vorkommen nennt. Dauerzellartige Bildungen beschreibt Geitler (1925) auch bei der Chamaesiphonale *Siphonema*. Schließlich seien noch die Dauerkeimlinge des eigenartigen, vielleicht den Chamaesiphonalen entfernt verwandten *Cyanophanon* (Geitler 1955, 1957) erwähnt, die anscheinend nach dem Auskeimen aus den „Stäbchen", den Sporen dieser Alge unter Verdickung der Membran das Wachstum sistieren können, aber nicht müssen. Vielleicht ein „Lag"-Phänomen, das unter bisher nicht näher definiertes Bedingungen eintritt.

Auch die *Dauerzellen sensu stricto*, die nur bei Hormogonalen vorkommen, schwellen bei ihrer Bildung unter Reservestoffspeicherung mehr oder weniger stark an und verdicken ihre Membran, wobei deren Außenschicht, welche vermutlich der Scheide der Trichome entspricht, in ähnlicher Weise strukturiert sein kann wie die Exine von Pollen und gelegentlich Farbstoffe einlagert. Bei Planktonformen kann sie blasenartig erweitert sein. Die stark lichtbrechende, relativ dünne Innenschicht ist immer farblos; darauf folgt eine sehr dünne Eigenmembran des Protoplasten. Die elektronenmikroskopische Untersuchung der

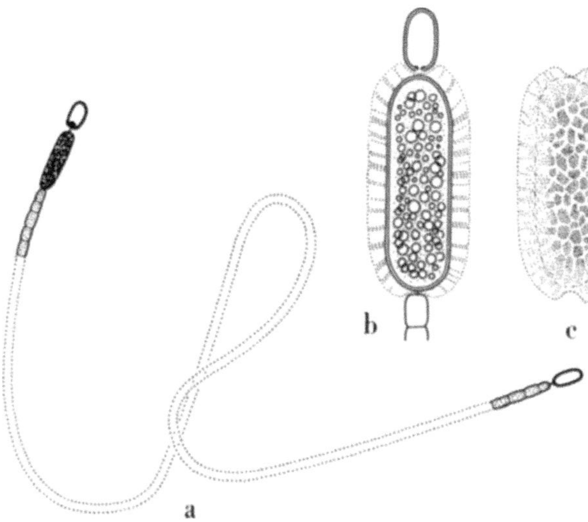

Abb. 6a—c. *Cylindrospermum alatosporum.* a Fadenaufbau, Terminalheterocysten, an die linke anschließend entwickelt sich eine Dauerspore, am rechten Ende würde eine solche später entstehen. b Reife Spore, stärker vergrößert. (a nach FRITSCH 1951, b, c nach GEITLER 1960.)

Anabaena cylindrica-Spore (WILDON und MERCER 1963) bestätigt den lichtoptischen Befund. Es gibt jedoch auch vierschichtige Dauersporenmembranen (*Nostoc elipsosporum*, GEITLER und RUTTNER 1935). SKUJA (1956) fand bei *Anabaena*-Arten eine (z.B. *A. planctonica*) oder zwei (*A. flos-aquae*) vorgebildete subterminale Rißstellen, an denen sich die Spore bei der Keimung öffnet. Die Membran ist cellulosefrei. Der Chromidialapparat der Dauerzellen ist relativ chromatinarm (POLJANSKI und PETRUSCHEWSKY), wohl, weil er ohne entsprechende DNS-Vermehrung einen vergrößerten Raum einnimmt. Der übrige plasmatische Inhalt ist bei *Anabaena* demjenigen der vegetativen Zelle ähnlich, jedoch treten gegenüber diesen charakteristische große Granen neu auf (WILDON und MERCER), die möglicherweise den Cyanophycinkörnchen des Lichtmikroskopes entsprechen.

Die Form der Dauerzellen (Abb. 1, 2, 6, 7, 8, 10b) weicht von jener der vegetativen Zellen durch erhöhte Länge, Breite oder beides, sowie häufig durch weitere Formunterschiede ab. Für ihre Lage im Fadenverband gibt es gattungsspezifische Besonderheiten. So entstehen sie bei *Nostoc*-Arten aus der Mehrzahl der Fadenzellen (Abb. 2f). Bei anderen Gattungen sind sie ± auf bestimmte Fadenabschnitte beschränkt, und zwar entweder als Nachbarzellen zu Heterocysten — besitzen diese interkalare Lage wie bei *Anabaena* (Abb. 7), so können sie ihnen beidseitig benachbart sein, liegen sie endständig wie bei *Cylindrospermum* (Abb. 6), so kann natürlich nur eine proximale Dauerzelle hinzutreten — oder sie entstehen im

Gegenteil entfernt von Heterocysten, mit denen sie bei *Anabaenopsis* z.B. alternieren können (Abb. 8) und welche sie hier auch insofern ersetzen, als sie wie diese in Paaren auftreten. Auch bei manchen *Anabaena*-Arten findet man sie an Stellen, an denen unter „normalen" Verhältnissen eine Heterocyste gebildet würde. Vielfach zieht die Bildung einer Dauerspore in Art einer „Ansteckung" die Umwandlung der benachbarten vegetativen Zellen in weitere Dauersporen nach sich (Abb. 7b).

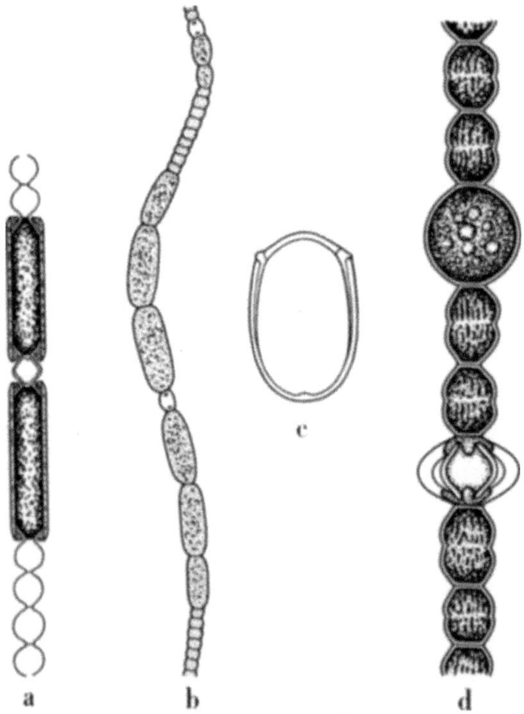

Abb. 7a—d. Dauersporen und Heterocysten von *Anabaena*-Arten. a *A. lapponica*, b *A. oscillarioides*, c Dauerspore mit zweiteiliger Wand von *A. scheremetievi*, d *A. planctonica*, Heterocyste zeigt nach Quellung im Konservierungsgemisch Struktur der Außenschicht. (a nach SKUJA aus GEITLER 1932, b nach FRÉMY 1930, c, d nach SKUJA 1956.)

Heterocysten. Mehr den Eindruck toter als lebender Zellen machen viele Heterocysten, da ihre plasmatischen Inhalte homogen wirken und wohl in hohem Quellungszustand vorliegen und da ihre Farbstoffe gewöhnlich bis auf Carotinoidreste abgebaut sind; doch bleiben bei tropischen Arten offenbar häufig auch die Chromoproteide erhalten (FRITSCH 1951). Nucleinsäuren und Eiweiß gehen zum Teil verloren. Die Heterocysten schrumpfen daher beim Trocknen stärker als vegetative Zellen. Selbst die Mineralsalze werden ziemlich vollständig abgeleitet (FOGG 1951). Doch bleiben Volutin-(Polyphosphat-?) Granula manchmal erhalten (TISCHER 1957). Auch liegen nach DRAWERT und TISCHER (1956) sowie TISCHER (1957) in den reifen, nicht in den heranwachsenden Heterocysten Orte sowohl erhöhter Reduktionsfähigkeit (Tetrazoliumchlorid, Redoxfarbstoffe) als auch erhöhten Oxydationsvermögens (Nadi-Reaktion), die wieder verschwinden, wenn die Heterocysten altern. Dies würde mit den Annahmen von CANNABAEUS (1929) übereinstimmen, die mit allerdings unzulänglicher Begründung in den Heterocysten Enzymspeicher sahen. Auch bei den Heterocysten wird die Membran verstärkt und liegt in einer ± gallertartigen äußeren, einer kräftigen, stark lichtbrechenden „Cellulose"schicht (Chlorzink- bzw. Schwefelsäure-Jod-Reaktion, wird aber nicht bei allen Heterocysten erhalten, s. FRITSCH 1951) und einer direkt nicht erkennbaren, an das Plasma angrenzenden sehr dünnen Eigenmembran vor (BRAND 1901, KOHL 1903, KLEIN 1915, GEITLER). Die Heterocystenmembran ist außen meist glatt (Abb. 1, 2, 9, 10, 11); selten nur treten wie bei *Anabaena planctonica* (SKUJA 1956) kompliziertere Membranstrukturen auf (Abb. 7d). Die inneren Ausmündungen der eine Verbindung mit den Nachbarzellen herstellenden

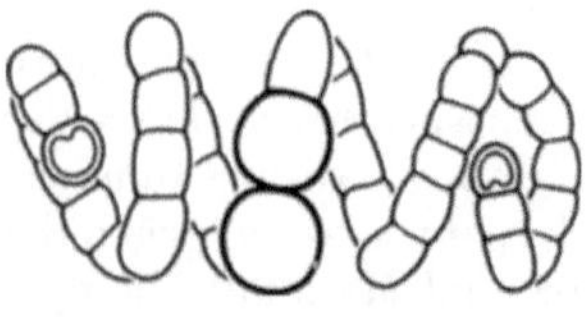

Abb. 8. Faden von *Anabaenopsis milleri* mit Terminalheterocysten und gepaarten Dauersporen. (Nach WORONICHIN aus GEITLER 1932.)

Tüpfelkanäle springen jedoch häufig kraterartig verstärkt in das Zellinnere vor und sind von innen häufig durch ein „Cyanophycinkörnchen" (GEITLER 1960), den „Verschlußkörper" KOHLS (1903), verschlossen. FOGG (1951) hält die Substanz für verschieden von Cyanophycin. Sie gibt zwar wie dieses starke Sakaguchi-Reaktion auf Arginin, löst sich aber nicht in verdünnter Säure.

Elektronenoptisch erweisen sich die Heterocysten (CHAPMAN und SALTON 1962, WILDON und MERCER 1963, SCHNEPF 1964) entgegen dem lichtoptischen Eindruck als reich an — vermutlich stark hydratisiertem — Plasma mit zahlreichen, allerdings blasenartig aufgetriebenen Thylakoiden, deren Lamellen häufig verzweigt sind. Gegenüber der vegetativen Zelle treten neue Granatypen auf. Von den drei Membranschichten ist die mittlere am kräftigsten. Sie hat auch den Hauptanteil an der Verstärkung des Tüpfelrandes; doch ist unmittelbar in Tüpfelnähe und in den Tüpfelkanal hinein auch die (innere) Eigenmembran recht dick (Abb. 11). Die „unit-membrane" des Plasmalemma kleidet auch den Tüpfelkanal aus und läuft an dessen Ende parallel zu derjenigen der vegetativen Nachbarzelle. Sie ist mit dieser durch die gemeinsame Eigenmembran querende Plasmodesmen-artige Fäden verbunden. Gelegentlich ist der Kanal durch einen osmophilen Pfropf verstopft, der wohl dem Verschlußkörper KOHLS (1903) entspricht. Erst an freigewordenen Heterocysten desorganisiert die Innenstruktur.

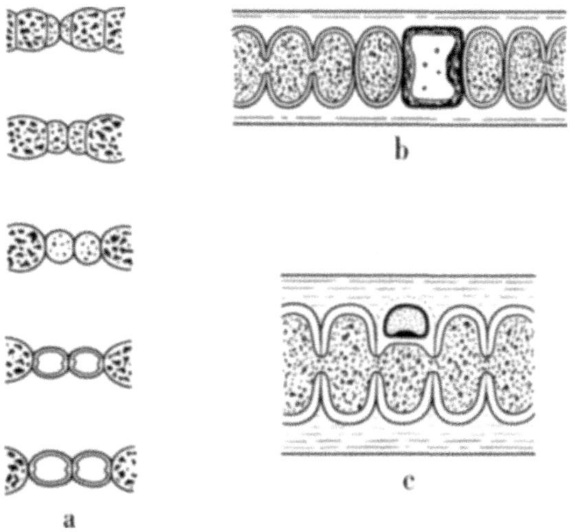

Abb. 9a—c. Entstehung eines Heterocystenpaares bei *Anabaenopsis arnoldii*, die Mutterzelle des Paares hatte sich äqual geteilt; die Mutterzellen der Einzelheterocysten teilen sich inäqual. Heterocysten zweitüpfelig. b Interkalarheterocyste, c Seitenast-(End-)Heterocyste von *Stigonema ocellatum* var. *braunii*. (a nach APTEKARJ aus GEITLER 1932, b, c nach FRITSCH 1951.)

Die Heterocysten zeigen sich als zweitüpfelig (Abb. 1, 7), wenn sie interkalar, oder als polar gebaut und eintüpfelig, wenn sie terminal am Faden stehen (Abb. 6, 9c, 10b); beides kann an der gleichen Pflanze vorkommen (Abb. 2e, 9b, c, 10). *Brachytrichia* bildet an Verzweigungsstellen sogar dreitüpfelige Heterocysten aus (IYENGAR u. DESIKACHARY 1954). Liegen „terminale" Heterocysten im Fadeninnern, so weist dies darauf hin, daß an ihrer tüpfelfreien Seite eine Unterbrechung des Fadenzusammenhanges stattgefunden hat, wie dieses meist geschieht, wenn die Nachbarzelle bereits eine Heterocyste war; dagegen nicht oder nicht immer, wenn sie gleichzeitig zur Heterocyste wird. Im übrigen weichen sie in Größe und Form meist weniger von vegetativen Zellen ab als Dauerzellen.

Heterocysten entstehen 1. einzeln, 2. in Paaren oder 3. in Ketten. Im letzten Falle werden mehrere Einzelheterocysten nacheinander gebildet, womit dieser auf Fall 1 zurückgeführt ist. Dabei kann die älteste Heterocyste interkalar oder endständig sein; die Folgeheterocysten werden wegen der Plasmaunterbrechung an der der älteren Heterocyste zugewendeten Seite immer eintüpfelig. Die Heterocyste ist mindestens gelegentlich Ergebnis einer inäqualen Zellteilung, welche die kleinere Zelle zur Sonderzelle determiniert; die Heterocystenpaare von *Anabaenopsis* entstehen entsprechend durch ein Paar von inäqualen Teilungen, deren Mutterzelle einer äqualen Teilung entstammt (Abb. 9a). Dabei besitzen nur die

Heterocysten von *A. arnoldii* (Aptekari 1926, zit. nach Geitler) zwei Tüpfel (Abb. 9a), jene aller anderen Arten (s. z.B. Pandey und Mitra 1962a), obwohl aus Schwesterzellen entstanden, einen Tüpfel; offenbar reißt im letzten Falle die Verbindung zwischen den Heterocysten sehr früh ab.

Abb. 10a u. b.

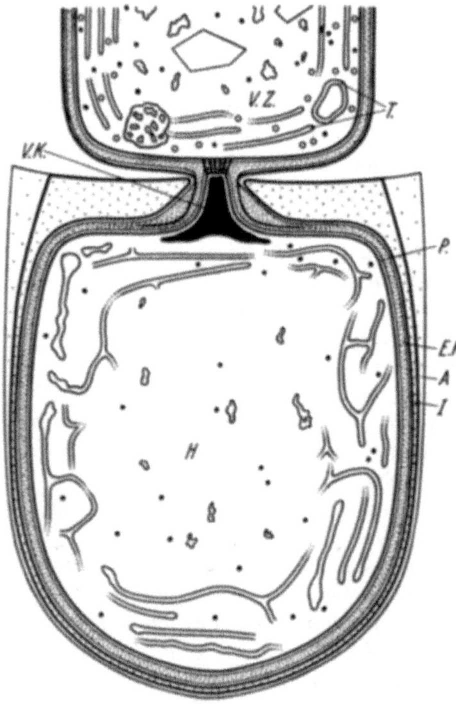

Abb. 11.

Abb. 10a u. b. *Nodularia spumigera*. a Faden mit äquidistanten Heterocysten. b Dauersporen, Interkalar- und Terminalheterocysten. (Nach Fritsch 1951.)

Abb. 11. Schemabild einer Heterocyste nach übermikroskopischen Aufnahmen. *H* Heterocyste; *V.Z.* Vegetative Zelle; *A* Außenschicht; *I* Innenschicht, „Cellulose" führend; *E.M.* Eigenmembran; *P.* Plasmalemma; *V.K.* Verschlußkörper; *T.* Thylakoide. (Nach Wildon und Mercer 1963.)

3. Entwicklungsphysiologie der Entstehung von Dauerzellen und Heterocysten

Allgemeine Bedingungen.

Die Bildung der Dauerzellen ist allgemein leichter zu beeinflussen als diejenige von Heterocysten. Sie fehlt in jungen Kulturen und tritt erst beim Altern auf (Glade 1914, Harder 1917b). Durch rechtzeitiges Übertragen der Fäden auf frische Nährböden kann sie aber häufig ganz verhindert werden. Glade erhält die Dauersporen von *Cylindrospermum*-Arten bei Erschöpfung der Nährlösung und deutet seine Versuche unter Zuhilfenahme derjenigen über die Dauerzellenkeimung in der Art, daß entweder Stickstoff- oder Phosphormangel induzierender Faktor sei. Ähnlich beobachtete Harder die Dauersporen von *Anabaena variabilis* und *Cylindrospermum muscicola* nur unter N-Mangelbedingungen. Bei den Stickstoffbindern *Cylindrospermum licheniforme*, *Anabaena oscillatorioides* und *A. variabilis* stellte Demeter (1955) auf den an gebundenem Stickstoff freien Böden gegenüber den N-haltigen Kulturen eine starke Vermehrung der Sporen fest, während der gleiche Faktor bei *Cyl. alatosporum* (N_2-Binder) nur umgestaltend auf die Form der Dauersporen wirkte, deren Menge jedoch nicht änderte. Bei

Anabaena cylindrica (FOGG 1944) war dagegen gebundener Stickstoff zur Dauersporenbildung notwendig, obwohl diese Art Stickstoff fixiert.

Die Erfahrungen WOLKS (1965b) an einem anderen Stamm von *Anabaena cylindrica* sind nicht im Widerspruch mit dieser Feststellung. Dieser prüfte eine Anzahl von Faktoren systematisch auf ihren Einfluß auf die Sporenbildung. Die Sporen werden in Ketten größerer oder geringerer Länge angelegt, welche den interkalaren Heterocysten unmittelbar benachbart sind. Im allgemeinen ruft P-Mangel, nicht aber N-Mangel die Sporenbildung hervor. Voraussetzung für reichliche Sporenproduktion sind geeignete Lichtintensität (400—1600 Lux) und Temperatur, wobei Temperaturerhöhung (bis 30^0) die Sporenbildung fördert, aber oberhalb 24^0 nur bei genügend reichlicher Ca-Versorgung (2,5 mM) ohne Schaden ertragen wird. Hohe Impfmenge ist günstig. Das pH muß um 8 liegen: zur Pufferung erweisen sich die Dipeptide Alanyl-Glycin und Alanyl-Alanin als günstig, während andere Di- und Tripeptide giftig waren. Weiter erhöht Na-Acetat die Zahl sporenbildender Heterocysten-Zentren und Ca-Gluconat die Anzahl der Sporen pro Sporenkette. Nicht klar wurde jedoch, warum gelegentlich auf alten Normalmedien trotz genügender P-Versorgung die Sporulation eintrat und sie auf P-Mangelmedien hin und wieder spärlich war.

Salzlösungen hemmen in höheren Konzentrationen allgemein die Ausbildung von Dauerzellen (DEMETER). Über die Folgen der Zugabe verschiedener Gifte berichten DEMETER und RENNER (1957). In den auslösenden Bedingungen ergibt sich bei manchen Arten eine Ähnlichkeit zu denen für die Heterocysten, so bei *Cylindrospermum licheniforme* (DEMETER). Darauf deutet auch in gewissen Fällen die äquivalente Verteilung beider Elemente in den Fäden (s. S. 609, 610).

Die Heterocysten finden sich in natürlichen Algenherkünften in artspezifisch weitgehend bestimmter Form und Anordnung und bedeuten daher dem Taxonomen ein wichtiges Merkmal zur Charakterisierung. Dennoch erwies sich ihr Auftreten in einigen Fällen als abhängig von Außenbedingungen. DEMETER (1956) und DEMETER und RENNER (1960) berichten über den Einfluß von Salzen (s. auch CANABAEUS 1929) und Giftstoffen. Vertreter aus beiden Stoffgruppen können die Heterocysten vermehren oder auch mehr oder weniger unterdrücken. Zum Beispiel wird bereits durch 1% NaCl bei den meisten der untersuchten Arten (s. oben) die Bildung von Heterocysten wie Dauerzellen gehemmt, bei *Anabaena oscillaraioides*, einer nach den Literaturangaben (GEITLER 1932) keineswegs halophilen Art, aber gefördert.

Wichtiger erscheinen die Beziehungen zum Stickstoff einerseits, zum Licht andererseits. Wie manche Dauersporen werden die Heterocysten bei einigen Arten durch N-Mangel begünstigt, so bei *Scytonema javanicum* und *Sc. spec.*, bei *Cylindrospermum licheniforme* und besonders *Anabaena apospora*, wo sich auch ihre Größe erhöht (DEMETER), sowie bei *A. cylindrica* (FOGG 1944, 1949) — und sie werden entsprechend durch N-Überschuß unterdrückt. PANDEY und MITRA (1962b) erzielten bei 6 Cyanophyceen aus 5 Gattungen in Medien, allerdings recht hohen Nitratgehaltes (Abb. 13), sogar völliges Ausbleiben der Heterocystenbildung. Die Zahlen der Abbildung wurden dadurch gewonnen, daß bei Erscheinen der ersten Heterocysten der Reststickstoff in der Kulturflüssigkeit bestimmt wurde; sie zeigen eine sehr unterschiedliche Stickstofftoleranz der einzelnen Arten. Da auch bei den höchsten Konzentrationen das Kation keinen oder geringen Einfluß besitzt, scheint wirklich das Nitrat der entscheidende Faktor zu sein.

Andererseits kann auch Lichtmangel die Heterocystenbildung unterdrücken, wie bei „*Scytonema polymorpha*“ (JAAG 1945), selbst dort, wo die Beleuchtung vegetatives Wachstum noch zuläßt. Der Entwurf des sehr langfristigen Versuchs läßt aber nicht erkennen, ob das Licht durch die infolge der erhöhten Produktion gesteigerte Stickstoffzehrung wirkte oder nicht. Dieser naheliegende Zusammenhang scheint in den Versuchen von LAZAROFF und VISHNIAC mit *Nostoc muscorum* nicht zu bestehen. Die Alge wächst auf Glucose-Medium im Dunkeln, wenn auch langsam, mit „aseriate morphology" (s. S. 605) und ist dabei frei von Heterocysten. Bereits Beleuchtung solchen Materials in Dosierungen, die auf Mineralböden noch kein Wachstum zulassen — 10 Lux und 50 Lux (Dauerlicht, Leuchtstoffröhre) genügt — um Hormogonien und danach in diesen Heterocysten entstehen zu lassen. Das gleiche geschieht, wenn ich die Autoren recht verstehe, auch im Dunkeln unter dem Einfluß eines Extraktes von lichtkultivierter Alge (s. S. 605). Die Produktionen sind so gering, daß der Lichteinfluß kaum über die Stickstoffzehrung wirken kann. Die Autoren betrachten die Heterocystenbildung allerdings als Folge der Wiederingangsetzung eines „Entwicklungscyclus" der Alge. Doch scheint unter Hinweis auf die Bemerkung von S. 605 und die gerade geschilderten Erfahrungen mit anderen Blaualgen nach dem gegenwärtigen Stand der Kenntnisse geringe Notwendigkeit zu bestehen, die Heterocysten anders als durch äußere (aber *nicht* innere) Bedingungen beeinflußt zu sehen.

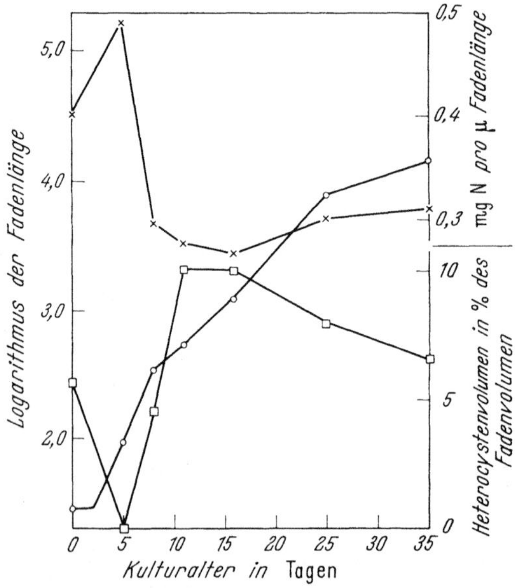

Abb. 12. Änderung von: ○ Fadenlänge in µ/mm³ der Kultur; □ Heterocystenvolumen pro Hundert Fadenvolumen; ×: $10^9 \times$ mg N/µ Fadenlänge während der Kultur von *Anabaena cylindrica* in einem an gebundenem Stickstoff freien Medium. (Nach FOGG 1944.)

Eine eingehendere Analyse der Abhängigkeit der Heterocystenbildung vom Stickstoff bei dem N-Binder *Anabaena cylindrica* lieferte FOGG (1944, 1949). Wird die Abhängigkeit der Heterocystenzahl vom N-Gehalt des Algenmaterials während des Wachstums einer Kultur mit N_2 als alleiniger Stickstoffquelle dargestellt (Abb. 12), so ergibt sich eine klare inverse Korrelation zwischen beiden Größen. Das Minimum der Heterocystenbildung fällt auf den Beginn der exponentiellen Phase des Wachstums. Tabelle 1 läßt aber erkennen, daß die Korrelation bei Abwandlung der N-Quelle nicht mehr besteht. NH_3-Stickstoff wirkt ausgesprochen hemmend auf die Heterocystenbildung, während NO_3^- in der gleichen Menge angeboten sich nahezu indifferent verhält. Dies legt nahe, daß es auf eine bestimmte N-Fraktion ankommt: es dürfte sich um NH_3 oder eine ihr metabolisch nahestehende Stickstoffverbindung handeln. Unwahrscheinlich ist dabei eine Wirkung der Ammonium-Base über die Reaktion des Zellinhaltes. Wurden weitere N-haltige Verbindungen herangezogen, so bewirkt Nitrat (geringfügig) wie auch Glykokoll und Asparagin eine anfängliche, dann aber bald vor-

Tabelle 1. *Heterocystenhäufigkeit von Anabaena cylindrica bei verschiedenen Stickstoffquellen.* (Nach FOGG 1949.)

Grundmedium ohne gebundenen Stickstoff plus	Heterocysten-häufigkeit	N-Gehalt % Trockengewicht
—	3,89 ± 0,13	8,14 ± 0,31
0,002 M KNO_3	3.47 ± 0.17	6.87 ± 0.14
0,002 M NH_4Cl	0.42 ± 0,16	7,10 ± 0.37

übergehende (bei Nitrat unvollständige) Senkung der Heterocystenhäufigkeit. FOGG erklärt dies mit anfänglicher Bildung und Anhäufung von Ammonium aus den Stickstoffverbindungen, die danach durch adaptive Einregulierungen des Stoffwechsels zurückgeht.

Eine Förderung der Heterocystenbildung durch Succinat oder Glucose dürfte auf der Verwendung dieser Verbindungen bzw. ihrer Metaboliten als Aminierungssubstrate beruhen, als welche sie die störende N-Fraktion entfernen. Eine gleichartige Wirkung von erhöhten Phosphatgaben könnte durch Ankurbelung der Glykolyse und weiter in ähnlicher Weise wie oben erklärt werden. Die Wirkung von teilweisem Lichtentzug bei einem an gebundenem Stickstoff freien Nährboden ist weniger leicht zu deuten. Bei niedrigerer Beleuchtung (260 Lux) sinkt die Heterocystenhäufigkeit zunächst und bis weit in die Phase exponentiellen Wachstums hinein auf nahe Null ab, erreicht dann aber fast das Doppelte gegenüber den bei stärkerer Beleuchtung (1300 und 3100 Lux) erhaltenen Werten. Gemäß Hypothese ist der zweite Anstieg vielleicht durch den geringeren Energiegehalt der Zellen infolge verminderter Photosynthese und dadurch bewirkte Abschwächung der N-Fixierung zu erklären. Fehlen des zur Stickstoff-Bindung notwendigen Molybdän steigert entsprechend die relative Heterocystenzahl ebenfalls. Der Mechanismus der Hemmwirkung des Ammoniumions ist ungeklärt. Daß jedoch die Sonderstellung dieser Form des Stickstoffs nicht in einfacher Weise zu verallgemeinern ist, scheinen die Versuche von PANDY und MITRA zu zeigen, in denen (Abb. 13) Ammoniak die Heterocystenbildung mehrerer Arten zwar stärker inhibiert als Nitrat, aber bei seinem guten Permeationsvermögen doch auffallend wenig stärker. Bei *Gloeotrichia natans* können die Heterocysten durch Ammoniumchlorid ganz unterdrückt werden (FOGG 1949).

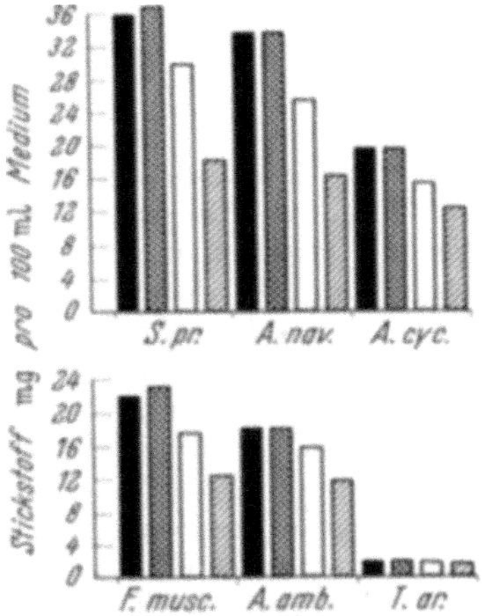

Abb. 13. Stickstoffkonzentration des Mediums, bei der die Heterocystenbildung beginnt für: *S. pr.*: *Scytonema praegnans*; *A. nav.*: *Anabaena naviculoides*; *A. cyc.*: *Anabaena cycadearum*; *F. musc.*: *Fisherella muscicola*; *A. amb.*: *Anabaenopsis ambigua*; *T. ar.*: *Tolypothrix arenicola*, wenn Stickstoffgabe als ■: KNO_3, ▩: $Mg(NO_3)_2$, □: NH_4NO_3, ▨: $(NH_4)_2SO_4$. (Nach PANDEY und MITRA 1962.)

Zum Mechanismus der Spontandifferenzierung von Heterocysten und Dauerzellen

Die linearen Muster der gegenseitigen Anordnung von Heterocysten und Dauerzellen (s. S. 609, 610, weitere Einzelheiten bei FRITSCH 1951) sind Ergebnis der beide Arten von Sonderzellen differenzierenden Vorgänge, und ihre Mannigfaltigkeit spiegelt die artspezifisch unterschiedlichen Ansprüche der Sonderzellen an die auslösenden Bedingungen wieder. Das Muster für den einzelnen Organismus muß modifizierbar sein. Da die Heterocysten früher da sind (s. z. B. WOLK 1965b) als die zu ihnen in bestimmte Lagebeziehung tretenden Dauerzellen, ist ihre Anlegung zuerst zu besprechen.

Die Heterocysten von *Anabaena cylindrica* finden sich in nicht sehr regelmäßigen Abständen im Faden. Diese Abstände sind, wie S. 614 geschildert, mit dem NH_4^+-Gehalt des Mediums positiv korreliert. Zwischen zwei alte Heterocysten des sich verlängernden Fadens wird dann zu geeigneter Zeit eine neue interpoliert. Die Ursache der Determinierung dieser Zelle zur Heterocyste sieht FOGG (1949) darin, daß durch den Stickstoffverbrauch während des Wachstums fluktuierende Gradienten von Ammonium (oder der diesem nahestehenden eigentlichen Hemmsubstanz, S. 614) im Faden entstehen. Eine Zelle wird induziert, wenn sie in ihrem individuellen Wachstumscyclus eine sensible Phase erreicht hat und wenn gleichzeitig ihr Gehalt an Hemmstoff unter einen kritischen Wert sinkt. Die Differenzierung wäre dann nur insofern von bereits vorhandenen Heterocysten abhängig, als diese die vegetativen Nachbarn einseitig gegen Stickstoffentzug abschirmen und dadurch vegetativ halten könnten, was auch Vorstellungen von DE PUYMALY (1957) entsprechen würde. Sehr regelmäßige gegenseitige Abstände der Heterocysten, wie sie bei *Nodularia* (FRITSCH 1951, GEITLER 1960) gefunden werden (Abb. 10), könnten auf zusätzlichen Sperreffekten (s. z. B. BÜNNING 1957) durch bereits vorhandene Heterocysten beruhen. Ähnlich ist vielleicht die häufig auch bei *Anabaena cylindrica* gefundene Endstellung der Heterocysten an Sporenkeimlingen oder Hormogonien zu erklären, was nahelegt, daß derartige Effekte auch bei den interkalaren Heterocysten der Art beteiligt sein möchten. Wie sich hier die Befunde von DRAWERT und TISCHER (1956) und TISCHER (1957) über erhöhte Fermentaktivität im Redoxsystem der Heterocysten einfügen, bleibt abzuwarten.

Die *Lage der Dauersporen* ist artspezifisch von derjenigen der Heterocysten und voneinander abhängig, wofür S. 609 bereits Beispiele gegeben wurden. Diese Muster werden erwartungsgemäß durch Außenbedingungen modifiziert. Nach CANNABAEUS (1929) entstehen die Dauerzellen bei solchen Keimlingen von *Anabaena cylindrica*, welche bald wieder in Dauersporen übergehen, entfernt von Heterocysten sogar in deren Abwesenheit, im Gegensatz zu erwachsenen Fäden, wo sie gewöhnlich unmittelbare Nachbarn der Heterocysten sind. *Anabaena variabilis* bildet gewöhnlich die Dauerzellen in Ketten und entfernt von Heterocysten; bei O_2-Mangel (Einleiten von H_2, wobei Gasvacuolen in den Zellen auftreten) entstehen sie im Gegensatz dazu einzeln und sind von zwei Heterocysten unmittelbar eingeschlossen. Direkte und beidseitige Nachbarstellung zu Heterocysten ist in der Gattung *Anabaena* besonders häufig. Beim Wolkschen Stamm von *Anabaena cylindrica* (WOLK 1965b) liegen sie in dieser Art entweder einzeln oder in Ketten von zwei oder seltener drei und vier Zellen (ähnlich wie Abb. 7b); Gluconat-Zugabe ruft sehr starke Verlängerung dieser Ketten (bis auf sieben und mehr Dauersporen) hervor (s. auch DEMETER 1956). Andererseits können aber aus unbekannten Gründen und selten eine oder zwei vegetative Zellen zwischen Heterocyste und Spore von *A. cylindrica* eingeschoben sein (ein Zustand, der z. B. bei *A. hasallii*, CANNABAEUS 1929, die Regel ist), und in Gegenwart von Gluconat treten auch von Heterocysten durch weitere Abstände isolierte Sporenstränge auf. Die Nachbarschaftslage von Dauerzellen zu Heterocysten läßt sich möglicherweise mit einer Abgabe induzierender Stoffe durch die Heterocyste erklären, wie dies in allgemeiner Form von FRITSCH (1951) und FOGG (1951) nahegelegt wird. Für einen besonderen Metabolismus der Heterocysten sprechen die S. 610 angegebenen allgemeinen Charakteristika dieser Zellen und z. B. die Befunde von DRAWERT und TISCHER (1956) und TISCHER (1957) über ihre erhöhte Fermentaktivität, für Stoffabgaben nach außen die Beobachtung TISCHERs, daß die Heterocysten Bakterien anlocken, für solche in die Nachbarzellen die Untersuchungen FOGGs (1951) über die Histochemie der Heterocysten und für ihre Weitergabe durch Diffusion oder Leitung die Bilder von Sporenketten mit proximalwärts steigender Altersordnung, wie sie sich bei *Anabaena oscillarioides* (Abb. 7b) oder *A. cylindrica* (WOLK 1965b) finden. Im letzteren Falle sind aber häufig die Sporen einer Kette auch in untereinander gleichem Reifezustand; Kettenverlängerungen, nachdem die Sporen als solche erkennbar geworden sind, scheinen in keinem der beiden Fälle nachträglich vorzukommen. Man kann daraus schließen, daß 1. der Akt der Determinierung ohne zunächst sichtbare Manifestation erfolgte (WOLK), 2. daß die sequentiellen Determinierungen im Vergleich zur Reifungsdauer der Sporen rasch vor sich gehen, und 3. vielleicht, daß die Heterocysten als Induktor nur in diesem kurzen Zeitraum „wirksam" sind.

Liegen andererseits bei einer Blaualge die Dauerzellen von den Heterocysten entfernt, so könnte dies auf ähnliche Bildungsbedingungen für beide Zellsorten oder (bzw. und damit) einen Sperreffekt der letzteren auf die ersteren hinweisen. Begründete Aussagen dazu sind einer zukünftigen entwicklungsphysiologischen Analyse vorbehalten.

4. Die Keimung von Heterocysten und Dauersporen.

Nach ersten Beobachtungen von BRAND (1901) und etwas zweifelhaften von SPRATT (1911) konnten GEITLER (1921) und später andere (Literatur s. S. 608) in größerem Umfang Keimlinge von *Nostoc*-Heterocysten und solchen anderer Arten erhalten (Abb. 14). Nachdem die vegetativen Zellen einer Aufsammlung von *Nostoc commune* zugrunde gegangen waren, keimten die resistenteren Heterocysten bei Übertragung in Nährlösung wieder aus (GEITLER). Auch FOGG (1951) berichtet von gelegentlichen Keimungen seiner *Anabaena cylindrica*-Heterocysten auf succinathaltigem Medium. Konsequente Kulturversuche zur Heterocystenkeimung stellte WOLK (1965a) mit seinem Stamm der gleichen Art und im Anschluß an die Erfahrungen FOGGS (1944, 1949) mit der Heterocysten-*Bildung* an. Da Heterocysten bei Stickstoff-, insbesondere NH_4^+-Mangel bevorzugt gebildet werden, sollte Restitution von Ammoniumsalzen zur Keimung führen. Tatsächlich war dies in entsprechenden Versuchen bei 3—10% der Heterocysten der Fall. In der Natur scheint die Heterocystenkeimung aber keine überwältigende Rolle zu spielen (DE PUYMALY 1957, FRITSCH 1951).

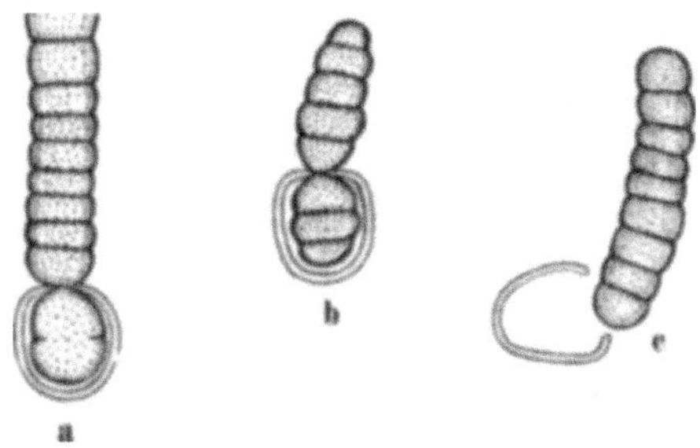

Abb. 14a—c. Keimung der Heterocysten von *Gloeotrichia raciborskii*. (Nach DESIKACHARI 1956, verändert.)

Die Keimung der Dauersporen geschieht 1. durch Verquellung der Membran (Beispiel: *Nostoc ellipsosporum*, GEITLER und RUTTNER 1935/36), 2. durch Aufreißen der Sporenwand (Beispiel: *Nostoc punctiforme*, HARDER 1917b; Abb. 2a), 3. durch Öffnen mittels präformierten (Abb. 7c) Deckels (Beispiel: *Anabaena flos-aquae*, SKUJA 1956). Der Inhalt tritt dann als Hormogonium aus.

Die Bedingungen der Keimung von Dauerzellen wurden von GLADE (1914) und später sehr eingehend von HARDER (1917b) untersucht. Der erstere kommt, allerdings auf Grund nur weniger Versuche, zu dem plausiblen Schluß, daß bei hinreichenden sonstigen Bedingungen die Keimung durch die Gegenwart desjenigen Ions begünstigt werde, dessen Fehlen ihre Bildung auslöste. In destilliertem Wasser jedenfalls (HARDER) erfolgt die Keimung nicht (*Cylindrospermum muscicola, Nostoc punctiforme, Anabaena variabilis*). Die Bedeutung der Salze geht daraus hervor, daß die Sporen — vom Einsetzen der auf anorganischem Substrat zur Keimung notwendigen Beleuchtung an gerechnet — rascher keimen, wenn sie einige Tage auf Nähragar gelegen haben, als wenn sie sogleich nach dem Übertragen belichtet wurden. Licht wirkt durch die gegebene Menge. Die Keimung kommt auf anorganischem Medien unabhängig von der Beleuchtungsstärke, wenn diese nur 12 Lux nicht wesentlich unterschreitet, nach etwa 1000 Lux-Std zustande. Unterschiede der Lichtfarbe waren im Rahmen der damaligen

Methodik ohne Einfluß. O_2 ist notwendig. Das Licht kann jedoch ganz oder teilweise durch organische C-Quellen, völlig durch Rohrzucker in seiner Wirkung auf die Keimung ersetzt werden. Erhöhung der Temperatur auf 30° erhöht dabei die Keimprozente stark. Im Licht ist Rohrzucker jedoch ohne Einfluß. Eine obligate Ruheperiode der Spore besteht nicht.

Wolk (1965a) erreicht die Keimung der spezifisch durch niedriges Phosphat-Niveau ausgelösten Dauersporen von *Anabaena cylindrica* auf Phosphat-haltigen Medien und diejenige der Heterocysten, welche infolge von NH_4^+-Mangel entstanden, durch Restitution von Ammoniumsalz. Er vermerkt daher, die Dauersporen seien bestimmt, die Art über Phosphor-Mangelperioden, die Heterocysten dagegen *Anabaena* bei Stickstoffmangel zu retten.

Literatur.

Beck, S.: Licht- und elektronenmikroskopische Untersuchungen an einer sporenbildenden Cyanophycee aus dem Formenkreis von *Pleurocapsa fuliginosa* Hauck. Flora (Jena) **153**, 194—216 (1963). — Böcher, T. W.: *Dichothrix gelatinosa sp.*, its structure and resting organs. Danske Vidnsk. Selsk. Biol. Skr. **4** (4), 1—14 (1946). — Borzi, A.: Note alla morfologia e biologia delle Alghe Ficocromacee. Nuovo G. bot. ital. **10**, 236—288 (1878). — Brand, F.: Der Formenkreis von *Gloeocapsa alpina*. Bot. Zbl. **83**, 224—236 (1900). — Bemerkungen über Grenzzellen und spontan rote Inhaltskörper der Cyanophyceen. Ber. dtsch. bot. Ges. **19**, 152—159 (1901). — Bünning, E.: Polarität und inäquale Teilung des pflanzlichen Protoplasten. Protoplasmatologia, (Wien) VII **9**a, 1—86 (1957). — Bunt, J. S.: Blue green algae. Growth. Nature (Lond.) **192**, 1274—1275 (1961).

Canabaeus, L.: Über die Heterocysten und Gasvakuolen der Blaualgen. Pflanzenforschung H. 13, 1—48 (1929). — Chapman, J. A., and M. R. J. Salton: A study of several blue-green algae in the electron microscope. Arch. Microbiol. **44**, 311—322 (1962). — Cholnoky, B. v.: Beiträge zur Kenntnis der Hormocystenbildung. Beih. Bot. Zbl. **53** A, 26—33 (1935). — Claus, O.: Re-evaluation of the genus *Gomontiella*. Rev. Algologique, N. S. **5**, 103—110 (1960).

Demeter, O.: Über Modifikationen bei Cyanophyceen. Arch. Mikrobiol. **24**, 105—133 (1956). — Demeter, O., u. O. Renner: Über Modifikationen bei Cyanophyceen. II. Über die Wirkung von Giften. Planta (Berl.) **54**, 195—209 (1960). — Desikachary, T. V.: Germination of heterocysts in two members of the *Rivulariaceae*, *Gloeotricha raciborskii* Wolosz. and *Rivularia mangini* Frémy. J. Indian bot. Soc. **25**, 11—17 (1956). — Douin, R.: Sur le polymorphisme de l'*Anabaena cycadeae* Reinke. Rev. gén. bot. **63**, 418—423 (1956). — Drawert, H., u. I. Tischer: Über Redox-Vorgänge bei Cyanophyceen unter besonderer Berücksichtigung der Heterocysten. Naturwissenschaften **43**, 132 (1956).

Feldmann, J., et G. Feldmann: Observations sur les genres *Dermocarpa*, *Dermocarpella (Cyanophyceae)*. Öst. bot. Z. **100**, 505—514 (1953). — Fogg, G. E.: Growth and heterocyst production in *Anabaena cylindrica* Lemm. New Phytologist **43**, 164—175 (1944). — Growth and heterocyst production in *Anabaena cylindrica*. II. In relation to carbon and nitrogen metabolism. Ann. Bot., N. S. **13**, 241—259 (1949). — Growth and heterocyst production in *Anabaena cylindrica* Lemm. III. The cytology of heterocysts. Ann. Bot., N. S. **15**, 23—35 (1951). — The comparative physiology and biochemistry of the blue green algae. Bact. Rev. **20**, 148—165 (1956). — Frémy, P.: Les Myxophycées de l'Afrique Équatoriale Française. Arch. Bot. **3**, Mem. 2, 1 (1930). — Friedmann, I.: *Chroococcidiopsis kashaii* sp. n. and the genus *Chroococcidiopsis*. (Studies on cave algae from Israel. III.) Öst. bot. Z. **108**, 354—367 (1961). — Fritsch, F. E.: Interrelations and classification of the *Myxophyceae*. New Phytologist **41**, 134—147 (1942). — The heterocyst: a botanical enigma. Proc. Linnean Soc. Lond. **1949/50**, 194—211 (1951).

Gardner, N. L.: Cytological studies in *Cyanophyceae*. Univ. California Publ. Bot. **2**, 237—296 (1906). — New pacific coast marine algae. II. Univ. California Publ. Bot. **6**, 429—454 (1917). — Geitler, L.: Kleine Mitteilungen über Blaualgen. Öst. bot. Z. **70**, 158—167 (1921). — Versuch einer Lösung des Heterocystenproblems. S.-B. Akad. Wiss. Wien, math.-nat. Kl. **130**, 223—245 (1921). — Neue und wenig bekannte Blaualgen. Ber. dtsch. bot. Ges. **40**, 283—287 (1922). — Über neue und wenig bekannte Balualgen aus der Gruppe der Chamaesiphonaceen. Arch. Protistenk. **45**, 413—418 (1925). — *Cyanophyceae*. Rabenhorsts Kryptogamenflora, Bd. 14. Leipzig: Akademische Verlagsgesellschaft 1932. — *Schizophyta:* Klasse Schizophyceen. A. Engler u. K. Prantl, Die natürlichen Pflanzenfamilien, 2. Aufl., Bd. 1 b. Berlin: Duncker & Humblot 1942. — Lebensweise, Fortpflanzung und Wachstumsvorgänge bei *Caryophanum mirabile* n. gen. n. spec., einem merkwürdigen blaualgenähnlichen Organismus. Öst. bot. Z. **102**, 235—272 (1955). — Zur morphologischen Variation und Ökologie auf kleinstem Raum von *Cyanophanum mirabile*. Öst. bot. Z. **104**, 100—114 (1957). — Schizophyceen. In: K. LINSBAUER, Handbuch der Pflanzenanatomie, 2. Aufl., Bd. VI/1. Berlin: Bornträger 1960. — GEITLER, L., u. F. RUTTNER: Die Cyanophyceen der Deutschen Limnologischen Sunda-Expedition. Arch. Hydrobiol., Suppl.-Bd. **14**, 308—483 (1935). — GLADE, K.: Zur Kenntnis der Gattung *Cylindrospermum*. Beitr. Biol. Pflanzen **12**, 295—346 (1914).

HARDER, R.: Ernährungsphysiologische Untersuchungen an Cyanophyceen. Z. Bot. **9**, 145—242 (1917a). — Über die Beziehung des Lichtes zur Keimung von Cyanophyceensporen. Jb. wiss. Bot. **58**, 237—294 (1917b). — HOLLERBACH, M. M.: Sur la morphologie de *Tolypothrix elenkinii* Holl. dans les conditions d'habituation naturelle et de laboratoire. Arch. Russe Protistol. **6**, 158—178 (1928). — HESS, U.: Über die hydraturabhängige Entwicklung und die Austrocknungsresistenz von Cyanophyceen. Arch. Mikrobiol. **44**, 189—218 (1962). — HOLM-HANSEN, O.: Viability of lyophilized algae. Canad. J. Bot. **42**, 127 (1964). — HUBER, J., et F. JADIN: Sur une nouvelle algue perforante d'eau douce. J. Bot. **6**, 278—286 (1892).

IYENGAR, M. O. P., and T. V. DESIKACHARY: A systematic account of some marine *Myxpohyceae* of the south indian coast. J. Madras Univ. **16**, 37—63 (1944). — Occurrence of three pored heterocysts in *Brachytrichia balani*. Curr. Sci. **23**, 180—181 (1954).

JAAG, O.: Untersuchungen über die Variabilität einer Blaualge unter dem Einfluß verschieden starker Beleuchtung. Verh. naturforsch. Ges. Basel **56**, 8—40 (1945). — Untersuchungen über die Vegetation und Biologie der Algen des nackten Gesteins in den Alpen, im Jura und im schweizerischen Mittelland. Beitr. Kryptogamenfl. Schweiz **13** (3), 1—560 (1945). — JOST, M.: Ultrastruktur von *Oscillatoria rubescens*. Arch. Mikrobiol. **50**, 211—245 (1965).

KLEIN, G.: Zur Chemie der Zellhäute der Cyanophyceen. S.-B. Akad. Wiss. Wien, math.-nat. Kl. **124**. 129—145 (1915). — KOHL, F. G.: Über die Organisation und Physiologie der Cyanophyceenzelle. Jena: Gustav Fischer 1903. — KUMAR, H. D.: Apparent genetic recombination in a blue green alga. Nature (Lond.) **196**. 1121—1122 (1962).

LAZAROFF, N., and J. SCHIFF: Developmental photoinduction of the blue green alga *Nostoc muscorum*. Science **137**. 603—604 (1962). — LAZAROFF, N., and W. VISHNIAC: Effect of light on the developmental cycle of *Nostoc muscorum*, a filamentous blue-green alga. J. gen. Microbiol. **25**, 365—374 (1961). — The participation of filament anastomosis in the developmental cycle of *Nostoc muscorum*, a blue-green alga. J. gen. Microbiol. **28**, 203—210 (1962). — The relationship of cellular differentiation to colonial morphogenesis of the blue-green alga, *Nostoc muscorum*. A. J. gen. Microbiol. **35**, 447—457 (1964).

PALIK, P.: Beiträge zur Fortpflanzung der *Gomontiella subtubulosa* Theodoresco und ein neues Vorkommen der Spezies. Nova Hedwigia **3**. 79—84 (1961). — PANDEY, D. C., and A. K. MITRA: On the morphology and life-history of a new species of *Anabaenopsis* — *A. ambigua* n. sp. Nova Hedwigia **4**, 345—350 (1962a). — Production of heterocysts in certain *Myxophyceae* as influenced by the concentration of nitrogen in the medium. Naturwissenschaften **49**. 89—90 (1962b). — PASCHER, A.: Über die Teilungsvorgänge bei der neuen Blaualge *Endonema*. Jb. wiss. Bot. **70**, 329—347 (1929). — POLJANSKI, G., u. G. PETRUSCHEWSKY: Zur Frage der Struktur der Cyanophyceenzelle. Arch. Protistenk. **67**, 11—45 (1929). — PUYMALY, A. DE: Les heterocystes des algues bleues. Leur nature et leur rôle. Botaniste **46**, 209—270 (1957).

SCHNEPF, E.: Zur Feinstruktur von *Geosiphon pyriforme*. Ein Versuch zur Deutung cytoplasmatischer Membranen und Kompartimente. Arch. Mikrobiol. **49**, 112—131 (1964). — SCHWABE, G.: Zur autotrophen Vegetation in ariden Böden. Öst. bot. Z. **107**, 281—309 (1960). — Über den thermobionten Kosmopoliten *Mastigocladus laminosus* Cohn. Schweiz. Z. Hydrol. **22**, 759—792 (1962). — SKUJA, H. (1956): Siehe Lit. E I Chrysophyceae. — SPRATT, E.: Some observations on the life history of *Anabaena cycadeae*. Ann. Bot. **25**, 369 (1911). — STEINECKE, FR.: Das Auskeimen alter Heterocysten bei *Calothrix Weberi*. Bot. Arch. **34**, 153—160 (1934). — SUBA-CLAUS, E.: Teratological forms among *Oscillatoriaceae* and their possible phylogenetic significance. Nova Hedwigia **9**, 104—130 (1965).

TISCHER, I.: Über die granulären Einschlüsse und das Reduktions-Oxydationsvermögen der Cyanophyceen. Arch. Mikrobiol. **27**, 400—428 (1957). — THURET, M. G.: Observations sur la reproduction de quelques Nostochinées. Mém. Soc. Imp. Sci. Nat. Cherbourg **5**, 19—32 (1857).

VARMA, A. K., and A. R. MITRA: On the life-history and mode of perennation of *Myxosarcina spectabilis* Geitler and Ruttner var. *dceolorata* n. var. Nova Hedwigia **4**, 351—358 (1962).

WATANABE, A.: Some devices for preserving blue-green algae in viable state. J. gen. Microbiol. **5**, 153—157 (1959). — WHITTON, B. A.: Effect of deep-freeze treatment on blue green algae cultures. Brit. Phycol. Bull. **2**, 177—178 (1962). — WILDON, D. C., and F. V. MERCER: The ultrastructure of the heterocyst and akinete of the blue-green algae. Arch. Mikrobiol. **47**, 19—31 (1963). — WOLK, C. P.: Heterocyst germination under defined conditions. Nature (Lond.) **205**, 201—202 (1965a). — Control of sporulation in a blue-green alga. Develop. Biol. (1965b, im Druck).

C. Euglenophyceae, Chloromonadophyceae und Cryptomonadophyceae.

Diese drei, vorwiegend in der monadoiden Organisationsform vertretenen Algenreihen können wegen ihrer konvergenten Eigenschaften (s. aber Chadefaud A 1960) gemeinsam behandelt werden; über die beiden letzten liegt experimentelles Material fast nicht vor.

Sexualität ist nur bei Euglenophyceen (Literatur bei Leedale 1962) und einer Chloromonade (Subrahmanyan 1954) bekannt oder wahrscheinlich gemacht worden.

Während der Zellteilung oder in Reaktion auf ungünstige Bedingungen werden häufig Gallerten oder auch festere Hüllen — meist wohl pektischer Natur — ausgeschieden, deren Material, wie Klebs (1883, 1886) zuerst erkannte, zum Teil oder ganz aus dem Inhalt von intra- oder subpelliculär gelegenen Schleimkörpern stammt. Über diese „corps mucifères" liegen Untersuchungen vor allem französischer Autoren (P. A. Dangeard 1901, 1929, 1939; P. Dangeard 1928) vor; nach Chadefaud (A 1960) sind diese Körper durch Übergänge mit Trichocysten verbunden, doch bestehen Meinungsverschiedenheiten über die Abgrenzung zwischen den beiden Organellensorten (Hollande 1942, 1952; Hovasse 1945). Der Grund für die Differenzen liegt wohl darin, daß zwei Termini nicht ausreichen, um die existierende Mannigfaltigkeit an solchen Körpern zu beschreiben. So sind nach elektronenmikroskopischen Untersuchungen die Schleimkörper von *Euglena spirogyra* längliche Bläschen mit Ausführgang, aber ohne besondere Differenzierungen des Inhaltes (Leedale 1964, Leedale, Meeuse und Pringsheim 1965), während diejenigen der Cryptomonaden, ob an der Körperoberfläche und klein (Schleimkörper, Hollande 1952) oder den Schlund pflasternd und groß (Trichocysten, Hollande), als komplex, aber einheitlich im Bau erscheinen; sie werden vermutlich von den Dictyosomen aus gebildet (s. z. B. Anderson 1962, Joyon 1963); wegen ihrer bisher singulären Struktur bezeichnet Anderson diese letzteren als Ejektosomen. Doch ließ sich der Inhalt auch der länglichen Schleimkörper einiger Euglenen färberisch differenzieren (Grassé und Poisson 1933, Chadefaud 1939). Den Trichocysten der Ciliaten ähnliche Gebilde kommen jedoch unter den hier behandelten Flagellaten nur bei Chloromonaden (Rhaphidiophyceen (!), Chadefaud 1960) vor. Nur scheinen sie nicht von dem System des Cinetidenapparates ausgebildet zu werden, wie dies für Ciliaten (Chatton, Lwoff und Lwoff 1931) wohl gilt. Bei *Gonyostomum* (Hovasse 1945) und *Chattonella* (Biecheler 1936a) treten sie anscheinend neben rundlichen Schleimkörpern auf; *Vacuolaria* aber führt nur die letzteren (P. A. Dangeard 1939, Poisson und Hollande 1943). Alle die genannten Körper können ihren Inhalt bei Chemikalienzusatz mehr oder weniger heftig und mehr oder weniger gleichzeitig ausstoßen. Nur die Euglenen mit runden Schleimkörpern, wie *E. pisciformis*, *E. viridis* und *E. polymorpha* sind nicht imstande, Schleim auf chemische Reize hin zu liefern. Hier ist es auch unklar, ob die Körper Artefakte ihres färberischen Nachweises sind oder nicht (Pringsheim 1956). Doch können auch diese Arten unter entsprechenden Außenbedingungen Schleimhüllen liefern (Pringsheim), ein Widerspruch, der aufzuklären bleibt.

a) Varianten der Zellteilung.

Da bereits der Zellteilungsmodus in allen 3 Gruppen genotypisch und, was hier wichtiger, auch modifikatorisch veränderlich ist, soll auf ihn eingegangen werden.

Bei *Euglenen*[1] kann die Zellteilung stattfinden (DANGEARD 1901) 1. als Längsteilung (Abb. 15b) in der Bewegung ⟨Schizotomie⟩ (z. B. *Astasia*, KLEBS 1883), 2. als Längsteilung nach Geißelabwurf, ohne wesentliche Formänderungen (*Euglena spirogyra*, KLEBS l. c., DANGEARD l. c.), wobei dann häufig die sich teilenden Zellen in leichte Gallertausscheidungen eingehüllt sind, und 3. als (Längs-) Teilung im mehr oder weniger abgerundetem Zustand ⟨Schizogonie⟩, der dann meist durch stärkere Gallertausscheidungen gekennzeichnet ist. Die Schleimentwicklung dürfte von der konstitutionellen Ausbildung der Schleimkörper (HOLLANDE 1952), aber auch von Außenbedingungen (*Euglena viridis*, KLEBS l. c.) abhängig sein. Wenn die Zellen des 3. bzw. auch 2. Typus nach der Teilung, welche sich im allgemeinen nachts (s. z. B. LEEDALE 1959) vollzieht, nicht wieder ausschwärmen, so entstehen mehr oder weniger vergallertende „Palmellen" (Abb. 15d, h) — diese, falls derbhäutig, auch als „Teilungscysten" ⟨Übergang zur Cytotomie⟩ bezeichnet (ZUMSTEIN 1900, GÜNTHER 1928) — in Lagern, die erhebliche Größe erreichen können. Dabei können Spezialhüllen um die Einzelzellen deutlich sein oder nicht (s. z. B. DANGEARD 1901, KHAWKINE 1885, 1886, TANNREUTHER 1923). Bei *Euglena „gracilis"* treten nach TANNREUTHER (1923) alle 3 Teilungsarten alternativ auf, wobei allerdings die freie Teilung als Querteilung ausgebildet ist und ferner als 4. Teilungsform neben einer Längsteilung auch noch eine Querteilung im ruhenden Zustand, aber ohne die Formänderung der Zelle vorkommen soll; bei erstgenannter wird die Geißel während der Teilung abgeworfen, bei letzterer nicht, was alles sowohl am lebenden als auch an präpariertem Material gesehen wird; leider fehlen Angaben über die speziellen Bedingungen. Das Vorkommen von Querteilungen wird seit KLEBS (1883), sonst jedoch kaum mehr angegeben. Nur einmal wurde berichtet (Abb. 15e bis g), daß Palmellazellen unter Heranwachsen vielkernig werden und sich darauf erst succedan und zwar innerhalb der Hülle aufteilen (*Euglena gracilis*, KRICHENBAUER 1937), doch liefert andererseits die in *Cyclops* parasitierende und im Darm dieses Wirtes zu außerordentlicher Größe heranwachsenden *Astasia mobilis* nach PENARD (1921) und CHATTON und MONOD (1941, zit. nach HOLLANDE 1952), in Freiheit gesetzt, durch succedane Aufteilung bis 256 sehr kleine Flagellaten. Hier ist die Mutterzelle allerdings beweglich und einkernig, doch sind ähnliche Unterschiede zwischen der bei Reife mehrkernigen und einkernigen Gametangien von *Protococcalen* her bekannt. Die von TANNREUTHER geschilderte Möglichkeit verschiedener Teilungsmodi bei der gleichen Art, nämlich im beweglichen und ruhenden Zustand, ist mehrfach beschrieben worden, z. B. bei: *Phacus pleuronectes*

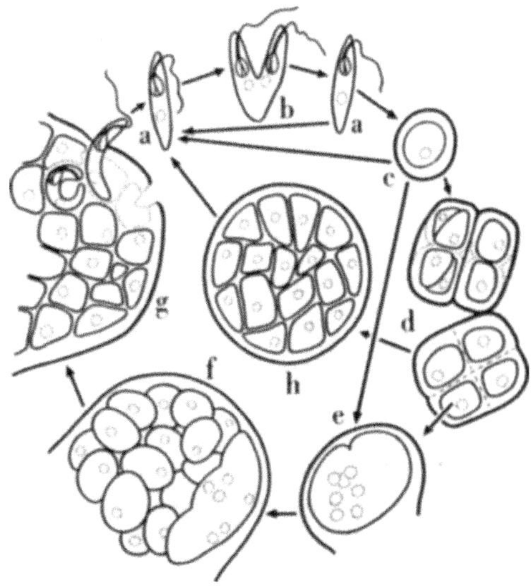

Abb. 15a—h. Entwicklungscyclus von *Euglena gracilis*. b Teilung im beweglichen Zustand. c Einzellige Palmellen. d Teilung im Palmellenzustand. e Vielkernige einzellige Palmelle, die sich in f simultan-succedan in einen Palmellenhaufen aus einkernigen Zellen aufteilt. g u. h Vielzellige Palmellen, die in g wieder ausschwärmen. (Nach KRICHENBAUER 1937.)

[1] Die Artnamen der Euglenen wurden im allgemeinen nicht korrigiert wiedergegeben, obwohl dieses in vielen Fällen mit Hilfe von PRINGSHEIM (1956) möglich gewesen wäre.

und *Trachelomonas hispida* von DANGEARD (1889a); *Euglena geniculata* von DANGEARD (1901); *Euglena levis* von BRABEZ (1941); *Euglena gracilis* von KRICHENBAUER (1937) und SKUJA (1948), allerdings ohne Angabe der Bedingungen. Nach PRINGSHEIM (1956) sind manche der angenommenen Teilungsstadien allerdings in Wirklichkeit Anomalien. ZUMSTEIN findet bei *E. gracilis* eine Teilung im „beweglichen Zustand" (genauer obige Teilungsform 2), in flüssigem Medium immer oder auf dünner (5%) Pepton-Gelatine nur im Licht, eine solche im Palmellastadium dagegen auf 5% Gelatine im Dunkeln und auf 20% Gelatine im Licht und im Dunkeln. Licht begünstigt hier also die Teilung in beweglichem Zustand. Nach KLEBS (1883) und den Versuchen von MAINX (1927a, b) mit einer großen Zahl von Arten verschiedener Neigung zur Palmellenbildung spielt aber nicht nur die Konsistenz, sondern auch die Konzentration des Mediums für diese Dinge eine Rolle. Ähnlich findet SCHÜLER (1910) bei der marinen *Euglena baltica* durch Erhöhung der NaCl-Konzentration bis auf 4% immer stärkere Anteile von Palmellenvermehrung, während die Teilungsrate sich nicht ändert.

Zu den Faktoren, welche eine Palmellenbildung begünstigen, gehören nach MAINX, neben hohen Nährstoffkonzentrationen, Temperaturextreme und Bedingungs*wechsel*, insbesondere auch plötzliche Änderungen der Wasserstoffionenkonzentration, die leicht Geißelabwürfe verursachen, während allmähliche Änderungen viel weniger von Einfluß sind; spezifisch sollen (freie) Essigsäure das Schwärmen von *E. gracilis* und Ca-Ionen dasjenige von *Astasia ocellata* fördern. Eine positive Wirkung des Lichtes auf das Schwärmen der Euglenen wurde von ZUMSTEIN (1900) festgestellt. Nach BOLTE (1920) werden jedoch nicht alle Arten durch Licht in den beweglichen Zustand versetzt, sondern die „photokinetisch negativen" wie zwei *Phacus*- und eine *Lepocinclis*-Art sogar in eine Lichtstarre. Ähnliches beobachtet MAINX bei gut ernährten Kulturen von mehreren *Euglena*-Arten, findet jedoch die Wirkung des Lichtes bei *E. gracilis* von der Vorbehandlung abhängig; werden Zellen in destilliertem Wasser im Dunklen gehalten, so tritt nach einigen Tagen Starre ein, die nun durch Licht revertiert wird. Da in diesen Versuchen nur (?) die Beweglichkeit, nicht der Teilungsmodus geprüft wurde, sei nur noch angemerkt, daß die Beweglichkeit (genauer die positive Phototaxis) bei *Euglena gracilis* an eine endogene, durch den Lichtwechsel steuerbare Rhythmik angekoppelt ist (s. z.B. POHL 1948, BRUCE 1960). MAINX weist, wie erwähnt, auf die Bedeutung von Störungen des bestehenden Zustandes für das Verhalten von Euglenen hin; so können auch die Ergebnisse von POPOFF und PASPALEFF (1924, 1925) verstanden werden. Diese Autoren machen durch eine Behandlung in „schwach mit Ammoniak versetztem" Wasser ihre *E. gracilis* unbeweglich und schließlich zu Palmellen (die sie fälschlich Cysten nennen). Erneute Beweglichkeit wird dann durch eine Behandlung mit einer ganzen Anzahl von anorganischen und organischen „Stimulantien" in jeweils verschiedenen optimalen Konzentrationen und Wirkungszeiten erreicht. Zusammenhänge zwischen Struktur und Wirkung dieser Mittel geben sich nicht zu erkennen. *Colacium vesiculosum* (s. z.B. DANGEARD 1901, MAINX 1927b) mit gestielten, meist epiphytischen Zellen ist eine im Formwechsel vorherrschend gewordene, spezialisierte Palmelle. Sie geht beim Stillegen der sie tragenden Kleinkrebse *(Cyclops)*, anscheinend also wieder auf Störungen der Außenbedingungen, in den Schwärmzustand über, setzt sich in Nährlösung aber bald erneut mit dem Vorderende fest und scheidet eine Hülle und vorn das Stielchen aus (MAINX). Sie kommt aber auch als unpolare Palmella vor, ein Zustand, in dem JOHNSON (1934) sie vornehmlich auch in flüssigen Medien erhielt; die gestielte *Colacium*-Form war aus unbekannten Gründen in seinen Kulturen selten.

Doch gibt es Euglenen, die nicht in Palmellazustände übergehen; es handelt sich um große Formen wie *E. olivacea*, *E. oblonga*, *E. velata* sowie *Astasia ocellata*; diese kontrahieren oder encystieren sich auf Agar, können sich aber nicht in diesem Zustand teilen.

Palmellenteilung existiert auch bei *Chloromonaden: Vacuolaria virescens* (DANGEARD 1910), anscheinend ohne Verlust der Geißel (POISSON u. HOLLANDE 1943) und *Vacuolaria viridis* (DANGEARD 1883b, FOTT 1935). *Hornellia* jedoch (SUBRAHMANYAN 1954), *Gonyostomum* (HOVASSE 1945) und *Chattonella* (BIECHELER 1936b) teilten sich im freien, begeißelten Zustand.

Ebenso besitzen die *Cryptomonaden* Längsteilung, frei oder nach Palmellenbildung (HOLLANDE 1942). *Phaeopalax marinus* (REINISCH 1911, PASCHER 1911) ist ein permanentes (?) Palmellastadium, das sich mit Schwärmern fortpflanzt; ähnlich vielleicht *Phaeococcus clementi* Borzi (PASCHER 1911).

b) Cysten.

Dauerzellen mit dickeren, meist gallertigen Membranen werden ebenfalls in allen drei Gruppen gebildet. Für die *Chloromonaden* sei auf PENNARD (1921), CIENKOWSKI (1870), DANGEARD (1889b, 1910), für die *Cryptomonaden* auf CIENKOWSKI (l. c.), DANGEARD (1889a, 1910), ALEXEIEFF (1911), HOLLANDE (1942) verwiesen. Sie sind, soweit man weiß, vegetativer Natur, nur bei der Chloromonade *Hornellia* (SUBRAHMANYAN 1954) und der farblosen Euglenale *Scytomonas* (DOBELL 1908) wurden Cysten sexueller Entstehung beschrieben. Die Cysten sind von dicken, geschichteten Gallertmembranen umgeben, die bei *Vacuolaria*

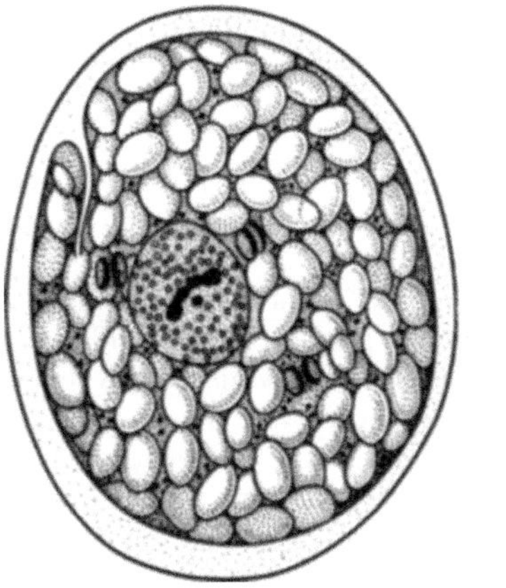

Abb. 16. Cyste der farblosen Euglenacee *Distigma proteus*. (Nach HOLLANDE 1952.)

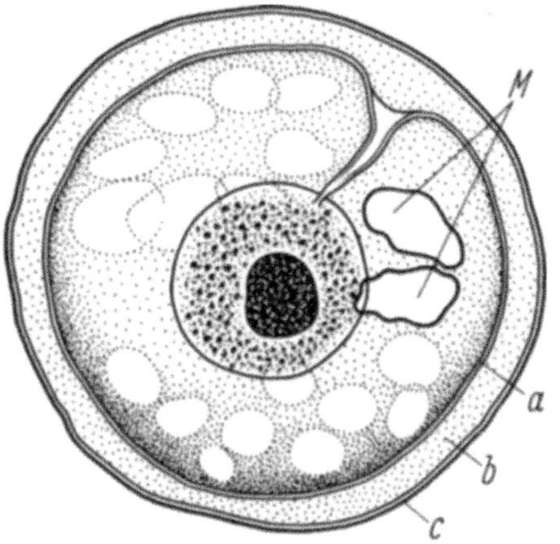

Abb. 17. Cyste der Cryptomonadine *Cryptomonas ovata*. *M* Maupas-Körper, *a*—*c* Schichten der Cystenwand. (Nach HOLLANDE 1942.)

virescens eine warzige (DANGEARD 1910), bei *Cryptomonas obovata* (SKUJA 1956) oder *Cryptomonas ovata* (HOLLANDE 1942) eine glatte aus dem Periplasten entstandene festere Membran umschließen (Abb. 17). Sie entstehen in alten Kulturen im ersten, unter „ungünstigen Lebensbedingungen" im zweiten Falle.

Bei den *Euglenophyceen* unterscheidet GÜNTHER (1928) seine Temporärcysten von den bereits besprochenen Teilungs- und den Dauercysten der anderen Autoren. Die *Temporärcysten* sind eine Anpassung an das Leben auf der Grenze von Schlamm gegen Luft. Die Zelle umgibt sich, ohne die Geißel zu verlieren, mit einer dünnen und festen, nur zum Substrat hin in einen Porus offenen Membran, die dem Körper anliegt, und teilt sich in diesem Zustand nicht. Die Bildungsbedingungen sind für *Euglena geniculata*, und *E. terricola* zwischen 3° und 25° durch einen mit der Temperatur ansteigenden (nicht näher definierten) Feuchtigkeitsbedarf gegeben. Die Temporärcysten sind beschränkt trocken-resistent; sie excystieren beim Übertragen in Wasser. Hier seien auch die langgestielten Cysten genannt, die bei *E. tuba* (CARTER 1869) anscheinend mit dem Stielchen der Wasseroberfläche ansitzen und bei *E. haematodes* an der Peripherie von Klumpen kugeliger Cysten vereinzelt vorkommen. Bei *E. pedunculata* (JOHNSON 1944) keimen sie unter Deckglasdruck (GOJDICS 1953), was näher beobachtet wurde. Alle drei Angaben beziehen sich nach PRINGSHEIM wahrscheinlich auf *E. sanguinea*.

Die *Dauercysten* sind bei *E. granulata* (SZABADOS 1950), an *Trachelomonas*-Gehäuse erinnernd, von urnenförmiger Gestalt und bei *E. chlamydophora* nach MAINX (1927a, b) durch Membranbildung des Flagellaten nach Verkürzung, aber nicht Abrundung von einer ähnlichen Form. Sie öffnen sich im letzten Falle mit einem Deckel; merkwürdigerweise entwickelt die gleiche Art noch eine Form runder Cysten mit geschichteter Membran, wie sie vielen Euglenen zukommt (Abb. 16). Die Gallertmembranen sind in einigen Fällen durch Einlagerung von Eisenoxydhydrat gefärbt. Die Cyste der *E. deses* ist zylindrisch (DANGEARD 1901).

Die Dauercysten entstehen nach langsamem Feuchtigkeitsentzug, durch Einschränkung der Bewegungsmöglichkeit (KLEBS 1883), unter Nährstoffmangel (ZUMSTEIN 1900, KHAWKINE 1885, 1886, DANGEARD 1901), Kältewirkung (*E. deses*, MAINX 1927), Verdunkelung in Nährstoffgegenwart oder in reinem Wasser, aber auch bei guter Nährstoffversorgung nach Störungen durch direkte Sonnenbestrahlung oder zu hohe Temperatur (KHAWKINE), ähnlich bei übermäßiger Vermehrung von Bakterien (DANGEARD l. c.). Die Cysten einiger „aerophiler" Arten dienen der Windverbreitung (BRABEZ 1941). Bei langsamem Austrocknen können einige Euglenen in Dauerstadien übergehen, ohne erkennbare Membranen auszuscheiden. *E. spirogyra* breitet sich blattartig aus und speichert Paramylon (KLEBS l. c.); *E. deses* var. *intermedia* sowie *E. granulata* und *E. velata* verhalten sich ähnlich; *Astasia margaritifera* und *Menoidium incurvum* verkürzen ihre Form nur etwas. Die *Phacus*-Arten sind durch ihre derbe Zellmembran allein geschützt (DANGEARD 1901).

Excystierung erfolgt bei *E. gracilis* (ZUMSTEIN, TANNREUTHER) nach Überführen in frische Nährlösung, Leitungswasser allein genügt nicht, und die Keimungsgeschwindigkeit ist in organischer gegenüber anorganischer Nährlösung erhöht. Farblose, in heterotropher Dunkelkultur encystierte *E. gracilis* keimte im Licht bereits ohne Nährstoffzugabe (ZUMSTEIN). Die Excystierung ist bei *Euglena gracilis* von einer Vierteilung des Protoplasten begleitet (ZUMSTEIN); *Astasia ocellata* (KHAWKINE) schlüpft direkt aus.

Literatur.

ALEXEIEFF, A.: Notes sur les Flagellés. Arch. zool. exp. et gén. **6**, 491—527 (1911). — ANDERSON, E.: A cytological study of *Chilomonas paramecium* with particular reference to the so-called trichocyst. J. Protozool. **9**, 380—395 (1962).

BIECHELER, B.: Sur une Chloromonadine nouvelle d'eau soumâtre. Arch. zool. exp. et gén. **56**, Notes et Rev. 79—83 (1936a). — Sur la division de la Chloromonadine *Chattonella subsalsa* Biecheler. C. R. Soc. Biol. (Paris) **123**, 1126 (1936b). — BOLTE, E.: Über die Wirkung von Licht und Kohlensäure auf die Beweglichkeit grüner und farbloser Schwärmzellen. Jb. wiss. Bot. **59**, 287—324 (1920). — BRABEZ, R.: Zur Kenntnis der Algenflora des Franzensbader und Sooser Thermenbereiches. Beih. Bot. Zbl. A **61**, 137—236 (1941). — BRUCE, V. G.: Environmental entrainment of circadian rhythms. Cold Spring Harb. Symp. quant. Biol. **25**, 23—48 (1960). — BÜTSCHLI, O.: Beiträge zur Kenntnis des Paramylon. Arch. Protistenk. **7**, 197—228 (1906).

CARTER, H. J.: Notes on filigerous green infusoria on the island of Bombay. Ann. a. Mag. Natur. History (4) **3**, 250—280 (1869). — CHADEFAUD, M.: Sur l'organisation d'*Euglena* *stellata* Mainx et sur la discrimination des Euglènes viridoides. Arch. zool. exp. et gén. **80**, 49—54 (1939). — Chatton, E., A. Lwoff et M. Lwoff: L'origine infaciliaire et la genèse de trichocystes et des trichites chez les Ciliés Foetteingeriidae. C. R. Acad. Sci. (Paris) **193**, 670—673 (1931). — Cienkowski, L.: Über Palmellaceen und einige Flagellaten. Arch. mikr. Anat. **6**, 421—438 (1870).

Dangeard, P.: Appareil mucifère et le vacuome chez les Euglènes. Annal. Protistol. **1**, 68—74 (1928). — Dangeard, P. A.: Recherches sur les Cryptomonadinae et les Euglénae. Botaniste **1**, 1—38 (1889a). — Mémoire sur les algues. Botaniste **1**, 127—174 (1889b). — Recherches sur les Eugléniens. Botaniste **8**, 97—357 (1901). — Études sur le développement et la structure des organisms inférieurs. V. Les algues inférieurs. Botaniste **11**, 181—227 (1910). — Remarques sur le vacuome des Eugléniens. Botaniste **21**, 327—332 (1929). — Nouvelles observations sur les *Vacuolaria virescens*. Botaniste **29**, 311—334 (1939). — Dobell, C. C.: The structure and life-history of *Copromonas subtilis*, nov. gen. & nova spec.: A contri) ι.on to our knowledge of the flagellates. Quart. J. micr. Sci. **52**, 75—120 (1908). — Droue.: .. and A. Cohen: The morphology of *Gonyostomum* semen from Woods Hole, Massachusetts. Biol. Bull. **68**, 422—439 (1935).

Fott, B.: Über den inneren Bau von *Vacuolaria viridis* (Dangeard) Senn. Arch. Protistenk. **84**, 242—250 (1935).

Goidics, M.: The genus *Euglena*. Madison: University of Wisconsin Press 1953. — Grassé, P. P., et R. Poisson: Nouvelles observations sur la cytologie des Euglènes. C. R. Acad. Sci. (Paris) **109**, 662—666 (1933). — Günther, F.: Über den Bau und die Lebensweise der Euglenen, besonders der Arten *Euglena terricola*, *geniculata*, *proxima*, *sanguinea* und *lucens* nov. spec. Arch. Protistenk. **60**, 511—590 (1928).

Hollande, A.: Etude cytologique et biologique de quelques Flagellates libres. Arch. zool. exp. et gén. **83**, 73—170 (1942). — Classe de Chloromonadines, Classe de Eugléniens, Classe de Cryptomonadines. In: P. Grassé: Traitée de Zoologie, vol. 1 (1). Paris: Masson & Cie. 1952. — Hovasse, R.: Contribution à l'étude des Chloromonadines: *Gonyostomum semen* Dies. Arch. zool. exp. et gén. **34**, 239—269 (1945). — Huber-Pestalozzi, G.: Das Phytoplankton der Binnengewässer. 4. Eugleninen. Stuttgart: E. Schweizerbart 1955.

Johnson, D. F.: Morphology and life history of *Colacium vesiculosum* Ehrenberg. Arch. Protistenk. **83**, 241—263 (1934). — Johnson, L. P.: *Euglenae* of Iowa. Trans. Amer. micr. Soc. **63**, 97—135 (1944). — Joyon, L.: Contribution à l'étude cytologique de quelques protozoaires flagellés. Ann. Fac. Sci. Univ. Clermont-Ferrand 22 Biol. Animale 1 (22), 1—96 (1963).

Khawkine, W.: Recherches biologiques sur l'*Astasia ocellata* n. sp. et l'*Euglena viridis* Ehrb. I. Ann. Sci. nat. Zool., Sér. VI, **19**, 1—48 (1885). — II. Ann. Sci. nat. Zool., Sér. VII, **1**, 9—376 (1886). — Klebs, G.: Über die Organisation einiger Flagellaten-Gruppen und ihre Beziehungen zu Algen und Infusorien. Untersuch. Bot. Inst. Tübingen **1** (2), 233—362 (1883). — Über die Organisation der Gallerte bei einigen Algen und Flagellaten. Unters. Bot. Inst. Tübingen **2**, 179—330 (1886). — Krichenbauer, H.: Beitrag zur Kenntnis der Morphologie und Entwicklungsgeschichte der Gattungen *Euglena* und *Phacus*. Arch. Protistenk. **90**, 88—122 (1937).

Leedale, G. F.: The time-scale of mitosis in the Euglenineae. Arch. Mikrobiol. **32**, 352—360 (1959). — The evidence for a meiotic process in the Euglenineae. Arch. Mikrobiol. **42**, 237—245 (1962). — Pellicle structure in *Euglena*. Brit. Phycol. Bull. 2 (5), 291—306 (1964). — Leedale, G. F., B. J. D. Meeuse, and E. G. Pringsheim: Structure and physiology of *Euglena spirogyra*. I and II. Arch. Mikrobiol. **50**, 68—102 (1965).

Mainx, F.: Beiträge zur Morphologie und Physiologie der Eugleninen. I. Morphologische Beobachtungen, Methoden und Erfolge der Reinkultur. Arch. Protistenk. **60**, 305—354 (1927a). — Beiträge zur Morphologie und Physiologie der Eugleninen. II. Untersuchungen über die Ernährung und Reizphysiologie. Arch. Protistenk. **60**, 355—414 (1927b). — Untersuchungen über den Einfluß der Außenfaktoren auf die phototaktische Stimmung. Arch. Protistenk. **68**, 105—176 (1929).

Pascher, A.: Über die Beziehungen der Cryptomonaden zu den Algen. Ber. dtsch. bot. Ges. **29**, 193—203 (1911). — Penard, E.: Studies on some flagellata. Proc. Acad. nat. Sci. (Philadelphia) **74**, 105—168 (1921). — Pohl, R.: Tagesrhythmus im phototaktischen Verhalten von *Euglena gracilis*. Z. Naturforsch. **3**b, 367—374 (1948). — Poisson, R., et A. Hollande: Considération sur la cytologie, la mitose et les affinités chez les Chloromonadines. Étude de *Vacuolaria virescens*. Ann. Sci. nat. Zool., Ser. XI, **5**, 147—160 (1943). — Popoff, M., u. G. Paspaleff: Experimentelle Zellstudien. IV. 1. Teil: Zur Physiologie des Encystierungsprozesses bei Protozoen. Encystierung und Stimulation. I. Zellstimulationsforsch. **1**, 39—56 (1924). — II. Zellstimulationsforsch. **1**, 129—149 (1924). — III. Zellstimulationsforsch. **1**, 391—412 (1925). — Pringsheim, E. G.: Contributions towards a monograph of the genus *Euglena*. Nova Acta Leopoldina, Abh., N.F. Nr 125, 1—168 (1956).

Reinisch, O.: Eine neue Phaeocapsacee. Ber. dtsch. bot. Ges. **29**, 77—83 (1911).

Schüler, J.: Über die Ernährungsbedingungen einiger Flagellaten des Meerwassers. Wiss. Meeresunters. Kiel. N.F. **9**, 359—365 (1910). — Skuja, H.: Taxonomie des Phytoplanktons einiger Seen in Upland (Schweden). Symbolae bot. Upsalensis **9** (3), 1—400 (1948). — Taxonomische und biologische Studien über das Phytoplankton in schwedischen Binnengewässern. Nova Acta roy. Soc. Sci. Upsalensis, Ser. IV, **16** (3), 1—404 (1956). — Subrahmanyan, R.: On the life history and ecology of *Hornellia marina* nov. gen. et spec. (Chloromonadineae), causing green discoloration of the sea and mortality among marine animals of the Malabar coast. Indian J. Fish. **1**, 182—202 (1954). — Szabados, M.: Adatok az *Euglena granulata* (Klebs) Lem. fejlödéséhez. Ann. Biol. Univ. Szeged **1**, 111—115 (1950).

Terenetz, C.: Beiträge zur Morphologie und Physiologie der *Euglena gracilis*. Jb. wiss. Bot. **51**, 435—514 (1912). — Tannreuther, G. W.: Nutrition and reproduction in *Euglena*. Arch. Entwickl.-Mech. Org. **52**, 367—383 (1923).

Weber, E.: Physiologische Untersuchungen an *Euglena olivacea*. S.-B. Öst. Akad. Wiss., math.-nat. Kl., Abt. I. **160**, 615—638 (1951).

Zumstein, H.: Zur Morphologie und Physiologie der *Euglena gracilis* Klebs. Jb. wiss. Bot. **34**, 149—198 (1900).

D. Dinophyta.

Der Stamm entwickelt seine Mannigfaltigkeit vorzüglich in der *monadoiden* Gestaltung und zeichnet sich durch sehr einfallsreiche Anpassungen insbesondere an die Erfordernisse planktischer Lebensweise aus. Daneben treten absonderliche *rhizopodiale* (unter diesen wohl große Teile der Radiolarien, die hier aber unberücksichtigt bleiben), weiter „*algenartige*" capsale, coccale und trichale Formen und schließlich durch die Gemeinsamkeit einer *parasitischen* Lebensweise vereinigte, in Vermehrungsart und Lebenscyclus aber reich variierende Typen auf.

Zur *Sexualität* der Dinophyten, um diese zunächst gegen die übrigen Fortpflanzungsarten abzugrenzen, ist immer noch wenig bekannt. Vollständig wurde ein Entwicklungsgang für *Glenodinium lubinensiforme* durch Diwald (1938) angegeben, und zwar als haplontischer Generationswechsel mit Isogamie und Hypnozygoten, aus denen vier individualisierte Gonen austreten. Doch werden die Resultate durch Schussnig (A 1960, S. 697) in Zweifel gezogen. Nunmehr ist aber der sexuelle Formwechsel des limnischen *Ceratium cornutum* im wesentlichen klargelegt, nachdem Skoczylas (1958) bereits Stadien der Meiosis beschrieben hatte. Nach heterogamer Flagellatenverschmelzung entsteht innerhalb des gepanzerten weiblichen Gameten, der Plasma und Kern des Androgameten aufnahm, und nach einer langen mit Wachstum verbundenen Schwärmperiode dieser Zygote die aus der Natur mehrfach beschriebene Cyste. Diese keimt mit einem Keimlingsschwärmer, welcher — der einzige Fall im Pflanzenreich — noch diploid ist und nach einiger Zeit erst unter Meiosis succedan vier haploide Zellen entstehen läßt (v. Stosch 1964, 1965). Das Stattfinden der Meiosis wurde inzwischen nicht nur cytologisch, sondern auch durch Tetradenanalyse genetisch nachgewiesen (v. Stosch, unveröff.). Marine *Ceratien* sind extremer heterogam, bilden aber keine Cysten; die Meiosis scheint in einer der vegetativen Zelle ähnlichen Zygote abzulaufen (darüber s. v. Stosch 1964). Die phagotrope marine *Oxyrrhis* ist isogam und verhält sich sonst ähnlich, nur wächst die gefräßige Zygote vor der Meiosis außerordentlich heran (v. Stosch, unveröff.). Gametenverschmelzungen wurden bei *Noctiluca* (Gross 1934) sowie in den parasitischen Gattungen *Dubosquella* und *Coccidinium* (s. Chatton 1952) gesehen. Auch bei den parasitischen Arten scheint aber die sexuelle Verwendung der Schwärmspore trotz häufigen Zoiden-Dimorphismus zu den Ausnahmen zu gehören (Cachon 1964).

a) Parasitische Dinophyten.

Um zunächst kurz auf die Parasiten einzugehen, so spielt sich bei ihnen der Lebenscyclus ab im Eindringen eines Dinophyten-Flagellaten (Dinospore) in den

Wirt oder seine Oberfläche, einem mehr oder weniger mächtigen Heranwachsen im einzelligen Zustand und einem schließlichen Zerfall in die Ausgangsform. Das Wachstum des „Trophonten" erfolgt entweder unter Vergrößerung des Kerns, der dabei unter Anhäufung nicht chromatischer Substanz haploid bleibt, in anderen Fällen polyploid wird, oder unter Polyenergidie durch Kernvermehrung. Der Zerfall führt simultan oder succedan, gelegentlich in Stufen, zu kleineren Einheiten, schließlich aber immer zur Dinospore herab. Da die Cyclen in Verbindung mit dem Wirt quasi autonom ablaufen und experimentelle Eingriffe bisher nicht vorgenommen wurden, genügt es, für sie auf CHATTON (1923, 1952) sowie auf CACHON (1964) zu verweisen.

Hierher ist wohl auch *Dissodinium lunula* zu ziehen, die gewöhnlich (s. z. B. OEHLKERS) als planktontische Dinococcale angesehen wird, deren kugelige Cysten jedoch, wenn sie im Plankton erkennbar werden — das geht bereits aus den Abbildungen DOGIELs (1906) hervor —, nahezu farblos sind und deren Plastiden sich, während der Bildung der Möndchencysten und danach der gymnodinoiden Schwärmer in diesen, fortschreitend vermehren und ergrünen. Da die Dinosporen, auch wenn die Möglichkeit zu etwaigen Kopulationen gegeben wird, zwar tagelang am Leben bleiben, aber sich nicht weiterentwickeln (v. STOSCH, unveröff.), dürften sie den durch Photosynthese motorisch autonom gemachten Schwärmzustand eines Parasiten darstellen, dessen Wirt freilich unbekannt ist.

b) Die monadoide Stufe.

1. Zellteilung, Gonidienbildung.

Die freilebenden Dinoflagellaten vermehren sich durch Zwei- oder gelegentlich auch succedane Vielfachteilungen und können daneben Cysten liefern, die Interessepunkt an dieser Stelle sein würden. Doch muß zunächst auf die Zellteilung eingegangen werden.

Dinoflagellaten können als *nackte* Formen oder mit meist kleingefelderter zarter Membran *gymnodinoid*, mit derber grobgefelderter aber nicht stukturierter Membran *glenodinoid* oder mit dem kräftigen ornamentierten *Plattenpanzer* ausgerüstet vorkommen, wie er etwa bei den Gattungen *Peridinium* und *Ceratium* bekannt ist, Bedeckungsformen, für welche LINDEMANN (1928, 1929) die Bezeichnungen (gleiche Reihenfolge) avalvat, gymnoid, prävalvat und valvat verwendet. Jede dieser Stufen vermag die höchste Organisationsform einer Art zu sein, diejenige, unter welcher diese im allgemeinen beschrieben ist; die avalvate Stufe kann bei allen drei höheren Formen als Entwicklungsstadium vorkommen und die prävalvate bei der valvaten Stufe, bei den beiden letztgenannten aber nicht die gymnoide Stufe mit ihrem abweichenden Panzerbau (LINDEMANN). Bei dem limnischen *Ceratium hirundinella* treten nach HUBER und NIPKOW (1922) diese Stufen, mit Ausnahme der gymnoiden, als Keimungsstadien der Cysten im Frühjahr nacheinander auf. Ebenso verhalten sich in der Gattung *Peridinium* (LEFÈVRE 1928) der Keimling einer Cyste oder das bei einer Häutung aus der gepanzerten Zelle ausschlüpfende *Gymnodinium* oder auch die nackten Schwärmer, die bei der „provenienten" Teilung (s. unten) entstehen. Ein durch Abwerfen des Panzers entstandenes *Gymnodinium* von *Peridinium aciculiferum* ist der Zustand, in welchem der Flagellat sich phagotroph ernähren kann (ENTZ und SEBESTYÉN 1935).

Nach LINDEMANN kann die Zellteilung in jeder dieser Stufen in Form einer Durchschnürung der ganzen Zelle inklusive etwaiger Hüllen unter *gleichzeitiger* Ergänzung der fehlenden Teile der Oberfläche bzw. der Hülle stattfinden. Die dazu für die prävalvate und insbesondere valvate Stufe gegebenen Bilder („prävalvate bzw. valvate Teilung“) erscheinen dem Referenten allerdings nicht überzeugend; es handelt sich zweifellos um Monstren, wie sie in Kulturen leicht durch Teilungshemmungen entstehen (siehe z. B. Nordli 1957). In der älteren Literatur wird häufig zu wenig zwischen einem aus einem Stadium erschlossenen und einem wirklich durchbeobachteten Ablauf unterschieden. Dann würden die valvaten und prävalvaten Teilungsformen fortfallen und hierher nur die

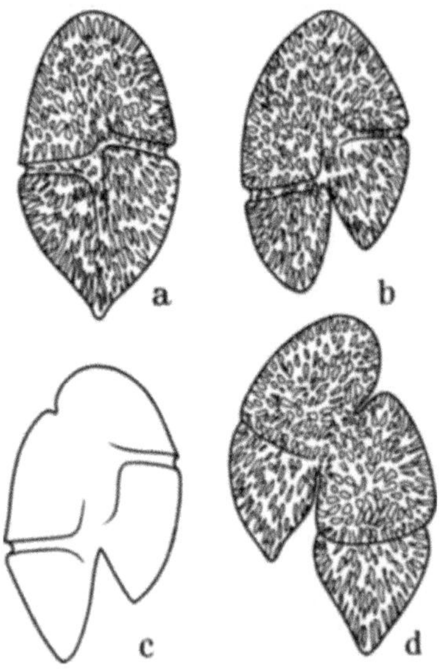

Abb. 18 a—d. Avalvate Teilung von *Gymnodinium fuscum*. (Nach Geitler 1928, aus Schiller.)

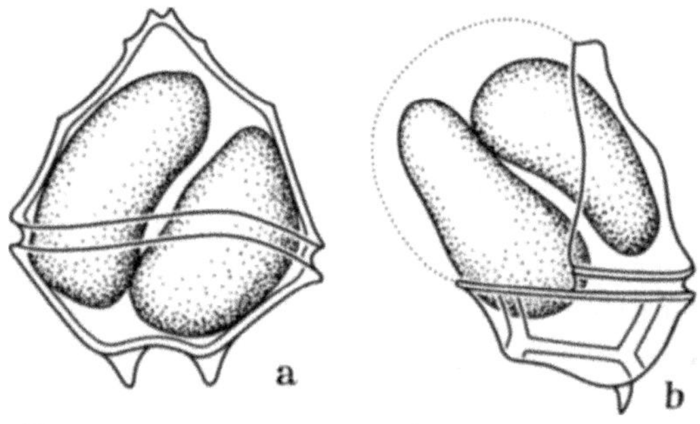

Abb. 19 a u. b. a Proveniente Teilung bei *Peridinium*. b Freiwerden der Zoosporen unter Absprengung eines Teiles des Panzers. (Nach Lefèvre 1928.)

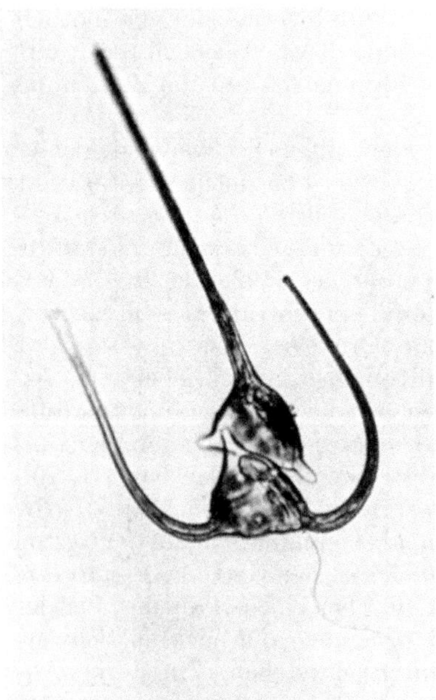

Abb. 20. *Ceratium horridum* in hemivalvater Teilung. Vordere Tochterzelle ergänzt nach der Schrägteilung ihre links hintere Panzerung, hintere Körperhälfte entsprechend die rechts vordere. (Original, Photo R. Schäfer.)

avalvate (Abb. 18) und vielleicht die *gymnoide Teilung* zu rechnen sein. Avalvat teilt sich z. B. regelmäßig (Entz 1926) das valvate *Peridinium borgei* in der Form seines durch Häutung befreiten *Gymnodinium*. Daher ist in beiden stärker gepanzerten Stufen als Zellteilung im engeren Sinne wohl nur der *hemivalvate* Modus wirksam (*Ceratium*, Dinophysideen, *Prorocentrum* nur dieser, bei *Glenodinium lubinensiforme* neben „provenienter“ Teilung), bei dem der Panzer schräg zur Längsachse entlang bestimmter Plattenfugen aufspringt und jede Zellhälfte einen Halbpanzer erhält, während die fehlende Hälfte von den neugebildeten Oberflächen ergänzt wird (Abb. 20). Alles dieses vermag im beweglichen oder unbeweglichen Zustand zu geschehen. Die Zelle kann sich ferner in *provenienter* Teilung innerhalb der Zellwand abrunden und sich dann nackt teilen. Die Tochterzellen werden danach frei (Abb. 19) und scheiden eine neue Schale aus. Hierbei kommt es häufig [Schütt (1895): *Pyrophacus* (Abb. 21), Woloszynska (1917): *Gymnodinium coronatum* (Abb. 22)] zu Mehrfachteilungen; es wäre überhaupt besser, statt von „provenienter Teilung“ von Zoosporen- oder Gonidienbildung (Schizogonie) zu sprechen.

Nicht selten behäutet sich, wie bei einigen marinen Peridineen oder bei *Glenodinium halli* (s. unten), der Protoplast vor der provenienten Teilung innerhalb der Mutterschale durch eine glatte Membran, in der nun die Teilung geschieht. *Peridinium acuminatum* soll dabei eine Wand einziehen und zuletzt aus jeder der beiden Teilsporangien ein *Gymnodinium* entlassen (SCHÜTT 1887). Durch Behäutung des Protoplasten unter Verlust der Schale, falls diese vorhanden, können sich runde Sporangien mit glatter, gallertiger oder fester Membran bilden. So entstehen bei *Gymnodinium fucorum* (DANGEARD 1938a) in derbhäutigen Sporangien bis zu 8 Gonidien.

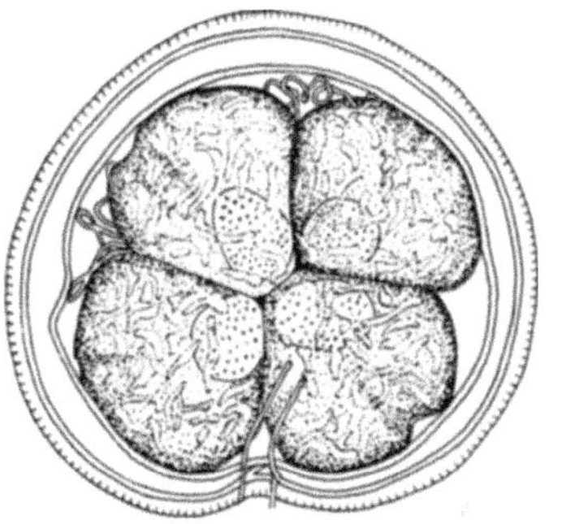

Abb. 21. *Pyrophacus horologicum* hat vier Zoosporen gebildet, Antapikalansicht. (Nach SCHÜTT 1895, verändert.)

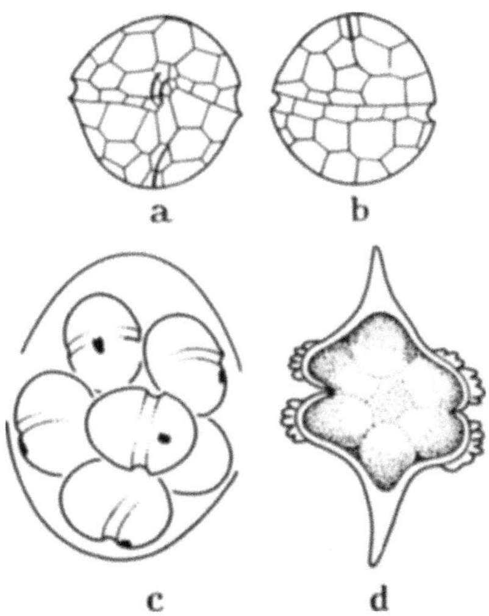

Abb. 22a—d. *Gymnodinium coronatum*. a Ventral-, b Dorsalansicht der vegetativen Zelle, c Zoosporenbildung, d Cyste. (Nach WOLOSZYNSKA 1917, aus SCHILLER.)

LINDEMANN u. a. zeigten nun, daß die vegetativen Teilungen auch eines Organismus je nach den Bedingungen dem einen oder dem anderen Modus gemäß ablaufen können. So teilt sich *Gymnodinium nivale* in (mineralischer) Nährlösung gymnoid, auf Agar aber provenient, d.h. nach Ablösung des Protoplasten von der Hülle. Bei dem kryophilen *Gymnodinium pascheri* kommen in der gleichen Population proveniente und avalvate Teilungen (SCHILLER 1954) nebeneinander vor und bei *Glenodinium halli* (FREUDENTHAL und LEE 1963) beide Varianten der provenienten Teilung, die eine am gepanzerten und beweglichen und die andere am durch Wiederbehäutung unbeweglichen Protoplasten; Phosphatreichtum des Mediums begünstigt hier anscheinend den ersten Vorgang. Die Umwandlung einer hemivalvaten Teilung des marinen *Peridinium triquetrum* in eine proveniente Teilung können BRAARUD u. PAPPAS (1951) noch während der Cytokinese erzielen; auf Zugabe von destilliertem Wasser werfen die Tochterzellen ihre Panzerhälften ab und nehmen ovale Form an. Schließlich seien noch Angaben LINDEMANNs (1929) wiedergegeben über jene Außenbedingungen (pH, pCa, Salzkonzentration, Torfextraktgabe), in denen die Teilungen von *Gymnodinium fuscum* entweder in freibeweglichem, nackt unbeweglichem oder in Gallerte eingehüllt unbeweglichem Zustand stattfinden (Tabelle 2). Nach Zusatz von starker Torflösung erfolgt die Vermehrung zwischen pH 7,3 und 6,2 in beiden Lösungen nur im unbeweglichen Zustand in Gallertlagern.

Auf die heteromorphen Teilungen einiger mariner *Ceratien*, bei denen aus der Normalform vereinfachte und kleinere Individuen, die sog. *truncata*-, *lineata*- und *lata*-Formen entstehen (LOHMANN 1908, APSTEIN 1911, KOFOID 1909, HASLE u. NORDLI 1951), braucht hier ebensowenig eingegangen zu werden, wie auf die mit einer „Amitose" des Kerns verbundenen „Knospungen" APSTEINs (l. c.) und

Tabelle 2. *Zustand und Teilungsform von Gymnodinium fuscum in Abhängigkeit von chemischen Bedingungen.* (Nach Lindemann 1929, vereinfacht.)

pH	3,1 mg-% Ca Salzkonzentration $5 \cdot 10^{-4}$		Kein Ca Salzkonzentration $25 \cdot 10^{-4}$ g/Liter ohne Torfextrakt
7,3	feine Gallerte		dickes Gallertlager
7,6	frei, beweglich	gute Vermehrung	frei, unbeweglich
6,2			
5,2	Gallerthöfe unbeweglich	ohne Vermehrung	feine Gallerte

Borgerts (1910); erstere sind Mikrogameten differenzierende Teilungen, letztere in der umgekehrten, also falschen Richtung gelesene Stadien der anisogamen Kopulation (v. Stosch 1964).

Unbeweglich und nackt sind auch die Zooxanthellen gewisser Medusen, die McLaughlin und Zahl (1957) in Reinkultur isolierten und deren Entwicklung Freudenthal (1962) studierte. Es handelt sich um ein *Gymnodinium*, bei dem die Tendenz zur Geißellosigkeit genetisch fixiert ist. In der gleichen abgerundeten Form, die seine Symbiose mit der Scyphomeduse *Cassiopeia* charakterisiert, wächst der Organismus auch in Kultur, kann sich aber hier unter nicht genauer umschriebenen Bedingungen mit einer dünnen Membran umhäuten und später aus diesen „Cysten" entweder in der nackten symbiontischen Form oder als *Gymnodinium* austreten oder auch vier Autosporen bilden. Schließlich können auch mehrere eingeißelige und vereinfachte Schwärmer entlassen werden, welche der Autor für Gameten hält, allerdings noch ohne zwingende Gründe. Die Zoosporen aus „Cysten" erscheinen vor allem in der exponentiellen Phase des Wachstums; das Ausschwärmen selbst wird durch Belichtung, möglicherweise in einer an den Lichtcyclus angekoppelten inneren Rhythmik induziert und zwar zu einem Maximum des Schwärmens 4—6 Std nach Lichtbeginn. Die Zoosporen gehen wieder in die Zooxanthellen-Form über, insbesondere in viskösen Medien. Eine Aplanospore entläßt die „Cyste" vor allem auf Agarboden, liefert aber mehrere Autosporen in alten Kulturen.

2. Cysten.

Was man bei Dinophyten als Cysten bezeichnet, ist sehr verschiedenartig, einmal in Form und Farbe, dann aber auch in der Entstehung und wahrscheinlich der entwicklungsgeschichtlichen Bedeutung. Cysten können gallertige Hüllen besitzen oder aber dünne bis kräftige feste Membranen. Beides kommt auch vereint vor, indem innerhalb der Membran des Flagellaten eine dünnwandige Cyste entsteht und Mutterzelle mit Cyste in derbe Gallerte eingebettet ist (*Peridinium borgei*, Entz 1926). Die Gallerte wird mindestens in einigen Fällen durch Ausstoßen von Trichocysten erzeugt (Biecheler 1952). Die feste Cystenmembran wird bei nackten Zellen von der Plasmaoberfläche ausgeschieden; der Vorgang ist z. B. durch Baumeister (1939) für ein *Cystodinium* geschildert worden. Bei beschalten Formen liegt die Cyste häufig innerhalb des Panzers (Abb. 23), der dadurch zum Epispor wird. Gallertcysten und Cysten mit fester Membran können übrigens nebeneinander vorkommen und unter ganz verschiedenen

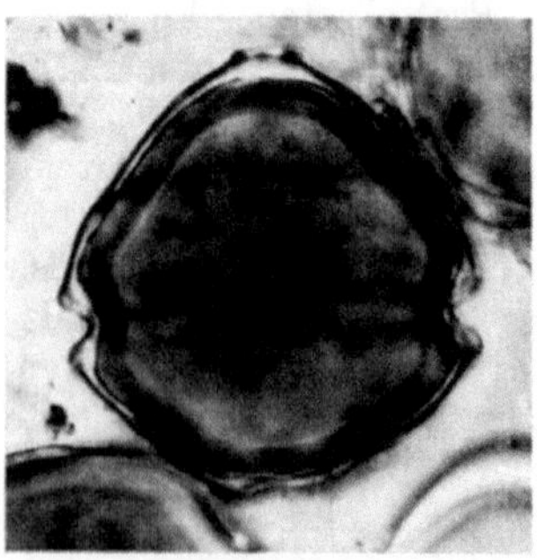

Abb. 23. *Peridinium lomnikii.* Mutterzelle mit Cyste, Mutterzellmembran wirkt als Epispor, Cyste selbst mit Exo- und Endospor. (Original.)

Bedingungen gebildet werden, wie bei *Peridinium borgei*, wo neben den bereits erwähnten innermembranären Cysten, mit denen die Art überwintert, „Gloeohypnodinien", d.h. mit geschichteten Gallertmembranen versehene, längliche oder nierenförmige Ruhezellen nach Abwerfen des Panzers entstehen. Die Cyste kann glatt (Abb. 23), warzig (Abb. 22d) oder bestachelt sein. Sie kann einer Kugel, einem Ellipsoid oder dem mehr oder weniger vereinfachten Ausguß des Panzers entsprechen wie bei den *Ceratium*-Arten des Süßwassers (Abb. 24) oder aber von einer von der Schwimmform her nicht verständlichen Gestalt — so die Halbmondcysten der *Cystodinien* oder einige der von WOLOSZYNSKA beschriebenen *Gymnodinium*-Cysten — sein (Abb. 22d). Ihre Wand ist zuweilen in ein äußeres, manchmal mehr oder weniger tiefgefärbtes Exospor und ein farbloses Endospor differenziert (DANGEARD 1938a, b).

Entwicklungsgeschichtliche Dignität der Cysten.

Die entwicklungsgeschichtliche Bedeutung der Cyste scheint in vielen Fällen die eines Behälters für die *Zellteilung* zu sein. Die Cytokinese findet dann nicht direkt am abgerundeten Protoplasten, sondern erst nach Ausscheidung einer Hülle statt, wie das SCHÜTT (1887) für einige marine, P. DANGEARD (1938a, b) u.a. für limnische Formen schilderten. Das würde zu den capsalen und coccalen Formen überleiten, auf die wir zurückkommen; der cystäre Zustand ist dann zum eigentlich vegetativen geworden.

Die Mehrzahl der Peridineencysten dürfte nach den Angaben ihrer Beschreiber vegetativen Dauerzuständen entsprechen (s. z.B. ENTZ 1926, ENTZ und SEBESTÝEN 1935, DANGEARD 1938a, b, NORDLI 1951, BRAARUD u. PAPPAS 1951).

Die Cysten der Süßwasserceratien sind aber mindestens zum Teil *Hypnozygoten*; für *Ceratium cornutum* wurde darüber auf S. 626 berichtet. Dafür, daß auch bei dem häufigeren *Ceratium hirundinella* sexuelle Cysten vorkommen, sprechen die Beobachtungen von HALL (1925), welcher in amerikanischem Material der Art reichlich zweikernige Cysten fand. Wenn dagegen ENTZ (1925) in Ungarn auch in jungen Cysten ganz überwiegend nur einen Kern beobachtete, könnte das damit zusammenhängen, daß die Kernverschmelzung sich bereits in einer Planozygote vollzog, wie dieses nach unseren Beobachtungen bei *C. cornutum* der Fall ist. Doch besteht natürlich auch die Möglichkeit, daß es sich bei ENTZ um Parthenozygoten handelt; selbst für *C. cornutum* ist ungewiß, ob solches nicht vielleicht vereinzelt vorkommt.

Bildungsbedingungen.

Die Bildungsbedingungen der Cysten sind wenig studiert. Bei den Süßwasser*ceratien* treten die Cysten vor allem am Ende der Vegetationsperiode im Herbst auf und überwintern dann (Lit. s. HUBER-PESTALOZZI 1951). Sie sind also wohl durch Temperatur-, Licht-, oder vielleicht auch Nährstoffaktoren bedingt. Die bestachelten Dauersporen des marinen *Gonyaulax polyedra* wurden (NORDLI 1951) nach dem Ende von „red tides" beobachtet, die durch diesen Organismus und ein *Ceratium* hervorgerufen waren und zu hohen Populationsdichten geführt hatten. BRAARUD und PAPPAS finden bei *Peridinium triquetrum* bestachelte Cysten in Kulturen, die längere Zeit bei 3° gestanden hatten, und glatte Cysten, möglicherweise Vorstufen zu ersteren, in schlecht wachsenden Kulturen. Umgekehrt entstehen die gehörnten Zygoten der Winterperidinee *Gymnodinium pascheri* bei langsamem Ansteigen der Temperatur im Frühjahr (SCHILLER 1955). Nährstoffmängel sind vielleicht verantwortlich für das Erscheinen von Cysten in alten Kulturen bei *Gonyaulax tamarensis* und *Protoceraticm reticulatum* (Braarud 1944) unter ähnlichen Bedingungen, und nur unter diesen erhielt Bursa (1959) dünnwandige, cystenartige Gebilde von *Prorocentrum*.

Reifung, Lebensdauer und Keimung der Cysten.

Über die Keimung der (vermutlich) vegetativen Cysten ist wenig bekannt. Offenbar kann der Inhalt direkt und als *Gymnodinium* schlüpfen, wie bei *Peridinium borgei* angegeben (Entz 1926), oder sich in zwei oder vier Zellen aufteilen

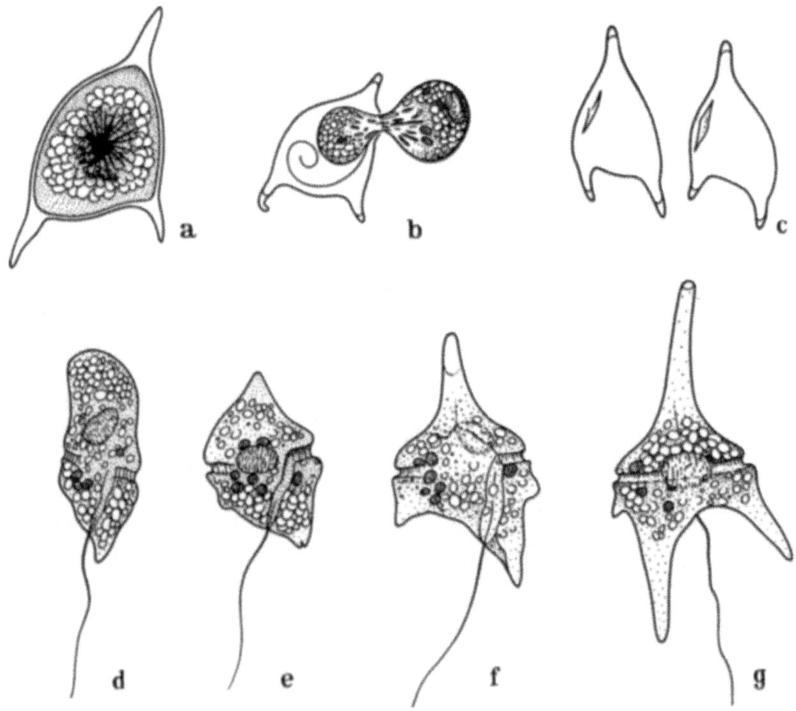

Abb. 24a—g. Cyste und Cystenkeimung von *Ceratium hirundinella*. a Cyste, b Keimling, c entleerte Cysten mit Keimungsriß, d Schwärmer im *Gymnodinium*stadium, 10 min nach Schlüpfen, e Gymnoceratium gegen Ende des Gymnodiniumstadiums, $^3/_4$ Std nach Schlüpfen, f u. g Präceratium ca. 4 Std nach Schlüpfen. Panzer noch zart, bis zur erwachsenen Form verlängern sich die Hörner noch weiter. (Nach Huber und Nipkow 1922.)

(*Gymnodinium fuscum*, *Discodinium poucheti*, Dangeard 1938a; *Gymnodinium pascheri*, Schiller 1955).

Dagegen sind wir über die vielleicht oder mit Sicherheit sexuell entstehenden Cysten der Süßwasserceratien in ihrer Reifung, Lebensdauer und Keimung gut unterrichtet, insbesondere durch die eingehenden Untersuchungen von Huber und Nipkow (1922, 1923) an *Ceratium hirundinella*. Das Material wurde aus limnischen Ablagerungen gewonnen, deren jahresperiodisch bedingte Bänderung die Altersbestimmung ermöglicht. Danach sind im Faulschlamm der Tiefe des Zürcher Sees bei gleichmäßig 4—5^0 eingelagerte Cysten bis zu $6^1/_2$ Jahren keimfähig, jedoch diejenigen von *Peridinium cinctum* $16^1/_2$ Jahre; Austrocknen und Gefrieren wurden von den Cysten nicht vertragen. Werden reife Cysten in Wasser von 17^0 suspendiert, so sondert sich innerhalb von 36—40 Std in ihnen ein apikales von einem antapikalen Plasma durch die Modellierung der Querfurche. 4 Std später schlüpft die Keimlingsform des „Gymnoceratiums" (Abb. 24b, d) und wandelt sich unter Ausformung kurzer Hörner, Abplattung des

Körpers und schließlich Ausscheidung des Panzers im Laufe der nächsten Stunden in das gedrungene „Präceratium“ um (Abb. 24e), das dann direkt in die endgültige *Ceratium*form (Abb. 24f, g) übergehen soll und damit teilungsfähig wird.

Die *Ceratium cornutum*-Cyste keimt ähnlich durch einen Riß in der Cystenwand (v. STOSCH 1965), also nicht durch Verquellung der Membran, wie SCHILLING (1891) dies beschrieb. Beim Aufreißen spielt eine starke Quellung des Endospor offenbar die Hauptrolle; dieses tritt auch als Hülle des jungen „Gymnoceratium“ mit aus (v. STOSCH, unveröff.). Der Keimlingsschwärmer wird dann frei und verwandelt sich in das durch den Bau von Vorder- und Hinterschale von der vegetativen „Sommerform“ abweichende Präceratium um, was FOLGNER (1899) bereits treffend beschrieb, und bleibt im Gegensatz zu dem für *C. hirundinella* Geschilderten auch auf dieser Stufe stehen. Das freischwimmende Präceratium macht bei 21° nach etwa 2 Tagen die erste an Meiosis I gebundene und seine Töchter führen nach weiteren 2—3 Tagen die zweite meiotische Teilung durch. Da die Tochterzellen eines *Ceratium* jeweils die eine Panzerhälfte der Mutterzelle erben, besitzen von den vier Gonen schließlich zwei die normale Panzerausrüstung und die beiden anderen ein „*Präceratium*“-Vorder- bzw. Hinterteil. Der Keimlingsschwärmer als Diplophase hat also eine gewisse Selbständigkeit, weshalb er besprochen wurde. Der Unterschied im Resultat seiner äußeren Entwicklung gegenüber den Zürcher *hirundinella*-Keimlingen könnte bedeuten, daß letztere nichtsexueller Entstehung gewesen waren.

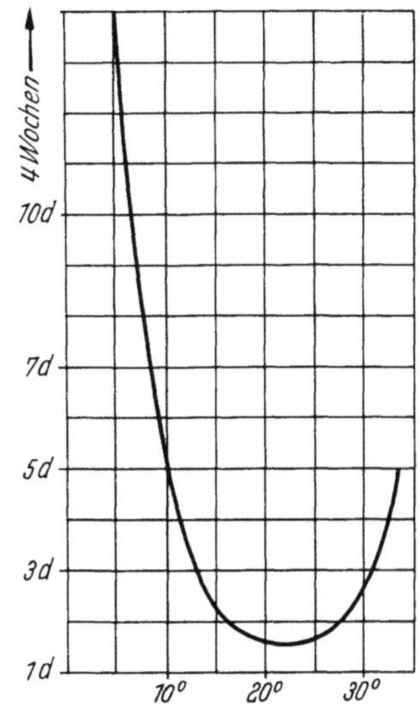

Abb. 25. Keimungsdauer bei keimungsreifen Cysten von *Ceratium hirundinella* in Abhängigkeit von der Temperatur. (Nach HUBER und NIPKOW 1923.)

Der Keimungsverlauf ist bei *C. hirundinella* nach HUBER und NIPKOW vom Alter der Cysten abhängig. Zunächst machen diese eine Reifung durch. $1^1/_2$ Monate nach der Bildung ist unter den oben angegebenen Bedingungen die Keimung verzögert (minimal 4 Tage): nach 3 Monaten wurde sie zwar bei einigen wenigen Cysten normal, gab aber viel Nachzügler; erst nach 7 Monaten sind die Cysten reif und keimen ziemlich einheitlich nach 48 Std. Mit steigendem Alter steigt dann auch die Keimungszeit wieder, und zwar nach 6 Jahren auf 6—8 Tage. Auch *Peridinium cinctum*-Cysten zeigen ähnliche Keimungsverzögerungen beim Altern. Über den Einfluß der Temperatur auf die Keimungsdauer unterrichtet Abb. 25, die insbesondere ein Optimum bei etwa 20° sowie eine starke Keimungsverzögerung bei 5°, also der Temperatur der Lagerstätte der Cyste, demonstriert.

Licht ist für die Keimung bei 18° bis zum fertigen *Ceratium* nicht erforderlich, doch soll Blaulicht (farbig lackierte Glasstürze) die Keimung etwas (auf 3 Tage), Grünlicht stark (auf 7 Tage) verzögern, während rotes und gelbes Licht ohne Einfluß sind.

Auch die Keimungszeit im Laboratorium erzeugter *cornutum*-Cysten (Hypnozygoten) ist eine Funktion der Reifungsdauer. Die bei 12° entstandenen Zygoten wurden nach einer Reifungszeit bei 3° (100 Lux) zur Keimung auf 21° und etwa 2000 Lux gebracht. Nach 21, bzw. 79, bzw. 156 Tagen Kältebehandlung keimten die Cysten (zu 50% oder mehr) nach 9, bzw. 3, bzw. 1 Tagen. Die Reifung ist auch hier lichtunabhängig, die Keimung (bei 21°) nach bisherigen Erfahrungen aber im Gegensatz zu *C. hirundinella* lichtbedürftig (v. STOSCH, unveröff.).

Nach der chemischen Seite wirkt destilliertes nicht anders als Standortwasser auf die Entwicklung von *Ceratium hirundinella* (Huber und Nipkow 1923). Schwache Natriumcarbonatlösungen (10^{-4} %) sind für die Keimung günstiger als die etwas sauren Klebs- und Knop-Lösungen. Kochsalz verzögert ab 0,1 % und führt gleichzeitig zu Formschäden; für diese und die morphogenetischen Wirkungen der übrigen Behandlungen muß auf die Originalarbeit verwiesen werden.

Abschließend ist zu sagen, daß nach diesen Experimenten die empirische Keimungsverhinderung am natürlichen Ort im Seeschlamm weder auf die Temperatur noch auf die Lichtverhältnisse allein zurückgeführt werden kann. Versuche der Verfasser, die auch Anaerobie allein als ungenügend zur Keimungsverhinderung (18°) zu zeigen schienen, sind als nicht mehr stichhaltig anzusehen, da bei ihnen Überschichtung des Mediums mit Paraffinöl zum Sauerstoffausschluß verwendet wurde, welches dafür aber bekanntlich nicht ausreicht.

Der Eintritt der Meiosis des Keimlingsschwärmers von *Ceratium cornutum* wird durch niedere Temperatur stark hinausgezögert. Die Existenzzeit des „Präceratium" beträgt dadurch bei 9° mindestens 7 Tage gegen 2 Tage bei 21° (v. Stosch, unveröff.).

c) Capsale, coccale und trichale Formen.

Im vegetativen Zustand liegen mit gallertigen oder festen Membranen umhäutete Formen vor, die sich mit Zoosporen, Autosporen oder durch Teilung und Neubehäutung innerhalb der Membran fortpflanzen. Interessanterweise kommen in einzelnen Gattungen behäutete mit gefelderter Membran versehene „glenodinoide" neben nackten „gymnodinoiden" Zoosporen vor. So hat *Cystodinium bataviense* die erste, *C. steinii* die zweite Schwärmerform (Klebs 1912). Von Stylodinien sollen *S. tarnum* und *S. polymorphum* (Baumeister 1943, 1957a, b) mit grob und spezifisch gefelderter Membran versehene Zoosporen besitzen, während diejenigen von *S. sphaera* nackt sind (Pascher 1944). Solche Formen wären nur *dann* als Cysten freilebender Peridineen, eben ihrer Schwärmer, aufzufassen, wenn letztere selbständig teilungsfähig sind. Der Beweis hierfür scheint aber in keinem Fall erbracht, auch bei denjenigen Cystodinien nicht, für welche Baumeister (1957a, b) überflüssigerweise die gerade durch diese Teilung definierte Gattung *Gymnocystodinium* aufstellt. Die Formen mit armierten Zoosporen zeigen besonders gut, daß man mit Pascher u. a. den coccalen Status bei Dinophyceen als von Cysten abgeleitet ansehen kann. Dieses würde dann auch für das mit gepanzerten Hemidinien schwärmende *Gloeodinium montanum* gelten, das zwar gewöhnlich (Schiller 1937) als capsal klassifiziert wird, aber auch durch den Membranbau (Klebs 1912) eher als coccal erscheint. Für Arten mit nackten Schwärmern kann man den gleichen Schluß ziehen; es ist dann eigenartig, daß bei dem hierhergehörigen *Stylodinium sphaera* (Pascher) innerhalb der Zellwand eine typische Cyste gebildet werden kann. Vergleichend morphologisch müßte sie als sekundär bezeichnet werden.

Besonders interessieren wieder diejenigen Fälle, die verschiedene Fortpflanzungsformen nebeneinander besitzen, wie Paschers (1927) *Dinastridium sexangulare*, bei dem die Vermehrung durch mehrere Zoosporen oder wahlweise durch Autosporen erfolgt (Abb. 26), oder *Tetradinium minus*, das sich ähnlich zu verhalten scheint. Dann wäre *Dinothrix* zu erwähnen, dessen Zoosporen entweder nur für kurze Zeit innerhalb der Zelle erscheinen, „um" die Zellteilung durchzuführen — die gymnodinoiden Töchter behäuten sich danach wieder innerhalb der Muttermembran, so daß der Fadenzusammenhalt gewahrt bleibt —, oder ungeteilt austreten und der Fortpflanzung dienen. Dagegen vermehrt sich *Gloeodinium* (KLEBS 1912) durch Zweiteilung ohne Schwärmerbildung so, daß die Tochter- und Enkelzellen durch die Membran der Mutterzelle zusammengehalten werden, und pflanzt sich außerdem nach KILLIAN (1924) und GEITLER (1928) durch hemidinoide Zoosporen fort: Aus den äußeren Membranschichten einer jungen Zelle schlüpft der Zellinhalt umhäutet aus, schwillt an und teilt sich: danach formen sich die Tochterzellen unter Ausstoßung von Ölreserven in die Dinosporen um, welche dann frei werden und für kurze Zeit schwärmen, ehe sie sich innerhalb ihres Panzers neu behäuten. KILLIAN sah dies am Naturmaterial vom August, das bei ihm direkt, während *Gloeodinium* bei GEITLER

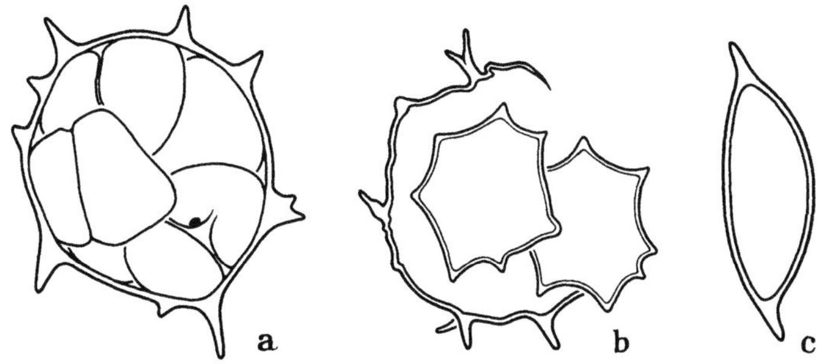

Abb. 26a—c. *Dinastridium sexangulare.* a Mutterzelle mit gymnodinoiden Zoosporen, diese können direkt ausschwärmen oder sich mit Membran umgeben (b) und danach entlassen werden. c Erwachsene Zelle, Seitenansicht. (Nach PASCHER 1927, aus SCHILLER.)

erst nach 14 Tagen im Laboratorium schwärmte. Wir fanden (unveröff.) bei einem aus bakterienfreien Agarkulturen (15°/1000 Lux) stammenden Material nach Übertragen in stark verdünnte Nährlösung Schwärmer im Bereich zwischen 6° und 24° (höhere und tiefere Temperaturen nicht geprüft), wobei die Schwärmerentlassung bei tieferen Temperaturen später einsetzte und länger andauerte (6°, 150 Lux: Einsatz nach 2 Tagen Dauer 9 Tage), gleichzeitig die Schwärmdauer der einzelnen Schwärmer größer war und bei den hohen Temperaturen früher einsetzte und nur kurz anhielt (24°, 150 Lux: Einsatz nach spätestens 2 Tagen, Dauer 2 Tage).

Literatur.

APSTEIN, C.: Biologische Studie über *Ceratium tripos* var. *subsalsa* Ostf. Wiss. Meeresunts. Abt. Kiel, N.F. **12**, 138—166 (1911).

BAUMEISTER, W.: Zur Kenntnis der Spitzzysten bildenden *Gymnodinien*. Int. Rev. Hydrobiol. **39**, 395—412 (1939). — Die Dinoflagellaten der Kreise Pfarrkirchen und Eggenfelden. 2. Arch. Protistenk. **96**, 344—363 (1943). — Zur Kenntnis der Dinophyceen des Seeoner Gebietes. Arch. Protistenk. **102**, 1—20 (1957a). — Neue Dinococcalen aus dem niederbayrischen Hügellande zwischen Isar und Inn. Arch. Protistenk. **102**, 21—43 (1957b). — BIECHELER, B.: Recherches sur les Péridiniens. Bull. biol. France et Belg., Suppl. **36**, 1—145 (1952). — BORGERT, A.: Kern- und Zellteilung bei marinen *Ceratium*-Arten. Arch. Protistenk. **20**, 1—46 (1910). — BRAARUD, T.: Morphological observations on marine dinoflagellate cultures (*Porella perforata, Goniaulax tamarensis, Protoceratium reticulatum*). Avh. Norske Vidensk. Akad. Oslo, math.-nat. Kl. **11**, 3—18 (1944). — BRAARUD, T., and I. PAPPAS: Experimental studies on the dinoflagellate *Peridinium triquetrum* (Ehrb.) Lebour. Avh. Norske Vidensk. Akad. Oslo, I. math.-nat. Kl. **1951** (2), 1—23 (1951). — BURSA, A.: The genus *Prorocentrum* Ehrenberg. Morphodynamics, protoplasmatic structures and taxonomy. Canad. J. Bot. **37**, 1—31 (1959).

CACHON, I.: Contribution à l'étude des Péridiniens parasites. Cytologie. Cycles evolutives. Thèses Fac. Sci. Univ. Paris, Ser. A No 4205, No d'ordre 5056, 1—158 (1964). — CHATTON, E.: Les Péridiniens parasites. Morphologie, reproduction, éthologie. Arch. zool. exp. et gén. **59**, 1—475 (1923). — Classe des Dinoflagélles ou Péridiniens. In: GRASSÉ, Traité de Zoologie, vol. 1 (1), p. 309—390. Paris: Masson & Cie. 1952.

Dangeard, P.: Mémoire sur la famille des Péridiniens. Botaniste **29**, 3—180 (1938a). — Second mémoire sur la famille des Péridiniens. Botaniste **29**, 267—308 (1938b). — Diwald, K.: Die ungeschlechtliche und geschlechtliche Fortpflanzung von *Glenodinium lubinensiforme* spec. nov. Flora (Jena) **132**, 174—192 (1938). — Dogiel, V.: Beiträge zur Kenntnis der Peridineen. Mitt. zool. Stat. Neapel **18**, 1—45 (1906).

Entz, G.: Über Cysten und Encystierung der Süßwasser-Ceratien. Arch. Protistenk. **51**, 131—183 (1925). — Zur Morphologie und Biologie von *Peridinium borgei* Lemmermann. Arch. Protistenk. **56**, 397—446 (1926). — Studien an Süßwasser-Ceratien. Arch. Protistenk. **58**, 344—440 (1927). — Über gehemmte Lebens- und Absterbeerscheinungen einiger Dinoflagellaten. Arb. 1. Abt. Ungar. Biol. Forsch.-Inst. **3**, 206—236 (1930). — Entz, G., u. O. Sebestyén: Morphologische, biologische und physicochemische Untersuchungen an *Peridinium aciculiferum* Lemmermann, mit besonderer Berücksichtigung der *Gymnodinium*-Form. Arb. 1. Abt. Ungar. Biol. Forsch.-Inst. **8**, 15—73 (1935/36).

Folgner, V.: Beiträge zur Kenntnis einiger Süßwasserperidineen. Öst. bot. Z. **49**, 81usf. (1899). — Freudenthal, H. D.: *Symbiodinium* gen. nov. and *Symbiodinium microadriaticum* sp. nov., a Zooxanthella: Taxonomy, life cycle, and morphology. J. Protozool. **9**, 45—52 (1962). — Freudenthal, H. D., and J. J. Lee: *Glenodinium halli* n. sp. and *Gyrodinium instriatum* n. sp., Dinoflagellates from New York waters. J. Protozool. **10**, 182—189 (1963).

Geitler, L.: Zwei neue Dinophyceenarten. Arch. Protistenk. **61**, 1—8 (1928). — Gross, F.: Zur Biologie. und Entwicklungsgeschichte von *Noctiluca miliaris*. Arch. Protistenk. **83**, 178—196 (1934).

Hall, R. P.: Mitosis in *Ceratium hirundinella*. Univ. California Publ. Zool. **28**, 29—64 (1925). — Hasle, G. R., and E. Nordli: From variation in *Ceratium fusus* and *tripos* populations in cultures and from the sea. Avh. Norske Vidensk. Akad. Oslo, I. math.-nat. Kl. **1951** (4), 1—25 (1951). — Huber, G., u. F. Nipkow: Experimentelle Untersuchungen über die Entwicklung von *Ceratium hirundinella* O. F. Müller. Z. Bot. **14**, 337—371 (1922). — Experimentelle Untersuchungen über Entwicklung und Formbildung von *Ceratium hirundinella* O. F. Müller. Flora (Jena) **116**, 114—215 (1923). — Huber-Pestalozzi, G.: Die Verbreitung der Süßwasserceratien auf der Erde. Verh. int. Ver. theor. ang. Limnol. **11**, 152—188 (1951).

Killian, Ch.: Le cycle evolutif du *Gloeodinium montanum* Klebs. Arch. Protistenk. **50**, 50—66 (1924). — Klebs, G. (1883): Siehe Lit. zu C. Euglenophyceae usw. — Über die Organisation und die systematische Stellung der Peridineen. Biol. Zbl. **4**, 704—710 (1885). — Über Flagellaten und algenähnliche Peridineen. Verh. nat. Ver. Heidelberg **11**, 367—451 (1912).— Kofoid, C. A.: Mutations in *Ceratium*. Bull. Mus. comp. Zool. Harvard **52**, 211—257 (1909).

Lefèvre, M.: Monographie des espèces d'eau douce du genre *Peridinium*. Arch. Bot. **2**, 1—210 (1928). — Lindemann, E.: Peridineae. In: Engler-Prantl, Die natürlichen Pflanzenfamilien. Bd. 2, S. 1—102. Leipzig: Wilhelm Engelmann 1928. — Experimentelle Studien über die Fortpflanzungserscheinungen der Süßwasserperidineen auf Grund von Reinkulturen. Arch. Protistenk. **68**, 1—104 (1929). — Lohmann, H.: Untersuchungen zur Feststellung des vollständigen Gehaltes des Meeres an Plankton. Wiss. Meeresunters. Abt. Kiel **10**, 129—370 (1908).

McLaughlin, I. I. A., and P. A. Zahl: Studies in marine biology. II. In vitro culture of zooxanthellae. Proc. Soc. exp. Biol. (N.Y.) **95**, 115—120 (1957).

Nordli, E.: Experimental studies on the ecology of *Ceratia*. Oikos 8, 201—265 (1957). — Resting spores in *Gonyaulax polyedra* Stein. Nytt Mag. Naturvidensk. **88**, 207—212 (1951).

Oehlkers, F.: Das Leben der Gewächse. Berlin-Göttingen-Heidelberg: Springer 1956.

Pascher, E.: Die braune Algenreihe aus der Verwandtschaft der Dinoflagellaten (Dinophyceen). Arch. Protistenk. **58**, 1—54 (1927). — Von einer neuen Dinococale *(Cystodinium phaseolus)* mit zwei verschiedenen Schwärmertypen. Arch. Protistenk. **63**, 241—254 (1928). — Über neue protococcoide, festsitzende Algengattungen aus der Verwandtschaft der Dinoflagellaten. Beih. Bot. Zbl. A **62**, 376—395 (1944).

Schiller, J.: Dinoflagellatae. Bd. 10/3. In: v. Rabenhorsts Kryptogamen-Flora. Leipzig: Akademische Verlagsanstalt 1937. — Über winterliche pflanzliche Bewohner des Wassers, Eises und des darauf liegenden Schneebreies. Öst. bot. Z. **101**, 235 (1954). — Untersuchungen an den planktischen Protophyten des Neusiedlersees 1950—1954, Teil I, Wiss. Arb. Burgenland, H. 9, 1—66 (1955). — Schilling, A. J.: Die Süßwasser-Peridineen. Flora (Jena) **74**, 220—299 (1891). — Schütt, F.: Über die Sporenbildung mariner Peridineen. Ber. dtsch. bot. Ges. **5**, 364—374 (1887). — Die Peridineen der Planktonexpedition. Erg. Plankton-exped. Humboldt-Stift. **4** (1895). — Skoczylas, O.: Über die Mitose von *Ceratium cornutum* und einigen anderen Peridineen. Arch. Protistenk. **103**, 193—228 (1958). — Stosch, H. A. v.: Zum Problem der sexuellen Fortpflanzung in der Peridineengattung *Ceratium*. Helgol. wiss. Meeresunters. **10** (1—4), 140—152 (1964). — Sexualität bei *Ceratium cornutum* (Dinophyta). Naturwissenschaften **52**, 112—113 (1965).

Woloszynska, J.: Neue Peridineenarten nebst Bemerkungen über den Bau der Hülle bei *Gymno-* und *Glenodinium*. Bull. Acad. Sci. Cracovie Sci., math.-nat. Ser. B **1917**, 114—122 (1917).

E. Chrysophyta.

Wenn wir von den *Phaeophyta* absehen, welche man heute in den Stamm der Chrysophyta einschließen sollte und welche hier nur aus praktischen Gründen — wie bisher üblich —von diesen getrennt behandelt werden, und wenn wir die *Silicoflagellaten* beiseite lassen, die weder überhaupt sicher einzureihen sind noch vegetative Fortpflanzung anders als Zweiteilung zu erkennen geben, so haben wir es in diesem Phylum mit den im Schwerpunkt ihrer Formbildung monadoiden *Chrysophyceen*, einer ihrer „coccalen" (besser wohl cystären) Sonderentwicklungen, den *Diatomeen* sowie den vorwiegend in Algenformen der verschiedenen Organisationstypen existierenden *Heterokonten* (Xanthophyceen) zu tun.

I. Chrysophyceae.

Bei den Chrysophyceen wird neuerdings eine wohl ziemlich grundsätzliche Unterscheidung in zwei Formenreihen (CONRAD 1941, KORNMANN 1955, BOURRELLY 1957, v. STOSCH 1958 und insbesondere PARKE 1961) sichtbar, von denen die eine die altbekannten, vorwiegend im Süßwasser verbreiteten *Euchrysophyceen* (Chrysophyceae, CHRISTENSEN A 1962) umfaßt mit Flimmergeißel, meist plus Peitschengeißel, häufig Kieselschuppen und endogenen Kieselcysten, und die andere die marinen *Haptophyceae* (CHRISTENSEN A 1962; *Prymnisiales*: v. STOSCH 1958) mit zwei isokonten oder subisokonten Peitschengeißeln (Abb. 33) und meist einem geißelähnlichen (daher früher als dritte Geißel aufgefaßt), aber im Feinbau verschiedenen zwischen den Geißeln liegenden Haptonema (PARKE, MANTON und CLARKE 1955, ferner die übrigen zitierten Arbeiten von PARKE u. Mitarb. und MANTON u. Mitarb.), das oft als Haftorgan fungiert. Diese letzteren besitzen, soweit nicht nackt, eine Körperbedeckung aus Schuppen organischer Substanz und, im Gegensatz zur vorigen Gruppe, von radialer oder zirkulärer Struktur, die bei den hierhergehörigen Coccolithophoriden verkalkt sind; Kieselsäure scheiden Haptophyceen offenbar niemals aus. Auch in den Farbstoffen bestehen zwischen den beiden Gruppen Unterschiede (DALES 1960).

Die monadoiden Formen vermögen einerseits vielfach in unbewegliche Zustände überzugehen, wobei die Geißeln meist (nicht z. B. bei manchen *Chromulina*-Arten) abgeworfen und häufig Gallerthüllen oder auch nur Gallertkappen ausgeschieden werden. Eine ursprüngliche Körperbedeckung aus Schuppen kann dabei erhalten bleiben. Andererseits werden zahlreiche Arten unter besonderen Umständen — wieder mit oder ohne Geißelverlust — amöboid wie z. B. *Ochromonas granularis* (DOFLEIN 1922) während der Zellteilung oder die Chrysochromulinen in alten Kulturen (s. z.B. PARKE, MANTON und CLARKE 1955). Die Bedingungskonstellationen für diese beiden Zustandsformen scheinen häufig einander ähnlich zu sein. Nur bei den Euchrysomonaden jedoch sind bis jetzt Arten bekannt, bei denen die amöboide Phase zur Hauptlebensform geworden ist (Rhizochrysidales).

Eine Monographie der Chrysomonaden verdanken wir BOURRELLY (1957).

a) Euchrysophyceae.

1. Vegetative Vermehrung.

Rhizopodiale Formen.

Unter jenen amöboiden Chrysomonaden, welche noch Schwärmerzustände ausbilden, vermögen sich *Chrysamoeba helvetica* (REVERDIN 1919) und die erwachsen polyenergide *Rhizochrysis* (PASCHER 1916) sowohl im begeißelten als auch unbegeißelten Zustand zu teilen, das letzte bei *Rhizochrysis* allerdings nur in der Form einer Plasmodienteilung. Die übrigen vermehren sich nach den Angaben nur als Amöben; bei der ein kelchartiges Gehäuse bewohnenden *Salpingorhiza* (KLUG 1936) schwärmt die eine Tochterzelle danach häufig als Monade ab, bei Beschädigung des Gehäuses beide Tochterzellen.

Flagellaten.

Die Monaden vermehren sich vorzugsweise durch Längsteilung im begeißelten Zustand (DOFLEIN 1922, 1923). Nur bei der mit Stielchen sessilen *Palatinella* ist durch LAUTERBORN (1906) und WISLOUCH (1910, 1914) eine Vermehrung durch Knospen beschrieben worden, welch letztere aus dem Rande des die Geißel umgebenden Tentakelkranzes ausgestülpt werden. Die „Knospungen" gehäusebewohnender Gattungen wie *Kephyrion*, *Pseudokephyrion*, *Stenocalyx* und ähnliche sind Zweiteilungen, modifiziert dadurch, daß der Protoplast teilweise aus dem Gehäuse austritt, sich unter Verlust oder Beibehaltung der Geißeln zerschnürt, und darauf die außenliegende Tochterzelle ein neues Gehäuse Mündung an Mündung mit dem alten ausscheidet (SCHILLER 1926b, SCHMID 1934, CONRAD 1931, 1942). Doch ist es nicht sicher, ob alle von den Autoren diesem Vorgang zugeordneten Zustände wirklich Teilungen oder eher Kopulationen angehören (FOTT 1959, Diskussion bei LUND 1960b und KRISTIANSEN 1963). Für *Kephyriopsis conica* und *K. cincta* nach den Abbildungen SCHILLERs und für *K. ornata* und *Kephyrion bacilliforme* nach den präzisen Angaben CONRADs (1931) ist es jedoch schwer, an dem Teilungscharakter solcher Paare zu zweifeln. Bei *Stenocalyx parvula* läßt nach SCHMID (1934) und bei einer weiteren *Stenocalyx*- und einer *Kephyrion*-Art nach KRISTIANSEN ein ähnlicher Vorgang das neue Gehäuse mit dem Hinterende zur Mündung des alten gewendet entstehen. Ganz ähnlich verhält sich *Chrysococcus cordiformis*, wo die Plasmaverbindung zur Tochterzelle nach der Teilung erhalten bleibt und die Entstehung eines antapikalen Porus im Tochtergehäuse verursacht (SKUJA 1956). In anderen Fällen, bei *Stenocalyx*, *Kephyrion*, *Pseudokephyrion*, *Kephyriopsis* (SCHILLER 1926b, 1929, SCHMID) und bei *Chrysococcus rufescens* (CONRAD 1926), sowie bei Arten von *Dinobryon* und *Hyalobryon* (z. B. WILLE 1887, GESSNER 1926), tritt die eine Tochter nackt aus und bildet frei oder bei den beiden letzteren Gattungen in größerer oder geringerer Nähe des Muttergehäuses, oft diesem selbst angeheftet, ein neues Gehäuse aus. Die Arten mit Schuppenpanzer, z. B. *Synura* (CONRAD 1926, SMITH A 1950), *Synuropsis* (SCHILLER 1926b), *Mallomonas* (CARTER 1937b, HARRIES 1953), *Chrysophaerella* (BOURRELLY 1957), teilen sich wohl gewöhnlich mit ihrer Hülle (Abb. 27), indem die fehlenden Schuppen während der Durchschnürung eingeschoben werden. Bei *Mallomonas* kann es aber auch wie bei den Formen mit festen Gehäusen zum Ausschwärmen einer oder beider Tochterzellen nach der Teilung im nackten (SCHILLER 1926b, REVERDIN 1919) oder (*M. anglica:* CARTER 1937b) im bereits wieder gepanzerten Zustand kommen. Man kann dann, bezogen auf die normale Fortpflanzungsart, von Zoosporenbildung sprechen. Hier sei erwähnt, daß gelegentlich auch Zellen der Kolonien von *Uroglenopsis europaea* (BOURRELLY: *Uroglena europaea*) nach zwei Teilungsschritten ausschwärmen (SCHILLER 1926b), während die Tochterflagellaten der normalen Bipartition im Verband bleiben; man sollte darauf achten, ob diese Kleinschwärmer nicht im Zusammenhang mit einer Sexualität stehen; *Uroglena* hat nämlich zweikernige Cysten.

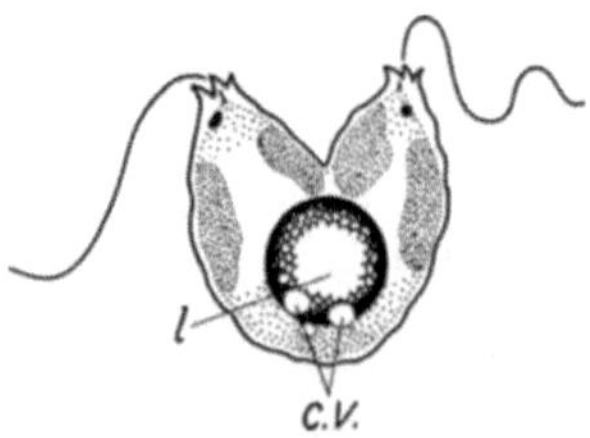

Abb. 27. *Mallomonas anglica.* Zellteilung unter gleichzeitiger Zerschnürung des Schuppenpanzers (*l* Leucosinvacuole; *c.v.* contractile Vacuolen). (Nach CARTER 1937b.)

Schließlich vermögen bei den verschiedensten gehäusebewohnenden oder hüllentragenden Formen die Zellen ungeteilt als Amöben (*Synura:* PASCHER 1912, *Mallomonas:* CONRAD 1914a) oder Monaden (z. B. *Synura:* PASCHER, *Mallomonas:* HARRIES 1953) auszutreten, auch im nackten Zustande Teilungen durchzuführen, oder die Flagellaten können unter Geißelverlust zu Amöben (*Synura:* PASCHER 1912) werden. Die beiden von PASCHER und CONRAD beschriebenen amöboiden Zustände können später unter Abrundung und Gallerte-Sekretion in Palmellen übergehen und in dieser Form ebenfalls Zellteilungen absolvieren; zuletzt treten sie aus der Gallerte als Schwärmer aus. Über die Bedingungen ist in allen diesen Fällen Näheres nicht bekannt.

Teilungen im unbeweglichen Zustand werden von zahlreichen monadoiden Arten angegeben. Bei *Ochromonas sociata* und *Chromulina hokeana* (PASCHER 1910) kommt es zur Gallertausscheidung und zur Bildung vorübergehend vier- bis achtzelliger Verbände, die hier noch beweglich bleiben und als Prototypen einer Koloniebildung von Monaden angesehen werden. Hier sei eingeschaltet, daß auch echt koloniale Formen unter gewissen Bedingungen (wie Einbringen vom Standort in das Laboratorium, aber auch in der Natur) vergallerten können. *Synura sphagnicola* (SKUJA 1956) stößt dabei die Eigenhüllen der Zellen ab, deren Schuppen an die Peripherie der Koloniegallerte verlagert werden. Die Geißeln ragen aus dieser Hülle heraus. Es resultiert, was EHRENBERG (A 1838) als *Syncrypta volvox* beschrieb, eine Gattung, die bis dahin anerkannt wurde. Bei anderen, insbesondere Arten von *Chromulina*, *Monochrysis*, *Ochromonas* sowie den oben erwähnten Beispielen aus den Genera *Mallomonas* und *Synura* können sich die Zellen, meist unter Geißelverlust, abrunden und Gallerte ausscheiden, in der Teilungen erfolgen (s. z.B. PASCHER 1912, DOFLEIN 1923, SKUJA 1948). *Monochrysis aphanaster* und *M. vesiculifera* teilen sich überhaupt nur noch in dieser Form, wobei (2)—4—(8—16) Töchter entstehen und danach ausschwärmen. Hiermit sind die Übergänge zu den capsalen Formen gegeben, deren Hauptzustand der unbewegliche ist (SKUJA 1948).

Kolonievermehrung bei Flagellaten.

Zunächst soll noch die Frage der Kolonievermehrung cönobialer Monaden berührt werden. Formen, welche die Verbindung zwischen den Zellen durch nicht lebendes Material herstellen, wie *Dinobryon*, wo Gehäuse, und *Uroglena*, bei der röhrenförmige Gallertstiele das Zusammenhalten besorgen, dürften eine Neugründung von Kolonien durch „Zoosporen", d. h. freiwerdende Einzelmonaden vollziehen. Bei solchen Arten aber, deren Zellen selbst *(Cyclonexis)* oder ihre basalen Enden (*Synura* und ähnlich, Abb. 28a) die Verbindung herstellen, kommt es daneben zu Teilungen der ganzen Kolonie. Der Ring von *Cyclonexis* — die Zellen liegen hier einreihig Flanke an Flanke (JANE 1940, HOVASSE 1949) — nimmt beim Heranwachsen der Kolonie ovale Form an, die langen Seiten nähern und berühren sich, nach HOVASSE unter sattelförmiger Durchbiegung der Kolonie, und es findet eine „Chiasmatypie", Bruch und Wiedervereinigung statt; zwei freie Tochterringe sind das Resultat. Es leuchtet ein, daß dieser Prozeß empfindlich ist und zu Fehlentwicklungen neigt. Bei *Synura* (STEIN 1878, SCHILLER 1926b), *Syncrypta* (JANE 1940, dort als *Synochromonas*), *Synochromonas perlata* (SKUJA 1956, bei BOURELLY als *Uroglena perlata*) sondern sich die vereinigten Hinterenden der Zellen auf ungeklärte Weise in zwei Zentren, womit die Kolonieteilung praktisch bereits vollzogen ist. Schwer zu verstehen ist dieser Modus bei der zuletzt genannten Art (Abb. 28b), weil die Zellen bei ihr durch kurze dichotom geteilte Gallertstiele zusammengehalten sind. Die Flagellaten der phagotrophen und farblosen *Anthophysa vegetans* sind ähnlich wie bei

Synura vereinigt. Ausreichende Fe- *und* Mn-Versorgung bei schwach oxidierenden Bedingungen vorausgesetzt, werden Einzelmonaden und die aus ihnen durch Teilung entstehenden Kolonien sessil und scheiden einen gemeinsamen aus Eisen-III- und höheren Mangen-Oxiden bestehenden Stiel aus. Derartige Kolonien pflegen sich mit nachfolgender jedesmaliger Gabelung des Stielchens als solche zu teilen. Neugründung gestielter Kolonien erfolgt von einzelnen sich anheftenden Schwärmzellen aus, die den Kolonieverband verlassen (PRINGSHEIM 1946, A 1963).

a b

Abb. 28a u. b. a Koloniezusammenhalt durch Plasmafortsätze der Einzelzellen bei *Synura*. (Nach SCHILLER 1926a.) b Kolonieteilung von *Synochromonas perlata*. (Nach SKUJA 1956.)

Capsale, coccale und trichale Formen.

BOURELLY (1957) beschreibt das Verhalten von *Chromulina woroniniana* in Reinkultur, welche in dem Maße, wie die Kulturlösung sich erschöpft, in den unbeweglichen Zustand übergeht, wobei die Zellen sich abrunden und unter Schleimausscheidung Lager bilden. Je zwei Tochterzellen bleiben durch die Gallerthülle der Mutterzelle vereinigt und behäuten sich in ihr durch eigene Polysaccharidausscheidung. Eine modifizierte Art dieses palmelloiden Zustandes sind die neustontischen Formen gewisser anderer *Chromulina* Arten, die mit einem Stielchen aus unbekanntem Material (LUND 1942, dort Literatur), oder auch ohne ein solches (VISCHER 1943), dem Oberflächenhäutchen aufsitzen und sich in diesem Zustand auch zwei- (oder mehrfach) teilen können. Je nachdem, ob die unbewegliche Phase die Geißel behält (*C. rosanoffii*, *C. aerophila*, Form A) oder nicht (*C. aerophila*, Form B), tritt beim Submergieren Beweglichkeit sofort oder später ein.

Wird das unbewegliche Stadium vorherrschend, so kommt man in den Bereich der bei Euchrysomonaden ziemlich gut vertretenen *capsalen* Organisation. Am höchsten entwickelt ist diese bei den bis mehrere Dezimeter langen verzweigten und mit Scheitelzelle wachsenden Kolonien von *Hydrurus* (KLEBS A 1936). Wird die Hülle der Zelle membranartig fest, wobei die behäuteten Zellen ihrerseits in Schleim eingebettet sein können, gelangt man zu den *coccalen* Formen. Fadenartige Reihung derbhäutiger Zellen in gemeinsamen Hüllen würde schließlich die *trichale* Organisation ergeben. Teilung in den Richtungen der Ebene *plakale* Formen. Die derben Membranen der letzten drei Stufen bestehen aus einem Stück; abweichend hiervon findet sich bei *Phaeoschizochlamys* (LEMMERMANN 1898) auf dem coccalen und bei *Nematochrysopsis* (CHADEFAUD 1947) auf dem trichalen Niveau ein zweischaliger Bau der Membranen, was für diese Formen cystäre Herkunft nahelegt, ähnlich wie man sie etwa bei den Diatomeen oder der Heterokonte *Tribonema* annehmen wird. Am höchsten unter den Trichalen steht *Chrysonephos lewisii* (TAYLOR 1951, 1952), die — pseudodichotom verzweigt — Längen bis zu 30 mm bei Fadenbreiten von 38 μ erreicht. Es ist allerdings nicht sichergestellt, ob die Form nicht zu den *Prymnesium*-Verwandten gehört.

Die normale Zellvermehrung erfolgt durch Zweiteilung oder succedane Vier- bis Achtfachteilung unter Neubehäutung der Tochterprotoplasten. Doch sind Mehrfachteilungen nicht gerade häufig. Nach dem Nomenklaturvorschlag von LUND (1962b) müssen wir hier überall von autosporiner Teilung sprechen. Bei den fadenbildenden Arten würden sich die „Sporen" immer in Zweizahl bilden, handelt es sich um Längsteilungen, wie bei *Phaeothamnion* und *Nematochrysis*, so schieben sie sich nachträglich voreinander. Nur bei *Nematochrysopsis* mit dem „H-Stück"-Bau ihrer Membran und möglicherweise bei *Chrysonephros*, deren Wandbildung noch nicht eingehend untersucht wurde, könnte man von „septater" Teilung (LUND) reden, bei welcher nicht vollständige Neubehäutung der Tochterzellen, sondern Neubildung zunächst nur eines gemeinsamen Teiles, der Membran des Septums, erfolgt.

Fortpflanzungsmöglichkeiten neben diesen das Wachstum vom Lager oder Thallus besorgenden Teilungen bestehen in einer Zoosporen- oder Autosporenbildung, in Isolierung der Zellen durch Verquellung der Membran und in endogener Kieselcystenentstehung, sowie in bestimmten spezialisierten, im nächsten Abschnitt behandelten Dauerorganen. Eine Gruppe trichal-placaler *(Phaeoplacales)* und coccaler *(Stichoglocales)* Gattungen sondert BOURELLY ab, da bei diesen bisher keine beweglichen Stadien gefunden wurden, was aber eines Tages geschehen könnte. Zoosporen entstehen gewöhnlich in vegetativen Zellen; nur bei zwei *Apistenoma*-Arten[1] („trichal"-verzweigt), *A. commutatum* (PASCHER 1925) und *A. submarinum* (MAGNE 1954), wurden vergrößerte und dadurch etwas spezialisierte Zoosporangien mit zahlreichen (bis 64 bzw. bis 16) Zoosporen beobachtet.

Bedingungen für die vegetative Fortpflanzung.

Schwärmsporen der atmophytischen Capsale *Ruttnera* (GEITLER 1943) entstehen beim Einbringen der Lager in Wasser innerhalb von einigen Stunden. Bei der marinen *Sarcinochrysis* (GEITLER 1930) kann Schwärmerbildung bereits durch geringfügige (um 5%) Hypotonisierung des Mediums (Schreiber-Seewasser) hervorgerufen werden. Eingehender wurde nur die Chrysocapsale *Hydrurus* durch KLEBS (A 1896) untersucht.

Diese Art kalter Gebirgsbäche wächst aktiv zwischen 2^0 und etwa 12^0, zeitweilig werden auch etwas höhere Temperaturen vertragen (s. auch BURSA 1934). Zoosporenbildung wird ausgelöst zwischen 1^0 und 16^0 durch einen Wechsel vom strömenden zum ruhenden Medium. Ein Temperatursprung ist dabei nicht erforderlich. Welches die wirksame Komponente der Situation „strömendes Medium" ist, bleibt fraglich; KLEBS selbst denkt an den höheren Sauerstoffgehalt. Doch kann es durchaus etwas anderes sein, da auch das Ionenmilieu in der Nähe der Pflanze, insbesondere im Hinblick auf die Nährstoffversorgung, stark von der Bewegung des Wassers abhängig sein muß. Für die letzte Deutung würde sprechen, daß der Effekt in „0,05%" Nährlösung (Knop?) gegenüber Wasser stark gemindert ist. Die ökologisch und morphologisch ähnliche *Celoniella* wird durch den gleichen Wechsel (PASCHER 1929) außerordentlich rasch zum Schwärmen gebracht.

[1] Apistonemen oder ähnlich primitiv trichale Formen gibt es vermutlich auch unter den Euchrysomonaden, nicht nur unter den Verwandten von *Prymnesium*.

2. Dauerstadien, Cysten und Sexualität.

Die Gruppe der Euchrysomonaden läßt sich durch ihre stark verkieselten Ampullencysten endoplasmatischen Ursprunges kennzeichnen.

Pseudocysten, Resistenz.

Es sei aber zunächst auf einige nicht diesem Typus folgende Dauerorgane hingewiesen, die jene bei ihren Besitzern ersetzen. Pascher (1931 b) schilderte *Pseudocysten* bei den capsalen, Gallertlager bildenden Arten *Chalcopyxis tetraspora* und *Chrysotilos ferrea*, welche Einzelzellen — bei der ersten Form auch kleinere Zellgruppen — mit exogenen, vererzten und dadurch gefärbten zweischaligen Kapseln umhüllen. Beide Teile der Membran sind hier im Gegensatz zu den echten Cysten von gleichem Material und annähernd gleich groß (Abb. 29a). Bei der Keimung öffnen sich diese Schalen und entlassen bereits wieder kleine Gallertkolonien (Abbildung 29b). Die Plasmodien von *Myxochysis* (Pascher 1916) umgeben sich in der wärmeren Jahreszeit mit einer ebenfalls Eisen und Manganoxide enthaltenden Kruste und sitzen dann in flachen Kissen dem Substrat auf. Der Inhalt zerfällt darauf in einkernige Portionen, die später als Amöben oder als Flagellaten austreten oder schließlich ihrerseits innerhalb der Plasmodienhülle aus einem Stück bestehende, jener ähnlich beschaffene Eigenschalen bilden; die braunen Sporen werden frei und keimen mit je einem Flagellaten. Das Ganze erinnert an die Plasmodiocarpien der Myxomyceten. Schließlich sei noch auf die Verbreitungseinrichtungen der capsalen atmophytischen Gattung *Gloeochrysis* (Pascher 1931 c) hingewiesen. In Trockenperioden erhalten die Zellen der auf feuchten Moorböden lebenden krustigen Lager sehr starkwandige feste Gallertmembranen, die sich nun — offenbar infolge hygroskopischer Bewegungen des ganzen Lagers oder innerhalb desselben — voneinander lösen (Abb. 29c, d). Das Freiwerden wird in diesem Falle durch die derbe eisen- und salzinkrustierte Außenschicht des Lagers erleichtert, welche beim Eintrocknen zerreißt, sich gebietsweise schüsselförmig wölbt und die Sporen dabei exponiert. In anderen Fällen bekommen nicht die ganzen Lager diese gefärbte Membran, sondern Zellpakete in ihnen; diese Schichten mit ihrer nicht klebrigen Oberfläche wirken dann als Trennungseinrichtung. Durch beide Methoden sind Mechanismen für eine Windverbreitung geschaffen, die weiterhin die plasmatische Eigenschaft der Trockenresistenz zur Voraussetzung hat. Diese erklärt wohl auch, daß morphologisch differenziertere Dauerzustände hier und bei der ebenfalls atmophytischen *Ruttnera* (Geitler 1943), wie bei übrigens zahlreichen luftlebenden Algen (s. z. B. Brabez 1942), fehlen.

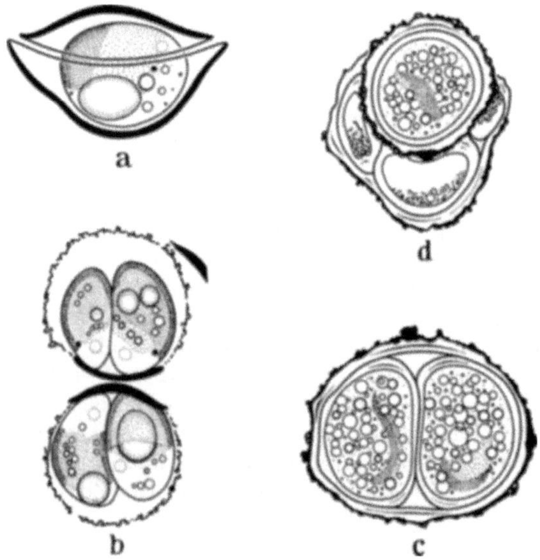

Abb. 29a—d. a Pseudocyste von *Chrysotila ferrea* und b deren Keimung. (Nach Pascher 1931 b.) c, d Verbreitungskörper von *Geochrysis turfosa*. (Nach Pascher 1931 c.)

Cysten.

Die in der Form und Ornamentierung mannigfaltig abgewandelten, grundsätzlich aber gleichartig gebauten Cysten entstehen endogen, wie SCHERFFEL (1911), DOFLEIN (1922, 1923), CONRAD (1927a, b), SKUJA (1956), BOURRELLY (1957) besonders eingehend schilderten. Dazu rundet sich die meist an Fett

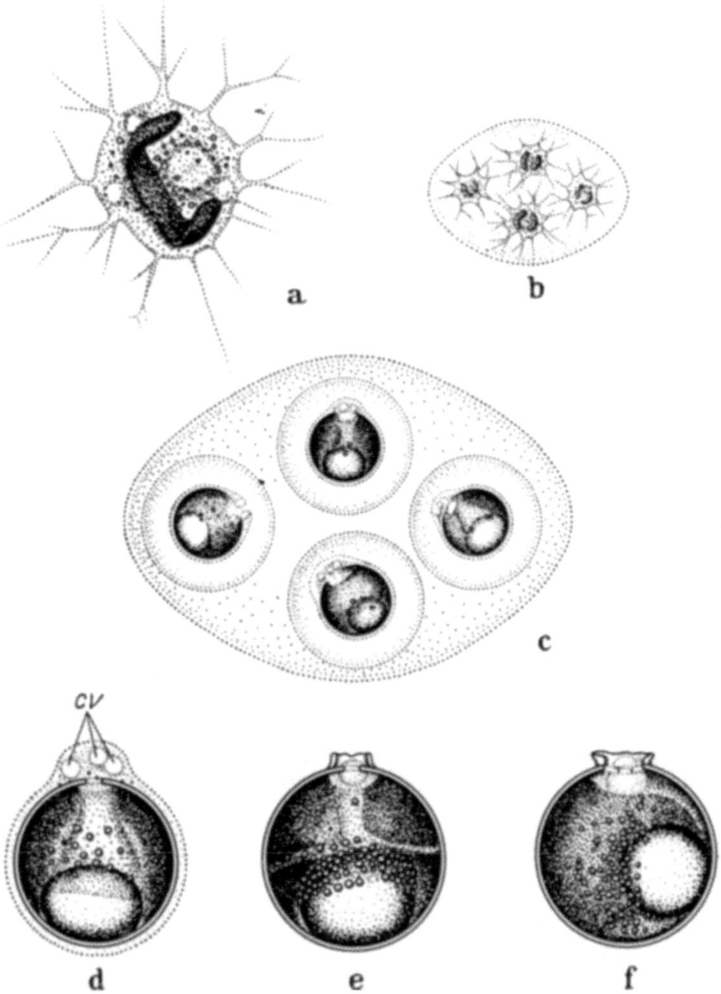

Abb. 30a—f. Cystenbildung von *Rhizochrysis tetragena*. a Amöbe, b cystogenes Gallertlager, c die Cysten in Bildung begriffen, d Cystenbildung (*cv* pulsierende Vacuolen), e, f reife Cysten. (Nach SKUJA 1956.)

und Leukosin reiche Mutterzelle unter Geißelverlust ab, sezerniert aus den Schleimkörpern des Periplasten eine Schleimhülle und unterlegt diese danach manchmal durch eine feste Cellulosemembran (z. B. *Chrysamoeba*, SCHERFFEL (1911); *Amphichrysis*, KORSHIKOV (1929); LUND (1962a); *Dinobryon*, *Pseudokephyrion*, s. z. B. BOURRELLY (1957)]. Darauf wird im Innern des Zellkörpers die kugelrunde oder ovale Kieselschale abgeschieden, die Plastiden, Kern, Stigma und meist die Reservestoffe umhüllt. In der Nähe des Vacuolenapparates, der im Periplasma weiterpulsiert, bleibt ein Porus frei. Dieser verstärkt sich häufig durch einen mehr oder weniger kompliziert gebauten außen aufgesetzten Hals (Abb. 30).

Nach Doflein soll bei *Ochromonas*-Arten der Halsteil und damit die Cyste selbst ganz geschlossen sein, später dann die oberste Kappe des Halskegels abgesprengt und damit die Öffnung hergestellt werden, was aber von anderen Autoren anscheinend nicht mehr gesehen wurde. Das außerhalb der Hülle verbliebene Periplasma stellt danach die Ornamentierung der Cystenaußenwand in Form von Warzen, Stacheln oder Leisten her (Abb. 31); die Mündung der Ampulle wird durch einen Stopfen aus Membransubstanz verschlossen, welcher mehr oder weniger stark, meist wohl schwächer als die Cyste selbst, verkieselt ist. Der Stopfen wird bei der Keimung in einigen Fällen aufgelöst, in anderen herausgedrückt.

Sexualität. Die Diskussion der entwicklungsgeschichtlichen Valenz der Cysten erfordert ein Eingehen auf den Sexualvorgang. Kopulationen wurden zuerst von Skuja (1950) an dem in kelchartigen Gehäusen einzeln lebenden *Dinobryon borgei*

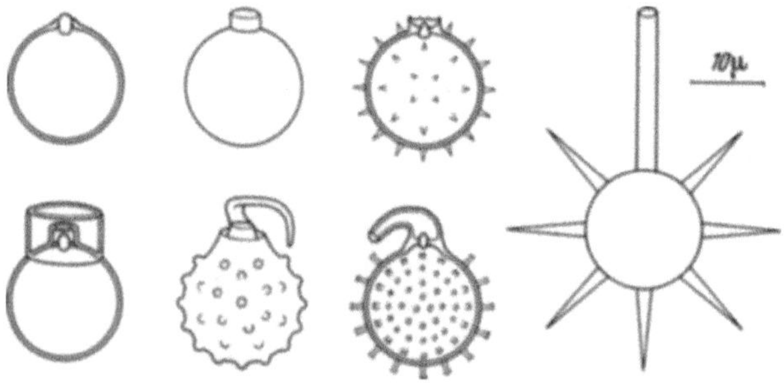

Abb. 31. Cystenformen von Arten der Gattung *Uroglena*. (Nach Bourrelly 1957.)

genauer verfolgt. Die Geißeln zweier Flagellaten umwinden sich, bringen dadurch die Gehäuse Mündung an Mündung, wonach Austreten und Verschmelzung der Plasmen im Raum zwischen den Gehäusen, gefolgt von einer Cystenbildung, stattfindet. Die Cyste, der die Gehäuse der Gameten noch anhängen, ist hier also Zygote. In der Folgezeit wurde nach einer Aufstellung von Lund (1960b, s. auch Kristiansen 1963) sexuelle Cysten bei weiteren elf gehäusebewohnenden Arten gesehen; sie sind, weil durch die beiden Gehäuse der Gameten markiert, leicht zu erkennen. Skuja wies auch die beiden Kerne in der Zygote von *Dinobryon borgei* nach, während Mack (1953a) in älteren Zygoten von *Kephyrion* bereits nurmehr einen Kern sah. Parke (1949) gibt für die marine *Chromulina pleiades* zweikernige Palmellen, möglicherweise sexueller Entstehung, an.

Zweikernige Cysten sind im übrigen häufiger beobachtet worden: so von Prowazek (1903) bei *Heterochromonas vivipara*, von Dangeard (1910) bei *Anthophysa*, von Korshikov (1929) bei *Synochromonas pallida* und *Uroglena volvox* (bei letzterer ebenso Geissbühler 1933 und Krieger 1930, s. dazu auch Geitler 1935), von Geitler (1935) bei *Dinobryon cylindricum* und *D. divergens*; letzteres wurde besonders eingehend untersucht. Hovasse (1943) sah sie bei *Heterochromonas beauchampi*, Mack (1951) bei *D. sertularia* var. *vindobonensis*, *D. divergens*, *Uroglena volvox*, *U. notabilis* sowie *Synura uvella* und Mack (1954) bei *Uroglena botrys* und *Amphichrysis compressa*. In diesem letzten Falle wurden 7 Wochen nach ihrer Bildung die ersten und 3 Wochen später *nur* noch einkernige Cysten beobachtet, so daß hier wohl eine Verschmelzung der Geschlechtskerne erfolgt war.

In den meisten der im vorigen Absatz genannten Fälle wurde von den Autoren angenommen, daß die Zweikernigkeit durch eine Mitose in der Präcyste zustande gekommen sei, woraus dann eine Autogamie *(Amphichrysis)* resultieren könnte. Prowazek und Geissbühler, welche für ihre Objekte an isogame

Allogamie dachten, hatten diese Annahme nicht zweifelsfrei belegt. Doch sind wohl bei kaum einem Objekt allogame, insbesondere extreme anisogame Prozesse völlig auszuschließen. Suspekt in dieser letzten Richtung erscheint insbesondere *Anthophysa* (DANGEARD 1910). Andererseits bedeutet, worauf GEITLER (1935, s. auch MACK 1951) als interessante Möglichkeit hinweist, eine Zweikernigkeit der jungen Cyste nicht unbedingt und in jedem Falle, daß später eine Kernfusion folgen muß; sie könnte auch Zwischenzustand einer in der Frühentwicklung der Cyste eingetretenen Mitose sein, deren Cytokinese dann bis zur Cystenkeimung hinausgeschoben wäre. Dazu würde das Verhalten von *Mallomonas* passen, deren Cysten zunächst einkernig sind (KRIEGER 1930, CONRAD 1927b, CARTER 1937b), in noch unfertigem Zustand (CONRAD: *M. pyriformis* und *M. mirabilis*) oder etwas später (CARTER: *M. anglica*) aber durch Mitose (CONRAD) zweikernig, schließlich *(M. anglica)* auch mehr als zweikernig sein können. Was daraus weiter wird, ist nicht bekannt. Es könnte sich also um eine Verschiebung der von GEITLER angenommenen präcystären Mitose in die Cyste handeln und dann im Sinne der von ihm diskutierten (aber nicht für sehr wahrscheinlich gehaltenen) Möglichkeit um Vorwegnahme von Teilungen bei der Keimung oder auch um das Vorspiel zu einer späteren autogamen Kernfusion. Jedenfalls läßt der Fall *Mallomonas* Vorsicht bei der Beurteilung zweikerniger Cysten angezeigt erscheinen. Einen Nachweis mit hohem Wahrscheinlichkeitswert für eine Karyogamie haben wir bisher nur bei der *Amphichrysis* MACKs.

Parthenogenetische Cysten. *Dinobryon borgei* (SKUJA 1950) und *Stenocalyx inconstans* (SKUJA 1956) bilden neben sexuellen Cysten auch solche, die vor der Mündung ungepaarter Gehäuse liegen und als parthenogenetischer Entstehung angesehen werden. Wichtig erscheint, daß sie durch geringere Größe von den sexuellen Cysten verschieden sind. Im Gegensatz dazu findet der gleiche Autor bei *Kephyrion boreale* (wo bisher Sexualität nicht bekannt ist) Cysten innerhalb des Gehäuses, die dann als vegetativ angesprochen werden.

Vegetative Cystenbildung. Schließlich sind Cysten häufig vom Beginn ihrer Bildung an einkernig. Wir nennen, abgesehen von dem erwähnten *Mallomonas*, die von DOFLEIN (1922, 1923) untersuchten Arten von *Ochromonas* und *Chromulina*, *Mallomonas tonsurata* (KRIEGER 1930), *Chromulina maxima*, in der Entstehung von MACK (1951) genau verfolgt, *Hydrurus* (MACK 1951, 1953b) und das ihm in der Cystenform ähnliche *Phaeodermatium* (GEITLER 1927). Vorkommen sexueller Cysten neben asexuellen scheinen, wenn man von den Parthenogenese-Fällen (s. oben) absieht, nicht bekannt zu sein.

Bedeutung und Bildungsbedingungen der Cysten. Die Cyste jeder Entstehungsform enthält die sie bildende Art während der für sie ungünstigen Zeit am Standort und mag weiter, insbesondere bei lagerbildenden Pflanzen, auch Mittel für eine passive Verbreitung sein. Ihre Bildung schließt die Vegetationsperiode der betreffenden Art sehr häufig ab und ist damit von deren ökologischen Ansprüchen abhängig. Bei den kryophilen Chrysomonaden dürfte häufig die Temperaturerhöhung im Frühjahr auslösend wirken, so bei *Hydrurus* (KLEBS A 1896); wärmeliebende Arten wie *Uroglena europaea* (SCHILLER 1926b) andererseits reagieren möglicherweise auf die Abkühlung des Wassers (unter 4^0) im Herbst. In weiteren Fällen wirken vielleicht auch durch Nährstoffmangel hervorgerufene Photosyntheseüberschüsse, wofür die Anreicherung von Fett und Leukosin bei der Cystenbildung sprechen könnte. Beide Bedingungen mögen selbstverständlich verknüpft, der Mangel etwa durch temperaturbedingte Impassen im Stoffwechsel verursacht sein. Experimentelles Material liegt kaum vor. Die Cysten der farblosen und phagotrophen *Anthophysa* entstehen mit klonspezifisch unterschiedlicher Leichtigkeit in alten Kulturen (PRINGSHEIM 1946). Jene von *Ochromonas granularis* (DOFLEIN 1922) überstehen eine Austrocknung und scheinen durch diese zur Keimung stimuliert zu werden.

Cystenkeimung wurde selten beobachtet: SCHERFFEL (1911) bei *Chrysamoeba radians:* ein oder zwei Amöben als Keimlinge; KRIEGER (1930) bei *Dinobryon:* ein Schwärmer; statt monadoider können aus *Dinobryon*-Cysten nach PASCHER (1943) auch amöboide kleine Protoplasten austreten; DOFLEIN (1922) bei *Chromulina friburgensis:* ein Schwärmer; ebenso SKUJA (1948, 1956) bei *Chrysococcus*-Arten, wo der Keimschwärmer durch den antapikalen Porus seines Gehäuses mit der anomalerweise ebenfalls bipolar-porigen Cyste im Zusammenhang bleibt; allerdings wurde hier wohl nicht die Keimung selbst, sondern nur ihr Ergebnis gesehen. Bei *Chromulina pascheri* (HOFFENDER 1913) sollen sich zuletzt zwei oder vier Protoplasten in der Cyste befinden. Bei *Phaeothamnium borzianum* schlüpfen zwei oder vier Schwärmer (PASCHER 1925). Alle diese Angaben beruhen auf Gelegenheitsbeobachtungen. Die experimentelle Keimungsauslösung bei den Cysten von *Anthophysa*, welche leicht im Laboratorium gewonnen werden können, gelang ebensowenig (DANGEARD 1910, PRINGSHEIM 1946) wie diejenige von Cysten phototropher Arten natürlicher Herkunft (MACK 1951).

b) Haptophyceae.

Diese zunächst auf das isokonte *Prymnesium* (MASSART 1920, CONRAD 1941) gegründete Gruppe hat in den letzten Jahren ein besonderes Gewicht dadurch bekommen, daß — außer einer größeren Zahl von lichtmikroskopisch meist nackten, elektronenoptisch aber mit nicht-mineralischen Schuppen bedeckten, vorwiegend marinen Monaden — die *Coccolithineae* (v. STOSCH 1955, 1958, PARKE 1961, MANTON und LEEDALE 1963a) sich als hierhergehörig erwiesen haben. Diese hoch organisierten mit verkalkten Plättchen (Coccolithen) sehr mannigfaltiger Gestalt (Abb. 34—36) armierten Flagellaten umfassen selbst wieder coccale (bzw. capsale) und trichale Zustände. Daneben enthält die Gruppe *capsale* (*Phaeocystis*, KORNMANN 1955, PARKE 1961) und „*trichale*“ Formen, denn es gibt anscheinend selbständige, nicht mit Coccolithineen in Zusammenhang stehende *Apistonema*-Herkünfte mit Haptophyceen-Schwärmern (v. STOSCH, unveröff.).

Wegen der größeren Zuverlässigkeit der Kulturverfahren für Seewasseralgen und des gestiegenen Interesses am marinen Plankton wissen wir über die Entwicklungsgeschichte dieser Gruppe heute mehr als über die häufiger studierten Euchrysophyceen.

1. Bewegliche und unbewegliche Phase; Generationswechsel.

Bei den besser, d. h. in Kulturen untersuchten Formen dieser Gruppe ist nach PARKE (1961) eine bewegliche von einer unbeweglichen Phase zu unterscheiden; die Einteilung verwendet zwar ein recht äußerliches Merkmal, erscheint aber wegen des geringen Wissens über den Kernphasenwechsel im Augenblick die einzig mögliche. Entwicklungsgeschichtlich sind diese Phasen sicher verschiedenwertig. Wenn wir SCHWARZ (1932) folgen, dient die bewegliche Form der haplontischen *Ochrosphaera* nur als kurzlebige Zoospore oder als Gamet der im übrigen haploiden sessilen Phase. Bei *Coccolithus pelagicus* würde, wenn überhaupt Kernphasenwechsel existiert, sehr wahrscheinlich die unbewegliche *Coccolithus*-Phase die diploide, die bewegliche und mit ganz anders gestalteten Kalkschuppen versehene *Crystallolithus*-Phase (Abb. 34a, b) dagegen die haploide sein (PARKE und ADAMS 1960). Bei *Syracosphaera carterae*, bei der cytologische

Untersuchungen vorliegen (RAYNS 1962, v. STOSCH et al., unveröff.), ist umgekehrt die monadoide coccolithophore Phase die haploide und die algenartige *Apistonema*-Stufe die diploide (Abb. 32). Bei den übrigen Formen wissen wir über den Generationswechsel zu wenig, um Aussagen machen zu können; dieses gilt auch für die Masse der pelagischen Coccolithophoriden. Unter diesen ist nur für *Coccolithus fragilis* aus Naturbeobachtungen ein komplexer Lebenscyclus angegeben worden (BERNARD 1948); wir kommen auf ihn zurück (s. auch Abb. 36).

Eine Anzahl dieser Formen besitzt grundsätzlich *unbegrenzte* vegetative Vermehrung in beiden Zuständen, so *Phaeocystis*, *Coccolithus pelagicus* und die Coccolithophoriden mit *Apistonema*-Phase. Solche Phasen wären als selbständige

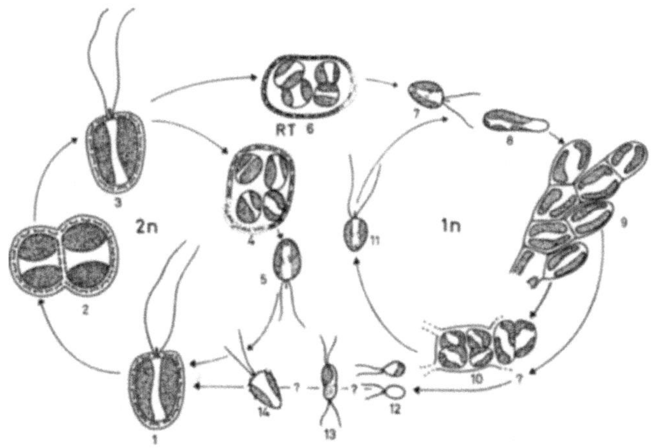

Abb. 32. Lebenscyclus von *Syracosphaera carterae*. *1*, *3* Schwärmer der beweglichen *Syracosphaera*-Phase und *2* dessen Teilung, *4* vegetative Zelle wird zum Zoosporangium und entläßt diploide Zoosporen *5* oder zum Meiosporangium *6*. Die Meiospore *7* keimt (*8*) zur haploiden *Apistonema*phase *9*. Letztere liefert unter Zerfallen des Thallus *10*, entweder pro Zelle vier haploide Zoosporen oder auf nicht genauer bekanntem Wege sehr kleine Gameten *12*, welche (*13*) vermutlich grundsätzlich isogam verschmelzen und deren vermutliche Zygote *14* dann erneut eine Coccolithenhülle sezerniert. (Original, v. STOSCH et al.)

Generationen zu bezeichnen; doch sind, soweit bekannt, die „nackten“, zu 2—4 in der Mutterzelle entstehenden Schwärmer von *Ochrosphaera* überhaupt nicht frei teilungsfähig und keimen direkt wieder zu Palmellen, oder sie können — dann in etwas größerer Zahl (bis 6) in der Mutterzelle gebildet — kopulieren und eine palmelloide Zygote liefern, welche nach Reduktionsteilung vier gewöhnliche Zoosporen entläßt (SCHWARZ 1932). Bei *Isochrysis*, *Dicrateria* (PARKE 1949) und *Chrysochromulina*-Arten (PARKE, MANTON und CLARKE 1955) dagegen ist die unbewegliche Phase anscheinend nur begrenzt vermehrungsfähig. Sie liefert, soweit die Beobachtungen gehen, je Zelle nur vier Autosporen, die später je einen Schwärmer entlassen. Die Mutterzelle der vier Autosporen entsteht bei *Isochrysis* und *Dicrateria* wahrscheinlich, bei *Chrysochromulina* möglicherweise in einem Sexualakt. Die vier Sporen wären dann Meiosporen.

Die unbewegliche Phase in Vermehrung und Fortpflanzung.

Die Zellen der unbeweglichen Phase sind bei *Phaeocystis* nackt und liegen in der Wand einer großen Gallertblase (zuletzt KORNMANN 1955), oder sie können mit mehr oder weniger fester Membranhülle versehen sein. Dieserart leben sie frei bei *Chrysochromulina* (s. z. B. PARKE, MANTON und CLARKE 1855) und zum Teil bei *Ochrosphaera* (MAGNE 1952) oder aber durch Gallerte oder Membranen der Mutterzellen zu vieren vereinigt bei *Dicrateria* und *Isochrysis* (PARKE 1949) sowie *Ochrosphaera* (SCHWARZ 1932). Die unbeweglichen Zellen des *Coccolithus pelagicus* sind frei, rundlich und mit verkalkten Placolithen (Abb. 34b) gepanzert (PARKE und ADAMS 1960). Schließlich kann die unbewegliche Form einen primitiven trichalen Typus (Abb. 32, *9*) repräsentieren: *Pleurochrysis scherffelii* (PRINGSHEIM 1955, PARKE 1961), *Syracosphaera carterae* (mehrere Lokalrassen, v. STOSCH 1955, PARKE 1961, VALKANOV 1962). Die sessil gewordenen nackten Meiozoosporen aus der beweglichen Phase scheiden derbe geschichtete Membranen aus; diese Zellen strecken sich und entwickeln Polarität, sie wachsen durch Zweiteilung (Furchung und Neubehäutung) unter Verzweigung zu gedrungenen bäumchenförmigen Thalli heran, die bei *Syracosphaera carterae* Stamm List über 1 mm Durchmesser erreichen können. Jedoch herrscht in diesen *Apistonema*-

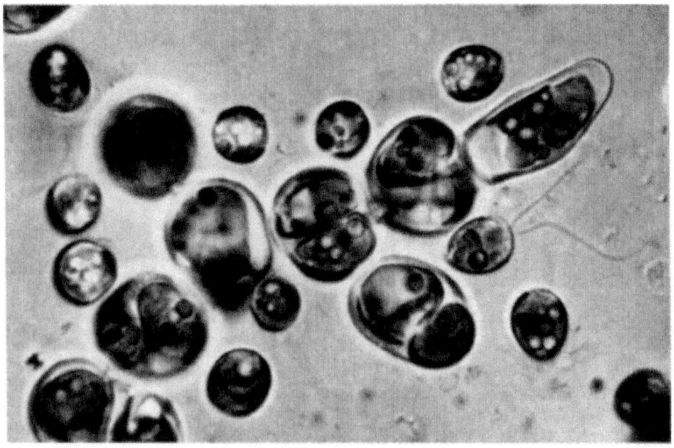

Abb. 33. *Syrocasphaera carterae. Apistonema*-Thallus zerfällt in Zoosporen. Unter der Thallusspitze rechts isokonte Zoospore mit Haptonema zwischen den Geißeln. (Original, Photo R. SCHÄFER.)

Phasen abhängig von den Kulturbedingungen außerordentliche Polymorphie; nach PARKE (1961) können Formen erhalten werden, die marinen Gliedern der früher beschriebenen Gattungen (s. z.B. ANAND 1937a, b), *Apistonema, Chrysonema, Thallochrysis* (s. auch VALKANOV), *Nematochrysis, Chrysosphaera* und *Chrysotila* ähnlich oder vielleicht mit ihnen identisch sind. Im letzten Falle bestehen sie aus Palmellen mit polar gestreckten, geschichteten Schleimhüllen, im vorletzten aus coccalen bis capsalen runden Einzelzellen. Diese benthischen Zustände besiedeln feste oder bewegliche Substrate im Küstengebiet (s. z. B. ANAND 1937, MAGNE 1954, 1957).

Auch bei den übrigen Formen geschieht die *Vermehrung* durch Zweiteilung. Die „Palmellen" von *Ochrosphaera*, bei denen häufig zwei Cytokinesen einander unmittelbar folgen, können dabei nur von häutigen Membranen umgeben sein (MAGNE 1952) oder auch Coccolithen tragen (SCHWARZ 1932). *Coccolithus pelagicus* schnürt in zwei kurz aufeinanderfolgenden Teilungsschritten die Zellen samt Coccolithen-tragender Hülle durch.

Vegetative Fortpflanzung innerhalb der unbeweglichen Phase ist für *Ochrosphaera* in der Form von nackten(?) Zoosporen bekannt, die einzeln in der Mutterzelle entstehen (SCHUSSNIG 1930). Die Apistonemen von *Syracosphaera carterae* Stamm List entlassen nach entsprechender Behandlung epidemieartig Zoosporen aus nahezu jeder Zelle (Abb. 32, *10*; Abb. 33). Dabei teilt sich der Inhalt des Zoosporangiums succedan in vier Zellen, die durch Membranverquellung frei werden und später zu jungen Apistonemen keimen (v. STOSCH et al., unveröff.).

Die bewegliche Phase, ihre Vermehrung und Fortpflanzung.

Die monadoide Phase ist bei *Isochrysis*, *Dicrateria* und *Diacronema* nackt, besitzt bei *Prymnesium*, *Phaeocystis*, *Chrysochromulina* und vermutlich bei *Ochrosphaera* aus zarten Schuppen organischer Natur zusammengesetzte und lichtmikroskopisch häufig nicht erkennbare Hüllen (PARKE 1961, SCHUSSNIG 1930, PRAUSER 1958, sowie die Arbeiten von PARKE et al. und MANTON et al.) oder schließlich bei den Coccolithophoriden mit Coccolithen durchsetzte Umhüllungen. Der Bau der letzteren ist nur bei *Crystallolithus* (Abb. 34a), der beweglichen Phase von *Coccolithus pelagicus*, näher untersucht. Hier bestehen sie aus einer von mehreren Lagen Schuppen aus organischer Substanz unterlegten äußeren Haut; einige von diesen Schuppen, die Coccolithen, scheinen auf der Außenseite die charakteristischen Kalkkörper zu tragen; doch sind die Beziehungen zwischen Haut, Schuppe und Kalkkörper noch unklar (MANTON und LEEDALE 1963b). Die Coccolithen existieren in großer artspezifischer Formenfülle (s. z. B. DEFLANDRE 1952, BRAARUD 1954, HALLDAL und MARKALI 1955, GAARDER und MARKALI 1956, GAARDER 1962). Das Haptonema ist noch nicht überall nachgewiesen.

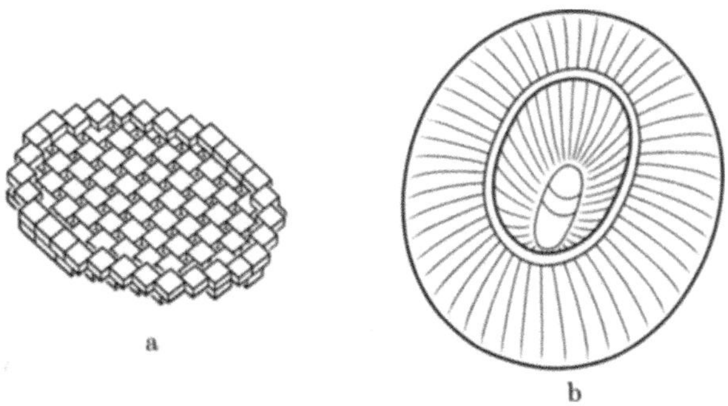

Abb. 34a u. b. Coccolithen von *Coccolithus pelagicus*; a der beweglichen *Crystallithus*-, b der unbeweglichen *Coccolithus*-Phase. Coccolith a ist viel kleiner als b, da a gegenüber b dreimal vergrößert. (a nach GAARDER und MARKALI 1956, b nach MURRAY und BLACKMAN, aus SCHILLER 1930, verändert.)

Die Vermehrung in der monadoiden Phase geschieht durch die übliche Flagellaten-Längsteilung unter gleichzeitiger Durchschnürung etwa vorhandener Hüllen, so bei den nicht verkalkten Formen. Nur bei einigen *Chrysochromulina*-Arten wie *Chr. minor* wurde als Ausnahme eine simultane Vierfachteilung mitsamt der Hülle beobachtet (PARKE, MANTON und CLARKE 1955), nachdem sich vorher der Geißelapparat vervierfacht und die Plastidenzahl verdoppelt hatte. Ähnliche Längsteilungen erfolgen auch bei gepanzerten Coccolithineen-Monaden, und zwar als Cytokinesen unter gleichzeitiger Durchschnürung der Hülle (*Hymenomonas roseola*, KLEBS 1893, SCHILLER 1926a, 1930) oder unter Einschnürung der Hülle und Bildung einer „Zellplatte" (nach Trennung der Plasmen) aus zwei Lagen im Plasmainnern vorgebildeter Coccolithen (Abb. 32, *2*), welche den Anschluß an die Ringfurche der Muttermembran finden; darauf „Spaltung" der neugebildeten „Membran" (CARTER 1937a, CHODAT und ROSSILO 1925, BRAARUD und FAGERLAND 1946, v. STOSCH u. Mitarb., unveröff.). Bei *Crystallolithus* endlich verlassen Zoosporen noch nackt die Mutterhülle und scheiden dann neue Crystallolithen aus, wobei manchmal der Abwurf der alten Schale erst nach Neubepanzerung der Töchter „gelingt". Ähnlich verhalten sich wohl *Pontosphaera huxlei* (neben Zweiteilung), *Syracosphaera pulchra* und wahrscheinlich die Arten mit sehr fester Kalkhülle (SCHILLER 1930).

Eine ähnliche Befreiung von vier nackten Schwärmern aus der Coccolithenhülle kann *Syracosphaera carterae* Stamm List als vegetative Fortpflanzung unter Beibehaltung der Phase neben der geschilderten Vermehrung durch Zweiteilung durchführen (Abb. 32, *4*). Diese Zoosporangien können bisher morphologisch nicht sicher von den ganz ähnlichen Meiosporangien (Abb. 32, *6*) unterschieden werden, liefern jedoch im Isolationsversuch im Unterschied zu diesen wieder Panzerflagellaten (v. Stosch et al., unveröff.). Bei anderen Coccolithineen wie *Syracosphaera mediterranea* und *Calyptosphaera oblonga* wurden (Schiller 1930) Stadien mit zahlreichen nackten Schwärmern gesehen, ihre Dignität im Formwechsel ist nicht bekannt (s. auch *Coccolithus fragilis*, weiter unten).

Die Übergänge zwischen den Phasen.

Die Übergänge beweglich zu unbeweglich oder umgekehrt vollziehen sich häufig unter Vermittlung besonderer Fortpflanzungszellen. Bei *Chrysochromulina* gehen der behäuteten Form amöboide Stadien voraus (Parke, Manton und Clarke 1955). *Crystallolithus* verliert Hülle und Geißeln, die nackten Palmellen wachsen (nach einem Sexualprozeß ?) heran und scheiden die Placolithen der *Coccolithus*-Form aus; auf der anderen Seite werden beschalte *Crystallolithus*-Monaden zu vieren (RT ?) in der Coccolithus-Mutterzelle gebildet (Parke und Adams 1960). *Syracosphaera carterae*-Flagellaten (Abb. 32) produzieren nach Verquellung der Hülle unter Meiosis vier nackte Schwärmer (v. Stosch 1955, Valkanov 1962, v. Stosch et al., unveröff.), welche austreten und als Apistonemen keimen. Diese letzteren können, wie eben geschildert, massenhaft Zoosporen (Abb. 33) freisetzen. Unter deren Keimlingen treten neben reichlich Apistonemen meist wenig, unter besonderen Umständen aber reichlich *Syracosphaera*-Monaden auf. Da diese, wie S. 647 angegeben, diploid, die Algenphase aber haploid ist, muß dabei eine Verdoppelung der Chromosomenzahl — man wird erwarten durch eine Gamie — eintreten. Unter den Zoosporen fanden sich nur Kleinschwärmer nicht genau bekannter Entstehung mit einem Plastiden oder auch ohne Plastiden, und zwar reichlich nur in den Kulturen, welche später viele Monaden lieferten. Man kann vermuten, daß es sich um Gameten handelt. Es wurden auch mehrfach viergeißelige Stadien gefunden, die wie Fusionspaare der Kleinschwärmer aussehen, sich aber unter der Beobachtung nicht weiterentwickelten (Abb. 32, *13*). Außerdem wurden mutmaßliche ältere Zygoten beobachtet, sie waren durch Größe und die Zweizahl der Plastiden den Zoosporen ähnlich, begannen aber während der Beobachtung Coccolithen auszuscheiden. Valkanov (1962) erhielt in seinen Kulturen der gleichen Art offenbar reichlicher Syracosphaeren aus der *Apistonema*-Phase, auch sah er möglicherweise Kopulationsstadien.

Phaeocystis als capsale Alge, die übrigens eine sehr merkwürdige Form von Kolonieteilung besitzt (Kornmann), liefert drei Arten von Zoosporen: Makro-, normale und Mikrozoosporen. Die ersten und die zweiten keimen zu neuen *Phaeocystis*-Kolonien, die Makrozoosporen unter Umständen nach Teilungen im beweglichen Zustand. Die Mikrozoosporen, von epidemieartig spontanem Auftreten, bilden eine selbständige Generation; wegen der geringen Zellgröße könnte man vermuten, daß sie haploid sind.

2. Cysten.

Das Bild wird kompliziert durch die Existenz von Cysten. Angaben über das Vorhandensein von Kieselcysten in dieser Gruppe (Carter 1937a, Conrad 1941, Parke 1949) bedürfen wohl der Überprüfung (Parke in literis). Die Coccolithophoriden liefern entweder homogen verkalkte Cysten in Ampullenform: *Pontosphaera steueri* (Kamptner 1937, 1941) oder solche ohne Öffnung und mit einer aus Coccolithen zusammengesetzten Hülle: *Rhabdothorax (Rhabdosphaera) erinaceus* (Kamptner 1937, 1958), *Rhabdosphaera subopaca* (Bernard 1938), *Coccolithus fragilis* (Bernard 1948). Bei den letzten Arten sind die Elemente der Cystenhülle (Rhabdolithen) von den Coccolithen der vegetativen Zellen

(Placolithen) stark verschieden (Abb. 36, *2*; Abb. 35a). Auch die *Ochrosphaera*-Cysten sind in ihren an Thecosphaeriden oder Discosphaeriden erinnernden Kalkkörpern denjenigen der vegetativen Zelle ganz unähnlich (PARKE 1961). Allerdings ist in den Naturfunden mit Täuschungen durch Parasiten zu rechnen; so gibt KAMPTNER für *Rhabdothorax* Stärke in der Cyste an, etwas für die gewöhnlich Leucosin speichernden Chrysophyten Unwahrscheinliches. Die Cysten von *Coccolithus fragilis* entlassen offenbar einen einzelnen arttypischen Flagellaten (BERNARD 1948), in denjenigen von *Rhabdosphaera subopaca* wurden vier unbegeißelte, zur Gattung *Coccolithus* gehörige Sporen beobachtet (BERNARD 1939).

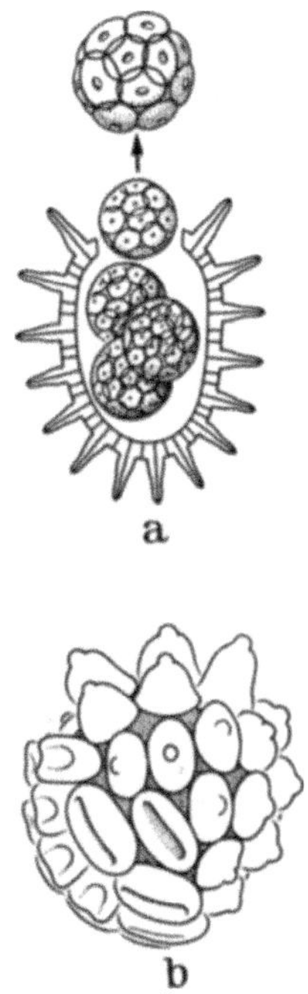

Abb. 35a u. b. a *Rhabdosphaera subopaca*, keimende Cyste. b Mosaikhülle aus Coccolithen von *Calyptrosphaera quadridentata* und *Acanthosphaera robusta*. (a nach BERNARD, b nach KAMPTNER, beide aus DEFLANDRE 1952.)

3. Pleomorphismus.

Ein solcher Pleomorphismus zwischen Cyste und Cysteninhalt könnte, da die Beobachtungen sich auf Naturmaterial beziehen, auch durch Parasitismus erklärt werden. Absolute Klärung kann nur der Kulturversuch bringen. Doch ist das Vorkommen zweier verschiedener Coccolithen-Typen bei der gleichen Art nach Entdeckung (PARKE und ADAMS) des entwicklungsgeschichtlichen Pleomorphismus in Kulturen von *Coccolithus pelagicus* nicht mehr gravierend. Hier wäre weiterhin an die Naturfunde von Hüllen mit Mosaikcharakter (s. z. B. KAMPTNER 1941) zu denken (Abb. 35b). *Syracosphaera wettsteini* kann kombiniert sein mit *Zygosphaera tuberculata*, *Anthosphaera robusta* mit *Calyptrosphaera quadridentata*, die letztere mit *Syracosphaera histrica* und diese letzte mit *Syracosphaera gracillima*. KAMPTNER diskutiert für solche Fälle einen Bastardcharakter der Zellen, es wäre aber wohl eher an die Möglichkeit alternativer Modifikation im Sinne der Pleomorphismen zu denken. Von einem Individuum zu anderen existiert Derartiges ja zweifellos: warum sollen solche Umstimmungen zu einer anderen Formbildung dann nicht in der Coccolithen-Sekretion des Individuums selbst eintreten, ähnlich wie dieses für die beiden zu verschiedener Zeit gebildeten Schalen einer Diatomeenzelle vielfach beobachtet wurde (s. S. 668). Umstimmungen in der Formbildung sind auch nicht zwangsläufig an einen Kernphasenwechsel gebunden, was viele Beispiele, darunter das eben Genannte und etwa die Regeneration des Moossporophyten mit Protonema als gametophytischer Wuchsform zeigen. Mosaikhüllen und Cysten der Coccolithophoriden können dem Planktologen, dessen Arbeit relativ arm an relevanten Ergebnissen für die Lebenscyclen und damit die Taxonomie ist, ein Mittel geben, aus der Fülle der Formen das genetisch Zusammengehörige verläßlich aufzufinden. Die Kulturmethode wird bei empfindlichen Arten noch auf lange Zeit versagen. Die bisherige Coccolithen-Taxonomie der Kalkflagellaten, sowenig entbehrlich sie ist, hat — das ergibt sich aus Kulturversuch und Freilandbeobachtung — nurmehr formalen Wert.

4. Der „theoretische" Cyclus von *Coccolithus fragilis*.

Nicht einbezogen wurden bisher die Arbeiten von BERNARD (Zusammenfassung 1948) an *Coccolithus fragilis*, weil sie, an Aufsammlungen gewonnen, bestenfalls Wahrscheinlichkeitsschlüsse vortragen. Doch stützen sie sich einerseits auf ein enormes Material an Beobachtungen dieser Allerweltsform des

Mittelmeeres, erscheinen andererseits nach den neuen experimentellen Befunden als möglich. Wir referieren (die in Klammern gesetzten Zahlen beziehen sich auf die Figuren von Abb. 36): Es gibt bei dieser Art einen Küsten- und einen Hochseecyclus, der erste repräsentiert durch einen oblongen, zur Teilung befähigten und mit zahlreichen kleinen und freien Coccolithen armierten Flagellaten (*11*), der zweite durch einen kugeligen Flagellaten (*1*) mit einer Hülle aus wenigen großen und verbackenen Kalkkörpern. Beide Arten begeißelter Zellen können unter ungünstigen Bedingungen in ihrem Innern mit einer dicken, unregelmäßig gestalteten Wand versehene Autosporen ausbilden, welche im ersten Falle in großer Zahl (*13*) entstehen und schwarz inkrustiert (*14*), im zweiten zu 4 bis 64

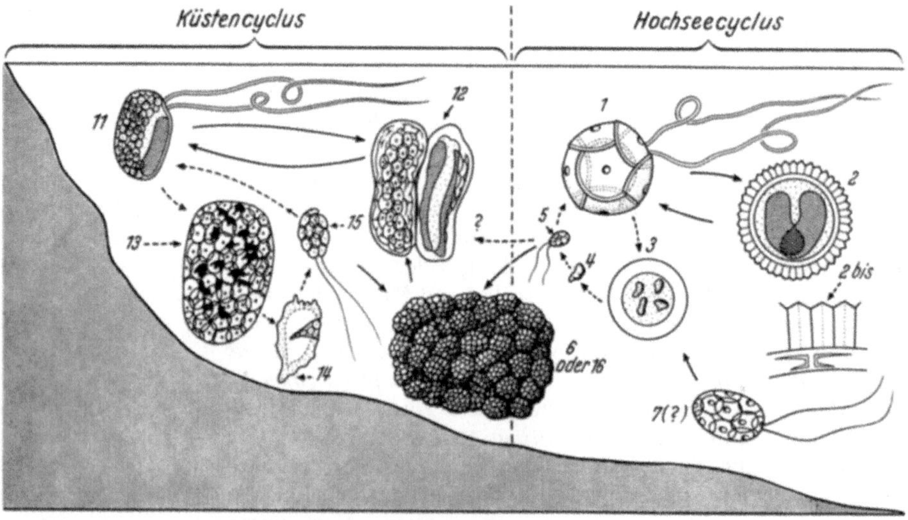

Abb. 36. Theoretischer Cyclus von *Coccolithus fragilis*, links Küsten-, rechts Hochseecyclus. Erklärungen im Text. (Nach BERNARD 1948.)

gebildet werden (*3*) und orangefarben sind (*4*). Beide Arten Sporen werden frei und entlassen gleichartige, zunächst nackte und später gepanzerte Kleinflagellaten (*15* bzw. *5*), die sich reichlich teilen und zu den entsprechenden großen Flagellaten (*11, 1*) heranwachsen oder aber beide durch Teilung ansehnliche Klumpen von unbeweglichen gepanzerten Zellen (*16*, *6*) entwickeln können. Diese Produkte (*15* und *16*) verbinden Hochsee- und Litoralcyclus. Der große nereitische Flagellat (*11*) vermag außerdem durch Abscheidung von Gallerte im Innern des Panzers Palmellen (*12*) hervorzubringen, die ihrerseits zum Flagellatenzustand zurückkehren oder sich mit schwarzen Sporen füllen können. Der ozeanische Flagellat (*1*) dagegen bildet bei raschen Bedingungsänderungen Rhabdolithen tragende Cysten (*2*) aus, die einen einzigen Flagellaten der Ausgangsform (*1*) zurückliefern. In Kulturen ließen sich diese Cyclen wegen der Empfindlichkeit der Art nicht verifizieren, sie bleiben natürlich problematisch.

5. Bedingungen für die Phasenübergänge.

Die unbeweglichen Zustände von *Isochrysis*, *Dicrateria*, einigen Chrysochromulinen und von *Coccolithus pelagicus* entstehen in alten Kulturen (PARKE 1949, PARKE et al. 1955); das gilt auch für die normalerweise palmelloide *Ochrosphaera*, welche nach SCHWARZ (1932) ausschließlich in jungen, in Teilung befindlichen Kulturen (spärlich) *schwärmt*. Nur im Herbst trat bei Einbringen alten Materials in frische Lösung „Gametangien“-Bildung (s. oben), gefolgt von vollständigem Ausschwärmen, ein. Nicht kopulierende „Gameten“ entwickeln sich parthenogenetisch zu gewöhnlichen Palmellen. PARKE (1961) fand jedoch bei dem von ihr untersuchten Stamm diese Gameten nicht. Eigene Versuche (v. STOSCH et al., unveröff.), die Bedingungen für die Phasenübergänge *Syracosphaera carterae* näher zu definieren, schlugen im ganzen fehl. Lediglich in synthetischem Seewasser (voll supplementiert) und in natürlichem Medium unter Vitamin B_1-Mangel (Vitamin B_1 neben Cobalamin bei Stamm List essentiell) traten in Kulturen der diploiden Monade reichlich nackte Schwärmer auf, welche sich unter diesen Umständen nicht weiterentwickelten. Aus unbekannten Gründen entstehen aber in ganz jungen Kulturen gelegentlich reichlich Meiocysten und aus ihnen *Apistonema*-Keimlinge (Abb. 32, *6—9*). Ein von BRAARUD erhaltener Stamm der Alge macht diesen Übergang nur äußerst selten. Reichliche Meiosporenbildung mit vorwiegender Direktkeimung der Meiosporen zur *Apistonema* unter Überspringung des beweglichen Stadiums scheint VALKANOV (1962) bei seinem bulgarischen Stamm der Alge „auf Objektträgern“ (halbaerophytisch) erreicht zu haben. Der umgekehrte Übergang *(Apistonema—Syracosphaera)* spielt sich bei VALKANOVS Form in 5–6 Wochen alten Kulturen anläßlich des Ausschwärmens der *Apistonema* ab. Wir können dieses Ausschwärmen zu nahezu 100% erzielen, wenn wir längere Zeit (mindestens 2 Wochen) bei Schwachlicht (100—200 Lux) gehaltene Apistonemen unter Lösungswechsel in stärkere Beleuchtung (700—2000 Lux, alles 15°) bringen. Die Schwärmdauer sinkt dann mit steigender Lichtintensität. Unter solchen Umständen entstehen aus der Mehrzahl der Schwärmer, die sich damit als Zoosporen erweisen, Apistonemen und nur in geringerer Zahl Kleinschwärmer sowie später Syracosphaeren(s. S. 648, 650). Die Ausbeute an den letzteren erheblich zu verbessern, gelang einige Male in Farblichtversuchen, die sich jedoch später nicht reproduzieren ließen.

Die normale Zoosporenform von *Phaeocystis* (s. S. 650) tritt nach künstlicher Zerlegung der Kolonie in die Einzelzellen auf, Makrozoosporen unter nicht faßbaren Bedingungen spontan, die Mikrozoosporen nur im Herbst beim Übertragen alter *Phaeocystis* in frische Lösung. Diese einzellig-bewegliche Phase vermehrt sich bei häufigem Übertragen stabil, geht aber bei erhöhtem Lichtgenuß in den *Phaeocystis*-Zustand zurück (KORNMANN).

Bei sehr geringen Nitratgehalten (weniger als 15 µg/Liter) wurden vor Monaco von *Coccolithus fragilis* nur noch Cysten im Plankton gefunden, dagegen erfolgte bei sehr niedrigen Phosphatgehalten (unter 0,1 µg/Liter P_2O_5) sogar noch Cystenkeimung.

Über die hydrographischen Bedingungen für das Auftreten der übrigen Stadien des Cyclus dieses Flagellaten muß auf das Original (BERNARD 1948) verwiesen werden.

c) Parthenogenese und Apogamie bei Chrysophyceen.

Für die Euchrysophyceen sei hierzu auf das auf S. 645 Gesagte und für die Haptophyceen auf S. 650 hingewiesen.

Literatur.

ANAND, P. L.: A taxonomic study of the algae of the british chalk-cliffs. J. Bot. 75, Suppl. 2, 1—51 (1937a). — An ecological study of the algae of the british chalk-cliffs. Part I and Part II. J. Ecol. 25, 153—188, 344—367 (1937b).

BERNARD, FR.: Coccolithophoridés nouveaux ou peu connus observés à Monaco en 1938. Notes et Rev. 81, 33—44 (1938). — Recherches sur les Coccolithophoridés. I. Principales espèces du plancton à Monaco. Bull. Inst. Oceanogr. No 767, 1—19 (1939). — Recherches sur le cycle du *Coccolithus fragilis* Lohm. Flagellé dominat les mers chaudes. J. conseil. Conseil permanent intern. exploration mer. **15**, 177—188 (1948). — BOURRELLY, P.: Recherches sur les Chrysophycées. Rev. Algol. Mem. Hors-Ser., No 1, 412 p. Paris 1957. — BRAARUD, T.: Studiet av planktonalger i electronmikroskop. Blyttia **2**, 102—108 (1954). — BRAARUD, T., u. E. FAGERLAND: *Syracosphaera carterae* n. sp. Videns Akad. Avh., I. math.-nat. Kl. **2**, 3—10 (1946). — BRABEZ, R.: Zur Kenntnis der Algenflora des Franzensbader und Sooser Thermenbereichs. Beih. Bot. Zbl. **61** A, 137—236 (1942). — BURSA, A.: *Hydrurus foetidus* Kirch. in der Polnischen Tatra. I. Ökologie, Morphologie, II. Phenologie. Bull. Inst. Acad. Polon. Sci. Let. **1934**, 69—84, 113—131 (1934).

CARTER, N.: New or interesting algae from brackish water. Arch. Protistenk. **90**, 1—68 (1937a). — *Pseudomallonomas anglica:* A new british flagellate. New Phytologist **36**, 57—63 (1937b). — CHADEFAUD, M.: Une nouvelle Chrysophycée marine filamenteuse: *Nematochrysopsis roscoffensis* n. g. n. sp. Bull. Soc. bot. France **94**, 237—243 (1947). — CHODAT, R., et A. R. ROSSILO: Sur une Coccolithophoridée d'eau douce. C. R. Soc. Physique Hist. natur. **42**, 51—60 (1925). — CONRAD, W.: Contribution à l'étude des Flagellates. Arch. Protistenk. **34**, 79—94 (1914a). — Contribution à l'étude des Flagellates. II. „*Thallochrysis pascheri*" n. g. n. sp., type d'une famille nouvelle. Ann. biol. lacustre **7**, 153—154 (1914b). — Recherches sur les Flagellates de nos eaus saumâtres. 2. Partie: Chrysomonadines. Arch. Protistenk. **56**, 167—231 (1926). — Le genre *Microglena* Ehrenberg. Arch. Protistenk. **60**, 415—439 (1927a). — Essay d'une monographie des genres *Mallomonas* Perty and *Pseudomallomonas* Chodat (1920). Arch. Protistenk. **59**, 423—505 (1927b). — Sur les Coccolithophoracées d'eau douce. Arch. Protistenk. **63**, 58—66 (1928). — Recherches sur les Flagellates de Belgique. I. A. Chrysomonadines, B. Volvocales. Mem. Mus. roy. Hist. natur. Belg. **8** (47), 1—65 (1931). — Notes protistologiques. III. Chrysomonadines intéressantes du nannoplankton saumâtre. Bull. Mus. roy. Hist. natur. Belg. **14** (29), 1—7 (1938). — Notes protistologiques. X. Sur le schorre de Lilloo. Bull. Mus. roy. Hist. natur. Belg. **15** (41), 1—18 (1939). — Notes protistologiques. XXI. Sur les Chrysomonadines à trois fouets. Aperçu synoptique. Bull. Mus. roy. Hist. natur. Belg. **17** (45), 1—16 (1941). — Notes protistologiques. XXVI. À propos des genres *Kephyrion* et *Pseudokephyrion*. Bull. Mus. roy. Hist. natur. Belg. **18**, 1—6 (1942).

DALES, R. P.: On the pigments of the *Chrysophyceae*. J. Mar. Biol. Ass. U. Kingd. **39**, 693—699 (1960). — DANGEARD, P. A.: Études sur le développement et la structure des organismes inférieurs. Botaniste **11**, 1—227 (1910). — Mémoire sur l'*Apistonema submarinum* sp. nov. et considérations générales sur la structure des Protozoaires et des Protophytes. Botaniste **26**, 235—240 (1934). — DEFLANDRE, G.: Classe des Coccolithophoridées. In: P. GRASSÉ, Traité Zool., vol. 1/1, p. 439—470. Paris: Masson & Cie. 1952. — DOFLEIN, FR.: Studien zur Naturgeschichte der Protozoen. IX. *Rhizochrysis*, eine Übergangsform unter den niederen Protozoen. Zool. Jb., Abt. Anat. u. Ontog. **40**, 383—420 (1917). — Untersuchungen über Chrysomonaden. I. *Ochromonas granularis* Dofl. II. Über *Chrysamoeba radians* Klebs. Arch. Protistenk. **44**, 149—205 (1922). — III. Arten von *Chromulina* und *Ochromonas* aus dem badischen Schwarzwald und ihre Cystenbildung. IV. Über einige aus dem Schwarzwald stammende, dort noch nicht bekannte oder neue Chrysomonadinen. Arch. Protistenk. **46**, 267—327 (1923).

FOTT, B.: Zur Frage der Sexualität bei Chrysomonaden. Nova Hedwigia **1**, 115—127 (1959).

GAARDER, K. R.: Electron microscope studies on holococcolithophorids. Nytt Mag. Bot. **10**, 35—51 (1962). — GAARDER, K. R., and J. MARKALI: On the coccolithophorid *Crystallolithus hyalinus* n. gen. n. sp. Nytt Mag. Bot. **5**, 1—5 (1956). — GEISSBÜHLER, H.: Grundlagen zu einer Algenflora einiger oberthurgauischer Moore. Thurg. naturforsch. Ges. **29**. 3—65 (1933). — GEITLER, L.: Zwei neue Chrysophyceen und eine neue Syncyanose aus dem Lunzer Untersee. Arch. Protistenk. **56**, 291—293 (1926). — Die Schwärmer und Kieselcysten von *Phaeodermatium rivulare* mit einer Berichtigung über *Phaeochrysis*. Arch. Protistenk. **58**, 272—280 (1927). — Ein grünes Filarplasmodium und andere neue Protisten. Arch. Protistenk. **69**, 615—636 (1930). — Über zweikernige Zysten von *Dinobryon divergens*. Öst. bot. Z. **84**, 282—286 (1935). — Neue luftlebige Algen aus Wien. Öst. bot. Z. **91**, 49—51 (1942a). — Zur Kenntnis der Bewohner des Oberflächenhäutchens einheimischer Gewässer. Biol. generalis (Wien) **16**, 450—475 (1942b). — Eine neue atmophytische Chrysophycee. *Ruttnera spectabilis*, nov. gen., nova spec. Int. Rev. Hydrobiol. **43**, 100—109 (1943). — GESSNER, F.: Die Gattung *Dinobryon* in phylogenetischer Betrachtung. Ber. dtsch. bot. Ges. **51**, 8—11 (1933).

HALLDAL, P., and J. MARKALI: Electron microscope studies on Coccolithophorids from the Norwegian sea, the Gulf Stream and the Mediterranean. Avh. Norske Vidensk. Akad. Oslo. I. math.-nat. Kl. **1**, 5—30 (1955). — HARRIS, K.: A contribution to our knowledge of *Mallomonas*. J. Linnean Soc. Bot. **55**, 88—102 (1953). — HOFFENDER, H.: Eine neue koloniebildende Chrysomonade. Arch. Protistenk. **29**, 293—307 (1913). — HOLLANDE, A.: Classe des Chrysomonades. In: P. GRASSÉ, Traité Zool., vol. 1/1, p. 471—570. Paris: Masson & Cie. 1952. — HOVASSE, R.: Contribution à l'étude des Flagellates libres. *Oicomonas beauchampi* sp. nov., Protomonadine et Chrysomonade. Arch. Zool. exp. et gén. **83**, 47—53 (1943). — Contribution à l'étude des Chrysomonadines. Botaniste **34**, 243—271 (1949).

JANE, F. W.: Two new Chrysophycean flagellates, *Cyclonexis erinus* and *Synochromonas elaeochus*. Proc. Linnean Soc. Lond. **152**, 298—309 (1940).

KAMPTNER, E.: Über Dauersporen bei marinen Coccolithineen. S.-B. Akad. Wiss. Wien, math.-nat. Kl., Abt. 1, **146**, 67—76 (1937). — Die Coccolithineen der Südwestküste von Istrien. Ann. Naturhist. Mus. Wien **51**, 54—149 (1941). — Betrachtungen zur Systematik der Kalkflagellaten. Arch. Protistenk. **103**, 54—116 (1958). — KLEBS, G.: Flagellatenstudien. I und II. Z. wiss. Zool. **55**, 265—351, 353—445 (1893). — KLUG, G.: *Salpingorhiza*, gen. nov. eine neue Craspedomonadaceengattung. Arch. Protistenk. **87**, 117—122 (1936). — KORNMANN, P.: Beobachtungen an *Phaeocystis*-Kulturen. Helgoländer wiss. Meeresunters. **5**, 218—233 (1955). — KORSHIKOV, A. A.: Studies on the Chrysomonads. I. Arch. Protistenk. **67**, 253—290 (1929). — On some new or little known Flagellates. Arch. Protistenk. **95**, 22—44 (1941). — KRIEGER, W.: Untersuchungen über Plankton-Chrysomonaden. Die Gattung *Mallomonas* und *Dinobryum* in monographischer Bearbeitung. Bot. Arch. (Lpz.) **29**, 257—329 (1930). — KRISTIANSEN, J.: Sexual and asexual reproduction in *Kephyrion* and *Stenocalyx* (Chrysophyceae). Bot. Tidsskr. **59**, 244—254 (1963).

LACKEY, B.: Scioto river forms of *Chrysococcus*. Amer. Midld Naturalist **20**, 619—623 (1938). — Some new flagellates from the Woods Hole area. Amer. Midld. Naturalist **23**, 463—471 (1940). — LAUTERBORN, R.: Eine neue Chrysomonadinen-Gattung (*Palatinella cystophora* nov. sp.). Zool. Anz. **30**, 423—428 (1906). — LUND, J. W. G.: Contributions to our knowledge of British *Chrysophyceae*. New Phytologist **41**, 274—292 (1942). — Some new or rare *Chrysophyceae* from the English Lake District. Hydrobiologia **16**, 97—108 (1960a). — New or rare British *Chrysophyceae*. 3. New records and observations on sexuality and classification. New Phytologist **59**, 349—360 (1960b). — A genus new to Britain: *Amphichysis* Korsh. Brit. Phycol. Bull. **2**, 116—120 (1962a). — Unsolved problems in the classification of the non-motile *Chrysophyceae* with reference to those in parallel groups. Preslia **34**, 140—146 (1962b).

MACK, BR.: Morphologische und entwicklungsgeschichtliche Untersuchungen an Chrysophyceen. Öst. bot. Z. **98**, 249—279 (1951). — Über einige epiphytische Chrysophyceen. Öst. bot. Z. **99**, 396—401 (1952). — Untersuchungen an Chrysophyceen. I—III. Öst. bot. Z. **100**, 147—152 (1953a). — Untersuchungen an Chrysophyceen. IV. Öst. bot. Z. **100**, 579—582 (1953b). — Untersuchungen an Chrysophyceen. V—VII. Öst. bot. Z. **101**, 64—73 (1954). — MAGNE, M. F.: Sur la cytologie d'une Chrysophycée marine vivant en aquarium, *Ochrosphaera neapolitana* Schussnig. Rev. gén. Bot. **59**, 231—240 (1952). — Les Chrysophycées marines de la Station Biologique de Roscoff. Rev. gén. Bot. **61**, 389—416 (1954). — Les conditions écologiques favorables aux Chrysophycées marines benthiques. Coll. int. Centre nat. Rech. sci. **81**, 91—100 (1957). — MANTON, I., and G. F. LEEDALE: Further observations on the fine structure of *Chrysochromulina ericina* Parke and Manton. J. Mar. Biol. Ass. U. Kingd. **41**, 145—155 (1961a). — Further observations on the fine structure of *Chrysochromulina minor* and *C. kappa* with special reference to the pyrenoids. J. Mar. Biol. Ass. U. Kingd. **41**, 519—525 (1961b). — Observations on the micro-anatomy of *Crystallolithus hyalinus* Gaarder and Markali. Arch. Mikrobiol. **47**, 115—136 (1963a). — Observations on the fine structure of *Prymnesium parvum* Carter. Arch. Mikrobiol. **45**, 285—303 (1963b). — MANTON, I., and M. PARKE: Further observations on small green flagellates with special reference to possible relatives of *Chromulina pusilla* Butcher. J. Mar. Biol. Ass. U. Kingd. **39**, 275—298 (1960). — Preliminary observations on scales and their mode of origin in *Chrysochromulina polylepis* sp. nov. J. Mar. Biol. Ass. U. Kingd. **42**, 565—578 (1962). — MASSART, J.: Recherches sur les organismes inférieurs. VIII. Sur la motilité des Flagellates. Acad. roy. de Belgique, Bull. Cl. Sci. Ser. V, **6**, 116—143 (1920).

PARKE, M.: Studies on marine flagellates. J. Mar. Biol. Ass. U. Kingd. **28**, 255—286 (1949). — Some remarks concerning the class *Chrysophyceae*. Brit. Phycol. Bull. **2**, 47—55 (1961). — PARKE, M., and I. ADAMS: The motile (*Crystallolithus hyalinus* Gaarder and Markali) and non-motile phases in the life-history of *Coccolithus pelagicus* (Wallich) Schiller. J. Mar. Biol. Ass. U. Kingd. **39**, 263—274 (1960). — PARKE, M., J. W. G. LUND, and I. MANTON: Observations on the biology and fine structure of the type species of *Chrysochromulina* (*C. parva* Lackey) in the English Lake District. Arch. Mikrobiol. **42**, 333—352 (1962). — PARKE, M., and I. MANTON: Studies on marine flagellates. VI. *Chrysochromulina pringsheimii* sp. nov. J. Mar. Biol. Ass. U. Kingd. **42**, 391—404 (1962). — PARKE, M., I. MANTON, and B. CLARKE: Studies on marine flagellates. II. Three new species of *Chrysochromulina*. J. Mar. Biol. Ass. U. Kingd. **34**, 579—609 (1955). — III. Three further species of *Chrysochromulina*. J. Mar. Biol. Ass. U. Kingd. **35**, 387—414 (1956). — IV. Morphology and microanatomy of a new species of *Chrysochromulina*. J. Mar. Biol. Ass. U. Kingd. **37**, 209—228 (1958). — Morphology and microanatomy of *Chrysochromulina strobilus* sp. nov. J. Mar.

Biol. Ass. U. Kingd. **38**, 168—188 (1959). — PASCHER, A.: Über einige Fälle vorübergehender Koloniebildung bei Flagellaten. Ber. dtsch. bot. Ges. **28**, 339—350 (1910). — Über Rhizopoden und Palmellenstadien bei Flagellaten. Arch. Protistenk. **25**, 153—200 (1912). — Fusionsplasmodien bei Flagellaten und ihre Bedeutung für die Ableitung der Rhizopoden von den Flagellaten. Arch. Protistenk. **37**, 31—64 (1916). — Die braune Algenreihe der Chrysophyceen. Arch. Protistenk. **52**, 489—564 (1925). — Über die Beziehungen zwischen Lagerform und Standortsverhältnissen bei einer Gallertalge (Chrysocapsale). Arch. Protistenk. **68**, 637—668 (1929). — Eine neue braune Fadenalge des Süßwassers (Chrysophycee). Arch. Protistenk. **73**, 60—72 (1931a). — Über eigenartige zweischalige Dauerstadien bei zwei tetrasporalen Chrysophyceen (Chrysocapsalen). Arch. Protistenk. **73**, 73—103 (1931b). — Eine braune, aerophile Gallertalge und ihre Einrichtungen für die Verbreitung durch den Wind. Beih. Bot. Zbl. **47**, Abt. I, 325—345 (1931c). — Über die Verbreitung endogener bzw. endoplasmatisch gebildeter Sporen bei den Algen. Beih. Bot. Zbl. A **49**, 293—308 (1932). — Rhizopodiale Chrysophyceen. Arch. Protistenk. **93**, 329—349 (1940a). — Filarplasmodiale Ausbildungen bei Algen. Arch. Protistenk. **94**, 293—309 (1940b). — Zur Kenntnis verschiedener Ausbildungen der planktontischen Dinobryen. Int. Rev. Hydrobiol. **43**, 110—123 (1943). — PASCHER, A., u. W. VLK: Zur Kenntnis der Chrysophyceen des salzhaltigen Flachmoores Hrabanow bei Lissa a. E. Lotos 88, 163—177 (1941/42). — PRAUSER, H.: *Diacronema vlkianum*, eine neue Chrysomonade. Arch. Protistenk. **103**, 117—128 (1958). — PRINGSHEIM, E. G.: On iron flagellates. Phil. Trans. B **232**, 311—327 (1946). — Kleine Mitteilungen über Flagellaten und Algen. I. Algenartige Chrysophyceen in Reinkultur. Arch. Mikrobiol. **21**, 401—410 (1955). — PROWAZEK, S.: Flagellatenstudien. Arch. Protistenk. **2**, 195—212 (1903).

RAYNS, D. G.: Alternation of generation in a coccolithophorid, *Cricosphaera carterae* (BRAARUD and FAGERL.) BRAARUD. J. Mar. Biol. Ass. U. Kingd. **42**, 481—484 (1962). — REVERDIN, L.: Étude phytoplanktonique, expérimentale et descriptive, des eaux du Lac de Genève. Arch. Sci. physiol., V. Per. **1**, 302—345, 403—450 (1919).

SCHERFFEL, A.: Beitrag zur Kenntnis der Chrysomonadinen. Arch. Protistenk. **22**, 299—344 (1911). — Über die Cyste von *Monas*. Arch. Protistenk. **48**, 187—195 (1924). — Beitrag zur Kenntnis der Chrysomonaden. Arch. Protistenk. **57**, 331—361 (1927). — SCHILLER, J.: Über Fortpflanzung, geißellose Gattungen und Nomenklatur der Coccolithophoraceen nebst Mitteilungen über die Kopulation bei *Dinobryon*. Arch. Protistenk. **53**, 326—342 (1926a). — Der thermische Einfluß und die Wirkung des Eises auf die planktischen Herbstvegetationen in den Altwässern der Donau bei Wien. Arch. Protistenk. **56**, 1—62 (1926b). — Neue Chryso- und Cryptomonaden aus Altwässern der Donau bei Wien. Arch. Protistenk. **66**, 436—458 (1929). — *Coccolithineae*. In: RABENHORSTs Kryptogamenflora, Bd. 10/2, S. 88—266. Leipzig: Akademische Verlagsgesellschaft 1930. — Neue Mikrophyten aus dem Neusiedler See und benachbarter Gebiete. Öst. bot. Z. **99**, 100—117 (1952). — SCHMID, G.: Die Chrysomonadengattungen *Kephyrion*, *Pseudokephyrion*, *Kephyriopsis* und *Stenocalyx* in Gewässern bei Wien. Öst. bot. Z. **83**, 161—172 (1934). — SCHUSSNIG, B.: *Ochrosphaera neapolitana* n. gen. et sp., eine neue Chrysomonade mit Kalkhülle. Öst. bot. Z. **79**, 164—169 (1930). — Über einige neue Protophyten aus der Adria. Arch. Protistenk. **93**, 327—330 (1940). — SCHWARZ, E.: Beiträge zur Entwicklungsgeschichte der Protophyten. IX. Der Formwechsel von *Ochrosphaera neapolitana*. Arch. Protistenk. **77**, 434—462 (1932). — SKUJA, H.: Taxonomie des Phytoplanktons einiger Seen in Uppland, Schweden. Symbolae bot. Upsalienses **9** (3), 1—399 (1948). — Körperbau und Reproduktion bei *Dinobryum borgei* Lem. Sv. bot. Tidskr. **44**, 96—107 (1950). — Taxonomische und biologische Studien über das Phytoplankton schwedischer Binnengewässer. Nova Acta R. Soc. Sci. upsal. Ser. IV. **16** (3), 1—401 (1956). — STEIN, FR.: Der Organismus der Infusionsthiere. Leipzig 1878. — STOSCH, H. A. v.: Ein morphologischer Phasenwechsel bei einer Coccolithophoride. Naturwissenschaften **42**, 423 (1955). — Der Geißelapparat einer Coccolithophoride. Naturwissenschaften **45**, 140—141 (1958). — STOSCH, H. A. v., G. DERSCH-SIEBERT, R. SCHÄFER u. V. LUCAN: Zum Entwicklungszyklus der Coccolithophoride *Syracosphaera carterae*, Stamm List. Unveröffentlicht.

TAYLOR, WM. R.: Structure and reproduction of *Chrysophaeum lewisii*. Hydrobiologia **3**, 123—130 (1951). — The algal genus *Chrysophaeum*. Bull. Torrey bot. Club **79**, 79 (1952).

VALKANOV, A.: Die Entwicklung von *Hymenomonas coccolithophora* Conrad. Rev. Algol.. N. S. **6**. 220—225 (1962). — VISCHER, W.: Über die Goldalge *Chromophyton rosanoffii* Woronin. Ber. schweiz. bot. Ges. **53**, 91—101 (1943).

WILLE, N.: Algologische Mitteilungen. Jb. wiss. Bot. **18**. 425—518 (1887). — WISLOUCH. S. M.: *Palatinella cytophora* Lauterborn fo. mihi *minor* und *Synura reticulata* Lemm.. zwei für Rußland neue Chrysomonaden. Arb. Bot. Lab. Nedz. Haueninst., St. Petersburg **22** (1910). Zit. nach BOURRELLY. — Sur les Chrysomonadines des environs de Petrograd. J. Mikrobiol. (Petrograd) **1**, 250--278 (1914). — WORONIN, M.: *Chromophyton rosanoffii*. Bot. Ztg **38**, 624—631. 641—648 (1880).

II. Diatomeen.

Die Kieselalgen, welche man als cystäre Sonderentwicklung am Chrysophytenstamm auffassen muß, vermehren sich gewöhnlich durch die Zweiteilung ihrer Zellen. Die letztere ist kompliziert durch den Schachtelbau der Membran an sich sowie seine Folgeerscheinung, das Absinken der mittleren Zellgröße in jedem Teilungsschritt. Diese Größenabnahme ist im allgemeinen begleitet von allometrischen Proportionsänderungen — z. B. verringern schlanke Pennatenzellen weniger die Dicke (Transapikalmaß) als die Breite (Valvarachse) — und vom Ausfallen von Strukturelementen, sei es durch Verminderung der Zahl, sei es durch Totalverlust (z. B. fehlt die Zentralarea von *Coscinodiscus perforatus* bei sehr schmalen Zellen). Beides ist Crux der Taxonomie. Dieses Absinken macht dann periodische Zellvergrößerungen notwendig, welche im allgemeinen durch Bildung von „Auxosporen" erreicht werden, ein Zusammenhang, den man als Pfitzer-MacDonaldsches Entwicklungsgesetz (s. GEITLER 1932) bezeichnet. Bei den pennaten Diatomeen weiß man schon lange, daß die Auxospore eine aus einer Syngamie entspringende Zygote oder besser ihr Entwicklungsprodukt ist; bei den zentrischen Diatomeen ist der gleiche Nachweis jünger (v. STOSCH 1951a, b, 1954, 1956, GEITLER 1952a, v. STOSCH und DREBES 1964), auch haben wir hier im Gegensatz zu der Isogamie der meisten Pennatae eine Oogamie, deren Spermien früher unter dem Namen Mikrosporen als Isogameten gedeutet wurden (KARSTEN 1928).

Die Entwicklung der Zygote verläuft nach einem artgemäß geregelten Ritual, Kurzwort für eine „Embryonalentwicklung" der Zelle, eine Abfolge von streng normierten Formbildungsprozessen, und führt auf die spezifische und streng normierte Ausgangsform der vegetativen Vermehrung zurück (siehe z. B. v. STOSCH 1962, 1965, v. STOSCH und DREBES 1964). Kurz formuliert wird die erste Stufe dieser Formbildung von der Zygotenhülle (Auxosporenmembran) getragen, die — bei *zentrischen Diatomeen* von Anfang an verkieselt — zu einer der endgültigen Zelle umschriebenen Form, häufig in ausgesprochen differentiellem Wachstum, führt. Bei *Pennaten* wird die Befruchtungsmembran funktionell meist sehr früh durch eine weitere Membranbildung, das ebenfalls verkieselte Perizonium, ersetzt, welches durch seinen besonderen Aufbau das stark anisometrische Auswachsen zu schlanken Formen zu stützen imstande ist; Ausnahmen und unsichere Fälle bleiben unberücksichtigt. Der Wechsel von der „rohen" zur endgültigen Form geschieht im wesentlichen durch zwei gezielte Plasmolysen, von denen die erste die Oberfläche für die sogleich erfolgende Ausscheidung der Epivalva, die zweite gegenüberliegende für jene der Hypovalva der Erstlingszelle schafft. Jeder der formbildenden Plasmolysen geht eine metagame Mitose voraus (GEITLER 1952b), d. h. eine *acytokinetische* Kernteilung, deren einer Tochterkern anschließend abortiert. Es handelt sich um eine entwicklungsmechanische Besonderheit der Diatomeen: die Schalenbildungen scheinen immer an Mitosen geknüpft (GEITLER 1953a, b; 1963).

Neben diesen Auxosporenbildungen infolge einer äußeren Befruchtung gibt es Fälle, wo eine Gamie nicht erkennbar wird, ohne daß das Ritual bei der Bildung der Erstlingszelle verlorengeht. Über sie wird im Kapitel Parthenogenese und Apogamie am Schluß des Abschnittes berichtet.

a) Sprunghafte vegetative Zellgrößenänderungen.

Neben diesen durch die Befruchtung oder wenigstens durch die spezifischen Formbildungsprozesse gekennzeichneten echten Auxosporen gibt es rein vegetative sprunghafte *Zellvergrößerungen*, die ungeregelt verlaufen, und die gelegentlich mit

Auxosporen verwechselt wurden. Daneben existieren am Rande des Abnormen auch sprunghafte *Zellverkleinerungen*. Beide Vorgänge sind häufige Artefakte des Kulturexperiments, kommen aber auch in der Natur vor, und der erste von ihnen ist in einem Einzelfall anscheinend zum Normalvorgang der Arterhaltung geworden.

Zu einer *vegetativen Zellvergrößerung* tritt nach Auseinanderweichen der Schalen durch Überdehnung der Zelle der Protoplast ganz oder teilweise aus, schwillt mehr oder weniger an und sezerniert neue vergrößerte Schalen. Da der offenbar an den sexuellen Ablauf gebundene entwicklungsmechanische Apparat fehlt, entstehen gegenüber der arteigenen Norm anomale Zellformen, und da der Teilungsmechanismus der Zelle prismatische Gürtelkonfigurationen nach sich zieht, Klone mit ins-

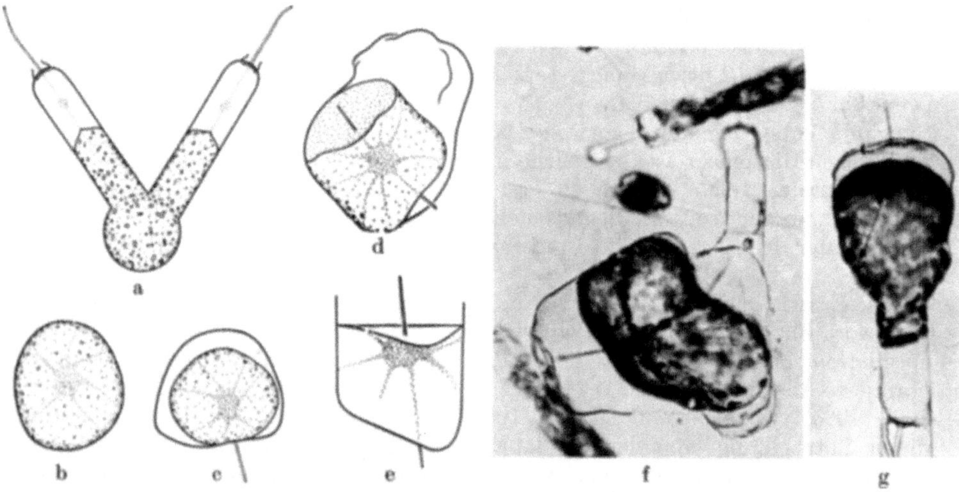

Abb. 37a—g. *Ditylum Brightwellii*. Vegetative Zellvergrößerung. a Plasmainhalt tritt am geöffneten Gürtel der Zelle als Blase aus und bildet in b—e neue Schalen. f Sehr unregelmäßig voranschreitende, g Auxosporen-ähnlich aussehende Zellvergrößerung. (a—e nach GROSS 1937, f, g Original.)

besondere anomaler Valvaransicht (d.h. Querschnittsform). Die meisten der in Kultur sich vegetativ vergrößernden Arten, die wir kennen, haben daneben auch die übliche Auxosporenbildung. Zuerst genauer beschrieben wurde der Vorgang von GROSS (1937, 1940), der ihn in Kulturen von *Ditylum brightwelli* regelmäßig beobachtete, was leicht verifiziert werden kann (v. STOSCH 1965), ihn aber irrtümlich für eine Auxosporenbildung hielt. Abb. 37d, f zeigt dazu insbesondere, daß der Prozeß der Wiederbeschalung durch Versuch und Irrtum vor sich geht. Die Zelle experimentiert mit Schalen-Neubildungen, ändert ihre Gestalt dabei auch mittels formbildender Plasmolysen, bis schließlich einmal eine Schale „paßt“ und die Gegenschale ausgeschieden werden kann. Eine der Auxosporenmembran der *Centrales* oder dem Perizonium der Pennaten vergleichbare Hülle um den jungen Protoplasten fehlt allgemein, höchstens unverkieselte „Pektin“-Membranen können gelegentlich festgestellt werden. In anderen Fällen tritt bei *Ditylum* (s. auch Abb. 37f, g) und weiteren Diatomeen der Protoplast anfänglich oder endgültig nur partiell aus der Mutterschale aus (v. STOSCH 1965). *Ditylum* hat „eigentlich“ regulär bi-, tri- oder quadriradiate Gestalt, dagegen ist die Querschnittsform der Pseudoauxosporen — das gilt auch für andere Arten — immer extrem ungeometrisch. Nach Ausweis aufgefundener Zellformen (s. insbesondere GRUNOW in VAN HEURCK 1885) scheint auch in der Natur, genauer in den westeuropäischen Gewässern, die vegetative Zellvergrößerung der wichtigste oder vorwiegende

Modus für die Größenerneuerung zu sein. Die entstandenen Querschnittsformen werden allerdings häufig, insbesondere bei Arten mit dünnen Schalen während der weiteren Zellteilungen eines etablierten Klones normalisiert. So entstehen bei *Stephanopyxis turris* bei auf Kieselsäuremangelwirkungen folgenden Regenerationen Klone ovalen Querschnittes, die im Laufe weniger Zellteilungsgenerationen auf die normale runde Form zurückregulieren (v. STOSCH 1965). Bei der etwas massiver gebauten *Stephanopyxis palmeriana* gelingt Ähnliches schon viel schwerer (DREBES unveröff.). Nun ist der Kreiszylinder einsichtig die Gleichgewichtsform, auf die Turgor und Oberflächenspannung hinarbeiten werden. Merkwürdigerweise kommen Ditylen aber, allerdings erst im Laufe recht zahlreicher Zellgenerationen, von der anfänglich ganz unregelmäßigen auf eine annähernd regulär tetra- oder triradiate Form zurück, was durch einfache Modellvorstellungen nicht mehr zu erklären ist, also besondere ,,Formbildungskräfte" der Zelle anzunehmen zwingt. Dieses Phänomen kann weiterhin verständlich machen, warum die in der Natur gefundenen *Ditylum*zellen nicht so irregulär gebaut sind, wie man nach vorstehender Schilderung über ihre Entstehung erwarten möchte.

Die *Methoden*, durch welche vegetative Vergrößerungen experimentell erreicht werden, sind, von der physiologischen Reaktionsnorm der Art abhängig, recht verschieden. Ganze Gruppen wie die *Coscinodiscus*arten und ihre Verwandten haben nach den jetzigen Erfahrungen überhaupt nicht die Fähigkeit, auf Protoplastenaustritte hin Schalen zu regenerieren. Wo letztere aber gegeben ist, hängt die vegetative Zellvergrößerung vom Austreten des Protoplasten ab, und für dieses gilt nach der Seite der *inneren Bedingungen* die Regel, daß es um so leichter erfolgt, je schmäler im arteigenen Größenbereich eine Zelle ist. Für diese Gesetzmäßigkeit wurde von GROSS (1937) ein schöner Beleg gegeben. Verschieden breite Klone von *Ditylum* zeigten mit dem Altern der Kultur, d.h. unter den durch die steigende Zellzahl bewirkten Veränderungen des Mediums um so früher Zellaustritte, je schmäler sie waren. So benötigten 43 μ breite Zellen 20 Tage Kulturdauer, 10 μ breite aber nur noch 3 Tage; Zwischengrößen liegen gestaffelt intermediär. Die notwendigen Veränderungen des Mediums sind also bei schmalen Zellen gering, müssen aber bei breiten größer sein. Zellen von 43 μ Breite haben, wenn sie Austritte zeigen, praktisch das Medium erschöpft. Die freien Protoplasten regenerieren daher anschließend nicht, schmale Zellen jedoch treten viel früher aus und ihre Regenerate wachsen danach kräftig weiter, da sie noch fähig sind, in einem Medium zu wachsen, das die schmale Ausgangsform bereits hemmt. Kausal scheint der Protoplastenaustritt nämlich wesentlich bedingt durch eine Inkongruenz zwischen Plasmawachstum und Zellteilung.

Nach der Seite der *äußeren Bedingungen* geben wir Beispiele, die zugleich die unterschiedliche Reaktionsweise zweier Idiotypen demonstrieren mögen. So kann man *Achnanthes longipes* (v. STOSCH 1942, 1965) durch Manganmangel zu epidemischen Zellaustritten bringen und bei Behebung des Mangels danach zur Regeneration. Kieselsäuremangel dagegen ruft übersteigertes Längenwachstum der Zellen hervor, Austritte sind selten. Auf der anderen Seite zeigt *Stephanopyxis* unter Manganmangel ebenfalls totale Protoplastenaustritte, regeneriert aber bei Restitution schlecht oder gar nicht. Kieselsäuremangel dagegen bewirkt Überdehnung und partielle Austritte, die nach Kieselsäuregabe reichlich vergrößerte Klone entstehen lassen.

Die Methode ist experimentell von Wichtigkeit, da sie ermöglicht, die Genotypen von Kulturstämmen zu halten, die durch die Umkombinationen verlorengehen, welche der Sexualakt vor der Auxosporenbildung erwarten läßt und praktisch bringt.

Abrupte *Zellverkleinerungen* können ebenfalls in der Natur beobachtet (Lit. Locker 1950) und bei einigen Arten in Kulturmaterial gefunden (Geitler 1932) oder erzwungen werden (v. Stosch 1942, 1965). Auch hier besteht ein Interesse für den Experimentator, weil es ihm erlaubt, Auxosporen-Nachkommenschaften anders als durch den langwierigen natürlichen Vorgang auf geringe Größen herunterzubringen, wie sie z.B. die Voraussetzung für erneute Sexualprozesse sind. Von den Methoden, Verkleinerungen zu erreichen, wird hier diejenige erwähnt, welche zu Inäqualteilungen führt. Alterierte Zellen legen gelegentlich die Teilungswände so fern von der Zellmitte und so schräg an, daß die kleinere Tochterzelle dreieckig wird und nicht mehr an die eine Längswand grenzt. Ist sie imstande, sich weiter zu vermehren, so etabliert sie einen schmalen Klon (Geitler 1932, Fig. 52; v. Stosch 1965). Bei *Achnanthes* läßt sich die erforderliche physiologische Situation durch SiO_2-Entzug schaffen, der die Teilungstätigkeit empfindlich stört. Nach Restitution erhält man die verkleinerten Klone. So kann man in einer einzigen Kulturschale von einer frischen Auxospore aus Nachkommenschaften erhalten, die klein genug sind, wieder den Sexualprozeß durchzuführen.

b) Kolonieteilung.

Da wir nur die gewöhnlichen Zellteilungen aus dem jetzigen Stoff ausschlossen, wären hier die Vorgänge zu bringen, welche die Trennung der Zellverbände bewirken. Bei Verbänden mit Verkittung ihrer Zellen durch Gallertverbindungen, wie denjenigen von *Stephanopyxis* (v. Stosch und Drebes 1964), werden sie anscheinend durch die mit dem Alter fortschreitende Quellung der Kittsubstanz bewirkt. Tatsächlich gehören die Kolonien dieser intrakolonial meist synchron teilenden Art im dynamischen Gleichgewicht der Kultur, je nach dessen Bedingungen, den Größenklassen 8, 16 oder 32 Zellen an.

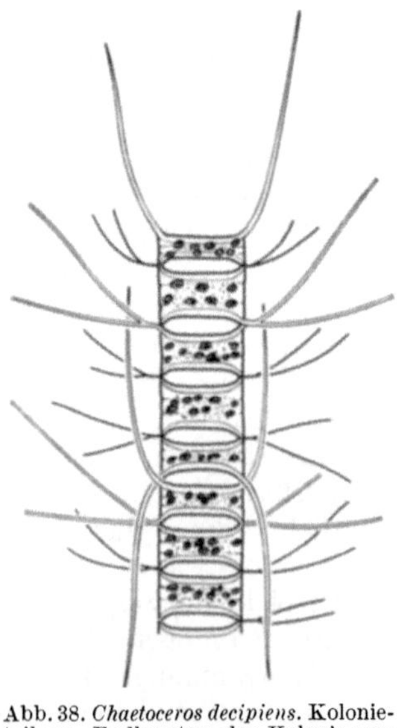

Abb. 38. *Chaetoceros decipiens*. Kolonieteilung. Endborsten der Kolonie von den übrigen verschieden. Gerade hat die zur Koloniespaltung führende heterovalvate Teilung stattgefunden. (Nach Cupp 1943.)

In anderen Fällen, so insbesondere bei den Gattungen *Bacteriastrum* und *Chaetoceros*, außerdem bei manchen Melosiren (z.B. *M. granulata*: O. Müller 1904) sind die Endzellen der Linearkolonien äußerlich von denjenigen im Innern des Verbandes verschieden, wodurch die Kolonie eine Abgeschlossenheit ihrer Gestalt und damit eine Individialität höherer Ordnung erhält (Abb. 38). Da nur *Chaetoceros* etwas genauer untersucht wurde (Schütt 1888), sei dieser hier allein geschildert. Die Zellen der Gattung, Zylinder mit elliptischer Grundfläche, tragen an den vier „Ecken" lang ausgezogene Hörner. Diese sprießen nach der Zellteilung an den neuen Schalen der Tochterzellen aus und verwachsen dort, wo sie sich paarweise kreuzen, unter Ausscheidung von Kieselsäure unlöslich, dadurch auch die Zellen selbst vereinigend. Soll nun eine Kolonieteilung erfolgen, so müssen zwei Schwesterzellen bei ihrer Entstehung andersartige Schalen mit nicht verwachsenden Borsten ausscheiden (Abb. 38). Die erzeugenden Teilungen haben wir (v. Stosch und Drebes 1964) als „*heterovalvat*" bezeichnet. Sie spielen in der Entwicklungsmechanik von Diatomeen neben inäqualen Cytokinesen und acytokinetischen Mitosen eine recht große Rolle. Die Endborsten sind in den meisten derartigen Fällen gröber als die normalen und abweichend ausgeformt. Es versteht sich, daß bei Synchronie der Teilungen sowie Symmetrie in bezug auf die Längsachse des Verbandes die Kolonien ungerade Zellzahlen aufweisen sollten. Dann müßte nämlich in jedem Teilungscyclus die zentrale Stelle der Kolonie eine heterovalvate Trennteilung durchführen. Darüber, was wirklich geschieht, fehlen Beobachtungen. Unbekannt ist auch, wie die sich differenzierende Zelle „erfährt", daß sie die alternative Teilungsform ins Spiel zu bringen hat.

Nicht untersucht sind ferner die Wirkungen von äußeren und inneren Bedingungen auf die Häufigkeit und Verteilung dieses Vorganges, welche im Zusammenspiel die tatsächlich beobachteten Änderungen der Koloniegrößen regeln müßten. Doch liegt für *Sceletonema*, deren Schalen durch Kränze auch an den Vereinigungsstellen verkieselter Borsten (HELMCKE u. KRIEGER 1953) verbunden sind und deren Kolonien daher Trennstellen besitzen müssen, eine Beobachtung von BENTLEY 1958 vor. In der Natur sowie in bakterienhaltigen Kulturen auf Grundlage natürlichen Seewassers umfassen die Verbände bis 50 Zellen. Bei Reinkulturen in synthetischen Medien erreichen sie höchstens drei Zellen. Die Koloniegröße, bei der eine Trennung eingeleitet wird, ist also stark herabgesetzt. Zusatz von Indolylessigsäure kann diese Größe teilweise, nämlich bis auf im Durchschnitt acht Zellen, normalisieren.

c) Dauerzustände und Dauerzellen.

1. Dauerzustände.

Zunächst sei darauf hingewiesen, daß vegetative Zellen der aerophytischen, auf Felsen und Rinde wachsenden und der im Boden vorkommenden Diatomeen in ausgetrocknetem Zustand lange Zeit weiterleben können (siehe z.B. BRISTOL 1919, 1920, BEGER 1927, LUND 1945, 1946, BREDEMÜHL 1947, BOCK 1963). Andererseits vermögen vegetative (NIPKOW 1950: *Melosira islandica* ssp. *helvetica*, *M. varians*, *Stephanodiscus hantzschii*, *Cymatopleura solea*) oder von solchen nur durch besondere Schalendicke verschiedene Zellen (LUND 1954: *Melosira italica*, ssp. *subarcta*, *M. granulata*; SKABITSCHEWSKY 1929: *M. baikalensis*) in der Tiefe der Seen den Winter oder im sauerstofffreien Schlamm noch länger zu überdauern. Die Zellen sind dann sehr ölreich und die Plastiden abgerundet. Wie bei dem hohen endgültigen Fettgehalt das Absinken der Zellen bewerkstelligt wird, ist unbekannt. NIPKOW konnte von *Melosira helvetica* noch Material aus fünfjährigen, von *Stephanodiscus* selbst bis zwölfjährigen Schlammschichten zum Keimen bringen. LUND erhielt *M. italica* lebend aus Sedimenten, deren Alter auf 16 Jahre geschätzt wurde und in Flaschen, in welchen die Diatomee zusammen mit Faulschlamm gefüllt wurde, ein spärliches Überleben nach 3 Jahren. Seenschlamm mit *M. ambigua*, *M. granulata* und einmal *Fragilaria construens* konnte analog mindestens 10 Monate gelagert werden, ohne daß alle Zellen abstarben.

2. Dauerzellen.

Dauerzellen (Cysten, Statocysten, Dauersporen, Statosporen, resting cells, endocysts) im eigentlichen Sinne, d.h. von den vegetativen durch Form, die geringere Größe und Dickwandigkeit unterschiedene Zellen, treten, abgesehen von wenigen Pennales, vorzugsweise bei zentrischen Diatomeen der nereitischen Bereiche des Meeres und selten aus dem Süßwasser auf. Entwicklungsgeschichtlich sind die Cysten verschiedenartige Gebilde, deren Genese im einzelnen nur in wenigen Fällen ausreichend untersucht wurde. Diese Lücken der Kenntnis erschweren eine rationale Klassifizierung nach den Entstehungsweisen, so daß man sich stellenweise auf plausible Annahmen über die letzteren stützen muß.

Bildung und System.

Modi der Membranbildung bei Dauerzellen. Da sich die Dauerspore von der vegetativen Zelle vor allem durch ihre Membran unterscheidet, wäre deren Genese von Interesse. Wir besprechen diese am Beispiel einiger Dauersporenentwicklungen. Nach der Zellteilung erfolgt die Bildung der neuen Valven auf den durch die cytokinetische *Durchfurchung* des Protoplasten *freigelegten Oberflächen.* Die Teilung kann dabei, bezogen auf die beiden Tochterzellen und deren neue Membranhälften, *äqual* oder *inäqual* sein. Bei der hemiendogenen Dauersporenbildung von *Stephanopyxis* (v. Stosch und Drebes 1964) ist beides vereinigt (Abb. 39). In

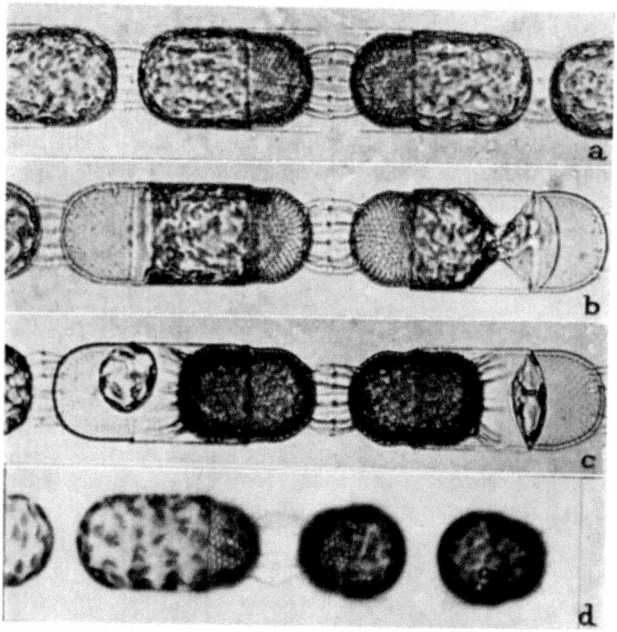

Abb. 39a—d. Entwicklung und Keimung der Dauerzellen von *Stephanopyxis turris.* a Rechtes Zellpaar in der auf die heterovalvate Teilung folgenden Streckungsphase, b gleiches Zellpaar kurz vor (links) bzw. während (rechts) Inäqualteilung, c fertiges Sporenpaar: am rechten „Richtungskörper" ist zarte Kieselmembran erkennbar. d Keimung zweier Dauersporen. (a—c Original, d Photo G. Drebes.)

äqualer, aber heterovalvater Teilung einer vegetativen Zelle entstehen zwei von der normalen abweichend gebaute verdickte Schalen, die Epivalven des Dauersporenpaares (Abb. 39a), in die hinein sich nun vor der nächsten Teilung die Hauptmenge von Plasma und Plastiden verschiebt (Abb. 39b). Dadurch wird diese Teilung sowohl in bezug auf die von ihr erzeugten Zellen als auch auf deren neue Membranen inäqual und trennt nun die große Dauerspore von einer kleinen plastidenarmen oder -freien richtungskörperartigen und später abortierenden Zelle. Die erste enthält eine der bereits vorhandenen Valva in der Form ähnliche, aber nicht gleiche, verstärkte, die letztere eine zarte, uhrglasförmige Schale (Abb. 39c). Wir kennen Fälle (Geitler 1953b, v. Stosch 1958) von inäqualen Teilungen bei Diatomeen, wo — wie gelegentlich auch bei *Stephanopyxis* — bei der nicht privilegierten Zelle die Schalenbildung ganz unterbleibt, und es scheint, daß in einem der beiden Modi der Sporenbildung von *Thalassiosira bioculata* (Meunier 1910) beide Schalen auf diese Art entstehen.

Eine weitere Art der Genese von Dauerzellen-Membranen ähnelt der Schalenbildung der Erstlingszelle in der Auxospore, indem jede Valva nach einer *acyto-*

kinetischen Mitose (nachgewiesen nur für *Chaetoceros teres*, v. STOSCH und DREBES 1964) und der an diese gebundenen formbildenden Plasmolyse auf dem Rückzugsmeniscus des Protoplasten ausgeschieden wird (Abb. 40a, b, d). Bei *Chaetoceros lorenzianus* verlängern sich wie bei *Ch. teres* und den meisten anderen Arten der Gattung, nach einer Zellteilung (Abb. 40a, b), die zwei Dauersporeninitiatorzellen erzeugt, die neu gebildeten Unterschalen dieser Mutterzellen stark, was SCHÜTT (1888) zuerst beschrieb (s. auch besonders MEUNIER 1910, 1913—1915). Dann löst sich das Plasma an den Oberschalen ab und kontrahiert sich unter Mitose, worauf an der freigelegten Plasmaoberfläche die topfförmige Oberschale der Dauerzelle ausgeschieden wird; danach hebt sich auch im anderen Zellteil das Plasma unter erneuter Mitose ab und sezerniert nun die Unterschale der Spore (Abb. 40a, b, d). Letzte stellt einen in den Topf eingepaßten Deckel mit zylindrischem Rand dar (Abb. 40e).

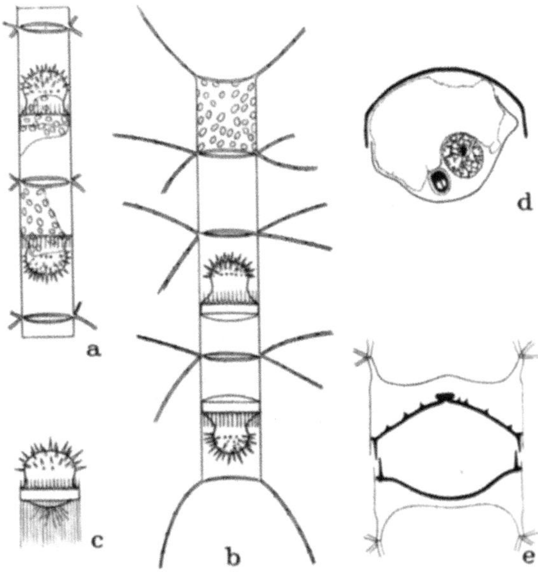

Abb. 40a—e. *Chaetoceros*-Dauersporen. a—c *Ch. lorenzianus*. a Zwei Schwesterzellen haben sich durch Verlängerung ihrer gemeinsamen Schalen stark (vgl. Fig. b oben) gestreckt und diesen gegenüber nach Kontraktion ihres Plasmas die Oberschalen der Dauersporen abgeschieden, in die nun das Plasma hineinwandert. b Dauersporen fertiggestellt, oben vegetative Zelle, c isolierte Dauerspore. (a—c nach MEUNIER 1913, umgezeichnet.) d *Ch. teres* (nahe verwandt mit voriger) Dauerspore mit Kern und zweitem pyknotischen Kern, der erste pyknotische Kern ist bereits resorbiert. e *Ch. messanensis*, Dauerspore teilweise geöffnet. (d, e Original.)

Die Frage, wieweit die Entstehung der Kieselfortsätze oder -borsten, welche die Ober- oder auch die Unterschale der Dauersporen tragen kann (Abb. 40c, e), mit dem Vorhandensein eines extramembranösen Plasmas (SCHÜTT 1888, 1899, 1900) zusammenhängt, wurde in neuerer Zeit nicht behandelt; sie muß erwähnt werden, da sich bei ihrer Bejahung das soeben gegebene Bild einer Membranentstehung an freien Oberflächen modifizieren müßte. Dann würde nicht alles Plasma aus den sich entleerenden Zellteilen zurückweichen, sondern noch Reste desselben in Form Hechtscher Fäden oder von Überbleibseln des Wandbelages erhalten bleiben. Auch wir haben derartiges bei dem mit *Chaetoceros* nahe verwandten *Bacteriastrum* gesehen. Die Sporenschale würde dann nicht eindeutig Oberflächenbildung sein. Nach den Erfahrungen der Elektronenmikroskopiker (siehe z.B. REIMANN, LEWIN und VOLCANI 1965) wird die Kieselsäure nicht an einer freien Oberfläche, sondern innerhalb vesiculärer Bildungen ausgeschieden. Vesikel aber können im Prinzip beliebige Lagen einnehmen.

Gewissermaßen zwischen dem für die *Stephanopyxis*-Hypovalva und dem von *Chaetoceros* liegt jener *dritte Typus der Membranbildung*, den MEUNIER (1910) bei der *Dauer*sporenbildung von *Thalassiosira gravida* schildert. Die Dinge beginnen wie bei *Chaetoceros* mit der Erstellung von zwei Mutterzellen. In jeder von diesen tritt eine etwas inäquale Plasmaspaltung mit ebenem Spalt auf; die kleinere Zelle liegt, bezogen auf das Paar, außen. Die Spaltfläche wölbt sich nun unter Autolyse (und Resorption ?) des Plasmas der kleineren Zelle in diese hinein vor und verkieselt als Oberschale der Dauerspore. Die Hypovalva der Spore entsteht dann auf der anderen Seite in grundsätzlich ähnlicher Weise, so daß schließlich ein Paar von bikonvexen Dauersporen vorliegt. Auch dieses Ergebnis entspricht dem Verhalten von *Chaetoceros*. Bemerkenswert ist nun, daß die beiden benachbarten Restzellen nach MEUNIER keine Kerne erhalten, die Plasmaspaltung also nicht wie üblich auf eine Teilungsspindel bezogen wird. Aus der Analogie zu den *Chaetoceros*-Dauersporen und den Auxosporen wird es wahrscheinlich, daß die Kerne auch hier geteilt werden, die Tochterkerne aber nicht in die Restzellen gelangen, doch gibt MEUNIERs Schilderung dafür keine Anhaltspunkte. Da hier eine Inäqualteilung in eine kernlose und in eine kernhaltige Zelle angegeben wird, wäre eine Nachuntersuchung von großem Interesse.

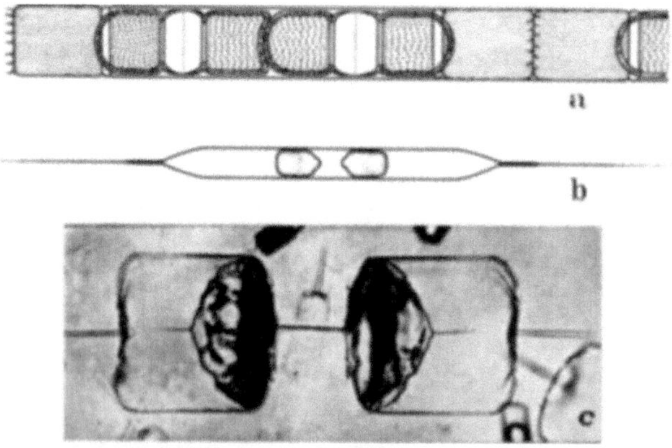

Abb. 41a—c. Dauersporentypen. a *Melosira italica*. Hälfte einer Vierergruppe von Dauersporen. Erklärung s. Text. (Nach O. MÜLLER 1906.) b *Rhizosolenia setigera*, Mutterzelle mit Dauersporen. Die einander zugekehrten Epivalven der Dauersporen sind becherförmig und tragen nur flach gewölbte Deckel (diese auf Figur nicht abgegrenzt). (Nach GRAN, aus CLEVE-EULER 1951.) c *Ditylum brightwellii*. Paar von Dauersporen in Kultur entstanden. Man beachte, daß bereits die einander zugekehrten Epivalven der Sporen von den normalen Valven etwas in der Form abweichen, also heterovalvat sind. (Photo G. DERSCH.)

Bemerkt sei noch, daß die Valven der Dauersporen fast immer ohne Vermittlung eines Gürtels aneinandergrenzen. Dadurch kommt es bei *Caetoceros teres*, *Ch. messanensis* und *Rhizosolenia setigera* (v. STOSCH, unveröff.) dazu, daß die Schalenmäntel sich übergreifen (Abb. 40e), also die Rolle des Gürtels übernehmen, wofür bei vegetativen Diatomeenschalen kein Beispiel bekannt ist. Nur an der Dauerspore von *Rhizosolenia polydactyla Castracane* = *Rh. styliformis* var. *latissima* Brightwell ist ein wohl ausgebildeter Gürtelteil vorhanden (VAN HEURCK 1909).

Klassifizierung der Dauerzellen. Solange die Modi der Schalenentstehung nicht überall genau bekannt sind, ist es notwendig, die Diatomeendauersporen pragmatisch zu klassifizieren. Daher sei eine *Schale* dann als *exogener Entstehung* bezeichnet, wenn sie nach Fertigstellung der Spore nicht in der Mutterzelle eingeschlossen bleibt, sonst als *endogen*:

1. Sporenepivalva exogen

a) Sporen entstehen in Gruppen von vieren:

Melosira italica-Typ (Abb. 41a): Durch eine äquivalvate Teilung zwei Tochterzellen, die sich jede mit vom Gruppenzentrum weg stark nach außen gebogenen Spaltflächen heterovalvat weiterteilen. Die an diesen Flächen entstehenden Valven sind Oberschalen je zweier formverschiedener Schwestercysten, deren Hypovalven dann endogen gebildet werden (O. MÜLLER 1906).

b) Sporen in Gruppen von zweien:

α) Sporenhypovalva entsteht durch eine inäquale Cytokinese:

Stephanopyxis-Typ (s. oben, Abb. 39). Bekannt von *Stephanopyxis turris*, *St. palmeriana* (HUSTEDT 1930, CUPP 1943) und *Melosira arctica* (GRAN 1900).

β) Sporenhypovalva entsteht acytokinetisch:

Ditylum-Typ: *Ditylum brightwelli* (GRAN u. ANGST 1931, CUPP, v. STOSCH 1965) (s. Abb. 41c), *Chaetoceros didymus* (MEUNIER 1913) und wenige andere

Arten der Gattung, *Thalassiosira bioculata* p.p., *Th. fallax* (MEUNIER 1910); *Detonula confervacea* (GRAN 1908), *Fragilaria oceanica*, *Achnanthes taeniata* (beide MEUNIER 1910, HUSTEDT 1959).

2. Beide Sporenmembranen endogen.

a) Sporen in Gruppen zu zweien:

α) Teilung der Gruppenmutterzelle liefert Epivalven *der Spore: Rhizosolenia setigera*-Typ (Abb. 41 b). Hierher *Rhizosolenia setigera* (HENSEN 1887, CLEVE 1897) sowie anscheinend *Cerataulina bergonii* (HUSTEDT 1930).

β) Teilung der Gruppenmutterzelle liefert Hypovalva der Sporenmutterzelle: *Chaetoceros teres*-Typ (Abb. 40). Hier Mehrzahl der *Chaetoceros*- und *Bacteriastrum*-Arten (MEUNIER 1910, 1915), *Rhizosolenia styliformis* und *Fragilaria groenlandica* (MEUNIER 1910), *Schroederella schroederi* (BERGON 1902) und *Thalassiosira gravida* (s. o.).

b) Sporen einzeln.

α) In vegetativen Zellen: *Attheya zachariasi*-Typ[1]: Hierher alternativ *Chaetoceros*-Arten des Typ 3, wie *Ch. compressum* Lauder sens. Hustedt (MEUNIER 1913), *Ch. messanensis* (v. STOSCH, unveröff.) sowie vielleicht *Thalassiosira bioculata* p.p. (MEUNIER 1910), *Podosira glacialis* (GRAN 1908, s. aber MEUNIER 1910) und *Attheya zachariasi* - *Rhizosolenia eriensis* (SCHULZ 1929). *Rhizosolenia styliformis* var. *latissima* (s. o.) hätte wegen der Ausbildung eines Sporengürtels eine Sonderstellung.

β) In der Auxospore: *Leptocylindrus*-Typ: Hierher *Leptocylindrus danicus* (GRAN 1915) (Abb. 42a), *Chaetoceros eibenii* (PAVILLARD 1921). Die Erstlingsschalen der Auxospore sind die Valven der Dauerspore.

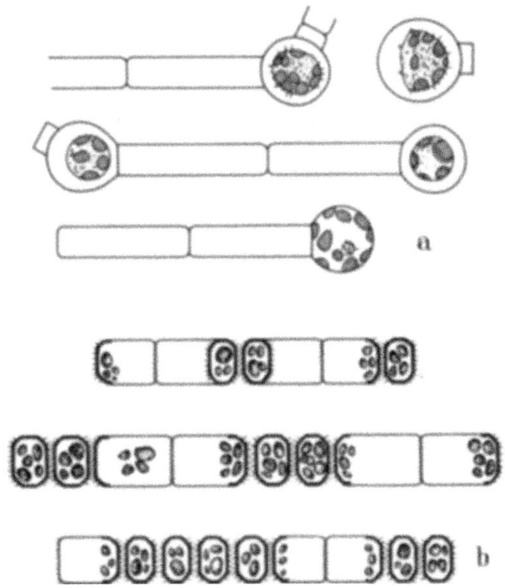

Abb. 42a u. b. a *Leptocylindrus danicus*. Dauersporen treten als Erstlingszellen in der Auxospore auf. (Nach GRAN, aus HUSTEDT 1930.) b *Detonula confervacea*. Entstehung von Dauerzellengruppen durch scheitelzellähnliche Teilungen der in der Gruppe distalen(langen) Zellen. Die Initiale ganz links im mittleren Faden wird noch (mindestens) eine weitere Dauerspore liefern, die drei anderen in gleichem Faden können nur noch die begonnene Cyste vollenden. Oberer Faden Mitte lieferte nur eine Spore je Mutterzelle. (Nach GRØNTVED 1956.)

Einen sehr interessanten Sonderfall bietet die bei 1 b β eingereihte Sporenbildung von *Detonula confervaceae*. Nach Untersuchungen von GRØNTVED (1956)[2] entstehen in den dänischen Gewässern nicht Auxosporenpaare, sondern Auxosporenketten (Abb. 42 b). In einer Entwicklung, die wie jene von *Stephanopyxis* beginnt, scheidet jede Sporenmutterzelle im zweiten Teilungsschritt des Paares zwei gleichartige und zwar Dauersporenschalen aus; die im Paar proximale komplettiert als Hypovalva die Dauerspore, die distale ist Epivalva einer zweiten Dauerspore. Die außen im Paar liegenden Zellen, die den Richtungskörpern von *Stephanopyxis* entsprechen, teilen sich nun scheitelzellenartig in der gleichen Weise weiter, bis zuletzt eine acytokinetische Schalenbildung sie entleert und

[1] Nur mit Vorbehalt als Typenvertreter genannt, da bei den solitären Arten ohne Verwendung von Kulturen nicht sicher festzustellen, ob die Sporen-Mutterzellen gepaart entstehen oder, wie hier angenommen, nicht.

[2] Herr GRØNTVED stellte freundlicherweise mehrfach Material der Art in Sporenbildung zur Verfügung, das seine Beobachtungen zu bestätigen gestattete.

damit den Vorgang unterbricht. Geschieht dieses Letzte bereits beim ersten Teilungsschritt der „Scheitelzelle", so erhält man die von Gran beobachtete Entwicklung nach dem *Ditylum*-Typ. Der Modus erscheint deshalb wichtig, weil er zum Verständnis des Pleomorphismus der Diatomeen beitragen könnte (s. u.).

Doppelschalen. Es herrscht keine Einstimmigkeit darüber, ob die „Doppelschalenbildungen" (Kratikularzustände) mancher Süßwasserdiatomeen zu den Dauerzellen zu rechnen sind. Die Sporenschalen wären dann in Dicke und Struktur, im Gegensatz zur Form, nicht sehr wesentlich von denjenigen vegetativer Zellen verschieden. Sie enthalten häufig Reservestoffe und treten in Depressionszuständen der Alge am Ende der Vegetationsperiode auf (Geitler 1953b, dort Lit.; Geitler 1963). Nach Pfitzer (1871) vermögen sie auch wieder zu vegetativen Zellen auszuwachsen. Sie werden in ähnlich artspezifisch festgelegter Teilungsfolge gebildet wie die eigentlichen Dauersporen. So würden die hierher gehörigen Bildungen von *Meridion circulare* (Geitler 1963) dem *Ditylum*-, diejenigen von *Eunotia pectinalis* (Geitler 1953a, b, 1963) dem *Chaetoceros*-Typ zuzurechnen sein. Nur entstehen die endogenen Schalen beider nicht acytokinetisch, sondern durch Inäqualteilungen extremer Art (s. auch v. Stosch und Drebes 1964). Nicht zu verwechseln hiermit sind die *Häutungen* (Schalenabwurf und Neubildung), die unter gewissen Umständen bei dazu disponierten Arten, wie *Ditylum* und *Achnanthes*, wiederholt und als mehr oder weniger pathologische Erscheinungen auftreten.

Entwicklungsphysiologie der Dauerzellen.

Bildung. *In der Natur* entstehen Dauersporen häufig am Ende der Vegetationsperiode der sie erzeugenden Art (Schütt 1888, Hart 1942, Braarud 1945). Bei der kryophilen *Detonula* geschieht dies nach Grøntved (1956) bei der Erwärmung des Wassers im Frühjahr. Sie sinken danach ab (Schütt) und stellen ihrem Namen entsprechend wohl das Einsaatmaterial für die nächste Planktonblüte. Weitere Angaben zur Ökologie könnten aus der ausgedehnten planktologischen Literatur zweifellos gewonnen werden.

In Kultur sind Dauersporen mehrfach erhalten worden: Von Gross (1937) bei *Chaetoceros didymus*, von Braarud (1945) bei *Ch. curvisetus*, *Ch. affinis*, von Guillard und Ryther (1962) bei der kryophilen *Detonula confervacea* an der oberen Grenze (10°) des Existenzbereiches und von uns (v. Stosch und Drebes 1964, v. Stosch 1965 u. unveröff.) bei *Stephanopyxis turris*, *St. palmeriana*, *Chaetoceros compressus*, *Ch. teres*, *Ch. Eibenii* (Typ 2b β, in den Auxosporen!), *Rhizosolenia setigera* und *Ditylum brightwelli*.

Experimentelles zur Dauersporenentstehung. Wie jeder Entwicklungsprozeß hängt die Fähigkeit, Dauersporen zu bilden, vom Idiotypus einerseits und im Zusammenspiel mit diesem von inneren und äußeren Bedingungen andererseits ab.

Die Betonung des *Genotypus* ist mehr als eine Banalität, weil wir einerseits auch innerhalb einer Art mit Differenzen der Neigung, Dauersporen zu bilden, zu rechnen haben. Selbst die sexuellen Nachkommen eines Klones können, obwohl äußerlich und in der Vitalität gleichartig, sehr unterschiedliche Tendenz besitzen, Cysten zu liefern, Eigenschaften, die sich mindestens bei *Stephanopyxis* während der Weiterkultur halten (*Stephanopyxis turris*, v. Stosch 1965; *Rhizosolenia setigera*, v. Stosch unveröff.). Andererseits kann auch für nahe verwandte Arten die Abhängigkeit der Dauerzellenbildung von der das Leben der Diatomeen beherrschenden inneren Bedingung, der Zellbreite, durchaus verschieden sein. Dafür kennen wir ein Beispiel.

Die Fähigkeit, Sexualzellen (und damit Auxosporen) zu entwickeln, kommt bei Diatomeen allgemein nur Zellen aus einem mehr oder weniger breiten Bezirk im unteren Teil des arteigenen Größenbereichs zu. Bei *Stephanopyxis turris* vikariieren nun in an sich „einleuchtender" Weise die Bereiche für die Bildung der Dauerzellen als vegetativer und der Auxosporen als generativer Fortpflanzungsart, wenigstens weitgehend (Abb. 43). Nur relativ breite Zellen sind zu ersterer fähig. An der Grenze der eingezeichneten Bereiche nimmt die Neigung zur Sporenbildung ab. Bei der wärmeliebenden *Stephanopyxis palmeriana* (der Klon stammt aus den Gewässern von Tokio) jedoch decken sich sexueller und Dauerzellenbereich weitgehend. Nur Zellen mittlerer und geringer Breite sind zur Cystenbildung imstande (DREBES 1965).

Die Fähigkeit zur Dauersporenbildung ist, wie diejenige zur Sexualisierung, keine Funktion des Zellalters als solchen, sondern nur abhängig von der Zellbreite. So gewinnen schmale Zellen (etwa 15 μ) von *Stephanopyxis turris* die Möglichkeit, die Sonderzellen zu differenzieren, sofort wieder, wenn sie auf genügend hohe Breite vegetativ vergrößert (s. S. 659) wurden. Ein durch letzteren Prozeß periodisch regenerierter Klon entwickelt noch nach mehr als 10 Jahren normale Dauerzellen.

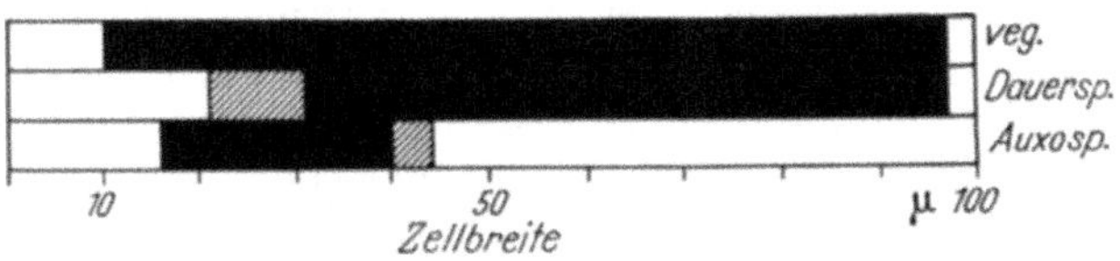

Abb. 43. Existenzbereiche für vegetative Zellen, Dauersporen und Auxosporen von *Stephanopyxis turris*. In den schraffierten Bereichen erfolgt Bildung erschwert. (Nach V. STOSCH und DREBES 1964.)

Um zu den *Außenbedingungen* überzugehen, werden Dauerzellen bei *Stephanopyxis turris* noch bei 3° gebildet; die obere Grenze liegt normalerweise zwischen 18° und 21° (sexuelle Fortpflanzung etwas über 21°, vegetative Vermehrung höher). Jedoch wird durch tiefere Temperaturen (15°) unter Lichtbedingungen (100 Lux), die, wie sogleich erläutert wird, die Sporenbildung nicht mehr zulassen, in einigen Tagen (Minimum 1 Tag) ein Zustand induziert, aus dem heraus die Bildung auch noch bei 21° (nicht mehr 24°) eintritt. Versuche mit verschiedener Lichtintensität bei 15° ergaben als untere Grenze für die Sporenbildung 100 Lux (Tageslänge wie in folgendem 16:8 Std). Diese Beleuchtung entspricht unter den Kulturbedingungen aber nahezu dem „ökologischen Kompensationspunkt", bei ihr finden entsprechend kaum noch vegetativ Teilungen statt. Da die Dauerzellenbildung (s. S. 662) zwei Zellteilungen erfordert, ist diese Grenze zu erwarten. Ist jedoch die erste (heterovalvate) Teilung abgelaufen, so kann die Inäqualteilung auch im Dunkeln stattfinden. Von 200 Lux setzt die Bildung der Sporen bis zu einem Optimum bei 800 Lux (einer relativ geringen Beleuchtungsstärke) ein, um dann bei höheren Lichtmengen rasch wieder abzusinken und bei 5000 Lux sehr gering zu werden. Was die Tageslänge betrifft, so wirkte Dauerlicht bei 15° stark hemmend auf die Entwicklung der Sporen; letztere fand dann nur noch bei geringer Beleuchtungsstärke (200 Lux) statt. Beleuchtungen im 10- und 8 Std-Tag sind dagegen mindestens nicht hinderlich.

Die Keimung der Dauersporen. Sie wurde von BRAARUD (1945) in Kultur erzielt, als er aus dem Plankton stammende Sporen von *Chaetoceros compressus*, *Ch. curvisetus* und *Ch. debilis* in Nährlösung brachte. Dauersporen von *Stephanopyxis turris*, welche in 12° entstanden, beginnen, 3 Wochen später auf 24° überführt, am nächsten Tag zu keimen, gleichgültig ob sie in Dunkelheit oder im Licht gehalten werden. Bei frisch gebildeten Dauerzellen ist die Keimungslatenzzeit jedoch auf 4 Tage verlängert, so daß eine relative Keimruhe existiert. Die Reservestoffe der Spore reichen zur Produktion vierzelliger Keimlinge (bei Dunkelheit) aus (v. STOSCH und DREBES 1964).

Zur Keimung weichen die beiden Valven der Dauerspore auseinander. Sie werden nach einer Zellteilung zu den Epivalven des Keimzellpaares (Abb. 39d) und später zu den Endschalen der Keimlingskolonie. Der Keimungsverlauf für *Chaetoceros* wurde von Lohmann (1920, zit. nach Michailowa) und Michailowa (1962) beschrieben. Hier werden die Sporenschalen bei oder nach der Keimung abgeworfen und nur von der Epitheca der Keimlingszelle Borsten entwickelt. Auch die *Rhizosolenia setigera*-Keimlingszelle macht sich von den Sporenschalen frei (v. Stosch unveröff.).

d) Pleomorphismus.

1. Abgrenzung.

Weil in den einen Dimorphismus erzeugenden Vorgängen von den bisherigen gestaltlich abweichende Zellen entstehen, welche sich als eigene „Phase" möglicherweise veränderter ökologischer Valenz, aber ungeänderten Genotypus weitervermehren, sei der Pleomorphismus der Diatomeen hier behandelt. Die in Ketten gebildeten Sporen von *Detonula* (S. 665) würden jedoch keine Phase darstellen. Das wäre erst der Fall, wenn sie sich ohne Eigenschaftswechsel selbst vermehren könnten, wofür es keinen Anhaltspunkt gibt.

Pleomorphismus ist bis auf die Ausnahme *Phaeodactylon* (Wilson 1942, Lewin et al. 1958, Lewin 1958, dort weitere Literatur) nur von Naturmaterial bekannt und besteht darin, daß sich zunächst zu verschiedenen Arten, Gattungen oder sogar Familien gerechnete Diatomeen als zusammengehörig erweisen. Bei konserviertem Material — die schönsten Beispiele stammen aus dem Plankton von Antarktis und Arktis — erkennt man dies durch Zellen, welche zwei verschiedene Formtypen in ihren beiden Schalen oder deren Teilen vereinigen, also heterovalvat bzw. heterothecat sind. Ausschließen muß man jedoch

1. die Fälle, in welchen nur quantitative Verschiedenheiten im Verkieselungszustand vorliegen und

2. diejenigen, bei denen die Neubildungen nicht teilungsfähige Phasen liefern, sondern wie bei Dauersporen (s. S. 661) auf einzelne Zellen beschränkt sind oder wo sie Gametangien darstellen.

Zu 1. wissen wir heute (v. Stosch 1942, Jørgensen 1955, Lewin 1957), daß die Schalendicke mancher Diatomeen vom Angebot an Kieselsäure und vermutlich von weiteren Faktoren, vor allem wohl der Wachstumsgeschwindigkeit abhängt (Lund 1954). Dickschaligkeit führt eo ipso zur Verstärkung und Verdeutlichung von Strukturen, was an sich noch kein Grund wäre, von Pleomorphismus zu sprechen. Erst das Hinzukommen von Unterschieden in der Form der Schalen oder in der Geometrie ihrer Strukturen würde dies erlauben. Der Pleomorphismus der Melosiren (O. Müller 1903, 1904, 1906) wird neben kleineren Formmerkmalen vor allem auf Unterschiede in den gegenseitigen Abständen und der Gestalt der Poren bezogen, die vom Entdecker wohl manchmal etwas überhöht gesehen wurden (Hustedt 1930), als solche aber existieren, z.B. bei *M. granulata*, *M. nyassensis* (Hustedt) und minder deutlich bei *M. islandica var. helvetica* (Cleve-Euler 1912). Doch ist *Melosira italica* var. *subarctica* nach Lund (1954) in der Englischen Seenplatte nicht pleomorph, sondern wandelt nur die Schalendicke, wohl aber nach O. Müller (1906) im Material von Island.

Unter Punkt 2. fällt mindestens überwiegend das von Müller (1906) als Pleomorphismus von *Melosira islandica* Vorgestellte. Die von der grob strukturierten Normalform abweichende, fein punktierte „Phase" ist, wie aus den Angaben über die Schalenlängen und über die Abhängigkeit dieses „Dimorphismus" von der Zellbreite (v. Stosch 1951), sowie aus dem vermerkten Zusammenhang mit der Auxosporenbildung hervorgeht, nichts als die Gesamtheit der männlich differenzierten Zellen, der Spermatogonien. Auch bei der „Pleomorphie“ von *Biddulphia anthropomorpha* van Heurck, welche sechs „Arten“ dieses Autors miteinander verbindet (MANGIN 1914, 1915, dort als *B. polymorpha* (Grunow) Wolle bezeichnet; HENDEY 1937; s. auch HEIDEN und KOLBE 1928), handelt es sich vielleicht um entwicklungsgeschichtliche Stadien, etwa eine allerdings sehr merkwürdige Entstehung von Dauersporen. Auch HART (1942) vertritt diese Ansicht. Jedenfalls wird nicht sicher, auch bei MANGIN 1915 nicht, ob mehr als eine dieser Formen als selbständige Phase auftritt.

2. Vorkommen.

Häufig finden sich in Dimorphismen grobe dauerzellenartige und zarte Formen vereinigt, wie bei dem Übergang der fo. *hiemalis* von *Rhizosolenia hebetata* in die

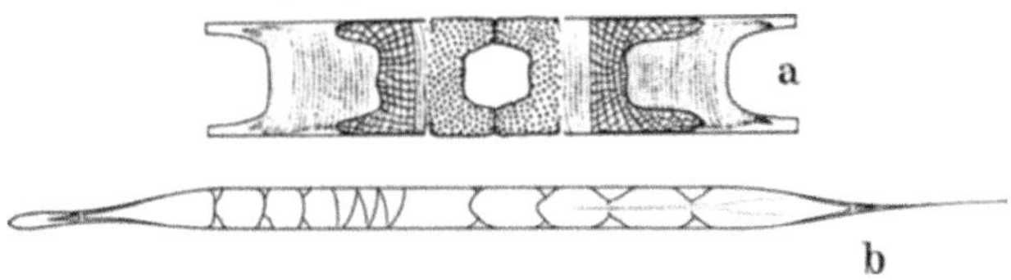

Abb. 44a u. b. Pleomorphismus bei Diatomeen. a *Eucampia balaustium*. Die Entstehung der *Eucampia*-Phase aus der *Moelleria*-Phase. Die grobe Maschenstruktur der beiden endogenen Schalen entspricht ihrem jungen, noch schwach verkieselten Zustand, später wird die Punkte-Struktur der in der Mitte liegenden, zuerst gebildeten Schalen erreicht. Weitere Erklärungen s. Text. (Nach MANGIN 1915.) b *Rhizosolenia hebetata*-Zelle mit den Phasen *semispina* (rechts) und *hiemalis* (links), die außer in den Calyptren auch im Gürtelbau verschieden sind. (Nach GRAN 1904.)

fo. *semispina* (GRAN 1904, MEUNIER 1910), die erste mit recht massiver und am Ende stumpf-keuliger Valva (Abb. 44b), die zweite zarter und in einen langen Dorn auslaufend. Ähnlich verhält sich ein von PAVILLARD (1937) beschriebener Fall aus der gleichen Gattung, sowie die mit *Moelleria antarctica* dimorph zusammenhängende *Eucampia balaustium* (Lit. s. S. 670) und ferner die pleomorphen Melosiren, die meist in zwei „Phasen“ auftreten, von denen jedoch *M. islandica* var. *helvetica* nach CLEVE-EULER (l.c.) in drei Phasen verschiedener Porendichte und Membrandicke vorkommt. Unterschiede im Verkieselungsgrad treten auch auf bei den drei im übrigen durch ihre Gürtelbandstrukturen charakterisierten Phasen, welche HENDEY (1937) bei *Dactyliosolen antarcticus*, sowie den fünf durch eine ganze Reihe von Merkmalen differierenden Typen, welche er bei *Corethron criophilum* unterscheidet und die als Phasen ebenso viele ältere Arten

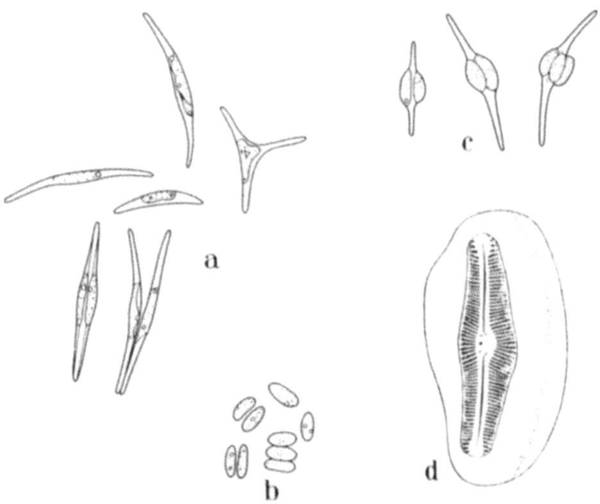

Abb. 45a—d. Pleomorphismus von *Phaeodactylon tricornutum*. a Die Spindelform, dabei eine triradiate Zelle, b die ovale Form, c Entstehung der ovalen aus der Spindelform. (a—c nach WILSON 1942.) d Ovale Form: Die zarte Kohlenhydrat-Membran trägt auf der einen ihrer beiden Schalen eine verkieselte Valva von *Cymbella*-Struktur. [Nach elektronenmikroskopischen Aufnahmen von LEWIN et al. (1959) und LEWIN (1959) gezeichnet.]

ersetzen; in beiden Fällen kommen wieder Mischformen vor, welche Schalen zweier Formzustände vereinigen (s. auch Manguin 1960, Heiden u. Kolbe). Verschiedene Familien bzw. Subfamilien sind in heterovalvaten Zellen vertreten, die, in zwei Exemplaren aufgefunden, *Asteromphalus roperianus* mit *Coscinodiscus lineatus* identifizieren lassen (Wood 1959).

Die einzige dem Referenten bekant gewordene nachweislich *pleomorphe Pennate* ist *Phaeodactylon tricornutum*, bei der die systematische Zugehörigkeit bis in die jüngste Zeit ungeklärt war (Lit. S. 668). Bei ihr treten zwei morphologische Phasen auf (Abb. 45), die möglicherweise jede eine eigene ökologische Position einnehmen: nämlich eine bis 35 μ lange und etwa 3 μ breite „Normalform" von Gestalt einer leicht gebogenen Spindel (Abb. 45a), früher mit *Nitzschia closterium* verwechselt (Lewin 1958), sowie ein zweiter ovaler etwa 7 × 3 μ messender Typus (Abb. 45b). Beide besitzen zarte und übereinandergreifende Theken ohne Zeichnung und nur die ovale Form trägt an einer dieser Theken, nicht der anderen — singuläres Faktum — eine strukturierte Valva aus Kieselsäure, einer winzigen *Cymbella* ähnlich — also auch mit Raphe, jedoch ohne Gürtel — und scheidet außerdem allein eine Kohlenhydratschleimhülle aus. Die Spindelform ist unbeweglich (Lewin et al. 1958), die „Hemicymbella" beweglich. Zur biradiaten Spindelform tritt noch eine triradiate Nebenform hinzu, die hier außer Betracht bleiben kann, obwohl sie der Art den Namen gab. Die Radiation wird nur durch den Mechanismus der Zellteilung weitergegeben, hat darüber hinaus keine besonderen Form- oder Strukturmerkmale und ist daher nicht mit den Phasen vergleichbar.

3. Phasenübergänge.

Die Entstehung der neuen Phase geschieht wohl in den meisten Fällen — anders sind die immer wieder gefundenen Zellen mit zwei ganz verschiedenen Schalen kaum zu verstehen — durch heterovalvate Teilungen. Bei Formen mit sehr lang anhaltendem Streckungswachstum, wie *Dactyliosolen antarcticus* Castr. sens. Hendey, zeigt sich die Formänderung jedoch bereits in der Struktur der nacheinander abgeschiedenen Zwischenbänder des Gürtels scharf (Hendey) oder auch unter Vermittlung einer Zwischenzone an (Heide und Kolbe). Leider beruht die einzige Studie, welche bei der leicht zugänglichen polymorphen Gattung *Melosira* (Cleve-Euler 1912) der Frage der Genese der neuen Phase nachgeht, auf unrichtigen Vorstellungen über die entwicklungsmechanischen Möglichkeiten der Diatomeenzelle. Von den zahlreichen Naturbeobachtungen bleibt daher zur Rekonstruktion der Übergangsprozesse nur der Fall der kettenbildenden *Eucampia balaustium*, wie er nach den sich ergänzenden Angaben mehrerer Arbeiten über antarktisches Plankton wahrscheinlich wird. Das Resultat kann hier nur gegeben, nicht näher diskutiert werden (Karsten 1905, van Heurck 1909, Mangin 1914, 1915, Heiden und Kolbe 1928, Hart 1934, 1942, Hendey 1937, Manguin 1960). Die grobschalige *Eucampia balaustium*-Phase entsteht aus der zartwandigen *Moelleria* durch eine heterovalvate Teilung (Abb. 44a), gefolgt von beiderseitigen „endogenen" Schalenbildungen[1], wobei zu beoachten ist, daß die Fortsätze dieser neuen Schalen spitzlich zulaufen, die beiden dickschaligen Zellen also ihrerseits in gemäßigter Form heterovalvat sind. Es entsteht ein Zwilling, der in seiner Genese ganz einem Dauersporenpaar vom Typus *Ditylum* (Abb. 41c) bzw. *Stephanopyxis* (Abb. 39) entspricht und dessen Zellen nun unter dem Muster der beiden ersten Schalen des Paares weiterteilend Kolonien der grobstrukturierten Phase erzeugen. Bei dieser würde es sich also vergleichend morphologisch um eine vegetativ gewordene Dauerspore handeln, womit keine phylogenetische Spekulation beabsichtigt ist.

[1] Endogen im Sinne unserer Definition (S. 664).

Einigermaßen sicher erscheint, daß in der *balaustium*-Form eine Phase von allerdings recht begrenzter Vermehrungsfähigkeit, nicht aber eine Dauersporenbildung nach dem *Detonula*-Typ vorliegt. Der umgekehrte Übergang wird, wie Abb. 7 und 7a der Tafel 11 bei KARSTEN (1905) zeigten, durch eine einfache, heterovalvate Teilung eingeleitet und verläuft damit ähnlich wie die Keimung der *Stephanopyxis*-Dauerzellen. Allerdings sind die erstgebildeten Schalen in der Struktur ihrer Valven, nicht aber in der Entwicklung der Gürtel noch *balaustium*-ähnlich.

Bei dem einzigen in Kultur untersuchten Polymorphismus, jenem von *Phaeodactylon*, scheint die Genese der massiver verkieselten Phase auf dem Wege einer endosporenartigen Kontraktion des Plasmas der Mutterzelle, gefolgt von Zellteilungen, der so entstandenen ovalen Form und Abwerfen der alten Membran zu geschehen (LEWIN et al.), nicht, wie WILSON glaubte, durch inäquale „exogene“ Teilungen der spindelförmigen Zelle (Abb. 45c). Der Übergang oval—spindlig vollzieht sich dagegen anscheinend durch das Auswachsen der ovalen Form an den Polen, das sich (gegen den Widerstand der Membran ?) über mehrere Teilungsgenerationen hinzieht und erst ein Ende findet, wenn mit einer rund 30 μ langen Spindelgestalt die neue Gleichgewichtslage erreicht ist (WILSON). Einem Herabteilen scheinen die Zellen nicht unterworfen zu sein. Der Beginn jenes Wachstums dürfte aber in einer sprunghaften Eigenschaftsänderung (begleitet vom Verlust der *Cymbella*-Schale ?) jeweils *einer* Mutterzelle seine Ursache haben. Sexuelle Prozesse, auch eine Auxosporenbildung (*sens. strict.* s. v. STOSCH 1962, 1965) wurde bisher bei der Art nicht bekannt.

4. Entwicklungsphysiologische Dignität.

Die entwicklungsphysiologische Stellung des Pleomorphismus der Diatomeen dürfte im Bereich der Dauermodifikationen zu suchen sein, auch wenn seine Verursachung durch Außenfaktoren nicht erwiesen ist. Mutationen, als Genmutationen, an die O. MÜLLER bei *Melosira* zunächst dachte (1903, 1904), bis er (1906) erkannte, daß die Änderungen *geläufig* reversibel sind, dürften schon wegen der Diploidie der Diatomeen kaum in Betracht zu ziehen sein. Doch können bei *Phaeodactylon* die Phasenübergänge — wohl beide Folgen von einmaligen Elementar-ereignissen an einer Zelle — als solche bisher offenbar nicht dirigiert werden (LEWIN et al. und J. C. LEWIN pers. Mitt.). Außenbedingungen würden dann nur als Selektionsfaktoren wirken.

5. Ökologisches.

Über die ökologische Bedeutung des Pleomorphismus bestehen nur Vermutungen. Bei *Phaeodactylon* erzeugt das geschilderte Phänomen eine Ambivalenz der Zelle, welche den Organismus, im Experiment wenigstens, in der gestreckten Form für flüssige Medien und in der gedrungenen beweglichen Phase für feste Substrate geeignet macht (LEWIN et al.). Die gröberen Schalen der ozeanischen *Rhizosolenia hebetata* wurden von GRAN (1904), das gleiche bei der antarktisch-nereitischen *Eucampia balaustium* durch KARSTEN (1905) als Winter- bzw. Kaltwasserformen angesehen. Später änderte sich die Lage im letzten Falle insofern, als MANGIN (1914, 1915) die Identität von *Moelleria antarctica* mit der Sommerform von *Eucampia balaustium* nachwies (Abb. 44a), was KARSTEN nicht oder nicht ganz klar erkannt hatte, sich im übrigen aber in dieser seiner Version der Auffassung KARSTENs anschließt. Ihm folgen HEIDEN und KOLBE, HENDEY und insbesondere HART mit seiner großen Erfahrung. Von *Melosira islandica* var. *helvetica*, die nach CLEVE-EULER (1912) im wesentlichen in drei Typen steigender Schalendicke und Grobporigkeit existiert, kommt im Mälarsee in Schweden die grobe Form der Wachstumsruhe oder -verlangsamung des Frühjahrs, die nächst feinere der pelagischen Tiefenpopulation des Sommers und die feinste und am meisten dünnschalige der oberflächennahen Hochproduktion des Herbstes zu: doch würde auch diese Untersuchung einer Vertiefung und Ausdehnung bedürfen.

e) Parthenogenese und Apogamie bei Diatomeen.

Auxosporen sind durch die S. 657 beschriebenen Prozesse bei ihrer Entstehung aus der Zygote, oder wie sich zeigen wird, auch der Azygote definiert; im letzten Falle kann der Vergleich mit verwandten allogamen Arten angezeigt sein. Sie entwickeln sich in überwiegendem Maße als Folgen einer sexuellen Zellverschmelzung. Wie die Zusammenstellungen bei Geitler (1932 und 1957) und Patrik (1954) zeigen, gibt es aber daneben einige Fälle der Auxosporenbildung, der eine Gametenverschmelzung nicht vorangeht; diese können heute vermehrt werden.

1. Centrales.

Beginnen wir mit den Centrales, so sei vorausgeschickt, daß die allogamen Verwandten der bisher involvierten Arten eine Oogonentwicklung nach dem *Melosira*-Typus besitzen: Das Oogon ist eine durch Umsteuerung einer vegetativen Zelle entstehende Meiocyte. Nach jedem Teilungsschritt der Meiosis degeneriert ein Kern, so daß das reife in den Oogonschalen eingeschlossene Ei einen funktionierenden haploiden und zwei pyknotische Kerne erhält. In der männlichen Entwicklung werden zunächst durch verkürzende und „depauperierende“ Teilungen kurze und plastidenarme Spermatogone differenziert, die nach Meiosis vier chromatophorenfreie Spermien liefern. Ein Spermium bzw. sein Kern dringt in das Ei ein, das nach der Karyogamie unter teilweiser Sprengung der Oogonschalen zur Auxospore anschwillt. *Melosira varians* ist wie alle bis heute untersuchten Centrales monözisch. Beim Herabteilen tritt sie erst in ein weibliches, dann in ein zwittriges und zuletzt bei ganz schmalen Zellen in ein rein männliches Stadium ein (v. Stosch 1951a).

Im Unterschied zum eben Geschilderten entsteht das Oogon von *Melosira nummuloides* nicht direkt aus der vegetativen Zelle, sondern diese macht noch eine differenzierende und inäquale Mitose durch (Rieth 1953, Erben 1959), ganz wie die Mutterzelle einer einzelnen Dauerspore von *Stephanopyxis*, deren größere Tochter nun in die Meiosis eintritt und zum Oogon wird. Erben zeigte nun, daß eine marine Herkunft (Helgoland) der Alge nach der Cytologie der Eientwicklung autogam ist: Die Tochterkerne der zweiten meiotischen Teilung bleiben beide erhalten und scheinen dann zu verschmelzen; eine Anwesenheit von Spermakernen im Ei wurde nicht beobachtet. Dennoch werden die für die Gattung normalen Spermatogone ausgebildet (dafür sprechen auch die von Schreiber 1931 für das gleiche Objekt gegebenen Bilder, nicht aber sein Text), diese entlassen auch auf die übliche Weise gebildete Spermien (Beobachtungen an drei Stämmen aus List/Sylt/Nordsee, v. Stosch u. Mitarb.). Auch unsere Lebendbeobachtungen (Drebes unveröff.) konnten niemals ein Eindringen des Spermiums in das Oogon feststellen. Die Nordseestämme sind also in aller Wahrscheinlichkeit in der angegebenen Weise autogam, entwickeln dennoch unter außerordentlicher Verschwendung an Zellen Spermien. Ein von Rieth im Leben studierter Binnenlandklon der gleichen Art (Salzstellen bei Artern/Mitteldeutschland) besitzt bei sonst gleichen äußeren Abläufen der Oogon- und Auxosporenbildung keine Spermatogone, würde also wesentlich „rentabler“ arbeiten. *Melosira moniliformis* var. *octogona* (Benennung nach Hustedt 1930, ist aber wohl selbständige Art) aus dem Mittelmeer würde sich hier anschließen. Sie liefert in Kultur Auxosporen aus Mutterzellen, welche hier direkt durch Umwandlung vegetativer Zellen entstehen, ohne daß Spermatogone auftreten (DREBES unveröff.). Von der bodenbewohnenden marinen *Melosira sulcata* wurden unter Umständen, die bei anderen zentrischen Grunddiatomeen Sexualität auslösten, Stadien der Meiosis im Oogon und Auxosporen gesehen, ohne daß Spermatogone auftraten (v. STOSCH 1951b).

Auch die Erfahrungen mit mehreren Klonen von *Actinoptychus undulatus* aus der Nordsee mahnen zur Vorsicht gegen eine Schlußweise, welche aus dem Vorkommen von Auxosporen und Spermien allein bereits eine sexuelle Bildung der ersteren zu postulieren versuchte. Die Entwicklung verläuft bis auf zahlreiche Details wie bei *Melosira varians*, auch die Folge weibliches, zwittriges und rein männliches Stadium im Verlauf der Größenabnahme ist vorhanden. Da sich aber Oogonien sowohl des weiblichen Stadiums als auch — nach rechtzeitiger Isolierung — der zwittrigen Phase in hohem Prozentanteil (bis 75%) zu Auxosporen entwickeln, ist die Art wenigstens fakultativ amiktisch. Tatsächlich wurde auch nur einmal das Eindringen eines Spermiums in das Oogon verfolgt, aber cytologisch niemals die bei allogamen Arten leicht auffindbaren Verschmelzungsstadien zwischen Eikern und Spermakern gesehen. Leider konnte die cytologische Untersuchung den Mechanismus der Parthenogenese nicht zweifelsfrei klären. Die Meiosis im Oogon verläuft bis zum Ende von RT I normal, es degeneriert ein Kern. Danach aber setzen nur 25% der Oogone die Entwicklung unter Bildung des zweiten pyknotischen Kernes fort. Diese Eier wären dann haploid. Da zwar eingedrungene männliche Kerne, aber keine Karyogamien gefunden wurden, müßten solche Eier abortieren, was mit der oben gegebenen Zahl in Übereinstimmung wäre. Die beobachteten 75% Auxosporen müßten dann aus den Oogonien mit nur einem pyknotischen Kern entstehen: wie kehren sie zur Diploidie zurück? Es wird vermutet, daß im Anschluß an die erste meiotische Teilung eine Centromeren-Spaltung und dadurch Chromosomenverdopplung eintritt. Doch wurde nur ein Stadium gefunden, das diese Annahme cytologisch stützt (BEHRE und v. STOSCH unveröff.). Es ist in jedem Falle auch hier merkwürdig, daß der ganze komplizierte männliche Apparat und zuletzt sogar ein vom Standpunkt der Erhaltung der Art und seiner selbst gänzlich hoffnungsloses männliches Stadium weiter erstellt wird.

2. Pennatae.

Unter den pennaten Diatomeen, die GEITLER (1932, 1957), zum Teil bereits mit Fragezeichen, als hierher gehörig aufführte, fallen heute die beiden *Rhabdonema*-Arten (v. STOSCH 1958, 1962) und weiter *Grammatophora marina* (MAGNE-SIMON 1960, 1962) als allogam (oogam bzw. morphologisch anisogam) fort. Der Besprechung der übrigen Fälle müssen wir wieder eine Bemerkung über den Sexualprozeß bei Fremdbefruchtern vorausschicken. Bei den allogamen pennaten Diatomeen paaren sich zwei Mutterzellen, die bis auf diejenigen der drei letzgenannten Arten vegetativen Zellen ähnlich sehen, und entwickeln sich — wir benützen die Gliederung GEITLERs (1957) — wie folgt weiter.

Bei Typ I (Normaltyp) entstehen durch Cytokinese nach dem ersten Schritt der Meiosis je zwei Gameten. In diesen geht im allgemeinen einer der Tochterkerne des zweiten meiotischen und acytokinetisch verlaufenden Teilungsschrittes zugrunde, so daß die Gameten endgültig je einen funktionierenden und einen abortierenden Kern enthalten, worauf sie allogam kopulieren (Abb. 47a—d). Morphologisch sind die Gameten, abgesehen von den Tabellariaceen (s.o.), einander in allen Fällen gleich.

Bei Typ II (reduzierter Typ) entsteht durch Inäqualteilung oder Acytokinese im ersten meiotischen Schritt ein Gamet mit zuletzt einem funktionierenden und einem bzw. zwei abortierten Kernen.

Bei beiden Typen kopulieren die Gameten verschiedener Gametangien paarweise. Sie können sich dabei, art- oder rassenspezifisch festgelegt, in physiologischer Isogamie gleichartig oder auch in Anisogamie verschieden stark bewegen. Die *Rhabdonema*-Arten unterscheiden sich durch extreme Differenzierung bereits der Gametangien und demzufolge auch der Gameten in den beiden Geschlechtern, wobei im übrigen *Rh. arcuatum* in beiden Geschlechtern dem Typ I folgt, während bei *Rh. adriaticum* die Spermatogonien sich nach Typ I, die Oogonien nach Typ II entwickeln. Die schwächer anisogame *Grammatophora* verhält sich in beiden Geschlechtern nach Typ II.

Fällt nun die Paarung oder doch die Allogamie zwischen den beiden Mutterzellen fort, so leiten sich ab:

Typ III (Automixis) von Typ I;

Typ IV (Apomixis) wahrscheinlich allgemein von Typ II.

Wir behandeln nur die einigermaßen gesicherten Fälle.

Typ III Automixis und das Problem der Geschlechtsbestimmung.

Kopulieren die Gameten einer Mutterzelle nach dem Normaltyp nicht mit denjenigen eines Partners, so können sie entweder A *pädogam* verschmelzen oder es fällt B bereits die die Gameten separierende Cytokinese aus, wodurch schließlich ein Doppelgamet mit zwei funktionierenden und zwei abortierten Kernen gedanklich entstehen sollte und merkwürdigerweise auch entsteht, dessen überlebende Kerne auch *autogam* fusionieren. Da A und B sich aus beobachtungstechnischen Gründen nicht immer sicher unterscheiden lassen, im Einzelfall auch alternativ realisiert sein mögen, behandeln wir hier beide Fälle, obwohl III A noch Gametenfusion besitzt und daher nach der eingangs getroffenen Festlegung auszuschließen wäre. Das angedeutete, bei dem Übergang von A zu B entstehende Problem ist folgendes: Bei Diatomeen sind gewöhnlich zwei Kerne in der gleichen Zelle nicht nebeneinander existenzfähig; das zeigt z.B. das Verhalten der überschüssigen Gametenkerne nach der zweiten meiotischen Teilung in Typ I, II und III A, ferner die Vorgänge bei den metagamen Mitosen (s. S. 657) und bei der endogenen Cystenbildung von *Chaetoceros* (s. S. 663). In allen diesen Fällen geht der überschüssige Kern unter Pyknose zugrunde. Dieser Mechanismus muß nach der ersten meiotischen Kernteilung autogamer Gameten abgeschaltet sein, obwohl er nach der zweiten Teilung der gegebenen Darstellung gemäß wieder funktioniert. Doch ist die Frage so komplex und so wenig abschließend darzustellen, daß wir uns mit dem Hinweis auf Diskussionen bei Geitler (1956) und v. Stosch und Drebes (1964) begnügen müssen.

Näher ist aber auf das von Geitler mehrfach (1951 a, 1953 a, 1957) diskutierte Problem des Ortes der Geschlechtsbestimmung einzugehen. Es tritt auf, wenn man versucht, eine die allogamen und pädogamen und autogamen Vorgänge umfassende Erklärung zu finden. Die mit Gallertstielen sessilen Gomphonemen führen ihre allogamen Kopulationen nach einer Paarbildung, bei der die eine Mutterzelle sessil bleibt und die zweite, häufig auch schmälere aktiv zuwandert, und nach darauffolgender Gallertausscheidung um das Paar herum derart durch, daß jedes Gametangium einen Gameten entsendet und einen empfängt. Reagieren bei *Gomphonema parvulum* die Gameten der einen Mutterzelle nicht, so kommt es, fakultativ also, zur Pädogamie der beiden Gameten in der zweiten. Während *G. constrictum* var. *typica* ebenfalls allogam ist (Geitler 1951), verhält sich ihre Varietät *capitata* (Abb. 46) obligatorisch pädogam (Geitler 1952 a) mit der wichtigen Abweichung gegenüber *G. parvulum*, daß bei ihr zwar „Kopulations“-Gallerte ausgeschieden wird, aber die Gametangienpaarung unterbleibt. Nebenbei bemerkt, verlaufen die pädogamen Kopulationen von *Gomphonema* allem Anschein nach isogam, so daß die Anisogamie bei Fremdbefruchtung vielleicht nur eine scheinbare ist. Sie könnte durch die schwierigen Raumverhältnisse innerhalb der Kopulationsgallerte bedingt und damit nur Folge der Paarung der Mutterzellen sein.

Versucht man nun eine Deutung der Tatsachen, so legt das Verhalten der Mutterzellen eine Geschlechtsbestimmung in der Diplophase mit der Folge physiologischer oder sogar morphologischer Anisogametangie, dasjenige der Pädogameten in Widerspruch dazu haplophasische Determination nahe. Bei der obligatorisch pädogamen *Gomphonema constricta* var. *capitata* besteht dieser Widerspruch nicht, da die Gametangienpaarung ausfällt; die Geschlechtsbestimmung würde hier dem Anschein nach in der Haplophase eintreten. Besteht man auf einer für beide Species einheitlichen Erklärung, so wäre dieses auch für *Gomphonema parvulum* anzunehmen und damit sowohl das Wanderungsverhalten der Gameten, des allogamen wie auch die Kopulation des pädogamen Falles verständlich gemacht. Dann wäre aber eine Begründung dafür nötig, daß bei Allogamie niemals die Gameten der gleichen Mutterzelle miteinander kopulieren. Es müßte Selbststerilität herrschen, die bei den Pädogamien etwa infolge Alterung der Gameten durchbrochen würde. Die Gametangienpaarung läge dann, wenig befriedigend, außerhalb der eigentlichen Sexualprozesse.

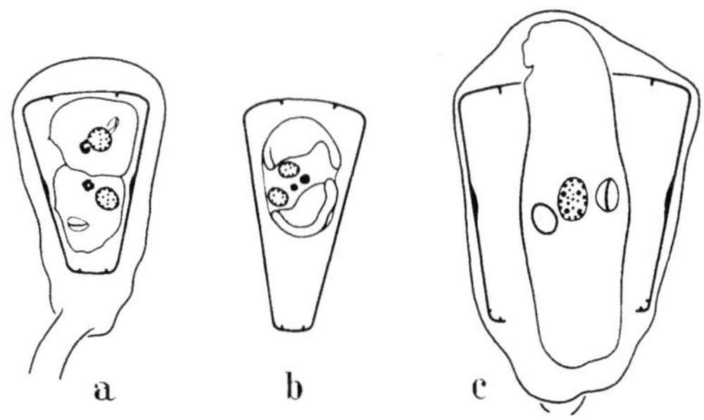

Abb. 46a—c. Pädogamie von *Gomphonema constrictum* var. *capitata*. Die beiden, je einen funktionierenden und je einen pyknotischen Kern enthaltenden Gameten einer nicht gepaarten Mutterzelle (a) kopulierten (b); die Zygote wächst nach Kernverschmelzung zur Auxospore aus (c). (Nach GEITLER 1952.)

Die Dinge werden nicht klarer durch die Beobachtungen an *Synedra ulna*. Hier wandern ausschließlich beide Gameten der einen allogamen Mutterzelle, die der anderen empfangen. Die Geschlechtsbestimmung ist also — nach dem hier wahrscheinlich nicht durch Gallerten beeinflußten Verhalten der Gameten — klar diplophasisch, bei (fakultativer) Parthenogenese aber erfolgt dennoch Zygotenbildung aus den Gameten *einer* Mutterzelle (GEITLER 1939).

Bei *Denticula tenuis* schließlich herrscht an einem Standort (I) in Lunz (Österreich) isogame Allogamie (Abb. 47a—g), auf einem anderen Standort (II) bei einer morphologisch gleichen Form Autogamie (nach Ausfall der Cytokinese im Gametangium, wie oben bereits besprochen), ohne vorherige Gametangienpaarung. Der Unterschied im Verhalten ist vermutlich Folge von Außenbedingungen (doch kann eine genotypische Verschiedenheit der Populationen nicht völlig ausgeschlossen werden). Eine morphologisch etwas abweichende Rasse aus Schladming verhält sich ganz wie Lunz II. Beim allogamen Prozeß wäre wegen der Gametangienpaarung wieder diplophasische Geschlechtsbestimmung anzunehmen. Bleibt die Paarung aus, so geschieht zweierlei: erstens ist die Cytokinese unterdrückt und zweitens wird die Geschlechtsdetermination erst bei den Gameten, hier ihren Kernen, erfolgen müssen. Da bei *Denticula* der Paarungsausfall greifbare morphologische Folgen hat, wird die Realität auch der physiologischen Änderung plausibel. Man kann den letzten Zusammenhang auf die Fälle *Gomphonema* und *Synedra* verallgemeinern: Ausfallende Paarung verschiebt die sonst diplophänotypische Geschlechtsdeterminierung auf die Haplophase oder „vorgesehene" Verschiebung der Determination verhindert die Paarung. Aber beides macht im Falle *G. parvulum*, wo Paarung und Pädogamie gleichzeitig bestehen, Schwierigkeiten, die man jedoch vielleicht durch

Zusatzannahmen beseitigen könnte. Doch bleibt eine weitere Möglichkeit: Grundsätzlich haben die besprochenen Objekte *diplophänotypische* Geschlechtsbestimmung; wenn aber, wie im Falle von *Gomphonema parvulum* und *Synedra ulna* durch Fehlen des Partners legitime Kopulationsmöglichkeiten nicht bestehen, kann die bereits festgelegte Sexualtendenz des einen Gameten umgestimmt werden. Zusatzannahme wäre dann nur noch, daß, wie oben bereits angedeutet, „die physiologische Anisogamie“ der allogamen Gomphonemen durch die Raumverhältnisse in der Kopulationsgallerte vorgetäuscht ist.

Für die letzte Auffassung sprechen nun zwei Momente: 1. existiert bei anderen, und zwar relativ ursprünglichen Pennaten — wie übrigens bei allen Centrales —

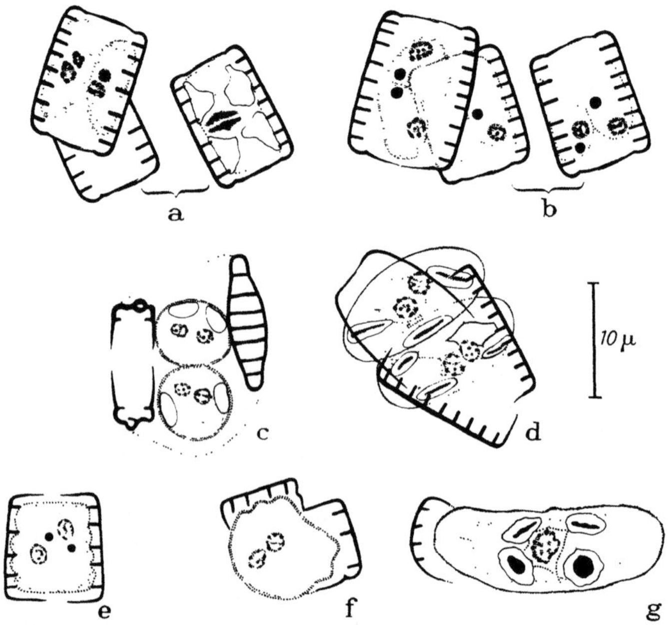

Abb. 47a—g. a—d Entwicklung von *Denticula tenuis* im Normaltyp der Auxosporenbildung. a Kopulationspaar, dessen linke Mutterzelle fertige Gameten besitzt, während sich rechte in Metaphase I befindet. b Beide Mutterzellen des Paares mit reifen Gameten, diese treten (c) aus der Mutterschale aus und kopulieren. Zygoten (d) wachsen danach zu Auxosporen heran (Kernverschmelzung erst etwas später). e—g Autogame Entwicklung. e die Mutterzelle ungepaart, die Gameten werden nicht individualisiert, so daß resultierende Zelle die Kernausrüstung zweier normaler Gameten enthält. f Junge Zygote tritt aus und wächst (g) zur Auxospore heran, Karyogamie gerade abgeschlossen. (Nach Geitler 1953, umgezeichnet.)

mit Sicherheit diplophasische Geschlechtsbestimmung, 2. kann wenigstens bei Gametangien die schon erfolgte Geschlechtsausprägung revertiert werden. Ad 1. Bei *Rhabdonema arcuatum* gibt es (v. Stosch 1962, unveröff.) bei modifikatorischer Geschlechtsbestimmung äußerlich von den vegetativen Zellen wie von den Oogonien durch Größe, Struktur und Zellinhalt unterschiedene Spermatogonien, die sogar in der differenzierten Form teilungsfähig sind und die in einer natürlichen Population immer kürzer und im Mittel, aber nicht im Einzelfall, auch *schmäler* (geringere Apikallänge) als die Oogonien ausfallen würden. Die Messungen Geitlers (1951d) an *Gomphonema*-Gametangienpaaren ergeben bezüglich der Breite für einige Arten das Gleiche, beweisen also wohl, daß bei diesen diplophasische Determination vorliegt. Die ebenfalls (in diesem Falle genotypisch) diplophasisch determinierte *Rh. adriaticum* würde aber in der „randomisierten“ Population — obwohl zwar beträchtliche Unterschiede zwischen den Geschlechtern in der Länge — wahrscheinlich überhaupt keine Unterschiede in der Breite geben, worauf es hier allein ankommt. Wenn GEITLER das gleiche bei anderen *Gomphonema*-Arten (oder -Rassen) findet, spricht das also keineswegs gegen eine diplophasische Geschlechtsdetermination bei diesen. Ad 2. Die Spermatogone von *Rhabdonema arcuatum* (s. o.) entstehen aus vegetativen Zellen spontan und irreversibel bei tiefer Temperatur (6°), sind unter diesen Bedingungen aber zu unbegrenzter Vermehrung fähig. Ein differenziertes Spermatogon dedifferenziert nur nach Temperaturerhöhung und kehrt zum vegetativen Zustand zurück. Seine Nachkommen aber geben unter erneuter Kultur bei 6° neben Spermatogonien auch Oogonien. Wenn eine Geschlechtsumkehr bei ausdifferenzierten Gametangien möglich ist, warum sollte sie bei Gameten nicht stattfinden können? Übrigens wurde kurzzeitliche Geschlechtsumkehr von (physiologisch) anisogamen Pilzgameten von KÖHLER (1956) bei *Synchytrium* beschrieben, auch die *Vaucheria*-Versuche F. v. WETTSTEINS (s. S. 691) werden ähnliches zeigen.

In der Gattung *Cymbella* sind die Möglichkeiten von Allogamie bei mehreren Arten, obligater Pädogamie bei *C. aspera* (GEITLER 1956), von Autogamie und/oder Pädogamie bei *C. sinuata* (GEITLER 1958c) und sicherer Autogamie bei *C. ventricosa* (GEITLER 1922, 1958c) verwirklicht. Pädogam ist nach älteren Beobachtungen (z. B. KARSTEN 1897) auch *Achnanthes subsessilis*. Obligate Autogamie findet sich noch bei *Amphora normanni* (GEITLER 1928) und *Synedra vaucheriae* (GEITLER 1958c), doch könnte in beiden Fällen auch Pädogamie vorliegen.

Apomixis.

Alle zu Typ IV gerechneten Prozesse entwickeln aus einer Mutterzelle eine Auxospore. Daß dabei Autogamie nicht erfolgt, und wie die Kernvorgänge in „Gameten" wirklich laufen, ist nur bei *Cocconeis placentula* var. *lineata* aus Lunz (Österreich) sichergestellt; doch kann man aus Analogiegründen annehmen, daß auch die anderen von GEITLER (1927 und später, zuletzt 1958a und b) untersuchten parthenogenetischen Formen der Art sich ähnlich verhalten.

Ehe wir auf diese eingehen, sei Anderes kurz zusammengestellt. *Cymatopleura solea* und *C. eliptica*: Scheinpaarung zweier Mutterzellen, direkte Entwicklung der „Gameten" zu Auxosporen. Voraus geht in jeder Mutterzelle vermutlich eine vegetative und acytokinetische Mitose, deren einer Tochterkern abortiert (KARSTEN 1900); *Surirella gemma* ebenso, doch cytologisch nicht untersucht (KARSTEN 1928, LIEBISCH 1929). *Bacillaria paradoxa* (KARSTEN 1899) und *Nitzschia* spec. (GEITLER 1932): eine ungepaarte Mutterzelle entwickelt eine Auxospore.

Sehr wichtig sind die langjährigen Untersuchungen GEITLERS (l. c.) an Varietäten und ihren Standortrassen von *Cocconeis placentula*. Die allogamen Formen der Art bilden je Mutterzelle nach Paarbildung und Gallertausscheidung in der als extreme Inäqualteilung ausgebildeten Meiosis I einen winzigen Richtungskörper und den Gameten. Die Zusammenführung der Geschlechtszellen geschieht danach isogam oder anisogam mittels Kopulationsschlauch. Von den acht untersuchten, gut charakterisierten morphologischen Varietäten sind var. *lineata* und „Var. II" sowie var. *euglypta* mit einer von drei Rassen und var. *klinorhapis* mit einer von zwei (drei) Rassen obligat parthenogenetisch; Sexualitätsverlust ist also überraschend häufig. Bei einigen allogamen Formen kommt Parthenogenese als seltene Ausnahme vor. In den meisten Fällen geht der parthenogenetischen Entwicklung eine „Scheinpaarung" von zwei Mutterzellen voraus, und danach ein mehr oder weniger einsinniges Eintreten in die Meiosis, was den Charakter eines rudimentären Sexualvorganges deutlich macht. Daneben kommt allerdings Auxosporenbildung in Einzelzellen vor. Bei der Reifung folgt auf eine meioseähnliche Prophase eine „Diakinese" mit ungepaarten Chromosomen, dann weiterer äußerer

Verlauf wie beim allogamen Gameten. Eine meist durch Paarbildung angestoßene Pseudomeiosis hat also eine diploide Azygote geliefert, die nun direkt zur Auxospore heranwächst.

Besonders interessieren im jetzigen Zusammenhang die Ausnahmen, Gamien bei Apomikten, bzw. Apomixis bei allogamen Formen, doch sind diese selten. Jedenfalls sehen die Auxosporen aus Ausnahme-Kopulation der gewöhnlich parthenogenetischen var. *lineata* normal aus, über die Entwicklungsfähigkeit ist nichts bekannt. Umgekehrt entwickeln sich in der allogamen Rasse von var. *klinorhaphis* einzelne Paare oder auch Zellen, wie vermutet wird in diploider Parthenogenese, weiter; das gleiche gilt für die allogame var. *euglyptoides*. Ergänzend noch die Bemerkung, daß bei einer anderen sexuellen Rasse der var. *klinorhaphis* gelegentlich nach mißlungenen Kopulationsversuchen in wahrscheinlich haploider Parthenogenese Auxosporen entstehen, die aber wohl nicht entwicklungsfähig sind (Geitler 1927).

3. Die Bedingungen.

Nach der Seite der Bedingungen für Parthenogenese und Apogamie bei Diatomeen wären hier in erster Linie diejenigen von Interesse, welche von der Allogamie zu Ersatzvorgängen führen und umgekehrt. Hier existiert nur das Beispiel von *Denticula tenuis* (s. S. 675), bei welcher das Ausbleiben von Gametangienpaarung mit Apogamie als Konsequenz gegenüber sonstiger Allogamie wahrscheinlich nur durch Standortverhältnisse bedingt war. Der entscheidende Faktor konnte nicht ermittelt werden, war aber anscheinend nicht die Temperatur.

Konstitutionelle, nicht allogame Auxosporenbildungen dürften vom vegetativen Zustand aus unter ähnlichen Bedingungen erfolgen, wie vergleichbare allogame. Größeres, experimentelles Material liegt allein für *Melosira nummuloides* vor (Bruckmayer-Berkenbusch 1955, Erben 1959). Das Größengesetz ist in der von der allogamen *Melosira varians* her bekannten Form (s. S. 672) auch hier gültig. Die Breite der Auxosporenmutterzellen (AMZ) ist nach oben (max. 24 μ) und unten (min. 12 μ) hin begrenzt, darunter existieren schmale Zellen (bis herab auf 6 μ), die nicht mehr Auxosporen — wohl aber Spermatogonien (Ref.) — liefern können. Die aktuellen Grenzen sind naturgemäß keine Konstanten, sondern ihre Lage wird durch die auslösenden Bedingungen beeinflußt und Bruckmayer-Berkenbusch hat demgemäß die Lage der oberen Grenze für die Auxosporenbildung — praktisch wird der Breitenmittelwert der AMZ bestimmt — als Maß für die Stärke des auslösenden Reizes benutzt. Die Wirkung der Zellgröße erklärt sie durch eine Verschiebung der Kern-Plasma-Relation. Eine solche liegt zweifellos vor, da das Volumen des Kerns beim Schmälerwerden etwas steigt, dasjenige der Zelle aber abnimmt (den von ihr dazu in ihrer Tabelle 10 angegebenen Zahlen liegen allerdings unrichtige Berechnungen zugrunde). Nach der Seite der physikalischen Faktoren können die Algen im ganzen Temperaturbereich zwischen 12 und 25° sowohl wachsen als Auxosporen bilden. Bei den tiefen Temperaturen entstehen letztere noch aus relativ breiten Zellen, nach höheren Temperaturen zu sinkt diese Breite ab. Auch nahe dem Kompensationspunkt für die Zellteilung liegende Beleuchtungen (50 Lux, 10:14 Std-Tag, 21°) drücken die Breite der AMZ, welche aber bereits bei 200 Lux maximal wird und nach der relativ geringen Beleuchtung von 1500 Lux hin durch die Konkurrenz der Sporenbildung mit dem Wachstum wieder stark abnimmt. Doch ist außer der Intensität auch die Dauer der Beleuchtung von Bedeutung, das Maximum der Breiten für AMZ liegt bei 1500 Lux (15:9 Std-Tag, 21°). Bei 24° und Dauerlicht wird die Auxosporenbildung zugunsten des Wachstums ganz aufgegeben, andererseits wirkt auch bei Dauerlicht geringe Beleuchtung (z. B. 250 Lux, 24°) für die Auxosporenbildung günstig. Es ist also durch ein entsprechendes Licht- und Temperaturregime möglich, die Auxosporenbildung bei anhaltendem Wachstum ganz zu unterdrücken; nicht das gleiche gelang ERBEN durch N- und P-Mangel oder -Überschuß, sowie Zugabe einiger Stoffwechselgifte. Von chemischen Agentien fördert Erdextrakt, ebenso eine Anzahl von Aminosäuren, ohne daß deren C-Gerüst von spezifischer Bedeutung erscheint; doch ist die Wirkung der ersteren nicht auf ihren Gehalt an letzteren zurückzuführen (BRUCKMAYER-BERKENBUSCH). Übrigens bleibt auch unter günstigsten Sporulationsbedingungen ein Teil der Zellen vegetativ. Welcher Faktor in einer an sich sporulationsbereiten Kultur eine Zelle zur AMZ determiniert, ist noch unbekannt. Biochemische Unterschiede (Aminosäuren, Monosaccharide, beide frei und nach Hydrolyse) zwischen ihrer Größe nach zu Auxosporenbildung befähigten und nicht befähigten Subklonen der Art waren nicht festzustellen (ERBEN).

Literatur.

BRAARUD, T.: Experimental studies on marine plankton diatoms. Avh. Norske Vidensk Akad. Oslo, I. math.-nat. Kl. **1945** (10), 1—16 (1945). — BEGER, H.: Beiträge zur Ökologie und Soziologie der luftlebigen (atmophytischen) Kieselalgen. Ber. dtsch. bot. Ges. **45**, 385—407 (1927). — BENTLEY (1958): Siehe Lit. Heterokonten. — BERGON, P.: Études sur la flore diatomique du bassin d'Arcachon, et des parages de l'Atlantique voisins de cette station. Soc. Sci. Arcachon Stat. Biol. **6**, 39—112 (1902). — Les processus de division, de rejeunissement de la cellule et de sporulation chez *Biddulphia mobiliensis* Bailey. Bull. Soc. bot. France **54**, 327—358 (1907). — BOCK, W.: Diatomeen extrem trockener Standorte. Nova Hedwigia **5**, 199—254 (1963). — BREDEMÜHL, I.: Über die Verbreitung der Erddiatomeen. Arch. Mikrobiol. **14**, 407—449 (1947). — BRISTOL, B. M.: On the retention of vitality by algae from old stored soils. New Phytologist **18**, 92—107 (1919). — On the algal flora of some desiccated English soils. Ann. Bot. **34**, 35—80 (1920). — BRISTOL-ROACH, B. M.: On the algae of some normal English soils. J. agric. Sci. **17**, 563—588 (1927). — BRUCKMAYER-BERKENBUSCH, H.: Die Beeinflussung der Auxosporenbildung von *Melosira nummuloides* durch Außenfaktoren. Arch. Protistenk. **100**, 183—211 (1955).

CLEVE, P. T.: Report on the phytoplankton collected on the expedition by H.M.S. "Research" 1896. 15th Ann. Rept. Fish. Board Scotland, Pt. III, 1897. — CLEVE-EULER, A.: Das Bacillariaceenplancton in Gewässern bei Stockholm. II. Zur Morphologie und Biologie einer pleomorphen *Melosira*. Arch. Hydrobiol. **7**, 119—139, 230—260 (1912). — Die Diatomeen von Schweden und Finnland. Kgl. Sv. Vetenskapsakad. Hd., Ser. IV, **2** (1) (1951); **3** (3) (1952); **4** (1) (1953); **4** (5) (1953); **5** (4) (1955). — CUPP, E.: Marine plankton diatoms of the West Coast of North America. Bull. Scripps Inst. Oceanogr. **5**, 1—238 (1943).

DENFFER, D. v.: Die planktische Massenkultur pennater Grunddiatomeen. Arch. Mikrobiol. **14**, 159—202 (1949). — DREBES, G.: Zur Entwicklungsgeschichte und Entwicklungsphysiologie der zentrischen Diatomee *Stephanopyxis turris* (Grev. et Arn.) Ralfs. Diss. Marburg 1965.

ERBEN, K.: Untersuchungen über Auxosporenentwicklung und Meioseauslösung an *Melosira nummuloides* (Dillw.) C. A. Agardh. Arch. Protistenk. **104**, 165—210 (1959).

GEITLER, L.: Somatische Teilung, Reduktionsteilung, Kopulation und Parthenogenese bei *Cocconeis placentula*. Arch. Protistenk. **59**, 506—549 (1927). — Autogamie bei *Amphora*. Öst. bot. Z. **77**, 81—91 (1928). — Der Formwechsel der pennaten Diatomeen (Kieselalgen). Arch. Protistenk. **78**, 1—226 (1932). — Gameten- und Auxosporenentwicklung von *Synedra ulna* im Vergleich mit anderen pennaten Diatomeen. Planta (Berl.) **30**, 551—566 (1939). — Zur Kenntnis der Rassenbildung und des Kopulationsverhaltens der Diatomee *Cocconeis placentula* und ihres Epiphytismus. Bot. Not. (Lund) **1948**, 84—92 (1948). — Beiträge zur Kenntnis der Auxosporenbildung pennater Diatomeen. Öst. bot. Z. **96**, 467—472 (1949a). — Die Auxosporenbildung von *Nitzschia sigmoidea* und die Geschlechtsbestimmung bei den Diatomeen. Portugal. Acta Biol., Ser. A, Goldschmidt Vol., 79—87 (1949b). — Die Stellungen der Kopulationspartner und der Auxosporen bei *Gomphonema*-Arten. Akad. Wiss. Lit. Mainz. Abh. math.-nat. Kl. **1951**, 217—225 (1951). — Untersuchungen über Kopulation und Auxosporenbildung pennater Diatomeen. I. Automixis bei *Gomphonema constrictum* var. *capitata*. Öst. bot. Z. **99**, 376—384 (1952a). — Oogamie, Mitose, Meiosis und metagame

Teilung bei der zentrischen Diatomee *Cyclotella*. Öst. bot. Z. **99**, 507—520 (1952b). — Allogamie und Autogamie bei der Diatomee *Denticula tenuis* und die Geschlechtsbestimmung der Diatomeen. Öst. bot. Z. **100**, 331—352 (1953a). — Abhängigkeit der Membranbildung von der Zellteilung bei Diatomeen und differentielle Teilungen im Zusammenhang mit der Bildung der Innenschalen. Planta (Berl.) **43**, 75—82 (1953b). — Das Auftreten zweier obligater metagamer Mitosen ohne Zellteilung während der Bildung der Erstlingsschalen bei Diatomeen. Ber. dtsch. bot. Ges. **66**, 222—227 (1953c). — Automixis, Geschlechtsbestimmung und Pyknose von Gonenkernen bei *Cymbella aspersa*. Planta (Berl.) **47**, 359—373 (1956). — Die sexuelle Fortpflanzung der pennaten Diatomeen. Biol. Rev. **32**, 261—295 (1957). — Selektive Paarung und gegenseitige Beeinflussung der Kopulationspartner bei der Diatomee *Cocconeis*. Planta (Berl.) **51**, 584—599 (1958a). — Fortpflanzungsbiologische Eigentümlichkeiten von *Cocconeis* und Vorarbeiten zu einer systematischen Gliederung von *Cocconeis placentula* nebst Beobachtungen an Bastarden. Öst. bot. Z. **105**, 350—379 (1958b). — Notizen über Rassenbildung, Fortpflanzung, Formwechsel und morphologische Eigentümlichkeiten bei pennaten Diatomeen. Öst. bot. Z. **105**, 408—442 (1958c). — Alle Schalenbildungen der Diatomeen treten als Folge von Zell- oder Kernteilungen auf. Ber. dtsch. bot. Ges. **75**, 393—396 (1963). — Gran, H. H.: Diatomaceae from the ice-floes and plankton of the Arctic Ocean. Norweg. North Polar Exped. 1893—1896. Sci. Res. ed. by Fr. Nansen. Kristiania 1900. — Die Diatomeen der arktischen Meere. I. Die Diatomeen des Planktons. In: Fauna Arctica, Bd. III/3. Jena: Gustav Fischer1904. — Diatomeen in Nordisches Plankton. Botanischer Teil, S. 1—146. Kiel u. Leipzig: Lipsius & Tischer 1908. — The plankton production in the North European waters in the spring (1912). Bull. planktol. ann. 1912, 146 p. Copenhague: A. F. Høst 1915. — Gran, H. H., and E. C. Angst: Plankton diatoms of the Puget Sound. Publ. Puget Sound Biol. Sta., Univ. Wash. **7**, 417—516 (1931). — Grøntved, J.: Planktological contributions. Taxonomical studies in some danish coastal localities. Meddr. Danm. Fisk- og Havunders., N. S. **1** (12), 3—13 (1956). — Gross, F.: On the life histories of some marine plankton diatoms. Phil. Trans. B **228**, 1—47 (1937). — The development of isolated resting spores into auxospores in *Ditylum Brightwellii*. J. Mar. Biol. Ass. U.Kingd. **24**, 375—380 (1940). — Guillard, R. R. L., and J. R. Ryther: Studies of marine planktonic diatoms .I. *Cyclotella nana* Hustedt and *Detonula confervacea* (Cleve) Gran. Canad. J. Microbiol. 8, 229—239 (1962).

Hart, T. J.: On the phytoplankton of the South-West Atlantic and the Bellinghausen Sea, 1929—1931. Discovery Repts 8, 1—268 (1934). — Phytoplankton periodicity in antarctic surface waters. Discovery Repts **21**, 261—356 (1942). — Heiden, H., u. R. W. Kolbe: Die marinen Diatomeen. Deutsche Südpolarexpedition 1901—1903. Berlin u. Leipzig: de Gryuter 8, 450—715 (1928). — Helmcke, I. G., u. W. Krieger: Diatoemenschalen im elektronenmikroskopischen Bild. Berlin-Wilmersdorf: Transmare Photo 1953. — Hendey, N. I.: The plankton diatoms of the southern seas. Discovery Repts **16**, 151—364 (1937). — Hensen, V.: Über die Bestimmung des Planktons. Ber. Kom. Wiss. Unters. dtsch. Meere (1882—1886). Berlin 1887. — Heurck, H. van: Synopsis des Diatomées de Belgique. Anvers 1885. — Diatomées. Exped. antarctique Belge **6**, 1—126 (1909). — Hustedt, F.: Die Kieslagen. Bd. 7 von Rabenhorsts Kryptogamenflora. Leipzig: Akademische Verlagsgesellschaft. I. Teil 1930. II. Teil 1959.

Jørgensen, E. G.: Variations in the silica content of diatoms. Physiol. Plantarum (Kbh.) 8, 840—845 (1955).

Karsten, G.: Untersuchungen über Diatomeen. Flora (Jena) **82**, 286—296 (1896). — Untersuchungen über Diatomeen. Flora (Jena) **83**, 33—53, 203—222 (1897). — Die Diatomeen der Kieler Bucht. Wiss. Meeresunters. Kiel, N. F. **4**, 7—205 (1899). — Die Auxosporenbildung der Gattungen *Cocconeis, Surirella* und *Cymatopleura*. Flora (Jena) **87**, 253—283 (1900). — Das Phytoplankton des Antarktischen Meeres. Wiss. Erg. Dtsch. Tiefsee-Exp. Valdivia. Jena: Gustav Fischer 1905. — Bacillariophyta. In: Engler-Prantl, Natürliche Pflanzenfamilien, 2. Aufl., Bd. 2. Leipzig: Wilhelm Engelmann 1928. — Köhler, E.: Zur Kenntnis der Sexualität bei *Synchytrium*. Ber. dtsch. bot. Ges. **69**, 121—127 (1956).

Lewin, J. C.: Silicon metabolism in diatoms. IV. Growth and frustule formation in *Navicula pelliculosa*. Canad. J. Microbiol. **3**, 427—433 (1957). — The taxonomic position of *Phaeodactylon tricornutum*. J. gen. Microbiol. **18**, 427—432 (1958). — Lewin, J. C., R. A. Lewin, and D. E. Philpott: Observations on *Phaeodactylon tricornutum*. J. gen. Microbiol. **18**, 418—426 (1958). — Liebisch, W.: Experimentelle und kritische Untersuchungen über die Pektinmembran der Diatomeen unter besonderer Berücksichtigung der Auxosporenbildung und der Kratikularzustände. Z. Bot. **22**, 1—65 (1929). — Locker, Fr.: Beiträge zur Kenntnis des Formwechsels der Diatomeen an Hand von Kulturversuchen. Öst. bot. Z. **97**, 322—332 (1950). — Lund, J. W. G.: Observations on soil algae. I. The ecology, size and taxonomy of british soil diatoms. New Phytologist **44**, 196—219 (1945). — Observations on soil algae. II. The biology, size and taxonomy of British soil diatoms. New Phytologist **45**, 56—110 (1946). — The seasonal cycle of the plancton diatom, *Melosira italica* (Ehr). Kütz subsp. *subarctica* O. Müller. J. Ecol. **42**, 151—173 (1954).

Magne-Simon, M. F.: Note sur le processus de l'auxosporulation chez une diatomée marine, *Grammatophora marina* (Lyngb) Kütz. C. R. Acad. Sci. (Paris) **251**, 3040—3042 (1960). — L'auxosporulation chez une Tabellariacée marine *Grammatophora marina* (Lyngb.) Kütz (Diatomée). Cahiers Biol. Mar. **3**, 79—89 (1962). — Mangin, L.: Sur le polymorphisme de certains Diatomées de l'Antarctique. C. R. Acad. Sci. (Paris) **159**, 344—350 (1914). — Phytoplancton de l'Antarctique. Deuxième Exped. Antarctique Francaise (1905—1910). Sci. Nat. Doc. Scient. 1915. — Manguin, E.: Les Diatomées de la Terre Adélie. Campagne du «Commandant Charcot» 1949—1950. Ann. Sci. Bot. natur. (Paris), Ser. XII, **1960**, 123—363 (1960). — Meunier, A.: Microplancton des Mers de Barents et de Kara. In: Campagne arctique de 1907. Bruxelles: Ch. Bulens 1910. — Microplankton de la Mer Flamande. Mem. Mus. roy. hist. Nat. Belg. **7** (2, 3) (1913—1915). — Michailova, N. F.: On the germination of resting spores of *Chaetoceros lauderi* Ralfs. Dokl. Akad. Nauk SSSR. **143**, 741—742 (1962). — Müller, O.: Sprungweise Mutationen bei Melosiren. Ber. dtsch. bot. Ges. **21**, 326—333 (1903). — Bacillariaceen aus dem Nyassalande. Zweite Folge. Englers Bot. Jb. **34**, 256—301 (1904). — Pleomorphismus, Auxosporen und Dauersporen bei *Melosira*-Arten. Jb. wiss. Bot. **43**, 49—88 (1906).

Nipkow, F.: Ruheformen planktonischer Kieselalgen im geschichteten Schlamm des Züricher Sees. Schweiz. Z. Hydrol. **12**, 263—270 (1950). — Die Auxosporenbildung bei *Fragilaria crotonensis* im Plankton des Zürichsees. Schweiz. Z. Hydrol. **15**, 302—310 (1953).

Patrick, R.: Sexual reproduction in diatoms, p. 82—99. In: Wenrich, Sex in microorganisms. Amer. Ass. Advanc. Sci. Washington 1954. — Pavillard, J.: Sur la reproduction du *Chaetoceros Eibenii* Meunier. C. R. Acad. Sci. (Paris) **172**, 469—471 (1921). — Hybridation (ou mutation) chez les Diatomées pélagiques du genre *Rhizosolenia*. C. R. Acad. Sci. (Paris) **204**, 896—898 (1937). — Pfitzer, E.: Untersuchungen über Bau und Entwicklung der Bacillariaceen (Diatomeen). Hansteins Bot. Abh. Bonn **1**, 1—189 (1871).

Reimann, B. E. F., J. C. Lewin, and B. E. Volcani: Studies on the biochemistry and fine structure of the silica shell formation in diatoms. I. The structure of the cell wall of *Cylindrotheca fusiformis* Reimann and Lewin. J. Cell Biol. **24**, 39—55 (1965). — Rieth, A.: Zur Auxosporenbildung bei *Melosira nummuloides*. Flora (Jena) **140**, 205—208 (1953).

Schreiber, E.: Reinkulturversuche und experimentelle Auxosporenbildung bei *Melosira nummuloides*. Arch. Protistenk. **13**, 331—345 (1931). — Schütt, F.: Über die Diatomeengattung *Chaetoceros*. Bot. Z. **46**, 161—170 (1888). — Zentrifugales Dickenwachstum der Membran und extramembranöses Plasma. Jb. wiss. Bot. **33**, 594—689 (1899). — Zentrifugale und simultane Membranverdickungen. Jb. wiss. Bot. **35**, 470—534 (1900). — Schulz, P.: Zellteilung und Dauersporenbildung zweier Diatomeengattungen. Bot. Arch. **24**, 505—524 (1929). — Skabitchewsky, A. P.: Über die Biologie von *Melosira baicalensis* (K. Meyer) Wisl. Russ. gidrobiol. Zh. **8**, 93—114 (1929). Zit. nach Lund. — Stosch, H. A. v.: Form und Formwechsel der Diatomee *Achnantes longipes* in Abhängigkeit von der Ernährung. Mit besonderer Berücksichtigung der Spurenstoffe. Ber. dtsch. bot. Ges. **60**, 2—15 (1942). — Entwicklungsgeschichtliche Untersuchungen an zentrischen Diatomeen. I. Die Auxosporenbildung von *Melosira varians*. Arch. Mikrobiol. **16**, 101—135 (1951a). — Zur Entwicklungsgeschichte zentrischer Meeresdiatomeen. Naturwissenschaften **38** (8), 191—192 (1951b). — Die Oogamie von *Biddulphia mobiliensis* und die bisher bekannten Auxosporenbildungen bei den Centrales. Com. VIII[e] Congr. internat. Bot. Sect. **17**, 58—68 (1954). — Pennate Diatomeen. Z. Bot. **43**, 89—99 (1955). — Entwicklungsgeschichtliche Untersuchungen an zentrischen Diatomeen. II. Geschlechtszellenreifung, Befruchtung und Auxosporenbildung einiger grundbewohnender Biddulphiaceen der Nordsee. Arch. Mikrobiol. **23**, 327—365 (1956). — Kann die oogame Arhaphidee *Rhabdonema adriaticum* als Bindeglied zwischen den beiden großen Diatomeengruppen angesehen werden? Ber. dtsch. bot. Ges. **71**, 241—249 (1958). — Über das Perizonium der Diatomeen. Vortr. Gesamtgeb. Bot. 1. Beitr. Phys. Morph. Algen. Stuttgart 1962, S. 43—52. — Manipulierung der Zellgröße von Diatomeen im Experiment. Phycologia **5**, 21—44 (1965). — Stosch, H. A. v., u. G. Drebes: Entwicklungsgeschichtliche Untersuchungen an zentrischen Diatomeen. IV. Die Planktondiatomee *Strphanopyxis turris* — ihre Behandlung und Entwicklungsgeschichte. Helgol. Wiss. Meeresunters. **11**, 209—257 (1964).

Wilson, P. D.: The triradiate and other forms of *Nitzschia Closterium* (Ehrenberg) W. Smith, f. *minutissima* of Allen and Nelson. J. Mar. Biol. Ass. U. Kingd. **26**, 235—270 (1942). — Wood, F. J. F.: An unusual diatom from the Antarctic. Nature (Lond.) **184**, 1962—1963 (1959).

III. Heterokontae (Xanthophyceae).

Durch die Eigenschaft der Heterokonten, Chlorophyceengenera in erstaunlichem Maße in Konvergenztypen zu imitieren, entsteht ein Filz von Formen, dessen Entwirrung auch heute noch nicht beendet ist. Wichtigste Differentialmerkmale zwischen diesen beiden Gruppen sind Begeißelung (Feinbau: Manton et al. 1952) der Schwärmer und Biochemie der Reservestoffe sowie Plastidenpigmente. Die Parallelität beschränkt sich nicht auf eine Anzahl ähnlicher Gestalten, sondern schließt auch nahezu alle wichtigen Organisationsformen ein. Und wenn die Heterokonten in der monadalen, trichalen, siphonocladalen und siphonalen Stufe ärmer vertreten sind als die Grünalgen — der Schwerpunkt liegt auf den capsalen und insbesondere coccalen Ausbildungen —, so besitzen sie sehr merkwürdige amöboide und plasmodiale Formen in der Gruppe der Rhizochrysideen, eine Organisationsstufe, welche den Chlorophyceen fehlt. Auch durch die Seltenheit, mit der sexuelle Fortpflanzung auftritt oder zur Beobachtung kommt, unterscheiden sich die beiden Gruppen. Die beiden letztgenannten Momente mögen zu der besonderen Vielgestaltigkeit der vegetativen Fortpflanzung bei den Heterokonten beigetragen haben.

a) Die Fortpflanzungsarten.

1. Die einzelligen und die trichalen Formen.

Sicht man von den erst in neuester Zeit als hierher gehörig erkannten Vaucherien (Chadefaud 1951, Koch 1951) als sehr hoch spezialisierten Vertretern der siphonalen Stufe ab, so kann man die mannigfaltigen Fortpflanzungsarten der Heterokonten dank der ordnenden Hände Paschers (1939) und neuerdings Ettls (1956) bis zu einem Grade überschauend behandeln und sagen, daß die niederen Lebensformen der Heterokonten sich insofern in der Fortpflanzung der höheren widerspiegeln, als sie und ihre eigenen Fortpflanzungsweisen integrale Bestandteile des Modalitätenvorrates jener sind und von ihnen nur „zögernd“ aufgegeben werden. So haben die *monadalen Formen* Zweiteilung und die Fähigkeit, zweischalige Cysten zu bilden, welche Schwärmer entlassen, sowie die Tendenz, in den amöboiden Zustand überzugehen, was man als Ausdruck einer Fortpflanzungstätigkeit betrachten kann. Die *Rhizochlorideen* teilen sich als Amöben, liefern aber gelegentlich Schwärmer; ihre zweischaligen Cysten (Abb. 48) keimen mit Schwärmern oder Amöben und bei den mehrkernigen plasmoidalen Formen tritt neu eine Riesencyste mit nicht unterteilter Membran auf, die sich später bei den siphonalen Gestaltungen wiederfindet. Die *Heterokapsineen* mit nackten, unbeweglichen Zellen als Hauptphase teilen sich in dieser Form, haben aber außerdem eine monadoide Phase samt deren Fortpflanzungsarten zur Verfügung, dazu einen ihrer eigenen Stufe entsprechenden, von Ettl als Hemiautospore bezeichneten

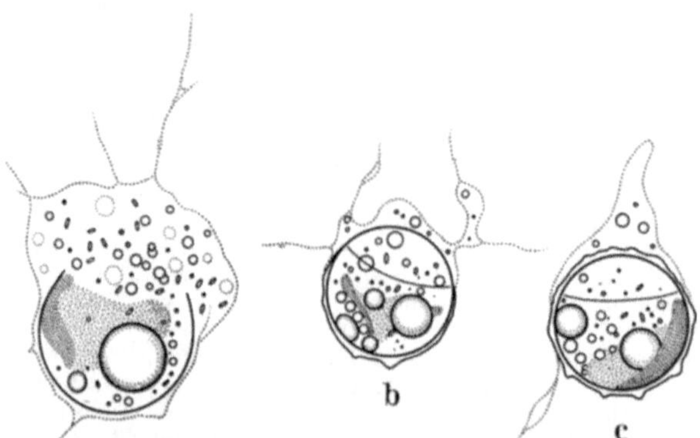

Abb. 48a—c. Endoplasmatische Cysten bei Heterokonten. a *Myxochloris (Rhizochloridales)*, b, c *Chloromeson (Heterochloridales)*. (Alles nach Pascher 1939.)

(Abb. 49) gonidialen Fortpflanzungskörper, dem die Geißeln und manchmal das Stigma, nicht jedoch eine pulsierende Vacuole fehlen. Bei den *coccalen Heterokonten* treten Zoosporen, Hemiautosporen und zweischalige Autosporen oder solche mit ungeteilter Membran auf. Für *Chlorocloster* wird eine in der Form einer Chrysomonadencyste ähnliche Dauerzelle beschrieben, deren Wand jedoch aus organischer Substanz besteht (CLÉMENÇON 1965). In der trichalen Stufe schließlich, wie sie durch *Tribonema* repräsentiert ist, findet sich nahezu der ganze Katalog von Fortpflanzungsarten der Ordnung.

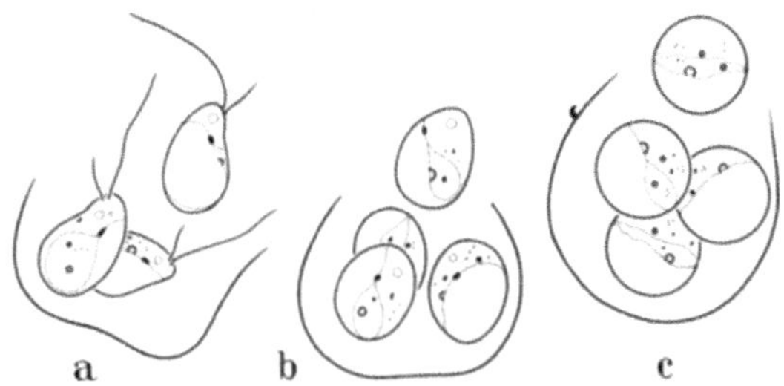

Abb. 49a—c. Ungeschlechtliche Vermehrung von Heterokonten. a Zoosporen, b Hemiautosporen, c Autosporen. (Nach ETTL 1954.)

Andererseits zeichnen die Dauerorgane der niederen Stufe die vegetative Entwicklung der höheren Formen vielfach vor. So entspricht der coccale sich mit Gonidien vermehrende sowie der sich von ihm ableitende trichale Typus, soweit mit zweischaligen Membranen ausgerüstet, wohl der zweiteiligen Cyste der nackten Formen. Die Heterococcalen mit einheitlichen Membranen und Zweiteilung (neben Gonidienbildung) wie *Chlorellidium* und die mit ihnen wohl zusammenhängenden pseudotrichalen *Heterococcus*-Arten (VISCHER 1937) mögen sich dagegen direkt von monadalen, vielleicht auch capsalen Formen herleiten. Die erwähnten vielkernigen Plasmodiencysten finden ihre Entsprechung in der heterosiphonalen Stufe.

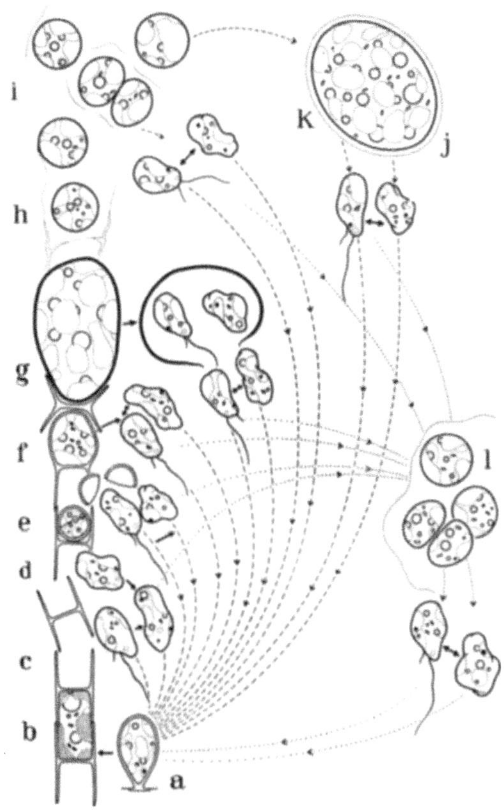

Abb. 50a—l. Fortpflanzungsarten von *Tribonema*. Erklärungen im Text. (Nach PASCHER 1939.)

PASCHERS Schema der *Fortpflanzung von Tribonema* als Vertreter der trichalen Gestaltung schildert daher auch die meisten der erwähnten Fortpflanzungsarten der einkernigen Heterokonten (Abb. 50). Es zeigt (Abb. 50c, d) Zoosporenbildung mit austretenden monadoiden oder amöboiden Gonidien. Doch können diese sich als Autosporen (Abb. 51a) früh mit zarter Membran umgeben und nun entweder mit Schwärmern — auch direkt als Faden — keimen oder aber zu dickwandigen Aplanosporen mit Reservestoffen und zweiteiliger Membran (Abb. 50e) werden. Weiter kann sich der ungeteilte Inhalt der Fadenzelle auch in eine dünnwandige oder dickwandige zweischalige Spore umwandeln (Abb. 50f), deren Membran sich unter

Umständen verstärkt und dunkel färbt (Abb. 51 b); die Keimung erfolgt hier wie bei den kleinen Sporen. Schließlich können vielkernige Riesencysten (Abb. 50 g) ebenfalls mit brauner Membran durch Heranwachsen des sich behäutenden Inhaltes einer Fadenzelle entstehen. Ihre Membran ist nicht unterteilt. Akineten (Abb. 51 c, d) bilden sich durch Verdickung der Zellwand der Fadenzelle selbst. Schließlich verquillt unter Umständen der ganze Faden und seine abgerundeten Zellinhalte liefern ein palmellaartiges Lager (Abb. 50 i), wie es ähnlich durch Verschleimung der Wände von Zoosporenkeimlingen auch direkt entstehen kann (Abbildung 50 l). Eine Palmella-Zelle vermag stark anzuschwellen (Abb. 50 k) und wiederum mehrkernig zu werden.

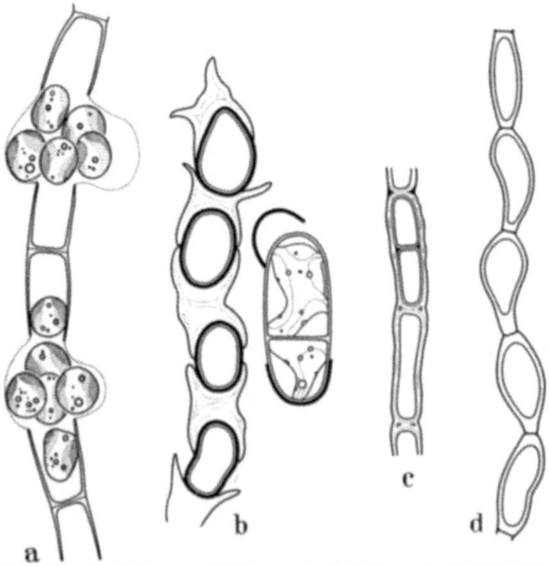

Abb. 51 a—d. *Tribonema* bildet a nach Aufteilung des Zellinhaltes zartwandige Autosporen, b unter Umhäutung des ganzen Protoplasten mit dicker, gefärbter Membran Aplanosporen, durch Verstärkung der gewöhnlichen Zellmembran *(Tr. subtilissima)* c Fadenakineten, d Einzelakineten. (Alle nach Pascher 1939.)

Die *zweischaligen Cysten* der Heterokonten haben wie die zweischaligen Zellen der Heterococcalen häufig verkieselte Membranen. Von den monadoiden und amöboiden Zuständen werden sie ähnlich den Ampullencysten der Chrysomonaden endoplasmatisch ausgeschieden (Abb. 48). Über ihre Entstehung bei den höheren Stufen ist wenig bekannt. Ihre Membranhälften sind aber weit weniger ungleich als bei Chrysomonaden, oft fast gleich groß und ähneln in dieser Hinsicht mehr den Dauersporen der Kieselalgen. Über die physiologischen Bedingungen, die diese Mannigfaltigkeit der Fortpflanzungsarten beherrschen, wissen wir kaum etwas.

2. Heterosiphonales.

Besonders sei noch auf die Heterosiphonalen in ihrer Fortpflanzung eingegangen, über die relativ eingehend experimentell gearbeitet wurde. Diese sind entweder wie *Botrydium* hapaxante, holocarpische und mit sich verzweigendem Rhizoidstamm versehene Blasen oder ausdauernde eucarpe und mehrfach wurzelnde Schlauchsysteme: *Asterosiphon*[1] (Dangeard 1941, Stahl 1879, Rieth 1962) und *Vaucheria*-Arten.

Bei *Botrydium* wächst bis zu einer artspezifischen Endgröße ihrer assimilierenden Blasen die Neigung, zur Fortpflanzung zu schreiten, was, genügende Feuchtigkeit vorausgesetzt, mit heterokonten Schwärmern erfolgt. Die übrigen Fortpflanzungsarten können durch progressive Hemmungen der Schwärmerbildung verstanden werden, die im ganzen vom Grade und der Geschwindigkeit der Austrocknung des Standorts abhängen, aber auch durch das Alter der Pflanzen und im Bereich der Gattung idioplasmatisch (Vischer 1938) reguliert

[1] *Asterosiphon*, durch Dangeard aufgestellt, wurde vorher für ein Stadium von *Vaucheria* gehalten.

werden. Die Schwärmzellen können durch einkernige Autosporen ersetzt sein; weiterhin wird aber bereits die Zerklüftung des Plasmas mehr oder weniger stark unterdrückt, wodurch zahlreiche wenigkernige oder wenige vielkernige und umhäutete Teilstücke entstehen. Durch starke Membranverdickung mögen sich aus diesen „Aplanosporen“ bzw. „Sporocysten“ — so nennt man die einkernigen

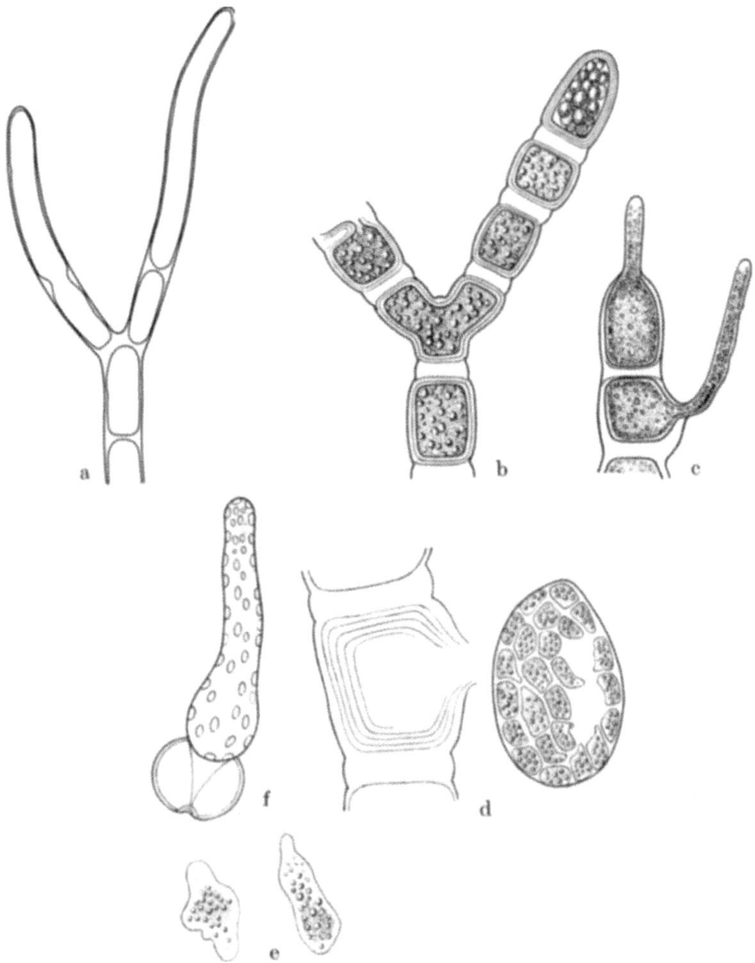

Abb. 52a—f. *Asterosiphon*, Fortpflanzung. a Thallusende beginnt sich zu kammern. (Nach DANGEARD 1941.) b Zoosporangien, oben mit Aplanosporen, c desgleichen das ganze Sporangium mit Keimschlauch keimend, d Entleerung des „Amöbosporangiums“, Inhalt zunächst noch in Blase eingeschlossen, e die befreiten Amöbosporen, f aus diesen entstandene Cysten keimend. (b—f nach STAHL 1879, umgezeichnet.)

bzw. die vielkernigen Dauerorgane — entwickeln. Sporocysten bilden sich vor allem auch in reifen, auf Agar unter Fortpflanzungsdruck stehenden, aber nicht mit überschüssigem Wasser versorgten Blasen (VISCHER l.c.). Bleibt schließlich der Blaseninhalt ungeteilt, so hätte man die „Makrocyste“ (PASCHER 1939, MILLER 1927). Manche Botrydien verlagern ihr Plasma in das Rhizoidsystem, wo es in „Rhizocysten“ zerlegt wird, von denen im Falle von *B. tuberosum* je eine die ampullenartig angeschwollenen Rhizoidspitzen ausfüllt. Alle unbeweglichen Stadien vermögen mit Keimschlauch zu keimen oder insbesondere aus den Cysten Schwärmer zu entlassen. Schwärmzellen jeglicher Art keimen nur auf festem

Substrat mit Keimschlauch zu einer Blase, runden sich sonst zu *Chromulina*-ähnlich der Wasseroberfläche aufsitzenden Bläschen, also einer neustonischen Phase. Letztere mögen wieder ausschwärmen oder auf der Oberfläche gespreitet das Phänomen der plötzlichen Massenansiedlungen von Botrydien auf trocken fallenden Ufern hervorrufen.

Die Sporangien der *schlauchförmigen Heterosiphonalen* werden bei *Vaucheria* einzeln, bei *Asterosiphon* (DANGEARD 1941) serienweise an Faden- oder Astenden durch Membranen abgetrennt (Abb. 52a). Im letzten Falle wachsen sie entweder (DE PUYMALY 1922) direkt (Abb. 52c) oder, nachdem sie als mehr oder weniger dickwandige Akineten geschlüpft sind, zu Keimschläuchen aus (Abb. 52d) oder endlich entlassen sie einkernige Gonidien (Abb. 52b). Diese (STAHL 1879, DE PUYMALY, DANGEARD) sind entweder amöboid (Abb. 52e), behäuten sich aber nach einiger Zeit oder tun dies bei mangelnder Feuchtigkeit als Aplanosporen direkt. In beiden Fällen können sie entweder unmittelbar zu *Asterosiphon*-Fäden keimen oder nachdem sie sich unter anhaltendem Wassermangel zu starkwandigen Cysten verdickten. Die letzteren erweisen sich bei der Keimung als zweischalig (Abb. 52e). RIETH (1962) findet in seinen Kulturen die Amöben und die dickwandigen Sporen nicht. Bei *Vaucheria* (Abb. 53) wird gewöhnlich der Inhalt des Zoosporangiums als polyvalenter Schwärmer, Coenozoospore mit zahlreichen Kernen und den dazugehörigen biflagellaten Cinetiden entlassen (neuere Beschreibung und Feinbau, GREENWOOD et al. 1957, GREENWOOD 1959) oder er schlüpft als Coenoaplanospore. Schließlich gibt es Fälle, wo das Sporangium, nach Abwurf durch eine Trennstelle oder nicht, direkt keimt. Die Entscheidungen Aplanospore—Zoospore (SMITH A1955) oder Zoospore—Sporangienkeimung (siehe z. B. WEBER 1958) werden, wie unten besprochen, in manchen Fällen durch Außenbedingungen bestimmt. Bei *Vaucheria pachyderma* soll eine Umwandlung des Fadeninhalts in einkernige Mikroaplanosporen vorkommen (SMITH 1944).

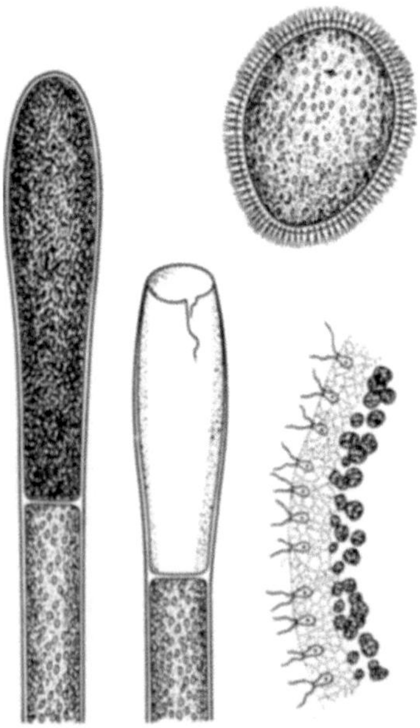

Abb. 53a—c. Synzoospore von *Vaucheria sessilis*. a Coenosporangium, b Coenozoospore ist entlassen, c Teil eines Schnittes durch Zoospore, jeder Kern ist mit einem zwei Geißeln tragenden Basalapparat verbunden. (a, b nach GÖTZ 1897, aus STRASSBURGER, c nach STRASSBURGER 1880.)

b) Experimentelle Untersuchungen zur Fortpflanzung.

1. Die Kultur von Heterokonten.

Kulturen sind in verhältnismäßig geringem Umfange überhaupt und seltener unter einigermaßen kontrollierten Bedingungen durchgeführt worden. Die wichtigsten Arbeiten an niederen Heterokonten werden anschließend genannt, da sie immer auch im folgenden nicht verwertete Angaben zur Fortpflanzung enthalten (s. auch VISCHER in PASCHER 1939): GERNECK 1907 (*Heterococcus*-Arten), SCHRAMM 1914 (*Botrydiopsis, Botrydium*), CHODAT (1913) (*Characiopsis, Bumilleria, Heterothrix, Tribonema, Heterococcus*), POULTON 1925 (*Vischeria, Characiopsis, Heterococcus, Tribonema*), HAWLITSCHKA 1932 (*Tribonema*-Arten), VISCHER 1932 (*Mischococcus*), 1936, 1937 (*Chlorellidium, Bumilleriopsis, Heterothrix, Hetero-*

coccus), 1938 (*Botrydium*-Arten), 1945 (*Pleurochloris-*, *Bumilleriopsis-*, *Heterothrix-*, *Heterococcus*-Arten, *Chloridella neglecta*, *Nephrodiella brevis*, *Botrydiopsis alpina*, *Vischeria punctata*), MOEWUS 1940 *(Botrydium)*, BELCHER u. FOGG 1958 *(Tribonema)*, MILLER u. FOGG 1957, 1958 *(Monodus)*, BELCHER und MILLER 1960 *(Tribonema, Bumilleriopsis, Chlorellidium, Botrydiopsis, Polyedriella)*, CLÉMENÇON 1965 *(Chlorocloster)*. Auf die sehr eingehenden, wenn auch mit Rohkulturen durchgeführten Arbeiten von KLEBS (A 1896) an *Tribonema*, *Bumilleria*, *Botrydium* und *Vaucheria*, von MILLER (1927) an *Botrydium* und auf einige andere wird noch eingegangen.

2. Die Alternative sexuelle—vegetative Fortpflanzung.

Nach MOEWUS (1940)[1] sind die Gonidien dreier untersuchter Herkünfte von *Botrydium* bei monözischer Geschlechtsverteilung in zwei, bei diözischer in einem Falle fähig zu kopulieren. Kommt die Kopulation nicht zustande, so verhalten sie sich als Zoosporen und keimen „parthenogenetisch". Da die Kopulationen an die Temperaturgrenzen, 1—14°, bzw. 8—27°, bzw. 5—24° für die drei Stämme gebunden sind, findet außerhalb dieser Grenzen vegetative Vermehrung statt. Die Temperaturansprüche werden im übrigen von mendelnden Genen bestimmt. Unklar bleibt, ob physiologische und morphologische Unterschiede zwischen Zoosporen und Gameten bestehen.

Für *Vaucheria* stellten die Arbeiten von KLEBS (A 1896) verhältnismäßig klare Alternativen in den Bedingungen für die Determination von sexueller Fortpflanzung einerseits, vegetativer anderseits heraus (s. auch HUSTEDE 1957):

Bedingungen für die Bildung von	
Zoosporen	*Gametangien*
Auch bei Dunkelheit	Nur im Licht
Starklicht hemmt	Starklicht fördert (Dauerlicht, HUSTEDE)
Nährsalze notwendig (wenigstens bei Vorkultur)	Nährsalze hemmen
Zucker ohne Wirkung	Zucker fördert
Nur im Wasser	Flüssiges Wasser hemmend (nicht absolut)
Ab 0,04 mm Hg O_2 (Wachstum ab 0,04 mm Hg O_2)	Ab 0,1 mm Hg O_2 (normal erst bei 1,5 mm Hg O_2)
Latenzzeit gering (ca. 1^d)	Latenzzeit groß (ca. 4^d HUSTEDE)

Beide Prozesse sind in ihren Temperaturansprüchen ähnlich und im strömenden Wasser verhindert, ohne daß im letzten Falle die eigentlich wirkende Ursache klar erkannt wurde. Dieses Ergebnis wird durch LEAGUE und GREULACH (1955), v. DENFFER und HUSTEDE (1955) sowie HUSTEDE (1957) im Grundsatz bestätigt. Die von KLEBS ermittelten Zusammenhänge bewirken, daß bei Überimpfen in frisches Medium die zunächst einsetzende Zoosporenbildung nach 14 Tagen durch reines Wachstum, dieses schließlich durch sexuelle Fortpflanzung abgelöst wird. Das Aufhören der Zoosporenbildung ist einer Veränderung der Nährlösung durch die Alge zuzuschreiben; v. DENFFER und HUSTEDE vermuten eine Hemmstoffausscheidung durch die Alge als Agens. Jedoch wurde eine „negative" Ursache nicht ausgeschlossen.

Diese Hemmung wird nun durch Tryptophan und einige seiner Abbauprodukte bereits in geringer Konzentration (ab 10^{-7} g/l) überspielt, derart, daß sich die

[1] Trotz der erwiesenen Unzuverlässigkeit der Angaben dieses Autors (s. RENNER 1958) sei die Arbeit kurz referiert.

ersten Glieder der aus Tryptophan durch Photolyse entstehenden Sequenz von Stoffen, soweit geprüft, als förderlich, die letzten als schädlich erweisen (Tabelle 3). Der Umschlagpunkt in der Photolyseserie liegt hinter Indolylglyoxylsäure.

Tabelle 3. *Wirkung von Stoffen aus der photolytischen Zerfallsserie des Tryptophans sowie einiger verwandter Verbindungen (die letzten drei) auf die Zoosporen- bzw. Gametenbildung von Vaucheria sessilis: + fördernd, — hemmend, 0 ohne Einfluß.* (Nach Angaben Hustedes.)

	Tryptophan	Tryptamin	Tryptophol	Indol-acet-aldehyd	Indol-essig-säure	Indol-glyoxyl-säure	Indol-aldehyd	Indol-carbon-säure	Indol-	Indol-propion-säure	Indol-acet-amid	Indol-aceto-nitrit
Zoosp.	+	+	+			+ (?)	—	—	—	+	0	0
Sexorg.	—			—	—		+	+	+			

Das antagonistische Verhalten beider Fortpflanzungsarten gegen Außenbedingungen setzt sich also gegenüber den Tryptophaniden fort. Auf die sexuelle Fortpflanzung wirken die Anfangsglieder der Umwandlungsreihe hemmend, die Endglieder wieder ab Indolaldehyd fördernd. Zieht man die Ergebnisse von Klebs heran, so heißt dies, daß die Veränderungen, die Licht am Tryptophan hervorruft, zu Produkten führen, welche, an der Pflanze geprüft, die gleichen Reaktionen auslösen, wie sie Lichtgenuß entsprechender Dauer direkt bewirkt. Da Indolderivate in Algen synthetisiert werden (Bentley 1958, 1960), besteht damit die Möglichkeit, daß diese Stoffe die Hormone einer Licht-vermittelten Umsteuerung von der vegetativen zur generativen Fortpflanzung darstellen. Fraglich ist noch, ob der Sporulationshemmstoff dieser Reihe angehört. Dann könnte man nunmehr die Alternativwirkung der Außenfaktoren auf die Frage nach der Wirkung der Wuchsstoffe eingeengt sehen. Wie groß die Reichweite eines solchen Mechanismus sein würde, ist nicht bekannt, doch sei auf die Beobachtungen Hustedes bei *Stigeoclonium* (s. S. 736) und *Bryopsis halymeniae* (1963) hingewiesen. Im letzten Falle wird allerdings zwischen zwei Formen der vegetativen Fortpflanzung differenziert. Die Ergebnisse über die Beeinflussung der Geschlechtsbestimmung durch die Indolderivate können hier unberücksichtigt bleiben.

3. Auslösung von Zoosporen und Aplanosporen.

Zoosporen.

Bei landlebenden Heterokonten wie *Botrydium* (Klebs A 1896, Miller 1927, Vischer 1938, Moewus 1940), *Asterosiphon* (de Puymaly 1924), *Vaucheria*, *Bumilleria* (Klebs l. c.) werden Zoosporen durch Fluten des Thallus hervorgerufen, wobei im Falle der beiden letztgenannten ein Lichtentzug die Reaktion verstärkt. Ähnlich verhält sich Material aus Agarkulturen von *Heterothrix* (s. auch James 1935) und *Heterococcus* (Vischer 1936) beim Überführen in flüssiges Medium. Doch führt *Heterococcus* (mehrere Arten, Vischer 1937) in jungen Agarkulturen den Zerfall auch durch, wenn Lichtwechsel, nicht jedoch, wennDauerlicht herrscht. Da bloße Verdunkelung ähnlich wirkt, ist anzunehmen, daß es sich nicht um einen photoperiodischen Effekt, sondern um einen einfachen Reizvorgang handelt. *Botrydium* liefert auch ein Beispiel für den Einfluß der Vorbehandlung, wenn von *Anfang* an in Nährlösung kultivierte Blasen nicht automatisch in der Flüssigkeit sporulieren, sondern wieder erst auf Verdunklung hin. Schwierig sind die Zoosporen mancher Tribonemen zu erhalten. Durch Verdunklung konnte Klebs (s. Gerneck 1907) sie nur bei günstigen inneren Bedingungen hervorrufen. Solches Material reagiert aber bereits im Licht auf Zugabe von Inulin in 1% und von Glykosiden wie Amygdalin in 0,5% Lösung; diese Stoffe scheinen nicht in erheblichem Maße assimiliert zu werden, im Gegensatz zu einfachen Zuckern, die aber nicht auslösend wirken. Es bleibt unklar, was die Grundlage dieser „spezifischen" Wirkungen war, die im Dunkeln auch reaktionsträges Material stimulieren können. Verdunklung erweist sich, wie bereits in der Einleitung (s. S. 600) und auf S. 688 angeführt, als besonders wirksam auch auf die Zoosporenbildung von *Vaucheria*, ebenso von *Bumilleria*.

In der Einleitung wurde auch bereits darauf hingewiesen, daß neben einer Verminderung der Nährsalzkonzentration (Ersatz der 0,2% Knop-Lösung durch Wasser: was dabei wirksam, osmotische Veränderung oder Mangel ?), die rasch zur Sporenbildung führt — wofür wieder *Vaucheria*, aber auch *Bumilleria* Beispiele sind — bei verschiedenen (und etwas verschieden reagierenden) *Vaucheria*-Arten auch Steigerung der Konzentration der Knop-Lösung (auf 0,6 bzw. 1%) die Zoosporenbildung auslösen kann (*Vaucheria repens* und besonders *V. clavata*).

Schließlich sei noch erwähnt, daß *Bumilleria* auf eine Temperaturerniedrigung (von 13—17° auf 5—6°) mit Zoosporenentlassung antwortet.

Als konstitutionell „hapaxanthe" Pflanzen gehen die *Botrydium*-Arten bei der Reife, d.h. der Erreichung einer artspezifischen Endgröße ihrer Coeloblasten ohne merkliche äußere Einwirkungen zur Fortpflanzung über; dennoch ist wohl eine Steuerung des Vorganges durch Außenbedingungen anzunehmen, da die Thalli einer Kultur weitgehend gleichzeitig fruchten (VISCHER 1938; die Versuche wurden anscheinend nicht bei konstanter Temperatur durchgeführt). Bereits während dieser Entwicklung ändert sich die Reaktion auf Submergieren. In älteren Thalli entstehen statt Zoosporen Aplanosporen. Auch von den verzweigt trichalen und aerophilen *Heterococcus*-Arten sind durch ein spontanes Zerfallen des Thallus in Zoo- oder Aplanosporen auf Agar charakterisierte, an eine Reifung erinnernde Vorgänge beschrieben worden (VISCHER 1936, 1937, 1945). Die Sporenbildung tritt schließlich auch dann auf, wenn Verdunkelungsreize (s.o.) fehlen, auch hier bei einer artspezifischen Thallusgröße. Es wäre bei solchen terrestrischen Formen denkbar, daß die mit dem Wachstum fortschreitende Entfernung einer Zelle vom Substrat und damit Änderung ihrer trophischen Situation die „innere" Ursache der Reifung bildete. Doch wurden etwaige Einflüsse der Verarmung usw. des Substrates durch das Wachstum auf diese Vorgänge nicht explizit untersucht.

Aplanosporen.

Unter ungünstigen Bedingungen (feuchte Kammer) lassen die Zoosporangien von *Vaucheria sessilis* ihren Inhalt als amöboide Masse austreten, die sich später behäutet; in stark bakterienhaltigen (Submers- ?) Kulturen unterbleibt auch die Entlassung und die Sporen keimen im Zoosporangium (WEBER 1958), ebenso wie bei der genotypisch Aplanosporen tragenden *V. geminata* (KLEBS A 1896). Die Bildungsbedingungen der Aplanosporen bei Arten dieses letzten Typus stehen nach KLEBS (l.c.) in starkem Gegensatz zu denjenigen der Zoosporen. *V. geminata* aus fließendem Wasser reagiert nicht auf Überführung in stehendes Wasser, erzeugt aber Aplanosporen in Massen, wenn sie auf feuchtem Boden trockener Luft ausgesetzt wird; unter den gleichen Bedingungen treten auch Sexualorgane auf. Der oben für *V. sessilis* beschriebene Antagonismus zwischen vegetativer und generativer Vermehrung existiert hier anscheinend nicht in der gleichen Form.

Ähnlich wie bei der Bildung der Sexualorgane der zoosporinen Arten scheint dabei auch Nährsalzmangel förderlich zu sein. So treten in geringer Wassermenge auch submers Aplanosporen auf, ebenso in Zuckerlösung. Die einzelnen Arten (*V. uncinata*, *V. racemosa*, *V. geminata*) verhalten sich aber etwas verschieden. Die marine, Aplanosporen entlassende *V. piloboloides* ließ sich andererseits durch keines der Mittel, welche hier für Süßwasserformen genannt wurden, sondern nur durch Minderung der Salzkonzentration zur vegetativen Fortpflanzung bringen (Ernst 1904).

Angaben über Aplanosporen bei *Tribonema* und ihren Verwandten finden sich bei Hawlitschka (1932) und Klebs.

4. Periodizität der Sporenentlassung.

Wie bei *Oedogonium* (s. S. 714) und einigen Rotalgen (Sagromski[1]) ist die Befreiung der Zoosporen (Klebs l. c., Rieth 1959) oder Aplanosporen (Ernst l. c.) von *Vaucheria* sowie der Amoebosporen von *Asterosiphon* (de Puymaly 1924) tageszeitlich geregelt. Durch Versuche mit *Vaucheria sessilis* zeigte Rieth, daß diese Periodik hier im Dauerdunkel und im Dauerlicht weitergeht, vom Einsatzpunkt der konstanten Bedingungen unabhängig ist, und durch invertierten Lichtwechsel nach einer Übergangsperiode in den neuen Rhythmus gezwungen werden kann. Die Eigenperiode in Dunkelheit ist zwischen 14° und 17° temperaturunabhängig; also Eigenschaften, die für die Existenz einer „inneren Uhr" bei *Vaucheria* sprechen.

5. Die Keimung der vegetativen Fortpflanzungskörper.

Bei Heterokonten erfolgt diese, soweit untersucht, unter geeigneten Bedingungen ohne Schwierigkeiten und Verzögerungen, nur bei Aplanosporen (*Vaucheria geminata*, Klebs) und Cysten von *Asterosiphon* (de Puymaly l. c.) treten Keimungshemmungen auf. Zoosporen von *Botrydium* keimen nicht in destilliertem Wasser, wohl aber in Nährlösungen (Klebs). Über Polaritätsverhältnisse bei der Keimung der *Vaucheria*-Spore vgl. Weber (1958).

d) Parthenogenese bei Heterokonten.

Sexuelle Vermehrung scheint mit einiger Sicherheit nur bei *Pleurochlonis terrestris*(Fritsch u. John 1942), *Tribonema* (Scherffel 1901), *Botrydium* (Moewus 1940; die Angaben von Rosenberg 1930 sind nach Pascher 1939 wohl irrtümlich) und der oogamen *Vaucheria* bekannt zu sein. Nach Angaben von Moewus dienen die Schwärmer der Blasen von *Botrydium* an sich als Gameten, können sich (sind die Kopulationsbedingungen nicht erfüllt oder fehlt der Partner) auch wie Zoosporen, d. h. in haploider Parthenogenese (Apomiktosis) verhalten und zu neuen Blasen auskeimen.

Die sexuelle Fortpflanzung von *Vaucheria* ist nach der cytologischen Seite immer noch nicht voll geklärt. Die coenocytischen und polyenergiden Fäden entwickeln an Seiten-, oder auch Hauptästen Oogonien und Antheridien, beide werden durch eine Wand abgegrenzt, und das Oogon ist schließlich durch Abwanderung (Oltmanns 1895, Heidinger 1908, Couch 1932) oder/und Vernichtung überschüssiger Kerne (Moreau 1913, Couch) einkernig und wird durch Spermien befruchtet. Die Hypnozygote ist von derber Membran umschlossen und die Meiosis spielt sich nach Hanatschek (1932) bei ihrer Keimung ab. Daß dabei wirklich alle vier Gonenkerne erhalten bleiben, erscheint nach Ergebnissen Rieths (1961) über die Vererbung von Geschlechtsfaktoren fraglich, wenn auch nicht unmöglich.

Verschiedentlich wurde Auswachsen von noch nicht abgegrenzten Oogonien-Anlagen zu vegetativen Fäden beobachtet (Lit. Winkler A 1942, zuletzt Rieth 1954), die als De- bzw. Umdeterminierungen aufzufassen sind und hier unberücksichtigt bleiben. Wenn aber vom Mutterfaden schon durch die Wand abgegrenzte Oogonien zum Austreiben gebracht werden, so müßte man von Parthenogenese oder in der verfeinerten Nomenklatur Winklers (1942) von Apomiktose sprechen, weil hier die Gamophase direkt wieder zur Gamophase wird. Die Aus-

[1] Zuletzt: Biol. Zbl. **80**, 681—692 (1961).

triebe aus Oogonien waren in v. WETTSTEINS (1920) Versuchen mit den Monözisten *V. hamata* und *V. sessilis* nämlich vegetative Fäden, die allerdings sehr bald wieder Sexualorgane beider Geschlechter anlegten. v. WETTSTEIN dienten diese Versuche vor allem für den Nachweis, daß selbst die ausgebildete Sexualzelle eines Monözisten noch beide Geschlechtstendenzen verkörpere. Wie RIETH (1954) und SCHUSSNIG (A 1960) bemerkten, wäre dieser Nachweis — wie übrigens auch der Name Apomiktosis — an die Voraussetzung gebunden, daß nur der ,,weibliche" Kern des Oogons die Regeneratkerne liefere. Dieses scheint nach den Erfahrungen der oben angeführten Autoren für das individualisierte Oogon gegeben, wenn der Nachweis auch im vorliegenden Falle nicht besonders geführt worden ist. Die Oogone wurden durch Anstich nach vorhergehender Plasmolyse zum Austreiben angeregt. DE WILDEMAN (1896) erhielt Ähnliches aber auch spontan. Wirkliche Parthenogenese (im Sinne WINKLERS) würde nur dann vorliegen, wenn ein unbefruchtetes Oogon eine Parthenozygote (also den Zygobionten) entwickeln würde. Dieses geschieht wohl allein bei *Vaucheria macrocarpa* (IWANOFF 1899); hier öffnen sich die Antheridien niemals, dennoch nehmen die Oogonien alle Charaktere von Oosporen an, letztere keimen jedoch nicht; die Art pflanzt sich durch Aplanosporen fort. Bei einem isoliert gezogenen weiblichen Klon von *V. dichotoma* jedoch gingen die Oogonien immer zugrunde (RIETH 1961).

Literatur.

BELCHER, J. H., and G. E. FOGG: Studies on the growth of *Xanthophyceae* in pure culture. III. *Tribonema aequale* Pascher. Arch. Mikrobiol. **30**, 17—22 (1958). — BELCHER, J. H., and J. D. A. MILLER: Studies on the growth of *Xanthophyceae* in pure culture. IV. Nutritional types amongst the *Xanthophyceae*. Arch. Mikrobiol. **36**, 219—228 (1960). — BENTLEY, J. A.: Role of plant hormones in algal metabolism and ecology. Nature (Lond.)**181**, 1499—1502 (1958). — Plant hormones in marine phytoplancton and sea water. J. mar. Biol. Ass. U. Kingd. **39**, 433—444 (1960). — BORZI, A.: Studi algologici I. Palermo 1883. — Studi algologici II. Palermo 1895.

CHADEFAUD, M.: Les Vauchéries sont des Xanthophycées. Bull. Soc. bot. France **98**, 210—211 (1951). — CHODAT, R.: Monographie d'algues en culture pure. Materiaux pour la flore cryptogamique de la Suisse **4**, fasc. 2 (1913). — CLEMENÇON, H.: Kultur und Entwicklung von *Chlorocloster engadinensis* Vischer *(Xanthophyceae)*. Arch. Mikrobiol. **51**, 199—212 (1965). — COUCH: Gametogenesis in *Vaucheria*. Bot. Gaz. **94**, 272—296 (1932).

DANGEARD, P.: Le nouveau genre *Asterosiphon* et sa place systématique. Botaniste **31**, 271—292 (1941). — DAVIS, B. M.: Oogenesis in *Vaucheria*. Bot. Gaz. **38**, 81—98 (1904). — DENFFER, D. v., u. H. HUSTEDE: Wuchsstoffbedingte Umstimmung von der sexuellen zur vegetativen Entwicklung bei *Vaucheria sessilis*. Flora (Jena) **142**, 489—492 (1955).

ERNST, A.: Siphoneenstudien. III. Zur Morphologie und Physiologie der Fortpflanzungszellen der Gattung *Vaucheria*. Beih. Bot. Zbl. **16**, 367—382 (1904). — ETTL, H.: Ein Beitrag zur Systematik der Heterokonten. Bot. Not. (Lund.) **109**, 411—445 (1956).

FRITSCH, F. E., and R. P. JOHN: An ecological and taxonomic study of the algae of british soils. II. Consideration of the species observed. Ann. Bot., N.F. **6**, 371—395 (1942).

GERNECK, R.: Zur Kenntnis der niederen Chlorophyceen. Beih. Bot. Zbl. **21**, Abt. 2, 221—290 (1907). — GREENWOOD, A. D.: Observations on the structure of the zoospores of *Vaucheria*. II. J. exp. Bot. **10**, 55—68 (1959). — GREENWOOD, A. D., J. MANTON, and B. CLARKE: Observations on the structure of the zoospores of *Vaucheria*. J. exp. Bot. **7**, 71—87 (1957).

HANATSCHEK, E.: Beiträge zur Entwicklungsgeschichte der Protophyten. X. Der Phasenwechsel bei der Gattung *Vaucheria*. Arch. Protistenk. **78**, 479—513 (1932). — HAWLITSCHKA, E.: Die Gattung *Tribonema*. Pflanzenforsch. H. 15. Jena: Gustav Fischer 1932. — HEIDINGER, W.: Die Entwicklung der Sexualorgane bei *Vaucheria*. Ber. dtsch. bot. Ges. **26**, 314—363 (1908). — HUSTEDE, H.: Untersuchungen über die stoffliche Beeinflussung der Entwicklung von *Stigeoclonium falclandicum* und *Vaucheria sessilis* durch Tryptophanabkömmlinge. Biol. Zbl. **76**, 555—595 (1957). — Bot. Marina **6**, 134—142 (1963).

IWANOFF, L.: Über neue Arten von Algen und Flagellaten usw. Bull. Soc. Imp. Natur. Moscou 423—449 (1899—1900). Zit. nach WINKLER.

JAMES, E. J.: An investigation of the algal growth in some naturally occurring soils. Beih. Bot. Zbl. A **53**, 519—553 (1935).

Koch, W. J.: A study of the motile cells of *Vaucheria*. J. Elisha Mitchel Sci. Soc. **67**. 123—131 (1951). — Kolkwitz, R.: Zur Ökologie und Systematik von *Botrydium granulatum*. Ber. dtsch. bot. Ges. **44**, 533—540 (1926).

League, E. A., and V. A. Greulach: Effects of day length and temperature on the reproduction of *Vaucheria sessilis*. Bot. Gaz. **117**, 45—51 (1955).

Manton, J., B. Clarke, A. D. Greenwood, and E. A. Flint: Further observations on the structure of plant cilia, etc. J. exp. Bot. **3**, 204—214 (1952). — Miller, J. A. D., and G. E. Fogg: Studies on the growth of *Xanthophyceae* in pure culture. I. Arch. Mikrobiol. **28**. 1—16 (1957). — II. The relations of *Monodus subterraneus* to organic substances. Arch. Mikrobiol. **30**, 1—16 (1958). — Miller, V.: Untersuchungen über die Gattung *Botrydium*. Ber. dtsch. bot. Ges. **45**, 151—170 (1927). — Moewus, F.: Über die Sexualität von *Botrydium granulatum*. Biol. Zbl. **60**, 485—498 (1940). — Moreau, F.: Recherches sur la reproduction des Mucorinées et de quelques autres Thallophytes. Botaniste **13**, 1—127 (1913). — Mundie. S. R.: Cytology and life history of *Vaucheria geminata*. Bot. Gaz. **87**, 397—410 (1929).

Oltmanns, F.: Über die Entwicklung der Sexualorgane bei *Vaucheria*. Flora (Jena) **80**, 388—420 (1895).

Pascher, A.: Heterokonten, Bd. 11 v. Rabenhorst's Kryptogamen-Flora. Leipzig: Akademische Verlagsgesellschaft 1939. — Poulton, E. M.: Études sur les Heterokontes. Univ. de Génève. Fac. Sci. Ser. X, **9**. 1—95 (1925). — Puymaly, A. de: Reproduction de *Vaucheria* par zoospores amiboides. C. R. Acad. Sci. (Paris) **174**, 824—827 (1922). — Recherches sur les algues vertes aeriennes. Bordeaux: Delberel 1924.

Renner, O.: Auch etwas über F. Moewus, *Forsythia* und *Chlamydomonas*. Z. Naturforsch. **13**b, 399—403 (1958). — Rieth, A.: Beobachtungen zur Entwicklungsgeschichte einer *Vaucheria* der Sektion *Woroninia*. Flora (Jena) **142**, 156—183 (1954). — Zur Kenntnis halophiler Vaucherien. I. Flora (Jena) **143**, 127—160 (1956a). — Zur Kenntnis halophiler Vaucherien. II. Flora (Jena) **143**, 281—294 (1956b). — Periodizität beim Ausschlüpfen der Schwärmsporen von *Vaucheria sessilis*. Flora (Jena) **147**, 35—42 (1959). — Eine diözische Rasse und experimentelle Kreuzung bei *Vaucheria dichotoma* (L.) Agh. Kurze Mitt. Forschst. Limnol. DAW Berlin 1961, 501—502 (1961). — Über *Gongrosira dichotoma* Kütz und *Asterosiphon* Dang. I. Gibt es ein „*Gongrosira*-Stadium" bei *Vaucheria*? Limnologica (Berl.) **1** (3) 197—210 (1962a). — Das Akinetenstadium von *Vaucheria*. Mber. Dtsch. Akad. Wiss. Berlin **4** (8), 519—522 (1962b). — Rosenberg, M.: Die geschlechtliche Fortpflanzung von *Botrydium*. Öst. bot. Z. **79**, 289 (1930).

Scherffel, A.: Kleine Beiträge zur Phylogenie einiger Gruppen niederer Organismen. Bot. Ztg **8**, 143—158 (1901). — Schramm, J. R.: Some pure culture methods in algae. Ann. Missouri bot. Gard. **1**, 23—45 (1914). — Smith, G. M.: Microaplanospores of *Vaucheria*. Farlowia **1**, 387—389 (1944). — Stahl, E.: Über die Ruhezustände der *Vaucheria geminata*. Bot. Ztg **37**, 129—137 (1879).

Vischer, W.: Experimentelle Untersuchungen (Gallertbildung) mit *Mischococcus sphaerocephalus* Vischer. Arch. Protistenk. **76**, 257—273 (1932). — Über Heterokonten und Heterokonten-ähnliche Grünalgen (*Bumilleriopsis, Heterothrix, Heterococcus* und *Dictyococcus, Muriella)*. Ber. schweiz. bot. Ges. **45**, 372—410 (1936). — Über einige Heterokonten *(Heterococcus, Chlorellidium)* und ihren Polymorphismus. Ber. schweiz. bot. Ges. **47**, 225—250 (1937). — Zur Kenntnis der Gattung *Botrydium* Wal. Ber. schweiz. bot. Ges. **48**, 538—561 (1938). — Heterokonten aus alpinen Böden, speziell dem schweizerischen Nationalpark. Erg. wiss. Unters. Schweiz. Nationalpark, N.F. **1**, 481—511 (1945).

Weber, W.: Zur Polarität von *Vaucheria*. Z. Bot. **46**, 141—198 (1958). — Wettstein. F. v.: Künstliche haploide Parthenogenese bei *Vaucheria* und die geschlechtliche Tendenz ihrer Keimzellen. Ber. dtsch. bot. Ges. **38**, 260—266 (1920). — Wildeman, E. de: Notes sur quelques *Vaucheria*. Notarisia **11**, 92—95 (1896).

F. Phaeophyta.

a) Vegetative Fortpflanzung.

1. Einleitung.

Die Kenntnisse der Fortpflanzungsverhältnisse der Braunalgen stützen sich weitgehend auf Beobachtungen an Freilandmaterial. Selbst wenn Beobachtungen an Laboratoriumskulturen vorliegen, sind sie nur in den wenigsten Fällen unter wohldefinierten reproduzierbaren Bedingungen angestellt worden. Außerdem ermangeln die meisten Beobachtungen cytologischer Angaben, so daß z. B. über die mit den Fortpflanzungsvorgängen einhergehenden Kernphasenverhältnisse häufig nur Vermutungen angestellt werden können. Dieser Mangel ist bedauerlich, und es ist zu hoffen, daß die beträchtlichen Fortschritte in den Kulturmethoden mariner Organismen zu einer vermehrten Bearbeitung mariner Algen mit Hilfe von Laboratoriumskulturen beitragen werden.

FRITSCH (1945) gab eine erschöpfende Zusammenfassung der über Braunalgen vorliegenden Kenntnisse. Weiter soll hier noch auf das Werk von PAPENFUSS (1951) hingewiesen werden.

Wir wollen im folgenden zunächst eine Übersicht über die vegetativen Fortpflanzungsweisen geben, dann einige physiologische Fragen berühren und schließlich die Rolle der vegetativen Fortpflanzung im Entwicklungscyclus an einigen Beispielen betrachten.

2. Die verschiedenen vegetativen Fortpflanzungsweisen.

SCHUSSNIG (1960) unterscheidet zwischen ungeschlechtlicher Fortpflanzung oder Reproduktion und vegetativer Vermehrung oder Propagation. Unter Reproduktion werden dabei alle Fortpflanzungsweisen verstanden, die sich auf den Typus einer Fortpflanzungscyste der Flagellaten zurückführen lassen, während die Propagation alle übrigen Fortpflanzungsweisen verschiedenartigster Natur umfaßt.

Tabelle 4. *Übersicht über die vegetativen Fortpflanzungsweisen der Braunalgen.*

	durch	Bildungsort	Vorkommen
Reproduktion	Schwärmsporen ⟨Planogonidien⟩	plurilokuläres Sporangium ⟨Angiothomus⟩	*Ectocarpales*[1] *Sphacelariales*
		unilokuläres Sporangium ⟨Gonidangium⟩	*Ectocarpales* *Sphacelariales* *Cutleriales* *Sporochnales* *Desmarestiales* *Laminariales*
	Aplanosporen ⟨Aplanogonidien⟩	plurilokuläres, unilokuläres Sporangium	*Acinetospora* *Zonaria*
	Tetrasporen ⟨Tetragonidien⟩	Tetrasporangium ⟨Tetragonidangium⟩	*Dictyotales*
Propagation	Monosporen	Monosporangium	*Tilopteridales* *Acinetospora*
	Brutkörper		*Sphacelariales*
	„Brutkörner"		*Sphacelaria olivacea*
	Akineten		*Tilopteris*
	Stolonen		*Sphacelariales* *Acinetospora*
	Fragmentation		*Laminaria*-Gametophyten *Halopteris* *Fucales*

[1] Im Sinne von FRITSCH (1945).

3. Plurilokuläre Sporangien.

Diese Organe werden durch eine mehr oder weniger große Anzahl von Zellteilungen aus einem Fadenabschnitt oder aus einer einzelnen Ausgangszelle gebildet. Jeder Teilungsschritt wird von einer Wandbildung gefolgt, so daß ein gekammerter Behälter entsteht (Abb. 54A), aus dessen Loculi im reifen Zustand je ein Schwärmer entlassen wird. Die Anzahl der Schwärmer pro Sporangium beträgt bei *Ectocarpus* über 800 (MÜLLER 1965b), kann sicher aber noch weit größere Zahlen erreichen. Die Kernteilungen bei der Bildung der plurilokulären Sporangien sind Mitosen, die Schwärmer besitzen also dieselbe Chromosomenzahl wie die Mutterpflanze. Bei manchen Formen (z. B. *Ectocarpus secundus*) kommen plurilokuläre Sporangien mit verschieden großer Fächerung vor. Es kann sich dabei um die Gametangien anisogamer Arten (Mega- und Mikrosporangien) oder auch um vegetative Vermehrungsorgane („Meiosporangien") handeln (Fritsch 1945)[1]. Im Rahmen der vegetativen Vermehrung haben wir hier nur die Fälle zu betrachten, in denen die plurilokulären Sporangien als Vermehrungsorgane der diploiden oder haploiden Phase dienen. Die Unterscheidung zwischen „Sporangien" und „Gametangien" ist jedoch häufig schwierig, da die Kopulationsbereitschaft potentieller Gameten starken Schwankungen unterliegen kann und diese sich außerdem ohne Kopulation entwickeln können (Knight 1931, S. 326; Kornmann 1957). Die Entleerung der plurilokulären Sporangien findet im allgemeinen durch eine apikale Öffnung statt, durch die die Schwärmer einzeln austreten, nachdem die Innenwände der einzelnen Loculi teilweise aufgelöst worden sind. Bei manchen Sphacelariales entweicht jedoch jeder Schwärmer unabhängig durch eine Öffnung in der Außenwand des Loculus direkt nach außen.

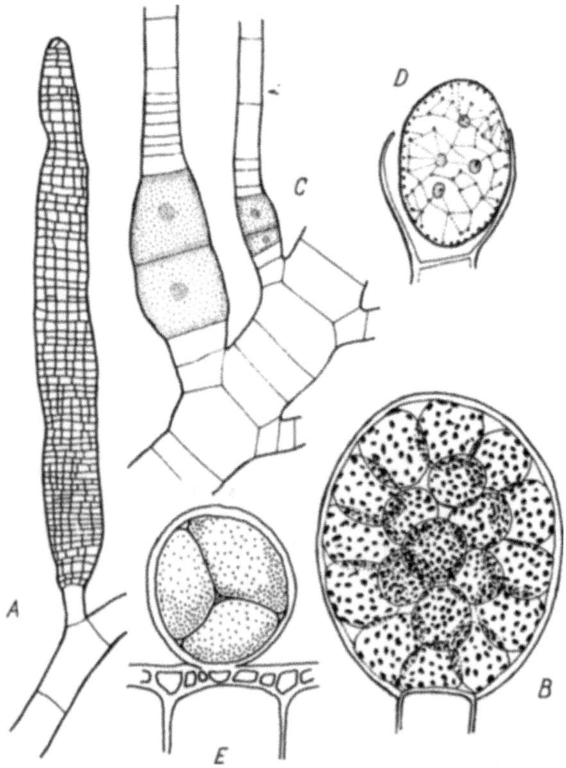

Abb. 54 A—E. Verschiedene Fortpflanzungsorgane von Braunalgen. A *Ectocarpus divergens*, plurilokuläres Sporangium, B *Acinetospora pusilla*, unilokuläres Sporangium, C *Tilopteris Mertensii*, Monosporangien mit einkernigen Monosporen, D *Haplospora globosa*, Monosporangium mit vierkerniger Monospore, E *Dictyota dichotoma*, Tetrasporangium. (A aus Kornmann 1957, B aus Schmidt 1940, C nach Bornet, aus Papenfuss 1951, D nach Reinke, aus Papenfuss 1951, E nach Thuret und Bornet, aus Papenfuss 1951.)

4. Unilokuläre Sporangien.

Unilokuläre Sporangien werden aus einzelnen Zellen eines Fadenverbandes, freistehenden oder zu vielen in Sori angeordneten Anlagen gebildet. Sie sind dadurch gekennzeichnet, daß die einzelnen Schwärmer nicht wie im plurilokulären Sporangium durch Wandbildungen voneinander getrennt sind (Abb. 54B). In einer Reihe von Fällen wurde festgestellt, daß die beiden ersten Kernteilungen in der Sporangienanlage meiotisch sind. Es gibt allerdings Abweichungen von

[1] Da keiner der in Frage kommenden Fälle vollständig aufgeklärt ist, sollte man sich zur Zeit auf deskriptive Bezeichnungen beschränken und den Ausdruck „Meiosporangien" vermeiden. Man könnte beispielsweise Makro-, Meso- und Mikrosporangien unterscheiden. Ist die Funktion der einzelnen Organe bekannt, so ließen sich z. B. Makro-(Gyno-)Gametangien ⟨Gynothomen⟩, Mikro-(Andro-)Gametangien ⟨Androthomen⟩ und geschlechtlich neutrale Sporangien ⟨Angiothomen⟩ unterscheiden.

dieser Regel, in denen eindeutig nachgewiesen ist, daß die Reduktion der Chromosomenzahl unterbleibt (*Ectocarpus siliculosus*, MÜLLER 1965b). Ähnliche Verhältnisse sind durch Ergebnisse von Kulturversuchen wahrscheinlich gemacht worden bei *Pylaiella rupincola* (KYLIN 1937), *Giffordia* (KORNMANN 1954), *Chordaria* (KORNMANN 1962a) und *Adenocystis* (NAYLOR 1955). Die aus dem unilokulären Sporangium entlassenen Schwärmer sind in der Regel asexuell, sie entwickeln sich zu haploiden Pflanzen. Bei *Ectocarpus siliculosus* von der englischen Küste sollen jedoch nach KNIGHT (1931) aus dem unilokulären Sporangium Gameten entlassen werden. Sowohl der Ausfall der Reduktionsteilung als auch die sofortige Kopulation der Schwärmer aus unilokulären Sporangien führt zum Verschwinden der Haplophase, vorausgesetzt, daß es sich um unilokuläre Sporangien auf diploiden Pflanzen handelt. SCHREIBER (1930) stellte fest, daß bei der Reduktionsteilung im unilokulären Sporangium von *Laminaria* die genotypische Geschlechtsbestimmung stattfindet. Sichere Fälle modifikatorischer Geschlechtsbestimmung diözischer Formen sind nicht bekannt. Die Zahl der Schwärmer, die pro Sporangium gebildet werden, ist in der Regel eine Potenz von 2, sie liegt bei vielen Arten bei 16 oder 32, kann jedoch bei *Ectocarpus* 256 beträchtlich überschreiten (MÜLLER 1965b). Die Entleerung geschieht durch eine apikale oder seitliche Öffnung, durch die im Gegensatz zu den plurilokulären Sporangien die Schwärmer nicht einzeln entweichen. Der gesamte Sporangieninhalt wird vielmehr als Ganzes mehr oder weniger ruckartig ausgestoßen. Die Schwärmer sind dabei noch in eine schleimige Masse eingehüllt, aus der sie sich allmählich befreien und damit ihre Beweglichkeit erlangen (KNIGHT 1931, SCHREIBER 1930). Die dem Sporangium anliegenden Paraphysen sollen nach SCHREIBER bei *Laminaria* eine Rolle beim Entleerungsmechanismus spielen. Bei *Ectocarpus* können sich jedoch völlig freipräparierte Sporangien ohne Schwierigkeit entleeren (MÜLLER 1965b).

5. Tetrasporangien.

Das Tetrasporangium (Abb. 54E) kann als homolog zum unilokulären Sporangium betrachtet werden (SCHUSSNIG 1960). Dies ist klar erkennbar bei den *Dictyotales*, wo neben Arten mit vier Tetrasporen pro Sporangium (*Dictyota*) auch Arten mit acht Sporen vertreten sind (*Zonaria*), für die dann der allgemeinere Ausdruck Aplanosporen verwendet wird. Bei der Ausbildung der Tetra- bzw. Aplanosporangien findet die Reduktionsteilung statt. Für *Dictyota* ist nachgewiesen, daß dabei die genotypische Geschlechtsbestimmung erfolgt (SCHREIBER 1935). Als Abweichung von dieser normalen Entwicklung wurde in einigen Fällen beobachtet, daß bereits die Tetrasporenmutterzelle auf der Mutterpflanze auskeimen kann (KNIEP 1928), oder die Tetrasporenmutterzelle als „Monospore“ ausgestoßen wird und auskeimt (ROBINSON 1932). Ob diese Keimlinge zu diploiden oder haploiden Pflanzen heranwachsen, wurde bisher allerdings noch in keinem Falle verfolgt. Es gibt jedoch eine Reihe von Freilandbeobachtungen, aus denen hervorgeht, daß an vielen Standorten die Tetrasporophytengeneration stark überwiegt. Es ist deshalb durchaus möglich, daß dies durch eine zusätzliche, außerhalb des regulären Generationswechsels liegende Vermehrungsweise bewirkt wird.

6. Monosporangien.

Als Monosporangien werden Behälter bezeichnet, aus denen eine einzelne, unbewegliche Spore (Monospore) entlassen wird. Solche Bildungen kommen bei Mitgliedern der Ordnung *Tilopteridales* vor, jedoch auch bei der von KORNMANN (1953) zu den *Ectocarpales* gestellten *Acinetospora*. Es gibt Fälle, in denen vierkernige Monosporen angetroffen werden (Abb. 54D), bei denen cytologische Beobachtungen für eine vorangegangene Reduktionsteilung sprechen (*Haplospora globosa*, NIENBURG 1923). Nähere Einzelheiten sind jedoch nicht bekannt. In anderen Fällen, z.B. *Acinetospora* (SCHMIDT 1940), *Tilopteris*, *Scaphospora* (KYLIN 1933) sind die Monosporen einkernig und werden als akzessorische, den Brutknospen vergleichbare Bildungen betrachtet (Abb. 54C).

7. Aplanosporangien.

Die unilokulären Sporangien (Aplanosporangien) mancher *Dictyotales* wurden bereits erwähnt. In manchen Fällen (z.B. *Acinetospora*, KORNMANN 1953) werden jedoch auch aus plurilokulären Sporangien statt beweglicher Schwärmer unbewegliche Aplanosporen entlassen. Die Fähigkeit zur Ausbildung begeißelter Schwärmer scheint dabei nicht vollständig verlorengegangen zu sein, denn zeitweise können aus denselben Sporangien auch normal bewegliche Schwärmer ausgestoßen werden.

8. Hypnocysten, Brutkörper, Adventivthalli.

In einer Reihe von Fällen dienen besonders ausdifferenzierte Thallusabschnitte der vegetativen Vermehrung. Bei *Tilopteris* werden nach SAUVAGEAU (1928) an absterbenden Thalli einzelne Zellen zu „Hypnocysten". Diese Strukturen sollen ein Dauerstadium darstellen und sich bei günstigen Bedingungen wieder zu ganzen Pflanzen entwickeln können. FRITSCH (1945) nennt diese Bildungen „Akineten". Bei den *Sphacelaria*-Arten werden in vielgestaltiger Weise Brutkörper (propagules) ausgebildet (Abb. 55A). Es handelt sich dabei um modifizierte Seitenzweige, die sich nach ihrer Ausbildung an einer vorgebildeten Stelle ablösen und zu einer neuen Pflanze heranwachsen. Bei manchen *Sphacelaria*-Arten überwiegt diese Vermehrung durch Brutknospen bei weitem die durch Schwärmer (FRITSCH 1945); PRINGSHEIM (1873) berichtet über „Brutkörner" bei *Sphacelaria olivacea*. Es handelt sich dabei um traubenförmige Anhäufungen ruhender Zweiginitialen, die der vegetativen Vermehrung dienen sollen. Bei *Zonaria Farlowii* (HAUPT 1932) entstehen am Thallus junge Adventivthalli aus vegetativen Zellen (Abb. 55C). Diese Pflänzchen können sich ablösen und zu neuen Individuen heranwachsen. Bei *Spathoglossum* (*Dictyotales*) kommen nach TAKAMATSU (1939) eigenartige „Brutkörper" vor: an einzelnen an der Thallusoberfläche entspringenden Fäden bilden sich mehrzellige Körper aus, die sich von der Mutterpflanze loslösen können. An diesen Brutkörpern werden noch auf der Mutterpflanze Tetrasporen gebildet.

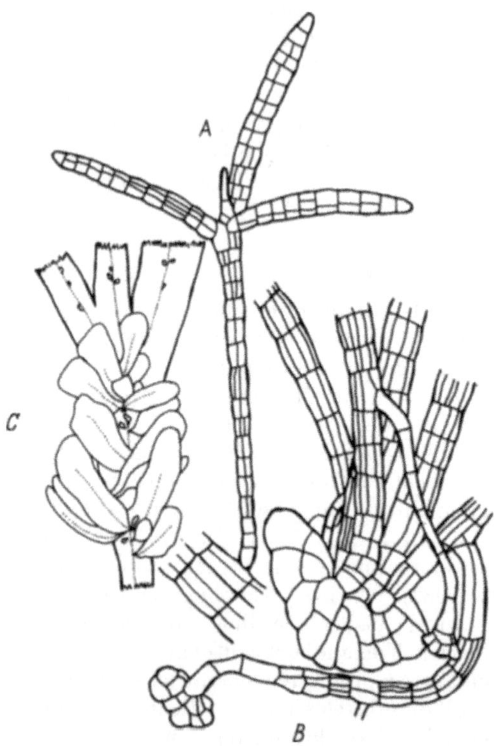

Abb. 55A—C. A *Sphacelaria cirrhosa*, Brutkörper, B *Sphacelaria cirrhosa*, Stolonenbildung, C *Zonaria Farlowii*, Adventivthalli. (A, B aus TAYLOR 1957, C aus HAUPT 1932.)

9. Haftscheiben, Stolonen.

Viele Algen können sich mit Hilfe ihrer Haftscheiben auf der Unterlage ausbreiten und an neuen Stellen den aufrechten Teil der Pflanze ausbilden (FRITSCH 1945). Vielfach heften sich auch zu Stolonen umgebildete Seitenäste bei erneutem Kontakt mit der Unterlage fest und bilden einen neuen Basalteil (Abb. 55B), aus dem dann wieder die aufrechten Teile der Pflanze hervorgehen (z.B. *Cladostephus*, FRITSCH 1945; *Phyllitis*, KYLIN 1933; *Acinetospora*, KORNMANN 1953).

10. Vegetative Vermehrung durch Fragmentation.

Für die starke Regenerationsfähigkeit der Braunalgen gibt es eine Reihe von eindrucksvollen Beispielen. SCHREIBER (1930) konnte *Laminaria*-Gametophyten, von einer einzigen Schwärmspore ausgehend, durch häufiges Zerteilen zu einer unbeschränkt großen Anzahl von Pflänzchen vermehren. HAMEL (1938) berichtet über das Auswachsen abgebrochener Thallusteile von *Halopteris* zu ganzen Pflanzen. Unter den *Fucales* gibt es einige Fälle, bei denen sich große Bestände ausschließlich durch Regenerationsprozesse erhalten. Hierher gehört z.B. *Fucus mytili*, bei dem NIENBURG (1932) feststellte, daß Zerreißen älterer Thallusstücke und Auswachsen der Bruchstücke die einzige Vermehrungsweise der Art darstellen. Oogonien werden zwar ausgebildet, aber nicht entleert, männliche Pflanzen konnten niemals gefunden werden. Ähnlich verhalten sich die eupelagischen *Sargassum*-Arten. Auch hier findet lebhafteste vegetative Vermehrung ohne spezifische Vermehrungsorgane statt (GESSNER 1955).

11. Einige Bemerkungen zur Physiologie der Fortpflanzung.

Ausbildung der Sporangien.

Das Auftreten bestimmter Fortpflanzungsorgane ist in sehr vielen Fällen an bestimmte Jahreszeiten gebunden (z.B. FUNK 1957). Genauer untersuchte Fälle zeigen, daß es sich dabei um eine direkte Koppelung an jahresperiodisch variierende Außenfaktoren handeln kann. KORNMANN (1962b) fand plurilokuläre Sporangien bei *Elachista* bevorzugt bei tiefer Temperatur. MÜLLER (1962) stellte fest, daß *Ectocarpus siliculosus* bei 13° C nur unilokuläre, bei 19° nur plurilokuläre Sporangien ausbildet. Andere Faktoren, wie Lichtgenuß, Salzkonzentration und Alter der Pflanzen spielten eine geringere Rolle. Pflanzen verschiedener Standorte können sich dabei unterschiedlich verhalten (MÜLLER 1963). Die unilokulären Sporangien von *Zonaria* werden nach HAUPT (1932) in einzelnen Schüben synchronisiert angelegt, und zwar in enger zeitlicher Koppelung an den lunaren Cyclus. Es handelt sich dabei möglicherweise um eine ähnliche steuernde Wirkung des Vollmondlichtes wie dies bei den Gametophyten von *Dictyota* der Fall ist (MÜLLER 1962).

Entleerung der Sporangien.

Aus einer Reihe von Beobachtungen geht hervor, daß die Bereitschaft zur Entleerung der reifen Sporangien tagesperiodischen Schwankungen unterworfen ist. KYLIN (1933) und HARTMANN (1934) z.B. sahen die Entleerung der plurilokulären Sporangien von *Ectocarpus* stets vormittags. Nach KORNMANN (1953) entleeren sich die Sporangien von *Acinetospora* in der Mitte der Nacht. Zeitweiliges Trockenliegen und Wiedereintauchen der Thalli in Seewasser scheint in vielen Fällen (z.B. *Pylaiella* und *Ectocarpus*) einen Reiz darzustellen, der die Entleerung der Sporangien auslöst (KYLIN 1933, MÜLLER 1964). Diese Reaktion dürfte eine Anpassung an die besonderen ökologischen Verhältnisse der Gezeitenzone darstellen. Nach Papenfuss (1934/35) bewirkt auch ein Wechsel der Kulturlösung Ausschüttung der Schwärmer bei *Ectocarpus*. Über die in Abständen von 4—5 Tagen wiederkehrende besonders intensive Schwärmerproduktion einiger Laminariales berichtet Suto (1952). Die Entleerung der Aplanosporangien von *Zonaria* ist zeitlich an den lunaren Cyclus gekoppelt. Bemerkenswert ist in diesem Fall, daß es sich nicht um die Ausstoßung von Geschlechtszellen handelt, wie dies bei allen übrigen bei Braunalgen bekannten lunarperiodischen Erscheinungen der Fall ist.

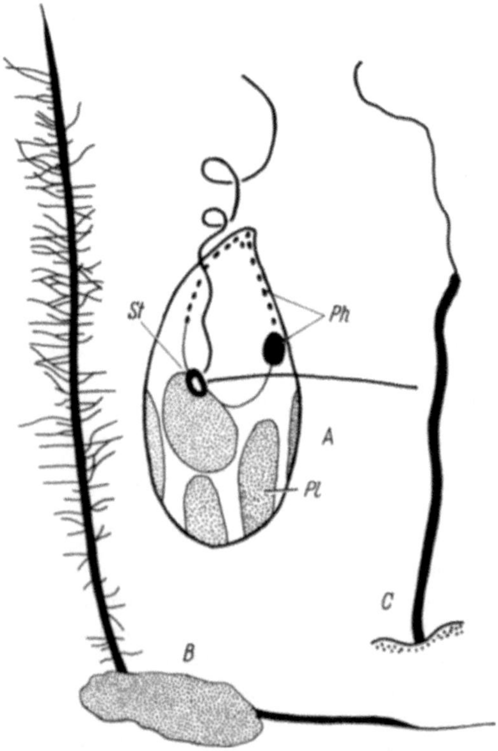

Abb. 56 A—C. A *Myriotricha densa*, Schwärmer aus plurilokulärem Sporangium. *Ph* Physoden; *St* Stigma; *Pl* Plastiden. B *Pylaiella litoralis*, Schwärmer aus unilokulärem Sporangium mit flimmertragender Vordergeißel. C *Sorocarpus uvaeformis*, Vordergeißel mit Akronema. (A aus Chadefaud 1948, B nach einer elektronen-mikroskopischen Aufnahme aus Manton und Clarke 1951, C aus Müller 1965a.)

Die Schwärmer der Braunalgen.

Die beweglichen Fortpflanzungszellen der Braunalgen sind nach einem sehr einheitlichen Schema gebaut (Abb. 56A). Neben einem oder mehreren Plastiden ist in der Regel ein Augenfleck vorhanden, der jedoch auch fehlen kann (z. B. bei *Acinetospora*, Kornmann 1953). Manton (1957) gibt eine ausführliche Beschreibung der Feinstruktur der Schwärmer von *Scytosiphon*. Die Schwärmer besitzen zwei Geißeln, von denen die nach vorne gerichtete längere nach elektronenmikroskopischen Untersuchungen von Manton und Clarke (1951) mit Flimmern besetzt ist (Abbildung 56B). An der Vordergeißel von *Ectocarpus* und *Sorocarpus* fand Müller (1965a) einen langen peitschenschnurartigen Fortsatz (Akronema, Abb. 56C). Die nach hinten gerichtete, in der Regel kürzere Geißel trägt keine Flimmern, jedoch ein Akronema.

Über Phototaxis bei Braunalgenschwärmern liegen eine größere Anzahl von Beobachtungen vor. Da jedoch meist nur über Ansammlung am lichtzu- oder abgewandten Teil eines Tropfens oder einer Kulturschale berichtet wird, sind präzise Angaben nicht möglich. Sicher ist jedoch, daß sowohl positive als auch negative Phototaxis vorkommen kann. Auch Umschlagen von positiver zu negativer Phototaxis in derselben Schwärmerpopulation (Kornmann 1954) sowie indifferentes Verhalten zur Lichtrichtung scheint möglich zu sein (Kornmann 1953). Manche Beobachtungen sprechen dafür, daß Belichtung für die Beweglichkeit der Schwärmer notwendig ist (Kornmann 1953). Suto (1948) konnte bei den Schwärmern einiger Laminariaceen positive Thigmotaxis und Rheotaxis feststellen. Die Mitwirkung eines Berührungsreizes beim Festsetzen von Schwärmern auf der Unterlage (Thigmokinesis) wurde von Müller (1964) gezeigt. Chemotaktisches Verhalten von ungeschlechtlichen Schwärmern wurde bisher nicht eindeutig beobachtet, jedoch ist es möglich, daß sie beim Festsetzen in Gruppen eine Rolle spielt. Häufig wird auch ein Festsetzen der Schwärmer an der Gefäßwand in Höhe der Wasser-Luft-Grenze beobachtet (z. B. KYLIN 1918), wofür möglicherweise Chemotaxis mitverantwortlich ist. Vielleicht ist Chemotaxis auch bei der Realisierung der Wirtsspezifität mancher Epiphyten im Spiel. In diesem Fall dürften aber auch andere Faktoren, z. B. lokale Oberflächenspannungsverhältnisse (BETH und LINSKENS 1964) beteiligt sein. SUTO (1950) gibt die Schwimmgeschwindigkeit verschiedener Laminariaceen-Schwärmer mit ca. 200 μ/sec. an. Er kommt zu der Feststellung, daß die Wirksamkeit der Eigenbewegung der Schwärmer weit hinter dem Verbreitungseffekt der Wasserbewegung zurücktritt. Dies schließt allerdings nicht aus, daß eine stetige gerichtete (z. B. lichtgesteuerte) Auf- oder Abwärtsbewegung wirkungsvoll die Verbreitung beeinflussen kann. Die Dauer der Schwärmerphase ist bei *Laminaria hyperborea* temperaturabhängig (KAIN 1964). Bewegungslos gewordene Schwärmer sollen mehrere Tage lang schwebend bleiben und dabei die ersten Keimungsstadien durchführen können. *Laminaria*-Schwärmer können sich noch nach 60 Tagen Aufenthalt in Dunkelheit zu lebensfähigen Gametophyten entwickeln.

Weiterentwicklung der festgesetzten Fortpflanzungszellen.

Die Fortpflanzungszellen der Braunalgen entwickeln sich im allgemeinen nach vollendeter Anheftung auf der Unterlage ohne Unterbrechung weiter. Die beweglichen Schwärmer werden nach Verlust der Geißeln und Ausbildung einer festen Wand zur Embryospore. SCHUSSNIG (1960) beschreibt den Keimungsvorgang aus-

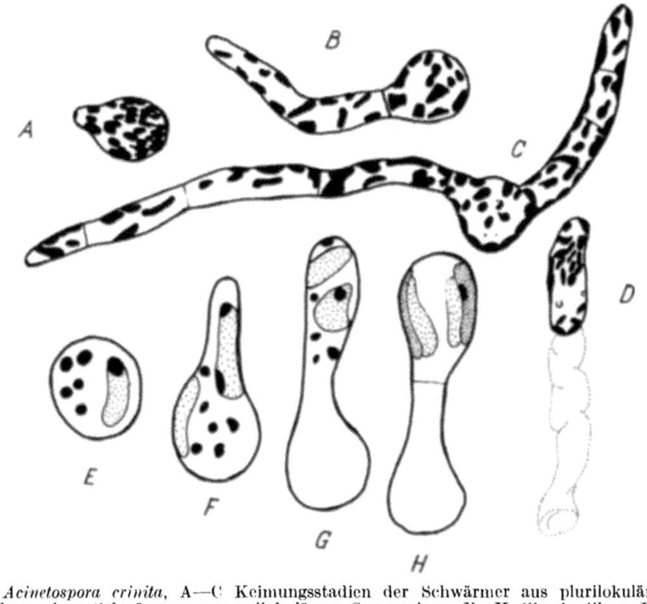

Abb. 57 A—H. *Acinetospora crinita*, A—C Keimungsstadien der Schwärmer aus plurilokulären Sporangien, D amöboide Phase eines Schwärmers aus unilokulärem Sporangium. E—H *Chorda filum*, Keimungsstadien von Schwärmern aus unilokulären Sporangien. (A—D aus KORNMANN 1953, E—H aus KYLIN 1918.)

führlich. Häufig ist ein merkwürdiges Keimungsverhalten anzutreffen: der gesamte lebende Inhalt der Embryospore wandert in den Keimschlauch ein, trennt sich durch eine Wand von der leeren Embryosporenhülle ab und beginnt erst dann mit der Bildung der eigentlichen Keimpflanze (Abb. 57 E—H). In vielen Fällen wird bereits bei der ersten Zellteilung eine Rhizoidzelle von der den aufrechten Teil der Pflanze bildenden Thalluszelle abgetrennt (Abb. 57 A—C). Bei den Tetrasporen der *Dictyotales* wird diese Polarität durch Licht induziert: die erste Wand wird senkrecht zur Lichtrichtung angelegt und trennt nach der lichtabgewandten Seite eine Rhizoidanlage ab (Robinson 1932). Bei *Acinetospora* (Kornmann 1953) zeigen die festgesetzten Schwärmer lebhafte amöboide Bewegung und hinterlassen dabei eine an der Unterlage haftende röhrenförmige Ausscheidung (Abb. 57 D). Nach Naylor (1955) findet die Keimung von *Adenocystis* nur im Licht statt. Die bei der Keimung gebildeten Zellfäden können positiv oder negativ phototropisch reagieren (Papenfuss 1934/35, Naylor 1956, Kornmann 1953).

Heteroblastie.

Bei Kulturversuchen an einigen Braunalgen wurde beobachtet, daß Schwärmer desselben Ursprungs eine grundlegend verschiedene Entwicklungsrichtung einschlagen können. Diese Erscheinung wurde zuerst von Sauvageau (1924) beschrieben und als Heteroblastie bezeichnet. (Diese Bezeichnung stimmt nicht mit dem von Goebel 1928 verwendeten Begriff „heteroblastische Entwicklung“ überein!) So können sich z. B. die Schwärmer aus den unilokulären Sporangien von *Chordaria*, die keinen Kernphasenwechsel im Entwicklungscyclus besitzt, direkt zu *Chordaria*-Pflanzen entwickeln oder aber zu einer Zwerggeneration, aus deren plurilokulären Sporangien dann wieder asexuelle *Chordaria*-Pflanzen hervorgehen (Kornmann 1962a). Bei *Giffordia* tritt nach Kornmann (1954) Heteroblastie an zwei Stellen im Entwicklungsgeschehen auf, nämlich bei den Schwärmern aus unilokulären Sprangien der *Giffordia* und denen aus plurilokulären Sporangien einer Zwerggeneration. In manchen Fällen ist die Entwicklungsrichtung, die die einzelnen Keimlinge einschlagen werden, bereits an morphologischen Merkmalen der Embryosporen zu erkennen (Kylin 1933, Abb. 58). Über die die Heteroblastie bewirkenden Faktoren gibt es nur wenige Angaben: Kylin (1933) meint, daß es für die künftige Entwicklungsrichtung entscheidend sei, ob sich die Schwärmer am Boden der Kulturschale oder an der Wasseroberfläche angesiedelt haben. In einem weiteren Fall von Heteroblastie beobachtete Naylor (1955), daß das Entwicklungsverhalten der entstehenden Keimlinge in Kulturgefäßen mit fließendem Kulturmedium deutlich verschoben war gegenüber dem Verhalten in Gefäßen mit stehendem Medium.

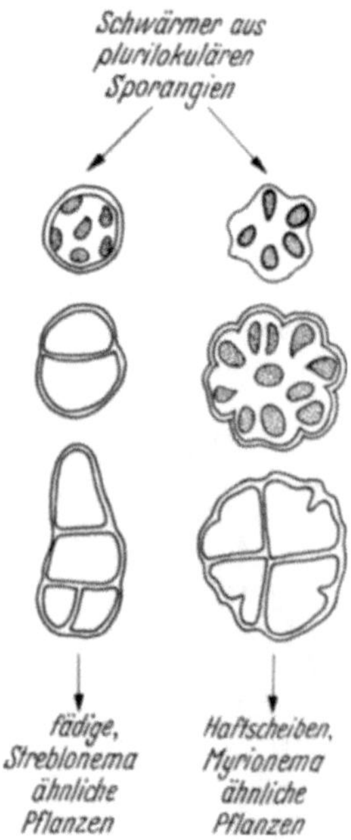

Abb. 58. *Eudesme zosterae*. Heteroblastie der Schwärmer aus unilokulären Sporangien. (Nach Kylin 1933.)

12. Die Bedeutung der vegetativen Fortpflanzung im Generationswechsel der Braunalgen.

Unter Generationswechsel versteht Winkler (1942) „den Wechsel zwischen zwei oder mehr Entwicklungszuständen, die sich bei sonstiger morphologischer Gleichheit oder Verschiedenheit durch die Art ihrer Fortpflanzung unterscheiden, wobei aus den Fortpflanzungszellen oder -körpern wieder dieselbe oder eine andere Entwicklungsstufe des Cyclus entstehen kann“. Dieser Fall ist bei den meisten Ordnungen der Phaeophyceen verwirklicht und wir wollen im Folgenden einige in neuerer Zeit relativ gut durchgearbeitete Beispiele anführen.

Kornmann (1956) berichtete über Kulturversuche mit einer *Ectocarpus confervoides*-Form (Abb. 59 A). Der diploide Sporophyt kann sich mit Hilfe plurilokulärer Sporangien selbständig vermehren. Aus den Schwärmern der unilokulären Sporangien dieses Sporophyten entstehen die anders gestalteten Gametophyten. Die auf diesen in plurilokulären Gametangien gebildeten Gameten können kopulieren und aus der Zygote entsteht wieder der diploide Sporophyt. Nicht kopulierte Gameten können sich parthenogenetisch zu Pflanzen mit Sporophyten-Wuchsform entwickeln (haploide Sporophyten), die wiederum imstande sind, sich mit Hilfe von Schwärmern aus plurilokulären Sporangien selbständig zu vermehren. Einzelne Fadenbereiche der haploiden Sporophyten können jedoch zum Gametophyten-Habitus umschlagen (Aposporie) und die plurilokulären Behälter dieser Bereiche liefern wieder Gameten.

Noch kompliziertere Verhältnisse fand Müller (1965b) bei *Ectocarpus siliculosus* aus Neapel (Abb. 59B). Auch hier liegen haploide und diploide Sporophyten

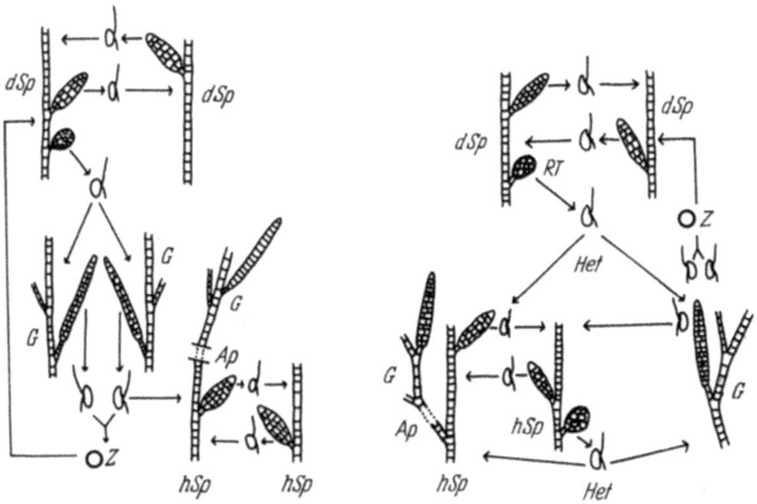

A B

Abb. 59A u. B. Schematische Darstellung der Entwicklungsmöglichkeiten von A *Ectocarpus confervoides*. (Nach Kornmann 1956.). B *Ectocarpus siliculosus* aus Neapel. (Nach Müller 1965b.) *Ap* Aposporie; *dSp* diploider Sporophyt; *G* Gametophyt; *Het* Heteroblastie der Schwärmer aus unilokulären Sporangien; *hSp* haploider Sporophyt; *RT* Reduktionsteilung; *Z* Zygote.

vor, die sich beide durch Schwärmsporen aus plurilokulären Sporangien selbständig vermehren können. Während im unilokulären Sporangium des diploiden Sporophyten die Reduktionsteilung abläuft, fällt sie im unilokulären Sporangium des haploiden Sporophyten aus. Außerdem tritt bei den Schwärmern aus unilokulären Sporangien der haploiden und diploiden Sporophyten Heteroblastie auf, d.h. sie können sich jeweils entweder zu haploiden Sporophyten oder Gametophyten entwickeln. Gametophyten können auch apospor auf haploiden Sporophyten entstehen. Die Entwicklung haploider Sporophyten aus Schwärmern der unilokulären Sporangien kann nach Winkler (1942) als Apogonosis bezeichnet werden.

Cavassilas (1963) beschäftigte sich mit der Aufklärung des Lebenscyclus von *Levringia brasiliensis* (Abb. 60). Es liegt hier ein komplizierter Wechsel heteromorpher Generationen vor: die Schwärmer aus den plurilokulären Sporangien des als typische *Levringia* bekannten Sporophyten entwickeln sich zu mikroskopisch kleinen männlichen und weiblichen „prothallischen Gametophyten". Die Schwärmer aus den unilokulären Sporangien dieser Gametophyten kopulieren und ergeben eine weitere mikroskopisch kleine Generation, die plurilokuläre Sporangien ausbildet und sich damit selbständig vermehren kann. Nach mehreren Generationen solcher „Plethysmothalli" treten „Protonemen" auf, aus deren vegetativen Zellen wieder die makroskopischen *Levringia*-Thalli hervorgehen. Die gesamte Folge der Plethysmothalli-Generationen kann übersprungen werden, indem sich aus der Zygote sogleich das Protonema-Stadium entwickelt. Das *Levringia*-Stadium ist nur während kurzer Zeit des Jahres aufzufinden, die übrige Zeit wird am Standort mit Hilfe der verschiedenen Zwergstadien überbrückt. SAUVAGEAU (1932) beschäftigte sich ausführlich mit solchen periodisch erscheinenden Braunalgen. Die makroskopische Form bezeichnet er als Delophyt, die Gesamtheit der mikroskopisch kleinen Stadien als Adelophyten. Ist der Gametophyt mikroskopisch klein und kann sich nicht selbständig vermehren, so wird er als Prothallium bezeichnet. Können sich die Zwergstadien selbständig vegetativ vermehren, so werden sie Plethysmothalli genannt. Schließlich nennt SAUVAGEAU die Zwergstadien, die aus vegetativen Zellen direkt zum Delophyten führen, ohne sich selbständig vermehren zu können, Protonemen. FRITSCH (1945) bezweifelt den Wert dieser Terminologie und schlägt vor, lediglich den Terminus „Plethysmothallus" für selbständig vermehrungsfähige Zwergstadien des diploiden Sporophyten beizubehalten.

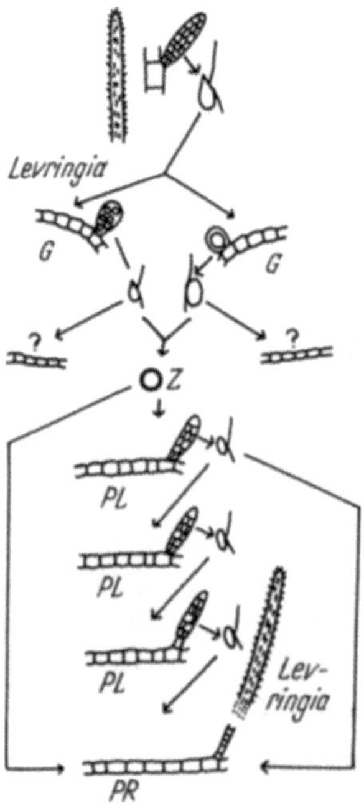

Abb. 60. Entwicklungsgang von *Levringia brasiliensis*. (Nach CAVASSILAS 1963.) *G* Gametophyt; *PL* Plethysmothallus; *PR* Protonema; *Z* Zygote; (?) parthenogenetisch entstandene Keimlinge unbekannter Natur.

An diesen wenigen Beispielen ist deutlich zu erkennen, daß bei den niederen Braunalgen (*Ectocarpales*) nicht von obligatem Generationswechsel gesprochen werden kann, wie das bei den höher entwickelten Formen mit stark ausgeprägter Anisogamie oder Oogamie der Fall ist (*Cutleriales*, *Sporochnales*, *Demarestiales*, *Laminariales*, *Dictyotales*). Das Durchlaufen einer sexuellen Phase ist bei den niederen Formen nicht obligatorisch in den Entwicklungsgang eingeschaltet, sondern vegetative Vermehrungsweisen herrschen vor. Nach den wenigen bisher vorliegenden experimentellen Befunden kann man annehmen, daß an vielen Stellen des Entwicklungsverlaufs labile Stadien eingeschaltet sind, an denen Außenfaktoren (mit häufig periodisch schwankender Natur) steuernd eingreifen können. Auf diese Weise ist denkbar, daß sich in manchen Fällen trotz vielfältiger potentieller Entwicklungsmöglichkeiten ein cyclischer Entwicklungsverlauf einspielt. Abschließend sei noch auf die von FELDMANN (1952) und KORNMANN (1954) vertretene Vorstellung hingewiesen, daß „heteromorphe Partner eines Entwicklungscyclus sich getrennt und zu selbständigen Formen entwickelt haben könnten". Als Beispiel für ein Zwischenstadium auf diesem Weg könnte man *Ectocarpus divergens* betrachten (KORNMANN 1957), bei dem zwei morphologisch verschiedene Phasen sich durch Schwärmer aus plurilokulären Sporangien selbständig vermehren können. Ein Übergang von einer Phase zur anderen ist nur noch in einer Richtung möglich, nämlich über die Schwärmer aus gelegentlich auftretenden unilokulären Sporangien. Ist die andere Phase einmal erreicht, so findet nur noch vegetative Vermehrung dieser Phase statt und ein Übergang zur Ausgangsform ist nicht mehr möglich. KORNMANN (1954) hält es für „durchaus wahrscheinlich, daß viele der im Freien vorkommenden mikroskopischen Braunalgen selbständig gewordene Partner ehemaliger oder noch bestehender Lebenscyclen sind".

b) Apogamie und Parthenogenese bei Braunalgen.

1. Parthenogenese isogamer Braunalgen.

Die Entwicklung unbefruchteter Gameten kommt bei den isogamen *Ectocarpales* und *Sphacelariales* häufig vor. Allerdings ist die eindeutige Unterscheidung von Gameten und ungeschlechtlichen Fortpflanzungszellen im Einzelfalle schwierig. Als Beispiel sei hierfür nur der häufig untersuchte *Ectocarpus siliculosus* aus Neapel angeführt. Nach HARTMANN (1934) kann die Kopulation zeitweise ohne ersichtlichen Grund ausbleiben. Tageszeit oder Alter der Gameten kann die Kopulationsbereitschaft beeinflussen (KNIGHT 1931). Der Gametencharakter der Schwärmer einzelner Pflanzen kann unterschiedlich stark ausgeprägt sein (HARTMANN 1934). Es ist deshalb durchaus möglich, daß die verschiedensten Abstufungen zwischen Gameten und ungeschlechtlichen Schwärmern vorkommen. In das Bild dieser labilen Geschlechtsausprägung paßt auch die Beobachtung, daß bei *Ectocarpus siliculosus* an anderen Standorten bereits die Schwärmer aus unilokulären Sporangien als Gameten fungieren können (KNIGHT 1931, SCHUSSNIG und KOTHBAUER 1934). Bei den morphologisch isogamen, physiologisch anisogamen Arten *Cladostephus spongiosus* (SCHREIBER 1931) und *Ectocarpus siliculosus* (HARTMANN 1934) sind sowohl die „weiblichen" als auch die „männlichen" Gameten ohne Befruchtung entwicklungsfähig. Für letzteren Fall wird dann statt Parthenogenesis gelgentlich die Bezeichnung Ephebogenesis verwendet. Nur in wenigen Fällen ist die Natur der aus unbefruchteten Gameten entstehenden Pflanzen bekannt. Bei *Ectocarpus* sind zwei eindeutige Fälle von Parthenogenese im engeren Sinne (WINKLER 1942) festgestellt worden: unbefruchtete Gameten von *E. confervoides* (KORNMANN 1956) und *E. siliculosus* aus Neapel (MÜLLER 1965) entwickeln sich zu haploiden Pflanzen mit Sporophytencharakter. Dagegen gehen aus unbefruchteten Gameten von *E. siliculosus* aus der Nordsee nach KORNMANN (1957) wieder Geschlechtspflanzen hervor, es liegt also Apomiktosis vor. KORNMANN ist dabei der Ansicht, „daß die überwiegende Mehrzahl der zusammengebrachten Schwärmer als geschlechtlich neutral anzusehen ist".

2. Parthenogenese oogamer Braunalgen.

Die Entwicklung unbefruchteter Eier ist bei einigen Braunalgenordnungen beobachtet worden. Ihre Bedeutung scheint jedoch gering zu sein, da die Keimlinge in den meisten untersuchten Fällen frühzeitig abstarben. In keinem Fall wurden bisher parthenogenetisch entstandene fruktifizierende Pflanzen erhalten.

Die Fähigkeit zu parthenogenetischer Entwicklung scheint in den verschiedenen Gruppen unterschiedlich stark ausgeprägt zu sein. SCHREIBER (1930) erhielt aus unbefruchteten Eiern von *Laminaria* Pflanzen vom Sporophytenhabitus. Parthenokeimlinge von verschiedenen *Dictyotales* sind öfters beobachtet worden, die Pflanzen gingen jedoch frühzeitig zugrunde (KNIEP 1928). Bei *Fucus*-Eiern ist Parthenogenese nur nach experimentellem Eingriff (Einwirkung organischer Säuren und hypertonischer Lösungen) möglich (OVERTON 1913). Auch kernlose Eifragmente beginnen sich zu entwickeln, wenn sie durch ein Spermatozoid „befruchtet" werden (Merogonie, WINKLER 1901).

3. Cutleria.

Bei den anisogamen *Cutleria*-Arten treten im Entwicklungsverlauf normalerweise zwei heteromorphe Phasen auf (*Cutleria*-Stadium als Gametophyt, *Aglaozonia*-Stadium als Sporophyt). Die regelmäßige Aufeinanderfolge von Gametophyt und Sporophyt kann jedoch an verschiedenen Stellen durchbrochen sein. So können sich nach den Beobachtungen älterer Autoren (s. Kniep 1928) die unbefruchteten weiblichen Gameten sowohl zu Sporophyten als auch zu Gametophyten entwickeln. Der erstere Fall müßte nach Winkler als Parthenogenese, der letztere als Apomiktosis bezeichnet werden. Weiterhin können aus unbefruchteten Eiern aber auch Zwischenformen entstehen, Keimlinge, die Merkmale beider Generationen besitzen. Ähnlich vielfältige Entwicklungsmöglichkeiten besitzen die Schwärmer aus den unilokulären Sporangien des Sporophyten. Schlösser (1935) fand, daß parthenogenetische Entwicklung der Eier nur nach experimentellem Eingriff möglich ist. Die Parthenokeimlinge wachsen langsamer als die Zygotenkeimlinge, sie bilden undifferenzierte Zellhaufen, aus denen fädige Abschnitte hervorgehen können, die Gametangien bilden. Schlösser hält diese Bildungen und auch die von anderen Autoren beschriebenen Zwischenformen für „Nebenfruchtformen" des Gametophyten. Werden die Parthenokeimlinge während der ersten Zellteilung zentrifugiert, so entstehen typische *Aglaozonia*-Keimlinge, die Schlösser für diploid hielt. Er zog daraus den Schluß, daß bei *Cutleria* Kernphasen- und Formwechsel obligat gekoppelt sind. Hartmann (1949/50) fand am selben Standort wie Schlösser (Neapel) spontane parthenogenetische Keimung der *Cutleria*-Eier. Experimentelle Eingriffe waren dabei nicht notwendig.

4. Apogamie.

Apogamie, die Entstehung des Zygobionten aus vegetativen Zellen des Gamobionten, ist bei Braunalgen nicht mit Sicherheit beobachtet. Bei *Carpomitra* (*Sporochnales*, Sauvageau 1926) und *Arthrocladia* (*Desmarestiales*, Sauvageau 1931) gehen die Sporophyten aus kolbenartig verdickten Zellen des Gametophyten hervor. Die Natur dieser Zellen ist jedoch unklar, es ist nicht auszuschließen, daß es sich um Oogonien handelt. Falls diese Möglichkeit zuträfe, würde Parthenogenese vorliegen.

Literatur.

Beth, K., u. H. F. Linskens: Physiologie epiphytischer Algen. Naturwiss. Rdsch. **17**, 254—257 (1964).

Cavassilas, Y.: Observations sur le développement et la reproduction de *Levringia brasiliensis* (Mont.) B. Joly (Phéophycée, Chordariale). Rev. gén. Bot. **70**, 621—632 (1963). — Chadefaud, M.: Sur la morphologie des zoospores des Phéophycées. C. R. Acad. Sci. Paris **227**, 1258—1260 (1948).

Feldmann, J.: Les cycles de reproduction des Algues et leurs rapports avec la phylogénie. Rev. Cytol. Biol. Végét. **13**, 1—49 (1952). — Fritsch, F. E.: The structure and reproduction of the algae II. Cambridge 1945. — Funk, G.: Fruktifikationszeiten der Meeresalgen bei Neapel. Pubbl. Staz. Zool. Napoli **29**, 126—138 (1957).

Gessner, F.: Hydrobotanik. I. Berlin 1955. — Goebel, K.: Organographie der Pflanzen. I. Allgemeine Organographie. Jena 1928.

Hamel, G.: Phéophycées de France, Paris 1931—1939. — Hartmann, M.: Untersuchungen über die Sexualität von *Ectocarpus siliculosus*. Arch. Protistenk. **83**, 110—153 (1934). — Beiträge zur Kenntnis der Befruchtung und Sexualität mariner Algen. Pubbl. Staz. zool. Napoli **22**, 120—128 (1949/50). — Haupt, A. W.: Structure and development of *Zonaria Farlowii*. Amer. J. Bot. **19**, 239—254 (1932).

Kain, J. M.: Aspects of the biology of *Laminaria hyperborea*. III. Survical and growth of gametophytes. J. Mar. Biol. Ass. U.K. **44**, 415—433 (1964). — Kniep, H.: Die Sexualität der niederen Pflanzen. Jena: 1928. — Knight, M.: Studies in the Ectocarpaceae. II. The life history and cytology of *Ectocarpus siliculosus* Dillw. Trans. Roy. Soc. Edinb. **56**, 307—332 (1931). — Kornmann, P.: Der Formenkreis von *Acinetospora crinita* (Carm.) nov. comb. Helgol. Wiss. Meeresuntersuch. **4**, 205—224 (1953). — *Giffordia fuscata* (Zan.) Kuck. nov. comb., eine Ectocarpacee mit heteromorphen, homophasischen Generationen. Helgol. Wiss., Meeresuntersuch. **5**, 41—52 (1954). — Über die Entwicklung einer *Ectocarpus confervoides*-Form. Pubbl. Staz. zool. Napoli **28**, 32—43 (1956). — Artspezifische Entwicklungsgänge in der Gattung *Ectocarpus*. Helgol. Wiss. Meeresuntersuch. **6**, 84—99 (1957). — Die Entwicklung von *Chordaria flagelliformis*. Helgol. Wiss. Meeresuntersuch. 8, 276—279 (1962a). —

Plurilokuläre Sporangien bei *Elachista fucicola*. Helgol. Wiss. Meeresuntersuch. 8, 293—297 (1962b). — KYLIN, H.: Studien über die Entwicklungsgeschichte der Phaeophyceen. Sv. bot. Tidskr. **12**, 1—64 (1918). — Über die Entwicklungsgeschichte der Phaeophyceen. Kgl. Fysiogr. Sällsk. Hdl., N. F. Avd. 2 **29**, 3—102 (1933). — Bemerkungen über die Entwicklungsgeschichte einiger Phaeophyceen. Kgl. Fysiogr. Sällsk. Hdl. **33**, 1—34 (1937).

MANTON, I.: Observations with the electron microscope on the internal structure of the zoospore of a brown alga. J. exp. Bot. 8, 249—303 (1957). — MANTON, I., and B. CLARKE: Electron microscope observations on the zoospores of *Pylaiella* and *Laminaria*. J. exp. Bot. **2**, 242—246 (1951). — MÜLLER, D. G.: Über jahres- und lunarperiodische Erscheinungen bei einigen Braunalgen. Botanica Marina **4**, 140—155 (1962). — Die Temperaturabhängigkeit der Sporangienbildung bei *Ectocarpus siliculosus* von verschiedenen Standorten. Pubbl. Staz. zool. Napoli **33**, 310—314 (1963). — Die Beteiligung eines Berührungsreizes beim Festsetzen von Algenschwärmern auf dem Substrat. Z. Bot. **52**, 193—198 (1964). — Bemerkung zum Bau der Geißeln von Braunalgenschwärmern. Naturwissenschaften **52**, 311 (1965a). — Untersuchungen zur Entwicklungsgeschichte der Braunlage *Ectocarpus siliculosus* aus Neapel. Planta **68**. 57—68 (1966).

NAYLOR, M.: The life history of *Adenocystis utricularis* (Bory) H. et H. Trans. Roy. Soc. New Zealand **83**, 295—301 (1955). — A note on the life history of *Scytothamnus australis* H. et H. New Phytologist **55**, 386—388 (1956). — NIENBURG, W.: Zur Entwicklungsgeschichte der Helgoländer *Haplospora*. Ber. dtsch. bot. Ges. **41**, 211—217 (1923). — *Fucus Mytili* spec. nov. Ber. dtsch. bot. Ges. **50**a. 28—41 (1932).

OVERTON, J. B.: Artificial parthenogenesis in *Fucus*. Science, N.S. **37**, 841—844 (1913).

PAPENFUSS, G. F.: Alternation of generations in *Ectocarpus siliculosus*. Bot. Gaz. **96**. 421—446 (1934/35). — Phaeophyta. In: SMITH, Manual of phycology. Waltham, Mass. 1951. — PRINGSHEIM, N. (1873): Zit. in FRITSCH 1945.

ROBINSON, W.: Observations on the development of *Taonia atomaria*. Ann. Bot. **46**. 113—120 (1932).

SAUVAGEAU, C.: Sur quelques exemples d'hétéroblastie dans le développement des Algues phéosporées. C. R. Acad. Sci. (Paris) **179**, 1576—1579 (1924). — (1926): Zit. in FRITSCH 1945. — (1928): Zit. in G. HAMEL: Phéophycées de France, fasc. II. 1935. — (1931): Zit. in FRITSCH 1945. — (1932): Le Pléthysmothalle. Arch. Zool. exp. et gén. **74**, 155—168. — SCHLÖSSER, L. A.: Zur Entwicklungsphysiologie des Generationswechsels von *Cutleria*. Biol. Zbl. **55**, 198—208 (1935). — SCHMIDT, P.: Über *Acinetospora pusilla* (Bornet) Sauvageau. ihr Vorkommen in der Helgoländer Algenflora und neue Kulturergebnisse. Z. Bot. **35**, 358—384 (1940). — SCHREIBER, E.: Untersuchungen über Parthenogenese, Geschlechtsbestimmung und Bastardierungsvermögen bei *Laminaria*. Planta (Berl.) **12**, 331—353 (1930). — Über die geschlechtliche Fortpflanzung der Sphacelariales. Ber. dtsch. bot. Ges. **49**, 235—240 (1931). — Über Kultur und Geschlechtsbestimmung von *Dictyota dichotoma*. Planta (Berl.) **24**, 266—275 (1935). — SCHUSSNIG, B.: Handbuch der Protophytenkunde. II. Jena 1960. — SCHUSSNIG, B., u. E. KOTHBAUER: Der Phasenwechsel von *Ectocarpus siliculosus*. Öst. bot. Z. **83**, 81—97 (1934). — SUTO, S.: Shedding, swimming and fixing of the zoospores in some species of Laminariaceae. Contr. Centr. Fish. Stat. Japan Tokyo **25**, 123—128 (1948). — Studies on shedding, swimming and fixing of the spores of seaweeds. Bull. Jap. Soc. Sci. Fish. **16**. 1—9 (1950). — On shedding of zoospores in some algae of Laminariaceae. II. Bull. Jap. Soc. Sci. Fish. **18**, 1—5 (1952).

TAKAMATSU, M. (1939): Zit. in B. SCHUSSNIG, Systematik und Phylogenie der Pflanzen. Fortschr. Bot. **17**, 186 (1955). — TAYLOR, W. R.: Marine algae of the northeastern coast of North America. Ann Arbor, Mich. 1957.

WINKLER, H. (1901): Zit. in F. OLTMANNS, Morphologie und Biologie der Algen, 2. Aufl., Bd. 3. Jena 1922. — Über den Biontenwechsel und die Abweichungen von seinem normalen Verlauf. Planta (Berl.) **33**, 1—90 (1942).

G. Chlorophyta.

I. Chlorophyceae

(mit Anhang e: *Pedinomonadales*).

Die Chlorophyceen zeigen morphologisch und entwicklungsgeschichtlich die fortschreitende Umbildung von Flagellaten zu unbeweglichen behäuteten Zellen, die sich auf verschiedenste Weise weiterentwickelt haben. Wie schon GEITLER (1934) betonte, handelt es sich bei der Algenzelle im Prinzip um die Ruhezelle eines Flagellaten (passend als „eingesperrter Flagellat“ bezeichnet). Während bei den Flagellaten die Ruhezelle nur als vorübergehendes Stadium in den Entwicklungscyclus eingeschaltet ist, stellt sie bei den Algen die Hauptphase der vegetativen Entwicklung dar. Die mit einem Geißelapparat versehenen *Volvocales* lassen sich mehr oder weniger deutlich mit den höheren Organisationsstufen der Chlorophyceen in Zusammenhang bringen (Pascher 1918a, 1931). Diese Organisationsstufen sind durch unbewegliche, von fester Membran umgebene Zellen gekennzeichnet. Mit diesen Tatsachen in engem Zusammenhang steht auch die Organisation der Zellfäden, bei denen die in einzelnen Teilungsfolgen entstandenen Zellen durch die Membranen der Mutterzellen polar und fadenartig zusammengehalten werden. Das trifft auch für die verzweigten und kompliziert gebauten Formen der *Chaetophorales* zu. Daß die hochdifferenzierten Algen eine potentiell monadoide Organisation besitzen, zeigt das Vorkommen von Zoosporen, bei deren Bildung, soweit mehrere in einer Mutterzelle entstehen, endogene Teilungen ohne Membranbildung ablaufen (Geitler 1934). Manchmal nimmt der Protoplast ohne jede Teilung unmittelbar die monadoide Organisation an und tritt als Flagellatenzelle aus. Diese Zusammenhänge werden auch von der Tatsache nicht berührt, daß manche Algen die monadoide Phase völlig verloren haben oder daß diese bisher noch nicht beobachtet wurde.

Gemäß der großen Formenfülle der Chlorophyceen und der Entwicklungstendenz von den monadoiden Typen bis zu komplizierten Verbänden und zum differenzierten Thallus ist die vegetative Fortpflanzung mannigfaltig. Insbesondere kann man sowohl morphologisch als auch entwicklungsgeschichtlich beobachten, daß sich die monadoiden Organisationstypen in der Fortpflanzung der übrigen Organisationsstufen widerspiegeln und dann allmählich in celluläre Fortpflanzungsgebilde übergehen (Pascher 1927, Schussnig 1960, Ettl 1958, 1961). Die beweglichen nackten Flagellaten vermehren sich primär durch Längsteilung der Mutterzelle, ebenso die behäuteten *Volvocales* und die von ihnen abgeleiteten Gallertzustände. So ist das Grundphänomen der Fortpflanzung bei den Chlorophyceen die Zellteilung (vgl. Hartmann 1929). Der Gepflogenheit, nur die Vielfachteilung als Fortpflanzung anzusehen, die Zweiteilung dagegen als Wachstum zu bezeichnen, kann bei den Chlorophyceen nicht völlig gefolgt werden. Man muß die verschiedenen zwischen Zweiteilung und Vielfachteilung vorhandenen Zusammenhänge in Betracht ziehen.

Bei den unbeweglichen und mit fester Membran versehenen Algenzellen erfolgt die Fortpflanzung häufig durch Zoosporen. Der Umweg über die Zoosporenbildung wird bei manchen Organisationstypen, vor allem bei den *Chlorococcales*, allmählich dadurch verlassen (Pascher 1927, Geitler 1934, Ettl 1961), daß bei der Fortpflanzung der Geißelapparat nicht mehr ausgebildet wird und auch die übrigen Merkmale der monadoiden Organisation abklingen. So werden die Tochterzellen schließlich zu Autosporen. Dabei gibt es eine Menge von Übergängen, welche die Zoosporenbildung mit der Autosporenbildung und nicht zuletzt auch mit der Fadenbildung der trichalen Organisation verbinden.

Außer diesen wichtigsten Fortpflanzungsarten, die wie ein roter Faden die Hierarchie der Organisationstypen durchziehen, existieren weitere Formen vegetativer Fortpflanzung. Viele davon können zugleich auch als Dauerzustände oder Dauerzellen aufgefaßt werden — Aplanosporen, Akineten, Palmellen, Thallusfragmente u.ä. Die heterogene Herkunft und Beschaffenheit letzterer gestattet jedoch eine Besprechung nur in Form einer Aufzählung.

a) Bewegliche Fortpflanzungszellen (Zoosporen-Typus).

1. Nackte Volvocales (Tochterzellenbildung).

Die vegetative Fortpflanzung wird bei diesen in sehr charakteristischer Weise vollzogen. Es handelt sich um eine ausgesprochene Längsteilung des Zellindi-

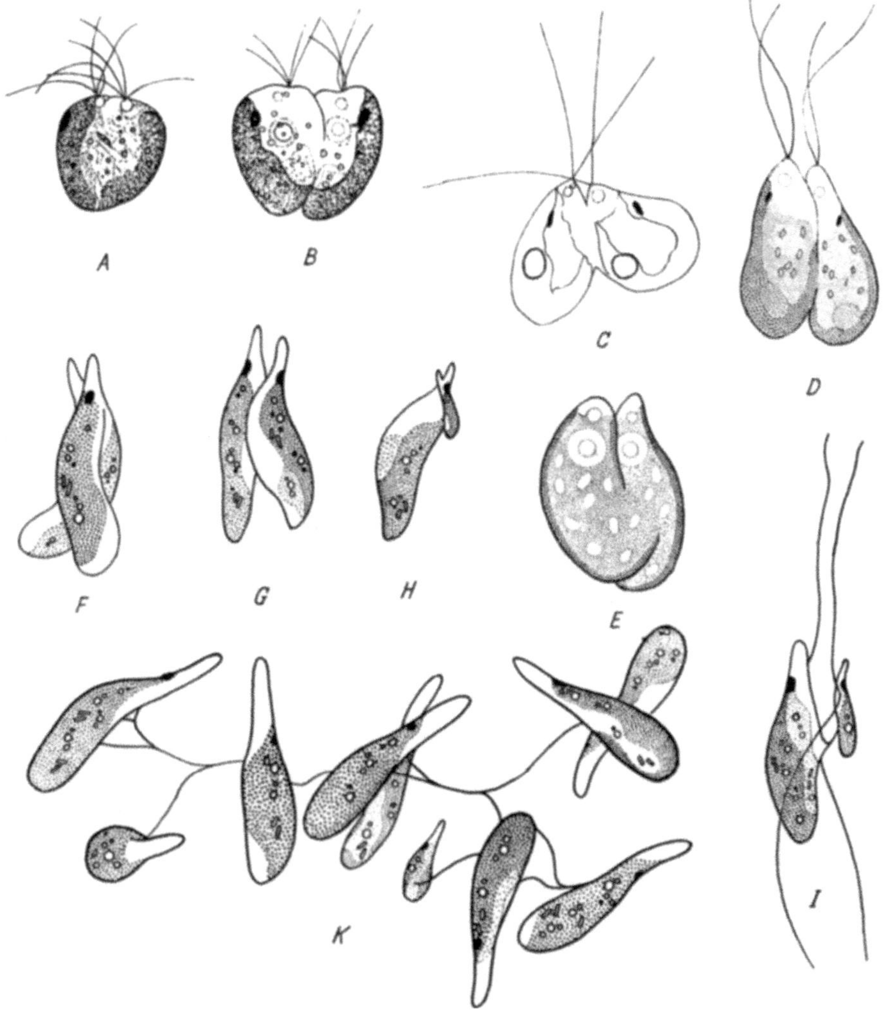

Abb. 61 A—K. Vegetative Fortpflanzung nackter *Volvocales*. A, B Teilungsverlauf von *Tetrachloris carterioides* (B fortgeschritteneres Stadium), C *Dunaliella lateralis*, D *Kremastochloris conus*, E *Dunaliella paupera*, F—I *Dangeardinella saltatrix*. F, G Teilung in gleiche Tochterzellen; H, I sehr ungleiche Teilung (Sprossung), K durch Plasmabrücken zusammenhängende Kolonien mit ungleich großen Zellen. (A—C nach PASCHER und JAHODA, D—I nach PASCHER.)

viduums (Schizotomie nach SCHUSSNIG 1960). Während oder kurz nach der Kernteilung, und nachdem sich auch der Chromatophor gespalten hat, das Pyrenoid geteilt wurde und pulsierende Vacuolen und Stigma verdoppelt sind, spaltet sich der Protoplast der Länge nach durch (Abb. 61A—G). Der Periplast und der Geißelapparat werden aufgeteilt, so daß die Tochterzellen jede die Hälfte der Geißeln und des Periplasten von dem Mutterindividuum mitbekommen (PASCHER 1927, SCHUSSNIG 1960, ETTL 1965a). Im Gegensatz zu den Chlamydomonadaceen (s. u.)

sind daher die Geißeln einer Polyblephariden-Zelle ontogenetisch ungleichwertig. Die Teilung kann dabei auf verschiedene Weise modifiziert sein. Manche Formen teilen sich inäqual, so daß die eine Tochterzelle beträchtlich größer wird als die andere (Abb. 61 H, I). Sogar sprossungsartige Vorgänge kommen vor, oder die Tochterzellen werden nach erfolgten Teilungen in Gruppen zusammengehalten (Pascher 1930). Dieses geschieht bei *Dangeardinella* durch Plasmaverbindungen, es sei aber betont, daß es sich nicht um Kolonien im gleichen engeren Sinne wie bei Volvocalen handelt (Abb. 61 K). Die Längsteilung erfolgt auch in Zellen, die sich zeitweilig in Gallertstadien befinden. Weil hier kein Zoosporangium gebildet wird, können — obwohl Ähnliches vorliegt — die Teilprodukte nicht als Zoosporen bezeichnet werden, besser ist es, einfach von Tochterzellenbildung zu sprechen.

2. Behäutete Volvocales.

Die vegetative Fortpflanzung ist hier ebenso wie bei den nackten Formen ein Ergebnis der Protoplastenteilung, die zwar modifiziert wird, aber im Prinzip einer Längsteilung beibehalten bleibt (Korschikoff 1938, Pascher 1927). Der bedeutendste Unterschied besteht darin, daß der Protoplast nicht zugleich mit seiner Membran, sondern innerhalb der erhaltenbleibenden Zellhülle geteilt wird. Pro Mutterzelle werden dann je nach der Zahl der sukzedanen Teilungswellen zwei oder mehrere Teilprodukte entstehen, die in Form von Zoosporen die Hülle der ursprünglichen Mutterzelle verlassen. Jede der Tochterzellen bildet vorher die Hülle und im Unterschied zu den nackten Formen auch den ganzen Geißelapparat neu aus (Pascher 1927, Schussnig 1960, Ettl 1965a). Die Zoosporenbildung erfolgt entweder im beweglichen oder im unbeweglichen Zustand, je nachdem, ob die alten Geißeln vor oder während der Teilung abgeworfen wurden. Im ersten Fall kann sich die zum Zoosporangium gewordene Mutterzelle noch lange mit ihren alten Geißeln bewegen, obzwar die Tochtermonaden fertig und mit eigenen Geißeln versehen sind (Abb. 62 K).

Die Teilung des Protoplasten bleibt, wie aus der Lage seiner Organellen zu erkennen, im Prinzip immer als Längsteilung erhalten. Doch kommt es oft noch vor der Teilung zu einer Verlagerung des Protoplasten innerhalb der Membran, so daß die Teilungsebene mit der Längsachse der vegetativen Zelle einen Winkel von bis zu 90° bildet. Dadurch wird eine Querteilung vorgetäuscht. So kommen bei *Chlamydomonas seriata* alle vier Tochterzellen quer und in einer Reihe hintereinander zu liegen (Abb. 62 B). Nach Pascher (1943b) ist die Querlagerung der Teilprotoplasten eine wesentliche phylogenetische Voraussetzung für die Bildung von Zellfäden, denn im Zellfaden erfolgt die Teilung quer zur Längsachse. Häufiger lagern sich die Teilprodukte im Zoosporangium in tetraëdrische Anordnungen um (Abb. 63 B). Auch andere Lagen kommen vor.

Bei den behäuteten *Volvocales* kann man von *Zoosporangien* sprechen, weil die beweglichen jungen Tochterzellen, Zoosporen, sukzedan und meist in der Mehrzahl innerhalb der Mutterzelle entstehen (Schussnig 1960). Die jungen Tochterzellen stellen, solange sie nicht ausgewachsen sind, Zoosporen dar. Nach Pascher (1927) verhält sich eine solche Zoospore zu einer ausgewachsenen Volvocale im Prinzip wie z. B. eine Zoospore von *Ulothrix* zum *Ulothrix*-Faden selber. In der Benennung Zoospore liegt nur der Begriff eines beweglichen Fortpflanzungsgebildes im Gegensatz zum erwachsenen Individuum. Man muß aber auch mit Schussnig übereinstimmen, der rein deskriptiv den Namen Zoosporangium beibehält, aber auf den Wesensunterschied zwischen solchen Zoosporangien der *Volvocales* und einem Sporangium der coccalen und anderen Chlorophyceen hinweist. Bei den Volvocalen entstehen die Zoosporen nicht infolge eines simultanen merogonen Zerfalls des Cytoplasten, sondern durch sukzedane, stark zusammengedrängte Längsteilungen innerhalb der Mutterhülle.

Im allgemeinen werden die Zellhüllen der Zoosporen bei den Volvocalen schon früh innerhalb der Mutterzelle (Ettl 1965c) gebildet (Abb. 63A); bei *Chlamydomonas koprophilos, Ch. insana, Carteria simplicissima* (Pascher 1932), jedoch verlassen die Zoosporen die Mutterzelle nackt und behäuten sich später. Auch die Zoosporen der Phacotaceen sind unbeschalt, wenn das Gehäuse der Mutterzelle gesprengt wird; so bei *Pteromonas*, wo sie allerdings noch in einer Gallertblase

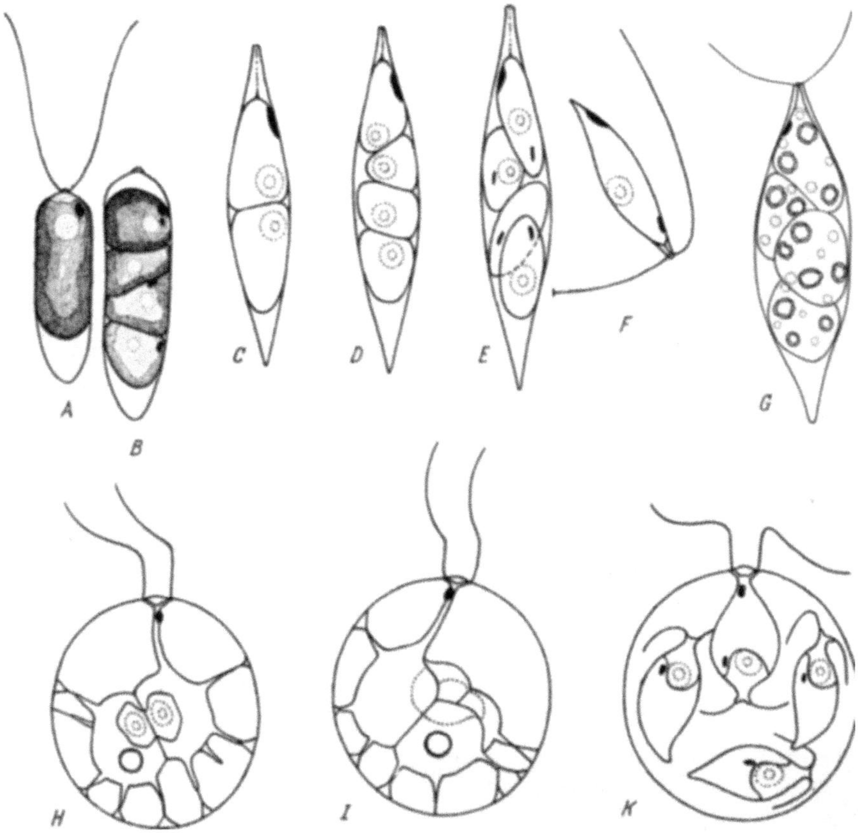

Abb. 62 A—K. Vegetative Fortpflanzung behäuteter *Volvocales*. A Vegetative Zelle von *Chlamydomonas seriata*; B Teilungsstadium derselben Art mit quer gelagerten und hintereinander gereihten Teilprotoplasten. C—G Verhalten des Stigmas während der Protoplastenteilung von *Chlorogonium euchlorum*. C Zweierteilung, das alte Stigma bleibt im vorderen Teilprotoplasten erhalten; D Viererteilung, vorderer Protoplast noch mit dem alten Stigma; E Bildung der Zoosporen. In der vorderen ist das alte Stigma neben dem neuen noch sichtbar, F die vordere Zoospore nach dem Freiwerden; G Viererteilung mit eingezeichneten Organellen in den Teilprotoplasten. H—K Zoosporenbildung von *Haematococcus capensis*; H Zweierteilung; I Viererteilung — man beachte das alte Stigma, das sich während der Teilung an der Geißelbasis des alten Kinetoms aufhält; K bei den fertigen Zoosporen hat die vordere vorübergehend zwei Geißelapparate und Stigmata. Der alte Geißelapparat und das alte Stigma gehen zugrunde. (A, B nach Pascher, C—F nach Geitler, G nach Ettl, H—K nach Pocock.)

eingeschlossen bleiben (Abb. 63C) und diese erst verlassen, wenn sie selbst Gehäuse ausgeschieden haben (Pascher 1927, Ettl 1964b).

Eine besondere Art der Zoosporenbildung (Abb. 62C—K) kommt bei *Haematococcus* und *Chlorogonium* vor (Geitler 1934, Pocock 1960, Schussnig 1960). Wir folgen der Beschreibung von Pocock an *Haematococcus capensis*. Die Vorbereitung zur Teilung beginnt mit der Gestaltsänderung des Protoplasten (Abb. 62H—K). In das sich dünn ausziehende Vorderende hinein wandert das Stigma bis in die Nähe der Geißelbasis. Da der Protoplast mit letzterer Verbindung behält und die Flagellen tätig bleiben, kann er sich nicht längs teilen. Vielmehr erfolgt eine Umgruppierung der Organellen mit Drehung des Chromatophoren. Die farblose Zone, die auch den Kern enthält, markiert das neue Vorderende, in bezug auf welches nunmehr eine Teilung des Protoplasten der Länge nach abläuft. Nach den Teilungen bleibt einer der Tochterprotoplasten noch mit den alten Geißeln verbunden, doch ist diese Stelle der

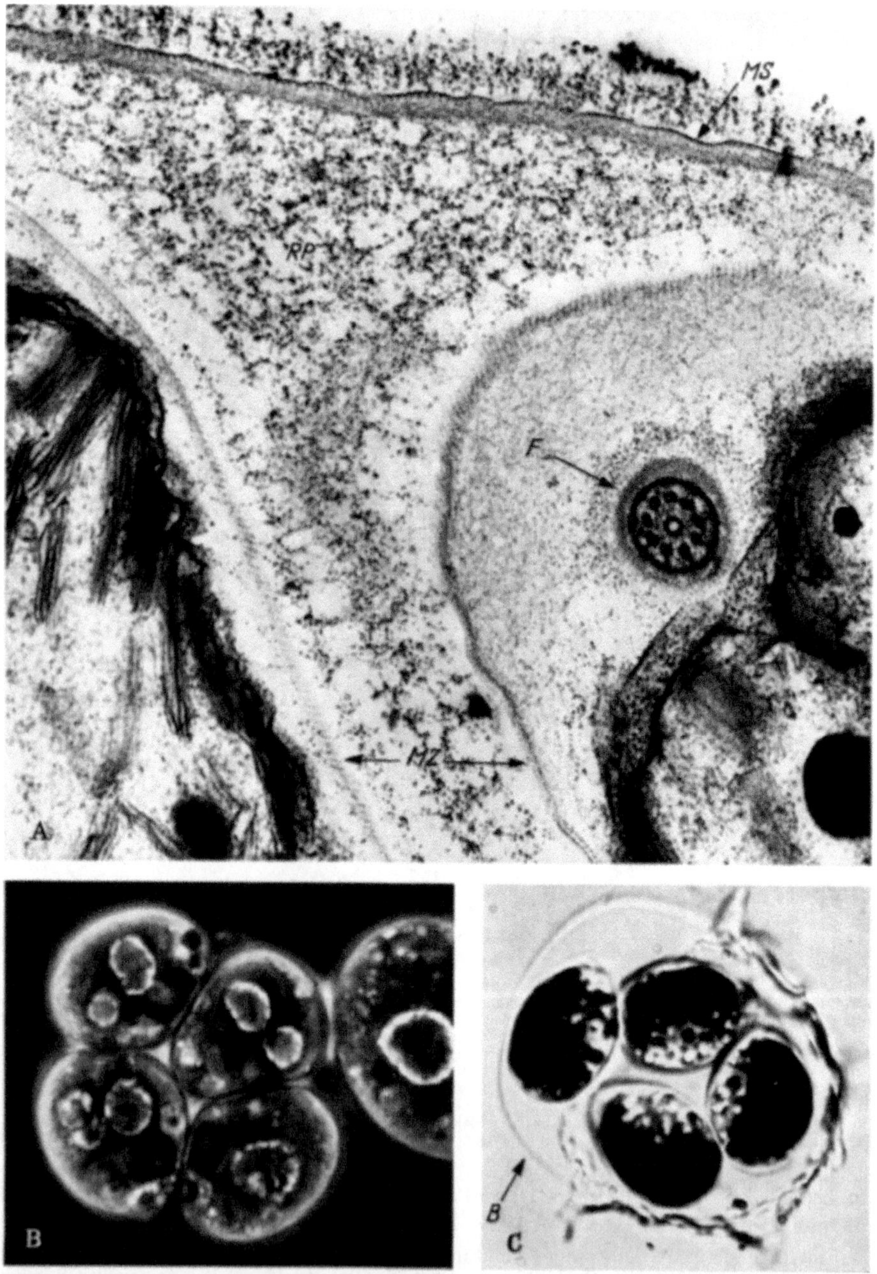

Abb. 63 A—C. A Zoosporenbildung bei *Chloromonas rosae*. Die Ultrastruktur eines Teiles des Sporangiums, das sowohl die Membran des Sporangiums (*MS*) als auch die neuen Membranen der Zoosporen (*MZ*) erkennen läßt, die durch ihre Struktur deutlich hervorkommen. Bemerkenswert ist auch der Querschnitt der Papille und Geißel (*F*) einer der Zoosporen und das Restplasma in dem Sporangium (*RP*). 38000× vergrößert. B Zoosporenbildung von *Chlamydomonas noctigama*. Tetraëdrisch angeordnete Zoosporen mit völlig ausgebildeten Organellen. 1500× vergrößert. C Teilprotoplasten von *Pteromonas aculeata* nach vollendeter Teilung in der Gallertblase (*B*) eingeschlossen. Sie sind noch nackt und ohne Gehäuse. 1500× vergrößert. (A nach einem Präparat und Photo von Manton; B, C nach Ettl.)

neuen Polarität entsprechend zum Hinterende geworden (nach der Polarität des Protoplasten). Am neuen Vorderende ist ein zweites Geißelpaar entstanden, so daß zeitweilig ein Kern zwei Geißelapparate dirigiert. Langsam trennt sich auch diese Zoospore vom alten Geißelapparat.

Die vegetative Fortpflanzung der kolonienbildenden *Volvocales* geht auf dasselbe Prinzip zurück wie bei den einzelligen. Doch ordnen sich die Zoosporen meist in einer der Mutterkolonie entsprechenden Zellenzahl bereits innerhalb des

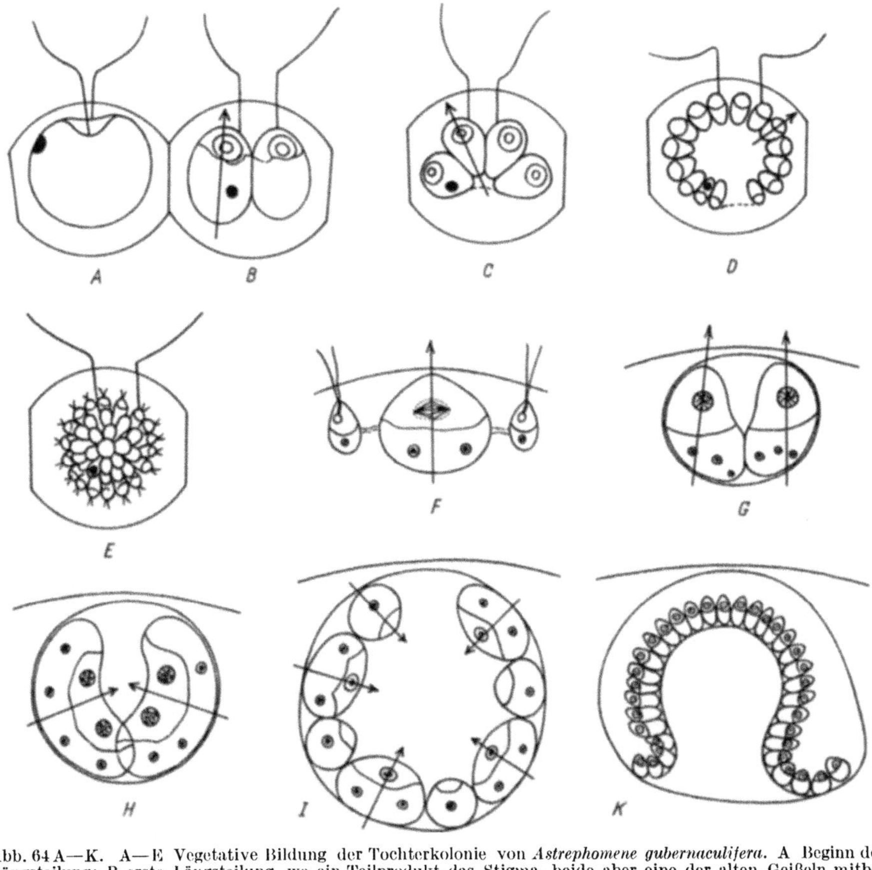

Abb. 64 A—K. A—E Vegetative Bildung der Tochterkolonie von *Astrephomene gubernaculifera*. A Beginn der Längsteilung; B erste Längsteilung, wo ein Teilprodukt das Stigma, beide aber eine der alten Geißeln mitbekommen; C Viererteilung; D die fast fertige Tochterkolonie; E fertige Tochterkolonie mit entwickelten neuen Geißeln. Der Pfeil kennzeichnet die Polarität der neuen Zellen. Man beachte, daß keine Umstülpung stattfindet. Das Stigma und der Geißelapparat der Mutterzelle bleiben bis zur völligen Ausbildung der Tochterkolonien erhalten. F—K Vegetative Bildung der Tochterkolonie bei *Volvox aureus*. F Beginn der Längsteilung; G vollzogene Längsteilung der sporogenen Zelle; H weitere Teilung der Protoplasten, wobei das Vorderende nach innen zu liegen kommt; I hohlkugelige Anordnung, wo die Vorderenden der Teilprotoplasten nach innen gerichtet sind; K Umstülpung, bei der die Protoplasten mit ihren Vorderenden nach außen gerichtet werden. Die Pfeile zeigen auch hier die Polarität. (A—E nach POCOCK, F—K nach ZIMMERMANN aus PASCHER, umgezeichnet.)

Zoosporangiums zu neuen Kolonien an. Die morphologisch einfacheren Spondylomoraceen (PASCHER 1927, KORSCHIKOFF 1938), sowie *Gonium* besitzen einander gleichwertige Zellen; bei den höher entwickelten Formen kommt es aber außerdem noch zu einer Differenzierung der vegetativen Zellen in bezug zur Fortpflanzungsmöglichkeit. Schon bei *Eudorina* und *Pandorina* geschieht das teilweise, indem die Zellen am morphologischen Vorderende der Kolonie in der Fortpflanzung gegenüber den anderen Zellen verzögert sind. *Eudorina illinoisensis* ist in der Differenzierung so weit fortgeschritten, daß die vorderen vier Zellen unfähig sind sich fortzupflanzen (s. PASCHER 1927). Bei der vielzelligen *Pleodorina californica* ist die Hälfte aller Zellen kleiner und für eine Fortpflanzung völlig funktionslos geworden. Ähnliches gilt für *Pleodorina sphaerica* (IYENGAR und RAMANATHAN 1951), doch vergrößern sich die Zellen der hinteren Koloniehälfte erst, wenn die Kolonien ein bestimmtes Alter erreichen. Schon früher unterscheiden sich die prospektiv somatischen Zellen der erstgenannten Art durch vergrößerte Stigmen. Bei der Gattung *Volvox* hat die Differenzierung ihren Höhepunkt erreicht. Die meisten Zellen sind somatisch und die Fortpflanzung ist nur auf wenige Zellen beschränkt, die an ihrer besonderen Größe häufig schon an jungen, noch in der Mutterkolonie befindlichen Tochterkolonien erkennbar sind.

Die Tochterkolonien entstehen selbst bei den differenzierten Volvocaceen (Abb. 64 F—K) durch eine sukzedane Längsteilung des Protoplasten der zur Fortpflanzung bestimmten Zelle (PASCHER 1927, KORSCHIKOFF 1940, IYENGAR 1951). Bis zum 8-Zellen-Stadium haben sie die Form einer napfartigen Platte, wobei die Geißelpole auf deren konkaver Seite liegen und letztere in der Mutterkolonie nach außen gekehrt ist. Schließt sich der Napf im 32—128-Zellen-Stadium (bis auf eine kleine Öffnung) zur Hohlkugel, so sind die Geißelpole entsprechend nach innen gerichtet und die Zellen müssen eine Drehung um 180° durchmachen, um in die endgültige Lage zu kommen. Dieses geschieht durch eine Umstülpung (Inversion), welche bei *Pandorina*, *Eudorina*, *Platydorina*, *Pleodorina* und *Volvox* beobachtet wurde. Nach der Umstülpung ist die Orientierung der Zellen definitiv, und die Tochterkolonien werden direkt oder bei großen Kolonien nach einer Wachstumsperiode frei. Bei *Astrephomene* erfolgt die Teilung sukzedan in einer Ebene, wobei es später zu einer Krümmung, aber *nach außen*, kommt und schließlich die Tochterkolonien gebildet werden (Abb. 64 A—E). Eine Inversion ist hier deshalb nicht mehr nötig (POCOCK 1956).

3. Celluläre Chlorophyceen.

Bei den unbeweglichen, von einer Membran umgebenen Chlorophyceen erfolgt die vegetative Fortpflanzung im Großteil unter Zurückgreifen auf das Flagellatenstadium, indem vorübergehend bewegliche Zoosporen vom Aussehen der charakteristischen *Volvovales*-Organisation gebildet werden. Der Typus der *Chlorophyceen-Zoospore* ist innerhalb eines gewissen Spielraumes (Abb. 65) konstant und entspricht einer *Volvocales*-Zelle. Die Zoosporen sind multiradiate, ei- oder birnförmige oder auch kugelige bis langgestreckte Zellen, deren verjüngtes Vorderende meist farblos ist und die Geißeln trägt. Ausnahmsweise werden etwas asymmetrische oder dorsiventral gebaute Zoosporen gebildet (*Hormidium*, *Coleochaete*). Jede Zoospore, mit Ausnahme der farblosen Formen, besitzt einen großen oder mehrere kleine Chromatophoren und oft auch Pyrenoide. Die meisten Zoosporen haben auch ein mehr oder weniger deutliches Stigma, das zweifellos in Beziehung zur Phototaxis steht. Das Stigma entsteht in der Regel als Neubildung und wird erst in den fertigen Zoosporen sichtbar. Letztere sind meist nackt, bei manchen *Tetrasporales* und *Chlorococcales* kommen jedoch auch behäutete, *Chlamydomonas*-artige Zoosporen vor. Im Innern der Zoosporen liegt noch der Zellkern und an der Geißelbasis oder in der Zelle zerstreut sind bei Süßwasserarten pulsierende Vacuolen nachweisbar. Auch die Art der Geißelinsertion entspricht nahezu überall der bei den *Volvocales* vorkommenden (REICH 1926). Wie bei diesen haben die meisten Zoosporen (PETROVÁ 1930, PASCHER 1939a) zwei oder vier jeweils zu Paaren geordnete Geißeln. (Näheres über Zoosporen s. auch PRINTZ 1927, PASCHER 1927, FRITSCH 1935, SMITH 1950.)

MANTON (1964) hat die Ultrastruktur der Zoosporen einer *Stigeoclonium*-Art untersucht. An den äußerst präzisen Abbildungen sieht man die Anordnung aller Organellen, die mit derjenigen einer *Chlamydomonas*-Zelle (mit Ausnahme der Membran) eine überraschende Übereinstimmung aufweisen (Abb. 66a, b).

Nur die relativ großen stephanokonten Zoosporen der *Oedogoniaceae* und die ähnlichen der *Derbesiaceae* weichen morphologisch vom Typus der üblichen Chlorophyceen-Zoosporen ab. Nach SCHUSSNIG (1960) handelt es sich bei beiden um Synzoosporen, deren Plurivalenz im ersten Falle allein aus der Vielzahl der monodynamischen Kinetiden erschlossen wird; im zweiten könnte man außerdem noch die coenocytische Herkunft der Spore anführen. Diese geht aus einem ursprünglich mehrkernigen, später durch Kernabort einkernigen Plasma-

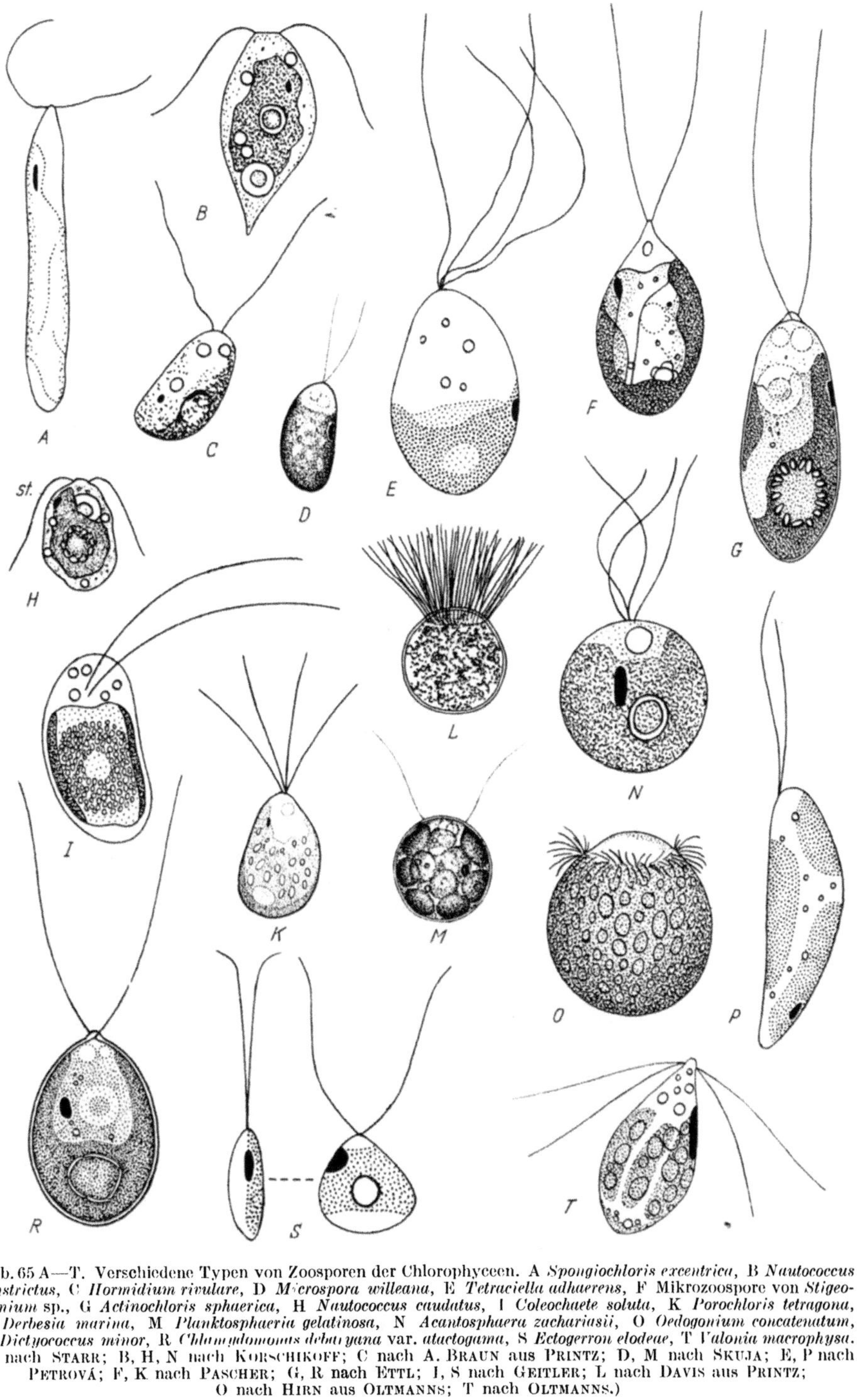

Abb. 65 A—T. Verschiedene Typen von Zoosporen der Chlorophyceen. A *Spongiochloris excentrica*, B *Nautococcus constrictus*, C *Hormidium rivulare*, D *Microspora willeana*, E *Tetraciella adhaerens*, F Mikrozoospore von *Stigeoclonium* sp., G *Actinochloris sphaerica*, H *Nautococcus caudatus*, I *Coleochaete soluta*, K *Porochloris tetragona*, L *Derbesia marina*, M *Planktosphaeria gelatinosa*, N *Acantosphaera zachariasii*, O *Oedogonium concatenatum*, P *Dictyococcus minor*, R *Chlamydomonas debaryana* var. *atactogama*, S *Ectogerron elodeae*, T *Valonia macrophysa*. (A nach STARR; B, H, N nach KORSCHIKOFF; C nach A. BRAUN aus PRINTZ; D, M nach SKUJA; E, P nach PETROVÁ; F, K nach PASCHER; G, R nach ETTL; I, S nach GEITLER; L nach DAVIS aus PRINTZ; O nach HIRN aus OLTMANNS; T nach OLTMANNS.)

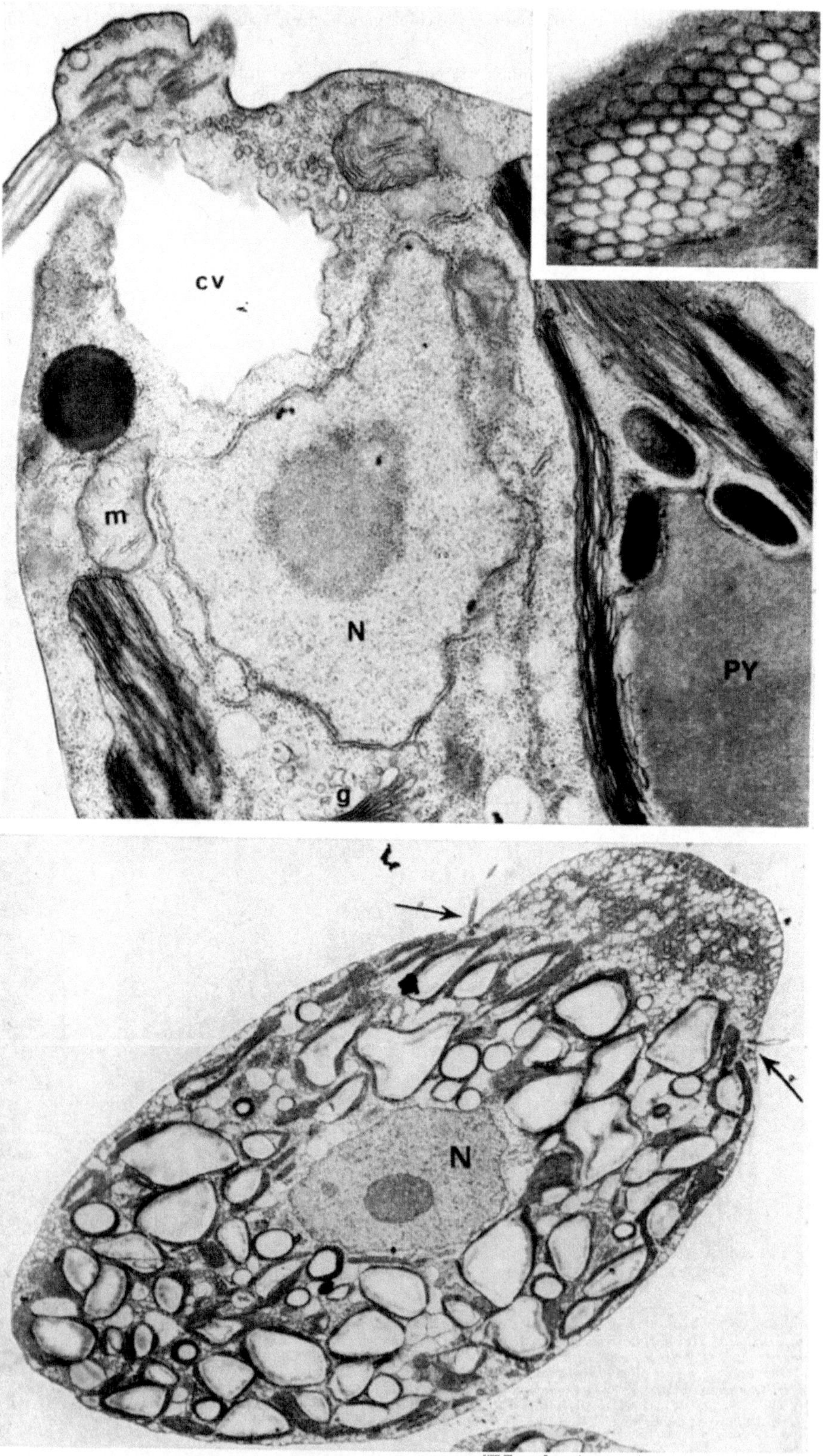

Abb. 66. (Legende s. S. 715.)

bereich hervor, und ist auch während der Entstehung des Geißelringes einkernig. Die Ultrastruktur der *Oedogonium*-Zoospore wurde von HOFFMANN und MANTON (1962) untersucht. Die Zoosporen haben ungefähr 120 Geißeln, die zu einem Geißelband angeordnet sind und durch einen Fibrillenring zusammengehalten werden. Die Struktur des Zellkörpers (Kern, Chromatophor und andere Komponenten) entspricht jedoch im allgemeinen dem der üblichen Zoosporen (Abb. 66c).

Bei den einfachsten Formen kann sich der Zellinhalt der vegetativen Zellen ohne vorherige Teilung des Protoplasten in eine Zoospore umwandeln (z. B. *Chlorangium*, *Asterococcus*). Auch bei manchen hoch entwickelten und differenzierten Fadenalgen kann der ganze Zellinhalt in eine Zoospore umgebildet werden. Beim Großteil aller Chlorophyceen geht jedoch der Zoosporenbildung eine Protoplastenteilung der sporogenen Zelle voran. Je nach der Organisation und Differenzierung der betreffenden Algenzelle wird die Zoosporenbildung durch verschiedene Teilungstypen eingeleitet. Von modifizierter Schizotomie bei einfachen *Tetrasporales* angefangen über Schizogonie, Cytogonie, Coenogonie bis zur Cytotomie komplizierter Zellindividuen und Zellverbände (s. SCHUSSNIG 1960). Die Zoosporen sind oft schon innerhalb der Muttermembran lebhaft beweglich, bei anderen dagegen beginnt die Bewegung erst, wenn die Zoosporen ins Freie gelangt sind. Sie werden in verschiedenster Weise aus den Zoosporangien entleert, wobei sie entweder einzeln oder in kleinen Gruppen, oder auch alle auf einmal das Sporangium verlassen.

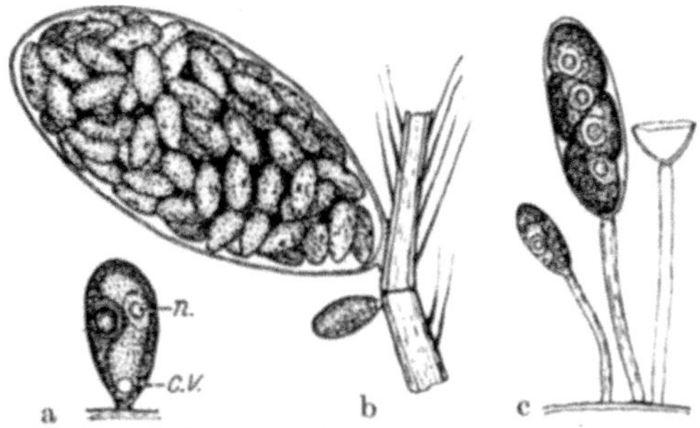

Abb. 67a—c. Zoosporenbildung von *Chlorangium epizooticum*. a Junge vegetative Zelle; b sehr großes Zoosporangium mit vielen Zoosporen; c eine langgestielte Form, zum Teil in Zoosporenbildung, rechts entleertes Sporangium. (Nach KORSCHIKOFF.)

Tetrasporales. Weil die tetrasporalen Chlorophyceen, wenn auch im vegetativen Zustand unbeweglich, den typischen Monadenbau beibehalten, ist ein Übergang zu beweglichen Zoosporen einfach (PASCHER 1927, ETTL 1958). So können sich die vegetativen Zellen direkt und durch Geißelbildung allein in Zoosporen umwandeln. Bei epizootischen Formen (z. B. *Chlorangium*, *Chlorangiopsis*) erfolgt dies äußerst rasch, binnen weniger Minuten, nach Änderung der Temperatur oder des Mediums oder auf Absterben des Wirtes hin. In den meisten Fällen entstehen die Zoosporen jedoch erst nach erfolgter Protoplastenteilung. Ihre Zahl hängt dann von der Größe und vom Alter der Mutterzelle ab. So geben junge Zellen von *Hypnomonas* nur acht, ältere 16 und die ältesten bis 64 Zoosporen. Sporangien von *Nautococcus* können mitunter sogar bis 124 Zoosporen bilden (KORSCHIKOFF 1926). Stark in der Größe veränderliche Zoosporangien bilden *Chlorangium*-Arten (KORSCHIKOFF 1932, ETTL 1958); besonders bei *Chl. epizooticum* ist die Zahl der Zoosporen sehr groß, nicht selten werden einige Hundert gebildet (Abb. 67). Die Teilung beginnt auch bei den *Tetrasporales* immer als eine Längsteilung des Protoplasten. Die Teilprodukte können sich aber während der Teilung auf verschiedenste Weise drehen (Abb. 68). Wenn sich diese Vorgänge während weiterer Teilungen wiederholen, erhält man eine größere Anzahl von Teilstücken unübersichtlicher Anordnung. Dies geschieht besonders bei den *Chlorangiales*. Im Prinzip aber bleibt die Teilung dieselbe wie bei den behäuteten *Volvocales* (GEITLER 1925, SCHUSSNIG 1960). Die *vielzelligen* in Gallertverbänden lebenden *Tetrasporales* bilden Zoosporen nur unter gewissen Umständen, da die Zellteilung dem Wachstum des Verbandes dient.

Abb. 66a—c. Ultrastruktur der Chlorophyceen-Zoosporen. a Längsschnitt durch das Vorderende einer Zoospore von *Stigeoclonium*. Man erkennt die wichtigsten Organellen — den Geißelapparat am Vorderende, eine der pulsierenden Vacuolen (*CV*), den Kern (*N*), den Chromatophor mit Pyrenoid (*PY*), Mitochondrien (*m*) und Golgi-Apparat (*g*). Typischer Bau einer monadoiden Chlorophyceen-Zelle. 20000× vergrößert. b Tangentialschnitt durch das Stigma derselben Zoospore. Die wabenartige Struktur der Pigmentkammern ist deutlich. 40000× vergrößert. c Längsschnitt durch eine Zoospore von *Oedogonium*, wo der Kern klar hervorkommt, umgeben vom Chromatophor, in dem reichlich Stärke vorhanden ist. Sichtbar ist auch der Geißelring unter dem farblosen haubenartigen Vorderende. 3000× vergrößert. (a, b nach MANTON, c nach HOFFMANN und MANTON.)

Chlorococcales. Eine coccale Zelle zeigt ähnlich wie die vegetative Cyste der Flagellaten keine Merkmale der Monadenorganisation, sondern stellt einen Zellzustand für sich dar. Auch die reproduktive Phase weicht von der der Monaden ab. Sie wird von der keimenden Flagellatencyste abgeleitet (Cytogonie — Schussnig 1960). Es sei betont, daß die Fortpflanzung der *Chlorococcales* niemals durch Zweiteilung, sondern nur durch simultane oder sukzedane Aufteilung des Inhaltes der vermehrungsreifen Zellen geschieht (Abb. 69). Das Ergebnis dieses multiplen Spaltungsvorganges ist eine merogene Ausbildung von Zoosporen in verschiedener Anzahl, im Minimalfall als Folge von zwei aufeinanderfolgenden Kernteilungen. Bei dieser Art der Fortpflanzung läßt sich keine bestimmte Polarität der Teilprodukte in bezug auf das Achsensystem des Mutterindividuums mehr erkennen. Die Polarität der Teilprodukte wird erst nach Vollendung der Zoosporen deutlich, wenn Geißelapparat, Stigma und pulsierende Vacuolen erschienen sind und das Vorderende der Zoospore markieren. Die Zoosporen sind zunächst immer nackt — das Sporangium wird niemals durch Membranen gefächert — und behäuten sich entweder erst nach dem Austreten oder bereits, aber sekundär, im Sporangium und verlassen letzteres dann als *Chlamydomonas*-ähnliche Schwärmzellen. So nicht selten bei einfachen Chlorococcalen.

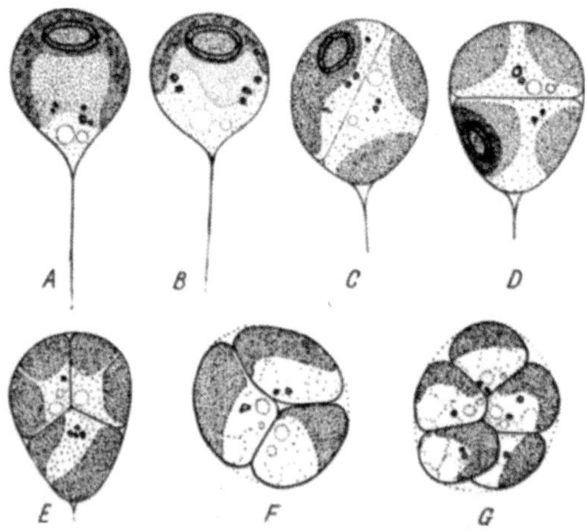

Abb. 68 A—G. Zoosporenbildung von *Stylosphaeridium stipitatum*. A, B Vegetative Zellen, an denen die Monadenorganisation zu erkennen ist; C einleitende Zweiteilung mit Querdrehung des Protoplasten; D vollendete Querdrehung; E, F weitere Teilung mit tetraëdrisch angeordneten Protoplasten; G vorgeschrittene Zoosporenbildung. (Nach Geitler aus Pascher.)

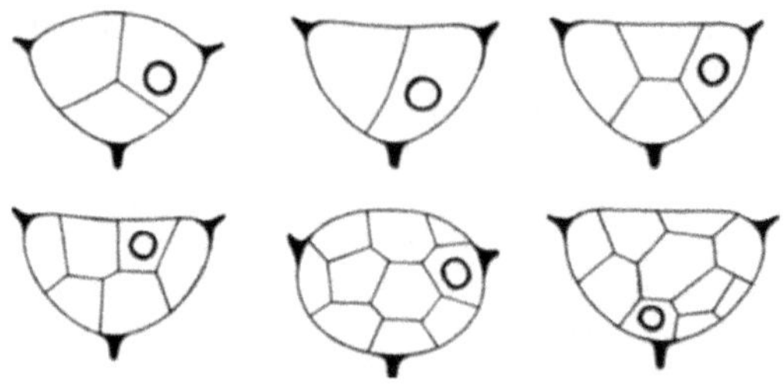

Abb. 69. *Sorastrum spinulosum*. Spaltung des Protoplasten vor der Gonidienbildung. Das alte Pyrenoid bleibt hier in einem der Teilprotoplasten erhalten. (Nach Geitler.)

Starr (1955) teilt die Zoosporen der *Chlorococcales* ein je nachdem, ob sie innerhalb der Mutterhülle die Membran ausbilden und das Sporangium behäutet verlassen als *Chlamydomonas*-Typus oder ob sie nackt schlüpfen und erst nach der Schwärmzeit als unbewegliche Keimlinge die Membran bilden als *Protosiphon*-Typus. Eigenartiges Verhalten behäuteter Zoosporen von *Planophila sphagnothermalis* beschrieb Pascher (1939a). Die Membran der Zoosporen wird nicht immer, vielleicht niemals zur Membran der vegetativen Zellen. Der Protoplast der Zoospore schlüpft nämlich aus der ursprünglichen Membran aus, um sich neu zur unbeweglichen Zelle zu behäuten.

Die Teilungsvorgänge selbst laufen verschiedenartig und Geitler (1924) unterscheidet zwei Hauptgruppen von Chlorococcalen. Eine, bei dem die Zoosporen (aber auch Autosporen!) *simultan*, durch Zerklüftung des vielkernigen Plasmas entstehen, wobei das ungeteilte Pyrenoid aufgelöst und in den Tochterzellen neu gebildet wird. Die andere, bei welcher durch *sukzedane* Teilung und unter Erhaltenbleiben des Pyrenoides die Tochterzellen entstehen. Auch bei den *Chlorococcales* können die Sporangien eine beträchtliche Größe erreichen und eine große Anzahl von Zoosporen bilden. So enthält der Flechtensymbiont *Myrmecia reticulata* (Tschermak-Woess 1951) und ebenso die alpine Bodenalge *M. incisor* (Reisigl 1964) in vergrößerten Sporangien nach streng sukzedaner Teilung bis 128 Zoosporen.

Ulothrichales. Im allgemeinen entstehen bei dieser Algengruppe die Zoosporen in Sporangien von demselben Aussehen wie die vegetativen Zellen (Abb. 71), doch kommt es bei morphologisch und entwicklungsgeschichtlich spezialisierten Arten zu einer Lokalisierung und einer gestaltlichen Differenzierung der zoosporenbildenden Zellen. So können die Zoosporangien nur an den Endzellen entstehen oder sich auch, wo eine strenge Trennung zwischen Hauptsproß und Seitenästen besteht (z. B. *Draparnaldia*), nur aus den Zellen der Äste entwickeln. Bei *Fritschiella* werden die Zoosporen merkwürdigerweise nur in den Zellen des prostraten Systems gebildet (Singh 1941), wogegen bei *Didymosporangium* (Abb. 72A) spezialisierte Zoosporangien durch zwei sukzedane Teilungen aus den mittleren

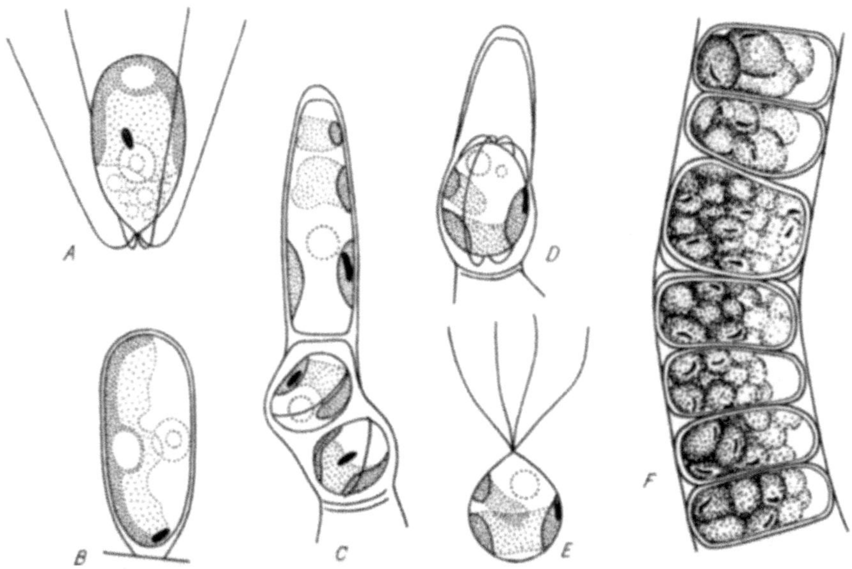

Abb. 70 A—F. Zoosporenbildung trichaler Chlorophyceen. A Eine viergeißelige Zoospore von *Ulothrix* sp. vor dem Festsetzen (Vorderende nach unten gerichtet). Der Chromatophor kleidet das Hinterende muldenförmig aus, hier das Pyrenoid, ungefähr in der Mitte der Augenfleck, darunter der Kern und im Vorderende mehrere pulsierende Vacuolen (vgl. mit Abb. 66a). B mit dem Vorderende festgesetzte Zoospore, die bereits eine Membran gebildet hat (Keimling). Der Zellinhalt hat sich gegenüber A um 90° gedreht. Durch die Drehung wird bei der ersten Teilung und bei den weiteren die Längsteilung des Protoplasten beibehalten, obwohl sie quer zur Fadenachse verlaufen. Die Polaritätsverhältnisse sind auf den Protoplasten einzubeziehen. C Zoosporenbildung im Astende von *Chaetophora pisiformis*. Oberste Zelle in Beginn der Umwandlung des Inhaltes in eine Zoospore (man beachte das Stigma und die pulsierende Vacuole). Untere Zelle mit zwei Zoosporen vor dem Entschlüpfen; D obere Zelle von C um 10 min später. Der Zellinhalt hat sich unter Kontraktion in eine Zoospore umgewandelt. E entschlüpfte Zoospore; F Zoosporenanlagen (Mikrozoosporen) bei *Ulothrix* sp. (A—E nach Geitler umgezeichnet, F nach Gross umgezeichnet.)

Zellen der Fäden entstehen. Bei *Sporocladus* (Abb. 73b) andererseits bilden sich die Zoosporangien durch Anschwellung seitlicher oder terminaler Zellen zu gedrungen keulenförmiger Gestalt. Auch bei *Gongrosira*, *Chloroclonium* und *Epibolium* (Abb. 72C—E, 73a) entstehen Zoosporangien durch Anschwellen der äußersten Zweigzellen (Printz 1927, West und Fritsch 1927, Smith 1950). Die tropische Gattung *Cephaleuros*, die epiphytisch oder parasitisch an Phanerogamen lebt, hat spezialisierte Sporangien. Vor allem jene Arten, die parasitisch in Blattgeweben leben, bilden lange Träger, die aus dem Gewebe herauskommen und an denen dann die Sporangien entstehen (Oltmanns 1922).

Derartig morphologisch und physiologisch spezialisierte Zoosporangien sind allgemein auch bei den atmophytischen Trentepohliaceen verbreitet (Flint 1959, Printz 1927). Es handelt sich vor allem um *Hakensporangien*, die für *Trentepohlia* geschildert seien (Abb. 72F, G). Diese sind elliptisch-eiförmige Zoosporangien, die sich vom Thallus auf mehr oder weniger hakenförmig gekrümmten, inhaltsarmen Trägerzellen erheben. Die Hakensporangien sind anfangs einkernig und werden bei trockenem Wetter von der Trägerzelle abgeworfen, was zwei ringförmige Membranverdickungen zwischen den Sporangien und Trägerzellen bewirken. Der Inhalt wird später mehrkernig und zuletzt in mehrere Zoosporen aufgeteilt. Erst nach Benetzung mit Niederschlagswasser erfolgt die endgültige Ausbildung der Zoosporen und deren Freiwerden mittels einer schnabelartigen Entleerungspapille. Das Ausschwärmen der Zoosporen findet also getrennt von der Mutterpflanze statt.

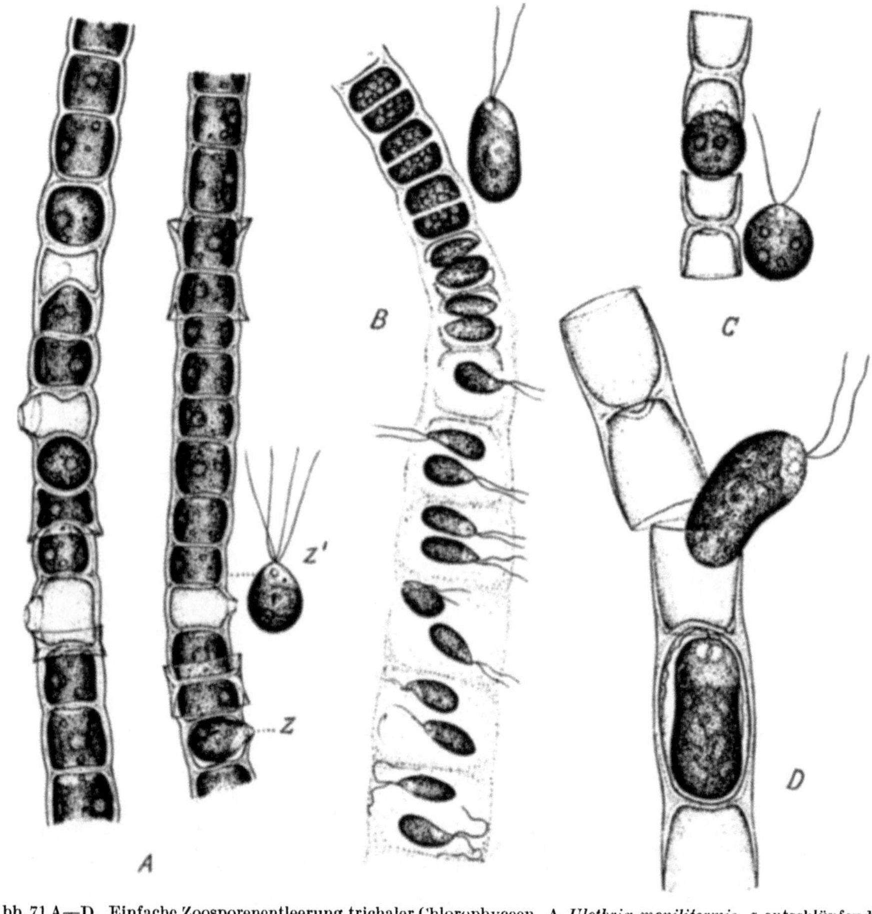

Abb. 71 A—D. Einfache Zoosporenentleerung trichaler Chlorophyceen. A *Ulothrix moniliformis*, *z* entschlüpfende Zoospore; *z'* Zoospore, die bereits das Sporangium verlassen hat. An den leeren Sporangien ist die Öffnung deutlich, durch die sich die Zoosporen hinauszwängen müssen. B *Microspora quadrata*, die Entleerung erfolgt durch Verschleimung der Membran; C *Microspora lauterbornii*. D *Microspora willeana*, bei den letzteren werden die Zoosporen durch den Zerfall der zweiteiligen Membran (H-Stücke) frei. (Nach Skuja.)

Es sei noch angeführt, daß bei gewissen Fadenalgen (*Ulothrix*, *Stigeoclonium* u.ä.), sogar bei ein und derselben Art, zweierlei Zoosporen vorkommen — *Makro-* und *Mikrozoosporen*. Sie unterscheiden sich sowohl durch ihre Größe als auch häufig durch die Anzahl (zwei bzw. vier) der Geißeln. Nach Juller (1937) keimen die Makrozoosporen von *Stigeoclonium* direkt, während sich die Mikrozoosporen zuerst zu Cysten abrunden und mit einer dicken Membran umgeben, worauf sie nach kürzerer oder auch längerer Zeit zu Zellfäden auswachsen. Auch in physiologischer Hinsicht sind beide Zoosporensorten verschieden, so zeigen sie z.B. eine unterschiedliche phototaktische Empfindlichkeit. Ob jedoch alle beobachteten Schwärmzellen neutrale Zoosporen sind, die direkt keimen, bleibt fraglich. Pascher (1906) beobachtete bei *Stigeoclonium nudiusculum* viergeißelige Makrozoosporen und viergeißelige Mikrozoosporen. Gross (1931) berichtet für *Ulothrix* über viergeißelige Zoosporen, die etwas in der Größe variieren, und über zweigeißelige Gameten. Nach Reich (1926) gibt es bei *Stigeoclonium flagelliferum* viergeißelige Makro- und Mikrozoosporen, daneben zweigeißelige „Mikrozoosporen“, die er als Gameten auffaßt, wenn sie auch unkopuliert keimen können. SINGH (1941, 1942) und andere Autoren vertreten jedoch die Ansicht, daß außer viergeißeligen auch zweigeißelige agame Mikrozoosporen vorkommen (*Fritschiella, Draparnaldiopsis*).

Über den Verlauf der Zoosporenbildung wurde mehrmals berichtet, sowohl nach lebendem Material als auch nach cytologischen Präparaten. SINGH (1942) beschreibt ihn eingehend an lebenden Zellen von *Draparnaldiopsis*. Vor der Fortpflanzung wandeln sich die sporogenen Zellen durch Anschwellen in Sporangien

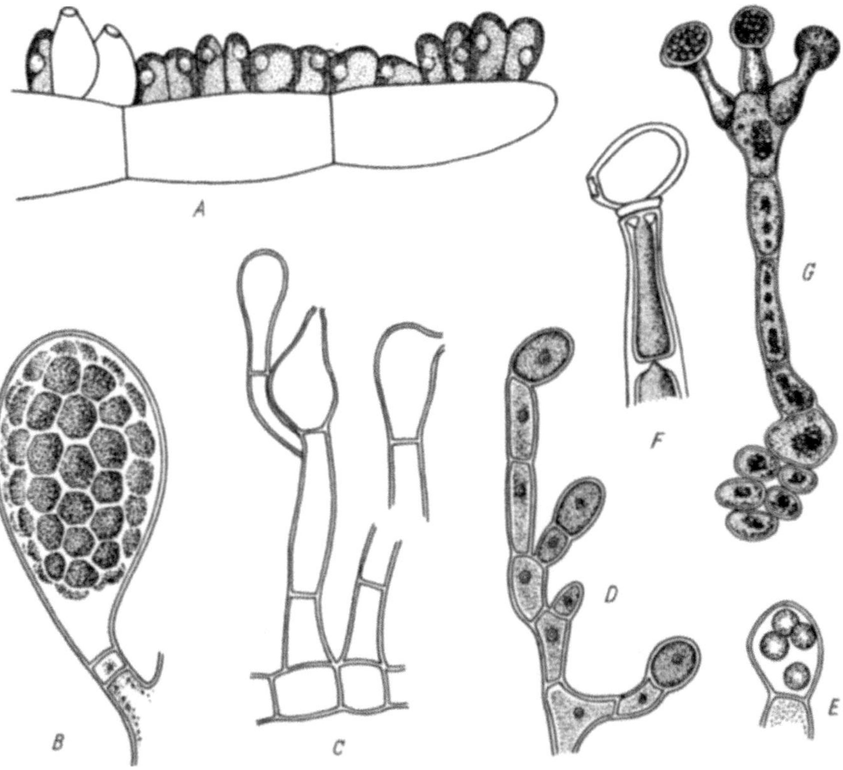

Abb. 72 A—G. Die Gestalt differenzierter Zoosporangien bei trichalen Chlorophyceen. A *Didymosporangium repens*, ein Teil der Alge in Zoosporangienbildung. Links zwei schon entleerte Zoosporangien. B *Derbesia marina*, Zoosporangium mit Zoosporenanlagen. C *Gongrosira stagnalis* mit einem reifen und zwei entleerten Zoosporangien. D *Epibolium dermaticola*, Zweigspitzen mit jungen Zoosporangien. E reifes Zoosporangium derselben Alge. F Hakensporangium von *Trentepohlia annulata* (Membranring deutlich). G Hakensporangium von *Trentepohlia umbrina*. (A nach LAMBERT aus PRINTZ; B nach KUCKUCK aus PRINTZ; C nach WEST und FRITSCH; D, E nach PRINTZ; F nach BRAND aus FOTT; G nach OLTMANNS aus FOTT.)

um, wobei der Ringchromatophor topfförmig wird. Dann löst sich der Protoplast von der Membran, rundet sich ab und wird weiter durch Umgestaltung und das Entstehen des Stigma, der pulsierenden Vacuolen und schließlich der Geißeln zur Zoospore. Besonders hübsch schildert GEITLER (1934) den Verlauf bei *Ulothrix* (Abb. 70 C—E). JULLER (1937) erkennt als erstes Anzeichen der Makrozoosporenbildung von *Stigeoclonium subspinosum* die Entstehung des Augenfleckes. An einem Vertreter der gleichen Gattung beobachtete REICH (1926) die Umwandlung der vegetativen Zellen in Zoosporen an gefärbten Präparaten. Die Basalkörner des Geißelapparates sollen von einem aus dem Karyosom austretenden färbbaren Körper gebildet werden. Die Cytologie der Kinetombildung ähnelt derjenigen der *Volvocales*, da auch bei diesen ein aus dem Karyosom austretendes, chromophiles Körnchen an der Geißelbildung beteiligt ist. Nach lebendem Material von *Ulothrix* berichtet Gross (1931), daß die Zoosporen, wenn in genügender Anzahl vorhanden (Mikrozoosporen), im Sporangium in einer Hohlkugel angeordnet liegen, die an einer Stelle offen ist (Abb. 70 F).

Oedogoniales. Die relativ großen stephanokonten Zoosporen entstehen bei Oedogoniaceen einzeln in jeder Zelle. Die Neigung zur Zoosporenbildung ist nach Mainx (1931) bei den einzelnen Arten verschieden stark. Ist der physiologische Zustand für die Zoosporenbildung günstig, so können bis zur völligen Auflösung der Fäden Zoosporen ausgelöst werden. Wie schon Klebs (1896) festgestellt hat, geschieht dies bei den festsitzenden Arten vom apikalen Ende und bei den flottierenden Arten von beiden Enden her. Auch bei *Bulbochaete* vermag sich der Inhalt jeder Zelle, mit Ausnahme der Haftzelle, in eine einzige Zoospore umzuwandeln (Cook 1962). Das hyaline Vorderende der zukünftigen Zoospore erscheint in der Mitte der Mutterzelle als ein farbloser Fleck, worauf sich der Inhalt abrundet. Die Geißeln kommen schon vor dem Ausschwärmen zum Vorschein. Dieser Prozeß ist jedoch keine einfache Verjüngung des Protoplasten, sondern ein kompliziertes Geschehen. Falls während der Festsetzung und Keimung der Zoosporen die Bedingungen zur Zoosporenbildung persistieren, kann der Inhalt der einzelligen Keimlinge durch Abwerfen eines apikalen Membrandeckels von neuem als Zoospore ausschlüpfen. Dies wiederholt sich manchmal (Mainx 1931).

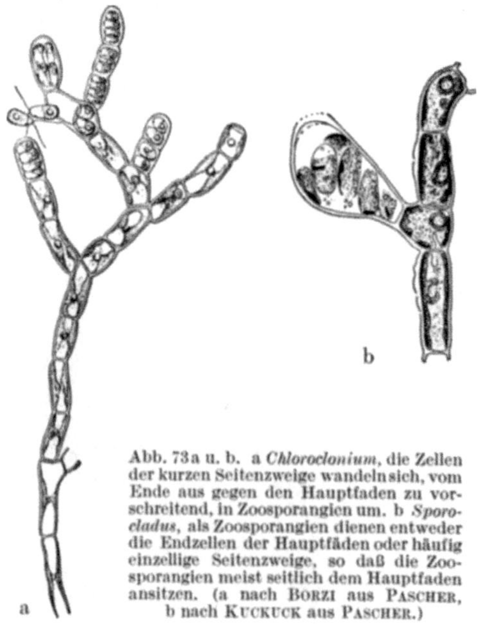
a b

Abb. 73a u. b. a *Chloroclonium*, die Zellen der kurzen Seitenzweige wandeln sich, vom Ende aus gegen den Hauptfaden zu vorschreitend, in Zoosporangien um. b *Sporocladus*, als Zoosporangien dienen entweder die Endzellen der Hauptfäden oder häufig einzellige Seitenzweige, so daß die Zoosporangien meist seitlich dem Hauptfaden ansitzen. (a nach Borzi aus Pascher, b nach Kuckuck aus Pascher.)

Coenoblastische und siphonoblastische Formen. Für diese zur Zeit der Fortpflanzung polyenergiden Chlorophyceen ist charakteristisch, daß die Fortpflanzung auf dem Wege einer segregativen Plasmotomie geschieht, die über ungleich große mehrkernige Teilstücke des Cytoplasten bis zur schizotomen Individualisierung einkerniger Zoosporen führt. Einen einfachen Typus stellt die Fortpflanzung von *Actinochloris* dar (Korschikoff 1953, Ettl 1964c). Die großen, erwachsenen Zellen sind polyenergid, mit 32—64 Kernen. Die Zahl der Kerne ergibt auch die Zahl der einkernigen Zoosporen. Hier sei aber betont, daß auch mittelgroße Zellen mit 8 oder 16, in seltenen Fällen mit nur 4 Kernen unter gewissen Umständen Zoosporen bilden können. Die vegetative Fortpflanzung erfolgt durch simultane Protoplastenteilung, wobei jeder Kern einen Teil des Protoplasten an sich zieht. Die Zoosporen sind behäutet, völlig dem *Chlamydomonas*-Typus entsprechend. Als abgeleitete Form des Coenoblasten ist der Siphonoblast anzusehen, der durch wandständige Lage des Cytoplasten und eine im Innern vorhandene große Saftvacuole charakterisiert ist. Auch hier sind die Zellen polyenergid und jeder Kern wirkt physiologisch-dynamisch auf einen gewissen Umkreis des Plasmas. Bei der Bildung einkerniger Zoosporen erfolgt die Individualisierung der Energiden (Schussnig 1954). Am einfachsten ist es bei *Characiosiphon rivularis*, wo die Energiden schon in den vegetativen Individuen als nackte, gegenseitig abgeplattete und parietal angeordnete Protoplasten mit Chromatophor und Pyrenoid individualisiert sind (IYENGAR 1954) und sich während der Fortpflanzung direkt in Zoosporen umwandeln.

Komplizierter ist der Teilungsvorgang bei den oberirdischen Blasen an *Protosiphon*, dessen Zellinhalt bei Anwesenheit von liquidem Wasser normalerweise völlig in zweigeißelige Zoosporen zerfällt. Das Gebilde wird zu einem Zoosporangium. Vorher kommt es jedoch zu einer Protoplastenkontraktion, wobei das Volumen bis auf ein Drittel zusammenschrumpft. Dann wird das Cytoplasma durch fortschreitende segregative Plasmotomie nach und nach in Zoosporenanlagen zerteilt. Die Oberfläche des ursprünglich glatten Protoplasten wird dann — an einem Pol beginnend und weiter über den ganzen Protoplasten fortschreitend —

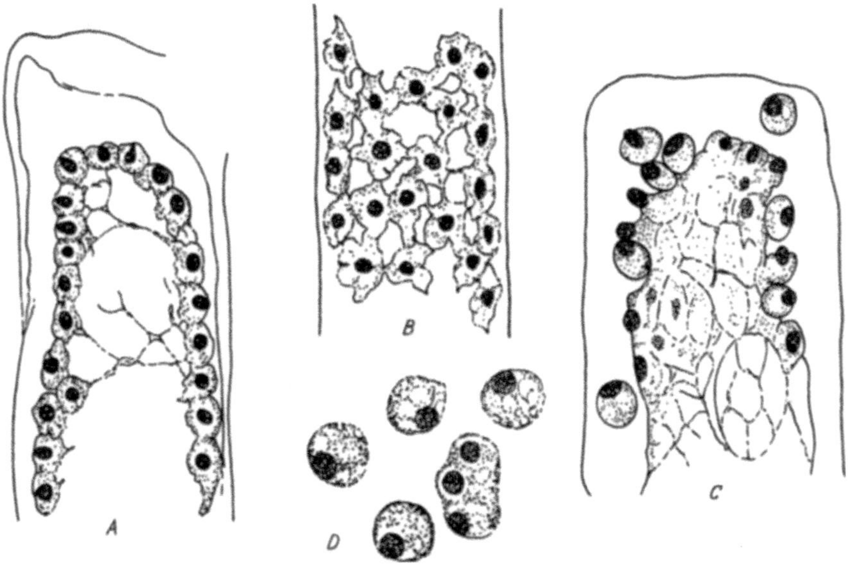

Abb. 74 A—D. Segregativer Zerfall des Siphonoblasten bei *Cladophora glomerata* während der Zoosporenbildung. A Im Längsschnitt; B dasselbe, aber als Ansicht von der Fläche. Beide Abbildungen zeigen den Beginn der Fertilisierung. C Stadium der Fertilisierung vor der Zoosporenbildung mit einzelnen losgelösten Zoosporenanlagen; D abgetrennte Zoosporenanlagen, spätes Stadium der Zoosporenbildung. Rechts eine dreikernige Anlage. (Nach CZEMPYREK umgezeichnet.)

gelappt und gefurcht. Nach BOLD (1933) ist der letzte Teilungsschritt der so entstandenen individualisierten, einkernigen Zoosporenanlagen eine schizotome Längsspaltung, wodurch erst die definitiven Zoosporen gebildet werden.

Bei *Cladophora* geben sich die Zellen, in denen es zur Zoosporenbildung kommt, durch starke Kernvermehrung zu erkennen. Der Beginn des sporogenen Zustandes ist charakteristisch und zeigt im Vergleich mit den vegetativen Zellen besondere Wesenszüge (CZEMPYREK 1930, SCHUSSNIG 1960). In den vegetativen Zellen ist die Zahl der zu einem Zeitpunkt teilenden Kerne immer beschränkt; bei den sporogenen Zellen nimmt dagegen die Zahl der durch Teilungen vermehrten Kerne unverhältnismäßig stark zu. Gleichzeitig wird das im Wandbelag der Zelle ausgespannte Chromatophorennetzwerk dichtmaschiger und zieht sich gegen die Zellkerne hin zusammen, was zu einem charakteristischen Aussehen der sporogenen Zellen führt. Mit dem Abschluß der Kernvermehrung rückt dann der Zellinhalt etwas von der Membran ab, und die zu den Kernen gehörigen Plasmaportionen werden durch Kontraktion des Plasmas deutlich (Abb. 74). Sie liegen in einfacher Schicht an der Oberfläche der zentralen Vacuole. Schließlich lösen sich die einzelnen Anlagen ganz voneinander.

Die Zoosporangien der Derbesiaceen entstehen als kurze Seitentriebe, die keulen- oder kugelförmig anschwellen und sich durch Doppelwände vom Hauptfaden abgrenzen (Abb. 72 B). In jedem Zoosporangium werden 8—20 stephanokonte Zoosporen gebildet (PRINTZ 1927). Bei den Arten mit heteromorphem Generationswechsel handelt es sich dabei sehr wahrscheinlich um Meiosporangien.

4. Modifizierte Zoosporen.

Hemizoosporen. Mit demselben Bau wie die Zoosporen, mit typischer monadoider Organisation und Geißelapparat. Sie sind zwar beweglich, aber nur *innerhalb* des Zoosporangiums oder in einer durch dieses gebildeten Gallertblase, und

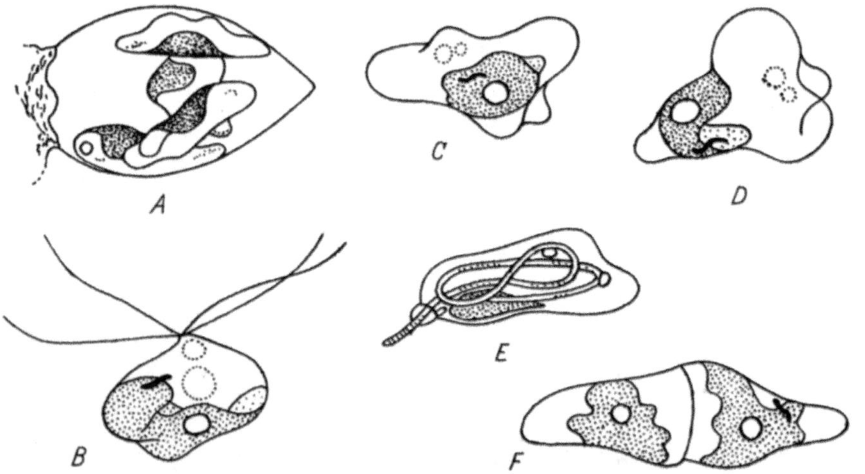

Abb. 75 A—F. Amöboide Zoosporen. A Einzelzelle der Kolonie von *Marthea tetras* mit vier amöboiden, in der Mutterhülle herumkriechenden Teilprotoplasten; B—F amöboide Zoosporen von Chaetophoraceen, B metabole Zoospore; C, D amöboid gewordene Zoosporen ohne Geißeln; E animalische Ernährung; F zweizelliger Keimling, der aus einer amöboiden Zoospore herangewachsen ist. (Alle Abbildungen nach PASCHER umgezeichnet.)

wandeln sich innerhalb des Sporangiums, ohne es zu verlassen, in das unbewegliche vegetative Stadium um. Hemizoosporen kommen vor allem bei Tetrasporalen und Chlorococcalen vor, vereinzelt auch bei Fadenalgen. PETROVÁ (1931) beobachtete sie bei *Dictyococcus minor*, wo sie sich schon innerhalb der Mutterhülle in behäutete vegetative Zellen umwandeln. KORSCHIKOFF (1953) hat Hemizoosporen bei mehreren Chlorococcalen festgestellt (z. B. bei *Hypnomonas*, *Nautococcus*, *Chlorophysema*, *Asterococcus*, *Characium* u. a.). Besondere Arten von Hemizoosporen werden bei den Hydrodictyaceen gebildet. Eingehend wird der Vorgang an *Sorastrum* von GEITLER (1924) dargestellt. Die Hemizoosporen werden in diesem Fall zwar aus der Mutterhülle entleert, bleiben aber in einer Gallertblase eingeschlossen. In dieser Blase bleiben sie einige Zeit in zitternder Bewegung, um schließlich ihre Eigenbewegung völlig aufzugeben, nachdem sie sich zu einer jungen Kolonie vereinigt haben. Erst die fertige Kolonie verläßt die Blase. Bei *Hydrodictyon* geht die Rückbildung des Geißelapparates noch weiter. RIETH (1952) fand bei *Sphaeroplea wilmani* zweigeißelige Hemizoosporen. Ihre bewegliche Phase ist sehr kurz, so daß sie in wenigen Minuten innerhalb des Sporangiums keimen. Es wird ausdrücklich betont, daß keine Parthenogenesis vorliegt.

Amöboide Stadien der Zoosporen. Bei den Chlorophyceen treten gelegentlich statt flagellatenähnlichen Zoosporen amöboide Gebilde auf, welche die Funktion der Zoosporen übernehmen (Abb. 75). Solche amöboide Stadien wurden von PASCHER (1909, 1915, 1918a, 1927) mehrmals beschrieben, sogar bei hoch entwickelten Fadenalgen. Auf amöboide Zoosporen bei *Draparnaldia* hat schon KLEBS aufmerksam gemacht. Die Makrozoosporen mancher Chaetophoraceen bilden Protoplasmaklumpen von unregelmäßiger Form mit plumpen Pseudopodien; der schüsselförmige Chromatophor bleibt erhalten, ebenso wie Pyrenoide, Stigma und pulsierende Vacuolen. Sie kriechen amöboid längere Zeit umher, um sich dann direkt in einen Keimling umzuwandeln. Solche amöboide Stadien entstehen entweder aus entschlüpften normalen Makrozoosporen oder können als solche sofort den Faden verlassen, ohne vorerst mit Geißeln versehen zu sein. Änliche Metabolie zeigen auch Zoosporen anderer Algen (PASCHER 1915, 1918a). Amöboide Zoosporen von *Stigeoclonium* bleiben länger beweglich als die begeißelten. Sie zeigen auch lebhafte animalische Ernährung, wobei die Aufnahme kleiner Organismen wie bei echten Amöben erfolgt.

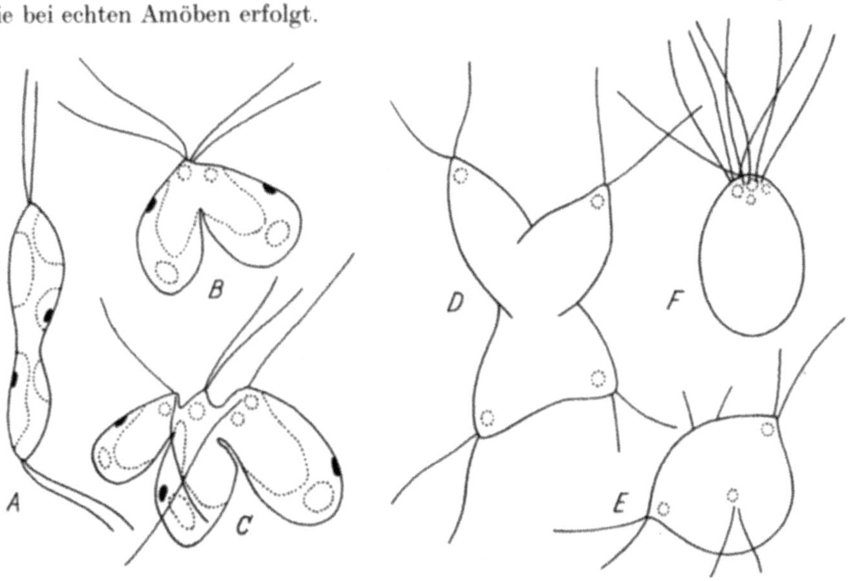

Abb. 76 A—F. Vielfachschwärmer und Synzoosporen. A Unvollständig getrennte Zoosporen von *Dictyococcus minor*; B, C ähnliche Gebilde von *Chlamydomonas*; D verschmolzene Zoosporen, die aus der Zygote von *Sphaeroplea cambrica* hervorkommen; E Vielfachschwärmer fast synzoosporenartig miteinander verbunden; F vier aus der keimenden Zygote austretende Zoosporen völlig zu einer Synzoospore verbunden. Diese ist ausgesprochen polar gebaut, einer normalen Zoospore nahekommend. (A vereinfacht nach PETROVÁ, B, C Original, D—F nach PASCHER.)

Vielfachschwärmer und Synzoosporen. Als solche bezeichnet man bewegliche Fortpflanzungszellen, die gewöhnlich vielkernig sind und meist auch vervielfachte Organellen und Geißelpaare aufweisen. Da im allgemeinen das Verhältnis der Kerne zu den Geißelpaaren und Organellen (Stigma, pulsierende Vacuolen) das gleiche sein dürfte wie bei den einkernigen Zoosporen, können sie als verschmolzene oder durch Teilungsverzögerungen und Teilungshemmungen des Sporangiuminhaltes entstandene Zoosporen aufgefaßt werden (PRINTZ 1927, PASCHER 1929, 1939b). Die verschmolzenen Zoosporen können in verschiedenster Weise aufgebaut sein — von einfachen Vielfachschwärmern bis zu extrem gebauten Synzoosporen. Fast bei allen Organisationsstufen der Chlorophyceen können sie gelegentlich festgestellt werden. Die Synzoosporen von *Follicularia* haben recht verschiedene Größe, enthalten 1—5 Pyrenoide, 1—5 Zellkerne und 1—5 Geißelpaare. Sie werden durch ein unvollkommenes Zerspalten des Protoplasten gebildet, in diesem Fall einer polyenergiden Zelle. Gelegentliche Vielfachschwärmer sind auch bei *Cladophora* und *Protosiphon* bekannt, wenn die Zoosporenanlagen nicht vollständig getrennt werden.

Pascher (1939b) beobachtete bei der Keimung der Zygoten von *Sphaeroplea*, daß nach vollzogener Reduktionsteilung, statt wie normal vier zweigeißeligen Zoosporen, verbundene und nicht vollständig getrennte Zoosporen hervorkamen (Abb. 76). So entstehen Doppel- oder Vielfachzoosporen. Da die Vorderenden mit ihren Geißelpaaren oft etwas vorspringen, so ist die Zahl der verbundenen Zoosporen leicht erkennbar. Manchmal jedoch sieht so ein Vielfachschwärmer den Einzelzoosporen zum Verwechseln ähnlich. Er stellt somit eine extrem polar gebaute Synzoospore dar, die nur an der Zahl der Vacuolensysteme und der Kerne den multiplen Charakter erkennen läßt und sich wieder dem Bauplan der Einzelzoospore nähert.

b) Unbewegliche Fortpflanzungszellen (Autosporen-Typus).

Den zoosporinen Chlorophyceen steht eine andere Reihe gegenüber, deren Entwicklungsgang wesentlich vereinfacht ist. Es werden hier ebenfalls innerhalb der Zelle mehrere Teilstücke des Protoplasten gebildet, doch wandeln sich diese nicht erst in Zoosporen um. Sie bleiben unbeweglich, behäuten sich und werden zu kleinen vegetativen Zellen. Diese Fortpflanzungszellen werden *Autosporen* genannt. Sie nehmen noch innerhalb des Sporangiums die für die gegebene Art charakteristische Zellform an (Pascher 1918, Ettl 1958). In der Fortentwicklung der unbeweglichen Algentypen führt die Umwandlung der Zoosporen über Zwischenstufen zu schließlich unbeweglichen geißellosen und behäuteten Autosporen.

Dieser Übergang läßt sich in schöner Reihe verstehen. Bei *Chlorococcum*, *Hypnomonas*, *Characium* u.ä. dienen der vegetativen Fortpflanzung bewegliche und das Sporangium verlassende Zoosporen (neben diesen auch Autosporen). Bei *Dictyococcus*, *Sorastrum*, *Pediastrum* u.ä sind es nur mehr ganz kurze Zeit bewegliche, das Sporangium nicht verlassende Hemizoosporen. Die Teilprotoplasten von *Marthea* haben noch Stigma und pulsierende Vacuolen, doch keine Geißeln und nur mehr kurzdauernde amöboide Bewegung in der Mutterzelle (Pascher 1918a). Als weiterer Schritt können Hemiautosporen betrachtet werden (s. unten). Diese sind unbeweglich, haben aber sonst Merkmale der Monadenorganisation. Sie werden häufig bei einfachen *Chlorococcales*, aber auch bei einigen Fadenalgen gebildet. Bei den autosporinen Formen sind die Fortpflanzungszellen ohne jedes Merkmal der Flagellaten und den vegetativen Zellen völlig ähnlich. In phylogenetischer Hinsicht sind die Zoosporen als die ursprünglicheren Gebilde aufzufassen, während die ihres Geißelapparates verlustig gegangenen Autosporen sekundär entfaltet wurden. Eine große Menge coccaler Typen bildet außer Autosporen auch Zoosporen, und zwar bei ein und derselben Art (z.B. *Chlorococcum infusionum*, *Borodinella polytetras*, *Planctococcus sphaerocystiformis* u.a.). Unter bestimmten nicht näher erklärbaren, sicher aber zum Teil endogen bedingten Umständen, kann jede Zelle dieser Formen anstatt unbeweglichen Autosporen auch begeißelte Zoosporen erzeugen. Die Fähigkeit. auch Zoosporen zu bilden, wahrscheinlich am Ende einer innenphysiologisch bestimmten Periode, ist genetisch verankert. Bei *Eremosphaera* treten in den Spindelpolen der die Fortpflanzung einleitenden Kernteilungen Centrosomen auf, obwohl sie sich nur durch Autosporen fortpflanzt. Sie können als Rest eines früher vorhandenen Geißelapparates aufgefaßt werden (Schussnig 1960), denn Centrosomen sind bei den zoosporinen Chlorococcalen nicht selten. Übrigens wurden in letzter Zeit auch bei typisch autosporinen Chlorococcalen Zoosporen gefunden, sogar bei solchen, wo man von einer ausschließlichen Fortpflanzung durch Autosporen überzeugt war. So hat Starr (1954a, b) bei *Planktosphaeria gelatinosa* außer Autosporen auch Zoosporen festgestellt, die in einer Anzahl von 2—64 gebildet werden. Auch bei *Tetraëdron bitridens* treten beiderlei Sporen als Fortpflanzungsgebilde nebeneinander auf. Das Überraschendste war jedoch die Feststellung von Zoosporen bei *Scenedesmus obliquus* durch Trainor (1963). Wenn die Schwärmzellen hier auch nur vereinzelt vorkamen, so ist dieser Fund von großem entwicklungsgeschichtlichem Interesse.

1. Hemiautosporen.

Wie schon erwähnt, sind die Hemiautosporen unbewegliche Fortpflanzungszellen, bei denen von der Monadenorganisation noch die pulsierende Vacuole, oft auch das Stigma vorhanden ist, die sich aber sonst wie Autosporen verhalten (Ettl 1958, 1961). Schon ältere Autoren legten den Hemiautosporen entwicklungsgeschichtlich besonderen Wert zu, bezeichneten sie jedoch als Autosporen mit Stigma und pulsierenden Vacuolen bzw. mit flagellatenähnlicher Polarität. Petrová (1930) beschreibt ihre Bildung bei *Tetraciella*, wo sie gelegentlich die Autosporen ersetzen und pulsierende Vacuolen führen; auch Pascher (1932) hat

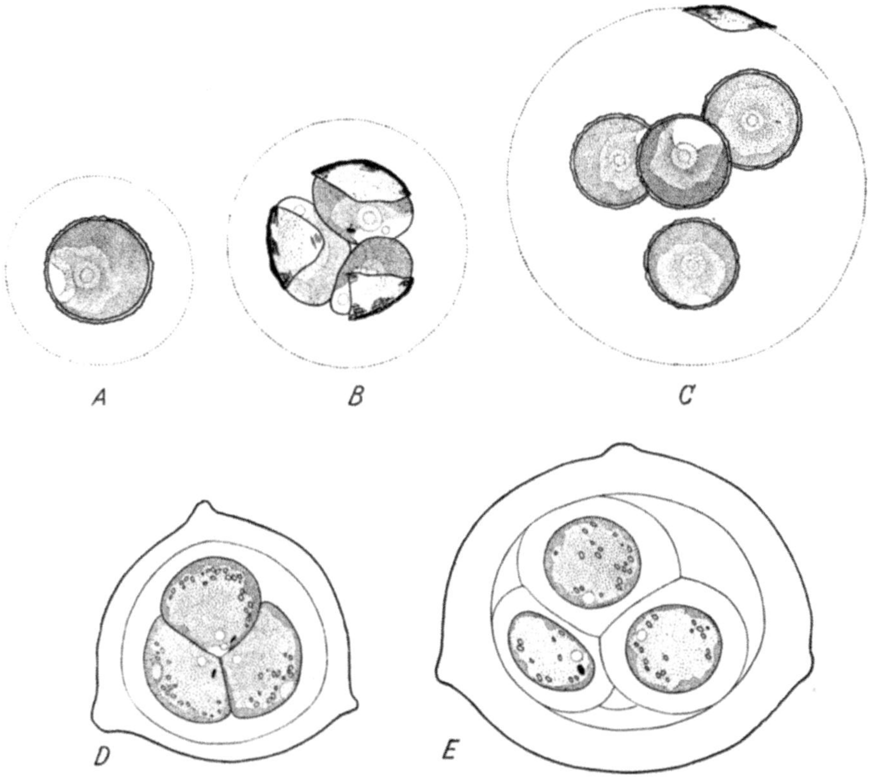

Abb. 77 A—E. Hemiautosporen bei Chlorophyceen. A—C *Thorakochloris tetras*. A Einzelzelle mit derber und spröder Membran und Gallerthülle; B Teilungsstadien mit tetraëdrisch angeordneten Hemiautosporen. Stigma und pulsierende Vacuolen sichtbar; C junge vegetative Zellen in der Gallerthülle der Mutterzelle. D Teilungsstadium von *Trigonidium galea* mit Hemiautosporen; E weiter fortgeschrittenes Stadium derselben Alge. Die Teilprotoplasten behalten lange die pulsierenden Vacuolen. (Nach PASCHER.)

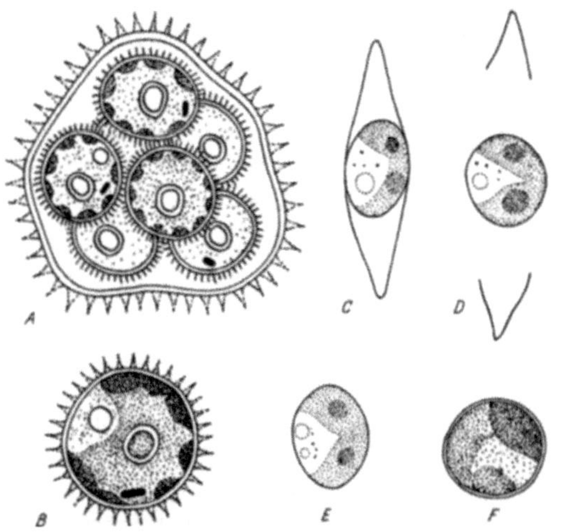

Abb. 78 A—F. A Sporangium von *Trochiscia aciculifera* mit Hemiautosporen; B eine einzelne Hemiautospore mit stacheliger Membran; Stigma und pulsierende Vacuole deutlich; C—F Aplanosporenbildung über Hemiautosporen, C Kontraktion des Protoplasten. D das Freiwerden der Hemiautospore, E einzelne Hemiautospore, F Aplanospore, die nach Behäutung der nackten Hemiautospore entstanden ist. (A, B nach KORSCHIKOFF, C—F nach LUND.)

sie bei typisch coccalen Formen gefunden (Abb. 77). So vollzieht sich bei *Trigonidium galea* die Fortpflanzung durch simultane Bildung von vier Teilprotoplasten, die tetraëdrisch gelagert werden. Diese Teilstücke besitzen einen großen topfförmigen Chromatophor, zwei pulsierende Vacuolen und in den meisten Fällen ein deutliches Stigma. Sie wandeln sich unter Rückbildung des Stigmas und der pulsierenden Vacuolen direkt in vegetative Zellen um. Ähnlich verhält sich *Thorakochloris* (Abb. 77A—C), und bei einer besonderen Form von *Asterococcus superbus* ist diejenige durch Hemiautosporen die einzige Fortpflanzungsart des Organismus (Ettl 1964a). Die Hemiautosporen müssen nicht immer nackt sein, sondern können noch innerhalb des Sporangiums von einer eigenen Membran umgeben werden und sich damit noch mehr den Autosporen nähern, wie dies von Korschikoff (1953) für verschiedene Tetrasporalen und Chlorococcalen und von Starr (1955) bei kugeligen Chlorococcaceen angegeben wurde. *Trochiscia aciculifera* endlich hat Hemiautosporen mit der bestachelten Membran und der Gestalt der vegetativen Zellen (Abb. 78A, B), die trotzdem eine Zeitlang Stigma und pulsierende Vacuolen erkennen lassen.

Bildung von Aplanosporen, die über Hemiautosporen führt, beschreibt Lund (1956) bei *Elakatothrix gelatinosa* var. *aplanospora* (Abb. 78C—F). Jede Zelle kann eine Hemiautospore bilden, wobei sich der Protoplast von der Membran abhebt, die nackte mit pulsierender Vacuole versehene Spore durch Zerfall der Membran frei wird und zuletzt unter Vacuolenverlust eine relativ dicke Membran ausscheidet. Ähnliche Sporen werden auch bei vielen Arten der Ulothrichaceen und Chaetophoraceen gefunden (Printz 1927).

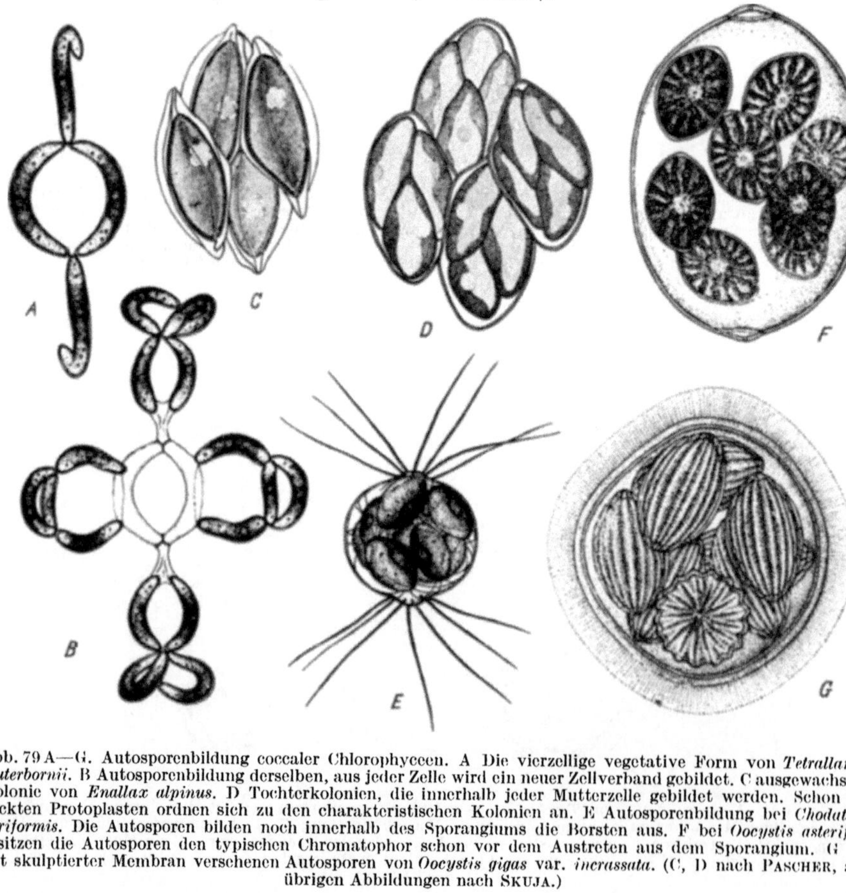

Abb. 79A—G. Autosporenbildung coccaler Chlorophyceen. A Die vierzellige vegetative Form von *Tetrallantos lauterbornii*. B Autosporenbildung derselben, aus jeder Zelle wird ein neuer Zellverband gebildet. C ausgewachsene Kolonie von *Enallax alpinus*. D Tochterkolonien, die innerhalb jeder Mutterzelle gebildet werden. Schon die nackten Protoplasten ordnen sich zu den charakteristischen Kolonien an. E Autosporenbildung bei *Chodatella citriformis*. Die Autosporen bilden noch innerhalb des Sporangiums die Borsten aus. F bei *Oocystis asterifera* besitzen die Autosporen den typischen Chromatophor schon vor dem Austreten aus dem Sporangium. G die mit skulptierter Membran versehenen Autosporen von *Oocystis gigas* var. *incrassata*. (C, D nach Pascher, alle übrigen Abbildungen nach Skuja.)

2. Autosporen.

Diese nehmen die Gestalt der vegetativen Zellen an und es fehlt hier jedes Merkmal einer Monadenorganisation (ETTL 1961). Sie werden aber auf dieselbe Art wie die Zoosporen gebildet, wobei die Teilung im Sporangium simultan oder sukzedan vor sich geht. Daß die Autosporen eigentlich schon fertige kleine vegetative Zellen sind, kommt besonders dort zum Vorschein, wo die Membran skulptiert ist oder irgendwelche Fortsätze oder Stacheln trägt (z. B. *Bohlinia*, *Franceia*, *Coelastrella* u. ä. — Abb. 79). Sehr schön ist die Autosporenbildung an *Keriochlamys* zu verfolgen (Abb. 80). Die vegetativen Zellen haben eine dicke, mit eigenartiger wabenförmiger Struktur versehene Membran (PASCHER 1943a). Die Autosporen sind zuerst zart behäutet, nehmen aber noch innerhalb der Mutterhülle deren Gestalt und eigenartige Membranstruktur an. Unter gewissen Bedingungen können die Autosporen durch das Sporangium lange Zeit paketartig zusammengepreßt werden, dann unterbleibt die Wabenbildung an der Membran der Autosporen und auch die Gestalt entspricht nicht der normalen. Alles wird erst nach dem Freiwerden nachgeholt. Auch in anderen Fällen kommt es bei begrenztem Raum zur Verformung der jungen Sporen. So werden bei *Oocystis* sukzedan vier tetraëderförmige Tochterzellen gebildet (TSCHERMAK 1942) und bei *Jaagichlorella geometrica* entstehen (REISIGL 1964) gewöhnlich durch streng geometrische und winkelgenaue Teilung vier Tochterprodukte, die sich nach Auflösung der Sporangiumhülle abrunden; werden die tetraëdrischen Sporen aber weiter zerlegt, so kommt es durch die Aneinanderpressung zu einer Deformierung, welche die Teilprodukte in der Aufsicht wie Fünfecke aussehen läßt. Normalerweise nehmen jedoch die Autosporen, auch bei kolonienbildenden Formen, schon in der Mutterzelle die Morphologie des vegetativen Zustandes an.

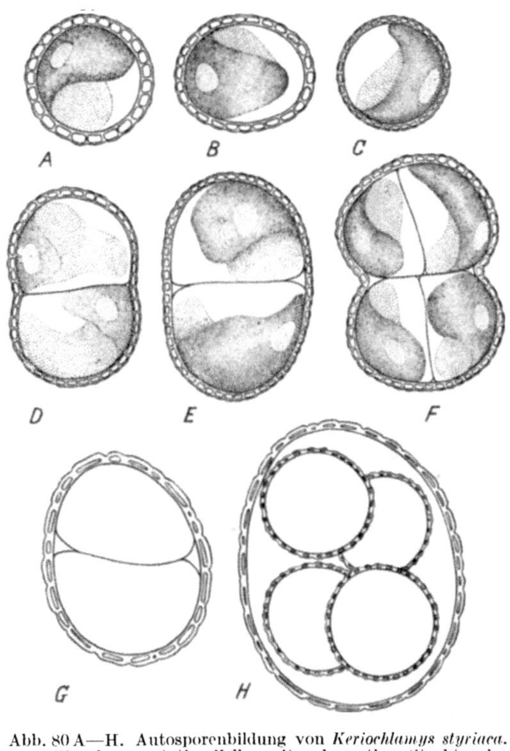

Abb. 80 A—H. Autosporenbildung von *Keriochlamys styriaca*. A—C Einzelne vegetative Zellen mit wabenartiger Struktur der Membran; D—F Teilungsstadien, bei F zwei Teilungen so rasch aufeinander folgend, daß sich die Teilprotoplasten erster Ordnung gar nicht vor der neuen Teilung behäuteten; G Teilungsstadium, wo die Mutterhülle gedehnt wird; H innerhalb der gedehnten Mutterhülle liegen die vier Autosporen vollständig behäutet mit Hülle von deutlicher Struktur. (Nach PASCHER.)

3. Autosporine Fadenbildung.

Als spezialisierte Autosporenbildung kann die Zellfadenentstehung der Ulothrichalen betrachtet werden. Obwohl dieser Vorgang vor allem dem Wachstum der Fäden dient, soll er hier kurz erwähnt werden, weil mit dem Wachstum der Fäden auch die Fragmentation oder die Bildung vegetativer Fortpflanzungsorgane verbunden ist. Nach PASCHER (1927) bleiben in diesem Fall die Autosporen durch die Membran der Mutterzelle zusammengehalten, ohne daß sie austreten, lagern sich in bestimmter Weise polar übereinander, um sich dann wieder zu teilen und wieder gleich orientierte Autosporen zu bilden. So entsteht ein System hintereinandergereihter, in die sukzedan gebildeten Membranen eingeschachtelter Zellen, der Zellfaden (Abb. 81). Die gewisse Selbständigkeit der Protoplasten in den Zellfäden demonstrieren *Protococcus*, *Ulothrix*, *Microspora* besonders gut. PASCHERS Ansichten finden in den Untersuchungen von GROSS (1931) und auch anderer

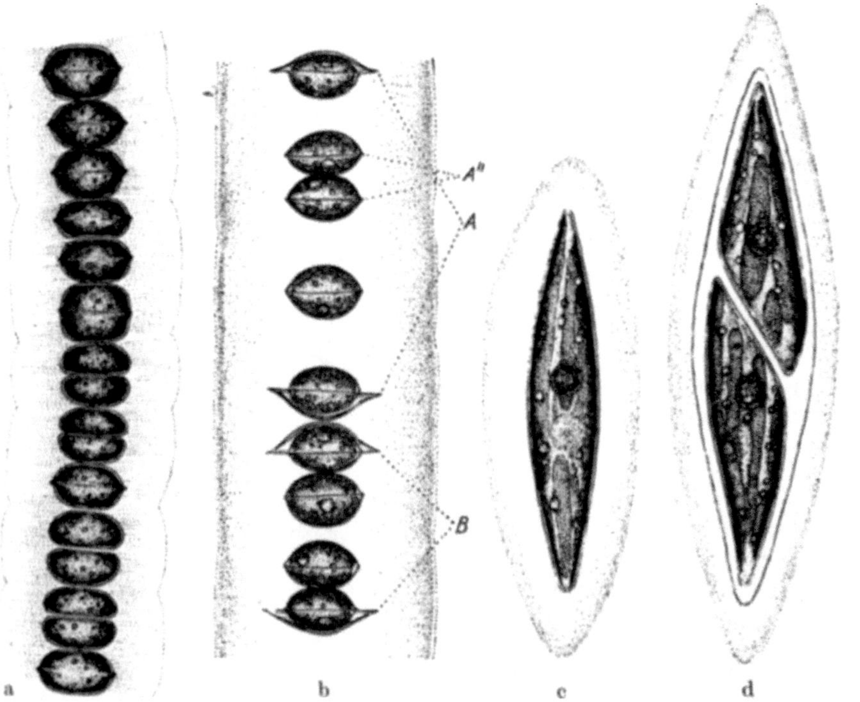

Abb. 81a—d. Modifizierte Autosporenbildung (Fadenbildung). a Vegetativer Faden von *Radiofilum mesomorphum*. Die Selbständigkeit der einzelnen Zellen tritt deutlich hervor; b das Wachstum des Fadens wird durch Autosporenbildung bewirkt, wobei sich die Autosporen in einer Reihe hintereinander anordnen. Die Buchstaben *A* und *B* zeigen die Reste der alten Mutterhüllen. c vegetative Zelle von *Elakatothrix viridis*; d zwei Autosporen von derselben Alge, die durch die alte Mutterhülle zusammengehalten werden. (Nach SKUJA.)

Autoren an Ulothrichalen, wo die Teilung des Cytoplasten sowohl beim ersten Teilungsschritt als auch bei den nachfolgenden als eine zentripetale Spaltung des Plasmas vor sich geht, eine wesentliche Stütze. Die Autosporenbildung ist ihrem Wesen nach immer noch erkennbar, sie hat aber den äußeren Grad der Rückbildung bis auf zwei Teilprodukte erreicht (SCHUSSNIG 1960).

Das Prinzip der Bildung ganzer neuer Membranen nach jeder Teilung wird jedoch bald von einem anderen Vorgang verdrängt. Nach der Plasmateilung übernimmt jede Tochterzelle die halbe Membran der Mutterzelle und bildet bloß die fehlende Hälfte neu. Das wird bei vielen Formen in verschiedenster Weise modifiziert. Im Extrem wird nur eine Querwand gebildet, was wahrscheinlich für die meisten höheren Algen kennzeichnend ist (GEITLER 1934). Fließende Übergänge unter diesen Vorgängen, bei denen die Tochterzellen eine gewisse Selbständigkeit behalten, erlauben keine strenge Absonderung der Autosporenbildung von der „vegetativen Teilung“ (PRINTZ 1927, HINDÁK 1962).

c) Andere Formen vegetativer Fortpflanzung.

Außer Zoo- und Autosporen (Gonidien) gibt es weitere Zellen und Organe verschiedener Herkunft, die der vegetativen Fortpflanzung dienen. Es handelt sich um ein- bis mehrzellige Gebilde, die in den verschiedensten Organisationsstufen der Chlorophyceen zur Ausprägung gelangten und teils nur gelegentlich, teils aber als ein für bestimmte Typen charakteristisches Merkmal auftreten. Häufig erfolgt diese Fortpflanzung durch *Fragmentation* einzelner Kolonien oder Fäden, die mehr oder weniger zufällig, aber auch durch innere und äußere Faktoren bestimmt wird. Ebenfalls häufig können *Dauerzustände* oder Dauerzellen der Fortpflanzung dienen, wenn eine Vermehrung des Inhaltes vorkommt. Hierher gehören auch die sog. Vermehrungsakineten (PRINTZ 1927), weiter Sprossen der *Bryopsidaceae* oder abgerissene Thallusteile der *Caulerpaceae*, die ein erstaunliches Regenerationsvermögen besitzen und als Fortpflanzungsorgane im weitesten Sinne aufgefaßt werden können.

1. Dauerzustände (Palmellen und Gloeocysten).

Ein Großteil aller Chlorophyceen verschiedener Organisationsstufen ist imstande, gallertumhüllte Zustände zu liefern (PASCHER 1927, FRITSCH 1935, SMITH 1950). Sie werden sowohl bei den monadoiden *Volvocales* als auch bei den coccalen

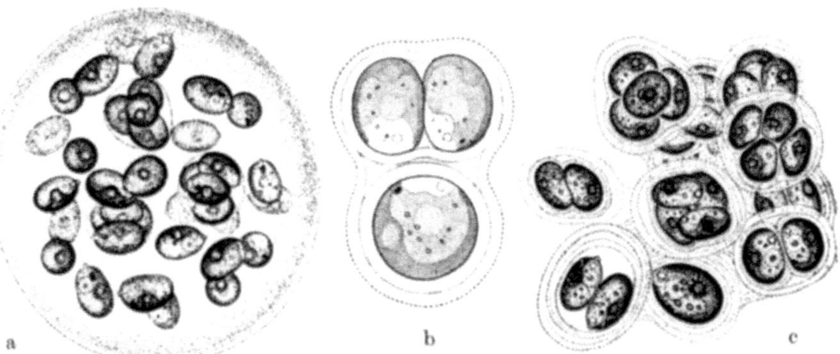

Abb. 82a—c. Dauerzustände bei Chlorophyceen. a Pallmelloides Gallertlager von *Chlamydomonas quiescens*; b Gloeocyste von *Kremastochloris conus*, dadurch entstanden, daß sich abgelöste Zellen abrunden, mit einer Gallerthülle umgeben und sich innerhalb dieser Gallerthülle weiterteilen, wobei jede Tochterzelle wieder neue Gallertschichten bildet; c Gloeocyste von *Chlamydomonas gloeopara*. (a, c nach SKUJA, b nach PASCHER.)

und trichalen Formen gebildet, indem durch Teilungen nach zwei oder drei Richtungen in mehr oder weniger dicke Gallerte eingebettete, einzelne Zellen oder Zellhaufen entstehen. Solche gallertartige Dauerzustände sind widerstandsfähiger als die gewöhnlichen vegetativen Zellen und mit ihrer Hilfe können die Algen ungünstige Bedingungen überstehen. Sehr häufig werden sie auch durch Kulturbedingungen, vor allem auf festen Nährböden, hervorgerufen. Die Gallertzustände können unter Normalbedingungen oder durch bisher unbekannte Faktoren wieder in die vegetative Form übergehen, was meist über Zoosporen verläuft. Es sei betont, daß innerhalb der Gallerte oft eine reichliche Vermehrung der Individuen erfolgt. Gallertstadien, bei denen die Gallerte deutlich geschichtet ist, werden als *Gloeocysten* (Abb. 82b, c) bezeichnet, wogegen solche ohne Schichtung *Palmellen* (Abb. 82a) genannt werden.

2. Dauerzellen (Aplanosporen und Akineten).

Nicht selten kommt es bei den Chlorophyceen vor, daß einzelne Zellen sich mit dicker Membran umgeben, unter gleichzeitiger reichlicher Speicherung von Reservestoffen im Zellinhalt. Dauerzellen treten sowohl bei monoenergiden als auch bei polyenergiden Algen in Erscheinung. Diese können sich alsbald weiterentwickeln oder auch in ein mehr oder weniger langes Ruhestadium eintreten. Sie sind Umwandlungsprodukte normaler vegetativer Zellen und entstehen meist durch ungünstige Außenbedingungen. Die Dauerzellen können sich in zweierlei Weise ausbilden. Bei nackten Flagellatenzellen (vor allem Zoosporen) rundet sich der Protoplast unter Aufgabe der Geißeln, der pulsierenden Vacuolen und des Stigmas ab und umgibt sich mit einer derben Membran. Falls die Zelle schon behäutet ist, tritt entweder der Inhalt vorher als nackte Zoospore oder Hemiauto-

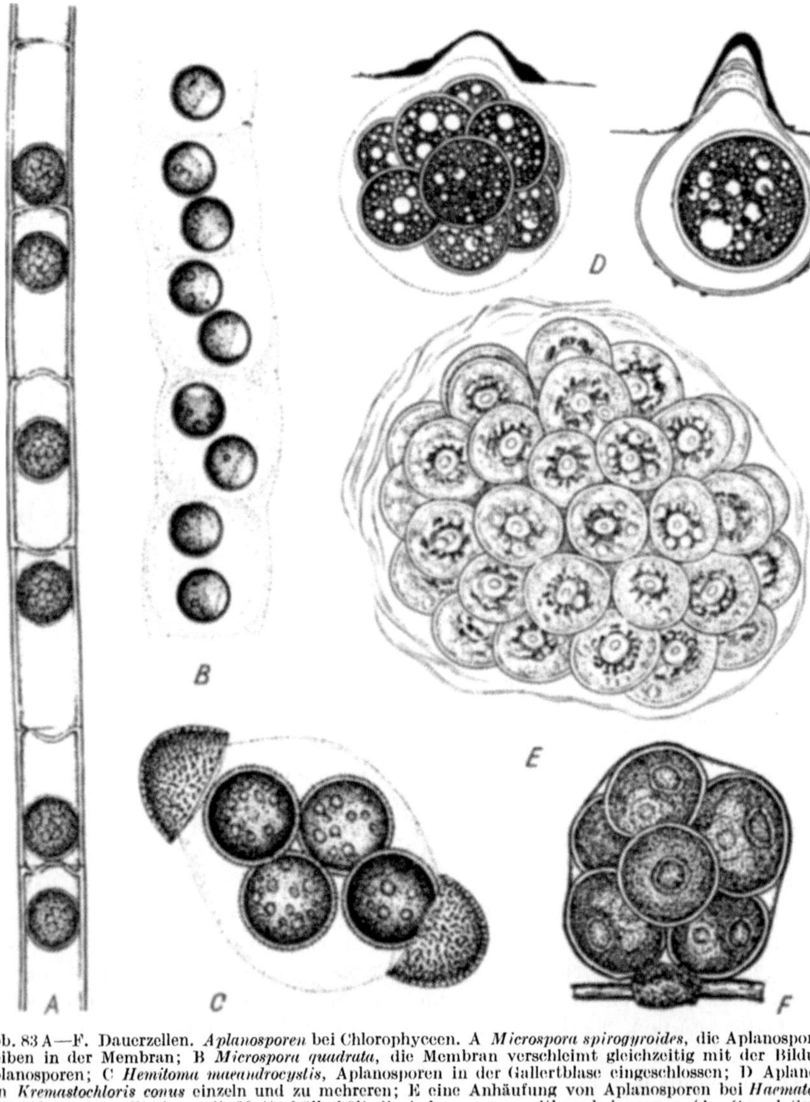

Abb. 83 A—F. Dauerzellen. *Aplanosporen* bei Chlorophyceen. A *Microspora spirogyroides*, die Aplanosporen verbleiben in der Membran; B *Microspora quadrata*, die Membran verschleimt gleichzeitig mit der Bildung der Aplanosporen; C *Hemitoma maeandrocystis*, Aplanosporen in der Gallertblase eingeschlossen; D Aplanosporen von *Kremastochloris conus* einzeln und zu mehreren; E eine Anhäufung von Aplanosporen bei *Haematococcus*; F *Chlorangium ellipticum*, die Mutterhülle hält die Aplanosporen zeitlang beisammen. (A—C nach Skuja, D nach Pascher, E nach Wollenweber aus Pascher, F nach Korschikoff.)

spore aus, um sich dann in die Dauerzelle umzuwandeln, oder die vegetative Zelle wandelt sich direkt um, wobei der Protoplast von der Membran abgehoben wird und eine neue dicke Membran bildet (Abb. 83). Die alte Mutterhülle nimmt in diesen Fällen an der Bildung der Dauerzelle *nicht* teil. Diese Art von Dauerzellen wird als *Aplanospore* bezeichnet (Pascher 1927). Nach einer Ruhezeit keimt sie unter Aufquellen direkt oder nach Teilung ihres Inhaltes in Zoosporen aus. Eine andere Art von Dauerzellen sind *Akineten*. Sie werden nur bei behäuteten Zellen gebildet, und zwar in der Weise, daß die Membran der Mutterzelle mit für die Bildung der Dauerzellwand verwendet wird (Abb. 85).

Bei *Haematococcus* sind die *Aplanosporen* Dauerzustände und Fortpflanzungsgebilde zugleich. Sie können dadurch keimen, daß Zoosporen austreten oder auch, daß sich ihr Inhalt zwei- oder mehrmals teilt, wobei die Teilprodukte wieder zu neuen Aplanosporen werden (PASCHER 1927). Bei Wiederholung entstehen ganze Haufen von Aplanosporen (Abb. 83 E). Nach PASCHER trifft das zu, wenn die Bedingungen zwar für die Fortpflanzung ausreichen, der normalen Erhaltung beweglicher Stadien jedoch nicht günstig sind. *Akineten* treten vor allem bei solchen Volvocalen auf, die sich in kurzen Zeiträumen abwechselnd encystieren und wieder beweglich werden. Viele Kenntnisse über Dauerzellen und bemerkenswerte, ihrer Bildung vorangehende Teilungen bei Chlamydomonaden haben Untersuchungen SKUJAS (1948, 1956) gebracht. Akineten besonderer Art werden bei *Chlamydomonas rapa* gebildet (Abb. 84), bei welcher die vegetativen Zellen direkt in Dauerzellen übergehen, aber unter ungewöhnlichem Größenwachstum, so daß der Durchmesser der Akineten die Länge der vegetativen Zellen fast um das Zweifache übertrifft (ETTL 1965 c). Sehr aufklärend sind die von CONRAD (1950) gegebenen schematischen Darstellungen der Unterschiede zwischen Akineten und Aplanosporen bei *Coccomonas*. Bei Fadenalgen werden Dauerzellen entweder zu mehreren oder einzeln aus jeder Zelle gebildet. Oft kommt es vor, daß Aplanosporen noch innerhalb des Fadens, in der Mutterhülle, durch direkte Umwandlung der Mikro- oder Makrozoosporen erzeugt werden (PASCHER 1906, PRINTZ 1927). Bei differenzierten Fadenalgen ist die Produktion von Dauerzellen genauso wie die Zoosporenbildung an die reproduktiven Zellen der Seitenäste gebunden. *Fritschiella* bildet entsprechend ihre Akineten in der pseudoparenchymatischen Sohle, die sich auflöst, nachdem die Zellmembranen verdickt wurden (RANDHAWA 1939).

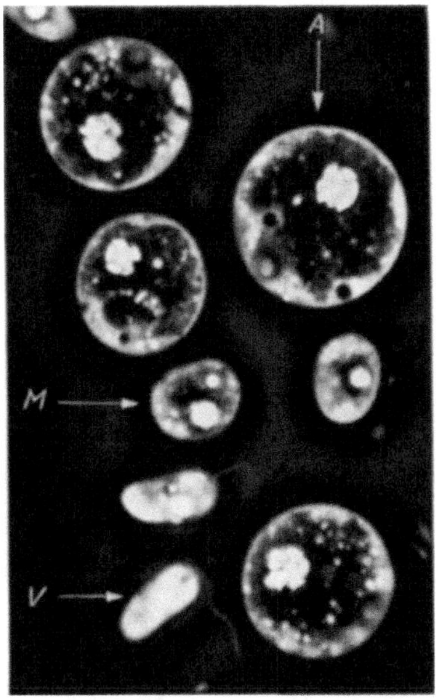

Abb. 84. Akinetenbildung bei *Chlamydomonas rapa*. *V* Vegetative Zelle; *M* Anfangsstadium der Akinetenbildung; *A* Akinete 1500 × vergrößert. (Nach ETTL.)

Die einfachsten Formen von *Vermehrungsakineten* findet man bei *Hormidium* und gewissen *Ulothrix*-Arten, wo einzelne Zellen sich abrunden und vom Mutterfaden ablösen. Der Vorgang kann mitunter von einer Umwandlung des Fadens in mächtige Gallerte begleitet sein (PRINTZ 1927). Eine besondere Art von Akinetenbildung kommt bei *Acrosiphonia*, *Spongomorpha* und *Chaetomorpha* vor. Die Rhizoiden bilden manchmal pseudoparenchymatische Gewebe, in deren Zellen sich Reservestoffe anhäufen, und wenn die übrigen Teile des Thallus absterben, bleiben diese Rhizoiden lebendig und wachsen später in neue Pflanzen aus. Bei *Pithophora* schwellen die zur Akinetenbildung bestimmten terminalen oder interkalaren Zellen an ihren oberen Enden an (Abb. 85 A). Der größere Teil des Inhaltes sammelt sich hier und wird durch eine Wand abgegrenzt, worauf das Gebilde von einer dicken Membran umgeben wird. Dies scheint die einzige Fortpflanzungsart dieser Alge zu sein. Durch Zerfall des cönoblastischen Inhaltes in rundliche Teilstücke und Umhüllung derselben mit einer derben, oft mehrschichtigen Membran entstehen vielkernige reservestoffreiche Dauerzellen in den Schläuchen und Rhizoiden von *Protosiphon* und *Bryopsis* (SCHUSSNIG 1960).

3. Fragmentation.

Eine einfache Art vegetativer Fortpflanzung erfolgt bei mehrzelligen Chlorophyceen durch Fragmentation des Thallus in Einzelzellen oder Zellkomplexe. Oft geschieht das ohne sichtbare Änderung in der Beschaffenheit der vegetativen Zellen (Heering 1914, Printz 1927, Iyengar 1951). Die Trennung kommt durch Auflösung der verbindenden Zellwand zustande. Der Zerfall von *Hormidium*-

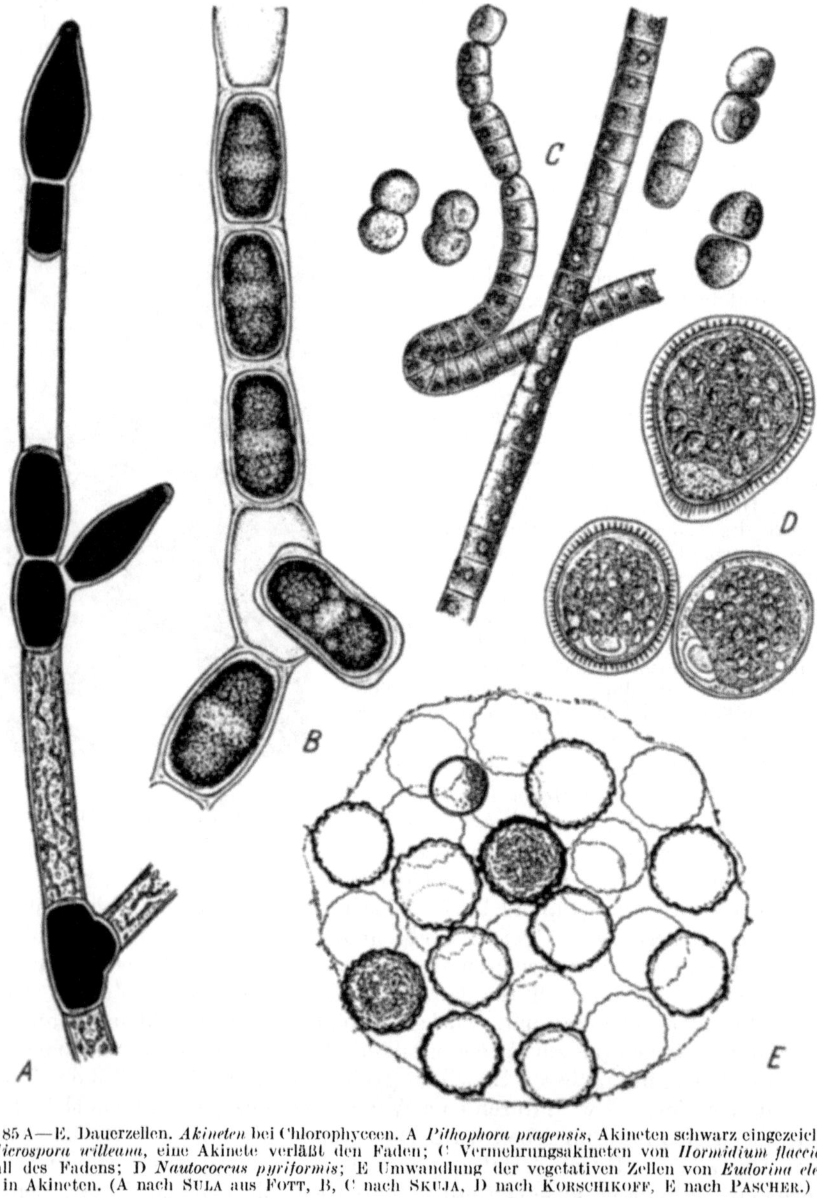

Abb. 85 A—E. Dauerzellen. *Akineten* bei Chlorophyceen. A *Pithophora pragensis*, Akineten schwarz eingezeichnet; B *Microspora willeana*, eine Akinete verläßt den Faden; C Vermehrungsakineten von *Hormidium flaccidum*, Zerfall des Fadens; D *Nautococcus pyriformis*; E Umwandlung der vegetativen Zellen von *Eudorina elegans* in Akineten. (A nach Sula aus Fott, B, C nach Skuja, D nach Korschikoff, E nach Pascher.)

Fäden in einzelne Zellen ist altbekannt (Abb. 85C). Bei *Stichococcus* ist diese Fortpflanzung der einzig beobachtete Vermehrungsmodus. Auch bei den Prasiolaceen ist die Fragmentation des Thallus die meist verbreitete Fortpflanzungsart (Fritsch 1935). Einige *Ulva*-Arten können aus den sekundären Rhizoiden durch Sprossung neue Tochterpflanzen bilden. Bei den Blastosporaceen geschieht die Fortpflanzung im Großteil durch losgelöste Thallusstücke, die mehr oder weniger regelmäßig abgetrennt werden, ähnlich auch bei Valoniaceen. Die Seitentriebe der Bryopsidaceen können sich von der Stammzelle lösen, im Wasser umhertreiben und, wenn sie an ein Substrat gelangen, zu neuen Individuen direkt heranwachsen. Typische Brutkörper kommen bei einigen *Codium*-Arten vor, die ähnlich wie Gametangien an den Blasen entstehen und durch Membranpfropfen von diesen getrennt werden (PRINTZ 1927, FRITSCH 1935).

PASCHER (1944) berichtet über conidienartige Sporen bei gewissen Trentepohliaceen. Am weitesten fortgeschritten erscheint darin die Gattung *Pirula* zu sein (Abb. 86). Es handelt sich um kurze, mehr- bis vielzellige Fäden, deren Endzellen sich kopfartig vergrößern und ablösen können. Die Fadenzellen wachsen aber auch seitlich schief aus und bilden durch Teilungen kleine Äste, deren Zellen sich voneinander ablösen und conidienartige Ketten bilden. Leider ist diese Alge wenig untersucht.

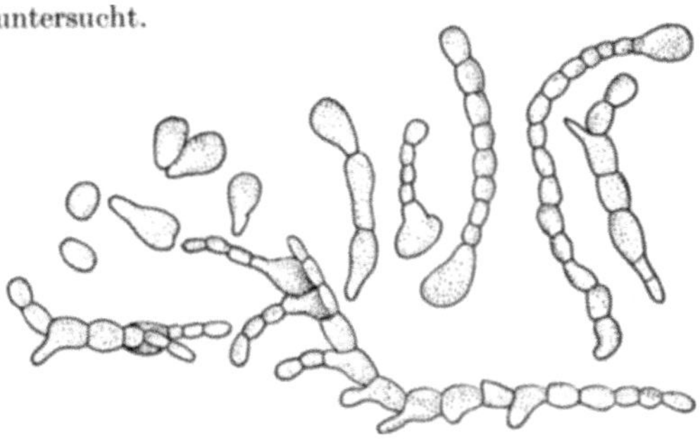

Abb. 86. Conidienartige Sporen bei *Pirula*. Links solche abgefallene Sporen. Unten ein vielzelliger Faden, dessen Zellen seitlich auswachsen. Aus diesen seitlich auswachsenden Zellen entwickeln sich Ketten von Zellen, die sich ablösen und zu Conidiensporen werden. Links unten ein kurzer, verzweigter Faden, dessen eine Zelle im Begriffe ist auszuwachsen, während die beiden rechten Zellen bereits seitliche Conidienketten gebildet haben. (Nach SNOW aus PASCHER.)

d) Auslösung von Fortpflanzungszellen.

Die Darlegung der ungeschlechtlichen Fortpflanzung wäre unvollkommen, wenn nicht auch die Bedingungen, unter denen diese Prozesse eintreten oder gefördert werden, erwähnt würden.

Von den Außenfaktoren, welche die Auslösung der Fortpflanzung bewirken oder fördern, seien vor allem folgende erwähnt: Licht-Dunkel-Wechsel, Temperaturänderung, Konzentration und Zusammensetzung der Nährlösung, Acidität, Wasserströmung und Benetzung. Nur selten wirkt ein einziger Faktor allein, meist kommt es zum Einfluß mehrerer gemeinsam. Natürlich hängt die Neigung zur Fortpflanzung auch von inneren physiologischen Faktoren ab. Die endogen bewirkten Zustände sind also ebenso wichtig. Besonders schön wurde das von MAINX (1931) an *Oedogonium* dargelegt. Maßgebend für die Wirksamkeit der die Zoosporenbildung auslösenden Außenfaktoren sind der Genotypus und der innere physiologische Zustand der Alge. Bei den stark zur Zoosporenbildung neigenden Arten kann man in allen Stadien durch Außenfaktoren Zoosporen auslösen. Die weniger dazu neigenden Arten lassen sich dagegen in gewissen physiologischen Zuständen überhaupt nicht zur Zoosporenbildung zwingen. Ist aber der innere physiologische Zustand für die Zoosporenbildung günstig, so wird durch bestimmte Außenbedingungen bei allen Arten reichliche Zoosporenbildung ausgelöst. Ein günstiger Zustand herrscht bei *Oedogonium* vor allem bei alten, mit Stärke überfüllten Zellen. Ungünstig ist im allgemeinen der Zustand gut wachsender und gesunder Zellen mit wenig Stärke und lebhaft gefärbtem Chromatophor.

1. Licht-Dunkel-Wechsel.

Als einer der wichtigsten Faktoren der Auslösung vegetativer Fortpflanzung kann wohl Licht-Dunkel-Wechsel angesehen werden. Die Gonidienbildung erfolgt meist in der Nacht (Dunkelperiode), wobei die Entleerung bei Einbruch des Tageslichtes (oder bei Einsetzen der Lichtperiode) stattfindet. Es ist allgemein bekannt, daß durch Aufbewahren der Algen im Dunkeln die Entleerung der Gonidien verzögert wird. Singh (1942) beobachtete bei *Draparnaldiopsis*, daß die Zoosporen früh am Morgen um 6.00 h entstehen und der Prozeß der Zoosporenbildung schon um 9.30 h aufhört. Wenn die Kulturen im Dunkeln gehalten wurden, verzögerte sich die Zoosporenentleerung um 2—3 Std. Auch Geitler (1960) berichtet über ähnliche Entleerung der Zoosporen bei *Coleochaete*, die in den Morgenstunden erfolgt, nachdem die Reifung der Sporangien in den späten Nachtstunden abgelaufen ist. Auslösend auf die Entleerung wirkt hier das Licht, denn durch Verdunkelung läßt sich der Austritt sogar um mehrere Stunden verschieben und durch Belichtung zu einem beliebigen Zeitpunkt auslösen. Das gilt allgemein auch für andere Organisationstypen der Chlorophyceen. Reich (1926) stellte bei einer *Stigeoclonium*-Art fest, daß keine Zoosporenbildung erfolgte, wenn die Algen ununterbrochen beleuchtet wurden.

Vorläufige Versuche an *Bulbochaete hiloensis* zeigten, daß die Entscheidung zwischen Gameten- und Zoosporenbildung vom Licht-Dunkel-Rhythmus abhängig ist. Bei einer Lichtintensität von 1100, 2200 und 5500 Lux und einem Rhythmus von 16:8 Licht-Dunkel wird die Bildung der Androgameten gefördert und beschleunigt. Wogegen bei derselben Lichtintensität, aber einem Rhythmus von 12:12 die Gametenbildung ausfällt und die Zoosporenbildung einsetzt (Cook 1962). Die Zoosporenbildung von *Stigeoclonium falklandicum* wird durch hohe Lichtintensität (2000 Lux) gehemmt, besonders bei langdauernder Einwirkungszeit (Hustede 1957). Die Sporulationshemmung ist durch einen Hemmstoff bedingt, der besonders bei hoher Lichtintensität gebildet wird. Durch die Überführung der Alge in frische Nährlösung soll sich die Sporulationshemmung beseitigen lassen.

Die Fortpflanzung erfolgt in Kulturen vielfach rhythmisch, wenn die untersuchten Algen durch Beleuchtungswechsel gesteuert werden. Die Einflüsse eines Licht-Dunkel-Wechsels auf die Fortpflanzung, vor allem auf die Zellteilung einzelliger Formen, aber auch auf die Gonidienbildung mehrzelliger Komplexe, werden zur Erzielung komplett synchroner Kulturen verwendet. Es wurden dabei mehrere Methoden ausgearbeitet, um durch künstlich hervorgerufenen Licht-Dunkel-Wechsel die Fortpflanzung an jedem Tag nur zu einem ganz bestimmten Zeitpunkt einzuleiten. Einer fortpflanzungsfreien Zeit folgt die schubartige Freisetzung der Fortpflanzungszellen (Einzelheiten darüber s. in Hase u.a. 1957, 1958, Pirson 1962, Pirson und Lorenzen 1958, Lorenzen 1957, Soeder 1960 u.a.). Die Algen einer Kultur in ihrer Entwicklung und Fortpflanzung zu synchronisieren, ermöglicht eine exakte Untersuchung physiologischer Vorgänge und auch der Fortpflanzung selbst (vor allem wurden *Chlorella* und *Chlamydomonas*-Arten bearbeitet). Das hat den Vorteil, daß Millionen von Individuen gleichen Entwicklungsalters zur Verfügung stehen. Eine Synchronisation durch Licht-Dunkel-Wechsel ist bei den Chlorophyceen mit relativ großer Schärfe zu erhalten. Tagesrhythmisch geregelte Entleerung der Zoosporen bei *Oedogonium* erhielt Bühnemann (1955a) ebenfalls durch Beleuchtungswechsel, doch reagieren die einzelnen Fadenzellen dabei unabhängig voneinander, jede Zoospore wird aus einer hinsichtlich der Sporenbildung als Einzelindividuum anzusprechenden Fadenzelle gebildet. Die Sporenbildung muß in jeder Fadenzelle einzeln induziert werden. Dabei hat das Alter der Kultur keinen Einfluß auf die Rhythmik der Sporenbildung als solche und in weitem Temperaturbereich beeinflussen nur die Beleuchtungsverhältnisse das Sporulationsverhalten der Algen. Die Sporulationsrhythmik wird durch eine „endogene Tagesrhythmik der sporogenen Stimmung" in den Einzelzellen verursacht (Bühnemann 1955a).

2. Temperaturänderung.

Erwärmung des Milieus fördert die Fortpflanzung, doch darf die Temperatur die optimale Grenze für die gegebene Art nicht überschreiten. Manchmal wirken auch niedrige Temperaturen auslösend (Kälteschock). Wirksam ist jedoch allgemein eine plötzliche Temperaturänderung (WINKLER 1934). Gefördert wird sowohl die Zoosporen- als auch die Autosporenbildung. Die Zoosporen von *Protosiphon* werden bei einer Temperatur von 4—6° C in 24—48 Std gebildet, wogegen bei optimaler Temperatur von 23—26° C diese schon binnen 2,5—3 Std entstehen (KLEBS, OLTMANNS). Periodischer zwölfstündiger Temperaturwechsel mit einer Temperaturdifferenz von 10° C (15°:25°) und 2,5° C (25°—27,5°) wirken ähnlich wie Lichtwechsel steuernd auf die in ihnen ablaufende Sporenbildung bei *Oedogonium* (BÜHNEMANN 1955c). Die Sporulationsmaxima liegen zu Beginn der Phasen höherer Temperatur. Temperaturwechsel führt ähnlich wie Licht-Dunkel-Wechsel zur Synchronisation von Kulturen (SOEDER 1960). Die Temperatur beeinträchtigt auch die Schwärmzeit der Zoosporen. So bewegen sich Zoosporen bei einer Temperatur von 3—4° C bis 48 Std, wogegen dieselben bei einer Temperatur von 12—15° C nur 7—9 Std umherschwärmen.

3. Konzentration und Zusammensetzung der Nährlösung.

Änderungen in der Konzentration und Zusammensetzung der Nährlösungen, in denen die betreffenden Chlorophyceen kultiviert werden, führen oft zur Auslösung der Fortpflanzung. So setzt manchmal intensive Fortpflanzung ein, wenn die Algen aus alter Nährlösung in ein frisches Medium oder in gewöhnliches Leitungswasser bzw. in destilliertes Wasser gebracht werden. Ein andermal genügt eine Änderung der Konzentration derselben Nährlösung. COOK (1962) und MAINX (1931) fanden bei *Bulbochaete* und *Oedogonium*, daß Zoosporen in Kulturen immer dann gebildet wurden, wenn die Algenfäden in frische Nährlösung gebracht wurden. Besonders erfolgt das bei alten, mit Stärke überfüllten Fäden. Durch Zusatz von frischer Nährlösung oder Wasser zu nicht zu alten Kulturen wandeln sich viele Zellen von *Pleurastrum* über Nacht zu Zoosporangien um (VISCHER 1933). Bei *Dictyococcus minor* können nach Überimpfung der Zellen von frischen Agarplatten in verdünnte Erddekokt-Lösung bereits in den ersten Tagen reichliche Zoosporen beobachtet werden (PETROVA 1931). Bei *Stigeoclonium* entwickeln sich zahlreiche Zoosporen in Nährlösungen geringerer Konzentration (z. B. Benecke-Lösung 0,01—0,005%). Ähnlich, wenn auch schwächer, wirkt gewöhnliches Leitungswasser. Als wirksamer Faktor muß in diesen Fällen die Veränderung des betreffenden Zustandes als solchem angesehen werden, da weder reines Wasser noch konzentrierte Nährlösung auf die Dauer die Fortpflanzung auslösen können.

Änderungen in der Zusammensetzung der Nährlösungen oder das Vorhandensein verschiedener organischer Verbindungen können den Vorgang der Fortpflanzung intensiv fördern, aber auch umgekehrt hemmen. TRAINOR (1963) gelang es, bei *Scenedesmus obliquus* Zoosporen zu bekommen, wenn die Zellen in einem Medium *ohne Stickstoff* gezüchtet wurden. Er benützte BEIJERINCKs Basalmedium ohne Ammoniumnitrat. In 3 Tage alten Kulturen kamen die Zoosporen zum Vorschein. Wogegen das Fehlen von $FeCl_3$, $MgSO_4$ oder K_2HPO_4 die Zoosporenbildung nicht stimulieren konnte. Autosporen wurden jedoch weiterhin gebildet. REISIGL (1964) beobachtete an einigen alpinen Bodenalgen (vor allem an *Rhopalocystis cucumis*), daß die Zoosporenbildung durch Glucosezusatz gefördert wird; Ähnliches VISCHER (1933) an *Stigeoclonium helveticum* (Optimum bei 1%). Auch Zusatz von Rohrzucker kann bei gewissen Arten die Fortpflanzung fördern. Bemerkenswert ist noch die Feststellung von TSCHERMAK-WOESS (1951), daß die als Flechtensymbiont vorkommende Alge *Myrmecia reticulata* Zoosporen nur dann bildet, wenn sie von der Einwirkung des Pilzes befreit ist. Offenbar werden den Sporangiumzellen durch den Pilz die zum Aufbau von Zoosporen nötigen Substanzen und Reservestoffe entzogen. In Symbiose mit dem Pilz werden nur Autosporen gebildet.

Die fördernde Wirkung der β-Indolyl-Essigsäure auf die Zoosporenbildung von Fadenalgen ist erwiesen (HUSTEDE 1957). Den Einfluß verschiedener physiologisch bedeutsamer Stoffe (β-Indolyl-Essigsäure, ATP, Riboflavin) auf die rhythmische Sporenbildung von *Oedogonium* hat BÜHNEMANN (1955b) untersucht. Die benutzten Substanzen waren hier ohne Einfluß auf die Sporulationsrhythmik, sie veränderten jedoch die Sporulationsintensität.

4. Acidität.

Es besteht kein Zweifel, daß die H-Ionenkonzentration die Fortpflanzung beeinflussen kann (vgl. Arbeiten CZURDAs und ULEHLAs). Nach ULEHLA soll sie einer der wichtigsten der Faktoren sein, welche als Folgeerscheinung anderer Änderungen die Fortpflanzung hervorrufen. Ein Übergang des pH von 7,8 auf 7,2 soll *Cladophora* zur Zoosporenbildung zwingen (WINKLER 1934). Bei *Oedogonium* wirkt die Verschiebung der H-Ionenkonzentration als auslösender Faktor (MAINX 1931). Die Makrozoosporen von *Draparnaldiopsis* sollen bei niedrigeren pH-Werten entstehen als die Mikrozoosporen. Bei Planktonformen hat eine Änderung der Acidität manchmal eine Massenproduktion bestimmter Arten zur Folge oder kann umgekehrt zur Hemmung führen.

5. Wasserströmung und Benetzung.

Viele Chlorophyceen, vor allem trichale Formen, die in rasch fließenden Gewässern gedeihen (Arten von *Ulothrix*, *Stigeoclonium*, *Draparnaldia*), bilden Zoosporen, wenn sie aus strömendem ins stehende Wasser gebracht werden. KLEBS war der Ansicht, daß das sauerstoffreiche strömende Wasser das Wachstum der Fäden fördert, wogegen beim Übergang ins stehende sauerstoffärmere Wasser die Zoosporenbildung einsetzt. Und das auch dann, wenn eine Temperaturänderung oder Änderung der Nährstoffe vermieden wird. Meist wird jedoch dieser Faktor mit Temperaturänderung kombiniert, vor allem beim Einsammeln der Algen aus kälteren strömenden Gewässern. Die Zoosporenbildung unter solchen Umständen kann manchmal aber ausfallen (SMITH 1950). Bei gewissen *Stigeoclonium*-Arten konnten keine Zoosporen erhalten werden, wenn sie aus stark strömenden Quellen in ein Sammelgefäß gebracht wurden.

Algen feuchter Stellen oder aerophile Formen (z.B. *Trentepohlia*) werden zur starken Zoosporenbildung angeregt, wenn diese Algen in liquides Wasser kommen. Ebenso bilden *Protosiphon* und *Hormidium* reichlich Zoosporen, wenn sie mit Wasser übergossen werden. Das gilt allgemein auch bei solchen Chlorophyceen, die in Reinkulturen auf festen Nährböden kultiviert werden. Die Übertragung des Algenmaterials von Agarplatten in flüssige Nährlösungen oder in gewöhnliches Leitungswasser stimuliert häufig die vegetative Fortpflanzung, vor allem die Zoosporenbildung. So konnte STARR (1954a) die Bildung von Zoosporen bei *Planktosphaeria gelatinosa* dadurch bewirken, daß er die Zellen von Agar ins destillierte Wasser überführte. Schon 12—24 Std später erfolgte die Zoosporenbildung, obzwar sonst bei dieser Art nur Autosporen entstehen. Bei *Planophila sphagnothermalis* werden Autosporen gebildet, wenn die Zellen an feuchten Moorstichen (aerophil) wachsen. Wenn genügend Feuchtigkeit vorhanden ist, treten statt Autosporen Zoosporen aus (PASCHER 1939a). Durch die Überführung von festen Nährböden in eine flüssige Nährlösung oder ins Wasser werden natürlich auch die Autosporen in ihrer Bildung gefördert. Entscheidend dabei scheint der plötzliche Übergang zu sein.

e) Anhang: Pedinomonadales.

Als Anhang zur Fortpflanzung der Chlorophyceen sei noch die der *Pedinomonadales* geschildert. Diese weichen durch ihren spezifischen Zellbau und modifizierte Fortpflanzung vom üblichen Chlorophyceen-Typus ab. CHRISTENSEN (1962) hat sie auf Grund dessen sogar als zwei selbständige Klassen — die *Loxophyceae* und *Prasinophyceae* — abgesondert. Es ist ganz sicher, daß diese Organismen, die im Großteil nur in der Monadenorganisation auftreten, von den *Volvocales* sensu stricto getrennt werden müssen. Auf der anderen Seite kann man jedoch die Zugehörigkeit zu den Chlorophyceen nicht abstreiten (ETTL 1965b). Schon das Vorhandensein beider Chlorophylle (*a* und *b*) und echter Stärke, ebenso wie die innerhalb des Chloroplasten eingebetteten Pyrenoide weisen darauf hin. Nach bisherigen Kenntnissen des Zellbaues und der Feinstruktur (vgl. MANTON und ETTL 1965, MANTON, RAYNS, ETTL und PARKE 1965, ETTL und MANTON 1964) werden zwei Unterordnungen geführt (ETTL 1965b), die auch in ihrem Teilungsmodus verschieden sind.

Pedinomonadineae sind durch den asymmetrischen Zellbau charakterisiert, der nicht nur an den vegetativen Zellen, sondern auch während der Fortpflanzung deutlich hervortritt. Im Prinzip handelt es sich immer um eine modifizierte und durch die Asymmetrie des Organismus beeinflußte Teilung, die eingehend an *Pedinomonas* (ETTL und MANTON 1964) und *Heteromastix* (ETTL 1967) untersucht wurde. Aplanosporen und Gallertstadien kommen seltener vor.

Die Teilung bei *Pedinomonas* beginnt an der Stelle, wo der Augenfleck liegt (Abb. 87). Dort wird die Zelle vorerst leicht, dann aber tief eingebuchtet. Man bemerkt zugleich eine leichte Dehnung der Zelle, wobei die Mitose verläuft. Danach geht die Teilung rasch weiter. Die Einbuchtung wird immer tiefer, und die Zelle wird von der Seite, wo der Augenfleck liegt, eingeschnürt. Die Einschnürung erfolgt einseitig und die ganze Zelle wird um den Augenfleck bogig gekrümmt. Danach werden die neue Geißel und pulsierende Vacuole de novo gebildet. Während des weiteren Verlaufes teilen sich erst Pyrenoid und Chloroplast, danach dann der Augenfleck und der Rest der Zelle. Der Vorgang verläuft ziemlich rasch binnen 15—20 min. Auffallend ist die Richtung der Teilung. Wird die Zellpolarität auf die Geißelinsertion bezogen, so kann man von einer schräg verlaufenden Schizotomie mit einseitiger Durchschnürung sprechen.

Bei *Heteromastix* (Abb. 88a—c) wird hingegen die Cytokinese durch die Teilung des Pyrenoids eingeleitet (MANTON, RAYNS, ETTL und PARKE 1965, ETTL 1966), welche mit dem Proteinkörper beginnt, während die gemeinsame Stärkeschale noch eine Zeitlang erhalten bleibt. Erst nach der Trennung der Pyrenoide voneinander findet auch die Einschnürung des Chloroplasten statt. Man kann zugleich auch eine leichte Einschnürung des Protoplasten beobachten, die auch hier einseitig, und zwar von der basalen Seite aus erfolgt. Nach beendeter Teilung des Chloroplasten entsteht der zweite Augenfleck de novo auf der entgegengesetzten Seite. Die ganze Zelle ist in diesem Stadium der Breite nach etwas gestreckt und ihre einseitige Einschnürung schreitet weiter fort. Von den Geißeln der Mutterzelle erhält der eine Tochterflagellat die längere, der andere die kürzere. Die jeweils fehlende Geißel wird neu gebildet, indem sich recht spät im Teilungsprozeß zwei Stummel zwischen den alten Geißeln und von deren Basen jeweils ausgehend vorschieben und dann rasch die endgültige Länge erreichen. Ähnliches Verhalten während der Teilung haben PARKE und RAYNS (1964) bei *Nephroselmis gilva* beobachtet.

Bei den *Pyramimonadineae* muß man die trichter- oder muldenförmige Vertiefung am Vorderende der Zellen berücksichtigen, durch die sie von den typischen

Chlorophyceen-Flagellaten abweichen (CHADEFAUD 1941, 1947, 1960). Dieser spezialisierte Bau spiegelt sich auch in der Zellteilung, vor allem was die Auf-

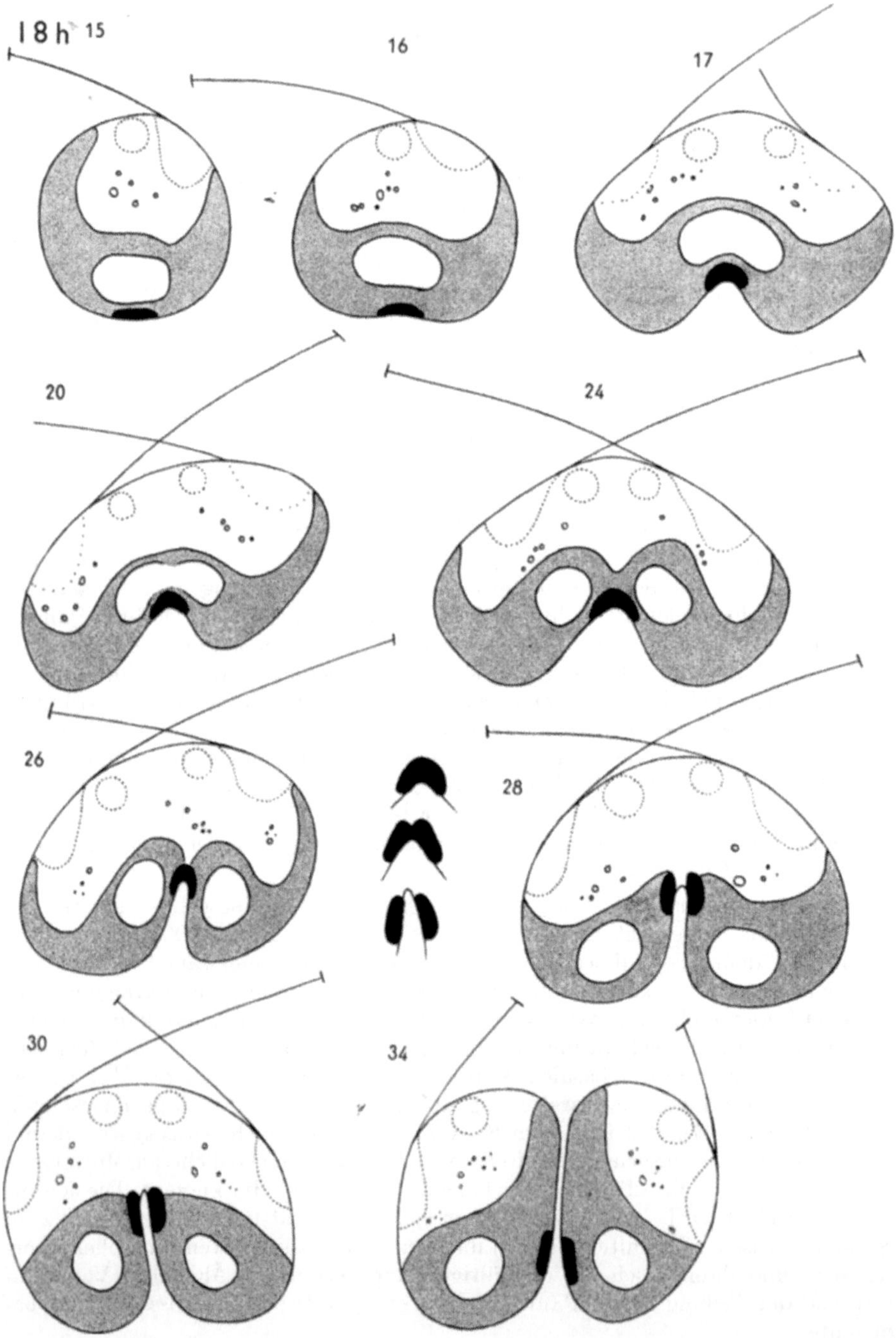

Abb. 87. Der Verlauf der Zellteilung von *Pedinomonas minor*. Die Zahlen geben die Zeitabstände in Minuten an. Man beachte den Beginn der Teilung am Augenfleck. Die Zelle wird nur von einer Seite aus eingebuchtet und durchgriffen. In der Mitte ist die Teilung des Augenflecks dargestellt. (Nach ETTL und MANTON.)

teilung der schlundartigen Vertiefung auf die Tochterzellen betrifft. Bei *Pyramimonas* gilt als erstes Teilungszeichen die Verdoppelung des Augenfleckes (BRETSCHNEIDER 1925). Kurz danach wird der Chloroplast mit Pyrenoid durch einen Längsspalt in zwei Hälften geteilt. Ziemlich früh findet die Bildung der neuen Geißeln statt, so daß das Mutterindividuum schon lange vor der Cytokinese 8-geißelig ist. Oft erfolgt Durchschnürung der Zelle von hinten her rascher als von

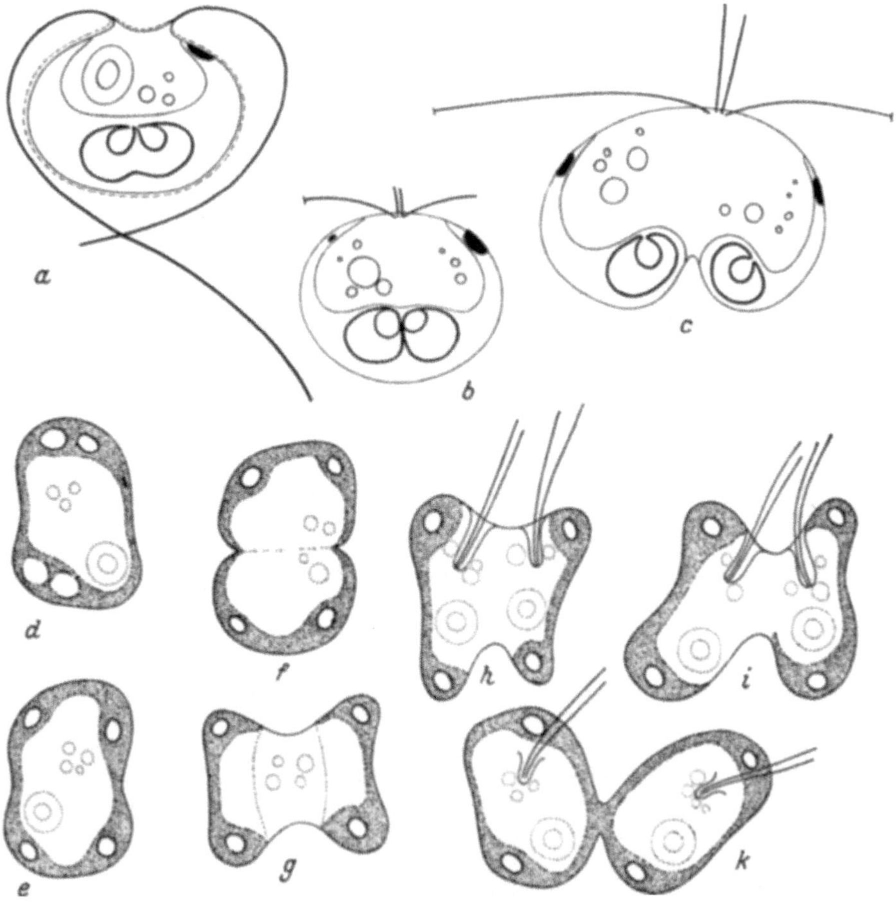

Abb. 88a—k. a—c Teilung von *Heteromastix rotunda*; a frühes Teilungsstadium mit geteiltem Proteinkörper des Pyrenoides; b Teilungsstadium mit geteiltem Pyrenoid, die neuen Geißeln wachsen zwischen den alten heran und es wird auch der neue Augenfleck gebildet; c letzter Teilungsschritt, wo der Protoplast durchschnürt wird (nach MANTON, RAYNS, ETTL und PARKE). d—k Teilung von *Mesostigma viride*; d frühes Teilungsstadium mit beginnender Pyrenoidbildung; e die Pyrenoide wandern in entgegengesetzte Richtungen; f der Protoplast wird geteilt, wobei die schlundartige Vertiefung verschwunden ist; g Protoplastenteilung; h, i späte Teilungsstadien, wobei schon zwei abgesonderte Vertiefungen mit Geißeln vorhanden sind; k nach vollendeter Teilung trennen sich die Tochterzellen. (Nach MANTON und ETTL.)

vorne. Der Periplast ist in elektronenoptischen Aufnahmen mit regelmäßigen Systemen von Schuppen bedeckt. Nach PASCHER (1927) hebt er sich bei der Teilung manchmal etwas ab, um sich schließlich doch noch durchzuteilen. Auch bei *Mesostigma viride* wurde dasselbe Verhalten des Periplasten beobachtet (MANTON und ETTL 1965). Hier ist letzterer besonders steif und von einem dicken Schuppenpanzer umgeben. Die Teilung dieser Art ist auf Abb. 88d—k wiedergegeben. Man beachte die Verdoppelung der Organellen und das Verschwinden der trichterartigen Vertiefung während der Teilung. Die Verdoppelung der Geißeln und die Deformation der schlundartigen Vertiefung vor der Protoplastenteilung wurden auch elektronenmikroskopisch beobachtet.

Bei den beweglichen Pyramimonaden kommen auch Gallertstadien mit leichter *Gloeocystis*-artiger Schichtung vor. In der Gallerte findet eine reichliche Vermehrung statt (Pascher 1932). Auch Aplanosporen werden gebildet, die durch einfache Kontraktion der Zellen zustande kommen, der jedoch eine weitgehende Metabolie vorangeht.

Es wurde festgestellt, daß manche *Pyramimonas*-artigen Flagellaten in ihrem Entwicklungscyclus unbewegliche Aplanosporen (Cysten) bilden, die der Grünalge *Halosphaera* Schmitz entsprechen (Manton, Oates und Parke 1963). Die „Cyste" entsteht aus einem besonders kurzgeißeligen Monadentypus nach Abrundung unter Verlust der Geißeln und der Ausscheidung einer festen äußeren Membran mit Sporopollenin-artigen Eigenschaften, der beim weiteren Wachstum eine quellungsfähige innere Membran angelagert wird. Sie wächst bei 15° von einer Ausgangsgröße von 14—18 μ je nach Lichtbedingungen in 20—45 Tagen bis zur Entlassung einer neuen Generation beweglicher Zellen heran. Von dieser Zeit scheint aber nur etwa die Hälfte dem Größenwachstum selbst zuzukommen, das bei *Halosphaera viridis* zu Kugeln bis zu 800 μ Durchmesser führt. Dabei vermehren sich Plastiden und Pyrenoide sehr stark, während der Kern bis oder fast bis zur Erreichung der Endgröße der Zelle ungeteilt bleibt. Erst dann setzen Mitosen ein. Das Plasma bleibt mindestens häufig bis zum Achtkernstadium einheitlich. Später und nach weiteren Mitosen verteilt es sich in Form einkerniger Plasmaballen, „Rosetten", an der Zelloberfläche. Bei *H. viridis* ist deren Zahl 256 oder 512. Die Rosetten teilen sich cytokinetisch auf, wobei nach dem zweiten Teilungscyclus die Geißeln entstehen. Schließlich ist die ganze Blase von Flagellaten erfüllt. Die innere Membran schlüpft nun aus, nimmt unter weiteren Teilungen der Flagellaten erheblich an Größe zu und entläßt ihren Inhalt schließlich durch einen Riß. Die Flagellaten können sich frei vermehren, wobei die Nachkommenschaft einer Cyste nach 14—21 Tagen wieder vollständig zur Cystenbildung schreitet oder nur zum Teil und ein weiterer Anteil des Restes dann im Abstand von 14—21 Tagen usf. Das Ausschlüpfen der äußeren aus der inneren Membran scheint in der Natur ebenfalls periodisch und mehr oder weniger synchron zu erfolgen (Parke u. Adams 1961, Parke u. den Hartog 1965, s. auch Braarud u. Nordli 1963). Zweifellos existiert hier eine eigene bewegliche Phase; ob man die „Cysten" als deren Aplanosporen oder als eigene unbewegliche Phase ansprechen will, ist etwas Geschmacksache.

Festsitzende Formen, wie *Prasinocladus* mit seinen bäumchenförmigen Kolonien und ähnliche, vermehren sich auch durch Zellteilung. Weil diese aber innerhalb einer Hülle geschieht (die Tochterzellen besitzen bei *Prasinocladus* auch Geißeln), kann von Zoosporenbildung gesprochen werden. Die Zoosporen sind vom Typus *Pyramimonas* bzw. *Platymonas* (Chadefaud 1947, Proskauer 1950) und entstehen entweder direkt aus den vegetativen Zellen durch Umwandlung oder erst nach Teilung derselben (Printz 1927, Fritsch 1935). Von der Fortpflanzung muß die Häutung der vegetativen Zellen unterschieden werden. Jede Zelle kann sich, ohne daß es zu einer Fortpflanzung kommt, mit einer neuen Membran umgeben (Proskauer 1950, Parke und Manton 1965). Durch wiederholte Häutung kommt es zu ineinander geschachtelten Systemen alter Membranen. Derselbe Vorgang wird auch bei *Platymonas* angegeben (Skuja 1948 u.a.).

Die Zoosporenbildung von *Prasinocladus ascus* findet am Morgen statt (Chihara 1963). Schon vor dem Ausschwärmen werden die Zoosporen in der Membran beweglich. Nach dem Ausschwärmen bewegen sie sich nur kurze Zeit. Aus den Keimlingen hervorgegangene Pflanzen bilden wieder Zoosporen erst am 13. Tag nach ihrer Keimung.

f) Parthenogenese und Apogamie bei Chlorophyceen.

Sexualisierte Individuen der Chlorophyceen, die keinen Partner finden, können sich oft parthenogenetisch zu neuen vegetativen Zellen weiterentwickeln (SCHUSSNIG 1960). Am einfachsten verläuft dieser Vorgang dort, wo Hologamie vorkommt, da die Gameten völlig den vegetativen Zellen gleichen (viele Arten der Gattung *Chlamydomonas*). In stärker abgeleiteten Fällen (z. B. bei der homothallischen und isogamen *Chlamydomonas paupera*) kann es aber vorkommen, daß nicht kopulierte Gameten einer parthenogenetischen Fortentwicklung unfähig sind und absterben (PASCHER 1931). Es liegt eine obligate Scheidung zwischen Gameten und agamen Zellen vor. Ähnliches Verhalten fand auch GEITLER (1931) bei *Tetraspora*. Die Unterschiede zwischen vegetativer Entwicklung agamer Zellen und Parthenogenese derselben Art können im einfachsten Falle äußerlich völlig verwischt werden. Doch lassen sich Gameten und Zoosporen dann noch physiologisch unterscheiden (WIESE 1965). Bei Formen, die Isogameten bilden, ist Parthenogenese bei beiden Geschlechtern recht häufig (also auch Androgenese — GEITLER 1934).

Auf die parthenogenetisch sich weiterentwickelnden Gameten haben besonderen Einfluß auch äußere Faktoren. So kann z.B. Temperaturerhöhung im richtigen Moment eine parthenogenetische Entwicklung hervorrufen. Auch spezielle Nährlösungen, Giftwirkungen usw. können Gameten zu selbständiger Entwicklung veranlassen (OLTMANNS 1923). Die entstandenen Parthenosporen zeigen bei gewissen Arten eine äußerliche Ähnlichkeit mit den Zygoten, bei anderen sind sie wieder verschieden (PRINTZ 1927). So sind z.B. bei *Protosiphon* die Parthenosporen dünnwandig, während die sternförmigen Zygoten eine derbe Membran haben. Dazu kommen noch physiologische Verschiedenheiten — so keimt die Parthenospore direkt, während die Zygote vorerst eine Ruhezeit durchmacht.

Es gibt eine ganze Menge von Angaben, nach denen sich Gameten, ohne daß es zur Befruchtung kommt, entweder direkt in vegetative Pflanzen umwandeln oder zygotenähnliche Ruhestadien bilden können. Nach SMITH (1950) kann schwerlich jede der diesbezüglichen Angaben als Parthenogenese betrachtet werden, weil sie oft in Mischkulturen oder im Freilandmaterial gewonnen wurden. Nur Reinkulturen, die von einzeln isolierten Zellen hervorkommen, können dieses Problem einwandfrei lösen. Es seien nun einige Beispiele für die verschiedenen Gruppen angeführt.

1. Volvocales.

Parthenogenese ist bei diesen keine Seltenheit. Schon KLEBS hat in seinen bekannten Untersuchungen gezeigt, daß bei der Herstellung von Bedingungen, welche die Kopulation hemmen, eine vegetative Teilung der Gameten von *Chlamydomonas media* einsetzt. Das wurde sowohl durch Verdunklung der im Stadium der Gametenbildung befindlichen Kulturen (der Kopulationsprozeß ist bei den *Volvocales* ans Licht gebunden) als auch durch Zugabe einer 1%igen Knopschen Nährlösung verursacht. Soweit die männlichen Gameten nicht zu weit differenziert sind, können sie sich auch vegetativ weiterentwickeln (Androgenese). In vielen Fällen kann kaum entschieden werden, ob eine Parthenospore oder eine vegetative Cyste vorliegt, weil sich vegetative Zellen und Gameten morphologisch kaum unterscheiden lassen (STREHLOW 1929). SCHULZE (1927) untersuchte Parthenogenese bei *Chlorogonium*. Wenn ein einziger Klon in Gametenbildung übergeht, so besteht keine Kopulationsmöglichkeit. Die Gameten von *Chlorogonium euchlorum* weisen aber nicht die sonst so weit verbreitete Entwicklungshemmung auf. Sie wachsen allmählich ohne Zwischenschaltung eines Ruhestadiums parthenogenetisch heran und werden zu normalen vegetativen Individuen, um dann erneut in die Gametenbildung einzutreten.

Von der getrenntgeschlechtlichen *Eudorina* wurden Klone isoliert. Ist der Klon männlich, so erschöpfen sich die Kolonien in der Spermatozoidbildung und gehen zugrunde. Die weiblichen Gameten, die im Zellverband verbleiben, können indessen, wenn die Befruchtung ausbleibt, leicht wieder zu vegetativer Teilung gebracht werden und Tochterzellen erzeugen (KNIEP 1928, MAINX 1929). Die Eizellen von *Volvox* sind fähig, wenn die Befruchtung ausbleibt, durch vegetative Teilung zu einer neuen Kolonie zu werden, ohne daß ein Ruhestadium eingeschaltet wird (KLEIN, JANET). MAINX (1929) fand an Klonkulturen von *Volvox aureus*, daß die Eizellen hier fast ausnahmslos zu Parthenosporen heranreifen, die in Größe, Gestalt, Reservestoffen, Membranbeschaffenheit und in der Art ihrer Reifung vollkommen den Zygoten gleichen.

2. Tetrasporales und Chlorococcales.

Bei den *Tetrasporales* liegen nur sehr wenige Angaben über Parthenogenese vor. So kommt sie bei *Tetraspora* vor, wo sich die Gameten ohne Kopulation mit einer Membran umgeben (PRINTZ 1927, GEITLER 1931). Auch bei den *Chlorangiales* wurde die Umwandlung von Gameten in vegetative Stadien beobachtet. Nicht viel besser bekannt sind die Verhältnisse bei den *Chlorococcales*, wenn auch *Protosiphon* und *Chlorococcum* in dieser Hinsicht eingehender untersucht wurden (KNIEP 1928, FRITSCH 1935). Die Gameten von *Protosiphon* entwickeln sich nach KLEBS parthenogenetisch weiter, wenn die Temperatur bei 25—27° C liegt, wogegen sie bei niedrigeren Temperaturen leicht kopulieren. Auch Änderungen in den Nährlösungen wirken ähnlich. Dazu gesellt sich Verdunklung usw.

3. Ulothrichales.

Sowohl die Keimung von Zygoten als auch die der Parthenosporen von *Ulothrix zonata* erfolgen nach KLEBS gleichartig, indem beide nach Sprengung der derben Membran sich in einige unbewegliche Zellen teilen, die direkt zu neuen Fäden auswachsen. Nur sollen bei Parthenosporen zwei, bei Zygoten hingegen vier Zellen während der Keimung entstehen (OLTMANNS 1922). Bei *Cylindrocapsa* bleiben unbefruchtete Eizellen grün, teilen sich oft in 2—4 Tochterzellen und wachsen direkt in neue Fäden aus (PRINTZ 1927). Die ohne Kopulation zur Ruhe kommenden Gameten von *Stigeoclonium subspinosum* sind leicht durch ihre geringere Größe von den Zygoten zu unterscheiden, außerdem sind auch bei letzteren fast stets die beiden Augenflecke sehr deutlich zu beobachten. Die parthenogenetischen Keimlinge sind dünner und weniger grün, scheinen aber später abzusterben (JULLER 1937). Bei den Chaetophoraceen scheint Parthenogenese ziemlich weit verbreitet zu sein, ohne daß die Bedingungen ihrer Auslösung genauer untersucht wären. So ist sie bekannt bei *Leptosira, Endoclonium, Entoderma, Draparnaldia* u.a. (KNIEP 1928). Die Oosphaere der Aphanochaetaceen kann ohne Befruchtung keimen. Sie bildet ein wenigzelliges Pflänzchen, das bald wieder Sexualorgane erzeugt.

Sehr gut untersucht sind in dieser Hinsicht die Ulvaceen. So kann bei *Enteromorpha* eine parthenogenetische Entwicklung vorkommen, an der bei *E. clathrata* nur die weiblichen Gameten beteiligt sind. Bei zwei Rassen von *E. compressa* scheint die Fortpflanzung durch Parthenogameten überhaupt vorherrschend zu sein (SCHUSSNIG 1960). Nach FÖYN (1934a) bleiben von den nichtkopulierten Gameten wenige von Hunderten am Leben und keimen aus. Das Wachstum geht jedoch während der ersten Tage ziemlich langsam vor sich, nimmt aber dann rasch zu, bis die Pflanzen zur Zoosporenbildung übergehen können. Das Aussehen der durch Parthenogenese entstandenen Pflanzen ist wie das der aus Zygoten und aus Zoosporen entstandenen. Die cytologischen Untersuchungen der parthenogenetisch entstandenen Pflanzen haben das interessante Ergebnis gebracht, daß die Pflanzen diploid sind. Nach MOEWUS (1938) sollen bei Ulvaceen folgende Fälle der Parthenogenese möglich sein:

a) Fehlen der Fähigkeit zur Parthenogenese in beiden Geschlechtern.

b) Fehlen der Parthenogenese in einem Geschlecht. Die eine Sorte von Gameten entwickelt sich zu normalen Gametophyten, die Gameten der anderen sterben ab.

c) Normale Parthenogenese, die Gameten beider Geschlechter entwickeln sich zu normalen Gametophyten.

d) Parthenogenese unter Aufregulierung der Chromosomenzahl.

4. Oedogoniales.

Parthenogenese ist bei *Oedogoniales* nicht selten. Bei *Oedogonium* können sich Oogonien in vereinzelten Fällen parthenogenetisch weiterentwickeln (Abb. 89). Bei *Bulbochaete pygmaea* wurde typische Fadenbildung, ausgehend von einem unbefruchteten Ei, beobachtet. Ähnliche Angaben wurden von mehreren Autoren geliefert. Die Tatsache, daß bei vielen *Oedogonium*-Arten Antheridien nicht gefunden wurden und den Oogonien die Befruchtungsöffnung fehlt, hat zur Vermutung geführt, daß das Vorkommen von Azygoten und deren parthenogenetische Entwicklung nicht ganz selten ist. Gelegentlich wurde auch vegetative Entwicklung von Androsporen (das sind die Kleinschwärmer, welche die Zwergmännchen hervorbringen) beobachtet (Kniep 1928). Wenn man *Bulbochaete*-Fäden, die Sexualzellen bilden, in Bedingungen versetzt, welche die Zoosporenproduktion fördern, wandeln primäre Oogonien ihren Inhalt in große vegetative Zoosporen um. Bei Oogonien, welche die zweite differenzierende Teilung hinter sich, also eine fortgeschrittene Spezialisierung erreicht haben, kommt es nicht mehr zur Zoosporenbildung, sondern sie wachsen direkt in einen vegetativen Faden aus (Cook 1962).

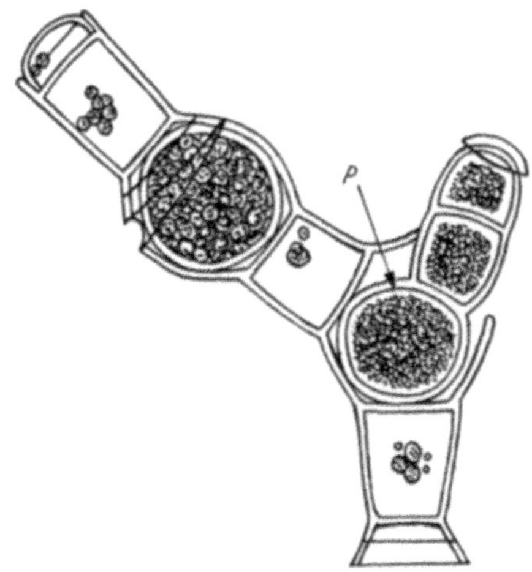

Abb. 89. *Oedogonium* mit parthenogenetisch entwickeltem Ei (*P*). (Nach Pringsheim aus Kniep umgezeichnet.)

Literatur.

Bliding, C.: Über Sexualität und Entwicklung bei der Gattung *Enteromorpha*. Sv. bot. Tidskr. **27**, 233—256 (1933). — Bold, H. C.: The life history and cytology of *Protosiphon botryoides*. Bull. Torrey Bot. Club **60**, 241—299 (1933). — Braarud, T., and E. Nordli: Reproduction and size variation in *Halosphaera viridis* of northern waters. Nytt Mag. Bot. **10**, 131—136 (1963). — Bretschneider, L. H.: *Pyramimonas utrajectina* spec. nov., eine neue *Polyblepharididae*. Arch. Protistenk. **53**, 124—130 (1925). — Brunnthaler, J.: *Protococcales* in Paschers Süßwasserflora Deutschlands, Österreichs und der Schweiz, H. 5. Jena: Gustav Fischer 1915. — Bühnemann, F.: Die rhythmische Sporenbildung von *Oedogonium cardiacum* Wittr. Biol. Zbl. **74**, 1—54 (1955a). — Das endodiurnale System der *Oedogonium*zelle II. Der Einfluß von Stoffwechselgiften und anderen Wirkstoffen. Biol. Zbl. **74**, 691—705 (1955b). — Das endodiurnale System der *Oedogonium*zelle III. Über den Temperatureinfluß. Z. Naturforsch. **10**b, 305—310 (1955c).

Chadefaud, M.: Sur l'organisation et la position systématique des flagellés du genre *Pyramidomonas*. Rev. Sci., Paris, Année **79**, 113—114 (1941). — Étude sur l'organisation de deux *Volvocales* sédentaires marines, *Prasinocladus lubricus* et *Chlorodendron subsalsum*. Rev. Sci., Paris, Année **85**, 862—865 (1947). — Les végétaux non vasculaires (Cryptogamie). In: M. Chadefaud et L. Emberger, Traité de Botanique, Systématique, vol. 1. Paris: Masson & Cie. 1960. — Chihara, M.: The life-history of *Prasinocladus ascus* as found in Japan, with special reference to the systematic position of the genus. Phycologia **3**, 19—28 (1963). — Conrad, W.: Flagellates nouveaux ou peu connus. Arch. Protistenk. **70**, 657—680 (1930). — Cook, P. W.: Growth and reproduction of *Bulbochaete hiloensis* in unialgal culture. Trans. Amer. micr. Soc. **81**, 384—395 (1962). — CHRISTENSEN, T.: Alger in BÖCHER, LANGE et SØRENSEN, Botanik, Bd. 2, Nr. 2. København: Munksgaard 1962. — CZEMPYREK, H.: Beitrag zur Kenntnis der Schwärmerbildung bei der Gattung *Cladophora*. Arch. Protistenk. **72**, 433—452 (1930). — CZURDA, V.: Experimentelle Analyse der kopulationsauslösenden Bedingungen bei Mikroorganismen. Beih. bot. Zbl. **51** (I), 711—762 (1933).

ETTL, H.: Einige Bemerkungen zur Systematik der Ordnung *Chlorangiales* Pascher. Algol. Studien. Prag: Verlag CSAV 1958. — Über pulsierende Vakuolen bei Chlorophyceen. Flora (Jena) **151**, 88—98 (1961). — Über eine besondere Form von *Asterococcus superbus* und deren systematische Stellung. Öst. bot. Z. **111**, 354—365 (1964a). — Die Morphologie einiger *Pteromonas*-Arten. Nova Hedwigia **8**, 323—331 (1964b). — Der Zellbau und die systematische Stellung der Alge *Actinochloris* Korsch. Protoplasma (Wien) **59**, 298—309 (1964c). — Beitrag zur Kenntnis der Morphologie der Gattung *Chlamydomonas* Ehrb. Arch. Protistenk. **108**, 271—430 (1965a). — Über systematische Gliederung kleiner Chlorophyceen. Nova Hedwigia **10**, 516—525 (1965b). — Untersuchungen an Flagellaten. Öst. bot. Z. **112**, 701—745 (1965c). — Der Verlauf der Zellteilung bei zwei marinen *Heteromastix*-Arten. Bot. marina (im Druck) (1967). — ETTL, H., u. I. MANTON: Die feinere Struktur von *Pedinomonas minor* Korsch. Nova Hedwigia **8**, 421—451 (1964).

FLINT, E. A.: The occurrence of zoospores in *Physolinum* Printz. New Phytologist **58**, 267—270 (1959). — FÖYN, B.: Lebenszyklus, Cytologie und Sexualität der Chlorophycee *Cladophora suhriana* Kütz. Arch. Protistenk. **83**, 1—56 (1934a). — Lebenszyklus und Sexualität der Chlorophycee *Ulva lactuca* L. Arch. Protistenk. **83**, 154—177 (1934b). — FRITSCH, F. E.: The structure and reproduction of the algae, vol. I. Cambridge: University Press 1935.

GEITLER, L.: Die Entwicklungsgeschichte von *Sorastrum spinulosum* und die Phylogenie der *Protococcales*. Arch. Protistenk. **47**, 440—447 (1924). — Beiträge zur Kenntnis der Flora ostholsteiner Seen. Arch. Protistenk. **52**, 603—611 (1925). — Untersuchungen über das sexuelle Verhalten von *Tetraspora lubrica*. Biol. Zbl. **51**, 173—187 (1931). — Grundriß der Cytologie. Berlin: Gebr. Bornträger 1934. — *Eremotyle affixa*, n. gen., n. sp., eine neue Protococcale (Chlorophycee). Öst. bot. Z. **99**, 244—252 (1952). — Zur Entwicklungsgeschichte und Chromatophorenbewegung von Coleochaeten. Öst. bot. Z. **107**, 409—424 (1960). — GROSS, I.: Entwicklungsgeschichte, Phasenwechsel und Sexualität bei der Gattung *Ulothrix*. Arch. Protistenk. **73**, 206—234 (1931).

HARTMANN, M.: Fortpflanzung und Befruchtung als Grundlage der Vererbung. In: Handbuch der Vererbungswissenschaften, Bd. I/A. Berlin: Gebr. Bornträger 1929. — Allgemeine Biologie. Stuttgart: Gustav Fischer 1953. — HASE, E., Y. MORIMURA, S. MIHARA, and H. TAMIYA: The role of sulfur in the cell division of *Chlorella*. Arch. Mikrobiol. **32**, 87—95 (1958). — HASE, E., Y. MORIMURA, and H. TAMIYA: Some data on the growth physiology of *Chlorella* studied by the technique of synchronous culture. Arch. Biochem. **69**, 149—165 (1957). — HEERING, W.: *Ulothrichales* in PASCHERS Süßwasserflora Deutschlands, Österreichs und der Schweiz, H. 6. Jena: Gustav Fischer 1914. — HINDÁK, F.: Systematische Revision der Gattungen *Fusola* Snow und *Elakatothrix* Wille. Preslia **34**, 277—292 (1962). — HOFFMAN, L., and I. MANTON: Observations on the fine structure of the zoospore of *Oedogonium cardiacum* with special reference to the flagellar apparatus. J. exp. Bot. **13**, 443—449 (1962). — HUSTEDE, H.: Untersuchungen über die stoffliche Beeinflussung der Entwicklung von *Stigeoclonium falklandicum* und *Vaucheria sessilis* durch Tryptophanabkömmlinge. Biol. Zbl. **76**, 555—595 (1957).

IYENGAR, M. O. P.: *Chlorophyceae*. In: G. M. SMITH et al., Manual of phycology. Waltham, Mass.: Chronica Botanica Co. 1951. — On the asexual and sexual reproduction of *Characiosiphon rivularis* Iyengar. J. Indian bot. Soc. **33**, 148—151 (1954). — IYENGAR, M. O. P., and K. R. RAMANATHAN: On the structure and reproduction of *Pleodorina sphaerica* Iyengar. Phytomorphology **1**, 215—224 (1951).

JANET, CH.: Le *Volvox*. I. Limoges 1912. — JULLER, E.: Der Generations- und Phasenwechsel bei *Stigeoclonium subspinosum*. Arch. Protistenk. **89**, 55—93 (1937).

KLEIN, L.: Vergleichende Untersuchungen über Morphologie und Biologie der Fortpflanzung bei der Gattung *Volvox*. Ber. naturforsch. Ges. Freiburg **5**, 1—92 (1890). — KNIEP, H.: Die Sexualität der niederen Pflanzen. Jena: Gustav Fischer 1928. — KORSCHIKOFF, A. A.: *Protochlorinae*, eine neue Gruppe der grünen Flagellaten. Arch. russ. Protistenk. **2**, 148—169 (1923). — On some new organisms from the groups *Volvocales* and *Protococcales*, and on the genetic relations of these groups. Arch. Protistenk. **55**, 439—503 (1926). — Studies in the *Vacuolatae*. I. Arch. Protistenk. **78**, 557—612 (1932). — *Volvocineae* in der Übersicht ukrainischer Algen, Bd. 4. Kijew 1938 [Ukrainisch]. — A note on *Gonium sacculiferum* Scherffel. J. Bot. Acad. Sci. U.S.S.R. 1940, 289—296 (1940). — *Protococcineae* in der Übersicht ukrainischer Algen, Bd. 5. Kijew 1953 [Ukrainisch].

LORENZEN, H.: Synchrone Zellteilung von *Chlorella* bei verschiedenen Licht-Dunkel-Wechseln. Flora (Jena) **144**, 473—496 (1957). — Periodizität von Nuklealreaktion und Kernteilung in *Chlorella*. Ber. dtsch. bot. Ges. **71**, 89—97 (1958). — LORENZEN, H., u. H. G. RUPPEL: Versuche zur Gliederung des Entwicklungsverlaufes der *Chlorella*-Zelle. Planta (Berl.) **54**, 394—403 (1960). — LUND, J. W. G.: On certain planktonic palmelloid green algae. J. Linnean Soc. London, Bot. **40**, 593—613 (1956).

MAINX, F.: Über die Geschlechtsverteilung bei *Volvox aureus*. Arch. Protistenk. **67**, 205—214 (1929). — Physiologische und genetische Untersuchungen an Oedogonien. Z. Bot. **24**, 481—527 (1931). — MANTON, I.: Observations on the fine structure of the zoospore and young germeling of *Stigeoclonium*. J. exp. Bot. **15**, 399—411 (1964). — MANTON, I., and H. ETTL: Observations on the fine structure of *Mesostigma viride* Lauterborn. J. Bot. **60**, 175—184 (1965). — MANTON, I., K. OATES, and M. PARKE: Observations on the fine structure of the *Pyramimonas* stage of *Halosphaera* and preliminary observations on three species of *Pyramimonas*. J. Mar. Biol. Ass. U. K. **43**, 225—238 (1963). — MANTON, I., D. G. RAYNS, H. ETTL, and M. PARKE: Further observations on green flagellates with scaly flagella: the genus *Heteromastix* Korshikov. J. Mar. Biol. Ass. U. K. **45**, 241—255 (1965). — METZNER, J.: A morphological and cytological study of a new form of *Volvox* I. Bull. Torrey Bot. Club **72**, 86—113 (1945). — MOEWUS, F.: Die Sexualität und der Generationswechsel der Ulvaceen und Untersuchungen über die Parthenogenese der Gameten. Arch. Protistenk. **91**, 357—441 (1938).

OLTMANNS, F.: Morphologie und Biologie der Algen, 2. Aufl., Bd. 1 u. 3. Jena: Gustav Fischer 1922/23.

PARKE, M., and I. ADAMS: The *Pyramimonas*-like motile stage of *Halosphaera viridis* Schmitz. Bull. Res. Coun. Israel **10**, 94—100 (1961). — PARKE, M., and I. DEN HARTOG-ADAMS: Three species of *Halosphaera*. J. Mar. Biol. Ass. U. K. **45**, 537—557 (1965). — PARKE, M., and I. MANTON: Preliminary observations on the fine structure of *Prasinocladus marinus*. J. Mar. Biol. Ass. U. K. **45**, 525—536 (1965). — PARKE, M., and D. G. RAYNS: Studies on marine flagellates VII. *Nephroselmis gilva* sp. nov. and some allied forms. J. Mar. Biol. Ass. U. K. **44**, 209—217 (1964). — PASCHER, A.: Über die Reproduktion bei *Stigeoclonium nudiusculum* und bei *Stigeoclonium* spec. Arch. Hydrobiol. **1**, 433—438 (1906). — Über merkwürdige amoeboide Stadien bei einer höheren Grünalge. Ber. dtsch. bot. Ges. **27**, 143—150 (1909). — Animalische Ernährung bei Grünalgen. Ber. dtsch. bot. Ges. **33**, 427—442 (1915). — Amoeboide Stadien bei einer *Protococcaceae* nebst Bemerkungen über den primitiven Charakter nicht festsitzender Algenformen. Ber. dtsch. bot. Ges. **36**, 253—260 (1918a). — Von einer allen Algenreihen gemeinsamen Entwicklungsregel. Ber. dtsch. bot. Ges. **36**, 390—409 (1918b). — *Volvocales* in PASCHERs Süßwasserflora Deutschlands, Österreichs und der Schweiz, H. 4. Jena: Gustav Fischer 1927. — Beiträge zur allgemeinen Zellehre. I. Doppelzellige Flagellaten und Parallelentwicklungen zwischen Flagellaten und Algenschwärmern. Arch. Protistenk. **68**, 261—304 (1929). — Über eine Fall von Widerstreit zwischen Zellform und Koloniebildung. Arch. Protistenk. **70**, 467—490 (1930). — Systematische Übersicht über die mit Flagellaten in Zusammenhang stehenden Algenreihen und Versuch einer Einreihung dieser Algenstämme in die Stämme des Pflanzenreiches. Beih. Bot. Zbl. **48** (II), 317—332 (1931). — Drei neue Protococcalengattungen. Arch. Protistenk. **76**, 409—419 (1932). — Zur Kenntnis einer Protococcale aus den salzhaltigen Mooren Franzensbads. Bot. Notiser **1939**, 158—168 (1939a). — Über gelegentliche Synzoosporenbildungen bei Algen und über die Verbreitung synzoosporer Organisationen. Beih. Bot. Zbl. **59** A, 389—408 (1939b). — Alpine Algen I. Neue Protococcalengattungen aus den Uralpen. Beih. Bot. Zbl. **62** A, 175—196 (1943a). — Beiträge zur Morphologie der ungeschlechtlichen und geschlechtlichen Vermehrung der Gattung *Chlamydomonas*. Beih. Bot. Zbl. **62** A, 197—220 (1943b). — Über Conidien-artige Sporen bei grünen Algen. Beih. Bot. Zbl. **62** A, 360—375 (1944). — PETROVÁ, J.: Eine neue festsitzende Protococcalengattung (*Tetraciella* nov. gen.). Arch. Protistenk. **71**, 550—566 (1930). — Die vermeintliche Heterokonte „*Botrydiopsis*" *minor* — eine Chlorophycee. Beih. Bot. Zbl. **48** (I), 221—228 (1931). — PIRSON, A.: Synchronanzucht von Algen im Licht-Dunkel-Wechsel. Vorträge aus dem Gesamtgebiet der Botanik, N.F. Nr **1**, 178—186 (1962). — PIRSON, A., u. H. LORENZEN: Photosynthetische Sauerstoffentwicklung von *Chlorella* nach Synchronisation durch Licht-Dunkel-Wechsel. Naturwissenschaften **45**, 497 (1958). — POCOCK, M. A.: Two multicellular motile green algae, *Volvulina* Playfair and *Astrephomene*, a new genus. Trans. roy. Soc. S. Afr. **34**, 103—127 (1956). — *Haematococcus* in Southern Africa. Trans. roy. Soc. S. Afr. **36**, 5—55 (1960). — PRINTZ, H.: *Chlorophyceae*. In: ENGLER-PRANTL, Natürliche Pflanzenfamilien, 2. Aufl., Bd. 3. Leipzig: Wilhelm Engelmann 1927. — PROSKAUER, J.: On *Prasinocladus*. Amer. J. Bot. **37**, 59—66 (1950).

RANDHAWA, M. S.: A note on cyst formation in *Fritschiella tuberosa* Iyengar. Arch. Protistenk. **92**, 131—136 (1939). — REICH, K.: Zur Kenntnis der Entwicklungsgeschichte und Cytologie von *Stigeoclonium*. Arch. Protistenk. **53**, 435—458 (1926). — REISIGL, H.: Zur Systematik und Ökologie alpiner Bodenalgen. Öst. bot. Z. **111**, 402—499 (1964). — RIETH, A.: Über die vegetative Vermehrung bei *Sphaeroplea Wilmani* Fritsch et Rich. Flora (Jena) **139**, 28—38 (1952).

SCHREIBER, E.: Zur Kenntnis der Physiologie und Sexualität höherer *Volvocales*. Z. Bot. **17**, 337—376 (1925). — SCHULZE, B.: Zur Kenntnis einiger *Volvocales*. Arch. Protistenk. **58**, 508—576 (1927). — SCHUSSNIG, B.: Grundriß der Protophytologie. Jena: Gustav Fischer 1954. — Handbuch der Protophytenkunde, Bd. 2. Jena: Gustav Fischer 1960. — SINGH, R. N.: On some phases in the life history of the terrestrial alga, *Fritschiella tuberosa* Iyengar, and its autecology. New Phytologist **40**, 170—182 (1941). — Reproduction in *Draparnaldiopsis indica* Bharadwaja. New Phytologist **41**, 262—273 (1942). — SKUJA, H.: Taxonomie des Phytoplanktons einiger Seen in Uppland, Schweden. Symbolae bot. upsalien. **9**/3, 1—399 (1948). — Taxonomische und biologische Studien über das Phytoplankton schwedischer Binnengewässer. Nova Acta Reg. Soc. Sci. upsalien., Ser. IV, **16**, 1—404 (1956). — SMITH, G. M.: The fresh-water algae of the United States, 2. ed. New York: McGraw Hill Book Co. 1950. — SOEDER, C. J.: Synchronisierte Mikroorganismen. Umschau **60**, 660—663 (1960). — STARR, R. C.: Reproduction by zoospores in *Planktosphaeria gelatinosa* G. M. Smith. Hydrobiologica **6**, 392—396 (1954a). — Reproduction by zoospores in *Tetraëdron bitridens*. Amer. J. Bot. **41**, 17—20 (1954b). — A comparative study of *Chlorococcum meneghini* and other spherical, zoospore-producing genera of the *Chlorococcales*. Indiana Univ. Publ. Sci., Ser. 20 (1955). — STREHLOW, K.: Über die Sexualität einiger *Volvocales*. Z. Bot. **21**, 625—692 (1929).

TRAINOR, F. R.: Zoospores in *Scenedesmus obliquus*. Science **142**, 1673—1674 (1963). — TSCHERMAK, E.: Über Vierteilung und succedane Autosporenbildung als gesetzmäßigen Vorgang, dargestellt an *Oocystis*. Arch. wiss. Bot. **32**, 585—595 (1942). — TSCHERMAK-WOESS, E.: Über wenig bekannte und neue Flechtengonidien. II. Öst. bot. Z. **98**, 412—419 (1951).

ULEHLA, V.: Ein Beitrag zur experimentellen Ökologie der Süßwasseralgen. [Tschechisch.] Studia Mendeliana, Brünn 1923, 229—253 (1923).

VISCHER, W.: Über einige kritische Gattungen und die Systematik der *Chaetophorales*. Beih. Bot. Zbl. **51** (I), 1—100 (1933).

WEST, G. S., and F. E. FRITSCH: A treatise on the British freshwater algae. Cambridge: Cambridge University Press 1927. — WIESE, L.: On sexual agglutination and mating-type substances (gamones) in isogamous heterothallic *Chlamydomonads*. I. Evidence for the identity of the gamones with the surface components responsible for sexual flagellar contact. J. Phycology **1**, 46—54 (1965). — WINKLER, H.: Fortpflanzung der Gewächse: Physiologie der Fortpflanzung. In: Handwörterbuch der Naturwissenschaften, 2. Aufl., Bd. 4, S. 435—450. Jena: Gustav Fischer 1934.

II. Conjugatophyceae.

Die Conjugaten sind dadurch charakterisiert, daß ihre Fortpflanzung vor allem eine sexuelle ist, die wir als *Konjugation* bezeichnen und die durch amöboide Gameten erfolgt. Eine vegetative Fortpflanzung mittels beweglicher Zoosporen, wie wir sie bei den Chlorophyceen kennen, fehlt hier. Auch die Bildung unbeweglicher Gonidien wurde nicht festgestellt und so kommt als vegetative Fortpflanzung eigentlich nur die Teilung einzelliger Formen und der Zerfall bei fadenförmigen *Zygnemales* in Frage. Außerdem kann man zur vegetativen Fortpflanzung auch die Bildung von Aplanosporen und Akineten (Chlamydosporen) rechnen (OLTMANNS 1922, PRINTZ 1927, FRITSCH 1935, SMITH 1950). Die vegetative Fortpflanzung der Conjugaten währt so lange, als die kopulationsauslösenden Bedingungen nicht eintreffen. Unter ausreichenden Lichtverhältnissen, geeigneter O_2- und CO_2-Tension, richtiger osmotischer Konzentration der Nährlösung und innerhalb eines mittleren Temperaturbereiches kommt es zum Eintritt sexueller Fortpflanzung, wenn ein bestimmter physiologischer Zustand, die Kopulationsdisposition, gegeben ist (CZURDA 1933). Wenn diese Disposition und die auslösenden Faktoren nicht vorliegen, vermehren sich die Conjugaten vegetativ.

a) Teilung.

Bei den einzelligen *Desmidiales* und *Mesotaeniales* erfolgt die vegetative Fortpflanzung durch einfache Zellteilung (FRITSCH 1935, FOTT 1959), zu der alle Individuen befähigt sind. Wo die Zellmembran aus zwei Stücken besteht, erfolgt bei der Teilung ein Auseinanderweichen der Membranhälften, worauf der Protoplast durchteilt wird. Nach der Kernteilung bildet sich ein irisblendenartig schließendes Diaphragma, das die sich teilende Zelle trennt. Jede Tochterzelle gleicht aber vorerst nur in einer Zellhälfte der Mutterzelle. Die voneinander weichenden Halbzellen bilden an der Stelle des Isthmus Fortsätze, mit denen die Tochterzellen verbunden bleiben. Die Fortsätze vergrößern sich und nehmen bald die Gestalt der fehlenden Halbzellen an (Abb. 90). Später dringt in diese auch der Chloroplast ein. Die beiden Tochterzellen bleiben miteinander meist so lange verbunden, bis die Formgestaltung der neuen Halbzellen vollendet ist. Zuweilen erfolgen aber die Teilungen so rasch nacheinander, daß bis zum Eintritt der neuen Teilung die jüngste Hälfte noch nicht Form und Skulptur der Mutterzelle angenommen hat, so daß Formen von einfacherem Aussehen entstehen. Die Membran der beiden Zellhälften ist immer ungleich alt, was besonders bei Arten mit eiseninkrustierten Membranen in Erscheinung tritt, bei denen die älteren Hälften stets dunkler sind. Die ganze Teilung ist in den Einzelheiten kompliziert und vollzieht sich in gewissen Abänderungen. Bei manchen Arten der Gattung *Penium* und *Closterium* werden Gürtelbänder gebildet, die sich in Form von ein bis zwei röhrenförmigen Membransegmenten zwischen die Membranhälften legen (Printz 1927, Fott 1959).

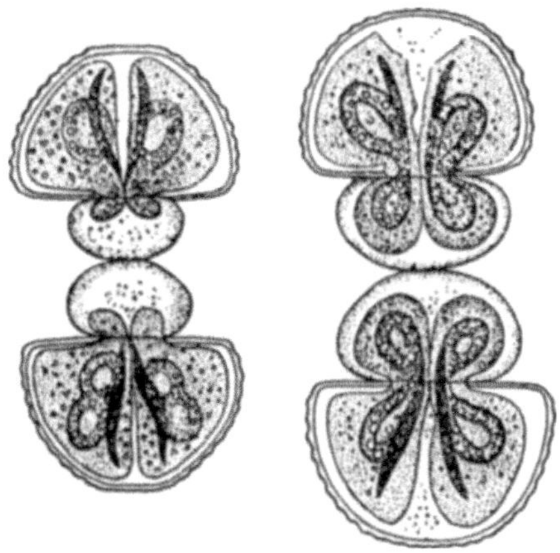

Abb. 90. Zellteilung von *Cosmarium sp.* (Nach De Bary aus Fott.)

b) Zerfall von Fäden.

Die Fäden der *Zygnemales* zerbrechen allgemein sehr leicht und bei vielen Arten sind spezielle Einrichtungen vorhanden, welche diese Fragmentation erleichtern. Wenn die Fäden eine gewisse Länge erreicht haben, zerfallen sie in zwei oder mehrere Stücke; oft kommt es unter bestimmten Bedingungen sogar zur völligen Isolierung der Einzelzellen. Durch die speziellen Einrichtungen wird der rapide Zerfall nur erleichtert. Ob der Faden in größere oder kleinere Teile oder auch in Einzelzellen zerfällt, immer haben diese Prozesse für die Fortpflanzung eine große Bedeutung. Denn jedes Teilstück und jede isolierte Zelle kann wieder zu einem neuen Faden auswachsen. Die Zerbrechung der Fäden kann entweder durch einen einfachen Mechanismus bedingt werden (z.B. *Mougeotia*) oder ein komplizierter Vorgang sein (manche *Spirogyra*-Arten) — vgl. Oltmanns 1922, Printz 1927, Smith 1950.

Der gewöhnliche Zerfall ist wahrscheinlich mit einer Änderung der chemischen Natur der Mittellamelle zweier benachbarter Zellen verbunden (Tiffany 1924). Die reifen Querwände zeigen gewöhnlich eine mehr oder weniger komplexe Struktur, die in Beziehung zur Fragmentation steht (Fritsch 1935). Es wurden aber bisher keine bestimmten Beziehungen zwischen Außenbedingungen und der Veränderung der Mittellamellen, die zum Zerfall führt, erkannt. Feststellen ließ sich, daß die Veränderung bei stärkerem Rückgang der Zellteilung, bei manchen *Zygnemales* unter bestimmten Kulturbedingungen nahezu regelmäßig erhalten werden kann (Czurda 1937). Der innere Zellzustand läßt weder kurz vor noch kurz nach dem Fadenzerfall eine bestimmte Veränderung erkennen. In neue

Lebensbedingungen übertragen, teilen sich die Zellen bald und neue Fäden werden gebildet.

Der eigentliche Mechanismus des Zerfalles wird dem Zerreißen der lateralen Membranen zugeschrieben, was mehr durch den verschiedenen Turgor in den benachbarten Zellen verursacht wird (SMITH 1950, OLTMANNS 1922, LLOYD 1926). Das Schema des Zerfalles ist für die dünnwandigen Arten von *Mougeotia* gegeben (Abb. 91 b), wo der Vorgang sehr einfach vor sich geht. Bei Spirogyren, vor allem, wo doppelte Querwände vorkommen, ist der Vorgang komplizierter. Diesen Querwänden sind kurze doppelwandige Faltenringe aufgesetzt, die erhalten bleiben, solange die Zellen im Verband sind. Bei Fadenzerfall werden sie durch den Turgor

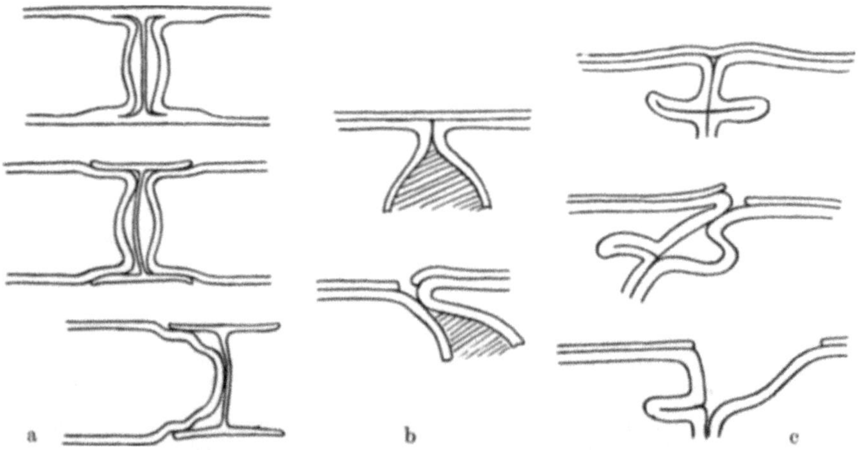

Abb. 91 a—c. Mechanismus des Fadenzerfalles bei *Zygnemales*. a *Spirogyra colligata*, Struktur des Septums und die Art des Fadenzerfalles mit Bildung von H-Stücken; b einfacher Zerfall von *Mougeotia sp.*; c *Spirogyra weberi*, Verdoppelung der Querwand und Trennung der benachbarten Zellen. (a nach HODGETTS, b und c nach LLOYD.)

der Zellen nach vorhergehender Lockerung der Cuticularschicht handschuhfingerartig vorgestülpt (Abb. 91 c). Dabei zerreißt die laterale Membran. Der Zerfall bei größeren und mit dicker Membran versehenen *Spirogyra*-Arten erfolgt die Trennung nach SMITH (1950) durch eine Absonderung von H-Stücken (Abb. 91 a) in der Nähe der Querwand. CZURDA (1937) hingegen hält diese H-förmigen Verbindungsstücke für Zellwandabsonderungen degenerierter Zellmembranen. Als Ursache des Fadenzerfalles wird ein unterschiedlicher Turgor benachbarter Zellen angesehen (LLOYD 1926). Wenn der Turgor der einen Zelle sinkt, so wird die Querwand in diese durch den höheren Druck der benachbarten Zelle unter gleichzeitiger Aufquellung der Mittellamelle (LLOYD spricht von Gallerte) eingedrückt und dann die gemeinsame laterale Membran zerrissen. Die Fragmentation kann auch durch Wasserströmung oder Wassertiere, die an den Algenrasen von diesen zehren, bewirkt werden. Der rapide Zerfall von Fäden der Algenwatten von Zygnemalen zeigt, daß es sich um eine recht wirksame Art der vegetativen Fortpflanzung handelt.

c) Aplanosporen.

Asexuelle Sporen wurden mehrmals bei den Conjugaten beobachtet und beschrieben. Doch hat man dieses Stadium gewöhnlich nicht in den Lebenscyclus eingereiht. Sie entstehen in Fäden, wo kein Anzeichen einer Kopulation besteht, also asexuell. In den meisten Fällen löst sich der Protoplast von der Muttermembran, rundet sich ab und umgibt sich mit einer neuen dicken Membran innerhalb der alten Mutterhülle. Manchmal ist diese Sporenmembran ebenso ornamentiert wie die der Zygoten. Weil sie auf dieselbe Weise entstehen wie die ähnlichen Dauerzellen der Chlorophyceen, ist der Name *Aplanospore* hier angebracht. Sie werden vor allem bei den *Zygnemales* gebildet, wurden aber auch bei den *Desmidiales* festgestellt (HUBER-PESTALOZZI 1927). STARR (1958) fand in Kulturen von *Closterium* Aplanosporen, deren Bildung durch das Auseinanderbrechen der Membranhälften der Mutterzelle eingeleitet wurde. An der Stelle des Bruches bildet sich Gallerte, die den heraustretenden Protoplasten umhüllt. Sobald der Protoplast herausgetreten ist, wandelt er sich in eine kugelige Masse um, die zwischen den leeren Membranhälften zu liegen kommt. Die Spore reift, indem sie sich mit einer neuen dicken Membran umgibt; ihr Zellinhalt wird dann undeutlich. STARR beobachtete auch ihre Keimung. Erstes Anzeichen der Keimung ist das Auftreten eines einzigen grünen Chromatophors in der Spore. Durch einen Riß in der Sporenmembran tritt dann der Protoplast heraus und wandelt sich wieder in eine vegetative Zelle um.

Bei den *Zygnemales* wurden Aplanosporen öfters gesehen. Hier zieht sich der Plasmainhalt der Zelle fast kugelförmig zusammen, umgibt sich mit einer neuen, derben Membran und speichert Reservestoffe. Die alten Membranen bleiben noch lange erhalten und verketten die Aplanosporen zu rosenkranzartigen Gebilden (WEST und STARKEY). Nach CZURDA (1937) sollten die in der Literatur als Aplanosporen benannten Gebilde nur Parthenosporen sein. Da aber die Zellen in diesen Fällen keine Kopulationskanäle bilden, so haben sie sich nicht als Geschlechtszellen differenziert; die in ihnen gebildeten Vermehrungszellen sind daher als Aplanosporen aufzufassen (vgl. OLTMANNS 1922, PRINTZ 1927, GEITLER 1958, 1959). Diese Aplanosporen sind keimungsfähig wie die Zygoten. Bei *Zygnema* und *Spirogyra* wurde auch die Keimung der Aplanosporen innerhalb des Mutterfadens gesehen (Abb. 92). Bemerkenswert scheint dabei der Befund, daß einer der Keimlinge noch innerhalb der Mutterzelle abermals einen Dauerzustand ausbildete.

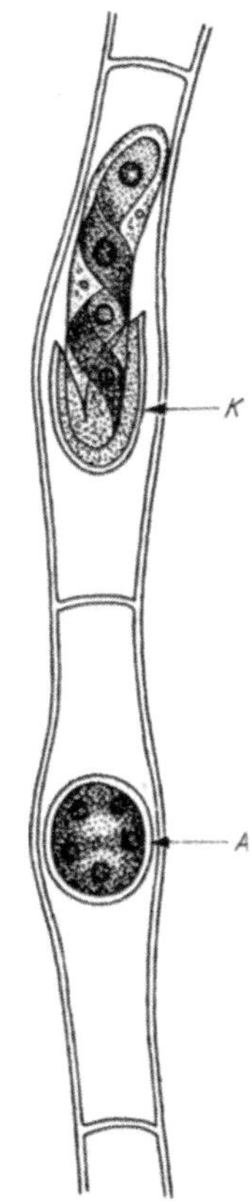

Abb. 92. Vegetative Aplanosporen von *Spirogyra mirabilis*. *A* Eine Aplanospore, die innerhalb der Zellwand durch Kontraktion des Protoplasten gebildet wurde; *K* noch innerhalb des Zellfadens keimende Aplanospore. (Nach KLEBS umgezeichnet.)

d) Akineten (Chlamydosporen).

Auch hier sind als Akineten solche Dauerzellen aufzufassen, die wie bei den Chlorophyceen aus vegetativen Zellen entstanden sind, indem sich die ursprüngliche Membran verstärkt hat. Manche *Zygnemales* bilden, ohne ihren Zellinhalt zu kontrahieren, Akineten, die sich in bekannter Weise durch Speicherung von Reservestoffen und durch erhebliche Verdickung der Membran auszeichnen (OLTMANNS 1922). Bei *Zygnema pectinatum* wird die derbe Membran sogar braun wie bei Zygoten. Diese Akineten entstehen beim Austrocknen des Wassers am Standort, aber auch bei niedriger Temperatur. Arten von *Mougeotia*, die in stark beschatteten Wasserlachen wachsen oder in Tümpeln der alpinen oder arktischen Region vorkommen, haben in den Fäden isolierte Zellen, die eine dicke Membran bekommen. Ähnliche Akineten wurden in Fäden von *Spirogyra*-Arten schnell fließender Gewässer oder bei erdbewohnenden Arten von *Zygnema* gefunden

(SMITH 1950). In diesem Zustand können sie selbst unter ungünstigen Lebensverhältnissen überdauern. In einigen Fällen werden in einem Faden viele Akineten gebildet, die rosenkranzartig verbunden bleiben (PRINTZ 1927).

In Kulturen treten Akineten (Chlamydosporen) bei den Zygnemalen unter ungünstigen Bedingungen auf. Akineten bei *Desmidiales* hat LHOTSKÝ (1948) beobachtet. Er hat bei alternden, in ihrem Wachstum gehemmten Zellen von *Closterium moniliferum* „Chlamydosporen" festgestellt. Diese entstanden aus dem Inhalt einer gehemmten Halbzelle und schieden innerhalb derselben eine dicke, glatte und farblose Membran aus (SCHUSSNIG 1960). Bei der Keimung dieser Dauersporen setzt die normale Zellvermehrung wieder ein. STARR (1958) hingegen hält die von LHOTSKÝ angegebenen Chlamydosporen für degenerierte Zellen, die in ungünstigen Kulturbedingungen entstehen. Bei der Akinetenbildung bilden die vegetativen Zellen gewöhnlich innerhalb ihrer Zellmembran eine zweite sehr dicke, skulptierte und gefärbte Wand aus, in welcher mehrere unregelmäßig verlaufende Rißlinien ähnlich wie bei Zygoten oder Aplanosporen angelegt werden.

Nach CZURDA (1937) können als Akineten auch weitgegangene Degenerationszustände gewisser Arten angesehen werden, die sich durch eine außerordentlich starke Verdichtung der Wand ohne Verfärbung und Skulptierung auszeichnen. Die Zellen führen viel Stärke, die oft vollkommen durch Fett ersetzt wird. In diesem Zustand vermögen die Zellen in feuchter und trockener Umgebung wochen- und monatelang entwicklungsfähig zu bleiben.

e) Parthenogenese und Apogamie bei Conjugaten.

Wenn bei den Conjugaten aus irgendeinem Grund der normale Kopulationsverlauf gestört oder gehemmt wird, so daß die Verschmelzung der beiden Gameten unterbleibt oder unterbrochen wird, so kommt es häufig zur Ausbildung von

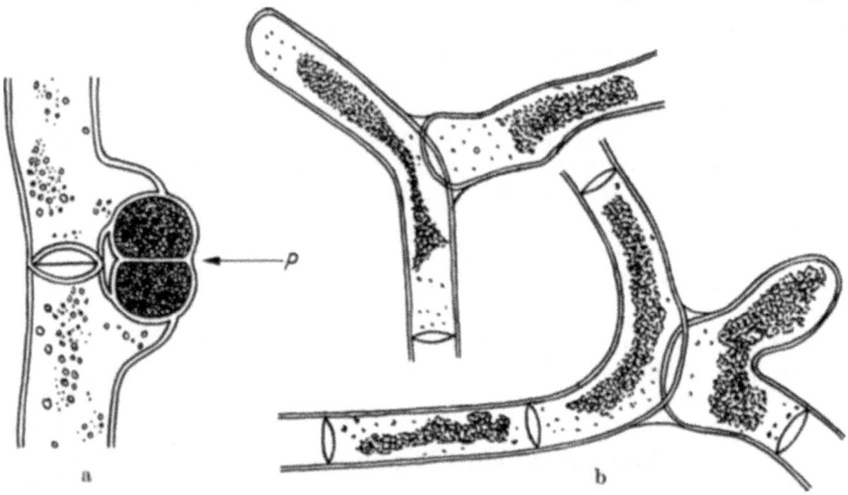

Abb. 93a u. b. a Zwei Azygoten von *Mougeotia mirabilis*. Bei der Seitenkopulation kommt es öfters vor, daß die die Gameten trennende Scheidewand nicht aufgelöst wird, dann umgibt sich jeder der beiden Gameten mit einer festen Membran und wird zur Azygote. b Apogame Konjugation von *Mougeotia genuflexa*. (a nach WITTROCK, b nach SCHUSSNIG umgezeichnet.)

Zygoten-ähnlichen Parthenosporen (Abb. 93a). Bei *Spirogyra*-Arten geschieht diese recht häufig und bei *Spirogyra mirabilis* ist die Parthenosporenbildung die Regel. Daß sowohl männliche als auch weibliche Protoplasten der Parthenogenese (resp. Androgenese) unterliegen, geht unzweifelbar daraus hervor, daß häufig zwei Zellen eine Kopulationsbrücke bilden, die Inhalte aber nicht miteinander verschmelzen, sondern sich abrunden und mit Membranen umgeben (KNIEP 1928). Bei *Spirogyra* unterscheiden sich Parthenosporen meist weder im Aussehen noch in der Weiterentwicklung nennenswert von den Zygoten. KLEBS konnte Parthenogenese bei *Spirogyra* herbeiführen, wenn er die Fäden im richtigen Moment in

6%ige Zucker- oder 1%ige Nährlösung überführte. Dann entstanden neben einigen normalen Zygoten, durch einfache Kontraktion von Gametenmutterzellen, mit derber Membran umgebene Parthenosporen, die zwar etwas empfindlicher sind als die Zygoten, aber doch im übrigen wie diese keimen.

Auch bei *Zygnema* ist die Azygotenbildung nicht selten. So kommen nach TRANSEAU bei *Z. cylindricum* nur Azygoten vor. Häufig tritt sie auch bei *Z. cruciatum* ein. Hier kann auch die sog. „*vegetative Konjugation*" bei *Mougeotia pulchella* angeführt werden, bei der wohl Bildung von Papillen und Kopulationsbrücken mit Fusion der Plasmakörper darin, jedoch keine Kernverschmelzung erfolgt. Etwas anders verhält sich *Mougeotia genuflexa*, bei der häufig knieförmige Kopulationsbilder angetroffen werden (Abb. 93b). Weil aber CZURDA selbst bei Untersuchungen eines sehr umfangreichen Materials niemals reife Zygoten feststellen konnte, wird man diese Art wohl für apogam halten können (SCHUSSNIG 1960). Ein schönes Beispiel einer Apogamie bei einer *Mougeotia* gibt auch GEITLER (1930). Die Art bildet Kopulationsknie und Kopulationspapillen zwischen Zellen gepaarter Fäden, aber die Differenzierung von Gameten sowie das Durchbrechen der Kopulationskanäle unterbleibt. Auch Azygoten entstehen nicht. Der Vorgang ließ sich in der Kultur reproduzieren. Es ist ein völliger Verlust der Sexualität eingetreten. Apogamie wurde bei Conjugaten wiederholt beschrieben.

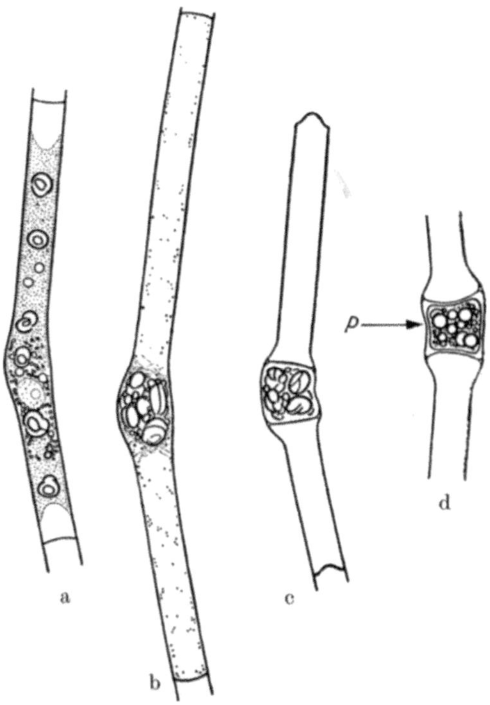

Abb. 94a—d. Azygotenbildung von *Mougeotia transeaui*. a Inhalt noch unkontrahiert mit einem Kern, einem Chromatophor und sechs Pyrenoiden, b fortgeschrittenes Stadium der Zusammenballung, c junge Azygote, d reife Azygote mit Öltropfen. (Nach GEITLER.)

Seltsame Bildung von Parthenosporen bei *Mougeotia heterogama* beobachtete GEITLER (1958). Die Parthenosporen gehen aus unterbrochener Kopulation hervor und entstehen gewöhnlich dadurch, daß der eine Partner, in der Regel der männliche, abstirbt und dies fast immer infolge einer Pilzinfektion. Weibliche Azygoten sind daher viel häufiger als männliche. Doch bei der Bildung männlicher Azygoten wurde niemals ein Absterben des weiblichen Partners beobachtet, sondern dessen Inhalt war unkontrahiert, also mehr oder weniger vegetativ geblieben. Bei Azygoten einer anderen *Mougeotia*-Art stellte GEITLER (1959) auffallende polare Kappen und ein braunes Mesospor fest. In den jüngeren Stadien der Kniebildung und Anschwellung, wie auch in den späteren der Kontraktion und Querwandbildung läßt sich immer nur ein einziger Zellkern beobachten, obwohl meistens zwei Chromatophoren vorhanden sind, wodurch der Eindruck einer Zygotenbildung erweckt wird (Abb. 94). Im Unterschied zu Fällen von Azygotenbildung mancher anderer Arten, bei denen ein Paar nötig ist, bedarf es bei *M. transeaui* keines Partners und die Azygoten entstehen regelmäßig in ungepaarten Fäden. Das ganze erfolgt innerhalb der Fäden in deutlichen „Blöcken", die durch vegetative Fadenabschnitte getrennt sind. Die Erscheinung läßt sich möglicherweise mit den in vegetativen Fäden nachgewiesenen Blöcken von mitotisch aktiven und inaktiven Zellen in Beziehung setzen (vgl. TSCHERMAK-WOESS und HASITSCHKA-JENSCHKE 1957). Das Gemeinsame in beiden Fällen besteht darin, daß die Fäden in bestimmter, wenn auch durch Außenfaktoren im weitesten Sinne modifizierter Weise der Länge nach physiologisch inhomogen sind.

Bei den *Desmidiales* ist Parthogenese eine relativ seltene Erscheinung. Trotzdem konnte Klebs auch an Desmidiaceen wie *Closterium lunula* und *Cosmarium botrytis* durch helle Beleuchtung unter Zuführung von Zucker Parthenosporen hervorbringen.

Literatur.

Benecke, W.: Mechanismus und Biologie des Zerfalles der Conjugatenfäden in die einzelnen Zellen. Jb. wiss. Bot. **32**, 453—476 (1898).

Cohn, F.: Zur Lehre vom Wachstum der Pflanzenzelle. Nova Acta Leopoldina **22**, 509—540 (1850). — Czurda, V.: *Zygnemales.* In: Paschers Süßwasserflora Mitteleuropas, H. 9. Jena: Gustav Fischer 1932. — (1933) s. L. Chlorophyceae. — *Conjugatae.* In: Linsbauers Handbuch der Pflanzenanatomie, 2. Abt., Bd. 6/2 B:b. Berlin: Gebr. Bornträger (1937).

Fott, B.: Algenkunde. Jena: Gustav Fischer 1959. — Fritsch, F. E. (1935) s. Lit. Chlorophyceae.

Geitler, L.: Über Apomixis bei *Mougeotia.* Arch. Protistenk. **70**, 307—312 (1930). — Sexueller Dimorphismus bei einer Konjugate (*Mougeotia heterogama* n. sp.). Öst. bot. Z. **105**, 301—322 (1958). — Bildung von Azygoten in Blöcken bei *Mougeotia transeaui.* Öst. bot. Z. **106**, 431—439 (1959).

Hodgetts, W. J.: A new species of *Spirogyra.* Ann. Bot. **34**, 519—524 (1920). — Huber-Pestalozzi, G.: Über Aplanosporenbildung bei einigen Desmidiaceen. Arch. Hydrobiol. **18**, 651 (1927).

Kolkwitz, R., u. H. Krieger: *Zygnemales.* In: Rabenhorsts Kryptogamenflora, Bd. 13. Abt. 2. Leipzig: Akademische Verlagsgesellschaft 1941.

Lhotský, O.: The production of chlamydospores by *Closterium moniliferum* (Bory) Ehrb. Stud. bot. Cechoslovaca **9**, 155 (1948). — Lloyd, F. E.: Cell disjunction in *Spirogyra.* Papers Michigan Acad. Sci. **6**, 275—287 (1926).

Oltmanns, F. (1922—1923) s. Lit. Chlorophyceae.

Printz, H. (1927) s. Lit. Chlorophyceae.

Schussnig, B. (1960) s. Lit. Chlorophyceae. — Smith, G. M. (1950) s. Lit. Chlorophyceae. — Starr, R. C.: Asexual spores in *Closterium didymotocum* Ralfs. New Phytologist **57**, 187—190 (1958).

Tiffany, L. H.: A physiological study of growth and reproduction among certain green algae. Ohio J. Sci. **24**, 65—98 (1924). — Transeau, E. N.: Notes on the *Zygnemales.* Ohio J. Sci. **16**, 17—31 (1915). — The genus *Debarya.* Ohio J. Sci. **25**, 193—199 (1925). — Tschermak-Woess, E., u. G. Hasitschka-Jenschke: Über die Teilungsfolge in den Fäden von zwei Grünalgen. Öst. bot. Z. **104**, 578—582 (1957).

West, G. S., and C. B. Starkey: A contribution to the cytology and life history of *Zygnema* etc. New Phytologist **14**, 194—205 (1915).

III. Charophyceae.

Der vegetativen Fortpflanzung dienen bei den Characeen Stengel- und Wurzelknöllchen, auch Bulbillen genannt. Sie sind lediglich Modifizierungen und Spezialisierungen der normalen Knoten, deren Entstehung hier darum kurz geschildert sei: Alle Descendenten der einschneidigen Scheitelzelle teilen sich inäqual in je eine prospektive Knotenzelle und in eine Internodialzelle (Abb. 95e). Letztere wird „somatisch“ und wächst zu einer enormen Länge heran, verbunden mit Endopolyploidisierung und Kernfragmentation (s. Hasitschka-Jentschke 1960). Erstere bleibt grundsätzlich teilungsfähig und „generativ“. Die Knotenzelle teilt sich nach dem Schema der Abb. 95 in (zunächst zwei) zentrale Knotenzellen, die mit den Internodien die Achse bilden, und in meist sechs periphere Zellen. Diese werden zu Scheitelzellen und bilden einen Wirtel lateraler Kurztriebe, die sog. „Blätter“, die aus einer beschränkten Folge von Knoten und Internodialzellen bestehen. Das erste Segment wird jeweils zum Basalknoten eines „Blattes“. Die peripheren Zellen dieser Basalknoten bleiben wiederum teilungsfähig. Der Basalknoten des ältesten Blattes (u_1 in Abb. 95) bringt einen pseudoaxillären Langtrieb hervor, der wie die Hauptachse gebaut ist. Bei einigen Nitellen wird ein zweiter Langtrieb an der Basis des nächstältesten Blattes gebildet

(u_2 in Abb. 95 — GIESENHAGEN 1902). Weiterhin entwickeln sich die Berindung der Internodien und die Stipularkränze bei den Charen aus Abkömmlingen dieser peripheren Basalknotenzellen. An unterirdischen Sproßknoten entstehen aus diesen außerdem verzweigte Rhizoiden, die sich ebenfalls in vereinfachte Knoten und Internodien gliedern (Abb. 100).

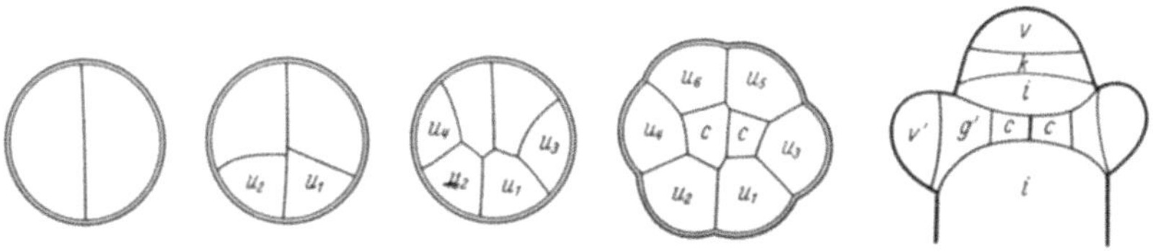

Abb. 95a—e. Entwicklung eines Sproßknotens von *Chara aspera* (= *Chara globularis* Thuill. var. *aspera*). a—d Querschnitte, e Längsschnitt durch Sproßspitze. u_1—u_6 periphere Knotenzellen in Reihenfolge der Entstehung; *c* zentrale Knotenzellen; *v* Scheitelzelle des Sprosses; *v'* Scheitelzelle eines lateralen Kurztriebes; *k* Knotenzelle; *i* Internodialzelle; *g* erstes Segment einer Scheitelzelle. (Nach GIESENHAGEN 1902; ca. 340×.)

a) Vegetative Vermehrung durch Sproßknoten.

An den Knoten behalten stets noch einige periphere Zellen embryonalen Charakter. Von diesen besitzen einzelne als ruhende Vegetationspunkte die Anlage, akzessorische Seitensprosse zu bilden; andere können nach weiteren Teilungen adventive Sprosse und Rhizoiden hervorbringen. Die normalen Knoten besitzen auf diese Weise bereits ein vegetatives Reproduktionsvermögen, das bei der intakten Pflanze einer korrelativen Hemmung durch die Vegetationspunkte der Haupt- und Seitenachsen unterliegt (BESSENICH 1923, GOEBEL 1918). Zur Aufhebung dieser Hemmung kommt es, wenn eine Pflanze der Sproßspitzen an Haupt- und Seitentrieben beraubt oder ein Knoten durch Absterben der benachbarten Internodien isoliert wird, was in der Natur z.B. durch Frost oder Tierfraß geschieht oder auch im Experiment vollzogen werden kann (AGARDH 1826, PRINGSHEIM 1862, RICHTER 1894, GIESENHAGEN 1902, DEBSKI 1898, MÜLLER 1907, BESSENICH 1923, DREW 1926). Vereinzelt ist es sogar gelungen, aus Blattknoten Adventivbildungen zu erhalten, nachdem die Blätter unter anderem durch Verwundung der Hemmung durch die embryonalen Anlagen des Basalknotens entzogen wurden (GIESENHAGEN 1902, DREW 1926, WALLNER 1932, GILLET 1956). Auch Wachstumshemmung der Haupt- und Seitensprosse durch erniedrigte Temperatur (ANDERSON u. LOMMARSON 1958 bei *Chara zeylanica* Willd.) und Verdunkelung (WALLNER 1932 bei verschiedenen Arten) kann die teilungsfähigen Knotenzellen zur Bildung von Ersatztrieben und Rhizoiden anregen. Zum gleichen Ergebnis führt bei intakten Pflanzen das Bedecken der Knoten mit Erde (RICHTER 1894 bei *Chara fragilis* Desv. = *Ch. globularis* Thuill.)[1].

Die *akzessorischen Sprosse*, die bei der vegetativen Reproduktion eines isolierten Stengelknotens zunächst vorwiegend an der Basis des Seitensprosses aus ruhenden Vegetationspunkten entstehen, unterscheiden sich nicht von normalen Langtrieben. Daneben treten in der weiteren Entwicklung die von PRINGSHEIM (1862) als ,,nacktfüßige Zweige" und ,,Zweigvorkeime" beschriebenen Triebe auf. Die *nacktfüßigen Zweige* sind Langtriebe mit vereinfachten Basalknoten und daher unvollständiger oder fehlender Berindung am ersten Internodium (Abbildung 96a—c). Die unberindeten Nitellen bilden entsprechende Triebe, bei denen der vereinfachte Bau der ersten Knoten nur in der Ausbildung der Kurztriebwirtel zum Ausdruck kommt (DREW 1926). Allgemein kann man hier im Gegen-

[1] Neuere Namen nach WOOD und IMAHORI (1965) jeweils in Klammern hinzugefügt.

satz zu den Zweigvorkeimen also nur von sekundären oder tertiären Langtrieben sprechen. Die *Zweigvorkeime* (engl.: secondary protonema) sind Kurztriebe, die mit dem aus einer Oospore entstehenden Vorkeim (engl.: protonema) völlig übereinstimmen, wie schon Agardh (1826) und Pringsheim (1862) feststellten.

Abb. 96.

Abb. 97.

Abb. 96. Austreibendes Sproßknöllchen von *Chara delicatula* var. *bulbifera* (= *Chara globularis* Thuill. var. *virgata*) mit vier „nacktfüßigen Zweigen" in verschiedenen Entwicklungsstadien (a—d), einem Zweigvorkeim (e = secondary protonema) und zahlreichen Rhizoiden. (Nach Kuczewski 1906; 7 ×.)

Abb. 97. Zweigvorkeim (= secondary protonema) von *Chara aspera* (= *Chara globularis* Thuill. var. *aspera*). *s* Vorkeimspitze; *l* Langtrieb; *w* Wurzelknoten mit Rhizoiden und adventivem Sproß. (Nach Giesenhagen 1902; 19 ×.)

Sie besitzen einen mehr oder weniger rudimentär ausgebildeten Blattquirl, der von der Vorkeimspitze meist nur wenig überragt wird (Abb. 97), und einen Wurzelknoten, dem Rhizoiden und zuweilen auch weitere Sprosse entspringen. Erst aus dem Knoten des Blattquirls geht ein Langtrieb hervor, der wie beim Oosporenkeimling zur eigentlichen Pflanze heranwächst (Abb. 97, *l*). Die nacktfüßigen Zweige bzw. sekundären Langtriebe entstehen offenbar vorwiegend akzessorisch aus ruhenden Vegetationspunkten. Die Zweigvorkeime dagegen sollen nur aus undifferenzierten embryonalen Zellen gebildet werden (RICHTER 1894, DREW 1926, GILLET 1956). Letztere können auch unmittelbar aus jungen Rhizoiden hervorgehen (NORDSTEDT u. WAHLSTEDT 1875, DEBSKI 1898, GIESENHAGEN 1902). Bei Nitellen treten sie nicht so häufig auf (PRINGSHEIM 1862, MIGULA 1897, RICHTER 1894, DREW 1926).

b) Bulbillen.

Bei den Sproß- und Wurzelknöllchen wird das vegetative Reproduktionsvermögen normaler Knoten durch Vermehrung der embryonalen Zellen verstärkt. Stärkehaltige Zellen sorgen für die beim Austreiben benötigten Reservestoffe. Letztere sind größer (bei den Wurzelknöllchen oft sehr groß) und besitzen fragmentierte Kerne (KUCZEWSKI 1906, MCNICOL 1907), was auf ihre Homologie mit Internodialzellen schließen läßt. Die fertigen Bulbillen können sehr unregelmäßig im Aufbau erscheinen, unterliegen aber alle der Gesetzmäßigkeit, welche die Entwicklung der Characeen beherrscht (GIESENHAGEN 1902). Die so als Brutorgane spezialisierten Knoten sind gegenüber den normalen Knoten in besonderer Weise geeignet, nach Absterben der Internodien von einer Vegetationsperiode zur anderen zu überdauern. Bei *Tollypellopsis stelligera* (Bauer) Migula (= *Nitellopsis obtusa* [Desv. in Lois.] J. Gr.) übernehmen die Stengel- und Wurzelknöllchen als Ausgleich für die stark beschränkte Fruktifikation den Hauptanteil der Fortpflanzung (STROEDE 1931).

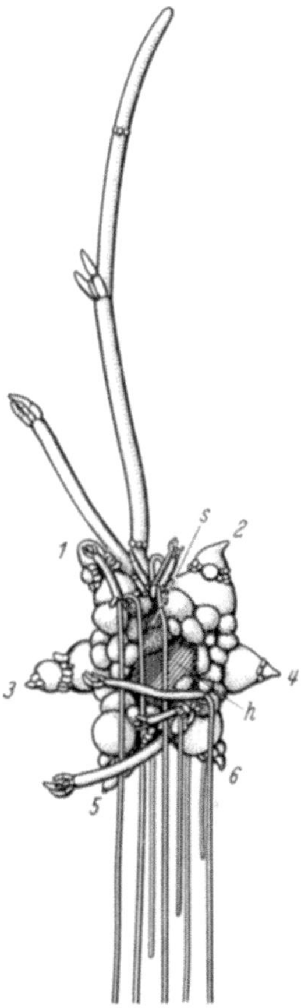

Abb. 98. Austreiben des Stengelknöllchen von *Chara stelligera* (= *Nitellopsis obtusa* J. Gr.). *1*—*6* rudimentäre Blätter; *h* und *s* Ansatz von Haupt- und Seitenachse. (Nach GIESENHAGEN 1902; 8×.)

1. Sproßknöllchen.

Ein kontinuierlicher Übergang von normalen Stengelknoten zu unterirdischen Sproßknöllchen ist bei *Chara delicatula* var. *bulbifera* zu finden (GIESENHAGEN 1902 u. KUCZEWSKI 1906; = *Chara globularis* Thuill. var. *virgata* f. *virgata*). Die Blätter werden mehr und mehr rudimentär. Die Entwicklung verläuft aber auch bei den Knöllchen regelmäßig bis zur Bildung der Blattbasalknoten, denen die kegelförmigen Scheitelzellen direkt aufsitzen können (KUCZEWSKI 1906). Häufig bilden hier, wie auch bei *Chara aspera* (vgl. S. 756), die ersten Knoten des Seitensprosses mit dem Knöllchen des Hauptsprosses einen zusammenhängenden Komplex.

Die sternförmigen Bulbillen an unterirdischen Sprossen von „*Chara stelligera*" (= *Tolypellopsis stelligera* Migula = *Nitellopsis obtusa* [Desv. in Lois.] J. Gr.) zeigen in ihren Strahlen die rudimentären Blattanlagen noch ausgeprägter (Abb. 98; GIESENHAGEN 1902).

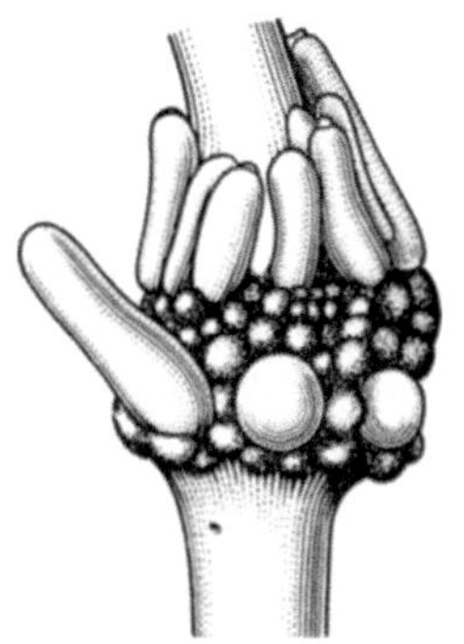

Abb. 99. Unterirdische Sproßknoten von *Chara baltica* (= *Chara hispida* L. var. *baltica*) mit rudimentärer Blatt und Rindenbildung. (Nach GIESENHAGEN 1902; 25×.)

Eine stärkere Metamorphose der Stengelknoten findet man in den Knöllchen von *Chara baltica* (GIESENHAGEN 1902; = *Chara hispida* L. var. *baltica*). Schon die einfachen

Knoten der unterirdischen Sprosse bilden nur rudimentäre Blätter und Rindenzellen aus (Abb. 99) und besitzen stärkehaltige Zellen. Sie sind nicht minder zur Bildung neuer Sprosse befähigt als die Knöllchen, bei denen die Blattanlagen — mehrere oder vereinzelt — zu zapfenförmigen Vorsprüngen auswachsen.

Das Austreiben der Sproßknöllchen und Knoten hat GIESENHAGEN (1902) bei „*Chara stelligera*" (= *Nitellopsis obtusa*, vgl. S. 755) in vivo verfolgt. Die Stärkekörner in den Speicherzellen lösen sich auf, und es beginnt eine lebhafte Plasmarotation. Die Rhizoiden und Sproßanlagen — nacktfüßige Zweige wie auch Zweigvorkeime — entspringen nur auf der Oberseite des Knöllchens, besonders an der Basis des Seitensprosses (Abb. 98). Nach MIGULA (1897) sollen hier die Knöllchenstrahlen direkt zu Seitensprossen austreiben, was von GIESENHAGEN jedoch nicht bestätigt wird. DREW (1926) zählte bei *Chara hispida* L., bei der die einfachen Stengelknoten der Überwinterung dienen (MÜLLER 1907), bis zu 45 Triebe an einem Knoten.

2. Wurzelknöllchen.

Die Wurzelknöllchen hat GIESENHAGEN (1902) bei *Chara aspera* ausführlich beschrieben (= *Chara globularis* Thuill. var. *aspera*). Er fand sie besonders im Herbst und im Winter als weiße Kügelchen oder Spindeln einzeln oder in Gruppen von zwei, drei oder vier an Verzweigungsknoten der Rhizoiden. Sie entstehen aus den vier Primärzellen des Knotens, die normalerweise zu weiteren Rhizoiden auswachsen (Abb. 100). Von den vier Primärzellen werden vorzugsweise die beiden unteren zu größeren, die beiden oberen dagegen meist nur zu kleineren Knöllchen (Abb. 101 a).

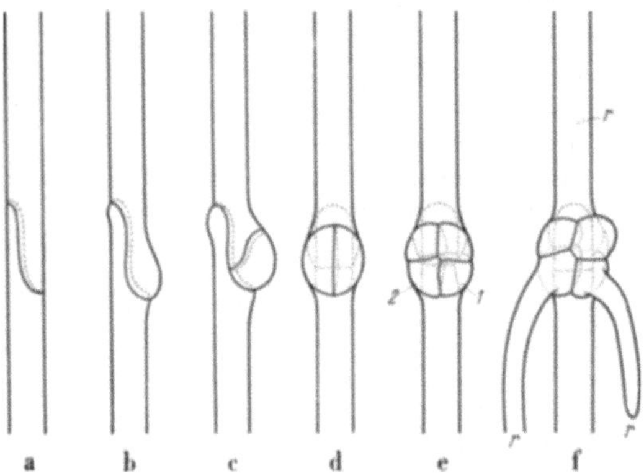

Abb. 100. Entwicklung eines Rhizoidknotens von *Chara aspera*. *r* Rhizoiden. (Nach GIESENHAGEN 1902; 40×.)

Die Bevorzugung der unteren Zellen entspricht der Teilungsfolge, die der in einer jungen Knotenzelle des Stengels homolog ist (vgl. Abb. 95a u. b und 100d u. e), wo der Seitensproß aus Zelle 1 und das größte Blatt aus der Zelle 2 hervorgehen. Die mit Stärke angefüllten Knöllchen besitzen eine sehr dicke Wand und erscheinen zunächst als einzellige Gebilde. Basal wie auch apikal befindet sich jedoch in die dicke Zellwand eingesenkt je eine Gruppe von Zellen, deren Anordnung und Entstehung mit der eines Stengelknotens übereinstimmt (Abb. 101 b). Wie beim Sproßknoten sind auch hier nur die peripheren Zellen noch teilungsfähig. Sie können zu Rhizoiden auswachsen oder Zweigvorkeime bilden (Abb. 101 c). Die Knöllchenzellen selbst sind als Internodialzellen zu betrachten. Aus den peripheren Zellen des basalen Knotens entstehen zuweilen noch weitere spindelförmige Knöllchen (Abb. 101 b). Die Teilzellen eines Verzweigungsknotens, die nicht zu Knöllchen werden, teilen sich zu maulbeerartigen Zellhaufen, die keine Stärke besitzen und embryonalen Charakter behalten. Einige davon können gleich zu Rhizoiden auswachsen und andere bilden beim Austreiben der Bulbillen nacktfüßige Zweige, während die Knöllchen selbst nur Zweigvorkeime hervorbringen.

Ähnlich wie bei *Chara aspera* sind die Wurzelknöllchen von *Lychnothamnus macropogon* (BRAUN und NORDSTEDT 1882 und GIESENHAGEN 1902; = *Lampro-*

thamnium papulosum [Wallr.] J. Gr. var. *papulosum*) und von *Lamprothamnus alopecuroides* (GIESENHAGEN 1902; = *Lamprothamnium papulosum*) aufgebaut.

Dabei können — wie vereinzelt auch bei *Chara aspera* (Abb. 101 b) — aus den peripheren Knotenzellen an der Basis der primären zahlreiche sekundäre Knöllchen entstehen, welche meist kleiner sind als jene. Sie stehen bei *Lychnothamnus* in Gruppen von 12 und mehr. Bei *Lamprothamnus alopecuroides* können die aus den basalen und apikalen Zellen des primären Knöllchens entspringenden

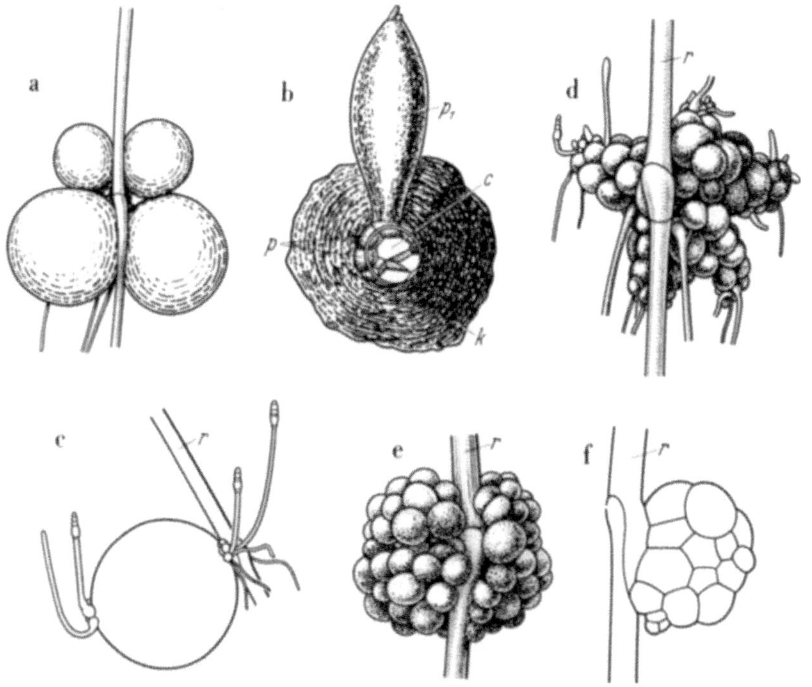

Abb. 101 a—f. Wurzelknöllchen von a—c *Chara aspera*; a aus den vier Primärzellen eines Rhizoidknotens hervorgegangene Knöllchen, b basaler Knoten eines Einzelknöllchens (*k* Ausschnitt der Knöllchenwand; *p* periphere Knotenzelle, von denen p_1 zu einem sekundären Knöllchen ausgewachsen ist; *c* zentrale Knotenzelle); c austreibendes Knöllchen; d *Chara baltica*; e und f *Chara fragifera* in zwei Ansichten. *r* Rhizoiden. (Nach GIESENHAGEN 1902; a 12,5 ×, b 50 ×, c 10 ×, d und e 16 ×, f 27 ×.)

Rhizoiden sekundäre und diese auf die gleiche Weise tertiäre Knöllchen bilden, die allerdings beträchtlich kleiner sind.

Die vielzelligen Wurzelknöllchen bei *Chara baltica* (GIESENHAGEN 1902; = *Chara hispida* L. var. *baltica*) besitzen zuweilen zapfenförmige Vorsprünge (Abb. 101 d), die durch eine zentrale Reihe größerer Zellen und sternförmige Anordnung ebenfalls als metamorphosierte Blätter zu erkennen sind. Durch eine abnorme erste Teilung des Wurzelgelenks entwickeln sie sich oft allseitig und sind dann kaum von Stengelknöllchen zu unterscheiden, bringen aber im Gegensatz zu letzteren keine nacktfüßigen Zweige, sondern nur Zweigvorkeime hervor.

Maulbeeartige, vielzellige Wurzelknöllchen gibt es bei *Chara fragifera* (Abb. 101 e und f; GIESENHAGEN 1902; = *Chara globularis* Thuill. var. *globularis* f. *fragifera*). Die gesetzmäßige Anordnung und Entstehung der Zellen soll hier im Schnitt an der Größe der Stärkekörner zu erkennen sein.

Bei den Wurzelknöllchen von *Chara delicatula* var. *bulbifera* (vgl. S. 755), die ebenso wie die von *Chara aspera* (S. 756 f.) entstehen, wurden keine Sproßbildungen beobachtet (Giesenhagen 1902, Kuczewski 1906). Sie sind offenbar reine Speicherorgane.

Die Wurzelknöllchen dienen als vegetative Fortpflanzungskörper nicht immer nur der Überwinterung, sondern fördern zuweilen auch die Verbreitung in derselben Vegetationsperiode. Bei *Nitella syncarpa* (Thuill.) Chev. schwillt der Wurzelknoten des Vorkeims bereits „zu einer zelligen Kugel von bedeutendem Umfang an“ (Ernst 1901), der neben Rhizoiden bis zu 26 (nach Braun sogar 50) Sprosse entspringen. „Einige dieser Sprosse legen sich nieder und bilden in einiger Entfernung von der Mutterpflanze ähnliche an Wurzeln und Sprossen reiche Anschwellungen“ (A. Braun 1849, nach Ernst 1901). Über Wurzelknöllchen und geschwollene basale Achselknoten bei *Nitellopsis bulbifera* berichten Carl de Donterberg (1960) und Tindall (1965).

Die Ausstattung mit vegetativen Fortpflanzungsorganen ist bei den Characeen nicht obligatorisch. *Chara canescens* z.B. (vgl. S. 758f.) bildet keinerlei Bulbillen und ist offenbar auch nicht zu vegetativer Fortpflanzung durch die normalen Stengelknoten befähigt (A. Braun 1856 und A. Ernst 1918). Manche Arten besitzen nur Wurzelknöllchen wie z.B. *Chara aspera* (S. 756), andere nur Stengelknöllchen wie z. B. *Chara stelligera* (S. 755) und wieder andere beides wie z.B. *Chara fragifera* (S. 757).

c) Parthenogenese.

Parthenogenese ist nur bei der diöcischen *Chara crinita* Wallroth (= *Chara canescens* Desv. u. Lois.) nachgewiesen. An den meisten Standorten tritt diese Art nur in weiblichen Exemplaren auf. Da die Pflanzen streng einjährig sind und sich nicht wie andere Arten vegetativ durch überwinternde Bulbillen fortpflanzen, worauf schon A. Braun (1856) hinwies, erhalten und vermehren sie sich an diesen Standorten offenbar nur durch Parthenosporen. Männliche Pflanzen sind nur von wenigen Standorten bekannt (Literatur bei Ernst 1918). Sie gehören einer amphimiktischen Form an, die Filarszky (nach Ernst 1918) an verschiedenen Stellen bei Budapest zum Teil mit der parthenogenetischen vergesellschaftet fand.

Den experimentellen Nachweis der Parthenogenese erbrachten De Bary (1872, 1875), Filarszky (1893, nach Ernst 1918) und Migula (1897, S. 357) an isoliert kultivierten weiblichen Pflanzen. Filarszky erhielt über 5 Jahre parthenogenetische Nachkommen aus einer Kultur. Migula beobachtete außerdem, daß die Keimkraft der Parthenosporen im Vergleich zu den Zygosporen der sexuellen Form nach einem Jahr erheblich nachließ. Aus den Parthenosporen gehen immer wieder nur weibliche parthenogenetische Pflanzen hervor.

A. Ernst (1918) stellte an dem Budapester Material fest, daß die parthenogenetische Form gegenüber der amphimiktischen diploid ist ($2n = 24$) und die Reduktionsteilung bei der Keimung der Oospore ausfällt. In Kulturen entwickelten sich fast alle Oogonien dieser Pflanzen zu Sporen, während bei isoliert wachsenden weiblichen Pflanzen der amphimiktischen Rasse die Sporenbildung ausblieb. Eine künstliche Auslösung der Parthenogenese bei haploiden Weibchen versuchte Ernst mit einigen in der Zoologie bewährten Methoden. Dabei führte nur die Anwendung von hyper- und hypotonischen Lösungen zu „einigen Resultaten“. Dennoch sei mit Ernst darauf hingewiesen, daß diöcische Characeen sehr geeignete Objekte für die Fortsetzung solcher Experimente sind.

Bei den Oogonien der parthenogenetischen Form unterbleibt die Bildung des Oogonienhalses und das Auseinanderweichen der Hüllschläuche in diesem Bereich. Dies dient bei der amphimiktischen Form zur Erleichterung der Befruchtung und wird besonders deutlich, wenn die Oogonien nicht unmittelbar nach der Reifung befruchtet werden (ERNST 1918).

Über die Entstehung der diploid-parthenogenetischen Rasse von *Chara crinita* liegen nur Vermutungen vor, die ausführlich von ERNST (1918) und WINKLER (1920) diskutiert werden. Ersterer hält die Entstehung durch (Aniso-)Autodiploidie infolge unterdrückter Reduktionsteilung für ausgeschlossen, da sonst, ähnlich wie in einigen Fällen bei getrenntgeschlechtlichen Moosen (MARCHAL 1909, 1911), Zwitter entstehen müßten. Er glaubt an die Bastardierung einer haploiden weiblichen *Chara crinita* mit dem Männchen einer nicht nachgewiesenen Art. Die Bastardierung betrachtet er zugleich als Ursache von Diploidie und parthenogenetischem Verhalten. (Hierauf baut er seine Hypothese von der „Bastardierung als Ursache der Apogamie im Pflanzenreich" auf.) Mit dieser Hypothese setzt er aber auch eine Epistasie der Weiblichkeitsfaktoren voraus, von ihm als „Metromorphie" bezeichnet.

Dem ist mit WINKLER (1920) entgegenzuhalten, daß es weder für die Bastardnatur des Apomikten einen Nachweis gibt noch dafür, daß eine Diploidisierung zur Parthenogenese und zum Ausfall der Meiosis führen muß. WINKLER selbst hält die Entstehung durch (Iso-)Autodiploidisierung einer weiblichen *Chara crinita* für wahrscheinlich und gibt dafür zwei mögliche Wege an. Er umgeht damit die Annahme der Epistasie des weiblichen Geschlechts, die allerdings heute nicht mehr a priori abgelehnt werden kann. Für die parthenogenetische Fortpflanzungsweise gibt er keine zwingende Erklärung.

Literatur.

AGARDH, C. A.: Über die Anatomie und den Kreislauf der Charen. Nova Acta Acad. Caesar. Leop. Carol. Nat. Cur. **13**, 113—162 (1826). — ANDERSON, R. G., and R. C. LOMMARSON: Some effects of temperature on growth of *Chara zeylanica* Willd. Butler Univ. bot. Stud. **13**. 113—120 (1958).

BESSENICH, K.: Über Beziehungen zwischen dem Vegetationspunkt und dem übrigen Pflanzenkörper bei *Chara*. Jb. wiss. Bot. **62**, 214—243 (1923). — BRAUN, A.: Übersicht der schweizerischen Characeen. Neue Denkschr. schweiz. Ges. Naturw. **10**, 1—23 (1849). — Über Parthenogenese bei Pflanzen. Abh. Kgl. Akad. Wiss. Berlin **1856**, 311—376 (1856). — BRAUN, A., u. O. NORDSTEDT: Fragmente einer Monographie der Characeen. Nach den hinterlassenen Manuskripten A. BRAUNS, herausgeg. von O. NORDSTEDT. Abh. Kgl. Akad. Wiss. Berlin **1882**, 1—211 (1882).

CARL DE DONTERBERG, C. C.: Une nueva especie de *Nitellopsis hallada* en la Argentina. Comm. Mus. Argentino Cienc. Natur. **1**, 3—10 (1960). — CLAVAUD, A.: Note sur les organes hypogés des Characées. Bull. Soc. bot. Fr. **10**, 137—148 (1863).

DE BARY, A.: Zur Keimungsgeschichte der Characeen. Bot. Ztg **33**, 377—385, 393—401 409—420 (1875). — DEBSKI, B.: Weitere Beobachtungen an *Chara fragilis* Desv. Jb. wiss. Bot. **32**, 635—670 (1898). — DREW, K. M.: The "leaf" of *Nitella opaca* Ag. and adventitious branch development from it. Ann. Bot. **40**, 321—348 (1926). — DURIEU DE MAISONEUVE, M. CH.: Nouvelles observations sur les bulbilles des Characées. Bull. Soc. bot. Fr. **7**, 627—633 (1860).

ERNST, A.: Über Pseudo-Hermaphroditismus und andere Mißbildungen der Oogonien von *Nitella syncarpa* (Thuill.) Kützing. Flora (Jena) **88**, 1—36 (1901). — Bastardierung als Ursache der Apogamie im Pflanzenreich. Jena: Gustav Fischer 1918.

GIESENHAGEN, K.: Untersuchungen über die Characeen. Marburg 1902 (Sonderausgabe der in Flora (Jena) **82**, 381—433, **83**, 160—202, **85**, 19—64 erschienenen Arbeiten). — GILLET, C.: Recherches expérimentales sur la morphogénèse de *Chara vulgaris* L. Lejeunia **20**. 5—14 (1956). — GOEBEL, K.: Zur Organographie der Characeen. Flora (Jena) **110**. 344—387 (1918).

HASITSCHKA-JENTSCHKE, G.: Beiträge zur Karyologie von Characeen. Öst. bot. Z. **107**, 228—240 (1960).

KUCZEWSKI, O.: Morphologische und biologische Untersuchungen an *Chara delicatula* f. *bulbillifera* A. Braun. Beih. bot. Zbl. **20**, 25—75 (1906).

MARCHAL, ÉL. et ÉM.: Aposporie et sexualité chez les Mousses. II, III. Bull. Acad. r. Belg. Cl. Sci. **12**, 1249—1288 (1909); **9/10**, 750—778 (1911). — MCNICOL, M.: The bulbils and proembryo of *Lamprothamnus alopecuroides* A. Braun. Ann. Bot. **21**, 61—70 (1907). — Migula, W.: Die Characeen Deutschlands, Österreichs und der Schweiz. In: Rabenhorsts Kryptogamenflora, Bd. 5. Leipzig: E. Kummer 1897. — Müller, A.: Beiträge zur Kenntnis der *Chara hispida* Linné und *Chara foetida* Braun. Diss. Zürich-München 1907.

Nordstedt, O., u. L. J. Wahlstedt: Über die Keimung der Characeen. Flora (Jena) **58**, 94—95 (1875).

Pringsheim, N.: Über die Vorkeime und die nacktfüßigen Zweige der Charen. Jb. wiss. Bot. **3**, 294—324 (1862).

Richter, J.: Über Reaktionen der Characeen auf äußere Einflüsse. Flora (Jena) **78**, 399—423 (1894).

Stroede, W.: Ökologie der Characeen. Diss. Berlin 1931.

Tindall, D. R., T. Sawa, and A. T. Hotchkiss: *Nitellopsis bulbillifera* in North America. J. Phycol. **1**, 147—150 (1965).

Wallner, F.: Die Beeinflussung der Gametangienbildung bei den Characeen. Flora (Jena) **126**, 249—293 (1932). — Winkler, H.: Verbreitung und Ursache der Parthenogenese im Pflanzen- und Tierreiche. Jena: Gustav Fischer 1920. — Wood, R. D., and K. Imahori: A revision of the Characeae. Weinheim: Cramer 1965.

H. Rhodophyta.

Bei den meisten Rotalgen tritt neben der für sie charakteristischen oogamen geschlechtlichen Fortpflanzung noch eine ungeschlechtliche Fortpflanzung auf, die bei einer Anzahl von Formen die einzige (bisher bekannte) Vermehrungsweise ist. Die Mittel dieser ungeschlechtlichen Fortpflanzung sind entweder Mono-, Para- und Tetrasporen, spezifische und unbegeißelte Einzelzellen also, die zum Teil in den Generations- und Kernphasenwechsel eingeschaltet sind, oder, in selteneren Fällen, aus einer Anzahl von vegetativen Zellen gebildete Brutkörper (Propagulen, Gemmen). Eine weitere Möglichkeit vegetativer Vermehrung besteht in der Fragmentation des Thallus.

a) Fortpflanzung durch Sporen.

1. Autosporen.

Die niederen Bangioideen *(Porphyridium, Chroothece)* vermehren sich ausschließlich durch wiederholte Zweiteilung. Dabei werden um die Tochterzellen neue Membranen gebildet, die Zellen bleiben durch die gedehnte Membran der Mutterzelle noch eine Zeitlang, oft durch zwei Teilungsgenerationen, miteinander verbunden. Bei *Rhodospora sordida* erfolgt die Fortpflanzung entweder ebenfalls durch eine solche (endogene) Zweiteilung oder es werden durch mehrere zueinander senkrechte Teilungen 4—8 (—16) Autosporen gebildet, die durch Aufreißen der Mutterzellmembran frei werden (Geitler 1927). Dieser Fall zeigt, daß zwischen Zweiteilung und Autosporenbildung kein prinzipieller Unterschied besteht; die Zweiteilung kann nämlich als Bildung zweier Autosporen aufgefaßt werden.

2. Monosporen.

Sie entstehen in Einzahl in den Monosporangien. Der mit Reservesubstanz angereicherte und oft infolge verstärkter Chromatophorenbildung kräftig rot gefärbte Inhalt der Sporangienzelle wird als nackte, cilienlose, anfangs häufig amöboide Spore entleert und umgibt sich nach einiger Zeit mit einer Membran. Die Monosporen keimen im allgemeinen ohne Ruheperiode aus. Die Vermehrung durch Monosporen ist auf die Bangioideae und, von wenigen Ausnahmen abgesehen, auf die Florideenordnung Nemalionales beschränkt.

Bei den *Bangioideae* treten Monosporen in zahlreichen Gattungen auf. Bei *Porphyridium*, das sich durch wiederholte Zweiteilung vermehrt, hat Vischer (1935) das gelegentliche Vorkommen nackter, amöboid beweglicher Monosporen festgestellt. In den verzweigten Fäden von *Asterocytis ramosa* können die Monosporen sowohl in den Endzellen wie auch in den Gliederzellen gebildet werden (SCHMITZ 1894, WILLE 1900). Nach Ausbildung der Membran keimt die Spore direkt zu einem neuen Faden aus. Ähnlich liegen die Verhältnisse bei *Goniotrichum elegans*.

Deutlicher gegenüber den vegetativen Zellen differenzierte Monosporangien besitzen *Erythrotrichia, Erythrocladia* und *Porphyropsis*. Hier sammelt sich in einer der oberen äußeren Ecken der vegetativen Zellen reichlich Plasma. Durch eine schräge, einwärts gebogene Wand wird die Ecke dann als kleinere Sporangienzelle abgegrenzt (Abb. 102). Bei *Erythrotrichia* kann sich der Vorgang an ein und derselben Zelle mehrfach wiederholen, indem nach dem Ausschlüpfen der Monospore der Raum von der inhaltsärmeren vegetativen Zelle wieder ausgefüllt und eine neue Sporangienzelle an Stelle der alten gebildet wird. Die ausgetretenen Monosporen führen, obwohl nicht amöboid, ruckweise Bewegungen aus, deren Ursache aber unbekannt ist (ROSENVINGE 1927). In ganz ähnlicher Weise werden die Monosporen bei *Rhodochaete* gebildet (THURET in BORNET 1892), einer Gattung, die wie *Compsopogon* eine unsichere Stellung zwischen den beiden Klassen der Rotalgen einnimmt. MAGNE (1960a, b) konnte indessen zeigen, daß der Monosporenbildung von *Rhodochaete* die Fusion einer bisher unbekannten kleinen Spore mit der vegetativen Zelle, von der das Monosporangium abgegliedert wird, vorausgeht. Diese kleinen Sporen, die nur etwa $^1/_3$ so dick wie die Monosporen sind, werden in derselben Weise wie die Monosporen erzeugt. Da nach der Zellverschmelzung eine Kernfusion erfolgt, muß man wohl mit MAGNE die kleinere Spore als Spermatium und die Monospore THURETs als Karpospore deuten. Da die inäquale Teilung jedoch mitotisch ist, entsteht die Frage nach dem Ort der Meiose und die Vermutung, daß ein bisher noch unentdeckter Phasen- und Generationswechsel existiert.

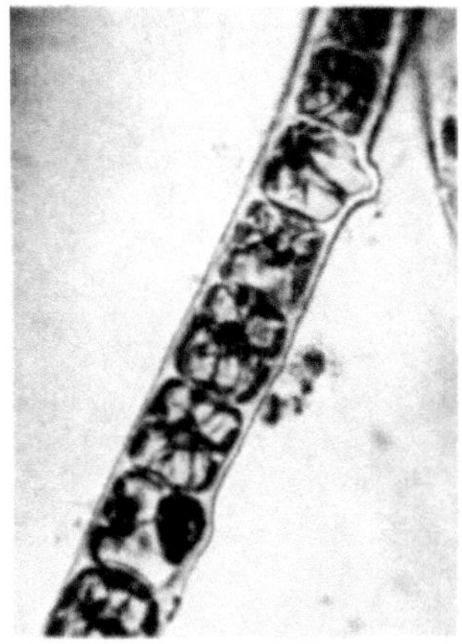

Abb. 102. *Erythrotrichia carnea*, unten: Monosporangium, oben: Monospore bereits ausgetreten, Vorwölbung bezeichnet die Austrittsstelle. 700×. (Original.)

Bei *Compsopogon* stellt THAXTER (1900) einen Sporendimorphismus fest. Bei der Bildung der sog. Makrospore gliedern Rindenzellen schräg nach außen inhaltsreiche Nebenzellen ab, deren ganzer Inhalt eine Monospore bildet. Nach deren Austritt füllt die Mutterzelle, in Analogie zu *Erythrotrichia*, den Hohlraum erneut aus. THAXTERs Mikrosporen entstehen in größerer Zahl in Sori, die nach Abgliederung einer Nebenzelle durch deren wiederholte Teilungen entstehen. Jede Zelle dieser Gruppe erzeugt eine Mikrospore, deren Entwicklung der Autor allerdings nicht beobachten konnte; sie entsprechen nach THAXTER möglicherweise Spermatien. Nach Ansicht SKUJAs (1938) stellen die Mikrosporen die normalen Fortpflanzungsorgane und die Makrosporangien lediglich in der Kultur auftretende, gehemmte Zweigansätze dar. Eine Untersuchung von NICHOLS (1964a) in Kultur scheint aber zu zeigen, daß eine scharfe Trennung zwischen beiden Sporensorten, was die Größe und auch das Resultat der Keimung betrifft, nicht existiert. Die Monosporen werden 1. wie oben geschildert, inäqual, 2. aus dem ganzen Inhalt einer Rindenzelle oder 3. nach dessen Aufteilung in „Mikroaplanosporen", 4. in Sori am einreihigen Faden oder 5. in hervorgewölbten Sori am berindeten Faden gebildet. Leider fehlt eine genügend eingehende Darstellung der Entstehung dieser Sporenformen. FLINT (1947) beschreibt eine geschlechtliche Fortpflanzung dieser Gattung, was aber wohl weiterer Untersuchungen bedarf.

Bei der kürzlich entdeckten Gattung *Boldia* (Herndon 1964) liefert die keimende Monospore ein fädiges Protonema, an dem sich Basalscheiben und aus diesen die saccaten Thalli entwickeln (Nichols 1964b). Protonema und Thalli vermögen Monosporen zu bilden.

Mit einer gewissen Unsicherheit behaftet sind die Angaben über die Fortpflanzung der Gattung *Kyliniella*, die etwas an *Rhodochaete* erinnert. Skuja (1926) beschreibt den Austritt eines ganzen Zellinhalts in Form einer Monospore bei *K. latvica*, während Flint (1953) die Emission männlicher Gameten und deren Verschmelzung mit benachbarten Zellen beobachtete. Weiter fand er abweichend gestaltete Pflanzen, deren Zellen in zwei oder vier „Segmente" geteilt waren. Magne ist geneigt, letztere für Tetrasporangien zu halten; man muß aber wohl Drew (1956) zustimmen, die zur Klärung der Fortpflanzungsverhältnisse weitere Studien für notwendig hält.

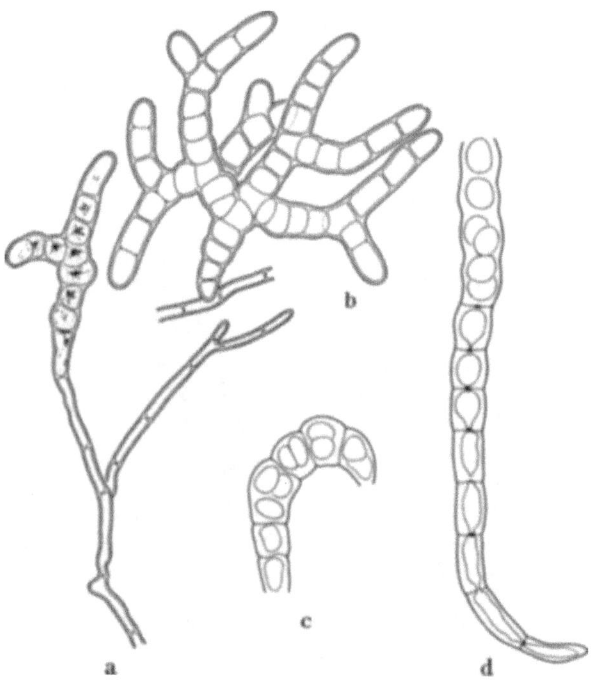

Abb. 103a—d. Conchosporangien und Conchosporen der *Conchocelis*-Phase von *Porphyra*. a, b Die breiten verzweigten Conchosporangienäste an der dünnfädigen *Conchocelis* (*Porphyra* spec.) 175 ×. c, d Conchosporen werden nach Zweiteilung des Inhaltes der Einzelsporangien und Auflösung der Querwände zwischen den sporenbildenden Zellen durch den entstandenen Schlauch entlassen (*P. tenera*). (a, b nach Kornmann, c, d nach Tseng und Chang.)

Drew (1956) klassifiziert die bei den Bangioideae vorkommenden Sporen in drei Typen. Typ 1: Monosporen, die in differenzierten Sporangien erzeugt werden; das Monosporangium wird durch eine inäquale Teilung abgegrenzt. Diese Art der Monosporenbildung kommt in typischer Weise bei den Erythropeltidaceen (z.B. *Erythrotrichia*) vor. Typ 2: Monosporen, die in undifferenzierten Zellen entstehen. Typ 3: Eine Anzahl von Sporen entsteht aus einer Mutterzelle durch sukzessive Teilungen. Diesen Sporenformen könnte man als vierten Typ die Conchosporen anreihen.

Sowohl die flächenhaft entwickelte *Porphyra* als auch die fädige, im Alter meist polystiche *Bangia* alternieren mit einer in der Natur Kalkschalen bewohnenden und ausdauernden Zwerggeneration, der verzweigt-zartfädigen *Conchocelis*. Der Cyclus ist für beide Gattungen in den letzten zwölf Jahren in der Kultur nachgewiesen worden (Literatur s. Drew 1956, Dixon 1963b). An den *Conchocelis*-Pflanzen entstehen unter geeigneten Bedingungen Sporangienfäden (Abb. 103a, b), die breiter als die vegetativen Fäden sind, streckenweise mehr als einreihig werden und außerdem zu ausschließlich Sporangien enthaltenden Verzweigungssystemen zusammentreten können (plantlets, Drew 1954). Bei der Reife scheint sich der Inhalt der Zellen im Sporangiumfaden noch einmal zu teilen (Tseng und Chang 1955a), danach werden die Wände zwischen den Zellen aufgelöst und der entstandene Schlauch, dessen Spitze aus der Kalkschale heraustritt, entleert die „Conchosporen" durch eine Öffnung ins Außenmedium (Abb. 103c, d). In Kultur können diese Vorgänge auch an freien, nicht endophytischen Thalli geschehen. Daneben gibt es bei *Porphyra tenera*- (IWASAKI und MATSUDAIRA 1963) und bei *Bangia atropurpourea-Conchocelis* (BELCHER 1960) auch aus Fadenendzellen entlassene Monosporen (Typ 2). An *Bangia* und *Porphyra* selbst können ganze Zellinhalte ebenfalls als Monosporen austreten, noch häufiger aber bilden sich Sporen nach Typ 3 (Aufteilung des Inhaltes in acht und mehr Zellen)[1]. Diese werden meist als Karposporen aufgefaßt. Doch ist die Befruchtung der Karpogone durch die ebenfalls nach Typ 3 entstehenden kleineren Spermatien mit Ausnahme vielleicht von *Porphyra tenera* (TSENG und CHANG 1955b) und *P. linearis* (s. z. B. MAGNE 1952) nicht mit völliger Sicherheit nachgewiesen worden (DREW 1956); außerdem besteht die Wahrscheinlichkeit, daß in der Natur amiktische Cyclen vorkommen, wobei die Spermatien dann funktionslos wären: *Porphyra purpurea* (KRISHNAMURTHY 1959) und *P. umbilicalis* (CONWAY 1964a, b). Ziemlich undurchsichtig sind die Dinge bei *Bangia*, weil hier auf dem gleichen Thallus beide Sporenformen entstehen, und man, da dieser selbst mehrreihig werden kann, zwischen beiden Sporenformen kaum sicher zu unterscheiden vermag. Aus diesen und aus rein cytologischen Gründen ist das Verhältnis der Kernphasen zwischen *Bangia* und *Porphyra* einerseits und *Conchocelis* andererseits nicht befriedigend geklärt. So finden TSENG und CHANG (1955b) bei *P. tenera* die Meiosis während der ersten beiden Teilungen des Karpogons nach der Befruchtung, welche bei *P. linearis* (MAGNE 1952) im Gegensatz dazu nur Mitosen sind; die zugehörigen *Conchocelis* wären also im ersten Falle haploid, im zweiten diploid. Die Monosporen von *Porphyra* wie auch von *Conchocelis* repetieren im allgemeinen die gleiche Phase, doch wurden *Porphyra*-Monosporen beobachtet, welche zur *Conchocelis* auskeimten (DREW 1954, KORNMANN 1961a). Die Sporen nach Typ 3 von *Porphyra* sowie die Conchosporen der *Conchocelis* liefern im allgemeinen die andere Phase. Doch kommen in beiden Fällen Ausnahmen vor (*Porphyra:* CONWAY 1964a, b; *Conchocelis:* IWASAKI und MATSUDAIRA 1963). Sporen, welche *Conchocelis* hervorbringen, keimen einseitig, solche mit *Porphyra*- bzw. *Bangia*-Keimlingen bipolar aus.

Nach Untersuchungen in Japan (KUROGI 1953, 1959, KUROGI und HIRANO 1956) beeinflussen Außenfaktoren wie Temperatur, Austrocknung, Salinität, Lichtintensität und insbesondere die Photoperiode das Wachstum, die Sporangienbildung und die Sporenemission der *Conchocelis*-Phase. Unter Kurztagbedingungen (8—11 Std) und geeigneter Temperatur entstehen an den *Conchocelis*-Fäden Monosporangien, in denen Sporen gebildet werden, welche zu *Porphyra*-Pflanzen auswachsen. Im Dauerlicht dagegen wird nur die *Conchocelis*-Phase reproduziert. IWASAKI (1961) sowie IWASAKI u. MATSUDAIRA (1963) registrierten an den unter Dauerlicht gehaltenen *Conchocelis*-Fäden verschiedenartige Reproduktionsorgane, die von den die *Porphyra*-Generation produzierenden Organen verschieden waren, nämlich (a) blasenartige, rundliche Zellen und (b) mehrzellige, erdbeerähnliche Organe. Aus (a) werden Einzelsporen entlassen, die sich zu *Conchocelis* entwickeln; aus (b) werden durch Auflösung des Organs die inneren Zellen frei und entwickeln sich unter Keimschlauchbildung ebenfalls zu *Conchocelis*-Filamenten. Je nach den herrschenden photoperiodischen Bedingungen ist dann auch die Gestalt der aus (a) und (b) hervorgehenden Filamente verschieden: Im Langtag (14 Std) entwickelt sich die dünnfädige, eigentliche *Conchocelis*, im Kurztag (9 Std) entwickeln sich die dickeren, massiveren „plant-

[1] Für die Frage, ob es unter diesen Sporangien zwei verschiedene oder nur einen Typus gibt, sei auf DREW (1956) und CONWAY (1964a, b) hingewiesen.

lets". — Das Mißlingen der Versuche von Drew u.a., den vollständigen Lebenscyclus von *Porphyra* und *Bangia* in der Kultur zu erhalten, ist wahrscheinlich darauf zurückzuführen, daß die gewählten lichtperiodischen Bedingungen zwar die Bildung von Sporen, die die *Conchocelis*-Phase reproduzieren, ermöglichten, nicht aber die Bildung *Porphyra*-produzierender Sporen. — Keine Korrelation besteht hingegen zwischen den Phasen des Lebenscyclus von *Porphyra* und der Temperatur, wie man nach der Saisonbedingtheit der Phasen (s. unten) vielleicht annehmen würde. Beide Phasen gedeihen im Temperaturbereich von 7 bis 21° C (Iwasaki 1961).

Bei den Helgoländer *Porphyra*-Arten *leucosticta, umbilicalis* und *purpurea* scheinen auffällige Abhängigkeiten der Reproduktion von der Tageslänge jedoch nicht zu existieren und eher die Temperatur der Kulturen die größere Rolle zu spielen (Kornmann 1961a, b).

Diese Fortpflanzungsverhältnisse sind von ökologischem Interesse, in Ostasien haben sie sogar ökonomische Bedeutung im Hinblick auf den Anbau von *Porphyra tenera*. Hier geht die Ernte der an den Küsten auf Reisbündeln kultivierten *Porphyra*-Pflanzen den ganzen Winter über vor sich. Im Frühjahr werden vor dem Absterben der Thalli Karposporen produziert, die zu *Conchocelis* auskeimen. Diese im Langtag wachsende Sommergeneration reproduziert sich vermutlich durch Monosporen, möglicherweise auch rein vegetativ, und erzeugt unter den Kurztagbedingungen im Herbst dann Conchosporen, aus denen wieder die *Porphyra*-Generation hervorgeht. Der Übergang zwischen den Phasen vollzieht sich ziemlich genau zur Zeit der Äquinoktien. In anderen Regionen sind die *Porphyra*-Pflanzen allerdings das ganze Jahr hindurch zu finden.

Nakazawa (1958) experimentierte mit den Monosporen der *Conchocelis*-Phase von *Porphyra tenera* mit dem Ziel, die prädeterminierte Polarität der Sporen vor der Keimung zu demonstrieren. Er erzielte mit Janusgrün B eine Differentialfärbung an der Spore in der Art, daß sich die basale Zone, mit der sich die Spore am Substrat festsetzt, rötlich färbt und die distale Zone, der präsumtive Apikalpol, blau. Bevor die inhärente Polarität sich morphologisch manifestiert, ist sie also an der ungekeimten Spore cytochemisch nachweisbar.

Auch für die Monosporen von *Bangia* ist die Annahme begründet, daß sie einer präformierten Polarität unterliegen. Jedenfalls läßt sich die Keimungspolarität nicht durch Außenfaktoren wie z.B. das Licht induzieren. Dagegen kann an einreihigen *Bangia*-Fäden gezeigt werden, daß erstens die Achsen der Monosporenkeimlinge den Querachsen der Sporangienzellen entsprechen, Keimlingsachse und Längsachse des *Bangia*-Thallus also senkrecht zueinander stehen, und daß zweitens die Polaritätsachsen der Keimlinge sogar über längere Abschnitte des Thallus parallel und gleichsinnig orientiert sind (Weber, unveröff.).

Unter den Ordnungen der *Florideen* ist die ungeschlechtliche Vermehrung durch Monosporen nur bei den Nemalionales von größerer Bedeutung. Die Monosporen sind hier akzessorische Keimzellen, welche die haploide oder auch die diploide Phase reproduzieren, und nicht in den Kernphasenwechsel eingeschaltet. Die Sporangien von *Acrochaetium* sitzen den Fäden seitlich an oder stellen die Endzellen kurzer Seitenzweige dar, auf denen sie auch in kleinen Gruppen zusammenstehen können. Bei *Colaconema* (Kylin 1944) stehen sie lateral an den vegetativen Zellen in Haufen von zwei bis sechs, in denen sowohl die terminalen wie die interkalaren Zellen Monosporen erzeugen. Einige *Acrochaetium*-Arten erzeugen sowohl Mono- wie Tetrasporen, die gemeinsam auf den Geschlechtspflanzen auftreten, zuweilen auch auf ungeschlechtlichen Pflanzen. In diesen Fällen ist es unsicher, ob bei der Tetrasporenbildung eine Reduktionsteilung stattfindet; vielleicht handelt es sich nur um vegetativ aufgeteilte Monosporangien. Andererseits mahnen die Untersuchungen MAGNEs (1961, 1964a, b) und v. STOSCHs (1965) an verschiedenen Nemalionales zur Vorsicht; viele solcher *Acrochaetium*-Formen könnten sich als die sporophytische Phase von Nemalionales mit heteromorph-heterophasischem Generationswechsel erweisen (s. auch weiter unten). Bei den meisten *Batrachospermum*-Arten ist die Vermehrung durch Monosporen auf die Jugendstadien (*Chantransia*-Stadien oder Pseudochantransien) beschränkt, doch treten Monosporen bei den entwickelten Pflanzen von *B. sporulans* und *B. vagum* auf, bei denen die geschlechtliche Vermehrung fast ganz fortgefallen ist. Die Monosporangien stehen hier in derselben Weise wie Spermatangien zu zweit oder dritt an den Enden der Kurztriebe, und es sind auch Übergänge zwischen beiden Keimzellbehältern festgestellt worden. Das Fehlen von Monosporen (auch am Vorkeim) wird bei *Lemanea* ausgeglichen durch vegetative Vermehrung (s. unten), während bei den ebenfalls zu den Nemalionales zählenden Thoreaceen die Monosporenbildung die einzige bisher bekannte Fortpflanzungsweise ist. Monosporenbildung wird weiterhin noch von den Gattungen *Scinaia*, *Helminthora* und *Liagora* berichtet. Die Monosporen der Nemalionales scheinen generell in derselben Weise wie deren Karposporen zu keimen, nämlich durch Anlage eines Keimschlauchs.

Unter Nicht-Nemalionales beschreibt SVEDELIUS (1914) das Vorkommen von Monosporangien auf weiblichen Gametophyten von *Nitophyllum punctatum*. Die Sporangien stehen in Gruppen um rudimentäre Cystocarpien, was auf eine gewisse Korrelation zwischen Sporangienbildung und Hemmung der Cystocarpbildung schließen läßt. Der Kern der Sporangienmutterzelle teilt sich mehrfach, alle Kerne bis auf einen degenerieren. Der ganze Inhalt des Sporangiums tritt als einzelne Spore aus. Bei der Bildung dieser Sporen wird der primär einschichtige Thallus mehrschichtig. — Bei *Polysiphonia violacea* und *Griffithsia globifera* wurden ungeteilte Sporangien auf männlichen Gametophyten beobachtet, bei *Polysiphonia urceolata* auf weiblichen Gametophyten mit reifen Cystocarpien. Nach BAUCH (1937) tritt Monosporenbildung schließlich noch bei einigen Corallinaceen auf. Ob die bei *Monospora pedicellata* beschriebenen ungeteilten Sporangien, die dieselbe Position wie die Tetrasporangien einnehmen, als Monosporangien betrachtet werden dürfen, erscheint nach der Art und Weise ihrer Ablösung und Keimung zweifelhaft (CHEMIN 1928).

Es sei darauf hingewiesen, daß auch die Karposporen der Florideen, die die Fortpflanzungszellen des Karposporophyten darstellen, bis auf wenige Ausnahmen Monosporen sind. Sie entstehen in Einzahl entweder in den Endzellen der Gonimoblastfäden, so z.B. bei den Nemalionales, oder auch interkalar, z.B. bei manchen Cryptonemiales.

3. Sporen aus mehrsporigen Sporangien.

Tetrasporen. Die Reduktionsteilung findet bei den Rhodophyten entweder unmittelbar nach der Befruchtung bei den ersten Teilungen der Zygote statt (Bangioideae [?], Teil der Nemalionales; s. aber MAGNE 1964, sowie auch weiter unten), oder sie ist durch Einschaltung einer selbständigen, dem Gametophyten meist gleich gestalteten, diploiden Phase, des Tetrasporophyten, bis zur Tetrasporenbildung hinausgeschoben (übrige Ordnungen der Florideen). Im ersten Fall bildet das Karpogon einen haploiden Gonimoblasten, der haploide Karposporen erzeugt, aus denen wieder Geschlechtspflanzen entstehen. Es handelt sich um zwei somatisch verschiedene Generationen von gleicher nucleärer, nämlich haploider Phase (haplobiontischer Typ). Im zweiten Fall entstehen auf den Geschlechtspflanzen nach der Befruchtung diploide Gonimoblasten. Die in ihnen erzeugten diploiden Karposporen entwickeln sich zu Tetrasporophyten, auf denen unter Reduktionsteilung Tetrasporen gebildet werden. Hier folgen drei somatisch verschiedene Generationen aufeinander, von denen der Gametophyt haploid, die beiden anderen aber diploid sind (diplobiontischer Typ). Gegenüber diesen beiden von Svedelius (1915) aufgestellten Typen sind jedoch so viele Abweichungen bekannt, daß inzwischen andere Typensysteme aufgestellt worden sind (z.B. von Drew 1955), die diesen Abweichungen Rechnung tragen. Darüber hinaus ist es durch Untersuchungen von Svedelius (1942) an *Galaxaura*, Feldmann und Feldmann 1942 an Bonnemaisoniaceen und durch die Arbeiten Magnes (1961, 1964a, b), v. Stoschs (1965) u.a. zweifelhaft geworden, ob die Haplobiontie für die Nemalionales überhaupt noch als charakteristisch gelten kann, ja Magne vertritt sogar die Ansicht, daß es haplobiontische Rhodophyceen im herkömmlichen Sinne möglicherweise nicht gibt. Tatsächlich führten dessen karyologische Studien zu dem überraschenden Ergebnis, daß die Karposporen von *Lemanea rigida*, *Nemalion helminthoides* und *Scinaia furcellata* diploid sind und diploide Keimlinge aus ihnen hervorgehen. v. Stosch gelang es neuerdings (1965), die bei Magne noch offen gebliebene Frage nach der Natur des Sporophyten dieser Algen im Falle von *Liagora farinosa* zu beantworten: er erhielt aus den Karposporen *Acrochaetium*-artige Sporophyten mit Mono- und Tetrasporangien; während die Monosporen die Phase reproduzieren, der sie entstammen, erzeugen die unter Meiose gebildeten Tetrasporen auf dem Weg über ein protonemaartiges Vorstadium normale *Liagora*-Thalli.

Mit einiger Sicherheit würden dann nur noch Batrachospermen haploide Karposporen besitzen, soweit man dies aus Kulturexperimenten Yoshidas (1959) schließen kann. In diesen lieferten Karposporen des *Batrachospermum* und Monosporen der *Chantransia* bei etwas unterschiedlicher Keimung gleichartig *Chantransia*-Vorkeime, welche dann *Batrachospermum*-Thalli als Knospen entwickelten.

Im allgemeinen stehen die Tetrasporangien an den Enden kurzer Seitenzweige, nur in seltenen Fällen interkalar. Das ist besonders deutlich an monosiphonen, reich verzweigten, aber lockeren Sproßsystemen wie bei *Callithamnion* und *Antithamnion* zu sehen, aber auch in Thalli von kompakter Struktur ist unschwer zu erkennen, daß sie terminal an kleinen Seitenzweigen in der Rindenschicht stehen. Die Verteilung der Sporangien über den Thallus unterliegt starken Unterschieden. Bei lockeren, monosiphonen Formen sind sie meist ziemlich gleichmäßig über die Sprosse verteilt, oft jedoch auch bei kompakten Thalli. Wo sie an bestimmten Stellen gruppiert sind und pustelartige Erhebungen bilden, spricht man von Nemathezien. Fertile Zweige, an denen die Sporangien gehäuft auftreten, und die sich auch morphologisch von den sterilen Zweigen unterscheiden, heißen Stichidien.

Die Tetrasporangien sind vergrößerte Zellen mit dichtem plasmatischen Inhalt. Je nach dem Modus der Teilung sind die vier Sporen im Sporangium verschiedenartig angeordnet: 1. In den paarig (cruciat) geteilten Sporangien entsteht zunächst eine Querwand, dann in beiden Hälften je eine Längswand; die beiden Längswände liegen entweder in einer Ebene oder stehen senkrecht zueinander (Abb. 104a). 2. In den quergeteilten Sporangien wird die keulenförmig gestreckte Mutterzelle durch drei parallele Querwände meist simultan in vier übereinanderliegende Zellen zerlegt (Abb. 104b). 3. In den tetraedrisch geteilten Sporangien entstehen die Sporen simultan durch Zerlegung der kugeligen Mutterzelle mittels radialer Wände (Abb. 104c). — Ebenso wie die Position der Sporangien im Thallus in den engeren Verwandtschaftskreisen festliegt und daher systematisch verwertet werden kann, stimmt meist auch der Teilungsmodus der Sporangien in größeren Gruppen überein. So treten in den Ordnungen Nemalionales und Gelidiales nur cruciat geteilte Sporangien auf, für viele Cryptonemiales und Gigartinales ist die Querteilung charakteristisch, während die meisten Ceramiales durch tetraedrische Teilung gekennzeichnet sind. — Die reife Sporangienmutterzelle ist gewöhnlich einkernig. Außer beim cruciaten Teilungstypus, der von den beiden anderen stärker abweicht und sich in der Entwicklung mehr den Monosporangien nähert, gehen die Kernteilungen den Zellteilungen vorauf. Die Kerne nehmen ihre definitive Lage ein, bevor die Trennungslamellen zentripetal eingezogen werden. Die jungen Sporangien sind vacuolisiert, die Chromatophoren sind undeutlich und bekommen erst bei der Keimung wieder scharfe Konturen.

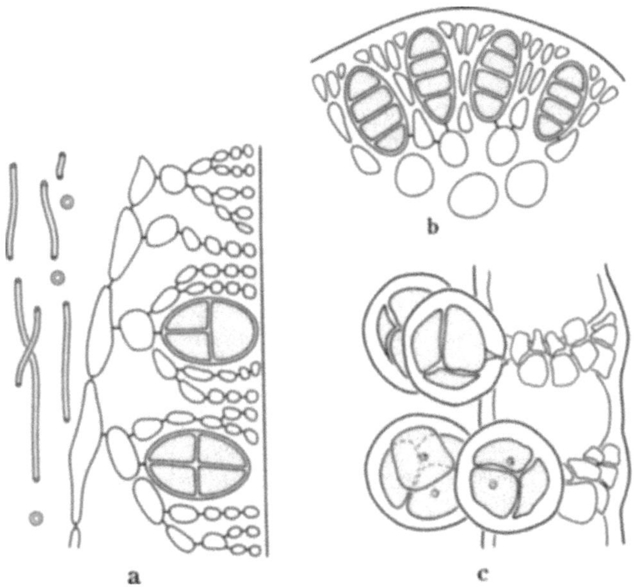

Abb. 104a—c. Teilungstypen bei den Tetrasporangien der Florideen: a paarig (bei *Nemastoma Laingii*), b quergeteilt (bei *Hypnea musciformis*), c tetraedrisch (bei *Spyridia filamentosa*). (a und b nach KYLIN, c nach HOMMERSAND.)

Im allgemeinen ist bei den diplobiontischen Florideen mit der Tetrasporenbildung die Reduktionsteilung verknüpft. Es sind jedoch Florideen bekannt, die sich nur durch Tetrasporen fortpflanzen, bei deren Bildung eine Reduktion ausgeschlossen werden muß, z.B. bei den *Rhodochorton*-Arten (sofern sich nicht ein entwicklungsgeschichtlicher Zusammenhang mit anderen Gattungen herausstellt, indem sie sich als deren Sporophyten erweisen). SVEDELIUS (1935) konnte nachweisen, daß *Lomentaria rosea* keinen Generationswechsel besitzt und nur Tetrasporen erzeugt, welche ohne Reduktionsteilung gebildet werden. Bei *Rhodymenia palmata* sind weibliche Pflanzen unbekannt, obwohl WESTBROOK (1928) behauptet, daß Meiosis in den Tetrasporangien stattfindet. MAGNE (1964b) zählte 14 Chromosomen sowohl bei den männlichen Gametophyten als auch bei den Tetrasporophyten, was wohl die Meiosis ausschließt. Bei *Lithophyllum corallinae* findet nach SUNESON (1950) in einigen Sporangien die Reduktion statt, in anderen nicht, so daß ein Individuum sowohl haploide wie diploide Sporen erzeugen kann. Nach DIXON scheint das auch auf einige *Gelidium*-Arten von England zuzutreffen. Vielleicht erklärt sich durch Ausfall der Reduktionsteilung auch der Befund von HASSINGER-HUIZINGA (1952) an *Callithamnion corymbosum*, daß gelegentlich aus Tetrasporen wieder Tetrasporophyten entstehen.

Bemerkenswert ist das Vorkommen von Tetrasporangien an den Gonimoblasten von *Liagora tetrasporifera* (Boergesen 1927) und *Helminthocladia Hudsoni* (Feldmann 1939). Sie stehen hier an Stelle der Karposporen und werden deshalb auch „Tetrakarposporen" genannt. Die Homologie ist um so naheliegender, als die meisten *Liagora*-Arten gewöhnliche Karposporen erzeugen. Wenn sich erweisen sollte, daß die Reduktionsteilung mit der Tetrasporenbildung zusammenfällt, könnte man nach Fritsch in diesen beiden Fällen ein Beispiel für die Verschiebung der Reduktionsteilung und damit einen möglichen Schritt in der Evolution der diplobiontischen Florideen erblicken. Nach den neuesten Untersuchungen v. Stoschs (1965) an *Liagora farinosa*, die eine selbständige, wenn auch zwerghafte sporophytische Phase besitzt, läge aber wohl die entgegengesetzte Annahme näher, daß es sich bei den Tetrakarposporen um extrem reduzierte Sporophyten handelt.

Von zahlreichen Gattungen ist ein saisonbedingter Wechsel zwischen Tetrasporophyten und Geschlechtspflanzen bekannt. So findet man die tetrasporentragende Generation von *Acrochaetium* in der kalten Jahreszeit, die geschlechtliche Generation in den warmen Monaten. Ganz entsprechend treten bei dem in der Ostsee häufigen *Ceramium corticatulum* Geschlechtspflanzen zu einem ganz bestimmten Zeitpunkt des Sommers auf, in der übrigen Jahreszeit nur Tetrasporenpflanzen, die sich wahrscheinlich durch Parasporen (s. unten) selbst reproduzieren (Svedelius 1927). An Kulturen von *Bonnemaisonia hamifera/Trailliella intricata* ist der jahreszeitliche Rhythmus, den J. und G. Feldmann (1942) auch für *Bonnemaisonia asparagoides/Hymenoclonium serpens* angeben, unter kontrollierten Bedingungen nachgewiesen worden. Wie unter natürlichen Bedingungen bildet der Gametophyt im Sommer und der Tetrasporophyt im Spätherbst Fortpflanzungsorgane aus (Koch 1950). Versuche, im Laboratorium schon im Frühjahr oder Sommer unter abgeänderten, dem Spätherbst entsprechenden Bedingungen Tetrasporangienbildung hervorzurufen, mißlangen. Allerdings kann sich jede der beiden Generationen durch Fragmente propagieren und sich also überall dort ansiedeln, wo zusagende ökologische Faktoren herrschen. — Häufig findet auch in einem bestimmten Biotop nur die eine Generation einer Art die ihr günstigen Lebensbedingungen und die andere Generation kann zu keiner Zeit des Jahres gefunden werden, während in anderen Meeresteilen die Verhältnisse umgekehrt sind.

Es ist oft nicht leicht, zwischen Tetra- und Monosporen scharf zu unterscheiden. Das gemeinsame Vorkommen beider Sporangientypen sowohl auf den Geschlechtspflanzen wie auch auf der ungeschlechtlichen Generation von *Acrochaetium*-Arten macht es unwahrscheinlich, daß hier bei der Tetrasporenbildung die Reduktionsteilung eintritt. Fritsch (A 1945) möchte die Tetrasporangien in diesen Fällen bloß als geteilte Monosporangien betrachten und bringt auch die bei einigen *Acrochaetium*-Arten auftretenden Bisporen und Polysporen damit in Verbindung.

Die austretenden Tetrasporen sind spezifisch schwerer als Wasser und sinken zu Boden. Durch ausgeschiedenen Schleim heften sie sich am Substrat fest. Quellung dieses Schleims ist einerseits wahrscheinlich die Ursache der gelegentlich zu beobachtenden (nicht amöboiden, s. aber v. Stosch 1965) Bewegungen, und andererseits verhindert sie bei gewissen Ceramiales die unmittelbare gegenseitige Berührung der Sporen auf dem Substrat. Die Membranbildung wird anscheinend durch den Kontaktreiz der festen Unterlage angeregt. Ein Agar-Substrat wird dabei nur bedingt als feste Unterlage empfunden. In Agar eingebettete Sporen bilden keine feste Zellmembran aus (Weber 1960a). Die Sporen keimen ohne Ruheperiode aus, auch bei völliger Dunkelheit. Nach den unterschiedlichen

Keimungsweisen lassen sich verschiedene Keimlingstypen aufstellen, die den Verwandtschaftsverhältnissen entsprechen (KILLIAN 1914, KYLIN 1917, CHEMIN 1937): 1. Der von den Ceramiaceen und Rhodomelaceen (auch *Bangia*; s. oben) repräsentierte „aufrechte" (besser: bipolare) Typus, der durch eine inäquale Primärteilung gekennzeichnet ist; 2. der durch eine äquale Primärteilung charakterisierte Haftscheibentypus; 3. der Keimschlauchtypus der Nemalionales. Dabei darf die Bezeichnung „aufrechter" Typus nur als Ausdruck der ausgeprägten Polarität der Keimlinge mit ihrem extrem verschiedenen Apikal- und Rhizoidpol, nicht aber gleichzeitig als Kennzeichnung der Orientierung zum Substrat aufgefaßt werden. Das Substrat scheint nämlich bei diesem Keimungstypus ohne Einfluß auf die Richtung der ersten Teilung zu sein. Dagegen steht bei den zum Haftscheibentypus gehörenden Sporenkeimlingen von *Chondrus crispus* die erste Teilungswand senkrecht zum (festen) Substrat, wird aber unabhängig von der Schwerkraftwirkung angelegt. Außerdem beeinflussen sich in unmittelbarem Kontakt stehende *Chondrus*-Sporen in der Weise, daß die Teilungswände der Nachbarsporen an der Kontaktstelle aufeinanderstoßen (WEBER 1960a). Abgesehen von diesen induzierenden Wirkungen des Substrats und der Nachbarsporen ist die Polarisierbarkeit der Rotalgensporen durch Außenfaktoren äußerst gering. Ihnen ist eine inhärente Polarität eigen, die sich bei der Keimung der in Agar eingebetteten Tetraden in der tangentialen Orientierung der Polaritätsachse manifestiert. Allerdings können Außeneinflüsse, z. B. Einwirkungen von nahe gelegenen Thallusstücken, über den innerhalb der festgelegten Achse anfangs noch labilen Richtungssinn der Polarität entscheiden (WEBER 1960b).

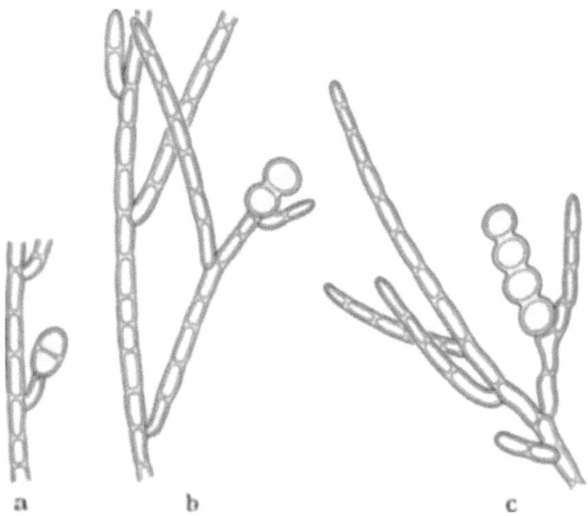

Abb. 105a—c. *Rhodochorton rothii*, a Bisporen, b und c Parasporen. (Nach KLAVESTAD.)

Sporentetraden keimen gelegentlich als Ganzes aus. Das ist anscheinend dann der Fall, wenn die Sporangien entleert werden, bevor die Sporen genügend isoliert sind. Solche a priori mehrzelligen Keimlinge entwickeln dementsprechend mehrere Rhizoide nach verschiedenen Seiten. Daneben gibt es andere, die aus noch ungeteilten Sporangien hervorgehen. Sie unterscheiden sich strukturell außer durch ihre enorme Größe nicht von normalen Keimlingen (WEBER 1960b).

An verschiedenen Arten ist die Keimung der Tetrasporen in situ beobachtet worden. Während die Sporen von *Falkenbergia* dabei je einen Keimling ergeben (FELDMANN 1942), beschreibt OSTERHOUT (1896) das Auswachsen des ganzen Sporangieninhalts als eine einzige kompakte Bildung bei *Agardhiella, Cystoclonium* und *Gracilaria*. Dieselbe Erscheinung ist von *Champia* und *Lomentaria* bekannt. Neuerdings haben DIXON (1960) dies bei verschiedenen *Ceramium*-Arten und TOKIDA u. YAMAMOTO (1965) bei *Pachymeniopsis* beobachtet. Die Verbindung des Keimlingsgewebes mit dem der Mutterpflanze ist hier oft so innig, daß es schwierig ist, zwischen einem adventiven Seitenzweig unbegrenzten Wachstums und dem Produkt einer solchen „inneren" Keimung zu unterscheiden. In den Fällen aber, wo statt der kontinuierlichen Berindung nur regelmäßige

Rindengürtel zwischen je zwei benachbarten Zentralzellen ausgebildet sind, ist die in-situ-Keimung unschwer erkennbar, die Keimlinge lassen sich leicht ablösen.

Bisporen. Eine Anzahl von Florideen erzeugt Sporangien mit nur zwei Sporen (Abb. 105a). Unter den Ceramiaceen sind sie besonders häufig bei *Crouania attenuata* und *Callithamnion furcellariae* (ROSENVINGE 1909); bei der letzteren treten sie an Individuen auf, denen Tetrasporangien fehlen; bei der ersteren besteht sogar ein jahreszeitlicher Wechsel zwischen Tetrasporen und Bisporen erzeugenden Pflanzen. Besonders charakteristisch ist Bisporenbildung für viele Corallinaceen. BAUCH (1937) unterscheidet obligat und fakultativ bisporische und tetrasporische Arten; bei einigen der letzteren treten bisporische und tetrasporische Rassen auf, die jeweils bestimmte Biotope bewohnen. Nach FOSLIE (1905) kommen bisporische Arten vornehmlich in kälteren Meeren vor und fehlen in den Tropen. Die Bisporen sind entweder einkernig (*Lithothamnion polymorphum*) oder zweikernig (*Lithophyllum incrustans*). In einkernigen Bisporen sind die Kerne doppelt so groß wie die Tetrasporenkerne derselben Art, was daher rühren mag, daß die zweite Teilung unterdrückt oder bis zur Keimung hinausgeschoben ist. Die Homologie von Bisporen und Tetrasporen liegt nahe, ist aber schließlich nur cytologisch nachzuweisen. SUNESON (1950), der die Bisporenbildung bei zwei *Lithophyllum*-Arten untersucht hat, fand bei *L. litorale* nur Bisporangien mit uninucleären Sporen, welche ohne Reduktionsteilung gebildet werden. *L. corallinae* erzeugt sowohl binucleäre als auch uninucleäre Bisporen.

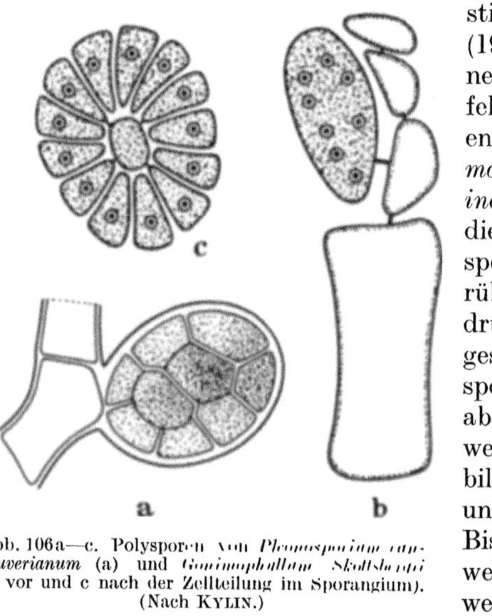

Abb. 106a—c. Polysporen von *Pleonosporium vancouverianum* (a) und *Gonimophyllum Skottsbergii* (b vor und c nach der Zellteilung im Sporangium). (Nach KYLIN.)

SUNESON vermutet, daß hier die Bildung der binucleären Bisporen unter Meiosis vonstatten geht und betont, daß es weniger von Bedeutung ist, ob Bi- oder Tetrasporen gebildet werden, als vielmehr, ob eine Reduktionsteilung stattfindet oder nicht.

Polysporen. Polysporen entstehen zu mehr als vier in besonderen Behältern, den Polysporangien. Sie sind bei den Ceramiaceen häufig und treten außerdem bei *Gonimophyllum* (Abb. 106b, c) und *Chylocladia* auf. Die Polysporangien von *Gonimophyllum* entstehen aus einkernigen Anlagen. Der Kern der Initiale teilt sich in 30—50 Tochterkerne auf, um die herum sich dann durch Aufteilung des Plasmas entsprechend viele radial gestellte Sporen bilden, die mit einer zentral liegenden Zelle in Tüpfelverbindung stehen. KYLIN (1956) nimmt als sicher an, daß bei der ersten Kernteilung der Initiale die Reduktionsteilung stattfindet und die haploiden Sporen Geschlechtspflanzen hervorbringen. Die Polysporen scheinen hier wie auch bei *Pleonosporium* (Abb. 106a) die Stelle der Tetrasporen einzunehmen. Bei *Spermothamnion snyderae* existiert ein normaler Generationswechsel, die ungeschlechtliche Generation erzeugt ausschließlich Polysporangien (DREW 1937). Deren Initialen enthalten mehrere diploide Kerne, die sich simultan meiotisch teilen, so daß die Polysporen in Vierergruppen auftreten. Auch hier unterliegt die Homologie mit Tetrasporangien wohl keinem Zweifel.

Bei den meisten Ceramiaceen treten die Polysporangien jedoch zusammen mit Tetrasporangien auf. Sie kommen als rundliche Körper z. B. bei verschiedenen

Ceramium-Arten, bei *Callithamnion* und *Spermothamnion* vor und haben dieselbe Position wie die Tetrasporangien, bei *Ceramium* allerdings mit dem Unterschied, daß sie den Rindengürteln außen aufsitzen, während die Tetrasporangien in diese eingebettet sind. Bei *Plumaria elegans* nehmen sie die Spitzen der Endverzweigungen ein. Nach DREW (1939) treten sie bei dieser Art auf besonderen triploiden Individuen (neben Tetrasporen unbekannten Schicksals) auf, während haploide Geschlechtspflanzen und diploide ungeschlechtliche Pflanzen einen normalen diplobiontischen Lebenscyclus bilden. Die Meinungen, ob auch in diesen Fällen die genannten Bildungen noch als den Tetrasporangien homologe Polysporangien anzusehen sind, gehen auseinander. Während FRITSCH (A 1945) ihre Homologie mit Tetrasporangien für gesichert hält, rechnen DREW (1939) und KYLIN (1956) sie schon zu den Parasporen.

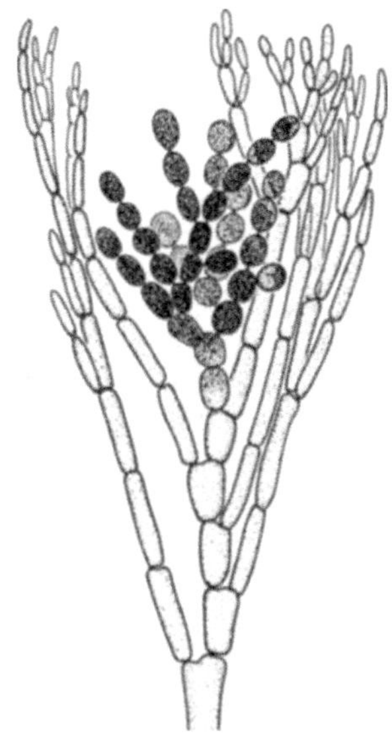

Abb. 107. *Seirospora Griffithsiana*, Zweigstück mit Parasporen. (Nach OLTMANNS.)

4. Parasporen.

Bei *Seirospora Griffithsiana* und anderen Arten der Gattung (für die ja solche „Seirosporen" genannten Parasporen charakteristisch sind) können die Zellen in den Enden jüngerer Zweigbüschel reihenweise stark anschwellen und sich in dickwandige, einkernige Sporen umbilden (Abb. 107). Die Zellen lösen sich bei der Reife voneinander ab, der Inhalt tritt aus der Membran heraus und keimt wie eine Tetraspore. Auf denselben Individuen können auch Bi- und Tetrasporangien erscheinen. An den Zweigspitzen von *Ceramium strictum* und *C. Deslongchampsii* treten die Parasporen in polysporangienartigen Haufen auf. Es ist unwahrscheinlich, daß die Parasporangien überhaupt genetisch mit echten Sporen zusammenhängen. Wahrscheinlich handelt es sich um eine Umwandlung vegetativer Zellen (nach Art der Akineten anderer Algengruppen ?).

DIXON (1960) ist der Überzeugung, daß sehr oft Gallbildungen für Poly- oder Parasporangien gehalten werden und daß es sich bei den in der Literatur berichteten Fällen in Wirklichkeit zumeist um solche Gallen handelt.

b) Sonstige Fortpflanzungskörper.

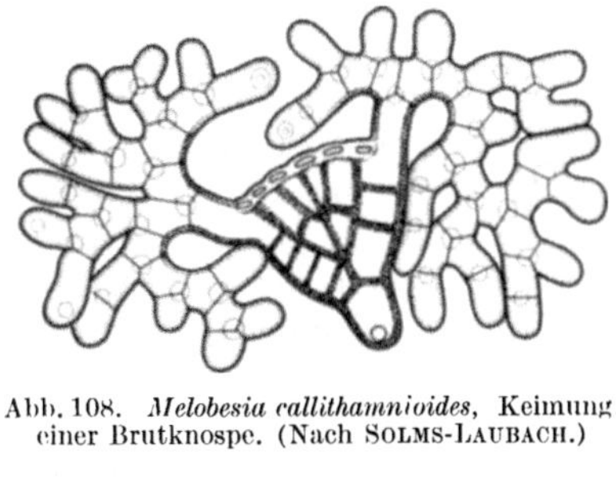

Abb. 108. *Melobesia callithamnioides*, Keimung einer Brutknospe. (Nach SOLMS-LAUBACH.)

Brutkörper. Spezielle Brutkörper (Propagulen, Gemmen) sind nur von wenigen Rotalgen bekannt. SOLMS-LAUBACH (1881) beschreibt solche Bildungen, die an die bekannten Brutknospen der Sphacelarien erinnern, für *Melobesia callithamnioides*. Hier entstehen sie an den Enden kurzer Fäden, welche sich über den epiphytisch wachsenden, krustenartigen Thallus erheben, als dreieckige Zellflächen mit dicken Zellwänden. Nach ihrer Ablösung wachsen sie zu neuen epiphytischen Thalli aus (Abb. 108). Über ähnliche Vermehrungskörper berichtet BORNET (1892) bei *Polysiphonia furcellata* und *P. atrorubescens*. Nach OKAMURA (1903) dienen bei *Chondria crassicaulis* kurze, mit Stärke angefüllte Ästchen als Brutknospen.

Akineten. Sehr selten scheinen unter den Rotalgen Akineten ausgebildet zu werden. Bisher ist offenbar nur der Fall von *Asterocytis ramosa* bekannt (WILLE 1900, ROSENVINGE 1909, BOERGESEN 1927). Die membranbekleideten Akineten stellen hier Dauersporen dar. Ihre Wände entstehen durch Umbildung der inneren Schichten der Zellwände des *Asterocytis*-Fadens.

c) Vegetative Vermehrung (sensu stricto).

Von größerer Bedeutung für die Vermehrung der Rotalgen ist ihre Fähigkeit der vegetativen Fortpflanzung im engeren Sinne, d.h. der Reproduktion neuer Individuen aus isolierten Stücken ihres Vegetationskörpers. Die vegetative Vermehrung kann zur Vermehrung durch spezielle Reproduktionsorgane hinzutreten, kann diese aber auch, zumindest zu bestimmten Jahreszeiten, völlig ersetzen. Es läßt sich sogar eine gewisse Korrelation zwischen Reproduktionsorganen und vegetativer Vermehrungsfähigkeit feststellen: bei Formen mit ausgeprägtem vegetativen Reproduktionsvermögen tritt die Ausbildung spezieller Fortpflanzungszellen in den Hintergrund. *Bornetia secundiflora* bildet Cystocarpien und Tetrasporen nur in den temperierten Zonen (Mittelmeer, Kalifornien), während in den kälteren Meeren eine Vermehrung nur auf rein vegetativem Wege möglich ist (CHEMIN 1928).

Die vegetative Fortpflanzungsfähigkeit hat natürlich ein Regenerationsvermögen zur Voraussetzung. TOBLER (1906) und besonders CHEMIN (1928) haben bei einer Anzahl von Arten die Regenerationsfähigkeit experimentell geprüft und dadurch Anhaltspunkte für die Art und Weise der vegetativen Vermehrung gewonnen. Als Beispiel für den Regenerationsverlauf bei einer monosiphonen Rhodophycee seien die Vorgänge bei *Griffithsia corallina* (nach CHEMIN) geschildert. Bei dieser Alge können einzelne Zellen neue Pflanzen regenerieren. Der Zusammenhalt der Zellen ist sehr locker und kann, insbesondere in jüngeren Thallusabschnitten, durch äußeren Eingriff leicht gelöst werden, z.B. durch die Bewegung des Wassers. Sogar im stagnierenden Wasser des Kulturgefäßes erfolgt spontan die Fragmentation der Zweige. Zunächst entsteht an der Basis des isolierten Thallusstücks bzw. der Zelle ein Hauptrhizoid in der Verlängerung der Achse, später treten Lateralrhizoide hinzu. Am Scheitel werden danach neue Gliederzellen gebildet. Der Fortsetzung der Entwicklung geht also die Festheftung am Substrat des Fragments mittels Rhizoidbildung voraus. Die Rhizoide können sogar schon ohne experimentellen Eingriff vor der Fragmentation gebildet werden, wenn eine interkalare Zelle degeneriert, wenn also der Bruch schon physiologisch, aber noch nicht mechanisch vollzogen ist. Auf diese Weise findet im natürlichen Milieu die vegetative Vermehrung statt. Die Regenerationsfähigkeit ist in den terminalen Partien mit den noch nicht adulten Zellen am größten. Bei der Regeneration wird eine deutliche Zellpolarität manifest, indem Rhizoide stets am basalen Pol, neue Thalluszellen am apikalen Pol der Zelle entstehen. Diese Polarität ist auch gegen Inversionen der Zelle invariant. Allerdings krümmen sich bei inverser Orientierung die Rhizoide positiv geotropisch nach unten.

Bei *Antithamnion plumula*, dessen Regenerationsfähigkeit in der starken Neigung zur Bildung von Adventivrhizoiden zum Ausdruck kommt (Abb. 109), nehmen diese ihren Ursprung von Zweigen zweiter und dritter Ordnung (CHEMIN 1928). Bei berindeten Florideen gehen diese Rhizoide vorwiegend aus den Berindungsfäden bzw. -zellen hervor, während bei polysiphonem Aufbau die Rhizoide den Perizentralen entspringen. Demgemäß entstehen sie an *Ceramium*-Thalli büschelweise gehäuft an den Knoten, an Polysiphonien (Abb. 110) dagegen über den Thallus verteilt, jeweils basal an den Perizentralzellen. Bei einseitiger Beleuchtung von Polysiphonia-Thallusstücken werden die Rhizoide nur an der lichtabgewandten Seite angelegt. Je mehr Adventivrhizoide an den Fragmenten entstehen, um so geringer ist die Bildung von Basalrhizoiden (CHEMIN 1928). Die Regeneration des Sprosses geht bei berindeten oder polysiphonen Florideen zumeist von den zentralen Stamm- bzw. Siphonalzellen aus (OLTMANNS 1895, MASSART 1898, CHEMIN 1928).

Abb. 109. *Antithamnion plumula*. Rhizoidbildung an den Zweigen nach mehrtägigem Aufenthalt im Kulturgefäß. 90 ×. (Original.)

Die Leichtigkeit, mit der an fädigen Rotalgen die Rhizoidbildung hervorgerufen werden kann, ist natürlich im Zusammenhang mit ihrer Gefährdung durch starke Wasserbewegung zu sehen. Die geringere Zerreißfestigkeit wird kompensiert durch stärkeres Regenerationsvermögen, das im Dienst der vegetativen Vermehrung steht.

TOBLER (1906) beobachtete an basalwärts verschmälerten Ästchen von *Polysiphonia urceolata* und *P. variegata* Rhizoidbildung vor der Ablösung und betrachtet die Ästchen als Vermehrungskörper. Damit ergäbe sich ein gleitender Übergang zwischen vegetativer Vermehrung durch Fragmentation des Thallus und Ausbildung spezieller Brutkörper.

Weit geringer ist die Regenerationsfähigkeit der Rotalgen mit massivem Thallus und demgemäß auch die Bedeutung der Vermehrung durch Fragmentation. In ökologischer Betrachtungsweise erscheint sie allerdings auch weniger notwendig, da diese Algen meist mechanisch resistenter sind. Bei ihnen können auf Kontaktreize hin die oberflächlichen Zellen Haftorgane ausbilden, die auch die Form von Häkchen annehmen können (z. B. *Cystoclonium purpurascens*). Durch Thallusstücke, die mittels solcher Haftorgane an anderen Gegenständen befestigt sind und aus dem Thallus herausgerissen werden, kann eine geringfügige Vermehrung stattfinden.

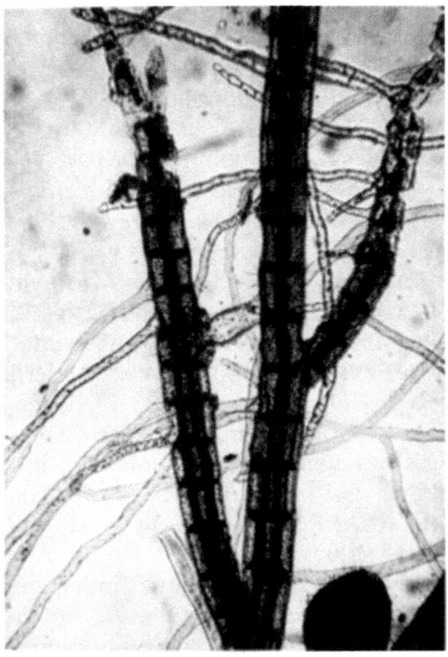

Abb. 110. *Polysiphonia urceolata*, Thallusstücke mit Rhizoidbildung, oben rechts beginnender Thalluszerfall. Rhizoidbildung nur an der linken Seite infolge einseitiger Belichtung von rechts. 75 ×. (Original.)

Einzelbeispiele für Vermehrung durch Thallusfragmentation, sowohl infolge Außeneinwirkungen wie spontan, sind in großer Zahl bekannt. So gehen bei verschiedenen *Ceramium*-Arten nach Beobachtungen von Oltmanns (1895), Massart (1898), Tobler (1906) und Chemin (1928) aus abgerissenen Stücken des Sprosses neue Thalli hervor, indem zunächst aus den Corticalzellen herauswachsende Adventivrhizoide die Anheftung am Substrat bewirken und dann aus der Achse die Regeneration einsetzt. Papenfuss (1937) berichtet über vegetative Vermehrung bei *Claudea multifida* und *Vanvoorstia spectabilis*, Sundene (1962) bei *Antithamnion boreale*, Chemin (1929) und Koch (1950) bei *Bonnemaisonia hamifera*, Iwasaki (1961) bei *Conchocelis*. Sie ist ebenso von großer Bedeutung für die Ausbreitung der *Chantransia*-Stadien von *Batrachospermum* und *Lemanea*. — Die Teilbarkeit und Regenerationsfähigkeit vieler Rotalgen hat man sich bei ihrer Kultur zunutze gemacht. Bonnemaisoniaceen, Rhodomelaceen, Ceramiaceen lassen sich weitgehend zerstückeln und auf diese Weise leicht vermehren.

d) Parthenogenese und Apogamie.

Eine apogame oder parthenogenetische Entwicklung ist bei den Florideen mit Sicherheit noch nicht nachgewiesen. Die Schwierigkeit des Nachweises der Parthenogenese resultiert aus der Kleinheit der männlichen Reproduktionsorgane. Bei vielen Gattungen und Arten sind bisher keine männlichen Organe gefunden worden. Dixon (1963a) weist jedoch mit Recht darauf hin, daß hierin noch kein Beweis ihres vollständigen Fehlens erblickt werden kann, und vielfach sind sie dann auch nach fortgesetztem Suchen entdeckt worden. Bei gewissen Gattungen wie z.B. *Batrachospermum*, deren Trichogyne nach der Fusion mit einem Spermatium persistiert, kann die Befruchtung leicht beobachtet werden. Bei den meisten Rhodophyten dagegen bildet sich die Trichogyne gleich nach der Befruchtung zurück, der Nachweis der Befruchtung bzw. der Parthenogenese wird dadurch außerordentlich erschwert. Zur Sicherung der Vermutungen über Fälle von Parthenogenese sollten daher cytologische Untersuchungen oder Kulturversuche unternommen werden.

Zu den Florideen, bei denen parthenogenetische Entwicklung angenommen wird, zählen *Platoma Bairdii* (Kuckuck 1912), *Ptilota plumosa* (Davis 1896), *Furcellaria fastigiata* und *Petrocelis Hennedyi* (Rosenvinge 1909). Die Behauptung von Davis bezüglich *Ptilota* wird jedoch von Phillips (1897) zurückgewiesen und auch von Kylin (1956) nicht akzeptiert. Die Angaben bezüglich der anderen Arten beruhen ebenfalls nicht auf direktem, d.h. experimentellem oder cytologischem Erweis, sondern auf dem vermeintlichen Fehlen männlicher Reproduktionsorgane und bedürfen daher einer exakten Nachprüfung.

Die Einordnung der Fortpflanzung nichtsexueller Porphyren unter den Bangiales hängt davon ab, ob man annimmt, daß es sexuelle Species der Gattung gibt. Tut man dies, so wären ohne nachweisbare Befruchtung entstandene „Karpo"-Sporen nach Typ III (s. S. 762) als parthenogenetisch zu bezeichnen. Hus (1902) konnte bei den 11 Arten der nordamerikanischen Westküste keine Anzeichen für Sexualität finden und Rosenvinge (1909) beobachtete nicht befruchtete Individuen von *P. linearis* und *P. purpurea* aus Dänemark in Karposporenbildung. Auf neuere Befunde ähnlicher Art wurde auf S. 763 hingewiesen.

Literatur.

Bauch, R.: Die Entwicklung der Bisporen der Corallinaceen. Planta (Berl.) **26**, 365—390 (1937). — Belcher, J. H.: Culture studies of *Bangia atropurpurea* (Roth) Ag. New Phytologist **59**, 367—373 (1960). — Boergesen, F.: Marine algae from the Canary Islands. Biol. méd. **6**, 1—97 (1927). — Bornet, E.: Les algues de P.K.A. Schousboe. Mém. soc. sci. nat. (Cherbourg) **28**, 165—376 (1892).

CHEMIN, E.: Multiplication végétative et dissémination chez quelques Algues Floridées. Trav. Stat. biol. Roscoff **7**, 1—61 (1928). — L'*Asparagopsis hamifera* (Hariot) Okamura et son mode de multiplication. Rev. Alg. **4**, 29—42 (1929). — Le développement des spores chez les Rhodophycées. Rev. gén. Bot. **49**, 205—234, 300—327, 352—371, 424—448, 478—535 (1937). — CONWAY, E.: Autecological studies of the genus *Porphyra*. I. The species found in Britain; II. *Porphyra umbilicalis*. Brit. Phycol. Bull. **2**, 342—363 (1964a, b).

DAVIS, B. M.: Development of the procarp and cystocarp in the genus *Ptilota*. Bot. Gaz. (Chicago) **22**, 353—378 (1896). — DIXON, P. S.: Studies on marine algae of the British Isles: The genus *Ceramium*. J. mar. biol. Assoc. U. Kingd. **39**, 331—374 (1960). — Variation and speciation in marine Rhodophyta. Syst. Ass. Publ. **5**, 51—62 (1963a). — The Rhodophyta: Some aspects of their biology. Oceanogr. Mar. Biol. Ann. Rev. **1**, 177—196 (1963b). — DREW, K. M.: *Spermothamnium Snyderae* Farlow, a floridean alga bearing polysporangia. Ann. Bot., N.S. **1**, 463—476 (1937). — An investigation of *Plumaria elegans* (Bonnem.) Schmitz with reference to triploid plants bearing parasporangia. Ann. Bot., N.S. **3**, 347—367 (1939). — Studies in the Bangioideae. I. Observations on *Bangia fuscopurpurea* (Dillw.) Lyngb. in culture. Phytomorphology **2**, 38—51 (1952). — Studies in the Bangioideae. III. The life history of *Porphyra umbilicalis* (L.) Kütz. var. *laciniata* (Lightf.) J. Ag. Ann. Bot., N.S. **18**, 181—211 (1954). — Life histories in the algae. With special reference to the Chlorophyta, Phaeophyta and Rhodophyta. Biol. Rev. **30**, 343—390 (1955). — Reproduction in the Bangiophycidae. Botan. Rev. **22**, 553—611 (1956).

FELDMANN, J.: Une Némalionale à carpotetraspores: *Helminthocladia Hudsoni* (C. Ag.) J. Ag. Bull. Soc. Histoire natur. Afrique N. Alger **30**, 87—97 (1939). — FELDMANN, J., et G. FELDMANN: Recherches sur les Bonnemaisoniacées et leur alternance de générations. Ann. Sci. Nat., Ser. **11**, 3, 75—175 (1942). — FLINT, L. H.: Studies of freshwater red algae. Amer. J. Bot. **34**, 125—131 (1947). — *Kyliniella* in America. Phytomorphology **3**, 76—80 (1953). — FOSLIE, M.: Remarks on northern *Lithothamnia*. Norsk Vid. Selsk. Skr. Trondhjem **1905** (3), 1—138 (1905).

GEITLER, L.: *Rhodospora sordida* nov. gen. et n. sp., eine neue Bangiacee des Süßwassers. Österr. bot. Z. **76**, 25—28 (1927).

HASSINGER-HUIZINGA, H.: Generationswechsel und Geschlechtsbestimmung bei *Callithamnion corymbosum* (Sm.) Lyngb. Arch. Protistenk. **98**, 91—124 (1952). — HERNDON, W. R.: *Boldia*: A new rhodophycean genus. Amer. J. Bot. **51**, 575—581 (1964). — HUS, H. T. A.: An account of the species of *Porphyra* found on the Pacific Coast of North America. Proc. Calif. Acad. Sci. III (Botany) **2**, 173—240 (1902).

IWASAKI, H.: The life-cycle of *Porphyra tenera* in vitro. Biol. Bull. **121**, 173—187 (1961). — IWASAKI, H., and C. MATSUDAIRA: Observations on the ecology and reproduction of free-living *Conchocelis* of *Porphyra tenera*. Biol. Bull. **124**, 268—276 (1963).

KILLIAN, K.: Über die Entwicklung einiger Florideen. Z. Bot. **6**, 209—278 (1914). — KLAVESTAD, N.: Paraspores in *Rhodochorton rothii* Naeg. Nytt Mag. Bot. **5**, 61—62 (1957). — KOCH, W.: Entwicklungsgeschichtliche und physiologische Untersuchungen an Laboratoriumskulturen der Rotalge *Trailliella intricata* Batters (Bonnemaisoniaceae). Arch. Mikrobiol. **14**, 635—660 (1950). — KORNMANN, P.: Die Entwicklung von *Porphyra leucosticta* im Kulturversuch. Helg. Wiss. Meeresuntersuchungen **8**, 167—175 (1961a). — Zur Kenntnis der *Porphyra*-Arten von Helgoland. Helg. Wiss. Meeresuntersuchungen **8**, 176—192 (1961b). — KRISHNAMURTHY, V.: Cytological investigations on *Porphyra umbilicalis* (L.) Kütz. var. *laciniata* (Lightf.) J. Ag. Ann. Bot., N.S. **23**, 147—176 (1959). — KUCKUCK, P.: Über *Platoma Bairdii* (Farl.) Kuck. Wiss. Meeresunters. Helgoland N.F. **5**, 187—208 (1912). — KUROGI, M.: Studies of the life-history of *Porphyra*. I. The germination and development of carpospores. Bull. Tohoku Reg. Fish. Lab. **2**, 67—103 (1953). — Influences of light on the growth and maturation of *Conchocelis*-thallus of *Porphyra*. I. Effect of photoperiod on the formation of monosporangia and liberation of monospores. Bull. Tohoku Reg. Fish. Lab. **15**, 33—42 (1959). — KUROGI, M., and K. HIRANO: Influence of water temperature on the growth, formation of monosporangia and monospore-liberation in the *Conchocelis*-phase of *Porphyra tenera* Kjellm. Bull. Tohoku Reg. Fish. Lab. **8**, 45—61 (1956). — KYLIN, H.: Über die Keimung der Florideensporen. Ark. Bot. **14**, No. 22 (1917). — Die Rhodophyceen der schwedischen Westküste. Lunds Univ. Årsskr., N.F., Avd. 2, 40 (2), 1—104 (1944). — Die Gattungen der Rhodophyceen. Lund: Gleerups 1956.

MAGNE, F.: La structure du noyau et le cycle nucléaire chez le *Porphyra linearis* Grev. C. R. Acad. Sci. (Paris) **234**, 986—988 (1952). — Sur l'existence d'une reproduction sexuée chez le *Rhodochaete parvula* Thuret. C. R. Acad. Sci. (Paris) **251**, 1554—1555 (1960a). — Le *Rhodochaete parvula* Thuret (Bangioidée) et sa reproduction sexuée. Cah. Biol. Mar. **1**, 407—420 (1960b). — Sur la caryologie de deux Rhodophycées considérées jusqu'ici comme à cycle cytologique entièrement haplophasique. C. R. Acad. Sci. (Paris) **252**, 4023—4024 (1961). — Les Rhodophycées à cycle haplophasique existent-elles? 4th Int. Seaweed Symp. 1961, Proc., p. 112—116 (1964a). — Recherches caryologiques chez les Floridées (Rhodophycées). Cah. Biol. Mar. **5**, 461—671 (1964b). — Massart, J.: La cicatrisation chez les végétaux. Mém. Acad. roy. Méd. Belg. **57**, No 1 (1898).

Nakazawa, S.: The predetermined polarity in *Porphyra* monospores shed from *Conchocelis*-thalli. Botan. Mag. (Tokyo) **71**, 144—150 (1958). — Nichols, H. W.: Culture and developmental morphology of *Compsopogon coeruleus*. Amer. J. Bot. **51**, 180—189 (1964a). — Developmental morphology and cytology of *Boldia erythrosiphon*. Amer. J. Bot. **51**, 653—659 (1964b).

Okamura, K.: On the vegetative multiplication of *Chondria crassicaulis* Harv. and its systematic position. Bot. Mag. (Tokyo) **17**, 1—5 (1903). — Oltmanns, F.: Notizen über die Cultur- und Lebensbedingungen der Meeresalgen. Flora (Jena) **80**, 38—55 (1895). — Osterhout, W. J. V.: On the life-history of *Rhabdonia tenera* J. Ag. Ann. Bot. **10**, 403—427 (1896).

Papenfuss, G. F.: The structure and reproduction of *Claudea multifida*, *Vanvoorstia spectabilis* and *V. coccinea*. Symbolae Botan. Upsalienses **2** (4), 1—66 (1937). — Phillips, R. W.: On the development of the cystocarp in Rhodymeniales. Ann. Bot. **11**, 347—368 (1897).

Rosenvinge, L. K.: The marine algae of Denmark. Danske Vid. Selsk. Skr. VII, Mat.-nat. Afd. **7** (1909—1931). — On mobility in the reproductive cells of the Rhodophyceae. Bot. Tidsskr. **40**, 72—79 (1927).

Schmitz, F.: Kleinere Beiträge zur Kenntnis der Florideen. V. Nuova Notarisia **5**, 705—720 (1894). — Skuja, H.: Eine neue Süßwasser-Bangiacee, *Kyliniella latvica* n. g., n. sp. Acta Hort. Bot. Univ. Latv. **1**, 1—6 (1926). — Untersuchungen über die Rhodophyceen des Süßwassers. III. *Batrachospermum Breutelii* Rbh. und seine Brutkörper. Arch. Protistenk. **80**, 357—366 (1933). — Die Süßwasser-Rhodophyceen der Deutschen Limnologischen Sunda-Expedition. Arch. Hydrobiol., Suppl. **15**, 603—637 (1938). — Solms-Laubach, H.: Die Corallinenalgen des Golfes von Neapel und der angrenzenden Meeresabschnitte. Fauna u. Flora d. Golfes v. Neapel **4** (1881). — Stosch, H. A. v.: The sporophyte of *Liagora farinosa* Lamour. Brit. Phyc. Bull. **2**, 486—496 (1965). — Sundene, O.: Reproduction and morphology in strains of *Antithamnion boreale* originating from Spitsbergen and Scandinavia. Skr. Norske Vid. Akad. I. Mat.-Nat. Kl., N.S. **5**, 1—19 (1962). — Suneson, S.: The cytology of the bispore formation in two species of *Lithophyllum* and the significance of the bispore in the Corallinaceae. Bot. Not. **1950**, 429—450. — Svedelius, N.: Über Sporen an Geschlechtspflanzen von *Nitophyllum punctatum*. Ber. dtsch. bot. Ges. **32**, 106—116 (1914). — Zytologisch-entwicklungsgeschichtliche Studien über *Scinaia furcellata*, ein Beitrag zur Frage der Reduktionsteilung der nicht tetrasporenbildenden Florideen. Nov. Acta Soc. Sci. Upsal., Ser. IV **4** (4), 1—55 (1915). — The seasonal alternation of generations of *Ceramium corticulatum* in the Baltic. Nov. Acta Soc. Sci. Upsal., vol. Extraord. 3—28 (1927). — *Lomentaria rosea*, eine Floridee ohne Generationswechsel, nur mit Tetrasporenbildung ohne Reduktionsteilung. Ber. dtsch. bot. Ges. **53**, 19—26 (1935). — Zytologisch-entwicklungsgeschichtliche Studien über *Galaxaura*, eine diplobiontische Nemalionales-Gattung. Nov. Acta Soc. Sci. Upsal., Ser. IV **13**, 5—154 (1942).

Thaxter, R.: Note on the structure and reproduction of *Compsopogon*. Bot. Gaz. **29**, 259—267 (1900). — Tobler, F.: Zerfall und Reproduktionsvermögen des Thallus einer Rhodomelacee. Ber. dtsch. bot. Ges. **20**, 357—365 (1902). — Über Vernarbung und Wundreiz an Algenzellen. Ber. dtsch. bot. Ges. **21**, 291—300 (1903). — Über Regeneration und Polarität sowie verwandte Wachstumsvorgänge bei *Polysiphonia* und anderen Algen. Jb. wiss. Bot. **42**, 461—502 (1906). — Tokida, J., and H. Yamamoto: Syntagmatic germination of tetraspores in *Pachymeniopsis yendoi*. Phycologia **5**, 15—20 (1965). — Tseng, C. K., and T. J. Chang: Studies on *Porphyra*. I. Life history of *Porphyra tenera* Kjellm. Acta Bot. Sinica **3**, 287—302 (1954). — Studies on *Porphyra*. II. On the *Conchocelis*-phase of *Porphyra tenera* Kjellm., and its Conchospores. Acta Bot. Sinica **4**, 27—46 (1955a). — Studies on *Porphyra*. III. Sexual reproduction of *Porphyra*. Acta Bot. Sinica **4**, 153—166 (1955b).

Vischer, W.: Zur Morphologie, Physiologie und Systematik der Blutalge, *Porphyridium cruentum* Naegeli. Verh. naturforsch. Ges. Basel **46**, 66—103 (1935).

Weber, W.: Keimungsphysiologische Untersuchungen an *Chondrus crispus*. Z. Bot. **48**, 251—257 (1960a). — Über die Sporenkeimung einiger Ceramiales. Bot. Mar. **2**, 182—188 (1960b). — Westbrook, M. A.: Contributions to the cytology of tetrasporic plants of *Rhodymenia palmata* (L.) Grev. and some other Florideae. Ann. Bot. **42**, 149—172 (1928). — Wille, N.: Algologische Notizen. I—IV. Nyt Mag. Naturvid. **38**, 1—27 (1900).

Yoshida, T.: Life-cycle of a species of *Batrachospermum* found in Northern Kyushu, Japan. Jap. J. Bot. **17**, 29—42 (1959).

Vegetative Fortpflanzung bei Pilzen.

Von

Georg Sörgel.

Neben der geschlechtlichen und der ungeschlechtlichen Vermehrung gibt es bei einer nicht geringen Zahl von Pilzen noch eine Fortpflanzung durch vegetative Zellen. Bei einigen Pilzen ist überhaupt nur diese Art der Vermehrung bekannt. Zu der vegetativen Fortpflanzung rechnet man außer *Sproßmycel* die Bildung von *Chlamydosporen*, von *Sklerotien* und von *Bulbillen*. Über die Bedingungen, unter denen diese Organe entstehen, ist im Vergleich zu unseren Kenntnissen über die Bildung von Gameten oder Sporen noch recht wenig bekannt.

1. Die Bildung von Sproßmycel.

Die Besonderheit des Sproßmycels besteht darin, daß sich an einer als Mutterzelle fungierenden runden oder ovalen Zelle meist mehrere Auswüchse bilden, die wieder zur normalen Größe heranwachsen. An solchen Tochterzellen pflegt sich dieser Sprossung genannte Vorgang mehrere Male zu wiederholen, so daß oft vielfach verzweigte Zellketten entstehen. Die Einzelzellen lösen sich leicht ab und werden so zum Ausgangspunkt neuer Individuen. Dadurch werden diese Zellen zu einer Art von ungeschlechtlichen Sporen, von denen sie sich jedoch durch ihre Bildungsweise unterscheiden. Sie sind in älteren Untersuchungen nicht selten mit ihnen verwechselt worden. Die Einzelzellen werden auch oft *Oidien* oder *Sproßconidien* genannt.

Typisch ist dieses Sproßmycel für die Hefepilze im weitesten Sinne. Als Beispiele seien genannt unter den Ascomyceten die Gattungen *Hansenula*, *Pichia* und *Saccharomyces*, unter den Basidiomyceten die Gattungen *Bullera* und *Sporobolomyces* und unter den Fungi imperfecti die Gattungen *Candida*, *Cryptococcus* und *Torulopsis*.

Verlängern sich die Einzelzellen und bleiben sie dauernd im Zusammenhang, so spricht man von einem Pseudomycel. Die Fähigkeit, Pseudomycel zu bilden, ist in vielen Fällen von Außenfaktoren abhängig.

Am eingehendsten wurde die Frage, unter welchen Bedingungen Sproß- und unter welchen Pseudomycel gebildet wird, bei einigen human- bzw. tierpathogenen Pilzen untersucht.

Ricketts (1901) beobachtet, daß *Blastomyces dermatitidis* in zwei morphologischen Formen vorkommt. Hamburger (1907) gelingt der Nachweis, daß für das Zustandekommen der einen oder anderen Wuchsform die Temperatur eine wichtige Rolle spielt: bei Zimmertemperatur entsteht Pseudomycel, bei höherer Temperatur dagegen Sproßmycel. Michelson (1928) stellt fest, daß bei den Temperaturen 37°, 41° und 45° der Pilz nur in Form von Sproßmycel wächst. In dieser Form findet er sich auch im Körper vor; das Pseudomycel stellt die saprophytische Entwicklungsphase dar. Dagegen zieht De Monbreun (1935) aus seinen Versuchen den Schluß, daß bei 37° nur auf eiweißreichen Nährböden Sproßmycel entsteht, auf eiweißarmen dagegen Pseudomycel. Levine und Ordal (1946) finden, daß unabhängig von der Reaktion des Nährbodens bei Zimmertemperatur und bei 33° Pseudomycel, bei 35—37° aber Sproßmycel gebildet wird. Auf einem Pepton-Glucose-Nährboden liegt nach diesen Autoren das Optimum für die Bildung von Pseudomycel bei 31°, für die Bildung von Sproßmycel bei 35°. Auf einem Mangelnährboden (Glucose und Ammonsulfat) ist das Optimum für beide Wuchsformen etwas erhöht: es liegt bei 33° bzw. bei 37°. Ausgedehnte Untersuchungen von Salvin (1949) zeigen, daß weder Kohlenhydrate noch Aminosäuren für die Ausbildung des Sproßmycels an sich von Bedeutung sind, sondern daß es dabei ausschließlich auf die Temperatur ankommt. Allerdings hängt das Ausmaß der Bildung von der Zusammensetzung des Nährbodens ab. Vitamine sind für diesen Vorgang ohne Bedeutung.

Aus diesen und ähnlichen Untersuchungen ergibt sich eindeutig, daß unabhängig von anderen Faktoren zwischen 33° und 35° der Übergang der einen Mycelform in die andere erfolgt.

Blastomyces dermatitidis kann demnach im lebenden Organismus überhaupt nur in Form von Sproßmycel vorkommen. Dasselbe trifft auch für folgende ebenfalls *dimorphe* Pilze zu: *Paracoccidioides brasiliensis*, *Histoplasma capsulatum* und *Sporotrichum schenckii*. Nach Nickerson und Edwards (1949) und Dowding (1950) bewirkt auch bei *Paracoccidioides brasiliensis* eine Temperatur von 37° die Entstehung von Sproßmycel. Bei *Histoplasma capsulatum* wird nach DeMonbreun (1934) nur auf Blut- oder Serumagar bei Temperaturen von 34—37° Sproßmycel gebildet; nach Salvin (1949) jedoch nur dann, wenn Verbindungen mit Sulfid- oder Sulfhydrilgruppen (einschließlich Biotin) im Nährmedium vorhanden sind. Bei *Sporotrichum schenckii* liegen nach Salvin (1947) die Verhältnisse noch etwas komplizierter: die Bildung von Sproßmycel hängt außer von der Temperatur noch von der Reaktion und der Konsistenz des Nährbodens und vom CO_2-Gehalt ab. Hervorzuheben ist weiterhin, daß zwischen Sproß- und Pseudomycel tiefgreifende physiologische Unterschiede vorhanden sind. So wurde von Nickerson und Edwards (1949) festgestellt, daß der Sauerstoffverbrauch (bezogen auf gleiches Trockengewicht) des Sproßmycels im Mittel 5,6mal so groß ist wie beim Pseudomycel. Nach Mariat (1959) variiert das Verhältnis von Ribonucleinsäure zu Desoxyribonucleinsäure bei *Sporotrichum schenckii* bei Sproßmycel von 0,71—1,40 und beträgt bei Pseudomycel 9,28.

Auch die Entomophthorales bilden im Insektenkörper Sproßmycel. Nach Sawyer (1921) entsteht dieses bei *Empusa sp.* und *Entomophthora pseudococci* auch auf Czapek-Agar.

Für *Candida albicans* berichten Linossier und Roux (1890), daß die Verschiedenheit der Form von der Verschiedenheit der Kohlenstoffquellen abhängt. Die Ergebnisse von Zikes (1920), daß bei 6° und 15° „zumeist nur langgestreckte, septierte Fäden“ und bei 42° „nur mehr Kurzsprosse“ gebildet werden, sowie die gegenteilige Meinung von Talice (1930), daß eine Temperatur von 37° für die Pseudomycelbildung besonders günstig sei, konnten von anderer Seite in dieser Form nicht bestätigt werden. Nach Fineman (1931) wird durch gärfähige Kohlenstoffverbindungen die Bildung von Sproßmycel begünstigt. Burkholder (1943) und Burkholder und Moyer (1943) finden bei der Kultur von *Candida albicans* auf einem Glucosenährboden nur Sproßmycel. Nickerson und van Rij (1949) prüfen die Wirkung von schwefelhaltigen Verbindungen. Sie finden unter anderem, daß Cystein am stärksten, weniger Glutathion und noch weniger Thioglycollat die Bildung von Pseudomycel verhindern. Begünstigt wird die Pseudomycelbildung dagegen durch Kobalt, Bor und Penicillin. Ihre Wirkung wird durch Cystein wieder gehemmt. Nickerson und Mankowski (1953) bestätigen den günstigen Einfluß von Glucose auf die Sproßmycelbildung, stellen aber außerdem fest, daß Polysaccharide plus Cystein die Bildung von Sproßmycel fördern. Sie kommen zu der Ansicht, daß die Sproßmycelbildung abhängig ist von dem Vorhandensein einer gut assimilierbaren C-Verbindung und außerdem von S-haltigen Verbindungen, welche die zur Aufrechterhaltung der Zellteilung erforderlichen SH-Gruppen liefern können. Sie betonen jedoch, daß dies nur ein Glied in der Kette der sich abspielenden Vorgänge sein kann. NICKERSON, TABER und FALCONE (1956) finden, daß Selenite und Tellurite eine Förderung der Zellteilung bewirken. JILLSON und NICKERSON (1948) zeigen, daß *Trichophyton rubrum* Stoffwechselprodukte bildet, welche verhindern, daß sich aus dem Sproßmycel von *Candida albicans* wieder Pseudomycel bildet. Es gelingt ihnen, aus Kulturfiltraten von *Trichophyton rubrum* zwei derartige Stoffe zu isolieren, einen hitzestabilen, wasser- und acetonlöslichen und einen hitzelabilen, wasser-, aber nicht acetonlöslichen. LANGERON und GUERRA (1939) führen die Bildung von Pseudomycel auf den einander zugewandten Seiten von dicht nebeneinander liegenden Impfstreifen ebenfalls auf Stoffwechselprodukte zurück.

Auch bei dem Basidiomyceten *Itersonilia perplexans* hängt nach OLIVE (1952) die Fähigkeit, Sproßmycel zu bilden, von der Zusammensetzung des Nährbodens ab.

Ob die Sproßmycelbildung der pathogenen Pilze als Anpassung an die parasitische Lebensweise aufzufassen ist, wie DUNCAN (1948) annimmt, mag dahingestellt bleiben. Fest steht jedenfalls, daß es andere Pilze und zwar Pflanzenparasiten gibt, bei denen die Verhältnisse gerade umgekehrt liegen: nur bei saprophytischer Entwicklung wird Sproßmycel gebildet! Zu diesen Pilzen gehört *Taphrina deformans*, *Ustilago zeae*, *Ustilago violacea*, *Ustilago panici-frumentacei* sowie *Doassansia sagittariae*. Daneben gibt es aber noch eine Reihe weiterer Pilze, in deren Entwicklungsgang gelegentlich Sproßmycel beobachtet wurde, ohne daß im einzelnen geklärt ist, unter welchen Bedingungen es auftritt. Von diesen Pilzen gibt bereits ZOPF (1890) eine Aufstellung, zu der bis heute nur noch wenige Beispiele hinzugefügt werden konnten. Sproßmycel bilden unter anderem die Gattungen *Bulgaria*, *Tympanis* und *Sphaerulina* unter den Ascomyceten, *Exobasidium vaccinii* und mehrere Arten der Gattung *Tremella* unter den Basidiomyceten sowie *Torula* und *Pullularia pullulans* unter den Fungi imperfecti. Lediglich bei der Gattung *Mucor* unter den Phycomyceten sind die Bedingungen näher untersucht worden.

Bereits 1857 beobachtet BAIL, daß *Mucor racemosus* in einer hefeartigen Form vorkommen kann. Dabei können sich auch hier die Einzelzellen voneinander lösen und zu neuem Sproßmycel auswachsen. Wenn BAIL noch glaubt, daß eine Umwandlung von *Mucor* zu echter Hefe erfolgen kann, so weist REESS (1870) nach, daß sich bei *Mucor mucedo* und *racemosus* nur dann Sproßmycel entwickeln kann, wenn sie untergetaucht in gärfähigen Flüssigkeiten wachsen. BREFELD (1873) ist noch der Ansicht, daß dies nur dann eintritt, wenn sich das Medium allmählich mit CO_2 sättigt. PASTEUR (1876) findet, daß sich der Pilz anfänglich bei genügend Sauerstoff noch normal entwickelt und erst bei Mangel Zucker vergärt und dabei in Sproßmycel übergeht. VAN TIEGHEM (1875) beobachtet, daß auch bei *Mucor circinelloides* und *M. spinosus* Sproßmycel gebildet werden kann. Dasselbe findet BAINIER (1884) bei *Mucor fragilis* und *Mucor erectus* und WEHMER (1900) kann diesen Arten noch *Mucor rouxianus* und *M. javanicus* hinzufügen. Weitere Angaben über die Bedingungen, die die Ausbildung von Sproßmycel zur Folge haben, verdanken wir KLEBS (1928). Nach ihm hemmt der allmählich einsetzende Sauerstoffmangel das Längenwachstum, wobei die CO_2-Vermehrung kaum wesentlich in den Prozeß eingreift. Hefeartig sprossende Zellen entstehen erst bei längerer Dauer der Gärung.

Bei anderen Pilzen können manchmal auch ungewöhnliche Bedingungen die Bildung von Sproßmycel bewirken. So berichtet Wolf (1955), daß *Fusarium oxysporum var. nicotianae* in Schüttelkulturen in Form von Sproßmycel wächst. Unsere Kenntnisse über die Bedingungen der Sproßmycelbildung sind noch zu gering, als daß sich schon allgemeingültige Gesetzmäßigkeiten erkennen ließen. Allerdings hat Nickerson (1948) diese Erscheinungen folgendermaßen zu deuten versucht: Er vergleicht die Umwandlung von Sproß- zu Pseudomycel mit Vorgängen bei Bakterien, bei denen unter besonderen Bedingungen sich aus der typischen Bakterienform lange Fäden entwickeln. In beiden Fällen wird die Zellteilung, nicht aber das Wachstum unterbunden. Vieles spricht dafür, daß es sich in beiden Organismengruppen um die gleichen Enzymsysteme handelt.

2. Die Bildung von Chlamydosporen.

Als Chlamydosporen bezeichnet man einzellige, meist runde — bei dichter Lagerung auch abgeplattete — dickwandige Dauerorgane. Sie zeichnen sich durch einen hohen Gehalt an Reservestoffen aus. Im allgemeinen sind sie dunkel gefärbt. Bei den Saprolegniales werden sie meist Gemmen genannt. Fälschlicherweise spricht man auch oft bei den Brandpilzen von Chlamydosporen.

Nach Reess (1870) entstehen die Chlamydosporen bei *Mucor mucedo* und *M. racemosus* bei ausreichender Ernährung nach der Sporangienbildung, bei ungenügender Ernährung jedoch schon vorher. Nach Klebs (1928) ist die Temperatur der entscheidende Faktor für ihre Bildung: das Optimum liegt bei 28—30°. Nach seinen Untersuchungen sind die Bedingungen für die Chlamydosporen- und Sporangienbildung gerade entgegengesetzt. Herrscht optimale Temperatur für die Bildung der Chlamydosporen, wird die Bildung der Sporangien fast vollkommen unterdrückt und umgekehrt. Auf Peptonnährböden bilden sich bei Zimmertemperatur überhaupt keine Chlamydosporen aus. Günstig für ihre Bildung sind Trauben- und Rohrzuckerlösungen.

Nach Klebs (1898) und Coker (1923) entstehen bei den Saprolegniales die Chlamydosporen leicht dann, wenn gut ernährtes Mycel in eine schwache Lösung von Natriumchlorid oder Tricalciumphosphat übertragen wird oder auch bei Erschöpfung der Nährstoffe.

Nach Nickerson und Mankowski (1953) ist bei *Candida albicans* die Chlamydosporenbildung als Ergebnis von Bedingungen aufzufassen, bei denen die Zellteilung nicht nur gehemmt ist, sondern auch gleichzeitig ein geringes Zellwachstum stattfindet. Bei Glykogenzusatz tritt eine Förderung ein, allerdings bei verschiedenen Herkünften unterschiedlich stark. Ebenso ist Aminopterin, ein Folsäure-Antagonist, für die Entstehung von Chlamydosporen besonders günstig. Dagegen führt ein Nährboden mit Polysacchariden und Cystein zu ihrer Unterdrückung. Nach Ingram (1955) fördert ebenfalls bei *Candida albicans* ein Medium mit wenig C und N die Bildung der Chlamydosporen.

Nach Venkat Ram (1952) läßt sich bei *Fusarium solani* durch Hinzufügen von bestimmten Bakterien zu den Kulturen die Chlamydosporenbildung erheblich steigern. Besonders stark wirkt eine nicht näher bestimmte Art, bei der einer Verdopplung der Bakterienzahl eine Verdopplung der Chlamydosporen gleichkommt. Zurückgeführt wird diese Wirkung auf antibiotisch wirkende Stoffe. Zur Hervorbringung dieses Effekts genügt bereits eine Zugabe von Kulturfiltraten.

Bei *Thielaviopsis basicola* ist nach Stover (1956) die Chlamydosporenbildung abhängig vom Thiamingehalt des Nährbodens: ohne Thiamin werden sie nicht gebildet. Auf Erddekokt entstehen nur wenig. Stimulierend wirken außerdem schwefelhaltige Aminosäuren. Verschiedene Herkünfte (pigmentierte und nicht pigmentierte Formen) verhalten sich nach diesen Untersuchungen allerdings nicht ganz gleich.

Bei *Mycosphaerella pinodes* konnte SÖRGEL (1953) zeigen, daß die Chlamydosporenbildung in starkem Maße von der Reaktion des Nährbodens abhängig ist: je alkalischer das Medium ist, desto mehr werden gebildet. So waren nach bestimmten Versuchen bei p_H 8,0 rund 18mal, bei p_H 7,0 rund 14mal und bei p_H 6,0 rund 4mal soviel Chlamydosporen gebildet werden wie bei p_H 5,0. Dagegen nimmt die Größe bei steigender Alkalität ab: bei p_H 5,0 beträgt ihre Größe im Mittel rund 146 μ^2, bei p_H 6,5 rund 115 μ^2 und bei p_H 8,0 nur noch 52 μ^2. Parallel zu dieser Größenabnahme der Sporen geht eine Abnahme der Länge und Breite der einzelnen Hyphenzellen.

Aus den wenigen Untersuchungen, die über die Bedingungen, unter denen die Chlamydosporen gebildet werden, geht hervor, daß zwar eine Reihe von physikalischen und chemischen Faktoren wirksam ist, daß sich allgemeine Gesetzmäßigkeiten vorläufig aber noch nicht ergeben haben. Manches spricht dafür. daß der Eintritt extremer Bedingungen oft ihrer Bildung günstig ist.

3. Die Bildung von Sklerotien.

Als Sklerotien bezeichnet man vielzellige Gebilde von etwa 0,5 mm bis 1 cm Größe. Meist sind sie dunkel gefärbt. Es besteht oft eine Differenzierung in ein inneres großzelliges und ein äußeres kleinzelliges Plektenchym mit verdickten Wänden. Die Sklerotien sind ausgezeichnet durch einen hohen Gehalt an Kohlenhydraten und oft auch an Öl. Nach manchen Autoren sollen sie den Rhizomorphen mancher Pilze homolog sein, wofür neben ihrer ähnlichen Bildungsweise auch ihre ähnlichen physiologischen Eigenschaften sprechen. Außer verschiedenen physikalischen Faktoren sind vor allem chemische Faktoren, welche ihre Bildung beeinflussen, untersucht worden.

Temperatur.

Grundsätzlich bilden sich die Sklerotien auch bei den Temperaturen, die für vegetatives Wachstum günstig sind. Das Optimum für ihre Entstehung liegt bei den einzelnen Pilzarten etwas verschieden. So stellt DANA (1931) für *Phymatotrichum omnivorum* einen Bereich von 21—32° fest, in dem die Sklerotienbildung vor sich geht. Nach ROGERS (1939) erstreckt sich dieser Bereich von 11—37°. Beide Autoren finden ein Optimum von 27°. PAGE (1956) gibt für *Botrytis squamosa* einen Bereich von 17—27° mit einem Optimum von 22° an. Nach SANFORD (1956) ist für die Sklerotienbildung von *Rhizoctonia solani* 16° günstiger als 24°. MILTHORPE (1941) und TOWNSEND (1957) geben für *Sclerotium rolfsii* einen Bereich von 25—30° an. Bei *Sclerotium cepivorum* geht dieser Bereich von 20—25° (TOWNSEND 1957). Nach TOWNSEND (1957) liegt das Optimum für *Sclerotinia gladioli* bei 25°, für *Rhizoctonia solani* bei 20°. Bei *Sclerotium rolfsii* geht nach ABEYGUNAWARDENA und WOOD (1957) der Temperaturbereich von 10—40° mit einem Optimum bei 30°.

Temperaturwechsel führt nach PAGE (1956) zu einer Anlage der Sklerotien in Ringen.

Von der Temperatur wird aber auch die Größe der Sklerotien beeinflußt. So werden nach RAMSEY (1925) bei *Sclerotinia intermedia*, *S. libertiana* und *S. minor* bei tieferer Temperatur die Sklerotien größer als bei höherer. CHIVERS (1929) kann dies für *Sclerotinia intermedia* bestätigen. Bei *Sclerotinia minor* findet er allerdings das Gegenteil: bei 9° sind die Sklerotien 0,75 mm, bei 18° 1,19 mm, bei 21° 1,51 mm und bei 25° 1,85 mm (Durschchnittswerte von 6 bzw. 7 Herkünften).

Auch die Farbe ist temperaturabhängig: bei niederen Temperaturen sind die Sklerotien heller gefärbt als bei höheren. Dies stellen Rogers (1939) für *Phymatotrichum omnivorum* und Abeygunawardena und Wood (1957) für *Sclerotium rolfsii* fest.

Licht.

Nach Paul (1929) werden bei *Botrytis cinerea* in der Dunkelheit grundsätzlich mehr Sklerotien gebildet. Auf Nährböden, die für das vegetative Wachstum günstig sind, sind die im Dunkeln gebildeten Sklerotien auch größer als die im Licht entstandenen. Nach Townsend (1957) werden bei *Sclerotium rolfsii* im Dauerlicht etwas mehr Sklerotien als im Dunkeln gebildet. Bei *Rhizoctonia solani* wird die Anzahl der Sklerotien durch die Dunkelheit nicht beeinflußt, wohl aber werden sie im Dunkeln eher angelegt. Die Reife geht aber wieder im Licht schneller vonstatten. Nach Abeygunawardena und Wood (1957) werden bei *Sclerotium rolfsii* im Dauerlicht viermal soviel gebildet und haben außerdem ein um 30% höheres Gewicht. Heim, Brack, Kobel, Hofmann und Cailleux (1958) finden, daß die Sklerotienbildung bei *Psilocybe mexicana* mengen- und gewichtsmäßig in der Dunkelheit größer als im Licht ist.

Ein Wechsel von Hell und Dunkel führt bei *Botrytis cinerea* nach Page (1956) zur Anlage der Sklerotien in Ringen.

Nach Page (1956) liegt das Optimum für die Bildung der Sklerotien bei *Botrytis cinerea* bei 370000—434000 erg/sec/cm^2; bei 500000 erg/sec/cm^2 entstehen nur noch ganz wenig.

Feuchtigkeit.

Aus den Versuchen von Paul (1929) über die Wirkung der Luftfeuchtigkeit geht hervor, daß bei *Botrytis cinerea* bei hoher Luftfeuchte viel große Sklerotien entstehen. Beim Kultivieren über 10% H_2SO_4 entstehen sie noch in ziemlicher Menge, über 20% H_2SO_4 werden aber nur noch wenige gebildet. Rogers (1939) findet, daß bei *Phymatotrichum omnivorum* zwischen 15% und 30% Bodenfeuchtigkeit Sklerotien entstehen; das Optimum für die Mycel- und Sklerotienentwicklung liegt zwischen 25% und 30%.

p_H.

Nach Allington (1936) ist die Sklerotienbildung wie das Wachstum bei *Rhizoctonia solani* bei p_H 7,0 optimal. Nach Pape (1937) liegt bei *Sclerotinia trifoliorum* das Optimum für die Sklerotienbildung bei p_H 5,5—6,5. Gewichtsmäßig werden bei p_H 5,5 am meisten gebildet. Bei p_H 8,2 sind die Sklerotien klein, bei p_H 2,6 entstehen keine mehr. Bei saurer Reaktion sind sie größer als bei alkalischer. Bei schwach saurer Reaktion werden sie am schnellsten reif, bei alkalischer und stark saurer Reaktion langsam. Nach Townsend (1957) begünstigt saures Medium die Sklerotienbildung verschiedener Pilze. Bei *Rhizoctonia solani* jedoch werden sie noch bei einem Anfangs-p_H 8,8 gebildet. Nach Abeygunawardena und Wood (1957) ist bei *Sclerotium rolfsii* zwischen p_H 2,4 und 4,2 das Optimum für Wachstum und Sklerotienbildung: bei p_H 2,8 wurden nach ihren Versuchen 1810, bei p_H 4,2 noch 1320 gebildet; das Gewicht betrug bei p_H 2,8 247 mg, bei p_H 4,2 269 mg. Bei p_H 1,9 und p_H 8,6 entstehen keine Sklerotien mehr.

Gase.

Nach Page (1956) werden bei *Botrytis cinerea* nur bei ungehindertem Luftzutritt Sklerotien gebildet: in verschlossenen Kulturgefäßen entstehen keine. Nach Pape (1937) findet bei *Sclerotinia trifoliorum* bei Zugabe von 0,5—5% CO_2 reichliche Bildung statt, bei 10% ist sie geringer und bei 20% hört sie vollkommen auf.

Ernährungsfaktoren.

Mehrere Autoren haben sich immer wieder mit der Wirkung verschiedenartiger Ernährung auf die Bildung der Sklerotien befaßt. Dabei wird fast stets betont, daß natürliche Substrate den künstlich zusammengesetzten erheblich überlegen sind. Schon BREFELD (1891) findet, daß Brot ein besonders gutes Medium für die Bildung von Sklerotien ist. Ebenso geeignet sind aber auch z.B. Kartoffeln, Rüben, Getreidekörner sowie Stengel und Blätter verschiedener Pflanzen. Auch Nährböden aus solchen natürlichen Substraten sind oft besonders geeignet. So fördert nach ASTHANA (1947) 20%iger Zwiebelagar die Sklerotienbildung von *Sclerotium cepivorum*; Tulpenagar ist besonders geeignet für die Sklerotien von *Sclerotium tuliparum*. Dieser ist allerdings auch für *Sclerotium cepivorum* günstig.

Von Kohlenhydraten wird ganz allgemein die günstige Wirkung von Glucose hervorgehoben, z.B. bei *Botrytis cinerea* nach PAUL (1929), bei *Phymatotrichum omnivorum* nach EZEKIEL, TAUBENHAUS und FUDGE (1934), bei *Sclerotinia trifoliorum* nach PAPE (1937), bei *Botrytis cinerea* nach PAGE (1956), bei *Botrytis cinerea*, *Botrytis allii*, *Sclerotinia gladioli*, *Sclerotium rolfsii*, *Sclerotium cepivorum* und *Rhizoctonia solani* nach TOWNSEND (1957) und bei *Sclerotium rolfsii* nach ABEYGUNAWARDENA und WOOD (1957). Weniger günstig wirken Maltose und Mannose; im allgemeinen entstehen keine Sklerotien auf Nährböden mit Galactose, Fructose, Lactose, Rhamnose, Xylose sowie auf Mannit und Glycerin. Einzelne Ausnahmen kommen aber immer wieder vor. So werden z.B. bei *Sclerotium rolfsii* nach TOWNSEND (1957) auf Mannit viel Sklerotien gebildet. Während auf Inulin meist keine Sklerotien entstehen, reagiert nach TOWNSEND (1957) *Sclerotium cepivorum* positiv. Saccharose wirkt bei einzelnen Pilzen fördernd, bei anderen jedoch nicht.

Ebenso unterschiedlich ist auch die Wirkung der einzelnen Aminosäuren. Sehr gut wird im allgemeinen die Wirkung von Asparagin beurteilt, z.B. von EZEKIEL, TAUBENHAUS und FUDGE (1934) bei *Phymatotrichum omnivorum*, von PAPE (1937) bei *Sclerotinia trifoliorum*, von DEMETRIADES (1953) bei *Sclerotinia sclerotiorum* und von ABEYGUNAWARDENA und WOOD (1953) bei *Sclerotium rolfsii*. Ebenso förderlich ist bei den meisten Pilzen auch Harnstoff; allerdings findet PAPE (1937) bei *Sclerotinia trifoliorum* das Gegenteil. Auch Pepton fördert die Sklerotienbildung. Ungeeignet sind bei *Sclerotinia sclerotiorum* nach DEMETRIADES (1953) Cystein, Cystin, Methionin, Tryptophan und Valin. HIGGINS (1927) ist der Meinung, daß organischer Stickstoff für die Sklerotienbildung von *Sclerotium rolfsii* erforderlich ist.

Günstig für die Sklerotienbildung von *Rhizoctonia solani* ist nach ALLINGTON (1936) hoher Stickstoff- und niedriger Kohlenstoffgehalt des Nährbodens.

Vielfach wird auch die fördernde Wirkung von Ammonsalzen hervorgehoben, z.B. von EZEKIEL, TAUBENHAUS und FUDGE (1934) bei *Phymatotrichum omnivorum*, von PAPE (1937) bei *Sclerotinia sclerotiorum*. Ähnlich gut ist die Wirkung von Nitraten, z.B. bei *Rhizoctonia solani* (ALLINGTON 1936), während Sulfate weniger geeignet sind.

Fehlt Magnesium oder Schwefel oder beide zusammen, so findet nach DEMETRIADES (1953) bei *Sclerotinia sclerotiorum* keine Bildung von Sklerotien statt. Weniger behindert ist die Mycelausbildung.

Für all diese Ernährungsversuche ist ferner bezeichnend, daß sich einzelne Herkünfte der Pilze oft etwas verschieden verhalten. Im allgemeinen handelt es sich dabei jedoch nur um quantitative Unterschiede; in einzelnen Fällen zeichnen sich bestimmte Herkünfte durch eine völlig andere Reaktion auf irgendeinen Nährstoff aus.

ROBBINS und KAVANAGH (1938) stellen für *Sclerotium delphinii* und *Sclerotium rolfsii* fest, daß bei Vorhandensein von Pyrimidin und Thiazol im Nährboden zahlreiche Sklerotien gebildet werden. Ist Pyrimidin allein vorhanden, geht die Entwicklung nicht nur langsamer vor sich, sondern es entstehen auch weniger. Ist aber nur Thiazol vorhanden, so bilden sich überhaupt keine Sklerotien aus. Von LILLY und BARNETT (1948) wurde diese Vitaminwirkung bei mehreren Pilzen genauer untersucht. Nach diesen Autoren bildet *Martinia panamaensis* nur bei Thiaminzufuhr Sklerotien; eine andere Herkunft desselben Pilzes bildet sie ohne Thiamin erst nach sehr langer Zeit aus (61 Tage). Auch bei *Sclerotinia minor* steigert Vitaminzufuhr die Sklerotienbildung. Dabei haben vier verschiedene Vitamine zusammen eine stärkere Wirkung als Biotin allein. Bei *Botryotinia convoluta* verhalten sich die einzelnen Herkünfte verschieden: eine benötigt kaum, eine andere etwas stärkere Vitaminzufuhr. *Stromatinia smilacina* benötigt für die Sklerotienbildung offenbar keine zusätzlichen Gaben. Nach TOWNSEND (1957) hat Aneurin bei *Sclerotium rolfsii* eine fördernde Wirkung. Verschieden hohe Gaben haben keine Wirkung auf die Zahl der gebildeten Sklerotien, wohl aber auf die Zahl der ausreifenden Sklerotien.

Das unterschiedliche Verhalten der einzelnen Herkünfte wird so erklärt, daß diejenigen Pilze, welche die für die Sklerotienbildung erforderlichen Vitamine selbst synthetisieren können, unabhängig von einer Zufuhr von außen sind. Je nachdem, in welchem Maße die Bildung von Vitaminen erfolgt, um so weniger oder um so mehr benötigen sie zusätzlich (LILLY und BARNETT 1948).

BROOK und CHESTERS (1957) geben an, daß die Sklerotienbildung von *Botrytis cinerea* vollkommen unterdrückt wird durch 2,3,5,6-tetrachlor-1-nitrobenzen und 2,3,4,6-tetrachlor-1-nitrobenzen, nicht aber durch das 2,3,4,5-Isomere.

Aus den vorliegenden Untersuchungen über die Bedingungen, die zur Sklerotienbildung führen, geht hervor, daß nur dann Sklerotien in größerer Menge gebildet werden können, wenn die Bedingungen zum vegetativen Wachstum günstig sind. In vielen Fällen, z.B. bei *Psilocybe mexicana* nach HEIM, BRACK, KOBEL, HOFMANN und CAILLEUX (1958) sind sogar die Ansprüche für die Sklerotienbildung noch größer als für das Mycelwachstum. Bemerkenswert in diesem Zusammenhange sind die Ergebnisse von TOWNSEND (1957). Danach sind für die Sklerotienbildung drei verschiedene Phasen zu unterscheiden: ihre Anlage, ihre weitere Entwicklung und ihr Ausreifen. Für diese drei Abschnitte müssen unterschiedliche Bedingungen erfüllt sein. Für die Anlage der Sklerotien ist in erster Linie ein besonders großer Anfangs-C-Gehalt erforderlich, in zweiter Linie kommt es erst auf den N-Gehalt an. Wenn die Sklerotien aber zur vollen Reife gelangen sollen, dürfen die Nährstoffe nicht vollkommen erschöpft sein. Die Reife kann aber erst dann beginnen, wenn im Mycelwachstum ein Stillstand eingetreten ist.

4. Die Bildung von Bulbillen.

Unter Bulbillen versteht man mehrzellige, braun gefärbte Organe, die in ihrer Morphologie zwischen Sklerotien und vielzelligen Sporen stehen. Sie kommen bei Ascomyceten *(Melanospora)*, bei Basidiomyceten *(Corticium)* und den Fungi Imperfecti vor *(Papulospora)*. Von diesen Pilzen gibt HOTSON (1912, 1917) eine ausführliche Darstellung. Sie ist jedoch rein beschreibender Natur. Über die Bedingungen, unter denen diese Bulbillen entstehen, ist offenbar nichts bekannt.

Literatur.

ABEYGUNAWARDENA, D. V. W., and R. K. S. WOOD: Factors affecting the germination of sclerotia and mycelial growth of *Sclerotium rolfsii* Sacc. Trans. Brit. mycol. Soc. **40**, 221—231 (1957). — ALLINGTON, W. B.: Sclerotial formation in *Rhizoctonia solani* as affected by nutritional and other factors. Phytopathology **26**, 831—844 (1936). — ASTHANA, R. P.: Studies on sclerotium-forming fungi. I. *Sclerotium cepivorum* Berk. and *S. tuliparum* Klebahn. Part 1. Cultural studies. Proc. Indian Acad. Sci. **26**, 93—107 (1947).

BAIL, TH.: Über Hefe. Flora (Jena) **40**, 417—430, 433—444 (1857). — BAINIER, G.: Nouvelles observations sur les zygospores des Mucorinées. Ann. Sci. nat., Sér. VI, Bot. **19**, 200—216 (1884). — BREFELD, O.: *Mucor racemosus* und Hefe, nebst einigen Bemerkungen zur Systematik der Pilze. Flora (Jena) **56**, 385—400 (1873). — Untersuchungen aus dem Gesamtgebiete der Mykologie, H. 10. Münster 1891. — BROOK, M., and C. G. C. CHESTERS: The growth of *Botrytis cinerea* Pers., *Fusarium caeruleum* (Lib.) Sacc., and *Phoma foveata* Foister in the presence of tetrachloronitrobenzene isomers. Ann. appl. Biol. **45**, 498—505 (1957). — BURKHOLDER, P. R.: Vitamin deficiencies in yeasts. Amer. J. Bot. **30**, 206—311 (1943). — BURKHOLDER, P. R., and D. MOYER: Vitamin deficiencies of fifty yeasts and molds. Bull. Torrey bot. Club **70**, 372—377 (1943).

CHIVERS, A. H.: A comparative study of *Sclerotinia minor* Jagger and *Sclerotinia intermedia* Ramsey in culture. Phytopathology **19**, 301—309 (1929). — COKER, W. C.: The Saprolegniaceae. Chapel Hill N. C. 1923.

DANA, B. F.: Soil cultures for the laboratory production of sclerotia in *Phymatotrichum omnivorum*. Phytopathology **21**, 551—556 (1931). — DEMETRIADES, S. D.: Études sur la biologie du *Sclerotinia sclerotiorum* (Lib.) Massee. III. L'action du magnesium et du soufre sur le développement du champignon et la formation de ses sclérotes. Ann. Inst. phyt. Benaki **7**, 15—20 (1953). — Études sur la biologie du *Sclerotinia sclerotiorum* (Lib.) Massee. IV. L'utilisation des diverses sources d'azote. Ann. Inst. phyt. Benaki **7**, 27—35 (1953). — DE MONBREUN, W. A.: Experimental chronic cutaneous blastomycosis in monkeys. Arch. Derm. Syph. (Chicago) **31**, 831—854 (1935). — The cultivation and cultural characteristics of DARLING's *Histoplasma capsulatum*. Ann. J. trop. Med. **14**, 93—125 (1934). — DOWDING, E. S.: *Histoplasma* and brazilian *Blastomyces*. Mycologia **42**, 668—679 (1950). — DUNCAN, J. T.: The epidemiology of fungus diseases. Trans. ryo. Soc. trop. Med. Hyg. **42**, 207—216 (1948).

EZEKIEL, W., J. T. TAUBENHAUS and J. F. FUDGE: Nutritional requirements of the root-rot fungus *Phymatotrichum omnivorum*. Plant Physiol. **9**, 187—216 (1934).

FINEMAN, B. C.: A study of the thrush parasite. J. infect. Dis. **28**, 185—206 (1921).

HAMBURGER, W. W.: A comparative study of four strains of organisms isolated from four cases of generalized blastomycosis. J. infect. Dis. **4**, 201—209 (1907). — HEIM, R., A. BRACK, H. KOBEL, A. HOFMANN et R. CAILLEUX: Déterminisme de la formation des carpophores et des sclérotes dans la culture du *Psilocybe mexicana* Heim, agaric hallucinogène du Mexique, et mise en évidence de la psilocybine et de la psilocine. Rev. de Mycol. **23**, 106—113 (1958). — HIGGINS, B. B.: Physiology and parasitism of *Sclerotium rolfsii* Sacc. Phytopathology **17**, 417—448 (1927). — HOTSON, J. W.: Culture studies of fungi producing bulbils and similar propagative bodies. Proc. Amer. Acad. Arts and Sci. **48**, 227—306 (1912). — Notes on bulbiferous fungi with a key to described species. Bot. Gaz. **64**, 265—284 (1917).

INGRAM, M.: An introduction to the biology of yeasts. London 1955.

JILLSON, O. F., and W. J. NICKERSON: Mutual antagonism between pathogenic fungi. Inhibition of dimorphism in *Candida albicans*. Mycologia **40**, 369—385 (1948).

KLEBS, G.: Die Bedingungen der Fortpflanzung bei einigen Algen und Pilzen. Jena 1928.

LANGERON, M., et P. GUERRA: Orientation de la filamentisation des champignons levuriformes cultivés sur lames gélosées. Ann. Parasit. hum. comp. **17**, 580—589 (1939). — LEVINE, S., and Z. ORDAL: Factors influencing the morphology of *Blastomyces dermatitidis*. J. Bact. **52**, 687—694 (1946). — LILLY, V. G., and H. L. BARNETT: Growth rates, vitamin deficiencies and sclerotia formation by some Sclerotiniaceae. Proc. W. Va. Acad. Sci. **20**, 69—74 (1948). — LINOSSIER, G., et G. ROUX: Recherches biologiques sur le champignon du muguet. Arch. Méd. exp. **2**, 222—253 (1890).

MARIAT, M. F.: Sur la teneur en acides nucléiques des formes levure et mycélienne de *Sporotrichum schenckii*. C. R. Acad. Sci. **248**, 3468—3469 (1959). — MICHELSON, I. D.: Blastomycosis, a pathologic and bacteriologic study. J. Amer. med. Ass. **91**, 1871—1876 (1928). — MILTHORPE, F. L.: Studies on *Corticium rolfsii* (Sacc.) Curzi (*Sclerotium rolfsii* Sacc.). I. Cultural characters and perfect stage. II. Mechanism of parasitism. Proc. Linnean Soc. **66**, 65—75 (1941).

NICKERSON, W. J.: Enzymatic control of cell division in microorganisms. Nature (Lond.) **162**, 4111, 241—245 (1948). — NICKERSON, W. J., and G. A. EDWARDS: Studies on the physiologic bases of morphogenesis in fungi. 1. The respiratory metabolism of dimorphic pathogenic fungi. J. gen. Phys. **33**, 41—55 (1949). — NICKERSON, W. J., and Z. MANKOWSKI: Role of nutrition in the maintenance of the yeast-shape in *Candida*. Amer. J. Bot. **28**, 617—621 (1953). — NICKERSON, W. J., W. A. TABER and G. FALCONE: Physiological bases of morphogenesis in fungi. 5. Effect of selenite and tellurite on cellular division of yeastlike fungi.

Canad. J. Microbiol. **2**, 575—584 (1956). — NICKERSON, W. J., and N. J. W. VAN RIJ: The effect of sulfhydryl compounds, penicillin, and cobalt on the cell division mechanism of yeasts. Biochim. biophys. Acta **3**, 461—475 (1949).

OLIVE, L. S.: Studies on the morphology and cytology of *Itersonilia perplexans* Derx. Bull. Torrey bot. Club **79**, 126—138 (1952).

PAGE, O. T.: The influence of light and other environmental factors on mycelial growth and sclerotial production by *Botrytis squamosa*. Canad. J. Bot. **34**, 881—890 (1956). — PAPE, R.: Beiträge zur Biologie und Bekämpfung des Kleekrebses (*Sclerotinia trifoliorum* Erikss.). Arb. biol. Reichsanst. Land- u. Forstw. Berlin-Dahlem **22**, 159—247 (1937). — PASTEUR, M. L.: Études sur la bière. Paris 1876. — PAUL, W. R. C.: A comparative morphological and physiological study of a number of strains of *Botrytis cinerea* Pers., with special reference to their virulence. Trans. Brit. mycol. Soc. **14**, 118—134 (1929).

RAMSEY, G. B.: *Sclerotinia* species causing decay of vegetables under transit and market conditions. J. agric. Res. **31**, 597—632 (1925). — REESS, M.: Botanische Untersuchungen über die Alkoholgärungspilze. Leipzig 1870. — RICKETTS, H. T.: Oidiomycosis (blastomycosis) of the skin and its fungi. J. med. Res. **6**, 373—547 (1901). — ROBBINS, W. J., and F. KAVANAGH: Vitamin B_1 or its intermediates and the growth of certain fungi. Amer. J. Bot. **25**, 229—236 (1938). — ROGERS, C. H.: Relation of moisture and temperature to growth of the cotton root-rot fungus. J. agric. Res. **58**, 701—709 (1939).

SALVIN, S. B.: Multiple budding in *Sporotrichum schenckii* Matruchot. J. invest. Derm. **69**, 315—320 (1947). — Phase-determining factors in *Blastomyces dermatitidis*. Mycologia **31**, 311—319 (1949). — Cysteine and related compounds in the growth of the yeastlike phase of *Histoplasma capsulatum*. J. infect. Dis. **84**, 275—283 (1949). — SANFORD, G. B.: Factors influencing formation of sclerotia by *Rhizoctonia solani*. Phytopathology **46**, 281—284 (1956). — SAWYER, W. H.: Observations on some entomogenous members of the Entomophthoraceae in artificial culture. Amer. J. Bot. **16**, 87—121 (1921). — SÖRGEL, G.: Über den Entwicklungsgang von *Mycosphaerella pinodes* (Berk. et Blox.) Stone. II. Der Einfluß der Wasserstoffionenkonzentration auf die Ausbildung der Fortpflanzungsorgane. Arch. Mikrobiol. **19**, 372—397 (1953). — STOVER, R. H.: Effect of nutrition on growth and chlamydospore formation in brown and gray cultures of *Thielaviopsis basicola*. Canad. J. Bot. **34**, 459—472 (1956).

TALICE, R. V.: Sur la filamentisation des monilia. Ann. Parasit. **8**, 394—410 (1930). — TOWNSEND, B. B.: Nutritional factors influencing the production of sclerotia by certain fungi. Ann. Bot. **21**, 153—166 (1957). — TIEGHEM, TH. VAN: Nouvelles recherches sur les Mucorinées. Ann. Sci. nat., Sér. VI **17** (1873).

VENKAT RAM, C. S.: Soil bacteria and chlamydospore formation in *Fusarium solani*. Nature (Lond.) **170**, 4334, 889 (1952).

WEHMER, C.: Die chinesische Hefe und der sogenannte *Amylomyces*. Zbl. Bact., II. Abt. **6**, 353—365 (1900). — WOLF, F. T.: Nutrition and metabolism of the tobacco wilt *Fusarium*. Bull. Torrey bot. Club **82**, 343—354 (1955).

ZIKES, H.: Über den Einfluß der Temperatur auf verschiedene Funktionen der Hefe. Teil II. Zbl. Bact., II. Abt. **50**, 385—410 (1920).

Vegetative Fortpflanzung bei Spermatophyten.

Von

Hans Weber.

Mit 12 Abbildungen.

I. Allgemeines.

Ähnlich wie im Bereich der niederen Pflanzen kann auch bei Spermatophyten der vegetativen Fortpflanzung eine bedeutende Rolle für die Erhaltung und Verbreitung der Arten zukommen. Dies gilt keineswegs allein für zahlreiche in Kultur befindliche Gewächse, es trifft ebenso für eine Fülle von Wildpflanzen zu, von denen manche sogar vorwiegend oder gar ausschließlich auf diese Form der Reproduktion angewiesen sind. Es sei nur an solche diöcischen Arten erinnert, von denen das eine Geschlecht in einem bestimmten Areal fehlt, wie *Elodea canadensis*, die in Europa nur durch weibliche Exemplare vertreten ist, oder wie *Hydrilla verticillata*, die sich ähnlich verhält (vgl. Möbius 1940, S. 29). Auch von *Stratiotes aloides* sind die Geschlechter oft auf verschiedene Areale verteilt (Glück 1906, S. 93). Wo selbststerile Individuen nur vereinzelt vorkommen, stellt die vegetative Reproduktion ebenfalls das einzige Mittel zur Klonbildung dar, wie es Lövkvist (1956, S. 41) etwa für arktische Sippen von *Cardamine pratensis* ausführt. In den Heiden und Mooren Lapplands konnten zahlreiche Arten allein durch vegetative Ausbreitung ein größeres Areal erobern, besonders solche, die aus benachbarten Waldgebieten in die klimatisch ungünstigeren Fjelde vordrangen, wo sie nicht mehr zum Samenansatz gelangen (Söyrinki 1938, S. 208ff.). Ähnliches gilt für *Acorus calamus* in Mittel- und Nordeuropa, der hier keine keimfähigen Samen zu erzeugen scheint. In tropischen Gebieten können z.B. die selten blühenden Bambuseen binnen kurzer Zeit große Flächen besiedeln, allein durch ihre vegetative Ausbreitung (vgl. Abb. 1).

Wir verstehen hier unter vegetativer Fortpflanzung die unmittelbare Bildung von selbständigen Tochterindividuen an bestimmten Vegetationsorganen einer Mutterpflanze, wie Sproßachse, Blatt oder Wurzel, d.h. also von Individuen, die nicht aus einer Zygote hervorgehen und somit abseits der Samenbildung entstehen. Parthenogenetische Entwicklungsvorgänge scheiden wir aus unserer Betrachtung aus. Diese erfolgen im Bereich der Blüte und stehen gewissermaßen zwischen sexueller und asexueller Reproduktion. Sie werden in einem besonderen Abschnitt dieses Handbuches behandelt.

Entscheidend für die vegetative Fortpflanzung ist es, daß die Tochtertriebe sich zu gegebener Zeit von ihrem Mutterorgan lösen, sei es durch aktive Abtrennung oder durch Zerfall der Ausgangsgewebe, und daß sie zur Bildung sproßbürtiger Wurzeln befähigt sind, die ihre Weiterentwicklung gewährleisten. Die wenigen Fälle wurzelloser Spermatophyten wie *Wolffia*, *Utricularia* und *Ceratophyllum* können dabei außer acht gelassen werden. Die sproßbürtige Radikation ist auch Vorbedingung für das Gedeihen jeglicher Stecklingskulturen, wie sie insbesondere für dikotyle Holzgewächse, aber auch für manche krautigen Pflanzen üblich sind. Doch sprechen wir hierbei besser von einer vegetativen *Vermehrung*, da es sich um regenerative Vorgänge handelt, die in der Natur im allgemeinen nicht spontan auftreten. Eine scharfe Grenze zwischen Fortpflanzung und Vermehrung kann und soll damit freilich nicht gezogen werden, zumal die meisten Autoren beide Begriffe nebeneinander im gleichen Sinne verwenden. Über Möglichkeiten und Grenzen von Stecklingskulturen existiert eine umfangreiche, wenn auch weit zerstreute Literatur, über die SWINGLE (1940, S. 301; 1952, S. 1) einen Überblick gegeben hat. Zum großen Teil handelt es sich dabei um die Frage der Beeinflußbarkeit der Wurzelbildung durch Applikation von Wuchsstoffen, die in unserem Zusammenhang nicht näher zu erörtern ist. Hingewiesen sei lediglich darauf, daß Radikation nur bei solchen Pflanzen bzw. Organen künstlich ausgelöst oder gefördert werden kann, die auch auf natürliche Weise unter geeigneten Bedingungen zur Wurzelbildung befähigt sind.

Im einzelnen sind die Formen der vegetativen Fortpflanzung höchst mannigfaltig und in ihren morphologischen Grundzügen schon vielfach beschrieben. Es sei nur auf die neueren zusammenfassenden Werke von GOEBEL (1928, 1933) und von TROLL (1937, 1943) verwiesen, die unter Auswertung des bis dahin vorliegenden Schrifttums eine Fülle von Beobachtungen wiedergeben. Einen geschlossenen Überblick über jene Phänomene hat zuletzt MÖBIUS (1940) vermittelt, der sich zum Teil auf die älteren Darstellungen von KERNER VON MARILAUN (1898) stützt. Wenn wir so über die morphologisch-anatomischen Verhältnisse heute ein recht vollständiges und anschauliches Bild besitzen, so steckt die physiologische Erforschung der betreffenden Erscheinungen noch in den Anfängen. Über die entwicklungsphysiologischen Voraussetzungen ist trotz einer Vielzahl von Arbeiten nur wenig bekannt, die Frage nach der Korrelation zwischen sexuellen und asexuellen Prozessen keineswegs gelöst, wenn auch Ansätze zu ihrer Beantwortung verschiedentlich gemacht werden konnten.

In unserer Darstellung soll zunächst ein Überblick über die Mannigfaltigkeit der Erscheinungen gegeben werden, wobei einige wenige Beispiele näher erläutert werden müssen. Im Anschluß daran wird auf die Physiologie der vegetativen Fortpflanzung einzugehen sein.

II. Morphologisch-anatomische Grundlagen.

1. Vegetative Fortpflanzung durch Sproßverzweigung.

Sproßverzweigung mit anschließender Isolierung der einzelnen Seitenorgane ist ein wesentliches Mittel der vegetativen Fortpflanzung. Im einfachsten Fall erkennen wir bei monopodial oder sympodial wachsenden Achsensystemen, insbesondere bei plagiotropen Rhizomen, eine normale Seitenastbildung. Durch Absterben und Verrottung der rückwärtigen Sproßabschnitte können die Seitentriebe verselbständigt werden und nunmehr zu neuen Individuen heranwachsen, wobei diese das Verhalten der Mutterpflanze wiederholen. Namentlich unter den monokotylen Pflanzen finden wir zahllose Beispiele (*Iris, Acorus, Butomus, Polygonatum, Convallaria, Bambusa, Saccharum, Agropyrum repens* u. v. a.) dafür, aber auch eine Reihe von Dikotylen (*Nuphar, Nymphaea, Anemone nemorosa* u. a.) verhält sich so (Abb. 1). Daß ein derartiger „Rhizomzerfall" nicht immer ein passiver Vorgang sein muß, sondern durch Differenzierung besonderer Trennungszonen vorbereitet werden kann, hat zuletzt NOLTE (1954, S. 224) für *Rumex* gezeigt. Besonders wirkungsvoll erweisen sich die sog. Kriechsprosse bzw. Legehalme, die neben normalen Rhizomen gelegentlich bei *Phragmites* auftreten. Dies sind horizontal wachsende Triebe, die eine Länge bis zu 20 m erreichen und an jedem Knoten, deren bis zu 80 gezählt wurden, einen orthotropen Achselsproß erzeugen können, der sich basal zu bewurzeln vermag (vgl. WEBER 1950, S. 231 und die dort zitierte Literatur sowie MÜLLER-STOLL 1952, S. 618). Auf diese Weise tragen sie wesentlich zur Verlandung mancher Gewässerzonen bei.

Im Gegensatz zu den Kriechhalmen von *Phragmites* verfügen die seitlichen *Ausläufer* oder *Stolonen*, die bei zahlreichen Arten auftreten, nur über eine geringere Zahl von gestreckten Internodien, im Extremfall nur über ein einziges, das sog. Hypopodium, wie dies aus Abb. 2 für *Saxifraga flagellaris* ersichtlich ist. Dieses wächst jedoch zu einer beträchtlichen Länge heran. Für alle Ausläufer gilt, daß sie nach einer plagiotropen Wachstumsphase, die unter- oder oberirdisch verlaufen kann, wiederum zur Bildung eines orthotropen Triebes bzw. — bei Rosettenpflanzen — einer Blattrosette schreiten. Die im Übergangsbereich stets vorliegende Hemmung der Internodienstreckung ist von einer allgemeinen Achsenerstarkung begleitet, wozu man Abbildung 3, *I* vergleiche. An den erstarkenden Teilen aber setzt bevorzugt Wurzelbildung ein, so daß nach Absterben des fadenförmigen Sproßabschnittes die junge Tochterpflanze für sich allein lebensfähig ist. Die Ausläuferbildung bewirkt also in der Tat eine „Fortpflanzung", indem sie die Besiedlung neuer, von der Mutterpflanze entfernter Räume ermöglicht. Ganz ähnlich wie *Saxifraga flagellaris* verhält sich die Erdbeerpflanze (*Fragaria vesca*), nur verfügen deren Stolonen über zwei stark ver-

Abb. 1. *Guadua angustifolia.* Durch Rhizomverzweigung sind zahlreiche Seitentriebe entstanden, die nach kurzer plagiotroper Wachstumsphase zu orthotropem Wuchs übergehen.

längerte Internodien. Zur weiteren Orientierung sei auf die sehr eingehende Darstellung dieses Beispieles bei TROLL (1937, S. 146; 1954, S. 135) verwiesen.

Nicht selten schließen Stolonen ihr Wachstum mit der Bildung von knollenartigen Speicherorganen ab, die erst nach einer Ruheperiode zur Weiterentwicklung gelangen[1]. Dies ist insbesondere bei unterirdischen Ausläufern der Fall,

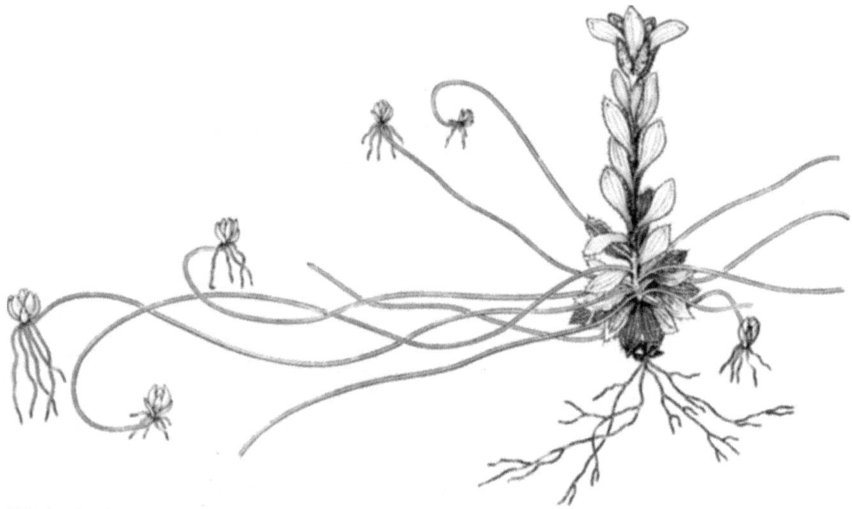

Abb. 2. *Saxifraga flagellaris.* Blühende Pflanze mit Ausläufern. Diese stellen basale Seitensprosse dar. (Nach WARMING.)

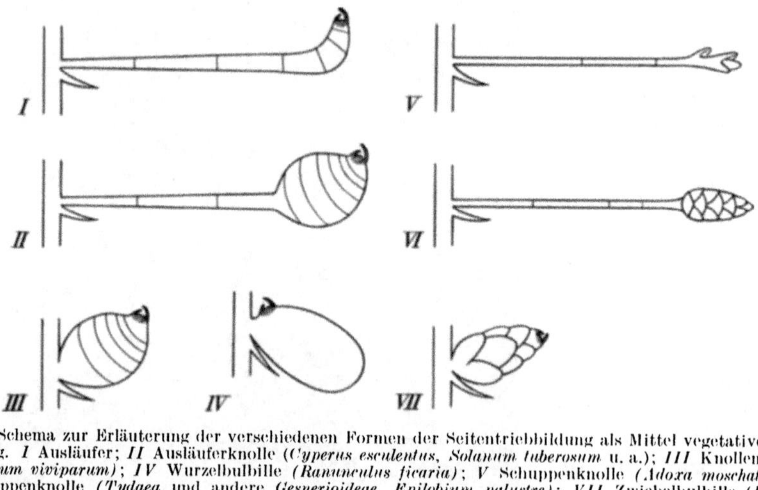

Abb. 3. Schema zur Erläuterung der verschiedenen Formen der Seitentriebbildung als Mittel vegetativer Fortpflanzung. *I* Ausläufer; *II* Ausläuferknolle (*Cyperus esculentus, Solanum tuberosum* u. a.); *III* Knollenbulbille (*Polygonum viviparum*); *IV* Wurzelbulbille (*Ranunculus ficaria*); *V* Schuppenknolle (*Adoxa moschatellina*); *VI* Schuppenknolle (*Tydaea* und andere *Gesnerioideae, Epilobium palustre*); *VII* Zwiebelbulbille (*Dentaria bulbifera* u. a.).

so z. B. bei der Kartoffelpflanze (*Solanum tuberosum*) oder bei der Erdmandel (*Cyperus esculentus*, Abb. 4). Hier ist also die Achsenerstarkung des Endabschnittes der Stolonen extrem (Abb. 3, *II*)[2]. Weniger ausgeprägt ist sie bei der

[1] Vgl. hierzu auch H. WEBER: Morphologisch-anatomische Grundlagen der Speicherung. Handbuch der Pflanzenphysiologie, Bd. VI, S. 817ff.

[2] Daß es bei der Kartoffelpflanze unter geeigneten Bedingungen auch an oberirdischen Sproßteilen zur Knollenbildung kommen kann, hat neuerdings THIJN (1959, S. 95) gezeigt. Man vergleiche hierzu die älteren Ausführungen von VÖCHTING (1899).

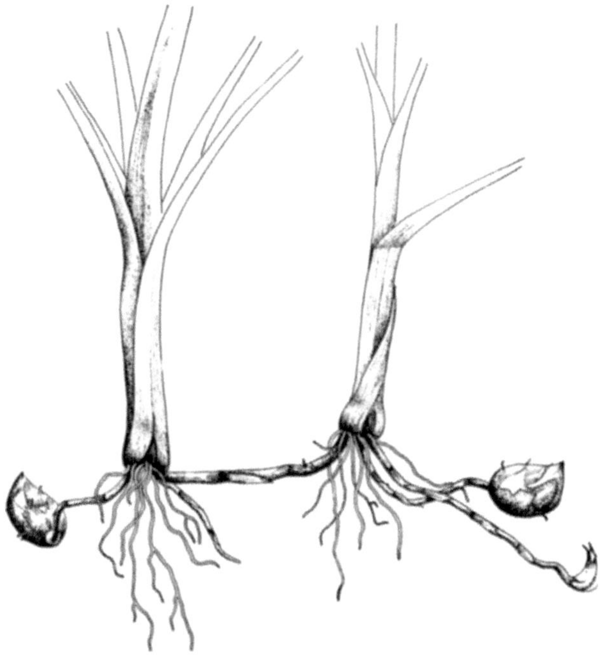

Abb. 4. *Cyperus esculentus.* Sproßsystem mit Ausläuferknollen.

zierlichen *Adoxa moschatellina*, wo die Speicherfunktion außer von der Achse von schuppenförmigen persistierenden Niederblättern übernommen wird (Abb. 3, *V*

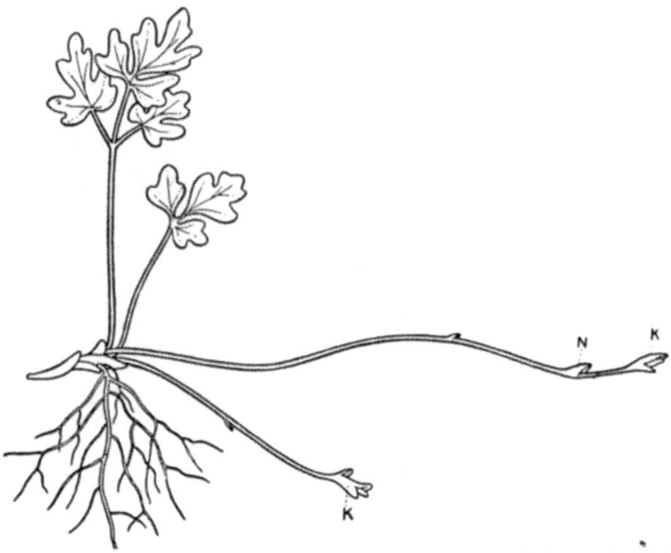

Abb. 5. *Adoxa moschatellina.* Pflanze mit Schuppenknollen (*K*), die am Ende von Ausläufern entstehen. *N* Niederblätter.

und Abb. 5). Bei stärkerer Häufung solcher Blattorgane kommen Bilder zustande, wie sie für verschiedene Gesneriaceen bekannt sind, etwa für die von RAUNKIAER (1934, S. 87) näher beschriebenen *Isoloma pictum* und *Tydaea lindeniana* (Abb. 6a, vgl. auch Abb. 3, *VI*). Auch das schon von IRMISCH (1847, S. 5) daraufhin untersuchte *Epilobium palustre* verhält sich so. Unterbleibt jegliche Achsenstreckung, so entstehen zwiebelartige Bildungen, wie sie von *Allium*-Arten und von zahlreichen anderen monokotylen Pflanzen zur Genüge bekannt sind, die

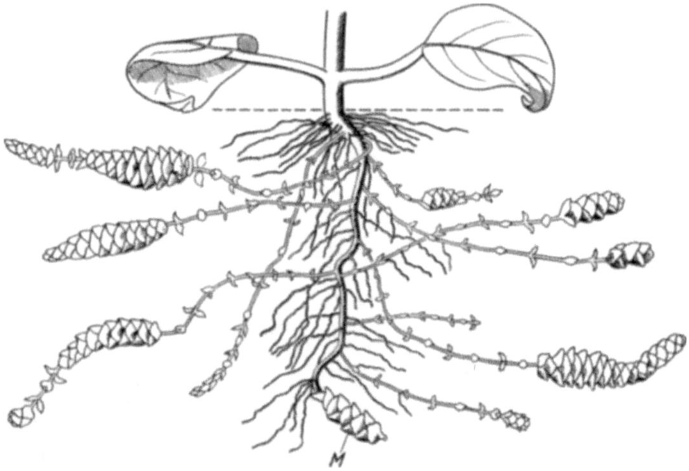

Abb. 6a. *Tydaea lindeniana*. Basaler Teil eines Sproßsystems, dessen Ausläufer mit Schuppenknollen enden. Der dargestellte Muttersproß ist selbst aus einer Schuppenknolle (*M*) hervorgegangen. (Nach Raunkiaer.)

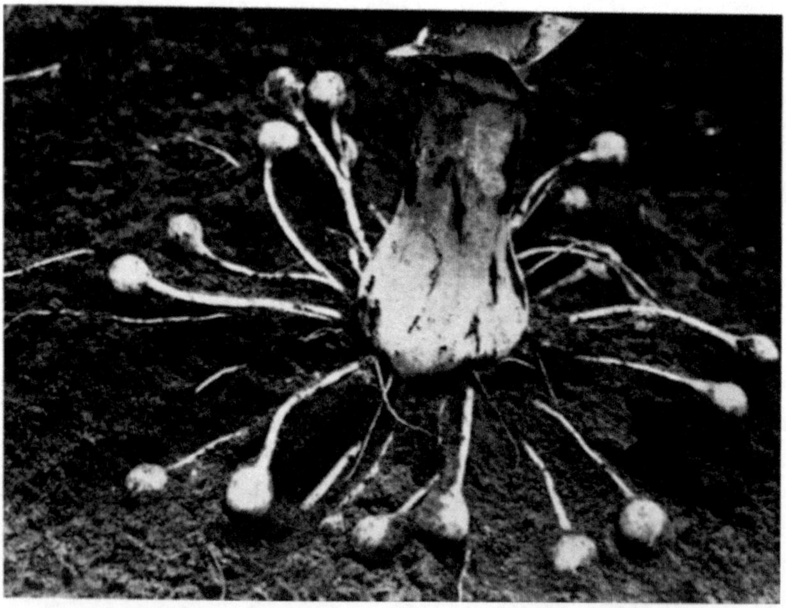

Abb. 6b. *Allium ampeloprasum*. Bildung von Tochterzwiebeln an der Mutterpflanze. (Nach Galil).

aber auch bei Dikotylen vorkommen, so z.B. bei südafrikanischen *Oxalis*-Arten (Hildebrand 1884). Recht interessant in dieser Hinsicht ist das mediterrane *Allium ampeloprasum*, von dem zwei erblich fixierte Formen existieren: eine, deren Tochterzwiebeln in der Achsel ihrer Tragblätter verharren, und eine andere, bei der die Achselprodukte durch Verlängerung der basalen Internodien von der Mutterpflanze entfernt werden (Zohary 1938, S. 35; Galil 1965, S. 135). Man vergleiche hierzu Abb. 6b.

Ähnliche Knollen- bzw. Zwiebelbildungen können jedoch auch an oberirdischen Achsen auftreten, wo sie gewöhnlich als Brutknospen oder Bulbillen bezeichnet werden. Als solche findet man sie sogar bevorzugt im Inflorescenzbereich, so unter anderen bei *Lilium bulbiferum* und einigen weiteren *Lilium*-Arten oder auch bei der Crucifere *Dentaria bulbifera (Cardamine bulbifera)*, wo sie die Blüten zuweilen vollständig ersetzen können (Abb. 7). Ähnlich zeigt *Polygonum viviparum* in der Inflorescenz an Stelle von Blüten leicht abfallende Bulbillen, an denen vor allem der Achsenanteil infolge seiner Verdickung hervortritt (Achsenbulbillen im Gegensatz zu den vorher genannten Zwiebelbulbillen, vgl. Abb. 3, *III*). *Agave*-, *Fourcroya*- und *Allium*-Arten liefern weitere Beispiele für Brutknospenbildung im Blütenstand. Bei letzteren entstehen nach HELM (1953, S. 288) die Zwiebelbulbillen stets im basalen Abschnitt der Inflorescenz, sofern diese heterogen ist (z. B. *Allium scorodoprasum*). Es kann aber auch, wie bei *Allium vineale var. compactum*, der Fall eintreten, daß die Blütenbildung gänzlich unterdrückt wird, so daß dann reine „Bulbillenstände" vorliegen. Als solche faßt GOEBEL (1931, S. 203) in seinem Überblick über „Infloreszenzen mit vegetativen Vermehrungsorganen" auch die Brutknospen tragenden Sprosse der Aracee *Remusatia vivipara* auf, die später MÖBIUS (1935, S. 123) eingehender beschrieben hat. Schließlich sei noch auf die sog. „viviparen" Formen verschiedener Gramineen hingewiesen, wie *Poa alpina vivipara* und *Poa bulbosa vivipara*, in deren Blütenständen gleichfalls an Stelle von Blüten bzw. Ährchen Bulbillen entstehen. ULBRICH (1928, S. 204) hat weitere Beispiele für ein solches Verhalten zusammengestellt. Im einzelnen vergleiche man die Darstellungen bei GOEBEL (1931, S. 199ff.), TROLL (1937, S. 563ff.) und MÖBIUS (1940, S. 49ff.) sowie die dort angeführte Literatur. Eine eingehende Analyse der Prolifikationserscheinungen, die im Bereich der Blütenstände von *Chlorophytum comosum* (Liliaceae) vorliegen, hat jüngst TROLL (1962, S. 59) gegeben. Hier kehren die Inflorescenzscheitel vollständig zu vegetativem Wachstum zurück, was zugleich mit der Wiedererlangung der Fähigkeit zu sproßbürtiger Radikation verbunden ist.

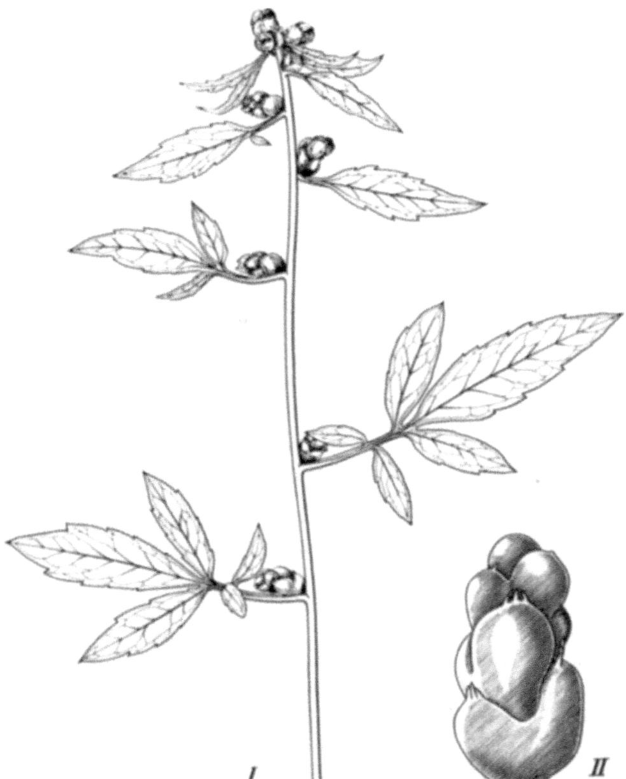

Abb. 7. *Dentaria bulbifera (Cardamine bulbifera)*. *I* Lufttrieb mit Brutknospen; *II* einzelne Zwiebelbulbille.

Durch den Besitz von Bulbillen zeichnet sich weiter *Ranunculus ficaria* aus, bei dem sie in den Achseln von Laubblättern stehen. Doch wird der Hauptteil der Knolle hier von einer kugel- oder walzenförmigen Wurzel gebildet, die der winzigen Achselknospe entspringt (Abb. 3, *IV*). Ganz ähnlich liegen die Verhältnisse bei *Aconitum napellus*. Hier bewurzeln sich die Achselknospen der basalen Niederblätter, und zwar geht aus ihnen je eine kräftige rübenartige Wurzel hervor, die sich jedoch, im Gegensatz zu *Ranunculus ficaria*, frühzeitig verzweigt.

Zum Schluß sei noch kurz auf die Turionen (auch als Gemmae, Hibernacula oder Winterknospen bezeichnet) eingegangen, die sich bei einer größeren Zahl von Wasserpflanzen vorfinden und die sehr eingehend von Glück (1906, S. 83ff.) studiert worden sind. Dabei handelt es sich um terminale oder seitenständige Knospen, die gegen Ende der Vegetationsperiode als zwiebelartige Organe entstehen und oft eigentümlich gebaute Blätter besitzen. Sie lösen sich von der Mutterpflanze ab, zumindest in einzelnen Fällen durch Vermittlung eines besonderen Trennungsgewebes (z. B. *Brasenia peltata* nach Chrysler 1938, S. 277), sinken zu Boden und wachsen, im allgemeinen nach einer Ruhezeit, wieder zu neuen Pflanzen aus. In der Regel werden sie alljährlich in großer Zahl erzeugt; sie stellen ein hervorragendes Mittel zur vegetativen Fortpflanzung dar, was gerade für Wassergewächse von Bedeutung ist, weil bei diesen Samenbildung und Keimung nicht selten äußerst erschwert sind. In diesem Zusammenhang sei auch an die wenig blühenden Lemnaceen erinnert, von denen ebenfalls turionenartige Sproßglieder bekannt sind (vgl. Landolt 1957, S. 388).

2. Knospenbildung an Blättern.

Aus der gärtnerischen Praxis ist es seit langem bekannt, daß aus Blättern oder Blattstücken bestimmter Gewächse, wenn diese als Stecklinge behandelt werden, neue Pflanzen hervorgehen. So entstehen z. B. an der Basis von Blattabschnitten von *Sansevieria zeylanica* junge Triebe, die sich bewurzeln und rasch weiterentwickeln (Abb. 8, *I*). Ähnliches beobachtet man an Blättern von *Peperomia* (Abb. 8, *II*) oder *Begonia rex*, wenn diese zuvor mit Einschnitten versehen werden. Die fleischigen Blätter einer Reihe von Crassulaceen bringen, nachdem sie abgeworfen sind, an ihrer Basis Sproßregenerate hervor. Besonders bekannt geworden ist in dieser Hinsicht *Sedum stahlii*, dessen sukkulente Blattorgane schon nach einer leichten Erschütterung der Pflanze abbrechen und hernach alsbald je eine Knospe erzeugen, die auf Kosten des Wasservorrates des Mutterblattes austreibt (Abb. 8, *III*). Aber auch Blattorgane, die noch mit der Mutterpflanze in Verbindung stehen, können zur Sproßbildung befähigt sein, so etwa bei der Liliacee *Drimiopsis kirkii*, hier aber nur dann, wenn die Blätter durch Einstiche oder Querbrüche vorher verletzt worden sind (Drawert 1952, S. 400; Lenski 1958, S. 586). Goebel (1908, S. 152) erzielte Knospenbildung auch am unverletzten Blatt von *Begonia rex*, wenn sämtliche Vegetationspunkte der Pflanze entfernt wurden, doch zeigen andere Begonien, wie *B. sinuata*, *B. prolifera* und *B. phyllomaniaca* diese Erscheinungen auch unter natürlichen Bedingungen (Goebel 1902, S. 427). Allerdings lösen sich die so entstandenen Sprosse nicht vom Mutterorgan ab. In all diesen Fällen geht der Knospenbildung eine Dedifferenzierung von Dauergewebe voraus, sei es, daß zunächst ein Callus gebildet wird, oder daß, wie es schon Hansen (1881, S. 45) für *Begonia* nachweisen konnte, und wie es auch für *Drimiopsis* zutrifft, die Entwicklung von einer einzelnen Epidermiszelle der Blattspreite ausgeht.

Einer solchen regenerativen Sproßentwicklung stehen die Fälle gegenüber, in denen es zu einer Art Brutknospenbildung kommt, derjenigen vergleichbar, die im vorhergehenden Abschnitt für Sproßachsen geschildert wurde. Hier werden stets die Tochtersprosse bereits an den Blättern angelegt, wenn diese noch mit der Mutterpflanze verbunden sind. Wie TROLL (1939, S. 1119), der diesen Fragen ein umfangreiches Kapitel gewidmet hat, im einzelnen ausführt, bestehen enge Zusammenhänge zwischen der Wachstumsverteilung im Blatt und der Knospenanlegung. Sprossungen sind nämlich an solchen Stellen zu erwarten,

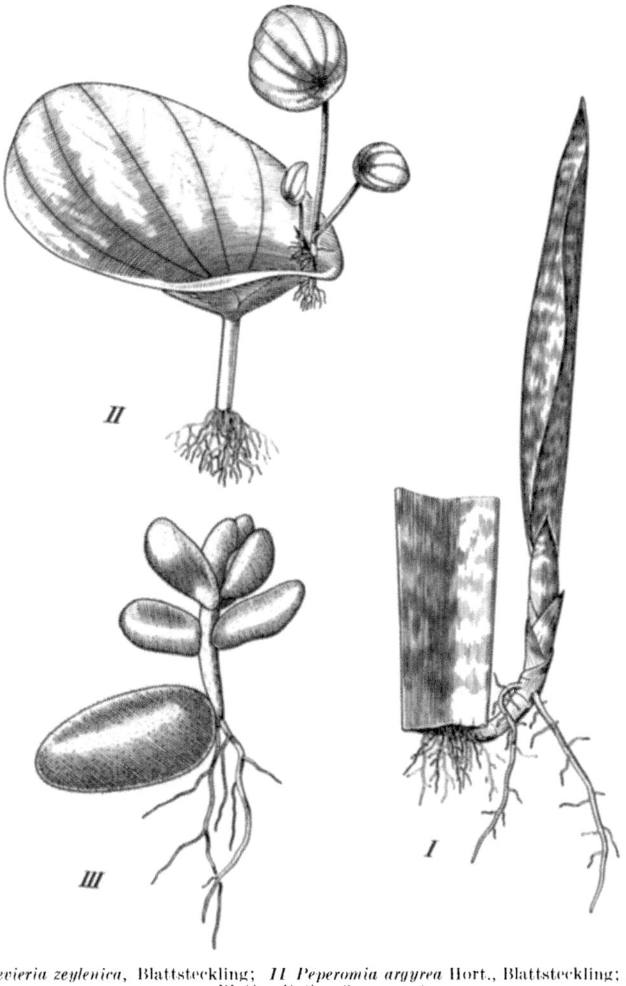

Abb. 8. *I Sansevieria zeylenica*, Blattsteckling; *II Peperomia argyrea* Hort., Blattsteckling; *III Sedum stahlii*, Blatt mit Sproßregenerat.

wo im Verlauf der Blattentwicklung der embryonale Charakter am längsten beibehalten wird. Während bei Farnblättern dies im allgemeinen für den Spitzenteil zutrifft, vollzieht sich das Wachstum der Angiospermenblätter in der Regel basal-interkalar und marginal. Dem entspricht Knospenbildung an der Spreitenbasis und am Blattrand. Aber auch an Blattstielen (z. B. *Pinellia tuberifera*) können Knospen auftreten.

Ein Beispiel für Sproßentstehung an der Spreitenbasis liefert *Nymphaea stellata var. bulbifera* (Abb. 9, *I*), an der Basis von Blattfiedern neben anderen *Cardamine*-Arten insbesondere *Cardamine pratensis*. Auch in diesem letzteren

Fall gehen die Knospen aus einem Meristemkomplex hervor, der erhalten bleibt, wenn das übrige Blattgewebe in den Dauerzustand eintritt (RIEHM 1904, S. 284). Ähnliches gilt für die schon oft untersuchte Knospenbildung am Rand der *Bryophyllum*-Blätter (die umfangreiche, bis 1947 vorliegende Literatur ist bei PEETERS zusammengestellt). Bei *Br. daigremontianum* und *Br. tubiflorum* treiben die Sprosse, die sehr frühzeitig an den Blättern angelegt werden, schon an der Mutterpflanze aus und vermögen sogar hier schon einige Wurzeln zu bilden (Abb. 10). Fallen sie ab, so können sie sich sogleich weiterentwickeln. Bei anderen Arten,

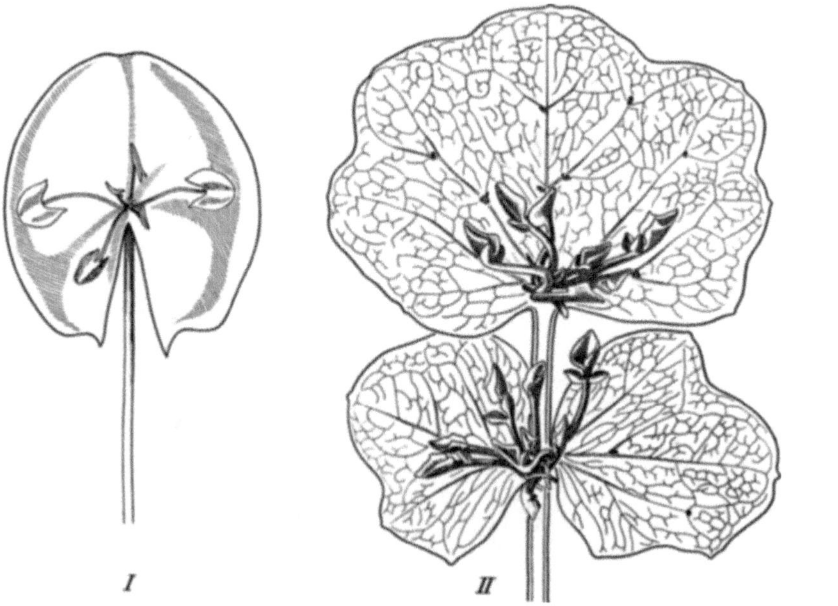

Abb. 9. *I Nymphaea stellata var. bulbifera*. Einzelnes Blatt mit austreibender Knospe. *II Cardamine pratensis*. Blatt mit Sproßbildung an der Basis der Fiedern. An Nervenverzweigungen sind weitere Knospen entstanden, die jedoch ruhend sind. (*I* nach TROLL, *II* nach GOEBEL.)

wie *Br. calycinum* und *Br. proliferum*, werden die Brutknospen erst sichtbar, wenn sich die Blätter vom Stamm abgelöst haben und geeignete Bedingungen das Sprossen begünstigen.

In allen Fällen aber gehen bei *Bryophyllum* die Bulbillen aus embryonalen Gewebekomplexen hervor, die zwischen den Randzähnen der Spreite erhalten bleiben und die zu sog. Knospenträgern (TROLL 1939, S. 1141) heranwachsen, an deren Enden sich die Tochterindividuen entwickeln. Und zwar scheint deren Anlegung im Innern der Knospenträger zu erfolgen (YARBROUGH 1932, CLAMP 1934, JOHNSON 1934, GÖTZ 1953). Bevor der junge Trieb ins Freie gelangt, muß er folglich erst das umgebende Blattgewebe durchbrechen. Dafür spricht auch die Beobachtung von RACK (1953, S. 9), daß an den Stellen, wo Brutknospen gerade sichtbar werden, Zerreißungen der Epidermis nachweisbar sind. So deutet die von RACK veröffentlichte Abb. 6 auf subepidermalen Ursprung der Knospen hin, was mit den Ausführungen von FREELAND (1933, S. 20) in Widerspruch steht, der exogene Entstehung angibt. Subepidermal erfolgt nach WIELING (1952, S. 427) auch die Anlegung der blattbürtigen Sprosse, die zuweilen an alternden Blättern von Tomatenpflanzen auftreten. Diese lösen sich jedoch nicht ab und sind daher für die vegetative Fortpflanzung bedeutungslos.

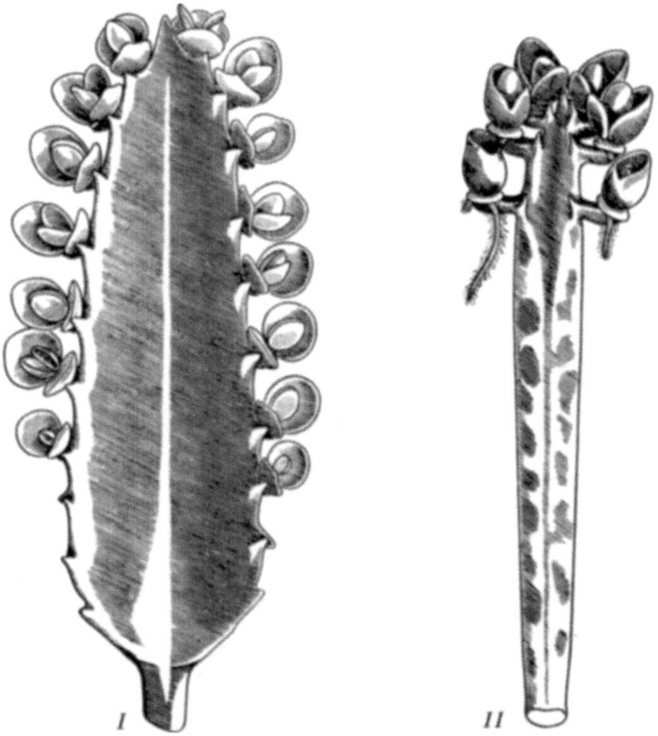

Abb. 10. *I Bryophyllum daigremontianum*; *II Bryophyllum tubiflorum*. Blätter mit Brutknospen. Diese sitzen auf Knospenträgern, vor denen die Randzähne des Blattes sichtbar sind.

3. Sproßbildung an Wurzeln.

Wie das Blatt vermag auch die Wurzel in den Dienst der Fortpflanzung zu treten, was wiederum auf ihrer Fähigkeit beruht, in zahlreichen Fällen Knospen zu erzeugen. Für einzelne Beispiele ist dieses Phänomen schon lange bekannt. WITTROCK (1884) hat als erster einen zusammenfassenden Überblick gegeben, dem die eingehenden Darstellungen von BEIJERINCK (1886), RAUH (1937) und TROLL (1943, S. 2301 ff.) gefolgt sind. Danach sind es vor allem in oberflächennahen Bodenschichten horizontal wachsende Seitenwurzeln von dikotylen Pflanzen (sog. Ausläuferwurzeln), die derartige Sprossungen zuweilen in großer Zahl — und dann oft in reihenartiger Anordnung, den Xylemstrahlen folgend — hervorbringen. Als Beispiel sei *Hippophaë rhamnoides* angeführt (Abb. 11). Ähnlich verhalten sich *Coronilla varia* und *Peltaria alliacea*. Auch der Meerrettich *(Armoracia rusticana)* sei genannt, dessen zahlreich auftretende Wurzelknospen für die Erhaltung der Art sorgen, was hier besonders bedeutsam ist, weil die Kulturformen nicht zum Samenansatz gelangen (vgl. TROLL 1959, S. **673**). Weithin bekannt durch ihre Wurzelsprossung sind ferner *Convolvulus arvensis*, *Cirsium arvense*, *Sonchus arvensis*, *Sium latifolium*, *Linaria vulgaris*, *Euphorbia cyparissias*, *Rumex acetosella* u.v.a. Jedoch ist zu bemerken, daß nicht bei allen Arten mit subkotyledonarer Sproßbildung jene Triebe als Mittel der Fortpflanzung angesehen werden können, nämlich dann nicht, wenn ihnen das Vermögen fehlt, sich sproßbürtig zu bewurzeln. In solchen Fällen bleiben die Wurzelsprosse nur so lange am Leben wie die Mutterpflanze selbst, sie stellen also lediglich Bereicherungstriebe dar. Beispiele dafür sind unter anderen *Euphorbia cyparissias* und *Rumex acetosella*. Hypokotylbürtige Sprosse scheinen niemals Wurzeln zu bilden.

Was die Anlegung der wurzelbürtigen Knospen betrifft, so erfolgt diese stets, ähnlich wie die Ausgliederung der Seitenwurzeln, akropetal, derart also, daß die jüngsten Sprosse der Wurzelspitze am nächsten liegen. Oft, jedoch keineswegs immer, bestehen enge Beziehungen zur Lokalisation der Seitenwurzeln insofern, als die Wurzelknospen diesen stark genähert sind und gleichsam aus ihrer Achsel hervorgehen. Die Bildung erfolgt endogen und zwar im allgemeinen in meristematischen Bezirken (Pericambium, Phellogen). Ausnahmen von dieser

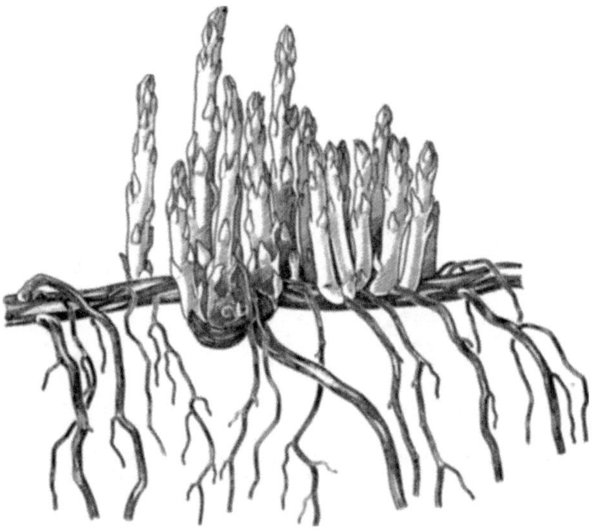

Abb. 11. *Hippophaë rhamnoides*. Wurzelstück mit Sproßbildung.

Regel machen *Aristolochia clematitis* (BEIJERINCK 1886, S. 109), *Bryophyllum proliferum* (OSSENBECK 1927, S. 381) und vor allem die Podostemonaceen (WARMING 1881, 1901; GOEBEL 1933, S. 1483; RAUH 1937, S. 534). Bei all diesen werden kleine Komplexe von Rindenzellen dedifferenziert und zu Knospenanlagen umgebildet. Exogene Anlegung ist nur von hypokotylbürtigen Sprossen bekannt, wie sie unter anderen bei *Linum*-Arten reichlich vorkommen. Bemerkenswert ist die Bildung der Hypokotylsprosse bei *Alliaria officinalis*. Sie sollen stets im Zusammenhang mit einem endogenen Wurzelprimordium angelegt werden, an dem sie exogen entstehen. CHAMPAGNAT, MARICHAL u. VINCENT (1962, S. 32), die diese Verhältnisse näher studiert haben, möchten diese Anlagen als „Adventivembryonen" bezeichnen.

Abweichend von dieser lateralen Knospenbildung erscheint das Auftreten von terminalen bzw. akrogenen Wurzelsprossen. Hier erfolgt die Sproßdifferenzierung an der äußersten Wurzelspitze, die dabei eine eigenartige Veränderung erfährt. Besonders bekannt geworden für ein solches Verhalten ist *Neottia nidus-avis* (WARMING 1874 u.a.), deren truppweises Vorkommen in der Natur auf das Auswachsen derartiger Wurzelknospen zurückzuführen ist. Die saprophytische Pflanze besitzt ein kurzes Rhizom, das mit unverzweigten, nestartig verschlungenen sproßbürtigen Wurzeln versehen ist (Abb. 12, *I*). Deren Spitzen schwellen, namentlich im Alter, köpfchenartig an, wie es Abb. 12, *II*—*V* zeigt. Die Auftreibungen sind aber nichts anderes als akrogene Knospen, die die Wurzelhaube gesprengt haben, sich nun ihrerseits frühzeitig sproßbürtig bewurzeln und alsbald von der absterbenden Mutterpflanze isoliert werden. Unter den Orchideen ist eine derartige Wurzelsproßbildung noch von *Listera cordata* (BRUNDIN 1895/96 u.a.), *Pogonia ophioglossoides* (HOLM 1900 u.a.) und *Taeniophyllum* (DOCTERS VAN LEEUWEN 1920) bekannt geworden, sie findet sich aber auch bei einigen Pteridophyten, so in den Gattungen *Ophioglossum*, *Platycerium* und *Diplazium*. Es ist nicht daran zu zweifeln, daß sie in den genannten Fällen von erheblicher Bedeutung für die Fortpflanzung ist. Daß die Sprossung hier akrogen verläuft, ist allem Anschein nach darauf zurückzuführen, daß das noch bildungsfähige Meristem allein auf die Wurzelspitze beschränkt ist, was schon daraus hervorgeht, daß es bei den aufgeführten Orchideen niemals zur Bildung von Seitenwurzeln kommt.

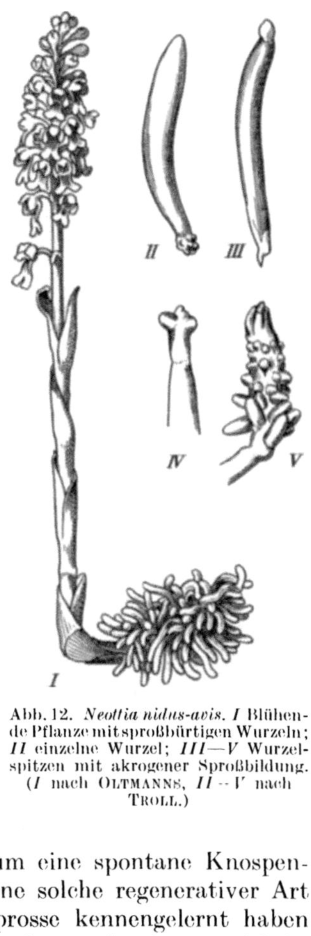

Abb. 12. *Neottia nidus-avis*. *I* Blühende Pflanze mit sproßbürtigen Wurzeln; *II* einzelne Wurzel; *III*—*V* Wurzelspitzen mit akrogener Sproßbildung. (*I* nach OLTMANNS, *II*—*V* nach TROLL.)

Auch bei Holzgewächsen ist Wurzelsprossung weit verbreitet. Für strauchige Wuchsformen wurde schon *Hippophaë rhamnoides* genannt. Doch gehören auch *Corylus*, *Berberis*, *Cydonia*, *Cornus sanguinea* u.a. zu dieser Gruppe. Baumartige Vertreter sind unter anderen *Prunus*-Arten, *Ulmus campestris*, *Robinia pseudacacia*, *Ailanthus glandulosa* und *Gymnocladus canadensis*. Auch auf die Gattung *Populus* ist hier hinzuweisen. Doch bleiben z.B. bei *Populus tremula*, ebenso wie bei der Mehrzahl der übrigen genannten Bäume, die stets vorhandenen Wurzelknospen im allgemeinen ruhend; sie treiben erst dann aus, wenn der Stamm gefällt wird bzw. abstirbt. Der Haupttrieb wirkt hier offensichtlich hemmend auf die Entwicklung der Wurzelsprosse (vgl. auch die Entwicklung der Blattbulbillen bei *Bryophyllum calycinum* und *Br. proliferum*, S. 796). Auf dem Vorhandensein solcher Organreserven beruht auch die Möglichkeit, aus Wurzelstecklingen neue Pflanzen heranzuziehen, wovon in der Praxis weitgehend Gebrauch gemacht wird (vgl. OLBRICH 1899; KRÜSSMANN 1935; v. WETTSTEIN 1940, S. 92; SCHRÖCK 1958, S. 71).

Wenn es sich bei den geschilderten Beispielen um eine spontane Knospenbildung an den Wurzeln handelt, so steht dieser eine solche regenerativer Art gegenüber, wie wir sie ähnlich für blattbürtige Sprosse kennengelernt haben (S. 794). Knospen werden in diesem Fall nur dann angelegt und weiterentwickelt, wenn die betreffenden Wurzeln bereits von der Mutterpflanze getrennt worden sind oder Verletzungen erlitten haben. TROLL (1943, S. 2304) und RAUH (1937, S. 530) haben die für ein solches Verhalten bekannten Fälle zusammengestellt. Nicht selten handelt es sich dabei um Arten, die eine pfahl- bzw. rübenartige Primärwurzel besitzen, an der die Knospen auftreten, so etwa *Oenothera biennis*, *Symphytum ottomanum*, *Taraxacum officinale* u.a. Auch *Falcaria vulgaris* kann als Beispiel hinzugefügt werden (WEBER 1953, S. 40). Allerdings scheinen die sich so entwickelnden Triebe nur selten Wurzeln schlagen zu können, so daß sie lediglich das Individuum erhalten, für eine vegetative Fortpflanzung dagegen kaum Bedeutung haben.

III. Zur Physiologie der vegetativen Fortpflanzung.

Goebel (1908, S. 141; 1928, S. 561) vertrat die Ansicht, daß es sich bei der vegetativen Reproduktion weitgehend oder gar ausschließlich um korrelative Wachstumsprozesse handelt. Im Hinblick auf solche *Bryophyllum*-Arten, deren Blattbulbillen sich normalerweise nicht entfalten, so lange die Blätter mit der Mutterpflanze verbunden sind, führt er aus, dies sei „bedingt durch den Zusammenhang des Blattes mit den Sproß- und Wurzelvegetationspunkten, deren Entfernung oder Inaktivierung ein Austreiben der blattbürtigen Knospen auch dann, wenn das Blatt an der Sproßachse bleibt, herbeiführt. Ebenso wirkt auch eine Unterbrechung der das Blatt mit der Sproßachse verbindenden Leitungsbahnen. Hier kann man annehmen, daß die Hemmung darauf beruht, daß Nährstoffe oder spezifische, das Wachstum bedingende Substanzen aus dem Blatte, in dem sie gebildet werden, ständig in die Sproß- und Wurzelvegetationspunkte abwandern, während sie bei Unterbrechung der Leitbahnen usw. den blattbürtigen Knospen zugute kommen und deren Austreiben bedingen". Indes haben zahlreiche spätere Untersuchungen (unter anderen Hartsema 1926, Behre 1929, Meyer 1953, Götz 1953, Kröner 1955) gezeigt, daß das Problem recht komplexer Natur ist. Insbesondere wurde man auf eine Reihe von Faktoren aufmerksam, die eindeutig (wenigstens in den untersuchten Fällen) die vegetative Reproduktion auslösen, fördern oder hemmen können, so vor allem Temperatur und Photoperiodismus. Wenn sich deren Wirkungen unter natürlichen Bedingungen auch weitgehend überlagern können (vgl. z.B. Vegis 1955), so sollen sie hier doch getrennt voneinander besprochen werden.

1. Einfluß der Temperatur.

Was zunächst den Temperaturfaktor anbelangt, so lassen schon pflanzengeographische Beobachtungen vermuten, daß Kälte die vegetative Reproduktion fördert. Recht gut sind wir in dieser Hinsicht über die nordfinnischen Fjelde und Moore unterrichtet, für die Söyrinki (1938, S. 227) nachweisen konnte, daß ein beträchtlicher Teil der besiedelnden Arten, darunter auch bestandsbildende, sich im wesentlichen vegetativ fortpflanzt. Bevorzugt sind dies aber solche Gewächse, die dem silvinen Element angehören, also aus südlicheren Gebieten vorgedrungen sind. Dabei nimmt im allgemeinen, wie dies auch Kujala (1926, S. 136) und Linkola (1930, S. 166) schon festgestellt haben, die generative Fortpflanzung der betreffenden Arten in zunehmendem Maße ab. Fries (1869, S. 151) betont die Bedeutung der vegetativen Ausbreitung für die höheren Pflanzen auf den Bäreninseln, Hart (1880, S. 74) und Gelting (1934, S. 309) weisen auf ihre Häufigkeit in Grönland hin. Von der in der Arktis verbreiteten *Nardosmia frigida* (Compositae) berichtet schon Kerner (1898, S. 408; vgl. auch Möbius 1940, S. 69), daß sie nur an der Südgrenze ihres Areals blüht, weiter im Norden sich dagegen durch unterirdische Ausläufer fortpflanzt. Ebenso soll *Adenostyles* in den Alpen unterhalb der Waldgrenze reichlich fruchten, in höheren Lagen dagegen nicht mehr blühen, sondern dort nur Ausläufer bilden. Ausgesprochen alpine Arten jedoch zeigen nach Söyrinki (1938, S. 228) nur in den wenigsten Fällen vegetative Reproduktion. Sie sind an die extremen Bedingungen ihrer exponierten Standorte angepaßt und gelangen hier zur vollen Blüh- und Samenreife. Es ist bezeichnend, daß solche nivalen Formen, sowohl in den skandinavischen Gebirgen als auch in den Alpen, vorwiegend diploid sind. Von diploiden Pflanzen aber wissen wir, daß sie im allgemeinen wenig zur vegetativen Reproduktion neigen (vgl. S. 804 sowie die bei Tischler 1956, S. 694

zitierte Literatur). Freilich gibt es auch hier einzelne Arten mit vegetativer Ausbreitung, wie etwa die der Schneetälchenflora angehörende *Salix herbacea*.

Kälteeinwirkung vor dem Schossen ist nach Schwarzenbach (1953, S. 96; 1956, S. 204) auch notwendig, wenn bei *Poa alpina var. vivipara* die Bildung der Bulbillen im Inflorescenzbereich ausgelöst werden soll. Für *Bryophyllum calycinum* haben schon Child und Bellamy (1920, S. 266) berichtet, daß eine mehrere Tage währende Unterkühlung des Blattstieles sich günstig auf das Austreiben der Brutknospen auswirkt. Sicherlich liegen hier ganz ähnliche Verhältnisse vor, wie sie sonst für Zwiebelpflanzen bekannt sind. So wird nach neueren Untersuchungen von Chaudhri, Bünning und Haupt (1956, S. 219; vgl. auch die dort angeführte Literatur) bei *Allium cepa* die Zwiebelbildung bei niedrigen Temperaturen (5° C) gefördert. Ähnliches scheint für die Knollenbildung der Kartoffel, wenigstens bei bestimmten Sorten, zu gelten (Vöchting 1906, S. 87; Goebel 1908, S. 112; Bodlaender 1960, S. 69; 1963, S. 199 u.a.). Nach Went (1959, S. 277) bleibt die Wirkung von niedrigen Temperaturen auch noch in den beiden folgenden Generationen erhalten. Allerdings kommt in all diesen Fällen noch eine Beeinflussung durch die Tageslänge hinzu, von der unten die Rede sein soll. Auch für die Bildung der S. 794 erwähnten Turionen der Wasserpflanzen scheint Temperaturerniedrigung weitgehend auslösend zu wirken (Glück 1906, S. 103; Ringel-Suessenguth 1922, S. 27), vielleicht neben Nährstoff- und Wassermangel sowie unzureichender Belichtung (Goebel 1891, S. 360). Dem stehen neuerdings die Beobachtungen von Vegis (1955, S. 24) entgegen, der bei *Hydrocharis* die Winterknospenbildung gerade bei höheren Temperaturen feststellte, unter der Voraussetzung freilich, daß „die tägliche Beleuchtung nicht zu lang war". Tiefe Temperaturen und niedrige Lichtintensität fördern auch die Knollenbildungen, die Marsden (1958, S. 1480) in den Blattachseln von *Helianthus tuberosus* (Topinambur) erzielen konnte, wenn die Pflanzen in einem sterilen Kokosmilch-Medium mit Casein-Hydrolysat gehalten wurden.

2. Einfluß des Photoperiodismus.

Sehr eingehend ist der Einfluß des Photoperiodismus auf die Bulbillenbildung bei *Bryophyllum* untersucht worden. Während Freeland (1933, S. 467) noch glaubt, daß die Tageslänge ohne wesentlichen Einfluß sei, haben zahlreiche spätere Autoren (v. Denffer 1941, Johnson und Kingsbury 1950, Resende 1951, Meyer 1953, Götz 1953, Kröner 1955, Sironval 1956 u.a.) übereinstimmend festgestellt, daß Langtagbedingungen vorliegen müssen, wenn es zur Ausbildung der Tochtertriebe an der Pflanze kommen soll. Deren Anlagen an den Brutknospenträgern sind zwar immer frühzeitig nachweisbar (vgl. S. 796), ihre Weiterentwicklung aber ist von einer zumindest vorübergehenden Langtageinwirkung abhängig. Dem entspricht auch eine Begünstigung der schon oben erwähnten Zwiebelbildung bei *Allium cepa* im Langtag. Für *Bryophyllum* glaubt Resende (1959, S. 3), daß der Langtag die Hemmung aufhebt, die vom terminalen Vegetationspunkt auf die Bulbillenentwicklung über den Wuchsstoff-Antiwuchsstoff-Spiegel ausgeübt wird.

Auch für andere Pflanzen ist die photoperiodische Abhängigkeit der vegetativen Vermehrung nachgewiesen worden. So konnte Hartmann (1947, S. 419) für *Fragaria vesca* zeigen, daß in einem 10 Std-Rhythmus (10 Std Licht, 10 Std Dunkelheit) die Ausbildung von Ausläufern unterbleibt, solche dagegen im Wechsel von 14stündiger Belichtung und 14stündiger Dunkelheit wohl entwickelt werden. Bei *Begonia evansiana* ist die Bildung von oberirdischen Knollen an Kurztag gebunden, die der unterirdischen Knollen soll von der Photoperiode unabhängig sein (ESASHI u. NAGAO 1958, S. 81; man vgl. auch die Literaturzitate bei KROH 1965, S. 219).

Seit den Untersuchungen von GARNER und ALLARD (1923) bemühten sich zahlreiche Autoren um die Klärung der photoperiodischen Einwirkungen auf die Kartoffelpflanze (*Solanum tuberosum*). Wenn sich die einzelnen Kultursorten auch verschieden verhalten können, so zeichnen sich doch einige allgemeinere Züge ab. Entscheidend ist nach GROSCH (1956, S. 318) die Tagesdauer bei der Keimung der Knollen. Bei der Kulturkartoffel wird „die Bildung der Tochterknollen im Kurztag beschleunigt und auch frühzeitig abgeschlossen. Im Langtag erfolgt sie zögernd und nimmt erst gegen Ende der Vegetationsperiode stärker zu, um dann aber die Knollenmasse im Kurztag meist zu überschreiten" (KRUG 1957, S. 35; vgl. auch die hier zusammengestellte Literatur). STEINECK (1956, S. 196) konnte bei verschiedenen Sorten unter Langtagbedingungen zwar reichliche Bildung von Stolonen erzielen, die sich aber als „wild" erwiesen und vielfach ohne Knollenansatz aus dem Boden herauswuchsen. Allerdings können die jeweils herrschenden Temperaturverhältnisse diese Beziehungen undeutlich machen, wie es STELZNER und TORKA (1940, S. 233) näher ausgeführt haben. Bei der Wildart *Solanum chacoense* gelang es ihnen sogar, durch Temperaturveränderungen die Tageslängenreaktion umzukehren.

Neben der Belichtungsdauer dürfte aber auch die Lichtintensität einen nicht unwesentlichen Einfluß auf die Bildung der Stolonen bzw. Knollen ausüben, worauf für *Solanum tuberosum* unter anderen WERNER (1941, S. 349; 1942, S. 257) sowie POHJAKALLIO (1951, S. 153; 1953, S. 140) und BODLAENDER (1960, S. 69; 1963, S. 199) hingewiesen haben. Wenn BODLAENDER sich für eine Förderung der Knollenbildung durch hohe Lichtintensitäten ausspricht, so stimmen die meisten älteren Autoren darin überein, daß es niedrige Intensitäten sind, die sich günstig auf den Knollenertrag auswirken. Man vergleiche hierzu auch die bereits oben zitierten Befunde für *Helianthus tuberosus* (MARSDEN). Diese Tatsache hat schon GOEBEL (1908, S. 108) unter Hinweis auf ältere Literatur und unter Einbeziehung weiterer Beispiele wie *Adoxa moschatellina*, *Scrophularia nodosa*, *Circaea intermedia* und *Aconitum napellus* hervorgehoben und die Ausläuferbildung im übrigen auf besondere Ernährungsbedingungen zurückgeführt. Auch die Beobachtungen von GLÜCK (1906, S. 103 und 144) über die Bildung der Turionen sprechen für die geschilderten Auffassungen. Bei der rhizogenen Sproßbildung dagegen scheint in einigen Fällen ein fördernder Einfluß einer stärkeren Belichtung auf die Knospenentwicklung vorzuliegen, so nach KLEBS (1903, S. 51) an den Wurzeln von *Ajuga reptans* und nach RAUH (1937, S. 503, 512 und 513) bei *Nasturtium amphibium*, *Alyssum saxatile*, *Bryophyllum proliferum* und *Silene acaulis*. Nähere Untersuchungen über diese Erscheinungen stehen jedoch noch aus. Was die Bulbillenbildung an *Bryophyllum*-Blättern anlangt, so vergleiche man dazu die Ausführungen bei KRÖNER (1955, S. 408).

3. Wuchsstoffwirkung.

Die mitgeteilten Untersuchungen zeigen die große Bedeutung, die der Temperatur und dem Photoperiodismus für die Auslösung bzw. Hemmung der Reproduktion im vegetativen Bereich zukommen. Sie zeigen aber auch deutlich, daß Organneubildungen immer nur dort auftreten können, wo die Möglichkeit zu ihrer Entstehung in der Gesamtgestalt der betreffenden Individuen gegeben ist, in deren allgemeinen Entwicklungsrhythmus sie sich einfügen. Die stofflichen Prinzipien, die zu ihrer Entfaltung führen, sind uns noch unbekannt. Die Ergebnisse, die die Wuchsstoff-Forschung im Hinblick auf die hier erörterten Probleme erzielt hat, sind uneinheitlich, zum Teil widerspruchsvoll und stets auf Einzelfälle beschränkt. Eine kurze Übersicht über die bisherige Literatur hat jüngst LENSKI (1958, S. 583) veröffentlicht. Es sei hinzugefügt, daß GÖTZ (1953, S. 469) an jungen *Bryophyllum*-Blättern durch Auftragen von Wuchsstoffpaste (β-Indolyl-3-essigsäure) keine Brutknospenstimulation erreichen konnte, ebensowenig wie LENSKI (1958, S. 601) an den Blättern von *Drimiopsis kirkii*. Bei *Lilium harrisii* wurde durch Auftragen von 3% β-Indolyl-Essigsäure in Pastenform auf den dekapitierten Sproß die Bildung von Bulbillen in den Achseln der 2—3 obersten Blätter ausgelöst (BEAL 1938, S. 881). Für *Solanum tuberosum* berichtet VAN SCHREVEN (1956, S. 75ff.), daß die Knollenbildung durch Zufuhr von Wuchsstoffen beschleunigt und die Zahl der Tochterknollen durch Zugabe von Vitamin B_1 und C erhöht werden konnte. GREGORY (1956, S. 281) scheint es erstmalig gelungen zu sein, bei der Kartoffel den die Knollenentwicklung induzierenden Faktor durch Pfropfung auf ein anderes Individuum zu übertragen. Entsprechende Versuche mit *Bryophyllum* waren ergebnislos geblieben (GÖTZ 1953, S. 468; KRÖNER 1955, S. 438). KRÖNER möchte die Existenz eines übertragbaren „Langtag-Prinzips" bei *Bryophyllum* sogar ausschließen. In neueren Arbeiten (CHAPMAN 1958, S. 215; OKAZAWA 1960, S. 121; OKAZAWA u. CHAPMAN 1962, S. 413) wird betont, daß bei der Kartoffelpflanze das Knollenbildung auslösende Stimulans in den jungen Blättern unter Kurztagbedingungen erzeugt und vorwiegend basipetal abgeleitet wird. Diese Auffassung wird durch Pfropfversuche erhärtet, die MADEC u. PERENNEC (1959, S. 22) sowie OKAZAWA u. CHAPMAN (1963, S. 623) durchgeführt haben. Sie pfropften u. a. Tomatenreiser auf Kartoffelunterlagen mit dem Ergebnis, daß die Knollenbildung unterblieb. Die letztgenannten Autoren vermuten für diesen Fall, daß nicht allein das knollenerzeugende Stimulans ausfällt, sondern daß darüber hinaus vom Tomatenreis eine hemmende Substanz (vermutlich Gibberelline) erzeugt wird, welche die Knollenbildung unterdrückt. Die gleichen Ergebnisse zeigten sich bei Verwendung von Reisern von *Solanum melongena*.

Daß das Auswachsen vorhandener Knospenanlagen in manchen Fällen durch Anwendung sog. Frühtreibmittel stimuliert werden kann, ist eine alte gärtnerische Erfahrung. Hier sei noch auf Versuche von KAKESITA (1928; 1930/31) hingewiesen, der bei *Bryophyllum calycinum* Brutknospenentwicklung durch Warmbadbehandlung, durch Einwirkung von Wasserstoff sowie durch Injektion von Brenztraubensäure, Aceton, Acetaldehyd und Äthylalkohol anregen konnte. HARIG (1931, S. 58 und 68) erzielte bei *Bryophyllum crenatum* außer durch Warmbad Erfolge durch Begasung der Pflanzen mit Blausäure, bei *Cardamine pratensis* außerdem durch Behandlung mit Äther, Chloroform, Dichloräthylen und Wasserstoff. Alle diese Mittel blieben dagegen erfolglos bei *Begonia rex*, an deren Blättern der Knospenbildung eine Dedifferenzierung von Epidermiszellen vorausgehen muß. Leuchtgas und Äthylen ließ DOSTÁL (1941, S. 437) auf *Circaea*- und *Scrophularia*-Sprosse einwirken, mit dem Ergebnis, daß an Stelle von Laub- und Inflorescenztrieben Ausläufer bzw. Knollen zur Entwicklung angeregt wurden.

Es ist mit Sicherheit anzunehmen, daß die genannten Agenzien tief in den Stoffwechsel der behandelten Organe eingreifen und so die vorher gehemmten induzierenden Prinzipien wirksam werden lassen. Doch können diese Fragen im Rahmen unserer Darstellung nicht erörtert werden. Zur Theorienbildung vergleiche man die Ausführungen bei BÜNNING (1953, S. 66ff.) sowie die Darstellungen in Band XIV—XVI dieses Handbuches.

4. Vegetative Fortpflanzung und Karyologie.

Hingewiesen sei noch darauf, daß Formen mit vegetativer Fortpflanzung sich häufig durch Polyploidie und damit oft auch ganz allgemein durch hohe Resistenz auszeichnen. TISCHLER (1939, S. 420) hat als einer der ersten auf diese Zusammenhänge aufmerksam gemacht und unter anderen *Ranunculus ficaria* und *Acorus calamus* als Beispiele angeführt. Von der erstgenannten Pflanze scheinen diploide Formen nur selten vorzukommen und dann stets als sexuelle Rassen, die keine Brutknospen hervorbringen. Viel weiter sind die Bulbillen tragenden Individuen verbreitet, die kaum zum Samenansatz gelangen; sie haben sich in den meisten Fällen als tetraploid erwiesen. Für *Acorus calamus*, der sich in der nördlichen Hemisphäre ausschließlich vegetativ fortzupflanzen scheint, hat WULFF (1940, S. 478) triploiden Chromosomensatz festgestellt. Recht eingehend hat sich unter anderen GUSTAFSSON (1935, 1948) mit diesen Fragen beschäftigt; er fand, daß unter Uferpflanzen sowie Unkräutern mit asexueller Reproduktion sich auffallend viele Polyploide befinden. Am Beispiel einiger Hydrocharitaceen haben SHARMA und BHATTACHARYYA (1956, S. 121) sich zu diesem Problem geäußert. Hochpolyploid sind auch die arktischen Sippen von *Cardamine pratensis* mit vegetativer Fortpflanzung. Diese ist um so ausgeprägter, je höher der Chromosomensatz ist, unter der Voraussetzung freilich, daß die betreffenden Pflanzen an genügend feuchten Standorten wachsen (LÖVKVIST 1956). Im übrigen vergleiche man die Literaturhinweise bei TISCHLER (1954, S. 266) und dessen umfangreiche Ausführungen über die Beziehungen der Karyologie zur Ökologie und Geobotanik (TISCHLER 1956, S. 635ff.) sowie eine neue zusammenfassende Darstellung von DAWSON (1962).

5. Korrelationen zwischen generativer und vegetativer Fortpflanzung.

Auf Korrelationen, die zwischen Samenbildung und vegetativer Fortpflanzung bestehen, wurde im Vorhergehenden wiederholt hingewiesen. Die einzelnen Arten verhalten sich darin verschieden. Es gibt solche, die unter natürlichen Bedingungen sowohl fruktifizieren als auch mehr oder weniger gleichzeitig vegetative Reproduktionsorgane erzeugen, und Arten, bei denen der eine Fortpflanzungsmodus zugunsten des anderen vollständig unterdrückt ist. Zwischen beiden Gruppen vermitteln Übergänge (KERNER 1898, S. 406 und 538; MÖBIUS 1897, S. 135; 1940, S. 66). Die Frage, warum die eine Art etwa durch Bulbillenbildung ausgezeichnet ist und die andere nicht, bleibt ungelöst. Nur soviel kann gesagt werden, daß zahlreiche Gewächse, die über beiderlei Fortpflanzungsmöglichkeiten verfügen, die eine oder die andere da bevorzugt verwirklichen können, wo die äußeren Bedingungen dies zulassen oder erzwingen. Licht- und Temperaturreize spielen dabei eine hervorragende Rolle. MÖBIUS (1940, S. 65) hat angenommen, daß „der Pflanze ein gewisses Quantum von Auxin zu Gebote steht", das entweder für die Blütenbildung oder für die vegetative Fortpflanzung eingesetzt werden kann, je nach den herrschenden Umweltfaktoren. Dies ist nur eine Hypothese. Es ist aber durchaus möglich, daß solche gegenläufig gesteuerten Prozesse stattfinden und daß ein und derselbe Wuchsstoff Blütenbildung *oder* Sproßentwicklung aktivieren kann. Für *Bryophyllum* haben dies RESENDE u. VIANA (1952, S. 74) und KRÖNER (1955, S. 410) wahrscheinlich gemacht. Bei der Kartoffelpflanze scheint zwischen dem Ausmaß der Blütenbildung und dem zeitlichen Verlauf der Knollenbildung „in gewissen Grenzen eine entgegengesetzte Beziehung zu bestehen" (GROSCH 1956, S. 305). Auch für *Fragaria* glaubt HARTMANN (1947, S. 419), daß Blütenentwicklung und Ausläuferbildung sich im allgemeinen „antagonistisch“ verhalten, d.h. Bedingungen, die das eine begünstigen, versuchen das andere zu hemmen. Für die Erdbeerpflanze hat neuerdings GUTTRIDGE (1959, S. 351) diese Fragen wieder aufgenommen.

Für die Erhaltung der Art ist damit jede Vorsorge getroffen, zumal vegetative Fortpflanzungseinheiten, wie Bulbillen, Zwiebeln, Knollen, Turionen, Rhizomstücke usw. im allgemeinen mit genügend Reservestoffen ausgestattet sind, die ihnen bei der Keimung zur Verfügung stehen. Auch in anderer Hinsicht verhalten sich viele von ihnen wie Samen. So machen z.B. die Bulbillen von *Dentaria bulbifera* eine Nachreifeperiode durch, die derjenigen von Samen entspricht (KÖNEMANN 1956, S. 516). Selbst verbreitungsbiologisch bestehen mancherlei Parallelen, worüber man die Darstellung von KERNER (1898, S. 538) vergleiche. Was schließlich die gelegentlich geäußerte Auffassung betrifft, daß bei andauernder und ausschließlicher vegetativer Reproduktion im Laufe der Zeit eine Schädigung oder Degeneration eintreten könne, so erscheint diese Meinung heute als unbegründet. Jedenfalls gibt es keinerlei zwingende Belege, die dafür sprechen. Solche Fragen sind schon eingehend erörtert worden (MÖBIUS 1897, S. 23; JOST 1923, S. 174; BÜNNING 1953, S. 255), so daß dieser Hinweis genügen mag.

Literatur.

Die mit * versehenen Angaben sind zusammenfassende Darstellungen, die ihrerseits zahlreiche Literaturhinweise enthalten.

BEAL, J. M.: Histological responses of three species of *Lilium* to indoleacetic acid. Bot. Gaz. **99**, 881—911 (1938). — BEHRE, K.: Physiologische und cytologische Untersuchungen über *Drosera*. Planta (Berl.) **7**, 208—306 (1929). — BEIJERINCK, M. W.: Beobachtungen und Betrachtungen über Wurzelknospen und Nebenwurzeln. Natuurk. Verh. Kgl. Akad. Amsterdam, Deel 25, **1887**. Auch in Verzamelde Geschriften **2**. Delft 1921. — BODLAENDER, K. B. A.: De invloed van de temperatuur op de ontwikkeling van de aardappel. Jaarb. Inst. biol. scheik. Onderz. Wageningen **1960**, 69—83. — Influence of temperature, radiation and photoperiod on development and yield. Proc. 10th Easter School Agric. Sci. Nottingham **1963**, 199—210. — BRUNDIN, J. A. Z.: Über Wurzelsprosse bei *Listera cordata* L. Bihang till Kgl. Svenska Vet.-Akad. Handl. **21**. Afd. III, Nr 12 (1895/96). — *BÜNNING, E.: Entwicklungs- und Bewegungsphysiologie der Pflanze. 3. Aufl. Berlin-Göttingen-Heidelberg: Springer 1953.

CHAMPAGNAT, M., J. MARICHAL et C. VINCENT: Mode d'initation de bourgeons hypocotylaires chez *Alliaria officinalis* L. Bull. Soc. bot. France, Mémoires **1962**, 32—44. — CHAPMAN, H. W.: Tuberization in the potato plant. Physiol. Plant. **11**, 215—224 (1958). — CHAUDHRI, J. J., E. BÜNNING u. W. HAUPT: Über die thermoperiodische Beeinflussung der Zwiebelbildung bei *Allium cepa* L. Beitr. Biol. Pflanzen **32**, 219—224 (1956). — CHILD, C. M., and A. W. BELLAMY: Physiological isolation by low temperature in *Bryophyllum*. Bot. Gaz. **70**, 249—267 (1920). — CHRYSLER, M. A.: The winter buds of *Brasenia*. Bull. Torrey bot. Club **65**, 277—283 (1938). — CLAMP, G.: Leaf development and vegetative propagation in *Kalanchoë tubiflora*. Trans. bot. Soc. Edinb. **31**, 327—338 (1934).

DAWSON, G. W. P.: An introduction to the cytogenetics of polyploids. Oxford 1962. — DENFFER, D. v.: Photoperiodische Beeinflußbarkeit von Habitus und Sukkulenz bei einigen Crassulaceenarten. Jb. wiss. Bot. **89**, 543—573 (1941). — DOCTERS VAN LEEUWEN, W.: On the vegetative propagation of two species of *Taeniophyllum* from Java. Ann. Jardin bot. Buitenzorg **31**, 46 (1920). — DOSTAL, R.: Über die Wirkung von Leuchtgas und Äthylen auf die Pflanzenmorphogenese und Produktion. Ber. dtsch. bot. Ges. **59**, 437—453 (1941). — DRAWERT, H.: Über das Auftreten blattbürtiger Adventivsprosse bei *Drimiopsis kirkii* Bak. nach Verwundung. Ber. dtsch. bot. Ges. **65**, 400—402 (1952).

ESASHI, Y., and M. NAGOA: Studies on the formation and sprouting of aerial tubers in *Begonia evansiana* ANDR. I. Photoperiodic conditions for tuberization. Sci. Rep. Tôhoku Univ. **24**, 81—88 (1958).

FREELAND, R. O.: Some morphological and physico-chemical changes accompanying proliferation of *Bryophyllum* leaves. Amer. J. Bot. **20**, 467—480 (1933). — FRIES, TH. M.: Om Beeren-Islands fanerogam-vegetation. Öfvers. Kgl. Vet-.Akad. Förhandl. **1869**, 121—144.

GALIL, J.: Vegetative dispersal of *Allium ampeloprasum* L. I. Vegetative reproduction. Israel J. Bot. **14**, 135—140 (1965). — GARNER, W., and N. A. ALLARD: Further studies in photoperiodism, the response of the plant to relative length of day and night. J. agric. Res. **23**, 871—920 (1923). — GELTING, P.: Studies on the vascular plants of East Greenland between Franz Joseph Fjord and Drove Bay (Lat. 73°15'—76°20' N). Meddel. om Grønland **101** (2), 1—337 (1934). — GLÜCK, H.: Biologische und morphologische Untersuchungen über Wasser- und Sumpfgewächse, Teil 2. Jena: Gustav Fischer 1906. — GOEBEL, K.: Pflanzenbiologische Schilderungen, Teil 2. Marburg: Elwert'sche Verlagsbuchhandlung 1891. — Regeneration im Pflanzenreich. Biol. Zbl. **22**, 304 (1902). — Einleitung in die experimentelle Morphologie der Pflanzen. Leipzig u. Berlin: Teubner 1908. — *Organographie der Pflanzen, Teil 1, 3. Aufl., Jena: Gustav Fischer 1928; Teil 3, 3. Aufl. Jena: Gustav Fischer 1933. — Blütenbildung und Sproßgestaltung. Jena: Gustav Fischer 1931. — GÖTZ, O.: Über die Brutknospenentwicklung der Gattung *Bryophyllum* im Langtag und Kurztag. Z. Bot. **41**, 445—482 (1953). — GRAHAM, R. J. D., and L. B. STEWART: Regeneration from roots of *Anchusa italica*. Trans. bot. Soc. Edinb. **29**, 333—334 (1927). — GREGORY, L. E.: Some factors for tuberization in the potato plant. Amer. J. Bot. **43**, 281—288 (1956). — GROSCH, H. G.: Weitere photoperiodische Versuche an Kulturkartoffeln. Z. Acker- u. Pflanzenbau **101**, 301—320 (1956). — GUSTAFSSON, A.: The importance of the apomicts for plant geography. Bot. Not. **1935**, 325—330. — Polyploidy, life-form and vegetative reproduction. Hereditas (Lund) **34**, 1—22 (1948). — GUTTRIDGE, C. G.: Evidence for a flower inhibitor and vegetative growth promotor in the strawberry. Ann. Bot. (Lond.), N.S. **23**, 351—360 (1959).

HANSEN, A.: Vergleichende Untersuchungen über Adventivbildungen bei den Pflanzen. Abh. Senckenberg. naturf. Ges. **12**, 147—198 (1881). — HARIG, A.: Untersuchungen über die experimentelle Beeinflußbarkeit von Wachstumsvorgängen bei vegetativer Fortpflanzung und Regeneration. Planta (Berl.) **15**, 43—104 (1931). — HART, H. C.: On the Botany of British Polar Expedition of 1875—1876. J. Bot., N. S. **9** (1880). Zit. nach SÖYRINKI 1938. — HARTMANN, H. T.: Some effects of temperature and photoperiod on flower formation and runner production in the Strawberry. Plant Physiol. **22**, 407—420 (1947). — HARTSEMA, A. M.: Anatomische und experimentelle Untersuchungen über das Auftreten von Neubildungen an den Blättern von *Begonia rex*. Rec. Trav. bot. néerl. **23**, 305 (1926). — HELM, J.: Die Entwicklung der Brutzwiebeln in den Infloreszenzen von *Allium vineale* L. und *Allium scorodoprasum* L. Flora (Jena) **140**, 288—297 (1953). — HILDEBRAND, F.: Die Lebensverhältnisse der *Oxalis*-Arten. Jena: Gustav Fischer 1884. — HOLM, TH.: *Pogonia ophioglossoides* Nutt., a morphological and anatomical study. Amer. J. Sci. and Arts, Ser. IV **9**, 13 (1900).

IRMISCH, TH.: Über die Stolonen von *Epilobium palustre* und von einigen anderen Arten desselben Genus. Bot. Ztg **5**, 5 (1847).

JOHNSON, E. L., and D. B. KINGSBURY: *Kalanchoë tubiflora* as influenced by radiation. Univ. Colorado Stud., Ser. Biol. **1**, 1—14 (1950). — JOHNSON, M. A.: The origin of the foliar pseudo-bulbils in *Kalanchoë daigremontiana*. Bull. Torrey bot. Club **61**, 355—365 (1934). — JOST, L.: Pflanzenphysiologie. Bd. II: Formwechsel und Ortswechsel, 4. Aufl. Jena: Gustav Fischer 1923.

KAKESITA, K.: Studies on regeneration in *Brypohyllum*. Jap. J. Bot. **4**, 27—35 (1928). — Experimental studies on the regeneration in *Bryophyllum calycinum*. Jap. J. Bot. **5**, 219—252 (1930/31). — KERNER VON MARILAUN, A.: Pflanzenleben, 2. Aufl., Bd. 2. Leipzig u. Wien: Bibliographisches Institut 1898. — KLEBS, K.: Willkürliche Entwicklungsänderungen bei Pflanzen. Jena: Gustav Fischer 1903. — KÖNEMANN, B.: Über den Stoffwechsel von Brutknospen im Laufe der Nachreife. (Untersuchungen an *Dentaria bulbifera*.) Planta (Berl.) **46**, 516—544 (1955/56). — KRIJTHE, N.: Influence of temperature and light on the early potato „Eersteling" under storage conditions. Versl. Landbouwk. Onderz. **54**, 1—40 (1948). — KRÖNER, E.: Experimentelle Beiträge zum Photoperiodismus der vegetativen Vermehrung der Gattung *Kalanchoë*. Flora (Jena) **142**, 400—465 (1955). — KROH, M.: Physiologie der Fortpflanzung und Sexualität. In Fortschr. Bot. **27**, 206—222 (1965). — KRÜSSMANN, G.: Die Vermehrung der Gehölze. Art, Zeit und Technik. Berlin: Paul Parey 1935. — KRUG, H.: Das photoperiodische Verhalten der Kartoffel *(Solanum tuberosum)*. Eine Übersicht. Angew. Bot. **31**, 29—44 (1957). — KUJALA, V.: Untersuchungen über die Waldvegetation in Süd- und Mittelfinnland. I. Zur Kenntnis des ökologisch-biologischen Charakters der Pflanzenarten unter spezieller Berücksichtigung der Bildung von Pflanzenvereinen. A. Gefäßpflanzen. Communic. ex inst. quaest. forest. Finlandiae editae **10**, 1—154 (1926).

LANDOLT, E.: Physiologische und ökologische Untersuchungen an Lemnaceen. Ber. schweiz. bot. Ges. **67**, 271—410 (1957). — LENSKI, J.: Über die Brutzwiebelbildung auf den Blättern von *Drimiopsis kirkii* Bak. Planta (Berl.) **50**, 579—621 (1958). — LINKOLA, A.: Über das Vorkommen von Samenkeimlingen bei Pollakanthen in den natürlichen Pflanzengesellschaften. Ann. bot. Soc. zool.-bot. fenn. „Vanamo" **11**, 150—172 (1930). — LÖVKVIST, B.: The *Cardamine pratensis* complex. Outlines of its cytogenetics and taxonomy. Symbolae bot. Upsalienses **14** (2), 1—131 (1956).

MADEC, P., and P. PERENNEC: Le rôle respectif du feuillage et du tuberculemere dans la tubérisation de la pomme de terre. Europ. Potato J. **2**, 22—49 (1959). — MARSDEN, RAY

M. P. F.: Formation of Jerusalem Artichoke Tubers in sterile culture. Nature (Lond.) **181**, 1480—1482 (1958). — MEYER, G.: Über photoperiodische Beeinflußbarkeit der Brutpflanzen-Bildung bei *Bryophyllum tubiflorum*. Z. Bot. **41**, 247—256 (1953). — *MÖBIUS, M.: Beiträge zur Lehre von der Fortpflanzung der Gewächse. Jena: Gustav Fischer 1897. — Über Brutknospen bei Araceen. Englers bot. Jb. **67**, 123—131 (1935). — *Die vegetative Vermehrung der Pflanzen. Jena: Gustav Fischer 1940. — MORTON, J. K.: The chromosome numbers of the british Menthae. Watsonia **3**, 244—252 (1956). — MÜLLER-STOLL, W. R.: Über die Entstehung von kriechenden Schilfsprossen auf Dünensand. Biol. Zbl. **71**, 618—626 (1952).

NAYLOR, E.: The morphology of regeneration in *Bryophyllum calycinum*. Amer. J. Bot. **19**, 32—40 (1932). — NOLTE, M.: Rhizomzerfall mit vegetativer Vermehrung bei *Rumex*. Bot. Jb. Systematik usw. **76**, 224—250 (1954).

OKAZAWA, Y.: Studies on the occurence of natural gibberellin and its effects on the tuber formation of potato plant. Proc. Crop. Sci. Soc. Japan **28**, 129 (1959). — Studies on the relation between the tuber formation of potato and its natural gibberellin content. Proc. Crop. Sci. Soc. Japan **29**, 121 (1960). — OKAZAWA, Y., and H. W. CHAPMAN: Regulation of tuber formation in the potato plant. Physiol. Plant. **15**, 413—419 (1962). — OLBRICH, ST.: Vermehrung und Schnitt der Obstgewächse. Stuttgart 1899.

PEETERS, E.: Le bourgeonnement épiphylle chez *Bryophyllum daigremontianum*. Cellule **51**, 193—258 (1947). — POHJAKALLIO, O.: On the effect of the intensity of light and length of day on the energy economy of certain cultivated plants. Acta agric. scand. **1**, 153—175 (1951). — On the effect of day-length on the yield of potato. Physiol. Plantarum (Cph.) **6**, 140—149 (1953).

RACK, K.: Untersuchungen über die Bedeutung der Verwundung und über die Rolle von Wuchsstoffen beim bakteriellen Pflanzenkrebs. Phytopath. Z. **21**, 1—44 (1953). — *RAUH, W.: Die Bildung von Hypokotyl- und Wurzelsprossen und ihre Bedeutung für die Wuchsformen der Pflanzen. Nova Acta Leopoldina, N. F. **4**, 393—553 (1937). — Über die Verzweigung ausläuferbildender Sträucher mit besonderer Berücksichtigung ihrer Beziehungen zu den Stauden. Hercynia (Halle/S.) **1**, 187—231 (1938). — Über die Vermehrungsweise von *Hyacinthus fastigiatus* Bert. Bot. Archiv **41**, 86—92 (1940). — RAUNKIAER, C.: De danske Blomsterplanters Naturhistorie, Bd. 1. Kopenhagen: Gyldendalske Boghandels Forlag 1895—1899. — The life forms of plants and statistical plant geography. Oxford: Clarendon Press 1934. — RESENDE, F.: Acção morfogenética do frio e do fotoperiodismo no „Habitus" vegetativo de *Bryophyllum diagremontianum*. Bol. Soc. portug. Ciê. nat., Sér. II **3**, Fasc. II, 256—264 (1951). — RESENDE, F., and M. J. VIANA: Some effects of synthetic auxins and antiauxin of the floral state. Bol. Soc. portug. Ciê. nat., Sér. II **4**, 74—88 (1952). — RESENDE, F.: Über die Brutknospenentwicklung bei *Bryophyllum* und ihre Vererbung. Ber. dtsch. bot. Ges. **72**, 3—10 (1959). — RIEHM, E.: Beobachtungen an isolierten Blättern. Z. Naturw. (Organ d. naturw. Ver. f. Sachsen u. Thüringen zu Halle a. S.) **77**, 311 (1904). — RINGEL-SUESSENGUTH, M.: Über Ruheorgane bei einigen Wasserpflanzen und Lebermoosen. Flora (Jena) **115**, 27—58 (1922).

SCHICK, R.: Der Einfluß der Tageslänge auf die Knollenbildung der Kartoffel. Züchter **3**, 365 (1931). — SCHREVEN, D. A. VAN: On the physiology of tuber formation in potatoes. I.—V. Plant and Soil (The Hague) 8, 49—104 (1956). — SCHRÖCK, O.: Die Graupappel und ihre vegetative Vermehrung. Züchter **28**, 71—80 (1958). — SCHWARZENBACH, F. H.: Die Abhängigkeit der Bulbillenbildung bei *Poa alpina vivipara* von Periodismus und Frost. Experientia (Basel) **9**, 96 (1953). — Die Beeinflussung der Viviparie bei einer grönländischen Rasse von *Poa alpina* L. durch den jahreszeitlichen Licht- und Temperaturwechsel. Ber. schweiz. bot. Ges. **66**, 204—223 (1956). — SHARMA, A. K., and B. BHATTACHARYYA: A study of the cytology of four members of the Hydrocharitaceae as an aid to trace the lines of evolution. Phyton (Argent.) **6**, 121—132 (1956). — SIRONVAL, C.: Action of day-length upon the formation of adventitious buds in *Bryophyllum tubiflorum* Harv. Nature (Lond.) **178**, 1357—1358 (1956). — SÖYRINKI, N.: Studien über die generative und vegetative Vermehrung der Samenpflanzen in der alpinen Vegetation Petsamo-Lapplands. I. Allgemeiner Teil. Ann. bot. Soc. zool.-bot. fenn. „Vanamo" **11**, Nr 1, 1—311 (1936). — STEINECK, O.: Tageslänge und Knollenbildung bei Kultursorten der Kartoffel. Z. Pflanzenzüchtung **36**, 197—213 (1956). — STELZNER, G., u. M. TORKA: Tageslänge, Temperatur und andere Umweltfaktoren in ihrem Einfluß auf die Knollenbildung der Kartoffel. Züchter **12**, 233—237 (1940). — STINGL, G.: Über regenerative Neubildungen an isolierten Blättern phanerogamer Pflanzen. Flora (Jena) **99**, 178—192 (1909). — *SWINGLE, CH. F.: Regeneration and vegetative propagation. Bot. Rev. **6**, 301—355 (1940). — *Regeneration and vegetative propagation. II. Bot. Rev. **18**, 1—13 (1952).

THIJN, G. A.: Above-ground tuber formation of potatoes. Euphytica 8, 95—97 (1959). — TISCHLER, G.: Die Beziehungen chromosomaler Rassenunterschiede zur Ökologie der Pflanzen. Forsch. u. Fortschr. **15**, 420—421 (1939). — *Allgemeine Pflanzenkaryologie.

Ergänzungsband Angewandte Pflanzenkaryologie. 2. Lief. Berlin: Bornträger 1954; 4. Lief., fortgeführt von H. D. WULFF. Berlin: Bornträger 1956. — TRÉCUL, A.: Sur l'origine des bourgeons adventifs. Ann. Sci. nat. Bot. 8, 268—295 (1847). — *TROLL, W.: Vergleichende Morphologie der höheren Pflanzen, Bd. I, Teil 1. Berlin: Bornträger 1937; Bd. I, Teil 2, Berlin: Bornträger 1939; Bd. I, Teil 3, Berlin: Bornträger 1943. — Praktische Einführung in die Pflanzenmorphologie, Teil 1: Der vegetative Aufbau. Jena: Gustav Fischer 1954. — Allgemeine Botanik. Ein Lehrbuch auf vergleichend-biologischer Grundlage, 3. Aufl. Stuttgart: Ferdinand Enke 1959. — TROLL, W.: Über die „Prolificität" von *Chlorophytum comosum*. Neue Hefte zur Morphologie (Weimar) **4**, 9—68 (1962).

ULBRICH, E.: Biologie der Früchte und Samen (Karpobiologie). Berlin 1928.

VEGIS, A.: Über den Einfluß der Temperatur und der täglichen Licht-Dunkel-Periode auf die Bildung der Ruheknospen, zugleich ein Beitrag zur Entstehung des Ruhezustandes. Symbolae bot. Upsalienses **14**, 1—175 (1955). — VÖCHTING, H.: Zur Physiologie der Knollengewächse. Studien über vicariierende Organe am Pflanzenkörper. Jb. wiss. Bot. **34**, 1—148 (1899). — Über die Keimbildung der Kartoffelknollen. Bot. Ztg **1903**, 87.

WARMING, E.: Om Rödderne hos *Neottia nidus avis* Lin. Videnskab. Meddel. naturhist. Foren. Kjöbenhavn **1874**, 26. — Familien Podostemonaceae. I.—VI. Kgl. Danske Vidensk. Selsk. Skr., 6. Raekke, naturvidensk. og math. Afd. Kopenhagen 1881, 1882, 1888, 1892, 1899, 1901. — Om Skudbygning, Overvintring og Foryngelse. Naturhist. Forenings Festskrift. Kopenhagen 1891. — Om Jordudløbere. Kgl. Danske Vidensk. Selsk. Skr., 8. Raekke, naturvidensk. og math. Afd. Kopenhagen 1918. — WEBER, E.: Entwicklungsgeschichtliche Untersuchungen über die Gattung *Allium*. Bot. Archiv **25**, 1—43 (1929). — WEBER, H.: Gramineenstudien. III. Neue Beobachtungen über die Kriechsprosse von *Phragmites communis* Trinius. Biol. Zbl. **69**. 323—334 (1950). — Die Bewurzelungsverhältnisse der Pflanzen. Freiburg: Herder 1953. — WENT, F. W.: Effects of environment of parent and grandparent generations on tuber production by potatoes. Amer. J. Bot. **46**, 277—282 (1959). — WERNER, H. O.: Flower and berry production by potatoes as influenced by two light intensities and two midwinter planting dates. Amer. Potato J. **18**, 349—355 (1941). — Relations of length of photoperiod and intensity of supplemental light to the production of flowers and berries in the greenhouse by several varieties of potatoes. J. agric. Res. **64**, 257—274 (1942). — WETTSTEIN, W. v.: Die vegetative Vermehrung von Aspen *(Populus tremula)*. Züchter **12**, 92—95 (1940). — WIELING, U.: Morphologische und anatomische Untersuchungen über die phyllogenen Sprossungen bei *Solanum lycopersicum* L. Biol. Zbl. **71**, 415—437 (1952). — *WITTROCK, V. B.: Über Wurzelsprossen bei krautartigen Gewächsen, mit besonderer Rücksicht auf ihre verschiedene biologische Bedeutung. Bot. Zbl. **17**, 227 (1884). — WULFF, H. D.: Über die Ursache der Sterilität des Kalmus (*Acorus Calamus* L.). Planta (Berl.) **31**, 478—491 (1940).

YARBROUGH, J. A.: Anatomical and developmental studies of the folian embryos of *Bryophyllum calycinum*. Amer. J. Bot. **19**, 443—453 (1932).

ZOHARY, M.: On the vegetative reproduction of some oriental geophytes. Palest. J. Bot. (Jerus. Series) **1**, 35—38 (1938).

Namenverzeichnis. — Author Index.

Die *kursiv* gesetzten Seitenzahlen beziehen sich auf die Literatur.

Page numbers in *italics* refer to the bibliography.

Abel, W. O. 352, *365*
Åberg, E. 573, *583*
Abeygunawardena, D. V. W., u. R. K. S. Wood 781, 782, 783, *784*
Abraham, A., u. P. M. Mathew 59, 61, *101*
— u. C. A. Ninan 549, *549*
— s. Kumar, L. S. S. 60, *105*
Adams, I. s. Parke, M. 637, 646, 648, 650, 651, *655*, 740, *745*
Adams, J. 384, *399*
Addicott, F. T. 387, *399*
Adelberg, E. A., u. S. N. Burns 13, 14, *26*
— u. J. Pittard 6, *26*
— s. Bouck, N. 22, 23, 24, *27*
— s. Driskell-Zamenhof, P. J. 13, 14, *27*
— s. Horiuchi, K. *28*
— s. Jacob, F. 16, *28*
— s. Pittard, J. *29*
— s. Taylor, A. L. 9, *29*
Aderson, T. F. 10, *26*
Afoldi, L. F., F. Jakob, E. L. Wollman u. R. Mazé 15, *26*
Afzelius, B. M. 374, *399*
— G. Erdtman u. F. S. Sjöstrand 374, *399*
Afzelius, K. 571, 578, 579, *583*
Agardh, C. H. 753, 754, *759*
Ahlström, L. s. v. Euler, H. 378, 379, 382, 383, 384, *400*
Ahrens 35
Akemine, T. 578, *583*
Åkerberg, E. 575, 576, 577, *583*
— u. S. Bingefors 575, *583*
— u. A. Nygren 576, *583*
— s. Grazi, F. 566, 577, *588*
— s. Nygren, A. 576, *593*
Åkerlund, E. 77, *101*
Akkoç, E. 89, *101*
Alexeieff, A. 623, *624*
Alexopoulos, C. J., u. S. H. Sun 326, *338*
Allard, H. A. s. Garner, W. W. 178, 191, 194, *205*
Allard, N. A. s. Garner, W. 802, *805*
Allen, C. E. 35, *49*, 58, 69, 91, 94, 95, *101*, 324, *338*
Allen, R. F. 325, 336, *338*
Allington, W. B. 782, 783, *784*
Allsopp, A., u. G. C. Mitra 242, *254*
Almgård, G. 568, 577, *583*
— s. Nygren, A. 563, 575, 576, *593*
Alston, H. A. G. 543, *549*
Amati, P. 15, *26*
Amelung, B. 373, *399*
Ames, Dodge, Drayton u. Zickler 46
Ames, L. M. 128, 129, *131*
Amici, J. B. 384, 393, *399*
Anand, P. L. 648, *653*
Anderson, D. 574, *583*
Anderson, E. 620, *624*
Anderson, L. E. 438, *443*
Anderson, R. F. s. Hesseltine, C. W. 296, *318*
Anderson, R. G., u. R. C. Lommarson 753, *759*
Anderson, R. J. *399*
— u. W. L. Kulp 382, 383, 384, *399*
Anderson, Th. F. 315, *317*
— u. R. Mazé 315, *317*
— E. L. Wollman u. F. Jacob 314, 315, *317*
Andersson, A. 569, 571, *583*
Andrus, C. F. 325, 336, *338*
Angst, E. C. s. Gran, H. H. 664, *680*
Anhäusser, H. 387, 388, *399*
— s. Kühlwein, H. 383, *401*
Anhder, L. D. s. Peterson, C. E. 188, 199, *210*
Anikiev, V. V., u. E. N. Goroščenko 372, *399*
Anraku, N., u. J. Tomizawa 364, *365*
Anthony, S., u. H. V. Harlan 397, *399*
Aoshima, K. 328, *338*
Apple, J. L. s. Dukes, P. D. 291, *317*
Apstein, C. 629, *635*
Aptekarj 611, 612
Araratjan, A. G. 59, *101*
Arber, A. 161, *202*
Archibald, E. E. A. 558, 570, *583*
Arens, P. 147, 148, *150*
Aristotle 257, *278*
Armstrong, J. M. 576, *583*
Arnal, C. 163, 181, 197, 200, *202*
Arnold, C. G. 501, 503, *503*
Arnold, J. D. 330, 333, *338*
Ar-Rushdi, Abbas, H. 95, *101*
Arruda, L. F. de s. Moreira, S. 581, *592*
Arsan 95
Asai, K. s. Namikawa, I. 182, *209*
Asbeck, F. 380, *399*
Ashworth, D. 336, *338*
Asthana, R. P. 783, *785*
Atal, C. K. 178, 199, *202*
Atkinson, G. F. 164, 192, 195, *202*
Atsmon, D., u. E. Galun 185, 193, *202*
— — u. K. M. Jakob 185, *202*
— s. Galun, E. 184, 185, *205*, *213*
Auclair, J. L., u. C. A. Jamieson 381, *399*
Auerbach, L. 258, *278*
Augustin, R. 380, *399*
Austin, C. R. *4*
Autenrieth, H. F. 195, *202*
Awada, M. 168, 193, 194, *202*, *203*
— u. W. Ikeda 168, 193, *203*
Awakian u. Jastreb *445*
Avery, P. 372, *399*
Aymard, M. 182, *203*

Babcock, E. B. 571, *583*
— u. G. L. Stebbins jr. 571, *583*
— s. Hughes, M. B. 509, *527*
Baccarani, M. 167, *203*
Bacchi, O. 581, *583*
Backus, M. P. 226, *233*
Baer, K. E. v. 258, *278*
Bail, Th. 78, *101*, 779, *785*
Baillaud, L. s. Courtot, V. 159, 197, *204*
Bainier, G. 779, *785*

Baker, E. P. s. Luig, N. H. 452, *477*
Baker, H. G. 68, *101*, 170, 171, 184, 196, *203*, 570, 579, *583*
Bali, P. N. 522, *525*
Balls, W. L. 456, *476*
Bally, W. 571, *583*
Bambacioni-Mezzetti, V. 578, *583*
Banbury, G. H. 217, 218, *233*, 296, *317*
Bandoni, R. J. 228, *233*, 335, *338*
Banerjee, S. 336, *338*
— u. K. R. Samadder 329, *338*
Banerji, S. N. s. Datta, R. M. 525, *526*
Barksdale, A. W. 218, 220, 222, 223, 231, *233*, 292, 294, 295, *317*
— M. J. Carlile u. L. Machlis 292, *317*
— u. McMorris, Trevor 223
Barlow, B. A. 570, *583*
Barnett, H. G., V. G. Lilly u. R. F. Krause 296, *317*
Barnett, H. L. 335, 336, *338*
— s. Lilly, V. G. 784, *785*
Baron, L. S., W. F. Carey u. W. M. Spilman *26*
— u. S. Falkow *27*
— W. M. Spilman u. W. F. Carey *27*
— s. Brinton, C. C. 10, *27*
— s. Falkow, S. 25, *27*
— s. Wohlhieter, J. A. *30*
Barski, G., S. Sorieul u. F. Cornefert 260, *278*
Barthelmess, A. 142, *150*
Baruah, H. K. s. Raghavan, V. 387, *403*
Basak, S. L. s. Jain, H. K. 351, *365*
Basarman, M. s. Heilbronn, A. 53, *103*
Bashaw, E. C. 575, *583*
— u. B. J. Hoff 566, *583*
— u. E. C. Holt 575, *583*
— u. D. J. Hutchison 575, *583*
— s. Bennett, H. W. *584*
— s. Fisher, W. D. 574, 575, *587*
— s. Hutchison, D. J. 575, *590*
Bateman, A. J. 457, *476*, 507, 509, 525, *525*
Bateson, W., u. A. F. Gairdner 96, *101*
Bathurst, N. O. 381, 382, *399*
Battaglia, E. 430, *443*, 552, 553, 571, 572, 573, *583*, *584*
Batygina, T. B. s. Gerassimova-Navashina, H. *445*
Batygina, T. P., O. A. Dolgova u. S. N. Korobova 524, *525*
Bauch, R. 228, *233*, 336, 337, *338*, 765, 770, *774*
— s. Kniep, H. 134, *151*
Bauer, L. 136, 141, 142, 143, 144, 145, 146, 147, 148, *150*, 245, 246, 247, 248, 249, 250, 251, 252, 253, *254*
Baumeister, W. 630, 634, *635*
Baur, E. 66, 67, *101*, *503*
Beadle, G. W. 359, *365*
— u. B. McClinstock 372, *399*
Beal, J. M. 803, *805*
Beatus, R. 509, *525*
Beck, S. 607, *618*
Beck, W. A., u. R. A. Joly 387, *399*
Becker 196, *203*
Becker, E. L. s. Nakamura, G. R. 381, *403*
Becker, T. 337, *338*
Beckwirth, A. M. 336, *338*
Beermann, W. *101*
Beger, H. 661, *679*
Béguinot, A. 172, 197, *203*
Behre 673
Behrke, K. 800, *805*
Beijerinck 735
Beijerinck, M. W. 797, 798, *805*
Belajeff 149
Belar, C. 77, *101*
Belcher, J. H. 763, *774*
— u. G. E. Fogg 687, *691*
— u. J. D. A. Miller 687, *691*
Bell, P. R. 252, *254*, 412, *443*
— u. K. Mühlethaler 412, *443*
— u. B. M. Richards 250, *254*
— s. Mühlethaler, K. 412, *444*
Bellamy, A. W. s. Child, C. M. 801, *805*
Bellartz, S. 380, 381, 382, *399*, 524, *525*
Belling, J. 351, 363, *365*
— s. Blakeslee, A. F. 245, *254*
Belowitzkaja, N. A., u. E. J. Grestschuchin 82, 83, *102*
Bemis, W. P. 462, *476*
— u. G. B. Wilson 59, 88, *102*
Benčat, F. 182, *203*
Benecke, W. 599, *602*, *752*
Benjamin, R. K., u. L. Shanor 127, *131*
Bennett, H. W., u. E. C. Bashaw *584*
Bennett, M. C., u. F. W. Cope 516, 517, *525*
Benoni, H. 502, *503*
— s. Schildknecht, H. 394, *404*, 438, *445*, 502, *504*
Bensaude, M. 331, *338*
Bentley 661, *679*
Bentley, J. A. 688, *691*
Bergener, A. D. s. Blakeslee, A. F. 245, *254*
Berger, A. 568, *584*
Berger, X. 154, *203*, 581, *584*
Bergman, B. 553, 554, 555, 571, 572, *584*
Bergner, A. D. 572, *584*
— s. Blakeslee, A. F. 36, *49*
Bergon, P. 665, *679*
Bernard, Fr. 647, 650, 651, 652, 653, *653*
Bernhard, Ch. s. Ernst, A. 570, *587*
Berrah, G., u. W. A. Konetzka 23, *27*
Berthold, G. 113, *119*, 237, *254*, 307, 309, *317*
Bertrand, G. 383, *399*
Bessenich, K. 753, *759*
Bessey, E. A. 173, *203*
Beth, K. 578, *584*
— u. H. F. Linskens 699, *704*
Betherik, O. s. Todd, F. E. 382, 383, 384, *405*
Betts, E. M. 326, *338*
— u. S. L. Meyer 326, *338*
Beyerle, R. 252, *254*
Bhaskaran, K. 25, *27*
— u. S. S. Iyer 25, *27*
Bhattacharjya, S. S., u. H. F. Linskens 525
Bhattacharyya, B. s. Sharma, A. K. 804, *807*
Biecheler, B. 620, 623, *624*, 630, *635*
Biggs, R. 327, 328, *338*
Bilge, E. 53, 95, *102*
Billen, D. s. Hewitt, R. *28*
Billings, F. H. 59, 66, *102*, 570, *584*
Bingefors, S. s. Åkerberg, E. 575, *583*
Bingham, W. E., S. L. Krugman u. E. F. Estermann 380, *399*
Biourge, P. 370, 390, *399*
Bird, R. s. Lark, K. G. *28*
Bishop, H. 124, 125, *131*, 219, *233*
Bistis, G. 224, 225, 226, 231, *233*, 297, *317*
— u. L. S. Olive 225
— u. J. R. Raper 224, *233*
Biswas, K. C. 155, *203*
Bitter, G. 87, *102*
Björling, K. 130, *131*
Black, C. A. 174, 193, *203*
Blackburn, K. B. 58, 59, 61, 72, *102*
— u. J. W. Heslop-Harrison 58, 59, 78, 80, *102*
Blackman s. Murray 649
Blakeslee, A. F. 31, *49*, 125, 126, *131*, 216, 217, *233*, 259, *279*, 296, *317*, 372, *399*
— J. Belling, M. E. Farnham u. D. A. Bergener 245, *254*

Blakeslee, A. F,, J. L. Cartledge, D. S. Welch u. A. D. Bergner 36, *49*
— u. H. E. Warmke 87, *102*
— s. Buchholz, J. T. 458, 465, *476*
— s. Rappaport, J. 524, *529*
— s. Rietsema, J. 524, *529*
— s. Satina, S. 148, *151*, 296, *319*
— s. Warmke, H. E. 61, 62, 72, 74, *109*
Blaringhem, L. 162, 192, 195, *203*
Blavet, M. A. 190, 194, *203*
Bliding, C. 237, *254*, *743*
Bliss, D. E. 325, *338*
Blondel, B. s. Rietsema, J. 524, *529*
Blotevogel, W. s. Dohrn, M. 198, *204*
Blumer, S. s. Schopfer, W. H. 171, 195, 196, *211*
Bobiloff-Preisser, W. 391, *399*
Bøcher, T. W. 556, 571, 573, 579, 582, *584*
Bock, W. 661, *679*
Bocquet, G., u. M. Stroun 172, 194, *203*
Bocsa, I. 83, *102*
Bodlaender, K. B. A. 801, 802, *805*
Bodmer, H. 515, *525*
Böcher, T. W. 608, *618*
Böös, G. 579, *584*
Boergensen 240
Boergesen, F. 768, 772, *774*
Bogdan, A. V. 574, *584*
Bogdanowa, E. N. *445*
Bold, H. C. 721, *743*
Bolte, E. 622, *624*
Bolton 548
Bolton, A. s. Raeber, J. G. 95, *107*
Bond, D. A. s. Holden, H. H.W. 509, *527*
Bonhoeffer, F., u. A. Gierer 19, *27*
Bonner, J. 360, *365*
Bonner, J. T. 241, *254*
Bonnet, O. T. 163, *203*
Boodle, C. *150*
Boost, C. s. Ludwig, W. 450, 460, *477*
Bopp 147
Bopp, M. 274, *279*
Bopp-Hassenkamp, G. 435, *443*
Bordage, E. 168, 195, *203*
Borgaonkar, D. S. 573, *584*
— u. J. M. J. de Wet 567, 573, 574, *584*
— J. R. Harlan u. J. M. J. de Wet 567, 573, 574, *584*
— u. A. P. Singh 567, 574, *584*
Borgaonkar, D. S. s. De Wet, J. M. J. 573, 574, *586*
— s. Harlan, J. R. 573, *589*
Borgert, A. 630, *635*
Bornet, E. 761, 771, *774*
— s. Thuret, G. 309, *320*, 694
Bornhagen, H. 142, 147, 148, *150*, 249, *254*
Borodina, N. A. s. Nekrasov, V. I. 181, *209*
Borriss, H., u. H. Krolop 387, *399*
Borthwick, H. A., u. N. J. Scully 174, 176, 193, 194, *203*
Bory de St. Vincent 195
Borzi, A. 603, *618*, *691*, 720
Bose, S., u. J. Janick 59, 89, *102*
Bose, S. R. 328, 329, *338*
Bossé, G. G. 166, 195, *203*
Bothe, F. 333, *338*
Bouck, N., u. E. A. Adelberg 22, 23, 24, *27*
Boulter, M. E. s. Burnett, J. H. 335, *338*
Bourrelly, P. 637, 638, 639, 640, 641, 643, 644, *654*
Bowden, W. M. 571, *584*
Bower, F. O. 149, *150*, 531, 545, *549*, 551, 556, *584*
Braarud, T. 632, *635*, 649, 653, *654*, 666, 667, *679*
— u. E. Fagerland 649, *654*
— u. E. Nordli 740, *743*
— u. I. Pappas 629, 631, *635*
Brabec, F. 53, *102*, 260, *279*
Brabez, R. 622, 624, *624*, 642, *654*
Brack, A. s. Heim, R. 782, 784, *785*
Braithwaite, A. F. 538, *549*
Brand, F. 608, 610, 617, *618*, 719
Branscheidt, P. 384, 387, *399*, 426, *443*
Brantley, B. B., u. G. F. Warren 184, 188, 192, 194, *203*
Braun, A. 139, *150*, 177, 181, *203*, 561, 570, 571, *584*, 713, 758, *759*
— u. O. Nordstedt 756, *759*
Braunitzer, G. s. Förster, H. *49*, 301, 302, *318*
— s. Plempel, M. 218, *234*, 296, *319*, 323, *341*
Bredemann, G., K. Garber, P. Harteck u. K. A. Suhr 396, 397, *399*
— — W. Huhnke u. R. v. Sengbusch 83, 85, *102*
Bredemühl, I. 661, *679*
Brefeld, O. 248, *254*, 779, 783, *785*
Breider, H. 169, *203*
— u. H. Scheu 90, *102*
Brenman, M. S. J., u. J. Doyle *443*
Brenner, S. s. Jacob, F. 19, 20, 22, 23, 24, *28*
Breslavetz, L. 61, *102*
Breslavec (Breslavetz), L. P. 154, 173, 175, 177, 194, 195, *203*
— u. E. Zaurov 154, 173, 174, 175, 194, *203*
Bretschneider, L. H. 739, *743*
Breuer, K. 521, *525*
Brewbaker, J. L. 507, 521, *525*
— u. B. H. Kwak 394, *399*
— u. S. T. Natarajan 521, *525*
— u. N. Shapiro 521, *525*
Bridges, C. B. 55, 69, 91, *102*
Brieger, F. 322, *338*, 481, 500, 501, *503*, 506, 507, 514, 515, 517, 521, *525*
Brieger, F. G. 447, 462, *476*
— G. E. Tidbury u. H. P. Tseng 462, *476*
Brink, R. A. 266, *279*, 384, 385, 387, 394, *399*, 447, 454, 460, 461, 471, *476*
— u. C. R. Burnham 466, *476*
— u. D. C. Cooper 525, *526*
— u. J. H. MacGillivray 460, 461, *476*
— s. Cooper, D. C. 573, *586*
Brinton, C. C. 10, *27*
— u. L. S. Baron 10, *27*
— A. Buzzell u. M. A. Lauffer 10, *27*
— u. M. J. Stone *27*
Brinton jr., C. C. 10, *27*
— P. Gemski jr. u. J.Carnahan 10, 17, 18, *27*
— — S. Falkow u. C. S. Baron *27*
— s. Wohlhieter, J. A. *30*
Bristol, B. M. 661, *679*
Bristol-Roach, B. M. *679*
Bristow, J. M. 252, *254*
Brittingham, W. H. 576, *584*
Britton, D. M. s. Tryon, Alice F. 534, 536, 539, 541, 542, 543, *550*
Brix, K. s. Gildenhuys, P. J. 575, *588*
Brock, D. T. 312, *317*
Brodie, H. J. 334, 335, *338*
— s. Smith, A. H. 334, *342*
— s. Vandendries, R. 328, 329, 332, 334, *343*
Brook, M., u. C. G. C. Chesters 784, *785*
Brookbank, J. W., u. M. R. Heisler 313, *317*

Brooks, Margaret Hoover s. Harlan, J. R. 573, 574, *589*
Brooks, M. H. *584*
— s. Harlan, J. R. 573, *589*
Brough, P., u. M. H. Taylor 419. *443*
Brown 142
Brown, A. M. 336, *338*
Brown, C. A. 328, *338*
Brown, E. D. W. 252, *254*
Brown, S. W. 370, *399*
— u. D. Zohary 351, *365*
Brown, W. L. 575, 576, *584*
Brown, W. V. 569, 573, 574, 575, 577, *585*
— u. W. H. P. Emery 558, 573, 574, 575, 577, *585*
— s. Emery, W. H. P. 575, *587*
Brubaker, F. s. Holman, R. M. 395, *401*
Bruce, V. G. 622, *624*
Bruchmann, H. 137, 145, 149, *150*, 548, 549, *549*
Bruckmayer-Berkenbusch, H. 678, 679, *679*
Brundin, J. A. Z. 799, *805*
Brunnthaler, J. *743*
Brunotte, C. 197, *203*
Brunswik, H. 130, *131*, 330, 331, *338*
Bryan, A. M. s. Nasatir, M. 380, *403*
Bryan, G. S., u. R. J. Evans 417, *443*
Bryant 574
Bubar, J. S. 506, *526*
Buchholz, J. T., u. A. F. Blakeslee 458, 465, *476*
Buchner, E. 491, *503*
Buchtien, O. 142, *150*
Bühnemann 598, *602*
Bühnemann, F. 734, 735, 736, *743*
Bünning, E. 248, *254*, 394, *400*, 599, *602*, 616, *618*, 803, 805, *805*
— u. P. Müller 598, *602*
— s. Chaudhri, J. J. 801, *805*
Bütschli, O. 258, *279*, *624*
Buijtenen, J. P. van u. D. W. Einspahr 59, 78, *102*
Buis, R. 163, *203*
Buisman, C. 325, *338*
Bukovac, M. J., u. S. H. Wittwer 188, 199, *203*
— s. Phatak, S. C. *213*
— s. Wittwer, S. H. 187, 188, 199, *213*
Bullen, M. R. s. Grant, W. F. 525, *527*
Buller, A. H. R. 130, *132*, 248, *254*, 323, 331, 333, *338*
Bunt, J. S. 605, *618*
Burck, W. 384, *400*
Burgeff, H. 35, 126, *132*, 214, 216, 217, *233*, 249, *254*, 295, 296, *317*, 323, *338*
Burk, E. F. 183, 194, *204*
Burkholder, P. R. 778, *785*
— u. D. Moyer 778, *785*
Burkill, J. H. 165, 172, 193, 197, *204*
Burlingame, L. L. 425, 426, 428, *443*
Burnett, J. H. 38, *49*, 122, 123, 128, *132*, 286, 296, *317*
— u. M. E. Boulter 335, *338*
Burnham, C. R. 462, *476*, *478*
— s. Brink, R. A. 466, *476*
Burns, S. N. s. Adelberg, E. A. 13, 14, *26*
Burrous, J. W. s. Luria, S. E. *29*
Bursa, A. 632, *635*, 641, *654*
Burton, G. W. 575, *585*
— u. I. Forbes 575, *585*
— u. Ian Forbes jr. 575, *585*
— u. B. J. Hoff 575
Butenandt, A. 198, *204*
— u. H. Jakobi 198, *204*
Butler. L. F. 327, *338*
Buzzell, A. s. Brinton, C. C. 10, *27*

Cachon, I. 626, 627, *635*
Cailleux, R. s. Heim, R. 782, 784, *785*
Cairn, J. 19, *27*
Cairns, J. 259, *279*
Callan s. Taylor, J. H. 354
Camefort, H. 427, *443*
Camerarius, R. J. 262, *279*, 384, *400*
Cameroni, R. s. Tappi, G. 379, *405*
Camp, W. H. 164, 197, *204*
Campbell, A. 21, *27*
Campbell, D. H. *150*
Canabaeus, L. 610, 613, 616, *618*
Capot, L. s. Eymé, J. 410, *443*
Carano, E. 558, 572, 573, *585*
Carber, R. J. s. Hansson, A. A. *589*
Carey, W. F. s. Baron, L. S. *26*
— s. Falkow, S. 25, *27*
Carl de Donterberg, C. C. 758, *759*
Carlile, M. J., u. L. Machlis 291, *317*
— s. Barksdale, A. W. 292, *317*
Carnahan. H. L., u. D. Helen Hill 577, *585*
Carnahan, J. s. Brinton jr. C. C. 10, 17, 18, *27*
Carniel, C. 370, *400*
Caro, L. s. Gross, J. D. 24, *28*
Caro, L. G., u. M. Schnös 316, *317*
Carpenter, P. L. s. Hunt, D. E. 313, *318*
Carr, D. J., u. Eng Kok Ng 575, *585*
Carr, P. H. s. O'Kelley, J. 390, *403*
Carter, H. J. 624, *624*
Carter, N. 237, *254*, 638, 645, 649, 650, *654*
Cartledge, J. L. s. Blakeslee, A. F. 36, *49*
Caruso, C. s. Savelli, R. 387, *404*
Casas-Builla, U. s. Gomez-Campo, C. 54, *103*
Caspari, E. *102*
Catarino, F. M. *204*, 272, 273, 275, *279*
Catcheside, D. G. 572, *585*
Cavalli-Sforza, L. L. 7, 8, *27*
— u. J. L. Jinks 8, 9, *27*
— J. Lederberg u. E. M. Lederberg 7, *27*
Cavalli, L. L. s. Lederberg, J. 7, *29*
Cavassilas, Y. 701, 702, *704*
Cayley, D. M. 326, *339*
Celarier, R. P. 573, *585*
— u. J. R. Harlan 573, 574, *585*
— K. L. Mehra u. M. L. Wulf *585*
— s. Harlan, J. R. 573, 574, 577, *589*
— s. Mehra, K. L. *591*
Cesca, G. *585*
Chadefaud, M. *602*, 620. *624*, 640, *654*, 682, *691*, 698, *704*, 738, 740, *743*
Chamberlain, C. J. 155. *204*, 347, *365*, 415, 418, 419, 420, 421, 430, *443*
Champagnat, M., J. Marichal u. C. Vincent 798, *805*
Chandler, C. s. Stout, A. B. 516, 521, *529*
Chandley, A. C. s. Grell, R. F. 352, *365*
Chang, T. J. s. Tseng, C. K. 762, 763, *776*
Chao, L. F. 455, 460, 463, *476*
Chapman, H. W. 803, *805*
— s. Okazawa, Y. 803, *807*
Chapman, J. A., u. M. R. J. Salton 611, *618*
Chapman, V. s. Riley, R. 350, *366*
Chase, S. S. *585*
Chatton, E. 626, 627, *635*
— A. Lwoff u. M. Lwoff 620, *625*
— u. Monod 621
Chaudhri, J. J., E. Bünning u. W. Haupt 801, *805*

Chedda, H. R., u. J. R. Harlan 573, *585*
Chemin, E. 765, 769, 772, 773, 774, *775*
Chen s. Jennings 43, 44
Chesters, C. G. C. s. Brook, M. 784, *785*
Chevallier, M. R. s. Luzzati, D. 15, *29*
Chiarugi, A. 571, *585*
— u. E. Francini 578, *585*
Chifflot, M. 163, *204*
Chihara, M. 740, *743*
Child, C. M., u. A. W. Bellamy 801, *805*
Chivers, A. H. 781, *785*
Chodat, R. 686, *691*
— u. A. R. Rossilo 649, *654*
Cholnoky, B. v. 608, *618*
Chopra, R. N. s. Sachar, R. C. 569, *594*
Choudhri, R. S., u. R. Krishan 162, 163, 192, 194, *204*
Chow, C. H. 330, 331, *339*
Chowdhury s. Mehra, P. 541, *550*
Christ, B. 509, 510, 521, *526*
Christen, H. R. 581, *585*
Christensen, J. J. 337, *339*
— s. Stakman, E. C. 337, *342*
Christensen, T. *602*, 637, 737, *744*
Christoff, M. 572, *585*
— u. M. A. Chrichoff 572, *585*
— u. G. Papasova 580, *585*
— u. A. Popoff 572, *585*
Christoff, M. A. 576, *585*
— s. Christoff, M. 572, *585*
Chrysler, M. A. 794, *805*
Church, A. H. 237, *254*
Cienkowski, L. 623, *625*
Citarella, R. V. s. Falkow, S. 16, 21, *27*
— s. Wohlhieter, J. A. *30*
Clamp, G. 796, *805*
Clark, F. J. 372, *400*
Clark, H. E. s. Freiberg, S. R. 380, *400*
Clarke, B. s. Greenwood, A. D. 686, *691*
— s. Manton, I. 698, *705*
— s. Manton, J. 682, *692*
— s. Parke, M. 637, 647, 649, 650, 652, *655*
Claus, O. 603, *618*
Clausen, J. 567, 576, 580, *585*
— P. Grun, W. M. Hiesey u. M. Nobs 567, *585*
— — A. Nygren u. M. Nobs 567, *585*
— W. M. Hiesey, P. Grun u. M. Nobs 567, 575, 576, *585*
— — u. M. A. Nobs 576, 575, 576, 577, *585*
Clausen, J., D. Keck u. W. M. Hiesey 567, 576, 580, *585*, *586*
— — — u. P. Grun 567, 568, *586*
— — — — A. Nygren u. u. M. Nobs 567, *586*
— M. A. Nobs u. W. M. Hiesey 567, 576, *586*
— s. Cox. L. M. 567, 575, 577, *586*
— s. Watson, P. J. 577, *596*
Clausen, T. s. Pratt, C. 566, 581, *593*
Claussen, P. 245, *254*
Clavaud, A. *759*
Cleland, R. E. s. Renner, O. 458, 472, *478*
Clemençon, H. 683, 687, *691*
Cleve, P. T. 665, *679*
Cleve-Euler, A. 664, 668, 669, 670, 671, *679*
Cleveland, L. R. 37, *49*, 121, *132*, 353, *365*
Clifford, H. T. *586*
Clint, H. B. 238, *254*
Clowes, R. C. 15, *27*
— u. D. Rowley 7, *27*
Cobb-Blanchard, F. 481, *503*
Coe, G. E. 569, *586*
Cohen, A. s. Drouet, F. *625*
Cohen, L. W., u. R. W. Siegel 284, *317*
Cohn, F. *752*
Coker, W. C. 124, *132*, 426, *443*, 780, *785*
Coleman, A. 600, *602*
Coleman, J. R., u. M. J. Moses 348, *365*
Collins, G. N., u. J. H. Kempton 461, *476*
Comolli, R. s. Maccararo, G. A. 315, 316, *318*
Conrad, K. 201, *204*
— u. K. Mothes 201, *204*
Conrad, W. 637, 638, 639, 643, 645, 646, 650, *654*, 731, *743*
Conway, E. 763, *775*
Cook, A. H., u. I. A. Elvidge 289, *317*
— — u. Sir I. Heilbronn 309, *317*
Cook, P. W. 720, 734, 735, 743, *743*
Cook, St. A. *4*
Cooper, D. C. 347, 356, *365*, 370, *400*, 437, *443*, 525, *526*
— u. R. A. Brink 573, *586*
— s. Brink, R. A. 525, *526*
Cooper, W. C. 384, 387, *400*
— s. Dandlikar, W. 387, *400*
Cope, F. W. 516, *526*
— s. Bennett, M. C. 516, 517, *525*
Copeland, E. B. 532, *549*
Corbett, M. K. s. Edwardson, J. R. 196, *204*
Cornefert, F. s. Barski, G. 260, *278*
Cornu, M. 196, *204*
Correns, C. 32, 43, *49*, 51, 52, 53, 55, 56, 57, 61, 69, 72, 74, 77, 78, 81, 93, 94, 99, 101, *102*, 134, 135, 136, 138, 139, 140, 141, 146, 147, 149, *150*, 153, 154, 157, 160, 161, 164, 166, 167, 168, 169, 170, 173, 175, 182, 183, 184, 190, 192, 193, 194, 195, 196, 197, 200, *204*, 274, 447, 448, 451, 452, 453, 454, 460, 463, 470, 471, *476*, 480, 481, 484, *503*, 517, 518, *526*
Couch 690, *691*
Couch, J. N. 124, *132*, 219, 223, *233*
Coulon, E. M. s. Nebel, B. R. 348, *366*
Coulon, J. de 575, *586*
Courtot, V., u. L. Baillaud 159, 197, *204*
Cox, L. M., u. J. Clausen 567, 575, 577, *586*
Crăciun, O. s. Preda, V. 163, *210*
Craigie, J. N. 324, 336, *339*
Crane, M. B., u. W. J. C. Lawrence 91, *102*, *526*
— u. P. T. Thomas 581, *586*
Crawford, E. M., u. R. F. Gesteland 10, 11, *27*
Creighton, H. B., u. B. McClintock 351, *365*
Crescini, Fr. 82, *102*
Crew, F. A. E. *102*
Crick, F. H. C. s. Watson, J. D. 349, *367*
Crouse, H. V. 363, *365*
Crowe, L. K. 507, 509, *526*
— s. Lewis, D. 93, *105*, 507, 524, *528*
Cuevas, N. Y. s. Juliano, I. B. 569, *590*
Cummins, G. B. 336, *339*
Cupp, E. 660, 664, *679*
Currence, T. M. 94, *102*, *204*
— s. Peirce, L. C. 90, *107*
Curtis, W. M. 578, *586*
Cuzin, F. s. Jacob, F. 23, 24, *28*
Czaja, A. Th. 137, 142, 146, 149, *150*
Czapek, F. 373, 380, *400*
Czapik, R. 580, *586*
— s. Skalinska, M. 580, *595*
Czempyrek, H. 721, *744*
Czurda, V. 111, *119*, 599, 600, *602*, 736, *744*, 746, 747, 748, 749, 750, 751, *752*

Dadds 538
Dade, H. A. 325, *339*
Dales, R. P. 637, *654*
Dahl, A. O. s. Fernandes-Morán, H. 374, *400*
Dahmen, M. s. Moutschen, J. 87, *106*
D'Amato, F. 578, 579, *586*
Dana, B. F. 781, *785*
Dana, S. K. s. Datta, R. M. 525, *526*
Dandlikar, W., W. C. Cooper u. H. P. Traub 387, *400*
Dangeard, P. *602*, 620, 625, 629, 631, 632, *636*, 684, 685, 686, *691*
— u. J. Eymé *443*
Dangeard, P. A. 620, 621, 622, 623, 624, *625*, 644, 645, 646, *654*
Daniel, L., u. E. Váróczy 385, *400*
Danielson, L. L. 186, *204*
Daris, G. K. s. Storey, W. B. 168, *212*
Dark, S. O. S. 62, *102*
Darlington, C. D. 62, 63, 64, *102*, 348, 349, 350, 351, 353, 359, 363, 364, *365*
— u. E. K. Janaki-Ammal 569, 570, 572, 574, 577, 578, 579, *586*
— u. A. P. Wylie *586*
Darrow, G. M. s. Waldo, G. F. 581, *596*
Darwin, C. 507, 517, 522, *526*
Datta, R. M., S. K. Dana u. S. N. Banerji 525, *526*
— u. S. K. Sen 524, *526*
— s. Patel, G. I. 524, *529*
Davey, A. J., u. C. M. Gibson 180, 193, 197, *204*
Davidson, R. W. s. Shear, C. L. 327, *342*
Davies, D. R., u. E. T. Wall 524, *526*
Davis 713
Davis, B. M. *691*, 774, *775*
Dawid, W. s. Plempel, M. 218, *234*, 297, *319*
Dawson, G. W. P. 349, *365*, 804, *805*
De, D. N. 349, 358
Dean, H. L. *102*
De Bary, A. 218, 219, *233*, 531, 532, 747, 758, *759*
Debski, B. 753, 755, *759*
Deflandre, G. 649, 651, *654*
DeLisle, D. G. 574, *586*
Dellingshausen, M. v., u. K. Stern 181, *204*
Delpino, F. 507, *526*
DeMaggio, A. E. 247, 248, *254*
— u. R. H. Wetmore 248, *254*
DeMaggio, A. E. s. Wetmore, R. H. 250, 252, *256*
Demereç, M. 464, *476*, 524, *526*
Demeter, O. 612, 613, 616, *618*
— u. O. Renner 613, *618*
Demetriades, S. D. 783, *785*
DeMonbreun, W. A. 777, 778, *785*
Dempsey, E. s. Rhoades, M. M. 462, *478*
Dempster, G. s. Duguid, J. P. 10, *27*
Denffer, D. v. *679*, 801, *805*
— u. H. Hustede 112, *119*, 599, 687, *691*
Derenne, P. 62, *102*
Dermen, H. 580, *586*
Dersch, G. 664
Dersch-Siebert, G. s. v. Stosch, H. A. 647, *656*
Derx, H. G. 325, *339*
Desai, Sharda 581, *586*
Desikachary, T. V. 608, 617, *618*
— s. Iyengar, M. O. P. 608, 611, *619*
Dettori, R., G. A. Maccacaro u. G. L. Piccinin 11, *27*
— s. Maccacaro, G. A. *29*
Devreux, M., G. Vallaeys, P. Pochet u. A. Gilles 509, *526*
Dewald, C. L., u. J. R. Harlan 573, 574, *586*
DeWet, J. M. J. 573, 575, *586*
— u. D. S. Borgaonkar 573, 574, *586*
— — u. W. L. Richardson 573, 574, *586*
— u. J. R. Harlan 573, *586*
— u. M. L. Higgins 573, *586*
— K. L. Mehra u. D. S. Borgaonkar 574, *586*
— u. W. L. Richardson 573, *586*
— u. B. D. Scott 573, *586*
— u. A. P. Singh 573, *586*
— s. Borgaonkar, D. S. 567, 573, 574, *584*
— s. Harlan, J. R. 573, 574, *589*
DeWitt, W., u. D. R. Helinski 15, *27*
Diannelidis, T. 477, *586*
Dianowa, V. s. Poddubnaja-Arnoldi, V. 573, *593*
Dickinson, S. 337, *339*
Dickson, H. 331, *339*
Diels 532
Diers, L. 412, *443*
Digby, L. s. Farmer, J. B. 538, 544, 548, *549*
Dijkman, M. J. 397, *400*
Dikshit, N. N. 168, 195, *204*
Dimock, A. W. 326, *339*
Dittmann, Chr. 491, *503*
Diwald, K. 626, *636*
Dixon, P. S. 762, 767, 769, 771, 774, *775*
D'Mello Joyce s. Mahabale, T. S. 548, *550*
Dmitrieva, A. M. s. Poljakow, I. M. 442, *446*
Dobell, C. C. 623, *625*
Docters van Leeuwen, W. 799, *805*
Dodds, K. S. *586*
— u. N. W. Simmonds 578, *586*
Dodge s. Ames 46
Dodge, B. O. 224, *233*, 326, *339*
— s. Shear, C. L. 326, 336, *342*
Döpp, W. 137, 143, 144, 147, 149, *150*, 252, 531, 532, 533, 534, 535, 536, 537, 538, 541, 544, *549*
Doflein, Fr. 637, 638, 639, 643, 644, 645, 646, *654*
Dogiel, V. 627, *636*
Dohrn, M., W. Faure, H. Poll u. W. Blotevogel 198, *204*
Dolgova, O. A. s. Batygina, T. P. 524, *525*
D'Oliveira, B. 336, *339*
Donk, M. van der s. Linskens, H. F. 520, 523, *528*
Dorsey, M. J. 370, *400*
Dorst, H. J. M. van s. van Koot, Y. 188, 196, *207*
Dostal, R. 803, *805*
Douin, R. *618*
Dove, W. F., u. J. J. Weigle *27*
Dowding, E. S. 326, *339*, 778, *785*
Downie, D. G. 416, *443*
Dowrick, G. J. 581, *586*
Doyle, J. s. Brenman, M. S. J. *443*
Drawert, H. 794, *805*
— u. I. Tischer 610, 616, *618*
Drayner, J. M. 509, *526*
Drayton, F. D., u. J. W. Groves 127, *132*
Drayton, F. L. 326, *339*
— u. J. W. Groves 326, *339*
Drayton s. Ames 46
Drebes, G. 659, 667, 672, 673, *679*
— s. v. Stosch, H. A. 657, 660, 662, 663, 667, 672, 674, *681*
Dressler, O. 59, 88, 89, *102*
Drew, K. M. 238, *254*, 753, 755, 756, *759*, 762, 763, 764, 766, 770, 771, *775*
— u. I. Friedmann *49*
Driga, J. E. 82, *102*
Driskell-Zamenhof, P. J., u. E. A. Adelberg 13, 14, *27*
Drouet, F., u. A. Cohen *625*
Druery, Ch. T. 551, 556, *586*

Duara, B. N., u. G. L. Stebbins jr. *587*
Dubrovina, A. V. 187, *204*
Düsing, C. 157, 192, 200, *204*
Duffield, J. W., u. A. G. Snow 397, *400*
Duguid, J. P., u. R. Gillies 10, *27*
— I. W. Smith, G. Dempster u. P. N. Edmonds 10, *27*
Duich, J. M., u. H. B. Musser 577, *587*
Dukes, P. D., u. J. L. Apple 291, *317*
Dulberger, R. 515, 516, *526*
Duncan, E. N. s. Pate, J. B. 202, *210*
Duncan, J. T. 779, *785*
Duncan, R. E. 252, *254*, 545, 547, *549*
Dunstan, R. T. 90, *103*
Dupler, A. W. 425, 426, *443*
Durand, B. 88, *103*
Durieu de Maisoneuve, M. Ch. *759*
Duvick, D. N. 94, *103*

East, E. M. 507, 518, 519, 521, 522, *526*
— u. A. J. Mangelsdorf 517, *526*
— u. J. B. Park *526*
— u. S. H. Yarnell 521, *526*
— s. Mangelsdorf, A. J. 100, *106*
Eaton, F. M. 202, *204*
Ebell, U. s. Oehlkers, F. 98, *107*
Eberle, Paul s. Oehlkers, F. 352, *366*
Ebersold, W. T. 351, *365*
Echols, R. M., u. F. Mergen 385, 387, *400*
Eddins, A. H. 337, *339*
Eddy, A. A. 313, *317*
— u. A. D. Rudin 313, *317*
Edel'štejn, V. I., u. A. N. Paponov 163, 192, *204*
Edman, G. 551, 579, *587*
Edmond, J. B. 186, 192, 194, *204*
Edmonds, H. W. s. Sax, K. 347, *366*
Edmonds, P. N. s. Duguid, J.P. 10, *27*
Edwards, G. A. s. Nickerson, W. J. 778, *785*
Edwardson, J. R., u. M. K. Corbett 196, *204*
Eggertson, E. 329, *339*
Eguchi, T., u. H. Ichikawa 69, *103*
Ehlers, H. 384, 385, *400*
Ehrenberg 639
Ehrenberg, L. 579, *587*
Ehrenberg, S. G. *602*
Einset, J. 566, 581, *587*
— s. Pratt, C. 566, 581, *593*
Einspahr, D. W. s. van Buijtenen, J. P. 59, 78, *102*
Eklundh-Ehrenberg, C., H. v. Euler u. G. Hevesy 442, *445*
Ekman, E. 571, *587*
El-Ani, A. S. 130, *132*
Elden, H. van s. Hofmeyr, J. D. J. *104*
Ellenhorn u. Swetosarowa *445*
Ellis, J. R., u. J. Janick 59, 66, 89, *103*
Elser, E., u. J. Ganzmüller 380, 381, 382, 383, *400*
Elsmann 148
Elvehjem, C. A. s. Kitzes, G. 383, *401*
Elvidge, I. A. s. Cook, A. H. 289, 309, *317*
Emecz, T. I. 372, *400*
Emerson, R. 123, 124, *132*
— u. C. M. Wilson 124, *132*
Emerson, R. A. 87, 96, *103*, 447. 461, 462, 464, *476*
Emerson, S. 453, 474, *476*, 511, 512, *526*
Emery, W. H. P. 574, 577, *587*
— u. W. V. Brown 575, *587*
— s. Brown, W. V. 558, 573, 574, 575, 577, *585*
Emmerling, M. H. 449, *476*
Emsweller, S. L., u. N. W. Stuart 523, *526*
— u. Uring 523
— s. Levan, A. 578, *591*
— s. Uhring, J. 578, *596*
Engelbert, V. 575, *587*
Eng Kok Ng s. Carr, D. J. 575, *585*
Engler, A. 433, *443*
— u. E. Gilg 160, 197, *204*
Entz, G. 628, 630, 631, 632, *636*
— u. O. Sebestyén 627, 631, *636*
Ephrussi, B., u. M. C. Weiss 260, *279*
Erben, K. 672, 678, 679, *679*
Erdelský, K., u. R. Herich 174, *204*
Erdtman, G. 373, *400*
— s. Afzelius, B. M. 374, *399*
Eresen, H. s. Reimann-Philipp, R. 59, 90, *107*
Erickson, E. O. 370, *400*
Erickson, R. O. 369, *400*
Erlandsson, S. 573, *587*
Erlanson u. Hermann 59
Erlenmeyer, H., u. M. Geiger-Huber 171, 196, *204*
Ernst, A. *103*, 142, 514, 523, 524, *526*, 531, 539, *549*, 569, 579, *587*, 690, *691*, 758, 759, *759*
— u. Ch. Bernhard 570, *587*
Erwin, N. S. s. Matney, T. S. 9, *29*
Esashi, Y., u. M. Nagoa 802, *805*
Esau, K. 572, *587*
Eschenberger, F. 467, *477*
Eschrich, W. 369, 390, *400*
Esser, K. 324, *339*, 385, *400*, 515, 516, 521, *526*
— u. R. Kuenen 121, 122, 123, 127, 128, *132*, 261, *279*, 293, 296, *317*, 323, 324, *339*
— u. J. R. Raper 324, *339*
— u. J. Straub 379, *400*, 516, *526*
— s. Linskens, H. F. 390, 394, *402*, 512, *528*
Estermann, E. F. s. Bingham, W. E. 380, *399*
Ettl, H. 682, 683, *691*, 706, 707, 708, 709, 710, 713, 715, 720, 724, 726, 727, 731, 737, *744*
— u. I. Manton 737, 738, *744*
— s. Manton, I. 737, 739, *745*
Eue, L. 521, *527*
Euler, H. v. 379, 381, *400*
— L. Ahlström, B. Högberg u. J. Petterson 378, 379, 382, 383, 384, *400*
— u. J. von Euler 381, *400*
— L. Heller u. K. G. Högberg 381, 382, *400*
— s. Eklundh-Ehrenberg, C. 442, *445*
Euler, J. von s. v. Euler, H. 381, *400*
Evans, A. M. 524, *527*
Evans, R. J. s. Bryan, G. S. 417, *443*
Eymé, J. 410, *443*
— u. L. Capot 410, *443*
— s. Dangeard, P. *443*
Ezekiel, W., J. T. Taubenhaus u. J. F. Fudge 783, *785*

Fähnrich, P. 386, *400*
Fagerland, E. s. Braarud, T. 649, *654*
Fagerlind, F. 430, 432, 437, *443*, *444*, 553, 554, 570, 572, 573, 582, *587*
Falcone, G. s. Nickerson, W. J. 779, *785*
Falkow, S., u. L. S. Baron *27*
— R. V. Citarella, J. A. Wohlhieter u. T. Watanabe 16, 21, *27*
— J. Marmur, W. F. Carey, W. M. Spilman u. L. S. Baron 25, *27*
— J. A. Wohlhieter, R. V. Citarella u. L. S. Baron *27*
— s. Baron, L. S. *27*
— s. Brinton jr., C. C. *27*

Falkow, S. s. Wohlhieter, J. A. *30*
Famintzin, A., u. D. S. Przybytek 384, *400*
Fantani, A. A. s. Olive, L. S. 326, *431*
Fargie, B. s. Holloway, B. W. *28*
Farkas, G. L. s. St. Gimesi, W. 360, *367*
Farlow, 531
Farmer, J. B., u. L. Digby 538, 544, 548, *549*
— u. J. L. Williams 309, 310, *317*
Farmer, R. E. *103*
Farnham, M. E. s. Blakeslee, A. F. 245, *254*
Farooq, M. 155, *205*
Farquharson, L. I. 577, *587*
Faruqi, S. A. 573, *587*
Faure, W. s. Dohrn, M. 198, *204*
Favarger 59
Favre-Duchartre, M. 414, 415, 416, 417, 418, 419, 420, 426, 427, 428, 429, 430, *444*
Fawcett, D. W. 348, *365*
Feiginson *445*
Fejér, S. O. 95, *103*
Feldman, M. 350, *365*
— T. Mello-Sampayo u. E. R. Sears 261, *279*, 350, *365*
Feldmann, G. s. Feldmann, J. *618*, 766, 768, *775*
Feldmann, J. 240, *254*, 702, *704*, 768, 769, *775*
— u. G. Feldmann *618*, 766, 768, *775*
Fenell, D. J. s. Raper, K. B. 325, *342*
Ferguson, M. C. 429, *444*
Ferguson-Hunt, F. 377, 378, 379, *400*
Fernald, M. L. 571, *587*
Fernandes-Morán, H., u. A. O. Dahl 374, *400*
Fields, W. G., u. J. Maniotis 326, *339*
Figdor, W. 178, *205*
Filarszky 758
Fineman, B. C. 778, *785*
Finn, W. 155, *205*
Firbas, H. 385, *400*
Fisch, C. 174, 193, *205*
Fischer, F. G., u. G. Werner 223, *233*, 285, 291, 292, 293, *317*
Fischer, G. W. 325, 337, *339*
— s. Holton, C. S. 337, *340*
Fischer, H. 373, *400*
Fisher, K. W. 22, *27*
Fisher, W. D., E. C. Bashaw u. D. C. Holt 574, 575, *587*
Flint, E. A. 717, *744*
— s. Manton, J. 682, *692*
Flint, L. H. 761, 762, *775*
Flor, H. H. 337, *339*
Flory 59
Flovik, K. 562, 571, 574, 576, 582, *587*
Focke, W. O. 551, *587*
Förster, H. 33, 36, *49*, 303, *317*
— u. L. Wiese 32, 33, 34, *49*, 229, *233*, 282, 300, 301, 302, 303, 306, *318*
— — u. G. Braunitzer *49*, 301, 302, *318*
Föyn, B. 742, *744*
Föyn, B. R. 237, 246, *254*
Fogg, G. E. 608, 610, 611, 613, 614, 615, 616, 617, *618*
— s. Belcher, J. H. 687, *691*
— s. Miller, J. A. D. 687, *692*
Folgner, V. 633, *636*
Forbes, I. s. Burton, G. W. 575, *585*
Forbes jr., Ian s. Burton, G. W. 575, *585*
Forenbacher, A. 580, *587*
Foslie, M. 770, *775*
Foster 60
Fott, B. *602*, 623, *625*, 638, *654*, 719, 732, 746, 747, *752*
Fox, R. A. 325, *339*
Francini, E. s. Chiarugi, A. 578, *585*
Fredericq, P. 14, 15, *27*
Freeland, R. O. 796. 801, *805*
Freiberg, S. R., u. H. E. Clark 380, *400*
Freksa, F. H. F. 260, *279*
Frémy, P. 604, 608, 610, *618*
Frenyo, W. s. St. Gimesi, W. 360, *367*
Freudenthal, H. D. 630, *636*
— u. J. J. Lee 629, *636*
Frey-Wyssling, A. 390, *400*
Fricke, B. s. Menke, W. 412, *444*
Friedmann, I. 41, *49*, 309, 310, *318*, 607, *618*
— s. Drew, K. M. *49*
Fries, N. 328, 329, 330, 335, *339*
— u. L. Jonasson 329, *339*
Fries, Th. C. E. 571, *587*
Fries, Th. M. 800, *805*
Fritsch, F. E. 235, 237, 238, 239, 240, 242, 243, *255*, 307, *318*, 597, *602*, 603, 608, 609, 610, 611, 612, 615, 616, 617, *618*, 693, 694, 696, 697, 702, *704*, 712, 729, 732, 733, 740, 742, *744*, 746, 747, *752*, 768
— u. R. P. John 690, *691*
— s. West, G. S. 717, 719, *746*
Frolova, G. D. 181, *205*
Frost, H. B. 558, 581, *587*
Fuchs, A. 433, *444*
Fudge, J. F. s. Ezekiel, W. 783, *785*
Fürnkranz, D. 573, *587*
Fultz, J. L. 574, *588*
Funk, G. 697, *704*
Funke, G. L. 330, 332, *339*
Furusato, K. 558, 559, 581, *588*
Fusakawa, T. s. Watanabe, T. 15, 16, *29*, *30*

Gaarder, K. R. 649, *654*
— u. J. Markali 649, *654*
Gadd, I. s. Gustafsson, Å. 566, *588*
Gärtel, W. 385, *400*
Gaertner, C. F. 169, 195, *205*
Gärtner, C. F. von 487, *504*
Gäumann, E. 38, 46, *49*, 123, 127, 128, *132*, 323, 324, *339*
Gairdner, A. F. s. Bateson, W. 96, *101*
Galán, F. 38, *49*, 54, 95, *103*, 274, 275, *279*
Galen 258, *279*
Galil, J. 792, *805*
Galun, E. 101, *103*, 184, 186, 187, 188, 194, 195, 198, 199, *205*, 275, *279*
— u. D. Atsmon 184, 185, *205*
— Sh. Izhar u. D. Atsmon *213*
— Y. Jung u. A. Lang 188, 198, 199, *205*, 275, *279*
— s. Atsmon, D. 185, 193, *202*
— s. Shifriss, O. 186, *212*
Ganesan, A. T., u. G. Lederberg 20, *27*
Ganong, W. F. 570, *588*
Ganzmüller, J. s. Elser, E. 380, 381, 382, 383, *400*
Garay, A., u. F. Sági 171, 201, *205*
Garber, E. D. *588*
Garber, K. s. Bredemann, G. 83, 85, *102*, 396, 397, *399*
Gardella, C. 524, *527*
Gardner, E. J. 572, *588*
— s. Powers, L. 572, *593*
Gardner, N. L. 607, *618*
Garner, W., u. N. A. Allard 802, *805*
Garner, W. W., u. H. A. Allard 178, 191, 194, *205*
Gates, R. R. *103*
Geddes, P., u. J. A. Thompson 160, *205*
Geerts, J. M. 501, *504*
Geiger-Huber, M. s. Erlenmeyer, H. 171, 196, *204*
Geissbühler, H. 644, *654*
Geitler, L. 41, *49*, 116, 117, 118, *119*, 154, *205*, 284, 306, 307, *318*, 577, *588*, 602, 603, 604, 606, 607, 608, 609, 610, 611, 612, 613, 616, 617, *618*, 628, 635, *636*, 641, 642, 644, 645,

654, 657, 660, 662, 666, 672, 673, 674, 675, 676, 677, 678, *679*, 705, 706, 709, 713, 715, 716, 717, 719, 722, 728, 734, 741, 742, *744*, 749, 751, *752*, 760, *775*
Geitler, L., u. F. Ruttner 609, 617, *619*
Gelin, O. 571, *588*
Gelting, P. 800, *806*
Gemski jr., P. s. Brinton jr., C. C. 10, 17, 18, *27*
Gentcheff, G. 572, *588*
— u. A. Gustafsson 441, *444*, 572, 579, 580, *588*
George jr., W. L. s. Shifriss, O. 101, *108*
Georgevitsch, P. 544, *549*
Gerassimova-Navashina, E. N. *588*
Gerassimova-Navashina, H. 440, 441, *444*, *588*
— u. T. B. Batygina *445*
Gerassimowa-Nawaschina, J. N. s. Nawaschin, M. S. *446*
Gerhardt, E. *150*
Gerisch, G. 241, *255*
Gerloff, J. 304, *318*
Gerneck, R. 686, 688, *691*
Gerstel, D. U. 349, *365*, 509, *527*, 572, *588*
— B. L. Hammond u. C. Kidd 572, *588*
— u. W. M. Mishanec 572, *588*
— u. Mary E. Riner 572, *588*
— u. L. L. Phillips 349, *365*
— s. Phillips, L. L. 349, *366*
Gessner, F. 638, *654*, 697, *704*
Gesteland, R. F. s. Crawford, E. M. 10, 11, *27*
Ghisa, A. s. Preda, V. 179, 198, *210*
Giard, A. 179, 192, 196, *205*
Gibson, C. M. s. Davey, A. J. 180, 193, 197, *204*
Gierer, A. s. Bonhoeffer, F. 19, *27*
Giesenhagen, K. 753, 754, 755, 756, 757, 758, *759*
Gildenhuys, P. J., u. K. Brix 575, *588*
Gilg, E. s. Engler, A. 160, 197, *204*
Gilles, A. s. Devreux, M. 509, *526*
Gillet, C. 753, 755, *759*
Gillies, R. s. Duguid, J. P. 10, *27*
Gilman, J. C. s. Welch, A. W. 326, *343*
Gilmore, K. A. 334, *339*
Girou de Buzareingues, C. 173, 174, *205*
Glade, K. 603, 612, 617, *619*
Glendinning, D. R. 516, *527*
Glenk, H. O. 394, 500, *504*
Glück, H. 139, *150*, 787, 794, 801, 802, *806*
Gluschtschenko, I. E. *445*
Goebel, K. 135, 136, 138, 139, 140, 143, 144, 145, 146, 149, *150*, 155, 163, *205*, 240, 241, 244, 245, 250, 252, 253, 254, *255*, 407, 408, 409, 418, 422, 424, 425, 441, *444*, 531, 541, 549, *549*, 700, *704*, 753, *759*, 788, 793, 794, 795, 798, 800, 801, 802, *806*
Götz 686
Götz, O. 796, 800, 801, 803, *806*
Goff, E. S. 395, *400*
Goidics, M. 624, *625*
Goldschmidt 35
Goldschmidt, R. 70, 71, 82, 94, 99, 101, *103*
Goldschmidt, R. B. 349, *365*
Goldschmidt, V. 337, *339*
Goldsmiht, E. P. s. Matney, T. S. 9, *29*
Golubinskij, I. N. (Golubinsky) 387, *400*
Gomez-Campo, C., u. U. Casas-Builla 54, *103*
Gordon, W. L. 326, *340*
Goroščenko, E. N. s. Anikiev, V. V. 372, *399*
Górska-Brylas, A. 381, *401*
Gotter s. Harte, C. 458
Gottschalk, W. *103*
Gould, F. W. 574, *588*
— s. Kapadia, Z. L. 574, *590*
— s. Mohamed, A. H. 574, *591*
Gräbner, F. 160, *205*
Graham, J. S. D. s. Stanley, R. G. 394, *405*
Graham, R. J. D., u. L. B. Stewart *806*
Gran, H. H. 664, 665, 669, 671, *680*
— u. E. C. Angst 664, *680*
Granick, E. B. 569, *588*
Grant, W. F. 55, 80, *103*
— M. R. Bullen u. D. de Nettancourt 525, *527*
Grassé, P. P., u. R. Poisson 620, *625*
Grau, W. 244, *255*
Grazi, F., M. Umaerus u. E. Åkerberg 566, 577, *588*
— s. Hjelmqvist, H. 563, *589*
Grebinskij, S. O., u. R. P. Rolik 399, *401*
Greenwood, A. D. 686, *691*
— J. Manton u. B. Clarke 686, *691*
— s. Manton, J. 682, *692*
Gregor, M. J. F. 325, *340*
Gregory, F. G., u. O. N. Purvis 161, *205*
Gregory, L. E. 803, *806*
Gregory, W. C. 360, *365*
Greguss, P. 134, *150*
Greis, H. 34, 46, *49*, 128, 130, *132*, 141, 142, *150*, 327, *340*
Grell, K. G. 45, *49*
Grell, R. F., u. A. C. Chandley 352, *365*
Gress, M. 490, 491, 493, *504*
Grestschuchin, E. J. s. Belowitzkaja, N. A. 82. 83, *102*
Greulach, V. A. s. League, E. A. 687, *692*
Griffiths, E. 326, *340*
Griggs, W. H., G. H. Vansell u. B. T. Iwakiri 397, 398, *401*
— — u. J. F. Reinhardt 397, *401*
Grinvald, C., E. Mavromati u. I. Gutenmacher 202, *205*
Grischko, N. N. 83, 85, *103*
Grömmer, J. s. Nielsen, N. 381, 382, 383, *403*
Grøntved, J. 665, 666, *680*
Grosch, H. G. 802, 804, *806*
Gross, D. s. Scaife, J. *29*
Gross, F. 626, *636*, 658, 659, 666, *680*
Gross, I. 717, 718, 720, 728, *744*
Gross, J. 6, *27*
Gross, J. D., u. L. Caro 24, *28*
Groves, J. W. s. Drayton, F. D. 127, *132*, 326, *339*
Grümmer, G. 387, *401*
Grun, P. 564, 576, *588*
— u. E. L. Triplett 568, 576, *588*
— s. Clausen, J. 567, 568, 575, 576, *585*
Grunow 658
Gudjónsson, G. 579, *588*
— s. Sörensen, Th. 563, 564, 573, *595*
Günther, E. 523, *527*
Günther, F. 621. 623, *625*
Guerra, P. s. Langeron, M. 779, *785*
Guillard, R. R. L., u. J. R. Ryther 666, *680*
Gupta, P. K. s. Roy, R. P. 575, *594*
Gurgel, J. T. A. s. Moreira, S. 581, *592*
Gustafsson, Ake 547, 548, *549*
Gustafsson, Å. 552, 555, 561, 566, 572, 573, 580, *588*, 804, *806*
— u. I. Gadd 566, *588*
— u. A. Nygren 555, 572, *588*
— s. Gentcheff, G. 441, *444*, 572, 579, 580, *588*

Gutenmacher, I. s. Grinvald, C. 202, *205*
Guttridge, C. G. 805, *806*
Gwynne-Vaughan, H. C. I. 326, *340*
— u. H. S. Williamson 326, *340*

Haagen-Smit, A. J. s. Raper, J. R. 292, *319*
Haberlandt, F. 174, 192, 193, *205*
Haberlandt, G. 572, *588*
Haeckel, A. 381, *401*
Häfliger, E. 579, *588*
Hämmerling, J. 113, *119*
— s. Hartmann, M. 113, *119*
Haertl s. Hoar 577
Haga, T. 59, *103*
Hagerup, O. *103*, 574, 578, 579, *588*, *589*
Hagerup, O. K. 155, *205*
Hagman, M. 519, 522, 524, 525, *527*
— u. L. Mikkola 524, *527*
Haigh, J. C. s. Harland, S. C. 447, 455, *477*
Hair, J. B. 573, *589*
Håkansson, A. 59, 80, *103*, 560, 561, 575, 577, 578, 580, *589*
— u. Albert Levan 577, *589*
Hakura, A., N. Otsuji u. V. Hirota 11, *28*
Hall, R. P. 631, *636*
Hall, W. C. 189, 192, 194, *205*
Halldal, P., u. J. Markali 649, *654*
Haller, Albrecht v. 154
Hallqvist, C. 455, *477*
Halsted, B. D. 174, 192, *205*
Hamant 145
Hamant, Cl. 410, *444*
Hamburger, W. W. 777, *785*
Hamel, G. 697, *704*
Hammond, B. L. s. Gerstel, D. U. 572, *588*
Hammond, M. R. s. Sparrow, A. H. 356, *367*
Hanatschek, E. 690, *691*
Hanna, G. C. s. Rick, C. M. 90, *107*, 160, *210*
Hanna, W. F. 331, 337, *340*
— u. W. Popp 337, *340*
— s. Popp, A. 337, *341*
Hanow, R. s. Nicolaisen, N. *106*, 173, *210*
Hansen, A. 794, *806*
Hansen, H. N., u. W. C. Snyder 129, 130, *132*
Hanson, A. A., u. F. V. Juska 566, 577, *589*
Hansson, A. A., R. J. Carber u. W. M. Myers *589*
Harada, K., M. Suzuki, H. Kameda u. S. Mitsuhashi 16, *28*
Haran, N. 573, *589*
Harder, R. 130, *132*, 248, *255*, 270, 603, 604, 605, 606, 612, 617, *619*
Harig, A. 803, *806*
Harlan, J. R. 558, 574, *589*
— M. H. Brooks u. R. P. Celarier *589*
— — D. S. Borgaonkar u. J. M. J. DeWet 573, *589*
— u. R. P. Celarier 573, 574, 577, *589*
— — W. L. Richardson, Margaret Hoover Brooks u. M. L. Mehra 573, 574, *589*
— u. J. M. J. DeWet 573, 574, *589*
— s. Borgaonkar, D. S. 567, 573, 574, *584*
— s. Celarier, R. P. 573, 574, *585*
— s. Chedda, H. R. 573, *585*
— s. Dewald, C. L. 573, 574, *586*
— s. DeWet, J. M. J. 573, *586*
— s. Snyder, L. A. 574, *595*
Harlan, H. V. s. Anthony, S. 397, *399*
Harland, S. C., J. C. Haigh u. J. L. Lochrie 447, 455, *477*
Harmsen, L. 582, *589*
Harnack, W. 334, *340*
Harris, K. 638, 639, *654*
Harrison, G. J. s. Kearney, T. H. 503, *504*
Hart, H. C. 800, *806*
Hart, T. J. 666, 669, 670, 671, *680*
Harte, C. 450, 452, 453, 455, 460, 462, 463, 465, 466, 467, 472, 473, 474, *477*
— u. Gotter 458
Harteck, P. s. Bredemann, G. 396, 397, *399*
Hartman-Dick, U., u. W. Müller-Stoll 393, *401*
Hartmann, H. T. 801, 804, *806*
Hartmann, K. M. 303, *318*
Hartmann, M. *4*, 31, 32, 33, 35, 36, 38, 42, 43, 46, *49*, 94, 99, 101, *103*, 110, 111, 112, 113, 114, 117, *119*, 123, 124, 125, 126, 127, *132*, 134, 136, 140, 141, *150*, 153, *205*, 215, 229, *233*, 235, 237, 250, *255*, 259, 260, 261, 264, 268, 274, *279*, 282, 283, 287, 302, 304, 307, 309, *318*, 321, 322, 323, *340*, 531, *550*, *602*, 697, 703, 704, *704*, 706, *744*
— u. J. Hämmerling 113, *119*
— u. O. Schartau 283, *318*
Hartog-Adams, I. den s. Parke, M. 740, *745*
Hartsema, A. M. 800, *806*
Hartshorne, J. W. 57, *103*
Hartung, M. E. 575, 576, *589*
Hase, E., Y. Morimura, S. Mihara u. H. Tamiya 734, *744*
— — u. H. Tamiya 734, *744*
Hashimoto, K. s. Ito, H. 188, *206*
Hasitschka-Jentschke, G. *759*
— s. Tschermak-Woess, E. 751, *752*
Haskell, G. 581, *589*
— u. J. Hill 581, *589*
— u. N. N. Tun 581, *589*
Haskell, G. M. 581, *589*
Hasle, G. R., u. E. Nordli 629, *636*
Hassan, A., u. M. H. Aboul-Wafa 383, *401*
Hassinger-Huizinga, H. 37, *49*, 114, *119*, 767, *775*
Hatano, K. 381, *401*
Haupt 145
Haupt, A. W. 429, *444*, 696, 697, *704*
Haupt, W. s. Chaudhri, J. J. 801, *805*
Haustein, E. *103*, 485, 486, *504*
Hawker 38
Hawker, E. L. 124, *132*
Hawker, L. E. 245, *255*
Hawlitschka, E. 686, 690, *691*
Hayase, H. 385, *401*
— u. Y. Hiraizumi 385, *401*
Hayes, Doris M. 538, 540, *550*
Hayes, W. 6, 7, 8, 9, 15, 16, 17, 18, 19, *28*
— s. Wollman, E. L. 314, 315, 316, *320*
Hayman 575
Hecht, A. 522, *527*
— s. Kumar, S. 520, 523, *527*
Hedberg, O. 574, *589*
Heering, W. 732, *744*
Heiden, H., u. R. W. Kolbe 669, 670, 671, *680*
Heidinger, W. 690, *691*
Heilbronn, A. 53, 95, *103*, 546, *550*
— u. M. Basarman 53, *103*
— u. N. Yakar 100, *103*, 154, *205*
Heilbronn, I. s. Cook, A. H. 309, *317*
Heim, R., A. Brack, H. Kobel, A. Hofmann u. R. Cailleux 782, 784, *785*
Heimans, J. 438, *444*
Heinen, W., u. H. F. Linskens 510, *527*
— s. Linskens, H. F. 381, *402*, 510, *528*
Heinricher, E. 140, *150*

Heisler, M. R. s. Brookbank, J. W. 313, *317*
Heitmüller, H. H., u. G. H. Melchior 159, 195, *205*
— s. Melchior, G. H. 158, 195, *208*
Heitz, E. 34, 35, *49*, 77, 141, *150*, 572, *589*
Heldmaier, C. 330, 334, *340*
Helinski, D. R. s. DeWitt, W. 15, *27*
Heller, L. s. v. Euler, H. 381, 382, *400*
Hellmers u. Machlis 393
Hellström, N. 373, *401*
Helm, J. 793, *806*
Helmcke, I. G., u. W. Krieger 661, *680*
Henderson, S. A. 352, *365*
Hendey, N. I. 669, 670, 671, *680*
Hensen, V. 665, *680*
Hensz, R. A., u. H. C. Mohr 189, 199, 202, *205*
Herbert, W. 507, *527*
Herbst 599
Heribert-Nilsson, M. 79, *103*
Heribert-Nilsson, N. 447, 448, 449, 452, 453, 454, 455, 465, 470, 474, *477*, 480, 481, *504*, 573, *589*
Herich, R. *103*, 174, 178, 199, *205*
— s. Erdelský, K. 174, *204*
Hermann s. Erlanson 59
Hernandez, A. R. s. Snyder, L. A. 575, *595*
Herndon, W. R. 762, *775*
Herophilos 258, *279*
Herrmann, S. 158, 195, 197, *206*
Hertwig, D. 258, *279*
Hertwig, G., u. P. Hertwig 72, 76, 77, *103*
Hertwig, P. s. Hertwig, G. 72, 76, 77, *103*
Heslop-Harrison, J. *103*, 156, 157, 160, 162, 164, 170, 171, 174, 177, 184, 186, 190, 193, 194, 197, 198, 199, 200, 201, *206*, 274, 275, *279*, 369, 377, *401*
— u. Y. Heslop-Harrison 88, 101, *103*, 169, 170, 174, 177, 178, 179, 194, 195, 198, 199, *206*, 273, *279*
— s. Knox, R. B. 563, 574, *590*
Heslop-Harrison, J. W. 59, 80, *103*, 181, 196, *206*
— s. Blackburn, K. B. 58, 59, 78, 80, *102*
Heslop-Harrison, Y. 581, *589*
— s. Heslop-Harrison, J. 88, 101, *103*, 169, 170, 174, 177, 178, 179, 194, 195, 198, 199, *206*, 273, *279*
Hess, D. 183, 184, 201, *206*
Hess, O. 261, *279*
Hess, U. 608, *619*
Hesse, C. O. s. King, J. R. 396, 398, *401*
Hesseltine, C. W., u. R. F. Anderson 296, *318*
Heun, Alphonse L. 537, 538, 540, *550*
Heurck, H. van 658, 664, 670, *680*
Hevesy, G. s. Eklundh-Ehrenberg, C. 442, *445*
Hewitson, W. M. s. Nickerson, N. H. 163, 198, *210*
Hewitt, R., u. D. Billen *28*
Heydak, M. H., u. L. S. Palmer 383, *401*
Heyer, F. 163, 181, 184, 185, 192, 193, 195, 196, 200, *206*
Heyl, F. W. 379, 380, 382, 383, *401*
Hickman, C. J. s. Royle, D. J. 291, *319*
Hiemenz, G. 442, *446*, 460, 470, *477*, 503, *504*
Hieronymus 143
Hieronymus, G., u. R. Sadebeck 549, *550*
Hiesey, W. M. 576, *589*
— s. Clausen, J. 567, 568, 575, 576, 577, 580, *585*
— s. Clausen, J. C. 567, 576, *586*
Higgins 238
Higgins, B. B. 783, *785*
Higgins, J. E. 168, 195, *206*
— u. V. Holt 168, 195, *206*
Higgins, M. L. s. DeWet, J. M. J. 573, *586*
Hildebrand 373
Hildebrand, F. 160, 179, 181, 182, 190, 192, 193, 197, 200, *206*, 507, *527*, 792, *806*
Hill, Helen D. s. Carnahan, H. L. 577, *585*
Hill, J. s. Haskell, G. 581, *589*
Hillyer, I. G. s. Wittwer, S. H. 187, 189, 198, 199, *213*
Hindák, F. 728, *744*
Hiorth, G. 100, *104*, 467, *477*, 481, 482, *504*
Hiraizumi, Y. s. Hayase, H. 385, *401*
Hirano, K. s. Kurogi, M. 763, *775*
Hirasé 418
Hirata, K. 58, 82, *104*
Hirayoshi, I. s. Kihara, H. 57, 58, 65, *104*
Hirn 713
Hirota, V. s. Hakura, A. 11, *28*
Hirota, Y. 13, *28*
— u. T. Iijima 12, *28*
— K. Nishimura, F. Osrkov u. I. Orskov *28*
Hirota, Y., u. P. H. A. Sneath 13, *28*
— s. Nishimura, Y. *29*
Hirsch, H. E. 34, *49*, 130, *132*
Hirst, G. K. *318*
Hjelmqvist, H. 579, 580, *589*
— u. F. Grazi 563, *589*
Hoar u. Haertl 577
Hodgetts, W. J. 748, *752*
Högberg, B. s. v. Euler, H. 378, 379, 382, 383, 384, *400*
Högberg, K. G. s. v. Euler, H. 381, 382, *400*
Hoff, B. J. s. Bashaw, E. C. 566, *583*
— s. Burton, C. W. 575
Hoffender, H. 646, *654*
Hoffman, L., u. I. Manton 715, *744*
Hoffman, L. R. 287, *318*
Hoffmann 35, 36, 136, 142, 145, 147, 148
Hoffmann, A. *255*
Hoffmann, H. 169, 192, 197, *206*
Hoffmann, I. s. Knapp, E. 33, 38, *50*
Hoffmann, W. 63, 82, 83, 85, 86, 87, *104*
— u. E. Knapp *104*
Hofmann, A. s. Heim, R. 782, 784, *785*
Hofmann, H. s. Weygand, F. 382, 383, *405*
Hofmeister 138
Hofmeister, L. 380, *401*
Hofmeister, W. 235, *255*, 262, *279*
Hofmeyr, J. D. J. 68, 90, 91, *104*
— u. H. van Elden *104*
— u. P. J. C. Oberholzer 581, *590*
Hofschneider 11
Holdaway, F. G. s. Storey, W. B. 168, *212*
Holden, H. H. W., u. D. A. Bond 509, *527*
Holdsworth, M. 169, 194, *206*
Holferty, G. M. 145, *151*
Holland, I. B. *28*
Hollande, A. 620, 621, 623, *625*, *654*
— s. Poisson, R. 620, 623, *625*
Hollerbach, M. M. 608, *619*
Holliday, R. 364, *365*
Holloway, B. W. 25, *28*
— u. B. Fargie *28*
— u. P. A. Jennings 25, *28*
Holm, Th. 799, *806*
Holm-Hansen, O. 608, *619*
Holman, R. M., u. F. Brubaker 395, *401*
Holmgren, I. 555, 572, *590*

Holmström, B. s. Nielsen, N. 381, *403*
Holt, D. C. s. Fisher, W. D. 574, 575, *587*
Holt, E. C. s. Bashaw, E. C. 575, *583*
Holt, V. s. Higgins, J. E. 168, 195, *206*
Holton, C. S. 228, *233*, 324, 337, *340*
— u. G. W. Fischer 337, *340*
Holuby, J. I. 82, *104*
Hommersand 767
Honma, S. s. Phatak, S. C. *213*
Horiuchi, K., u. E. A. Adelberg *28*
Horn, C. L. 577, *590*
Hotchkiss, A. T. s. Tindall, D. R. 758, *760*
Hotson, J. W. 784, *785*
Hotta, Y., M. Ito u. H. Stern 349, 354, 360, *365*
— u. H. Stern 358, *365*
Hou-Liu, S. Y. 573, *590*
Houwink, A. L., u. W. van Iterson 10, *28*
— u. P. A. Roelofsen 390, *401*
Hovasse, R. 620, 623, *625*, 639, 644, *655*
Hovin, A. W. 202, *206*
Howarth, S. s. Ozeki, H. 15, *29*
Howlett, F. S. 183, 192, 193, 194, *206*
Hubbard, J. 563, 573, *590*
Huber, G., u. F. Nipkow 627, 632, 633, 634, *636*
Huber, H. 488, 494, *504*
Huber, J., u. F. Jadin *619*
Huber, P. s. Schoch-Bodmer, H. 393, *404*
Huber-Pestalozzi, G. *625*, 631, *636*, 749, *752*
Hubert 195
Hüttig, W. 34, *49*, 128, *132*, 141, 142, *151*, 337, *340*
Huf-Brinckmann, M., u. H. F. Linskens 507, *527*
Huggler, K. s. Zetschke, F. 375, *406*
Hughes, M. B., u. E. B. Babcock 509, *527*
Hughes, W. S. s. Taylor, J. H. 353, *367*
Huhnke, W., Chr. Jordan, H. Neuer u. R. v. Sengbusch 83, *104*, 175, 177, 193, 194, 195, 197, *206*
— s. Bredemann, G. 83, 85, *102*
Hulewicz, D. 174, *206*
Hull, E. D. 570, *590*
Humphrey, J. E. 124, *132*
Humphrey, L. M. s. Sax, K. 353, *367*
Hunt, D. E., u. P. L. Carpenter 313, *318*
Hunziker, H. R. 580, *590*
— s. Rutishauser, A. 555, *594*
Hus, H. T. A. 774, *775*
Husfeld, B. 90, *104*
Huskins, C. L. 353, *365*
Hustede, H. 112, 113, *119*, 125, *132*, 687, 688, *691*, 734, 736, *744*
— s. v. Denffer, D. 112, *119*, 599, 687, *691*
Hustedt, F. 664, 665, 668, 672, *680*
Hutchinson, A. H. 430, *444*
Hutchison, D. J., u. E. C. Bashaw 575, *590*
— s. Bashaw, E. C. 575, *583*
Hutner, S. H., u. L. Provasoli 303, 307, *318*
Huyskes, J. A., u. J. Sneep 90, *104*
Hwang, S. W. 329, *340*
Hyde, B. B. s. Paliwal, R. L. 95, *107*
Hylmö, B. 173, *206*

Ichikawa, H. s. Eguchi, T. 69, *103*
Iijima, T. 15, *28*
— s. Hirota, Y. 12, *28*
Iizuka, M. s. Janick, J. *104*
Iizuka, W., u. J. Janick 59, 66, 89, *104*
Ikeda, W. s. Awada, M. 168, 193, *203*
Ikeno, S. 572, *590*
Iltis, H. 163, 196, *206*
Imahori, K. s. Wood, R. D. 753, *760*
Imai, Y. 69, *104*
Ingram, M. 780, *785*
Ippen, K. A., u. R. C. Valentine 12, *28*
— s. Wendt, L. W. 12, *30*
Irmisch, Th. 791, *806*
Irmscher, E. 167, 192, 197, *206*
Isenbeck, K. 337, *340*
Ito, H., u. T. Kato 188, *206*
— — K. Hashimoto u. T. Saito 188, *206*
— u. T. Saito 188, *206*, *207*
— s. Hotta, Y. 349, 354, 360, *365*
Ivanov, M. A. 245, *255*
Iwakiri, B. T. s. Griggs, W. H. 397, 398, *401*
Iwanami, Y. 393, 394, *401*
Iwanoff, L. 691, *691*
Iwasaki, H. 763, 764, 774, *775*
— u. C. Matsudaira 763, *775*
Iyengar, M. O. P. 712, 721, 732, *744*
— u. T. V. Desikachary 608, 611, *619*
— u. K. R. Ramanathan 712, *744*
Iyer, C. P. A., u. G. S. Randhawa *213*
Iyer, S. S. s. Bhaskaran, K. 25, *27*
Izhar, Sh. s. Galun, E. *213*
Izmailow, R. s. Jankun, A. 579, *590*

Jaag, O. 608, 613, *619*
Jachimsky, H. *151*
Jacob, F., u. E. A. Adelberg 16, *28*
— u. S. Brenner 19, 20, 22, 23, 24, *28*
— — u. F. Cuzin 23, 24, *28*
— A. Lwoff, A. Siminovitch u. E. L. Wollman *28*
— P. Schaeffer u. E. L. Wollman 13, *28*
— L. Siminovitch u. E. L. Wollman 15, *28*
— u. E. L. Wollman *4*
— s. Anderson, Th. F. 314, 315, *317*
— s. Lavallé, R. 13, *29*
— s. Wollman, E. L. 10, *30*, 314, 315, 316, *320*
Jacob, L., u. E. L. Wollman 6, 7, 8, 9, 12, 17, 18, 19, 21 *28*
Jacobsen P. 58 62 64 65, *104*
Jacques-Félix, H. 161, 193, *207*
Jadin, F. s. Huber, J. *619*
Jagannath, D. R., B. W. X. Ponnaiya u. W. S. Raman 574, *590*
Jahoda 707
Jain, H. K., u. S. L. Basak 351, *365*
Jakob, F. s. Afoldi, L. F. 15, *26*
Jakob, K. M. s. Atsmon, D. 185, *202*
Jakobi, H. s. Butenandt, A. 198, *204*
Jakowlew, M. S. s. Nawaschin, M. S. *446*
Jakuschkina, N. L. 388, *401*
James, E. J. 688, *691*
Jamieson, C. A. s. Auclair, J. L. 381, *399*
Jana, M. K. *213*
Janaki-Ammal, E. K. s. Darlington, C. D. 569, 570, 572, 574, 577, 578, 579, *586*
Jane, F. W. 639, *655*
Janet, Ch. 742, *744*
Janick, J. 89, *104*
— u. M. Iizuka *104*
— u. E. C. Stevenson 59, 88, 89, *104*, 172, 194, *207*
— s. Bose, S. 59, 89, *102*
— s. Ellis, J. R. 59, 66, 89, *103*
— s. Iizuka, M. 59, 66, 89, *104*
— s. Mahoney, D. L. 88, *106*

Jankun, A., u. R. Izmailow 579, *590*
Jansen 142
Janssens, F. A. 350, *365*
Jastreb s. Awakian *445*
Jeffrey, E. C. 578, *590*
Jenkins, J. A. s. Stebbins jr., G. L. 571, *595*
Jennings u. Chen 43, 44
Jennings, P. A. s. Holloway, B. W. 25, *28*
Jensen 397
Jensen, W. A. 440, *444*
Jerchel, D. s. Kuhn, R. 215, *233*
Jillson, O. F., u. W. J. Nickerson 779, *785*
Jiménez, J. R., u. O. E. Nelson 465, *477*
Jinks, J. L. s. Cavalli-Sforza, L. L. 8, 9, *27*
Johannsen, W. 480, *504*.
Johansen, D. 578, *590*
John, R. P. s. Fritsch, F. E. 690, *691*
Johnson, A. 578, *590*
Johnson, A. G. 59, 95, *104*
Johnson, B. L. *590*
Johnson, D. F. 622, *625*
Johnson, E. L., u. D. B. Kingsbury 801, *806*
Johnson, I. J. s. Whiteley, J.R. 524, *530*
Johnson, L. P. 624, *625*
Johnson, L. P. V. 397, *401*
Johnson, M. A. 796, *806*
Johnsson, H. 59, 78. *104*
Johri, B. M., u. B. Tiagi 154, 155, *207*
— u. I. K. Vasil 368, 395, *401*
— s. Maheshwari, P. 431, 434, *444*
Jollos, V. 113, *119*, 229, *233*, 301, *318*
Joly, R. A. s. Beck, W. A. 387, *399*
Jonasson, L. s. Fries, N. 329, *339*
Jones 544
Jones, D. F. 87, 95, 96, *104*, 454, 456, 457, 461, 464, 465, 468, *477*, 481, *504*
— s. Mangelsdorf, P. C. 461, 462, *477*
Jones, H. A. s. Robbins, W. W. 90, *107*
Jones, K. L. 191, 194, 195, *207*
Jones, R. F. s. Wiese, L. 303, *320*
Jones, W. W. s. Storey, W. B. 168, *212*
Jordan, Chr. s. Huhnke, W. 83, *104*, 175, 177, 193, 194, 195, 197, *206*
Jørgensen, C. A. 61, *104*
Jørgensen, E. G. 668, *680*
Joshi, A. C. 161, 195, 197, *207*
Jost, L. 384, 385, *401*, 509, 517, 518, *527*, 805, *806*
Jost, M. 604, *619*
Joyet-Lavergne, Ph. 148, *151*
Joyon, L. 620, *625*
Juel, O. 555, 571, *590*
Jüssen, F. J. s. Schürhoff, P.N. 569, *594*
Juhl, H. 576, *590*
Juhl-Noodt, H. 576, *590*
Julén, Gösta 565, 566, 576, *590*
Juliano, I. B., u. N. Y. Cuevas 569, *590*
Juliano, J. B. 569, *590*
Julin, E. 579, *590*
Juller, E. 240, *255*, 718, 719, 742, *744*
Jung, Y. s. Galun, E. 188, 198, 199, *205*, 275, *279*
Juska, F. V. s. Hanson, A. A. 566, 577, *589*

Kachroo, P. 144, *151*
Käfer, E. s. Pontecorvo, K. 264, *280*
Kämmerling, H. 337, *340*
Kämpfer, E. 395, *401*
Kaienburg, A. L. 394, *401*, 500, *504*
Kain, J. M. 699, *704*
Kakesita, K. 803, *806*
Kakizaki, Y. 509, 521, *527*
— u. T. Kasai 522, *527*
Kameda, H. s. Harada, K. 16, *28*
Kamo 59
Kamptner, E. 650, 651, *655*
Kana, K., N. S. Ranga Swamy u. P. Maheshwari 525, *527*
Kapadia, Z. L., u. F. W. Gould 574, *590*
Kapoor 316
Kappert, H. 90, 94, *104*, 442, *446*, 573, *590*
Karrer, P., u. E. Leumann 378, *401*
Karschon, R. 159, *207*
Karsten, G. 657, 670, 671, 677, *680*
Kasai, T. s. Kakizaki, Y. 522, *527*
Kashyap, S. R. 138, 142, *151*
Katayama, Y., u. S. Shida 69, *104*
Kato, T. s. Ito, H. 188, *206*
Kaufert, F. 334, *340*
Kauffman, C. H. 219, *233*
Kaufmann, B. P. 348, 352, *365*
Kavanagh, F. s. Robbins, W. J. 784, *786*
Kawamura, E. 329, 336, *340*
Kawano, S. 574, *590*
Kayano, H. 351, *365*
Kearney, T. H., u. G. J. Harrison 503, *504*
Keck, D. s. Clausen, J. 567, 568, 576, 580, *585*, *586*
Kehl, H. 217, *233*
Keitt, G. W., u. D. H. Palmiter 325, *340*
— s. Langford, M. H. 325, *340*
Kellenberger, G., M. L. Zichichi u. J. J. Weigle 364, *365*
Kemper, W. 130, *132*
Kempton, J. H. 461, *477*
— s. Collins, G. N. 461, *476*
Keng, H. *104*
Kerner, A. 571, *590*
Kerner von Marilaun, A. 788, 800, 804, 805, *806*
Khan, R. 430, *444*
Khawkine, W. 621, 624, *625*
Kidd, C. s. Gerstel, D. U. 572, *588*
Kiellander, C. L. 576, *590*
Kiermayer, O. 183, 191, 199, *207*
Kiesel, A., u. B. Rubin 380, 382, 383, *401*
Kiesselbach, T. A., u. F. Peterson 461, *477*
Kihara, H. 58, 60, 63, 64, *104*
— u. I. Hirayoshi 57, 58, 65, *104*
— u. T. Ono 58, 65, 69, *104*
— u. Y. Yamamoto 65, 70, *105*
Kikudome, G. 462, *478*
Killian, Ch. 635, *636*
Killian, K. 769, *775*
Kimball, R. F. 44, *50*
Kimber, G. s. Riley, R. 350, *366*
King, E. D. s. Sax, K. 363, *367*
King, J. R., u. C. O. Hesse 396, 398, *401*
Kingsbury, D. B. s. Johnson, E. L. 801, *806*
Kirby, R. S. 325, *340*
Kircheisen, G. 266, *279*
Kirk, L. E., u. T. M. Stevenson 509, *527*
Kitzes, G., H. A. Schuette u. C. A. Elvehjem 383, *401*
Kiyohara s. Sinoto, Y. 59
Klavestad, N. 769, *775*
Klebs 307, 687, 688, 689, 690, 720, 723, 735, 736, 741, 742, 749, 750
Klebs, G. 111, 112, *119*, 142, *151*, 166, 192, 196, 200, *207*, 598, 599, 600, *602*, 620, 621, 622, 624, *625*, 634, 635, *636*, 640, 641, 645, 649, *655*, 779, 780, *785*
Klebs, K. 802, *806*
Klein, G. 610, *619*
Klein, L. 742, *744*
Klug, G. 638, *655*
Knapp, E. 58, *105*, 245, *255*, 309, *318*

Knapp, E., u. I. Hoffmann 33, 58, *50*
— s. Hoffmann, W. *104*
Kniep, H. 34, 46, 47, *50*, 110, *119*, 123, 127, 130, *132*, 134, 237, 250, *255*, 327, 328, 329, 330, 331, 332, 333, 334, 335, 337, *340*, *602*, 695, 703, 704, *704*, 742, 743, *744*, 750
— u. R. Bauch 134, *151*
Knight, A. 507, *527*
Knight, L. I. 384, *401*
Knight, M. 238, 244, *255*, 694, 695, 703, *704*
Knight, R., u. H. H. Rogers 516, *527*
Knight, Th. A. 184, 194, *207*
Knoll, F. 372, 373, *401*
Knott, J. E. 172, *207*
Knox, R. B., u. J. Heslop-Harrison 563, 574, *590*
Knuth, P. 507, *527*
Kny 137
Kobel, H. s. Heim, R. 782, 784, *785*
Koch, W. 579, *590*, 768, 774, *775*
Koch, W. J. 682, *692*
Köhler, D. 83, 85, 101, *105*, 157, 173, 176, 178, 193, 199, 200, 201, *207*
— u. A. Lang 201, *207*
Köhler, E. 677, *680*
Köhler, F. 217, *233*
Köhler, K. 37, *50*, 113, 115, 118, *119*, 121, *132*, 229, *233*, 285, 298, 299, 304, 307, 308, *318*
Kölreuter 72
Koelreuter, J. G. 384, *401*, 507, *527*
Könemann, B. 805, *806*
Koepchen, W. s. Schwemmle, J. 450, 467, *478*, 484, 491, 502, *505*
Koessler, J. H. 383, *401*
Kofoid, C. A. 629, *636*
Kohl, F. G. 610, 611, *619*
Kojan, S. s. Sparrow, A. H. 360, *367*
Kolbe, R. W. s. Heiden, H. 669, 670, 671, *680*
Kolkwitz, R. *692*
— u. H. Krieger *752*
Kolsow *445*
Konetzka s. Roeser 23
Konetzka, W. A. s. Berrah, G. 23, *27*
Konokotine, A. s. Nadson, G. 127, *132*
Koopmans, A. 99, 100, *105*
Koot, Y. van, u. H. J. M. van Dorst 188, 196, *207*
Kornmann, P. 237, 239, 243, 246, *255*, 637, 646, 647, 650, 653, *655*, 694, 695, 696, 697, 698, 699, 700, 701, 702, 703, *704*, 762, 763, 764, *775*
Korobova, S. N. s. Batygina, T. P. 524, *525*
Koroleva, V. A. 573, *590*
Korschikoff, A. A. 708, 711, 712, 713, 715, 720, 722, 725, 726, 730, 732, *744*
Korshikov, A. A. 643, 644, *655*
Kosan 360
Kostoff, D. 372, *401*
Kothbauer, E. s. Schussnig, B. 238, *256*, 703, *705*
Kotte 285, 287, 309 310
Krafczyk H. 217 *233*
Krause, O. 582, *590*
Krause, R. F. s. Barnett, H. G. 296, *317*
Kribben, F. J. 186, 198, *207*
— s. Laibach, F. 187, 189, 198, 200, *207*, 269, 270, 275, *279*
Krichenbauer, H. 621, 622, *625*
Krieger, H. s. Kolkwitz, R. *752*
Krieger, W. 644, 645, 646, *655*
— s. Helmcke, I. G. 661, *680*
Krijthe, N. *806*
Krishan, R. s. Choudhri, R. S. 162, 163, 192, 194, *204*
Krishnamurthy, V. 763, *775*
Kristiansen, J. 638, 644, *655*
Kröner, E. 800, 801, 802, 803, 804, *806*
Kroh, M. 509, 510, 511, 521, 525, *527*, 802. *806*
Krolop, H. s. Borriss, H. 387, *399*
Krüssmann, G. 799, *806*
Krug, C. A. 581, *590*
Krug, H. 802, *806*
Krugman, S. L. s. Bingham, W. E. 380, *399*
Kuckuck 719, 720
Kuckuck, H. 89, *105*
Kuckuck, P. 113, *119*, 248, *255*, 774, *775*
Kuczewski, O. 754, 755, 758, *759*
Kudrjavcev, V. A. 372, *401*
Kühlwein, H. 388, *401*
— u. H. Anhäusser 393, *401*
Kühn, A. 260, *279*, 322, *340*
Kühner, R. 330, 332, *340*
— u. P. Terra 333, *340*
Kuenen, R. s. Esser, K. 121, 122, 123, 127, 128, *132*, 261, *279*, 293, 296, *317*, 323, 324, *339*
Küstenmacher, M. 373, *402*
Küster, E. 393, *402*
Kütz 608
Kuhn 283
Kuhn, E. 55, 87, 92, 95, 100, *105*, 177, 179, 195, *207*, 385, 387, *402*
Kuhn, R. 379, *402*
— u. I. Löw 379, 382, *402*
— F. Moewus u. D. Jerchel 215, *233*
Kujala, V. 800, *806*
Kulp, W. L. s. Anderson, R. J. 382, 383, 384, *399*
Kumar, H. D. 602, *619*
Kumar, L. S. S., A. Abraham u. V. K. Srinivasan 60, *105*
— u. S. Vishveshwaralah 60, 80, *105*
Kumar, S., u. A. Hecht 520, 523, *527*
Kunieda, H., u. S. Suto 309, *318*
Kunze, G. 68, *105*, 448, *477*
Kurita 60
Kurogi, M. 763, *775*
— u. K. Hirano 763, *775*
Kurtz, E. B. s. Nitsch, J. P. 186, 189, 194, 198, *210*
Kurtz jr., E. B., u. J. L. Liverman 372, *402*
Kusano, S. 230, *233*
Kušnerenko, S. V. s. Minina, E. G. 186, 195, 198, *209*
Kwack, B. H. 522, 523, *527*
— s. Brewbaker, J. L. 394, *399*
Kylin, H. 695, 696, 697, 699, 700, *705*, 764, 767, 769, 770, 771, 774, *775*

Lackey, B. *655*
Laczyńska-Hulewiczowa, T. *105*, 176, 194, *207*
Lagerberg, T. *151*
Laibach, F. 164, 165, 179, 187, 194, 197, 198, 200, *207*, 269, 270, *279*, 383, 387, *402*
— u. F. J. Kribben 187, 189, 198, 200, *207*, 269, 270, 275, *279*
Lal, M. 245, 250, 251, 252, *255*
Lamb, I. M. 324, 336, *340*
Lambert, J. 161, 196, *207*
La Moine d. Anhder s. Peterson, C. E. 101, *107*
Lamoure, D. 331, *340*
Lampert 719
Lamprecht, H. 503, *504*
Land, W. J. G. 425, 430, *444*
Landmann, W. s. Steffen, K. 435, *445*
Landolt, E. 794, *806*
Landspersky, N. 499, *504*
Lang, A. 157, *207*, 519, *527*
— s. Galun, E. 188, 198, 199, *205*, 275, *279*
— s. Köhler, D. 201, *207*
Lang, W. H. 249, 252, *255*, 545, 546, *550*

Lange, A. H. 91, *105*, 168, 193, 194, 195, 199, *207*
Lange, M. 330, *340*
Langendorff, H. 460, *477*
Langeron, M., u. P. Guerra 779, *785*
Langford, M. H., u. G. W. Keitt 325, *340*
Lark, C., u. K. G. Lark *28*
— s. Lark, K. G. *29*
Lark, K. G., u. R. Bird *28*
— u. C. Lark *29*
— s. Lark, C. *28*
— s. Pritchard, R. H. 20, *29*
Larsen, P., u. S. M. Tung 383, 388, *402*
Laskowska, R. 57, *105*
Lata, M. s. Negi, V. 430, *444*
Lauffer, M. A. s. Brinton, C. C. 10, *27*
Laurent, É. 173, 174, 179, 193, *208*
Lausten, O. s. Winge, Ø. *109*
Lauterborn, R. 638, *655*
Lavallé, R., u. F. Jacob 13, *29*
Lawrence, W. E. 563, 574, *590*
Lawrence, W. J. C. 57, *105*, 457, *477*
— s. Crane, M. B. 91, *102*, *526*
Lawson, A. A. 429, *444*
Lawton, Elva 539, 546, *550*
Lazarenko, A. S. 245, 250, 251, *255*
— C. T. Pashuk u. E. N. Lesnyak 250, *255*
Lazaroff, N., u. J. Schiff 605, *619*
— u. W. Vishniac 603, 604, 605, 606, 614, *619*
League, E. A., u. V. A. Greulach 687, *692*
Lebedeff, G. A. 71, *105*
Lebek, G. 19, *29*
Lechner, K. 485, *504*
Lecocq, M. *213*
Lederberg, E. M. s. Cavalli-Sforza, L. L. 7, *27*
— s. Lederberg, J. 7, *29*
— s. Makela, P. H. 25, *29*
— s. Morse, M. L. *29*
Lederberg, G. s. Ganesan, A. T. 20, *27*
Lederberg, H. s. Sneath, P. H. A. 19, *29*
Lederberg, J. 6, 12, *29*, 259, *280*, 314, 315, *318*
— L. L. Cavalli u. E. M. Lederberg 7, *29*
— s. Cavalli-Sforza, L. L. 7, *27*
— s. Makela, P. H. 25, *29*
— s. Morse, M. L. *29*
— s. Sneath, P. H. 316, *319*
— s. Tatum, E. L. 314, *320*
Lee, C. L. 419, *444*
Lee, J. J. s. Freudenthal, H. D. 629, *636*
Leedale, G. F. 620, 621, *625*
— B. J. D. Meeuse u. E. G. Pringsheim 620, *625*
— s. Manton, I. 646, 649, *655*
Leeuwenhoek, A. van 257, 258, *280*
Lefévre, M. 627, 628, *636*
Lehmann, E. 507, *527*
Leitgeb, H. 533, *550*
Leman, A. 77, *105*
LeMinor 316
Lemmermann 640
Lendner, A. 330, *340*
Lenski, J. 794, 803, *806*
Lenz, F. 448, 460, *477*
Leplae, E. 179, *208*
Lerch, G. s. Müller-Stoll, W. R. 390, *403*
Lerche, W. 306, *318*
Leroy, J. F. 569, *590*
Lesley, M. M., u. J. W. Lesley 95, *105*
Lesnyak, E. N. s. Lazarenko, A. S. 250, *255*
Lester, D. T. 78, *105*
Leuchtmann, G. 488, *504*
Leumann, E. s. Karrer, P. 378, *401*
Leupold, U. 38, *50*, 126, 127, *132*, 323, *340*
Levan, A. *366*, 577, 578, *590*
— u. S. L. Emsweller 578, *591*
— u. A. Müntzing 264, 276, *280*
Levan, Albert s. Håkansson, A. 577, *589*
Levi, J. D. 228, *234*, 294, 304, *318*, 323, *341*
Levine, S., u. Z. Ordal 777, 778, *785*
Levring, T. R. 307, 308, 309, 310, *318*
Lewin, J. C. 668, 669, 670, 671, *680*
— R. A. Lewin u. D. E. Philpott 668, 669, 670, 671, *680*
— s. Reimann, B. E. F. 663, *681*
Lewin, R. A. 32, 33, *50*, 240, *255*, 282, 303, 304, 305, *318*
— u. J. O. Meinhart 304, *318*
— s. Lewin, J. C. 668, 669, 670, 671, *680*
Lewis, D. 94, *105*, 322, *341*, 379, *402*, 507, 508, 513, 514, 516, 519, 521, 522, 523, 524, *527*, *528*
— u. L. K. Crowe 93, *105*, 507, 524, *528*
— s. Mäkinen, Y. L. A. 520, *528*
Lewitzky, G. A. 160, 192, *208*
Leydecker 174, 192, *208*
Lhotský, O. 570, *752*
Lichte, H. F. 388, *402*
Lidforss, B. 384, *402*, 566, *591*
Liebisch, W. 677, *680*
Lilienfeld, F. 55, *105*
Li, Lai-Yung 575, 577, *591*
Liljefors, A. 581, *591*
Lillo, M. 172, 197, *208*
Lilly, V. G., u. H. L. Barnett 784, *785*
— s. Barnett, H. G. 296, *317*
Lima-deFaria, A. 353, 360, *366*
Limberk, J. *105*, 177, 178, 193, 195, *208*
Limpricht, K. G. 134, *151*
Lind, E. M. 240, *255*
Lindberg 135
Lindegren, C. C. 126, *132*, 351, *366*
— u. G. Lindegren 126, *132*
Lindegren, G. s. Lindegren, C. C. 126, *132*
Lindemann, E. 627, 629, 630, *636*
Lindsay, R. H. 59, 60, *105*
Linkola, A. 800, *806*
Linnaeus 262
Linnaeus, C. 551, 562, *591*
Linnert, G. 93, *105*
Linossier, G., u. G. Roux 778, *785*
Linskens, H. F. 260, 264, 266, 269, 270, *280*, 368, 369, 370, 373, 380, 382, 383, 395, *402*, 506, 512, 513, 514, 519, 520, 522, *528*
— u. Kl. Esser 390, 394, *402*, 512, 528
— u. W. Heinen 381, *402*, 510, *528*
— u. F. Resende 271, *280*
— u. J. Schrauwen 380
— J. A. M. Schrauwen u. M. van der Donk 520, 523, *528*
— u. J. Tupý 382, *402*, 513, *528*
— s. Beth, K. 699, *704*
— s. Bhattacharjya, S. S. *525*
— s. Heinen, W. 510, *527*
— s. Huf-Brinckmann, M. 507, *527*
— s. Mühlethaler, K. 390, 391, *403*, 512, *528*
— s. van der Pluijm, J. E. 520, *529*
— s. Resende, F. 96, *107*, 272, 275, 276, *281*
— s. Stanley, R. G. 381, *405*
Lissouba, P., J. Mousseau, G. Rizet u. J. L. Rossignol 364, *366*
List, H. 236, *255*
Liverman, J. L. s. Kurtz jr., E. B. 372, *402*
— s. Nitsch, J. P. 186, 189, 194, 198, *210*

Lloyd, F. E. 748, *752*
Lobanov, G. A. 387, *402*
Lochrie, J. L. s. Harland, S. C. 447, 455, *477*
Locker, Fr. 660, *680*
Loeb, T. 11, 17, *29*
— u. N. D. Zinder 11, *29*
Loehwing, W. F. 157, *208*
Loertzer, B. 479, 480, 488, 496, 502, *504*
Löve, Á. 58, 63, 65, 68, 80, 81, *105*, 565, 566, 577, 579, *591*
— u. D. Löve 58, 59, 80, *105*, 171, 173, 195, 198, *208*, 575, 576, 579, 582, *591*
— u. N. Sarkar 58, 63, 80, 81, *106*
Löve, D. 77, *105*
— s. Löve, Á. 58, 59, 80, *105*, 171, 173, 195, 198, *208*, 575, 576, 579, 582, *591*
Lövkvist, B. 787, 804, *806*
Löw, I. s. Kuhn, R. 379, 382, *402*
Loewe, S., u. E. Spohr 198, *208*
Loewus, F. s. Stanley, R. G. 386, 390, *405*
Lohmann 668
Lohmann, H. 629, *636*
Lohmeyer, H. s. Schwemmle, J. 485, *505*
Lommarson, R. C. s. Anderson, R. G. 753, 759
Longley, A. E. 449, 462, 467, *477*, 580, 581, *591*
Longman, K. A., u. P. F. Wareing 181, 197, *208*
Longo, B. 582, *591*
Loomis, N. H., C. F. Williams u. M. M. Murphy *106*
Lopriore, G. 373, *402*
Lorbeer, G. 35, 36, 38, *50*, *106*, 141, *151*, 249, *255*
Lorenz, F. 335, *341*
Lorenzen, H. 734, *744*
— u. H. G. Ruppel *745*
— s. Pirson, A. 734, *745*
Lorz, A. P. 59, *106*
Loutit, J. S. s. Pittard, J. *29*
Loveless, M. H., u. S. N. Timasheff 380, *402*
Lucan, V. s. v. Stosch, H. A. 647, *656*
Ludwig, (F.) 192, *208*
Ludwig, W., u. C. Boost 450, 460, *477*
Ludwigs, K. 138, 142, 143, *151*, 408, *444*
Luecke, R. W. s. Sarkar, B. C. R. 381, *404*
Lüning, K. G. *106*
Luerssen 139, 140
Luig, N. H. 452, *477*
— K. S. McWhirter u. E. P. Baker 452, *477*
Lund, H. A. 383, *402*
Lund, J. W. G. 638, 640, 641, 643, 644, *655*, 661, 668, *680*, 725, 726, *745*
— s. Parke, M. 637, 649, *655*
Lundén, R. 373, *402*
— s. Nielsen, N. 381, 382, 383, *403*
Lundquist, A. 519, 521, *528*
Luria, S. E. *29*
— u. J. W. Burrous *29*
Luzzati, D., u. M. R. Chavallier 15, *29*
Lwoff, A., u. P. Schaeffer 322, *341*
— s. Chatton, E. 620, *625*
— s. Jacob, F. *28*
Lwoff, M. s. Chatton, E. 620, *625*
Lyon, H. L. 413, *444*

Maaløe, O. 19, *29*
MacAllister, F. 578, *591*
Maccacaro, G. A. *29*, 316, *318*
— u. R. Comolli 315, 316, *318*
— u. R. Dettori *29*
— s. Dettori, R. 11, *27*
MacDonald, J. A. 328, 329, *341*
MacGillivray, J. H. s. Brink, R. A. 460, 461, *476*
Machlis, L. 215, 223, *234*, 288, 290, 291, *319*
— W. H. Nutting, M. W. Williams u. H. Rapoport 218, *234*
— u. E. Rawitscher-Kunkel 394, *402*
— s. Barksdale, A. W. 292, *317*
— s. Carlile, M. J. 291, *317*
— s. Mascarenhas, J. P. 394, *402*
— s. Rawitscher-Kunkel, E. 216, 226, 232, *234*, 287, *319*
Machlis s. Hellmers 393
Mack, Br. 644, 645, 646, *655*
Mackay, E. 58, 82, *106*
Macrae, R. 334, *341*
— s. Mounce, I. 328, 329, *341*
— s. Nobles, M. K. 324, 327, 328, 329, 330, *341*
— s. Vanterpool, T. C. 329, *343*
Madec, P., u. P. Perennec 803, *806*
Mäckel, H. G. 138, 150, *151*
Maekawa, T. *106*, 164, 174, 176, 192, 195, 197, *208*
Mäkinen, Y. L. A., u. D. Lewis 520, *528*
Magne, F. 761, 762, 763, 765, 766, 767, *775*
Magne, M. F. 641, 647, 648, *655*
Magne-Simon, M. F. 673, *681*
Magnin, A. 161, 170, 196, *208*
Mahabale, T. S., u. Joyce D'Mello 548, *550*
Maheshwari, P. 368, *402*, 431, 434, *442*
— u. B. M. Johri 431, 434, *444*
— u. V. Vasil 430, *444*
— s. Kana, K. 525, *527*
Mahoney, D. L., J. Janick u. E. C. Stevenson 88, *106*
Mailing, H. *106*
Mainx, F. 89, *106*, 111, *119*, 230, *234*, 286, *319*, 622, 624, *625*, 720, 733, 735, 736, 742, *745*
Majsuradze, N. I. 581, *591*
Majumder, P. K. s. Singh, R. N. 168, *212*
Makela, P. H. 25, *29*
— J. Lederberg u. E. M. Lederberg 25, *29*
Malecka, J. 573, *591*
Malhotra, R. C. 161, *208*
Mallik, P. C., R. K. Sahay u. D. L. Singh 167, 198, *208*
Mamporija, F. D. 581, *591*
Manaresi, A. 398, *402*
Manarte, M. s. Resende, F. 264, 269, *281*
Mangelsdorf, A. J., u. E. M. East 100, *106*
— s. East, E. M. 517, *526*
Mangelsdorf, P. C. 453, 454, *477*, 481, *504*
— u. D. F. Jones 461, 462, *477*
— u. R. G. Reeves 525, *528*
Mangenot, G. 426, *444*
Mangin, L. 369, 395, *402*, 669, 670, 671, *681*
Manguin, E. 670, *681*
Maniotis, J. s. Fields, W. G. 326, *339*
Mankowski, Z. s. Nickerson, W. J. 778, 780, *785*
Mann, L. K. 191, 194, *208*
Manoiloff, E. O. 148, *151*, 283, *319*
Manton, Irene 531, 532, 534, 537, 538, 539, 541, 545, 547, *550*
Manton, I. 245, 252, *255*, 352, 353, *366*, 411, *444*, 698, *705*, 710, 712, *745*
— u. B. Clarke 698, *705*
— u. H. Ettl 737, 739, *745*
— u. G. F. Leedale 646, 649, *655*
— K. Oates u. M. Parke 740, *745*
— u. M. Parke 649, *655*
— D. G. Rayns, H. Ettl u. M. Parke 737, 739, *745*
— u. W. A. Slegde 532, 538, 539, 541, *550*
— u. S. Walker 545, 546, 547, *550*
— s. Ettl, H. 737, 738, *744*
— s. Hoffman, L. 715, *744*

Manton, I. s. Parke, M. 637, 647, 649, 650, 652, *655*, 740, *745*
Manton, J., B. Clarke, A. D. Greenwood u. E. A. Flint 682, *692*
— s. Greenwood, A. D. 686, *691*
Marchal, Él. u. Ém. 759, *759*
— u. Em. Marchaz 146, *151*
Marchaz, Em. s. Marchal, El. 146, *151*
Mariat, M. F. 778, *785*
Marichal, J. s. Champagnat, M. 798, *805*
Markali, J. s. Gaardner, K. R. 649, *654*
— s. Halldal, P. 649, *654*
Marklund, G. 579, *591*
— u. A. Rousi 579, *591*
Marmur, J. s. Falkow, S. 25, *27*
Marquardt, H. 100, *106*
Marquardt, P., u. G. Vogg 381, *402*
Marschall, C. C. 140, *151*
Marsden-Jones, E. M., u. W. B. Turrill 579, *591*
Marsden, Ray, M. P. F. 801, 802, *806*, *807*
Martens, J. L. 163, *208*
Martens, P. 323, *341*
— u. R. Vandendries 334, *341*
Martin, E. M. 130, *132*
Mascarenhas, J. P., u. L. Machlis 394, *402*
Massart, J. 646, *655*, 773, 774, *776*
Masters, M. T. 155, *208*
Mather, K. 350, 351, *366*, *528*
Mathew, P. M. s. Abraham, A. 59, 61, *101*
Mathias, W. T. 238, *255*
Matney, T. S., E. P. Goldsmith, N. S. Erwin u. R. N. Scrogg 9, *29*
Matsudaira, C. s. Iwasaki, H. 763, *775*
Matsuura, H., u. T. Suto 573, *591*
Mattfeld, J. 551, *591*
Matzke, E. B. 167, 171, 172, 192, 195, *208*
Mauritzon, J. 570, 582, *591*
Maurizio, A. 124, *132*
Mauz 173
Mavromati, E. s. Grinvald, C. 202, *205*
Mazé, R. s. Afoldi, L. F. 15, *26*
— s. Anderson, Th. F. 315, *317*
McClintock, B. 350, *366*
— s. Beadle, G. W. 372, *399*
— s. Creighton, H. B. 351, *365*
— s. Rhoades, M. M. 350, *366*
McClung, C. E. *106*
McDonald, C. C. 572, *591*
McGuire, D. C. 396, *402*
— u. C. M. Rick 525, *528*
McLaughlin, I. I. A., u. P. A. Zahl 630, *636*
McMaster, R. D. s. Taylor, J.H. 347, 354, 356, 363, *367*
McMorris, Trevor s. Barksdale, A. W. 223
McNicol, M. 755, *759*
McPhee, H. C. 82, *106*, 175, 177, 194, 195, *208*
McVeigh 199
McVey, G. R. 159, *208*
McWilliam, J. R. 524, 525, *528*
— u. F. Mergen 429, *444*
McWhirter, K. S. s. Luig, N. H. 452, *477*
Mechanik, F. Ja. 187, 198, *208*
Mechelke, F. 370, *402*
Medina, D. M., u. C. H. T. Mendes Conagin 509, *528*
Meehan, Th. 158, 169, 191, 192, *208*
Meeuse, B. J. D. s. Leedale, G. F. 620, *625*
Mehra, K. L. 574, *591*
— u. R. P. Celarier *591*
— K. N. Subramanyam u. M. S. Swaminathan 573, 574, *591*
— s. Celarier, R. P. *585*
— s. DeWet, J. M. J. 574, *586*
— s. Singh, A. P. 574, *594*
Mehra, M. L. s. Harlan, J. R. 573, 574, *589*
Mehra, P. 532, *550*
— u. Chowdhury 541, *550*
Mehra, P. N. 430, *444*
— u. S. C. Verma 532, *550*
Meinhart, J. O. s. Lewin, R. A. 304, *318*
Melchior, G. H. 158, 159, 195, *208*
— u. H. H. Heitmüller 158, 195, *208*
— s. Heitmüller, H. H. 159, 195, *205*
Meletti, P. 161, 194, 202, *208*
Mello-Sampayo, T. s. Feldman, M. 261, *279*, 350, *365*
Mendes, A. J. T. 437, 441, *444*, *591*
Mendes Conagin, C. H. T. s. Medina, D. M. 509, *528*
Menke, W., u. B. Fricke 412, *444*
Menzel, M. Y. 344, 350, *366*
— u. J. M. Price 348, *366*
Menziani, E. s. Tappi, G. 379, *405*
Mercer, F. V. s. Wildon, D. C. 609, 611, 612, *619*
Mergen, F. s. Echols, R. M. 385, 387, *400*
— s. McWilliams, J. R. 429, *444*
Meselson, M., u. J. J. Weigle 364, *366*
Messeri, A. 577, *591*
Mettenius, G. 140, *151*
Metz, Ch. B. 282, 284, 304, *319*
Metzner, J. *745*
Meunier, A. 662, 663, 664, 665, 669, *681*
Meurman, O. 59, 60, 61, 63, *106*
Meyen, F. J. F. 373, *402*
Meyer, D. 545, 546, 547, *550*
Meyer, G. 348, *366*, 800, 801, *807*
Meyer, G. F. 261, *280*
Meyer, S. L. s. Betts, E. M. 326, *338*
Michaelis, P. 96, *106*
Michailova, N. F. 668, *681*
Michailowa, P. W. s. Poljakow, I. M. 503, *504*
Michelson, I. D. 777, *785*
Miduno, T. 578, *591*
Miedzyrzecki, Ch. s. Simonet, M. 570, *594*
Miège, J. 180, 190, 194, *208*
Migula, W. 755, 756, 758, *760*
Mihara, S. s. Hase, E. 734, *744*
Miki, H. 394, *402*
Miki-Hirosige, H. 394, *402*
Mikkelsen, V. M. 372, *402*
Mikkola, L. s. Hagman, M. 524, *527*
Millar, P. R. 336, *341*
Miller, E. C. 163, *208*
Miller, J. A. D., z. G. E. Fogg 687, *692*
— s. Belcher, J. H. 687, *691*
Miller, V. 685, 687, 688, *692*
Milthorpe, F. L. 781, *785*
Minina, E. G. 157, 163, 182, 185, 186, 192, 197, 198, *209*
— u. S. V. Kušnerenko 186, 195, 198, *209*
— u. L. G. Tylkina 186, 187, 198, *209*
Mishanec, W. M. s. Gerstel, D. U. 572, *588*
Mitchell, H. K. 350, *366*
— s. Wagner, R. P. 259, *281*
Mitchell, W. D., u. S. H. Wittwer 101, *106*, 187, 188, 199, *209*
Mitra, A. K. s. Pandey, D. C. 612, 613, 615, *619*
Mitra, A. R. s. Varma, A. K. 603, 607, 608, *619*
Mitra, G. C. s. Allsopp, A. 242, *254*
Mitra, S. 352, 362, 363, 364, *366*
Mitsuhashi, S. s. Harada, K. 16, *28*

Mitter, J. H. 324, *341*
Mittmann, G. 325, *341*
Miyake, A. 284, *319*
Miyake, K. s. Shibata, K. 582, *594*
Modilevsky, J. S. *442*
Moebius, M. 377, 378, 379, *402*, 787, 788, 793, 800, 804, 805, *807*
Moewus 33, 34, 36, 38
Moewus, F. 118, *119*, 141, 142, *151*, 229, 232, *234*, 259, *280*, 282, 301, 306, *319*, 351, *366*, 379, *402*, 516, *528*, 687, 688, 690, *692*, 742, *745*
— s. Kuhn, R. 215, *233*
Moffett, A. A. 154, *209*, 581, *591*
Mohamed, A. H., u. F. W. Gould 574, *591*
Mohl, H. v. 157, *209*, 384, *403*
Mohr, H. 245, *255*
Mohr, H. C. s. Hensz, R. A. 189, 199, 202, *205*
Mol, W. E. de 153, 154, *209*
Molisch, H. 384, 385, 395, *403*
Molliard, M. 155, 173, 174, 175, 179, 184, 194, 196, *209*
Monod s. Chatton, E. 621
Monroy, A. *4*
Montemartini, I. 570, *592*
Montgomery, H. B. S. 328, *341*
Monzani, A. s. Tappi, G. 379, *405*
Moor, H. 393
Moore, J. F. 202, *209*
Moore, R. H. 163, 202, *209*
Moreau, F. 690, *692*
Moreira, S., J. T. A. Gurgel u. L. F. de Arruda 581, *592*
Morel, G., u. R. H. Wetmore 545, *550*
— s. Wetmore, R. H. 250, 252, *256*
Morel, M. 252, *255*
Morel, M. G., u. R. H. Wetmore 252, *256*
Morgan 55
Morgando, A. 396, *403*
Morimura, Y. s. Hase, E. 734, *744*
Moritz, O. 513, 514, 519, *528*
Morse, M. L., E. M. Lederberg u. J. Lederberg *29*
Morton, J. K. *807*
Moses, M. J. 348, *366*
— u. J. H. Taylor 347, 354, 356, *366*
— s. Coleman, J. R. 348, *365*
— s. Sparrow, A. H. 354, 363, *367*
Mosig, G. 521, *528*
Mosko, M. s. Robbins, K. C. 382, *404*
Mostovoj, K. *106*
Mothes, K. s. Conrad, K. 201, *204*
Mottier, D. M. *151*, 545, *550*
Mounce, I. 329, 330, 331, 334, *341*
Mousseau, J. s. Lissouba, P. 364, *366*
— u. R. Macrae 328, 329, *341*
Moutschen 35, 145, 147
Moutschen, J., u. M. Dahmen 87, *106*
Moyer, D. s. Burkholder, P. R. 778, *785*
Mügge, E. 118, *119*
Mühldorf, A. *403*
Mühlethaler, K. 369, 373, 375, 376, 377, 378, *403*
— u. P. R. Bell 412, *444*
— u. H. F. Linskens 390, 391, *403*, 512, *528*
— s. Bell, P. R. 412, *443*
Müller, A. 753, 756, *760*
Müller, D. G. 238, *256*, 693, 695, 697, 698, 701, 703, *705*
Müller, F. 507, *528*
Müller, H. 160, 192, *209*, 507, 509, *529*
Müller, K. 162, 195, *209*
Müller, O. 660, 664, 668, 671, *681*
Müller, P. s. Bünning, E. 598, *602*
Müller, R. 383, *403*
Müller-Stoll, W. 388, *403*
— s. Hartman-Dick, U. 393, *401*
Müller-Stoll, W. R. 789, *807*
— u. G. Lerch 390, *403*
Müntzing, A. 59, 78, *106*, 556, 575, 576, 580, *592*
— u. G. Müntzing 580, *592*
— u. A. Nygren 575, *592*
— s. Levan, A. 264, 276, *280*
Müntzing, G. s. Müntzing, A. 580, *592*
Muir, R. M 383, *403*
Mukerji, S. K. 180, *209*
Mukherjee, S. K. *592*
Muller, H. J. 72, *106*
Mulligan, G. A. 573, *592*
Mullins, J. T., u. J. R. Raper 223, *234*
Mundie, S. R. *692*
Murata, Y. s. Nakaya, R. 15, *29*
Murbeck, Sv. 579, *592*
Murgai, P. 569, *592*
— s. Sachar, R. C. 569, *594*
Murneek, A. E. 165, 192, 195, *209*
Murphy, M. M. s. Loomis, N. H. *106*
Murphy, P. A. 123, *132*
— s. Pethybridge, G. H. 123, *133*
Murray, M. J. 54, 80, 94, *106*
Murray u. Blackman 649
Murrill, W. A. 430, *444*
Musser, H. B. s. Duich, J. M. 577, *587*
Myers, W. M. 574, *592*
— s. Hansson, A. A. *589*

Nadson, G., u. A. Konokotine 127, *132*
Nadson, G. A., u. G. S. Philippov 127, *132*
Näf, U. 143, 144, *151*
Näg 608
Nagai, I. 142, 143, *151*
Nagai, N. 124, *132*
Nagata 19
Nagoa, M. s. Esashi, Y. 802, *805*
Naithani, S. P. 153, 154, *209*
Nakajima, G. 59, 60, 65, 66, *106*
Nakamura, A. s. Nakaya, R. 15, *29*
— s. Rownd, R. 16, *29*
Nakamura, G. R., u. E. L. Becker 381, *403*
Nakaya, R., A. Nakamura u. Y. Murata 15, *29*
— s. Rownd, R. 16, *29*
Nakazawa, S. 764, *776*
Namikawa, I., M. Sisa u. K. Asai 182, *209*
Nannfeldt, J. A. 558, 574, 575, 576, *592*
Narang, N. 430, *444*
Narayan, K. N. 575, *592*
Narbel-Hofstetter, M. 601, *602*
Nasatir, M., A. M. Bryan u. S. D. Rodenberg 380, *403*
Natarajan, S. T. s. Brewbaker, J. L. 521, *525*
Naudin, Ch. 262, *280*
Naugol'nych, V. N. 187, *209*
Naundorf, G. 516, 517, *529*
Nawaschin, M. S., J. N. Gerassimowa-Nawaschina u. M. S. Jakowlew *446*
Naylor, A. W. 163, 202, *209*
Naylor, E. *807*
Naylor, F. L. 191, 194, *209*
Naylor, M. 695, 700, *705*
Nebel, B. R. 398, *403*
— u. E. M. Coulon 348, *366*
— u. M. L. Ruttle 397, *403*
Neger, F. W. *151*
Negi, V., u. M. Lata 430, *444*
Negodi, G. 91, *106*
Negrul, A. M. 90, *106*
Neidle, E. K. 191, 193, 194, *209*
Nekrasov, V. I., L. S. Vartazarova u. N. A. Borodina 181, *209*
Nelson, C. H. 176, 194, *209*
Nelson, O. E. s. Jiménez, J. R. 465, *477*
Nelson jr., O. E. 464, *477*
Nelson, P. M., u. E. C. Rossman 163, 199, *209*

Nelson, T. C. 314, 316, *319*
Němec, B. 153, *209*
Nettancourt, D. de s. Grant, W. F. 525, *527*
Neuer, H., u. R. v. Sengbusch 83, *106*, 177, 195, *209*
— s. Huhnke, W. 83, *104*, 175, 177, 193, 194, 195, 197, *206*
Neumann, K. 370, *403*
Neve, R. A. 58, *106*
Neves 561
Newton, D. E. 130, *132*, 330, 331, *341*
Nichols, H. W. 761, 762, *776*
Nickerson, N. H. 163, 199, *210*
— u. W. M. Hewitson 163, 198, *210*
Nickerson, W. J. 780, *785*
— u. G. A. Edwards 778, *785*
— u. Z. Mankowski 778, 780, *785*
— W. A. Taber u. G. Falcone 779, *785*
— u. N. J. W. van Rij 778, *786*
— s. Jillson, O. F. 779, *785*
Nicolaisen, N., u. R. Hanow *106*, 173, *210*
Nielsen, E. L. 576, *592*
— u. D. C. Smith 576, *592*
Nielsen, N. 383, 396, *403*
— J. Grömmer u. R. Lundén 381, 382, 383, *403*
— u. B. Holmström 381, *403*
Nienburg, W. 696, 697, *705*
Nigtevecht, G. van 57, 67, 68, 76, 77, 78, 88, *106*
Niklas, L. s. Wettstein, W. 372, *405*
Nilsson, H. 430, *444*
Nilsson-Ehle, H. 78, *106*
Ninan, C. A. 549, *550*
— s. Abraham, A. 549, *549*
Nipkow, F. 661, *681*
— s. Huber, G. 627, 632, 633, 634, *636*
Nishimura, K. s. Hirota, Y. *28*
Nishimura, Y., u. Y. Hirota *29*
Nisikado, Y., u. K. Yamauti 331, *341*
Nisioka, T., u. N. Takahashi 572, *592*
Nitsch, J. P. 275, *280*
— E. B. Kurtz, J. L. Liverman u. F. W. Went 186, 189, 194, 198, *210*
Noack, K. 561, 577, *592*
Noack, R. 167, 168, 194, 197, 200, *210*
Noble, M. 328, *341*
Nobles, M. K. 327, 329, 330, *341*
— R. Macrae u. B. P. Tomlin 324, 327, 328, 329, 330, *341*
Nobs, M. s. Clausen, J. 567, 575, 576, *585*
Nobs, M. A. s. Clausen, J. 567, 575, 576, 577, *585*
— s. Clausen, J. C. 567, 576, *586*
Noda, S. 351, *366*
Noguchi, Y. 88, *106*
Noll, F. 171, 174, *210*, 260, *280*
Nolte, M. 789, *807*
Nordhagen 543
Nordli, E. s. Braarud, T. 740, *743*
— s. Hasle, G. R. 629, *636*
Nordlin, E. 628, 631, *636*
Nordstedt, O., u. L. J. Wahlstedt 755, *760*
— s. Braun, A. 756
Norén, C. O. 422, *444*
Norstog, K. 575, *592*
Nürnberg-Krüger, U. 100, *106*, 196
Nunes, M. A. *213*, 274, 275, *280*
Nutting, W. H. s. Machlis, L. 218, *234*
Nygren, A. 55, 76, *106*, 554, 555, 556, 557, 558, 562, 565, 571, 574, 575, 576, 577, *592*, *593*
— u. E. Åkerberg 576, *593*
— u. G. Almgård 563, 575, 576, *593*
— s. Åkerberg, E. 576, *583*
— s. Clausen, J. 567, *585*
— s. Gustafsson, Å. 555, 572, *588*
— s. Müntzing, A. 575, *592*
Nypels 179, *210*

Oates, K. s. Manton, I. 740, *745*
Oberholzer, P. C. J. s. Hofmeyr, J. D. J. 581, *590*
Oehler, E. 573, *593*
Oehlkers, F. 96, 97, 98, *107*, 183, 194, 197, *210*, 244, 246, *256*, 369, 372, *403*, 452, 453, 473, 474, *477*, 627, *636*
— u. U. Ebell 98, *107*
— u. Paul Eberle 352, *366*
Oelke, J. 509, *529*
Oikawa, K. 331, 332, 334, *341*
Oka, H. T. 572, *593*
Okabe, S. 552, 553, 572, 582, *593*
Okada, M. s. Watanabe, T. 11, *30*
Okamoto, M. s. Sears, E. R. 350, *367*
Okamura, K. 771, *776*
Okazawa, Y. 803, *807*
— u. H. W. Chapman 803, *807*
O'Kelley, J. 393, *403*
— u. P. H. Carr 390, *403*
O'Kelley, J. C. 385, *403*
Okonuki, K. 381, *403*
Olbrich, St. 799, *807*
Oldén, E. J. 580, *593*
Olive, L. S. 364, *366*, 779, *786*
— u. A. A. Fantani 326, *341*
— s. Bistis, G. 225
Olmo, H. P. 396, 398, *403*
Olson, O. E. 325, *341*
Oltmanns, F. 307, 308, *319*, *602*, 690, *692*, 713, 717, 719, 735, 741, 742, *745*, 746, 747, 748, 749, *752*, 771, 773, 774, *776*, 799
Oltmanns, Fr. 113, 118, *119*
Ono, T. 58, 65, 66, 70, 71, 74, *107*
— u. N. Shimotomai 70, *107*
— s. Kihara, H. 58, 65, 69, *104*
Ordal, Z. s. Levine, S. 777, 778, *785*
Orr, M. Y. 570, *593*
Ørskov, F. s. Hirota, Y. *28*
— s. Ørskov, I. *29*, 316, *319*
Ørskov, I., u. F. Ørskov *29*, 316, *319*
— s. Hirota, Y. *28*
Orth, H. 138, 144, *151*
Osawa, J. 553, 558, 559, 581, *593*
Ossenbeck 789
Ostenfeld, C. H. 556, 572, *593*
Osterhout, W. J. V. 769, *776*
Otsuji, N. s. Hakura, A. 11, *28*
Overton, J. B. 703, *705*
Ozeki, H. *29*
— u. S. Howarth 15, *29*

Pace, L. 560, 569, *593*
Page, O. T. 781, 782, 783, *786*
Page, W. M. 326, *341*
Paixão-Correia, C. A. da 182, *210*
Palik, P. 603, *619*
Paliwal, R. L., u. B. B. Hyde 95, *107*
Palmer, E. J. 580, *593*
Palmer, L. S. s. Heydak, M. H. 383, *401*
— s. Vivino, A. E. 378, 383, 384, *405*
Palmiter, D. H. s. Keitt, G. W. 325, *340*
Palumbo, R. F. 381, *403*
Pandey, D. C., u. A. K. Mitra 612, 613, 615, *619*
Pandey, K. K. 516, 517, 520, 521, 522, *529*
Pankow, H. 373, *403*
Papasova, G. s. Christoff, M. 580, *585*
Papazian, H. P. 48, *50*
Pape, R. 782, 783, *786*
Papenfuss, G. F. 693, 694, 698, 700, *705*, 774, *776*
Paponov, A. N. s. Edel'štejn, V. I. 163, 192, *204*
Pappas, I. s. Braarud, T. 629, 631, *635*

Park, J. B. s. East, E. M. *526*
Parke, M. 637, 644, 646, 647, 648, 650, 651, 652, 653, *655*
— u. I. Adams 637, 646, 648, 650, 651, *655*, 740, *745*
— u. I. den Hartog-Adams 740 *745*
— J. W. G. Lund u. I. Manton 637, 649, *655*
— u. I. Manton 637, 649, *655*, 740, *745*
— — u. B. Clarke 637, 647, 649, 650, 652, *655*
— u. D. G. Rayns 737, *745*
— s. Manton, I. 649, *655*, 737, 739, 740, *745*
Parnell, F. R. 460, *478*
Parsons, P. A. *107*
Partanen, C. R. 266, *280*
— s. Steeves, T. A. 252, *256*
Pascher, A. 230, *234*, 238, *256*, 285, 307, 310, *319*, 607, *619*, 623, *625*, 637, 639, 641, 642, 646, *656*, 682, 683, 684, 685, 686, 690, *692*, 706, 707, 708, 709, 711, 712, 713, 715, 716, 718, 720, 722, 723, 724, 725, 726, 727, 729, 730, 731, 732, 733, 736, 739, 740, 741, *745*
— u. W. Vlk *656*
Pascher, E. 634, 635, *636*
Pashuk, C. T. s. Lazarenko, A. S. 250, *255*
Paspaleff, G. s. Popoff. M, 622, *625*
Passecker, F. 395, *403*
Pasteur, M. L. 779, *786*
Pate, J. B., u. E. N. Duncan 202, *210*
Patel 60
Patel, G. I., u. R. M. Datta 524, *529*
Paton, J. B. 381, 384, *403*
Patrick, R. 672, *681*
Patterson, Mary Thomasine 532, *550*
Paul, W. R. C. 782, 783, *786*
Pavillard, J. 665, 669, *681*
Peacock, A. S., u. U. Weidmann *602*
Pease, D. 326, *341*
Peeters, E. 796, *807*
Peirce, L. C., u. T. M. Currence 90, *107*
Penard, E. 621, 623, *625*
Percival, M. S. 372, *403*
Perennec, P. s. Madec, P. 803, *806*
Perkins, D. D. 351, *366*
Perry, B. A. 573, *593*
Perry, T. O. s. Schmitt, D. 516, *529*
Peter, J. 570, *593*
Petersen, O. G. 169, 197, *210*
Peterson, C. E., u. L. D. Anhder 188, 199, *210*
— u. LaMoine d. Anhder 101, *107*
Peterson, F. s. Kiesselbach, T. A. 461, *477*
Pethybridge, G. H., u. P. A. Murphy 123, *133*
Petit, J. 82, *107*, 175, 177, 194, *210*
Peto, F. H. 59, 79, *107*
Petrov, D. F., u. D. A. Tukan 55, *107*
Petrová, J. 712, 713, 722, 723, 724, 735, *745*
Petrovskaya, T. P. s. Zinger, N. V. 573, *596*
Petrovskaya-Baranova, T. P., u. N. V. Zinger 573, *593*
Petrů, E., u. R. Řetovský 163, *210*
— u. Z. Šesták 163, *210*
Petruschewsky, G. s. Poljanski, G. 609, *619*
Petterson, J. s. v. Euler, H. 378, 379, 382, 383, 384, *400*
Pfeffer, W. 598, *602*
Pfeiffer 412
Pfeiffer, N. E. 379, 396, 397, *403*
Pfitzer, E. 666, *681*
Pfitzer, P. 569, *593*
Pfundt, M. 385, 395, *403*
Phatak, S. C., S. H. Wittwer, S. Honma u. M. J. Bukovac *213*
Philippov, G. S. s. Nadson, G. A. 127, *132*
Phillips, L. L., u. D. U. Gerstel 349, *366*
— s. Gerstel, D. U. 349, *365*
Phillips, R. W. 774, *776*
Philpott, D. E. s. Lewin, J. C. 668, 669, 670, 671, *680*
Piccinin, G. L. s. Dettori, R. 11, *27*
Pierson, R. H. 336, *341*
Pijl, L. van der 578, *593*
Pillay, T. P. 130, *133*
Pinto-Lopes, J. 331, *341*
— s. Quintanilha, A. 324, *342*
Pirson, A. 734, *745*
— u. H. Lorenzen 734, *745*
Pittard, J., u. E. A. Adelberg *29*
— J. S. Loutit u. E. A. Adelberg *29*
— s. Adelberg, E. A. 6, *26*
Plaut, W. S. 354, *366*
Plempel, M. 215, 217, 218, *234*, 295, 296, 297, *319*, 323, *341*
— u. G. Braunitzer 218, *234*, 296, *319*, 323, *341*
— u. W. Dawid 218, *234*, 297, *319*
Pluijm, J. E. van der, u. H. F. Linskens 520, *529*
Plumier, W. 571, *593*
Pochet, P. s. Devreux, M. 509, *526*
Pocock, M. 709, 712, *745*
Poddubnaja-Arnoldi, V. A. *443*, *445*, *446*
— s. Zinger, N. V. 573, *596*
Poddubnaja-Arnoldi, V., u. V. Dianowa 573, *593*
Poddubnaja-Arnoldi, W. A. 554, 571, 573, *593*
Pöllnitz, K. v. 266, *280*
Pohjakallio, O. 802, *807*
Pohl, F. 373, *403*
Pohl, R. 387, 388, *403*, 622, *625*
Poir 563
Poisson, R., u. A. Hollande 620, 623, *625*
— s. Grassé, P. P. 620, *625*
Poljakow, I. M. 442, *446*
— u. A. M. Dmitrieva 442, *446*
— u. P. W. Michailowa 503, *504*
Poljanski, G., u. G. Petruschewsky 609, *619*
Poll, H. s. Dohrn, M. 198, *204*
Pollister s. Woods 347
Polunina, N. N. s. Zinger, N. V. 573, *596*
Pond, V. s. Sparrow, A. H. 360, *367*
Ponnaiya, B. W. X. s. Jagannath, D. R. 574, *590*
Pontecorvo, G. 264, *280*, 364, *366*
Pontecorvo, K. 260, 261, 264, *280*
— u. E. Käfer 264, *280*
Popoff, A. 580, *593*
— s. Christoff, M. 572, *585*
Popoff, M., u. G. Paspaleff 622, *625*
Popp, A., u. W. F. Hanna 337, *341*
Popp, W. s. Hanna, W. F. 337, *340*
Porsch, O. 169, 195, *210*
Porsild, M. P. 571, *593*
Poulton, E. M. 686, *692*
Powers, L. 572, *593*
— u. E. J. Gardner 572, *593*
— u. R. C. Rollins 572, *593*
Prantl, K. 142, 143, 149, *151*
Prat, H. 158, 197, *210*
Pratt, C., u. J. Einset 566, 581, *593*
— — u. T. Clausen 566, 581, *593*
Prauser, H. 649, *656*
Preda, V., u. O. Crăciun 163, *210*
— u. A. Ghisa 179, 198, *210*
Prell, H. H. *29*

Prensky, W. s. Wimber, D. E. 349, 354, *367*
Present *445*
Preuss-Herzog, G. s. Renner, O. 500, *504*
Price, J. M. s. Menzel, M. Y. 348, *366*
Pringsheim, E. G. *602*, 620, 621, 622, 624, *625*, 640, 646, 648, *656*
— s. Leedale, G. F. 620, *625*
Pringsheim, N. 238, 248, *256*, 695, *705*, 753, 754, 755, *760*
Printz, H. 712, 713, 717, 719, 722, 723, 726, 728, 729, 731, 732, 733, 740, 741, 742, *745*, 746, 747, 749, 750, *752*
Pritchard, F. J. 176, 195, *210*
Pritchard, R. H., u. K. G. Lark 20, *29*
Proskauer, J. 740, *745*
Provasoli, L. s. Hutner, S. H. 303, 307, *318*
Prowazek, S. 644, *656*
Prud'homme, N. 261, *280*
Pruzsinsky, S. 397, 398, *403*
Przybytek, D. S. s. Famintzin, A. 384, *400*
Purvis, O. N. s. Gregory, F. G. 161, *205*
Puymaly, A. de 604, 608, 616, 617, *619*, 686, 688, 690, *692*

Quinones, J. A. s. Shifriss, O. 101, *108*
Quintanilha, A. 46, 47, *50*, 263, 264, *280*, 324, 331, 332, 333, 334, *341*
— u. J. Pinto-Lopes 324, *342*
— L. Quintanilha u. A. Vasermanis 330, 331, 332, 333, 334, *342*
Quintanilha, L. s. Quintanilha, A. 330, 331, 332, 333, 334, *342*

Rack, K. 796, *807*
Raeber, J. G., u. A. Bolton 95, *107*
Raghavan, T. S. *593*
Raghavan, V., u. K. H. Baruah 387, *403*
Raghuvanshi, S. S. 581, *593*
Rainio, A. J. 79, *107*
Ram, M. 154, *210*
Raman, W. S. s. Jagannath, D. R. 574, *590*
Ramanathan, K. R. s. Iyengar, M. O. P. 712, *744*
Ramsey, G. B. 781, *786*
Rancken, G. 576, *593*
Randhawa, G. S. s. Iyer, C. P. A. *213*
Randhawa, M. S. 731, *745*
Rangaswami, G. s. Satyanarayana, G. 190, 198, *211*
Ranga Swamy, N. S. s. Kana, K. 525, *527*
Rao, V. S. 155, *210*, 421
Raper, J. 282, *319*
Raper, J. R. 33, 38, 47, 48, *50*, 122, 123, 124, 125, 128, 131, *133*, 215, 219, 220, 221, 223, 224, 226, 230, 231, *234*, 292, 293, 301, *319*, 324, 334, *342*
— u. A. J. Haagen-Smit 292, *319*
— s. Bistis, G. 224, *233*
— s. Esser, K. 324, *339*
— s. Mullins, J. T. 223, *234*
Raper, K. B., u. D. J. Fenell 325, *342*
Rapoport, H. s. Machlis, L. 218, *234*
Rappaport, J., S. Satina u. A. F. Blakeslee 524, *529*
Rappoport 291
Rauh, W. 797, 798, 799, 802, *807*
Raunkiaer, C. 791, 792, *807*
Rawitscher-Kunkel, E., u. L. Machlis 216, 226, 232, *234*, 287, *319*
— s. Machlis, L. 394, *402*
Rayns, D. G. 647, *656*
— s. Manton, I. 737, 739, *745*
— s. Parke, M. 737, *745*
Redemann, C. T. 383, *403*
Reess, M. 779, 780, *785*
Reeves 14
Reeves, R. G. s. Mangelsdorf, P. C. 525, *528*
Rehm, S. 183, 189, 199, *210*
Reich, K. 712, 718, 719, 734, *745*
Reichel, L., u. M. Wallis 296, *319*
Reichle u. Snider 326
Reimann, B. E. F., J. C. Lewin u. B. E. Volcani 663, *681*
Reimann-Philipp, R. 59, 64, *107*, 456, *478*
— B. Zilm u. H. Eresen 59, 90, *107*
Reiner 89, *107*
Reinhardt, J. F. s. Griggs, W. H. 397, *401*
Reinisch, O. 623, *625*
Reinke 694
Reinöhl, F. 172, 193, *210*
Reisigl, H. 716, 727, 735, *745*
Renner, O. 118, *120*, 277, *280*, 282, *319*, 385, 393, *403*, 448, 451, 452, 458, 465, 472, *478*, 480, 481, 482, 500, *504*, 551, *593*, 687, *692*
— u. R. E. Cleland 458, 472, *478*
— u. G. Preuss-Herzog 500, *504*
Renner, O. s. Demeter, O. 613
Resende, F. 157, 166, 183, 193, 194, 199, 200, *210*, 264, 266, 269, 270, 271, 272, 273, 274, 275, 276, 277, *280*, 801, *807*
— u. H. F. Linskens 96, *107*, 272, 275, 276, *281*
— u. M. Manarte 264, 269, *281*
— u. M. J. Viana 101, *107*, 166, 183, 199, *210*, 269, 271, 274, *281*, 804, *807*
— u. J. W. Warden 266, 269, *281*
— s. Linskens, H. F. 271, *280*
Řetovský, R. s. Petrů, E. 163, *210*
Reusch, J. D. H. 575, *593*
Reverdin, L. 637, 638, *656*
Reznik, H. 379, *404*, 516, *529*
Rhoades, M. M. 96, *107*, 359, *366*, 448, 449, 462, *478*
— u. B. McClintock 350, *366*
— u. E. Dempsey 462, *478*
— u. V. H. Rhoades 453, *478*
Rhoades, V. H. s. Rhoades, M. M. 453, *478*
Richards, B. M. s. Bell, P. R. 250, *254*
Richardson, C. W. 55, 100, *107*
Richardson, W. L. s. DeWet, J. M. J. 573, 574, *586*
— s. Harlan, J. R. 567, 573, 574, *589*
Richey, F. D., u. G. F. Sprague 162, 192, 195, *210*
Richter, A. 8, 9, *29*
Richter, C. M. 493, 494, *504*
Richter, F. I. 384, *404*
Richter, J. 753, 755, *760*
Richter, S. 373, *404*
Richter-Landmann, W. 435, 439, 440, *444*
— s. Steffen, K. *445*
Rick, C. M., u. G. C. Hanna 90, *107*, 160, *210*
— s. McGuire, D. C. 525, *528*
Ricketts, H. T. 777, *786*
Ridi, M. S., u. A. Wafa 383, 384, *404*
Riede, W. 163, 181, 193, 195, *210*
Riehm, E. 796, *807*
Rieth 598, *602*
Rieth, A. 112, *120*, 672, *681*, 684, 686, 690, 691, *692*, 722, *745*
Rietsema, J., B. Blondel, S. Satina u. A. F. Blakeslee 524, *529*
— S. Satina u. A. F. Blakeslee 524, *529*
Rij, N. J. W. van s. Nickerson, W. J. 778, *786*
Riley, H. P. 509, *529*

Riley, R. 350, *366*
— u. V. Chapman 350, *366*
— — u. G. Kimber 350, *366*
Riner, E. Mary s. Gerstel, D. U. 572, *588*
Ringel-Suessenguth, M. 801, *807*
Ris, H. 352, *366*
Rittinghaus, P. 395, *404*, *529*
Rizet, G. 326, *342*
— s. Lissouba, P. 364, *366*
Robak, H. 327, 329, 330, *342*
Robbins, K. C., A. Samuel u. M. Mosko 382, *404*
Robbins, W. J., u. F. Kavanagh 784, *786*
Robbins, W. W., u. H. A. Jones 90, *107*
Roberts, C. s. Winge, Ö. 126, *133*
Roberts, R. H. *213*
— u. B. E. Struckmeyer *210*
Robinson, W. 695, 700, *705*
Robyn, G. s. Vandendries, R. 331, *343*
Rodenberg, S. D. s. Nasatir, M. 380, *403*
Rodenhiser, H. A. 336, 337, *342*
Röbbelen, G. 524, *529*
Roelofsen, P. A. s. Houwink, A. L. 390, *401*
Roemer, Th. 395, *404*
Roeser u. Konetzka 23
Roger, N. s. Zinder, N. D. 11, *30*
Rogers, C. H. 781, 782, *786*
Rogers, H. H. s. Knight, R. 516, *527*
Rogers, L. M. 409, 412, *445*
Roggen, H. P. J. R. 524, *529*
Rolik, R. P. s. Grebinskij, S. O. 399, *401*
Rollins, R. C. 572, *593*
— s. Powers, L. 572, *593*
Ronsdorf, L. 217, *234*
Rosa, J. T. 69, *107*, 172, *211*
Rosen, W. G. 394, *404*
Rosenberg, M. 690, *692*
Rosenberg, O. 531, 533, *550*, 553, 555, 556, 571, 572, *594*
Rosendahl, Gertraud 537, 538, *550*
Rosenvinge, L. K. 761, 770, 772, 774, *776*
Rossen u. Westergaard 364
Rossignol, J. L. s. Lissouba, P. 364, *366*
Rossilo, A. R. s. Chodat, R. 649, *654*
Rossman, E. C. s. Nelson, P. M. 163, 199, *209*
Rottmann, W. 545, *550*
Rousi, A. 579, *594*
— s. Marklund, G. 579, *591*
Routien, J. B. 330, 331, 333, 334, *342*
Roux, G. s. Linossier, G. 778, *785*
Rowell, J. B. 337, *342*
Rowley, D. s. Clowes, R. C. 7, *27*
Rowley, J. R. 370, *404*
Rownd, R., R. Nakaya u. A. Nakamura 16, *29*
Roy, R. P., A. P. Singh u. P. K. Gupta 575, *594*
— u. B. M. B. Sinha 532, 541, *550*
Royle, D. J., u. C. J. Hickman 291, *319*
Rozanova, M. A. 579, *594*
Roze, E. 196, *211*
Rubin, B. s. Kiesel, A. 380, 382, 383, *401*
Rudenko, Ch. Ju. 177, 195, *211*
Rudenko, F. E. 174, *211*
Rudin, A. D. s. Eddy, A. A. 313, *317*
Rudloff, F. C. 449, *478*
Rumberg, I. *151*
Runquist, E. W. 78, *107*
Ruppel, H. G. s. Lorenzen, H. *745*
Rutishauser, A. 555, 560, 561, 579, 580, *594*
— u. H. R. Hunziker 555, *594*
Ruttle, M. L. s. Nebel, B. R. 397, *403*
Ruttner, F. s. Geitler, L. 609, 617, *619*
Ryan 33
Ryan, F. J. 229, *234*
Rychlewski, J. 575, *594*
Ryther, J. R. s. Guillard, R. R. L. 666, *680*

Sabinin, D. A. 163, 192, *211*
Saccardo, P. A. 174, *211*
Sachar, R. C. 154, 155, *211*
— u. R. N. Chopra 569, *594*
— u. P. Murgai 569, *594*
Sachs, J. 214, *234*
Sadebeck, R. 531, *550*
— s. Hieronymus, G. 549, *550*
Saez, F. *107*
Sager, R. 322, *342*
Sági, F. s. Garay, A. 171, 201, *205*
Sagromsky, H. 383, *404*
Sahay, R. K. s. Mallik, P. C. 167, 198, *208*
Saikowskaja *445*
Saito, T. s. Ito, H. 188, *206*, *207*
Saito, Y. 158, 159, 195, 198, *211*
Sakai, K. *594*
Salter, S. J. 155, *211*
Salton, M. R. J. s. Chapman, J. A. 611, *618*
Salvin, S. B. 778, *786*
Samadder, K. R. s. Banerjee, S. 329, *338*
Sampson, D. R. 511, 525, *529*
Samuel, A. s. Robbins, K. C. 382, *404*
Sandsten, E. P. 395, *404*
Sandt, W. 155, *211*
Sanford, G. B. 781, *786*
Sansome, E. 223, *234*
Sansome, F. W. 78, *107*, 195, *211*
Santos, J. K. 58, 59, 61, *107*
Sanwal, M. 430, *445*
Sarbadhikari, P. C. 544, *550*
Sarkar, B. C. R., S. H. Wittwer, R. W. Luecke u. H. M. Sell 381, *404*
Sarkar, N. s. Löve, Á. 58, 63, 80, 81, *106*
Sartoris, G. B. 398, *404*
Sarvella, P. 349, *366*
Sass, J. E. 331, 332, 333, *342*
Sassen, M. M. A. 374, 380, 389, 390, 393, *404*
Satina, S., u. A. F. Blakeslee 148, *151*, 296, *319*
— s. Rappaport, J. 524, *529*
— s. Rietsema, J. 524, *529*
Satyanarayana, G., u. G. Rangaswami 190, 198, *211*
Sauerland, Heinz 352, *366*
Sauvageau, C. 237, 238, 248, 250, 253, *256*, 309, *319*, 601, *602*, 696, 700, 702, 704, *705*
Savelli, R. 189, *211*, 387, *404*
— u. C. Caruso 387, *404*
— u. N. Soster 176, 195, *211*
Savine 177
Savitsky, V. F. 94, *107*
Šavrov, L. A. 161, *211*
Sawa, T. s. Tindall, D. R. 758, *760*
Sawyer, W. H. 778, *786*
Sax 507
Sax, H. J. 580, *594*
— u. K. Sax 347, *366*
Sax, K. 350, *366*, 580, 581, *594*
— u. H. W. Edmonds 347, *366*
— u. L. M. Humphrey 353, *367*
— u. E. D. King 363, *367*
— s. Sax, H. J. 347, *366*
Scaife, J., u. D. Gross *29*
Scaramella Petri, P., u. P. Strigoli *213*
Schade, C. 399, *404*
Schäfer, R. 628, 647
— s. v. Stosch, H. A. 647, *656*
Schaeffer, P. s. Jacob, F. 13, *28*
— s. Lwoff, A. 322, *341*
Schaffner, J. H. *107*, 160, 162, 164, 173, 174, 175, 176, 178, 192, 193, 194, 195, 197, *211*
Schartau, O. s. Hartmann, M. 283, *318*

Schaudinn, F. *319*
Schellenberg, G. 135, 142, 144, *151*
Scherffel, A. 643, 646, *656*, 690, *692*
Scheu, H. s. Breider, H. 90, *102*
Schick, R. *807*
Schiemann, E. 55, *107*
Schiff, J. s. Lazaroff, N. 605, *619*
Schifferdecker, Ilse 536, 537, 539, *550*
Schildknecht, H., u. H. Benoni 394, *404*, 438, 445, 502, *504*
Schiller, J. 237, *256*, 628, 629, 631, 632, 634, 635, *636*, 638, 639, 640, 645, 649, 650, *656*
Schilling, A. J. 633, *636*
Schleip, W. 448, 460, *478*
Schlenker, G. 180, *211*
Schlimbach, H. 569, *594*
Schlösser, K. 512, 513, *529*
Schlösser, L. A. 237, 238, 250, *256*, 704, *705*
Schlumberger, O. 408, *445*
Schmelzeisen, W. *151*
Schmid, G. 638, *656*
Schmidt, P. 694, 696, *705*
Schmitt, D., u. T. O. Perry 516, *529*
Schmitz, F. 761, *776*
Schmucker, Th. 385, 387, *404*
Schnarf, K. 368, 370, *404*, *443*, 501, *504*, 572, 573, *594*
Schneider, G. 394, *404*, 484, 497, 502, *504*
Schnepf, E. 611, *619*
Schnös, M. s. Caro, L. G. 316, *317*
Schoch, M. 570, *594*
Schoch-Bodmer, H. 372, 373, 378, 380, 384, 389, 390, *404*, 516, *529*
— u. P. Huber 393, *404*
Scholz, E. 169, *211*
Schopfer, W. H. 171, 195, 196, *211*
— u. S. Blumer 171, 195, 196, *211*
Schramm, J. R. 686, *692*
Schramm, R. 162
Schratz, E. 138, 143, 146, 148, 149, *151*
Schraudolf, H. 144, *151*
Schrauwen, J. s. Linskens, H. F. 380
Schrauwen, J. A. M. s. Linskens, H. F. 520, 523, *528*
Schreiber, E. 308, 309, *319* 672, *681*, 695, 697, 703, *705*, *746*
Schreven, D. A. van 803, *807*
Schröck, O. 799, *807*
Schüler, J. 622, *626*
Schürhoff, P. N. 570, *594*
— u. F. J. Jüssen 569, *594*
Schütt, F. 628, 629, 631, *636*, 660, 663, 666, *681*
Schuette, H. A. s. Kitzes, G. 383, *401*
Schütz, Georg 482, 483, 486, 488, 493 *504*
Schultze, O. 143, *152*
Schulz, A. 68, *108*, 163, 167, 168, 181, 182, 184, 197, *211*, *212*
Schulz, P. 665, *681*
Schulze, B. 111, *120*, 741, *746*
Schulze, K. L. 113, *120*
Schussnig, B. 236, 237, 238, 240, 241, 242, 248, 250, *256*, 597, *602*, 626, 648, 649, *656*, 691, 693, 695, 699, *705*, 706, 707, 708, 709, 712, 715, 716, 720, 721, 724, 728, 731, 741, 742, *746*, 750, 751, *752*
— u. E. Kothbauer 238, *256*, 703, *705*
Schuster, J. 159, *212*, 419, *445*
Schwabe, G. 606, *619*
Schwanitz, F. 372, *404*
Schwartz, D. 96, *108*, 462, 464, *478*
Schwarz, E. 646, 647, 648, 652, *656*
Schwarzenbach 563, 575
Schwarzenbach, F. H. 379, 384, 387, *404*, 801, *807*
Schwarzenbach, M. 249, *256*
Schweizer 147, 148
Schwemmle, J. 457, 467, *478*. 479, 480, 482, 483, 484, 486, 488, 490, 491, 492, 493, 494, 495, 496, 497, 498, 499, 500, 502, 503, *504*
— u. Mitarb. 479, *505*
— u. W. Koepchen 450, 467, *478*, 484, 491, 502, *505*
— u. H. Lohmeyer 485, *505*
— u. R. Simon 488, 489, 492, 493, 502, *505*
Schwerdtfeger, G. 154, *212*
Scott, B. D. s. DeWet, J. M. J. 573, *586*
Scott, G. W. 189, *212*
Scrogg, R. N. s. Matney, T. S. 9, *29*
Scully, N. J. s. Borthwick, H. A. 174, 176, 193, 194, *203*
Sears, E. R. 508, 509, 516, *529*
— u. M. Okamoto 350, *367*
— s. Feldman, M. 261, *279*, 350, *365*
Sebestýen, O. s. Entz, G. 627, 631, *636*
Sedgwick, P. G. 421, *445*
Seiler, J. 36, *50*, 117, *120*
Seiler, J. u.a. 36, *50*
Seitz, W. 79, *108*
Sell, H. M. s. Sarkar, B. C. R. 381, *404*
Sen, S. K. s. Datta, R. M. 524, *526*
Sengbusch, R. v. 82, 83, 84, 85, 90, *108*
— s. Bredemann, G. 83, 85, *102*
— s. Huhnke, W. 83, *104*, 175, 177, 193, 194, 195, 197, *206*
— s. Neuer, H. 83, *106*, 177, 195, *209*
Sernander, R. 551, *594*
Serra, J. A. 259, 260, *281*
Šesták, Z. s. Petrů, E. 163, *210*
Shama Kumari, K. 575, *594*
Shanor, L. s. Benjamin, R. K. 127, 131
Shapiro, N. s. Brewbaker, J. L. 531, *525*
Sharma, A. K., u. B. Bhattacharyya 804, *807*
Sharma, D. K. s. Singh, R. N. 168, *212*
Shattuck, C. H. 145, *152*
Shear, C. L., u. R. W. Davidson 327, *342*
— u. B. O. Dodge 326, 336, *342*
Sherwood, W. A. 224, *234*
Shibata, K., u. K. Miyake 582, *594*
Shibukawa, J. s. Ushirozawa, K. 398, *405*
Shida, S. s. Katayama, Y. 69, *104*
Shifriss, O. 101, *108*, 180, 194 197, 199, *212*
— u. E. Galun 186, *212*
— W. L. George jr. u. J. A. Quinones 101, *108*
Shimakura, K. 359, *367*
Shimotomai, N. 580, *594*
— s. Ono T. 70, *107*
Showalter A. M. 141, *152*
Shull, G. H. 56, 66, 67, 72, 76, 77, *108*, 467, *478*
Shumway, C. P. s. Tyler, L. J. 337, *342*
Siebold, C. Th. E. v. 551, *594*
Siegel, R. W. s. Cohen, L. W. 284, *317*
Signoret, J. 266, *281*
Silvers, S. D. *29*
Simon, R. s. Schwemmle, J. 488, 489, 492, 493, 502, *505*
Simonet, M., u. Ch. Miedzyrzecki 570, *594*
Siminovitch, A. s. Jacob, F. *28*
Siminovitch, L. s. Jacob, F. 15, *28*
Simmonds, N. W. s. Dodds, K. S. 578, *586*
Singh, A. P. 574, *594*
— u. K. L. Mehra 574, *594*
— s. Borgaonkar, D. S. 567, 574, *584*
— s. DeWet, J. M. J. 573, *586*
— s. Roy, R. P. 575, *594*

Singh, D. L. s. Mallik, P. C. 167, 198, *208*
Singh, R. N. 717, 719, 734, *746*
— P. K. Majumder u. D. K. Sharma 168, *212*
Singh, R. P. 155, *212*
Singh, S. P. 155, *212*
Sinha, B. M. B. s. Roy, R. P. 532, 541, *550*
Sinoto, Y. 58, 59, 60, 65, 66, *108*
— u. Kiyohara 59
Sirks, M. J. 447, 452, 453, 460, 466, *478*, 506. *529*
Sironval, C. 801, *807*
Sisa, M. s. Namikawa, I. 182, *209*
Sitte, P. 373, 375, 376, 377, *404*
Sjöstrand, F. S. s. Afzelius, B. M. 374, *399*
Skabitchewsky, A. P. 661, *681*
Skalinska, M. 562, 575, 576, *595*
— u. R. Czapik 580, *595*
Skarżyński, B. 198, *212*
Skawinska, R. 572, *595*
Skoczylas, O. 626, *636*
Skolko, A. J. 327, *342*
Skovstedt, A. 65, *108*, 582, *595*
Skuja, H. 304, *319*, 609, 610, 617, *619*, 622, 623, *626*, 638, 639, 640, 643, 644, 645, 646, *656*, 713, 718, 726, 728, 729, 730, 731, 732, 740, *746*, 761, 762, *776*
Slegde, W. A. s. Manton, I. 532, 538, 539, 541, *550*
Sleumer, H. O. 337, *342*
Sloover-Colinet, A. de 414, *445*
Smith 638
Smith, A. H., u. H. J. Brodie 334, *342*
Smith, B. W. 58, 59, 61, 63, 65, 71, 80, 81, *108*, 575, *595*
— u. M. Th. Smith 65, *108*
Smith, D. C. s. Nielsen, E. L. 576, *592*
Smith, E. C. 181, *212*
Smith, F. L. 452, *478*
Smith, G. L. 580, *595*
Smith, G. M. 237, 238, 239, 240, 243, *256*, *602*, 686, *692*, 712, 717, 729, 736, 741, *746*, 746, 747, 748, 750, *752*
Smith, I. W. s. Duguid, J. P. 10, *27*
Smith, J. 558, *595*
Smith, M. Th. s. Smith, B. W. 65, *108*
Smith, P. F. 387, *404*
Smith, R. W. 425, *445*
Smith, S. G. 351, *367*
Smith, S. M., u. B. A. D. Stocker 15, *29*
Sneath, P. H., u. J. Lederberg 316, *319*
Sneath, P. H. A. 9, *29*
Sneath, P. H. A. u. H. Lederberg 18, *29*
— s. Hirota, Y. 13, *28*
Sneep, J. 90, *108*
— s. Huyskes, J. A. 90, *104*
Snider s. Reichle 326
Snow 733
Snow, A. G. s. Duffield, J. W. 397, *400*
Snyder, L. A. 575, *595*
— u. J. R. Harlan 574, *595*
— A. R. Hernandez u. H. E. Warmke 575, *595*
Snyder, W. C. s. Hansen, H. N. 129, 130, *132*
Soeder, C. J. 734, 735, *746*
Söderberg, E. 572, 582, *595*
Sörensen s. Westergård 576
Sörensen, Th. 573, 579, *595*
— u. G. Gudjónsson 563, 564, 573, *595*
Sörgel, G. 237, 248, 253, *256*, 781, *786*
Söyrinki, N. 787, 800, *807*
Sokolowska-Kulczycka, A. 572, *595*
Solms-Laubach, H. 771, *776*
Sonneborn, T. M. 44, *50*
Sorieul, S. s. Barski, G. 260, *278*
Sosa, A., u. C. Sosa-Bourdouil 383, *404*
Sosa-Bourdouil, C. 381, *404*
— s. Sosa, A. 383, *404*
Sost, H. 245, 248, *256*
Soster, N. s. Savelli, R. 176, 195, *211*
Souèges, E. C. R. 432, *445*
Spada, A. s. Tappi, G. 379, *405*
Spallanzani 173
Sparrow, A. H. 353, *367*
— u. M. R. Hammond 356, *367*
— M. J. Moses u. R. Steele 354, 363, *367*
— V. Pond u. S. Kojan 360, *367*
— u. Rhoda Sparrow 347, *367*
Sparrow, Rhoda s. Sparrow, A. H. 347, *367*
Spessard, E. A. 111, *120*
Spilman, W. M. s. Baron, L. S. *26*
— s. Falkow, S. 25, *27*
Spohr, E. s. Loewe, S. 198, *208*
Sprague, G. F., u. A. Tavčar 95, *108*
— s. Richey, F. D. 162, 192, 194, *210*
Sprecher, A. 173, 174, 193, *212*
Sprecher, M. A. 577, *595*
Sprengel, C. K. 384, *404*, 507, *529*
Springer 148
Springer, E. 245, 250, 251, *256*
Spratt, E. 608, 617, *619*
Srinivasan, V. K. s. Kumar, L. S. S. 60, *105*
Srivastava, R. K. 430, *445*
Stahl, E. 248, *256*, 684, 685, 686, *692*
Stakman, E. C., u. J. J. Christensen 337, *342*
Stamps 59
Stanley, R. G. 368, 394, *404*
— u. H. F. Linskens 381, *405*
— u. F. Loewus 386, 390, *405*
— u. A. W. G. Yee 382, *405*
— L. C. T. Young u. J. S. D. Graham 394, *405*
Starkey, C. B. s. West, G. S. 749, *752*
Starr, R. C. *50*, 713, 716, 724, 726, 736, *746*, 749, 750, *752*
Staudt, G. 55, *108*
Stebbins, G. L., u. J. A. Jenkins 571, *595*
Stebbins jr., G. L. 552, 555, 571, 573, *595*
— s. Babcock, E. B. 571, *583*
— s. Duara, B. N. *587*
Stecher, M. 521, *529*
Steele, R. s. Sparrow, A. H. 354, 363, *367*
Steeves, T. A., I. M. Sussex u. C. R. Partanen 252, *256*
— s. Whittier, D. P. 252, *256*, 542, *550*
Steffen, K. 373, 394, *405*, 431, 432, 434, 435, 436, 438, 439, *445*
— u. W. Landmann 435, *445*
— u. W. Richter-Landmann *445*
Steil, W. N. 252, *256*, 410, *445*, 531, 532, 538, 544, *550*
Stein, Fr. 639, *656*
Stein, R. J. 306, *319*
Steineck, O. 802, *807*
Steinecke, Fr. 608, *619*
Stelzner, G., u. M. Torka 802, *807*
Stenar, H. 574, 578, *595*
Stern, C. 261, *281*, 351, *367*
Stern, H. s. Hotta, Y. 349, 354, 358, 360, *365*
Stern, K. s. v. Dellingshausen, M. 181, *204*
Stevenson, E. C. s. Janick, J. 59, 88, 89, *104*, 172, 194, *207*
— s. Mahoney, D. L. 88, *106*
Stevenson, T. M. s. Kirk, L. E. 509, *527*
Stewart, L. B. s. Graham, R. J. D. *806*
St. Gimesi, W. Frenyo u. G. L. Farkas 360, *367*
Stifter 34, 36
Stingl, G. *807*
Stocker, B. A. D. s. Smith, S. M 15, *29*

Stoeckenius, W. s. Zinder, N. D. 11, *30*
Stokey, A. 531, 532, 538, 542, *550*
Stone, M. J. s. Brinton, C. C. *27*
Storey, W. B. 60, 90, 91, *108*
— W. W. Jones, G. K. Daris u F. G. Holdaway 168, *212*
Stosch, H. A. v. 115, 116, *120*, *256*, 599, 626, 627, 630, 633, 634, *636*, 637, 646, 647, 648, 649, 651, 653, *656*, 657, 658, 659, 660, 662, 664, 665, 666, 668, 671, 672, 673, 676, *681*, 765, 766, 768, *776*
— G. Dersch-Siebert, R. Schäfer u. V. Lucan 647, *656*
— u. G. Drebes 657, 660, 662, 663, 666, 667, 672, 674, *681*
Stout, A. B. 165, 182, *212*, 506, 523, *529*
— u. C. Chandler 516, 521, *529*
Stover, R. H. 325, *342*, 780, *786*
Stow, I. 153, 154, *212*
Strand, M. s. Valentine, R. C. 12, *29*
Strandhede, Sven-Olov 264, *281*
Strasburger, E. 82, 87, 91, *108*, 113, *120*, 140, *152*, 155, 163, 170, 173, 174, 175, 177, 179, 195, 196, *212*, 258, 264, 266, *281*, 309, *319*, 373, 376, *405*, 413, *445*, 454, *478*, 515, 517, *529*, 547, 548, *550*
Strasburger, Ed. 558, 569, 571, 573, 578, 581, *595*, 686
Straub, J. 352, *367*, 511, 516, 518, 519, 521, 522, 523, *529*
— s. Esser, K. 379, *400*, 516, *526*
Streetman, L. J. 575, *595*
Strehlow, K. 111, *120*, 741, *746*
Strigoli, P. s. Scaramella Petri, P. *213*
Stroede, W. 755, *760*
Stroun, M. s. Bocqet, G. 172, 194, *203*
Struckmeyer, B. E. s. Roberts, R. H. *210*
Stuart, N. W. s. Emsweller, S. L. 523, *526*
Stubbe, H. 257, 262, *281*
Stumm, C. 236, 237, 245, *256*
Sturtevant, A. H. 55, 71, *108*
Suba-Claus, E. 603, *619*
Subba Rao, A. M. 578, *595*
Subramanyam, K. N. s. Mehra, K. L. 573, 574, *5891*
Subrahmanyan, R. 117, *120*, 309, *319*, 620, 623, *626*
Sueoka, N. s. Yoshikawa, H. 19, *30*
Suessenguth, K. 579, *595*
Sugimoto, Y. 69, 88, *108*
Sugiyama, S., u. T. Sutô 173, *212*
— s. Suto, T. 59, 69, 89, *108*
Suhr, K. A. s. Bredemann, G. 396, 397, *399*
Sukačev, V. N. 570, *595*
Sula 732
Sun, S. H. s. Alexopoulos, C. J. 326, *338*
Sundene, O. 774, *776*
Suneson, C. A. 202, *212*
Suneson, S. 767, 770, *776*
Suomalainen, E. 601, *602*
Surikov, I. M. 521, *529*
Sussex, I. M. s. Steeves, T. A. 252, *256*
Suto, S. 698, 699, *705*
— s. Kunieda, H. 309, *318*
Sutô, T. 154, *212*, 578, *595*
— u. S. Sugiyama 59, 69, 89, *108*
— s. Matsuura, H. 573, *591*
— s. Sugiyama, S. 173, *212*.
Suzuki, M. s. Harada, K. 16, *28*
Svedelius, N. 238, 240, 244, 250, *256*, 765, 766, 767, 768, *776*
Swaminathan, M. S. 521, 524, *529*
— s. Mehra, K. L. 573, 574, *591*
Swamy, G. L. 560, 578, 579, *595*
Swamy, R. 581, *595*
Swanson, C. P. 353, *367*
Swetosarowa s. Ellenhorn *445*
Swift, H. 354, *367*
Swingle, Ch. F. 788, *807*
Szabados, M. 624, *626*
Szymkiewicz, D. 190, *212*

Tabata, M. 462
Taber, W. A. s. Nickerson, W. J. 779, *785*
Tabushi, J. 89, *108*
Täckholm, G. 551, *595*
Tahara, M. 572, *595*
Takahashi, H. s. Tano, S. 382, *405*
Takahashi, N. s. Nisioka, T. 572, *592*
Takamatsu, M. 696, *705*
Takano, T. s. Watanabe, T. 16, *30*
Takao, A. 427, *445*
Takats, Stephen T. 347, 356, *367*
Takenaka, Y. 63, 65, 71, 87, *108*
Talice, R. V. 778, *786*
Tamiya, H. s. Hase, E. 734, *744*
Tanaka, K. 393, *405*
Tanaka, R. 523, *530*
Tannreuther, G. W. 621, 624, *626*
Tano, S., u. H. Takahashi 382, *405*
Tapke, V. F. 337, *342*
Tappi, G. 378, 379, *405*
— u. E. Menziani 379, *405*
— u. A. Monzani 379, *405*
— A. Spada u. R. Cameroni 379, *405*
Tatebe, T. 509, 521, *530*
Tatum, E. L. 6, *29*
— u. J. Lederberg 314, *320*
Taubenhaus, J. T. s. Ezekiel, W. 783, *785*
Tavčar, A. s. Sprague, G. F. 94, *108*
Taylor, A. L., u. E. A. Adelberg 9, *29*
Taylor, J. H. 347, 348, 349, 351, 352, 353, 354, 355, 356, 357, 360, 361, 364, *367*
— u. Callan 354
— u. R. D. McMaster 347, 354, 356, 363, *367*
— u. S. H. Taylor 357, 358, *367*
— P. S. Woods u. W. S. Hughes 353, *367*
— s. Moses, M. J. 347, 354, 356, *366*
Taylor, M. H. s. Brough, P. 419, *443*
Taylor, N. W. 312, *320*
Taylor, S. H. s. Taylor, J. H. 357, 358, *367*
Taylor, W. R. 696, *705*
Taylor, Wm. R. 641, *656*
Ter-Awanesjan, D. W. *446*
Terenetz, C. *626*
Terra, P. 327, 328, 329, 330, 331, 332, 333, 334, 335, *342*
— s. Kühner, R. 333, *340*
Thaxter, R. 127, *133*, 761, *776*
Theophrastus 262, *281*
Thijn, G. A. 790, *807*
Thomas, P. T. s. Crane, M. B. 581, *586*
Thomas, R. G. 88, *109*, 176, 179, 194, *212*
Thomas, W. I. s. Whiteley, J. R. 524, *530*
Thompson, A. E. 172, 193, 194, *212*
Thompson, J. A. s. Geddes, P. 160, *205*
Thompson, K. F. *530*
Thompson, R. C. 432, *445*
Thren, R. 337, *342*
Thuret, G. 761
Thuret, G. 237, *256*, 309, *320*
— u. E. Bornet 309, *320*, 694
Thuret, M. G. *619*
Tiagi, B. s. Johri, B. M. 154, 155, *207*
Tibeau, M. E. 174, 192, *212*
Tidbury, G. E. s. Brieger, F. G. 462, *476*
Tiedjens, V. A. 185, 186, 192, 193, 194, 195, *212*

Tieghem, Th. van 779, *786*
Tiffany, L. H. 111, *120*, 747, *752*
Timasheff, S. N. s. Loveless, M. H. 380, *402*
Tindall, D. R., T. Sawa u. A. T. Hotchkiss 758, *760*
Tinline, R. D. 325, *342*
Tinney, F. W. 576, *595*
Tischer, I. 608, 610, 616, *619*
— s. Drawert, H. 610, 616, *618*
Tischler, G. 385, *405*, 578, *595*, 800, 804, *807*
Tiwary, N. K. 578, *595*
Tobler, F. 772, 773, 774, *776*
Todd, F. E., u. O. Betherik 382, 383, 384, *405*
Tokida, J., u. H. Yamamoto 769, *776*
Tomizawa, J. s. Anraku, N. 364, *365*
Tomlin, B. P. s. Nobles, M. K. 324, 327, 328, 329, 330, *341*
Torka, M. s. Stelzner, G. 802, *807*
Tournois, J. 175, 178, 193, 194, 197, 200, *212*
Townsend, B. B. 781, 782, 783, 784, *786*
Toxopeus, H. J. 581, *595*
Trainor, F. R. 724, 735, *746*
Transeau, E. N. 751, *752*
Traub, H. P. 581, *595*
— s. Dandlikar, W. 387, *400*
Trécul, A. *808*
Treskow, M. v. 164, 192, *212*
Treub, M. 137, *152*
Triplett, E. L. 576, *595*
— s. Grun, P. 568, 576, *588*
Troll, W. 373, *405*, 788, 790, 793, 795, 796, 797, 799, *808*
Tryon, Alice F., u. D. M. Britton 534, 536, 539, 541, 542, 543, *550*
Tryon, Rolla M. 532, 549, *550*
Tsao, T. H. 394, *405*
Tsao, Tsung-Hsun 438, *445*
Tschermak, E. 727, *746*
Tschermak-Woess, E. 573, *596*, 716, 735, *746*
— u. G. Hasitschka-Jemschke 751, *752*
Tseng, C. K., u. T. J. Chang 762, 763, *776*
Tseng, H. P. s. Brieger, F. G. 462, *476*
Tsubo, Y. 289, 290, 303, 305, 307, *320*
Tukan, D. A. s. Petrov, D. F. 55, *107*
Tulasne, Ch. s. Tulasne, L. R. 163, 170, 196, *212*
Tulasne, L. R., u. Ch. Tulasne 163, 170, 196, *212*
Tulecke, W. R. 397, 399, *405*

Tun, N. N. s. Haskell, G. 581, *589*
Tung, S. M. s. Larsen, P. 383, 388, *402*
Tupý, J. 512, 513, *530*
— s. Linskens, H. F. 382, *402*, 513, *528*
Turbin *445*
Turesson, B. s. Turesson, G. 572, *596*
Turesson, G. 576, 579, *596*
— u. B. Turesson 572, *596*
Turesson, K. 277, *281*
Turrill, W. B. 573, *596*
— s. Marsden-Jones, E. M. 579, *591*
Tyler, L. J. 337, *342*
— u. C. P. Shumway 337, *342*
Tylkina, L. G. s. Minina, E. G. 186, 187, 198, *209*
Tysdal, H. M. *596*

Ubisch, G. v. 67, 91, *109*, 166, 196, *212*, 467, *478*
Uhring, J., u. S. L. Emsweller 578, *596*
Ulbrich, E. 793, *808*
Ulehla, V. 736, *746*
Umaerus, M. s. Grazi, F. 566, 577, *588*
Uphof, J. C. Th. 167, *212*
Urbańska, K. 571, *596*
Urbańska-Worytkiewicz, K. 571, *596*
Uring s. Emsweller, S. L. 523
Ushirozawa, K., u. J. Shibukawa 398, *405*
Ustinova, E. I. 163, *212*

Vaheeduddin, S. 337, *342*
Valentine, R. C., u. M. Strand 12, *29*
— s. Ippen, K. A. 12, *28*
— s. Wendt, L. W. 12, *30*
— s. Zinder, N. D. 11, *30*
Valkanov, A. 648, 650, 653, *656*
Valleau, W. D. 55, 90, *109*
Vallaeys, G. s. Devreux, M. 509, *526*
Vandendries, R. 130, *133*, 327, 329, 330, 331, 332, 333, 334, 335, *342*, *343*
— u. H. J. Brodie 328, 329, 332, 334, *343*
— u. G. Robyn 331, *343*
— s. Martens, P. 334, *341*
Van Iterson, W. s. Houwink, A. L. 10, *28*
Vansell, G. H. s. Griggs, W. H. 397, 398, *401*
Vanterpool, T. C., u. R. Macrae 329, *343*
Varma, A. K., u. A. R. Mitra 603, 607, 608, *619*

Váróczy, E. s. Daniel, L. 385, *400*
Vartazarova, L. S. s. Nekrasov, V. I. 181, *209*
Vasermanis, A. s. Quintanilha, A. 330, 331, 332, 333, 334, *342*
Vasil, I. K. 360, *367*, 385, *405*
— s. Johri, B. M. 368, 395, *401*
Vasil, V. 430, *445*
— s. Maheshwari, P. 430, *444*
Vazart, B. 160, 197, *212*
Vazart, Bernard *443*
Vazart, J. 434, 441, *445*
Vegis 291
Vegis, A. 800, 801, *808*
Venkat Ram, C. S. 780, *786*
Venkatarayan, S. V. 173, *212*
Verall, A. F. 328, *343*
Verkaik, C. 217, *234*
Verma, S. C. s. Mehra, P. N. 532, *550*
Viana, M. J. s. Resende, F. 101, *107*, 166, 183, 199, *210*, 269, 271, 274, *281*, 804, *807*
Vicari, H. s. Zetschke, F. 375, 377, *406*
Vignoli, L. 569, *596*
Vincent, C. s. Champagnat, M. 798, *805*
Vinson, C. G. 381, *405*
Vischer, W. 640, *656*, 683, 684, 685, 686, 688, 689, *692*, 735, *746*, 760, *776*
Vishniac, W. s. Lazaroff, N. 603, 604, 605, 606, 614, *619*
Vishveshwaralah, S. s. Kumar, L. S. S. 60, 80, *105*
Visser, T. 379, 385, 387, 388, 395, 397, 398, *405*, 516, *530*
Vivino, A. E., u. L. S. Palmer 378, 383, 384, *405*
Vlk, W. s. Pascher, A. *656*
Vöchting, H. 179, 183, 193, 195, *212*, 790, 801, *808*
Voeller, B. 144, *152*
Vogg, G. s. Marquardt, P. 381, *402*
Volcani, B. E. s. Reimann, B. E. F. 663, *681*
Vries, H. de 455, 467, *478*, 481, 487, 502, *505*
Vuillemin, P. 170, 196, *213*

Wafa, A. s. Ridi, M. S. 383, 384, *404*
Wafa-Aboul, M. H. s. Hassan, A. 383, *401*
Wagenitz, G. 372, 375, *405*
Wagner, E. 90, *109*
Wagner, R. P., u. H. K. Mitchell 259, *281*
Wahlstedt, L. J. s. Nordstedt, O. 755, *760*

Walderdorf, M. 385, *405*
Waldo, G. F., u. G. M. Darrow 581, *596*
Walker, D. s. Wittmann, G. 374, *405*
Walker, E. 138, 143, *152*
Walker, S. s. Manton, I. 545, 546, 547, *550*
Walker, T. G. 532, 539, 540, 541, *550*
Wall, E. T. s. Davies, D. R. 524, *526*
Wallis, M. s. Reichel, L. 296, *319*
Wallner, F. 753, *760*
Ward, M. 252, *256*
Warden, J. W. s. Resende, F. 266, 269, *281*
Wareing, P. F. 158, 195, 197, *213*
— s. Longman, K. A. 181, 197, *208*
Warming 424
Warming, E. 798, *808*
Warmke, H. E. 58, 73, 74, 75, 76, 78, *109*, 575, *596*
— u. A. F. Blakeslee 61, 62, 72, 74, *109*
— s. Blakeslee, A. F. 87, *102*
— s. Snyder, L. A. 575, *595*
Warren, G. F. s. Brantley, B. B. 184, 188, 192, 194, *203*
Wassilzowa *445*
Watanabe, A. 608, *619*
Watanabe, K. *405*
Watanabe, T. 15, *29*
— u. T. Fusakawa 15, 16, *29*
— — u. T. Takano 16, *30*
— u. M. Okada 11, *30*
— s. Falkow, S. 16, *27*
Waterkeyn, L. 430, *445*
Watson, J. D., u. F. H. C. Crick 349, *367*
Watson, P. J. 581, *596*
— u. J. Clausen 577, *596*
Watson, W. C. R. 581, *596*
Weber, E. *626*, *808*
Weber, H. 789, 790, 799, *808*
Weber, J. M. 569, 581, *596*
Weber, W. 686, 689, 690, *692*, 764, 768, 769, *776*
Webber, H. J. 416, 419, *445*
Wehmer, C. 779, *786*
Weidel, W. 259, 265, *281*, 314, *320*
Weidmann, U. s. Peacock, A. S. *602*
Weidner-Rauh, E. 488, 493, 497, 501, *505*
Weigle, J. J. s. Dove, W. F. *27*
— s. Kellenberger, G. 364, *365*
— s. Meselson, M. 364, *366*
Weiling, F. 260, *281*
Weinland, H. 370, *405*
Weiss, M. C. s. Ephrussi, B. 260, *279*
Weiss, P. 282, *320*
Welch, A. W., u. J. C. Gilman 326, *343*
Welch, D. S. s. Blakeslee, A. F. 36, *49*
Welzel, G. 370, 371, 372, *405*
Wendt, L. W., K. A. Ippen u. R. C. Valentine 12, *30*
Wenrick, D. H. *4*
Went, F. A. F. C. 393, *405*
Went, F. W. 162, 194, *213*, 801, *808*
— s. Nitsch, J. P. 186, 189, 194, 198, *210*
Werber, H. J. *596*
Werfft, R. 380, 399, *504*
Werner, G. s. Fischer, F. G. 223, *233*, 285, 291, 292, 293, *317*
Werner, H. O. 802, *808*
Werth, E. 171, *213*
West, G. S., u. F. E. Fritsch 717, 719, *746*
— u. C. B. Starkey 749, *752*
Westbrook, M. A. 767, *776*
Westergård u. Sörensen 576
Westergaard, M. 38, *50*, 58, 59, 60, 61, 62, 63, 64, 66, 68, 72, 73, 74, 75, 76, 77, 78, 85, 86, 88, 91, 92, 94, 99, *109*, 170, 171, 195, 196, *213*, 268, 274, 275, *281*
Westergaard s. Rossen 364
Weston, E. W., *109* 179, 199, *213*
Wetmore, R. H., A. E. DeMaggio u. G. Morel 250, 252, *256*
— s. DeMaggio, A. E. 248, *254*
— s. Morel, G. 545, *550*
— s. Morel, M. G. 252, *256*
Wettstein, D. v. 252, *256*
Wettstein, F. v. 32, 35, *50*, 51, 94, 96, *109*, 112, *120*, 141, 142, 144, 145, 146, 147, 148, *152*, 170, *213*, 245, 250, 251, *256*, 266, *281*, 677, 691, *692*
Wettstein, R. 137, 138, 139, 140, *152*, 241, *256*, 324, *343*
Wettstein, W. v. 799, *808*
— u. L. Niklas 372, *405*
Weygand, F., u. H. Hofmann 382, 383, *405*
Wheeler, H. E. 47, *50*
Whetzel, H. H. 326, *343*
Whitaker, T. W. 184, *213*
White 360
White, M. J. D. *602*
Whitehouse, A. L. K. 49, *50*
Whitehouse, H. L. K. 121, 122, 123, 128, *133*, 324, *343*, 364, *367*
Whiteley, J. R., W. I. Thomas u. I. J. Johnson 524, *530*
Whittier, D. P. 252, *256*, *542*, *550*
— u. T. A. Steeves 252, *256*, 542, *550*
Whitton, B. A. 608, *619*
Wicherley, P. R. 574, 575, *596*
Wickerham, L. J. 310, 311, 313, *320*
Widder, F. 235, *256*
Wieling, U. 796, *808*
Wiese, L. 36, 38, *50*, 741, *746*
— u. R. F. Jones 303, 304, *320*
— s. Förster, H. 32, 33, 34, *49*, 229, *233*, 282, 300, 301, 302, 303, 306, *318*
Wiger, J. 570, *596*
Wilbois, A. 306, *320*
Wildeman, E. de 691, *692*
Wildon, D. C., u. F. V. Mercer 609, 611, 612, *619*
Wilkie, D. 412, *445*
Wilkinsson 59
Wille, N. 638, *656*, 761, 772, *776*
Williams, C. F. s. Loomis, N. H. *106*
Williams, J. L. s. Farmer, J. B. 309, 310, *317*
Williams, L. J. 291, 307, 308, 309, *320*
Williams, M. M. 113, *120*
Williams, M. W. s. Machlis, L. 218, *234*
Williamson, H. S. s. Gwynne-Vaughan, H. C. I. 326, *340*
Willis, I. C. 165, *213*
Wilson, C. M. s. Emerson, R. 124, *132*
Wilson, E. B. 57, 60, 61, *109*
Wilson, G. B. s. Bemis, W. P. 59, 88, *102*
Wilson, M. 145, *152*
Wilson, P. D. 668, 669, 671, *681*
Wimber, D. E., u. W. Prensky 349, 354, *367*
Wineland, G. O. 326, *343*
Winge, Ö. 45, *50*
— u. C. Roberts 126, *133*
Winge, Ø. 58, 59, 61, 62, 64, 65, 67, 68, 72, 76, 77, 86, *109*
— u. O. Lausten *109*
Winkler, H. *256*, 260, 276, 277, 278, *281*, 531, *550*, 551, *596*, 600, 601, *602*, 690, 691, 700, 701, 703, 704, *705*, 735, 736, *746*, 759, *760*
Wislouch, S. 240, *256*
Wislouch, S. M. 638, *656*
Witsch, H. v. 191, 194, 199, *213*, *281*
Witschi, E. *109*
Wittmann, G., u. D. Walker 374, *405*
Wittrock 750

Wittrock, V. B. 168, 197, *213*, 797, *808*
Wittwer, S. H., u. M. J. Bukovac 187, 188, 199, *213*
— u. I. G. Hillyer 187, 189, 198, 199, *213*
— s. Bukovac, M. J. 188, 199, *203*
— s. Mitchell, W. D. 101, *106*, 187, 188, 199, *209*
— s. Phatak, S. C. *213*
— s. Sarkar, B. C. R. 381, *404*
Wodehouse, R. P. 373, *405*
Wohlhieter, J. A., C. C. Brinton jr. u. L. S. Baron *30*
— S. Falkow, R. V. Citarella u. L. S. Baron *30*
— s. Falkow, S. 16, *27*
Wolf, F. T. 780, *786*
Wolk, C. P. 608, 613, 615, 616, 617, 618, *619*
Wollenweber 730
Wollersheim 138, 143, 146, 148, 149
Wollman, E. L., u. F. Jacob 10, *30*
— — u. W. Hayes 314, 315, 316, *320*
— s. Afoldi, L. F. 15, *26*
— s. Anderson, Th. F. 314, 315, *317*
— s. Jacob, F. *4*, 13, 15, *28*
— s. Jacob, L. 6, 7, 8, 9, 12, 17, 18, 19, 21, *28*
Woloszynska, J. 628, 629, 631, *636*
Wood, F. J. F. 670, *681*
Wood, R. D., u. K. Imahori 753, *760*
Wood, R. K. S. s. Abeygunawardena, D. V. W. 781, 782, 783, *784*
Woods u. Pollister 347
Woods, P. S. s. Taylor, J. H. 353, *367*
Woodworth, R. H. 570, *596*
Woronichin 610
Woronin, H. 532, 533, 541, 542, 544, *550*
Woronin, M. *656*
Wuist, E. D. *152*
Wulf, M. L. s. Celarier, R. P. *585*
Wulff, H. D. 368, *405*, 804, *808*
Wunderlich, R. 370, *406*
Wylie, A. P. s. Darlington, C. D. *586*
Wylie, R. B. 438, *445*

Yakar, N. s. Heilbronn, A. 100, *103*, 154, *205*
Yakushina, N. I. 383, *406*
Yamada, I. 58, 82, *109*
Yamaha, G. 369, 370, *406*
Yamamoto, H. s. Tokida, J. 769, *776*
Yamamoto, Y. 58, 70, 71, *109*
— s. Kihara, H. 65, 70, *105*
Yamauti, K. s. Nisikado, Y. 331, *341*
Yampolsky, C. 87, *109*, 154, 155, 179, 195, *213*
— u. H. Yampolsky 52, *109*, 157, *213*
Yampolsky, H. s. Yampolsky, C. 52, *109*, 157, *213*
Yarbrough, J. A. 796, *808*
Yarnell, S. H. s. East, E. M. 521, *526*
Yarwood, C. E. 325, *343*
Yasuda, S. 512, 518, 522, *530*
Yee, A. W. G. s. Stanley, R. G. 382, *405*
Yen, H. C. 327, 329, 330, 331, 332, 333, 334, 335, *343*
Yong, D. P. *596*
Yoshida, T. 766, *776*
Yoshikawa, H., u. N. Sueoka 19, *30*
Young, L. C. T. s. Stanley, R. G. 394, *405*
Youngner, V. B. 202, *213*, 563, 576, *596*
Yuasa, A. 410, 411, 413, *445*
Yuichi, S. 101, *109*

Zachow, F. 442, *446*
Zahl, P. A. s. McLaughlin, I. I. A. 630, *636*
Zanoni, G. 387, *406*
Zarubin, A. F. 577, *596*
Zattler, F. 330, 334, *343*
Zaurov, E. s. Breslavec, L. P. 154, 173, 174, 175, 194, *203*
Zeijlemaker, F. C. J. 394, *406*
Zeilinga, A. E. 580, *596*
Zen, S. 351, *367*
Zetschke, F. 373, 375, *406*
— u. K. Huggler 375, *406*
— u. H. Vicari 375, 377, *406*
Zichichi, M. L. s. Kellenberger, G. 364, *365*
Zickler, H. 226, *234*, 297, *320*, 326, *343*
Zickler s. Ames 46
Ziegler, A. 90, *109*
Zielinski, F. 409, *445*
Zikes, H. 778, *786*
Zillig, H. 337, *343*
Zilm, B. s. Reimann-Philipp, R. 59, 90, *107*
Zimmermann 711
Zimmermann, W. 243, *256*
Zinder, N. D., R. C. Valentine, N. Roger u. W. Stoeckenius 11, *30*
— s. Loeb, T. 11, *29*
Zinger, N. V., V. A. Poddubnaja-Arnoldi, T. P. Petrovskaya u. N. N. Polunina 573, *596*
— s. Petrovskaya-Baranova, T. P. 573, *593*
Zirkle, C. 395, *406*
Zitek, R. v. 488, *505*
Zohary, D. 351, *367*
— s. Brown, S. W. 351, *365*
Zohary, M. 792, *808*
Zolovivh u. a. 398
Zopf 779
Zoschke, U. 59, *109*, 173, *213*
Zschokke 90
Zumstein, H. 621, 622, 624, *626*
Zych, C. C. 524, *530*
Zycha, H. 326, *343*

Sachverzeichnis.

(Deutsch-Englisch.)

Bei gleicher Schreibweise in beiden Sprachen sind die Stichworte jeweils einfach aufgeführt. Ä, Ö, Ü sind wie Ae, Oe, Ue eingereiht.

Abnormal blasser Typus, *abnormal pale type* 67
Abstandswirkungen, stoffliche, beim Kopulationsvorgang, *long distance effect in copulation* 297
Abweichungen vom zu erwartenden Mendelschen Spaltungsverhältnis, *segregation, preferential* 449
Acrasin, *acrasine* 241
Acrosom-Kontakt, *acrosomal reaction* 309
active site 19
adaptive Enzyme, Synthese, *adaptive enzymes, synthesis* 311
Adventiv-Embryonie, *adventitious embryony* 558—560
Änderungen im Spiralisierungs-Cyclus, *variations in coiling cycle* 352—353
Affinität, genetische Grundlagen bei der selektiven Befruchtung, *affinity, genetical background in selective fertilization* 489—492
—, geschlechtliche, *sexual affinity* 229
—, sexuelle, *affinity, sexual* 517
— zwischen Samenanlagen und Pollenschläuchen, *between ovules and pollen tubes* 485—489
Agamospermie, die verschiedenen Typen der, *agamospermy, various types* 551
Agglutination, *agglutination* 32
— bei der Befruchtung, *in fertilization* 285
—, geschlechtsspezifische, *sex specific* 312
— bei *Hansenula, of Hansenula* 310—311
— bei Hefen, *in yeasts* 310—313
—, Rolle der Ca-Ionen, *role of Ca ions* 313
—, sexuelle, *sexual* 311
— von Geißeln, *of flagellae* 304
Agglutinationsstoffe, *agglutination substances* 306
— bei *Tetraspora, in Tetraspora* 306
Agglutinationswirkungen, *agglutination effects* 297—316
AG-Komplex, *AG complex* 51, 98f., 101, 140
Aglazonia-Stadium, *Aglazonia stage* 237f., 244, 703
Agamospermie, Definition, *agamospermy, definition* 551
Algen, Heterogenität des Begriffs, *algae, heterogenity of definition* 598
Allogamie bei Diatomeen, *allogamy in diatoms* 116
Allogamie, isogame, bei Diatomeen, *isogamic, in diatoms* 675
Allomyces-Hormone, *Allomyces hormones* 290f.
Alternative sexuelle-vegetative Fortpflanzung bei den Heterokonten, *alternative sexual-vegetative reproduction in Heterokontae* 687f.
Alterung des Pollens, Änderungen in der Pigmentausstattung, *aging of pollen, changes of pollen pigment in connection with* 379
Androgamone, Definition, *androgamones, definition* 283
Androhermaphroditen, *androhermaphrodites* 74, 78
Andro-Intersexualität, *andro intersexuality* 79
angustifolia-Faktor, *angustifolia factor* 66
Anheftung von Zellen, *attachment of cells* 297—298
Anisogametie, *anisogamety* 283—284
Anlagerung des Spermakernes an den Eikern, *attaching of the sperm nucleus to the egg nucleus* 439—440
Antheren, Einbaustudien während der Mikrosporen-Entwicklung, *anthers, incorporation during microsporogenesis* 355—358
—, Überlebenschancen, *potential for surviving* 360
—, Wachstumskurve, *growth* 369
Antherenbildung, *anthers, induction of* 75
Antheren-Entwicklung, Einfluß der Lichtbedingungen auf, *anther development, influence of light conditions* 193
—, — von 2-Thiouracil, *of 2-thiouracil* 184
—, — von TJBS, *of TIBA* 183
Antherenkultur, *anthers, culture of excised* 352, 359—361
Antherensterilität, *sterility, anther* 170
Antheridien, Öffnung der, *antheridia, opening of* 407
Antigen, spezifisches, an der Bakterienoberfläche, *antigen, specific, on the bacteria surface* 316
Antikörper, geschlechtsspezifische, *antibodies, sex specific* 304
Antiserum bei der Gruppenbildung von Algengameten, *antiserum in group formation of algal gametes* 304
Aplanosporen, *aplanospores* 748f.

Aplanosporen, Auslösung der, bei Heterokonten, *aplanospores, induction of, in Heterokontae* 689f.
Apogamie, *apogamy* 278
— bei den Algen, *in algae* 600—601
— bei den Braunalgen, *in brown algae* 703, 704
— bei den Chlorophyceen, *in Chlorophyceae* 741—743
— bei den Conjugaten, *in Conjugales* 750—752
—, Definition, *definition* 530
— bei Diatomeen, *in diatoms* 672—677
—, Entstehung durch Bastardierung, *origin by hybridization* 539—541
—, — durch Mutation, *by mutation* 541
—, — durch Umweltfaktoren, *induction by environmental factors* 542
—, experimentell induzierte, *experimentally induced* 250—252, 542
— bei Farnen, *in ferns* 532ff.
—, genetische Voraussetzungen, *genetical conditions* 250—252
— bei Moosen, *in mosses* 245
—, obligate 532ff.
—, —, Bedeutung der, *importance* 542—544
—, —, Entstehung, *origin* 539—542
—, physiologische Voraussetzungen, *physiological conditions* 251
— bei den Rhodophyten, *in Rhodophyta* 774
Apogonose, *apogonosis* 279
Apomeiose, *apomeiosis* 279
apomiktische Arten, Artbildung, Biotypen und Formen, *apomictic species, species formation, biotype and form* 562—568
— Löwenzahn-Arten, abweichende Nachkommenschaft, *Taraxacum species, aberrant offspring* 564
— Pflanzen, *plants* 561
apomiktisches System in der Gattung *Rubus*, *apomictic system in Rubi* 566
Apomixis 276, 277
— bei den Archegoniaten, *in bryophytes and ferns* 530ff.
— bei den Angiospermen, *in angiosperms* 551—582
—, Arten mit bekannter oder vermuteter (Tabelle), *species with known or suspected (table)* 569—582
—, Definition 530, 531, 551, 601
— bei Diatomeen, *in diatoms* 116f,. 674, 677—678
—, experimentelle Kontrolle der, *experimental control* 563
—, experimenteller Nachweis, *experimental control* 563
—, fakultative, *facultative* 563, 568
—, Liste der Arten bei denen bekannt, *list of species* 599ff.
Aposporie, *apospory* 248ff., 279, 544ff., 556—558, 601
Archegonien, Öffnung der, *archegonia, opening of* 409
Artbastarde, *interspecific hybrids* 567
— von *Poa*, Zahl der Nachkommen, *of Poa, number of progenics* 568
Artkreuzungen, *species crosses* 80, 96
—, Solanum, *in Solanum* 100
Ascusbildung bei Hefen, *ascus formation in yeast* 313
A-Substanz bei Pteridium, *A substance in Pteridium* 143f.
Asynapsis 360
Atmungsquotient von Pollenschläuchen, *respiration quotient of pollen tubes* 513
Atmungsverhalten von Griffeln während des Pollenschlauchwachstums, *respiration in styles during pollen tube growth* 513
„Attrappen" der Befruchtung, *"imitations" of fertilization* 121
— für die Meiose, *of the meiotic division* 121
Aurea-Faktor, *aurea factor* 67
Ausläufer, *runners* 561
Auslösung von ungeschlechtlichen Fortpflanzungszellen bei den Chlorophyceen, *induction of asexual reproduction in Chlorophyceae* 733—736
Austausch von Erbmaterial, *genetic exchange* 2, 5
Autogamie bei Amphora, *autogamy in Amphora* 677
Automixis bei Diatomeen, *in diatoms* 116, 674—677
Autopolyploidisierung, *autopolyploidisation* 759
Autosomen, *autosomes* 60, 64
Autosporen, *autospores* 724—728, 780
—, Einfluß der Temperatur auf die Bildung von, *influence of temperature on formation of* 735
Auxinspiegel, Beeinflussung durch Virusinfektion, *auxin level, influence of virus infection on* 188
Auxosporen bei Diatomeen, *auxospores in diatoms* 657f.
—, Entstehung durch haploide Parthenogenese, *development by haploid parthenogenesis* 678
—, Entwicklung von Gameten zu, *development of gametes into* 677
— und Sexualakt bei Diatomeen, *and sexual process in diatoms* 659, 672
Auxosporenbildung, abgeleitete Typen, *auxospore formation, derived types* 116
—, Bedingungen für, *auxospores, conditions for formation* 678—679
— bei Denticula, *in Denticula* 676
Auxosporen-Membran bei Diatomeen, *auxospore membrane, diatoms* 657
Auxosporenmutterzelle bei Diatomeen, *auxospore mother cell in diatoms* 678
Azygotenbildung bei Mougeotia, *azygote formation in Mougeotia* 751

Bakterielle Chromosomen, *bacterial chromosomes* 5, 13
—, Methode der Replikation, *mode of replication* 19
Bakteriophagen, geschlechtsspezifische, *bacteriophages, sex specific* 11
—, Wechselwirkung von, *interaction of* 11

Balance zwischen Kerngenen und Plasmon, *balance genome-plasmone* 92
Balance-Theorie nach BRIDGES, *balance therory of* BRIDGES 91
Bastardierung zwischen pseudogamen Arten, *hybridization among pseudogamous species* 563, 566
Bastard-Keimlinge bei Citrus-Kreuzungen, *hybrid seedlings in Citrus crosses* 558
Befruchtung, *fertilization* 2, 3, 258
— bei den Angiospermen, *in angiosperms* 430—442
— bei Archegoniaten und Blütenpflanzen, *in embryophyta* 407
— bei Bakterien, *in bacteriae* 313—316
—, chemische Grundlagen, *chemical background* 283ff.
— bei oogamen Formen, *in oogamic forms* 307—310
—, partielle, bei Bakterien, *partial, in bacteriae* 315
—, Phasen der, *phases of* 240
—, physiologische Vorgänge bei der, *physiological processes during* 283—285
— bei Prasiola, *in Prasiola* 309—310
—, Selektion bei der, *selection during* 449
—, selektive, *selective* 447, 479—502
Befruchtungsmembran, *fertilization membrane* 285, 308
Befruchtungsprozeß bei Bakterien, Physiologie des, *mating process in bacteria, physiology of* 16
Befruchtungstypen, *fertilization types* 440
Befruchtungsvorgang bei Oedogonium, *fertilization process in Oedogonium* 287
Befruchtungsvorgänge bei Algen, Übersicht, *fertilization processes in algae, survey* 285
— bei Pilzen, Übersicht, *in fungi, survey* 286
Beschneidung, Beeinflussung des Geschlechtes durch, *resection, influence on sex* 177, 195
Bestäubung, *pollination* 453
— bei Coniferen, *in Gymnosperms* 421f.
Bestäubungstropfen, *pollination fluid* 414f., 421f.
Beweglichkeit der Gameten, *mobility of gametes* 298
Bipolarität, sexuelle, *sexuality, bipolar* 31, 32
bisexuelle Potenz, *bisexual potency* 123, 124
Bisporen, *bispores* 769
„Blüh-Grad" von Blüten, *floral degree of flowers* 272, 273
Blühimpuls, *impulse, flowering* 268, 274
Blüh-Realisation, *flower-realisation* 268, 269
Blühzustand, Grad des, *flowering, degree of* 269
Blüte, *flower* 271ff.
—, sterile, durch Infektion mit Ustilago, *sterile flowers, by infection with Ustilago* 161, 163, 166, 170, 184
—, vegetative 271
Blütenausprägung, *floral realisation* 274
Blütenbildung, *flower formation* 159
—, beteiligte Gene, *formation of flower, genes involved in* 274
Blütenbildung, optimale Tageslänge, *flower development, optimal day length* 193f.
Blütenentwicklung, genetische Kontrolle, *flower development, genetical control* 272
Blütenfarbe, geschlechtsgebundene Vererbung der, *flower colours, sex linked heredity of* 68
Blütenveränderung, *flower variation* 272
Bor-Säure, Einfluß auf Pollen-Keimung und -Schlauchwachstum, *boric acid, influence on pollen germination and pollen tube growth* 385—387
BOUCK-BRENNER-Modell, BOUCK-BRENNER *model* 23, 24
Brutkörper in Rhodophyten, *gemmae* in *Rhodophyta* 771
Bryokinin, *bryokinin* 247
Bryonia-Versuche, CORRENS', *Bryonia experiments*, CORRENS' 52
Bulbillen, *bulbills* 561, 755—758, 784, 793

Calluswuchs, *callus growth* 253
Certationsversuche, *certation experiments* 56
chemotaktische Anlockung der Gameten bei Chlamydomonas, *chemotactic attraction of gametes in Chlamydomonas* 288
— Stoffe bei Fucus, *substances, in Fucus* 309
— — bei Oedogonium, *in Oedogonium* 286
Chemotaxis 241, 284, 285ff., 310
—, phobische, bei Fucus, *phobic, in Fucus* 287
chemotropische Anlockung bei Ginkgo und Cycas, *chemotropical attration in Ginkgo and Cycas* 419
— Anziehung der antheridialen Hyphen, *chemotropic attraction of antheridial hyphae* 222
— Kontrolle des Wachstums der Trichogyne, *control of trichogynal growth* 226
— Wirkung von Aminosäuren-Gemischen, *activity of amino acid mixtures* 223
— Wirkungen bei Hefestämmen, *effect in yeasts* 294
Chemotropismus, *chemotropism* 284, 291
—, Einfluß von Calcium-Ionen, *influence of Ca ions* 394
— der Pollenschläuche, *of pollen tubes* 431, 432, 437f.
— der Pollenschläuche im Griffel, *of pollen tubes in the style* 394
— bei Saprolegnia-Arten, *in Saprolegnia species* 223
— und selektive Befruchtung, *in connection with selective fertilization* 502
— vegetativer Hyphen bei Saprolegnia, *between vegetative hyphae of Saprolegnia* 292f.
Chiasmata-Bildung, *chiasmata, formation of* 350—352
„Chiasmatypie", "*chiasmatypy*" 639
Chimären, das Problem der, *chimaera, problem of* 276
Chitingehalt in den Membranen von Pilzen, *chitin content of the membranes of fungi* 241

Chlamydosporen, *chlamydospores* 749, 780f.
Chromosom, bakterielles, *chromosome, bacterial* 313
Chromosomenabweichungen, *chromosome aberrations* 362
Chromosomenbrüche, *chromosome breaks* 361
Chromosomen-Konstitution und Geschlecht, *chromosome constitution and sex* 75
Chromosomenpaarung, *chromosomes, pairing of* 260, 282, 538
—, genetische Kontrolle der, *pairing of chromosomes, genetic control of* 350
Chromosomensatz, Verdoppelung, Einfluß auf Entwicklung, *chromosome set, doublication, influence on development* 245
Chromosomentransfer, *transfer of chromosomes* 259
—, Mechanismus des, *chromosome transfer, mechanism of* 22
Chromosomenverdoppelung, *chromosomes, duplication of* 19, 536
Chromosomenzahl bei apogamen Farnen, *chromosome number in apogamic ferns* 539
— bei Poa-Bastarden, *in Poa hybrids* 568
— der Nachkommenschaft von Poa irrigata, *of offspring of Poa irrigata* 566
—, Variationen der, bei apomiktischen Arten, *variation in, in apomictic species* 565, 566
— und vegetative Fortpflanzung, *and vegetative propagation* 804
Colchicin-Behandlung, *colchicine treatment* 72
Colicine (Antibiotica), *colicins* 14
— produzierende Stämme, *colicinogenic strains* 14, 15
Conchosporen, *conchospores* 762
cross over suppressor 91
Cutinase 510
—, Aktivierung der, bei Inkompatibilitätsreaktion auf der Narbe, *activation of, in relation to incompatibility reaction on the stigma* 511
Cutleria-Pflanze, vegetative, *Cutleria plants, vegetative* 237f.
Cutleria-Stadium, *Cutleria stage* 703
Cyanophyceen, Fortpflanzung bei den, *cyanophyta, reproduction in* 602—619
Cyclus, theoretischer, von Coccolithus fragilis, *cycle, theoretical, of Coccolithus fragilis* 651—652
Cysten, Bildung der, bei den Chloromonaden, Cryptomonaden und Euglenophyceen, *cysts, formation, in Chloromonadene, Cryptomonadene and Euglenophyceae* 623f.
—, Bildungsbedingungen, *conditions for formation* 624, 631, 645
—, Dignität der, *significance* 631
— bei den Dinophyten, *in Dinophyta* 630—635
— —, Keimung der, *germination of* 632—634
— bei den Haptophyceen, *in Haptophyceae* 646
—, Keimung der, bei den Chrysophyten, *germination of, in Chrysophyta* 646
—, sexuelle Entstehung, *of sexual origin* 623
Cysten und Sexualität bei den Chrysophyten, *cysts and sexuality in Chrysophyta* 642—645
—, zweikernige, *binuclear* 644
Cystenformen in der Gattung Uroglena, *cysts, various form in the genus Uroglena* 644

Dauerlicht, Einfluß auf Geschlechtsverteilung, *continuous light, influence on sex rate* 161
Dauermodifikation, *modification, persistent* 146
Dauersporentypen bei Diatomeen, *endocysts in diatoms, various types* 664
Dauerzellen, *endocysts* 661
— bei Diatomeen, *cysts in diatoms* 661
— —, Klassifizierung, *resting cells in diatoms, classification* 664—665
—, Entwicklungsphysiologie bei Diatomeen *in diatoms, developmental physiology* 666—668
— bei den Heterokonten, *cysts, in Heterokontae* 682
—, Keimung bei Diatomeen, *endocysts, germination of, in diatoms* 662, 667
Dauerzustände bei Diatomeen, *resting cells in diatoms* 661
Deckschicht des Makroprothalliums bei Cycas, *jacket layer in the macroprothallium in Cycas* 417
Dedifferenzierung, *dedifferentiation* 803
Degeneration als Folge andauernder vegetativer Reproduktion, *as a consequence of vegetative propagation* 805
Detergentia, Einfluß auf Receptorgruppen an der Spermienoberfläche, *detergents, influence on receptor sites on sperm surface* 309
—, Wirkung von, auf Befruchtung, *effect of, on copulation* 307
Determination des Gametophyten, *of gametophyte* 235ff.
—, labile 251, 253
—, schrittweise, eines spezifischen Charakters, *stepwise, of a specific character* 253
— des Sporophyten, *of sporophyte* 235ff.
—, stabile, *stable* 253
Determinationsfaktoren, *determination factors* 244, 251
— bei Generationswechsel, *in life cycles* 238
— bei Sporen, *in spores* 238
Determinationsmuster der Entwicklung, *determination pattern of development* 238
Diatomeen, Vermehrung der, *diatoms, reproduction* 657—659
„Dichtsaateffekt" auf die Geschlechtsausprägung, *crowding effect on sex expression* 143
Differenzierung, bisexuelle, *differentiation, bisexual* 128
— der Inflorescenz, *of inflorescence* 268
—, sexualspezifische, *sex specific* 32
—, sexuelle, von Kernen, *sex differentiation, of nuclei* 130

Differenzierung, Umschlagen der, *differentiation, switch in* 253
—, unisexuelle, *unisexual* 130
— vegetativer Zellen bei den Volvocales, *of vegetative cells in Volvocales* 711
— der zoosporenbildenden Zellen, *of zoospore forming cells* 717
Differenzierungsmuster, *differentiation pattern* 48
Differenzierungsstoffwechsel, *differentiation, metabolism of* 245
dikaryotische Phase, Bildung der, *dikaryotic phase, formation of* 240
dikaryotischer Zustand, *dikaryotic condition* 130, 131
Dimorphismus, *dimorphism* 148
Dinophyten, parasitische, *Dinophyta, parasitic* 626
Diöcie, *dioecism* 34, 83ff.
diploide Phase, Begriff, *diploid phase, term* 264
diplophänotypische Geschlechtsbestimmung, *diplophenotypic sex determination* 114
Diplosporie, Terminologie, *diplospory, terminology* 552
DNS-Replikation, *DNA replication* 23
DNS-Synthese, *DNA synthesis* 22, 23, 24
— während der meiotischen Prophase, *during meiotic prophase* 364
Doppelbestäubung, *pollination, double* 522
doppelte Befruchtung bei Pinus, *double fertilization in pine* 430
Drosophila-Rumex-Typ der Geschlechtsbestimmung, *Drosophila-Rumex-type of sex determination* 275
Drosophila-Typus, *Drosophila type* 69

Eiapparat bei den Angiospermen, *egg apparatus in angiosperms* 435
Eigallerte, *egg jelly* 307
Eindringen, aktives, der Spermatozoiden bei Cycas und Ginkgo, *penetration of spermatozoids in Cycas and Ginkgo* 419f.
— von Phagenteilchen, *of phage particles* 314
— der Pollenschläuche in die Eizelle, bei Coniferen, *of the pollen tubes into the egg cell in pine* 428
eingeschlechtliche Blüten, *unisexual flowers* 275
Ein-Gen-Mutation, *one-gene-mutation* 237
Eiwasser, chemotaktische Anlockung der Spermatozoiden von Fucus, *egg water, chemotactic attraction of spermatozoids of Fucus* 309
— von Seeigeln, *of sea urchins* 309
Eizelloberfläche, *egg surface* 413
Eliminierung, selektive, von Gameten, *elimination, selective, of gametes* 93
Embryonen, Zahl der, per Samen bei Citrus-Varietäten, *embryos, number of, per seed in Citrus varieties* 558
Embryobildung, fraktionierte, *embryo formation, fractionated* 244
Embryoentwicklung, *embryo development* 524
Embryosack bei den Angiospermen, *embryo sac in angiosperms* 435—437
—, Bildung des, *formation* 552—555, 556—558
—, —, durch semiheterotypische Teilung, *semiheterotypic division* 553
end-to-end-Paarung, *end-to-end pairing* 62, 64
endogene Tagesrhythmik der sporogenen Stimmung bei Oedogonium, *endogenic clock of sporulation in Oedogonium* 734
Endopolyploidisierung, *endopolyploidization* 752
Endosperm, Chromosomenzahlen im, *chromosome numbers* 561
Endosperm-Entwicklung in agamospermen Arten, *endosperm development in agamospermous species* 558, 560—561
—, Störung der, *inhibition of* 525
Endosporen bei den Blaualgen, Entwicklung der, *endospores in cyanophyta, development of* 606—607
Entblätterung, Einfluß auf Geschlechtsausprägung, *defoliation, influence on sex expression* 164
Entdeterminierung, *dedetermination* 244
Entleerung der Sporangien, tagesperiodische Schwankungen, *release of sporangia contents, daily fluctuations* 897
Entobacteriaceae 24f.
Entwicklung der Eizelle in Cycas, *development of the megaspore in Cycas* 416f.
—, phänotypische Veränderungen, *phenotypic variation* 267
Entwicklungsrhythmus, *rhythm of development* 802
Entwicklungstendenzen, mögliche, *developmental tendency, possible* 245
Entwicklungscyclus, *life cycle* 34
— einer Mutante von Allomyces arbuscula, *of an Allomyces arbuscula mutant* 236
— von Codiolum, Abhängigkeit von Temperatur, *of Codiolum, in dependence of temperature* 238
— von Ectocarpus, *of Ectocarpus* 701
— von Levringia, *of Levringia* 701f.
Enzyme, geschlechtsbestimmende, *sex determining enzymes* 200
Epi-Hypostasie, genetische Wechselwirkung bei, *epi-hypostasy type, genetic interaction* 274
epi-hypostatisches Wechselspiel, *epi-hypostatic interplay* 275—276
Epiphytismus bei Algen, *epiphytism, in algae* 899
Episomen, *episomes* 5
Epistasie von Weiblichkeitsfaktoren, *epistasy of female factors* 759
Epistasie-Hypostasie-Mechanismus zwischen Genen, die für Gynaeceum und Androeceum verantwortlich sind, *epistasy-hypostasy mechanism between genes, responsible for gynaecium and androecium* 273
Erhaltung der Art, *preservation of species* 805

Erholungsperiode gepaarter Gameten bei Chlamydomonas, *recovery of paired gametes in Chlamydomonas* 305—306
„Erschöpfung" der weiblichen Tendenz, *"exhaustion" of female tendency* 142
Euglena gracilis, Entwicklungscyclus, *life cycle* 621
Exine 373
extrasexuelle Vorgänge, *extra sexual processes* 121

F-Duktion, *F-duction* 21
F-Faktor, *F-factor* 6, 7, 8, 9, 10, 11, 12, 13
—, Übertragung des, *transfer of* 20
F^--Mutanten, *F^- mutants* 7
F-pili-Phage 10, 12, 17
F-Replikon, Integration des bakteriellen und des, *F replicon, integration of bacterial and* 21
Fallenreaktion von Gameten, *trap reaction of gametes* 285
Fernreaktionen, *telemorphic reactions* 220
Fernwirkungen bei der Befruchtung, *long distance effects in fertilization* 285
Fertilität der Antheren, *fertility of anthers* 75
— bei Bakterien, Faktoren für, *in bacteria, factors effecting* 12ff.
— der Intersexe, *of intersexes* 70
— des Pollens, veränderte, *of pollen, changed* 88
Fertilitätsfaktor, *fertility factor* 6
Festheften des männlichen Gameten, *attachment of male gamete* 309
Festsetzen von Sporen, Einfluß der Oberflächenspannung, *adhesion of spores, influence of surface tension* 899
Flagellenkontakt bei der Kopulation von Algen, *contact of flagella in copulation of algae* 299—306
Formbildungsstoffe, *morphoregulators* 274
Formwechsel, sexueller, von Ceratium cornutum, *life cycle, sexual, in Ceratium cornutum* 626
Fortpflanzung, apomiktische, *reproduction, apomictic* 3
— bei den Braunalgen, Physiologie der, *in brown algae, physiology of* 672—702
—, geschlechtliche, *sexual* 3, 4
— bei den Heterokonten, *in Heterokontae* 682—691
—, sexuelle, phylogenetische Entwicklung der, *sexual, phylogenetic development of* 51
— bei Tribonema, *in Tribonema* 683
—, ungeschlechtliche, *asexual* 3, 4
Fortpflanzungskörper mit Dauersporencharakter bei den Blaualgen, *reproductive organs with the character of permanent spore in Cyanophaty* 608—618
Fortpflanzungssysteme bei Pilzen, *reproduction systems in fungi* 122
Fortpflanzungsverhältnisse bei den Ectocarpales, *reproduction in Ectocarpales* 238
Fortpflanzungsverhalten, *reproduction behaviour* 241
Fortpflanzungsweisen, vegetative, bei den Braunalgen, *reproduction, vegetative, various types, in brown algae* 693ff.
Fragmentation von Kolonien, *of colonies* 729, 731—733
Fremdbefruchtung, *cross fertilization* 82
Fruchtbarkeit, *fertility* 2
Fruchtkörperbildung bei den Acrasieen, *fruit body formation in Acrasieae* 285
Fucus-Ei, *Fucus egg* 307
Fusion, somatogene, von Hyphen, *somatogame, of hyphae* 130

Gametangien, *gametangia* 265
Gametangien-Befruchtung, *gametangiogamy* 284
Gametangiengruppierung, Wirkung von Tryptophan-Abkömmlingen, *grouping of gametangia, influence of tryptophane derivates* 113
Gameten, *gametes* 258
—, Beweglichkeit, *mobility* 298
—, Selektion der, *selection of* 450
—, Vegetative Entwicklung der, *vegetative development* 237—238
Gametenbildung, *gametes, formation* 236
— bei Vaucheria, durch photolytische Zerfallsprodukte des Tryptophans, *gamete formation, in Vaucheria, by photolytic products of tryptophane* 688
Gameten-Filtrate, Wirkung in vitro, *gamete filtrates, effect in vitro* 288
Gametenkopulation, *copulation of gametes* 283
Gametenverschmelzung, *gamete fusion* 298ff.
— bei *Noctiluca, in Noctiluca* 628
Gametophyten, Elimination der, Tendenz zur, *gametophytes, elimination of, tendency to* 238
Gamon von Chlamydomonas, Molekülgewicht, *gamone, of Chlamydomonas, M.W.* 302
—, niedermolekulares, bei Hefe, *low molecular, in yeast* 294
Gamon-Produktion, Wirkungsspektrum, *gamone production, wavelength dependence* 303
Gamone (Befruchtungsstoffe), *gamones* 32, 33, 215, 218, 260, 261, 282f., 311
— bei Achlya, *in Achlya* 292
—, agglutinierende, *agglutinating* 306
— bei Bakterien, *in bacteriae* 315—316
—, Definition 283
—, Geschlechtsspezifität, *sex specificity* 312
—, Glykoproteide als, *glycoproteids as* 303
—, Proteinkomponenten, *protein constitutions* 312
Geißel, Bedeutung der, bei der Gruppenbildung, *flagellum, importance of, in pairing process* 303
Gemischtgeschlechtlichkeit, *sexuality, mixed* 136
—, Tendenz zur, *tendency to* 112
Gen-Austausch, *gene exchange* 101

Gen-Balance-System, *gene balance system* 93
Gen-System, *gene system* 92
Gene, additive, bei der Kontrolle der natürlichen Wachstumsprozesse, *genes, additive, controlling native growth regulation* 272
Generation, Begriff, *term* 265
Generationen, Abfolge der, *generations, succession of* 264
Generationsdauer, *generation time* 158
Generationswechsel, *alternation of generations (see: life cycles)* 3, 39—42, 235ff., 241, 262—268, 275, 279
—. Abhängigkeit von Temperaturwechsel, *life cycles, dependence on temperature* 239
—. allgemeine Prinzipien, *general principles* 257ff.
— bei Braunalgen, *of Phaeophyta* 266, 700—702
—. diphasischer, diphasic 239
—. Entstehung, origin 235ff.
—. phylogenetische Gesichtspunkte, *phylogenetic aspects* 235
—, veränderte Stoffwechselansprüche beim, *changed metabolism of* 244
Generationswechseltypen, *life cycles, types* 240, 242
genetische Information, Übertragung der, *genetic information, transfer of* 2, 5
genetisches Material, Übertragung des, *genetic material, transfer of* 16, 17
Genokopie, *genocopy* 273
genotypische/phänotypische Geschlechtsbestimmung, Übergang, *genotypic/phenotypic sex determination, transition* 114
Geschlecht, *sex* 2, 31
—, der Begriff, *concept of* 257
—, Elimination, unterschiedliche, *elimination, differential* 160
—, Evolution 101
— bei haploiden Pflanzen, *of haploid plants* 89
—, Nachweis bei höheren Pflanzen, *in higher plants, existence of* 262
—, weibliches, Manifestation des, Abhängigkeit von Außenbedingungen, *female, manifestation of, dependence upon external conditions* 137
Geschlechter, Reihenfolge der, *sexes, sequence of* 136
—, Verhältnis der, *sex ratio* 52
Geschlechterverhältnis, in tetraploiden und aneuploiden Formen, *sex ratio, in tetraploid and aneuploid forms* 78
Geschlechterverteilung, *sex distribution* 135
geschlechtliche Affinität, Determination, mittels diffusibler Sekrete, *sexual affinity, determination of, by diffusible secretions* 232
— Anziehung, Determination der, *affinity, determination* 228—233
— Bipolarität, *bipolarity* 274
— Fortpflanzung, allgemeine Prinzipien, *reproduction, general principles* 257ff.
— Wechselwirkung, Muster der, *interaction, pattern of* 219ff., 226
Geschlechtlichkeit, Zeichen der, *sexuality, sign of* 265
Geschlechtsaufspaltung, *sex segregation* 79, 82
Geschlechtsausbildung, Änderung der, *sex development, change of* 94
—, verschiedene experimentell erhaltene Typen bei Melandrium, *sex expression, different experimental types in Melandrium* 73
Geschlechtsausprägung, *sex expression* 79, 82, 97, 100, 274
—, Änderung, *change* 182
—, Änderung der, durch Pfropfungen, *changing of, by transplantation* 171, 177
— bei Angiospermen, *in angiosperms* 274
— bei Bastarden, *in hybrids* 183
—, Einfluß von Gibberellinsäurebehandlung auf, *influence of gibberellic acid treatment* 166, 178, 180, 183, 187ff.
—, — der Größe und des Alters der Samen, *of size and age of seeds* 174
—, — von Kinetin-Behandlung, *of kinetin treatment* 166
—, — von Kohlenmonoxydbehandlung, *CO atmosphere* 177, 179f, 187
—, — der mineralischen Düngung, *of mineral supply* 163ff., 167, 172, 174, 189
—, — der Wetterbedingungen, *of weather conditions* 168, 180, 190
—, — der Zellgröße auf, *of cell size on* 117
—, Temperaturwirkung auf, *temperature effects on* 162
Geschlechtsbestimmung, *sex determination* 39, 113ff., 179, 267
— bei Ascobolus, *in Ascobolus* 231
—, auslösende Faktoren, bei Moosen, *determination of, in bryophytes* 140
— bei Diatomeen, *in diatoms* 115
— bei pennaten Diatomeen, *in pennate diatoms* 674—677
—, diplogenotypische, *diplogenotypic* 51ff.
— in der Diplophase, *during diplophase* 116
— in der Gattung Melandrium, *in Melandrium* 170f.
—, genetische Faktoren bei der, *genetical factors in* 140ff.
—, genotypische, *genotypic* 128
—, haplogenotypische, *haplogenotypic* 31, 42
— bei Hefen, *in yeast* 126
—, Modifikation der, im Sporophyten, *modification of, in sporophytes* 155ff.
—, modifikative, Beeinflussung durch Wuchsstoffe, *phenotypic, influence of growth hormones* 198f.
—, — Einfluß des Alters der Pflanzen, *influence of age of the plants* 196f.
—, —, — der Behandlung mit tierischen Geschlechtshormonen, *of treatment with sex hormones* 198
—, —, — des Bodenwassergehaltes, *of soil water relation* 193
—, —, — edaphischer Faktoren, *of edaphic factors* 192ff.
—, —, — der Lichtbedingungen, *of light conditions* 193

Geschlechtsbestimmung, modifikative Einfluß, der Luftfeuchtigkeit, *sex determination, phenotypic, influence of humidity* 193
—, —, — von Parasitenbefall, *of infection* 196
—, —, — der Temperatur, *of temperature* 194
—, —, — der Transplantation, *of transplantation* 194f.
—, —, Interpretation, *theoretical interpretation* 199ff.
—, —, Pflanzenarten mit, species with 157ff.
—, modifikatorische, bei Dikotyledonen, *phenotypic, in Dicotyledonae* 164—192
—, —, bei Gymnospermen, *in Gymnosperms* 158—159
—, —, bei Monokotyledonen, *in Monocotyledonae* 160—164
—, —, bei Spermatophyten, *in spermatophytes* 153ff.
—, —, im Sporophyten, *in sporophytes* 155ff.
—, nicht-genetische Faktoren bei der, *non-genetic factors in* 142
—, phänotypische, bei Algen, *phenotypic, in algae* 110ff.
—, —, bei Archegoniaten, *in archegoniatae* 134ff.
—, —, bei Pilzen, *in fungi* 121ff.
—, Varianten der, *variations in* 35
—, Wirkung von Wuchsstoffen, *influence of growth hormones* 143
Geschlechtsbestimmungs-Mechanismus, *sex determination mechanism* 65
— bei Melandrium, *in Melandrium* 61
— —, Schema des, *scheme of* 76
Geschlechtscharakter, *sex character* 51
—, komplementärer, *complementary* 36
Geschlechtschromosomen, *sex chromosomes* 52, 57ff., 63, 89, 101
—, heteromorphe, *heteromorphic* 58—60, 89, 267—268
—, —, möglicher Entstehungsmechanismus, *possible mechanism of the origin* 275, 276
—, homomorphe, *homomorphic* 66
—, Paarungsverhalten der, *pairing behaviour* 77
Geschlechtschromosomen-Komplex, *sex chromosomes complex* 64
Geschlechtschromosomen-Mechanismus, *sex chromosomes mechanism* 64
Geschlechtschromosomen-Typen, multiple, *sex chromosome types, multiple* 65
Geschlechtsdetermination durch Hormone, *sex determination by hormones* 230
—, Stabilität der, *stability of* 146
Geschlechtsdifferenzierung, *sex differentiation* 36
— bei E. coli, *in E. coli* 6
—, Muster der, bei Algen, *pattern of, in algae* 229
Geschlechtsdimorphismus, *sexual dimorphism* 82, 201
Geschlechtsduktion, *sex duction* 21
Geschlechtsentwicklung bei Achlya, *sexual progression in Achlya* 221—226
Geschlechtsexpression, *sex expression* 265
— und Blühstadium, *and flowering* 268ff.
Geschlechtsfaktoren, *sexual factors* 30, 51ff., 90, 93, 125
—, Lokalisation der, *localisation of* 93
—, multiple Allelie der, *multiple allely of* 90
—, im Plasma lokalisierte, *plasmatic* 94
—, Übertragung von, *transfer of* 18
—, Wirkung von, *effect of* 93
Geschlechtsgene, *sex genes* 75, 140
Geschlechtshormone, *sexual hormones* 215ff.
— bei Achlya, *in Achlya* 221-224
— bei Bombardia lunata, *in Bombardia lunata* 226
— bei Oedogonium, *in Oedogonium* 226, 232
Geschlechtsmerkmale, sekundäre, *sex characters, secondary* 86, 87, 147—149
Geschlechtsorgane, *sex organs* 257, 265
— in Blüten, *in flowers* 272f.
—, Differenzierung, *differentiation* 268, 275
—, — der, bei Saprolegniaceen, *in Saprolegniaceae* 124
—, Entwicklung der, *development of* 117
—, Hemmung der, *inhibition of* 155
—, Induktion der, *induction of* 214ff.
—, koordinierte Entwicklung, *coordinated development* 216—228
—, Verteilung der, *distribution of* 136
Geschlechtspaarung, *pairing, sexual* 261
Geschlechtspartner, chemische Polarität, *sexual cells, chemical polarity* 316
Geschlechtsprozesse, chemische Regulation der, *sexual processes, chemical regulation* 282ff.
—, hormonale Kontrolle, bei Hefe, *hormonal control in Saccharomyces* 228
Geschlechtsreaktion, Koordination der, *sexual reaction, coordination* 219
—, Reihenfolge der Stadien, *sexual reaction, sequence of stages* 219
Geschlechtsrealisation, Trigger für die, *sex, realisation of, trigger for* 275
Geschlechts-Realisatoren, *sex realisators* 92
Geschlechtsstoffe, diffusible, bei niederen Pflanzen, *sexual substances, diffusible, in lower plants* 232
Geschlechtstendenz, *sex tendency* 54, 98
—, Ausprägung der, *expression of* 169
Geschlechtstheorie, nach HARTMANN, *sexuality, theory of, following* HARTMANN 117
Geschlechtstypen, *sex types* 90
Geschlechtsumkehr, Tendenz zur, *sex change, tendency to* 165
Geschlechtsunterschiede, *sex differences* 31, 32
—, Kriterien, bei Diatomeen, *criteria for, in diatoms* 117
Geschlechtsveränderung, Einfluß der Chromosomen-Verdoppelung, *sex, change of, influence of chromosome duplication* 88
Geschlechtsvererbung, polyfaktorielle, *sex inheritance, polyfactors* 88

Geschlechtsverhältnis, *sex ratio* 54, 56, 87, 88, 93, 94, 114
—, Einfluß der Behandlung mit α-Naphtylessigsäure, *influence of NAA treatment* 167, 170, 177, 180, 183, 187, 190
—, — der Düngung, *of mineral nutrition* 174
—, — der Kulturbedingungen, *of culture conditions* 182
—, — der Lichtintensität, *of light intensity* 182
—, — der Tageslänge, *of day length* 169f., 174ff., 178f.
—, — der Temperatur, *of temperature* 176, 185
—, jährliche Änderung, *yearly change of* 169, 186
—, normales, *normal* 63
—, Verschiebung des, *change of* 57, 69, 91, 143, 454
Geschlechtsverhalten bei Moosen, *sex behaviour, in mosses* 147
Geschlechtsverteilung, *sex distribution* 92
gestaltbildende Prozesse, *morphogenetic processes* 241
Gestaltprinzip, *morphogenetic principles* 241
Gestaltunterschiede zwischen den Generationen, *morphological differences between the generations* 243f.
Gewebebildung, *tissue formation* 242
Giberelline, *gibberellins* 803
Gibberellinsäure, Antagonisten, *gibberellic acid, antagonists* 201
—, Behandlung mit, *treatment with* 159
Gibberellinsäurebehandlung, Einfluß auf Geschlechtsausprägung, *gibberellic acid treatment, influence on sex expression* 166
Gonenkonkurrenz, *competition of gones* 447—476
—, entwicklungsgeschichtliche Grundlagen, *developmental background* 450f.
—, genetische Grundlagen, *genetical background* 448—451
— und genetische Wechselwirkungen, *and genic interactions* 474
— in der Samenanlage, *in the ovary* 471—475
—, Variabilität der, *variability of* 472
—, Zusammenhang mit der Selbststerilität, *certation, relation to self sterility* 468f.
Gonidienbildung, *gonidia, formation of* 627
Gradient in der Blüte, *floral gradient* 271
Griffel, Atmungsverhalten während des Pollenschlauchwachstums, *styles, respiration during pollen tubes passing* 513
—, Gradient der Pollenschlauchhemmung, *gradient of inhibition within* 518
—, Pollenschlauchwachstum im Innern des, *pollen tube growth within* 457f.
—, Reduktion des, durch Kurztagbehandlung, *reduction, by SD treatment* 169
Größen-Variation, bei Diatomeen, *size variation, in diatoms* 114
Größenverschiebung der Generationen, *shift in size of the generations* 244
Gruppenbildung bei den Braunalgen, *group formation brown algae* 308
Gruppenbildung der Gameten, *clumping of gametes* 283, 285
Gynogamon 303
—, Definition 283
Gyno-Intersexualität, *gyno intersexuality* 79

Häufigkeit Weibchen/Männchen, *frequency, male-female* 53
haploide Phase, Begriff, *haploid phase, term* 264
Hefen, Genetik der, *yeast, genetics of* 126
Hemiautosporen, *hemiautospores* 724—726
Hemizoosporen, *hemizoospores* 722
Hemmungsfaktor, *inhibitor* 88
Herbst-Fertilität, *end-season-fertility* 522
Hermaphroditismus, funktioneller, *hermaphroditism, functional* 215
Heteroblastie, *heteroblasty* 700
Heterochromatin-Gehalt, *heterochromatin content* 58
Heterocysten, Einfluß der Stickstoffkonzentration im Medium, *heterocysts, influence of nitrogen concentration* 615
—, Entstehung der, bei Anabaenopsis, *heterocysts, development of, in Anabaenopsis* 611
—, Entwicklungsphysiologie der Entstehung, *heterocysts, physiology of development* 612—617
—, Feinstruktur, *fine structure* 612
—, Keimung, *germination* 617—618
—, Mechanismus der Spontandifferenzierung, *mechanisms of spontaneous differentiation* 615
—, relative Zahl der, Molybdän-Bedarf, *relative number, influence of molybdenum* 615
Heterogametie, *heterogamety* 52, 55, 87
Heterokonten, alternative sexuelle-vegetative Fortpflanzung, *Heterocontae, alternative sexual-vegetative reproduction* 687f.
—, Fortpflanzung, *reproduction* 682—691
Heteromorphie von X- und Y-Chromosomen, *heteromorphy of X and Y chromosomes* 64
Heterosis 544
Heterosomen, *heterosomes* 57
Heterothallie, *heterothally* 49
Hfr-DNS, *Hfr DNA* 23
Hitzebehandlung, Einfluß auf Geschlechtsbestimmung, *heat treatment, influence on sex expression* 165
Homogametie, *homogamety* 55
homo-heterothallische Reaktionen bei Achlya, *homo-heterothallic reactions in Achlya* 231
Homozygotie-Heterozygotie-Mechanismus, *homozygoty heterozygoty mechanism* 52
Hormogonien, Entwicklung der, *hormogonia, development of* 603—606
Hormon A, chemische Analyse, *hormone A, chemistry* 223
Hormon A-Komplex, Achlya, *hormone A-complex, Achlya* 221, 222
Hormon B, Achlya, *hormone B, Achlya* 222
Hormon C, Achlya, *hormone C, Achlya* 222

Hormon D, Achlya, *hormone D, Achlya* 222
hormonaler Koordinationsmechanismus bei Achlya, *hormonal coordinating mechanism, in Achlya* 219—223
— — bei Ascobolus, *in Ascobolus* 224
Hormonproduktion männlicher Mycelien bei Achlya, *hormone production of male mycelia in Achlya* 294
Hybridengewinnung, *hybrids, production of* 524
Hyphenfusion, *hyphal fusion* 284

IES, *IAA* 272, 803
IES-Oxydase, *IAA oxydase* 88
IES-Oxydase-Aktivität, geschlechtsabhängige, *IAA oxydase activity, sex dependent* 171
Immunitäts-Theorie der Inkompatibilität, *immune theory of incompatibility* 518—519
Immunreaktion zwischen Pollenschlauch und Leitgewebe, *immunological reaction, between pollen tubes and style* 512—513
Incompatibilität s. Inkompatibilität (Fußnote S. IX)
Induktion von Konjugationsschläuchen, bei Hefe, *induction of conjugation tubes in Saccharomyces cerevisiae* 228
Inflorescenz, *inflorescence* 268—271
—, blütenlose, *without flowers* 269, 270
—, endständige, *terminal* 162
—, vegetative 270, 271
Initialzonen bei der Gewebebildung, *initial zones, in tissues* 242
Initiator 19
Inkompatibilität, *incompatibility* 35, 131
— bei Ascobolus, *in Ascobolus* 231
—, heterogenische, *heterogenic* 324
—, homogenische, *homogenic* 324—337
—, interspezifische, *interspecific* 506, 524—525
— bei Phanerogamen, *in tracheophyta* 506—525
— bei Pilzen, *in fungi* 224f.
— bei den Thallophyten, *in thallophytes* 321ff.
—, unilaterale, interspezifische, *unilateral, interspecific* 524
—, Verbreitung, *occurrence* 322ff., 507f.
Inkompatibilitätsfaktoren, *incompatibility factors* 33, 35, 127, 131
Inkompatibilitätsreaktion, *incompatibility reaction* 511
—, Einfluß von Bestrahlung auf, *influence of radiation* 523
—, — von Wuchsstoffen, *of growth hormones* 523
—, Isoantikörper, *isoantibodies* 519
—, Lokalisierung der, *localization* 520
— bei polyploiden Pflanzen, *in polyploids* 521
— in der Samenanlage, *in the ovule* 516—517
—, serologische Spezifität, *serological specificity* 519
Inkompatibilitätsreaktion, Temperaturabhängigkeit, *incompatibility reaction, temperature dependence* 519, 522f.
—, Überwindung der, *break down of* 520—523
Inkompatibilitäts-Theorien, *incompatibility, theories* 517ff.
„innere Uhr“ bei Vaucheria, *internal clock, in Vaucheria* 690
Instabilität, sexuelle, *instability, sexual* 126
Intersexe, *intersexes* 71, 79, 82, 87
—, primäre, *primary* 94
Intersexualität, *intersexuality* 175
Intersexualitätsgrad, *intersexuality, degree of* 70
intersexuelle Einzelblüten, *flowers, intersexual* 181
Intersexueller Charakter, *intersex character* 78
Intine 373
Isoagglutination 306
—, in konzentrierten Filtraten von Algen-Gameten, *in concentrated filtrates of gametes of algae* 300—302
isogame Befruchtung bei Algen, *isogamie fertilization in algae* 298—307
Isogametie, *isogamety* 283—284
Isogamie, *isogamy* 258

Jacob-Brenner-Modell, Jacob-Brenner *model* 22, 23
Jacob-Monod-Modell, Jacob-Monod *model* 100
Jugendformen, *juvenile forms* 266

Kallus, gametophytischer, *callus, gametophytic* 251—252
Kalluswuchsform, *callus growth form* 253
Karposporen, *carpospores* 238
Karyogamie, *caryogamy* 240, 264
— (Kernverschmelzung) bei den Angiospermen, *karyogamy in angiosperms* 439—441
—, verzögerte, Ursachen für, *retarded, causes of* 240
Karyologie und Ökologie, *karyology and ecology* 804
Kastration, physiologische, *emasculation, physiological* 202
Keimung, spontane parthenogenetische, von Cutleria-Eiern, *germination, spontaneous parthenogenetic, of Cutleria eggs* 704
— der Zoosporen bei Ulothrix, *of zygotes in Ulothrix* 742
— der Zygote, *of zygote* 240
Keimungsbedingungen für Pollen, *germination conditions, pollen* 384—385
Keimzellen, allmähliche Reduktion der Zahl der, *sexual cells, reduction of the number of* 236—237
—, Qualität der, *germ cells, quality* 235—237
Kernfragmentation, *nucleus fragmentation* 752
Kernfusion, *nuclear fusion* 282
Kern-Plasma-Relation, bei Diatomeen, *nucleus-plasma-relation, in diatoms* 678

Kern-Plasma-System, *nuclear plasma system* 93, 100
Kernteilung, acytokinetische, bei Diatomeen, *nuclear division, acytokinetic, in diatoms* 657, 662f.
Kernverschmelzung, *nuclear fusion* 285, 286
— bei Ginkgo, *in Ginkgo* 420
— bei Pinus, *in pine* 429
Kernwanderung, *nuclear migration* 286
Kieselsäureausscheidung, bei Diatomeen, *secretion of silicate, in diatoms* 660
Kieselsäuremangelwirkungen, *deficiency, silicate, effect of* 659
Kinetombildung, Cytologie der, *kinetom formation, cytology of* 719
Klebrigkeit der Geißel, *stickiness of the flagellum* 304
Klon-Kulturen von Diatomeen, *clonal cultures of diatoms* 114
Klon-Selektionstheorie der Inkompatibilität, *clonal selection theory of incompatibility* 520
Knollenbildung, *tuber formation* 790, 803
Knospenbestäubung, *bud pollination* 521, 522
Knospenbildung an Blättern, *bud formation in leaves* 794—797
— an Wurzeln, *in roots* 797—799
Kolonieteilung bei Diatomeen, *colonies, division of, in diatoms* 660
Kolonievermehrung bei Flagellaten, *reproduction of colonies in Flagellatae* 639—641
Kompatibilität, *compatibility* 35
Kompatibilitätsgruppen, *compatibility groups* 130
Konjugation 258
—, bakterielle, bei anderen Bakterien als E. coli, *bacterial, other than E. coli* 24ff.
— bei Bakterien, *in bacteria* 259
— bei den Conjugaten, *in Conjugatophyceae* 748
—, molekulare, *molecular* 259
Konjugationsschläuche bei Hefe, *conjugation tubes, in Saccharomyces cerevisiae* 228
Konkurrenz, *competition* 56
— in Pollenschlauch und Samenanlage, *in pollen tube and ovary* 474f.
— unter den S-Genen, *of S genes* 521
Konkurrenzversuche, *competition experiments* 72
Konstruktionsprinzipien, morphogenetische, *construction principles, morphogenetic* 240f.
Kontakt, Kinetik des effektiven, bei Bakterien, *contact, kinetics of the, in bacteriae* 314f.
Koordinationsmechanismus, hormonaler, bei der Entwicklung von Geschlechtsorganen, *coordinating mechanism, hormonal, in development of sexual organs* 220
Koordinierung, zeitliche, bei der Befruchtung der Coniferen, *timing of events during fertilization of pine* 427
Koppelungsgruppen, *linkage groups* 9
Koppelungsgruppen, ringförmige, bei E. coli, *circular linkage group of E. coli* 18
Kopulation 130, 258
— bei Ascobolus, *in Ascobolus* 297
— von Bakterien, *of bacteriae* 314
— bei Bombardia, *in Bombardia* 297
— von Gameten bei Dinobryon, *between gametes of Dinobryon* 304
—, illegitime, bei Hefen, *illegal, in yeast* 126
—, innerhalb des Klons, *within a clone* 117
—, isogame 117
—, pädogame, *paedogamic* 127
—, Physiologie der, *physiology of* 43
Kopulationshäufigkeit, *copulation, frequency* 35
Kopulationsvorgänge, *copulation processes* 32, 214
Kopulationszone bei Mucor, *copulation zone in Mucor* 294
Korrelation zwischen generativer und vegetativer Fortpflanzung, *correlation between sexual and vegetative reproduction* 804
— zwischen sexuellen und asexuellen Prozessen, *between sexual and asexual processes* 788
Korrelationserscheinungen zwischen den sexuellen Tendenzen, *correlation between sexual tendencies* 144f.
korrelative Beziehung zwischen Bestäubung, Pollenschlauchwachstum und Entwicklung der weiblichen Fortpflanzungsorgane, *correlation between pollination, pollen tube growth and development of female organs* 441
Kreuzungs-Inkompatibilität, *cross incompatibility* 506, 523—524
Kreuzungssterilität, *cross sterility* 523
Kreuzungstyp, tetrapolarer, *mating types, tetrapolar* 46
Kulturbedingungen bei Heterokonten, *culture conditions in Heterokontae* 686f.

lac Operon 21
Lagerungsbedingungen, Einfluß auf die Lebensfähigkeit des Pollens, *storage conditions, influence on viability of pollen* 397—398
Langtag-Prinzip, übertragbares, *long-day principle, transferable* 803
Lebenscyclus von Syracospaera carterae, *life cycle of Syracospaera carterae* 647
Lebensfähigkeit, Pollen, *viability, pollen* 395ff.
— des Pollens, Einfluß der Temperatur, *viability of pollen, influence of temperature* 396f.
Leptotän-Chromosomen, Doppelstrang, *leptotene chromosomes, double strands* 363
Letalfaktoren, *lethal factors* 93
Lichteinfluß auf die Keimung von Cysten, *light, influence on germination of cysts* 633, 635, 646
Lichtintensität, Einfluß auf vegetative Fortpflanzung, *light intensity, influence on vegetative propagation* 802

Lichtintensität, reduzierte, Beeinflussung der Geschlechtsausprägung durch, *light intensity, reduced, influence on sex expression* 166, 172
„Liebesleben", *sex phenomena* 259
Locus für die Geschlechtsfaktor-Affinität, *sex factor affinity locus (sfa)* 25
Luftfeuchtigkeit, Einfluß auf Lebensfähigkeit des Pollens, *humidity, influence on viability of pollen* 395f.
Lygaeus-Typus, *Lygaeus type* 61
Lysogenese, *lysogenesis* 2

männlich blühende Pflanzen mit weiblichem Habitus, *male plants with female habit* 83f.
männliche Populationen, *male populations* 81
Männlichkeitsfaktor, plasmatischer, *male factor, plasmatic* 92
Männlichkeitsfaktoren, *male factors* 81
Mannane, *mannans* 313
Meiocysten, *meiocysts* 653
Meiose, *meiosis* 1, 61, 63, 236, 243, 264, 275, 344—364, 369, 532, 626, 630, 761, 764
—, Abgabe von Kernmaterial, *extrusion of nuclear material* 347
—, Abweichungen der, *aberrations* 359—363
—, —, induziert durch ionisierende Strahlen, *induced by ionizing radiations* 361—363
—, Determination der, *determination of* 239, 246f.
—, einheitliches Verhalten der Centromeren homologer Chromosomen, *unitary behaviour of the centromeres of homologous chromosomes* 353f.
—, lange Prophase, *long prophase* 347
—, Paarung der homologen Chromosomen, *pairing of homologous chromosomes* 347—350
—, Synchronisierung in Antheren, *synchronisation in anthers* 370
—, Synthese der Nucleinsäuren und Proteine, *synthesis of nucleic acids and proteins* 354—358
—, unregelmäßige, in diplosporen Arten, *irregular, in diplosporous species* 556
—, unterdrückte, *suppressed* 759
—, unterscheidende Kennzeichen, *meiotic division, distinguishing features* 344—358
—, Verhalten der Tapetumzellen, *behaviour of tapetal cells* 355f.
—, Verschiebung der, *shift of* 239
meioseähnliche Prophase bei Diatomeen, *meiosis-like prophase, in diatoms* 677
Meiose-Chromosomen, Spiralisierung-Charakteristika, *meiotic chromosomes, coiling characteristics* 352
Meiose-Theorien, *meiosis, theories* 363f.
Meiosewellen in Antheren, *meiotic waves in anthers* 370
Meiosporen-Bildung, *meiospores, formation of* 653
Melandrium-Typus, *Melandrium type* 72ff., 87
Membranbildung bei den Flagellaten, *membrane formation in Flagellatae* 624
Mendelsches Spaltungsverhältnis, Abweichungen vom zu erwartenden, *segregation, preferential* 449
Meromixis 314, 315
Merozygote 21
Metromorphie, *metromorphy* 759
Mikrosporen, Keimung der, bei Ginkgo, *microspores, germination, in Ginkgo* 415
Mikrosporenentwicklung, Einbauprozesse in Nucleinsäuren und Proteine, *microgametogenesis, incorporation into nucleic acids and proteins* 355ff.
Mikrosporenmutterzelle, *microspore mother cell* 154
Mischbildungen zwischen Gametophyt und Sporophyt, *mixture between gametophyte and sporophyte* 253—264
Mischmycel, *mycelium, mixed* 253
Mitose bei Diatomeen, *mitosis in diatoms* 657
Modifikation, korrelative, bei der Blütenbildung, *modification, correlative, in flower formation* 270
Modifikator-Gene für die S-Gene, *modificator genes of S genes* 521
Monöcie, *monoecism* 34, 83ff.
Monosporangium von Erythrotrichia, *of Erythrotrichia* 761
Monosporen, *monospores* 765—780
morphogenetische Prinzipien, *morphogenetic principles* 240ff.
morphologische Differenzierung im Zusammenhang mit der Fortpflanzung, *morphological differentiation in connection with reproduction* 286
multiple Allelie der Geschlechtsfaktoren, *allely of sex factors* 90
Mutanten von Chlamydomonas, *mutations in Chlamydomonas* 305
—, eingeschlechtliche, *mutants, monosexual* 128
Mutualreaktion zwischen keimenden Pollenkörnern, *mutual reaction between pollen grains* 387

Narbencuticula, *stigmatic cuticle* 509
Narbenoberfläche, *stigma surface* 521
Narbenschleim, *stigmatic exudate* 385, 522
NES, *NAA* 273
„Neuter"-Pflanzen, "*neuter*" *plants* 95
„Neuters" (Pflanzen ohne Blüten), "*neuters*" *(plants without flowers)* 81
Nondisjunktion, *nondisjunction* 34
Nucellarembryogenie, *nucellar embryogenesis* 277

Oberfläche der Bakterien, *bacterial surface* 9, 10
Oberflächenkomponenten, aktive, *surface, active sites* 312
Oberflächenkontakt bei Bakterien, *surface contact, in bacteriae* 314
Obturator 431—433
Öffnung der Pollenschläuche bei Cycas und Ginkgo, *opening of the pollen tubes in Cycas and Ginkgo* 418

Ontogenese, Determination in der, *ontogenesis determination during* 245ff.
Ontogenieabschnitte, *ontogeny, phases* 235f.
Operator-Regulator-System 100
Organbildung, *organ formation* 242
Organcharakter, geschlechtlicher, *sexual character of organs* 145
Organisationsstufe, *organisation level* 241
osmotischer Schock, Wirkung der Gewebekultur, *osmotic shock, effect of culture* 361
Ovarialstoffe, *ovarial substances* 518
Ovarialtumoren, *ovarial tumors* 524

Paarung, *pairing* 63
— bei Diatomeen, *in diatoms* 675
— von kompatiblen Zellen, *of compatible cells* 16
—, Spezifität der, Hypothese basierend auf dem Vorkommen von einfachen DNS-Strängen, *specificity of, based on the occurrence of single stranded DNA* 349
Paarungsstoffe, *mating substances* 316
Paarungsstrukturen („Cores"), *pairing-structures ("cores")* 261
Paarungstypen, *mating types* 6, 43, 44, 45, 46, 48, 49, 122, 127, 129
Paarungstypsubstanzen, *mating type substances* 284
— bei Paramaecium, *in Paramaecium* 304
Paarungsverhältnisse bei Diatomeen, *pairing condition in diatoms* 116
Pädogamie, *paedogamy* 127
— bei Cymbella, *in Cymbella* 677
— bei Gomphonema, *in Gomphonema* 675
Palmella-Bildung, Faktoren für, *palmella formation, factors concerning* 622
Panspermie, Theorie der, *panspermy, theory of* 257
parasexuelle Erscheinungen, *parasexual phenomena* 121
— Cyclen, *cycles* 2
Parasporen, *paraspores* 769
Parthenogameten, *parthenogametes* 237, 600f.
Parthenogenese, *parthenogenesis* 277f.
— bei den Charophyceen, *in Charophyceae* 758—759
— bei den Chlorophyceen, *in Chlorophyceae* 741—743
— bei den Conjugalen, *in Conjugales* 750—752
—, Definition 530f.
— bei Diatomeen, *in diatoms* 672—677
—, diploide, *diploid* 117
—, haploide, *haploid* 551
— bei Heterokonten, *in Heterokontae* 690f.
— bei den Rhodophyten, *in Rhodophyta* 774
— bei den Spermatophyten, *in spermatophytes* 787
parthenogenetische Entwicklung, *parthenogenetic development* 237
PEA (Phenäthylalkohol), Inhibitor für die DNS-Synthese, *PEA (phenethyl alcohol), inhibitor of DNA synthesis* 23

Pektin-Synthese, Einfluß der Borsäure im Pollenschlauch, *pectin synthesis, influence of boric acid in pollen tubes* 386
Pfropfung, Beeinflussung des Geschlechtes durch, *grafting, sex exchange by* 177
Phasenübergang bei den Haptophyceen, Bedingungen für, *phase transition, conditions for, in Haptophyceae* 652—653
Phasenwechsel, *alternation of phases* 262ff.
Photoperiodismus, *photoperiodism* 801—802
Phototaxis bei Braunalgen, *in brown algae* 698
—, positive, bei Euglena gracilis, *in Euglena gracilis* 622
Physiologie der vegetativen Fortpflanzung bei den Spermatophyten, *physiology of vegetative reproduction in spermatophytes* 800ff.
„pili" (Oberflächen-Auswüchse), *(surface outgrowths)* 10
pili (charakteristische Fortsätze) an der Bakterienoberfläche, *on the surface of bactria* 316
Pistil, Ausbildung des, *development of* 77
Plasma-Brücken bei verschmelzenden Bakterien, *cytoplasmatic bridges in conjugating bacteria* 259
plasmatische Inkompatibilität, *plasmatic incompatibility* 524
plasmodiale Organisation, *plasmodial organisation* 241
Plasmodienbildung, *plasmodia, formation of* 241
Plasmogamie, *plasmogamy* 240, 264, 286
— (Plasmaverschmelzung) bei den Angiospermen), *in angiosperms* 438—439
Plasmon, Beteiligung des, bei Artkreuzungen, *plasmone, participation of, in species crosses* 96
Plasmon-Mutationen, *plasmone mutations* 94
Plasmonwirkung bei der Gonenkonkurrenz, *plasmone influence on certation* 467
Pleomorphismus der Diatomeen, *pleomorphism in diatoms* 668—672
— zwischen Cysten und Cysteninhalt, *between cyst and cyst content* 651
Polarität, lichtinduzierte, bei Tetrasporen von Dictyota, *polarity, light induced, on tetraspores of Dictyota* 899
—, sexuelle, *sexual* 282
— der Sporen, prädeterminierte, *of spores, predetermination of* 764
Pollen 56, 368—399
—, Abnahme der Keimfähigkeit, Ursachen für, *decrease of viability, reasons* 396—399
—, Alterungsversuche mit, *aging experiments* 57, 78
—, anorganische Substanzen, *inorganic material* 383
—, Behandlung mit Alkohol, *treatment with alcohol* 56
—, Bildung des, *formation* 555f., 558
—, —, durch verschiedene Typen abweichender meiotischer Teilung, *by various types of irregular meiotic division* 555

Pollen, Biochemie des, *biochemistry* 373—384
—, Entwicklung des Tapetums, *tapetum development* 369—370
—, Farbstoffe, *pigments* 377—380, 516
—, —, Funktion, *function* 379—380
—, Fermente, *enzymes* 380—381, 513
—, —, Abnahme der Aktivität während der Alterung, *decrease of activity with aging* 381
—, —, Zunahme der Aktivität während der Keimung, *increase of activity during germination* 381
—, gegenseitige Beeinflussung, *interaction of species* 387—388
—, Gehalt an freien Aminosäuren, *content of free amino acids* 381 f.
—, Inhaltsstoffe, *chemical substances* 380—384
—, Kohlenhydrate, *carbohydrates* 382 f.
—, Kolloide, Quellung der, *colloids, swelling* 388
—, Lebensfähigkeit, *viability* 395 ff.
—, —, Einfluß der Aufbewahrungsbedingungen, *storage conditions* 395—397
—, Lipide, *lipids* 383
—, Morphologie, *morphology* 373
—, Nucleinsäuren, *nucleic acids* 382
—, Protein, Aminosäurenzusammensetzung, *proteins, amino acid composition* 380
—, —, Änderungen der Zusammensetzung während der Entwicklung, *protein pattern, changes during development* 380, 513
—, Verlust der Keimfähigkeit, *loss of germination capacity* 379, 385, 397—399
—, Verwendung von altem, zur Bestäubung, *use of aged, in fertilization* 57
—, Wandbildung, *wall formation* 369
—, Cellulosegehalt, *cellulose content* 375
Pollenausschüttung, *pollen, shedding* 372
Pollenbildung, Beeinflussung durch genetische und Milieu-Faktoren, *pollen formation, interference genetical and external* 370, 372
— im Embryosack, *in the embryo sac* 154
Pollenentwicklung, *pollen grains, development* 154, 368 ff.
—, Einfluß der Lichtbedingungen auf, *pollen development, influence of light conditions* 193
—, gestörte, *abnormal* 80
—, Hemmung der, *pollen formation, inhibition of* 161
Pollenfertilität, *pollen fertility* 90, 167
Pollengemische, *pollen mixture* 456 f.
Pollen-Größe, *pollen size* 378
Pollen-Hormon, *pollen hormone* 387
Pollenkammer bei Cycas und Ginkgo, *pollen chamber in Cycas and Ginkgo* 414
Pollenkeimung, *pollen germination* 384—394, 524
—, Abhängigkeit von der Zahl der Pollenkörner in einem bestimmten Volumen des Keimmediums, *dependence upon the number of grains in a given volume* 387—388
Pollenkeimung, Einfluß von Borsäure, *pollen germination, influence of boric acid* 385—387
—, Hemmung der, bei der Inkompatibilität, *inhibition during incompatibility reaction* 509—511
— bei Pinus, *in pine* 423
—, in vitro 384—388, 458
—, in vivo 393, 394
Pollenkitt, *pollen cement* 372—373
Pollenkontrolle, sporophytische, *pollen control, sporophytic* 93
Pollenkorn, Größenklassen, *pollen grain size* 515
Pollenkorn-Größe, Abhängigkeit vom Alter der Blüte, *pollen grain, size, dependence from age of flower* 377
Pollenkornmitose, *pollen grain mitosis* 434
—, Verteilung der Plasmapartikel, *distribution of cytoplasm particles* 435
Pollenmutterzellen, *pollen mother cells* 154
Pollenschlauch, *pollen tube* 388—393
—, Chemotropismus, *chemotropism* 394
—, Eindringen in den Embryosack, *penetration into the embryo sac* 436—438, 515
—, Ernährung, *nutrition* 393—394, 512
— -Griffel-Wechselwirkung, *-style interaction* 469
—, Hemmung bei Heterostylen, *inhibition in heterostylic plants* 514—516
—, Hydrotropismus, *hydrotropism* 394
—, Kallosepfropfen, *callose plugs* 390—391
—, Öffnung der Spitze, *opening of the tip* 437
—, plasmatische Feinstruktur, *plasmatic fine structure* 392—393
—, Resorption von Griffelmaterial, *of style material* 393, 512—513
—, Spitze, *tip* 389, 511
—, Transformierung der Reservestoffe, *transformation of reserve material* 389
—, Wachstum, *growth* 457—459
—, — bei Coniferen, *development in pine* 423
—, —, Deutung von Versuchsresultaten, *growth, interpretation of experimental results* 458 f.
—, —, differentielles, *differential* 447
—, —, Hemmung des, *inhibition* 469, 511—516
—, —, —, bei Heterostylen, *in heterostylic plants* 514—516
—, —, —, bei Homostylen, *in homostylic plants* 512—514
—, —, und Vitalität der nächsten Generation, *and vitality of the next generation* 470 f
—, Wachstumsgeschwindigkeit nach Selbstbestäubung, *growth rate, after self pollination* 511
—, Wachstumsleistung, *growth performance* 393
—, Wand, Oberfläche, *wall, surface* 520
—, Wandaufbau, *wall* 390
Pollenschläuche, Kernteilung in, *pollen tubes, nuclear division in* 434
—, stoffwechselphysiologische Wechselwirkung mit dem Leitgewebe, *metabolic interaction with transmitting tissue* 512 f.

Pollensterilität, *pollen sterility* 96, 202
—, plasmatisch bedingte, *plasmatic* 94
Pollenwand (Sporoderm), *pollen wall* 373—377
—, Nomenklatur, *nomenclature* 373—375
Polyploide bei Gametophyten, *polyploids in gametophytes* 266
polyploide Abweicher bei apomiktischen Arten, *polyploid aberrants in apomictic species* 563, 564
Polyploidenreihe, *polyploidy, series of* 62
Polyploidie, *polyploidy* 804
Polyploidiegrad bei apogamen Farnen, *polyploidy in apogame ferns* 539
Polyspermie, *polyspermy* 310
— bei den Angiospermen, *in angiosperms* 442
— bei den Moosen, *in mosses* 413
Polysporen, *polyspores* 770f.
Potenz, bisexuelle, *potency, bisexual* 49, 101
Protandrie, *protandry* 36
Proteinaseeinwirkung bei der Verschmelzung der Zellkörper, *protease effect in fusion process* 300
Proteinsynthese während der Meiose, *protein synthesis during meiosis* 357—358
Protenor-Typus, *protenor type* 60
Protogynie, *protogyny* 36
Protoplastenteilung bei den Volvocales, *division of the protoplast in Volvocales* 708
Pseudofertilität, *pseudo fertility* 522
Pseudogamie, Definition, *pseudogamy* 551
Pseudomonas aeruginosa 25
Pseudomixis 538
Pseudozygoten-Bildung, *pseudozygote formation* 278

Radikation, *radication* 786
Realisator 54, 70, 92
—, Stärke, Zahlenwert für, *strength of, index* 70
—, weiblicher, *female* 86
Realisatorallel, *realisator allele* 83
Realisator-Mechanismus, *realisator mechanism* 101
Realisatorsysteme, *realisator systems* 43, 85
Realisatoren, *realisators (realizers)* 274
—, ausbalancierte, *balanced* 99
—, Wirkung der, *effect of* 141
Realisatorgene, *realisator genes* 114, 117
Reduktion der Chromosomenzahl, *chromatic reduction* 275
Reduktionsteilung (s. auch Meiose), *meiosis* 31, 34, 39, 51, 134
— in Gametangien, *in gametangia* 113
Regenerate, Mischcharakter, *regenerates, mixed character* 253
Regeneration 146
— aus Farnwedeln, *from fern leaves* 253
— des Laubmoos-Sporogons, *of sporogones in Musci* 249
— freier Protoplasten, *free protoplast* 659
— von Sporogonen, *of sporogones* 147
Regenerationsexperimente an geschlechtlich determinierten Zellen, *regeneration experiments in sexual determined cells* 146
Rekombination, genetische, *recombination, genetical* 261
Rekombinationssysteme, *recombination systems* 26
Replikator, *replicator* 19
Replikon, *replicon* 13, 19
Replikon-Modell, *replicon model* 10, 20, 21
Repression der Struktur-Gene, *of structural genes* 274
Repressor 261
Reproduktion im vegetativen Bereich, *reproduction in the vegetative region* 802
Restitutionskernbildung, *nucleus, restitution, formation of* 533, 536
—, physiologische Ursachen, *physiological background* 538
RNS-Synthese während der Meiose, *RNA synthesis during meiosis* 356—357
Röntgenmutante von Allomyces, *x ray mutation of Allomyces* 245
Rückkreuzungspopulationen, *back cross populations* 90
Ruhestadium, *resting stage* 240

Samenanlage, Anzahl, *ovule number* 273
Samenanlagen, unterschiedlicher Zustand in einem Fruchtknoten, *ovules, differential condition of, in one ovary* 498—500
Samenbildung, *seed formation* 804
SAT-Chromosom, *SAT chromosome* 61
Sauerstoffpartialdruck, Einfluß auf die Lebensfähigkeit des Pollens, *oxygen tension, influence on viability of pollen* 396f.
Scheitelzelle, *apical cell* 247, 251
Schizotomie, *schizotomy* 707
Schwärmer der Braunalgen, Bauschema, *spores of brown algae, construction* 698
Seeigel-Ei, *sea urchin egg* 307
Seeigel-Eier, Homologie zwischen Fucus und, *sea urchin eggs, homology between Fucus and* 309
Sekretion, spezifische, *secretion, specific* 219
Selbstbefruchtung bei apomiktischen Arten, *self-fertilization in apomictic species* 563, 565
Selbst-Inkompatibilität, *self incompatibility* 469, 506, 508ff.
—, genetische Systeme, *genetic systems* 508
—, Überwindung der, *breakdown of* 520—523
Selbstreproduktion, sporogoninduzierender Faktor, *replication, sporogone inducing factor* 251
Selbststerilität, *self sterility* 117, 128f.
—, Zusammenhang mit der Gonenkonkurrenz, *relation to certation* 468
Selbst-Sterilitäts-Faktoren, *self sterility factors* 93
Selbststerilitätsreaktion, *self sterility reaction* 518
Selektionsprinzipien, *selection principles* 244
selektive Befruchtung, *selective fertilization* 447, 479—502
—, materielle Grundlagen, *physiological background* 500ff.

selektive Befruchtung, Reaktionsgeschwindigkeit, *selective fertilization, reaction speed* 483—485
Sex-Duktion, *sex duction* 21
Sexualcharakter, morphologischer, *sex character, morphological* 48
Sexualforschung, Geschichte der, *sex, history of* 257ff.
Sexualgene, *sex genes* 90
Sexualhormone, tierische, Einfluß auf Geschlechtsausprägung bei Pflanzen, *sexual hormones, animal, influence on sex expression in plants* 171, 173, 179
Sexualität, *sexuality* 2, 122f.
— bei den höheren Ascomyceten, *in higher Ascomycetes* 224
—, Ausprägung der, *sex expression* 131
— bei Bakterien, *sexuality in bacteria* 14
—, Begriffsysteme, *terminology of* 128
—, chemische Differenzierung, *chemical differentiation* 300
— bei den Dinophyten, *in Dinophyta* 626
— bei den Euglenophyta, *in Euglenophyta* 620
— und Inkompatibilität, Verhältnis von, *and incompatibility, relation between* 43
—, multipolare, *multipolar* 31
—, relative 36, 37, 38, 125, 261
—, —, Hypothese der, *hypothesis of* 117
—, somatische, *somatic* 260
Sexualitätserscheinungen bei Pilzen, *sexual phenomena in fungi* 131
Sexualitätstheorie, *sex theory* 31
Sexualorgane bei Ceramiales, *sexual organs in Ceramiales* 238
—, diploide, *diploid* 238
—, Verlust der, *loss of* 127
Sexualphysiologie, *sex, physiology of* 257
Sexualreaktion bei Mucor mucedo, *sexual reaction in Mucor mucedo* 296
— bei Mucorineen, *in Mucorineae* 294—297
Sexualstoffe, *sexual hormones* 215—233
Sexualvorgang bei Achlya, hormonale Kontrolle, *sexual process in Achlya, hormonal control* 293
Sexualvorgänge, *sexual processes* 39
— bei Pilzen, *in fungi* 122ff.
sexuelle Phase, determinierender Faktor der Schalengröße bei Diatomeen, *sexual phase influence of cell size in diatoms* 116
sexuelle Potenz, *sexual potency* 33
— Vorgänge bei Bakterien, *processes in bacteriae* 314
sexuelles Verhalten von Sporen, *sexual behaviour of spores* 238
S-Gen, Inaktivierung des, *S gene, inactivation* 521
—, multiple Allelie, *allely* 521
Sirenin, *sirenin* 218, 285ff., 288, 291
—, Molekülgewicht, *molecular weight* 291
Sklerotien, *sclerotia* 781—784
Somatogamie, *somatogamy* 46, 260, 261
Spaltungszahlen, abnorme, Ursachen für, *segregations, abnormal, reasons for* 449—450
Spermatozoiden, Bau der, bei den Moosen und Farnen, *spermatozoids, morphology of, in mosses and ferns* 410
—, Eindringen in den Archegoniumhals bei Farnen, *penetration of the neck canal of the archegonium in ferns* 412
—, Erguß der, in die Archegoniumkammer bei Cycas, *release of, in the archegonial chamber in Cycas* 418
—, frei bewegliche, bei Ginkgo und Cycas, *free motility, in Ginkgo and Cycas* 415f.
—, Weg der, zu den Archegonien, *pathway of, to the archegonia* 412
Spermazellen, Entstehung der, bei Coniferen, *sperm cells, formation of, in pine* 424—426
Spermienfalle, *sperm trap* 287
spezifischer Kontakt zwischen Zellen, *specific contact between cells* 284
Sphaerocarpos 267—268
Spiralisierung der Geißeln, *spiralisation of flagellae* 300
Spiralisierungscyclus, Änderungen im, *coiling cycle, variations in* 352—353
Sporangien, *sporangia* 265
Sporenbildung, *sporulation* 129
— bei Hefen, *in yeast* 126
—, Verlust der, bei apogamen Farnen, *spore formation, loss of, in ferns* 544
Sporenentlassung, Periodizität der, bei Heterokonten, *spore release, periodicity, in Heterokontae* 690
Sporenkeimung bei apogamen Farnen, *spore germination in apogamic ferns* 537
Sporenmembran bei Diatomeen, elektronenmikroskopische Untersuchungen, *spore membrane, in diatoms, electron microscopical observations* 663
Sporenmutterzellen, *spore mother cells* 533—538
Sporen-Tetrade, Aufspaltung in der, *spore tetrad, segregation* 267
Sporentypen bei den Bangioideae, *spores, various types, in Bangioideae* 762—763
Sporogonbildung, apogame, bei Farnen, *sporogone formation, apogamic, in ferns* 252
— in vitro 251
— bei Moosen, *in mosses* 246ff.
—, stofflicher Faktor, *chemical factors* 251
Sporophyll, intersexuelles, *sporophyll, intersexual* 156
Sporophyt, *sporophyte* 240
Sporophyten, Determination der, *sporophytes, determination* 248
—, —, Stabilitätsgrad der, *degree of stability* 248—250
—, Konstruktionselemente, Determination der, *construction elements, determination* 247
Sporophyten-Charakter, Determination des, schrittweise, *sporophyte characters, determination of, stepwise* 247
Sporophyten-Status, Determination des, *sporophyte status, determination* 249—250

Sporophytenstufe bei Bryophyten, *sporophytic level in bryophytes* 138
sporophytische Generationen, *sporophytic generations* 266
Sporopollenin, *sporopollenine* 375
S-Proteine, *S proteins* 520
Stecklinge, Geschlechtsbeeinflussung, *cutting, sex of* 181
Sterilität, männliche, *sterility, male* 97
— im männlichen Geschlecht, Beeinflussung durch NES-Behandlung, *in male sex, influence of NAA* 183
—, plasmatisch bedingte, *plasmatic* 96
Sterilitätsfaktoren, *sterility factors* 43, 47, 96
— beim Mais, *in corn* 95
Sterilitätsphänomene, *sterility phenomena* 561
Strukturformel des Gamons von Mucor, *structure, chemical, of the gamone in Mucor* 296
Struktur-Heterozygotie, *structural heterozygoty* 62
Subletalfaktoren, *sublethal factors* 93
Subtilisin, *subtilisin* 300
suprasexuelle Vorgänge, *supra sexual processes* 121
Synergiden, Rolle der, *synergids, role of* 434f., 436, 437
Syngamie, *syngamy* 316
Synzoospore von Vaucheria, *in Vaucheria* 686
Synzoosporen, *synzoospores* 723

Tapetum 355ff.
Teildetermination, *determination, partial* 253
Teilungen, endogene, ohne Membranbildung, bei den Chlorophyceen, *division, endogenous, without membrane formation, in Chlorophyceae* 706
Telemorphosen, *telemorphosis* 286
Termone, *termones* 33, 215
— bei Chlamydomonas, *in Chlamydomonas* 229, 232
Tetrapolarität, *tetrapolarity* 127
Tetrasporangien bei den Rotalgen, *tetrasporangia in Rhodophyceae* 768
Tetrasporen, *tetraspores* 765—771
—, Bildung der, *formation of* 240
Tetrasporophyten, *tetrasporophytes* 238
Thigmotropismus, *thigmotropism* 286
Tochterkolonien, vegetative, Bildung der, bei Astrephomene, *daughter colonies, vegetative, formation of, in Astrephomene* 711
Transduktion, *transduction* 2, 314
—, rekombinative, *recombinative* 260
Transfer von genetischem Material, Kinetik des, *of genetic material, kinetics of* 17, 19
Transformation 2, 314
Translokationschromosom, isobranchiales, bei Spinat, *translocation chromosome, isobranchial, of Spinacia* 89
Translokationstypen, *translocation types* 66
Tropfenbildung an der Geißelspitze von Chaetomorpha, *drop formation on the tip of the flagellum in Chaetomorpha* 301

Übergang vom Gametophyten zum Sporophyten, *transition of gametophyte to sporophyte* 533
— vom Wasser- zum Land-Leben, *from water to land* 244
Übertragung von Chromosomen, *chromosomal transfer* 26
— männlicher Plastiden durch Spermatozoid bei Farnen, *transfer of plastids by sperms in ferns* 413
Überweibchen, *female, super* 98
Umdetermination des Keimzell-Charakters, *redifferentiation of the germ cell character* 237
Umstimmung, sexuelle, *sex determination, change of* 154
Umstülpung der Kolonien bei den Volvocales, *inversion of colonies in Volvocales* 712
ungeschlechtliche Fortpflanzung, allgemeine Prinzipien, *asexual reproduction, general principles* 257ff.
— — bei den Rhodophyten, *in Rhodophyta* 760—774
— Vermehrung, *asexual reproduction* 276f.
— —, Typen, *types of* 276
Ungeschlechtlichkeit, *asexuality* 266, 277
Unterdrücker-Locus, *suppressor locus* 96
Unverträglichkeit (Inkompatibilität), *incompatibility* 2

variegata-Typus, *variegata type* 68
Varietätenbildung bei apogamen Farnen, *varieties, formation of, in apogamic ferns* 544
vegetative Determination der Sporen, unterschiedliche, *of spores, differential* 238f.
— Fortpflanzung bei den Braunalgen, *propagation, in brown algae* 692—702
— — bei den Charophyceen, *in Charophyceae* 752—759
— — und Karyologie, *and karyology* 804
— — bei den Pilzen, *in fungi* 777—784
— — bei den Spermatophyten, *in spermatophytes* 787
— — der Volvocales, *of Volvocales* 707
— Teilungen, *divisions* 629
— Tendenz, *tendency* 147
— — des weiblichen Geschlechtes, *of the female sex* 145
— Vermehrung, *propagation* 277
— — bei den Algen, *in algae* 596, 601
— — bei den Angiospermen, *in angiosperms* 561—563
— — der Chrysophyceen, *of Chrysophyceae* 637—640, 641
— —, Definition 551, 597, 601
— — bei den Rhodophyten, *in Rhodophyta* 772—774
— — bei höheren Pflanzen, *in higher plants* 561ff.
Verbrauchstheorie der Inkompatibilität, *consumption theory of incompatibility* 519
Vereinigung der Gameten, *fusion of gametes* 116

Vereinigung, sexuelle, von Myxoflagellaten, *fusion, sexual, of myxoflagellatae* 241
Vererbung, geschlechtsgebundene, *heredity, sex linked* 66ff.
Vermehrung, *multiplication* 1
—, geschlechtliche, *reproduction, sexual* 268
—, vegetative, *propagation, vegetative* 52
Verschmelzung der Gameten, *fusion of gametes* 264
— von Myxamoeben, *of myxamoebae* 241
— der Zellkörper, *of cell body* 300
Verschmelzungsmuster bei Brandpilzen, *conjugation pattern in Ustilago* 228
Verteilung der Makro- und Mikro-Sporen, *distribution of macro- and microspores* 139
Verweiblichung männlicher Gametophyten, *embryo sac-like structure* 153
Verzweigung, *ramification* 268
Vibrio cholerae 25
Virus-Infektion, *virus infection* 314
Vitalität und Pollenschlauchwachstum, *vitality and pollen tube growth* 470f.
Vitamin B, *vitamin B* 803
Vitamin C, *vitamin C* 803
Viviparie, *vivipary* 551, 561
— bei Gräsern, *in grasses* 562f.

wachstumsfördernde Stoffe, *growth promotors* 272
wachstumshemmende Stoffe, *growth inhibitors* 272
Wachstumsprozesse, korrelative, growth *processes, correlative* 800
Wachstumsregulatoren, *growth regulators* 274
Wechselwirkungsreaktion zwischen Pollenschläuchen und den Zellen des Griffel-Leitgewebes, *mutual reaction between pollen tube and cells of the transmitting tissue of the style* 519
weiblich blühende Pflanzen mit männlichem Habitus, *female plants with male habit* 83f.
Weiblichkeitsfaktor, homozygote, *female factor, homozygote* 92
Weiblichkeitsfaktoren, *female factors* 81
Weiblichkeits-Tendenz, *female tendency* 98
Weiblichkeits-Unterdrücker, *suppressor, female* 75, 77
Wuchsstoff, Wirkung auf Blätter, *growth promotor, action of, on leaves* 273
Wuchsstoff-Antiwuchsstoff und Blühzustand, *auxin antiauxin and floral state* 804
Wuchsstoff-Gleichgewicht, additive Genregulation der, *growth hormone balance, additive genes regulating the* 275
Wuchsstoff-Haushalt, Einfluß auf die Geschlechtsbestimmung, *hormone balance, influence on sex determination* 275
Wuchsstoffe, *auxins* 804
—, Einfluß auf Blühzustand, *growth regulators, influence on flowering* 269
—, natürliche, Einfluß auf das Blühverhalten, *auxin, native, influence of level, on the floral degree* 271
Wuchsstoffwirkung auf die vegetative Fortpflanzung bei den Spermatophyten, *growth hormones effect of, on vegetative propagation in spermatophytes* 802—803

XO-Typus, *XO type* 60
XY-Intersexe, *XY intersexes* 77
XY-Typus, *XY type* 61
XY_1Y_2-Typus, XY_1Y_2 *type* 64

Zelle, *cell* 1
Zellfusion, *cell fusion* 282
Zellkerne, Größenunterschiede zwischen den männlichen und weiblichen, bei Ginkgo und Bowenia, *nuclei, size differences, between male and female, in Ginkgo and Bowenia* 420f.
Zell-Kontakt, Herstellung des, *cellular contact, establishment of* 16, 17
Zellkontakte, *cell contact* 282, 297ff.
Zelloberfläche, physiologische Bereitschaft, *cell surface, physiological competence* 315
—, Veränderungen der, bei Bakterien, *change of, in bacteriae* 315
Zellteilung bei Cosmarium, *cell division in Cosmarium* 747
— bei den Dinophyten, *in Dinophyta* 627—630
— bei Heteromastix, *of Heteromastix* 739
— bei Pedinomonas, *of Pedinomonas* 738
—, Varianten der, bei Euglena, *variations, in Euglena* 620—623
Zellvergrößerung, bei Diatomeen, *cell size, increase, in diatoms* 657ff.
Zellverkleinerung, bei Diatomeen, *cell size, reduction, in diatoms* 658, 660
Zellverschmelzung, *cellular fusion* 285
Zerfall von Zellfäden bei den Conjugaten, *fragmentation of colonies, in Conjugatophyceae* 747f.
— —, Mechanismus bei den Zygnemalen, *mechanisms in Zygnemales* 748
Zertation, *certation* 447ff.
—, Einfluß der Pollenmenge, *influence of amount of pollen* 452
—, — des Abschneidens des Griffels, *of shortening of the style* 452ff.
—, — von Umweltbedingungen, *of environmental conditions* 454—456
—, experimentelle Untersuchungen, *experimental investigations* 451—471
—, Genphysiologie der, *gene physiology of* 463—471
—, Reaktionsschema, *reaction scheme* 469f.
Zoosporangien, Bildung der, bei den Ulotrichales, *zoosporangia formation, in Ulotrichales* 717
— bei den Volvocales, *in Volvocales* 708
Zoosporen, Auslösung von, bei Heterokonten, *zoospores, induction of, in Heterokontae* 688f.
—, Bildung der, bei Chlorangium, *formation of, in Chlorangium* 715

Zoosporen, Bildung der, bei den Oedogoniales, *zoospores, formation of, in Oedogoniales* 720
—, —, bei Stylosphaeridium, *in Stylosphaeridium* 716
—, —, Einfluß des Chromatophorennetzwerks bei Chladophora, *influence of the chromatophore, in Chladophora* 721
—, —, — des Säuregrades, *of acidity* 738
— der Chlorophyceen, *of Chlorophyceae* 707—724
—, modifizierte, *modified* 722—724
—, Ultrastruktur, bei Oedogonium, *ultrastructure, of Oedogonium* 714
—, —, bei Stigeoclonium-Arten, *of Stigeoclonium species* 712
—, verschiedene Typen bei den Chlorophyceen, *various types, in Chlorophyceae* 713
Zoosporenbildung, *zoospore formation* 706ff.
— bei Chloromonas, *in Chloromonas* 710
Zoosporenentleerung bei Chlorophyceen, *zoospores, release of, in Chlorophyceae* 718
Zoosporen-Membran, *zoospore membrane* 710
Zurückschlagen in den vegetativen Zustand, *regression to the vegetative state* 272
Zweigeschlechtlichkeit, bipolare, *potency, bisexual* 33, 35
Zweihäusigkeit, *dioecism* 267
—, Entstehung der, aus der Einhäusigkeit, *dioecism, origin of, from monoecism* 275, 276
Zwergmännchen, *nannocytes* 238
—, Anheftung der, *androspore, attachment of* 227f
— bei Oedogonium, *nannandrium in Oedogonium* 227, 286
Zwergpflanzen, *dwarf plants* 238, 239
Zwergvorkeim von Chara, *protonema, secondary, in Chara* 754
Zwitter, *intersex* 36, 37
Zwittrigkeit, *monoecism* 101
Zygophoren induzierende Substanzen, *zygophore-inducing substance* 218ff., 232
— bei Mucor, *zygophores in Mucor* 295
Zygotenbildung, *zygote formation* 51
—, Kinetik der, *kinetics of* 19
Zygotenprodukte, Morphogenese der, *zygote products, morphogenesis of* 244
Zygotropismus, *zygotropism* 217f., 286

Subject Index.

(English-German.)

Where English and German spellings of a word are identical, the *italicized* (German) entry is omitted.
Ä, Ö, Ü are taken as Ae, Oe, Ue.

A substance in Pteridium, *A-Substanz bei Pteridium* 143f.
abnormal pale type, *abnormal blasser Typus* 67
acrasine, *Acrasin* 241
acrosomal reaction, *Acrosom-Kontakt* 309
activation of cutinase in relation to incompatibility reaction on the stigma, *Aktivierung der Cutinase bei Inkompatibilitätsreaktion auf der Narbe* 511
active site 19
adaptive enzymes, synthesis of, *adaptive Enzyme, Synthese von* 311
adhesion of spores, influence of surface tension, *Festsetzen von Sporen, Einfluß der Oberflächenspannung* 899
adventitious embryogeny, *Adventiv-Embryonie* 558—560
affinity, genetical background, in selective fertilization, *Affinität, genetische Grundlagen, bei der selektiven Befruchtung* 489—492
— between ovules and pollen tubes, *zwischen Samenanlagen und Pollenschläuchen* 485, 489
—, sexual, *sexuelle* 517
agamospermy, definition, *Agamospermie* 551
—, various types, *die verschiedenen Typen der* 551
AG complex, *AG-Komplex* 51, 98f., 101, 140
agglutination 32
— effects, *Agglutinationswirkungen* 297—316
— in fertilization, *bei der Befruchtung* 285
— of flagellae, *von Geißeln* 304
— of Hansenula, *bei Hansenula* 310—311
—, role of Ca ions, *Rolle der Ca-Ionen* 313
—, sex specific, *geschlechtsspezifische* 312
—, sexual, *sexuelle* 311
— substances, *Agglutinationsstoffe* 306
— — in Tetraspora, *bei Tetraspora* 306
— in yeasts, *bei Hefen* 310—313
aging of pollen, changes of pollen pigment in connection with, *Alterung des Pollens, Änderungen in der Pigmentausstattung* 379
Aglazonia stage, *Aglazonia-Stadium* 237f., 244, 703
algae, heterogenity of definition, *Algen, Heterogenität des Begriffs* 598
allogamy in diatoms, *Allogamie bei Diatomeen* 116
—, isogamic, in diatoms, *isogame, bei Diatomeen* 675
Allomyces hormones, *Allomyces-Hormone* 290f.
alternation of generation (see: life cycles), *Generationswechsel* 262—268
— — in brown algae, *bei den Braunalgen* 700—702
— of phases, *Phasenwechsel* 262ff.
alternative sexual-vegetative reproduction in Heterokontae, *alternative sexuelle-vegetative Fortpflanzung bei den Heterokonten* 687f.
androgamones, definition, *Androgamone* 283
androhermaphrodites, *Androhermaphroditen* 74, 78
andro intersexuality, *Andro-Intersexualität* 79
androspore, attachment of, *Zwergmännchen, Anheftung der* 227f.
angustifolia factor, *angustifolia-Faktor* 66
anisogamety, *Anisogametie* 283—284
anther development, influence of light conditions, *Antheren-Entwicklung, Einfluß der Lichtbedingungen auf* 193
— —, — of 2-thiouracil, *von 2-Thiouracil* 184
— —, — of TIBA, *von TJBS* 183
antheridia, opening of, *Antheridien, Öffnung der* 407
anthers, culture of excised, *Antherenkultur* 352, 359—361
—, fertility of, *Antheren, Fertilität der* 75
—, growth, *Anthere, Wachstumskurve* 369
—, incorporation during microsporogenesis, *Einbaustudien während der Mikrosporen-Entwicklung* 355—358
—, induction of, *Bildung* 75
—, potential for surviving, *Überlebenschancen* 360
antibodies, sex specific, *Antikörper, geschlechtsspezifische* 304
antigen, specific, of the bacteria surface, *Antigen, spezifisches, an der Bakterienoberfläche* 316
antiserum in group formation of algal gametes, *Antiserum bei der Gruppenbildung von Algengameten* 304

apical cell, *Scheitelzelle* 247, 251
aplanospores, *Aplanosporen* 748f.
—, induction of, in Heterokontae, *Aplanosporen, Auslösung der, bei Heterokonten* 689f.
apogamy, *Apogamie* 278
— in algae, *bei den Algen* 600—601
— in brown algae, *bei den Braunalgen* 703, 704
— in Chlorophyceae, *bei den Chlorophyceen* 741—743
— in Conjugales, *bei den Conjugaten* 750—752
—, definition 530
— in diatoms, *bei Diatomeen* 672—677
—, experimental induction, *experimentelle Induktion* 250—252, 542
— in ferns, *bei Farnen* 532ff.
—, genetical conditions, *genetische Voraussetzungen* 250—252
—, induction by environmental factors, *Entstehung durch Umweltfaktoren* 542
— in mosses, *bei Moosen* 245
—, obligate 532ff.
—, —, importance, *Bedeutung der* 542—544
—, —, origin, *Entstehung* 539—542
—, origin by hybridization, *Entstehung durch Bastardierung* 539—541
—, — by mutation, *durch Mutation* 541
—, physiological conditions, *physiologische Voraussetzungen* 251
— in Rhodophyta, *bei den Rhodophyten* 774
apogonosis, *Apogonose* 279
apomeiosis, *Apomeiose* 279
apomictic plants, *apomiktische Pflanzen* 561
— species, species formation, biotype and form, *Arten, Artbildung, Biotypen und Formen* 562—568
— system in Rubi, *apomiktisches System in der Gattung Rubus* 566
— Taraxacum species, aberrant offspring, *apomiktische Löwenzahn-Arten, abweichende Nachkommenschaft* 564
apomixis 276f.
— in angiosperms, *bei den Angiospermen* 551—582
— in bryophytes and ferns, *bei den Archegoniaten* 530ff.
—, definition 530, 531, 551, 601
— in diatoms, *bei Diatomeen* 116f., 674, 677—678
—, experimental control, *experimentelle Kontrolle der* 563
—, facultative, *fakultative* 563, 568
—, list of species, *Liste der Arten bei denen bekannt* 599ff.
—, species with known or suspected (table), *Arten mit bekannter oder vermuteter (Tabelle)* 569—582
apospory, *Aposporie* 248ff., 276f., 279, 544ff., 556—558, 601
archegonia, opening of, *Archegonien, Öffnung der* 409
ascus formation in yeast, *Ascusbildung bei Hefen* 313
asexual reproduction, *ungeschlechtliche Vermehrung* 276f.
— —, general principles, *Fortpflanzung, allgemeine Prinzipien* 257ff.
— — in Rhodophyta, *bei den Rhodophyten* 760—774
— —, types of, *Typen* 276
asexuality, *Ungeschlechtlichkeit* 266, 277
asynapsis 360
attachment of cells, *Anheftung von Zellen* 297—298
— of male gamete, *Festheften des männlichen Gameten* 309
attaching of the sperm nucleus to the egg nucleus, *Anlagerung des Spermakernes an den Eikern* 439—440
aurea factor, *Aurea-Faktor* 67
autogamy in Amphora, *Autogamie bei Amphora* 677
automixis in diatoms, *bei Diatomeen* 116, 674—677
autopolyploidisation, *Autopolyploidisierung* 759
autosomes, *Autosomen* 60, 64
autospores, *Autosporen* 724—728, 780
—, influence of temperature on formation of, *Einfluß der Temperatur auf die Bildung von* 735
auxin-antiauxin and floral state, *Wuchsstoff-Antiwuchsstoff und Blühzustand* 804
auxin level, influence of virus infection on, *Auxinspiegel, Beeinflussung durch Virusinfektion* 188
—, native, influence of level on the floral degree, *Wuchsstoffe, natürliche, Einfluß auf das Blühverhalten* 271
auxins, *Wuchsstoffe* 804
auxospore formation, derived types, *Auxosporen-Bildung, abgeleitete Typen* 116
— — in Denticula, *bei Denticula* 676
— membrane in diatoms, *Auxosporen-Membran bei Diatomeen* 657
— mother cell in diatoms, *Auxosporenmutterzelle bei Diatomeen* 678
auxospores and sexual process in diatoms, *Auxosporen und Sexualakt bei Diatomeen* 659, 672
—, conditions for formation, *Auxosporenbildung, Bedingungen für* 678—679
—, development by haploid parthenogenesis, *Auxosporen, Entstehung durch haploide Parthenogenese* 678
—, — of gametes into, *Entwicklung von Gameten zu* 677
— in diatoms, *bei Diatomeen* 657ff.
azygote formation in Mougeotia, *Azygotenbildung bei Mougeotia* 751

back cross populations, *Rückkreuzungspopulationen* 90
bacterial chromosomes, *bakterielle Chromosomen* 5, 13
— —, mode of replication, *Bakterien-Chromosom, Methode der Replikation* 19
— surface, *Oberfläche der Bakterien* 9, 10

bacteriophages, interaction of, *Bakteriophagen, Wechselwirkung von* 11
—, sex specific, *geschlechtsspezifische* 11
balance genome-plasmone, *zwischen Kerngenen und Plasmon* 92
— theory of BRIDGES, *Balance-Theorie nach* BRIDGES 91
bisexual potency, *bisexuelle Potenz* 101, 123, 124
bispores, *Bisporen* 769
boric acid, influence on pollen germination and pollen tube growth, *Bor-Säure, Einfluß auf Pollen-Keimung und -Schlauchwachstum* 385—387
bryokinine, *Bryokinin* 247
Bryonia experiments, CORRENS', *Bryonia-Versuche*, CORRENS' 52
BOUCK-BRENNER model, BOUCK-BRENNER-*Modell* 23, 24
bud formation in leaves, *Knospenbildung an Blättern* 794—796
— — in roots, *an Wurzeln* 797—799
— pollination, *Knospenbestäubung* 521, 522
bulbills, *Bulbillen* 561, 755—758, 784, 793

callus, gametophytic, *Callus, gametophytischer* 251—252
— growth, *Calluswuchs* 253
— — form, *Calluswuchsform* 253
carpospores, *Karposporen* 238
caryogamy, *Karyogamie* 264
cell, *Zelle* 1
— contact, *Zellkontakt* 282, 297ff.
— division in Cosmarium, *Zellteilung bei Cosmarium* 747
— — in Dinophyta, *bei den Dinophyten* 627—630
— — of Heteromastix, *bei Heteromastix* 739
— — of Pedinomonas, *bei Pedinomonas* 738
— —, variations, in Euglena, *Varianten der, bei Euglena* 620—623
— fusion, *Zellfusion* 282, 285
— size, increase, in diatoms, *Zellvergrößerung bei Diatomeen* 657f.
— —, reduction, in diatoms, *Zellverkleinerung bei Diatomeen* 658, 660
— surface, change of, in bacteriae, *Zelloberfläche, Veränderungen der, bei Bakterien* 315
— —, physiological, competence *physiologische Bereitschaft* 315
cellular contact, establishment of, *Zell-Kontakt, Herstellung des* 16, 17
certation 447ff.
—, experimental investigations, *experimentelle Untersuchungen* 451—471
— experiments, *Certationsversuche* 56, 69
—, gen physiology of, *Genphysiologie der* 463—471
—, influence of amount of pollen, *Einfluß der Pollenmenge* 452
—, — of environment conditions, *von Umweltbedingungen* 454—456
—, — of shortening of the style, *des Abschneidens des Griffels* 452ff.
certation, reaction scheme, *Reaktionsschema* 469f.
—, relation to self sterility, *Gonenkonkurrenz, Zusammenhang mit der Selbststerilität* 468f.
chemotactic attraction of gametes in Chlamydomonas, *chemotaktische Anlockung der Gameten bei Chlamydomonas* 288
— substances in Fucus, *Stoffe bei Fucus* 309
— — in Oedogonium, *bei Oedogonium* 286
chemotaxis 241, 284, 285ff., 310
—, phobic, in Fucus, *phobische, bei Fucus* 287
chemotropic activity of amino acid mixtures, *chemotropische Wirkung von Aminosäuren-Gemischen* 223
— attraction of antheridial hyphae, *Anziehung der antheridialen Hyphen* 222
— control of trichogynal growth, *Kontrolle des Wachstums der Trichogyne* 226
— effect in yeasts, *Wirkungen bei Hefestämmen* 294
chemotropical attraction in Ginkgo und Cycas, *chemotropische Anlockung bei Ginkgo und Cycas* 419
chemotropism, *Chemotropismus* 284, 291
— in connection with selective fertilization, *und selektive Befruchtung* 502
—, influence of Ca-ions, *Einfluß von Calcium-Ionen* 394
— of pollen tubes in the style, *der Pollenschläuche im Griffel* 394
— of pollen tubes, *der Pollenschläuche* 431, 432, 437f.
— in Saprolegnia species, *bei Saprolegnia-Arten* 223
— between vegetative hyphae of Saprolegnia, *vegetativer Hyphen bei Saprolegnia* 292f.
chiasmata, formation of, *Chiasmata-Bildung* 350—352
"chiasmatypy", „*Chiasmatypie*" 639
chimaera, problem of, *Chimären, das Problem der* 276
chitin content of the membranes of fungi, *Chitingehalt in den Membranen von Pilzen* 241
chlamydospores, *Chlamydosporen* 749, 780f.
chromosomal transfer, *Übertragung von Chromosomen* 26
chromosome aberrations, *Chromosomenabweichungen* 362
—, bacterial, *Chromosom, bakterielles* 313
— breaks, *Chromosomenbrüche* 361
— constitution and sex, *Chromosomen-Konstitution und Geschlecht* 75
— number in apogamic ferns, *Chromosomenzahl bei apogamen Farnen* 539
— — and vegetative propagation, *und vegetative Fortpflanzung* 804
— — in Poa hybrids, *bei Poa-Bastarden* 568
— — of offspring of Poa irrigata, *der Nachkommenschaft von Poa irrigata* 566
— —, variation in, in apomictic species, *Variationen der, bei apomiktischen Arten* 565, 566

chromosome replication, *Chromosomen-Verdoppelung* 19
— set, duplication, influence on development, *Chromosomensatz, Verdoppelung, Einfluß auf Entwicklung* 245
— transfer, mechanism of, *Chromosomen-Transfer, Mechanismus des* 22
chromosomes, duplication of, *Chromosomenverdoppelung* 536
—, pairing of, *Chromosomenpaarung* 538
circular linkage group of E. coli, *Koppelungsgruppe, ringförmige, bei E. coli* 18
clonal cultures of diatoms, *Klon-Kulturen von Diatomeen* 114
— selection theory of incompatibility, *Klon-Selektionstheorie der Inkompatibilität* 520
clumping of gametes, *Gruppenbildung der Gameten* 283, 285
colchicine treatment, *Colchicin-Behandlung* 72
colicinogenic strains, *Colicine produzierende Stämme* 14, 15
colicins, *Colicine (Antibiotica)* 14
colonies, division of, in diatoms, *Kolonieteilung bei Diatomeen* 660
compatibility, *Kompatibilität* 35
— groups, *Kompatibilitätsgruppen* 130
competition, *Konkurrenz* 56
— experiments, *Konkurrenzversuche* 72
— of gones, *Gonenkonkurrenz* 447—476
—, developmental background, *entwicklungsgeschichtliche Grundlagen* 450f.
— — in the ovary, *in der Samenanlage* 471—475
— —, variability of, *Variabilität der* 472
— — and genic interactions, *und genetische Wechselwirkungen* 474
— in pollen tube and ovary, *Konkurrenz in Pollenschlauch und Samenanlage* 474f.
— of S genes, *Konkurrenz unter den S-Genen* 521
conchospores, *Conchosporen* 762
conjugation, *Konjugation* 258
— in bacteria, *bei Bakterien* 259
—, bacterial, ohter than E. coli, *bakterielle, bei anderen Bakterien als E. coli* 24ff.
— in Conjugatophyceae, *bei den Conjugaten* 748
— pattern in Ustilago, *Verschmelzungsmuster bei Brandpilzen* 228
— tubes in Saccharomyces cerevisiae, *Konjugationsschläuche bei Hefe* 228
construction principles, morphogenetic, *Konstruktionsprinzipien, morphogenetische* 240f.
consumption theory of incompatibility, *Verbrauchstheorie der Inkompatibilität* 519
contact, kinetics of the, in bacteriae, *Kontakt, Kinetik des effektiven, bei Bakterien* 314f.
— of flagella in copulation of algae, *Flagellenkontakt bei der Kopulation von Algen* 299—306
continuous light, influence on sex rate, *Dauerlicht, Einfluß auf Geschlechtsverteilung* 161
coordinating mechanism, hormonal, in development of sexual organs, *Koordinationsmechanismus, hormonaler, bei der Entwicklung von Geschlechtsorganen* 220
copulation, *Kopulation* 130, 258
— in Ascobolus, *bei Ascobolus* 297
— of bacteriae, *von Bakterien* 314
— in Bombardia, *bei Bombardia* 297
— within a clone, *innerhalb des Klons* 117
—, frequency, *Kopulationshäufigkeit* 35
—, physiology of, *Physiologie der* 43
— processes, *Kopulationsvorgänge* 32
— of gametes, *Gametenkopulation* 283
— between gametes of Dinobryon, *von Gameten bei Dinobryon* 304
— of gamontes, *Gamontenkopulation* 284
—, isogamic 117
—, paedogamic, *pädogame* 127
—, illegal, in yeast, *illegitime, bei Hefen* 126
— zone in Mucor, *Kopulationszone bei Mucor* 294
copulatory process, *Kopulationsprozeß* 214
correlation between pollination, pollen tube growth and development of female organs, *korrelative Beziehung zwischen Bestäubung, Pollenschlauch-Wachstum und Entwicklung der weiblichen Fortpflanzungsorgane* 441
— between sexual and asexual processes, *zwischen sexuellen und asexuellen Prozessen* 788
— between sexual and vegetative reproduction, *zwischen generativer und vegetativer Fortpflanzung* 804
— between sexual tendencies, *Korrelationserscheinungen zwischen den sexuellen Tendenzen* 144f.
cross fertilization, *Fremdbefruchtung* 82
cross incompatibility, *Kreuzungs-Inkompatibilität* 506, 523—524
cross sterility, *Kreuzungssterilität* 523
cross over suppressor 91
crowding effect on sex expression, *„Dichtsaateffekt" auf die Geschlechtsausprägung* 143
culture conditions in Heterokontae, *Kulturbedingungen bei Heterokonten* 686f.
cutinase 510
Cutleria plants, vegetative, *Cutleria-Pflanze* 237f.
Cutleria stage, *Cutleria-Stadium* 703
cutting, sex of, *Stecklinge, Geschlechtsbeeinflussung* 181
cyanophyta, reproduction in *Cyanophyceen, Fortpflanzung bei den* 602—619
cycle, theoretical, of Coccolithus fragilis, *Cyclus, theoretischer, von Coccolithus fragilis* 651—652
cysts, *Dauerzustände* 729—731
—, binuclear, *Cysten, zweikernige* 644
—, conditions for formation, *Bildungsbedingungen* 624, 631, 645
— in diatoms, *Dauerzellen bei Diatomeen* 661
— in Dinophyta, *Cysten bei den Dinophyten* 630—635

cysts in Dinophyta, germination of, *Cysten bei den Dinophyten, Keimung der* 632—634
—, formation, in Chloromonadeae, Cryptomonadeae and Euglenophyceae, *Bildung der, bei den Chloromonaden, Cryptomonaden und Euglenophyceen* 623f.
—, germination of, in Chrysophyta, *Keimung der, bei den Chrysophyten* 646
— in Haptophyceae, *bei den Haptophyceen* 646
— in Heterokontae, *Dauerzellen bei den Heterokonten* 682
— of sexual origin, *Cysten, sexuelle Entstehung* 623
— and sexuality in Chrysophyta, *und Sexualität bei den Chrysophyten* 642—645
—, significance, *Dignität der* 631
—, various forms in the genus Uroglena, *Cystenformen in der Gattung Uroglena* 644
cytoplasmatic bridges in conjugating bacteria, *Plasma-Brücken bei verschmelzenden Bakterien* 259

daughter colonies, formation of, in Astrephomene, *Tochterkolonien, vegetative Bildung der, bei Astrephomene* 711
dedetermination, *Entdeterminierung* 244
dedifferentiation, *Dedifferenzierung* 803
deficiency, silicate, effect of, *Kieselsäuremangelwirkungen* 659
defoliation, influence on sex expression, *Entblätterung, Einfluß auf Geschlechtsausprägung* 164
degeneration as a consequence of vegetative propagation, *als Folge andauernder vegetativer Reproduktion* 805
detergents, effect of, on copulation, *Detergentia, Wirkung von, auf Befruchtung* 307
—, influence on receptor sites on sperm surface, *Einfluß auf Receptorgruppen an der Spermienoberfläche* 309
determination of gametophyte, *des Gametophyten* 235ff.
—, labile 251, 253
—, partial, *Teildetermination* 253
— of sporophyte, *des Sporophyten* 235ff.
—, stabile 253
—, stepwise, of a specific character, *schrittweise, eines spezifischen Charakters* 253
— factors, *Determinationsfaktoren* 244, 251
— — in life cycles, *bei Generationswechsel* 238
— — in spores, *bei Sporen* 238
— pattern of development, *Determinationsmuster der Entwicklung* 238
development of the megaspore in Cycas, *Entwicklung der Eizelle in Cycas* 416f.
—, phenotypic variation, *phänotypische Veränderungen* 267
developmental tendency, possible, *Entwicklungstendenzen, mögliche* 245
diatoms, reproduction, *Diatomeen, Vermehrung der* 657—679
differentiation, bisexual, *Differenzierung, bisexuelle* 128
— of inflorescence, *der Infloreszenz* 258
differentiation, metabolism of, *Differenzierungsstoffwechsel* 245
—, sex specific, *Differenzierung, sexualspezifische* 32
—, switch in, *Umschlagen der* 253
—, unisexual, *unisexuelle* 130
— of vegetative cells in Volvocales, *vegetativer Zellen bei den Volvocales* 711
— of zoospore forming cells, *der zoosporenbildenden Zellen* 717
— pattern, *Differenzierungsmuster* 48
dikaryotic condition, *dikaryotischer Zustand* 130, 131
— phase, formation of, *dikaryotische Phase, Bildung der* 240
dimorphism, *Dimorphismus* 148
—, sexual, *Geschlechtsdimorphismus* 82
Dinophyta, parasitic, *Dinophyten, parasitische* 626
dioecism, *Diöcie (Zweihäusigkeit)* 34, 83ff., 267
—, origin of, from monoecism, *Zweihäusigkeit, Entstehung der, aus der Einhäusigkeit* 275, 276
diplogenotypic sex determination, *Geschlechtsbestimmung, diplogenotypische* 51ff.
diploid phase, term, *diploide Phase, Begriff* 264
diplophenotypic sex determination, *diplophänotypische Geschlechtsbestimmung* 114
diplospory, terminology, *Diplosporie, Terminologie* 552
distribution of macro- and micro-spores, *Verteilung der Makro- und Mikro-Sporen* 139
division, endogenous, without membrane formation, in Chlorophyceae, *Teilungen, endogene, ohne Membranbildung, bei den Chlorophyceen* 706
— of the protoplast, in Volvocales, *Protoplastenteilung bei den Volvocales* 708
DNA replication, *DNS-Replikation* 23
DNA synthesis, *DNS-Synthese* 22, 23, 24
— during meiotic prophase, *während der meiotischen Prophase* 364
double fertilization in pine, *doppelte Befruchtung bei Pinus* 430
drop formation on the tip of the flagellum in Chaetomorpha, *Tropfenbildung an der Geißelspitze von Chaetomorpha* 301
Drosophila type, *Drosophila-Typus* 69
Drosophila-Rumex type of sex, *Drosophila-Rumex-Typ der Geschlechtsbestimmung* 275
dwarf male, Oedogonium, *„Zwergmännchen" Oedogonium* 227
— plants, *Zwergpflanzen* 238, 239

egg apparatus in angiosperms, *Eiapparat bei den Angiospermen* 435
egg jelly, *Eigallerte* 307
egg surface, *Eizelloberfläche* 413
egg water, chemotactic attraction of spermatozoids of Fucus, *Eiwasser, chemotaktische Anlockung der Spermatozoiden von Fucus* 309
— of sea urchins, *von Seeigeln* 309

elimination, selective, of gametes, *Eliminierung, selektive, von Gameten* 93
emasculination, physiological, *Kastration, physiologische* 202
embryo development, *Embryoentwicklung* 524
embryo formation, fractionated, *Embryobildung, fraktionierte* 244
embryo sac in angiosperms, *Embryosack bei den Angiospermen* 435—437
—, formation, *Bildung des* 552—555, 556—558
— —, division, semiheterotypic, *durch semiheterotypische Teilung* 553
— -like structure, *Verweiblichung männlicher Gametophyten* 153
embryos, number of, per seed in Citrus varieties, *Embryonen, Zahl der, per Samen bei Citrus-Varietäten* 558
endocysts, *Dauerzellen* 661
— in diatoms, various types, *Dauersporentypen bei Diatomeen* 664
—, germination of, in diatoms, *Dauerzellen, Keimung, bei Diatomeen* 662, 667
endogenic clock of sporulation in Oedogonium, *endogene Tagesrhythmik der sporogenen Stimmung bei Oedogonium* 734
endopolyploidization, *Endopolyploidisierung* 752 .
endosperm, chromosome numbers, *Chromosomenzahlen im* 561
— development, inhibition of, *Endospermentwicklung, Störung der* 525
— — in agamospermous species, *in agamospermen Arten* 558, 560—561
endospores in cyanophyta, development of, *Endosporen bei den Blaualgen, Entwicklung der* 606—607
end-season-fertility, *Herbst-Fertilität* 522
end-to-end pairing, *end-to-end-Paarung* 62, 64
Entobacteriaceae 24f.
enzymes, sex determinating, *Enzyme, geschlechtsbestimmende* 200
epi-hypostasy type, genetic interaction, *Epi-Hypostasie, genetische Wechselwirkung bei* 274
epi-hypostatic interplay, *epi-hypostatisches Wechselspiel* 275—276
epiphytism in algae, *Epiphytismus bei Algen* 899
episomes, *Episomen* 5
epistasy-hypostasy mechanism between genes, responsible for gynaecium and androecium, *Epistasie-Hypostasie-Mechanismus zwischen Genen, die für Gynaeceum und Androeceum verantwortlich sind* 273
epistasy of female factors, *Epistasie von Weiblichkeitsfaktoren* 759
Euglena gracilis, life cycle, *Entwicklungscyclus* 621
"exhaustion" of female tendency, *„Erschöpfung" der weiblichen Tendenz* 142
exine 373
extra sexual processes, *extrasexuelle Vorgänge* 121

F-duction, *F-Duktion* 21
F-factor, *F-Faktor* 6, 7, 8, 9, 10, 11, 12, 13
—, transfer of, *Übertragung des* 20
F-mutants, *F-Mutanten* 7
F-pili phage, *F-pili Phage* 10, 12, 17
F replicon, integration of bacterial and, *F-Replikon, Integration des bakteriellen und des* 21
fecundation, *Befruchtung* 258
female factor, homozygote, *Weiblichkeitsfaktor, homozygote* 92
female factors, *Weiblichkeitsfaktoren* 81
female plants with male habit, *weiblich blühende Pflanzen mit männlichem Habitus* 83f.
female, super, *Überweibchen* 98
female tendency, *Weiblichkeits-Tendenz* 98
fertility, *Fruchtbarkeit* 2
— in bacteria, factors effecting, *Fertilität bei Bakterien, Faktoren für* 12ff.
— factor, *Fertilitätsfaktor* 6
— of intersexes, *Fertilität der Intersexe* 70
— of pollen, changed, *des Pollens, veränderte* 88
fertilization, *Befruchtung* 2, 3
— in angiosperms, *bei den Angiospermen* 430—442
— in bacteriae, *bei Bakterien* 313—316
—, chemical background, *chemische Grundlagen* 283ff.
— in embryophyta, *bei Archegoniaten und Blütenpflanzen* 407
— membrane, *Befruchtungsmembran* 285, 308
— in oogamic forms, *bei oogamen Formen* 307—310
—, partial, in bacteriae, *partielle, bei Bakterien* 315
—, phases of, *Phasen der* 240
—, physiological processes during, *physiologische Vorgänge bei der* 283—285
— in Prasiola, *bei Prasiola* 309—310
— process in Oedogonium, *Befruchtungsvorgang bei Oedogonium* 287
— processes in algae, survey, *Befruchtungsvorgänge bei Algen, Übersicht* 285
— — in fungi, survey, *bei Pilzen, Übersicht* 286
—, selection during, *Selektion bei der* 449
—, selective, *selektive* 447, 479—502
— types, *Befruchtungstypen* 440
flagellum, importance of in pairing process, *Geißel, Bedeutung der bei der Gruppenbildung* 303
floral degree of flowers, *„Blüh-Grad" von Blüten* 272, 273
— gradient, *Gradient in der Blüte* 271
— realisation, *Blütenausprägung* 274
flower, *Blüte* 271ff.
— colours, sex linked heredity of, *Blütenfarbe, geschlechtsgebundene Vererbung der* 68
— development, genetical control, *Blütenentwicklung, genetische Kontrolle* 272
— —, optimal day length, *Blütenbildung, optimale Tageslänge* 193f.

flower formation, *Blütenbildung* 159
— realisation, *Blüh-Realisation* 268, 269
— variation, *Blütenveränderung* 272
—, vegetative, *Blüte, vegetative* 271
flowering, degree of, *Blühzustand, Grad des* 269
— impulse, *Blühimpuls* 274
flowers, intersexual, *intersexuelle Einzelblüten* 181
foliar embryos, *Knospenbildung an Blättern* 795—797
formation of flower, genes involved in, *Blütenbildung, beteiligte Gene* 274
fragmentation of colonies, *von Kolonien* 729, 731—733
— — in Conjugatophyceae, *Zerfall von Zellfäden bei den Conjugaten* 747f.
— —, mechanisms in Zygnemales, *Mechanismus bei den Zygnemalen* 748
free protoplast, regeneration, *Regeneration freier Protoplasten* 659
frequency, male/female, *Häufigkeit Weibchen/Männchen* 53
fruit body formation in Acrasieae, *Fruchtkörperbildung bei den Acrasieen* 285
Fucus egg, *Fucus-Ei* 307
fusion of cell bodies, *Verschmelzung der Zellkörper* 300
— of gametes, *der Gameten* 116, 264
— of myxamoebae, *von Myxamöben* 241
—, sexual, of myxoflagellatae, *Vereinigung, sexuelle, von Myxoflagellaten* 241
—, sogatogame, of hyphae, *somatogene, von Hyphen* 130

gametangia, *Gametangien* 265
gametangiogamy, *Gametangien-Befruchtung* 284
gamete filtrates, effect in vitro, *Gameten-Filtrate, Wirkung in vitro* 288
— formation, in Vaucheria, by photolytic products of tryptophane, *Gametenbildung bei Vaucheria, durch photolytische Zerfallsprodukte des Tryptophans* 688
— fusion, *Gametenverschmelzung* 298ff.
— — in Noctiluca, *bei Noctiluca* 628
gametes, *Gameten* 258
—, formation, *Gametenbildung* 236
—, selection of, *Gameten, Selektion der* 450
—, vegetative development, *vegetative Entwicklung der* 237—238
—, vegetative development, *Entwicklung der* 237—238
gametophytes, elimination of, tendency to, *Gametophyten, Elimination der, Tendenz zur* 238
gamones, *Gamone (Befruchtungsstoffe)* 32, 33, 215, 218, 260, 261, 282ff., 311
— in Achlya, *Gamone bei Achlya* 292
—, agglutinating, *agglutienierende* 306
— in bacteriae, *bei Bakterien* 315—316
— of Chlamydomonas, M. W., *von Chlamydomonas, Molekülgewicht* 302
—, definition 283
—, glycoproteids as, *Glykoproteide als* 303
gamones, low molecular, in yeast, *Gamone, niedermolekulare, bei Hefe* 294
— production, wave length, dependence, *Produktion, Wirkungsspektrum* 303
—, protein constitutions, *Proteinkomponenten* 312
—, sex specificity, *Geschlechtsspezifität* 312
gemmae in Rhodophyta, *Brutkörper in Rhodophyten* 771
gene balance system, *Gen-Balance-System* 93
— exchange, *Gen-Austausch* 101
gene system, *Gen-System* 92
generation, term, *Begriff* 265
generations, succession of, *Generationen, Abfolge der* 264
generation time, *Generationsdauer* 158
genes, additive, controling native growth regulation, *Gene, additive, bei der Kontrolle der natürlichen Wachstumsprozesse* 272
genetic exchange, *Austausch von Erbmaterial* 2, 5
— information, transfer of, *genetische Information, Übertragung der* 2, 5
— material, transfer of, *genetisches Material, Übertragung des* 16, 17
genocopy, *Genokopie* 273
genotypic/phenotypic sex determination, transition, *genotypische/phänotypische Geschlechtsbestimmung, Übergang* 114
germ cells, quality, *Keimzellen, Qualität der* 235—237
germination conditions, pollen, *Keimungsbedingungen für Pollen* 384—385
—, spontanous parthenogenetic, of Cutleria eggs, *Keimung, spontane parthenogenetische, von Cutleria-Eiern* 704
— of zygote, *der Zygote* 240
— — in Ulothrix, *bei Ulothrix* 742
gibberellic acid, antagonists, *Gibberellinsäure, Antagonisten* 201
— —, treatment with, *Behandlung mit* 159
— — —, influence on sex expression, *Gibberellinsäurebehandlung, Einfluß auf Geschlechtsausprägung* 166
gibberellins, *Gibberelline* 803
gone competition, genetical background, *Gonenkonkurrenz, genetische Grundlagen* 448—451
gonidia, formation of, *Gonidienbildung* 627
grafting, sex change by, *Pfropfung, Beeinflussung des Geschlechtes durch* 177
group formation brown algae, *Gruppenbildung bei den Braunalgen* 308
grouping of gametangia, influence of tryptophane derivates, *Gametangiengruppierung, Wirkung von Tryptophan-Abkömmlingen* 113
growth hormone balance, additive genes regulating the, *Wuchsstoff-Gleichgewicht, additive Genregulation des* 275
— hormones, effect of on vegetative propagation in spermatophytes, *Wuchsstoffwirkung auf die vegetative Fortpflanzung bei den Spermatophyten* 802—803

growth inhibitors, *wachstumshemmende Stoffe* 272
— processes, correlative, *Wachstums-prozesse, korrelative* 800
— promotor, action of, on leaves, *Wuchsstoff, Wirkung auf Blätter* 273
— —, *wachstumsfördernde Stoffe* 272
— regulators, *Wachstumsregulatoren* 274
— —, influence on flowering, *Wuchsstoffe, Einfluß auf Blühzustand* 269
— —, ratio, *Wachstumsregulatoren, Verhältnis* 272
gynogamone, *Gynogamon* 303
gynogamones, definition, *Gynogamone, Definition* 283
gyno intersexuality, *Gyno-Intersexualität* 79

haploid phase, term, *haploide Phase, Begriff* 264
heat treatment, influence on sex expression, *Hitzebehandlung, Einfluß auf Geschlechtsbestimmung* 165
hemiautospores, *Hemiautosporen* 724—726
hemizoospores, *Hemizoosporen* 722
heredity, sex linked, *Vererbung, geschlechtsgebundene* 66ff.
hermaphroditism, functional, *Hermaphroditismus, funktioneller* 215
heteroblasty, *Heteroblastie* 700
heterochromatin content, *Heterochromatin-Gehalt* 58
heterocysts, development of, in Anabaenopsis, *Heterocysten, Entstehung der, bei Anabaenopsis* 611
—, fine structure, *Feinstruktur* 612
—, germination, *Keimung* 617—618
—, influence of nitrogen concentration, *Einfluß der Stickstoffkonzentration im Medium* 615
—, mechanisms of spontaneous differentiation, *Mechanismus der Spontandifferenzierung* 615
—, physiology of development, *Entwicklungsphysiologie der Entstehung* 612—617
—, relative number, influence of molybdenum, *Zahl der, Molybdän-Bedarf* 615
heterogamety, *Heterogametie* 52, 55, 87
Heterokontae, reproduction in, *Fortpflanzung bei den Heterokonten* 682—691
heteromorphy of X and Y chromosomes, *Heteromorphie von X- und Y-Chromosomen* 64
heterosis 544
heterosomes, *Heterosomen* 57
heterothaly, *Heterothallie* 49
Hfr DNA, *Hfr-DNS* 23
homo-heterothallic reactions in Achlya, *homo-heterothallische Reaktionen bei Achlya* 231
homozygoty heterozygoty mechanism, *Homozygotie-Heterozygotie-Mechanismus* 52
hormogonia, development of, *Hormogonien, Entwicklung der* 603—606
hormonal coordinating mechanism in Achlya, *hormonaler Koordinationsmechanismus bei Achlya* 219—223
— — — in Ascobolus, *bei Ascobolus* 224
hormone A, chemistry, *Hormon A, chemische Analyse* 223
hormone A-complex, Achlya, *Hormon A-Komplex, Achlya* 221, 222
hormone B, Achlya, *Hormon B, Achlya* 222
hormone C, Achlya, *Hormon C, Achlya* 222
hormone D, Achlya, *Hormon D, Achlya* 222
hormone balance, influence on sex determination, *Wuchsstoff-Haushalt, Einfluß auf die Geschlechtsbestimmung* 275
— production of male mycelia in Achlya, *Hormonproduktion männlicher Mycelien bei Achlya* 294
humidity, influence on viability of pollen, *Luftfeuchtigkeit, Einfluß auf Lebensfähigkeit des Pollens* 395f.
hybrid seedlings in Citrus crosses, *Bastard-Keimlinge bei Citrus-Kreuzungen* 558
hybrids, production of, *Hybridengewinnung* 524
hybridization, among pseudogamous species, *Bastardiierung, zwischen pseudogamen Arten* 563, 566
hyphal fusion, *Hyphenfusion* 284

IAA, *IES* 272, 803
IAA oxydase, *IES-Oxydase* 88
—, activity, sex dependent, *IES-Oxydase-Aktivität, geschlechtsabhängige* 171
"imitations" of fertilization, *„Attrappen" der Befruchtung* 121
— of the meiotic division, *für die Meiose* 121
immune theory of incompatibility, *Immunitäts-Theorie der Inkompatibilität* 518—519
immunological reaction, between pollen tubes and style, *Immunreaktion zwischen Pollenschlauch und Leitgewebe* 512—513
impulse, flowering, *Blühimpuls* 268
incompatibility, *Unverträglichkeit (Inkompatibilität)* 2, 35, 131
— in Ascobolus, *Inkompatibilität bei Ascobolus* 231
— factors, *Inkompatibilitätsfaktoren* 33, 35, 127, 131
— in fungi, *Inkompatibilität bei Pilzen* 224f.
—, heterogenic, *heterogenische* 324
—, homogenic, *homogenische* 324—337
—, interspecific, *interspezifische* 506, 524—525
—, occurrence, *Verbreitung* 322ff., 507f.
— reaction, *Inkompatibilitätsreaktion* 511
— —, break down of, *Überwindung der* 520—523
— —, influence of growth hormones, *Einfluß von Wuchsstoffen* 523
— —, — of radiation, *von Bestrahlung auf* 523
— —, isoantibodies, *Isoantikörper* 519
—, localization, *Lokalisierung der* 520

incompatibility localization, in the ovule, *Inkompatibilität, Lokalisierung in der Samenanlage* 516—517
— — in polyploids, *bei polyploiden Pflanzen* 521
— —, serological specificity, *serologische Spezifität* 519
— —, temperature dependence, *Temperaturabhängigkeit* 519, 522 f.
— in thallophytes, *bei den Thallophyten* 321 ff.
—, theories, *Inkompatibilitäts-Theorien* 517 ff.
— in tracheophyta, *Inkompatibilität bei Phanerogamen* 506—525
induction of asexual reproduction in Chlorophyceae, *Auslösung von ungeschlechtlichen Fortpflanzungszellen bei den Chlorophyceen* 733—736
— of conjugation tubes in Saccharomyces cerevisiae, *Induktion von Konjugationsschläuchen bei Hefe* 228
inflorescence, *Inflorescenz* 268—271
— without flowers, *blütenlose* 269, 270
—, terminal, *endständige* 162
—, vegetative 270, 271
inhibitor, *Hemmungsfaktor* 88
initial zones, in tissues, *Initialzonen bei der Gewebebildung* 242
initiator 19
instability, sexual, *Instabilität, sexuelle* 126
internal clock in Vaucheria, *„innere Uhr" bei Vaucheria* 690
intersex 36, 37, 87
— character, *intersexueller Charakter* 78
—, primary, *primäre* 94
intersexes, *Intersexe* 71, 79, 82
intersexuality, *Intersexualität* 175
—, degree of, *Intersexualitätsgrad* 70
interspecific hybrids, *Artbastarde* 567
— — of Poa, number of progenics, *Artbastarde von Poa, Zahl der Nachkommen* 568
intine 373
inversion of colonies in Volvocales, *Umstülpung der Kolonien bei den Volvocales* 712
isoagglutination 306
— in concentrated filtrates of gametes of algae, *in konzentrierten Filtraten von Algen-Gameten* 300—302
isogame fertilization in algae, *Befruchtung bei Algen* 298—307
isogamety, *Isogametie* 283—284
isogamy, *Isogamie* 258

jacket layer, in the macroprothallium in Cycas, *Deckschicht des Makroprothalliums bei Cycas* 417
Jacob-Brener model, Jacob-Brenner-*Modell* 22. 23
Jacob-Monod model, Jacob-Monod-*Modell* 100
juvenile forms, *Jugendformen* 266

karyogamy, *Karyogamie* 240
— in angiosperms, *(Kernverschmelzung) bei den Angiospermen* 439—441
—, retarded, causes of, *verzögerte, Ursachen für* 240
karyology and ecology, *Karyologie und Ökologie* 804
kinetom formation, cytology of, *Kinetombildung, Cytologie der* 719

lac operon 21
leptotene chromosomes, double strands, *Leptotän-Chromosomen, Doppelstrang* 363
lethal factors, *Letalfaktoren* 93
life cycles, *Generationswechsel* 3, 34, 39—42, 235f., 241, 244, 262—268, 275, 279
— — of an Allomyces arbuscula mutant, *Entwicklungscyclus einer Mutante von Allomyces arbuscula* 236
— —, changed metabolism of, *Generationswechsel, veränderte Stoffwechselansprüche beim* 244
— — of Codiolum, in dependence on temperature, *Entwicklungscyclus von Codiolum, Abhängigkeit von Temperatur* 238
— —, diphasic, *Generationswechsel, diphasischer* 239
— —, dependence on temperature, *Abhängigkeit von Temperaturwechsel* 239
— —, of Ectocarpus, *Entwicklungscyclus von Ectocarpus* 701
— —, general principles, *Generationswechsel, allgemeine Prinzipien* 257 ff.
— — of Levringia, *Entwicklungscyclus von Levringia* 701 f.
— —, origin, *Generationswechsel, Entstehung des* 235 ff.
— — of Phaeophyta, *bei Braunalgen* 266
— —, phylogenetic aspects, *phylogenetische Gesichtspunkte* 235
— —, sexual, in Ceratium cornutum, *Formwechsel, sexueller, von Ceratium cornutum* 626
— — of Syracospaera carterae, *Lebenscyclus von Syracospaera carterae* 647
— —, types, *Generationswechseltypen* 240, 242
light, influence on germinations of cysts, *Lichteinfluß auf die Keimung von Cysten* 633, 635, 646
— intensity, influence on vegetative propagation, *Lichtintensität, Einfluß auf vegetative Fortpflanzung* 802
— —, reduced, influence on sex expression, *Lichtintensität, reduzierte, Beeinflussung der Geschlechtsausprägung durch* 166, 172
linkage groups, *Koppelungsgruppen* 9
long-day principle, transferable, *Langtag-Prinzip, übertragbares* 803
long distance effect in copulation, *Abstandswirkungen, stoffliche, beim Kopulationsvorgang* 297
— — — in fertilization, *Fernwirkungen, bei der Befruchtung* 285
Lygaeus type, *Lygaeus-Typus* 61
lysogenesis, *Lysogenese* 2

male factor, *Männlichkeitsfaktor* 81
— —, plasmatic, *plasmatischer* 92
— plants with female habit, *männlich blühende Pflanzen mit weiblichem Habitus* 83f.
— populations, *männliche Populationen* 81
mannans, *Mannane* 313
mating process in bacteria, physiology of, *Befruchtungsprozess bei Bakterien, Physiologie des* 16
mating substances, *Paarungsstoffe* 316
— types, *Paarungstypen* 6, 43, 44, 45, 46, 48, 49, 122, 127, 129
— — type substances, *Paarungstypsubstanzen* 284
— — — — in Paramaecium, *bei Paramaecium* 304
— types, tetrapolar, *Kreuzungstyp, tetrapolarer* 46
meiocysts, *Meiocysten* 653
meiosis, Meiose 1, 31, 34, 39, 61, 63, 243, 236, 264, 275, 344—364, 369, 532, 626, 630, 761, 764
—, aberrations, *Abweichungen von* 359—363
—, — induced by ionizing radiations, *Abweichungen von, induziert durch ionisierende Strahlen* 361—363
—, behavior of tapetal cells, *Verhalten der Tapetumzellen* 355f.
—, determination, *Determination der* 239, 246f.
—, extrusion of nuclear material, *Abgabe von Kernmaterial* 347
— in gametangia, *Reduktionsteilung in Gametangien* 113
—, irregular, in diplosporous species, *Meiose, unregelmäßige, in diplosporen Arten* 556
—, long prophase, *lange Prophase* 347
—, pairing of homologous chromosomes, *Paarung der homologen Chromosomen* 347—350
—, shift of, *Verschiebung der* 239
—, suppressed, *unterdrückte* 759
—, synchronisation, in anthers, *Synchronisierung, in Antheren* 370
—, synthesis of nucleic acids and proteins, *Synthese der Nucleinsäuren und Proteine* 354—358
—, theories, *Meiose-Theorien* 363f.
—, unitary behaviour of the centromeres of homologous chromosomes, *einheitliches Verhalten der Centromeren homologer Chromosomen* 353f.
meiosis-like prophase, in diatoms, *meioseähnliche Prophase bei Diatomeen* 677
meiospores, formation of, *Meiosporen-Bildung* 653
meiotic chromosomes, coiling characteristics, *Meiose-Chromosomen, Spiralisierungs-Charakteristika* 352
— division, distinguishing features, *Meiose, unterscheidende Kennzeichen* 344—358
— waves in anthers, *Meiosewellen in Antheren* 370
Melandrium type, *Melandrium-Typus* 72ff., 87
membrane formation in Flagellatae, *Membranbildung bei den Flagellaten* 624
meromixis 314, 315
merozygote 21
metromorphy, *Metromorphie* 759
microgametogenesis, incorporation into nucleic acids and proteins, *Mikrosporenentwicklung, Einbauprozesse in Nucleinsäuren und Proteine* 355ff.
microspore mother cell, *Mikrosporenmutterzelle* 154
microspores, germination, in Ginkgo, *Mikrosporen, Keimung der, bei Ginkgo* 415
mitosis in diatoms, *Mitose bei Diatomeen* 657
mixture between gametophyte and sporophyte, *Mischbildungen zwischen Gametophyt und Sporophyt* 253—254
mobility of gametes, *Beweglichkeit der Gameten* 298
modification, correlative, in flower formation, *Modifikation, korrelative, bei der Blütenbildung* 270
—, persistent, *Dauermodifikation* 146
modificator genes, of S genes, *Modifikator-Gene für die S-Gene* 561
molecular conjugation, *Konjugation, molekulare* 259
monoecism, *Monöcie* 34, 83ff., 101
monosporangium of Erythrotrichia, *Monosporangium von Erythrotrichia* 761
monospores, Monosporen 780—765
morphogenetic principles, *morphogenetische Prinzipien* 240ff.
— processes, *gestaltbildende Prozesse* 241
morphological differences between the generations, *Gestaltunterschiede zwischen den Generationen* 243f.
morphological differences between the generations, *Gestaltunterschiede zwischen den Generationen* 243f
— differentiation, in connection with reproduction, *morphologische Differenzierung im Zusammenhang mit der Fortpflanzung* 286
morphoregulators, *Formbildungsstoffe* 274
multiple allely of sex factors, *Allelie der Geschlechtsfaktoren* 90
multiplication, *Vermehrung* 1
mutants, monosexual, *Mutanten, eingeschlechtliche* 128
mutations in Chlamydomonas, *Mutanten von Chlamydomonas* 305
mutual reaction of pollen grains, *Mutualreaktion zwischen keimenden Pollenkörnern* 387
— — between pollen tube and cells of the transmitting tissue of the style, *Wechselwirkungsreaktion zwischen Pollenschläuchen und den Zellen des Griffel-Leitgewebes* 519
mycelium, mixed, *Mischmycel* 253

NAA, *NES* 273
nannandrium, Oedogonium, „*Zwergmännchen*", *Oedogonium* 227, 238, 286
"neuters" (plants without flowers), „*Neuters*" *(Pflanzen ohne Blüten)* 81, 95
nondisjunction, *Nondisjunktion* 34
nucellar embryony, *Nucellarembryogenie* 277
nuclear division, acytokinetic, in diatoms, *Kernteilung, acytokinetische, bei Diatomeen* 657, 662f.
nuclear fusion, *Kernfusion* 282, 285, 286
— in Ginkgo, *Kernverschmelzung bei Ginkgo* 420
— in pine, *bei Pinus* 429
nuclear migration, *Kernwanderung* 286
nuclear plasma system, *Kern-Plasma-System* 93, 100
nuclei, size differences between male and female in Ginkgo and Bowenia, *Zellkerne, Größenunterschiede zwischen den männlichen und weiblichen, bei Ginkgo und Bowenia* 420f.
nucleus fragmentation, *Kernfragmentation* 752
nucleus-plasma-relation in diatoms, *Kern-Plasma-Relation bei Diatomeen* 678
nucleus, restitution, formation of, *Restitutionskernbildung* 533, 536
—, —, —, physiological background, *physiologische Ursachen* 538

obturator 431—433
one-gene-mutation, *Ein-Gen-Mutation* 237
ontogenesis, determination during, *Ontogenese, Determination in der* 245ff.
ontogeny, phases, *Ontogenieabschnitte* 235f.
opening of the pollen tubes in Cycas and Ginkgo, *Öffnung der Pollenschläuche bei Cycas and Ginkgo* 418
operator regulator system, *Operator-Regulator-System* 100
organ formation, *Organbildung* 242
organisation level, *Organisationsstufe* 241
osmotic shock, effect of culture, *osmotischer Schock, Wirkung der Gewebekultur* 361
ovarial substances, *Ovarialstoffe* 518
ovarial tumors, *Ovarialtumoren* 524
ovule number, *Samenanlage* 273
ovules, differential condition of, in one ovary, *Samenanlagen, der unterschiedliche Zustand, in einem Fruchtknoten* 498—500
oxygen tension, influence on viability of pollen, *Sauerstoffpartialdruck, Einfluß auf die Lebensfähigkeit des Pollens* 396f.

paedogamy, *Pädogamie* 127
— in Cymbella, *bei Cymbella* 677
— in Gomphonema, *bei Gomphonema* 675
pairing, *Paarung* 63
— of chromosomes, *Chromosomenpaarung* 260, 282
— —, genetic control of, *genetische Kontrolle der* 350
— of compatible cells, *Paarung von kompatiblen Zellen* 16
pairing condition in diatoms, *Paarungs-Bedingung bei Diatomeen* 116
— in diatoms, *Paarung bei Diatomeen* 675
—, sexual, *Geschlechtspaarung* 261
—, specificity of, based on the occurrence of single stranded DNA, *Paarung, Spezifität der, Hypothese basierend auf dem Vorkommen von einfachen DNS-Strängen* 349
pairing-structures ("cores"), *Paarungsstrukturen („Cores")* 261
palmella formation, factors concerning, *Palmella-Bildung, Faktoren für* 622
panspermy, theory of, *Panspermie, Theorie der* 257
parasexual cycles, *parasexuelle Cyclen* 2
— phenomena, *Erscheinungen* 121
paraspores, *Parasporen* 769
parthenogametes, *Parthenogameten* 237, 600f.
parthenogenesis, *Parthenogenese* 277f.
— in brown algae, *Aopgamie bei den Braunalgen* 703f.
— in Charophyceae, *Parthenogenese bei den Charophyceen* 758—759
— in Chlorophyceae, *bei den Chlorophyceen* 741—743
— in Conjugales, *bei den Conjugalen* 750—752
—, definition 530, 551
— in diatoms, *bei Diatomeen* 672—677
—, diploid, *diploide* 117
—, haploid, *haploide* 551
— in Heterokontae, *bei Heterokonten* 690f.
— in Rhodophyta, *bei den Rhodophyten* 774
— in spermatophytes, *bei den Spermatophyten* 787
parthenogenetic development, *parthenogenetische Entwicklung* 237
PEA (phenethyl alcohol), inhibitor of DNA synthesis, *PEA (Phenäthylalkohol), Inhibitor für die DNS-Synthese* 23
pectin synthesis, influence of boric acid in pollen tubes, *Pektin-Synthese, Einfluß der Borsäure im Pollenschlauch* 386
penetration of phage particles, *Eindringen von Phagenteilchen* 314
— of the pollen tubes into the egg cell in pine, *der Pollenschläuche in die Eizelle bei Coniferen* 428
— of spermatozoids in Cycas and Ginkgo, *aktives, der Spermatozoiden bei Cycas und Ginkgo* 419f.
phase transition, conditions for, in Haptophyceae, *Phasenübergang bei den Haptophyceen, Bedingungen für* 652—653
photoperiodism, *Photoperiodismus* 801—802
phototaxis in brown algae, *bei Braunalgen* 698
—, positive, in Euglena gracilis, *bei Euglena gracilis* 622
"pili" (surface outgrowths), „*pili*" *(Oberflächen-Auswüchse)* 10
pili on the surface of bacteria, *(charakteristische Fortsätze) an der Bakterienoberfläche* 316
pistil, development of, *Ausbildung des* 77

plasmatic incompatibility, *plasmatische Inkompatibilität* 524
plasmodia, formation of, *Plasmodienbildung* 241
plasmodial organisation, *plasmodiale Organisation* 241
plasmogamy, *Plasmogamie* 240, 264, 286
— in angiosperms, *(Plasmaverschmelzung) bei den Angiospermen* 438—439
plasmon influence on certation, *Plasmonwirkung bei der Gonenkonkurrenz* 467
— mutations, *Plasmon-Mutationen* 94
—, participation of, in species crosses, *Beteiligung des, bei Artkreuzungen* 96
pleomorphism between cyst and cyst content, *Pleomorphismus zwischen Cysten und Cysteninhalt* 651
— in diatoms, *der Diatomeen* 668—672
polarity, light induced, on tetraspores of Dictyota, *Polarität, lichtinduzierte, bei Tetrasporen von Dictyota* 899
— of spores, predetermination of, *der Sporen, prädeterminierte* 764
—, sexual, *sexuelle* 282
pollen 56, 368—399
—, aging of, *Alterungsversuche* 57, 78
—, biochemistry, *Biochemie des* 373—384
—, carbohydrates, *Kohlenhydrate* 382f.
—, cellulose content, *Cellulosegehalt* 375
— cement, *Pollenkitt* 372—373
— chamber in Cycas and Ginkgo, *Pollenkammer bei Cycas und Ginkgo* 414
—, chemical substances, *Inhaltsstoffe* 380—384
—, colloids, swelling, *Kolloide, Quellung der* 388
—, content of free amino acids, *Gehalt an freien Aminosäuren* 381f.
— control, sporophytic, *Pollenkontrolle, sporophytische* 93
—, decrease of viability, reasons, *Abnahme der Keimfähigkeit, Ursachen für* 396—399
— development, *Pollenentwicklung* 154
— —, abnormal, *gestörte* 80
— —, influence of light conditions, *Einfluß der Lichtbedingungen auf* 193
—, enzymes, *Fermente* 380—381, 513
—, —, decrease of activity with aging, *Abnahme der Aktivität während der Alterung* 381
—, —, increase of activity during germination, *Zunahme der Aktivität während der Keimung* 381
— fertility, *Pollenfertilität* 90, 167
—, formation, *Bildung des* 555f., 558
— —, by various types of irregular meiotic division, *durch verschiedene Typen abweichender meiotischer Teilung* 555
— —, inhibition of, *Pollenentwicklung, Hemmung der* 161
— —, interference, genetical and external, *Pollenbildung, Beeinflussung durch genetische und Milieufaktoren* 370, 372
— — in the embryo sac, *im Embryosack* 154
pollen germination, *Pollenkeimung* 384—394, 524
— —, dependence from the number of grains in a given volume, *Abhängigkeit von der Zahl der Pollenkörner in einem bestimmten Volumen des Keimmediums* 387—388
— —, influence of boric acid, *Einfluß von Borsäure* 385—387
— —, inhibition, during incompatibility reaction, *Hemmung der, bei der Inkompatibilität* 509—511
— —, in vitro 384—388, 458
— —, in vivo 393, 394
— — in pine, *bei Pinus* 423
— grains, development, *Pollenentwicklung* 368ff.
— grain mitosis, *Pollenkornmitose* 434
— — —, distribution of cytoplasm particles, *Verteilung der Plasmapartikel* 435
— — size, *Pollenkorn, Größenklassen* 515
— —, —, dependence upon age of flower, *Pollenkorn-Größe, Abhängigkeit vom Alter der Blüte* 377
— hormone, *Pollen-Hormon* 387
—, inorganic material, *anorganische Substanzen* 383
—, interaction of species, *gegenseitige Beeinflussung* 387—388
—, lipids, *Lipide* 383
—, loss of germination capacity, *Verlust der Keimfähigkeit* 379, 385, 397—399
— mixtures, *Pollengemische* 456f.
—, morphology, *Morphologie* 373
— mother cells, *Pollenmutterzellen* 154
—, nucleic acids, *Nucleinsäuren* 382
—, pigments, *Farbstoffe* 377—380, 516
—, —, function, *Funktion* 379—380
—, protein pattern, changes during development, *Protein, Änderungen der Zusammensetzung während der Entwicklung* 380, 513
—, proteins, amino acid composition, *Protein, Aminosäurenzusammensetzung* 380
—, shedding, *Pollenausschüttung* 372
— size, *Pollen-Größe* 78
— sterility, *Pollensterilität* 96, 202
— —, plasmatic, *plasmatisch bedingte* 94
—, tapetum development, *Entwicklung des Tapetums* 369—370
—, treatment with alcohol, *Behandlung mit Alkohol* 56
— tube, *Pollenschlauch* 388—393
— —, callose plugs, *Callosepfropfen* 390—391
— —, chemotropism, *Chemotropismus* 394
— —, development in pine, *Pollenschlauch-Entwicklung bei Coniferen* 423
— —, entrance into the embryo sac, *Eintritt des, in den Embryosack* 436—438
— —, growth, *Wachstum* 457—459
— —, —, differential, *differentielles* 447
— —, —, inhibition, *Hemmung des* 469, 511—516
— —, —, —, in heterostylic plants, *bei Heterostylen* 514—516
— —, —, —, in homostylic plants, *bei Homostylen* 512—514

pollen tube, growth, interpretation of experimental results, *Pollenschlauch-Wachstum, Deutung von Versuchsresultaten* 458f.
— —, growth performance, *Wachstumsleistung* 393
— —, growth rate, after self pollination, *Wachstumsgeschwindigkeit nach Selbstbestäubung* 511
— —, —, and vitality of the next generation, *und Vitalität der nächsten Generation* 470f.
— —, hydrotropism, *Hydrotropismus* 394
— —, inhibition in heterostylic plants, *Hemmung bei Heterostylen* 514—516
— —, metabolic interaction with transmitting tissue, *stoffwechselphysiologische Wechselwirkung mit dem Leitgewebe* 512f.
— —, nuclear division in, *Kernteilung in den* 434
— —, nutrition, *Ernährung* 393—394, 512
— —, opening of the tip, *Öffnung der Spitze* 437
— —, penetration into the embryo sac, *Eindringen in den Embryosack* 515
— —, plasmatic fine structure, *plasmatische Feinstruktur* 392—393
— —, resorption of style material, *von Griffelmaterial* 393, 512—513
— — -style interaction, *-Griffel-Wechselwirkung* 469
— —, tip, *Spitze* 389, 511
— —, transformation of reserve material, *Transformierung der Reservestoffe* 389
— —, wall, *Wandaufbau* 390
— —, —, surface, *Wand, Oberfläche* 520
—, use of aged, in fertilization, *Verwendung von altem, zur Bestäubung* 57
—, viability, *Lebensfähigkeit* 395ff.
—, —, storage conditions, *Einfluß der Aufbewahrungsbedingungen* 395—397
— wall, *Pollenwand (Sporoderm)* 373—377
—, — formation, *Wandbildung* 369
— —, nomenclature, *Pollenwand, Nomenklatur* 373—375
pollination, *Bestäubung* 453
—, double, *Doppelbestäubung* 522
— fluid, *Bestäubungstropfen* 414f., 421f.
— in Gymnosperms, *bei Coniferen* 421f.
polyploid aberrants in apomictic species, *polyploide Abweicher bei apomiktischen Arten* 563, 564
polyploids in gametophyte, *Polyploide bei Gametophyten* 266
polyploidy, *Polyploidie* 804
— in apogame ferns, *Polyploidiegrad bei apogamen Farnen* 539
—, series of, *Polyploidenreihe* 62
polyspermy, *Polyspermie* 310
— in angiosperms, *bei den Angiospermen* 442
— in mosses, *bei Moosen* 413
polyspores, *Polysporen* 770f.
potency, bisexual, *Zweigeschlechtlichkeit, bipolare* 33, 35, 49
preservation of species, *Erhaltung der Art* 805
proliferation of the ovary, *Durchwachsung des Fruchtknotens* 161
propagation, vegetative, *Vermehrung* 52
protandry, *Protandrie* 36
protease effect in fusion process, *Proteinaseeinwirkung bei der Verschmelzung der Zellkörper* 300
protein synthesis during meiosis, *Proteinsynthese während der Meiose* 357—358
protenor type, *Protenor-Typus* 60
protogyny, *Protogynie* 36
protonema, secondary, in Chara, *Zwergvorkeim von Chara* 754
pseudo fertility, *Pseudofertilität* 522
pseudogamy, definition, *Pseudogamie* 551
pseudomixis 538
Pseudomonas aeruginosa 25
pseudozygote formation, *Pseudozygoten-Bildung* 278

radication, *Radikation* 786
ramification, *Verzweigung* 268
realisator allele 83
—, female, *weiblicher* 86
— genes, *Realisatorgene* 114, 117
— mechanism, *Realisator-Mechanismus* 101
—, strength of, index, *Stärke, Zahlenwert für* 70
— systems, *Realisatorsysteme* 43, 85
realisators (realizers), *Realisatoren* 54, 70, 92, 274
—, balanced, *ausbalancierte* 99
—, effect of, *Wirkung der* 141
recombination, genetical, *Rekombination, genetische* 261
— systems, *Rekombinationssysteme* 26
recovery of paired gametes in Chlamydomonas, *Erholungsperiode gepaarter Gameten bei Chlamydomonas* 305—306
redifferentiation of the germ cell character, *Umdetermination des Keimzell-Charakters* 237
reduction division (see also: meiosis), *Reduktionsteilung (siehe: Meiose)* 51, 134
regenerates, mixed character, *Regenerate, Mischcharakter* 253
regeneration 146
— experiments in sexual determined cells, *Regenerationsexperimente an geschlechtlich determinierten Zellen* 146
— from fern leaves, *aus Farnwedeln* 253
— of sporogones, *von Sporogonen* 147
— — in Musci, *des Laubmoos-Sporogons* 249
regression to the vegetative state, *Zurückschlagen in den vegetativen Zustand* 272
release of spores, day periodical, *Entleerung von Sporangien, tagesperiodische Schwankungen* 897
replication, sporogone inducing factor, *Selbstreproduktion, sporogoninduzierender Faktor* 251
replicator, *Replikator* 19
replicon, *Replikon* 13, 19
— model, *Replikon-Modell* 19, 20, 21
repression of structural genes, *der Struktur-Gene* 274

repressor 261
reproduction, *Fortpflanzung* 1
—, apomictic, *apomiktische* 3
—, asexual, *ungeschlechtliche* 3, 4
— behaviour, *Fortpflanzungsverhalten* 241
— in brown algae, physiology of, *bei den Braunalgen, Physiologie der* 697—702
— of colonies in Flagellatae, *Kolonievermehrung bei Flagellaten* 639—641
— in Ectocarpales, *Fortpflanzungsverhältnisse bei den Ectocarpales* 238
— in the vegetative region, *Reproduktion im vegetativen Bereich* 802
— in Tribonema, *Fortpflanzung bei Tribonema* 683
—, sexual, *Vermehrung, geschlechtliche* 3, 4, 268
—, —, phylogenetic development of, *Fortpflanzung, sexuelle, phylogenetische Entwicklung der* 51
— systems in fungi, *Fortpflanzungssysteme bei Pilzen* 122
—, vegetative, various types, in Phaeophyta, *Fortpflanzungsweisen, vegetative, bei den Braunlagen* 693ff.
reproductive organs with the character of permanent spores, in cyanophyta, *Fortpflanzungskörper mit Dauersporencharakter bei den Blaulagen* 608—618
resection, influence on sex, *Beschneidung, Beeinflussung des Geschlechtes durch* 177, 195
respiration quotient of pollen tubes, *Atmungsquotient von Pollenschläuchen* 513
— in styles during pollen tube growth, *Atmungsverhalten von Griffeln während des Pollenschlauchwachstums* 513
resting cells in diatoms, *Dauerzustände bei Diatomeen* 661
— —, classification, *Dauerzellen bei Diatomeen, Klassifizierung* 664—665
— —, developmental physiology, *Entwicklungsphysiologie* 666—668
resting stage, *Ruhestadium* 240
RNA synthesis during meiosis, *RNS-Synthese während der Meiose* 356—357
runners, *Ausläufer* 561
rhythm of development, *Entwicklungsrhythmus* 802

SAT chromosome, *SAT-Chromosom* 61
schizotomy, *Schizotomie* 707
sea urchin egg, *Seeigel-Ei* 307
— —, homology between Fucus and, *Seeigel-Ei, Homologie zwischen Fucus- und* 309
secretion of silicate in diatoms, *Kieselsäureausscheidung bei Diatomen* 660
—, specific, *Sekretion, spezifische* 219
seed formation, *Samenbildung* 804
segregation, preferential, *Abweichungen vom zu erwartenden Mendelschen Spaltungsverhältnis* 449
—, abnormal, reasons for, *Spaltungszahlen, abnorme, Ursachen für* 449—450
selection principles, *Selektionsprinzipien* 244
selective fertilization, *Befruchtung, selektive* 447, 479—502
— —, physiological background, *selektive Befruchtung, materielle Grundlagen* 500ff.
— —, reaction speed, *Reaktionsgeschwindigkeit* 483—485
self-fertilization, in apomictic species, *Selbstbefruchtung bei apomiktischen Arten* 563, 565
self incompatibility, *Selbstimkompatibilität* 469
— —, break down of, *Überwindung der* 520—523
— —, genetic systems, *genetische Systeme* 508
self sterility, *Selbststerilität* 117, 128f.
— — factors, *Selbst-Sterilitäts-Faktoren* 93
— — reaction, *Selbststerilitätsreaktion* 518
— —, relation to certation, *Selbststerilität, Zusammenhang mit der Gonenkonkurrenz* 468
sex, *Geschlecht* 2, 31
— behaviour, in mosses, *Geschlechtsverhalten bei Moosen* 147
—, change of, influence of chromosome duplication, *Geschlechtsveränderung, Einfluß der Chromosomen-Verdoppelung* 88
— change, tendency to, *Geschlechtsumkehr, Tendenz zur* 165
— character, *Geschlechtscharakter* 51
— —, complementary, *komplementärer* 36
— —, morphological, *Sexualcharakter, morphologischer* 48
— —, secondary, *Geschlechtsmerkmale, sekundäre* 86, 87, 147—149
— chromosomes, *Geschlechtschromosomen* 52, 57ff., 63, 89, 101
— — complex, *Geschlechtschromosomen-Komplex* 64
— —, heteromorphic, *Geschlechtschromosomen, heteromorphe* 58—60, 89, 267—268
— —, heteromorphic, *Heteromorphie des* 89
— — heteromorphy, possible mechanism of the origin, *heteromorphe, möglicher Entstehungsmechanismus* 275, 276
— —, homomorphic, *homomorphe* 66
— — mechanism, *Geschlechtschromosomen-Mechanismus* 64
— —, pairing behaviour, *Geschlechtschromosomen, Paarungsverhalten der* 77
— — types, multiple, *Geschlechtschromosomen-Typen, multiple* 65
—, concept of, *Geschlecht, der Begriff* 2, 257
— development, change of, *Geschlechtsausbildung der* 94
— determining enzymes, *Enzyme, geschlechtsbestimmende* 200
— determination, *Geschlechtsbestimmung* 39, 113ff., 267
— — in Ascobolus, bei Ascobolus 231
— — in bryophytes, *auslösende Faktoren, bei Moosen* 140

sex determination, change of, *Geschlechts-Umstimmung* 154, 179
— — in diatoms, *Geschlechtsbestimmung bei Diatomeen* 115
— — during diplophase, *in der Diplophase* 116
— —, genetical factors in, *genetische Faktoren bei der* 140ff.
— —, genotypic, *genotypische* 128
— —, haplogenotypic, *haplogenotypische* 31, 42
— — by hormones, *durch Hormone* 230
— —, influence of growth hormones, *Wirkung von Wuchsstoffen* 143
— — mechanism, *Geschlechtsbestimmungs-Mechanismus* 65
— — —, genetical background of the, *genetische Zusammensetzung des* 69
— — — in Melandrium, *bei Melandrium* 61
— — — in Melandrium, scheme of, *bei Melandrium, Schema des* 76
— — in Melandrium, *in der Gattung Melandrium* 170f.
— —, modification of, in sporophytes, *Modifikation der, im Sporophyten* 155ff.
— —, non-genetic factors in, *nichtgenetische Faktoren bei der* 142
— — in pennate diatoms, *bei pennaten Diatomen* 674—677
— —, phenotypic, in algae, *phänotypische, bei Algen* 110ff.
— —, —, in archegoniatae, *phänotypische, bei Archegoniaten* 134ff.
— —, —, in Dicotyledonae, *modifikative, bei Dikotyledonen* 164—192
— —, —, in fungi, *phänotypische, bei Pilzen* 121ff.
— —, —, in gymnosperms, *modifikative, bei Gymnospermen* 158—159
— —, —, influence of age of the plants, *Einfluß des Alters der Pflanzen* 196f.
— —, —, — of edaphic factors, *edaphischer Faktoren* 192ff.
— —, —, — of growth hormones, *Beeinflussung durch Wuchsstoffe* 198f.
— —, —, — of humidity, *Einfluß der Luftfeuchtigkeit* 193
— —, —, — of infection, *von Parasitenbefall* 196
— —, —, — of light conditions, *der Lichtbedingungen* 193
— —, —, — of soil water relation, *des Bodenwassergehaltes* 193
— —, —, — of temperature, *der Temperatur* 194
— —, —, — of transplantation, *der Transplantation* 194f.
— —, —, — of treatment with sex hormones, *der Behandlung mit tierischen Geschlechtshormonen* 198
— —, —, in Monocotyledonae, *bei Monokotyledonen* 160—164
— —, — in sporophytes, *modifikatorische, im Sporophyten* 155ff.
sex determination, phenotypic, in spermatophytes, *Geschlechtsbestimmung, phänotypische, bei Spermatophyten* 153ff.
— —, —, species with, *modifikative, Pflanzenarten mit* 157ff.
— —, —, theoretical interpretation, *Interpretation* 199ff.
— — stability of, *Stabilität des* 146
— —, variations in, *Varianten der* 35
— — in yeast, *bei Hefen* 126
sex differences, *Geschlechtsunterschiede* 31, 32
— —, criteria for, in diatoms, *Kriterien, bei Diatomeen* 117
— differentiation, *Geschlechtsdifferenzierung* 36
— — of nuclei, *Differenzierung, sexuelle, von Kernen* 130
— distribution, *Geschlechtsverteilung* 92, 135
— duction, *Geschlechtsduktion* 21
— elimination, differential, *Geschlecht, Elimination, unterschiedliche* 160
— evolution, *Geschlechts, Evolution des* 101
— expression, *Geschlechtsausprägung* 79, 82, 97, 100, 131, 265, 274
— — in angiosperms, *bei Angiospermen* 274
— —, change, *Änderung* 182
— —, changing of, by transplantation, *Änderung der, durch Pfropfungen* 171, 177
— —, different experimental types in Melandrium, *Geschlechtsausbildung, verschiedene experimentell erhaltene Typen bei Melandrium* 73
— — and flowering, *Geschlechtsexpression und Blühstadium* 268ff.
— — in hybrids, *Geschlechtsausprägung bei Bastarden* 183
— —, influence of cell size on, *Einfluß der Zellgröße auf* 117
— —, — CO atmosphere, *von Kohlenmonoxydbehandlung* 177, 179f. 187
— —, — of gibberellic acid treatment, *von Gibberellinsäurebehandlung auf* 166, 178, 180, 183, 187ff.
— —, — of kinetin treatment, *von Kinetin-Behandlung* 166
— —, — of mineral supply, *der mineralischen Düngung* 163ff., 167, 172, 174, 189
— —, — of size and age of seeds, *der Größe und des Alters der Samen* 174
— —, — of weather conditions, *der Wetterbedingungen* 168, 180, 190
— —, temperature effect on, *Temperaturwirkung auf* 162
— factor affinity locus (sfa), *Lokus für die Geschlechtsfaktor-Affinität* 25
— factors, *Geschlechtsfaktoren* 30, 51ff., 90, 93
— —, effect of, *Wirkung von* 93
— —, localisation of, *Lokalisation der* 93
— —, multiple allely of, *Allelie der* 90
— —, plastic, *im Plasma lokalisierte* 94
— —, transfer of, *Übertragung von* 18
—, female, manifestation of, dependence upon external conditions, *Geschlecht, weibliches, Manifestation des, Abhängigkeit von Außenbedingungen* 137

sex genes, *Geschlechtsgene* 75, 90, 140
— of haploid plants, *Geschlecht bei haploiden Pflanzen* 89
—, in higher plants, existence of, *Geschlechts, Nachweis des, bei höheren Pflanzen* 262
—, history of, *Sexualforschung, Geschichte der* 257ff.
— inheritance, polyfactors, *Geschlechtsvererbung, polyfaktorielle* 88
— organs, *Geschlechtsorgane* 257, 265
— —, development of, *Entwicklung der* 117
— —, differentiation, *Differenzierung* 268, 275
— —, —, in Saprolegniaceae, *bei Saprolegniaceen* 124
— —, distribution of, *Verteilung der* 136
— —, in flowers, *in Blüten* 272f.
— —, inhibition of, *Hemmung der* 155
— phenomena, „*Liebesleben*" 259
—, physiology of, *Sexualphysiologie* 257
— processes, *Sexualvorgänge* 39
— ratio, *Geschlechtsverhältnis* 52, 54, 56, 87, 88, 93, 94, 114
— —, change of, *Verschiebung des* 57, 143
— —, influence of culture conditions, *der Kulturbedingungen* 182
— —, — of day length, *der Tageslänge* 169f., 174ff., 178f.
— —, — of light intensity, *der Lichtintensität* 182
— —, — of mineral nutrition, *der Düngung* 174
— —, — of NAA treatment, *der Behandlung mit α-Naphthylessigsäure* 167, 170, 177, 180, 183, 187, 190
— —, — of temperature, *der Temperatur* 176, 185
— —, normal, *normales* 63
— —, shift of, *Verschiebung des* 69, 454
— —, in tetraploid and aneuploid forms, *in tetraploiden und aneuploiden Formen* 78
— —, yearly change of, *jährliche Änderungen* 169, 186
—, realisation of, trigger for, *Geschlechtsrealisation, Trigger für die* 275
— realisators, *Geschlechts-Realisatoren* 92
— segregation, *Geschlechtsaufspaltung* 79, 82
— shift, *Geschlechtsverschiebung* 91
— tendency, *Geschlechtstendenz* 54, 98
— theory, *Sexualitätstheorie* 31
— types, *Geschlechtstypen* 90
sexes, sequence of, *Geschlechter, Reihenfolge der* 136
sexual affinity, *Affinität, geschlechtliche* 229
— —, determination, *geschlechtliche Anziehung, Determination der* 228—233
— —, — of by diffusible secretions, *Affinität, Determination, mittel diffusibler Sekrete* 232
— behaviour of spores, *Sexuelles Verhalten von Sporen* 238
— bipolarity, *geschlechtliche Bipolarität* 274
— cells, chemical polarity, *Geschlechtspartner, chemische Polarität* 316
sexual cells, reduction of the number of, *Geschlechtspartner, Keimzellen, allmähliche Reduktion der Zahl der* 236—237
— character of organs, *Organcharakter, geschlechtlicher* 145
— differentiation in E. coli, *Geschlechtsdifferenzierung bei E. coli* 6
— —, pattern of, in algae, *Muster der, bei Algen* 229
— dimorphism, *Geschlechtsdimorphismus* 201
— factors, *Geschlechtsfaktoren* 125
— hormones, *Sexualstoffe* 215—233
— — in Achlya, *Geschlechtshormone bei Achlya* 221—224
— —, animal, influence on sex expression in plants, *Sexualhormone, tierische, Einfluß auf Geschlechtsausprägung bei Pflanzen* 171, 173, 179
— — in Bombardia lunata, *Geschlechtshormone, bei Bombardia lunata* 226
— — in Oedogonium, *bei Oedogonium* 226
— — in Oedogonium development, *bei Entwicklung von Oedogonium* 226, 232
— interaction, pattern of, *geschlechtliche Wechselwirkung, Muster der* 219ff., 226
— organs in Ceramiales, *Sexualorgane bei Ceramiales* 238
— —, coordinated development, *Geschlechtsorgane, koordinierte Entwicklung* 216—228
— —, diploid, *Sexualorgane, diploide* 238
— —, induction of, *Geschlechtsorgane, Induktion der* 214ff.
— —, loss of, *Sexualorgane, Verlust der* 127
— phase, influence of cell size, in diatoms, *sexuelle Phase, determinierender Faktor der Schalengröße bei Diatomeen* 116
— phenomena, in fungi, *Sexualitätserscheinungen bei Pilzen* 131
— potency, *sexuelle Potenz* 33
— process in Achlya, hormonal control, *Sexualvorgang bei Achlya, hormonale Kontrolle* 293
— processes in bacteriae, *sexuelle Vorgänge bei Bakterien* 314
— —, chemical regulation, *Geschlechtsprozesse, chemische Regulation der* 282ff.
— — in fungi, *Sexualvorgänge bei Pilzen* 122ff.
— —, hormonal control in Saccharomyces, *Geschlechtsprozesse, hormonale Kontrolle bei Hefen* 228
— progression in Achlya, *Geschlechtsentwicklung bei Achlya* 221—226
— reaction, coordination, *Geschlechtsreaktion, Koordination der* 219
— — in Mucorineae, *Sexualreaktion bei Mucorineen* 294—297
— —, sequence of stages, *Geschlechtsreaktion, Reihenfolge der Stadien* 219
— reproduction, general principles, *geschlechtliche Fortpflanzung allgemeine Prinzipien* 257ff.
— substances, diffusible, in lower plants, *Geschlechtsstoffe, diffusible, bei niederen Pflanzen* 232

sexual tendency, expression of, *Geschlechtstendenz, Ausprägung der* 169
sexuality, *Sexualität* 2, 122f.
— in bacteria, *bei Bakterien* 14
—, bipolar, *Bipolarität, sexuelle* 31, 32
—, chemical differentiation, *Sexualität, chemische Differenzierung* 300
— in Dinophyta, *bei den Dinophyten* 626
— in Euglenophyta, *bei den Euglenophyta* 620
— in higher Ascomycetes, *bei den höheren Ascomyceten* 224
— and incompatibility, relation between, *und Inkompatibilität, Verhältnis von* 43
—, mixed, *Gemischtgeschlechtlichkeit* 136
—, —, tendency to, *Tendenz zur* 112
—, multipolar, *Sexualität, multipolare* 31
—, relative 36, 37, 38, 125, 261
—, —, hypothesis of, *Hypothese der* 117
—, sign of, *Geschlechtlichkeit, Zeichen der* 265
—, somatic, *somatische* 260
—, terminology of, *Begriffsysteme* 128
—, theory of, following HARTMANN, *Geschlechtstheorie nach* HARTMANN 117
S gene, inactivation, *S-Gen, Inaktivierung des* 521
—, multiple allely, *S-Gene, multiple Allelie* 521
shift in size of the generations, *Größenverschiebung der Generationen* 244
sirenin, *Sirenin* 218, 285ff., 288
—, M.W., *Molekülgewicht* 291
size variation, in diatoms, *Größen-Variation bei Diatomeen* 114
sclerotia, *Sklerotien* 781—784
somatogamy, *Somatogamie* 46, 260, 261
species crosses, *Artkreuzungen* 80, 96
— — in Solanum, *Solanum-* 100
specific contact between cells, *spezifischer Kontakt zwischen Zellen* 284
sperm cells, formation of, in pine, *Spermazellen, Entstehung der, bei Coniferen* 424—426
— trap, *Spermium-Falle* 287
spermatozoids, free motility, in Cycas and Ginkgo, *Spermatozoiden, frei bewegliche, bei Ginkgo und Cycas* 415f.
—, morphology of, in mosses and ferns, *Bau der, bei den Moosen und Farnen* 410
—, pathway of to the archegonia, *Weg der, zu den Archegonien* 412
—, penetration of the neck canal of the archegonium in ferns, *Eindringen in den Archegoniumhals bei Farnen* 412
—, release of, in the archegonial chamber in Cycas, *Erguß der, in die Archegoniumkammer bei Cycas* 418
Sphaerocarpos 267—268
spiralisation of flagellae, *Spiralisierung der Geißeln* 300
sporangia, *Sporangien* 265
spore formation, loss of, in ferns, *Sporenbildung, Verlust der, bei apogamen Farnen* 544
spore germination in apogamic ferns, *Sporenkeimung bei apogamen Farnen* 537
— membrane in diatoms, electron microscopical observations, *Sporenmembran bei Diatomeen, elektronenmikroskopische Untersuchungen* 663
— mother cells, *Sporenmutterzellen* 533—538
— release, periodicity, in Heterokontae, *Sporenentlassung, Periodizität der, bei Heterokonten* 690
— tetrad, segregation, *Sporen-Tetrade, Aufspaltung in der* 267
spores of brown algae, construction, *Schwärmer der Braunalgen, Bauschema* 698
—, various types in Bangioideae, *Sporentypen bei den Bangioideae* 762—763
sporogone formation, apogame, in ferns, *Sporogonbildung, apogame, bei Farnen* 252
— —, chemical factors, *stofflicher Faktor* 251
— — in vitro 251
— — in mosses, *bei Moosen* 246ff.
sporophyll, intersexual, *Sporophyle, intersexuelles* 156
sporophyte, *Sporophyt* 240
— characters, determination of, stepwise, *Sporophyten-Charakter, Determination des, schrittweise* 247
sporophytes, construction elements, determination, *Sporophyten, Konstruktionselemente, Determination der* 247
—, determination 248
sporophytic generations, *sporophytische Generationen* 266
— level in bryophytes, *Sporophytenstufe bei Bryophyten* 138
sporophyte status, determination, *Sporophyten-Status, Determination des* 249—250
sporophytes, determination, degree of stability, *Sporophyten, Determination, Stabilitätsgrad der* 248—250
sporopollenin, *Sporopollenin* 375
sporulation, *Sporenbildung* 129
— in yeast, *bei Hefen* 126
S proteins, *S-Proteine* 520
sterile flowers by infection with Ustilago, *sterile Blüten durch Infektion mit Ustilago* 161, 163, 166, 170, 184
sterility, anther, *Antherensterilität* 170
— factors, *Sterilitätsfaktoren* 43, 47, 96
— — in corn, *beim Mais* 95
—, male, *Sterilität, männliche* 97
— in male sex, influence of NAA, *Sterilität im männlichen Geschlecht, Beeinflussung durch NES-Behandlung* 183
— phenomena, *Sterilitätsphänomene* 561
—, plasmatic, *Sterilität, plasmatisch bedingte* 96
stickiness of the flagellum, *Klebrigkeit der Geißel* 304
stigma surface, *Narbenoberfläche* 521
stigmatic cuticle, *Narbencuticula* 509
— exudate, *Narbenschleim* 385, 522

storage conditions, influence on viability of pollen, *Lagerungsbedingungen, Einfluß auf die Lebensfähigkeit des Pollens* 397—398
structural heterozygoty, *Struktur-Heterozygotie* 62
structure chemical of the gamone in Mucor, *Strukturformel des Gamons von Mucor* 296
style, gradient of inhibition within, *Griffel, Gradient der Pollenschlauchhemmung* 518
—, pollen tube growth within, *Pollenschlauchwachstum im Innern des* 457f.
— reduction by SD treatment, *Reduktion des, durch Kurztagbehandlung* 169
styles, respiration during pollen tube growth, *Griffel, Atmungsverhalten während des Pollenschlauchwachstums* 513
sublethal factors, *Subletalfaktoren* 93
subtilisin, *Subtilisin* 300
suppressor, female, *Weiblichkeits-Unterdrücker* 75, 77
— locus, *Uunterdrücker-Locus* 96
supra sexual processes, *suprasexuelle Vorgänge* 121
surface, active sites, *Oberflächenkomponenten, aktive* 312
— contact in bacteriae, *Oberflächenkontakt bei Bakterien* 314
synergids, role of *Synergiden, Rolle der* 434f., 436, 437
syngamy, *Syngamie* 316
synzoospores, *Synzoosporen* 723
synzoospore in Vaucheria, *von Vaucheria* 686

tapetum 355ff.
telemorphic reactions, *Fernreaktionen* 220
telemorphosis, *Telemorphosen* 286
termones, *Termone* 33, 215
— in Chlamydomonas, *bei Chlamydomonas* 229, 232
tetrapolarity, *Tetrapolarität* 127
tetrasporangia in Rhodophyceae, *Tetrasporangien bei den Rotalgen* 768
tetraspores, *Tetrasporen* 765—771
—, formation of, *Bildung der* 240
tetrasporophytes, *Tetrasporophyten* 238
thigmotropism, *Thigmotropismus* 286
timing of events during fertilization of pine, *Koordinierung, zeitliche, bei der Befruchtung der Coniferen* 427
tissue formation, *Gewebebildung* 242
transduction, *Transduktion* 2, 314
—, recombinative, *rekombinative* 260
transfer of chromosomes, *Chromosomentransfer* 259
— of genetic material, kinetics of, *von genetischem Material, Kinetik des* 17, 19
— of plastids by sperms, in ferns, *Übertragung männlicher Plastiden durch Spermatozoid bei Farnen* 413
transformation 2, 314
transition of gametophyte to sporophyte, *Übergang vom Gametophyten zum Sporophyten* 533
— from water to land, *vom Wasser- zum Land-Leben* 244
translocation chromosome, isobranchial, of Spinacia, *Translokationschromosom, isobranchiales, bei Spinat* 89
— types, *Translokationstypen* 66
trap reaction of gametes, *Fallenreaktion von Gameten* 285
tuber formation, *Knollenbildung* 790, 803

unisexual flowers, *eingeschlechtliche Blüten* 275

variations in coiling cycle, *Änderungen im Spiralisierungs-Cyclus* 352—353
variegata type, *variegata-Typus* 68
varieties, formation of, in apogame ferns, *Varietätenbildung bei apogamen Farnen* 544
vegetative determination of spores, differential, *der Sporen, unterschiedliche* 238f.
— divisions, *Teilungen* 629
— propagation, *Vermehrung* 277
— —, definition, *Vermehrung, definition* 551, 597, 601
— — in algae, *bei den Algen* 516, 601
— — in angiosperms, *bei den Angiospermen* 561—563
— — in brown algae, *Fortpflanzung bei den Braunalgen* 692—702
— — in Charophyceae, *bei den Charophyceen* 752—759
— — in higher plants, *Vermehrung bei höheren Pflanzen* 561ff.
— — in Rhodophyta, *bei den Rhodophyten* 772—774
— reproduction of Chrysophyceae, *Vermehrung der Chrysophyceen* 637—640, 641
— — in fungi, *Fortpflanzung bei den Pilzen* 777—784
— — and karyology, *und Karyologie* 804
— —, physiology of, in spermatophytes, *Physiologie der vegetativen Fortpflanzung bei den Spermatophyten* 800ff.
— — in spermatophytes, *Fortpflanzung bei den Spermatophyten* 787
— — of Volvocales, *der Volvocales* 707
— tendency, *Tendenz* 147
— — of the female sex, *des weiblichen Geschlechtes* 145
viability, pollen, *Lebensfähigkeit* 395ff.
—, —, influence of temperature, *Einfluß der Temperatur* 396f.
Vibrio cholerae 25
virus infection, *Virus-Infektion* 314
vitality and pollen tube growth, *Vitalität und Pollenschlauchwachstum* 470f.
vitamin B, *Vitamin B* 803
vitamin C, *Vitamin C* 803
vivipary, *Viviparie* 551, 561
— in grasses, *bei Gräsern* 562f.

x ray mutation of Allomyces, *Röntgen-mutante von Allomyces* 245
XO types, *XO-Typus* 60
XY intersexes, *XY-Intersexe* 77
XY type, *XY-Typus* 61
XY_1Y_2 type, XY_1Y_2*-Typus* 64

yeasts, genetics of, *Hefen, Genetik der* 126

zoosporangia formation in Ulotrichales, *Zoosporangien, Bildung der, bei den Ulotrichales* 717
— in Volvocales, *bei den Volvocales* 708
zoospore formation, *Zoosporenbildung* 706ff.
— — in Chloromonas, *bei Chloromonas* 710
— membranes, *Zoosporen-Membran* 710
zoospores of Chlorophyceae, *Zoosporen der Chlorophyceen* 707—724
— in Chlorophyceae, release of, *Zoosporenentleerung bei Chlorophyceen* 718
—, formation of, in Chlorangium, *Zoosporen, Bildung der, bei Chlorangium* 715
—, —, influence of acidity, *Einfluß des Säuregrades* 738
zoospores, formation, influence of the chromatophore, in Chladophora, *Zoosporen, Bildung der, Einfluß des Chromatophorennetzwerkes bei Chladophora* 721
—, —, in Oedogoniales, *bei den Oedogoniales* 720
—, —, in Stylosphaeridium, *bei Stylosphaeridium* 716
—, induction of, in Heterokontae, *Auslösung von, bei Heterokonten* 688f.
—, modified, *modifizierte* 722—724
—, ultrastructure, of Oedogonium, *Ultrastruktur, bei Oedogonium* 714
—, —, of Stigeoclonium species, *bei Stigeoclonium*-Arten 712
—, various types, in Chlorophyceae, *verschiedene Typen bei den Chlorophyceen* 713
zygophore-inducing substance, *Zygophoren induzierende Substanz* 218ff., 232
zygophores in Mucor, *Zygophoren bei Mucor* 295
zygote formation, *Zygotenbildung* 51
— —, kinetics of, *Kinetik der* 19
zygote products, morphogenesis of, *Zygotenprodukte, Morphogenese der* 244
zygotropism, *Zygotropismus* 217f., 286

GPSR Compliance

The European Union's (EU) General Product Safety Regulation (GPSR) is a set of rules that requires consumer products to be safe and our obligations to ensure this.

If you have any concerns about our products, you can contact us on ProductSafety@springernature.com

In case Publisher is established outside the EU, the EU authorized representative is:

Springer Nature Customer Service Center GmbH
Europaplatz 3
69115 Heidelberg, Germany

Batch number: 10620876

Printed by Printforce, the Netherlands